See the Difference

SHOWCASE EXAMPLES

Showcase Examples introduce key topics and provide "how-to" instruction by walking **step-by-step** through the problem-solving process. The left and middle "columns" can be thought of as the instructor's voice offering an explanation (left column) and then summarizing information (middle column) during a classroom lecture.

"I found the [Showcase] step-by-step examples to be an outstanding feature of this text...This is a great feature for students who need to see every step, as they are learning something new." – Elizabeth Morrison, Valencia Community College

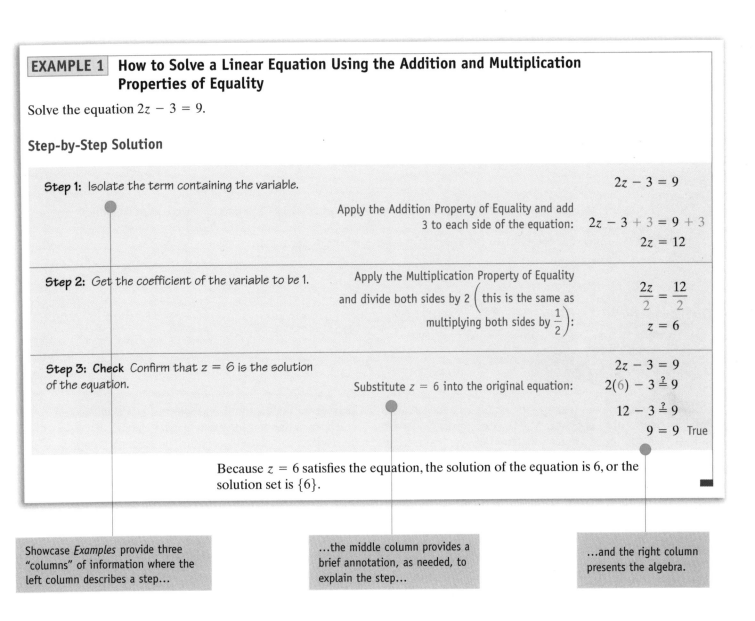

EXAMPLE 1 **How to Solve a Linear Equation Using the Addition and Multiplication Properties of Equality**

Solve the equation $2z - 3 = 9$.

Step-by-Step Solution

Step 1: Isolate the term containing the variable.

$$2z - 3 = 9$$

Apply the Addition Property of Equality and add 3 to each side of the equation: $2z - 3 + 3 = 9 + 3$

$$2z = 12$$

Step 2: Get the coefficient of the variable to be 1.

Apply the Multiplication Property of Equality and divide both sides by 2 $\left(\text{this is the same as multiplying both sides by } \frac{1}{2}\right)$: $\frac{2z}{2} = \frac{12}{2}$

$$z = 6$$

Step 3: Check Confirm that $z = 6$ is the solution of the equation.

$$2z - 3 = 9$$

Substitute $z = 6$ into the original equation: $2(6) - 3 \stackrel{?}{=} 9$

$$12 - 3 \stackrel{?}{=} 9$$

$$9 = 9 \quad \text{True}$$

Because $z = 6$ satisfies the equation, the solution of the equation is 6, or the solution set is $\{6\}$.

Showcase *Examples* provide three "columns" of information where the left column describes a step...

...the middle column provides a brief annotation, as needed, to explain the step...

...and the right column presents the algebra.

QUICK CHECK EXERCISES AS HOMEWORK

Quick Check exercises can serve as immediate reinforcement, review, or a "warm-up" to the section exercise set. By assigning *Quick Check* exercises, you can be assured that students will have access to a detailed, step-by-step Example that is similar to the assignment.

Additional Resources to Help Students Succeed

Student Study Pack

A single, easy-to-use package, available bundled with your textbook or by itself, for purchase through your bookstore. This package contains the following resources to help you succeed:

Student Solutions Manual

- Solutions to the odd-numbered section exercises
- Solutions to the Quick Check exercises
- Solutions to the Preparing for This Section, Putting the Concepts Together (mid-chapter review), Chapter Review, Chapter Test, and Cumulative Review exercises

Prentice Hall Math Tutor Center

- Staffed by qualified math instructors who provide students with tutoring on examples and odd-numbered exercises from the textbook. Tutoring is available via toll-free telephone, toll-free fax, email, or the Internet.

CD Lecture Series

- Perfect for review of a section or a specific topic, these mini-lectures cover the key concepts from each section of the text in approximately 10-15 minutes.
- Includes fully worked-out solutions to the exercises marked with a CD video icon.

Online Homework and Tutorial Resources

MyMathLab MyMathLab

MyMathLab is a series of text specific, easily customizable, online courses for Prentice Hall textbooks in mathematics and statistics. MyMathLab is powered by CourseCompass™—Pearson Education's online teaching and learning environment—and by MathXL®—our online homework, tutorial, and assessment system. MyMathLab gives instructors the tools they need to deliver all or a portion of their course online, whether students are in a lab setting or working from home. MyMathLab provides a rich and flexible set of course materials, featuring free-response exercises that are algorithmically generated for unlimited practice and mastery. Students can also use online tools, such as video lectures, animations, and a multimedia textbook, to independently improve their understanding and performance. MyMathLab is available to qualified adopters. For more information, visit our Web site at *www.mymathlab.com* or contact your Prentice Hall sales representative. (MyMathLab must be set up and assigned by your instructor.)

MathXL® www.mathxl.com MathXL

MathXL is a powerful online homework, tutorial, and assessment system that accompanies the text. With MathXL, instructors can create, edit, and assign online homework and tests using algorithmically generated exercises correlated to your textbook. All student work is tracked in MathXL's online gradebook. Students can take chapter tests in MathXL and receive personalized study plans based on their test results. The study plan diagnoses weaknesses and links students directly to tutorial exercises for the objectives they need to study and retest. Students can also access supplemental animations and video clips directly from selected exercises. MathXL is available to qualified adopters. For more information, visit our Web site at *www.mathxl.com* or contact your Prentice Hall sales representative for a product demonstration. (MathXL must be set up and assigned by your instructor.)

Annotated Instructor's Edition

Elementary Algebra

Michael Sullivan, III
Joliet Junior College

Katherine R. Struve
Columbus State Community College

Upper Saddle River, New Jersey 07458

Executive Editor: Paul Murphy
Editor in Chief: Christine Hoag
Executive Project Manager: Ann Heath
Production Editor: Barbara Mack
Senior Managing Editor: Linda Mihatov Behrens
Executive Managing Editor: Kathleen Schiaparelli
Media Project Manager: Audra J. Walsh
Media Production Editor: John Cassar
Managing Editor, Digital Supplements: Nicole M. Jackson
Manufacturing Buyer: Maura Zaldivar
Manufacturing Manager: Alexis Heydt-Long
Director of Marketing: Patrice Jones
Senior Marketing Manager: Kate Valentine
Marketing Assistant: Jennifer de Leeuwerk
Development Editor: Don Gecewicz
Editor in Chief, Development: Carol Trueheart
Editorial Assistant: Abigail Rethore
Project Manager/Class Testing: Dawn Nuttall
Art Director: Jonathan Boylan
Interior Designers: Wanda España, Mary Siener
Cover Designer: Wanda España
Art Editor: Thomas Benfatti
Creative Director: Juan R. López
Director of Creative Services: Paul Belfanti
Director, Image Resource Center: Melinda Reo
Manager, Rights and Permissions: Zina Arabia
Manager, Visual Research: Beth Brenzel
Manager, Cover Visual Research & Permissions: Karen Sanatar
Image Permission Coordinator: Richard Rodrigues
Photo Researcher: Teri Stratford
Cover Photo: © Celia Pearson/Pearson Photography
Art Studios: Precision Graphics, Laserwords
Compositor: Interactive Composition Corporation

Pearson Prentice Hall
Pearson Education, Inc.
Upper Saddle River, New Jersey 07458

Printed in the United States of America

10 9 8 7 6 5 4 3 2 1

ISBN 0-13-146811-1
ISBN 0-13-146766-2 (Student Edition)

Pearson Education LTD., *London*
Pearson Education Australia PTY, Limited, *Sydney*
Pearson Education Singapore, Pte. Ltd
Pearson Education North Asia Ltd, *Hong Kong*
Pearson Education Canada, Ltd., *Toronto*
Pearson Educación de Mexico, S.A. de C.V.
Pearson Education—Japan, *Tokyo*
Pearson Education Malaysia, Pte. Ltd

To my father, Michael Sullivan, an unbelievable mentor, to the memory of my Mother, who is missed dearly, and to Stefan and Zofia Wilk, who helped in so many ways.

- Michael Sullivan

To my husband, Dan Struve, for his encouragement and support.

- Katherine R. Struve

About the Authors

With training in mathematics, statistics, and economics, Michael Sullivan, III has a varied teaching background that includes 15 years of instruction in both high school and college-level mathematics. He is currently a full-time professor of mathematics at Joliet Junior College. Michael has numerous textbooks in publication, including an Introductory Statistics series, and a Precalculus series, which he writes with his father, Michael Sullivan.

Michael believes that his experiences writing texts for college-level math and statistics courses give him a unique perspective as to where students are headed once they leave the developmental mathematics tract. This experience is reflected in the philosophy and presentation of his developmental text series. When not in the classroom or writing, Michael enjoys spending time with his three children, Michael, Kevin, and Marissa, and playing golf. Now that his two sons are getting older, he has the opportunity to do both at the same time!

Kathy Struve has been a classroom teacher for nearly 25 years, first at the high school level, and, for the past 13 years, at Columbus State Community College. Kathy emphasizes classroom diversity: diversity of age, learning styles, and previous learning success. She is aware of the challenges of teaching mathematics at a large, urban community college, where students have varied mathematics backgrounds, and may enter college with a high level of mathematics anxiety.

Kathy served as Lead Instructor of the Developmental Algebra sequence at Columbus State where she developed curriculum and provided leadership to adjunct faculty in implementing graphing calculator technology in the classroom. She has authored classroom activities at the Elementary Algebra, Intermediate Algebra, and College Algebra levels. In her spare time Kathy enjoys biking, hiking, traveling, and reading British detective mysteries.

Contents

PART III Developing Algebraic Skills Using Two Unknowns

A Word about Textbook Design and Student Success

As students and instructors have related in Prentice Hall focus groups and market research surveys, developmental math textbooks should not look "cluttered" or "busy." A busy design can distract a student from what is most important in the text. It can also heighten math anxiety.

As a result of this research, the design of this text is understated and focused on the most important pedagogical elements. Students and instructors helped us to identify the primary elements of this text, which are central to student success. They include:

- Exercise Sets
- Examples and *Quick Check* exercises (practice problems)
- Rules, Property, and Definition boxes
- Study Aids: *In Words, Work Smart,* and *Work Smart: Study Skills*

As you will notice, these primary features are the most prominent elements in the design. We have made every attempt to ensure that these components are the features to which the eye is drawn. The remaining features, the secondary elements, blend into the "fabric" or "grain" of the overall design.

Our thanks go to all of the students and instructors who helped us develop the design of this text. Their feedback proved invaluable in helping us to make the right decisions. We are confident the design of this text will be both practical and engaging as it serves its educational and learning purposes.

Sincerely,

Paul Murphy

Executive Editor
Developmental Mathematics
Prentice Hall

Kate Valentine

Senior Marketing Manager
Developmental Mathematics
Prentice Hall

Preface

The Elementary Algebra course serves a diverse group of students. Some of them have never been exposed to algebra, while others have been introduced to the material but have not yet grasped all the concepts. Still other students have succeeded in the course some time ago and need a refresher. Not only do the backgrounds of students vary with regard to their mathematical abilities but students' motivation, study skills, and reading skills also range considerably.

This diversity makes teaching Elementary Algebra challenging. It is imperative that a new text recognize the diversity of the classroom and address the array of needs of the many students.

Elementary Algebra introduces students to the logic and precision of mathematics. We expect our students to leave the course with an appreciation of this precision as well as the power of mathematics. Our students have to understand that the concepts we teach in the course form the basis for future mathematics courses. Once they have a conceptual understanding of algebra, students recognize that the material is not a series of unconnected topics. Instead, they see a story in which each new chapter builds on concepts learned in previous chapters.

To reinforce this idea, we remind the students of a helpful fact—mathematics is about taking a problem and reducing it to another problem that we have already seen. Taking a problem and reducing it to parts that are easier to solve helps students to see the forest for the trees (and, to carry the metaphor further, prevents them from feeling that they are lost in the woods).

In short, to address the many needs of today's Elementary Algebra students, we established the following as our goals when we began to write this text:

- Provide the student with a strong conceptual foundation in mathematics through a clear, distinct, comprehensive presentation of the topics.
- Present a variety of study aids and tips so the student quickly comes to view the text as a useful and reliable tool that can increase their success in the course.
- Offer comprehensive exercise sets that build students' skills, show various intriguing applications of mathematics, begin to build up mathematical thinking, and reinforce mathematical concepts.
- Provide students with ample opportunity to see the connections among the various topics learned in the course.

A Strong Foundation Through an Innovative Approach

One of the exciting aspects of creating a new text is that we found ourselves before a blank canvas that provided us with the opportunity to rethink what has become the conventional presentation of the course material. In particular, we had the opportunity to reconsider the organization and location of topics within the text.

To help students gradually master the core concepts of algebra, we organized our Elementary Algebra textbook into the following parts:

Part I: Foundations

Part II: Developing Algebraic Skills Using One Unknown

Part III: Developing Algebraic Skills Using Two Unknowns

We created what we call our ***one-unknown, two-unknown*** approach. Working with expressions and solving equations in one variable is covered comprehensively (including a thorough coverage of quadratic equations in one variable) before we present equations in two variables.

Part I: Foundations

Here we review operations on real numbers and algebraic expressions. Part I gives an instructor a chance to focus on potential trouble spots and to fill gaps in student knowledge before continuing to more complicated algebraic ideas.

Part II: Developing Algebraic Skills Using One Unknown

Relying on one unknown, we extend the fundamental skills by solving equations in one variable and problem solving; simplifying and factoring polynomials; simplifying rational and radical expressions; and solving quadratic equations.

Part III: Developing Algebraic Skills Using Two Unknowns

Once the students have mastered operations with one unknown, we introduce equations in two variables such as graphing linear equations, solving systems of linear equations, and functions. We believe that moving from a single unknown to two unknowns will be much simpler for the student than moving back and forth between operations with one unknown and operations with two unknowns.

In summary, the ***one-unknown, two-unknown approach*** will help students:

- Gradually master core algebraic skills and concepts
- Develop better critical thinking skills through a logical progression of the material
- Provide a more seamless transition to Intermediate Algebra

Developing an Effective Text for Use In and Out of the Classroom

Given the hectic lives led by most students, coupled with the anxiety and trepidation with which they approach this course, an outstanding developmental mathematics text must provide pedagogical support that makes the text valuable to students as they study and do assignments. Pedagogy must be presented within a framework that teaches students how to study math; pedagogical devices must also address what students see as the "mystery" of mathematics—and solve that mystery.

To encourage students and to clarify the material, we developed a set of pedagogical features that help students develop good study skills, garner an understanding of the connections between topics, and work smarter in the process. The pedagogy used in this text is based upon the more than 40 years of classroom teaching experience that the authors bring to this text.

Examples are often the determining factor in how valuable a textbook is to a student. Students look to Examples to provide them with guidance and instruction when they need it most—the times when they are away from the instructor and the classroom. We have developed several Example formats in an attempt to provide superior guidance and instruction for the students. The formats include:

Innovative *Sullivan/Struve Examples*

The innovative *Sullivan/Struve Example* has a two-column format in which annotations are provided to the **left** of the algebra, rather than the right, as is the practice in most texts. Because we read from **left-to-right,** placing the annotation on the left will make more sense to the student. It becomes clear that the annotation describes what we are about to do instead of what was just done. The annotations may be thought of as the teacher's voice offering clarification immediately before writing the solution on the board. Consider the following:

EXAMPLE 1 **Multiplying Rational Numbers (Fractions)**

Find the product: $\frac{2}{9} \cdot \left(-\frac{15}{19}\right)$

Solution

We begin by rewriting the rational number $-\frac{15}{19}$ as $\frac{-15}{19}$. Then we multiply the numerators and multiply the denominators.

$$\frac{2}{9} \cdot \left(\frac{-15}{19}\right) = \frac{2 \cdot (-15)}{9 \cdot 19}$$

Write the numerator and the denominator as the product of prime factors: $= \frac{2 \cdot 3 \cdot (-5)}{3 \cdot 3 \cdot 19}$

Showcase Examples

Showcase Examples are used strategically to introduce key topics or important problem-solving techniques. These examples provide "how-to" instruction by offering a guided, step-by-step approach to solving a problem. Students can then immediately see how each of the steps is employed. We remind students that the *Showcase Example* is meant to provide "how-to" instruction by including the words "how to" in the example title.

The *Showcase Example* has a three-column format in which the left column describes a step, the middle column provides a brief annotation, as needed, to explain the step, and the right column presents the algebra. With this format, the left and middle columns can be thought of as the instructor's voice offering an explanation (left) and then summarizing information (middle) during a lecture.

EXAMPLE 1 **How to Solve a Linear Equation Using the Addition and Multiplication Properties of Equality**

Solve the equation $2z - 3 = 9$.

Step-by-Step Solution

Step 1: *Isolate the term containing the variable.*		$2z - 3 = 9$
	Apply the Addition Property of Equality and add 3 to each side of the equation:	$2z - 3 + 3 = 9 + 3$
		$2z = 12$
Step 2: *Get the coefficient of the variable to be 1.*	Apply the Multiplication Property of Equality and divide both sides by 2 $\left(\text{this is the same as multiplying both sides by } \frac{1}{2}\right)$:	$\frac{2z}{2} = \frac{12}{2}$
		$z = 6$

Quick Check Exercises

Placed at the conclusion of every Example, the *Quick Check* exercises provide students with an opportunity for immediate reinforcement. By working the problems that mirror the example just presented, students get instant feedback and gain confidence in their understanding of the concept. All the answers to *Quick Check* exercises are provided in the back of the text. We think that the *Quick Check* exercises will make the text more accessible and encourage students to read, consult, and use the text regularly.

Study Skills and Student Success

We have included study skills and student success as regular themes throughout this text starting with *Section 1.1: Success in Mathematics*. In addition to this dedicated section that covers many of the basics that are essential to success in any math course, we have included several recurring study aids that appear in the margin. These features were designed to anticipate the student's needs and to provide immediate help—as if the teacher were looking over his or her shoulder. These margin features include *In Words; Work Smart;* and *Work Smart: Study Skills.*

Section 1.1: Success in Mathematics focuses the student on the basics of study skills, including what to do during the first week of the semester; what to do before, during, and after class; how to use the text effectively; and how to prepare for an exam.

In Words helps to address the difficulty that students have in reading mathematically precise definitions and theorems by explaining them in plain English.

Work Smart provides "tricks of the trade" hints, tips, reminders, and alerts. It also identifies some common errors to avoid and helps students work more efficiently.

Work Smart: Study Skills reminds students of study skills that will help them to succeed at various points in the course. Attention to these practices will help them to become better, more proficient, learners.

Test Preparation and Student Success

The Chapter Tests in this text and the companion Chapter Test Prep Video CD have been designed to help students make the most of their valuable study time.

Chapter Test
In preparation for their classroom test, students should take the practice test to make sure they understand the key topics in the chapter. The exercises in the Chapter Tests have been crafted to reflect the level and types of exercises a student is likely to see on a classroom test.

Chapter Test Prep Video CD
Packaged with each new copy of the text, the Chapter Test Prep Video CD provides students with help at the critical juncture when they are studying for a test. The CD Video presents step-by-step solutions to the exact exercises found in each of the book's Chapter Tests. Easy video navigation allows students to access instantly the worked-out solutions to the exercises they want to study or review.

Superior Exercise Sets: Paired with Purpose

Students learn algebra by doing algebra. The superior end-of-section exercise sets in this text provide students with ample practice of both procedures and concepts. The exercises are paired and present problem types with every possible derivative.

The exercises also present a gradual increase in difficulty level. The early, basic exercises keep the student's focus on as few "levels of understanding" as possible. The later or higher-numbered exercises are "multi-task" (or Mixed Practice) exercises where students are required to utilize multiple skills, concepts, or problem-solving techniques.

Throughout the textbook, the exercise sets are grouped into seven categories—some of which appear only as needed:

1. **Concepts and Vocabulary** exercises are fill-in-the-blank, true/false, and open-ended questions that test a student's understanding of the vocabulary and concepts presented within the section. We have found that students do not succeed if they do not become familiar with the basic vocabulary of mathematics.
2. **Building Skills** exercises are drill problems that develop the student's understanding of the procedures and skills in working with the methods presented in the section. Often these exercises can be linked back to a single example in the section.
3. **Mixed Practice** exercises are also drill problems, but they offer a comprehensive assessment of the skills learned in the section by asking problems that relate to more than one concept or objective.

4. **Applying the Concepts** exercises are problems that allow students to see the relevancy of the material learned within the section. Problems in this category are either situational problems that use material learned in the section to solve "real-world" problems or they are problems that ask a series of questions to enhance a student's conceptual understanding of the mathematics presented in the section.
5. **Extending the Concepts** exercises can be thought of as problems that go beyond the basics. Within this block of problems an instructor will find a variety of problems to sharpen students' critical-thinking skills.
6. Finally, we also include some coverage of the **Graphing Calculator** when we reach the two-unknown part of the book (Chapter 8) where it may be incorporated more logically. Because instructors differ widely in their views about the appropriateness of including graphing technology at this or any other course level, we made an effort to make graphing technology entirely optional and placed it exclusively in the chapters that cover solving equations with two unknowns. When appropriate, technology exercises are included at the close of a section's exercise set and examples using the technology precede the problems. The graphing calculator exercises are, therefore, self-contained so that they do not interrupt the discussion of the algebra.

How It All Fits Together: The Big Picture

Another important role of the pedagogy in this text is to help students see and understand the connection between the mathematical topics being presented. Several section-opening and margin features help to reinforce connections:

The Big Picture: Putting It Together (Chapter Opener)
This feature is based on how we start each chapter in the classroom—with a quick sketch of what we plan to cover. Before tackling a chapter, we tie concepts and techniques together by summarizing material covered previously and then relate these ideas to material we are about to discuss. It is important for students to understand that content truly builds from one chapter to the next. We find that students need to be reminded that the familiar operations of addition, subtraction, multiplication, and division are being applied to different or more complex objects.

Preparing for This Section
As part of this building process, we think it is important to remind students of specific material that they will need from earlier in the course to be successful within a given section. The *Preparing for*... feature that begins each section not only provides a list of prerequisite skills that a student should understand before tackling the content of a new section but also offers a short quiz to test students' preparedness. Answers to the quiz are provided as a footnote on the same page, and a cross-reference to the material in the text is provided so that the student can remediate when necessary.

Putting the Concepts Together (Mid-Chapter Review)
Each chapter has a group of exercises at the appropriate point in the chapter, entitled *Putting the Concepts Together*. These exercises serve as a review—synthesizing material introduced up to that point in the chapter. The exercises in these mid-chapter reviews are carefully chosen to assist students in seeing the "big picture."

Cumulative Review
Learning algebra is a building process and building involves considerable reinforcement. The cumulative review exercises at the end of each odd-numbered chapter, starting with Chapter 3, help students to reinforce and solidify their knowledge by revisiting concepts and using them in context. This way, studying for the final exam should be fairly easy.

In Closing

When we started writing this textbook, we discussed what improvements we could make in organization and in a textbook's approach to fundamental concepts. We also wanted to enhance staples such as examples and problems, and we "re-engineered" selected pedagogical features to make them truly useful. After writing and rewriting, and reading many thoughtful reviews from instructors, we focused on the following features of this Elementary Algebra text to help students get a better sense of the "big picture" and be more successful in this course:

- The one-unknown, two-unknown approach enables students to gradually master the core concepts of algebra more systematically through a clear, comprehensive presentation of the topics.
- The **innovative *Sullivan/Struve Examples*** and ***Showcase Examples*** provide students with superior guidance and instruction when they need it most—when they are away from the instructor and the classroom.
- The ***Quick Check*** exercises provide students with immediate reinforcement and instant feedback to determine their understanding of the concepts presented in the examples.
- We developed each of the margin features such as ***In Words, Work Smart,*** and ***Work Smart: Study Skills*** with the goals of improving study skills, making the textbook easier to navigate, and increasing student success.
- ***Exercise Sets: Paired with Purpose***—The exercise sets are structured to assess student understanding of vocabulary, concepts, drill, problem solving, and applications. The exercise sets are graded in difficulty level to build confidence and to enhance students' mathematical thinking.
- ***Putting the Concepts Together*** and ***Synthesis Review*** help students see the big picture and provide a structure for learning each new concept and skill in the course.
- The organization of this text provides a distinct transition from skill-based Elementary Algebra to the functions-based Intermediate Algebra approach employed by this author team. In doing so, the Sullivan/Struve algebra texts afford the instructor with the opportunity to minimize overlap between Elementary and Intermediate Algebra courses.

Instructor and Student Resources

The following resources are available to help instructors and students use this text more effectively.

Instructor Resources

Annotated Instructor's Edition (0-13-146811-1)

Instructor Solutions Manual (0-13-146813-8)

Instructor's Resource Manual with Tests (0-13-146814-6)

CD Lecture Series—Lab Pack (0-13-227684-4)

TestGen (0-13-146819-7)

- Enables instructors to build, edit, print, and administer tests.
- Features a computerized bank of questions developed to cover all text objectives.
- Available on dual-platform Windows/Macintosh CD-Rom.

MyMathLab Instructor Version (0-13-1478982)

MyMathLab is a series of text specific, easily customizable, online courses for Prentice Hall textbooks in mathematics and statistics. MyMathLab is powered by Course Compass™—Pearson Education's online teaching and learning environment—and by MathXL®—our online homework, tutorial, and assessment system.

MathXL® Instructor Version (0-13-1478958) www.mathxl.com

MathXL is a powerful online homework, tutorial, and assessment system that accompanies the text. With MathXL, instructors can create, edit, and assign online homework and tests using algorithmically generated exercises correlated to your textbook.

Student Resources

Student Solutions Manual (0-13-146822-7)

- Solutions to the odd-numbered section exercises.
- Solutions to the Quick Check exercises.
- Solutions to the Preparing for This Section, Putting the Concepts Together (mid-chapter review), Chapter Review, Chapter Test, and Cumulative Review exercises.

Prentice Hall Math Tutor Center (0-13-064604-0)

- Staffed by qualified math instructors who provide students with tutoring on examples and odd-numbered exercises from the textbook.
- Tutoring is available via toll-free telephone, toll-free fax, e-mail, or the Internet.
- White board technology allows tutors and students to see problems worked while they "talk" in real time over the Internet during tutoring sessions.

Elementary Algebra Student Study Pack (0-13-204412-9)

The Student Study Pack includes:

- CD Lecture Series
- Student Solutions Manual
- Prentice Hall Math Tutor Center access code

Chapter Test Prep Video CD—Standalone (0-13-134608-3)

- Includes fully worked-out solutions to every problem from each Chapter Test in the text.

MathXL® Tutorial on CD (0-13-219679-4)

- Provides algorithmically generated practice exercises that correlate to exercises at the end of sections.
- Every exercise is accompanied by an example and a guided solution; selected exercises include a video clip.
- The software recognizes student errors and provides feedback. It can also generate printed summaries of students' progress.

Interact Math® Tutorial Web Site www.interactmath.com

Get practice and tutorial help online! This interactive tutorial Web site provides algorithmically generated practice exercises that correlate directly to the exercises in your textbook.

Acknowledgments

Textbooks are written by authors but evolve through the efforts of many people. We would like to extend our thanks to the following individuals for their important contributions to the project. From Prentice Hall: Paul Murphy, who saw the vision of this text from its inception and made it happen; Kate Valentine and Patrice Jones for their innovative marketing ideas; Ann Heath for her dedication, enthusiasm, and attention to detail (quite honestly, Ann was the cement of the project); Chris Hoag for her support and encouragement; Dawn Nuttall for her perseverance, publishing acumen, and attention to detail with the class testing effort; Barbara Mack for her attention to detail throughout production; Jonathan Boylan and the design team for the attractive and functional design; Thomas Benfatti for his attentive eye in overseeing the creation of literally thousands of pieces of art; Maura Zaldivar for coordinating the scheduling of this project with the compositor and the printer; Linda Behrens for her watchful eye and management over countless production details; and finally, the Prentice Hall sales team, for their confidence and support of our books.

We would like to thank two people who contributed greatly to the development of this book. Thanks to Don Gecewicz for his ability to manage and maintain a single voice as well as his talent to make us think about each sentence. Thanks also go to Janet Mazzarella for her perspective, teaching philosophies, and numerous contributions to the text and its supplement package.

We would like to offer special thanks to a number of instructors who helped us with this project, including: Kelly Jade Hammer for help with solutions and to Cindy Trimble and Ondine Parker for their attention to detail in preparing and accuracy checking the answer manuscript and final solutions manuals; Darren Wiberg for his thorough accuracy review of the final manuscript and for lending his teaching style and caring to the CD Lecture Series; Kimberly Neuburger for her participation in the Chapter Prep Test Video CD; and Andreana Grimaldo and Denise Robichaud for their creativity with Chapter Activities. We would also like to thank Rafiq Ladhani, Jon Stockdale, and Fred Landwehr for their dedication to accuracy in checking the art, examples, and answers; and Sarah Streett and Jenny Crawford for attention to detail and consistency in accuracy checking revised text pages and answers.

We offer many thanks to all the instructors from across the country who participated in reviewer conferences and focus groups, and reviewed and/or class-tested some aspect of the manuscript. Their insights and ideas form the backbone of this text. Hundreds of instructors contributed their time, energy, and ideas to help us shape this text. We will attempt to thank them all here. We apologize for any omissions. *Note:* At the time this book went to press, more class tests were being secured. Our thanks also go to those instructors who tested the manuscript with their students after the printing deadline.

Class Testers

Marwan Abu-Sawwa, *Florida Community College—Jacksonville*
Mary Lou Baker, *Columbia State Community College*
Donna Beatty, *Ventura College*
Becky Bradshaw, *Lake Superior College*
Tim Britt, *Jackson State Community College*
Beverly Broomell, *SUNY Suffolk*
Hien Bui, *Hillsborough Community College—Dale Mabry*
Elena Catoiu, *Joliet Junior College*
John Close, *Salt Lake Community College*
Shirley Davis, *South Plains College*
Erica Egizio, *Joliet Junior College*
Sanford Geraci, *Broward Community College*
Susan Grody, *Broward Community College*
Pete Herrera, *Southwestern College*
Becky Hubiak, *Tidewater Community College—Virginia Beach*
Sally Jackman, *Richland College*
Nancy Johnson, *Broward Community College*
Mike Kirby, *Tidewater Community College—Virginia Beach*
Carla Kulinsky, *Salt Lake Community College*
Lynn Marecek, *Santa Ana College*
Janet Mazzarella, *Southwestern College*
Michael McComas, *Marshall University*
Judy Meckley, *Joliet Junior College*
Ron Moore, *Florida Community College—Jacksonville*
Hossein Navid-Tabrizi, *Houston Community College*
Charlotte Newsom, *Tidewater Community College—Virginia Beach*
Charles Odion, *Houston Community College*
Eugenia Peterson, *Daley College*
Elise Price, *Tarrant County Community College*
RB Pruitt, *South Plains College*
William Radulovich, *Florida Community College—Jacksonville*
Pavlov Rameau, *Miami Dade Community College—Wolfson*
Nancy Ressler, *Oakton Community College*
George Rhys, *College of the Canyons*
Togba Sapolucia, *Houston Community College*
Gisela Spieler-Persad, *Rio Hondo Community College*
Patrick Stevens, *Joliet Junior College*
Jennifer Strehler, *Oakton Community College*
Katalin Szucs, *East Carolina University*
Jo Tucker, *Tarrant County Community College*
Richard Watkins, *Tidewater Community College*

Reviewers

Darla Aguilar, *Pima State University*
Grant Alexander, *Joliet Junior College*
Philip Anderson, *South Plains College*
Mary Lou Baker, *Columbia State Community College*
Bill Bales, *Rogers State*
Tony Barcellos, *American River College*
John Beachy, *Northern Illinois University*
David Bell, *Florida Community College—Jacksonville*
Sandy Berry, *Hinds Community College*
Lori Braselton, *Georgia Southern University*
Beverly Broomell, *Suffolk Community College*
Joanne Brunner, *Joliet Junior College*
Connie Buller, *Metropolitan Community College*
Annette Burden, *Youngstown State University*
James Butterbach, *Joliet Junior College*
Marc Campbell, *Daytona Beach Community College*
Elena Catoiu, *Joliet Junior College*
Nancy Chell, *Anne Arundel Community College*
John Close, *Salt Lake Community College*
Bobbi Cook, *Indian River Community College*
Carlos Corona, *San Antonio College*
Faye Dang, *Joliet Junior College*
Vivian Dennis-Monzingo, *Eastfield College*
Alvio Dominguez, *Miami Dade Community College—Wolfson*
Karen Driskell, *South Plains College*
Brenda Dugas, *McNeese State University*
Doug Dunbar, *Okaloosa-Walton Junior College*
Laura Dyer, *Southwestern Illinois State University*
Bill Echols, *Houston Community College—Northwest*
Erica Egizio, *Joliet Junior College*
Jason Eltrevoog, *Joliet Junior College*
Nancy Eschen, *Florida Community College—Jacksonville*
Mike Everett, *Santa Ana College*
Phil Everett, *Ohio State University*
Scott Fallstrom, *Shoreline Community College*
Betsy Farber, *Bucks County Community College*
Fitzroy Farqharson, *Valencia Community College—West*

Dorothy French, *Community College of Philadelphia*
Donna Gerken, *Miami Dade Community College—Kendall*
Adrienne Goldstein, *Miami Dade Community College—Kendall*
Marion Graziano, *Montgomery County Community College*
Susan Grody, *Broward Community College*
Tom Grogan, *Cincinnati State University*
Barbara Grover, *Salt Lake Community College*
Shawna Haider, *Salt Lake Community College*
Margaret Harris, *Milwaukee Area Technical College*
Teresa Hasenauer, *Indian River Community College*
Mary Henderson, *Okaloosa-Walton Junior College*
Celeste Hernandez, *Richland College*
Bob Hervey, *Hillsborough Community College—Dale Mabry*
Teresa Hodge, *Broward Community College*
Sandee House, *Georgia Perimeter College*
Becky Hubiak, *Tidewater Community College—Virginia Beach*
John Jarvis, *Utah Valley State College*
Steven Kahn, *Anne Arundel Community College*
Linda Kass, *Bergen Community College*
Donna Katula, *Joliet Junior College*
Mohammed Kazemi, *University of North Carolina—Charlotte*
Doreen Kelly, *Mesa Community College*
Mike Kirby, *Tidewater Community College—Virginia Beach*
Keith Kuchar, *College of Dupage*
Carla Kulinsky, *Salt Lake Community College*
Julie Labbiento, *Leigh Carbon Community College*
Kathy Lavelle, *Westchester Community College*
Deanna Li, *North Seattle Community College*
Brian Macon, *Valencia Community College—West*
Jim Matovina, *Community College of Southern Nevada*
Jean McArthur, *Joliet Junior College*
Mikal McDowell, *Cedar Valley College*
Lee McEwen, *Ohio State University*
Angela McNulty, *Joliet Junior College*
Debbie McQueen, *Fullerton College*
Judy Meckley, *Joliet Junior College*
Lynette Meslinsky, *Erie Community College—City Campus*
Kausha Miller, *Lexington Community College*
Chris Mizell, *Okaloosa Walton Junior College*
Jim Moore, *Madison Area Technical College*
Ronald Moore, *Florida Community College—Jacksonville*
Elizabeth Morrison, *Valencia Community College—West*
Roya Namavar, *Rogers State*
Hossein Navid-Tabrizi, *Houston Community College*
Carol Nessmith, *Georgia Southern University*
Kim Neuburger, *Portland Community College*
Larry Newberry, *Glendale Community College*
Elsie Newman, *Owens Community College*
Charlotte Newsome, *Tidewater Community College*
Charles Odion, *Houston Community College*
Viann Olson, *Rochester Community and Technical College*
Linda Padilla, *Joliet Junior College*
Carol Perry, *Marshall Community and Technical College*
Faith Peters, *Miami Dade Community College—Wolfson*
Philip Pina, *Florida Atlantic University*
Carol Poos, *Southwestern Illinois University*
William Radulovich, *Florida Community College—Jacksonville*
David Ray, *University of Tennessee—Martin*
Michael Reynolds, *Valencia Community College—West*
George Rhys, *College of the Canyons*
Jorge Romero, *Hillsborough Community College—Dale Mabry*
David Ruffato, *Joliet Junior College*
Carol Rychly, *Augusta State University*
David Santos, *Community College of Philadelphia*
Doug Smith, *Tarrant Community College*
Gisela Spieler-Persad, *Rio Hondo Community College*
Raju Sriram, *Okaloosa-Walton Junior College*
Patrick Stevens, *Joliet Junior College*
Bryan Stewart, *Tarrant Community College*
Elizabeth Suco, *Miami Dade Community College—Wolfson*
Katalin Szucs, *East Carolina University*
KD Taylor, *Utah Valley State College*
Suzanne Topp, *Salt Lake Community College*
Suzanne Trabucco, *Nassau Community College*
Jo Tucker, *Tarrant Community College*
Bob Tuskey, *Joliet Junior College*
Mary Vachon, *San Joaquin Delta College*
Carol Walker, *Hinds Community College*
Kim Ward, *Eastern Connecticut State University*
Natalie Weaver, *Daytona Beach Community College*
Darren Wiberg, *Utah Valley State College*
Rachel Wieland, *Bergen Community College*
Christine Wilson, *Western Virginia University*
Brad Wind, *Miami Dade Community College—North*
Roberta Yellott, *McNeese State University*
Steve Zuro, *Joliet Junior College*

Additional Acknowledgments

We also would like to extend thanks to our colleagues at Joliet Junior College and Columbus State Community College, who provided encouragement, support, and the teaching environment where the ideas and teaching philosophies in this text were developed.

Michael Sullivan, III
Katherine R. Struve

CHAPTER

1 Operations on Real Numbers and Algebraic Expressions

Did you know that the arrangement of seeds in a sunflower can be described using a sequence of numbers called the Fibonacci sequence? See Problem 121 in Section 1.3.

OUTLINE

The Big Picture: Putting It Together

Welcome to Elementary Algebra! This course is taken by a diverse group of individuals. Some of you may never have taken an algebra course, while others may have taken algebra at some time in the past. In any case, we have written this text with both groups in mind.

The first chapter of the text serves as a review of the topics we would learn in an arithmetic course. The material is presented with an eye on the future, which is algebra. This means that we will slowly build our discussion so that the shift from arithmetic to algebra is painless. Take care to study the methods used in this section because these same methods will be used again in later chapters.

1.1 Success in Mathematics

OBJECTIVES

1. What to Do the First Week of the Semester
2. What to Do Before, During, and After Class
3. How to Use the Text Effectively
4. How to Prepare for an Exam

Let's start by having a frank discussion about the "big picture" goals of the course and how this book can help you to be successful at mathematics. The first "big picture" goal of the class is to develop algebraic skills and gain an appreciation for the power of algebra and mathematics. But there is also a second "big picture" goal. By studying mathematics, we develop our sense of logic and exercise the part of our brain that deals with logical thinking. The examples and problems that appear throughout the text are like the crunches that we do in a gym to exercise our body. The goal of running or walking is to get from point A to point B, so doing fifty crunches on a mat does not accomplish this goal, but crunches do make our upper bodies, backs, and heart stronger when we need to run or walk.

In Words
Doing crunches doesn't solve the problem of running a race, but they are truly effective at conditioning your body.

Logical thinking can assist us in solving difficult everyday problems, so solving algebra problems "builds the muscles" in the part of our brain that performs logical thinking. So, when you are studying algebra, and getting frustrated with the amount of work that needs to be done, and you say, "My brain hurts," remember the phrase that we all use in the gym, "No pain, no gain."

Another phrase to keep in mind is, "Success Breeds Success." Mathematics is everywhere. You already are successful at doing some everyday mathematics. With practice, you can take your initial successes and become even more successful. Have you ever done any of the following everyday activities?

- Compare the price per ounce of different sizes of jars of peanut butter or jam.
- Leave a tip at a restaurant.
- Figure out how many calories your bowl of breakfast cereal gives you.
- Take an opinion survey along with many other people.
- Measure the distances between cities as you plan your summer vacation.
- Order the appropriate number of gallons of paint to cover the walls of a room that you are renovating.
- Buy a car and take out a car loan with interest.
- Double a cookie recipe.
- Change American dollars for Canadian dollars.
- Fill up a basketball or soccer ball with air (balls are spheres, after all).
- Coach a Little League team (scores, statistics, catching, and throwing all involve math).
- Check the percentages of saturated and unsaturated fats in a chocolate bar.

We just listed twelve of the many everyday mathematical activities, and you may do five or ten in a single day! The everyday mathematics that you already know is the foundation for your success in this course.

1 What to Do the First Week of the Semester

You have enrolled in an Elementary Algebra course. The first week of the semester gives you the opportunity to prepare for a successful course. Here are the things that you should do:

1. **Pick a good seat.** As you enter the classroom for the first time, choose a seat that gives you a good view of the room. Sit close enough to the front so that you can easily see the board and hear the professor.
2. **Read the syllabus to learn about your instructor and the course.** Be sure to take note of your instructor's name, office location, e-mail address, telephone number, and office hours. Also, pay attention to any additional help that can be found on campus such as tutoring centers, videos in the library, software, on-line tutorials, and so on. Make sure that you fully understand all of the instructor's policies for

the class. This includes the policy on absences, missed exams or quizzes, and homework. Ask questions.

3. **Learn the names of some of your classmates and exchange contact information.** One of the best ways to learn math is through group study sessions. Try to create time each week to study with your classmates. Knowing how to get in contact with classmates is also useful if you ever miss class because you can obtain the assignment for the day.
4. **Budget your time.** Most students have a tendency to "bite off more than they can chew." To help with time management, consider the following general rule for studying mathematics: You should plan on studying *at least* two hours outside of class for each hour in class. So, if you enrolled in a four-hour math class, you should set aside at least eight hours each week to study for the course. If this is not your only course, you will have to set aside time for other courses as well. Consider your work schedule and personal life when creating your budget as well.

Work Smart: Study Skills
Plan on studying two hours outside of class for each hour in class every week.

2 What to Do Before, During, and After Class

Now that the semester is underway, we present the following ideas for what to do before, during, and after each class meeting. While these suggestions may sound overwhelming, we guarantee that by following them, you will be successful in mathematics (and other courses). Also, you will find that studying for exams becomes much easier by following this plan.

Work Smart: Study Skills
Take a few minutes to plan out the next academic term.

Before Class Begins

1. Make sure you are mentally prepared for class. This means that your mind should be alert and ready to concentrate for the entire class period. (Invest in a cup of coffee and eat lots of protein for breakfast!)
2. Read the section or sections that will be covered in the upcoming class meeting.
3. Based upon your reading, prepare a list of questions. Jot them down. In many cases, your questions will be answered through the lecture. You can then ask any questions that are not answered completely.

During Class

1. Arrive early enough to prepare your mind and material for the lecture.
2. Stay alert. Do not doze off or daydream during class. It will be very difficult to understand the lecture when you "return to class."
3. Take thorough notes. It is normal not to get certain topics the first time that you hear them through the lecture. However, this does not mean that you throw your hands up in despair. Rather, continue to write your class notes.
4. You can ask questions when appropriate. Do not be afraid to ask questions. In fact, instructors love when students ask questions, for two reasons. First, we know as teachers that if one student has a question, there are many more in class with the same question. Second, by asking questions, you are teaching the teacher what topics cause difficulty.

Work Smart: Study Skills
Be sure to ask questions during class.

After Class

1. Reread (and possibly rewrite) your class notes. In our experience as students, we were amazed how often confusion that existed during class went away after studying our in-class notes later when we had more time to absorb the material.
2. Reread the section. This is an especially important step. Once you have heard the lecture, the section will make more sense and you will understand much more.

Work Smart: Study Skills
The reason for homework is to build your skill and confidence. Don't skip assignments.

3. Do your homework. **Homework is not optional.** There is an old Chinese proverb that says,

   I hear . . . and I forget
   I see . . . and I remember
   I do . . . and I understand

   This proverb applies to any situation in life in which you want to succeed. Would a pianist expect to be the best if she didn't practice? The only way you are going to learn algebra is by doing algebra. Remember: Success breeds success.
4. And don't forget, when you get a problem wrong, try to figure out why you got the problem wrong. If you can't discover your error, be sure to ask for help.
5. If you have questions, visit your professor during office hours. You can also ask someone in your study group or go to the tutoring center on campus, if available.

Math Courses: No Brain Freezes Here!

Learning algebra is a building process. Learning is the art of making connections between thousands of neurons (specialized cells) in the brain. Memory is the ability to reactivate these neural networks—it is a conversation among neurons.

Math isn't a mystery. You already know some math. But you do have to practice what you know and expand your knowledge. Why? The brain contains thousands of neurons. Through repeated practice, a special coating forms that allows the signals to travel faster and reduces interference. The cells "fire" more quickly and connections are made faster and with less effort. Practice forms the pathways that allow us to retrieve concepts and facts at test time. Remember those crunches, which are a way of making your body more robust and nimble—learning does the same to your brain.

Have We Mentioned Asking Questions?

To move information from short-term memory to long-term memory, we need to think about the information, comprehend its meaning, and ask questions about it.

3 How to Use the Text Effectively

When we sat down to write this text, we knew based upon experience from teaching our own students that students typically do not read their mathematics text. Rather than saying to you, "Ah, but our book is different—it can be read!," we decided to accept how students study math.

Students usually go through the following steps:

1. Attend the lecture and watch the instructor do some problems on the board. Perhaps work some problems in class.
2. Go home and work on the homework assignment.
3. After each problem, check the answer in the back of the text. If right, move on, but if wrong, go back and see where the solution went wrong.
4. Maybe, the mistake can be identified, but if not, go to the class notes or try to find a similar example in the text. With a little luck, a similar example can be found and the student can determine where the solution went wrong in the problem.
5. If not, mark the problem and ask about it in the next class meeting, which leads us back to step 1.

Work Smart: Study Skills
Learn what the different features of this book are designed to do. Decide which ones you may need the most.

So with this model in mind, we started to develop this text so that there is more than one way to extract the information you need from it.

All of the features have been included in the text to help you succeed. We list each feature in the order they appear and briefly explain its purpose and how it can be used to help you succeed in this course:

Preparing for This Section: Warming Up

Immediately after the title of the section, each section (after Section 1.1) begins with a short "readiness quiz." The readiness quiz asks questions about material that was presented earlier in the course that is needed for the upcoming section. You should take the readiness quiz to be sure that you understand the material that the new section will be based on. Answers to the readiness quiz appear as footnotes on the page of the quiz. Check your answers. If you get a problem wrong, or don't know how to do a problem, go back to the section listed and review the material.

Objectives: A "Road Map" through the Course

To the left of the readiness quiz, we present a list of objectives to be covered in the section. If you follow the objectives, you will get a good idea of the section's "big picture"—the important concepts, techniques, and procedures. Once you have completed your homework for the section, you should be able to answer "yes" to the statement, "You should be able to . . . " for each objective.

The objectives are numbered. (See the numbered headline at the beginning of this section.) When we begin discussing a particular objective, the number appears in the left-hand column of the text.

Examples: Where to Look for Information

You look to examples to provide you with guidance and instruction when you need it most—when you are away from the instructor and the classroom. With this in mind, we have developed two example formats.

Step-by-Step Examples have a three-column format where the left column describes a step, the middle column provides a brief explanation of the step, and the right column presents the algebra. With this format, the left and middle columns can be thought of as your instructor's voice during a lecture. *Step-by-Step Examples* are used to introduce key topics or important problem-solving strategies. They are meant to provide easy-to-understand, practical instructions by including the words "how to" in the examples' headline.

Annotated Examples have a two-column format in which explanations are provided to the left of the algebra. Because we read from left to right, placing the explanation on the left clearly describes what we are about to do in the order that we will do it. Again, the annotations can be thought of as your instructor's voice right before he or she writes the solution on the board.

Quick Check: Practice for the Examples

After a concept has been presented, we will provide between 1 and 4 (occasionally up to 8 for some concepts) problems to solve. The answers to these problems are provided in the back of the book, so that you can quickly check your answers. This feature allows you to get immediate feedback about whether or not you understand the concept presented. The Quick Checks usually follow each example in the text. This feature came from our belief that students use the examples as a template for solving problems, so we decided to place the Quick Checks just after an example. If you can do the Quick Check problems, you will be prepared for the end-of-section exercises, which are a comprehensive review of the material in the section.

In Words: Math in Everyday Language

Have you ever been given a math definition in class and said, "What in the world does that mean?" As teachers, we have heard that from our students. So we added the "In Words" feature, which takes definitions that are in their mathematical form and restates them in everyday language. These boxes will help you to understand the language of mathematics better.

Work Smart

These are "tricks of the trade" that can be used to help you solve problems. They also show alternative approaches to solving problems. Yes, there is more than one way to solve a math problem!

Work Smart—Study Skills

Working smart also means studying smart. We provide tips throughout the text to help you understand the study skills required for success in this and other mathematics courses.

Chapter Review

The chapter review is arranged section by section. For each section, we list key concepts, key terms, and objectives. For each objective, we provide the examples from the text, along with page references, that illustrate the objective. Also, for each objective, we list the problems in the review exercises that test your understanding. If you get a problem wrong, use this feature to help you to identify how to work the problem.

Chapter Test

We have included a chapter test. Once you think that you are prepared for the exam, take the chapter test. If you do well on the chapter test, chances are you will do well on your in-class exam. Be sure to take the chapter test under the conditions that you will face in class. If you are unsure how to solve a problem in the chapter test, watch the Chapter Test Prep Video CD, which is a video of an instructor solving each problem in the chapter test.

Cumulative Review: Reinforcing Your Knowledge

As we mentioned, learning algebra is a building process. Building involves a lot of reinforcement. To do so, we provide cumulative reviews at the end of every odd-numbered chapter starting with Chapter 3. Do these cumulative reviews after each chapter test, so that you are always refreshing your memory—making those neurons do their calisthenics. This way, studying for the final exam should be fairly easy.

4 How to Prepare for an Exam

The following steps are time-tested suggestions to help you prepare for an exam.

Step 1: Revisit your homework and the chapter review problems: Beginning about one week before your exam, start to redo your homework assignments. If you don't understand a topic, be sure to seek out help. You should also work the problems given in the chapter review. The problems are keyed to the objectives in the course. If you get a problem wrong, identify the objective and examples that illustrate the objective. Then review this material and try the problem in the chapter review again. If you get the problem wrong again, seek out help.

Step 2: Test yourself: A day or two before the exam, take the chapter test under test conditions. Be sure to check your answers. If you got any problems wrong, determine why you got them wrong and remedy the situation. Don't forget about the Chapter Test Prep Video CD, which is a video of an instructor solving each problem in the chapter test.

Work Smart: Study Skills
Do not "cram" for an exam by pulling an "all nighter."

Step 3: Follow these rules as you train: Be sure to arrive early at the location of the exam. Prepare your mind for the exam. Also, be sure that you are well rested. Don't try to pull "all-nighters." If you need to study all night long for an exam, then your time management is poor and you should rethink how you are using your time or whether you have enough time set aside for the course.

Step 4: **Relax, and work on the exam thoughtfully:** While taking the exam, be sure to read the instructions. Show all your work, and be neat so that your instructor can follow your work and find your solution. Also, taking an exam is not a race, so there is no reason to turn your exam in early. If you finish early, go over each problem and check your answers.

1.1 Exercises

Answers will vary.

1. Why do you want to be successful in mathematics? Are your goals positive or negative? If you stated your goal negatively ("Just get me out of this course!"), can you restate it positively?

2. Name three activities in your daily life that involve the use of math (for instance, scoring a game of poker, bridge, or pinochle, operating your computer, or reading a credit-card bill).

3. What is your instructor's name?

4. What are your instructor's office hours? Where is your instructor's office?

5. Does your instructor have an e-mail address? If so, what is it?

6. Does your class have a Web site? Do you know how to access it? What information is located on the Web site?

7. Are there tutors available for this course? If so, where are they located? When are they available?

8. Name two other students in your class. What is their contact information? When can you meet with them to study?

9. List some of the things that you should do before class begins.

10. List some of the things that you should do during class.

11. List some of the things that you should do after class.

12. What is the point of the Chinese proverb on page 4?

13. What is the "readiness quiz"? How should it be used?

14. Name three features that appear in the margins. What is the purpose of each of them?

15. How should the chapter review material be used?

16. How should the chapter test be used?

17. How should the cumulative review be used?

18. List the four steps that should be followed when preparing for an exam.

19. Use the chart below to help manage your time. Be sure to fill in time allocated to various activities in your life including school, work and leisure.

	Monday	Tuesday	Wednesday	Thursday	Friday	Saturday	Sunday
7 A.M.							
8 A.M.							
9 A.M.							
10 A.M.							
11 A.M.							
Noon							
1 P.M.							
2 P.M.							
3 P.M.							
4 P.M.							
5 P.M.							
6 P.M.							
7 P.M.							
8 P.M.							
9 P.M.							

20. What is the Chapter Test Prep Video CD?

1.2 The Number Systems and the Real Number Line

OBJECTIVES

1. Classify Numbers
2. Plot Points on the Real Number Line
3. Use Inequalities to Order Real Numbers
4. Compute the Absolute Value of a Real Number

Preparing for the Number Systems and the Real Number Line

Before getting started, take the following readiness quiz. If you get a problem wrong, go to the section cited and review the material.

1. Write $\frac{5}{8}$ as a decimal. [Appendix, Section A.2, pp. A9–A10]

2. Write $\frac{9}{11}$ as a decimal. [Appendix, Section A.2, pp. A9–A10]

The goal of this section is to discuss the *real number system.* We use the real numbers every day in our lives, so it is an idea that you are already familiar with. In short, real numbers are numbers that we use to count or measure things: For instance, there might be 25 students in your class. Your car might get 18.4 miles per gallon. You might have a $130 debt.

As we proceed through the course, we will be dealing with various types of numbers. The kinds of numbers that we deal with are organized in *sets.* A **set** is a collection of objects. For example, we can identify the students enrolled in Elementary Algebra at your college as a set. The collection of numbers 0, 1, 2, 3, 4, 5, 6, 7, 8, and 9 may also be identified as a set. If we let A represent this set of numbers, then we can write

$$A = \{0, 1, 2, 3, 4, 5, 6, 7, 8, 9\}$$

In this notation, braces { } are used to enclose the objects, or **elements,** in the set. When a set has no elements in it, we say that the set is an **empty set.** Empty sets are denoted by the symbol ∅ or { }.

Work Smart

The use of the word "real" to describe numbers leads us to question "Are there 'nonreal' numbers?" The answer is "yes." We use the word "imaginary" to describe nonreal numbers. Imaginary does not mean that these numbers are made up, however. Imaginary numbers have many interesting applications in areas such as biology and the development of high-definition antennas. We will discuss imaginary numbers in Section 7.5. For most of the text, we concentrate on real numbers.

Preparing for...Answers **1.** 0.625 **2.** 0.8181... or $0.\overline{81}$

Classroom Example ➔
Write the set that represents the vowels in the word "mathematics."

Answer:

$\{a, e, i\}$

EXAMPLE 1 Writing a Set

Write the set that represents the vowels.

Solution

The vowels are a, e, i, o, and u. If we let V represent this set, then

$$V = \{a, e, i, o, u\}$$

QUICK ✓

1. Write the set that represents the first 4 positive, odd numbers.
2. Write the set that represents the states that begin with the letter A.
3. Write the set that represents the states that begin with the letter Z.

1 Classify Numbers

Teaching Tip
The Quick Check exercises are designed to help get students "into" the book. They can be assigned as homework so that students know exactly what example can be used as a reference. They also can be used for classroom exercises.

We will develop the real number system by looking at the history of numbers. The first types of numbers that humans worked with are called the *natural numbers* or *counting numbers.*

DEFINITION

The **natural numbers,** or **counting numbers,** are the numbers in the set $\{1, 2, 3, \dots\}$. The three dots, called *ellipsis,* indicate that the pattern continues indefinitely.

Figure 1
The natural numbers.

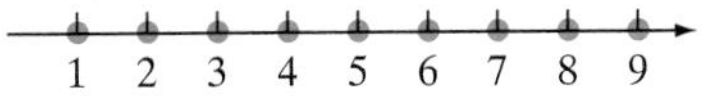

1 2 3 4 5 6 7 8 9

As their name implies, the counting numbers are often used to count things. For example, we can count the number of cars that arrive at a Wendy's drive-thru between 12 noon and 1:00 P.M. We can represent the counting numbers graphically using a number line. See Figure 1. The arrow on the right is used to indicate the direction in which the numbers increase.

Since we do not count the number of cars waiting in the drive-thru by saying, "zero, one, two, three . . . ," zero is not a natural, or counting, number. When we add the number 0 to the set of counting numbers, we get the set of *whole numbers.*

DEFINITION

The **whole numbers** are the numbers in the set $\{0, 1, 2, 3, \dots\}$.

Figure 2
The whole numbers.

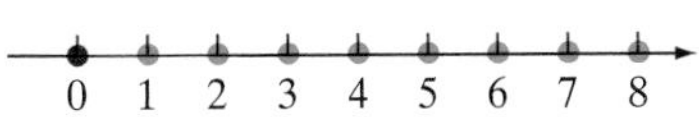

Figure 2 represents the whole numbers on the number line. Notice that the set of natural numbers, $\{1, 2, 3, \dots\}$ is included in the set of whole numbers.

By expanding the numbers to the left of zero on the number line, we have the set called *integers.*

DEFINITION

The **integers** are the numbers in the set $\{\dots, -3, -2, -1, 0, 1, 2, 3, \dots\}$.

Figure 3
The integers.

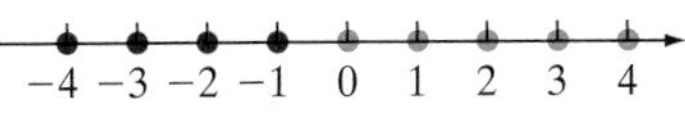

Figure 3 represents the integers on the number line. Notice that the whole numbers and natural numbers are included in the set of integers.

Integers are useful in many situations. For example, we could not discuss temperatures above 0°F (positive counting numbers) and temperatures below 0°F (negative counting numbers) without integers. A debt of 300 dollars can be represented as an integer as -300 dollars.

Work Smart: Study Skills
Do you need a refresher on fractions, decimals, or percents? If so, go to Appendix A and review your skills.

How can you represent a part of a whole? For example, how do you represent a part of last night's leftover pizza or part of a dollar? To address this problem, we enlarge our number system to include *rational numbers*.

DEFINITION

A **rational number** is a number that can be expressed as a fraction (or quotient) of two integers. That is, a rational number is a number that can be written in the form $\frac{p}{q}$, where p and q are integers. However, q cannot equal zero.

Work Smart
Remember, all integers are also rational numbers. For example $42 = \frac{42}{1}$.

Examples of rational numbers are $\frac{2}{5}, \frac{5}{2}, \frac{0}{8}, -\frac{7}{9}$, and $\frac{31}{4}$. Because $\frac{p}{1} = p$ for any integer p, it follows that all integers are also rational numbers. For example, 7 is an integer, but it is also a rational number because it can be written as $\frac{7}{1}$. We illustrate this idea below.

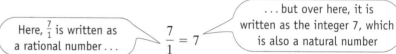

Here, $\frac{7}{1}$ is written as a rational number . . . $\frac{7}{1} = 7$ . . . but over here, it is written as the integer 7, which is also a natural number

Teaching Tip
Remind students that a decimal that terminates ends. A repeating decimal is one that repeats in a pattern.

In addition to representing rational numbers as fractions, we can also represent rational numbers in decimal form as either a repeating decimal or a terminating decimal. Table 1 shows various rational numbers in fraction form and decimal form.

Table 1

Fraction Form of Rational Number	Decimal Form of Rational Number	Terminating or Repeating Decimal
$\frac{7}{2}$	3.5	Terminating
$\frac{5}{11}$	$0.454545\ldots = 0.\overline{45}$	Repeating
$-\frac{3}{8}$	-0.375	Terminating
$-\frac{23}{21}$	$-1.095238095238\ldots = -1.\overline{095238}$	Repeating

For example, the repeating decimal $0.\overline{3}$ and the terminating decimal 0.27 are rational numbers because they can be converted to fractions (see Section A.2). The repeating decimal $0.\overline{3}$ is equivalent to the fraction $\frac{1}{3}$. The terminating decimal 0.27 is equivalent to the fraction $\frac{27}{100}$.

Decimals that neither terminate nor repeat are called irrational numbers. This means that irrational numbers cannot be written as the quotient of two integers. An example of an irrational number is the symbol π, whose value is approximately 3.141592. Another example of an irrational number is the symbol $\sqrt{2}$, whose value is approximately 1.41421.

Teaching Tip
Some students may have heard the expression "nonrepeating, nonterminating decimal" applied to irrational numbers.

DEFINITION

An **irrational number** is a number that cannot be written as the quotient of two integers. That is, it cannot be expressed as a terminating decimal or a repeating decimal.

Now that we have defined the set of rational numbers and the set of irrational numbers, we are ready for a formal definition of the set of *real numbers.*

DEFINITION
The set of rational numbers combined with the set of irrational numbers is called the set of **real numbers.**

Figure 4 shows the relationship among the various types of numbers. Notice the oval that represents the whole numbers surrounds the oval that represents the natural numbers. This means the whole numbers include all the natural numbers.

Figure 4

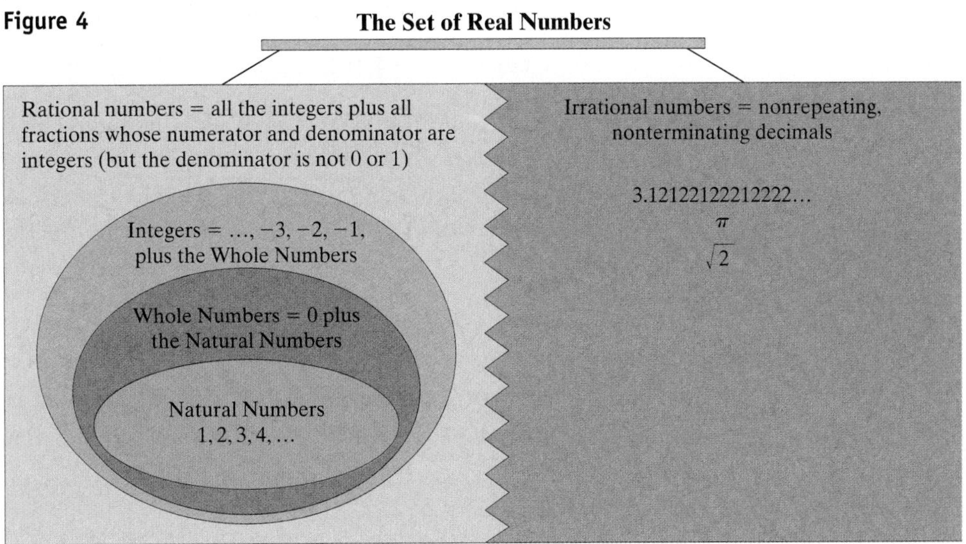
The Set of Real Numbers
Rational numbers = all the integers plus all fractions whose numerator and denominator are integers (but the denominator is not 0 or 1)
Integers = ..., −3, −2, −1, plus the Whole Numbers
Whole Numbers = 0 plus the Natural Numbers
Natural Numbers
1, 2, 3, 4, ...
Irrational numbers = nonrepeating, nonterminating decimals
3.12122122212222...
π
√2

Summary of the Set of Real Numbers

Classification	Elements of the Set	Description
Natural numbers (also called counting numbers)	$\{1, 2, 3, \ldots\}$	
Whole numbers	$\{0, 1, 2, 3, \ldots\}$	Natural numbers and 0
Integers	$\{\ldots, -3, -2, -1, 0, 1, 2, 3, \ldots\}$	Positive and negative natural numbers and 0
Rational numbers	A number that can be written in the form $\frac{p}{q}$ where p and q are integers. However, q cannot equal zero. Examples: $3, 7, -11, -\frac{4}{5}, 7.\overline{16}, 4.125$	Any number that can be written as the quotient of two integers with the denominator not equal to 0 Terminating decimals or nonterminating, repeating decimals
Irrational numbers	Numbers that **cannot** be written as the quotient of two integers with a nonzero denominator Examples: 1.1121231234 . . . , 3.14159 . . . , 1.8975314253648607 . . .	Nonrepeating, nonterminating decimals
Real numbers	Set of rational numbers combined with the set of irrational numbers.	Rational numbers (any number that can be written as the quotient of two integers with the denominator not equal to 0; terminating decimals or nonterminating, repeating decimals) and Irrational numbers (nonrepeating, nonterminating decimals.)

Classroom Example Use the directions in Example 2 with the set

$\left\{7, -\frac{4}{9}, -2, 0, -5.131131113\ldots, 7.\overline{538}, 26.6868\ldots\right\}$

Answer:

(a) 7

(b) 0, 7

(c) −2, 0, 7

(d) $7, -\frac{4}{9}, -2, 0, 7.\overline{538}, 26.6868\ldots$

(e) −5.131131113...

(f) all the numbers are real

EXAMPLE 2 Classifying Numbers in a Set

List the numbers in the set

$$\left\{9, -\frac{2}{7}, -4, 0, -4.010010001\ldots, 3.\overline{632}, 18.3737\ldots\right\}$$

that are

(a) Natural numbers **(b)** Whole numbers **(c)** Integers
(d) Rational numbers **(e)** Irrational numbers **(f)** Real numbers

Solution

(a) 9 is the only natural number

(b) 0 and 9 are the whole numbers

(c) 9, −4, and 0 are the integers

(d) $9, -\frac{2}{7}, -4, 0, 3.\overline{632}$, and $18.3737\ldots$ are the rational numbers.

(e) $-4.010010001\ldots$ is the only irrational number because the decimal does not repeat, nor does it terminate.

(f) All the numbers listed are real numbers. Real numbers consist of rational numbers together with irrational numbers.

Teaching Tip Differentiating between categories of real numbers is difficult for students. You may want to use a chart to check off the category into which a number falls. See Problems 55–62 in Exercise Set 1.2.

QUICK ✓ *List the numbers in the set* $\left\{\frac{11}{5}, -5, 12, 2.\overline{76}, 0, \pi, \frac{18}{4}\right\}$ *that are*

4. Natural numbers **5.** Whole numbers **6.** Integers
7. Rational numbers **8.** Irrational numbers **9.** Real numbers

2 Plot Points on the Real Number Line

In Words We can think of the real number line as the graph of the set of all real numbers.

Look back at Figure 3. Notice that there are gaps between the integers plotted on the number line. These gaps are filled in with the real numbers that are not integers.

To construct a **real number line,** pick a point on a line somewhere in the center, and label it *O*. This point, called the **origin,** corresponds to the real number 0. See Figure 5.

Figure 5
The real number line.

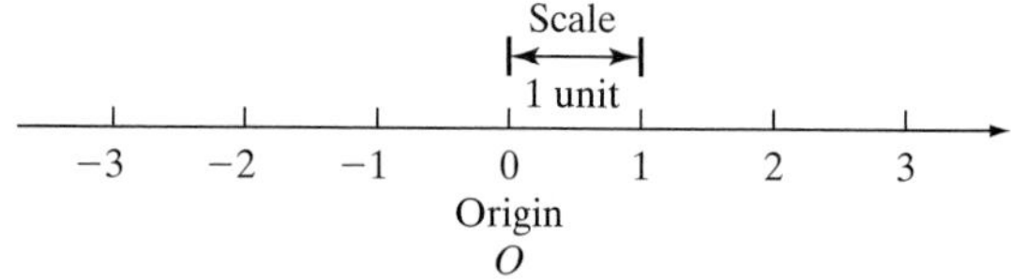

The point 1 unit to the right of *O* corresponds to the real number 1. The distance between 0 and 1 determines the **scale** of the number line. For example, the point associated with the number 2 is twice as far from *O* as 1 is. Notice that an arrowhead on the right end of the line indicates the direction in which the numbers increase. Points to the left of the origin correspond to the real numbers −1, −2, and so on.

DEFINITION

The real number associated with a point *P* is called the **coordinate** of *P*.

Classroom Example ➤
On the real number line, label the points with coordinates 0, 5, −3, 1.5 and $-\frac{5}{2}$.

Answer:

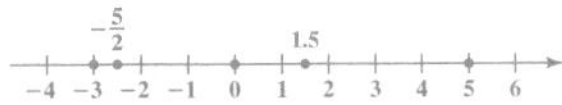

EXAMPLE 3 Plotting Points on the Real Number Line

On the real number line, label the points with coordinates 0, 6, −2, 2.5, $-\frac{1}{2}$.

Solution

We draw a real number line and then plot the points. See Figure 6. Notice that 2.5 is midway between 2 and 3. Also notice that $-\frac{1}{2}$ is midway between −1 and 0.

Figure 6

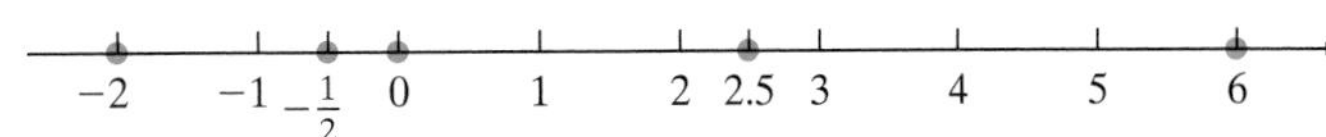

QUICK ✓

10. On the real number line, label the points with coordinates 0, 3, −2, $\frac{1}{2}$, and 3.5.

The real number line consists of three classes (or categories) of real numbers, as shown in Figure 7.

Figure 7

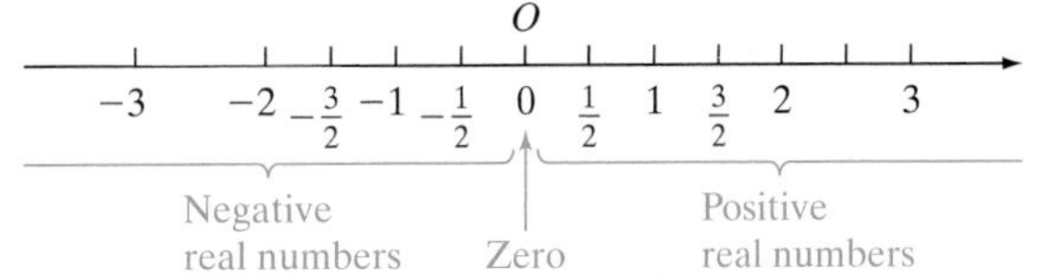

- The **negative real numbers** are the coordinates of points to the left of the origin, O.
- The real number **zero** is the coordinate of the origin denoted O.
- The **positive real numbers** are the coordinates of points to the right of the origin, O.

The **sign** of a number refers to whether the number is a positive real number or a negative real number. For example, the sign of −4 is negative while the sign of 100 is positive.

3 Use Inequalities to Order Real Numbers

An important property of the real number line follows from the fact that given two numbers (points) a and b, either a is to the left of b, denoted $a < b$, a is the same as b, denoted $a = b$, or a is to the right of b, denoted $a > b$. See Figure 8.

If a is either less than or equal to b, we write $a \leq b$. Similarly, $a \geq b$ means that a is either greater than or equal to b. Collectively, the symbols $<, >, \leq,$ and $\geq$ are called **inequality symbols.** The "arrowhead" in an inequality always points to the smaller number. For $3 < 5$, the "arrowhead" points to the 3.

Note that $a < b$ and $b > a$ mean the same thing. For example, it does not matter whether we write $2 < 3$ (2 is to the left of 3) or $3 > 2$ (3 is to the right of 2).

Figure 8

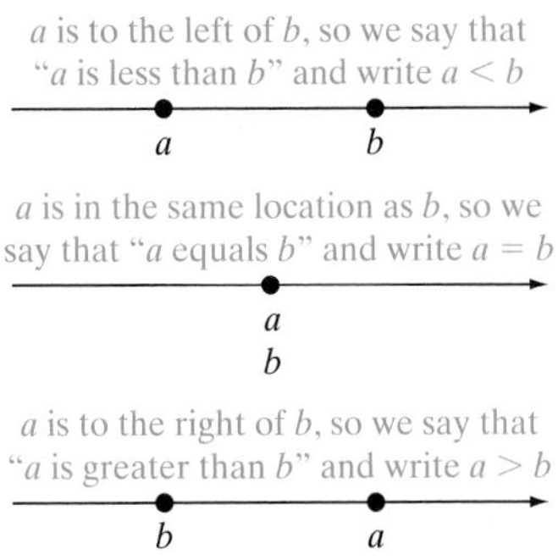

a is to the left of b, so we say that "a is less than b" and write $a < b$

a b

a is in the same location as b, so we say that "a equals b" and write $a = b$

a
b

a is to the right of b, so we say that "a is greater than b" and write $a > b$

b a

Teaching Tip
Point out that when a is to the left of b, then a is less than b, or $a < b$. When a is to the right of b, $a > b$.

Classroom Example

(a) \$2 is less than \$8, so $2 < 8$.

(b) Owing your friend \$10 is not as bad as owing your friend \$20, so $-10 > -20$.

(c) $1.5 \le \frac{3}{2}$ because $1.5 = \frac{3}{2}$.

(d) $\frac{7}{12} < \frac{5}{8}$ because $\frac{14}{24} < \frac{15}{24}$. 14 parts out of 24 is less than 15 parts out of 24.

Work Smart

Write fractions with a common denominator or change fractions to decimals to compare the location of the numbers on the number line.

EXAMPLE 4 Using Inequality Symbols

(a) We know that having \$3 is less than having \$7 or having 3 apples is fewer than having 7 apples. Using the real number line, we say $3 < 7$ because the coordinate 3 lies to the left of the coordinate 7 on the real number line.

(b) Being \$2 in debt is not as bad as being \$5 in debt, so $-2 > -5$. Using the real number line, $-2 > -5$ because the coordinate -2 lies to the right of the coordinate -5 on the real number line.

(c) $2.7 > \frac{5}{2}$ because $\frac{5}{2} = 2.5$ and $2.7 > 2.5$.

(d) $\frac{5}{6} > \frac{4}{5}$ because $\frac{5}{6} = \frac{25}{30}$ and $\frac{4}{5} = \frac{24}{30}$. Having 25 parts out of 30 is more than having 24 parts out of 30. We could also write $\frac{5}{6} = 0.8\overline{3}$ and $\frac{4}{5} = 0.8$. Because $0.8\overline{3}$ is greater than 0.8, the coordinate $\frac{5}{6}$ lies to the right of the coordinate $\frac{4}{5}$ on the real number line.

QUICK ✓ *Replace the question mark by <, >, or =, whichever is correct.*

11. 2 ? 9 **12.** -5 ? -3 **13.** $\frac{4}{5}$? $\frac{1}{2}$

14. $\frac{4}{7}$? 0.5 **15.** $\frac{4}{3}$? $\frac{20}{15}$

If you look carefully at the results of Example 4, you should notice that the direction of the inequality symbol always points to the smaller number.

Based upon the discussion so far, we conclude that

$a > 0$	is equivalent to	a is positive
$a < 0$	is equivalent to	a is negative

We sometimes read $a > 0$ by saying that "a is positive." If $a \ge 0$, then either $a > 0$ or $a = 0$, and we may read this as "a is nonnegative" or "a is greater than or equal to zero."

4 Compute the Absolute Value of a Real Number

The real number line can be used to describe the concept of *absolute value*.

In Words

Think of absolute value as the number of units you must count to get from 0 to a number. The absolute value of a number can never be negative because it represents a distance.

DEFINITION

The **absolute value** of a number a, written $|a|$, is the distance from 0 to a on the real number line.

For example, because the distance from 0 to 3 on the real number line is 3, the absolute value of 3, $|3|$, is 3. Because the distance from 0 to -3 on the real number line is 3, the absolute value of -3, $|-3|$, is 3. See Figure 9.

Teaching Tip

When we say that absolute value cannot be negative, remind students that it may be 0.

Figure 9

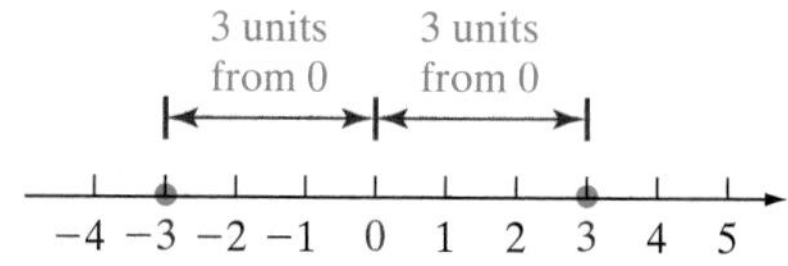

Classroom Example
Evaluate each of the following:
(a) $|-5|$ (b) $|-2.8|$
(c) $|7|$ (d) $|0|$

Answer:
(a) 5 (b) 2.8
(c) 7 (d) 0

EXAMPLE 5 Computing Absolute Value

Evaluate each of the following:

(a) $|-7|$ **(b)** $|-1.5|$ **(c)** $|6|$ **(d)** $|0|$

Solution

(a) $|-7| = 7$ because the distance from 0 to -7 on the real number line is 7.

(b) $|-1.5| = 1.5$ because the distance from 0 to -1.5 on the real number line is 1.5.

(c) $|6| = 6$ because the distance from 0 to 6 on the real number line is 6.

(d) $|0| = 0$ because the distance from 0 to 0 on the real number line is 0.

QUICK ✓ *Evaluate each expression.*

16. $|-15|$ **17.** $\left|-\frac{3}{4}\right|$

1.2 Exercises

For Extra Help: Student Solutions Manual, CD Video, PH Math/Tutor Center, MathXL Tutorials on CD, MathXL®, MyMathLab

Concepts and Vocabulary

In Problems 1–3, fill in the blanks.

1. Real numbers that can be represented with a terminating decimal are called _____ numbers.

2. Real numbers that cannot be represented as a ratio of *integers* are called _____ numbers.

3. The distance from zero to a point on the number line is called the _____ _____ of the number.

In Problems 4–6, answer True or False to each statement.

4. When comparing two real numbers, the number that is farther to the right on the number line is always the greater of the two.

5. Every integer is a rational number.

6. 0 is neither positive nor negative.

7. Investigate the history of the number zero. Where and when was zero first used?

8. Write a definition of "irrational number" in your own words. Describe the characteristics to look for when placing numbers into this set.

Building Skills

In Problems 9–14, write each set.

9. A is the set of *whole* numbers less than 5.

10. B is the set of *natural* numbers less than 25.

11. D is the set of *natural* numbers less than 5.

12. C is the set of *integers* between -6 and 4, not including -6 or 4.

13. E is the set of even *natural* numbers between 4 and 15, not including 4 or 15.

14. F is the set of odd *natural* numbers less than 1.

1. rational
2. irrational
3. absolute value
4. True 5. True 6. True
7. Answers may vary.
8. Answers may vary.
9. $A = \{0, 1, 2, 3, 4\}$
10. $B = \{1, 2, 3, 4, \dots, 24\}$
11. $D = \{1, 2, 3, 4\}$
12. $C = \{-5, -4, -3, -2, -1, 0, 1, 2, 3\}$
13. $E = \{6, 8, 10, 12, 14\}$
14. $F = \{\ \}$ or $\varnothing$

15. 3
16. 3, 0
17. $-4, 3, 0$
18. $-4, 3, -\frac{13}{2}, 0$
19. 2.303003000...
20. All numbers listed
21. All numbers listed
22. $-4.2, 3.\overline{5}, \frac{5}{5} = 1$
23. π
24. $\frac{5}{5} = 1$
25. $\frac{5}{5} = 1$
26. $\frac{5}{5} = 1$
27.

28.
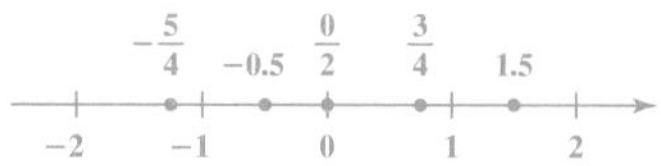
29. 12 30. 8 31. 4 32. 7
33. $\frac{3}{8}$
34. $\frac{13}{9}$
35. 2.1
36. 3.2
37. True
38. False
39. True
40. True
41. True
42. False
43. False
44. False
45. $<$
46. $>$
47. $>$
48. $<$
49. $>$
50. $=$
51. $=$
52. $>$
53. (a)
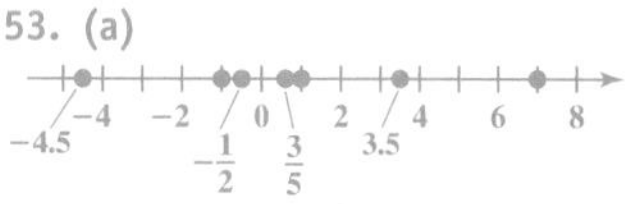
(b) $-4.5, -1, -\frac{1}{2}, \frac{3}{5}, 1, 3.5, |-7| = 7$
(c) i. $-1, 1, |-7| = 7$
ii. All numbers listed
54. (a)
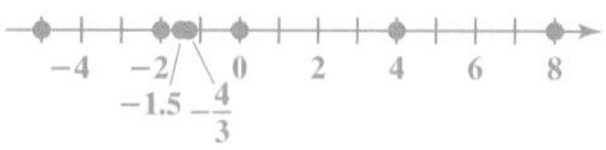
(b) $-\frac{15}{3} = -5, -2, -1.5, -\frac{4}{3}, 0, |-4| = 4, 8$
(c) i. $-\frac{15}{3} = -5, -2, 0, 8$
ii. All numbers listed

In Problems 15–20, use the set $\left\{-4, 3, \frac{-13}{2}, 0, 2.303003000\ldots\right\}$. *List all the elements that are...*

15. natural numbers
16. whole numbers
17. integers
18. rational numbers
19. irrational numbers
20. real numbers

In Problems 21–26, use the set $\left\{-4.2, 3.\overline{5}, \pi, \frac{5}{5}\right\}$. *List all the elements that are...*

21. real numbers
22. rational numbers
23. irrational numbers
24. integers
25. whole numbers
26. natural numbers

In Problems 27 and 28, plot the points in each set on a real number line.

27. $\left\{0, \frac{3}{3}, -1.5, -2, \frac{4}{3}\right\}$
28. $\left\{\frac{3}{4}, \frac{0}{2}, -\frac{5}{4}, -0.5, 1.5\right\}$

In Problems 29–36, evaluate each expression.

29. $|-12|$ **30.** $|-8|$ **31.** $|4|$ **32.** $|7|$
33. $\left|-\frac{3}{8}\right|$ **34.** $\left|-\frac{13}{9}\right|$ **35.** $|-2.1|$ **36.** $|-3.2|$

In Problems 37–44, determine whether the statement is True or False.

37. $-2 > -3$ **38.** $0 < -5$ **39.** $-6 \leq -6$ **40.** $-3 \geq -5$
41. $\frac{3}{2} = 1.5$ **42.** $4.7 = 4.\overline{7}$ **43.** $\pi = 3.14$ **44.** $\frac{1}{3} = 0.33$

In Problems 45–52, replace the ? with the correct symbol: $>$, $<$, $=$.

45. $-1 \; ? \; 0$ **46.** $-8 \; ? \; -8.5$ **47.** $\frac{5}{8} \; ? \; \frac{6}{11}$ **48.** $\frac{5}{12} \; ? \; \frac{2}{3}$
49. $\frac{6}{13} \; ? \; 0.46$ **50.** $\frac{5}{11} \; ? \; 0.\overline{45}$ **51.** $|-7| \; ? \; |7|$ **52.** $\left|-\frac{3}{4}\right| \; ? \; \frac{3}{5}$

Mixed Practice

In Problems 53 and 54, (a) plot the points on the real number, (b) write the numbers in ascending order, (c) list the numbers that are (i) integers, (ii) rational numbers.

53. $\left\{\frac{3}{5}, -1, -\frac{1}{2}, 1, 3.5, |-7|, -4.5\right\}$
54. $\left\{8, -2, |-4|, -1.5, -\frac{4}{3}, 0, -\frac{15}{3}\right\}$

Applying the Concepts

In Problems 55–62, place a ✓ in the box if the given number belongs to the set.

		Natural	Whole	Integers	Rational	Irrational	Real
55.	-100			✓	✓		✓
56.	0		✓	✓	✓		✓
57.	-10.5				✓		✓
58.	π					✓	✓
59.	$\frac{75}{25}$	✓	✓	✓	✓		✓
60.	4	✓	✓	✓	✓		✓
61.	7.56556555...					✓	✓
62.	$6.\overline{45}$				✓		✓

63. True
64. False
65. False
66. False
67. True
68. True
69. True
70. True
71. True
72. True
73. Irrational numbers
74. Whole numbers
75. Real numbers
76. Rational numbers
77. 0
78. ∅
79. True
80. False
81. True
82. False
83. {7, 8, 9, 10, 11, 12, 13, 14, 15}
84. {10, 11, 12}
85. {10, 11, 12, 13, 14, 15}
86. {11, 12, 13, 14, 15}
87. {11, 12}
88. {7, 8, 9, 10, 11, 12, 13, 14, 15}
89. {2, 4, 6, 8, 10}
90. {60, 62, 64, ..., 80}
91. (a) {1}, {2}, {3}, {4}, {1, 2}, {1, 3}, {1, 4}, {2, 3}, {2, 4}, {3, 4}, {1, 2, 3}, {1, 2, 4}, {1, 3, 4}, {2, 3, 4}, {1, 2, 3, 4}, ∅
(b) 16
92. (a) {*a*}, {*b*}, {*c*}, {*a*, *b*}, {*a*, *c*}, {*b*, *c*}, {*a*, *b*, *c*}, ∅
(b) 8
(c) 2^n

In Problems 63–72, determine whether the statement is True or False.

63. Every *whole* number is also an *integer*.

64. Every decimal number is a *rational* number.

65. There are numbers which are both *rational* and *irrational.*

66. 0 is a positive number.

67. Every *natural* number is also a *whole* number.

68. Every *integer* is also a *real* number.

69. Every terminating decimal is a *rational* number.

70. Some numbers in the form $\frac{p}{q}$, $q \neq 0$ are *integers.*

71. 0 is a nonnegative *integer*.

72. -1 is a nonpositive *integer.*

In Problems 73–78, give the name of the set or elements of the set that matches the following description:

73. nonterminating and nonrepeating decimals

74. nonnegative integers

75. the set of rational numbers combined with the set of irrational numbers

76. terminating or repeating decimals

77. numbers which are both nonnegative and nonpositive

78. numbers which are both negative and positive

Extending the Concepts

If every element of set A is also an element of set B, we say A is a **subset** *of B and we write $A \subseteq B$. In Problems 79–82, use this definition and following sets to answer True or False to each statement.*

$$X = \{a, b, c, d, e\} \quad Y = \{c, e\} \quad Z = \{c, e, f\}$$

79. $Y \subseteq X$ **80.** $Z \subseteq Y$

81. $Y \subseteq Z$ **82.** $Z \subseteq X$

The **intersection** *of two sets is the set that contains the elements common to both A and B and is written $A \cap B$. The* **union** *of two sets is the set of all elements that are in either A or B and is written $A \cup B$. In Problems 83–86, write the elements of each set, using sets A, B, and C below.*

$$A = \{7, 8, 9, 10, 11, 12\} \quad B = \{10, 11, 12, 13, 14, 15\} \quad C = \{11, 12, 13, 14, 15\}$$

83. $A \cup B$ **84.** $A \cap B$ **85.** $B \cup C$

86. $B \cap C$ **87.** $A \cap C$ **88.** $A \cup C$

89. If $A = \{\text{even integers}\}$ and $B = \{\text{whole numbers less than } 11\}$, find $A \cap B$.

90. If $X = \{48, 49, 50, \dots\}$ and $Y = \{60, 62, 64, \dots 80\}$, find $X \cap Y$.

91. When writing subsets, it is important to be orderly when creating the list. Think of a pattern and then answer the following:

(a) List all possible subsets of set Z where $Z = \{1, 2, 3, 4\}$.
(b) How many subsets did you find?

92. Use the set $M = \{a, b, c\}$ to answer the following:

(a) List all possible subsets of M.
(b) How many subsets did you find?
(c) Determine a rule for finding the number of subsets of a set that has n elements.

1.3 Adding, Subtracting, Multiplying, and Dividing Integers

OBJECTIVES

1. Add Integers
2. Determine the Additive Inverse of a Number
3. Subtract Integers
4. Multiply Integers
5. Divide Integers

Preparing for Adding, Subtracting, Multiplying, and Dividing Integers

Before getting started, take the following readiness quiz. If you get a problem wrong, go back to the section cited and review the material.

1. Write $\frac{16}{36}$ as a fraction in lowest terms. [Appendix, Section A.1, pp. A5–A6]

In this section, we perform addition, subtraction, multiplication, and division, called **operations,** on integers. The symbols used in algebra for addition, subtraction, multiplication, and division are +, −, ·, and /, respectively. The words used to describe the results of these four operations are **sum, difference, product,** and **quotient.** Table 2 summarizes these ideas.

Table 2

Operation	Symbols	Words
Addition	$a + b$	Sum: a plus b
Subtraction	$a - b$	Difference: a minus b
Multiplication	$a \cdot b$, $(a) \cdot b$, $a \cdot (b)$, $(a) \cdot (b)$, ab, $(a)b$, $a(b)$, $(a)(b)$	Product: a times b
Division	a/b or $\frac{a}{b}$	Quotient: a divided by b

In algebra, we avoid using the multiplication sign × used in arithmetic. Instead, when two expressions are placed next to each other without an operation symbol, as in ab, or in parentheses, as in $(a)(b)$, we know that the expressions are to be multiplied.

A **mixed number** is a whole number followed by a fraction. We do not use mixed numbers in algebra. When mixed numbers are used, addition is understood. For example, $3\frac{2}{5}$ means $3 + \frac{2}{5}$.

When you see a mixed number, rewrite it as a fraction. Do you remember how to do this? To write $3\frac{2}{5}$ as a fraction, we multiply the whole number 3 by the denominator 5 and obtain 15. Then add this result to the numerator 2 and obtain 17. This result is the numerator of the fraction. The denominator remains 5. So

$$3\frac{2}{5} = \frac{17}{5}$$

Work Smart

Do not use mixed numbers in algebra.

In algebra, using mixed numbers becomes confusing because we interpret the lack of an operation symbol between two terms to mean multiplication. To avoid confusion, write $3\frac{2}{5}$ as 3.4 or as $\frac{17}{5}$.

1 Add Integers

Adding Integers with the Same Sign Using a Number Line

We will use the real number line to discover a pattern for adding integers. When we add a positive integer, we move to the right on the number line, and when we add a negative integer, we move to the left on the number line.

Remember, the *sign* of a number refers to whether the number is positive or negative. For example, the sign of 4 is positive, while the sign of −12 is negative. We will first consider adding integers with the same sign.

Preparing for...Answer **1.** $\frac{4}{9}$

Classroom Example Find the sum: $7 + 2$

Answer: 9

EXAMPLE 1 Adding Two Positive Integers Using a Number Line

Find the sum: $5 + 3$

Solution

We begin at 5 on the number line and move 3 spaces to the right, so $5 + 3 = 8$. See Figure 10.

Figure 10

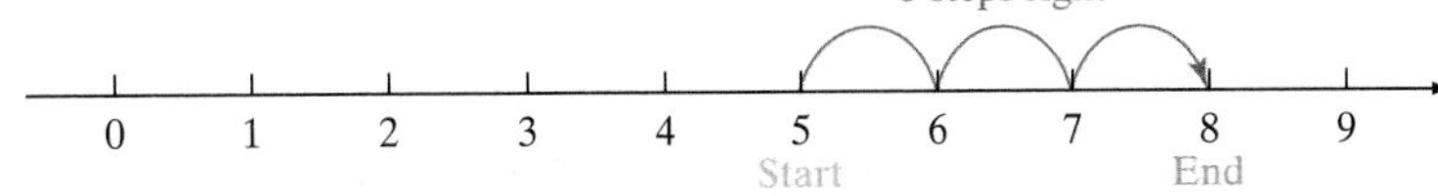

Classroom Example Find the sum: $-5 + (-3)$

Answer: -8

Teaching Tip Students may think of positive numbers as gains and negative numbers as losses. For example, if on one day you lose \$2 on a stock in the stock market and the next day you lose \$1 on that same stock, you have lost \$3. So $-2 + (-1) = -3$.

EXAMPLE 2 Adding Two Negative Integers Using a Number Line

Find the sum: $-7 + (-4)$

Solution

We begin at -7 on the number line and move 4 spaces to the left, so $-7 + (-4) = -11$. See Figure 11.

Figure 11

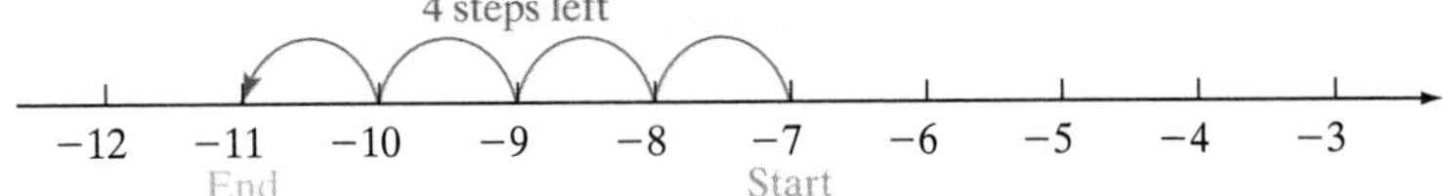

QUICK ✓ *Use a number line to find each sum.*

1. $8 + 19$ **2.** $-3 + (-5)$ **3.** $8 + 6$

4. $-3 + (-4)$ **5.** $-5 + (-12)$

Adding Integers with Different Signs Using a Number Line

We now consider the sum of two integers with different signs.

Classroom Example Find the sum: $-6 + 2$

Answer: -4

EXAMPLE 3 Adding Integers with Different Signs Using a Number Line

Find the sum: $-5 + 3$

Solution

We begin at -5 and then move 3 units to the right. From Figure 12 we see that $-5 + 3 = -2$.

Figure 12

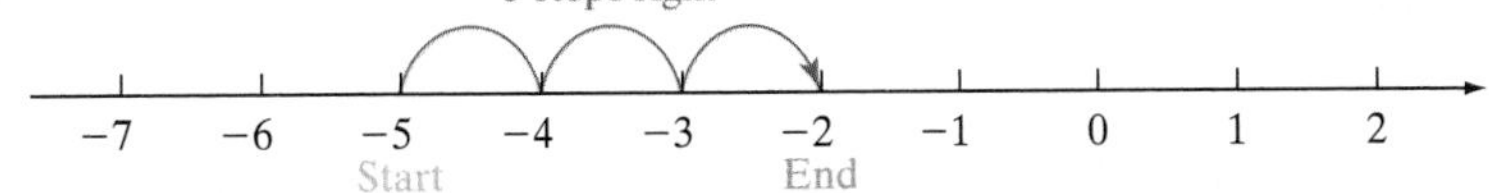

Classroom Example Find the sum: $6 + (-2)$

Answer: 4

EXAMPLE 4 Adding Integers with Different Signs Using a Number Line

Find the sum: $7 + (-4)$

Solution

We begin at 7 and move 4 spaces to the left. We see that $7 + (-4) = 3$. See Figure 13.

Figure 13

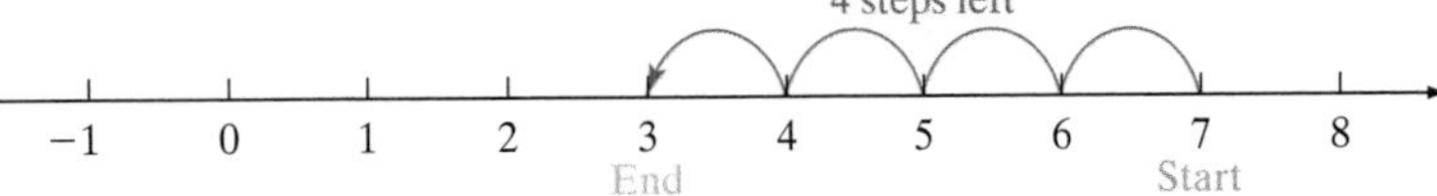

QUICK ✓ *Use a number line to find each sum.*

6. $-1 + 4$ **7.** $3 + (-4)$ **8.** $-8 + 10$ **9.** $-8 + 4$

10. $-17 + (-3)$ **11.** $-12 + 6$ **12.** $15 + (-5)$

Adding Integers Using Absolute Value

Did you discover a pattern for adding integers from Examples 1–4? When we add integers with the same sign (both positive or both negative), we add the absolute values of the integers being added and attach the common sign. When we add integers with different signs (one positive and one negative), we subtract the smaller absolute value from the larger absolute value and attach the sign of the integer having the larger absolute value.

Classroom Example
Find $-12 + (-9)$ using absolute value.
Answer: -21

EXAMPLE 5 How to Add Real Numbers with the Same Sign Using Absolute Value

Find $-16 + (-24)$ using absolute value.

Step-by-Step Solution

Step 1: Add the absolute values of the two integers.

We have $|-16| = 16$ and $|-24| = 24$. So

$$16 + 24 = 40$$

Step 2: Attach the common sign, either positive or negative.

Both integers are negative in the original problem, so

$$-16 + (-24) = -40$$

Classroom Example
Find $-28 + 17$ using absolute value.
Answer: -11

EXAMPLE 6 How to Add Real Numbers with Different Signs Using Absolute Value

Find $-31 + 16$ using absolute value.

Step-by-Step Solution

Step 1: Subtract the smaller absolute value from the larger absolute value.

We have $|-31| = 31$ and $|16| = 16$. The smaller absolute value is 16, so we compute

$$31 - 16 = 15$$

Step 2: Attach the sign of the integer with the larger absolute value.

The larger absolute value is 31, which was a negative number in the original problem. So, the sum is negative. Therefore,

$$-31 + 16 = -15$$

Work Smart

When we add integers with the same sign, we add the absolute values of the integers and attach the common sign. When we add integers with different signs, we subtract the smaller absolute value from the larger absolute value and attach the sign of the integer having the larger absolute value.

SUMMARY: Steps to Add Two Nonzero Integers Using Absolute Value

To add integers with the same sign (both positive or both negative),

Step 1: Add the absolute values of the two integers.

Step 2: Attach the common sign, either positive or negative.

To add integers with different signs (one positive and one negative),

Step 1: Subtract the smaller absolute value from the larger absolute value.

Step 2: Attach the sign of the integer with the larger absolute value.

QUICK ✓ *Use absolute value to find each sum.*

13. $-11 + 7$ **14.** $5 + (-8)$ **15.** $-8 + (-16)$ **16.** $-94 + 38$

2 Determine the Additive Inverse of a Number

What is $3 + (-3)$? What is $10 + (-10)$? What is $-143 + 143$? Of course, the answer to all three of these questions is 0! In fact, these results are true in general.

Teaching Tip

Point out that the concept of the opposite, or additive inverse, of a number will be used in many ways in the course, especially when solving equations.

ADDITIVE INVERSE PROPERTY

For any real number a other than 0, there is a real number $-a$, called the **additive inverse,** or **opposite,** of a, having the following property.

$$a + (-a) = -a + a = 0$$

Work Smart

The additive inverse of a, $-a$, is sometimes called the *negative* of a. Be careful when using this term because it suggests that the opposite is a negative number, which may not be true! For example, the additive inverse of -11 is 11, a positive number.

Any two numbers whose sum is zero are additive inverses, or opposites.

EXAMPLE 7 Finding an Additive Inverse

(a) The additive inverse of 9 is -9 because $9 + (-9) = 0$.

(b) The additive inverse of -12 is $-(-12) = 12$ because $-12 + 12 = 0$.

Classroom Example

(a) The additive inverse of 6 is -6 because $6 + (-6) = 0$.

(b) The additive inverse of -15 is $-(-15) = 15$ because $-15 + 15 = 0$.

Notice from Example 7(b) that $-(-a) = a$ for any real number a.

QUICK ✓ *Determine the additive inverse of the given real number.*

17. 7 **18.** $\frac{3}{7}$ **19.** -21 **20.** $-\frac{8}{5}$ **21.** -5.75

3 Subtract Integers

Now that we understand addition and have the ability to determine the additive inverse of a number, we can proceed to subtract integers. Good news! There is not a new set of rules to follow for subtracting integers. We simply rewrite the subtraction problem as an addition problem and use the addition rules that were presented earlier.

From arithmetic, we write $10 - 6 = 4$. Using the additive inverse, we can write $10 - 6 = 4$ as $10 + (-6) = 4$.

DEFINITION

The **difference** $a - b$, read "a minus b" or a less b," is defined as

$$a - b = a + (-b)$$

In words, to subtract b from a, add the "opposite" of b to a.

Classroom Example
Compute each difference:
(a) $-16 - 43$
(b) $9 - (-12)$
(c) $35 - 81$
Answer: (a) -59 (b) 21 (c) -46

So, subtracting b from a is the same as adding the additive inverse (opposite) of b to a.

EXAMPLE 8 How to Subtract Integers

Compute the difference: $-18 - (-40)$

Step-by-Step Solution

Step 1: Change the subtraction problem to an equivalent addition problem. $-18 - (-40) = -18 + 40$

Step 2: Find the sum. $= 22$

In Words
The problem in Example 8 is read "negative eighteen minus negative 40" not "negative 18 minus minus 40."

SUMMARY: Steps to Subtract Nonzero Integers

1. Change the subtraction problem to an equivalent addition problem using $a - b = a + (-b)$.
2. Find the sum.

QUICK ✓ *Find the value of each expression.*

22. $59 - (-21)$ **23.** $-32 - 146$ **24.** 17 minus 35
25. -382 subtracted from -2954

For the remainder of this text, the direction **evaluate** will mean to find the numerical value of an expression. When we evaluate a numerical expression in which there is both addition and subtraction, change all subtraction to addition. Then add left to right.

Classroom Example
Evaluate: $12 - 36 + 40$
Answer: 16

EXAMPLE 9 Evaluating an Expression Containing Three Integers

Evaluate the expression: $10 - 18 + 25$

Solution

$$10 - 18 + 25 = 10 + (-18) + 25$$

Add from left to right: $= -8 + 25$

$= 17$

Work Smart
When adding and subtracting more than two numbers, add in order from left to right.

EXAMPLE 10 Evaluating an Expression Containing Four Integers

Evaluate the expression: $162 - (-46) + 80 - 274$

Classroom Example
Evaluate: $154 - (-12) + 73 + (-123)$
Answer: 116

Solution

$$162 - (-46) + 80 - 274 = 162 + 46 + 80 + (-274)$$

Add from left to right: $= 208 + 80 + (-274)$

$= 288 + (-274)$

$= 14$

QUICK ✓ *Evaluate each expression.*

26. $8 - 13 + 5 - 21$

27. $-27 - 49 + 18$

28. $3 - (-14) - 8 + 3$

29. $-825 + 375 - (-735) + 265$

4 Multiply Integers

A quick review of the language of multiplication is needed: In the multiplication statement $9 \cdot 2 = 18$, the numbers 9 and 2 are called **factors** and the number 18 is the **product.**

$$\underbrace{9}_{\text{factor}} \cdot \underbrace{2}_{\text{factor}} = \underbrace{18}_{\text{product}}$$

When we first learned how to multiply natural numbers, we were told that we can think of multiplication as repeated addition. For example, $3 \cdot 4$ is equivalent to adding 4 three times. That is,

$$3 \cdot 4 = \underbrace{4 + 4 + 4}_{\substack{\text{Add 4} \\ \text{three times}}} = 12$$

Notice that the product of two positive factors produces a positive product. We knew that from arithmetic! But what is the product $3 \cdot (-4)$?

$$3 \cdot (-4) = \underbrace{-4 + (-4) + (-4)}_{\substack{\text{Add } -4 \\ \text{three times}}} = -12$$

We conclude that the product of two real numbers with *different signs* is negative.

What about the product of two negative numbers? Consider the pattern shown below.

$$-4 \cdot 3 = -12$$
$$-4 \cdot 2 = -8$$
$$-4 \cdot 1 = -4$$
$$-4 \cdot 0 = 0$$

Teaching Tip
Advise students to summarize: When multiplying factors with like signs, the product is positive. When multiplying factors with unlike signs, the product is negative.

Notice that each time the second factor decreases by 1, the product increases by four. Assuming this pattern continues, we would have

$$-4 \cdot -1 = 4$$
$$-4 \cdot -2 = 8$$

The pattern suggests that the product of two negative numbers is a positive number.

In Words
The sign of the product is *positive* if the signs of the two factors are the *same* and *negative* if the signs of the two factors are *different.*

RULES OF SIGNS FOR MULTIPLYING TWO INTEGERS

1. If we multiply two positive integers, the product is positive.
2. If we multiply one positive integer and one negative integer, the product is negative.
3. If we multiply two negative integers, the product is positive.

Classroom Example
Find each product:

(a) $3(-5) = -15$
(b) $-7(4) = -28$
(c) $(-6)(-7) = 42$
(d) $-15(30) = -450$

EXAMPLE 11 Multiplying Integers

(a) $2(-4) = -8$

(b) $-6(5) = -30$

(c) $(-7)(-8) = 56$

(d) $-25(18) = -450$

QUICK ✓ *Find the product.*

30. $-3(7)$ **31.** $13(-4)$ **32.** $5 \cdot 16$ **33.** $-9(-12)$ **34.** $(-13)(-25)$

Find the Product of Several Integers

When we find the product of more than two integers, we multiply in order, left to right.

Classroom Example ➤
Find the product: $3 \cdot (-5) \cdot 6$
Answer: -90

EXAMPLE 12 Multiplying Three Integers

Find the product: $3 \cdot (-4) \cdot 7$

Solution

Multiply left to right: $$3 \cdot (-4) \cdot 7 = -12 \cdot 7 = -84$$

Classroom Example ➤
Find the product: $-3 \cdot (-5) \cdot 2 \cdot (-6)$
Answer: -180

EXAMPLE 13 Multiplying Four Integers

Find the product: $-8 \cdot (-1) \cdot 4 \cdot (-5)$

Solution

Multiply left to right: $$-8 \cdot (-1) \cdot 4 \cdot (-5) = 8 \cdot 4 \cdot (-5) = 32 \cdot (-5) = -160$$

Work Smart
If we multiply an *even* number of negative factors the product is *positive.*

If we multiply an *odd* number of negative factors the product is *negative.*

QUICK ✓ *Find the product.*

35. $-3 \cdot 9 \cdot (-4)$ **36.** $(-3) \cdot (-4) \cdot (-5) \cdot (-6)$

(5) Divide Integers

When we divide, the numerator is called the **dividend,** the denominator is called the **divisor,** and the answer is called the **quotient.**

$$\text{dividend} \rightarrow \frac{28}{7} \leftarrow \text{divisor} = 4 \leftarrow \text{quotient} \quad \text{or} \quad \underbrace{28}_{\text{dividend}} \div \underbrace{7}_{\text{divisor}} = \underbrace{4}_{\text{quotient}}$$

Two nonzero numbers are *reciprocals* if their product is 1. Reciprocal is another word for the term *multiplicative inverse.*

In Words
Any two numbers whose product is 1 are called multiplicative inverses or reciprocals of each other.

MULTIPLICATIVE INVERSE (RECIPROCAL) PROPERTY

For each *nonzero* real number a, there is a real number $\frac{1}{a}$, called the **multiplicative inverse** or **reciprocal** of a, having the following property:

$$a \cdot \frac{1}{a} = \frac{1}{a} \cdot a = 1 \quad a \neq 0$$

Classroom Example ➤
(a) The multiplicative inverse or reciprocal of 4 is $\frac{1}{4}$.
(b) The multiplicative inverse or reciprocal of -7 is $-\frac{1}{7}$.

EXAMPLE 14 Finding the Multiplicative Inverse (or Reciprocal) of an Integer

(a) The multiplicative inverse or reciprocal of 5 is $\frac{1}{5}$.

(b) The multiplicative inverse or reciprocal of -8 is $-\frac{1}{8}$.

QUICK ✓ *Find the reciprocal of each integer.*

37. 6 **38.** −2

We use the idea behind the multiplicative inverse to define division of real numbers.

DEFINITION

If b is a nonzero real number, the **quotient** $\frac{a}{b}$, read as "a divided by b" or "the **ratio** of a to b," is defined as

$$\frac{a}{b} = a \cdot \frac{1}{b} \quad \text{if } b \neq 0$$

For example, $\frac{40}{8} = 40 \cdot \frac{1}{8}$ and $\frac{12}{7} = 12 \cdot \frac{1}{7}$. Because division can be represented as multiplication, the same rules of signs that apply to multiplication also apply to division.

In Words

These are the same rules as before . . . a positive divided by a positive is positive; positive divided by negative is negative, and so on.

RULES OF SIGNS FOR DIVIDING TWO INTEGERS

1. If we divide two positive integers, the quotient is positive. That is, $\frac{+a}{+b} = \frac{a}{b}$.
2. If we divide one positive integer and one negative integer, the quotient is negative. That is, $\frac{-a}{b} = \frac{a}{-b} = -\frac{a}{b}$.
3. If we divide two negative integers, the quotient is positive. That is, $\frac{-a}{-b} = \frac{a}{b}$.

Finding the quotient of two integers is identical to writing a fraction in lowest terms.

Classroom Example ➤
Find each quotient:

(a) $\frac{-75}{15}$ (b) $225 \div -25$

Answer: (a) −5 (b) −9

EXAMPLE 15 Finding the Quotient of Two Integers

Find each quotient:

(a) $\frac{-90}{20}$ **(b)** $200 \div (-5)$

Solution

(a)

$$\frac{-90}{20} = \frac{-9 \cdot 10}{2 \cdot 10}$$

Divide out the common factor: $= \frac{-9 \cdot \cancel{10}}{2 \cdot \cancel{10}}$

$$= \frac{-9}{2}$$

$$= -\frac{9}{2}$$

(b)

$$200 \div (-5) = \frac{200}{-5}$$

$$= \frac{40 \cdot 5}{-1 \cdot 5}$$

Divide out the common factor: $= \frac{40 \cdot \cancel{5}}{-1 \cdot \cancel{5}}$

$$= -40$$

QUICK ✓ *Find the quotient.*

39. $\dfrac{20}{-4}$ **40.** $\dfrac{707}{-101}$ **41.** $-63 \div -7$ **42.** $\dfrac{-54}{4}$

1.3 Exercises

1. sum 2. difference
3. product 4. False
5. False 6. False
7. Answers may vary.
8. Answers may vary.
9. 15
10. 10
11. 4
12. 8
13. 4
14. 6
15. −19
16. −18
17. 21
18. 17
19. −328
20. −213
21. −11
22. 2
23. 43
24. 72
25. 325
26. 34
27. −125
28. −7
29. 11
30. 12
31. −8
32. −7
33. −28
34. −24
35. 54
36. 32
37. 0
38. 0
39. −41
40. −18
41. −31
42. 71
43. 172
44. 98
45. 40
46. 63
47. −56
48. −63
49. 0
50. 0
51. 144
52. 110
53. −126
54. −896
55. −90
56. −192
57. 210
58. 144
59. 120
60. −72

For Extra Help:

Student Solutions Manual

CD Video

PH Math/Tutor Center

MathXL Tutorials on CD

MathXL®

MyMathLab

Concepts and Vocabulary

In Problems 1–3, fill in the blanks.

1. The answer to an addition problem is called the ________.

2. The answer to a subtraction problem is called the ________.

3. The answer to a multiplication problem is called the ________.

In Problems 4–6, answer True or False to each statement.

4. When adding two negative numbers, we subtract their absolute values.

5. The subtraction problem $-3 - 10$ is equivalent to $3 + (-10)$.

6. The quotient of two negative numbers is negative.

✎ **7.** Write a sentence or two that justifies the fact that the product of two negative integers is positive. You may use an example.

✎ **8.** Explain how a subtraction problem can be written as an equivalent addition problem. Explain how $42 \div 4$ can be written as a multiplication problem.

Building Skills

In Problems 9–24, find the sum.

9. $8 + 7$ **10.** $6 + 4$ **11.** $-5 + 9$ **12.** $-4 + 12$

13. $9 + (-5)$ **14.** $13 + (-7)$ **15.** $-11 + (-8)$ **16.** $-13 + (-5)$

17. $-16 + 37$ **18.** $-32 + 49$ **19.** $-119 + (-209)$ **20.** $-145 + (-68)$

21. $-14 + 21 + (-18)$ **22.** $(-13) + 37 + (-22)$

23. $74 + (-13) + (-23) + 5$ **24.** $-34 + 46 + (-12) + 72$

In Problems 25–28, determine the additive inverse of each real number.

25. -325 **26.** -34 **27.** 125 **28.** 7

In Problems 29–44, find the difference.

29. $23 - 12$ **30.** $35 - 23$ **31.** $9 - 17$ **32.** $12 - 19$

33. $-20 - 8$ **34.** $-15 - 9$ **35.** $13 - (-41)$ **36.** $14 - (-18)$

37. $-36 - (-36)$ **38.** $-15 - (-15)$ **39.** $0 - 41$ **40.** $0 - 18$

41. $-93 - (-62)$ **42.** $46 - (-25)$ **43.** $86 - (-86)$ **44.** $49 - (-49)$

In Problems 45–60, find the product.

45. $5 \cdot 8$ **46.** $7 \cdot 9$ **47.** $8(-7)$ **48.** $9(-7)$

49. $(0)(-21)$ **50.** $-21 \cdot 0$ **51.** $(-48)(-3)$ **52.** $(-22)(-5)$

53. $(-42)3$ **54.** $(-128)7$ **55.** $-5 \cdot 6 \cdot 3$ **56.** $-6 \cdot 4 \cdot 8$

57. $-10(3)(-7)$ **58.** $-8(2)(-9)$

59. $(-2)(4)(-1)(3)(5)$ **60.** $(-3)(-4)(6)(-1)$

61. $\frac{1}{8}$ 62. $\frac{1}{10}$ 63. $-\frac{1}{4}$ 64. $-\frac{1}{3}$
65. 1 66. $\frac{1}{2}$ 67. 5 68. 4 69. 7
70. 9 71. −15 72. −24 73. $\frac{7}{2}$
74. $\frac{5}{4}$ 75. $-\frac{10}{7}$ 76. $-\frac{20}{11}$ 77. $\frac{35}{4}$
78. $\frac{20}{3}$ 79. −72 80. −105
81. 60
82. 68
83. 171
84. 1463
85. −15
86. 6
87. −42
88. 37
89. $\frac{15}{4}$ 90. $\frac{7}{2}$ 91. 40
92. −52
93. −238
94. 213
95. $28 + (-21) = 7$
96. $32 + (-64) = -32$
97. $-21 - 47 = -68$
98. $-16 - (-85) = 69$
99. $-12 \cdot 18 = -216$
100. $32 \cdot (-8) = -256$
101. $-36 \div (-108)$ or $\frac{-36}{-108} = \frac{1}{3}$
102. $-40 \div 100$ or $\frac{-40}{100} = -\frac{2}{5}$
103. −3.25 points
104. −14°
105. −6 yards
106. $125,000
107. −$48
108. 12,368
109. 14 miles
110. 48 yard-marker
111. $655
112. $352

In Problems 61–66, find the multiplicative inverse (or reciprocal) of each number.

61. 8 **62.** 10 **63.** −4
64. −3 **65.** 1 **66.** 2

In Problems 67–78, divide.

67. $10 \div 2$ **68.** $36 \div 9$ **69.** $\dfrac{-56}{-8}$ **70.** $\dfrac{-63}{-7}$

71. $\dfrac{-45}{3}$ **72.** $\dfrac{-144}{6}$ **73.** $\dfrac{35}{10}$ **74.** $\dfrac{20}{16}$

75. $\dfrac{60}{-42}$ **76.** $\dfrac{120}{-66}$ **77.** $\dfrac{-105}{-12}$ **78.** $\dfrac{-80}{-12}$

Mixed Practice

In Problems 79–94, evaluate the expression.

79. $-4 \cdot 18$ **80.** $7 \cdot (-15)$ **81.** $-16 - (-76)$ **82.** $87 - 19$

83. $-9 \cdot (-19)$ **84.** $7 \cdot 209$ **85.** $\dfrac{120}{-8}$ **86.** $\dfrac{-156}{-26}$

87. $-98 + 56$ **88.** $103 + (-66)$ **89.** $\dfrac{75}{|-20|}$ **90.** $\dfrac{|-42|}{12}$

91. $|-14| + |-26|$ **92.** $|-10| + (-62)$
93. $|-389| - 627$ **94.** $|-193| - (-20)$

In Problems 95–102, write each expression using mathematical symbols. Then evaluate the expression.

95. the sum of 28 and −21 **96.** the sum of 32 and −64
97. −21 minus 47 **98.** −16 minus −85
99. −12 multiplied by 18 **100.** 32 multiplied by −8
101. −36 divided by −108 **102.** −40 divided by 100

Applying the Concepts

In Problems 103–108, write the positive or negative number for each of the following.

103. Stock The price of IBM stock fell by 3.25 points.

104. Temperature The current temperature in Juneau, Alaska is 14° below zero.

105. Chargers Football The Chargers lost 6 yards on the play.

106. Profit Mark's BMW dealership showed a profit of $125,000 this quarter.

107. Checking Account Leila's checking account is now overdrawn by $48.

108. Census The number of people in Brian's home town grew by 12,368.

In Problems 109–116, write each problem as a numerical expression and evaluate.

109. Hiking Loren and Richard went on a hiking trip. They walked 8 miles to the base of Snow Creek Falls where they set up camp. They went another 3 miles to see the Falls and then returned to their campsite. How many miles did they walk that day?

110. Football The St. Louis Rams took over possession of the ball on their own 15 yard line. The following plays occurred: QB sack, loss of 7 yards; Marshall Faulk ran for 14 yards; Isaac Bruce caught a pass for 26 yards. What yard-marker is the ball on now?

111. Bank Balance When Martha balanced her checkbook, she had $563 in the account. Then the following transactions occurred: She wrote a check to Home Depot for $46; deposited $233; wrote a check to Vons for $63; and wrote a check to Petco for $32. What is Martha's new balance?

112. Checkbook Balance Josie began the month with $399 in her bank account. She deposited her paycheck of $839. She paid her $69 telephone bill, the electric bill for $78, and rent of $739. How much does she have left for spending money?

113. No; −125 cases
114. 26,285 feet
115. 25,725 feet
116. 14,775 feet
117. −3, −5
118. 6, −4
119. −12, 2
120. −15, −3
121. (a) 1, 2, 1.5, $1.\overline{6}$, 1.6, 1.625, 1.615, 1.619, 1.618, . . .
(b) 1.618
(c) Answers may vary.

113. Warehouse Inventory The warehouse began the month with 725 cases of soda. During the month the following occurred: 120 cases were shipped out, 590 cases were shipped out, and a delivery to the warehouse of 310 cases. Does the warehouse have enough stock on hand to fill an order for 450 cases of soda? What is the difference between what they have and what has been requested?

114. Altitude of an Airplane A pilot leveled off his airplane at 35,000 feet at the beginning of the flight. The following adjustments were made during the trip: gained 4290 feet, dropped 10,400 feet and then dropped 2605 feet. At what altitude is the plane currently flying?

115. Distance An airplane flying at 25,350 feet is directly over a submarine that is 375 feet below sea level. What is the distance between a person in the plane and a person in the submarine?

116. Elevation The highest point in California is Mt. Whitney at an elevation of 14,495 feet and the lowest point is in Death Valley at 280 feet below sea level. What is the maximum difference in elevation between two people in California?

Extending the Concepts

117. Find two integers whose sum is −8 and whose product is 15.

118. Find two integers whose sum is 2 and whose product is −24.

119. Find two integers whose sum is −10 and whose product is −24.

120. Find two integers whose sum is −18 and whose product is 45.

121. The Fibonacci Sequence The Fibonacci sequence is a famous sequence of numbers that were discovered by Leonardo Fibonacci of Pisa. The numbers in the sequence are 1, 1, 2, 3, 5, 8, 13, 21, 34, 55, . . . where each term after the second term is the sum of the two preceding terms.
(a) Form fractions of consecutive terms in the sequence. Find the decimal approximation to $\frac{1}{1}, \frac{2}{1}, \frac{3}{2}, \frac{5}{3}$, and so on.
(b) What number does the ratio get close to? This number is called the **golden ratio** and has application in many different areas.
(c) Research Fibonacci numbers and cite three different applications.

1.4 Adding, Subtracting, Multiplying, and Dividing Rational Numbers Expressed as Fractions and Decimals

OBJECTIVES

1. Multiply Rational Numbers Expressed as Fractions
2. Divide Rational Numbers Expressed as Fractions
3. Add and Subtract Rational Numbers Expressed as Fractions
4. Add, Subtract, Multiply, and Divide Rational Numbers Expressed as Decimals

Preparing for Adding, Subtracting, Multiplying, and Dividing Rational Numbers Expressed as Fractions and Decimals

Before getting started, take the following readiness quiz. If you get a problem wrong, go to the section cited and review the material.

1. Find the least common denominator of $\frac{5}{12}$ and $\frac{3}{16}$. [Appendix, Section A.1, pp. A4–A5]

2. Rewrite $\frac{4}{5}$ as an equivalent fraction with a denominator of 30. [Appendix, Section A.1, pp. A3–A5]

Preparing for...Answers **1.** LCD = 48 **2.** $\frac{4}{5} = \frac{24}{30}$

Now that we are comfortable with operations on integers, we expand our skill set so that we can perform operations on rational numbers. Remember, rational numbers can be expressed either as fractions or as a terminating decimal or a nonterminating, repeating decimal. We begin with operations on rational numbers expressed as fractions and end the section with the operations on rational numbers in decimal form.

Work Smart
Remember, any integer can be written as a fraction with 1 in the denominator. For example, $7 = \frac{7}{1}$.
All the skills we learn in this section apply to integers as well.

Finding the quotient of two integers is identical to writing a fraction in lowest terms. So, the direction to "find the quotient $\frac{-90}{20}$" is the same as "write $\frac{-90}{20}$ in lowest terms." For example, the rational number $\frac{-90}{20}$ is written in lowest terms as $-\frac{9}{2}$.

QUICK ✓ *Write each rational number in lowest terms.*

1. $\frac{-4}{14}$ **2.** $-\frac{18}{30}$ **3.** $\frac{24}{-4}$

1 Multiply Rational Numbers Expressed as Fractions

In Words
When finding the product of two or more fractions, multiply the numerators together. Then multiply the denominators together. Write the fraction in lowest terms, if necessary.

The following property tells us how to multiply two rational numbers when they are in fractional form:

MULTIPLYING FRACTIONS

$$\frac{a}{b}\cdot\frac{c}{d} = \frac{a\cdot c}{b\cdot d}, \text{ where } b \text{ and } d \neq 0$$

The rules of signs that apply to integers also apply to rational numbers: The product of two positive rational numbers is positive; the product of a positive rational number and a negative rational number is negative, and so on.

Classroom Example ➤
Find the product: $\frac{5}{8}\cdot\left(-\frac{20}{7}\right)$

Answer: $-\frac{25}{14}$

EXAMPLE 1 Multiplying Rational Numbers (Fractions)

Find the product: $\frac{2}{9}\cdot\left(-\frac{15}{19}\right)$

Solution

We begin by rewriting the rational number $-\frac{15}{19}$ as $\frac{-15}{19}$. Then we multiply the numerators and multiply the denominators.

$$\frac{2}{9}\cdot\left(\frac{-15}{19}\right) = \frac{2\cdot(-15)}{9\cdot 19}$$

Write the numerator and the denominator as the product of prime factors: $= \frac{2\cdot 3\cdot(-5)}{3\cdot 3\cdot 19}$

Divide out common factors: $= \frac{2\cdot \cancel{3}\cdot(-5)}{\cancel{3}\cdot 3\cdot 19}$

$$= \frac{2\cdot(-5)}{3\cdot 19}$$

Perform the multiplication: $= -\frac{10}{57}$

Teaching Tip
Remind students that when we divide out common factors, the quotient of those factors is 1.

QUICK ✓ *Find each product, and write in lowest terms if possible.*

4. $\frac{3}{4}\cdot\frac{9}{8}$ **5.** $\frac{-5}{7}\cdot\frac{56}{15}$ **6.** $\frac{12}{45}\cdot\left(-\frac{18}{20}\right)$ **7.** $-\frac{25}{75}\cdot\left(-\frac{9}{4}\right)$ **8.** $\frac{7}{3}\cdot\frac{1}{14}\cdot\left(-\frac{9}{11}\right)$

(2) Divide Rational Numbers Expressed as Fractions

To divide rational numbers, we must know how to determine the reciprocal of a rational number. We introduced the term *reciprocal* in Section 1.3 by saying that two numbers are reciprocals, or multiplicative inverses, if the product of the numbers is 1. This definition applies to any non-zero real number. So, for example,

$\frac{3}{2}$ and $\frac{2}{3}$ are reciprocals because $\frac{3}{2}\cdot\frac{2}{3} = 1$ 9 and $\frac{1}{9}$ are reciprocals because $9\cdot\frac{1}{9} = 1$

$-\frac{4}{7}$ and $-\frac{7}{4}$ are reciprocals because $-\frac{4}{7}\cdot\left(-\frac{7}{4}\right) = 1$

QUICK ✓ *Find the reciprocal of each number.*

9. 12 **10.** $\frac{7}{5}$ **11.** $-\frac{1}{4}$ **12.** $-\frac{31}{20}$

We divide rational numbers by rewriting the division as an equivalent multiplication problem, according to the following.

Classroom Example Find the quotient: $\frac{5}{12} \div \frac{25}{18}$

Answer: $\frac{3}{10}$

DIVIDING RATIONAL NUMBERS EXPRESSED AS FRACTIONS

$$\frac{a}{b} \div \frac{c}{d} = \frac{a}{b}\cdot\frac{d}{c} = \frac{a\cdot d}{b\cdot c}, \quad \text{where } b, c, d \neq 0$$

EXAMPLE 2 How to Divide Rational Numbers (Fractions)

Find the quotient: $\frac{3}{10} \div \frac{12}{25}$

Step-by-Step Solution

Step 1: Write the equivalent multiplication problem.

$$\frac{3}{10} \div \frac{12}{25} = \frac{3}{10}\cdot\frac{25}{12}$$

Step 2: Write the product in factored form and divide out common factors.

$$= \frac{3\cdot 5\cdot 5}{5\cdot 2\cdot 4\cdot 3}$$

$$= \frac{\cancel{3}\cdot \cancel{5}\cdot 5}{\cancel{5}\cdot 2\cdot 4\cdot \cancel{3}}$$

$$= \frac{5}{2\cdot 4}$$

Step 3: Multiply the remaining factors.

$$= \frac{5}{8}$$

So $\frac{3}{10} \div \frac{12}{25} = \frac{5}{8}$.

QUICK ✓ *Find the quotient.*

13. $\frac{5}{7} \div \frac{7}{10}$ **14.** $-\frac{9}{12} \div \frac{14}{7}$ **15.** $\frac{8}{35} \div \left(\frac{-1}{10}\right)$ **16.** $-\frac{18}{63} \div \left(-\frac{54}{35}\right)$

3 Add and Subtract Rational Numbers Expressed as Fractions

Figure 14

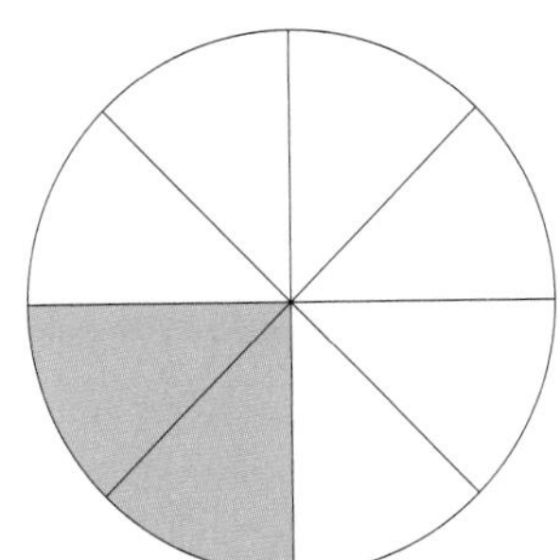

Consider Figure 14, which shows a circle divided into 8 equal parts. We have two regions shaded, each of which represents the fraction $\frac{1}{8}$. Together, the two shaded regions make up $\frac{1}{4}$ of the circle. This implies that

$$\frac{1}{8}+\frac{1}{8}=\frac{1+1}{8}$$
$$=\frac{2}{8}$$
$$=\frac{1}{4}$$

Or, suppose Bobby has \$0.25 and his grandma gives him \$0.50. How much does he now have? Of course, \$0.25 + \$0.50 = \$0.75 or $\frac{3}{4}$ of a dollar. Because $0.25 = \frac{1}{4}$ and $0.50 = \frac{1}{2}$, we can determine Bobby's good fortune using fractions:

$$\frac{1}{4}+\frac{1}{2}=\frac{1}{4}+\frac{2}{4}$$
$$=\frac{3}{4}$$

Based on these results, we might conclude that to add fractions with the same denominators, we add the numerators and write the result over the common denominator. This conclusion is correct. Also, because we can write any subtraction problem as an equivalent addition problem, we have the following methods for adding and subtracting rational numbers.

In Words
When adding or subtracting fractions with a common denominator, you add or subtract the numerators and retain the denominator.

ADDING OR SUBTRACTING RATIONAL NUMBERS (FRACTIONS) WITH THE SAME DENOMINATOR

$$\frac{a}{c}+\frac{b}{c}=\frac{a+b}{c} \quad \text{where } c \neq 0 \qquad \frac{a}{c}-\frac{b}{c}=\frac{a-b}{c}=\frac{a+(-b)}{c} \quad \text{where } c \neq 0$$

Classroom Example Find the sum and write in lowest terms, if necessary.

$$-\frac{5}{12}+\frac{11}{12}$$

Answer: $\frac{1}{2}$

EXAMPLE 3 Adding Rational Numbers (Fractions) with the Same Denominator

Find the sum and write in lowest terms, if necessary: $-\frac{1}{8}+\frac{3}{8}$

Solution

$$-\frac{1}{8}+\frac{3}{8}=\frac{-1}{8}+\frac{3}{8}$$

Write the numerators as a sum over the common denominator: $=\frac{-1+3}{8}$

Add the numerators: $=\frac{2}{8}$

Factor 8: $=\frac{1\cdot 2}{4\cdot 2}$

Divide out the 2s: $=\frac{1\cdot \cancel{2}}{4\cdot \cancel{2}}$

$$=\frac{1}{4}$$

So $-\frac{1}{8}+\frac{3}{8}=\frac{1}{4}$.

Teaching Tip
Prompt students to remember that we add numerators ONLY, not denominators. So

$$-\frac{1}{8}+\frac{3}{8}\neq\frac{2}{16}$$

Classroom Example

Find the difference and write in lowest terms, if necessary:

$$\frac{13}{10} - \frac{5}{10}$$

Answer: $\frac{4}{5}$

EXAMPLE 4 Subtracting Rational Numbers (Fractions) with the Same Denominator

Find the difference and write in lowest terms, if necessary: $\frac{9}{16} - \frac{3}{16}$

Solution

$$\frac{9}{16} - \frac{3}{16} = \frac{9-3}{16}$$

Rewrite as an addition problem: $= \frac{9+(-3)}{16}$

Add the numerators: $= \frac{6}{16}$

Factor 6; Factor 16: $= \frac{3 \cdot 2}{8 \cdot 2}$

Divide out the 2s: $= \frac{3 \cdot \cancel{2}}{8 \cdot \cancel{2}}$

$$= \frac{3}{8}$$

So $\frac{9}{16} - \frac{3}{16} = \frac{3}{8}$.

Work Smart: Study Skills
Notice the title of Example 5: "How to Add Rational Numbers with Unlike Denominators." This three-column example provides a guided, step-by-step approach to solving a problem so you can immediately see each of the steps. Cover the third column and try to work the example yourself. Then look at the entries in the third column and check your solution. Was it correct?

QUICK ✓ *Find the sum or difference, and write in lowest terms if necessary.*

17. $-\frac{9}{10} - \frac{3}{10}$ **18.** $\frac{8}{11} + \frac{2}{11}$ **19.** $-\frac{18}{35} + \frac{3}{35}$ **20.** $\frac{19}{63} - \frac{10}{63}$

Adding and Subtracting Rational Numbers (Fractions) with Unlike Denominators

Work Smart: Study Skills
Do you need a refresher on finding the LCD? If so, go to the Appendix on pages A4–A5 and review your skills.

How do we add rational numbers with different denominators? We must find the least common denominator of the two rational numbers. The **least common denominator (LCD)** is the smallest number that each denominator has as a common multiple.

EXAMPLE 5 How to Add Rational Numbers (Fractions) with Unlike Denominators

Find the sum: $\frac{5}{6} + \frac{3}{8}$

Classroom Example

Find the sum: $\frac{2}{3} + \frac{1}{4}$

Answer: $\frac{11}{12}$

Step-by-Step Solution

Step 1: Find the least common denominator of the denominators.

Write each denominator as the product of prime factors, arranging like factors vertically:

$$6 = 2 \quad\quad \cdot 3$$
$$8 = 2 \cdot 2 \cdot 2$$

Find the product of each of the prime factors the greatest number of times it appears in any factorization:

$$\text{LCD} = 2 \cdot 2 \cdot 2 \cdot 3 = 24$$

Step 2: Write each rational number with the denominator found in Step 1.

Use $1 = \frac{4}{4}$ to change the denominator 6 to 24, and use $1 = \frac{3}{3}$ to change the denominator 8 to 24:

$$\frac{5}{6} + \frac{3}{8} = \frac{5}{6} \cdot \frac{4}{4} + \frac{3}{8} \cdot \frac{3}{3}$$
$$= \frac{20}{24} + \frac{9}{24}$$

Step 3: Add the numerators and write the result over the common denominator.

$$= \frac{20 + 9}{24}$$
$$= \frac{29}{24}$$

Step 4: Write in lowest terms. In this case, the rational number is already in lowest terms.

$$\text{Thus, } \frac{5}{6} + \frac{3}{8} = \frac{29}{24}.$$

EXAMPLE 6 How to Subtract Rational Numbers (Fractions) with Unlike Denominators

Find the difference: $-\frac{9}{14} - \frac{1}{6}$

Classroom Example

Find the difference: $-\frac{5}{14} - \frac{5}{6}$

Answer: $-\frac{25}{21}$

Step-by-Step Solution

Step 1: Find the least common denominator of the denominators.

Write each denominator as the product of prime factors, aligning like factors vertically:

$$14 = 2 \cdot 7$$
$$6 = 2 \quad \cdot 3$$

Find the product of each of the prime factors the greatest number of times it appears in any factorization:

$$\text{LCD} = 2 \cdot 7 \cdot 3$$
$$= 42$$

Step 2: Write each rational number with the denominator found in Step 1.

Use $1 = \frac{3}{3}$ to change the denominator 14 to 42, and use $1 = \frac{7}{7}$ to change the denominator 6 to 42:

$$-\frac{9}{14} - \frac{1}{6} = \frac{-9}{14} \cdot \frac{3}{3} - \frac{1}{6} \cdot \frac{7}{7}$$
$$= \frac{-27}{42} - \frac{7}{42}$$

Step 3: Subtract the numerators and write the result over the common denominator.

$$= \frac{-27 - 7}{42}$$
$$= \frac{-34}{42}$$

Step 4: Write in lowest terms.

Factor −34 and 42:

$$\frac{-34}{42} = \frac{2 \cdot (-17)}{2 \cdot 21}$$

Write in lowest terms by dividing out like factors:

$$= \frac{\cancel{2} \cdot (-17)}{\cancel{2} \cdot 21}$$
$$= -\frac{17}{21}$$

Teaching Tip
Suggest that students align like factors vertically to more easily find the LCD.

$$\text{We see that } -\frac{9}{14} - \frac{1}{6} = -\frac{17}{21}.$$

SUMMARY: Steps to Add or Subtract Rational Numbers (Fractions) with Unlike Denominators

Step 1: Find the LCD of the rational numbers.

Step 2: Write each rational number with the LCD.

Step 3: Add or subtract the numerators and write the result over the common denominator.

Step 4: Write the result in lowest terms, if possible.

QUICK ✓ *Find each sum or difference, and write in lowest terms if necessary.*

21. $\frac{5}{12} - \frac{5}{18}$ **22.** $\frac{3}{14} + \frac{10}{21}$ **23.** $-\frac{23}{6} + \frac{7}{12}$ **24.** $\frac{3}{5} + \left(-\frac{4}{11}\right)$

Remember, the direction "evaluate" means to find the value of the expression, so the direction "evaluate $4 - \frac{2}{3}$" means to find the difference between 4 and $\frac{2}{3}$.

Classroom Example ➤
Evaluate and write in lowest terms if necessary: $3 - \frac{5}{8}$

Answer: $\frac{19}{8}$

EXAMPLE 7 Evaluating an Expression Containing Rational Numbers

Evaluate and write in lowest terms if necessary: $4 - \frac{2}{3}$

Solution

The key is to remember that $4 = \frac{4}{1}$.

$$4 - \frac{2}{3} = \frac{4}{1} - \frac{2}{3}$$

Rewrite each fraction with LCD = 3: $= \frac{4}{1} \cdot \frac{3}{3} - \frac{2}{3}$

$$= \frac{12}{3} - \frac{2}{3}$$

$$= \frac{10}{3}$$

QUICK ✓ *Evaluate and write in lowest terms, if necessary.*

25. $-2 + \frac{7}{16}$ **26.** $6 + \left(\frac{-9}{4}\right)$

4 Add, Subtract, Multiply, and Divide Rational Numbers Expressed as Decimals

Adding and Subtracting Decimals

To add or subtract decimals, we arrange the numbers in a column with the decimals aligned. Then add or subtract the digits in the like place values, and place the decimal point in the answer directly below the decimal point in the problem.

Classroom Example ➤
Evaluate each expression:
(a) $6.19 + 2.7 + 4.069$
(b) $47.2 - 3.87$

Answer:
(a) 12.959
(b) 43.33

EXAMPLE 8 Adding and Subtracting Decimals That Are the Same Sign

Evaluate each expression:

(a) $2.93 + 7.2 + 3.026$ **(b)** $76.4 - 4.95$

Solution

(a) Add zeros as placeholders:

$$\begin{array}{r} 2.930 \\ 7.200 \\ +3.026 \\ \hline 13.156 \end{array}$$

Work Smart
A whole number has an implied decimal point. For example,

$74 = 74.000$

(b) Add a zero as a placeholder:

$$\begin{array}{r} \overset{5\ 13\ 10}{76.40} \\ -4.95 \\ \hline 71.45 \end{array}$$

Classroom Example Evaluate each expression:

(a) $68.91 - (-32.614)$

(b) $-34.98 + 20.76$

Answer:

(a) 101.524

(b) -14.22

EXAMPLE 9 Adding and Subtracting Decimals That Are Different Signs

Evaluate each expression:

(a) $100.32 - (-32.015)$ **(b)** $-23.03 + 18.49$

Solution

(a) Remember, $a - (-b) = a + b$, so we can write $100.32 - (-32.015)$ as $100.32 + 32.015$.

$$\begin{array}{r} 100.320 \\ +32.015 \\ \hline 132.335 \end{array}$$

(b) To add real numbers with different signs (one positive and one negative), we subtract the smaller absolute value from the larger absolute value and attach the sign of the larger absolute value. Because $|-23.03| = 23.03$ and $|18.49| = 18.49$, we compute $23.03 - 18.49$ and attach a negative sign to the difference.

$$\begin{array}{r} 23.03 \\ -18.49 \\ \hline 4.54 \end{array}$$

So $-23.03 + 18.49 = -4.54$.

QUICK ✓ *Find the sum or difference.*

27. $9.67 + 11.344$ **28.** $81.96 - 17.39$ **29.** $14.95 + 7.118 + 0.3$

30. $345.67 - 8.0912$ **31.** $-180.782 + 100.3 + 9.07$ **32.** $-74.28 - 14.832$

Multiplying Decimals

The rules for multiplying decimals come from the rules for multiplying rational numbers written as fractions. For example,

$$\underbrace{-0.7}_{\text{1 decimal place}} \times \underbrace{0.03}_{\text{2 decimal places}} = \frac{-7}{10} \times \frac{3}{100} = \frac{-21}{1000} = \underbrace{-0.021}_{\text{3 decimal places}}$$

Notice that there are three digits to the right of the decimal point in the answer. This is equal to the sum of the number of digits to the right of the decimal point in the factors.

Notice also that the rules of signs that we learned in Section 1.3 apply to decimals as well. We use this result to demonstrate how to multiply decimals.

Classroom Example Find the product:

(a) 4.73×5.2

(b) -2.14×0.03

Answer:

(a) 24.596

(b) -0.0642

EXAMPLE 10 Multiplying Decimals

Find the product:

(a) 3.43×2.6 **(b)** -3.17×0.02

Solution

(a)

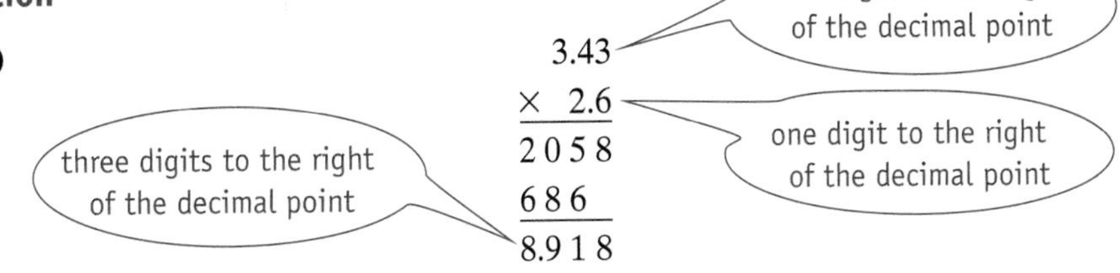

(b)

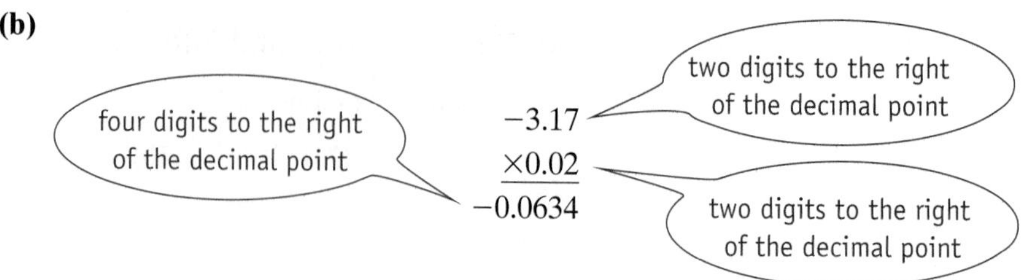

We summarize the procedure for multiplying decimals.

Work Smart
The number of digits to the right of the decimal point in the product is the *sum* of the number of digits to the right of the decimal point in the factors.

SUMMARY: Steps to Multiply Decimals

Step 1: Multiply the factors as if they were whole numbers.

Step 2: Count the total number of digits to the right of the decimal point in the factors.

Step 3: Place the decimal point in the product so that it contains the *sum* of the number of digits to the right of the decimal point in the factors.

QUICK ✓ *Find the product.*

33. 23.9×0.2 **34.** 9.1×7.24 **35.** -3.45×0.03

36. $257 \times (-3.5)$ **37.** $-0.03 \times (-0.45)$ **38.** 9.9×0.002

Dividing Decimals

First, we review vocabulary. In the division problem $2\overline{)7.94}$ with quotient 3.97, the number 2 is called the *divisor,* 7.94 is called the *dividend,* and 3.97 is the *quotient.* You see from this example that when we divide by a whole number, we place the decimal point in the quotient directly above the decimal point in the dividend. In algebra, we typically write this division problem as $\frac{7.94}{2} = 3.97$.

Teaching Tip
Students may be familiar with the method of dividing decimals that "moves" the decimal point in the divisor. Explain that the reason the decimal point is "moved" is that it is being transformed into a whole number.

To divide decimals, we want the divisor to be a whole number, so we multiply the dividend and the divisor by a power of 10 that will make the divisor a whole number. Then divide as though we were working with whole numbers. The decimal point in the quotient lies directly above the decimal point in the dividend.

Classroom Example ➤
Divide:

(a) $17.68 \div 13.6$

(b) $\frac{0.02484}{-0.27}$

Answer:

(a) 1.3

(b) −0.092

EXAMPLE 11 Dividing Decimals

(a) Divide: $\frac{22.26}{15.9}$ **(b)** Divide: $\frac{0.03724}{-0.38}$

Solution

(a) Since the divisor 15.9 is fifteen and nine-tenths, we multiply $\frac{22.26}{15.9}$ by $\frac{10}{10}$ to make the divisor a whole number. $\frac{22.26}{15.9} \cdot \frac{10}{10} = \frac{222.6}{159}$. Now we divide.

$$\begin{array}{r} 1.4 \\ 159\overline{)222.6} \\ \underline{159} \\ 63\ 6 \\ \underline{63\ 6} \\ 0 \end{array}$$

So $\frac{22.26}{15.9} = 1.4$.

(b) Because we have a positive number divided by a negative number, the quotient will be negative. Now, perform the division $\frac{0.03724}{0.38}$. Once we obtain this quotient, we will make it negative. The divisor 0.38 is thirty-eight hundredths, so we multiply $\frac{0.03724}{0.38}$ by $\frac{100}{100}$ to make the divisor a whole number. $\frac{0.03724}{0.38} \cdot \frac{100}{100} = \frac{3.724}{38}$. Now we divide.

$$\begin{array}{r} 0.098 \\ 38\overline{)3.724} \\ \underline{3\,42} \\ 304 \\ \underline{304} \\ 0 \end{array}$$

Therefore, $\frac{0.03724}{-0.38} = -0.098$.

We generalize the process for dividing decimals.

In Words
To divide decimals, change the divisor to a whole number and divide.

Steps to Divide Decimals

Step 1: Multiply the dividend and divisor by a power of ten that will make the divisor a whole number.

Step 2: Divide as though working with whole numbers.

Step 3: Place the decimal point in the quotient above the decimal point in the dividend.

QUICK ✓ *Find the quotient.*

39. $\frac{18.25}{73}$ **40.** $\frac{1.0032}{0.12}$ **41.** $\frac{-4.2958}{45.7}$ **42.** $\frac{0.1515}{-5.05}$

1.4 Exercises

1. reciprocals, multiplicative inverses
2. least common denominator
3. $\frac{a-b}{c}$
4. False
5. False
6. False

For Extra Help: Student Solutions Manual  CD Video  PH Math/Tutor Center  MathXL Tutorials on CD  MathXL® MyMathLab

Concepts and Vocabulary

In Problems 1–3, fill in the blank.

1. Two numbers are __________, or __________ __________, if the product of the numbers is 1.

2. The __________ __________ __________ is the smallest number that each denominator has as a common multiple.

3. $\frac{a}{c} - \frac{b}{c} = \frac{\_\_ - \_\_}{\_\_}, c \neq 0$

In Problems 4–6, answer True or False to each statement.

4. To add two rational expressions with the same denominator, we add the numerators and add the denominators.

5. $\frac{2}{3} \cdot \frac{5}{3} = \frac{10}{3}$

6. The reciprocal of $\frac{2}{3}$ is $-\frac{2}{3}$.

7. Answers may vary.
8. Answers may vary.
9. $\frac{2}{3}$ 10. $\frac{3}{5}$ 11. $-\frac{19}{9}$ 12. $-\frac{9}{4}$
13. $-\frac{1}{2}$ 14. $-\frac{8}{9}$ 15. $\frac{4}{5}$ 16. $\frac{7}{9}$
17. $-\frac{3}{5}$ 18. $-\frac{2}{3}$ 19. $\frac{2}{3}$ 20. $\frac{12}{13}$
21. $\frac{12}{25}$ 22. $\frac{5}{12}$ 23. -25 24. -27
25. $-\frac{2}{3}$ 26. $-\frac{8}{5}$ 27. 8 28. $\frac{5}{9}$
29. $\frac{31}{3}$ 30. $\frac{5}{2}$ 31. $\frac{6}{11}$ 32. $\frac{9}{28}$
33. $\frac{5}{3}$ 34. $\frac{4}{9}$ 35. $-\frac{1}{5}$ 36. $-\frac{1}{8}$
37. $\frac{5}{6}$ 38. 1 39. $-\frac{1}{9}$ 40. $-\frac{1}{16}$
41. $-\frac{2}{3}$ 42. $-\frac{40}{27}$ 43. $\frac{9}{11}$
44. $\frac{4}{3}$ 45. -12 46. $-\frac{27}{7}$
47. 32 48. 18 49. $\frac{3}{2}$ 50. 2
51. $\frac{1}{2}$ 52. 2 53. 2 54. 3
55. $\frac{1}{3}$ 56. $\frac{25}{8}$ 57. $\frac{5}{2}$ 58. $\frac{4}{3}$
59. $-\frac{13}{12}$ 60. $-\frac{16}{15}$ 61. $\frac{1}{4}$
62. $\frac{3}{8}$ 63. $\frac{9}{5}$ 64. $\frac{53}{9}$ 65. $-\frac{1}{6}$
66. $\frac{25}{18}$ 67. $\frac{33}{40}$ 68. $-\frac{203}{60}$
69. $\frac{139}{36}$ 70. $-\frac{53}{48}$ 71. $-\frac{7}{18}$
72. $-\frac{7}{8}$ 73. -6.5 74. -6.2
75. 8.4 76. 43.2 77. 55.92
78. 42.21

7. What does it mean for a rational number to be written in lowest terms?

8. Write an example that illustrates why $\frac{1}{2} + \frac{1}{2} = 1$.

Building Skills

In Problems 9–20, write each rational number in lowest terms.

9. $\frac{14}{21}$ **10.** $\frac{9}{15}$ **11.** $\frac{38}{-18}$ **12.** $\frac{81}{-36}$

13. $-\frac{22}{44}$ **14.** $-\frac{24}{27}$ **15.** $\frac{32}{40}$ **16.** $\frac{49}{63}$

17. $\frac{-15}{25}$ **18.** $\frac{-34}{51}$ **19.** $\frac{150}{225}$ **20.** $\frac{144}{156}$

In Problems 21–32, find the product, and write in lowest terms, if necessary.

21. $\frac{6}{5} \cdot \frac{2}{5}$ **22.** $\frac{7}{8} \cdot \frac{10}{21}$ **23.** $\frac{5}{-2} \cdot 10$ **24.** $\frac{3}{-7} \cdot 63$

25. $-\frac{3}{2} \cdot \frac{4}{9}$ **26.** $-\frac{5}{2} \cdot \frac{16}{25}$ **27.** $-\frac{22}{3} \cdot \left(-\frac{12}{11}\right)$ **28.** $-\frac{60}{75} \cdot \left(-\frac{25}{36}\right)$

29. $5 \cdot \frac{31}{15}$ **30.** $9 \cdot \frac{5}{18}$ **31.** $\frac{3}{4} \cdot \frac{8}{11}$ **32.** $\frac{4}{7} \cdot \frac{9}{16}$

In Problems 33–36, find the reciprocal of each number.

33. $\frac{3}{5}$ **34.** $\frac{9}{4}$ **35.** -5 **36.** -8

In Problems 37–48, find the quotient, and write in lowest terms, if necessary.

37. $\frac{4}{9} \div \frac{8}{15}$ **38.** $\frac{1}{2} \div \frac{3}{6}$ **39.** $-\frac{1}{3} \div 3$ **40.** $-\frac{1}{4} \div 4$

41. $\frac{5}{6} \div \left(-\frac{5}{4}\right)$ **42.** $\frac{4}{3} \div \left(-\frac{9}{10}\right)$ **43.** $\frac{36}{28} \div \frac{22}{14}$ **44.** $\frac{44}{63} \div \frac{11}{21}$

45. $-8 \div \left(\frac{2}{3}\right)$ **46.** $-3 \div \frac{7}{9}$ **47.** $-8 \div \left(-\frac{1}{4}\right)$ **48.** $-3 \div \left(-\frac{1}{6}\right)$

In Problem 49–72, find the sum or difference and write in lowest terms, if necessary.

49. $\frac{3}{4} + \frac{3}{4}$ **50.** $\frac{6}{11} + \frac{16}{11}$ **51.** $\frac{9}{8} - \frac{5}{8}$

52. $\frac{12}{5} - \frac{2}{5}$ **53.** $\frac{6}{7} - \left(-\frac{8}{7}\right)$ **54.** $\frac{2}{3} - \left(-\frac{7}{3}\right)$

55. $-\frac{5}{3} + 2$ **56.** $-\frac{7}{8} + 4$ **57.** $6 - \frac{7}{2}$

58. $3 - \frac{5}{3}$ **59.** $-\frac{4}{3} + \frac{1}{4}$ **60.** $-\frac{2}{5} + \left(-\frac{2}{3}\right)$

61. $\frac{7}{5} + \left(-\frac{23}{20}\right)$ **62.** $\frac{3}{4} + \left(-\frac{3}{8}\right)$ **63.** $\frac{7}{15} - \left(-\frac{4}{3}\right)$

64. $\frac{8}{3} - \left(-\frac{29}{9}\right)$ **65.** $\frac{8}{15} - \frac{7}{10}$ **66.** $\frac{17}{6} - \frac{13}{9}$

67. $-\frac{33}{10} - \left(-\frac{33}{8}\right)$ **68.** $-\frac{29}{6} - \left(-\frac{29}{20}\right)$ **69.** $\frac{19}{12} - \left(-\frac{41}{18}\right)$

70. $\frac{13}{12} - \frac{35}{16}$ **71.** $-\frac{2}{3} + \left(-\frac{5}{9}\right) + \frac{5}{6}$ **72.** $-\frac{1}{2} + \frac{3}{8} + \left(-\frac{3}{4}\right)$

In Problems 73–92, perform the indicated operation(s).

73. $-10.5 + 4$ **74.** $-13.2 + 7$ **75.** $-(-3.5) + 4.9$

76. $-(-32.9) + 10.3$ **77.** $39.1 - (-16.82)$ **78.** $29.23 - (-12.98)$

79. $-5.21 - (-6.7)$ **80.** $-4.94 - (-3.87)$ **81.** $45 - 2.45$

82. $32 - 5.68$ **83.** 4.3×5.8 **84.** 3.1×10.9

85. 0.075×120 **86.** 0.065×340 **87.** $\dfrac{136.08}{5.6}$

88. $\dfrac{332.59}{7.9}$ **89.** $\dfrac{25.48}{0.052}$ **90.** $\dfrac{48}{0.03}$

91. $-1.25 - (-0.6) + 1.6$ **92.** $-5.82 - (-2.9) + (-2.74)$

Mixed Practice

In Problems 93–124, evaluate and write in lowest terms, if necessary.

93. $-\dfrac{5}{6} + \dfrac{7}{15}$ **94.** $-\dfrac{8}{9} + \left(-\dfrac{16}{21}\right)$ **95.** $-\dfrac{10}{21} \cdot \dfrac{14}{5}$

96. $\dfrac{24}{5} \cdot \left(-\dfrac{35}{4}\right)$ **97.** $\dfrac{3}{8} \div \left(-\dfrac{9}{16}\right)$ **98.** $\dfrac{-12}{7} \div \dfrac{4}{-21}$

99. $-\dfrac{5}{12} + \dfrac{2}{12}$ **100.** $-\dfrac{4}{9} + \dfrac{1}{9}$ **101.** $-\dfrac{2}{7} - \dfrac{17}{5}$

102. $-\dfrac{3}{4} - \dfrac{1}{5}$ **103.** $-8.7 - (-10.3)$ **104.** $-4.63 - (-12.9)$

105. $\dfrac{1}{12} + \left(-\dfrac{5}{28}\right)$ **106.** $\dfrac{3}{16} + \left(-\dfrac{7}{40}\right)$ **107.** -12.03×4.2

108. $34.2 \times (-8.43)$ **109.** $36 \cdot \left(-\dfrac{4}{9}\right)$ **110.** $-\dfrac{8}{3} \cdot 15$

111. $-27 \div \dfrac{9}{5}$ **112.** $-24 \div \dfrac{8}{7}$ **113.** $3.62 - 10.2$

114. $4.75 - 6.2$ **115.** $\dfrac{-145.518}{18.42}$ **116.** $\dfrac{-297.078}{22.17}$

117. $\dfrac{12}{7} - \dfrac{17}{14} - \dfrac{48}{21}$ **118.** $\dfrac{9}{4} - \dfrac{21}{6} - \dfrac{11}{8}$

119. $54.2 - 18.78 - (-2.5) + 20.47$ **120.** $90.3 - 100.9 - (-34.26) + 32.95$

121. $400 \times 25.8 \times 0.003$ **122.** $500 \times 12.4 \times 0.02$

123. $-\dfrac{11}{12} - \left(-\dfrac{1}{6}\right) + \dfrac{7}{8}$ **124.** $\dfrac{8}{15} - \left(-\dfrac{7}{9}\right) + \dfrac{2}{3}$

Applying the Concepts

125. Watching TV If Rachel spends $\frac{1}{8}$ of her life watching TV, how many hours of TV does she watch in one week?

126. Halloween Candy Henry decided to make $\frac{2}{3}$ oz. bags of candy for treats at Halloween. If he bought 16 oz. of candy, how many bags will he have to give away?

127. Biology Class Susan's biology class begins with 36 students. If $\frac{2}{3}$ will finish the course and $\frac{3}{4}$ of those get a passing grade, how many students will pass Susan's biology class this term?

128. Pizza Time Joyce and Ramie bought a pizza. Joyce ate $\frac{2}{5}$ of the pizza and Ramie ate $\frac{1}{9}$ of what was left. What fraction of the pizza remains uneaten?

129. Overdrawn Maria's checking account was overdrawn by \$43.29. The bank charged overdraft fees amounting to \$25.50. What is the balance in Maria's checking account now?

79. 1.49
80. -1.07
81. 42.55
82. 26.32
83. 24.94
84. 33.79
85. 9
86. 22.1
87. 24.3
88. 42.1
89. 490
90. 1600
91. 0.95
92. -5.66
93. $-\frac{11}{30}$
94. $-\frac{104}{63}$
95. $-\frac{4}{3}$
96. -42
97. $-\frac{2}{3}$
98. 9
99. $-\frac{1}{4}$
100. $-\frac{1}{3}$
101. $-\frac{129}{35}$
102. $-\frac{19}{20}$
103. 1.6
104. 8.27
105. $-\frac{2}{21}$
106. $\frac{1}{80}$
107. -50.526
108. -288.306
109. -16
110. -40
111. -15
112. -21
113. -6.58
114. -1.45
115. -7.9
116. -13.4
117. $-\frac{25}{14}$
118. $-\frac{21}{8}$
119. 58.39
120. 56.61
121. 30.96
122. 124
123. $\frac{1}{8}$
124. $\frac{89}{45}$
125. 21 hours
126. 24 bags
127. 18 students
128. $\frac{8}{15}$
129. $-\$68.79$

130. 39.3°
131. \$2.40
132. −\$174.81

130. Temperature The temperature in Yellowstone National Park fell from 25.6° above zero to 13.7° below zero. How much did the temperature drop?

131. Stock Prices The price per share of Intel stock has been up and down lately. On Monday it rose 2.75; on Tuesday it rose 0.87; on Wednesday it dropped 1.12; on Thursday it rose 0.52; and on Friday it fell 0.62. What was the net change in Intel's stock price per share for the week?

132. Bank Balance Henry started the month with \$43.68 in his checking account. During the month the following transactions occurred: He deposited his paycheck of \$929.30; and he wrote checks for rent \$650, phone \$33.49, credit card \$229.50, cable service \$75.50, and groceries \$159.30. How much does he have in his account now?

Extending the Concepts

Problems 133–136 use the following definition.

If P and Q are two points on a real number line with coordinates a and b, respectively, the **distance between P and Q**, denoted by $d(P, Q)$, is

$$d(P, Q) = |b - a|$$

133. 13.2
134. 15.1
135. $\frac{86}{15}$
136. $\frac{29}{6}$

133. Find the distance between the points P and Q on the real number line if $P = -9.7$ and $Q = 3.5$.

134. Find the distance between the points P and Q on the real number line if $P = -12.5$ and $Q = 2.6$.

135. Find the distance between the points P and Q on the real number line if $P = -\frac{13}{3}$ and $Q = \frac{7}{5}$.

136. Find the distance between the points P and Q on the real number line if $P = -\frac{5}{6}$ and $Q = 4$.

PUTTING THE CONCEPTS TOGETHER (SECTIONS 1.2–1.4)

1. (a) $-12, -\frac{14}{7} = -2, 0, 3$
(b) $-12, -\frac{14}{7} = -2, -1.25, 0, 3, 11.2$
(c) $\sqrt{2}$
(d) All the numbers in the set are real numbers.
2. <
3. −11
4. −65
5. −27
6. 27
7. −5.5
8. −10
9. −100 10. 72
11. −5

These problems cover important concepts from Sections 1.2 to 1.4. We designed these problems so that you can review the chapter so far and show your mastery of the concepts. Take time to work these problems before proceeding with the next section. The answers to these problems are located at the back of the text starting on page AN-1.

1. Use the set $\left\{-12, -\frac{14}{7}, -1.25, 0, \sqrt{2}, 3, 11.2\right\}$. List all of the elements that are:

(a) integers **(b)** rational numbers
(c) irrational numbers **(d)** real numbers

2. Replace the ? with the correct symbol >, <, =: $\frac{1}{8}$? 0.5

In Problems 3–23, perform the indicated operation. Reduce to lowest terms if necessary.

3. $17 + (-28)$ **4.** $-23 + (-42)$ **5.** $18 - 45$

6. $3 - (-24)$ **7.** $-18 - (-12.5)$ **8.** $(-5)(2)$

9. $25(-4)$ **10.** $(-8)(-9)$ **11.** $\frac{-35}{7}$

12. 16
13. −9 14. −3
15. $\frac{31}{5}$ 16. $\frac{31}{36}$
17. $-\frac{17}{36}$ 18. $\frac{9}{5}$
19. $-\frac{1}{28}$ 20. 0
21. 10.76 22. 7.646
23. 1.46 24. 22.232

12. $\frac{-32}{-2}$ **13.** $27 \div -3$ **14.** $-\frac{4}{5} - \frac{11}{5}$

15. $7 - \frac{4}{5}$ **16.** $\frac{7}{12} + \frac{5}{18}$ **17.** $-\frac{5}{12} - \frac{1}{18}$

18. $\frac{6}{25} \cdot 15 \cdot \frac{1}{2}$ **19.** $\frac{2}{7} \div (-8)$ **20.** $\frac{0}{-8}$

21. $3.56 - (-7.2)$ **22.** $18.946 - 11.3$ **23.** $62.488 \div 42.8$

24. $(7.94)(2.8)$

1.5 Properties of Real Numbers

OBJECTIVES

1. Understand and Use the Identity Properties of Addition and Multiplication
2. Understand and Use the Commutative Properties of Addition and Multiplication
3. Understand and Use the Associative Properties of Addition and Multiplication
4. Understand the Multiplication and Division Properties of 0

Preparing for Properties of Real Numbers

Before getting started, take the following readiness quiz. If you get a problem wrong, go to the section cited and review the material.

1. Find the sum: $12 + 3 + (-12)$ [Section 1.3, pp. 18–21]

2. Find the product: $\frac{3}{4} \cdot 11 \cdot \frac{4}{3}$ [Section 1.4, p. 29]

This section is dedicated to presenting properties of real numbers. A property in mathematics is a rule that is always true. These properties will be used throughout this course and future math courses, so it is extremely important that you understand these properties and know how to use them.

1 Understand and Use the Identity Properties of Addition and Multiplication

The real number 0 has an interesting property. It turns out that 0 is the only number that when added to any real number a results in the same real number a.

In Words
The word "identity" comes from a Latin word that means "the same" or "alike." The Identity Property of Addition means adding zero to a real number keeps the number the same.

IDENTITY PROPERTY OF ADDITION

For any real number a,

$$0 + a = a + 0 = a$$

That is, the sum of any number and 0 is that number. We call 0 the **additive identity.**

The real number 1 also has an interesting property. Recall that the expression $5 \cdot 3$ is equivalent to adding 5 three times. That is, $5 \cdot 3 = 5 + 5 + 5$. Therefore, $5 \cdot 1$ means to add 5 once, so, $5 \cdot 1 = 5$. This result is true in general.

MULTIPLICATIVE IDENTITY

For any real number a,

$$a \cdot 1 = 1 \cdot a = a$$

That is, the product of any number and 1 is that number. We call 1 the **multiplicative identity.**

Preparing for...Answers **1.** 3 **2.** 11

We also use the multiplicative identity throughout our math careers to create new expressions that are equivalent to previous expressions. For example, the expressions

$$\frac{4}{5} \quad \text{and} \quad \frac{4}{5}\cdot\frac{3}{3}$$

are equivalent because $\frac{3}{3} = 1$.

Conversion

A practical use of the multiplicative identity is conversion. **Conversion** is changing the units of measure (such as inches or pounds) from one measure to a different measure. For example, we might change a length from inches to feet or a weight from ounces to pounds.

Classroom Example Marcello measures his brick patio and finds that it is 159 inches long. How many feet long is the patio?

Answer: The patio is $\frac{53}{4} = 13\frac{1}{4}$ feet or 13 feet 3 inches long.

Teaching Tip Just as dividing out identical numbers produces a quotient of 1, dividing out identical units of measure also produces a quotient of 1.

EXAMPLE 1 Converting from Inches to Feet

Janice measures her family room and finds that its length is 184 inches long. How many feet long is Janice's family room? (*Note:* 12 inches = 1 foot)

Solution

The trick in doing conversions is to make sure that the units of measure that you are trying to remove get divided out and the new unit of measure remains. In this problem, we want the inches to divide out with feet remaining. Because 12 inches equals 1 foot, multiplying 184 inches by $\frac{1 \text{ foot}}{12 \text{ inches}}$ is like multiplying by one.

$$184 \text{ inches} = 184 \text{ inches}\cdot\overbrace{\frac{1 \text{ foot}}{12 \text{ inches}}}^{\text{Multiplying by 1}}$$

The inches "divide out": $= \frac{184}{12}\text{feet}$

$184 = 2\cdot2\cdot2\cdot23;\ 12 = 2\cdot2\cdot3$: $= \frac{2\cdot2\cdot2\cdot23}{2\cdot2\cdot3}\text{feet}$

Divide out common factors: $= \frac{\not{2}\cdot\not{2}\cdot2\cdot23}{\not{2}\cdot\not{2}\cdot3}\text{feet}$

$= \frac{46}{3}\text{feet}$

Work Smart

$$\begin{array}{r} 15 \\ 3\overline{)46} \\ \underline{-3} \\ 16 \\ \underline{-15} \\ 1 \end{array}$$

So 184 inches equals $\frac{46}{3}$ feet. Because 46 divided by 3 is 15 with a remainder of 1 (Why? See the Work Smart), $\frac{46}{3}$ feet is equivalent to $15\frac{1}{3}$ feet. Since $\frac{1}{3}\text{foot}\cdot\frac{12 \text{ inches}}{1 \text{ foot}} = 4$ inches, we have that $\frac{46}{3}$ feet is equivalent to 15 feet, 4 inches. ■

QUICK ✓ *Convert each measurement to the indicated unit of measurement.*

1. 96 inches = ? feet [1 foot = 12 inches]

2. 500 minutes = ? hours [60 minutes = 1 hour]

3. 88 ounces = ? pounds [16 ounces = 1 pound]

2 Understand and Use the Commutative Properties of Addition and Multiplication

We illustrate another property of real numbers in the next example.

Classroom Example
(a) $3 + 10 = 13$
$10 + 3 = 13$
so $3 + 10 = 10 + 3$
(b) $4 \cdot 9 = 36$
$9 \cdot 4 = 36$
so $4 \cdot 9 = 9 \cdot 4$

EXAMPLE 2 Illustrating the Commutative Properties

(a) $4 + 7 = 11$ and
$7 + 4 = 11$ so
$4 + 7 = 7 + 4$

(b) $3 \cdot 8 = 24$ and
$8 \cdot 3 = 24$ so
$3 \cdot 8 = 8 \cdot 3$

The results of Example 2 are true in general.

COMMUTATIVE PROPERTIES OF ADDITION AND MULTIPLICATION

If a and b are real numbers, then

$$a + b = b + a \quad \text{and} \quad a \cdot b = b \cdot a$$

In Words
The **Commutative Property** of real numbers states that the order in which we add or multiply real numbers does not affect the final result.

We add real numbers from left to right. We multiply real numbers from left to right. The **Commutative Property** allows us to write $3 + 5$ as $5 + 3$ or $3 \cdot 5$ as $5 \cdot 3$ without affecting the value of the expression. Why is this important? As the next example illustrates, by rearranging addition or multiplication problems, some expressions become a lot easier to evaluate.

Classroom Example
Evaluate the expression: $15 + 4 + (-15)$
Answer: 4

EXAMPLE 3 Using the Commutative Property of Addition

Evaluate the expression: $18 + 3 + (-18)$

Solution

$$18 + 3 + (-18) = 18 + (-18) + 3$$

Add $18 + (-18)$: $= 0 + 3$

Add $3 + 0$: $= 3$

It is also true here that if we add in order, left to right, we also obtain the sum $18 + 3 + (-18) = 21 + (-18) = 3$. Notice that rearranging the numbers made the problem easier!

Teaching Tip
Have students think of real life situations in which the commutative property is true. Then think of other situations in which it is not true. Is putting on your socks and putting on your shoes commutative? Is putting on you shirt and putting on your shoes commutative?

Does subtraction obey the Commutative Property? In other words, does $3 - 14 = 14 - 3$? Because $3 - 14 = 3 + (-14) = -11$ while $14 - 3 = 14 + (-3) = 11$, we see that **subtraction is not commutative.**

QUICK ✓ *Use the Commutative Property of Addition to find the sum of the real numbers.*

4. $(-8) + 22 + 8$ **5.** $\frac{8}{15} + \frac{3}{20} + \left(-\frac{8}{15}\right)$ **6.** $2.1 + 11.98 + (-2.1)$

Classroom Example
Find each product:
(a) $-15 \cdot 9 \cdot \left(-\frac{4}{5}\right)$
(b) $1000 \cdot 470.8 \cdot 0.001$
Answer:
(a) 108 (b) 470.8

EXAMPLE 4 Using the Commutative Property of Multiplication

Find each product.

(a) $-27 \cdot 7 \cdot \left(-\frac{2}{9}\right)$ **(b)** $100 \cdot 307.5 \cdot 0.01$

Solution

(a) $-27 \cdot 7 \cdot \left(-\frac{2}{9}\right) = -27 \cdot \left(-\frac{2}{9}\right) \cdot 7$

$= -\overset{3}{\cancel{27}} \cdot \left(-\frac{2}{\underset{1}{\cancel{9}}}\right) \cdot 7$

$= -3 \cdot (-2) \cdot 7$

$= 6 \cdot 7$

$= 42$

(b) $100 \cdot 307.5 \cdot 0.01 = 100 \cdot 0.01 \cdot 307.5$

$= 1 \cdot 307.5$

$= 307.5$

QUICK ✓ *Find the product of the real numbers.*

7. $-8 \cdot (-13) \cdot \left(-\frac{3}{4}\right)$ **8.** $\frac{5}{22} \cdot \frac{18}{331} \cdot \left(-\frac{44}{5}\right)$ **9.** $100{,}000 \cdot 349 \cdot 0.00001$

Work Smart
Neither division nor subtraction is commutative.

Does division obey the Commutative Property? That is, does $a \div b = b \div a$? To see the answer, we ask does $8 \div 2 = 2 \div 8$? Because $8 \div 2 = \frac{8}{2} = 4$, but $2 \div 8 = \frac{2}{8} = \frac{1}{4}$, we conclude that **division is not commutative.**

3 Understand and Use the Associative Properties of Addition and Multiplication

Sometimes **grouping symbols,** such as parentheses ( ), brackets [ ], or braces { } are used to indicate that the operation within the grouping symbols is to be performed first. For example, $3 \cdot (8 + 3)$ indicates that we should first add 8 and 3 and then multiply this sum by 3.

Earlier we mentioned that addition is to be performed from left to right. We also stated that multiplication should be performed from left to right. But does the order in which we add (or multiply) three or more numbers matter? Let's see.

Classroom Example
Use Example 5 as a classroom example.

EXAMPLE 5 Illustrating the Associate Properties

(a) $2 + (8 + 6) = 2 + 14 = 16$ and $(2 + 8) + 6 = 10 + 6 = 16$

so

$$2 + (8 + 6) = (2 + 8) + 6$$

(b) $-4 \cdot (9 \cdot 2) = -4 \cdot (18) = -72$ and $(-4 \cdot 9) \cdot 2 = -36 \cdot 2 = -72$

so

$$-4 \cdot (9 \cdot 2) = (-4 \cdot 9) \cdot 2$$

Examples 4(a) and (b) illustrate the **Associative Properties of Addition and Multiplication.**

Teaching Tip
Point out that the arrangement of the numbers doesn't change in the Associative Property—the *grouping* changes.

ASSOCIATIVE PROPERTIES OF ADDITION AND MULTIPLICATION

If a, b, and c are real numbers, then

$$a + (b + c) = (a + b) + c = a + b + c$$
$$a \cdot (b \cdot c) = (a \cdot b) \cdot c = a \cdot b \cdot c$$

Classroom Example ➤
Use the Associative Property of Addition to evaluate $17 + 962 + (-962)$.

Answer: 17

EXAMPLE 6 Using the Associative Property of Addition

Use the Associative Property to evaluate $23 + 453 + (-453)$.

Solution

There are three numbers to be added. However, we see that the second and third numbers are additive inverses, so we use the Associative Property of Addition and insert parentheses around $453 + (-453)$ and perform this operation first.

$$\begin{aligned} 23 + 453 + (-453) &= 23 + (453 + (-453)) \\ &= 23 + 0 \\ &= 23 \end{aligned}$$

Classroom Example ➤
Use the Associative Property of Multiplication to evaluate $-\frac{4}{13} \cdot \frac{5}{9} \cdot \frac{27}{10}$.

Answer: $-\frac{6}{13}$

EXAMPLE 7 Using the Associative Property of Multiplication

Use the Associative Property to evaluate $-\frac{3}{11} \cdot \frac{9}{4} \cdot \frac{8}{3}$.

Solution

There are three factors. If we look carefully at the problem, we notice that the second and third factors have common factors that can be divided out. So, we use the Associative Property of Multiplication and insert parentheses around $\frac{9}{4} \cdot \frac{8}{3}$ and perform this operation first.

$$\begin{aligned} -\frac{3}{11} \cdot \frac{9}{4} \cdot \frac{8}{3} &= -\frac{3}{11} \cdot \left(\frac{9}{4} \cdot \frac{8}{3}\right) \\ &= -\frac{3}{11} \cdot \frac{\overset{3}{\cancel{9}}}{\underset{1}{\cancel{4}}} \cdot \frac{\overset{2}{\cancel{8}}}{\underset{1}{\cancel{3}}} \\ &= -\frac{3}{11} \cdot (3 \cdot 2) \\ &= -\frac{3}{11} \cdot 6 \\ &= -\frac{18}{11} \end{aligned}$$

QUICK ✓ *Use the Associative Property to evaluate each expression.*

10. $14 + 101 + (-101)$

11. $14 \cdot \frac{1}{5} \cdot 5$

12. $-34.2 + 12.6 + (-2.6)$

13. $\frac{19}{2} \cdot \frac{4}{38} \cdot \frac{50}{13}$

4 Understand the Multiplication and Division Properties of 0

We know that when we multiply any real number a by 1, the product is a. Now let's look at multiplication by zero.

MULTIPLICATION PROPERTY OF ZERO

For any real number a, the product of a and 0 is always 0; that is,

$$a \cdot 0 = 0 \cdot a = 0$$

We now introduce some division properties of the number 0.

Work Smart

Division by zero is not allowed. That is, 0 cannot be used as a divisor.

DIVISION PROPERTIES OF ZERO

For any nonzero real number a,

1. The quotient of 0 and a is 0. That is, $\frac{0}{a} = 0$.

2. The quotient of a and 0 is **undefined.** That is, $\frac{a}{0}$ is undefined.

Teaching Tip

Strongly emphasize the difference between expressions such as $\frac{0}{8} = 0$ and $\frac{8}{0}$, undefined.

To see why these statements are true, consider the following. When we divide, we can check the quotient by multiplication. For example, $\frac{12}{4} = 3$ because $4 \cdot 3 = 12$. In the same way, $\frac{0}{4} = 0$ because $4 \cdot 0 = 0$. However, what is the value of $\frac{12}{0}$? To determine this quotient, we should be able to determine a real number such that $0 \cdot \square = 12$. But since the product of 0 and every real number is 0, there is no replacement value for $\square$.

Classroom Example

Find the quotient:

(a) $\frac{17}{0}$ (b) $\frac{0}{11}$

Answer:

(a) undefined

(b) 0

EXAMPLE 8 Using Zero as a Divisor and a Dividend

Find the quotient:

(a) $\frac{23}{0}$ **(b)** $\frac{0}{17}$

Solution

(a) $\frac{23}{0}$ is undefined because 0 is the divisor.

(b) $\frac{0}{17} = 0$ because 0 is the dividend.

QUICK ✓ *Tell if the quotient is zero or undefined.*

14. $\frac{0}{22}$ **15.** $\frac{-11}{0}$ **16.** $-\frac{0}{5}$ **17.** $\frac{5678}{0}$

We conclude this section with a summary of the properties of addition, multiplication, and division.

Work Smart

The Commutative Property changes **order** and the Associative Property changes **grouping.**

SUMMARY: Properties of Addition

Identity Property of Addition

For any real number a, $0 + a = a + 0 = a$.

Commutative Property of Addition

If a and b are real numbers, then $a + b = b + a$.

Additive Inverse Property

For any real number a, $a + (-a) = -a + a = 0$.

Associative Property of Addition

If a, b, and c are real numbers, then $a + (b + c) = (a + b) + c$.

SUMMARY: Properties of Multiplication and Division

Commutative Property of Multiplication

If a and b are real numbers, then $a \cdot b = b \cdot a$.

Multiplication Property of Zero

For any real number a, the product of a and 0 is always 0; that is, $a \cdot 0 = 0 \cdot a = 0$.

Multiplicative Identity

$a \cdot 1 = 1 \cdot a = a$ for any real number a.

Associative Property of Multiplication

If a, b, and c are real numbers, then $a \cdot (b \cdot c) = (a \cdot b) \cdot c$.

Multiplicative Inverse Property

$$a \cdot \frac{1}{a} = \frac{1}{a} \cdot a = 1 \quad \text{provided that } a \neq 0$$

Division Properties of Zero

For any nonzero number a,

1. The quotient of 0 and a is 0. That is, $\frac{0}{a} = 0$.

2. The quotient of a and 0 is undefined. That is, $\frac{a}{0}$ is undefined.

1.5 Exercises

1. Identity Property; Addition
2. Multiplicative Identity
3. undefined
4. False
5. True
6. True
7. Answers may vary.
8. Answers may vary.
9. Additive Inverse Property
10. Commutative Property of Multiplication
11. Multiplicative Identity Property
12. Commutative Property of Addition

For Extra Help:  Student Solutions Manual  CD Video  PH Math/Tutor Center  MathXL Tutorials on CD MathXL® MyMathLab

Concepts and Vocabulary

In Problems 1–3, fill in the blank.

1. The _________ _________ of _________ states that for any real number a, $0 + a = a + 0 = a$.

2. Because $a \cdot 1 = 1 \cdot a = a$ for any real number a, we call 1 the _________ _________.

3. The quotient of a and 0 is _________.

In Problems 4–6, answer True or False to each statement.

4. Division is commutative.

5. Addition is associative.

6. The Commutative Property of Addition says that changing the order of an addition problem will not change the sum.

7. How is the Identity Property of Addition related to the Additive Inverse Property? How is the Multiplicative Identity Property related to the Multiplicative Inverse Property?

8. In your own words, explain the Associative Property of Addition. Explain the key elements to look for when identifying this property and when you would use it to evaluate an expression.

Skill Building

In Problems 9–20, state the property of real numbers that is being illustrated.

9. $16 + (-16) = 0$

10. $4 \cdot 63 \cdot \frac{1}{4} = 4 \cdot \frac{1}{4} \cdot 63$

11. $\frac{3}{4}$ is equivalent to $\frac{3}{4} \cdot \frac{5}{5}$

12. $4 + 5 + (-4)$ is equivalent to $4 + (-4) + 5$

13. Multiplicative Inverse Property
14. Additive Inverse Property
15. Additive Inverse Property
16. Associative Property of Multiplication
17. Commutative Property of Multiplication
18. Multiplicative Inverse Property
19. Commutative Property of Addition
20. Multiplicative Property of Zero
21. 156 inches
22. $43\frac{1}{3}$ yards = 43 yards, 1 foot
23. 45 meters
24. 59 meters
25. $10\frac{1}{2}$ gallons
26. $14\frac{1}{2}$ gallons
27. $11\frac{1}{4}$ pounds
28. $7\frac{1}{2}$ pounds
29. $4\frac{1}{2}$ hours = 4 hours, 30 min.
30. $6\frac{1}{4}$ hours = 6 hours, 15 min.
31. 29
32. 59
33. 18
34. 28
35. -65
36. -72
37. 347
38. 593
39. -90
40. 13
41. undefined
42. 0
43. -34
44. 2
45. 0
46. 1
47. 0
48. 30
49. $-\frac{20}{3}$
50. $-\frac{21}{16}$
51. \$203.16
52. \$31.72
53. Answers may vary.
54. Answers may vary.
55. Answers may vary.
56. $-3-(4-10)$
57. $-6-(4+10)$
58. $-15+10-(4-8)$
59. $25-(6-10)-1$
60. 44 feet per second
61. $58\frac{2}{3}$ feet per second

13. $12\cdot\frac{1}{12}=1$

14. $-236+236=0$

15. $34.2+(-34.2)=0$

16. $(4\cdot5)\cdot7=4\cdot(5\cdot7)$

17. $\frac{2}{3}\cdot\left(-\frac{12}{43}\right)\cdot\frac{3}{2}=\frac{2}{3}\cdot\frac{3}{2}\cdot\left(-\frac{12}{43}\right)$

18. $\frac{5}{12}\cdot\frac{12}{5}=1$

19. $5.23+4.98+(-5.23)=5.23+(-5.23)+4.98$

20. $16.4\cdot0=0$

In Problems 21–30, convert each measurement to the indicated unit of measurement. Use the following conversions:

1 foot = 12 inches   3 feet = 1 yard   1 gallon = 4 quarts
100 centimeters = 1 meter   16 ounces = 1 pound

21. 13 feet to inches

22. 130 feet to yards

23. 4500 centimeters to meters

24. 5900 centimeters to meters

25. 42 quarts to gallons

26. 58 quarts to gallons

27. 180 ounces to pounds

28. 120 ounces to pounds

29. 16,200 seconds to hours

30. 22,500 seconds to hours

Mixed Practice

In Problems 31–50, evaluate each expression by using the properties of real numbers.

31. $54+29+(-54)$

32. $46+59+(-46)$

33. $\frac{9}{5}\cdot\frac{5}{9}\cdot18$

34. $\frac{4}{9}\cdot\frac{9}{4}\cdot28$

35. $-25\cdot13\cdot\frac{1}{5}$

36. $36\cdot(-12)\cdot\frac{1}{6}$

37. $347+456+(-456)$

38. $593+306+(-306)$

39. $\frac{9}{2}\cdot\left(-\frac{10}{3}\right)\cdot6$

40. $\frac{13}{2}\cdot\frac{8}{39}\cdot\frac{39}{4}$

41. $\frac{7}{0}$

42. $\frac{0}{100}$

43. $100(-34)(0.01)$

44. $4000(0.5)(0.001)$

45. $569.003\cdot0$

46. $104\cdot\frac{1}{104}$

47. $\frac{45}{3902}+\left(-\frac{45}{3902}\right)$

48. $30\cdot\frac{4}{4}$

49. $-\frac{5}{44}\cdot\frac{80}{3}\cdot\frac{11}{5}$

50. $\frac{7}{48}\cdot\left(-\frac{21}{4}\right)\cdot\frac{12}{7}$

Applying the Concepts

51. Balancing the Checkbook Alberto's checking account balance at the start of the month was \$321.03. During the month, he wrote checks for \$32.84, \$85.03, and \$120.56. He also deposited a check for \$120.56. What is Alberto's balance at the end of the month?

52. Stock Price Before the opening bell on Monday, a certain stock was priced at \$32.04. On Monday the stock was up \$0.54, on Tuesday it was down \$0.32, and on Wednesday it was down \$0.54. What was the closing price of the stock on Wednesday?

53. In your own words, explain why 0 does not have a multiplicative inverse.

54. Why does $2(4\cdot5)$ not equal $(2\cdot4)\cdot(2\cdot5)$?

55. Why does $\frac{0}{4}=0$? Why is $\frac{4}{0}$ undefined?

Extending the Concepts

In Problems 56–59, insert parentheses to make the statement true.

56. $-3-4-10=3$

57. $-6-4+10=-20$

58. $-15+10-4-8=-1$

59. $25-6-10-1=28$

60. Convert 30 miles per hour to feet per second. (*Note:* 1 mile = 5280 feet)

61. Convert 40 miles per hour to feet per second. (*Note:* 1 mile = 5280 feet)

1.6 Exponents and the Order of Operations

OBJECTIVES

1. Evaluate Exponential Expressions
2. Apply the Rules for Order of Operations

Preparing for Exponents and the Order of Operations

Before getting started, take this readiness quiz. If you get a problem wrong, go back to the section cited and review the material.

1. Find the sum: $9 + (-19)$ [Section 1.3, pp. 18–21]

2. Find the difference: $28 - (-7)$ [Section 1.3, pp. 21–23]

3. Find the product: $-7 \cdot \frac{8}{3} \cdot 36$ [Section 1.4, p. 29]

4. Find the quotient: $\frac{100}{-15}$ [Section 1.3, pp. 24–26]

1 Evaluate Exponential Expressions

Suppose that we wanted to multiply 2 eight times. We would write this as

$$2 \cdot 2 \cdot 2 \cdot 2 \cdot 2 \cdot 2 \cdot 2 \cdot 2$$

That's a lot of writing! To reduce the amount of writing that is needed to represent this repeated multiplication, we introduce some new notation. Using this new notation, called **exponential notation,** we would write $2 \cdot 2 \cdot 2 \cdot 2 \cdot 2 \cdot 2 \cdot 2 \cdot 2$ as 2^8. The number 2 is called the **base** and the number 8 is called the **exponent.**

EXPONENTIAL NOTATION

If n is a natural number and a is a real number, then

$$a^n = \underbrace{a \cdot a \cdot a \cdot \cdots \cdot a}_{n \text{ factors}}$$

where a is called the **base** and the natural number n is called the **exponent** or **power.** The exponent tells the number of times the base is used as a factor.

An expression written in the form a^n is said to be in **exponential form.** The expression 6^2 is read "six squared," 8^3 is read "eight cubed," and the expression 11^4 is read "eleven to the fourth power." In general, we read a^n as "a to the nth power."

Classroom Example
Write each expression in exponential form:
(a) $7 \cdot 7 \cdot 7 \cdot 7 \cdot 7 \cdot 7$
(b) $(-8) \cdot (-8) \cdot (-8)$
Answer:
(a) 7^6 (b) $(-8)^3$

EXAMPLE 1 Writing a Numerical Expression in Exponential Form

Write each expression in exponential form.

(a) $5 \cdot 5 \cdot 5$ **(b)** $(-4) \cdot (-4) \cdot (-4) \cdot (-4) \cdot (-4) \cdot (-4)$

Solution

(a) The expression $5 \cdot 5 \cdot 5$ contains three factors of 5, so $5 \cdot 5 \cdot 5 = 5^3$.

(b) The expression $(-4) \cdot (-4) \cdot (-4) \cdot (-4) \cdot (-4) \cdot (-4)$ contains 6 factors of -4, so $(-4) \cdot (-4) \cdot (-4) \cdot (-4) \cdot (-4) \cdot (-4) = (-4)^6$.

QUICK ✓ *Write each expression in exponential form.*

1. $11 \cdot 11 \cdot 11 \cdot 11 \cdot 11$ **2.** $7 \cdot 7 \cdot 7 \cdot 7 \cdot 7 \cdot 7 \cdot 7 \cdot 7$ **3.** $(-2) \cdot (-2) \cdot (-2)$

To evaluate an exponential expression, we write the expression in **expanded form.** For example, 2^8 in expanded form would be $2 \cdot 2 \cdot 2 \cdot 2 \cdot 2 \cdot 2 \cdot 2 \cdot 2$. We then perform the multiplication.

Preparing for...Answers **1.** -10 **2.** 35 **3.** -672 **4.** $-\frac{20}{3}$

Classroom Example
Evaluate each exponential expression:
(a) 5^4 (b) $\left(\frac{3}{4}\right)^3$
Answer:
(a) 625 (b) $\frac{27}{64}$

EXAMPLE 2 Evaluating an Exponential Expression

Evaluate each exponential expression:

(a) 6^4 **(b)** $\left(\frac{5}{3}\right)^5$

Solution

(a) $6^4 = 6 \cdot 6 \cdot 6 \cdot 6$
$= 1296$

(b) $\left(\frac{5}{3}\right)^5 = \left(\frac{5}{3}\right)\left(\frac{5}{3}\right)\left(\frac{5}{3}\right)\left(\frac{5}{3}\right)\left(\frac{5}{3}\right)$
$= \frac{5 \cdot 5 \cdot 5 \cdot 5 \cdot 5}{3 \cdot 3 \cdot 3 \cdot 3 \cdot 3}$
$= \frac{3125}{243}$

Classroom Example
Evaluate each exponential expression:
(a) $(-3)^5$ (b) -3^5
Answer:
(a) -243 (b) -243

EXAMPLE 3 Evaluating an Exponential Expression—Odd Exponent

Evaluate each exponential expression:

(a) $(-5)^3$ **(b)** -5^3

Solution

(a) $(-5)^3 = (-5) \cdot (-5) \cdot (-5)$
$= -125$

(b) $-5^3 = -(5 \cdot 5 \cdot 5)$
$= -125$

Classroom Example
Evaluate each exponential expression:
(a) $(-3)^4$ (b) -3^4.
Answer:
(a) 81 (b) -81

EXAMPLE 4 Evaluating an Exponential Expression—Even Exponent

Evaluate each exponential expression:

(a) $(-5)^4$ **(b)** -5^4

Solution

(a) $(-5)^4 = (-5) \cdot (-5) \cdot (-5) \cdot (-5)$
$= 625$

(b) $-5^4 = -(5 \cdot 5 \cdot 5 \cdot 5)$
$= -625$

Work Smart
There is a difference between finding $(-5)^4$ and -5^4.
The parentheses around (-5) tell us to use four factors of -5. However, in the expression -5^4, we use 5 as a factor four times and then multiply the result by -1. We could also read -5^4 as "Take the opposite of the quantity 5^4."

QUICK ✓ *Evaluate each exponential expression.*

4. 2^4 **5.** $(-8)^2$ **6.** $\left(-\frac{1}{6}\right)^3$

7. $(0.9)^2$ **8.** -2^4 **9.** $(-2)^4$

2 Apply the Rules for Order of Operations

Suppose you wish to evaluate the mathematical expression $3 \cdot 5 + 4$. Do you multiply first, then add to get $15 + 4 = 19$ *or* add first, then multiply to get $3 \cdot 9 = 27$?

Because $3 \cdot 5$ is equivalent to $5 + 5 + 5$, we have

$$3 \cdot 5 + 4 = 5 + 5 + 5 + 4$$
$$= 19$$

In Words
Multiply first, then add.

Based on this, **whenever the two operations of addition and multiplication appear in the same expression, the multiplication operation is always performed first, followed by the addition operation.**

Because any division problem can be written as a multiplication problem, we perform division before addition as well. Also, because any subtraction problem can be written as an addition problem, we perform multiplication and division before addition and subtraction.

Classroom Example Evaluate each expression:

(a) $12 + 3 \cdot -5$ (b) $2 + 15 \div 5 \cdot 4$

Answer:

(a) -3 (b) 14

EXAMPLE 5 Finding the Value of an Expression Containing Multiplication and Addition

Evaluate each expression:

(a) $11 + 2 \cdot (-6)$ **(b)** $7 + 12 \div 3 \cdot 5$

Solution

(a) Multiply first: $11 + 2 \cdot (-6) = 11 + (-12)$

Add: $= -1$

(b) Multiply/divide left to right: $7 + 12 \div 3 \cdot 5 = 7 + 4 \cdot 5$

Multiply: $= 7 + 20$

$= 27$

QUICK ✓ *Evaluate each expression.*

10. $1 + 7 \cdot 2$ **11.** $-11 \cdot 3 + 2$ **12.** $18 + 3 \div \left(-\frac{1}{2}\right)$

13. $9 \cdot 4 - 5$ **14.** $\frac{15}{2} \div (-5) - \frac{3}{2}$

Parentheses

If we want to add two numbers first and then multiply, we use parentheses and write $(3 + 5) \cdot 4$. In other words, **the computation in any expression in parentheses is always done first.**

Classroom Example Evaluate each expression:

(a) $(7 + 2) \cdot 4$

(b) $\left(\frac{3}{4} - \frac{7}{4}\right)\left(\frac{13}{5} + \frac{2}{5}\right)$

Answer:

(a) 36 (b) -3

EXAMPLE 6 Finding the Value of an Expression Containing Parentheses

Evaluate each expression:

(a) $(5 + 3) \cdot 2$ **(b)** $\left(\frac{3}{2} - \frac{5}{2}\right)\left(\frac{7}{3} + \frac{2}{3}\right)$

Solution

(a) $(5 + 3) \cdot 2 = 8 \cdot 2$

$= 16$

(b) $\left(\frac{3}{2} - \frac{5}{2}\right)\left(\frac{7}{3} + \frac{2}{3}\right) = \left(-\frac{2}{2}\right)\left(\frac{9}{3}\right)$

$= (-1)(3)$

$= -3$

QUICK ✓ *Evaluate each expression.*

15. $(8 \cdot 2) + 3$ **16.** $(2 - 9) \cdot (5 + 4)$ **17.** $\left(\frac{6}{7} + \frac{8}{7}\right) \cdot \left(\frac{11}{8} + \frac{5}{8}\right)$

Work Smart
The division bar (also called the *vinculum*) acts like parentheses. The word "vinculum" means "a link." So the division bar "ties things together."

The Division Bar

Another type of problem involves dividing numerical expressions that contain a division bar. When we divide numerical expressions in this form, we agree to treat the terms above and below the division bar as if they were in parentheses. To evaluate the numerical expression $\frac{3+5}{9+7}$, we see that

$$\frac{3+5}{9+7} = \frac{(3+5)}{(9+7)} = \frac{8}{16} = \frac{\cancel{8}\cdot 1}{\cancel{8}\cdot 2} = \frac{1}{2}$$

Classroom Example Evaluate each expression:

(a) $\frac{9 \cdot 3}{5 + 11 \cdot 2}$ (b) $\frac{2 + 3 \div \frac{1}{4}}{-8 \cdot 2 + 9}$

Answer:

(a) 1 (b) −2

EXAMPLE 7 Finding the Value of an Expression That Contains a Division Bar

Evaluate each expression:

(a) $\frac{7 \cdot 3}{3 + 9 \cdot 2}$ **(b)** $\frac{1 + 7 \div \frac{1}{5}}{-6 \cdot 2 + 8}$

Solution

(a) Multiply: $\frac{7 \cdot 3}{3 + 9 \cdot 2} = \frac{21}{3 + 18}$

Add: $= \frac{21}{21}$

$= 1$

(b) Write division as multiplication: $\frac{1 + 7 \div \frac{1}{5}}{-6 \cdot 2 + 8} = \frac{1 + 7 \cdot 5}{-6 \cdot 2 + 8}$

Multiply: $= \frac{1 + 35}{-12 + 8}$

Add: $= \frac{36}{-4}$

$= \frac{9 \cdot \cancel{4}}{-1 \cdot \cancel{4}}$

$= -9$

QUICK ✓ *Evaluate each expression.*

18. $\frac{2 + 5 \cdot 6}{-3 \cdot 8 - 4}$ **19.** $\frac{(12 + 14) \cdot 2}{13 \cdot 2 + 13 \cdot 5}$ **20.** $\frac{4 + 3 \div \frac{1}{7}}{2 \cdot 9 - 3}$

In Words
When there is more than one set of grouping symbols in an expression, evaluate the innermost grouping symbol first, and work outward.

Embedded Grouping Symbols

We now introduce more grouping symbols. Grouping symbols include parentheses ( ), brackets [ ], and braces { }, and absolute value symbols, | |. They are used to group numbers and mathematical expressions together so that operations within the grouping symbols are performed first. **When multiple pairs of grouping symbols exist and are nested inside one another, we evaluate the information in the innermost grouping symbol first and work our way outward.**

Classroom Example Evaluate each expression:

(a) $2 \cdot [4 \cdot (3 + 5) - 7]$

(b) $\left[5 + \left(\frac{3}{4} \cdot (-8)\right)\right] \cdot 2$

Answer:

(a) 50 (b) −2

EXAMPLE 8 Finding the Value of an Expression Containing Grouping Symbols

Evaluate each expression:

(a) $2 \cdot [3 \cdot (6 + 3) - 7]$ **(b)** $\left[4 + \left(\frac{2}{3} \cdot (-9)\right)\right] \cdot 3$

Solution

(a) Perform the operation in parentheses first: $2 \cdot [3 \cdot (6 + 3) - 7] = 2 \cdot [3 \cdot 9 - 7]$

Perform the operations in brackets, multiply first: $= 2 \cdot [27 - 7]$
$= 2 \cdot [20]$
$= 40$

(b) Perform the operation in parentheses first: $\left[4 + \left(\frac{2}{3} \cdot (-9)\right)\right] \cdot 3 = [4 + (-6)] \cdot 3$

Perform the operation in brackets: $= -2 \cdot 3$
$= -6$

QUICK ✓ *Evaluate each expression.*

21. $4 \cdot [2 \cdot (3 + 7) - 15]$ **22.** $2 \cdot \{4 \cdot [26 - (9 + 7)] - 15\} - 10$

In Words
Evaluate exponents before multiplication.

When do we evaluate exponents in the order of operations? Consider the expression $2 \cdot 4^3$. Do we multiply first and then evaluate the exponent to obtain $2 \cdot 4^3 = 8^3 = 512$, or do we evaluate the exponent first and then multiply to obtain $2 \cdot 4^3 = 2 \cdot 64 = 128$? Because $2 \cdot 4^3 = 2 \cdot 4 \cdot 4 \cdot 4 = 128$, we **evaluate exponents before multiplication.**

Classroom Example
Evaluate each expression:
(a) $3 + 5(-2)^2$
(b) $\frac{3 \cdot 4^2 - 10}{2(2 - 11)}$

Answer:
(a) 23 (b) $-\frac{19}{9}$

EXAMPLE 9 Finding the Value of an Expression Containing Exponents

Evaluate each of the following:

(a) $2 + 7(-4)^2$ **(b)** $\frac{2 \cdot 3^2 + 4}{3(2 - 6)}$

Solution

(a) Evaluate the exponent: $2 + 7(-4)^2 = 2 + 7 \cdot 16$
Multiply: $= 2 + 112$
Add: $= 114$

(b) Evaluate the exponent: $\frac{2 \cdot 3^2 + 4}{3(2 - 6)} = \frac{2 \cdot 9 + 4}{3(-4)}$

Find products: $= \frac{18 + 4}{-12}$

Add terms in numerator: $= \frac{22}{-12}$

Write in lowest terms: $= \frac{\cancel{2} \cdot 11}{\cancel{2} \cdot -6}$

$= -\frac{11}{6}$

QUICK ✓ *Evaluate each of the following:*

23. $\frac{7 - 5^2}{2}$ **24.** $3(7 - 3)^2$ **25.** $\frac{(-3)^2 + 7(1 - 3)}{3 \cdot 2 + 5}$ **26.** $2 + 5 \cdot 3^2 - \frac{3}{2} \cdot 2^2$

We now summarize the rules for performing operations on mathematical expressions.

ORDER OF OPERATIONS

Perform all operations within *grouping symbols* first. When an expression has nested grouping symbols, begin within the innermost pair of grouping symbols and work outward.

Step 1: Evaluate expressions containing exponents.

Step 2: Perform *multiplication and division* in the order they occur, working from *left to right.*

Step 3: Perform *addition and subtraction* in the order they occur, working from *left to right.*

Classroom Example
Evaluate:
$12 + 4(3^2 - 15) + 2^3$

Answer: -4

EXAMPLE 10 How to Evaluate an Expression Using Order of Operations

Evaluate: $18 + 7(2^3 - 26) + 5^2$

Step-by-Step Solution

Step 1: We evaluate the expression in the parentheses first. In the parentheses, evaluate the expression containing the exponent first.

$18 + 7(2^3 - 26) + 5^2 = 18 + 7(8 - 26) + 25$

Evaluate 8 − 26 in the parentheses: $= 18 + 7(-18) + 25$

Step 2: Perform multiplication and division in the order they occur, working from left to right.

Multiply 7(−18): $= 18 - 126 + 25$

Step 3: Perform addition and subtraction in the order in which they occur, working from left to right.

$= -108 + 25$

$= -83$

Classroom Example
Evaluate:
$$\left(\frac{6 - 3^2}{12 - 2\cdot 4}\right)^2$$

Answer: $\frac{9}{16}$

EXAMPLE 11 Evaluating a Numerical Expression Using Order of Operations

Evaluate: $\left(\dfrac{2^3 - 6}{10 - 2\cdot 3}\right)^2$

Solution

Evaluate exponential expression inside the parentheses: $\left(\dfrac{2^3 - 6}{10 - 2\cdot 3}\right)^2 = \left(\dfrac{8 - 6}{10 - 2\cdot 3}\right)^2$

Multiply inside the parentheses: $= \left(\dfrac{8 - 6}{10 - 6}\right)^2$

Add/subtract inside the parentheses: $= \left(\dfrac{2}{4}\right)^2$

$= \left(\dfrac{1\cdot \cancel{2}}{2\cdot \cancel{2}}\right)^2$

Divide out common factors: $= \left(\dfrac{1}{2}\right)^2$

Evaluate exponential expression: $= \dfrac{1}{4}$

QUICK ✓ *Evaluate each expression.*

27. $\dfrac{(4-10)^2}{2^3-5}$ **28.** $-3[(-4)^2 - 5(8-6)]^2$ **29.** $\dfrac{(2.9+7.1)^2}{5^2-15}$

1.6 Exercises

1. base
2. exponent
3. grouping symbols
4. False
5. False
6. False
7. Answers may vary.
8. Answers may vary.
9. 5^2
10. 4^5
11. $\left(\frac{3}{5}\right)^3$
12. $(-2)^3$
13. 64
14. 64
15. 64
16. −64
17. 1000
18. 16
19. $\frac{27}{64}$
20. $\frac{625}{16}$
21. 2.25
22. 0.0016
23. −9
24. −625
25. −1
26. −1
27. 0
28. 1
29. $\frac{1}{64}$
30. $-\frac{243}{32}$
31. $-\frac{1}{27}$
32. $\frac{9}{16}$
33. 14
34. 36
35. −3
36. −27
37. 2500
38. 40
39. 160
40. 80
41. 20
42. 5
43. 4
44. 12
45. $\frac{3}{5}$ 46. $\frac{4}{9}$ 47. −1
48. −1 49. 42
50. 23 51. 5
52. 13.5 53. −4
54. −6 55. 115 56. 58

For Extra Help:

Student Solutions Manual

CD Video

PH Math/Tutor Center

MathXL Tutorials on CD

MathXL®

MyMathLab

Concepts and Vocabulary

In Problems 1–3, fill in the blanks.

1. In the expression 3^5, 3 is called the ________.

2. In the expression 3^5, 5 is called the ________.

3. ( ) and [ ] are called ________ ________.

In Problems 4–6, answer True or False to each statement.

4. When evaluating an expression with more than one operation, always multiply before dividing.

5. The square of 4 is 2.

6. To evaluate -3^3, multiply $-3 \cdot -3 \cdot -3$

7. What does it mean to square a number? What does it mean to cube a number? In general, what does it mean to raise a real number to the nth power?

8. Explain the difference between -3^2 and $(-3)^2$. Identify the distinguishing characteristics between the two problems, and explain how you correctly identify which number to use as the base.

Building Skills

In Problems 9–12, write in exponential form.

9. $5 \cdot 5$ **10.** $4 \cdot 4 \cdot 4 \cdot 4 \cdot 4$ **11.** $\frac{3}{5} \cdot \frac{3}{5} \cdot \frac{3}{5}$ **12.** $(-2)(-2)(-2)$

In Problems 13–32, evaluate each exponential expression.

13. 8^2 **14.** 4^3 **15.** $(-8)^2$ **16.** $(-4)^3$

17. 10^3 **18.** 2^4 **19.** $\left(\frac{3}{4}\right)^3$ **20.** $\left(\frac{5}{2}\right)^4$

21. $(1.5)^2$ **22.** $(0.04)^2$ **23.** -3^2 **24.** -5^4

25. -1^{20} **26.** $(-1)^{19}$ **27.** 0^4 **28.** 1^6

29. $\left(-\frac{1}{2}\right)^6$ **30.** $\left(-\frac{3}{2}\right)^5$ **31.** $\left(-\frac{1}{3}\right)^3$ **32.** $\left(-\frac{3}{4}\right)^2$

In Problems 33–64, evaluate each expression.

33. $2 + 3 \cdot 4$ **34.** $12 + 8 \cdot 3$ **35.** $-5 \cdot 3 + 12$ **36.** $-3 \cdot 12 + 9$

37. $100 \div 2 \cdot 50$ **38.** $50 \div 5 \cdot 4$ **39.** $156 - 3 \cdot 2 + 10$ **40.** $86 - 4 \cdot 3 + 6$

41. $(2+3) \cdot 4$ **42.** $(7-5) \cdot \frac{5}{2}$ **43.** $8 \div 4 \cdot 2$ **44.** $4 \div 7 \cdot 21$

45. $\dfrac{4+2}{2+8}$ **46.** $\dfrac{5+3}{3+15}$ **47.** $\dfrac{14-6}{6-14}$ **48.** $\dfrac{15-7}{7-15}$

49. $13 - [3 + (-8)4]$ **50.** $12 - [7 + (-6)3]$

51. $(-8.75 - 1.25) \div (-2)$ **52.** $(-11.8 - 15.2) \div (-2)$

53. $4 - 2^3$ **54.** $10 - 4^2$ **55.** $15 + 4 \cdot 5^2$ **56.** $10 + 3 \cdot 2^4$

57. -5
58. $-\frac{49}{2}$
59. $-\frac{65}{4}$
60. $-\frac{217}{4}$
61. -24
62. 24
63. $-\frac{13}{12}$
64. $\frac{5}{24}$
65. 0.5
66. 5.2
67. 12
68. 27
69. $-\frac{1}{2}$
70. $\frac{9}{2}$
71. $\frac{3}{4}$
72. $-\frac{1}{11}$
73. 1
74. $\frac{8}{3}$
75. 24
76. -72
77. $\frac{133}{8}$
78. $-\frac{9}{4}$
79. $\frac{2}{5}$
80. $\frac{11}{20}$
81. 3
82. 2
83. $\frac{3}{2}$
84. $\frac{1}{2}$
85. $\frac{64}{27}$
86. $\frac{1}{15}$
87. $-\frac{1}{6}$
88. -2
89. $2^3 \cdot 3^2$
90. $3^3 \cdot 5^2$
91. $2^4 \cdot 3$
92. $2^3 \cdot 5^2$
93. $(4 \cdot 3 + 6) \cdot 2$
94. $4 \cdot (7 - 4^2)$
95. $(4 + 3) \cdot (4 + 2)$
96. $6 - (4 + 3 - 1)$
97. $(6 - 4) + (3 - 1)$
98. $[4 + 3 \cdot (2 - 1)] \cdot 6$
99. \$514.93
100. \$3590
101. 603.19 in.^{2}

57. $-2^3 + 3^2 \div (2^2 - 1)$

58. $-5^2 + 3^2 \div (3^2 + 9)$

59. $12 \div 6(-2)^3 - \left(\frac{1}{2}\right)^2$

60. $42 \div 21 \cdot (-3)^3 - \left(\frac{1}{2}\right)^2$

61. $-2 \cdot [5 \cdot (9 - 3) - 3 \cdot 6]$

62. $3 \cdot [6 \cdot (5 - 2) - 2 \cdot 5]$

63. $\left(\frac{4}{3} + \frac{5}{6}\right)\left(\frac{2}{5} - \frac{9}{10}\right)$

64. $\left(\frac{3}{4} + \frac{1}{2}\right)\left(\frac{2}{3} - \frac{1}{2}\right)$

Mixed Practice

In Problems 65–88, evaluate each expression.

65. $-2.5 + 4.5 \div 1.5$

66. $7.2 - 10.4 \div 5.2$

67. $4 + 2 \cdot (6 - 2)$

68. $3 + 6 \cdot (9 - 5)$

69. $\dfrac{12 - 16 \div 4 + (-24)}{16 \cdot 2 - 4 \cdot 0}$

70. $\dfrac{6 + 15 \div 3 + 16}{6 + 10 \cdot 0}$

71. $\frac{1}{2} + \frac{3}{4} \cdot \left[-2 \cdot \left(\frac{1}{4} + \frac{5}{12}\right) + \frac{5}{3}\right]$

72. $\frac{2}{5} + \frac{4}{11} \cdot \left[-3 \cdot \left(\frac{1}{5} + \frac{7}{20}\right) + \frac{3}{10}\right]$

73. $\dfrac{5^2 - 10}{3^2 + 6}$

74. $\dfrac{12(2)^3}{4^2 + 4 \cdot 5}$

75. $|6 \cdot (5 - 3^2)|$

76. $-6 \cdot (2 + |2 \cdot 3 - 4^2|)$

77. $\frac{81}{8} + \frac{13}{4} \div \frac{1}{2}$

78. $\frac{5}{12} \div \frac{1}{3} - \frac{7}{2}$

79. $\frac{-7}{20} + \frac{3}{8} \div \frac{1}{2}$

80. $-\frac{4}{5} + \frac{3}{10} \div \frac{2}{9}$

81. $\dfrac{21 - 3^2}{1 + 3}$

82. $\dfrac{5 + 3^2}{2 + 5}$

83. $\frac{3}{4} \cdot \left[\frac{5}{4} \div \left(\frac{3}{8} - \frac{1}{8}\right) - 3\right]$

84. $\left[\frac{9}{10} \div \left(\frac{2}{5} + \frac{1}{5}\right) + \frac{7}{2}\right] \cdot \frac{1}{10}$

85. $\left(\frac{4}{3}\right)^3 - \left(\frac{1}{2}\right)^2 \cdot \left(\frac{8}{3}\right) + 2 \div 3$

86. $\frac{1}{18} \cdot \frac{46}{5} - \left(\frac{2}{3}\right)^2$

87. $\dfrac{5^2 - 3^3}{|4 - 4^2|}$

88. $\dfrac{3 \cdot 2^3 - 2^2 \cdot 12}{3 + 3^2}$

Applying the Concepts

In Problems 89–92, express each number as the product of prime factors. Write the answer in exponential form.

89. 72 **90.** 675 **91.** 48 **92.** 200

In Problems 93–98, insert grouping symbols so that the expression has the desired value.

93. $4 \cdot 3 + 6 \cdot 2$ results in 36

94. $4 \cdot 7 - 4^2$ results in -36

95. $4 + 3 \cdot 4 + 2$ results in 42

96. $6 - 4 + 3 - 1$ results in 0

97. $6 - 4 + 3 - 1$ results in 4

98. $4 + 3 \cdot 2 - 1 \cdot 6$ results in 42

99. Cost of a TV The total amount paid for a flat screen television that costs \$479, plus state tax of 7.5% is found by evaluating the expression $479 + 0.075(479)$. Evaluate this expression rounded to the nearest cent.

100. Manufacturing Cost Evaluate the expression $3000 + 6(100) - \frac{100^2}{1000}$ to find the weekly production cost of manufacturing 100 calculators.

△**101. Surface Area** The surface area of a right circular cylinder whose radius is 6 inches and height is 10 inches is given approximately by $2 \cdot 3.1416 \cdot 6^2 + 2 \cdot 3.1416 \cdot 6 \cdot 10$. Evaluate this expression rounded to two decimal places.

102. 113.10 cubic centimeters
103. \$1060.90
104. \$6230.91
105. 115.75°
106. 39°

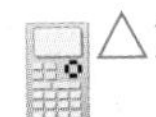

△ **102. Volume of a Cone** The volume of a cone whose radius is 3 centimeters and whose height is 12 centimeters is given approximately by $\frac{1}{3} \cdot 3.1416 \cdot 3^2 \cdot 12$. Evaluate this expression rounded to two decimal places.

103. Investing If \$1000 is invested at 3% annual interest and remains untouched for 2 years, the amount of money that is in the account after 2 years is given by the expression $1000(1 + 0.03)^2$. Evaluate this expression, rounded to the nearest cent.

104. Investing If \$5000 is invested at 4.5% annual interest and remains untouched for 5 years, the amount of money that is in the account after 5 years is given by the expression $5000(1 + 0.045)^5$. Evaluate this expression, rounded to the nearest cent.

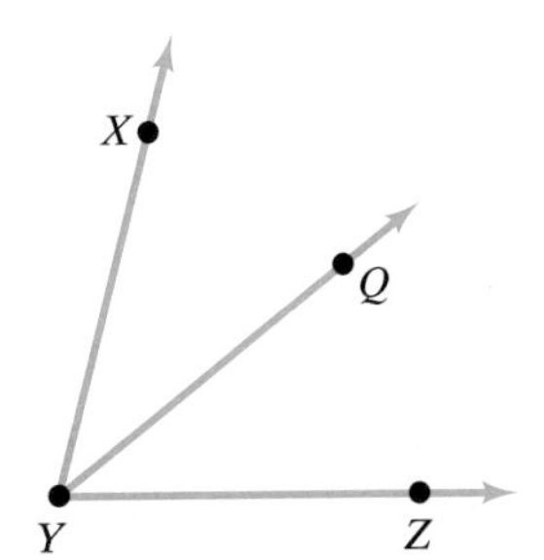

Extending the Concepts

The Angle Addition Postulate from geometry states that the measure of an angle is equal to the sum of the measures of its parts. Refer to the figure. Use the Angle Addition Postulate to answer Problems 105 and 106.

△ **105.** If $\angle XYQ = 46.5°$ and $\angle QYZ = 69.25°$, find the measure of $\angle XYZ$.

△ **106.** If $\angle QYZ = 18°$ and $\angle XYZ = 57°$, find the measure of $\angle XYQ$.

1.7 Simplifying Algebraic Expressions

OBJECTIVES

1. Evaluate Algebraic Expressions
2. Identify Like Terms and Unlike Terms
3. Use the Distributive Property
4. Simplify Algebraic Expressions by Combining Like Terms

Preparing for Simplifying Algebraic Expressions

Before getting started, take this readiness quiz. If you get a problem wrong, go back to the section cited and review the material.

1. Find the sum: $-3 + 8$ [Section 1.3, pp. 18–21]

2. Find the difference: $-7 - 8$ [Section 1.3, pp. 21–23]

3. Find the product: $-\frac{4}{3}(27)$ [Section 1.4, p. 29]

What is algebra? The word "algebra" is derived from the Arabic word, *al-jabr,* which means "restoration." Today algebra means more. According to the American Heritage Dictionary, **algebra** is a branch of mathematics in which symbols, usually letters of the alphabet, represent numbers or members of a set. These symbols are used to represent quantities and to express general relationships that hold for all members of the set. In this course, the set of numbers referred to in the definition is the set of real numbers.

1 Evaluate Algebraic Expressions

In arithmetic, we work with numbers. As stated in the definition, in algebra, we use letters such as x, y, a, b, and c to represent numbers.

DEFINITION

When a letter represents any number from a set of numbers, it is called a **variable.**

The set of numbers that we use in this textbook is the set of real numbers.

DEFINITION

A **constant** is either a fixed number, such as 5, or a letter or symbol that represents a fixed number.

For example, in Einstein's Theory of Relativity, $E = mc^2$, E and m are variables that represent total energy and mass, respectively, while c is a constant that represents the speed of light (299,792,458 meters per second).

Preparing for...Answers **1.** 5 **2.** −15 **3.** −36

DEFINITION

An **algebraic expression** is any combination of variables, constants, grouping symbols, and mathematical operations such as addition, subtraction, multiplication, division, and exponents.

Some examples of algebraic expressions are

$$x - 5, \quad \frac{1}{2}x, \quad 2x - 7, \quad x^2 + 3 \quad \text{and} \quad \frac{x - 1}{x + 1}$$

Recall that a variable represents a number from a set of numbers. One of the procedures we perform on algebraic expressions is *evaluating an algebraic expression.*

DEFINITION

To **evaluate an algebraic expression,** substitute the numerical value for each variable into the expression and simplify the result.

Classroom Example Evaluate each expression for the given value of the variable:

(a) $3x + 2$ for $x = 6$

(b) $n^2 - 3n + 2$ for $n = -2$

Answer: (a) 20 (b) 12

EXAMPLE 1 Evaluating an Algebraic Expression

Evaluate each expression for the given value of the variable.

(a) $2x + 5$ for $x = 8$ **(b)** $a^2 - 2a + 4$ for $a = -3$

Solution

(a) We substitute 8 for x in the expression $2x + 5$:

$$\begin{aligned} 2(8) + 5 &= 16 + 5 \\ &= 21 \end{aligned}$$

(b) We substitute -3 for a in $a^2 - 2a + 4$:

$$\begin{aligned} (-3)^2 - 2(-3) + 4 &= 9 + 6 + 4 \\ &= 19 \end{aligned}$$

Classroom Example Use the expression in Example 2, but evaluate $4.5x + 2.50y$ for $x = 62$ and $y = 40$.

Answer: $379

EXAMPLE 2 An Algebraic Expression for Revenue

The expression $4.50x + 2.50y$ represents the total amount of money, in dollars, received at a school play where x represents the number of adult tickets sold and y represents the number of student tickets sold. Evaluate $4.50x + 2.50y$ for $x = 50$ and $y = 82$. Interpret the result.

Solution

We substitute 50 for x and 82 for y in the expression $4.50x + 2.50y$.

$$4.50(50) + 2.50(82) = 225 + 205 = 430$$

So $430 was collected by selling 50 adult tickets and 82 student tickets.

QUICK ✓ *Evaluate each expression for the given value of the variable.*

1. $-3k + 5$ for $k = 4$ **2.** $\frac{5}{4}t - 6$ for $t = 12$ **3.** $-2y^2 - y + 8$ for $y = -2$

4. The Amadeus Coffee Shop wishes to blend two types of coffee to create a breakfast blend. They will mix x pounds of a mild coffee that sells for $7.50 per pound with y pounds of a robust coffee that sells for $10.00 per pound. An algebraic expression that represents the value of the breakfast blend, in dollars, is $7.50x + 10y$. Evaluate this expression for $x = 8$ and $y = 16$.

2 Identify Like Terms and Unlike Terms

Many algebraic expressions consist of the sum or difference of terms.

DEFINITION

A **term** is a constant or the product or quotient of a constant and one or more variables raised to powers.

Classroom Example Identify the terms in the following algebraic expressions.

(a) $3x^3 + 7y^2 - 4w + 6$

(b) $6 - \frac{a}{5} + 3b$

Answer:

(a) $3x^3$; $7y^2$; $-4w$; 6

(b) 6; $-\frac{a}{5}$; $3b$

EXAMPLE 3 Identifying the Terms in an Algebraic Expression

Identify the terms in the following algebraic expressions.

(a) $4a^3 + 5b^2 - 8c + 12$ **(b)** $\frac{x}{4} - 7y + 8z$

Solution

(a) The algebraic expression $4a^3 + 5b^2 - 8c + 12$ can be written as

$$4a^3 + 5b^2 + (-8c) + 12$$

so the terms are $4a^3$, $5b^2$, $-8c$, and 12. There are a total of 4 terms in the algebraic expression.

(b) The algebraic expression $\frac{x}{4} - 7y + 8z$ has three terms: $\frac{x}{4}$, $-7y$, and $8z$.

QUICK ✓

Identify the terms in each algebraic expression.

5. $5x^2 + 3xy$

6. $9ab - 3bc + 5ac - ac^2$

7. $\frac{2mn}{5} - \frac{3n}{7}$

8. $\frac{m^2}{3} - 8$

DEFINITION

The **coefficient** of a term is the numerical factor of the term.

For example, the coefficient of $7x$ is 7; the coefficient of $-2x^2y$ is -2. For terms that have no number as a factor, such as mn, the coefficient is 1. The coefficient of $-y$ is -1 since $-y = -1 \cdot y$. If a term consists of just a constant, the coefficient is the number itself. We say the coefficient of 14 is 14.

Classroom Example Determine the coefficient of each term:

(a) $\frac{3}{2}ab^3$ (b) $-\frac{n}{5}$

Answer:

(a) $\frac{3}{2}$ (b) $-\frac{1}{5}$

EXAMPLE 4 Determining the Coefficient of a Term

Determine the coefficient of each term:

(a) $\frac{1}{2}xy^2$ **(b)** $-\frac{t}{12}$

Solution

(a) The coefficient of $\frac{1}{2}xy^2$ is $\frac{1}{2}$.

(b) The coefficient of $-\frac{t}{12}$ is $-\frac{1}{12}$ because $-\frac{t}{12}$ can be written as $-\frac{1}{12} \cdot t$.

Classroom Example
Determine the coefficient of each term:
(a) mn^2 (b) 16
Answer:
(a) 1 (b) 16

EXAMPLE 5 Determining the Coefficient of a Term

Determine the coefficient of each term:

(a) ab^3 **(b)** 12

Solution

(a) The coefficient of ab^3 is 1 because ab^3 can be written as $1 \cdot ab^3$.

(b) The coefficient of 12 is 12 because the coefficient of a constant is the number itself.

QUICK ✓ *Determine the coefficient of each term.*

9. $2z^2$ **10.** xy **11.** $-b$ **12.** 5 **13.** $-\frac{2}{3}z$ **14.** $\frac{x}{6}$

Sometimes we have to rewrite algebraic expressions by combining like terms.

Work Smart
Like terms can have different coefficients, but they cannot have different variables or different exponents on those variables.

DEFINITION
Terms that have the same variable factor(s) with the same exponent(s) are called **like terms.**

For example, $3x^2$ and $-7x^2$ are like terms because the variable x is raised to the second power, but $3x^2$ and $-7x^3$ are not like terms because the variable x is raised to the second power in $3x^2$ but to the third power in the term $-7x^3$. Constant terms are like terms: -9 and 3 are like terms.

Classroom Example
Classify the following pairs of terms as *like* or *unlike*.
(a) $4a^2$ and $-7a^2$
(b) $7mn$ and $\frac{4}{3}m^3n$
(c) 3 and 7
Answer:
(a) like
(b) unlike
(c) like

EXAMPLE 6 Classifying Terms as Like or Unlike

Classify the following pairs of terms as *like* or *unlike:*

(a) $2p^3$ and $-5p^3$ **(b)** $7kr$ and $\frac{1}{4}k^2r$ **(c)** 5 and 8

Solution

(a) $2p^3$ and $-5p^3$ are *like* terms. They have the same variable p raised to the same power, 3.

(b) $7kr$ and $\frac{1}{4}k^2r$ are *unlike* terms. Although they both have the variables k and r, the variable k is raised to the first power in $7kr$ and to the second power in $\frac{1}{4}k^2r$.

(c) 5 and 8 are *like* terms. They are both constants.

QUICK ✓ *Tell if the terms are like or unlike.*

15. $-\frac{2}{3}p^2; \frac{4}{5}p^2$ **16.** $\frac{m}{6}; 4m$ **17.** $3a^2b; -2ab^2$ **18.** $8a; 11$

3 Use the Distributive Property

Now let's illustrate another property of real numbers. The Distributive Property of real numbers will be used throughout this course and in future courses.

THE DISTRIBUTIVE PROPERTY

If a, b, and c are real numbers, then

$$a \cdot (b + c) = a \cdot b + a \cdot c$$
$$(a + b) \cdot c = a \cdot c + b \cdot c$$

That is, multiply each of the terms inside the parentheses by the factor on the outside.

Because $b - c = b + (-c)$, it is also true that $a(b - c) = a \cdot b - a \cdot c$. One of the purposes of the Distributive Property is to remove parentheses from an algebraic expression.

Classroom Example ➤
Use the Distributive Property to remove the parentheses.

(a) $2(a + 7)$ (b) $-\frac{1}{4}(12x - 16)$

Answer:

(a) $2a + 14$ (b) $-3x + 4$

EXAMPLE 7 Using the Distributive Property to Remove Parentheses

Use the Distributive Property to remove the parentheses.

(a) $3(x + 5)$ **(b)** $-\frac{1}{3}(6x - 12)$

Solution

(a) To use the Distributive Property, we multiply each term in the parentheses by 3:

$$3(x + 5) = 3 \cdot x + 3 \cdot 5$$
$$= 3x + 15$$

(b) Multiply each term in the parentheses by $-\frac{1}{3}$:

$$-\frac{1}{3}(6x - 12) = -\frac{1}{3} \cdot 6x - \left(-\frac{1}{3}\right) \cdot 12$$
$$= -2x + 4$$

Work Smart

The long name for the Distributive Property is The Distributive Property of Multiplication over Addition. This name helps to remind us that we do not distribute across multiplication. For example,

$$6x(5xy) \neq 6x \cdot 5x \cdot 6x \cdot y$$

QUICK ✓ *Use the Distributive Property to remove the parentheses.*

19. $6(x + 2)$ **20.** $-5(x + 2)$ **21.** $-2(k - 7)$ **22.** $(8x + 12)\frac{3}{4}$

4 Simplify Algebraic Expressions by Combining Like Terms

An algebraic expression that contains the sum or difference of like terms may be simplified using the Distributive Property "in reverse." When we use the Distributive Property to add coefficients of like terms we say that we are **combining like terms.**

EXAMPLE 8 Using the Distributive Property to Combine Like Terms

Combine like terms:

(a) $2x + 7x$ **(b)** $x^2 - 5x^2$

Classroom Example Simplify each algebraic expression by combining like terms.
(a) $4b + 9b$
(b) $b^2 - 3b^2$

Answer:
(a) $13b$
(b) $-2b^2$

Solution

(a) $2x + 7x = (2 + 7)x$
$= 9x$

(b) $x^2 - 5x^2 = (1 - 5)x^2$
$= -4x^2$

Look carefully at the results of Example 8. Notice that when we combine like terms, we add the coefficients of the like terms and keep the variables and exponents the same.

QUICK ✓ *Combine like terms.*

23. $3x - 8x$

24. $-5x^2 + x^2$

25. $-7x - x + 6 - 3$

26. $4x - 12x - 3 + 17$

Sometimes we must rearrange the terms in an algebraic expression using the Commutative Property of Addition.

Classroom Example Simplify the algebraic expression by combining like terms:
(a) $3a + 6b - 8a + 5b$
(b) $\frac{3}{4}y + \frac{5}{3}z - \frac{1}{4}y + \frac{1}{3}z - 7$

Answer:
(a) $-5a + 11b$
(b) $\frac{1}{2}y + 2z - 7$

EXAMPLE 9 Combining Like Terms Using the Commutative Property

Combine like terms: $4x + 5y + 12x - 7y$

Solution

Rearrange terms: $4x + 5y + 12x - 7y = 4x + 12x + 5y - 7y$
Use the Distributive Property "in reverse": $= (4 + 12)x + (5 - 7)y$
Combine like terms: $= 16x + (-2y)$
Write the answer in simplest form: $= 16x - 2y$

QUICK ✓ *Combine like terms.*

27. $3a + 2b - 5a + 7b - 4$

28. $(5ac + 2b) + (7ac - 5a) + (-b)$

29. $5ab^2 + 7a^2b + 3ab^2 - 8a^2b$

30. $\frac{4}{3}rs - \frac{3}{2}r^2 + \frac{2}{3}rs - 5$

Often, we need to first remove parentheses by using the Distributive Property before we can combine like terms. Recall that the rules for order of operations of real numbers place multiplication before addition or subtraction. In this section, the direction **simplify** will mean to remove all parentheses and combine like terms.

Classroom Example Simplify the algebraic expression $4 - 2(3x + 5) + 7x - 6$

Answer: $x - 12$

EXAMPLE 10 Combining Like Terms Using the Distributive Property

Simplify the algebraic expression: $3 - 4(2x + 3) + 5x - 1$

Solution

Use the Distributive Property to remove parentheses: $3 - 4(2x + 3) + 5x - 1 = 3 - 8x - 12 + 5x - 1$
Rearrange terms using the Commutative Property of Addition: $= -8x + 5x + 3 - 12 - 1$
Combine like terms: $= -3x - 10$

Work Smart
Remember that multiplication comes before subtraction. In the first step of Example 10, do not compute $3 - 4$ first to obtain $-1(2x + 3)$.

QUICK ✓ *Simplify each expression.*

31. $3x + 2(x - 1) - 7x + 1$

32. $m + 2n - 3(m + 2n) - (7 - 3n)$

33. $2(a - 4b) - (a + 4b) + b$

34. $\frac{1}{2}(6x + 4) - \frac{1}{3}(12 - 9x)$

Work Smart
You know that an expression has been simplified when all like terms have been combined and all arithmetic operations on constants have been performed.

We summarize the steps for simplifying an algebraic expression below.

SUMMARY: Simplifying an Algebraic Expression

Step 1: Remove any parentheses using the Distributive Property.

Step 2: Combine any like terms.

1.7 Exercises

For Extra Help: Student Solutions Manual | CD Video | PH Math/Tutor Center | MathXL Tutorials on CD | MathXL® | MyMathLab

Concepts and Vocabulary

In Problems 1–3, fill in the blanks.

1. The ________ Property is used to remove parentheses in an expression.

2. ________ ________ are terms having the same variable with the same exponents.

3. The coefficient of the expression $-xy^2$ is ________.

In Problems 4–6, answer True or False to each statement.

4. The coefficient of $\frac{x}{8}$ is 1.

5. The terms $-3x^2$ and $9x^2$ are unlike terms.

6. The algebraic expression $\frac{3x + 5y}{4}$ contains 2 terms.

7. Explain why the sum $2x^2 + 4x^2$ is *not* equivalent to $6x^4$. What is the correct answer? Choose any value for x and then compare the results for each of these expressions.

8. Use $x = 4$ and $y = 5$ to answer parts (a), (b), and (c).
(a) Evaluate $x^2 + y^2$.
(b) Evaluate $(x + y)^2$.
(c) Are the results the same? Explain your response.

Building Skills

In Problems 9–12, for each expression, identify the terms and then name the coefficient of each term.

9. $2x^3 + \frac{x^2}{4} - x + 6$

10. $3m^4 - m^3n^2 + \frac{5n}{7} - 1$

11. $z^2 + \frac{2y}{3}$

12. $t^3 - \frac{t}{4}$

In Problems 13–26, evaluate each expression using the given value of the variables.

13. $2x + 5$ for $x = 4$

14. $3x + 7$ for $x = 2$

15. $x^2 + 3x - 1$ for $x = 3$

16. $n^2 - 4n + 3$ for $n = 2$

17. $4 - k^2$ for $k = -5$

18. $-2p^2 + 5p + 1$ for $p = -3$

19. $\frac{5x}{y} + y^2$ for $x = 8, y = 10$

20. $m^2 - \frac{3n}{m}$ for $m = 2, n = 4$

21. $\frac{9x - 5y}{x + y}$ for $x = 3, y = 2$

22. $\frac{3y + 2z}{y - z}$ for $y = 4, z = -3$

23. $(x + 3y)^2$ for $x = 3, y = 4$

24. $(a - 2b)^2$ for $a = 1, b = -2$

25. $x^2 + 9y^2$ for $x = 3, y = 4$

26. $a^2 - 4b^2$ for $a = 1, b = -2$

1. Distributive
2. Like terms
3. -1
4. False
5. False
6. True
7. Answers may vary.
8. (a) 41 (b) 81 (c) No; answers may vary.
9. $2x^3, \frac{x^2}{4}, -x, 6; 2, \frac{1}{4}, -1, 6$
10. $3m^4, -m^3n^2, \frac{5n}{7}, -1; 3, -1, \frac{5}{7}, -1$
11. $z^2, \frac{2y}{3}; 1, \frac{2}{3}$
12. $t^3, -\frac{t}{4}; 1, -\frac{1}{4}$
13. 13
14. 13
15. 17
16. -1
17. -21
18. -32
19. 104
20. -2
21. $\frac{17}{5}$
22. $\frac{6}{7}$
23. 225
24. 25
25. 153
26. -15

27. unlike
28. unlike
29. like
30. like
31. like
32. like
33. unlike
34. unlike
35. $3m + 6n$
36. $12s + 6t$
37. $18n^2 + 12n - 6$
38. $18a^4 - 12a^2 + 6$
39. $-x + y$
40. $-5k + 5n$
41. $-4x + 3y$
42. $-6.4a - 4.8b$
43. $3x$
44. $3k$
45. $6z$
46. $3m$
47. $10m + 10n$
48. $6x + 9y$
49. $2.2x^7$
50. $3.8n^4 - n^2$
51. $10y^6$
52. $-5p^5$
53. $-6w - 12y + 13z$
54. $6m - 9n + 8p$
55. $-3k + 15$
56. $21 - 4z$
57. $4n - 8$
58. $9m - 6$
59. $-4n + 20$
60. $27r + 81$
61. $\frac{5}{6}x$
62. $\frac{13}{10}y$
63. $-\frac{11}{2}$
64. $-21x + 21$
65. $-3.5x - 6$
66. $-7.4x + 13$
67. $6x - 0.06$
68. $5.1x + 1.42$
69. 32
70. -30
71. 27
72. -75
73. 0
74. 27
75. -13
76. 60
77. -7
78. -4
79. -12
80. 8
81. 44
82. $\frac{135}{2}$
83. $-\frac{3}{2}$
84. $-\frac{1}{2}$
85. 36
86. 1
87. \$78.70
88. \$71.60

In Problems 27–34, determine if the terms are like or unlike.

27. $8x$ and 8 **28.** $11p$ and 11 **29.** 54 and -21 **30.** -13 and 38

31. $12b$ and $-b$ **32.** $6a^2$ and $-3a^2$ **33.** r^2s and rs^2 **34.** x^2y^3 and y^2x^3

In Problems 35–42, use the Distributive Property to remove the parentheses.

35. $3(m + 2n)$ **36.** $3(4s + 2t)$ **37.** $(3n^2 + 2n - 1)6$ **38.** $(6a^4 - 4a^2 + 2)3$

39. $-(x - y)$ **40.** $-5(k - n)$ **41.** $(8x - 6y)(-0.5)$ **42.** $(16a + 12b)(-0.4)$

In Problems 43–68, simplify each expression by combining like terms, if possible.

43. $5x - 2x$ **44.** $14k - 11k$

45. $4z - 6z + 8z$ **46.** $9m - 8m + 2m$

47. $2m + 3n + 8m + 7n$ **48.** $x + 2y + 5x + 7y$

49. $0.3x^7 + x^7 + 0.9x^7$ **50.** $1.7n^4 - n^2 + 2.1n^4$

51. $-3y^6 + 13y^6$ **52.** $-7p^5 + 2p^5$

53. $-(6w + 12y - 13z)$ **54.** $-(-6m + 9n - 8p)$

55. $5(k + 3) - 8k$ **56.** $3(7 - z) - z$

57. $7n - (3n + 8)$ **58.** $18m - (6 + 9m)$

59. $-6(n - 3) + 2(n + 1)$ **60.** $-9(7r - 6) + 9(10r + 3)$

61. $\frac{2}{3}x + \frac{1}{6}x$ **62.** $\frac{3}{5}y + \frac{7}{10}y$

63. $\frac{1}{2}(8x + 5) - \frac{2}{3}(6x + 12)$ **64.** $\frac{1}{5}(60 - 15x) + \frac{3}{4}(12 - 24x)$

65. $2(0.5x + 9) - 3(1.5x + 8)$ **66.** $3(0.2x + 6) - 5(1.6x + 1)$

67. $3.2(x + 1.6) + 1.4(2x - 3.7)$ **68.** $1.8(x + 2.5) + 1.1(3x - 2.8)$

Mixed Practice

In Problems 69–80, (a) evaluate the expression for the given value(s) of the variable(s) before combining like terms, (b) simplify the expression by combining like terms and then evaluate the expression for the given value(s) of the variable(s). Compare your results.

69. $5x + 3x; x = 4$ **70.** $8y + 2y; y = -3$

71. $-2a^2 + 5a^2; a = -3$ **72.** $4b^2 - 7b^2; b = 5$

73. $4z - 3(z + 2); z = 6$ **74.** $8p - 3(p - 4); p = 3$

75. $5y^2 + 6y - 2y^2 + 5y - 3; y = -2$ **76.** $3x^2 + 8x - x^2 - 6x; x = 5$

77. $\frac{1}{2}(4x - 2) - \frac{2}{3}(3x + 9); x = 3$ **78.** $\frac{1}{5}(5x - 10) - \frac{1}{6}(6x + 12); x = -2$

79. $3a + 4b - 7a + 3(a - 2b); a = 2, b = 5$

80. $-4x - y + 2(x - 3y); x = 3, y = -2$

Applying the Concepts

In Problems 81–86, evaluate each expression using the given value of the variables.

81. $\frac{1}{2}h(b + B); h = 4, b = 5, B = 17$ **82.** $\frac{1}{2}h(b + B); h = 9, b = 3, B = 12$

83. $\frac{a - b}{c - d}; a = 6, b = 3, c = -4, d = -2$ **84.** $\frac{a - b}{c - d}; a = -5, b = -2, c = 7, d = 1$

85. $b^2 - 4ac; a = 7, b = 8, c = 1$ **86.** $b^2 - 4ac; a = 2, b = 5, c = 3$

87. **Renting a Truck** The cost of renting a truck from Hamilton Auto Rental is \$59.95 per day plus \$0.15 per mile. The expression $59.95 + 0.15m$ represents the cost of renting a truck for one day and driving it m miles. Evaluate $59.95 + 0.15m$ for $m = 125$.

88. **Renting a Car** The cost of renting a compact car for one day from CMH Auto is \$29.95 plus \$0.17 per mile. The expression $29.95 + 0.17m$ represents the total daily cost. Evaluate the expression $29.95 + 0.17m$ for $m = 245$.

89. \$4819
90. \$884
91. (a) $8w - 8$ (b) 32 yards
92. (a) $4l - 22$ (b) 38 meters
93. \$228.88
94. \$201.75
95. Answers may vary; $-3x^2 + 7x - 3$
96. $153.9x + 1626.80$

89. **Ticket Sales** The Center for Science and Industry sells adult tickets for \$12 and children's tickets for \$7. The expression $12a + 7c$ represents the total revenue from selling a adult tickets and c children's tickets. Evaluate the algebraic expression $12a + 7c$ for $a = 156$ and $c = 421$.

90. **Ticket Sales** A community college theatre group sold tickets to a recent production. Student tickets cost \$5 and nonstudent tickets cost \$8. The algebraic expression $5s + 8n$ represents the total revenue from selling s student tickets and n nonstudent tickets. Evaluate $5s + 8n$ for $s = 76$ and $n = 63$.

△ 91. **Rectangle** The width of a rectangle is w yards and the length of the rectangle is $(3w - 4)$ yards. The perimeter of the rectangle is given by the algebraic expression $2w + 2(3w - 4)$.
(a) Simplify the algebraic expression $2w + 2(3w - 4)$.
(b) Determine the perimeter of a rectangle whose width w is 5 yards.

△ 92. **Rectangle** The length of a rectangle is l meters and the width of the rectangle is $(l - 11)$ meters. The perimeter of the rectangle is given by the algebraic expression $2l + 2(l - 11)$.
(a) Simplify the expression $2l + 2(l - 11)$.
(b) Determine the perimeter of a rectangle whose length l is 15 meters.

93. **Finance** Novella invested some money in two investment funds. She placed s dollars in stocks that yield 5.5% annual interest and b dollars in bonds that yield 3.25% annual interest. Evaluate the expression $0.055s + 0.0325b$ for $s = \$2950$ and $b = \$2050$. Round your answer to the nearest penny.

94. **Finance** Jonathan received an inheritance from his grandparents. He invested x dollars in a Certificate of Deposit that pays 2.95% and y dollars in an off-shore oil drilling venture that is expected to pay 12.8%. Evaluate the algebraic expression $0.0295x + 0.128y$ for $x = \$2500$ and $y = \$1000$.

Extending the Concepts

95. Explain how you can simplify this algebraic expression (cleverly!!) using the Distributive Property – in reverse! $2.75(-3x^2 + 7x - 3) - 1.75(-3x^2 + 7x - 3)$.

96. Simplify by using the Distributive Property in reverse:
$11.23(7.695x + 81.34) + 8.77(7.695x + 81.34)$.

CHAPTER 1 ACTIVITY: THE MATH GAME

Focus: Performing order of operations and simplifying expressions
Time: 10 minutes
Group size: 2–4

- The instructor will announce when the groups may begin solving the problems below.
- When your group has completed all of the problems, ask the instructor to check the answers. The instructor will tell you how many answers are correct, but not which ones.
- The first group to complete all of the problems correctly will win a prize, as determined by the instructor.

1. -20
2. $-2/5$
3. -36
4. $-9/2$
5. $-13x - 5$
6. $4x^2 + 3x + 12$

1. Evaluate: $-8 \div 2^2 \cdot 6 + (-2)^3$

2. Evaluate: $\dfrac{6(-3) + 4^2}{25 + 4(-9 + 4)}$

3. Evaluate: $x^3 - x^2$ for $x = -3$

4. Evaluate: $\dfrac{(x + 2y)^2}{xy}$ for $x = 1, y = -2$

5. Simplify: $-2(4x + 3) - (5x - 1)$

6. Simplify: $\dfrac{3}{4}(8x^2 + 16) - 2x^2 + 3x$

CHAPTER 1 REVIEW

Section 1.2 The Number Systems and the Real Number Line		
KEY CONCEPTS	**KEY TERMS**	
• $a < b$ means a is to the left of b on the real number line • $a = b$ means a and b are in the same position on the real number line • $a > b$ means a is to the right of b on the real number line • $\|a\|$ is the distance from 0 to a on the real number line	Set Elements Empty set Natural numbers Counting numbers Whole numbers Integers Rational number Irrational number Real numbers	Real number line Origin Scale Coordinate Negative real numbers Zero Positive real numbers Sign Inequality symbols Absolute value
YOU SHOULD BE ABLE TO . . .	**EXAMPLE**	**REVIEW EXERCISES**
① Classify numbers (p. 9)	Example 2	1–10, 27, 28
② Plot points on the real number line (p. 12)	Example 3	11, 12
③ Use inequalities to order real numbers (p. 13)	Example 4	13–16, 21–26
④ Compute the absolute value of a real number (p. 14)	Example 5	17–20, 25, 26

1. $A = \{0, 1, 2, 3, 4, 5, 6\}$
2. $B = \{1, 2, 3\}$
3. $C = \{-2, -1, 0, 1, 2, 3, 4, 5\}$
4. $D = \{-2, -1, 0, 1, 2, 3\}$
5. $\frac{9}{3}$, 11
6. 0, $\frac{9}{3} = 3$, 11
7. −6, 0, $\frac{9}{3} = 3$, 11
8. −6, −3.25, 0, $\frac{9}{3} = 3$, 11, $\frac{5}{7}$
9. 5.030030003...
10. All numbers listed

11., 12.

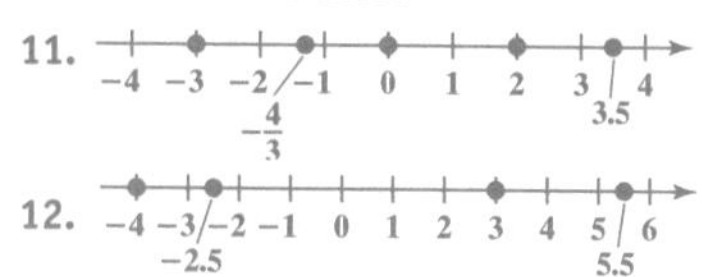

13. False 14. True
15. True 16. True
17. $-\frac{1}{2}$ 18. 7
19. −6 20. −8.2
21. = 22. <
23. > 24. >
25. > 26. <
27. Answers may vary.
28. natural numbers

In Problems 1–4, write each set.

1. A is the set of whole numbers less than 7.

2. B is the set of natural numbers less than or equal to 3.

3. C is the set of integers greater than -3 and less than or equal to 5.

4. D is the set of integers greater than or equal to -2 and less than 4.

In Problems 5–10, use the set $\left\{-6, -3.25, 0, 5.030030003\ldots, \frac{9}{3}, 11, \frac{5}{7}\right\}$. *List all the elements that are*

5. natural numbers **6.** whole numbers

7. integers **8.** rational numbers

9. irrational numbers **10.** real numbers

11. Plot the points $\left\{-3, -\frac{4}{3}, 0, \frac{4}{2}, 3.5\right\}$ on a real number line.

12. Plot the points $\left\{-4, -\frac{5}{2}, \frac{6}{2}, 5.5\right\}$ on a real number line.

In Problems 13–16, determine whether the statement is True or False.

13. $-3 > -1$ **14.** $5 \le 5$ **15.** $-5 \le -3$ **16.** $\frac{1}{2} = 0.5$

In Problems 17–20, evaluate each expression.

17. $-\left|\frac{1}{2}\right|$ **18.** $|-7|$ **19.** $-|-6|$ **20.** $-|-8.2|$

In Problems 21–26, replace the ? with the correct symbol: >, <, =.

21. $\frac{1}{4}$? 0.25 **22.** -6 ? 0 **23.** 0.83 ? $\frac{3}{4}$ **24.** -2 ? -10

25. $|-4|$? $|-3|$ **26.** $\frac{4}{5}$? $\left|-\frac{5}{6}\right|$

27. Explain the difference between a rational number and an irrational number. Be sure that your explanation includes a discussion of terminating decimals and nonterminating decimals.

28. What do we call the set of positive integers?

Section 1.3 Adding, Subtracting, Multiplying, and Dividing Integers

KEY CONCEPTS	KEY TERMS
• **Rules of Signs for Multiplying Two Integers** **1.** If we multiply two positive integers, the product is positive. **2.** If we multiply one positive integer and one negative integer, the product is negative. **3.** If we multiply two negative integers, the product is positive. • **Rules of Signs for Dividing Two Integers** **1.** If we divide two positive integers, the quotient is positive. That is, $\frac{+a}{+b} = \frac{a}{b}$ **2.** If we divide one positive integer and one negative integer, the quotient is negative. That is, $\frac{-a}{b} = \frac{a}{-b} = -\frac{a}{b}$ **3.** If we divide two negative integers, the quotient is positive. That is, $\frac{-a}{-b} = \frac{a}{b}$	Operations Sum Difference Product Quotient Mixed number Additive Inverse Opposite Evaluate Factors Dividend Divisor Multiplicative Inverse Reciprocal Ratio Golden ratio

YOU SHOULD BE ABLE TO . . .	EXAMPLE	REVIEW EXERCISES
(1) Add integers (p. 18)	Examples 1 through 6	29–36, 41–42, 59–60, 63, 69–70
(2) Determine the additive inverse of a number (p. 21)	Example 7	57–58
(3) Subtract integers (p. 21)	Examples 8 through 10	37–42, 61, 62, 64, 69–71
(4) Multiply integers (p. 23)	Examples 11 through 13	43–48, 65–66, 72
(5) Divide integers (p. 24)	Example 15	49–56, 67–68

In Problems 29–56, perform the indicated operation.

29. $-2 + 9$ **30.** $6 + (-10)$ **31.** $-23 + (-11)$

32. $-120 + 25$ **33.** $-|-2 + 6|$ **34.** $-|-15| + |-62|$

35. $-110 + 50 + (-18) + 25$ **36.** $-28 + (-35) + (-52)$

37. $-10 - 12$ **38.** $18 - 25$ **39.** $-11 - (-32)$

40. $0 - (-67)$ **41.** $34 - 18 + 10$ **42.** $-49 - 8 + 21$

43. $-6(-2)$ **44.** $4(-10)$ **45.** $13(-86)$

46. -19×423 **47.** $(11)(13)(-5)$ **48.** $(-53)(-21)(-10)$

49. $\frac{-20}{-4}$ **50.** $\frac{60}{-5}$ **51.** $\frac{|-55|}{11}$

52. $-\left|\frac{-100}{4}\right|$ **53.** $\frac{120}{-15}$ **54.** $\frac{64}{-20}$

55. $\frac{-180}{54}$ **56.** $\frac{-450}{105}$

In Problems 57 and 58, determine the additive inverse of each number.

57. 13 **58.** -45

29. 7 30. −4 31. −34
32. −95 33. −4 34. 47
35. −53 36. −115 37. −22
38. −7 39. 21 40. 67
41. 26 42. −36
43. 12 44. −40
45. −1118 46. −8037
47. −715 48. −11,130
49. 5 50. −12
51. 5 52. −25
53. −8 54. $-\frac{16}{5}$
55. $-\frac{10}{3}$
56. $-\frac{30}{7}$
57. −13
58. 45

In Problems 59–68, write the expression using mathematical symbols, and then evaluate the expression.

59. $-43 + 101 = 58$
60. $45 + (-28) = 17$
61. $-10 - (-116) = 106$
62. $74 - 56 = 18$
63. $13 + (-8) = 5$
64. $-60 - (-10) = -50$
65. $-21 \cdot (-3) = 63$
66. $54 \cdot (-18) = -972$
67. $-34 \div (-2)$ or $\frac{-34}{-2} = 17$
68. $-49 \div 14$ or $\frac{-49}{14} = -\frac{7}{2}$
69. 26 yards
70. $-3°F$
71. 24°F
72. 87 points

59. -43 plus 101

60. 45 plus -28

61. -10 minus -116

62. 74 minus 56

63. the sum of 13 and -8

64. the difference between -60 and -10

65. -21 multiplied by -3

66. 54 multiplied by -18

67. -34 divided by -2

68. -49 divided by 14

69. Football Clinton Portis had three possessions of the football within the first few minutes of the game. On his first possession he gained 20 yards, on his second possession he lost 6 yards, and on his third possession he gained 12 yards. What was his total yardage?

70. Temperature On a winter day in Detroit, Michigan, the temperature was 10°F in the morning. The temperature rose 12°F in the afternoon, and then fell 25°F by midnight. What was the temperature at midnight in Detroit?

71. Temperature One day in Bismarck, North Dakota, the high temperature was 6°F above zero and the low temperature was 18°F below zero. What was the difference between the high and the low temperature on that day in Bismarck?

72. Test Score Ms. Rosen awards 5 points for each correct multiple choice question and awards 8 points for each correct free response question. On one of Ms. Rosen's tests, Sarah got 11 multiple choice questions correct and 4 free response questions correct. What was Sarah's test score?

Section 1.4 Adding, Subtracting, Multiplying, and Dividing Rational Numbers Expressed as Fractions and Decimals

KEY CONCEPTS	KEY TERM
• **Multiplying Fractions** $\frac{a}{b} \cdot \frac{c}{d} = \frac{a \cdot c}{b \cdot d}$, where b and $d \neq 0$ • **Dividing Fractions** $\frac{a}{b} \div \frac{c}{d} = \frac{a}{b} \cdot \frac{d}{c} = \frac{a \cdot d}{b \cdot c}$, where $b, c, d \neq 0$ • **Adding or Subtracting Fractions with the Same Denominator** $\frac{a}{c} + \frac{b}{c} = \frac{a+b}{c}$, where $c \neq 0$; $\frac{a}{c} - \frac{b}{c} = \frac{a-b}{c} = \frac{a+(-b)}{c}$, where $c \neq 0$ • **Adding or Subtracting Fractions with the Unlike Denominators** **Step 1:** Find the LCD of the fractions. **Step 2:** Find equivalent fractions with the LCD by multiplying by a factor of 1. **Step 3:** Add or subtract the numerators and write the result over the common denominator. **Step 4:** Simplify the result.	Least common denominator

YOU SHOULD BE ABLE TO . . .	EXAMPLE	REVIEW EXERCISES
(1) Multiply rational numbers expressed as fractions (p. 29)	Example 1	77–80, 110
(2) Divide rational numbers expressed as fractions (p. 30)	Example 2	81–84
(3) Add and subtract rational numbers expressed as fractions (p. 31)	Examples 3 through 7	85–96, 111
(4) Add, subtract, multiply, and divide rational numbers expressed as decimals (p. 34)	Examples 8 through 11	97–109, 112

In Problems 73–76, write each rational number in lowest terms.

73. $\frac{32}{64}$ **74.** $-\frac{27}{81}$ **75.** $\frac{-100}{150}$ **76.** $\frac{35}{-25}$

In Problems 77–96, perform the indicated operation. Write in lowest terms if necessary.

77. $\frac{2}{3}\cdot\frac{15}{8}$ **78.** $-\frac{3}{8}\cdot\frac{10}{21}$ **79.** $\frac{5}{8}\cdot\left(-\frac{2}{25}\right)$ **80.** $5\cdot\left(-\frac{3}{10}\right)$

81. $\frac{24}{17}\div\frac{18}{3}$ **82.** $-\frac{5}{12}\div\frac{10}{16}$ **83.** $-\frac{27}{10}\div 9$ **84.** $20\div\left(-\frac{5}{8}\right)$

85. $\frac{2}{9}+\frac{1}{9}$ **86.** $-\frac{6}{5}+\frac{4}{5}$ **87.** $\frac{5}{7}-\frac{2}{7}$ **88.** $\frac{7}{5}-\left(-\frac{8}{5}\right)$

89. $\frac{3}{10}+\frac{1}{20}$ **90.** $\frac{5}{12}+\frac{4}{9}$ **91.** $-\frac{7}{35}-\frac{2}{49}$ **92.** $\frac{5}{6}-\left(-\frac{1}{4}\right)$

93. $-2-\left(-\frac{5}{12}\right)$ **94.** $-5+\frac{9}{4}$

95. $-\frac{1}{10}+\left(-\frac{2}{5}\right)+\frac{1}{2}$ **96.** $-\frac{5}{6}-\frac{1}{4}+\frac{3}{24}$

In Problems 97–108, perform the indicated operation.

97. $30.3 + 18.2$ **98.** $-43.02 + 18.36$ **99.** $201.37 - 118.39$

100. $-35.1 - 18.64$ **101.** $(-0.04)(-2.01)$ **102.** $(87.3)(-2.98)$

103. $\frac{69.92}{3.8}$ **104.** $-\frac{1.08318}{0.042}$ **105.** $12.5 - 18.6 + 8.4$

106. $-13.5 + 10.8 - 20.2$ **107.** $12.9 \times 1.4 \times (-0.3)$ **108.** $2.4 \times 6.1 \times (-0.05)$

109. Checking Account Lee had a balance of $256.75 in her checking account. Lee wrote a check for $175.68 on Wednesday and wrote a check for $180.00 on Thursday. What is her checking account balance now? Is Lee's account overdrawn?

110. Super Bowl Party Jarred had 36 friends at his Super Bowl XXXVIII party. Two-thirds of his friends wanted the Carolina Panthers to win. How many of Jarred's friends wanted the Panthers to win the Super Bowl?

111. Ribbon Cutting Tara has a piece of ribbon that is 15 inches long. If she cuts off a $3\frac{1}{2}$ inch piece of the ribbon, what is the length of the piece that remains?

112. Buying Clothes While shopping at her favorite store, Sierra bought 5 sweaters. If the sweaters cost $35 each and sales tax is 6.75% of the net price (net price = price × quantity), how much did Sierra spend on the clothes?

73. $\frac{1}{2}$ 74. $-\frac{1}{3}$ 75. $-\frac{2}{3}$
76. $-\frac{7}{5}$ 77. $\frac{5}{4}$ 78. $-\frac{5}{28}$
79. $-\frac{1}{20}$ 80. $-\frac{3}{2}$
81. $\frac{4}{17}$ 82. $-\frac{2}{3}$
83. $-\frac{3}{10}$ 84. -32
85. $\frac{1}{3}$ 86. $-\frac{2}{5}$
87. $\frac{3}{7}$ 88. 3
89. $\frac{7}{20}$ 90. $\frac{31}{36}$
91. $-\frac{59}{245}$ 92. $\frac{13}{12}$
93. $-\frac{19}{12}$ 94. $-\frac{11}{4}$
95. 0
96. $-\frac{23}{24}$
97. 48.5
98. −24.66
99. 82.98
100. −53.74
101. 0.0804
102. −260.154
103. 18.4
104. −25.79
105. 2.3
106. −22.9
107. −5.418
108. −0.732
109. −$98.93
110. 24 friends
111. $\frac{23}{2}$ or $11\frac{1}{2}$ inches
112. $186.81

Section 1.5 Properties of Real Numbers

KEY CONCEPTS	KEY TERMS
• **Identity Property of Addition** For any real number a, $0 + a = a + 0 = a$. • **Commutative Property of Addition** If a and b are real numbers, then $a + b = b + a$. • **Additive Inverse Property** For any real number a, $a + (-a) = -a + a = 0$. • **Associative Property of Addition** If a, b, and c are real numbers, then $a + (b + c) = (a + b) + c$. • **Commutative Property of Multiplication** If a and b are real numbers, then $a \cdot b = b \cdot a$. • **Multiplication Property of Zero** For any real number a, the product of a and 0 is always 0; that is, $a \cdot 0 = 0 \cdot a = 0$. • **Multiplicative Identity** $a \cdot 1 = 1 \cdot a = a$ for any real number a. • **Associative Property of Multiplication** If a, b, and c are real numbers, then $a \cdot (b \cdot c) = (a \cdot b) \cdot c$. • **Multiplicative Inverse Property** $a \cdot \frac{1}{a} = \frac{1}{a} \cdot a = 1$ provided that $a \neq 0$ • **Division Properties of Zero** For any nonzero number a, **1.** The quotient of 0 and a is 0. That is, $\frac{0}{a} = 0$. **2.** The quotient of a and 0 is undefined. That is, $\frac{a}{0}$ is undefined.	Additive Identity Multiplicative Identity Conversion Commutative Property Grouping symbols Associative Property Undefined

YOU SHOULD BE ABLE TO . . .	EXAMPLE	REVIEW EXERCISES
(1) Understand and use the Identity Properties of Addition and Multiplication (p. 41)	Example 1	114, 115, 118–120, 122, 125–130, 138–140
(2) Understand and use the Commutative Properties of Addition and Multiplication (p. 43)	Examples 2 through 4	116, 117, 121, 125–128, 131–132, 135–136, 141–142
(3) Understand and use the Associative Properties of Addition and Multiplication (p. 44)	Examples 5 through 7	113, 124
(4) Understand the Multiplication and Division Properties of 0 (p. 45)	Example 8	123, 133–134, 137

113. Associative Property of Multiplication
114. Multiplicative Inverse Property
115. Multiplicative Inverse Property
116. Commutative Property of Multiplication
117. Commutative Property of Multiplication
118. Additive Inverse Property
119. Identity Property of Addition
120. Identity Property of Addition

In Problems 113–124, state the property of real numbers that is being illustrated.

113. $(5 \cdot 12) \cdot 10 = 5 \cdot (12 \cdot 10)$

114. $20 \cdot \frac{1}{20} = 1$

115. $\frac{8}{3} \cdot \frac{3}{8} = 1$

116. $\frac{5}{3} \cdot \left(-\frac{18}{61}\right) \cdot \frac{3}{5} = \frac{5}{3} \cdot \frac{3}{5} \cdot \left(-\frac{18}{61}\right)$

117. $9 \cdot 73 \cdot \frac{1}{9} = 9 \cdot \frac{1}{9} \cdot 73$

118. $23.9 + (-23.9) = 0$

119. $36 + 0 = 36$

120. $-49 + 0 = -49$

121. Commutative Property of Addition
122. Multiplicative Identity Property
123. Multiplication Property of Zero
124. Associative Property of Addition
125. 29 126. 99
127. 18 128. 121
129. 3.4 130. 5.3
131. -33 132. 6
133. undefined
134. 0
135. -334
136. 2
137. 0
138. 1
139. 0
140. 130
141. $-\frac{5}{3}$
142. $-\frac{150}{13}$

121. $23 + 5 + (-23)$ is equivalent to $23 + (-23) + 5$

122. $\frac{7}{8}$ is equivalent to $\frac{7}{8} \cdot \frac{3}{3}$ **123.** $14 \cdot 0 = 0$

124. $-5.3 + (5.3 + 2.8) = (-5.3 + 5.3) + 2.8$

In Problems 125–142, evaluate each expression, if possible, by using the properties of real numbers.

125. $144 + 29 + (-144)$ **126.** $76 + 99 + (-76)$

127. $\frac{19}{3} \cdot 18 \cdot \frac{3}{19}$ **128.** $\frac{14}{9} \cdot 121 \cdot \frac{9}{14}$

129. $3.4 + 42.56 + (-42.56)$ **130.** $5.3 + 3.6 + (-3.6)$

131. $\frac{9}{7} \cdot \left(-\frac{11}{3}\right) \cdot 7$ **132.** $\frac{13}{5} \cdot \frac{18}{39} \cdot 5$ **133.** $\frac{7}{0}$

134. $\frac{0}{100}$ **135.** $1000(-334)(0.001)$ **136.** $400(0.5)(0.01)$

137. $43{,}569{,}003 \cdot 0$ **138.** $154 \cdot \frac{1}{154}$ **139.** $\frac{3445}{302} + \left(-\frac{3445}{302}\right)$

140. $130 \cdot \frac{42}{42}$ **141.** $-\frac{7}{48} \cdot \frac{20}{3} \cdot \frac{12}{7}$ **142.** $\frac{9}{8} \cdot \left(-\frac{25}{13}\right) \cdot \frac{48}{9}$

Section 1.6 Exponents and the Order of Operations

KEY CONCEPTS	KEY TERMS
• **Exponential Notation** If n is a natural number and a is a real number, then $a^n = \underbrace{a \cdot a \cdot a \cdot \ldots \cdot a}_{n \text{ factors}}$ where a is called the base and the natural number n is called the exponent or power. • **Rules for Order of Operations** Perform all operations within *grouping symbols* first. When an expression has multiple grouping symbols, begin with the innermost pair of grouping symbols and work outward. **Step 1:** Evaluate expressions containing *exponents.* **Step 2:** Perform *multiplication and division* in the order in which they occur, working from *left to right.* **Step 3:** Perform *addition and subtraction* in the order in which they occur, working from *left to right.*	Exponential notation Base Exponent Power Exponential form Expanded form

YOU SHOULD BE ABLE TO . . .	EXAMPLE	REVIEW EXERCISES
(1) Evaluate exponential expressions (p. 49)	Examples 1 through 4	143–152
(2) Apply the rules for order of operations (p. 50)	Examples 5 through 11	153–160

143. 3^4 144. $\left(\frac{2}{3}\right)^3$
145. $(-4)^2$ 146. $(-3)^3$
147. 125 148. 32
149. 81 150. -64
151. -81
152. $\frac{1}{64}$
153. -4
154. -76

In Problems 143–146, write in exponential form.

143. $3 \cdot 3 \cdot 3 \cdot 3$ **144.** $\frac{2}{3} \cdot \frac{2}{3} \cdot \frac{2}{3}$ **145.** $(-4)(-4)$ **146.** $(-3)(-3)(-3)$

In Problems 147–152, evaluate each expression.

147. 5^3 **148.** 2^5 **149.** $(-3)^4$

150. $(-4)^3$ **151.** -3^4 **152.** $\left(\frac{1}{2}\right)^6$

In Problems 153–160, evaluate each expression.

153. $-2 + 16 \div 4 \cdot 2 - 10$ **154.** $-4 + 3[2^3 + 4(2 - 10)]$

155. 206
156. 32
157. 2
158. $\frac{1}{2}$ 159. $\frac{6}{5}$ 160. $\frac{7}{5}$

155. $(12 - 7)^3 + (19 - 10)^2$

156. $5 - (-12 \div 2 \cdot 3) + (-3)^2$

157. $\dfrac{2 \cdot (4 + 8)}{3 + 3^2}$

158. $\dfrac{3 \cdot (5 + 2^2)}{2 \cdot 3^3}$

159. $\dfrac{6 \cdot [12 - 3 \cdot (5 - 2)]}{5 \cdot [21 - 2 \cdot (4 + 5)]}$

160. $\dfrac{4 \cdot [3 + 2 \cdot (8 - 6)]}{5 \cdot [14 - 2 \cdot (2 + 3)]}$

Section 1.7 Simplifying Algebraic Expressions

KEY CONCEPT	KEY TERMS
• **Distributive Property** If $a, b,$ and c are real numbers, then $a \cdot (b + c) = a \cdot b + a \cdot c$ and $(a + b) \cdot c = a \cdot c + b \cdot c$	Algebra Variable Constant Algebraic expression Evaluate an algebraic expression Term Coefficient Like terms Combining like terms Simplify

YOU SHOULD BE ABLE TO . . .	EXAMPLE	REVIEW EXERCISES
(1) Evaluate algebraic expressions (p. 57)	Examples 1 and 2	161–164, 179
(2) Identify like terms and unlike terms (p. 59)	Examples 3 through 6	165–170
(3) Use the Distributive Property (p. 61)	Example 7	174–178
(4) Simplify algebraic expressions by combining like terms (p. 61)	Examples 8 through 10	171–178

161. 21
162. −18
163. −729
164. −3
165. $3x^2$, $-x$, 6; 3, −1, 6
166. $2x^2y^3$, $-\frac{y}{5}$; 2, $-\frac{1}{5}$
167. like
168. unlike
169. unlike
170. like
171. $-3x$
172. $-4x - 15$
173. $-4.1x^4 + 0.3x^3$
174. $-3x^4 + 6x^2 + 12$
175. $18 - x$
176. $4x - 18$
177. $9x - 4$
178. −1
179. \$98.70

In Problems 161–164, evaluate each expression using the given values of the variables.

161. $x^2 - y^2$ for $x = 5, y = -2$

162. $x^2 - 3y^2$ for $x = 3, y = -3$

163. $(x + 2y)^3$ for $x = -1, y = -4$

164. $\dfrac{a - b}{x - y}$ for $a = 5, b = -10, x = -3, y = 2$

In Problems 165 and 166, for each expression, identify the terms and then name the coefficient of each term.

165. $3x^2 - x + 6$

166. $2x^2y^3 - \dfrac{y}{5}$

In Problems 167–170, determine if the terms are like or unlike.

167. $4xy^2, -6xy^2$ **168.** $-3x, 4x^2$ **169.** $-6y, -6$ **170.** $-10, 4$

In Problems 171–178, simplify each algebraic expression.

171. $4x - 6x - x$

172. $6x - 10 - 10x - 5$

173. $0.2x^4 + 0.3x^3 - 4.3x^4$

174. $-3(x^4 - 2x^2 - 4)$

175. $20 - (x + 2)$

176. $-6(2x + 5) + 4(4x + 3)$

177. $5 - (3x - 1) + 2(6x - 5)$

178. $\dfrac{1}{6}(12x + 18) - \dfrac{2}{5}(5x + 10)$

179. Moving Van The cost of renting a moving van for one day is \$19.95 plus \$0.25 per mile. The expression $19.95 + 0.25m$ represents the total cost of renting the truck for one day and driving m miles. Evaluate the expression $19.95 + 0.25m$ for $m = 315$.

CHAPTER 1 TEST

Remember to use your Chapter Test Prep Video CD to see fully worked-out solutions to any of these problems you would like to review.

Note to Instructor: A special file in TestGen provides algorithms specifically matched to the problems in this Chapter Test for easy-to-replicate practice or assessment purposes.

1. $\frac{1}{3}$
2. $\frac{9}{4}$
3. $-\frac{320}{3}$
4. -102
5. -12.16
6. -20
7. undefined
8. -14
9. 55
10. (a) 6
 (b) 0, 6
 (c) -2, 0, 6
 (d) $-2, -\frac{1}{2}, 0, 2.5, 6$
 (e) none
 (f) All those listed.
11. $<$
12. $=$
13. -7
14. 15
15. -102
16. -343
17. $-16x - 28$
18. $-4x^2 + 5x + 4$
19. \$531.85
20. $4x + 10$

In Problems 1–9, perform the indicated operation. Write in lowest terms if necessary.

1. $\frac{4}{15} - \left(-\frac{2}{30}\right)$ **2.** $\frac{21}{4} \cdot \frac{3}{7}$ **3.** $-16 \div \frac{3}{20}$

4. $14 - 110 - (-15) + (-21)$ **5.** $-14.5 + 2.34$

6. $(-4)(-1)(-5)$ **7.** $16 \div 0$

8. -6 subtracted from -20 **9.** -110 divided by -2

10. Use the set $\left\{-2, -\frac{1}{2}, 0, 2.5, 6\right\}$. List all of the elements that are:

(a) natural numbers **(b)** whole numbers **(c)** integers
(d) rational numbers **(e)** irrational numbers **(f)** real numbers

In Problems 11 and 12, replace the ? with the correct symbol $>$, $<$, or $=$.

11. $-|-14|$? -12 **12.** $\left|-\frac{2}{5}\right|$? 0.4

In Problems 13–15, evaluate each expression.

13. $-16 \div 2^2 \cdot 4 + (-3)^2$

14. $\dfrac{4(-9) - 3^2}{25 + 4(-6 - 1)}$

15. $8 - 10[6^2 - 5(2 + 3)]$

16. Evaluate $(x - 2y)^3$ for $x = -1$ and $y = 3$.

In Problems 17 and 18, simplify each algebraic expression.

17. $-6(2x + 5) - (4x - 2)$

18. $\frac{1}{2}(4x^2 + 8) - 6x^2 + 5x$

19. Bank Account Latoya started with \$675.15 in her bank account. She wrote a check for \$175.50, withdrew \$78.00 in cash, and made a deposit of \$110.20. How much money does Latoya have in her bank account now?

20. Perimeter The length of a rectangle is 5 feet more than its width. The algebraic expression $2(x + 5) + 2x$ represents the perimeter of the rectangle. Simplify the expression $2(x + 5) + 2x$.

CHAPTER

2 Equations and Inequalities in One Variable

You plan to remodel your bathroom and you've chosen 1 foot-by-1 foot ceramic tiles to cover the floor. The bathroom is 7 feet 6 inches long and 8 feet 2 inches wide. How many tiles do you need to cover the floor of your bathroom? If each tile costs \$6, how much will it cost to tile your floor? Suppose that the store selling the tile offers a discount of 10% on orders over \$350. Does your order qualify for the discount?

You can solve everyday problems such as these using mathematical formulas and models. See Problem 79 in Section 2.4.

OUTLINE

The Big Picture: Putting It Together

In Chapter 1, we reviewed arithmetic skills that will be needed throughout the course. We also introduced algebraic expressions and discussed how to simplify and evaluate algebraic expressions.

In this chapter we dive into the discussion of algebra. The word "algebra" is derived from the Arabic word, *al-jabr*. The word *al-jabr* means "restoration." This is a reference to the fact that, if a number is added to one side of an equation, then it must also be added to the other side in order to "restore" the equality. While algebra now means a whole lot more than "restoration," we will concentrate on the "restoration" part of algebra in this chapter.

2.1 Linear Equations: The Addition and Multiplication Properties of Equality

OBJECTIVES

1. Determine If a Number Is a Solution of an Equation
2. Use the Addition Property of Equality to Solve Linear Equations
3. Use the Multiplication Property of Equality to Solve Linear Equations

Preparing for Linear Equations: The Addition and Multiplication Properties of Equality

Before getting started, take this readiness quiz. If you get a problem wrong, go back to the section cited and review the material.

1. Determine the additive inverse of 3. [Section 1.3, p. 21]

2. Determine the multiplicative inverse of $-\frac{4}{3}$. [Section 1.3, pp. 24–25]

3. Evaluate: $\frac{2}{3}\left(\frac{3}{2}\right)$ [Section 1.4, p. 29]

4. Use the Distributive Property to simplify $-4(2x + 3)$. [Section 1.7, p. 61]

1 Determine If a Number Is a Solution of an Equation

We begin with a definition.

Teaching Tip
This chapter covers a lot of material. Consider dividing the exam into two parts. Part I could cover Sections 2.1–2.4. Part II could cover Sections 2.5–2.8.

DEFINITION

A **linear equation in one variable** is an equation that can be written in the form $ax + b = c$, where a, b, and c are real numbers and a does not equal 0.

Examples of linear equations (in the variable x) are

$$x - 5 = 8 \qquad \frac{1}{2}x - 7 = \frac{3}{2}x + 8 \qquad 0.2(x + 5) - 1.5 = 4.25 - (x + 3)$$

The algebraic expressions in the equation are called the **sides** of the equation. For example, in the equation $\frac{1}{2}x - 7 = \frac{3}{2}x + 8$, the algebraic expression $\frac{1}{2}x - 7$ is the **left side** of the equation and the algebraic expression $\frac{3}{2}x + 8$ is the **right side** of the equation.

An equation may be true or false. For example, the equation $x - 5 = 8$ is true if the variable x is replaced by the number 13, but the equation $x - 5 = 8$ is false if the variable x is replaced by the number 2.

Because replacing x by 13 in the equation $x - 5 = 8$ results in a true statement, we say that 13 is a *solution* of the equation $x - 5 = 8$. We also say that $x = 13$ *satisfies* the equation.

DEFINITION

The **solution** of a linear equation is the value or values of the variable that make the equation a true statement. The set of all solutions of an equation is called the **solution set.** We sometimes say that the solution **satisfies** the equation.

We use set notation to indicate the solution set of an equation. For example, because $x = 13$ satisfies the equation $x - 5 = 8$, we say that the solution set is $\{13\}$.

To determine whether a number satisfies an equation we replace the variable with the number and find out whether the left side of the equation equals the right side of the equation. If it does, we have a true statement and the replacement value is a solution of the equation.

Preparing for...Answers **1.** -3 **2.** $-\frac{3}{4}$ **3.** 1 **4.** $-8x - 12$

Classroom Example
Determine if the given value of the variable is a solution of the equation $2x + 3 = 17$.

(a) $x = -3$ (b) $x = 7$

Answer:

(a) no (b) yes

EXAMPLE 1 Determine Whether a Number Is a Solution of an Equation

Determine if the given value of the variable is a solution of the equation $4x + 7 = 19$.

(a) $x = -2$ **(b)** $x = 3$

Solution

(a)

$$4x + 7 = 19$$

Replace x with -2: $4(-2) + 7 \stackrel{?}{=} 19$

Simplify: $-8 + 7 \stackrel{?}{=} 19$

$-1 = 19$ False

Since the left side does not equal the right side when we replace x by -2, $x = -2$ is *not* a solution of the equation.

(b)

$$4x + 7 = 19$$

Replace x with 3: $4(3) + 7 \stackrel{?}{=} 19$

Simplify: $12 + 7 \stackrel{?}{=} 19$

$19 = 19$ True

Since the left side equals the right side when we replace x by 3, $x = 3$ is a solution of the equation.

QUICK ✓ *Determine if the given number is a solution of the equation.*

1. $a - 4 = -7; a = -3$

2. $\frac{1}{2} + x = 10; x = \frac{21}{2}$

3. $3x - (x + 4) = 8; x = 6$

4. $-9b + 3 + 7b = -3b + 8; b = -3$

2 Use the Addition Property of Equality to Solve Linear Equations

We solve linear equations by writing a series of steps that result in the equation

$$x = \textit{a number}$$

One method for solving equations algebraically requires that a series of *equivalent equations* be developed from the original equation until a solution results.

Teaching Tip
The Addition Property of Equality given in the text can be expanded to say for real numbers a, b, and c, $a = b$ if and only if $a + c = b + c$.

DEFINITION

Two or more equations that have precisely the same solutions are called **equivalent equations.**

We form equivalent equations using mathematical properties that transform the original equation into a new equation that has the same solution. The first property is called the *Addition Property of Equality*.

In Words
The Addition Property of Equality says that whatever you add to one side of the equation, you must also add to the other side.

ADDITION PROPERTY OF EQUALITY

The **Addition Property of Equality** states that for real numbers a, b, and c,

$$\text{if } a = b, \text{ then } a + c = b + c$$

Also, because $a - b$ is equivalent to $a + (-b)$, the Addition Property of Equality can be used to add a real number to each side of the equation, or to subtract a real number from each side of the equation.

Classroom Example Solve the linear equation $x - 5 = 12$.

Answer: $\{17\}$

Because the goal in solving a linear equation is to get the variable by itself with a coefficient of 1, we say that we want to **isolate the variable.**

EXAMPLE 2 How to Use the Addition Property of Equality to Solve a Linear Equation

Solve the linear equation $x - 6 = 11$.

Step-by-Step Solution

Since the coefficient of the variable x in this equation is already 1, we just need to "get x by itself."

Step 1: Isolate the variable x on the left side of the equation.

$$x - 6 = 11$$

Add 6 to each side of the equation: $x - 6 + 6 = 11 + 6$

Step 2: Simplify the left and right sides of the equation.

Apply the Additive Inverse Property, $a + (-a) = 0$: $x + 0 = 17$

Apply the Additive Identity Property, $a + 0 = a$: $x = 17$

Step 3: Check Verify that $x = 17$ is the solution.

$$x - 6 = 11$$

Replace 17 for x in the original equation to see if a true statement results: $17 - 6 \stackrel{?}{=} 11$

$11 = 11$ True

Because $x = 17$ satisfies the original equation, the solution is 17, or the solution set is $\{17\}$.

Classroom Example Solve the linear equation $n + \frac{4}{3} = \frac{6}{5}$

Answer: $\left\{-\frac{2}{15}\right\}$

EXAMPLE 3 Using the Addition Property of Equality to Solve a Linear Equation

Solve the linear equation $x + \frac{5}{2} = \frac{1}{4}$.

Solution

$$x + \frac{5}{2} = \frac{1}{4}$$

Subtract $\frac{5}{2}$ from each side of the equation: $x + \frac{5}{2} - \frac{5}{2} = \frac{1}{4} - \frac{5}{2}$

$a + (-a) = 0$; LCD $= 4$: $x = \frac{1}{4} - \frac{5}{2} \cdot \frac{2}{2}$

$$x = \frac{1}{4} - \frac{10}{4}$$

$$x = -\frac{9}{4}$$

Check Verify that $x = -\frac{9}{4}$ is the solution.

$$x + \frac{5}{2} = \frac{1}{4}$$

$$-\frac{9}{4} + \frac{5}{2} \stackrel{?}{=} \frac{1}{4}$$

$$-\frac{9}{4} + \frac{10}{4} \stackrel{?}{=} \frac{1}{4}$$

$$\frac{1}{4} = \frac{1}{4} \quad \text{True}$$

Because $x = -\frac{9}{4}$ satisfies the original equation, the solution is $-\frac{9}{4}$, or the solution set is $\left\{-\frac{9}{4}\right\}$.

QUICK ✓ *Solve each equation using the Addition Property of Equality.*

5. $x - 11 = 21$

6. $y + 7 = 21$

7. $-8 + a = 4$

8. $12 + c = -3$

9. $z - \frac{2}{3} = \frac{5}{3}$

10. $p + \frac{5}{4} = \frac{1}{4}$

11. $w - \frac{1}{4} = \frac{3}{8}$

12. $\frac{5}{4} + x = \frac{1}{6}$

Classroom Example ➤
Use Example 4, but the total cost including sales tax of \$1.71 was \$30.21. Find the price p of the CD set by solving $p + 1.71 = 30.21$.

Answer: \$28.50

EXAMPLE 4 How Much Is the CD Set?

The total cost for purchasing a two-CD set including a sales tax of \$1.56 was \$27.56. To find the price p of the CD set before tax, solve the equation $p + 1.56 = 27.56$ for p.

Solution

$$p + 1.56 = 27.56$$

Subtract 1.56 from each side of the equation: $p + 1.56 - 1.56 = 27.56 - 1.56$

Apply the Additive Inverse Property, $a + (-a) = 0$: $p + 0 = 26$

Apply the Additive Identity Property, $a + 0 = a$: $p = 26$

Since $p = 26$, the two-CD set cost \$26 before tax.

QUICK ✓

13. The total cost for a new car, including tax, title, and dealer preparation charges of \$1472.25 is \$13,927.25. To find the price p of the car before the extra charges, solve the equation $p + 1472.25 = 13,927.25$ for p.

3 Use the Multiplication Property of Equality to Solve Linear Equations

Teaching Tip
The Multiplication Property of Equality given in the text can be expanded to say for real numbers a, b, and c, $c \neq 0$, $a = b$ if and only if $ac = bc$.

A second property that allows us to create an equivalent equation is called the *Multiplication Property of Equality*.

In Words
The Multiplication Property of Equality says that when you multiply one side of an equation by a nonzero quantity, you must also multiply the other side by the same nonzero quantity.

MULTIPLICATION PROPERTY OF EQUALITY

The **Multiplication Property of Equality** states that for real numbers a, b, and c, where c does not equal 0,

if $a = b$, then $ac = bc$

Let's see how to use the Multiplication Property of Equality to solve an equation in the next example.

EXAMPLE 5 How to Solve a Linear Equation Using the Multiplication Property of Equality

Solve the equation $5x = 30$.

Step-by-Step Solution

Step 1: *Get the coefficient of the variable x to be 1.*

$$5x = 30$$

Apply the Multiplication Property of Equality and multiply each side of the equation by $\frac{1}{5}$: $\frac{1}{5}(5x) = \frac{1}{5}(30)$

Step 2: *Simplify the left and right sides of the equation.*

Regroup factors using the Associative Property of Multiplication: $\left(\frac{1}{5} \cdot 5\right)x = \frac{1}{5}(30)$

Apply the Multiplicative Inverse Property, $a \cdot \frac{1}{a} = 1$: $1 \cdot x = 6$

Apply the Multiplicative Identity, $1 \cdot a = a$: $x = 6$

Step 3: Check *Verify the solution.*

$$5x = 30$$

Replace $x = 6$ in the original equation: $5(6) \stackrel{?}{=} 30$

$30 = 30$ True

Because $x = 6$ satisfies the original equation, the solution is 6, or the solution set is $\{6\}$.

Classroom Example
Solve the linear equation $4x = 28$.

Answer: $\{7\}$

Work Smart
We multiply when the coefficient of the variable is a fraction, and we divide when the coefficient is an integer.

In Step 1 of Example 5, we multiplied both sides of the equation by $\frac{1}{5}$ to get the coefficient of x to equal 1. Instead of multiplying by $\frac{1}{5}$, we could also divide both sides of the equation by 5 because dividing by 5 is the same as multiplying by the reciprocal of 5, $\frac{1}{5}$.

$$5x = 30$$

Divide each side of the equation by 5: $\frac{5x}{5} = \frac{30}{5}$

Simplify: $x = 6$

Which approach do you prefer?

Classroom Example
Solve the linear equation $-6z = 15$.

Answer: $\left\{-\frac{5}{2}\right\}$

EXAMPLE 6 Solving a Linear Equation Using the Multiplication Property of Equality

Solve the equation $-4n = 18$.

Solution

$$-4n = 18$$

Multiply each side of the equation by $-\frac{1}{4}$: $-\frac{1}{4}(-4n) = -\frac{1}{4} \cdot 18$

Regroup: $\left(-\frac{1}{4} \cdot -4\right)n = -\frac{18}{4}$

Simplify: $n = -\frac{9}{2}$

Check Verify the solution by replacing $n = -\frac{9}{2}$ in the original equation to see if a true statement results.

$$-4n = 18$$

$$-4\left(-\frac{9}{2}\right) \stackrel{?}{=} 18$$

Divide out the common factor: $$\overset{-2}{\cancel{-4}}\left(-\frac{9}{\underset{1}{\cancel{2}}}\right) \stackrel{?}{=} 18$$

$$18 = 18$$

Because $n = -\frac{9}{2}$ satisfies the original equation, the solution is $-\frac{9}{2}$, or the solution set is $\left\{-\frac{9}{2}\right\}$.

QUICK ✓ *Solve each equation using the Multiplication Property of Equality.*

14. $8p = 16$ **15.** $-7n = 14$ **16.** $6z = 15$ **17.** $-12b = 28$

Classroom Example ➤
Solve the equation $15 = \frac{5}{4}n$.

Answer: $\{12\}$

EXAMPLE 7 Solving a Linear Equation with a Fraction as a Coefficient

Solve the equation $12 = \frac{2}{3}x$.

Solution

Work Smart
When the variable is on the right side of an equation, we isolate the variable in the same way we did when it was on the left side of the equation.

$$12 = \frac{2}{3}x$$

Multiply both sides of the equation by $\frac{3}{2}$, the reciprocal of $\frac{2}{3}$: $$\frac{3}{2}(12) = \frac{3}{2}\left(\frac{2}{3}x\right)$$

Use the Associative Property of Multiplication: $$\frac{3}{2}(12) = \left(\frac{3}{2}\cdot\frac{2}{3}\right)x$$

Apply the Multiplicative Inverse Property, $a \cdot \frac{1}{a} = 1$: $$18 = (1)x$$

Apply the Multiplicative Identity Property, $1 \cdot a = a$: $$18 = x$$

Check Verify the solution by replacing $x = 18$ in the original equation to see if a true statement results.

Teaching Tip
You may want to tell students that $18 = x$ is equivalent to $x = 18$ because of the Symmetric Property of Real Numbers, which states for real numbers a and b, if $a = b$, then $b = a$.

$$12 = \frac{2}{3}x$$

$$12 \stackrel{?}{=} \frac{2}{3}(18)$$

$$12 = 12 \quad \text{True}$$

Because $x = 18$ satisfies the original equation, the solution is 18, or the solution set is $\{18\}$.

QUICK ✓ *Solve each equation using the Multiplication Property of Equality.*

18. $\frac{4}{3}n = 12$ **19.** $\frac{7}{3}k = -21$ **20.** $15 = -\frac{9}{2}z$

Classroom Example Solve the equation $-\frac{10}{3} = -\frac{5}{6}k$.

Answer: {4}

EXAMPLE 8 Solving a Linear Equation with Fractions

Solve the equation $\frac{4}{5} = -\frac{2}{15}p$.

Solution

$$\frac{4}{5} = -\frac{2}{15}p$$

Multiply both sides by $-\frac{15}{2}$: $$-\frac{15}{2}\cdot\frac{4}{5} = -\frac{15}{2}\cdot\left(-\frac{2}{15}p\right)$$

Simplify: $$-\frac{\overset{3}{\cancel{15}}}{\underset{1}{\cancel{2}}}\cdot\frac{\overset{2}{\cancel{4}}}{\underset{1}{\cancel{5}}} = \left(-\frac{15}{2}\cdot-\frac{2}{15}\right)p$$

$$-6 = p$$

Check Let $p = -6$ in the original equation to see if a true statement results.

$$\frac{4}{5} = -\frac{2}{15}p$$

Let $p = -6$: $$\frac{4}{5} \stackrel{?}{=} -\frac{2}{15}(-6)$$

Simplify: $$\frac{4}{5} \stackrel{?}{=} -\frac{2}{\underset{5}{\cancel{15}}}(-\overset{2}{\cancel{6}})$$

$$\frac{4}{5} = \frac{4}{5} \quad \text{True}$$

The solution is -6, or the solution set is $\{-6\}$.

Working with fractional coefficients can be tricky. The three equations $\frac{-x}{3} = 7$, $-\frac{1}{3}x = 7$, and $\frac{x}{-3} = 7$ are all equivalent. Do you see why? In the case of $\frac{-x}{3} = 7$, we can write $\frac{-x}{3}$ as $\frac{-1x}{3}$, which is equivalent to $-\frac{1}{3}x$. In the case of $\frac{x}{-3} = 7$, we can write $\frac{x}{-3}$ as $\frac{1x}{-3}$, which is equivalent to $-\frac{1}{3}x$.

Work Smart

Whenever an equation is in the form $ax = b$, if a is an integer, either multiply by the reciprocal of a, or divide by a.

$$2x = 7 \qquad \frac{1}{2}(2x) = \frac{1}{2}\cdot 7 \qquad x = \frac{7}{2}$$

$$5x = -\frac{15}{2} \qquad \frac{1}{5}(5x) = \frac{1}{5}\left(-\frac{15}{2}\right) \qquad x = -\frac{3}{2}$$

$$-3x = 72 \qquad \frac{-3x}{-3} = \frac{72}{-3} \qquad x = -24$$

If a is a noninteger rational number, multiply by the reciprocal of a.

$$\frac{4}{5}x = -16 \qquad \frac{5}{4}\left(\frac{4}{5}x\right) = \frac{5}{4}(-16) \qquad x = \frac{5}{\cancel{4}}\cdot\frac{\overset{-4}{\cancel{-16}}}{1} \qquad x = -20$$

QUICK ✓ *Solve each equation using the Multiplication Property of Equality.*

21. $\frac{3}{8}b = \frac{9}{4}$ **22.** $-\frac{4}{9} = \frac{-t}{6}$ **23.** $\frac{1}{4} = -\frac{7}{10}m$

2.1 Exercises

For Extra Help:  Student Solutions Manual  CD Video  PH Math/Tutor Center MathXL Tutorials on CD MathXL® MyMathLab

1. Addition Property of Equality
2. equivalent equations
3. subtract 9
4. False
5. False
6. True
7. Answers may vary.
8. Answers may vary.
9. Answers may vary.
10. The reasoning is correct; however, the answer is not correct; Answers may vary.
11. Yes
12. No
13. No
14. No
15. Yes
16. Yes
17. Yes
18. No
19. {20}
20. {10}
21. {−12}
22. {−4}
23. {19}
24. {19}
25. {−13}
26. {−15}
27. {2}
28. $\left\{\frac{1}{2}\right\}$
29. $\left\{\frac{1}{4}\right\}$
30. $\left\{\frac{1}{10}\right\}$

Concepts and Vocabulary

In Problems 1–3, fill in the blanks.

1. The ________ ________ ______ ________ says that whatever number you add to one side of an equation you must also add to the other side.

2. Two or more equations that have precisely the same solution are called ________ ________.

3. To solve the equation $p + 9 = -11$, we ________ ______ (from/to) each side of the equation.

In Problems 4–6, answer True or False to each statement.

4. The equation $3x^2 - 5x = 9$ is a linear equation.

5. To find the solution to the equation $\frac{4}{5}y = 16$, we subtract $\frac{4}{5}$ from each side of the equation.

6. The solution to $-\frac{3}{11}b = 12$ is $b = -44$.

7. Explain what is meant by finding the *solution* to an equation.

8. Consider the equation $3z = \frac{15}{4}$. You could either multiply both sides of the equation by $\frac{1}{3}$ or you could divide both sides by 3. Explain which operation you would choose and why you think it is easier.

9. Explain in your own words the difference between an algebraic expression and an equation. Write an equation involving the algebraic expression $x - 10$. Then solve the equation.

10. A classmate suggests that to solve the equation $4x = 2$, you must divide by 4 and the result will be $x = 2$. Explain why this reasoning is or is not correct.

Building Skills

In Problems 11–18, determine if the given value is a solution to the equation. Answer Yes or No.

11. $3x - 1 = 5; x = 2$ **12.** $4t + 2 = 16; t = 3$

13. $4 - (m + 2) = 3(2m - 1); m = 1$ **14.** $3(x + 1) - x = 5x - 9; x = -3$

15. $8k - 2 = 4; k = \frac{3}{4}$ **16.** $-15 = 3x - 16; x = \frac{1}{3}$

17. $r + 1.6 = 2r + 1; r = 0.6$ **18.** $3s - 6 = 6s - 3.4; s = -1.2$

In Problems 19–34, solve the equation using the Addition Property of Equality. Be sure to check your solution.

19. $x - 9 = 11$ **20.** $y - 8 = 2$ **21.** $x + 4 = -8$ **22.** $r + 3 = -1$

23. $12 = n - 7$ **24.** $13 = u - 6$ **25.** $-8 = x + 5$ **26.** $-2 = y + 13$

27. $x - \frac{2}{3} = \frac{4}{3}$ **28.** $x - \frac{1}{8} = \frac{3}{8}$ **29.** $z + \frac{1}{2} = \frac{3}{4}$ **30.** $n + \frac{3}{5} = \frac{7}{10}$

31. $\left\{\frac{19}{24}\right\}$ 32. $\left\{\frac{13}{24}\right\}$
33. $\{-6.1\}$ 34. $\{-7.5\}$
35. $\{5\}$ 36. $\{6\}$
37. $\{-4\}$ 38. $\{-5\}$
39. $\left\{\frac{7}{2}\right\}$ 40. $\left\{\frac{15}{2}\right\}$
41. $\left\{-\frac{5}{2}\right\}$ 42. $\left\{-\frac{5}{2}\right\}$
43. $\{21\}$ 44. $\{12\}$
45. $\{121\}$ 46. $\{30\}$
47. $\left\{\frac{3}{5}\right\}$ 48. $\left\{\frac{3}{2}\right\}$
49. $\left\{-\frac{1}{3}\right\}$ 50. $\left\{-\frac{3}{5}\right\}$
51. $\{9\}$ 52. $\{14\}$
53. $\left\{-\frac{4}{9}\right\}$ 54. $\left\{-\frac{5}{9}\right\}$
55. $\left\{-\frac{5}{3}\right\}$ 56. $\left\{-\frac{11}{4}\right\}$
57. $\{2\}$ 58. $\{-3\}$
59. $\{-3\}$ 60. $\{-3\}$
61. $\left\{\frac{2}{3}\right\}$ 62. $\left\{\frac{9}{5}\right\}$
63. $\{-6\}$ 64. $\{-9\}$
65. $\{19\}$ 66. $\{-495\}$
67. $\{283\}$ 68. $\{5\}$
69. $\{-50\}$ 70. $\{-36\}$
71. $\{41.1\}$ 72. $\{-444.9\}$
73. $\left\{\frac{20}{3}\right\}$ 74. $\left\{-\frac{15}{4}\right\}$
75. $\{-4\}$ 76. $\{-8\}$
77. $\{-12\}$ 78. $\{-20\}$
79. $\left\{\frac{1}{2}\right\}$ 80. $\left\{\frac{4}{5}\right\}$
81. $\left\{\frac{3}{16}\right\}$ 82. $\left\{\frac{23}{4}\right\}$
83. $\left\{-\frac{5}{4}\right\}$ 84. $\left\{-\frac{1}{6}\right\}$
85. \$18,499.80 86. \$799
87. \$68 88. \$1469

31. $\frac{5}{12} = x - \frac{3}{8}$ **32.** $\frac{3}{8} = y - \frac{1}{6}$ **33.** $w + 3.5 = -2.6$ **34.** $z + 4.9 = -2.6$

In Problems 35–56, solve the equation using the Multiplication Property of Equality. Be sure to check your solution.

35. $5c = 25$ **36.** $8b = 48$ **37.** $-7n = 28$ **38.** $-8s = 40$

39. $4k = 14$ **40.** $4z = 30$ **41.** $-6w = 15$ **42.** $-8p = 20$

43. $\frac{5}{3}a = 35$ **44.** $\frac{4}{3}b = 16$ **45.** $-\frac{3}{11}p = -33$ **46.** $-\frac{6}{5}n = -36$

47. $\frac{6}{5} = 2x$ **48.** $\frac{9}{2} = 3b$ **49.** $5y = -\frac{5}{3}$ **50.** $4r = -\frac{12}{5}$

51. $\frac{1}{2}m = \frac{9}{2}$ **52.** $\frac{1}{4}w = \frac{7}{2}$ **53.** $-\frac{3}{8}t = \frac{1}{6}$ **54.** $\frac{3}{10}q = -\frac{1}{6}$

55. $\frac{5}{24} = \frac{-y}{8}$ **56.** $\frac{11}{36} = -\frac{t}{9}$

Mixed Practice

In Problems 57–84, solve the equation. Be sure to check your solution.

57. $n - 4 = -2$ **58.** $m - 6 = -9$ **59.** $b + 12 = 9$

60. $c + 4 = 1$ **61.** $2 = 3x$ **62.** $9 = 5y$

63. $-4q = 24$ **64.** $-6m = 54$ **65.** $-39 = x - 58$

66. $-637 = c - 142$ **67.** $-18 = -301 + x$ **68.** $-46 = -51 + q$

69. $\frac{x}{5} = -10$ **70.** $\frac{z}{3} = -12$ **71.** $m - 56.3 = -15.2$

72. $p - 26.4 = -471.3$ **73.** $-40 = -6c$ **74.** $-45 = 12x$

75. $14 = -\frac{7}{2}c$ **76.** $12 = -\frac{3}{2}n$ **77.** $\frac{3}{4} = -\frac{x}{16}$

78. $\frac{5}{9} = -\frac{h}{36}$ **79.** $x - \frac{5}{16} = \frac{3}{16}$ **80.** $w - \frac{7}{20} = \frac{9}{20}$

81. $-\frac{3}{16} = -\frac{3}{8} + z$ **82.** $\frac{5}{2} = -\frac{13}{4} + y$

83. $\frac{5}{6} = -\frac{2}{3}z$ **84.** $-\frac{4}{9} = \frac{8}{3}b$

Applying the Concepts

85. New Car The total cost for a new car, including tax, title, and dealer preparation charges of \$1562.35, is \$20,062.15. To find the price of the car without the extra charges, solve the equation $y + 1562.35 = 20{,}062.15$, where y represents the price of the car without the extra charges.

86. New Kayak The total cost for a new kayak is \$862.92, including sales tax of \$63.92. To find the cost of the kayak without tax, solve the equation $k + 63.92 = 862.92$, where k represents the cost of the kayak.

87. Discount The cost of a sleeping bag has been discounted by \$17, so that the sale price of the bag is \$51. Find the original price of the sleeping bag, p, by solving the equation $p - 17 = 51$.

88. Discount The cost of a computer has been discounted by \$239, so that the sale price is \$1230. Find the original price of the computer, c, by solving the equation $c - 239 = 1230$.

89. 12
90. \$12
91. 0.18 or 18%
92. 0.12 or 12%
93. $x = 48 - \lambda$
94. $x = 25 + \beta$
95. $x = \frac{14}{\theta}$
96. $x = \frac{2}{5\psi}$
97. $\lambda = \frac{50}{9}$
98. $\beta = 8.81$
99. $\theta = -\frac{6}{7}$
100. $\psi = -\frac{31}{298}$

89. Eating Out Rebecca purchased several "Happy Meals" at McDonald's for her child's play group. Each Happy Meal costs \$4 and she spent a total of \$48 on the food. To determine the number of Happy Meals, h, she purchased, solve the equation $4h = 48$.

90. Paperback Books Anne bought 3 paperback books to read on the flight to Europe. She paid \$36 for the books (without sales tax). To find the price, p, of each book solve the equation $3p = 36$.

91. Interest Suppose you have a credit card debt of \$3,000. Last month, the bank charged you \$45 interest on the debt. The solution to the equation $45 = \frac{3000}{12} \cdot r$ represents the annual interest rate, r, on the credit card. Find the annual interest rate on the credit card.

92. Interest Suppose you have a credit card debt of \$4,000. Last month, the bank charged you \$40 interest on the debt. The solution to the equation $40 = \frac{4000}{12} \cdot r$ represents the annual interest rate, r, on the credit card. Find the annual interest rate on the credit card.

Extending the Concepts

93. Let λ represent some real number. Solve the equation $x + \lambda = 48$ for x.

94. Let β represent some real number. Solve the equation $x - \beta = 25$ for x.

95. Let θ represent some real number except 0. Solve the equation $14 = \theta x$ for x.

96. Let ψ represent some real number except 0. Solve the equation $\frac{2}{5} = \psi x$ for x.

97. Find the value of λ in the equation $x + \lambda = \frac{16}{3}$ so that the solution is $-\frac{2}{9}$.

98. Find the value of β in the equation $x - \beta = -13.6$ so that the solution is -4.79.

99. Find the value of θ in the equation $-\frac{3}{4} = \theta x$ so that the solution is $\frac{7}{8}$.

100. Find the value of ψ in the equation $-0.31 = \psi x$ so that the solution is 2.98.

2.2 Linear Equations: Using the Properties Together

OBJECTIVES

1. Apply the Addition and the Multiplication Properties of Equality to Solve Linear Equations
2. Combine Like Terms and Apply the Distributive Property to Solve Linear Equations
3. Solve a Linear Equation with the Variable on Both Sides of the Equation
4. Use Linear Equations to Solve Problems

Preparing for Linear Equations: Using the Properties Together

Before getting started, take this readiness quiz. If you get a problem wrong, go back to the section cited and review the material.

1. Simplify by combining like terms: $6 - (4 + 3x) + 8$ [Section 1.7, pp. 61–63]
2. Evaluate the expression $2(3x + 4) - 5$ for $x = -1$. [Section 1.7, pp. 57–58]

1 Apply the Addition and the Multiplication Properties of Equality to Solve Linear Equations

In the last section, we solved equations such as $x + 3 = 7$ and $\frac{1}{2}z = 8$, which required that we use either the Addition Property or the Multiplication Property of Equality, but not both. We now consider equations such as $2x - 3 = 7$ and $\frac{1}{2}z + 6 = 18$, which

Preparing for...Answers **1.** $10 - 3x$ **2.** -3

require using both the Addition and Multiplication Properties of Equality to solve for the variable. For example, the equation $2x + 3 = 7$ can be read as "Two *times* a number x *plus* three equals seven." We must undo both the multiplication and addition to find the solution of the equation.

Classroom Example
Solve the equation $5a - 8 = 12$.

Answer: $\{4\}$

EXAMPLE 1 How to Solve a Linear Equation Using the Addition and Multiplication Properties of Equality

Solve the equation $2z - 3 = 9$.

Step-by-Step Solution

Step 1: Isolate the term containing the variable.

$$2z - 3 = 9$$

Apply the Addition Property of Equality and add 3 to each side of the equation: $2z - 3 + 3 = 9 + 3$

$$2z = 12$$

Step 2: Get the coefficient of the variable to be 1.

Apply the Multiplication Property of Equality and divide both sides by 2 $\left(\text{this is the same as multiplying both sides by } \frac{1}{2}\right)$:

$$\frac{2z}{2} = \frac{12}{2}$$

$$z = 6$$

Step 3: Check Confirm that $z = 6$ is the solution of the equation.

$$2z - 3 = 9$$

Substitute $z = 6$ into the original equation: $2(6) - 3 \stackrel{?}{=} 9$

$$12 - 3 \stackrel{?}{=} 9$$

$$9 = 9 \text{ True}$$

Because $z = 6$ satisfies the equation, the solution of the equation is 6, or the solution set is $\{6\}$.

Classroom Example
Solve the equation $\frac{5}{4}x + 2 = 17$.

Answer: $\{12\}$

EXAMPLE 2 Solving a Linear Equation Using the Addition and Multiplication Properties of Equality

Solve the equation $\frac{3}{2}p + 3 = 12$.

Solution

$$\frac{3}{2}p + 3 = 12$$

Subtract 3 from both sides of the equation: $\frac{3}{2}p + 3 - 3 = 12 - 3$

Simplify: $\frac{3}{2}p = 9$

Multiply both sides of the equation by $\frac{2}{3}$: $\frac{2}{3}\left(\frac{3}{2}p\right) = \frac{2}{3}(9)$

Simplify: $p = 6$

Check Let $p = 6$ in the original equation to verify the solution.

$$\frac{3}{2}p + 3 = 12$$

Let $p = 6$: $\frac{3}{2}(6) + 3 \stackrel{?}{=} 12$

Simplify: $9 + 3 \stackrel{?}{=} 12$

$12 = 12$ True

Because $p = 6$ satisfies the equation, the solution is $p = 6$, or the solution set is $\{6\}$.

QUICK ✓ *Solve each equation.*

1. $5x - 4 = 11$

2. $8 - 5r = -2$

3. $\frac{2}{3}k - 4 = 8$

4. $-\frac{3}{2}n + 2 = -\frac{1}{4}$

2 Combine Like Terms and Apply the Distributive Property to Solve Linear Equations

Often, we must combine like terms before we can use the Addition or Multiplication Properties of Equality.

Classroom Example ➤ Solve the equation $4x + 9 + 2x = -9$.

Answer: $\{-3\}$

EXAMPLE 3 Combining Like Terms to Solve a Linear Equation

Solve the equation $2x - 6 + 3x = 14$.

Solution

$$2x - 6 + 3x = 14$$

Combine like terms: $5x - 6 = 14$

Add 6 to each side of the equation: $5x - 6 + 6 = 14 + 6$

$5x = 20$

Divide both sides by 5: $\frac{5x}{5} = \frac{20}{5}$

$x = 4$

Check Verify that $x = 4$ is the solution.

$$2x - 6 + 3x = 14$$

Substitute 4 for x in the original equation: $2(4) - 6 + 3(4) \stackrel{?}{=} 14$

$8 - 6 + 12 \stackrel{?}{=} 14$

$14 = 14$

Since $x = 4$ results in a true statement, the solution of the equation is 4, or the solution set is $\{4\}$.

QUICK ✓ *Solve the equation.*

5. $7b - 3b + 3 = 11$

6. $-3a + 4 + 4a = 13 - 27$

7. $6c - 2 + 2c = 18$

8. $-12 = 5x - 3x + 4$

When an equation contains parentheses, we use the Distributive Property to eliminate parentheses before we use the Addition or Multiplication Properties of Equality.

Classroom Example Solve the equation $3(4y + 1) - 4 = -13$.

Answer: $\{-1\}$

EXAMPLE 4 Solve a Linear Equation Using the Distributive Property

Solve the equation $4(2x + 3) - 7 = -11$.

Solution

$$4(2x + 3) - 7 = -11$$

Apply the Distributive Property to remove parentheses: $8x + 12 - 7 = -11$

Combine like terms: $8x + 5 = -11$

Subtract 5 from each side of the equation: $8x + 5 - 5 = -11 - 5$

$$8x = -16$$

Divide both sides by 8: $\frac{8x}{8} = \frac{-16}{8}$

$$x = -2$$

Check Verify that $x = -2$ is the solution.

$$4(2x + 3) - 7 = -11$$

Substitute -2 for x in the original equation: $4[2(-2) + 3] - 7 \stackrel{?}{=} -11$

$$4(-4 + 3) - 7 \stackrel{?}{=} -11$$

$$4(-1) - 7 \stackrel{?}{=} -11$$

$$-4 - 7 \stackrel{?}{=} -11$$

$$-11 = -11 \quad \text{True}$$

Because $x = -2$ results in a true statement, the solution of the equation is -2, or the solution set is $\{-2\}$.

QUICK ✓ *Solve the equation.*

9. $2(y + 5) - 3 = 11$

10. $\frac{1}{2}(4 - 6x) + 5 = 3$

11. $4 - (6 - x) = 11$

12. $8 + \frac{2}{3}(2n - 9) = 10$

3 Solve a Linear Equation with the Variable on Both Sides of the Equation

When solving a linear equation, our goal is to get the terms that contain the variable on one side of the equation and the constants on the other side.

Classroom Example Solve the equation $8a - 3 = 4a + 7$.

Answer: $\left\{\frac{5}{2}\right\}$

EXAMPLE 5 Solve a Linear Equation with a Variable on Both Sides of the Equation

Solve the equation $9y - 5 = 5y + 9$.

Solution

$$9y - 5 = 5y + 9$$

Subtract $5y$ from each side of the equation: $9y - 5 - 5y = 5y + 9 - 5y$

$$4y - 5 = 9$$

Add 5 to each side of the equation: $4y - 5 + 5 = 9 + 5$

$$4y = 14$$

Divide both sides by 4: $\frac{4y}{4} = \frac{14}{4}$

Simplify: $y = \frac{7}{2}$

Work Smart
When an equation is one in which the variable appears on both sides of the equation, work with the variable expression first. Once you have the expression containing the variable isolated, then add or subtract the constant to get it on the other side.

Check Confirm that $y = \frac{7}{2}$ is the solution.

$$9y - 5 = 5y + 9$$

Substitute $\frac{7}{2}$ for y in the original equation: $9\left(\frac{7}{2}\right) - 5 \stackrel{?}{=} 5\left(\frac{7}{2}\right) + 9$

$$\frac{63}{2} - 5 \stackrel{?}{=} \frac{35}{2} + 9$$

$$\frac{63}{2} - \frac{10}{2} \stackrel{?}{=} \frac{35}{2} + \frac{18}{2}$$

$$\frac{53}{2} = \frac{53}{2} \quad \text{True}$$

Because $y = \frac{7}{2}$ results in a true statement, the solution of the equation is $\frac{7}{2}$, or the solution set is $\left\{\frac{7}{2}\right\}$.

QUICK ✓ *Solve the equation.*

13. $3x + 4 = 5x - 8$

14. $10m + 3 = 6m - 11$

We now summarize the steps that you should follow to solve an equation in one variable. Not all steps may be necessary to solve every equation, but you should use this summary as a guide.

SUMMARY: Steps for Solving an Equation in One Variable

Step 1: Remove any parentheses using the Distributive Property.

Step 2: Combine like terms on each side of the equation.

Step 3: Use the Addition Property of Equality to get the terms with the variable on one side of the equation and the constants on the other side.

Step 4: Use the Multiplication Property of Equality to get the coefficient of the variable term to be 1.

Step 5: Check the solution to verify that it satisfies the original equation.

Classroom Example
Solve the equation
$2(b - 5) + 4b = 3 - (b - 1)$.
Answer: $\{2\}$

EXAMPLE 6 Solving a Linear Equation in One Variable Using the Steps

Solve the equation $2(z - 4) + 3z = 4 - (z + 2)$.

Teaching Tip
Continue to emphasize the goal of isolating the variable on one side of the equation and the constant on the other side.

Step-by-Step Solution

Step 1: Remove any parentheses using the Distributive Property.

$$2(z - 4) + 3z = 4 - (z + 2)$$
$$2z - 8 + 3z = 4 - z - 2$$

Step 2: Combine like terms on each side of the equation.

$$5z - 8 = 2 - z$$

Step 3: Use the Addition Property of Equality to get the terms with the variable on one side of the equation and the constants on the other side.

Add z to both sides of the equation: $5z - 8 + z = 2 - z + z$

Simplify: $6z - 8 = 2$

Add 8 to both sides of the equation: $6z - 8 + 8 = 2 + 8$

Simplify: $6z = 10$

Step 4: Use the Multiplication Property of Equality to get the coefficient of the variable term to be 1.

Divide both sides of the equation by 6: $\frac{6z}{6} = \frac{10}{6}$

Simplify: $z = \frac{5}{3}$

Step 5: Check the solution to verify that it satisfies the original equation.

We leave the check to you.

The solution to the equation is $z = \frac{5}{3}$, or the solution set is $\left\{\frac{5}{3}\right\}$.

QUICK ✓ *Solve the equation.*

15. $-9x + 3(2x - 3) = -10 - 2x$

16. $3 - 4(p + 5) = 5(p + 2) - 12$

4 Use Linear Equations to Solve Problems

We continue to solve problems that are modeled by algebraic equations.

Classroom Example Use Example 7, but Alejandro earns \$625 before taxes and works 45 hours that week. Solve the equation $40w + 5(2w) = 625$.

Answer:

Alejandro earns \$12.50 per hour.

Teaching Tip
Take time to discuss how the equation in Example 7 is developed from the words of the problem.

EXAMPLE 7 How Much Does Alejandro Make in an Hour?

Alejandro works as an engineer for a hospital. His contract calls for him to earn double time on all hours worked in excess of 40 hours for any given week. One week, Alejandro worked 46 hours and earned \$1326 before taxes. To determine Alejandro's hourly wage, w, solve the equation $40w + 6(2w) = 1326$.

Solution

$$40w + 6(2w) = 1326$$

Simplify: $40w + 12w = 1326$

Combine like terms: $52w = 1326$

Divide both sides of the equation by 52: $\frac{52w}{52} = \frac{1326}{52}$

Simplify: $w = 25.5$

Alejandro makes \$25.50 per hour.

QUICK ✓

17. Marcella works at a clothing store. Whenever she works more than 40 hours in a week, she gets paid twice her regular hourly wage of \$10 per hour. One week, Marcella earned \$640 before taxes. To determine how many hours, h, Marcella worked, solve the equation $400 + 20(h - 40) = 640$ for h.

2.2 Exercises

For Extra Help: Student Solutions Manual  CD Video  PH Math/Tutor Center  MathXL Tutorials on CD MathXL® MyMathLab

1. add 8
2. Distributive
3. add x
4. False
5. True
6. False
7. Answers may vary.
8. Answers may vary.
9. Answers may vary.
10. Answers may vary.
11. $\{1\}$
12. $\{2\}$
13. $\{-2\}$
14. $\{-1\}$
15. $\{-3\}$
16. $\{-3\}$
17. $\left\{\frac{3}{2}\right\}$
18. $\left\{\frac{5}{3}\right\}$
19. $\{-3\}$
20. $\{-1\}$
21. $\{12\}$
22. $\{8\}$
23. $\{4\}$
24. $\{25\}$
25. $\{-2\}$
26. $\{-8\}$

Concepts and Vocabulary

In Problems 1–3, fill in the blanks.

1. To solve the equation $2x - 8 = 41$, the first step is to __________ _____ to both sides of the equation.

2. To solve the equation $3 - 2(7x + 1) + 8x = 12$, first use the __________ Property to remove the parentheses.

3. To solve the equation $3x = -x - 19$, the first step is to __________ _____ to each side of the equation.

In Problems 4–6, answer True or False to each statement.

4. We use the Distributive Property to simplify $3(4x)$.

5. When solving the equation $z + 6z = 35$, we first combine z and $6z$ and obtain $7z = 35$.

6. It is not necessary to check solutions to equations.

7. Explain the difference between $6x - 2(x + 1)$ and $6x - 2(x + 1) = 6$. In general, what is the difference between an algebraic expression and an algebraic equation?

8. In your own words, explain the Addition and Multiplication Properties of Equality. In the Multiplication Property of Equality, why do you think we cannot multiply both sides of the equation by 0?

9. A classmate begins to solve the equation $7x + 3 - 2x = 9x - 5$ by adding $2x$ in the following manner:

$$\begin{array}{rl} 7x + 3 - 2x = & 9x \quad - 5 \\ +2x & +2x \end{array}$$

Will this lead to the correct solution? Why or why not? Write an explanation telling the steps you would use to solve this equation.

10. A *corollary* is a rule or theorem that is closely related to a previous rule. Write a corollary to the Addition Property of Equality and title it the Subtraction Property of Equality. Write a corollary to the Multiplication Property of Equality, called the Division Property of Equality. What restrictions would you place on the Division Property of Equality? Why do you think these properties were not included in the text?

Building Skills

In Problems 11–46, solve the equation. Check your solution.

11. $3x + 4 = 7$

12. $5t + 1 = 11$

13. $2y - 1 = -5$

14. $6z - 2 = -8$

15. $-3p + 1 = 10$

16. $-4x + 3 = 15$

17. $8y + 3 = 15$

18. $6z - 7 = 3$

19. $5 - 2z = 11$

20. $1 - 3k = 4$

21. $\frac{2}{3}x + 1 = 9$

22. $\frac{5}{4}a + 3 = 13$

23. $\frac{7}{2}y - 1 = 13$

24. $\frac{1}{5}p - 3 = 2$

25. $3x - 7 + 2x = -17$

26. $5r + 2 - 3r = -14$

27. $\{-5\}$
28. $\{-3\}$
29. $\{-8\}$
30. $\{-2\}$
31. $\{-5\}$
32. $\{-2\}$
33. $\{-8\}$
34. $\{-5\}$
35. $\{3\}$
36. $\{2\}$
37. $\left\{-\frac{7}{2}\right\}$
38. $\left\{\frac{1}{3}\right\}$
39. $\{7\}$
40. $\{2\}$
41. $\{-18\}$
42. $\{-7\}$
43. $\left\{\frac{7}{2}\right\}$
44. $\left\{\frac{7}{4}\right\}$
45. $\left\{-\frac{27}{16}\right\}$
46. $\left\{-\frac{15}{2}\right\}$
47. $\{2\}$
48. $\{4\}$
49. $\left\{-\frac{3}{4}\right\}$
50. $\left\{-\frac{10}{7}\right\}$
51. $\left\{\frac{1}{3}\right\}$
52. $\left\{\frac{1}{2}\right\}$
53. $\left\{\frac{1}{16}\right\}$
54. $\left\{\frac{17}{42}\right\}$
55. $\left\{\frac{15}{2}\right\}$
56. $\left\{-\frac{14}{3}\right\}$
57. $\left\{\frac{1}{2}\right\}$
58. $\left\{-\frac{1}{4}\right\}$
59. $\{-7\}$
60. $\{9\}$
61. $\left\{\frac{51}{7}\right\}$
62. $\left\{-\frac{143}{4}\right\}$
63. McDonald's: 23 g; Burger King: 27 g
64. 34.5 g
65. width: $\frac{13}{3}$ or $4\frac{1}{3}$ feet; length: $\frac{32}{3}$ or $10\frac{2}{3}$ feet

27. $2k - 7k - 8 = 17$

28. $2b + 5 - 8b = 23$

29. $2(x + 1) = -14$

30. $3(t - 4) = -18$

31. $-3(2 + r) = 9$

32. $-5(6 + z) = -20$

33. $2x + 9 = x + 1$

34. $7z + 13 = 6z + 8$

35. $2t - 6 = 3 - t$

36. $3 + 8x = 21 - x$

37. $14 - 2n = -4n + 7$

38. $6 - 12m = -3m + 3$

39. $-3(5 - 3k) = 6k + 6$

40. $-4(10 - 7x) = 3x + 10$

41. $2(2x + 3) = 3(x - 4)$

42. $3(5 + x) = 2(2x + 11)$

43. $-3(2y + 3) - 1 = -4(y + 6) + 2y$

44. $-5(b + 2) + 3b = -2(1 + 5b) + 6$

45. $9(6 + a) + 33a = 10a$

46. $5(12 - 3w) + 25w = 2w$

Mixed Practice

In Problems 47–62, solve the equation. Check your solution.

47. $-5x + 11 = 1$

48. $-6n + 14 = -10$

49. $4m + 5 = 2$

50. $7x + 1 = -9$

51. $-2(3n - 2) = 2$

52. $-5(2n - 3) = 10$

53. $\frac{1}{4} = \frac{3}{8} - 2x$

54. $\frac{6}{7} = -\frac{5}{14} + 3x$

55. $2y + 36 = 6 + 6y$

56. $7a - 26 = 13a + 2$

57. $\frac{1}{2}(-4k + 28) = 6 + 14k$

58. $\frac{2}{3}(9a - 12) = -6a - 11$

59. $-\frac{5}{2}(x + 6) + \frac{3}{2}x = -8$

60. $\frac{4}{3}a - \left(\frac{7}{3}a + 6\right) = -15$

61. $8[4 - 6(x - 1)] + 5[(2x + 3) - 5] = 18x - 338$

62. $3[10 - 4(x - 3)] + 2[(3x + 6) - 2] = 2x + 360$

Applying the Concepts

63. Burger King A Burger King chicken salad contains 4 more grams of fat than a McDonald's Crispy Chicken California Cobb salad. Find the number of grams of fat in each salad if there are 50 grams of fat in the two salads by solving the equation $x + (x + 4) = 50$, where x represents the number of fat grams in the McDonald's Crispy Chicken California Cobb salad and $x + 4$ represents the number of fat grams in the Burger King chicken salad. (*Source: Newsweek,* May 16, 2003)

64. Taco Bell A Taco Bell Express taco salad contains 3.5 fewer fat grams than a Wendy's mandarin chicken salad. Find the number of grams of fat in the Wendy's mandarin chicken salad if there are 65.5 grams of fat in the two salads by solving the equation $x + (x - 3.5) = 65.5$, where x represents the number of grams of fat in a Wendy's salad and $x - 3.5$ represents the number of fat grams in the Taco Bell salad. (*Source: Newsweek,* May 16, 2003)

△ **65. Dog Run** The length of a rectangular dog run is two feet more than twice the width, w, and the perimeter of the dog run is 30 feet. Solve the equation $2w + 2(2w + 2) = 30$ to find the width, w, of the dog run. Then find the length, $2w + 2$, of the dog run.

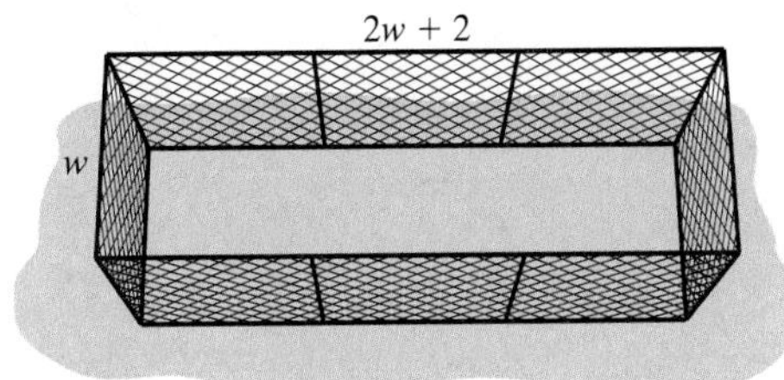

66. 8 yards
67. \$8
68. \$8.75
69. Yes
70. 68 in., 70 in., 72 in.
71. {2.47}
72. {23.2}
73. {−5.6}
74. {6.37}
75. $\frac{20}{3}$
76. $-\frac{14}{5}$
77. $-\frac{5}{4}$
78. $-\frac{1}{4}$

△ **66. Garden** The width of a rectangular garden is one yard more than one-half the length, L. The perimeter of the garden is 26 yards. Find the length, L, of the garden by solving the equation $2L + 2\left(\frac{1}{2}L + 1\right) = 26$.

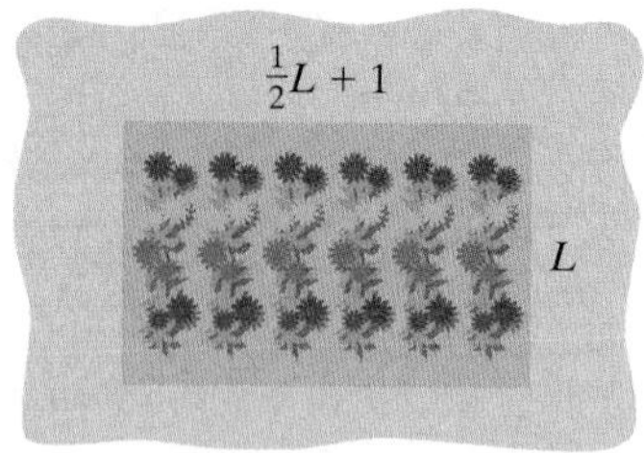

67. Overtime Pay Jennifer worked 44 hours last week, including 4 hours of overtime, and earned \$368. She is paid at a rate of 1.5 times her regular hourly rate for overtime hours. Solve the equation $40x + 4(1.5x) = 368$ to find her regular hourly pay rate, x.

68. Overtime Pay Juan worked a total of 50 hours last week and earned \$498.75. He earned 1.5 times his regular hourly rate for 6 hours, and double his hourly rate for 4 holiday hours. Solve the equation $40x + 6(1.5x) + 4(2x) = \$498.75$, where x is Juan's regular hourly rate.

△ **69. Hanging Wallpaper** Becky purchased a remnant of 42 feet of a wallpaper border to hang in a rectangular bedroom. She knows that one wall of the bedroom is 5 feet longer than the other wall. Let x represent the length of the shorter wall. Assuming the length of the shorter wall is 8 feet, does Becky have enough wallpaper to hang the border? Use the equation $2x + 2(x + 5) = 42$ to answer the question.

△ **70. Perimeter of a Triangle** The perimeter of a triangle is 210 inches. If the sides are made up of 3 consecutive even integers, find the lengths of each of the 3 sides by solving the equation $x + (x + 2) + (x + 4) = 210$, where x represents the length of the shortest side.

Extending the Concepts

In Problems 71–74, use a calculator to solve the equation and round your answer to the indicated place.

71. $3(36.7 - 4.3x) - 10 = 4(10 - 2.5x) - 8(3.5 - 4.1x)$ to the nearest hundredth

72. $12(2.3 - 1.5x) - 6 = -3(18.4 - 3.5x) - 6.1(4x + 3)$ to the nearest tenth

73. $3.5\{4 - [6 - (2x + 3)] + 5\} = -18.4$ to the nearest tenth

74. $9\{3 - [4(2.3z - 1)] + 6.5\} = -406.3$ to the nearest hundredth

In Problems 75–78, determine the value of d to make the statement true.

75. In the equation $3d + 2x = 12$, the solution is -4.

76. In the equation $5d - 2x = -2$, the solution is -6.

77. In the equation $\frac{2}{3}x - d = 1$, the solution is $-\frac{3}{8}$.

78. In the equation $\frac{2}{5}x + 3d = 0$, the solution is $\frac{15}{8}$.

2.3 Solving Linear Equations Involving Fractions and Decimals; Classifying Equations

OBJECTIVES

1. Use the Least Common Denominator to Solve a Linear Equation Containing Fractions
2. Solve a Linear Equation Containing Decimals
3. Classify Linear Equations as Identity, Conditional, or Contradiction
4. Use Linear Equations to Solve Problems

Preparing for Solving Linear Equations Involving Fractions and Decimals; Classifying Equations

Before getting started, take this readiness quiz. If you get a problem wrong, go back to the section cited and review the material.

1. Find the LCD of $\frac{3}{5}$ and $\frac{3}{4}$. [Appendix, Section A.1, pp. A4–A5]

2. Find the LCD of $\frac{3}{8}$ and $-\frac{7}{12}$. [Appendix, Section A.1, pp. A4–A5]

1 Use the Least Common Denominator to Solve a Linear Equation Containing Fractions

Work Smart
A review of finding the least common denominator (LCD) can be found on pages A4–A5 in the Appendix, Section A.1.

Sometimes equations are easier to solve if they do not contain fractions or decimals. To solve a linear equation containing fractions, we can multiply each side of the equation by the least common denominator (LCD) to clear the equation of fractions. Recall that the LCD is the smallest number that each denominator has as a common multiple. The Multiplication Property of Equality allows us to multiply both sides of the equation by the LCD.

Work Smart
Although removing fractions is not required to solve an equation, it frequently makes the arithmetic easier.

EXAMPLE 1 How to Solve a Linear Equation That Contains Fractions

Solve the linear equation: $\frac{1}{2}x + \frac{2}{3}x = \frac{14}{3}$

Classroom Example
Solve the equation $\frac{4}{3}x - \frac{7}{4}x = -\frac{5}{4}$.

Answer: $\{3\}$

Step-by-Step Solution

Before we follow the steps given in Section 2.2 on page 88, we will eliminate the fractions by multiplying both sides of the equation by the LCD. The LCD of 2 and 3 is 6, so we multiply both sides of the equation by 6.

$$6\left(\frac{1}{2}x + \frac{2}{3}x\right) = 6\left(\frac{14}{3}\right)$$

Now we can follow Steps 1–5 from the summary in Section 2.2 to solve the equation.

Step 1: Apply the Distributive Property to remove parentheses.

$$6\left(\frac{1}{2}x + \frac{2}{3}x\right) = 6\left(\frac{14}{3}\right)$$

Use the Distributive Property: $6\left(\frac{1}{2}x\right) + 6\left(\frac{2}{3}x\right) = 6\left(\frac{14}{3}\right)$

$$3x + 4x = 28$$

Step 2: Combine like terms.

$$7x = 28$$

Step 3: Use the Addition Property of Equality to get the terms with the variable on one side of the equation and the constants on the other side.

This step is not necessary because the variable is already on one side of the equation in this problem.

Preparing for...Answers **1.** 20 **2.** 24

Step 4: *Get the coefficient of the variable to be 1.* Divide both sides of the equation by 7:

$$\frac{7x}{7} = \frac{28}{7}$$
$$x = 4$$

Step 5: *Confirm that $x = 4$ is the solution of the equation.*

$$\frac{1}{2}x + \frac{2}{3}x = \frac{14}{3}$$

Substitute 4 for x in the original equation: $\frac{1}{2}(4) + \frac{2}{3}(4) \stackrel{?}{=} \frac{14}{3}$

$$2 + \frac{8}{3} \stackrel{?}{=} \frac{14}{3}$$
$$\frac{6}{3} + \frac{8}{3} \stackrel{?}{=} \frac{14}{3}$$
$$\frac{14}{3} = \frac{14}{3} \quad \text{True}$$

The solution of the equation is 4, or the solution set is $\{4\}$.

Teaching Tip
Point out that the goal in eliminating fractions or decimals is to transform the equation into an equation that students already know how to solve.

QUICK ✓ *Solve each equation by multiplying by the LCD.*

1. $\frac{2x}{5} - \frac{x}{4} = \frac{3}{2}$
2. $\frac{5}{6}x + \frac{1}{9} = -\frac{1}{6}x - \frac{1}{6}$

Classroom Example ➔
Solve the equation
$\frac{3n + 5}{2} = 4 + \frac{4n + 2}{4}$.

Answer: $\{4\}$

EXAMPLE 2 Solve a Linear Equation Using the LCD

Solve the linear equation: $\frac{7n + 5}{8} = 2 + \frac{3n + 15}{10}$

Solution

Because the equation contains fractions, we multiply both sides of the equation by the LCD. Because the LCD of 8 and 10 is 40, we multiply both sides of the equation by 40.

Work Smart
When clearing fractions from an equation, first count the number of terms in the equation. You must have that many products when you apply the Distributive Property. So in Example 2, remember to multiply the "2" by 40 also.

$$\frac{7n + 5}{8} = 2 + \frac{3n + 15}{10}$$

Multiply both sides by the LCD, 40: $40\left(\frac{7n + 5}{8}\right) = 40\left(2 + \frac{3n + 15}{10}\right)$

Use the Distributive Property to multiply all terms of the equation by the LCD: $40\left(\frac{7n + 5}{8}\right) = 40(2) + 40\left(\frac{3n + 15}{10}\right)$

Divide out common factors: $\overset{5}{\cancel{40}}\left(\frac{7n + 5}{\underset{1}{\cancel{8}}}\right) = 40(2) + \overset{4}{\cancel{40}}\left(\frac{3n + 15}{\underset{1}{\cancel{10}}}\right)$

$$5(7n + 5) = 40(2) + 4(3n + 15)$$

Use Distributive Property to remove parentheses: $35n + 25 = 80 + 12n + 60$

Combine like terms: $35n + 25 = 140 + 12n$

Isolate n: $35n + 25 - 12n = 140 + 12n - 12n$

$$23n + 25 = 140$$
$$23n + 25 - 25 = 140 - 25$$
$$23n = 115$$

Divide both sides by 23: $\frac{23n}{23} = \frac{115}{23}$

$$n = 5$$

Does $n = 5$ satisfy the original equation? Let's see.

Check

$$\frac{7n+5}{8} = 2 + \frac{3n+15}{10}$$
$$\frac{7(5)+5}{8} \stackrel{?}{=} 2 + \frac{3(5)+15}{10}$$
$$\frac{35+5}{8} \stackrel{?}{=} 2 + \frac{15+15}{10}$$
$$\frac{40}{8} \stackrel{?}{=} 2 + \frac{30}{10}$$
$$5 \stackrel{?}{=} 2 + 3$$
$$5 = 5 \quad \text{True}$$

The solution $n = 5$ checks, so the solution is 5, or the solution set is $\{5\}$.

QUICK ✓ *Solve each equation by multiplying by the LCD.*

3. $\frac{a}{3} - \frac{1}{3} = -5$

4. $\frac{3x-3}{4} - 1 = \frac{3}{5}x$

2 Solve a Linear Equation Containing Decimals

When decimals occur in linear equations, we can clear the decimal using the same techniques that we used to clear the equation of fractions. That is, we multiply both sides of the equation by a power of 10 so that the decimal is cleared. For example, $0.8x$ is equivalent to $\frac{8}{10}x$, so we multiply by 10 to clear the decimal from the equation. Because 0.34 is equivalent to $\frac{34}{100}$, we multiply by 100 to clear the decimal from the equation.

Teaching Tip
A quick discussion of place value may be beneficial at this point.
$0.3 = \frac{3}{10}$,
$0.07 = \frac{7}{100}$, etc.

Classroom Example ➤
Solve the equation $0.4z - 6 = 12$.
Answer: $\{45\}$

EXAMPLE 3 Solving a Linear Equation with a Decimal Coefficient

Solve the equation: $0.3x - 4 = 11$

Solution

We want to clear the decimal from the equation. We can do this by multiplying both sides of the equation by 10. Do you see why? $0.3 = \frac{3}{10}$, so multiplying by the LCD $= 10$ will clear the decimal.

Work Smart
The equation $0.3x - 4 = 11$ is equivalent to

$$\frac{3}{10}x - 4 = 11$$

and may be solved by multiplying both sides of the equation by 10 to clear the fraction.

	$0.3x - 4 = 11$
Multiply both sides of the equation by 10:	$10 \cdot (0.3x - 4) = 10 \cdot 11$
Distribute:	$10 \cdot 0.3x - 10 \cdot 4 = 110$
	$3x - 40 = 110$
Add 40 to both sides of the equation:	$3x = 150$
Divide both sides of the equation by 3:	$x = 50$

Check We verify our solution by replacing $x = 50$ in the original equation to see if a true statement results.

$$\begin{aligned} 0.3x - 4 &= 11 \\ 0.3(50) - 4 &\stackrel{?}{=} 11 \\ 15 - 4 &\stackrel{?}{=} 11 \\ 15 &= 15 \quad \text{True} \end{aligned}$$

The solution of the equation $0.3x - 4 = 11$ is 50, or the solution set is $\{50\}$.

It's not a requirement to clear fractions or decimals before solving an equation. Here is another way to solve $0.3x - 4 = 11$.

$$\begin{aligned} 0.3x - 4 &= 11 \\ \text{Add 4 to each side:} \quad 0.3x - 4 + 4 &= 11 + 4 \\ 0.3x &= 15 \\ \text{Divide each side by 0.3:} \quad \frac{0.3x}{0.3} &= \frac{15}{0.3} \\ x &= 50 \end{aligned}$$

Which method do you prefer?

QUICK ✓ *Solve each equation by clearing the decimal.*

5. $0.2z = 20$

6. $0.15p - 2.5 = 5$

Classroom Example ➤ Solve the equation $p + 0.06p = 296.8$.
Answer: $\{280\}$

EXAMPLE 4 Solving a Linear Equation Containing Decimals

Solve the equation: $p + 0.08p = 129.6$

Solution

We can clear the decimals by multiplying both sides of the equation by 100. Do you see why? $0.08 = \frac{8}{100}$ and $129.6 = \frac{1296}{10}$, so multiplying by the LCD $= 100$ will clear the decimals. However, before we multiply both sides of the equation by 100, we will first combine like terms on the left-hand side of the equation.

Teaching Tip
Remind students that when the coefficient of the variable is not written, it's understood to be 1.

$$\begin{aligned} p + 0.08p &= 129.6 \\ p = 1 \cdot p: \quad 1p + 0.08p &= 129.6 \\ \text{Combine like terms:} \quad 1.08p &= 129.6 \\ \text{Multiply both sides of the equation by 100:} \quad 100 \cdot 1.08p &= 100 \cdot 129.6 \\ \text{Simplify:} \quad 108p &= 12{,}960 \\ \text{Divide both sides of the equation by 108:} \quad \frac{108p}{108} &= \frac{12{,}960}{108} \\ \text{Simplify:} \quad p &= 120 \end{aligned}$$

We leave the check to you. The solution of the equation $p + 0.08p = 129.6$ is 120, or the solution set is $\{120\}$.

QUICK ✓ *Solve each equation by first clearing the decimal.*

7. $p + 0.05p = 52.5$

8. $c - 0.25c = 120$

Classroom Example ➤
Solve the equation
$0.04x + 0.06(10{,}000 - x) = 480$.
Answer: {6000}

EXAMPLE 5 Solving a Linear Equation That Contains Decimals

Solve the equation: $0.05x + 0.08(10{,}000 - x) = 680$

Solution

Before we eliminate the decimals, we will distribute the 0.08.

$$0.05x + 0.08(10{,}000 - x) = 680$$
$$0.05x + 800 - 0.08x = 680$$
Combine like terms: $$-0.03x + 800 = 680$$
Subtract 800 from both sides: $$-0.03x = -120$$
Multiply both sides of the equation by 100 to eliminate the decimal: $$100(-0.03x) = 100(-120)$$
$$-3x = -12{,}000$$
Divide both sides by -3: $$x = 4000$$

We leave the check to you. The solution to the equation $0.05x + 0.08(10{,}000 - x) = 680$ is 4000, or the solution set is $\{4000\}$.

QUICK ✓ *Solve each equation.*

9. $0.36y - 0.5 = 0.16y + 0.3$

10. $0.12x + 0.05(5000 - x) = 460$

3 Classify Linear Equations as Identity, Conditional, or Contradiction

All of the linear equations we have solved so far have had a single solution. This one value of the variable made the equation a true statement, while all other values of the variable make the equation false. We give these types of equations a special name.

DEFINITION

A **conditional equation** is an equation that is true for some values of the variable and false for other values of the variable.

For example, the equation

$$x + 5 = 11$$

is a conditional equation because it is true when $x = 6$ and false for every other real number x. All the equations that we have studied to this point have been conditional equations.

There are equations that are false for all values of the variable.

DEFINITION

A **contradiction** is an equation that is false for every replacement value of the variable.

For example, the equation

$$2x + 3 = 7 + 2x$$

is a contradiction because it is false for any replacement value for x. Contradictions are identified through the process of creating equivalent equations. For example, if we subtract $2x$ from both sides of $2x + 3 = 7 + 2x$, we obtain $3 = 7$, which is clearly false. **Contradictions have no solution, so the solution set is the empty set, written as { } or ∅.**

Work Smart
Do not write the empty set as {∅}.

Classroom Example Solve the equation $4y - (7y + 2) = 9y - 3(4y - 5)$.

Answer: $\emptyset$ or $\{\ \}$

EXAMPLE 6 Solving a Linear Equation That Is a Contradiction

Solve the equation: $3y - (5y + 4) = 12y - 7(2y - 1)$

Solution

$$3y - (5y + 4) = 12y - 7(2y - 1)$$

Use the Distributive Property to remove parentheses: $3y - 5y - 4 = 12y - 14y + 7$

Combine like terms: $-2y - 4 = -2y + 7$

Isolate y: $-2y + 2y - 4 = -2y + 2y + 7$

$$-4 = 7$$

Work Smart
The term "contradiction" describes the type of equation. The solution set is $\emptyset$.

The last statement states that $-4 = 7$. This is a false statement, so the equation is a contradiction. The solution set is $\emptyset$ or $\{\ \}$.

Some equations are true for all real numbers for which the equation is defined.

DEFINITION
An **identity** is an equation that is satisfied for all values of the variable for which both sides of the equation are defined.

An example of an identity is the equation

$$2(x - 5) + 1 = 5x - (9 + 3x)$$

This equation is an identity because any real number x makes the equation a true statement. Just as with contradictions, identities are recognized through the process of creating equivalent equations. For example, if we use the Distributive Property and combine like terms in the equation $2(x - 5) + 1 = 5x - (9 + 3x)$, we obtain $2x - 9 = 2x - 9$, which is true for all replacement values for x. **The solution set of linear identities is the set of all real numbers.**

Classroom Example $4(x + 2) = 6x - (2x - 5) + 3$

Answer: The set of all real numbers

EXAMPLE 7 Solving a Linear Equation That Is an Identity

Solve the equation: $2(x + 5) = 4x - (2x - 10)$

Solution

$$2(x + 5) = 4x - (2x - 10)$$

Use the Distributive Property to remove parentheses: $2x + 10 = 4x - 2x + 10$

Combine like terms: $2x + 10 = 2x + 10$

Isolate x: $2x - 2x + 10 = 2x - 2x + 10$

$$10 = 10$$

Teaching Tip
Remind students that when they obtain the true statement $10 = 10$ it does not mean that the solution is $x = 10$.

The last statement states that $10 = 10$. This is a true statement for all real numbers x. The solution set is the set of all real numbers.

QUICK ✓ *Solve the equation and state the solution set.*

11. $3(x + 4) = 4 + 3x + 18$

12. $\frac{1}{3}(6x - 9) - 1 = 6x - [4x - (-4)]$

13. $-5 - (9x + 8) + 23 = 7 + x - (10x - 3)$

14. $\frac{3}{2}x - 8 = x + 7 + \frac{1}{2}x$

Teaching Tip
Expand on this summary. A conditional equation results in $x = a$ *number*. When the variable is eliminated and a false statement results, the solution set is the empty set. When the variable is eliminated and a true statement results, the solution set is all real numbers.

SUMMARY:

- A *conditional* equation is true for some values of the variable and false for others.
- A *contradiction* is false for all values of the variable.
- An *identity* is true for all of the permitted values of the variable.

Classroom Example
(a) $-(4a + 5) + 2a = 8a - 5 - 10a$
(b) $3(2 - 4n) + 1 = -\frac{1}{2}(24n - 6)$
Answer:
(a) the set of all real numbers
(b) no solution; $\emptyset$ or $\{\ \}$

EXAMPLE 8 Classifying a Linear Equation

Solve the equation $2(2 + a) - 12a = -a + 5 - 9a$. Tell if the equation is a contradiction, an identity, or a conditional equation.

Solution

$$2(2 + a) - 12a = -a + 5 - 9a$$

Use the Distributive Property to remove parentheses: $4 + 2a - 12a = -a + 5 - 9a$

Combine like terms: $4 - 10a = 5 - 10a$

Isolate a: $4 - 10a + 10a = 5 - 10a + 10a$

$$4 = 5$$

The last statement states that $4 = 5$. This is a false statement, so the equation is a contradiction. The solution set is $\emptyset$ or $\{\ \}$.

QUICK ✓ *Solve the equation and tell if each equation is a contradiction, an identity, or a conditional equation.*

15. $2(x - 7) + 8 = 6x - (4x + 2) - 4$

16. $\frac{4(7 - x)}{3} = x$

17. $\frac{1}{2}(4x - 6) = 6\left(\frac{1}{3}x - \frac{1}{2}\right) + 4$

18. $4(5x - 4) + 1 = -2 + 20x$

4 Use Linear Equations to Solve Problems

We continue to solve problems that are modeled by algebraic equations.

Classroom Example
Use Example 9 with the amount of interest equal to \$140. Solve $0.02x + 0.04(5000 - x) = 140$.
Answer: You have \$3000 to invest in CDs.

EXAMPLE 9 Solving a Problem from Finance

You have \$5000 to invest, and your financial advisor advises you to put part of the money in a Certificate of Deposit (CD) that earns 2% simple interest compounded annually, and the rest in bonds that earn 4% simple interest compounded annually. To determine the amount you should invest in the CD to earn \$170 interest at the end of one year, solve the equation $0.02x + 0.04(5000 - x) = 170$, where x is the amount of money invested in CDs.

Solution

We want to solve the equation for x, the amount of money invested in CDs.

Teaching Tip
Take time to discuss how the equation is developed from the words of the problem.

$$0.02x + 0.04(5000 - x) = 170$$

Use the Distributive Property to remove parentheses: $0.02x + 200 - 0.04x = 170$

Combine like terms: $-0.02x + 200 = 170$

Isolate x: $-0.02x + 200 - 200 = 170 - 200$

$$-0.02x = -30$$

Multiply both sides by 100 to eliminate the decimal: $-2x = -3000$

Divide both sides by -2: $x = 1500$

Since $x = 1500$, you must invest \$1500 in Certificates of Deposit to earn \$170 in interest at the end of one year.

QUICK ✓ *Solve the equation for the unknown quantity.*

19. Janet Majors invested part of her lottery winnings in a savings account that pays 4% annual interest, and \$250 more than that in a mutual fund that pays 6% annual interest. Her total interest was \$65. To determine the amount she invested in the savings account, solve the equation $0.04x + 0.06(x + 250) = 65$, where x represents the amount invested in the savings account.

2.3 Exercises

1. conditional equation
2. least common denominator
3. 100
4. True
5. False
6. False
7. Answers may vary.
8. No, there is an error. All terms must be multiplied by 3; Answers may vary.
9. $\left\{\frac{9}{2}\right\}$
10. $\left\{-\frac{7}{3}\right\}$
11. $\{-2\}$
12. $\{6\}$
13. $\{-6\}$
14. $\{3\}$

Concepts and Vocabulary

In Problems 1–3, fill in the blanks.

1. A(n) __________ __________ is an equation that is true for some values of the variable and false for other values of the variable.

2. To clear fractions from an equation, we multiply each side of the equation by the __________ __________ __________.

3. To clear decimals from the equation $0.25x + 5 = 7 - 0.3x$, we multiply both sides of the equation by __________.

In Problems 4–6, answer True or False to each statement.

4. A linear equation can have one solution, no solution, or infinitely many solutions.

5. An equation that is satisfied for every choice of the variable for which both sides of the equation are defined is called a contradiction.

6. To solve the equation $\frac{1}{10}x + 8 = \frac{1}{5}x - 3$, we multiply both sides of the equation by 5 to clear fractions.

7. A student solved the equation $3(x + 8) = \frac{1}{2}(6x + 4)$ and wrote the answer $24 = 2$. The instructor did not give full credit for this answer. Explain the student's error and determine the correct solution.

8. When solving the equation $\frac{2}{3}x - 5 = \frac{1}{2}x$, a student decided to multiply both sides by the LCD and wrote $6 \cdot \frac{2}{3}x - 5 = \frac{1}{2}x \cdot 6$. This resulted in the next line $4x - 5 = 3x$. Is this correct? Explain the steps necessary to finish by this technique and then suggest another list of steps that would arrive at the correct solution.

Building Skills

In Problems 9–44, solve the equation. Check your solution.

9. $\frac{2k - 1}{4} = 2$

10. $\frac{3a + 2}{5} = -1$

11. $\frac{3x + 2}{4} = \frac{x}{2}$

12. $\frac{2x - 3}{5} = \frac{3x}{10}$

13. $\frac{1}{5}x + \frac{3}{2} = \frac{3}{10}$

14. $\frac{3}{2}n - \frac{4}{11} = \frac{91}{22}$

15. $\left\{\frac{2}{3}\right\}$ 16. $\left\{\frac{44}{9}\right\}$
17. {30} 18. {20}
19. {−4} 20. {5}
21. {50} 22. {30}
23. {4.8} 24. {2}
25. {150} 26. {250}
27. {6} 28. {8}
29. $\left\{\frac{25}{8}\right\}$ 30. $\left\{\frac{8}{3}\right\}$
31. {−20} 32. $\left\{-\frac{72}{25}\right\}$
33. {1} 34. $\left\{-\frac{9}{2}\right\}$
35. {−12} 36. {8}
37. {60} 38. {−4}
39. {5} 40. {1}
41. {2} 42. {1}
43. {75} 44. {450}
45. contradiction; ∅ or { }
46. contradiction; ∅ or { }
47. identity; The set of all real numbers
48. identity; The set of all real numbers
49. conditional equation; $\left\{-\frac{1}{2}\right\}$
50. conditional equation; $\left\{\frac{3}{2}\right\}$
51. contradiction; ∅ or { }
52. contradiction; ∅ or { }
53. contradiction; ∅ or { }
54. contradiction; ∅ or { }
55. identity; The set of all real numbers
56. identity; The set of all real numbers
57. $\left\{-\frac{3}{4}\right\}$ 58. $\left\{\frac{9}{8}\right\}$
59. identity; The set of all real numbers
60. identity; The set of all real numbers
61. $\left\{-\frac{1}{3}\right\}$ 62. $\left\{-\frac{7}{2}\right\}$
63. {−20} 64. {−4}
65. $\left\{\frac{7}{2}\right\}$ 66. {−10}
67. contradiction; ∅ or { }
68. contradiction; ∅ or { }
69. {−2} 70. {−30}
71. {3} 72. {−11}
73. contradiction; ∅ or { }
74. contradiction; ∅ or { }
75. {−4} 76. {−5}
77. {39} 78. {15}
79. {0} 80. {0}

15. $\frac{-2x}{3} + 1 = \frac{5}{9}$ **16.** $\frac{3m}{8} - 1 = \frac{5}{6}$ **17.** $0.4w = 12$

18. $0.3z = 6$ **19.** $-1.3c = 5.2$ **20.** $-1.7q = -8.5$

21. $1.05p = 52.5$ **22.** $1.06z = 31.8$ **23.** $p + 1.5p = 12$

24. $2.5a + a = 7$ **25.** $p + 0.05p = 157.5$ **26.** $p + 0.04p = 260$

27. $\frac{a}{4} - \frac{a}{3} = -\frac{1}{2}$ **28.** $\frac{3}{2}b - \frac{4}{5}b = \frac{28}{5}$ **29.** $\frac{5}{4}(2a - 10) = -\frac{3}{2}a$

30. $\frac{2}{3}(6 - x) = \frac{5x}{6}$ **31.** $\frac{y}{10} + 3 = \frac{y}{4} + 6$ **32.** $\frac{p}{8} - 1 = \frac{7p}{6} + 2$

33. $\frac{4x - 9}{3} + \frac{x}{6} = \frac{x}{2} - 2$ **34.** $\frac{3x + 2}{4} - \frac{x}{12} = \frac{x}{3} - 1$

35. $0.3x + 2.3 = 0.2x + 1.1$ **36.** $0.7y - 4.6 = 0.4y - 2.2$

37. $0.65x + 0.3x = x - 3$ **38.** $0.5n - 0.35n = 2.5n + 9.4$

39. $3 + 1.5(z + 2) = 3.5z - 4$ **40.** $5 - 0.2(m - 2) = 3.6m + 1.6$

41. $0.02(2c - 24) = -0.4(c - 1)$ **42.** $0.3(6a - 4) = -0.10(2a - 8)$

43. $0.15x + 0.10(250 - x) = 28.75$ **44.** $0.03t + 0.025(1000 - t) = 27.25$

In Problems 45–56, solve each equation. State whether the equation is a contradiction, an identity, or a conditional equation.

45. $4z - 3(z + 1) = 2(z - 3) - z$ **46.** $4(y - 2) = 5y - (y + 1)$

47. $6q - (q - 3) = 2q + 3(q + 1)$ **48.** $-3x + 2 + 5x = 2(x + 1)$

49. $9a - 5(a + 1) = 2(a - 3)$ **50.** $7b + 2(b - 4) = 8b - (3b + 2)$

51. $\frac{4x - 9}{6} - \frac{x}{2} = \frac{x}{6} + 3$ **52.** $\frac{2m + 1}{4} - \frac{m}{6} = \frac{m}{3} - 1$

53. $\frac{5z + 1}{5} = \frac{2z - 3}{2}$ **54.** $\frac{2y - 7}{4} = \frac{3y - 13}{6}$

55. $\frac{q}{3} + \frac{4}{5} = \frac{5q + 12}{15}$ **56.** $\frac{2x}{3} + \frac{x + 3}{12} = \frac{3x + 1}{4}$

Mixed Practice

In Problems 57–82, solve each equation.

57. $-3(2n + 4) = 10n$ **58.** $-15(z - 3) = 25z$

59. $-2x + 5x = 4(x + 2) - (x + 8)$ **60.** $5m - 3(m + 1) = 2(m + 1) - 5$

61. $-6(x - 2) + 8x = -x + 10 - 3x$ **62.** $3 - (x + 10) = 3x + 7$

63. $\frac{3}{4}x = \frac{1}{2}x - 5$ **64.** $\frac{1}{3}x = 2 + \frac{5}{6}x$

65. $\frac{1}{2}x + 2 = \frac{4x + 1}{4}$ **66.** $\frac{x}{2} + 4 = \frac{x + 7}{3}$

67. $0.3p + 2 = 0.1(p + 5) + 0.2(p + 1)$ **68.** $1.6z - 4 = 2(z - 1) - 0.4z$

69. $-0.7x = 1.4$ **70.** $0.2a = -6$

71. $\frac{3(2y - 1)}{5} = 2y - 3$ **72.** $\frac{4(2n + 1)}{3} = 2n - 6$

73. $0.6x - 0.2(x - 4) = 0.4(x - 2)$ **74.** $0.3x - 1 = 0.5(x + 2) - 0.2x$

75. $\frac{3x - 2}{4} = \frac{5x - 1}{6}$ **76.** $\frac{x - 1}{4} = \frac{x - 4}{6}$

77. $0.3x + 2.6x = 5.7 - 1.8 + 2.8x$ **78.** $0.3(z - 10) - 0.5z = -6$

79. $\frac{3}{2}x - 6 = \frac{2(x - 9)}{3} + \frac{1}{6}x$ **80.** $\frac{2x - 3}{4} + 5 = \frac{3(x + 3)}{4} - \frac{x}{2} + 2$

81. $\left\{\frac{2}{3}\right\}$
82. $\left\{-\frac{4}{15}\right\}$
83. -1.70
84. 4
85. -13.2
86. -4.67
87. \$50
88. 5 hours
89. \$18,000
90. \$1200
91. \$8.50
92. \$28
93. \$95
94. \$48
95. 15 quarters
96. 10 nickels
97. 6 units

81. $\frac{2}{3}\left[4 - \left(\frac{x}{2} + 6\right) - 2x\right] + 3 = \frac{5x}{6}$

82. $\frac{1}{2}\left[3 - \left(\frac{2x}{3} - 1\right) + 3x\right] = \frac{-4x + 1}{3} + 1$

In Problems 83–86, solve the equation and round to the indicated place.

83. $2.8x + 13.754 = 4 - 2.95x$ to the nearest hundredth

84. $-4.88x - 5.7 = 2(-3.41x) + 1.2$ to the nearest whole number

85. $x - \{1.5x - 2[x - 3.1(x + 10)]\} = 0$ to the nearest tenth

86. $-3x - 2\{4 + 3[x - (1 + x)]\} = 12$ to the nearest hundredth

Applying the Concepts

87. Sales Tax The price of a pair of jeans including sales tax of 6% is \$53. To find the price p of the jeans, solve the equation $1.06p = 53$.

88. Gardening Bob Adams rented a rototiller for x hours at a cost of \$7.50 per hour. Bob paid a bill of \$37.50. Find the number of hours he rented the tiller by solving the equation $7.50x = 37.50$.

89. Purchasing a Car The total cost (including 6% sales tax) for the purchase of an automobile was \$19,080. To determine the cost of the auto before the sales tax was added, solve the equation $x + 0.06x = 19{,}080$, where x represents the cost of the car before taxes.

90. Purchasing a Kayak The total cost, including 5.5% sales tax, for the purchase of a kayak was \$1266. Find the price of the kayak, k, before sales tax, by solving the equation $k + 0.055k = 1266$.

91. Hourly Pay Bob recently received a 4% pay increase. His hourly wage is now \$8.84. To determine his hourly wage, w, before the 4% pay raise, solve the equation $w + 0.04w = 8.84$ for w.

92. Hourly Pay A union representing airline employees recently agreed to a 6% cut in hourly wage in order to help the airline avoid filing for bankruptcy. A baggage handler will now earn \$26.32 per hour. To find the baggage handler's hourly wage, w, before the pay cut solve the equation $w - 0.06w = 26.32$.

93. CD Player Tamara purchased a CD player at a "25% off" sale for \$71.25. To determine the original price, p, of the CD player, solve the equation $p - 0.25p = 71.25$.

94. Team Sweatshirt Your favorite college has logo sweatshirts on sale for \$33.60. The sweatshirt has been marked down by 30%. To find the original price of the sweatshirt, x, solve the equation $x - 0.30x = 33.60$.

95. Piggy Bank Celeste saves dimes and quarters in a piggy bank. She opened the bank and discovered that she had \$7.05 and the number of dimes was 3 more than twice the number of quarters. To find the number of quarters, q, solve the equation $0.25q + 0.10(2q + 3) = 7.05$.

96. Clean Car Pablo cleaned out his car and found nickels and quarters in the car seats. He found \$4.25 in change and noticed that the number of quarters was 5 less than twice the number of nickels. Solve the equation $0.05n + 0.25(2n - 5) = 4.25$ to find n, the number of nickels Pablo found.

△**97. Comparing Perimeters** The two rectangles shown below have the same perimeter. Solve the equation $2x + 2(x + 3) = 2\left(\frac{1}{2}x\right) + 2(x + 6)$ for x, the width of the first rectangle.

98. 10 units
99. $12,000
100. $75,000
101. Answers may vary.

△ **98. Comparing Perimeters** The square and the rectangle shown have the same perimeter. Solve the equation $4x = 2\left(\frac{1}{2}x\right) + 2(x + 5)$ for x, the length of the side of the square.

99. Paying Your Taxes You are single and just determined that you paid $1442.50 in federal income taxes for 2004. The solution to the equation $1442.50 = 0.15(x - 7150) + 715$ represents your adjusted gross income in 2004. Determine your adjusted gross income in 2004. (*Source:* Internal Revenue Service)

100. Paying Your Taxes You are married and just determined that you paid $12,225 in federal income taxes in 2004. The solution to the equation $12{,}225 = 0.25(x - 58{,}100) + 8000$ represents the adjusted gross income of you and your spouse in 2004. Determine the adjusted gross income of you and your spouse in 2004. (*Source:* Internal Revenue Service)

Extending the Concepts

101. Make up a linear equation that has one solution. Make up a linear equation that has no solution. Make up a linear equation that is an identity. Comment on the differences and similarities in making up each equation.

2.4 Evaluating Formulas and Solving Formulas for a Variable

OBJECTIVES

1. Evaluate a Formula
2. Solve a Formula for a Variable

Preparing for Evaluating Formulas and Solving Formulas for a Variable

Before getting started, take this readiness quiz. If you get a problem wrong, go back to the section cited and review the material.

1. Evaluate the expression $2L + 2W$ for $L = 7$ and $W = 5$. [Section 1.7, pp. 57–58]

2. Round the expression 0.5873 to hundredths place. [Section A.2, p. A9]

1 Evaluate a Formula

Teaching Tip
Because Chapter 2 is a rather long chapter with a lot of concepts, you may consider giving an exam at the end of this section. Then, give a second exam covering Sections 2.5–2.8.

In this section, we use *formulas* to solve mathematical problems.

DEFINITION

A mathematical **formula** is an equation that describes how two or more variables are related.

For example, a formula for the area of a rectangle is *area* = *length* · *width*, or $A = l \cdot w$. You use mathematical formulas every day, many times without even realizing it. For instance, if you plan to paint your apartment, you must find the surface area of the walls to compute the amount of paint you will purchase. To determine the number of gallons of gasoline you can afford to put in your car, you may also use a formula, as illustrated in the next example.

Classroom Example
Use Example 1, except you have $20 to purchase gas at $2.50 per gallon.
Answer: 8 gallons

EXAMPLE 1 Evaluating a Formula

The number of gallons of gasoline that you put in your car is found by using the formula $\frac{C}{p} = n$, where C is the total cost, p is the price per gallon, and n is the

Preparing for...Answers **1.** 24 **2.** 0.59

number of gallons. How many gallons of gasoline can you purchase for \$10.00 if gas costs \$2.50 per gallon?

Solution

$$\frac{C}{p} = n$$

Replace C with cost, \$10, and p with price per gallon, \$2.50: $\frac{\$10.00}{\$2.50} = n$

Evaluate the numerical expression: $4 = n$

You can purchase 4 gallons for \$10.

QUICK ✓ *Evaluate each formula for the unknown quantity.*

1. Your best friend, who is on spring break in Europe, e-mailed you that the temperature in Paris today is 15° Celsius. Use the formula $F = \frac{9}{5}C + 32$, where F is degrees Fahrenheit and C is degrees Celsius, to find the approximate temperature in degrees Fahrenheit.
2. The size of a dress purchased in Europe is different from one purchased in the U.S. The equation $c = a + 30$ gives the European (Continental) dress size c in terms of the size in the U.S., a. Find the Continental dress size that corresponds to a size 10 dress in the United States.

Classroom Example ➔
Use Example 2, with the original price of the jeans equal to \$30.

Answer:

The sale price is \$22.50.

EXAMPLE 2 Evaluating a Formula Containing a Percent

The formula $S = P - 0.25P$ gives the sale price S of an item which originally cost P dollars and that was reduced by 25%. Find the sale price of a pair of jeans that originally cost \$40.00.

Solution

$$S = P - 0.25P$$

Replace P with the price of the jeans, \$40: $S = 40 - 0.25(40)$

Evaluate the numerical expression: $S = 40 - 10$

$$S = 30$$

The sale price of the jeans is \$30.

QUICK ✓ *Evaluate each formula for the unknown quantity.*

Teaching Tip
Remind students to label units in their answers.

3. The formula $E = 250 + 0.05S$ is a formula for the earnings, E, of a salesman who receives \$250 per week plus 5% commission on all weekly sales, S. Find the earnings of a salesman who had weekly sales of \$1250.
4. The formula $N = p + 0.06p$ models the new population, N, of a town whose current population, p, is given, if the town expects population growth of 6% next year. Find the new population of a town whose current population is 5600 persons.

The Simple Interest Formula

Interest is money paid for the use of money. The total amount borrowed is called the **principal.** The principal can be in the form of a loan (an individual borrows from the bank) or a deposit (the bank borrows from the individual). The **rate of interest,**

expressed as a percent, is the amount charged for the use of the principal for a given period of time, usually on yearly basis.

SIMPLE INTEREST FORMULA

If an amount of money, P, called the **principal** is invested for a period of t years at an annual interest rate r, expressed as a decimal, the amount of interest I earned is

$$I = Prt$$

Interest earned according to this formula is called **simple interest.**

Classroom Example ➤
Use Example 3 with \$1500 invested in the CD.

Answer: Janice earned \$18 in interest; the new total is \$1518.

EXAMPLE 3 Evaluating Using the Simple Interest Formula

Janice invested \$500 in a two-year Certificate of Deposit (CD) paying 4% simple interest. Find the amount of interest Janice will earn in 6 months. Also find the total amount of money Janice will have at the end of 6 months.

Solution

Work Smart
Be sure to express t in years when using the simple interest formula.

We see that the amount of money Janice invested, P is \$500. The interest rate r is $4\% = 0.04$. Because 6 months is $\frac{1}{2}$ a year, we have that $t = \frac{1}{2}$.

$$I = Prt$$

Replace $P = \$500$, $r = 0.04$, and $t = \frac{1}{2}$: $I = 500 \cdot 0.04 \cdot \frac{1}{2}$

Evaluate the numerical expression: $I = 10$

Janice earned \$10 on her investment. At the end of 6 months, Janet will have $\$500 + \$10 = \$510$.

QUICK ✓ *Find the unknown quantity.*

5. Bill invested his \$2500 Virginia Lottery winnings in an 8-month Certificate of Deposit that earns 3% simple interest. Find the amount of interest Bill's investment will earn at the end of 8 months. Also find the total amount of money bill will have at the end of 8 months.

Geometry Formulas

Let's review a few common terms from geometry.

DEFINITIONS

The **perimeter** is the sum of the lengths of all the sides of a figure.

The **area** is the amount of space enclosed by a two-dimensional figure measured in units squared.

The **surface area** of a solid is the sum of the areas of the surfaces of a three-dimensional figure.

The **volume** is the amount of space occupied by a three-dimensional figure measured in units cubed.

The **radius** r of a circle is the line segment that extends from the center of the circle to any point on the circle.

The **diameter** of a circle is any line segment that extends from one point on the circle through the center to a second point on the circle. The diameter is two times the length of the radius, $d = 2r$.

In circles, we use the term **circumference** to mean the perimeter.

Formulas from geometry are useful in solving many types of problems. We list some of these formulas in Table 1.

Teaching Tip
This may be a good time to discuss the difference between perimeter, area, surface area, and volume. You may use the floor or the walls of your classroom to point out the difference between these concepts.

Table 1

Plane Figures		Formulas
Square	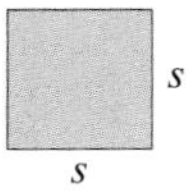 s, s	**Area:** $A = s^2$ **Perimeter:** $P = 4s$
Rectangle	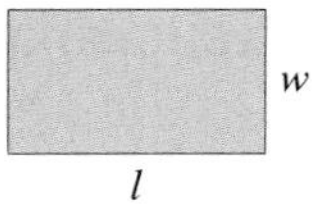 w, l	**Area:** $A = lw$ **Perimeter:** $P = 2l + 2w$
Triangle	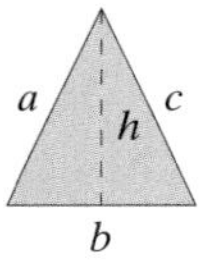 a, c, h, b	**Area:** $A = \frac{1}{2}bh$ **Perimeter:** $P = a + b + c$
Trapezoid	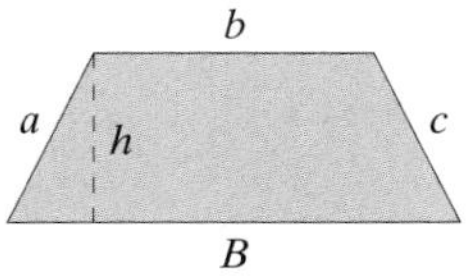 b, a, h, c, B	**Area:** $A = \frac{1}{2}h(B + b)$ **Perimeter:** $P = a + b + c + B$
Parallelogram	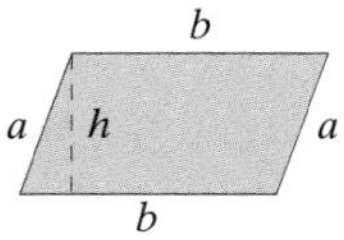 b, a, h, a, b	**Area:** $A = bh$ **Perimeter:** $P = 2a + 2b$
Circle	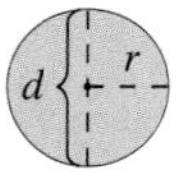 r, d	**Area:** $A = \pi r^2$ **Circumference:** $C = 2\pi r = \pi d$
Solids		**Formulas**
Cube	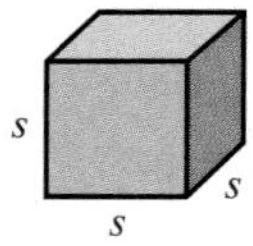 s, s, s	**Volume:** $V = s^3$ **Surface Area:** $S = 6s^2$
Rectangular Solid	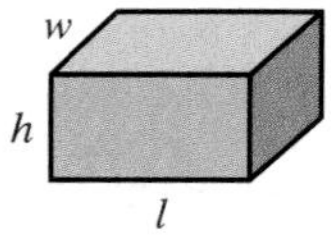 w, h, l	**Volume:** $V = lwh$ **Surface Area:** $S = 2lw + 2lh + 2wh$
Sphere	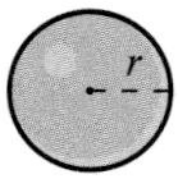 r	**Volume:** $V = \frac{4}{3}\pi r^3$ **Surface Area:** $S = 4\pi r^2$
Right Circular Cylinder	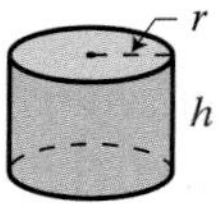 r, h	**Volume:** $V = \pi r^2 h$ **Surface Area:** $S = 2\pi r^2 + 2\pi rh$
Cone	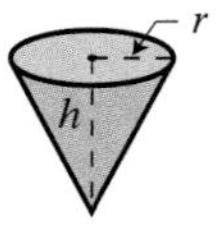 r, h	**Volume:** $V = \frac{1}{3}\pi r^2 h$

Classroom Example
Use Example 4 with the garden 6.75 feet long and 3.5 feet wide.

Answer:

20.5 feet of fencing are needed.

Figure 1

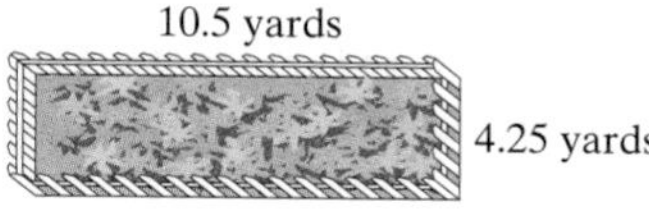

EXAMPLE 4 Evaluating a Formula for the Perimeter of a Rectangle

The number of yards of fencing that must be purchased to fence a rectangular garden is given by $P = 2l + 2w$, where P is the perimeter of the garden, l is the length of the garden and w is its width. Find the number of yards of fencing that must be purchased to enclose a garden that is 10.5 yards long and 4.25 yards wide as illustrated in Figure 1.

Solution

$$P = 2l + 2w$$

Replace l with 10.5 and w with 4.25: $P = 2(10.5) + 2(4.25)$

Evaluate the numerical expression: $P = 21 + 8.5$

$$P = 29.5$$

We find that 29.5 yards of fencing must be purchased.

QUICK ✓ *Evaluate the formula for the unknown quantity.*

6. Find the area of a trapezoid whose height is 4.5 inches, $B = 9$ inches, and $b = 7$ inches. Use $A = \frac{1}{2}h(B + b)$.

Sometimes we need to use more than one geometry formula in order to solve a problem.

Classroom Example
Use Example 5 with pool diameter 20 feet with a rectangular yard that is 100 feet by 60 feet.

Answer:

(a) Area is 5686 sq feet

(b) Cost is $1421.50 installed

EXAMPLE 5 Finding the Area of a Lawn and the Cost of Sod

A circular swimming pool whose diameter is 30 feet is to be installed in a rectangular yard that is 100 feet by 60 feet. Once the pool is installed, grass is to be installed on the remaining land. See Figure 2.

(a) Determine the area of land that is to receive grass.

(b) If sod costs $0.25 per square foot installed, what will be the cost of the lawn?

Figure 2

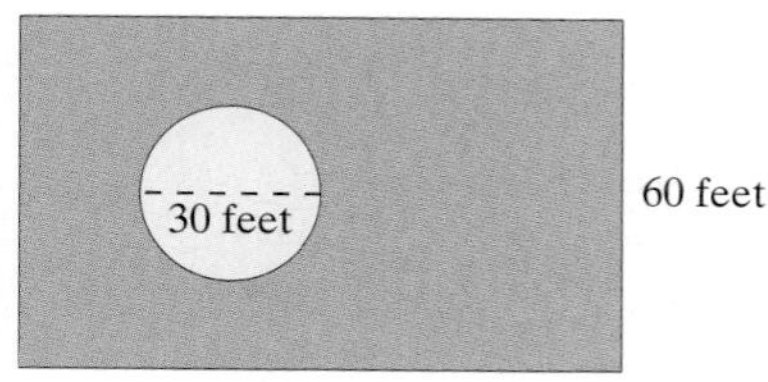

Solution

(a) The area of a rectangle is $A = lw$ and the area of a circle is $A = \pi r^2$. We find the area of the remaining lawn by subtracting the area of the circle from the area of the lawn.

$$\text{Area of remaining lawn} = \text{Area of rectangle} - \text{Area of circle}$$
$$= lw - \pi r^2$$

The length l of the lawn is 100 feet and the width is 60 feet. The radius of the circle is one-half its diameter, so the radius is 15 feet. Substituting into the formulas, we obtain

$$\text{Area of remaining lawn} = (100\text{ feet})(60\text{ feet}) - \pi(15\text{ feet})^2$$
$$= 6000\text{ feet}^2 - 225\pi\text{ feet}^2$$

$\pi \approx 3.14159$: $\approx 5293\text{ feet}^2$

There is approximately 5293 square feet of land that needs sod.

(b) We determine the cost of sod by multiplying the square footage by the cost per square foot.

$$\text{Cost for sod} = 5293\text{ square feet} \cdot \frac{\$0.25}{1\text{ square foot}}$$
$$= \$1323.25$$

The sod will cost $1323.25.

Teaching Tip
Area is expressed in square units because you multiply units (i.e., feet · feet = feet²). Perimeter is expressed as a linear quantity.

In Words
The word "per" means for each one. For example, the sod costs $0.25 for each square foot installed. When you see the word "per," think fraction. So $0.25 per square foot is represented as

$$\frac{\$0.25}{1\text{ square foot}}$$

It is worth noting in Example 5(b) that the units "square feet" divided out, so that the answer is measured in dollars. Because the answer is in the appropriate units, we have more confidence that our answer is correct.

QUICK ✓ *Solve for the unknown quantity.*

7. A circular water feature whose diameter is 4 feet is to be installed in a rectangular garden that is 20 feet by 10 feet. Once the water feature is installed, grass is to be installed on the remaining land.

(a) Determine the area, to the nearest square foot, of land that is to receive grass.

(b) If sod costs \$0.25 per square foot installed, what will be the cost of the lawn?

Geometry formulas can even be used to make us savvy consumers.

Classroom Example ➔
Use Example 6 with an 18″ pizza that costs \$16.99 and a 12″ pizza that costs \$9.99.

Answer: The 18″ pizza is the "better deal." Large is \$0.07 per square inch and small is \$0.09 per square inch.

EXAMPLE 6 Determining the Better Buy

At Mamma da Vinci's Pizza Parlor, a 16″ pizza costs \$14.99 and a 12″ pizza costs \$9.99. Which is the "better" buy?

Solution

The "better" buy will be the pizza that costs less per square inch. Our plan is: (1) find the area of each pizza and then (2) find the price per square inch of each pizza. We must consider, too, that a 16″ pizza has a *diameter* of 16 inches so that the radius is 8″ and a 12″ pizza has a *diameter* of 12 inches so that the radius is 6″.

1. Find the area of each pizza.

Since pizzas are circular, we compute the area of the pizza using the formula for the area of a circle.

	Area of the 16″ pizza:	Area of the 12″ pizza:
	$A = \pi r^2$	$A = \pi r^2$
Replace r with its given value:	$A = \pi \cdot 8^2$	$A = \pi \cdot 6^2$
Evaluate the numerical expression:	$A \approx 201.06$ sq in.	$A \approx 113.10$ sq in.

2. Find the price per square inch of each pizza.

The price per square inch can be found by dividing the price of the pizza by the number of square inches of area.

price per square inch of large pizza: $\dfrac{\$14.99}{201.06} = \$0.075$

price per square inch of medium pizza: $\dfrac{\$9.99}{113.10} = \$0.088$

The price per square inch of the large pizza is about 7.5¢, while the price per square inch of the medium pizza is about 8.8¢. The large pizza is the "better" buy. ■

QUICK ✓ *Evaluate each formula for the unknown quantity.*

8. A homeowner plans to construct a circular brick pad for his barbeque grill. The diameter of the brick pad is to be 6 feet. Find, to the nearest hundredth of a square foot, the area of the barbeque pad.

9. An extra large (18″) pizza at Dante's Pizza costs \$16.99 and a small (9″) pizza costs \$8.99. Which is the better buy?

---

2 Solve a Formula for a Variable

The expression "solve for a variable" means to isolate the variable with a coefficient of 1 on one side of the equation and all other variables and constants, if any, on the other side by forming equivalent equations. For example, in the formula for the area of a rectangle, $A = lw$, the formula is solved for A because A is by itself with a coefficient of 1 on one side of the equation while all other variables are on the other side.

The steps that we follow when solving formulas for a certain variable are identical to those that we followed when solving equations.

Solve for x:	$15 = \frac{3}{2}x$	**Solve for h:**	$A = \frac{1}{2}bh$
Multiply both sides of the equation by 2 to clear the fraction.	$2(15) = 2\left(\frac{3}{2}x\right)$	Multiply both sides of the equation by 2 to clear the fraction.	$2(A) = 2\left(\frac{1}{2}bh\right)$
	$30 = 3x$		$2A = bh$
Divide by 3 to isolate the variable x.	$\frac{30}{3} = \frac{3x}{3}$	Divide by b to isolate the variable h.	$\frac{2A}{b} = \frac{bh}{b}$
	$10 = x$		$\frac{2A}{b} = h$

Classroom Example Solve the formula $P = 2l + 2w$ for l.

Answer: $l = \frac{P - 2w}{2}$

EXAMPLE 7 How to Solve a Formula for a Variable

Solve the formula $P = 2l + 2w$ for w. *Note:* $P = 2l + 2w$ is the formula for the perimeter of a rectangle.

Step-by-Step Solution

Step 1: Isolate the term containing the variable.

Subtract $2l$ from both sides of the equation to isolate the term $2w$:

$$P = 2l + 2w$$
$$P - 2l = 2l + 2w - 2l$$
$$P - 2l = 2w$$

Step 2: Get the coefficient of the variable to be 1.

Divide both sides of the equation by 2:

$$\frac{P - 2l}{2} = \frac{2w}{2}$$
$$\frac{P - 2l}{2} = w$$

Work Smart

The Symmetric Property states that if $a = b$, then $b = a$.

We can write $\frac{P - 2l}{2} = w$ as $w = \frac{P - 2l}{2}$ using the *Symmetric Property*. We have solved the equation $P = 2l + 2w$ for w because w is isolated with a coefficient of 1 on one side of the equation and all other terms are on the other side of the equation.

QUICK ✓ *Solve for the specified variable.*

10. To convert from degrees Celsius to degrees Fahrenheit, we use the formula $F = \frac{9}{5}C + 32$. Solve this formula for C.

11. The formula $A = 2\pi rh + 2\pi r^2$ represents the surface area A of a right circular cylinder whose radius is r and height is h. Solve the formula for h.

This next example illustrates a skill that will be extremely important when we study graphing lines in Chapter 8.

Classroom Example Solve the equation $4x + 3y = 12$ for y.

Answer: $y = \frac{12 - 4x}{3}$

EXAMPLE 8 Solving for a Variable in a Formula

Solve the equation $3x + 2y = 6$ for y.

Solution

We wish to solve for y. That is, we want to get y isolated on one side of the equation and the variable x and constants on the other side. We begin by isolating the term containing y on the left side of the equation.

$$3x + 2y = 6$$

Subtract $3x$ from both sides of the equation to isolate the term containing y: $3x + 2y - 3x = 6 - 3x$

Simplify both sides of the equation: $2y = 6 - 3x$

Divide both sides by 2: $\frac{2y}{2} = \frac{6 - 3x}{2}$

Simplify: $y = \frac{6 - 3x}{2}$

Teaching Tip

Remind students that $\frac{A + B}{C} = \frac{A}{C} + \frac{B}{C}$.

So when we solve the equation $3x + 2y = 6$ for y, we obtain the equation $y = \frac{6 - 3x}{2}$. This equation may also be written in the form $y = 3 - \frac{3}{2}x$ or $y = -\frac{3}{2}x + 3$ by dividing 2 into each term in the numerator. Do you see why these three equations are equivalent?

QUICK ✓ *Solve each equation for the indicated variable.*

12. $x + 2y = 7$; for y

13. $5x - 3y = 15$; for y

14. $\frac{3}{4}a + 2b = 7$; for b

15. $3rs + \frac{1}{2}t = 12$ for t.

Classroom Example ➤

The formula $A = P + Prt$ is used to compute the amount A in a savings account that pays simple interest where P is the principal, r is the interest rate (expressed as a decimal), and t is time in years.

(a) Solve the equation for t.

(b) Find the number of years that it took a principal of \$500 to grow to \$575 at 5% simple interest compounded annually.

Answer:

(a) $t = \frac{A - P}{Pr}$

(b) 3 years

EXAMPLE 9 Solving a Formula for a Specified Variable and Evaluating the Formula

The amount of profit, P, earned by a manufacturer is given by the formula $P = R - C$, where R represents the manufacturer's revenue and C represents the manufacturer's costs.

(a) Solve the equation for R, the manufacturer's revenue.

(b) Find the amount of revenue if a manufacturer has a \$12,500 profit and \$6000 in costs.

Solution

(a) We wish to solve the equation for R, so we must isolate R on one side of the equation.

$$P = R - C$$

Add C to both sides of the equation: $P + C = R - C + C$

Simplify: $P + C = R$

Thus, $R = P + C$.

(b) Evaluate $R = P + C$ if $P = \$12,500$ and $C = \$6000$.

$$R = P + C$$

Replace P with 12,500 and C with 6000: $R = 12,500 + 6000$

Evaluate: $R = 18,500$

The manufacturer has revenue of \$18,500 when it has \$12,500 profit and \$6000 in costs.

QUICK ✓ *Solve each equation for the indicated variable and evaluate.*

16. **(a)** Solve the formula $d = rt$ for t, where d represents distance, r represents an average speed, and t represents time.
    (b) Suppose that a family drives from Columbus, Ohio, to Raleigh, North Carolina, a distance of 550 miles. Their average speed for the trip is 60 mph. How long will it take the family to reach their destination?
17. **(a)** Solve the simple interest formula $I = Prt$ for t.
    (b) Find the number of years that \$1000 must be invested at 7% annual interest to earn \$35 interest. **Hint:** Convert the percent to a decimal.

2.4 Exercises

1. formula
2. multiply; 4
3. subtract; from
4. True
5. False
6. False
7. Answers may vary.
8. Answers may vary.
9. Rogan 1099.14 feet; Nurek 984.3 feet
10. 1066.8 meters
11. \$19.20

Concepts and Vocabulary

In Problems 1–3, fill in the blanks.

1. A mathematical ________ is a statement that describes how two or more variables are related.
2. To solve the formula $\frac{a + b + 2c}{4} = G$ for a, we first ________ both sides of the equation by ________.
3. To solve the formula $P = 2l + 2w$ for w, we first ________ $2l$ (to/from) both sides of the equation.

In Problems 4–6, answer True or False to each statement.

4. Solving $d = rt$ for t results in $t = \frac{d}{r}$.
5. The perimeter of a rectangle is found by multiplying length and width.
6. Solving $x - y = 6$ for y results in $y = -x + 6$.
7. A student solved the equation $x + 2y = 6$ for y and obtained the result $y = \frac{-x + 6}{2}$. Another student solved the same equation for y and obtained the result $y = -\frac{1}{2}x + 3$. Are both solutions correct? Explain why or why not.
8. Make up an example of a linear equation whose coefficients are integers. Make up a similar example where the coefficients are constants (letters). Solve both and then explain which steps in the solution are alike and which are different.

Building Skills

In Problems 9–28, substitute the given values into the formula and then evaluate to find the unknown quantity. Label units in the answer. If the answer is not exact, round your answer to the nearest hundredth.

9. **Height of a Dam** The formula $f = 3.281m$ converts a length in meters m to feet f. The world's highest dams, the Rogun and the Nurek, are both in Tajikistan. The Rogun is 335 meters high and the Nurek is 300 meters high. Convert these heights to feet.
10. **Length of a Bridge** The formula $m = 0.3048f$ converts length in feet f to length in meters m. The George Washington Bridge in New York City is 3500 feet long. How long is the George Washington Bridge in meters?
11. **Buying a CD Player** The formula $S = P - 0.20P$ gives the sale price S of an item which originally cost P dollars that was reduced by 20%. Find the sale price of a CD player that originally cost \$24.00.

12. \$617.50
13. \$650
14. \$834
15. 20°C
16. 50°F
17. \$3
18. \$150
19. (a) 50 units (b) 144 square units
20. (a) 104 units (b) 640 square units
21. (a) 36.2 meters (b) 70 square meters
22. (a) $2\frac{1}{2}$ miles (b) $\frac{3}{8}$ square miles
23. (a) 36 units (b) 81 square units
24. (a) 14 units (b) 12.25 square units

12. **Buying a Computer** The formula $S = P - 0.35P$ gives the sale price S of an item which originally cost P dollars that was reduced by 35%. Find the sale price of a computer that originally cost \$950.00.

13. **Salesperson's Earnings** The formula $E = 500 + 0.15S$ is a formula for the earnings, E, of a salesperson who receives \$500 per week plus 15% commission on all sales, S. Find the earnings of a salesperson who had weekly sales of \$1000.

14. **Salesperson's Earnings** The formula $E = 750 + 0.07S$ is a formula for the earnings, E, of a salesperson who receives \$750 per week plus 7% commission on all sales, S. Find the earnings of a salesperson who had weekly sales of \$1200.

15. **Planning a Trip** A businessperson is planning a trip from Tokyo to San Diego, California, in March. She learns that the average high temperature in San Diego in March is 68°. Use the formula $C = \frac{5}{9}(F - 32)$ to convert 68° Fahrenheit to degrees Celsius.

16. **Planning a Trip** An English Literature professor plans to take twenty students on a study-abroad trip to London in March. He learns that the average daytime temperature in London in March is 10 degrees Celsius. Use the formula $F = \frac{9}{5}C + 32$ to convert 10 degrees Celsius to degrees Fahrenheit.

17. **Lottery Earnings** Therese invested her \$200 West Virginia Lottery winnings in a 6-month Certificate of Deposit (CD) that earns 3% simple interest. Use the formula $I = Prt$ to find the amount of interest Therese's investment will earn.

18. **Investing an Inheritance** Christopher invested part of his \$5000 inheritance from his grandmother in a 9-month Certificate of Deposit that earns 4% simple interest. Use the formula $I = Prt$ to find the amount of interest Christopher's investment will earn.

△ 19. Find (a) the perimeter and (b) the area of the rectangle.

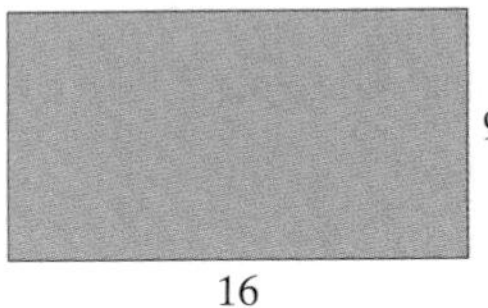

△ 20. Find (a) the perimeter and (b) the area of the rectangle.

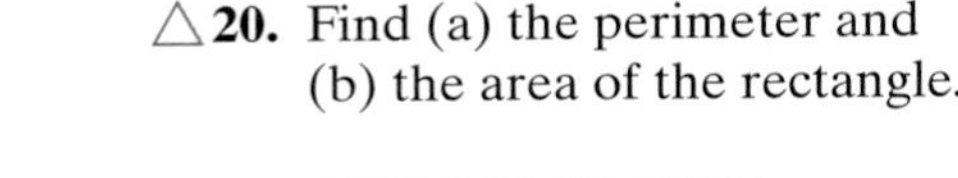

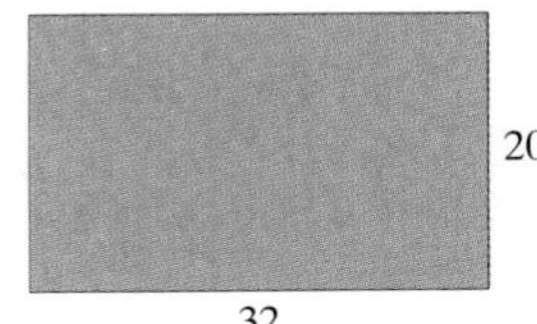

△ 21. Find (a) the perimeter and (b) the area of the rectangle.

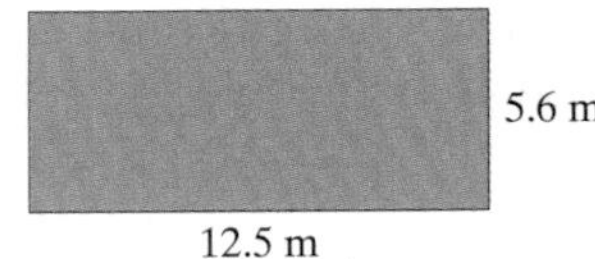

△ 22. Find (a) the perimeter and (b) the area of the rectangle.

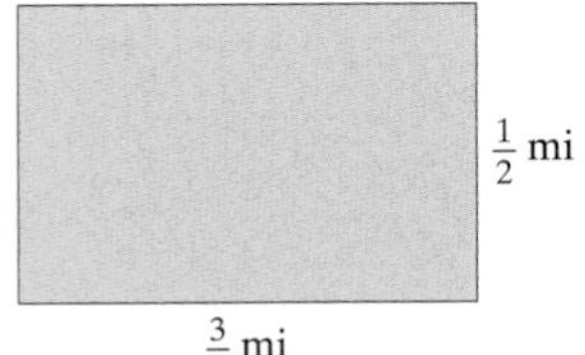

△ 23. Find (a) the perimeter and (b) the area of the square.

△ 24. Find (a) the perimeter and (b) the area of the square.

25. (a) 31.4 cm (b) 78.5 cm^2
26. (a) 17.58 yards (b) 24.62 square yards
27. 68.44 square inches
28. 19.63 square kilometers
29. $t = \frac{d}{r}$
30. $w = \frac{A}{l}$
31. $d = \frac{C}{\pi}$
32. $m = \frac{F}{v^2}$
33. $r = \frac{I}{Pt}$
34. $W = \frac{v}{LH}$
35. $h = \frac{2A}{b}$
36. $B = \frac{3V}{h}$
37. $a = P - b - c$
38. $b = S - a - c$
39. $t = \frac{A - P}{Pr}$
40. $w = \frac{P - 2l}{2}$
41. $b = \frac{2A}{h} - B$
42. $h = \frac{S}{2\pi} - r^2$
43. $y = -3x + 12$
44. $y = 2x + 18$
45. $y = 2x - 5$
46. $y = 2x - 3$
47. $y = \frac{-4x + 13}{3}$
48. $y = -\frac{5}{6}x + 3$
49. $y = 3x - 12$
50. $y = \frac{4}{15}x - 2$
51. (a) $C = R - P$ (b) \$450
52. (a) $R = P + C$ (b) \$6000
53. (a) $t = \frac{I}{Pr}$ (b) 2 years
54. (a) $r = \frac{I}{Pt}$ (b) 3%
55. (a) $m = \frac{2K}{v^2}$ (b) 16
56. (a) $x = Z\sigma + \mu$ (b) 130
57. (a) $C = \frac{5}{9}(F - 32)$ (b) 15°

△ **25.** Find (a) the circumference and (b) the area of the circle with radius $r = 5$ cm. Use $\pi \approx 3.14$.

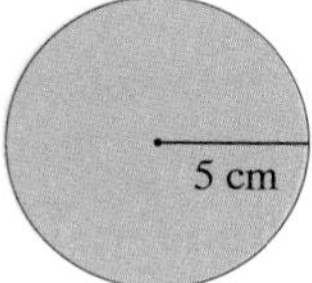

△ **26.** Find (a) the circumference and (b) the area of the circle with radius $r = 2.8$ yards. Use $\pi \approx 3.14$.

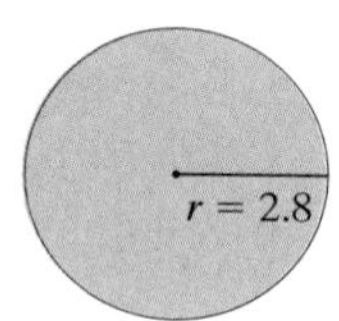

△ **27. Area of a Circle** Find the area of a circle A when $\pi \approx \frac{22}{7}$ and $r = \frac{14}{3}$ inches.

△ **28. Area of a Circle** Find the area of a circle A when $\pi \approx 3.14$ and $r = 2.5$ km.

In Problems 29–42, solve each formula for the stated variable.

29. $d = rt$; solve for t

30. $A = lw$; solve for w

31. $C = \pi d$; solve for d

32. $F = mv^2$; solve for m

33. $I = Prt$; solve for r

34. $v = LWH$; solve for W

35. $A = \frac{1}{2}bh$; solve for h

36. $V = \frac{1}{3}Bh$; solve for B

37. $P = a + b + c$; solve for a

38. $S = a + b + c$; solve for b

39. $A = P + Prt$; solve for t

40. $P = 2l + 2w$: solve for w

41. $A = \frac{1}{2}h(B + b)$; solve for b

42. $S = 2\pi(h + r^2)$; solve for h

In Problems 43–50, solve for y.

43. $3x + y = 12$

44. $-2x + y = 18$

45. $10x - 5y = 25$

46. $12x - 6y = 18$

47. $4x + 3y = 13$

48. $5x + 6y = 18$

49. $\frac{1}{2}x - \frac{1}{6}y = 2$

50. $\frac{2}{3}x - \frac{5}{2}y = 5$

Applying the Concepts

In Problems 51–64, (a) solve for the indicated variable, and then (b) find the value of the unknown quantity. When given, label units in the answer.

51. Profit = Revenue − Cost: $P = R - C$
(a) Solve for C. **(b)** Find C when $P = \$1200$ and $R = \$1650$.

52. Profit = Revenue − Cost: $P = R - C$
(a) Solve for R. **(b)** Find R when $P = \$4525$ and $C = \$1475$.

53. Simple Interest: $I = Prt$
(a) Solve for t. **(b)** Find t when $I = \$42$, $P = \$525$, and $r = 4\%$.

54. Simple Interest: $I = Prt$
(a) Solve for r. **(b)** Find r when $I = \$225$, $P = \$5000$, and $t = 1.5$ years.

55. Physics Formula: $K = \frac{1}{2}mv^2$
(a) Solve for m. **(b)** Find m when $K = 8192$ and $v = 32$.

56. Statistics Formula: $Z = \frac{x - \mu}{\sigma}$
(a) Solve for x. **(b)** Find x when $Z = 2$, $\mu = 100$, and $\sigma = 15$.

57. Fahrenheit/Celsius Temperature Conversion: $F = \frac{9}{5}C + 32$
(a) Solve for C. **(b)** Find C when $F = 59°$.

58. (a) $m = \frac{y - 5}{x}$ (b) -2
59. (a) $r = \frac{A - P}{Pt}$ (b) 0.04 or 4%
60. (a) $t = \frac{A - P}{Pr}$ (b) 1.5 years
61. (a) $h = \frac{V}{\pi r^2}$ (b) 5 mm
62. (a) $h = \frac{V}{\pi r^2}$ (b) 12 in.
63. (a) $b = \frac{2A}{h}$ (b) 18 ft
64. (a) $h = \frac{2A}{b + B}$ (b) 9 cm
65. 1834.05
66. 1333.62
67. (a) $h = \frac{S - 2\pi r^2}{2\pi r}$
(b) 1.25 inches
68. (a) $h = \frac{V}{\pi r^2}$ (b) 10 inches
69. medium 12″
70. large 16″
71. (a) 12 hours (b) \$336

58. Algebra: $y = mx + 5$
(a) Solve for m. **(b)** Find m when $x = 3$ and $y = -1$.

59. New Amount = Principal + Interest: $A = P + Prt$
(a) Solve for r. **(b)** Find r when $A = \$540$, $P = \$500$, and $t = 2$.

60. New Amount = Principal + Interest: $A = P + Prt$
(a) Solve for t. **(b)** Find t when $A = \$249$, $P = \$240$, and $r = 2.5\% = 0.025$.

△ **61. Volume of a Right Circular Cylinder:** $V = \pi r^2 h$
(a) Solve for h. **(b)** Find h when $V = 320\pi$ mm^3 and $r = 8$ mm.

△ **62. Volume of a Right Circular Cylinder:** $V = \pi r^2 h$
(a) Solve for h. **(b)** Find h when $V = 972\pi$ in.^{3} and $r = 9$ in.

△ **63. Area of a Triangle:** $A = \frac{1}{2}bh$
(a) Solve for b. **(b)** Find b when $A = 45$ ft and $h = 5$ ft

△ **64. Area of a Trapezoid:** $A = \frac{1}{2}h(b + B)$
(a) Solve for h. **(b)** Find h when $A = 99$ cm^2, $b = 19$ cm, and $B = 3$ cm.

65. Energy Expenditure Basal energy expenditure (E) is the amount of energy required to maintain the body's normal metabolic activity such as respiration, maintenance of body temperature, and so on. For males, the Basal energy expenditure is given by the formula

$$E = 66.67 + 13.75W + 5H - 6.76A$$

where W is the weight of the male (in kilograms), H is the height of the male (in centimeters), and A is the age of the male. Determine the Basal energy expenditure of a 37-year-old male who is 178 cm (5 feet, 10 inches) tall and weighs 82 kg (180 pounds).

66. Energy Expenditure See Problem 65. The Basal energy expenditure for females is given by

$$E = 665.1 + 9.56W + 1.85H - 4.68A$$

Compute the Basal energy expenditure of a 40-year-old female who is 168 cm tall (5 feet, 6 inches) and weighs 57 kg (125 pounds).

△ **67. Soup Can** The formula $S = 2\pi rh + 2\pi r^2$ gives the surface area S of a right circular cylinder whose radius is r and height is h.
(a) Solve the formula for h.
(b) Find the height of a right circular cylinder whose surface area is 8.25π square inches and radius is 1.5 inches.

△ **68. Cylinders** The volume V of a right circular cylinder is given by the formula $V = \pi r^2 h$, where r is the radius and h.
(a) Solve the formula for h.
(b) Find the height of a right circular cylinder whose volume is 90π cubic inches and whose radius is 3 inches.

△ **69. Grocery Store** You are standing at the freezer case at your local grocery store trying to decide which is the "better" buy: a medium (12″) pizza for \$9.99 or 2 small pizzas (8″) that are on special for \$4.49 each. Which should you choose to get the best deal?

70. Pizza for Dinner Mama Mimi's Take and Bake Pizzeria is running a special on southwestern-style pizzas: a large 16″ pizza for \$13.99 or two small 8″ pizzas for \$12.99. Which should you choose to get the best deal?

71. Taking a Trip Jason drives a truck as an independent contractor. He bills himself out at \$28 per hour. Suppose Jason has a contract that calls for him to leave a dock

72. (a) 5 hours (b) $160
73. 50.5 square inches
74. 76.5 cm^2
75. 96π $cm^3 \approx 301.59$ cm^3
76. $(24 + 2\pi)ft^2 \approx 30.28$ ft^2
77. (a) $I = \frac{D - P}{0.03} + 137{,}300$
(b) $158,000

at 9:00 A.M. and travel 600 miles to a warehouse. Jason has driven this route many times and figures that he can travel at an average speed of 50 miles per hour.

(a) Using the formula $d = rt$, where d is the distance traveled, r is the average speed, and t is the time spent traveling, determine how long Jason expects the trip to take.

(b) How much money does Jason expect to earn from this contract?

72. Taking a Trip Messai drives a truck as an independent contractor. He bills himself out at $32 per hour. Messai has a contract that calls for him to leave a dock at 8:00 A.M. and travel 145 miles to a warehouse. At the warehouse, he will wait while the truck is loaded (this takes 2 hours) and then return to his original dock. Messai has driven this route many times and figures that he can travel at an average speed of 58 miles per hour.

(a) Using the formula $d = rt$, where d is the distance traveled, r is the average speed, and t is the time spent traveling, determine how long Messai expects the roundtrip to take. Exclude the time Messia waits for the truck to be loaded.

(b) How much money does Messai expect to earn from this contract driving his truck?

△ **73. Area of a Region** Find the area of the figure below.

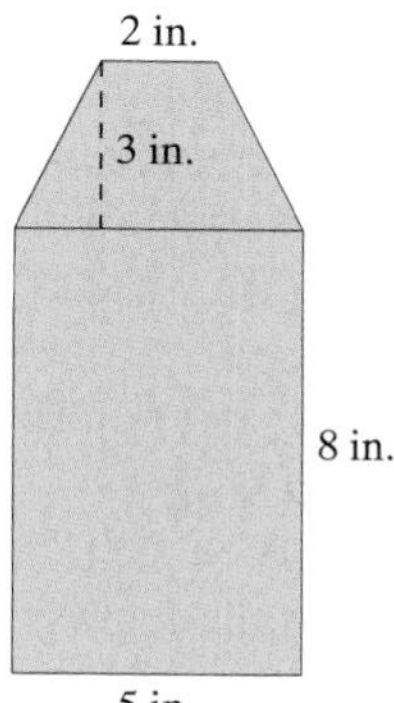

△ **74. Area of a Region** Find the area of the figure below.

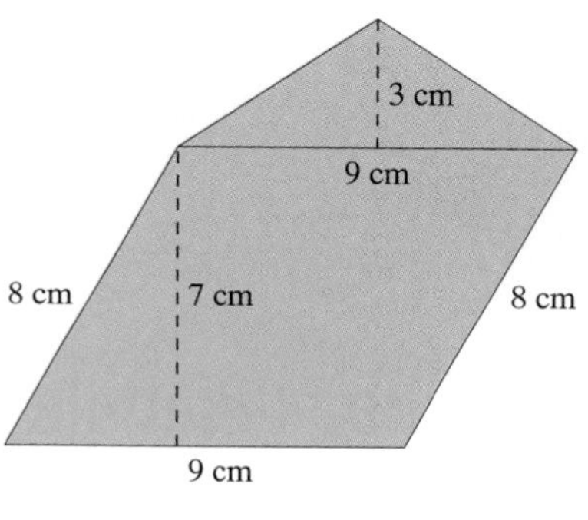

△ **75. Ice Cream Cone** Find the amount of ice cream in a cone if the radius of the cone is 4 cm and its height is 10 cm. The ice cream fully fills the cone and the hemisphere of ice cream on the top has a radius of 4 cm.

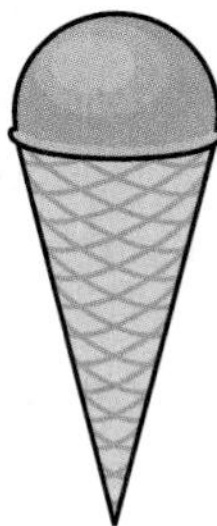

△ **76. Window** Find the area of the window given that the upper portion is a semicircle:

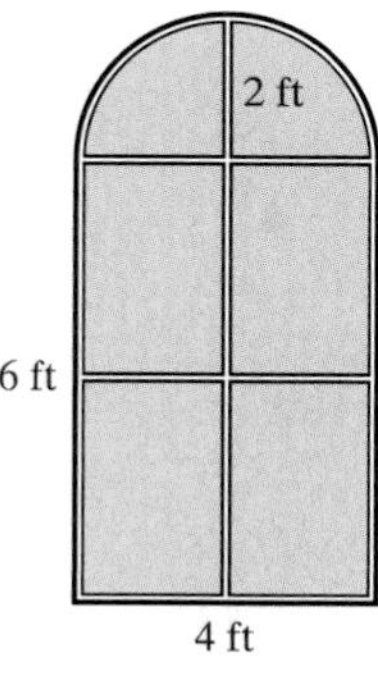

77. Federal Taxes According to the tax code in 2002, a married couple that earns over $137,300 per year filing a joint income tax return is subject to having their itemized deductions reduced. The formula $P = D - 0.03(I - 137{,}300)$ can be used to determine the permitted deductions P, where D represents the amount of deductions from Schedule A and I represents the couple's adjusted gross income.

(a) Solve the formula for I.

(b) Determine the adjusted gross income of a married couple filing a joint return whose allowed deductions were $15,200 and Schedule A deductions were $15,821.

78. (a) $d = \frac{6G - a - b - 2c}{2}$ (b) 92
79. (a) 62 (b) $372 (c) Yes
80. 2 gallons
81. (a) 4948 ft^2 (b) $1237
82. (a) 2912 ft^2 (b) 6 pallets (c) $576

78. Computing a Grade Jose's art history instructor uses the equation $G = \frac{a + b + 2c + 2d}{6}$ to compute her students' semester grade. The variables a and b represent the grades on two tests, c represents the grade on a research paper, d represents the final exam grade, and G represents the student's average.

(a) Solve the equation for d, Jose's final exam grade.

(b) A final average of 84 will earn Jose a B in the course. Compute the grade Jose must make on his final exam in order to earn a B for the semester, if he scored 78 and 74 on his tests, and 84 on his research paper.

△ **79. Remodel a Bathroom** You plan to remodel your bathroom and you've chosen 1 foot-by-1 foot ceramic tiles for the floor. The bathroom is 7 feet 6 inches long and 8 feet 2 inches wide.

(a) How many tiles do you need to cover the floor of your bathroom?

(b) Each tile costs $6. How much will it cost to tile your floor?

(c) The store from which you purchase the tile offers a discount of 10% on orders over $350. Does your order qualify for the discount?

△ **80. Painting a Room** A gallon of paint can cover about 500 square feet. Find the number of gallon containers of paint that must be purchased to paint two coats on each wall of a rectangular room measuring 8 feet by 12 feet, with a 10-foot ceiling. *Note:* You cannot purchase a partial can of paint!

△ **81. Landscaping a Back Yard** A circular swimming pool whose diameter is 24 feet is to be installed in a rectangular yard that is 60 feet by 90 feet. Once the pool is installed, grass is to be installed on the remaining land.

(a) Determine the area of land that is to receive grass. Round your answer to the nearest foot. Use $\pi \approx 3.14159$.

(b) If sod costs $0.25 per square foot installed, what will be the cost of the lawn?

△ **82. Landscaping a Back Yard** A rectangular swimming pool whose dimensions are 12 feet by 24 feet is to be installed in a rectangular yard that is 80 feet by 40 feet. Once the pool is installed, grass is to be installed on the remaining land.

(a) Determine the area of land that is to receive grass. Round your answer to the nearest foot. Use $\pi \approx 3.14159$.

(b) A pallet of sod covers 500 square feet. How many pallets of sod are required?

(c) Each pallet of sod costs $96. What is the cost of the sod?

Extending the Concepts

83. (a) 1080 $in.^2$ (b) 7.5 ft^2
84. (a) 216 ft^2 (b) 24 yd^2
85. Multiply by $\frac{1 \text{ ft}^2}{144 \text{ in.}^2}$.
86. Multiply by $\frac{9 \text{ ft}^2}{1 \text{ yd}^2}$.

△ **83.** A rectangle has length 5 feet and width 18 inches.

(a) What is the area in square inches? **(b)** What is the area in square feet?

△ **84.** A rectangle has length 9 yards and width 8 feet.

(a) What is the area in square feet? **(b)** What is the area in square yards?

△ **85.** Determine a formula for converting square inches to square feet.

△ **86.** Determine a formula for converting square yards to square feet.

PUTTING THE CONCEPTS TOGETHER (SECTIONS 2.1–2.4)

These problems cover important concepts from Sections 2.1 to 2.4. We designed these problems so that you can review the chapter so far and show your mastery of the concepts. Take time to work these problems before proceeding with the next section. The answers to these problems are located at the back of the text starting on page AN-3.

1. (a) Yes (b) No

1. Determine if the given value of the variable is a solution of the equation

$$4 - (6 - x) = 5x - 8$$

(a) $x = \frac{3}{2}$ **(b)** $x = -\frac{5}{2}$

2. (a) No (b) Yes
3. $\left\{-\frac{2}{3}\right\}$
4. $\{-40\}$
5. $\{-6\}$
6. $\{3\}$
7. $\{-6\}$
8. $\left\{-\frac{7}{3}\right\}$
9. $\left\{-\frac{57}{2}\right\}$
10. $\{25\}$
11. $\{6\}$
12. $\left\{\frac{74}{5}\right\}$ or $\{14.8\}$
13. contradiction; $\emptyset$ or { }
14. identity; $\{x \mid x \text{ is any real number}\}$
15. \$5000
16. (a) $b = \frac{2A}{h} - B$ (b) 6 in.
17. (a) $h = \frac{V}{\pi r^2}$ (b) 13 in.
18. $y = -\frac{3}{2}x + 7$

2. Determine if the given value of the variable is a solution of the equation

$$\frac{1}{2}(x - 4) + 3x = x + \frac{1}{2}$$

(a) $x = -4$ **(b)** $x = 1$

In Problems 3–14, solve the equation and check the solution.

3. $x + \frac{1}{2} = -\frac{1}{6}$ **4.** $-0.4m = 16$ **5.** $14 = -\frac{7}{3}p$ **6.** $8n - 11 = 13$

7. $\frac{5}{2}n - 4 = -19$ **8.** $-(5 - x) = 2(5x + 8)$

9. $7(x + 6) = 2x + 3x - 15$ **10.** $-7a + 5 + 8a = 2a + 8 - 28$

11. $-\frac{1}{2}(x - 6) + \frac{1}{6}(x + 6) = 2$ **12.** $0.3x - 1.4 = -0.2x + 6$

13. $5 + 3(2x + 1) = 5x + x - 10$ **14.** $3 - 2(x + 5) = -2(x + 2) - 3$

15. Investment You have \$7500 to invest and your financial advisor suggests that you put part of the money in a Certificate of Deposit that earns 2.4% simple interest and the remainder in bonds that earn 4% simple interest. To determine the amount you should invest in the CD to earn \$220 interest at the end of one year, solve the equation $0.024x + 0.04(7500 - x) = 220$, where x represents the amount of money invested in CDs.

16. Area of a trapezoid: $A = \frac{1}{2}h(B + b)$

(a) Solve for b. **(b)** Find b when $A = 76$ in.^{2}, $h = 8$ in., and $B = 13$ in.

17. Volume of a right circular cylinder: $V = \pi r^2 h$

(a) Solve for h. **(b)** Find h when $V = 117\pi$ sq. in., and $r = 3$ in.

18. Solve the equation $3x + 2y = 14$ for y.

2.5 Introduction to Problem Solving: Direct Translation Problems

OBJECTIVES

1. Translate English Phrases to Algebraic Expressions
2. Translate English Sentences to Equations
3. Build Models for Solving Direct Translation Problems

Preparing for Introduction to Problem Solving: Direct Translation Problems

Before getting started, take this readiness quiz. If you get a problem wrong, go back to the section cited and review the material.

1. Solve the equation: $x + 34.95 = 60.03$ [Section 2.1, pp. 76–78]
2. Solve the equation: $x + 0.25x = 60$ [Section 2.2, p. 86]

1 Translate English Phrases to Algebraic Expressions

One of the neat features of mathematics is that the symbols we use allow us to express English phrases briefly and consistently. An algebraic expression is similar to an English phrase. For example, the English phrase "5 more than a number x" is represented algebraically as $x + 5$.

There are certain words or phrases in English that easily translate into mathematical symbols. Table 2 lists various English words or phrases and their corresponding math symbol.

Preparing for...Answers **1.** {25.08} **2.** {48}

Table 2 Math Symbols and the Words They Represent

Add (+)	Subtract (−)	Multiply (·)	Divide (/)
sum	difference	product	quotient
plus	minus	times	divided by
greater than	subtracted from	of	per
more than	less	twice	ratio
exceeds by	less than	double	
in excess of	decreased by	half	
added to	fewer		
increased by			
combined			
altogether			

EXAMPLE 1 Writing English Phrases Using Math Symbols

Express each English phrase using mathematical symbols.

(a) The sum of 2 and 5.
(b) The difference of 12 and 7.
(c) The product of −3 and 8.
(d) The quotient of 10 and 2.
(e) 9 less than 15.
(f) A number z decreased by 11.
(g) Three times the sum of a number x and 8.

Solution

(a) Because we are talking about a sum, we know to use the + symbol, so "The sum of 2 and 5" is represented mathematically as $2 + 5$.

(b) Because we are talking about a difference, we know to use the − symbol, so "The difference of 12 and 7" is represented mathematically as $12 - 7$.

(c) "The product of −3 and 8" is represented as $-3 \cdot 8$.

(d) "The quotient of 10 and 2" is represented mathematically as $\frac{10}{2}$.

(e) "9 less than 15" is represented mathematically as $15 - 9$.

(f) "A number z decreased by 11" is represented algebraically as $z - 11$.

(g) "Three times the sum of a number x and 8" is represented as an algebraic expression as $3(x + 8)$.

Classroom Example
(a) The sum of 5 and 8.
(b) The difference of 25 and 12.
(c) The product of −2 and 13.
(d) The quotient of 24 and 6.
(e) 7 less than 12.
(f) A number b decreased by 9.
(g) Four times the sum of a number n and 3.

Answer:
(a) $5 + 8$
(b) $25 - 12$
(c) $-2 \cdot 13$
(d) $\frac{24}{6}$
(e) $12 - 7$
(f) $b - 9$
(g) $4(n + 3)$

Work Smart
When translating from English to math, try some specific examples. For example, to translate "a number z decreased by 11," pick specific values of z, as in "16 decreased by 11," which would be $16 - 11$ or 5. So "z decreased by 11" is $z - 11$.

In Example 1(g), we know the mathematical representation of the phrase is $3(x + 8)$ rather than $3x + 8$ because the phrase "three times the sum" means to multiply the sum of the two numbers by 3. The English phrase that would result in $3x + 8$ might be "the sum of three times a number and 8." Do you see the difference?

QUICK ✓ *Express each English phrase using mathematical symbols.*

1. The sum of 5 and 17.
2. The product of −2 and 6.
3. The quotient of 25 and 3.
4. The difference of 7 and 4.
5. Twice a less 2.
6. Three plus the quotient of z and 4.

Classroom Example ➤ Use Example 2.

EXAMPLE 2 Translate from an English Phrase to an Algebraic Expression

Write an algebraic expression for each problem.

(a) The Raiders scored p points in a football game. The Packers scored 12 more points than the Raiders. Write an algebraic expression for the number of points the Packers scored.

(b) A lumberman cuts a 50-foot log into 2 pieces. One piece is t feet long. Express the length of the second piece as an algebraic expression in t.

(c) The number of quarters in a drink machine is two fewer than the number of dimes, d, in the machine. Write an expression for the number of quarters as an algebraic expression in d.

Solution

(a) The phrase "more than" implies addition. The Packers scored $p + 12$ points in the football game.

(b) The log is 50 feet long. If the lumberman cuts one piece 20 feet long, then the other piece must be $50 - 20 = 30$ feet long. In general, if the lumberman cuts one piece that is t feet long, then the remaining piece must be $50 - t$ feet long.

(c) The phrase "less than" implies subtraction. Two less than the number of dimes is represented algebraically as $d - 2$. ∎

QUICK ✓ *Translate each phrase to an algebraic expression.*

7. Terry earned z dollars last week. Anne earned \$50 more than Terry last week. Write an algebraic expression for Anne's earnings in terms of z.
8. Melissa paid x dollars for her college math book and \$15 less than that for her college sociology book. Express the cost of her sociology book in terms of x.
9. Tim raided his piggy bank and found he had 75 dimes and quarters. Tim has d dimes. Express the number of quarters in his piggy bank as an algebraic expression in d.

Classroom Example ➤ Use Example 3.

EXAMPLE 3 Translate from an English Phrase to an Algebraic Expression

Write an algebraic expression for each problem.

(a) The number of student tickets sold to a play is five fewer than four times the number n of nonstudent tickets sold. Write an algebraic expression for the number of nonstudent tickets sold in terms of n.

Teaching Tip
Try using specific numerical examples to reinforce the notion that variables represent quantities.

(b) The height of a full-grown maple tree is ten feet more than three times the height h of a sapling. Write an algebraic expression for the height of the full-grown tree in terms of h.

(c) A pedestrian bridge over a river is f feet long. Further downstream, another pedestrian bridge over the same river is three feet less than twice the length of the first bridge. Write an algebraic expression for the length of the second bridge in terms of f.

Solution

(a) The phrase "fewer than" implies subtraction. The number of student tickets sold is five fewer than four times the number of nonstudent tickets sold. So the algebraic expression is 4 times the number of nonstudent tickets sold minus 5, or $4n - 5$.

(b) The full-grown tree is ten feet more than three times the height of the sapling, so we use the phrase "height of sapling times three plus ten" to obtain $3h + 10$ feet as the height of the full-grown tree.

(c) The second bridge is three feet less than twice the length of the first bridge, so we say "twice the length of the first bridge minus 3 feet" to obtain $2f - 3$ feet as the length of the second bridge.

QUICK ✓ *Translate each phrase to an algebraic expression.*

10. The width of a platform is 2 feet less than three times the length, l. Express the width of the platform as an algebraic expression in terms of l.

11. T.J. has quarters and dimes in his piggy bank. The number of quarters is three more than twice the number of dimes. T.J. has q quarters. Express the number of dimes as an algebraic expression in terms of q.

12. The number of blue M&Ms in a bowl is five less than three times the number of brown M&Ms, b, in the bowl. Express the number of blue M&Ms in the bowl as an expression in terms of b.

2 Translate English Sentences to Equations

In Words
English phrase is to algebraic expression as English sentence is to algebraic equation.

Work Smart
We learned in Section 2.1 that an equation is a statement in which two algebraic expressions are equal.

Nearly every word problem that we do in algebra requires some type of translation. Learning to speak the language of math is the same as learning to speak any language. Now that we have the ability to translate English phrases to algebraic expressions, we will extend the idea to translating English sentences to algebraic equations.

In English, a complete sentence must contain a subject and a verb, so expressions or "phrases" are not complete sentences. For example "Beats me!" is an expression, but it is not a complete sentence because it does not contain a subject. The expression "5 more than a number x" does not contain a verb and therefore is not a complete sentence either. The statement "5 more than a number x is 18" is a complete sentence because it contains a subject and a verb. Because this is a complete sentence, we can translate it into a mathematical statement. In mathematics, statements can be represented symbolically through equations.

In English, statements can be true or false. For example, "The moon is made of green cheese" is a false statement, while "The sky is blue" is a true statement. Mathematical statements can be true or false as well—we called them conditional equations.

Table 3 provides a summary of words that typically translate into an equal sign.

Table 3 Words That Translate into an Equal Sign

is	yields	are	equals
was	gives	results in	is equal to
is equivalent to			

Classroom Example
Translate each of the following sentences into an equation. Do not solve the equation.
(a) Eight more than a number y is 19.
(b) Five times the sum of a number n and 4 is 12.
(c) The difference of x and 3 equals the quotient of x and 2.

Answer:
(a) $y + 8 = 19$
(b) $5(n + 4) = 12$
(c) $x - 3 = \frac{x}{2}$

Notice that the words that translate into an equal sign are all verbs. So, the equal sign in an equation acts like a verb in a sentence.

Let's look at some examples where we translate English sentences into equations.

EXAMPLE 4 Translating English Sentences into Equations

Translate each of the following sentences into an equation. Do not solve the equation.

(a) Five more than a number x is 20.

(b) Four times the sum of a number z and 3 is 15.

(c) The difference of x and 5 equals the quotient of x and 2.

Solution

(a) 5 more than a number x is 20

$$\underbrace{x + 5}_{\text{5 more than a number } x} \; \underbrace{=}_{\text{is}} \; \underbrace{20}_{20}$$

(b) Because the expression reads "Four times the sum," we first need to determine the sum and then multiply this result by 4.

Four times the sum of a number z and 3 is 15

$$4(z + 3) = 15$$

Work Smart

The English sentence "The sum of four times a number z and 3 is 15" would be expressed mathematically as $4z + 3 = 15$. Do you see how this differs from Example 4(b)?

(c) The difference of x and 5 equals the quotient of x and 2

$$x - 5 = \frac{x}{2}$$

QUICK ✓ *Translate each English statement into an equation. Do not solve the equation.*

13. The product of 3 and y is equal to 21.

14. The sum of 3 and x is equivalent to the product of 5 and x.

15. The difference of x and 10 equals the quotient of x and 2.

16. Three less than a number y is five times y.

An Introduction to Problem Solving and Mathematical Models

Every day we encounter various types of problems that must be solved. **Problem solving** is the ability to use information, tools, and our own skills to achieve a goal. For example, suppose 4-year-old Kevin wants a glass of water, but he is too short to reach the sink. Kevin has a problem. To solve the problem, he finds a step stool and pulls it over to the sink. He uses the step stool to climb on the counter, opens the kitchen cabinet and pulls out a cup. He then crawls along the counter top, turns on the faucet, fills the cup, and proceeds to drink the water. Problem solved!

Of course, this is not the only way that Kevin could solve the problem. Can you think of any other solutions? Just as there are various approaches to solving life's everyday problems, there are many ways to solve problems using mathematics. However, regardless of the approach, there are always some common aspects in solving any problem. For example, regardless of how Kevin ultimately ends up with his cup of water, someone must get a cup from the cabinet and someone must turn on the faucet.

One of the purposes of learning algebra is to be able to solve certain types of problems. To solve these problems, we will need techniques that can help us translate the verbal description of the problem into an equation that can be solved. The process of taking a verbal description of a problem and developing a mathematical equation that can be used to solve the problem is **mathematical modeling.**

Mathematical modeling begins with a problem. The problem is summarized as a verbal description. The verbal description is then translated into the language of mathematics. This translation results in an equation that can be solved (the mathematical problem). The solution must be checked against the mathematical problem (the equation) and the verbal description. This entire process is called the **modeling process.** We call the equation that is developed the **mathematical model.** See Figure 3.

Not all models are mathematical. In general, a **model** is a way of using graphs, pictures, small-scale reproductions, equations, or even verbal descriptions to represent a real-life situation. Because the world is an extremely complex place, we often need to simplify information when we develop a model. For example, a map is a model of our road system. Maps don't show all the details of the system such as trees, buildings, or potholes, but they do a good job of describing how to get from point A to point B.

Figure 3

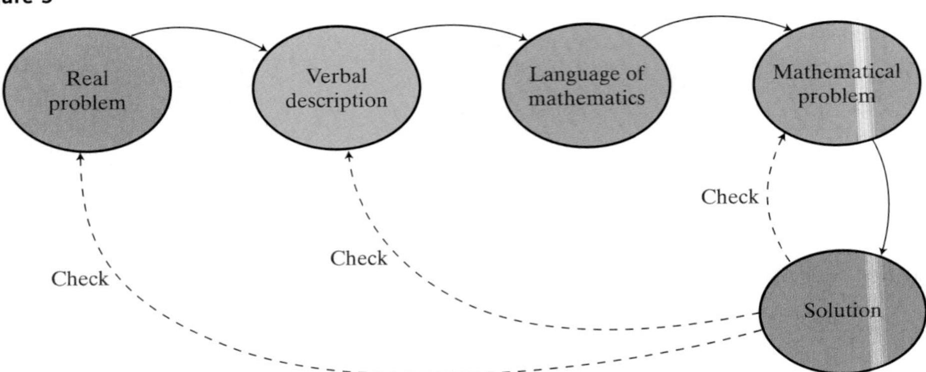

Mathematical models are similar in that we often make assumptions regarding our world in order to make the mathematics more manageable.

It is difficult to give a step-by-step approach for solving problems because each problem is unique in some way. However, because there are common links to many types of problems, we can categorize problems. In this text, we will present five categories of problems.

Five Categories of Problems

1. **Direct Translation**—problems in which we must translate from English into the language of mathematics by using key words in the verbal description.
2. **Geometry**—problems in which the unknown quantities are related through geometric formulas.
3. **Mixtures**—problems in which two or more quantities are combined in some fashion.
4. **Uniform Motion**—problems in which an object travels at a constant speed.
5. **Work Problems**—problems in which two or more entities join forces to complete a job.

We will present strategies for solving categories of problems throughout the text. In this section, we will concentrate on direct translation problems.

Regardless of the type of problem we solve, there are certain steps that should always be followed to assist in solving the problem. Below we provide you with a series of steps that should be followed when developing any mathematical model. As we proceed through this course, and in future courses, you will use the techniques that you have studied in this course to solve more complicated problems, but the approach remains the same.

Steps for Solving Problems with Mathematical Models

Step 1: Identify What You Are Looking For Read the problem very carefully, perhaps two or three times. Identify the type of problem. Identify the information that is given, and the information that we wish to learn from the problem. It is fairly typical that the last sentence in the problem indicates what it is we wish to solve for in the problem.

Step 2: Give Names to the Unknowns Assign variables to the unknown quantities in the problem. Choose a variable that is representative of the unknown quantity it represents. For example, use t for time.

Step 3: Translate the Problem into the Language of Mathematics Read the problem again. This time, after each sentence is read, determine if the sentence can be translated into a mathematical statement or expression in terms of the variables identified in Step 2. It is often helpful to create a table, chart, or figure. When you have finished reading the problem, if necessary, combine the mathematical statements or expressions into an equation that can be solved.

Step 4: Solve the Equation(s) Found in Step 3 Solve the equation for the variable and then answer the question posed by the original problem.

Step 5: Check the Reasonableness of Your Answer Check your answer to be sure that it makes sense. If it does not, go back and try again.

Step 6: Answer the Question Write your answer in a complete sentence.

Let's review each of these steps, one at a time.

- **Identify** Carefully read the problem. Reading a verbal description of a problem is not like reading a spy novel. You may need to read the problem three or four times. You may not know how to solve the problem while reading it, but you should get a sense of which of the five categories the problem falls into, what information you are given, and what you are being asked to do.
- **Name** Reread the problem and assign variables to the unknowns. You should write the name of each variable and what it represents. You will use this to check your final answer.
- **Translate** In this step, you develop a model (equation) that mathematically describes the problem. Be sure to use the guidelines presented in each category of problem. We will present these guidelines shortly.
- **Solve** the equation. This is generally the easy part. Most students say, "I could solve the problem, if I could find the right equation."
- **Check** Checking your answer can be difficult because you can make two types of errors while setting up or solving the problem. One type of error occurs if you correctly translate the problem into a model but then make an error solving the equation. A second type of error occurs if you misinterpret the problem and develop an incorrect model. The solution you obtain may still satisfy your model, but it probably will not be the solution to the original problem. We can check for this type of error by determining whether the solution is reasonable. Does your answer make sense? Always be sure that you are answering the question that is being asked.
- **Answer the question** identified in the problem in words.

Work Smart
Remember that you can use any letter to represent the unknown(s) when you make your model. Choose a letter that reminds you what it represents. For example, use t for time.

3 Build Models for Solving Direct Translation Problems

Let's look at a "direct translation" problem. Remember, these are problems that can be set up by reading the problem and using everyday language to translate the verbal description into a mathematical equation.

Classroom Example ➤
Use Example 5.

EXAMPLE 5 Solving a Direct Translation Problem

For the 2001–2002 season, the price of a ticket to a New York Knicks NBA game was \$11 more than twice the price of a ticket to a Minnesota Timberwolves NBA game. If you buy a ticket to a Knicks game and a ticket to a Timberwolves game, and the total cost of the two tickets is \$128, what does each of the tickets cost?

Solution

Step 1: Identify This is a direct translation problem. We can obtain an equation from the words of the problem. We want to know the price of a ticket to a NY Knicks basketball game and the price of a ticket to a Minnesota Timberwolves game.

Work Smart
It is helpful to assign the variable to the quantity that you know **least** about.

Step 2: Name We know that the price of a NY Knicks ticket was \$11 more than twice the price of a ticket to a Minnesota Timberwolves game. We will let t represent the price of a ticket to a Timberwolves game. Then $2t + 11$ represents the price of a NY Knicks ticket.

Step 3: Translate Since we know that the total cost for a ticket to both games is \$128, we use the equation

$$\underbrace{\text{price of ticket to Timberwolves game}}_{t} \text{ plus } \underbrace{\text{price of ticket to Knicks game}}_{2t + 11} \text{ equals } \underbrace{\text{total cost}}_{128}$$

$$t + 2t + 11 = 128$$

Step 4: Solve We now solve the equation

$$t + (2t + 11) = 128$$

Combine like terms: $3t + 11 = 128$

Subtract 11 from each side of the equation: $3t + 11 - 11 = 128 - 11$

$$3t = 117$$

Divide each side by 3: $\dfrac{3t}{3} = \dfrac{117}{3}$

$$t = 39$$

We have found $t = 39$. Recall that t represents the price of a ticket to a Timberwolves game. The price of a ticket to a Knicks game is \$11 more than twice the price of a ticket to a Timberwolves game. So $2t + 11 = 2(39) + 11 = \$89$. It costs \$89 for a ticket to a NY Knicks game.

Teaching Tip
Remind students to determine if the answer makes sense in context. Would a ticket to an NBA basketball game cost $-\$20$? No! If the answer doesn't make sense, go back and look for a mistake.

Step 5: Check Is the total cost of a ticket to a Timberwolves game and a Knicks game \$128? $\$39 + \$89 = \$128$, so our answers are correct.

Step 6: Answer The price of a ticket to a Minnesota Timberwolves game is \$39 and the price of a ticket to a NY Knicks game is \$89.

QUICK ✓ *Translate the problem to an algebraic equation and solve the equation for the unknowns.*

17. Sean and Connor decide to buy a pizza. The pizza costs \$15 and they decide to split the cost based upon how much pizza each eats. Connor eats two-thirds of the amount that Sean eats, so Connor pays two-thirds of the amount that Sean pays. How much does each pay?

Consecutive Integer Problems

Recall that an *integer* is a member of the set $\{\ldots, -3, -2, -1, 0, 1, 2, 3, \ldots\}$. An *even integer* is an integer that is divisible by two. For example, 16 and 124 are even integers. An *odd integer* is an integer that is not even. For example, 9 and 17 are odd integers. Consecutive integers differ by one, so if n represents the first integer, $n + 1$ represents the second integer, $n + 2$ represents the third integer, and so on. Consecutive *even* integers, such as 14 and 16, differ by two so if n represents the first even integer, $n + 2$ represents the second even integer and $n + 4$ represents the third even integer. Consecutive *odd* integers also differ by 2 (41 and 43, for example), so if n represents the first odd integer, $n + 2$ represents the second odd integer, and $n + 4$ represents the third odd integer. See Figure 4.

Work Smart
Examples of consecutive even integers are

16, 18, 20
78, 80, 82
752, 754, 756

Examples of consecutive odd integers are

9, 11, 13
31, 33, 35
623, 625, 627

Teaching Tip
You may need to use several numerical examples to help students understand the algebraic notation n, $n + 1$, $n + 2$.

Figure 4

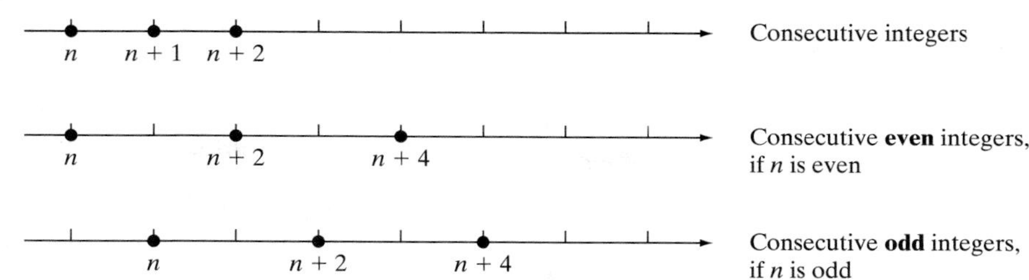

Classroom Example
The sum of three consecutive even integers is 174. Find the integers.

Answer: The numbers are 56, 58, and 60.

EXAMPLE 6 Solve a Direct Translation Problem: Consecutive Integers

The sum of three consecutive even integers is 324. Find the integers.

Solution

Step 1: Identify This is a direct translation problem. We are looking for three consecutive even integers, and we know that their sum is 324. Examples of consecutive even integers are 6, 8, 10, ...

Step 2: Name We will let n represent the first even integer, so $n + 2$ is the next even integer, and $n + 4$ is the third even integer.

Step 3: Translate We know that the sum of the three consecutive even integers is 324. We also know that the word "sum" translates to "addition" and the word "is" translates to "equals." So our equation is

$$\underbrace{n}_{\text{first even integer}} + \underbrace{n + 2}_{\text{second even integer}} + \underbrace{n + 4}_{\text{third even integer}} = \underbrace{324}_{\text{sum}}$$

Step 4: Solve We solve the equation.

$$n + (n + 2) + (n + 4) = 324$$

Combine like terms: $3n + 6 = 324$

Subtract 6 from each side of the equation: $3n + 6 - 6 = 324 - 6$

$$3n = 318$$

Divide each side by 3: $\dfrac{3n}{3} = \dfrac{318}{3}$

$$n = 106$$

Since $n = 106$, is the first even integer, the remaining even integers are 108 and 110.

Step 5: Check These numbers are all even integers, so we know that they are possible solutions. Does $106 + 108 + 110 = 324$? Yes! We know we have the correct answer.

Step 6: Answer The three consecutive even integers are 106, 108, and 110.

QUICK ✓ *Translate the problem to an algebraic equation and solve the equation for the unknowns.*

18. The sum of three consecutive even integers is 270. Find the integers.

Classroom Example
A 14-foot piece of lumber is cut into three pieces: The longest piece is 4 feet longer than the middle-sized piece, and the shortest piece is 2 feet shorter than the middle-sized piece. Find the length of each piece of lumber.

Answer: 8 ft, 4 ft, 2 ft

EXAMPLE 7 Solve a Direct Translation Problem: Piece Lengths

A carpenter is building a shelving system and cuts a 14-foot length of cherry shelving into three pieces. The second piece is twice as long as the first, and the third piece is 2 feet longer than the first. Find the length of each piece of cherry shelving.

Work Smart
Sometimes these problems are called "the whole equals the sum of the parts" problems because the values of the "parts" must sum to the value of the "whole."

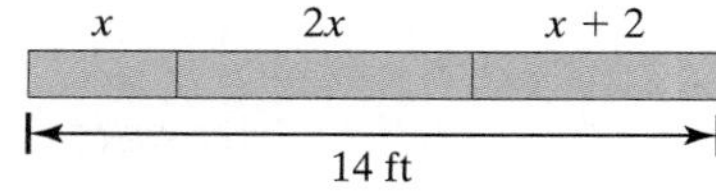

Solution

Step 1: Identify This is a direct translation problem. We are looking for the length of each piece of shelving. We know that the board is 14 feet long.

Step 2: Name Because the length of the second and third pieces are described in terms of the length of the first piece, we will let x represent the length of the first piece of shelving. The second piece is twice as long as the first piece, so we let $2x$ represent the length of the second piece. Because the third piece is 2 feet longer than the first piece, its length is $x + 2$.

Step 3: Translate We know that the lengths of the three pieces must total 14 feet, so our equation is

$$\underbrace{x}_{\text{length of first piece}} + \underbrace{2x}_{\text{length of second piece}} + \underbrace{x+2}_{\text{length of third piece}} = \underbrace{14}_{\text{total length}}$$

Step 4: Solve We solve the equation.

$$x + 2x + (x + 2) = 14$$

Combine like terms: $$4x + 2 = 14$$

Subtract 2 from each side of the equation: $$4x + 2 - 2 = 14 - 2$$

$$4x = 12$$

Divide each side by 4: $$\frac{4x}{4} = \frac{12}{4}$$

$$x = 3$$

Since $x = 3$ feet is the length of the first shelf, the second piece of shelving is $2x = 2 \cdot 3 = 6$ feet and the third piece is $x + 2 = 5$ feet long.

Step 5: Check Is the sum of the three pieces of cherry shelving equal to 14 feet? Does $3 + 6 + 5 = 14$? Yes! We know we have the correct answer.

Step 6: Answer The three pieces of shelving are 3 feet, 6 feet, and 5 feet long.

QUICK ✓ *Translate the problem to an algebraic equation and solve.*

19. A 76-inch length of ribbon is to be cut into three pieces. The longest piece is to be 24 inches longer than the shortest piece, and the third piece is to be half the length of the longest piece. Find the length of each piece of ribbon.

Classroom Example ➤
A total of $17,500 is to be invested, some in CDs and $4000 less than that in bonds. How much is to be invested in each type of investment?

Answer: $10,750 in CDs and $6750 in bonds.

EXAMPLE 8 Investment Decisions

A total of $25,000 is to be invested, some in bonds and some in certificates of deposit (CDs). The amount invested in CDs is to be $8000 less than the amount invested in bonds. How much is to be invested in each type of investment?

Solution

Step 1: Identify We want to know the amount invested in each type of investment.

Step 2: Name Let b represent the amount invested in bonds.

Step 3: Translate The equation that we will use is

$$\text{Amount invested in bonds} + \text{Amount invested in CDs} = \text{Total investment}$$

Suppose we invested $18,000 in bonds; then the amount in CDs will be $8000 less than this amount, or $10,000. In general, b is the amount invested in bonds, so $b - 8000$ represents the amount invested in CDs. Our total investment is $25,000. Substituting into the above equation, we have

$$\underbrace{b}_{\text{Amount invested in bonds}} + \underbrace{b - 8000}_{\text{Amount invested in CDs}} = \underbrace{25{,}000}_{\text{Total investment}}$$

Step 4: Solve We now solve the equation

$$b + (b - 8000) = 25{,}000$$

Combine like terms: $2b - 8000 = 25{,}000$

Add 8000 to each side of the equation: $2b = 33{,}000$

Divide both sides by 2: $b = 16{,}500$

Teaching Tip
Ensure that students answer the question posed in the problem. In this case, the question is to find the amount in each investment. Also, encourage students to verify the reasonableness of their answer. For example, a negative investment in bonds does not make sense.

Step 5: Check If we invest \$16,500 in bonds, then the amount invested in CDs should be \$8000 less than this amount or \$8500. The total investment should then be \$25,000. Since the amount invested in bonds plus the amount invested in CDs is \$16,500 + \$8500 = \$25,000, the answer checks.

Step 6: Answer Invest \$16,500 in bonds and \$8500 in CDs.

QUICK ✓ *Translate the problem to an algebraic equation and solve.*

20. A total of \$18,000 is to be invested, some in stocks and some in bonds. If the amount invested in bonds is twice that invested in stocks, how much is invested in each category?

Classroom Example
EZ Rider Truck Rental charges \$50 per day plus \$0.25 per mile to rent a small truck. Trucks-R-Us charges \$35 per day plus \$0.30 per mile for the same model truck. For how many miles will the costs of renting the two trucks be the same?

Answer: 300 miles

EXAMPLE 9 Choosing a Long-Distance Carrier

MCI has a long-distance phone plan that charges \$2.00 a month plus \$0.09 per minute of usage. Sprint has a long-distance phone plan that charges \$3.50 a month plus \$0.07 per minute of usage. For how many minutes of long distance calls will the costs for the two plans be the same? (*Source:* MCI and Sprint)

Solution

Step 1: Identify This is a direct translation problem. We are looking for the number of minutes for which the two plans cost the same.

Step 2: Name Let m represent the number of long distance minutes used in the month.

Step 3: Translate The monthly fee for MCI is \$2.00 plus \$0.09 for each minute used. So, if one minute is used, the fee is $2.00 + 0.09(1) = 2.09$ dollars. If two minutes are used, the fee is $2.00 + 0.09(2) = 2.18$ dollars. In general, if m minutes are used, the monthly fee is $2.00 + 0.09m$ dollars. Similar logic results in the monthly fee for Sprint being $3.50 + 0.07m$ dollars. We want to know the number of minutes for which the cost for the two plans will be the same, which means we need to solve

$$\text{Cost for MCI} = \text{Cost for Sprint}$$
$$2.00 + 0.09m = 3.50 + 0.07m$$

Step 4: Solve

Subtract 2.00 from both sides: $0.09m = 1.50 + 0.07m$

Subtract $0.07m$ from both sides: $0.02m = 1.50$

Divide both sides by 0.02: $m = 75$

Step 5: Check We believe the cost of the two plans will be the same if 75 minutes are used. The cost of MCI's plan will be $2.00 + 0.09(75) = \$8.75$. The cost of Sprint's plan will be $3.50 + 0.07(75) = \$8.75$. They are the same!

Step 6: Answer the Question The cost of the two plans will be the same if 75 minutes are used.

QUICK ✓

21. **Truck Rentals** You need to rent a moving truck. You have identified two companies that rent trucks. EZ-Rental charges \$30 per day plus \$0.15 per mile. Do It Yourself Rental charges \$15 per day plus \$0.25 per mile. For how many miles will the cost of renting be the same?

2.5 Exercises

For Extra Help: Student Solutions Manual, CD Video, PH Math/Tutor Center, MathXL Tutorials on CD, MathXL®, MyMathLab

Concepts and Vocabulary

In Problems 1 and 2, fill in the blanks.

1. Letting variables represent unknown quantities and then expressing relationships among the variables in the form of equations is called __________ __________.
2. In English we use sentences. To translate a sentence into a mathematical statement we use __________.

In Problems 3 and 4, answer True or False to each statement.

3. If the first of three consecutive odd integers is represented by the variable n, the second consecutive odd integer would be $n + 1$ and the third odd integer would be $n + 3$.
4. Suppose a total of \$10,000 is to be invested in stocks and bonds. If we let b represent the amount invested in bonds, then $b - 10{,}000$ represents the amount invested in stocks.
5. How is mathematical modeling related to problem solving? Why do we make assumptions when creating mathematical models?
6. What is the difference between an algebraic expression and an equation? How is each related to phrases and English statements?
7. Two students write an equation to solve a word problem with consecutive odd integers. One student assigns the variables as $n - 1, n + 1, n + 3$, where n is an even integer. A second student uses $n, n + 2, n + 4$, where n is an odd integer. Which student is correct? Will the value for n be the same for both? Make up a problem that can be solved in more than one way and explain how the variables were assigned.
8. Using the algebraic expression $3x + 5$, make up a problem that uses the direction evaluate. Using the same algebraic expression, make up a problem that uses the direction solve.

Building Skills

In Problems 9–26, translate each phrase to an algebraic expression. Let x represent the unknown number.

9. the sum of −5 and a number
10. a number increased by 32.3
11. the product of a number and $\frac{2}{3}$
12. the product of −2 and number
13. half of a number
14. double a number
15. a number less −25
16. 8 less than a number
17. the quotient of a number and 3
18. the quotient of −14 and a number

1. mathematical modeling
2. equations
3. False
4. False
5. Answers may vary.
6. Answers may vary.
7. Answers may vary.
8. Answers may vary.
9. $-5 + x$
10. $x + 32.3$
11. $x\left(\frac{2}{3}\right)$ or $\frac{2}{3}x$
12. $-2x$
13. $\frac{1}{2}x$
14. $2x$
15. $x - (-25)$
16. $x - 8$
17. $\frac{x}{3}$
18. $\frac{-14}{x}$

19. $x + \frac{1}{2}$
20. $\frac{4}{5}x$
21. $6x + 9$
22. $4x + 21$
23. $2(13.7 + x)$
24. $\frac{1}{2}x - 50$
25. $2x + 31$
26. $2x + 45$
27. $x + 15 = -34$
28. $43 + x = -72$
29. $35 = 3x - 7$
30. $49 = 2x - 3$
31. $\frac{x}{-4} + 5 = 36$
32. $\frac{x}{-6} - 15 = 30$
33. $2[x + 6] = x + 3$
34. $2[x + 5] = x + 7$
35. Braves: r; Clippers: $r + 5$
36. Indians: r; Blue Jays: $r - 3$
37. Bill's amount: b; Jan's amount: $b + 0.55$
38. Ralph's amount: R; Beryl's amount: $3R + 0.25$
39. Janet's share: j; Kathy's share: $200 - j$
40. Juan's share: j; Emilio's share: $1500 - j$
41. number adults: a; number children: $1433 - a$
42. number paid tickets: p; number promotion tickets: $12{,}765 - p$
43. 83
44. −16
45. −14
46. −16
47. 54, 55, 56
48. 25, 27, 29
49. Verrazano-Narrows Bridge: 4260 ft; Golden Gate Bridge: 4200 ft
50. Taipei Tower: 106 stories; Sears Tower: 110 stories
51. $11,215
52. $269

19. $\frac{1}{2}$ more than a number

20. $\frac{4}{5}$ of a number

21. 9 more than 6 times a number

22. 21 more than 4 times a number

23. twice the sum of 13.7 and a number

24. 50 less than half of a number

25. the sum of twice a number and 31

26. the sum of twice a number and 45

In Problems 27–34, translate each statement into an equation. Let x represent the unknown number. DO NOT SOLVE.

27. The sum of a number and 15 is −34.

28. The sum of 43 and a number is −72.

29. 35 is 7 less than triple a number.

30. 49 is 3 less than twice a number.

31. The quotient of a number and −4, increased by 5, is 36.

32. The quotient of a number and −6, decreased by 15, is 30.

33. Twice the sum of a number and 6 is the same as 3 more than the number.

34. Twice the sum of a number and 5 is the same as 7 more than the number.

In Problems 35–42, choose a variable to represent one quantity. State what that quantity represents and then express the second quantity in terms of the first.

35. The Columbus Clippers scored 5 more runs than the Richmond Braves.

36. The Toronto Blue Jays scored 3 fewer runs than the Cleveland Indians.

37. Jan has $0.55 more in her piggy bank than Bill.

38. Beryl has $0.25 more than 3 times the amount Ralph has.

39. Janet and Kathy will share the $200 grant.

40. Juan and Emilio will share the $1500 lottery winnings.

41. There were 1433 visitors to the Arts Center Spring show. Some were adults and some were children.

42. There were 12,765 fans at a recent NBA game. Some held paid admission tickets and some held special promotion tickets.

Applying the Concepts

43. Number Sense The sum of a number and −12 is 71. Find the number.

44. Number Sense The difference between a number and 13 is −29. Find the number.

45. Number Sense 25 less than twice a number is −53. Find the number.

46. Number Sense The sum of 13 and twice a number is −19. Find the number.

47. Consecutive Integers The sum of three consecutive integers is 165. Find the numbers.

48. Consecutive Integers The sum of three consecutive odd integers is 81. Find the numbers.

49. Bridges The longest bridge in the United States is the Verrazano-Narrows Bridge. The second-longest bridge in the United States is the Golden Gate Bridge, which is 60 feet shorter than the Verrazano-Narrows Bridge. The combined length of the two bridges is 8460 feet. Find the length of each bridge.

50. Towers The tallest towers in the world (those having the most stories) are the Sears Tower in Chicago, IL and the Taipei 101 Tower in Taipei, Taiwan. The Taipei Tower has 4 fewer stories than the Sears Tower. The two buildings together have 216 stories. Find the number of stories in each tower.

51. Buying a Motorcycle The total price for a new motorcycle is $11,894.79. The tax, title, and dealer preparation charges amount to $679.79. Find the price of the motorcycle before the extra charges.

52. Buying a Desk The total price for a new desk is $285.14, including sales tax of $16.14. Find the original cost of the desk.

53. CD: \$8500; Bonds: \$11,500
54. Sean: \$6500; George: \$3500
55. Stocks: \$20,000; Bonds: \$12,000
56. Stocks: \$24,000; Bonds: \$16,000
57. Smart Start: 2 g; Go Lean: 8 g
58. \$12.45
59. \$29,140
60. \$26
61. 150 miles
62. 240 minutes
63. 2000 pages
64. 80 vacuums
65. Jensen: \$36,221; Maureen: \$35,972

53. **Finance** A total of \$20,000 is to be invested, some in bonds and some in certificates of deposit (CDs). The amount invested in bonds is to be \$3000 greater than the amount invested in CDs. How much is to be invested in each type of investment?

54. **Finance** A total of \$10,000 is to be divided between Sean and George. George is to receive \$3000 less than Sean. How much will each receive?

55. **Investments** Suppose that your Aunt May has left you an unexpected inheritance of \$32,000. You have decided to invest the money rather than blow it on frivolous purchases. Your financial advisor has recommended that you diversify by placing some of the money in stocks and some in bonds. Based upon current market conditions, she has recommended that the amount in bonds should equal three-fifths of the amount invested in stocks. How much should be invested in stocks? How much should be invested in bonds?

56. **Investments** Jack and Diane have \$40,000 to invest. Their financial advisor has recommended that they diversify by placing some of the money in stocks and some in bonds. Based upon current market conditions, he has recommended that the amount in bonds should equal two-thirds of the amount invested in stocks. How much should be invested in stocks? How much should be invested in bonds?

57. **Cereal** A serving of Kashi Go Lean Crunch cereal contains 4 times the amount of fiber as a serving of Kellogg's Smart Start Cereal. If you eat a serving of each cereal, you will consume 10 g of dietary fiber. Find the amount of dietary fiber in each cereal.

58. **Books** A paperback edition of a book costs \$12.50 less than the hardback edition of the book. If you purchase one of each book, you will pay \$37.40. Find the cost of the paperback edition of the book.

59. **Income** On a joint income tax return, Elizabeth Morrell's adjusted gross income was \$2549 more than her husband Dan's adjusted gross income. Their combined adjusted gross income was \$55,731. Find Elizabeth Morrell's adjusted gross income.

60. **Spring Break** Allison went shopping to prepare for her Spring Break trip. Her bathing suit cost \$8 more than a pair of shorts, and a T-shirt cost \$2 less than the shorts. Find the cost of the bathing suit if Allison spent \$60 on the items, before sales tax.

61. **Truck Rentals** You need to rent a moving truck. You have identified two companies that rent trucks. EZ-Rental charges \$35 per day plus \$0.15 per mile. Do It Yourself Rental charges \$20 per day plus \$0.25 per mile. For how many miles will the cost of renting be the same?

62. **Cellular Telephones** You need a new cell phone for emergencies only. Company A charges \$12 per month plus \$0.10 per minute, while Company B charges \$0.15 per minute with no monthly service charge. For how many minutes will the monthly cost be the same?

63. **Comparing Printers** Samuel is trying to decide between two laser printers, one manufactured by Hewlett-Packard, the other by Brother. Both have similar features and warranties, so price is the determining factor. The Hewlett-Packard costs \$200 and printing costs are approximately \$0.03 per page. The Brother costs \$240 and printing costs are approximately \$0.01 per page. How many pages need to be printed for the cost of the two printers to be the same?

64. **Comparing Job Offers** Hans has just been offered two sales jobs selling vacuums. The first job offer is a base monthly salary of \$2000 plus a commission of \$50 for each vacuum sold. The second job offer is a base monthly salary of \$1200 plus a commission of \$60 for each vacuum sold. How many vacuums must be sold for the two jobs to pay the same salary?

65. **Adjusted Gross Income** On a joint income tax return, Jensen Beck's adjusted gross income was \$249 more than his wife Maureen's adjusted gross income. Their combined adjusted gross income was \$72,193. Find Jensen and Maureen Beck's adjusted gross income.

66. lantern: \$45; cookware: \$75; cook stove: \$79
67. 5, 6, 7, and 8
68. 85
69. Answers may vary.
70. Answers may vary.
71. Answers may vary.
72. Answers may vary.
73. 20°, 40°, 120°
74. 19°, 58°, 103°

66. Camping Trip Jaime Juarez purchased some new camping equipment. He spent \$199 on a cookware set, a lantern, and a cook stove. The cookware set cost \$30 more than the lantern, and the cook stove cost \$34 more than the lantern. Find the cost of each item.

67. Baseball Games The Columbus Comets played a double-header and won both games. The scores of each of the two games were consecutive integers, and a total of 26 runs were scored. Find the number of runs scored in each of the two baseball games.

68. Computing Grades Going into the final exam, which will count as two tests, Brooke has test scores of 80, 83, 71, 61, and 95. What score does Brooke need on the final exam in order to have an average score of 80?

Extending the Concepts

In Problems 69–72, write a problem that would translate into the given equation.

69. $10x = 370$ **70.** $5n + 10 = 170$ **71.** $\frac{x + 74}{2} = 80$ **72.** $n + n + 2 = 98$

△ **73. Angles** The sum of the measures of the three angles in a triangle is 180 degrees. The measure of the smallest angle of a triangle is half the measure of the second angle. The measure of the third angle is 40° more than 4 times the measure of the first. Find the measure of each angle.

△ **74. Angles** The sum of the measures of the three angles in a triangle is 180 degrees. The measure of one angle of a triangle is one degree more than three times the measure of the smallest angle. The measure of the third angle is 13° less than twice the measure of the second angle. Find the measure of each angle.

2.6 Problem Solving: Direct Translation Problems Involving Percent

OBJECTIVES

1. Solve Direct Translation Problems Involving Percent
2. Model and Solve Direct Translation Problems from Business Involving Percent

Preparing for Problem Solving: Direct Translation Problems Involving Percent

Before getting started, take the following readiness quiz. If you get a problem wrong, go back to the section cited and review the material.

1. Write 45% as a decimal. [Appendix, Section A.2, p. A11]

2. Write 0.2875 as a percent. [Appendix, Section A.2, pp. A11–A12]

Teaching Tip
Sections 2.6 and 2.7 are optional and can be omitted without loss of continuity.

1 Solve Direct Translation Problems Involving Percent

Percent means "divided by 100" or "per hundred." We use the symbol % to denote percent, so 45% means 45 out of 100 or $\frac{45}{100}$ or 0.45. In applications involving percents, we often encounter the word "of," as in 20% of 100. The word "of" translates into "multiplication" in mathematics, so 20% of 100 means $0.20 \cdot 100 = 20$.

Classroom Example ➤
A number is 45% of 60. Find the number.
Answer: 27

EXAMPLE 1 Solving an Equation Involving Percent

A number is 35% of 40. Find the number.

Solution

Step 1: Identify We want to know the unknown number.

Step 2: Name Let n represent the number.

Preparing for...Answers **1.** 0.45 **2.** 28.75%

Step 3: Translate We translate the words of the problem:

$$\underbrace{\text{a number}}_{n} \ \underbrace{\text{is}}_{=} \ \underbrace{35\%}_{0.35} \ \underbrace{\text{of}}_{\cdot} \ \underbrace{40}_{40}$$

The equation we want to solve is $n = 0.35(40)$.

Step 4: Solve We now solve the equation.

$$n = 0.35(40)$$

Multiply: $n = 14$

Step 5: Check Check the multiplication: $0.35(40) = 14$.

Step 6: Answer 14 is 35% of 40.

QUICK ✓ *Find the number.*

1. A number is 89% of 900. Find the number.
2. A number is 3.5% of 72. Find the number.
3. A number is 150% of 24. Find the number.
4. A number is $8\frac{3}{4}\%$ of 40. Find the number.

Classroom Example → The number 120 is what percent of 600?

Answer: 20%

EXAMPLE 2 Solving an Equation Involving Percent

The number 240 is what percent of 800?

Solution

Step 1: Identify We want to know the percentage.

Step 2: Name Let x represent the percent.

Step 3: Translate We translate the words of the problem:

$$\underbrace{240}_{240} \ \underbrace{\text{is}}_{=} \ \underbrace{\text{what percent}}_{x} \ \underbrace{\text{of}}_{\cdot} \ \underbrace{800?}_{800}$$

The equation we will solve is $240 = 800x$.

Step 4: Solve We now solve the equation.

$$240 = 800x$$

Divide each side by 800: $\dfrac{240}{800} = \dfrac{800x}{800}$

$$0.3 = x$$

Since we are finding a percent and our answer is a decimal, we must change 0.3 to a percent by moving the decimal point two places to the right: $0.30 = 30\%$.

Step 5: Check Is 240 equal to 30% of 800? Does $(0.30)(800) = 240$? Yes!

Step 6: Answer The number 240 is 30% of 800.

QUICK ✓ *Find the percent.*

5. The number 8 is what percent of 20?
6. The number 15 is what percent of 40?
7. The number 12.3 is what percent of 60?
8. The number 44 is what percent of 40?

Classroom Example
28 is 40% of what number?
Answer: 70

EXAMPLE 3 Solving an Equation Involving Percent

42 is 35% of what number?

Solution

Step 1: Identify We want to know a number.

Step 2: Name Let x represent the number.

Step 3: Translate We translate the words of the problem:

$$\underbrace{42}_{42}\ \underbrace{\text{is}}_{=}\ \underbrace{35\%}_{0.35}\ \underbrace{\text{of}}_{\cdot}\ \underbrace{\text{what number?}}_{x}$$

The equation we will solve is $42 = 0.35x$.

Step 4: Solve We now solve the equation.

$$42 = 0.35x$$

Divide each side by 0.35: $$\frac{42}{0.35} = \frac{0.35x}{0.35}$$

$$120 = x$$

Step 5: Check Is 35% of 120 equal to 42? Because $(0.35)(120) = 42$, the answer is correct.

Step 6: Answer 42 is 35% of 120.

QUICK ✓ *Find the number.*

9. 14 is 28% of what number?

10. 111 is 74% of what number?

11. 14.8 is 18.5% of what number?

12. 102 is 136% of what number?

Classroom Example
In an election for school board, the winner received 55% of the votes cast. The winner received 6600 votes. How many votes were cast in the school board election?
Answer: 12,000 votes

EXAMPLE 4 Educational Attainment of U.S. Residents

In 2003, the number of U.S. residents 25 years of age or older was approximately 185,000,000. If 30% of all U.S. residents 25 years of age or older are high school graduates, determine the number of U.S. residents 25 years of age or older in 2003 who were high school graduates. (*Source:* U.S. Census Bureau)

Solution

Step 1: Identify We want to know the number of U.S. residents 25 years of age or older who are high school graduates.

Step 2: Name Let x represent the number of high school graduates.

Step 3: Translate We know that 30% of U.S. residents 25 years of age or older are high school graduates. We also know that in the year 2003 there were approximately 185,000,000 U.S. residents 25 years of age or older. We translate the words of the problem:

$$\underbrace{30\%}_{0.30}\ \underbrace{\text{of}}_{\cdot}\ \underbrace{\text{U.S. residents}}_{185{,}000{,}000}\ \underbrace{\text{are}}_{=}\ \underbrace{\text{high school graduates}}_{x}$$

Our equation is $(0.30)(185{,}000{,}000) = x$.

Step 4: Solve We solve the equation.

$$(0.30)(185{,}000{,}000) = x$$

Multiply: $$55{,}500{,}000 = x$$

Step 5: Check We can recheck our arithmetic: Because $0.30 \cdot 185{,}000{,}000 = 55{,}500{,}000$, the answer is correct.

Step 6: Answer In the year 2003, the number of U.S. residents 25 years of age or older were high school graduates was 55,500,000.

QUICK ✓

13. In 2003, the number of U.S. residents 25 years of age or older was approximately 185,000,000. If 17% of all U.S. residents 25 years of age or older have bachelor's degrees, determine the number of U.S. residents 25 years of age or older in 2003 who have bachelor's degrees.

2 Model and Solve Direct Translation Problems from Business Involving Percent

Now let's look at direct translation problems that involve percents. Typically, "percent problems" involve discounts or mark-ups that businesses use in determining their prices. They may also include finding the cost of an item including sales tax, as shown in Example 5.

Classroom Example ➤
You just purchased a new jacket. The price of the jacket, including sales tax of 7%, was $48.15. How much did the jacket cost excluding the sales tax?

Answer: The jacket cost $45.

EXAMPLE 5 Finding the Cost of an Item Excluding Sales Tax

You just purchased a new pair of jeans. The price of the jeans, including sales tax of 6% was $41.34. How much did the jeans cost excluding the sales tax?

Solution

Step 1: Identify We want to know the price of the jeans before sales tax.

Step 2: Name Let p represent the price of the jeans before sales tax.

Work Smart
To change a percent to a decimal, move the decimal point two places to the left.

Step 3: Translate The algebraic expression for new cost is computed by adding the price of the jeans before sales tax p and the amount of tax. The amount of tax is 6% of the price of jeans before sales tax.

$$\underbrace{\text{original cost}}_{p} + \underbrace{\text{amount of tax}}_{0.06p}$$

So the final cost of the jeans is represented by the algebraic expression $p + 0.06p$. We can now construct our equation.

Teaching Tip
Use a numerical example to show that the sales tax is a percentage of the *original cost*. Students sometimes (incorrectly!) multiply 0.06 by the total cost and subtract that amount to obtain the original cost.

$$\underbrace{\text{original cost}}_{p} + \underbrace{\text{amount of tax}}_{0.06p} = \underbrace{\text{total cost}}_{41.34}$$

Step 4: Solve We solve the equation.

$$p + 0.06p = 41.34$$

Combine like terms. Remember that the coefficient of p is 1.

$$1p + 0.06p = 41.34$$

$$1.06p = 41.34$$

Divide each side of the equation by 1.06.

$$\frac{1.06p}{1.06} = \frac{41.34}{1.06}$$

$$p = 39$$

The jeans cost $39.

Step 5: Check If the jeans cost $39, then the jeans plus the 6% sales tax on $39 amounts to $\$39 + (0.06)(39) = \$39 + \$2.34 = \$41.34$, so our answer is correct.

Step 6: Answer The jeans cost $39 before the sales tax.

QUICK ✓

14. As a reward for being named "Teacher of the Year," Janet received a 2.5% pay raise. If Janet's current salary is $39,000, determine Janet's new salary.

15. Suppose that you just purchased a used car. The price of the car including 7% sales tax was $7811. What was the price of the car excluding sales tax?

Another type of percent problem involves discounts or mark-ups that businesses use in determining their prices. When dealing with percents and the price of goods, it is helpful to remember the following:

$$\text{Original Price} - \text{Discount} = \text{Sale Price}$$
$$\text{Wholesale Price} + \text{Markup} = \text{Selling Price}$$

Classroom Example
A hardware store marks down a table saw by 25%. The sale price of the saw is $90. What was the original price?

Answer: $120 was the original price.

EXAMPLE 6 Markdown

Suppose you just learned that a local clothing store is going out of business and that all merchandise is marked down by 40%. The sale price of a jacket is $108. What was the original price?

Solution

Step 1: Identify This is a direct translation problem. We are looking for the original price of a jacket that was marked down by 40%, and we know that the sale price is $108.

Teaching Tip
Point out that a percentage is always a percent of something. For example, in Example 6 the equation is

$$p - 0.40p = 108$$

not

$$p - 0.40 = 108$$

Step 2: Name Let p represent the original price of the jacket.

Step 3: Translate We know that the original price minus the amount of discount will give us the sale price. We also know that the sale price was $108, so

$$p - \text{discount} = 108$$

"Marked down by 40%" means that the discount is 40% off of the original price, so discount is represented by the expression $0.40p$. Substituting into the equation $p - \text{discount} = 108$, we obtain the equation

$$p - 0.40p = 108$$

Step 4: Solve the equation.

$$p - 0.40p = 108$$

Combine like terms:
$$1p - 0.40p = 108$$
$$0.60p = 108$$

Divide each side by 0.60:
$$\frac{0.60p}{0.60} = \frac{108}{0.60}$$
$$p = 180$$

Remember, p represents the original price, so the original price of the jacket was $180.

Step 5: Check If the original price of the jacket was $180, then the discount would be $0.40(180) = \$72$. Subtracting $72 from the original price of $180 results in a sale price of $108. This answer agrees with the information in the problem.

Step 6: Answer The original price of the jacket was $180.

QUICK ✓

16. Suppose that a gas station marks its gasoline up 80%. If the gas station charges $2.25 per gallon of 87 octane gasoline, what does it pay for the gasoline?

17. A furniture store marks recliners down by 25%. The sale price, excluding the sales tax, is $494.25. Find the original price of each recliner.

2.6 Exercises

For Extra Help: Student Solutions Manual CD Video PH Math/Tutor Center MathXL Tutorials on CD MathXL® MyMathLab

1. 0.032
2. Markup
3. True
4. True
5. Answers may vary.
6. No; explanations may vary.
7. 80
8. 42.5
9. 14
10. 15
11. 4.8
12. 13.5
13. 210
14. 305
15. 72
16. 2000
17. $8\frac{1}{3}$
18. 200
19. 40%
20. 20%
21. 7.5%
22. 16%
23. 200%
24. 75%
25. $54
26. $8000
27. $102,000

Concepts and Vocabulary

In Problems 1 and 2, fill in the blanks.

1. Expressed as a decimal, 3.2% = ________.

2. Wholesale Price + ________ = Selling Price

In Problems 3 and 4, answer True or False to each statement.

3. A computer is marked down 20% and sells for $1225. An equation to calculate the original price is $0.8x = 1225$.

4. Original Price − Discount = Sale Price.

5. The sales tax rate is 6%. Explain why $1.06x$ will correctly calculate the total purchase price of any item that sells for x dollars.

6. An item is reduced by 10% and then this is reduced by another 20%. Is this the same as reducing the item by 30%? Explain why or why not.

Building Skills

In Problems 7–24, find the unknown in each percent question.

7. What is 50% of 160?

8. What is 85% of 50?

9. 7% of 200 is what number?

10. 75% of 20 is what number?

11. What number is 16% of 30?

12. What number is 150% of 9?

13. 31.5 is 15% of what number?

14. 40% of what number is 122?

15. 60% of 120 is what number?

16. 45% of what number is 900?

17. 10 is 120% of what number?

18. 11 is 5.5% of what number?

19. What percent of 60 is 24?

20. 15 is what percent of 75?

21. 1.5 is what percent of 20?

22. 4 is what percent of 25?

23. What percent of 300 is 600?

24. What percent of 16 is 12?

Applying the Concepts

25. Sales Tax The sales tax in Delaware County, Ohio, is 6%. The total cost of purchasing a tennis racket, including tax, is $57.24. Find the cost of the tennis racket before sales tax.

26. Sales Tax The sales tax in Franklin County, Ohio, is 5.75%. The total cost of purchasing a used Honda Civic, including sales tax, is $8460. Find the cost of the car before sales tax.

27. Pay Cut Todd works for a computer firm, and in 2004 he earned a salary of $120,000. Recently, Todd was required to take a 15% pay cut. Find Todd's salary after the pay cut.

28. $22.57
29. $25,000
30. $200,000
31. $68
32. $15,550
33. $570
34. $41.48
35. $400
36. $1170
37. winner: 530; loser: 318
38. 320
39. $285,700
40. 85 full-page; 170 half-page
41. 597
42. 40
43. 32.4 million
44. 28.75 million

28. **Pay Raise** MaryBeth works from home as a graphic designer. Recently she raised her hourly rate by 5% to cover increased costs. Her new hourly rate is $23.70. Find MaryBeth's previous hourly rate.

29. **Bad Investment** After Mrs. Fisher lost 9% of her investment, she had $22,750. What was Mrs. Fisher's original investment?

30. **Good Investment** Perry just learned that his house increased in value by 4% over the past year. The value of the house is now $208,000. What was the value of the home one year ago?

31. **Bookstore Purchase** The bookstore at Marietta College had a one-day-only 25%-off sale on all merchandise, except for computer software and textbooks. You purchased pens, notebooks, and a book on study skills for $51. What was the price of the merchandise before the discount?

32. **Hailstorm** Toyota Town had a 15%-off sale on cars that had been damaged in a hailstorm. A new Toyota truck is on sale for $13,217.50. What was the price of the truck before the hailstorm discount?

33. **Business: Discount Pricing** A wool suit, discounted by 30% for a clearance sale, has a price tag of $399. What was the suit's original price?

34. **Business: Marking up the Price of Books** A college bookstore marks up the price that it pays the publisher for a book by 35%. If the selling price of a book is $56.00, how much did the bookstore pay for the book?

35. **Furniture Sale** A furniture store discounted a dining room table by 40%. The discounted price of the dining room table was $240. Determine the original price of the dining room table.

36. **Vacation Package** The Liberty Travel Agency advertised a 3-night vacation package in Jamaica for 30% off the regular price. The sale price of the package is $819. How much is the vacation package before the 30% off sale? (*Source: New York Times,* 6/29/03)

37. **Voting** In an election for school president, the loser received 60% of the winner's votes. If 848 votes were cast, how many did each receive?

38. **Voting** On a committee consisting of Republicans and Democrats, there are twice as many Republicans as Democrats. If 30% of the Republicans and 20% of the Democrats voted in favor of a bill and there were 160 yes votes, how many people are on the committee?

39. **Commission** Melanie receives a 3% commission on every house she sells. If she received a commission of $8571, what was the value of the house she sold?

40. **Commission** Mario collects a commission for bringing in advertisers to his magazine company. He receives 8% on $450 full-page ads and 5% for $300 half-page ads. If he sells twice as many half-page as full-page ads and his commission was $5610, how many of each type did he bring in?

41. **Voting** In the 2000 presidential election, George Bush received 48.85% of Florida's votes and Al Gore received 48.84% of the votes. If 5,963,070 votes were cast in Florida, how many votes separated the two candidates?

42. **Grades** If 15% of Grant's astronomy class received an A, how many students were in his astronomy class if 6 students earned A's this term?

43. **Bachelors** Based on data obtained from the U.S. Census Bureau, 30% of the 108 million males aged 15 years or older have never married. How many males aged 15 years or older have never married?

44. **Census Data** Based on data obtained from the U.S. Census Bureau, 25% of the 115 million females aged 15 years or older have never married. How many females aged 15 years or older have never married?

45. 70.2%
46. 68.1%
47. 24.6%
48. 6.1%
49. $24
50. 13%
51. 28.6%
52. 15.9%
53. (a) $31,314.38 (b) 43.7%
54. 23.67%
55. 21.2%

According to the U.S. Census Bureau, the types of households are changing. The table below shows the number of family households, single-occupant households, and other nonfamily households in 1990 and in 2000. In Problems 45–48, use the information found in the following table to answer each question.

	1990		2000	
	Number	**Percent**	**Number**	**Percent**
Households	91,947,410	100.0	105,480,101	100.0
Family households	64,517,947		71,787,347	
Single-occupant households	22,580,420		27,230,075	
Other nonfamily households	4,849,043		6,462,679	

SOURCE: *United States Census Bureau,* Census 2000.

45. Find, to the nearest tenth of a percent, the percent of family households in the United States in 1990.

46. Find, to the nearest tenth of a percent, the percent of family households in the United States in 2000.

47. Find, to the nearest tenth of a percent, the percent of single-occupant households in the United States in 1990.

48. Find, to the nearest tenth of a percent, the percent of nonfamily households in the United States in 2000.

Extending the Concepts

49. Discount Pricing Suppose that you are the manager of a clothing store and have just purchased 100 shirts for $12 each. After 1 month of selling the shirts at the regular price, you plan to have a sale giving 25% off the original selling price. However, you still want to make a profit of $6 on each shirt at the sale price. What should you price the shirts at initially to ensure this?

In Problems 50–55, find the percent increase or decrease. The percent increase or percent decrease is defined as $\frac{\text{amount of change}}{\text{original amount}} \times 100\%$.

50. Population Growth The population in a small fishing town grew from 2500 to 2825. Find the percent increase.

51. Gas Mileage The gas mileage on your VW van decreases due to the heavy weight of extra passengers and gear. With the extra weight your van gets 15 mpg and without the passengers and gear it gets 21 mpg. To the nearest tenth, what is your percent decrease in gas mileage with extra passengers and gear?

52. Teaching Salaries The highest average teaching salary in 2000–01 was $53,507 in Connecticut. You decided to move to Oregon where the average teaching salary was $44,988. After the move, to the nearest tenth, what percent decrease in salary do you expect to take?

53. Car Depreciation A new car decreases in value by 25% each year. At *www.bmwusa.com* Misha priced his convertible M3 at $55,670.

(a) After two years, what will the car be worth?

(b) To the nearest tenth of a percent, after 2 years, what is the overall decrease in the value of the car?

54. Gas Prices When gas prices went up from $1.69 to $2.09 per gallon, to the nearest hundredth, what was the percent increase?

55. Gas Prices Last week Danivan bought gas for $1.89 per gallon. This week gas is selling for $2.29 per gallon. To the nearest tenth, what is the percent increase in gas prices?

2.7 Problem Solving: Geometry and Uniform Motion

OBJECTIVES

1. Set Up and Solve Complementary and Supplementary Angle Problems
2. Set Up and Solve Angles of a Triangle Problems
3. Use Geometry Formulas to Solve Problems
4. Set Up and Solve Uniform Motion Problems

Preparing for Problem Solving: Geometry and Uniform Motion

Before getting started, take the following readiness quiz. If you get a problem wrong, go back to the section cited and review the material.

1. Solve: $q + 2q - 30 = 180$ [Section 2.2, p. 86]

2. Solve: $30w + 20(w + 5) = 300$ [Section 2.2, p. 87]

In this section we continue to solve problems using the six-step method introduced in Section 2.5.

1 Set Up and Solve Complementary and Supplementary Angle Problems

We begin by defining *complementary* and *supplementary angles.*

DEFINITION

Two angles whose measures sum to 90° are called **complementary angles.** Each angle is called the *complement* of the other.

Figure 5

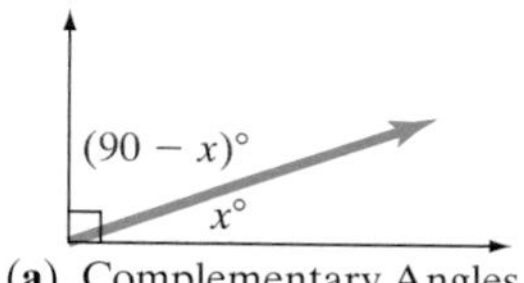

(a) Complementary Angles

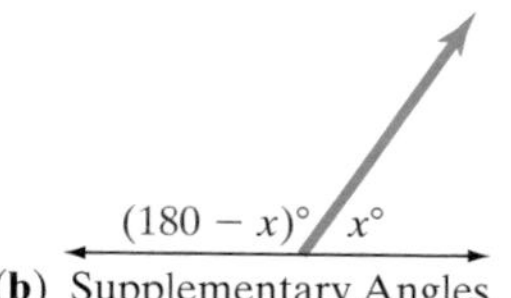

(b) Supplementary Angles

For example, the angles shown in Figure 5(a) are complements because their sum is 90°. Notice the use of the symbol ⌉ to show the 90° angle.

DEFINITION

Two angles whose measures sum to 180° are called **supplementary angles.** Each angle is called the *supplement* of the other.

The angles shown in Figure 5(b) are supplementary. We use the notation $m\angle A$ to say "the measure of angle A."

Classroom Example Find the measure of two complementary angles such that the larger angle is 26° more than the smaller angle.

Answer: 32° and 58°

EXAMPLE 1 Solving a Complementary Angle Problem

Find the measure of two complementary angles such that the measure of the larger angle is 6° greater than twice the measure of the smaller angle.

Solution

Step 1: Identify This is a complementary angle problem. We are looking for the measure of two angles whose sum is 90°.

Step 2: Name We know least about the measure of the smaller angle so we let x represent the measure of the smaller angle.

Step 3: Translate The measure of the larger angle is 6° more than twice the measure of the smaller angle, so $2x + 6$ represents the measure of the larger angle.

We use the formula

$$m\angle A + m\angle B = 90$$

$$\underbrace{x}_{\text{measure of angle A}} + \underbrace{(2x + 6)}_{\text{measure of angle B}} = 90$$

Preparing for...Answers **1.** $\{70\}$ **2.** $\{4\}$

Step 4: Solve the equation.

$$x + (2x + 6) = 90$$

Combine like terms: $3x + 6 = 90$

Subtract 6 from each side of the equation: $3x = 84$

Divide each side by 3: $x = 28$

Step 5: Check The measure of the smaller angle, $\angle A$, is 28°. The measure of the larger angle, $\angle B$, is $(2x + 6)$ degrees: $2(28) + 6 = 56 + 6 = 62°$. Is the sum of the measures of these angles 90°? $28° + 62° = 90°$. Our answer is correct.

Step 6: Answer The two complementary angles measure 28° and 62°.

QUICK ✓ *Solve each problem for the unknown angle measures.*

1. Find two complementary angles such that the measure of the larger angle is 12° more than the measure of the smaller angle.
2. Find two supplementary angles such that the measure of the larger angle is 30° less than twice the measure of the smaller angle.

Figure 6

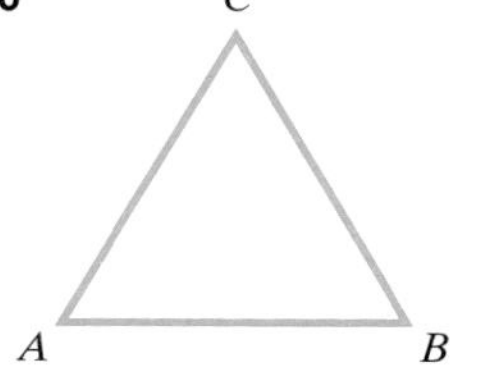

2 Set Up and Solve Angles of a Triangle Problems

An important fact from geometry is that the sum of the measures of the interior angles of a triangle is 180°. In the triangle in Figure 6, we have three angles, A, B, and C. Once again, we use the notation $m\angle A$ to represent the measure of an angle. In a triangle whose angles are A, B, and C, we have $m\angle A + m\angle B + m\angle C = 180°$.

Classroom Example The measure of the smallest angle of a triangle is 40° less than the middle angle and the measure of the largest angle is 25° larger than the middle angle. Find the measure of each angle of the triangle.

Answer: 25°, 65°, 90°

EXAMPLE 2 Solving Sum of Angles of a Triangle Problem

The measure of the largest angle of a triangle is 20° more than twice the measure of the smallest angle, and the measure of the second angle is 10° more than twice the measure of the smallest angle. Find the measure of each angle of the triangle.

Solution

Step 1: Identify This is an "angles of a triangle" problem. We know that the sum of the measures of the interior angles of a triangle is 180°.

Step 2: Name We know least about the measure of the smallest angle so we let x represent the measure of the smallest angle.

Step 3: Translate The measure of the largest angle is 20° more than twice the measure of the smallest angle, so $2x + 20$ represents the measure of the largest angle. The measure of the second angle is 10° more than twice the measure of the smallest angle so we let $2x + 10$ represent the measure of the second angle. Using the formula $m\angle A + m\angle B + m\angle C = 180°$, we have

$$\underbrace{x}_{\text{measure of angle A}} + \underbrace{(2x + 10)}_{\text{measure of angle B}} + \underbrace{(2x + 20)}_{\text{measure of angle C}} = 180°$$

Step 4: Solve the equation.

$$x + (2x + 10) + (2x + 20) = 180°$$

Combine like terms: $5x + 30 = 180°$

Subtract 30 from each side of the equation: $5x = 150°$

Divide each side by 5: $x = 30°$

Step 5: Check 30 degrees is the measure of the smallest angle, $\angle A$. The largest angle, $\angle C$, has a measure of $(2x + 20)$ degrees: $2(30°) + 20 = 60 + 20 = 80°$. The second angle, $\angle B$, has a measure of $(2x + 10)$ degrees: $2(30°) + 10 = 60 + 10 = 70°$. Is the sum of the measures of these angles 180°? Because $30° + 70° + 80° = 180°$, our answer is correct.

Step 6: Answer The measures of the angles of the triangle are 30°, 70°, and 80°.

QUICK ✓ *Solve for the unknown angle measures.*

3. The measure of the smallest angle of a triangle is one-third the measure of the largest angle. The measure of the second angle is 65° less than the measure of the largest angle. Find the measure of the angles of the triangle.

---

3 Use Geometry Formulas to Solve Problems

Recall from Section 2.4, that the perimeter of a figure is the sum of the lengths of its sides.

Classroom Example The perimeter of a rectangular swimming pool is 64 feet and the length is 8 feet more than the width. Find the dimensions of the pool.

Answer: length = 20 feet and width = 12 feet

EXAMPLE 3 Solving a Perimeter Problem

The perimeter of the rectangular swimming pool shown in Figure 7 is 80 feet. If the length is 10 feet more than the width, find the length and the width of the pool.

Figure 7

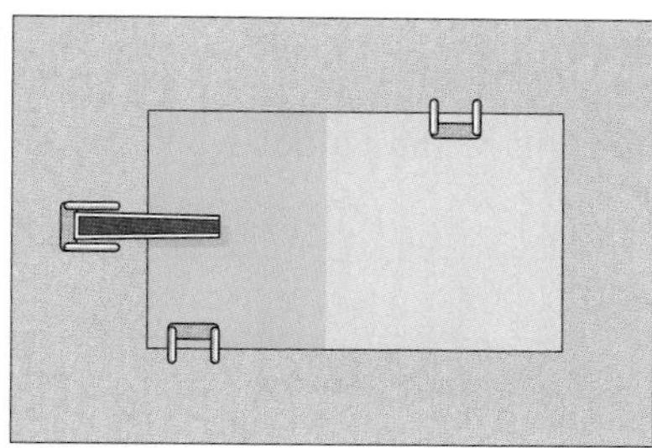

Solution

Step 1: Identify This is a perimeter problem. We want to find the length and the width of the pool, given the perimeter. We know that the perimeter of a rectangle is the sum of the measures of the sides.

Step 2: Name Let w represent the width of the pool.

Step 3: Translate The length of the pool is 10 feet more than the width, so the length of the pool is $w + 10$. The formula for perimeter of a rectangle is $P = 2l + 2w$, where l represents the length of the rectangle and w represents the width. So we have

$$\underbrace{2(w+10)}_{\substack{2l \\ 2\cdot\text{length}}} + \underbrace{2(w)}_{\substack{2w \\ 2\cdot\text{width}}} = \underbrace{80}_{\substack{P \\ \text{perimeter}}}$$

Step 4: Solve the equation.

$$2(w + 10) + 2(w) = 80$$

Use the Distributive Property: $2w + 20 + 2w = 80$

Combine like terms: $4w + 20 = 80$

Subtract 20 from each side of the equation: $4w = 60$

Divide each side by 4: $w = 15$

Step 5: Check The width of the pool is $w = 15$ feet, so the length is $w + 10 = 15 + 10 = 25$ feet. We need to see if the perimeter of the pool is 80 feet. Does $15 + 15 + 25 + 25 = 80$? Yes! Our answer is correct.

Step 6: Answer The length of the rectangular pool is 25 feet and the width of the pool is 15 feet.

Work Smart

In solving problems dealing with geometric figures, it is helpful to draw a picture.

QUICK ✓ *Solve for the dimensions of the rectangle.*

4. The perimeter of a small rectangular garden is 9 feet. If the length is twice the width, find the width and length of the garden.

Recall, the area of a plane (two-dimensional) figure is the number of square units that the figure contains, such as square feet, square inches or square yards.

Classroom Example The area of a trapezoid is 105 square inches. The height is 7 inches and the longer base is 8 inches more than the shorter base. Find the length of each base of the trapezoid.

Answer: 11 inches and 19 inches

EXAMPLE 4 Solving an Area Problem

A garden in the shape of a trapezoid between a sidewalk and curb has an area of 18 square feet. The height is 3 feet and the shorter base is 2 feet less than the length of the longer base. Find the length of each base of the trapezoid. See Figure 8.

Figure 8

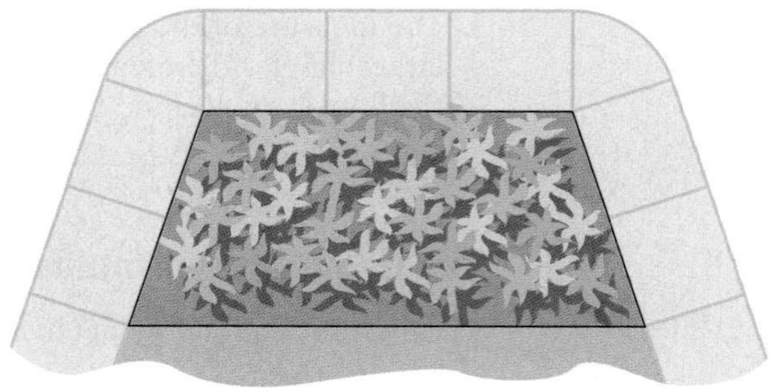

Solution

Step 1: Identify This problem is about the area of a trapezoid. The formula for the area of a trapezoid is $A = \frac{1}{2}h(B + b)$, where h is the height, B is the length of the longer base, and b is the length of the shorter base. We are given the area and the height of the trapezoid.

Step 2: Name We know that one base is 2 feet shorter than the other. Let B represent the length of the longer base.

Step 3: Translate Since one base is 2 feet shorter than the other base, and B represents the length of the longer base, then $B - 2$ represents the length of the shorter base. Using the area of a trapezoid formula, we replace the values we know for A, h, B, and b.

$$A = \frac{1}{2}h(B + b)$$

$$\underbrace{18}_{\text{area}} = \frac{1}{2} \cdot \underbrace{3}_{\text{height}} \; (\underbrace{B}_{B} + \underbrace{B - 2}_{b})$$

Step 4: Solve the equation.

$$18 = \frac{1}{2} \cdot 3(B + B - 2)$$

Combine like terms in the parentheses: $18 = \frac{1}{2} \cdot 3(2B - 2)$

Multiply by 2 to clear fractions: $2[18] = 2\left[\frac{1}{2} \cdot 3(2B - 2)\right]$

$$36 = 3(2B - 2)$$

Use the Distributive Property: $36 = 6B - 6$

Add 6 to each side of the equation: $42 = 6B$

Divide each side by 6: $7 = B$

Work Smart
Instead of distributing the 3 in

$$36 = 3(2B - 2)$$

we could divide both sides by 3. Try it! Which approach do you prefer?

Step 5: Check The longer base is 7 feet long. The smaller base is $B - 2 = 7 - 2 = 5$ feet long. Is the area of the trapezoidal garden 18 square feet? Because $\frac{1}{2} \cdot 3(7 + 5) = \frac{1}{2} \cdot 3(12) = 18$, the answers 7 feet and 5 feet are correct.

Step 6: Answer The lengths of the two bases are 5 feet and 7 feet.

QUICK ✓ *Solve for the unknown quantity.*

5. The surface area of a rectangular box is 62 square feet. If the length of the box is 3 feet and the width is 2 feet, find the height of the box.

4 Set Up and Solve Uniform Motion Problems

Objects that move at a constant velocity (speed) are said to be in **uniform motion.** When the average speed of an object is known, it can be interpreted as its constant velocity. For example, a car traveling at an average speed of 45 miles per hour is in uniform motion. An object traveling down an assembly line at a constant speed is also in uniform motion.

In Words
The uniform motion formula states that distance equals rate times time.

UNIFORM MOTION FORMULA

If an object moves at an average speed r, the distance d covered in time t is given by the formula

$$d = rt$$

We will use a chart to set up uniform motion problems as shown in Table 4.

Teaching Tip
Point out that "rate" is another term for speed. Do a numerical example: If you drive 150 miles in 3 hours, what is your rate (speed)?

Table 4

	Rate	·	Time	=	Distance
Object #1					distance 1
Object #2					distance 2

Rate, time, and distance must be expressed in corresponding units. For example, if rate (speed) is stated in miles per hour, then distance must be in miles and time must be in hours. If rate is measured in kilometers per minute, then distance is kilometers and time is minutes.

Classroom Example
Two planes are traveling toward each other when an air traffic controller sees they may be on a collision course. The faster plane is flying at 300 mph and the slower plane is flying at 100 mph. The planes are 100 miles apart. How long does the traffic controller have to prevent a collision?

Answer: $\frac{1}{4}$ hour

EXAMPLE 5 Solve a Uniform Motion Problem for Time

Bob and Karen drove from Atlanta, Georgia, to Durham, North Carolina, a distance of 390 miles, to attend a family reunion. Their average rate of speed for the first part of the trip was 60 miles per hour. Due to road construction, their average rate of speed for the remainder of the trip was 45 miles per hour. How long did they travel at 45 miles per hour if they drove 3 hours longer at 60 mph than at 45 mph?

Solution

Step 1: Identify This is a uniform motion problem. We wish to know the number of hours Bob and Karen drove at 45 mph.

Step 2: Name Let t represent the number of hours Bob and Karen drove at 45 miles per hour. Since they traveled 3 hours longer at 60 mph, $t + 3$ represents the number of hours driven at 60 mph.

Step 3: Translate We will set up Table 5, listing the information that we know.

Table 5

	Rate (in mph)	·	Time	=	Distance
First part of trip	60		$t + 3$		$60(t + 3)$
Second part of trip	45		t		$45t$
Total					390

We summarize the information from Table 5. We know that the total distance that Bob and Karen traveled is 390 miles, so we state that

$$\underbrace{60(t+3)}_{\text{distance traveled at 60 mph}} + \underbrace{45t}_{\text{distance traveled at 45 mph}} = \underbrace{390}_{\text{total distance}}$$

Step 4: Solve We wish to solve for t: $60(t + 3) + 45t = 390$

Use the Distributive Property: $60t + 180 + 45t = 390$

Combine like terms: $105t + 180 = 390$

Subtract 180 from both sides: $105t + 180 - 180 = 390 - 180$

$105t = 210$

Divide both sides by 105: $\dfrac{105t}{105} = \dfrac{210}{105}$

$t = 2$

Work Smart

The variable time in the formula $d = rt$ stands for the number of hours traveled, not time of day.

Step 5: Check It appears that Bob and Karen drove for 2 hours at 45 mph. So they drove $t + 3 = 2 + 3 = 5$ hours at 60 miles per hour. Let's see if 2 hours driven at 45 mph plus 5 hours driven at 60 mph equals a distance of 390 miles. Because 2 hrs. · 45 mi./hr. + 5 hrs. · 60 mi./hr. = 90 miles + 300 miles = 390 miles, our answer checks!

Step 6: Answer the Question Bob and Karen drove for 2 hours at 45 mph. ■

Classroom Example ➤

One hiker starts up a 15-mile-long trail on Grandfather Mountain at the same time another hiker begins the trek down the same trail. The speed of the hiker going up the trail is 1 mph slower than the hiker going downhill. If the hikers meet on the trail in 3 hours, what is the speed of each hiker?

Answer: Speed of hiker going uphill is 2 mph and speed of hiker going downhill is 3 mph.

EXAMPLE 6 Solve a Uniform Motion Problem for Rate

Two groups of friends took a canoe trip down Big Darby Creek. The first group left Dan's Canoe Livery at 12 noon. One-half hour later, the second group left Dan's Canoe Livery, traveling at an average speed that was 0.75 miles per hour faster than the first group. At 2:30 P.M. the second group caught up to the first group. How fast was each group paddling?

Solution

Step 1: Identify This is a uniform motion problem. We wish to know the rate of speed that each group is paddling.

Step 2: Name Let r represent the rate that the first group paddled. The second group's rate of paddling was 0.75 miles per hour greater than the first group, so we let $r + 0.75$ represent the rate of the second group.

Step 3: Translate Notice that we are given a specific time that the first group left, 12 noon. The second group left $\frac{1}{2}$ hour later than the first group. The variable time in our formula represents number of hours traveled. So the first group traveled for $2\frac{1}{2}$ (or 2.5) hours (12 noon until 2:30 P.M.), but the second group only traveled for 2 hours. We're now ready to fill in the chart as shown in Table 6.

Table 6

	Rate (in mph)	·	Time	=	Distance
First group	r		2.5		$2.5r$
Second group	$r + 0.75$		2		$2(r + 0.75)$

We summarize the information from the chart. We know that the second group caught up to the first group, so the total distance the two groups traveled was the same.

$$\underbrace{2.5r}_{\text{distance traveled by first group}} = \underbrace{2(r + 0.75)}_{\text{distance traveled by second group}}$$

Step 4: Solve We wish to solve for r:

$$2.5r = 2(r + 0.75)$$

Use the Distributive Property:

$$2.5r = 2r + 1.5$$

Subtract $2r$ from both sides:

$$2.5r - 2r = 2r - 2r + 1.5$$

$$0.5r = 1.5$$

Divide both sides by 0.5:

$$\frac{0.5r}{0.5} = \frac{1.5}{0.5}$$

$$r = 3$$

Step 5: Check It appears that the first group paddled at 3 miles per hour and the second group paddled at $r + 0.75 = 3 + 0.75 = 3.75$ miles per hour. Let's see if the distance traveled by each group is the same. Does 3 miles per hour · 2.5 hours = 2 hours · 3.75 miles per hour? $(3)(2.5) = 7.5$ miles and $(2)(3.75)$ also equals 7.5 miles, so our answers are correct.

Step 6: Answer the Question The first group paddled at 3 miles per hour and the second group paddled at 3.75 miles per hour.

QUICK ✓

6. Two bikers, José and Luis, start at the same point at the same time and travel in opposite directions. José's average speed is 5 miles per hour more than that of Luis, and after 3 hours the bikers are 63 miles apart. Find the average speed of each biker.

7. Tanya, a long-distance runner, runs at an average speed of 8 miles per hour. Two hours after Tanya leaves your house you leave in your car and follow the same route. If your average speed is 40 miles per hour, how long will it be before you catch up to Tanya? How far will each of you be from your home?

2.7 Exercises

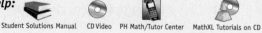

For Extra Help: Student Solutions Manual | CD Video |  PH Math/Tutor Center |  MathXL Tutorials on CD | MathXL® | MyMathLab

1. 90
2. 180
3. True
4. False
5. False
6. Answers may vary.

Concepts and Vocabulary

In Problems 1 and 2, fill in the blanks.

1. Complementary angles are angles whose measures sum to _________ degrees.
2. The sum of the measures of the angles of a triangle is _________ degrees.

In Problems 3–5, answer True or False to each statement.

3. The perimeter of a square is $4s$, where s is the length of a side of the square.
4. The perimeter of a rectangle can be found by multiplying the length of rectangle by the width of the rectangle.
5. When using $d = rt$ to calculate the distance traveled, it is not necessary to travel at a constant speed.
6. When setting up a uniform motion problem, you wrote $65t + 40t = 115$. Your classmate wrote $65t - 40t = 115$. Write a word problem for each of these equations, and explain the keys to recognizing the difference between the two types.

7. 20°, 70°
8. 28°, 62°
9. 50°, 130°
10. 76°, 104°
11. 60°, 75°, 45°
12. 29°, 58°, 93°
13. 42°, 44°, 94°
14. 32°, 65°, 83°
15. 47.5°, 132.5°
16. 36°, 144°
17. 52.5°, 37.5°
18. 57.5°, 32.5°
19. 44°, 46°
20. 58°, 60°, 62°
21. 18°, 72°, 90°
22. 30°, 50°, 100°
23. l = 32 feet; w = 12 feet

Building Skills

In Problems 7–14, find the value of **x** *and then identify the measure of each of the angles.*

△**7.**

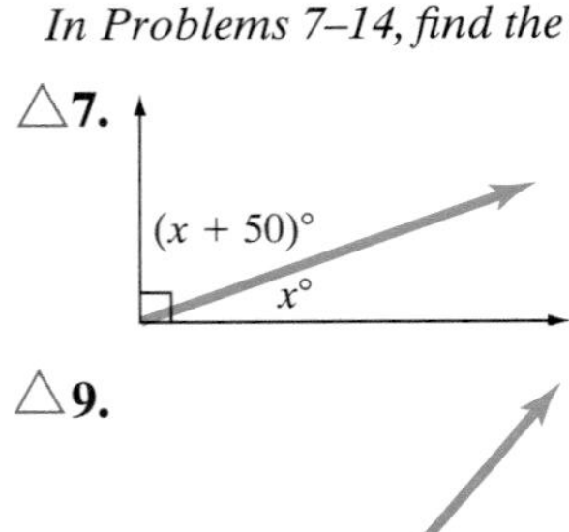

△**8.**

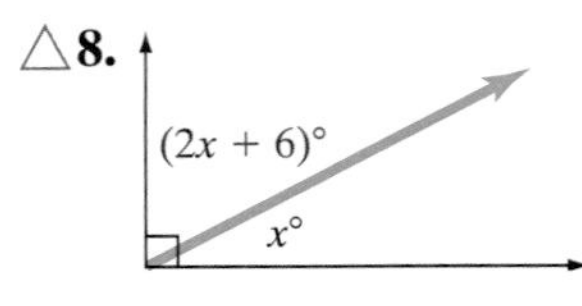

△**9.**

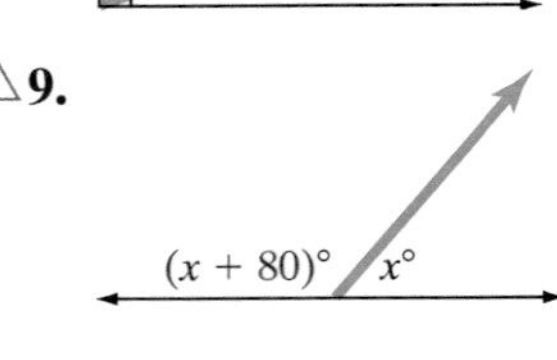

△**10.**

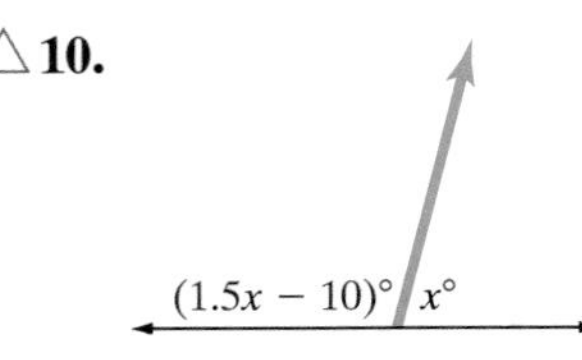

△**11.**

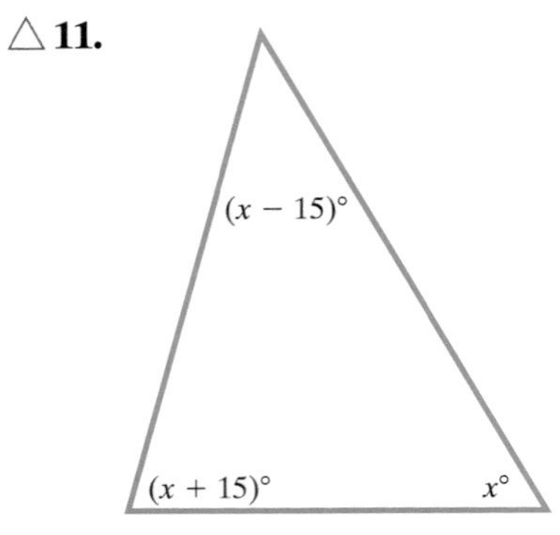

△**12.**

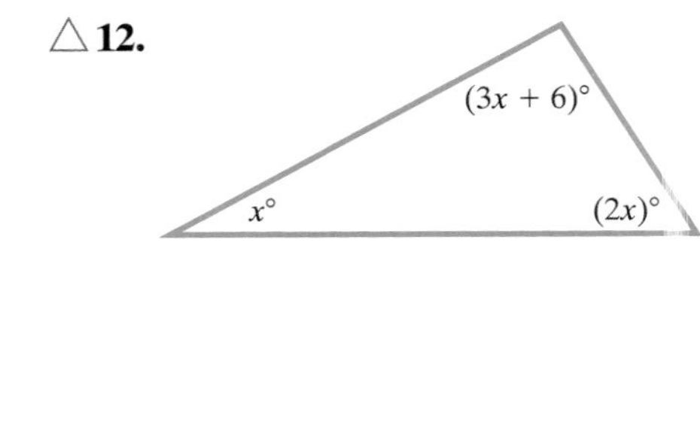

△**13.**

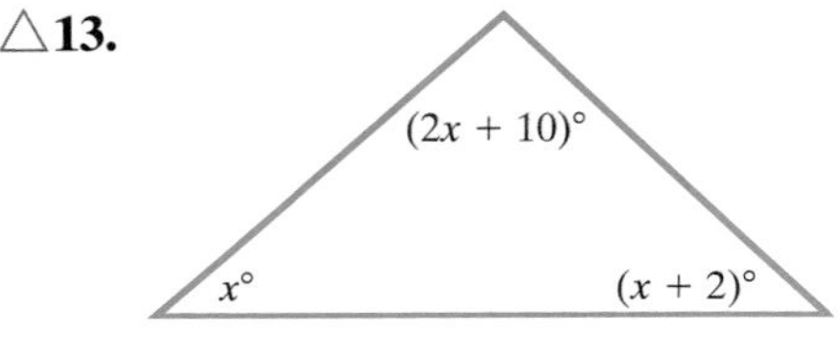

△**14.**

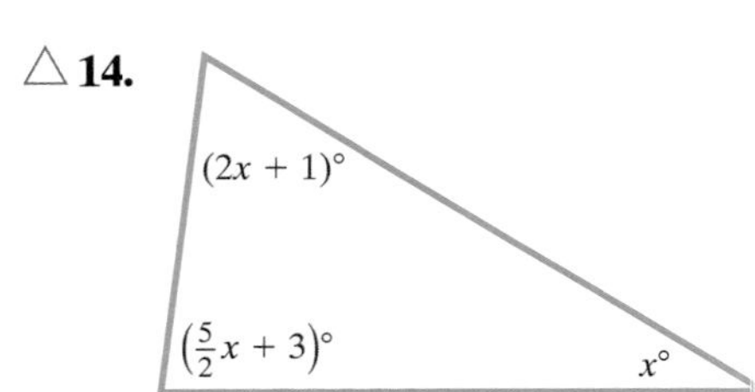

△**15.** Find two supplementary angles such that the measure of the first angle is 10° less than three times the measure of the second.

△**16.** Find two supplementary angles such that the measure of the first angle is four times the measure of the second.

△**17.** Find two complementary angles such that the measure of the first angle is 15° more than the measure of the second.

△**18.** Find two complementary angles such that the measure of the first angle is 25° less than the measure of the second.

△**19.** The measures of two complementary angles are consecutive even integers. Find the measure of each angle.

△**20.** The measures of the angles of a triangle are consecutive even integers. Find the measure of each angle.

△**21.** In a triangle, the second angle measures four times the first. The measure of the third angle is 18° more than the second. Find the measures of the three angles.

△**22.** In a triangle, the second angle measures 20° more than the first. The measure of the third angle is twice the second. Find the measures of the three angles.

△**23.** The length of a rectangle is 8 ft longer than twice the width. If the perimeter is 88 ft, find the length and width of the rectangle.

24. $l = 24$ meters; $w = 2$ meters
25. 42 ft by 84 ft
26. 31 m by 93 m
27. 26.5 in.
28. 39 cm
29. 49°, 49°, 82°
30. 46°, 46°, 88°
31. (a) $62t$ (b) $68t$
(c) $62t + 68t$
(d) $62t + 68t = 585$
32. (a) $72t$ (b) $66t$
(c) $72t - 66t$
(d) $72t - 66t = 45$
33. $528(t + 10) = 880t$

△ **24.** The width of a rectangle is 10 m less than half of the length. If the perimeter is 52 meters, find the length of each side of the rectangle.

△ **25.** A rectangular field is divided into 2 squares of the same size and shape. If it takes 294 yards of fencing to enclose the field and divide the field into the two parcels, find the dimensions of the field. See the figure.

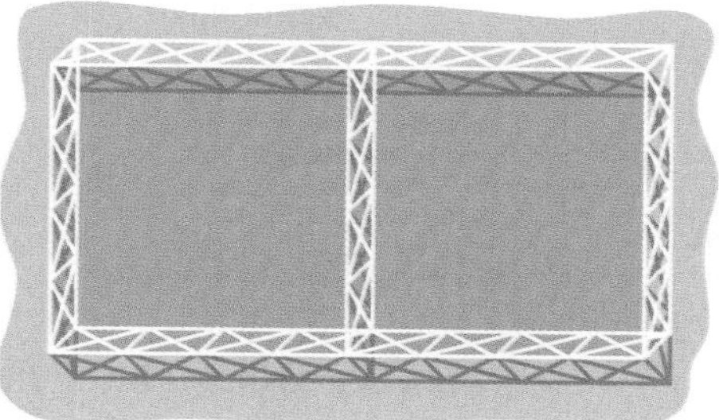

△ **26.** A rectangular field has been divided so that the length of one of the parcels is twice the other. The smaller parcel is a square and the larger parcel is a rectangle. If it takes 279 m of fencing to enclose the field and divide the two parcels, find the dimensions of the field.

△ **27.** An **isosceles triangle** has exactly two sides that are equal in length (*congruent*). If the base (the third side) measures 45 inches and the perimeter is 98 inches, find the length of the two congruent sides, called *legs*.

△ **28.** An isosceles triangle has a base of 17 cm. If the perimeter is 95 cm, find the length of each of the legs.

△ **29.** In an isosceles triangle, the base angles (angles opposite the two congruent legs) are equal in measure (*congruent*). Find the measures of the angles of an isosceles triangle in which the third angle (called the *vertex angle*) has a measure that is 16° less than twice the measures of the base angles.

△ **30.** In an isosceles triangle, the measure of the third angle is 4 degrees less than twice the measures of the base angles. Find the measure of each of the angles of the triangle.

31. Two cars leave Chicago, one traveling east and the other west. The car going east is traveling at 62 mph and the car going west is traveling at 68 mph. How long before they are 585 miles apart?

(a) Write an algebraic expression for the distance traveled by the car going east.
(b) Write an algebraic expression for the car going west.
(c) Write an algebraic expression for the total distance traveled by the two cars.
(d) Write an equation to answer the question.

32. Two trains leave Albuquerque, traveling the same direction on parallel tracks. One train is traveling at 72 mph and the other is traveling at 66 mph. How long before they are 45 miles apart?

(a) Write an algebraic expression for the distance traveled by the faster train.
(b) Write an algebraic expression for the slower train.
(c) Write an algebraic expression for the difference in distance between the two.
(d) Write an equation to answer the question.

In Problems 33 and 34, fill in the table from the information given. Then write the equation that will solve the problem. DO NOT SOLVE.

33. Martha is running in her first marathon. She can run at a rate of 528 ft per min. Ten minutes later her mom starts the same course, running at a rate of 880 ft per min. How long before Mom catches up to Martha?

	Rate	•	Time	=	Distance
Martha	?		?		?
Mom	?		?		?

34. $2x + 3(x - 10) = 580$
35. length = 14 in.; width = 4 in.
36. 4 ft
37. 40 ft
38. (a) width = 9 ft; length = 18 ft
(b) 432 ft^2 (c) 48 yd^2
39. 40 ft; 32 ft
40. 12 ft
41. (a) length = 11 ft; width = 19 ft
(b) 209 ft^2
42. length = 54 in.; width = 36 in.
43. 13 hours
44. eastbound: 18 mph; westbound: 22 mph
45. fast car: 60 mph; slow car: 48 mph
46. freight: 240 mph; passenger: 480 mph
47. freeway: 4.5 hours; 2-lane: 1.5 hours

34. A 580-mile trip in a small plane took a total of 5 hours. The first two hours were flown at one rate and then the plane encountered a head wind and was slowed by 10 mph. Find the rate for each portion of the trip.

	Rate	•	Time	=	Distance
Beginning of trip	?		?		?
Rest of the trip	?		?		?
Total			?		?

Applying the Concepts

△ **35. Rectangle** The width of a rectangle is 3 inches less than one-half the length. Find the length and the width of the rectangle if the perimeter of the rectangle is 36 inches.

△ **36. Garden** The length of a rectangular garden is 9 feet. If 26 feet of fencing are required to fence the garden, find the width of the garden.

△ **37. Billboard** A billboard along a highway has a perimeter of 110 feet. Find the length of the billboard if its height is 15 feet.

△ **38. Buying Wallpaper** Erika is buying wallpaper for her bedroom. She remembers that the perimeter of the room is 54 ft and that the room is twice as long as it is wide.

(a) Find the dimensions of the room.
(b) If the walls are 8 ft high, how many square feet of wallpaper does she need to buy?
(c) Erika arrives at the decorating store and finds that wallpaper is sold by the square yard. How many square yards of wallpaper does Erika need to buy?

△ **39. Back Yard** Bob's back yard is in the shape of a trapezoid with height of 60 feet. The shorter base is 8 feet shorter than the longer base, and the area of the back yard is 2160 square feet. Find the length of each base of the trapezoidal yard.

△ **40. Buying Fertilizer** Melinda has to buy fertilizer for a flower garden in the shape of a right triangle. If the area of the garden is 54 square feet and the base of the garden measures 9 feet, find the height of the triangular garden.

△ **41. Garden** The perimeter of a rectangular garden is 60 yards. The width of the garden is three yards less than twice the length.

(a) Find the length and width of the garden.
(b) What is the area of the garden?

△ **42. Table** The Jacksons are having a custom rectangular table made for a small dining area. The length of the table is 18 inches more than the width, and the perimeter is 180 inches. Find the length and the width of the table.

43. Boats Two boats leave a port at the same time, one going north and the other traveling south. The northbound boat travels 16 mph faster than the southbound boat. If the southbound boat is traveling at 47 mph, how long will it be before they are 1430 miles apart?

44. Cyclists Two cyclists leave a city at the same time, one going east and the other going west. The west-bound cyclist bikes at 4 mph faster than the east-bound cyclist. After 5 hours they are 200 miles apart. How fast is the east-bound cyclist riding?

45. Road Trip Two cars leave a city on the same road, one driving 12 mph faster than the other. After 4 hours, the car traveling at the faster speed stops for lunch. After 4 hours and 30 minutes, the car traveling at the slower speed stops for lunch. Assuming that the person in the faster car is still eating lunch, the cars are now 24 miles apart. How fast is each car driving?

46. Passenger and Freight Trains Two trains leave a city on parallel tracks, traveling the same direction. The passenger train is going twice as fast as the freight train. After 45 minutes, the trains are 180 miles apart. Find the speed of each train.

47. Down the Highway A 360-mile trip began on a freeway in a car traveling at 62 mph. Once the road became a 2-lane highway, the car slowed to 54 mph. If the total trip took 6 hours, find the time spent on each type of road.

48. $2\frac{1}{2}$ hours
49. 6 mph
50. 20 miles
51. $x = 3$
52. $x = 10$
53. $x = 15$
54. $x = 16.5$
55. $x = 12$
56. $x = 60$

48. River Trip Max lives on a river, 30 miles from town. Max travels downstream (with the current) at 20 mph. Returning upstream (against the current) his progress is at 12 mph. If the total trip to town and back took 4 hours, how long did he spend returning from town?

49. Walking and Jogging Carol knows that when she jogs along her neighborhood greenway, she can complete the route in 10 minutes. It takes 30 minutes to cover the same distance when she walks. If her jogging rate is 4 mph faster than her walking rate, find the speed at which she jogs.

50. Trip to School Dien drives to school at 40 mph. Five minutes $\left(\frac{1}{12}\text{hour}\right)$ after he left home, his mother sees that he forgot his homework and leaves to take it to him, driving 48 mph. If they arrive at school at the same time, how far away is the school?

Extending the Concepts

Parallel lines are lines in the same plane that never intersect (think of railroad tracks going infinitely far out into space). A line that cuts two parallel lines is called a *transversal*. The transversal forms 8 different angles that are related in following ways:

Corresponding angles are equal in measure.

Alternate interior angles are equal in measure.

Interior angles on the same side of the transversal are supplementary.

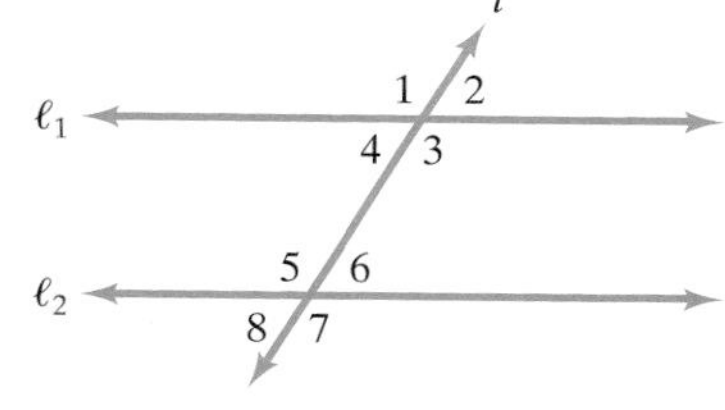

In the figure shown, lines l_1 and l_2 are parallel ($l_1 \parallel l_2$) and the transversal is labeled t. In this figure, there are 4 pairs of corresponding angles:

$$\angle 1 \text{ and } \angle 5,\ \angle 2 \text{ and } \angle 6,\ \angle 3 \text{ and } \angle 7,\ \angle 4 \text{ and } \angle 8$$

There are 2 pairs of alternate interior angles: $\angle 3$ and $\angle 5$, $\angle 4$ and $\angle 6$.

There are 2 pairs of interior angles on the same side of the transversal: $\angle 3$ and $\angle 6$, $\angle 4$ and $\angle 5$.

In Problems 51–56, given $l_1 \parallel l_2$, use the appropriate properties from geometry to solve for x.

△ **51.**

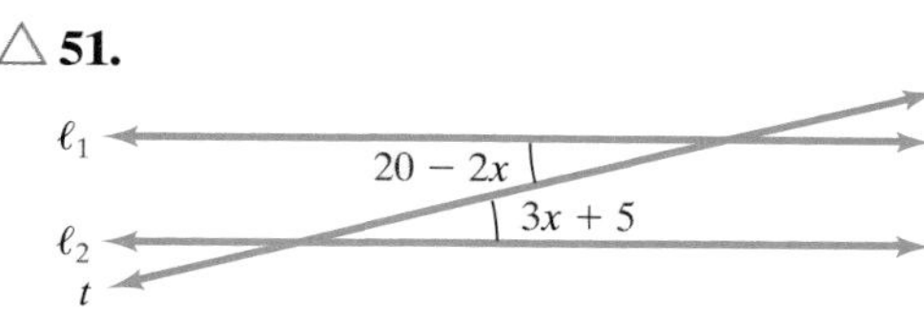

△ **52.**

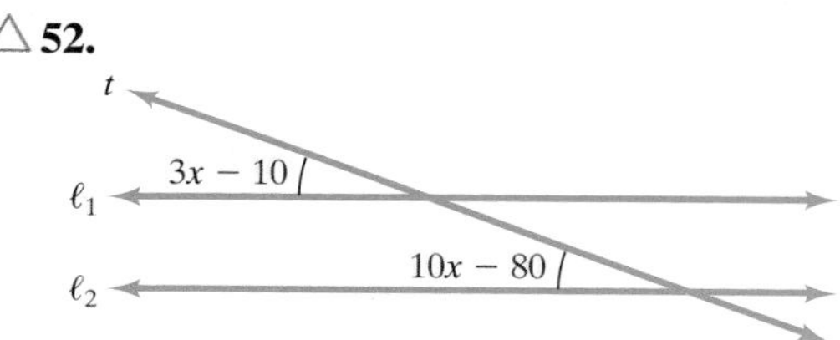

△ **53.**

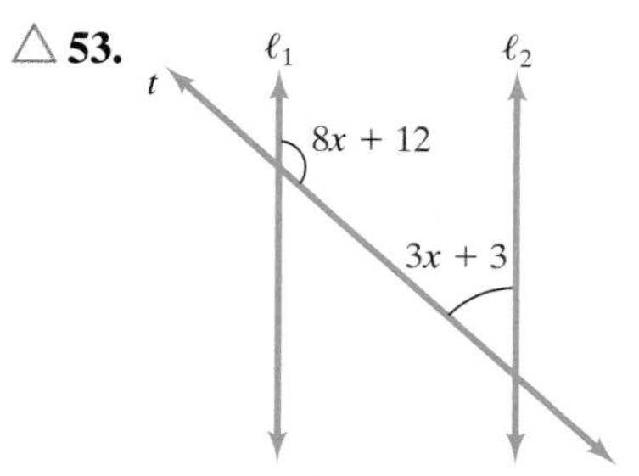

△ **54.**

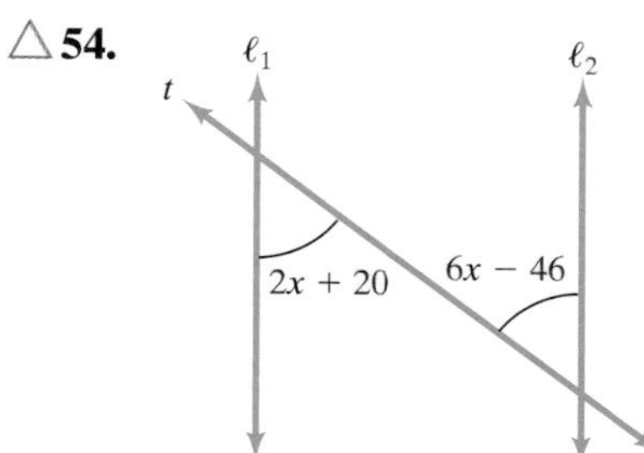

△ **55.**

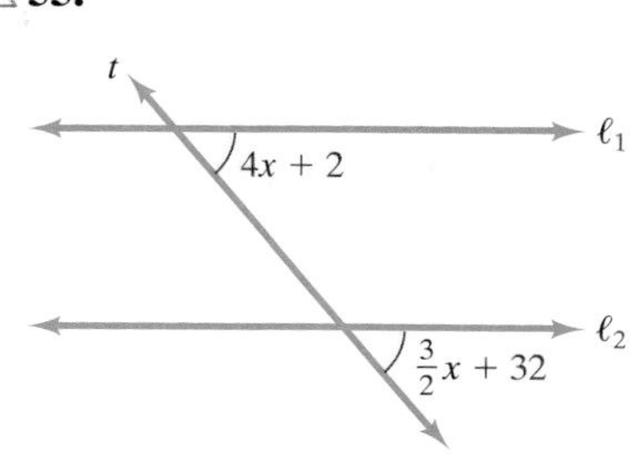

△ **56.**

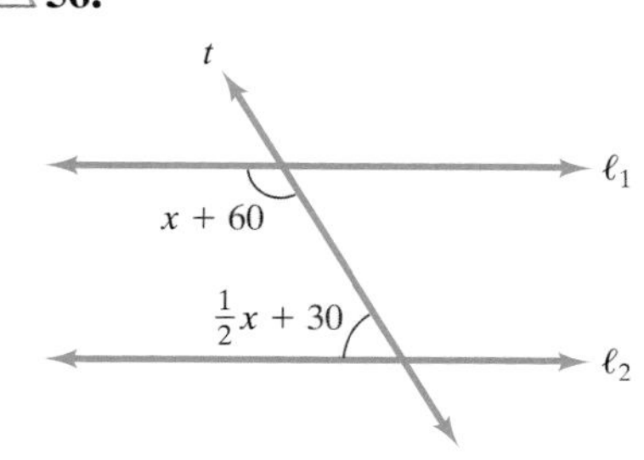

2.8 Solving Linear Inequalities in One Variable

OBJECTIVES

1. Graph Inequalities on the Real Number Line
2. Use Interval Notation
3. Solve Linear Inequalities Using Properties of Inequalities
4. Model Inequality Problems

Preparing for Solving Linear Inequalities in One Variable

Before getting started, take this readiness quiz. If you get a problem wrong, go back to the section cited and review the material.

In Problems 1–4, replace the question mark by <, >, or = to make the statement true.

1. $4 \ ? \ 19$ **2.** $-11 \ ? \ -24$ **3.** $\frac{1}{4} \ ? \ 0.25$ **4.** $\frac{5}{6} \ ? \ \frac{4}{5}$ [Section 1.2, pp. 13–14]

The goal of this section is to solve linear inequalities in one variable.

DEFINITION

A **linear inequality in one variable** is an inequality that can be written in the form

$$ax + b < c \quad \text{or} \quad ax + b \le c \quad \text{or} \quad ax + b > c \quad \text{or} \quad ax + b \ge c$$

where a, b, and c are real numbers and $a \neq 0$.

Examples of linear inequalities in one variable, x, are

$$x + 2 > 7 \qquad \frac{1}{2}x + 4 \le 9 \qquad 5x > 0 \qquad 3x + 7 \ge 8x - 3$$

Before we discuss methods for solving linear inequalities, we will present three ways of representing inequalities. One of the methods for representing an inequality is through set-builder notation. However, set-builder notation can be somewhat cumbersome, so we present a more streamlined way to represent an inequality using *interval notation.* Inequalities can also be represented graphically on the real number line.

1 Graph Inequalities on the Real Number Line

Inequalities that contain one inequality symbol are called **simple inequalities.** For example, the simple inequality

$x > 2$ means the set of all real numbers x greater than two

Since there are infinitely many real numbers that are greater than 2, we cannot list each real number greater than 2. Instead we use **set-builder notation** to express the inequality in written form. For example, the set of all real numbers greater than two is represented in set-builder notation as

$$\{ x \mid x > 2\}$$

The set of all x ↑ such that ↑ x is greater than 2 ↑

Representing an inequality on a number line is called graphing the inequality, and the picture is called the **graph of the inequality.** For example, we graph $\{x|x > 2\}$ on the real number line by shading the portion of the number line that is to the right of 2. The number 2 is called an **endpoint.** We use a parenthesis to indicate that the number 2 is *not* included in the solution set. See Figure 9.

Figure 9
$x > 2$

−4 −3 −2 −1 0 1 2 3 4

Preparing for...Answers **1.** < **2.** > **3.** = **4.** >

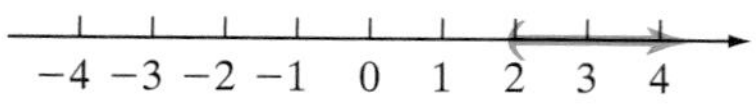

Teaching Tip
The parentheses and bracket notation may be unfamiliar to students who are used to open and closed circles. Point out that when graphing on a number line, parentheses correspond to an open circle, and brackets correspond to a closed circle.

To graph the inequality $x \geq 2$, we also shade the number line to the right of 2, but this time we use a bracket on the endpoint (to indicate that 2 is included in the solution set). See Figure 10.

Figure 10
$x \geq 2$

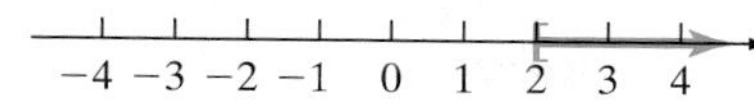

In a similar way, we graph the inequality $x < 4$ in Figure 11 and $x \leq 4$ in Figure 12.

Figure 11
$x < 4$

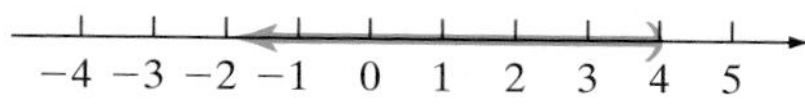

Figure 12
$x \leq 4$

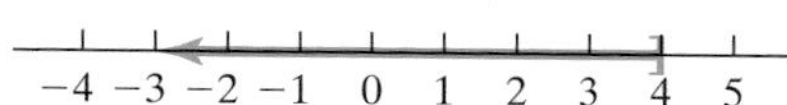

Classroom Example
Graph each inequality on the real number line.
(a) $x > 3$
(b) $x \leq -2$
Answer:
(a) $x > 3$

(b) $x \leq -2$

EXAMPLE 1 Graphing Simple Inequalities on the Real Number Line

Graph each inequality on the real number line.

(a) $x > 5$ **(b)** $x \leq -4$

Solution

(a) The inequality $x > 5$ represents all real numbers greater than 5. Because the number 5 is not included in this set, we draw a parenthesis on the 5 and shade to the right. See Figure 13.

Figure 13
$x > 5$

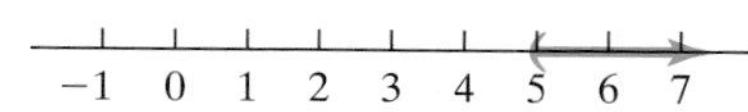

(b) The inequality $x \leq -4$ is all real numbers less than or equal to negative four. Because -4 is included in the set, we place a bracket on the number -4 and shade to the left. See Figure 14.

Work Smart
The inequalities $x > 5$ and $5 < x$ have the same graph. Do you see why?

Figure 14
$x \leq -4$

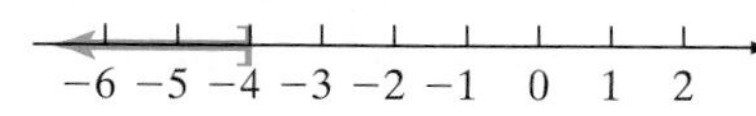

QUICK ✓ *Graph each inequality on the real number line.*

1. $n \geq 8$ **2.** $a < -6$ **3.** $x > -1$ **4.** $p \leq 0$

2 Use Interval Notation

In addition to using set-builder notation to represent inequalities, we can use *interval notation*. To use interval notation, we need to introduce a new symbol.

The symbol ∞ (read "infinity") is not a real number; it is used to indicate that there is no right endpoint on an inequality. The symbol $-\infty$ (read as "minus infinity" or "negative infinity") also is not a real number and means that there is no left endpoint on an inequality. Using the symbols ∞ and $-\infty$, we can define five intervals for simple inequalities:

Work Smart
The symbols ∞ and $-\infty$ are never included as endpoints because they are not real numbers. So, we use parentheses when $-\infty$ or ∞ are endpoints.

INTERVALS INCLUDING ∞

$[a, \infty)$	consists of all real numbers x for which $x \geq a$
(a, ∞)	consists of all real numbers x for which $x > a$
$(-\infty, a]$	consists of all real numbers x for which $x \leq a$
$(-\infty, a)$	consists of all real numbers x for which $x < a$
$(-\infty, \infty)$	consists of all real numbers x (or $-\infty < x < \infty$)

For example, we now have three ways to represent all real numbers greater than 2.

SET-BUILDER NOTATION	NUMBER LINE GRAPH	INTERVAL NOTATION
$\{x \mid x > 2\}$	−4 −3 −2 −1 0 1 2 3 4	$(2, \infty)$

Other simple inequalities can be represented using interval notation as well. We represent inequalities involving $\leq$ and $\geq$ graphically using brackets, [and]. For example, interval notation for $x \leq 1$ is $(-\infty, 1]$.

Table 7 summarizes interval notation, inequality notation, and their graphs.

Work Smart
Intervals are always read from left to right. For example $(5, \infty)$ represents all real numbers greater than 5 because 5 is the left endpoint. But $(5, \infty)$ as an inequality can be represented as $x > 5$ or $5 < x$.

Table 7

Set-Builder Notation	Interval Notation	Graph
$\{x \mid x \geq a\}$	$[a, \infty)$	a
$\{x \mid x > a\}$	(a, ∞)	a
$\{x \mid x \leq a\}$	$(-\infty, a]$	a
$\{x \mid x < a\}$	$(-\infty, a)$	a
$\{x \mid x \text{ is a real number}\}$	$(-\infty, \infty)$	

Classroom Example ➔
Write each inequality using interval notation.
(a) $x > 5$
(b) $x \leq -2$
(c)

(d)
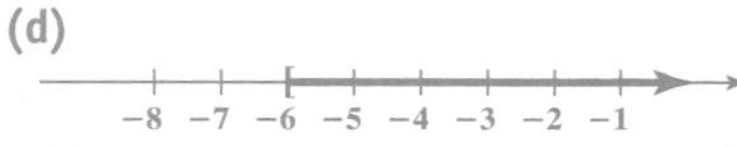

Answer:
(a) $(5, \infty)$
(b) $(-\infty, -2]$
(c) $(-\infty, 2.5)$
(d) $[-6, \infty)$

EXAMPLE 2 Writing an Inequality in Interval Notation

Write each inequality using interval notation.

(a) $x > -4$ **(b)** $x \leq 8$

(c)
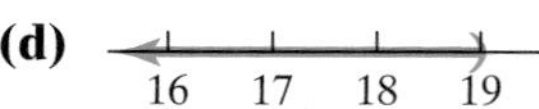

(d) 16 17 18 19

Solution

Remember, if the inequality contains $<$ or $>$, use a parenthesis, (or). If the inequality contains $\leq$ or $\geq$, use a bracket, [or].

(a) $(-4, \infty)$ **(b)** $(-\infty, 8]$ **(c)** $[1.75, \infty)$ **(d)** $(-\infty, 19)$ ■

QUICK ✓ *Write the inequality in interval notation.*

5. $x \geq -3$

6. $x < 12$

7. 1.0 1.5 2.0 2.5 3.0

8. 115 120 125 130 135

3 Solve Linear Inequalities Using Properties of Inequalities

Teaching Tip
The Addition Property of Inequality parallels the Addition Property of Equality.

In Words
The Addition Property of Inequality states that the direction of the inequality does not change when the same quantity is added to each side of the inequality.

To **solve an inequality** means to find all replacement values of the variable for which the statement is true. These values are called **solutions** of the inequality. The set of all solutions is called the **solution set.** As with equations, one method for solving a linear inequality is to replace it by a series of equivalent inequalities until an inequality with an obvious solution, such as $x > 2$, is obtained.

Two inequalities that have exactly the same solution set are called **equivalent inequalities.** We obtain equivalent inequalities by applying some of the same operations as those used to find equivalent equations.

Consider the inequality $3 < 8$. If we add 2 to both sides of the inequality, the left side becomes 5 and the right side becomes 10. Since $5 < 10$, we see that adding the same quantity to both sides of an inequality does not change the sense, or direction, of the inequality. This result is called the **Addition Property of Inequality.**

ADDITION PROPERTY OF INEQUALITY
For real numbers a, b, and c

If $a < b$, then $a + c < b + c$

If $a > b$, then $a + c > b + c$

Classroom Example
Solve the linear inequality and express the solution set using set-builder notation and interval notation. Graph the solution set.

$4n + 3 > 3n + 7$

Answer: $\{n|n > 4\}$; $(4, \infty)$;

−2 −1 0 1 2 3 4 5 6

The Addition Property of Inequality also holds true for subtracting a real number from both sides of an inequality, since $a - b$ is equivalent to $a + (-b)$. In other words, subtracting a quantity from both sides of an inequality does not change the sense of the inequality.

EXAMPLE 3 How to Solve an Inequality Using the Addition Property of Inequality

Solve the linear inequality $3y + 5 \leq 2y + 8$ and express the solution set using set-builder notation and interval notation. Graph the solution set.

Step-by-Step Solution

Step 1: Get the expressions containing variables on the left side of the inequality.

$$3y + 5 \leq 2y + 8$$

Apply the Addition Property of Inequality and subtract $2y$ from both sides:

$$3y + 5 - 2y \leq 2y + 8 - 2y$$
$$y + 5 \leq 8$$

Step 2: Isolate the variable y on the left side.

Apply the Addition Property of Inequality and subtract 5 from both sides:

$$y + 5 - 5 \leq 8 - 5$$
$$y \leq 3$$

The solution set using set-builder notation is $\{y|y \leq 3\}$. The solution set using interval notation is $(-\infty, 3]$. The solution set is graphed in Figure 15.

Figure 15

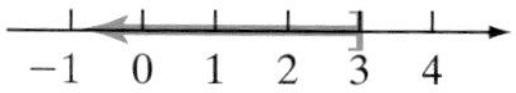

−1 0 1 2 3 4

QUICK ✓ *Find the solution of the linear inequality and express the solution set in set-builder notation or interval notation. Graph the solution set.*

9. $5n - 4 > 11$

10. $-2x + 3 < 7 - 3x$

11. $5n + 8 \leq 4n + 4$

12. $3(4x - 8) + 12 > 11x - 13$

We've seen what happens when we add a real number to both sides of an inequality. Let's look at two examples from arithmetic to see if we can figure out what happens when we multiply or divide both sides of an inequality by a non-zero constant.

Classroom Example

(a) Express the inequality that results by multiplying both sides of the inequality $-3 < 6$ by 2.

(b) Express the inequality that results by dividing both sides of the inequality $12 > 8$ by 4.

Answer:

(a) $-6 < 12$

(b) $3 > 2$

EXAMPLE 4 Multiplying or Dividing an Inequality by a Positive Number

(a) Express the inequality that results by multiplying both sides of the inequality $-2 < 5$ by 3.

(b) Express the inequality that results by dividing both sides of the inequality $18 > 14$ by 2.

Solution

(a) We begin with

$$-2 < 5$$

Multiplying both sides by 3 results in the numbers -6 and 15 on each side of the inequality, so we have

$$-6 < 15$$

(b) We begin with

$$18 > 14$$

Dividing both sides by 2 results in the numbers 9 and 7 on each side of the inequality, so we have

$$9 > 7$$

Based on the results of Example 4, we see that multiplying (or dividing) both sides of an inequality by a positive real number does not affect the sense, or direction, of the inequality.

Classroom Example

(a) Express the inequality that results by multiplying both sides of the inequality $-3 < 6$ by -2.

(b) Express the inequality that results by dividing both sides of the inequality $12 > 8$ by -4.

Answer:

(a) $6 > -12$

(b) $-3 < -2$

EXAMPLE 5 Multiplying or Dividing an Inequality by a Negative Number

(a) Express the inequality that results by multiplying both sides of the inequality $-2 < 5$ by -3.

(b) Express the inequality that results by dividing both sides of the inequality $18 > 14$ by -2.

Solution

(a) We begin with

$$-2 < 5$$

Multiplying both sides by -3 results in the numbers 6 and -15 on each side of the inequality, so we have

$$6 > -15$$

(b) We begin with

$$18 > 14$$

Dividing both sides by -2 results in the numbers -9 and -7 on each side of the inequality, so we have

$$-9 < -7$$

Teaching Tip

The Multiplication Property of Inequality parallels the Multiplication Property of Equality EXCEPT when multiplying or dividing by a negative.

In Example 5, we see that multiplying (or dividing) both sides of an inequality by a negative real number produces an inequality that has the opposite sense, or direction, of the original inequality. The results of Examples 4 and 5 lead us to the **Multiplication Properties of Inequality.**

In Words

The Multiplication Property of Inequality states that if you multiply both sides of an inequality by a positive number, the inequality symbol remains the same, but if you multiply by a negative number, the inequality symbol reverses.

MULTIPLICATION PROPERTIES OF INEQUALITY

Let a, b, and c be real numbers.

If $a < b$, and if $c > 0$, then $ac < bc$.
If $a > b$, and if $c > 0$, then $ac > bc$.

If $a < b$, and if $c < 0$, then $ac > bc$.
If $a > b$, and if $c < 0$, then $ac < bc$.

Classroom Example
Solve each linear inequality and state the solution set using set-builder notation and interval notation. Graph the solution set.

(a) $9x > -45$

(b) $-6x \geq 18$

Answer:

(a) $\{x \mid x > -5\}$; $(-5, \infty)$;

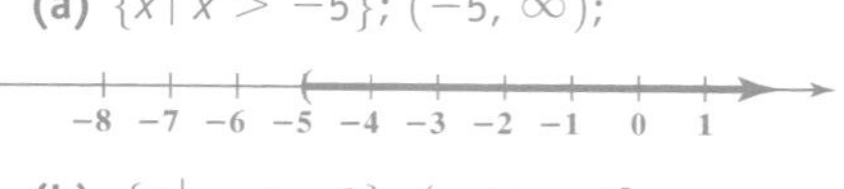

(b) $\{x \mid x \leq -3\}$; $(-\infty, -3]$;

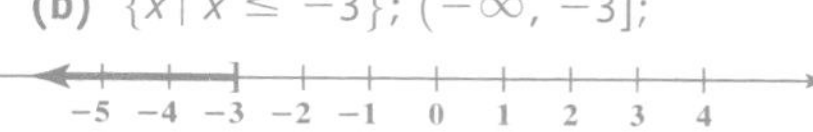

EXAMPLE 6 Solving a Linear Inequality Using the Multiplication Properties of Inequality

Solve each linear inequality and state the solution set using set-builder notation and interval notation. Graph the solution set.

(a) $7x > -35$ **(b)** $-4x \geq 24$

Solution

(a)

$$7x > -35$$

Divide both sides of the inequality by 7: $$\frac{7x}{7} > \frac{-35}{7}$$

$$x > -5$$

The solution set using set-builder notation is $\{x|x > -5\}$. The solution set using interval notation is $(-5, \infty)$. The graph of the solution set is shown in Figure 16.

Figure 16

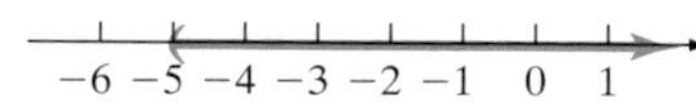

(b)

$$-4x \geq 24$$

Divide both sides of the inequality by -4. Remember to reverse the inequality symbol! $$\frac{-4x}{-4} \leq \frac{24}{-4}$$

$$x \leq -6$$

Teaching Tip
Note the difference to students:

$3x > -15$ is equivalent to $x > -5$

but

$-3x > -15$ is equivalent to $x < 5$

The solution set using set-builder notation is $\{x|x \leq -6\}$. The solution set using interval notation is $(-\infty, -6]$. The graph of the solution set is shown in Figure 17.

Figure 17

−8 −7 −6 −5 −4 −3 −2 −1 0

QUICK ✓ *Find the solution of the linear inequality and express the solution set in set-builder notation or interval notation. Graph the solution set.*

13. $6k < -36$ **14.** $2n \geq -5$ **15.** $-\frac{3}{2}k > 12$ **16.** $-\frac{4}{3}p \leq -\frac{4}{5}$

We are now ready to solve inequalities using both the Addition and Multiplication Properties of Inequalities. In each solution, we isolate the variable on the left side of the inequality, so the inequality is easier to read. Just remember that

$$a < x \text{ is equivalent to } x > a$$
$$\text{and} \quad a > x \text{ is equivalent to } x < a$$

In general, if the two sides of the inequality are interchanged, the direction of the inequality reverses.

EXAMPLE 7 How to Solve an Inequality Using Both the Addition and Multiplication Properties of Inequalities

Solve the inequality $4(x + 1) - 2x < 8x - 26$. State the solution set using set-builder notation and interval notation. Graph the solution set.

Step-by-Step Solution

Step		
Step 1: Remove parentheses.		$4(x + 1) - 2x < 8x - 26$
	Use the Distributive Property:	$4x + 4 - 2x < 8x - 26$
Step 2: Combine like terms on each side of the inequality.		$2x + 4 < 8x - 26$
Step 3: Get the variable expressions on the left side of the inequality and the constants on the right side.	Subtract 8x from both sides:	$2x + 4 - 8x < 8x - 26 - 8x$
		$-6x + 4 < -26$
	Subtract 4 from both sides:	$-6x + 4 - 4 < -26 - 4$
		$-6x < -30$
Step 4: Get the coefficient of the variable term to be one.	Divide both sides by −6: Remember to reverse the inequality symbol!	$\frac{-6x}{-6} > \frac{-30}{-6}$
		$x > 5$

Classroom Example Solve the inequality $5(2x - 1) - 6x < 11x - 19$ and state the solution set using set-builder notation and interval notation. Graph the solution set.

Answer: $\{x|x > 2\}$; $(2, \infty)$;

The solution set using set-builder notation is $\{x|x > 5\}$, or using interval notation, $(5, \infty)$. The graph of the solution is given in Figure 18.

Figure 18

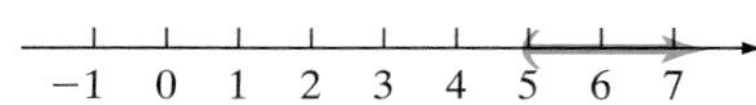

We can gather evidence to support that our solution to an inequality is correct in the same way we check the solution of an equation. We substitute a value for the variable that is in the solution set into the original inequality and see if we obtain a true statement. If we obtain a true statement, then we have evidence our solution is correct. Be warned, however, this does not prove your solution is correct. The check for Example 7 is shown below.

Check Since the solution of the inequality $4(x + 1) - 2x < 8x - 26$ is any real number greater than 5, let's replace x by 10.

$$4(x + 1) - 2x < 8x - 26$$

Replace x with 10: $$4(10 + 1) - 2(10) \stackrel{?}{<} 8(10) - 26$$

$$4(11) - 20 \stackrel{?}{<} 80 - 26$$

Perform the arithmetic: $$24 < 54 \quad \text{True}$$

$24 < 54$ is a true statement, so $x = 10$ is in the solution set. We have some evidence that our solution set, $(5, \infty)$, is correct.

QUICK ✓ *Find the solution of the linear inequality and express the solution set using set-builder notation or interval notation. Graph the solution set.*

17. $3x - 7 > 14$

18. $-4n - 3 < 9$

19. $2x - 6 < 3(x + 1) - 5$

20. $-4(x + 6) + 18 \geq -2x + 6$

Classroom Example ➤
Solve the inequality $\frac{1}{3}(x - 4) \geq \frac{5}{6}(2x - 3)$ and state the solution set using set-builder notation and interval notation. Graph the solution set.

Answer:

$\left\{x \middle| x \leq \frac{7}{8}\right\}; \left(-\infty, \frac{7}{8}\right];$

Teaching Tip
In Example 8, suggest that students try subtracting $2x$ from both sides of the equations and solve. Show that the result, $-\frac{11}{4} \geq x$, is equivalent to $x \leq -\frac{11}{4}$.

EXAMPLE 8 Solving a Linear Inequality Containing Fractions

Solve the inequality $\frac{1}{2}(x - 4) \geq \frac{3}{4}(2x + 1)$. Express the solution using set-builder notation and interval notation. Graph the solution set.

Solution

To clear the inequality of fractions, we multiply both sides of the inequality by 4, the least common denominator of $\frac{1}{2}$ and $\frac{3}{4}$.

$$\frac{1}{2}(x - 4) \geq \frac{3}{4}(2x + 1)$$

$$4 \cdot \frac{1}{2}(x - 4) \geq 4 \cdot \frac{3}{4}(2x + 1)$$

$$2(x - 4) \geq 3(2x + 1)$$

Use the Distributive Property: $$2x - 8 \geq 6x + 3$$

Subtract $6x$ from both sides: $$2x - 8 - 6x \geq 6x + 3 - 6x$$

$$-4x - 8 \geq 3$$

Add 8 to both sides: $$-4x - 8 + 8 \geq 3 + 8$$

$$-4x \geq 11$$

Divide both sides by -4 and remember to reverse the inequality symbol! $$\frac{-4x}{-4} \leq \frac{11}{-4}$$

$$x \leq -\frac{11}{4}$$

The solution of the inequality $\frac{1}{2}(x - 4) \geq \frac{3}{4}(2x + 1)$ is $\left\{x \middle| x \leq -\frac{11}{4}\right\}$, or using interval notation, $\left(-\infty, -\frac{11}{4}\right]$. The graph of the solution is given in Figure 19.

Figure 19

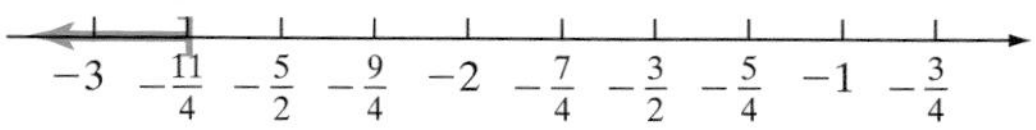

QUICK ✓ *Find the solution of the linear inequality and express the solution set using set-builder notation or interval notation. Graph the solution set.*

21. $\frac{1}{2}(x + 2) > \frac{1}{5}(x + 17)$

22. $\frac{4}{3}x - \frac{2}{3} \leq \frac{4}{5}x + \frac{3}{5}$

There are inequalities that are true for all values of the variable and inequalities that are false for all values of the variable. We present these special cases now.

Classroom Example Solve the linear inequality $2(3x - 2) + 7 > 10x - (4x + 5)$ and state the solution set using set-builder notation and interval notation. Graph the solution set.

Answer:

$\{x \mid x \text{ is a real number}\}$; $(-\infty, \infty)$.

EXAMPLE 9 Solving an Inequality for Which the Solution Set Is All Real Numbers

Solve the inequality $3(x + 4) - 5 > 7x - (4x + 2)$. Express the solution using set-builder notation and interval notation. Graph the solution set.

Solution

$$3(x + 4) - 5 > 7x - (4x + 2)$$

Use the Distributive Property: $$3x + 12 - 5 > 7x - 4x - 2$$

$$3x + 7 > 3x - 2$$

Subtract $3x$ from each side: $$3x + 7 - 3x > 3x - 2 - 3x$$

$$7 > -2$$

Since 7 is always greater than -2, the solution to this inequality is all real numbers. The solution set is $\{x|x \text{ is any real number}\}$. In interval notation, the solution is $(-\infty, \infty)$. Figure 20 shows the graph of the solution set.

Figure 20

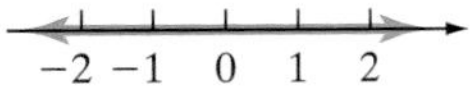

Classroom Example Solve the linear inequality $6\left(1 - \frac{1}{3}x\right) + 5x \le 3x - 4$ and state the solution set using set-builder notation and interval notation, if possible. Graph the solution set.

Answer: The solution is the empty set $\emptyset$ or $\{\ \}$.

EXAMPLE 10 Solving an Inequality for Which the Solution Set Is the Empty Set

Solve the inequality $8\left(\frac{1}{2}x - 1\right) + 2x \le 6x - 10$. Express the solution using set-builder notation and interval notation, if possible. Graph the solution set.

Solution

$$8\left(\frac{1}{2}x - 1\right) + 2x \le 6x - 10$$

Use the Distributive Property: $$4x - 8 + 2x \le 6x - 10$$

$$6x - 8 \le 6x - 10$$

Subtract $6x$ from each side: $$6x - 8 - 6x \le 6x - 10 - 6x$$

$$-8 \le -10$$

The statement $-8 \le -10$ is a false statement. Therefore, there is no solution to this inequality. The solution is the empty set $\emptyset$ or $\{\ \}$. Figure 21 shows the graph of the solution set on a real number line.

Figure 21

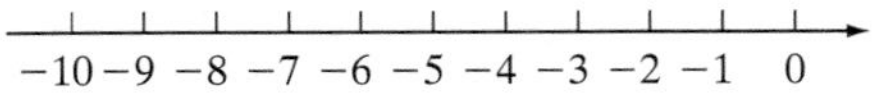

QUICK ✓ *Find the solution of the linear inequality and express the solution set using set-builder notation and interval notation, if possible. Graph the solution set.*

23. $-2x + 7(x - 5) \le 6x + 32$ **24.** $-x + 7 - 8x \ge 2(8 - 5x) + x$

25. $\frac{3}{2}x + 5 - \frac{5}{2}x < 4x - 3(x + 1)$ **26.** $0.8x + 3.2(x + 4) \ge 2x + 12.8 + 3x - x$

4 Model Inequality Problems

Work Smart

The context of the words "less than" is important! 3 **less than** x is $x - 3$, but 3 **is less than** x is $3 < x$.

When solving word problems that are modeled by inequalities, you need to look for key words that indicate the type of inequality symbol that should be used. Some key words and phrases are listed in Table 8.

Table 8

Word or Phrase	Inequality Symbol
at least	$\geq$
no less than	$\geq$
more than	$>$
greater than	$>$
no more than	$\leq$
at most	$\leq$
fewer than	$<$
less than	$<$

When we solve applications involving linear inequalities, we use the same steps for setting up applied problems that we introduced in Section 2.5, on page 122.

Classroom Example A plumber charges a flat fee of $40 plus $27.50 per hour for a job. How many hours does the plumber need to work to make more than $700?

Answer: more than 24 hours

EXAMPLE 11 A Handyman's Fee

A handyman charges a flat fee of $60 plus $22 per hour for a job. How many hours does this handyman need to work to make more than $500?

Solution

Step 1: Identify This is a direct translation problem. We want to know the number of hours the handyman must work to make more than $500.

Step 2: Name Let h represent the number of hours that the handyman must work.

Step 3: Translate We know that the sum of the flat fee that the handyman charges plus the hourly charge must exceed $500. So

$$\underbrace{\text{flat fee}}_{60}\ \underbrace{\text{plus}}_{+}\ \underbrace{\text{hourly wage}}_{22}\ \underbrace{\text{times}}_{\cdot}\ \underbrace{\text{no. hours worked}}_{h}\ \underbrace{\text{more than}}_{>}\ \underbrace{500}_{500}$$

Step 4: Solve the inequality.

$$60 + 22h > 500$$

Subtract 60 from each side and simplify: $$60 - 60 + 22h > 500 - 60$$

$$22h > 440$$

Divide both sides by 22: $$\frac{22h}{22} > \frac{440}{22}$$

$$h > 20$$

So the handyman must work more than 20 hours to earn more than $500.

Step 5: Check If the handyman works 23 hours (for example), will he earn more than $500? Since $60 + 22(23) = 566$ is greater than 500, we have evidence that our answer is correct.

Step 6: Answer The handyman must work more than 20 hours to earn more than $500.

QUICK ✓

27. A worker in a large apartment complex uses an elevator to move supplies. The elevator has a weight limit of 2000 pounds. The worker weighs 180 pounds and each box of supplies weighs 91 pounds. Find the maximum number of boxes of supplies the worker can move on the elevator.

2.8 Exercises

For Extra Help: Student Solutions Manual | CD Video | PH Math/Tutor Center | MathXL Tutorials on CD | MathXL® | MyMathLab

1. solve
2. Addition Property of Inequality
3. $\{x|x \text{ is any real number}\}$
4. False
5. True
6. True
7. A left parenthesis is used when the endpoint is not included. A left bracket is used when the endpoint is included.
8. The inequality symbol is reversed when you multiply or divide each side of the inequality by a negative number.
9. $x \geq 16{,}000$
10. $x \leq 120{,}000$
11. $x \leq 20{,}000$
12. $x \geq 250$
13. $x > 12{,}000$
14. $x < 25$
15. $x > 0$
16. $x \geq 0$
17. $x \leq 0$
18. $x < 0$

Concepts and Vocabulary

In Problems 1–3, fill in the blanks.

1. To _________ an inequality means to find the set of all replacement values of the variable for which the statement is true.

2. The __________ _________ ____ _______ states that the direction, or sense, of each inequality remains the same when a positive number is added to each side of the inequality.

3. When you solve an inequality and the result is $-2 \leq 7$, the solution set is _________.

In Problems 4–6, answer True or False to each statement.

4. The solution to the inequality $-\frac{1}{2}x > 9$ is $\{x|x < -18\}$. This solution is written in interval notation as $(-18, -\infty)$.

5. When dividing both sides of a simple inequality by a negative number, reverse the direction of the inequality symbol.

6. When solving an inequality, if you obtain the result $7 < 1$, then there is no solution.

7. When graphing an inequality, when is a left parenthesis used? When is a left bracket used?

8. Explain the circumstances in which the direction of the inequality symbol is reversed when solving a simple inequality.

Building Skills

In Problems 9–18, write the given statement using inequality symbols. Let **x** *represent the unknown quantity.*

9. Karen's salary this year will be at least \$16,000.

10. Bob's salary this year will be at most \$120,000.

11. There will be at most 20,000 fans at the Cleveland Indians game today.

12. The cost of a new lawnmower is at least \$250.

13. The cost to remodel a kitchen is more than \$12,000.

14. There are fewer than 25 students in your math class on any given day.

15. x is a positive number

16. x is a nonnegative number

17. x is a nonpositive number

18. x is a negative number

In Problems 19–26, graph each inequality on a number line, and write each inequality in interval notation.

19. $x > 2$ **20.** $n > 5$ **21.** $x \le -1$

22. $x \le 6$ **23.** $z \ge -3$ **24.** $x \ge -2$

25. $x < 4$ **26.** $y < -3$

In Problems 27–32, use interval notation to express the inequality shown in each graph.

27. −1 0 1 2

28. −2 −1 0 1 2 3

29. −2 −1 0 1 2 3

30. −2 −1 0 1 2 3

31. −2 −1 0 1 2

32. −2 −1 0 1 2 3

In Problems 33–52, solve the inequality and express the solution set in set-builder notation or interval notation. Graph the solution set on the real number line.

33. $x + 1 < 5$ **34.** $x + 4 \le 3$

35. $x - 6 \ge -4$ **36.** $x - 2 < 1$

37. $3x \le 15$ **38.** $4x > 12$

39. $-5x < 35$ **40.** $-7x \ge 28$

41. $3x - 7 > 2$ **42.** $2x + 5 > 1$

43. $3x - 1 \ge 3 + x$ **44.** $2x - 2 \ge 3 + x$

45. $1 - 2x \le 3$ **46.** $2 - 3x \le 5$

47. $-2(x + 3) < 8$ **48.** $-3(1 - x) > x + 8$

49. $4 - 3(1 - x) \le 3$ **50.** $8 - 4(2 - x) \le -2x$

51. $\frac{1}{2}(x - 4) > x + 8$ **52.** $3x + 4 > \frac{1}{3}(x - 2)$

In Problems 53–62, solve the inequality and express the solution set in set-builder notation or interval notation, if possible. Graph the solution set on the real number line.

53. $4(x - 1) > 3(x - 1) + x$ **54.** $2y - 5 + y < 3(y - 2)$

55. $5(n + 2) - 2n \le 3(n + 4)$ **56.** $3(p + 1) - p \ge 2(p + 1)$

57. $2n - 3(n - 2) < n - 4$ **58.** $4x - 5(x + 1) \le x - 3$

19. 0 2 $(2, \infty)$
20. 0 5 $(5, \infty)$
21. −1 0 $(-\infty, -1]$
22. 0 6 $(-\infty, 6]$
23. −3 0 $[-3, \infty)$
24. −2 0 $[-2, \infty)$
25. 0 4 $(-\infty, 4)$
26. −3 0 $(-\infty, -3)$
27. $(-\infty, 2)$ 28. $[-1, \infty)$
29. $\varnothing$ or { }
30. $(-\infty, 3]$
31. $(-\infty, \infty)$
32. $(-2, \infty)$
33. $\{x \mid x < 4\}$; $(-\infty, 4)$ 0 4
34. $\{x \mid x \le -1\}$; $(-\infty, -1]$ −1 0
35. $\{x \mid x \ge 2\}$; $[2, \infty)$ 0 2
36. $\{x \mid x < 3\}$; $(-\infty, 3)$ 0 3
37. $\{x \mid x \le 5\}$; $(-\infty, 5]$ 0 5
38. $\{x \mid x > 3\}$; $(3, \infty)$ 0 3
39. $\{x \mid x > -7\}$; $(-7, \infty)$ −7 0
40. $\{x \mid x \le -4\}$; $(-\infty, -4]$ −4 0
41. $\{x \mid x > 3\}$; $(3, \infty)$ 0 3
42. $\{x \mid x > -2\}$; $(-2, \infty)$ −2 0
43. $\{x \mid x \ge 2\}$; $[2, \infty)$ 0 2
44. $\{x \mid x \ge 5\}$; $[5, \infty)$ 0 5
45. $\{x \mid x \ge -1\}$; $[-1, \infty)$ −1 0
46. $\{x \mid x \ge -1\}$; $[-1, \infty)$ −1 0
47. $\{x \mid x > -7\}$; $(-7, \infty)$ −7 0
48. $\left\{x \mid x > \frac{11}{2}\right\}$; $\left(\frac{11}{2}, \infty\right)$ 0 $\frac{11}{2}$

49. $\left\{x \mid x \leq \frac{2}{3}\right\}$; $\left(-\infty, \frac{2}{3}\right]$

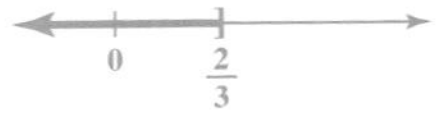

50. $\{x \mid x \leq 0\}$; $(-\infty, 0]$

51. $\{x \mid x < -20\}$; $(-\infty, -20)$

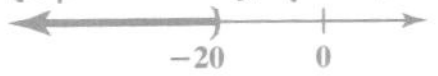

52. $\left\{x \mid x > -\frac{7}{4}\right\}$; $\left(-\frac{7}{4}, \infty\right)$

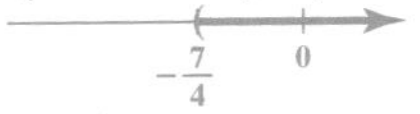

53. $\varnothing$ or { }

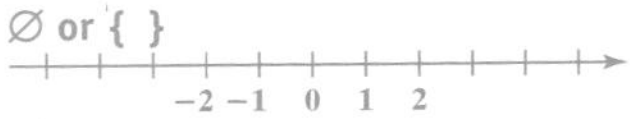

54. $\varnothing$ or { }

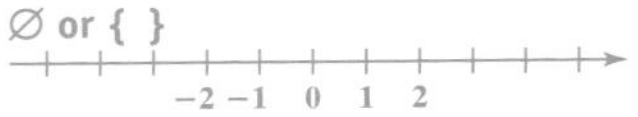

55. $\{n \mid n$ is any real number$\}$; $(-\infty, \infty)$

56. $\{p \mid p$ is any real number$\}$; $(-\infty, \infty)$

57. $\{n \mid n > 5\}$; $(5, \infty)$

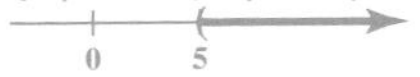

58. $\{x \mid x \geq -1\}$; $[-1, \infty)$

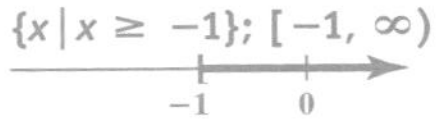

59. $\{w \mid w$ is any real number$\}$; $(-\infty, \infty)$

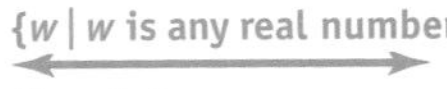

60. $\varnothing$ or { }

61. $\left\{y \mid y < -\frac{3}{2}\right\}$; $\left(-\infty, -\frac{3}{2}\right)$

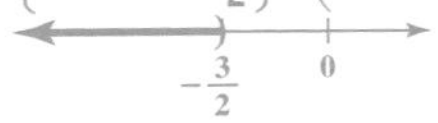

62. $\{x \mid x$ is any real number$\}$; $(-\infty, \infty)$

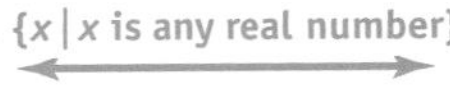

63. $\{x \mid x > 4\}$; $(4, \infty)$

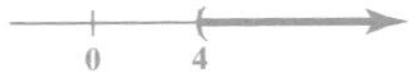

64. $\{x \mid x \leq -9\}$; $(-\infty, -9]$

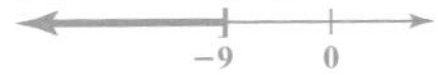

65. $\left\{x \mid x < \frac{3}{4}\right\}$; $\left(-\infty, \frac{3}{4}\right)$

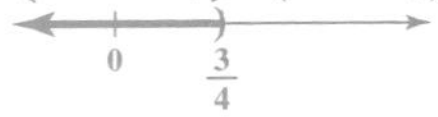

66. $\left\{x \mid x < -\frac{5}{6}\right\}$; $\left(-\infty, -\frac{5}{6}\right)$

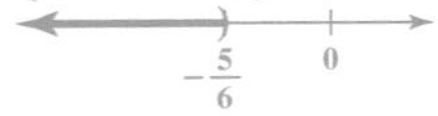

67. $\{x \mid x$ is any real number$\}$; $(-\infty, \infty)$

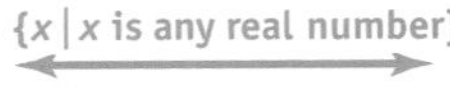

68. $\{x \mid x$ is any real number$\}$; $(-\infty, \infty)$

69. $\{a \mid a < -1\}$; $(-\infty, -1)$

70. $\left\{b \mid b \geq -\frac{9}{2}\right\}$; $\left[-\frac{9}{2}, \infty\right)$

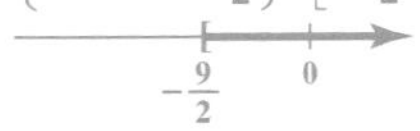

71. $\{n \mid n$ is any real number$\}$; $(-\infty, \infty)$

72. $\varnothing$ or { }

73. $\left\{x \mid x \geq \frac{4}{3}\right\}$; $\left[\frac{4}{3}, \infty\right)$

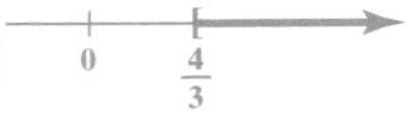

74. $\{x \mid x \geq 12\}$; $[12, \infty)$

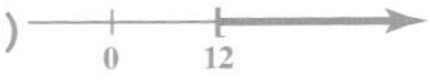

59. $4(2w - 1) \geq 3(w + 2) + 5(w - 2)$

60. $3q - (q + 2) > 2(q - 1)$

61. $3y - (5y + 2) > 4(y + 1) - 2y$

62. $8x - 3(x - 2) \geq x + 4(x + 1)$

Mixed Practice

In Problems 63–80, solve the inequality and express the solution set in set-builder notation or interval notation, if possible. Graph the solution set on the real number line.

63. $-1 < x - 5$

64. $6 \geq x + 15$

65. $-\frac{3}{4}x > -\frac{9}{16}$

66. $-\frac{5}{8}x > \frac{25}{48}$

67. $3(x + 1) > 2(x + 1) + x$

68. $5(x - 2) < 3(x + 1) + 2x$

69. $-4a + 1 > 9 + 3(2a + 1) + a$

70. $-5b + 2(b - 1) \leq 6 - (3b - 1) + 2b$

71. $n + 3(2n + 3) > 7n - 3$

72. $2k - (k - 4) \geq 3k + 10 - 2k$

73. $\frac{x}{2} \geq 1 - \frac{x}{4}$

74. $\frac{x}{3} \geq 2 + \frac{x}{6}$

75. $\frac{x + 5}{2} + 4 > \frac{2x + 1}{3} + 2$

76. $\frac{3z - 1}{4} + 1 \leq \frac{6z + 5}{2} + 2$

77. $-5z - (3 + 2z) > 3 - 7z$

78. $2(4a - 3) \leq 5a - (2 - 3a)$

79. $1.3x + 3.1 < 4.5x - 15.9$

80. $4.9 + 2.6x < 4.2x - 4.7$

Applying the Concepts

81. Auto Rental A car can be rented from Certified Auto Rental for \$55 per week plus \$0.18 per mile. How many miles can be driven if you have at most \$280 to spend for weekly transportation?

82. Truck Rental A truck can be rented from Acme Truck Rental for \$80 per week plus \$0.28 per mile. How many miles can be driven if you have at most \$100 to spend on truck rental?

83. Final Grade Yvette has scores of 72, 78, 66, and 81 on her algebra tests. Find the minimum score she can make on the final exam in order to pass the course with at least 360 points. The final exam counts as two test grades.

84. Final Grade To earn an A in Mrs. Smith's elementary statistics class, Elizabeth must earn at least 540 points. Thus far, Elizabeth has earned scores of 85, 83, 90, and 96. The final exam counts as two test grades. How many points does Elizabeth have to score on the final exam to earn an A?

85. Calling Plan Imperial Telephone has a long-distance calling plan that has a monthly fee of \$10 and a charge of \$0.03 per minute used. Mayflower Communications has a monthly fee of \$6 and a fee of \$0.04 per minute. For how many minutes is Imperial Telephone the cheaper plan?

86. Commission A recent college graduate had an offer of a sales position that pays \$15,000 per year plus 1% of all sales.

(a) Write an expression for the total annual salary based on sales of S dollars.

(b) For what total sales amount will the college graduate earn in excess of \$150,000 annually?

87. Borrowing Money The amount of money that a lending institution will allow you to borrow mainly depends on the interest rate and your annual income. The equation $L = 2.98I - 76.11$ describes the amount of money, L, that a bank will lend at an interest rate of 7.5% for 30 years, based upon annual income, I. For what annual income, I, will a bank lend at least \$150,000? (*Source: Information Please Almanac*)

88. Advertising A marketing firm found that the equation $S = 2.1A + 224$ describes the amount of sales S of a product depends on A, the amount spent on advertising the product. Both S and A are measured in thousands of dollars. For what amount, A, is the sales of a product at least 350 thousand dollars?

75. $\{x \mid x < 25\}$; $(-\infty, 25)$

76. $\left\{z \mid z \geq -\frac{5}{3}\right\}$; $\left[-\frac{5}{3}, \infty\right)$

77. $\varnothing$ or $\{\ \}$

78. $\{a \mid a \text{ is any real number}\}$; $(-\infty, \infty)$

79. $\{x \mid x > 5.9375\}$; $(5.9375, \infty)$

80. $\{x \mid x > 6\}$; $(6, \infty)$

81. at most 1250 miles
82. at most 71 miles 83. 32
84. at least 186 points
85. more than 400 minutes
86. (a) $15{,}000 + 0.01S$ (b) \$13,500,000
87. greater than \$50,361.11
88. at least \$60,000 89. at least 74
90. 2 packages, 17 letters
91. $\{x \mid -33 < x < -14\}$
92. $\{x \mid -4 \leq x \leq 63\}$
93. $\{x \mid -4 \leq x \leq 6\}$
94. $\{x \mid -4.5 < x < 13.5\}$
95. $\{x \mid -2 \leq x < 9\}$ 96. $\{x \mid -1 < x \leq 7\}$
97. $\{x \mid -8 \leq x < 4\}$ 98. $\{x \mid 12 < x \leq 18\}$

Extending the Concepts

89. Grades In your Economics 101 class, you have scores of 68, 82, 87, and 89 on the first four of five tests. To earn a grade of B or higher, the average of the first five test scores must be greater than or equal to 80. Find the minimum score that you need on the last test to earn a B.

90. Delivery Service A messenger service charges \$10 to make a delivery to an address. In addition, each letter delivered costs \$3 and each package delivered costs \$8. If there are 15 more letters than packages delivered to this address, what is the maximum number of items that can be delivered for \$85?

Compound inequalities are solved using the properties of inequality. Study the example below and then solve Problems 91–98. State the solution in set-builder notation.

$$\begin{aligned} -15 &\leq 2x - 1 < 37 \\ -15 + 1 &\leq 2x - 1 + 1 < 37 + 1 \\ \frac{-14}{2} &\leq \frac{2x}{2} < \frac{38}{2} \\ -7 &\leq x < 19 \end{aligned}$$

91. $-3 < x + 30 < 16$

92. $-41 \leq x - 37 \leq 26$

93. $-6 \leq \frac{3x}{2} \leq 9$

94. $-4 < \frac{8x}{9} < 12$

95. $-7 \leq 2x - 3 < 15$

96. $4 < 3x + 7 \leq 28$

97. $4 < 6 - \frac{x}{2} \leq 10$

98. $-1 \leq 5 - \frac{x}{3} < 1$

CHAPTER 2 ACTIVITY: PASS TO THE RIGHT

1. (a) 6 (b) $\frac{-72}{7}$ (c) -12 (d) $\frac{-70}{19}$

6. (a) $(-27, \infty)$ (b) $\left[\frac{4}{9}, \infty\right)$ (c) $(-31, \infty)$ (d) $\left[\frac{5}{9}, \infty\right)$

Focus: Solving linear equations and inequalities as a group.

Time: 20–30 minutes

Group size: 3–4

1. Each member of the group should choose one of the following four equations and write it down on a piece of paper. Do not begin to solve.

(a) $\frac{x + 2}{2} - \frac{5x - 12}{6} = 1$ **(b)** $-\frac{1}{9}(x + 27) + \frac{1}{3}(x + 3) = x + 6$

(c) $\frac{x + 3}{3} - \frac{2x - 12}{9} = 1$ **(d)** $-\frac{1}{2}(x + 6) + \frac{1}{7}(x + 7) = x + 3$

2. Each member of the group should pass their equation to the person on their right. This member should perform the first step in solving the equation. When you are done, pass your paper to the next group member on your right.

3. Upon receipt of the equation, each group member should check the previous member's work and then perform the next step. If an error is found, discuss and correct the error.

4. Continue passing the problems until all equations have been solved.

5. As a group, discuss the results.

6. If time permits, repeat this activity except choose one of the following inequalities. Express your final answer using interval notation.

(a) $\frac{1}{4}(2x + 12) > \frac{3}{8}(x - 1)$ **(b)** $\frac{3x + 1}{10} - \frac{1 + 6x}{5} \leq -\frac{1}{2}$

(c) $\frac{1}{2}(2x + 14) > \frac{3}{4}(x - 1)$ **(d)** $\frac{3x + 1}{21} - \frac{1 + 4x}{7} \leq -\frac{1}{3}$

CHAPTER 2 REVIEW

Section 2.1 Linear Equations: The Addition and Multiplication Properties of Equality

KEY CONCEPTS	KEY TERMS
• **Linear Equation in One Variable** An equation equivalent to one of the form $ax + b = c$, where a, b, and c are real numbers and $a \neq 0$. • **Addition Property of Equality** For real numbers a, b, and c, if $a = b$, then $a + c = b + c$. • **Multiplication Property of Equality** For real numbers a, b, and c where $c \neq 0$, if $a = b$, then $ac = bc$.	Linear equation Sides Left Side Right Side Solution Solution set Satisfies Equivalent equations

YOU SHOULD BE ABLE TO . . .	EXAMPLE	REVIEW EXERCISES
(1) Determine if a number is a solution of an equation (p. 75)	Example 1	1–4
(2) Use the Addition Property of Equality to solve linear equations (p. 76)	Examples 2 through 4	5–10, 15, 16, 19
(3) Use the Multiplication Property of Equality to solve linear equations (p. 78)	Examples 5 through 8	11–14, 17, 18, 20

1. No
2. No
3. No
4. Yes
5. {16}
6. {20}
7. {−16}
8. {−7}
9. {−95}
10. {50}
11. {24}
12. {80}
13. {−6}
14. {5}
15. $\left\{-\frac{1}{3}\right\}$
16. $\left\{\frac{3}{8}\right\}$
17. {4}
18. {5}
19. \$20,100
20. \$2.55

In Problems 1–4, determine if the given value is the solution to the equation. Answer Yes or No.

1. $3x + 2 = 7; x = 5$

2. $5m - 1 = 17; m = 4$

3. $6x + 6 = 12; x = \frac{1}{2}$

4. $9k + 3 = 9; k = \frac{2}{3}$

In Problems 5–18, solve the equation. Check your solution.

5. $n - 6 = 10$

6. $n - 8 = 12$

7. $x + 6 = -10$

8. $x + 2 = -5$

9. $-100 = m - 5$

10. $-26 = m - 76$

11. $\frac{2}{3}y = 16$

12. $\frac{1}{4}x = 20$

13. $-6x = 36$

14. $-4x = -20$

15. $z + \frac{5}{6} = \frac{1}{2}$

16. $m - \frac{1}{8} = \frac{1}{4}$

17. $1.6x = 6.4$

18. $1.8m = 9$

19. Discount The cost of a new Honda Accord has been discounted by \$1200, so that the sale price is \$18,900. Find the original price of the Honda Accord, p, by solving the equation $p - 1200 = 18{,}900$.

20. Coffee While studying for an exam, Randi drank coffee at a local coffee house. She bought 3 cups of coffee for a total of \$7.65. Find the cost of each cup of coffee, c, by solving the equation $3c = 7.65$.

Section 2.2 Linear Equations: Using the Properties Together		
KEY CONCEPT		**KEY TERM**
• Steps for solving an equation in one variable page 88.		Distributive Property
YOU SHOULD BE ABLE TO . . .	**EXAMPLE**	**REVIEW EXERCISES**
1 Apply the Addition and Multiplication Properties of Equality to solve linear equations (p. 84)	Examples 1 and 2	21–24
2 Combine like terms and apply the Distributive Property to solve linear equations (p. 86)	Examples 3 and 4	25–30
3 Solve a linear equation with the variable on both sides of the equation (p. 87)	Examples 5 and 6	31–34
4 Use linear equations to solve problems (p. 89)	Example 7	35, 36

21. $\{-4\}$
22. $\{4\}$
23. $\{9\}$
24. $\{-21\}$
25. $\{-4\}$
26. $\{-6\}$
27. $\{4\}$
28. $\{-5\}$
29. $\{6\}$
30. $\{-6\}$
31. $\left\{\frac{4}{3}\right\}$
32. $\{5\}$
33. $\{2\}$
34. $\{7\}$
35. 14 years
36. width = 19 yards; length = 29 yards

In Problems 21–34, solve the equation. Check your solution.

21. $5x - 1 = -21$

22. $-3x + 7 = -5$

23. $\frac{2}{3}x + 5 = 11$

24. $\frac{5}{7}x - 2 = -17$

25. $-2x + 5 + 6x = -11$

26. $3x - 5x + 6 = 18$

27. $2m + 0.5m = 10$

28. $1.4m + m = -12$

29. $-2(x + 5) = -22$

30. $3(2x + 5) = -21$

31. $5x + 4 = -7x + 20$

32. $-3x + 5 = x - 15$

33. $4(x - 5) = -3x + 5x - 16$

34. $4(m + 1) = m + 5m - 10$

35. Ages Skye is 4 years older than Beth. The sum of their ages is 24. Find Skye's age by solving the equation $x + x + 4 = 24$, where x represents Beth's age and $x + 4$ represents Skye's age.

36. Parking Lot The length of a rectangular parking lot is 10 yards longer than the width, w, and the perimeter of the parking lot is 96 yards. Solve the equation $2w + 2(w + 10) = 96$ to find the width, w, of the parking lot. Then find the length of the parking lot, $w + 10$.

Section 2.3 Solving Linear Equations Involving Fractions and Decimals; Classifying Equations	
KEY CONCEPTS	**KEY TERMS**
• A **conditional equation** is an equation that is true for some values of the variable and false for other values of the variable. • An equation that is false for every replacement value of the variable is called a **contradiction.** • An equation that is satisfied for every choice of the variable for which both sides of the equation are defined is called an **identity.**	Least common denominator (LCD) Identity Conditional equation Contradiction

YOU SHOULD BE ABLE TO . . .	EXAMPLE	REVIEW EXERCISES
1 Use the least common denominator to solve a linear equation containing fractions (p. 93)	Examples 1 and 2	37–40, 45, 46
2 Solve a linear equation containing decimals (p. 95)	Examples 3 through 5	41–44, 47, 48
3 Classify linear equations as identity, conditional, or contradiction (p. 97)	Examples 6 through 8	49–54
4 Use linear equations to solve problems (p. 99)	Example 9	55, 56

37. $\left\{-\frac{35}{12}\right\}$
38. $\left\{-\frac{62}{3}\right\}$
39. $\{-2\}$
40. $\left\{\frac{6}{5}\right\}$
41. $\left\{\frac{3}{2}\right\}$
42. $\{3\}$
43. $\{4\}$
44. $\{-1\}$
45. $\left\{-\frac{7}{2}\right\}$
46. $\{-3\}$
47. $\{58\}$
48. $\{-10.5\}$
49. contradiction; ∅ or { }
50. contradiction; ∅ or { }
51. conditional equation; {0}
52. conditional equation; {0}
53. identity; The set of all real numbers
54. identity; The set of all real numbers
55. \$15.75
56. 3 dimes

In Problems 37–48, solve the equation. Check your solution.

37. $\frac{6}{7}x + 3 = \frac{1}{2}$

38. $\frac{1}{4}x + 6 = \frac{5}{6}$

39. $\frac{n}{2} + \frac{2}{3} = \frac{n}{6}$

40. $\frac{m}{8} + \frac{m}{2} = \frac{3}{4}$

41. $1.2r = -1 + 2.8$

42. $0.2x + 0.5x = 2.1$

43. $1.2m - 3.2 = 0.8m - 1.6$

44. $0.3m + 0.8 = 0.5m + 1$

45. $\frac{1}{2}(x + 5) = \frac{3}{4}$

46. $-\frac{1}{6}(x - 1) = \frac{2}{3}$

47. $0.1(x + 80) = -0.2 + 14$

48. $0.35(x + 6) = 0.45(x + 7)$

In Problems 49–54, determine if the equation is a contradiction, an identity, or a conditional equation. State the solution of the equation.

49. $4x + 2x - 10 = 6x + 5$

50. $-2(x + 5) = -5x + 3x + 2$

51. $-5(2n + 10) = 6n - 50$

52. $8m + 10 = -2(7m - 5)$

53. $10x - 2x + 18 = 2(4x + 9)$

54. $-3(2x - 8) = -3x - 3x + 24$

55. T-Shirt Al purchased a tie-dyed T-shirt at a "20% off" sale for \$12.60. What was the original price, p, of the shirt? Solve the equation $p - 0.20p = 12.60$.

56. Couch Cushions Juanita was cleaning under her couch cushions and found some nickels and dimes. She found \$0.55 in change, with the number of nickels one less than twice the number of dimes. Solve the equation $0.10x + 0.05(2x - 1) = 0.55$ to find x, the number of dimes Juanita found.

Section 2.4 Evaluating Formulas and Solving Formulas for a Variable

KEY CONCEPTS	KEY TERMS
• **Simple interest formula** $I = Prt$, I represents the amount of interest, P is the principal, r is the rate of interest, and t is time (expressed in years). • **Geometry Formulas (page 106)**	Formula Interest Principal Rate of interest

YOU SHOULD BE ABLE TO . . .	EXAMPLE	REVIEW EXERCISES
1 Evaluate a formula (p. 103)	Examples 1 through 6	57–60, 69, 70
2 Solve a formula for a variable (p. 108)	Examples 7 through 9	61–66, 67, 68

57. 48 in.^{2}
58. 64 cm
59. $\frac{3}{2}$ yards
60. 15 mm
61. $H = \frac{V}{LW}$
62. $P = \frac{I}{rt}$
63. $W = \frac{S - 2LH}{2L + 2H}$
64. $M = \frac{p - mv}{V}$
65. $y = \frac{-2x + 10}{3}$
66. $x = \frac{14 + 7y}{6}$
67. (a) $P = \frac{A}{(1 + r)^t}$
(b) $2238.65
68. (a) $h = \frac{A - 2\pi r^2}{2\pi r}$
(b) 5 cm
69. $11.25
70. $\frac{9}{4}\pi$ ft^2 $\approx$ 7.1 ft^2

In Problems 57–60, substitute the given values into the formula and then simplify to find the unknown quantity. Label units in the answer.

57. area of a rectangle: $A = lw$; Find A when $l = 8$ inches and $w = 6$ inches.

58. perimeter of a square: $P = 4s$; Find P when $s = 16$ cm.

59. perimeter of a rectangle: $P = 2l + 2w$; Find w when $P = 16$ yards and $l = \frac{13}{2}$ yards.

60. circumference of a circle: $C = \pi d$; Find C when $d = \frac{15}{\pi}$ mm.

In Problems 61–66, solve each formula for the stated variable.

61. $V = LWH$; solve for H.

62. $I = Prt$; solve for P.

63. $S = 2LW + 2LH + 2WH$; solve for W.

64. $\rho = mv + MV$; solve for M.

65. $2x + 3y = 10$; solve for y.

66. $6x - 7y = 14$; solve for x.

67. Finance The formula $A = P(1 + r)^t$ can be used to find the future value A of a deposit of P dollars in an account that earns an annual interest rate r (expressed as a decimal) after t years.

(a) Solve the formula for P.

(b) How much would you have to deposit today in order to have $3000 in 6 years in a bank account that pays 5% annual interest? Round your answer to the nearest penny.

68. Cylinders The surface area A of a right circular cylinder is given by the formula $A = 2\pi rh + 2\pi r^2$, where r is the radius and h is the height.

(a) Solve the formula for h.

(b) Determine the height of a right circular cylinder whose surface area is 72π square centimeters and whose radius is 4 centimeters.

69. Christmas Bonus Samuel invested his $500 Christmas bonus in a 9-month Certificate of Deposit that earns 3% simple interest. Use the formula $I = Prt$ to find the amount of interest Samuel's investment will earn.

70. Coffee Table Find the area of a circular coffee table top with a diameter of 3 feet.

Section 2.5 Introduction to Problem Solving: Direct Translation Problems

KEY CONCEPTS	KEY TERMS
• **Six steps for solving problems with mathematical models** **Identify** what you are looking for **Name** the unknown(s) **Translate** to a mathematical equation **Solve** the equation **Check** the answer **Answer** the question in a complete sentence	Problem solving Mathematical modeling Modeling process Mathematical model Model Direct translation

YOU SHOULD BE ABLE TO . . .	EXAMPLE	REVIEW EXERCISES
1 Translate English phrases to algebraic expressions (p. 117)	Examples 1 through 3	71–76; 83–86
2 Translate English sentences to equations (p. 120)	Example 4	77–82
3 Build models for solving direct translation problems (p. 123)	Examples 5 through 9	87–90

71. $x - 6$
72. $x - 8$
73. $-8x$
74. $\frac{x}{10}$
75. $2(6 + x)$
76. $4(5 - x)$
77. $6 + x = 2x + 5$
78. $6x - 10 = 2x + 1$
79. $x - 8 = \frac{1}{2}x$
80. $\frac{6}{x} = 10 + x$
81. $4(2x + 8) = 16$
82. $5(2x - 8) = -24$
83. Sarah's age: s; Jacob's age: $s + 7$
84. Consuelo's speed: c; José's speed: $2c$
85. Max's amount: m; Irene's amount: $m - 6$
86. Victor's amount: v; Larry's amount: $350 - v$
87. 153 pounds
88. 12, 13, 14
89. Juan: \$11,000; Roberto: \$9000
90. 100 miles

In Problems 71–76, translate each phrase to an algebraic expression. Let x represent the unknown number.

71. the difference between a number and 6

72. eight subtracted from a number

73. the product of -8 and a number

74. the quotient of a number and 10

75. twice the sum of 6 and a number

76. four times the difference of 5 and a number

In Problems 77–82, translate each statement into an equation. Let x represent the unknown number. DO NOT SOLVE.

77. The sum of 6 and a number is equal to twice the number increased by 5.

78. The product of 6 and a number decreased by 10 is one more than double the number.

79. Eight less than a number is the same as half of the number.

80. The ratio of 6 to a number is the same as the number added to 10.

81. Four times the sum of twice a number and 8 is 16.

82. Five times the difference between double a number and 8 is -24.

In Problems 83–86, choose a variable to represent one quantity. State what the quantity represents, and then express the second quantity in terms of the first.

83. Jacob is seven years older than Sarah.

84. José runs twice as fast as Consuelo.

85. Irene has \$6 less than Max.

86. Victor and Larry will share \$350.

87. Losing Weight Over the past year Lee Lai lost 28 pounds. Find Lee Lai's weight one year ago if her current weight is 125 pounds.

88. Consecutive Integers The sum of three consecutive integers is 39. Find the integers.

89. Finance A total of \$20,000 is to be divided between Roberto and Juan, with Roberto to receive \$2000 less than Juan. How much will each receive?

90. Truck Rentals You need to rent a moving truck. You identified two companies that rent trucks. ABC-Rental charges \$30 per day plus \$0.15 per mile. U-Do-It Rental charges \$15 per day plus \$0.30 per mile. For how many miles will the cost of renting be the same?

Section 2.6 Problem Solving: Direct Translation Problems Involving Percent

KEY CONCEPTS

- Percent means divided by 100 or per hundred.
- Total Cost = Original Cost + Sales Tax
- Original Price − Discount = Sale Price
- Wholesale Price + Markup = Selling Price

YOU SHOULD BE ABLE TO . . .	EXAMPLE	REVIEW EXERCISES
(1) Solve direct translation problems involving percent (p. 131)	Examples 1 through 4	91–94
(2) Model and solve direct translation problems from business involving percent (p. 134)	Examples 5 and 6	95–100

91. 5.2
92. 60
93. 13%
94. 50
95. \$18.50
96. \$30
97. \$40
98. \$200
99. \$125,000
100. winner: 500 votes; loser: 400 votes
101. 10°, 80°
102. 80°, 100°
103. 30°, 60°, 90°
104. 50°, 45°, 85°
105. length = 31 in.; width = 8 in.
106. length = 28 cm; width = 7 cm
107. (a) length = 20 ft; width = 40 ft (b) 800 ft^2
108. 50 ft; 40 ft
109. 5 hours 110. 65 mph

In Problems 91–94, find the unknown in each percent question.

91. What is 6.5% of 80?

92. 18 is 30% of what number?

93. 16 is what percent of 120?

94. 110% of what number is 55?

95. Sales Tax The sales tax in Florida is 6%. The total cost of purchasing a leotard, including tax, was \$19.61. Find the cost of the leotard before sales tax.

96. Tutoring Mei Ling is a private tutor. She raised her hourly fee by 8.5% to cover her traveling expenses. Her new hourly rate is \$32.55. Find Mei Ling's previous hourly fee.

97. Business: Discount Pricing A sweater, discounted by 70% for an end of the year clearance sale, has a price tag of \$12. What was the sweater's original price?

98. Business: Mark Up A clothing store marks up the price that it pays for a suit 80%. If the selling price of a suit is \$360, how much did the store pay for the suit?

99. Salary Tanya earns \$500 each week plus 2% of the value of the computers she sells each week. If Tanya wishes to earn \$3000 this week, what must be the total value of the computers she sells?

100. Voting In an election for school president, the loser received 80% of the winner's votes. If 900 votes were cast, how many did each receive?

Section 2.7 Problem Solving: Geometry and Uniform Motion

KEY CONCEPTS	KEY TERMS
• Complementary angles are two angles whose measures sum to 90°. • Supplementary angles are two angles whose measures sum to 180°. • The sum of the measures of the angles of a triangle is 180°. • **Uniform Motion** If an object moves at an average speed r, the distance d covered in time t is given by the formula $d = rt$.	Complementary angles Supplementary angles Uniform motion

YOU SHOULD BE ABLE TO . . .	EXAMPLE	REVIEW EXERCISES
(1) Set up and solve complementary and supplementary angle problems (p. 139)	Example 1	101, 102
(2) Set up and solve angles of a triangle problems (p. 140)	Example 2	103, 104
(3) Use geometry formulas to solve problems (p. 141)	Examples 3 and 4	105–108
(4) Set up and solve uniform motion problems (p. 143)	Examples 5 and 6	109, 110

△ **101. Complementary Angles** Find two complementary angles such that the measure of the first angle is 20° more than six times the second.

△ **102. Supplementary Angles** Find two supplementary angles such that the measure of the first angle is 60° less than twice the second.

△ **103. Triangles** In a triangle, the measure of the second angle is twice the first. The measure of the third angle is 30° more than the second. Find the measures of the three angles.

△ **104. Triangles** In a triangle, the measure of the second angle is 5° less than the first. The measure of the third angle is 5° less than twice the second. Find the measures of the three angles.

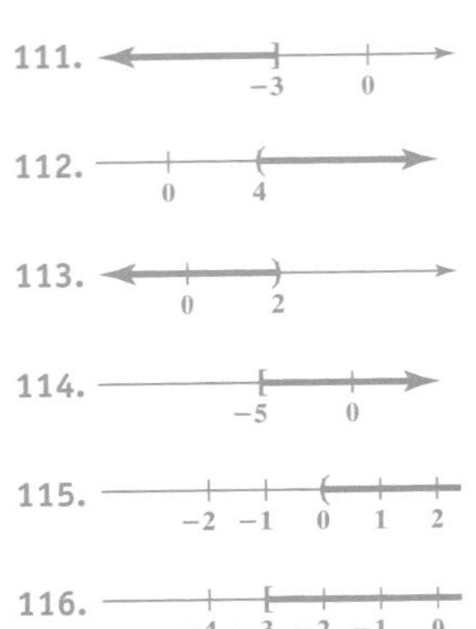

117. $(-\infty, -4)$
118. $[7, \infty)$
119. $[2, \infty)$
120. $(-\infty, 3)$
121. $\left\{x \mid x < -\frac{13}{2}\right\}$; $\left(-\infty, -\frac{13}{2}\right)$

$-\frac{13}{2}$ 0

△ **105. Rectangle** The length of a rectangle is 15 inches longer than twice its width. If the perimeter is 78 inches, find the length and width of the rectangle.

△ **106. Rectangle** The length of a rectangle is four times its width. If the perimeter is 70 cm, find the length and width of the rectangle.

△ **107. Garden** The perimeter of a rectangular garden is 120 feet. The width of the garden is twice the length.
(a) Find the length and width of the garden.
(b) What is the area of the garden?

△ **108. Back Yard** Yvonne's back yard is in the shape of a trapezoid with height of 80 feet. The shorter base is 10 feet shorter than the longer base, and the area of the back yard is 3600 square feet. Find the length of each base of the trapezoidal yard.

109. Boating Two motorboats leave the same dock at the same time traveling in the same direction. One boat travels at 18 miles per hour and the other travels at 25 miles per hour. In how many hours will the motorboats be 35 miles apart?

110. Train Station Two trains leave a station at the same time. One train is traveling east at 10 miles per hour faster than the other train, which is traveling west. After 6 hours, the two trains are 720 miles apart. At what speed did the faster train travel?

Section 2.8 Solving Linear Inequalities in One Variable

KEY CONCEPTS

- A linear inequality is of the form $ax + b < c$, $ax + b \leq c$, $ax + b > c$ or $ax + b \geq c$, where a, b, and c are real numbers and $a \neq 0$.
- **Addition Property of Inequalities**
  For real numbers a, b, and c
  If $a < b$, then $a + c < b + c$.
  If $a > b$, then $a + c > b + c$.
- **Multiplication Properties of Inequalities**
  For real numbers a, b, and c
  If $a < b$, and if $c > 0$, then $ac < bc$.
  If $a > b$, and if $c > 0$, then $ac > bc$.
  If $a < b$, and if $c < 0$, then $ac > bc$.
  If $a > b$, and if $c < 0$, then $ac < bc$.

Set-Builder Notation	Interval Notation	Graph
$\{x \mid x \geq a\}$	$[a, \infty)$	a
$\{x \mid x > a\}$	(a, ∞)	a
$\{x \mid x \leq a\}$	$(-\infty, a]$	a
$\{x \mid x < a\}$	$(-\infty, a)$	a
$\{x \mid x$ is a real number$\}$	$(-\infty, \infty)$	

KEY TERMS

Linear inequality
Simple inequality
Set-builder notation
Graph of an inequality
Endpoint
Interval
Solve an inequality
Equivalent inequalities

YOU SHOULD BE ABLE TO . . .	EXAMPLE	REVIEW EXERCISES
1 Graph inequalities on the real number line (p. 150)	Example 1	111–116
2 Use interval notation (p. 151)	Example 2	117–120
3 Solve linear inequalities using properties of inequalities (p. 153)	Examples 3 through 10	121–128
4 Model inequality problems (p. 159)	Example 11	129, 130

122. $\left\{x \mid x \geq -\frac{7}{3}\right\}$; $\left[-\frac{7}{3}, \infty\right)$

123. $\left\{x \mid x \geq -\frac{4}{5}\right\}$; $\left[-\frac{4}{5}, \infty\right)$

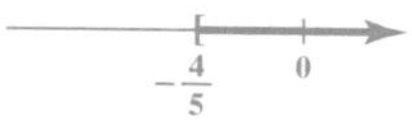

124. $\{x \mid x > -12\}$; $(-12, \infty)$

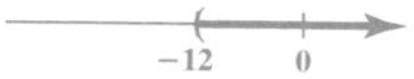

125. $\varnothing$ or { } 126. $\varnothing$ or { }

126. $\{x \mid x$ is any real number$\}$; $(-\infty, \infty)$

127. $\{x \mid x > 3\}$; $(3, \infty)$

128. $\left\{x \mid x < -\frac{38}{5}\right\}$; $\left(-\infty, -\frac{38}{5}\right)$

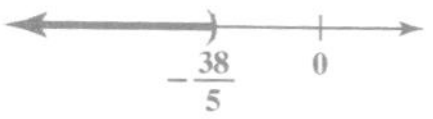

129. at most 65 miles

130. more than 150

In Problems 111–116, graph each inequality on a number line.

111. $x \leq -3$ **112.** $x > 4$ **113.** $m < 2$

114. $m \geq -5$ **115.** $0 < n$ **116.** $-3 \leq n$

In Problems 117–120, write the inequality in interval notation.

117. $x < -4$ **118.** $x \geq 7$

119. 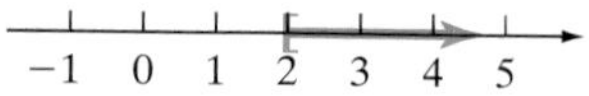

120. 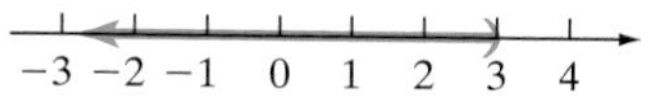

In Problems 121–128, solve the inequality and express the solution in set-builder notation or interval notation if possible. Graph the solution set on the real number line.

121. $4x + 3 < 2x - 10$

122. $3x - 5 \geq -12$

123. $-4(x - 1) \leq x + 8$

124. $6x - 10 < 7x + 2$

125. $-3(x + 7) > -x - 2x$

126. $4x + 10 \leq 2(2x + 7)$

127. $\frac{1}{2}(3x - 1) > \frac{2}{3}(x + 3)$

128. $\frac{5}{4}x + 2 < \frac{5}{6}x - \frac{7}{6}$

129. Moving The Rent-A-Moving-Van Company charges a flat rate of \$19.95 per day plus \$0.20 for each mile driven. How many miles can be driven if a customer can afford to spend at most only \$32.95?

130. Bowling Travis is bowling three games in a tournament. In the first game, his score was 148. In the second game, his score was 155. What score must Travis get in the third game for his tournament average to be greater than 151?

CHAPTER 2 TEST

Remember to use your Chapter Test Prep Video CD to see fully worked-out solutions to any of these problems you would like to review.

Note to Instructor: A special file in TestGen provides algorithms specifically matched to the problems in this Chapter Test for easy-to-replicate practice or assessment purposes.

1. $\{-17\}$
2. $\left\{-\frac{4}{9}\right\}$
3. $\{4\}$
4. $\left\{\frac{10}{13}\right\}$
5. $\left\{\frac{5}{8}\right\}$
6. $\{5\}$
7. $\varnothing$ or { }
8. all real numbers
9. (a) $l = \frac{V}{wh}$ (b) 9 in.
10. (a) $y = -\frac{2}{3}x + 4$ (b) $y = -\frac{4}{3}$
11. $6(x - 8) = 2x - 5$

In Problems 1–8, solve the equation. Check your solution.

1. $x + 3 = -14$

2. $-\frac{2}{3}m = \frac{8}{27}$

3. $5(2x - 4) = 5x$

4. $-2(x - 5) = 5(-3x + 4)$

5. $-\frac{2}{3}x + \frac{3}{4} = \frac{1}{3}$

6. $-0.6 + 0.4y = 1.4$

7. $8x + 3(2 - x) = 5(x + 2)$

8. $2(x + 7) = 2x - 2 + 16$

In Problems 9 and 10, (a) solve for the indicated variable, and then (b) find the value of the unknown quantity. Label units in the answer.

9. Volume of a rectangular solid: $V = lwh$

(a) Solve for l.

(b) Find l when $V = 540$ in.^{3}, $w = 6$ in., and $h = 10$ in.

10. Equation of a line: $2x + 3y = 12$

(a) Solve for y.

(b) Find y when $x = 8$.

11. Translate the following statement into an equation: Six times the difference between a number and 8 is equal to 5 less than twice the number. DO NOT SOLVE.

12. 60
13. 15, 16, 17
14. 10 in., 24 in., 26 in.
15. 3.5 hours
16. shorter piece is 5 feet; longer is 16 feet
17. $36
18. $\{x \mid x \leq 6\}$; $(-\infty, 6]$

0 6

19. $\left\{x \,\middle|\, x > \frac{5}{4}\right\}$; $\left(\frac{5}{4}, \infty\right)$

0 $\frac{5}{4}$

20. at most 200 minutes

12. 18 is 30% of a number. Find the number.

13. Consecutive Integers The sum of three consecutive integers is 48. Find the integers.

14. Trading Spaces On the show *Trading Spaces,* designer Vern Yip constructs a triangular art piece for a bedroom. The length of the longest side is two inches longer than the length of the middle side. The shortest side is 14 inches shorter than the middle side. If the perimeter of the art piece is 60 inches, what is the length of each side?

15. Buses Kimberly and Clay leave a concert hall at the same time traveling in buses going in opposite directions. Kimberly's bus travels at 40 mph and Clay's bus travels at 60 mph. In how many hours will Kimberly and Clay be 350 miles apart?

16. Construction A carpenter cuts an oak board 21 feet long into two pieces. The longer board is 1 foot longer than three times the shorter one. Find the length of each board.

17. New Backpack Sherry purchased a new backpack for her daughter. The backpack was on sale for 20% off the regular price. If Sherry paid $28.80 for the backpack without sales tax, what was the original price?

In Problems 18 and 19, solve the inequality and express the solution in set-builder notation or interval notation. Graph the solution set on the real number line.

18. $3(2x - 5) \leq x + 15$

19. $-6x - 4 < 2(x - 7)$

20. Cell Phone Danielle's cell phone plan has a $30 monthly fee and an extra charge of $0.35 a minute. For how many minutes can Danielle use her cell phone so that the monthly bill is at most $100?

CHAPTER

3 Exponents and Polynomials

Music producers follow trends in consumer taste. Did you know that the number of music videos sold in the United States reached its peak in 1998, and then began to decline? The polynomial $-0.425t^2 + 4.57t + 8.31$ describes the number of music videos sold, in millions, since 1993, where t represents the number of years since 1993. A recording company may want to use this model to decide how many music videos to produce in a given year. See Problem 108 in Section 3.1, p. 183.

OUTLINE

The Big Picture: Putting It Together

The first two chapters of the text were devoted to reviewing and developing fundamental skills with real numbers, algebraic expressions, and equations. We will use these skills often as we proceed through the course. In particular, we will describe how new topics and skills relate to those that we already know.

This chapter introduces polynomials. Polynomial expressions, such as the one given in the chapter opener, are used to describe many situations in business, consumer affairs, the physical and biological sciences, and even the entertainment industry. In this chapter, we will learn how to add, subtract, multiply, and divide polynomial expressions. You should pay attention to the techniques for performing these operations and how they are related to the techniques for adding, subtracting, multiplying, and dividing real numbers. Why? If we think of algebra as an extension of arithmetic skills, then comprehension and understanding of the material will come more quickly and easily.

3.1 Adding and Subtracting Polynomials

OBJECTIVES

1. Define Monomial and Determine the Degree of a Monomial
2. Define Polynomial and Determine the Degree of a Polynomial
3. Simplify Polynomials by Combining Like Terms
4. Evaluate Polynomials

Preparing for Adding and Subtracting Polynomials

Before getting started, take the following readiness quiz. If you get a problem wrong, go back to the section cited and review the material.

1. What is the coefficient of $-4x^5$? [Section 1.7, pp. 59–60]

2. Combine like terms: $-3x + 2 - 2x - 6x - 7$ [Section 1.7, pp. 61–62]

3. Use the Distributive Property to remove the parentheses: $-4(x - 3)$ [Section 1.7, p. 61]

4. Evaluate the expression $5 - 3x$ for $x = -2$. [Section 1.7, p. 58]

Recall from Section 1.7 that a term is a number or the product of a number and one or more variables raised to a power. The numerical factor of a term is the coefficient. A constant is a single number, such as 2 or $-\frac{3}{2}$. For example, consider Table 1, where some algebraic expressions are given. Note how the terms and coefficients of each algebraic expression are identified.

Table 1

Algebraic Expression	Terms	Coefficients
$3x + 2$	$3x$, 2	3, 2
$4x^2 - 7x + 5 = 4x^2 + (-7x) + 5$	$4x^2$, $-7x$, 5	4, -7, 5
$9x^2 + 7y^3$	$9x^2$, $7y^3$	9, 7

Notice in the algebraic expression $4x^2 - 7x + 5$ given in Table 1 that we first rewrite it as a sum, $4x^2 + (-7x) + 5$, to identify the terms and coefficients.

1 Define Monomial and Determine the Degree of a Monomial

In this chapter, we study *polynomials*. Polynomials are made up of terms that are **monomials.**

Work Smart

The prefix "mono" means "one." For example, a monorail has one rail. So a monomial in one variable is either a constant or a single term with a variable in it.

DEFINITION

A **monomial** in one variable is the product of a constant and a variable raised to a whole number (0, 1, 2, ...) power. A monomial in one variable is of the form

$$ax^k$$

where a is a constant, x is a variable, and k is a whole number.

In Words

The degree of a monomial can be thought of as the number of times the variable occurs. For example, because $2x^4 = 2 \cdot x \cdot x \cdot x \cdot x$, the degree of $2x^4$ is 4.

In a monomial of the form ax^k, where $a \neq 0$, we call k the **degree** of the monomial. The degree of a nonzero constant is zero. The number 0 has no degree.

Classroom Example

Determine the coefficient and degree of each monomial:

(a) $3y^5$ (b) $-\frac{3}{2}n^2$ (c) $-p$
(d) b^4 (e) 6

Answer:

(a) 3; 5 (b) $-\frac{3}{2}$; 2 (c) -1, 1
(d) 1; 4 (e) 6; 0

EXAMPLE 1 Monomials

	MONOMIAL	COEFFICIENT	DEGREE
(a)	$2x^4$	2	4
(b)	$-\frac{7}{4}x^2$	$-\frac{7}{4}$	2
(c)	$-x = -1 \cdot x$	-1	1
(d)	$x^3 = 1 \cdot x^3$	1	3
(e)	8	8	0

Now let's look at some expressions that are not monomials.

Preparing for...Answers **1.** -4 **2.** $-11x - 5$ **3.** $-4x + 12$ **4.** 11

Classroom Example Examples of expressions that are not monomials:

(a) $7a^{2/3}$ (b) $-3y^{-5}$

EXAMPLE 2 Examples of Expressions That Are Not Monomials

(a) $5x^{\frac{1}{2}}$ is not a monomial because the exponent of the variable x is $\frac{1}{2}$ and $\frac{1}{2}$ is not a whole number.

(b) $2x^{-4}$ is not a monomial because the exponent of the variable x is -4 and -4 is not a whole number.

QUICK ✓ *Determine whether the expression is a monomial. For those that are monomials, determine the coefficient and degree.*

1. $12x^6$ **2.** $3x^{-3}$ **3.** 10 **4.** $n^{\frac{1}{3}}$

So far, we have discussed only monomials in one variable. A monomial may contain more than one variable factor, such as $ax^m y^n$, where a is a constant (called the coefficient), x and y are variables, and m and n are whole numbers. An example of a monomial with more than one variable factor is $-4x^2y^3$. The **degree of the monomial** $ax^m y^n$ is the sum of the exponents, $m + n$.

Classroom Example State the degree and the coefficient of these monomials in more than one variable:

(a) $-2a^2b^5$ (b) $9xy^3$

Answer:

(a) degree $= 7$; coefficient is -2

(b) degree $= 4$; coefficient is 9

EXAMPLE 3 Monomials in More than One Variable

(a) $-4x^3y^4$ is a monomial in x and y of degree $3 + 4 = 7$. The coefficient is -4.

(b) $10ab^5$ is a monomial in a and b of degree $1 + 5 = 6$. The coefficient is 10.

QUICK ✓ *Determine whether the expression is a monomial. For those that are monomials, determine the coefficient and degree.*

5. $3x^5y^2$ **6.** $-2m^3n$ **7.** $4ab^{1/2}$ **8.** $-xy$

2 Define Polynomial and Determine the Degree of a Polynomial

We now turn our attention to polynomials.

DEFINITION

A **polynomial** is a monomial or the sum of monomials.

We give special names to certain polynomials. For example, a polynomial with exactly one term is a monomial; a polynomial that has two different monomials is called a **binomial;** and a polynomial that contains three different monomials is called a **trinomial.** So

Work Smart
The prefix "bi" means "two" as in bicycle. The prefix "tri" means "three" as in tricycle. The prefix "poly" means "many."

$-14x$ is a polynomial	but more specifically	$-14x$ is a monomial
$2x^3 - 5x$ is a polynomial	but more specifically	$2x^3 - 5x$ is a binomial
$-x^3 - 4x + 11$ is a polynomial	but more specifically	$-x^3 - 4x + 11$ is a trinomial
$3x^2 + 6xy - 2y^2$ is a polynomial	but more specifically	$3x^2 + 6xy - 2y^2$ is a trinomial

We use the term *monomial* to describe a polynomial with a *single* term, and the term *polynomial* to describe the sum of two or more monomials.

A polynomial is in **standard form** if it is written with the terms in descending order according to degree. The **degree of a polynomial** is the highest degree of all the terms of the polynomial. Remember, the degree of a nonzero constant is 0 and the number 0 has no degree.

Classroom Example ➤
Examples of polynomials and their degree:
(a) $5a^3 - 3a^2 + 8a - 1$; degree = 3
(b) $5 - 6x + x^2$; degree = 2
(c) $m^3n^2 - 2mn^3 + 3m^2n$; degree = 5
(d) $x^3y - 5x^4y^2 + 2$; degree = 6;
(e) 5; degree = 0
(f) 0; no degree

EXAMPLE 4 Examples of Polynomials

	POLYNOMIAL	DEGREE
(a)	$7x^3 - 2x^2 + 6x + 4$	3
(b)	$3 - 8x + x^2 = x^2 - 8x + 3$	2
(c)	$-7x^4 + 24$	4
(d)	$x^3y^4 - 3x^3y^2 + 2x^3y$	7
(e)	$p^2q - 8p^3q^2 + 3 = -8p^3q^2 + p^2q + 3$	5
(f)	6	0
(g)	0	No Degree

Although we have been using x to represent the variable, other letters may also be used.

$7t^4 + 14t^2 + 8$ is a polynomial (in t) of degree 4
$-5z^2 + 3z - 6$ is a polynomial (in z) of degree 2
$8y^3 - y^2 + 2y - 12$ is a polynomial (in y) of degree 3

Remember, polynomials consist of monomials. So, if any term in an algebraic expression is not a monomial, then the algebraic expression is not a polynomial.

Classroom Example ➤
Algebraic expressions that are not polynomials:
(a) $3x^{-2} - 6x + 2$
(b) $\frac{3}{x^2}$
(c) $5b^2 - 3b + 11b^{1/2}$

EXAMPLE 5 Algebraic Expressions That Are Not Polynomials

(a) $4x^{-2} - 5x + 1$ is not a polynomial because the exponent on the first term, -2, is not a whole number.

(b) $\frac{4}{x^3}$ is not a polynomial because there is a variable expression in the denominator of the fraction.

(c) $4z^2 + 9z - 3z^{1/2}$ is not a polynomial because the exponent on the third term, $-3z^{1/2}$, is not a whole number.

QUICK ✓ *Determine whether the algebraic expression is a polynomial. For those that are polynomials, determine the degree.*

9. $-4x^3 + 2x^2 - 5x + 3$ **10.** $2m^{-1} + 7$ **11.** $\frac{-1}{x^2 + 1}$

12. $5n^3 - \frac{3}{2}n^2 + 7n^5 - 6$ **13.** $5p^3q - 8pq^2 + pq$

3 Simplify Polynomials by Combining Like Terms

In Section 1.7, we learned how to combine like terms. To simplify a polynomial means to perform all indicated operations and combine like terms. One operation we perform on polynomials is addition. To add polynomials, combine the like terms of the polynomials.

Classroom Example ➤
Find the sum:
$(-4x^3 + 7x^2 + 5x - 3) + (x^3 - 8x^2 + 5)$
Answer: $-3x^3 - x^2 + 5x + 2$

EXAMPLE 6 Simplifying Polynomials: Addition

Find the sum: $(-5x^3 + 6x^2 + 2x - 7) + (3x^3 + 4x + 1)$

Solution

We can find the sum using either horizontal addition or vertical addition.

Horizontal Addition:

The idea here is to combine like terms.

$$(-5x^3 + 6x^2 + 2x - 7) + (3x^3 + 4x + 1) = -5x^3 + 6x^2 + 2x - 7 + 3x^3 + 4x + 1$$

Rearrange terms: $= -5x^3 + 3x^3 + 6x^2 + 2x + 4x - 7 + 1$

Use the Distributive Property: $= (-5 + 3)x^3 + 6x^2 + (2 + 4)x + (-7 + 1)$

Simplify: $= -2x^3 + 6x^2 + 6x - 6$

Work Smart
Remember, like terms have the same variable and the same exponent on the variable.

Vertical Addition:

The idea here is to line up like terms in each polynomial vertically and then add the coefficients.

$$\begin{array}{r} -5x^3 + 6x^2 + 2x - 7 \\ \underline{3x^3 \qquad\quad + 4x + 1} \\ -2x^3 + 6x^2 + 6x - 6 \end{array}$$

QUICK ✓ *Add the polynomials using either horizontal or vertical addition.*

14. $(9x^2 - x + 5) + (3x^2 + 4x - 2)$

15. $(4z^4 - 2z^3 + z - 5) + (-2z^4 + 6z^3 - z^2 + 4)$

Adding polynomials in two variables is handled the same way as adding polynomials in a single variable, that is, by adding like terms.

Classroom Example ➤
Find the sum:
$(5x^2y - 3xy + 12xy^2) + (x^2y + 12xy - 5xy^2)$
Answer: $6x^2y + 9xy + 7xy^2$

EXAMPLE 7 Simplifying Polynomials in Two Variables: Addition

Find the sum: $(6a^2b - 4ab + 11ab^2) + (a^2b + 9ab - 3ab^2)$

Solution

To save space, we only present the horizontal format.

$$(6a^2b - 4ab + 11ab^2) + (a^2b + 9ab - 3ab^2) = 6a^2b - 4ab + 11ab^2 + a^2b + 9ab - 3ab^2$$

Rearrange terms: $= 6a^2b + a^2b + 11ab^2 - 3ab^2 - 4ab + 9ab$

Use the Distributive Property: $= (6 + 1)a^2b + (11 - 3)ab^2 + (-4 + 9)ab$

Simplify: $= 7a^2b + 8ab^2 + 5ab$

QUICK ✓ *Simplify by adding the polynomials.*

16. $(7x^2y + x^2y^2 - 5xy^2) + (-2x^2y + 5x^2y^2 + 4xy^2)$

Work Smart
Taking the opposite of each term following the subtraction symbol is the same as multiplying each term of the polynomial by -1.

We can subtract polynomials using either the horizontal or vertical approach as well. However, the first step in subtracting polynomials requires that we remember the method for subtracting two real numbers. Remember,

$$a - b = a + (-b)$$

So, to subtract one polynomial from another, we add the opposite of each term in the polynomial following the subtraction sign and then combine like terms.

Classroom Example ➤
Find the difference:
$(7k^3 + 5k^2 - 6) - (-2k^3 + 8k^2 - 7k + 2)$
Answer: $9k^3 - 3k^2 + 7k - 8$

EXAMPLE 8 Simplifying Polynomials: Subtraction

Find the difference: $(6z^3 + 2z^2 - 5) - (-3z^3 + 9z^2 - z + 1)$

Solution

Horizontal Subtraction:

First, we write the problem as an addition problem and determine the opposite of each term following the subtraction sign.

$$(6z^3 + 2z^2 - 5) - (-3z^3 + 9z^2 - z + 1) = (6z^3 + 2z^2 - 5) + (3z^3 - 9z^2 + z - 1)$$

Remove parentheses: $= 6z^3 + 2z^2 - 5 + 3z^3 - 9z^2 + z - 1$

Rearrange terms: $= 6z^3 + 3z^3 + 2z^2 - 9z^2 + z - 5 - 1$

Combine like terms: $= 9z^3 - 7z^2 + z - 6$

Vertical Subtraction:

We line up like terms vertically,

$$\begin{array}{r} 6z^3 + 2z^2 \qquad - 5 \\ -(-3z^3 + 9z^2 - z + 1) \\ \hline \end{array}$$

then change the sign of each coefficient of the second polynomial, and add.

$$\begin{array}{r} 6z^3 + 2z^2 \qquad - 5 \\ +3z^3 - 9z^2 + z - 1 \\ \hline 9z^3 - 7z^2 + z - 6 \end{array}$$

QUICK ✓ *Simplify by subtracting the polynomials.*

17. $(7x^3 - 3x^2 + 2x + 8) - (2x^3 + 12x^2 - x + 1)$

18. $(6y^3 - 3y^2 + 2y + 4) - (-2y^3 + 5y + 10)$

Classroom Example ➤
Find the difference:
$(2a^2b + ab - 7ab^2) - (5a^2b - 4ab + 6ab^2)$
Answer: $-3a^2b + 5ab - 13ab^2$

EXAMPLE 9 Simplifying Polynomials in Two Variables: Subtraction

Find the difference: $(2p^2q + pq + 5pq^2) - (3p^2q - 6pq + 9pq^2)$

Solution

$$(2p^2q + pq + 5pq^2) - (3p^2q - 6pq + 9pq^2) = (2p^2q + pq + 5pq^2) + (-3p^2q + 6pq - 9pq^2)$$

Remove parentheses: $= 2p^2q + pq + 5pq^2 - 3p^2q + 6pq - 9pq^2$

Rearrange terms: $= 2p^2q - 3p^2q + 5pq^2 - 9pq^2 + pq + 6pq$

Simplify: $= -p^2q - 4pq^2 + 7pq$

QUICK ✓ *Subtract the polynomials.*

19. $(9x^2y + 6x^2y^2 - 3xy^2) - (-2x^2y + 4x^2y^2 + 6xy^2)$

Classroom Example ➤
Perform the indicated operations:
$(5x^2 + 2xy - y^2) - (9x^2 - 2y^2) + (7x^2 + 5xy)$
Answer: $3x^2 + 7xy + y^2$

EXAMPLE 10 Simplifying Polynomials

Perform the indicated operations: $(3x^2 - 2xy + y^2) - (7x^2 - y^2) + (3xy - 5y^2)$

Solution

$$(3x^2 - 2xy + y^2) - (7x^2 - y^2) + (3xy - 5y^2) = 3x^2 - 2xy + y^2 - 7x^2 + y^2 + 3xy - 5y^2$$

Rearrange terms: $= 3x^2 - 7x^2 - 2xy + 3xy + y^2 + y^2 - 5y^2$

Simplify: $= -4x^2 + xy - 3y^2$

QUICK ✓ *Perform the indicated operations.*

20. $(3a^2 - 5ab + 3b^2) + (4a^2 - 7b^2) - (8b^2 - ab)$

---

4 Evaluate Polynomials

To **evaluate** a polynomial, we substitute the given number for the value of the variable and simplify, just as we did in Section 1.7.

Classroom Example ➤
Evaluate the polynomial $3x^3 - 2x^2 + x - 1$ for (a) $x = 2$; (b) $x = -4$
Answer:
(a) 17 (b) -229

EXAMPLE 11 Evaluating a Polynomial

Evaluate the polynomial $2x^3 - 5x^2 + x - 3$ for

(a) $x = 2$ **(b)** $x = -3$

Work Smart
In each example apply the exponent before you multiply.

Solution

(a) $2x^3 - 5x^2 + x - 3 = 2(2)^3 - 5(2)^2 + 2 - 3$
$= 16 - 20 + 2 - 3$
$= -5$

(b) $2x^3 - 5x^2 + x - 3 = 2(-3)^3 - 5(-3)^2 + (-3) - 3$
$= -54 - 45 - 3 - 3$
$= -105$

QUICK ✓

21. Evaluate the polynomial $-2x^3 + 7x + 1$ for
(a) $x = 0$ **(b)** $x = 5$ **(c)** $x = -4$

Classroom Example ➤
Evaluate the polynomial $3x^2y + 2x^2y^2 - 5xy$ for $x = -2$ and $y = -1$.
Answer: -14

EXAMPLE 12 Evaluating a Polynomial in Two Variables

Evaluate the polynomial $2a^2b + 3a^2b^2 - 4ab$ for $a = -3$ and $b = -1$.

Solution

Let $a = -3$ and $b = -1$ in $2a^2b + 3a^2b^2 - 4ab$.

$$\begin{aligned} 2a^2b + 3a^2b^2 - 4ab &= 2(-3)^2(-1) + 3(-3)^2(-1)^2 - 4(-3)(-1) \\ &= 2(9)(-1) + 3(9)(1) - 4(-3)(-1) \\ &= -18 + 27 - 12 \\ &= -3 \end{aligned}$$

QUICK ✓

22. Evaluate the polynomial $-2m^2n + 3mn - n^2$ for
(a) $m = -2$ and $n = -4$ **(b)** $m = 1$ and $n = 5$

Classroom Example ➤
Use Example 13 with $x = 250$.
Answer: 3750; The revenue from selling 250 clocks per month is \$3750.

EXAMPLE 13 How Much Revenue?

The monthly revenue (in dollars) from selling x clocks is given by the polynomial $-0.3x^2 + 90x$. Evaluate the polynomial for $x = 200$ and describe the result in practical terms.

Solution

The variable x represents the number of clocks sold in a month. When we evaluate the polynomial $-0.3x^2 + 90x$ for $x = 200$, we are finding the revenue (in dollars) when 200 clocks are sold in a month. To find the revenue we replace x by 200 in the expression $-0.3x^2 + 90x$.

$$\begin{aligned} -0.3x^2 + 90x &= -0.3(200)^2 + 90(200) \\ &= -0.3(40{,}000) + 18{,}000 \\ &= -12{,}000 + 18{,}000 \\ &= 6000 \end{aligned}$$

According to the model, the revenue from selling 200 clocks in a month is \$6000.

QUICK ✓ *Evaluate the polynomial for the given value.*

23. The polynomial $40x - 0.2x^2$ represents the monthly revenue (in dollars) achieved by selling x wristwatches. Find the monthly revenue for selling 75 wristwatches by evaluating the polynomial $40x - 0.2x^2$ for $x = 75$.

3.1 Exercises

For Extra Help:  Student Solutions Manual  CD Video  PH Math/Tutor Center  MathXL Tutorials on CD · MathXL® · MyMathLab

Concepts and Vocabulary

In Problems 1–3, fill in the blanks.

1. A __________ in one variable is the product of a number and a variable raised to a whole number power.

2. The implied coefficient of a term such as x^2 or z is __________.

3. The degree of the polynomial $3x^5 - 7x + 1$ is __________.

In Problems 4–6, answer True or False to each statement.

4. $3x + 4 + 6x$ is an example of a trinomial.

5. The degree of the polynomial $3ab^2 - 4a^2b^3 + \frac{6}{5}a^2b^2$ is 5.

6. The degree of the polynomial $-7y^3 + 2y^2 - \frac{1}{2}$ is 5.

7. When adding polynomials you may use either a horizontal or vertical format. Explain which format you prefer and what you view as its advantages.

8. Use the two polynomials $3x - 5$ and $2x + 3$ to make up three different problems. In the directions to the first problem use the word "evaluate"; in the second problem use the direction "solve"; and in the third use "simplify."

Building Skills

In Problems 9–20, determine whether the given expression is a monomial (Yes or No). For those that are monomials, state the coefficient and degree.

9. $\frac{1}{2}y^3$ **10.** $-3x^2$ **11.** $\frac{x^2}{7}$ **12.** $4m^{101}$

13. z^{-6} **14.** $\frac{1}{y^7}$ **15.** $12mn^4$ **16.** $-x^6y$

17. $\frac{3}{n^2}$ **18.** y **19.** 4 **20.** $\frac{2}{3}$

In Problems 21–24, write the polynomial in standard form.

21. $4x - 5 + x^3$ **22.** $6 - 2m - m^2$

23. $-3y + 2 - y^4 + y^2$ **24.** $2t^5 + 6t - t^3$

In Problems 25–44, determine whether the algebraic expression is a polynomial (Yes or No). If it is a polynomial, write the polynomial in standard form, determine its degree, and state if it is a monomial, binomial or trinomial. If it is a polynomial with more than 3 terms, identify the expression as polynomial.

25. $6x^2 - 10$ **26.** $4x + 1$ **27.** $\frac{-20}{n}$ **28.** $\frac{1}{x}$

29. $3y^{1/3} + 2$ **30.** $8m - 4m^{1/2}$ **31.** $\frac{1}{8}$ **32.** 32

33. $3t^2 - \frac{1}{2}t^4 + 6t$ **34.** $4x^7 - 3x^3 - 1$ **35.** $7x^{-1} + 4$ **36.** $4y^{-2} + 6y - 1$

37. $5z^3 - 10z^2 + z + 12$ **38.** $p^5 - 3p^4 + 7p + 8$

39. $3x^2y^2 + 2xy^4 + 4$ **40.** $4mn^3 - 2m^2n^3 + mn^8$

41. $4x^3 - 3y^5$ **42.** $-6s^{10} + 4t^3$

43. $4pqr + 2p^2q + 3pq^{1/4}$ **44.** $-2xyz^2 + 7x^3z - 8y^{1/2}z$

1. monomial
2. 1
3. 5
4. False
5. True
6. False
7. Answers may vary.
8. Answers may vary.
9. Yes; coefficient $\frac{1}{2}$; degree 3
10. Yes; coefficient -3; degree 2
11. Yes; coefficient $\frac{1}{7}$; degree 2
12. Yes; coefficient 4; degree 101
13. No
14. No
15. Yes; coefficient 12; degree 5
16. Yes; coefficient -1; degree 7
17. No
18. Yes; coefficient 1; degree 1
19. Yes; coefficient 4; degree 0
20. Yes, coefficient $\frac{2}{3}$; degree 0
21. $x^3 + 4x - 5$
22. $-m^2 - 2m + 6$
23. $-y^4 + y^2 - 3y + 2$
24. $2t^5 - t^3 + 6t$
25. Yes; $6x^2 - 10$; degree 2; binomial
26. Yes; $4x + 1$; degree 1; binomial
27. No
28. No
29. No
30. No
31. Yes; $\frac{1}{8}$; degree 0; monomial
32. Yes; 32; degree 0; monomial
33. Yes; $-\frac{1}{2}t^4 + 3t^2 + 6t$; degree 4; trinomial
34. Yes; $4x^7 - 3x^3 - 1$; degree 7; trinomial
35. No
36. No
37. Yes; $5z^3 - 10z^2 + z + 12$; degree 3; polynomial
38. Yes; $p^5 - 3p^4 + 7p + 8$; degree 5; polynomial
39. Yes; $2xy^4 + 3x^2y^2 + 4$; degree 5; trinomial
40. Yes; $mn^8 - 2m^2n^3 + 4mn^3$; degree 9; trinomial
41. Yes; $-3y^5 + 4x^3$; degree 5; binomial
42. Yes; $-6s^{10} + 4t^3$; degree 10; binomial
43. No
44. No

In Problems 45–58, add the polynomials. Express your answer in standard form.

45. $(4x - 3) + (3x - 7)$

46. $(-13z + 4) + (9z - 10)$

47. $(-4m^2 + 2m - 1) + (2m^2 - 2m + 6)$

48. $(x^2 - 2) + (6x^2 - x - 1)$

49. $(p - p^3 + 2) + (6 - 2p^2 + p^3)$

50. $(4r^4 + 3r - 1) + (2r^4 - 7r^3 + r^2 - 10r)$

51. $(2y - 10) + (-3y^2 - 4y + 6)$

52. $(3 - 12w^2) + (2w^2 - 5 + 6w)$

53. $\left(\frac{1}{2}p^2 - \frac{2}{3}p + 2\right) + \left(\frac{3}{4}p^2 + \frac{5}{6}p - 5\right)$

54. $\left(\frac{3}{8}b^2 - \frac{3}{5}b + 1\right) + \left(\frac{5}{6}b^2 + \frac{2}{15}b - 1\right)$

55. $(5m^2 - 6mn + 2n^2) + (m^2 + 2mn - 3n^2)$

56. $(4a^2 + ab - 9b^2) + (-6a^2 - 4ab + b^2)$

57. $\begin{array}{r} 4n^2 - 2n + 1 \\ \underline{+(-6n + 4)} \end{array}$

58. $\begin{array}{r} 8x^3 \qquad + 2x + 2 \\ \underline{+(-3x^2 - 4x - 2)} \end{array}$

In Problems 59–72, subtract the polynomials. Express your answer in standard form.

59. $(3x - 10) - (4x + 6)$

60. $(-7t + 3) - (-2t - 4)$

61. $(12x^2 - 2x - 4) - (-2x^2 + x + 1)$

62. $(3x^2 + x - 3) - (x^2 - 2x + 4)$

63. $(y^3 - 2y + 1) - (-3y^3 + y + 5)$

64. $(m^4 - 3m^2 + 5) - (3m^4 - 5m^2 - 2)$

65. $(3y^3 - 2y) - (2y + y^2 + y^3)$

66. $(2x - 4x^3) - (-3 - 2x + x^3)$

67. $\left(\frac{5}{3}q^2 - \frac{5}{2}q + 4\right) - \left(\frac{1}{9}q^2 + \frac{3}{8}q + 2\right)$

68. $\left(\frac{7}{4}x^2 - \frac{5}{8}x - 1\right) - \left(\frac{7}{6}x^2 + \frac{5}{12}x + 5\right)$

69. $(-4m^2n^2 - 2mn + 3) - (4m^2n^2 + 2mn + 10)$

70. $(4m^2n - 2mn - 4) - (10m^2n - 6mn - 3)$

71. $\begin{array}{r} 6x - 3 \\ \underline{-(10x + 2)} \end{array}$

72. $\begin{array}{r} 6n^2 - 2n - 3 \\ \underline{-(4n + 7)} \end{array}$

In Problems 73–80, evaluate the polynomial for each of the given value(s).

73. $2x^2 - x + 3$
(a) $x = 0$
(b) $x = 5$
(c) $x = -2$

74. $-x^2 + 10$
(a) $x = 0$
(b) $x = -1$
(c) $x = 1$

75. $7 - x^2$
(a) $x = 3$
(b) $x = -\frac{5}{2}$
(c) $x = -1.5$

76. $2 + \frac{1}{2}n^2$
(a) $n = 4$
(b) $n = 0.5$
(c) $n = -\frac{1}{4}$

45. $7x - 10$
46. $-4z - 6$
47. $-2m^2 + 5$
48. $7x^2 - x - 3$
49. $-2p^2 + p + 8$
50. $6r^4 - 7r^3 + r^2 - 7r - 1$
51. $-3y^2 - 2y - 4$
52. $-10w^2 + 6w - 2$
53. $\frac{5}{4}p^2 + \frac{1}{6}p - 3$
54. $\frac{29}{24}b^2 - \frac{7}{15}b$
55. $6m^2 - 4mn - n^2$
56. $-2a^2 - 3ab - 8b^2$
57. $4n^2 - 8n + 5$
58. $8x^3 - 3x^2 - 2x$
59. $-x - 16$
60. $-5t + 7$
61. $14x^2 - 3x - 5$
62. $2x^2 + 3x - 7$
63. $4y^3 - 3y - 4$
64. $-2m^4 + 2m^2 + 7$
65. $2y^3 - y^2 - 4y$
66. $-5x^3 + 4x + 3$
67. $\frac{14}{9}q^2 - \frac{23}{8}q + 2$
68. $\frac{7}{12}x^2 - \frac{25}{24}x - 6$
69. $-8m^2n^2 - 4mn - 7$
70. $-6m^2n + 4mn - 1$
71. $-4x - 5$
72. $6n^2 - 6n - 10$
73. (a) 3 (b) 48 (c) 13
74. (a) 10 (b) 9 (c) 9
75. (a) -2 (b) $\frac{3}{4}$ (c) 4.75
76. (a) 10 (b) 2.125 (c) $\frac{65}{32}$

77. 45
78. 4
79. -328
80. $-\frac{3}{2}$
81. $3t - 3$
82. $-12x^2 - 2$
83. $3x^2 - 3x - 3$
84. $7m^2 - 5m - 4$
85. $9xy^2 + 1$
86. $-13xy - 7$
87. $-y^2 + 15y + 1$
88. $-11z^2 + z + 11$
89. $\frac{7}{3}q^2 + \frac{5}{3}$
90. $\frac{9}{10}t^2 - \frac{4}{15}t - 3$
91. $13d^2 + d + 10$
92. $-2x - 5$
93. $6a^2 + 5a - 3$
94. $2b^2 + 7b - 5$
95. $-2x^2 - 2xy + y^2$
96. $10st^4 + 4s^2t^2 - 2t^4$
97. $-5x + 12$
98. $-x^2 - 4x + 4$
99. $3x^2 + 4x - 3$
100. $-4x + 4$
101. $-14x^2 + 4x - 13$
102. $3n^2 + 12n - 1$
103. $-3x + 5$
104. $-a - 8b$
105. 26 ft
106. 80 ft

77. $-x^2y + 2xy^2 - 3$ for $x = 2$ and $y = -3$

78. $-2ab^2 - 2a^2b - b^3$ for $a = 1$ and $b = -2$

79. $st + 2s^2t + 3st^2 - t^4$ for $s = -2$ and $t = 4$

80. $m^2n^2 - mn^2 + 3m^2 - 2$ for $m = \frac{1}{2}$ and $n = -1$

Mixed Practice

In Problems 81–96, perform the indicated operations. Express your answer in standard form.

81. $(7t - 3) - 4t$

82. $6x^2 - (18x^2 + 2)$

83. $(5x^2 + x - 4) + (-2x^2 - 4x + 1)$

84. $(4m^2 - m + 6) + (3m^2 - 4m - 10)$

85. $(2xy^2 - 3) + (7xy^2 + 4)$

86. $(-14xy + 3) - (-xy + 10)$

87. $(4 + 8y - 2y^2) - (3 - 7y - y^2)$

88. $(9 + 2z - 6z^2) - (-2 + z + 5z^2)$

89. $\left(\frac{5}{6}q^2 - \frac{1}{3}\right) + \left(\frac{3}{2}q^2 + 2\right)$

90. $\left(\frac{7}{10}t^2 - \frac{5}{12}t\right) + \left(\frac{3}{15}t^2 + \frac{3}{20}t - 3\right)$

91. $14d^2 - (2d - 10) - (d^2 - 3d)$

92. $3x - (5x + 1) - 4$

93. $(4a^2 - 1) + (a^2 + 5a + 2) - (-a^2 + 4)$

94. $(2b^2 + 3b - 5) - (b^2 - 4b + 1) + (b^2 + 1)$

95. $(x^2 - 2xy - y^2) - (3x^2 + xy - y^2) + (xy + y^2)$

96. $(2st^4 - 3s^2t^2 + t^4) + (7s^2t^2 - 3t^4 + 8st^4)$

97. Find the sum of $3x + 10$ and $-8x + 2$.

98. Find the sum of $4x^2 - 2x - 3$ and $-5x^2 - 2x + 7$.

99. Find the difference of $-x^2 + 2x + 3$ and $-4x^2 - 2x + 6$.

100. Find the difference of $4x - 3$ and $8x - 7$.

101. Subtract $14x^2 - 2x + 3$ from $2x - 10$.

102. Subtract $-4n - 8$ from $3n^2 + 8n - 9$.

103. What polynomial should be added to $3x - 5$ so that the sum is zero?

104. What polynomial should be added to $2a + 7b$ so that the sum is $a - b$?

Applying the Concepts

105. Height of a Ball The height above ground (in feet) of a ball dropped from the top of a 30-foot tall building after t seconds is given by the polynomial $-16t^2 + 30$. What is the height of the ball after $\frac{1}{2}$ second?

106. Height of a Ball The height above ground (in feet) of a ball tossed upward after t seconds is given by the polynomial $-16t^2 + 32t + 100$. What is the height of the ball after 2.5 seconds?

107. (a) \$22,579.65 (b) \$43,793.85
108. 1.95 million
109. (a) $-2.125x^2 + 105x$ (b) \$1250
110. (a) $-0.001x^2 + 5.2x - 3000$ (b) \$2760
111. (a) $10x$ (b) \$500 (c) \$26,000

107. Income The bar graph shown represents the average per-capita income (in dollars) for residents of the United States by age in 2000. The polynomial $-50.27a^2 + 4426.16a - 40803.75$ can be used to approximate the average income I, where a is the age of the individual.

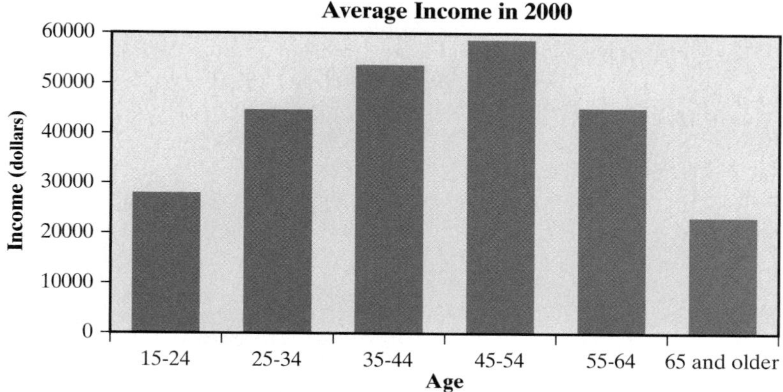

(a) Use the polynomial to estimate the average income of an 18-year-old in 2000.
(b) Use the polynomial to estimate the average income of a 60-year-old in 2000.

108. Selling Music Videos The polynomial $-0.425t^2 + 4.57t + 8.31$ describes the number (in millions) of music videos sold since 1993, where t represents the number of years since 1993. Use the polynomial to estimate the number of music videos sold in 2005.

109. Manufacturing Calculators The revenue (in dollars) from manufacturing and selling x calculators in a day is given by the polynomial $-2x^2 + 120x$. The cost of manufacturing and selling x calculators in a day is given by the polynomial $0.125x^2 + 15x$.

(a) In business, profit is equal to revenue minus costs. Write a polynomial that represents the profit from manufacturing and selling x calculators in a day.
(b) Find the profit if 20 calculators are produced and sold each day.

110. Manufacturing DVDs The revenue (in dollars) from manufacturing and selling x DVDs each day is given by the polynomial $-0.001x^2 + 6x$. The cost of manufacturing and selling x DVDs each day is given by the polynomial $0.8x + 3000$.

(a) Write a polynomial that expresses the profit from manufacturing and selling x DVDs each day.
(b) Find the profit if 1600 DVDs are produced and sold in a day.

111. Lawn Service Marissa started a lawn-mowing business in Tampa, Florida, in which she has x lawns on her route. Her costs are \$5 per lawn for fertilizer and \$10 per lawn for labor.

(a) If she charges \$25 per lawn, write a polynomial that represents her weekly profit.
(b) If Marissa mows 50 lawns in a week, what is her weekly profit?
(c) If Marissa mows 50 lawns per week, what is her annual profit?

112. Skateboard Business Kyle owns a skateboard shop. Kyle charges \$40 for each skateboard he sells. The cost for manufacturing each skateboard is \$12. In addition, Kyle has weekly costs of \$504 regardless of how many skateboards he manufactures.

112. (a) $28x - 504$ (b) \$756 (c) 18 skateboards
113. $16x - 12$
114. $4x^2 + 8x - 10$
115. $7x + 18$
116. $4x^2 - 8 + 2\pi$
117. $2x - 5$
118. $6x^2 + 12x + 4$
119. $-x^2 - x + 15$
120. $3y^2 + 13y - 1$
121. $2p^2 + 10p + 16$
122. $-2n^2$
123. $y^3 - 2y^2 + 11y - 14$
124. $-y^2 + 7y - 13$

(a) If Kyle manufactures and sells x skateboards in a week, write a polynomial that represents his weekly profit.
(b) What is Kyle's profit if he sells 45 skateboards per week?
(c) How many skateboards does Kyle need to sell in order to break even (profit = \$0)?

△ *In Problems 113–116, write the polynomial that represents the perimeter of each figure.*

113.

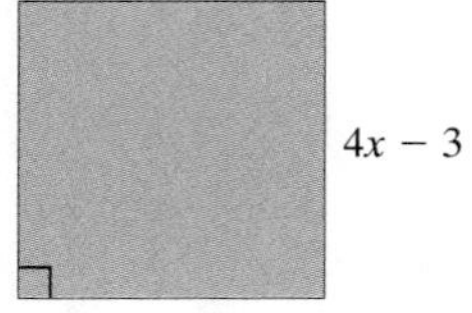

114.

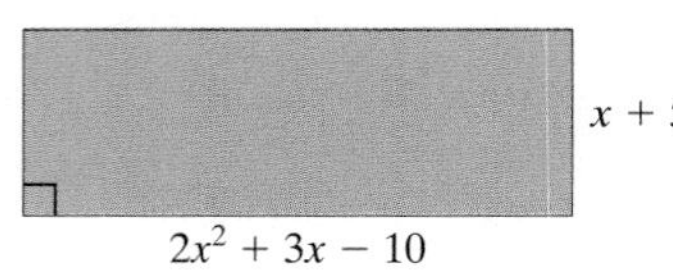

115.

10
2x + 1
3x − 4

116.

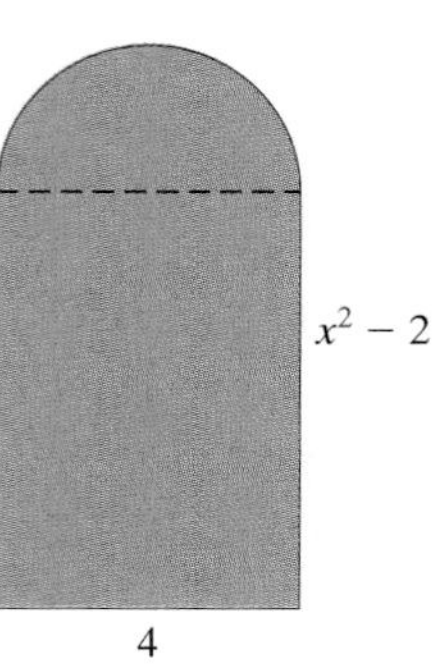

In Problems 117 and 118, write the polynomial that represents the unknown length.

117. 3x − 10; x − 5; ?

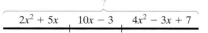

118. ?; 2x² + 5x; 10x − 3; 4x² − 3x + 7

Extending the Concepts

In Problems 119–124, simplify each of the following.

119. $(2x^2 - 3x + 1) + (x^2 + 9) - (4x^2 - 2x - 5)$
120. $(y^2 + 6y - 2) + (4y^2 - 9) - (2y^2 - 7y - 10)$
121. $(p^2 + 25) - (3p^2 - p + 4) - (-4p^2 - 9p + 5)$
122. $(n^2 + 4n - 5) - (2n^2 - 3n + 1) + (-n^2 - 7n + 6)$
123. $(4y - 3) + (y^3 - 8) - (2y^2 - 7y + 3)$
124. $(y^3 - 1) + (y^2 - 9) - (y^3 + 2y^2 - 7y + 3)$

3.2 Multiplying Monomials: The Product and Power Rules

OBJECTIVES

1. Simplify Exponential Expressions Using the Product Rule
2. Simplify Exponential Expressions Using the Power Rule
3. Simplify Exponential Expressions Containing Products
4. Multiply a Monomial by a Monomial

Preparing for...Answers **1.** 64 **2.** $12 - 15x$

Preparing for Multiplying Monomials: The Product and Power Rules

Before getting started, take the following readiness quiz. If you get a problem wrong, go back to the section cited and review the material.

1. Evaluate: 4^3 [Section 1.6, pp. 49–50]
2. Use the Distributive Property to simplify: $3(4 - 5x)$ [Section 1.7, p. 61]

In Section 1.6, we gave a definition for raising a real number to a natural number exponent. Recall that if a is a real number and n is a natural number, then the symbol a^n means that we should use a as a factor n times:

$$a^n = \underbrace{a \cdot a \cdot \ldots \cdot a}_{n \text{ factors}}$$

(exponent: n; base: a)

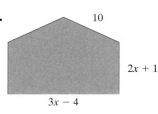

Work Smart
The natural numbers are 1, 2, 3, ...

For example,

$$4^3 = \underbrace{4 \cdot 4 \cdot 4}_{\text{3 factors}}$$

or

$$y^4 = \underbrace{y \cdot y \cdot y \cdot y}_{\text{4 factors}}$$

In the notation a^n we call a the base and n the power or exponent. We read a^n as "a raised to the power n" or "a raised to the nth power." We usually read a^2 as "a squared" and a^3 as "a cubed." The expression 4^3 is called **exponential form** and the expression $4 \cdot 4 \cdot 4$ is called **expanded form.**

1 Simplify Exponential Expressions Using the Product Rule

We can discover several general rules for simplifying expressions with natural number exponents. The first rule that we introduce is used when multiplying two exponential expressions that have the same base. Consider the following:

$$\underset{\text{Same base}}{x^2 \cdot x^4} = \underbrace{(x \cdot x)}_{\text{2 factors}}\underbrace{(x \cdot x \cdot x \cdot x)}_{\text{4 factors}} = \underbrace{x \cdot x \cdot x \cdot x \cdot x \cdot x}_{\text{6 factors}} = \underset{\text{Same base}}{x^{\overset{\text{Sum of powers 2 and 4}}{6}}}$$

The following rule generalizes the result shown above.

In Words
The product of two exponential expressions with the same base is that base with the sum of the exponents as the power.

PRODUCT RULE FOR EXPONENTS

If a is a real number, and m and n are natural numbers, then

$$a^m \cdot a^n = a^{m+n}$$

Classroom Example Evaluate each expression:
(a) $3^2 \cdot 3^3$
(b) $(-2)^3(-2)$
Answer:
(a) $3^5 = 243$
(b) $(-2)^4 = 16$

Teaching Tip
A common student error is $2^2 \cdot 2^4 = 4^6$. Explain why this answer is incorrect.

EXAMPLE 1 Using the Product Rule to Evaluate Numerical Expressions with Exponents

Evaluate each expression:

(a) $2^2 \cdot 2^4$ **(b)** $(-3)^2(-3)$

Solution

(a)
$$2^2 \cdot 2^4 = 2^{2+4} = 2^6 = 64$$

(b)
$$(-3)^2(-3) = (-3)^{2+1} = (-3)^3 = -27$$

Classroom Example Simplify:
(a) $n^3 \cdot n^5$
(b) $a^4 \cdot a \cdot a^2$
Answer:
(a) n^8
(b) a^7

EXAMPLE 2 Using the Product Rule to Simplify Algebraic Expressions with Exponents

Simplify the following expressions:

(a) $t^4 \cdot t^7$ **(b)** $a^6 \cdot a \cdot a^4$

Solution

(a)
$$t^4 \cdot t^7 = t^{4+7} = t^{11}$$

(b)
$$a^6 \cdot a \cdot a^4 = a^6 \cdot a^1 \cdot a^4 = a^{6+1+4} = a^{11}$$

Classroom Example ➔
Simplify the expression: $a \cdot a^2 \cdot b^7$
Answer: a^3b^7

EXAMPLE 3 Using the Product Rule to Simplify Algebraic Expressions with Exponents

Simplify the expression: $m^3 \cdot m^5 \cdot n^9$

Solution

To use the Product Rule for Exponents, the bases must be the same.

$$m^3 \cdot m^5 \cdot n^9 = m^{3+5} \cdot n^9$$
$$= m^8n^9$$

An expression such as $a^5 \cdot b^2$ is in simplest form because the bases are different.

QUICK ✓ *Evaluate or simplify each expression, if possible.*

1. $3^2 \cdot 3$ **2.** $(-5)^2(-5)^3$ **3.** $c^6 \cdot c^2$ **4.** $y^3 \cdot y \cdot y^5$ **5.** $a^4 \cdot a^5 \cdot b^6$

(2) Simplify Exponential Expressions Using the Power Rule

Another law of exponents applies when an exponential expression containing a power is itself raised to a power.

$$(3^2)^4 = \underbrace{3^2 \cdot 3^2 \cdot 3^2 \cdot 3^2}_{\text{4 factors}} = \underbrace{\underbrace{(3 \cdot 3)}_{\text{2 factors}} \cdot \underbrace{(3 \cdot 3)}_{\text{2 factors}} \cdot \underbrace{(3 \cdot 3)}_{\text{2 factors}} \cdot \underbrace{(3 \cdot 3)}_{\text{2 factors}}}_{2 \cdot 4 = 8 \text{ factors}} = 3^8$$

We have the following result:

In Words
If an exponential expression contains a power raised to a power, keep the base and multiply the powers.

POWER RULE FOR EXPONENTS

If a is a real number and m, n are natural numbers, then

$$(a^m)^n = a^{m \cdot n}$$

Classroom Example ➔
Simplify each expression and write the answer in exponential form:
(a) $(3^2)^4$ (b) $(p^3)^5$
Answer:
(a) 3^8 (b) p^{15}

Teaching Tip
Note to students the difference between $(2^3)^2$ and $2^3 \cdot 2^2$.

EXAMPLE 4 Using the Power Rule to Simplify Exponential Expressions

Simplify each expression. Write the answer in exponential form.

(a) $(2^3)^2$ **(b)** $(z^4)^2$

Solution

(a) $(2^3)^2 = 2^{3 \cdot 2}$
$= 2^6$

(b) $(z^4)^2 = z^{4 \cdot 2}$
$= z^8$

Classroom Example ➔
Simplify each expression. Write the answer in exponential form:
(a) $[(-2)^2]^3$ (b) $[(-x)^3]^7$
Answer:
(a) $(-2)^6$ (b) $(-x)^{21}$

EXAMPLE 5 Using the Power Rule to Simplify Exponential Expressions

Simplify each expression. Write the answer in exponential form.

(a) $[(-4)^2]^3$ **(b)** $[(-n)^3]^5$

Solution

(a) $[(-4)^2]^3 = (-4)^{2 \cdot 3}$
$= (-4)^6$

(b) $[(-n)^3]^5 = (-n)^{3 \cdot 5}$
$= (-n)^{15}$

Teaching Tip
Remind students that if they forget the Power Rule, they can change to the Product Rule by writing the product in expanded form (i.e., $(z^4)^2 = z^4 \cdot z^4 = z^{4+4} = z^8$).

QUICK ✓ *Simplify each expression. Write the answer in exponential form.*

6. $(2^2)^4$ **7.** $[(-3)^3]^2$ **8.** $(b^2)^5$ **9.** $(z^4)^5$

3 Simplify Exponential Expressions Containing Products

The next law of exponents deals with raising a product to a power. Consider the following:

$$(x \cdot y)^3 = (x \cdot y) \cdot (x \cdot y) \cdot (x \cdot y) = (x \cdot x \cdot x) \cdot (y \cdot y \cdot y) = x^3 \cdot y^3$$

The following rule generalizes this result.

In Words
When we raise a product to a power, we raise each factor to the power. (Remember, each of the numbers or variables in a multiplication problem is a factor.)

PRODUCT TO A POWER RULE FOR EXPONENTS
If a, b are real numbers and n is a natural number, then

$$(a \cdot b)^n = a^n \cdot b^n$$

Classroom Example ➤
Simplify each expression.
(a) $(2a^3)^4$
(b) $(-4m^2n)^3$
Answer:
(a) $16a^{12}$
(b) $-64m^6n^3$

Teaching Tip
Stress the importance of raising **each factor** in the product to the indicated power.

EXAMPLE 6 Using the Product to a Power Rule to Simplify Exponential Expressions

Simplify each expression:

(a) $(3b^2)^4$ **(b)** $(-2a^2b)^3$

Solution

Each expression contains the product of factors, so we use the Product to a Power Rule.

(a) $(3b^2)^4 = (3)^4(b^2)^4$
Evaluate 3^4: $= 81b^8$

(b) $(-2a^2b)^3 = (-2)^3(a^2)^3(b)^3$
Evaluate $(-2)^3$: $= -8a^{2 \cdot 3}b^3$
$= -8a^6b^3$

QUICK ✓ *Simplify each expression.*

10. $(2n)^3$ **11.** $(6y)^2$ **12.** $(-5x^4)^3$ **13.** $(-7a^3b)^2$

Let's summarize the three rules that we use when we find products of monomials.

Work Smart: Study Skills
Are you having trouble remembering the rules for multiplying exponents? Try making flash cards with the property on the front side, and an example or two on the back side. Also, write in words what each rule means and when it should be applied. This is a good study strategy that may apply to other topics.

RULES FOR MULTIPLYING MONOMIALS

- Product Rule for Exponents
  If a is a real number and m and n are natural numbers, then $a^m \cdot a^n = a^{m+n}$.
- Power Rule for Exponents
  If a is a real number and m and n are natural numbers, then $(a^m)^n = a^{m \cdot n}$.
- Product to a Power Rule for Exponents
  If a and b are real numbers and n is a natural number, then $(a \cdot b)^n = a^n \cdot b^n$.

4 Multiply a Monomial by a Monomial

To multiply two monomials, we multiply the coefficients of the monomials and use the Product Rule for Exponents to multiply the variable expressions.

Classroom Example
Multiply and simplify:
(a) $(3y^3)\cdot(7y^4)$
(b) $(5z^2)\cdot(-4z)$

Answer:
(a) $21y^7$
(b) $-20z^3$

EXAMPLE 7 Multiplying a Monomial by a Monomial: One Variable

Multiply and simplify:

(a) $(5x^2)(6x^4)$ **(b)** $(2p^3)(-5p^2)$

Solution

In each problem, we use the Commutative Property to rearrange factors

(a) $$(5x^2)(6x^4) = 5\cdot 6\cdot x^2\cdot x^4 = 30x^{2+4} = 30x^6$$

(b) $$2p^3(-5p^2) = 2\cdot(-5)\cdot p^3\cdot p^2 = -10p^{3+2} = -10p^5$$

Classroom Example
Multiply and simplify:
$(-4a^7b)\cdot(-6a^4b^5)$
Answer: $24a^{11}b^6$

EXAMPLE 8 Multiplying a Monomial by a Monomial: Two Variables

Multiply and simplify: $(-3x^2y)(-7x^5y^3)$

Solution

$$(-3x^2y)(-7x^5y^3) = -3\cdot(-7)\cdot x^2\cdot x^5\cdot y\cdot y^3 = 21\cdot x^{2+5}\cdot y^{1+3} = 21x^7y^4$$

QUICK ✓ *Multiply and simplify.*

14. $(2a^6)(4a^5)$

15. $(3p^2)(-4p^5)$

16. $(3m^2n^4)(-6mn^5)$

17. $\left(\frac{4}{3}xy^2\right)\left(\frac{1}{2}x^2y\right)$

3.2 Exercises

For Extra Help:

Student Solutions Manual

CD Video

PH Math/Tutor Center

MathXL Tutorials on CD

MathXL® MyMathLab

Concepts and Vocabulary

In Problems 1–3, fill in the blanks.

1. $(a^m)^n =$ ______.

2. In the exponential expression $10x^2$, x is called the ______ and 2 is called the ________ or __________.

3. The expression $y\cdot y\cdot y\cdot y$ is written in ________ ________.

In Problems 4 and 5, answer True or False to each statement.

4. $3^4\cdot 3^2 = 9^6$

5. $(ab^3)^2 = ab^6$

6. Provide a justification for the product rule for exponential expressions.

7. Provide a justification for the power rule for exponential expressions.

8. Provide a justification for the product to a power rule for exponential expressions.

1. a^{mn}
2. base; power; exponent
3. expanded form
4. False
5. False
6. Answers may vary.
7. Answers may vary.
8. Answers may vary.

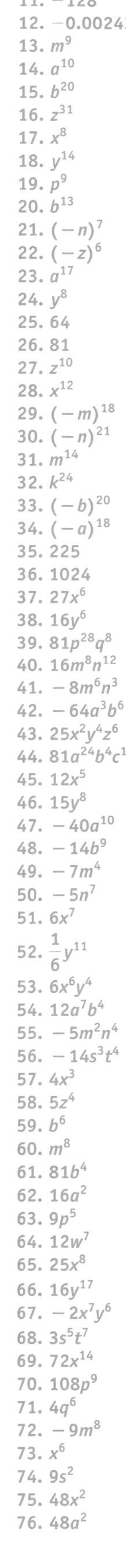

9. 1024
10. 81
11. -128
12. -0.00243
13. m^9
14. a^{10}
15. b^{20}
16. z^{31}
17. x^8
18. y^{14}
19. p^9
20. b^{13}
21. $(-n)^7$
22. $(-z)^6$
23. a^{17}
24. y^8
25. 64
26. 81
27. z^{10}
28. x^{12}
29. $(-m)^{18}$
30. $(-n)^{21}$
31. m^{14}
32. k^{24}
33. $(-b)^{20}$
34. $(-a)^{18}$
35. 225
36. 1024
37. $27x^6$
38. $16y^6$
39. $81p^{28}q^8$
40. $16m^8n^{12}$
41. $-8m^6n^3$
42. $-64a^3b^6$
43. $25x^2y^4z^6$
44. $81a^{24}b^4c^{16}$
45. $12x^5$
46. $15y^8$
47. $-40a^{10}$
48. $-14b^9$
49. $-7m^4$
50. $-5n^7$
51. $6x^7$
52. $\frac{1}{6}y^{11}$
53. $6x^6y^4$
54. $12a^7b^4$
55. $-5m^2n^4$
56. $-14s^3t^4$
57. $4x^3$
58. $5z^4$
59. b^6
60. m^8
61. $81b^4$
62. $16a^2$
63. $9p^5$
64. $12w^7$
65. $25x^8$
66. $16y^{17}$
67. $-2x^7y^6$
68. $3s^5t^7$
69. $72x^{14}$
70. $108p^9$
71. $4q^6$
72. $-9m^8$
73. x^6
74. $9s^2$
75. $48x^2$
76. $48a^2$

Building Skills

In Problems 9–44, simplify each expression.

9. $4^2 \cdot 4^3$ **10.** $3 \cdot 3^3$ **11.** $(-2)^3(-2)^4$ **12.** $(-0.3)^2(-0.3)^3$

13. $m^4 \cdot m^5$ **14.** $a^3 \cdot a^7$ **15.** $b^9 \cdot b^{11}$ **16.** $z^8 \cdot z^{23}$

17. $x^7 \cdot x$ **18.** $y^{13} \cdot y$ **19.** $p \cdot p^2 \cdot p^6$ **20.** $b^5 \cdot b \cdot b^7$

21. $(-n)^3(-n)^4$ **22.** $(-z)(-z)^5$ **23.** $a^5 \cdot a^{10} \cdot a^2$ **24.** $y \cdot y^2 \cdot y^5$

25. $(2^3)^2$ **26.** $(3^2)^2$ **27.** $(z^2)^5$ **28.** $(x^3)^4$

29. $[(-m)^9]^2$ **30.** $[(-n)^7]^3$ **31.** $(m^2)^7$ **32.** $(k^8)^3$

33. $[(-b)^4]^5$ **34.** $[(-a)^6]^3$ **35.** $(3 \cdot 5)^2$ **36.** $(8 \cdot 4)^2$

37. $(3x^2)^3$ **38.** $(4y^3)^2$ **39.** $(-3p^7q^2)^4$ **40.** $(-2m^2n^3)^4$

41. $(-2m^2n)^3$ **42.** $(-4ab^2)^3$ **43.** $(-5xy^2z^3)^2$ **44.** $(-3a^6bc^4)^4$

In Problems 45–56, multiply the monomials.

45. $(4x^2)(3x^3)$ **46.** $(5y^6)(3y^2)$ **47.** $(10a^3)(-4a^7)$

48. $(7b^5)(-2b^4)$ **49.** $(-m^3)(7m)$ **50.** $(-n^6)(5n)$

51. $\left(\frac{4}{5}x^4\right)\left(\frac{15}{2}x^3\right)$ **52.** $\left(\frac{3}{8}y^5\right)\left(\frac{4}{9}y^6\right)$ **53.** $(2x^2y^3)(3x^4y)$

54. $(6a^2b^3)(2a^5b)$ **55.** $\left(\frac{1}{4}mn^3\right)(-20mn)$ **56.** $\left(\frac{2}{3}s^2t^3\right)(-21st)$

Mixed Practice

In Problems 57–72, simplify each expression completely.

57. $x^2 \cdot 4x$ **58.** $z^3 \cdot 5z$ **59.** $(b^3)^2$

60. $(m^2)^4$ **61.** $(-3b)^4$ **62.** $(-4a)^2$

63. $(-6p)^2\left(\frac{1}{4}p^3\right)$ **64.** $(-8w)^2\left(\frac{3}{16}w^5\right)$ **65.** $(5x)^2(x^2)^3$

66. $(4y)^2(y^3)^5$ **67.** $(x^2y)^3(-2xy^3)$ **68.** $(s^2t^3)^2(3st)$

69. $(-3x)^2(2x^4)^3$ **70.** $(-2p^3)^2(3p)^3$

71. $\left(\frac{4}{5}q\right)^2(-5q)^2\left(\frac{1}{4}q^2\right)$ **72.** $\left(\frac{2}{3}m\right)^2(-3m)^3\left(\frac{3}{4}m^3\right)$

Applying the Concepts

△ **73. Cubes** Suppose the length of a side of a cube is x^2. Find the volume of the cube in terms of x.

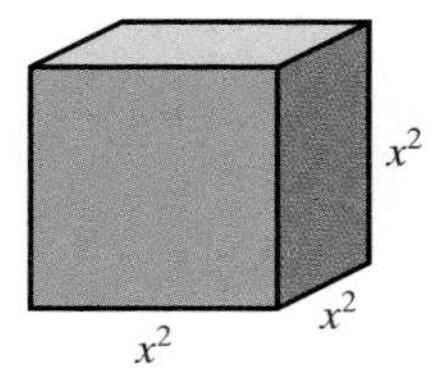

△ **74. Squares** Suppose that the length of a side of a square is $3s$. Find the area of the square in terms of s.

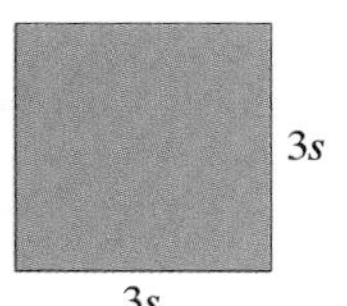

△ **75. Rectangle** Suppose that the width of a rectangle is $4x$ and its length is $12x$. Write an algebraic expression for the area of the rectangle in terms of x.

△ **76. Rectangle** Suppose that the width of a rectangle is $6a$ and its length is $8a$. Write an algebraic expression for the area of the rectangle in terms of a.

77. $16x^{14}$
78. $45y^{13}$
79. $-\frac{24}{5}m^{17}n^9$
80. $-108a^9b^{19}$

Extending the Concepts

In Problems 77–80, simplify each expression.

77. $(3x)^2(-2x)^4\left(\frac{1}{3}x^4\right)^2$

78. $(5y)^3(-3y)^2\left(\frac{1}{5}y^4\right)^2$

79. $-3(-5mn^3)^2\left(\frac{2}{5}m^5n\right)^3$

80. $-2(-4a^3b^2)^2\left(\frac{3}{2}ab^5\right)^3$

3.3 Multiplying Polynomials

OBJECTIVES

1. Multiply a Polynomial by a Monomial
2. Multiply Two Binomials Using the Distributive Property
3. Multiply Two Binomials Using the FOIL Method
4. Multiply the Sum and Difference of Two Terms
5. Square a Binomial
6. Multiply a Polynomial by a Polynomial

Preparing for Multiplying Polynomials

Before getting started, take the following readiness quiz. If you get a problem wrong, go back to the section cited and review the material.

1. Use the Distributive Property to simplify: $2(4x - 3)$ [Section 1.7, p. 61]
2. Find the product: $(3x^2)(-5x^3)$ [Section 3.2, p. 188]
3. Find the product: $(9a)^2$ [Section 3.2, p. 187]

1 Multiply a Polynomial by a Monomial

Now that we know how to multiply two monomials, we can extend our discussion to the product of a monomial and a polynomial. If we were asked to simplify $3(2x + 1)$, we would use the Distributive Property to obtain $3(2x) + 3(1)$. In general, when we multiply a polynomial by a monomial, we use the following property.

Teaching Tip
Stress the importance of multiplying polynomials. Students will need strong multiplication skills to check their answers to factoring polynomials, seen in the next chapter.

EXTENDED FORM OF THE DISTRIBUTIVE PROPERTY

$$a(b + c + \cdots + z) = a \cdot b + a \cdot c + \cdots + a \cdot z$$

where $a, b, c, \ldots, z$ are real numbers.

In Words
The Extended Form of the Distributive Property says to multiply each term in parentheses by a.

Classroom Example Multiply and simplify:
$4x^3(x^2 - 3x + 2)$
Answer: $4x^5 - 12x^4 + 8x^3$

EXAMPLE 1 Multiplying a Monomial by a Trinomial Using the Extended Form of the Distributive Property

Multiply and simplify: $2x^2(x^2 + 3x + 5)$

Solution

We are multiplying a monomial by a trinomial, so we use the Extended Form of the Distributive Property and multiply $2x^2$ by each term in parentheses.

$$\begin{aligned} 2x^2(x^2 + 3x + 5) &= 2x^2(x^2) + 2x^2(3x) + 2x^2(5) \\ &= 2x^4 + 6x^3 + 10x^2 \end{aligned}$$

Classroom Example Multiply and simplify:
$-4ab(2a + 3ab + 5a^2)$
Answer: $-8a^2b - 12a^2b^2 - 20a^3b$

EXAMPLE 2 Multiplying a Monomial by a Trinomial

Multiply and simplify: $-2xy(3x^2 + 5xy + 2y^2)$

Solution

Use the Extended Form of the Distributive Property.

$$\begin{aligned} -2xy(3x^2 + 5xy + 2y^2) &= -2xy(3x^2) + (-2xy)(5xy) + (-2xy)(2y^2) \\ &= -6x^3y - 10x^2y^2 - 4xy^3 \end{aligned}$$

Preparing for...Answers **1.** $8x - 6$ **2.** $-15x^5$ **3.** $81a^2$

Classroom Example Multiply and simplify:

$$\left(\frac{3}{2}y^2 + 8y + \frac{4}{3}\right)\frac{7}{4}y^2$$

Answer: $\frac{21}{8}y^4 + 14y^3 + \frac{7}{3}y^2$

Work Smart

Example 3 is an algebraic **expression** so you do not clear fractions.

EXAMPLE 3 Multiplying a Trinomial by a Monomial

Multiply and simplify: $\left(\frac{4}{3}z^2 + 8z + \frac{1}{4}\right)\frac{1}{2}z^3$

Solution

$$\left(\frac{4}{3}z^2 + 8z + \frac{1}{4}\right)\frac{1}{2}z^3 = \frac{4}{3}z^2\left(\frac{1}{2}z^3\right) + 8z\left(\frac{1}{2}z^3\right) + \frac{1}{4}\left(\frac{1}{2}z^3\right)$$

$$= \frac{2}{3}z^5 + 4z^4 + \frac{1}{8}z^3$$

QUICK ✓ *Multiply and simplify each product.*

1. $3x(x^2 - 2x + 4)$ **2.** $-a^2b(2a^2b^2 - 4ab^2 + 3ab)$ **3.** $\left(5n^3 - \frac{15}{8}n^2 - \frac{10}{7}n\right)\frac{3}{5}n^2$

2 Multiply Two Binomials Using the Distributive Property

We now discuss how to multiply two binomials. To help understand the process, we review multiplication of two-digit numbers. Suppose we want to find 32×14. We proceed as follows: Multiply 2 by 4 and obtain 8, multiply 3 by 4 and obtain 12, so that $32 \times 4 = 128$. Now, multiply 2 by 1 and obtain 2, multiply 3 by 1 and obtain 3, so that $32 \times 1 = 32$. We add vertically to obtain the product, 448.

$$\begin{array}{rcl} & 32 & \\ & \times 14 & \\ 4 \times 3 \rightarrow & 128 & \leftarrow 4 \times 2 \\ 1 \times 3 \rightarrow & 32 & \leftarrow 1 \times 2 \\ & 448 & \end{array}$$

Now suppose we want to multiply $(3x + 2)$ and $(x + 4)$. That is, we want to find

$$\begin{array}{r} 3x + 2 \\ \times \quad x + 4 \\ \hline \end{array}$$

We proceed in exactly the same way as we did when multiplying two-digit numbers. Multiply 2 by 4, multiply $3x$ by 4, so that $(3x + 2) \cdot 4 = 12x + 8$. Multiply 2 by x, multiply $3x$ by x, so that $(3x + 2) \cdot x = 3x^2 + 2x$. Now add vertically and obtain the product, $3x^2 + 14x + 8$.

$$\begin{array}{rcl} & 3x + 2 & \\ & \times \quad x + 4 & \\ 4 \cdot 3x \longrightarrow & 12x + 8 & \leftarrow 4 \cdot 2 \\ x \cdot 3x \rightarrow & 3x^2 + \quad 2x & \leftarrow x \cdot 2 \\ & 3x^2 + 14x + 8 & \end{array}$$

Multiplying two binomials uses exactly the same procedures that is used when multiplying two-digit numbers!

The multiplication that we just went through is known as *vertical multiplication*. We can also multiply using *horizontal multiplication*. Horizontal multiplication requires repeated use of the Distributive Property. For example, to find $(3x + 2)(x + 4)$, we distribute $3x + 2$ to x and then distribute $3x + 2$ to 4.

$$(3x + 2)(x + 4) = (3x + 2)x + (3x + 2) \cdot 4$$

Distribute x, distribute 4: $= 3x \cdot x + 2 \cdot x + 3x \cdot 4 + 2 \cdot 4$

$$= 3x^2 + 2x + 12x + 8$$

$$= 3x^2 + 14x + 8$$

This is the same result that we obtained using vertical multiplication. Let's do a couple more examples using both horizontal and vertical multiplication.

Classroom Example Multiply and simplify: $(a + 5)(a - 3)$

Answer: $a^2 + 2a - 15$

EXAMPLE 4 Multiplying Two Binomials

Find the product: $(x + 3)(x - 8)$

Solution

Vertical Multiplication

$$\begin{array}{r} x + 3 \\ \times\ x - 8 \\ \hline -8x - 24 \leftarrow -8(x+3) \\ x^2 + 3x \leftarrow x(x+3) \\ \hline x^2 - 5x - 24 \end{array}$$

Horizontal Multiplication

We distribute the first binomial to each term in the second binomial.

$$(x + 3)(x - 8) = (x + 3)x + (x + 3)(-8)$$
$$= x^2 + 3x - 8x - 24$$

Combine like terms: $= x^2 - 5x - 24$

In either case, $(x + 3)(x - 8) = x^2 - 5x - 24$.

Classroom Example Find the product: $(2x - 3)(3x + 7)$

Answer: $6x^2 + 5x - 21$

EXAMPLE 5 Multiplying Two Binomials

Find the product: $(3x + 4)(2x - 5)$

Solution

Vertical Multiplication

$$\begin{array}{r} 3x + 4 \\ \times\ 2x - 5 \\ \hline -15x - 20 \leftarrow -5(3x+4) \\ 6x^2 + 8x \leftarrow 2x(3x+4) \\ \hline 6x^2 - 7x - 20 \end{array}$$

Horizontal Multiplication

We distribute the first binomial to each term in the second binomial.

$$(3x + 4)(2x - 5) = (3x + 4)(2x) + (3x + 4)(-5)$$
$$= 3x \cdot 2x + 4 \cdot 2x + 3x \cdot (-5) + 4 \cdot (-5)$$
$$= 6x^2 + 8x - 15x - 20$$

Combine like terms: $= 6x^2 - 7x - 20$

In either case, $(3x + 4)(2x - 5) = 6x^2 - 7x - 20$.

QUICK ✓ *Find the product.*

4. $(x + 7)(x + 2)$ **5.** $(2n + 1)(n - 3)$ **6.** $(3p - 2)(2p - 1)$

3 Multiply Two Binomials Using the FOIL Method

Work Smart

The FOIL method is a form of the Distributive Property, and FOIL can be used ***only*** when we multiply binomials.

When multiplying two binomials, there is a second method referred to as the **FOIL method**. FOIL stands for First, Outer, Inner, Last. "First" means that we are multiplying the *first* terms in each binomial, "Outer" means we are multiplying the *outside* terms of each binomial, "Inner" means we are multiplying the *innermost* terms of each binomial, and "Last" means we are multiplying the *last* terms of each binomial. The FOIL method is illustrated below. For comparison the horizontal method is illustrated as well.

The FOIL Method	Horizontal Multiplication Using the Distributive Property
First, Last, Inner, Outer F O I L $(ax + b)(cx + d) = ax \cdot cx + ax \cdot d + b \cdot cx + b \cdot d$	$(ax + b)(cx + d) = (ax + b)cx + (ax + b)d$ $= ax \cdot cx + b \cdot cx + ax \cdot d + b \cdot d$ Rearrange terms: $= ax \cdot cx + ax \cdot d + b \cdot cx + b \cdot d$

The FOIL method does not give different results than the Distributive Property. Instead, it is a memory device to help when multiplying two binomials.

Classroom Example ➤
Find the product: $(x + 2)(x + 6)$
Answer: $x^2 + 8x + 12$

EXAMPLE 6 Using the FOIL Method to Multiply Two Binomials

Find the product: $(y + 3)(y + 5)$

Solution

First Last, Inner, Outer

$$\begin{aligned}(y + 3)(y + 5) &= \overset{F}{(y)(y)} + \overset{O}{(y)(5)} + \overset{I}{(3)(y)} + \overset{L}{(3)(5)} \\ &= y^2 + 5y + 3y + 15 \\ &= y^2 + 8y + 15\end{aligned}$$

Classroom Example ➤
Find the product: $(n + 4)(n - 2)$
Answer: $n^2 + 2n - 8$

EXAMPLE 7 Using the FOIL Method to Multiply Two Binomials

Find the product: $(a + 1)(a - 3)$

Solution

First Last, Inner, Outer

$$\begin{aligned}(a + 1)(a - 3) &= \overset{F}{(a)(a)} + \overset{O}{(a)(-3)} + \overset{I}{(1)(a)} + \overset{L}{(1)(-3)} \\ &= a^2 - 3a + a - 3 \\ &= a^2 - 2a - 3\end{aligned}$$

QUICK ✓ *Find the product.*

7. $(x + 3)(x + 4)$ **8.** $(y + 5)(y - 3)$ **9.** $(a - 1)(a - 5)$

Classroom Example ➤
Find the product: $(2z + 3)(z - 5)$
Answer: $2z^2 - 7z - 15$

EXAMPLE 8 Using the FOIL Method to Multiply Two Binomials

Find the product: $(3a + 1)(a - 4)$

Solution

First Last, Inner, Outer

$$\begin{aligned}(3a + 1)(a - 4) &= \overset{F}{(3a)(a)} + \overset{O}{(3a)(-4)} + \overset{I}{(1)(a)} + \overset{L}{(1)(-4)} \\ &= 3a^2 - 12a + a - 4 \\ &= 3a^2 - 11a - 4\end{aligned}$$

Classroom Example ➤
Find the product: $(4x - 3y)(7x - 2y)$
Answer: $28x^2 - 29xy + 6y^2$

EXAMPLE 9 Using the FOIL Method to Multiply Two Binomials

Find the product: $(3a - 4b)(5a - 2b)$

Solution

$$\begin{aligned}(3a - 4b)(5a - 2b) &= \overset{F}{(3a)(5a)} + \overset{O}{(3a)(-2b)} - \overset{I}{(4b)(5a)} - \overset{L}{(4b)(-2b)} \\ &= 15a^2 - 6ab - 20ab + 8b^2 \\ &= 15a^2 - 26ab + 8b^2\end{aligned}$$

QUICK ✓ *Find the product.*

10. $(2x + 3)(x + 4)$ **11.** $(3y + 5)(2y - 3)$

12. $(2a + 1)(3a - 4)$ **13.** $(5x + 2y)(3x - 4y)$

4 Multiply the Sum and Difference of Two Terms

Certain binomials lead to products that result in patterns. For this reason, we call these products **special products**. The first special product we discuss is a product of the form $(a - b)(a + b)$.

Classroom Example ➤
Find the product: $(n - 4)(n + 4)$
Answer: $n^2 - 16$

EXAMPLE 10 Finding a Product of the Form $(a - b)(a + b)$

Find the product: $(x - 5)(x + 5)$

Solution

$$\begin{aligned}(x - 5)(x + 5) &= \overset{F}{(x)(x)} + \overset{O}{(x)(5)} - \overset{I}{(5)(x)} - \overset{L}{(5)(5)} \\ &= x^2 + 5x - 5x - 25 \\ &= x^2 - 25\end{aligned}$$

We call products of the form $(a - b)(a + b)$ "the sum and difference of two terms." Do you see why?

In Example 10, did you notice that the outer product, $5x$, and the inner product, $-5x$, were opposites? This result is not a coincidence. The sum of the outer product and the inner product of two binomials in the form $(a - b)(a + b)$ is *always* zero, so the product $(a - b)(a + b)$ is the *difference* $a^2 - b^2$. We have the following formula based upon the results of Example 10.

PRODUCT OF THE SUM AND DIFFERENCE OF TWO TERMS

$$(a - b)(a + b) = a^2 - b^2$$

Classroom Example ➤
Find each product:
(a) $(3x - 2)(3x + 2)$
(b) $(5a - 3b)(5a + 3b)$
Answer:
(a) $9x^2 - 4$
(b) $25a^2 - 9b^2$

EXAMPLE 11 Finding the Product of the Sum and Difference of Two Terms

Find each product:

(a) $(2x + 5)(2x - 5)$ **(b)** $(4x - 3y)(4x + 3y)$

Solution

(a) $$\begin{aligned}(2x + 5)(2x - 5) &= (2x)^2 - 5^2 \\ &= 4x^2 - 25\end{aligned}$$

(b) $$\begin{aligned}(4x - 3y)(4x + 3y) &= (4x)^2 - (3y)^2 \\ &= 16x^2 - 9y^2\end{aligned}$$

QUICK ✓ *Find each product.*

14. $(a - 4)(a + 4)$ **15.** $(3w + 7)(3w - 7)$
16. $(6b - 2c)(6b + 2c)$ **17.** $(x - 2y^3)(x + 2y^3)$

5 Square a Binomial

Another special product involves squaring a binomial. For example, since $(x + 3)^2 = (x + 3)(x + 3)$, we use FOIL and obtain

$$(x + 3)^2 = (x + 3)(x + 3) = x^2 + 3x + 3x + 9 = x^2 + 6x + 9$$

Did you notice that the outer product and the inner product are the same, namely $3x$? Study Table 2 to discover the pattern.

Table 2

$(n + 5)^2 =$	$(n + 5)(n + 5) =$	$n^2 + 5n + 5n + 25 =$	$n^2 + 2(5n) + 25 =$	$n^2 + 10n + 25$
$(2a - 3)^2 =$	$(2a - 3)(2a - 3) =$	$4a^2 - 6a - 6a + 9 =$	$4a^2 - 2(6a) + 9 =$	$4a^2 - 12a + 9$
$(n + 5p)^2 =$		$n^2 + 5np + 5np + 25p^2 =$	$n^2 + 2(5np) + 25p^2 =$	$n^2 + 10np + 25p^2$
$(4b - 9)^2 =$			$16b^2 - 2(4b)(9) + 81 =$	$16b^2 - 72b + 81$

Teaching Tip
Explain that there is no "Distributive Property" for exponents.

$$(5 + x)^2 \neq 5^2 + x^2$$

Work Smart

$$(x + y)^2 \neq x^2 + y^2$$
$$(x - y)^2 \neq x^2 - y^2$$

Whenever you feel the urge to perform an operation that you're not quite sure about, try it with actual numbers. For example, does

$$(3 + 2)^2 = 3^2 + 2^2?$$

NO! So

$$(x + y)^2 \neq x^2 + y^2$$

The results of Table 2 lead to the following.

$$(a + b)^2 = \underbrace{a^2}_{(\text{first term})^2} + \underbrace{2ab}_{2(\text{product of terms})} + \underbrace{b^2}_{(\text{second term})^2}$$

$$(a - b)^2 = \underbrace{a^2}_{(\text{first term})^2} - \underbrace{2ab}_{2(\text{product of terms})} + \underbrace{b^2}_{(\text{second term})^2}$$

SQUARES OF BINOMIALS

$$(a + b)^2 = a^2 + 2ab + b^2$$
$$(a - b)^2 = a^2 - 2ab + b^2$$

The trinomials $a^2 + 2ab + b^2$ and $a^2 - 2ab + b^2$ are called **perfect square trinomials**.

EXAMPLE 12 How to Find a Product of the Form $(a + b)^2$

Find the product: $(y + 7)^2$

Classroom Example
Find the product: $(x + 5)^2$
Answer: $x^2 + 10x + 25$

Step-by-Step Solution

Step 1: Use the $(a + b)^2$ pattern.

$$\underbrace{(a + b)^2}_{(y + 7)^2} = \underbrace{a^2}_{(y)^2} + \underbrace{2ab}_{2(y)(7)} + \underbrace{b^2}_{7^2}$$

Step 2: Find each product.

$$= y^2 + 14y + 49$$

Classroom Example
Find the product: $(7 - k)^2$
Answer: $49 - 14k + k^2$

EXAMPLE 13 Finding a Product of the Form $(a - b)^2$

Find the product: $(3 - r)^2$

Solution

Use the $(a - b)^2$ pattern: $(3 - r)^2 = 3^2 - 2(3)(r) + r^2$

$$= 9 - 6r + r^2$$

QUICK ✓ *Find each product.*

18. $(z - 9)^2$ **19.** $(p + 1)^2$ **20.** $(4 - a)^2$ **21.** $(w + y)^2$

Classroom Example
Find each product:
(a) $(7z - 2)^2$ (b) $(3x + 8y)^2$

Answer:
(a) $49z^2 - 28z + 4$
(b) $9x^2 + 48xy + 64y^2$

EXAMPLE 14 Finding Products of the Form $(a + b)^2$ or $(a - b)^2$

Find each product:

(a) $(9p - 4)^2$ **(b)** $(2x + 5y)^2$

Work Smart

If you can't remember the formulas for a perfect square, don't panic! Simply use the fact that $(x + a)^2 = (x + a)(x + a)$ and then FOIL. The same logic applies to perfect squares of the form $(x - a)^2$.

Solution

(a) $(9p - 4)^2 = (9p)^2 - 2(9p)4 + 4^2$
$= 81p^2 - 72p + 16$

(b) $(2x + 5y)^2 = (2x)^2 + 2(2x)(5y) + (5y)^2$
$= 4x^2 + 20xy + 25y^2$

QUICK ✓ *Find each product.*

22. $(3z - 4)^2$ **23.** $(5p + 1)^2$ **24.** $(4a - 5)^2$ **25.** $(2w + 7y)^2$

For convenience, we present a summary of binomial products.

SUMMARY OF BINOMIAL PRODUCTS

- When multiplying two binomials, we can use the Distributive Property.

$$\begin{aligned}(3x + 5)(4x - 1) &= (3x + 5)(4x) + (3x + 5)(-1)\\ &= 12x^2 + 20x - 3x - 5\\ &= 12x^2 + 17x - 5\end{aligned}$$

- When multiplying two binomials, we can use the FOIL pattern.

$$\begin{aligned}(3x + 5)(4x - 1) &= 3x(4x) + 3x(-1) + 5(4x) + 5(-1)\\ &= 12x^2 - 3x + 20x - 5\\ &= 12x^2 + 17x - 5\end{aligned}$$

- When finding a product in the form $(a - b)(a + b)$, we use the product of the sum and difference of two terms, $a^2 - b^2$.

$$\begin{aligned}(7a + 2b)(7a - 2b) &= (7a)^2 - (2b)^2\\ &= 49a^2 - 4b^2\end{aligned}$$

- When squaring a binomial, $(a + b)^2 = a^2 + 2ab + b^2$ and $(a - b)^2 = a^2 - 2ab + b^2$. The trinomial product is called a perfect square trinomial.

$$\begin{aligned}(3x + 5)^2 &= (3x)^2 + 2(3x)(5) + (5)^2\\ &= 9x^2 + 30x + 25\end{aligned}$$

$$\begin{aligned}(4y - 3)^2 &= (4y)^2 - 2(4y)(3) + (3)^2\\ &= 16y^2 - 24y + 9\end{aligned}$$

6 Multiply a Polynomial by a Polynomial

When we find the product of two polynomials, we make repeated use of the Extended Form of the Distributive Property. We can use either a horizontal or vertical format. In addition, it is a good idea to write each polynomial in standard form.

Classroom Example ➤
Find the product $(5x + 2)(x^2 + 3x - 1)$ using horizontal multiplication.
Answer: $5x^3 + 17x^2 + x - 2$

EXAMPLE 15 Multiplying Polynomials Using Horizontal Multiplication

Find the product: $(2x + 3)(x^2 + 5x - 1)$

Solution

We distribute $2x + 3$ to each term in the trinomial.

$$\begin{aligned}(2x + 3)(x^2 + 5x - 1) &= (2x + 3)x^2 + (2x + 3)\cdot 5x + (2x + 3)\cdot(-1)\\ &= 2x(x^2) + 3(x^2) + 2x(5x) + 3(5x) + 2x(-1) + 3(-1)\\ \text{Rearrange terms:} \quad &= 2x^3 + 3x^2 + 10x^2 + 15x - 2x - 3\\ \text{Combine like terms:} \quad &= 2x^3 + 13x^2 + 13x - 3\end{aligned}$$

QUICK ✓ *Find the product using horizontal multiplication.*

26. $(x - 2)(x^2 + 2x + 4)$ **27.** $(3y - 2)(y^2 + 2y + 4)$

The best way to use vertical multiplication to multiply two polynomials is to place the polynomial with more terms on top, align terms of the same degree, and then multiply. Make sure both polynomials are written in standard form.

Classroom Example Find the product $(5x + 2)(x^2 + 3x - 1)$ using vertical multiplication.

Answer: $5x^3 + 17x^2 + x - 2$

EXAMPLE 16 Multiplying a Binomial and a Trinomial Using Vertical Multiplication

Find the product: $(2x + 3)(x^2 + 5x - 1)$

Solution

We arrange the polynomials vertically by placing the trinomial on top since it has more terms than the binomial. Now we multiply the trinomial by the 3 in the binomial and then multiply the trinomial by the $2x$ in the binomial.

$$\begin{array}{rr} & x^2 + 5x - 1 \\ & \times \quad 2x + 3 \\ \hline 3(x^2+5x-1): & 3x^2 + 15x - 3 \\ 2x(x^2+5x-1): & 2x^3 + 10x^2 - 2x \quad\quad \\ \hline \text{Add vertically:} & 2x^3 + 13x^2 + 13x - 3 \end{array}$$

QUICK ✓ *Find the product using vertical multiplication.*

28. $(x - 2)(x^2 + 2x + 4)$

29. $(3y - 2)(y^2 + 2y + 4)$

Classroom Example Find the product:

$$(-3x)(2x - 3)(x + 7)$$

Answer: $-6x^3 - 33x^2 + 63x$

Teaching Tip

The process for finding the product of three polynomials is parallel to the process for finding the product of three numbers:

$$\frac{1}{2}(8)(11) = 4(11) = 44$$

so

$$\frac{1}{2}x(8x)(11x + 2) =$$
$$4x^2(11x + 2) =$$
$$44x^3 + 8x^2$$

EXAMPLE 17 Multiplying Three Polynomials

Find the product: $2x(x - 4)(3x - 5)$

Solution

To find the product of three polynomials, multiply any two factors and then multiply that product by the remaining factor. We'll start by using the Distributive Property to multiply $2x$ and $(x - 4)$.

$$\begin{aligned} 2x(x - 4)(3x - 5) &= (2x^2 - 8x)(3x - 5) \\ \text{FOIL:} \quad &= 2x^2\cdot 3x - 2x^2\cdot 5 - 8x\cdot 3x - 8x\cdot(-5) \\ &= 6x^3 - 10x^2 - 24x^2 + 40x \\ \text{Combine like terms:} \quad &= 6x^3 - 34x^2 + 40x \end{aligned}$$

QUICK ✓ *Find the product.*

30. $-2a(4a - 1)(3a + 5)$

31. $\frac{3}{2}(5x - 2)(2x + 8)$

3.3 Exercises

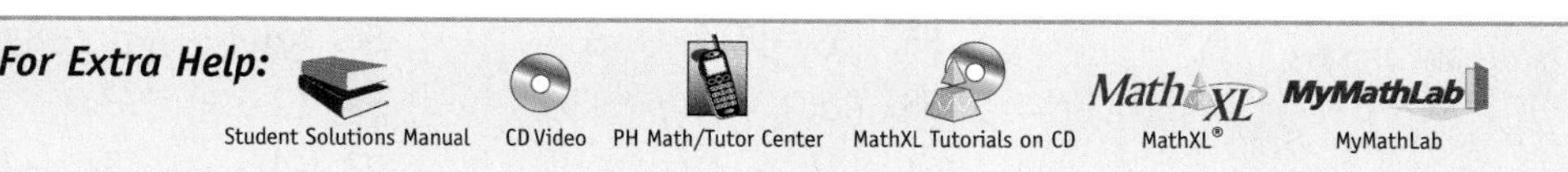

Concepts and Vocabulary

In Problems 1–3, fill in the blanks.

1. The product $(a - b)(a + b)$ is equal to __________.
2. When using the FOIL method to multiply two binomials, the letter I indicates that the __________ terms should be multiplied.
3. $x^2 + 2xy + y^2$ is referred to as a __________ __________ __________.

1. $a^2 - b^2$
2. innermost
3. perfect square trinomial

4. False
5. False
6. False
7. Yes; $x = 0, y = 0$
8. $(20 - 1)(20 + 1) = 20^2 - 1^2 = 399$
9. $6x^2 - 10x$
10. $6m^2 - 21m$
11. $-7b^2 + 35b$
12. $-15b^2 - 45b$
13. $2n^2 - 3n$
14. $9b^2 - 3b$
15. $12n^4 + 6n^3 - 15n^2$
16. $8w^3 + 12w^2 - 20w$
17. $12m^3n - 4m^2n^2 + 2mn^3$
18. $54x^3y + 12x^2y^2 - 18xy^3$
19. $4x^4y^2 - 3x^3y^3$
20. $14r^3s + 6r^2s^3$
21. $x^2 + 5x + 6$
22. $x^2 + 10x + 21$
23. $q^2 - 13q + 42$
24. $n^2 - 9n + 20$
25. $y^2 - y - 20$
26. $m^2 + 3m - 18$
27. $x^4 + 4x^2 + 3$
28. $x^4 - 7x^2 + 10$
29. $42 - 13x + x^2$
30. $8 - 6y + y^2$
31. $-n^2 + 4n - 3$
32. $-x^2 + 14x - 45$
33. $3a^2 + 5a + 2$
34. $6m^2 - 7m + 2$
35. $10u^2 + 17uv + 6v^2$
36. $3a^2 - 11ab + 6b^2$
37. $4n^2 - 9p^2$
38. $16x^2 - 9y^2$
39. $x^3 + x^2 - 5x - 2$
40. $6a^3 - 17a^2 - 4a + 3$
41. $2y^3 - 12y^2 + 19y - 3$
42. $4m^3 + 3m + 2$
43. $2x^3 - 7x^2 + 4x + 3$
44. $8y^3 - 2y^2 - 13y - 3$
45. $6p^3 - 13p^2 + 4p + 3$
46. $m^3n^2 - 2m^2n^2 - 3m^2n + 7mn - 3$
47. $2b^3 + 2b^2 - 24b$
48. $5a^3 + 25a^2 - 30a$
49. $-x^3 + 9x$
50. $-4k^3 - 16k^2 + 84k$
51. $2x^4 - 5x^3 + 10x^2 - 11x - 4$
52. $2a^4 + 5a^3 + 9a^2 + 15a + 18$
53. $10y^5 + 3y^4 + 4y^3 + 3y^2 + 2y + 2$
54. $-2m^5 + m^4 - 8m^3 - 7m - 4$
55. $x^2 - 4x - 21$
56. $p^2 - 3p - 40$
57. $z^2 - 7z - 30$
58. $p^2 + p - 90$
59. $6x^2 + 7x - 3$
60. $12x^2 - 5x - 2$
61. $7k^2 + 39k - 18$
62. $12r^2 - 23r + 5$
63. $14x^2 + 41xy + 15y^2$
64. $8m^2 - 10mn - 3n^2$
65. $10a^2 - ab - 2b^2$
66. $18r^2 + 51rs + 35s^2$
67. $x^2 - 9$ 68. $y^2 - 49$
69. $4z^2 - 25$ 70. $36r^2 - 1$
71. $16x^4 - 1$
72. $9a^4 - 4$ 73. $4x^2 - 9y^2$
74. $64a^2 - 25b^2$ 75. $x^2 - \frac{1}{9}$

In Problems 4–6, answer True or False to each statement.

4. The product $(x - y)(x^2 + 2xy + y^2)$ can be found using the FOIL method.

5. $(x - y)^2 = x^2 - y^2$

6. The product of a binomial and a binomial is always a trinomial.

7. Do you think there are values of x and y such that $(x + y)^2 = x^2 + y^2$? Use some numerical values for x and y to explain your reasoning.

8. Explain how the difference of squares can be used to calculate $19 \cdot 21$.

Building Skills

In Problems 9–20, use the Distributive Property to find each product.

9. $2x(3x - 5)$ **10.** $3m(2m - 7)$

11. $-7b(b - 5)$ **12.** $-5b(3b + 9)$

13. $\frac{1}{2}n(4n - 6)$ **14.** $\frac{3}{5}b(15b - 5)$

15. $3n^2(4n^2 + 2n - 5)$ **16.** $4w(2w^2 + 3w - 5)$

17. $2mn(6m^2 - 2mn + n^2)$ **18.** $6xy(9x^2 + 2xy - 3y^2)$

19. $(4x^2y - 3xy^2)x^2y$ **20.** $(7r + 3s^2)2r^2s$

In Problems 21–54, find each product.

21. $(x + 2)(x + 3)$ **22.** $(x + 3)(x + 7)$

23. $(q - 6)(q - 7)$ **24.** $(n - 4)(n - 5)$

25. $(y - 5)(y + 4)$ **26.** $(m - 3)(m + 6)$

27. $(x^2 + 3)(x^2 + 1)$ **28.** $(x^2 - 5)(x^2 - 2)$

29. $(7 - x)(6 - x)$ **30.** $(2 - y)(4 - y)$

31. $(1 - n)(n - 3)$ **32.** $(x - 5)(9 - x)$

33. $(3a + 2)(a + 1)$ **34.** $(2m - 1)(3m - 2)$

35. $(5u + 6v)(2u + v)$ **36.** $(3a - 2b)(a - 3b)$

37. $(2n - 3p)(2n + 3p)$ **38.** $(4x + 3y)(4x - 3y)$

39. $(x - 2)(x^2 + 3x + 1)$ **40.** $(3a - 1)(2a^2 - 5a - 3)$

41. $(2y^2 - 6y + 1)(y - 3)$ **42.** $(2m^2 - m + 2)(2m + 1)$

43. $(2x - 3)(x^2 - 2x - 1)$ **44.** $(4y + 1)(2y^2 - y - 3)$

45. $(3p^2 - 2p - 1)(2p - 3)$ **46.** $(m^2n - 2mn + 1)(mn - 3)$

47. $2b(b - 3)(b + 4)$ **48.** $5a(a + 6)(a - 1)$

49. $-\frac{1}{2}x(2x + 6)(x - 3)$ **50.** $-\frac{4}{3}k(k + 7)(3k - 9)$

51. $(x^2 - x + 4)(2x^2 - 3x - 1)$ **52.** $(2a^2 - a + 6)(a^2 + 3a + 3)$

53. $(5y^3 - y^2 + 2)(2y^2 + y + 1)$ **54.** $(2m^2 - m + 4)(-m^3 - 2m - 1)$

In Problems 55–66, use the FOIL method to find each product.

55. $(x - 7)(x + 3)$ **56.** $(p + 5)(p - 8)$ **57.** $(z - 10)(z + 3)$

58. $(p + 10)(p - 9)$ **59.** $(2x + 3)(3x - 1)$ **60.** $(3z - 2)(4z + 1)$

61. $(7k - 3)(k + 6)$ **62.** $(4r - 1)(3r - 5)$ **63.** $(7x + 3y)(2x + 5y)$

64. $(2m - 3n)(4m + n)$ **65.** $(2a - b)(5a + 2b)$ **66.** $(3r + 5s)(6r + 7s)$

In Problems 67–90, find the special products.

67. $(x - 3)(x + 3)$ **68.** $(y + 7)(y - 7)$ **69.** $(2z + 5)(2z - 5)$

70. $(6r - 1)(6r + 1)$ **71.** $(4x^2 + 1)(4x^2 - 1)$ **72.** $(3a^2 + 2)(3a^2 - 2)$

73. $(2x - 3y)(2x + 3y)$ **74.** $(8a - 5b)(8a + 5b)$ **75.** $\left(x - \frac{1}{3}\right)\left(x + \frac{1}{3}\right)$

76. $y^2 - \frac{4}{81}$
77. $x^6 - y^4$
78. $4x^4 - 9z^6$
79. $x^2 - 4x + 4$
80. $x^2 + 8x + 16$
81. $x^2 + 4xy + 4y^2$
82. $9x^2 - 12xy + 4y^2$
83. $4a^2 - 12ab + 9b^2$
84. $25x^2 + 20xy + 4y^2$
85. $16x^2 + 24xy + 9y^2$
86. $4p^2 - 28pq + 49q^2$
87. $25a^2 - 20ab^2 + 4b^4$
88. $49m^4 + 28m^2n^3 + 4n^6$
89. $x^2 + x + \frac{1}{4}$
90. $y^2 - \frac{2}{3}y + \frac{1}{9}$
91. $3x^3 - 6x^2 + 1$
92. $-7y^3 + 4y^2 - 8y + 3$
93. $-5x^7 + 4x^6 - 6x^5$
94. $-18y^2 + 3y - 2$
95. $5x^3 + 21x^2 + 2x$
96. $12a^4 + b^4$
97. $-3w^3 + 3w^2 + 36w$
98. $5y^3 + 10y^2 - 75y$
99. $3a^3 + 24a^2 + 48a$
100. $2m^3 - 12m^2 + 18m$
101. $2n^2 + 6n$
102. $2s^2 - 12s$
103. $24ab$
104. $40ab$
105. $x^2 - 0.6x + 0.09$
106. $n^2 + 0.4n + 0.04$
107. $\frac{4}{9}b^2 - \frac{25}{81}$
108. $\frac{16}{25}b^2 - \frac{4}{49}$
109. $-x^2 + 3x + 2$
110. $-2a^2 - 5a + 13$
111. $4x^2 + 4x + 1$
112. $9x^2 - 12xy + 4y^2$
113. $x^3 - 3x^2 + 3x - 1$
114. $8a^3 + 12a^2b + 6ab^2 + b^3$
115. $2x^2 - 9$
116. $2x^2 + 5x - 6$
117. $2x^2 + 7x - 15$
118. $3x^2 + 32x + 45$
119. $x^2 + 10x + 25$
120. $x^2 - 6x + 9$
121. $9x^2 - 25$
122. $4x^2 - 9$
123. $2x^2 + x - 1$
124. $14x^2 - 27x + 9$

76. $\left(y + \frac{2}{9}\right)\left(y - \frac{2}{9}\right)$ **77.** $(x^3 - y^2)(x^3 + y^2)$ **78.** $(2x^2 - 3z^3)(2x^2 + 3z^3)$

79. $(x - 2)^2$ **80.** $(x + 4)^2$ **81.** $(x + 2y)^2$

82. $(3x - 2y)^2$ **83.** $(2a - 3b)^2$ **84.** $(5x + 2y)^2$

85. $(4x + 3y)(4x + 3y)$ **86.** $(2p - 7q)(2p - 7q)$ **87.** $(5a - 2b^2)^2$

88. $(7m^2 + 2n^3)^2$ **89.** $\left(x + \frac{1}{2}\right)^2$ **90.** $\left(y - \frac{1}{3}\right)^2$

Mixed Practice

In Problems 91–110, perform the indicated operation.

91. $2x(x^2 - 3x + 2) + (x^3 - 4x + 1)$ **92.** $-4y^2(2y - 1) + (y^3 - 8y + 3)$

93. $-\frac{1}{2}x^2(10x^5 - 6x^4 + 12x^3) + (x^3)^2$ **94.** $-\frac{1}{3}(27y^2 - 9y + 6) - (3y)^2$

95. $7x^2(x + 3) - 2x(x^2 - 1)$ **96.** $2(3a^4 + 2b^4) - 3(b^4 - 2a^4)$

97. $-3w(w - 4)(w + 3)$ **98.** $5y(y + 5)(y - 3)$

99. $3a(a + 4)^2$ **100.** $2m(m - 3)^2$

101. $(n + 3)(n - 3) + (n + 3)^2$ **102.** $(s + 6)(s - 6) + (s - 6)^2$

103. $(a + 6b)^2 - (a - 6b)^2$ **104.** $(2a + 5b)^2 - (2a - 5b)^2$

105. $(x - 0.3)^2$ **106.** $(n + 0.2)^2$

107. $\left(\frac{2}{3}b - \frac{5}{9}\right)\left(\frac{2}{3}b + \frac{5}{9}\right)$ **108.** $\left(\frac{4}{5}b + \frac{2}{7}\right)\left(\frac{4}{5}b - \frac{2}{7}\right)$

109. $(x + 1)^2 - (2x + 1)(x - 1)$ **110.** $(a + 3)^2 - (a + 4)(3a - 1)$

111. Square $2x + 1$. **112.** Square $3x - 2y$.

113. Find the cube of $x - 1$. **114.** Find the cube of $2a + b$.

115. Subtract $x - 6$ from the product of $x + 3$ and $2x - 5$.

116. Add $2x + 3$ to the product of $x + 3$ and $2x - 3$.

Applying the Concepts

In Problems 117–124, find the area of the shaded region.

△ **117.**

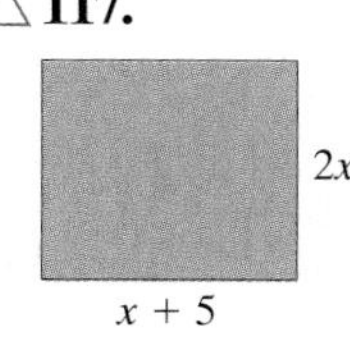

△ **118.**

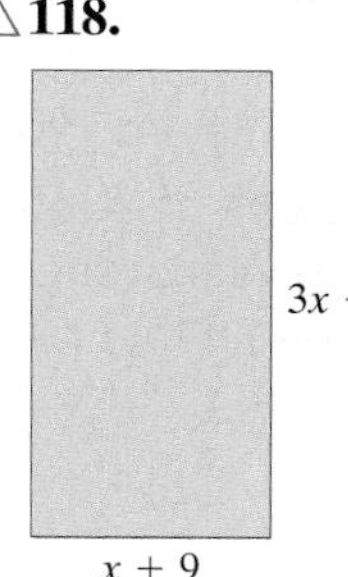

△ **119.**

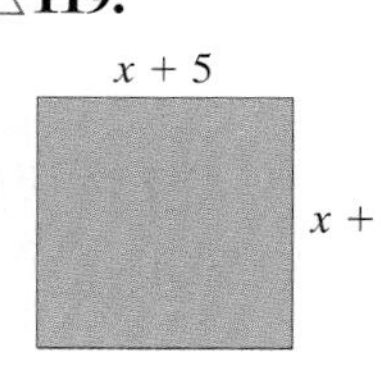

△ **120.**

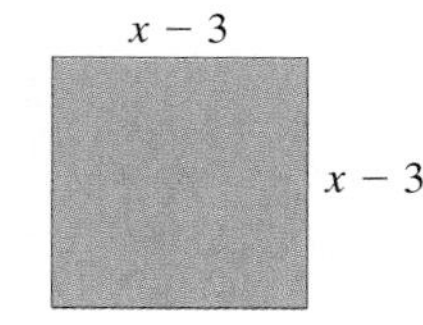

△ **121.**

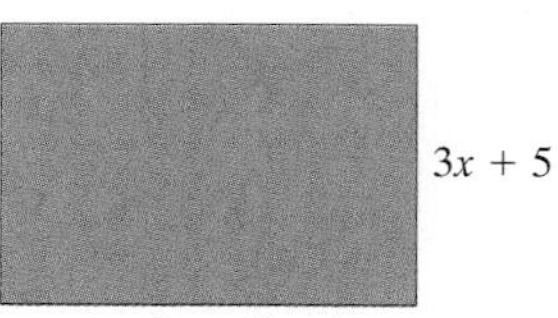

△ **122.**

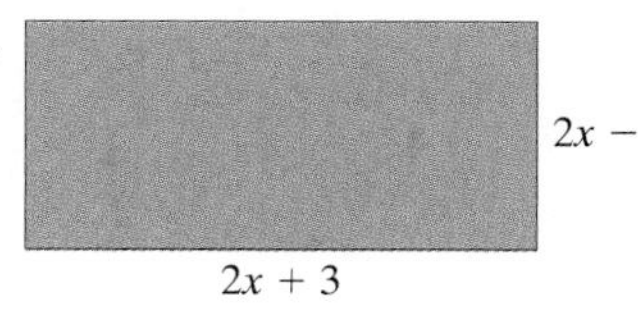

△ **123.**

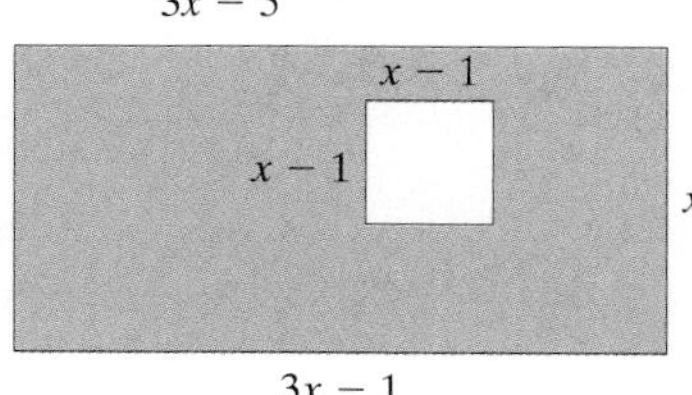

△ **124.**

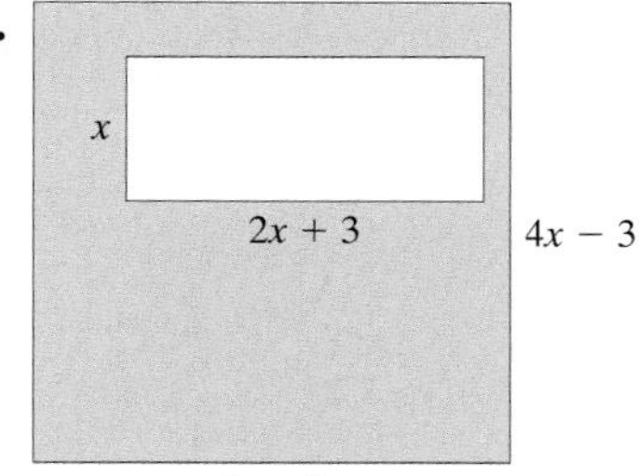

125. $x^2 + 12x + 20$
126. $12x^2 + 17x + 6$
127. $x^2 + 3x + 2$
128. $x^2 + 4x$
129. $40.8 + 0.4x$
130. $120{,}000 - 600$
131. $\pi x^2 + 4\pi x + 4\pi$ feet
132. $4\pi y^2 - 12\pi y + 9\pi$ meters
133. (a) a^2, ab, ab, b^2
(b) $a^2 + 2ab + b^2$
(c) length: $a + b$, width: $a + b$, area: $(a + b)^2$
134. $4x^2 + 56x + 195$ square inches
135. $x^3 + 2x^2 - 4x - 8$
136. $x^3 + 9x^2 + 27x + 27$
137. $x^2 + 14x + 49$
138. $9 - x^2 - 2xy - y^2$
139. $x^2 - 2xy + y^2 - 9$
140. $x^2 + 8x - 33$
141. $x^3 + 6x^2 + 12x + 8$
142. $y^3 - 9y^2 + 27y - 27$
143. $z^4 + 12z^3 + 54z^2 + 108z + 81$
144. $m^4 - 8m^3 + 24m^2 - 32m + 16$

In Problems 125 and 126, find an algebraic expression for the area of the rectangle by finding the sum of the four interior rectangles. Then find the area of the rectangle by multiplying the width and height. Compare the two expressions. How is multiplying a binomial related to finding the area of the rectangle?

△**125.**

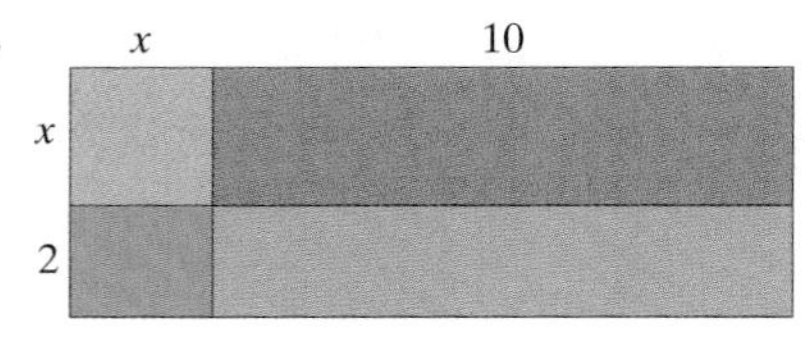

△**126.**

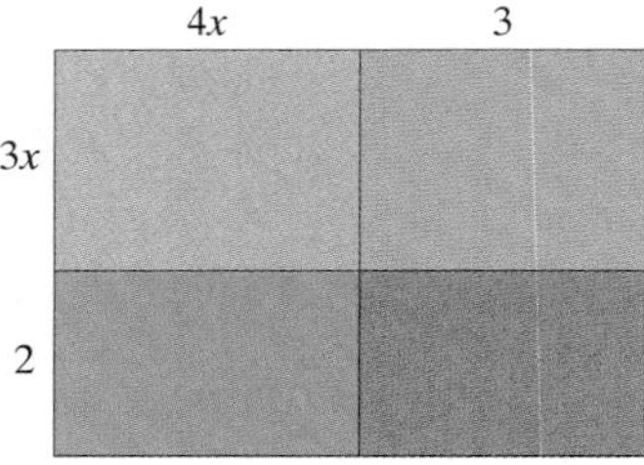

127. Consecutive Integers If x represents the first of three consecutive integers, write a polynomial that represents the product of the next two consecutive integers.

128. Consecutive Integers If x represents the first of three consecutive odd integers, write a polynomial that represents the product of the first and the third integer.

129. Sales Tax In a certain city, the sales tax rate has been increased by 2%. If the original sales tax rate was $x\%$, write a polynomial that represents the total cost of a \$40 purchase after the new sales tax has been imposed.

130. Property Tax In a certain city, the property tax rate has been decreased by 0.5%. If the original tax rate was $x\%$, write a polynomial that represents the amount of property tax on a \$120,000 home after the decrease in property tax rate.

△**131. Area of a Circle** Write a polynomial for the area of a circle with radius $x + 2$ feet.

△**132. Area of a Circle** Write a polynomial for the area of a circle with radius $2y - 3$ meters.

△**133. Perfect Square** Why is the expression $(a + b)^2$ called a perfect square? Consider the figure to the right.

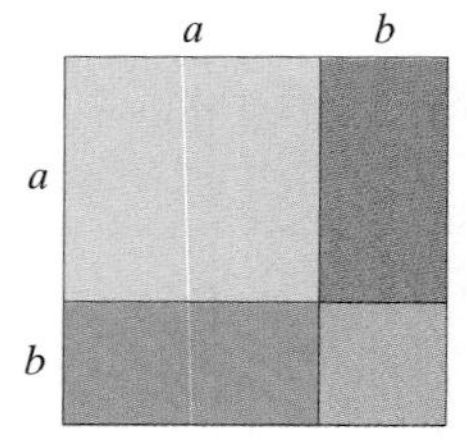

(a) Find the area of each of the four quadrilaterals.
(b) Use the result from part (a) to find the area of the entire region.
(c) Find the length and width of the entire region in terms of a and b. Use this result to find the area of the entire region. What do you notice?

△**134. Picture Frame** A photo is 13″ by 15″ and is bordered by a matting that is x inches wide. Write the polynomial that calculates the area of the entire framed picture.

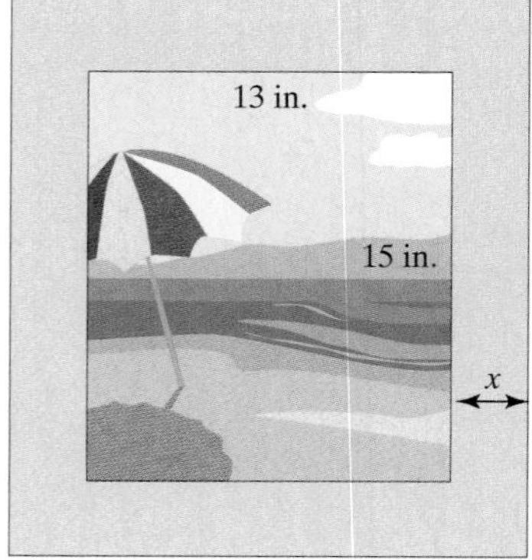

Extending the Concepts

In Problems 135–144, find the product.

135. $(x - 2)(x + 2)^2$

136. $(x + 3)^3$

137. $[(x + 4) + 3]^2$

138. $[3 - (x + y)][3 + (x + y)]$

139. $[(x - y) + 3][(x - y) - 3]$

140. $[(a + 4) - 7][(a + 4) + 7]$

141. $(x + 2)^3$

142. $(y - 3)^3$

143. $(z + 3)^4$

144. $(m - 2)^4$

3.4 Dividing Monomials: The Quotient Rule and Integer Exponents

OBJECTIVES

1. Simplify Exponential Expressions Using the Quotient Rule
2. Simplify Exponential Expressions Using the Quotient to a Power Rule
3. Simplify Exponential Expressions Using Zero as an Exponent
4. Simplify Exponential Expressions Involving Negative Exponents
5. Simplify Exponential Expressions Using the Laws of Exponents

Preparing for Dividing Monomials: The Quotient Rule and Integer Exponents

Before getting started, take the following readiness quiz. If you get a problem wrong, go back to the section cited and review the material.

1. Find the product: $(3a^2)^3$ [Section 3.2, page 187]

2. Evaluate: $\left(\frac{2}{3}\right)^2$ [Section 1.6, page 49–50]

3. Find the reciprocal: (a) 5 (b) $-\frac{6}{7}$ [Section 1.4, page 30]

Before we can turn our attention to dividing polynomials, we must first discuss dividing monomials.

1 Simplify Exponential Expressions Using the Quotient Rule

To find a general rule for the quotient of two exponential expressions, consider the following:

$$\frac{y^6}{y^2} = \frac{\overbrace{y \cdot y \cdot y \cdot y \cdot \not{y} \cdot \not{y}}^{6 \text{ factors}}}{\underbrace{\not{y} \cdot \not{y}}_{2 \text{ factors}}} = \underbrace{y \cdot y \cdot y \cdot y}_{4 \text{ factors}} = y^4$$

We might conclude from this result that

$$\frac{y^6}{y^2} = y^{6-2} = y^4$$

This result is true in general.

In Words
When dividing two exponential expressions with a common base, subtract the exponent in the denominator from the exponent in the numerator. Then write the base to that power.

QUOTIENT RULE FOR EXPONENTS

If a is a real number and if m and n are natural numbers, then

$$\frac{a^m}{a^n} = a^{m-n} \qquad \text{if } a \neq 0$$

Classroom Example
Simplify each expression:

(a) $\frac{4^5}{4^2}$ (b) $\frac{y^7}{y^3}$

Answer:

(a) $4^3 = 64$ (b) y^4

Teaching Tip

Emphasize to students that $\frac{4^5}{4^2} \neq 1^3$.

Students can write out the quotient in expanded form to see why:

$$\frac{4^5}{4^2} = \frac{4 \cdot 4 \cdot 4 \cdot 4 \cdot 4}{4 \cdot 4} = 4^3$$

EXAMPLE 1 Using the Quotient Rule to Simplify Expressions

Simplify each expression:

(a) $\frac{6^5}{6^3}$ **(b)** $\frac{n^6}{n^2}$

Solution

(a)
$$\frac{6^5}{6^3} = 6^{5-3}$$
$$= 6^2$$
Evaluate 6^2: $= 36$

(b)
$$\frac{n^6}{n^2} = n^{6-2}$$
$$= n^4$$

Preparing for...Answers **1.** $27a^6$ **2.** $\frac{4}{9}$ **3.** **(a)** $\frac{1}{5}$ **(b)** $-\frac{7}{6}$

Classroom Example Simplify each expression:

(a) $\frac{16z^5}{12z^3}$

(b) $\frac{-35a^7b^3}{25a^6b}$

Answer:

(a) $\frac{4}{3}z^2$

(b) $-\frac{7}{5}ab^2$

EXAMPLE 2 Using the Quotient Rule to Simplify Expressions

Simplify each expression:

(a) $\frac{25b^9}{15b^6}$ **(b)** $\frac{-24x^6y^4}{10x^4y}$

Solution

(a)
$$\frac{25b^9}{15b^6} = \frac{25}{15} \cdot \frac{b^9}{b^6}$$

Divide out like factors; apply the Quotient Rule:
$$= \frac{\cancel{5} \cdot 5}{\cancel{5} \cdot 3} \cdot b^{9-6}$$
$$= \frac{5}{3}b^3$$

(b)
$$\frac{-24x^6y^4}{10x^4y} = \frac{-24}{10} \cdot \frac{x^6}{x^4} \cdot \frac{y^4}{y}$$

Divide out like factors; apply the Quotient Rule:
$$= \frac{-12 \cdot \cancel{2}}{5 \cdot \cancel{2}} \cdot x^{6-4} \cdot y^{4-1}$$
$$= -\frac{12}{5}x^2y^3$$

QUICK ✓ *Simplify each expression.*

1. $\frac{3^7}{3^5}$ **2.** $\frac{y^9}{y^3}$ **3.** $\frac{14c^6}{10c^5}$ **4.** $\frac{-21w^4z^8}{14w^3z}$

2 Simplify Exponential Expressions Using the Quotient to a Power Rule

Now let's look at a quotient raised to a power:

$$\left(\frac{3}{2}\right)^4 = \left(\frac{3}{2}\right) \cdot \left(\frac{3}{2}\right) \cdot \left(\frac{3}{2}\right) \cdot \left(\frac{3}{2}\right) = \frac{\overbrace{3 \cdot 3 \cdot 3 \cdot 3}^{\text{4 factors}}}{\underbrace{2 \cdot 2 \cdot 2 \cdot 2}_{\text{4 factors}}} = \frac{3^4}{2^4}$$

We are led to the following result:

In Words
When a quotient is raised to a power, both the numerator and the denominator are raised to the indicated power.

QUOTIENT TO A POWER RULE FOR EXPONENTS

If a, b are real numbers and n is a natural number, then

$$\left(\frac{a}{b}\right)^n = \frac{a^n}{b^n} \quad \text{if } b \neq 0$$

Classroom Example Simplify the expression $\left(\frac{a}{2}\right)^3$.

Answer: $\frac{a^3}{8}$

EXAMPLE 3 Using the Quotient to a Power Rule to Simplify an Expression

Simplify the expression $\left(\frac{z}{3}\right)^2$.

Solution

This expression is a quotient, so we apply the Quotient to a Power Rule, raising both the numerator and the denominator to the indicated power.

$$\left(\frac{z}{3}\right)^2 = \frac{z^2}{3^2}$$

Evaluate 3^2: $= \dfrac{z^2}{9}$

Classroom Example ➤ Simplify the expressions:

(a) $\left(-\frac{3x}{y}\right)^2$ (b) $\left(\frac{2m^3}{n^2}\right)^4$

Answer:

(a) $\frac{9x^2}{y^2}$ (b) $\frac{16m^{12}}{n^8}$

EXAMPLE 4 Using the Quotient to a Power Rule to Simplify Expressions

Simplify the expressions:

(a) $\left(-\dfrac{2y}{z}\right)^3$ **(b)** $\left(\dfrac{3a^2}{b^3}\right)^4$

Solution

(a) $\left(-\dfrac{2y}{z}\right)^3 = \dfrac{(-2y)^3}{z^3}$

Use the Power Rule, $(ab)^n = a^n b^n$: $= \dfrac{(-2)^3(y)^3}{z^3}$

Evaluate $(-2)^3$: $= \dfrac{-8y^3}{z^3}$

(b) $\left(\dfrac{3a^2}{b^3}\right)^4 = \dfrac{(3a^2)^4}{(b^3)^4}$

Use the Power Rule, $(ab)^n = a^n b^n$: $= \dfrac{3^4(a^2)^4}{(b^3)^4}$

Evaluate $(3)^4$: $= \dfrac{81a^{2\cdot 4}}{b^{3\cdot 4}}$

$= \dfrac{81a^8}{b^{12}}$

QUICK ✓ *Simplify each expression.*

5. $\left(\dfrac{p}{2}\right)^4$ **6.** $\left(\dfrac{b}{3}\right)^3$ **7.** $\left(\dfrac{a^3}{b^4}\right)^2$ **8.** $\left(-\dfrac{2a^2}{b^4}\right)^3$

3 Simplify Exponential Expressions Using Zero as an Exponent

Now that we have fully discussed exponential expressions with natural number (or positive integer) exponents, we will extend the definition of exponential expressions to integer exponents. We begin with raising a real number to the 0 power.

DEFINITION OF ZERO EXPONENT

If a is a nonzero real number (that is, $a \neq 0$), we define

$$a^0 = 1$$

The reason that $a^0 = 1$ is based upon the Quotient Rule for Exponents. That is,

$$\frac{a^n}{a^n} = a^{n-n} = a^0. \text{ In addition, } \frac{a^n}{a^n} = \frac{\overbrace{a \cdot a \cdot \ldots \cdot a}^{n \text{ factors}}}{\underbrace{a \cdot a \cdot \ldots \cdot a}_{n \text{ factors}}} = 1. \text{ Therefore, } a^0 = 1.$$

Classroom Example
(a) 2^0 (b) n^0 (c) $(3x)^0$
(d) $18c^0$ (e) -6^0
Answer:
(a) 1 (b) 1 (c) 1
(d) 18 (e) −1

EXAMPLE 5 Using Zero as an Exponent

(a) $3^0 = 1$ **(b)** $x^0 = 1$ **(c)** $(4y)^0 = 1$

(d) $8w^0 = 8 \cdot 1 = 8$ **(e)** $-5^0 = -1 \cdot 5^0 = -1$

Work Smart
In Example 5(d), the exponent 0 applies *only* to the factor w.

QUICK ✓ *Simplify each expression. When they occur, assume variables are nonzero.*

9. 10^0 **10.** $4z^0$ **11.** $-p^0$

4 Simplify Exponential Expressions Involving Negative Exponents

Now let's look at exponents that are negative integers. Suppose that we wanted to simplify $\frac{z^3}{z^5}$. If we use the Quotient Rule for Exponents, we obtain

$$\frac{z^3}{z^5} = z^{3-5} = z^{-2}$$

We could also simplify this expression directly by dividing like factors:

$$\frac{z^3}{z^5} = \frac{\cancel{z} \cdot \cancel{z} \cdot \cancel{z}}{\cancel{z} \cdot \cancel{z} \cdot \cancel{z} \cdot z \cdot z} = \frac{1}{z^2}$$

This implies that $z^{-2} = \frac{1}{z^2}$. Considering this result, we define a raised to a negative integer power as follows:

DEFINITION OF NEGATIVE EXPONENT

If n is a positive integer and if a is a nonzero real number (that is, $a \neq 0$), then we define

$$a^{-n} = \frac{1}{a^n}$$

Teaching Tip
You may need to take a few minutes to remind students of the difference between the reciprocal of a number, used in negative exponents, and the opposite of a number.

Remember, the reciprocal of a nonzero number a is $\frac{1}{a}$. For example, the reciprocal of 7 is $\frac{1}{7}$ and the reciprocal of $-\frac{5}{8}$ is $-\frac{8}{5}$. Whenever we encounter a negative exponent, we should think, "take the reciprocal of the base."

Classroom Example
Simplify:
(a) x^{-2} (b) 3^{-2}
Answer:
(a) $\frac{1}{x^2}$ (b) $\frac{1}{9}$

EXAMPLE 6 Simplifying Exponential Expressions Containing Negative Integer Exponents

(a) $x^{-4} = \frac{1}{x^4}$ **(b)** $2^{-3} = \frac{1}{2^3} = \frac{1}{8}$

Classroom Example
Simplify each expression:
(a) $(-4)^{-2}$ (b) $(-a)^{-5}$
Answer:
(a) $\frac{1}{16}$ (b) $-\frac{1}{a^5}$

EXAMPLE 7 Simplifying Exponential Expressions Containing Negative Integer Exponents

Simplify each expression:

(a) $(-5)^{-2}$ **(b)** $(-n)^{-9}$

Solution

(a) $(-5)^{-2} = \dfrac{1}{(-5)^2}$

Evaluate: $= \dfrac{1}{25}$

(b) $(-n)^{-9} = \dfrac{1}{(-n)^9}$

$= \dfrac{1}{(-1)^9 \cdot n^9}$

Evaluate: $= -\dfrac{1}{n^9}$

Classroom Example Simplify: $9^{-1} + 3^{-2}$

Answer: $\frac{2}{9}$

EXAMPLE 8 Simplifying the Sum of Exponential Expressions Containing Negative Integer Exponents

Simplify: $2^{-2} + 4^{-1}$

Solution

$$2^{-2} + 4^{-1} = \frac{1}{2^2} + \frac{1}{4^1}$$

Evaluate: $= \dfrac{1}{4} + \dfrac{1}{4}$

Find the sum: $= \dfrac{2}{4}$

Write in lowest terms: $= \dfrac{1}{2}$

QUICK ✓ *Simplify each expression. Write answers with only positive exponents.*

12. z^{-5} **13.** $(-p)^{-4}$ **14.** $-x^{-3}$ **15.** $(-2)^{-4}$

16. -7^{-2} **17.** $(-7)^{-2}$ **18.** $4^{-1} - 2^{-3}$

Teaching Tip
You may want to make this argument first with a numerical expression:

$$\frac{1}{2^{-3}} = \frac{1}{\frac{1}{2^3}} = \frac{1}{\frac{1}{8}}$$
$$= 1 \div \frac{1}{8}$$
$$= 1 \cdot 8 = 8$$

How might we simplify $\dfrac{1}{a^{-n}}$? Because

$$\frac{1}{a^{-n}} = \frac{1}{\frac{1}{a^n}} = 1 \cdot \frac{a^n}{1} = a^n$$

we have the following result.

> If a is a real number and n is an integer, then
> $$\frac{1}{a^{-n}} = a^n \qquad \text{if } a \neq 0$$

Classroom Example Simplify:

(a) $\dfrac{3}{4^{-2}}$ (b) $\dfrac{5}{b^{-3}}$

Answer:

(a) 48 (b) $5b^3$

Work Smart

$$\frac{6}{2^{-3}} = \frac{6}{\frac{1}{8}}$$
$$= 6 \div \frac{1}{8}$$
$$= 6 \cdot 8$$

EXAMPLE 9 Simplifying Quotients Containing Negative Integer Exponents Using $\frac{1}{a^{-n}} = a^n$

Simplify:

(a) $\dfrac{6}{2^{-3}}$ **(b)** $\dfrac{7}{n^{-2}}$

Solution

(a) $\dfrac{6}{2^{-3}} = 6 \cdot 2^3 = 6 \cdot 8 = 48$ **(b)** $\dfrac{7}{n^{-2}} = 7 \cdot n^2 = 7n^2$

QUICK ✓ *Simplify each numerical expression.*

19. $\dfrac{1}{3^{-2}}$ **20.** $\dfrac{1}{v^{-1}}$ **21.** $\dfrac{1}{-10^{-2}}$ **22.** $\dfrac{5}{2z^{-2}}$

Classroom Example Simplify: $\left(\frac{4}{3}\right)^{-2}$

Answer: $\frac{9}{16}$

EXAMPLE 10 Simplifying Quotients Containing Negative Integer Exponents

Simplify: $\left(\dfrac{3}{2}\right)^{-3}$

Solution

$$\left(\frac{3}{2}\right)^{-3} = \frac{1}{\left(\frac{3}{2}\right)^{3}}$$

Use $\left(\frac{a}{b}\right)^{n} = \frac{a^n}{b^n}$:

$$= \frac{1}{\frac{3^3}{2^3}}$$

$$= \frac{1}{\frac{27}{8}}$$

$$= 1 \cdot \frac{8}{27}$$

$$= \frac{8}{27}$$

Work Smart

This "shortcut" says to simplify a quotient to a negative exponent, first take the reciprocal, then raise that expression to the positive exponent power.

The following shortcut is based upon the results of Example 10:

If a and b are real numbers and n is an integer, then

$$\left(\frac{a}{b}\right)^{-n} = \left(\frac{b}{a}\right)^{n} \quad \text{if } a \neq 0, b \neq 0$$

For example, $\left(\dfrac{3}{2}\right)^{-3} = \left(\dfrac{2}{3}\right)^{3} = \dfrac{8}{27}$.

Classroom Example Simplify: $\left(-\frac{y^2}{5}\right)^{-2}$

Answer: $\frac{25}{y^4}$

EXAMPLE 11 Simplifying Quotients Involving Negative Exponents

Simplify: $\left(-\dfrac{x^3}{4}\right)^{-2}$

Solution

$$\left(-\frac{x^3}{4}\right)^{-2} = \left(-\frac{4}{x^3}\right)^{2}$$

Use $\left(\frac{a}{b}\right)^{n} = \frac{a^n}{b^n}$:

$$= \frac{(-4)^2}{(x^3)^2}$$

$$= \frac{16}{x^6}$$

QUICK ✓ *Simplify each expression.*

23. $\left(\dfrac{7}{8}\right)^{-1}$ **24.** $\left(-\dfrac{1}{4}\right)^{-3}$ **25.** $\left(\dfrac{3a}{5}\right)^{-2}$ **26.** $\left(-\dfrac{2}{3n^4}\right)^{-3}$

5 Simplify Exponential Expressions Using the Laws of Exponents

We now summarize the Laws of Exponents where the exponents are integers.

THE LAWS OF EXPONENTS

If a and b are real numbers and if m and n are integers, then, assuming the expression is defined,

Rule		Examples
Product Rule	$a^m \cdot a^n = a^{m+n}$	$7^3 \cdot 7 = 7^4$; $x^2 \cdot x^5 = x^7$
Power Rule	$(a^m)^n = a^{m \cdot n}$	$(3^4)^2 = 3^8$; $(x^3)^2 = x^6$
Product to Power Rule	$(a \cdot b)^n = a^n \cdot b^n$	$(x^3y)^4 = x^{12}y^4$
Quotient Rule	$\frac{a^m}{a^n} = a^{m-n}$, if $a \neq 0$	$\frac{9^5}{9^3} = 9^2$; $\frac{y^{11}}{y^8} = y^3$
Quotient to Power Rule	$\left(\frac{a}{b}\right)^n = \frac{a^n}{b^n}$, if $b \neq 0$	$\left(\frac{3}{4}\right)^2 = \frac{9}{16}$; $\left(\frac{2x}{y^2}\right)^3 = \frac{8x^3}{y^6}$
Zero Exponent Rule	$a^0 = 1$, if $a \neq 0$	$3^0 = 1$; $(5m)^0 = 1$ $5m^0 = 5 \cdot 1 = 5$
Negative Exponent Rules	$a^{-n} = \frac{1}{a^n}$, if $a \neq 0$ $\frac{1}{a^{-n}} = a^n$, if $a \neq 0$	$2^{-3} = \frac{1}{8}$; $b^{-4} = \frac{1}{b^4}$ $\frac{1}{t^{-5}} = t^5$
Quotient to a Negative Power Rule	$\left(\frac{a}{b}\right)^{-n} = \left(\frac{b}{a}\right)^n$, if $a \neq 0$, $b \neq 0$	$\left(\frac{5}{k}\right)^{-2} = \left(\frac{k}{5}\right)^2 = \frac{k^2}{25}$

Let's do some examples where we use one or more of the rules listed above. To evaluate or simplify these exponential expressions, ask yourself the questions in the following display.

USING THE RULES OF EXPONENTS

- Does the exponential expression contain numerical expressions? If so, *evaluate* them. ("Evaluate" means "find the value.")
- Is the expression a product of monomials? If so, use the Product Rule:
$$a^m \cdot a^n = a^{m+n}$$
- Is the expression a quotient (division problem)? If so, use the Quotient Rule:
$$\frac{a^m}{a^n} = a^{m-n}$$
- Do you see a quantity raised to a power? If so, use one of the Power Rules:
$$(a^m)^n = a^{m \cdot n}, (a \cdot b)^n = a^n \cdot b^n, \text{ or } \left(\frac{a}{b}\right)^n = \frac{a^n}{b^n}.$$
- Is there a negative exponent in the expression? Remember, a negative in an exponent means "take the reciprocal of the base." Plus, an expression is not simplified if it contains negative exponents.
- Are you required to raise a quantity to the zero power? Remember, $a^0 = 1$ for $a \neq 0$.

Teaching Tip
Remind students that in a quotient such as $-3a$, $-3a \neq \frac{a}{-3}$; nor does $-3a = \frac{1}{3a}$. A negative **exponent** means "take the reciprocal."

EXAMPLE 12 How to Simplify Expressions Using the Product Rule and the Negative Exponent Rule

Simplify $\left(\frac{3}{2}a^3b^{-2}\right)(-12a^{-4}b^5)$. Write the answer with only positive exponents.

Classroom Example Simplify: $\left(\frac{4}{3}x^2y^{-3}\right)(-9x^{-4}y^5)$ Write answer with only positive exponents.

Answer: $\frac{-12y^2}{x^2}$

Step-by-Step Solution

Step 1: Rearrange factors.

$$\left(\frac{3}{2}a^3b^{-2}\right)(-12a^{-4}b^5) = \left(\frac{3}{2}\cdot(-12)\right)(a^3\cdot a^{-4})(b^{-2}\cdot b^5)$$

Step 2: Find each product.

Evaluate $\left(\frac{3}{2}\right)\cdot(-12)$. Apply the Product Rule to variable factors:

$$= -18a^{3+(-4)}b^{-2+5}$$
$$= -18a^{-1}b^3$$

Step 3: Simplify. Write the product so that the exponents are positive.

Apply the Negative Exponent Rule:

$$= -18\cdot\frac{1}{a}\cdot b^3$$
$$= \frac{-18b^3}{a}$$

Work Smart

$\frac{-18b^3}{a} \neq \frac{b^3}{18a}$

Do you know why?

QUICK ✓ *Simplify each expression. Write the answers with only positive exponents.*

27. $(-4a^{-3})(5a)$

28. $\left(-\frac{2}{5}m^{-2}n^{-1}\right)\left(-\frac{15}{3}mn^0\right)$

EXAMPLE 13 How to Simplify Expressions Using the Quotient Rule and the Negative Exponent Rule

Simplify $-\frac{27ab^4}{18a^3b}$. Write the answer with only positive exponents.

Classroom Example Simplify $-\frac{25a^5b}{15ab^4}$. Write answer with only positive exponents.

Answer: $-\frac{5a^4}{3b^3}$

Step-by-Step Solution

Step 1: Write the quotient as the product of factors.

Use $-\frac{a}{b} = \frac{-a}{b}$:

$$-\frac{27ab^4}{18a^3b} = \frac{-27}{18}\cdot\frac{a}{a^3}\cdot\frac{b^4}{b}$$

Step 2: Find each quotient.

Simplify $\frac{-27}{18}$. Use the Quotient Rule.

$$= \frac{-3}{2}\cdot a^{1-3}\cdot b^{4-1}$$
$$= -\frac{3}{2}\cdot a^{-2}\cdot b^3$$

Step 3: Simplify. Write the quotient so that the exponents are positive.

Apply the Negative Exponent Rule:

$$= -\frac{3}{2}\cdot\frac{1}{a^2}\cdot b^3$$
$$= -\frac{3b^3}{2a^2}$$

QUICK ✓ *Simplify. Write answers with only positive exponents.*

29. $\dfrac{16m^2}{2m^7}$ **30.** $-\dfrac{16a^4b^{-1}}{12ab^{-4}}$ **31.** $\dfrac{45x^{-2}y^{-2}}{35x^{-4}y}$

Classroom Example ➤ Simplify $\left(\dfrac{36x^{-3}y}{30xy^{-2}}\right)^{-2}$. Write answer with only positive exponents.

Answer: $\dfrac{25x^8}{36y^6}$

EXAMPLE 14 Simplifying Expressions Using the Power Rule and the Negative Exponent Rule

Simplify $\left(\dfrac{25a^{-2}b}{10ab^{-1}}\right)^{-2}$. Write the answer with only positive exponents.

Solution

Because there are like factors inside the parentheses, we first perform operations within parentheses before we apply the Power Rule.

$$\left(\frac{25a^{-2}b}{10ab^{-1}}\right)^{-2} = \left(\frac{25\cdot b\cdot b}{10\cdot a^2\cdot a}\right)^{-2}$$

Simplify $\dfrac{25}{10}$; use $a^m a^n = a^{m+n}$: $= \left(\dfrac{5b^2}{2a^3}\right)^{-2}$

Use $\left(\dfrac{a}{b}\right)^{-n} = \left(\dfrac{b}{a}\right)^n$: $= \left(\dfrac{2a^3}{5b^2}\right)^2$

Use $\left(\dfrac{a}{b}\right)^n = \dfrac{a^n}{b^n}$: $= \dfrac{(2a^3)^2}{(5b^2)^2}$

Use $(ab)^n = a^nb^n$ in the numerator and denominator: $= \dfrac{4a^6}{25b^4}$

QUICK ✓ *Simplify. Write the answers with only positive exponents.*

32. $(3y^2z^{-3})^{-2}$ **33.** $\left(\dfrac{2wz^{-3}}{7w^{-1}}\right)^2$

Classroom Example ➤ Simplify:

$$(5a^{-2}b)\left(\frac{2a^{-2}b^3}{3b}\right)^2$$

Write answer with only positive exponents.

Answer: $\dfrac{20b^5}{9a^6}$

EXAMPLE 15 Simplifying Exponential Expressions Using Exponent Rules

Simplify: $(6c^{-2}d)\left(\dfrac{3c^{-3}d^4}{2d}\right)^2$

Write the answer with only positive exponents.

Solution

Since the quotient in the parentheses is messy, we will simplify it first.

$$(6c^{-2}d)\left(\frac{3c^{-3}d^4}{2d}\right)^2 = \frac{6c^{-2}d}{1}\left(\frac{3c^{-3}d^{4-1}}{2}\right)^2$$

Simplify; use $a^{-n} = \dfrac{1}{a^n}$: $= \dfrac{6d}{c^2}\left(\dfrac{3d^3}{2c^3}\right)^2$

Use $\left(\frac{a}{b}\right)^n = \frac{a^n}{b^n}$ and $(ab)^n = a^n b^n$: $= \frac{6d}{c^2}\left(\frac{3^2(d^3)^2}{2^2(c^3)^2}\right)$

Use $(a^m)^n = a^{m \cdot n}$: $= \frac{6d}{c^2}\left(\frac{9d^6}{4c^6}\right)$

Simplify; rearrange factors: $= 6 \cdot \frac{9}{4} \cdot \frac{d \cdot d^6}{c^2 \cdot c^6}$

Evaluate the product $6 \cdot \frac{9}{4}$;
Use $a^m a^n = a^{m+n}$: $= \frac{27}{2} \cdot \frac{d^{1+6}}{c^{2+6}}$

$= \frac{27d^7}{2c^8}$

QUICK ✓ *Simplify the expression. Write answers with only positive exponents.*

34. $\left(\frac{-6p^{-2}}{p}\right)(3p^8)^{-1}$ **35.** $(-25k^5r^{-2})\left(\frac{2}{5}k^{-3}r\right)^2$

3.4 Exercises

1. a^{m-n}
2. 1
3. $\frac{1}{a^n}$
4. True
5. False
6. False
7. Answers may vary.
8. Answers may vary.
9. 27
10. 256
11. 16
12. 1000
13. x^9
14. x^6
15. a^{12}
16. a^{21}
17. $4y^3$
18. $\frac{1}{3}a^3$
19. $-\frac{2}{3}m^7$
20. $-\frac{3}{2}xy$

Concepts and Vocabulary

In Problems 1–3, fill in the blanks.

1. $\frac{a^m}{a^n} =$ __________ provided that $a \neq 0$.

2. Any expression with a base (other than zero) raised to the zero power has the value of __________.

3. $a^{-n} =$ __________ provided that $a \neq 0$.

In Problems 4–6, answer True or False to each statement.

4. When simplifying an exponential expression, it is possible to apply different exponent laws in different orders and still obtain the correct result.

5. $(-a)^n \cdot (-a)^n$ will always represent a positive number.

6. $\frac{6^{10}}{6^4} = 1^6$

7. Explain two different approaches to simplify $\left(\frac{x^8}{x^2}\right)^3$. Which do you prefer?

8. Use the Quotient Rule and the expression $\frac{a^n}{a^n}$, $a \neq 0$, to explain why $a^0 = 1$, $a \neq 0$.

Building Skills

In Problems 9–24, use the Quotient Rule to simplify.

9. $\frac{3^5}{3^2}$ **10.** $\frac{4^{10}}{4^6}$ **11.** $\frac{2^{23}}{2^{19}}$ **12.** $\frac{10^5}{10^2}$

13. $\frac{x^{15}}{x^6}$ **14.** $\frac{x^{20}}{x^{14}}$ **15.** $\frac{a^{13}}{a}$ **16.** $\frac{a^{22}}{a}$

17. $\frac{16y^4}{4y}$ **18.** $\frac{9a^4}{27a}$ **19.** $\frac{-16m^{10}}{24m^3}$ **20.** $\frac{-36x^2y^5}{24xy^4}$

21. $2m^8n^2$ 22. $3rs$
23. $\frac{11}{8}b^2$ 24. $\frac{3}{4}b^5$ 25. $\frac{27}{8}$ 26. $\frac{16}{81}$
27. $\frac{x^{15}}{27}$ 28. $\frac{49}{y^4}$ 29. $\frac{x^{20}}{y^{28}}$ 30. $-\frac{a^{15}}{b^{50}}$
31. $\frac{49a^4b^2}{c^6}$ 32. $\frac{16m^4n^8}{q^{12}}$
33. 1 34. -1 35. -1 36. 1
37. 18 38. 10
39. $\frac{1}{3}$ 40. $\frac{1}{16}$ 41. $\frac{1}{1000}$ 42. $\frac{1}{5}$
43. $\frac{1}{m^2}$ 44. $\frac{1}{k^2}$ 45. $\frac{4}{a^2}$ 46. $\frac{5}{b^3}$
47. $-\frac{4}{y^3}$ 48. $-\frac{7}{z^5}$ 49. $\frac{11}{18}$ 50. $-\frac{1}{16}$
51. 2 52. 9 53. $\frac{25}{4}$ 54. $\frac{8}{27}$
55. $\frac{z^2}{3}$ 56. $\frac{p^4}{16}$ 57. $-\frac{m^6}{8n^3}$ 58. $-\frac{27b^6}{125}$
59. 16 60. 36 61. $6x^4$ 62. $4b^3$
63. 4 64. 9
65. $\frac{1}{8}$ 66. $\frac{1}{10}$
67. 81 68. 3125
69. $\frac{1}{x^9}$ 70. $\frac{1}{b^{18}}$
71. x^{13} 72. $\frac{1}{a^{10}}$
73. $\frac{4}{x^3}$ 74. $\frac{4}{m^4}$
75. $-\frac{3z^3}{2x^3}$ 76. $-\frac{4a^{15}}{9b^{18}}$
77. $\frac{4a}{5b}$ 78. $\frac{x^4}{y^4}$
79. $\frac{1}{4x^3y^9}$ 80. $-\frac{42}{c^4}$
81. $-\frac{1}{a^{12}}$ 82. $-\frac{1}{a^{10}}$
83. $\frac{27}{m^6}$ 84. $\frac{8}{z^3}$ 85. $\frac{x^4y^6}{9}$
86. $-\frac{a^3b^2c}{4}$ 87. $\frac{2}{p^7}$ 88. $\frac{7}{a^5}$
89. 12 90. $-\frac{x^3}{6y^5}$ 91. $\frac{2x^8}{3}$
92. $\frac{4}{3x^8}$ 93. $\frac{16y}{27x^{18}}$ 94. $\frac{16b^{11}}{a^{10}}$
95. $\frac{9y}{2z^3}$ 96. $\frac{t^{10}}{324s^5}$ 97. $216x^3y^6$
98. $\frac{9a^{21}}{2}$ 99. $\frac{5ab^7}{3}$
100. $\frac{x^6y^{14}}{4}$ 101. $\frac{2x^6}{5}$
102. $\frac{a^{10}}{80b^{10}}$

21. $\dfrac{-12m^9n^3}{-6mn}$ **22.** $\dfrac{-15r^9s^2}{-5r^8s}$ **23.** $\dfrac{22b^3c^7}{16bc^7}$ **24.** $\dfrac{39ab^6}{52ab}$

In Problems 25–32, use the Quotient to a Power Rule to simplify.

25. $\left(\dfrac{3}{2}\right)^3$ **26.** $\left(\dfrac{4}{9}\right)^2$ **27.** $\left(\dfrac{x^5}{3}\right)^3$ **28.** $\left(\dfrac{7}{y^2}\right)^2$

29. $\left(-\dfrac{x^5}{y^7}\right)^4$ **30.** $\left(-\dfrac{a^3}{b^{10}}\right)^5$ **31.** $\left(\dfrac{7a^2b}{c^3}\right)^2$ **32.** $\left(\dfrac{2mn^2}{q^3}\right)^4$

In Problems 33–38, use the Zero Exponent Rule to simplify.

33. 3^0 **34.** -100^0 **35.** $-\left(\dfrac{1}{2}\right)^0$ **36.** $\left(\dfrac{2}{5}\right)^0$

37. $18 \cdot 2^0$ **38.** $14^0 \cdot 3^0 \cdot 10$

In Problems 39–62, use the Negative Exponent Rules to simplify. Write answers with only positive exponents.

39. 3^{-1} **40.** 4^{-2} **41.** 10^{-3} **42.** 5^{-1}

43. m^{-2} **44.** k^{-2} **45.** $4a^{-2}$ **46.** $5b^{-3}$

47. $-4y^{-3}$ **48.** $-7z^{-5}$ **49.** $2^{-1} + 3^{-2}$ **50.** $4^{-2} - 2^{-3}$

51. $\left(\dfrac{1}{2}\right)^{-1}$ **52.** $\left(\dfrac{1}{3}\right)^{-2}$ **53.** $\left(\dfrac{2}{5}\right)^{-2}$ **54.** $\left(\dfrac{3}{2}\right)^{-3}$

55. $\left(\dfrac{3}{z^2}\right)^{-1}$ **56.** $\left(\dfrac{4}{p^2}\right)^{-2}$ **57.** $\left(-\dfrac{2n}{m^2}\right)^{-3}$ **58.** $\left(-\dfrac{5}{3b^2}\right)^{-3}$

59. $\dfrac{1}{4^{-2}}$ **60.** $\dfrac{1}{6^{-2}}$ **61.** $\dfrac{6}{x^{-4}}$ **62.** $\dfrac{4}{b^{-3}}$

Mixed Practice

In Problems 63–102, simplify. Write answers with only positive exponents.

63. $2^5 \cdot 2^{-3}$ **64.** $3^8 \cdot 3^{-6}$ **65.** $2^{-7} \cdot 2^4$ **66.** $10^{-4} \cdot 10^3$

67. $\dfrac{3}{3^{-3}}$ **68.** $\dfrac{5^4}{5^{-1}}$ **69.** $\dfrac{x^6}{x^{15}}$ **70.** $\dfrac{b^{24}}{b^{42}}$

71. $\dfrac{x^{10}}{x^{-3}}$ **72.** $\dfrac{a^{-4}}{a^6}$ **73.** $\dfrac{8x^2}{2x^5}$ **74.** $\dfrac{24m^4}{6m^8}$

75. $\dfrac{-27xy^3z^4}{18x^4y^3z}$ **76.** $\dfrac{8a^{15}b^2}{-18b^{20}}$ **77.** $\dfrac{-12a^2b^3c^6}{-15ab^4c^6}$

78. $(4x^{-6}y^3)(4^{-1}x^{10}y^{-7})$ **79.** $(3x^2y^{-3})(12^{-1}x^{-5}y^{-6})$ **80.** $(-14c^2d^4)(3c^{-6}d^{-4})$

81. $(-a^4)^{-3}$ **82.** $(-x^{-2})^5$ **83.** $(3m^{-2})^3$

84. $(2z^{-1})^3$ **85.** $(-3x^{-2}y^{-3})^{-2}$ **86.** $(-4a^{-3}b^{-2}c^{-1})^{-1}$

87. $2p^{-4} \cdot p^{-3} \cdot p^0$ **88.** $7a^{-2} \cdot a^0 \cdot a^{-3}$

89. $(-16a^3)(-3a^4)\left(\dfrac{1}{4}a^{-7}\right)$ **90.** $(-3x^{-4}y)(2x^{-3}y^{-6})\left(\dfrac{1}{36}x^{10}\right)$

91. $\dfrac{8x^2 \cdot x^7}{12x^{-3} \cdot x^4}$ **92.** $\dfrac{32x^{-3} \cdot x^2}{24x^2 \cdot x^5}$ **93.** $(2x^{-3}y^{-2})^4(3x^2y^{-3})^{-3}$

94. $(4a^{-2}b)^3(2a^2b^{-4})^{-2}$ **95.** $\left(\dfrac{y}{2z}\right)^{-3} \cdot \left(\dfrac{3y^2}{4z^3}\right)^2$ **96.** $\left(\dfrac{s}{4t^2}\right)^3 \cdot \left(\dfrac{3s^2}{2t^4}\right)^{-4}$

97. $(4x^2y)^3 \cdot \left(\dfrac{2x}{3y}\right)^{-3}$ **98.** $(2a^5b^2)^3 \cdot \left(\dfrac{4a^{-2}b^3}{3a}\right)^{-2}$ **99.** $\dfrac{(5a^{-3}b^2)^2}{a^{-4}b^{-4}}(15a^{-3}b)^{-1}$

100. $\left(\dfrac{2x^{-4}y^{-3}}{4xy^3}\right)^{-2}(4x^2y^{-1})^{-2}$ **101.** $\dfrac{4x^{-3}(2x^3)}{20(x^{-3})^2}$ **102.** $\dfrac{(-2a^{-3}b^2)^{-4}}{5a^2b^2c^0}$

103. $x^3 + 6x^2 + 12x + 8$ cubic feet
104. $8x^3 - 12x^2 + 6x - 1$ cubic meters
105. $3x^3$ cubic meters
106. $36n^3$ cubic yards
107. $V = \frac{\pi d^2 h}{4}$
108. πh^3 or πr^3
109. $x^5 + 3x^4 - 3x^3 - 9x^2$ cubic cm
110. $4x^4 + 6x^3 - 4x^2$ cubic ft
111. $9\pi x^3$ square units
112. $\frac{\pi x^2}{16}$ m^2
113. $\frac{1}{x^n}$
114. $\frac{1}{x^{7a+6}}$
115. $x^{6n-2}y^2$
116. $\frac{1}{a^{3n-3}b^{6m-6}}$
117. $\frac{x^{6a^2}y^{3ab}}{z^{3ac}}$
118. $\frac{z^{8ac}}{x^{2a^2}y^{4ab}}$
119. $\frac{9}{8a^{10n}}$
120. $\frac{1}{y^{3a}}$

Applying the Concepts

△ **103. Volume of a Cube** Write the polynomial in simplified form that represents the volume of a cube if each side has a length of $x + 2$ ft, where *volume* = *side*^{3}.

△ **104. Volume of a Cube** Write the polynomial in simplified form that represents the volume of a cube if each side has a length of $2x - 1$ m, where *volume* = *side*^{3}.

△ **105. Volume of a Box** The volume of a rectangular solid is given by the equation $V = l \cdot w \cdot h$. Find the volume of a shipping box whose dimensions are $l = x$ m, $w = x$ m, and $h = 3x$ m.

△ **106. Volume of a Box** The volume of a rectangular solid is given by the equation $V = l \cdot w \cdot h$. Find the volume of a shoe box whose dimensions are $l = 3n$ yards, $w = 2n$ yards, and $h = 6n$ yards.

△ **107. Volume of a Cylinder** The volume of a right circular cylinder is given by the equation $V = \pi r^2 h$, where r is the radius of the circular base of the cylinder and h is its height. Rewrite this equation in terms of the diameter, d, of the base of the cylinder.

△ **108. Volume of a Cylinder** The volume of a right circular cylinder is given by the equation $V = \pi r^2 h$, where r is the radius of the circular base of the cylinder and h is its height. Write an algebraic expression for the volume of a cylinder in which the radius of the circular base is equal to the height of the cylinder.

△ **109. Volume of a Trough** A trough is in the shape of a triangular prism. The volume of a prism is found by multiplying the area of the triangle by the height of prism (the distance between the two triangular bases). Use this fact to find the polynomial which represents the volume of the trough whose dimensions (in centimeters) are shown to the right.

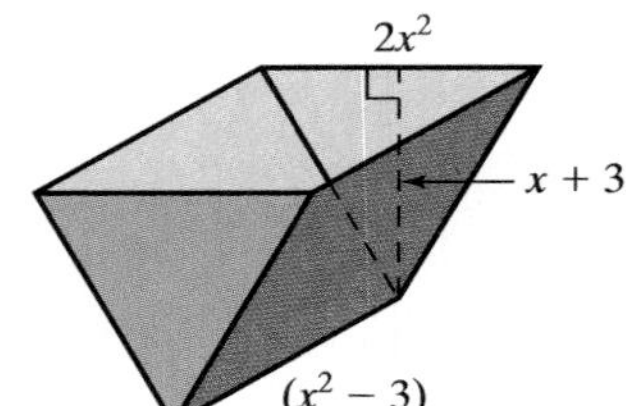

△ **110. Volume of a Trough** The triangular prism shown to the right has the following dimensions: $x + 2$ ft is the length of the base of the triangle and $4x^2$ ft is the height of the triangle. The length of the prism is $2x - 1$ ft. Write the polynomial which represents the volume of the trough.

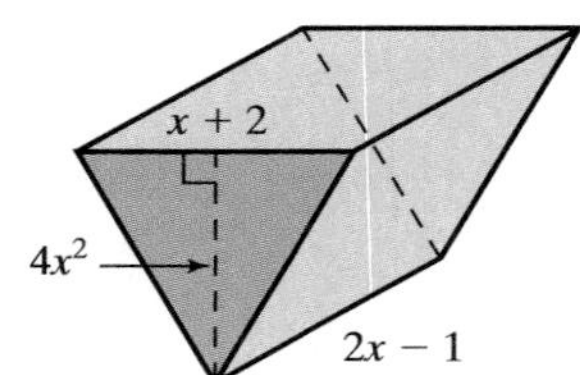

△ **111. Making Buttons** How much fabric is required to cover x buttons in the shape of a circle whose radius is $3x$?

△ **112. Making a Tablecloth** How much fabric is required to make a round tablecloth if the diameter of the tablecloth is $\left(\frac{1}{2}x\right)$ meters?

Extending the Concepts

In Problems 113–120, simplify. Write answers with only positive exponents.

113. $\frac{x^{2n}}{x^{3n}}$

114. $\frac{(x^{a-2})^3}{(x^{2a})^5}$

115. $\left(\frac{x^n y^m}{x^{4n-1}y^{m+1}}\right)^{-2}$

116. $\left(\frac{a^n b^{2m}}{ab^2}\right)^{-3}$

117. $(x^{2a}y^b z^{-c})^{3a}$

118. $(x^{-a}y^{-2b}z^{4c})^{2a}$

119. $\frac{(3a^n)^2}{(2a^{4n})^3}$

120. $\frac{y^a}{y^{4a}}$

PUTTING THE CONCEPTS TOGETHER (SECTIONS 3.1–3.4)

1. Yes; degree 6; trinomial
2. (a) 0 (b) −4 (c) 2
3. $8x^4 - 4x^2 + 14$
4. $5x^2y + 3xy + 2y^2$
5. $-6m^3n^2 + 2m^2n^4$
6. $5x^2 - 17x - 12$
7. $8x^2 - 26xy + 21y^2$
8. $25x^2 - 64$
9. $4x^2 + 12xy + 9y^2$
10. $6a^3 + 22a^2 - 40a$
11. $16m^4 + 4m^3 + 10m^2 - 20m - 24$
12. $-8x^8z^9$
13. $-\frac{15}{mn}$
14. $-3a^3b^2$
15. $\frac{1}{2x^2y^2}$
16. $\frac{y^4}{16z^6}$
17. $\frac{8}{27r^6}$
18. $\frac{q^{56}}{r^{40}t^{64}}$
19. $\frac{1}{128x^6y^2}$
20. $-\frac{48y^6}{x^{12}}$

These problems cover important concepts from Sections 3.1 to 3.4. We designed these problems so that you can review the chapter so far and show your mastery of the concepts. Take time to work these problems before proceeding with the next section. The answers to these problems are located at the back of the text starting on page AN-6.

1. Determine whether the following algebraic expression is a polynomial (Yes or No). If it is a polynomial, state the degree and then state if it is a monomial, binomial, or trinomial: $6x^2y^4 - 8x^5 + 3$.

2. Evaluate the polynomial $-x^2 + 3x$ for the given values: (a) 0 (b) −1 (c) 2.

In Problems 3–11, perform the indicated operation.

3. $(6x^4 - 2x^2 + 7) - (-2x^4 - 7 + 2x^2)$

4. $(2x^2y - xy + 3y^2) + (4xy - y^2 + 3x^2y)$

5. $-2mn(3m^2n - mn^3)$

6. $(5x + 3)(x - 4)$

7. $(2x - 3y)(4x - 7y)$

8. $(5x + 8)(5x - 8)$

9. $(2x + 3y)^2$

10. $2a(3a - 4)(a + 5)$

11. $(4m + 3)(4m^3 - 2m^2 + 4m - 8)$

In Problems 12–20, simplify each expression. Write answers with only positive exponents.

12. $(-2x^7y^0z)(4xz^8)$

13. $(5m^3n^{-2})(-3m^{-4}n)$

14. $\frac{-18a^8b^3}{6a^5b}$

15. $\frac{16x^7y}{32x^9y^3}$

16. $(4y^{-2}z^3)^{-2}$

17. $\left(\frac{3}{2}r^2\right)^{-3}$

18. $\left(\frac{q^{-6}rt^5}{qr^{-4}t^{-3}}\right)^{-8}$

19. $(4x^{-3}y^4)^{-2}\cdot(2x^4y^{-2})^{-3}$

20. $\left(\frac{2x^3y^{-3}}{4x^{-1}y^0}\right)^{-4}\cdot(-3x^4y^{-6})$

3.5 Dividing Polynomials

OBJECTIVES

① Divide a Polynomial by a Monomial

② Divide a Polynomial by a Binomial

Preparing for Dividing Polynomials

Before getting started, take the following readiness quiz. If you get a problem wrong, go back to the section cited and review the material.

1. Find the quotient: $\frac{24a^3}{6a}$ [Section 3.4, pp. 201–202]
2. Find the quotient: $\frac{-49x^3}{21x^4}$ [Section 3.4, p. 208]
3. Find the product: $3x(7x - 2)$ [Section 3.3, pp. 190–191]
4. Tell the degree of $4x^2 - 2x + 1$. [Section 3.1, pp. 175–176]

Work Smart

A polynomial is a monomial or the sum of monomials.

We have now presented a discussion of adding, subtracting, and multiplying polynomials. All that's left to discuss is polynomial division! We begin with dividing a polynomial by a monomial. Recall, when we compute a quotient using long division, such as $15\overline{)345}$ with quotient 23, the number 15 is called the *divisor,* the number 345 is called the *dividend,* and the number 23 is called the *quotient.* We use the same terminology with polynomial division.

Preparing for...Answers **1.** $4a^2$ **2.** $-\frac{7}{3x}$ **3.** $21x^2 - 6x$ **4.** 2

1 Divide a Polynomial by a Monomial

Dividing a polynomial by a monomial requires the Quotient Rule for Exponents. For convenience, we repeat the rule below.

QUOTIENT RULE FOR EXPONENTS

If a is a real number and m and n are integers, then

$$\frac{a^m}{a^n} = a^{m-n}, \quad a \neq 0$$

In Words
To divide a polynomial by a monomial, divide each of the terms of the polynomial numerator (dividend) by the monomial denominator (divisor).

Remember, to add two rational numbers, the denominators must be the same. When we have the same denominator, we add the numerators and write the result over the common denominator. For example, $\frac{3}{11} + \frac{4}{11} = \frac{3+4}{11} = \frac{7}{11}$. When dividing a polynomial by a monomial, we reverse this process. So, if A, B, and C are monomials, we can write $\frac{A+B}{C}$ as $\frac{A}{C} + \frac{B}{C}$. We can extend this result to polynomials with three or more terms quite easily.

Classroom Example ➤
Divide and simplify: $\frac{12y^3 + 16y^2}{4y}$
Answer: $3y^2 + 4y$

EXAMPLE 1 Dividing a Binomial by a Monomial

Divide and simplify: $\frac{9x^3 - 21x^2}{3x}$

Solution

We divide each term in the numerator by the denominator.

$$\frac{9x^3 - 21x^2}{3x} = \frac{9x^3}{3x} - \frac{21x^2}{3x}$$

Use the Quotient Rule, $\frac{a^m}{a^n} = a^{m-n}$: $= \frac{9}{3}x^{3-1} - \frac{21}{3}x^{2-1}$

Simplify: $= 3x^2 - 7x$ ■

Classroom Example ➤
Divide and simplify:
$\frac{21n^4 - 18n^3 + 12n^2}{3n^2}$
Answer: $7n^2 - 6n + 4$

EXAMPLE 2 Dividing a Trinomial by a Monomial

Divide and simplify: $\frac{12p^4 + 24p^3 + 4p^2}{4p^2}$

Solution

We divide each term in the numerator by the denominator.

$$\frac{12p^4 + 24p^3 + 4p^2}{4p^2} = \frac{12p^4}{4p^2} + \frac{24p^3}{4p^2} + \frac{4p^2}{4p^2}$$

Use the Quotient Rule, $\frac{a^m}{a^n} = a^{m-n}$: $= \frac{12}{4}p^{4-2} + \frac{24}{4}p^{3-2} + \frac{4}{4}\cdot\frac{p^2}{p^2}$

Simplify: $= 3p^2 + 6p + 1$ ■

Work Smart

$$\frac{12p^4 + 24p^3 + 4p^2}{4p^2} \neq \frac{12p^4 + 24p^3 + \cancel{4p^2}}{\cancel{4p^2}} \neq 12p^4 + 24p^3$$

You must divide *each* term in the numerator by the monomial in the denominator.

Classroom Example Divide and simplify:
$$\frac{9x^2y^2 + 27x^2y - 10xy^2}{3x^2y^2}$$
Answer: $3 + \frac{9}{y} - \frac{10}{3x}$

EXAMPLE 3 Dividing a Trinomial by a Monomial

Divide and simplify: $\dfrac{8a^2b^2 - 6a^2b + 5ab^2}{2a^2b^2}$

Solution

$$\frac{8a^2b^2 - 6a^2b + 5ab^2}{2a^2b^2} = \frac{8a^2b^2}{2a^2b^2} - \frac{6a^2b}{2a^2b^2} + \frac{5ab^2}{2a^2b^2}$$

Use the Quotient Rule, $\frac{a^m}{a^n} = a^{m-n}$:
$$= \frac{8}{2}a^{2-2}b^{2-2} - \frac{6}{2}a^{2-2}b^{1-2} + \frac{5}{2}a^{1-2}b^{2-2}$$

Simplify:
$$= 4a^0b^0 - 3a^0b^{-1} + \frac{5}{2}a^{-1}b^0$$

Use the Negative Exponent Rule, $a^{-n} = \frac{1}{a^n}$:
$$= 4 \cdot 1 \cdot 1 - 3 \cdot 1 \cdot \frac{1}{b} + \frac{5}{2} \cdot \frac{1}{a} \cdot 1$$
$$= 4 - \frac{3}{b} + \frac{5}{2a}$$

QUICK ✓ *Find the quotient.*

1. $\dfrac{10n^4 - 20n^3 + 5n^2}{5n^2}$

2. $\dfrac{12k^4 - 18k^2 + 5}{2k^2}$

3. $\dfrac{x^4y^4 + 8x^2y^2 - 4xy}{4x^3y}$

2 Divide a Polynomial by a Binomial

The procedure for dividing a polynomial by a binomial is similar to the procedure for dividing two integers. Although this procedure should be familiar to you, we review it in the following example.

Classroom Example Divide 328 by 12.
Answer: $27\frac{1}{3} = 27 + \frac{1}{3}$

EXAMPLE 4 Dividing an Integer by an Integer Using Long Division

Divide 579 by 16.

Solution

```
                 36 <— quotient
divisor —> 16)579 <— dividend
              48   <— 3 · 16 = 48 (subtract)
              99      bring down the 9
              96 <— 6 · 16 = 96 (subtract)
               3 <— remainder
```

So $\dfrac{579}{16} = 36\dfrac{3}{16}$.

We can always check our work after completing a long division problem by multiplying the quotient by the divisor and adding this product to the remainder. The result should be the dividend. That is,

$$(\text{Quotient})(\text{Divisor}) + \text{Remainder} = \text{Dividend}$$

For example, we can check the results of Example 4 as follows:

$$(36)(16) + 3 = 576 + 3 = 579$$

One other comment regarding Example 4: we wrote the solution as $\frac{579}{16} = 36\frac{3}{16}$. Remember, the mixed number $36\frac{3}{16}$ means $36 + \frac{3}{16}$. So, the answer is written in the form

$$\text{Quotient} + \frac{\text{Remainder}}{\text{Divisor}}$$

Classroom Example
Find the quotient when $x^2 + 13x + 42$ is divided by $x + 6$.
Answer: $x + 7$

Now let's go over an example that introduces "how to" divide a polynomial by a binomial using long division.

EXAMPLE 5 How to Divide a Polynomial by a Binomial Using Long Division

Find the quotient when $x^2 + 10x + 21$ is divided by $x + 7$.

Step-by-Step Solution

To divide two polynomials, we first write each polynomial in standard form (descending order of degree). The dividend is $x^2 + 10x + 21$ and the divisor is $x + 7$.

Step 1: Divide the highest degree term of the dividend, x^2, by the highest degree term of the divisor, x. Enter the result over the term x^2.

$$x + 7\overline{)x^2 + 10x + 21} \quad \text{quotient: } x \leftarrow \frac{x^2}{x} = x$$

Step 2: Multiply x by $x + 7$. Be sure to vertically align like terms.

$$\begin{array}{r} x \\ x + 7\overline{)x^2 + 10x + 21} \\ \underline{x^2 + 7x} \end{array} \leftarrow x(x + 7) = x^2 + 7x$$

Step 3: Subtract $x^2 + 7x$ from $x^2 + 10x + 21$.

Teaching Tip
Remind students of the definition of subtraction: $a - b = a + (-b)$.

$$\begin{array}{r} x \\ x + 7\overline{)x^2 + 10x + 21} \\ \underline{-(x^2 + 7x)} \\ 3x + 21 \end{array} \leftarrow (x^2 + 10x + 21) - (x^2 + 7x) = 3x + 21$$

Step 4: Repeat Steps 1 – 3 treating $3x + 21$ as the dividend.

$$\begin{array}{r} x + 3 \\ x + 7\overline{)x^2 + 10x + 21} \\ \underline{-(x^2 + 7x)} \\ 3x + 21 \\ \underline{-(3x + 21)} \\ 0 \end{array} \quad 3 \leftarrow \frac{3x}{x} = 3, \quad (3x + 21) - (3x + 21) = 0$$

The quotient is $x + 3$ and the remainder is 0.

Step 5: Verify by showing that (Quotient)(Divisor) + Remainder = Dividend.

$$(x + 3)(x + 7) + 0 = x^2 + 10x + 21$$

The product checks, so $x^2 + 10x + 21$ divided by $x + 7$ is $x + 3$. We can express this division using fractions, as follows:

$$\frac{x^2 + 10x + 21}{x + 7} = x + 3$$

EXAMPLE 6 How to Divide a Polynomial by a Binomial Using Long Division

Find the quotient when $6x^2 + 9x - 10$ is divided by $2x - 1$.

Classroom Example
Find the quotient when $4x^2 - 4x - 9$ is divided by $2x + 3$.

Answer: $2x - 5 + \frac{6}{2x + 3}$

Step-by-Step Solution

Each polynomial is in standard form. The dividend is $6x^2 + 9x - 10$ and the divisor is $2x - 1$.

Step 1: Divide the highest degree term of the dividend, $6x^2$, by the highest degree term of the divisor, $2x$. Enter the result over the term $6x^2$.

$$\begin{array}{r} 3x \\ 2x - 1\overline{)6x^2 + 9x - 10} \end{array} \qquad \frac{6x^2}{2x} = 3x$$

Step 2: Multiply $3x$ by $2x - 1$. Be sure to vertically align like terms.

$$\begin{array}{r} 3x \\ 2x - 1\overline{)6x^2 + 9x - 10} \\ \underline{6x^2 - 3x} \end{array} \qquad 3x(2x - 1) = 6x^2 - 3x$$

Step 3: Subtract $6x^2 - 3x$ from $6x^2 + 9x - 10$.

$$\begin{array}{r} 3x \\ 2x - 1\overline{)6x^2 + 9x - 10} \\ \underline{-(6x^2 - 3x)} \\ 12x - 10 \end{array} \qquad 6x^2 + 9x - 10 - (6x^2 - 3x) = 12x - 10$$

Step 4: Repeat Steps 1–3 treating $12x - 10$ as the dividend.

Teaching Tip
Most student errors in polynomial division occur during the subtraction step. Suggest that students write the product in parentheses and show the subtraction symbol, as is done in the examples.

$$\begin{array}{r} 3x + 6 \\ 2x - 1\overline{)4x^2 + 9x - 10} \\ \underline{-(6x^2 - 3x)} \\ 12x - 10 \\ \underline{-(12x - 6)} \\ -4 \end{array} \qquad \begin{array}{l} \frac{12x}{2x} = 6 \\ 6(2x - 1) = 12x - 6 \\ (12x - 10) - (12x - 6) \\ = 12x - 12x - 10 + 6 = -4 \end{array}$$

Because -4 is a lower degree than the divisor, $2x - 1$, the process ends. The quotient is $3x + 6$ and the remainder is -4.

Step 5: Check We verify that (Quotient)(Divisor) + Remainder = Dividend.

$$(3x + 6)(2x - 1) + (-4) = 6x^2 - 3x + 12x - 6 + (-4)$$

Combine like terms: $= 6x^2 + 9x - 10$

Work Smart
You know that you're finished dividing when the degree of the remainder is less than the degree of the divisor.

The product checks, so our answer is correct. So

$$\frac{6x^2 + 9x - 10}{2x - 1} = 3x + 6 + \frac{-4}{2x - 1}.$$

QUICK ✓ *Find the quotient by performing long division.*

4. $\frac{x^2 - 3x - 40}{x + 5}$

5. $\frac{2x^2 - 5x - 12}{2x + 3}$

6. $\frac{4x^2 + 17x + 21}{x + 3}$

Classroom Example
Simplify by performing long division:
$$\frac{7 + 2b^4 + 5b^2}{b^2 + 2}$$

Answer: $2b^2 + 1 + \frac{5}{b^2 + 2}$

EXAMPLE 7 Dividing Two Polynomials Using Long Division

Simplify by performing long division: $\frac{8 - 9x + 2x^2 + 12x^3 + 5x^5}{x^2 + 3}$

Solution

Do you notice that the dividend is not written in standard form (that is, the dividend is not written in descending order of degree)? Also, when we write the dividend in standard form, do you notice that there is no x^4 term? When a term is missing, its coefficient is 0, so we rewrite the division problem as follows:

$$\frac{5x^5 + 0 \cdot x^4 + 12x^3 + 2x^2 - 9x + 8}{x^2 + 3}.$$

$$\begin{array}{rl}
 & \quad 5x^3 \qquad\quad - 3x + 2 \\
x^2 + 3 \overline{)} & 5x^5 + 0x^4 + 12x^3 + 2x^2 - 9x + 8 \\
 & \underline{-(5x^5 \qquad + 15x^3)} \leftarrow 5x^3(x^2 + 3) \\
 & -3x^3 + 2x^2 - 9x + 8 \leftarrow 5x^5 + 12x^3 + 2x^2 - 9x + 8 - (5x^5 + 15x^3) = -3x^3 + 2x^2 - 9x + 8 \\
 & \underline{-(-3x^3 \qquad - 9x)} \leftarrow -3x(x^2 + 3) \\
 & 2x^2 \qquad + 8 \leftarrow (-3x^3 + 2x^2 - 9x + 8) - (-3x^3 - 9x) = 2x^2 + 8 \\
 & \underline{-(2x^2 \qquad + 6)} \leftarrow 2(x^2 + 3) \\
 & 2 \leftarrow \text{Remainder}
\end{array}$$

The quotient is $5x^3 - 3x + 2$ and the remainder is 2. We now check our work.

Check (Quotient)(Divisor) + Remainder = Dividend

$$\begin{aligned}(5x^3 - 3x + 2)(x^2 + 3) + 2 &= 5x^5 + 15x^3 - 3x^3 - 9x + 2x^2 + 6 + 2 \\ &= 5x^5 + 12x^3 + 2x^2 - 9x + 8\end{aligned}$$

Our answer checks, so $\dfrac{8 - 9x + 2x^2 + 12x^3 + 5x^5}{x^2 + 3} = 5x^3 - 3x + 2 + \dfrac{2}{x^2 + 3}$.

QUICK ✓ *Simplify by performing long division.*

7. $\dfrac{x + 1 - 3x^2 + 4x^3}{x + 2}$ **8.** $\dfrac{2x^3 + 3x^2 + 10}{2x - 5}$ **9.** $\dfrac{4x^3 - 3x^2 + x + 1}{x^2 + 2}$

3.5 Exercises

For Extra Help:

Student Solutions Manual

CD Video

PH Math/Tutor Center

MathXL Tutorials on CD

MathXL®

MyMathLab

Concepts and Vocabulary

In Problems 1–3, fill in the blanks.

1. standard
2. $\frac{4x^4}{2x}$; $\frac{8x^2}{2x}$
3. quotient; remainder; dividend
4. True
5. False
6. False

1. To begin a polynomial division problem, write the divisor and the dividend in ________ form.

2. The first step to simplify $\dfrac{4x^4 + 8x^2}{2x}$ would be to rewrite $\dfrac{4x^4 + 8x^2}{2x}$ as ________ + ________.

3. To check the result of long division, multiply the ________ and the divisor and add this result to the ______. If correct, this result will be equal to the ________.

In Problems 4–6, answer True or False to each statement.

4. The division problem $\dfrac{3x^2 - 4}{x}$ can be written as $\dfrac{3x^2}{x} - \dfrac{4}{x}$.

5. $\dfrac{1}{x - 3} = \dfrac{1}{x} - \dfrac{1}{3}$

6. $\dfrac{x^2 - 2}{x + 2} = x - 1$

7. Answers may vary.
8. Answers may vary.
9. $2x - 1$ 10. $x - 2$
11. $3a + 9 - \frac{1}{a^2}$ 12. $2m + 1 - \frac{1}{2m^2}$
13. $\frac{n^4}{5} - \frac{2n^2}{5} - 1$ 14. $x - 3 + \frac{2}{x}$
15. $\frac{5r^2}{3} - 3$ 16. $\frac{4}{5}a - \frac{3}{5} + \frac{2}{5a}$
17. $-x^2 + 3x - \frac{9}{x^2}$
18. $-\frac{7}{6} - \frac{7}{2p} + \frac{1}{2p^2}$
19. $\frac{3}{8} + \frac{z^2}{2} - \frac{z}{4}$ 20. $-1 + 3y^2 - \frac{16y^3}{5}$
21. $x + \frac{3x^2}{2} + \frac{2}{3x}$
22. $\frac{2}{x^2} + \frac{7}{2} - \frac{3x^2}{2} + 3x$
23. $-7x + 5y$ 24. $-7x - 4y$
25. $-\frac{6y}{x} + 15$ 26. $-\frac{3y^2}{x^2} - 5$
27. $-\frac{5}{a} - \frac{2c^2}{a^2b}$ 28. $-\frac{2}{mn} + \frac{3m}{n}$
29. $x - 7$ 30. $x + 12$
31. $x - 5$ 32. $x + 8$
33. $x^2 + 6x - 3$ 34. $x^2 - 7x + 2$
35. $x^3 - 3x^2 + 6x - 2$
36. $x^3 - x^2 + 1$
37. $x^2 - 2x + 3 + \frac{5}{x + 1}$
38. $x^2 - 4x + 3 + \frac{-2}{x - 3}$
39. $2x + 3$ 40. $2x - 7$
41. $x^2 + 6x + 7 + \frac{16}{x - 2}$
42. $x^2 - 3x + 4 + \frac{2}{x + 1}$
43. $2x^3 + 5x^2 + 9x - 4 + \frac{-17}{x - 4}$
44. $3x^3 - 2x^2 + x + 5 + \frac{-3}{x + 3}$
45. $2x - 6$ 46. $4x + 5$
47. $x^2 + 4x - 3 + \frac{2}{2x - 1}$
48. $3a^2 + 2a + 4 + \frac{5}{4a + 1}$
49. $x - 4 + \frac{-4}{5 + x}$
50. $x - 7 + \frac{7}{-9 + x}$
51. $2x - 1 + \frac{6}{1 + 2x}$
52. $3x - 2 + \frac{-10}{2 + 3x}$
53. $x + 2 + \frac{2x - 4}{x^2 - 2}$
54. $16x^4 + 12x^2 + 9$
55. $x^2 + 3x - 6$
56. $3x^3 - x - 1$

7. Explain how to divide polynomials when the divisor is a monomial and then when the divisor is a binomial. Which procedures are the same and which are different?

8. The first steps of a division problem are written below. Describe what has occurred. Are there any potential errors with this presentation? What would you recommend this student do to improve his or her chances of obtaining the correct answer?

$$\begin{array}{r} 2x^2 \\ x^2 - 3 \overline{) \; 2x^4 - 3x^3 + 2x + 1} \\ \underline{-(2x^4 - 6x^2)} \end{array}$$

Building Skills

In Problems 9–28, divide and simplify.

9. $\frac{4x^2 - 2x}{2x}$ **10.** $\frac{3x^3 - 6x^2}{3x^2}$ **11.** $\frac{9a^3 + 27a^2 - 3}{3a^2}$

12. $\frac{16m^3 + 8m^2 - 4}{8m^2}$ **13.** $\frac{5n^5 - 10n^3 - 25n}{25n}$ **14.** $\frac{5x^3 - 15x^2 + 10x}{5x^2}$

15. $\frac{15r^5 - 27r^3}{9r^3}$ **16.** $\frac{16a^5 - 12a^4 + 8a^3}{20a^4}$ **17.** $\frac{3x^7 - 9x^6 + 27x^3}{-3x^5}$

18. $\frac{7p^4 + 21p^3 - 3p^2}{-6p^4}$ **19.** $\frac{3z + 4z^3 - 2z^2}{8z}$ **20.** $\frac{-5y^2 + 15y^4 - 16y^5}{5y^2}$

21. $\frac{6x^2 + 9x^3 + 4}{6x}$ **22.** $\frac{4 + 7x^2 - 3x^4 + 6x^3}{2x^2}$ **23.** $\frac{14x - 10y}{-2}$

24. $\frac{35x + 20y}{-5}$ **25.** $\frac{12y - 30x}{-2x}$ **26.** $\frac{21y^2 + 35x^2}{-7x^2}$

27. $\frac{25a^3b^2c + 10a^2bc^3}{-5a^4b^2c}$ **28.** $\frac{16m^2n^3 - 24m^4n^3}{-8m^3n^4}$

In Problems 29–56, find the quotient using long division.

29. $\frac{x^2 - 4x - 21}{x + 3}$ **30.** $\frac{x^2 + 18x + 72}{x + 6}$ **31.** $\frac{x^2 - 9x + 20}{x - 4}$

32. $\frac{x^2 + 4x - 32}{x - 4}$ **33.** $\frac{x^3 + 4x^2 - 15x + 6}{x - 2}$ **34.** $\frac{x^3 - x^2 - 40x + 12}{x + 6}$

35. $\frac{x^4 - x^3 + 10x - 4}{x + 2}$ **36.** $\frac{x^4 - 2x^3 + x^2 + x - 1}{x - 1}$ **37.** $\frac{x^3 - x^2 + x + 8}{x + 1}$

38. $\frac{x^3 - 7x^2 + 15x - 11}{x - 3}$ **39.** $\frac{2x^2 - 7x - 15}{x - 5}$ **40.** $\frac{2x^2 + 5x - 42}{x + 6}$

41. $\frac{x^3 + 4x^2 - 5x + 2}{x - 2}$ **42.** $\frac{x^3 - 2x^2 + x + 6}{x + 1}$

43. $\frac{2x^4 - 3x^3 - 11x^2 - 40x - 1}{x - 4}$ **44.** $\frac{3x^4 + 7x^3 - 5x^2 + 8x + 12}{x + 3}$

45. $\frac{6x^2 - 28x + 30}{3x - 5}$ **46.** $\frac{12x^2 - 25x - 50}{3x - 10}$ **47.** $\frac{2x^3 + 7x^2 - 10x + 5}{2x - 1}$

48. $\frac{12a^3 + 11a^2 + 18a + 9}{4a + 1}$ **49.** $\frac{-24 + x^2 + x}{5 + x}$ **50.** $\frac{-16x + 70 + x^2}{-9 + x}$

51. $\frac{4x^2 + 5}{1 + 2x}$ **52.** $\frac{9x^2 - 14}{2 + 3x}$

53. $\frac{x^3 + 2x^2 - 8}{x^2 - 2}$ **54.** $\frac{64x^6 - 27}{4x^2 - 3}$

55. $\frac{18 + x^4 - 9x^2 + 3x^3 - 9x}{x^2 - 3}$ **56.** $\frac{x^2 - 1 + 6x^4 - x - 2x^3}{2x^2 + 1}$

57. $a^2 + a - 30$
58. $-3m^2 + 8m$
59. $-x^2 - x - 6$
60. $21x^4 + 6x^3 - 9x^2$
61. $3x + 1$
62. $x^2 + 3x + \frac{5}{2x} - \frac{1}{2x^2}$
63. $-2ab + 2b^2$
64. $\frac{a^8b}{2}$
65. $x - 2 + \frac{3}{x}$
66. $b^2 - b - 6$
67. $\frac{2}{3}x^7y^8z^{12}$
68. $6ab + 9ab^2 - 12a^2b$
69. $n^2 - 6n + 9$
70. $3x - 1$
71. $-10x^5 - 10x^3 + 40x^6$
72. $x^3 - x^2 + 18x$
73. $6pq - 2q^2 - p^2$
74. $4x^2 - 7x + 10$
75. $4n^2 + 10nm - 6m^2$
76. $6x^2 - 11xy + 3y^2$
77. $x - 3 + \frac{-10}{x - 5}$
78. $x - 6 + \frac{-4}{x + 2}$
79. $x^3 + 6x^2 + 4x - 5$
80. $x^3 - 6x^2 + 11x - 12$
81. $15rs^2 - 4r^2s$
82. $x^2 + 12x + 36$
83. $-3n - \frac{8}{3}m + 4$
84. $-3x^5 + 6x^3 - 3x^2$
85. $2x^6 - 4x^5 + 6x^4$
86. $n^2 - 2 + \frac{7n}{2} - \frac{1}{2n}$
87. $-x + \frac{2}{x^2} - \frac{4}{x}$
88. $1 + \frac{-5}{x + 2}$
89. $-\frac{1}{2x} - \frac{2}{x^2} + \frac{1}{x^3}$
90. $\frac{144}{x^2}$
91. x
92. $(x - 6)$ in.
93. $z + 3$
94. $6a + 4$
95. $4x + 4$
96. $(x + 1)$ ft

Mixed Practice

In Problems 57–86, perform the indicated operation.

57. $(a - 5)(a + 6)$

58. $(7n + 3m^2 + 4m) + (-6m^2 - 7n + 4m)$

59. $(2x - 8) - (3x + x^2 - 2)$

60. $3x^2(7x^2 + 2x - 3)$

61. $\dfrac{6x^2 - 1 - x}{2x - 1}$

62. $\dfrac{2x^4 - 6x^3 - 5x + 1}{-2x^2}$

63. $(2ab + b^2 - a^2) + (b^2 - 4ab + a^2)$

64. $\dfrac{3a^4b^{-2}}{6a^{-4}b^{-3}}$

65. $\dfrac{x^2 - 2x + 3}{x}$

66. $(b - 3)(b + 2)$

67. $\dfrac{18x^3y^{-4}z^6}{27x^{-4}y^{-12}z^{-6}}$

68. $(4ab + 6ab^2) - (12a^2b - 2ab - 3ab^2)$

69. $(n - 3)^2$

70. $\dfrac{3 + 6x^2 - 11x}{2x - 3}$

71. $(x^3 + x - 4x^4)(-10x^2)$

72. $(2x^3 + 6x) + (12x - x^2 - x^3)$

73. $(2pq - q^2) + (4pq - p^2 - q^2)$

74. $(4x^2 - 6x + 3) - (x - 7)$

75. $(4n - 2m)(n + 3m)$

76. $(2x - 3y)(3x - y)$

77. $\dfrac{x^2 - 8x + 5}{x - 5}$

78. $\dfrac{x^2 - 4x - 16}{x + 2}$

79. $(x^2 + x - 1)(x + 5)$

80. $(x^2 - 2x + 3)(x - 4)$

81. $(7rs^2 - 2r^2s) - (2r^2s - 8rs^2)$

82. $(x + 6)^2$

83. $\dfrac{-9mn^2 - 8m^2n + 12mn}{3mn}$

84. $(x^4 - 2x^2 + x)(-3x)$

85. $2x^4(x^2 - 2x + 3)$

86. $\dfrac{2n^3 - 4n + 7n^2 - 1}{2n}$

Applying the Concepts

87. Find the quotient of $(3x^4 - 6x + 12x^2)$ and $-3x^3$.

88. Divide $x - 3$ by $x + 2$.

89. Divide the sum of $x^2 + 3x - 1$ and $x - 1$ by $-2x^3$.

90. Divide x^2 into the square of the difference of $x - 9$ and $x + 3$.

△ **91. Volume of a Box** The volume of a rectangular solid is $x^3 - 5x^2 + 6x$. One side measures $x - 2$ and the other measures $x - 3$. What is the measure of the third side?

△ **92. Area of a Rectangle** A rectangle has area $x^2 + 2x - 48$ square inches. If one side measures $x + 8$ inches, what is the measurement of the other side?

△ **93. Area of a Rectangle** If the area of a rectangle is $z^2 + 6z + 9$ and the length is $z + 3$, what is the width?

△ **94. Area of a Triangle** If the area of a triangle is $6a^2 - 5a - 6$ and the length of the base is $2a - 3$, what is the height?

△ **95. Area of a Triangle** If the area of a triangle is $6x^3 - 2x^2 - 8x$ and the height is $3x^2 - 4x$, what is the length of the base?

△ **96. Volume of a Box** The volume of a rectangular solid is $x^3 + 2x^2 - x - 2$ square feet. One side measures $x + 2$ feet and the other measures $x - 1$ feet. What is the measure of the third side?

97. \$182.11 per computer
98. \$73.88 per camera
99. 6
100. $2x$

97. Average Cost The average cost of manufacturing x computers per day is given by $\dfrac{0.004x^3 - 0.8x^2 + 180x + 5000}{x}$. Rewrite this quotient as the sum of four expressions by dividing x into each term in the numerator. Use this result to determine the average cost of manufacturing $x = 140$ computers in a day.

98. Average Cost The average cost of manufacturing x digital cameras per day is given by $\dfrac{0.0024x^3 - 0.4x^2 + 46x + 4000}{x}$. Rewrite this quotient as the sum of four expressions by dividing x into each term in the numerator. Use this result to determine the average cost of manufacturing $x = 90$ cameras in a day.

Extending the Concepts

In Problems 99 and 100, determine the value of the missing term so that the remainder is zero.

99. $\dfrac{(6x^2 - 13x + ?)}{2x - 3}$ **100.** $\dfrac{(3x^2 + ? - 5)}{x - 1}$

3.6 Applying Exponent Rules: Scientific Notation

OBJECTIVES

1. Convert Decimal Notation to Scientific Notation
2. Convert Scientific Notation to Decimal Notation
3. Use Scientific Notation to Multiply and Divide

Preparing for Applying Exponent Rules: Scientific Notation

Before getting started, take the following readiness quiz. If you get a problem wrong, go back to the section cited and review the material.

1. Find the product: $(3a^6)(4.5a^4)$ [Section 3.2, p. 188]

2. Find the product: $(7n^3)(2n^{-2})$ [Section 3.4, pp. 207–208]

3. Find the quotient: $\dfrac{3.6b^9}{0.9b^{-2}}$ [Section 3.4, pp. 207–209]

Did you know that *Star Wars: Episode 1—The Phantom Menace* had box office revenues of \$922,800,000? Did you know that the mass of a dust particle is 0.000000000753 kg? These numbers are difficult to write and difficult to read, so we use exponents to rewrite them.

1 Convert Decimal Notation to Scientific Notation

We use two types of notation to write numbers: decimal notation and scientific notation. Decimal notation is the notation we commonly see when we read the newspaper or a magazine. The numbers \$922,800,000 and 0.000000000753 kg are written in decimal notation. When we write a number in scientific notation we express that number as the product of two factors: One factor is a number between 1 and 10, including 1 but not including 10, and the other factor is an integer power of 10.

DEFINITION

When a number has been written as the product of a number x, where $1 \le x < 10$, and a power of 10, it is said to be written in **scientific notation.** That is, a number is written in scientific notation when it is of the form

$$x \times 10^N$$

where

$$1 \le x < 10 \text{ and } N \text{ is an integer}$$

Preparing for...Answers **1.** $13.5a^{10}$ **2.** $14n$ **3.** $4b^{11}$

Notice in the definition, $x < 10$. That's because when $x = 10$, we have 10^1, a power of ten.

For example, in scientific notation,

Box Office Revenue for *Star Wars: The Phantom Menace* $= \$9.228 \times 10^8$ dollars

Mass of a dust particle $= 7.53 \times 10^{-10}$ kilograms

In Words

When you have a number greater than or equal to 1, use $10^{\text{positive exponent}}$. When you have a number between 0 and 1, use $10^{\text{negative exponent}}$.

Steps to Convert from Decimal Notation to Scientific Notation

To change a positive number into scientific notation:

Step 1: Count the number N of decimal places that the decimal point must be moved in order to arrive at a number x, where $1 \leq x < 10$.

Step 2: If the original number is greater than or equal to 1, the scientific notation is $x \times 10^N$. If the original number is between 0 and 1, the scientific notation is $x \times 10^{-N}$.

Classroom Example
Write 648 in scientific notation.
Answer: 6.48×10^2

EXAMPLE 1 How to Convert from Decimal Notation to Scientific Notation

Write 5283 in scientific notation.

Step-by-Step Solution

For a number to be in scientific notation, the decimal must be moved so that there is a single nonzero digit to the left of the decimal point. All remaining digits must appear to the right of the decimal point.

Step 1: The "understood" decimal point in 5283 follows the 3. Therefore, we will move the decimal to the left until it is between the 5 and the 2. Do you see why? This requires that we move the decimal $N = 3$ places. 5.2.8.3.

Step 2: The original number is greater than 1, so we write 5283 in scientific notation as 5.283×10^3

Classroom Example
Write 0.062 in scientific notation.
Answer: 6.2×10^{-2}

EXAMPLE 2 How to Convert from Decimal Notation to Scientific Notation

Write 0.054 in scientific notation.

Step-by-Step Solution

Step 1: Because 0.054 is less than 1, we shall move the decimal point to the right until it is between the 5 and the 4. This requires that we move the decimal $N = 2$ places. 0.05.4

Step 2: The original number is between 0 and 1, so we write 0.054 in scientific notation as 5.4×10^{-2}

QUICK ✓ *Write each number in scientific notation.*

1. 432 **2.** 10,302 **3.** 5,432,000 **4.** 0.093 **5.** 0.0000459 **6.** 0.00000008

2 Convert Scientific Notation to Decimal Notation

Now we are going to convert a number from scientific notation to decimal notation. Study Table 3 to discover the pattern.

Table 3

Scientific Notation	Product	Decimal Notation	Location of Decimal Point
3.69×10^2	3.69×100	369	moved 2 places to the right
3.69×10^1	3.69×10	36.9	moved 1 place to the right
3.69×10^0	3.69×1	3.69	didn't move
3.69×10^{-1}	3.69×0.1	0.369	moved 1 place to the left
3.69×10^{-2}	3.69×0.01	0.0369	moved 2 places to the left

We present the following steps for converting a number from scientific notation to decimal notation.

Steps to Convert a Number from Scientific Notation to Decimal Notation

Step 1: Determine the exponent on the number 10.

Step 2: If the exponent is positive, then move the decimal N decimal places to the right. If the exponent is negative, then move the decimal $|N|$ decimal places to the left.

Classroom Example
Write 2.7×10^4 in decimal notation.

Answer: 27,000

EXAMPLE 3 How to Convert from Scientific Notation to Decimal Notation

Write 2.3×10^3 in decimal notation.

Step-by-Step Solution

Step 1: Determine the exponent on the number 10. — The exponent on the 10 is 3.

Step 2: Since the exponent is positive, we move the decimal point 3 places to the right. Notice we add zeros to the right of 3, as needed. — 2.300

So, $2.3 \times 10^3 = 2300$.

Classroom Example
Write 3.8×10^{-4} in decimal notation.

Answer: 0.00038

EXAMPLE 4 How to Convert from Scientific Notation to Decimal Notation

Write 4.57×10^{-5} in decimal notation.

Step-by-Step Solution

Step 1: Determine the exponent on the number 10. — The exponent on the 10 is -5.

Step 2: Since the exponent is negative, we move the decimal 5 places to the left. — Add zeros to the left of the original decimal point. 0.0000457

So, $4.57 \times 10^{-5} = 0.0000457$.

Work Smart

In Example 3, notice that

$$2.3 \times 10^3 = 2.3 \times 1000 = 2300$$

In Example 4, notice that

$$4.57 \times 10^{-5} = 4.57 \times 0.00001 = 0.0000457$$

QUICK ✓ *Write each number in decimal notation.*

7. 3.1×10^2 **8.** 9.01×10^{-1} **9.** 1.7×10^5 **10.** 7×10^0 **11.** 8.9×10^{-4}

3 Use Scientific Notation to Multiply and Divide

The Laws of Exponents make it relatively straightforward to multiply and divide numbers that are written in scientific notation. The two Laws of Exponents that we make use of are the Product Rule, $a^m \cdot a^n = a^{m+n}$, and the Quotient Rule, $\frac{a^m}{a^n} = a^{m-n}$. We will use these laws where the base is 10 as follows:

$$10^m \cdot 10^n = 10^{m+n} \quad \text{and} \quad \frac{10^m}{10^n} = 10^{m-n}$$

Classroom Example Perform the indicated operation. Express the answer in scientific notation.

$(1.5 \times 10^2) \cdot (3 \times 10^3)$

Answer: 4.5×10^5

EXAMPLE 5 Multiplying Using Scientific Notation

Perform the indicated operation. Express the answer in scientific notation.

$$(3 \times 10^2) \cdot (2.5 \times 10^5)$$

Solution

$$(3 \times 10^2) \cdot (2.5 \times 10^5) = (3 \cdot 2.5) \times (10^2 \cdot 10^5)$$

Multiply $3 \cdot 2.5$; use the Product Rule for Exponents on the base 10 factors: $= 7.5 \times 10^7$

Classroom Example Perform the indicated operation and express the answer in scientific notation.

(a) $(3 \times 10^{-3}) \cdot (8 \times 10^7)$

(b) $(1.2 \times 10^{-3}) \cdot (9 \times 10^{-1})$

Answer:

(a) 2.4×10^5

(b) 1.08×10^{-3}

EXAMPLE 6 Multiplying Using Scientific Notation

Perform the indicated operation. Express the answer in scientific notation.

(a) $(4 \times 10^{-2}) \cdot (6 \times 10^8)$ **(b)** $(3.2 \times 10^{-3}) \cdot (4.8 \times 10^{-4})$

Solution

(a) $(4 \times 10^{-2}) \cdot (6 \times 10^8) = (4 \cdot 6) \times (10^{-2} \cdot 10^8)$

Multiply $4 \cdot 6$; use the Product Rule: $= 24 \times 10^6$

Convert 24 to scientific notation: $= (2.4 \times 10^1) \times 10^6$

Apply the Product Rule for Exponents: $= 2.4 \times 10^7$

(b) $(3.2 \times 10^{-3}) \cdot (4.8 \times 10^{-4}) = (3.2 \cdot 4.8) \times (10^{-3} \cdot 10^{-4})$

Multiply $3.2 \cdot 4.8$; use the Product Rule: $= 15.36 \times 10^{-7}$

Convert 15.36 to scientific notation: $= (1.536 \times 10^1) \times 10^{-7}$

Apply the Product Rule for Exponents: $= 1.536 \times 10^{-6}$

Teaching Tip Point out that the result from multiplication using scientific notation initially may not be in scientific notation. See Example 6(b) as an illustration.

QUICK ✓ *Perform the indicated operation. Express the answer in scientific notation.*

12. $(3 \times 10^4) \cdot (2 \times 10^3)$ **13.** $(2 \times 10^{-2}) \cdot (4 \times 10^{-1})$

14. $(5 \times 10^{-4}) \cdot (3 \times 10^7)$ **15.** $(8 \times 10^{-4}) \cdot (3.5 \times 10^{-2})$

Classroom Example
Perform the indicated operation and express the solution in scientific notation.

(a) $\frac{12 \times 10^7}{3 \times 10^3}$

(b) $\frac{1.8 \times 10^2}{9 \times 10^{-2}}$

Answer:

(a) 4×10^4

(b) 2×10^3

EXAMPLE 7 Dividing Using Scientific Notation

Perform the indicated operation. Express the answer in scientific notation.

(a) $\frac{6 \times 10^6}{2 \times 10^2}$ **(b)** $\frac{2.4 \times 10^4}{3 \times 10^{-2}}$

Solution

(a) Divide $\frac{6}{2}$; use the Quotient Rule for Exponents:

$$\frac{6 \times 10^6}{2 \times 10^2} = \frac{6}{2} \times \frac{10^6}{10^2} = 3 \times 10^4$$

(b) Divide $\frac{2.4}{3}$; use the Quotient Rule for Exponents:

$$\frac{2.4 \times 10^4}{3 \times 10^{-2}} = \frac{2.4}{3} \times \frac{10^4}{10^{-2}} = 0.8 \times 10^{4-(-2)}$$

Convert 0.8 to scientific notation:

$$= (8 \times 10^{-1}) \times 10^6 = 8 \times 10^5$$

QUICK ✓ *Perform the indicated operation. Express the answer in scientific notation.*

16. $\frac{8 \times 10^6}{2 \times 10^1}$ **17.** $\frac{2.8 \times 10^{-7}}{1.4 \times 10^{-3}}$ **18.** $\frac{3.6 \times 10^3}{7.2 \times 10^{-1}}$ **19.** $\frac{5 \times 10^{-2}}{8 \times 10^2}$

Classroom Example
Use Example 8. Find the circulation of *USA Today* in the month of July 2003, when *USA Today* was published 23 times.

Answer:

5.06×10^7 newspapers = 50,600,000

EXAMPLE 8 Finding Newspaper Circulation

In 2003, *USA Today* had a daily circulation of 2.2×10^6 newspapers. What was the circulation of *USA Today* in the month of April when *USA Today* was published 21 times? (*Note*: *USA Today* does not publish on weekends.) Express the answer in scientific notation and in decimal notation.(*Source: Information Please* Almanac)

Solution

To find the total monthly circulation for the newspaper in April, we multiply the number of daily copies by the number of publication days. There are 21 publication days in April.

Let's first change 21 to scientific notation: $21 = 2.1 \times 10^1$. So we have

$$(2.2 \times 10^6) \cdot (2.1 \times 10^1) = (2.2 \cdot 2.1) \times (10^6 \cdot 10^1)$$
$$= 4.62 \times 10^7$$
$$= 46{,}200{,}000 \text{ newspapers.}$$

A total of 46,200,000 copies of *USA Today* were circulated in April 2003.

QUICK ✓

20. Saudi Arabia produced nearly 8.4×10^6 barrels of oil per day in August 2003. How many barrels of oil did Saudi Arabia produce in August 2003? Express the solution in scientific notation and decimal notation.

3.6 Exercises

For Extra Help: Student Solutions Manual | CD Video | PH Math/Tutor Center | MathXL Tutorials on CD | MathXL® | MyMathLab

1. scientific notation
2. left
3. positive
4. True
5. False
6. True
7. Answers may vary.
8. Answers may vary.
9. 3×10^5
10. 4.21×10^8
11. 6.4×10^7
12. 8×10^9
13. 5.1×10^{-4}
14. 1×10^{-7}
15. 1×10^{-9}
16. 2.83×10^{-5}
17. 8.007×10^9
18. 4.01×10^8
19. 3.09×10^{-5}
20. 2.03×10^{-4}
21. 6.2×10^2
22. 8×10^0
23. 4×10^0
24. 1.2×10^2
25. 420,000
26. 375
27. 100,000,000
28. 6,000,000
29. 0.0039
30. 0.0000061
31. 0.4
32. 0.0005
33. 3760
34. 0.49
35. 0.0082
36. 540,000
37. 0.00006
38. 0.005123
39. 7,050,000
40. 700,000,000
41. 5×10^{-4}
42. 1,000,000,000
43. 142,000
44. 6×10^9
45. 0.008
46. 3.5×10^{-3}
47. 6.9×10^4
48. 0.4
49. 6.415×10^9
50. 2.94975×10^8
51. 1.46×10^9
52. 3.67×10^{11}

Concepts and Vocabulary

In Problems 1–3, fill in the blanks.

1. A number written as 3.2×10^{-6} is said to be written in _____ _____.

2. To write 3.2×10^{-6} in decimal notation, move the decimal point in 3.2 six places to the _____.

3. When writing 47,000,000 in scientific notation, the power of 10 will be _____. (positive or negative).

In Problems 4–6, answer True or False to each statement.

4. To convert 2.4×10^3 to decimal notation, move the decimal three places to the right.

5. When a number is expressed in scientific notation, it is expressed as the product of a number x, $0 \le x < 1$, and a power of 10.

6. It is useful to do calculations with scientific notation when using very large or very small numbers.

7. Explain the advantages of using scientific notation. Are there any disadvantages?

8. Explain why scientific notation is used to perform calculations that involve multiplying and dividing but not adding and subtracting.

Building Skills

In Problems 9–24, write each number in scientific notation.

9. 300,000 **10.** 421,000,000 **11.** 64,000,000 **12.** 8,000,000,000

13. 0.00051 **14.** 0.0000001 **15.** 0.000000001 **16.** 0.0000283

17. 8,007,000,000 **18.** 401,000,000 **19.** 0.0000309 **20.** 0.000201

21. 620 **22.** 8 **23.** 4 **24.** 120

In Problems 25–40, write each number in decimal notation.

25. 4.2×10^5 **26.** 3.75×10^2 **27.** 1×10^8 **28.** 6×10^6

29. 3.9×10^{-3} **30.** 6.1×10^{-6} **31.** 4×10^{-1} **32.** 5×10^{-4}

33. 3.76×10^3 **34.** 4.9×10^{-1} **35.** 8.2×10^{-3} **36.** 5.4×10^5

37. 6×10^{-5} **38.** 5.123×10^{-3} **39.** 7.05×10^6 **40.** 7×10^8

In Problems 41–48, if a number is given in scientific notation, write it in decimal notation. If it is written in decimal notation, write it in scientific notation.

41. 0.0005 **42.** 1×10^9 **43.** 1.42×10^5 **44.** 6,000,000,000

45. 8×10^{-3} **46.** 0.0035 **47.** 69,000 **48.** 4×10^{-1}

In Problems 49–54, a number is given in decimal notation. Write the number in scientific notation.

49. World Population At the beginning of 2005, the population of the world was approximately 6,415,000,000 persons.

50. United States Population At the beginning of 2005, the population of the United States was approximately 294,975,000 persons.

51. Stock Exchange Daily Volume The average daily volume of the New York Stock Exchange in the year 2004 was 1.46 billion shares. (*Source:* New York Stock Exchange)

52. Stock Exchange Total Volume The total volume of shares traded on the New York Stock Exchange in 2004 was approximately 367 billion shares. (*Source:* New York Stock Exchange)

53. 3×10^{-5} mm
54. 7.5×10^{-3} mm
55. 2,158,000
56. 3,302,000,000
57. 0.000000000000001
58. 0.000000000753
59. 0.00225
60. 0.000000385
61. 3×10^{9}
62. 2.4×10^{-8}
63. 8.4×10^{-3}
64. 1×10^{4}
65. 1.05×10^{-3}
66. 1.8×10^{6}
67. 3×10^{8}
68. 5×10^{-2}
69. 2.5×10^{1}
70. 4×10^{5}
71. 1.8×10^{8}
72. 1.2×10^{-8}
73. 8×10^{8}
74. 1.1×10^{-6}
75. 9×10^{15}
76. 2×10^{11}
77. 3×10^{-14}
78. 4.5×10^{-16}
79. 2.11×10^{5} miles
80. 1.702×10^{8} km
81. 4.87×10^{8} km
82. 1.987×10^{8} mi
83. 4.56×10^{8} km
84. $\$8.2 \times 10^{8}$

53. Smallpox Virus The diameter of a smallpox virus is 0.00003 mm.

54. Human Blood Cell The diameter of a human blood cell is 0.0075 mm.

In Problems 55–60, a number is given in scientific notation. Write the number in decimal notation.

55. Farms In the year 2002, there were 2.158×10^6 farms in the United States.

56. Crude Oil Imports In the year 2002, the United States imported 3.302×10^9 barrels of crude oil.

57. Time A femtosecond is equal to 1×10^{-15} second.

58. Dust Particle The mass of a dust particle is 7.53×10^{-10} kg.

59. Vitamin A One-A-Day vitamin pill contains 2.25×10^{-3} grams of zinc.

60. Water Molecule The diameter of a water molecule is 3.85×10^{-7} m.

Mixed Practice

In Problems 61–78, perform the indicated operations. Express your answer in scientific notation.

61. $(2 \times 10^6)(1.5 \times 10^3)$

62. $(3 \times 10^{-4})(8 \times 10^{-5})$

63. $(1.2 \times 10^0)(7 \times 10^{-3})$

64. $(4 \times 10^7)(2.5 \times 10^{-4})$

65. $(7 \times 10^{-6})(1.5 \times 10^2)$

66. $(9 \times 10^{-1})(2 \times 10^6)$

67. $\dfrac{9 \times 10^4}{3 \times 10^{-4}}$

68. $\dfrac{6 \times 10^3}{1.2 \times 10^5}$

69. $\dfrac{2 \times 10^{-3}}{8 \times 10^{-5}}$

70. $\dfrac{4.8 \times 10^7}{1.2 \times 10^2}$

71. $\dfrac{5.4 \times 10^6}{3 \times 10^{-2}}$

72. $\dfrac{1.44 \times 10^{-3}}{1.2 \times 10^5}$

73. $\dfrac{56{,}000}{0.00007}$

74. $\dfrac{0.000275}{2{,}500}$

75. $\dfrac{300{,}000 \times 15{,}000{,}000}{0.0005}$

76. $\dfrac{24{,}000{,}000{,}000}{0.00006 \times 2{,}000}$

77. $\dfrac{0.00072}{3{,}000 \times 8{,}000{,}000}$

78. $\dfrac{0.00075 \times 0.0000003}{500{,}000}$

Applying the Concepts

In 2003 and 2004, two NASA rovers landed on Mars. Landing a rover such as NASA's Opportunity or Spirit on the Martian surface required several course corrections as it traveled from Earth to Mars.

79. The rovers were launched from unmanned *Boeing Delta II* launch vehicles once out of our atmosphere and out of the grasp of Earth's gravity. The trajectory the rovers were originally launched on was designed to miss Mars by 211,000 miles. Write 211,000 in scientific notation.

80. The rovers traveled in an indirect path to Mars. The distance from Earth to Mars on the day the *Spirit* rover landed was 170,200,000 kilometers. Write 170,200,000 in scientific notation.

81. However, after traveling around Earth to gather speed and including the course corrections *Spirit* had to make to ensure a successful landing, the total distance traveled by the rover was 487,000,000 kilometers. Write 487,000,000 in scientific notation.

82. When the second NASA rover, *Opportunity,* landed on Mars in January 2004, the distance between the Earth and Mars was 198,700,000 miles. Write the distance between Earth and Mars in January 2004 in scientific notation.

83. Factoring in all of *Opportunity*'s course corrections and extra distance traveled, that craft traveled 456,000,000 kilometers. Write the total distance that *Opportunity* traveled in scientific notation.

84. The two-lander *NASA* project cost \$820,000,000. Write \$820,000,000 in scientific notation.

Once on Mars, the rovers looked for geological evidence that life may have existed at some point. NASA is planning future missions to Mars to look for a biological footprint of life.

85. Future missions are searching only for certain molecules present in Martian soil or ice. The tests will include the use of molecules engineered here on Earth that

85. 2.5×10^{-8} M
86. 2.14×10^{-10} M
87. $\approx 1.65 \times 10^{3}$ lb/person
88. $\approx 1.14 \times 10^{1}$ barrels/person
89. 1.116×10^{7} miles
90. 6.762×10^{4} feet
91. 2.5×10^{-4} m
92. 6.04×10^{-8} m
93. 8×10^{-10} m
94. 4×10^{-2} m
95. 7.15×10^{-8} m
96. 2×10^{-4} m
97. $1.2348\pi \times 10^{-14}$ m^3
98. $5.625\pi \times 10^{-28}$ m^3
99. $2.88\pi \times 10^{-25}$ m^3
100. $1.679616\pi \times 10^{-12}$ m^3

will give us a visible response if the molecules being sought after are found. These engineered molecules are expensive and difficult to create, however, so very dilute solutions will be used. Scientists express the concentration of solutions in molarity, abbreviated M. The solutions of engineered molecules in Martian experiments will be on the order of 0.000000025 M. Write 0.000000025 M in scientific notation.

86. This dilute solution will be so sensitive that it is able to detect these molecules of interest down to a concentration of less than 0.000000000214 M. Write 0.000000000214 M in scientific notation.

87. **Garbage** In 2000, the total waste generated in the United States was 4.638×10^{11} pounds. Also in 2000, the United States population was 2.81×10^{8} people. Determine the garbage per capita (per person) in the United States in the year 2000.

88. **Fossil Fuels** In 2002, the United States imported 3.3×10^{9} barrels of oil. The United States population in 2002 was 2.89×10^{8} people. Determine the per capita number of barrels of oil imported into the United States in 2002.

89. **Speed of Light** Light travels at the rate of 1.86×10^{5} miles per second. How far does light travel in one minute (6.0×10^{1} seconds)?

90. **Speed of Sound** Sound travels at the rate of 1.127×10^{3} feet per second . How far does sound travel in one minute (6.0×10^{1} seconds)?

Extending the Concepts

Scientists often need to measure very small things, such as cells. They use the following units of measure:

millimeter (mm) $= 1 \times 10^{-3}$ *meter*	*micron (μm)* $= 1 \times 10^{-6}$ *meter*
nanometer (nm) $= 1 \times 10^{-9}$ *meter*	*picometer (pm)* $= 1 \times 10^{-12}$ *meter*

Write the following measurements in meters using scientific notation:

91. 250 μm **92.** 60.4 nm **93.** 800 pm

94. 40 mm **95.** 71.5 nm **96.** 200 μm

Assume a cell is in the shape of a sphere. Given that the volume of a sphere is $V = \frac{4}{3}\pi r^3$, *find the volume of a cell whose radius is given. Express the answer as a multiple of π in cubic meters.*

97. 21 μm **98.** 0.75 nm **99.** 6 nm **100.** 108 μm

CHAPTER 3 ACTIVITY: WHAT IS THE QUESTION?

Focus: Using exponent rules, using scientific notation, and performing operations with polynomials

Time: 15–20 minutes

Group size: 2 or 4

In this activity you will work as a team to solve eight multiple-choice questions. However these questions are different from most multiple-choice questions. You are given the answer to a problem and must determine which of the multiple-choice options has the correct question for the given answer.

Before beginning the activity, decide how you will approach this task as a team. For example:

- If there are 2 members on your team, one member will always examine choices (a) and (b) and the other will always examine choices (c) and (d).
- If there are 4 members on your team, one member will always examine choice (a), another member will always examine choice (b), and so on.

1. d
2. b
3. c
4. c
5. c
6. a
7. d
8. c

1. The answer is $-3x^2 - 10x$. What is the question?
(a) Simplify: $2x - 3x(x^2 + 4)$
(b) Find the quotient: $(6x^3 - 20x^2) \div (-2x)$
(c) Simplify: $-3x(x^2 + 3) + 1$
(d) Find the quotient: $(-6x^4 - 20x^3) \div (2x^2)$

2. The answer is $-10x^4y^8$. What is the question?
(a) Simplify: $(-5x^2y^4)^2$
(b) Simplify: $(5x^3y^3)(-2xy^5)$
(c) Simplify: $\dfrac{-30x^{-2}y^6}{3x^2y^{-2}}$
(d) Simplify: $(5x^2y^4)(-2x^2y^2)$

3. The answer is $12x^2 - 16x - 3$. What is the question?
(a) Multiply: $(6x - 1)(2x + 3)$
(b) Divide: $(24x^3 - 32x^2 + 6x) \div (2x)$
(c) Multiply: $(6x + 1)(2x - 3)$
(d) Simplify: $(14x^2 - x + 2) - (2x^2 + 16x + 5)$

4. The answer is $x^2 + 5x + 6$. What is the question?
(a) Find the product: $(x + 6)(x - 1)$
(b) Simplify: $2x^2 + 7x + 9 - (x^2 - 2x - 3)$
(c) Find the product: $(x + 2)(x + 3)$
(d) Simplify: $(x + 6)^2$

5. The answer is 3. What is the question?
(a) What is the name of the variable in $16z^2 + 3z - 5$?
(b) What is the degree of the polynomial $2mn + 6m - 3$?
(c) How many terms are in the polynomial $2mn + 6m - 3$?
(d) What is the coefficient of b in the polynomial $3a^2b - 9a + 5b$?

6. The answer is 43. What is the question?
(a) Evaluate: $4x^2 - 2x + 1$ for $x = -3$
(b) Evaluate: $2x^2 + 3x - 4$ for $x = -2$
(c) Evaluate: $-x^2 - 5x + 1$ for $x = -1$
(d) Evaluate: $x^2 + 4x - 8$ for $x = -5$

7. The answer is $\dfrac{2x^5}{y^5}$. What is the question?
(a) Simplify: $(6x^{-4}y)^0\left(\dfrac{12x^{-1}y^{-2}}{x^{-4}y^3}\right)$
(b) Simplify: $(6x^{-3}y^0)^{-1}\left(\dfrac{12xy^{-3}}{x^{-2}y^2}\right)$
(c) Simplify: $(6x^2y^0)^{-1}\left(\dfrac{12x^{-2}y^3}{xy^{-4}}\right)$
(d) Simplify: $(6x^{-2}y^0)^{-1}\left(\dfrac{12x^0y^{-2}}{x^{-3}y^3}\right)$

8. The answer is 2.5×10^{-8}. What is the question?
(a) Find the quotient: $\dfrac{2 \times 10^3}{8 \times 10^{-4}}$
(b) Find the product: $(5 \times 10^{-4}) \cdot (5 \times 10^{-4})$
(c) Find the quotient: $\dfrac{2 \times 10^{-3}}{8 \times 10^4}$
(d) Find the product: $(5 \times 10^{12}) \cdot (5 \times 10^{-4})$

CHAPTER 3 REVIEW

<table>
<tr><th colspan="3">Section 3.1 Adding and Subtracting Polynomials</th></tr>
<tr><th colspan="2">KEY CONCEPTS</th><th>KEY TERMS</th></tr>
<tr><td colspan="2">• In a monomial in the form ax^k, k is the degree of the monomial.
• The degree of a polynomial is the highest degree of all the terms of the polynomial.</td><td>Monomial
Degree of a monomial
Polynomial
Binomial
Trinomial
Standard form
Degree of a polynomial</td></tr>
<tr><th>YOU SHOULD BE ABLE TO . . .</th><th>EXAMPLE</th><th>REVIEW EXERCISES</th></tr>
<tr><td>1 Define monomial and determine the degree of a monomial (p. 174)</td><td>Examples 1 through 3</td><td>1–4</td></tr>
<tr><td>2 Define polynomial and determine the degree of a polynomial (p.175)</td><td>Examples 4 and 5</td><td>5–10</td></tr>
<tr><td>3 Simplify polynomials by combining like terms (p. 176)</td><td>Examples 6 through 10</td><td>11–16</td></tr>
<tr><td>4 Evaluate polynomials (p. 178)</td><td>Examples 11 through 13</td><td>17–20</td></tr>
</table>

In Problems 1–4, determine whether the given expression is a monomial (Yes or No). For those that are monomials, state the degree and the coefficient.

1. $4x^3$ **2.** $6x^{-3}$ **3.** $m^{1/2}$ **4.** mn^2

In Problems 5–10, determine whether the algebraic expression is a polynomial (Yes or No). If it is a polynomial, state the degree and then state if it is a monomial, binomial, or trinomial.

5. $4x^6 - 4x^{1/2}$ **6.** $\frac{3}{x} - \frac{1}{x^2}$ **7.** 6 **8.** $3x^3 - 4xy^4$

9. $-2x^5y - 7x^4y + 7$ **10.** $\frac{1}{2}x^3 + 2x^{10} - 5$

In Problems 11–14, perform the indicated operation.

11. $(6x^2 - 2x + 1) + (3x^2 + 10x - 3)$ **12.** $(-7m^3 - 2mn) + (8m^3 - 5m + 3mn)$

13. $(4x^2y + 10x) - (5x^2y - 2x)$

14. $(3y^2 - yz + 3z^2) - (10y^2 + 5yz - 6z^2)$

15. Find the sum of $-6x^2 + 5$ and $4x^2 - 7$.

16. Subtract $20y^2 - 10y + 5$ from $-18y + 10$.

In Problems 17–20, evaluate the polynomial for the given value(s).

17. $3x^2 - 5x$
(a) $x = 0$
(b) $x = -1$
(c) $x = 2$

18. $-x^2 + 3$
(a) $x = 0$
(b) $x = -1$
(c) $x = \frac{1}{2}$

19. $x^2y + 2xy^2$ for $x = -2$ and $y = 1$

20. $4a^2b^2 - 3ab + 2$ for $a = -1$ and $b = -3$

1. Yes; degree 3; coefficient 4
2. No
3. No
4. Yes; degree 3; coefficient 1
5. No
6. No
7. Yes; degree 0; monomial
8. Yes; degree 5; binomial
9. Yes; degree 6; trinomial
10. Yes; 10; trinomial
11. $9x^2 + 8x - 2$
12. $m^3 + mn - 5m$
13. $-x^2y + 12x$
14. $-7y^2 - 6yz + 9z^2$
15. $-2x^2 - 2$
16. $-20y^2 - 8y + 5$
17. (a) 0 (b) 8 (c) 2
18. (a) 3 (b) 2 (c) $\frac{11}{4}$ 19. 0 20. 29

Section 3.2 Multiplying Monomials: The Product and Power Rules

KEY CONCEPTS	KEY TERMS
• **Product Rule for Exponents** If a is a real number and m, n are natural numbers, then $a^m \cdot a^n = a^{m+n}$ • **Power Rule for Exponents** If a is a real number and m, n are whole numbers greater than 0, then $(a^m)^n = a^{m \cdot n}$ • **Product to a Power Rule for Exponents** If a, b are real numbers and n is a whole number, then $(a \cdot b)^n = a^n \cdot b^n$	Base Power Exponent

YOU SHOULD BE ABLE TO . . .	EXAMPLE	REVIEW EXERCISES
1 Simplify exponential expressions using the Product Rule (p. 185)	Examples 1 through 3	21, 22, 25, 26
2 Simplify exponential expressions using the Power Rule (p. 186)	Examples 4 and 5	23, 24, 27, 28
3 Simplify exponential expressions containing products (p. 187)	Example 6	29–32
4 Multiply a monomial by a monomial (p. 188)	Examples 7 and 8	33–40

In Problems 21–32, simplify each expression.

21. $6^2 \cdot 6^5$ **22.** $\left(-\frac{1}{3}\right)^2\left(-\frac{1}{3}\right)^3$ **23.** $(4^2)^6$ **24.** $[(-1)^4]^3$

25. $x^4 \cdot x^8 \cdot x$ **26.** $m^4 \cdot m^2$ **27.** $(r^3)^4$ **28.** $(m^8)^3$

29. $(4x)^3(4x)^2$ **30.** $(-2n)^3(-2n)^3$ **31.** $(-3x^2y)^4$ **32.** $(2x^3y^4)^2$

In Problems 33–40, multiply.

33. $3x^2 \cdot 5x^4$ **34.** $-4a \cdot 9a^3$ **35.** $-8y^4 \cdot (-2y)$ **36.** $12p \cdot (-p^5)$

37. $\frac{8}{3}w^3 \cdot \frac{9}{2}w$ **38.** $\frac{1}{3}z^2 \cdot \left(-\frac{9}{4}z\right)$ **39.** $(3x^2)^3 \cdot (2x)^2$ **40.** $(-4a)^2 \cdot (5a^4)$

21. 279,936 22. $-\frac{1}{243}$ 23. 16,777,216 24. 1 25. x^{13} 26. m^6 27. r^{12} 28. m^{24} 29. $1024x^5$ 30. $64n^6$ 31. $81x^8y^4$ 32. $4x^6y^8$ 33. $15x^6$ 34. $-36a^4$ 35. $16y^5$ 36. $-12p^6$ 37. $12w^4$ 38. $-\frac{3}{4}z^3$ 39. $108x^8$ 40. $80a^6$

Section 3.3 Multiplying Polynomials

KEY CONCEPTS	KEY TERMS
• **Extended form of the Distributive Property** $a(b + c + \cdots + z) = a \cdot b + a \cdot c + \cdots + a \cdot z$ where $a, b, c, \ldots, z$ are real numbers. • **FOIL Method for multiplying two binomials** $(ax + b)(cx + d) = \overset{F}{ax \cdot cx} + \overset{O}{ax \cdot d} + \overset{I}{b \cdot cx} + \overset{L}{b \cdot d}$ • **Product of the Sum and Difference of Two Terms** $(a - b)(a + b) = a^2 - b^2$ • **Squares of Binomials** $(a + b)^2 = a^2 + 2ab + b^2$, $(a - b)^2 = a^2 - 2ab + b^2$	FOIL method Special products Sum and difference of two terms Difference of two squares Squares of binomials Perfect square trinomial

YOU SHOULD BE ABLE TO . . .	EXAMPLE	REVIEW EXERCISES
1 Multiply a polynomial by a monomial (p. 190)	Examples 1 through 3	41, 42
2 Multiply two binomials using the Distributive Property (p. 191)	Examples 4 and 5	43–46
3 Multiply two binomials using the FOIL method (p. 192)	Examples 6 through 9	49–54
4 Multiply the sum and difference of two terms (p. 194)	Examples 10 and 11	55, 56, 59, 60, 63, 64
5 Square a binomial (p. 194)	Examples 12 through 14	57, 58, 61, 62, 65, 66
6 Multiply a polynomial by a polynomial (p. 196)	Examples 15 through 17	47, 48

41. $-8x^5 + 6x^4 - 2x^3$
42. $2x^7 + 4x^6 - x^4$ 43. $6x^2 - 7x - 5$
44. $4x^2 - 5x - 6$ 45. $x^2 - 3x - 40$
46. $w^2 + 9w - 10$
47. $24m^3 - 22m^2 + 7m - 3$
48. $8y^5 + 12y^4 + 4y^3 + 6y^2 - 6y - 9$
49. $x^2 + 8x + 15$
50. $2x^2 - 17x + 8$
51. $6m^2 + 17m - 14$
52. $48m^2 - 26m - 4$
53. $21x^2 + 5xy - 6y^2$
54. $20x^2 + 7xy - 3y^2$
55. $x^2 - 16$ 56. $4x^2 - 25$
57. $4x^2 + 12x + 9$ 58. $49x^2 - 28x + 4$
59. $9x^2 - 16y^2$ 60. $64m^2 - 36n^2$
61. $25x^2 - 20xy + 4y^2$
62. $4a^2 + 12ab + 9b^2$
63. $x^2 - 0.25$ 64. $r^2 - 0.0625$
65. $y^2 + \frac{4}{3}y + \frac{4}{9}$ 66. $y^2 - y + \frac{1}{4}$

In Problems 41–48, multiply.

41. $-2x^3(4x^2 - 3x + 1)$ **42.** $\frac{1}{2}x^4(4x^3 + 8x^2 - 2)$ **43.** $(3x - 5)(2x + 1)$

44. $(4x + 3)(x - 2)$ **45.** $(x + 5)(x - 8)$ **46.** $(w - 1)(w + 10)$

47. $(4m - 3)(6m^2 - m + 1)$ **48.** $(2y + 3)(4y^4 + 2y^2 - 3)$

In Problems 49–54, use the FOIL method to find each product.

49. $(x + 5)(x + 3)$ **50.** $(2x - 1)(x - 8)$ **51.** $(2m + 7)(3m - 2)$

52. $(6m - 4)(8m + 1)$ **53.** $(3x + 2y)(7x - 3y)$ **54.** $(4x - y)(5x + 3y)$

In Problems 55–66, find the special products.

55. $(x - 4)(x + 4)$ **56.** $(2x + 5)(2x - 5)$ **57.** $(2x + 3)^2$

58. $(7x - 2)^2$ **59.** $(3x + 4y)(3x - 4y)$ **60.** $(8m - 6n)(8m + 6n)$

61. $(5x - 2y)^2$ **62.** $(2a + 3b)^2$ **63.** $(x - 0.5)(x + 0.5)$

64. $(r + 0.25)(r - 0.25)$ **65.** $\left(y + \frac{2}{3}\right)^2$ **66.** $\left(y - \frac{1}{2}\right)^2$

Section 3.4 Dividing Monomials: The Quotient Rule and Integer Exponents

KEY CONCEPTS

- **Quotient Rule for Exponents**
  If a is a nonzero real number and if m and n are integers, then $\frac{a^m}{a^n} = a^{m-n}$.
- **Definition of Zero as an Exponent**
  If a is a nonzero real number (that is, $a \neq 0$), we define $a^0 = 1$.
- **Quotient to a Power Rule for Exponents**
  If a, b are real numbers and n is an integer, then $\left(\frac{a}{b}\right)^n = \frac{a^n}{b^n}$ if $b \neq 0$.

  If n is negative or 0, then a cannot be 0.
- **Definition of a Negative Exponent**
  If n is a positive integer and if a is a nonzero real number (that is, $a \neq 0$), then we define
  $a^{-n} = \frac{1}{a^n}$ and $\frac{1}{a^{-n}} = a^n$ if $a \neq 0$.
- **Quotient to a Negative Power**

  If a and b are real numbers and n is an integer, then $\left(\frac{a}{b}\right)^{-n} = \left(\frac{b}{a}\right)^n$ if $a \neq 0, b \neq 0$.

YOU SHOULD BE ABLE TO . . .	EXAMPLE	REVIEW EXERCISES
① Simplify exponential expressions using the Quotient Rule (p. 201)	Examples 1 and 2	67–70, 75, 76
② Simplify exponential expressions using the Quotient to a Power Rule (p. 202)	Examples 3 and 4	77–80
③ Simplify exponential expressions using zero as an exponent (p. 203)	Example 5	71–74
④ Simplify exponential expressions involving negative exponents (p. 204)	Examples 6 through 11	81–86
⑤ Simplify exponential expressions using the Laws of Exponents (p. 207)	Examples 12 through 15	87–92

67. 36 68. $\frac{1}{343}$ 69. x^4 70. $\frac{1}{x^8}$
71. 1 72. -1 73. 1 74. -1
75. $\frac{5x^2}{2y^3}$ 76. $\frac{x^2}{3y^8}$ 77. $\frac{x^{15}}{y^{10}}$
78. $\frac{343}{x^6}$ 79. $\frac{8m^6n^3}{p^{12}}$
80. $\frac{81m^4n^8}{p^{20}}$ 81. $-\frac{1}{25}$
82. 64 83. $\frac{81}{16}$ 84. 27
85. $\frac{7}{12}$ 86. $\frac{1}{8}$
87. $\frac{2x^3y^5}{3}$ 88. $\frac{3}{7xy^{10}}$
89. $\frac{9m^4}{16}$ 90. $\frac{64}{9m^6n^6}$
91. $\frac{8r^4s^6}{9}$ 92. $\frac{32}{3r^{10}s^{15}}$

In Problems 67–92, simplify. Write answers with only positive exponents.

67. $\frac{6^5}{6^3}$ **68.** $\frac{7}{7^4}$ **69.** $\frac{x^{16}}{x^{12}}$ **70.** $\frac{x^3}{x^{11}}$

71. 5^0 **72.** -5^0 **73.** $m^0, m \neq 0$ **74.** $-m^0, m \neq 0$

75. $\frac{25x^3y^7}{10xy^{10}}$ **76.** $\frac{3x^4y^2}{9x^2y^{10}}$ **77.** $\left(\frac{x^3}{y^2}\right)^5$ **78.** $\left(\frac{7}{x^2}\right)^3$

79. $\left(\frac{2m^2n}{p^4}\right)^3$ **80.** $\left(\frac{3mn^2}{p^5}\right)^4$ **81.** -5^{-2} **82.** $\frac{1}{4^{-3}}$

83. $\left(\frac{2}{3}\right)^{-4}$ **84.** $\left(\frac{1}{3}\right)^{-3}$ **85.** $2^{-2} + 3^{-1}$ **86.** $4^{-1} - 2^{-3}$

87. $\frac{16x^{-3}y^4}{24x^{-6}y^{-1}}$ **88.** $\frac{15x^0y^{-6}}{35xy^4}$ **89.** $(2m^{-3}n)^{-4}(3m^{-4}n^2)^2$

90. $(4m^{-6}n^0)^3(3m^{-6}n^3)^{-2}$ **91.** $\left(\frac{3rs^{-1}}{4s^2}\right)^{-2} \cdot (2r^{-6}t^0)^{-1}$ **92.** $(6r^4s^{-3})^2 \cdot \left(\frac{3r^4s}{2r^{-2}s^{-2}}\right)^{-3}$

Section 3.5 Dividing Polynomials

KEY CONCEPTS	KEY TERMS
• If A, B, and C are monomials, $\frac{A+B}{C} = \frac{A}{C} + \frac{B}{C}$. • (Quotient)(Divisor) + Remainder = Dividend	Divisor Dividend Quotient Remainder

YOU SHOULD BE ABLE TO . . .	EXAMPLE	REVIEW EXERCISES
1 Divide a polynomial by a monomial (p. 214)	Examples 1 through 3	93–98
2 Divide a polynomial by a binomial (p. 215)	Examples 4 through 7	99–104

93. $6x^5 - 4x^4 + 5$
94. $3x^4 + 5x^2 - 6x$
95. $4n^3 + 1 - \frac{5}{2n^4}$
96. $6n - 4 - \frac{16}{5n^2}$
97. $-\frac{p^3}{8} - \frac{1}{4} + \frac{1}{2p^2}$
98. $-\frac{p^2}{2} + 1 - \frac{3}{2p^2}$
99. $4x - 7$ 100. $x + 6$
101. $x^2 + 7x + 5 + \frac{6}{x-1}$
102. $2x^2 - 3x - 12 + \frac{-16}{x-2}$
103. $x^2 - 2x + 4$
104. $3x^2 + 15x + 77 + \frac{378}{x-5}$

In Problems 93–104, divide each of the following.

93. $\frac{36x^7 - 24x^6 + 30x^2}{6x^2}$ **94.** $\frac{15x^5 + 25x^3 - 30x^2}{5x}$ **95.** $\frac{16n^8 + 4n^5 - 10n}{4n^5}$

96. $\frac{30n^6 - 20n^5 - 16n^3}{5n^5}$ **97.** $\frac{2p^8 + 4p^5 - 8p^3}{-16p^5}$ **98.** $\frac{3p^4 - 6p^2 + 9}{-6p^2}$

99. $\frac{8x^2 - 2x - 21}{2x + 3}$ **100.** $\frac{3x^2 + 17x - 6}{3x - 1}$ **101.** $\frac{6x^2 + x^3 - 2x + 1}{x - 1}$

102. $\frac{-6x + 2x^3 - 7x^2 + 8}{x - 2}$ **103.** $\frac{x^3 + 8}{x + 2}$ **104.** $\frac{3x^3 + 2x - 7}{x - 5}$

Section 3.6 Applying Exponent Rules: Scientific Notation

KEY CONCEPTS	KEY TERMS
• **Definition of Scientific Notation** A number is written in scientific notation when it is of the form $x \times 10^N$, where $1 \leq x < 10$ and N is an integer. • **Convert from Decimal Notation to Scientific Notation** To change a positive number into scientific notation: **Step 1:** Count the number N of decimal places that the decimal point must be moved to arrive at a number x, where $1 \leq x < 10$. **Step 2:** If the original number is greater than or equal to 1, the scientific notation is $x \times 10^N$. If the original number is between 0 and 1, the scientific notation is $x \times 10^{-N}$. • **Convert from Scientific Notation to Decimal Notation** **Step 1:** Determine the exponent on the number 10. **Step 2:** If the exponent is positive, then move the decimal N decimal places to the right. If the exponent is negative, then move the decimal $\lvert N \rvert$ decimal places to the left.	Decimal notation Scientific notation

YOU SHOULD BE ABLE TO . . .	EXAMPLE	REVIEW EXERCISES
(1) Convert decimal notation to scientific notation (p. 221)	Examples 1 and 2	105–110
(2) Convert scientific notation to decimal notation (p. 223)	Examples 3 and 4	111–116
(3) Use scientific notation to multiply and divide (p. 224)	Examples 5 through 8	117–122

105. 2.7×10^7 106. 1.23×10^9
107. 6×10^{-5} 108. 3.05×10^{-6}
109. 3×10^0 110. 8×10^0
111. 0.0006 112. 0.00125
113. 613,000 114. 80,000
115. 0.37 116. 54,000,000
117. 6×10^3 118. 4.2×10^{-8}
119. 2×10^2 120. 2×10^8
121. 4×10^{15} 122. 2×10^2

In Problems 105–110, write in scientific notation.

105. 27,000,000 **106.** 1,230,000,000 **107.** 0.00006

108. 0.00000305 **109.** 3 **110.** 8

In Problems 111–116, write in decimal notation.

111. 6×10^{-4} **112.** 1.25×10^{-3} **113.** 6.13×10^5

114. 8×10^4 **115.** 3.7×10^{-1} **116.** 5.4×10^7

In Problems 117–122, perform the indicated operations. Express your answer in scientific notation.

117. $(1.2 \times 10^{-5})(5 \times 10^8)$ **118.** $(1.4 \times 10^{-10})(3 \times 10^2)$ **119.** $\dfrac{2.4 \times 10^{-6}}{1.2 \times 10^{-8}}$

120. $\dfrac{5 \times 10^6}{25 \times 10^{-3}}$ **121.** $\dfrac{200{,}000 \times 4{,}000{,}000}{0.0002}$ **122.** $\dfrac{1{,}200{,}000}{0.003 \times 2{,}000{,}000}$

CHAPTER 3 TEST

Remember to use your Chapter Test Prep Video CD to see fully worked-out solutions to any of these problems you would like to review.

Note to instructor: A special file in TestGen provides algorithms specifically matched to the problems in this Chapter Test for easy-to-replicate practice or assessment purposes.

1. Yes; degree 5; binomial
2. (a) 5
(b) 21
(c) 26
3. $7x^2y^2 - 6x - 3y$

1. Determine whether the algebraic expression $6x^5 - 2x^4$ is a polynomial (Yes or No). If it is a polynomial, state the degree and then state if it is a monomial, binomial, or trinomial.

2. Evaluate the polynomial $3x^2 - 2x + 5$ for the given values:

(a) $x = 0$ **(b)** $x = -2$ **(c)** $x = 3$

In Problems 3–11, perform the indicated operation.

3. $(3x^2y^2 - 2x + 3y) + (-4x - 6y + 4x^2y^2)$

4. $10m^3 + m^2 - 6$
5. $-6x^5 + 18x^4 - 15x^3$
6. $2x^2 - 3x - 35$
7. $4x^2 - 28x + 49$
8. $16x^2 - 9y^2$
9. $6x^3 + x^2 - 25x + 8$
10. $2x - \frac{8}{3} + \frac{3}{x^3}$
11. $3x^2 - 11x + 33 + \frac{-94}{x+3}$
12. $-12x^4y^6$
13. $\frac{2m^3}{3n^5}$
14. $\frac{n^{24}}{m^{30}}$
15. $\frac{x^{22}}{y^{14}}$
16. $\frac{8m^7}{n^7}$
17. 1.2×10^{-5}
18. 210,100
19. 3.57×10^4
20. 2×10^{-7}

4. $(8m^3 + 6m^2 - 4) - (5m^2 - 2m^3 + 2)$

5. $-3x^3(2x^2 - 6x + 5)$

6. $(x - 5)(2x + 7)$

7. $(2x - 7)^2$

8. $(4x - 3y)(4x + 3y)$

9. $(3x - 1)(2x^2 + x - 8)$

10. $\frac{6x^4 - 8x^3 + 9}{3x^3}$

11. $\frac{3x^3 - 2x^2 + 5}{x + 3}$

In Problems 12–16, simplify each expression. Write answers with only positive exponents.

12. $(4x^3y^2)(-3xy^4)$

13. $\frac{18m^5n}{27m^2n^6}$

14. $\left(\frac{m^{-2}n^0}{m^{-7}n^4}\right)^{-6}$

15. $(4x^{-3}y)^{-2}(2x^4y^{-3})^4$

16. $(2m^{-4}n^2)^{-1} \cdot \left(\frac{16m^0n^{-3}}{m^{-3}n^2}\right)$

17. Write 0.000012 in scientific notation.

18. Write 2.101×10^5 in decimal notation.

In Problems 19 and 20, perform the indicated operation. Express your answer in scientific notation.

19. $(2.1 \times 10^{-6}) \cdot (1.7 \times 10^{10})$

20. $\frac{3 \times 10^{-4}}{15 \times 10^2}$

CUMULATIVE REVIEW CHAPTERS 1–3

1. (a) $\{\sqrt{25}\}$
   (b) $\{0, \sqrt{25}\}$
   (c) $\left\{-6, -\frac{4}{2}, 0, \sqrt{25}\right\}$
   (d) $\left\{-6, -\frac{4}{2}, 0, 1.4, \sqrt{25}\right\}$
   (e) $\{\sqrt{7}\}$
   (f) $\left\{-6, -\frac{4}{2}, 0, 1.4, \sqrt{7}, \sqrt{25}\right\}$
2. $-\frac{4}{9}$
3. -19
4. $6x^3 + 5x^2 - 3x$
5. $-18x + 8$
6. $\{1\}$
7. $4(x - 5) = 2x + 10$
8. \$640

1. Use the set $\left\{-6, -\frac{4}{2}, 0, 1.4, \sqrt{7}, \sqrt{25}\right\}$. List all of the elements that are

(a) natural **(b)** whole **(c)** integers
(d) rational **(e)** irrational **(f)** real

In Problems 2 and 3, evaluate each expression.

2. $-\frac{1}{2} + \frac{2}{3} \div 4 \cdot \frac{1}{3}$

3. $2 + 3[3 + 10(-1)]$

In Problems 4 and 5, simplify each algebraic expression.

4. $6x^3 - (-2x^2 + 3x) + 3x^2$

5. $-4(6x - 1) + 2(3x + 2)$

6. Solve: $-2(3x - 4) + 6 = 4x - 6x + 10$

7. Translate the following statement into an equation. DO NOT SOLVE. Four times the difference of a number and 5 is equal to 10 more than twice the number.

8. Paycheck Kathy's monthly paycheck from her part-time job working at an electronics store totaled \$659.20. This amount included a 3% raise over her previous month's earnings. What were Kathy's monthly earnings before the 3% raise?

9. 45 mi/h
10. $x < -3$ or $(-\infty, -3)$;
11. $-\frac{1}{2}$
12. $6x^3 - 3x^2 + 7x$
13. $28m^2 - 13m - 6$
14. $9m^2 - 4n^2$
15. $49x^2 + 14xy + y^2$
16. $4m^3 - 19m + 15$
17. $\frac{2}{x} + \frac{1}{y}$
18. $x^2 - 3x + 9$
19. $-24n^4$
20. $-\frac{5n^8}{2m^2}$
21. $\frac{1}{64x^6y^{24}z^{12}}$
22. $\frac{1}{2x^{12}y^3}$
23. 6.05×10^{-5}
24. 2,175,000
25. 7.14×10^5

9. Driving Cheyenne and Amber live 306 miles apart. They start driving toward each other and meet in three hours. If Cheyenne drives 12 miles per hour faster than Amber, find Amber's driving speed.

10. Solve and graph the following inequality: $-5x + 2 > 17$

11. Evaluate $\frac{x^2 - y^2}{z}$ when $x = 3$, $y = -2$, and $z = -10$.

In Problems 12–18, perform the indicated operation.

12. $(4x^2 + 6x) - (-x + 5x^2) + (6x^3 - 2x^2)$

13. $(4m - 3)(7m + 2)$

14. $(3m - 2n)(3m + 2n)$

15. $(7x + y)^2$

16. $(2m + 5)(2m^2 - 5m + 3)$

17. $\frac{14xy^2 + 7x^2y}{7x^2y^2}$

18. $\frac{x^3 + 27}{x + 3}$

In Problems 19–22, simplify the expression. Write your answers with only positive exponents.

19. $(4m^0n^3)(-6n)$

20. $\frac{25m^{-6}n^{-2}}{-10m^{-4}n^{-10}}$

21. $\left(\frac{2xy^4}{z^{-2}}\right)^{-6}$

22. $(x^4y^{-2})^{-4} \cdot \left(\frac{6x^{-4}y^3}{3y^{-8}}\right)^{-1}$

23. Write 0.0000605 in scientific notation.

24. Write 2.175×10^6 in decimal notation.

25. Perform the indicated operation. Write your answer in scientific notation.

$$(3.4 \times 10^8)(2.1 \times 10^{-3})$$

CHAPTER

4 Factoring Polynomials

The size of a television set is described by its diagonal measure. For example, suppose you want to purchase a 50-inch big-screen TV before the World Series. The TV measures 40 inches across the bottom, but how do you know if your TV cabinet will be big enough for this large TV? To find the height of a 50-inch television, we use the Pythagorean Theorem. See Problem 33 in Section 4.7 on page 297.

OUTLINE

The Big Picture: Putting It Together

In Chapter 3, we learned how to multiply polynomials. We began by multiplying a monomial and a trinomial using the Distributive Property. We then learned how to multiply two binomials, and, in general, how to multiply two polynomials. In this chapter, we reverse the process. That is, we want to write a polynomial as a product. This process is called *factoring*.

In Chapter 2, we solved linear (first-degree) equations such as $2x + 5 = 8$. In this chapter, we discuss how factoring can be used to solve equations such as $2x^2 + 7x + 3 = 0$. The approach requires that we rewrite $2x^2 + 7x + 3$ as the product of two polynomials of degree 1. This ultimately leads us to solving two linear equations—something we already know how to do! This is one of the goals of algebra: Simplify a problem until it becomes a problem you already know how to solve!

Factoring is important for solving equations, but also will play a major role in Chapters 5, 6, and 7, so be sure to work hard to learn the factoring techniques presented in this chapter.

4.1 Greatest Common Factor and Factoring by Grouping

OBJECTIVES

1. Find the Greatest Common Factor of Two or More Expressions
2. Factor Out the Greatest Common Factor in Polynomials
3. Factor Polynomials by Grouping

Preparing for Greatest Common Factor and Factoring by Grouping

Before getting started, take the following readiness quiz. If you get a problem wrong, go back to the section cited and review the material.

1. Write 48 as the product of prime numbers. [Section A.1, p. A1]
2. Distribute: $2(5x - 3)$ [Section 1.7, p. 61]
3. Find the product: $(2x + 5)(x - 3)$ [Section 3.3, pp. 191–193]

Consider the following products:

$$5 \cdot 3 = 15$$
$$5(y + 5) = 5y + 25$$
$$(3x - 1)(x + 5) = 3x^2 + 14x - 5$$

The expressions on the left side are called **factors** of the expression on the right side. For example, $3x - 1$ and $x + 5$ are factors of $3x^2 + 14x - 5$. In the last chapter, we learned how to multiply factors to obtain a product. For example, we learned how to multiply expressions such as $3x - 1$ and $x + 5$ to obtain $3x^2 + 14x - 5$.

In this chapter, we learn how to obtain the factors of a polynomial such as $3x^2 + 14x - 5$. That is, we learn how to write $3x^2 + 14x - 5$ as $(3x - 1)(x + 5)$.

In Words
Factoring is "undoing" multiplication.

DEFINITION
To **factor** a polynomial means to write the polynomial as a product.

The process of factoring reverses the process of multiplying, as shown below.

Multiplication →

Factored form → $3(x - 7) = 3x - 21$ ← Product

← Factoring

1 Find the Greatest Common Factor of Two or More Expressions

If you look carefully at the illustration above, you will notice that 3 is the largest number that divides evenly into both $3x$ and 21 in the expression $3x - 21$. For this reason, 3 is the greatest common factor of $3x - 21$.

DEFINITION
The **greatest common factor (GCF)** of a list of polynomials is the largest expression that divides evenly into all the polynomials.

Writing $3x - 21$ as $3(x - 7)$ is referred to as factoring out the greatest common factor. But how can we find the GCF? The following example shows us how.

Preparing for...Answers **1.** $3 \cdot 2^4$ **2.** $10x - 6$ **3.** $2x^2 - x - 15$

EXAMPLE 1 How to Find the GCF of a List of Numbers

Find the GCF of 12 and 18.

Step-by-Step Solution

Step 1: Write each number as the product of prime factors.	$12 = 2 \cdot 2 \cdot 3$ $18 = 2 \quad \cdot 3 \cdot 3$
Step 2: Determine the common prime factors.	The common factors are 2 and 3.
Step 3: Find the product of the common factors found in Step 2. This number is the GCF.	The GCF is $2 \cdot 3 = 6$.

Work Smart

Remember, a prime number is a number greater than 1 that has no factors other than itself and 1. For example, 3, 7, and 13 are prime numbers while $4\ (= 2 \cdot 2)$, $12\ (= 2 \cdot 2 \cdot 3)$, and $35\ (= 5 \cdot 7)$ are not prime.

Also, it is helpful to align the common factors vertically.

Classroom Example

Find the greatest common factor (GCF) of the terms.

(a) 16, 24 **(b)** 27, 36, 54

Answer:

(a) 8 **(b)** 9

EXAMPLE 2 Finding the GCF of a List of Numbers

Find the GCF of 24, 40, and 72.

Solution

We write each number as the product of prime factors.

$$24 = 4 \cdot 6 = 2 \cdot 2 \cdot 2 \cdot 3$$
$$40 = 4 \cdot 10 = 2 \cdot 2 \cdot 2 \cdot \quad 5$$
$$72 = 8 \cdot 9 = 2 \cdot 2 \cdot 2 \cdot 3 \cdot 3$$

Because all three numbers contain three factors of 2, the GCF is $2 \cdot 2 \cdot 2 = 8$.

Notice the greatest common factor in Example 2 could be written as 2^3. The exponent 3 represents the number of times the factor 2 appears in the factorization of each number.

QUICK ✓ *Find the GCF of each list of numbers.*

1. 32, 40 **2.** 15, 25 **3.** 12, 45 **4.** 21, 35, 84

What if we want to find the greatest common factor among two or more expressions that contain variables? The approach to finding the GCF is the same as it is for numbers. Consider the terms x^3, x^5, and x^6. We can write each of these terms as the product of factors of x as follows:

$$x^3 = x \cdot x \cdot x$$
$$x^5 = x \cdot x \cdot x \cdot x \cdot x$$
$$x^6 = x \cdot x \cdot x \cdot x \cdot x \cdot x$$

Work Smart

The GCF of a variable factor is the lowest power of that variable.

Each of the terms contains three factors of x, so the greatest common factor is x^3. It is no coincidence that the exponent of the GCF is 3, the smallest exponent of the terms x^3, x^5, and x^6. This approach to finding the GCF for variable expressions will work in general. We illustrate how to find the GCF of two expressions in the next example.

Classroom Example ➤ Find the greatest common factor (GCF) of $12n^3$ and $18n^2$.

Answer: $6n^2$

EXAMPLE 3 **Finding the Greatest Common Factor of Two Expressions**

Find the greatest common factor (GCF) of $8y^4$, $12y^2$.

Solution

Step 1: The coefficients are 8 and 12. We need to determine the GCF of 8 and 12.

$$8 = 2 \cdot 2 \cdot 2$$
$$12 = 2 \cdot 2 \quad \cdot 3$$

The GCF of the coefficients is $2 \cdot 2 = 4$.

Step 2: The variable expressions are y^4 and y^2. For each variable, determine the smallest exponent that each variable is raised to.

The GCF of y^4 and y^2 is y^2.

Step 3: Find the product of the common factors found in Steps 1 and 2. This expression is the GCF.

The GCF is $4y^2$.

Classroom Example ➤ Find the greatest common factor (GCF) of the terms.

(a) $6x^3$, $12x^2$, $15x$

(b) $14x^3y^4$, $28x^2y^3$, $42xy^2$

Answer:

(a) $3x$ (b) $7xy^2$

EXAMPLE 4 **Finding the Greatest Common Factor of Three Expressions**

Find the GCF of the expressions:

(a) $3x^3, 9x^2, 21x$ **(b)** $10x^5y^4, 15x^2y^3, 25x^3y^5$

Solution

We determine the GCF of the coefficients and then determine the variable expression with the smallest exponent. The product of these two factors will be the GCF.

(a) The coefficients, 3, 9, and 21, written as the product of prime numbers are

Factor the coefficients as a product of primes:
$$3 = 3$$
$$9 = 3 \cdot 3$$
$$21 = 3 \cdot \quad 7$$

The GCF of the coefficients is 3.

The variable expressions are x^3, x^2, and x. The smallest exponent is 1, so the GCF of the variable expressions is x. Therefore, the GCF of $3x^3$, $9x^2$, $21x$ is $3x$.

(b) The coefficients are 10, 15, and 25. We write these coefficients as the product of prime numbers.

Factor the coefficients as a product of primes:
$$10 = 5 \cdot 2$$
$$15 = 5 \cdot \quad 3$$
$$25 = 5 \cdot \qquad 5$$

The GCF of the coefficients is 5.

The GCF of x^5, x^2, and x^3 is x^2. The GCF of y^4, y^3, and y^5 is y^3. Therefore, the GCF of $10x^5y^4$, $15x^2y^3$, $25x^3y^5$ is $5x^2y^3$.

QUICK ✓ *Find the greatest common factor (GCF) of the terms.*

5. $14y^3, 35y^2$ **6.** $6z^3, 8z^2, 12z$ **7.** $4x^3y^5, 8x^2y^3, 24xy^4$

The greatest common factor can be a binomial, as illustrated by the following example.

Classroom Example ➤
Find the GCF of each pair of expressions:
(a) $2(x - 7)$ and $5(x - 7)$
(b) $3(n + 2)(n - 4)$ and $6(n - 4)^2$
Answer:
(a) $x - 7$ (b) $3(n - 4)$

EXAMPLE 5 The GCF as a Binomial

Find the greatest common factor of each pair of expressions.

(a) $3(x - 1)$ and $8(x - 1)$ **(b)** $2(z + 3)(z + 5)$ and $4(z + 5)^2$

Solution

(a) There is no common factor between the coefficients, 3 and 8. However, each expression has $x - 1$ as a factor, so the GCF of $3(x - 1)$ and $8(x - 1)$ is $x - 1$.

(b) The GCF between 2 and 4 is 2. The GCF between $(z + 3)(z + 5)$ and $(z + 5)^2$ is $z + 5$. The GCF of the expressions $2(z + 3)(z + 5)$ and $4(z + 5)^2$ is $2(z + 5)$.

QUICK ✓ *Find the greatest common factor (GCF) of each pair of expressions.*

8. $7(2x + 3)$ and $-4(2x + 3)$ **9.** $9(k + 8)(3k - 2)$ and $12(k - 1)(k + 8)^2$

SUMMARY: Steps to Find the Greatest Common Factor of Two or More Expressions

Step 1: Find the GCF of the coefficients of each variable expression.

Step 2: For each variable expression common to all the terms, determine the smallest exponent that the variable expression is raised to.

Step 3: Find the product of the common factors found in Steps 1 and 2. This expression is the GCF.

2 Factor Out the Greatest Common Factor in Polynomials

The first step in factoring any polynomial is to look for the greatest common factor. Once the GCF is identified, we use the Distributive Property "in reverse" to factor the polynomial as shown below.

$$ab + ac = a(b + c) \quad \text{or} \quad ab - ac = a(b - c)$$

When we use this method, we say that we "factor out" the greatest common factor. Example 6 illustrates the method.

EXAMPLE 6 How to Factor Out the Greatest Common Factor in a Polynomial

Classroom Example ➤
Factor $7z^2 - 14z$ by factoring out the greatest common factor.
Answer: $7z(z - 2)$

Factor $2x - 10$ by factoring out the greatest common factor.

Step-by-Step Solution

Step 1: Find the GCF.	GCF = 2
Step 2: Rewrite each term as the product of the GCF and remaining factor.	$2x - 10 = 2(x) - 2(5)$

(continued)

Step 3: *Factor out the GCF.*	$= 2(x - 5)$
Step 4: Check	$2(x - 5) = 2(x) - 2(5)$ $= 2x - 10$

So $2x - 10 = 2(x - 5)$.

Work Smart
The Distributive Property comes in handy again for factoring out the GCF. Note how it is used in the check step, too.

Below we summarize the steps to follow when factoring out the greatest common factor.

Steps to Factor a Polynomial Using the Greatest Common Factor

Step 1: Identify the greatest common factor (GCF) of the terms that make up the polynomial.

Step 2: Rewrite each term as the product of the GCF and the remaining factor.

Step 3: Use the Distributive Property "in reverse" to factor out the GCF.

Step 4: Use the Distributive Property to verify that the factorization is correct.

Classroom Example Factor the binomial $8p^4 + 24p^3$ by factoring out the greatest common factor.
Answer: $8p^3(p + 3)$

EXAMPLE 7 Factoring Out the Greatest Common Factor in a Binomial

Factor the binomial $9z^3 + 36z^2$ by factoring out the greatest common factor.

Solution

The greatest common factor between 9 and 36 is 9. The greatest common factor between z^3 and z^2 is z^2. Therefore, the GCF is $9z^2$.

Rewrite each term as the product of the GCF and remaining factor: $9z^3 + 36z^2 = 9z^2(z) + 9z^2(4)$

Factor out the GCF: $= 9z^2(z + 4)$

Check $9z^2(z + 4) = 9z^2(z) + 9z^2(4)$
$= 9z^3 + 36z^2$

So $9z^3 + 36z^2 = 9z^2(z + 4)$.

Classroom Example Factor the trinomial $4a^2b^2 - 10ab^3 + 18a^3b^4$ by factoring out the greatest common factor.
Answer: $2ab^2(2a - 5b + 9a^2b^2)$

EXAMPLE 8 Factoring Out the Greatest Common Factor in a Trinomial

Factor the trinomial $6a^2b^2 - 8ab^3 + 18a^3b^4$ by factoring out the greatest common factor.

Solution

The GCF of $6a^2b^2 - 8ab^3 + 18a^3b^4$ is $2ab^2$. We now rewrite each term as the product of the GCF and the remaining factor.

$$6a^2b^2 - 8ab^3 + 18a^3b^4 = 2ab^2(3a) - 2ab^2(4b) + 2ab^2(9a^2b^2)$$

Factor out the GCF: $= 2ab^2(3a - 4b + 9a^2b^2)$

Check $2ab^2(3a - 4b + 9a^2b^2) = 2ab^2(3a) - 2ab^2(4b) + 2ab^2(9a^2b^2)$
$= 6a^2b^2 - 8ab^3 + 18a^3b^4$

So $6a^2b^2 - 8ab^3 + 18a^3b^4 = 2ab^2(3a - 4b + 9a^2b^2)$.

QUICK ✓ *Factor each polynomial by factoring out the greatest common factor.*

10. $5z^2 - 30z$

11. $12p^2 + 30p^4$

12. $16y^3 - 12y^2 + 4y$

13. $6m^4n^2 + 18m^3n^4 - 22m^2n^5$

When the coefficient of the term of highest degree is negative, we factor the negative out of the polynomial as part of the GCF.

Classroom Example Factor $-5y^3 + 10y$ by factoring out the greatest common factor.

Answer: $-5y(y^2 - 2)$

EXAMPLE 9 Factoring Out a Negative as Part of the GCF

Factor $-7a^3 + 14a$ by factoring out the greatest common factor.

Solution

This binomial is written in standard form. Since the coefficient on the highest-degree term, $-7a^3$, is negative, we factor the negative out as part of the GCF. So we use $-7a$ as the greatest common factor.

$$-7a^3 + 14a = -7a(a^2) + (-7a)(-2)$$

Factor out GCF: $$= -7a(a^2 - 2)$$

Check

$$-7a(a^2 - 2) = -7a(a^2) + (-7a)(-2)$$
$$= -7a^3 + 14a$$

So $-7a^3 + 14a = -7a(a^2 - 2)$.

QUICK ✓ *Factor out the greatest common factor.*

14. $-4y^2 + 8y$

15. $-6a^3 + 12a^2 - 3a$

Sometimes the greatest common factor is a binomial.

Classroom Example Factor out the greatest common binomial factor.

$4x(x - 3) + 5(x - 3)$

Answer: $(x - 3)(4x + 5)$

EXAMPLE 10 Factoring Out a Binomial as the Greatest Common Factor

Factor out the greatest common binomial factor: $5x(x - 2) + 3(x - 2)$

Solution

Do you see that $x - 2$ is common to both terms? The GCF is the binomial $x - 2$.

$$5x(x - 2) + 3(x - 2) = 5x(x - 2) + 3(x - 2)$$

Factor out $x - 2$: $$= (x - 2)(5x + 3)$$

Check

$$(x - 2)(5x + 3) = (x - 2)5x + (x - 2)3$$
$$= 5x(x - 2) + 3(x - 2)$$

So $5x(x - 2) + 3(x - 2) = (x - 2)(5x + 3)$.

Teaching Tip When factoring out a GCF that is a binomial, students may incorrectly want to write the answer to Example 10 as $(x - 2)^2(5x + 3)$. Have students multiply out the terms. Does $(x - 2)^2(5x + 3) = 5x(x - 2) + 3(x - 2)$? No!

QUICK ✓ *Factor out the greatest common factor.*

16. $2a(a - 5) + 3(a - 5)$

17. $7z(z + 5) - 4(z + 5)$

3 Factor Polynomials by Grouping

Sometimes a common factor does not occur in every term of the polynomial. If a polynomial contains four terms, it may be possible to find a GCF of the first two terms and a different GCF of the second two terms. When this happens, the common factor can be

Work Smart
Try factoring by grouping when a polynomial contains four terms.

factored out of each group of terms. This technique is called **factoring by grouping,** as is illustrated in the next example.

EXAMPLE 11 How to Factor by Grouping

Classroom Example
Factor by grouping: $5x + 5y + ax + ay$
Answer: $(x + y)(5 + a)$

Factor by grouping: $3x - 3y + ax - ay$

Step-by-Step Solution

Step 1: Group terms with common factors.	$3x - 3y + ax - ay = (3x - 3y) + (ax - ay)$
Step 2: In each grouping, factor out the GCF.	$= 3(x - y) + a(x - y)$
Step 3: Factor out the common factor, $x - y$, that remains.	$= (x - y)(3 + a)$
Step 4: Check	FOIL: $(x - y)(3 + a) = 3x + ax - 3y - ay$ Rearrange terms: $= 3x - 3y + ax - ay$

So $3x - 3y + ax - ay = (x - y)(3 + a)$.

Based upon Example 11, we have the following steps for factoring by grouping.

Work Smart
We could have written the answer to Example 11 as $(3 + a)(x - y)$. Do you know why?

Steps to Factor a Polynomial by Grouping

Step 1: Group the terms with common factors.
Step 2: In each grouping, factor out the greatest common factor (GCF).
Step 3: Factor out the common factor that remains.
Step 4: Check your work by finding the product of the factors.

QUICK ✓ *Factor by grouping.*

18. $4x + 4y + bx + by$

Classroom Example
Factor by grouping:
$2ax - 2ay - 3x + 3y$
Answer: $(2a - 3)(x - y)$

EXAMPLE 12 Factoring by Grouping

Factor by grouping: $5x - 5y - 4bx + 4by$

Solution

First, we group terms with common factors.

$$5x - 5y - 4bx + 4by = (5x - 5y) + (-4bx + 4by)$$

Factor out the common factor in each group: $= 5(x - y) + (-4b)(x - y)$

Factor out the common factor that remains: $= (x - y)(5 - 4b)$

Work Smart
Be careful when working with signs:
$-4b(x - y) = -4bx + 4by$

Check Use FOIL: $(x - y)(5 - 4b) = 5x - 4bx - 5y + 4by$

Rearrange terms: $= 5x - 5y - 4bx + 4by$

So $5x - 5y - 4bx + 4by = (x - y)(5 - 4b)$.

Teaching Tip
Suggest that students group different pairs of terms in Example 12. Does every combination factor?
Also, consider using the classroom example for Example 12 by arranging the terms as $2ax + 3y - 2ay - 3x$ so students see some rearrangement may be necessary before factoring by grouping.

Work Smart
Sometimes it will be necessary to rearrange terms before factoring by grouping.

In Example 12, we could have rearranged terms using the Commutative Property of Addition to write the problem $5x - 5y - 4bx + 4by$ as $5x - 4bx - 5y + 4by$ and factored as follows.

$$5x - 4bx - 5y + 4by = x(5 - 4b) - y(5 - 4b)$$
$$= (5 - 4b)(x - y)$$

Notice that the answer contains the same factors, just written in reverse order. But $(5 - 4b)(x - y) = (x - y)(5 - 4b)$ by the Commutative Property of Multiplication, so we know that the answer is correct.

QUICK ✓ *Factor by grouping.*

19. $6az - 2a - 9bz + 3b$

20. $8b + 4 - 10ab - 5a$

Whenever we encounter a factoring problem, the first thing we should always do is look for a common factor. The next example illustrates this idea.

Classroom Example
Factor by grouping:
$4x^3 - 8x^2 + 6x^2y - 12xy$
Answer: $2x(2x + 3y)(x - 2)$

Work Smart
Whenever factoring, always look for a GCF first.

EXAMPLE 13 Factoring by Grouping

Factor: $3x^3 + 12x^2 - 6x - 24$

Solution

Do all four terms contain a common factor? Yes!! There is a GCF of 3, so we factor it out.

$$3x^3 + 12x^2 - 6x - 24 = 3(x^3 + 4x^2 - 2x - 8)$$

Now we group terms with common factors.

$$3(x^3 + 4x^2 - 2x - 8) = 3[(x^3 + 4x^2) + (-2x - 8)]$$

Factor out the common factor in each group: $= 3[(x^2(x + 4) + (-2)(x + 4)]$

Factor out the common factor that remains: $= 3(x + 4)(x^2 - 2)$

Check Use FOIL: $3(x + 4)(x^2 - 2) = 3(x^3 - 2x + 4x^2 - 8)$

Distribute the 3: $= 3x^3 - 6x + 12x^2 - 24$

Rearrange terms: $= 3x^3 + 12x^2 - 6x - 24$

So $3x^3 + 12x^2 - 6x - 24 = 3(x + 4)(x^2 - 2)$.

QUICK ✓ *Factor by grouping.*

21. $3z^3 + 12z^2 + 6z + 24$

22. $2n^4 + 2n^3 - 4n^2 - 4n$

4.1 Exercises

For Extra Help: Student Solutions Manual | CD Video | PH Math/Tutor Center | MathXL Tutorials on CD | MathXL® | MyMathLab

Concepts and Vocabulary

In Problems 1–4, fill in the blanks.

1. The largest expression that divides evenly into a set of numbers is called the _______ _______ _______.

2. To _______ a polynomial means to write the polynomial as a product.

1. greatest common factor
2. factor

3. Distributive
4. grouping
5. True
6. False
7. False
8. False
9. Answers may vary.
10. Answers may vary.
11. 2
12. 7
13. 1
14. 1
15. 5
16. 12
17. 26
18. 13
19. 4
20. 7
21. 18
22. 15
23. x^2
24. y
25. $7x$
26. $4a^2$
27. m^2n^2
28. xy
29. $15ab^2$
30. $13xy$
31. $2abc^2$
32. xyz
33. 3
34. $(x + y)$
35. $2(x - 4)^2$
36. $3(a - b)$
37. $6(x - 3)^2$
38. $3(2a - 1)^2$
39. $4x^3$
40. $4x^2$
41. $-3s$
42. $-5y^3z$
43. mn^5
44. $3yz$
45. $6(2x - 3)$
46. $3(a + 2)$
47. $5x^2y(1 - 3xy)$
48. $4a^3b^2(2 + 3a^2)$
49. $3x(x^2 + 2x - 1)$
50. $5x^2(x^2 + 2x - 5)$
51. $3(3m^5 - 6m^3 - 4m^2 + 27)$
52. $5z^2(1 + 2z^2 - 3z - 9z^3)$
53. $15ab^2(ab^2 - 4b + 3a^2)$
54. $3rs(4r^2s + 1 - 2s^3)$
55. $(x - 3)(x - 5)$
56. $(a - 5)(a + 6)$
57. $(x - 1)(x^2 + y^2)$
58. $(b + 2)(a^2 - b)$
59. $(4x + 1)(x^2 + 2x + 5)$
60. $(s^2 - 1)(s^2 + 4s + 7)$
61. $-x(3x - 4)$
62. $-3b(3b + 2a)$
63. $-5x(x^2 - 2x + 3)$
64. $-2(y^2 - 5y + 7)$
65. $-4z(3z^2 - 4z + 2)$
66. $-2n(11n^3 - 9n - 7)$
67. $-5(3b^3 + b - 2)$
68. $-2m^2(12m^2 + 8m - 7)$

3. When we factor a polynomial using the GCF, we use the _______ Property in reverse.

4. When a polynomial contains four terms try factoring by _______.

In Problems 5–8, answer True or False to each statement.

5. The first step in any factoring problem is to look for the GCF of the polynomial.

6. There is more than one correct factorization of any given polynomial.

7. Every polynomial containing four terms can be factored by grouping.

8. We factor $(2x + 1)(x - 3) + (2x + 1)(2x + 7)$ as $(2x + 1)^2(3x + 4)$.

9. In your own words, write a list of steps for finding the greatest common factor and then write a second list of steps for factoring the GCF from a polynomial.

10. Describe how to factor a negative from a polynomial. What types of errors might happen during this process?

Building Skills

In Problems 11–38, find the greatest common factor, GCF, of each group of expressions.

11. 8, 6 **12.** 49, 35 **13.** 15, 14 **14.** 6, 55

15. 15, 20 **16.** 24, 36 **17.** 78, 104 **18.** 65, 91

19. 12, 28, 48 **20.** 35, 42, 63 **21.** 36, 54, 72 **22.** 60, 75, 135

23. x^{10}, x^2, x^8 **24.** y^3, y^5, y **25.** $7x, 14x^3$ **26.** $8a^4, 20a^2$

27. m^3n^2, m^2n^2 **28.** xy^2, x^4y **29.** $45a^2b^3, 75ab^2c$ **30.** $26xy^2, 39x^2y$

31. $4a^2bc^3, 6ab^2c^2, 8a^2b^2c^4$ **32.** $2x^2yz, xyz^2, 5x^3yz^2$

33. $3(x - 1)$ and $6(x + 1)$ **34.** $8(x + y)$ and $9(x + y)$

35. $2(x - 4)^2$ and $4(x - 4)^3$ **36.** $6(a - b)$ and $15(a - b)^3$

37. $12(x + 2)(x - 3)^2$ and $18(x - 3)^2(x - 2)$

38. $15(2a - 1)^2(2a + 1)$ and $18(2a - 1)^3(2a + 3)^2$

In Problems 39–44, identify the missing factor to make the statement true.

39. $12x^6 = 3x^3 \cdot ?$ **40.** $40x^5 = 10x^3 \cdot ?$ **41.** $9s^2t^2 = -3st^2 \cdot ?$

42. $-5y^4z^2 = yz \cdot ?$ **43.** $15m^3n^5 = 15m^2 \cdot ?$ **44.** $-120z^6y = -40z^5 \cdot ?$

In Problems 45–60, factor the GCF from the polynomial.

45. $12x - 18$ **46.** $3a + 6$

47. $5x^2y - 15x^3y^2$ **48.** $8a^3b^2 + 12a^5b^2$

49. $3x^3 + 6x^2 - 3x$ **50.** $5x^4 + 10x^3 - 25x^2$

51. $9m^5 - 18m^3 - 12m^2 + 81$ **52.** $5z^2 + 10z^4 - 15z^3 - 45z^5$

53. $15a^2b^4 - 60ab^3 + 45a^3b^2$ **54.** $12r^3s^2 + 3rs - 6rs^4$

55. $(x - 3)x - (x - 3)5$ **56.** $a(a - 5) + 6(a - 5)$

57. $x^2(x - 1) + y^2(x - 1)$ **58.** $(b + 2)a^2 - (b + 2)b$

59. $x^2(4x + 1) + 2x(4x + 1) + 5(4x + 1)$ **60.** $s^2(s^2 - 1) + 4s(s^2 - 1) + 7(s^2 - 1)$

In Problems 61–68, factor out the GCF using a negative sign.

61. $-3x^2 + 4x$ **62.** $-9b^2 - 6ab$ **63.** $-5x^3 + 10x^2 - 15x$

64. $-2y^2 + 10y - 14$ **65.** $-12z^3 + 16z^2 - 8z$ **66.** $-22n^4 + 18n^2 + 14n$

67. $10 - 5b - 15b^3$ **68.** $14m^2 - 16m^3 - 24m^4$

69. $(x + 3)(y + 4)$
70. $(x + a)(x + 2)$
71. $(y + 1)(z - 1)$
72. $(m - 3)(n + 2)$
73. $(x - 1)(x^2 + 2)$
74. $(z + 4)(z^2 + 3)$
75. $(2t - 1)(t^2 - 1)$
76. $(x - 1)(x^2 - 1)$
77. $(2t - 1)(t^3 - 1)$
78. $(3z - 4)(2y - 3)$
79. $4(y - 5)$
80. $3(z - 7)$
81. $7m(4m^2 + m + 9)$
82. $5x(2x^2 - 3x + 1)$
83. $6m^2n(2mnp - 3)$
84. $4st(st^2 - 6)$
85. $3(p + 3)(3p + 1)$
86. $(3x - 2)(5x - 9)$
87. $3(2x - y)(3a - 2b)$
88. $(3x - 2y)(2m + n)$
89. $(x - 2)(x - 2)$ or $(x - 2)^2$
90. $(2b - 1)(a + 1)$
91. $3x(5x - 2)(x^2 + 2)$
92. $2(a - 2)(a^2 + 4)$
93. $-3x(x^2 - 2x + 3)$
94. $-4z^2(2z^2 + 3z - 7)$
95. $-4b(3 - 4b)$
96. $4a(5a - 2)$
97. $(4y + 3)(3x - 2)$
98. $-3x(4m + 1)(z - 2)$
99. $\frac{1}{3}x^2\left(x - \frac{2}{3}\right)$
100. $\frac{1}{4}p^3(3p - 1)$
101. Answers may vary.
102. Answers may vary.
103. Answers may vary.
104. Answers may vary.
105. $-2t(8t - 75)$
106. $-16(t^2 - 5t - 3)$
107. $150(140 - p)$
108. $-4p(p - 1000)$
109. $4x^3(2x^2 - 7)$
110. $3n(6n^2 - 5n^2 + 2)$
111. $2\pi r(r + 4)$

In Problems 69–78, factor by grouping.

69. $xy + 3y + 4x + 12$ **70.** $x^2 + ax + 2a + 2x$ **71.** $yz + z - y - 1$

72. $mn - 3n + 2n - 6$ **73.** $x^3 - x^2 + 2x - 2$ **74.** $z^3 + 4z^2 + 3z + 12$

75. $2t^3 - t^2 - 2t + 1$ **76.** $x^3 - x^2 - x + 1$

77. $2t^4 - t^3 - 2t + 1$ **78.** $6yz - 8y - 9z + 12$

Mixed Practice

In Problems 79–100, factor each polynomial.

79. $4y - 20$ **80.** $3z - 21$

81. $28m^3 + 7m^2 + 63m$ **82.** $10x^3 - 15x^2 + 5x$

83. $12m^3n^2p - 18m^2n$ **84.** $4s^2t^3 - 24st$

85. $(2p - 1)(p + 3) + (7p + 4)(p + 3)$ **86.** $(3x - 2)(4x + 1) + (3x - 2)(x - 10)$

87. $18ax - 9ay - 12bx + 6by$ **88.** $6xm + 3xn - 4ym - 2yn$

89. $(x - 2)(x - 3) + (x - 2)$ **90.** $(a + 2)(2b - 1) - (2b - 1)$

91. $15x^4 - 6x^3 + 30x^2 - 12x$ **92.** $2a^3 - 4a^2 + 8a - 16$

93. $-3x^3 + 6x^2 - 9x$ **94.** $-8z^4 - 12z^3 + 28z^2$

95. $-12b + 16b^2$ **96.** $-8a + 20a^2$

97. $12xy + 9x - 8y - 6$ **98.** $-12mxz - 3xz + 24mx + 6x$

99. $\frac{1}{3}x^3 - \frac{2}{9}x^2$ **100.** $\frac{3}{4}p^4 - \frac{1}{4}p^3$

Applying the Concepts

101. Write any trinomial that has $-4xy$ as a GCF.

102. Write any binomial that has $3x^3$ as a GCF.

103. Write any binomial that has $(x - 1)$ as a GCF.

104. Write any trinomial that has $(y + 2)$ as a GCF.

105. Height of a Toy Rocket The height of a toy rocket after t seconds when fired straight up with an initial speed of 150 feet per second from the ground is given by the polynomial $-16t^2 + 150t$. Write the polynomial $-16t^2 + 150t$ in factored form.

106. Height of a Ball The height of a ball after t seconds when thrown straight up with an initial speed of 80 feet per second from a height of 48 feet is given by the polynomial $-16t^2 + 80t + 48$. Write the polynomial $-16t^2 + 80t + 48$ in factored form.

107. Selling Calculators A manufacturer of calculators found that the number of calculators sold at a price of p dollars is given by the polynomial $21{,}000 - 150p$. Write $21{,}000 - 150p$ in factored form.

108. Revenue A manufacturer of a gas clothes dryer has found that the revenue (in dollars) from selling clothes dryers at a price of p dollars is given by the expression $-4p^2 + 4000p$. Write $-4p^2 + 4000p$ in factored form.

△ **109. Area of a Rectangle** A rectangle has an area of $8x^5 - 28x^3$ square feet. Write $8x^5 - 28x^3$ in factored form.

△ **110. Area of a Parallelogram** A parallelogram has an area of $18n^4 - 15n^3 + 6n$ square cm. Write $18n^4 - 15n^3 + 6n$ in factored form.

△ **111. Surface Area** The surface area of a cylindrical can whose radius is r inches and height is 4 inches is given by $S = 2\pi r^2 + 8\pi r$ square inches. Express the surface area in factored form.

112. $2\pi r(r + 8)$
113. $x - 2$
114. $3x + 2$
115. $4x^{2n} + 5$
116. $x^2 + 2x^{4n}$
117. $3x^3 - 4x^2 + 2$
118. $x^2 - 3x + 2$
119. $x^4 - 3x + 2$
120. $2x^6 + x - 1$
121. $\frac{3}{5}x^3 - \frac{1}{2}x + 1$
122. $\frac{2}{5}x^2 - \frac{1}{3}x + 1$

△ **112. Surface Area** The surface area of a cylindrical can whose radius is r inches and height is 8 inches is given by $S = 2\pi r^2 + 16\pi r$ square inches. Express the surface area in factored form.

In Problems 113 and 114, the area of the polygon is given. Write a polynomial that represents the missing length.

△ **113.**

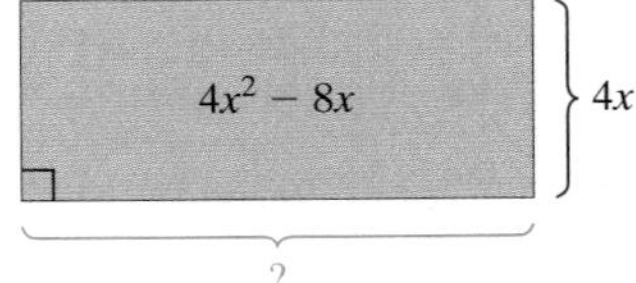

△ **114.**

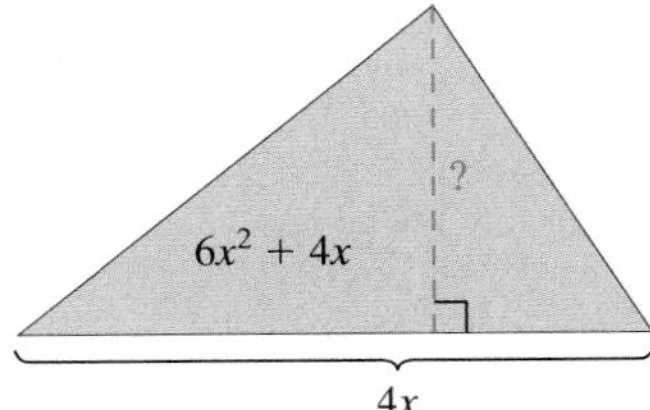

Extending the Concepts

In Problems 115–122, find the missing factor.

115. $8x^{3n} + 10x^n = 2x^n \cdot ?$

116. $3x^{2n+2} + 6x^{6n} = 3x^{2n} \cdot ?$

117. $3 - 4x^{-1} + 2x^{-3} = x^{-3} \cdot ?$

118. $1 - 3x^{-1} + 2x^{-2} = x^{-2} \cdot ?$

119. $x^2 - 3x^{-1} + 2x^{-2} = x^{-2} \cdot ?$

120. $2x^2 + x^{-3} - x^{-4} = x^{-4} \cdot ?$

121. $\frac{6}{35}x^4 - \frac{1}{7}x^2 + \frac{2}{7}x = \frac{2}{7}x \cdot ?$

122. $\frac{2}{15}x^3 - \frac{1}{9}x^2 + \frac{1}{3}x = \frac{1}{3}x \cdot ?$

4.2 Factoring Trinomials of the Form $x^2 + bx + c$

OBJECTIVES

1. Factor Trinomials of the Form $x^2 + bx + c$
2. Factor Out the GCF, Then Factor $x^2 + bx + c$

Preparing for Factoring Trinomials of the Form $x^2 + bx + c$

Before getting started, take the following readiness quiz. If you get a problem wrong, go back to the section cited and review the material.

1. Find a pair of factors of 18 whose sum is 11. [Section A.1, p. A1]
2. Find a pair of factors of -24 whose sum is -2. [Section A.1, p. A1]
3. Determine the coefficients of $3x^2 - x - 4$. [Section 1.7, pp. 59–60]
4. Find the product: $-5p(p + 4)$ [Section 3.3, pp. 190–191]
5. Find the product: $(z - 1)(z + 4)$ [Section 3.3, pp. 191–193]

In this section, we are going to factor trinomials that are of degree 2. Because the word **quadratic** means "relating to a square," these trinomials are called *quadratic trinomials.*

DEFINITION

A **quadratic trinomial** is a polynomial of the form $ax^2 + bx + c$, $a \neq 0$ where a represents the coefficient of the squared (second-degree) term, b represents the coefficient of the linear (first-degree) term, and c represents the constant.

When the trinomial is written in standard form (descending order of degree) the coefficient of the squared term is also called the **leading coefficient.** We begin by looking at quadratic trinomials where the leading coefficient, a, is 1. Examples of quadratic trinomials whose leading coefficient is 1 are

$$x^2 + 4x + 3 \qquad a^2 + 4a - 21 \qquad p^2 - 11p + 18$$

1 Factor Trinomials of the Form $x^2 + bx + c$

The idea behind factoring a second-degree trinomial of the form $x^2 + bx + c$ is to see whether it can be written as the product of two first-degree polynomials.

Preparing for...Answers **1.** 9 and 2 **2.** -6 and 4 **3.** 3, -1, -4 **4.** $-5p^2 - 20p$ **5.** $z^2 + 3z - 4$

For example,

Multiplication →

Factored form → $(x + 3)(x - 7) = x^2 - 4x - 21$ ← Product

← Factoring

The factors of $x^2 - 4x - 21$ are $x + 3$ and $x - 7$. Notice the following:

$$x^2 - 4x - 21 = (x + 3)(x - 7)$$

The sum of -7 and 3 is -4.

The product of -7 and 3 is -21.

In general, if $x^2 + bx + c = (x + m)(x + n)$, then $mn = c$ and $m + n = b$. We illustrate how to factor a trinomial of the form $x^2 + bx + c$ in the following example.

EXAMPLE 1 How to Factor a Trinomial of the Form $x^2 + bx + c$, where c Is Positive

Factor: $x^2 + 7x + 12$

Classroom Example

Factor: $x^2 + 8x + 12$

Answer: $(x + 6)(x + 2)$

Step-by-Step Solution

Step 1: When we compare $x^2 + 7x + 12$ with $x^2 + bx + c$, we see that $b = 7$ and $c = 12$. We are looking for factors of $c = 12$ whose sum is $b = 7$. We begin by listing all factors of 12 and computing the sum of these factors.

Integers Whose Product Is 12	1, 12	2, 6	3, 4	−1, −12	−2, −6	−3, −4
Sum	13	8	7	−13	−8	−7

We can see that $3 \cdot 4 = 12$ and $3 + 4 = 7$, so $m = 3$ and $n = 4$.

Step 2: We write the trinomial in the form $(x + m)(x + n)$.

$$x^2 + 7x + 12 = (x + 3)(x + 4)$$

Step 3: Check

Use FOIL: $(x + 3)(x + 4) = x^2 + 4x + 3x + 3(4) = x^2 + 7x + 12$

So $x^2 + 7x + 12 = (x + 3)(x + 4)$.

We summarize the steps for factoring a trinomial of the form $x^2 + bx + c$ below.

Work Smart

Because multiplication is commutative, the order in which we list the factors does not matter.

Steps to Factor a Trinomial of the Form $x^2 + bx + c$

Step 1: Find the pair of integers whose product is c and whose sum is b. That is, determine m and n such that $mn = c$ and $m + n = b$.

Step 2: Write $x^2 + bx + c = (x + m)(x + n)$.

Step 3: Check your work by multiplying the binomials using the FOIL method.

In Example 1, notice that the coefficient of the middle term is positive and the constant is positive. **If the coefficient of the middle term and the constant are both positive, then m and n must both be positive.**

Classroom Example

Factor: $n^2 - 10n + 21$

Answer: $(n - 7)(n - 3)$

EXAMPLE 2 Factoring a Trinomial of the Form $x^2 + bx + c$, where c Is Positive

Factor: $p^2 - 11p + 24$

Solution

We first identify $b = -11$ and $c = 24$. We are looking for factors of $c = 24$ whose sum is $b = -11$. We begin by listing all factors of 24 and computing the sum of these factors.

Integers Whose Product Is 24	1, 24	2, 12	3, 8	4, 6	−1, −24	−2, −12	−3, −8	−4, −6
Sum	25	14	11	10	−25	−14	−11	−10

Work Smart

The sum of two even numbers is even, the sum of two odd numbers is even, the sum of an odd and even number is odd. In Example 2, the coefficient of the middle term is odd (−11). So one of the factors of 24 will be odd, and the other will be even.

Use this result on the remaining examples to reduce the list of possible factors.

We can see that $-3 \cdot (-8) = 24$ and $-3 + (-8) = -11$, so $m = -3$ and $n = -8$.

Write the trinomial in the form $(p + m)(p + n)$: $p^2 - 11p + 24 = (p + (-3))(p + (-8))$
$= (p - 3)(p - 8)$

Check

Use the FOIL pattern: $(p - 3)(p - 8) = p^2 - 8p - 3p + (-3)(-8)$
$= p^2 - 11p + 24$

So $p^2 - 11p + 24 = (p - 3)(p - 8)$.

In Example 2, notice that the coefficient of the middle term is negative and the constant is positive. **If the coefficient of the middle term is negative and the constant is positive, then m and n must both be negative.**

QUICK ✓ *Factor each trinomial.*

1. $y^2 + 9y + 20$ **2.** $w^2 + 10w + 24$ **3.** $q^2 - 5q + 6$ **4.** $z^2 - 9z + 14$

Classroom Example
Factor: $r^2 + 2r - 15$
Answer: $(r + 5)(r - 3)$

EXAMPLE 3 Factoring a Trinomial of the Form $x^2 + bx + c$, where c Is Negative

Factor: $z^2 + 3z - 28$

Solution

We see that $b = 3$ and $c = -28$. We are looking for factors of $c = -28$ whose sum is $b = 3$. We begin by listing all factors of −28 and computing the sum of these factors. Because c is negative, we know that one of the factors must be positive and the other negative.

Integers Whose Product Is −28	−1, 28	−2, 14	−4, 7	1, −28	2, −14	4, −7
Sum	27	12	3	−27	−12	−3

We can see that $-4 \cdot 7 = -28$ and $-4 + 7 = 3$, so $m = -4$ and $n = 7$.

Write the trinomial in the form $(z + m)(z + n)$: $z^2 + 3z - 28 = (z + (-4))(z + 7)$
$= (z - 4)(z + 7)$

Check

Use the FOIL pattern: $(z - 4)(z + 7) = z^2 + 7z - 4z + (-4)(7)$
$= z^2 + 3z - 28$

Thus, $z^2 + 3z - 28 = (z - 4)(z + 7)$.

In Example 3, notice that the coefficient of the middle term is positive and the constant is negative. **If the constant is negative, then m and n must have opposite signs.** In addition, **if the coefficient of the middle term is positive, then the factor with the larger absolute value must be positive.**

Classroom Example ➤ Factor: $x^2 - 3x - 28$

Answer: $(x - 7)(x + 4)$

EXAMPLE 4 Factoring a Trinomial of the Form $x^2 + bx + c$, where c Is Negative

Factor: $z^2 - 7z - 18$

Solution

To factor the expression, what do we want to find? We need factors of $c = -18$ whose sum is $b = -7$. We begin by listing all factors of -18 and computing the sum of these factors.

Integers Whose Product Is −18	$-1, 18$	$-2, 9$	$-3, 6$	$1, -18$	$2, -9$	$3, -6$
Sum	17	7	3	−17	−7	−3

We can see that $2 \cdot (-9) = -18$ and $2 + (-9) = -7$, so $m = 2$ and $n = -9$.

Write the trinomial in the form $(z + m)(z + n)$: $z^2 - 7z - 18 = (z + 2)(z + (-9))$
$= (z + 2)(z - 9)$

Check

Use FOIL: $(z + 2)(z - 9) = z^2 - 9z + 2z + 2(-9)$
$= z^2 - 7z - 18$

So $z^2 - 7z - 18 = (z + 2)(z - 9)$. ■

In Example 4, notice that the coefficient of the middle term is negative and the constant is also negative. **If the coefficient of the middle term is negative and the constant is negative, then m and n must have opposite signs and the factor with the larger absolute value must be negative.**

QUICK ✓ *Factor each trinomial.*

5. $y^2 - 2y - 15$ **6.** $w^2 + w - 12$ **7.** $q^2 - 4q - 12$ **8.** $n^2 + 5n - 6$

Table 1 summarizes the four possibilities for factoring a quadratic trinomial in the form $x^2 + bx + c$.

Teaching Tip
Review this summary table with students. After reviewing the entries, present four more examples for which students must distinguish the patterns for themselves.

Table 1

Form	Signs of m and n	Example
$x^2 + bx + c$, where b and c are both positive	m and n are both positive.	$x^2 + 3x + 2 = (x + 2)(x + 1)$
$x^2 + bx + c$, where b is negative and c is positive	m and n are both negative.	$a^2 - 7a + 12 = (a - 4)(a - 3)$
$x^2 + bx + c$, where b is positive and c is negative	m and n are opposite in sign and the factor with the larger absolute value is positive.	$y^2 + 2y - 24 = (y + 6)(y - 4)$
$x^2 + bx + c$, where b is negative and c is negative	m and n are opposite in sign and the factor with the larger absolute value is negative.	$b^2 - 4b - 21 = (b - 7)(b + 3)$

DEFINITION

A polynomial that cannot be written as the product of two other polynomials (other than 1 or -1) is said to be a **prime polynomial.**

Classroom Example Show that $z^2 + 8z + 3$ is prime.

Answer: There are no factors of 3 whose sum is 8.

EXAMPLE 5 Identifying a Prime Trinomial

Show that $y^2 + 10y + 12$ is prime.

Solution

We are looking for factors of $c = 12$ whose sum is $b = 10$. Because both b and c are positive, we know that m and n must both be positive, so we only list positive factors of 12 and compute the sum of these factors.

Integers Whose Product Is 12	1, 12	2, 6	3, 4
Sum	13	8	7

There are no factors of 12 whose sum is 10. Therefore, $y^2 + 10y + 12$ is prime.

QUICK ✓ *Factor the trinomial. If the trinomial cannot be factored, state that it is prime.*

9. $c^2 - 5c + 8$

10. $q^2 + 4q - 45$

Teaching Tip Emphasize that the technique used for factoring a polynomial in two variables is identical to factoring a polynomial in one variable.

If a trinomial has more than one variable, we take the same approach as that used for trinomials with one variable. So trinomials of the form

$$x^2 + bxy + cy^2$$

will factor as

$$(x + my)(x + ny)$$

where

$$mn = c \quad \text{and} \quad m + n = b$$

Classroom Example Factor: $x^2 + 8xy + 15y^2$

Answer: $(x + 5y)(x + 3y)$

EXAMPLE 6 Factoring Trinomials with Two Variables

Factor: $p^2 + 4pq - 21q^2$

Solution

The trinomial $p^2 + 4pq - 21q^2$ will factor as $(p + mq)(p + nq)$, where $mn = -21$ and $m + n = 4$. We are looking for factors of $c = -21$ whose sum is $b = 4$. We list factors of -21, but because c is negative and b is positive, we know the factor of -21 with the larger absolute value will be positive.

Integers Whose Product Is −21	−1, 21	−3, 7
Sum	20	4

We can see that $-3 \cdot 7 = -21$ and $-3 + 7 = 4$, so $m = -3$ and $n = 7$.

Write the trinomial in the form $(p + mq)(p + nq)$:
$$p^2 + 4pq - 21q^2 = (p + (-3)q)(p + 7q)$$
$$= (p - 3q)(p + 7q)$$

Check

Use the FOIL pattern: $$(p - 3q)(p + 7q) = p^2 + 7pq - 3pq - 21q^2$$
$$= p^2 + 4pq - 21q^2$$

We conclude that $p^2 + 4pq - 21q^2 = (p - 3q)(p + 7q)$.

QUICK ✓ *Factor each trinomial.*

11. $x^2 + 9xy + 20y^2$ **12.** $m^2 + mn - 42n^2$

Classroom Example → Factor: $6n + n^2 - 16$
Answer: $(n + 8)(n - 2)$

EXAMPLE 7 Factoring a Trinomial Not Written in Standard Form

Factor: $2w + w^2 - 8$

Solution

Do you notice that the trinomial is not in standard form $ax^2 + bx + c$? We will first write it in standard form (descending order of degree) and then factor.

Rearrange terms: $$2w + w^2 - 8 = w^2 + 2w - 8$$

We now look for factors of $c = -8$ whose sum is $b = 2$. Since the coefficient of w is positive and the constant term is negative, we know that the factor with the larger absolute value will be positive.

Integers Whose Product Is −8	−1, 8	−2, 4
Sum	7	2

We see that $-2 \cdot 4 = -8$ and $-2 + 4 = 2$, so $m = -2$ and $n = 4$. The factors are $(w - 2)(w + 4)$.

Check

Use the FOIL pattern: $$(w - 2)(w + 4) = w^2 + 4w - 2w + (-2)(4)$$
$$= w^2 + 2w - 8$$

So $w^2 + 2w - 8 = (w - 2)(w + 4)$.

QUICK ✓ *Completely factor each trinomial.*

13. $-56 + n^2 + n$ **14.** $y^2 + 35 - 12y$

Teaching Tip
Remind students that they should always look for a GCF first, and to include the GCF in the final factored form.

2 Factor Out the GCF, Then Factor $x^2 + bx + c$

Some algebraic expressions can be factored as trinomials in the form $x^2 + bx + c$ after we factor out a greatest common factor.

Classroom Example → Factor: $3a^3 - 9a^2 - 120a$
Answer: $3a(a + 5)(a - 8)$

EXAMPLE 8 Factoring Trinomials with a Common Factor

Factor: $3u^3 - 21u^2 - 90u$

Solution

Did you notice that there is a GCF of $3u$ in the trinomial? We factor the $3u$ out:

$$3u^3 - 21u^2 - 90u = 3u(u^2 - 7u - 30)$$

We now concentrate on factoring the trinomial in parentheses, $u^2 - 7u - 30$. We are looking for factors of $c = -30$ whose sum is $b = -7$. Because c is negative and b is negative, the factor of -30 with the larger absolute value will be negative.

Integers Whose Product Is −30	1, −30	2, −15	3, −10	5, −6
Sum	−29	−13	−7	−1

We can see that $3(-10) = -30$ and $3 + (-10) = -7$, so $m = 3$ and $n = -10$.

Write the trinomial in the form $(u + m)(u + n)$: $u^2 - 7u - 30 = (u + 3)(u + (-10))$
$= (u + 3)(u - 10)$

So we have

$$3u^3 - 9u^2 - 90u = 3u(u^2 - 7u - 30)$$
$$= 3u(u + 3)(u - 10)$$

Work Smart
We could have checked the result of Example 8 as follows:
$3u(u + 3)(u - 10)$
$= (3u^2 + 9u)(u - 10)$
$= 3u^3 - 30u^2 + 9u^2 - 90u$
$= 3u^3 - 21u^2 - 90u$

Check

Use FOIL: $3u(u + 3)(u - 10) = 3u(u^2 - 10u + 3u - 30)$
Combine like terms: $= 3u(u^2 - 7u - 30)$
Distribute: $= 3u^3 - 21u^2 - 90u$

So $3u^3 - 21u^2 - 90u = 3u(u + 3)(u - 10)$.

We say that a polynomial is **factored completely** if each factor in the final factorization is prime. For example, $3x^2 - 6x - 45 = (x - 5)(3x + 9)$ is not factored completely because the binomial $3x + 9$ has a common factor of 3. However, $3x^2 - 6x - 45 = 3(x - 5)(x + 3)$ is factored completely.

QUICK ✓ *Factor each trinomial completely.*

15. $4m^2 - 16m - 84$ **16.** $3z^3 + 12z^2 - 15z$

As we stated in the last section, sometimes the leading coefficient may be negative. If this is the case, we factor out the negative to make our factoring lives easier.

Classroom Example
Factor completely: $-y^2 - 3y + 28$
Answer: $-1(y + 7)(y - 4)$

EXAMPLE 9 Factoring Trinomials with a Leading Coefficient That Is Negative

Factor completely: $-w^2 - 5w + 24$

Solution

Work Smart
It's easier to factor a trinomial in standard form when the leading coefficient is positive.

We notice that the leading coefficient is -1. It is easier to factor a trinomial when the leading coefficient is positive, so we will use -1 as the GCF and rewrite $-w^2 - 5w + 24$ as

$$-w^2 - 5w + 24 = -1(w^2 + 5w - 24)$$

To factor $w^2 + 5w - 24$, we look for two integers whose product is -24 and whose sum is 5. Since we have a negative product, but a positive sum, we know that one factor will be positive, and the other negative. Further, we know that the factor of -24 with the larger absolute value will be positive.

Integers Whose Product Is −24	−1, 24	−2, 12	−3, 8	−4, 6
Sum	23	10	5	2

We factor $w^2 + 5w - 24$ as $(w - 3)(w + 8)$. But remember that we already factored out the greatest common factor, -1, so

$$\begin{aligned} -w^2 - 5w + 24 &= -1(w^2 + 5w - 24) \\ &= -1(w - 3)(w + 8) \end{aligned}$$

QUICK ✓ *Factor each trinomial completely.*

17. $-w^2 - 3w + 10$ **18.** $-2a^2 - 8a + 24$

4.2 Exercises

For Extra Help: Student Solutions Manual  CD Video  PH Math/Tutor Center  MathXL Tutorials on CD  MathXL® MyMathLab

1. quadratic trinomial
2. negative
3. opposite
4. False
5. False
6. False
7. Answers may vary.
8. Answers may vary.
9. $(x + 2)(x + 3)$
10. $(p + 6)(p + 1)$
11. $(m + 3)(m + 6)$
12. $(n + 2)(n + 10)$
13. $(x - 3)(x - 12)$
14. $(z - 4)(z - 9)$
15. $(a - 2)(a - 6)$
16. $(b - 4)(b - 3)$
17. $(x - 4)(x + 3)$
18. $(y + 1)(y - 9)$
19. $(x - 4)(x + 5)$
20. $(a - 4)(a + 10)$
21. $(z - 3)(z + 15)$
22. prime
23. prime
24. $(x + 1)(x - 12)$
25. $(x - 2y)(x - 3y)$
26. $(x - 2y)(x - 12)$
27. $(r - 2s)(r + 3s)$
28. $(p - 2q)(p + 7q)$
29. $(x - 4y)(x + y)$
30. $(x + 12y)(x - 3y)$
31. prime
32. $(m + 4n)(m + 12n)$
33. $3(x + 2)(x - 1)$
34. $5(x^2 + 2)(x^2 + 4)$
35. $3a(a - 3)(a - 5)$
36. $4p^2(p - 2)(p + 1)$
37. $5z(x - 8)(x + 4)$
38. $8z^2(x - 2)(x - 5)$
39. $-2(y - 2)^2$
40. $-3(x + 1)(x + 5)$

Concepts and Vocabulary

In Problems 1–3, fill in the blanks.

1. A ________ ________ is a polynomial of the form $ax^2 + bx + c, a \neq 0$.

2. When factoring $x^2 - 10x + 24$, the sign of the factors of 24 must both be ________.

3. When factoring $x^2 - 10x - 24$, the factors of 24 must have ________ signs.

In Problems 4–6, answer True or False to each statement.

4. $(x - 3)(3x + 6)$ is factored completely.

5. A trinomial will always factor as a product of two binomials.

6. $4 + 4x + x^2$ has a leading coefficient of 4.

7. The answer key to your algebra exam said the factored form of a trinomial is $(x - 1)(x - 2)$. You have $(1 - x)(2 - x)$ on your paper. Is your answer marked correct or incorrect? Explain your reasoning.

8. In the trinomial $x^2 + bx + c$, both b and c are negative. Explain how to factor the trinomial. Make up trinomial and then use your rules to factor it.

Building Skills

In Problems 9–32, factor each trinomial completely. If the trinomial cannot be factored, say it is prime.

9. $x^2 + 5x + 6$ **10.** $p^2 + 7p + 6$ **11.** $m^2 + 9m + 18$

12. $n^2 + 12n + 20$ **13.** $x^2 - 15x + 36$ **14.** $z^2 - 13z + 36$

15. $a^2 - 8a + 12$ **16.** $b^2 - 7b + 12$ **17.** $x^2 - x - 12$

18. $y^2 - 8y - 9$ **19.** $x^2 + x - 20$ **20.** $a^2 + 6a - 40$

21. $z^2 + 12z - 45$ **22.** $t^2 + 2t - 38$ **23.** $x^2 - x - 32$

24. $x^2 - 11x - 12$ **25.** $x^2 - 5xy + 6y^2$ **26.** $x^2 - 14xy + 24y^2$

27. $r^2 + rs - 6s^2$ **28.** $p^2 + 5pq - 14q^2$ **29.** $x^2 - 3xy - 4y^2$

30. $x^2 + 9xy - 36y^2$ **31.** $z^2 + 7zy + y^2$ **32.** $m^2 + 16mn + 48n^2$

In Problems 33–40, factor each trinomial completely by factoring out the GCF first and then factoring the resulting trinomial.

33. $3x^2 + 3x - 6$ **34.** $5x^2 + 30x + 40$ **35.** $3a^3 - 24a^2 + 45a$

36. $4p^4 - 4p^3 - 8p^2$ **37.** $5x^2z - 20xz - 160z$ **38.** $8x^2z^2 - 56xz^2 + 80z^2$

39. $-2y^2 + 8y - 8$ **40.** $-3x^2 - 18x - 15$

41. $(x - 6)(x + 3)$
42. $(x - 5)(x - 10)$
43. $(x + 8)(x - 4)$
44. $(x + 15)(x - 5)$
45. $x(x + 5)(x - 3)$
46. $x(x - 10)(x + 2)$
47. $(a + 4b)(a - 7b)$
48. $(x - 12y)(x + 3y)$
49. prime
50. prime
51. $(k - 5)(k + 4)$
52. $(x - 7)(x + 5)$
53. $(x - 5y)(x + 6y)$
54. $(p - 5q)(p - q)$
55. $(st - 3)(st - 5)$
56. $(xy + 1)(xy + 2)$
57. $-3c(c - 2)(c + 1)$
58. $-2z(z + 4)(z - 3)$
59. $-(x + 3)(x - 2)$
60. $-(r + 6)^2$
61. prime
62. prime
63. $n^2(n - 6)(n + 5)$
64. $x(x - 8)(x + 2)$
65. $(m + 5n)(m - 3n)$
66. $(x + 3y)(x + 4y)$
67. $(a + 2)(b - 4)$
68. $(2x - y)(5a - 1)$
69. $(s + 5)(s + 7)$
70. $(x + 5)^2$
71. $3xy(x^2 - 2x - 2)$
72. $3pq^2(p^2 - 2p + 5)$
73. prime
74. prime
75. $(ab - 6)(ab - 2)$
76. $(xy - 6)(xy + 3)$
77. $-x(x - 14)(x + 2)$
78. $-3r(r - 1)^2$
79. $(y - 6)^2$
80. $(x - 9)(x - 6)$
81. $(2r - 3)(3s - 2)$
82. $(3a - 2)(4b + 1)$
83. $-2x(x - 3)(x - 6)$
84. $-5m(2n - m)^2$
85. $-7xy(3x^2 + 2y)$
86. $-4xy(3x - 2y^2)$
87. $(x - a)(x + b)$
88. $(2p - 3q)(2m + 3n)$
89. $-16(t + 1)(t - 5)$
90. $-16(t - 2)(t + 1)$
91. $(x + 6)$ and $(x + 3)$
92. $(x + 4)$ and $(x + 2)$
93. $(x + 5)$ and $(x - 3)$
94. $(x + 6)$ and $(x + 4)$
95. $-7, -5, 5, 7$
96. $-9, -3, 3, 9$
97. $-20, -4, 4, 20$
98. $-13, -8, -7, 7, 8, 13$
99. 1
100. 2
101. 6, 10, 12
102. 5, 8, 9

In Problems 41–46, write the trinomial in standard form (descending order of degree) and then completely factor.

41. $-3x + x^2 - 18$ **42.** $50 - 15x + x^2$ **43.** $4x - 32 + x^2$
44. $-75 + x^2 + 10x$ **45.** $2x^2 + x^3 - 15x$ **46.** $x^3 - 20x - 8x^2$

Mixed Practice

In Problems 47–88, factor each polynomial completely. If the polynomial cannot be factored, say that it is prime.

47. $a^2 - 3ab - 28b^2$ **48.** $x^2 - 9xy - 36y^2$ **49.** $x^2 + x + 6$
50. $t^2 - 8t - 7$ **51.** $k^2 - k - 20$ **52.** $x^2 - 2x - 35$
53. $x^2 + xy - 30y^2$ **54.** $p^2 - 6pq + 5q^2$ **55.** $s^2t^2 - 8st + 15$
56. $x^2y^2 + 3xy + 2$ **57.** $-3c^3 + 3c^2 + 6c$ **58.** $-2z^3 - 2z^2 + 24z$
59. $-x^2 - x + 6$ **60.** $-r^2 - 12r - 36$ **61.** $g^2 - 4g + 21$
62. $x^2 - x + 6$ **63.** $n^4 - 30n^2 - n^3$ **64.** $-16x + x^3 - 6x^2$
65. $m^2 + 2mn - 15n^2$ **66.** $x^2 + 7xy + 12y^2$ **67.** $ab + 2b - 4a - 8$
68. $10ax - 5ay - 2x + y$ **69.** $35 + 12s + s^2$ **70.** $25 + 10x + x^2$
71. $3x^3y - 6x^2y - 6xy$ **72.** $3p^3q^2 - 6p^2q^2 + 15pq^2$ **73.** $n^2 - 9n - 45$
74. $x^2 - 6x - 42$ **75.** $a^2b^2 - 8ab + 12$ **76.** $x^2y^2 - 3xy - 18$
77. $-x^3 + 12x^2 + 28x$ **78.** $-3r^3 + 6r^2 - 3r$ **79.** $y^2 - 12y + 36$
80. $x^2 - 15x + 54$ **81.** $6rs - 9s - 4r + 6$ **82.** $12ab - 8b + 3a - 2$
83. $-36x + 18x^2 - 2x^3$ **84.** $-20mn^2 + 20m^2n - 5m^3$
85. $-21x^3y - 14xy^2$ **86.** $-12x^2y + 8xy^3$
87. $x^2 - ax + bx - ab$ **88.** $4mp - 6mq + 6np - 9nq$

Applying the Concepts

89. Punkin Chunkin In a recent Punkin Chunkin contest, the height of a pumpkin after t seconds was given by the trinomial $-16t^2 + 64t + 80$ feet. Write this polynomial in factored form.

90. Punkin Chunkin In a recent Punkin Chunkin contest, the height of a pumpkin after t seconds was given by the trinomial $-16t^2 + 16t + 32$ feet. Write this polynomial in factored form.

△ **91. A Rectangular Field** The trinomial $x^2 + 9x + 18$ square meters represents the area of a rectangular field. Find two binomials that represent the length and width of the field.

△ **92. Another Rectangular Field** The trinomial $x^2 + 6x + 8$ square yards represents the area of a rectangular field. Find two binomials that represent the length and width of the field.

△ **93. Area as a Polynomial** The area of a triangle is given by the trinomial $\frac{1}{2}x^2 + x - \frac{15}{2}$. Find algebraic expressions for the base and height of the triangle. (**Hint:** Factor out $\frac{1}{2}$ as a common factor.)

△ **94. Triangle Trinomial** The area of a triangle is given by the trinomial $\frac{1}{2}x^2 + 5x + 12$. Find algebraic expressions for the base and height of the triangle. (**Hint:** Factor out $\frac{1}{2}$ as a common factor.)

Extending the Concepts

In Problems 95–98, find all possible values of b so that the trinomial is factorable.

95. $x^2 + bx + 6$ **96.** $x^2 + bx - 10$ **97.** $x^2 + bx - 21$ **98.** $x^2 + bx + 12$

In Problems 99–102, find all possible positive values of c so that the trinomial is factorable.

99. $x^2 - 2x + c$ **100.** $x^2 - 3x + c$ **101.** $x^2 - 7x + c$ **102.** $x^2 + 6x + c$

4.3 Factoring Trinomials of the Form $ax^2 + bx + c, a \neq 1$

OBJECTIVES

1. Factor $ax^2 + bx + c, a \neq 1$ Using Trial and Error
2. Factor $ax^2 + bx + c, a \neq 1$ Using Grouping

Preparing for Factoring Trinomials of the Form $ax^2 + bx + c, a \neq 1$

Before getting started, take the following readiness quiz. If you get a problem wrong, go back to the section cited and review the material.

1. List the prime factorization of 24. [Section A.1, p. A1]

2. Determine the coefficients of $5x^2 - 3x + 7$. [Section 1.7, pp. 59–60]

3. Find the product: $(2x + 7)(3x - 1)$ [Section 3.3. pp. 191–193]

In this section, we factor quadratic trinomials of the form $ax^2 + bx + c$ in which the leading coefficient, a, is not 1. Examples of quadratic trinomials of the form $ax^2 + bx + c, a \neq 1$ are

$$2x^2 + 3x + 1 \qquad 5y^2 - y - 4 \qquad 10z^2 - 7z + 6$$

Before we begin to factor these quadratic trinomials, let's review multiplication of binomials using FOIL.

Factored Form	F O I L	Polynomial
$(2x + 3)(x + 4) =$	$2x^2 + 8x + 3x + 12 =$	$2x^2 + 11x + 12$
$(3x - 4)(x + 7) =$	$3x^2 + 21x - 4x - 28 =$	$3x^2 + 17x - 28$
$(5a - 2b)(3a - b) =$	$15a^2 - 5ab - 6ab + 2b^2 =$	$15a^2 - 11ab + 2b^2$

To factor a trinomial in the form $ax^2 + bx + c, a \neq 1$, we reverse the FOIL multiplication process. That is, we will be given a trinomial such as $2x^2 + 11x + 12$, and we will write it as the product $(2x + 3)(x + 4)$. To factor trinomials in this form, we have two methods that can be used.

1. Trial and error
2. Factoring by grouping

There are pros and cons to both methods. We will point out these pros and cons as we proceed. We start with factoring by trial and error.

1 Factor $ax^2 + bx + c, a \neq 1$ Using Trial and Error

The idea behind using trial and error is to list various binomials and use FOIL to find their product until the combination of binomials is found that results in the trinomial to be factored. While this method may sound haphazard, experience and logic will play a role in minimizing the number of possibilities that you must try before a factored form is found. The following example illustrates the method.

Preparing for...Answers **1.** $2^3 \cdot 3$ **2.** $5, -3, 7$ **3.** $6x^2 + 19x - 7$

EXAMPLE 1 How to Factor $ax^2 + bx + c, a \neq 1$ Using Trial and Error

Classroom Example
Factor: $2x^2 + 7x + 3$
Answer: $(2x + 1)(x + 3)$

Factor: $2x^2 + 7x + 5$

Step-by-Step Solution

Step 1: List the possibilities for the first terms of each binomial whose product is ax^2.

We list possible ways of representing the first term, $2x^2$. Since 2 is a prime number, we have one possibility:

$$(2x + \_\_)(x + \_\_)$$

(continued)

Step 2: List the possibilities for the last terms of each binomial whose product is c.

The last term, 5, is also prime. It has the factors (1)(5) and $(-1)(-5)$. However, did you notice that the coefficient of x, 7, is positive? To produce a positive sum, $7x$, we must have two positive factors. So we do not include the factors $(-1)(-5)$ in our list.

Step 3: Write out all the combinations of factors found in Steps 1 and 2. Multiply the binomials until a product is found that equals the trinomial.

Possible Factorization of $2x^2 + 7x + 5$	Product
$(2x + 1)(x + 5)$	$2x^2 + 11x + 5$
$(2x + 5)(x + 1)$	$2x^2 + 7x + 5$

The second row is the factorization that works, so $2x^2 + 7x + 5 = (2x + 5)(x + 1)$.

Teaching Tip
Emphasize the process for factoring polynomials in which the leading coefficient is not one, compared to those in which the leading coefficient is one. The important numbers here are a, the leading coefficient, and c, the constant. In polynomials of the form $ax^2 + bx + c, a \neq 1$, we don't look for two numbers whose product is c whose sum is b.

We summarize the steps used in Example 1 below.

Steps to Factor $ax^2 + bx + c, a \neq 1$ Using Trial and Error: a, b, and c Have No Common Factors

Step 1: List the possibilities for the first terms of each binomial whose product is ax^2.

$$(\_x + \quad)(\_x + \quad) = ax^2 + bx + c$$

Step 2: List the possibilities for the last terms of each binomial whose product is c.

$$(\_x + \square)(\_x + \square) = ax^2 + bx + c$$

Step 3: Write out all the combinations of factors found in Steps 1 and 2. Multiply the binomials until a product is found that equals the trinomial.

QUICK ✓ *Factor each trinomial completely.*

1. $3x^2 + 5x + 2$ **2.** $7y^2 + 22y + 3$

Classroom Example
Factor completely: $5x^2 - 17x + 6$
Answer: $(5x - 2)(x - 3)$

EXAMPLE 2 Factoring $ax^2 + bx + c, a \neq 1$, Using Trial and Error

Factor completely: $7x^2 - 18x + 8$

Solution

We list possible ways of representing the first term, $7x^2$. Since 7 is a prime number, we have

$$(7x + \_)(x + \_)$$

Let's look at the last term, 8, and list its factors:

Factors of 8	
$1 \cdot 8$	$-1 \cdot -8$
$2 \cdot 4$	$-2 \cdot -4$

Now let's concentrate on the middle term, $-18x$. To produce a negative sum, $-18x$, from a positive product, 8, we must have two *negative* factors. So we do not include the factors $1 \cdot 8$ or $2 \cdot 4$ in our list. Now we list the possible combinations of factors.

Work Smart
When factoring using trial and error it is only necessary to find the sum of the outer and inner products to see which factorization works.

Possible Factorization	Product
$(7x - 1)(x - 8)$	$7x^2 - 57x + 8$
$(7x - 8)(x - 1)$	$7x^2 - 15x + 8$
$(7x - 2)(x - 4)$	$7x^2 - 30x + 8$
$(7x - 4)(x - 2)$	$7x^2 - 18x + 8$

We see that $7x^2 - 18x + 8 = (7x - 4)(x - 2)$.

QUICK ✓ *Factor each trinomial completely.*

3. $3x^2 - 13x + 12$ **4.** $5p^2 - 21p + 4$

Classroom Example
Factor completely: $6x^2 - x - 2$
Answer: $(3x - 2)(2x + 1)$

EXAMPLE 3 Factoring $ax^2 + bx + c, a \neq 1$ Using Trial and Error

Factor completely: $10x^2 - 13x - 3$

Solution

We list possible ways of representing the first term, $10x^2$.

$$(10x + \_\_\_)(x + \_\_\_)$$
$$(5x + \_\_\_)(2x + \_\_\_)$$

The last term, -3, has factors $-1 \cdot 3$ or $1 \cdot -3$.

We list the possible combinations of factors. We also highlight the sum of the "outer" and "inner" products.

Possible Factorization	Product
$(10x - 1)(x + 3)$	$10x^2 + 29x - 3$
$(10x - 3)(x + 1)$	$10x^2 + 7x - 3$
$(10x + 1)(x - 3)$	$10x^2 - 29x - 3$
$(10x + 3)(x - 1)$	$10x^2 - 7x - 3$
$(5x - 1)(2x + 3)$	$10x^2 + 13x - 3$
$(5x - 3)(2x + 1)$	$10x^2 - x - 3$
$(5x + 1)(2x - 3)$	$10x^2 - 13x - 3$
$(5x + 3)(2x - 1)$	$10x^2 + x - 3$

The row with the correct factorization is highlighted. So $10x^2 - 13x - 3 = (5x + 1)(2x - 3)$.

QUICK ✓ *Factor each trinomial completely.*

5. $2n^2 - 17n - 9$ **6.** $4b^2 - 5b - 6$

Factoring trinomials of the form $ax^2 + bx + c, a \neq 1$ can at first seem overwhelming. There are so many possibilities! Plus, the technique may at first seem haphazard. In fact, however, calling the technique "trial and error" is not quite truth in advertising because some thought is necessary to reduce the list of possible factors. To keep the list of possibilities small, here are some questions you should ask yourself before you begin to factor.

HINTS FOR USING TRIAL AND ERROR TO FACTOR $ax^2 + bx + c, a \neq 1$

- Is the constant c positive? If so, then the factors of the constant, c, must be the same sign as the coefficient of the middle term, b. For example,
$$2a^2 + 11a + 5 = (2a + 1)(a + 5)$$
$$2a^2 - 11a + 5 = (2a - 1)(a - 5)$$
- Is the constant c negative? If so, then the factors of c must have opposite signs. For example,
$$10b^2 + 19b - 15 = (5b - 3)(2b + 5)$$
$$10b^2 - 19b - 15 = (5b + 3)(2b - 5)$$
- If $ax^2 + bx + c$ has no common factor, then the binomials in the factored form cannot have common factors either.
- Is the value of b small? If so, then choose factors of a and factors of c that are close to each other. If the value of b is large, then choose factors of a and factors of c that are far from each other.
- Is the value of b correct, but has the wrong sign? Then interchange the signs in the binomial factors.

Classroom Example ➤
Factor completely:
$24y^2 - 14y - 3$
Answer: $(6y + 1)(4y - 3)$

EXAMPLE 4 Factoring $ax^2 + bx + c, a \neq 1$ Using Trial and Error

Factor completely: $18x^2 + 3x - 10$

Solution

Remember, the first step in any factoring problem is to look for common factors. There are no common factors in $18x^2 + 3x - 10$.

List all possible ways of representing $18x^2$: $(18x + \_\_)(x + \_\_)$
$(9x + \_\_)(2x + \_\_)$
$(6x + \_\_)(3x + \_\_)$

Let's look at the last term, -10, and list its factors:

Factors of -10	
$-10 \cdot 1$	$10 \cdot -1$
$-5 \cdot 2$	$5 \cdot -2$

Before we list the possible combinations of factors, we ask some questions. Is the coefficient of the middle term of the polynomial $18x^2 + 3x - 10$ small? Yes, it is $+3$. Because this middle term is positive and small, the binomial factors we list should have outer and inner products that sum to a positive, small number. Therefore, we will start with $(6x + \_\_)(3x + \_\_)$ and the factors $(2)(-5)$ and $(-2)(5)$. We do not use $6x + 2$ or $6x - 2$ as possible factors because there is a common factor of 2 in these binomials and the trinomial to be factored has no common factors.

Let's try $(6x - 5)(3x + 2)$.
$$(6x - 5)(3x + 2) = 18x^2 + 12x - 15x - 10$$
$$= 18x^2 - 3x - 10$$

Close! The only problem is that the middle term has the opposite sign of the one that we want. Therefore, let's flip-flop the signs of -5 and 2 in the binomial and try $(6x + 5)(3x - 2)$.
$$(6x + 5)(3x - 2) = 18x^2 - 12x + 15x - 10$$
$$= 18x^2 + 3x - 10$$

So $18x^2 + 3x - 10 = (6x + 5)(3x - 2)$. ■

The moral of the story in Example 4 is that the name *trial and error* is a bit misleading. With some thought, you won't have to choose binomial factors haphazardly "until the cows come home" provided you use the helpful hints given and some care.

QUICK ✓ *Factor each trinomial completely.*

7. $12x^2 + 17x + 6$ **8.** $12y^2 + 32y - 35$

Classroom Example ➤
Factor completely:
$24x^2 + 14x - 20$

Answer: $2(4x + 5)(3x - 2)$

Work Smart
Don't divide by the GCF because we have an algebraic expression, not an equation.

Teaching Tip
Point out that NOT taking out the GCF first will produce the same factorization, but may make the process of factoring more difficult. That's because one factor will contain a GCF, which still must be factored out at some point. In Example 5, if 2 isn't factored out first, $(12x + 10y)(4x - 3y)$ has the GCF of 2 in the first term.

Continue to remind students to include the GCF in the final factorization.

EXAMPLE 5 Factoring Trinomials with Two Variables

Factor completely: $48x^2 + 4xy - 30y^2$

Solution

Did you notice that there is a greatest common factor of 2? We first factor out this GCF and obtain the polynomial $2(24x^2 + 2xy - 15y^2)$. The trinomial in parentheses will factor in the form $24x^2 + 2xy - 15y^2 = (\_x + \_y)(\_x + \_y)$.

List all possible ways of representing $24x^2$:

$(24x + \_y)(x + \_y)$
$(12x + \_y)(2x + \_y)$
$(8x + \_y)(3x + \_y)$
$(6x + \_y)(4x + \_y)$

We list the factors of the last term, -15:

Factors of -15	
$-15 \cdot 1$	$15 \cdot -1$
$-5 \cdot 3$	$5 \cdot -3$

We do not use $24x + 3y$, $12x + 3y$, $3x + 3y$, or $6x + 3y$ as possible factors because there is a common factor in these binomials, and a common factor does not exist in the trinomial $24x^2 + 2xy - 15y$. Also, since the middle term has a small coefficient, we will start with $(6x + \_y)(4x + \_y)$ and the factors $(3)(-5)$ and $(-3)(5)$.

Let's try $(6x - 5y)(4x + 3y)$.

$$(6x - 5y)(4x + 3y) = 24x^2 - 2xy - 15y^2$$

Close! The only problem is that the middle term is the opposite sign that we want. Therefore, let's flip-flop the signs of -5 and 3 in the binomial and try $(6x + 5y)(4x - 3y)$.

$$\begin{aligned}(6x + 5y)(4x - 3y) &= 24x^2 - 18xy + 20xy - 15y^2 \\ &= 24x^2 + 2xy - 15y^2\end{aligned}$$

So $48x^2 + 4xy - 30y^2 = 2(6x + 5y)(4x - 3y)$. ∎

QUICK ✓ *Factor the trinomial completely.*

9. $90x^2 + 21xy - 6y^2$

Classroom Example ➤
Factor: $-5z^2 + 13z + 6$

Answer: $-1(5z + 2)(z - 3)$

Work Smart
It's easier to factor a trinomial in standard form when the leading coefficient is positive.

EXAMPLE 6 Factoring Trinomials with a Negative Leading Coefficient

Factor: $-14x^2 + 29x + 15$

Solution

While there are no common factors in $-14x^2 + 29x + 15$, notice that the coefficient of the squared term is negative. Factor -1 out of the trinomial to obtain

$$-14x^2 + 29x + 15 = -1(14x^2 - 29x - 15)$$

Now factor the expression in parentheses and obtain

$$\begin{aligned} -14x^2 + 29x + 15 &= -1(14x^2 - 29x - 15) \\ &= -1(7x + 3)(2x - 5) \end{aligned}$$

So $-14x^2 + 29x + 15 = -1(7x + 3)(2x - 5)$.

QUICK ✓ *Factor each trinomial completely.*

10. $-6y^2 + 23y + 4$

11. $-9x^2 - 21xy - 10y^2$

12. $-12x^2 - 10x + 8$

13. $-6x^2 - 3x + 45$

2 Factor $ax^2 + bx + c$, $a \neq 1$ Using Grouping

We now introduce a second method for factoring trinomials of the form $ax^2 + bx + c$, $a \neq 1$. The second method uses a factoring technique introduced in the last section, factoring by grouping. The next example illustrates the procedure.

Classroom Example
Factor: $3x^2 + 17x + 10$
Answer: $(3x + 2)(x + 5)$

EXAMPLE 7 How to Factor $ax^2 + bx + c$, $a \neq 1$ by Grouping

Factor: $3x^2 + 14x + 15$

Step-by-Step Solution

First, we notice that $3x^2 + 14x + 15$ has no common factors. Comparing $3x^2 + 14x + 15$ to $ax^2 + bx + c$, we find that $a = 3$, $b = 14$, and $c = 15$.

Step 1: Find the value of *ac*.

The value of $a \cdot c = 3 \cdot 15 = 45$.

Step 2: Find the pair of integers, *m* and *n*, whose product is *ac* and whose sum is *b*.

We want to determine the integers whose product is 45 and whose sum is 14. Because both 14 and 45 are positive, we only list the positive factors of 45.

Integers Whose Product Is 45	1, 45	3, 15	5, 9
Sum	46	18	14

Step 3: Write $ax^2 + bx + c$ as $ax^2 + mx + nx + c$.

Write $3x^2 + 14x + 15$ as $3x^2 + 9x + 5x + 15$

$14x = 9x + 5x$

Step 4: Factor the expression in Step 3 by grouping.

$$3x^2 + 9x + 5x + 15 = (3x^2 + 9x) + (5x + 15)$$

Common factor in 1st group: $3x$; common factor in 2nd group: 5: $= 3x(x + 3) + 5(x + 3)$

Factor out $x + 3$: $= (x + 3)(3x + 5)$

Step 5: Check Multiply out the factored form.

$$\begin{aligned} (x + 3)(3x + 5) &= 3x^2 + 5x + 9x + 15 \\ &= 3x^2 + 14x + 15 \end{aligned}$$

So $3x^2 + 14x + 15 = (x + 3)(3x + 5)$.

We summarize below the steps used in Example 7.

Steps to Factor $ax^2 + bx + c, a \neq 1$ by Grouping: a, b, and c Have No Common Factors

Step 1: Find the value of ac.

Step 2: Find the pair of integers, m and n, whose product is ac and whose sum is b.

Step 3: Write $ax^2 + bx + c = ax^2 + mx + nx + c$.

Step 4: Factor the expression in Step 3 by grouping.

Step 5: Multiply out the factored form to verify your answer.

Classroom Example ➤
Factor: $8x^2 + 2x - 3$
Answer: $(4x + 3)(2x - 1)$

EXAMPLE 8 Factoring $ax^2 + bx + c, a \neq 1$ by Grouping

Factor: $12x^2 - x - 1$

Solution

First, we see that there is no greatest common factor in the expression $12x^2 - x - 1$. Comparing $12x^2 - x - 1$ to $ax^2 + bx + c$, we see that $a = 12$, $b = -1$, and $c = -1$.

The value of $a \cdot c = 12(-1) = -12$. We want to determine the integers whose product is -12 and whose sum is -1. Because the product $a \cdot c$ is negative, -12, we know that one integer will be positive and the other negative. Since the value of b is negative ($b = -1$), we know the factor of -12 with the larger absolute value will be negative.

Teaching Tip
Use the Work Smart suggestion below to show students two ways to factor $12x^2 - x - 1$.

Integers Whose Product Is −12	1, −12	2, −6	3, −4
Sum	−11	−4	−1

The integers whose product is -12 and whose sum is -1 are 3 and -4.

Work Smart
In Example 8, we could have written $12x^2 - x - 1$ as $12x^2 - 4x + 3x - 1$ and obtained the same result.

Write $12x^2 - x - 1$ as $12x^2 + 3x - 4x - 1$, and factor by grouping.

$-x = 3x - 4x$

$$12x^2 + 3x - 4x - 1 = (12x^2 + 3x) + (-4x - 1)$$

Common factor in 1st group: $3x$; common factor in 2nd group: -1: $= 3x(4x + 1) - 1(4x + 1)$

Factor out $4x + 1$: $= (4x + 1)(3x - 1)$

Check

$$(4x + 1)(3x - 1) = 12x^2 - 4x + 3x - 1$$
$$= 12x^2 - x - 1$$

So $12x^2 - x - 1 = (4x + 1)(3x - 1)$.

QUICK ✓ *Factor each trinomial completely.*

14. $3x^2 - 2x - 8$

15. $10z^2 + 21z + 9$

Teaching Tip
Discuss the fact that if the product ac is large, factoring by grouping may be cumbersome. For example, in $12x^2 - 40x - 63$, $ac = -756$!

The advantage of factoring quadratic trinomials of the form $ax^2 + bx + c, a \neq 1$ by grouping is that it is algorithmic (that is, step by step). However, if the product $a \cdot c$ gets large, then there are a lot of factors of ac whose sum must be determined. This can get overwhelming. Under these circumstances, it may be better to try trial and error.

Classroom Example ➤
Factor: $20n^2 - 22n - 12$

Answer: $2(5n + 2)(2n - 3)$

EXAMPLE 9 Factoring $ax^2 + bx + c$, $a \neq 1$ by Grouping

Factor: $18x^2 - 33x - 30$

Solution

The first question we should ask is, "Is there is a greatest common factor in the expression $18x^2 - 33x - 30$?" Yes, there is a GCF of 3, so we first factor out the 3.

$$18x^2 - 33x - 30 = 3(6x^2 - 11x - 10)$$

Now we factor the remaining trinomial, $6x^2 - 11x - 10$, by grouping. We see that $a = 6$, $b = -11$, and $c = -10$. The value of $a \cdot c = 6(-10) = -60$.

Now, we want to determine the integers whose product is -60 and whose sum is -11.

Integers Whose Product Is −60	1, −60	2, −30	3, −20	4, −15	5, −12	6, −10
Sum	−59	−28	−17	−11	−7	−4

The integers whose product is -60 and whose sum is -11 are 4 and -15.
Write $6x^2 - 11x - 10$ as $6x^2 + 4x - 15x - 10$.

$-11x = 4x - 15x$

Now, factor by grouping.

$$6x^2 + 4x - 15x - 10 = (6x^2 + 4x) + (-15x - 10)$$

Common factor in 1st group: $2x$; common factor in 2nd group: -5: $= 2x(3x + 2) - 5(3x + 2)$

Factor out $3x + 2$: $= (3x + 2)(2x - 5)$

Check

$$\begin{aligned} 3(3x + 2)(2x - 5) &= 3(6x^2 - 15x + 4x - 10) \\ &= 3(6x^2 - 11x - 10) \\ &= 18x^2 - 33x - 30 \end{aligned}$$

So $18x^2 - 33x - 30 = 3(3x + 2)(2x - 5)$.

QUICK ✓ *Factor the trinomial completely.*

16. $24x^2 + 6x - 9$

17. $-10n^2 + 17n - 3$

Work Smart

Let's compare the two methods presented in this section, trial and error and factoring by grouping, by factoring $3x^2 + 10x + 8$. The first question to ask is: Is there a GCF? No, there's not, so let's continue.

Trial and Error

Step 1: The coefficient of x^2, 3, is prime, so we list the possibilities for the binomial factors: $(3x + \_\_)(x + \_\_)$.

Step 2: The last term, 8, is not prime. Its factors are (1)(8) or (2)(4) or (−1)(−8) or (−2)(−4). Since $c = 8$ is positive and the coefficient of the middle term is also positive, we'll consider only (1)(8) and (2)(4).

Grouping

Step 1: For the polynomial $3x^2 + 10x + 8$, $a = 3$ and $c = 8$; the value of ac is $3 \cdot 8 = 24$.

Step 2: The two integers whose product is $ac = 24$ whose sum is $b = 10$ are 6 and 4.

Step 3: Rewrite $3x^2 + 10x + 8$ as $3x^2 + 10x + 8 = 3x^2 + 6x + 4x + 8$.

(continued)

Trial and Error	**Grouping**
Step 3: The coefficient of the middle term is not large, so let's start by trying $(3x + 4)(x + 2)$ and $(3x + 2)(x + 4)$. We will multiply out $(3x + 4)(x + 2)$ first. $(3x + 4)(x + 2) = 3x^2 + 10x + 8$ It works! Therefore, $3x^2 + 10x + 8 = (3x + 4)(x + 2)$.	**Step 4:** Factor the expression $3x^2 + 6x + 4x + 8$ by grouping. $3x^2 + 6x + 4x + 8 = 3x(x + 2) + 4(x + 2) = (x + 2)(3x + 4)$ So $3x^2 + 10x + 8 = (x + 2)(3x + 4)$.

Because $(3x + 4)(x + 2)$ is equivalent to $(x + 2)(3x + 4)$, we see that both methods give the same result. Which method do you prefer?

4.3 Exercises

For Extra Help:  Student Solutions Manual  CD Video PH Math/Tutor Center  MathXL Tutorials on CD MathXL® MyMathLab

Concepts and Vocabulary

In Problems 1–3, fill in the blanks.

1. When factoring any polynomial, begin by determining whether each term contains a __________ __________ other than one.

2. When factoring $6x^2 + x - 1$ using grouping, $ac =$ __________ and $b =$ __________.

3. When factoring $ax^2 + bx + c$ and b is a small number, choose factors of the product ac that are __________ to each other.

In Problems 4–6, answer True or False to each statement.

4. The trinomial $12x^2 + 22x + 6$ is completely factored as $(4x + 6)(3x + 1)$.

5. To factor $2x^2 - 13x + 6$ using grouping, find factors whose product is 6 and sum is -13.

6. To check if a trinomial has been factored correctly, use FOIL to multiply the binomials and verify that the product is the original trinomial.

7. Describe when you would use the trial and error method and when you would use the grouping method to factor a trinomial. Make up two examples that demonstrate your reasoning.

8. How can you tell if a trinomial is not factorable? Make up an example to demonstrate your reasoning.

Building Skills

In Problems 9–36, factor completely. If a trinomial cannot be factored, say it is prime.

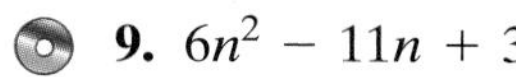

9. $6n^2 - 11n + 3$	**10.** $5w^2 + 13w - 6$	**11.** $4w^2 - 8w - 5$
12. $4x^2 - 9x + 4$	**13.** $5x^2 - 13x + 2$	**14.** $3x^2 + 16x - 12$
15. $4x^2 + 4x - 3$	**16.** $7a^2 + 37a + 10$	**17.** $27z^2 + 3z - 2$
18. $25t^2 + 5t - 2$	**19.** $9x^2 - 6x + 1$	**20.** $3x^2 - 17x + 10$
21. $2x^2 + 5x + 3$	**22.** $2x^2 - 7x + 3$	**23.** $6n^2 - 11n + 5$
24. $6y^2 - 5y - 6$	**25.** $12x^2 + 2xy - 4y^2$	**26.** $20t^2 + 23t + 8$
27. $4m^2 + 9m - 5$	**28.** $6x^2 - 14xy - 12y^2$	**29.** $6m^2 - 5m - 4$
30. $6m^2 - 23m - 4$	**31.** $-18n^2 + 39n - 18$	**32.** $-3x - 4 + 10x^2$
33. $12a^2 + 19a + 5$	**34.** $-5 - 9x + 18x^2$	
35. $2x^2 - 5x + 3$	**36.** $2x^2 + 7x + 3$	

1. common factor
2. $-6, 1$
3. close
4. False
5. False
6. True
7. Answers may vary.
8. Answers may vary.
9. $(2n - 3)(3n - 1)$
10. $(w + 3)(5w - 2)$
11. $(2w - 5)(2w + 1)$
12. prime
13. prime
14. $(x + 6)(3x - 2)$
15. $(2x - 1)(2x + 3)$
16. $(a + 5)(7a + 2)$
17. $(9z - 2)(3z + 1)$
18. $(5t + 2)(5t - 1)$
19. $(3x - 1)^2$
20. $(x - 5)(3x - 2)$
21. $(2x + 3)(x + 1)$
22. $(2x - 1)(x - 3)$
23. $(6n - 5)(n - 1)$
24. $(2y - 3)(3y + 2)$
25. $2(2x - y)(3x + 2y)$
26. prime
27. prime
28. $2(x - 3y)(3x + 2y)$
29. $(3m - 4)(2m + 1)$
30. $(m - 4)(6m + 1)$
31. $-3(3n - 2)(2n - 3)$
32. $(5x - 4)(2x + 1)$
33. $(4a + 5)(3a + 1)$
34. $(6x - 5)(3x + 1)$
35. $(2x - 3)(x - 1)$
36. $(2x + 1)(x + 3)$

37. $(3x - 4)(5x - 1)$
38. $x(3x - 5)(4x + 3)$
39. $-(a + 4)(4a - 3)$
40. $2(2x + 3)(3x - 4)$
41. $2(x - 2y)(5x + 6y)$
42. $(4n + 5)(3n - 2)$
43. $(3x - 4)(3x - 2)$
44. $2(x + 5)(9x - 1)$
45. $-3(4x + 3)(2x - 1)$
46. $-2(12z^2 - 9z - 1)$
47. $4x^2y^2(x - 2y - 1)$
48. $3xy(3x^2 + 2x + 1)$
49. $(x - 6y)(x - 7y)$
50. $(k + 7p)(k - 5p)$
51. $(m + 3n)(4m + n)$
52. $(4m - 3n)(m - 4n)$
53. $-(x - 12)(x + 2)$
54. $-(x - 7)(x + 4)$
55. $(n - 1)(n^2 + 1)$
56. $(a - x)(b - y)$
57. prime
58. prime
59. $6(2x + 5y)(2x - y)$
60. $(2a + 5b)(3a - 4b)$
61. $4(a + 2)(a - 2)$
62. $2(x + 1)(x - 1)$
63. $3(3a + 8b)(2a - b)$
64. $3y^2(2x + 7)(2x + 3)$
65. $(2ab^2 - c)(a + b)$
66. $y(2 + x^2)(3xy - 1)$
67. $-2x^2(2x + 5)(x - 1)$
68. $-3n(2n - 3)(3n + 1)$
69. $(2x + 7)$ m and $(3x - 4)$ m
70. $(2x + 1)$ cm and $(3x - 1)$ cm
71. $2x - 5$
72. $4x - 3$
73. $-2x(4x + 3)(3x - 5)$
74. $2x(2x - 3)(7x - 8)$
75. $(9z^2 + 8)(3z^2 + 2)$
76. $(3n^3 + 2)(5n^3 - 1)$
77. $(3x^n + 1)(x^n + 6)$
78. $(2x^n - 5)(x^n + 1)$
79. $(x^2 + 1)(2x - 7)(3x - 2)$
80. $(x - 1)(2x - 1)(5x + 2)$
81. ± 2 or ± 14
82. 13, 17, 23, or 43

Mixed Practice

In Problems 37–68, factor completely using any method you wish. If a polynomial cannot be factored, say it is prime.

37. $15x^2 - 23x + 4$ **38.** $12x^3 - 11x^2 - 15x$ **39.** $-13a + 12 - 4a^2$

40. $2x + 12x^2 - 24$ **41.** $10x^2 - 8xy - 24y^2$ **42.** $12n^2 + 7n - 10$

43. $8 - 18x + 9x^2$ **44.** $18x^2 + 88x - 10$ **45.** $-6x + 9 - 24x^2$

46. $-24z^2 + 18z + 2$ **47.** $4x^3y^2 - 8x^2y^3 - 4x^2y^2$ **48.** $9x^3y + 6x^2y + 3xy$

49. $x^2 - 13xy + 42y^2$ **50.** $k^2 + 2kp - 35p^2$ **51.** $4m^2 + 13mn + 3n^2$

52. $4m^2 - 19mn + 12n^2$ **53.** $10x - x^2 + 24$ **54.** $28 - x^2 + 3x$

55. $n^3 - n^2 + n - 1$ **56.** $ab - bx - ay + xy$ **57.** $6x^2 - 17x - 12$

58. $8x^2 + 14x - 15$ **59.** $48xy + 24x^2 - 30y^2$ **60.** $-20b^2 + 6a^2 + 7ab$

61. $4a(a + 2) - 8(a + 2)$ **62.** $2x(x + 1) - 2(x + 1)$

63. $18a^2 + 39ab - 24b^2$ **64.** $63y^3 + 60xy^3 + 12x^2y^3$

65. $2a^2b^2 - ac + 2ab^3 - bc$ **66.** $6xy^2 + 3x^3y^2 - 2y - x^2y$

67. $-6x^3 + 10x^2 - 4x^4$ **68.** $21n^2 - 18n^3 + 9n$

Applying the Concepts

△ **69. Area of a Triangle** A triangle has area described by the polynomial $3x^2 + \frac{13}{2}x - 14$ square meters. Find the base and height of the triangle. (**Hint:** The area of a triangle is $A = \frac{1}{2}bh$.)

△ **70. Area of a Rectangle** A rectangle has area described by the polynomial $6x^2 + x - 1$ square centimeters. Find the length and width of the rectangle. (**Hint:** The area of a rectangle is $A = lw$.)

71. Suppose that we know one factor of $6x^2 - 11x - 10$ is $3x + 2$. What is the other factor?

72. Suppose that we know one factor of $8x^2 + 22x - 21$ is $2x + 7$. What is the other factor?

Extending the Concepts

In Problems 73–80, factor completely.

73. $30x + 22x^2 - 24x^3$ **74.** $48x - 74x^2 + 28x^3$ **75.** $27z^4 + 42z^2 + 16$

76. $15n^6 + 7n^3 - 2$ **77.** $3x^{2n} + 19x^n + 6$ **78.** $2x^{2n} - 3x^n - 5$

79. $6x^2(x^2 + 1) - 25x(x^2 + 1) + 14(x^2 + 1)$

80. $10x^2(x - 1) - x(x - 1) - 2(x - 1)$

In Problems 81 and 82, find all possible integer values of b so that the polynomial is factorable.

81. $3x^2 + bx - 5$ **82.** $6x^2 + bx + 7$

4.4 Factoring Special Products

OBJECTIVES

1. Factor Perfect Square Trinomials
2. Factor the Difference of Two Squares
3. Factor the Sum or Difference of Two Cubes

Preparing for...Answers **1.** 25 **2.** -8 **3.** $125p^6$ **4.** $9z^2 + 12z + 4$ **5.** $16m^2 - 25$

Preparing for Factoring Special Products

Before getting started, take the following readiness quiz. If you get a problem wrong, go back to the section cited and review the material.

1. Evaluate: 5^2 [Section 1.6, pp. 49–50]

2. Evaluate: $(-2)^3$ [Section 1.6, pp. 49–50]

3. Find the product: $(5p^2)^3$ [Section 3.2, p. 187]

4. Find the product: $(3z + 2)^2$ [Section 3.3, pp. 194–196]

5. Find the product: $(4m + 5)(4m - 5)$ [Section 3.3, p. 194]

In Section 3.3, we presented three binomial products that were "special products," that is, special cases of the FOIL pattern. We briefly review two of them here.

Pattern	Product	Example
Squaring a binomial	$(a + b)^2 = a^2 + 2ab + b^2$ $(a - b)^2 = a^2 - 2ab + b^2$	$(3a + 5)^2 = (3a)^2 + 2(3a)(5) + 5^2$ $= 9a^2 + 30a + 25$
Product of the sum and difference of two terms	$(a + b)(a - b) = a^2 - b^2$	$(4z + 7)(4z - 7) = (4z)^2 - 7^2$ $= 16z^2 - 49$

Recall that $a^2 + 2ab + b^2$ and $a^2 - 2ab + b^2$ are called perfect square trinomials. We call $a^2 - b^2$ is called the difference of two squares.

In this section, we look at polynomials that can be categorized as having "special formulas" for factoring. Why would we want to learn a method for factoring using "special formulas"? If you recognize a polynomial as a perfect square trinomial or the difference of two squares, you'll be able to factor it quickly without using trial and error or the grouping method.

We begin with factoring perfect square trinomials.

Teaching Tip
Before introducing factoring perfect square trinomials, do a few examples of squaring a binomial to remind students of the pattern.

1 Factor Perfect Square Trinomials

Reversing the formulas $(a + b)^2 = a^2 + 2ab + b^2$ and $(a - b)^2 = a^2 - 2ab + b^2$, we obtain a method for factoring perfect square trinomials.

In Words
A perfect square trinomial is a trinomial in which the first term and the third term are perfect squares and the second term is either 2 times or −2 times the product of the expressions being squared in the first and third terms.

PERFECT SQUARE TRINOMIALS

$$a^2 + 2ab + b^2 = (a + b)^2$$
$$a^2 - 2ab + b^2 = (a - b)^2$$

Perfect square trinomials can be factored quickly when they are recognized. For a polynomial to be a perfect square trinomial, two conditions must be satisfied.

1. The first and last terms must be perfect squares. The perfect squares are $1^2 = 1, 2^2 = 4, 3^2 = 9$, and so on. Any variable raised to an even exponent is a perfect square. So, x^2, $x^4 = (x^2)^2$, $x^6 = (x^3)^2$ are all perfect squares. Examples of perfect squares are

49	because	$7^2 = 49$
121	because	$11^2 = 121$
$9x^2$	because	$(3x)^2 = 9x^2$
$25a^4$	because	$(5a^2)^2 = 25a^4$

2. The "middle term" must equal 2 times or −2 times the product of the expressions being squared in the first and last term.

Let's look at an example.

Classroom Example
Factor completely: $x^2 + 10x + 25$
Answer: $(x + 5)^2$

EXAMPLE 1 How to Factor Perfect Square Trinomials

Factor completely: $z^2 + 6z + 9$

Step-by-Step Solution

Step 1: Determine whether the first term and the third term are perfect squares.

The first term, z^2, is the square of z and the third term, 9, is the square of 3.

(continued)

Step 2: Determine whether the middle term is 2 times or −2 times the product of the expressions being squared in the first and last term.

The middle term, $6z$, is 2 times the product of z and 3.

Step 3: Use $a^2 + 2ab + b^2 = (a + b)^2$ to factor the expression.

$$\begin{array}{c} a^2 + 2 \cdot a \cdot b + b^2 \\ \downarrow \quad \downarrow \; \downarrow \quad \downarrow \\ z^2 + 6z + 9 = z^2 + 2 \cdot z \cdot 3 + 3^2 \end{array}$$

Factor as $(a + b)^2$ with $a = z$ and $b = 3$: $= (z + 3)^2$

So $z^2 + 6z + 9 = (z + 3)^2$.

Classroom Example ➤
Factor completely:
(a) $49n^2 - 28n + 4$
(b) $9a^2 + 30ab + 25b^2$

Answer:
(a) $(7n - 2)^2$
(b) $(3a + 5b)^2$

EXAMPLE 2 Factoring Perfect Square Trinomials

Factor completely:

(a) $4x^2 - 20x + 25$ **(b)** $9x^2 + 42xy + 49y^2$

Solution

(a) The first term, $4x^2$, is the square of $2x$, and the third term, 25, is the square of 5. The middle term, $-20x$, is -2 times the product of $2x$ and 5. We use $a^2 - 2ab + b^2 = (a - b)^2$ to factor the expression.

$$\begin{aligned} & \quad a^2 - 2 \cdot a \cdot b + b^2 \\ 4x^2 - 20x + 25 &= (2x)^2 - 2 \cdot 2x \cdot 5 + (5)^2 \\ &= (2x - 5)^2 \end{aligned}$$

Therefore, $4x^2 - 20x + 25 = (2x - 5)^2$.

(b) The first term, $9x^2$, is the square of $3x$, and the third term, $49y^2$, is the square of $7y$. The middle term, $42xy$, is 2 times the product of $3x$ and $7y$. We factor as $a^2 + 2ab + b^2 = (a + b)^2$.

$$\begin{aligned} & \quad a^2 + 2 \cdot a \cdot b + b^2 \\ 9x^2 + 42xy + 49y^2 &= (3x)^2 + 2 \cdot 3x \cdot 7y + (7y)^2 \\ &= (3x + 7y)^2 \end{aligned}$$

We see that $9x^2 + 42xy + 49y^2 = (3x + 7y)^2$.

QUICK ✓ *Factor each trinomial completely.*

1. $x^2 - 12x + 36$ **2.** $16x^2 + 40x + 25$ **3.** $9a^2 - 60ab + 100b^2$

Classroom Example ➤
Factor completely, if possible:
$a^2 - 8a + 64$

Answer: prime polynomial

EXAMPLE 3 Factoring a Trinomial

Factor completely, if possible: $b^2 - 10b + 36$

Solution

The first term, b^2, is the square of b. The third term, 36, is the square of 6. Is the middle term, $-10b$, -2 times the product of the expressions being squared in the first and last term?

$$-2(b)(6) = -12b \neq -10b.$$

So $b^2 - 10b + 36$ is not a perfect square trinomial. Can we factor $b^2 - 10b + 36$ using another strategy? The trinomial $b^2 - 10b + 36$ is in the form $x^2 + bx + c$, so we need two integers whose product is 36, whose sum is -10. We choose negative factors of 36, since the coefficient of the middle term is negative: The possibilities are $(-1)(-36)$, $(-2)(-18)$, $(-3)(-12)$, or $(-9)(-4)$. None of these factors sum to -10. We conclude that $b^2 - 10b + 36$ is prime.

QUICK ✓ *Factor each trinomial completely, if possible.*

4. $z^2 - 8z + 16$ **5.** $4n^2 + 12n + 9$ **6.** $y^2 + 13y + 35$

Classroom Example ➤
Factor completely: $18z^4 - 84z^2 + 98$
Answer: $2(3z^2 - 7)^2$

EXAMPLE 4 Factoring a Perfect Square Trinomial Having a GCF

Factor completely: $32m^4 - 48m^2 + 18$

Solution

Work Smart
m^4 is a square because $m^4 = (m^2)^2$.

Do you remember the first step in any factoring problem? It is to look for the GCF. Notice there is a common factor of 2 in the trinomial, so we factor it out.

$$32m^4 - 48m^2 + 18 = 2(16m^4 - 24m^2 + 9)$$

We now examine the trinomial $16m^4 - 24m^2 + 9$. The first term, $16m^4$, is the square of $4m^2$. The third term, 9, is the square of 3. The middle term, $-24m^2$, is -2 times the product of $4m^2$ and 3.

$$\begin{aligned} & \quad\; a^2 \quad - \quad 2 \cdot a \cdot b + b^2 \\ 16m^4 - 24m^2 + 9 &= (4m^2)^2 - 2 \cdot 4m^2 \cdot 3 + 3^2 \\ \text{Factor as } (a-b)^2\text{:} \quad &= (4m^2 - 3)^2 \end{aligned}$$

Work Smart
Don't forget the GCF that was factored out in the first step.

Therefore,

$$\begin{aligned} 32m^4 - 48m^2 + 18 &= 2(16m^4 - 24m^2 + 9) \\ &= 2(4m^2 - 3)^2 \end{aligned}$$

QUICK ✓ *Factor each trinomial completely.*

7. $4z^2 + 24z + 36$ **8.** $50a^3 + 80a^2 + 32a$ **9.** $50a^4 + 40a^2b + 8b^2$

Perfect square trinomials can also be factored using the methods introduced in the last section, but if you recognize the perfect square pattern, you won't have to use either trial and error or the grouping method.

2 Factor the Difference of Two Squares

In Words
The difference of two squares is just that! A perfect square minus a perfect square. This pattern is easy to recognize: Just look for two squares that have been subtracted.

Do you recall finding products such as $(x + 2y)(x - 2y)$ in Chapter 3? We found that $(x + 2y)(x - 2y) = x^2 - (2y)^2 = x^2 - 4y^2$. In general,

$$(a - b)(a + b) = a^2 - b^2$$

We will now reverse the multiplication process and find the factors of the difference of two squares.

Teaching Tip
Before introducing factoring the difference of two squares, do a few examples of multiplying binomials in the form $(a - b)(a + b)$ to remind students of the pattern.

DIFFERENCE OF TWO SQUARES

$$a^2 - b^2 = (a - b)(a + b)$$

Classroom Example
Factor completely:
(a) $y^2 - 100$
(b) $9a^2 - 16b^2$
Answer:
(a) $(y - 10)(y + 10)$
(b) $(3a - 4b)(3a + 4b)$

EXAMPLE 5 Factoring the Difference of Two Squares

Factor completely:

(a) $n^2 - 81$ **(b)** $16x^2 - 9y^2$

Solution

(a) We notice that $n^2 - 81$ is the difference of two squares, n^2 and $81 = 9^2$. So

$$\begin{aligned} & \quad\ \ a^2 \quad - \quad b^2 \\ & \quad\ \ \downarrow \qquad\quad \downarrow \\ n^2 - 81 &= (n)^2 - (9)^2 \\ a^2 - b^2 = (a - b)(a + b): \quad &= (n - 9)(n + 9) \end{aligned}$$

Work Smart
Remember that the Commutative Property of Multiplication tells us that the order of the factors in the answer doesn't matter:
$(a - b)(a + b) = (a + b)(a - b)$

(b) We notice that $16x^2 - 9y^2$ is the difference of two squares, $16x^2 = (4x)^2$ and $9y^2 = (3y)^2$. So

$$\begin{aligned} & \quad\ \ a^2 \quad - \quad b^2 \\ & \quad\ \ \downarrow \qquad\quad \downarrow \\ 16x^2 - 9y^2 &= (4x)^2 - (3y)^2 \\ a^2 - b^2 = (a - b)(a + b): \quad &= (4x - 3y)(4x + 3y) \end{aligned}$$

You can check the answer to any of these differences of two squares by using FOIL. To check Example 5(b),

Check $$\begin{aligned} (4x - 3y)(4x + 3y) &= 16x^2 + 12xy - 12xy - 9y^2 \\ &= 16x^2 - 9y^2 \end{aligned}$$

Our answer checks, so $16x^2 - 9y^2 = (4x - 3y)(4x + 3y)$.

QUICK ✓ *Factor completely.*

10. $z^2 - 25$ **11.** $81m^2 - 16n^2$ **12.** $16a^2 - \frac{4}{9}b^2$

Classroom Example
Factor completely:
(a) $16a^4 - 25$ (b) $n^4 - 81$
Answer:
(a) $(4a^2 - 5)(4a^2 + 5)$
(b) $(n - 3)(n + 3)(n^2 + 9)$

EXAMPLE 6 Factoring the Difference of Two Squares

Factor completely:

(a) $49k^4 - 100$ **(b)** $p^4 - 1$

Solution

(a) Because $49k^4 = (7k^2)^2$ and $100 = 10^2$, we have the difference of two squares. So

$$\begin{aligned} & \quad\ \ a^2 \quad - \quad b^2 \\ & \quad\ \ \downarrow \qquad\quad \downarrow \\ 49k^4 - 100 &= (7k^2)^2 - 10^2 \\ a^2 - b^2 = (a - b)(a + b): \quad &= (7k^2 - 10)(7k^2 + 10) \end{aligned}$$

So $49k^4 - 100 = (7k^2 - 10)(7k^2 + 10)$.

(b) The expressions p^4 and 1 are both perfect squares. $p^4 = (p^2)^2$ and $1 = 1^2$. So

$$\begin{aligned} & \quad\ \ a^2 \quad - \quad b^2 \\ & \quad\ \ \downarrow \qquad\quad \downarrow \\ p^4 - 1 &= (p^2)^2 - 1^2 \\ a^2 - b^2 = (a - b)(a + b): \quad &= (p^2 - 1)(p^2 + 1) \end{aligned}$$

But $p^2 - 1$ is a difference of two squares! So we factor again:

$$= (p + 1)(p - 1)(p^2 + 1)$$

So $p^4 - 1 = (p + 1)(p - 1)(p^2 + 1)$.

You may be asking yourself, "What about the sum of two squares—how does it factor?" The answer is that if a and b are real numbers, **the sum of two squares, $a^2 + b^2$, is prime and does not factor.** So, if a variable represents a real number, binomials such as $a^2 + 1$ or $4y^2 + 81$ are prime.

QUICK ✓ *Factor each polynomial completely.*

13. $36b^4 - c^2$ **14.** $100k^4 - 81w^2$ **15.** $x^4 - 16$

Classroom Example
Factor completely: $48a^2 - 27b^2$

Answer: $3(4a - 3b)(4a + 3b)$

EXAMPLE 7 Factoring the Difference of Two Squares Containing a GCF

Factor completely: $50x^2 - 72y^2$

Solution

What's the first thing we look for when we factor? A greatest common factor! The GCF of 50 and 72 is 2, so we proceed as follows:

Factor out GCF = 2: $50x^2 - 72y^2 = 2(25x^2 - 36y^2)$

Difference of two squares with $a = 5x$ and $b = 6y$: $= 2[(5x)^2 - (6y)^2]$

Factor as $(a - b)(a + b)$: $= 2(5x - 6y)(5x + 6y)$

So $50x^2 - 72y^2 = 2(5x - 6y)(5x + 6y)$.

QUICK ✓ *Factor each polynomial completely.*

16. $147x^2 - 48$ **17.** $-27a^3b + 75ab^3$

3 Factor the Sum or Difference of Two Cubes

Consider the following products:

$$(a + b)(a^2 - ab + b^2) = a^3 - a^2b + ab^2 + a^2b - ab^2 + b^3$$
$$= a^3 + b^3$$
$$(a - b)(a^2 + ab + b^2) = a^3 + a^2b + ab^2 - a^2b - ab^2 - b^3$$
$$= a^3 - b^3$$

These products show us that we can factor the sum or difference of two cubes as follows:

THE SUM OF TWO CUBES

$$a^3 + b^3 = (a + b)(a^2 - ab + b^2)$$

THE DIFFERENCE OF TWO CUBES

$$a^3 - b^3 = (a - b)(a^2 + ab + b^2)$$

Notice in the formulas that the sign between the cubes matches the sign in the first set of parentheses and that the sign in the middle of the trinomial factor must be opposite the sign between the cubes. Also remember that the perfect cubes are $1^3 = 1, 2^3 = 8, 3^3 = 27$, and so on. Any variable raised to a multiple of 3 is a perfect cube. So $x^3, x^6 = (x^2)^3, x^9 = (x^3)^3$, and so on are all perfect cubes.

Classroom Example ➤
Factor completely: $x^3 - 27$
Answer: $(x - 3)(x^2 + 3x + 9)$

EXAMPLE 8 Factoring the Sum or Difference of Two Cubes

Factor completely: $x^3 - 8$

Solution

We notice that we have the difference of two cubes, x^3 and $8 = 2^3$. We let $a = x$ and $b = 2$ in the factoring formula for the difference of two cubes.

$$a^3 - b^3 = (a - b)(a^2 + a\,b + b^2)$$
$$x^3 - 8 = x^3 - 2^3 = (x - 2)(x^2 + x(2) + 2^2)$$
$$= (x - 2)(x^2 + 2x + 4)$$

So $x^3 - 8 = (x - 2)(x^2 + 2x + 4)$.

Teaching Tip
You may suggest that students use vertical multiplication to verify the result.

Work Smart
The trinomial in the factored form of the sum or difference of two cubes will always be prime.

Classroom Example ➤
Factor completely: $27a^6 - 8b^3$
Answer:
$(3a^2 - 2b)(9a^4 + 6a^2b + 4b^2)$

EXAMPLE 9 Factoring the Sum or Difference of Two Cubes

Factor completely: $8m^3 + 125n^6$

Solution

Because $8m^3 = (2m)^3$ and $125n^6 = (5n^2)^3$, the expression $8m^3 + 125n^6$ is the sum of two cubes. We let $a = 2m$ and $b = 5n^2$ in the factoring formula for the sum of two cubes.

$$8m^3 + 125n^6 = (2m)^3 + (5n^2)^3$$
$$= (2m + 5n^2)[(2m)^2 - (2m)(5n^2) + (5n^2)^2]$$
$$= (2m + 5n^2)(4m^2 - 10mn^2 + 25n^4)$$

So $8m^3 + 125n^6 = (2m + 5n^2)(4m^2 - 10mn^2 + 25n^4)$.

QUICK ✓ *Factor each polynomial completely.*

18. $z^3 + 125$ **19.** $8p^3 - 27q^6$

Classroom Example ➤
Factor completely: $24x^4 - 375x$
Answer:
$3x(2x - 5)(4x^2 + 10x + 25)$

EXAMPLE 10 Factoring the Sum or Difference of Two Cubes Having a GCF

Factor completely: $250x^4 + 54x$

Solution

Remember, the first step in any factoring problem is to look for a common factor. Here, we have a common factor of $2x$.

$$250x^4 + 54x = 2x(125x^3 + 27)$$
$$(5x)^3 = 125x^3 \text{ and } 3^3 = 27: \quad = 2x[(5x)^3 + 3^3]$$
$$= 2x(5x + 3)(25x^2 - 15x + 9)$$

So $250x^4 + 54x = 2x(5x + 3)(25x^2 - 15x + 9)$.

QUICK ✓ *Factor each polynomial completely.*

20. $54a - 16a^4$ **21.** $-375b^3 + 3$

4.4 Exercises

For Extra Help:   Student Solutions Manual  CD Video   PH Math/Tutor Center  MathXL Tutorials on CD MathXL® MyMathLab

1. sum; two cubes
2. $(a - b)^2$
3. difference, two squares
4. True
5. False
6. False
7. Answers may vary.
8. Answers may vary.
9. $(x - 3)^2$
10. $(n - 1)^2$
11. $(x + 5)^2$
12. $(m + 6)^2$
13. $(2p - 1)^2$
14. $(3a - 2)^2$
15. $(4x + 3)^2$
16. $(4y - 9)^2$
17. $(x - 2y)^2$
18. $(2a + 5b)^2$
19. $(2z - 3)^2$
20. $(5k - 7)^2$
21. $(x - 11)(x + 11)$
22. $(x - 3)(x + 3)$
23. $(4x - 7y)(4x + 7y)$
24. $(6m - 5n)(6m + 5n)$
25. $(2x - 5)(2x + 5)$
26. $(5m - 3)(5m + 3)$
27. $(10n^4 - 9p^2)(10n^4 + 9p^2)$
28. $(6s^3 - 7t^2)(6s^3 + 7t^2)$
29. $(k - 2)(k + 2)(k^2 + 4)(k^4 + 16)$
30. $(a - 2)(a + 2)(a^2 + 4)$
31. $(5p^2 - 7q)(5p^2 + 7q)$
32. $(6b^2 - 11a)(6b^2 + 11a)$
33. $(1 + x)(1 - x + x^2)$
34. $(2v + 1)(4v^2 - 2v + 1)$
35. $(2x - 3y)(4x^2 + 6xy + 9y^2)$
36. $(4r - 5s)(16r^2 + 20rs + 25s^2)$
37. $(x^2 - 2y)(x^4 + 2x^2y + 4y^2)$
38. $(m^3 - 3n^2)(m^6 + 3m^3n^2 + 9n^4)$
39. $(3c + 4d^3)(9c^2 - 12cd^3 + 16d^6)$
40. $(5y + 3z^2)(25y^2 - 15yz^2 + 9z^4)$
41. prime
42. prime
43. $\left(x - \frac{1}{3}\right)^2$
44. $\left(x + \frac{1}{4}\right)^2$
45. $(4m + 5n)^2$
46. $(5pq - 8)^2$
47. $2(x - 3)^2$
48. $2(x + 5)^2$
49. $3a(4a + 3)^2$
50. $3n(2n - 3)^2$
51. $\left(9p - \frac{1}{2}\right)\left(9p + \frac{1}{2}\right)$
52. $\left(\frac{x}{12} - \frac{1}{5}\right)\left(\frac{x}{12} + \frac{1}{5}\right)$

Concepts and Vocabulary

In Problems 1–3, fill in the blanks.

1. The binomial $27x^3 + 64y^3$ is called the ________ of ________ ________.

2. $a^2 - 2ab + b^2 =$ ______.

3. $4m^2 - 81n^2$ is called the ________ of ________ ________ and factors into two binomials.

In Problems 4–6, answer True or False to each statement.

4. $x^2 + 9$ is prime.

5. $(x - 1)(x^2 - x + 1)$ can be factored further.

6. $4x^2 - 16y^2$ factors completely to $(2x - 4y)(2x + 4y)$.

7. In your own words, create a list of steps for factoring the sum or difference of two cubes.

8. You are asked to factor the polynomial $x^2 + 4$. What is your answer and what rationale would you give for this result?

Building Skills

In Problems 9–20, factor each perfect square trinomial completely.

9. $x^2 - 6x + 9$ **10.** $n^2 - 2n + 1$ **11.** $x^2 + 10x + 25$

12. $m^2 + 12m + 36$ **13.** $4p^2 - 4p + 1$ **14.** $9a^2 - 12a + 4$

15. $16x^2 + 24x + 9$ **16.** $16y^2 - 72y + 81$ **17.** $x^2 - 4xy + 4y^2$

18. $4a^2 + 20ab + 25b^2$ **19.** $4z^2 - 12z + 9$ **20.** $25k^2 - 70k + 49$

In Problems 21–32, factor each difference of two squares completely.

21. $x^2 - 121$ **22.** $x^2 - 9$ **23.** $16x^2 - 49y^2$

24. $36m^2 - 25n^2$ **25.** $4x^2 - 25$ **26.** $25m^2 - 9$

27. $100n^8 - 81p^4$ **28.** $36s^6 - 49t^4$ **29.** $k^8 - 256$

30. $a^4 - 16$ **31.** $25p^4 - 49q^2$ **32.** $36b^4 - 121a^2$

In Problems 33–42, factor each sum or difference of two cubes completely.

33. $1 + x^3$ **34.** $8v^3 + 1$ **35.** $8x^3 - 27y^3$

36. $64r^3 - 125s^3$ **37.** $x^6 - 8y^3$ **38.** $m^9 - 27n^6$

39. $27c^3 + 64d^9$ **40.** $125y^3 + 27z^6$ **41.** $16x^3 + 125y^3$

42. $25a^3 + 81b^6$

Mixed Practice

In Problems 43–84, factor completely. If the polynomial is prime, state so.

43. $x^2 - \frac{2}{3}x + \frac{1}{9}$ **44.** $x^2 + \frac{1}{2}x + \frac{1}{16}$ **45.** $16m^2 + 40mn + 25n^2$

46. $25p^2q^2 - 80pq + 64$ **47.** $18 - 12x + 2x^2$ **48.** $50 + 20x + 2x^2$

49. $48a^3 + 72a^2 + 27a$ **50.** $12n^3 - 36n^2 + 27n$ **51.** $81p^2 - \frac{1}{4}$

52. $\frac{x^2}{144} - \frac{1}{25}$ **53.** $x^4y^2 - x^2y^4$ **54.** $x^6y^2 - 16x^2y^2$

53. $x^2y^2(x - y)(x + y)$
54. $x^2y^2(x - 2)(x + 2)(x^2 + 4)$
55. $2t(t - 3)(t^2 + 3t + 9)$
56. $2x^2(2x - 3)(4x^2 + 6x + 9)$
57. $3s(s^2 + 2)(s^4 - 2s^2 + 4)$
58. $4(2a + b^2)(4a^2 - 2ab^2 + b^4)$
59. $4x(1 - x)(1 + x)$
60. $2x(x - 3)(x + 3)(x^2 + 9)$
61. $xy(x - y)(x + y)$
62. $3(2z - 1)(2z + 1)(4z^2 + 1)$
63. $(xy + 1)(x^2y^2 - xy + 1)$
64. $(2rs + t)(4r^2s^2 - 2rst + t^2)$
65. $(x^4 - 5y^5)(x^4 + 5y^5)$
66. $x^2(x - 15)(x + 15)$
67. $\left(x - \frac{y^2}{4}\right)\left(x^2 + \frac{xy^2}{4} + \frac{y^4}{16}\right)$
68. $\left(\frac{x}{5} + 2\right)\left(\frac{x^2}{25} - \frac{2}{5}x + 4\right)$
69. $2(x - 2)^2$
70. $3(x + 3)^2$
71. $-2n(2n - 1)^2(2n + 1)^2$
72. $-3x(2x^2 - 3)^2$
73. $-2x(1 + 2x)^2$
74. $2x^2(2x - 3)^2$
75. $(x^2 - y^3)(x^4 + x^2y^3 + y^6)$
76. $xy^3(x + 6)(x^2 - 6x + 36)$
77. prime
78. prime
79. $(x - 2)(x + 2)(2x + 5)$
80. $(z - 3)(z + 3)(4z + 3)$
81. $(2y + 5)(y - 4)(y + 4)$
82. $(m + 2)(m - 5)(m + 5)$
83. $(x - 2)(x + 2)(x^2 - 2x + 4)$
84. $(b - 4)(b - 3)(b^2 + 3b + 9)$
85. $2x + 5$
86. $3x - 1$
87. $(x - 3)(x + 1)$
88. $3\pi(2x + 3)$
89. $-3(2x - 1)$
90. $-4mp$
91. $x(x^2 - 3xy + 3y^2)$
92. $-1(z^2 - z - 1) \times (z^4 + z^3 + 2z^2 + 2z + 1)$
93. $(x - 2)(x + 4)$
94. $(x - 2)(x + 8)$
95. $(x + 1)(2a - 5)(a - 6)$
96. $(x - 1)^2(5a + 1)(5a - 2)$
97. $x(5x + 13)$
98. $(6a + 3b + 4)(6a + 3b - 2)$

55. $2t^4 - 54t$ **56.** $16x^5 - 54x^2$ **57.** $3s^7 + 24s$

58. $32a^3 + 4b^6$ **59.** $4x - 4x^3$ **60.** $2x^5 - 162x$

61. $x^3y - xy^3$ **62.** $48z^4 - 3$ **63.** $x^3y^3 + 1$

64. $8r^3s^3 + t^3$ **65.** $x^8 - 25y^{10}$ **66.** $x^4 - 225x^2$

67. $x^3 - \frac{y^6}{64}$ **68.** $\frac{x^3}{125} + 8$ **69.** $2x^2 - 8x + 8$

70. $3x^2 + 18x + 27$ **71.** $16n^3 - 32n^5 - 2n$ **72.** $36x^3 - 12x^5 - 27x$

73. $-2x - 8x^2 - 8x^3$ **74.** $18x^2 - 24x^3 + 8x^4$ **75.** $x^6 - y^9$

76. $x^4y^3 + 216xy^3$ **77.** $9x^2 + y^2$ **78.** $16a^2 + 49b^2$

79. $2x(x^2 - 4) + 5(x^2 - 4)$ **80.** $4z(z^2 - 9) + 3(z^2 - 9)$

81. $2y^3 + 5y^2 - 32y - 80$ **82.** $m^3 + 2m^2 - 25m - 50$

83. $x^4 - 2x^3 + 8x - 16$ **84.** $b^4 - 4b^3 - 27b + 108$

Applying the Concepts

△ **85. Area of a Square** The area of a square is given by the polynomial $4x^2 + 20x + 25$. What is an algebraic expression for the length of one side of the square? (**Hint:** The area of a square is $A = s^2$.)

△ **86. Area of a Square** The area of a square is given by the polynomial $9x^2 - 6x + 1$. What is an algebraic expression for the length of one side of the square?

In Problems 87 and 88, calculate the area of the shaded region and then write this polynomial in factored form.

△ **87.**

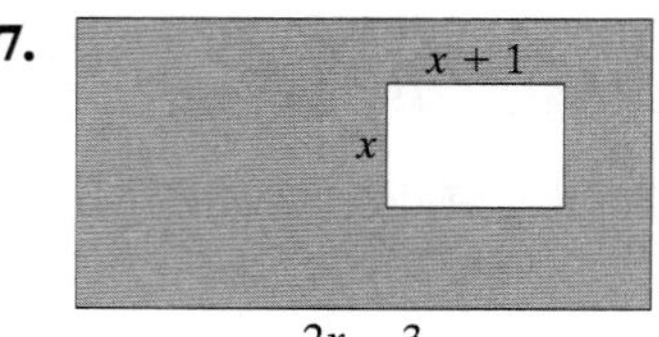

△ **88.**

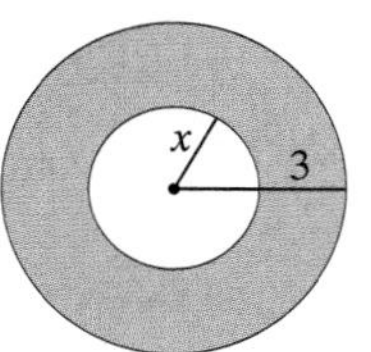

Extending the Concepts

In Problems 89–98, factor completely and then simplify, if possible.

89. $(x - 2)^2 - (x + 1)^2$ **90.** $(m - p)^2 - (m + p)^2$ **91.** $(x - y)^3 + y^3$

92. $(z + 1)^3 - z^6$ **93.** $(x + 1)^2 - 9$ **94.** $(x + 3)^2 - 25$

95. $2a^2(x + 1) - 17a(x + 1) + 30(x + 1)$

96. $25a^2(x - 1)^2 - 5a(x - 1)^2 - 2(x - 1)^2$

97. $5(x + 2)^2 - 7(x + 2) - 6$ **98.** $9(2a + b)^2 + 6(2a + b) - 8$

4.5 Summary of Factoring Techniques

OBJECTIVE

① Factor Polynomials Completely

① Factor Polynomials Completely

We have one objective in this section—to put together all the various factoring techniques we have discussed in Sections 4.1–4.4. Recall that we say that a polynomial is factored completely if each factor in the final factorization is prime. Here we present guidelines for factoring any polynomial.

Teaching Tip
The goal of this section is to give students practice recognizing and applying the factoring patterns. Take time to go over this summary of factoring strategies. Students may not have observed that counting the number of terms will yield a factoring plan.

Steps for Factoring

Step 1: Is there a greatest common factor? If so, factor out the greatest common factor (GCF).

Step 2: Count the number of terms.

Step 3: **(a)** 2 terms (Binomials)

- Is it the difference of two squares? If so,
$$a^2 - b^2 = (a - b)(a + b)$$
- Is it the difference of two cubes? If so,
$$a^3 - b^3 = (a - b)(a^2 + ab + b^2)$$
- Is it the sum of two cubes? If so,
$$a^3 + b^3 = (a + b)(a^2 - ab + b^2)$$

(b) 3 terms (Trinomials)

- Is it a perfect square trinomial? If so,
$$a^2 + 2ab + b^2 = (a + b)^2 \quad \text{or} \quad a^2 - 2ab + b^2 = (a - b)^2$$
- Is the coefficient of the square term 1? If so,
$$x^2 + bx + c = (x + m)(x + n) \text{ where } mn = c \text{ and } m + n = b$$
- Is the coefficient of the square term different from 1? If so,
  a. Use factoring by grouping.
  b. Use trial and error.

(c) 4 terms

- Use factoring by grouping.

Step 4: Check your work by multiplying out the factors.

Work Smart
Remember, if a and b are real numbers, the sum of two squares, $a^2 + b^2$ is prime.

EXAMPLE 1 How to Factor Completely

Factor completely: $2x^2 - 4x - 48$

Classroom Example
Factor completely: $3x^2 - 15x - 72$
Answer: $3(x - 8)(x + 3)$

Step-by-Step Solution

Step 1: Is there a GCF? Yes. We factor out the GCF, 2.

$$2x^2 - 4x - 48 = 2(x^2 - 2x - 24)$$

Step 2: Identify the number of terms in the polynomial in parentheses.

This polynomial has three terms in parentheses.

Step 3: We concentrate on the trinomial in parentheses, $x^2 - 2x - 24$. It is not a perfect square trinomial. The leading coefficient is 1, so we try $(x + m)(x + n)$, where $mn = c$ and $m + n = b$.

We need to find two factors of -24 whose sum is -2. Since $-6(4) = -24$ and $-6 + 4 = -2$, we have that $m = -6$ and $n = 4$.

$$2(x^2 - 2x - 24) = 2(x - 6)(x + 4)$$

Step 4: Check

$$\begin{aligned} \text{FOIL:} \quad 2(x - 6)(x + 4) &= 2(x^2 + 4x - 6x - 24) \\ &= 2(x^2 - 2x - 24) \\ \text{Distribute:} \quad &= 2x^2 - 4x - 48 \end{aligned}$$

So $2x^2 - 4x - 48 = 2(x - 6)(x + 4)$.

Classroom Example Factor completely: $49r^2 - 16s^2$

Answer: $(7r - 4s)(7r + 4s)$

QUICK ✓ *Factor each polynomial completely.*

1. $2p^2 + 8p - 90$

2. $-45x^2 + 3xy + 6y^2$

EXAMPLE 2 How to Factor Completely

Factor completely: $25p^2 - 9q^2$

Step-by-Step Solution

Step 1: Is there a GCF?	There is no GCF.
Step 2: Identify the number of terms in the polynomial.	There are two terms.
Step 3: Because the first term, $25p^2 = (5p)^2$, and the second term, $9q^2 = (3q)^2$, are both perfect squares, we have the difference of two squares.	$25p^2 - 9q^2 = ((5p)^2 - (3q)^2)$ $a^2 - b^2 = (a - b)(a + b)$: $= (5p - 3q)(5p + 3q)$
Step 4: Check	FOIL: $(5p - 3q)(5p + 3q) = 25p^2 + 15pq - 15pq - 9q^2$ Combine like terms: $= 25p^2 - 9q^2$

So $25p^2 - 9q^2 = (5p - 3q)(5p + 3q)$.

Classroom Example Factor completely: $162a^2 + 72ab + 8b^2$

Answer: $2(9a + 2b)^2$

QUICK ✓ *Factor each polynomial completely.*

3. $100x^2 - 81y^2$

4. $2ab^2 - 242a$

EXAMPLE 3 How to Factor Completely

Factor completely: $12x^2 + 36xy + 27y^2$

Step-by-Step Solution

Step 1: We notice that the coefficients 12, 36, and 27 are all multiples of 3, so we factor out the GCF of 3.	$12x^2 + 36xy + 27y^2 = 3(4x^2 + 12xy + 9y^2)$
Step 2: Identify the number of terms in the polynomial in parentheses.	There are three terms in the polynomial in parentheses.
Step 3: We concentrate on the polynomial in parentheses. Is it a perfect square trinomial? The first term is a perfect square, $4x^2 = (2x)^2$. The third term is also a perfect square, $9y^2 = (3y)^2$. The middle term is 2 times the product of $2x$ and $3y$. The polynomial in parentheses is a perfect square trinomial.	$3(4x^2 + 12xy + 9y^2) = 3[(2x)^2 + 2(2x)(3y) + (3y)^2]$ $a^2 + 2ab + b^2 = (a + b)^2$: $= 3(2x + 3y)^2$

Step 4: Check

$$\begin{aligned}3(2x + 3y)^2 &= 3(2x + 3y)(2x + 3y)\\ &= 3(4x^2 + 6xy + 6xy + 9y^2)\\ &= 3(4x^2 + 12xy + 9y^2)\\ &= 12x^2 + 36xy + 27y^2\end{aligned}$$

So $12x^2 + 36xy + 9y^2 = 3(2x + 3y)^2$.

QUICK ✓ *Factor each polynomial completely.*

5. $p^2 - 12pq + 36q^2$ **6.** $75x^2 + 90x + 27$

Classroom Example ➤
Factor completely: $250a^3 + 16$
Answer:
$2(5a + 2)(25a^2 - 10a + 4)$

EXAMPLE 4 Factoring Completely

Factor completely: $16m^3 + 54$

Solution

Notice that there is a greatest common factor of 2. Let's factor it out. The polynomial in the parentheses has two terms.

$$16m^3 + 54 = 2(8m^3 + 27)$$

Sum of two cubes with $a = 2m$ and $b = 3$: $= 2[(2m)^3 + 3^3]$

$a^3 + b^3 = (a + b)(a^2 - ab + b^2)$: $= 2[(2m + 3)((2m)^2 - (2m)(3) + (3)^2)]$

$= 2[(2m + 3)(4m^2 - 6m + 9)]$

Check $2[(2m + 3)(4m^2 - 6m + 9)] = 2[8m^3 - 12m^2 + 18m + 12m^2 - 18m + 27]$
$= 2[8m^3 + 27]$
$= 16m^3 + 54$

So $16m^3 + 54 = 2[(2m + 3)(4m^2 - 6m + 9)]$.

QUICK ✓ *Factor each polynomial completely.*

7. $125y^3 - 64$ **8.** $-24a^3 + 3b^3$

Classroom Example ➤
Factor completely:
$-2p^5 + 162p$
Answer:
$-2p(p - 3)(p + 3)(p^2 + 9)$

EXAMPLE 5 Factoring Completely

Factor completely: $-5x^5 + 80x$

Solution

We know that the first thing to look for is a greatest common factor. Because the leading coefficient is negative we factor out a negative as part of the GCF. The GCF is $-5x$.

$$-5x^5 + 80x = -5x(x^4 - 16)$$

There are two terms in the factor in parentheses. The terms x^4 and 16 are perfect squares since $x^4 = (x^2)^2$ and $16 = 4^2$. So $x^4 - 16$ is a difference of two squares:

Work Smart
Check each factor to determine if any factor can be factored again.

Difference of two squares with $a = x^2$ and $b = 4$: $-5x(x^4 - 16) = -5x(x^2 - 4)(x^2 + 4)$.

Difference of two squares with $a = x$ and $b = 2$: $= -5x(x - 2)(x + 2)(x^2 + 4)$

Check $-5x(x - 2)(x + 2)(x^2 + 4) = -5x(x^2 - 4)(x^2 + 4)$
$= -5x(x^4 - 16)$
$= -5x^5 + 80x$

So $-5x^5 + 80x = -5x(x - 2)(x + 2)(x^2 + 4)$.

QUICK ✓ *Factor each polynomial completely.*

9. $x^4 - 1$

10. $-36x^2y + 16y$

Classroom Example ➢
Factor completely:
$9x^3 - 6x^2 - 36x + 24$
Answer: $3(3x - 2)(x + 2)(x - 2)$

EXAMPLE 6 Factoring Completely

Factor completely: $4x^3 - 6x^2 - 36x + 54$

Solution

Is there a greatest common factor among these four terms? Yes! The GCF is 2. Also notice that there are four terms, so we attempt to factor by grouping.

Factor out the GCF of 2: $4x^3 - 6x^2 - 36x + 54 = 2(2x^3 - 3x^2 - 18x + 27)$

Factor by grouping: $= 2[(2x^3 - 3x^2) + (-18x + 27)]$

Factor out common factor in each group: $= 2[x^2(2x - 3) - 9(2x - 3)]$

Factor out $2x - 3$: $= 2(2x - 3)(x^2 - 9)$

$x^2 - 9$ is the difference of two squares: $= 2(2x - 3)(x - 3)(x + 3)$

Check

$$2(2x - 3)(x - 3)(x + 3) = 2(2x - 3)(x^2 - 9)$$
$$= 2(2x^3 - 18x - 3x^2 + 27)$$
$$= 4x^3 - 6x^2 - 36x + 54$$

So $4x^3 - 6x^2 - 36x + 54 = 2(2x - 3)(x - 3)(x + 3)$.

Work Smart: Study Skills

You know how to begin factoring a polynomial—look for the GCF. But do you know when a polynomial is factored completely? That is, do you know when to stop? Which of the following is not completely factored?

(a) $(x + 5)(2x + 6)$
(b) $(z - 3)(z + 3)(z^2 + 9)$
(c) $(2n - 5)(5n + 7)$
(d) $(4y - 3yz)(2y + 7yz)$

Did you recognize that (a) and (d) are not completely factored because each has a common factor in one of the binomials?

QUICK ✓ *Factor each polynomial completely.*

11. $2x^3 + 3x^2 + 4x + 6$

12. $6x^3 + 9x^2 - 6x - 9$

Classroom Example ➢
Factor completely:
$-5xy^2 + 10xy + 45x$
Answer: $-5x(y^2 - 2y - 9)$

EXAMPLE 7 Factoring Completely

Factor completely: $-4xy^2 + 4xy + 12x$

Solution

Notice that each term has a common factor of $-4x$. Factor out the GCF.

GCF $= -4x$: $-4xy^2 + 4xy + 12x = -4x(y^2 - y - 3)$

There are three terms in the polynomial in parentheses, so we concentrate on the polynomial in parentheses, $y^2 - y - 3$. It is not a perfect square trinomial, so we need to find two factors of -3 whose sum is -1. There are no such factors. Therefore, $y^2 - y - 3$ is prime.

Check $-4x(y^2 - y - 3) = -4xy^2 + 4xy + 12x$

So $-4xy^2 + 4xy + 12x = -4x(y^2 - y - 3)$.

QUICK ✓ *Factor each polynomial completely.*

13. $-3z^2 + 9z - 21$

14. $6xy^2 + 15x^3$

4.5 Exercises

1. perfect square trinomial, perfect squares
2. grouping
3. greatest common factor
4. False 5. False 6. True
7. Answers may vary.
8. Answers may vary.
9. $(x-10)(x+10)$
10. $(x-16)(x+16)$
11. $(t+3)(t-2)$
12. $(n+3)(n+2)$
13. $(x+y)(1+2a)$
14. $(x-2a)(y+3b)$
15. $(a-2)(a^2+2a+4)$
16. $(1-y)(1+y+y^2)(1+y^3+y^6)$
17. $(a-3b)(a+2b)$
18. $(x-6y)(x+5y)$
19. $(2x-7)(x+1)$
20. $(4n-1)(3n-5)$
21. $2(x-5y)(x+2y)$
22. $(2x+3y)(x-y)$
23. $(3-a)(3+a)$
24. $(5-m)(5+m)$
25. $(u-11)(u-3)$
26. $(x+6)(x-1)$
27. $(x-a)(y-b)$
28. $(2x-1)(x^2-9)$
29. $(w+2)(w+4)$
30. $(x-3)(x-5)$
31. $(6a-7b^2)(6a+7b^2)$
32. $25(2x-y)(2x+y)$
33. $(x+4m)(x-2m)$
34. $(s-3t)(s-8t)$
35. $(3xy-2)(2xy-3)$
36. $(3st-1)(2st+1)$
37. $(x+1)(x^2+1)$
38. $(x-3)(x^2+2)$
39. $6(2z^2+2z+3)$
40. $3(y+1)^2$
41. $(7c-1)(2c+3)$
42. $3(4x+5)(2x+3)$
43. $(3m+4n^2)(9m^2-12mn^2+16n^4)$
44. $(2x^2+5y)(4x^4-10x^2y+25y^2)$
45. $2j^2(j-1)(j+1)(j^2+1)$
46. $3m(2-m^2)(2+m^2)(4+m^4)$
47. $(4a-b)(2a+5b)$
48. $(5p-3q)(2p+5q)$
49. $2a(a^2+3)$
50. $4t^2(t^2+4)$
51. $3(2z-1)(2z+1)$
52. $2n(2n-3)(2n+3)$
53. prime 54. prime
55. $2a(2a+b)(4a^2-2ab+b^2)$
56. $3q(2p+3q)(4p^2-6pq+9q^2)$
57. $(pq+7)(pq-1)$
58. $(xy+12)(xy+8)$
59. $(s+2)^2(s-2)$
60. $(x+y)(x-4)(x+4)$
61. $-2x(3x+1)(2x-1)$
62. $-x(3x-2)(x+5)$
63. $(5v+2)(2v-1)$
64. $(6h-1)(3h+5)$
65. $-n^2(n-4)(n+1)$
66. $-2x(x+2)(x-1)$
67. $-2ab(2a^2-a+1)$
68. $-4xy(x^2-x+1)$

For Extra Help:  Student Solutions Manual  CD Video  PH Math/Tutor Center  MathXL Tutorials on CD MathXL® MyMathLab

Concepts and Vocabulary

In Problems 1–3, fill in the blanks.

1. $x^2 + 2xy + y^2$ is called a ________ ________ ________ because x^2 and y^2 are ________ ________ and the middle term is twice xy.
2. When factoring a polynomial with four terms, try factoring by ________.
3. The first step in any factoring problem is to look for the ________ ________ ________.

In Problems 4–6, answer True or False to each statement.

4. The polynomial $x^4 - 16 = (x^2+4)(x^2-4)$ is factored completely.
5. There is only one technique that will factor a given polynomial correctly.
6. $a(x-y) + x - y$ can be factored by grouping.
7. Describe the method of factoring you find most difficult. Write a polynomial and factor it using this method.
8. You are grading the paper of a student who factors the polynomial $x^2 + 4x - 3x - 12$ as follows:

$$(x^2+4x)(-3x-12) = x(x+4) - 3(x-4)$$
$$= (x-3)(x+4)$$

Did the student correctly factor the polynomial? What comments would you write on the student's paper?

Mixed Practice

In Problems 9–80, factor completely. If a polynomial cannot be factored, say it is prime.

9. $x^2 - 100$
10. $x^2 - 256$
11. $t^2 + t - 6$
12. $n^2 + 5n + 6$
13. $x + y + 2ax + 2ay$
14. $xy - 2ay + 3bx - 6ab$
15. $a^3 - 8$
16. $1 - y^9$
17. $a^2 - ab - 6b^2$
18. $x^2 - xy - 30y^2$
19. $2x^2 - 5x - 7$
20. $12n^2 - 23n + 5$
21. $2x^2 - 6xy - 20y^2$
22. $2x^2 + 4xy - 6y^2$
23. $9 - a^2$
24. $25 - m^2$
25. $u^2 - 14u + 33$
26. $x^2 + 5x - 6$
27. $xy - ay - bx + ab$
28. $2x^3 - x^2 - 18x + 9$
29. $w^2 + 6w + 8$
30. $x^2 - 8x + 15$
31. $36a^2 - 49b^4$
32. $100x^2 - 25y^2$
33. $x^2 + 2xm - 8m^2$
34. $s^2 - 11st + 24t^2$
35. $6x^2y^2 - 13xy + 6$
36. $6s^2t^2 + st - 1$
37. $x^3 + x^2 + x + 1$
38. $x^3 - 3x^2 + 2x - 6$
39. $12z^2 + 12z + 18$
40. $3y^2 + 6y + 3$
41. $14c^2 + 19c - 3$
42. $24x^2 + 66x + 45$
43. $27m^3 + 64n^6$
44. $8x^6 + 125y^3$
45. $2j^6 - 2j^2$
46. $48m - 3m^9$
47. $8a^2 + 18ab - 5b^2$
48. $10p^2 - 15q^2 + 19pq$
49. $2a^3 + 6a$
50. $4t^4 + 16t^2$
51. $12z^2 - 3$
52. $8n^3 - 18n$
53. $x^2 - x + 6$
54. $n^2 + 2n + 8$
55. $16a^4 + 2ab^3$
56. $24p^3q + 81q^4$
57. $p^2q^2 + 6pq - 7$
58. $x^2y^2 + 20xy + 96$
59. $s^2(s+2) - 4(s+2)$
60. $x^2(x+y) - 16(x+y)$
61. $-12x^3 + 2x^2 + 2x$
62. $-3x^3 - 13x^2 + 10x$
63. $10v^2 - 2 - v$
64. $27h - 5 + 18h^2$
65. $4n^2 - n^4 + 3n^3$
66. $4x - 2x^2 - 2x^3$
67. $-4a^3b + 2a^2b - 2ab$
68. $-4x^3y + 4x^2y - 4xy$

69. $-p(p-4)(p+3)$
70. $-3x(x+3)(x+1)$
71. $-8x(2x-3y)(2x+3y)$
72. $-3p^2(4p-5)(4p+5)$
73. $2(n-5)(n^2-3)$
74. $3x(2x+1)(x-2)(x+2)$
75. $4(4x-3)(x+1)$
76. $6(3x+4)(2x-1)$
77. $(3x^2+2)(x^2+4)$
78. $(7r^3+2)(2r^3+7)$
79. $-xy(2x-3y)(x+y)$
80. $a(2a-b)(3a-b)$
81. $x(x-5)(x-3)$
82. $n(n+6)(n-2)$
83. $2x$, $2x+1$, $x-3$
84. $3x$, $3x+2$, $2x-1$
85. $(2m-n+p)(2m-n-p)$
86. $(b+c-a)(b+c+a)$
87. $(x+y-z)(x-y+z)$
88. $(4x-y-4)(4x+y+4)$
89. $(x-y-4)(x-y-2)$
90. $(a+b-3)(a+b+2)$
91. $(x+y-a+b)(x+y+a-b)$
92. $(x-2m)^2-(y^2+2ny+4n^2)$

69. $12p - p^3 + p^2$ **70.** $-9x - 3x^3 - 12x^2$ **71.** $-32x^3 + 72xy^2$
72. $-48p^4 + 75p^2$ **73.** $2n^3 - 10n^2 - 6n + 30$
74. $6x^4 + 3x^3 - 24x^2 - 12x$ **75.** $16x^2 + 4x - 12$ **76.** $36x^2 + 30x - 24$
77. $14x^2 + 3x^4 + 8$ **78.** $14 + 53r^3 + 14r^6$ **79.** $-2x^3y + x^2y^2 + 3xy^3$
80. $6a^3 - 5a^2b + ab^2$

Applying the Concepts

81. Profit on Newspapers The revenue for selling x newspapers is given by the expression $x^3 + 8x$. The cost to produce x newspapers is $8x^2 - 7x$. If profit is calculated as revenue minus costs, write an expression, in factored form, that calculates the profit on x newspapers.

82. Profit on T-shirts Candy and her boyfriend produce silk-screened T-shirts. The cost to produce n T-shirts is given by the expression $12n - 2n^2$ and the revenue for selling the same number of T-shirts is $n^3 + 2n^2$. Write an expression, in factored form, that calculates the profit on n T-shirts.

△ **83. Volume of a Box** The volume of a box of nails is given by the formula $V = lwh$. The volume of the box is represented by the expression $4x^3 - 10x^2 - 6x$. Factor this expression to determine algebraic expressions for the dimensions of the box.

△ **84. Volume of a Box** The volume of a shoe box is given by the formula $V = lwh$. The volume of the box is represented by the expression $18x^3 + 3x^2 - 6x$. Factor this expression to determine algebraic expressions for the dimensions of the box.

Extending the Concepts

Although the most common pattern for factoring a polynomial with four terms is to group the first two terms and group the second two terms, it is not the only possibility. Here the polynomial has been written as three terms in the first group and a single term in the second group. Study the example and then try to find a grouping that will factor each polynomial.

$$\begin{aligned} x^2 + 2xy + y^2 - z^2 &= (x^2 + 2xy + y^2) - z^2 \\ &= (x + y)^2 - z^2 \\ &= (x + y + z)(x + y - z) \end{aligned}$$

In Problems 85–88, factor completely.

85. $4m^2 - 4mn + n^2 - p^2$ **86.** $b^2 + 2bc + c^2 - a^2$
87. $x^2 - y^2 + 2yz - z^2$ **88.** $16x^2 - y^2 - 8y - 16$

In Problems 89–92, see if you can extend this concept to find a creative way to group and then factor each of the following. You may need to rearrange the terms.

89. $x^2 - 2xy + y^2 - 6x + 6y + 8$ **90.** $a^2 + 2ab + b^2 - a - b - 6$
91. $x^2 + 2xy + y^2 - a^2 + 2ab - b^2$ **92.** $x^2 - 4mx + 4m^2 - y^2 - 2ny - 4n^2$

PUTTING THE CONCEPTS TOGETHER (SECTIONS 4.1–4.5)

These problems cover important concepts from Sections 4.1 to 4.5. We designed these problems so that you can review the chapter so far and show your mastery of the concepts. Take time to work these problems before proceeding with the next section. The answers to these problems are located at the back of the text starting on page AN-8.

1. $5x^2y$
2. $(x-4)(x+1)$
3. $(x^2-3)(x^4+3x^2+9)$
4. $(2x+1)(6x+5z)$
5. $(x+6y)(x-y)$
6. $(x+4)(x^2-4x+16)$
7. prime

1. Find the GCF of $10x^3y^4z$, $15x^5y$, and $25x^2y^7z^3$.

In Problems 2–16, factor each polynomial completely. If the polynomial cannot be factored, say it is prime.

2. $x^2 - 3x - 4$ **3.** $x^6 - 27$
4. $6x(2x + 1) + 5z(2x + 1)$ **5.** $x^2 + 5xy - 6y^2$
6. $x^3 + 64$ **7.** $4x^2 + 49y^2$

8. $3(x + 6y)(x - 2y)$
9. $4z^2(3z^3 - 11z - 6)$
10. prime
11. $m^2(m + 2)(4m - 3)$
12. $(5p - 2)(p - 3)$
13. $(2m + 5)(5m - 3)$
14. $6(3m - 1)(2m + 1)$
15. $(2m - 5)^2$
16. $(5x + 4y)(x - y)$
17. $S = 2\pi r(h + r)$
18. $h = 16t(3 - t)$

8. $3x^2 + 12xy - 36y^2$

9. $12z^5 - 44z^3 - 24z^2$

10. $x^2 + 6x - 5$

11. $4m^4 + 5m^3 - 6m^2$

12. $5p^2 - 17p + 6$

13. $10m^2 + 25m - 6m - 15$

14. $36m^2 + 6m - 6$

15. $4m^2 - 20m + 25$

16. $5x^2 - xy - 4y^2$

17. Surface Area The surface area of a right circular cylinder is given by the formula $S = 2\pi rh + 2\pi r^2$. Factor the right side of this formula.

18. Rocket Height A toy rocket is launched upward from the ground with an initial velocity of 48 feet per second. Its height, h, in feet, after t seconds, is given by the equation $h = 48t - 16t^2$. Factor the right side of this equation.

4.6 Solving Polynomial Equations by Factoring

OBJECTIVES

1. Solve Quadratic Equations Using the Zero-Product Property
2. Solve Polynomial Equations of Degree Three or Higher Using the Zero-Product Property

Preparing for Solving Polynomial Equations by Factoring

Before getting started, take the following readiness quiz. If you get a problem wrong, go back to the cited section and review the material.

1. Solve: $x + 5 = 0$ [Section 2.1, pp. 77–79]

2. Solve: $2(x - 4) - 10 = 0$ [Section 2.2, pp. 86–87]

3. Evaluate $2x^2 + 3x - 4$ when (a) $x = 2$ (b) $x = -1$. [Section 1.7, pp. 57–58]

In Sections 4.1–4.5, we learned how to factor polynomial expressions. A question that you may have been asking yourself is, "Why do I care about factoring? What good is it?" It turns out that there are many uses of factoring, but one use that we can present now is that factoring is essential for solving polynomial equations.

Work Smart
Remember, the degree of a polynomial is the value of the **largest exponent** on the variable. For example, the degree of $4x^3 - 9x^2 + 1$ is 3.

DEFINITIONS

A **polynomial equation** is any equation that contains only polynomial expressions. The **degree of a polynomial equation** is the degree of the polynomial expression in the equation.

Teaching Tip
Emphasize the difference between a linear equation (first degree) and a quadratic equation (second degree). Tell students that we use the skills acquired in solving linear equations to solve quadratic equations.

Some examples of polynomial equations are

$4x + 5 = 17$	$2x^2 - 5x - 3 = 0$	$y^3 + 4y^2 = 3y + 18$
Polynomial equation of degree 1	Polynomial equation of degree 2	Polynomial equation of degree 3

We learned how to solve polynomial equations of degree 1 back in Chapter 2. It seems logical that the next step is to learn how to solve polynomial equations of degree 2.

1 Solve Quadratic Equations Using the Zero-Product Property

Teaching Tip
Try this exercise before introducing the Zero-Product Property:

$\_\cdot\_ = 6$
$\_\cdot\_ = -12$
$\_\cdot\_ = 0$

If a polynomial equation can be solved using factoring, we make use of the following property.

THE ZERO-PRODUCT PROPERTY

If the product of two factors is zero, then at least one of the factors is 0. That is,

$$\text{if } ab = 0, \text{ then } a = 0 \text{ or } b = 0 \text{ or both } a \text{ and } b \text{ are } 0.$$

For example, if $2x = 0$, then either $2 = 0$ or $x = 0$. Since $2 \neq 0$, it must be that $x = 0$.

Preparing for...Answers **1.** $\{-5\}$ **2.** $\{9\}$ **3.** **(a)** 10 **(b)** -5

Classroom Example
Solve: $(x-6)(4x+3)=0$

Answer: $\left\{-\frac{3}{4}, 6\right\}$

EXAMPLE 1 Using the Zero-Product Property

Solve: $(x+4)(2x-5)=0$

Solution

We have the product of two factors, $x+4$ and $2x-5$, set equal to 0. According to the Zero-Product Property, the value of at least one of the factors must equal 0. Therefore, we set each of the factors equal to 0 and solve each equation separately.

$$x+4=0 \quad \text{or} \quad 2x-5=0$$

Subtract 4 from both sides: $x=-4$

Add 5 to both sides: $2x=5$

Divide both sides by 2: $\frac{2x}{2}=\frac{5}{2}$

$$x=\frac{5}{2}$$

Check

$x=-4$

$$(x+4)(2x-5)=0$$
$$(-4+4)(2(-4)-5)\stackrel{?}{=}0$$
$$0(-13)\stackrel{?}{=}0$$
$$0=0 \quad \text{True}$$

$x=\frac{5}{2}$

$$(x+4)(2x-5)=0$$
$$\left(\frac{5}{2}+4\right)\left(2\cdot\left(\frac{5}{2}\right)-5\right)\stackrel{?}{=}0$$
$$\left(\frac{13}{2}\right)(5-5)\stackrel{?}{=}0$$
$$\frac{13}{2}\cdot 0\stackrel{?}{=}0$$
$$0=0 \quad \text{True}$$

The solution set is $\left\{-4, \frac{5}{2}\right\}$.

QUICK ✓ *Use the Zero-Product Property to solve the equation.*

1. $x(x+3)=0$ **2.** $(x-2)(4x+5)=0$

The Zero-Product Property comes in handy when we need to solve quadratic equations.

Work Smart
In a quadratic equation, why can't a equal 0? Because if a were equal to zero, the equation would be a linear equation.

DEFINITION

A **quadratic equation** is an equation equivalent to one of the form

$$ax^2+bx+c=0$$

where a, b, and c are real numbers and $a \neq 0$.

Teaching Tip
Emphasize that the goal is to write a quadratic equation in standard form before it can be solved by factoring.

The following are examples of quadratic equations.

$$2x^2+5x-3=0 \qquad -6z^2+12z=0 \qquad y^2-25=0 \qquad p^2+12p=-36$$

Notice that the equation $y^2-25=0$ is a quadratic equation even though it is missing the "y" term. The equation $-6z^2+12z=0$ is a quadratic equation even though it is missing a constant term.

A quadratic equation is a specific type of a polynomial equation. The term "quadratic" means "of, or relating to, a square." There are many real-world situations that are modeled by these second-degree equations, such as problems involving revenue for selling x units of a good. Additionally, quadratic equations can be used to describe the height of a projectile over time.

Sometimes a quadratic equation is called a **second-degree equation** because the highest power of the variable in a quadratic equation is two.

In Words
A quadratic equation is in standard form if it is written in descending order of exponents, and is set equal to zero.

DEFINITION

A quadratic equation is said to be in **standard form** if it is written in the form $ax^2 + bx + c = 0$.

For example, the equation $2x^2 + x - 5 = 0$ is in standard form, while the equation $p^2 + 12p = -36$ is not in standard form. To place $p^2 + 12p = -36$ in standard form, we add 36 to both sides of the equation to obtain $p^2 + 12p + 36 = 0$.

When a quadratic equation is written in standard form, $ax^2 + bx + c = 0$, it may be possible to factor the expression $ax^2 + bx + c$ as the product of two first-degree polynomials. If it is possible to factor the trinomial, we can then use the Zero-Product Property to solve the quadratic equation.

Classroom Example
Solve: $x^2 + 3x + 2 = 0$
Answer: $\{-2, -1\}$

EXAMPLE 2 How to Solve a Quadratic Equation by Factoring

Solve: $x^2 - 4x - 21 = 0$

Step-by-Step Solution

Step 1: Is the equation in standard form? Yes, it is written in the form $ax^2 + bx + c = 0$.

$$x^2 - 4x - 21 = 0$$

Step 2: Factor the expression on the left side of the equation.

Two integers whose product is -21 and whose sum is -4 are -7 and 3.

$$(x + 3)(x - 7) = 0$$

Step 3: Set each factor to 0.

$$x + 3 = 0 \quad \text{or} \quad x - 7 = 0$$

Step 4: Solve each first-degree equation.

$$x = -3 \quad \text{or} \quad x = 7$$

Step 5: Check

$x = -3$:
$$x^2 - 4x - 21 = 0$$
$$(-3)^2 - 4(-3) - 21 \stackrel{?}{=} 0$$
$$9 + 12 - 21 \stackrel{?}{=} 0$$
$$0 = 0 \quad \text{True}$$

$x = 7$:
$$x^2 - 4x - 21 = 0$$
$$7^2 - 4(7) - 21 \stackrel{?}{=} 0$$
$$49 - 28 - 21 \stackrel{?}{=} 0$$
$$0 = 0 \quad \text{True}$$

The solution set is $\{-3, 7\}$.

Look at Step 3 in the solution to Example 2. To solve a quadratic equation, we wish to put the equation into a form that we already know how to solve. That is, we "transform" the quadratic equation $x^2 - 4x - 21 = 0$ into two linear equations, $x + 3 = 0$ or $x - 7 = 0$. We will present methods for solving $ax^2 + bx + c = 0$ when we cannot factor the expression $ax^2 + bx + c$ in Chapter 7.

Steps to Solve a Quadratic Equation by Factoring

Step 1: Write the quadratic equation in standard form, $ax^2 + bx + c = 0$.

Step 2: Factor the expression on the left side of the equation.

Step 3: Set each factor found in Step 2 equal to zero using the Zero-Product Property.

Step 4: Solve each first-degree equation for the variable.

Step 5: Be sure to check your answers by substituting into the *original* equation.

Classroom Example
Solve: $3t^2 - 14t = 5$

Answer: $\left\{-\frac{1}{3}, 5\right\}$

EXAMPLE 3 Solving a Quadratic Equation Not in Standard Form

Solve: $3x^2 + 5x = 14x$

Solution

We first write the equation in standard form, $ax^2 + bx + c = 0$.

$$3x^2 + 5x = 14x$$

Subtract $14x$ from both sides of the equation: $3x^2 - 9x = 0$

Factor: $3x(x - 3) = 0$

Set each factor to 0: $3x = 0$ or $x - 3 = 0$

Solve each first-degree equation: $x = 0$ or $x = 3$

Check Substitute $x = 0$ and $x = 3$ into the original equation.

$$3x^2 + 5x = 14x$$

$x = 0$: $3(0)^2 + 5(0) \stackrel{?}{=} 14(0)$

$0 = 0$ True

$$3x^2 + 5x = 14x$$

$x = 3$: $3(3)^2 + 5(3) \stackrel{?}{=} 14(3)$

$27 + 15 \stackrel{?}{=} 42$

$42 = 42$ True

The solution set is $\{0, 3\}$.

QUICK ✓ *Solve each quadratic equation by factoring.*

3. $p^2 - 6p + 8 = 0$
4. $2t^2 - 5t = 3$
5. $2x^2 + 3x = 5$
6. $z^2 + 20 = -9z$

Classroom Example
Solve: $2m^2 - 3m - 7 = 2m + 5$

Answer: $\left\{-\frac{3}{2}, 4\right\}$

EXAMPLE 4 Solving a Quadratic Equation Not in Standard Form

Solve: $3m^2 - 3m = 2 - 2m$

Solution

Once again, we first write the quadratic equation in standard form, $ax^2 + bx + c = 0$.

$$3m^2 - 3m = 2 - 2m$$

Add $2m$ to both sides and subtract 2 from both sides to obtain standard form: $3m^2 - 3m + 2m - 2 = 2 - 2 - 2m + 2m$

$$3m^2 - m - 2 = 0$$

Factor $3m^2 - m - 2$: $(3m + 2)(m - 1) = 0$

Set each factor to 0: $3m + 2 = 0$ or $m - 1 = 0$

Solve each first-degree equation: $3m = -2$ or $m = 1$

$$m = -\frac{2}{3}$$

Check Check the two solutions by substituting $m = -\frac{2}{3}$ and $m = 1$ in the original equation. We leave this to you.

The solution set is $\left\{-\frac{2}{3}, 1\right\}$.

QUICK ✓ *Solve each quadratic equation by factoring.*

7. $2k^2 - k - 4 = 1 + 2k$

8. $3x^2 + 9x = 4 - 2x$

Classroom Example
Solve: $(x - 1)(2x + 3) = 6x$

Answer: $\left\{-\frac{1}{2}, 3\right\}$

EXAMPLE 5 Solving a Quadratic Equation Not in Standard Form

Solve: $(2x + 5)(x - 3) = 6x$

Solution

$$(2x + 5)(x - 3) = 6x$$

Multiply using FOIL: $2x^2 - x - 15 = 6x$

Write in standard form: $2x^2 - 7x - 15 = 0$

Factor the polynomial: $(2x + 3)(x - 5) = 0$

Set each factor to 0: $2x + 3 = 0$ or $x - 5 = 0$

Solve each first-degree equation: $2x = -3$ or $x = 5$

$$x = -\frac{3}{2}$$

Work Smart
Do not attempt to solve $(2x + 5)(x - 3) = 6x$ by setting each factor to $6x$ as in $2x + 5 = 6x$ and $x - 3 = 6x$. The Zero-Product Property can only be applied when the product equals zero.

Check We leave it to you to substitute $x = -\frac{3}{2}$ and $x = 5$ into the original equation to verify the answer.

The solution set is $\left\{-\frac{3}{2}, 5\right\}$.

QUICK ✓ *Solve each quadratic equation by factoring.*

9. $(x - 3)(x + 5) = 9$

10. $(x + 3)(2x - 1) = 7x - 3x^2$

Classroom Example
Solve: $9a^2 + 25 = 30a$

Answer: $\left\{\frac{5}{3}\right\}$

EXAMPLE 6 Solving a Quadratic Equation Not in Standard Form

Solve: $4k^2 + 9 = -12k$

Solution

$$4k^2 + 9 = -12k$$

Write in standard form: $4k^2 + 12k + 9 = 0$

Factor the trinomial $a^2 + 2ab + b^2 = (a + b)^2$: $(2k + 3)^2 = 0$

Set each factor to 0: $2k + 3 = 0$ or $2k + 3 = 0$

Solve each first-degree equation: $2k = -3$ $\quad$ $2k = -3$

$$k = -\frac{3}{2} \qquad k = -\frac{3}{2}$$

Check Substitute $k = -\frac{3}{2}$ into the original equation to confirm that the solution is correct.

The solution set is $\left\{-\frac{3}{2}\right\}$.

Notice in Example 6 that the solution $k = -\frac{3}{2}$ occurred twice. When this occurs, the solution is called a **double root.**

QUICK ✓ *Solve the quadratic equation by factoring.*

11. $9p^2 + 16 = 24p$

Classroom Example ➔
Solve: $-2x^2 + 10x + 48 = 0$
Answer: $\{-3, 8\}$

EXAMPLE 7 Solving a Quadratic Equation Containing a GCF

Solve: $-3x^2 + 6x + 72 = 0$

Solution

$$-3x^2 + 6x + 72 = 0$$

Factor the polynomial, GCF = −3: $$-3(x^2 - 2x - 24) = 0$$

$$-3(x - 6)(x + 4) = 0$$

Set each factor to 0: $$-3 = 0 \quad \text{or} \quad x - 6 = 0 \quad \text{or} \quad x + 4 = 0$$

Solve each first-degree equation: $$x = 6 \qquad\qquad x = -4$$

The statement $-3 = 0$ is false. So the solutions are 6 and -4.

Check Substitute $x = 6$ and $x = -4$ into the original equation to check.

The solution set is $\{-4, 6\}$.

Another technique that we could have used in Example 7 is to use the Multiplication Property of Equality and first divide each term of the equation by -3. Then we would solve the quadratic equation $x^2 - 2x - 24 = 0$ and also obtain the solutions $x = -4$ or $x = 6$.

Classroom Example ➔
Use Example 9 with a building 180 feet tall with an initial velocity of 32 feet per second to find when the height of the ball will be 192 feet from the ground. The equation is $-16t^2 + 32t + 180 = 192$.
Answer: After 0.5 seconds and 1.5 seconds the ball will be 192 feet from the ground.

QUICK ✓ *Solve the quadratic equation by factoring.*

12. $4x^2 + 12x - 72 = 0$ **13.** $-2x^2 + 2x = -12$

EXAMPLE 8 Throwing a Ball from the Top of a Building

A ball is thrown vertically upward from the top of a building 96 feet tall with an initial velocity of 80 feet per second. Solve the equation $-16t^2 + 80t + 96 = 192$ to find the time t (in seconds) at which the ball is 192 feet from the ground. See Figure 1.

Figure 1

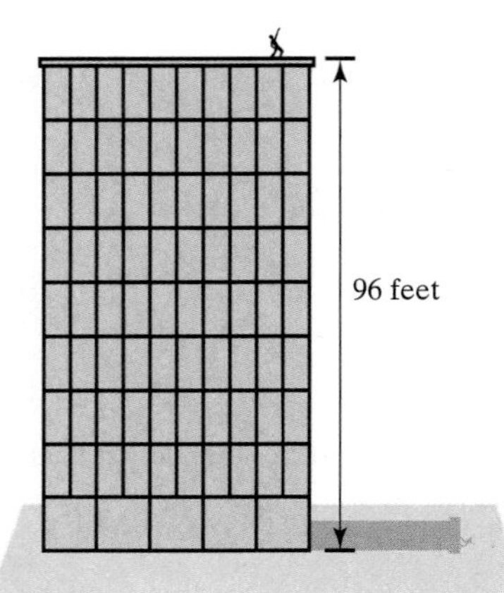

Solution

$$-16t^2 + 80t + 96 = 192$$

Write the quadratic equation in standard form: $$-16t^2 + 80t + 96 - 192 = 192 - 192$$

$$-16t^2 + 80t - 96 = 0$$

Factor the polynomial: $$-16(t^2 - 5t + 6) = 0$$

$$-16(t - 3)(t - 2) = 0$$

Set each factor to 0: $$-16 = 0 \quad \text{or} \quad t - 3 = 0 \quad \text{or} \quad t - 2 = 0$$

Solve each first-degree equation: $$t = 3 \quad \text{or} \quad t = 2$$

The statement $-16 = 0$ is false. The solutions are $t = 3$ or $t = 2$. After both 2 seconds and 3 seconds the ball will be 192 feet from the ground. Do you see why?

QUICK ✓

14. A toy rocket is shot directly up from the ground with an initial velocity of 80 feet per second. Solve the equation $-16t^2 + 80t = 64$ to find the time (in seconds) at which the toy rocket is 64 feet from the ground.

Classroom Example
Solve: $4n^3 - n^2 - 5n = 0$

Answer: $\left\{0, -1, \frac{5}{4}\right\}$

2 Solve Polynomial Equations of Degree Three or Higher Using the Zero-Product Property

The Zero-Product Property is also used to solve higher-degree polynomial equations.

EXAMPLE 9 How to Solve a Polynomial Equation of Degree Three or Higher

Solve: $18x^3 + 3x^2 - 6x = 0$

Step-by-Step Solution

Step 1: Write the equation in standard form.

$$18x^3 + 3x^2 - 6x = 0$$

Step 2: Factor the expression on the left side of the equation. Begin with the GCF, 3x.

$$3x(6x^2 + x - 2) = 0$$
$$3x(3x + 2)(2x - 1) = 0$$

Step 3: Set each factor to 0.

$3x = 0$ or $3x + 2 = 0$ or $2x - 1 = 0$

Step 4: Solve each first-degree equation.

$x = 0$; $3x = -2$, $x = -\frac{2}{3}$; $2x = 1$, $x = \frac{1}{2}$

Step 5: Check each solution by substituting $x = 0$, $x = -\frac{2}{3}$, and $x = \frac{1}{2}$ into the original equation.

We leave the check to you.

The solution set is $\left\{0, -\frac{2}{3}, \frac{1}{2}\right\}$.

Work Smart
Do not divide both sides of the equation by the factor $3x$. We can divide both sides by a constant, but not by a variable because the variable may have a value of zero.

QUICK ✓ *Solve each polynomial equation and state the solutions.*

15. $(4x - 5)(x^2 - 9) = 0$ **16.** $3x^3 + 9x^2 + 6x = 0$

4.6 Exercises

For Extra Help: Student Solutions Manual | CD Video | PH Math/Tutor Center | MathXL Tutorials on CD | MathXL® | MyMathLab

Concepts and Vocabulary

In Problems 1–3, fill in the blanks.

1. A __________ equation is an equation that can be written in the form $ax^2 + bx + c = 0$, where a, b, and c are real numbers and $a \neq 0$.
2. The Zero-Product Property states that if $ab = 0$, then either ________ or ________.
3. Quadratic equations are also known as __________-degree equations.

1. quadratic
2. $a = 0$, $b = 0$
3. second

4. False 5. False 6. True
7. Answers may vary.
8. Answers may vary.
9. linear 10. linear 11. quadratic
12. quadratic
13. $\{-4, 0\}$ 14. $\{-9, 0\}$
15. $\{-3, 9\}$ 16. $\{-8, 4\}$
17. $\left\{-\frac{1}{3}, 5\right\}$ 18. $\left\{-4, \frac{3}{4}\right\}$
19. $\left\{-\frac{3}{5}, \frac{2}{7}\right\}$ 20. $\left\{\frac{5}{6}, \frac{3}{8}\right\}$
21. $\{-1, 4\}$ 22. $\{-9, 7\}$
23. $\{-7, -2\}$ 24. $\{-3, 8\}$
25. $\left\{-\frac{1}{2}, 0\right\}$ 26. $\left\{0, \frac{2}{7}\right\}$
27. $\left\{-\frac{1}{2}, 2\right\}$ 28. $\left\{-\frac{7}{3}, 2\right\}$
29. $\{3\}$ 30. $\{-6\}$
31. $\{0, 6\}$
32. $\left\{0, \frac{5}{2}\right\}$
33. $\{-2, 3\}$ 34. $\{-2, 8\}$
35. $\{2, 9\}$ 36. $\{-3, 10\}$
37. $\left\{\frac{1}{4}, 1\right\}$ 38. $\left\{-\frac{3}{2}, \frac{1}{2}\right\}$
39. $\{-4, 6\}$
40. $\{-2, 1\}$
41. $\{-5, 10\}$
42. $\{-6, 4\}$
43. $\{-5, 1\}$
44. $\left\{\frac{5}{2}, -\frac{3}{2}\right\}$
45. $\{-3, 0, 2\}$
46. $\left\{-\frac{7}{3}, 0, 2\right\}$
47. $\{-3, -2, 2\}$
48. $\{-3, -2, 3\}$
49. $\left\{-\frac{3}{2}, 2, -2\right\}$
50. $\left\{-1, 1, \frac{3}{2}\right\}$
51. $\left\{-\frac{3}{5}, 4\right\}$
52. $\left\{-5, \frac{2}{3}\right\}$
53. $\{-4, 5\}$ 54. $\{5, 8\}$
55. $\{-5\}$ 56. $\{-3\}$
57. $\left\{-\frac{3}{4}, 7\right\}$ 58. $\left\{-\frac{2}{5}, 4\right\}$
59. $\{-4, 2\}$ 60. $\{-11, 1\}$
61. $\left\{-4, -\frac{1}{2}, 4\right\}$
62. $\left\{-2, \frac{1}{2}, 2\right\}$
63. $\{20\}$ 64. $\{7\}$ 65. $\{-10, 10\}$
66. $\{-7, 7\}$
67. $\left\{-3, -\frac{2}{3}, 5\right\}$ 68. $\left\{-5, \frac{3}{2}, 6\right\}$
69. $\left\{-\frac{1}{2}, \frac{3}{2}, 5\right\}$ 70. $\left\{-\frac{2}{3}, 1, \frac{11}{7}\right\}$
71. $\{0, 8\}$ 72. $\left\{-\frac{4}{3}, -1\right\}$
73. $\left\{-4, \frac{3}{2}\right\}$ 74. $\left\{-8, \frac{3}{4}\right\}$
75. $\left\{-6, \frac{1}{2}\right\}$ 76. $\left\{-\frac{2}{3}, 1\right\}$

In Problems 4–6, answer True or False to each statement.

4. $x(x - 3) = 4$ means that $x = 4$ or $x - 3 = 4$.

5. $3x - 6 + x^2 = 0$ is written in standard form.

6. $x^3 - 4x^2 - 12x = 0$ can be solved using the Zero-Product Property.

7. When solving polynomial equations, we always begin by writing the equation in standard form. Explain why this important.

8. Explain the difference between a quadratic polynomial and a quadratic equation.

Building Skills

In Problems 9–12, identify each as a linear equation or a quadratic equation.

9. $3(x + 4) - 1 = 5x + 2$ **10.** $2x + 1 - (x + 7) = 3x + 1$

11. $x^2 - 2x = 8$ **12.** $(x + 2)(x - 2) = 14$

In Problems 13–20, solve each equation using the Zero-Product Property.

13. $2x(x + 4) = 0$ **14.** $3x(x + 9) = 0$

15. $(n + 3)(n - 9) = 0$ **16.** $(a + 8)(a - 4) = 0$

17. $(3p + 1)(p - 5) = 0$ **18.** $(4z - 3)(z + 4) = 0$

19. $(5y + 3)(2 - 7y) = 0$ **20.** $(6m - 5)(3 - 8m) = 0$

In Problems 21–44, solve each quadratic equation by factoring.

21. $x^2 - 3x - 4 = 0$ **22.** $x^2 + 2x - 63 = 0$ **23.** $n^2 + 9n + 14 = 0$

24. $p^2 - 5p - 24 = 0$ **25.** $4x^2 + 2x = 0$ **26.** $14x - 49x^2 = 0$

27. $2x^2 - 3x - 2 = 0$ **28.** $3x^2 + x - 14 = 0$ **29.** $a^2 - 6a + 9 = 0$

30. $k^2 + 12k + 36 = 0$ **31.** $6x^2 = 36x$ **32.** $2x^2 = 5x$

33. $n^2 - n = 6$ **34.** $a^2 - 6a = 16$ **35.** $b^2 + 18 = 11b$

36. $m^2 - 30 = 7m$ **37.** $1 - 5m = -4m^2$ **38.** $4p - 3 = -4p^2$

39. $n(n - 2) = 24$ **40.** $p(p + 1) = 2$ **41.** $(x - 2)(x - 3) = 56$

42. $(x + 5)(x - 3) = 9$ **43.** $(c + 2)^2 = 9$ **44.** $(2a - 1)^2 = 16$

In Problems 45–50, solve each polynomial equation by factoring.

45. $2x^3 + 2x^2 - 12x = 0$ **46.** $3x^3 + x^2 - 14x = 0$

47. $y^3 + 3y^2 - 4y - 12 = 0$ **48.** $m^3 + 2m^2 - 9m - 18 = 0$

49. $2x^3 + 3x^2 = 8x + 12$ **50.** $-2x + 3 = 3x^2 - 2x^3$

Mixed Practice

In Problems 51–76, solve each equation. Be careful; the problems represent a mix of linear, quadratic, and third-degree polynomial equations.

51. $(5x + 3)(x - 4) = 0$ **52.** $(3x - 2)(x + 5) = 0$ **53.** $p^2 - p - 20 = 0$

54. $z^2 - 13z + 40 = 0$ **55.** $4w + 3 = 2w - 7$ **56.** $7y + 3 = 2y - 12$

57. $4a^2 - 25a = 21$ **58.** $5m^2 = 18m + 8$ **59.** $2a(a + 1) = a^2 + 8$

60. $2y(y + 5) = y^2 + 11$ **61.** $2x^3 + x^2 = 32x + 16$ **62.** $2n^3 + 4 = n^2 + 8n$

63. $4(b - 3) - 3b = 8$ **64.** $2(p - 3) = p + 1$ **65.** $y^2 + 5y = 5(y + 20)$

66. $3z^2 + 7z = 7(z + 21)$ **67.** $(a + 3)(a - 5)(3a + 2) = 0$

68. $(w - 6)(w + 5)(2w - 3) = 0$ **69.** $(2k - 3)(2k^2 - 9k - 5) = 0$

70. $(7m - 11)(3m^2 - m - 2) = 0$ **71.** $(w - 3)^2 = 9 + 2w$

72. $3(k + 2)^2 = 5k + 8$ **73.** $\frac{1}{2}x^2 + \frac{5}{4}x = 3$ **74.** $z^2 + \frac{29}{4}z = 6$

75. $8x^2 + 44x = 24$ **76.** $9q^2 = 3q + 6$

77. 1 sec, 3 sec
78. 1 sec, 3 sec
79. 1 sec
80. 3 sec
81. -4 and -3 or 3 and 4
82. -11 and -13 or 11 and 13
83. -12, -10, and -8, or 8, 10, and 12
84. 7, 9, and 11 or -13, -11, and -9
85. 15 by 17
86. 20 feet by 22 feet
87. 8 teams
88. 9 teams
89. $x(x - 3)(x + 5) = 0$
90. $n(n - 8)(n + 2) = 0$
91. $(z - 6)(z - 6) = 0$ or $(z - 6)^2 = 0$
92. $(x + 2)(x + 2) = 0$ or $(x + 2)^2 = 0$
93. $(a - 4)(2a + 1)(3a - 2) = 0$
94. $(a - 3)(3a - 1)(5a - 4) = 0$

Applying the Concepts

77. Tossing a Ball A ball is thrown vertically upward from the top of a building 80 feet tall with an initial velocity of 64 feet per second. Solve the equation $-16t^2 + 64t + 80 = 128$ to find the time t (in seconds) at which the ball is 128 feet from the ground.

78. Tossing a Ball A ball is thrown vertically upward from the ground with an initial velocity of 64 feet per second. Solve the equation $-16t^2 + 64t = 48$ to find the time t (in seconds) at which the ball is 48 feet from the ground.

79. Water Balloon A water balloon is dropped from a height of 40 feet. Solve the equation $-16t^2 + 40 = 24$ to find the time t (in seconds) at which the water balloon is 24 feet from the ground.

80. Flying Money A stunt-man dropped a bag containing fake money from a hot-air balloon at a height of 240 feet. Solve the equation $-16t^2 + 240 = 96$ to find the time t (in seconds) at which the bag was 96 feet from the ground.

81. Consecutive Integers The product of two consecutive integers is 12. Find the integers.

82. Consecutive Odd Integers The product of two consecutive odd integers is 143. Find the integers.

83. Consecutive Even Integers Find three consecutive even integers such that the product of the first and the third is 96.

84. Consecutive Odd Integers Find three consecutive odd integers such that the product of the second and the third is 99.

△ **85. Rectangle** The length and width of two sides of a rectangle are consecutive odd integers. The area of the rectangle is 255. Find the dimensions of the rectangle.

△ **86. Garden Area** The State University Landscape Club wants to establish a horticultural garden near the administration building. The length and width of the space that is available are consecutive even integers, and the area of the garden is 440 square feet. Find the dimensions of the garden.

The equation $N = \dfrac{t^2 - t}{2}$ *models the number of soccer games that must be scheduled in a league with t teams, when each team plays every other team exactly once. Use this equation to solve Problems 87 and 88.*

87. If a league has 28 games scheduled, how many teams are in the league?

88. If a league has 36 games scheduled, how many teams are in the league?

Extending the Concepts

89. Write a polynomial equation with integer coefficients that has $x = 0$, $x = 3$, and $x = -5$ as solutions.

90. Write a polynomial equation with integer coefficients that has $n = 0$, $n = 8$, and $n = -2$ as solutions.

91. Write a polynomial equation with integer coefficients that has $z = 6$ as a double root.

92. Write a polynomial equation with integer coefficients that has $x = -2$ as a double root.

93. Write a polynomial equation with integer coefficients that has $a = 4$, $a = -\dfrac{1}{2}$, and $a = \dfrac{2}{3}$ as solutions.

94. Write a polynomial equation with integer coefficients that has $a = 3$, $a = \dfrac{1}{3}$, and $a = \dfrac{4}{5}$ as solutions.

95. Student divided by a variable expression; $\left\{0, \frac{1}{3}\right\}$
96. Student should first write in standard form; $\{-2, 3\}$
97. $\{a, -b\}$
98. $\{-3a, 2a\}$
99. $\{0, 2a\}$
100. $\left\{\frac{3a}{2}, -\frac{5b}{2}\right\}$

95. A student solved a quadratic equation using the following procedure. Explain the student's error and then work the problem correctly.

$$15x^2 = 5x$$
$$\frac{15x^2}{x} = \frac{5x}{x}$$
$$15x = 5$$
$$x = \frac{5}{15} = \frac{1}{3}$$

96. A student solved a quadratic equation using the following procedure. Explain the student's error and then work the problem correctly.

$$(x-4)(x+3) = -6$$
$$x - 4 = -6 \quad \text{or} \quad x + 3 = -6$$
$$x = -2 \qquad\qquad x = -9$$

In Problems 97–100, solve for x in each equation.

97. $x^2 - ax + bx - ab = 0$

98. $x^2 + ax - 6a^2 = 0$

99. $2x^3 - 4ax^2 = 0$

100. $4x^2 - 6ax + 10bx - 15ab = 0$

4.7 Modeling and Solving Problems with Quadratic Equations

OBJECTIVES

1 Model and Solve Problems Involving Quadratic Equations

2 Model and Solve Problems Using the Pythagorean Theorem

Preparing for Modeling and Solving Problems with Quadratic Equations

Before getting started, take the following readiness quiz. If you get a problem wrong, go back to the section cited and review the material.

1. Evaluate: 15^2. [Section 1.6, pp. 49–50]

2. Solve: $x^2 - 5x - 14 = 0$ [Section 4.6, pp. 281–286]

Preparing for...Answers **1.** 225 **2.** $\{-2, 7\}$

The solutions to many applied problems require solving polynomial equations by factoring. For example, the height of a projectile over time, the area of a triangular sail, and the dimensions of a big-screen TV may be found by solving a polynomial equation.

1 Model and Solve Problems Involving Quadratic Equations

Let's begin by solving a quadratic equation to determine the time at which a projectile is at a certain height.

Classroom Example Use Example 1 with a cliff having a height of 180 feet.

(a) When will the height of the ball be 180 feet above sea level?

(b) When will the ball strike the water?

Answer:

(a) The ball will be at a height of 180 feet after 0 seconds and 2 seconds.

(b) The ball will strike the water after 4.5 seconds.

EXAMPLE 1 Projectile Motion

A ball is thrown off a cliff from a height of 240 feet above sea level. The height h of the ball above the water (in feet) at any time t (in seconds) can be modeled by the equation

$$h = -16t^2 + 32t + 240$$

See Figure 2.

Figure 2

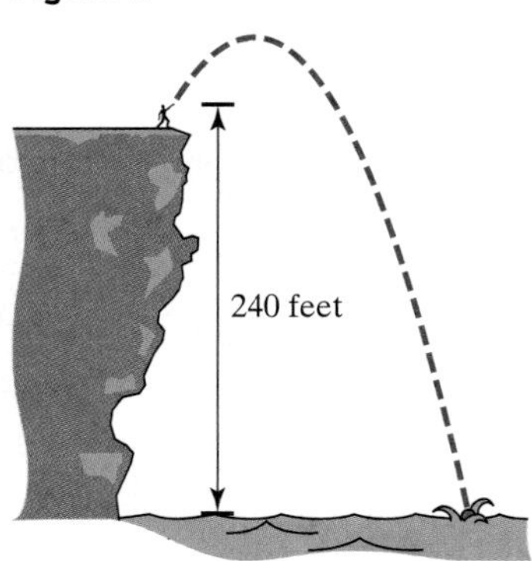

(a) When will the height of the ball be 240 feet above sea level?
(b) When will the ball strike the water?

Solution

(a) To determine when the height of the ball will be 240 feet above sea level, we let $h = 240$ and solve the resulting equation.

$$-16t^2 + 32t + 240 = 240$$

Subtract 240 from both sides: $$-16t^2 + 32t = 0$$

Factor out $-16t$: $$-16t(t - 2) = 0$$

Set each factor to 0: $$-16t = 0 \quad \text{or} \quad t - 2 = 0$$

Solve each equation: $$t = 0 \quad \text{or} \quad t = 2$$

The ball will be at a height of 240 feet the instant the ball leaves the child's hand and after 2 seconds of flight.

(b) The ball will strike the water at the instant its height is 0. So we let $h = 0$ and solve the resulting equation.

$$-16t^2 + 32t + 240 = 0$$

Factor out -16: $$-16(t^2 - 2t - 15) = 0$$

Factor: $$-16(t - 5)(t + 3) = 0$$

Set each factor to 0: $$-16 = 0 \quad \text{or} \quad t - 5 = 0 \quad \text{or} \quad t + 3 = 0$$

Solve each equation: $$t = 5 \quad \text{or} \quad t = -3$$

The equation $-16 = 0$ is false, and since t represents time, we discard the solution $t = -3$. Therefore, the ball will strike the water after 5 seconds.

QUICK ✓

1. A model rocket is fired straight up from the ground. The height h of the rocket (in feet) at any time t (in seconds) can be modeled by the equation $h = -16t^2 + 160t$.
   (a) When will the height of the rocket be 384 feet from the ground?
   (b) When will the rocket strike the ground?

In the next two examples, we will employ the problem-solving strategy first presented in Section 2.5.

Classroom Example ➤
The width of a rectangular room is 1 foot less than twice the length. Find the dimensions of the room if the area is 45 square feet.

Answer: 5 feet by 9 feet

EXAMPLE 2 Geometry: Area of a Rectangle

A carpet installer finds that the length of a rectangular hallway is 3 feet more than twice the width. If the area of the hallway is 44 square feet, what are the dimensions of the hallway? See Figure 3.

Figure 3

Solution

Step 1: Identify This is a geometry problem involving the area of a rectangle.

Step 2: Name Because we know less about the width of the hallway, we let w represent the width. The length of the hallway is 3 feet more than twice the width, so we let $2w + 3$ represent the length.

Step 3: Translate We are given that the area of the hallway is 44 square feet. We know that the area of a rectangle = (length)(width), so we have

$$\text{area} = (\text{length})(\text{width})$$

$$44 = (2w + 3)(w) \quad \text{The Model}$$

Step 4: Solve We now proceed to solve the equation. Do you recognize this equation as being a quadratic equation? The first step is to put the equation in standard form, $ax^2 + bx + c = 0$.

$$w(2w + 3) = 44$$

Distribute: $$2w^2 + 3w = 44$$

Subtract 44 from both sides: $$2w^2 + 3w - 44 = 0$$

Factor: $$(2w + 11)(w - 4) = 0$$

Set each factor to 0: $$2w + 11 = 0 \quad \text{or} \quad w - 4 = 0$$

Solve: $$2w = -11 \quad \text{or} \quad w = 4$$

$$\frac{2w}{2} = -\frac{11}{2}$$

$$w = -\frac{11}{2}$$

Step 5: Check Since w represents the width of the rectangular hallway, we discard the solution $w = -\frac{11}{2}$. If the width of the hallway is 4 feet, then the length would be $2w + 3 = 2(4) + 3 = 11$ feet. The area of a hallway that is 4 feet by 11 feet would be $4(11) = 44$ square feet. We have the right answer!

Step 6: Answer The dimensions of the hallway are 4 feet by 11 feet.

QUICK ✓

2. A rectangular plot of land has length that is 3 kilometers less than twice its width. If the area of the land is 104 square kilometers, what are the dimensions of the land?

Classroom Example The base of a triangle is 3 meters shorter than its height. What are the height and base of the triangle if its area is 44 square meters?

Answer: The base is 8 meters and the height is 11 meters.

EXAMPLE 3 Geometry: Area of a Triangle

The height of a triangle is 5 inches less than the length of the base, and the area of the triangle is 42 square inches. Find the height of the triangle.

Solution

Step 1: Identify This is a geometry problem involving the area of a triangle.

Step 2: Name The height of the triangle is 5 inches less than the base. We will let b represent the length of the base and $b - 5$ represent the height of the triangle. See Figure 4.

Figure 4

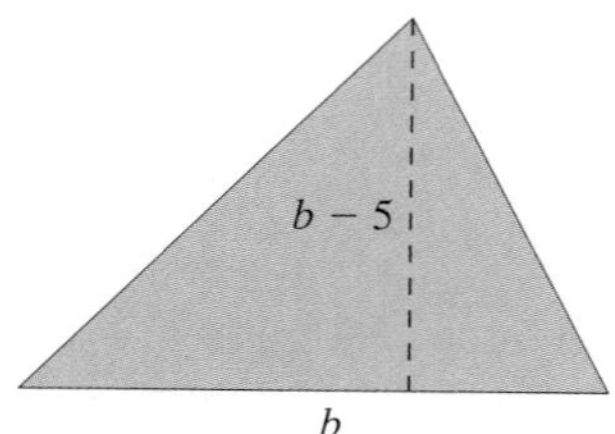

Step 3: Translate We know that the area of a triangle is given by the formula Area $= \frac{1}{2}$(base)(height). In addition, we are given the area of the triangle to be 42 square feet, so

$$\text{Area} = \frac{1}{2}(\text{base})(\text{height})$$

$$42 = \frac{1}{2}(b)(b - 5) \quad \text{The Model}$$

Step 4: Solve Our model is a quadratic equation, so we first put the equation in standard form, $ax^2 + bx + c = 0$.

Work Smart

Multiply only the $\frac{1}{2}$ by 2—don't multiply the factors b and $(b - 5)$ by 2 also.

$$42 = \frac{1}{2}(b)(b - 5)$$

Multiply by 2 to clear fractions: $$2(42) = 2\left[\frac{1}{2}(b)(b - 5)\right]$$

$$84 = b(b - 5)$$

Distribute: $84 = b^2 - 5b$

Subtract 84 from both sides: $0 = b^2 - 5b - 84$

Factor: $0 = (b - 12)(b + 7)$

Apply Zero-Product Property: $b - 12 = 0$ or $b + 7 = 0$

Solve: $b = 12$ or $b = -7$

Step 5: Check Since b represents the base of the triangle, we discard the solution $b = -7$. Do you see why? The base of the triangle is 12 inches, so the height is $b - 5 = 12 - 5 = 7$ inches. The area of a triangle that has a base of 12 inches and a height of 7 inches is $\frac{1}{2} \cdot 12 \cdot 7 = 42$ square inches. We have the right answer!

Step 6: Answer The height of the triangle is 7 inches.

QUICK ✓

3. The base of a triangular garden is 4 yards longer than the height, and the area of the garden is 48 square yards. Find the dimensions of the triangle.

2 Model and Solve Problems Using the Pythagorean Theorem

The Pythagorean Theorem is a statement about right triangles.

DEFINITIONS

A **right triangle** is one that contains a **right angle,** that is, an angle of 90°. The side of the triangle opposite the 90° angle is called the **hypotenuse;** the remaining two sides are called **legs.**

In Figure 5 we use c to represent the length of the hypotenuse and a and b to represent the lengths of the legs. Notice the use of the symbol ⌐ to show the 90° angle.

We now state the Pythagorean Theorem.

Figure 5

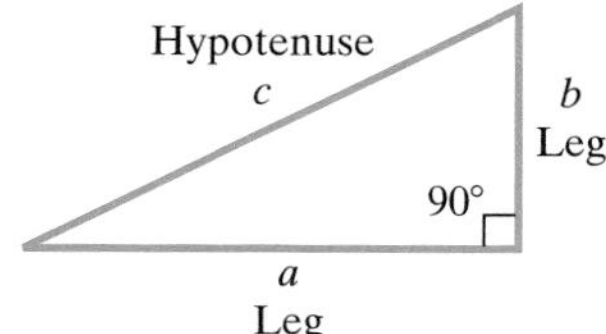

THE PYTHAGOREAN THEOREM

In a right triangle, the square of the length of the hypotenuse is equal to the sum of the squares of the lengths of the legs. That is, in the right triangle shown in Figure 5,

$$a^2 + b^2 = c^2 \quad \text{or} \quad \text{leg}^2 + \text{leg}^2 = \text{hypotenuse}^2$$

Classroom Example Use Example 4 with legs labeled x, $x + 4$, and the hypotenuse 20.

Answer: $x = 12$ and $x + 4 = 16$

EXAMPLE 4 Using the Pythagorean Theorem

Find the lengths of the sides of the right triangle in Figure 6 on page 294.

Solution

Figure 6 shows a right triangle, so we use the Pythagorean Theorem, $a^2 + b^2 = c^2$. The legs have lengths x and $x + 7$, and the hypotenuse has length 13. We substitute into the equation $a^2 + b^2 = c^2$.

Figure 6

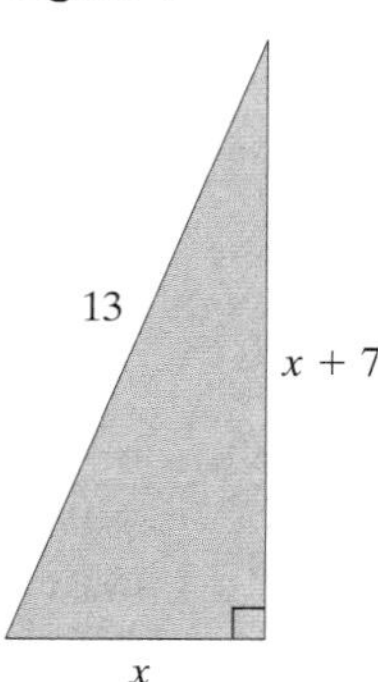

$$a^2 + b^2 = c^2$$

Substitute $a = x, b = x + 7, c = 13$: $x^2 + (x + 7)^2 = 13^2$

$(a + b)^2 = a^2 + 2ab + b^2$: $x^2 + x^2 + 14x + 49 = 169$

Combine like terms: $2x^2 + 14x + 49 = 169$

Subtract 169 from both sides: $2x^2 + 14x - 120 = 0$

Factor: $2(x^2 + 7x - 60) = 0$

$$2(x + 12)(x - 5) = 0$$

Set each factor = 0: $2 = 0$ or $x + 12 = 0$ or $x - 5 = 0$

Solve each equation: $x = -12$ or $x = 5$

The equation $2 = 0$ is false, and the solution $x = -12$ makes no sense because x represents the length of a leg of a right triangle. So $x = 5$ is the length of one leg of the triangle. The other leg is $x + 7 = 5 + 7 = 12$.

To check our solutions, let's replace a by 5 and b by 12 and see if our solutions satisfy the Pythagorean Theorem. Does $5^2 + 12^2 = 13^2$? Is $25 + 144 = 169$? Yes! Our answers are correct.

QUICK ✓

4. Find the length of each leg of the right triangle pictured below.

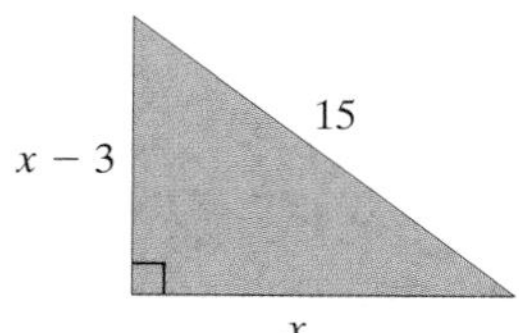

Classroom Example A rectangular plasma TV that measures 25 inches on the diagonal is 5 inches longer than it is tall. Will the TV fit in a media center that is 24 inches long?

Answer: Yes. The length of the TV is 20 inches.

EXAMPLE 5 Will the Television Fit in the Media Cabinet?

A rectangular 20-inch television screen (measured diagonally) is 4 inches wider than it is tall. Will this TV fit in your new media cabinet that is 20 inches wide?

Solution

Step 1: Identify This is a geometry problem involving the lengths of the sides of a right triangle.

Step 2: Name We know that the diagonal of the television screen (hypotenuse) is 20 inches long. Let x represent the height of the screen. The screen is 4 inches wider than it is tall, so $x + 4$ represents the width.

Step 3: Translate We know that the Pythagorean Theorem tells us the relationship between the lengths of the sides of a right triangle, so we use $a^2 + b^2 = c^2$.

$$a^2 + b^2 = c^2$$
$$x^2 + (x + 4)^2 = 20^2$$

Work Smart
Remember
$$(x + 4)^2 \neq x^2 + 16$$
$$(x + 4)^2 = x^2 + 8x + 16$$

Step 4: Solve We now proceed to solve the equation. The first step is to put the quadratic equation in standard form, $ax^2 + bx + c = 0$.

$$x^2 + (x + 4)^2 = 20^2$$
$$x^2 + x^2 + 8x + 16 = 400$$
$$2x^2 + 8x + 16 = 400$$

Subtract 400 from both sides: $2x^2 + 8x - 384 = 0$

Factor: $2(x^2 + 4x - 192) = 0$

Factor: $2(x + 16)(x - 12) = 0$

Set each factor to 0: $2 = 0$ or $x + 16 = 0$ or $x - 12 = 0$

Solve: $x = -16$ or $x = 12$

Step 5: Check The statement $2 = 0$ is false. Since x represents the length of the shorter leg of the triangle, we discard the solution $x = -16$. If the shorter leg of the triangle (height of the TV) is 12 inches, then the longer leg (width of the television) would be $x + 4 = 12 + 4 = 16$ inches. Does $12^2 + 16^2 = 20^2$? Since $144 + 256 = 400$, our answer is correct!

Step 6: Answer The television screen has dimensions 12 inches by 16 inches, so it will fit in the new cabinet.

QUICK ✓

5. A rectangular 10-inch television screen (measured diagonally) is 2 inches wider than it is tall. What are the dimensions of the TV screen?

4.7 Exercises

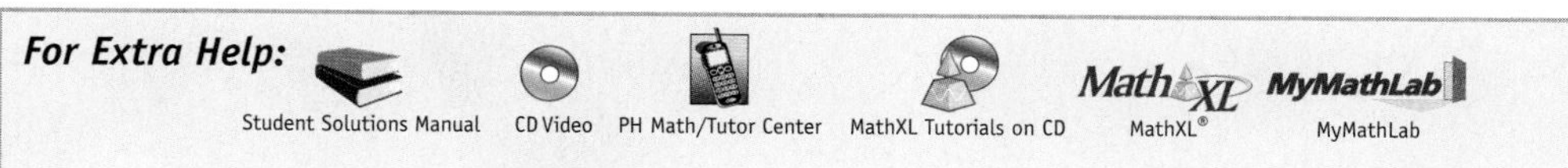

1. hypotenuse
2. legs, a, b
3. $a^2 + b^2 = c^2$
4. False
5. False
6. False
7. Answers may vary.
8. Answers may vary.
9. 4, 14
10. 10, 25
11. 2, 9
12. 14, 26

Concepts and Vocabulary

In Problems 1–3, fill in the blanks.

1. The ______ of a right triangle is always the side opposite the 90° angle.

2. The sides of a triangle other than the hypotenuse are called ______ and are labeled as ______ and ______.

3. The Pythagorean Theorem states that in a right triangle, if c is the hypotenuse and a, b are the legs, then ______.

In Problems 4–6, answer True or False to each statement.

4. If the length of the side of rectangle is 5 less than twice the width and the width is labeled x, the length should be labeled $5 - 2x$.

5. The difference of 25 and the square of a number is the same as the difference of 25 and a number squared.

6. The Pythagorean Theorem can be used to find the length of the side of any triangle.

7. In a projectile motion problem, you have two positive numbers that satisfy the word problem. Explain the meaning of each of these numbers and then give a case where you would have only one positive solution.

8. You have to give the class some steps to follow when solving a word problem. What steps would you include and what would you say about the numbers that might appear in the solution to your equation?

Building Skills

In Problems 9–12, use the given area to find the missing sides of the rectangle.

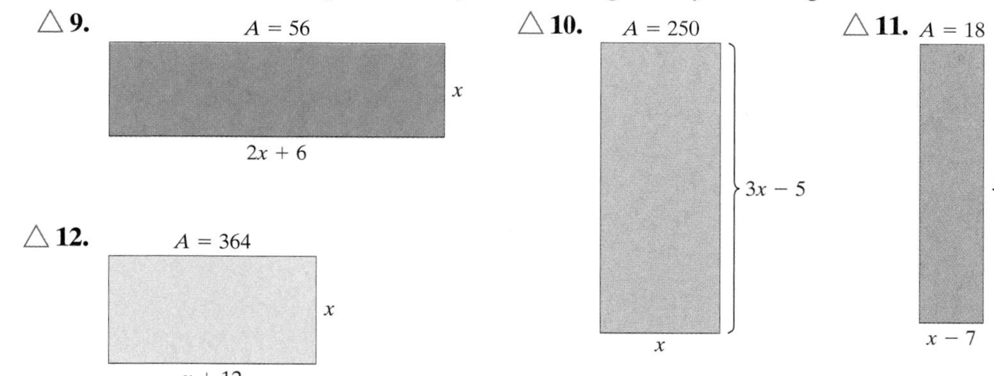

13. base = 26; height = 8
14. base = 6; height = 14
15. base = 18; height = 16
16. base = 12; height = 14
17. base = 13; height = 11
18. base = 11; height = 7
19. $B = 53$; $b = 43$
20. $B = 17$; $h = 5$; $b = 11$
21. 9, 12
22. 9, 12, 15

In Problems 13–16, use the given area to find the height and base of the triangle.

△ **13.**

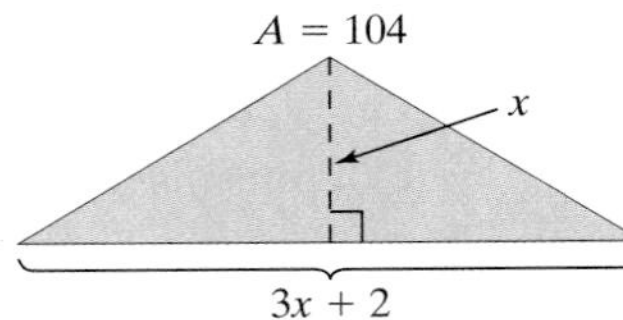

△ **14.**

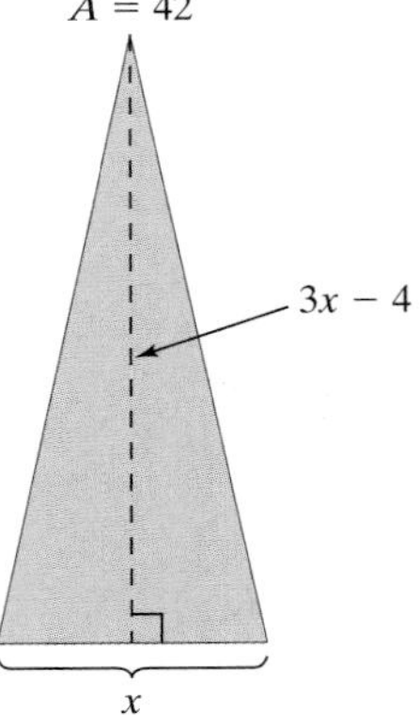

△ **15.**

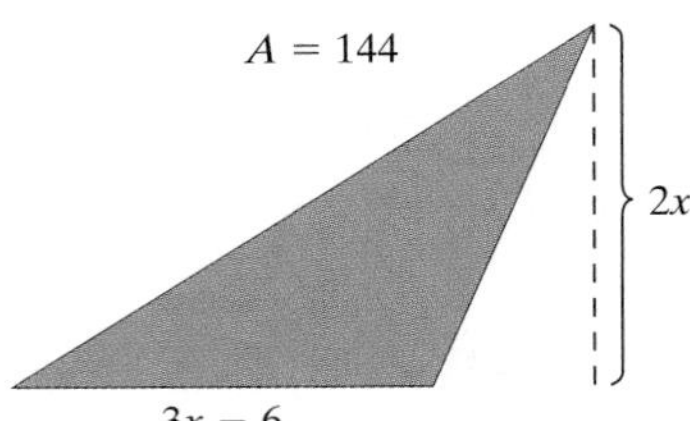

△ **16.**

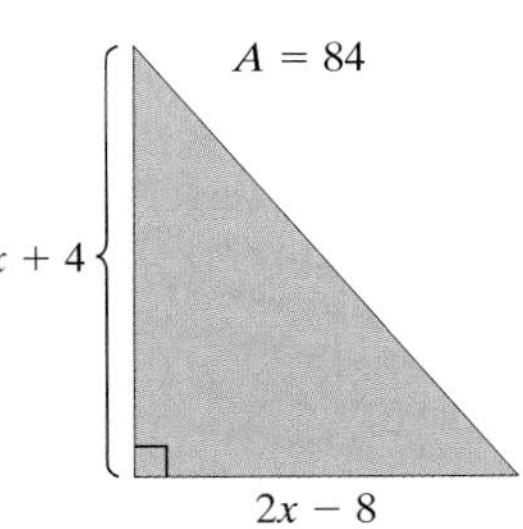

In Problems 17–20, use the given area to find the dimensions of the quadrilateral.

△ **17.**

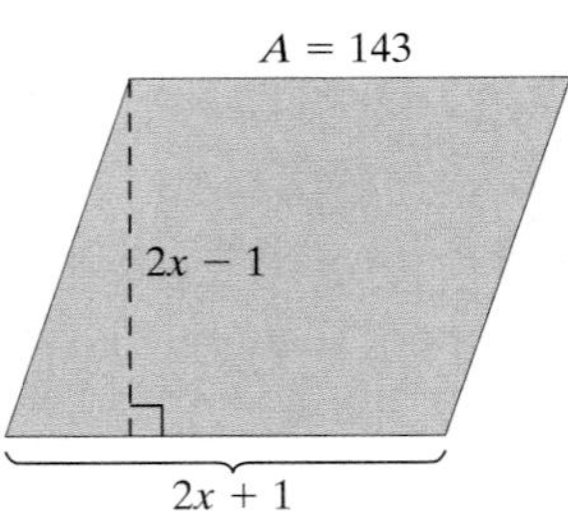

△ **18.**

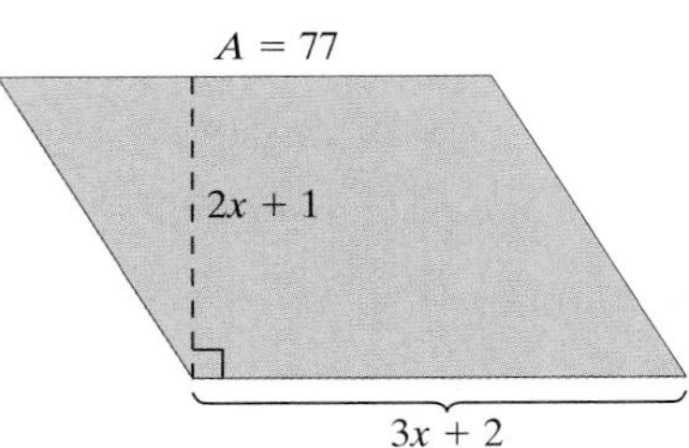

△ **19.**

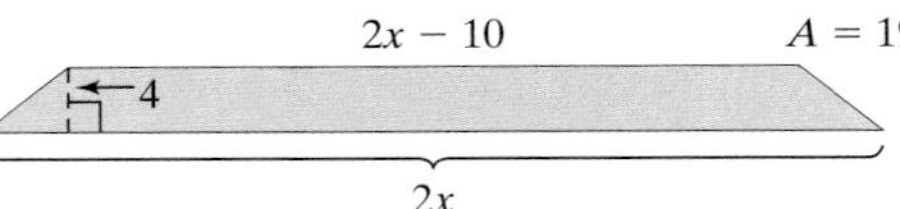

△ **20.**

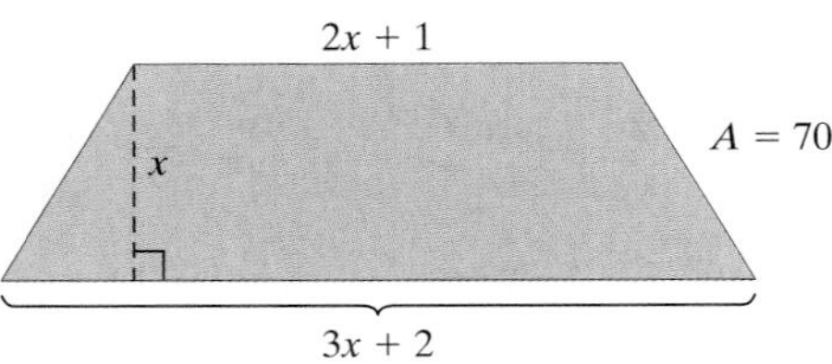

In Problems 21–24, use the Pythagorean Theorem to find the lengths of the sides of the triangle.

△ **21.**

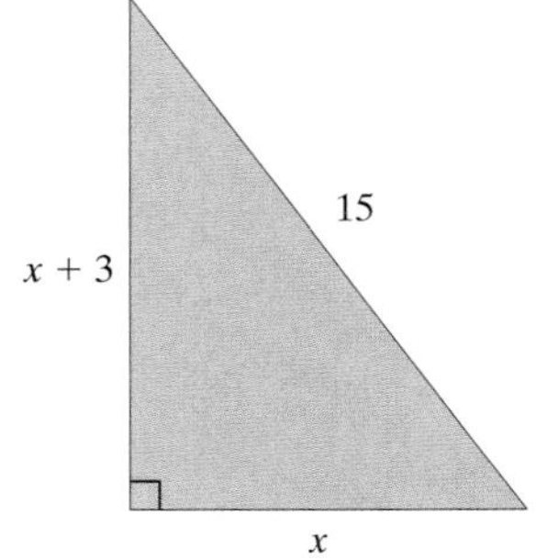

△ **22.**

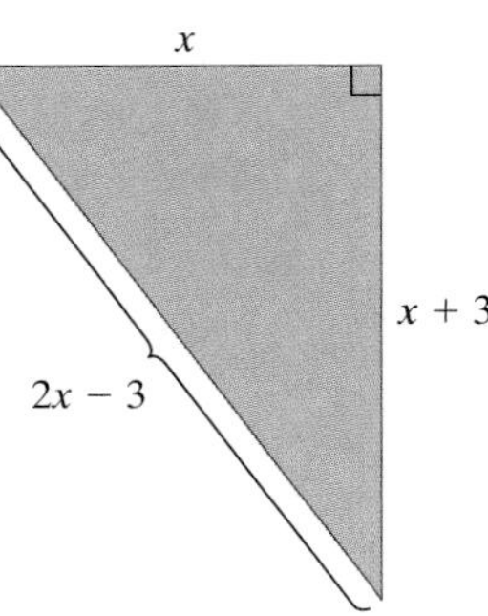

23. 12, 5
24. 15, 8, 17
25. 256, 252, 240, 220, 192, 156, 112, 60, 0 feet
26. 0, 28, 48, 60, 64, 60, 48, 28, 0 feet
27. 6 sec
28. 3 sec
29. width = 4 m; length = 12 m
30. width = 3 ft; length 7 ft
31. base = 6 ft; height = 18 ft
32. base = 8 m; height = 6 m
33. 30 inches
34. 21 in.
35. width = 12 mm; length = 25 mm
36. width = 13 yd; length = 8 yd
37. width = 4 cm; length = 25 cm
38. width = 4 in.; length = 9 in.

△ **23.**

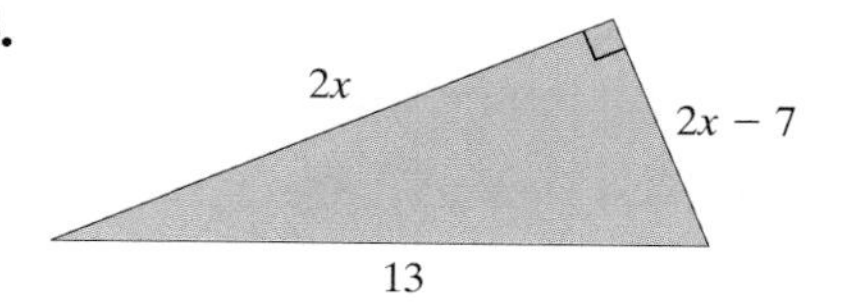

△ **24.**

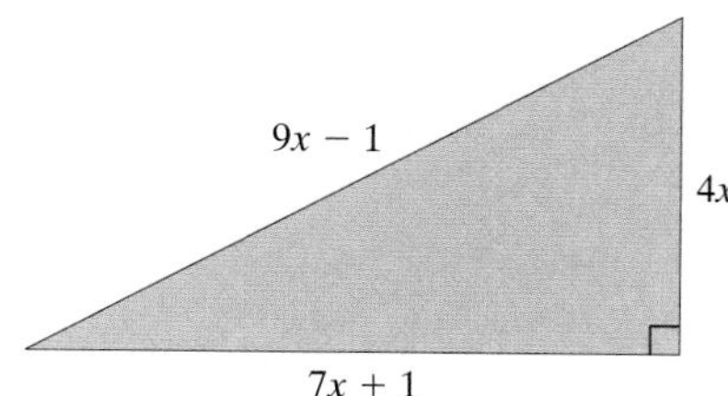

Applying the Concepts

25. Projectile Motion The height, h, of an object t seconds after it is dropped from a cliff 256 feet tall is given by the equation $h = -16t^2 + 256$. Suppose you dropped your glasses off the cliff. Fill in the table below to find the height of your glasses at each time, t.

Time, in Seconds	0	0.5	1	1.5	2	2.5	3	3.5	4
Height, in Feet									

26. Projectile Motion The height, h, of an object t seconds after it is propelled from ground level is given by the equation $h = -16t^2 + 64t$. Fill in the table below to find the height of the object at each time, t.

Time, in Seconds	0	0.5	1	1.5	2	2.5	3	3.5	4
Height, in Feet									

27. Projectile Motion If $h = -16t^2 + 96t$ represents the height of a rocket, in feet, t seconds after it was fired, when will the rocket hit the ground? (**Hint:** The rocket is on the ground when $h = 0$.)

28. Projectile Motion If $h = -16t^2 + 96t$ represents the height of a rocket, in feet, t seconds after it was fired, when will the rocket be 144 feet high?

△ **29. Rectanglular Room** The length of a rectangular room is 8 meters more than the width. If the area of the room is 48 square meters, find the dimensions of the room.

△ **30. Rectangular Room** The width of a rectangular room is 4 feet less than the length. If the area of the room is 21 square feet, find the dimensions of the room.

△ **31. Sail Boat** The sail on a sail boat is in the shape of a triangle. If the height of the sail is 3 times the length of base and the area is 54 square feet, find the dimensions of the sail.

△ **32. Triangle** The base of a triangle is 2 meters more than the height. If the area of the triangle is 24 square meters, find the base and height of the triangle.

33. Big-Screen TV Your big-screen TV measures 50 inches on the diagonal. If the front of the TV measures 40 inches across the bottom, find the height of the TV.

34. Big-Screen TV Hannah owns a 29-inch TV (that is, it measures 29 inches on the diagonal). If the television is 20 inches high, find the distance across the bottom.

△ **35. Dimensions of a Rectangle** The length of a rectangle is 1 mm more than twice the width. If the area is 300 square mm, find the dimensions of the rectangle.

△ **36. Playing Field Dimensions** The width of a rectangular playing field is 3 yards less than twice the length. If the area of the field is 104 square yards, what are its dimensions?

△ **37. Rectangle and Square** A rectangle and a square have the same area. The width of the rectangle is 6 cm less than the side of the square and the length of the rectangle is 5 cm more than twice the side of the square. What are the dimensions of the rectangle?

△ **38. Square and Rectangle** A rectangle and a square have the same area. The width of the rectangle is 2 in. less than the side of the square and the length of the rectangle is 3 in. less than twice the side of the square. What are the dimensions of the rectangle?

39. (a) width = 25 in.; length = 42 in.
(b) 28 in. by 45 in.
40. (a) width = 26 in.; length = 36 in.
(b) No
41. 6 ft
42. 5 ft by 12 ft
43. width = 7 ft; length = 14 feet
44. 4 m by 4 m

△ **39. Watching the World Series** David is an avid baseball fan and has purchased a plasma TV just in time to watch the World Series. The TV screen is 17 inches longer than it is wide and there is a $1\frac{1}{2}$ inch wide casing that surrounds the TV screen.

(a) David begins to install the TV and remembers that he was told that it measures 53 inches on the diagonal, including the casing. What are the dimensions of the TV screen?

(b) What size opening is required to fit the TV into David's entertainment center?

△ **40. Jasper's Big-Screen TV** Jasper purchased a new big-screen TV. The TV screen is 10 inches longer than it is wide and is surrounded by a casing that is 2 inches wide.

(a) Jasper lost his tape measure but sees on the box that the TV measures 50 inches on the diagonal, including the casing. What are the dimensions of the TV screen?

(b) Jasper knows that the size of the opening where he wants the TV installed is 29 by 42 inches. Will this TV fit into his space?

△ **41. How Tall Is the Pole?** A pole is supported by a 10-foot guy-wire that is attached to the ground. The distance from the pole to the point that wire attaches to the ground is 2 feet greater than the height of the pole. Find the height of the pole.

△ **42. Sailing on Lake Erie** A sail on a sailboat is in the shape of a right triangle. The longest side of the sail is 13 feet long and one side of the sail is 7 feet longer than the other. Find the dimensions of the sail.

Extending the Concepts

△ **43. Gardening** Beth has 28 feet of fence to enclose a small garden. One side of the garden lies along her house, so only three sides require fencing. Find the dimensions of the garden if she encloses 98 square feet of the garden with the fence.

△ **44. Volume of a Box** A rectangular solid has a square base and is 8 meters high. What are the dimensions of the base if the volume of the solid is 128 cubic meters? The volume of a rectangular solid is (length)(width)(height).

CHAPTER 4 ACTIVITY: WHICH ONE DOES NOT BELONG?

Focus: Factoring polynomials.

Time: 20–30 minutes

Group size: 3–4

Each group member should decide which polynomial from each row does not belong. Answers will vary. As a group, discuss each member's results. Each group member should be prepared to explain WHY that particular polynomial does not belong with the other three. Be creative!

1.–5. Answers will vary.

	A	B	C	D
1	$15x^2 - 10ax - 15xy + 10ay$	$xr - 3xs - ry + 3sy$	$2ax - 14bx - 2ay + 14by$	$3ab - 3ay - 3bx + 3xy$
2	$-x^2 - x + 6$	$3x^2 + 39x + 120$	$2x^2 - 12xy - 54y^2$	$7x^2 + 7x - 140$
3	$3x^2 - 16x + 21$	$6x^2 - 3x + 15$	$3x^2 - 13x - 56$	$3x^2 + x + 1$
4	$9a^2 + 30a + 25$	$16x^2 - 24x + 9$	$4x^2 - 12xy + 9y^2$	$18s^2 - 60s + 9$
5	$4x^2 + 13x + 10$	$16x^2 + 40xy + 25y^2$	$16x^2 - 25y^2$	$8x^2 - 2xy - 15y^2$

Chapter 4 Review

Section 4.1 Greatest Common Factor and Factoring by Grouping

KEY CONCEPTS	KEY TERMS
• **To find the greatest common factor of two or more expressions,** **Step 1:** Find the GCF of the coefficients of each variable expression. **Step 2:** For each variable expression, determine the smallest exponent that the variable expression is raised to. **Step 3:** Find the product of the common factors found in Steps 1 and 2. This expression is the GCF. • **To factor a polynomial using the greatest common factor,** **Step 1:** Identify the greatest common factor (GCF) of the terms that make up the polynomial. **Step 2:** Rewrite each term as the product of the GCF and remaining factor. **Step 3:** Use the Distributive Property "in reverse" to factor out the GCF. **Step 4:** Use the Distributive Property to verify that the factorization is correct.	Factors Greatest common factor Factoring by grouping

YOU SHOULD BE ABLE TO . . .	EXAMPLE	REVIEW EXERCISES
1 Find the greatest common factor (GCF) of two or more expressions (p. 238)	Examples 1 through 5	1–10
2 Factor out the greatest common factor in polynomials (p. 241)	Examples 6 through 10	11–16
3 Factor polynomials by grouping (p. 243)	Examples 11 through 13	17–20

1. 12 2. 27 3. 10
4. 4 5. x^2 6. m
7. $15ab^2$
8. $6x^3y^2z$
9. $2(2a + 1)^2$
10. $9(x - y)$
11. $-6a^2(3a + 4)$
12. $-3x(3x - 4)$
13. $5y^2z(3 + y^5 + 4y)$
14. $7xy(x^2 - 3xy + 2y^2)$
15. $(5 - y)(x + 2)$
16. $(a + b)(z + y)$
17. $(5m + 2n)(m + 3n)$
18. $(2x + y)(y + x)$
19. $(x + 2)(8 - y)$
20. $(y^2 + 1)(x - 3)$

In Problems 1–10, find the greatest common factor, GCF, of each group of expressions.

1. 24, 36 **2.** 27, 54 **3.** 10, 20, 30 **4.** 8, 16, 28

5. x^4, x^2, x^8 **6.** m^3, m, m^5 **7.** $30a^2b^4, 45ab^2$ **8.** $18x^4y^2z^3, 24x^3y^5z$

9. $4(2a + 1)^2$ and $6(2a + 1)^3$ **10.** $9(x - y)$ and $18(x - y)$

In Problems 11–16, factor the GCF from the polynomial.

11. $-18a^3 - 24a^2$ **12.** $-9x^2 + 12x$

13. $15y^2z + 5y^7z + 20y^3z$ **14.** $7x^3y - 21x^2y^2 + 14xy^3$

15. $x(5 - y) + 2(5 - y)$ **16.** $z(a + b) + y(a + b)$

In Problems 17–20, factor by grouping.

17. $5m^2 + 2mn + 15mn + 6n^2$ **18.** $2xy + y^2 + 2x^2 + xy$

19. $8x + 16 - xy - 2y$ **20.** $xy^2 + x - 3y^2 - 3$

Section 4.2 Factoring Trinomials of the Form $x^2 + bx + c$

KEY CONCEPT	KEY TERMS
• **To factor a trinomial of the form $x^2 + bx + c$, we use the following steps.** **Step 1:** Find the pair of integers whose product is c and whose sum is b. That is, determine m and n such that $mn = c$ and $m + n = b$. **Step 2:** Write $x^2 + bx + c = (x + m)(x + n)$. **Step 3:** Check your work by multiplying the binomials using the FOIL pattern.	Quadratic trinomial Leading coefficient Prime polynomial

YOU SHOULD BE ABLE TO . . .	EXAMPLE	REVIEW EXERCISES
1 Factor trinomials of the form $x^2 + bx + c$ (p. 248)	Examples 1 through 7	21–28
2 Factor out the GCF, then factor $x^2 + bx + c$ (p. 253)	Examples 8 and 9	29–34

21. $(x+3)(x+2)$
22. $(x+2)(x+4)$
23. $(x-7)(x+3)$
24. $(x+5)(x-2)$
25. prime 26. prime
27. $(x-3y)(x-5y)$
28. $(m+5n)(m-n)$
29. $-(p+6)(p+5)$
30. $-(y-5)(y+3)$
31. $3x(x^2+11x+12)$
32. $4(x+1)(x+8)$
33. $2(x-7y)(x+6y)$
34. $4y(y+5)(y-2)$

In Problems 21–34, factor completely. If the polynomial cannot be factored, say it is prime.

21. $x^2 + 5x + 6$ **22.** $x^2 + 6x + 8$ **23.** $x^2 - 21 - 4x$

24. $3x + x^2 - 10$ **25.** $m^2 + m + 20$ **26.** $m^2 - 6m - 5$

27. $x^2 - 8xy + 15y^2$ **28.** $m^2 + 4mn - 5n^2$ **29.** $-p^2 - 11p - 30$

30. $-y^2 + 2y + 15$ **31.** $3x^3 + 33x^2 + 36x$ **32.** $4x^2 + 36x + 32$

33. $2x^2 - 2xy - 84y^2$ **34.** $4y^3 + 12y^2 - 40y$

Section 4.3 Factoring Trinomials of the Form $ax^2 + bx + c,\ a \neq 1$

KEY CONCEPTS

- **Factoring $ax^2 + bx + c,\ a \neq 1$ using trial and error: a, b, and c have no common factors**

  Step 1: List the possibilities for the first terms of each binomial whose product is ax^2.

  $(\_x + \_)(\_x + \_) = ax^2 + bx + c$

  Step 2: List the possibilities for the last terms of each binomial whose product is c.

  $(\_x + \square)(\_x + \square) = ax^2 + bx + c$

  Step 3: Write out all the combinations of factors found in Steps 1 and 2. Multiply the binomials out until a product is found that equals the trinomial.

- **Factoring $ax^2 + bx + c,\ a \neq 1$ by grouping: a, b, and c have no common factors**

  Step 1: Find the value of ac.

  Step 2: Find the pair of integers, m and n, whose product is ac and whose sum is b.

  Step 3: Write $ax^2 + bx + c = ax^2 + mx + nx + c$.

  Step 4: Factor the expression in Step 3 by grouping.

  Step 5: Multiply out the factored form to verify your answer.

YOU SHOULD BE ABLE TO . . .	EXAMPLE	REVIEW EXERCISES
1 Factor $ax^2 + bx + c,\ a \neq 1$ using trial and error (p. 257)	Examples 1 through 6	35–44
2 Factor $ax^2 + bx + c,\ a \neq 1$ using grouping (p. 262)	Examples 7 through 9	35–44

35. $(5y-6)(y+4)$
36. $(y-7)(6y+1)$
37. $(2x-3)(x-1)$
38. $(2x+7)(3x+1)$ 39. prime
40. $(4m+3)(2m+3)$
41. $3m(m+n)(3m+7n)$
42. $2(7m+n)(m+n)$
43. $x(5x+2)(3x-1)$
44. $p^2(3p-1)(2p+1)$
45. $(2x-3)^2$ 46. $(x-5)^2$
47. $(x+3y)^2$ 48. prime

In Problems 35–44, factor completely using any method you wish. If a polynomial cannot be factored, say it is prime.

35. $5y^2 + 14y - 24$ **36.** $6y^2 - 41y - 7$ **37.** $-5x + 2x^2 + 3$

38. $23x + 6x^2 + 7$ **39.** $2x^2 - 7x - 6$ **40.** $8m^2 + 18m + 9$

41. $9m^3 + 30m^2n + 21mn^2$ **42.** $14m^2 + 16mn + 2n^2$ **43.** $15x^3 + x^2 - 2x$

44. $6p^4 + p^3 - p^2$

Section 4.4 Factoring Special Products

KEY CONCEPTS	KEY TERMS
• **Perfect Square Trinomials** $a^2 + 2ab + b^2 = (a + b)^2$ $a^2 - 2ab + b^2 = (a - b)^2$ • **Difference of Two Squares** $a^2 - b^2 = (a - b)(a + b)$ • **Sum or Difference of Two Cubes** $a^3 + b^3 = (a + b)(a^2 - ab + b^2)$ $a^3 - b^3 = (a - b)(a^2 + ab + b^2)$	Perfect square trinomial Difference of two squares Sum of two cubes Difference of two cubes

YOU SHOULD BE ABLE TO . . .	EXAMPLE	REVIEW EXERCISES
1 Factor perfect square trinomials (p. 267)	Examples 1 through 4	45–50
2 Factor the difference of two squares (p. 269)	Examples 5 through 7	51–56
3 Factor the sum or difference of two cubes (p. 271)	Examples 8 through 10	57–62

49. $2(2m + 1)^2$ 50. $2(m - 6)^2$
51. $(2x - 5y)(2x + 5y)$
52. $(7x - 6y)(7x + 6y)$ 53. prime
54. prime 55. $(x - 3)(x + 3)(x^2 + 9)$
56. $(x - 5)(x + 5)(x^2 + 25)$
57. $(m + 3)(m^2 - 3m + 9)$
58. $(m + 5)(m^2 - 5m + 25)$
59. $(3p - 2)(9p^2 + 6p + 4)$
60. $(4p - 1)(16p^2 + 4p + 1)$
61. $(y^3 + 4z^2)(y^6 - 4y^3z^2 + 16z^4)$
62. $(2y + 3z^2)(4y^2 - 6yz^2 + 9z^4)$

In Problems 45–62, factor completely. If a polynomial cannot be factored, say it is prime.

45. $4x^2 - 12x + 9$ **46.** $x^2 - 10x + 25$ **47.** $x^2 + 6xy + 9y^2$

48. $9x^2 + 24xy + 4y^2$ **49.** $8m^2 + 8m + 2$ **50.** $2m^2 - 24m + 72$

51. $4x^2 - 25y^2$ **52.** $49x^2 - 36y^2$ **53.** $x^2 + 25$

54. $x^2 + 100$ **55.** $x^4 - 81$ **56.** $x^4 - 625$

57. $m^3 + 27$ **58.** $m^3 + 125$ **59.** $27p^3 - 8$

60. $64p^3 - 1$ **61.** $y^9 + 64z^6$ **62.** $8y^3 + 27z^6$

Section 4.5 Summary of Factoring Techniques

KEY CONCEPT

Steps for Factoring

Step 1: Is there a greatest common factor? If so, factor out the greatest common factor (GCF).

Step 2: Count the number of terms.

Step 3: (a) 2 terms

- Is it the difference of two squares? If so,

  $a^2 - b^2 = (a - b)(a + b)$

- Is it the difference of two cubes? If so,

  $a^3 - b^3 = (a - b)(a^2 + ab + b^2)$

- Is it the sum of two cubes? If so,

  $a^3 + b^3 = (a + b)(a^2 - ab + b^2)$

(b) 3 terms

- Is it a perfect square trinomial? If so,

  $a^2 + 2ab + b^2 = (a + b)^2$ or $a^2 - 2ab + b^2 = (a - b)^2$

- Is the leading coefficient 1? If so,

  $x^2 + bx + c = (x + m)(x + n)$ where $mn = c$ and $m + n = b$

- Is the coefficient of the square term different from 1? If so,
  a. Use factoring by grouping
  b. Use trial and error

(c) 4 terms
- Use factoring by grouping

Step 4: Check your work by multiplying out the factors.

YOU SHOULD BE ABLE TO . . .	EXAMPLE	REVIEW EXERCISES
(1) Factor polynomials completely (p. 274)	Examples 1 through 7	63–78

63. $(5a - 2b)(3a^2 - 5b^2)$
64. $(4a - 3b)(3a + b)$
65. prime 66. $(x - 12y)(x + 2y)$
67. $x(x - 7)(x + 6)$
68. $3x^4(x - 7)(x - 3)$
69. $(3x + 1)(2x + 3)$
70. $(2z + 3)(5z - 3)$
71. $(3x + 2)(9x^2 - 6x + 4)$
72. $(2z - 1)(4z^2 + 2z + 1)$
73. $2(2y - 1)(y + 5)$
74. $xy^2(5x - 3)(x - 1)$
75. $(5k - 9m)(5k + 9m)$
76. $(x^2 - 3)(x^2 + 3)$
77. prime 78. prime

In Problems 63–78, factor completely. If a polynomial cannot be factored, say it is prime.

63. $15a^3 - 6a^2b - 25ab^2 + 10b^3$

64. $12a^2 - 9ab + 4ab - 3b^2$

65. $x^2 - xy - 48y^2$

66. $x^2 - 10xy - 24y^2$

67. $x^3 - x^2 - 42x$

68. $3x^6 - 30x^5 + 63x^4$

69. $6x^2 + 11x + 3$

70. $10z^2 + 9z - 9$

71. $27x^3 + 8$

72. $8z^3 - 1$

73. $4y^2 + 18y - 10$

74. $5x^3y^2 - 8x^2y^2 + 3xy^2$

75. $25k^2 - 81m^2$

76. $x^4 - 9$

77. $16m^2 + 1$

78. $m^4 + 25$

Section 4.6 Solving Polynomial Equations by Factoring

KEY CONCEPT	KEY TERMS
• **The Zero-Product Property** If the product of two factors is zero, then at least one of the factors is 0. That is, if $ab = 0$, then $a = 0$ or $b = 0$ or both a and b are 0.	Polynomial equation Degree of a polynomial equation Quadratic equation Second-degree equation Standard form

YOU SHOULD BE ABLE TO . . .	EXAMPLE	REVIEW EXERCISES
(1) Solve quadratic equations using the Zero-Product Property (p. 281)	Examples 1 through 8	79–88
(2) Solve polynomial equations of degree three or higher using the Zero-Product Property (p. 287)	Example 9	89–92

79. $\left\{\frac{3}{2}, 4\right\}$ 80. $\left\{-7, -\frac{1}{2}\right\}$
81. $\{-3, 15\}$ 82. $\{2, 5\}$
83. $\{0, -2\}$ 84. $\left\{-\frac{9}{2}, 0\right\}$
85. $\{-1, 3\}$ 86. $\{-3, -1\}$
87. $\{-3, -1\}$ 88. $\left\{-\frac{1}{2}, \frac{3}{2}\right\}$
89. $\{-14, 0, 3\}$ 90. $\{-3, 0, 6\}$
91. $\left\{-3, \frac{4}{3}, 3\right\}$ 92. $\left\{-\frac{5}{2}, -2\right\}$

In Problems 79–92, solve each equation by factoring.

79. $(x - 4)(2x - 3) = 0$

80. $(2x + 1)(x + 7) = 0$

81. $x^2 - 12x - 45 = 0$

82. $x^2 - 7x + 10 = 0$

83. $3x^2 + 6x = 0$

84. $4x^2 + 18x = 0$

85. $3x(x + 1) = 2x^2 + 5x + 3$

86. $2x^2 + 6x = (3x + 1)(x + 3)$

87. $5x^2 + 5 = -20x - 10$

88. $8x^2 - 10x = -2x + 6$

89. $x^3 = -11x^2 + 42x$

90. $-3x^2 = -x^3 + 18x$

91. $(3x - 4)(x^2 - 9) = 0$

92. $(2x + 5)(x^2 + 4x + 4) = 0$

Section 4.7 Modeling and Solving Problems with Quadratic Equations	
KEY CONCEPT	**KEY TERMS**
• **The Pythagorean Theorem** In a right triangle, the square of the length of the hypotenuse is equal to the sum of the squares of the lengths of the legs. That is, if a and b are the lengths of the legs and c is the length of the hypotenuse, then $\text{leg}^2 + \text{leg}^2 = \text{hypotenuse}^2$ or $c^2 = a^2 + b^2$.	Right triangle Right angle Hypotenuse Legs

YOU SHOULD BE ABLE TO . . .	EXAMPLE	REVIEW EXERCISES
① Model and solve problems involving quadratic equations (p. 290)	Examples 1 through 3	93–96
② Model and solve problems using the Pythagorean Theorem (p. 293)	Example 4 and 5	97, 98

93. 5 sec
94. 2 sec or 3 sec
95. 9 ft, 6 ft
96. 3 yd, 5 yd
97. 8 ft, 6 ft
98. 24 ft, 10 ft

93. If $h = -16t^2 + 80t$ represents the height of a jet of water from a geyser t seconds after the geyser erupts, when will the water hit the ground?

94. If $h = -16t^2 + 80t$ represents the height of the jet of water from a geyser, when will the jet of water be 96 feet high?

△**95. Tabletop** The length of a rectangular tabletop is 3 feet shorter than twice its width. If the area of the tabletop is 54 square feet, what are the dimensions of the tabletop?

△**96. Tarp** The length of a rectangular tarp is 1 yard shorter than twice its width. If the area of the tarp is 15 square yards, what are the dimensions of the tarp?

△**97. Right Triangle** The shorter leg of a right triangle is 2 feet shorter than the longer leg. The hypotenuse is 10 feet. How long is each leg?

△**98. Right Triangle** The shorter leg of a right triangle is 14 feet shorter than the longer leg. The hypotenuse is 26 feet. How long is each leg?

CHAPTER 4 TEST

Remember to use your Chapter Test Prep Video CD to see fully worked-out solutions to any of these problems you would like to review.

Note to instructor: A special file in TestGen provides algorithms specifically matched to the problems in this Chapter Test for easy-to-replicate practice or assessment purposes.

1. $4x^4y^2$
2. $(x + 3)(x - 3)(x^2 + 9)$
3. $9x(2x^2 - 3)(x^2 + 1)$
4. $(x - 7)(y - 4)$
5. $(3x + 5)(9x^2 - 15x + 25)$
6. $(y - 12)(y + 4)$
7. $(6m + 5)(m - 1)$
8. prime
9. $(x - 5)(4 + y)$
10. $3y(x - 7)(x + 2)$
11. prime
12. $3(x - 1)(x + 1)(x^2 + x + 1)(x^2 - x + 1)$
13. $3x(3x + 1)(x + 4)$
14. $(3m + 2)(2m + 1)$
15. $2(2m^2 - 3mn + 2)$
16. $(5x + 7y)^2$
17. $\left\{-\frac{1}{5}, -3\right\}$
18. $\{0, 2\}$
19. length = 7 in.; width = 5 in.
20. legs: 12 in., 5 in., hypotenuse: 13 in.

1. Find the GCF of $16x^5y^2$, $20x^4y^6$, and $24x^6y^8$.

In Problems 2–16, factor each polynomial completely. If the polynomial cannot be factored, say it is prime.

2. $x^4 - 81$

3. $18x^5 - 9x^3 - 27x$

4. $xy - 7y - 4x + 28$

5. $27x^3 + 125$

6. $y^2 - 8y - 48$

7. $6m^2 - m - 5$

8. $4x^2 + 25$

9. $4(x - 5) + y(x - 5)$

10. $3x^2y - 15xy - 42y$

11. $x^2 + 4x + 12$

12. $3x^6 - 3$

13. $9x^3 + 39x^2 + 12x$

14. $6m^2 + 7m + 2$

15. $4m^2 - 6mn + 4$

16. $25x^2 + 70xy + 49y^2$

In Problems 17 and 18, solve the equation by factoring.

17. $5x^2 = -16x - 3$

18. $5x^3 - 20x^2 + 20x = 0$

19. The length of a rectangle is 8 inches shorter than three times its width. If the area of the rectangle is 35 square inches, what is the length of the rectangle?

20. The hypotenuse of a right triangle is one inch longer than the longer leg. The shorter leg is 7 inches shorter than the longer leg. Find the length of all three sides of the triangle.

CHAPTER

5 Rational Expressions and Equations

To plan effectively for purchasing materials and hiring part-time college students, the president of a bicycle manufacturing company wants to know the average daily cost of manufacturing bicycles. A rational equation is used to model his average daily costs. See Example 7 in Section 5.7, page 364.

OUTLINE

The Big Picture: Putting It Together

In this chapter, we discuss rational expressions. A rational expression is the ratio of two polynomials (that is, a polynomial divided by another polynomial). The techniques that we will learn in this chapter are similar to the techniques we learned when dealing with fractions. In the Appendix, Section A.1, we learned how to write a fraction in lowest terms by factoring the numerator and denominator and dividing out like factors—this same approach is used on rational expressions. The ability to factor a polynomial (discussed in Chapter 4) plays a *huge* role in being able to simplify a rational expression, so make sure you are good at factoring.

In Section 1.4 we discussed how to add, subtract, multiply, and divide rational numbers expressed as fractions. These same skills apply to adding, subtracting, multiplying, and dividing rational expressions. However, the skills learned in Chapter 3 plus the factoring skills acquired in Chapter 4 will also be needed to perform these operations. Just remember, the methods that we use to perform operations on rational expressions are identical to those that we use on rational numbers. So we really aren't learning any new methods here—just new ways to apply the skills that we already have!

5.1 Simplifying Rational Expressions

OBJECTIVES

1. Evaluate a Rational Expression
2. Determine Undefined Values of a Rational Expression
3. Simplify Rational Expressions

Preparing for Simplifying Rational Expressions

1. Evaluate $\dfrac{3x + y}{2}$ for $x = -1$ and $y = 7$. [Section 1.7, p. 58]
2. Factor: $2x^2 + x - 3$ [Section 4.3, pp. 257–265]
3. Solve: $3x^2 - 5x - 2 = 0$ [Section 4.6, pp. 281–284]
4. Write $\dfrac{21}{70}$ in lowest terms. [Appendix, Section A.1, p. A5]
5. Divide: $\dfrac{x^3y^4}{xy^2}$ [Section 3.4, pp. 201–202]

Work Smart

The numbers $\frac{4}{3}, \frac{-5}{2}$, and 18 are examples of rational numbers.

Remember, a rational number is the quotient of two integers, where the denominator is not zero.

DEFINITION

A **rational expression** is the quotient of two polynomials. That is, a rational expression is written in the form $\dfrac{p}{q}$, where p and q are polynomials and $q \neq 0$.

Some examples of rational expressions are

(a) $\dfrac{x - 3}{4x + 1}$ **(b)** $\dfrac{x^2 - 4x - 12}{x^2 - 5}$ **(c)** $\dfrac{3a^2 + 7b + 2b^2}{a^2 - 2ab + 8b^2}$

Expressions (a) and (b) are rational expressions in one variable, x, while expression (c) is a rational expression in two variables, a and b.

1 Evaluate a Rational Expression

To **evaluate** a rational expression, we replace the variable with its assigned numerical value and perform the arithmetic, using the order of operations.

Classroom Example ➤

Evaluate $\dfrac{-4}{x + 3}$ for (a) $x = 7$ (b) $x = -5$.

Answer:

(a) $-\dfrac{2}{5}$ (b) 2

EXAMPLE 1 Evaluating a Rational Expression

Evaluate $\dfrac{-2}{x + 5}$ for (a) $x = -3$ (b) $x = 11$.

Solution

(a) Substitute -3 for x:

$$\frac{-2}{x+5} = \frac{-2}{-3+5} = \frac{-2}{2} = -1$$

(b) Substitute 11 for x:

$$\frac{-2}{x+5} = \frac{-2}{11+5} = \frac{-2}{16}$$

Divide out common factors:

$$= \frac{-1 \cdot \cancel{2}}{8 \cdot \cancel{2}} = -\frac{1}{8}$$

Preparing for...Answers **1.** 2 **2.** $(2x + 3)(x - 1)$ **3.** $\left\{-\frac{1}{3}, 2\right\}$ **4.** $\dfrac{3}{10}$ **5.** x^2y^2

Classroom Example ➢
Evaluate $\frac{a^2 + 4}{2a^2 + 3a - 5}$ for (a) $a = -2$. (b) $a = 2$.

Answer:

(a) $-\frac{8}{3}$ (b) $\frac{8}{9}$

EXAMPLE 2 Evaluating a Rational Expression of a Higher Degree

Evaluate $\frac{p^2 - 9}{2p^2 + p - 10}$ for (a) $p = 1$ (b) $p = -2$.

Solution

(a) Substitute 1 for p:
$$\frac{p^2 - 9}{2p^2 + p - 10} = \frac{(1)^2 - 9}{2(1)^2 + 1 - 10} = \frac{1 - 9}{2(1) + 1 - 10} = \frac{-8}{2 + 1 - 10} = \frac{-8}{-7} = \frac{8}{7}$$

(b) Substitute -2 for p:
$$\frac{p^2 - 9}{2p^2 + p - 10} = \frac{(-2)^2 - 9}{2(-2)^2 + (-2) - 10} = \frac{4 - 9}{2(4) - 2 - 10} = \frac{-5}{8 - 2 - 10} = \frac{-5}{-4} = \frac{5}{4}$$

QUICK ✓ *Evaluate the rational expression for (a)* $x = -5$ *(b)* $x = 3$.

1. $\frac{3}{5x + 1}$ **2.** $\frac{x^2 + 6x + 9}{x + 1}$

Classroom Example ➢
Evaluate (a) $\frac{3a - 4b}{5c - 2}$ for $a = 4$, $b = -1$, and $c = -2$; (b) $\frac{x + 2y}{y + 5}$ for $x = 4$, $y = -5$.

Answer:

(a) $-\frac{4}{3}$ (b) undefined

EXAMPLE 3 Evaluating a Rational Expression with More than One Variable

Evaluate:

(a) $\frac{2a + 4b}{3c - 1}$ for $a = 1, b = -2, c = 3$ **(b)** $\frac{w - 3v}{v + 3w}$ for $w = 2$ and $v = -6$

Solution

(a) We substitute 1 for a, -2 for b, and 3 for c.

$$\frac{2a + 4b}{3c - 1} = \frac{2(1) + 4(-2)}{3(3) - 1} = \frac{2 + (-8)}{9 - 1} = \frac{-6}{8}$$

Divide out common factors: $= \frac{\cancel{2} \cdot -3}{\cancel{2} \cdot 4} = -\frac{3}{4}$

(b) We substitute 2 for w and -6 for v.

$$\frac{w-3v}{v+3w} = \frac{2-3(-6)}{-6+3(2)} = \frac{2+18}{-6+6} = \frac{20}{0}$$

Division by 0 is undefined, so the expression $\frac{w-3v}{v+3w}$ is not defined for $w = 2$ and $v = -6$.

Teaching Tip
Remind students that division by 0 is undefined. That is, $\frac{20}{0}$ is undefined but $\frac{0}{20} = 0$.

QUICK ✓ *Evaluate the rational expression for the given values of the variable.*

3. $\frac{3m-5n}{p-4}$ for $m = 2, n = -1$, and $p = 6$

4. $\frac{5y+2}{2y-z}$ for $y = 3, z = 6$

2 Determine Undefined Values of a Rational Expression

Work Smart
A rational expression is undefined if the *denominator* equals 0. It is okay for the *numerator* to equal 0.

A rational expression is **undefined** for those values of the variable(s) that make the denominator zero. **We find the values for which a rational expression is undefined by setting the denominator equal to zero and solving for the variable.**

Classroom Example
Determine the value of x for which $\frac{3}{x+2}$ is undefined.

Answer: The expression is undefined for $x = -2$.

EXAMPLE 4 Determining the Values for Which a Rational Expression Is Undefined

Find the value of x for which the expression $\frac{2}{x+3}$ is undefined.

Solution

We want to find all values of x in the rational expression $\frac{2}{x+3}$ that cause $x + 3$ to equal 0.

Set the denominator equal to 0: $x + 3 = 0$

Subtract 3 from both sides: $x = -3$

So -3 causes the denominator, $x + 3$, to equal 0. Therefore, the expression $\frac{2}{x+3}$ is undefined for $x = -3$.

Classroom Example
Find the values of x for which $\frac{6}{x^2-2x-24}$ is undefined.

Answer: The expression is undefined for $x = -4$ or $x = 6$.

EXAMPLE 5 Determining the Values for Which a Rational Expression Is Undefined

Find the values of x for which the expression $\frac{8x}{x^2-2x-3}$ is undefined.

Solution

We want to find all values of x in the rational expression $\frac{8x}{x^2-2x-3}$ that cause the denominator, $x^2 - 2x - 3$, to equal 0.

Work Smart
The equation $x^2 - 2x - 3 = 0$ is quadratic because the highest exponent on the variable term is 2.

Set the denominator equal to 0: $x^2 - 2x - 3 = 0$

This is a quadratic equation, so we factor the polynomial: $(x-3)(x+1) = 0$

Set each factor equal to 0: $x - 3 = 0$ or $x + 1 = 0$

Solve each equation for x: $x = 3$ or $x = -1$

Since 3 or -1 cause the denominator to equal zero, the rational expression $\frac{8x}{x^2-2x-3}$ is undefined for $x = 3$ or $x = -1$.

Work Smart
In Example 5, the expression $\dfrac{8x}{x^2 - 2x - 3}$ is *not* undefined at $x = 0$ because 0 causes the numerator to equal 0, which is okay.

Check Let's verify our answer by evaluating $\dfrac{8x}{x^2 - 2x - 3}$ for each of these values.

If $x = 3$: $\dfrac{8(3)}{(3)^2 - 2(3) - 3} = \dfrac{24}{9 - 6 - 3} = \dfrac{24}{0}$ undefined

If $x = -1$: $\dfrac{8(-1)}{(-1)^2 - 2(-1) - 3} = \dfrac{-8}{1 + 2 - 3} = \dfrac{-8}{0}$ undefined

QUICK ✓ *Find the value(s) for which the rational expression is undefined.*

5. $\dfrac{3}{x + 7}$ **6.** $\dfrac{-4}{3n + 5}$ **7.** $\dfrac{8x}{x^2 + 2x - 3}$ **8.** $\dfrac{2k}{k^2 - 9}$

3 Simplify Rational Expressions

Remember that we write fractions in lowest terms by using the fact that $\dfrac{a \cdot c}{b \cdot c} = \dfrac{a}{b}$. For example, to write the fraction $\dfrac{18}{45}$ in lowest terms, we factor both 18 and 45 and divide out common factors.

$$\frac{18}{45} = \frac{2 \cdot \cancel{3} \cdot \cancel{3}}{\cancel{3} \cdot \cancel{3} \cdot 5} = \frac{2}{5}$$

To **simplify** a rational expression means to write the rational expression in the form $\dfrac{p}{q}$ where p and q are polynomials that have no common factors. We use the same ideas to simplify rational expressions as we do to write fractions in lowest terms.

SIMPLIFYING RATIONAL EXPRESSIONS

If p, q, and r are polynomials, then

$$\frac{p \cdot r}{q \cdot r} = \frac{p}{q} \quad \text{if } q \neq 0 \text{ and } r \neq 0$$

Work Smart
A rational expression is simplified if the numerator and the denominator share no common factor other than 1.

So to simplify a rational expression, factor the numerator, factor the denominator, and divide out common the factors.

EXAMPLE 6 How to Simplify a Rational Expression

Simplify: $\dfrac{7x + 14}{x^2 - 4}$, $x \neq -2, x \neq 2$

Classroom Example
Simplify
$\dfrac{3x + 12}{x^2 - 16}$, $x \neq -4, x \neq 4$

Answer: $\dfrac{3}{x - 4}$

Step-by-Step Solution

We include the restrictions $x \neq -2$, $x \neq 2$ because these values of x cause division by 0.

Step 1: Completely factor the numerator and the denominator.

$$\frac{7x + 14}{x^2 - 4} = \frac{7(x + 2)}{(x + 2)(x - 2)}$$

Step 2: Divide out common factors.

$$= \frac{7\cancel{(x + 2)}}{\cancel{(x + 2)}(x - 2)} = \frac{7}{x - 2}$$

So $\dfrac{7x + 14}{x^2 - 4} = \dfrac{7}{x - 2}$.

When we simplify a rational expression by dividing out common factors, we are changing the values of the variable that are not allowed as replacement values. In Example 6 we include the restriction $x \neq -2$, $x \neq 2$ because the expression $\frac{7x+14}{x^2-4}$ is equal to $\frac{7}{x-2}$ for all values except $x = -2$ and $x = 2$. When $x = 2$, both expressions are undefined and when $x = -2$, the original expression is also undefined. By not allowing x to equal -2 in both instances, the expressions remain equal. In general, we must restrict all values of the variable that are allowed in the *original* rational expression. For the remainder of the text, we shall not include the restrictions on the variable, but you should be aware that the restrictions are necessary to maintain equality.

We summarize the steps that you can use to simplify a rational expression.

Steps to Simplify a Rational Expression

Step 1: Completely factor the numerator and denominator of the rational expression.

Step 2: Divide out common factors.

Classroom Example Simplify:

(a) $\frac{4x-20}{6x^2-30x}$ (b) $\frac{x^2-4}{2x^2-3x-14}$

Answer:

(a) $\frac{2}{3x}$ (b) $\frac{x-2}{2x-7}$

Work Smart

Notice in Example 7(a) that when all the factors in the numerator divide out, we are left with a factor of 1, not 0!

EXAMPLE 7 Simplifying a Rational Expression

Simplify:

(a) $\frac{2x-10}{4x^2-20x}$ **(b)** $\frac{x^2-9}{2x^2-3x-9}$

Solution

(a)

Factor the numerator and denominator: $\frac{2x-10}{4x^2-20x} = \frac{2(x-5)}{4x(x-5)}$

Divide out common factors: $= \frac{\cancel{2}\cancel{(x-5)}}{\cancel{2}\cdot 2x\cancel{(x-5)}}$

$= \frac{1}{2x}$

(b)

Factor the numerator and denominator: $\frac{x^2-9}{2x^2-3x-9} = \frac{(x+3)(x-3)}{(2x+3)(x-3)}$

Divide out common factors: $= \frac{(x+3)\cancel{(x-3)}}{(2x+3)\cancel{(x-3)}}$

$= \frac{x+3}{2x+3}$

Work Smart

A common error students make is to divide out terms rather than to divide out factors. **When simplifying, we can only divide out common factors, not common terms.**

WRONG! $\frac{x+3}{x} = \frac{\cancel{x}+3}{\cancel{x}} = 3$ WRONG! $\frac{x^2-4x+3}{x+3} = \frac{x^2-4\cancel{x}+\cancel{3}}{\cancel{x}+\cancel{3}} = x^2-4$

If you aren't sure whether you can divide out, try the computation with numbers and see if it works.

For example, does $\frac{2+3}{2} = \frac{\cancel{2}+3}{\cancel{2}} = 3$? No!

QUICK ✓ *Simplify the rational expression.*

9. $\dfrac{3n + 12}{6n + 24}$ **10.** $\dfrac{10p + 5}{4p^2 - 1}$ **11.** $\dfrac{2z^2 + 6z + 4}{-4z - 8}$ **12.** $\dfrac{a^2 + 3a - 28}{2a^2 - a - 28}$

Classroom Example ➤
Simplify:

$$\frac{xn - xm + 2n - 2m}{x^2 + 4x + 4}$$

Answer: $\dfrac{n - m}{x + 2}$

Work Smart
Remember: you can simplify only after the numerator and denominator have been completely factored!

EXAMPLE 8 Simplifying a Rational Expression

Simplify: $\dfrac{ab + 3b - ac - 3c}{a^2 + 6a + 9}$

Solution

Factor the numerator and the denominator: $\dfrac{ab + 3b - ac - 3c}{a^2 + 6a + 9} = \dfrac{b(a + 3) - c(a + 3)}{(a + 3)^2}$

$$= \frac{(a + 3)(b - c)}{(a + 3)^2}$$

Divide out common factors: $= \dfrac{\cancel{(a + 3)}(b - c)}{\cancel{(a + 3)}(a + 3)}$

$$= \frac{b - c}{a + 3}$$

QUICK ✓ *Simplify the rational expression.*

13. $\dfrac{xz + x - yz - y}{z^2 - z - 2}$ **14.** $\dfrac{4k^2 + 4k + 1}{4k^2 - 1}$

Teaching Tip
Ask students which form a quotient of −1:

(a) $\dfrac{3 + 2x}{2x + 3}$ (b) $\dfrac{2 - z}{z - 2}$

(c) $\dfrac{m - 7}{m + 7}$

Work Smart

$\dfrac{x - 7}{x + 7} \neq -1$

and, because addition is commutative,

$\dfrac{x + 7}{7 + x} = 1$

In a rational expression, when a factor in the numerator and a factor in the denominator are identical, they form a quotient of 1. When a factor in the numerator is the *opposite* of the factor in the denominator, they form a quotient of −1.

$\dfrac{3}{-3} = -1$ 3 and −3 are opposites, so $\dfrac{3}{-3} = -1$

$\dfrac{-42}{42} = -1$ −42 and 42 are opposites, so $\dfrac{-42}{42} = -1$

$\dfrac{5 - x}{x - 5} = -1$ $\dfrac{5 - x}{x - 5} = \dfrac{-1(-5 + x)}{x - 5} = \dfrac{-1(x - 5)}{x - 5} = -1$

$\dfrac{4x - 7}{7 - 4x} = -1$ $\dfrac{4x - 7}{7 - 4x} = \dfrac{4x - 7}{-1(-7 + 4x)} = \dfrac{4x - 7}{-1(4x - 7)} = -1$

Classroom Example ➤
Simplify:

$$\frac{9 - x^2}{2x^2 - 7x + 3}$$

Answer: $\dfrac{-(3 + x)}{2x - 1}$

EXAMPLE 9 Simplifying a Rational Expression Containing Opposite Factors

Simplify: $\dfrac{4 - x^2}{2x^2 - x - 6}$

Solution

Factor the numerator and denominator: $\dfrac{4 - x^2}{2x^2 - x - 6} = \dfrac{(2 + x)(2 - x)}{(2x + 3)(x - 2)}$

Factor −1 from 2 − x: $= \dfrac{(2 + x)(-1)(-2 + x)}{(2x + 3)(x - 2)}$

Divide out common factors: $= \dfrac{(2 + x)(-1)\cancel{(x - 2)}}{(2x + 3)\cancel{(x - 2)}}$

$$= \frac{-(2 + x)}{2x + 3}$$

Work Smart
In Example 9, we could also write

$$\frac{4 - x^2}{2x^2 - x - 6} = \frac{-(2 + x)}{2x + 3}$$
$$= -\frac{2 + x}{2x + 3}$$
$$= -\frac{x + 2}{2x + 3}$$

QUICK ✓ *Simplify the rational expression.*

15. $\dfrac{7a - 7b}{b - a}$ **16.** $\dfrac{12 - 4x}{4x^2 - 13x + 3}$ **17.** $\dfrac{25z^2 - 1}{3 - 15z}$

5.1 Exercises

For Extra Help: Student Solutions Manual |  CD Video |  PH Math/Tutor Center |  MathXL Tutorials on CD | MathXL® | MyMathLab

Concepts and Vocabulary

In Problems 1–3, fill in the blanks.

1. The quotient of two polynomials is called a ________ ________.

2. When the denominator of any rational expression has a value of zero, the expression is said to be ________.

3. To ________ a rational expression means to write the rational expression in the form $\dfrac{p}{q}$ where p and q are polynomials that have no common factors.

In Problems 4–6, answer True or False to each statement.

4. $\dfrac{a + b}{a - b} = -1$ **5.** $\dfrac{x + y}{y + x} = 1$ **6.** $\dfrac{2n + 3}{2n + 6} = \dfrac{3}{6} = \dfrac{1}{2}, n \neq -3$

7. Is $x^2 - 3x + 1$ a rational expression? Explain your response.

8. Rational expressions that represent the same quantity are said to be *equivalent*. Are the rational expressions $\dfrac{x - y}{y - x}$ and $\dfrac{y - x}{x - y}$ equivalent? Explain how to determine whether two rational expressions are equivalent.

Building Skills

In Problems 9–18, evaluate each expression for the given values.

9. $\dfrac{x}{x - 5}$
(a) $x = 10$ **(b)** $x = -5$ **(c)** $x = 5$

10. $\dfrac{x}{x + 4}$
(a) $x = 8$ **(b)** $x = -4$ **(c)** $x = -6$

11. $\dfrac{2a - 3}{a}$
(a) $a = 0$ **(b)** $a = -3$ **(c)** $a = 9$

12. $\dfrac{2m - 1}{m}$
(a) $m = 0$ **(b)** $m = 1$ **(c)** $m = -1$

13. $\dfrac{x + 2}{x - 2}$
(a) $x = 4$ **(b)** $x = 2$ **(c)** $x = -2$

14. $\dfrac{a - 3}{a + 3}$
(a) $a = 5$ **(b)** $a = -3$ **(c)** $a = 3$

15. $\dfrac{x^2 - 2x}{x - 4}$
(a) $x = 3$ **(b)** $x = 2$ **(c)** $x = -3$

16. $\dfrac{a^2 - 2a}{a - 4}$
(a) $a = 5$ **(b)** $a = -1$ **(c)** $a = -4$

17. $\dfrac{x^2 - y^2}{2x - y}$
(a) $x = 2, y = 2$ **(b)** $x = 2, y = 4$ **(c)** $x = 1, y = 2$

18. $\dfrac{b^2 - a^2}{(a - b)^2}$
(a) $a = 3, b = 2$ **(b)** $a = 5, b = 4$ **(c)** $a = -2, b = -2$

In Problems 19–30, find the value(s) of the variable for which the rational expression is undefined.

19. $\dfrac{2 - 4x}{3x}$ **20.** $\dfrac{-x + 1}{5x}$ **21.** $\dfrac{3p}{p - 5}$ **22.** $\dfrac{5m^3}{m + 8}$

1. rational expression
2. undefined
3. simplify
4. False
5. True
6. False
7. Answers may vary.
8. Yes; Answers may vary.
9. (a) 2 (b) $\frac{1}{2}$ (c) undefined
10. (a) $\frac{2}{3}$ (b) undefined (c) 3
11. (a) undefined (b) 3 (c) $\frac{5}{3}$
12. (a) undefined (b) 1 (c) 3
13. (a) 3 (b) undefined (c) 0
14. (a) $\frac{1}{4}$ (b) undefined (c) 0
15. (a) -3 (b) 0 (c) $-\frac{15}{7}$
16. (a) 15 (b) $-\frac{3}{5}$ (c) -3
17. (a) 0 (b) undefined (c) undefined
18. (a) -5 (b) -9 (c) undefined
19. 0
20. 0
21. 5
22. -8

23. $\frac{3}{2}$ 24. $\frac{3}{4}$ 25. $-6, 6$ 26. $-5, 5$
27. $2, 5$ 28. $-2, 1$ 29. $-2, 0, 3$
30. $-4, -1, 0$ 31. $\frac{3}{x-2}$ 32. $\frac{1}{n+3}$
33. $\frac{(1+z)(1-z)}{3}$ 34. $\frac{2p+1}{2}$
35. $\frac{1}{p+2}$ 36. $\frac{1}{x-1}$ 37. -1
38. -1 39. $-2k$ 40. $-2v$
41. $\frac{x-1}{x+4}$ 42. $\frac{x-3}{x+2}$ 43. $\frac{5}{9}$ 44. $\frac{7}{9}$
45. $\frac{10}{11}$ 46. $\frac{38}{35}$ 47. $\frac{1}{2x^3}$ 48. $\frac{1}{3m^3}$
49. $\frac{12}{a^2}$ 50. $3ab^2$ 51. -3 52. -2
53. $\frac{b-5}{4}$ 54. $\frac{3}{p+4}$ 55. $\frac{x+3}{x-3}$
56. $\frac{x-1}{x+3}$ 57. $\frac{x-3}{x-5}$ 58. $\frac{x+2}{x-3}$
59. $-(x+y)$ 60. $-\frac{1}{b+a}$
61. $\frac{4}{a+b}$ 62. $\frac{5}{m+n}$
63. $\frac{x}{x-4}$ 64. $\frac{1}{m-1}$
65. $\frac{-(4+c)}{c-4}$ 66. $\frac{-(x+3)}{x-3}$
67. $\frac{2(x-2)}{3(x-5)}$ 68. $\frac{x+3}{2x-3}$
69. $\frac{-(x-3)}{x-2}$ 70. $\frac{-(x+1)}{x+5}$
71. $\frac{2}{t^2+9}$ 72. $\frac{x}{x^2-4}$
73. $\frac{-3}{w-1}$ 74. $-\frac{1}{a+2}$
75. (a) 5 mg/mL (b) 4 mg/mL
76. (a) $\frac{30}{1801}$ mg/mL (b) $\frac{60}{7201}$ mg/mL

23. $\frac{8}{3-2x}$ **24.** $\frac{12}{4a-3}$ **25.** $\frac{6z}{z^2-36}$ **26.** $\frac{5x}{25-x^2}$

27. $\frac{x}{x^2-7x+10}$ **28.** $\frac{2x^2}{x^2+x-2}$ **29.** $\frac{12x+5}{x^3-x^2-6x}$ **30.** $\frac{3h+2}{h^3+5h^2+4h}$

In Problems 31–42, simplify each rational expression. Assume that no variable has a value which results in a denominator with a value of zero.

31. $\frac{15}{5x-10}$ **32.** $\frac{3}{3n+9}$ **33.** $\frac{z-z^3}{3z}$ **34.** $\frac{6p^2+3p}{6p}$

35. $\frac{p-3}{p^2-p-6}$ **36.** $\frac{x-3}{x^2-4x+3}$ **37.** $\frac{2-x}{x-2}$ **38.** $\frac{4-z}{z-4}$

39. $\frac{2k^2-14k}{7-k}$ **40.** $\frac{2v^2-6v}{3-v}$ **41.** $\frac{x^2-1}{x^2+5x+4}$ **42.** $\frac{x^2-9}{x^2+5x+6}$

Mixed Practice

In Problems 43–74, simplify each rational expression. Assume that no variable has a value which results in a denominator with a value of zero.

43. $\frac{20}{36}$ **44.** $\frac{49}{63}$ **45.** $\frac{150}{165}$ **46.** $\frac{456}{420}$

47. $\frac{3x^2}{6x^5}$ **48.** $\frac{3m}{9m^4}$ **49.** $\frac{24a^5b}{2a^7b}$ **50.** $\frac{45a^2b^3}{15ab}$

51. $\frac{-3x-3y}{x+y}$ **52.** $\frac{-2a-2b}{a+b}$ **53.** $\frac{b^2-25}{4b+20}$ **54.** $\frac{3p-12}{p^2-16}$

55. $\frac{x^2-2x-15}{x^2-8x+15}$ **56.** $\frac{x^2+x-2}{x^2+5x+6}$ **57.** $\frac{x^3+x^2-12x}{x^3-x^2-20x}$ **58.** $\frac{x^3+3x^2+2x}{x^3-2x^2-3x}$

59. $\frac{x^2-y^2}{y-x}$ **60.** $\frac{a-b}{b^2-a^2}$ **61.** $\frac{4a-4b}{a^2-b^2}$ **62.** $\frac{5m+5n}{m^2+2mn+n^2}$

63. $\frac{x^2}{x^2-4x}$ **64.** $\frac{9m}{9m^2-9m}$ **65.** $\frac{16-c^2}{(c-4)^2}$ **66.** $\frac{9-x^2}{(x-3)^2}$

67. $\frac{4x^2-20x+24}{6x^2-48x+90}$ **68.** $\frac{2x^2+5x-3}{4x^2-8x+3}$ **69.** $\frac{6+x-x^2}{x^2-4}$ **70.** $\frac{5+4x-x^2}{x^2-25}$

71. $\frac{2t^2-18}{t^4-81}$ **72.** $\frac{x^3+4x}{x^4-16}$ **73.** $\frac{12w-3w^2}{w^3-5w^2+4w}$ **74.** $\frac{7a-a^2}{a^3-5a^2-14a}$

Applying the Concepts

75. Drug Concentration The concentration, C, in mg/mL, of a drug in a patient's bloodstream t minutes after an injection is given by the formula $C = \frac{50t}{t^2+25}$.

(a) Find the concentration in a patient 5 minutes after receiving an injection.
(b) Find the concentration in a patient 10 minutes after receiving an injection.

76. Drug Concentration The concentration, D, in mg/mL, of a drug in a patient's bloodstream t minutes after an injection is given by the formula $D = \frac{t}{2t^2+1}$.

(a) Find the concentration in a patient 30 minutes after receiving an injection.
(b) Find the concentration in a patient 60 minutes after receiving an injection.

77. Yes; BMI = 27.5
78. 23.5
79. \$15,600
80. \$12,320; Answers may vary
81. $\frac{(c^2 - c + 1)(c^2 + c + 1)}{c^2 + 1}$
82. $\frac{-(x^4 - x^2 + 1)}{x + 1}$
83. $\frac{(x^2 + 4)(x^2 - 3)}{(x + 2)(x^3 - 9)}$
84. $\frac{(n + 3)(n^2 - 8)}{(n - 4)(n^2 + 3)}$
85. $\frac{(t^2 + 4)(t + 2)}{t^2 - 2t + 4}$
86. $\frac{-(8x^3 + 9)}{8x^2 + 9}$

77. Body Mass Index BMI, or Body Mass Index, is used to determine if a person's weight is in a healthy range. The formula used to determine Body Mass Index is given by BMI $= \frac{k}{m^2}$, where k is the person's weight in kilograms and m is the height in meters. If a BMI range of 19 to 24.9 is considered healthy and 25 to 29.9 is considered overweight, should a person 2 meters tall weighing 110 kilograms start looking for a weight loss program?

78. Body Mass Index A formula to calculate the BMI when the weight, w, is in given in pounds and the height, h, is given in inches is given by BMI $= \frac{705w}{h^2}$. What is the BMI of a person weighing 120 pounds who is 5 feet tall?

79. Cost of a Car The average cost, in thousands of dollars, to produce Chevy Cobalts is given by the rational expression $\frac{0.2x^3 - 2.3x^2 + 14.3x + 10.2}{x}$, where x is the number of cars produced. If two Cobalts are produced, what is the average cost per car?

80. Cost of a Car Use the rational expression from Problem 79 to find the average cost of producing 10 Cobalts. Do you believe that the cost will continue to decrease as more cars are made?

Extending the Concepts

In Problems 81–86, simplify each rational expression.

81. $\frac{c^{12} - 1}{(c^4 - 1)(c^6 + 1)}$ **82.** $\frac{(1 - x)(x^6 + 1)}{x^4 - 1}$ **83.** $\frac{x^4 + x^2 - 12}{x^4 + 2x^3 - 9x - 18}$

84. $\frac{n^3 + 3n^2 - 8n - 24}{n^3 - 4n^2 + 3n - 12}$ **85.** $\frac{(t + 2)^3(t^4 - 16)}{(t^3 + 8)(t + 2)(t^2 - 4)}$ **86.** $\frac{24x^4 - 16x^3 + 27x - 18}{18 - 27x + 16x^2 - 24x^3}$

5.2 Multiplying and Dividing Rational Expressions

OBJECTIVES

1. Multiply Rational Expressions
2. Divide Rational Expressions

Preparing for Multiplying and Dividing Rational Expressions

Before getting started, take the following readiness quiz. If you get a problem wrong, go back to the section cited and review the material.

1. Find the product: $\frac{3}{14} \cdot \frac{28}{9}$ [Section 1.4, p. 29]

2. Find the reciprocal of $\frac{5}{8}$. [Section 1.4, p. 30]

3. Find the quotient: $\frac{12}{25} \div \frac{12}{5}$ [Section 1.4, p. 30]

4. Factor: $3x^3 - 27x$ [Section 4.4, pp. 269–271]

5. Simplify the rational expression: $\frac{3x + 12}{5x^2 + 20x}$ [Section 5.1, pp. 308–311]

1 Multiply Rational Expressions

The steps for multiplying rational expressions follow the same logic as the steps for multiplying rational numbers. For example,

$$\frac{2}{3} \cdot \frac{12}{5} = \frac{2 \cdot 12}{3 \cdot 5} = \frac{2 \cdot \cancel{3} \cdot 4}{\cancel{3} \cdot 5} = \frac{2 \cdot 4}{5} = \frac{8}{5}$$

Preparing for...Answers **1.** $\frac{2}{3}$ **2.** $\frac{8}{5}$ **3.** $\frac{1}{5}$ **4.** $3x(x + 3)(x - 3)$ **5.** $\frac{3}{5x}$

EXAMPLE 1 How to Multiply Rational Expressions

Multiply $\frac{x-3}{5} \cdot \frac{5x+35}{x^2-9}$. Simplify the result, if possible.

Classroom Example Multiply and simplify: $\frac{x+2}{7} \cdot \frac{7x+21}{x^2-4}$

Answer: $\frac{x+3}{x-2}$

Step-by-Step Solution

Step 1: Completely factor the polynomials in each numerator and denominator.

$$\frac{x-3}{5} \cdot \frac{5x+35}{x^2-9} = \frac{x-3}{5} \cdot \frac{5(x+7)}{(x+3)(x-3)}$$

Step 2: Multiply.

$$= \frac{5(x-3)(x+7)}{5(x+3)(x-3)}$$

Step 3: Divide out common factors in the numerator and denominator. The common factors are $x - 3$ and 5.

$$= \frac{\cancel{5}\,\cancel{(x-3)}(x+7)}{\cancel{5}(x+3)\cancel{(x-3)}}$$

$$= \frac{x+7}{x+3}$$

Teaching Tip Continue to remind students to divide factors, not terms.

Did you remember that $\frac{x+7}{x+3}$ cannot be simplified further? We divide *factors,* not terms, so $\frac{x+7}{x+3} \neq \frac{\cancel{x}+7}{\cancel{x}+3} \neq \frac{7}{3}$.

In Words To multiply two rational expressions, factor each polynomial, write the expression as a single fraction in factored form and then divide out common factors.

Steps to Multiply Rational Expressions

Step 1: Factor the polynomials in each numerator and denominator.

Step 2: Use the fact that if $\frac{a}{b}$ and $\frac{c}{d}$, $b \neq 0$, $d \neq 0$, are two rational expressions, then $\frac{a}{b} \cdot \frac{c}{d} = \frac{ac}{bd}$ to multiply the rational expressions.

Step 3: Divide out common factors in the numerator and denominator. Leave your answer in factored form.

Classroom Example Multiply. Simplify the result, if possible.

$\frac{5z+15}{2z} \cdot \frac{6}{z^2-2z-15}$

Answer: $\frac{15}{z(z-5)}$

EXAMPLE 2 Multiplying Rational Expressions

Multiply: $\frac{7y+28}{4y} \cdot \frac{8}{y^2+2y-8}$. Simplify the result, if possible.

Solution

Factor each polynomial: $\frac{7y+28}{4y} \cdot \frac{8}{y^2+2y-8} = \frac{7(y+4)}{4y} \cdot \frac{4 \cdot 2}{(y+4)(y-2)}$

Multiply: $= \frac{7(y+4) \cdot 4 \cdot 2}{4y(y+2)(y-2)}$

Divide out common factors, 4 and $y + 4$: $= \frac{7\cancel{(y+4)} \cdot \cancel{4} \cdot 2}{\cancel{4}y\cancel{(y+4)}(y-2)}$

$= \frac{7 \cdot 2}{y(y-2)}$

Multiply: $= \frac{14}{y(y-2)}$

QUICK ✓ *Multiply the rational expressions. Simplify the result, if possible.*

1. $\frac{p^2-9}{5} \cdot \frac{25p}{2p-6}$

2. $\frac{2x+8}{x^2+3x-4} \cdot \frac{7x-7}{6x+30}$

Classroom Example ➤
Find the product of $\frac{a^2 - a - 12}{a^2 + 2a - 3}$ and $\frac{a-1}{12-3a}$. Simplify, if possible.

Answer: $-\frac{1}{3}$

EXAMPLE 3 Multiplying Rational Expressions

Find the product of $\frac{p^2 - 9}{p^2 + 5p + 6}$ and $\frac{p+2}{6-2p}$. Simplify, if possible.

Solution

Factor each polynomial: $\frac{p^2 - 9}{p^2 + 5p + 6} \cdot \frac{p+2}{6-2p} = \frac{(p+3)(p-3)}{(p+3)(p+2)} \cdot \frac{p+2}{2 \cdot (-1)(p-3)}$

Multiply: $= \frac{(p+3)(p-3)(p+2)}{(p+3)(p+2) \cdot 2 \cdot (-1)(p-3)}$

Divide out common factors: $p+3$, $p-3$, and $p+2$: $= \frac{\cancel{(p+3)}\cancel{(p-3)}\cancel{(p+2)}}{\cancel{(p+3)}\cancel{(p+2)} \cdot 2 \cdot (-1)\cancel{(p-3)}}$

Multiply: $= -\frac{1}{2}$

Work Smart
$6 - 2p = 2(3 - p) = -2(p - 3)$

Work Smart
When all the factors in the numerator divide out. We are left with a factor of 1, *not* 0.

QUICK ✓ *Multiply the rational expressions. Simplify the result, if possible.*

3. $\frac{x^2 - 9}{x^2 - 25} \cdot \frac{2x - 10}{x - 3}$

4. $\frac{15a - 3a^2}{7} \cdot \frac{3 + 2a}{2a^2 - 7a - 15}$

Classroom Example ➤
Find the product and simplify:
$\frac{a^2 + 2ab + b^2}{6a^2 + 6ab} \cdot \frac{9a + 3b}{3a^2 + 4ab + b^2}$

Answer: $\frac{1}{2a}$

EXAMPLE 4 Multiplying Rational Expressions with Two Variables

Find the product and simplify: $\frac{m^2 - n^2}{10m^2 - 10mn} \cdot \frac{10m + 5n}{2m^2 + 3mn + n^2}$

Solution

Factor each polynomial: $\frac{m^2 - n^2}{10m^2 - 10mn} \cdot \frac{10m + 5n}{2m^2 + 3mn + n^2} = \frac{(m+n)(m-n)}{10m(m-n)} \cdot \frac{5(2m+n)}{(2m+n)(m+n)}$

Multiply: $= \frac{(m+n)(m-n) \cdot 5(2m+n)}{5 \cdot 2m(m-n)(2m+n)(m+n)}$

Divide out common factors: $m+n$, $m-n$, $2m+n$, and 5: $= \frac{\cancel{(m+n)}\cancel{(m-n)} \cdot \cancel{5}\cancel{(2m+n)}}{\cancel{5} \cdot 2m\cancel{(m-n)}\cancel{(2m+n)}\cancel{(m+n)}}$

Multiply: $= \frac{1}{2m}$

QUICK ✓ *Multiply the rational expressions. Simplify the result, if possible.*

5. $\frac{a^2 + 2ab + b^2}{3a + 3b} \cdot \frac{a - b}{a^2 - b^2}$

2 Divide Rational Expressions

Let's review the language we use in division. In the expression $\frac{20x^5}{3y} \div \frac{4x^2}{15y^5}$, $\frac{20x^5}{3y}$ is called the *dividend* and $\frac{4x^2}{15y^5}$ is called the *divisor*. The answer is called the *quotient*. We could also write $\frac{20x^5}{3y} \div \frac{4x^2}{15y^5}$ as $\frac{\frac{20x^5}{3y}}{\frac{4x^2}{15y^5}}$.

Teaching Tip
Before introducing division of rational expressions, have students practice finding the reciprocal of some algebraic expressions such as $\frac{3}{a}$, $\frac{n^2 - 1}{n^2 - 2n - 8}$, and $c + 2$.

The steps for finding the quotient of rational expressions are the same as the steps for finding the quotient of rational numbers. For example, to evaluate $\frac{2}{3} \div \frac{7}{9}$, we take the reciprocal of the divisor, $\frac{7}{9}$, and multiply.

reciprocal of divisor

$$\frac{2}{3} \div \frac{7}{9} = \frac{2}{3}\cdot\frac{9}{7} = \frac{2\cdot\cancel{3}\cdot 3}{\cancel{3}\cdot 7} = \frac{2\cdot 3}{1\cdot 7} = \frac{6}{7}$$

dividend

Classroom Example

Find the quotient: $\frac{x+2}{5x} \div \frac{2x+4}{15x^2}$

Answer: $\frac{3x}{2}$

EXAMPLE 5 How to Divide Rational Expressions

Find the quotient and simplify: $\frac{x+3}{2x-8} \div \frac{9}{4x}$

Step-by-Step Solution

Step 1: Multiply the dividend by the reciprocal of the divisor.

The reciprocal of $\frac{9}{4x}$ is $\frac{4x}{9}$: $\quad \frac{x+3}{2x-8} \div \frac{9}{4x} = \frac{x+3}{2x-8}\cdot\frac{4x}{9}$

Step 2: Completely factor the polynomials in each numerator and denominator.

$$= \frac{x+3}{2(x-4)}\cdot\frac{2\cdot 2\cdot x}{9}$$

Step 3: Multiply.

$$= \frac{(x+3)\cdot 2\cdot 2\cdot x}{2(x-4)\cdot 9}$$

Step 4: Divide out common factors in the numerator and denominator. The common factor is 2.

Simplify: $\quad = \frac{(x+3)\cdot\cancel{2}\cdot 2\cdot x}{\cancel{2}(x-4)\cdot 9}$

$$= \frac{2x(x+3)}{9(x-4)}$$

So $\frac{x+3}{2x-8} \div \frac{9}{4x} = \frac{2x(x+3)}{9(x-4)}$.

Next, we summarize the procedure for dividing rational expressions.

Steps to Divide Rational Expressions

Step 1: Multiply the dividend by the reciprocal of the divisor.

Step 2: Factor each polynomial in the numerator and denominator.

Step 3: Multiply.

Step 4: Divide out common factors in the numerator and denominator. Leave the remaining factors in factored form.

Classroom Example

Find the quotient:

$$\frac{a^2+2a-8}{2a^2+a-3} \div \frac{a^2+4a}{2a^2-a-6}$$

Answer: $\frac{(a-2)^2}{a(a-1)}$

EXAMPLE 6 Dividing Rational Expressions

Find the quotient and simplify: $\frac{y^2-9}{2y^2-y-15} \div \frac{3y^2+10y+3}{2y^2+y-10}$

Solution

Multiply the dividend by the reciprocal of the divisor: $\frac{y^2 - 9}{2y^2 - y - 15} \div \frac{3y^2 + 10y + 3}{2y^2 + y - 10} = \frac{y^2 - 9}{2y^2 - y - 15} \cdot \frac{2y^2 + y - 10}{3y^2 + 10y + 3}$

Factor each polynomial in the numerator and denominator: $= \frac{(y - 3)(y + 3)}{(2y + 5)(y - 3)} \cdot \frac{(2y + 5)(y - 2)}{(3y + 1)(y + 3)}$

Multiply: $= \frac{(y - 3)(y + 3)(2y + 5)(y - 2)}{(2y + 5)(y - 3)(3y + 1)(y + 3)}$

Divide out common factors: $y - 3$, $y + 3$, and $2y + 5$: $= \frac{\cancel{(y - 3)}\cancel{(y + 3)}\cancel{(2y + 5)}(y - 2)}{\cancel{(2y + 5)}\cancel{(y - 3)}(3y + 1)\cancel{(y + 3)}}$

$$= \frac{y - 2}{3y + 1}$$

QUICK ✓ *Find the quotient and simplify, if possible.*

6. $\frac{12}{x^2 - x} \div \frac{4x - 2}{x^2 - 1}$

7. $\frac{x^2 - 9}{x^2 - 16} \div \frac{x^2 - x - 12}{x^2 + x - 12}$

Classroom Example
Find the quotient:

$\frac{y^2 - 2y - 8}{y^2 + y} \div (y + 2)$

Answer: $\frac{y - 4}{y(y + 1)}$

EXAMPLE 7 Dividing Rational Expressions

Find the quotient and simplify: $\frac{a^2 + 3a + 2}{a^2 - 9} \div (a + 2)$

Solution

We write the divisor $a + 2$ as $\frac{a + 2}{1}$ and proceed as we did in Example 6.

$$\frac{a^2 + 3a + 2}{a^2 - 9} \div (a + 2) = \frac{a^2 + 3a + 2}{a^2 - 9} \div \frac{a + 2}{1}$$

$$= \frac{a^2 + 3a + 2}{a^2 - 9} \cdot \frac{1}{a + 2}$$

Factor: $= \frac{(a + 2)(a + 1)}{(a + 3)(a - 3)} \cdot \frac{1}{a + 2}$

Multiply: $= \frac{(a + 2)(a + 1)}{(a + 3)(a - 3)(a + 2)}$

Divide out common factors: $= \frac{\cancel{(a + 2)}(a + 1)}{(a + 3)(a - 3)\cancel{(a + 2)}}$

$$= \frac{a + 1}{(a + 3)(a - 3)}$$

QUICK ✓ *Find the quotient and simplify.*

8. $\frac{q^2 - 6q - 7}{q^2 - 25} \div (q - 7)$

As we stated earlier, we could write an expression such as $\frac{20x^5}{3y} \div \frac{4x^2}{15y^5}$ vertically as $\frac{\frac{20x^5}{3y}}{\frac{4x^2}{15y^5}}$. We use this form of division in the next example.

Classroom Example
Find the quotient and simplify:

$$\frac{\frac{16z}{7z+14}}{\frac{8z^2}{z^2-4}}$$

Answer: $\frac{2(z-2)}{7z}$

EXAMPLE 8 **Dividing Rational Expressions Written Vertically**

Find the quotient and simplify: $\dfrac{\frac{12x}{5x+20}}{\frac{4x^2}{x^2-16}}$

Solution

Multiply by the reciprocal of the divisor: $\dfrac{\frac{12x}{5x+20}}{\frac{4x^2}{x^2-16}} = \dfrac{12x}{5x+20}\cdot\dfrac{x^2-16}{4x^2}$

Factor each polynomial in the numerator and denominator: $= \dfrac{4\cdot 3\cdot x}{5(x+4)}\cdot\dfrac{(x-4)(x+4)}{4\cdot x\cdot x}$

Multiply: $= \dfrac{4\cdot 3\cdot x(x-4)(x+4)}{5(x+4)\cdot 4\cdot x\cdot x}$

Divide out common factors 4, x, and $x+4$: $= \dfrac{\cancel{4}\cdot 3\cdot \cancel{x}(x-4)\cancel{(x+4)}}{5\cancel{(x+4)}\cdot \cancel{4}\cdot \cancel{x}\cdot x}$

$= \dfrac{3(x-4)}{5x}$

QUICK ✓ *Find the quotient and simplify.*

9. $\dfrac{\frac{x+3}{x^2-4}}{\frac{4x+12}{7x^2+14x}}$

Classroom Example
Find the quotient and simplify:

$$\frac{3x-3y}{5z^2} \div \frac{6y-6x}{20z}$$

Answer: $-\frac{2}{z}$

EXAMPLE 9 **Dividing Rational Expressions with More Than One Variable**

Find the quotient and simplify: $\dfrac{5a-5b}{7c^2} \div \dfrac{10b-10a}{21c}$

Solution

Multiply by the reciprocal of the divisor: $\dfrac{5a-5b}{7c^2} \div \dfrac{10b-10a}{21c} = \dfrac{5a-5b}{7c^2}\cdot\dfrac{21c}{10b-10a}$

Factor each polynomial in the numerator and denominator: $= \dfrac{5(a-b)}{7\cdot c\cdot c}\cdot\dfrac{7\cdot 3\cdot c}{5\cdot 2\cdot(-1)(a-b)}$

Multiply: $= \dfrac{5(a-b)\cdot 7\cdot 3\cdot c}{7\cdot c\cdot c\cdot 5\cdot 2\cdot(-1)(a-b)}$

Divide out common factors, 5, 7, c, and $(a-b)$: $= \dfrac{\cancel{5}\cancel{(a-b)}\cdot \cancel{7}\cdot 3\cdot \cancel{c}}{\cancel{7}\cdot c\cdot \cancel{c}\cdot \cancel{5}\cdot 2\cdot(-1)\cancel{(a-b)}}$

$= \dfrac{3}{c\cdot 2\cdot(-1)}$

Multiply: $= -\dfrac{3}{2c}$

So $\dfrac{5a-5b}{7c^2} \div \dfrac{10b-10a}{21c} = -\dfrac{3}{2c}$.

QUICK ✓ *Find the quotient and simplify.*

10. $\dfrac{3m-6n}{5n} \div \dfrac{m^2-4n^2}{10mn}$

5.2 Exercises

For Extra Help:

Student Solutions Manual

CD Video

PH Math/Tutor Center

MathXL Tutorials on CD

MathXL® MyMathLab

1. quotient, product
2. reciprocal
3. factor, common factors
4. False
5. False
6. False
7. Answers may vary.
8. Answers may vary.
9. $\frac{16}{63x^6}$
10. $\frac{3}{2}$
11. $-\frac{5a^4}{8}$
12. $\frac{1}{8x^2}$
13. $\frac{x}{(x+1)^2}$
14. $\frac{2(n+2)}{3(n-2)}$
15. $p - 1$
16. $z + 2$
17. $\frac{18}{25z}$
18. $\frac{27}{7b^4}$
19. $-\frac{3m^5}{10}$
20. $\frac{3y^3}{10x^2}$
21. x
22. $\frac{9y(y-1)}{2}$
23. 1
24. $\frac{3}{2}$
25. $\frac{10}{3}$
26. $\frac{539}{676}$
27. 1

Concepts and Vocabulary

In Problems 1–3, fill in the blanks.

1. The answer to a division problem is called the _______ and the answer to a multiplication problem is called the _______.

2. To divide two rational expressions, rewrite the problem as an equivalent multiplication problem by finding the _______ of the divisor.

3. To multiply two rational expressions, first _______ each numerator and denominator and then divide out any _______ _______.

In Problems 4–6, answer True or False to each statement.

4. $\frac{3}{x} \cdot \frac{x}{9} = 3$ **5.** $\frac{8}{x-7} \cdot \frac{7-x}{16x} = \frac{1}{2x}$ **6.** $\frac{3}{x^2} \div \frac{x}{9} = \frac{x^2}{3} \cdot \frac{x}{9}$

7. To multiply two fractions, we can multiply and then simplify or multiply, factor, and then divide out common factors. Use $\frac{4}{15} \cdot \frac{25}{28}$ to demonstrate these two techniques. When multiplying rational expressions, why would it not be a good idea to first multiply the numerators and denominators and then simplify the resulting rational expression?

8. A unit fraction is any representation of one. For instance, both $\frac{x}{x}$ and $\frac{1 \text{ foot}}{12 \text{ inches}}$ are unit fractions. Since multiplying by one does not change the value of any real number, use this concept to explain how to convert 56 inches into feet.

Building Skills

In Problems 9–16, multiply and simplify the result, if possible.

9. $\frac{4x^2}{9x^5} \cdot \frac{12}{21x^3}$ **10.** $-\frac{8y}{12y^5} \cdot \left(-\frac{9y^6}{4y^2}\right)$ **11.** $-\frac{15a^2b^2}{18ab} \cdot \frac{12a^3}{16b}$

12. $\frac{5x^2y}{26y^3} \cdot \frac{13xy^2}{20x^5}$ **13.** $\frac{x^2 - x}{x^2 - x - 2} \cdot \frac{x-2}{x^2-1}$ **14.** $\frac{8n - 8}{n^2 - 3n + 2} \cdot \frac{n+2}{12}$

15. $\frac{p^2 - 1}{2p - 3} \cdot \frac{2p^2 + p - 6}{p^2 + 3p + 2}$ **16.** $\frac{z^2 - 4}{3z - 2} \cdot \frac{3z^2 + 7z - 6}{z^2 + z - 6}$

In Problems 17–24, divide and simplify the result, if possible.

17. $\frac{12z^2}{35} \div \frac{20z^6}{42z^3}$ **18.** $\frac{18a^2b}{5b^3} \div \frac{14a^3b^3}{15ab}$ **19.** $\frac{-\frac{3m^5}{8m^2}}{\frac{15m}{12m^3}}$ **20.** $\frac{-\frac{27xy^2}{60x^2y^4}}{-\frac{36x}{24y^5}}$

21. $\frac{x^2 - x}{x^2 - 1} \div \frac{x + 2}{x^2 + 3x + 2}$ **22.** $\frac{3y^2 + 6y}{8} \div \frac{y+2}{12y - 12}$

23. $\frac{(x+2)^2}{x^2 - 4} \div \frac{x^2 - x - 6}{x^2 - 5x + 6}$ **24.** $\frac{4z^2 + 12z + 9}{8z + 16} \div \frac{(2z+3)^3}{12z + 24}$

Mixed Practice

In Problems 25–60, perform the indicated operation and simplify.

25. $\frac{14}{9} \cdot \frac{15}{7}$ **26.** $\frac{7}{52} \cdot \frac{77}{13}$ **27.** $-\frac{20}{16} \div \left(-\frac{30}{24}\right)$

28. $-\frac{44}{5}$ 29. $-\frac{4}{11}$
30. $-\frac{4}{3}$ 31. $\frac{4}{3y(y-3)}$
32. $\frac{1}{(x+4)(2x-3)}$
33. $\frac{a}{3}$ 34. 1 35. $\frac{3x}{x-4}$
36. $\frac{2(x+5)(x+5)}{45x}$
37. $\frac{4-w}{(w+2)(w+4)}$ 38. -2
39. $-\frac{(x-y)^2}{x}$ 40. $r-3s$
41. $-\frac{1}{2}$ 42. $\frac{1}{9}$ 43. 1
44. $\frac{x+2}{x-2}$
45. $\frac{3(2n+3)(n-3)}{2(n+4)}$
46. $\frac{x^2+1}{(x+1)(x+1)}$
47. $\frac{x}{4y^2(x-2y)}$
48. $\frac{2x^3(x-3)}{(x+3)}$
49. $\frac{2a-b}{2a(a+b)}$
50. $-\frac{3(x-2y)}{2(2x-y)}$
51. $\frac{(t+3)(t-2)}{3}$
52. $-\frac{p-7}{(p+2)(p-2)}$
53. -1
54. $\frac{3}{2(2z+3)}$
55. $-\frac{x(x+3)}{(x+1)^2}$
56. $\frac{b-9}{b-3}$
57. $\frac{a+b}{2}$
58. $\frac{t(t+3)}{(t+4)(t-3)}$
59. $\frac{1}{3}$
60. $\frac{p+2pq+4p^2q^2}{p+2q}$
61. $\frac{1}{3(x-y)}$
62. $\frac{x}{14y}$
63. $\frac{9}{8}$
64. $\frac{2(x-3)^2}{x-9}$
65. $\frac{x-y}{(x+y)^3}$

28. $-\frac{60}{35} \div \frac{15}{77}$

29. $\frac{8}{11} \div (-2)$

30. $\frac{16}{3} \div (-4)$

31. $\frac{3y}{y^2 - y - 6} \cdot \frac{4y + 8}{9y^2}$

32. $\frac{x - 4}{12x - 18} \cdot \frac{6}{x^2 - 16}$

33. $\frac{4a + 8b}{a^2 + 2ab} \cdot \frac{a^2}{12}$

34. $\frac{m^2 - n^2}{3m - 3n} \cdot \frac{6}{2m + 2n}$

35. $\frac{3x^2 - 6x}{x^2 - 2x - 8} \div \frac{x - 2}{x + 2}$

36. $\frac{x + 5}{3} \div \frac{30x}{4x + 20}$

37. $\frac{(w - 4)^2}{4 - w^2} \div \frac{w^2 - 16}{w - 2}$

38. $\frac{a^2 - b^2}{b - a} \cdot \frac{2a + 2b}{a^2 + 2ab + b^2}$

39. $\frac{3xy - 2y^2 - x^2}{x + y} \cdot \frac{x^2 - y^2}{x^2 - 2xy}$

40. $\frac{2r^2 + rs - 3s^2}{r^2 - s^2} \cdot \frac{r^2 - 2rs - 3s^2}{2r + 3s}$

41. $\dfrac{\frac{2c - 4}{8}}{\frac{2 - c}{2}}$

42. $\dfrac{\frac{2x - 3}{3}}{6x - 9}$

43. $(x + 1) \cdot \frac{x - 6}{x^2 - 5x - 6}$

44. $(x - 3) \cdot \frac{x + 2}{x^2 - 5x + 6}$

45. $\frac{4n^2 - 9}{6n + 18} \cdot \frac{9n^2 - 81}{2n^2 + 5n - 12}$

46. $\frac{2x^2 - 5x + 3}{x^2 - 1} \cdot \frac{x^2 + 1}{2x^2 - x - 3}$

47. $\frac{x^2y}{2x^2 - 5xy + 2y^2} \div \frac{(2xy^2)^2}{2x^2y - xy^2}$

48. $\frac{(x - 3)^2}{8xy^2} \div \frac{x^2 - 9}{(4x^2y)^2}$

49. $\frac{2a^2 + 3ab - 2b^2}{a^2 - b^2} \cdot \frac{a^2 - ab}{2a^3 + 4a^2b}$

50. $\frac{3y^2 - 3x^2}{2x^2 + xy - y^2} \cdot \frac{3x - 6y}{6x - 6y}$

51. $\frac{3t^2 - 27}{t + 2} \cdot \frac{t^2 - 4}{9t - 27}$

52. $\frac{p^2 - 49}{p^2 - 5p - 14} \cdot \frac{p - 7}{14 - 5p - p^2}$

53. $\frac{(x + 2)^2}{x^2 - 4} \div \frac{-x^2 + x + 6}{x^2 - 5x + 6}$

54. $\frac{4z^2 + 12z + 9}{8z + 16} \div \frac{(2z + 3)^3}{12z + 24}$

55. $\frac{9 - x^2}{x^2 + 5x + 4} \div \frac{x^2 - 2x - 3}{x^2 + 4x}$

56. $\frac{1}{b^2 + b - 12} \div \frac{1}{b^2 - 5b - 36}$

57. $\dfrac{\frac{a^2 - b^2}{a^2 + b^2}}{\frac{4a - 4b}{2a^2 + 2b^2}}$

58. $\dfrac{\frac{t^2}{t^2 - 16}}{\frac{t^2 - 3t}{t^2 - t - 12}}$

59. $\frac{x^3 - 1}{x^4 - 1} \div \frac{3x^2 + 3x + 3}{x^3 + x^2 + x + 1}$

60. $\frac{p^3 - 8q^3}{p^2 - 4q^2} \div \frac{p^2 + 4pq + 4q^2}{(p + 2q)^2}$

Applying the Concepts

61. Find the product of $\frac{x^2 + 3xy + 2y^2}{x^2 - y^2}$ and $\frac{3x - 3y}{9x^2 + 9xy - 18y^2}$.

62. Find the product of $\frac{x}{6y^2}$ and $\frac{21x^2y}{(7x)^2}$.

63. Find the quotient of $\frac{x}{2y}$ and $\frac{(2xy)^2}{9xy^3}$.

64. Find the quotient of $\frac{x - 3}{2x + 6}$ and $\frac{x - 9}{4x^2 - 36}$.

65. Find $\frac{x - y}{x + y}$ squared divided by $x^2 - y^2$.

66. $\frac{a-b}{a+b}$
67. $\frac{4x}{(x-2)(x-3)}$
68. $\frac{x^2-3}{4x^2}$
69. $\frac{1}{x}$ square feet
70. $2x$ square meters
71. $\frac{2(x+2)}{x-3}$ square inches
72. $\frac{2}{(x+3)(x+2)}$ square centimeters
73. $\frac{(x-a)^2}{2(x+a)}$
74. $-(x^2+1)$
75. $\frac{1-x}{4}$
76. $\frac{(x-3y)^2}{x+3y}$
77. $\frac{2a^2}{3}$
78. $\frac{(2v-1)^3(1+2v+4v^2)}{(1-2v)^4}$
79. $\frac{(x+y)^2}{(x^2-xy+y^2)(x+2)}$
80. $\frac{(a-2)(a+3)}{8a^3(a+9)}$
81. $3p^5q^2(p^2+3pq+9q^2)(p-q)$
82. $\frac{x(x-1)}{(x+3)(x^2+9)}$
83. $6x+12$
84. x^2-4x+4

66. Find $(a-b)$ squared divided by a^2-b^2.

67. What is $\frac{3x-9}{2x+4}$ divided into $\frac{6x}{x^2-4}$?

68. What is $\frac{2x^2}{3}$ divided into $\frac{x^3-3x}{6x}$?

△ **69. Area of a Rectangle** Write an algebraic expression for the area of the rectangle.

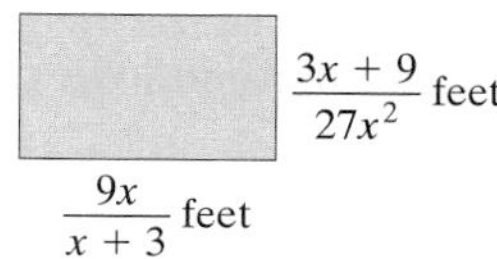

△ **70. Area of a Rectangle** Write an algebraic expression for the area of the rectangle.

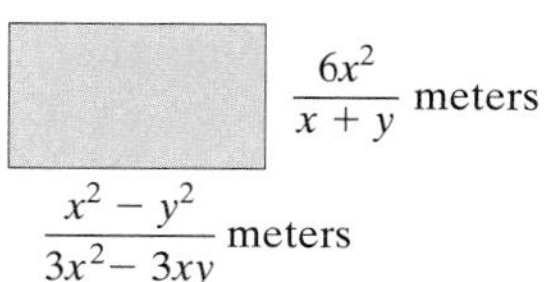

△ **71. Area of a Triangle** Write an algebraic expression for the area of the triangle.

$h = 4x^2 + 20x + 24$ inches

$b = \frac{1}{x^2-9}$ inches

△ **72. Area of a Triangle** Write an algebraic expression for the area of the triangle.

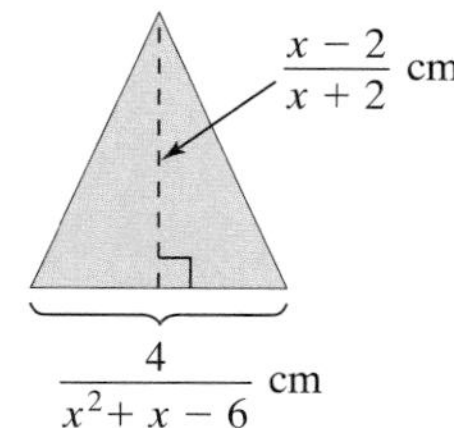

Extending the Concepts

In Problems 73–82, perform the indicated operation and simplify the result.

73. $\frac{xy-ay+xb-ab}{xy+ay-xb-ab} \cdot \frac{2xy-2ay-2xb+2ab}{4b+4y}$

74. $\frac{x^3-x^2+x-1}{x+1} \cdot \frac{x^2+2x+1}{1-x^2}$

75. $\frac{x^2+x-12}{x^2-2x-35} \div \frac{x+4}{x^2+4x-5} \div \frac{12-4x}{x-7}$

76. $\frac{x^2-6xy+9y^2}{x^2-4y^2} \div \frac{x-3y}{x^2-5xy+6y^2} \div \frac{x^2-9y^2}{x^2-xy-6y^2}$

77. $\frac{a^2-2ab}{2b-3a} \div \frac{3a^2-4ab-4b^2}{16a^2b^2-36a^4} \div (6a)$

78. $\frac{(2v-1)^6}{(1-2v)^5} \cdot \frac{1-8v^3}{4v^2-4v+1} \div (2v-1)$

79. $\frac{x^2+xy-3x-3y}{x^3+y^3} \cdot \frac{x^2+2xy+y^2}{x^2-x-6}$

80. $\frac{a^2+4a-12}{a^2+6a-27} \cdot \frac{a^2-9}{2a^2+12a} \div (4a^2)$

81. $\frac{p^3-27q^3}{9pq} \cdot \frac{(3p^2q)^3}{p^2-9q^2} \div \frac{1}{p^2+2pq-3q^2}$

82. $\frac{x^2-1}{x^4-81} \div \frac{x^2+1}{(x-3)^2} \cdot \frac{x^3+x}{x^2-2x-3}$

In Problems 83 and 84, find the missing expression.

83. $\frac{2x}{x^3-3x^2} \cdot \frac{x^2-x-6}{?} = \frac{1}{3x}$

84. $\frac{x^2+12x+36}{?} \div \frac{x^2+11x+30}{x^2+3x-10} = \frac{x+6}{x-2}$

5.3 Adding and Subtracting Rational Expressions with a Common Denominator

OBJECTIVES

1. Add Rational Expressions with a Common Denominator
2. Subtract Rational Expressions with a Common Denominator
3. Add or Subtract Rational Expressions with Opposite Denominators

Preparing for Adding and Subtracting Rational Expressions with a Common Denominator

Before getting started, take the following readiness quiz. If you get a problem wrong, go back to the section cited and review the material.

1. Write $\frac{12}{15}$ in lowest terms. [Appendix, Section A.1, p. A5]

2. Find each sum and write in lowest terms:

(a) $\frac{7}{5}+\frac{2}{5}$ **(b)** $\frac{5}{6}+\frac{11}{6}$ [Section 1.4, p. 31]

3. Find each difference and write in lowest terms:

(a) $\frac{7}{9}-\frac{5}{9}$ **(b)** $\frac{7}{8}-\frac{5}{8}$ [Section 1.4, p. 32]

4. Determine the additive inverse of 5. [Section 1.3, p. 21]

5. Simplify: $-(x-2)$ [Section 1.7, p. 61]

In the previous section, we learned how to multiply and divide rational expressions. We now learn how to add and subtract rational expressions.

1 Add Rational Expressions with a Common Denominator

Teaching Tip
You may wish to present two or three examples of adding rational numbers before introducing adding rational expressions.

The rules for adding rational expressions are the same as the rules for adding rational numbers. If the denominators of two rational expressions are the same, then we add the numerators, write the result over the common denominator, and simplify if possible. Two examples are worked below. In the left column, rational numbers are added; in the right column, rational algebraic expressions are added. Do you see that the steps are the same?

ADDING RATIONAL NUMBERS	ADDING RATIONAL EXPRESSIONS
$\frac{3}{5}+\frac{8}{5}=\frac{3+8}{5}=\frac{11}{5}$	$\frac{3}{5a}+\frac{8}{5a}=\frac{3+8}{5a}=\frac{11}{5a}, a \neq 0$
$\frac{7}{12}+\frac{11}{12}=\frac{18}{12}=\frac{\cancel{6}\cdot 3}{\cancel{6}\cdot 2}=\frac{3}{2}$	$\frac{7}{12z}+\frac{11}{12z}=\frac{18}{12z}=\frac{\cancel{6}\cdot 3}{\cancel{6}\cdot 2\cdot z}=\frac{3}{2z}, z \neq 0$

Throughout the section, we assume that a variable cannot take on any values that result in division by zero.

Preparing for...Answers **1.** $\frac{4}{5}$ **2. (a)** $\frac{9}{5}$ **(b)** $\frac{8}{3}$ **3. (a)** $\frac{2}{9}$ **(b)** $\frac{1}{4}$ **4.** -5 **5.** $2-x$

EXAMPLE 1 How to Add Rational Expressions with a Common Denominator

Find the sum and simplify, if possible: $\frac{5}{x+1}+\frac{3}{x+1}$

Classroom Example
Find the sum and simplify, if possible:
$\frac{4}{x+3}+\frac{6}{x+3}$
Answer: $\frac{10}{x+3}$

Step-by-Step Solution

The rational expressions in the sum $\frac{5}{x+1}+\frac{3}{x+1}$ have a common denominator, $x+1$.

Step 1: Add the numerators and write the result over the common denominator.

Use $\frac{a}{c}+\frac{b}{c}=\frac{a+b}{c}$: $\frac{5}{x+1}+\frac{3}{x+1}=\frac{5+3}{x+1}$

Combine like terms in the numerator: $=\frac{8}{x+1}$

Step 2: Simplify the rational expression by writing the rational expression in lowest terms. | The sum is already simplified.

So $\dfrac{5}{x+1} + \dfrac{3}{x+1} = \dfrac{8}{x+1}$.

Here are the steps we use to add rational expressions with a common denominator.

In Words
To add rational expressions with a common denominator, add the numerators and write the result over the common denominator. Then simplify, if necessary.

Steps to Add Rational Expressions

Step 1: Use the fact that if $\dfrac{a}{c}$ and $\dfrac{b}{c}$, $c \neq 0$, are two rational expressions, then

$$\frac{a}{c} + \frac{b}{c} = \frac{a+b}{c}$$

to add the rational expressions.

Step 2: Simplify the sum by writing the rational expression in lowest terms. This step will not always be necessary.

Classroom Example Find the sum and simplify the result, if possible.

$$\frac{a^2}{a+5} + \frac{5a}{a+5}$$

Answer: a

EXAMPLE 2 Adding Rational Expressions with a Common Denominator

Find the sum $\dfrac{x^2}{x+7} + \dfrac{7x}{x+7}$ and simplify the result, if possible.

Solution

Use $\dfrac{a}{c} + \dfrac{b}{c} = \dfrac{a+b}{c}$: $\quad \dfrac{x^2}{x+7} + \dfrac{7x}{x+7} = \dfrac{x^2+7x}{x+7}$

Factor the numerator: $\quad = \dfrac{x(x+7)}{x+7}$

Divide out like factors: $\quad = \dfrac{x}{1} = x$

QUICK ✓ *Find the sum and simplify the result, if possible.*

1. $\dfrac{1}{x-2} + \dfrac{3}{x-2}$

2. $\dfrac{2x+1}{x+1} + \dfrac{x^2}{x+1}$

Classroom Example Find the sum and simplify the result, if possible:

$$\frac{3a^2+a}{a^2-9} + \frac{2a-2a^2}{a^2-9}$$

Answer: $\dfrac{a}{a-3}$

EXAMPLE 3 Adding Rational Expressions in Which a Numerator Contains More Than One Term

Find the sum: $\dfrac{2x^2+x}{x^2-4} + \dfrac{x-x^2}{x^2-4}$. Simplify the result, if possible.

Solution

Use $\dfrac{a}{c} + \dfrac{b}{c} = \dfrac{a+b}{c}$: $\quad \dfrac{2x^2+x}{x^2-4} + \dfrac{x-x^2}{x^2-4} = \dfrac{2x^2+x+x-x^2}{x^2-4}$

Combine like terms in the numerator: $\quad = \dfrac{x^2+2x}{x^2-4}$

Factor the numerator and denominator: $\quad = \dfrac{x(x+2)}{(x+2)(x-2)}$

Divide out like factors: $\quad = \dfrac{x}{x-2}$

QUICK ✓ *Find the sum and simplify the result, if possible.*

3. $\dfrac{9x}{6x-5}+\dfrac{2x-3}{6x-5}$

4. $\dfrac{2x-2}{2x^2-7x-15}+\dfrac{5}{2x^2-7x-15}$

2 Subtract Rational Expressions with a Common Denominator

The steps for subtracting rational expressions are the same as the steps for subtracting rational numbers. If the denominators of two rational expressions are the same, then we subtract the numerators and write the result over the common denominator.

SUBTRACTING RATIONAL NUMBERS	SUBTRACTING RATIONAL EXPRESSIONS
$\dfrac{7}{3}-\dfrac{2}{3}=\dfrac{7-2}{3}=\dfrac{5}{3}$	$\dfrac{7}{3x}-\dfrac{2}{3x}=\dfrac{7-2}{3x}=\dfrac{5}{3x}$
$\dfrac{3}{10}-\dfrac{7}{10}=\dfrac{3-7}{10}=\dfrac{-4}{10}=\dfrac{-2\cdot 2}{5\cdot 2}=\dfrac{-2}{5}$	$\dfrac{3}{10b}-\dfrac{7}{10b}=\dfrac{3-7}{10b}=\dfrac{-4}{10b}=\dfrac{-2\cdot 2}{5\cdot 2\cdot b}=\dfrac{-2}{5b}$

Once again, we exclude values of the variable that result in division by zero.

EXAMPLE 4 How to Subtract Rational Expressions with a Common Denominator

Find the difference: $\dfrac{2n^2}{n^2-1}-\dfrac{2n}{n^2-1}$. Simplify the result, if possible.

Step-by-Step Solution

Step 1: Subtract the numerators and write the result over the common denominator.

Use $\dfrac{a}{c}-\dfrac{b}{c}=\dfrac{a-b}{c}$: $\dfrac{2n^2}{n^2-1}-\dfrac{2n}{n^2-1}=\dfrac{2n^2-2n}{n^2-1}$

Step 2: Simplify the rational expression by writing the rational expression in lowest terms.

Factor the numerator and denominator: $=\dfrac{2n(n-1)}{(n-1)(n+1)}$

Divide out like factors: $=\dfrac{2n}{n+1}$

Therefore, $\dfrac{2n^2}{n^2-1}-\dfrac{2n}{n^2-1}=\dfrac{2n}{n+1}$.

Classroom Example Find the difference and simplify the result, if possible:

$\dfrac{3k^2}{k^2-k-2}-\dfrac{6k}{k^2-k-2}$

Answer: $\dfrac{3k}{k+1}$

Here are the steps we use to subtract rational expressions with a common denominator.

In Words
To subtract rational expressions with a common denominator, subtract the numerators and write the result over the common denominator. Then simplify, if possible.

Steps to Subtract Rational Expressions with a Common Denominator

Step 1: Use the fact that if $\dfrac{a}{c}$ and $\dfrac{b}{c}$, $c \neq 0$, are two rational expressions, then

$$\frac{a}{c}-\frac{b}{c}=\frac{a-b}{c}$$

to subtract the rational expressions.

Step 2: Simplify the difference by writing the rational expression in lowest terms. This step will not always be necessary.

QUICK ✓ *Find the difference and simplify the result, if possible.*

5. $\dfrac{8y}{2y - 5} - \dfrac{6}{2y - 5}$

6. $\dfrac{10 + 3z}{6z} - \dfrac{7}{6z}$

When finding the difference between two rational expressions in which the numerator of the rational expression being subtracted contains more than one term, enclose the terms in the numerator that follow the subtraction sign in parentheses. This will remind you to distribute the minus sign across all the terms. The next example illustrates this point.

Classroom Example ➤
Find the difference and simplify the result:

$\dfrac{7x - 3}{x + 2} - \dfrac{3x - 11}{x + 2}$

Answer: 4

EXAMPLE 5 Subtracting Rational Expressions in Which a Numerator Contains More than One Term

Find the difference: $\dfrac{9x + 1}{x + 1} - \dfrac{6x - 2}{x + 1}$. Simplify the result.

Solution

First, we note that the rational expressions have the same denominator, $x + 1$.

Work Smart
Enclose the terms in the numerator that follow the subtraction sign in parentheses because this will remind you to distribute the minus sign across all the terms.

Notice the use of parentheses: $\dfrac{9x + 1}{x + 1} - \dfrac{6x - 2}{x + 1} = \dfrac{9x + 1 - (6x - 2)}{x + 1}$

Distribute the minus sign into the parentheses: $= \dfrac{9x + 1 - 6x + 2}{x + 1}$

Combine like terms: $= \dfrac{3x + 3}{x + 1}$

Factor the numerator: $= \dfrac{3(x + 1)}{x + 1}$

Divide out like factors: $= \dfrac{3}{1} = 3$

QUICK ✓ *Find the difference and simplify the result.*

7. $\dfrac{2x^2 - 5x}{3x} - \dfrac{x^2 - 13x}{3x}$

8. $\dfrac{3x^2 + 8x - 1}{x^2 - 3x - 28} - \dfrac{2x^2 + 2x - 9}{x^2 - 3x - 28}$

3 Add or Subtract Rational Expressions with Opposite Denominators

Suppose you were asked to find the sum $\dfrac{w^2 - 3w}{w - 4} + \dfrac{4}{4 - w}$. Although the denominators of these rational expressions are different, you might have noticed that they are additive inverses (opposites) of each other. Recall that $4 - w = -w + 4 = -1(w - 4)$. We use this idea to rewrite the sum so that each rational expression has the same denominator.

Classroom Example ➤
Find the sum and simplify the result:

$\dfrac{n^2 - 2n}{n - 4} + \dfrac{8}{4 - n}$

Answer: $n + 2$

EXAMPLE 6 Adding Rational Expressions with Opposite Denominators

Find the sum: $\dfrac{w^2 - 3w}{w - 4} + \dfrac{4}{4 - w}$. Simplify the result.

Solution

We rewrite the denominators so they are the same by factoring -1 from $4 - w$ in the second denominator.

Factor -1 from $4 - w$: $\dfrac{w^2 - 3w}{w - 4} + \dfrac{4}{4 - w} = \dfrac{w^2 - 3w}{w - 4} + \dfrac{4}{-1(w - 4)}$

Use $\dfrac{a}{-b} = \dfrac{-a}{b}$: $= \dfrac{w^2 - 3w}{w - 4} + \dfrac{-4}{w - 4}$

$= \dfrac{w^2 - 3w - 4}{w - 4}$

Factor: $= \dfrac{(w - 4)(w + 1)}{w - 4}$

Divide out like factors: $= \dfrac{w + 1}{1}$

$= w + 1$

So $\dfrac{w^2 - 3w}{w - 4} + \dfrac{4}{4 - w} = w + 1.$

QUICK ✓ *Find the sum and simplify the result.*

9. $\dfrac{3x}{x - 5} + \dfrac{1}{5 - x}$ **10.** $\dfrac{a^2 + 2a}{a - 7} + \dfrac{a^2 + 14}{7 - a}$

Classroom Example
Find the difference and simplify the result:

$$\frac{p^2}{p^2 - 81} - \frac{9p}{81 - p^2}$$

Answer: $\dfrac{p}{p - 9}$

EXAMPLE 7 Subtracting Rational Expressions with Opposite Denominators

Find the difference: $\dfrac{b^2 - 11}{b^2 - 25} - \dfrac{3b + 1}{25 - b^2}$. Simplify the result.

Solution

We must first rewrite the denominators so that they are the same. This is done by factoring -1 from $25 - b^2$.

$$\frac{b^2 - 11}{b^2 - 25} - \frac{3b + 1}{25 - b^2} = \frac{b^2 - 11}{b^2 - 25} - \frac{3b + 1}{-(b^2 - 25)}$$

$-\dfrac{a}{-b} = +\dfrac{a}{b}$: $= \dfrac{b^2 - 11}{b^2 - 25} + \dfrac{3b + 1}{b^2 - 25}$

Add numerators: $= \dfrac{b^2 - 11 + 3b + 1}{b^2 - 25}$

Combine like terms: $= \dfrac{b^2 + 3b - 10}{b^2 - 25}$

Factor: $= \dfrac{(b + 5)(b - 2)}{(b + 5)(b - 5)}$

Divide out like factors: $= \dfrac{b - 2}{b - 5}$

QUICK ✓ *Find the difference and simplify the result.*

11. $\dfrac{2n}{n^2 - 9} - \dfrac{6}{9 - n^2}$ **12.** $\dfrac{2k}{6k - 6} - \dfrac{9 + 4k}{6 - 6k}$

5.3 Exercises

For Extra Help: Student Solutions Manual  CD Video  PH Math/Tutor Center  MathXL Tutorials on CD MathXL® MyMathLab

1. numerators
2. $x + 3$
3. $a + b$; c
4. False
5. True
6. True
7. Incorrect; Answers may vary.
8. Answers may vary.
9. $\frac{7p}{4}$
10. $\frac{2m}{3}$
11. $2n$
12. $\frac{3 + y}{x}$
13. $\frac{2a - 1}{a}$
14. $2n$
15. $\frac{2(5c - 2)}{c - 1}$
16. $\frac{2}{3}$
17. 2
18. 4
19. $\frac{7x(1 + x)}{x + 2}$
20. $\frac{5p}{p - 1}$
21. 2
22. $2(x + 2)$
23. $2x$
24. $\frac{2n^2}{3}$
25. $x - 3$
26. $\frac{-x + 7}{8}$
27. $-\frac{x}{2}$
28. $2(x + 2)$
29. $-\frac{c + 3}{c}$
30. $2(x - 1)$
31. 3

Concepts and Vocabulary

In Problems 1–3, fill in the blanks.

1. When adding or subtracting rational expressions written with a common denominator, add or subtract the ________ only.

2. $\frac{3x + 1}{x - 4} - \frac{x + 3}{x - 4} = \frac{3x + 1 - (\quad)}{x - 4}$.

3. If $\frac{a}{c}$ and $\frac{b}{c}$, $c \neq 0$, are two rational expressions, then $\frac{a}{c} + \frac{b}{c} = \frac{\quad}{\quad}$.

In Problems 4–6, answer True or False to each statement.

4. $\frac{2x}{a} + \frac{4x}{a} = \frac{6x}{2a} = \frac{3x}{a}$

5. When adding or subtracting rational expressions the result should always be simplified.

6. $8 - x = -1(x - 8)$

7. Is the following correct or incorrect? Explain your reasoning.

$$\frac{x - 2}{x} - \frac{x + 4}{x} = \frac{x - 2 - x + 4}{x} = \frac{2}{x}$$

8. The denominators in the expression $\frac{4x}{x - 2} - \frac{2x}{2 - x}$ are opposites. Explain how to rewrite this difference so that the terms have a common denominator and then describe the steps to simplify the result.

Building Skills

In Problems 9–22, add the rational expressions and simplify the result, if possible.

9. $\frac{3p}{8} + \frac{11p}{8}$

10. $\frac{2m}{9} + \frac{4m}{9}$

11. $\frac{n}{2} + \frac{3n}{2}$

12. $\frac{3}{x} + \frac{y}{x}$

13. $\frac{4a - 1}{3a} + \frac{2a - 2}{3a}$

14. $\frac{4n + 1}{8} + \frac{12n - 1}{8}$

15. $\frac{8c - 3}{c - 1} + \frac{2c - 1}{c - 1}$

16. $\frac{4n}{6n + 27} + \frac{18}{6n + 27}$

17. $\frac{2x}{x + y} + \frac{2y}{x + y}$

18. $\frac{4}{2x + 1} + \frac{8x}{2x + 1}$

19. $\frac{7x}{x + 2} + \frac{7x^2}{x + 2}$

20. $\frac{4p - 1}{p - 1} + \frac{p + 1}{p - 1}$

21. $\frac{12a - 1}{2a + 6} + \frac{13 - 8a}{2a + 6}$

22. $\frac{3x^2 + 4}{3x - 6} + \frac{3x^2 - 28}{3x - 6}$

In Problems 23–36, subtract the rational expressions and simplify the result, if possible.

23. $\frac{11x}{3} - \frac{5x}{3}$

24. $\frac{5n^2}{6} - \frac{n^2}{6}$

25. $\frac{4x - 3}{3} - \frac{x + 6}{3}$

26. $\frac{x + 2}{8} - \frac{2x - 5}{8}$

27. $\frac{x^2 - x}{2x} - \frac{2x^2 - x}{2x}$

28. $\frac{3x + 3x^2}{x} - \frac{x^2 - x}{x}$

29. $\frac{2c - 3}{4c} - \frac{6c + 9}{4c}$

30. $\frac{7x}{3} - \frac{x + 6}{3}$

31. $\frac{x^2 - 6}{x - 2} - \frac{x^2 - 3x}{x - 2}$

32. $\frac{x-4}{x-1}$ 33. -1
34. $\frac{z+6}{z-1}$ 35. $x-1$
36. $\frac{n+3}{n+1}$ 37. $-\frac{1}{x}$
38. x 39. -1
40. 1 41. 0
42. $\frac{3n(n-2)}{2n-1}$ 43. $n-1$
44. $\frac{x(x+2)}{x^2+1}$ 45. $\frac{3}{v+3}$
46. $\frac{x-2}{x-1}$ 47. $\frac{x+1}{x-1}$ 48. $\frac{x+3}{x-3}$
49. $\frac{x^2+3}{x(x-3)}$ 50. $\frac{4}{x-2}$
51. $\frac{1}{a-5}$ 52. $\frac{1}{x-3}$ 53. 1
54. $\frac{4(1-2x)}{x-1}$ 55. $\frac{2(2p+3q)}{p-q}$
56. $\frac{2(a+2b)}{a-b}$ 57. $\frac{2p^2+2p+1}{p-1}$
58. $\frac{x^2+x+6}{x-1}$
59. $\frac{4b}{a-b}$
60. $2k$
61. -1
62. $\frac{x+3}{x+1}$
63. $\frac{2}{x-y}$
64. $\frac{m+n}{n-m}$
65. $2(x-1)$
66. $\frac{x^2-3x-2}{x^2-1}$
67. $\frac{3x+3y-2}{x-y}$
68. $\frac{-4s+t}{s-t}$
69. $\frac{p+2q}{p^2-6q^2}$
70. $\frac{n+3}{2n-3}$
71. $\frac{2g-7}{g-3}$
72. $\frac{6x+7y}{x^2-y^2}$
73. $12a$
74. $\frac{(n+2)(n+1)}{3n+2}$
75. $\frac{n-3}{n+1}$
76. $\frac{(p-2)(p-2)}{p-4}$
77. $\frac{4}{n}$
78. $\frac{-x-4}{2}$

32. $\dfrac{2x-3}{x-1} - \dfrac{x+1}{x-1}$ **33.** $\dfrac{2}{c^2-4} - \dfrac{c^2-2}{c^2-4}$ **34.** $\dfrac{2z^2+7z}{z^2-1} - \dfrac{z^2-6}{z^2-1}$

35. $\dfrac{2x^2+5x}{2x+3} - \dfrac{4x+3}{2x+3}$ **36.** $\dfrac{2n^2+n}{n^2-n-2} - \dfrac{n^2+6}{n^2-n-2}$

Mixed Practice

In Problems 37–72, perform the indicated operation and simplify the result, if possible.

37. $\dfrac{4}{7x} - \dfrac{11}{7x}$ **38.** $\dfrac{2x}{5} - \left(-\dfrac{3x}{5}\right)$ **39.** $\dfrac{5}{2x-5} - \dfrac{2x}{2x-5}$

40. $\dfrac{2b+1}{b+1} - \dfrac{b}{b+1}$ **41.** $\dfrac{2n^3}{n-1} - \dfrac{2n^3}{n-1}$ **42.** $\dfrac{3n^2}{2n-1} - \dfrac{6n}{2n-1}$

43. $\dfrac{n^2-3}{2n+3} + \dfrac{n^2+n}{2n+3}$ **44.** $\dfrac{5x}{x^2+1} + \dfrac{x^2-3x}{x^2+1}$ **45.** $\dfrac{3v-1}{v^2-9} - \dfrac{8}{v^2-9}$

46. $\dfrac{x^2-x}{x^2-1} - \dfrac{2}{x^2-1}$ **47.** $\dfrac{x^2}{x^2-1} + \dfrac{2x+1}{x^2-1}$ **48.** $\dfrac{x^2}{x^2-9} + \dfrac{6x+9}{x^2-9}$

49. $\dfrac{2x^2+3}{x^2-3x} - \dfrac{x^2}{x^2-3x}$ **50.** $\dfrac{4x}{x^2-4} + \dfrac{8}{x^2-4}$

51. $\dfrac{a}{a^2-3a-10} + \dfrac{2}{a^2-3a-10}$ **52.** $\dfrac{2x}{2x^2-x-15} + \dfrac{5}{2x^2-x-15}$

53. $\dfrac{n}{n-3} + \dfrac{3}{3-n}$ **54.** $\dfrac{4}{x-1} + \dfrac{8x}{1-x}$ **55.** $\dfrac{12q}{2p-2q} - \dfrac{8p}{2q-2p}$

56. $\dfrac{2a}{a-b} - \dfrac{4b}{b-a}$ **57.** $\dfrac{2p^2-1}{p-1} - \dfrac{2p+2}{1-p}$ **58.** $\dfrac{x^2+3}{x-1} - \dfrac{x+3}{1-x}$

59. $\dfrac{2a}{a-b} + \dfrac{2a-4b}{b-a}$ **60.** $\dfrac{4k^2}{2k-1} + \dfrac{2k}{1-2k}$ **61.** $\dfrac{3}{p^2-3} + \dfrac{p^2}{3-p^2}$

62. $\dfrac{x^2+6x}{x^2-1} + \dfrac{4x+3}{1-x^2}$ **63.** $\dfrac{2x}{x^2-y^2} - \dfrac{2y}{y^2-x^2}$ **64.** $\dfrac{m^2+2mn}{n^2-m^2} - \dfrac{n^2}{m^2-n^2}$

65. $\dfrac{3x^2}{x+1} - \dfrac{x^2+2}{x+1}$ **66.** $\dfrac{x^2}{x^2-1} - \dfrac{3x+2}{x^2-1}$ **67.** $\dfrac{3x-1}{x-y} + \dfrac{1-3y}{y-x}$

68. $\dfrac{2s-3t}{s-t} + \dfrac{6s-4t}{t-s}$ **69.** $\dfrac{2p-3q}{3p^2-18q^2} - \dfrac{9q+p}{18q^2-3p^2}$

70. $\dfrac{n+3}{2n^2-n-3} - \dfrac{-n^2-3n}{2n^2-n-3}$ **71.** $\dfrac{2g}{g-3} - \left(\dfrac{g+3}{g-3} - \dfrac{g-4}{g-3}\right)$

72. $\dfrac{7x-3y}{x^2-y^2} - \left(\dfrac{2x-3y}{x^2-y^2} - \dfrac{x+7y}{x^2-y^2}\right)$

Applying the Concepts

73. Find the sum of $\dfrac{7a^2}{a}$ and $\dfrac{5a^2}{a}$.

74. Find the sum of $\dfrac{3n}{3n+2}$ and $\dfrac{n^2+2}{3n+2}$.

75. Find the difference of $\dfrac{2n}{n+1}$ and $\dfrac{n+3}{n+1}$.

76. Find the difference of $\dfrac{p^2-4}{p-4}$ and $\dfrac{4p}{p-4}$.

77. Subtract $\dfrac{3}{2n}$ from $\dfrac{11}{2n}$.

78. Subtract $\dfrac{3x+4}{4}$ from $\dfrac{x-4}{4}$.

79. $\frac{x}{x+3}$
80. $\frac{5x+3}{x-3}$
81. $\frac{4x-2}{x+2}$
82. $\frac{x-2}{x+1}$
83. $\frac{2(6x-1)}{x}$ cm
84. $\frac{6n}{2n+1}$ yd
85. $\frac{-x(x-1)}{(3x-1)(x+2)}$
86. $\frac{7x-9}{(x-2)(x+2)}$
87. $\frac{-5x+12}{x-2}$
88. $\frac{-7a+5b}{a-b}$
89. $-4n-6$
90. $2n-4$

79. Find a rational expression which, when subtracted from $\frac{x^2}{x^2-9}$, gives a difference of $\frac{3x}{x^2-9}$.

80. Find a rational expression which, when subtracted from $\frac{6x}{x-3}$, gives a difference of one.

81. Find a rational expression which, when added to $\frac{-3x+4}{x+2}$, gives a sum of one.

82. Find a rational expression which, when added to $\frac{x-3}{x+1}$, gives a sum of $\frac{2x-5}{x+1}$.

△ **83. Perimeter of a Rectangle** Find the perimeter of the rectangle.

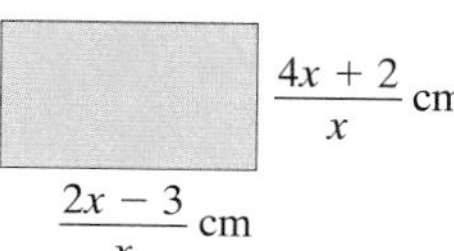

△ **84. Perimeter of a Rectangle** Find the perimeter of the rectangle.

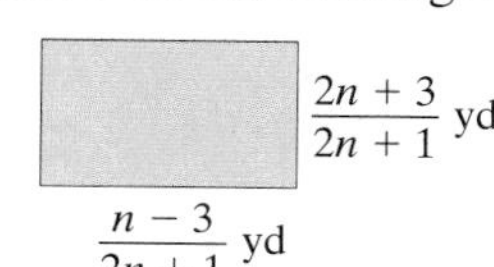

Extending the Concepts

In Problems 85–88, perform the indicated operations and simplify the result, if possible.

85. $\frac{x^2}{3x^2+5x-2} - \frac{x}{3x-1} \cdot \frac{2x-1}{x+2}$

86. $\frac{4}{x-2} \cdot \frac{x-2}{x+2} + \frac{3x-1}{x^2-4}$

87. $\frac{5x}{x-2} + \frac{2(x+3)}{x-2} - \frac{6(2x-1)}{x-2}$

88. $\frac{2a-b}{a-b} - \frac{6(a-2b)}{a-b} + \frac{3(2b+a)}{b-a}$

In Problems 89 and 90, find the missing expression.

89. $\frac{2n+1}{n-3} - \frac{?}{n-3} = \frac{6n+7}{n-3}$

90. $\frac{?}{n^2-1} - \frac{n-3}{n^2-1} = \frac{1}{n+1}$

5.4 Finding the Least Common Denominator and Forming Equivalent Rational Expressions

OBJECTIVES

1. Find the Least Common Denominator of Two or More Rational Expressions
2. Write a Rational Expression That Is Equivalent to a Given Rational Expression
3. Use the LCD to Write Equivalent Rational Expressions

Preparing for Finding the Least Common Denominator and Forming Equivalent Rational Expressions

Before getting started, take the following readiness quiz. If you get a problem wrong, go back to the section cited and review the material.

1. Write $\frac{5}{12}$ as a fraction with 24 as its denominator. [Section A.1, pp. A3–A4]

2. Find the least common denominator (LCD) of $\frac{4}{15}$ and $\frac{7}{25}$. [Section A.1, pp. A4–A5]

1 Find the Least Common Denominator of Two or More Rational Expressions

Work Smart

Don't confuse the LCD (least common denominator) with the GCF (greatest common factor)!

To add or subtract rational expressions the denominators must be the same. What if the denominators are different? We use the same ideas that we did for adding or subtracting rational numbers with unlike denominators: we rewrite each rational expression over the least common denominator.

Before we find the least common denominator (LCD) of rational expressions, let's review how to find the LCD of rational numbers.

Preparing for...Answers **1.** $\frac{10}{24}$ **2.** 75

EXAMPLE 1 How to Find the Least Common Denominator of Rational Numbers

Find the least common denominator (LCD) of $\frac{5}{6}$ and $\frac{11}{45}$.

Step-by-Step Solution

Step 1: Write each denominator as the product of prime factors.	$6 = 2 \cdot 3$ $45 = \quad 3 \cdot 3 \cdot 5$
Step 2: Write down the factor(s) that the denominators share. Then copy the remaining factors the greatest number of times that the factor occurs in a denominator.	The shared factor is 3 (indicated in blue). The remaining factors are 2, 3, and 5.
Step 3: Multiply the factors listed in Step 2. The product is the least common denominator (LCD).	The LCD is $3 \cdot 2 \cdot 3 \cdot 5 = 90$.

Classroom Example
Find the LCD of the rational numbers $\frac{5}{12}$ and $\frac{8}{15}$.

Answer: 60

In Example 1, we could have written the factorization as follows:

$$6 = 2 \cdot 3$$
$$45 = \quad 3^2 \cdot 5$$

To find the LCD, we notice that the factor 3 is written to the second power in the factorization of 45. We write 3^2 as part of the LCD because it is the "common factor to the highest power." We then write the factors that are not common, 2 and 5, as part of the LCD. The product of these factors gives us the least common denominator (LCD). So, the LCD is $3^2 \cdot 2 \cdot 5 = 90$.

Work Smart
It may helpful to align the factors vertically when finding the LCD.

Keep in mind that arithmetic is a special case of algebra, so the approach we just followed for finding the least common denominator of rational numbers will be the same approach we follow for finding the least common denominator of rational expressions.

DEFINITION
The **least common denominator (LCD)** of two or more rational expressions is the polynomial of least degree that is a multiple of each denominator in the rational expressions to be added or subtracted.

Classroom Example
Find the LCD of the rational expressions $\frac{7}{10x^2}$ and $\frac{2}{15x}$.

Answer: $30x^2$

Let's see how to find the least common denominator of rational expressions.

EXAMPLE 2 How to Find the Least Common Denominator of Rational Expressions with Monomial Denominators

Find the least common denominator of the rational expressions $\frac{3}{4x^3}$ and $\frac{5}{6x}$.

Step-by-Step Solution

Step 1: Express each denominator as the product of factors.	$4x^3 = 2^2 \quad \cdot x^3$ $6x = 2 \cdot 3 \cdot x$
Step 2: List the common factors raised to the highest power. Then list the factors that are not common.	• We list 2^2 because 2 is the highest power on the factor 2. • We list x^3 because 3 is the highest power on the factor x. • The factor that is not common is 3.

Step 3: Find the product of the factors written Step 2. This algebraic expression is the LCD.

$$\text{LCD} = 2^2 \cdot 3 \cdot x^3 = 12x^3$$

We summarize the steps to finding the least common denominator below.

Steps to Find the Least Common Denominator of Rational Expressions

Step 1: Factor each denominator completely. When factoring, write the factored form using powers. For example, write $x^2 + 4x + 4$ as $(x + 2)^2$.

Step 2: If factors are common except for their power, then list the factor with the highest power. That is, list each factor the greatest number of times that it appears. Then list the factors that are not common.

Step 3: The LCD is the product of the factors written in Step 2.

Classroom Example ➤ Find the LCD of the rational expressions $\frac{1}{6a^2b}$ and $\frac{2}{9ab^3}$.

Answer: $18a^2b^3$

Work Smart
List the factor with the highest power in the LCD.

EXAMPLE 3 Finding the Least Common Denominator of Rational Expressions with Monomial Denominators

Find the least common denominator of the rational expressions $\frac{1}{15xy^2}$ and $\frac{7}{18x^3y}$.

Solution

Express each denominator as the product of factors:

$$15xy^2 = 3 \cdot 5 \cdot x \cdot y^2$$
$$18x^3y = 2 \cdot 3^2 \cdot x^3 \cdot y$$

The product of each different factor the greatest number of times it appears is the LCD. So

$$\text{LCD} = 2 \cdot 3^2 \cdot 5 \cdot x^3 \cdot y^2 = 90x^3y^2$$

QUICK ✓ *Find the least common denominator of the rational expressions.*

1. $\frac{5}{8x^2y}$ and $\frac{1}{12xy^3}$

2. $\frac{1}{6a^3b^2}$ and $\frac{5}{21ab^3}$

Classroom Example ➤ Find the LCD of the rational expressions $\frac{5}{2a}$ and $\frac{7}{4a+4}$.

Answer: $4a(a + 1)$

Work Smart
The factor a in the denominator $3a$ is different from the a in the binomial factor $(a + 1)$. The a in $(a + 1)$ is a *term*, not a *factor*.

EXAMPLE 4 Finding the Least Common Denominator of Rational Expressions with Polynomial Denominators

Find the LCD of the rational expressions $\frac{7}{3a}$ and $\frac{5}{6a + 6}$.

Solution

Factor the denominators:

$$3a = 3 \cdot a$$
$$6a + 6 = 6(a + 1) = 2 \cdot 3 \cdot (a + 1)$$

The common factor is 3. The factors that are not common are 2, a, and $a + 1$. The product of these factors gives us the least common denominator.

$$\text{LCD} = 2 \cdot 3 \cdot a \cdot (a + 1) = 6a(a + 1)$$

Classroom Example Find the LCD of the rational expressions $\frac{5}{x^2 - x - 6}$ and $\frac{9}{x^2 + 3x + 2}$.

Answer: $(x + 1)(x + 2)(x - 3)$

EXAMPLE 5 Finding the Least Common Denominator of Rational Expressions with Polynomial Denominators

Find the LCD of the rational expressions $\frac{7}{x^2 - x - 2}$ and $\frac{3}{x^2 + 2x + 1}$.

Solution

Factor the denominators:
$$x^2 - x - 2 = (x - 2)(x + 1)$$
$$x^2 + 2x + 1 = (x + 1)^2$$

The factor $x + 1$ is common except that it is written to the second power in the factorization of $x^2 + 2x + 1$. Therefore, $(x + 1)^2$ is part of the LCD. The factor that is not common is $x - 2$. The product of these factors gives us the least common denominator.

$$\text{LCD} = (x + 1)^2(x - 2)$$

QUICK ✓ *Find the least common denominator of the rational expressions.*

3. $\frac{2}{15z}$ and $\frac{7}{5z^2 + 5z}$

4. $\frac{3}{x^2 + 4x - 5}$ and $\frac{1}{x^2 + 10x + 25}$

Classroom Example Find the LCD of the rational expressions $\frac{4}{n^2 - 9}$ and $\frac{8}{21 - 7n}$.

Answer: $-7(n - 3)(n + 3)$

EXAMPLE 6 Finding the Least Common Denominator with Opposite Factors

Find the LCD of the rational expressions $\frac{1}{x^2 - 25}$ and $\frac{9}{10 - 2x}$.

Solution

Factor the denominators:
$$x^2 - 25 = (x + 5)(x - 5)$$
$$10 - 2x = 2(5 - x)$$

It looks as though there are no common factors. But you should notice that the factors $x - 5$ and $5 - x$ are opposites. So we can write the factorization of $10 - 2x$ as follows:

$$10 - 2x = 2(5 - x)$$
Factor out -1:
$$= 2(-1)(x - 5)$$
$$= -2(x - 5)$$

Now, the common factor in each factorization is $x - 5$. The factors that are not common are -2 and $x + 5$. So the least common denominator is

$$\text{LCD} = -2(x + 5)(x - 5)$$

QUICK ✓ *Find the least common denominator of the rational expressions.*

5. $\frac{3}{21 - 3x}$ and $\frac{8}{x^2 - 49}$

2 Write a Rational Expression That Is Equivalent to a Given Rational Expression

Rational expressions which represent the same quantity are said to be *equivalent*. Now that we know how to find the LCD, we learn to write a rational expression as an equivalent rational expression with the LCD as its denominator. We can obtain equivalent rational expressions by multiplying the numerator and denominator of a rational expression by the same quantity as follows:

$$\frac{p}{q} = \frac{p}{q} \cdot 1 = \frac{p}{q} \cdot \frac{r}{r} = \frac{pr}{qr}$$

In other words we obtain equivalent rational expressions by multiplying the rational expression by 1, the multiplicative identity. Let's review how to do this using rational numbers.

Classroom Example Write $\frac{8}{3}$ as an equivalent fraction with a denominator of 12.

Answer: $\frac{32}{12}$

Teaching Tip Encourage students to show the value of "1" when writing equivalent fractions.

Classroom Example Write the rational expression $\frac{7}{3x^2 + 2x}$ as an equivalent rational expression with denominator $12x^2 + 8x$.

Answer: $\frac{28}{4x(3x + 2)}$

EXAMPLE 7 Forming Equivalent Rational Numbers

Write $\frac{2}{5}$ as an equivalent fraction with a denominator of 20.

Solution

We want to change from a denominator of 5 to a denominator of 20. In other words,

$$\frac{2}{5} \cdot \frac{?}{?} = \frac{\square}{20}$$

We know that $5 \cdot 4 = 20$, so we form the factor of $1 = \frac{4}{4}$.

$$\frac{2}{5} \cdot \frac{4}{4} = \frac{8}{20}$$

The equivalent fraction is $\frac{8}{20}$. ■

The approach that we use to form equivalent fractions is the same as the approach we use to form equivalent rational expressions.

EXAMPLE 8 How to Form an Equivalent Rational Expression

Write the rational expression $\frac{4}{5x^2 + x}$ as an equivalent rational expression with denominator $10x^2 + 2x$.

Step-by-Step Solution

Step 1: Write each denominator in factored form.	$5x^2 + x = x(5x + 1)$ $10x^2 + 2x = 2 \cdot x(5x + 1)$
Step 2: Determine the "missing factor(s)."	The new denominator has a factor of 2 that the denominator of the original rational expression is missing.
Step 3: Multiply the original rational expression by 1.	We multiply $\frac{4}{5x^2 + x}$ by $1 = \frac{2}{2}$. $\frac{4}{5x^2 + x} = \frac{4}{x(5x + 1)} \cdot \frac{2}{2}$
Step 4: Find the product. Leave the denominator in factored form.	$= \frac{8}{2x(5x + 1)}$

So $\frac{4}{5x^2 + x} = \frac{8}{2x(5x + 1)}$. Notice that $2x(5x + 1) = 10x^2 + 2x$, as required. ■

SUMMARY: Steps to Form Equivalent Rational Expressions

Step 1: Write each denominator in factored form.

Step 2: Determine the "missing factor(s)." That is, what factor(s) does the new denominator have that is missing from the original denominator?

Step 3: Multiply the original rational expression by $1 = \dfrac{\text{missing factor(s)}}{\text{missing factor(s)}}$.

Step 4: Find the product. Leave the denominator in factored form.

QUICK ✓ *Write each rational expression as an equivalent rational expression with the given denominator.*

6. $\dfrac{3}{4p^2 - 8p}$ with denominator $16p^3(p - 2)$.

7. $\dfrac{2}{x^2 + 5x + 6}$ with denominator $(x + 2)(x - 4)(x + 3)$.

3 Use the LCD to Write Equivalent Rational Expressions

The last skill that we need before we add or subtract rational expressions with unlike denominators is to find the LCD of two or more rational expressions and then write them as equivalent rational expressions.

Classroom Example

Find the LCD of the rational expressions $\dfrac{3}{x^2 - x - 6}$ and $\dfrac{5}{x^2 - 9}$. Then rewrite each rational expression with the LCD.

Answer: LCD $= (x + 2)(x - 3)(x + 3)$; $\dfrac{3}{x^2 - x - 6} = \dfrac{3x + 9}{(x + 2)(x - 3)(x + 3)}$

$\dfrac{5}{x^2 - 9} = \dfrac{5x + 10}{(x + 2)(x - 3)(x + 3)}$

EXAMPLE 9 Using the LCD to Write Equivalent Rational Expressions

Find the LCD of the rational expressions $\dfrac{4}{x^2 + 3x + 2}$ and $\dfrac{9}{x^2 - 4}$. Then rewrite each rational expression with the LCD.

Solution

Factor each denominator:

$$x^2 + 3x + 2 = (x + 1)(x + 2)$$
$$x^2 - 4 = (x + 2)(x - 2)$$

Find the product of the factors written in the previous step to obtain the LCD:

$$\text{LCD} = (x + 2)(x + 1)(x - 2)$$

Rewrite each rational expression with a denominator of $(x + 2)(x + 1)(x - 2)$. Multiply out the numerators, but leave the denominator in factored form.

$$\frac{4}{x^2 + 3x + 2} = \frac{4}{(x + 2)(x + 1)} \cdot \frac{(x - 2)}{(x - 2)} = \frac{4x - 8}{(x + 2)(x + 1)(x - 2)}$$

$$\frac{9}{x^2 - 4} = \frac{9}{(x - 2)(x + 2)} \cdot \frac{(x + 1)}{(x + 1)} = \frac{9x + 9}{(x - 2)(x + 2)(x + 1)}$$

QUICK ✓ *Find the least common denominator of each rational expression. Then rewrite each rational expression with the LCD.*

8. $\dfrac{5}{x^2 - 4x - 5}$ and $\dfrac{-3}{x^2 - 7x + 10}$

5.4 Exercises

For Extra Help: Student Solutions Manual | CD Video | PH Math/Tutor Center | MathXL Tutorials on CD | MathXL® | MyMathLab

1. least common denominator
2. $x + 3$
3. 1
4. False
5. False
6. True
7. Answers may vary.
8. Answers may vary.
9. 36
10. 24
11. 60
12. 72
13. $5x^2$
14. $49y^2$
15. $60x^3y^2$
16. $72x^2y^2$
17. $x(x + 1)$
18. $2(2x + 1)$
19. $2x + 1$
20. $y - 1$
21. $4(b - 3)$
22. $24(x - 3)$
23. $p(p + 1)(p - 2)$
24. $4(x - 3)^2(x + 3)$
25. $(r + 2)^2(r + 1)(r - 2)$
26. $(x - 1)^2(x + 1)$
27. $-(x - 4)$
28. $-(a - 3)$
29. $-2(x + 3)(x - 3)$
30. $-3(c - 7)(c + 7)$

Concepts and Vocabulary

In Problems 1–3, fill in the blanks.

1. The ________ ________ ________ of two or more rational expressions is the smallest polynomial that is a multiple of each denominator in the rational expressions to be added or subtracted.

2. If we wanted to rewrite the rational expression $\frac{2x + 1}{x - 1}$ with a denominator of $(x - 1)(x + 3)$, we would multiply the numerator and denominator of $\frac{2x + 1}{x - 1}$ by ________.

3. Any time we multiply a rational expression by ________, we form an equivalent rational expression.

In Problems 4–6, answer True or False to each statement.

4. To find the sum $\frac{3}{4a^2b} + \frac{5}{4ab}$, we begin by adding $3 + 5$.

5. In the expression $\frac{3}{4a^2b} + \frac{5}{4ab}$, the LCD is $16a^3b^2$.

6. The least common denominator of two rational expressions is unique.

7. If you were teaching the class how to write equivalent rational expressions with a common denominator, what steps would you list on the board for the class?

8. A member of your class says that $\frac{1}{(x + 3)(x + 2)}$ and $\frac{4}{(x + 2)}$ have an LCD of $(x + 3)(x + 2)^2$. This class member then rewrites the rational expressions as $\frac{x + 2}{(x + 3)(x + 2)^2}$ and $\frac{4(x + 3)(x + 2)}{(x + 3)(x + 2)^2}$. Use this result to explain why $(x + 3)(x + 2)^2$ is not the least common denominator.

Building Skills

In Problems 9–34, identify the LCD of the given rational expressions.

9. $\frac{5}{12}; \frac{4}{9}$

10. $\frac{7}{12}; \frac{3}{8}$

11. $\frac{4}{15}; \frac{3}{5}; \frac{1}{20}$

12. $\frac{1}{9}; \frac{3}{8}; \frac{7}{36}$

13. $\frac{14}{5x^2}; \frac{4}{5x}$

14. $\frac{3}{7y^2}; \frac{4}{49y}$

15. $\frac{7}{12xy^2}; \frac{4}{15x^3y}$

16. $\frac{5}{36xy^2}; \frac{1}{24x^2y}$

17. $\frac{3}{x}; \frac{4}{x + 1}$

18. $\frac{2x - 1}{2x + 1}; \frac{3x}{2}$

19. $\frac{4}{2x + 1}; 2$

20. $\frac{7}{y - 1}; 3$

21. $\frac{7}{2b - 6}; \frac{3b}{4b - 12}$

22. $\frac{2}{6x - 18}; \frac{3}{8x - 24}$

23. $\frac{6}{p^2 + p}; \frac{7}{p^2 - p - 2}$

24. $\frac{5}{2x^2 - 12x + 18}; \frac{4}{4x^2 - 36}$

25. $\frac{3}{r^2 + 4r + 4}; \frac{4}{r^2 - r - 2}$

26. $\frac{8}{x^2 - 1}; \frac{3}{x^2 - 2x + 1}$

27. $\frac{2}{x - 4}; \frac{3}{4 - x}$

28. $\frac{7}{3 - a}; \frac{1}{a - 3}$

29. $\frac{8}{x^2 - 9}; \frac{2}{6 - 2x}$

30. $\frac{-1}{c^2 - 49}; \frac{5}{21 - 3c}$

31. $\dfrac{1}{(x-1)(x+2)}; \dfrac{11x}{(1-x)(2+x)}$

32. $\dfrac{3z}{(z+5)(z-4)}; \dfrac{-z}{(4-z)(5+z)}$

33. $\dfrac{11}{p^2-p}; \dfrac{-2}{p^3-p^2}; \dfrac{8}{p^2-4p+3}$

34. $\dfrac{7}{4n^3-2n^2}; \dfrac{9}{8n^2-4n}; \dfrac{6}{4n^2-1}$

In Problems 35–44, write an equivalent rational expression with the given denominator.

35. $\dfrac{4}{x}$ with denominator $3x^3$

36. $\dfrac{7}{x}$ with denominator $2x^2$

37. $\dfrac{3+c}{a^2b^2c}$ with denominator $a^2b^2c^2$

38. $\dfrac{7a+1}{abc}$ with denominator a^2b^2c

39. $\dfrac{x-4}{x+4}$ with denominator x^2-16

40. $\dfrac{3}{x+2}$ with denominator x^2+5x+6

41. $\dfrac{3n}{2n+2}$ with denominator $6n^2-6$

42. $\dfrac{7a}{3a^2+9a+6}$ with denominator $6a^2+18a+12$

43. $4t$ with denominator $t-1$

44. 7 with denominator t^2+1

Mixed Practice

In Problems 45–66, identify the LCD and then write each as an equivalent rational expression with that denominator.

45. $\dfrac{7}{12}; \dfrac{-8}{9}$

46. $\dfrac{-4}{9}; \dfrac{-3}{8}$

47. $\dfrac{7}{15}; 2$

48. $\dfrac{5}{6}; 4$

49. $\dfrac{2x}{3y}; \dfrac{4}{9y^2}$

50. $\dfrac{4}{5n^2}; \dfrac{3}{7n}$

51. $\dfrac{3a+1}{2a^3}; \dfrac{4a-1}{4a}$

52. $\dfrac{2p^2-1}{6p}; \dfrac{3p^3+2}{8p^2}$

53. $\dfrac{2}{m}; \dfrac{3}{m+1}$

54. $\dfrac{x+2}{x}; \dfrac{x}{x+2}$

55. $\dfrac{y-2}{4y}; \dfrac{y}{8y-4}$

56. $\dfrac{3b}{7a}; \dfrac{2a}{7a+14b}$

57. $\dfrac{1}{x-1}; \dfrac{2x}{1-x}$

58. $\dfrac{2a}{a-2}; \dfrac{a}{2-a}$

59. $\dfrac{3x}{x-7}; \dfrac{-5}{7-x}$

60. $\dfrac{5}{m-6}; \dfrac{2m}{6-m}$

61. $\dfrac{4x}{x^2-4}; \dfrac{2}{x+2}$

62. $\dfrac{4a}{a^2-1}; \dfrac{7}{a+1}$

63. $\dfrac{x+1}{x^2-9}; \dfrac{x+2}{x^2-2x-3}$

64. $\dfrac{4n}{n^2-n-6}; \dfrac{2}{n^2+4n+4}$

65. $\dfrac{3}{x+4}; \dfrac{2x-1}{2x^2+7x-4}$

66. $\dfrac{3}{2n^2-7n+3}; \dfrac{2n}{n-3}$

Applying the Concepts

67. Painting a Room It takes an experienced painter x hours to paint a room alone; assuming that he works at a constant rate, he completes $\dfrac{1}{x}$ of the job per hour. It takes an apprentice 3 times as long to paint the same room as it takes the experienced painter, so the apprentice completes $\dfrac{1}{3x}$ of the job per hour. Identify the LCD of the two rational expressions that represent the rate of work of the experienced painter and the apprentice, and then write the equivalent fractions using this denominator.

31. $-(x-1)(x+2)$
32. $-(z-4)(z+5)$
33. $p^2(p-1)(p-3)$
34. $4n^2(2n-1)(2n+1)$
35. $\frac{12x^2}{3x^3}$ 36. $\frac{14x}{2x^2}$
37. $\frac{3c+c^2}{a^2b^2c^2}$ 38. $\frac{7a^2b+ab}{a^2b^2c}$
39. $\frac{(x-4)^2}{x^2-16}$ 40. $\frac{3x+9}{x^2+5x+6}$
41. $\frac{9n^2-9n}{6n^2-6}$
42. $\frac{14a}{6a^2+18a+12}$
43. $\frac{4t^2-4t}{t-1}$
44. $\frac{7t^2+7}{t^2+1}$ 45. $\frac{21}{36}; \frac{-32}{36}$
46. $\frac{-32}{72}; \frac{-27}{72}$
47. $\frac{7}{15}; \frac{30}{15}$ 48. $\frac{5}{6}; \frac{24}{6}$
49. $\frac{6xy}{9y^2}; \frac{4}{9y^2}$ 50. $\frac{28}{35n^2}; \frac{15n}{35n^2}$
51. $\frac{6a+2}{4a^3}; \frac{4a^3-a^2}{4a^3}$
52. $\frac{8p^3-4p}{24p^2}; \frac{9p^3+6}{24p^2}$
53. $\frac{2m+2}{m(m+1)}; \frac{3m}{m(m+1)}$
54. $\frac{(x+2)^2}{x(x+2)}; \frac{x^2}{x(x+2)}$
55. $\frac{2y^2-5y+2}{4y(2y-1)}; \frac{y^2}{4y(2y-1)}$
56. $\frac{3ab+6b^2}{7a^2+14ab}; \frac{2a^2}{7a^2+14ab}$
57. $\frac{-1}{-(x-1)}; \frac{2x}{-(x-1)}$
58. $\frac{-2a}{-(a-2)}; \frac{a}{-(a-2)}$
59. $\frac{-3x}{-(x-7)}; \frac{-5}{-(x-7)}$
60. $\frac{-5}{-(m-6)}; \frac{2m}{-(m-6)}$
61. $\frac{4x}{(x+2)(x-2)}; \frac{2x-4}{(x+2)(x-2)}$
62. $\frac{4a}{(a+1)(a-1)}; \frac{7a-7}{(a+1)(a-1)}$
63. $\frac{x^2+2x+1}{(x+3)(x-3)(x+1)}; \frac{x^2+5x+6}{(x+3)(x-3)(x+1)}$
64. $\frac{4n^2+8n}{(n+2)^2(n-3)}; \frac{2n-6}{(n+2)^2(n-3)}$
65. $\frac{6x-3}{(x+4)(2x-1)}; \frac{2x-1}{(x+4)(2x-1)}$
66. $\frac{3}{(n-3)(2n-1)}; \frac{4n^2-2n}{(n-3)(2n-1)}$
67. $3x; \frac{3}{3x}; \frac{1}{3x}$

68. $x(x + 2)$; $\frac{x + 2}{x(x + 2)}$; $\frac{x}{x(x + 2)}$
69. $(r - 4)(r + 4)$; $\frac{12r + 48}{(r - 4)(r + 4)}$; $\frac{12r - 48}{(r - 4)(r + 4)}$
70. $(r - 50)(r + 50)$; $\frac{500r + 25{,}000}{(r - 50)(r + 50)}$; $\frac{500r - 25{,}000}{(r - 50)(r + 50)}$
71. $(x + 2)(x - 2)(x^2 - 2x + 4)(x^2 + 2x + 4)$
72. $12(x - 1)(x + 1)(x^2 + x + 1)$
73. $4a^2b^2(a + b)(a - b)(a^2 + ab + b^2)$
74. $(3x - 2y)(x - 2y)(x + y)$

68. Mowing the Lawn It takes Mario 2 hours longer to mow the lawn than it takes it Marco. If Marco takes x hours, he completes $\frac{1}{x}$ of the lawn per hour. Since Mario requires 2 more hours, he completes $\frac{1}{x + 2}$ of the job per hour. Identify the LCD of the two rational expressions that represent the rate of work for Mario and Marco, and then write the equivalent fractions using this denominator.

69. Traveling on a River The time to complete a journey can be calculated by the formula $t = \frac{d}{r}$, where t is the time, d is the distance traveled, and r is rate. Suppose the time for a boat to travel a distance of 12 miles upstream on a river whose current is 4 miles per hour is given by $\frac{12}{r - 4}$. The time for the boat to travel 12 miles downstream on the same river is $\frac{12}{r + 4}$. Find the LCD of the two rational expressions and then write each as an equivalent rational expression using this denominator.

70. Traveling by Plane The time to complete a journey can be calculated by the formula $t = \frac{d}{r}$, where t is the time, d is the distance traveled, and r is rate. Suppose a plane flies 500 miles west into a headwind of 50 miles per hour in a time of $\frac{500}{r - 50}$ hours, where r is the speed of plane in still air. The plane flies 500 miles east with a tailwind of 50 miles per hour in a time of $\frac{500}{r + 50}$ hours. Find the LCD of the two rational expressions and then write each as an equivalent rational expression using this denominator.

Extending the Concepts

In Problems 71–74, identify the LCD of the rational expressions.

71. $\frac{4}{x^3 + 8}; \frac{1}{x^2 - 4}; \frac{5}{x^3 - 8}$

72. $\frac{x}{2x^3 - 2}; \frac{1}{3x^2 - 3}; \frac{2x + 1}{4x - 4}$

73. $\frac{a}{2a^4 - 2a^2b^2}; \frac{b}{4ab^2 + 4b^3}; \frac{ab}{a^3b - b^4}$

74. $\frac{5}{3x^2 + xy - 2y^2}; \frac{2}{x^2 - xy - 2y^2}$

5.5 Adding and Subtracting Rational Expressions with Unlike Denominators

OBJECTIVE

① Add and Subtract Rational Expressions with Unlike Denominators

Preparing for Adding and Subtracting Rational Expressions with Unlike Denominators

1. Find the least common denominator (LCD) of 15 and 24. [Section A.1, pp. A2–A5]
2. Factor: $2x^2 + 3x + 1$ [Section 4.3, pp. 257–265]
3. Simplify: $\frac{3x + 9}{6x}$ [Section 5.1, pp. 308–311]
4. Simplify: $\frac{1 - x^2}{2x - 2}$ [Section 5.1, pp. 308–311]
5. Find the sum: $\frac{5}{2x} + \frac{7}{2x}$ [Section 5.3, pp. 322–324]

Preparing for...Answers **1.** 120 **2.** $(2x + 1)(x + 1)$ **3.** $\frac{x + 3}{2x}$ **4.** $\frac{-(1 + x)}{2}$ **5.** $\frac{6}{x}$

1 Add and Subtract Rational Expressions with Unlike Denominators

We are now ready to add and subtract rational expressions with unlike denominators! Keep in mind that the steps for adding and subtracting rational expressions with unlike denominators parallel the steps for adding and subtracting rational numbers with unlike denominators.

Before we add and subtract rational expressions with unlike denominators, we provide an example to remind you how to add rational numbers with unlike denominators.

EXAMPLE 1 How to Add Rational Numbers with Unlike Denominators

Evaluate $\frac{1}{6} + \frac{9}{14}$. Be sure to write the sum in lowest terms.

Classroom Example

Evaluate $\frac{5}{12} + \frac{2}{15}$. Write the sum in lowest terms.

Answer: $\frac{11}{20}$

Step-by-Step Solution

Step 1: Find the least common denominator of 14 and 6.

$$6 = 2 \cdot 3$$
$$14 = 2 \cdot \quad 7$$
$$\text{LCD} = 2 \cdot 3 \cdot 7 = 42$$

Step 2: Write each fraction as an equivalent fraction, with denominator of 42.

Because $6 \cdot 7 = 42$, we use $\frac{7}{7} = 1$ to rewrite the denominator 6 as 42. Since $14 \cdot 3 = 42$, we use $\frac{3}{3} = 1$ to rewrite the denominator 14 as 42.

$$\frac{1}{6} \cdot \frac{7}{7} + \frac{9}{14} \cdot \frac{3}{3} = \frac{7}{42} + \frac{27}{42}$$

Step 3: Find the sum of the numerators, written over the common denominator.

$$= \frac{7 + 27}{42}$$
$$= \frac{34}{42}$$

Step 4: Simplify.

Factor 34 and 42: $= \frac{2 \cdot 17}{2 \cdot 21}$

$$= \frac{\cancel{2} \cdot 17}{\cancel{2} \cdot 21}$$

Divide out common factors: $= \frac{17}{21}$

Therefore, $\frac{1}{6} + \frac{9}{14} = \frac{17}{21}$.

Teaching Tip

When introducing adding rational expressions, you may want to set up two examples side by side: adding rational numbers with unlike denominators and adding rational expressions with unlike denominators. Encourage students to see the similarity in the process of adding numbers and expressions.

QUICK ✓ *Find each sum or difference. Write the sum or difference in lowest terms.*

1. $\frac{5}{12} + \frac{5}{18}$ **2.** $\frac{5}{6} + \frac{8}{3}$ **3.** $\frac{19}{6} - \frac{5}{12}$ **4.** $\frac{8}{15} - \frac{3}{10}$

Now that you've practiced adding and subtracting rational numbers, let's learn how to add and subtract rational expressions.

EXAMPLE 2 How to Add Rational Expressions with Unlike Monomial Denominators

Find the sum and simplify the result, if possible: $\frac{5}{6x^2} + \frac{4}{15x}$

Classroom Example
Find the sum and simplify the result, if possible: $\frac{3}{4y^3} + \frac{7}{18y}$

Answer: $\frac{27 + 14y^2}{36y^3}$

Step-by-Step Solution

Step 1: Find the least common denominator.

$$6x^2 = 2 \cdot 3 \cdot \quad x^2$$
$$15x = \quad 3 \cdot 5 \cdot x$$

The LCD is $2 \cdot 3 \cdot 5 \cdot x^2 = 30x^2$

Step 2: Rewrite each rational expression with the common denominator.

We multiply $\frac{5}{6x^2}$ by $\frac{5}{5}$ to get $30x^2$ in the denominator; we multiply $\frac{4}{15x}$ by $\frac{2x}{2x}$ to get $30x^2$ in the denominator:

$$\frac{5}{6x^2} + \frac{4}{15x} = \frac{5}{6x^2} \cdot \frac{5}{5} + \frac{4}{15x} \cdot \frac{2x}{2x}$$
$$= \frac{25}{30x^2} + \frac{8x}{30x^2}$$

Step 3: Add the rational expressions found in Step 2.

Add using $\frac{a}{c} + \frac{b}{c} = \frac{a+b}{c}$:

$$= \frac{25 + 8x}{24x^2}$$

Step 4: Simplify.

The rational expression is already simplified.

So $\frac{5}{6x^2} + \frac{4}{15x} = \frac{25 + 8x}{24x^2}$.

Work Smart
Notice in Step 2, when we rewrite the rational expression with the common denominator, we are multiplying by 1.

We summarize the steps for adding or subtracting rational expressions with unlike denominators.

Steps for Adding or Subtracting Rational Expressions with Unlike Denominators

Step 1: Find the least common denominator.

Step 2: Rewrite each rational expression with the common denominator.

Step 3: Add or subtract the rational expressions found in Step 2.

Step 4: Simplify the result.

QUICK ✓ *Find each sum and simplify the result, if possible.*

5. $\frac{1}{8a^3b} + \frac{5}{12ab^2}$

6. $\frac{1}{15xy^2} + \frac{7}{18x^3y}$

Classroom Example
Find the sum and simplify the result, if possible: $\frac{-2}{x+5} + \frac{3}{x+4}$

Answer: $\frac{x+7}{(x+4)(x+5)}$

EXAMPLE 3 Adding Rational Expressions with Unlike Polynomial Denominators

Find the sum and simplify the result, if possible: $\frac{-1}{x+3} + \frac{3}{x+2}$

Solution

First, we find that the least common denominator is $(x + 3)(x + 2)$ because the two denominators have no common factors. Now we need to rewrite each rational expression using the least common denominator. To do this, we multiply $\frac{-1}{x+3}$ by

$\frac{x+2}{x+2}$ to get $(x + 3)(x + 2)$ in the denominator and we multiply $\frac{3}{x+2}$ by $\frac{x+3}{x+3}$ to get $(x + 3)(x + 2)$ in the denominator.

Teaching Tip
Remind students that multiplying by $\frac{x+2}{x+2}$ or $\frac{x+3}{x+3}$ is an application of the Multiplicative Identity Property. The value of the expression doesn't change, only its appearance.

$$\frac{-1}{x+3} + \frac{3}{x+2} = \frac{-1}{x+3}\cdot\frac{x+2}{x+2} + \frac{3}{x+2}\cdot\frac{x+3}{x+3}$$

Multiply out the numerators; leave the denominators in factored form: $= \frac{-x-2}{(x+3)(x+2)} + \frac{3x+9}{(x+2)(x+3)}$

Add using $\frac{a}{c} + \frac{b}{c} = \frac{a+b}{c}$: $= \frac{-x-2+3x+9}{(x+3)(x+2)}$

Combine like terms: $= \frac{2x+7}{(x+3)(x+2)}$

There are no common factors, so $\frac{-1}{x+3} + \frac{3}{x+2} = \frac{2x+7}{(x+3)(x+2)}$.

QUICK ✓ *Find each sum and simplify the result, if possible.*

7. $\frac{5}{x-4} + \frac{3}{x+2}$ **8.** $\frac{-1}{n-3} + \frac{4}{n+1}$

Classroom Example ➤
Find the sum and simplify the results, if possible:

(a) $\frac{3}{x+2} + \frac{8-2x}{x^2-4}$

(b) $\frac{x+1}{x+2} + \frac{-4}{x^2-4}$

Answer:

(a) $\frac{1}{x-2}$ (b) $\frac{x-3}{x-2}$

EXAMPLE 4 Adding Rational Expressions with Unlike Polynomial Denominators

Find the sum and simplify the result: $\frac{4}{x+1} + \frac{16}{x^2-2x-3}$

Solution

Notice that the denominators are not the same. So, what's our first step? We need to find the least common denominator. We begin by factoring each denominator.

$$x + 1 = x + 1$$
$$x^2 - 2x - 3 = (x-3)(x+1)$$

The LCD is $(x - 3)(x + 1)$. Now we rewrite each rational expression with the least common denominator.

Teaching Tip
Remind students that before they can find the common denominator of two or more rational expressions, they may need to factor the denominator(s) of the given rational expressions.

$$\frac{4}{x+1} + \frac{16}{x^2-2x-3} = \frac{4}{x+1} + \frac{16}{(x-3)(x+1)}$$

Rewrite each rational expression with the common denominator: $= \frac{4}{x+1}\cdot\frac{x-3}{x-3} + \frac{16}{(x+1)(x-3)}$

Don't reduce $\frac{4(x-3)}{(x+1)(x-3)}$ or you'll be right back where you started!: $= \frac{4(x-3)}{(x+1)(x-3)} + \frac{16}{(x+1)(x-3)}$

Multiply out the numerator: $= \frac{4x-12}{(x+1)(x-3)} + \frac{16}{(x+1)(x-3)}$

Add using $\frac{a}{c} + \frac{b}{c} = \frac{a+b}{c}$: $= \frac{4x-12+16}{(x+1)(x-3)}$

Add: $= \frac{4x+4}{(x+1)(x-3)}$

Factor the numerator: $= \frac{4(x+1)}{(x+1)(x-3)}$

Divide out like factors: $= \frac{4}{x-3}$

Classroom Example
Find the difference and simplify the result, if possible: $\frac{7}{2a} - \frac{3}{a+5}$

Answer: $\frac{a+35}{2a(a+5)}$

QUICK ✓ *Find each sum and simplify the result, if possible.*

9. $\frac{2}{x-3} + \frac{-2}{x^2-5x+6}$ **10.** $\frac{-z+1}{z^2+7z+10} + \frac{2}{z+5}$ **11.** $\frac{1}{x^2+5x} + \frac{1}{x^2-5x}$

So far we have concentrated on adding rational expressions with unlike denominators. Now let's go over how to subtract rational expressions with unlike denominators.

EXAMPLE 5 How to Subtract Rational Expressions with Unlike Denominators

Find the difference and simplify the result, if possible: $\frac{8}{x} - \frac{2}{x+1}$

Step-by-Step Solution

Step 1: Find the least common denominator.

The LCD is $x(x+1)$

Step 2: Rewrite each rational expression with the common denominator. Multiply out the numerators, but leave the denominators in factored form.

Multiply $\frac{8}{x}$ by $\frac{x+1}{x+1}$ to get $x(x+1)$ in the denominator; multiply $\frac{2}{x+1}$ by $\frac{x}{x}$ to get $x(x+1)$ in the denominator.

$$\frac{8}{x} - \frac{2}{x+1} = \frac{8}{x}\cdot\frac{x+1}{x+1} - \frac{2}{x+1}\cdot\frac{x}{x}$$
$$= \frac{8(x+1)}{x(x+1)} - \frac{2\cdot x}{x(x+1)}$$
$$= \frac{8x+8}{x(x+1)} - \frac{2x}{x(x+1)}$$

Step 3: Subtract the rational expressions found in Step 2.

Subtract using $\frac{a}{c} - \frac{b}{c} = \frac{a-b}{c}$:

$$= \frac{6x+8}{x(x+1)}$$

Step 4: Simplify the result.

Factor the numerator:

$$= \frac{2(3x+4)}{x(x+1)}$$

Since there are no common factors, the expression is written in lowest terms. So $\frac{8}{x} - \frac{2}{x+1} = \frac{2(3x+4)}{x(x+1)}$.

QUICK ✓ *Find each difference and simplify the result, if possible.*

12. $\frac{-4}{5ab^2} - \frac{3}{4a^2b^3}$ **13.** $\frac{5}{x} - \frac{3}{x-4}$

Classroom Example
Find the sum and simplify, if possible: $\frac{3}{x^2-25} - \frac{4}{x-5}$

Answer: $\frac{-4x-17}{(x-5)(x+5)}$

EXAMPLE 6 Adding Rational Expressions with Unlike Denominators—Both Denominators Factor

Perform the indicated operation and simplify, if possible: $\frac{-2}{a^2-4a} + \frac{2}{4a-16}$

Solution

First, we factor each denominator to find the LCD.

$$a^2 - 4a = a\cdot(a-4)$$
$$4a - 16 = 4\cdot(a-4)$$

The LCD is $4a(a-4)$, so we must rewrite each denominator with a new denominator of $4a(a-4)$. We multiply $\frac{-2}{a(a-4)}$ by $\frac{4}{4}$ and we multiply $\frac{2}{4(a-4)}$ by $\frac{a}{a}$.

$$\frac{-2}{a(a-4)} + \frac{2}{4(a-4)} = \frac{-2}{a(a-4)} \cdot \frac{4}{4} + \frac{2}{4(a-4)} \cdot \frac{a}{a}$$

Multiply numerators: $= \frac{-8}{4a(a-4)} + \frac{2a}{4a(a-4)}$

Add numerators using $\frac{a}{c} + \frac{b}{c} = \frac{a+b}{c}$: $= \frac{2a-8}{4a(a-4)}$

Factor the numerator and denominator: $= \frac{2(a-4)}{2 \cdot 2a(a-4)}$

Divide out like factors: $= \frac{1}{2a}$

So $\frac{-2}{a^2-4a} + \frac{2}{4a-16} = \frac{1}{2a}$.

Work Smart
When all the factors in the numerator divide out, the remaining factor is 1, not 0.

Classroom Example
Perform the indicated operation and simplify, if possible:

(a) $\frac{4}{x^2-6x} + \frac{5}{2x-12}$

(b) $\frac{2}{y^2-4} - \frac{1}{y^2-2y}$

Answer:

(a) $\frac{8+5x}{2x(x-6)}$

(b) $\frac{1}{y(y+2)}$

EXAMPLE 7 Subtracting Rational Expressions with Unlike Denominators—Both Denominators Factor

Perform the indicated operation and simplify, if possible: $\frac{a-2}{a^2+5a+6} - \frac{2a-3}{3a^2+9a}$

Solution

First, we factor each denominator.

$$a^2 + 5a + 6 = (a+3)(a+2)$$
$$3a^2 + 9a = 3a(a+3)$$

The LCD is $3a(a+3)(a+2)$. Now we rewrite each expression with the common denominator.

$$\frac{a-2}{a^2+5a+6} - \frac{2a-3}{3a^2+9a} = \frac{a-2}{(a+3)(a+2)} \cdot \frac{3a}{3a} - \frac{2a-3}{3a(a+3)} \cdot \frac{a+2}{a+2}$$

Multiply the numerators: $= \frac{(a-2) \cdot 3a}{3a(a+3)(a+2)} - \frac{(2a-3)(a+2)}{3a(a+3)(a+2)}$

$= \frac{3a^2-6a}{3a(a+3)(a+2)} - \frac{2a^2+a-6}{3a(a+3)(a+2)}$

Write numerators over the common denominator; don't forget the parentheses: $= \frac{3a^2-6a-(2a^2+a-6)}{3a(a+3)(a+2)}$

Distribute the minus sign: $= \frac{3a^2-6a-2a^2-a+6}{3a(a+3)(a+2)}$

Combine like terms: $= \frac{a^2-7a+6}{3a(a+3)(a+2)}$

Factor the numerator: $= \frac{(a-6)(a-1)}{3a(a+3)(a+2)}$

There are no common factors, so $\frac{a-2}{a^2+5a+6} - \frac{2a-3}{3a^2+9a} = \frac{(a-6)(a-1)}{3a(a+3)(a+2)}$.

QUICK ✓ *Perform the indicated operation and simplify, if possible.*

14. $\dfrac{3}{(x-5)(x+4)} - \dfrac{2}{(x-5)(x-1)}$ **15.** $\dfrac{x-2}{x^2-3x} + \dfrac{x+3}{4x-12}$

Classroom Example
Find the sum and simplify, if possible:
$\dfrac{2}{3-b} + \dfrac{1}{b^2-9}$

Answer: $\dfrac{-2b-5}{(b-3)(b+3)}$

EXAMPLE 8 Adding Rational Expressions Containing Opposite Factors

Find the sum and simplify, if possible: $\dfrac{1}{1-n} + \dfrac{2n}{n^2-1}$

Solution

First, we factor each denominator:

$$1 - n = 1 - n$$
$$n^2 - 1 = (n+1)(n-1)$$

Do you see that the factors $n - 1$ and $1 - n$ are opposites? We factor -1 from $1 - n$ so that $1 - n = -1(-1 + n) = -1(n - 1)$. We rewrite the sum as

$$\frac{1}{1-n} + \frac{2n}{n^2-1} = \frac{1}{-1(n-1)} + \frac{2n}{(n+1)(n-1)}$$

Use $\frac{a}{-b} = \frac{-a}{b}$: $$= \frac{-1}{(n-1)} + \frac{2n}{(n+1)(n-1)}$$

In this form, we can see that the LCD is $(n-1)(n+1)$. Now we rewrite each expression with the common denominator.

Multiply $\frac{-1}{n-1}$ by $\frac{n+1}{n+1}$: $$= \frac{-1}{n-1}\cdot\frac{n+1}{n+1} + \frac{2n}{(n+1)(n-1)}$$

Multiply the numerators: $$= \frac{-1(n+1)}{(n+1)(n-1)} + \frac{2n}{(n+1)(n-1)}$$

Distribute the -1: $$= \frac{-n-1}{(n+1)(n-1)} + \frac{2n}{(n+1)(n-1)}$$

Write numerators over the common denominator: $$= \frac{-n-1+2n}{(n+1)(n-1)}$$

Combine like terms: $$= \frac{n-1}{(n+1)(n-1)}$$

Divide out like factors: $$= \frac{1}{n+1}$$

So $\dfrac{1}{1-n} + \dfrac{2n}{n^2-1} = \dfrac{1}{n+1}$.

Work Smart
Remember that when like factors divide, they form a quotient of 1. That's why

$$\frac{n-1}{(n+1)(n-1)}$$

simplifies to be

$$\frac{\cancel{n-1}^{1}}{(n+1)\cancel{(n-1)}_{1}} = \frac{1}{n+1}$$

QUICK ✓ *Find the sum and simplify, if possible.*

16. $\dfrac{7}{3p^2-3p} + \dfrac{5}{6-6p}$

Classroom Example
Find the sum and simplify, if possible:
$2 + \dfrac{3}{2x-5}$

Answer: $\dfrac{4x-7}{2x-5}$

EXAMPLE 9 Adding an Integer and a Rational Expression

Find the sum and simplify, if possible: $5 + \dfrac{2}{3x-4}$

Solution

First, note that $5 = \frac{5}{1}$. Let's rewrite the expression as $\frac{5}{1} + \frac{2}{3x - 4}$ and proceed. The LCD is $3x - 4$, so we rewrite the expression with the LCD by multiplying $\frac{5}{1}$ by $\frac{3x - 4}{3x - 4}$.

$$\frac{5}{1} + \frac{2}{3x - 4} = \frac{5}{1} \cdot \frac{3x - 4}{3x - 4} + \frac{2}{3x - 4}$$

Multiply the numerators: $= \frac{5(3x - 4)}{3x - 4} + \frac{2}{3x - 4}$

$= \frac{15x - 20}{3x - 4} + \frac{2}{3x - 4}$

Write numerators over the common denominator: $= \frac{15x - 20 + 2}{3x - 4}$

Combine like terms: $= \frac{15x - 18}{3x - 4}$

Factor the numerator: $= \frac{3(5x - 6)}{3x - 4}$

We see that $5 + \frac{2}{3x - 4} = \frac{3(5x - 6)}{3x - 4}$.

QUICK ✓ *Perform the indicated operation and simplify, if possible.*

17. $1 + \frac{3}{z - 5}$

18. $2 - \frac{3}{x - 1}$

Classroom Example ➤ Perform the indicated operations and simplify, if possible:

$\frac{3}{x + 2} + \frac{2}{2 - x} - \frac{5}{x^2 - 4}$

Answer: $\frac{x - 15}{(x - 2)(x + 2)}$

EXAMPLE 10 Adding and Subtracting Rational Expressions

Perform the indicated operations and simplify, if possible.

$$\frac{5}{n - 2} + \frac{5}{n + 2} - \frac{6}{n^2 - 4}$$

Solution

To find the LCD, each denominator must be factored completely. We see that $n^2 - 4 = (n - 2)(n + 2)$, so the LCD is $(n - 2)(n + 2)$.

$$\frac{5}{n - 2} + \frac{5}{n + 2} - \frac{6}{n^2 - 4} = \frac{5}{n - 2} + \frac{5}{n + 2} - \frac{6}{(n - 2)(n + 2)}$$

Rewrite each expression with the common denominator: $= \frac{5}{n - 2} \cdot \frac{n + 2}{n + 2} + \frac{5}{n + 2} \cdot \frac{n - 2}{n - 2} - \frac{6}{(n - 2)(n + 2)}$

Multiply the numerators: $= \frac{5n + 10}{(n - 2)(n + 2)} + \frac{5n - 10}{(n - 2)(n + 2)} - \frac{6}{(n - 2)(n + 2)}$

Write numerators over the common denominator: $= \frac{5n + 10 + 5n - 10 - 6}{(n - 2)(n + 2)}$

Combine like terms: $= \frac{10n - 6}{(n - 2)(n + 2)}$

Factor the numerator: $= \frac{2(5n - 3)}{(n - 2)(n + 2)}$

Thus, $\frac{5}{n - 2} + \frac{5}{n + 2} - \frac{6}{n^2 - 4} = \frac{2(5n - 3)}{(n - 2)(n + 2)}$.

QUICK ✓ *Perform the indicated operation and simplify, if possible.*

19. $\dfrac{1}{x+1} - \dfrac{2}{x^2-1} + \dfrac{3}{x-1}$

20. $\dfrac{3}{x} - \left(\dfrac{1}{x-2} - \dfrac{6}{x^2-2x}\right)$

5.5 Exercises

For Extra Help:

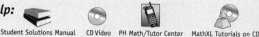

Student Solutions Manual

CD Video

PH Math/Tutor Center

MathXL Tutorials on CD

MathXL®

MyMathLab

1. least common denominator
2. lowest terms
3. denominator
4. False
5. True
6. True
7. Answers may vary.
8. Answers may vary.
9. $-\dfrac{5}{6}$ 10. $\dfrac{4}{3}$
11. $\dfrac{5}{3x}$ 12. $\dfrac{25+12x}{10x}$
13. $\dfrac{2a^2+7a-3}{(2a-1)(2a+1)}$
14. $\dfrac{x^2+3}{(x-1)(x+1)}$
15. $\dfrac{-2}{x-4}$ 16. $\dfrac{-1}{n-4}$
17. $\dfrac{2a^2-3a-12}{4a(a-5)}$
18. $\dfrac{x+6}{x+2}$ 19. $\dfrac{5}{x}$
20. $\dfrac{4}{x-4}$
21. $\dfrac{8}{75}$
22. $\dfrac{22}{105}$
23. $\dfrac{(m-4)(m+4)}{m}$

Concepts and Vocabulary

In Problems 1–3, fill in the blanks.

1. The first step in adding or subtracting rational expressions with unlike denominators is to determine the ________ ________ ________.

2. After adding or subtracting rational expressions, factor the numerator of the result to verify that the fraction is written in ________ ________.

3. To rewrite a rational expression with a least common denominator, multiply the numerator and ________ by the factor(s) missing from the LCD.

In Problems 4–6, answer True or False to each statement.

4. $\dfrac{3}{x} + \dfrac{6}{7x^2} = \dfrac{3+6}{7x^2}$

5. To add or subtract rational expressions, we write the rational expressions over a common denominator and then add or subtract the numerators.

6. $\dfrac{3}{x} + \dfrac{6}{-x} = \dfrac{3-6}{x}$

7. In your own words, explain how to add or subtract rational expressions with unlike denominators.

8. To find the product $\dfrac{1}{x-1} \cdot \dfrac{1}{x+1}$, your classmate decided that each fraction should be written with a common denominator and wrote the problem as $\dfrac{(x+1)}{(x-1)(x+1)} \cdot \dfrac{(x-1)}{(x+1)(x-1)}$. Is this a correct approach? Explain your response.

Building Skills

In Problems 9–20, find each sum and simplify, if possible.

9. $\dfrac{-4}{3} + \dfrac{1}{2}$

10. $\dfrac{7}{12} + \dfrac{3}{4}$

11. $\dfrac{2}{3x} + \dfrac{1}{x}$

12. $\dfrac{5}{2x} + \dfrac{6}{5}$

13. $\dfrac{a}{2a-1} + \dfrac{3}{2a+1}$

14. $\dfrac{2}{x-1} + \dfrac{x-1}{x+1}$

15. $\dfrac{3}{x-4} + \dfrac{5}{4-x}$

16. $\dfrac{7}{n-4} + \dfrac{8}{4-n}$

17. $\dfrac{a+3}{5a-a^2} + \dfrac{2a+1}{4a-20}$

18. $\dfrac{2x-6}{x^2-x-6} + \dfrac{x+4}{x+2}$

19. $\dfrac{2x+4}{x^2+2x} + \dfrac{3}{x}$

20. $\dfrac{3}{x-4} + \dfrac{x+4}{x^2-16}$

In Problems 21–32, find each difference and simplify, if possible.

21. $\dfrac{7}{15} - \dfrac{9}{25}$

22. $\dfrac{8}{21} - \dfrac{6}{35}$

23. $m - \dfrac{16}{m}$

24. $\dfrac{(3-x)(3+x)}{x}$ 25. $\dfrac{2(4x-3)}{(x-3)(x+3)}$
26. $\dfrac{-8x}{(x+2)(x-2)}$ 27. $\dfrac{6(x+1)}{(x+3)^2}$
28. $\dfrac{(x+2)^2}{(x+4)^2}$ 29. $\dfrac{-4(a-1)}{a+3}$
30. $\dfrac{x+3}{2(x-3)}$ 31. $\dfrac{x}{x-2}$
32. $\dfrac{-p^3+2p^2-1}{p(p+2)(p-1)}$ 33. $\dfrac{5}{4}$ 34. $\dfrac{-22}{3}$
35. $\dfrac{20-9y}{12y^2}$ 36. $\dfrac{5n-12}{4n^2}$
37. $\dfrac{6}{5x}$ 38. $\dfrac{7-6b^3}{4b^2}$
39. $\dfrac{7y+4x}{2x^2y^2}$ 40. $\dfrac{-4x^2+3}{xy}$
41. $\dfrac{-3}{2x+3}$ 42. $\dfrac{a(a-1)}{a-2}$
43. $\dfrac{2(n^2-2)}{n(n-2)}$ 44. $\dfrac{10(x-1)}{(x+2)(x-3)}$
45. $\dfrac{2x^2-5x+15}{x(x-3)}$ 46. $\dfrac{(x-4)(x+1)}{(x-2)(x-1)}$
47. $\dfrac{-1}{a(a-1)}$ 48. $\dfrac{2(-3y+4)}{5y(y+2)}$
49. $\dfrac{2n+5}{2n+1}$ 50. $\dfrac{3x+4}{x+3}$
51. $\dfrac{4x-5}{x-2}$ 52. $\dfrac{19}{6(x+1)}$
53. $\dfrac{3x-28}{x-9}$ 54. $\dfrac{x^2-5x+9}{2(2x-3)(x+1)}$
55. $\dfrac{4n^2-7n+9}{n^2(n-3)(n-1)}$
56. $\dfrac{-13x+6}{(x+6)(x-6)^2}$
57. $\dfrac{-13}{2a(a-2)}$ 58. $\dfrac{4}{3n(n-3)}$
59. $\dfrac{n^2-n-3}{-(n+2)(n-2)(n+3)}$
60. $\dfrac{a^2-5a-3}{2a^2(a-3)}$
61. $\dfrac{2x^2+x+1}{(x+1)(x-1)^2}$
62. $\dfrac{x^2+3x+5}{(x+2)^2(x-1)}$
63. $\dfrac{-3n+5}{(n+3)(n-2)}$
64. $\dfrac{7x+23}{(x+5)(x-3)(x+2)}$
65. $\dfrac{x}{x-2}$ 66. -1
67. $\dfrac{x-4}{x}$ 68. $\dfrac{x^3-11x+14}{(x-3)(x-2)(x+2)}$
69. $\dfrac{6}{m(m-2)}$ 70. $\dfrac{15}{w(w-3)}$
71. $\dfrac{a+2}{a(a-1)^2}$ 72. $\dfrac{2(3b+4)}{b(b+2)^2}$
73. $\dfrac{x^2+4}{(x+2)(x+3)}$

24. $\dfrac{9}{x} - x$ **25.** $\dfrac{x}{x-3} - \dfrac{x-2}{x+3}$ **26.** $\dfrac{x-2}{x+2} - \dfrac{x+2}{x-2}$

27. $\dfrac{x+2}{x+3} - \dfrac{x^2-x}{x^2+6x+9}$ **28.** $\dfrac{x}{x+4} - \dfrac{-4}{x^2+8x+16}$ **29.** $\dfrac{6}{2a+6} - \dfrac{4a-1}{a+3}$

30. $\dfrac{3x+7}{2x-6} - \dfrac{x+2}{x-3}$ **31.** $\dfrac{-3x-9}{x^2+x-6} - \dfrac{x+3}{2-x}$ **32.** $\dfrac{p+1}{p^2+2p} - \dfrac{p^2-p}{p^2+p-2}$

Mixed Practice

In Problems 33–72, perform the indicated operation and simplify, if possible.

33. $\dfrac{-3}{4} + 2$ **34.** $4 - \dfrac{34}{3}$ **35.** $\dfrac{5}{3y^2} - \dfrac{3}{4y}$ **36.** $\dfrac{5}{4n} - \dfrac{3}{n^2}$

37. $\dfrac{9}{5x} - \dfrac{6}{10x}$ **38.** $\dfrac{7}{4b^2} - \dfrac{3b^2}{2b}$ **39.** $\dfrac{7}{2x^2y} + \dfrac{8}{4xy^2}$ **40.** $\dfrac{-4x}{y} + \dfrac{3x}{x^2y}$

41. $\dfrac{2x}{2x+3} - 1$ **42.** $a + \dfrac{a}{a-2}$ **43.** $\dfrac{n}{n-2} + \dfrac{n+2}{n}$ **44.** $\dfrac{6}{x+2} + \dfrac{4}{x-3}$

45. $\dfrac{2x}{x-3} - \dfrac{5}{x}$ **46.** $\dfrac{x}{x-2} - \dfrac{2}{x-1}$ **47.** $\dfrac{a}{a^2} - \dfrac{1}{a-1}$ **48.** $\dfrac{4}{5y} - \dfrac{2}{y+2}$

49. $\dfrac{4}{2n+1} + 1$ **50.** $\dfrac{x-2}{x+3} + 2$

51. $\dfrac{-12}{3x-6} + \dfrac{4x-1}{x-2}$ **52.** $\dfrac{2}{4x+4} + \dfrac{8}{3x+3}$

53. $\dfrac{3x-1}{x} - \dfrac{9}{x^2-9x}$ **54.** $\dfrac{2x}{8x-12} - \dfrac{3}{2x+2}$

55. $\dfrac{4}{n^2-3n} - \dfrac{3}{n^3-n^2}$ **56.** $\dfrac{x-1}{x^2-36} - \dfrac{x}{x^2-12x+36}$

57. $\dfrac{5}{2a-a^2} - \dfrac{3}{2a^2-4a}$ **58.** $\dfrac{-1}{3n-n^2} + \dfrac{1}{3n^2-9n}$

59. $\dfrac{2n+1}{n^2-4} + \dfrac{3n}{6-n^2-n}$ **60.** $\dfrac{a-5}{2a^2-6a} - \dfrac{6}{12a^2-4a^3}$

61. $\dfrac{x}{x^2-1} + \dfrac{x+1}{x^2-2x+1}$ **62.** $\dfrac{x+1}{x^2+4x+4} + \dfrac{3}{x^2+x-2}$

63. $\dfrac{2n+1}{n+3} + \dfrac{7-2n^2}{n^2+n-6}$ **64.** $\dfrac{4}{x^2+2x-15} + \dfrac{3}{x^2-x-6}$

65. $\dfrac{-3x-9}{x^2+x-6} - \dfrac{x+3}{2-x}$ **66.** $\dfrac{2x-1}{1-x} - \dfrac{-x^2-2x}{x^2+x-2}$

67. $\dfrac{4}{x+2} + \dfrac{-5x-2}{x^2+2x} - \dfrac{3-x}{x}$ **68.** $\dfrac{x+1}{x-3} + \dfrac{x+2}{x-2} - \dfrac{x^2+3}{x^2-x-6}$

69. $\dfrac{2}{m+2} - \dfrac{3}{m} + \dfrac{m+10}{m^2-4}$ **70.** $\dfrac{7}{w-3} - \dfrac{5}{w} - \dfrac{2w+6}{w^2-9}$

71. $\dfrac{2}{a} - \left(\dfrac{2}{a-1} - \dfrac{3}{(a-1)^2}\right)$ **72.** $\dfrac{2}{b} - \left(\dfrac{2}{b+2} - \dfrac{2}{(b+2)^2}\right)$

Applying the Concepts

In Problems 73–80, perform the indicated operation, if possible. Addition, subtraction, multiplication, and division of rational expressions are included.

73. Find the quotient of $\dfrac{x-3}{x+2}$ and $\dfrac{x^2-9}{x^2+4}$.

74. $\frac{3x}{5}$

75. $\frac{x+3}{(x+1)(x-1)}$

76. $\frac{-1}{2(x+2)}$

77. $\frac{2x^2-x+10}{(x+3)(x-2)(x+2)}$

78. $x+4$

79. $\frac{2x^2-5x+3}{(x+6)(x-7)}$

80. $\frac{x-2}{x+2}$

81. $\frac{x-4}{8}$ square units

82. $\frac{7x-5}{4}$ units

83. $\frac{2x^2-9x-18}{6}$ square units

84. $\frac{6(3x-4)}{x-3}$ square units

85. $\frac{x^3-2x^2+4x+3}{x^2(x+1)(x-1)}$

86. $\frac{3x^3-5x^2+2x+1}{x^2(x-1)^2}$

87. $\frac{1}{a+b}$

88. $\frac{7a-b}{(a-b)(a+b)}$

89. $\frac{x^2+2x+5}{(x-4)(x+1)}$

90. $\frac{1}{m-n}$

91. $\frac{-x^3+6x^2-11x+10}{x-2}$

92. $\frac{3(3a+1)}{(a+2)(a-3)}$

74. Find the quotient of $\frac{x}{x+1}$ and $\frac{5}{3x+3}$.

75. Find the difference of $\frac{4x+6}{2x^2+x-3}$ and $\frac{x-1}{x^2-1}$.

76. Find the difference of $\frac{x+4}{2x^2-8}$ and $\frac{3}{4x-8}$.

77. Find the sum of $\frac{x+2}{x^2+x-6}$ and $\frac{x-3}{x^2+5x+6}$.

78. Find the sum of $\frac{x^2}{x-4}$ and $\frac{16}{4-x}$.

79. Find the product of $\frac{2x-3}{x+6}$ and $\frac{x-1}{x-7}$.

80. Find the product of $\frac{x^2-3x+2}{x^2+2x-3}$ and $\frac{x^2+x-6}{x^2-4}$.

△ **81. Area of a Trapezoid** Use the formula $A = \frac{1}{2}h(B+b)$ to find the area of the trapezoid shown:

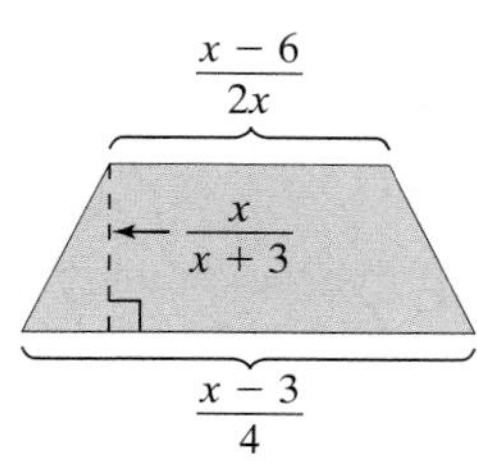

△ **82. Perimeter of a Rectangle** Find the perimeter of the rectangle:

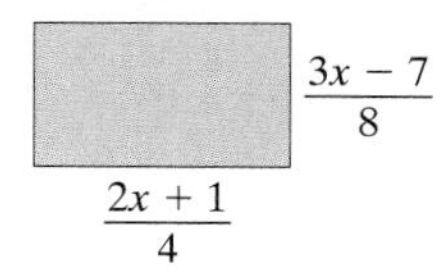

△ **83. Area of a Rectangle** Find the area of the rectangle:

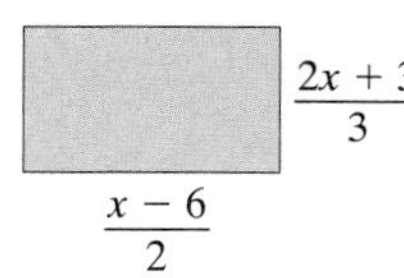

△ **84. Area of a Trapezoid** Find the area of the trapezoid:

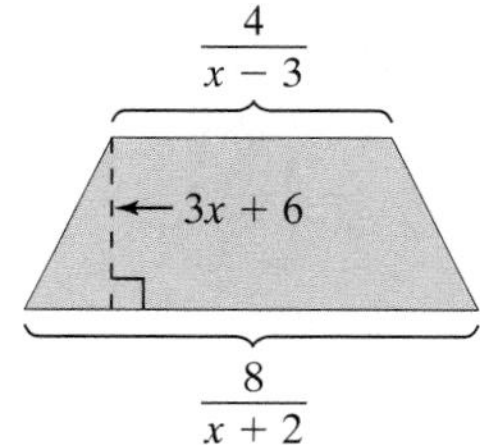

Extending the Concepts

In Problems 85–92, perform the indicated operations and simplify, if possible.

85. $\frac{1}{x} - \frac{2}{x^2+x} + \frac{3}{x^3-x^2}$

86. $\frac{x}{(x-1)^2} + \frac{2}{x} - \frac{x+1}{x^3-x^2}$

87. $\frac{2a+b}{a-b}\cdot\frac{2}{a+b} - \frac{3a+3b}{a^2-b^2}$

88. $\frac{2a+b}{a-b} \div \frac{a+b}{2} + \frac{3}{a+b}$

89. $\frac{x-3}{x-4} + \frac{x+2}{x-4}\cdot\frac{4}{x+1}$

90. $\frac{3}{m+n} - \frac{m-2n}{m-n}\cdot\frac{2}{m+n}$

91. $\frac{x+2}{x-2} - \frac{x-2}{x+2} \div \frac{1}{x^2-4}$

92. $\frac{a+3}{a-3} - \frac{a+3}{a-3}\cdot\frac{a^2-4a+3}{a^2+5a+6}$

5.6 Complex Rational Expressions

OBJECTIVES

1. Simplify a Complex Rational Expression by Simplifying the Numerator and Denominator Separately
2. Simplify a Complex Rational Expression Using the Least Common Denominator

Preparing for Complex Rational Expressions

Before getting started, take the following readiness quiz. If you get a problem wrong, go back to the section cited and review the material.

1. Factor: $6y^2 - 5y - 6$ [Section 4.3, pp. 257–265]

2. Find the quotient: $\dfrac{x + 3}{12} \div \dfrac{x^2 - 9}{15}$ [Section 5.2, pp. 315–318]

DEFINITION

A **complex rational expression** is a fraction in which the numerator and/or the denominator contains the sum or difference of rational expressions.

The following are examples of complex rational expressions.

$$\frac{3}{\frac{4}{x} - \frac{x}{3}} \qquad \frac{\frac{x + 1}{x + 4} - 1}{16} \qquad \frac{3 - \frac{1}{x}}{1 + \frac{1}{x}}$$

To **simplify** a complex rational expression means to write the rational expression in the form $\frac{p}{q}$, where p and q are polynomials that have no common factors. We can simplify complex rational expressions using one of two methods:

1. By simplifying the numerator and the denominator separately, or
2. By using the least common denominator.

The goal in each of these methods is to write the complex rational expression in the form $\frac{\text{rational expression}}{\text{rational expression}}$. Then, we can simplify the expression using the division techniques introduced in Section 5.2. To help refresh your memory regarding this technique, we present Example 1.

Classroom Example

Find the quotient: $\dfrac{\frac{4}{x + 3}}{\frac{8}{x^2 - 9}}$

Answer: $\dfrac{x - 3}{4}$

EXAMPLE 1 Dividing Rational Expressions

Find the quotient: $\dfrac{\frac{3}{x + 4}}{\frac{9}{x^2 - 16}}$

Solution

Notice the rational expression is in the form $\frac{\text{rational expression}}{\text{rational expression}}$, where the rational expression in the numerator is $\frac{3}{x + 4}$ and the rational expression in the denominator is $\frac{9}{x^2 - 16}$. To find the quotient, we multiply the rational expression in the numerator

Preparing for...Answers

1. $(3y + 2)(2y - 3)$ **2.** $\dfrac{5}{4(x - 3)}$

by the reciprocal of the rational expression in the denominator.

$$\frac{\frac{3}{x+4}}{\frac{9}{x^2-16}} = \frac{3}{x+4} \cdot \frac{x^2-16}{9}$$

Factor the numerator and denominator: $= \frac{3}{x+4} \cdot \frac{(x+4)(x-4)}{3 \cdot 3}$

Multiply: $= \frac{3(x+4)(x-4)}{(x+4) \cdot 3 \cdot 3}$

Divide out common factors: $= \frac{\cancel{3}\cancel{(x+4)}(x-4)}{\cancel{(x+4)} \cdot \cancel{3} \cdot 3}$

$$= \frac{x-4}{3}$$

QUICK ✓ *Simplify each expression.*

1. $\dfrac{\frac{2}{k+3}}{\frac{4}{k^2+4k+3}}$ **2.** $\dfrac{\frac{1}{n+3}}{\frac{8n}{2n+6}}$

Classroom Example

Simplify: $\dfrac{\frac{1}{3}+\frac{1}{y}}{\frac{y+3}{4}}$

Answer: $\dfrac{4}{3y}$

(1) Simplify a Complex Rational Expression by Simplifying the Numerator and Denominator Separately

As mentioned, there are two methods that can be used to simplify a complex rational expression. We'll start by simplifying the numerator and denominator separately. We call this Method 1.

EXAMPLE 2 How to Simplify a Complex Rational Expression (Method 1)

Simplify: $\dfrac{\frac{1}{5}+\frac{1}{x}}{\frac{x+5}{2}}$

Step-by-Step Solution

Step 1: Write the numerator of the complex fraction as a single rational expression.

The numerator of the complex rational expression is $\frac{1}{5}+\frac{1}{x}$. Its least common denominator is $5x$.

LCD = $5x$: $\frac{1}{5}+\frac{1}{x} = \frac{1}{5}\cdot\frac{x}{x} + \frac{1}{x}\cdot\frac{5}{5}$

$= \frac{x}{5x} + \frac{5}{5x}$

Use $\frac{a}{c}+\frac{b}{c}=\frac{a+b}{c}$: $= \frac{x+5}{5x}$

Step 2: Write the denominator of the complex rational expression as a single rational expression.

The denominator of the complex rational expression is $\frac{x+5}{2}$. It is already written as a single rational expression.

(continued)

Step 3: Rewrite the complex rational expression using the rational expressions determined in Steps 1 and 2.

$$\frac{\frac{1}{5}+\frac{1}{x}}{\frac{x+5}{2}} = \frac{\frac{x+5}{5x}}{\frac{x+5}{2}}$$

Step 4: Simplify the rational expression using the techniques for dividing rational expressions from Section 5.2.

Rewrite the division problem as a multiplication problem: $= \frac{x+5}{5x}\cdot\frac{2}{x+5}$

Multiply: $= \frac{(x+5)\cdot 2}{5x(x+5)}$

Divide out like factors: $= \frac{2}{5x}$

So $\frac{\frac{1}{5}+\frac{1}{x}}{\frac{x+5}{2}} = \frac{2}{5x}$.

We summarize the steps that you can use to simplify a complex rational expression.

Work Smart

First simplify (add or subtract) both the numerator and the denominator into a single rational expression before you take the reciprocal and multiply.

Teaching Tip

Emphasize that students must simplify (add or subtract) both the numerator and denominator into a single rational expression before they take the reciprocal and multiply.

Steps to Simplify a Complex Rational Expression by Simplifying The Numerator and Denominator Separately (Method 1)

Step 1: Write the numerator of the complex rational expression as a single rational expression.

Step 2: Write the denominator of the complex rational expression as a single rational expression.

Step 3: Rewrite the complex rational expression using the rational expressions determined in Steps 1 and 2.

Step 4: Simplify the rational expression using the techniques for dividing rational expressions from Section 5.2.

Classroom Example Simplify: $\frac{\frac{2}{a^2}+\frac{3}{a}}{\frac{1}{a^3}-\frac{4}{a}}$

Answer: $\frac{a(2+3a)}{(1+2a)(1-2a)}$

EXAMPLE 3 Simplifying a Complex Rational Expression (Method 1)

Simplify: $\frac{\frac{2}{y}-\frac{8}{y^3}}{\frac{2}{y^2}+\frac{1}{y}}$

Solution

We write the numerator and denominator of the complex rational expression as a single rational expression.

NUMERATOR:

LCD is y^3: $\frac{2}{y}-\frac{8}{y^3} = \frac{2}{y}\cdot\frac{y^2}{y^2}-\frac{8}{y^3}$

$= \frac{2y^2-8}{y^3}$

$= \frac{2(y^2-4)}{y^3}$

$= \frac{2(y+2)(y-2)}{y^3}$

DENOMINATOR:

LCD is y^2: $\frac{2}{y^2}+\frac{1}{y} = \frac{2}{y^2}+\frac{1}{y}\cdot\frac{y}{y}$

$= \frac{2}{y^2}+\frac{y}{y^2}$

$= \frac{2+y}{y^2}$

Now we rewrite the complex rational expression with the new numerator and denominator.

$$\frac{\frac{2}{y}-\frac{8}{y^3}}{\frac{2}{y^2}+\frac{1}{y}} = \frac{\frac{2(y+2)(y-2)}{y^3}}{\frac{2+y}{y^2}}$$

Multiply the numerator by the reciprocal of the denominator: $= \frac{2(y+2)(y-2)}{y^3} \cdot \frac{y^2}{2+y}$

Simplify by dividing out common factors: $= \frac{2\cancel{(y+2)}(y-2)}{\cancel{y^2} \cdot y} \cdot \frac{\cancel{y^2}}{\cancel{y+2}}$

$$= \frac{2(y-2)}{y}$$

QUICK ✓ *Simplify each expression.*

3. $\dfrac{\frac{1}{3}+\frac{1}{x}}{\frac{x+3}{2}}$

4. $\dfrac{\frac{3}{y}-1}{\frac{9}{y}-y}$

Classroom Example ➤ Find the quotient: $\dfrac{\frac{1}{x+3}+1}{x-\frac{4}{x+3}}$

Answer: $\dfrac{1}{x-1}$

EXAMPLE 4 Dividing Rational Expressions (Method 1)

Simplify: $\dfrac{\frac{1}{x+2}+1}{x-\frac{3}{x+2}}$

Solution

Write the numerator of the complex rational expression, $\frac{1}{x+2}+1$, as a single rational expression. The LCD is $(x+2)$.

$$\frac{1}{x+2}+1 = \frac{1}{x+2}+1\cdot\frac{x+2}{x+2}$$

Multiply: $= \frac{1}{x+2}+\frac{x+2}{x+2}$

Write the numerators over a common denominator: $= \frac{1+x+2}{x+2}$

Combine like terms: $= \frac{x+3}{x+2}$

Write the denominator of the complex rational expression, $x-\frac{3}{x+2}$, as a single rational expression. The LCD of the expression is $x+2$.

$$x-\frac{3}{x+2} = \frac{x}{1}\cdot\frac{x+2}{x+2}-\frac{3}{x+2}$$

Multiply: $= \frac{x^2+2x}{x+2}-\frac{3}{x+2}$

Write the numerators over a common denominator: $= \frac{x^2+2x-3}{x+2}$

Factor: $= \frac{(x+3)(x-1)}{x+2}$

Rewrite the complex rational expression using the numerator and denominator just found, and then simplify.

$$\frac{\frac{1}{x+2}+1}{x-\frac{3}{x+2}} = \frac{\frac{x+3}{x+2}}{\frac{(x+3)(x-1)}{x+2}}$$

Multiply the reciprocal of the denominator by the numerator: $= \frac{x+3}{x+2} \cdot \frac{x+2}{(x+3)(x-1)}$

Multiply: $= \frac{\cancel{(x+3)}\,\cancel{(x+2)}}{\cancel{(x+2)}\,\cancel{(x+3)}\,(x-1)}$

Divide like factors: $= \frac{1}{x-1}$

QUICK ✓ *Simplify each expression.*

5. $\dfrac{\frac{1}{3}+\frac{1}{x+5}}{\frac{x+8}{9}}$

6. $\dfrac{\frac{8}{y+3}-2}{y-\frac{4}{y+3}}$

② Simplify a Complex Rational Expression Using the Least Common Denominator

We now introduce a second method for simplifying complex rational expressions.

EXAMPLE 5 How to Simplify a Complex Rational Expression Using the Least Common Denominator (Method 2)

Simplify: $\dfrac{\frac{2}{y}-\frac{8}{y^3}}{\frac{2}{y^2}+\frac{1}{y}}$

Classroom Example

Simplify using Method 2: $\dfrac{\frac{1}{a^2}+\frac{2}{a}}{\frac{1}{a^3}-\frac{4}{a}}$

Answer: $\dfrac{a}{1-2a}$

Step-by-Step Solution

Step 1: Find the least common denominator among each denominator in the complex rational expression.

The denominators of the complex rational expression are y, y^3, and y^2. The least common denominator is y^3.

Step 2: Multiply both the numerator and denominator of the complex rational expression by the least common denominator found in Step 1.

Multiply the numerator and denominator by y^3: $\left(\dfrac{\frac{2}{y}-\frac{8}{y^3}}{\frac{2}{y^2}+\frac{1}{y}}\right)\cdot\left(\dfrac{y^3}{y^3}\right) = \dfrac{\frac{2}{y}\cdot y^3-\frac{8}{y^3}\cdot y^3}{\frac{2}{y^2}\cdot y^3+\frac{1}{y}\cdot y^3}$

Divide out common factors: $= \dfrac{\frac{2}{\cancel{y}}\cdot \cancel{y^3}y^2-\frac{8}{\cancel{y^3}}\cdot \cancel{y^3}1}{\frac{2}{\cancel{y^2}}\cdot \cancel{y^3}y+\frac{1}{\cancel{y}}\cdot \cancel{y^3}y^2}$

Teaching Tip

Students may want to short-cut Step 2 by not showing each of the products obtained using the Distributive Property. Encourage them to show each step, even though it takes up space on their paper!

Step 3: Simplify.

Multiply: $= \dfrac{2y^2 - 8}{2y + y^2}$

Factor: $= \dfrac{2(y^2 - 4)}{y(2 + y)}$

$= \dfrac{2(y + 2)(y - 2)}{y(y + 2)}$

Divide out like factors: $= \dfrac{2(y - 2)}{y}$

Teaching Tip
You may wish to show Method 1 and Method 2 worked side by side, using Example 4.

So $\dfrac{\dfrac{2}{y} - \dfrac{8}{y^3}}{\dfrac{2}{y^2} + \dfrac{1}{y}} = \dfrac{2(y - 2)}{y}$. This is the same result we obtained in Example 3.

Let's redo Example 4 using Method 2 so that we can compare the results.

Classroom Example Find the quotient: $\dfrac{\dfrac{1}{x + 3} + 1}{x - \dfrac{4}{x + 3}}$

Answer: $\dfrac{1}{x - 1}$

EXAMPLE 6 Simplifying a Complex Rational Expression Using the Least Common Denominator

Simplify: $\dfrac{\dfrac{1}{x + 2} + 1}{x - \dfrac{3}{x + 2}}$

Solution

The least common denominator is $x + 2$. So, we multiply the numerator and denominator by $(x + 2)(x + 1)$.

$$\frac{\dfrac{1}{x + 2} + 1}{x - \dfrac{3}{x + 2}} = \frac{\dfrac{1}{x + 2} + 1}{\dfrac{x}{1} - \dfrac{3}{x + 2}} \cdot \frac{x + 2}{x + 2}$$

Distribute the LCD to each term: $= \dfrac{\dfrac{1}{x + 2} \cdot (x + 2) + 1 \cdot (x + 2)}{\dfrac{x}{1} \cdot (x + 2) - \dfrac{3}{x + 2} \cdot (x + 2)}$

Divide out like factors: $= \dfrac{\dfrac{1}{\cancel{x + 2}} \cdot \cancel{(x + 2)} + x + 2}{\dfrac{x}{1} \cdot (x + 2) - \dfrac{3}{\cancel{x + 2}} \cdot \cancel{(x + 2)}}$

Distribute: $= \dfrac{1 + x + 2}{x^2 + 2x - 3}$

Combine like terms; factor: $= \dfrac{x + 3}{(x + 3)(x - 1)}$

Divide out like factors: $= \dfrac{1}{x - 1}$

This is the same result we obtained in Example 4.

Next, we summarize the steps you can use to simplify a complex rational expression using the least common denominator method.

Steps to Simplify a Complex Rational Expressions Using the Least Common Denominator (Method 2)

Step 1: Find the least common denominator among each denominator in the complex rational expression.

Step 2: Multiply both the numerator and denominator of the complex rational expression by the least common denominator found in Step 1.

Step 3: Simplify the rational expression, if possible.

QUICK ✓ *Simplify the complex rational expression using Method 2.*

7. $\dfrac{\dfrac{1}{x}+\dfrac{2}{y}}{\dfrac{2}{x}-\dfrac{1}{y}}$

5.6 Exercises

For Extra Help:

Student Solutions Manual | CD Video | PH Math/Tutor Center | MathXL Tutorials on CD | MathXL® | MyMathLab

Concepts and Vocabulary

In Problems 1–3, fill in the blanks.

1. A rational expression such as $\dfrac{\dfrac{x}{2}+\dfrac{5}{x}}{\dfrac{2x-1}{3}}$ is called a ________ ________ ________.

2. To ______ a complex rational expression means to write the rational expression in the form $\dfrac{p}{q}$, where p and q are polynomials that have no common factors.

3. To simplify $\dfrac{\dfrac{x+1}{6}}{\dfrac{x^2+3x+2}{3}}$ take the ______ of $\dfrac{x^2+3x+2}{3}$ and ______ it by $\dfrac{x+1}{6}$.

In Problems 4–6, answer True or False to each statement.

4. A complex rational expression is considered simplified if there is only one fraction bar and there are no factors common to both the numerator and denominator.

5. To simplify using Method 1, we rewrite $\dfrac{x-y}{\dfrac{1}{x}+\dfrac{2}{y}}$ as $\dfrac{x-y}{1}\cdot\left(\dfrac{x}{1}+\dfrac{y}{2}\right)$.

6. The expression $\dfrac{\dfrac{2}{x+1}}{\dfrac{1}{x-5}}$ is simplified.

7. State which method you prefer to use when simplifying complex rational expressions. State your reasons for choosing this method and then explain how to use this method to simplify $\dfrac{\dfrac{1}{x}+\dfrac{1}{y}}{\dfrac{1}{x^2}-\dfrac{1}{y^2}}$.

8. Explain the property of real numbers that is being applied in Method 2. Make up a complex rational expression and explain how to use Method 2 to simplify the expression.

1. complex rational expression
2. simplify
3. reciprocal, multiply
4. True
5. False
6. False
7. Answers may vary.
8. Multiplicative Identity; Answers may vary.

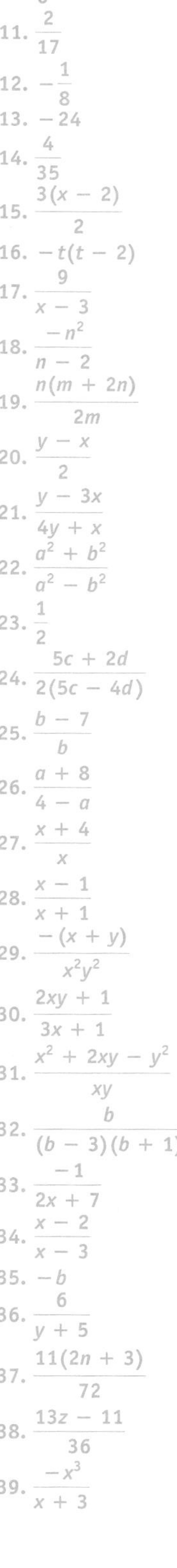

9. -5
10. $\frac{7}{6}$
11. $\frac{2}{17}$
12. $-\frac{1}{8}$
13. -24
14. $\frac{4}{35}$
15. $\frac{3(x-2)}{2}$
16. $-t(t-2)$
17. $\frac{9}{x-3}$
18. $\frac{-n^2}{n-2}$
19. $\frac{n(m+2n)}{2m}$
20. $\frac{y-x}{2}$
21. $\frac{y-3x}{4y+x}$
22. $\frac{a^2+b^2}{a^2-b^2}$
23. $\frac{1}{2}$
24. $\frac{5c+2d}{2(5c-4d)}$
25. $\frac{b-7}{b}$
26. $\frac{a+8}{4-a}$
27. $\frac{x+4}{x}$
28. $\frac{x-1}{x+1}$
29. $\frac{-(x+y)}{x^2y^2}$
30. $\frac{2xy+1}{3x+1}$
31. $\frac{x^2+2xy-y^2}{xy}$
32. $\frac{b}{(b-3)(b+1)}$
33. $\frac{-1}{2x+7}$
34. $\frac{x-2}{x-3}$
35. $-b$
36. $\frac{6}{y+5}$
37. $\frac{11(2n+3)}{72}$
38. $\frac{13z-11}{36}$
39. $\frac{-x^3}{x+3}$

Building Skills

In Problems 9–20, simplify the complex rational expressions using both Method 1 and Method 2. State which method you prefer for the problem.

9. $\dfrac{\frac{4}{3}+1}{\frac{8}{15}-1}$ **10.** $\dfrac{\frac{5}{9}+1}{\frac{7}{3}-1}$ **11.** $\dfrac{1-\frac{3}{4}}{\frac{1}{8}+2}$ **12.** $\dfrac{\frac{2}{3}-\frac{3}{4}}{\frac{1}{6}+\frac{1}{2}}$

13. $\dfrac{1}{\frac{5}{6}-\frac{7}{8}}$ **14.** $\dfrac{\frac{3}{7}-\frac{1}{5}}{2}$ **15.** $\dfrac{\frac{x^2}{12}-\frac{1}{3}}{\frac{x+2}{18}}$ **16.** $\dfrac{\frac{4}{t^2}-1}{\frac{t+2}{t^3}}$

17. $\dfrac{\frac{x+3}{}}{}$ $\dfrac{x+3}{\frac{x^2}{9}-1}$ **18.** $\dfrac{n+2}{\frac{4}{n^2}-1}$ **19.** $\dfrac{\frac{m}{2}+n}{\frac{m}{n}}$ **20.** $\dfrac{\frac{1}{x}-\frac{1}{y}}{\frac{2}{xy}}$

Mixed Practice

In Problems 21–36, simplify the complex rational expression using either Method 1 or Method 2.

21. $\dfrac{\frac{1}{x}-\frac{3}{y}}{\frac{4}{x}+\frac{1}{y}}$ **22.** $\dfrac{\frac{a}{b}+\frac{b}{a}}{\frac{a}{b}-\frac{b}{a}}$ **23.** $\dfrac{\frac{3}{m}+\frac{2}{m^2}}{\frac{6}{m}+\frac{4}{m^2}}$ **24.** $\dfrac{\frac{3c}{4}+\frac{3d}{10}}{\frac{3c}{2}-\frac{6d}{5}}$

25. $\dfrac{1-\frac{49}{b^2}}{1+\frac{7}{b}}$ **26.** $\dfrac{\frac{a}{2}+4}{2-\frac{a}{2}}$ **27.** $\dfrac{1+\frac{5}{x}}{1+\frac{1}{x+4}}$ **28.** $\dfrac{\frac{x}{x+1}}{1+\frac{1}{x-1}}$

29. $\dfrac{\frac{1}{x^2}-\frac{1}{y^2}}{x-y}$ **30.** $\dfrac{6x+\frac{3}{y}}{\frac{9x+3}{y}}$ **31.** $\dfrac{\frac{x}{x-y}+\frac{y}{x+y}}{\frac{xy}{x^2-y^2}}$ **32.** $\dfrac{\frac{b}{b+1}-1}{\frac{b+3}{b}-2}$

33. $\dfrac{\frac{2}{x+4}}{\frac{2}{x+4}-4}$ **34.** $\dfrac{1-\frac{4}{x^2}}{1-\frac{1}{x}-\frac{6}{x^2}}$ **35.** $\dfrac{\frac{b^2}{b^2-16}-\frac{b}{b+4}}{\frac{b}{b^2-16}-\frac{1}{b-4}}$ **36.** $\dfrac{\frac{-6}{y^2+5y+6}}{\frac{2}{y+3}-\frac{3}{y+2}}$

Applying the Concepts

37. Finding the Mean The arithmetic mean of a set of numbers is found by adding the numbers and then dividing by the number of entries on the list. Write a complex rational expression to find the arithmetic mean of the expressions $\frac{n}{6}$, $\frac{n+3}{2}$, and $\frac{2n-1}{8}$ and then simplify the complex rational expression.

38. Finding the Mean See Problem 37. Write a complex rational expression to find the arithmetic mean of the expressions $\frac{z}{2}$, $\frac{z-3}{4}$, and $\frac{2z-1}{6}$ and then simplify the complex rational expression.

△ **39. Area of a Rectangle** In a rectangle, the length can be found by dividing the area by the width. If the area of a rectangle is $\frac{2x+3}{x^2}-\frac{3}{x}$ and the width is $\frac{x^2-9}{x^5}$, write a complex rational expression to find the length and then simplify the complex rational expression.

40. $\dfrac{-1}{x^2(x-4)}$

41. (a) $R = \dfrac{R_1R_2}{R_2 + R_1}$ (b) $\dfrac{15}{4}$ ohms

42. (a) $R = \dfrac{R_1R_2R_3}{R_2R_3 + R_1R_3 + R_1R_2}$
(b) $\dfrac{60}{31}$ ohms

43. $\dfrac{2x+1}{x+1}$

44. $\dfrac{-1}{x-3}$

45. $\dfrac{5-2x}{2-x}$

46. $\dfrac{3x^2+2x+6}{x^2+2x+2}$

△ **40. Area of a Rectangle** In a rectangle, the length can be found by dividing the area by the width. If the area of a rectangle is $\dfrac{x-4}{x^3} - \dfrac{2}{x^2}$ and the width is $\dfrac{x^2-16}{x}$, write a complex rational expression to find the length and then simplify the complex rational expression.

41. Electric Circuits An electrical circuit contains two resistors connected in parallel, as shown in the figure. If the resistance of each is R_1 and R_2 ohms, respectively, then their combined resistance R is given by the formula

$$R = \dfrac{1}{\dfrac{1}{R_1} + \dfrac{1}{R_2}}$$

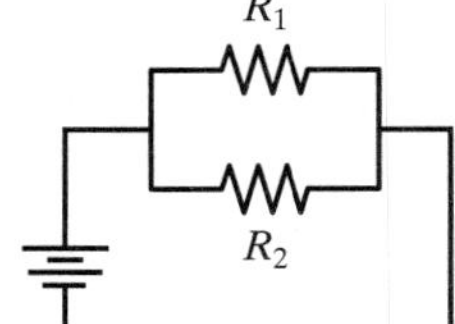

(a) Express R as a simplified rational expression.
(b) Evaluate the rational expression if $R_1 = 6$ ohms and $R_2 = 10$ ohms.

42. Electric Circuits An electrical circuit contains three resistors connected in parallel. If the resistance of each is R_1, R_2, and R_3 ohms, respectively, then their combined resistance is given by the formula

$$R = \dfrac{1}{\dfrac{1}{R_1} + \dfrac{1}{R_2} + \dfrac{1}{R_3}}$$

(a) Express R as a simplified rational expression.
(b) Evaluate the rational expression if $R_1 = 4$ ohms, $R_2 = 6$ ohms, and $R_3 = 10$ ohms.

Extending the Concepts

In Problems 43–46, write each expression in lowest terms.

43. $1 + \dfrac{1}{1 + \dfrac{1}{x}}$ **44.** $1 - \dfrac{1}{1 - \dfrac{1}{x-2}}$ **45.** $\dfrac{1}{1 - \dfrac{1}{2 - \dfrac{1}{3-x}}}$ **46.** $1 + \dfrac{2}{1 + \dfrac{2}{x + \dfrac{2}{x}}}$

PUTTING THE CONCEPTS TOGETHER (SECTIONS 5.1–5.6)

These problems cover important concepts from Sections 5.1 to 5.6. We designed these problems so that you can review the chapter so far and show your mastery of the concepts. Take time to work these problems before proceeding with the next section. The answers to these problems are located at the back of the text starting on page AN-11.

1. -35
2. (a) The expression is undefined for $a = 6$.
(b) The expression is undefined for $y = 0$ or $y = -4$.
3. (a) $\dfrac{a-4b}{2}$ (b) $\dfrac{-(x+2)}{x+1}$
4. The LCD is $8(a+2b)(a-4b)$.

1. Evaluate $\dfrac{a^3 - b^3}{a+b}$ when $a = 2$ and $b = -3$.

2. Find the values for which the following rational expressions are undefined:

(a) $\dfrac{-3a}{a-6}$ **(b)** $\dfrac{y+2}{y^2+4y}$

3. Simplify:

(a) $\dfrac{ax + ay - 4bx - 4by}{2x + 2y}$ **(b)** $\dfrac{x^2 + x - 2}{1 - x^2}$

4. Find the least common denominator (LCD) of the rational expressions:

$$\dfrac{5}{2a+4b}, \dfrac{-1}{4a+8b}, \dfrac{3ab}{8a-32b}$$

5. $\dfrac{7}{3x^2 - x} = \dfrac{35x}{5x^2(3x - 1)}$
6. $\dfrac{-2y^2}{y + 1}$
7. $\dfrac{2(m - 1)^2}{m^2}$
8. $\dfrac{2(y - 1)}{5(y - 3)}$
9. $4x + 1$
10. $\dfrac{1}{x + 1}$
11. $\dfrac{1}{2x - 1}$
12. $\dfrac{15m + 17}{(m + 3)(3m + 2)}$
13. $\dfrac{m}{(m + 5)(m + 4)}$
14. $-\dfrac{1}{m - 1}$
15. $\dfrac{a(24 + a)}{3(6 + a^2)}$

5. Write $\dfrac{7}{3x^2 - x}$ as an equivalent rational expression with denominator $5x^2(3x - 1)$.

In Problems 6–13, perform the indicated operations and simplify, if possible.

6. $\dfrac{y^2 - y}{3y} \cdot \dfrac{6y^2}{1 - y^2}$

7. $\dfrac{m^2 + m - 2}{m^3 - 6m^2} \cdot \dfrac{2m^2 - 14m + 12}{m + 2}$

8. $\dfrac{4y + 12}{5y - 5} \div \dfrac{2y^2 - 18}{y^2 - 2y + 1}$

9. $\dfrac{4x^2 + x}{x + 3} + \dfrac{12x + 3}{x + 3}$

10. $\dfrac{x^2}{x^3 + 1} - \dfrac{x - 1}{x^3 + 1}$

11. $\dfrac{-8}{2x - 1} - \dfrac{9}{1 - 2x}$

12. $\dfrac{4}{m + 3} + \dfrac{3}{3m + 2}$

13. $\dfrac{3m}{m^2 + 7m + 10} - \dfrac{2m}{m^2 + 6m + 8}$

14. Simplify the complex rational expression: $\dfrac{\dfrac{-1}{m + 1} - 1}{m - \dfrac{2}{m + 1}}$

15. Simplify the complex rational expression: $\dfrac{\dfrac{4}{a} + \dfrac{1}{6}}{\dfrac{3}{a^2} + \dfrac{1}{2}}$

5.7 Rational Equations

OBJECTIVES

1. Solve Equations Containing Rational Expressions
2. Solve for a Variable in a Rational Equation

Preparing for Rational Equations

Before getting started, take the following readiness quiz. If you get a problem wrong, go back to the section cited and review the material.

1. Solve: $3k - 2(k + 1) = 6$ [Section 2.2, pp. 86–87]

2. Factor: $3p^2 - 7p - 6$ [Section 4.3, pp. 257–265]

3. Solve: $8z^2 - 10z - 3 = 0$ [Section 4.6, pp. 281–287]

4. Find the values for which the expression $\dfrac{x + 4}{x^2 - 2x - 24}$ is undefined. [Section 5.1, pp. 307–308]

5. Solve for y: $4x - 2y = 10$ [Section 2.4, pp. 109–112]

1 Solve Equations Containing Rational Expressions

Up to this point, we have learned how to solve linear equations (Sections 2.1–2.3), quadratic equations (Section 4.6), and equations that contain polynomial expressions that can be factored (Section 4.6). We now introduce another type of equation.

DEFINITION

A **rational equation** is an equation that contains a rational expression.

Examples of rational equations are

$$\frac{3}{x + 4} = \frac{1}{x + 2} \quad \text{and} \quad \frac{4}{3} + \frac{7}{x - 4} = \frac{x - 1}{3x - 12}$$

Preparing for...Answers **1.** $\{8\}$ **2.** $(3p + 2)(p - 3)$ **3.** $\left\{-\frac{1}{4}, \frac{3}{2}\right\}$ **4.** $x = -4$ or $x = 6$ **5.** $y = 2x - 5$

Teaching Tip
Spend a few minutes in class describing the difference in the way the LCD is used in adding and subtracting algebraic expressions as contrasted to solving rational equations.

The goal in solving these equations is to use algebraic techniques to rewrite the equation until it becomes one that you already know how to solve, such as a linear or quadratic equation. In Chapter 2, you learned how to solve equations such as $\frac{3x}{4} - \frac{x}{2} = \frac{1}{2}$. Recall, one approach to solving an equation whose coefficients are fractions is to multiply each side of the equation by the least common denominator of the fractions, which will eliminate the fractions.

To solve an equation such as $\frac{8}{p} + \frac{1}{4p} = \frac{11}{8}$ we use the same process. Let's compare the approaches.

Classroom Example ➤
Solve:

(a) $\frac{x}{4} - \frac{2x}{3} = \frac{5}{6}$

(b) $\frac{2}{p} - \frac{4}{3p} = -\frac{1}{6}, \ p \neq 0$

Answer:
(a) $\{-2\}$ (b) $\{-4\}$

EXAMPLE 1 Solving a Rational Equation

Solve: **(a)** $\frac{3x}{4} - \frac{x}{2} = \frac{1}{2}$ **(b)** $\frac{8}{p} + \frac{1}{4p} = \frac{11}{8}, \quad p \neq 0$

Solution

Notice Example 1(a) is a linear equation with fractional coefficients. Example 1(b) is a rational equation because the equation contains a rational expression. As we solve these two equations, notice that the steps for solving the two equations are identical. They may also look familiar because they are the same steps that we used back in Chapter 2.

(a)
$$\frac{3x}{4} - \frac{x}{2} = \frac{1}{2}$$

Find the LCD: The LCD is 4.

Multiply each side of the equation by the LCD:
$$4\left(\frac{3x}{4} - \frac{x}{2}\right) = 4\left(\frac{1}{2}\right)$$

Distribute and divide out common factors:
$$4\left(\frac{3x}{4}\right) - 4\left(\frac{x}{2}\right) = 4\left(\frac{1}{2}\right)$$
$$3x - 2x = 2$$
$$x = 2$$

(b)
$$\frac{8}{p} + \frac{1}{4p} = \frac{11}{8}, \quad p \neq 0$$

Find the LCD: The LCD is $8p$.

Multiply each side of the equation by the LCD:
$$8p\left(\frac{8}{p} + \frac{1}{4p}\right) = 8p\left(\frac{11}{8}\right)$$

3.

Distribute and divide out common factors:
$$8p\left(\frac{8}{p}\right) + 8p\left(\frac{1}{4p}\right) = 8p\left(\frac{11}{8}\right)$$
$$64 + 2 = 11p$$

Solve the linear equation:
$$66 = 11p$$
$$6 = p$$

We leave the check up to you. The solution set of $\frac{3x}{4} - \frac{x}{2} = \frac{1}{2}$ is $\{2\}$. The solution set of $\frac{8}{p} + \frac{1}{4p} = \frac{11}{8}$ is $\{6\}$. ■

Did you notice the restriction that $p \neq 0$ for the equation $\frac{8}{p} + \frac{1}{4p} = \frac{11}{8}$? We include that condition because if p is replaced by the number 0, both $\frac{8}{p}$ and $\frac{1}{4p}$ become undefined. When solving a rational equation, always determine the value(s) of the variable that will cause the expression to be undefined.

You may recall from Chapter 2 that some equations in one variable have no solution. That's also true with rational equations.

Classroom Example
Solve:

$$\frac{-12}{x^2-9} = \frac{2}{x+3} - \frac{5}{x-3}$$

Answer: { } or ∅

EXAMPLE 6 Solving a Rational Equation with No Solution

Solve: $\frac{6}{z^2-1} = \frac{5}{z-1} - \frac{3}{z+1}$

Solution

$$\frac{6}{z^2-1} = \frac{5}{z-1} - \frac{3}{z+1}$$

Factor $z^2 - 1$: $$\frac{6}{(z-1)(z+1)} = \frac{5}{z-1} - \frac{3}{z+1}$$

Work Smart: Study Skills
Showing your work makes it easier for you to see how you get the extraneous solution. Don't skip steps to get the answer. Double check every step as you proceed through the problem to catch possible errors.

From the factored form of the denominator, we can see that $z \neq 1$ and $z \neq -1$. We also see that the LCD is $(z-1)(z+1)$. So we multiply both sides of the equation by the LCD, $(z-1)(z+1)$, and divide out common factors.

$$(z-1)(z+1) \cdot \frac{6}{(z-1)(z+1)} = (z-1)(z+1)\left(\frac{5}{z-1}\right) - (z-1)(z+1)\left(\frac{3}{(z+1)}\right)$$

$$\cancel{(z-1)(z+1)}\left(\frac{6}{\cancel{(z-1)(z+1)}}\right) = \cancel{(z-1)}(z+1)\left(\frac{5}{\cancel{z-1}}\right) - (z-1)\cancel{(z+1)}\left(\frac{3}{\cancel{(z+1)}}\right)$$

$$6 = 5(z+1) - 3(z-1)$$

Distribute: $6 = 5z + 5 - 3z + 3$

Combine like terms: $6 = 2z + 8$

Subtract 8 from both sides: $-2 = 2z$

Divide both sides by 2: $-1 = z$

Since z cannot equal -1 (because it results in division by 0), we reject $z = -1$ as a solution. Therefore, the equation $\frac{6}{z^2-1} = \frac{5}{z-1} - \frac{3}{z+1}$ has no solution. The solution set is { } or ∅.

Work Smart
The solution set { } is not the same as the solution set {0}.

QUICK ✓ *Solve the rational equation.*

8. $\frac{4}{y+1} = \frac{7}{y-1} - \frac{8}{y^2-1}$

Work Smart

When we solve a rational equation we multiply both sides of the equation by the LCD, but when we add or subtract rational expressions, we retain the LCD. Notice the difference in the directions: we *simplify* expressions and *solve* equations.

Simplify: $\frac{2}{x} - \frac{1}{6} + \frac{5}{2x} - \frac{1}{3}$

$$\frac{2}{x} - \frac{1}{6} + \frac{5}{2x} - \frac{1}{3} = \frac{2}{x}\cdot\frac{6}{6} - \frac{1}{6}\cdot\frac{x}{x} + \frac{5}{2x}\cdot\frac{3}{3} - \frac{1}{3}\cdot\frac{2x}{2x}$$
$$= \frac{12}{6x} - \frac{x}{6x} + \frac{15}{6x} - \frac{2x}{6x}$$
$$= \frac{12 - x + 15 - 2x}{6x}$$
$$= \frac{17 - 3x}{6x}$$

Solve: $\frac{2}{x} - \frac{1}{6} = \frac{5}{2x} - \frac{1}{3}$

$$6x\left(\frac{2}{x} - \frac{1}{6}\right) = 6x\left(\frac{5}{2x} - \frac{1}{3}\right)$$
$$6x\left(\frac{2}{x}\right) - 6x\left(\frac{1}{6}\right) = 6x\left(\frac{5}{2x}\right) - 6x\left(\frac{1}{3}\right)$$
$$6(2) - x = 3(5) - 2x$$
$$12 - x = 15 - 2x$$
$$12 - x + 2x = 15 - 2x + 2x$$
$$12 + x = 15$$
$$x = 3$$

Classroom Example
Use the equation from Example 7 with $\overline{C} = 240$.

Answer: The level of production is 40 bicycles or 125 bicycles.

EXAMPLE 7 An Application of Rational Equations: Average Daily Cost

Suppose that the average daily cost $\overline{C}$ of manufacturing x bicycles is given by the equation

$$\overline{C} = \frac{x^2 + 75x + 5000}{x}$$

Determine the number of bicycles for which the average daily cost will be \$225.

Solution

First, notice that x must be greater than 0. Do you know why? Since we want to know the level of production to obtain an average daily cost of \$225, we wish to solve the equation $\overline{C} = 225$.

$$\overline{C} = \frac{x^2 + 75x + 5000}{x}$$

$$225 = \frac{x^2 + 75x + 5000}{x}$$

Multiply both sides by x: $$x \cdot 225 = \frac{x^2 + 75x + 5000}{x} \cdot x$$

$$225x = x^2 + 75x + 5000$$

Subtract $225x$ from both sides: $$0 = x^2 - 150x + 5000$$

Factor: $$0 = (x - 50)(x - 100)$$

Zero-Product Property: $$x - 50 = 0 \quad \text{or} \quad x - 100 = 0$$

$$x = 50 \quad \text{or} \quad x = 100$$

The level of production for which the average daily cost will be \$225 is at 50 bicycles or 100 bicycles.

QUICK ✓

9. The concentration C of a certain drug in a patient's bloodstream t hours after injection is given by $C = \frac{50t}{t^2 + 25}$. When will the concentration of the drug be 4 milligrams per liter?

2 Solve for a Variable in a Rational Equation

Work Smart
The steps that we follow when solving formulas for a certain variable are identical to those that we followed when solving rational equations.

Recall from Section 2.4 that the expression "solve for the variable" means to get the variable by itself on one side of the equation with all other variables and constants, if any, on the other side. The steps that we follow when solving formulas for a certain variable are identical to those that we followed when solving rational equations. We compare solving the rational equation $5 = \frac{3}{1 + x}$ for x with solving the rational equation $P = \frac{A}{1 + r}$ for r.

	$5 = \frac{3}{1+x}$	$P = \frac{A}{1+r}$
Multiply by the LCD:	$(1+x)5 = (1+x)\frac{3}{1+x}$	$(1+r)P = (1+r)\frac{A}{1+r}$
Distribute:	$5 + 5x = 3$	$P + Pr = A$
Isolate the unknown:	$5 - 5 + 5x = 3 - 5$	$P - P + Pr = A - P$
	$5x = -2$	$Pr = A - P$
	$\frac{5x}{5} = \frac{-2}{5}$	$\frac{Pr}{P} = \frac{A - P}{P}$
	$x = -\frac{2}{5}$	$r = \frac{A - P}{P}$

You may recall from Chapter 2 that some equations in one variable have no solution. That's also true with rational equations.

Classroom Example Solve:

$$\frac{-12}{x^2-9}=\frac{2}{x+3}-\frac{5}{x-3}$$

Answer: { } or $\varnothing$

EXAMPLE 6 Solving a Rational Equation with No Solution

Solve: $\dfrac{6}{z^2-1}=\dfrac{5}{z-1}-\dfrac{3}{z+1}$

Solution

$$\frac{6}{z^2-1}=\frac{5}{z-1}-\frac{3}{z+1}$$

Factor z^2-1: $\dfrac{6}{(z-1)(z+1)}=\dfrac{5}{z-1}-\dfrac{3}{z+1}$

Work Smart: Study Skills
Showing your work makes it easier for you to see how you get the extraneous solution. Don't skip steps to get the answer. Double check every step as you proceed through the problem to catch possible errors.

From the factored form of the denominator, we can see that $z \neq 1$ and $z \neq -1$. We also see that the LCD is $(z-1)(z+1)$. So we multiply both sides of the equation by the LCD, $(z-1)(z+1)$, and divide out common factors.

$$(z-1)(z+1)\cdot\frac{6}{(z-1)(z+1)}=(z-1)(z+1)\left(\frac{5}{z-1}\right)-(z-1)(z+1)\left(\frac{3}{(z+1)}\right)$$

$$\cancel{(z-1)(z+1)}\left(\frac{6}{\cancel{(z-1)(z+1)}}\right)=\cancel{(z-1)}(z+1)\left(\frac{5}{\cancel{z-1}}\right)-(z-1)\cancel{(z+1)}\left(\frac{3}{\cancel{(z+1)}}\right)$$

$$6=5(z+1)-3(z-1)$$

Distribute: $6=5z+5-3z+3$

Combine like terms: $6=2z+8$

Subtract 8 from both sides: $-2=2z$

Divide both sides by 2: $-1=z$

Since z cannot equal -1 (because it results in division by 0), we reject $z=-1$ as a solution. Therefore, the equation $\dfrac{6}{z^2-1}=\dfrac{5}{z-1}-\dfrac{3}{z+1}$ has no solution. The solution set is { } or $\varnothing$.

Work Smart
The solution set { } is not the same as the solution set {0}.

QUICK ✓ *Solve the rational equation.*

8. $\dfrac{4}{y+1}=\dfrac{7}{y-1}-\dfrac{8}{y^2-1}$

Work Smart
When we solve a rational equation we multiply both sides of the equation by the LCD, but when we add or subtract rational expressions, we retain the LCD. Notice the difference in the directions: we *simplify* expressions and *solve* equations.

Simplify: $\dfrac{2}{x}-\dfrac{1}{6}+\dfrac{5}{2x}-\dfrac{1}{3}$

$$\frac{2}{x}-\frac{1}{6}+\frac{5}{2x}-\frac{1}{3}=\frac{2}{x}\cdot\frac{6}{6}-\frac{1}{6}\cdot\frac{x}{x}+\frac{5}{2x}\cdot\frac{3}{3}-\frac{1}{3}\cdot\frac{2x}{2x}$$

$$=\frac{12}{6x}-\frac{x}{6x}+\frac{15}{6x}-\frac{2x}{6x}$$

$$=\frac{12-x+15-2x}{6x}$$

$$=\frac{17-3x}{6x}$$

Solve: $\dfrac{2}{x}-\dfrac{1}{6}=\dfrac{5}{2x}-\dfrac{1}{3}$

$$6x\left(\frac{2}{x}-\frac{1}{6}\right)=6x\left(\frac{5}{2x}-\frac{1}{3}\right)$$

$$6x\left(\frac{2}{x}\right)-6x\left(\frac{1}{6}\right)=6x\left(\frac{5}{2x}\right)-6x\left(\frac{1}{3}\right)$$

$$6(2)-x=3(5)-2x$$

$$12-x=15-2x$$

$$12-x+2x=15-2x+2x$$

$$12+x=15$$

$$x=3$$

Classroom Example Use the equation from Example 7 with $\overline{C} = 240$.

Answer: The level of production is 40 bicycles or 125 bicycles.

EXAMPLE 7 An Application of Rational Equations: Average Daily Cost

Suppose that the average daily cost $\overline{C}$ of manufacturing x bicycles is given by the equation

$$\overline{C} = \frac{x^2 + 75x + 5000}{x}$$

Determine the number of bicycles for which the average daily cost will be \$225.

Solution

First, notice that x must be greater than 0. Do you know why? Since we want to know the level of production to obtain an average daily cost of \$225, we wish to solve the equation $\overline{C} = 225$.

$$\overline{C} = \frac{x^2 + 75x + 5000}{x}$$

$$225 = \frac{x^2 + 75x + 5000}{x}$$

Multiply both sides by x: $$x \cdot 225 = \frac{x^2 + 75x + 5000}{x} \cdot x$$

$$225x = x^2 + 75x + 5000$$

Subtract $225x$ from both sides: $$0 = x^2 - 150x + 5000$$

Factor: $$0 = (x - 50)(x - 100)$$

Zero-Product Property: $$x - 50 = 0 \quad \text{or} \quad x - 100 = 0$$

$$x = 50 \quad \text{or} \quad x = 100$$

The level of production for which the average daily cost will be \$225 is at 50 bicycles or 100 bicycles.

QUICK ✓

9. The concentration C of a certain drug in a patient's bloodstream t hours after injection is given by $C = \frac{50t}{t^2 + 25}$. When will the concentration of the drug be 4 milligrams per liter?

---

2 Solve for a Variable in a Rational Equation

Work Smart

The steps that we follow when solving formulas for a certain variable are identical to those that we followed when solving rational equations.

Recall from Section 2.4 that the expression "solve for the variable" means to get the variable by itself on one side of the equation with all other variables and constants, if any, on the other side. The steps that we follow when solving formulas for a certain variable are identical to those that we followed when solving rational equations. We compare solving the rational equation $5 = \frac{3}{1 + x}$ for x with solving the rational equation $P = \frac{A}{1 + r}$ for r.

	$5 = \frac{3}{1 + x}$	$P = \frac{A}{1 + r}$
Multiply by the LCD:	$(1 + x)5 = (1 + x)\frac{3}{1 + x}$	$(1 + r)P = (1 + r)\frac{A}{1 + r}$
Distribute:	$5 + 5x = 3$	$P + Pr = A$
Isolate the unknown:	$5 - 5 + 5x = 3 - 5$	$P - P + Pr = A - P$
	$5x = -2$	$Pr = A - P$
	$\frac{5x}{5} = \frac{-2}{5}$	$\frac{Pr}{P} = \frac{A - P}{P}$
	$x = -\frac{2}{5}$	$r = \frac{A - P}{P}$

Do you see that the steps for solving for the variable r in $P = \frac{A}{1 + r}$ are identical to the steps in solving for x in $5 = \frac{3}{1 + x}$?

EXAMPLE 8 How to Solve for a Variable in a Rational Equation

Solve $\frac{x}{1 + y} = z$ for y.

Step-by-Step Solution

We need to isolate the variable y. That is, we need to get y by itself on one side of the equation and the other variables on the other side of the equation.

Step 1: Determine the value(s) of the variable that results in any undefined rational expression in the rational equation.

$$y \neq -1$$

Step 2: Determine the least common denominator (LCD) of all the denominators.

The LCD is $1 + y$.

Step 3: Multiply both sides of the equation by the LCD and simplify the expression on each side of the equation.

$$\frac{x}{1 + y} = z$$

$$(1 + y)\left(\frac{x}{1 + y}\right) = z(1 + y)$$

$$x = z(1 + y)$$

Step 4: Solve the resulting equation for y.

Divide both sides by z: $\frac{x}{z} = \frac{z(1 + y)}{z}$

$$\frac{x}{z} = 1 + y$$

Subtract 1 from both sides: $\frac{x}{z} - 1 = 1 - 1 + y$

$$\frac{x}{z} - 1 = y$$

Classroom Example

Solve $\frac{a}{1 + b} = c$ for b.

Answer: $b = \frac{a - c}{c}$ or $b = \frac{a}{c} - 1$

Look back at Step 4, in which we divided both sides of the equation by z. Did you think about using the Distributive Property to simplify $x = z(1 + y)$? If you did, you are correct! Another way to complete solving $\frac{x}{1 + y} = z$ for y is as follows:

$$x = z(1 + y)$$

Use the Distributive Property: $x = z + zy$

Isolate the expression containing the variable y: $x - z = z - z + zy$

$$x - z = zy$$

Divide both sides by z: $\frac{x - z}{z} = \frac{zy}{z}$

$$\frac{x - z}{z} = y$$

The result that we get using this method produces the equation $y = \frac{x - z}{z}$, which is equivalent to the equation $y = \frac{x}{z} - 1$. Do you see why?

QUICK ✓ *Solve the equation for the indicated variable.*

10. Solve $R = \frac{4g}{x}$ for x

11. $S = \frac{a}{1 - r}$ for r

Classroom Example ➤ Solve $\frac{1}{x} = \frac{1}{y} + \frac{1}{z}$ for z.

Answer: $z = \frac{xy}{y - x}$

EXAMPLE 9 Solving for a Variable in a Rational Equation

Solve $\frac{1}{a} + \frac{1}{b} = \frac{1}{c}$ for b.

We can see that neither a, b, nor c can be equal to zero. Further, the LCD is abc.

$$\frac{1}{a} + \frac{1}{b} = \frac{1}{c}$$

Multiply both sides by the LCD: $$(abc)\left(\frac{1}{a} + \frac{1}{b}\right) = (abc)\left(\frac{1}{c}\right)$$

Distribute: $$(abc)\left(\frac{1}{a}\right) + (abc)\left(\frac{1}{b}\right) = (abc)\left(\frac{1}{c}\right)$$

Divide out common factors: $$(\not{a}bc)\left(\frac{1}{\not{a}}\right) + (a\not{b}c)\left(\frac{1}{\not{b}}\right) = (ab\not{c})\left(\frac{1}{\not{c}}\right)$$

$$bc + ac = ab$$

Work Smart

To solve for a variable that occurs on both sides of the equation, get all terms containing that variable on the same side of the equation. Then factor out the variable you wish to solve for.

To solve for b, we need to get both terms that contain the variable b on the same side of the equation. We'll subtract bc from both sides of the equation and continue.

Subtract bc from both sides: $$ac = ab - bc$$

Factor out b as a greatest common factor: $$ac = b(a - c)$$

Divide each side by $a - c$: $$\frac{ac}{a - c} = \frac{b(a - c)}{a - c}$$

$$\frac{ac}{a - c} = b$$

When we solve $\frac{1}{a} + \frac{1}{b} = \frac{1}{c}$ for b, we obtain the equation $b = \frac{ac}{a - c}$.

QUICK ✓ *Solve the equation for the indicated variable.*

12. $\frac{1}{f} = \frac{1}{p} + \frac{1}{q}$ for p

5.7 Exercises

For Extra Help:

Student Solutions Manual

CD Video

PH Math/Tutor Center

MathXL Tutorials on CD

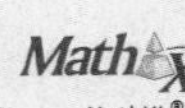

MathXL®

MyMathLab

Concepts and Vocabulary

In Problems 1–3, fill in the blanks.

1. least common denominator
2. extraneous solutions

1. To solve an equation that contains one or more rational expressions, multiply both sides of the equation by the __________ __________ __________ to form an equivalent equation without any rational expressions.

2. Apparent solutions that do not satisfy the original equation are called __________ __________.

3. undefined
4. False
5. True
6. True
7. Answers may vary.
8. Answers may vary.
9. {18}
10. {27}
11. {−3}
12. {−4}
13. {2}
14. {−2}
15. {6}
16. {−36}
17. $\left\{\frac{5}{2}\right\}$
18. $\left\{-\frac{3}{4}\right\}$
19. $\left\{\frac{31}{2}\right\}$
20. $\left\{\frac{12}{13}\right\}$
21. {−7}
22. {36}
23. $\left\{\frac{1}{2}\right\}$
24. $\left\{-\frac{3}{2}\right\}$
25. {7}
26. $\left\{-\frac{16}{9}\right\}$
27. { } or ∅
28. { } or ∅
29. {−3}
30. $\left\{-\frac{8}{3}\right\}$
31. $\left\{\frac{3}{4}\right\}$
32. {7}
33. $\left\{-\frac{5}{6}\right\}$
34. $\left\{\frac{3}{2}\right\}$
35. {−4}
36. $\left\{-\frac{11}{7}\right\}$

3. When solving an equation that contains rational expressions, extraneous solutions are values for which a rational expression in the equation is ________.

In Problems 4–6, answer True or False to each statement.

4. There are no values of x for which the rational expressions in the equation $\frac{3x+1}{5} = \frac{10}{x+2}$ are undefined.

5. Equivalent equations are different equations which have the same solution set.

6. A rational equation can have no solution.

7. Explain the difference between the direction *simplify* and *solve.* Make up a problem using each of these directions with the rational expressions $\frac{5}{x+1}$ and $\frac{6}{x-1}$.

8. Explain how using the LCD to solve an equation that contains rational expressions is different from using the LCD in a problem requiring adding and subtracting rational expressions with unlike denominators.

Building Skills

In Problems 9–36, solve each equation and state the solution set. Remember to state the values of the variable for which the expressions in each rational equation are undefined, if necessary.

9. $\frac{4}{x} = \frac{2}{9}$

10. $\frac{6}{x} = \frac{2}{9}$

11. $\frac{6}{x} + 5 = 3$

12. $\frac{12}{x} - 5 = -8$

13. $\frac{5}{3y} - \frac{1}{2} = \frac{5}{6y} - \frac{1}{12}$

14. $\frac{3}{z} - \frac{3}{2} = \frac{6}{z}$

15. $\frac{1}{3} + \frac{1}{x} = \frac{3}{x}$

16. $\frac{5}{6} + \frac{3}{x} = \frac{3}{4}$

17. $\frac{4}{x} - \frac{11}{5} = \frac{3}{2x} - \frac{6}{5}$

18. $\frac{6}{x} + \frac{2}{3} = \frac{4}{2x} - \frac{14}{3}$

19. $\frac{3}{2}m + 6 = 2m - \frac{7}{4}$

20. $\frac{4}{p} - \frac{5}{4} = \frac{5}{2p} + \frac{3}{8}$

21. $\frac{4}{x-1} = \frac{3}{x+1}$

22. $\frac{4}{x-4} = \frac{5}{x+4}$

23. $\frac{2}{x+2} + 2 = \frac{7}{x+2}$

24. $\frac{4}{x+1} + 2 = \frac{3}{x+1}$

25. $\frac{r-4}{3r} + \frac{2}{5r} = \frac{1}{5}$

26. $\frac{x-2}{4x} - \frac{x+2}{3x} = \frac{1}{6x} + \frac{2}{3}$

27. $\frac{2}{a-1} + \frac{3}{a+1} = \frac{-6}{a^2-1}$

28. $\frac{6}{t} - \frac{2}{t-1} = \frac{2-4t}{t^2-t}$

29. $\frac{1}{4-x} + \frac{2}{x^2-16} = \frac{1}{x-4}$

30. $\frac{4}{x-3} - \frac{3}{x-2} = \frac{4x+9}{x^2-5x+6}$

31. $\frac{3}{2t-2} - \frac{2t}{3t-3} = -4$

32. $\frac{2x+3}{x-1} - 2 = \frac{3x-1}{4x-4}$

33. $\frac{2}{5} + \frac{3-2a}{10a-20} = \frac{2a+1}{a-2}$

34. $\frac{1}{x-2} + \frac{2}{2-x} = \frac{5}{x+1}$

35. $\frac{6}{j^2-1} - \frac{4j}{j^2-5j+4} = -\frac{4}{j-1}$

36. $\frac{3}{2x+2} - \frac{5}{4x-4} = \frac{2x}{x^2-1}$

37. $\{-3, -2\}$ 38. $\left\{-\frac{7}{11}\right\}$
39. $\{-1\}$ 40. $\left\{-\frac{1}{3}, 2\right\}$
41. $\{-9, 4\}$ 42. $\left\{-\frac{1}{2}\right\}$
43. $\left\{-\frac{2}{3}, \frac{1}{2}\right\}$ 44. $\left\{-\frac{3}{2}, \frac{2}{3}\right\}$
45. { } or $\varnothing$ 46. { } or $\varnothing$
47. $\{-5, 5\}$ 48. $\{-3, 4\}$
49. $\{-1, 5\}$ 50. $\left\{\frac{2}{3}\right\}$
51. $\{-6\}$ 52. $\{3\}$
53. $\{-5\}$ 54. $\{2, -2, -6\}$
55. { } or $\varnothing$ 56. $\left\{\frac{2}{3}\right\}$
57. $y = \frac{2}{x}$ 58. $m = \frac{k}{4}$
59. $R = \frac{E}{I}$ 60. $R = \frac{D}{T}$
61. $b = \frac{2A}{h} - B$
62. $y = m(x - x_1) + y_1$
63. $y = \frac{x}{z} - 3$ 64. $b = \frac{a - 2c}{c}$
65. $S = \frac{RT}{T - R}$ 66. $j = \frac{4ik}{3k - 8i}$
67. $y = \frac{an - p}{am}$ 68. $c = \frac{ab}{b - a}$
69. $x = \frac{Ay}{y - A}$ 70. $x = \frac{ck - m}{cB}$
71. $y = \frac{xz}{2z - 6x}$ 72. $b = \frac{a - ay}{y}$
73. $x = \frac{cy}{y - 1}$ 74. $a = \frac{bX}{X - b}$
75. $\frac{4x + 5}{x(x + 5)}$ 76. $\frac{7}{x - 5}$
77. $\{-2, 3\}$ 78. $\{1, 3\}$
79. $\frac{x + 7}{2}$ 80. $\frac{4n^2 + 3n - 16}{n^2 - 4}$
81. $\left\{-\frac{1}{6}, 1\right\}$ 82. $\left\{-1, \frac{2}{9}\right\}$
83. $\frac{x^3 - x^2 + 4}{x(x + 2)(x - 2)}$
84. $\frac{3x^2 - 2x - 3}{(x - 1)(x + 1)}$

Mixed Practice

In Problems 37–56, solve each equation and state the solution set. Be sure to check for extraneous solutions.

37. $\frac{x}{x - 2} = \frac{3}{x + 8}$ **38.** $\frac{x + 5}{x - 7} = \frac{x - 3}{x + 7}$ **39.** $\frac{x}{x + 3} = \frac{6}{x - 3} + 1$

40. $\frac{3x^2}{x + 1} = 2 + \frac{3x}{x + 1}$ **41.** $\frac{-p}{p - 6} = \frac{p + 6}{5}$ **42.** $\frac{1}{n - 3} = \frac{3n - 1}{9 - n^2}$

43. $x = \frac{2 - x}{6x}$ **44.** $x = \frac{6 - 5x}{6x}$ **45.** $\frac{2x + 3}{x - 1} = \frac{x - 2}{x + 1} + \frac{6x}{x^2 - 1}$

46. $\frac{2x}{x + 4} = \frac{x + 1}{x + 2} - \frac{7x + 12}{x^2 + 6x + 8}$ **47.** $\frac{2x}{x + 3} - \frac{2x^2 + 2}{x^2 - 9} = \frac{-6}{x - 3} + 1$

48. $\frac{2}{b + 2} - \frac{5b + 6}{b^2 - b - 6} = \frac{-b}{b - 3}$ **49.** $\frac{5x}{2x - 3} = \frac{3x}{x - 1} - \frac{5}{2x^2 - 5x + 3}$

50. $\frac{2x + 1}{x^2 + 2x - 3} = \frac{x - 1}{x^2 + 5x + 6} + \frac{x + 1}{x^2 + x - 2}$

51. $\frac{x}{x^2 - 1} - \frac{x + 3}{x^2 - x} = \frac{-3}{x^2 + x}$ **52.** $\frac{2t}{t^2 + 2t + 1} + \frac{t - 1}{t^2 + t} = \frac{6t + 8}{t^3 + 2t^2 + t}$

53. $\frac{5}{x - 2} - \frac{2}{2 - x} = \frac{4}{x + 1}$ **54.** $\frac{x}{12} + \frac{1}{2} = \frac{1}{3x} + \frac{2}{x^2}$

55. $\frac{a}{2a - 2} - \frac{2}{3a + 3} = \frac{5a^2 - 2a + 9}{12a^2 - 12}$

56. $\frac{2x + 1}{x^2 + 2x - 3} = \frac{x - 1}{x^2 + 5x + 6} + \frac{x + 1}{x^2 + x - 2}$

In Problems 57–74, solve the equation for the indicated variable.

57. $x = \frac{2}{y}$ for y **58.** $4 = \frac{k}{m}$ for m **59.** $I = \frac{E}{R}$ for R

60. $T = \frac{D}{R}$ for R **61.** $h = \frac{2A}{B + b}$ for b **62.** $m = \frac{y - y_1}{x - x_1}$ for y

63. $\frac{x}{3 + y} = z$ for y **64.** $\frac{a}{b + 2} = c$ for b **65.** $\frac{1}{R} = \frac{1}{S} + \frac{1}{T}$ for S

66. $\frac{3}{i} - \frac{4}{j} = \frac{8}{k}$ for j **67.** $m = \frac{n}{y} - \frac{p}{ay}$ for y **68.** $\frac{1}{a} = \frac{1}{b} + \frac{1}{c}$ for c

69. $A = \frac{xy}{x + y}$ for x **70.** $B = \frac{k}{x} - \frac{m}{cx}$ for x **71.** $\frac{2}{x} - \frac{1}{y} = \frac{6}{z}$ for y

72. $y = \frac{a}{a + b}$ for b **73.** $y = \frac{x}{x - c}$ for x **74.** $X = \frac{ab}{a - b}$ for a

In Problems 75–84, simplify the expression or solve the equation.

75. $\frac{1}{x} + \frac{3}{x + 5}$ **76.** $\frac{2}{x - 5} + \frac{5}{x - 5}$ **77.** $x - \frac{6}{x} = 1$ **78.** $z + \frac{3}{z} = 4$

79. $\frac{3}{x - 1} \cdot \frac{x^2 - 1}{6} + 3$ **80.** $5 - \frac{n^2 - 3n - 4}{n^2 - 4}$ **81.** $2b - \frac{5}{3} = \frac{1}{3b}$

82. $3a + \frac{7}{3} = \frac{2}{3a}$ **83.** $\frac{x^2}{x^2 - 4} - \frac{1}{x}$ **84.** $\frac{x}{x + 1} + \frac{2x - 3}{x - 1}$

85. 1 hour and 9 hours
86. 1 hour and 3 hours
87. 40 or 125 bicycles
88. 80%
89. $k = 33$
90. $k = 6$

Applying the Concepts

85. Drug Concentration The concentration C of a drug in a patient's bloodstream in milligrams per liter t hours after ingestion is modeled by $C = \frac{40t}{t^2 + 9}$. When will the concentration of the drug be 4 milligrams per liter?

86. Drug Concentration The concentration C of a drug in a patient's bloodstream in milligrams per liter t hours after ingestion is modeled by $C = \frac{40t}{t^2 + 3}$. When will the concentration of the drug be 10 milligrams per liter?

87. Average Cost Suppose that the average daily cost $\overline{C}$ of manufacturing x bicycles is given by the equation $\overline{C} = \frac{x^2 + 75x + 5000}{x}$. Determine the level of production for which the average daily cost will be \$240.

88. Cost-Benefit Model Environmental scientists often use cost-benefit models to estimate the cost of removing a pollutant from the environment related to the percentage of pollutant removed. Suppose a cost-benefit model for the cost C (in millions of dollars) of removing x percent of the pollutants from Maple Lake is given by $C = \frac{25x}{100 - x}$. If the federal government budgets \$100 million to clean up the lake, what percent of the pollutants can be removed?

Extending the Concepts

89. For what value of k will the solution set of $\frac{4x + 3}{k} = \frac{x - 1}{3}$ be $\{2\}$?

90. For what value of k will the solution set of $\frac{1}{2} + \frac{3x}{k} = 1 + \frac{x}{3}$ be $\{1\}$?

5.8 Models Involving Rational Equations

OBJECTIVES

1. Model and Solve Ratio and Proportion Problems
2. Model and Solve Problems with Similar Figures
3. Model and Solve Work Problems
4. Model and Solve Uniform Motion Problems

Preparing for Models Involving Rational Equations

Before getting started, take the following readiness quiz. If you get a problem wrong, go back to the section cited and review the material.

1. Solve: $\frac{150}{r} = \frac{250}{r + 20}$ [Section 5.7, pp. 357–363]

1 Model and Solve Ratio and Proportion Problems

We begin with a definition.

DEFINITION

A **ratio** is the quotient of two numbers or two quantities. The ratio of two numbers a and b can be written as

$$a \text{ to } b \qquad \text{or} \qquad a{:}b \qquad \text{or} \qquad \frac{a}{b}$$

When solving algebraic problems, we write ratios as $\frac{a}{b}$. For example, the odds of winning the "Pick Three" Instant Ohio Lottery are 1 in 1000, so we write that ratio as $\frac{1}{1000}$. Generally, a ratio is written as a fraction reduced to lowest terms. If 8 oz of a cleaner are required for every 4 gallons of water, the ratio is $\frac{8 \text{ oz}}{4 \text{ gal}} = \frac{2 \text{ oz}}{1 \text{ gal}}$.

Preparing for...Answer **1.** $\{30\}$

A rational equation which involves two ratios is called a *proportion*.

DEFINITION

A **proportion** is an equation of the form $\frac{a}{b} = \frac{c}{d}$, where $b \neq 0$ and $d \neq 0$. We call a, b, c, and d the **terms** of the proportion. The terms b and c are called the **means** of the proportion and the terms a and d are called the **extremes** of the proportion.

We may solve a proportion using the same method that we used to solve a rational equation.

Classroom Example ➤

Solve the proportion $\frac{n}{8} = \frac{56}{64}$.

Answer: $\{7\}$

EXAMPLE 1 Solving a Proportion

Solve the proportion: $\frac{x}{5} = \frac{63}{105}$

Solution

Because $105 = 21 \cdot 5$, the LCD is 105. So we multiply both sides of the equation by 105.

$$\frac{x}{5} = \frac{63}{105}$$

$$105\left(\frac{x}{5}\right) = 105\left(\frac{63}{105}\right)$$

Divide out common factors: $21x = 63$

Divide both sides by 21: $x = 3$

The solution set is $\{3\}$.

Work Smart

The equation $\frac{3}{8} = \frac{7}{x} + 1$ is NOT a proportion. Do you see why?

Work Smart

Another method frequently used to solve a proportion is the method of cross multiplication. When this process is used, the same result is obtained as multiplying both sides of the equation by the LCD.

CROSS-MULTIPLICATION PROPERTY

If $\frac{a}{b} = \frac{c}{d}$, where $b \neq 0$ and $d \neq 0$, then $a \cdot d = b \cdot c$.

We can use this method to solve a proportion because for the proportion $\frac{a}{b} = \frac{c}{d}$, the LCD is bd. If we multiply both sides of the equation $\frac{a}{b} = \frac{c}{d}$ by the LCD, we obtain the result given by the Cross-Multiplication Property:

$$bd\left(\frac{a}{b}\right) = \left(\frac{c}{d}\right)bd$$

$$ad = bc$$

If we solve the proportion in Example 1, $\frac{x}{5} = \frac{63}{105}$, using cross multiplication, we obtain

$$\frac{x}{5} = \frac{63}{105}$$

Cross multiply: $105x = 5 \cdot 63$

$105x = 315$

Divide both sides by 105: $x = 3$

QUICK ✓ *Solve the proportion for the indicated variable.*

1. $\frac{x}{9} = \frac{-4}{3}$ **2.** $\frac{5}{y+4} = \frac{2}{y-1}$ **3.** $\frac{2p+1}{4} = \frac{p}{8}$ **4.** $\frac{6}{x^2} = \frac{2}{x}$

Classroom Example ➤
Use Example 2, but Laura spent 86 euros on a Venetian vase.

Answer: She spent \$107.97 on the vase.

EXAMPLE 2 Exchange Rates

Last summer Laura took a Study Abroad trip to Italy. When she returned, she examined her credit card bill and found that she spent 199 euros (€) on the digital camera she bought in Venice. The bill stated that \$1 US was equal to 0.796509 euros (€). How much did the camera cost in U.S. dollars?

Solution

Step 1: Identify We want to know the cost of the camera in U.S. dollars.

Step 2: Name We let c represent the cost of the camera in U.S. dollars.

Step 3: Translate We want to know how many dollars are equivalent to 199 euros (€). Since c represents the cost of the camera (in dollars) and we know 1 dollar equals 0.796509 euros, we set up the proportion using $\frac{\text{dollars}}{\text{euros}}$.

$$\frac{\$1}{0.796509\,€} = \frac{c}{199\,€} \quad \text{The Model}$$

Notice we include the units of measure in the model. This is done to keep track of the units. Our answer should be in dollars.

Work Smart

You could also set up the model using $\frac{\text{euros}}{\text{dollars}}$.

$$\frac{0.796509\,€}{\$1} = \frac{199\,€}{c}$$

Step 4: Solve We'll now solve the equation.

Cross multiply: $\$1 \cdot 199\,€ = [0.796509\,€]c$

Divide both sides by 0.796509: $\frac{\$1 \cdot 199\,€}{0.796509\,€} = \frac{[0.796509\,€]c}{0.796509\,€}$

Simplify using a calculator: $\$249.84 \approx c$

Step 5: Check Is this answer reasonable? First, we notice our answer is in dollar. This is encouraging. Second, the ratio 1 dollar to 0.796509 euros should be approximately equal to the ratio of \$249.84 to 199 euros. That is, $\frac{1}{0.796509}$ should equal $\frac{249.84}{199}$. Because $\frac{1}{0.796509} \approx 1.2555$ and $\frac{249.84}{199} \approx 1.2555$, our answer checks.

Step 6: Answer The camera that Laura purchased in Italy cost about \$249.84 U.S.

QUICK ✓

5. Taryn is reading the newest *Harry Potter* book. If she can read 54 pages in 60 minutes, how long will it take her to read the last 100 pages of the book?
6. An automobile manufacturer is advertising special financing on its autos. A buyer will have a monthly payment of \$16.67 for every \$1000 borrowed. If Clem borrows \$14,000, find his monthly payment.

Sometimes proportions can be used as a general model for economic behavior.

Classroom Example ➤
Use Example 3 with a 1300-square-foot house that sells for about \$97,500. How much should an agent price a 1700 square-foot house?

Answer: The house should be priced at \$127,500.

EXAMPLE 3 Model and Solve a Problem from Business

A real estate agent knows that a 1800-square-foot house in a particular neighborhood sold for \$150,000. How much should the real estate agent appraise a 2100-square-foot house in the same neighborhood?

Solution

Step 1: Identify We want to know the price of a 2100-square-foot home in the same neighborhood as an 1800-square-foot house that sold for \$150,000.

Step 2: Name Let p represent the price of the 2100-square-foot house.

Step 3: Translate The selling price of an 1800-square-foot house is \$150,000. We wish to know the selling price p of a 2100-square foot house. We use the ratio $\frac{\text{square feet}}{\text{selling price}}$ to set up the proportion

$$\frac{1800 \text{ ft}^2}{\$150{,}000} = \frac{2100 \text{ ft}^2}{p}$$

Step 4: Solve

Multiply by the LCD 150,000p: $\$150{,}000p\left(\frac{1800 \text{ ft}^2}{\$150{,}000}\right) = \$150{,}000p\left(\frac{2100 \text{ ft}^2}{p}\right)$

Divide like factors: $(1800 \text{ ft}^2)p = (\$150{,}000)(2100 \text{ ft}^2)$

Divide both sides by 1800: $p = \frac{(\$150{,}000)(2100 \text{ ft}^2)}{1800 \text{ ft}^2}$

$$p = \$175{,}000$$

Step 5: Check The answer is in dollars. Plus, the ratio $\frac{1800}{150{,}000} = 0.012$ equals the ratio $\frac{2100}{175{,}000} = 0.012$, so our answer checks.

Step 6: Answer We expect that the selling price of a 2100-square-foot house in this neighborhood will be about \$175,000.

QUICK ✓

7. On a map, $\frac{1}{4}$ inch represents a distance of 15 miles. According to the map, Springfield and Brookhaven are $3\frac{1}{2} = \frac{7}{2}$ inches apart. Find the number of miles between the two cities.

2 Model and Solve Problems with Similar Figures

You may be familiar with using proportions to solve problems involving *similar figures* from geometry.

DEFINITION

Two figures are **similar** if their corresponding angle measures are equal and their corresponding sides are proportional.

Figure 1

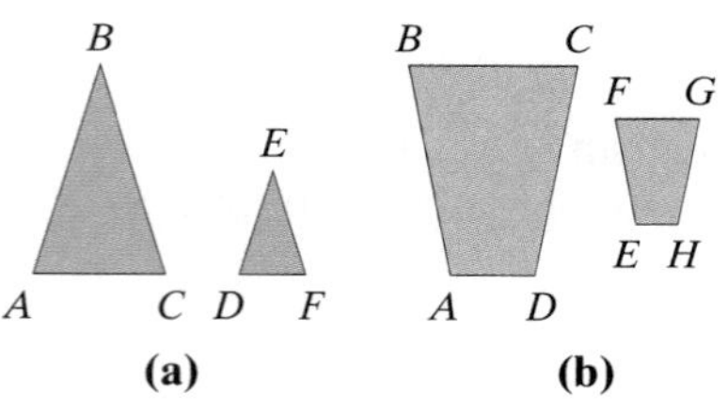

Figure 1 shows examples of similar figures.

In Figure 1(a), $\triangle ABC$ is similar to $\triangle DEF$ and in Figure 1(b), quadrilateral $ABCD$ is similar to quadrilateral $EFGH$. Because $\triangle ABC$ is similar to $\triangle DEF$, we know that the ratio of AB to AC equals the ratio of DE to DF. That is,

$$\frac{AB}{AC} = \frac{DE}{DF}$$

The principle that the ratios of corresponding sides are equal can be used to find unknown lengths in similar figures.

Teaching Tip
Note to students that the ratio $\frac{AB}{AC} = \frac{DE}{DF}$ is equivalent to $\frac{AB}{DE} = \frac{AC}{DF}$.

Classroom Example
Use Figure 2 in Example 4, with $AB = 35$, $AC = 14$, $DF = 10$ to find DE.

Answer: $DE = 25$

EXAMPLE 4 Solving a Problem with Similar Triangles

Find the length of side DE in triangle $\triangle DEF$, given that $\triangle ABC$ is similar to $\triangle DEF$ based on the triangles shown in Figure 2. All measurements are in inches.

Solution

We know that $\triangle ABC$ is similar to $\triangle DEF$. We also know that the corresponding sides are proportional, so $\frac{AB}{AC} = \frac{DE}{DF}$. We substitute the known values and solve for the unknown value, letting $x = DE$.

Figure 2

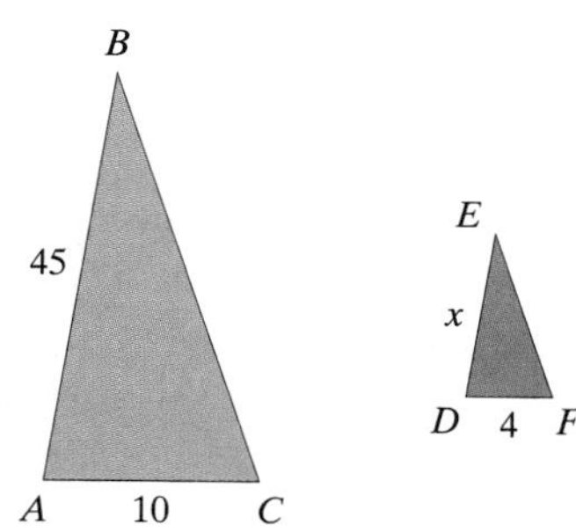

$$\frac{AB}{AC} = \frac{DE}{DF}$$

$AB = 45,\ AC = 10,\ DF = 4$: $$\frac{45}{10} = \frac{x}{4}$$

Cross multiply: $$10x = 180$$

Divide both sides by 10: $$x = 18$$

The length of side DE is 18 inches.

In Example 4, we could also have used the proportion $\frac{AB}{DE} = \frac{AC}{DF}$. Then we would have solved the proportion $\frac{45}{x} = \frac{10}{4}$.

QUICK ✓ *Find the length of side XY given that $\triangle MNP$ is similar to $\triangle XYZ$.*

8.

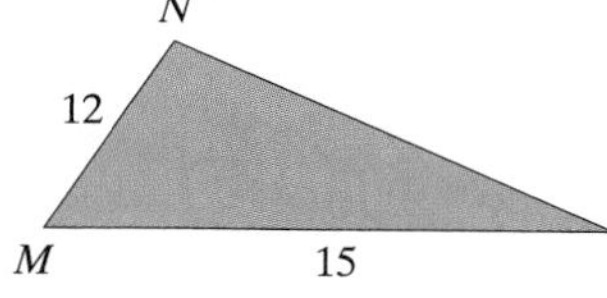

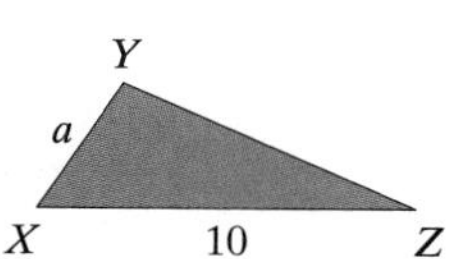

Classroom Example Use Example 5, and Figure 3 with the shadow of the tree—24 feet—and the shadow of the yardstick—2 feet.

Answer: The height of the tree is 36 feet.

EXAMPLE 5 Finding the Height of a Tree

A fifth-grade student is conducting an experiment to find the height of a tree in the schoolyard. The student measures the length of the tree's shadow and then immediately measures the length of the shadow that a yardstick forms. The tree's shadow measures 24 feet, and the yardstick's shadow measures 4 feet. Find the height of the tree.

Solution

The ratio of the length of the tree's shadow to the height of the tree equals the ratio of the length of the yardstick's shadow to the height of the yardstick. This is because the tree and its shadow form a triangle that is similar to the yardstick and its shadow. See Figure 3.

Figure 3

Let h represent the height of the tree and use 3 feet for the length of the yardstick. We set up the proportion problem as

$$\frac{24}{h} = \frac{4}{3} \quad \text{The Model}$$

Now we solve the proportion.

$$\frac{24}{h} = \frac{4}{3}$$

Cross multiply: $$4h = 72$$

Divide both sides by 4: $$h = 18$$

The height of the tree is 18 feet.

Work Smart

We could have also solved the proportion $\frac{h}{3} = \frac{24}{4}$. Do you see why?

QUICK ✓

9. A 60-foot-tall tree casts a shadow of 25 feet. At the same time of day, a man casts a shadow of 2.5 feet. How tall is the man?

3 Model and Solve Work Problems

Work Smart
Remember, when we model we make simplifying assumptions to make the math easier to deal with.

We are now going to solve work problems. These problems assume that jobs are performed at a **constant rate,** which means an individual works at the same pace throughout the entire job. While this assumption is reasonable for machines, it is not likely to be true for people simply because of the old phrase "too many chefs spoil the broth." Think of it this way—if you continually add more people to paint a room, the time to complete the job may decrease initially, but eventually the painters get in each other's way and the time to completion actually increases. While we could take this into account when modeling situations such as this, we will make the "constant rate" assumption for humans as well to keep the mathematics manageable.

The constant-rate assumption states that if it takes t units of time to complete a job, then $\frac{1}{t}$ of the job is done in 1 unit of time. For example, if it takes 5 hours to paint a room, then $\frac{1}{5}$ of the room should be painted in 1 hour.

To solve these work problems, we'll use the six-step process that was introduced in Section 2.5.

Classroom Example ➤
Use Example 6; however, Dan plants 1800 seedlings in 4 hours and Hala requires 6 hours to plant 1800 seedlings.

Answer: It takes 2.4 hours working together to plant 1800 seedlings.

EXAMPLE 6 Planting Seedlings

Dan, an experienced horticulturist, can plant 1000 seedlings in 2 hours. Hala, a student assistant, requires 5 hours to plant 1000 seedlings. How long would it take Dan and Hala working together to plant 1000 seedlings?

Solution

Step 1: Identify We want to know how long it will take for Dan and Hala to plant 1000 seedlings together.

Step 2: Name We let t represent the time (in hours) that it takes to plant the seedlings working together. Then in 1 hour they will complete $\frac{1}{t}$ of the job.

Step 3: Translate Since we know that Dan can finish the job in 2 hours, Dan will finish $\frac{1}{2}$ of the job in 1 hour. We know that Hala can finish the job in 5 hours, so Hala will finish $\frac{1}{5}$ of the job in 1 hour. Based on this information, we set up Table 1.

Work Smart
The part of the job completed per hour is the reciprocal of the number of hours to complete the job.

Table 1

	Number of Hours to Complete the Job	Part of the Job Completed per Hour
Dan	2	$\frac{1}{2}$
Hala	5	$\frac{1}{5}$
Together	t	$\frac{1}{t}$

We set up the model using the following logic:

$$\begin{pmatrix}\text{Part done by Dan}\\ \text{in 1 hour}\end{pmatrix} + \begin{pmatrix}\text{Part done by Hala}\\ \text{in 1 hour}\end{pmatrix} = \begin{pmatrix}\text{Part done together}\\ \text{in 1 hour}\end{pmatrix}$$

$$\frac{1}{2} \qquad + \qquad \frac{1}{5} \qquad = \qquad \frac{1}{t}$$

Step 4: Solve We now proceed to solve the equation.

$$\frac{1}{2} + \frac{1}{5} = \frac{1}{t}$$

Multiply both sides by the LCD, $10t$: $$10t \cdot \left(\frac{1}{2} + \frac{1}{5}\right) = 10t \cdot \frac{1}{t}$$

Distribute: $$10t \cdot \frac{1}{2} + 10t \cdot \frac{1}{5} = 10t \cdot \frac{1}{t}$$

Divide out common factors: $$5t + 2t = 10$$

Combine like terms: $$7t = 10$$

Divide both sides by 7: $$t = \frac{10}{7}$$

$$t \approx 1.428$$

Work Smart

We convert 0.428 hours to minutes by multiplying 0.428 by 60 minutes and obtain 24 minutes.

Step 5: Check It is always a good idea to make sure your answer is reasonable. We expect our answer to be greater than 0 but less than 2 (because it takes Dan 2 hours working by himself). Our answer of 1.428 hours or 1 hour, 24 minutes seems reasonable.

Step 6: Answer It will take Dan and Hala about 1 hour 24 minutes to plant 1000 seedlings.

QUICK ✓

10. It takes Molly 3 hours to shovel her driveway after a snowstorm working by herself. It takes her elderly neighbor 6 hours to shovel his driveway working by himself. How long would it take Molly and her neighbor to shovel the neighbor's driveway working together assuming each driveway is the same dimension?

Classroom Example ➢

Bob Johnson can paint a 1500-square-foot apartment in 20 hours. If his son Mark, a college student, helps him, they can paint a 1500-square-foot apartment in 12 hours. How long does it take Mark to paint the apartment alone?

Answer: It takes Mark 30 hours.

EXAMPLE 7 Mowing the Lawn

Sara can cut the grass in 3 hours working by herself. When Sara cuts the grass with her younger brother Brian, it takes 2 hours. How long would it take Brian to cut the grass if he worked by himself?

Solution

Step 1: Identify We want to know how long it will take for Brian to cut the grass by himself.

Step 2: Name We let t represent the time (in hours) that it takes Brian to cut the grass by himself. Then, in 1 hour he will complete $\frac{1}{t}$ of the job.

Step 3: Translate Since we know that Sara can finish the job in 3 hours, Sara will finish $\frac{1}{3}$ of the job in 1 hour. We also know that together the job is completed in 2 hours, so $\frac{1}{2}$ of the job is completed in 1 hour. We set up the model using the following logic:

$$\begin{pmatrix}\text{Part done by}\\\text{Sara in 1 hour}\end{pmatrix} + \begin{pmatrix}\text{Part done by}\\\text{Brian in 1 hour}\end{pmatrix} = \begin{pmatrix}\text{Part done}\\\text{together in 1 hour}\end{pmatrix}$$

$$\frac{1}{3} + \frac{1}{t} = \frac{1}{2}$$

Step 4: Solve We now proceed to solve the equation.

$$\frac{1}{3} + \frac{1}{t} = \frac{1}{2}$$

Multiply both sides by the LCD, $6t$: $$6t \cdot \left(\frac{1}{3} + \frac{1}{t}\right) = 6t \cdot \frac{1}{2}$$

Distribute: $$6t \cdot \frac{1}{3} + 6t \cdot \frac{1}{t} = 6t \cdot \frac{1}{2}$$

Divide out common factors: $$2t + 6 = 3t$$

Subtract $2t$ from each side: $$6 = t$$

Step 5: Check Is this answer correct? Does Brian do $\frac{1}{6}$ of the job per hour? Is $\frac{1}{3} + \frac{1}{6} = \frac{1}{2}$? Since $\frac{2}{6} + \frac{1}{6} = \frac{3}{6} = \frac{1}{2}$, the answer is correct.

Step 6: Answer It will take Brian 6 hours to mow the grass, working by himself.

QUICK ✓

11. It takes Leon 5 hours to seal his driveway working by himself. If Michael helps Leon, they can seal the driveway in 2 hours. How long would it take Michael to seal the driveway alone?

---

4 Model and Solve Uniform Motion Problems

Teaching Tip
You may need to introduce upstream and downstream motion. The total downstream speed of the boat is the speed of the boat in still water plus the speed of the current. Similarly, the total upstream speed is the speed of the boat in still water minus the speed of the current.

We first introduced uniform motion problems back in Section 2.7. Recall that uniform motion problems use the fact that distance equals rate times time, that is, $d = rt$. When modeling uniform motion problems that lead to rational equations, we usually end up using a variation on this model, $t = \frac{d}{r}$. These problems use the idea of constant rate, just like work problems. For example, a bicyclist may not ride for several hours at exactly the same speed, but we can assume that over the time period the average speed is constant. Again, the assumption makes the mathematics easier to deal with.

Classroom Example ➤
Barbara uses her canoe to travel on a river whose current is running at 1 mile per hour. After traveling 12 miles downstream, she turns around and makes the trip back upstream. She notices that she covered only 10 miles going upstream in the same amount of time that she covered 12 miles going downstream. How fast does Barbara paddle the canoe in still water?

Answer: 11 miles per hour

EXAMPLE 8 A Boat Trip on the Mighty Olentangy

The Olentangy River has a current of 2 miles per hour. A motorboat takes the same amount of time to go 48 miles downstream as it takes to go 36 miles upstream. What is the speed of the boat in still water?

Solution

Step 1: Identify This is a uniform motion problem. We wish to know the speed of the boat in still water.

Step 2: Name Let r represent the speed of the boat in still water.

Step 3: Translate The current pushes the boat when it is going downstream, so the total speed of the boat will be the speed of the boat in still water plus the speed of the current. We let $r + 2$ represent the speed of the boat downstream. Going upstream, the current slows the boat, so the total speed of the boat upstream will be $r - 2$. We set up Table 2.

The amount of time traveling downstream is the same as the amount of time spent traveling upstream, so

$$\frac{48}{r+2} = \frac{36}{r-2} \quad \text{The Model}$$

Work Smart

Remember, $d = r \cdot t$, so $t = \frac{d}{r}$.

Table 2

	Distance (miles)	Rate (miles per hour)	Time $= \frac{\text{Distance}}{\text{Rate}}$ (hours)
Downstream	48	$r + 2$	$\frac{48}{r+2}$
Upstream	36	$r - 2$	$\frac{36}{r-2}$

Step 4: Solve We wish to solve for r: $\frac{48}{r+2} = \frac{36}{r-2}$

Multiply both sides by the LCD, $(r + 2)(r - 2)$: $48(r - 2) = 36(r + 2)$

Distribute: $48r - 96 = 36r + 72$

Subtract $36r$ from both sides: $12r - 96 = 72$

Add 96 to both sides: $12r = 168$

Divide each side by 12: $\frac{12r}{12} = \frac{168}{12}$

$r = 14$

Step 5: Check Does the boat travel 48 miles downstream in the same time that it travels 36 miles upstream? Let's see. Going downstream, the boat travels at $14 + 2 = 16$ miles per hour, so it takes $\frac{48}{16} = 3$ hours to travel downstream. Going upstream, the boat travels at $14 - 2 = 12$ miles per hour, and it takes $\frac{36}{12} = 3$ hours to travel upstream. Our answer checks!

Step 6: Answer the Question The boat travels 14 miles per hour in still water.

QUICK ✓

12. A small airplane can travel 120 mph in still air. The plane can fly 700 miles with the wind in the same time that it can travel 500 miles against the wind. Find the speed of the wind.

Classroom Example ➤
Use Example 9 with distance to the waterfall = 15 miles, and total time = 8 hours.

Answer: Average speed biking from the waterfall is 5 mph.

EXAMPLE 9 The Ride Back through the Forest

Every weekend, you ride your bicycle on a forest preserve path. The path is 12 miles long and ends at a waterfall, at which point you relax and then make the trip back to the starting point. One weekend, you notice that because the ride to the waterfall is partially uphill, your cycling speed returning is 2 miles per hour faster than the speed riding to the waterfall. If the round trip takes 5 hours (excluding resting time), find the average speed cycling back from the waterfall.

Solution

Step 1: Identify This is a uniform motion problem. We wish to know the average speed returning from the waterfall.

Step 2: Name Let r represent your cycling speed when you are cycling to the waterfall. Then $r + 2$ represents the speed cycling back from the waterfall to the starting point.

Step 3: Translate We know that the distance to the waterfall is 12 miles. We set up Table 3.

Table 3

	Distance (miles)	Rate (miles per hour)	$\text{Time} = \dfrac{\text{Distance}}{\text{Rate}}$ (hours)
Going to the Waterfall	12	r	$\dfrac{12}{r}$
Returning Home	12	$r + 2$	$\dfrac{12}{r+2}$

We know that the total time spent cycling was 5 hours, so we set up the equation

$$\begin{pmatrix}\text{time cycling to} \\ \text{waterfall}\end{pmatrix} + \begin{pmatrix}\text{time cycling to} \\ \text{starting point}\end{pmatrix} = 5 \text{ hours}$$

$$\frac{12}{r} + \frac{12}{r+2} = 5 \quad \text{The Model}$$

Step 4: Solve We wish to solve for r:

$$\frac{12}{r} + \frac{12}{r+2} = 5$$

Multiply both sides by the LCD, $r(r + 2)$:
$$r(r+2)\left(\frac{12}{r} + \frac{12}{r+2}\right) = (5)r(r+2)$$

Distribute:
$$r(r+2)\left(\frac{12}{r}\right) + r(r+2)\left(\frac{12}{r+2}\right) = (5)r(r+2)$$

Divide out common factors:
$$(r+2)12 + r(12) = 5r(r+2)$$

Multiply and distribute:
$$12r + 24 + 12r = 5r^2 + 10r$$

Combine like terms:
$$24r + 24 = 5r^2 + 10r$$

Write the equation in standard form $ax^2 + bx + c = 0$:
$$0 = 5r^2 - 14r - 24$$

Factor the polynomial:
$$0 = (5r+6)(r-4)$$

Set each factor equal to zero:
$$5r + 6 = 0 \quad \text{or} \quad r - 4 = 0$$

Solve each equation:
$$5r = -6 \qquad r = 4$$
$$r = -\frac{6}{5}$$

We discard the solution $r = -\frac{6}{5}$ because r represents the cycling speed and speeds cannot be negative. So $r = 4$ miles per hour is the speed cycling to the waterfall. We are asked to find the speed cycling back to the starting point, $r + 2$, so $r + 2 = 4 + 2 = 6$ mph.

Step 5: Check Let's check our solution. We should find that the time cycling to the waterfall + the time returning to the starting point = 5 hours. Because $\dfrac{12 \text{ mi}}{4 \text{ mph}} + \dfrac{12 \text{ mi}}{6 \text{ mph}} = 3 \text{ hr} + 2 \text{ hr} = 5 \text{ hr}$, our answer is correct.

Step 6: Answer Your average speed cycling from the waterfall to the starting point is 6 mph.

QUICK ✓

13. To prepare for a half-marathon run, Sue ran 18 miles before she blistered her heel, and then she walked 1 additional mile. Her running speed was 12 times as fast as her walking speed. She was running and walking for 5 hours. Find Sue's running speed.

5.8 Exercises

For Extra Help: Student Solutions Manual | CD Video | PH Math/Tutor Center | MathXL Tutorials on CD | MathXL® | MyMathLab

1. extremes, means
2. similar
3. $d = rt$
4. False
5. True
6. False
7. Answers may vary.
8. Answers may vary.

Concepts and Vocabulary

In Problems 1–3, fill in the blanks.

1. In the proportion $\frac{x}{y} = \frac{3}{8}$, the terms x and 8 are called the ________ while y and 3 are called the ________.

2. In geometry, two figures are ________ if their corresponding angle measures are equal and their corresponding sides are proportional.

3. To solve uniform motion problems, we use the formula ________.

In Problems 4–6, answer True or False to each statement.

4. The following figures are similar.

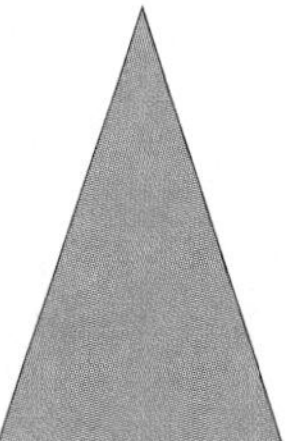

5. The following proportions all yield the same result.

(a) $\frac{x}{2} = \frac{4}{9}$ **(b)** $\frac{2}{x} = \frac{9}{4}$ **(c)** $\frac{x}{4} = \frac{2}{9}$ **(d)** $\frac{9}{2} = \frac{4}{x}$ **(e)** $\frac{4}{9} = \frac{x}{2}$

6. A student reads the following sentence from a word problem: *Barbara rows her boat in a stream that has a current of 1 mph. She travels 10 miles upstream and returns to her starting point 5 hours later*. She makes the following chart. The information is correctly placed for this part of the problem.

d	r	t
10	$r + 1$	5
10	$r - 1$	5

7. When modeling work problems, we assume that individuals work at a constant rate. When modeling uniform motion problems, we assume that individuals travel at a constant rate. Explain what these assumptions mean.

8. You are teaching the class how to solve the following problem: *Marcos can write a budget analysis in 8 hours, and Alonso can complete the same analysis in 6 hours. If they work together, how long will it take to finish the analysis?* Write an equation that could be used to find the answer to the question and then explain your justification for setting up the equation your way. If you can think of another way to do the problem, explain how to set it up differently and then explain why this logic will also solve the problem correctly.

9. $\{12\}$
10. $\{15\}$
11. $\left\{\frac{18}{7}\right\}$
12. $\left\{\frac{20}{9}\right\}$
13. $\{16\}$
14. $\left\{\frac{1}{6}\right\}$
15. $\left\{\frac{24}{5}\right\}$
16. $\{2\}$
17. $\left\{-\frac{4}{5}\right\}$
18. $\left\{-\frac{9}{7}\right\}$
19. $\{-2\}$
20. $\{-9\}$
21. $\left\{\frac{5}{7}\right\}$
22. $\left\{\frac{9}{2}\right\}$
23. $\{-3, 2\}$
24. $\{-4, -3\}$
25. $\{2, 5\}$
26. $\{-2, 6\}$
27. $\{-2, 9\}$
28. $\{-3, 6\}$
29. $XZ = 12$
30. $BC = 21$
31. $n = 3$; $ZY = 5$
32. $n = 1$; $BC = 3$
33. $x = 30$; $AB = 28$
34. $x = 2.5$; $HE = 2$
35. $2x$
36. $\frac{n}{2}$
37. $t - 3$
38. $p - 45$
39. $r - 2$
40. $r + 25$
41. $\frac{1}{5} + \frac{1}{3} = \frac{1}{t}$

Building Skills

In Problems 9–28, solve the proportion.

9. $\frac{9}{x} = \frac{3}{4}$
10. $\frac{x}{5} = \frac{12}{4}$
11. $\frac{4}{7} = \frac{2x}{9}$

12. $\frac{6}{5} = \frac{8}{3x}$
13. $\frac{6}{5} = \frac{x+2}{15}$
14. $\frac{5}{9} = \frac{2x+3}{6}$

15. $\frac{b}{b+6} = \frac{4}{9}$
16. $\frac{k}{k+3} = \frac{6}{15}$
17. $\frac{y}{y-10} = \frac{2}{27}$

18. $\frac{y}{y-3} = \frac{3}{10}$
19. $\frac{p+2}{4} = \frac{2p+4}{5}$
20. $\frac{n+6}{3} = \frac{n+4}{5}$

21. $\frac{2z-1}{z} = \frac{3}{5}$
22. $\frac{2k-3}{k} = \frac{4}{3}$
23. $\frac{2}{v^2-v} = \frac{1}{3-v}$

24. $\frac{n-2}{4n+7} = \frac{-2}{n+1}$
25. $\frac{10-x}{4x} = \frac{1}{x-1}$
26. $\frac{4}{x^2} = \frac{1}{x+3}$

27. $\frac{2p-3}{p^2+12p+6} = \frac{1}{p+4}$
28. $\frac{1}{2z-1} = \frac{z+4}{z^2+10z+14}$

In Problems 29–32, $\triangle ABC$ is similar to $\triangle XYZ$.

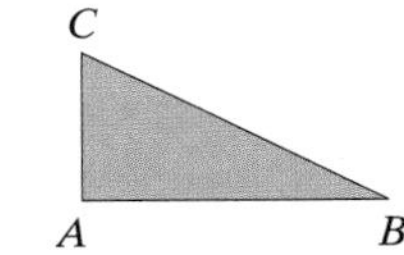

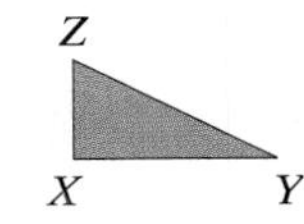

△ **29.** If $AB = 6$, $XY = 9$, $AC = 8$, find XZ.

△ **30.** If $XY = 8$, $YZ = 7$, $AB = 24$, find BC.

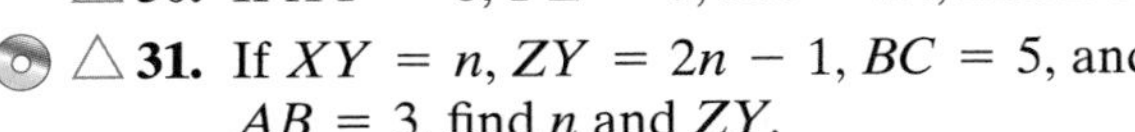

△ **31.** If $XY = n$, $ZY = 2n - 1$, $BC = 5$, and $AB = 3$, find n and ZY.

△ **32.** If $AB = n$, $BC = n + 2$, $XY = 5$, and $YZ = 15$, find n and BC.

In Problems 33 and 34, rectangle ABCD is similar to rectangle EFGH.

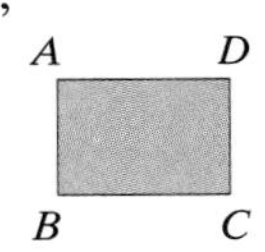

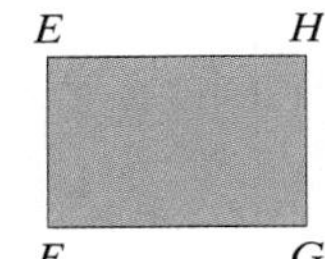

△ **33.** If $AB = x - 2$, $EF = 4$, $BC = 2x + 3$, and $FG = 9$, find x and AB.

△ **34.** If $DC = 6$, $HG = x - 1$, $DA = 8$, and $HE = 2x - 3$ find x and HE.

In Problems 35–40, write an algebraic expression that represents each phrase.

35. If Mariko worked x hours and Natalie worked twice as many hours as Mariko, write an algebraic expression that represents the number of hours Natalie worked.

36. If Brian worked for n days and Kellen worked for half as many days, write an algebraic expression that represents the number of days Kellen worked.

37. If David painted for 3 days less than Pierre and Pierre painted for t days, write an algebraic expression that represents the number of days David painted.

38. If Valerie danced for p minutes and Melody danced for 45 minutes less than Valerie, write an algebraic expression that represents the number of minutes Melody danced.

39. If the rate of the current of a stream is 2 mph and Joe can swim r mph in still water, write an algebraic expression that represents Joe's rate when he swims upstream.

40. If the wind is blowing at 25 mph and a plane flies at r mph in still air, write an algebraic expression that represents the rate the plane is traveling when it flies with the wind.

In Problems 41–46, write an equation that could be used to model each of the following. SET UP BUT DO NOT SOLVE THE EQUATION.

41. Painting Chairs Christina can paint a chair in 5 hours, and Victoria can paint a chair in 3 hours. How many hours will it take to paint the chair when the two girls work together?

42. $\frac{1}{6} + \frac{1}{t} = \frac{1}{2}$
43. $\frac{1}{b+4} + \frac{1}{b} = \frac{1}{7}$
44. $\frac{1}{n} + \frac{1}{2n} = \frac{1}{6}$
45. $\frac{4}{14-c} = \frac{7}{14+c}$
46. $\frac{10}{s+3} = \frac{4}{s-3}$
47. 145 miles
48. $15\frac{1}{3}$ cm
49. $8\frac{1}{3}$ lb
50. 5 lb
51. 37,245 rubles
52. 1600 pesos
53. 24 ft
54. 20 m
55. $1\frac{7}{8}$ hr
56. $2\frac{6}{11}$ hr
57. $3\frac{3}{7}$ minutes
58. 12 hours

42. Weeding the Yard Bill can weed the back yard in 6 hours. When he worked with Tamra, it only took 2 hours. How long would it take Tamra to weed the back yard if she worked alone?

43. Filling a Tank There are two different inlet pipes that can be used to fill a 3000-gallon tank. Pipe A takes 4 hours longer than Pipe B to fill the tank. With both pipes open, it takes 7 hours to fill the tank. How long would it take Pipe B alone to fill the tank?

44. Canning Peaches A factory that was producing canned peaches had an old canning machine. They decided to add a newer machine that could work twice as fast. With both machines on line, the plant could produce 10,000 cans in 6 hours. How long would it take to produce the same number of cans using only the newer machine?

45. Bob's River Trip Bob's boat travels at 14 mph in still water. Find the speed of the current if he can go 4 miles upstream in the same time that it takes to go 7 miles downstream.

46. Marty's River Trip A stream has a current of 3 mph. Find the speed of Marty's boat in still water if she can go 10 miles downstream in the same time it takes to go 4 miles upstream.

Applying the Concepts

47. Yasmine's Map On Yasmine's map, $\frac{1}{4}$ inch represents 10 miles. According to the map, Yellow Springs and Hillsboro are $3\frac{5}{8}$ inches apart. Find the number of miles between the two towns.

48. Beth's Map On Beth's map, 0.5 cm represents 60 miles. If it is 1840 miles from San Diego to New Orleans, how far is this distance on her map?

49. Making Bread If it takes 5 lb of flour to make 3 loaves of bread, how much flour is needed to make 5 loaves of bread?

50. Buying Candy If it costs $3.50 for 2 lb of candy, how much candy can be purchased for $8.75?

51. Comparing Rubles and Dollars $5 U.S. is worth approximately 143.25 Russian rubles. How many rubles can Hortencia purchase for $1300 U.S.?

52. Comparing Pesos and Pounds 50 Mexican pesos are worth approximately 2.5 British pounds. If Sean takes 80 pounds as spending money in Mexico City, how many pesos will he have?

53. Tree Shadow If a 6-ft man casts a shadow that is 2.5 ft long, how tall is a tree that has a shadow 10 ft long?

54. Telephone Pole Shadow A bush that is 4 m tall casts a shadow that is 1.75 m long. How tall is a telephone pole if its shadow is 8.75 m?

55. Cleaning the Math Building Josh can clean the math building on his campus in 3 hours. Ken takes 5 hours to complete the same building. If they work together, how long will it take them to finish cleaning this building?

56. Pruning Trees Martin can prune his fruit trees in 4 hours. His neighbor can prune the same trees for him in 7 hours. If they work together on this job, how long will take to prune the trees?

57. Retrieving Volleyballs After hitting practice for the Long Beach State volleyball team, Dyanne can retrieve all of the balls in the gym in 8 minutes. It takes Makini 6 minutes to retrieve all the balls. If they work together, how long will it take these two players to return the volleyballs and be ready to start the next round of hitting practice?

58. Stuffing Envelopes It took Kirsten 6 hours to stuff 3000 envelopes. When she worked with Brittany, it only took 4 hours to stuff the next 3000 envelopes. How long would it take Brittany working alone to stuff the 3000 envelopes?

59. $5\frac{1}{7}$ hr
60. 14.4 min
61. 15 hr
62. 40 hours
63. 24 hr
64. 35 sec
65. $\frac{14}{3}$ km per hour
66. 45 mph
67. 1 hr, 40 min
68. $\frac{15}{7}$ hr
69. 40 mph
70. 30 mph
71. 35 mph
72. 9 mph

59. **Painting City Hall** Three people were given the job of painting city hall. They divided the job into equal portions and decided they would each go home after completing their assigned portion. It took José 9 hours to paint his part and then he went home. It took Joaquín 12 hours to complete his part. The third person played around, never got started, and was promptly fired from the job. If José and Joaquín went back the second day and worked together to finish the undone portion, how long did it take them?

60. **Mowing the Grass** It takes 18 minutes to cut the grass in the outfield of the stadium. A new mower was purchased and with both mowers working, the same task takes 8 minutes. How long would it take the new mower to cut the outfield grass?

61. **Replacing Pipes** It takes an apprentice twice as long as the experienced plumber to replace the pipes under an old house. If it takes them 5 hours when they work together, how long would it take the apprentice alone?

62. **Making Care Packages** It takes Rocco 3 times as long as Traci to make 100 Red Cross care packages for refugees. If it takes them 10 hours when they work together to make 100 care packages, how long would it take Rocco working alone to make the same number?

63. **Filling the Pool** Using a single hose, Janet can fill a pool in 6 hours. The same pool can be drained in 8 hours by opening a drainpipe. If Janet forgets to close the drainpipe, how long would it take her to fill the pool?

64. **Jill's Bucket** Jill has a bucket with a hole in it. When the hole did not exist, Jill could fill the bucket in 30 seconds. With the hole, the bucket now empties in 210 seconds (3.5 minutes). How long will it take Jill to fill the bucket now that it has a hole in it?

65. **Travel by Boat** A boat can travel 10 km down the river in the same time it can go 4 km up the river. If the current in the river is 2 km per hour, how fast can the boat travel in still water?

66. **Travel by Plane** A small plane can travel 1000 miles with the wind in the same time it can go 600 miles against the wind. If the speed of the plane in still air is 180 mph, what is the speed of the wind?

67. **Iron Man Training** While training for an iron man competition, Tony bikes for 60 miles and runs for 15 miles. If his biking speed is 8 times his running speed and it takes 5 hours to complete the training, how long did he spend on his bike?

68. **Robin's Workout** Robin can run twice as fast as she walks. If she runs for 12 miles and walks for 8 miles, the total time to complete the trip is 5 hours. How many hours did she run?

69. **Driving during a Snowstorm** Claire and Chris were driving to their home in Milwaukee when they came upon a snowstorm. They drove at an average rate of 20 miles per hour slower for the last 60 miles during the snowstorm than they did for the first 90 miles. Find their average rate of speed during the last 60 miles.

70. **Driving during Road Construction** David and Lisa drove 150 miles through road construction at a certain average rate. By increasing their speed by 20 miles per hour, they traveled the next 250 miles without road construction in the same time they spent on the 150-mile leg of their trip. Find their average rate during the part of the trip that had road construction.

71. **Tough Commute** You have a 20-mile commute into work. Since you leave very early, the trip going to work is easier than the trip home. You can travel to work in the same time that it takes for you to make it 16 miles on the trip back home. Your average speed coming home is 7 miles per hour slower than your average speed going to work. What is your average speed going to work?

72. **A Bike Trip** A bicyclist rides his bicycle 12 miles up a hill and then 16 miles on level terrain. His speed on level ground is 3 miles per hour faster than his speed going uphill. The cyclist rides for the same amount of time going uphill and on level ground. Find his speed going uphill.

Extending the Concepts

In geometry, when a line is parallel to one side of a triangle, the intersection of the parallel line with the other two sides of the triangle will form a new triangle that is similar to the original triangle. In the figure below, $\overline{XY}$ is parallel to $\overline{BC}$, therefore $\triangle XAY$ is similar to $\triangle BAC$.

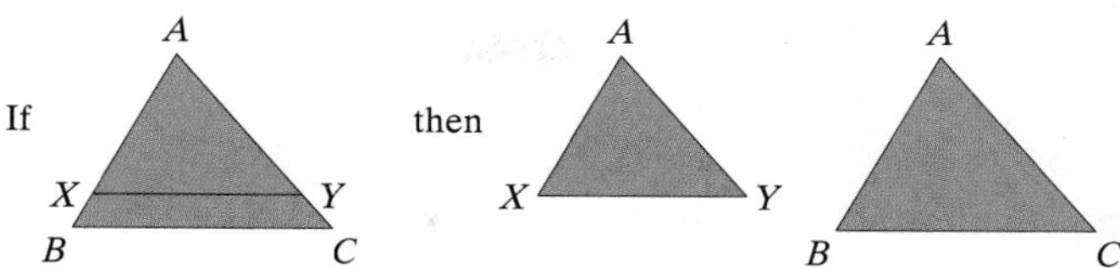

In Problems 73–78, solve for x.

73. $x = 6$
74. $x = 12$
75. $x = \frac{98}{5}$
76. $x = 24$
77. $x = \frac{72}{5}$
78. $x = 6$

△ **73.** $XA = 4, BA = 12, CA = 18$, and $YA = x$.

△ **74.** $XA = 4, BA = 10, CA = 30$, and $YA = x$.

△ **75.** $XA = 5, XB = 9, XY = 7$, and $BC = x$.

△ **76.** $XA = 3, XB = 15, XY = 4$, and $BC = x$.

△ **77.** $AY = 5, YC = 12, AX = 6$, and $XB = x$.

△ **78.** $AY = 4, YC = x, AX = x$, and $XB = 9$.

CHAPTER 5 ACTIVITY: CORRECT THE QUIZ

Focus: Performing operations with rational expressions and solving rational equations

Time: 15–20 minutes

Group size: 2

In this activity you will work as a team to grade the student quiz shown below. If an answer is correct, mark it correct. If an answer is wrong, mark it wrong and show the correct answer.

Once all of the quiz questions are graded, compute the final score for the quiz. Be prepared to discuss your results with the rest of the class.

1. $\frac{9}{4}$
2. $\frac{y^2}{x^2 - y^2}$
3. $\{2\}$
4. $\frac{2x+3}{5x+3}$
5. 1 hr, 12 min
6. $\frac{-(x+17)}{5x+13}$
7. $\frac{x^2+5x+20}{(x+5)^2(x+1)}$

Quiz score $= \frac{4}{7}$, or 57%

Student Quiz	
Name: *Ima Student*	Quiz Score: ________
(1) Multiply: $\frac{6x-12}{4x+8} \cdot \frac{3x+6}{2x-4}$	Answer: $\frac{9}{4}$
(2) Subtract: $\frac{xy}{x^2-y^2} - \frac{y}{x+y}$	Answer: $\frac{y^2}{x^2-y^2}$
(3) Solve: $\frac{5x}{x+1} = 2 + \frac{2x}{x+1}$	Answer: $\{2\}$
(4) Divide: $\frac{x^2+x-6}{5x^2-7x-6} \div \frac{3x^2+13x+12}{6x^2+17x+12}$	Answer: $\frac{2}{5}$
(5) Solve: Joe can mow his lawn in 2 hrs. Mike can mow the same lawn in 3 hrs. If they work together, how long will it take them to mow the lawn?	Answer: 1 hr, 12 min
(6) Simplify: $\dfrac{\frac{1}{x+5} - \frac{2}{x-7}}{\frac{4}{x-7} + \frac{1}{x+5}}$	Answer: $\frac{x+17}{5x+13}$
(7) Add: $\frac{x}{x^2+10x+25} + \frac{4}{x^2+6x+5}$	Answer: $\frac{x^2+5x+20}{(x+5)(x+1)}$

Chapter 5 Review

Section 5.1 Simplifying Rational Expressions

KEY CONCEPTS	KEY TERMS
• To determine undefined values of a rational expression, set the denominator equal to zero and solve for the variable. • **Simplifying a Rational Expression** First, completely factor the numerator and the denominator of the rational expression; then divide out common factors using the fact that if p, q, and r are polynomials, then $\frac{p \cdot r}{q \cdot r} = \frac{p}{q}$ if $q \neq 0$ and $r \neq 0$.	Rational expression Evaluate Undefined Simplify

YOU SHOULD BE ABLE TO . . .	EXAMPLE	REVIEW EXERCISES
① Evaluate a rational expression (p. 305)	Examples 1 through 3	1–4
② Determine undefined values of a rational expression (p. 307)	Examples 4 and 5	5–10
③ Simplify rational expressions (p. 308)	Examples 6–9	11–16

1. (a) $\frac{1}{2}$ (b) undefined (c) 1
2. (a) 0 (b) -1 (c) undefined
3. (a) undefined (b) 9 (c) 49
4. (a) -1 (b) undefined (c) 42
5. $\frac{7}{3}$
6. $\frac{1}{2}$
7. none
8. none
9. -10 or -2
10. -1 or 4
11. $\frac{4x^4}{5y^4}$
12. $\frac{3}{4x^2y^4}$
13. $\frac{3}{k+2}$
14. $-\frac{2}{x-3}$
15. $\frac{x+5}{2x-1}$
16. $\frac{3x-1}{2(x^2-2x+4)}$

In Problems 1–4, evaluate each expression for the given values.

1. $\frac{2x}{x+3}$
(a) $x = 1$ **(b)** $x = -3$ **(c)** $x = 3$

2. $\frac{3x}{x-4}$
(a) $x = 0$ **(b)** $x = 1$ **(c)** $x = 4$

3. $\frac{x^2 + 2xy + y^2}{x - y}$
(a) $x = 1$, $y = 1$ **(b)** $x = 2$, $y = 1$ **(c)** $x = -3$, $y = -4$

4. $\frac{2x^2 - x - 3}{x + z}$
(a) $x = 1$, $z = 1$ **(b)** $x = 1$, $z = -1$ **(c)** $x = 5$, $z = -4$

In Problems 5–10, find the value(s) of the variable for which the rational expression is undefined.

5. $\frac{3x}{3x-7}$ **6.** $\frac{x+1}{4x-2}$ **7.** $\frac{5}{x^2+25}$

8. $\frac{17}{4x^2+49}$ **9.** $\frac{5x+2}{x^2+12x+20}$ **10.** $\frac{3x-1}{x^2-3x-4}$

In Problems 11–16, simplify each rational expression, if possible. Assume that no variable has a value that results in a denominator with a value of zero.

11. $\frac{20x^5y}{25xy^5}$ **12.** $\frac{45x^2y^2}{60x^4y^6}$ **13.** $\frac{3k-21}{k^2-5k-14}$

14. $\frac{-2x-2}{x^2-2x-3}$ **15.** $\frac{x^2+8x+15}{2x^2+5x-3}$ **16.** $\frac{3x^2+5x-2}{2x^3+16}$

Section 5.2 Multiplying and Dividing Rational Expressions

KEY CONCEPTS

- **Multiplying Rational Expressions**

  If $\frac{a}{b}$ and $\frac{c}{d}$, $b \neq 0, d \neq 0$, are two rational expressions, then $\frac{a}{b} \cdot \frac{c}{d} = \frac{ac}{bd}$.

- **Dividing Rational Expressions**

  If $\frac{a}{b}$ and $\frac{c}{d}$, $b \neq 0, c \neq 0, d \neq 0$, are two rational expressions, then $\frac{\frac{a}{b}}{\frac{c}{d}} = \frac{a}{b} \div \frac{c}{d} = \frac{a}{b} \cdot \frac{d}{c} = \frac{ad}{bc}$.

YOU SHOULD BE ABLE TO ...	EXAMPLE	REVIEW EXERCISES
(1) Multiply rational expressions (p. 313)	Examples 1 through 4	17, 18, 21, 22, 25, 26
(2) Divide rational expressions (p. 315)	Examples 5 through 9	19, 20, 23, 24, 27, 28

17. $\frac{2m^2}{n^2}$ 18. $\frac{y^3}{6x}$
19. $\frac{14}{9n^2}$ 20. $\frac{1}{ab^7}$
21. $-\frac{5(x-2)}{x+3}$
22. $-\frac{4(x-3)(x+3)}{(x-9)(x-9)}$
23. $\frac{x-2}{4(x-3)}$ 24. $15x^3$
25. $\frac{x+5}{x+6}$ 26. $\frac{y+1}{y-3}$
27. $\frac{x^2}{(x+1)(x+9)}$
28. $\frac{y-2}{3y+1}$

In Problems 17–28, perform the indicated operations and simplify, if possible.

17. $\frac{12m^4n^3}{7} \cdot \frac{21}{18m^2n^5}$

18. $\frac{3x^2y^4}{4} \cdot \frac{2}{9x^3y}$

19. $\frac{10m^2n^4}{9m^3n} \div \frac{15mn^6}{21m^2n}$

20. $\frac{5ab^3}{3b^4} \div \frac{10a^2b^8}{6b^2}$

21. $\frac{5x - 15}{x^2 - x - 12} \cdot \frac{x^2 - 6x + 8}{3 - x}$

22. $\frac{4x - 24}{x^2 - 18x + 81} \cdot \frac{x^2 - 9}{6 - x}$

23. $\frac{\frac{x^2 - 4}{x^2 - 8x + 15}}{\frac{12x + 24}{3x - 15}}$

24. $\frac{\frac{5x^3 + 10x^2}{3x}}{\frac{2x + 4}{18x^2}}$

25. $\frac{3x^2 + 14x - 5}{x^2 + x - 30} \cdot \frac{x^2 - 2x - 15}{3x^2 + 8x - 3}$

26. $\frac{y^2 - 5y - 14}{y^2 - 2y - 35} \cdot \frac{y^2 + 6y + 5}{y^2 - y - 6}$

27. $\frac{x^2 - 9x}{x^2 + 3x + 2} \div \frac{x^2 - 81}{x^2 + 2x}$

28. $\frac{y^2 - 9}{2y^2 - y - 15} \div \frac{3y^2 + 10y + 3}{2y^2 + y - 10}$

Section 5.3 Adding and Subtracting Rational Expressions with a Common Denominator

KEY CONCEPT

- **Adding/Subtracting Rational Expressions**

  To add or subtract rational expressions, if $\frac{a}{c}$ and $\frac{b}{c}$, $c \neq 0$, then $\frac{a}{c} + \frac{b}{c} = \frac{a+b}{c}$ and $\frac{a}{c} - \frac{b}{c} = \frac{a-b}{c}$.

YOU SHOULD BE ABLE TO ...	EXAMPLE	REVIEW EXERCISES
(1) Add rational expressions with a common denominator (p. 322)	Examples 1 through 3	29–34
(2) Subtract rational expressions with a common denominator (p. 324)	Examples 4 and 5	35–38
(3) Add or subtract rational expressions with opposite denominators (p. 325)	Examples 6 and 7	39–42

29. $\frac{9}{x-3}$ 30. $\frac{10}{x+4}$ 31. m
32. $\frac{1}{m}$ 33. $\frac{1}{m-2}$ 34. $2m+3$
35. $\frac{1}{3m}$ 36. $\frac{2}{b}$
37. $y+7$
38. $\frac{1}{m-6}$
39. $-\frac{7}{x-y}$
40. $\frac{4}{a-b}$
41. $\frac{3x-5}{x^2-25}$
42. $\frac{x+4}{x-3}$

In Problems 29–42, perform the indicated operations and simplify, if possible.

29. $\frac{4}{x-3}+\frac{5}{x-3}$ **30.** $\frac{3}{x+4}+\frac{7}{x+4}$

31. $\frac{m^2}{m+3}+\frac{3m}{m+3}$ **32.** $\frac{1}{6m}+\frac{5}{6m}$

33. $\frac{-m+1}{m^2-4}+\frac{2m+1}{m^2-4}$ **34.** $\frac{2m^2}{m+1}+\frac{5m+3}{m+1}$

35. $\frac{11}{15m}-\frac{6}{15m}$ **36.** $\frac{15b}{2b^2}-\frac{11b}{2b^2}$

37. $\frac{2y^2}{y-7}-\frac{y^2+49}{y-7}$ **38.** $\frac{6m+5}{m^2-36}-\frac{5m-1}{m^2-36}$

39. $\frac{3}{x-y}+\frac{10}{y-x}$ **40.** $\frac{7}{a-b}+\frac{3}{b-a}$

41. $\frac{2x}{x^2-25}-\frac{x-5}{25-x^2}$ **42.** $\frac{x+5}{2x-6}-\frac{x+3}{6-2x}$

Section 5.4 Finding the Least Common Denominator and Forming Equivalent Rational Expressions

KEY CONCEPTS	KEY TERMS
• The steps to find the LCD of rational expressions are given on page 331. • The steps to form equivalent rational expressions are given on page 334.	Least common denominator (LCD) Equivalent rational expressions

YOU SHOULD BE ABLE TO . . .	EXAMPLE	REVIEW EXERCISES
① Find the LCD of two or more rational expressions (p. 329)	Examples 1 through 6	43–48
② Write a rational expression that is equivalent to a given rational expression (p. 332)	Examples 7 and 8	49–52
③ Use the LCD to write equivalent rational expressions (p. 334)	Example 9	53–58

43. $12x^4y^7$ 44. $60a^3b^4c^7$
45. $8a(a+2)$ 46. $5a(a+6)$
47. $4(x-3)(x+1)$
48. $x(x-7)(x+7)$
49. $\frac{6xy^6}{x^4y^7}$ 50. $\frac{11a^4b^3}{a^7b^5}$
51. $\frac{(x-1)(x+2)}{(x-2)(x+2)}$
52. $\frac{(m+2)(m-2)}{(m+7)(m-2)}$
53. $\frac{20x^2y}{24x^5}; \frac{21}{24x^5}$ 54. $\frac{12}{10a^3}; \frac{11a^2b}{10a^3}$
55. $\frac{-4x}{-(x-2)}; \frac{6}{-(x-2)}$
56. $\frac{-3}{-(m-5)}; \frac{-2m}{-(m-5)}$
57. $\frac{2m+4}{(m+7)(m-2)(m+2)}; \frac{m^2-m-2}{(m+7)(m-2)(m+2)}$
58. $\frac{n^2+3n-10}{n(n-5)(n+5)}; \frac{n^2}{n(n-5)(n+5)}$

In Problems 43–48, identify the LCD of the given rational expressions.

43. $\frac{6}{4x^2y^7}; \frac{8}{6x^4y}$ **44.** $\frac{3}{20a^3bc^4}; \frac{7}{30ab^4c^7}$ **45.** $\frac{11}{4a}; \frac{7a}{8a+16}$

46. $\frac{6}{5a}; \frac{17}{a^2+6a}$ **47.** $\frac{x+1}{4x-12}; \frac{3x}{x^2-2x-3}$ **48.** $\frac{11}{x^2-7x}; \frac{x+2}{x^2-49}$

In Problems 49–52, write an equivalent rational expression with the given denominator.

49. $\frac{6}{x^3y}$ with denominator x^4y^7 **50.** $\frac{11}{a^3b^2}$ with denominator a^7b^5

51. $\frac{x-1}{x-2}$ with denominator x^2-4 **52.** $\frac{m+2}{m+7}$ with denominator $m^2+5m-14$

In Problems 53–58, identify the LCD and then write each as an equivalent rational expression with that denominator.

53. $\frac{5y}{6x^3}; \frac{7}{8x^5}$ **54.** $\frac{6}{5a^3}; \frac{11b}{10a}$

55. $\frac{4x}{x-2}; \frac{6}{2-x}$ **56.** $\frac{3}{m-5}; \frac{-2m}{5-m}$

57. $\frac{2}{m^2+5m-14}; \frac{m+1}{m^2+9m+14}$ **58.** $\frac{n-2}{n^2-5n}; \frac{n}{n^2-25}$

Section 5.5 Adding and Subtracting Rational Expressions with Unlike Denominators

KEY CONCEPT

- The steps for adding or subtracting rational expressions with unlike denominators are given on page 339.

YOU SHOULD BE ABLE TO . . .	EXAMPLE	REVIEW EXERCISES
(1) Add and subtract rational expressions with unlike denominators (p. 338)	Examples 1 through 10	59–72

59. $\frac{4xz + 8y^4}{x^2y^3z}$ 60. $\frac{5xy^2 + x^2y}{10x^3y^3}$
61. $\frac{x^2 - 5x + 14}{(x + 7)(x - 7)}$
62. $\frac{2x^2 + 13x - 9}{(2x + 3)(2x - 3)}$
63. $\frac{-4x - 10}{x(x - 2)}$ 64. $\frac{2x^2 - 6x - 5}{x(x + 5)}$
65. $\frac{3x - 1}{4x + 1}$ 66. $\frac{3x - 4}{(x - 1)(x - 2)}$
67. $\frac{-(4m - 9)(m - 4)}{(m + 3)(m - 3)}$
68. $\frac{2m^2 + 5m - 2}{(m - 5)(m + 2)}$
69. $\frac{4}{m - 2}$ 70. $\frac{m + n}{m - n}$
71. $\frac{5x + 12}{x + 3}$ 72. $\frac{7x + 13}{x + 2}$

In Problems 59–72, perform the indicated operations and simplify, if possible.

59. $\frac{4}{xy^3} + \frac{8y}{x^2z}$

60. $\frac{x}{2x^3y} + \frac{y}{10xy^3}$

61. $\frac{x}{x + 7} + \frac{2}{x - 7}$

62. $\frac{4}{2x + 3} + \frac{x + 1}{2x - 3}$

63. $\frac{x + 5}{x} - \frac{x + 7}{x - 2}$

64. $\frac{3x}{x + 5} - \frac{x + 1}{x}$

65. $\frac{3x - 4}{4x + 1} + \frac{3x + 6}{4x^2 + 9x + 2}$

66. $\frac{7}{3x^2 + x - 4} + \frac{9x + 2}{3x^2 - 2x - 8}$

67. $\frac{m}{m^2 - 9} - \frac{4m - 12}{m + 3}$

68. $\frac{2m + 1}{m - 5} - \frac{4}{m^2 - 3m - 10}$

69. $\frac{3}{m - 2} - \frac{1}{2 - m}$

70. $\frac{m}{m - n} - \frac{n}{n - m}$

71. $4 + \frac{x}{x + 3}$

72. $7 - \frac{1}{x + 2}$

Section 5.6 Complex Rational Expressions

KEY CONCEPT	KEY TERMS
• There are two methods that can be used to simplify a complex rational expression. The steps for Method 1 are presented on page 350 while the steps for Method 2 are presented on page 354.	Complex rational expression Simplify

YOU SHOULD BE ABLE TO . . .	EXAMPLE	REVIEW EXERCISES
(1) Simplify a complex rational expression by simplifying the numerator and denominator separately (p. 349)	Examples 2 through 4	73–80
(2) Simplify a complex rational expression using the least common denominator (p. 352)	Examples 5 and 6	73–80

73. $-\frac{3}{23}$ 74. $\frac{18}{11}$
75. $\frac{2m^2 - 10m}{m^2 + 10}$
76. $\frac{6 + 4m^2}{6m - 5m^2}$
77. $\frac{1}{6}$ 78. $\frac{56}{9}$
79. $\frac{y + 8}{-y + 2}$
80. $\frac{5(y + 10)}{-5y - 22}$

In Problems 73–80, simplify the complex rational expressions.

73. $\dfrac{\frac{1}{2} - \frac{2}{3}}{\frac{4}{9} + \frac{5}{6}}$

74. $\dfrac{\frac{1}{4} + \frac{1}{2}}{\frac{5}{8} - \frac{1}{6}}$

75. $\dfrac{\frac{1}{5} - \frac{1}{m}}{\frac{1}{10} + \frac{1}{m^2}}$

76. $\dfrac{\frac{1}{m^2} + \frac{2}{3}}{\frac{1}{m} - \frac{5}{6}}$

77. $\dfrac{\frac{x}{4} - \frac{1}{2}}{\frac{3x}{2} - 3}$

78. $\dfrac{\frac{7x}{3} + 7}{\frac{3x + 9}{8}}$

79. $\dfrac{\frac{8}{y + 4} + 2}{\frac{12}{y + 4} - 2}$

80. $\dfrac{\frac{25}{y + 5} + 5}{\frac{3}{y + 5} - 5}$

Section 5.7 Rational Equations

KEY CONCEPTS	KEY TERMS
• The steps for solving any rational equation are given on pages 359–360. • To solve for a variable in a formula, get the variable by itself on one side of the equation.	Rational equation Extraneous solution

YOU SHOULD BE ABLE TO . . .	EXAMPLE	REVIEW EXERCISES
(1) Solve equations containing rational expressions (p. 357)	Examples 1 through 7	81–92
(2) Solve for a variable in a rational equation (p. 364)	Examples 8 and 9	93–96

81. $x = -5$ or $x = 0$
82. $x = -6$ or $x = 6$
83. $\{-4\}$ 84. $\left\{\frac{2}{3}\right\}$ 85. $\{5\}$
86. $\left\{\frac{14}{3}\right\}$
87. $\{38\}$
88. $\left\{\frac{9}{7}\right\}$
89. $\{\ \}$ or $\varnothing$
90. $\{\ \}$ or $\varnothing$
91. $\left\{-\frac{5}{2}\right\}$
92. $\{24\}$
93. $k = \frac{4}{y}$
94. $y = \frac{x}{6}$
95. $y = \frac{xz}{x - z}$
96. $z = \frac{xy}{x + y}$

In Problems 81 and 82, state the values of the variable for which the expressions in the rational equation are undefined.

81. $\dfrac{4}{x+5} + \dfrac{1}{x} = 3$ **82.** $\dfrac{2}{x^2 - 36} = \dfrac{1}{x+6}$

In Problems 83–92, solve each equation and list the solution(s) in a solution set. Be sure to check for extraneous solutions.

83. $\dfrac{2}{x} - \dfrac{3}{4} = \dfrac{5}{x}$ **84.** $\dfrac{4}{x} + \dfrac{3}{4} = \dfrac{2}{3x} + \dfrac{23}{4}$

85. $\dfrac{4}{m} - \dfrac{3}{2m} = \dfrac{1}{2}$ **86.** $\dfrac{3}{m} + \dfrac{5}{3m} = 1$

87. $\dfrac{m+4}{m-3} = \dfrac{m+10}{m+2}$ **88.** $\dfrac{m-6}{m+5} = \dfrac{m-3}{m+1}$

89. $\dfrac{2x}{x-1} - 5 = \dfrac{2}{x-1}$ **90.** $\dfrac{2x}{x-2} - 3 = \dfrac{4}{x-2}$

91. $\dfrac{1}{x+3} + \dfrac{1}{x-3} = \dfrac{-5}{x^2 - 9}$ **92.** $\dfrac{3}{x-5} - \dfrac{11}{x^2 - 25} = \dfrac{4}{x+5}$

In Problems 93–96, solve the equation for the indicated variable.

93. $y = \dfrac{4}{k}$ for k **94.** $6 = \dfrac{x}{y}$ for y **95.** $\dfrac{1}{x} + \dfrac{1}{y} = \dfrac{1}{z}$ for y **96.** $\dfrac{1}{x} + \dfrac{1}{y} = \dfrac{1}{z}$ for z

Section 5.8 Models Involving Rational Equations

KEY CONCEPT	KEY TERMS
• **Cross-Multiplication Property** If $\dfrac{a}{b} = \dfrac{c}{d}$, then $a \cdot d = b \cdot c$, b and $d \neq 0$.	Ratio Proportion Terms Means Extremes Cross multiplication Similar figures Constant rate

YOU SHOULD BE ABLE TO . . .	EXAMPLE	REVIEW EXERCISES
(1) Model and solve ratio and proportion problems (p. 369)	Examples 1 through 3	97–100, 103, 104
(2) Model and solve problems with similar figures (p. 372)	Examples 4 and 5	101, 102
(3) Model and solve work problems (p. 374)	Examples 6 and 7	105–108
(4) Model and solve uniform motion problems (p. 376)	Examples 8 and 9	109–112

97. {4}
98. {5}
99. {1}
100. {3}
101. $x = 15$
102. $x = 3.5$
103. 8 tanks
104. $26.25
105. $1\frac{1}{5}$ hours
106. $1\frac{7}{8}$ hours
107. 15 min
108. $9\frac{1}{3}$ hours
109. 9 mph
110. 55 mph
111. 70 mph
112. 45 mph

In Problems 97–100, solve the proportion.

97. $\dfrac{6}{4y+5} = \dfrac{2}{7}$ **98.** $\dfrac{2}{y-3} = \dfrac{5}{y}$ **99.** $\dfrac{y+1}{8} = \dfrac{1}{4}$ **100.** $\dfrac{6y+7}{10} = \dfrac{2y+9}{6}$

In Problems 101 and 102, use the similar triangles to solve for x.

101. 2, 10, 9

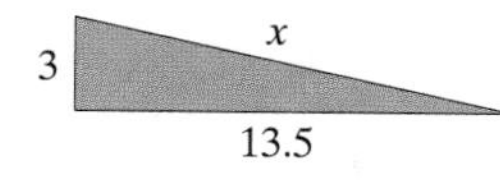

102.

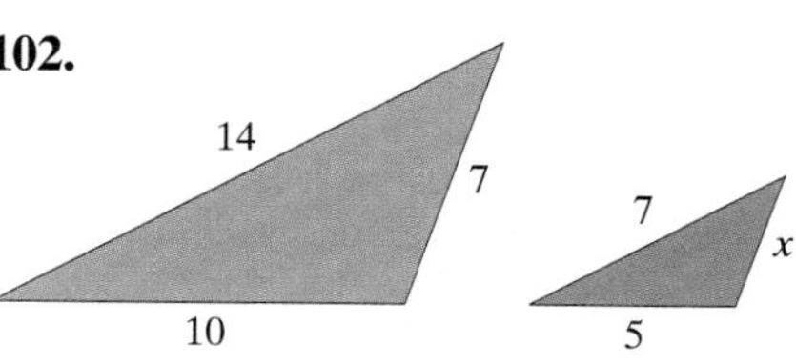

103. Cement A cement mixer uses 4 tanks of water to mix 15 bags of cement. How many tanks of water are needed to mix 30 bags of cement?

104. Pizza If 4 small pizzas cost $15.00, find the cost of 7 small pizzas.

105. Washing Lucille can wash the walls in 3 hours working alone, and Teresa can wash the walls in 2 hours. How long would it take them to wash the walls working together?

106. Carpet Fred can install the carpet in a room in 3 hours, but Barney needs 5 hours to install the carpet. How long would it take them to complete the carpet installation if they work together?

107. Dishes Working alone, it takes Jake 5 minutes longer to wash the dishes than it takes Adrienne when she washes the dishes alone. Washing the dishes together, Jake and Adrienne can finish the job in 6 minutes. How long does it take Jake to wash the dishes by himself?

108. Painting Working together, Donovan and Ben can paint a room in 4 hours. Working alone, it takes Donovan 7 hours to paint the room. How long would it take Ben to paint the room working alone?

109. Paddling Paul can paddle a kayak 15 miles upstream in the same amount of time it takes him to paddle a kayak 27 miles downstream. If the current is 2 mph, what is Paul's speed as he paddles downstream?

110. Cruising A cruise ship traveled for 275 miles with the current in the same amount of time it traveled 175 miles against the current. The speed of the current was 10 mph. What was the speed of the cruise ship as it traveled with the current?

111. Vacation On their vacation, a family traveled 135 miles by train and then traveled 855 miles by plane. The speed of the plane was three times the speed of the train. If the total time of the trip was 6 hours, what was the speed of the train?

112. Driving Tamika drove for 90 miles in the city. When she got on the highway, she increased her speed by 20 mph and drove for 130 miles. If Tamika drove a total of 4 hours, how fast did she drive in the city?

CHAPTER 5 TEST

Remember to use your Chapter Test Prep Video CD to see fully worked-out solutions to any of these problems you would like to review.

Note to instructor: A special file in TestGen provides algorithms specifically matched to the problems in this Chapter Test for easy-to-replicate practice or assessment purposes.

1. 2
2. $x = 5$ or $x = -2$
3. $-\dfrac{x+3}{2}$

1. Evaluate $\dfrac{3x - 2y^2}{6z}$ when $x = 2$, $y = -3$, and $z = -1$.

2. Find the values for which the following rational expression is undefined: $\dfrac{x+5}{x^2 - 3x - 10}$

3. Simplify: $\dfrac{x^2 - 4x - 21}{14 - 2x}$

4. $\frac{14}{5x^2}$
5. $\frac{25}{9}$
6. $\frac{x-y}{2x-3y}$
7. y
8. $\frac{x-5}{x+3}$
9. $\frac{-1}{y-z}$ or $\frac{1}{z-y}$
10. $\frac{2(x+3)(x-1)}{(x-2)(2x+1)}$
11. $\frac{x^2-5x-2}{(x+2)(x+3)(x-1)}$
12. $\frac{y-3}{3y}$
13. $\{-20, 5\}$
14. $\{2\}$
15. $x = \frac{yz}{y-z}$
16. $\{5\}$
17. $x = \frac{24}{5}$
18. \$33
19. 36 min
20. 4 mph

In Problems 4–11, perform the indicated operations and simplify, if possible.

4. $\dfrac{\frac{35x^6}{9x^4}}{\frac{25x^5}{18x}}$ **5.** $\dfrac{5x-15}{3x+9} \cdot \dfrac{5x+15}{3x-9}$ **6.** $\dfrac{\frac{2x^2-5xy-12y^2}{x^2+xy-20y^2}}{\frac{4x^2-9y^2}{x^2+4xy-5y^2}}$

7. $\dfrac{y^2}{y+3} + \dfrac{3y}{y+3}$ **8.** $\dfrac{x^2}{x^2-9} - \dfrac{8x-15}{x^2-9}$ **9.** $\dfrac{6}{y-z} + \dfrac{7}{z-y}$

10. $\dfrac{x}{x-2} + \dfrac{3}{2x+1}$ **11.** $\dfrac{2x}{x^2+5x+6} - \dfrac{x+1}{x^2+2x-3}$

12. Simplify the following complex rational expression: $\dfrac{\frac{1}{9} - \frac{1}{y^2}}{\frac{1}{3} + \frac{1}{y}}$

In Problems 13 and 14, solve the rational equations. Check for extraneous solutions.

13. $\dfrac{m}{5} + \dfrac{5}{m} = \dfrac{m+3}{4}$ **14.** $\dfrac{4}{x+3} + \dfrac{5}{x-6} = \dfrac{4x+1}{x^2-3x-18}$

15. Solve $\dfrac{1}{x} + \dfrac{1}{y} = \dfrac{1}{z}$ for x. **16.** Solve the proportion: $\dfrac{2}{y+1} = \dfrac{1}{y-2}$

17. Use the similar triangles to solve for x.

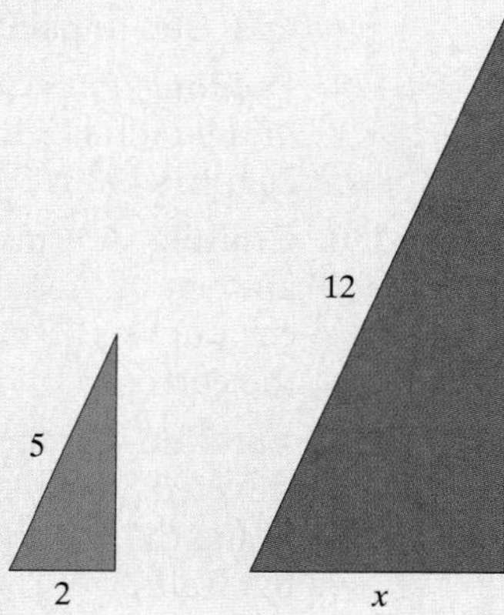

18. Barrettes If 8 hair barrettes cost \$22.00, how much would 12 hair barrettes cost?

19. Car Washing It takes Frank 18 more minutes than Juan to wash a car. If they can wash the car together in 12 minutes, how long does it take Frank to wash the car by himself?

20. Excursion On a vacation excursion, some tourists walked 7 miles on a nature path and then hiked 12 miles up a mountainside. The tourists walked 3 mph faster than they hiked. The total time of the excursion was 4 hours. At what rate did the tourists hike?

CUMULATIVE REVIEW CHAPTERS 1–5

1. -144
2. $9x - 11$
3. $\left\{\frac{19}{6}\right\}$
4. all real numbers
5. $\{4\}$
6. 7 feet
7. 44, 46, 48
8. $x \geq 1$

9. $9x^2 - 12xy + 4y^2$
10. $\frac{16}{a^{16}b^{12}}$
11. $\frac{27z^9}{y^{12}}$
12. $2b(4a - 7)(a + 6)$
13. $\left\{0, \frac{1}{6}, 5\right\}$
14. 10 cm
15. 0
16. $-\frac{3(x + 1)}{5x(2x - 3)}$
17. $\frac{(x - 2)(x + 1)}{4(x + 2)}$
18. $\frac{1}{x - 6}$
19. $\frac{17x + 30}{2(x - 2)(x + 2)}$
20. $\frac{4(10 - 3x)}{85x}$
21. $\{\ \}$ or $\varnothing$
22. $\{-16\}$
23. 10 inches
24. 9 hr
25. 7 mph

In Problems 1 and 2, simplify each expression.

1. $-6^2 + 4(-5 + 2)^3$

2. $3(4x - 2) - (3x + 5)$

In Problems 3–5, solve each equation.

3. $-3(x - 5) + 2x = 5x - 4$

4. $3(2x - 1) + 5 = 6x + 2$

5. $0.25x + 0.10(x - 3) = 0.05(22)$

6. Movie Poster In the entranceway of the Sawgrass Cinema, there is a large rectangular poster advertising the movie *Spiderman 2*. If the poster's length is 3 feet less than twice its width, and its perimeter is 24 feet, what is the length of the poster?

7. Integers The sum of three consecutive even integers is 138. Find the integers.

8. Solve and graph the following inequality: $2(x - 3) - 5 \leq 3(x + 2) - 18$

9. Multiply: $(3x - 2y)^2$

In Problems 10 and 11, simplify each expression. Write your answers with positive exponents only.

10. $\left(\frac{2a^5b}{4ab^{-2}}\right)^{-4}$

11. $(3x^0y^{-4}z^3)^3$

12. Factor completely: $8a^2b + 34ab - 84b$

13. Solve the equation using the Zero-Product Property: $6x^3 - 31x^2 = -5x$.

14. Rectangle The length of a rectangle is 2 cm longer than twice its width. If the area of the rectangle is 40 square centimeters, what is the length of the rectangle?

15. Find the value of $\frac{x + 5}{x^2 + 25}$ when $x = -5$.

In Problems 16–19, perform the indicated operations and simplify, if possible.

16. $\frac{3x + 3}{5x - 5x^2} \cdot \frac{2x^2 + x - 3}{4x^2 - 9}$

17. $\frac{x^2 - x - 2}{10} \div \frac{2x + 4}{5}$

18. $\frac{2x + 3}{x^2 - x - 30} - \frac{x - 2}{x^2 - x - 30}$

19. $\frac{15}{2x - 4} + \frac{x}{x^2 - 4}$

20. Simplify the following complex rational expression: $\dfrac{\frac{2}{x^2} - \frac{3}{5x}}{\frac{4}{x} + \frac{1}{4x}}$

In Problems 21 and 22, solve each equation.

21. $\frac{3}{x - 4} = \frac{5x + 4}{x^2 - 16} - \frac{4}{x + 4}$

22. $\frac{x - 5}{3} = \frac{x + 2}{2}$

23. Map On a city map, 4 inches represents 50 miles. How many inches would represent 125 miles?

24. Inventory It takes Trent 9 hours longer than Sharona to do a store's inventory. If they can finish the inventory in 6 hours working together, how long does it take Sharona to do the inventory working alone?

25. Marathon During a marathon, Francisco jogged for 35 miles, then walked for 6 miles. Francisco jogs 4 mph faster than he walks. If it took him 7 hours to finish the marathon, how fast was he jogging?

CHAPTER

6 Roots and Radicals

When driving on a gravel road, the distance d, in feet, to stop a car going S miles per hour can be calculated using the equation $S = \sqrt{15d}$. A driver traveling 25 mph on a gravel road hits his brakes to avoid a jackrabbit 40 feet away. Will he stop the car in time to avoid hitting the rabbit? See Problem 87 on page 439.

OUTLINE

The Big Picture: Putting It Together

Addition, subtraction, multiplication, and division of polynomial expressions were presented in Chapter 3. Chapter 4 presented the idea that polynomial multiplication can be "undone" by factoring. Throughout your mathematics career, you will learn a process and a method for "undoing" the process.

This chapter discusses the process of "undoing" integer exponents. The focus of Sections 6.1–6.5 is on "undoing" the squaring operation (raising real numbers to the second power). This technique involves a symbol called a radical. Once the idea of how radicals "undo" exponents is understood, the presentation shifts to addition, subtraction, multiplication, and division of radical expressions. After we have a complete understanding of simplifying radical expressions, we turn our attention to solving equations involving these radical expressions in Section 6.6. Finally, Section 6.7 deals with "undoing" positive integer exponents that are 3 or greater. Recall, in Chapter 3 we presented a discussion of raising a real number to integer exponents. In Section 6.7, we extend this idea by discussing how to raise a real number to rational exponents.

6.1 Introduction to Square Roots

OBJECTIVES

1. Evaluate Square Roots
2. Determine Whether a Square Root Is Rational, Irrational, or Not a Real Number
3. Find Square Roots of Variable Expressions

Preparing for Introduction to Square Roots

Before getting started, take the following readiness quiz. If you get a problem wrong, go to the section cited and review the material.

In Problems 1–3, use the set $\left\{-4, \frac{5}{3}, 0, \sqrt{2}, 6.95, 13, \pi\right\}$.

1. Which of the numbers are integers? [Section 1.2, p. 9]
2. Which of the numbers are rational numbers? [Section 1.2, p. 10]
3. Which of the numbers are irrational numbers? [Section 1.2, p. 10]
4. Evaluate: (a) $\left(\frac{3}{2}\right)^2$ (b) $(0.4)^2$ [Section 1.6, pp. 49–50]

Teaching Tip
Up to this point, students have worked with rational numbers only. This chapter deals with both rational and irrational numbers. You may wish to review the definitions of rational numbers and irrational numbers.

In Section 1.6, we introduced the concept of exponents. Exponents are used to indicate repeated multiplication. For example, 4^2 means $4 \cdot 4$, so $4^2 = 16$; $(-6)^2$ means $(-6) \cdot (-6)$, so $(-6)^2 = 36$. In this section, we will reverse the process of raising a number to the second power and ask questions such as, "What number, or numbers, when squared, give me 16?"

1 Evaluate Square Roots

If asked, "What is 5^2?", you would respond "25." If asked, "What is $(-5)^2$?", you would also respond "25." If instead you were asked, "What number, when squared, results in 25?", you would have to think for a while, but eventually you would respond, "Either −5 or 5."

In Words
Taking the square root of a number "undoes" raising a number to the second power.

DEFINITION
For any real numbers a and b, b is a **square root** of a if $b^2 = a$.

Put another way, the square roots of a are the numbers whose square give a.

For example, the square roots of 25 are −5 and 5 because $(-5)^2 = 25$ and $5^2 = 25$. The square roots of 144 are −12 and 12 because $(-12)^2 = 144$ and $12^2 = 144$.

Classroom Example
Find the square roots of
(a) 49 (b) 81.
Answer:
(a) 7 and −7 (b) 9 and −9

EXAMPLE 1 Finding the Square Roots of Numbers

Find the square roots of (a) 36 (b) 0

Solution

(a) 6 is a square root of 36 because $6^2 = 36$.
−6 is also a square root of 36 because $(-6)^2 = 36$.

(b) 0 is the only square root of 0 because only $0^2 = 0$. ■

Classroom Example
Find the square roots of
(a) $\frac{9}{100}$ (b) 0.25
Answer:
(a) $\frac{3}{10}$ and $-\frac{3}{10}$ (b) 0.5 and −0.5

Work Smart
When finding square roots of fractions, look at the numerator and denominator separately.

EXAMPLE 2 Finding the Square Roots of Numbers

Find the square roots of (a) $\frac{16}{49}$ (b) 0.09

Solution

(a) $-\frac{4}{7}$ is a square root of $\frac{16}{49}$ because $\left(-\frac{4}{7}\right)^2 = \frac{16}{49}$ and $\frac{4}{7}$ is also a square root of $\frac{16}{49}$ because $\left(\frac{4}{7}\right)^2 = \frac{16}{49}$.

Preparing for...Answers **1.** −4, 0, 13 **2.** $-4, \frac{5}{3}, 0, 6.95, 13$ **3.** $\sqrt{2}, \pi$ **4. (a)** $\frac{9}{4}$ **(b)** 0.16

(b) -0.3 is a square root of 0.09 because $(-0.3)^2 = 0.09$ and 0.3 is a square root of 0.09 because $(0.3)^2 = 0.09$.

QUICK ✓ *Find the square roots of each real number.*

1. 64 **2.** $\frac{25}{49}$ **3.** 0.36

Work Smart

Math is all about symbols and notation. These symbols allow us to express our thoughts using shorthand. So $\sqrt{100}$ means "give me the positive number whose square is 100." If you want the negative number whose square is 100, write $-\sqrt{100}$.

Notice in Examples 1 and 2 that every positive number has a positive square root and a negative square root. When we want only the positive square root of a number, we use the symbol $\sqrt{\ }$, called a **radical,** to denote the **principal square root,** or nonnegative (zero or positive) square root. For example, if we want the positive square root of 25, we would write $\sqrt{25} = 5$. We read $\sqrt{25} = 5$ as "the positive square root of 25 is 5." But what if we want the negative square root of a real number? In that case, we use the expression $-\sqrt{25} = -5$ to obtain the negative square root of 25.

PROPERTIES OF SQUARE ROOTS

- Every positive real number has two square roots, one positive and one negative.
- We use the symbol $\sqrt{\ }$, called a radical, to denote the nonnegative square root of a real number. The nonnegative square root is called the principal square root.
- The square root of 0 is 0. In symbols, $\sqrt{0} = 0$.
- The number under the radical is called the **radicand.** For example, the radicand in $\sqrt{25}$ is 25.

Classroom Example ➤
Evaluate each square root:
(a) $\sqrt{36}$ (b) $-\sqrt{121}$
Answer:
(a) 6 (b) -11

EXAMPLE 3 Evaluating Square Roots

Evaluate each square root:

(a) $\sqrt{64}$ **(b)** $-\sqrt{81}$

Solution

(a) The notation $\sqrt{64}$ means that we want the positive number whose square is 64. Because $8^2 = 64$, we have that

$$\sqrt{64} = 8$$

(b) The notation $-\sqrt{81}$ means that we want the negative number whose square is 81. So

$$-\sqrt{81} = -1 \cdot \sqrt{81} = -1 \cdot 9 = -9$$

In Words

Example 3(b) in words states, "the negative square root of 81 equals -9," or "the opposite of the principal square root of 81 equals -9."

Classroom Example ➤
Evaluate each square root:
(a) $\sqrt{\frac{25}{16}}$ (b) $-\sqrt{0.49}$
Answer:
(a) $\frac{5}{4}$ (b) -0.7

EXAMPLE 4 Evaluating Square Roots

Evaluate each square root:

(a) $\sqrt{\frac{1}{4}}$ **(b)** $-\sqrt{0.01}$

Solution

(a) $\sqrt{\frac{1}{4}} = \frac{1}{2}$ because $\left(\frac{1}{2}\right)^2 = \frac{1}{4}$.

(b) $-\sqrt{0.01} = -0.1$ because $0.1^2 = 0.01$.

Work Smart
The perfect squares of positive integers are

$1^2 = 1$
$2^2 = 4$
$3^3 = 9$
$4^2 = 16$
$5^2 = 25$
$6^2 = 36$
$7^2 = 49$

and so on.

A rational number is a **perfect square** if it is the square of a rational number. For example, 64, 81, $\frac{1}{4}$, and 0.01 are all perfect squares because their square roots are rational numbers. (Refer back to Examples 3 and 4.) We can think of perfect squares geometrically as in Figure 1, where we have a square whose area is 64 square units. The square root of the area, $\sqrt{64}$, gives us the length of each side of the square, 8 units.

Figure 1

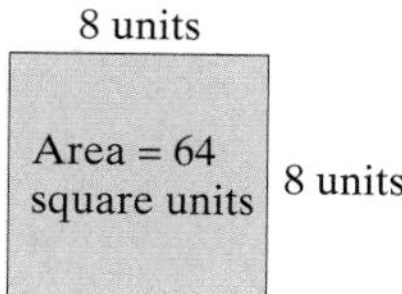

QUICK ✓ *Evaluate each square root.*

4. $\sqrt{100}$ **5.** $-\sqrt{9}$ **6.** $\sqrt{\frac{25}{49}}$ **7.** $\sqrt{0.36}$

Classroom Example ➤
Evaluate: $-2\sqrt{36}$
Answer: -12

EXAMPLE 5 Evaluating an Expression Containing Square Roots

Evaluate: $-4\sqrt{25}$

Solution

The expression $-4\sqrt{25}$ is asking us to find -4 times the positive square root of 25. So we first find the positive square root of 25 and then multiply this result by -4.

$$\begin{aligned}-4\sqrt{25} &= -4 \cdot 5\\ &= -20\end{aligned}$$

Teaching Tip
Remind students that just as $-4x$ indicates the product of the factors -4 and x, $-4\sqrt{25}$ also represents the product of the factors -4 and $\sqrt{25}$.

Classroom Example ➤
Evaluate each expression:
(a) $\sqrt{144} + \sqrt{25}$
(b) $\sqrt{144 + 25}$
(c) $\sqrt{49 - 4 \cdot 3 \cdot 2}$
Answer: (a) 17 (b) 13 (c) 5

EXAMPLE 6 Evaluating an Expression Containing Square Roots

Evaluate each expression:

(a) $\sqrt{9} + \sqrt{16}$ **(b)** $\sqrt{9 + 16}$ **(c)** $\sqrt{64 - 4 \cdot 7 \cdot 1}$

Solution

(a) $$\begin{aligned}\sqrt{9} + \sqrt{16} &= 3 + 4\\ &= 7\end{aligned}$$

(b) $$\begin{aligned}\sqrt{9 + 16} &= \sqrt{25}\\ &= 5\end{aligned}$$

(c) $$\begin{aligned}\sqrt{64 - 4 \cdot 7 \cdot 1} &= \sqrt{64 - 28}\\ &= \sqrt{36}\\ &= 6\end{aligned}$$

Work Smart
In Examples 6(a) and (b), notice that $\sqrt{9} + \sqrt{16} \neq \sqrt{9 + 16}$. In general,

$$\sqrt{a} + \sqrt{b} \neq \sqrt{a + b}$$

The radical acts like a grouping symbol, so always simplify the radicand before taking the square root.

QUICK ✓ *Evaluate each expression.*

8. $2\sqrt{36}$ **9.** $\sqrt{25 + 144}$ **10.** $\sqrt{25} + \sqrt{144}$ **11.** $\sqrt{25 - 4 \cdot 3 \cdot (-2)}$

---

(2) Determine Whether a Square Root Is Rational, Irrational, or Not a Real Number

Because there is no rational number whose square is 5, $\sqrt{5}$ is not a rational number. In fact, $\sqrt{5}$ is an *irrational* number. Remember, an irrational number is a number that cannot be written as the quotient of two integers.

Work Smart
The square roots of negative real numbers are not real.

What if we wanted to evaluate $\sqrt{-16}$? Because any positive real number squared is positive, any negative real number squared is also positive, and 0 squared is 0, there is no real number whose square is -16. We conclude: **Negative real numbers do not have square roots that are real numbers!**

The following comments regarding square roots are important.

MORE PROPERTIES OF SQUARE ROOTS

- The square root of a perfect square is a rational number.
- The square root of a positive rational number that is not a perfect square is an irrational number. For example, $\sqrt{20}$ is an irrational number because 20 is not a perfect square.
- The square root of a negative real number is not a real number. For example, $\sqrt{-2}$ is not a real number.

If needed, find decimal approximations for irrational square roots using a calculator.

Classroom Example Approximate $\sqrt{7}$ by writing it rounded to two decimal places.
Answer: $\sqrt{7} \approx 2.65$

EXAMPLE 7 Approximating Square Roots

Approximate $\sqrt{5}$ by writing it rounded to two decimal places.

Solution

We know that $\sqrt{4} = 2$ and $\sqrt{9} = 3$, so it seems reasonable to expect $\sqrt{5}$ to be between 2 and 3 since 5 is between 4 and 9. We use a calculator or Appendix B and find $\sqrt{5} \approx 2.23607$. Rounded to two decimal places, we have $\sqrt{5} \approx 2.24$.

QUICK ✓ *Express each square root as a decimal rounded to two decimal places.*

12. $\sqrt{35}$ **13.** $-\sqrt{6}$

Classroom Example Use Example 8 with (a) $\sqrt{19}$, (b) $\sqrt{256}$, (c) $\sqrt{-9}$.
Answer:
(a) irrational: $\sqrt{19} \approx 4.36$
(b) $\sqrt{256}$ is rational because $16^2 = 256$, so $\sqrt{256} = 16$.
(c) $\sqrt{-9}$ is not a real number.

EXAMPLE 8 Determining Whether a Square Root of an Integer Is Rational, Irrational, or Not a Real Number

Determine if each square root is rational, irrational, or not a real number. Then evaluate each real square root. For each square root that is irrational, express the square root as a decimal rounded to two decimal places.

(a) $\sqrt{51}$ **(b)** $\sqrt{169}$ **(c)** $\sqrt{-81}$

Solution

(a) $\sqrt{51}$ is irrational because 51 is not a perfect square. There is no rational number whose square is 51. Using a calculator or Appendix B, we find $\sqrt{51} \approx 7.14$.

(b) $\sqrt{169}$ is a rational number because $13^2 = 169$, so $\sqrt{169} = 13$.

(c) $\sqrt{-81}$ is not a real number. There is no real number whose square is -81.

Work Smart
$\sqrt{-81}$ is not a real number, but $-\sqrt{81}$ is because $-\sqrt{81} = -9$. Note the placement of the negative sign!

QUICK ✓ *Determine whether each square root is rational, irrational, or not a real number. Evaluate each square root that is rational. For each square root that is irrational, round your answer to two decimal places.*

14. $\sqrt{49}$ **15.** $\sqrt{71}$ **16.** $\sqrt{-25}$ **17.** $-\sqrt{16}$

3 Find Square Roots of Variable Expressions

What is $\sqrt{4^2}$? Because $4^2 = 16$, we have that $\sqrt{4^2} = \sqrt{16} = 4$. Based on this result, we might conclude that $\sqrt{a^2} = a$ for any real number a. Before we jump to this conclusion, let's consider $\sqrt{(-4)^2}$. Our "formula" says that $\sqrt{a^2} = a$, so we would think that $\sqrt{(-4)^2} = -4$, right? Wrong! $\sqrt{(-4)^2} = \sqrt{16} = 4$. So $\sqrt{4^2} = 4$ and $\sqrt{(-4)^2} = 4$. Regardless of whether the "a" in $\sqrt{a^2}$ is positive or negative, the result ends up being positive. So to say that $\sqrt{a^2} = a$ would not be correct. How can we fix our "formula"? In Section 1.2, we learned that $|a|$ will be a positive number if a is nonzero. From this, we have the following result:

In Words
The square root of a nonzero number squared will always be positive. The absolute value ensures this.

For any **real number** a,

$$\sqrt{a^2} = |a|$$

For example, $\sqrt{(-4)^2} = |-4| = 4$. We use this result when evaluating square roots that have variable expressions because it guarantees the result will be nonnegative regardless of whether the variable is negative, positive, or zero.

Classroom Example
Simplify each square root. The variable can be any real number.
(a) $\sqrt{(6z)^2}$ (b) $\sqrt{(n-2)^2}$
Answer:
(a) $6|z|$ (b) $|n-2|$

EXAMPLE 9 Finding Square Roots of Variable Expressions

(a) $\sqrt{(2x)^2} = |2x| = 2|x|$ for any real number x.

(b) $\sqrt{(a-4)^2} = |a-4|$ for any real number a.

QUICK ✓ *Simplify each square root. The variable represents any real number.*

18. $\sqrt{b^2}$ **19.** $\sqrt{(4p)^2}$ **20.** $\sqrt{(w+3)^2}$

If it is stated that $a \geq 0$, then the absolute value bars are not needed because $|a| = a$ when $a \geq 0$. So $\sqrt{a^2} = a$ when $a \geq 0$.

Classroom Example
Simplify each square root.
(a) $\sqrt{(7n)^2}$, $n \geq 0$
(b) $\sqrt{(t+5)^2}$, $t+5 \geq 0$
Answer:
(a) $7n$ (b) $t+5$

EXAMPLE 10 Finding Square Roots of Variable Expressions

Simplify each square root.

(a) $\sqrt{(3y)^2}$, $y \geq 0$ **(b)** $\sqrt{(b+3)^2}$, $b+3 \geq 0$

Solution

(a) $\sqrt{(3y)^2} = 3y$ because $y \geq 0$.

(b) $\sqrt{(b+3)^2} = b+3$ because $b+3 \geq 0$.

QUICK ✓ *Simplify each square root.*

21. $\sqrt{(5b)^2}$, $b \geq 0$ **22.** $\sqrt{(z-6)^2}$, $z-6 \geq 0$

23. $\sqrt{(2h-5)^2}$, $2h-5 \geq 0$

6.1 Exercises

For Extra Help:

Student Solutions Manual

CD Video

PH Math/Tutor Center

MathXL Tutorials on CD

MathXL®

MyMathLab

Concepts and Vocabulary

In Problems 1–3, fill in the blanks.

1. The _______ _______ of a are the numbers whose square is a.

2. In the expression $\sqrt{4x}$, $4x$ is called the _______.

3. When we want only the positive square root of a number, we use the symbol $\sqrt{\ }$, called a _______, to denote the _________ square root.

In Problems 4–6, answer True or False to each statement.

4. $-\sqrt{9}$ and $\sqrt{-9}$ represent the same number.

5. If a positive number is not a perfect square, then its square root is an irrational number.

6. For all possible positive values of a and b, $\sqrt{a+b} \neq \sqrt{a} + \sqrt{b}$.

7. You don't have a calculator handy but need an estimate for $\sqrt{20}$. Explain how you might find an approximate value for this number without a calculator.

8. Explain why, in the set of real numbers, the radicand of a square root cannot be negative. Make up an example to demonstrate why this is true.

Building Skills

In Problems 9–16, find the value(s) for each expression.

9. the square of 4

10. the square roots of 1

11. the square roots of 4

12. the square of 1

13. the square roots of $\frac{1}{9}$

14. the square of $\frac{9}{4}$

15. the square of $\frac{1}{9}$

16. the square roots of $\frac{9}{4}$

In Problems 17–36, find the exact value of each square root without a calculator.

17. $\sqrt{144}$ **18.** $\sqrt{4}$ **19.** $-\sqrt{9}$ **20.** $-\sqrt{1}$

21. $\sqrt{36}$ **22.** $\sqrt{25}$ **23.** $\sqrt{225}$ **24.** $\sqrt{169}$

25. $\sqrt{\frac{1}{121}}$ **26.** $\sqrt{\frac{1}{49}}$ **27.** $\sqrt{0.04}$ **28.** $\sqrt{0.09}$

29. $-6\sqrt{9}$ **30.** $-4\sqrt{4}$ **31.** $\sqrt{\frac{36}{81}}$ **32.** $\sqrt{\frac{25}{9}}$

33. $\sqrt{0.0016}$ **34.** $\sqrt{0.0049}$ **35.** $5\sqrt{49}$ **36.** $4\sqrt{16}$

In Problems 37–42, use a calculator or Appendix B to find the approximate value of the square root, rounded to the indicated place.

37. $\sqrt{8}$ to 3 decimal places

38. $\sqrt{12}$ to 3 decimal places

39. $\sqrt{30}$ to the nearest hundredth

40. $\sqrt{18}$ to the nearest hundredth

41. $\sqrt{57}$ to the nearest tenth

42. $\sqrt{42}$ to the nearest tenth

In Problems 43–54, tell if the square root is rational, irrational, or not a real number. If the square root is rational, find the exact value; if the square root is irrational, write the approximate value rounded to two decimal places.

43. $\sqrt{-4}$ **44.** $\sqrt{-100}$ **45.** $\sqrt{400}$ **46.** $\sqrt{900}$

47. $\sqrt{\frac{1}{4}}$ **48.** $\sqrt{\frac{49}{64}}$ **49.** $\sqrt{54}$ **50.** $\sqrt{24}$

51. $\sqrt{50}$ **52.** $\sqrt{12}$ **53.** $\sqrt{-16}$ **54.** $\sqrt{-64}$

1. square roots
2. radicand
3. radical, principal (or nonnegative)
4. False 5. True 6. True
7. Answers may vary.
8. Answers may vary.
9. 16 10. $-1, 1$
11. $-2, 2$ 12. 1
13. $-\frac{1}{3}, \frac{1}{3}$ 14. $\frac{81}{16}$
15. $\frac{1}{81}$ 16. $-\frac{3}{2}, \frac{3}{2}$
17. 12
18. 2
19. -3
20. -1
21. 6
22. 5
23. 15
24. 13
25. $\frac{1}{11}$
26. $\frac{1}{7}$
27. 0.2
28. 0.3
29. -18
30. -8
31. $\frac{6}{9}$
32. $\frac{5}{3}$
33. 0.04
34. 0.07
35. 35
36. 16
37. 2.828
38. 3.464
39. 5.48
40. 4.24
41. 7.5
42. 6.5
43. not a real number
44. not a real number
45. rational; 20
46. rational; 30
47. rational; $\frac{1}{2}$
48. rational; $\frac{7}{8}$
49. irrational; 7.35
50. irrational; 4.90
51. irrational; 7.07
52. irrational; 3.46
53. not a real number
54. not a real number

55. d
56. k
57. $|x - 9|$
58. $y - 16$
59. $2m - n$
60. $|p + q|$
61. 1.73
62. 1.41
63. 0.63
64. $\frac{2}{9}$
65. not a real number
66. not a real number
67. 10
68. 15
69. 14
70. 21
71. 6
72. −30
73. −8
74. 1
75. not a real number
76. not a real number
77. 8
78. 7
79. 6.63
80. 4.58
81. 5
82. 5
83. 6.71
84. 8.94
85. 25 ft
86. 16 m
87. 16 km
88. 20 in.
89. 7 m
90. 5 ft
91. 14 in.
92. 17 cm
93. 398 m
94. 1276 ft
95. 2
96. 3
97. 5
98. 2

In Problems 55–60, simplify each square root.

55. $\sqrt{d^2}, d \geq 0$

56. $\sqrt{k^2}, k \geq 0$

57. $\sqrt{(x-9)^2}$

58. $\sqrt{(y-16)^2}, y - 16 \geq 0$

59. $\sqrt{(2m-n)^2}, 2m - n \geq 0$

60. $\sqrt{(p+q)^2}$

Mixed Practice

In Problems 61–84, evaluate each expression. If the square root is irrational, write the approximate value rounded to two decimal places. If the square root is not a real number, so state.

61. $\sqrt{3}$

62. $\sqrt{2}$

63. $\sqrt{0.4}$

64. $\sqrt{\frac{4}{81}}$

65. $\sqrt{-2}$

66. $\sqrt{-10}$

67. $\sqrt{36 + 64}$

68. $\sqrt{81 + 144}$

69. $\sqrt{36} + \sqrt{64}$

70. $\sqrt{81} + \sqrt{144}$

71. $3\sqrt{4}$

72. $-10\sqrt{9}$

73. $3\sqrt{\frac{25}{9}} - \sqrt{169}$

74. $2\sqrt{\frac{9}{4}} - \sqrt{4}$

75. $\sqrt{3 - 19}$

76. $\sqrt{5 - 9}$

77. $\sqrt{4 - (4)(1)(-15)}$

78. $\sqrt{9 - (4)(1)(-10)}$

79. $\sqrt{100 - (4)(2)(7)}$

80. $\sqrt{121 - (4)(5)(5)}$

81. $\sqrt{7^2 - (4)(-2)(-3)}$

82. $\sqrt{13^2 - (4)(-6)(-6)}$

83. $\sqrt{(11 - 8)^2 + (11 - 5)^2}$

84. $\sqrt{(-3 - 5)^2 + (-5 - (-1))^2}$

Applying the Concepts

The area, A, of a square whose side has length s is given by $A = s^2$. We can calculate the length, s, of the side of a square as the positive square root of the area, A, using $s = \sqrt{A}$. In Problems 85–88, find the length of the side of the square whose area is given.

△ **85.** 625 square feet

△ **86.** 256 square meters

△ **87.** 256 square kilometers

△ **88.** 400 square inches

The area, A, of a circle whose radius is r is given by the formula $A = \pi r^2$. We can calculate the radius when the area is given using $r = \sqrt{\frac{A}{\pi}}$. In Problems 89–92, find the radius of the circle with the following areas.

△ **89.** 49π square meters

△ **90.** 25π square feet

△ **91.** 196π square inches

△ **92.** 289π square centimeters

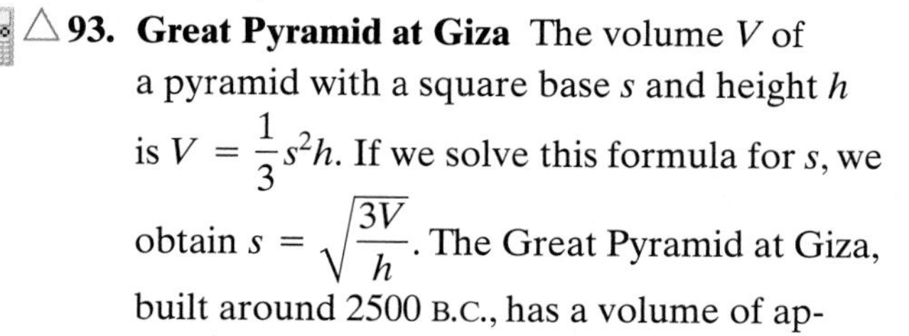

△ **93. Great Pyramid at Giza** The volume V of a pyramid with a square base s and height h is $V = \frac{1}{3}s^2h$. If we solve this formula for s, we obtain $s = \sqrt{\frac{3V}{h}}$. The Great Pyramid at Giza, built around 2500 B.C., has a volume of approximately 7,700,000 cubic meters. Find the length, to the nearest meter, of the side of the Great Pyramid if you know that the height is approximately 146 meters.

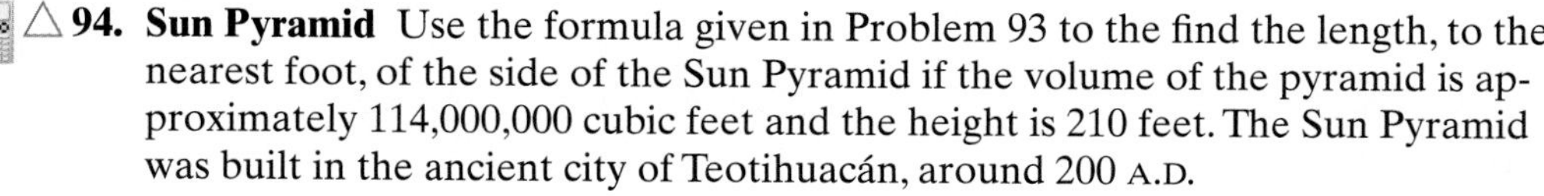

△ **94. Sun Pyramid** Use the formula given in Problem 93 to the find the length, to the nearest foot, of the side of the Sun Pyramid if the volume of the pyramid is approximately 114,000,000 cubic feet and the height is 210 feet. The Sun Pyramid was built in the ancient city of Teotihuacán, around 200 A.D.

Extending the Concepts

In Problems 95–98, simplify each expression.

95. $\sqrt{\sqrt{16}}$

96. $\sqrt{\sqrt{81}}$

97. $\sqrt{5 \cdot \sqrt{25}}$

98. $\sqrt{2 \cdot \sqrt{4}}$

6.2 Simplifying Square Roots

OBJECTIVES

1. Use the Product Rule to Simplify Square Roots of Constants
2. Use the Product Rule to Simplify Square Roots of Variable Expressions
3. Use the Quotient Rule to Simplify Square Roots of Constants and Variable Expressions

Preparing for Simplifying Square Roots

Before getting started, take the following readiness quiz. If you get a problem wrong, go to the section cited and review the material.

1. List the powers of 2 that are less than 100. [Section 1.6, p. 50]
2. List the powers of 3 that are less than 100. [Section 1.6, p. 50]
3. Simplify: $\sqrt{49}$ [Section 6.1, pp. 394–395]

1 Use the Product Rule to Simplify Square Roots of Constants

Work Smart

The perfect square integers are

$$1^2 = 1$$
$$2^2 = 4$$
$$3^3 = 9$$
$$4^2 = 16$$
$$5^2 = 25$$
$$6^2 = 36$$
$$7^2 = 49$$

and so on.

Recall that a number that is the square of a rational number is called a perfect square. For example, $1^2 = 1, 2^2 = 4, 3^2 = 9, \left(\frac{2}{3}\right)^2 = \frac{4}{9}$ are all perfect squares.

DEFINITION

A square root expression is **simplified** if the radicand does not contain any factors that are perfect squares.

Teaching Tip

Emphasize that $\sqrt{a}$ and $\sqrt{b}$ must be real numbers in order to apply the Product Rule.

For example, $\sqrt{50}$ is not simplified because 25 is a factor of 50 and 25 is a perfect square.

But how do we simplify a square root expression? Consider the following:

$$\sqrt{25 \cdot 4} = \sqrt{100} = 10$$
$$\sqrt{25} \cdot \sqrt{4} = 5 \cdot 2 = 10$$

Because both expressions equal 10, we might conclude that

$$\sqrt{25 \cdot 4} = \sqrt{25} \cdot \sqrt{4}$$

which suggests the following result:

In Words

$\sqrt{ab} = \sqrt{a} \cdot \sqrt{b}$ can be stated as follows: "The square root of a product equals the product of the square roots."

PRODUCT RULE OF SQUARE ROOTS

If $\sqrt{a}$ and $\sqrt{b}$ are real numbers, then

$$\sqrt{ab} = \sqrt{a} \cdot \sqrt{b}$$

Example 1 illustrates how to use this rule to simplify square root expressions.

Preparing for...Answers **1.** $2^1 = 2$, $2^2 = 4, 2^3 = 8, 2^4 = 16, 2^5 = 32, 2^6 = 64$ **2.** $3^1 = 3, 3^2 = 9, 3^3 = 27, 3^4 = 81$ **3.** 7

EXAMPLE 1 How to Use the Product Property to Simplify Square Roots

Simplify: $\sqrt{12}$

Classroom Example
Simplify: $\sqrt{75}$
Answer: $5\sqrt{3}$

Step-by-Step Solution

Step 1: Write the radicand as the product of factors, one of which is a perfect square. Write 12 as $4 \cdot 3$ since 4 is a factor of 12 and 4 is a perfect square: $\sqrt{12} = \sqrt{4 \cdot 3}$

Step 2: Write the radicand as the product of two or more radicals. Use $\sqrt{ab} = \sqrt{a} \cdot \sqrt{b}$: $= \sqrt{4} \cdot \sqrt{3}$

Step 3: Take the square root of any perfect square. $\sqrt{4} = 2$: $= 2\sqrt{3}$

So $\sqrt{12} = 2\sqrt{3}$.

We summarize the steps used to simplify the square root expression from Example 1 below.

Steps to Simplify a Square Root Expression

Step 1: Write the radicand as the product of factors, one of which is a perfect square. Look for the largest factor of the radicand that is a perfect square.

Step 2: Use the Product Rule of Square Roots to write the radicand as the product of two or more radicals.

Step 3: Take the square root of each perfect square.

Classroom Example ➤
Simplify:
(a) $\sqrt{28}$ (b) $\sqrt{54}$
Answer:
(a) $2\sqrt{7}$ (b) $3\sqrt{6}$

EXAMPLE 2 Using the Product Property to Simplify Square Roots

Simplify each of the following:

(a) $\sqrt{45}$ **(b)** $\sqrt{48}$

Solution

(a) Since 9 is a factor of 45 and 9 is a perfect square, we write 45 as $9 \cdot 5$.

$$\sqrt{45} = \sqrt{9 \cdot 5}$$

$$\sqrt{ab} = \sqrt{a} \cdot \sqrt{b}: \quad = \sqrt{9} \cdot \sqrt{5}$$

$$\sqrt{9} = 3: \quad = 3\sqrt{5}$$

(b) Since 16 is a factor of 48 and 16 is a perfect square, we write 48 as $16 \cdot 3$.

$$\sqrt{48} = \sqrt{16 \cdot 3}$$

$$\sqrt{ab} = \sqrt{a} \cdot \sqrt{b}: \quad = \sqrt{16} \cdot \sqrt{3}$$

$$\sqrt{16} = 4: \quad = 4\sqrt{3}$$

Work Smart
Sometimes there is more than one way to factor a number. Use the method that produces the **largest** perfect square.

Did you notice in Example 2(b) that 48 can be factored in more than one way? $48 = 6 \cdot 8$ and $48 = 12 \cdot 4$. Either one of these factorizations of 48 would also have produced $4\sqrt{3}$. For example,

$$\sqrt{48} = \sqrt{12 \cdot 4} = \sqrt{3 \cdot 4 \cdot 4} = 2 \cdot 2\sqrt{3} = 4\sqrt{3}$$

By not choosing the *largest* factor of a radicand that is a *perfect square,* you may make your mathematical life rather difficult. Be sure to Work Smart!

QUICK ✓ *Simplify the square root.*

1. $\sqrt{20}$ **2.** $\sqrt{50}$ **3.** $\sqrt{24}$ **4.** $\sqrt{72}$

Watch Out! The direction "simplify" does not mean find a decimal approximation. The simplified form of a radical is the exact value of the radical, not the decimal approximation.

Classroom Example ➤
Simplify:

$$\frac{6 - \sqrt{63}}{3}$$

Answer: $2 - \sqrt{7}$

EXAMPLE 3 Simplifying an Expression Involving a Square Root

Simplify: $\dfrac{4 - \sqrt{20}}{2}$

Work Smart

$\frac{4-\sqrt{20}}{2} \neq 2-\sqrt{20}$

Solution

We can simplify this expression using two different approaches.

METHOD 1:

4 is the largest perfect square factor of 20: $\frac{4-\sqrt{20}}{2} = \frac{4-\sqrt{4\cdot 5}}{2}$

Use $\sqrt{a\cdot b} = \sqrt{a}\cdot\sqrt{b}$: $= \frac{4-\sqrt{4}\cdot\sqrt{5}}{2}$

$= \frac{4-2\cdot\sqrt{5}}{2}$

Factor out the 2 in the numerator: $= \frac{2(2-\sqrt{5})}{2}$

Divide out the common factor: $= 2-\sqrt{5}$

METHOD 2:

4 is the largest perfect square factor of 20: $\frac{4-\sqrt{20}}{2} = \frac{4-\sqrt{4\cdot 5}}{2}$

Use $\sqrt{a\cdot b} = \sqrt{a}\cdot\sqrt{b}$: $= \frac{4-\sqrt{4}\cdot\sqrt{5}}{2}$

$= \frac{4-2\cdot\sqrt{5}}{2}$

Use $\frac{A-B}{C} = \frac{A}{C} - \frac{B}{C}$: $= \frac{4}{2} - \frac{2\cdot\sqrt{5}}{2}$

Divide out the common factor: $= 2-\sqrt{5}$

Teaching Tip
Simplifying an expression such as the one in Example 3 is an important skill for students to have when they solve quadratic equations in Chapter 7.

QUICK ✓ *Simplify the expression.*

5. $\frac{6+\sqrt{45}}{3}$

6. $\frac{-2+\sqrt{32}}{4}$

2 Use the Product Rule to Simplify Square Roots of Variable Expressions

In Section 6.1, we learned that

$$\sqrt{a^2} = a \text{ if } a \geq 0$$

We use this result to simplify square roots of variable expressions. Consider the following:

$$(a^1)^2 = a^2, \quad (a^2)^2 = a^4, \quad (a^3)^2 = a^6, \quad (a^4)^2 = a^8$$

and so on. Based on this, we conclude that if a is a nonnegative real number,

$$\sqrt{a^2} = a, \quad \sqrt{a^4} = \sqrt{(a^2)^2} = a^2, \quad \sqrt{a^6} = \sqrt{(a^3)^2} = a^3,$$

$$\text{and} \quad \sqrt{a^8} = \sqrt{(a^4)^2} = a^4$$

In other words, even powers of a variable expression are perfect squares.

Work Smart
Even powers of a variable expression are perfect squares.

Classroom Example
Simplify $\sqrt{49z^4}$. Assume the variable is a nonnegative real number.
Answer: $7z^2$

EXAMPLE 4 Simplifying a Square Root Containing a Variable to an Even Power

Simplify $\sqrt{16x^6}$. Assume the variable is a nonnegative real number.

Solution

$\sqrt{a\cdot b} = \sqrt{a}\cdot\sqrt{b}$: $\sqrt{16x^6} = \sqrt{16}\cdot\sqrt{x^6}$

$= \sqrt{16}\cdot\sqrt{(x^3)^2}$

Simplify using $\sqrt{a^2} = a$ for $a \geq 0$: $= 4x^3$

Classroom Example Simplify $\sqrt{27a^6b^8}$. Assume the variables are nonnegative real numbers.

Answer: $3a^3b^4\sqrt{3}$

EXAMPLE 5 Simplifying a Square Root Containing Variables to Even Powers

Simplify $\sqrt{50m^4n^6}$. Assume the variables are nonnegative real numbers.

Solution

We proceed as we did in Example 4.

25 is a factor of 50 and 25 is a perfect square: $\sqrt{50m^4n^6} = \sqrt{25 \cdot 2 \cdot m^4n^6}$

$25 = 5^2$; $m^4 = (m^2)^2$ and $n^6 = (n^3)^2$ are perfect squares: $= \sqrt{25m^4n^6 \cdot 2}$

$\sqrt{ab} = \sqrt{a} \cdot \sqrt{b}$: $= \sqrt{25m^4n^6} \cdot \sqrt{2}$

$\sqrt{25} = 5$; $\sqrt{m^4} = m^2$; $\sqrt{n^6} = n^3$: $= 5m^2n^3\sqrt{2}$

QUICK ✓ *Simplify the expression. Assume all variables represent nonnegative real numbers.*

7. $\sqrt{36y^8}$ **8.** $\sqrt{12n^6}$ **9.** $\sqrt{28a^2b^{10}}$ **10.** $\sqrt{45x^{12}}$

What if the variable expression in the radicand does not have an even exponent? We use the same strategy that we used when we simplified square root expressions when the radicand wasn't a perfect square: We factor the variable expression so that one of the factors is a perfect square.

Classroom Example Simplify. Assume the variables represent nonnegative real numbers.

(a) $\sqrt{25b^3}$ (b) $\sqrt{72x^4y^7}$

Answer:

(a) $5b\sqrt{b}$ (b) $6x^2y^3\sqrt{2y}$

EXAMPLE 6 Simplify a Square Root in Which the Variable Is Not a Perfect Square

Simplify each of the following. Assume the variables represent nonnegative real numbers.

(a) $\sqrt{36x^5}$ **(b)** $\sqrt{300m^4n^9}$

Solution

(a) x^4 is a perfect square: $\sqrt{36x^5} = \sqrt{36x^4 \cdot x}$

$\sqrt{ab} = \sqrt{a} \cdot \sqrt{b}$: $= \sqrt{36x^4} \cdot \sqrt{x}$

$\sqrt{36} = 6$; $\sqrt{x^4} = x^2$: $= 6x^2\sqrt{x}$

(b) 100 is a perfect square: m^4 and n^8 are perfect squares: $\sqrt{300m^4n^9} = \sqrt{100m^4n^8 \cdot 3n}$

$\sqrt{ab} = \sqrt{a} \cdot \sqrt{b}$: $= \sqrt{100m^4n^8} \cdot \sqrt{3n}$

$\sqrt{100} = 10$; $\sqrt{m^4} = m^2$; $\sqrt{n^8} = n^4$: $= 10m^2n^4\sqrt{3n}$

Work Smart

To simplify square roots that contain variable expressions, we can use the following "trick of the trade": Divide the exponent on the variable expression by 2. The quotient is the exponent on the variable expression outside the square root, the remainder is the exponent on the variable expression inside the square root. For example,

$$\sqrt{p^{15}} = p^7\sqrt{p} \text{ if } p \geq 0$$

because $15 \div 2 = 7$ with a remainder of 1.

QUICK ✓ *Simplify the expression. Assume the variables represent nonnegative real numbers.*

11. $\sqrt{25y^{11}}$ **12.** $\sqrt{121a^5}$ **13.** $\sqrt{32ab^9}$ **14.** $\sqrt{12x^7y^{13}}$

3 Use the Quotient Rule to Simplify Square Roots of Constants and Variable Expressions

Now consider the following:

$$\sqrt{\frac{64}{4}} = \sqrt{\frac{4 \cdot 16}{4}}$$
$$= \sqrt{16}$$
$$= 4$$
$$\frac{\sqrt{64}}{\sqrt{4}} = \frac{8}{2}$$
$$= 4$$

Because both expressions equal 4, we might conclude that

$$\sqrt{\frac{64}{4}} = \frac{\sqrt{64}}{\sqrt{4}}$$

which suggests the following result:

In Words

$\sqrt{\frac{a}{b}} = \frac{\sqrt{a}}{\sqrt{b}}$ means the square root of the quotient equals the quotient of the square roots.

QUOTIENT RULE OF SQUARE ROOTS

If $\sqrt{a}$ and $\sqrt{b}$ are real numbers and $b \neq 0$, then

$$\sqrt{\frac{a}{b}} = \frac{\sqrt{a}}{\sqrt{b}}$$

Earlier in the section we said that a square root expression is simplified if the radicand does not contain any factors that are perfect squares. We now add an additional requirement that must be satisfied for a square root to be simplified.

A square root is simplified when the radicand does not contain any fractions.

So $\sqrt{\frac{4}{81}}$ or $\sqrt{\frac{27}{3}}$ are not considered to be simplified. The Quotient Rule of Square Roots gives us the tools we need to simplify expressions such as these.

Classroom Example ➤
Simplify:

(a) $\sqrt{\frac{25}{121}}$ (b) $\sqrt{\frac{13}{100}}$

Answer:

(a) $\frac{5}{11}$ (b) $\frac{\sqrt{13}}{10}$

EXAMPLE 7 Simplifying Square Roots of Quotients Using the Quotient Rule

Simplify:

(a) $\sqrt{\frac{4}{81}}$ **(b)** $\sqrt{\frac{7}{36}}$

Solution

(a) $\sqrt{\frac{a}{b}} = \frac{\sqrt{a}}{\sqrt{b}}$: $\sqrt{\frac{4}{81}} = \frac{\sqrt{4}}{\sqrt{81}}$

Simplify: $= \frac{2}{9}$

(b) $\sqrt{\frac{a}{b}} = \frac{\sqrt{a}}{\sqrt{b}}$: $\sqrt{\frac{7}{36}} = \frac{\sqrt{7}}{\sqrt{36}}$

Simplify: $= \frac{\sqrt{7}}{6}$

EXAMPLE 8 Simplifying Square Roots of Quotients Involving Variables Using the Quotient Rule

Simplify: $\sqrt{\dfrac{25x^5}{y^6}}, x \geq 0, y > 0$

Solution

$$\sqrt{\frac{a}{b}} = \frac{\sqrt{a}}{\sqrt{b}}: \quad \sqrt{\frac{25x^5}{y^6}} = \frac{\sqrt{25x^5}}{\sqrt{y^6}}$$

$$x^4 \text{ is a perfect square:} \quad = \frac{\sqrt{25x^4 \cdot x}}{\sqrt{y^6}}$$

$$\sqrt{ab} = \sqrt{a} \cdot \sqrt{b}: \quad = \frac{\sqrt{25x^4} \cdot \sqrt{x}}{\sqrt{y^6}}$$

$$\text{Simplify each radical:} \quad = \frac{5x^2\sqrt{x}}{y^3}$$

QUICK ✓ *Find the quotient and simplify. Assume the variables represent nonnegative real numbers, and that the variable in the denominator is not zero.*

15. $\sqrt{\dfrac{64}{25}}$ **16.** $\sqrt{\dfrac{49a^3}{100}}$ **17.** $\sqrt{\dfrac{11}{16m^2}}$

Sometimes we perform the division in the radicand first and then simplify the radical. This approach is useful when dividing first results in a radicand that is a perfect square.

Classroom Example ➤
Simplify:

$\sqrt{\dfrac{45m^7}{5m^2}}, m > 0$

Answer: $3m^2\sqrt{m}$

EXAMPLE 9 Simplifying Square Roots of Quotients

Simplify: $\sqrt{\dfrac{28a^5}{7a^2}}, a > 0$

Solution

Look carefully at the radicand. Do you see that if we divide 28 by 7 we obtain 4, a perfect square? For this reason, we will perform the division in the radicand first.

$$\text{Divide radicand first:} \quad \sqrt{\frac{28a^5}{7a^2}} = \sqrt{4a^3}$$

$$a^3 = a^2 \cdot a: \quad = \sqrt{4a^2 \cdot a}$$

$$\sqrt{ab} = \sqrt{a} \cdot \sqrt{b}: \quad = \sqrt{4a^2} \cdot \sqrt{a}$$

$$\text{Simplify:} \quad = 2a\sqrt{a}$$

QUICK ✓ *Find the quotient and simplify. Assume the variable represents a nonnegative real number.*

18. $\sqrt{\dfrac{125}{5}}$ **19.** $\sqrt{\dfrac{144b^5}{16}}$

Teaching Tip
Emphasize this point with another example or two so that students can distinguish which method is most efficient for a given problem. For example, $\sqrt{\dfrac{288}{2m^4}}, m > 0$, compared to $\sqrt{\dfrac{15}{4p^2}}, p > 0$.

Based on the results of Examples 8 and 9, we have the following guidelines for simplifying radicals using the Quotient Rule for Radicals. **To simplify square roots, if the radicand has a numerator or denominator containing a perfect square, simplify it first. (See Example 8.) If performing the division in the radicand first results in a perfect square, do the division first. (See Example 9.)**

6.2 Exercises

For Extra Help:  Student Solutions Manual  CD Video  PH Math/Tutor Center  MathXL Tutorials on CD — MathXL® — MyMathLab

Concepts and Vocabulary

In Problems 1–3, fill in the blanks.

1. If $\sqrt{a}$ and $\sqrt{b}$ are real numbers, then $\sqrt{ab} =$ ________.

2. To simplify $\sqrt{50}$, write $\sqrt{\_\_ \cdot 2}$.

3. If $\sqrt{a}$ and $\sqrt{b}$ are real numbers and $b \neq 0$, then $\sqrt{\frac{a}{b}} =$ ________.

In Problems 4–6, answer True or False to each statement.

4. To simplify a square root expression, write the radicand as a product of any two factors.

5. If a square root expression has a radicand that does not have a factor that is a perfect square and does not contain any fractions, then it is simplified.

6. The Quotient Rule of Square Roots allows for the following:

$$\frac{\sqrt{100}}{4} = \sqrt{\frac{100}{4}} = \sqrt{25} = 5$$

7. Is the following simplified? If not, complete the problem by simplifying. Is there an alternate method for simplifying the quotient? If so, explain the steps you would use to simplify.

$$\sqrt{\frac{315}{5}} = \frac{\sqrt{315}}{\sqrt{5}} = \frac{\sqrt{9 \cdot 35}}{\sqrt{5}} = \frac{3\sqrt{35}}{\sqrt{5}}$$

8. Example 8 and Example 9 show two different approaches to simplifying a radical that contains a quotient. Explain what characteristics to look for in the radicand when deciding which method to use.

Building Skills

In Problems 9–28, simplify each square root, if possible.

9. $\sqrt{8}$ **10.** $\sqrt{27}$ **11.** $\sqrt{40}$ **12.** $\sqrt{32}$

13. $\sqrt{18}$ **14.** $\sqrt{12}$ **15.** $\sqrt{33}$ **16.** $\sqrt{21}$

17. $\sqrt{45}$ **18.** $\sqrt{50}$ **19.** $\sqrt{125}$ **20.** $\sqrt{200}$

21. $\sqrt{42}$ **22.** $\sqrt{30}$ **23.** $\sqrt{98}$ **24.** $\sqrt{80}$

25. $\sqrt{270}$ **26.** $\sqrt{180}$ **27.** $\sqrt{132}$ **28.** $\sqrt{243}$

In Problems 29–40, simplify each square root. Assume all variables represent nonnegative real numbers.

29. $\sqrt{x^{10}}$ **30.** $\sqrt{y^{12}}$ **31.** $\sqrt{n^{144}}$ **32.** $\sqrt{m^{100}}$

33. $\sqrt{9y^4}$ **34.** $\sqrt{25x^8}$ **35.** $\sqrt{24z^{14}}$ **36.** $\sqrt{32p^{16}}$

37. $\sqrt{49x^7}$ **38.** $\sqrt{81y^3}$ **39.** $\sqrt{75c^4d^2}$ **40.** $\sqrt{125s^2t^6}$

In Problems 41–60, simplify each square root. Assume all variables represent positive real numbers.

41. $\sqrt{\frac{25}{36}}$ **42.** $\sqrt{\frac{9}{16}}$ **43.** $\sqrt{\frac{1}{196}}$ **44.** $\sqrt{\frac{1}{361}}$

45. $\sqrt{\frac{11}{9}}$ **46.** $\sqrt{\frac{13}{4}}$ **47.** $\sqrt{\frac{y^{16}}{121}}$ **48.** $\sqrt{\frac{x^{10}}{36}}$

49. $\sqrt{\frac{3}{a^4}}$ **50.** $\sqrt{\frac{5}{b^2}}$ **51.** $\sqrt{\frac{x^5}{y^8}}$ **52.** $\sqrt{\frac{a^7}{b^6}}$

1. $\sqrt{a} \cdot \sqrt{b}$
2. 25
3. $\frac{\sqrt{a}}{\sqrt{b}}$
4. False
5. True
6. False
7. No, not simplified. Answers may vary.
8. Answers may vary.
9. $2\sqrt{2}$
10. $3\sqrt{3}$
11. $2\sqrt{10}$
12. $4\sqrt{2}$
13. $3\sqrt{2}$
14. $2\sqrt{3}$
15. $\sqrt{33}$
16. $\sqrt{21}$
17. $3\sqrt{5}$
18. $5\sqrt{2}$
19. $5\sqrt{5}$
20. $10\sqrt{2}$
21. $\sqrt{42}$
22. $\sqrt{30}$
23. $7\sqrt{2}$
24. $4\sqrt{5}$
25. $3\sqrt{30}$
26. $6\sqrt{5}$
27. $2\sqrt{33}$
28. $9\sqrt{3}$
29. x^5
30. y^6
31. n^{72}
32. m^{50}
33. $3y^2$
34. $5x^4$
35. $2z^7\sqrt{6}$
36. $4p^8\sqrt{2}$
37. $7x^3\sqrt{x}$
38. $9y\sqrt{y}$
39. $5c^2d\sqrt{3}$
40. $5st^3\sqrt{5}$
41. $\frac{5}{6}$
42. $\frac{3}{4}$
43. $\frac{1}{14}$
44. $\frac{1}{19}$
45. $\frac{\sqrt{11}}{3}$
46. $\frac{\sqrt{13}}{2}$
47. $\frac{y^8}{11}$
48. $\frac{x^5}{6}$
49. $\frac{\sqrt{3}}{a^2}$
50. $\frac{\sqrt{5}}{b}$
51. $\frac{x^2\sqrt{x}}{y^4}$
52. $\frac{a^3\sqrt{a}}{b^3}$

53. $\frac{2xy^2\sqrt{5y}}{z^2}$
54. $\frac{2ab^3\sqrt{6a}}{c}$
55. $\frac{\sqrt{15}}{x^2}$
56. $\frac{\sqrt{10}}{y^2}$
57. $\frac{5a}{2}$
58. $\frac{b^3}{2}$
59. $\frac{8n^2\sqrt{n}}{3}$
60. $\frac{9m^2\sqrt{m}}{2}$
61. $3\sqrt{7}$
62. $3\sqrt{6}$
63. $\frac{2\sqrt{3}}{5}$
64. $\frac{3\sqrt{2}}{7}$
65. $12a^3$
66. $15b^2$
67. $\frac{x^2\sqrt{x}}{3}$
68. $\frac{1}{2y^4}$
69. $2a^2bc^3\sqrt{3}$
70. $3xy^4z^4\sqrt{2}$
71. $q^2r^2\sqrt{15pq}$
72. $m^3p^3\sqrt{30np}$
73. $\frac{2x^3\sqrt{x}}{y^6}$
74. $\frac{9x\sqrt{x}}{y^2}$
75. $3a^{12}\sqrt{3a}$
76. $4x^{24}\sqrt{2x}$
77. 5
78. -1
79. $3 + \sqrt{2}$
80. $2 - \sqrt{3}$
81. $\frac{-4 - 9\sqrt{2}}{6}$
82. $\frac{-3 + 2\sqrt{3}}{4}$
83. $\frac{1 - \sqrt{2}}{2}$
84. $-1 + \sqrt{3}$
85. $x = 90\sqrt{2}$
86. $x = 60\sqrt{2}$
87. $x = 7\sqrt{5}$
88. $x = 5\sqrt{5}$
89. $x = 2\sqrt{34}$
90. $x = 6\sqrt{10}$
91. $2\sqrt{73}$
92. $4\sqrt{17}$
93. (a) $\frac{\sqrt{6}}{\pi}$ (b) 0.78

53. $\sqrt{\frac{20x^2y^5}{z^4}}$ **54.** $\sqrt{\frac{24a^3b^6}{c^2}}$ **55.** $\sqrt{\frac{105x^2}{7x^6}}$ **56.** $\sqrt{\frac{30y}{3y^5}}$

57. $\sqrt{\frac{100a^3}{16a}}$ **58.** $\sqrt{\frac{36b^{10}}{144b^4}}$ **59.** $\sqrt{\frac{64n^9}{9n^4}}$ **60.** $\sqrt{\frac{81m^{11}}{4m^6}}$

Mixed Practice

In Problems 61–84, simplify each expression. Assume all variables represent positive real numbers.

61. $\sqrt{63}$ **62.** $\sqrt{54}$ **63.** $\sqrt{\frac{12}{25}}$ **64.** $\sqrt{\frac{18}{49}}$

65. $\sqrt{144a^6}$ **66.** $\sqrt{225b^4}$ **67.** $\sqrt{\frac{10x^5}{90}}$ **68.** $\sqrt{\frac{6}{24y^8}}$

69. $\sqrt{12a^4b^2c^6}$ **70.** $\sqrt{18x^2y^8z^8}$ **71.** $\sqrt{15pq^5r^4}$ **72.** $\sqrt{30m^6np^7}$

73. $\sqrt{\frac{4x^7}{y^{12}}}$ **74.** $\sqrt{\frac{81x^3}{y^4}}$ **75.** $\sqrt{27a^{25}}$ **76.** $\sqrt{32x^{49}}$

77. $\frac{4 + \sqrt{36}}{2}$ **78.** $\frac{5 - \sqrt{100}}{5}$ **79.** $\frac{9 + \sqrt{18}}{3}$ **80.** $\frac{10 - \sqrt{75}}{5}$

81. $\frac{-4 - \sqrt{162}}{6}$ **82.** $\frac{-6 + \sqrt{48}}{8}$ **83.** $\frac{7 - \sqrt{98}}{14}$ **84.** $\frac{-6 + \sqrt{108}}{6}$

Applying the Concepts

△ *In Problems 85–92, use the following information. According to the Pythagorean Theorem, the length of the hypotenuse of a right triangle is $c = \sqrt{a^2 + b^2}$, where c is the length of the hypotenuse and a and b are the lengths of the other two sides. The equation $c = \sqrt{a^2 + b^2}$ can also be used to find the diagonal of a square or a rectangle. Find the missing length, x, and simplify the result.*

85.
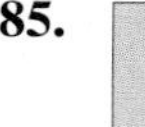
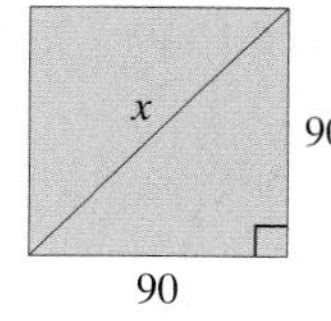

86.
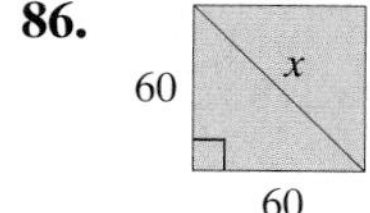

87.
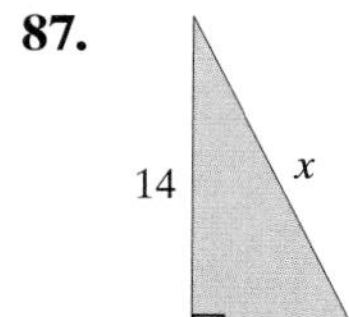

88.
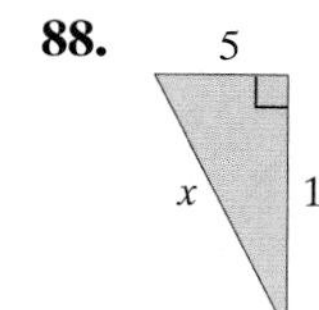

89.
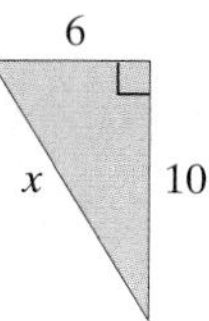

90.
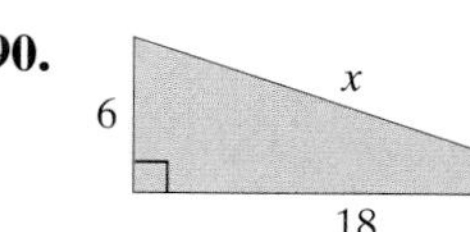

91.

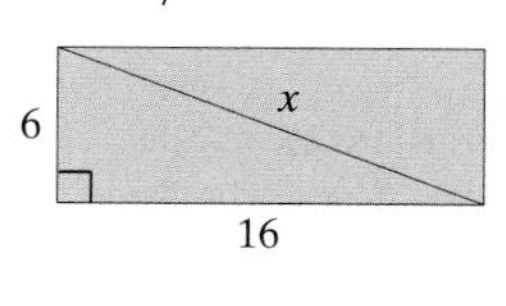

92.
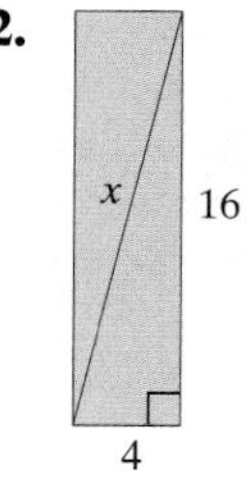

In Problems 93 and 94, use the following information. The frequency of a wave is the number of full oscillations it completes per unit of time. For instance, if you stood on a pier in the Pacific Ocean and counted the number of times that a wave went past you within a certain amount of time, you could calculate the wave frequency. Frequency is measured in oscillations per second. One oscillation per second is equal to one hertz, Hz. The frequency of an object attached to a spring can be calculated with the formula $f = \frac{1}{2\pi}\sqrt{\frac{k}{m}}$ cycles per second. In this formula, k is a constant that depends on the material that is oscillating and m stands for the mass of the object.

93. Suppose you place a 5-kg weight at the end of a spring.

(a) If the constant k is 120, what is the frequency of the vibration of the spring? Write your result as a simplified radical, leaving π in your answer.

(b) Approximate this value to the nearest hundredth.

94. (a) $\frac{5\sqrt{2}}{4\pi}$ (b) 0.56
95. x^{24}
96. a^{40}
97. $16s^{18n}$
98. $81p^{32m}$

94. Suppose you place a 16-kg weight at the end of a spring.
   (a) If the constant k is 200, what is the frequency of the vibration of a spring? Write your result as a simplified radical, leaving π in your answer.
   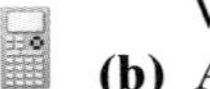
   (b) Approximate this value to the nearest hundredth.

Extending the Concepts

In Problems 95–98, find the missing expression. Assume all variables represent positive real numbers.

95. $\sqrt{?} = x^{12}$ 96. $\sqrt{?} = a^{20}$ 97. $\sqrt{?} = 4s^{9n}$ 98. $\sqrt{?} = 9p^{16m}$

6.3 Adding and Subtracting Square Roots

OBJECTIVES

1. Add and Subtract Square Root Expressions with Like Square Roots
2. Add and Subtract Square Root Expressions with Unlike Square Roots

Preparing for Adding and Subtracting Square Roots

Before getting started, take the following readiness quiz. If you get a problem wrong, go to the section cited and review the material.

1. Which of the following are like terms? [Section 1.7, p. 60]
   $3a$ and $5a$; $4m$ and $-6n$
2. Add: $(3x^2 - 2x + 4) + (x^2 - 4x - 1)$ [Section 3.1, pp. 176–177]
3. Subtract: $(3 - 5k + k^2) - (6 - 3k - 4k^2)$ [Section 3.1, pp. 177–178]
4. Simplify: $3x(x - 7)$ [Section 3.3, pp. 190–191]

In Section 1.7, we learned how to add algebraic expressions by combining like terms using the Distributive Property. In this section, we will learn how to use the Distributive Property to add radical expressions. We continue to assume that variables represent nonnegative real numbers.

1 Add and Subtract Square Root Expressions with Like Square Roots

In Words
Remember, like terms are terms that have the same variables and the same exponent on each variable. Like square roots have the same radicand.

DEFINITION
Square root expressions are **like square roots** if each square root has the same radicand.

For example, $5\sqrt{7}$ and $3\sqrt{7}$ are like square roots because they have the *same radicand,* 7. Additionally, $3\sqrt{2x}$ and $-5\sqrt{2x}$ are also like square roots. Do you see why? However, $\sqrt{3}$ and $\sqrt{11}$ are not like square roots.

To add or subtract square root expressions, we combine like square roots using the Distributive Property "in reverse." Compare the approach for adding square root expressions to the approach for combining like terms from Section 1.6.

Combining Like Terms	Adding Like Square Roots
$6x + 8x = (6 + 8)x$ $= 14x$	$6\sqrt{3} + 8\sqrt{3} = (6 + 8)\sqrt{3}$ $= 14\sqrt{3}$

So all we are doing here is using an old technique in a new way.

Preparing for...Answers 1. $3a$ and $5a$ 2. $4x^2 - 6x + 3$ 3. $5k^2 - 2k - 3$ 4. $3x^2 - 21x$

Classroom Example ➤
Add or subtract, as indicated.
(a) $7\sqrt{3} + 2\sqrt{3}$
(b) $9\sqrt{5} - 4\sqrt{5} + 8\sqrt{2}$
Answer:
(a) $9\sqrt{3}$
(b) $5\sqrt{5} + 8\sqrt{2}$

Work Smart
Remember that to add or subtract square roots, the square roots must have the same radicand.

EXAMPLE 1 Adding and Subtracting Square Roots with Like Square Roots

Add or subtract, as indicated.

(a) $5\sqrt{2} + 9\sqrt{2}$ **(b)** $2\sqrt{7} - 3\sqrt{7} + 9\sqrt{5}$

Solution

(a) These terms are like square roots because the radicand is 2.

Apply the Distributive Property "in reverse": $5\sqrt{2} + 9\sqrt{2} = (5 + 9)\sqrt{2}$
$= 14\sqrt{2}$

(b) The first two terms are like square roots, so we find their difference using the Distributive Property.

$$2\sqrt{7} - 3\sqrt{7} + 9\sqrt{5} = (2 - 3)\sqrt{7} + 9\sqrt{5}$$
$$= -\sqrt{7} + 9\sqrt{5}$$

Classroom Example ➤
Perform the indicated operations:
$-7\sqrt{13z} + 2\sqrt{13z} - \sqrt{13z}$
Answer: $-6\sqrt{13z}$

EXAMPLE 2 Adding and Subtracting Square Root Expressions with Radicands That Contain Variables

Perform the indicated operations: $4\sqrt{11p} + 9\sqrt{11p} - 5\sqrt{11p}$

Solution

Apply the Distributive Property "in reverse": $4\sqrt{11p} + 9\sqrt{11p} - 5\sqrt{11p} = (4 + 9 - 5)\sqrt{11p}$
$= 8\sqrt{11p}$

QUICK ✓ *Add or subtract, as indicated.*

1. $6\sqrt{3} + 8\sqrt{3}$
2. $7\sqrt{11} - 8\sqrt{11} + 2\sqrt{10}$
3. $3\sqrt{13} + 4\sqrt{6} - 11\sqrt{13} - 3\sqrt{6}$
4. $4\sqrt{5x} - 5\sqrt{5x} + 13\sqrt{5x}$

2 Add and Subtract Square Root Expressions with Unlike Square Roots

What if the square root expressions to be added or subtracted aren't like square roots? In this case, we attempt to simplify the square roots so they are like. Then we add or subtract.

Classroom Example ➤
Add: $\sqrt{18} + \sqrt{50}$
Answer: $8\sqrt{2}$

EXAMPLE 3 Adding Unlike Square Roots

Add: $\sqrt{12} + \sqrt{75}$

Solution

The radicands are different. However, when we simplify each square root we will have like square roots. Remember, to simplify, we begin by writing each square root as the product of factors, with one of the factors a perfect square.

4 and 25 are perfect squares: $\sqrt{12} + \sqrt{75} = \sqrt{4 \cdot 3} + \sqrt{25 \cdot 3}$
Simplify each radical: $= 2\sqrt{3} + 5\sqrt{3}$
Add like radicals: $= 7\sqrt{3}$

Work Smart
$\sqrt{a} + \sqrt{b} \neq \sqrt{a + b}$
For example,
$\sqrt{75} + \sqrt{12} \neq \sqrt{87}$

Classroom Example Subtract:

$3\sqrt{8} - \frac{1}{2}\sqrt{200}$

Answer: $\sqrt{2}$

EXAMPLE 4 Subtracting Unlike Square Roots

Subtract: $\frac{1}{2}\sqrt{32} - 2\sqrt{72}$

Solution

The radicands are different, so we simplify the square roots to obtain like square roots.

16 and 36 are perfect squares: $\frac{1}{2}\sqrt{32} - 2\sqrt{72} = \frac{1}{2}\sqrt{16\cdot 2} - 2\sqrt{36\cdot 2}$

Simplify each radical: $= \frac{1}{2}\cdot 4\sqrt{2} - 2\cdot 6\sqrt{2}$

$= 2\sqrt{2} - 12\sqrt{2}$

Add like radicals: $= -10\sqrt{2}$

Teaching Tip When these expressions are being simplified, remind students that they have a **product** of factors; that is,

$\frac{1}{2}\sqrt{32} = \frac{1}{2}\cdot\sqrt{32} = \frac{1}{2}\cdot 4\cdot\sqrt{2}$

QUICK ✓ *Add or subtract, as indicated.*

5. $\sqrt{50} + \sqrt{32}$ **6.** $\sqrt{48} - \sqrt{108}$ **7.** $\sqrt{45} + \sqrt{20} - \frac{1}{2}\sqrt{72}$

Classroom Example Subtract, assuming $x \geq 0$:

$3x\sqrt{20x} - 7\sqrt{5x^3}$

Answer: $-x\sqrt{5x}$

EXAMPLE 5 Subtracting Unlike Square Roots That Contain Variables

Assuming $x \geq 0$, find the difference: $x\sqrt{18} - 2\sqrt{128x^2}$

Solution

The radicands are different, so we simplify each square root and then combine like radicals.

9 and 64 are perfect squares: $x\sqrt{18} - 2\sqrt{128x^2} = x\sqrt{9\cdot 2} - 2\sqrt{64x^2\cdot 2}$

Simplify each radical: $= 3x\sqrt{2} - 2\cdot 8x\sqrt{2}$

$= 3x\sqrt{2} - 16x\sqrt{2}$

Distributive Property "in reverse": $= (3x - 16x)\sqrt{2}$

Combine like radicals: $= -13x\sqrt{2}$

QUICK ✓ *Add or subtract, as indicated.*

8. $\sqrt{48z} + \sqrt{108z} - \sqrt{12z}$ **9.** $\frac{1}{3}\sqrt{108a^3} - 2a\sqrt{75a}, a > 0$

6.3 Exercises

For Extra Help:

Student Solutions Manual

CD Video

PH Math/Tutor Center

MathXL Tutorials on CD

MathXL®

MyMathLab

Concepts and Vocabulary

In Problems 1–3, fill in the blanks.

1. To add or subtract square root expressions, the _______ must be the same.
2. $4\sqrt{3} + 7\sqrt{3} = (\_ + \_)\sqrt{3}$.
3. The coefficient of $\sqrt{2}$ is ________.

In Problems 4–6, answer True or False to each statement.

4. We say that square root expressions are like square roots if each square root has the same radicand.

1. radicands
2. 4, 7
3. 1
4. True

5. False 6. False
7. Answers may vary.
8. Answers may vary.
9. $2\sqrt{3}$ 10. $2\sqrt{2}$
11. $2a\sqrt{2a}$ 12. $2k\sqrt{11k}$
13. $\sqrt{15}+\sqrt{7}$
14. $\sqrt{3}+\sqrt{10}$
15. 0 16. 0
17. $-\sqrt{2}$
18. $-\sqrt{3}$
19. $-3\sqrt{6}-6\sqrt{11}$
20. $-11\sqrt{2}-13\sqrt{7}$
21. $2\sqrt{22}$
22. $-13\sqrt{30}$
23. $7\sqrt{14}$
24. $-2\sqrt{6}$
25. 0 26. 0
27. $30\sqrt{7}$
28. $33\sqrt{5}$
29. 0 30. 0
31. $-5\sqrt{y}$
32. $-3\sqrt{7}+4\sqrt{5}$
33. $\sqrt{6}+9\sqrt{5}$
34. $9\sqrt{x}$
35. $-13\sqrt{qr}$
36. $15\sqrt{xy}$
37. $9\sqrt{2x}-5\sqrt{2}$
38. $7\sqrt{3p}-4\sqrt{3}$
39. $-2n\sqrt{5n}$
40. $-9a\sqrt{2a}$
41. $-\sqrt{3}$
42. $-2\sqrt{2}$
43. $21\sqrt{15}$
44. $26\sqrt{3}$
45. $3\sqrt{5}$
46. $-15\sqrt{6}$
47. $18\sqrt{3}$
48. $8\sqrt{3}$
49. $11\sqrt{5}-4\sqrt{6}$
50. $-5\sqrt{7}-28$
51. $18\sqrt{2}-36\sqrt{3}$
52. $58\sqrt{3}+22\sqrt{2}$
53. $-\sqrt{6x}$ 54. $2\sqrt{5y}$
55. 0 56. 0
57. $5x\sqrt{5y}-5x\sqrt{5}$
58. $5y\sqrt{7x}-2y\sqrt{5x}$
59. $a^4\sqrt{10a}+6a\sqrt{3a}$
60. $n^4(9n^2-10)\sqrt{6}+4n^4\sqrt{6n}$
61. $7+13\sqrt{2}$ 62. $-5\sqrt{5}-1$
63. $-1+5\sqrt{3}$ 64. $2-4\sqrt{2}$
65. $\frac{7}{6}\sqrt{3}$ 66. $\frac{13}{12}\sqrt{5}$ 67. $-\frac{16}{5}\sqrt{5}$
68. $\frac{7}{15}\sqrt{3}$ 69. $\frac{9}{20}\sqrt{3}$ 70. $\frac{10}{21}\sqrt{2}$
71. $\frac{5}{6}\sqrt{15}$ 72. $\frac{11}{8}\sqrt{3}$ 73. $\frac{11}{3}\sqrt{6}$
74. $\frac{7}{3}\sqrt{3}$

5. $\sqrt{8}$ and $\sqrt{18}$ cannot be added because the radicands are different.
6. $\sqrt{5}+\sqrt{6}=\sqrt{5+6}$
7. Look at the problem below. Tell which terms are like terms and which terms are unlike terms. Then, tell if the problem has been correctly simplified. If not, explain how you would simplify the problem.
$$6-2\sqrt{x}-10+5\sqrt{x}=4\sqrt{x}-5\sqrt{x}=-\sqrt{x}$$
8. Give an example of two square roots that are like and two that are unlike. Make up a problem that requires adding or subtracting these four expressions. Explain how to simplify this problem.

Building Skills

In Problems 9–22, add or subtract the square root expressions and simplify, if possible.

9. $\sqrt{3}+\sqrt{3}$
10. $\sqrt{2}+\sqrt{2}$
11. $a\sqrt{2a}+a\sqrt{2a}$
12. $k\sqrt{11k}+k\sqrt{11k}$
13. $\sqrt{15}+\sqrt{7}$
14. $\sqrt{3}+\sqrt{10}$
15. $\sqrt{19}-\sqrt{19}$
16. $\sqrt{34}-\sqrt{34}$
17. $4\sqrt{2}-5\sqrt{2}$
18. $8\sqrt{3}-9\sqrt{3}$
19. $-3\sqrt{6}-6\sqrt{11}$
20. $-11\sqrt{2}-13\sqrt{7}$
21. $-\sqrt{22}-\left(-3\sqrt{22}\right)$
22. $-14\sqrt{30}-\left(-\sqrt{30}\right)$

Mixed Practice

In Problems 23–74, add or subtract, as indicated. Assume all variables represent positive real numbers.

23. $15\sqrt{14}-8\sqrt{14}$
24. $18\sqrt{6}-20\sqrt{6}$
25. $3\sqrt{2}-3\sqrt{2}$
26. $4\sqrt{5}-4\sqrt{5}$
27. $12\sqrt{7}+18\sqrt{7}$
28. $20\sqrt{5}+13\sqrt{5}$
29. $4\sqrt{6}-3\sqrt{6}-\sqrt{6}$
30. $5\sqrt{5}-\sqrt{5}-4\sqrt{5}$
31. $\sqrt{y}-6\sqrt{y}$
32. $9\sqrt{7}+4\sqrt{5}-12\sqrt{7}$
33. $15\sqrt{6}-14\sqrt{6}+9\sqrt{5}$
34. $8\sqrt{x}+\sqrt{x}$
35. $7\sqrt{qr}-20\sqrt{rq}$
36. $3\sqrt{yx}+12\sqrt{xy}$
37. $\sqrt{2x}+8\sqrt{2x}-5\sqrt{2}$
38. $13\sqrt{3p}-6\sqrt{3p}-4\sqrt{3}$
39. $2n\sqrt{5n}-4n\sqrt{5n}$
40. $6a\sqrt{2a}-15a\sqrt{2a}$
41. $\sqrt{12}-\sqrt{27}$
42. $\sqrt{8}-\sqrt{32}$
43. $6\sqrt{60}+3\sqrt{135}$
44. $2\sqrt{27}+5\sqrt{48}$
45. $3\sqrt{20}-3\sqrt{125}+4\sqrt{45}$
46. $-2\sqrt{96}+4\sqrt{24}-3\sqrt{150}$
47. $2\sqrt{75}+6\sqrt{12}-\sqrt{48}$
48. $-2\sqrt{243}+8\sqrt{48}-\sqrt{108}$
49. $4\sqrt{20}+\sqrt{45}-2\sqrt{24}$
50. $-3\sqrt{63}+2\sqrt{28}-4\sqrt{49}$
51. $5\sqrt{72}-4\sqrt{243}-\sqrt{288}$
52. $4\sqrt{300}+3\sqrt{108}+2\sqrt{242}$
53. $\sqrt{54x}-\sqrt{96x}$
54. $\sqrt{80y}-\sqrt{20y}$
55. $5\sqrt{32y^4}-4\sqrt{50y^4}$
56. $3\sqrt{12x^2}-2\sqrt{27x^2}$
57. $\sqrt{45x^2y}+\sqrt{20x^2y}-\sqrt{125x^2}$
58. $\sqrt{63xy^2}+\sqrt{28xy^2}-\sqrt{20xy^2}$
59. $2a^2\sqrt{40a^5}-a\sqrt{90a^7}+3\sqrt{12a^3}$
60. $3n^4\sqrt{54n^4}-2n\sqrt{150n^6}+2\sqrt{24n^9}$
61. $3+2\sqrt{8}+4+3\sqrt{18}$
62. $\sqrt{5}-5+4-3\sqrt{20}$
63. $-1-2\sqrt{27}+4\sqrt{48}-\sqrt{75}$
64. $2+3\sqrt{8}-\sqrt{32}-2\sqrt{18}$
65. $\frac{1}{2}\sqrt{3}+\frac{2}{3}\sqrt{3}$
66. $\frac{1}{4}\sqrt{5}+\frac{5}{6}\sqrt{5}$
67. $\frac{2}{5}\sqrt{20}-\frac{4}{3}\sqrt{45}$
68. $\frac{3}{5}\sqrt{27}-\frac{2}{3}\sqrt{12}$
69. $\sqrt{\frac{3}{25}}+\sqrt{\frac{3}{16}}$
70. $\sqrt{\frac{2}{9}}+\sqrt{\frac{2}{49}}$
71. $4\sqrt{\frac{15}{9}}-2\sqrt{\frac{15}{16}}$
72. $3\sqrt{\frac{3}{4}}-\frac{1}{2}\sqrt{\frac{3}{16}}$
73. $\frac{2}{3}\sqrt{\frac{18}{3}}+\frac{3}{2}\sqrt{\frac{120}{5}}$
74. $\frac{4}{3}\sqrt{\frac{18}{6}}+\frac{1}{2}\sqrt{\frac{72}{6}}$

75. $24\sqrt{3}$ cm
76. $48\sqrt{2}$ ft
77. $(18\sqrt{2} + 32\sqrt{3})$ inches
78. $48\sqrt{5}$ meters
79. 23.3 ft
80. 14.4 ft
81. $13 + 8\sqrt{3}$
82. $26\sqrt{2}$
83. Answers may vary.

Applying the Concepts

△ **75.** Find the perimeter of a square if it measures $3\sqrt{12}$ cm on each side.

△ **76.** Find the perimeter of a square if it measures $4\sqrt{18}$ ft on each side.

△ **77.** Find the perimeter of a rectangle if it measures $3\sqrt{18}$ inches on one side and $4\sqrt{48}$ inches on the other.

△ **78.** Find the perimeter of a rectangle if it measures $2\sqrt{45}$ meters on one side and $9\sqrt{20}$ meters on the other.

△ **79.** **Building a Shed** Alex is planning to strengthen the side of a wooden shed, which measures 10 feet long and 6 feet tall, by purchasing two metal bars and constructing a support in the shape of an X. If he buys one metal bar and cuts it into 2 pieces to make the X, what length of metal, to the nearest tenth of a foot, does he need to buy? [**Hint:** Use $c = \sqrt{a^2 + b^2}$, where $a = 10$ feet and $b = 6$ feet.]

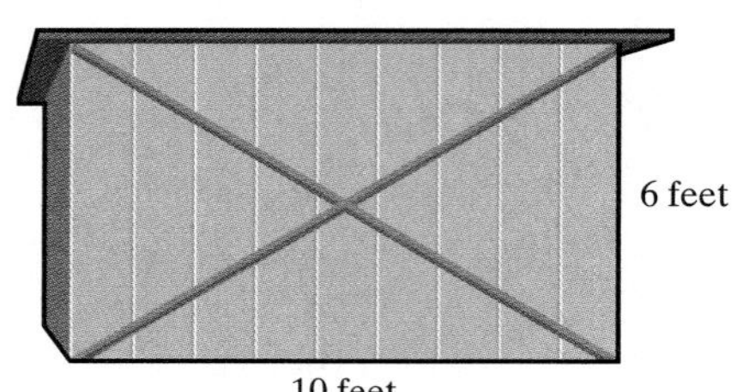

△ **80.** **Strengthening a Gate** Barry sees his neighbor, Alex (from Problem 79), working on his shed and thinks that he should use the same plan to strengthen his gate. If his gate measures 4 feet by 6 feet, how long, to the nearest tenth of a foot, is the metal bar that he needs to buy?

In Problems 81 and 82, find the perimeter of each figure.

81.

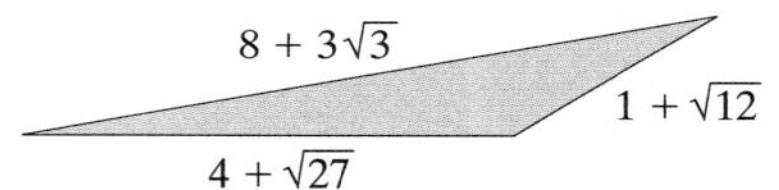

82.

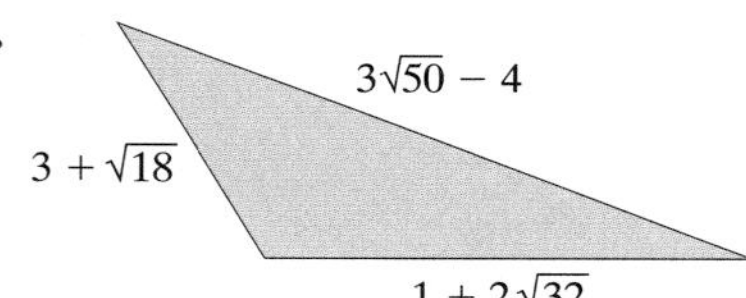

Extending the Concepts

The symbol indicating the square root of a number, $\sqrt{\ }$, did not become commonplace until the seventeenth century. In ancient times, Arab scholars thought of numbers that were perfect squares as growing out of a root in the ground. Later translations of Arabic work used the word radix, *meaning root, when dealing with these numbers. Today, mathematicians still use the phrase "extracting roots" when simplifying radicals.*

Over the centuries, many different symbols have been used to indicate the square root of a number. Scholars writing in Latin used ℞*, as it was derived from the word* radix. *During the sixteenth century, all of the following were used to indicate the square root of a number: ✓ √ √ √. The √ symbol appeared in the work of Christoff Rudolff in 1525 and is the basis of the square root symbol used today, although it took until the seventeenth century to become the standard notation.*

83. Math history has many interesting stories to tell. Go to the library or the Internet and research a person, notation, or concept. Some topics of interest might be the Pythagorean Theorem and the Pythagorean society, ancient numeration systems used by the Egyptians or Mayans, early algebra, prime numbers, or the origins of zero. Write a brief report on your findings.

6.4 Multiplying Expressions with Square Roots

OBJECTIVES

1. Find the Product of Square Roots Containing One Term
2. Find the Product of Square Roots Using the Distributive Property
3. Find the Product of Square Roots Using FOIL
4. Find the Product of Square Roots Using Special Products: $(a + b)^2$, $(a - b)^2$, and $(a + b)(a - b)$

Preparing for Multiplying Expressions with Square Roots

Before getting started, take the following readiness quiz. If you get a problem wrong, go to the section cited and review the material.

1. Find the product: $(3x)^2$ [Section 3.2, p. 187]

2. Find the product: $3(2x^2 - 4x + 1)$ [Section 3.3, pp. 190–191]

3. Find the product: $(2a + 3)(3a - 4)$ [Section 3.3, pp. 191–193]

4. Find the product: $(5b + 2)^2$ [Section 3.3, pp. 194–196]

5. Find the product: $(3y - 4)(3y + 4)$ [Section 3.3, p. 194]

In this section, you'll learn how to multiply square roots. You'll see that the techniques that you learned to find polynomial products are useful in finding products of square root expressions. Throughout the section assume that all variables are positive.

Teaching Tip

You may wish to do a quick review of multiplication examples with polynomials:

(a) $(2x)(5x) = 10x^2$

(b) $(7p)^2 = 7^2 \cdot p^2 = 49p^2$

(c) $5(z - 4) = 5z - 20$

(d) $(2x + 5)(x - 3) = 2x^2 - x - 15$

(e) $(2a - 3)^2 = 4a^2 - 12a + 9$

(f) $(3b - 4)(3b + 4) = 9b^2 - 16$

1 Find the Product of Square Roots Containing One Term

In Section 6.2, we learned how to use the Product Rule of Square Roots to simplify square roots. In this section, we will use the same property to find the product of square root expressions.

In Words

The Product Rule of Square Roots states that "the product of the square roots equals the square root of the product."

PRODUCT RULE OF SQUARE ROOTS

If $\sqrt{a}$ and $\sqrt{b}$ are real numbers, then

$$\sqrt{a} \cdot \sqrt{b} = \sqrt{ab}$$

EXAMPLE 1 Using the Product Rule to Multiply Square Roots

Find each product.

(a) $\sqrt{5} \cdot \sqrt{7}$ **(b)** $\sqrt{3} \cdot \sqrt{11}$ **(c)** $\sqrt{3} \cdot \sqrt{2a}$

Solution

(a) $\sqrt{5} \cdot \sqrt{7} = \sqrt{5 \cdot 7} = \sqrt{35}$

(b) $\sqrt{3} \cdot \sqrt{11} = \sqrt{3 \cdot 11} = \sqrt{33}$

(c) $\sqrt{3} \cdot \sqrt{2a} = \sqrt{3 \cdot 2a} = \sqrt{6a}$

Classroom Example Find each product:

(a) $\sqrt{3} \cdot \sqrt{13}$

(b) $\sqrt{7} \cdot \sqrt{2}$

(c) $\sqrt{5y} \cdot \sqrt{3}$

Answer:

(a) $\sqrt{39}$

(b) $\sqrt{14}$

(c) $\sqrt{15y}$

QUICK ✓ *Find each product.*

1. $\sqrt{3} \cdot \sqrt{5}$ **2.** $\sqrt{6} \cdot \sqrt{11}$ **3.** $\sqrt{5} \cdot \sqrt{6c}$ **4.** $\sqrt{2z} \cdot \sqrt{7}$

Sometimes after we multiply two radicands, the resulting square root can be simplified. Remember, a square root can be simplified if the radicand has a factor that is a perfect square. So, after multiplying, look for factors that are perfect squares.

EXAMPLE 2 Multiplying and Simplifying Square Root Expressions

Find the product and simplify, if possible.

(a) $\sqrt{3} \cdot \sqrt{12}$ **(b)** $\sqrt{6} \cdot \sqrt{15}$

Solution

(a) $\sqrt{a} \cdot \sqrt{b} = \sqrt{ab}$: $\sqrt{3} \cdot \sqrt{12} = \sqrt{3 \cdot 12} = \sqrt{36} = 6$

Classroom Example Find each product:

(a) $\sqrt{20} \cdot \sqrt{5}$ (b) $\sqrt{6} \cdot \sqrt{8}$

Answer:

(a) 10 (b) $4\sqrt{3}$

Preparing for...Answers **1.** $9x^2$ **2.** $6x^2 - 12x + 3$ **3.** $6a^2 + a - 12$ **4.** $25b^2 + 20b + 4$ **5.** $9y^2 - 16$

(b) $\sqrt{a}\cdot\sqrt{b} = \sqrt{ab}$: $\sqrt{6}\cdot\sqrt{15} = \sqrt{6\cdot 15}$
$= \sqrt{90}$
9 is largest factor of 90 that is a perfect square: $= \sqrt{9\cdot 10}$
$\sqrt{a\cdot b} = \sqrt{a}\cdot\sqrt{b}$: $= \sqrt{9}\cdot\sqrt{10}$
$= 3\sqrt{10}$

Classroom Example ➤
Find the product and simplify, if possible:
(a) $\sqrt{8z^3}\cdot\sqrt{3z}$
(b) $\sqrt{15k^5}\cdot\sqrt{5k^2}$
Answer:
(a) $2z^2\sqrt{6}$
(b) $5k^3\sqrt{3k}$

EXAMPLE 3 Multiplying and Simplifying Square Root Expressions Containing Variables

Find the product and simplify, if possible. Assume all variables represent nonnegative real numbers.

(a) $\sqrt{6p}\cdot\sqrt{2p^3}$ **(b)** $\sqrt{10m^2}\cdot\sqrt{5m^3}$

Solution

(a) $\sqrt{a}\cdot\sqrt{b} = \sqrt{ab}$: $\sqrt{6p}\cdot\sqrt{2p^3} = \sqrt{6p\cdot 2p^3}$
$= \sqrt{12p^4}$
4 is a perfect square factor of 12: $= \sqrt{4p^4\cdot 3}$
$\sqrt{a\cdot b} = \sqrt{a}\cdot\sqrt{b}$: $= \sqrt{4p^4}\cdot\sqrt{3}$
$\sqrt{4p^4} = 2p^2$: $= 2p^2\sqrt{3}$

(b) $\sqrt{a}\cdot\sqrt{b} = \sqrt{ab}$: $\sqrt{10m^2}\cdot\sqrt{5m^3} = \sqrt{50m^5}$
25 is the largest perfect square factor of 50: $= \sqrt{25m^4\cdot 2m}$
$\sqrt{a\cdot b} = \sqrt{a}\cdot\sqrt{b}$: $= \sqrt{25m^4}\cdot\sqrt{2m}$
$\sqrt{25} = 5$; $\sqrt{m^4} = m^2$: $= 5m^2\sqrt{2m}$

Classroom Example ➤
Multiply and simplify, if possible:
$\sqrt{8z^5}\cdot\sqrt{5z^6}$
Answer: $2z^5\sqrt{10z}$

EXAMPLE 4 Multiplying and Simplifying Expressions with Square Roots

Multiply and simplify, if possible: $\sqrt{2w^3}\cdot\sqrt{12w^6}$. Assume all variables represent nonnegative real numbers.

Solution

Sometimes more than one approach will lead to the correct solution.

METHOD 1:
$\sqrt{a}\cdot\sqrt{b} = \sqrt{ab}$: $\sqrt{2w^3}\cdot\sqrt{12w^6} = \sqrt{24w^9}$
$= \sqrt{4w^8\cdot 6w}$
$\sqrt{a\cdot b} = \sqrt{a}\cdot\sqrt{b}$: $= \sqrt{4w^8}\cdot\sqrt{6w}$
$\sqrt{4} = 2$; $\sqrt{w^8} = w^4$: $= 2w^4\sqrt{6w}$

METHOD 2:
Simplify first: $\sqrt{2w^3}\cdot\sqrt{12w^6} = \sqrt{w^2\cdot 2w}\cdot\sqrt{4w^6\cdot 3}$
$\sqrt{a\cdot b} = \sqrt{a}\cdot\sqrt{b}$: $= \sqrt{w^2}\cdot\sqrt{2w}\cdot\sqrt{4w^6}\cdot\sqrt{3}$
$\sqrt{w^2} = w$; $\sqrt{4w^6} = 2w^3$: $= w\sqrt{2w}\cdot 2w^3\sqrt{3}$
Multiply: $= 2w^4\sqrt{6w}$

Work Smart
Remember, for a radical to be simplified the exponent on the variable in the radicand must be less than 2.

Which approach do you like better?

QUICK ✓ *Find the product and simplify. Assume all variables represent nonnegative real numbers.*

5. $\sqrt{15b}\cdot\sqrt{5b}$ **6.** $\sqrt{24k}\cdot\sqrt{3k^3}$ **7.** $\sqrt{2z^3}\cdot\sqrt{10z^4}$

What happens when we square an expression containing a square root? Let's see.

Classroom Example ➤
Simplify:
(a) $(\sqrt{2})^2$ (b) $(-\sqrt{13})^2$
Answer:
(a) 2 (b) 13

EXAMPLE 5 Squaring a Square Root

Simplify:

(a) $(\sqrt{3})^2$ **(b)** $(-\sqrt{7})^2$

Work Smart
$(a)^2 = a \cdot a$ so $(\sqrt{a})^2 = \sqrt{a} \cdot \sqrt{a}$.

Solution

(a) $(\sqrt{3})^2 = \sqrt{3} \cdot \sqrt{3}$
$= \sqrt{9}$
$= 3$

(b) $(-\sqrt{7})^2 = (-\sqrt{7}) \cdot (-\sqrt{7})$
$= \sqrt{49}$
$= 7$

We can generalize the results of Example 5 as follows:

SQUARING A SQUARE ROOT

$(\sqrt{a})^2 = a$ and $(-\sqrt{a})^2 = a$ for any real number $a \geq 0$.

QUICK ✓ *Simplify each expression.*

8. $(\sqrt{11})^2$ **9.** $(-\sqrt{31})^2$

What if we have a product such as $(2\sqrt{5})(-4\sqrt{6})$? We can find this product using the Commutative Property of Multiplication, which allows us to multiply in any order without changing the product. We used this property when finding products of monomials. For example, $(2x)(-4y) = 2 \cdot (-4) \cdot x \cdot y = -8xy$, so

$$(2\sqrt{5})(-4\sqrt{6}) = 2 \cdot (-4) \cdot \sqrt{5} \cdot \sqrt{6} = -8\sqrt{30}$$

Classroom Example Find each product:
(a) $(2\sqrt{7})(5\sqrt{7})$
(b) $(3\sqrt{5})(-2\sqrt{10})$

Answer:
(a) 70 (b) $-30\sqrt{2}$

EXAMPLE 6 Multiplying Square Root Expressions

Find each product:

(a) $(3\sqrt{5})(2\sqrt{5})$ **(b)** $(7\sqrt{2})(-4\sqrt{12})$

Solution

(a) Rearrange factors: $(3\sqrt{5})(2\sqrt{5}) = 3 \cdot 2 \cdot \sqrt{5} \cdot \sqrt{5}$
$= 6 \cdot (\sqrt{5})^2$
$(\sqrt{a})^2 = a$: $= 6 \cdot 5$
$= 30$

(b) Rearrange factors: $(7\sqrt{2})(-4\sqrt{12}) = 7 \cdot -4 \cdot \sqrt{2} \cdot \sqrt{12}$
$\sqrt{a} \cdot \sqrt{b} = \sqrt{a \cdot b}$: $= -28(\sqrt{24})$
4 is the largest factor of 24 that is a perfect square: $= -28\sqrt{4 \cdot 6}$
$\sqrt{4 \cdot 6} = \sqrt{4} \cdot \sqrt{6} = 2\sqrt{6}$: $= -28 \cdot 2\sqrt{6}$
$= -56\sqrt{6}$

QUICK ✓ *Find the product.*

10. $(7\sqrt{3})(\sqrt{3})$ **11.** $(\sqrt{12})(-4\sqrt{3})$ **12.** $(2\sqrt{5})(-3\sqrt{10})$ **13.** $(-2\sqrt{11})^2$

2 Find the Product of Square Roots Using the Distributive Property

We just multiplied square root expressions in which one square root was multiplied by a second square root. Now we concentrate on multiplying square root expressions using the Distributive Property.

Classroom Example
Multiply and simplify:
(a) $3(2 + \sqrt{7})$
(b) $-\sqrt{3}(4 - 2\sqrt{3})$
Answer:
(a) $6 + 3\sqrt{7}$
(b) $-4\sqrt{3} + 6$

Work Smart
$8 + 2\sqrt{3} \neq 10\sqrt{3}$
Remember the order of operations—multiplication before addition.

EXAMPLE 7 Multiplying Square Roots Using the Distributive Property

Multiply and simplify:

(a) $2(4 + \sqrt{3})$ **(b)** $-\sqrt{2}(4 - 3\sqrt{2})$

Solution

(a) We use the Distributive Property and multiply each term in the parentheses by 2.

$$2(4 + \sqrt{3}) = 2 \cdot 4 + 2 \cdot \sqrt{3}$$
Multiply: $= 8 + 2\sqrt{3}$

(b) We use the Distributive Property and multiply each term in parentheses by $-\sqrt{2}$.

$$-\sqrt{2}(4 - 3\sqrt{2}) = -\sqrt{2} \cdot 4 - (-\sqrt{2})(3\sqrt{2})$$
$$= -4\sqrt{2} + 3 \cdot \sqrt{2} \cdot \sqrt{2}$$
Multiply radicals: $= -4\sqrt{2} + 3 \cdot 2$
$$= -4\sqrt{2} + 6$$

QUICK ✓ *Find the product.*

14. $-3(5 - \sqrt{7})$ **15.** $\sqrt{5}(3 - \sqrt{5})$
16. $\sqrt{2}(-4 + \sqrt{10})$ **17.** $-\sqrt{8}(\sqrt{2} - \sqrt{6})$

3 Find the Product of Square Roots Using FOIL

When we multiplied polynomials such as $(2x + 1)(x - 4)$, we used the **F**irst, **O**uter, **I**nner, **L**ast (FOIL) method as follows:

First Last Inner Outer

F O I L
$$(2x + 1)(x - 4) = 2x \cdot x + 2x \cdot (-4) + 1 \cdot x + 1 \cdot (-4)$$
$$= 2x^2 - 8x + x - 4$$
$$= 2x^2 - 7x - 4$$

We'll use the same process to multiply square root expressions such as $(7 + \sqrt{2})(6 + \sqrt{2})$.

Classroom Example
Find each product:
(a) $(5 + \sqrt{3})(2 + \sqrt{3})$
(b) $(7 - \sqrt{5w})(2 - \sqrt{5w})$
(c) $(3 - \sqrt{7})(4 + \sqrt{7})$
Answer:
(a) $13 + 7\sqrt{3}$
(b) $14 - 9\sqrt{5w} + 5w$
(c) $5 - \sqrt{7}$

EXAMPLE 8 Multiplying Square Roots Using FOIL

Find each product. Assume all variables represent nonnegative real numbers.

(a) $(7 + \sqrt{2})(6 + \sqrt{2})$ **(b)** $(3 - \sqrt{2a})(5 - \sqrt{2a})$

Solution

(a) We use the FOIL pattern to multiply.

F O I L
$$(7 + \sqrt{2})(6 + \sqrt{2}) = 7 \cdot 6 + 7\sqrt{2} + 6\sqrt{2} + \sqrt{2} \cdot \sqrt{2}$$
Simplify: $= 42 + 7\sqrt{2} + 6\sqrt{2} + 2$
Combine like terms: $= 44 + 13\sqrt{2}$

(b)

$$\begin{aligned} & \quad\ \text{F} \qquad\quad \text{O} \qquad\quad \text{I} \qquad\quad \text{L} \\ \left(3-\sqrt{2a}\right)\left(5-\sqrt{2a}\right) &= 3\cdot 5 - 3\left(\sqrt{2a}\right) - \sqrt{2a}\cdot 5 - \sqrt{2a}\left(-\sqrt{2a}\right) \\ \text{Multiply:} \quad &= 15 - 3\sqrt{2a} - 5\sqrt{2a} + \sqrt{4a^2} \\ \text{Combine like terms:} \quad &= 15 - 8\sqrt{2a} + 2a \end{aligned}$$

QUICK ✓ *Find the product.*

18. $\left(3+\sqrt{7}\right)\left(4+\sqrt{7}\right)$ **19.** $\left(7-\sqrt{5n}\right)\left(6+\sqrt{5n}\right)$ **20.** $\left(\sqrt{3}-1\right)\left(\sqrt{3}+5\right)$

Classroom Example ➤ Find the product: $\left(4+5\sqrt{2}\right)\left(2-7\sqrt{2}\right)$

Answer: $-62-18\sqrt{2}$

EXAMPLE 9 Multiplying Square Roots Using FOIL

Find the product: $\left(9+5\sqrt{10}\right)\left(1-3\sqrt{10}\right)$

Solution

Once again, use the FOIL method to multiply.

$$\begin{aligned} \left(9+5\sqrt{10}\right)\left(1-3\sqrt{10}\right) &= 9\cdot 1 - 9\left(3\sqrt{10}\right) + 5\sqrt{10}\cdot 1 + \left(5\sqrt{10}\right)\left(-3\sqrt{10}\right) \\ \text{Multiply:} \quad &= 9 - 27\sqrt{10} + 5\sqrt{10} - 15\left(\sqrt{10}\right)^2 \\ \text{Combine like terms:} \quad &= 9 - 22\sqrt{10} - 15\cdot 10 \\ \text{Multiply:} \quad &= 9 - 22\sqrt{10} - 150 \\ \text{Combine like terms:} \quad &= -141 - 22\sqrt{10} \end{aligned}$$

QUICK ✓ *Find the product.*

21. $\left(4-2\sqrt{3}\right)\left(3+5\sqrt{3}\right)$ **22.** $\left(2-3\sqrt{7}\right)\left(5+2\sqrt{7}\right)$

---

4 Find the Product of Square Roots Using Special Products: $(a+b)^2$, $(a-b)^2$, and $(a+b)(a-b)$

Teaching Tip
Continue to stress that the process for multiplying square roots parallels that for multiplying polynomials.

We can use our special products formulas first introduced in Section 3.3 to multiply radicals as well as polynomials. First, we are going to use the formulas for perfect square trinomials,

$$(a+b)^2 = a^2 + 2ab + b^2 \qquad \text{and} \qquad (a-b)^2 = a^2 - 2ab + b^2$$

Classroom Example ➤ **Find each product:**
(a) $\left(2+\sqrt{3}\right)^2$
(b) $\left(7-3\sqrt{2x}\right)^2$

Answer:
(a) $7+4\sqrt{3}$
(b) $49-42\sqrt{2x}+18x$

EXAMPLE 10 Multiplying Square Roots Using Perfect Square Trinomials

Find each product. Assume all variables represent nonnegative real numbers.

(a) $\left(3+\sqrt{5}\right)^2$ **(b)** $\left(4-2\sqrt{3y}\right)^2$

Solution

(a) We'll use $(a+b)^2 = a^2 + 2ab + b^2$ to find the product.

$$\begin{aligned} (a\ +\ b)^2 &= a^2 + 2\cdot a\cdot\ b\ +\ b^2 \\ \left(3+\sqrt{5}\right)^2 &= 3^2 + 2\cdot 3\cdot\sqrt{5} + \left(\sqrt{5}\right)^2 \\ &= 9 + 6\sqrt{5} + 5 \\ \text{Combine like terms:} \quad &= 14 + 6\sqrt{5} \end{aligned}$$

(b) We use $(a-b)^2 = a^2 - 2ab + b^2$ to find this product.

$$\begin{aligned} (a\ -\ b)^2 &= a^2 - 2\cdot a\cdot\ b\ +\ b^2 \\ \left(4-2\sqrt{3y}\right)^2 &= 4^2 - 2\cdot 4\cdot 2\sqrt{3y} + \left(2\sqrt{3y}\right)^2 \\ \left(2\sqrt{3y}\right)^2 = 2^2\left(\sqrt{3y}\right)^2 = 4\cdot 3y = 12y: \quad &= 16 - 16\sqrt{3y} + 12y \end{aligned}$$

QUICK ✓ *Find the product.*

23. $(8 + \sqrt{3})^2$ **24.** $(2 - \sqrt{7z})^2$ **25.** $(2 + 3\sqrt{5})^2$ **26.** $(-1 - 5\sqrt{2})^2$

Now let's look at products in the form of the Difference of Two Squares,

$$(a + b)(a - b) = a^2 - b^2$$

Classroom Example ➤
Find each product:
(a) $(8 - \sqrt{5})(8 + \sqrt{5})$
(b) $(4 + 3\sqrt{7})(4 - 3\sqrt{7})$
Answer:
(a) 59 (b) −47

EXAMPLE 11 Multiplying Square Roots Using the Difference of Two Squares

Find each product:

(a) $(5 + \sqrt{6})(5 - \sqrt{6})$ **(b)** $(2 - 3\sqrt{5})(2 + 3\sqrt{5})$

Solution

(a) First, notice that $(5 + \sqrt{6})(5 - \sqrt{6})$ is in the form $(a + b)(a - b)$.

$$\begin{aligned} (a + b)(a - b) &= a^2 - b^2 \\ (5 + \sqrt{6})(5 - \sqrt{6}) &= 5^2 - (\sqrt{6})^2 \\ &= 25 - 6 \\ &= 19 \end{aligned}$$

(b)

$$\begin{aligned} (a - b)(a + b) &= a^2 - b^2 \\ (2 - 3\sqrt{5})(2 + 3\sqrt{5}) &= 2^2 - (3\sqrt{5})^2 \\ (3\sqrt{5})^2 = 3^2 \cdot (\sqrt{5})^2 = 9 \cdot 5 = 45: \quad &= 4 - 45 \\ &= -41 \end{aligned}$$

Notice that the products found in Example 11 are integers. That is, there are no square roots in the product. Square root expressions such as $5 + \sqrt{6}$ and $5 - \sqrt{6}$ are called **conjugates.** When we multiply square root expressions that are conjugates, the result will never contain a square root. This result plays an important role in the next section.

QUICK ✓ *Find the product. Assume all variables represent nonnegative real numbers.*

27. $(6 + \sqrt{3})(6 - \sqrt{3})$ **28.** $(\sqrt{2} - 7)(\sqrt{2} + 7)$ **29.** $(1 + 4\sqrt{5n})(1 - 4\sqrt{5n})$

6.4 Exercises

For Extra Help: Student Solutions Manual  CD Video  PH Math/Tutor Center  MathXL Tutorials on CD · MathXL® · MyMathLab

Concepts and Vocabulary

In Problems 1–3, fill in the blanks.

1. To multiply $(3 - \sqrt{2})(\sqrt{3} + 5)$, we use the ________ method.
2. If a is a positive number, $(\sqrt{a})^2 =$ ________.
3. To multiply $4(3\sqrt{x} + 2)$, use the ________ Property.

In Problems 4–6, answer True or False to each statement.

4. If $\sqrt{a}$ and $\sqrt{b}$ are real numbers, then $\sqrt{a} \cdot \sqrt{b} = \sqrt{ab}$.
5. $(\sqrt{a})^2 = a$ and $(-\sqrt{a})^2 = a$ for any real number a.
6. $(\sqrt{a} + \sqrt{b})^2 = a + b$.

1. FOIL
2. a
3. Distributive
4. True
5. False
6. False

7. It is true that $(\sqrt{-4})^2 = -4$. However, we cannot use $(\sqrt{a})^2 = a$ to obtain this result because this property requires that $a \geq 0$. In addition, we cannot simplify $(\sqrt{-4})^2$ using $(\sqrt{a})^2 = \sqrt{a}\cdot\sqrt{a}$ because this also requires that $a \geq 0$. The student said that $(\sqrt{-4})^2 = -4$ is correct, but for the wrong reasons.
8. Answers may vary.
9. 3 10. 2 11. $7x$ 12. $5n$
13. $\sqrt{39}$
14. $\sqrt{10}$
15. $\sqrt{14a}$
16. $\sqrt{15x}$
17. $-2\sqrt{15}$
18. $-5\sqrt{6}$
19. $24\sqrt{3}$
20. $18\sqrt{15}$
21. $-9x^2$
22. $-10a^4$
23. $72\sqrt{6}$
24. $4q^3\sqrt{3q}$
25. $270\sqrt{6}$
26. $240\sqrt{3}$
27. $-45\sqrt{6}$
28. $-32\sqrt{15}$
29. $32x^2\sqrt{3x}$
30. $300n\sqrt{3n}$
31. $42y^7\sqrt{3}$
32. $30z^7\sqrt{7}$
33. 37
34. 42
35. $7x$
36. $6y$
37. $24 + 8\sqrt{5}$
38. $21\sqrt{2} - 15$
39. $12 - 3\sqrt{6}$
40. $4\sqrt{2} + 24$
41. $21\sqrt{5} + 4\sqrt{3}$
42. $5\sqrt{5} + 15\sqrt{3}$
43. $a\sqrt{2} + 2\sqrt{a}$
44. $b\sqrt{3} - 3\sqrt{b}$
45. $33 + 11\sqrt{3}$
46. $58 - 15\sqrt{2}$
47. $3 - 14\sqrt{x} + 8x$
48. $14 + 37\sqrt{y} + 5y$
49. $20 - 15\sqrt{2} + 8\sqrt{3} - 6\sqrt{6}$
50. $14 - 21\sqrt{2} + 6\sqrt{3} - 9\sqrt{6}$
51. $8x\sqrt{3} - 7x$
52. $-4x + 15x\sqrt{2}$
53. $19 + 8\sqrt{3}$
54. $13 - 6\sqrt{2}$
55. $9x - 12\sqrt{x} + 4$
56. $4x + 12\sqrt{x} + 9$
57. 1
58. -29
59. $x - 9y$
60. $160x - y$
61. $3m^3\sqrt{5}$
62. $6n^2\sqrt{3}$
63. $-5\sqrt{2}$
64. $21\sqrt{2} + 6\sqrt{3}$
65. $12\sqrt{2n}$
66. 0

7. A student in your class used the Property for Squaring Square Roots on page 415 to obtain the following result $(\sqrt{-4})^2 = -4$. You believe the correct answer is $\sqrt{-4}\cdot\sqrt{-4} = \sqrt{(-4)(-4)} = \sqrt{16} = 4$. Is either answer correct? Explain any limitations that need to be included for this problem and the use of this property.

8. Explain how to multiply each of the following problems. Tell what possible errors students might make when doing problems of this type.
(a) $2\cdot 3\sqrt{5}$ **(b)** $\sqrt{8}\cdot 3\sqrt{2}$

Building Skills

In Problems 9–36, find the product and simplify, if possible. Assume all variables represent nonnegative real numbers.

9. $\sqrt{3}\cdot\sqrt{3}$ **10.** $\sqrt{2}\cdot\sqrt{2}$ **11.** $\sqrt{7x}\cdot\sqrt{7x}$ **12.** $\sqrt{5n}\cdot\sqrt{5n}$
13. $\sqrt{3}\cdot\sqrt{13}$ **14.** $\sqrt{2}\cdot\sqrt{5}$ **15.** $\sqrt{2}\cdot\sqrt{7a}$ **16.** $\sqrt{5x}\cdot\sqrt{3}$
17. $-\sqrt{10}\cdot\sqrt{6}$ **18.** $-\sqrt{10}\cdot\sqrt{15}$ **19.** $\sqrt{24}\cdot\sqrt{72}$ **20.** $\sqrt{90}\cdot\sqrt{54}$
21. $-\sqrt{3x}\cdot\sqrt{27x^3}$ **22.** $-\sqrt{2a^2}\cdot\sqrt{50a^6}$ **23.** $3\sqrt{12}\cdot 4\sqrt{18}$
24. $\sqrt{6q^3}\cdot\sqrt{8q^4}$ **25.** $5\sqrt{12}\cdot 9\sqrt{18}$ **26.** $8\sqrt{15}\cdot 3\sqrt{20}$
27. $-\sqrt{45}\cdot 3\sqrt{30}$ **28.** $-4\sqrt{24}\cdot\sqrt{40}$ **29.** $2\sqrt{6x}\cdot 4\sqrt{8x^4}$
30. $10\sqrt{6n^2}\cdot 5\sqrt{18n}$ **31.** $7y\sqrt{2y^6}\cdot 3y^2\sqrt{6y^2}$ **32.** $2z^3\sqrt{3z^4}\cdot 5z\sqrt{21z^2}$
33. $(\sqrt{37})^2$ **34.** $(-\sqrt{42})^2$ **35.** $(-\sqrt{7x})^2$ **36.** $(\sqrt{6y})^2$

In Problems 37–60, find the product and simplify, if possible. Assume all variables represent nonnegative real numbers.

37. $4(6 + 2\sqrt{5})$ **38.** $3(7\sqrt{2} - 5)$
39. $\sqrt{6}(2\sqrt{6} - 3)$ **40.** $\sqrt{8}(2 + 3\sqrt{8})$
41. $\sqrt{3}(7\sqrt{15} + 4)$ **42.** $\sqrt{5}(5 + 3\sqrt{15})$
43. $\sqrt{a}(\sqrt{2a} + 2)$ **44.** $\sqrt{b}(\sqrt{3b} - 3)$
45. $(6 + \sqrt{3})(5 + \sqrt{3})$ **46.** $(8 - \sqrt{2})(7 - \sqrt{2})$
47. $(3 - 2\sqrt{x})(1 - 4\sqrt{x})$ **48.** $(7 + \sqrt{y})(2 + 5\sqrt{y})$
49. $(5 + 2\sqrt{3})(4 - 3\sqrt{2})$ **50.** $(7 + 3\sqrt{3})(2 - 3\sqrt{2})$
51. $(2\sqrt{3x} + \sqrt{x})(5\sqrt{x} - 2\sqrt{3x})$ **52.** $(7\sqrt{x} - 3\sqrt{2x})(2\sqrt{x} + 3\sqrt{2x})$
53. $(4 + \sqrt{3})^2$ **54.** $(3 - \sqrt{2})^2$
55. $(3\sqrt{x} - 2)^2$ **56.** $(2\sqrt{x} + 3)^2$
57. $(3 + 2\sqrt{2})(3 - 2\sqrt{2})$ **58.** $(4 - 3\sqrt{5})(4 + 3\sqrt{5})$
59. $(\sqrt{x} - 3\sqrt{y})(\sqrt{x} + 3\sqrt{y})$ **60.** $(4\sqrt{10x} + \sqrt{y})(4\sqrt{10x} - \sqrt{y})$

Mixed Practice

In Problems 61–90, perform the indicated operation and simplify, if possible. Assume all variables represent nonnegative real numbers.

61. $\sqrt{15m^5}\cdot\sqrt{3m}$ **62.** $\sqrt{18n^2}\cdot\sqrt{6n^2}$
63. $3\sqrt{2} - 4\sqrt{8}$ **64.** $7\sqrt{18} + 6\sqrt{3}$
65. $3(2\sqrt{2n} + \sqrt{8n})$ **66.** $5(\sqrt{27x} - 3\sqrt{3x})$

67. 32
68. 45
69. $a + 2\sqrt{ab} + b$
70. $p - 2\sqrt{pq} + q$
71. $27 + 2\sqrt{30}$
72. $200 - 180\sqrt{3}$
73. $m\sqrt{2} - m$
74. $x\sqrt{3} - x\sqrt{6}$
75. $1 + 2\sqrt{6}$
76. $1 - 2\sqrt{2}$
77. $-\dfrac{\sqrt{3}}{12}$
78. $\dfrac{8}{45}\sqrt{5}$
79. $9s - 12\sqrt{s} + 4$
80. $25 + 30\sqrt{t} + 9t$
81. -120
82. -756
83. $6a^2\sqrt{a}$
84. $4b^2\sqrt{b}$
85. $-\dfrac{9}{2}\sqrt{3}$
86. $\dfrac{7}{3}\sqrt{6}$
87. 4
88. -6
89. 9
90. 8
91. $126\sqrt{3}$ square units
92. $65\sqrt{2}$ square units
93. $112\sqrt{7}$ square units
94. $390\sqrt{2}$ square units
95. $(12 + 18\sqrt{3})$ square units
96. $\left(\dfrac{15}{2} + \dfrac{55}{2}\sqrt{2}\right)$ square units
97. 147π square meters
98. 864π square yards
99. 126π square inches
100. 84π square centimeters
101. $12\sqrt{30}$ square units
102. $8\sqrt{110}$ square units
103. $50\sqrt{14}$ square units
104. $24\sqrt{570}$ square units

67. $(4\sqrt{2})^2$

68. $(3\sqrt{5})^2$

69. $(\sqrt{a} + \sqrt{b})^2$

70. $(\sqrt{p} - \sqrt{q})^2$

71. $(3\sqrt{5} - 2\sqrt{6})(5\sqrt{5} + 4\sqrt{6})$

72. $(6\sqrt{15} + 2\sqrt{5})(3\sqrt{15} - 7\sqrt{5})$

73. $\sqrt{m}(\sqrt{2m} - \sqrt{m})$

74. $\sqrt{x}(\sqrt{3x} + \sqrt{6x})$

75. $\dfrac{2 + \sqrt{96}}{2}$

76. $\dfrac{3 - \sqrt{72}}{3}$

77. $\dfrac{\sqrt{3}}{6} - \dfrac{\sqrt{3}}{4}$

78. $\dfrac{\sqrt{5}}{9} + \dfrac{\sqrt{5}}{15}$

79. $(3\sqrt{s} - 2)^2$

80. $(5 + 3\sqrt{t})^2$

81. $-3\sqrt{5}\cdot 4\sqrt{20}$

82. $-6\sqrt{28}\cdot 3\sqrt{63}$

83. $\sqrt{12a}\cdot\sqrt{3a^4}$

84. $\sqrt{8b^2}\cdot\sqrt{2b^3}$

85. $\dfrac{\sqrt{27}}{2} - 3\sqrt{12}$

86. $\sqrt{54} - \dfrac{\sqrt{24}}{3}$

87. $(3 - \sqrt{5})(3 + \sqrt{5})$

88. $(2 + \sqrt{10})(2 - \sqrt{10})$

89. $\sqrt{3}(\sqrt{3} + \sqrt{12})$

90. $\sqrt{2}(\sqrt{18} + \sqrt{2})$

Applying the Concepts

The area A of a trapezoid is calculated with the formula $A = \frac{1}{2}h(B + b)$, *where h is the height, B is the length of one base, and b is the length of the other base see the figure to the right. In Problems 91–96, find the area of each trapezoid.*

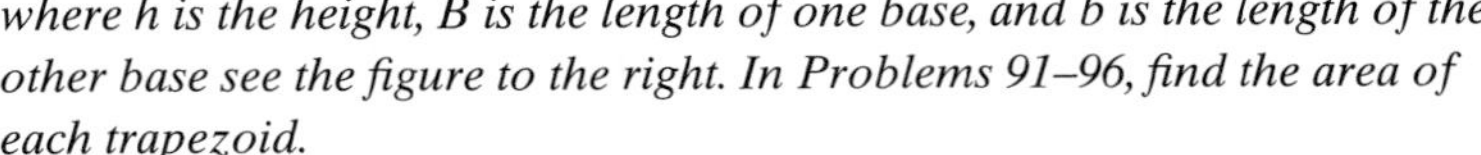

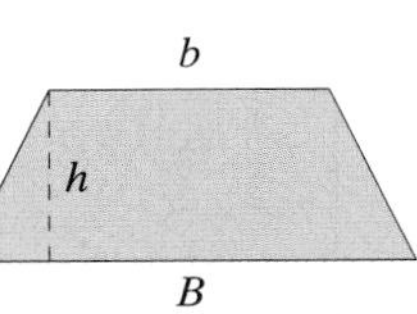

△ **91.** The height is $3\sqrt{6}$ and the bases have lengths $3\sqrt{32}$ and $10\sqrt{18}$.

△ **92.** The height is $2\sqrt{10}$ and the bases have lengths $3\sqrt{45}$ and $2\sqrt{20}$.

△ **93.** The height is $2\sqrt{14}$ and the bases have lengths $4\sqrt{128}$ and $6\sqrt{32}$.

△ **94.** The height is $5\sqrt{6}$ and the bases have lengths $4\sqrt{75}$ and $8\sqrt{48}$.

△ **95.** B is $\sqrt{27}$, h is 4, and b is $6 + 3\sqrt{12}$.

△ **96.** B is $\sqrt{98}$, h is 5, and b is $3 + 2\sqrt{8}$.

In Problems 97–100, use the formula $A = \pi r^2$ *to find the exact area of each circle. (Write your answer as an expression containing* π*.)*

△ **97.** Find the area of a circle if its radius is $7\sqrt{3}$ meters.

△ **98.** Find the area of a circle if its radius is $12\sqrt{6}$ yards.

△ **99.** Find the area of a circle if its diameter is $3\sqrt{56}$ inches.

△ **100.** Find the area of a circle if its diameter is $2\sqrt{84}$ centimeters.

Extending the Concepts

Heron's Formula, $A = \sqrt{s(s - a)(s - b)(s - c)}$, *can be used to find the area of a triangle when the lengths of the three sides are known. See the figure to the right. In this formula, a, b, and c are the lengths of the three sides of the triangle, and s represents the semiperimeter* $\left(s = \frac{1}{2}(a + b + c)\right)$. *In Problems 101–104, use Heron's Formula to find the exact area of each triangle. (This formula has always been credited to Heron of Alexandria, but it is possible that it is actually the work of Archimedes.)*

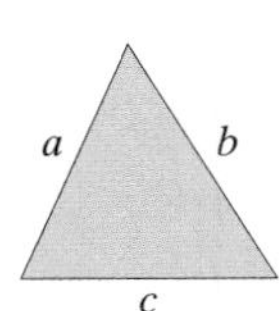

△ **101.** Find the area of a triangle with sides of lengths 11, 12, and 17.

△ **102.** Find the area of a triangle with sides of lengths 12, 21, and 31.

△ **103.** Find the area of a triangle with sides of lengths 15, 25, and 30.

△ **104.** Find the area of a triangle with sides of lengths 34, 42, and 68.

6.5 Dividing Expressions with Square Roots

OBJECTIVES

1. Find the Quotient of Two Square Roots
2. Rationalize a Denominator Containing One Term
3. Rationalize a Denominator Containing Two Terms

Preparing for Dividing Expressions with Square Roots

Before getting started, take the following readiness quiz. If you get a problem wrong, go to the section cited and review the material.

1. Find the quotient: $\dfrac{55x^8}{11x}$ [Section 3.4, pp. 201–202]

2. Find the quotient: $\sqrt{\dfrac{45x^7}{5x^3}}$ [Section 6.2, pp. 404–405]

Throughout this section, we assume that all variables represent positive real numbers.

1 Find the Quotient of Two Square Roots

Recall from Section 6.2 that we have two requirements for a square root to be simplified.

1. The radicand cannot contain any factors that are perfect squares.
2. The radicand cannot contain any fractions.

For example, neither $\sqrt{\dfrac{50x^6}{4x^4}}$ or $\sqrt{\dfrac{50x^3}{2x}}$ are simplified. We simplified expressions such as these using two methods back in Section 6.2.

Teaching Tip
Demonstrate Methods 1 and 2 in class to help students distinguish between the two approaches. For example Method 1 can be used to simplify $\sqrt{\dfrac{10}{36x^2}}$, while Method 2 can be used to simplify $\sqrt{\dfrac{24a^5}{6a}}$.

METHOD 1:

Write the square root of the quotient as the quotient of square roots and simplify.

$$\sqrt{\frac{50x^6}{4x^4}} = \frac{\sqrt{50x^6}}{\sqrt{4x^4}} = \frac{\sqrt{25x^6 \cdot 2}}{\sqrt{4x^4}} = \frac{5x^3\sqrt{2}}{2x^2}$$

METHOD 2:

Divide first, then take the square root.

$$\sqrt{\frac{50x^3}{2x}} = \sqrt{\frac{50}{2} \cdot x^{3-1}} = \sqrt{25x^2} = \sqrt{25} \cdot \sqrt{x^2} = 5x$$

We used Method 1 in Examples 7 and 8 of Section 6.2. This option works well when the denominator of the radicand is already a perfect square. We used Method 2 in Example 9 of Section 6.2. This option works well if the quotient of the radicand results in the denominator of the radicand becoming a perfect square. In both methods, the problem began as a single square root.

We now introduce a third (and final) requirement for a square root to be simplified.

3. No square roots may appear as denominators in a fraction.

Work Smart
The Quotient Rule states that $\sqrt{\dfrac{a}{b}} = \dfrac{\sqrt{a}}{\sqrt{b}}$.

When square roots appear in denominators $\left(\text{as in } \dfrac{\sqrt{18}}{\sqrt{50}} \text{ or } \dfrac{\sqrt{69x}}{\sqrt{12x}}\right)$, we can use the Quotient Rule "in reverse" and write the quotient of two square roots as the square root of a quotient, as in $\dfrac{\sqrt{a}}{\sqrt{b}} = \sqrt{\dfrac{a}{b}}$.

Classroom Example
Simplify:

(a) $\dfrac{\sqrt{20}}{\sqrt{45}}$ (b) $\dfrac{\sqrt{93z}}{\sqrt{27z}}$

Answer:

(a) $\dfrac{2}{3}$ (b) $\dfrac{\sqrt{31}}{3}$

EXAMPLE 1 Simplifying Quotients of Square Roots

Simplify:

(a) $\dfrac{\sqrt{18}}{\sqrt{50}}$ **(b)** $\dfrac{\sqrt{69x}}{\sqrt{12x}}$

Preparing for...Answers **1.** $5x^7$ **2.** $3x^2$

Solution

(a) The denominator contains a square root. We will use the Quotient Rule to simplify the expression.

$$\frac{\sqrt{a}}{\sqrt{b}} = \sqrt{\frac{a}{b}}: \quad \frac{\sqrt{18}}{\sqrt{50}} = \sqrt{\frac{18}{50}} = \sqrt{\frac{9 \cdot \cancel{2}}{25 \cdot \cancel{2}}}$$

$$= \sqrt{\frac{9}{25}} = \frac{3}{5}$$

(b) The denominator contains a square root. We will use the Quotient Rule to rewrite the expression.

$$\frac{\sqrt{69x}}{\sqrt{12x}} = \sqrt{\frac{69x}{12x}}$$

$$= \sqrt{\frac{23 \cdot \cancel{3} \cdot \cancel{x}}{4 \cdot \cancel{3} \cdot \cancel{x}}}$$

$$= \sqrt{\frac{23}{4}}$$

$$= \frac{\sqrt{23}}{\sqrt{4}}$$

$$= \frac{\sqrt{23}}{2}$$

QUICK ✓ *Simplify the quotient.*

1. $\dfrac{\sqrt{108}}{\sqrt{12}}$ **2.** $\dfrac{\sqrt{35y}}{\sqrt{5y}}$ **3.** $\dfrac{\sqrt{32}}{\sqrt{50}}$

Classroom Example Find the quotient:

$$\frac{\sqrt{27} - \sqrt{18}}{\sqrt{3}}$$

Answer: $3 - \sqrt{6}$

EXAMPLE 2 Dividing Square Root Expressions with More than One Term in the Numerator

Find the quotient: $\dfrac{\sqrt{8} - \sqrt{12}}{\sqrt{2}}$

Solution

We see that we have two terms in the numerator and a single term in the denominator, so we use the fact that $\dfrac{A + B}{C} = \dfrac{A}{C} + \dfrac{B}{C}$ to simplify.

$$\frac{A+B}{C} = \frac{A}{C} + \frac{B}{C}: \quad \frac{\sqrt{8} - \sqrt{12}}{\sqrt{2}} = \frac{\sqrt{8}}{\sqrt{2}} - \frac{\sqrt{12}}{\sqrt{2}}$$

$$\text{Quotient Rule:} \quad = \sqrt{\frac{8}{2}} - \sqrt{\frac{12}{2}}$$

$$\text{Divide:} \quad = \sqrt{4} - \sqrt{6}$$

$$= 2 - \sqrt{6}$$

QUICK ✓ *Find the quotient.*

4. $\dfrac{\sqrt{30} - \sqrt{75}}{\sqrt{5}}$

---

2 Rationalize a Denominator Containing One Term

Can you answer this question? Under what circumstances does a square root simplify to a rational number? Before reading on, try to answer the question. Here's the answer: If the radicand of a square root is a perfect square, the square root can be written as a rational number. For example, because 4 is a perfect square, we can write $\sqrt{4} = 2$, a rational number; because $\frac{25}{36}$ is a perfect square, we can write $\sqrt{\frac{25}{36}} = \frac{5}{6}$, a rational number. However, because 11 is not a perfect square, $\sqrt{11}$ is an irrational number.

We stated earlier in this section that no square roots may appear as denominators in a fraction. In Example 1(a) of this section, we were asked to simplify $\frac{\sqrt{18}}{\sqrt{50}}$. Although 50 is not a perfect square, we were able to simplify the expression (that is, we were able to eliminate the square root from the denominator) using the Quotient Rule. However, this method does not always work. For example, if we were asked to simplify $\frac{3}{\sqrt{2}}$, we would not be able to use the approach of Example 1 to remove the square root from the denominator. We now present a method that can be used to make denominators rational numbers.

DEFINITION

The process of rewriting a quotient in which the denominator contains a square root that is irrational as an equivalent quotient in which the denominator is rational is called **rationalizing the denominator.**

To rationalize the denominator, we need to make the radicand in the denominator a perfect square. For example, to simplify $\frac{3}{\sqrt{2}}$ we must make the denominator, $\sqrt{2}$, a rational number. How do we do this? A little thought might lead you to say, "multiply $\sqrt{2}$ by $\sqrt{2}$." The result would be $\sqrt{4}$, which equals 2, a rational number! Be careful though, because if you multiply the denominator by $\sqrt{2}$, you must also multiply the numerator by $\sqrt{2}$.

In general, multiply the denominator and numerator by a square root such that the radicand in the denominator becomes a perfect square. Since we are multiplying both the numerator and denominator by the same value, we are multiplying the radical expression by 1.

Work Smart
Multiply the denominator and numerator by a square root such that the radicand in the denominator becomes a perfect square.

EXAMPLE 3 How to Rationalize a Denominator Containing a Square Root

Rationalize the denominator of the expression $\frac{1}{\sqrt{5}}$.

Classroom Example
Rationalize the denominator of the expression $\frac{1}{\sqrt{7}}$.

Answer: $\frac{\sqrt{7}}{7}$

Step-by-Step Solution

Step 1: Determine the number that you multiply the radicand in the denominator by to obtain a perfect square.

Multiply $\frac{1}{\sqrt{5}}$ by $1 = \frac{\sqrt{5}}{\sqrt{5}}$: $\frac{1}{\sqrt{5}} = \frac{1}{\sqrt{5}} \cdot \frac{\sqrt{5}}{\sqrt{5}}$

Step 2: Multiply numerators; multiply denominators.

$= \frac{\sqrt{5}}{\sqrt{25}}$

Step 3: Simplify.

$= \frac{\sqrt{5}}{5}$

Work Smart
The process of rationalizing the denominator is the same method we use to write equivalent fractions. For example, $\frac{2}{3} \cdot \frac{3}{3} = \frac{6}{9}$.

Classroom Example ➤
Rationalize the denominator of each expression:

(a) $\frac{-4}{3\sqrt{5}}$ (b) $\sqrt{\frac{6}{20}}$

Answer:

(a) $\frac{-4\sqrt{5}}{15}$ (b) $\frac{\sqrt{30}}{10}$

Teaching Tip
Remind students that this is an application of the Multiplicative Identity Property—the same property that was used to rewrite rational expressions with a different denominator.

EXAMPLE 4 Rationalizing a Denominator Containing a Square Root

Rationalize the denominator of each expression:

(a) $-\frac{3}{2\sqrt{3}}$ **(b)** $\sqrt{\frac{3}{32}}$

Solution

(a) The denominator of the expression $-\frac{3}{2\sqrt{3}}$ has a radicand of 3. What can we multiply 3 by to obtain a perfect square? We multiply 3 by 3, because $3 \cdot 3 = 9$. Since the factor 2 in the denominator is already a rational number, it does not play a role in the process of rationalizing the denominator.

Multiply by $1 = \frac{\sqrt{3}}{\sqrt{3}}$: $-\frac{3}{2\sqrt{3}} = -\frac{3}{2\sqrt{3}} \cdot \frac{\sqrt{3}}{\sqrt{3}}$

Multiply numerators; multiply denominators: $= -\frac{3\sqrt{3}}{2\sqrt{9}}$

$= -\frac{3\sqrt{3}}{2 \cdot 3}$

Divide out the 3s: $= -\frac{\sqrt{3}}{2}$

(b) First, we write the square root as the quotient of two square roots. So $\sqrt{\frac{3}{32}} = \frac{\sqrt{3}}{\sqrt{32}}$. This denominator has a radicand of 32. We multiply the expression $\frac{\sqrt{3}}{\sqrt{32}}$ by $1 = \frac{\sqrt{2}}{\sqrt{2}}$ because $32 \cdot 2 = 64$, a perfect square.

$\frac{\sqrt{3}}{\sqrt{32}} = \frac{\sqrt{3}}{\sqrt{32}} \cdot \frac{\sqrt{2}}{\sqrt{2}}$

Multiply numerators; multiply denominators: $= \frac{\sqrt{6}}{\sqrt{64}}$

$= \frac{\sqrt{6}}{8}$

Work Smart
A second approach to simplifying the expression in Example 3(b) would be to first simplify $\sqrt{32}$ so that it does not contain any factors that are perfect squares.

$$\sqrt{\frac{3}{32}} = \frac{\sqrt{3}}{\sqrt{32}} = \frac{\sqrt{3}}{\sqrt{16 \cdot 2}} = \frac{\sqrt{3}}{4\sqrt{2}} = \frac{\sqrt{3}}{4\sqrt{2}} \cdot \frac{\sqrt{2}}{\sqrt{2}} = \frac{\sqrt{6}}{4\sqrt{4}} = \frac{\sqrt{6}}{4 \cdot 2} = \frac{\sqrt{6}}{8}$$

QUICK ✓ *Rationalize each denominator.*

5. $\frac{1}{\sqrt{2}}$ **6.** $-\frac{\sqrt{2}}{5\sqrt{3}}$ **7.** $\frac{5}{\sqrt{20}}$

3 Rationalize a Denominator Containing Two Terms

Recall from the previous section that expressions such as $5 + \sqrt{6}$ and $5 - \sqrt{6}$ are called conjugates.

Classroom Example ➤
Use Example 5 with the following:

(a) The conjugate of $2 + \sqrt{3}$ is $2 - \sqrt{3}$.

(b) The conjugate of $\sqrt{7} - 8$ is $\sqrt{7} + 8$.

(c) The conjugate of $-3 + 5\sqrt{2}$ is $-3 - 5\sqrt{2}$.

(d) The conjugate of $4\sqrt{3} + \sqrt{5}$ is $4\sqrt{3} - \sqrt{5}$.

EXAMPLE 5 Determining the Conjugate of a Square Root Expression

(a) The conjugate of $3 + \sqrt{5}$ is $3 - \sqrt{5}$.

(b) The conjugate of $\sqrt{7} - 5$ is $\sqrt{7} + 5$.

(c) The conjugate of $-2 + 3\sqrt{11}$ is $-2 - 3\sqrt{11}$.

(d) The conjugate of $2\sqrt{7} + \sqrt{6}$ is $2\sqrt{7} - \sqrt{6}$.

In Example 11(a) from the previous section we found the product $(5 + \sqrt{6})(5 - \sqrt{6})$ to be 19, a rational number. The product of two conjugates containing square roots will always be a rational number. To rationalize a denominator containing two terms, we multiply both the numerator and denominator of the quotient by the conjugate of the denominator. For example, if the quotient is $\frac{3}{2 + \sqrt{3}}$, we multiply both the numerator and the denominator by the conjugate of $2 + \sqrt{3}$, $2 - \sqrt{3}$. Since we are multiplying both the numerator and denominator by $2 - \sqrt{3}$, we are multiplying $\frac{3}{2 + \sqrt{3}}$ by 1, where $1 = \frac{2 - \sqrt{3}}{2 - \sqrt{3}}$.

Classroom Example ➤
Rationalize the denominator: $\frac{28}{4 - \sqrt{2}}$

Answer: $8 + 2\sqrt{2}$

EXAMPLE 6 How to Rationalize a Denominator Containing Two Terms

Rationalize the denominator: $\frac{4}{2 - \sqrt{6}}$

Step-by-Step Solution

Step 1: Multiply the numerator and denominator of the quotient by the conjugate of the denominator.

Multiply $\frac{4}{2 - \sqrt{6}}$ by $1 = \frac{2 + \sqrt{6}}{2 + \sqrt{6}}$: $\frac{4}{2 - \sqrt{6}} = \frac{4}{2 - \sqrt{6}} \cdot \frac{2 + \sqrt{6}}{2 + \sqrt{6}}$

Step 2: Multiply numerators; multiply denominators.

$$= \frac{4(2 + \sqrt{6})}{(2 - \sqrt{6})(2 + \sqrt{6})}$$

$(a + b)(a - b) = a^2 - b^2$ in the denominator: $= \frac{4(2 + \sqrt{6})}{(2)^2 - (\sqrt{6})^2}$

$$= \frac{4(2 + \sqrt{6})}{4 - 6}$$

$$= \frac{4(2 + \sqrt{6})}{-2}$$

Step 3: Simplify, if possible.

Find the quotient $\frac{4}{-2} = -2$: $= \frac{{}^{-2}\cancel{4}(2 + \sqrt{6})}{\cancel{-2}_1}$

$$= -2(2 + \sqrt{6})$$

Distribute the -2: $= -4 - 2\sqrt{6}$

Teaching Tip
Compare rationalizing when the denominator is a monomial versus when the denominator is a binomial. That is, $\frac{2}{\sqrt{3}}$ versus $\frac{2}{1 + \sqrt{3}}$.

Work Smart
Did you notice that in Example 6 we didn't use the Distributive Property in the numerator until after we had simplified the denominator? We kept the numerator in factored form so that we could easily simplify the expression.

QUICK ✓ *Rationalize the denominator.*

8. $\dfrac{2}{2+\sqrt{3}}$ **9.** $\dfrac{12}{3-\sqrt{5}}$

Classroom Example ➤
Rationalize the denominator:
$\dfrac{\sqrt{2}}{\sqrt{7}-\sqrt{3}}$

Answer: $\dfrac{\sqrt{14}+\sqrt{6}}{4}$

EXAMPLE 7 Rationalizing a Denominator Containing Two Terms

Rationalize the denominator: $\dfrac{\sqrt{2}}{\sqrt{x}+\sqrt{5}}$

Solution

We multiply the numerator and denominator by the conjugate of $\sqrt{x}+\sqrt{5}$, $\sqrt{x}-\sqrt{5}$.

Multiply by $1 = \dfrac{\sqrt{x}-\sqrt{5}}{\sqrt{x}-\sqrt{5}}$: $\dfrac{\sqrt{2}}{\sqrt{x}+\sqrt{5}} = \dfrac{\sqrt{2}}{\sqrt{x}+\sqrt{5}} \cdot \dfrac{\sqrt{x}-\sqrt{5}}{\sqrt{x}-\sqrt{5}}$

Multiply the numerators and denominators: $= \dfrac{\sqrt{2}(\sqrt{x}-\sqrt{5})}{(\sqrt{x})^2-(\sqrt{5})^2}$

$(a+b)(a-b)=a^2-b^2$ in the denominator: $= \dfrac{\sqrt{2}(\sqrt{x}-\sqrt{5})}{x-5}$

Distribute $\sqrt{2}$: $= \dfrac{\sqrt{2x}-\sqrt{10}}{x-5}$

Work Smart: Study Skills
Can you distinguish between the different types of problems involving division presented in this chapter? Make a list of examples, such as $\sqrt{\dfrac{100x^6}{4x^4}}$, $\dfrac{\sqrt{18}}{\sqrt{50}}$, $\dfrac{4+\sqrt{20}}{2}$, $\dfrac{\sqrt{8}-\sqrt{12}}{\sqrt{2}}$, $\dfrac{1}{\sqrt{5}}$, and $\dfrac{4}{2-\sqrt{6}}$. Can you simplify each one?

QUICK ✓ *Rationalize the denominator.*

10. $\dfrac{\sqrt{3}}{4+\sqrt{7}}$ **11.** $\dfrac{\sqrt{3}}{3-\sqrt{x}}$ **12.** $\dfrac{-\sqrt{2}}{\sqrt{3}+\sqrt{2}}$

We summarize the requirements for a square root to be simplified.

SUMMARY:
A square root is simplified if
- the radicand does not contain any factors that are perfect squares.
- the radicand does not contain any fractions.
- the denominator of a fraction contains no square roots.

6.5 Exercises

For Extra Help:  Student Solutions Manual  CD Video  PH Math/Tutor Center  MathXL Tutorials on CD MathXL® MyMathLab

Concepts and Vocabulary

In Problems 1–3, fill in the blanks.

1. rationalize the denominator
2. conjugates

1. To _________ _____ ________ means to rewrite a quotient in an equivalent form which has a rational number in the denominator.

2. If two binomials have the same terms but opposite signs between the terms, these binomials are called ________.

3. $\frac{\sqrt{2}}{\sqrt{2}}$; $\frac{1+\sqrt{2}}{1+\sqrt{2}}$
4. False
5. True
6. False
7. Neither is correct. The correct approach is $\frac{3}{\sqrt{2}+\sqrt{3}} \cdot \frac{\sqrt{2}-\sqrt{3}}{\sqrt{2}-\sqrt{3}}$.
8. Incorrect. The correct approach is $\frac{3+\sqrt{6}}{6+\sqrt{2}} \cdot \frac{6-\sqrt{2}}{6-\sqrt{2}}$.

3. To rationalize the denominator of $\frac{1}{\sqrt{2}}$, multiply the quotient by $1 =$ ______; to rationalize the denominator of $\frac{1}{1-\sqrt{2}}$, multiply the quotient by $1 =$ ______.

In Problems 4–6, answer True or False to each statement.

4. $\frac{9}{\sqrt{4}}$ requires multiplying by $1 = \frac{\sqrt{4}}{\sqrt{4}}$ to rationalize the denominator.

5. $\frac{\sqrt{18}+\sqrt{12}}{\sqrt{3}}$ can be simplified by rationalizing the denominator or by rewriting the quotient as $\frac{\sqrt{18}}{\sqrt{3}} + \frac{\sqrt{12}}{\sqrt{3}} = \sqrt{\frac{18}{3}} + \sqrt{\frac{12}{3}}$.

6. $-\sqrt{3}+2$ and $\sqrt{3}+2$ are conjugates.

7. The following question appeared on an exam: Rationalize the denominator of $\frac{3}{\sqrt{2}+\sqrt{3}}$. One student wrote $\frac{3}{\sqrt{2}+\sqrt{3}} \cdot \frac{\sqrt{2}+\sqrt{3}}{\sqrt{2}+\sqrt{3}}$. A second student wrote $\frac{3}{\sqrt{2}+\sqrt{3}} \cdot \frac{\sqrt{2}+\sqrt{3}}{\sqrt{2}-\sqrt{3}}$. Is either approach correct? If not, what is the correct approach?

8. There was another question on this exam that asked the student to rationalize the denominator of $\frac{3+\sqrt{6}}{6+\sqrt{2}}$. The paper you are grading has the following steps. Is this procedure correct or incorrect? Explain how you would rationalize the denominator of $\frac{3+\sqrt{6}}{6+\sqrt{2}}$.

$$\frac{3+\sqrt{6}}{6+\sqrt{2}} = \frac{3}{6} + \frac{\sqrt{6}}{\sqrt{2}} = \frac{1}{2} + \sqrt{3}$$

Building Skills

9. 4 10. 6 11. $-\frac{1}{5}$ 12. $-\frac{1}{3}$ 13. $3\sqrt{2}$ 14. $4\sqrt{3}$ 15. $2\sqrt{7}$ 16. $5\sqrt{6}$ 17. $-5\sqrt{y}$ 18. $-3x$ 19. 8 20. $7a$

In Problems 9–20, simplify the quotient. Assume all variables represent positive real numbers.

9. $\frac{\sqrt{96}}{\sqrt{6}}$ **10.** $\frac{\sqrt{108}}{\sqrt{3}}$ **11.** $-\frac{\sqrt{5}}{\sqrt{125}}$ **12.** $-\frac{\sqrt{10}}{\sqrt{90}}$

13. $\frac{\sqrt{54}}{\sqrt{3}}$ **14.** $\frac{\sqrt{144}}{\sqrt{3}}$ **15.** $\frac{\sqrt{196}}{\sqrt{7}}$ **16.** $\frac{\sqrt{300}}{\sqrt{2}}$

17. $\frac{-\sqrt{75y^2}}{\sqrt{3y}}$ **18.** $\frac{-\sqrt{18x^3}}{\sqrt{2x}}$ **19.** $\sqrt{\frac{128a^2b^5}{2a^2b^5}}$ **20.** $\sqrt{\frac{147a^3b^4}{3ab^4}}$

21. $\frac{3\sqrt{5}}{5}$ 22. $\frac{9\sqrt{11}}{11}$ 23. $\frac{10\sqrt{6}}{3}$ 24. $\frac{5\sqrt{21}}{7}$ 25. $-\frac{8\sqrt{2}}{3}$ 26. $-\frac{5\sqrt{3}}{2}$ 27. $-\frac{2\sqrt{x}}{x}$ 28. $-\frac{5\sqrt{y}}{y}$ 29. $\frac{9\sqrt{2y}}{2}$ 30. $\frac{3\sqrt{5x}}{5}$ 31. $\frac{\sqrt{3}}{2}$ 32. $\frac{5\sqrt{7}}{7}$ 33. $\frac{3\sqrt{2n}}{2n}$ 34. $\frac{2\sqrt{3p}}{3p^2}$ 35. $\frac{4\sqrt{3}}{9}$ 36. $\frac{\sqrt{30}}{4}$ 37. $\frac{\sqrt{6ab}}{21b}$ 38. $-\frac{\sqrt{10ab}}{4b}$ 39. $-\frac{7\sqrt{10}}{12}$ 40. $\frac{3\sqrt{15}}{20}$

In Problems 21–40, rationalize the denominator of each expression. Assume all variables represent positive real numbers.

21. $\frac{3}{\sqrt{5}}$ **22.** $\frac{9}{\sqrt{11}}$ **23.** $\frac{20}{\sqrt{6}}$ **24.** $\frac{15}{\sqrt{21}}$ **25.** $-\frac{16}{3\sqrt{2}}$

26. $-\frac{15}{2\sqrt{3}}$ **27.** $\frac{2}{-\sqrt{x}}$ **28.** $\frac{5}{-\sqrt{y}}$ **29.** $\frac{9y}{\sqrt{2y}}$ **30.** $\frac{3x}{\sqrt{5x}}$

31. $\frac{6}{\sqrt{48}}$ **32.** $\frac{10}{\sqrt{28}}$ **33.** $\frac{6n}{\sqrt{8n^3}}$ **34.** $\frac{4p}{\sqrt{12p^5}}$ **35.** $\frac{4\sqrt{2}}{3\sqrt{6}}$

36. $\frac{5\sqrt{3}}{2\sqrt{10}}$ **37.** $\frac{2\sqrt{a}}{7\sqrt{6b}}$ **38.** $-\frac{5\sqrt{a}}{2\sqrt{10b}}$ **39.** $-\frac{7\sqrt{5}}{2\sqrt{18}}$ **40.** $\frac{3\sqrt{3}}{2\sqrt{20}}$

41. $\frac{\sqrt{2}}{2}$ 42. $\frac{\sqrt{3}}{3}$
43. $-\frac{\sqrt{6}}{3}$ 44. $-\frac{\sqrt{21}}{7}$ 45. $\frac{\sqrt{30}}{10}$
46. $\frac{5\sqrt{6}}{6}$ 47. $\frac{\sqrt{7xy}}{7x^2y}$ 48. $\frac{\sqrt{6xy}}{2x^2}$
49. $2-\sqrt{3}$; 1 50. $4+\sqrt{5}$; 11
51. $\sqrt{6}+1$; 5 52. $\sqrt{10}-3$; 1
53. $-1+3\sqrt{14}$; -125
54. $-8-5\sqrt{21}$; -461
55. $\sqrt{p}-3\sqrt{q}$; $p-9q$
56. $\sqrt{x}+2\sqrt{y}$; $x-4y$
57. $\frac{3-\sqrt{5}}{4}$ 58. $\frac{6-\sqrt{2}}{34}$
59. $\frac{27+3\sqrt{3}}{26}$
60. $\frac{8+2\sqrt{10}}{3}$ 61. $\frac{24-4\sqrt{7}}{29}$
62. $-12\sqrt{3}+24$ 63. $\frac{3\sqrt{y}+18}{y-36}$
64. $\frac{8\sqrt{x}+16}{x-4}$ 65. $32\sqrt{2}+48$
66. $\frac{6\sqrt{3}+45}{71}$ 67. $\frac{6\sqrt{2}+\sqrt{42}}{15}$
68. $\frac{4\sqrt{3}+\sqrt{30}}{6}$ 69. $\frac{n\sqrt{n}-n}{n-1}$
70. $\frac{z\sqrt{z}-2z}{z-4}$ 71. $\frac{2\sqrt{x}-3x}{4-9x}$
72. $\frac{\sqrt{y}-2y}{1-4y}$ 73. $\frac{\sqrt{55}+\sqrt{30}}{5}$
74. $\frac{\sqrt{66}+\sqrt{110}}{-2}$ 75. $\frac{\sqrt{3}+1}{2}$
76. $\sqrt{2}+1$ 77. $\sqrt{3}-\sqrt{6}$
78. $3+2\sqrt{2}$ 79. $\frac{3\sqrt{5}}{10}-\frac{\sqrt{2}}{2}$
80. $\frac{2\sqrt{3}}{9}+\frac{\sqrt{10}}{6}$ 81. $7\sqrt{2}$
82. $6\sqrt{2}$ 83. $\frac{5\sqrt{7}}{98}$ 84. $\frac{3\sqrt{5}}{100}$
85. $2\sqrt{2x}$ 86. $2\sqrt{5y}$
87. $\frac{2\sqrt{3}}{3}$ 88. $\frac{\sqrt{10}}{5}$
89. $-8\sqrt{2}+6\sqrt{5}$ 90. 0
91. $4-2\sqrt{3}$ 92. $\frac{16+4\sqrt{5}}{11}$
93. $\frac{\sqrt{2}}{10}$ 94. $\frac{\sqrt{3}}{9}$
95. $11-6\sqrt{2}$ 96. $19+8\sqrt{3}$

In Problems 41–48, simplify each expression. Assume all variables represent positive real numbers.

41. $\sqrt{\frac{1}{2}}$ **42.** $\sqrt{\frac{1}{3}}$ **43.** $-\sqrt{\frac{10}{15}}$ **44.** $-\sqrt{\frac{6}{14}}$

45. $\sqrt{\frac{3}{4}}\cdot\sqrt{\frac{6}{15}}$ **46.** $\sqrt{\frac{5}{8}}\cdot\sqrt{\frac{20}{3}}$ **47.** $\sqrt{\frac{5x}{y}}\cdot\sqrt{\frac{1}{35x^4}}$ **48.** $\sqrt{\frac{3x}{y}}\cdot\sqrt{\frac{y^2}{2x^4}}$

In Problems 49–56, determine the conjugate of each expression. Then find the product of the expression and its conjugate.

49. $2+\sqrt{3}$ **50.** $4-\sqrt{5}$ **51.** $\sqrt{6}-1$ **52.** $\sqrt{10}+3$

53. $-1-3\sqrt{14}$ **54.** $-8+5\sqrt{21}$ **55.** $\sqrt{p}+3\sqrt{q}$ **56.** $\sqrt{x}-2\sqrt{y}$

In Problems 57–76, rationalize the denominator of each expression. Assume all variables represent positive real numbers.

57. $\frac{1}{3+\sqrt{5}}$ **58.** $\frac{1}{6+\sqrt{2}}$ **59.** $\frac{9}{9-\sqrt{3}}$ **60.** $\frac{4}{4-\sqrt{10}}$

61. $\frac{4}{\sqrt{7}+6}$ **62.** $\frac{12}{\sqrt{3}+2}$ **63.** $\frac{3}{\sqrt{y}-6}$ **64.** $\frac{8}{\sqrt{x}-2}$

65. $\frac{16}{-2\sqrt{2}+3}$ **66.** $\frac{9}{-2\sqrt{3}+15}$ **67.** $\frac{\sqrt{2}}{6-\sqrt{21}}$ **68.** $\frac{\sqrt{3}}{4-\sqrt{10}}$

69. $\frac{n}{\sqrt{n}+1}$ **70.** $\frac{z}{\sqrt{z}+2}$ **71.** $\frac{\sqrt{x}}{2+3\sqrt{x}}$ **72.** $\frac{\sqrt{y}}{1+2\sqrt{y}}$

73. $\frac{\sqrt{5}}{\sqrt{11}-\sqrt{6}}$ **74.** $\frac{\sqrt{22}}{\sqrt{3}-\sqrt{5}}$ **75.** $\frac{\sqrt{2}}{\sqrt{6}-\sqrt{2}}$ **76.** $\frac{\sqrt{3}}{\sqrt{6}-\sqrt{3}}$

Mixed Practice

In Problems 77–96, simplify the expression. Assume all variables represent positive real numbers.

77. $\frac{\sqrt{6}-\sqrt{12}}{\sqrt{2}}$ **78.** $\frac{\sqrt{27}+\sqrt{24}}{\sqrt{3}}$ **79.** $\frac{\sqrt{18}-\sqrt{20}}{2\sqrt{10}}$

80. $\frac{\sqrt{8}+\sqrt{15}}{3\sqrt{6}}$ **81.** $\frac{2}{3}\sqrt{18}+\frac{5}{2}\sqrt{8}$ **82.** $\frac{3}{5}\sqrt{50}+\frac{3}{4}\sqrt{32}$

83. $\frac{5}{14\sqrt{7}}$ **84.** $\frac{3}{20\sqrt{5}}$ **85.** $\frac{4x}{\sqrt{2x}}$

86. $\frac{10y}{\sqrt{5y}}$ **87.** $\sqrt{\frac{4}{3}}$ **88.** $\sqrt{\frac{2}{5}}$

89. $-4\sqrt{8}+3\sqrt{20}$ **90.** $-3\sqrt{12}+2\sqrt{27}$ **91.** $\frac{2}{2+\sqrt{3}}$

92. $\frac{4}{4-\sqrt{5}}$ **93.** $\frac{1}{\sqrt{50}}$ **94.** $\frac{1}{\sqrt{27}}$

95. $(3-\sqrt{2})^2$ **96.** $(4+\sqrt{3})^2$

Applying the Concepts

In Problems 97–104, use the following information.

An answer which contains a square root is said to be "the exact answer" when it contains a radical in simplified form and there are no square roots in the denominator of a fraction. Irrational numbers, such as some of the square roots we have seen in this chapter, do not have an exact decimal representation. Any decimal form of an answer containing irrational numbers is only an approximation. For each expression, do the following:

(a) *Use your calculator to find the approximate value of the expression, to four decimal places.*
(b) *Rationalize the denominator to find the exact value of this expression.*
(c) *Use your calculator to find the approximate value of (b).*
(d) *Compare your results.*

97. $\dfrac{1}{2+\sqrt{3}}$ **98.** $\dfrac{1}{3-\sqrt{5}}$ **99.** $\dfrac{5}{2\sqrt{3}-1}$ **100.** $\dfrac{2}{3\sqrt{5}+2}$

101. $\dfrac{\sqrt{8}}{\sqrt{2}-3}$ **102.** $\dfrac{\sqrt{12}}{\sqrt{3}-4}$ **103.** $\dfrac{\sqrt{6}}{3+5\sqrt{3}}$ **104.** $\dfrac{\sqrt{5}}{5+2\sqrt{2}}$

Extending the Concepts

In Problems 105–110, rationalize the denominator of each expression.

105. $\dfrac{1+\sqrt{3}}{1-\sqrt{3}}$ **106.** $\dfrac{2+\sqrt{2}}{2-\sqrt{2}}$ **107.** $\dfrac{\sqrt{6}-\sqrt{3}}{\sqrt{2}-\sqrt{3}}$ **108.** $\dfrac{\sqrt{5}-\sqrt{2}}{\sqrt{3}-\sqrt{2}}$

109. $\dfrac{2\sqrt{x}+\sqrt{y}}{\sqrt{x}+4\sqrt{y}}$ **110.** $\dfrac{\sqrt{x}+6\sqrt{y}}{3\sqrt{x}+\sqrt{y}}$

97. (a) 0.2679 (b) $2-\sqrt{3}$ (c) 0.2679 (d) They are the same.
98. (a) 1.3090 (b) $\dfrac{3+\sqrt{5}}{4}$ (c) 1.3090 (d) They are the same.
99. (a) 2.0291 (b) $\dfrac{10\sqrt{3}+5}{11}$ (c) 2.0291 (d) They are the same.
100. (a) 0.2297 (b) $\dfrac{6\sqrt{5}-4}{41}$ (c) 0.2297 (d) They are the same.
101. (a) -1.7836 (b) $-\dfrac{4+6\sqrt{2}}{7}$ (c) -1.7836 (d) They are the same.
102. (a) -1.5274 (b) $-\dfrac{6+8\sqrt{3}}{13}$ (c) -1.5274 (d) They are the same.
103. (a) 0.2101 (b) $\dfrac{5\sqrt{2}-\sqrt{6}}{22}$ (c) 0.2101 (d) They are the same.
104. (a) 0.2856 (b) $\dfrac{5\sqrt{5}-2\sqrt{10}}{17}$ (c) 0.2856 (d) They are the same.
105. $-2-\sqrt{3}$ 106. $3+2\sqrt{2}$
107. $-2\sqrt{3}-3\sqrt{2}+\sqrt{6}+3$
108. $\sqrt{15}+\sqrt{10}-\sqrt{6}-2$
109. $\dfrac{2x-7\sqrt{xy}-4y}{x-16y}$
110. $\dfrac{3x+17\sqrt{xy}-6y}{9x-y}$

PUTTING THE CONCEPTS TOGETHER (SECTIONS 6.1–6.5)

These problems cover important concepts from Sections 6.1 to 6.5. We designed these problems so that you can review the chapter so far and show your mastery of the concepts. Take time to work these problems before proceeding with the next section. The answers to these problems are located at the back of the text starting on page AN-13.

1. Determine if each square root is rational or irrational. Then evaluate the square root. If the square root is irrational, express the square root as a decimal rounded to two decimal places.

(a) $\sqrt{98}$ **(b)** $\sqrt{361}$

In Problems 2–6, simplify each of the following. Assume all variables represent positive real numbers.

2. $-2\sqrt{18}$ **3.** $\sqrt{32x^3y^6}$ **4.** $\sqrt{\dfrac{100x^3}{16x^5}}$

5. $\sqrt{121-4(6)(3)}$ **6.** $\dfrac{\sqrt{72m^6}}{\sqrt{8}}$

In Problems 7–16, perform the indicated operation and simplify. Assume all variables represent positive real numbers.

7. $5\sqrt{3}-\sqrt{12}$ **8.** $3\sqrt{50}-6\sqrt{8}$ **9.** $\sqrt{12}\cdot\sqrt{18}$

10. $(2\sqrt{7})(-3\sqrt{7})$ **11.** $6\sqrt{5}(2\sqrt{10}+3)$ **12.** $(6-\sqrt{3})(2-4\sqrt{3})$

13. $(3-2\sqrt{5})(3+2\sqrt{5})$ **14.** $(\sqrt{7}-2)^2$ **15.** $\dfrac{-4+\sqrt{28}}{2}$ **16.** $\dfrac{\sqrt{90x^2}}{\sqrt{5x^6}}$

In Problems 17 and 18, rationalize the denominator.

17. $\dfrac{8}{\sqrt{24}}$ **18.** $\dfrac{12}{\sqrt{3}+3}$

1. (a) irrational; $\sqrt{98}\approx 9.90$ (b) rational; $\sqrt{361}=19$
2. $-6\sqrt{2}$
3. $4xy^3\sqrt{2x}$
4. $\dfrac{5}{2x}$
5. 7
6. $3m^3$
7. $3\sqrt{3}$
8. $3\sqrt{2}$
9. $6\sqrt{6}$
10. -42
11. $60\sqrt{2}+18\sqrt{5}$
12. $24-26\sqrt{3}$
13. -11
14. $11-4\sqrt{7}$
15. $-2+\sqrt{7}$
16. $\dfrac{3\sqrt{2}}{x^2}$
17. $\dfrac{2\sqrt{6}}{3}$
18. $-2\sqrt{3}+6$

6.6 Solving Equations Containing Square Roots

OBJECTIVES

1. Determine Whether or Not a Number Is a Solution of a Radical Equation
2. Solve Equations Containing One Square Root
3. Solve Equations That Involve Two Square Roots
4. Solve Problems Modeled by a Radical Equation

Preparing for Solving Equations Containing Square Roots

Before getting started, take the following readiness quiz. If you get a problem wrong, go to the section cited and review the material.

1. Solve: $4x - 5 = 0$ [Section 2.2, pp. 84–86]

2. Solve: $2p^2 + 4p - 16 = 0$ [Section 4.6, pp. 281–286]

3. Find the product: $(2x - 3)^2$ [Section 3.3, pp. 194–196]

4. Find the product: $(\sqrt{x} + 2)^2$ [Section 6.4, p. 417]

When the variable in an equation occurs in a square root, the equation is called a **radical equation.** Examples of radical equations are

$$\sqrt{3x + 1} = 5 \qquad \sqrt{x - 2} - \sqrt{2x + 5} = 2$$

1 Determine Whether or Not a Number Is a Solution of a Radical Equation

A number is a **solution** to a radical equation if, when we substitute the number for the variable, a true statement results.

Classroom Example

(a) Determine whether $x = 9$ is a solution of $\sqrt{3x - 2} = 5$.

(b) Is $x = 4$ a solution of $\sqrt{3x + 4} = -4$?

Answer:

(a) yes (b) no

EXAMPLE 1 Determining Whether or Not a Number Is a Solution to a Radical Equation

(a) Determine whether $x = 5$ is a solution of the equation $\sqrt{3x + 1} = 4$.

(b) Is $x = 1$ a solution of $\sqrt{5x - 1} = -2$?

Solution

(a) To determine if $x = 5$ is a solution of $\sqrt{3x + 1} = 4$, we substitute 5 for x.

$$\sqrt{3x + 1} = 4$$

Substitute 5 for x: $\sqrt{3(5) + 1} \stackrel{?}{=} 4$

Multiply: $\sqrt{15 + 1} \stackrel{?}{=} 4$

$$\sqrt{16} \stackrel{?}{=} 4$$

$$4 = 4 \quad \text{True}$$

Since $x = 5$ produces a true statement, $x = 5$ is a solution of $\sqrt{3x + 1} = 4$.

(b) Substitute 1 for x in the equation $\sqrt{5x - 1} = -2$.

$$\sqrt{5x - 1} = -2$$

Substitute 1 for x: $\sqrt{5(1) - 1} \stackrel{?}{=} -2$

Multiply: $\sqrt{5 - 1} \stackrel{?}{=} -2$

$$\sqrt{4} \stackrel{?}{=} -2$$

$$2 = -2 \quad \text{False}$$

The value $x = 1$ gives a false statement. So $x = 1$ is not a solution of $\sqrt{5x - 1} = -2$.

QUICK ✓ *Determine if the number is a solution to the given equation.*

1. $\sqrt{6x + 4} = 8; x = 10$ **2.** $\sqrt{n + 2} = n; n = -1$ **3.** $\sqrt{2y + 11} = 3; y = -1$

Preparing for...Answers **1.** $\left\{\frac{5}{4}\right\}$ **2.** $\{-4, 2\}$ **3.** $4x^2 - 12x + 9$ **4.** $x + 4\sqrt{x} + 4$

Teaching Tip
Solving equations with radicals will involve solving both linear and quadratic equations. You may wish to have students work "Preparing for . . ." problems 1 and 2 before beginning the discussion of solving these equations.

2 Solve Equations Containing One Square Root

Remember when we solved quadratic equations? We used factoring to rewrite the quadratic equation as two linear equations. When we solved rational equations, we multiplied both sides of the equation by the least common denominator and rewrote the rational equation as a linear or quadratic equation. So, each time a problem came up, we used techniques to rewrite it as a problem we already know how to solve.

The same approach will be used to solve radical equations. We will rewrite the radical equation as a problem that we already know how to solve. To turn a radical equation into one we already know how to solve uses

$$(\sqrt{a})^2 = a$$

provided that $a > 0$. Examples 2 and 3 illustrate how to solve a radical equation.

Work Smart
Notice $(\sqrt{a})^2 = a$, but $\sqrt{a^2} = |a|$.

EXAMPLE 2 **How to Solve a Radical Equation Containing One Radical**

Solve: $\sqrt{x - 3} = 2$

Classroom Example
Solve: $\sqrt{2x - 3} = 5$
Answer: $\{14\}$

Step-by-Step Solution

Step 1: Isolate the radical.

$$\sqrt{x - 3} = 2$$

Step 2: Square both sides of the equation.

$$(\sqrt{x - 3})^2 = 2^2$$
$$x - 3 = 4$$

Step 3: Solve the equation that results.

Add 3 to both sides: $x - 3 + 3 = 4 + 3$
$$x = 7$$

Step 4: Check

$$\sqrt{x - 3} = 2$$
Let $x = 7$ in the original equation: $\sqrt{7 - 3} \stackrel{?}{=} 2$
$$\sqrt{4} \stackrel{?}{=} 2$$
$$2 = 2 \quad \text{True}$$

The solution set is $\{7\}$.

EXAMPLE 3 **How to Solve a Radical Equation Containing One Radical**

Solve: $\sqrt{4x + 1} - 2 = 3$

Classroom Example
Solve: $\sqrt{3x - 2} + 2 = 6$
Answer: $\{6\}$

Step-by-Step Solution

Step 1: Isolate the radical.

$$\sqrt{4x + 1} - 2 = 3$$
Add 2 to each side of the equation: $\sqrt{4x + 1} - 2 + 2 = 3 + 2$
$$\sqrt{4x + 1} = 5$$

Step 2: Square both sides of the equation.

$$(\sqrt{4x + 1})^2 = 5^2$$
$$4x + 1 = 25$$

(continued)

Step 3: *Solve the equation that results.*

Subtract 1 from both sides: $4x + 1 - 1 = 25 - 1$

$4x = 24$

Divide both sides of the equation by 4: $x = 6$

Step 4: Check

$$\sqrt{4x + 1} - 2 = 3$$

Let $x = 6$ in the original equation: $\sqrt{4(6) + 1} - 2 \stackrel{?}{=} 3$

$$\sqrt{25} - 2 \stackrel{?}{=} 3$$

$$5 - 2 \stackrel{?}{=} 3$$

$$3 = 3 \quad \text{True}$$

The solution set is $\{6\}$.

SUMMARY: Steps to Solve a Radical Equation Containing One Radical

Step 1: Isolate the radical. That is, get the radical by itself on one side of the equation and everything else on the other side.

Step 2: Square both sides of the equation. This will eliminate the radical from the equation.

Step 3: Solve the equation that results.

Step 4: Check your answer in the original equation.

QUICK ✓ *Solve the equation and check the solution.*

4. $\sqrt{y - 5} = 3$ **5.** $\sqrt{x - 4} + 4 = 7$ **6.** $2\sqrt{p} = 12$ **7.** $4\sqrt{t - 3} - 2 = 10$

In Section 5.7, we found that when we solve a rational equation, some apparent solutions to the equation do not satisfy the original equation. When we solve equations containing square roots, we also may find apparent solutions that do not satisfy the original equation. Apparent solutions are solutions that result from the process of solving the equation.

Work Smart

Extraneous solutions are likely to occur when one side of the equation contains a square root expression and the other side contains a variable expression.

DEFINITION

Apparent solutions that do not satisfy the original equation are called **extraneous solutions.**

We can identify extraneous solutions by substituting apparent solutions into the original equation. If the solution does not satisfy the equation, it is extraneous.

Classroom Example ➤

Solve: $\sqrt{x + 12} = x$

Answer: $\{4\}$

EXAMPLE 4 Solving a Radical Equation with an Extraneous Solution

Solve: $\sqrt{x + 6} = x$

Solution

$$\sqrt{x + 6} = x$$

The radical is already isolated, so we square both sides of the equation: $\left(\sqrt{x + 6}\right)^2 = x^2$

$$x + 6 = x^2$$

The resulting equation, $x + 6 = x^2$, is a quadratic equation (do you see why?), so we write it in standard form, $ax^2 + bx + c = 0$, and solve.

Subtract x and 6 from each side: $x - x + 6 - 6 = x^2 - x - 6$

Write in standard form: $0 = x^2 - x - 6$

Factor the polynomial: $0 = (x - 3)(x + 2)$

Set each factor equal to 0: $x - 3 = 0$ or $x + 2 = 0$

Solve the resulting equations: $x = 3$ or $x = -2$

Let's check these two apparent solutions.

Check

$x = 3$: Let $x = 3$ in the original equation:

$$\sqrt{x + 6} = x$$
$$\sqrt{3 + 6} \stackrel{?}{=} 3$$
$$\sqrt{9} \stackrel{?}{=} 3$$
$$3 = 3 \quad \text{True}$$

$x = 3$ checks so $x = 3$ is a solution of the equation.

$x = -2$: Let $x = -2$ in the original equation:

$$\sqrt{x + 6} = x$$
$$\sqrt{-2 + 6} \stackrel{?}{=} -2$$
$$\sqrt{4} \stackrel{?}{=} -2$$
$$2 = -2 \quad \text{False}$$

$x = -2$ does not check, so $x = -2$ is an extraneous solution.

The solution set is $\{3\}$.

QUICK ✓ *Solve the equation and check the solution.*

8. $\sqrt{2a + 3} = a$ **9.** $\sqrt{2p + 8} = p$

Classroom Example
Solve: $\sqrt{x + 5} = x - 1$
Answer: $\{4\}$

EXAMPLE 5 Using the Square of a Binomial to Solve a Radical Equation

Solve: $\sqrt{w} + 6 = w$

Solution

$$\sqrt{w} + 6 = w$$

Isolate the radical expression: $\sqrt{w} + 6 - 6 = w - 6$

$$\sqrt{w} = w - 6$$

Square both sides of the equation: $(\sqrt{w})^2 = (w - 6)^2$

$(a - b)^2 = a^2 - 2ab + b^2$: $w = w^2 - 12w + 36$

Work Smart

$(a - b)^2 \neq a^2 - b^2$
$(a - b)^2 = a^2 - 2ab + b^2$

We have a quadratic equation, so we write it in standard form and solve.

Subtract w from each side: $w - w = w^2 - 12w - w + 36$

Write in standard form: $0 = w^2 - 13w + 36$

Factor the polynomial: $0 = (w - 9)(w - 4)$

Set each factor equal to 0: $w - 9 = 0$ or $w - 4 = 0$

Solve the resulting equations: $w = 9$ or $w = 4$

Let's check these two apparent solutions.

Check

$w = 9$: Let $w = 9$ in the original equation.

$$\sqrt{w} + 6 = w$$
$$\sqrt{9} + 6 \stackrel{?}{=} 9$$
$$3 + 6 \stackrel{?}{=} 9$$
$$9 = 9 \quad \text{True}$$

$w = 9$ checks, so $w = 9$ is a solution of the equation.

$w = 4$: Let $w = 4$ in the original equation.

$$\sqrt{w} + 6 = w$$
$$\sqrt{4} + 6 \stackrel{?}{=} 4$$
$$2 + 6 \stackrel{?}{=} 4$$
$$8 = 4 \quad \text{False}$$

$w = 4$ does not check, so $w = 4$ is an extraneous solution.

The solution set is $\{9\}$.

QUICK ✓ *Solve the equation and check the solution.*

10. $\sqrt{m + 10} = m - 2$ **11.** $\sqrt{17 - 2x} + 1 = x$

Is it possible for a radical equation to have no solution? Yes!

Classroom Example Solve: $\sqrt{7t + 4} + 5 = 2$

Answer: $\{\ \}$ or $\varnothing$

EXAMPLE 6 Solving a Radical Equation with No Solution

Solve: $\sqrt{5p - 3} + 7 = 3$

Solution

$$\sqrt{5p - 3} + 7 = 3$$

Isolate the radical expression: $\sqrt{5p - 3} + 7 - 7 = 3 - 7$

$$\sqrt{5p - 3} = -4$$

Square both sides: $(\sqrt{5p - 3})^2 = (-4)^2$

$$5p - 3 = 16$$

$$5p = 19$$

$$p = \frac{19}{5}$$

Work Smart

Look closely at the equation

$$\sqrt{5p - 3} = -4.$$

Because the principal square root of a number cannot be negative, we know that there is no real solution to the equation $\sqrt{5p - 3} + 7 = 3$. So the solution set is $\varnothing$ or $\{\ \}$.

Check

$$\sqrt{5p - 3} + 7 = 3$$

$p = \frac{19}{5}$: $\sqrt{5\left(\frac{19}{5}\right) - 3} + 7 \stackrel{?}{=} 3$

$$\sqrt{19 - 3} + 7 \stackrel{?}{=} 3$$

$$\sqrt{16} + 7 \stackrel{?}{=} 3$$

$$11 = 3 \quad \text{False}$$

Because $p = \frac{19}{5}$ does not satisfy the original equation, it is extraneous. So the solution set is $\varnothing$ or $\{\ \}$.

QUICK ✓ *Solve the equation and check the solution.*

12. $\sqrt{3b - 2} + 8 = 5$

3 Solve Equations That Involve Two Square Roots

When a radical equation contains two square roots, we place one of the radicals on one side of the equation and the other radical on the other side of the equation. We then follow the same steps that we used when we solved a radical equation with one square root. That is, we square both sides of the equation and solve the resulting equation.

Classroom Example Solve:

$$\sqrt{x + 4} = \sqrt{3x - 8}$$

Answer: $\{6\}$

EXAMPLE 7 Solving a Radical Equation with Two Square Roots

Solve: $\sqrt{x + 9} = \sqrt{2x + 5}$

Solution

Notice that each side of the equation contains one radical expression. We square both sides of the equation and solve the resulting equation.

$$\sqrt{x+9} = \sqrt{2x+5}$$

Square both sides of the equation: $\left(\sqrt{x+9}\right)^2 = \left(\sqrt{2x+5}\right)^2$

$$x + 9 = 2x + 5$$

Subtract x from both sides: $x - x + 9 = 2x - x + 5$

$$9 = x + 5$$

Subtract 5 from both sides: $4 = x$

Check

$$\sqrt{x+9} = \sqrt{2x+5}$$

$x = 4$: $\sqrt{4+9} \stackrel{?}{=} \sqrt{2(4)+5}$

$$\sqrt{13} \stackrel{?}{=} \sqrt{8+5}$$

$$\sqrt{13} = \sqrt{13} \quad \text{True}$$

We see that $x = 4$ satisfies the equation $\sqrt{x+9} = \sqrt{2x+5}$, so the solution set is $\{4\}$.

Classroom Example
Solve:

$$3\sqrt{n} - \sqrt{6n+12} = 0$$

Answer: $\{4\}$

EXAMPLE 8 Solving a Radical Equation with Two Square Roots

Solve: $\sqrt{y+6} - 2\sqrt{y} = 0$

Solution

Notice that both radicals are on the same side of the equation. Our first step will be to add $2\sqrt{y}$ to both sides to get one radical expression on each side of the equation.

$$\sqrt{y+6} - 2\sqrt{y} = 0$$

$$\sqrt{y+6} - 2\sqrt{y} + 2\sqrt{y} = 0 + 2\sqrt{y}$$

$$\sqrt{y+6} = 2\sqrt{y}$$

Square both sides of the equation: $\left(\sqrt{y+6}\right)^2 = (2\sqrt{y})^2$

$(2\sqrt{y})^2 = 2^2 \cdot (\sqrt{y})^2 = 4y$: $y + 6 = 4y$

Subtract y from both sides: $y - y + 6 = 4y - y$

$$6 = 3y$$

Divide both sides by 3: $\frac{6}{3} = \frac{3y}{3}$

$$2 = y$$

Is $y = 2$ the solution of the equation $\sqrt{y+6} - 2\sqrt{y} = 0$? Let's check.

Check

$$\sqrt{y+6} - 2\sqrt{y} = 0$$

$y = 2$: $\sqrt{2+6} - 2\sqrt{2} \stackrel{?}{=} 0$

$$\sqrt{8} - 2\sqrt{2} \stackrel{?}{=} 0$$

$$2\sqrt{2} - 2\sqrt{2} \stackrel{?}{=} 0$$

$$0 = 0 \quad \text{True}$$

Since $y = 2$ checks, the solution set is $\{2\}$.

QUICK ✓ *Solve each equation and check the solution.*

13. $\sqrt{3x+1} = \sqrt{2x+7}$

14. $\sqrt{2w^2 - 3w - 4} = \sqrt{w^2 + 6w + 6}$

15. $2\sqrt{k+5} - \sqrt{8k+4} = 0$

We now discuss one more type of equation with two radicals. In these equations, it's necessary to square both sides of the equation twice to eliminate radicals. After we square the first time, a radical remains. We then follow the same steps for solving a radical equation with one radical.

Classroom Example ➤ Solve:

$$\sqrt{x} + 2 = \sqrt{x + 12}$$

Answer: $\{4\}$

EXAMPLE 9 Solving a Radical Equation—Squaring Twice

Solve: $\sqrt{x} + 1 = \sqrt{x + 5}$

Solution

There is a radical expression on each side of the equation, so we square both sides.

$$\sqrt{x} + 1 = \sqrt{x + 5}$$

Square both sides: $(\sqrt{x} + 1)^2 = \left(\sqrt{x + 5}\right)^2$

$(\sqrt{x} + 1)^2 = (\sqrt{x})^2 + 2(\sqrt{x})(1) + 1^2$
$= x + 2\sqrt{x} + 1$

Teaching Tip
Remind students to square the sides of the equation, not individual terms. That is, $(a + b)^2 = a^2 + 2ab + b^2$.

$$x + 2\sqrt{x} + 1 = x + 5$$

Isolate the remaining radical: $x - x + 2\sqrt{x} + 1 - 1 = x - x + 5 - 1$

$$2\sqrt{x} = 4$$

Square both sides: $(2\sqrt{x})^2 = 4^2$

$$4x = 16$$

Divide both sides of the equation by 4: $x = 4$

Check

$$\sqrt{x} + 1 = \sqrt{x + 5}$$

$x = 4$: $\sqrt{4} + 1 \stackrel{?}{=} \sqrt{4 + 5}$

$$2 + 1 \stackrel{?}{=} 3$$

$3 = 3$ True

Since $x = 4$ results in a true statement, the solution set is $\{4\}$.

QUICK ✓ *Solve the equation and check the solution.*

16. $\sqrt{x} = \sqrt{x + 9} - 1$

---

(4) Solve Problems Modeled by a Radical Equation

Let's see how square roots are used in real-world applications.

Classroom Example ➤ Use Example 10 with $P = \$2000$ and $r = 4\%$.

Answer: \$2163.20

EXAMPLE 10 How Much Money?

The annual rate of interest r (expressed as a decimal) required to have A dollars after 2 years from an initial deposit of P dollars is given by the equation $r = \sqrt{\frac{A}{P}} - 1$. Suppose you deposit \$1000 in an account that pays 5% annual interest. How much money will you have after two years?

Solution

We want to know the amount of money that will be in the account after 2 years. That is, we are solving for the value of A. We know that $P = 1000$, and that $r = 5\% = 0.05$.

$$r = \sqrt{\frac{A}{P}} - 1$$

Substitute $r = 0.05$, $P = 1000$:
$$0.05 = \sqrt{\frac{A}{1000}} - 1$$

Isolate the square root by adding 1 to both sides of the equation:
$$0.05 + 1 = \sqrt{\frac{A}{1000}} - 1 + 1$$

$$1.05 = \sqrt{\frac{A}{1000}}$$

Square both sides:
$$(1.05)^2 = \left(\sqrt{\frac{A}{1000}}\right)^2$$

$$1.1025 = \frac{A}{1000}$$

Multiply both sides by 1000:
$$1000(1.1025) = \left(\frac{A}{1000}\right)1000$$

$$1102.50 = A$$

There will be \$1102.50 in the account after 2 years.

QUICK 

17. A method to "curve" grades on an exam uses the equation $N = 10\sqrt{O}$, where N is the new grade and O is the original grade. (This method is sometimes referred to as a "square root curve.") Use this equation to find the original score on a test that has a new score of 70.

6.6 Exercises

1. radical equation
2. extraneous solutions
3. isolate
4. False
5. False
6. True
7. Incorrect; Answers may vary.

Concepts and Vocabulary

In Problems 1–3, fill in the blanks.

1. An equation such as $\sqrt{x+3} = 5$ is call a(n) ________ ________.

2. Apparent solutions that do not satisfy the original equation are called ________ ________.

3. To solve an equation such as $\sqrt{x} + 1 = 16$, the first step should be to ________ the radical on one side of the equal sign.

In Problems 4–6, answer True or False to each statement.

4. To solve $\sqrt{x} + 3 = \sqrt{2x+1}$, the first step is $\left(\sqrt{x}\right)^2 + (3)^2 = \left(\sqrt{2x+1}\right)^2$.

5. The first step in solving $x + \sqrt{x-3} = 5$ is to square both sides of the equation to remove the square root.

6. It is possible to follow all the rules of algebra and still get apparent solutions do not satisfy the original equation.

7. An exam asked to solve the radical equation $\sqrt{x-1} = 3 - \sqrt{x+6}$. The following steps appear on a student's paper. Is the student's work correct or incorrect? What property did the student apply to get the second line? If there is an error, explain the correct steps to solve the equation.

$$\sqrt{x-1} = 3 - \sqrt{x+6}$$
$$x - 1 = 9 - (x+6)$$

8. Principal square root of a real number cannot be negative.
9. Yes 10. Yes
11. No 12. No
13. Yes 14. No
15. No 16. Yes
17. {16}
18. {81}
19. {−4}
20. {−1}
21. {7}
22. {2}
23. {4}
24. {6}
25. {9}
26. {1}
27. ∅
28. ∅
29. {25}
30. {8}
31. {3, 6}
32. {2, 5}
33. {1}
34. {3}
35. ∅
36. ∅
37. {9}
38. {4, 6}
39. {6, 9}
40. {3, 4}
41. {9}
42. {8}
43. {6}
44. {−8}
45. {1}
46. $\left\{\frac{1}{2}\right\}$
47. {0}
48. ∅
49. {−5}
50. ∅
51. {5}
52. {18}
53. ∅
54. {1}
55. {6}
56. ∅
57. $\left\{\frac{13}{4}\right\}$
58. ∅
59. {23}
60. {4}
61. {20}
62. {8}
63. {−1}
64. {−1}
65. {16}
66. {81}
67. {2}
68. {3}
69. $\left\{\frac{1}{4}\right\}$
70. $\left\{\frac{1}{4}\right\}$
71. {−4, 7}
72. {−5, 7}
73. {4}
74. {3}

8. Explain why we know that the solution set of $\sqrt{2x-5} = -3$ is ∅ without solving the equation.

Building Your Skills

In Problems 9–16, determine if the number is a solution to the equation.

9. $\sqrt{6x+1} = 5; x = 4$
10. $\sqrt{4n-7} = 3; n = 4$
11. $\sqrt{5n-6} = -6; n = -6$
12. $\sqrt{3y-4} = -5; y = -7$
13. $\sqrt{p+6} = p+4; p = -2$
14. $\sqrt{p+6} = p+4; p = -5$
15. $\sqrt{2y+3} = y; y = -1$
16. $\sqrt{2y+3} = y; y = 3$

In Problems 17–42, solve the equation, and check the solution.

17. $\sqrt{a} = 4$
18. $\sqrt{x} = 9$
19. $\sqrt{5-x} = 3$
20. $\sqrt{3-y} = 2$
21. $5 = \sqrt{4x-3}$
22. $4 = \sqrt{3x+10}$
23. $3\sqrt{p} = 6$
24. $5\sqrt{a-2} = 10$
25. $2\sqrt{x} + 2 = 8$
26. $3\sqrt{n} + 3 = 6$
27. $3 = 6 + \sqrt{n}$
28. $10 = 23 + \sqrt{p}$
29. $\sqrt{2y-1} - 4 = 3$
30. $\sqrt{3y+1} - 3 = 2$
31. $\sqrt{9z-18} = z$
32. $\sqrt{7v-10} = v$
33. $y = \sqrt{4-3y}$
34. $y = \sqrt{24-5y}$
35. $n+1 = \sqrt{n^2-3}$
36. $p+3 = \sqrt{p^2-15}$
37. $\sqrt{3x-11} = x-5$
38. $\sqrt{2x-8} = x-4$
39. $\sqrt{31x+10} - 8 = x$
40. $\sqrt{17x+13} - 5 = x$
41. $6 + \sqrt{x} - x = 0$
42. $4 + \sqrt{2x} - x = 0$

In Problems 43–62, solve the equation and check the solution.

43. $\sqrt{3x-4} = \sqrt{x+8}$
44. $\sqrt{5x+1} = \sqrt{3x-15}$
45. $\sqrt{x^2+3x} = \sqrt{x^2+6x-3}$
46. $\sqrt{x^2+4} = \sqrt{x^2+8x}$
47. $2\sqrt{a+5} - \sqrt{7a+20} = 0$
48. $\sqrt{7a-6} - 2\sqrt{a-3} = 0$
49. $\sqrt{2x^2-10} - \sqrt{x^2-3x} = 0$
50. $\sqrt{2x^2-2x} - \sqrt{x^2-3} = 0$
51. $2\sqrt{n} = \sqrt{n^2-5}$
52. $\sqrt{y^2-36} = 4\sqrt{y}$
53. $\sqrt{2x} + 2 = \sqrt{2x-12}$
54. $\sqrt{x} + 3 = \sqrt{x+15}$
55. $\sqrt{z+3} = 1 + \sqrt{z-2}$
56. $\sqrt{p+13} = \sqrt{p-8} - 7$
57. $2\sqrt{t+9} = \sqrt{4t+3} + 3$
58. $2\sqrt{s-4} = 2 + \sqrt{4s}$
59. $\sqrt{x+2} + 2 = \sqrt{2x+3}$
60. $\sqrt{x} - 1 = \sqrt{2x-7}$
61. $\sqrt{3x+4} - \sqrt{4x+1} = -1$
62. $\sqrt{4x-7} - \sqrt{2x} = 1$

Mixed Practice

In Problems 63–78, solve the equation and check the solution.

63. $-2 = \sqrt{5x+6} - 3$
64. $-4 = \sqrt{2x+11} - 7$
65. $2\sqrt{x} - 8 = 0$
66. $3\sqrt{x} - 27 = 0$
67. $3x - 2 = \sqrt{9x^2-20}$
68. $2x - 1 = \sqrt{4x^2-11}$
69. $3 = \sqrt{w+12} - 2w$
70. $1 = \sqrt{3-3c} - 2c$
71. $\sqrt{a^2-3a+5} = \sqrt{2a^2-6a-23}$
72. $\sqrt{2n^2-5n-20} = \sqrt{n^2-3n+15}$
73. $\sqrt{2y+1} - 2 = 1$
74. $\sqrt{4z-3} - 6 = -3$

75. ∅
76. {1}
77. ∅
78. ∅
79. 11
80. 27
81. 3
82. 0 or $\frac{1}{4}$
83. (a) 8 seconds (b) $\frac{5\sqrt{3}}{2}$ seconds
84. (a) 2 seconds (b) 5 seconds
85. about 28 ft
86. 160 ft
87. (a) about 42 ft (b) Yes
88. (a) 240 ft (b) He avoids the big crash.

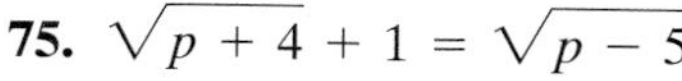

75. $\sqrt{p+4}+1=\sqrt{p-5}$

76. $\sqrt{p+8}-1=\sqrt{p+3}$

77. $\sqrt{11x-18}=-x$

78. $\sqrt{10x-21}=-x$

Applying the Concepts

79. Fun with Numbers The square root of twice the difference of a number and 3 is 4. Find the number.

80. Fun with Numbers The square root of 5 less than twice a number is 7. Find the number.

81. Fun with Numbers. Three times the square root of one more than an unknown number is the same as twice the square root of 3 more than double the unknown number. Find the number.

82. Fun with Numbers Twice the square root of one more than the square of an unknown number is the same as the square root of the sum of the unknown number and 4. Find the number.

Problems 83 and 84 use the following information. The time t that it takes an object to fall h feet is modeled by the formula

$$t=\sqrt{\frac{h}{16}}$$

83. (a) If a ball is dropped off a cliff that is 1024 feet tall, how long will it take to hit the ground?

(b) If a tomato is dropped from the top of a building that is 300 feet tall, exactly how long will it take to hit the ground?

84. (a) If a ball is dropped off a cliff that is 64 feet tall, how long will it take to hit the ground?

(b) If a tomato is dropped from the top of a building that is 400 feet tall, exactly how long will it take to hit the ground?

Problems 85 and 86 use the following information. When driving on asphalt, the distance in feet, d, to stop a car going S miles per hour can be calculated using the following formula:

$$S=\sqrt{22.5d}$$

85. Braking Distance Mark's car is going down a street made of asphalt. If he is driving at 25 mph and slams on the brakes, to the nearest foot, how far will the car skid before it comes to a stop?

86. Braking Distance Dale is driving his Hummer down a county highway made of asphalt in the desert. If he is going 60 mph and slams on the brakes, how far will the car skid before it comes to a stop?

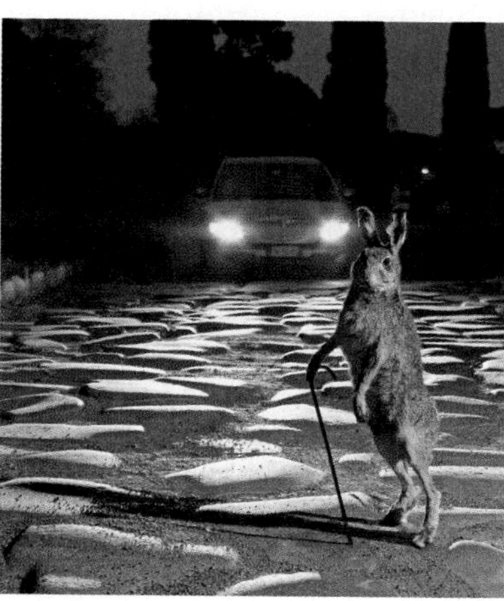

Problems 87 and 88 use the following information. When driving on gravel, the distance in feet, d, to stop a car going S miles per hour can be calculated using the following formula:

$$S=\sqrt{15d}$$

87. Braking Distance Mark gets off the asphalt street and drives onto a gravel road leading to his house, still going 25 mph, when he hits the brakes to avoid a jackrabbit.

(a) To the nearest foot, how far will the car skid before it comes to a stop?

(b) If Mark spotted the jackrabbit 40 feet away when he hit his brakes, will Mark hit the rabbit?

88. Braking Distance Dale decided to go off-road driving in his Hummer and turns off the pavement onto a gravel road at 60 mph. After driving a short time, he sees a huge ditch in the road and he hits his brakes.

(a) How far will the Hummer skid before it comes to a stop?

(b) Dale spotted the ditch 241 feet away when he hit the brakes. Does he land in the ditch or manage to avoid a big crash?

89. $A = \pi r^2$ 90. $g = \frac{v^2}{2h}$
91. $h = \frac{3V}{S^2}$ 92. $h = \frac{3V}{\pi r^2}$
93. $a = \frac{b^2 - b}{2}$ 94. $n = \frac{m - m^2}{5}$
95. $n = \frac{1}{2}m$ 96. $p = -\frac{5}{7}q$

Extending the Concepts

In Problems 89–96, solve for the indicated variable.

89. $\sqrt{\frac{A}{\pi}} = r$; solve for A

90. $v = \sqrt{2gh}$; solve for g

91. $S = \sqrt{\frac{3V}{h}}$; solve for h

92. $r = \sqrt{\frac{3V}{\pi h}}$; solve for h

93. $b = \sqrt{2a + b}$; solve for a

94. $m = \sqrt{m - 5n}$; solve for n

95. $2\sqrt{m - 2n} = \sqrt{4n - 2m}$; solve for n

96. $2\sqrt{3p + 5q} = \sqrt{5q - 9p}$; solve for p

6.7 Higher Roots and Rational Exponents

OBJECTIVES

1. Evaluate Higher Roots
2. Use Product and Quotient Rules to Simplify Higher Roots
3. Define and Evaluate Expressions of the Form $a^{1/n}$
4. Define and Evaluate Expressions of the Form $a^{m/n}$
5. Use Laws of Exponents to Simplify Expressions with Rational Exponents

Preparing for Higher Roots and Rational Exponents

Before getting started, take the following readiness quiz. If you get a problem wrong, go to the section cited and review the material.

1. Evaluate: (a) 2^4 (b) $(-4)^3$ [Section 1.6, p. 50]

2. Simplify: $x^3 \cdot x^5$ [Section 3.2, pp. 185–186]

3. Simplify: $\frac{y^5}{y}$ [Section 3.4, pp. 201–202]

4. Simplify: $(b^3)^4$ [Section 3.2, p. 186]

1 Evaluate Higher Roots

Earlier in this chapter, we commented that finding the square root of a number "reverses" the squaring process. In a similar way, we can reverse the process of raising numbers to other positive integer powers.

In Words

When you see the notation $\sqrt[n]{a} = b$, think to yourself, "Find a number b such that raising that number to the nth power gives me a."

DEFINITION

The **principal *n*th root of a number *a*,** symbolized by $\sqrt[n]{a}$, where $n \geq 2$ is an integer, is defined as follows:

$$\sqrt[n]{a} = b \quad \text{means} \quad a = b^n$$

- If $n \geq 2$ and even, then a and b must be greater than or equal to 0.
- If $n \geq 3$ and odd, then a and b can be any real number.

In the notation $\sqrt[n]{a}$, the integer n, $n \geq 2$, is called the **index.** If a radical is written without the index, it is understood that we mean the square root, so $\sqrt{a}$ represents the square root of a. If the index is 3, we call $\sqrt[3]{a}$ the **cube root** of a. If the index is even, then the radicand must be greater than or equal to 0. If the index is odd, then the radicand can be any real number. Do you know why? Since $\sqrt[n]{a} = b$ means $b^n = a$, if the index n is even, then $b^n \geq 0$ so $a \geq 0$. If we have an odd index n, then b^n can be any real number, so a can be any real number.

Before we evaluate nth roots, we list some perfect powers of 2, 3, 4, and 5 squares, cubes, and so on. See the table on the next page. Having this list will be a great help in finding roots in the examples that follow. Notice that perfect cubes can be negative, but perfect fourths cannot. Do you know why?

Preparing for...Answers **1. (a)** 16 **(b)** -64 **2.** x^8 **3.** y^4 **4.** b^{12}

Teaching Tip
Take time to go through this list of perfect squares, cubes, fourths, and fifths. Emphasize that knowing these numbers will be valuable in simplifying radical expressions.

Perfect Squares	Perfect Cubes	Perfect Fourths	Perfect Fifths
$1^2 = 1$	$(-2)^3 = -8$	$1^4 = 1$	$(-2)^5 = -32$
$2^2 = 4$	$(-1)^3 = -1$	$2^4 = 16$	$(-1)^5 = -1$
$3^2 = 9$	$1^3 = 1$	$3^4 = 81$	$1^5 = 1$
$4^2 = 16$ and so on	$2^3 = 8$ and so on	$4^4 = 256$ and so on	$2^5 = 32$ and so on

Classroom Example ➤
Use Example 1 to provide examples of evaluating higher roots.

Work Smart
$\sqrt[3]{-8} = -2$ because $(-2)^3 = 8$. But $\sqrt[4]{-16}$ is not a real number because there's no real number b such that $b^4 = -16$. Remember, we can take odd roots of any real number, but we can only take even roots of nonnegative real numbers.

EXAMPLE 1 Evaluating Higher Roots

(a) $\sqrt[3]{-1} = -1$ because $(-1)^3 = -1$

(b) $\sqrt[4]{16} = 2$ because $2^4 = 16$

(c) $\sqrt[5]{-32} = -2$ because $(-2)^5 = -32$

(d) $\sqrt[3]{\dfrac{1}{64}} = \dfrac{1}{4}$ because $\left(\dfrac{1}{4}\right)^3 = \dfrac{1}{64}$

(e) $\sqrt[4]{-256}$ is not a real number. There is no real number b such that $b^4 = -256$.

QUICK ✓ *Find each root, if possible.*

1. $\sqrt[5]{32}$ **2.** $\sqrt[3]{-27}$ **3.** $\sqrt[4]{-16}$ **4.** $\sqrt[6]{64}$

Remember in Section 6.1, we said that $\sqrt{a^2} = a$ provided that $a \geq 0$. The same type of rule applies to nth roots. For example, what is $\sqrt[3]{4^3}$? Here we are asking, what number, when raised to the 3rd power, gives me 4^3? The answer is 4, so $\sqrt[3]{4^3} = 4$. As another example, what is $\sqrt[4]{10^4}$? We are asking, what number, when raised to the 4th power, gives me 10^4? The answer is 10, so $\sqrt[4]{10^4} = 10$. We have the following general result.

In Words
Odd roots of positive numbers are positive and odd roots of negative numbers are negative. If the radicand is nonnegative, the even root is nonnegative.

SIMPLIFYING $\sqrt[n]{a^n}$

If $n \geq 2$ is an integer, then

$$\sqrt[n]{a^n} = a \quad \text{if } n \text{ is odd}$$
$$\sqrt[n]{a^n} = |a| \quad \text{if } n \text{ is even}$$

If $a \geq 0$, then $\sqrt[n]{a^n} = a$ for $n \geq 2$ is an integer.

Classroom Example ➤
Simplify:
(a) $\sqrt[5]{7^5}$ (b) $\sqrt[7]{x^7}$
(c) $\sqrt[4]{n^4}$ for $n \geq 0$ (d) $\sqrt[3]{(5a)^3}$
Answer:
(a) 7 (b) x (c) n (d) $5a$

EXAMPLE 2 Simplifying Expressions of the Form $\sqrt[n]{a^n}$

(a) $\sqrt[4]{5^4} = 5$ **(b)** $\sqrt[5]{x^5} = x$ **(c)** $\sqrt[6]{y^6} = y$ for $y \geq 0$ **(d)** $\sqrt[3]{(6z)^3} = 6z$

QUICK ✓ *Simplify each expression.*

5. $\sqrt[6]{7^6}$ **6.** $\sqrt[4]{w^4}$ for $w \geq 0$ **7.** $\sqrt[5]{(12p)^5}$

2 Use Product and Quotient Rules to Simplify Higher Roots

Remember that we simplified square roots such as $\sqrt{18}$ by looking for a factor of 18 that is a perfect square. In the same way, we can simplify radicals such as $\sqrt[3]{16}$ or $\sqrt[4]{32}$. To do this, we use the Product Rule of Radicals that applies for any integer index $n \geq 2$.

In Words

$\sqrt[n]{ab} = \sqrt[n]{a} \cdot \sqrt[n]{b}$ means "the product of the roots equals the root of the product."

PRODUCT RULE OF RADICALS

If $\sqrt[n]{a}$ and $\sqrt[n]{b}$ are real numbers and $n \geq 2$ is an integer, then

$$\sqrt[n]{ab} = \sqrt[n]{a} \cdot \sqrt[n]{b}$$

To simplify higher roots we follow the same steps we used to simplify square roots: We write the radicand as the product of two or more factors, at least one of which is a perfect power of the index, and then take the indicated root. For example, if we are simplifying a cube root, we want to factor the radicand so that one of the factors is a perfect cube.

Classroom Example

Simplify:

(a) $\sqrt[3]{32}$ (b) $\sqrt[4]{48y^4}, y \geq 0$

Answer:

(a) $2\sqrt[3]{4}$ (b) $2y\sqrt[4]{3}$

EXAMPLE 3 Simplifying Higher Roots Using the Product Rule

Simplify each of the following:

(a) $\sqrt[3]{54}$ **(b)** $\sqrt[4]{32x^4}, x \geq 0$

Solution

(a) 27 is a factor of 54 and 27 is a perfect cube, so we write 54 as $27 \cdot 2$: $\sqrt[3]{54} = \sqrt[3]{27 \cdot 2}$

$\sqrt[n]{ab} = \sqrt[n]{a} \cdot \sqrt[n]{b}$: $= \sqrt[3]{27} \cdot \sqrt[3]{2}$

$27 = 3^3$: $= \sqrt[3]{3^3} \cdot \sqrt[3]{2}$

$\sqrt[n]{a^n} = a$: $= 3\sqrt[3]{2}$

(b) 16 is a factor of 32 and 16 is a fourth power ($16 = 2^4$), so we write 32 as $16 \cdot 2$: $\sqrt[4]{32x^4} = \sqrt[4]{16x^4 \cdot 2}$

$\sqrt[n]{ab} = \sqrt[n]{a} \cdot \sqrt[n]{b}$: $= \sqrt[4]{16x^4} \cdot \sqrt[4]{2}$

$16x^4 = 2^4x^4 = (2x)^4$: $= \sqrt[4]{(2x)^4} \cdot \sqrt[4]{2}$

$\sqrt[n]{a^n} = a$: $= 2x\sqrt[4]{2}$

Teaching Tip

The same steps are used to simplify higher roots as were used with square roots. You may wish to work two examples side by side to show the similarities.

QUICK ✓ *Simplify the radical.*

8. $\sqrt[3]{-32}$ **9.** $\sqrt[3]{250n^3}$ **10.** $\sqrt[4]{81b^8}, b \geq 0$

The Quotient Rule of Radicals can also be generalized to any integer index $n \geq 2$.

In Words

$$\sqrt[n]{\frac{a}{b}} = \frac{\sqrt[n]{a}}{\sqrt[n]{b}}$$

means "the root of the quotient equals the quotient of the roots."

QUOTIENT RULE OF RADICALS

If $\sqrt[n]{a}$ and $\sqrt[n]{b}$ are real numbers, $b \neq 0$, and $n \geq 2$ is an integer, then

$$\sqrt[n]{\frac{a}{b}} = \frac{\sqrt[n]{a}}{\sqrt[n]{b}}$$

Classroom Example
Simplify:

(a) $\dfrac{\sqrt[3]{-108}}{\sqrt[3]{2}}$ (b) $\dfrac{\sqrt[4]{96y^5}}{\sqrt[4]{2y}}, y > 0$

Answer:

(a) $-3\sqrt[3]{2}$ (b) $2y\sqrt[4]{3}$

EXAMPLE 4 Simplifying Higher Roots Using the Quotient Rule

Simplify:

(a) $\dfrac{\sqrt[3]{-48}}{\sqrt[3]{3}}$ **(b)** $\dfrac{\sqrt[4]{64z^9}}{\sqrt[4]{2z}}, z > 0$

Solution

(a)

$\dfrac{\sqrt[n]{a}}{\sqrt[n]{b}} = \sqrt[n]{\dfrac{a}{b}}$: $\dfrac{\sqrt[3]{-48}}{\sqrt[3]{3}} = \sqrt[3]{\dfrac{-48}{3}}$

$= \sqrt[3]{-16}$

-8 is a factor of -16 and -8 is a perfect cube $(-8 = (-2)^3)$, so we write -16 as $-8 \cdot 2$: $= \sqrt[3]{-8 \cdot 2}$

$\sqrt[n]{ab} = \sqrt[n]{a} \cdot \sqrt[n]{b}$: $= \sqrt[3]{-8} \cdot \sqrt[3]{2}$

Simplify each radical: $= -2\sqrt[3]{2}$

(b)

$\dfrac{\sqrt[n]{a}}{\sqrt[n]{b}} = \sqrt[n]{\dfrac{a}{b}}$: $\dfrac{\sqrt[4]{64z^9}}{\sqrt[4]{2z}} = \sqrt[4]{\dfrac{64z^9}{2z}}$

$= \sqrt[4]{32z^8}$

16 is a factor of 32 and $16 = 2^4$: $= \sqrt[4]{16z^8 \cdot 2}$

$\sqrt[n]{ab} = \sqrt[n]{a} \cdot \sqrt[n]{b}$: $= \sqrt[4]{16z^8} \cdot \sqrt[4]{2}$

$\sqrt[4]{16} = 2$; $\sqrt[4]{z^8} = z^2$: $= 2z^2\sqrt[4]{2}$

QUICK ✓ *Find the product or quotient and simplify.*

11. $\dfrac{\sqrt[4]{324}}{\sqrt[4]{4}}$ **12.** $\dfrac{\sqrt[3]{512m^{10}}}{\sqrt[3]{4m}}, m \neq 0$

3 Define and Evaluate Expressions of the Form $a^{1/n}$

The world cannot easily be described using only integers, so it seems logical that we would want to extend the rules that apply to integer exponents to rational exponents.

Work Smart

A rational number is a number that can be written as a fraction in which the denominator is not zero.

We start by providing a definition for "a raised to the power $\dfrac{1}{n}$," where a is a real number and n is a positive integer. Remember, one of the Laws of Exponents states that $(a^m)^n = a^{m \cdot n}$. We can use this law to help us determine what $a^{1/n}$ means.

$a^2 = a \cdot a$ $(5^{1/2})^2 = 5^{\frac{1}{2}} \cdot 5^{\frac{1}{2}}$

$a^m \cdot a^n = a^{m+n}$: $= 5^{\frac{1}{2}+\frac{1}{2}}$

$= 5^1$

$= 5$

We also know that $(\sqrt{5})^2 = 5$, so it is reasonable to conclude that

$$5^{1/2} = \sqrt{5}$$

This suggests the following definition:

Teaching Tip

Be sure to note that the root must exist. For example, $(-25)^{1/2}$ is not a real number.

DEFINITION OF $a^{1/n}$

If a is a real number and $n \geq 2$ is an integer, then

$$a^{1/n} = \sqrt[n]{a}$$

provided $\sqrt[n]{a}$ exists.

Classroom Example
Evaluate each expression:
(a) $9^{1/2}$ (b) $(-8)^{1/3}$
Answer:
(a) 3 (b) -2

EXAMPLE 5 Evaluating Expressions of the Form $a^{1/n}$

Evaluate each of the following expressions:

(a) $16^{1/2}$ **(b)** $(-27)^{1/3}$

Solution

(a) $a^{1/2} = \sqrt{a}$: $16^{1/2} = \sqrt{16} = 4$ **(b)** $a^{1/3} = \sqrt[3]{a}$: $(-27)^{1/3} = \sqrt[3]{-27} = -3$

Classroom Example
Evaluate:
(a) $-49^{1/2}$ (b) $(-49)^{1/2}$
Answer:
(a) -7 (b) not a real number

EXAMPLE 6 Evaluating Expressions of the Form $a^{1/n}$

Evaluate each of the following expressions:

(a) $-81^{1/2}$ **(b)** $(-81)^{1/2}$

Solution

(a)
$$-81^{\frac{1}{2}} = -1 \cdot 81^{\frac{1}{2}}$$
$$a^{\frac{1}{2}} = \sqrt{a}: \quad = -\sqrt{81}$$
$$= -9$$

(b) $(-81)^{1/2} = \sqrt{-81}$ is not a real number because there is no real number whose square is -81.

QUICK ✓ *Evaluate each of the expressions, if possible.*

13. $49^{1/2}$ **14.** $(-64)^{1/3}$ **15.** $-100^{1/2}$ **16.** $(-100)^{1/2}$

4 Define and Evaluate Expressions of the Form $a^{m/n}$

We now look for a definition for $a^{m/n}$, where m and n are integers. In the expression $a^{m/n}$, m/n is reduced to lowest terms and $n \geq 2$. The definition we provide obeys all the laws of exponents presented earlier. For example,

$$a^{\frac{m}{n}} = a^{m \cdot \frac{1}{n}} = (a^m)^{\frac{1}{n}} = \sqrt[n]{a^m}$$

and

$$a^{\frac{m}{n}} = a^{\frac{1}{n} \cdot m} = \left(a^{\frac{1}{n}}\right)^m = (\sqrt[n]{a})^m$$

This suggests the following definition:

Work Smart
The expression $a^{m/n}$ is not a real number when a is a negative real number and n is an even integer.

DEFINITION OF $a^{m/n}$

If a is a real number and m/n is a rational number in lowest terms with $n \geq 2$, then

$$a^{m/n} = \sqrt[n]{a^m} = (\sqrt[n]{a})^m$$

provided that $\sqrt[n]{a}$ exists.

If an expression is written as a radical, we can rewrite it with rational exponents, using the definitions of rational exponents.

Classroom Example
Write each radical with a rational exponent.
(a) $\sqrt{n}$ (b) $\sqrt[3]{t^2}$ (c) $\sqrt[5]{r^3}$
Answer:
(a) $n^{1/2}$ (b) $t^{2/3}$ (c) $r^{3/5}$

EXAMPLE 7 Writing a Radical Expression with Rational Exponents

Write each radical with a rational exponent:

(a) $\sqrt{a}$ **(b)** $\sqrt[3]{z^2}$ **(c)** $\sqrt[4]{p^7}$

Solution

(a) $\sqrt{a} = a^{1/2}$ **(b)** $\sqrt[3]{z^2} = z^{2/3}$ **(c)** $\sqrt[4]{p^7} = p^{7/4}$

QUICK ✓ *Write each radical with rational exponents.*

17. $\sqrt[3]{n}$ **18.** $\sqrt[4]{d^3}$ **19.** $\sqrt[5]{y^8}$

Likewise, if an expression is written with rational exponents, we can write it in radical form using the definitions of rational exponents.

Classroom Example
Write each exponential expression as a radical:
(a) $m^{3/4}$ (b) $2y^{2/5}$ (c) $(5z)^{2/3}$

Answer:
(a) $\sqrt[4]{m^3} = (\sqrt[4]{m})^3$
(b) $2\sqrt[5]{y^2} = 2(\sqrt[5]{y})^2$
(c) $\sqrt[3]{(5z)^2} = (\sqrt[3]{5z})^2$

EXAMPLE 8 Writing an Expression with Rational Exponents as a Radical

Write each exponential expression as a radical:

(a) $b^{2/3}$ **(b)** $7x^{3/5}$ **(c)** $(7x)^{3/5}$

Solution

(a)
$$b^{2/3} = \sqrt[3]{b^2}$$
$$\sqrt[n]{a^m} = (\sqrt[n]{a})^m: \quad = (\sqrt[3]{b})^2$$

(b)
$$7x^{3/5} = 7\sqrt[5]{x^3}$$
$$\sqrt[n]{a^m} = (\sqrt[n]{a})^m: \quad = 7(\sqrt[5]{x})^3$$

(c)
$$(7x)^{3/5} = \sqrt[5]{(7x)^3}$$
$$\sqrt[n]{a^m} = (\sqrt[n]{a})^m: \quad = (\sqrt[5]{7x})^3$$

QUICK ✓ *Write each expression in radical form.*

20. $z^{4/5}$ **21.** $2n^{2/3}$ **22.** $(2n)^{2/3}$

When evaluating or simplifying $a^{m/n}$, either $\sqrt[n]{a^m}$ or $(\sqrt[n]{a})^m$ may be used. Use the one that makes simplifying the expression easier. Generally, taking the root first, as in $(\sqrt[n]{a})^m$, is easier.

Classroom Example
Evaluate each of the following:
(a) $9^{3/2}$ (b) $125^{2/3}$

Answer:
(a) 27 (b) 25

EXAMPLE 9 Evaluating Expressions of the Form $a^{m/n}$

Evaluate each of the following expressions.

(a) $16^{3/2}$ **(b)** $27^{2/3}$

Solution

(a) $16^{3/2} = (\sqrt{16})^3 = 4^3 = 64$

(b) $27^{2/3} = (\sqrt[3]{27})^2 = 3^2 = 9$

Work Smart
When simplifying expressions of the form $a^{m/n}$, it is typically easier to evaluate the radical first.

Classroom Example
Evaluate each expression, if possible:
(a) $-25^{3/2}$ (b) $(-8)^{4/3}$
(c) $(-100)^{5/2}$

Answer:
(a) -125 (b) 16
(c) not a real number

EXAMPLE 10 Evaluating Expressions of the Form $a^{m/n}$

Evaluate each expression, if possible:

(a) $-9^{3/2}$ **(b)** $(-64)^{4/3}$ **(c)** $(-25)^{\frac{7}{2}}$

Solution

(a)
$$\begin{aligned} -9^{3/2} &= -1 \cdot 9^{3/2} \\ &= -1 \cdot (\sqrt{9})^3 \\ &= -1 \cdot 3^3 \\ &= -1 \cdot 27 \\ &= -27 \end{aligned}$$

(b)
$$\begin{aligned} (-64)^{4/3} &= (\sqrt[3]{-64})^4 \\ &= (-4)^4 \\ &= 256 \end{aligned}$$

(c) $(-25)^{\frac{7}{2}}$ is not a real number because $(-25)^{\frac{7}{2}} = (\sqrt{-25})^7$ and $\sqrt{-25}$ is not a real number.

QUICK ✓ *Evaluate each expression.*

23. $25^{3/2}$ **24.** $64^{2/3}$ **25.** $-16^{3/4}$ **26.** $(-8)^{4/3}$ **27.** $(-49)^{5/2}$

If a rational exponent is negative, then we can use the rule for negative exponents given in Section 3.4 on page 204.

Teaching Tip
You may wish to review the negative exponent rule with integer exponents before introducing negative exponent rule with rational exponents.

DEFINITION: NEGATIVE-EXPONENT RULE

If m/n is a rational number and if a is a nonzero real number (that is, if $a \neq 0$), then

$$a^{-m/n} = \frac{1}{a^{m/n}} \text{ or } \frac{1}{a^{-m/n}} = a^{m/n}$$

Classroom Example Rewrite each expression with a positive exponent and simplify, if possible.
(a) $81^{-\frac{1}{2}}$ (b) $8^{-\frac{2}{3}}$ (c) $-16^{-\frac{3}{2}}$
Answer:
(a) $\frac{1}{9}$ (b) $\frac{1}{4}$ (c) $-\frac{1}{64}$

EXAMPLE 11 Evaluating Expressions with Negative Rational Exponents

Rewrite each of the following with a positive exponent and completely simplify, if possible.

(a) $25^{-\frac{1}{2}}$ **(b)** $27^{-\frac{2}{3}}$ **(c)** $-4^{-\frac{3}{2}}$

Solution

(a)
$$a^{-m/n} = \frac{1}{a^{m/n}}: \quad 25^{-\frac{1}{2}} = \frac{1}{25^{\frac{1}{2}}}$$
$$a^{1/2} = \sqrt{a}: \quad = \frac{1}{\sqrt{25}}$$
$$= \frac{1}{5}$$

(b)
$$a^{-m/n} = \frac{1}{a^{m/n}}: \quad 27^{-\frac{2}{3}} = \frac{1}{27^{\frac{2}{3}}}$$
$$a^{m/n} = (\sqrt[n]{a})^m: \quad = \frac{1}{(\sqrt[3]{27})^2}$$
$$= \frac{1}{3^2}$$
$$= \frac{1}{9}$$

Work Smart
A negative exponent doesn't mean a negative answer.

(c)
$$-4^{-\frac{3}{2}} = -1 \cdot 4^{-\frac{3}{2}}$$
$$a^{-m/n} = \frac{1}{a^{m/n}}: \quad = \frac{-1}{4^{\frac{3}{2}}}$$
$$a^{m/n} = (\sqrt[n]{a})^m: \quad = \frac{-1}{(\sqrt{4})^3}$$
$$= \frac{-1}{2^3}$$
$$= \frac{-1}{8} = -\frac{1}{8}$$

QUICK ✓ *Rewrite each of the following with positive exponents and completely simplify, if possible.*

28. $36^{-\frac{1}{2}}$ **29.** $8^{-\frac{4}{3}}$ **30.** $-25^{-\frac{3}{2}}$

Teaching Tip
Emphasize that these properties of exponents apply to rational exponents as well as integer exponents. The properties don't change, just the type of numbers in the exponents.

5 Use Laws of Exponents to Simplify Expressions with Rational Exponents

The Laws of Exponents presented in Chapter 3 applied to integer exponents. It turns out that these same laws apply to rational exponents as well. To refresh your memory, we present the laws on the next page.

LAWS OF EXPONENTS

If a and b are real numbers and if r and s are rational numbers, then assuming the expression is defined,

Product Rule: $a^r \cdot a^s = a^{r+s}$

Quotient Rule: $\dfrac{a^r}{a^s} = a^{r-s} = \dfrac{1}{a^{s-r}}$ if $a \neq 0$

Power Rule: $(a^r)^s = a^{r \cdot s}$

Product to a Power Rule: $(ab)^r = a^r \cdot b^r$

Negative-Exponent Rule: $a^{-r} = \dfrac{1}{a^r}$ if $a \neq 0$

The direction **simplify** shall mean the following:

- All the exponents are positive.
- The base only occurs once.
- There are no parentheses in the expression.
- There are no powers written to powers.

Classroom Example
Simplify:
(a) $5^{2/3} \cdot 5^{4/3}$ (b) $\dfrac{11^{1/3}}{11^{4/3}}$ (c) $(a^6)^{1/3}$

Answer:
(a) 25 (b) $\dfrac{1}{11}$ (c) a^2

EXAMPLE 12 Using Laws of Exponents to Simplify Expressions Involving Rational Exponents

Simplify each of the following:

(a) $3^{1/2} \cdot 3^{3/2}$ **(b)** $\dfrac{7^{2/3}}{7^{5/3}}$ **(c)** $(x^4)^{1/2}$

Solution

(a) $a^r \cdot a^s = a^{r+s}$: $3^{1/2} \cdot 3^{3/2} = 3^{\frac{1}{2}+\frac{3}{2}}$
$= 3^{4/2}$
$= 3^2$
$= 9$

(b) $\dfrac{a^r}{a^s} = a^{r-s}$: $\dfrac{7^{2/3}}{7^{5/3}} = 7^{\frac{2}{3}-\frac{5}{3}}$
$= 7^{-\frac{3}{3}}$
$= 7^{-1}$
$= \dfrac{1}{7}$

(c) $(a^r)^s = a^{r \cdot s}$: $(x^4)^{1/2} = x^{4 \cdot \frac{1}{2}}$
$= x^2$

Classroom Example
Simplify:
(a) $(8n^{1/2})^{2/3}$ (b) $\dfrac{a^{5/3} \cdot a^{-4/3}}{a^{-2/3}}$

Answer:
(a) $4n^{1/3}$ (b) a

EXAMPLE 13 Using Laws of Exponents to Simplify Expressions Involving Rational Exponents

Simplify each of the following:

(a) $\left(16x^{\frac{2}{3}}\right)^{\frac{3}{4}}$ **(b)** $\dfrac{n^{\frac{4}{3}} \cdot n^{-\frac{5}{3}}}{n^{-\frac{2}{3}}}$

Solution

(a) $(ab)^r = a^r \cdot b^r$: $\left(16x^{\frac{2}{3}}\right)^{\frac{3}{4}} = 16^{\frac{3}{4}} \cdot \left(x^{\frac{2}{3}}\right)^{\frac{3}{4}}$

$(a^r)^s = a^{r \cdot s}$: $= 16^{\frac{3}{4}} \cdot \left(x^{\frac{2}{3} \cdot \frac{3}{4}}\right)$

Simplify the exponent: $= 16^{\frac{3}{4}} \cdot \left(x^{\frac{2}{3} \cdot \frac{3}{4}}\right)$

$= 16^{\frac{3}{4}} \cdot \left(x^{\frac{1}{2}}\right)$

Write $16^{\frac{3}{4}}$ as a radical: $= \left(\sqrt[4]{16}\right)^3 \cdot x^{\frac{1}{2}}$

Simplify: $= 2^3 \cdot x^{\frac{1}{2}}$
$= 8x^{\frac{1}{2}}$

(b) $a^r \cdot a^s = a^{r+s}$: $\dfrac{n^{\frac{4}{3}} \cdot n^{-\frac{5}{3}}}{n^{-\frac{2}{3}}} = \dfrac{n^{\frac{4}{3}+\left(-\frac{5}{3}\right)}}{n^{-\frac{2}{3}}}$

$= \dfrac{n^{-\frac{1}{3}}}{n^{-\frac{2}{3}}}$

$= n^{-\frac{1}{3}-\left(-\frac{2}{3}\right)}$

$\dfrac{a^r}{a^s} = a^{r-s}$: $= n^{-\frac{1}{3}+\left(\frac{2}{3}\right)}$

$= n^{\frac{1}{3}}$

QUICK ✓ *Simplify each expression.*

31. $7^{-\frac{5}{12}} \cdot 7^{\frac{7}{12}}$ **32.** $\dfrac{11^{\frac{6}{5}}}{11^{\frac{3}{5}}}$ **33.** $\left(2^{\frac{3}{8}}\right)^{\frac{16}{3}}$ **34.** $\dfrac{a^{\frac{3}{2}} \cdot a^{\frac{5}{2}}}{a^{-\frac{3}{2}}}$

6.7 Exercises

For Extra Help: Student Solutions Manual | CD Video | PH Math/Tutor Center | MathXL Tutorials on CD | MathXL® | MyMathLab

1. radicand, index
2. $a^{1/n}$
3. even, negative
4. False
5. True
6. False
7. First answer incorrect; second answer correct; answers may vary.
8. Answers may vary.
9. 8^3 or 512
10. $\sqrt[3]{8}$ or 2
11. $\sqrt[3]{-1}$ or -1
12. $(-1)^3$ or -1
13. $\sqrt[4]{16}$ or 2
14. $(-2)^4$ or 16
15. $(-1)^4$ or 1
16. $\sqrt[4]{-1}$, not a real number
17. 3 18. 4
19. not a real number
20. not a real number
21. -5 22. -10
23. 5 24. 3
25. 0 26. 1
27. 3 28. 2
29. x 30. x

Concepts and Vocabulary

In Problems 1–3, fill in the blanks.

1. In the expression $\sqrt[n]{a}$, a is called the ________ and n is called the ________.

2. $\sqrt[n]{a} =$ ________ in exponential form.

3. The expression $\sqrt[n]{a}$ is not a real number when the index n is ________ and the radicand a is ________.

In Problems 4–6, answer True or False to each statement.

4. To evaluate $8^{2/3}$, it is best to begin by squaring 8.

5. 64 can be used as a factor to simplify square roots and cube roots.

6. $\sqrt[4]{32} = \sqrt[4]{16 \cdot 2} = 4\sqrt[4]{2}$

7. You are helping your friend study for an exam on radical expressions and find that your friend has simplified $\sqrt{-4} = -2$ and $-4^{1/2} = -2$. Are these answers correct or incorrect? Explain your reasoning.

8. Make up an example of a radical expression whose index is 3 and that can be simplified using 27 as a factor. Explain, in general, how to simplify cube roots. What potential errors might occur when simplifying a radical expression of this type?

Building Skills

In Problems 9–16, find the value of each expression.

9. the cube of 8 **10.** the cube root of 8

11. the cube root of -1 **12.** the cube of -1

13. the fourth root of 16 **14.** the fourth power of -2

15. the fourth power of -1 **16.** the fourth root of -1

In Problems 17–30, evaluate each expression.

17. $\sqrt[3]{27}$ **18.** $\sqrt[3]{64}$ **19.** $\sqrt[4]{-1}$ **20.** $\sqrt[4]{-81}$ **21.** $\sqrt[3]{-125}$

22. $\sqrt[3]{-1000}$ **23.** $\sqrt[4]{625}$ **24.** $\sqrt[4]{81}$ **25.** $\sqrt[6]{0}$ **26.** $\sqrt[6]{1}$

27. $\sqrt[5]{243}$ **28.** $\sqrt[5]{32}$ **29.** $\sqrt[3]{x^3}$ **30.** $\sqrt[4]{x^4}, x \geq 0$

In Problems 31–48, simplify each expression, if possible. All variables represent positive real numbers.

31. $\sqrt[3]{16}$ **32.** $\sqrt[3]{24}$ **33.** $\sqrt[4]{-64}$ **34.** $\sqrt[4]{-162}$ **35.** $\sqrt[3]{-40}$

36. $\sqrt[3]{-48}$ **37.** $\dfrac{\sqrt[4]{243}}{\sqrt[4]{3}}$ **38.** $\dfrac{\sqrt[4]{64}}{\sqrt[4]{4}}$ **39.** $-\sqrt[4]{32n^8}$ **40.** $-\sqrt[4]{48n^{12}}$

41. $\sqrt[4]{20{,}000}$ **42.** $\sqrt[4]{50{,}000}$ **43.** $-\sqrt[3]{-128}$ **44.** $-\sqrt[3]{-250}$ **45.** $\sqrt[3]{81x^3}$

46. $\sqrt[3]{54x^3}$ **47.** $\dfrac{\sqrt[3]{56b^5}}{\sqrt[3]{7b^2}}$ **48.** $\dfrac{\sqrt[3]{320w^7}}{\sqrt[3]{5w}}$

In Problems 49–54, write each radical with a rational exponent.

49. $\sqrt{c}$ **50.** $\sqrt[3]{s}$ **51.** $\sqrt[3]{x^2}$ **52.** $\sqrt{x^3}$ **53.** $\sqrt[4]{p^5}$ **54.** $\sqrt[4]{n^3}$

In Problems 55–60, write each expression as a radical expression.

55. $u^{2/3}$ **56.** $v^{3/5}$ **57.** $(2a)^{1/2}$ **58.** $3x^{1/3}$ **59.** $4x^{2/3}$ **60.** $(4x)^{3/4}$

In Problems 61–80, evaluate each expression, if possible.

61. $-16^{1/4}$ **62.** $-81^{1/4}$ **63.** $-8^{1/3}$ **64.** $-216^{1/3}$ **65.** $64^{3/2}$

66. $8^{2/3}$ **67.** $81^{-\frac{1}{2}}$ **68.** $4^{-\frac{1}{2}}$ **69.** $(-216)^{1/3}$ **70.** $(-8)^{1/3}$

71. $64^{2/3}$ **72.** $27^{4/3}$ **73.** $(-36)^{1/2}$ **74.** $(-25)^{1/2}$ **75.** $-9^{-\frac{3}{2}}$

76. $-16^{-\frac{3}{2}}$ **77.** $49^{-\frac{3}{2}}$ **78.** $9^{-\frac{5}{2}}$ **79.** $(-16)^{5/4}$ **80.** $(-81)^{3/4}$

In Problems 81–92, use Laws of Exponents to simplify each of the following. All variables represent positive real numbers.

81. $4^{\frac{2}{3}} \cdot 4^{\frac{4}{3}}$ **82.** $6^{\frac{12}{5}} \cdot 6^{-\frac{2}{5}}$ **83.** $\dfrac{6^{\frac{3}{2}}}{6^{\frac{7}{2}}}$ **84.** $\dfrac{5^{\frac{1}{3}}}{5^{\frac{4}{3}}}$

85. $(x^9)^{2/3}$ **86.** $(y^4)^{5/2}$ **87.** $(8n^{1/2})^{2/3}$ **88.** $(16y^{1/3})^{3/4}$

89. $\dfrac{x^{-1/4} \cdot x^{3/4}}{x^{1/4}}$ **90.** $\dfrac{x^{-3/5} \cdot x^{9/5}}{x^{2/5}}$ **91.** $\dfrac{(2x)^{-3/2}}{(2x)^{-1/2}}$ **92.** $\dfrac{(3x)^{5/4}}{(3x)^{-3/4}}$

Mixed Practice

In Problems 93–106, simplify each expression. Assume that all variables represent positive real numbers.

93. $-25^{\frac{3}{2}}$ **94.** $-16^{\frac{3}{2}}$ **95.** $\dfrac{6^{\frac{5}{2}}}{6^{\frac{7}{2}}}$ **96.** $\dfrac{5^{\frac{1}{3}}}{5^{\frac{4}{3}}}$ **97.** $-125^{-\frac{2}{3}}$

98. $-64^{-\frac{4}{3}}$ **99.** $4^{\frac{1}{2}} + 25^{\frac{3}{2}}$ **100.** $100^{\frac{1}{2}} - 4^{\frac{3}{2}}$ **101.** $\left(\sqrt[4]{25}\right)^2$ **102.** $\left(\sqrt[6]{27}\right)^2$

103. $\left(25^{\frac{3}{4}} \cdot 4^{-\frac{3}{4}}\right)^2$ **104.** $\left(36^{-\frac{1}{4}} \cdot 9^{\frac{3}{4}}\right)^{-2}$ **105.** $\dfrac{y^{-\frac{1}{3}} \cdot y^{-\frac{1}{3}}}{y^{\frac{4}{3}}}$ **106.** $\dfrac{n^{-\frac{5}{4}} \cdot n^{-\frac{1}{4}}}{n^{\frac{1}{2}}}$

Applying the Concepts

107. Evaluate $2x^{-\frac{3}{2}}$ for $x = 4$.

108. Evaluate $10x^{-\frac{2}{3}}$ for $x = 8$.

109. Evaluate $(27x)^{\frac{1}{3}}$ for $x = 8$.

110. Evaluate $(36x)^{\frac{1}{2}}$ for $x = 4$.

Extending the Concepts

In Problems 111–116, simplify the expression. Express your answer in radical form. All variables represent positive real numbers.

111. $x^{\frac{1}{2}} \cdot x^{-\frac{3}{4}}$ **112.** $x^{-\frac{3}{2}} \cdot x^{\frac{7}{6}}$ **113.** $\dfrac{(-8a^4)^{\frac{1}{3}}}{a^{\frac{5}{6}}}$

114. $\dfrac{(100a^3)^{\frac{1}{2}}}{(16a)^{\frac{3}{4}}}$ **115.** $\dfrac{\left(3x^{\frac{2}{3}}\right)^4}{x^{\frac{5}{6}}}$ **116.** $\dfrac{\left(2x^{\frac{2}{5}}\right)^3}{x^{\frac{7}{10}}}$

31. $2\sqrt[3]{2}$ 32. $2\sqrt[3]{3}$
33. not a real number
34. not a real number
35. $-2\sqrt[3]{5}$ 36. $-2\sqrt[3]{6}$
37. 3 38. 2
39. $-2n^2\sqrt[4]{2}$ 40. $-2n^3\sqrt[4]{3}$
41. $10\sqrt[4]{2}$ 42. $10\sqrt[4]{5}$
43. $4\sqrt[3]{2}$ 44. $5\sqrt[3]{2}$
45. $3x\sqrt[3]{3}$ 46. $3x\sqrt[3]{2}$
47. $2b$ 48. $4w^2$ 49. $c^{1/2}$
50. $s^{1/3}$ 51. $x^{2/3}$ 52. $x^{3/2}$
53. $p^{5/4}$ 54. $n^{3/4}$ 55. $\sqrt[3]{u^2} = \left(\sqrt[3]{u}\right)^2$
56. $\sqrt[5]{v^3} = \left(\sqrt[5]{v}\right)^3$ 57. $\sqrt{2a}$ 58. $3\sqrt[3]{x}$
59. $4\sqrt[3]{x^2} = 4\left(\sqrt[3]{x}\right)^2$ 60. $\sqrt[4]{64x^3}$
61. -2 62. -3 63. -2 64. -6
65. 512 66. 4
67. $\frac{1}{9}$ 68. $\frac{1}{2}$
69. -6 70. -2
71. 16 72. 81
73. not a real number
74. not a real number
75. $-\frac{1}{27}$ 76. $-\frac{1}{64}$
77. $\frac{1}{343}$ 78. $\frac{1}{243}$
79. not a real number
80. not a real number
81. 16 82. 36
83. $\frac{1}{36}$ 84. $\frac{1}{5}$ 85. x^6
86. y^{10} 87. $4\sqrt[3]{n}$ 88. $8\sqrt[4]{y}$
89. $x^{1/4}$ or $\sqrt[4]{x}$ 90. $x^{4/5}$ or $\sqrt[5]{x^4}$
91. $\frac{1}{2x}$ 92. $9x^2$ 93. -125
94. -64 95. $\frac{1}{6}$
96. $\frac{1}{5}$ 97. $-\frac{1}{25}$ 98. $-\frac{1}{256}$
99. 127 100. 2 101. 5
102. 3 103. $\frac{125}{8}$ 104. $\frac{2}{9}$
105. $\frac{1}{y^2}$ 106. $\frac{1}{n^2}$ 107. $\frac{1}{4}$
108. $\frac{5}{2}$ 109. 6 110. 12
111. $\frac{1}{\sqrt[4]{x}}$
112. $\frac{1}{\sqrt[3]{x}}$
113. $-2\sqrt{a}$
114. $\frac{5}{4}\sqrt[4]{a^3}$
115. $81x\sqrt[6]{x^5}$
116. $8\sqrt{x}$

CHAPTER 6 ACTIVITY: WORKING TOGETHER WITH RADICALS

Focus: As a group, perform operations on radicals.

Time: 20 minutes

Group size: 3–4

As a group, answer the following questions:

1. Assuming that $q = 81x^9y^2$, find: $\sqrt{q}$, $\sqrt[3]{q}$, $\sqrt[4]{q}$
2. Assuming that $x = a\sqrt[3]{125a^3b^4c^2}$
   $y = 10b\sqrt[3]{a^6bc^2}$, find: $x + y$, $x - y$, $x \cdot y$, $x \div y$
3. Assuming that $a = 2x^{\frac{5}{4}}y^{\frac{1}{3}}$
   $b = 6x^{\frac{2}{3}}y^{\frac{5}{2}}$
   $c = 4x^{\frac{1}{5}}y$, find: $a + c$, $a \cdot b$, $b \div c$, $\frac{ab}{c}$

1. $9x^4y\sqrt{x}$, $3x^3\sqrt[3]{3y^2}$, $3x^2\sqrt[4]{xy^2}$
2. $15a^2b\sqrt[3]{bc^2}$, $-5a^2b\sqrt[3]{bc^2}$, $50a^4b^2c\sqrt[3]{b^2c}$, $1/2$
3. $2x^{\frac{5}{4}}y^{\frac{1}{3}} + 4x^{\frac{1}{5}}y$, $12x^{\frac{23}{12}}y^{\frac{17}{6}}$, $\frac{3}{2}x^{\frac{7}{15}}y^{\frac{3}{2}}$, $3x^{\frac{103}{60}}y^{\frac{11}{6}}$

CHAPTER 6 REVIEW

Section 6.1 Introduction to Square Roots

KEY CONCEPTS	KEY TERMS
• The principal square root of any positive number is positive. • The principal square root of 0 is 0 because $0^2 = 0$. That is, $\sqrt{0} = 0$. • The square root of a perfect square is a rational number. • The square root of a positive rational number that is not a perfect square is an irrational number. • The square root of a negative real number is not a real number. • For any real number a, $\sqrt{a^2} = \lvert a\rvert$. If $a \geq 0$, $\sqrt{a^2} = a$.	Square root Radical Principal square root Radicand Perfect square

YOU SHOULD BE ABLE TO . . .	EXAMPLE	REVIEW EXERCISES
(1) Evaluate square roots (p. 393)	Examples 1 through 6	1–12
(2) Determine whether a square root is rational, irrational, or not a real number (p. 395)	Examples 7 and 8	13–16
(3) Find square roots of variable expressions (p. 397)	Examples 9 and 10	17, 18

In Problems 1 and 2, find the value of each expression.

1. the square roots of 4
2. the square roots of 81

In Problems 3–12, find the exact value of each expression.

3. $-\sqrt{1}$
4. $-\sqrt{25}$
5. $\sqrt{0.16}$
6. $\sqrt{0.04}$
7. $\frac{3}{2}\sqrt{\frac{25}{36}}$
8. $\frac{4}{3}\sqrt{\frac{81}{4}}$
9. $\sqrt{25 - 9}$
10. $\sqrt{169 - 25}$
11. $\sqrt{9^2 - (4)(5)(-18)}$
12. $\sqrt{13^2 - (4)(-3)(-4)}$

1. $-2, 2$
2. $-9, 9$
3. -1
4. -5
5. 0.4
6. 0.2
7. $\frac{5}{4}$
8. 6
9. 4
10. 12
11. 21
12. 11

In Problems 13–16, determine whether each square root is rational, irrational, or not a real number. Then evaluate the square root. For each square root that is irrational, use a calculator or Appendix B to round your answer to two decimal places.

13. rational; -3
14. irrational; -3.46
15. irrational; 3.74
16. not a real number
17. $|4x - 9|$
18. $|16m - 25|$

13. $-\sqrt{9}$ **14.** $-\frac{1}{2}\sqrt{48}$ **15.** $\sqrt{14}$ **16.** $\sqrt{-2}$

In Problems 17 and 18, simplify each square root. Assume that the variable can be any real number.

17. $\sqrt{(4x - 9)^2}$ **18.** $\sqrt{(16m - 25)^2}$

Section 6.2 Simplifying Square Roots

KEY CONCEPTS

- **Product Rule of Square Roots**
  If $\sqrt{a}$ and $\sqrt{b}$ are real numbers, then $\sqrt{ab} = \sqrt{a} \cdot \sqrt{b}$.
- **Quotient Rule of Square Roots**
  If $\sqrt{a}$ and $\sqrt{b}$ are nonnegative real numbers and $b \neq 0$, then $\sqrt{\frac{a}{b}} = \frac{\sqrt{a}}{\sqrt{b}}$.

YOU SHOULD BE ABLE TO . . .	EXAMPLE	REVIEW EXERCISES
(1) Use the Product Rule to simplify square roots of constants (p. 400)	Examples 1 through 3	19–24
(2) Use the Product Rule to simplify square roots of variable expressions (p. 402)	Examples 4 through 6	25–30
(3) Use the Quotient Rule to simplify square roots of constants and variable expressions (p. 404)	Examples 7 through 9	31–34

19. $2\sqrt{7}$ 20. $3\sqrt{5}$
21. $10\sqrt{2}$ 22. $5\sqrt{6}$
23. $-1 + \sqrt{2}$
24. $\frac{-1 + \sqrt{3}}{2}$
25. a^{18} 26. x^8
27. $4x^5$ 28. $7a^6$
29. $3n^4\sqrt{2n}$
30. $2y^{12}\sqrt{2y}$
31. $\frac{3}{x^4}$ 32. $\frac{2}{x^2}$
33. $\frac{9y^2}{5}$ 34. $\frac{6}{11n^4}$

In Problems 19–24, simplify each of the square roots, if possible.

19. $\sqrt{28}$ **20.** $\sqrt{45}$ **21.** $\sqrt{200}$

22. $\sqrt{150}$ **23.** $\frac{-2 + \sqrt{8}}{2}$ **24.** $\frac{-3 + \sqrt{27}}{6}$

In Problems 25–30, simplify each of the following. Assume that the variables represent nonnegative real numbers.

25. $\sqrt{a^{36}}$ **26.** $\sqrt{x^{16}}$ **27.** $\sqrt{16x^{10}}$

28. $\sqrt{49a^{12}}$ **29.** $\sqrt{18n^9}$ **30.** $\sqrt{8y^{25}}$

In Problems 31–34, find the quotient and simplify. Assume that variables represent positive real numbers.

31. $\sqrt{\frac{27}{3x^8}}$ **32.** $\sqrt{\frac{8}{2x^4}}$ **33.** $\sqrt{\frac{81y^5}{25y}}$ **34.** $\sqrt{\frac{36n}{121n^9}}$

Section 6.3 Adding and Subtracting Square Roots

KEY CONCEPT	KEY TERM
• To add or subtract square roots, the radicand must be the same.	Like square roots

YOU SHOULD BE ABLE TO . . .	EXAMPLE	REVIEW EXERCISES
(1) Add and subtract square root expressions with like square roots (p. 408)	Examples 1 and 2	35–40
(2) Add and subtract square root expressions with unlike square roots (p. 409)	Examples 3 through 5	41–50

35. $-2\sqrt{7}$ 36. $3\sqrt{11}$
37. $\sqrt{x}$ 38. $-\sqrt{n}$
39. $25a\sqrt{ab}$
40. $3x\sqrt{3xy}$
41. $5\sqrt{3}$ 42. $8\sqrt{2}$
43. $4+\sqrt{5}$ 44. $6-2\sqrt{3}$
45. $19n^2\sqrt{2n}$
46. $9a^2\sqrt{14a}$
47. $2\sqrt{2}+2\sqrt{3}$
48. $\frac{9}{2}\sqrt{6}+\frac{2}{3}\sqrt{3}$
49. $\frac{11}{20}\sqrt{3}$ 50. $-\frac{8}{5}\sqrt{2}$

In Problems 35–40, add or subtract as indicated. Assume all variables represent nonnegative real numbers.

35. $\sqrt{7}-3\sqrt{7}$ **36.** $4\sqrt{11}-\sqrt{11}$ **37.** $4\sqrt{x}-3\sqrt{x}$

38. $5\sqrt{n}-6\sqrt{n}$ **39.** $20a\sqrt{ab}+5a\sqrt{ba}$ **40.** $2x\sqrt{3xy}+x\sqrt{3yx}$

In Problems 41–50, add or subtract as indicated. Assume all variables represent nonnegative real numbers.

41. $-2\sqrt{12}+3\sqrt{27}$ **42.** $-4\sqrt{18}+5\sqrt{32}$ **43.** $4+2\sqrt{20}-\sqrt{45}$

44. $6-2\sqrt{27}+\sqrt{48}$ **45.** $2n\sqrt{8n^3}+5\sqrt{18n^5}$ **46.** $3\sqrt{56a^5}+a^2\sqrt{126a}$

47. $\frac{3}{4}\sqrt{32}+\frac{2}{3}\sqrt{27}-\frac{1}{2}\sqrt{8}$ **48.** $\frac{5}{2}\sqrt{24}+\frac{2}{9}\sqrt{27}-\frac{1}{6}\sqrt{54}$

49. $3\sqrt{\frac{3}{16}}-\frac{1}{2}\sqrt{\frac{12}{25}}$ **50.** $2\sqrt{\frac{2}{25}}-3\sqrt{\frac{8}{9}}$

Section 6.4 Multiplying Expressions with Square Roots

KEY CONCEPTS	KEY TERM
• If $\sqrt{a}$ and $\sqrt{b}$ are real numbers, then $\sqrt{a}\cdot\sqrt{b}=\sqrt{ab}$. • $(\sqrt{a})^2=a$ and $(-\sqrt{a})^2=a$ for any real number a, $a\ge 0$.	Conjugates

YOU SHOULD BE ABLE TO . . .	EXAMPLE	REVIEW EXERCISES
1 Find the product of square roots containing one term (p. 413)	Examples 1 through 6	51–60
2 Find the product of square roots using the Distributive Property (p. 415)	Example 7	61, 62
3 Find the product of square roots using FOIL (p. 416)	Examples 8 and 9	63–66
4 Find the product of square roots using special products: $(a+b)^2$, $(a-b)^2$, and $(a+b)(a-b)$ (p. 417)	Examples 10 and 11	67–72

51. $4\sqrt{6}$ 52. $4\sqrt{15}$
53. $7x^2\sqrt{6x}$
54. $6y^2\sqrt{5y}$
55. $-128\sqrt{10}$
56. -200
57. $8a$
58. $45n$
59. -24
60. -154
61. $3\sqrt{a}+3\sqrt{5a}$
62. $2x+2x\sqrt{2}$
63. $6+12\sqrt{3}+4\sqrt{6}+24\sqrt{2}$
64. $15+10\sqrt{6}-9\sqrt{3}-18\sqrt{2}$
65. $4\sqrt{y}+12-2y\sqrt{2}-6\sqrt{2y}$
66. $\sqrt{x}+9+3x\sqrt{3}+27\sqrt{3x}$

In Problems 51–60, find the product and simplify. Assume all variables represent nonnegative real numbers.

51. $\sqrt{12}\cdot\sqrt{8}$ **52.** $\sqrt{24}\cdot\sqrt{10}$ **53.** $\sqrt{14x^3}\cdot\sqrt{21x^2}$

54. $\sqrt{6y}\cdot\sqrt{30y^4}$ **55.** $-4\sqrt{20}\cdot 8\sqrt{8}$ **56.** $-2\sqrt{10}\cdot 5\sqrt{40}$

57. $(2\sqrt{2a})^2$ **58.** $(3\sqrt{5n})^2$

59. $4\sqrt{2}\cdot(-3\sqrt{2})$ **60.** $7\sqrt{11}\cdot(-2\sqrt{11})$

In Problems 61 and 62, find the product and simplify. Assume all variables represent nonnegative real numbers.

61. $\sqrt{3}(\sqrt{3a}+\sqrt{15a})$ **62.** $\sqrt{x}(\sqrt{4x}+\sqrt{8x})$

In Problems 63–66, find the product and simplify. Assume all variables represent nonnegative real numbers.

63. $(3+2\sqrt{6})(2+4\sqrt{3})$ **64.** $(5-3\sqrt{3})(3+2\sqrt{6})$

65. $(4-2\sqrt{2y})(\sqrt{y}+3)$ **66.** $(1+3\sqrt{3x})(\sqrt{x}+9)$

67. $11 - 6\sqrt{2}$
68. $21 - 8\sqrt{5}$
69. $11x + 6x\sqrt{2}$
70. $13n + 4n\sqrt{3}$
71. $9 - 25x$ 72. $4x - 16$

In Problems 67–72, find the product and simplify. Assume all variables represent nonnegative real numbers.

67. $(3 - \sqrt{2})^2$ **68.** $(4 - \sqrt{5})^2$ **69.** $(\sqrt{2x} + 3\sqrt{x})^2$

70. $(\sqrt{n} + 2\sqrt{3n})^2$ **71.** $(3 + 5\sqrt{x})(3 - 5\sqrt{x})$ **72.** $(2\sqrt{x} - 4)(2\sqrt{x} + 4)$

Section 6.5 Dividing Expressions with Square Roots

KEY CONCEPTS	KEY TERM
• To rationalize the denominator when the denominator contains a single square root, we multiply by a factor of 1, such that the product forms a radicand in the denominator that is a perfect square. • To rationalize a denominator containing two terms, we multiply the numerator and the denominator by the conjugate of the denominator.	Rationalizing the denominator

YOU SHOULD BE ABLE TO . . .	EXAMPLE	REVIEW EXERCISES
1 Find the quotient of two square roots (p. 421)	Examples 1 and 2	73–78
2 Rationalize a denominator containing one term (p. 422)	Examples 3 and 4	79–82
3 Rationalize a denominator containing two terms (p. 424)	Examples 5 through 7	85–88

73. $3\sqrt{6}$ 74. $3\sqrt{5}$ 75. $\frac{\sqrt{6}}{2x}$
76. $\frac{x\sqrt{15}}{6}$ 77. $\sqrt{5} + \sqrt{6}$
78. $\sqrt{6} - \sqrt{3}$ 79. $2\sqrt{7}$
80. $5\sqrt{5}$ 81. $\frac{\sqrt{6}}{4}$
82. $\frac{\sqrt{10}}{25}$
83. $-2 - 3\sqrt{5}; -41$
84. $-3 + 2\sqrt{7}; -19$
85. $4 + 2\sqrt{2}$
86. $\frac{6 - \sqrt{3}}{11}$
87. $2\sqrt{2} - 2$ 88. $\sqrt{3} + 1$

In Problems 73–78, simplify the quotient. Assume all variables represent nonnegative real numbers.

73. $\frac{\sqrt{108}}{\sqrt{2}}$ **74.** $\frac{\sqrt{135}}{\sqrt{3}}$ **75.** $\sqrt{\frac{6x}{4x^3}}$

76. $\sqrt{\frac{15x^4}{36x^2}}$ **77.** $\frac{\sqrt{10} + \sqrt{12}}{\sqrt{2}}$ **78.** $\frac{\sqrt{30} - \sqrt{15}}{\sqrt{5}}$

In Problems 79–82, rationalize the denominator of the expression.

79. $\frac{14}{\sqrt{7}}$ **80.** $\frac{25}{\sqrt{5}}$ **81.** $\frac{3}{2\sqrt{6}}$ **82.** $\frac{2}{5\sqrt{10}}$

In Problems 83 and 84, determine the conjugate of each expression. Then find the product of the expression and its conjugate.

83. $-2 + 3\sqrt{5}$ **84.** $-3 - 2\sqrt{7}$

In Problems 85–88, rationalize the denominator.

85. $\frac{4}{2 - \sqrt{2}}$ **86.** $\frac{3}{6 + \sqrt{3}}$ **87.** $\frac{\sqrt{12}}{\sqrt{3} + \sqrt{6}}$ **88.** $\frac{\sqrt{8}}{\sqrt{6} - \sqrt{2}}$

Section 6.6 Solving Equations Containing Square Roots

KEY CONCEPT	KEY TERMS	
• Steps to solve a radical equation containing one radical see p. 432	Radical equation Solution to a radical equation	Extraneous solution

YOU SHOULD BE ABLE TO . . .	EXAMPLE	REVIEW EXERCISES
1 Determine whether or not a number is a solution of a radical equation (p. 430)	Example 1	89, 90
2 Solve radical equations containing one square root (p. 431)	Examples 2 through 6	91–98
3 Solve equations that involve two square roots (p. 434)	Examples 7 through 9	99–102
4 Solve problems modeled by a radical equation (p. 436)	Example 10	103, 104

In Problems 89 and 90, determine if the number is a solution to the equation.

89. Yes

90. Yes

91. $\{-13\}$

92. $\{13\}$

93. $\{2\}$

94. $\{9\}$

95. $\{3, 5\}$

96. $\{6\}$

97. $\left\{\frac{1}{4}\right\}$

98. $\{-3\}$

99. $\{3\}$

100. $\{2\}$

101. $\varnothing$

102. $\varnothing$

103. 18π square cm

104. 6 in.

89. Is $x = 8$ a solution of $\sqrt{3x + 1} = 5$?

90. Is $x = 4$ a solution of $3\sqrt{x} - 9 = -3$?

In Problems 91–98, solve the equation and check the solution.

91. $\sqrt{3 - x} = 4$ **92.** $5 = \sqrt{2x - 1}$

93. $3\sqrt{2x} + 4 = 10$ **94.** $5\sqrt{x} - 3 = 12$ **95.** $x = \sqrt{8x - 15}$

96. $\sqrt{3x + 18} = x$ **97.** $n + 2 = \sqrt{n^2 + 5}$ **98.** $n + 3 = \sqrt{n^2 - 9}$

In Problems 99–102, solve the equation and check the solution.

99. $\sqrt{2x + 7} = \sqrt{5x - 2}$ **100.** $3\sqrt{a} - \sqrt{4a + 10} = 0$

101. $\sqrt{2x - 8} = 3 + \sqrt{2x + 1}$ **102.** $\sqrt{x - 6} = 2 - \sqrt{x + 10}$

△ **103.** Use the formula $r = \sqrt{\frac{A}{\pi}}$ where r represents the radius and A represents the area of a circle to find the exact area of a circle whose radius is $3\sqrt{2}$ cm.

△ **104.** The formula $r = \sqrt{\frac{3V}{\pi h}}$ can be used to calculate the volume of a cone. If the radius of the base of the cone, r, is 7 inches and the volume, V, is 98π cubic inches, find the height, h, of the cone.

Section 6.7 Higher Roots and Rational Exponents

KEY CONCEPTS	KEY TERMS
• The principal nth root of a number a, symbolized by $\sqrt[n]{a}$, where $n \geq 2$ is an integer, is defined as follows: $\sqrt[n]{a} = b$ means $a = b^n$. • If $n \geq 2$ and even, then a and b must be greater than or equal to 0. • If $n \geq 3$ and odd, then a and b can be any real number. • If $n \geq 2$ is an integer, then $\sqrt[n]{a^n} = a$ if n is odd $\quad \sqrt[n]{a^n} = \lvert a \rvert$ if n is even • If $\sqrt[n]{a}$ and $\sqrt[n]{b}$ are real numbers and $n \geq 2$ is an integer, then $\sqrt[n]{ab} = \sqrt[n]{a} \cdot \sqrt[n]{b}$. • If $\sqrt[n]{a}$ and $\sqrt[n]{b}$ are real numbers, $b \neq 0$, and $n \geq 2$ is an integer, then $\sqrt[n]{\frac{a}{b}} = \frac{\sqrt[n]{a}}{\sqrt[n]{b}}$. • If a is a real number and $n \geq 2$ is an integer, then $a^{1/n} = \sqrt[n]{a}$ provided that $\sqrt[n]{a}$ exists. • If a is a real number and m/n is a rational number in lowest terms with $n \geq 2$, then $a^{m/n} = \sqrt[n]{a^m} = (\sqrt[n]{a})^m$ provided that $\sqrt[n]{a}$ exists. • If m/n is a rational number and if a is a nonzero real number, then $a^{-\frac{m}{n}} = \frac{1}{a^{\frac{m}{n}}}$ or $\frac{1}{a^{-\frac{m}{n}}} = a^{\frac{m}{n}}$. • If a and b are real numbers and if r and s are rational numbers, then assuming the expression is defined, Product Rule: $a^r \cdot a^s = a^{r+s}$ Quotient Rule: $\frac{a^r}{a^s} = a^{r-s} = \frac{1}{a^{s-r}}$ if $a \neq 0$ Power Rule: $(a^r)^s = a^{r \cdot s}$ Power of a Product Rule: $(ab)^r = a^r b^r$ Negative-Exponent Rule: $a^{-r} = \frac{1}{a^r}$ if $a \neq 0$	Cube root Principal nth root Index

YOU SHOULD BE ABLE TO . . .	EXAMPLE	REVIEW EXERCISES
1 Evaluate higher roots (p. 440)	Examples 1 and 2	105, 106
2 Use Product and Quotient Rules to simplify higher roots (p. 442)	Examples 3 and 4	107–110
3 Define and evaluate expressions of the form $a^{1/n}$ (p. 443)	Examples 5 and 6	111, 112
4 Define and evaluate expressions of the form $a^{m/n}$ (p. 444)	Examples 7 through 11	113–120
5 Use Laws of Exponents to simplify expressions with rational exponents (p. 446)	Examples 12 and 13	121–124

105. -2
106. not a real number
107. $2x\sqrt[4]{3}$
108. $5x\sqrt[3]{5}$
109. $2n\sqrt[3]{3n}$
110. $3z\sqrt[4]{2z^2}$
111. -2
112. not a real number
113. $x^{3/4}$
114. $z^{1/3}$
115. $2\sqrt{x^3} = 2x\sqrt{x}$
116. $\sqrt[3]{3v}$
117. 9
118. 32
119. $-\frac{1}{64}$
120. $\frac{1}{8}$
121. 4
122. 36
123. $\frac{1}{x^{3/4}}$
124. $x^{1/5}$

In Problems 105 and 106, evaluate each expression, if possible.

105. $\sqrt[3]{-8}$ **106.** $\sqrt[4]{-16}$

In Problems 107–110, simplify each expression. Assume all variables represent positive real numbers.

107. $\sqrt[4]{48x^4}$ **108.** $\sqrt[3]{625x^3}$ **109.** $\frac{\sqrt[3]{48n^6}}{\sqrt[3]{2n^2}}$ **110.** $\frac{\sqrt[4]{324z^7}}{\sqrt[4]{2z}}$

In Problems 111 and 112, evaluate each expression, if possible.

111. $-16^{1/4}$ **112.** $(-16)^{1/4}$

In Problems 113 and 114, write each radical with a rational exponent.

113. $\sqrt[4]{x^3}$ **114.** $\sqrt[3]{z}$

In Problems 115 and 116, write each expression as a radical.

115. $2x^{3/2}$ **116.** $(3v)^{1/3}$

In Problems 117–120, evaluate each expression, if possible.

117. $27^{\frac{2}{3}}$ **118.** $16^{\frac{5}{4}}$ **119.** $-16^{-\frac{3}{2}}$ **120.** $32^{-\frac{3}{5}}$

In Problems 121–124, simplify each of the following. Assume all variables represent positive real numbers.

121. $2^{\frac{2}{3}} \cdot 2^{\frac{4}{3}}$ **122.** $6^{\frac{1}{2}} \cdot 6^{\frac{3}{2}}$ **123.** $\frac{x^{\frac{3}{4}} \cdot x^{-\frac{1}{4}}}{x^{\frac{5}{4}}}$ **124.** $\frac{x^{-\frac{1}{5}} \cdot x^{\frac{4}{5}}}{x^{\frac{2}{5}}}$

CHAPTER 6 TEST

Remember to use your Chapter Test Prep Video CD to see fully worked-out solutions to any of these problems you would like to review.

Note to instructor: A special file in TestGen provides algorithms specifically matched to the problems in this Chapter Test for easy-to-replicate practice or assessment purposes.

1. $-6\sqrt{3}$
2. $5xy\sqrt{2y}$
3. $\frac{4x}{3}$
4. 7
5. $\sqrt{2}$
6. $17\sqrt{3}$
7. $15\sqrt{3}$
8. $12\sqrt{3} - 4\sqrt{2}$
9. $19 - 8\sqrt{3}$
10. $2x\sqrt{3x}$
11. $2 - 10\sqrt{2}$
12. $\frac{-1 + 3\sqrt{2}}{2}$ 13. $\frac{6\sqrt{5}}{5}$
14. $-3\sqrt{2} + 6$

In Problems 1–4, simplify each of the following. Assume all variables represent positive real numbers.

1. $-3\sqrt{12}$ **2.** $\sqrt{50x^2y^3}$ **3.** $\sqrt{\frac{16x^4}{9x^2}}$ **4.** $\sqrt{25 - 4(3)(-2)}$

In Problems 5–12, perform the indicated operation and simplify. Assume all variables represent positive real numbers.

5. $3\sqrt{2} - \sqrt{8}$ **6.** $3\sqrt{27} + 2\sqrt{48}$ **7.** $\sqrt{45} \cdot \sqrt{15}$

8. $2\sqrt{2}(3\sqrt{6} - 2)$ **9.** $(4 - \sqrt{3})^2$ **10.** $\frac{\sqrt{36x^4}}{\sqrt{3x}}$

11. $(2 - 3\sqrt{2})(4 + \sqrt{2})$ **12.** $\frac{-3 + \sqrt{162}}{6}$

In Problems 13 and 14, rationalize the denominator.

13. $\frac{12}{\sqrt{20}}$ **14.** $\frac{6}{\sqrt{2} + 2}$

15. {5}
16. {5}
17. {7, −3}
18. $3x\sqrt[3]{4x^2}$
19. $\frac{1}{9}$
20. $3\sqrt{5} \approx 6.7$ feet

In Problems 15–17, solve each equation.

15. $3\sqrt{2x - 1} = 9$ **16.** $2\sqrt{x + 11} - 3 = x$ **17.** $\sqrt{x^2 - 3x} = \sqrt{x + 21}$

In Problems 18 and 19, simplify each expression.

18. $\sqrt[3]{108x^5}$ **19.** $27^{-\frac{2}{3}}$

20. Two legs of a right triangle measure 3 feet and 6 feet. Find the exact length of the hypotenuse and then find the decimal approximation to the nearest tenth of a foot. [**Hint:** $c = \sqrt{a^2 + b^2}$ where c is the length of the hypotenuse and a and b are the lengths of the legs.]

CHAPTER

7 Quadratic Equations

Many problems from the physical, behavioral, and social sciences require solving a quadratic equation. For example, there were 1314 traffic fatalities among highway workers in 1999, but this number has declined in recent years. The equation $N = -11x^2 + 75x + 1134$ models the number, N, of fatalities of private industry highway workers per year x years since 1995. According to the model, in which year were there 1000 highway worker fatalities? See Example 4 in Section 7.4.

OUTLINE

The Big Picture: Putting It Together

In Chapter 2, we solved linear, or first-degree, equations. In Chapter 4, we introduced quadratic equations, second-degree equations of the form $ax^2 + bx + c = 0$. All of the quadratic equations in Chapter 4 were solved by factoring the expression $ax^2 + bx + c$ and setting each factor equal to zero using the Zero-Product Property. But not all quadratic equations can be solved using factoring.

With our knowledge of radicals covered in Chapter 6, we now have the ability to develop additional techniques for solving any quadratic equation.

7.1 Solving Quadratic Equations Using the Square Root Property

OBJECTIVES

1 Solve Quadratic Equations Using the Square Root Property

2 Solve Problems Using the Pythagorean Theorem

Preparing for Solving Quadratic Equations Using the Square Root Property

Before getting started, take the following readiness quiz. If you get a problem wrong, go back to the section cited and review the material.

1. Simplify: (a) $\sqrt{25}$ (b) $\sqrt{48}$ [Section 6.1, p. 394; Section 6.2, pp. 400–401]

2. Simplify: $\dfrac{3 - \sqrt{72}}{6}$ [Section 6.2, pp. 401–402]

3. Factor: $z^2 - 8z + 16$ [Section 4.4, pp. 267–269]

In Chapter 4, you learned how to solve quadratic equations by factoring. A quadratic equation is one that can be written in the form $ax^2 + bx + c = 0$, where a, b, and c are real numbers and $a \neq 0$. For example, to solve the quadratic equation $x^2 - 2x - 8 = 0$, we used the following steps:

$$x^2 - 2x - 8 = 0$$

Factor the polynomial: $(x - 4)(x + 2) = 0$

Use the Zero-Product Property: $x - 4 = 0$ or $x + 2 = 0$

Solve the resulting equations: $x = 4$ or $x = -2$

The solution set of the equation is $\{-2, 4\}$.

However, some polynomials of the form $ax^2 + bx + c$ are not easily factored. Further, some polynomials of the form $ax^2 + bx + c$ cannot be factored over the integers at all! To solve quadratic equations where the quadratic expression cannot be factored, we need other methods. We will learn three more methods for solving quadratic equations. In this section, we discuss the technique that uses the *Square Root Property*.

1 Solve Quadratic Equations Using the Square Root Property

Suppose that we wanted to solve the quadratic equation

$$x^2 = p$$

where p is any real number. In words, this equation is saying, "Give me all numbers whose square is p." So if $p = 16$, then we would have the equation

$$x^2 = 16$$

which means we want "all numbers whose square is 16." There are two numbers whose square is 16, -4 and 4, so the solution set to the equation $x^2 = 16$ is $\{-4, 4\}$.

In general, we have the following method for solving equations of the form $x^2 = p$.

Teaching Tip

When we evaluate a square root such as $\sqrt{16}$, we take the principal square root. When we solve an equation, we include both the positive and the negative root.

THE SQUARE ROOT PROPERTY

If $x^2 = p$, where $p \geq 0$, then $x = \sqrt{p}$ or $x = -\sqrt{p}$.

When using the Square Root Property to solve an equation such as $x^2 = p$, we usually abbreviate the solutions as $x = \pm\sqrt{p}$, read "x equals plus or minus the square root of p." For example, the two solutions of the equation

$$x^2 = 16$$

are

$$x = \pm\sqrt{16}$$

and since $\sqrt{16} = 4$, we have

$$x = \pm 4$$

Work Smart

The Square Root Property is useful for solving equations of the form "some unknown squared equals a real number." To solve this equation, we take the square root of both sides of the equation. Because any positive real number has two square roots, don't forget the $\pm$ symbol!

Preparing for...Answers **1. (a)** 5 **(b)** $4\sqrt{3}$ **2.** $\dfrac{1 - 2\sqrt{2}}{2}$ or $\dfrac{1}{2} - \sqrt{2}$ **3.** $(z - 4)^2$

EXAMPLE 1 How to Solve a Quadratic Equation Using the Square Root Property

Solve the quadratic equation $x^2 - 32 = 0$.

Step-by-Step Solution

Did you notice that this equation *isn't* in the form $x^2 = p$? The first thing we must do is to write it in the form $x^2 =$ "a number."

Step 1: Isolate the expression containing the squared term.

$x^2 - 32 = 0$

Add 32 to each side: $x^2 = 32$

Step 2: Use the Square Root Property. Don't forget the $\pm$ symbol.

Use the Square Root Property: $x = \pm\sqrt{32}$

Simplify the radical: $x = \pm\sqrt{16 \cdot 2}$

$x = \pm 4\sqrt{2}$

Step 3: Isolate the variable, if necessary.

The variable is already isolated.

Step 4: Verify your solutions.

$x = -4\sqrt{2}$: $(-4\sqrt{2})^2 - 32 \stackrel{?}{=} 0$

$(a \cdot b)^2 = a^2 \cdot b^2$: $(-4)^2(\sqrt{2})^2 - 32 \stackrel{?}{=} 0$

$16 \cdot 2 - 32 \stackrel{?}{=} 0$

$0 = 0$ True

$x = 4\sqrt{2}$: $(4\sqrt{2})^2 - 32 \stackrel{?}{=} 0$

$(a \cdot b)^2 = a^2 \cdot b^2$: $(4)^2(\sqrt{2})^2 - 32 \stackrel{?}{=} 0$

$16 \cdot 2 - 32 \stackrel{?}{=} 0$

$0 = 0$ True

Classroom Example
Solve:
(a) $k^2 = 121$ (b) $y^2 = 8$
(c) $n^2 - 24 = 0$

Answer:
(a) $\{-11, 11\}$
(b) $\{-2\sqrt{2}, 2\sqrt{2}\}$
(c) $\{-2\sqrt{6}, 2\sqrt{6}\}$

The solution set is $\{-4\sqrt{2}, 4\sqrt{2}\}$.

If you have a quadratic equation in the form $x^2 = p$, where p is a perfect square, such as $y^2 = 25$, you could solve the equation using the Square Root Property or by factoring.

SQUARE ROOT PROPERTY:

$y^2 = 25$

Use the Square Root Property: $y = \pm\sqrt{25}$

$y = \pm 5$

$y = 5$ or $y = -5$

FACTORING:

$y^2 = 25$

Subtract 25 from both sides: $y^2 - 25 = 0$

Factor the difference of two squares: $(y - 5)(y + 5) = 0$

Zero-Product Property: $y - 5 = 0$ or $y + 5 = 0$

$y = 5$ or $y = -5$

Which approach do you prefer? We summarize the steps that are used to solve quadratic equations using the Square Root Property below.

Work Smart
To use the Square Root Property, the equation must be in the form "some unknown squared equals a number."

Steps to Solve a Quadratic Equation Using the Square Root Property

Step 1: Write the equation in the form $x^2 = p$. The coefficient on the squared term must be one.

Step 2: Use the Square Root Property, which states that if $x^2 = p$, then $x = \pm\sqrt{p}$. Don't forget the $\pm$ symbol.

Step 3: Isolate the variable, if necessary.

Step 4: Verify your solution(s).

QUICK ✓ *Solve the quadratic equation using the Square Root Property.*

1. $s^2 = 36$ **2.** $p^2 - 20 = 0$

Classroom Example ➤
Solve each quadratic equation:

(a) $2n^2 = 40$ (b) $\frac{4}{3}a^2 + 2 = 11$

Answer:
(a) $\{-2\sqrt{5}, 2\sqrt{5}\}$
(b) $\left\{-\frac{3\sqrt{3}}{2}, \frac{3\sqrt{3}}{2}\right\}$

EXAMPLE 2 Solving a Quadratic Equation in the Form $ax^2 = p$

Solve each quadratic equation:

(a) $3y^2 = 36$ **(b)** $\frac{9}{2}b^2 + 4 = 15$

Solution

(a)

$$3y^2 = 36$$

Divide both sides of the equation by 3: $$\frac{3y^2}{3} = \frac{36}{3}$$

$$y^2 = 12$$

Use the Square Root Property: $$y = \pm\sqrt{12}$$

Simplify the radical expression: $$y = \pm\sqrt{4 \cdot 3}$$

$$y = \pm 2\sqrt{3}$$

The solution set is $\{-2\sqrt{3}, 2\sqrt{3}\}$. We leave it to you to check the solutions.

(b)

$$\frac{9}{2}b^2 + 4 = 15$$

Subtract 4 from both sides: $$\frac{9}{2}b^2 + 4 - 4 = 15 - 4$$

$$\frac{9}{2}b^2 = 11$$

Multiply both sides by $\frac{2}{9}$: $$\frac{2}{9} \cdot \frac{9}{2}b^2 = \frac{2}{9} \cdot 11$$

Use the Square Root Property: $$b^2 = \frac{22}{9}$$

Don't forget the $\pm$ symbol: $$b = \pm\sqrt{\frac{22}{9}}$$

$\sqrt{\frac{a}{b}} = \frac{\sqrt{a}}{\sqrt{b}}$: $$b = \pm\frac{\sqrt{22}}{\sqrt{9}}$$

$$b = \pm\frac{\sqrt{22}}{3}$$

The solution set is $\left\{-\frac{\sqrt{22}}{3}, \frac{\sqrt{22}}{3}\right\}$. We leave it to you to check the solutions.

QUICK ✓ *Solve the quadratic equation using the Square Root Property.*

3. $3p^2 = 75$ **4.** $\frac{1}{2}z^2 + 1 = 10$

Sometimes we need to rationalize the denominator of the solution.

Classroom Example Solve the quadratic equation $5z^2 + 1 = 50$.

Answer: $\left\{-\frac{7\sqrt{5}}{5}, \frac{7\sqrt{5}}{5}\right\}$

Teaching Tip Continue to emphasize that the equation must be in the form "some unknown squared equals a number" in order to apply the Square Root Property.

EXAMPLE 3 Solving a Quadratic Equation in the Form $ax^2 = p$

Solve the quadratic equation $3x^2 - 25 = 0$.

Solution

$$3x^2 - 25 = 0$$

Add 25 to both sides: $3x^2 = 25$

Divide both sides by 3: $x^2 = \frac{25}{3}$

Use the Square Root Property: $x = \pm\sqrt{\frac{25}{3}}$

$\sqrt{\frac{a}{b}} = \frac{\sqrt{a}}{\sqrt{b}}$: $x = \pm\frac{\sqrt{25}}{\sqrt{3}}$

$$x = \pm\frac{5}{\sqrt{3}}$$

Rationalize the denominator: $x = \pm\frac{5}{\sqrt{3}} \cdot \frac{\sqrt{3}}{\sqrt{3}}$

$$x = \pm\frac{5\sqrt{3}}{3}$$

The solution set is $\left\{-\frac{5\sqrt{3}}{3}, \frac{5\sqrt{3}}{3}\right\}$. We leave it to you to check the solutions.

QUICK ✓ *Solve the quadratic equation using the Square Root Property.*

5. $x^2 = \frac{1}{2}$

6. $5x^2 - 16 = 0$

Classroom Example Solve each quadratic equation:

(a) $(x - 2)^2 = 25$

(b) $(3y - 3)^2 - 7 = 11$

Answer:

(a) $\{-3, 7\}$ (b) $\{1 - \sqrt{2}, 1 + \sqrt{2}\}$

Solving Quadratic Equations of the Form $(ax + b)^2 = p$

When solving a quadratic equation such as $(3y - 6)^2 - 2 = 25$, we isolate the entire expression containing the variable and then follow the steps for using the Square Root Property.

EXAMPLE 4 How to Solve a Quadratic Equation in the Form $(ax + b)^2 = p$

Solve: $(3y - 6)^2 - 2 = 25$

Step-by-Step Solution

Step 1: Isolate the expression containing the squared term.

$$(3y - 6)^2 - 2 = 25$$

Add 2 to each side: $(3y - 6)^2 - 2 + 2 = 25 + 2$

$$(3y - 6)^2 = 27$$

(continued)

Step 2: Use the Square Root Property. Don't forget the ± symbol.

Use the Square Root Property: $3y - 6 = \pm\sqrt{27}$

Simplify the radical expression: $3y - 6 = \pm\sqrt{9 \cdot 3}$

$$3y - 6 = \pm 3\sqrt{3}$$

Step 3: Isolate the variable, if necessary.

Add 6 to each side: $3y - 6 + 6 = \pm 3\sqrt{3} + 6$

Apply the Commutative Property of Addition to rearrange terms: $3y = 6 \pm 3\sqrt{3}$

Divide each side by 3: $\dfrac{3y}{3} = \dfrac{6 \pm 3\sqrt{3}}{3}$

$$y = \frac{6 \pm 3\sqrt{3}}{3}$$

Factor the numerator: $y = \dfrac{3(2 \pm \sqrt{3})}{3}$

Divide out common factors: $y = 2 \pm \sqrt{3}$

Work Smart

Be careful simplifying the radical expression $\dfrac{6 \pm 3\sqrt{3}}{3}$.

Step 4: Verify your solutions.

$y = 2 - \sqrt{3}$:

$$(3y - 6)^2 - 2 = 25$$
$$\left[3(2 - \sqrt{3}) - 6\right]^2 - 2 \stackrel{?}{=} 25$$
$$\left[6 - 3\sqrt{3} - 6\right]^2 - 2 \stackrel{?}{=} 25$$
$$\left[-3\sqrt{3}\right]^2 - 2 \stackrel{?}{=} 25$$
$$(ab)^2 = a^2b^2 \quad 9 \cdot 3 - 2 \stackrel{?}{=} 25$$
$$27 - 2 \stackrel{?}{=} 25$$
$$25 = 25 \quad \text{True}$$

$y = 2 + \sqrt{3}$:

$$(3y - 6)^2 - 2 = 25$$
$$\left[3(2 + \sqrt{3}) - 6\right]^2 - 2 \stackrel{?}{=} 25$$
$$\left[6 + 3\sqrt{3} - 6\right]^2 - 2 \stackrel{?}{=} 25$$
$$\left[3\sqrt{3}\right]^2 - 2 \stackrel{?}{=} 25$$
$$(ab)^2 = a^2b^2 \quad 9 \cdot 3 - 2 \stackrel{?}{=} 25$$
$$27 - 2 \stackrel{?}{=} 25$$
$$25 = 25 \quad \text{True}$$

The solution set is $\{2 - \sqrt{3}, 2 + \sqrt{3}\}$.

Teaching Tip

Remind students that when simplifying an expression such as $\dfrac{6 + 3\sqrt{3}}{3}$ in Step 3 of Example 4, either factor the numerator and divide out the common factor, or divide each term in the numerator by 3.

In Step 3 of Example 4, we added 6 to both sides of the equation to obtain $3y - 6 + 6 = \pm 3\sqrt{3} + 6$, or $3y = \pm 3\sqrt{3} + 6$. We then used the Commutative Property to obtain $3y = 6 \pm 3\sqrt{3}$. It is common practice to put rational numbers first, followed by square roots.

Another way to simplify the expression $\dfrac{6 \pm 3\sqrt{3}}{3}$ is to use $\dfrac{A + B}{C} = \dfrac{A}{C} + \dfrac{B}{C}$, and write $y = \dfrac{6}{3} \pm \dfrac{3\sqrt{3}}{3} = 2 \pm \sqrt{3}$. Notice that we obtain the same answer as when we factored the numerator and divided out common factors.

QUICK ✓ *Solve the quadratic equation using the Square Root Property.*

7. $(y - 3)^2 = 121$

8. $(5k + 1)^2 - 2 = 26$

Classroom Example

Solve:

(a) $x^2 - 10x + 25 = 49$

(b) $(2x + 5)^2 - 20 = 0$

Answer:

(a) $\{-2, 12\}$

(b) $\left\{\dfrac{-5 - 2\sqrt{5}}{2}, \dfrac{-5 + 2\sqrt{5}}{2}\right\}$

EXAMPLE 5 Solving a Quadratic Equation Containing a Perfect Square Trinomial

Solve: $x^2 + 6x + 9 = 7$

Solution

Did you notice that the left side of the equation $x^2 + 6x + 9 = 7$ is a perfect square trinomial? Remember that $a^2 + 2ab + b^2 = (a + b)^2$. We use this fact along with the Square Root Property to solve the equation.

$$x^2 + 6x + 9 = 7$$

Factor the trinomial $x^2 + 6x + 9$: $$(x + 3)^2 = 7$$

Use the Square Root Property: $$x + 3 = \pm\sqrt{7}$$

Subtract 3 from both sides: $$x = -3 \pm \sqrt{7}$$

The solution is $\{-3 - \sqrt{7}, -3 + \sqrt{7}\}$.

QUICK ✓ *Solve the quadratic equation using the Square Root Property.*

9. $4c^2 - 12c + 9 = 25$

10. $(x + 4)^2 - 9 = 15$

Classroom Example Solve the equation $(5x - 3)^2 = -8$.

Answer: no real solution

EXAMPLE 6 Solving a Quadratic Equation of the Form $(ax + b)^2 = p$, where p Is Negative

Solve: $(2x + 1)^2 = -10$

Solution

The equation is in the form $(ax + b)^2 = p$. It states the square of 2 times a number plus 1 is negative 10. But there is no real number whose square is negative, so there is no real solution to this equation.

QUICK ✓ *Solve the quadratic equation.*

11. $(n + 3)^2 = -4$

12. $(2x + 5)^2 + 8 = 7$

---

(2) Solve Problems Using the Pythagorean Theorem

Recall from Section 4.7 that the Pythagorean Theorem is a statement about *right triangles.* A right triangle is one that contains a right angle, that is, an angle of 90°. The side of the triangle opposite the 90° angle is the hypotenuse; the remaining two sides are the legs. In Figure 1, we use c to represent the length of the hypotenuse and a and b to represent the lengths of the legs. Notice the use of the symbol ⌐ to show the 90° angle.

Figure 1

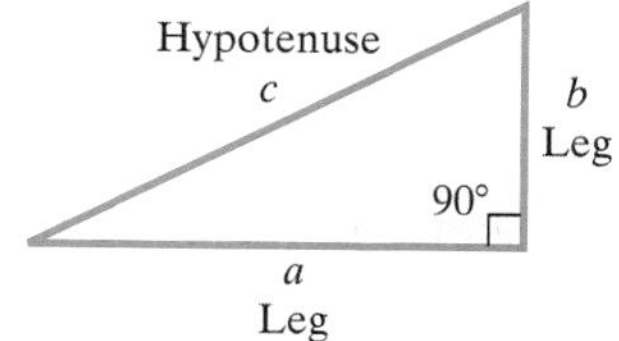

THE PYTHAGOREAN THEOREM

In a right triangle, the square of the length of the hypotenuse is equal to the sum of the squares of the lengths of the legs. That is, in the right triangle shown in Figure 1,

$$a^2 + b^2 = c^2$$

Classroom Example Use Example 7 with one leg = 4 inches and the other leg = 8 inches.

Answer: The hypotenuse has length = $4\sqrt{5}$ inches or approximately 8.9 inches.

EXAMPLE 7 Finding the Hypotenuse of a Right Triangle

In a right triangle, one leg is of length 4 inches and the other is of length 6 inches. Find both the exact length and the approximate (to the nearest tenth of an inch) length of the hypotenuse.

Solution

Figure 2 on p. 464 shows the right triangle with the lengths of the legs labeled and the unknown length of the hypotenuse labeled. Because the triangle is a right triangle, we

Figure 2

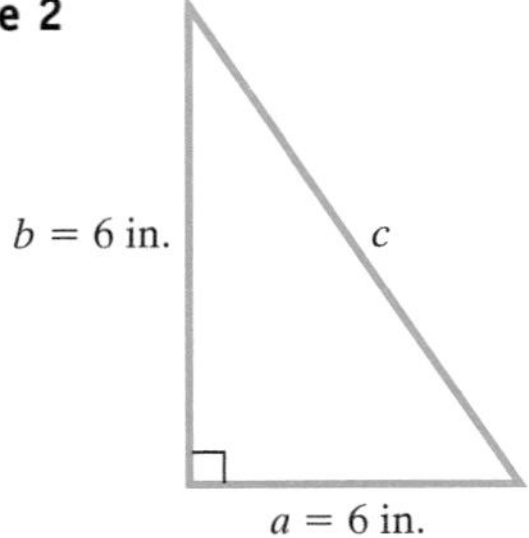

use the Pythagorean Theorem with $a = 4$ and $b = 6$ to find the length c of the hypotenuse. We know that

$$a^2 + b^2 = c^2$$

so that

$$4^2 + 6^2 = c^2$$
$$16 + 36 = c^2$$
$$52 = c^2$$

We now use the Square Root Property to find c, the length of the hypotenuse.

$$\pm\sqrt{52} = c$$
$$\pm\sqrt{4 \cdot 13} = c$$
$$\pm 2\sqrt{13} = c$$

Work Smart

Notice that the **exact** answer contains a radical, and the **approximate** answer is a decimal.

We use only the positive square root because c represents the length of a side of a triangle and a negative length does not make sense. The exact length of the hypotenuse is $2\sqrt{13}$ inches. Use a calculator or Appendix B to find that the approximate length of the hypotenuse is 7.2 inches. ■

Teaching Tip

You may need to discuss the difference between an exact answer and an approximate answer.

QUICK ✓ *The lengths of the legs of a right triangle are given. Find the exact length and the approximate (to the nearest tenth) length of the hypotenuse.*

13. $a = 5, b = 15$ **14.** $a = 6, b = 6$

Here's how we apply the Pythagorean Theorem in an everyday situation.

Classroom Example ➤

Use Example 8 with the 25-foot ladder placed 5 feet from the wall of the house.

Answer: The ladder reaches approximately 24.5 feet up the wall.

EXAMPLE 8 Spring Cleaning: Washing Windows

Bob wants to wash the windows of his house. He has a 25-foot ladder and places the base of the ladder 10 feet from the wall of his house. To the nearest tenth of a foot, how far up the wall will the ladder reach?

Solution

We can apply the problem-solving approach from Chapter 2 to solve this problem.

Figure 3

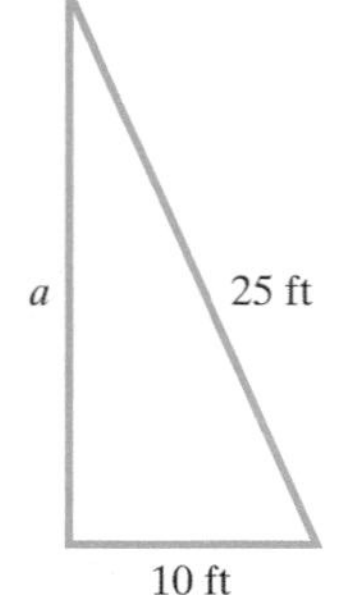

Step 1: Identify We wish to know how far up the wall the ladder reaches.

Step 2: Name We let a represent the height the ladder reaches on the side of the house.

Step 3: Translate To help visualize the problem, we make a sketch to represent the information we're given. See Figure 3. We can express the relation among the three sides of the right triangle using the Pythagorean Theorem.

$$a^2 + b^2 = c^2$$

Substitute $b = 10$ and $c = 25$: $a^2 + 10^2 = 25^2$

Step 4: Solve

$$a^2 + 100 = 625$$
$$a^2 = 525$$

Use the Square Root Property: $a = \pm\sqrt{525}$

Because a represents the height on the building, we are seeking the positive square root, $a = \sqrt{525} = 5\sqrt{21}$ ft. Using a calculator, we find $a \approx 22.9$ feet.

Step 5: Check Does $(\sqrt{525})^2 + 10^2 = 25^2$? Because, $525 + 100 = 625$, our answer checks.

Step 6: Answer The ladder reaches approximately 22.9 feet up the wall.

QUICK ✓

15. Stefan needs to clean his gutters. He has a 30-foot ladder. The manufacturer recommends that the base of the ladder be 8 feet from the wall. If the gutters are 25 feet above the ground, can Stefan use this ladder to clean his gutters?

7.1 Exercises

For Extra Help: Student Solutions Manual | CD Video | PH Math/Tutor Center | MathXL Tutorials on CD | MathXL® | MyMathLab

Concepts and Vocabulary

In Problems 1–3, fill in the blanks.

1. If $x^2 = p$, then $x =$ ________ or $x =$ ________.

2. In a right triangle, the longest side of the triangle is called the ________ and the other two sides are called the ________.

3. When using the Square Root Property to solve an equation in the form $(ax + b)^2 = p$, where p is a positive real number, you will have (how many) ______ unique solutions.

In Problems 4–6, answer True or False to each statement.

4. Only some quadratic equations can be solved by factoring.

5. The following is the correct process for simplifying the expression $\frac{4 \pm 2\sqrt{3}}{2}$.

$$\frac{4 + 2\sqrt{3}}{2} = \frac{6\sqrt{3}}{2} = 3\sqrt{3}; \frac{4 - 2\sqrt{3}}{2} = \frac{2\sqrt{3}}{2} = \sqrt{3}.$$

6. We use the Pythagorean Theorem to find the area of a right triangle.

7. You now know two different methods for solving quadratic equations: factoring and using the Square Root Property. Describe each method, explaining when it can and cannot be used. Make up two quadratic equations and solve one by factoring and one using the Square Root Property.

8. Consider the two quadratic equations, $x^2 + 36 = 0$ and $x^2 - 36 = 0$. Describe how to solve each equation. Explain the difference in the two solutions.

Building Skills

In Problems 9–60, solve each quadratic equation using the Square Root Property. Express radicals in simplest form.

9. $x^2 = 144$ **10.** $x^2 = 81$ **11.** $u^2 = 0$ **12.** $y^2 = 1$

13. $32 = q^2$ **14.** $18 = x^2$ **15.** $12 = x^2$ **16.** $48 = t^2$

17. $s^2 = \frac{4}{9}$ **18.** $x^2 = \frac{25}{4}$ **19.** $x^2 = \frac{4}{3}$ **20.** $d^2 = \frac{36}{5}$

21. $\frac{1}{2}r^2 = 16$ **22.** $\frac{1}{3}w^2 = 9$ **23.** $x^2 - 169 = 0$ **24.** $r^2 - 36 = 0$

Answers:
1. $\sqrt{p}; -\sqrt{p}$
2. hypotenuse; legs
3. two
4. True
5. False
6. False
7. Answers may vary.
8. The equation $x^2 + 36 = 0$ has no real solution. The solution set of $x^2 - 36 = 0$ is $\{-6, 6\}$.
9. $\{-12, 12\}$
10. $\{-9, 9\}$
11. $\{0\}$
12. $\{-1, 1\}$
13. $\{-4\sqrt{2}, 4\sqrt{2}\}$
14. $\{-3\sqrt{2}, 3\sqrt{2}\}$
15. $\{-2\sqrt{3}, 2\sqrt{3}\}$
16. $\{-4\sqrt{3}, 4\sqrt{3}\}$
17. $\left\{-\frac{2}{3}, \frac{2}{3}\right\}$
18. $\left\{-\frac{5}{2}, \frac{5}{2}\right\}$
19. $\left\{-\frac{2\sqrt{3}}{3}, \frac{2\sqrt{3}}{3}\right\}$
20. $\left\{-\frac{6\sqrt{5}}{5}, \frac{6\sqrt{5}}{5}\right\}$
21. $\{-4\sqrt{2}, 4\sqrt{2}\}$
22. $\{-3\sqrt{3}, 3\sqrt{3}\}$
23. $\{-13, 13\}$
24. $\{-6, 6\}$

25. $\{-5\sqrt{2}, 5\sqrt{2}\}$
26. $\{-2\sqrt{5}, 2\sqrt{5}\}$
27. no real solution
28. no real solution
29. $\left\{-\frac{1}{3}, \frac{1}{3}\right\}$ 30. $\left\{-\frac{1}{4}, \frac{1}{4}\right\}$
31. $\{-8, 8\}$ 32. $\{-10, 10\}$
33. $\{-6, 6\}$ 34. $\{-5, 5\}$
35. $\left\{-\frac{5\sqrt{3}}{3}, \frac{5\sqrt{3}}{3}\right\}$
36. $\left\{-\frac{4\sqrt{7}}{7}, \frac{4\sqrt{7}}{7}\right\}$
37. no real solution
38. no real solution
39. $\{-4\sqrt{2}, 4\sqrt{2}\}$
40. $\{-3\sqrt{5}, 3\sqrt{5}\}$
41. $\{-6, -2\}$ 42. $\{-9, -1\}$
43. $\left\{\frac{2}{3}, \frac{4}{3}\right\}$ 44. $\left\{-\frac{3}{2}, -\frac{1}{2}\right\}$
45. $\{-5-\sqrt{6}, -5+\sqrt{6}\}$
46. $\{-4-\sqrt{15}, -4+\sqrt{15}\}$
47. $\{6-2\sqrt{2}, 6+2\sqrt{2}\}$
48. $\{3-2\sqrt{3}, 3+2\sqrt{3}\}$
49. $\left\{-\frac{5}{3}, 1\right\}$ 50. $\{-1, 2\}$
51. $\left\{\frac{5-\sqrt{15}}{2}, \frac{5+\sqrt{15}}{2}\right\}$
52. $\left\{\frac{-2-\sqrt{21}}{5}, \frac{-2+\sqrt{21}}{5}\right\}$
53. $\{-3, 5\}$
54. $\{-9, 3\}$
55. $\left\{-\frac{1}{2}, -\frac{1}{4}\right\}$
56. $\left\{\frac{1}{4}, \frac{1}{2}\right\}$
57. $\left\{\frac{-2-3\sqrt{6}}{9}, \frac{-2+3\sqrt{6}}{9}\right\}$
58. $\left\{\frac{-2-3\sqrt{2}}{3}, \frac{-2+3\sqrt{2}}{3}\right\}$
59. no real solution
60. no real solution
61. $\{-9, 1\}$ 62. $\{-14, -2\}$
63. $\{-3, 4\}$ 64. $\left\{-\frac{2}{3}, 2\right\}$
65. $c = 10$ 66. $c = 13$
67. $c = 3\sqrt{2} \approx 4.24$
68. $c = 2\sqrt{5} \approx 4.47$
69. $a = 8\sqrt{2} \approx 11.31$
70. $a = 8$ 71. $\{4, 9\}$ 72. $\{6, 8\}$
73. $\left\{-1, \frac{1}{2}\right\}$ 74. $\left\{-\frac{1}{3}, 1\right\}$
75. $\{-2\sqrt{2}, 2\sqrt{2}\}$
76. $\{-2\sqrt{3}, 2\sqrt{3}\}$
77. $\{-\sqrt{15}, \sqrt{15}\}$ 78. $\{-6, 6\}$
79. $\left\{-\frac{\sqrt{15}}{5}, \frac{\sqrt{15}}{5}\right\}$
80. $\left\{-\frac{\sqrt{21}}{3}, \frac{\sqrt{21}}{3}\right\}$
81. no real solution
82. no real solution

25. $x^2 - 50 = 0$ **26.** $x^2 - 20 = 0$ **27.** $a^2 + 16 = 0$ **28.** $0 = h^2 + 4$

29. $27x^2 = 3$ **30.** $64x^2 = 4$ **31.** $\frac{1}{16}a^2 - 4 = 0$ **32.** $\frac{1}{4}m^2 - 25 = 0$

33. $65 = 2n^2 - 7$ **34.** $76 = 3v^2 + 1$ **35.** $3x^2 + 20 = 45$ **36.** $7x^2 + 8 = 24$

37. $2k^2 + 12 = 10$ **38.** $4 = 3z^2 + 10$ **39.** $2x^2 - 15 = 49$ **40.** $100 = 2x^2 + 10$

41. $4 = (x + 4)^2$ **42.** $16 = (x + 5)^2$ **43.** $27(x - 1)^2 = 3$ **44.** $8(v + 1)^2 = 2$

45. $\frac{(n + 5)^2}{3} = 2$ **46.** $\frac{(z + 4)^2}{5} = 3$ **47.** $\frac{2}{3}(x - 6)^2 = \frac{16}{3}$ **48.** $\frac{2}{15}(p - 3)^2 = \frac{8}{5}$

49. $\left(p + \frac{1}{3}\right)^2 = \frac{16}{9}$ **50.** $\left(a - \frac{1}{2}\right)^2 = \frac{9}{4}$ **51.** $\left(x - \frac{5}{2}\right)^2 = \frac{15}{4}$

52. $\left(x + \frac{2}{5}\right)^2 = \frac{21}{25}$ **53.** $(x - 1)^2 - 7 = 9$ **54.** $(x + 3)^2 - 12 = 24$

55. $(8k + 3)^2 - 5 = -4$ **56.** $(8g - 3)^2 - 2 = -1$ **57.** $30 = (9x + 2)^2 - 24$

58. $20 = (3c + 2)^2 + 2$ **59.** $(2b + 8)^2 - 12 = -24$ **60.** $(3x - 6)^2 - 40 = -48$

In Problems 61–64, solve each quadratic equation by first factoring the perfect square trinomial on one side of the equation and then using the Square Root Property.

61. $x^2 + 8x + 16 = 25$ **62.** $x^2 + 16x + 64 = 36$

63. $49 = 4w^2 - 4w + 1$ **64.** $16 = 9z^2 - 12z + 4$

In Problems 65–70, use the right triangle shown in the margin and find the missing length. Give exact answers and decimal approximations correct to two decimal places.

65. $a = 6, b = 8$ **66.** $a = 5, b = 12$ **67.** $a = 3, b = 3$

68. $a = 4, b = 2$ **69.** $b = 4, c = 12$ **70.** $b = 6, c = 10$

Mixed Practice

In Problems 71–94, solve each quadratic equation by either factoring or the Square Root Property. For each problem, choose the method that is most efficient.

71. $x^2 - 13x + 36 = 0$ **72.** $x^2 - 14x + 48 = 0$ **73.** $2m^2 = 1 - m$

74. $3x^2 = 2x + 1$ **75.** $2n^2 = 16$ **76.** $3q^2 = 36$

77. $3x^2 - 9 = 36$ **78.** $81 = 2x^2 + 9$ **79.** $0 = 15x^2 - 9$

80. $0 = 6t^2 - 14$ **81.** $2 = r^2 + 6$ **82.** $x^2 + 17 = 8$

83. $12 = x^2 + x$ **84.** $15 = n^2 + 2n$ **85.** $d^2 - 27 = 0$

86. $b^2 - 2 = 0$ **87.** $x^2 - 10x + 25 = 0$ **88.** $x^2 + 14x + 49 = 0$

89. $2(x + 2)^2 + 3 = 8$ **90.** $3(z - 1)^2 + 4 = 6$ **91.** $3x^2 + 4 = 6$

92. $8 = 2z^2 + 5$ **93.** $2x^2 - 5x - 12 = 0$ **94.** $2y^2 - 5y - 3 = 0$

In Problems 95–98, solve each equation.

95. $\sqrt{x^2 - 15} = 5$ **96.** $\sqrt{p^2 + 4} = 4$

97. $\sqrt{q^2 + 13} + 3 = 6$ **98.** $\sqrt{z^2 - 2} + 9 = 4$

83. $\{-4, 3\}$
84. $\{-5, 3\}$
85. $\{-3\sqrt{3}, 3\sqrt{3}\}$
86. $\{-\sqrt{2}, \sqrt{2}\}$
87. $\{5\}$
88. $\{-7\}$
89. $\left\{\frac{-4-\sqrt{10}}{2}, \frac{-4+\sqrt{10}}{2}\right\}$
90. $\left\{\frac{3-\sqrt{6}}{3}, \frac{3+\sqrt{6}}{3}\right\}$
91. $\left\{-\frac{\sqrt{6}}{3}, \frac{\sqrt{6}}{3}\right\}$
92. $\left\{-\frac{\sqrt{6}}{2}, \frac{\sqrt{6}}{2}\right\}$
93. $\left\{-\frac{3}{2}, 4\right\}$
94. $\left\{-\frac{1}{2}, 3\right\}$
95. $\{-2\sqrt{10}, 2\sqrt{10}\}$
96. $\{-2\sqrt{3}, 2\sqrt{3}\}$
97. no real solution
98. no real solution
99. $2\sqrt{10}$
100. $6\sqrt{2}$
101. (a) $2\sqrt{30}$ ft
(b) 11.0 ft
102. (a) 15 in.
(b) 15.00 in.
103. $20\sqrt{26}$ yd
104. $30\sqrt{101} \approx 301$ ft

Applying the Concepts

In Problems 99 and 100, find the length of the diagonal in each figure.

99.

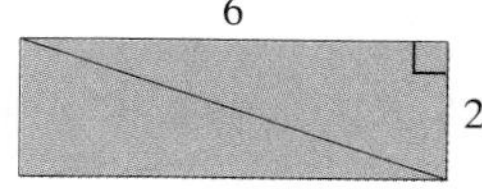

100.

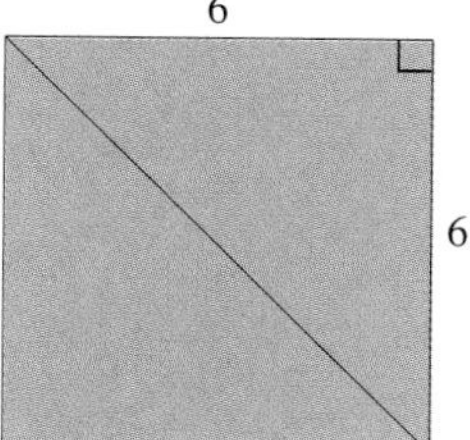

△ **101. Firefighter's Ladder** A firefighter has a 13-foot ladder. Suppose he places the ladder 7 feet from a building.

(a) Exactly how far up the building will his ladder reach?

(b) Approximate this height to the nearest tenth of a foot.

△ **102. Picture Frame** A rectangular picture frame measures 17 inches on the diagonal and is 8 inches high.

(a) Exactly how wide is this frame?

(b) Find the width of the picture frame to 2 decimal places.

103. Golf A golfer hits an errant tee shot that lands in the rough. The golfer finds that the ball is exactly 20 yards to the right of the 100-yard marker that indicates the distance to the center of the green, as shown in the figure. How far is the ball from the center of the green?

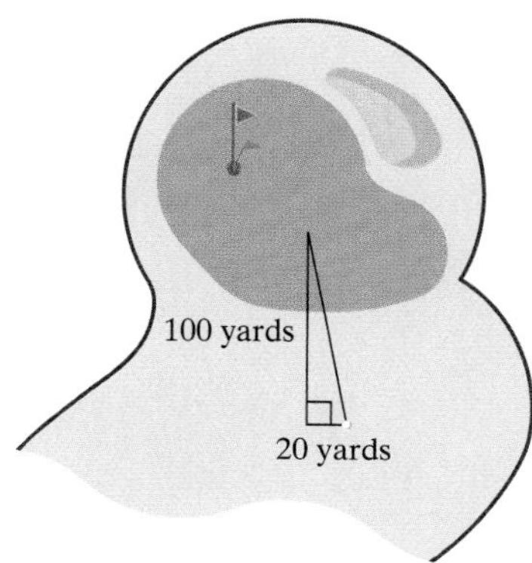

104. Baseball Ichiro Suzuki plays right field for the Seattle Mariners. He catches a fly ball 30 feet from the right field foul line, as indicated in the figure. How far is it to home plate? Round your answer to the nearest foot.

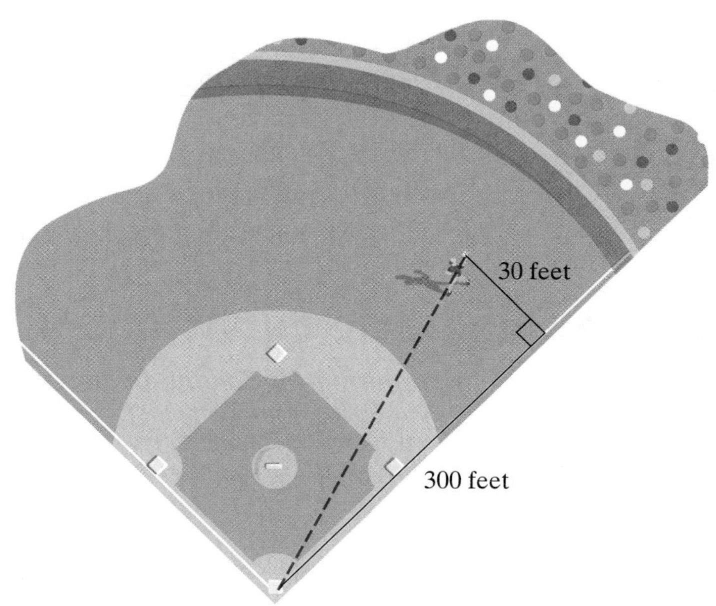

105. (a) 13.8 ft
(b) 13 ft 9.6 in.
106. (a) 17.5 ft
(b) 15 ft 6 in.
107. (a) −2.08 or 0.08
(b) 8%
108. (a) −2.05 or 0.05
(b) 5%
109. $x^2 - 16 = 0$
110. $x^2 - 81 = 0$
111. $x^2 - 6 = 0$
112. $x^2 - 15 = 0$
113. $x^2 - 45 = 0$
114. $x^2 - 28 = 0$

△ **105. Football Goal Post** The grounds keeper at the football stadium suspects that the goal post is beginning to lean slightly. He decides to determine if it is square by measuring the distance from the end of the cross bar to the point where the pole goes in the ground.

(a) How long should this distance be if the distance from the center of the cross bar to the end is 9.25 feet and the goal post is 10.25 feet high? Find the distance, to the nearest tenth of a foot.

(b) What is this distance in feet and inches?

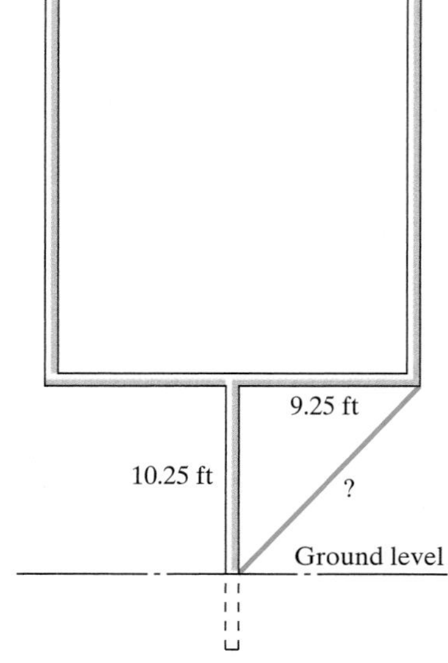

△ **106. Ham Radio Antenna** Bob is installing a new ham radio antenna in his back yard. The antenna is 16 feet high and he wants to place a guy wire 7 feet from the antenna for added support during windy weather.

(a) Find the length of the guy wire to the nearest tenth of a foot.

(b) What is this distance in inches and feet?

The formula for calculating the amount of money in an account earning interest after 2 years is given by the formula $A = P(1 + r)^2$, where A is the new amount, P is the principal, and r is the interest rate, written in decimal form. Use this formula in Problems 107 and 108.

107. Leticia deposited \$500 and had \$583.20 after two years.

(a) Use this information in the formula above to solve for r.

(b) What is the rate of interest paid on Leticia's account?

108. James deposited \$1200 and had \$1323 after two years.

(a) Use this information in the formula above to solve for r.

(b) What is the rate of interest paid on James's account?

Extending the Concepts

In Problems 109–114, find a quadratic equation that has the indicated solutions. Write the equation in standard form, $ax^2 + bx + c = 0, a \neq 0$.

109. The solutions to the equation are $x = 4$ and $x = -4$ $(x = \pm 4)$.

110. The solutions to the equation are $x = 9$ and $x = -9$ $(x = \pm 9)$.

111. The solutions to the equation are $x = \pm\sqrt{6}$.

112. The solutions to the equation are $x = \pm\sqrt{15}$.

113. The solutions to the equation are $x = \pm 3\sqrt{5}$.

114. The solutions to the equation are $x = \pm 2\sqrt{7}$.

7.2 Solving Quadratic Equations by Completing the Square

OBJECTIVES

1. Complete the Square in One Variable
2. Solve Quadratic Equations by Completing the Square

Preparing for Solving Quadratic Equations by Completing the Square

Before getting started, take the following readiness quiz. If you get a problem wrong, go back to the section cited and review the material.

1. Factor: $a^2 + 10a + 25$ [Section 4.4, pp. 267–269]

2. Solve: $(x + 3)^2 = 16$ [Section 7.1, pp. 458–462]

3. Find the quotient: $\dfrac{3x^2 - 5x + 4}{3}$ [Section 3.5, pp. 213–215]

Preparing for...Answers **1.** $(a + 5)^2$ **2.** $\{-7, 1\}$ **3.** $x^2 - \frac{5}{3}x + \frac{4}{3}$

You have now learned two methods for solving quadratic equations: factoring and using the Square Root Property. A third method for solving quadratic equations is called the method of *completing the square*. Before we illustrate how the method is used to solve a quadratic equation, we first discuss how to complete the square.

1 Complete the Square in One Variable

Teaching Tip
Tell students that completing the square is used in other ways in algebra. Completing the square is used to find the vertex of a parabola and the center of a circle, to name just two applications.

The idea behind the method of completing the square is to "adjust" the left side of a quadratic equation to make it a perfect square trinomial. Recall that perfect square trinomials are trinomials of the form

$$a^2 + 2ab + b^2 = (a + b)^2 \quad \text{or} \quad a^2 - 2ab + b^2 = (a - b)^2$$

For example, $x^2 + 8x + 16$ is a perfect square trinomial because $x^2 + 8x + 16 = (x + 4)^2$.

We "adjust" the equation by adding a "well-chosen" number to the left side of the equation to make it a perfect square trinomial. We also have to add this number to the right side. For example, to make $x^2 + 10x$ a perfect square we would add 25. But where does this 25 come from? Look at the coefficient on the first-degree term, 10. Notice that if we divide the coefficient by 2 and then square the result, we obtain 25.

OBTAINING A PERFECT SQUARE TRINOMIAL BY COMPLETING THE SQUARE

To complete the square on $x^2 + bx$, identify b, the coefficient of the first-degree term. Multiply this coefficient by $\frac{1}{2}$ and then square the result. Add this number to $x^2 + bx$ to obtain a perfect square trinomial.

Classroom Example ➤
Use Example 1 with
(a) $a^2 + 16a$
(b) $m^2 - 10m$
(c) $z^2 + 5z$
(d) $c^2 - \frac{4}{3}c$

Answer:
(a) add 64; $a^2 + 16a + 64 = (a + 8)^2$
(b) add 25; $m^2 - 10m + 25 = (m - 5)^2$
(c) add $\frac{25}{4}$; $z^2 + 5z + \frac{25}{4} = \left(z + \frac{5}{2}\right)^2$
(d) add $\frac{4}{9}$; $c^2 - \frac{4}{3}c + \frac{4}{9} = \left(c - \frac{2}{3}\right)^2$

EXAMPLE 1 Obtaining a Perfect Square Trinomial

Determine the number that must be added to each expression to make it a perfect square trinomial. Then factor the expression $a^2 + 2ab + b^2 = (a + b)^2$ or $a^2 - 2ab + b^2 = (a - b)^2$.

Start	Add	Result	Factored Form
$y^2 + 8y$	$\left(\frac{1}{2}\cdot 8\right)^2 = 4^2 = 16$	$y^2 + 8y + 16$	$(y + 4)^2$
$x^2 + 12x$	$\left(\frac{1}{2}\cdot 12\right)^2 = 6^2 = 36$	$x^2 + 12x + 36$	$(x + 6)^2$
$a^2 - 20a$	$\left(\frac{1}{2}\cdot(-20)\right)^2 = (-10)^2 = 100$	$a^2 - 20a + 100$	$(a - 10)^2$
$p^2 - 5p$	$\left(\frac{1}{2}\cdot(-5)\right)^2 = \left(-\frac{5}{2}\right)^2 = \frac{25}{4}$	$p^2 - 5p + \frac{25}{4}$	$\left(p - \frac{5}{2}\right)^2$
$n^2 + \frac{3}{2}n$	$\left(\frac{1}{2}\cdot\frac{3}{2}\right)^2 = \left(\frac{3}{4}\right)^2 = \frac{9}{16}$	$n^2 + \frac{3}{2}n + \frac{9}{16}$	$\left(n + \frac{3}{4}\right)^2$

Work Smart
To complete the square, take half the coefficient of b, the first-degree term, and square it. Then factor the polynomial as $\left(x + \frac{b}{2}\right)^2$, or $\left(x - \frac{b}{2}\right)^2$.

Did you notice in the factored form that the perfect square trinomial always factors so that

$$x^2 + bx + \left(\frac{b}{2}\right)^2 = \left(x + \frac{b}{2}\right)^2 \quad \text{or} \quad x^2 - bx + \left(\frac{b}{2}\right)^2 = \left(x - \frac{b}{2}\right)^2$$

The $\frac{b}{2}$ represents half the value of the coefficient of the first-degree term.

QUICK ✓ *Determine the number that must be added to the expression to make it a perfect square trinomial. Then factor the expression.*

1. $p^2 + 6p$ **2.** $z^2 - 14z$ **3.** $b^2 - 18b$ **4.** $w^2 + 3w$

Figure 4

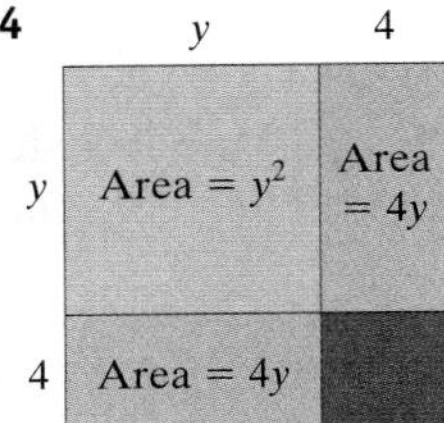

Are you wondering why we call making an expression a perfect square trinomial "completing the square"? Consider the expression $y^2 + 8y$ given in Example 1. We can geometrically represent this algebraic expression as shown in Figure 4. The green area is y^2 and each of the orange areas is $4y$ (for a total area of $8y$). But what is the area of the red region in order to make the square complete? The dimensions of the red region must be 4 by 4, so the area of the red region is 16. The area of the entire square region, $(y + 4)^2$, equals the sum of the area of the regions that make up the square: $y^2 + 4y + 4y + 16 = y^2 + 8y + 16$.

2 Solve Quadratic Equations by Completing the Square

We're now ready to learn how to use the method of completing the square to solve quadratic equations in the form $x^2 + bx + c = 0$. Notice that the coefficient on the square term is 1. Later in the section, we address how to complete the square if the coefficient of the square term is not 1.

Work Smart

While completing the square, we also use the Addition Property of Equality—whatever you add to one side of an equation, you must also add to the other side.

EXAMPLE 2 How to Solve a Quadratic Equation by Completing the Square

Solve: $p^2 + 10p + 21 = 0$

Classroom Example
Solve:
$n^2 + 8n + 3 = 12$
Answer: $\{-9, 1\}$

Step-by-Step Solution

Step 1: Rewrite $x^2 + bx + c = 0$ as $x^2 + bx = -c$.

$$p^2 + 10p + 21 = 0$$

Subtract 21 from both sides: $p^2 + 10p = -21$

Step 2: Complete the square on the expression $x^2 + bx$.

$\left(\frac{1}{2}\cdot 10\right)^2 = 5^2 = 25$, so add 25 to both sides of the equation: $p^2 + 10p + 25 = -21 + 25$

$$p^2 + 10p + 25 = 4$$

Step 3: Factor the perfect square trinomial on the left side of the equation.

$$(p + 5)^2 = 4$$

Step 4: Solve the equation using the Square Root Property.

Use the Square Root Property: $p + 5 = \pm\sqrt{4}$

$$p + 5 = \pm 2$$

$p + 5 = -2$ or $p + 5 = 2$

Subtract 5 from both sides: $p = -7$ or $p = -3$

Step 5: Verify your solution(s).

$p = -7$: $p^2 + 10p + 21 = 0$
$(-7)^2 + 10(-7) + 21 \stackrel{?}{=} 0$
$49 - 70 + 21 \stackrel{?}{=} 0$
$0 = 0$ True

$p = -3$: $p^2 + 10p + 21 = 0$
$(-3)^2 + 10(-3) + 21 \stackrel{?}{=} 0$
$9 - 30 + 21 \stackrel{?}{=} 0$
$0 = 0$ True

The solution set is $\{-7, -3\}$.

The following steps are used to solve quadratic equations by completing the square.

Steps to Solve a Quadratic Equation in the Form $x^2 + bx + c = 0$ by Completing the Square

Step 1: Rewrite $x^2 + bx + c = 0$ as $x^2 + bx = -c$ by subtracting (or adding) the constant from (to) both sides of the equation.

Step 2: Complete the square on the expression $x^2 + bx$ by making it a perfect square trinomial. Don't forget, whatever you add to the left side of the equation must also be added to the right side.

Step 3: Factor the perfect square trinomial on the left side of the equation.

Step 4: Solve the equation using the Square Root Property.

Step 5: Verify your solutions.

QUICK ✓ *Solve the quadratic equation by completing the square.*

5. $n^2 + 8n + 7 = 0$

6. $y^2 + 4y - 32 = 0$

The solutions in Example 2 were rational numbers. We know from solving quadratic equations using the Square Root Property that some quadratic equations have irrational solutions.

Classroom Example ➤
Solve: $y^2 + 6y = -1$

Answer:
$\{-3 - 2\sqrt{2}, -3 + 2\sqrt{2}\}$

EXAMPLE 3 Solving a Quadratic Equation Having Irrational Solutions

Solve: $a^2 + 8a = -4$

Solution

First, notice that the equation is already written in the form $x^2 + bx = -c$, so we complete the square on the left-hand side of the equation:

$$a^2 + 8a = -4$$

Add 16 to both sides since $\left(\frac{1}{2} \cdot 8\right)^2 = 4^2 = 16$: $$a^2 + 8a + 16 = -4 + 16$$

Factor: $$(a + 4)^2 = 12$$

Use the Square Root Property: $$a + 4 = \pm\sqrt{12}$$

Simplify the radical: $$a + 4 = \pm 2\sqrt{3}$$

$$a + 4 = -2\sqrt{3} \quad \text{or} \quad a + 4 = 2\sqrt{3}$$

Subtract 4 from both sides: $$a = -4 - 2\sqrt{3} \quad \text{or} \quad a = -4 + 2\sqrt{3}$$

Check $a = -4 - 2\sqrt{3}$:

$$a^2 + 8a = -4$$
$$(-4 - 2\sqrt{3})^2 + 8(-4 - 2\sqrt{3}) \stackrel{?}{=} -4$$
$$16 + 16\sqrt{3} + 12 - 32 - 16\sqrt{3} \stackrel{?}{=} -4$$
$$-4 = -4 \quad \text{True}$$

$a = -4 + 2\sqrt{3}$:

$$a^2 + 8a = -4$$
$$(-4 + 2\sqrt{3})^2 + 8(-4 + 2\sqrt{3}) \stackrel{?}{=} -4$$
$$16 - 16\sqrt{3} + 12 - 32 + 16\sqrt{3} \stackrel{?}{=} -4$$
$$-4 = -4 \quad \text{True}$$

The solution set is $\{-4 - 2\sqrt{3}, -4 + 2\sqrt{3}\}$.

QUICK ✓ *Solve the quadratic equation by completing the square.*

7. $z^2 + 8z = -9$ **8.** $b^2 - 10b = -19$

Classroom Example ➤
Solve:

$(x - 3)(x + 5) = 12$

Answer:
$\{-1 - 2\sqrt{7}, -1 + 2\sqrt{7}\}$

Work Smart
Do NOT solve $(x - 1)(x + 3) = 9$ by setting each factor equal to 9 as in

$x - 1 = 9$ or $x + 3 = 9$

Why not?

EXAMPLE 4 Solving a Quadratic Equation Not in Standard Form

Solve: $(x - 1)(x + 4) = 9$

Solution

What is the first thing that you notice about this equation? It's not in the form $x^2 + bx = -c$. Our first step will be to find the product $(x - 1)(x + 4)$ using FOIL.

$$(x - 1)(x + 4) = 9$$

Find the product using FOIL: $$x^2 + 3x - 4 = 9$$

Add 4 to both sides: $$x^2 + 3x - 4 + 4 = 9 + 4$$

$$x^2 + 3x = 13$$

We're now ready to complete the square.

Add $\frac{9}{4}$ to both sides since $\left(\frac{1}{2} \cdot 3\right)^2 = \left(\frac{3}{2}\right)^2 = \frac{9}{4}$: $$x^2 + 3x + \frac{9}{4} = 13 + \frac{9}{4}$$

$$x^2 + 3x + \frac{9}{4} = \frac{52}{4} + \frac{9}{4}$$

$$x^2 + 3x + \frac{9}{4} = \frac{61}{4}$$

Factor: $$\left(x + \frac{3}{2}\right)^2 = \frac{61}{4}$$

Use the Square Root Property: $$x + \frac{3}{2} = \pm\sqrt{\frac{61}{4}}$$

Subtract $\frac{3}{2}$ from both sides; use $\sqrt{\frac{a}{b}} = \frac{\sqrt{a}}{\sqrt{b}}$: $$x = -\frac{3}{2} \pm \frac{\sqrt{61}}{\sqrt{4}}$$

$$x = -\frac{3}{2} \pm \frac{\sqrt{61}}{2}$$

$$x = \frac{-3 - \sqrt{61}}{2} \quad \text{or} \quad x = \frac{-3 + \sqrt{61}}{2}$$

The solution set is $\left\{\frac{-3 - \sqrt{61}}{2}, \frac{-3 + \sqrt{61}}{2}\right\}$. We leave it to you to check the two solutions.

QUICK ✓ *Solve the quadratic equation by completing the square.*

9. $(n + 5)(n + 3) = 24$ **10.** $(a + 6)(a - 1) = 4$

As mentioned earlier, to use the method of completing the square to solve a quadratic equation, the quadratic equation must be in the form $x^2 + bx = -c$. In other words, the leading coefficient must be 1. What happens if the leading coefficient is not 1?

Classroom Example
Solve:

$$3c^2 - 4c - 6 = 0$$

Answer:

$$\left\{\frac{2 - \sqrt{22}}{3}, \frac{2 + \sqrt{22}}{3}\right\}$$

Work Smart
To solve a quadratic equation using completing the square, the coefficient of the squared term must be 1.

EXAMPLE 5 Solving a Quadratic Equation Whose Leading Coefficient Is Not 1

Solve: $2x^2 - 5x - 12 = 0$

Solution

First, we must get the coefficient of the squared term to be 1, so we divide each side of the equation by 2.

$$2x^2 - 5x - 12 = 0$$

Divide each side by 2: $$\frac{2x^2 - 5x - 12}{2} = \frac{0}{2}$$

$$x^2 - \frac{5}{2}x - 6 = 0$$

Add 6 to both sides: $$x^2 - \frac{5}{2}x = 6$$

Because $\left(\frac{1}{2} \cdot -\frac{5}{2}\right)^2 = \left(-\frac{5}{4}\right)^2 = \frac{25}{16}$, we add $\frac{25}{16}$ to both sides: $$x^2 - \frac{5}{2}x + \frac{25}{16} = 6 + \frac{25}{16}$$

Factor: $$\left(x - \frac{5}{4}\right)^2 = \frac{96}{16} + \frac{25}{16}$$

$$\left(x - \frac{5}{4}\right)^2 = \frac{121}{16}$$

Use the Square Root Property: $$x - \frac{5}{4} = \pm\sqrt{\frac{121}{16}}$$

$$x - \frac{5}{4} = \pm\frac{11}{4}$$

$$x - \frac{5}{4} = -\frac{11}{4} \quad \text{or} \quad x - \frac{5}{4} = \frac{11}{4}$$

Add $\frac{5}{4}$ to both sides: $$x = \frac{5}{4} - \frac{11}{4} \quad \text{or} \quad x = \frac{5}{4} + \frac{11}{4}$$

$$x = \frac{-6}{4} \quad \text{or} \quad x = \frac{16}{4}$$

$$x = -\frac{3}{2} \quad \text{or} \quad x = 4$$

We leave it to you to verify the solutions. The solution set is $\left\{-\frac{3}{2}, 4\right\}$.

QUICK ✓ *Solve the quadratic equation by completing the square.*

11. $2m^2 = m + 10$

12. $3b^2 - 4b - 2 = 0$

Do you recall from Section 7.1 that quadratic equations of the form $(ax + b)^2 = p$, where p is negative, have no real solution? We illustrate this point again in the next example.

Classroom Example ➤
Solve:

$k(k + 8) + 21 = 3$

Answer: no real solution; $\varnothing$ or $\{\ \}$

Teaching Tip
Some students may be familiar with complex solutions. Explain that we're interested only in real number solutions at this point.

Work Smart
No real solution does not mean *no solution.* We will learn how to find the solution to equations such as the one in Example 14 in Section 7.5.

EXAMPLE 6 Solving a Quadratic Equation Using Completing the Square

Solve: $x(x + 6) + 28 = 10$

Solution

What's the first step? Write the equation in the form $x^2 + bx = -c$.

$$x(x + 6) + 28 = 10$$

Use the Distributive Property: $x^2 + 6x + 28 = 10$

Subtract 28 from each side: $x^2 + 6x + 28 - 28 = 10 - 28$

$$x^2 + 6x = -18$$

Add 9 to both sides because $\left(\frac{1}{2}\cdot 6\right)^2 = 3^2 = 9$: $x^2 + 6x + 9 = -18 + 9$

$$x^2 + 6x + 9 = -9$$

Factor: $(x + 3)^2 = -9$

There is no real number whose square is -9, so there is no real solution to this equation.

QUICK ✓ *Solve the quadratic equation by completing the square.*

13. $n(n - 4) = -12$ **14.** $(p - 2)(p + 3) = 36$

7.2 Exercises

For Extra Help:

Student Solutions Manual

CD Video

PH Math/Tutor Center

MathXL Tutorials on CD

MathXL®

MyMathLab

1. 25 2. 2 3. $x + \frac{b}{2}$ 4. False
5. False 6. False 7. $\frac{b^2}{4a^2}$
8. Answers may vary. 9. 64; $(x + 8)^2$
10. 144; $(x + 12)^2$ 11. 36; $(z - 6)^2$
12. 324; $(p - 18)^2$
13. $\frac{4}{9}$; $\left(k + \frac{2}{3}\right)^2$ 14. $\frac{9}{49}$; $\left(y + \frac{3}{7}\right)^2$
15. $\frac{81}{4}$; $\left(y - \frac{9}{2}\right)^2$ 16. $\frac{25}{4}$; $\left(z - \frac{5}{2}\right)^2$
17. $\frac{1}{4}$; $\left(t + \frac{1}{2}\right)^2$ 18. $\frac{1}{4}$; $\left(w - \frac{1}{2}\right)^2$
19. $\frac{1}{16}$; $\left(s - \frac{1}{4}\right)^2$ 20. $\frac{1}{64}$; $\left(r + \frac{1}{8}\right)^2$
21. $\{-3, 7\}$ 22. $\{-10, 4\}$
23. $\{-2 - \sqrt{11}, -2 + \sqrt{11}\}$
24. $\{-1 - \sqrt{6}, -1 + \sqrt{6}\}$
25. $\left\{\frac{5 - \sqrt{65}}{2}, \frac{5 + \sqrt{65}}{2}\right\}$
26. $\{-1, 6\}$ 27. $\{-3, 6\}$ 28. $\{2, 3\}$
29. $\{3, 5\}$ 30. $\{2, 4\}$
31. no real solution 32. no real solution
33. $\{-1, 4\}$ 34. $\{1, 6\}$
35. $\{4 - 2\sqrt{7}, 4 + 2\sqrt{7}\}$
36. $\{3 - 3\sqrt{3}, 3 + 3\sqrt{3}\}$
37. $\left\{\frac{3 - 3\sqrt{13}}{2}, \frac{3 + 3\sqrt{13}}{2}\right\}$
38. $\left\{\frac{3 - \sqrt{89}}{2}, \frac{3 + \sqrt{89}}{2}\right\}$
39. $\{-5, -2\}$ 40. $\{-4, -1\}$
41. $\left\{\frac{-7 - \sqrt{249}}{2}, \frac{-7 + \sqrt{249}}{2}\right\}$

Concepts and Vocabulary

In Problems 1–3, fill in the blanks.

1. $x^2 + 10x +$ __________ is a perfect square trinomial.

2. To solve $2x^2 + 4x = 9$ by completing the square, the first step is to divide both sides of the equation by __________.

3. In general, $x^2 + bx + \left(\frac{b}{2}\right)^2 = (\_\_\_\_\_\_)^2$.

In Problems 4–6, answer True or False to each statement.

4. To solve $3x^2 - 12x = -5$ by completing the square, add 36 to both sides of the equation.

5. To solve $x^2 + 18x = 6$ by completing the square, add 9 to both sides of the equation.

6. The equation $(x - 1)(x + 5) = 10$ can be solved by solving $x - 1 = 10$ and $x + 5 = 10$.

7. Given the quadratic equation $x^2 + \frac{b}{a}x = \frac{c}{a}$, find the value that needs to be added to both sides of the equation to complete the square and describe each step of this process.

8. We now have three techniques to solve quadratic equations: factoring, using the Square Root Property, and completing the square. Which method is easiest to use when solving $x^2 + 3x - 208 = 0$ and why?

Building Skills

In Problems 9–20, determine the number that must be added to the expression to make it a perfect square trinomial. Then factor the expression.

9. $x^2 + 16x$ **10.** $x^2 + 24x$ **11.** $z^2 - 12z$

12. $p^2 - 36p$ **13.** $k^2 + \frac{4}{3}k$ **14.** $y^2 + \frac{6}{7}y$

15. $y^2 - 9y$ **16.** $z^2 - 5z$ **17.** $t^2 + t$

18. $w^2 - w$ **19.** $s^2 - \frac{s}{2}$ **20.** $r^2 + \frac{r}{4}$

In Problems 21–66, solve the quadratic equation using completing the square.

21. $x^2 - 4x = 21$ **22.** $x^2 + 6x = 40$ **23.** $x^2 + 4x = 7$

24. $x^2 + 2x = 5$ **25.** $p^2 - 5p - 10 = 0$ **26.** $u^2 - 5u - 6 = 0$

27. $m^2 - 3m - 18 = 0$ **28.** $z^2 - 5z + 6 = 0$ **29.** $r^2 - 8r + 15 = 0$

30. $p^2 - 6p + 8 = 0$ **31.** $x^2 + 2x + 24 = 0$ **32.** $x^2 - 8x + 18 = 0$

33. $x^2 = 3x + 4$ **34.** $r^2 = 7r - 6$ **35.** $a^2 = 8a + 12$

36. $n^2 = 6n + 18$ **37.** $27 = a^2 - 3a$ **38.** $20 = k^2 - 3k$

39. $-10 = x^2 + 7x$ **40.** $-4 = x^2 + 5x$ **41.** $n^2 + 7n - 50 = 0$

42. $p^2 + 7p - 48 = 0$ **43.** $10x = -26 - x^2$ **44.** $12x = -32 - x^2$

45. $3x^2 + 2x - 1 = 0$ **46.** $3x^2 - 4x - 4 = 0$ **47.** $4x^2 - 4x - 3 = 0$

48. $5x^2 + 8x - 4 = 0$ **49.** $2x^2 + 5x - 3 = 0$ **50.** $2a^2 + a = 3$

51. $3c^2 + 4c = 2$ **52.** $3x^2 = 4 - 2x$ **53.** $4x^2 = 5 - 4x$

54. $4n^2 + 12n + 1 = 0$ **55.** $4m^2 + 5m + 7 = 0$ **56.** $-9a = 2a^2 + 44$

57. $-16t + 4t^2 + 15 = 0$ **58.** $5d^2 + 12d = 9$ **59.** $4p^2 - 12p = -7$

60. $2r^2 - 5r + 1 = 0$ **61.** $4r^2 - 24r - 1 = 0$ **62.** $5 = 4n^2 - 32n$

63. $10 = 9a^2 - 54a$ **64.** $9c^2 - 36c - 7 = 0$ **65.** $2b^2 - 3b - 4 = 0$

66. $6x^2 - 13x + 5 = 0$

Mixed Practice

In Problems 67–94, solve the quadratic equation using any method you wish.

67. $(x + 2)(x - 4) = 16$ **68.** $(x - 8)(x - 1) = 1$ **69.** $4x^2 - 4x + 1 = 0$

70. $4x^2 + 12x + 9 = 0$ **71.** $a^2 + 6a - 3 = 0$ **72.** $n^2 + 10n - 15 = 0$

73. $3z^2 + 12 = 72$ **74.** $2y^2 + 15 = 51$ **75.** $3t^2 = 2 - 6t$

76. $2p^2 = 4p - 1$ **77.** $5x = 18 - x^2$ **78.** $x^2 + 3x = 6$

79. $h^2 = 9$ **80.** $k^2 = 36$ **81.** $(x + 2)^2 - 16 = 0$

82. $(x - 9)^2 - 4 = 0$ **83.** $n^2 + 10n = 24$ **84.** $n^2 - 13n = 30$

85. $y^2 - 18 = 0$ **86.** $z^2 - 48 = 0$ **87.** $2x^2 - 24 = 0$

88. $3x^2 - 24 = 0$ **89.** $12u + 2u^2 - 27 = 0$ **90.** $3v = 3v^2 - 10$

91. $r(3r - 11) = -6$ **92.** $t(2t + 9) = -4$ **93.** $(x + 6)(x - 4) = 9$

94. $(x - 2)(x - 3) = 8$

Applying the Concepts

95. Fun with Numbers The product of a number and one less than half of the number is 12. Find the two numbers that satisfy this condition.

96. Fun with Numbers The product of a number and one more than one-quarter of this number is 3. Find the two numbers that satisfy this condition.

97. Fun with Numbers The square of one more than one-third of a number is the same as twice the number. Find the two numbers that satisfy this condition.

98. Fun with Numbers The square of one less than one half of a number is the same as three more than twice the number. Find the two numbers that satisfy this condition.

42. $\left\{\frac{-7-\sqrt{241}}{2}, \frac{-7+\sqrt{241}}{2}\right\}$
43. no real solution
44. $\{-8, -4\}$
45. $\left\{-1, \frac{1}{3}\right\}$ 46. $\left\{-\frac{2}{3}, 2\right\}$
47. $\left\{-\frac{1}{2}, \frac{3}{2}\right\}$ 48. $\left\{-2, \frac{2}{5}\right\}$
49. $\left\{-3, \frac{1}{2}\right\}$ 50. $\left\{-\frac{3}{2}, 1\right\}$
51. $\left\{\frac{-2-\sqrt{10}}{3}, \frac{-2+\sqrt{10}}{3}\right\}$
52. $\left\{\frac{-1-\sqrt{13}}{3}, \frac{-1+\sqrt{13}}{3}\right\}$
53. $\left\{\frac{-1-\sqrt{6}}{2}, \frac{-1+\sqrt{6}}{2}\right\}$
54. $\left\{\frac{-3-2\sqrt{2}}{2}, \frac{-3+2\sqrt{2}}{2}\right\}$
55. no real solution
56. no real solution 57. $\left\{\frac{3}{2}, \frac{5}{2}\right\}$
58. $\left\{-3, \frac{3}{5}\right\}$
59. $\left\{\frac{3-\sqrt{2}}{2}, \frac{3+\sqrt{2}}{2}\right\}$
60. $\left\{\frac{5-\sqrt{17}}{4}, \frac{5+\sqrt{17}}{4}\right\}$
61. $\left\{\frac{6-\sqrt{37}}{2}, \frac{6+\sqrt{37}}{2}\right\}$
62. $\left\{\frac{8-\sqrt{69}}{2}, \frac{8+\sqrt{69}}{2}\right\}$
63. $\left\{\frac{9-\sqrt{91}}{3}, \frac{9+\sqrt{91}}{3}\right\}$
64. $\left\{\frac{6-\sqrt{43}}{3}, \frac{6+\sqrt{43}}{3}\right\}$
65. $\left\{\frac{3-\sqrt{41}}{4}, \frac{3+\sqrt{41}}{4}\right\}$
66. $\left\{\frac{1}{2}, \frac{5}{3}\right\}$ 67. $\{-4, 6\}$
68. $\left\{\frac{9-\sqrt{53}}{2}, \frac{9+\sqrt{53}}{2}\right\}$
69. $\left\{\frac{1}{2}\right\}$ 70. $\left\{-\frac{3}{2}\right\}$
71. $\{-3-2\sqrt{3}, -3+2\sqrt{3}\}$
72. $\{-5-2\sqrt{10}, -5+2\sqrt{10}\}$
73. $\{-2\sqrt{5}, 2\sqrt{5}\}$ 74. $\{-3\sqrt{2}, 3\sqrt{2}\}$
75. $\left\{\frac{-3-\sqrt{15}}{3}, \frac{-3+\sqrt{15}}{3}\right\}$
76. $\left\{\frac{2-\sqrt{2}}{2}, \frac{2+\sqrt{2}}{2}\right\}$
77. $\left\{\frac{-5-\sqrt{97}}{2}, \frac{-5+\sqrt{97}}{2}\right\}$
78. $\left\{\frac{-3-\sqrt{33}}{2}, \frac{-3+\sqrt{33}}{2}\right\}$
79. $\{-3, 3\}$ 80. $\{-6, 6\}$
81. $\{-6, 2\}$ 82. $\{7, 11\}$
83. $\{-12, 2\}$ 84. $\{-2, 15\}$
85. $\{-3\sqrt{2}, 3\sqrt{2}\}$ 86. $\{-4\sqrt{3}, 4\sqrt{3}\}$
87. $\{-2\sqrt{3}, 2\sqrt{3}\}$ 88. $\{-2\sqrt{2}, 2\sqrt{2}\}$
89. $\left\{\frac{-6-3\sqrt{10}}{2}, \frac{-6+3\sqrt{10}}{2}\right\}$

90. $\left\{\frac{3-\sqrt{129}}{6}, \frac{3+\sqrt{129}}{6}\right\}$
91. $\left\{\frac{2}{3}, 3\right\}$ 92. $\left\{-4, -\frac{1}{2}\right\}$
93. $\left\{-1-\sqrt{34}, -1+\sqrt{34}\right\}$
94. $\left\{\frac{5-\sqrt{33}}{2}, \frac{5+\sqrt{33}}{2}\right\}$
95. -4 or 6 96. -6 or 2
97. $6-3\sqrt{3}$ or $6+3\sqrt{3}$
98. $6-2\sqrt{11}$ or $6+2\sqrt{11}$
99. $15\sqrt{2}$ 100. $9\sqrt{2}$ 101. 3 and 6
102. 4 and 8
103. (a) 9 (b) $\{3, 6\}$ (c) $3+6=9$
104. (a) -5 (b) $\{-8, 3\}$
(c) $-8+3=-5$
105. (a) 2 (b) $\left\{1-\frac{\sqrt{6}}{2}, 1+\frac{\sqrt{6}}{2}\right\}$
(c) $1-\frac{\sqrt{6}}{2}+1+\frac{\sqrt{6}}{2}=2$
106. (a) -2
(b) $\left\{-1-\sqrt{3}, -1+\sqrt{3}\right\}$
(c) $-1-\sqrt{3}+\left(-1+\sqrt{3}\right)=-2$
107. product $=-\frac{3}{4}$; $\left\{-\frac{3}{2}, \frac{1}{2}\right\}$;
$\left(-\frac{3}{2}\right)\left(\frac{1}{2}\right)=-\frac{3}{4}$
108. product $=\frac{4}{3}$; $\left\{\frac{2}{3}, 2\right\}$; $\left(\frac{2}{3}\right)(2)=\frac{4}{3}$
109. product $=-7$; $\left\{1-2\sqrt{2}, 1+2\sqrt{2}\right\}$;
$\left(1-2\sqrt{2}\right)\left(1+2\sqrt{2}\right)=-7$
110. product $=-\frac{13}{2}$; $\left\{\frac{2-\sqrt{30}}{2}, \frac{2+\sqrt{30}}{2}\right\}$;
$\left(\frac{2-\sqrt{30}}{2}\right)\left(\frac{2+\sqrt{30}}{2}\right)=-\frac{13}{2}$
111. center $=(5, -3)$, radius $=2$
112. center $(-4, 2)$, radius $=1$
113. center $=(10, 8)$, radius $=\sqrt{142}$
114. center $=(-7, -3)$, radius $=\sqrt{70}$

△ **99. The Pythagorean Theorem** In a right triangle, the lengths of the two legs are equal. If the length of the hypotenuse is 30, find the length of the legs.

△ **100. The Pythagorean Theorem** In a right triangle, the lengths of the two legs are equal. If the length of the hypotenuse is 18, find the length of the legs.

△ **101. The Pythagorean Theorem** In a right triangle, the length of the hypotenuse is twice the length of the shorter leg. If the length of the longer leg is $\sqrt{27}$, find the lengths of the other two sides of the triangle.

△ **102. The Pythagorean Theorem** In a right triangle, the length of the hypotenuse is twice the length of the shorter leg. If the length of the longer leg is $\sqrt{48}$, find the lengths of the other two sides of the triangle.

Extending the Concepts

When a quadratic equation is written in the form $x^2+bx+c=0$, the sum of the two solutions of the equation is $-b$, the opposite of the coefficient of the linear term. In Problems 103–106, use this fact to (a) identify the sum of the solutions for each quadratic equation, (b) find the solutions of the quadratic equation by completing the square, and (c) verify that the sum of the solutions is $-b$.

103. $x^2-9x+18=0$ **104.** $x^2+5x-24=0$

105. $2x^2-4x-1=0$ **106.** $2x^2+4x-4=0$

When a quadratic equation is written in the form $x^2+bx+c=0$, the product of the two solutions is c, the constant term. In Problems 107–110, use this fact to identify the product of the solutions for each quadratic equation. Then solve the quadratic equation by completing the square and verify that the product of the solutions is c.

107. $4x^2+4x-3=0$ **108.** $3x^2-8x+4=0$

109. $2x^2=4x+14$ **110.** $2x^2-4x=13$

An equation of the form $(x-h)^2+(y-k)^2=r^2$ is the equation of a circle whose radius is r and center is (h, k). Use completing the square to write the following equations in this form. State the center and radius of each circle.

111. $x^2+y^2-10x+6y+30=0$ **112.** $x^2+y^2+8x-4y+19=0$

113. $x^2+y^2-20x-16y+22=0$ **114.** $x^2+y^2+14x+6y-12=0$

7.3 Solving Quadratic Equations Using the Quadratic Formula

OBJECTIVES

1. Solve Quadratic Equations Using the Quadratic Formula
2. Use the Discriminant to Determine Which Method to Use When Solving a Quadratic Equation

Preparing for Solving Quadratic Equations Using the Quadratic Formula

Before getting started, take the following readiness quiz. If you get a problem wrong, go back to the section cited and review the material.

1. Evaluate $-4+\sqrt{16-4(2)(1)}$. Give the exact value. [Section 6.2, pp. 400–402]

2. Evaluate $\frac{6+\sqrt{6^2-4(3)(-2)}}{6}$. Give the exact value. [Section 6.2, pp. 400–402]

3. Write each quadratic equation in standard form:
(a) $4c-c^2=12$ **(b)** $(n+2)(n-3)=6$ [Section 4.6, pp. 283–285]

4. Solve: $\frac{3}{x}+4=\frac{15}{x}$ [Section 5.7, pp. 357–363]

Preparing for...Answers **1.** $-4+2\sqrt{2}$ **2.** $\frac{3+\sqrt{15}}{3}$ or $1+\frac{\sqrt{15}}{3}$ **3. (a)** $c^2-4c+12=0$ **(b)** $n^2-n-12=0$ **4.** $\{3\}$

Currently, we know three methods for solving quadratic equations: (1) factoring, (2) using the Square Root Property, and (3) completing the square. Why do we need three methods? Because each method provides the "quickest" route to the solution, when used appropriately. For example, if the quadratic expression is easy to factor, factoring will get you to the solution more easily than the other two. If the equation is in the form $x^2 = p$ or $(ax \pm b)^2 = p$, using the Square Root Property is easiest. We use completing the square to solve the remaining quadratic equations. But the method of completing the square is tedious. So you may be asking yourself, "Is there an alternative to completing the square?" The answer, you'll be happy to know, is yes!

1 Solve Quadratic Equations Using the Quadratic Formula

Teaching Tip
Students may be unfamiliar with the derivation of a formula. Explain that formulas, such as the quadratic formula, don't just appear as if by magic. The quadratic formula is a generalization for solving a quadratic equation of the form $ax^2 + bx + c = 0$, using completing the square.

We can use the method of completing the square to obtain a general formula for solving the quadratic equation

$$ax^2 + bx + c = 0, \qquad a \neq 0$$

As we have seen, to complete the square, we first must get the constant c on the right-hand side of the equation.

$$ax^2 + bx = -c, \qquad a \neq 0$$

Since $a \neq 0$, we can divide both sides of the equation by a to get

$$x^2 + \frac{b}{a}x = -\frac{c}{a}$$

Now that the coefficient of x^2 is 1, we can complete the square on the left side. We do this by adding the square of $\frac{1}{2}$ of the coefficient of x to both sides of the equation. That is, we add

$$\left(\frac{1}{2} \cdot \frac{b}{a}\right)^2 = \frac{b^2}{4a^2}$$

to both sides of the equation. We now have

$$x^2 + \frac{b}{a}x + \frac{b^2}{4a^2} = -\frac{c}{a} + \frac{b^2}{4a^2}$$

or

$$x^2 + \frac{b}{a}x + \frac{b^2}{4a^2} = \frac{b^2}{4a^2} - \frac{c}{a}$$

We combine terms on the right-hand side by writing the right-hand side over a common denominator. The least common denominator (LCD) on the right-hand side is $4a^2$. So we multiply $-\frac{c}{a}$ by $1 = \frac{4a}{4a}$:

$$x^2 + \frac{b}{a}x + \frac{b^2}{4a^2} = \frac{b^2}{4a^2} - \frac{c}{a} \cdot \frac{4a}{4a}$$

$$x^2 + \frac{b}{a}x + \frac{b^2}{4a^2} = \frac{b^2}{4a^2} - \frac{4ac}{4a^2}$$

$$x^2 + \frac{b}{a}x + \frac{b^2}{4a^2} = \frac{b^2 - 4ac}{4a^2}$$

Work Smart
To factor any perfect square trinomial of the form $x^2 + bx + c$, we write $\left(x + \frac{b}{2}\right)^2$.

The expression on the left-hand side is a perfect square trinomial. We factor the left-hand side and obtain

$$\left(x + \frac{b}{2a}\right)^2 = \frac{b^2 - 4ac}{4a^2}$$

At this point, we will assume that $a > 0$ (you'll see why in a little while). This assumption does not compromise the results because if $a < 0$, we could multiply both sides of the equation $ax^2 + bx + c = 0$ by -1 to make a positive. With this assumption, we use the

Square Root Property and get

$$x + \frac{b}{2a} = \pm\sqrt{\frac{b^2 - 4ac}{4a^2}}$$

$\sqrt{\frac{a}{b}} = \frac{\sqrt{a}}{\sqrt{b}}$: $$x + \frac{b}{2a} = \pm\frac{\sqrt{b^2 - 4ac}}{\sqrt{4a^2}}$$

$\sqrt{4a^2} = 2a$ since $a > 0$: $$x + \frac{b}{2a} = \pm\frac{\sqrt{b^2 - 4ac}}{2a}$$

Subtract $\frac{b}{2a}$ from both sides: $$x = -\frac{b}{2a} \pm \frac{\sqrt{b^2 - 4ac}}{2a}$$

Write over a common denominator: $$x = \frac{-b \pm \sqrt{b^2 - 4ac}}{2a}$$

Work Smart
This step shows why $a > 0$. Without this assumption, $\sqrt{4a^2} = 2|a|$.

This gives us the *quadratic formula.*

THE QUADRATIC FORMULA
The solution(s) to the quadratic equation $ax^2 + bx + c = 0, a \neq 0$, are given by the **quadratic formula**

$$x = \frac{-b \pm \sqrt{b^2 - 4ac}}{2a}$$

In Words
The quadratic formula says that the solution to the equation $ax^2 + bx + c = 0$ is "the opposite of b plus or minus the square root of b squared minus $4ac$ all over $2a$."

EXAMPLE 1 How to Solve a Quadratic Equation Using the Quadratic Formula

Solve: $2x^2 + 11x + 15 = 0$

Classroom Example
Solve:
$3x^2 + 13x + 12 = 0$
Answer: $\left\{-3, -\frac{4}{3}\right\}$

Step-by-Step Solution

Step 1: Write the equation in standard form $ax^2 + bx + c = 0$ and identify the values of a, b, and c.

The equation is already in standard form.

$a = 2, b = 11, c = 15$: $$2x^2 + 11x + 15 = 0$$

Step 2: Substitute the values of a, b, and c into the quadratic formula.

Write the quadratic formula: $$x = \frac{-b \pm \sqrt{b^2 - 4ac}}{2a}$$

Substitute 2 for a, 11 for b, and 15 for c: $$x = \frac{-11 \pm \sqrt{11^2 - 4(2)(15)}}{2(2)}$$

Step 3: Simplify the expression found in Step 2.

$$= \frac{-11 \pm \sqrt{121 - 120}}{2(2)}$$

$$= \frac{-11 \pm \sqrt{1}}{4}$$

$$= \frac{-11 \pm 1}{4}$$

$$x = \frac{-11 - 1}{4} \quad \text{or} \quad x = \frac{-11 + 1}{4}$$

$$x = \frac{-12}{4} \quad \text{or} \quad x = \frac{-10}{4}$$

Reduce each rational number to lowest terms: $$x = -3 \quad \text{or} \quad x = -\frac{5}{2}$$

Work Smart: Study Skills
Notice in Step 2 we wrote out the quadratic formula. When solving homework problems, always write the quadratic formula as part of the solution to help you memorize the formula.

Step 4: *Verify your solutions.*

$$
\begin{aligned}
x = -3:\quad 2x^2 + 11x + 15 &= 0\\
2(-3)^2 + 11(-3) + 15 &\stackrel{?}{=} 0\\
2\cdot 9 - 33 + 15 &\stackrel{?}{=} 0\\
18 - 33 + 15 &\stackrel{?}{=} 0\\
0 &= 0 \quad \text{True}
\end{aligned}
$$

$$
\begin{aligned}
x = -\frac{5}{2}:\quad 2x^2 + 11x + 15 &= 0\\
2\left(-\frac{5}{2}\right)^2 + 11\left(-\frac{5}{2}\right) + 15 &\stackrel{?}{=} 0\\
2\cdot\frac{25}{4} - \frac{55}{2} + 15 &\stackrel{?}{=} 0\\
\frac{25}{2} - \frac{55}{2} + 15 &\stackrel{?}{=} 0\\
-\frac{30}{2} + 15 &\stackrel{?}{=} 0\\
0 &= 0 \quad \text{True}
\end{aligned}
$$

The solution set is $\left\{-3, -\frac{5}{2}\right\}$.

The following steps can be used to solve any quadratic equation using the quadratic formula.

Steps to Solve a Quadratic Equation Using the Quadratic Formula

Step 1: Write the equation in standard form $ax^2 + bx + c = 0$, and identify the values of a, b, and c.

Step 2: Substitute the values of a, b, and c into the quadratic formula.

Step 3: Simplify the expression found in Step 2.

Step 4: Verify your solution(s).

QUICK ✓ *Solve each equation using the quadratic formula.*

1. $3x^2 + 7x + 4 = 0$ **2.** $2x^2 + 13x + 6 = 0$

Classroom Example ➤
Solve: $8a^2 + 2a = 3$

Answer: $\left\{-\frac{3}{4}, \frac{1}{2}\right\}$

EXAMPLE 2 Solving a Quadratic Equation Using the Quadratic Formula

Solve: $6x^2 + 7x = 20$

Solution

To solve a quadratic equation using the quadratic formula, the quadratic equation must be in standard form, $ax^2 + bx + c = 0$. To express $6x^2 + 7x = 20$ in standard form, we subtract 20 from each side of the equation, and proceed as follows:

Teaching Tip
Note to students that we list values for *a, b,* and *c* *after* the quadratic equation is written in standard form.

$$
\begin{aligned}
6x^2 + 7x &= 20\\
6x^2 + 7x - 20 &= 20 - 20\\
\underset{a}{6}x^2 + \underset{b}{7}x \underset{c}{- 20} &= 0
\end{aligned}
$$

Work Smart
A common error is to accidentally write the quadratic formula as

$$x = -b \pm \frac{\sqrt{b^2 - 4ac}}{2a}$$

Be sure to extend the fraction bar under the entire numerator as in

$$x = \frac{-b \pm \sqrt{b^2 - 4ac}}{2a}$$

We see that $a = 6$, $b = 7$, and $c = -20$.

Write the quadratic formula: $x = \dfrac{-b \pm \sqrt{b^2 - 4ac}}{2a}$

Substitute $a = 6$, $b = 7$, $c = -20$: $x = \dfrac{-7 \pm \sqrt{7^2 - 4(6)(-20)}}{2(6)}$

$$= \frac{-7 \pm \sqrt{49 - (-480)}}{12}$$

$$= \frac{-7 \pm \sqrt{49 + 480}}{12}$$

$$= \frac{-7 \pm \sqrt{529}}{12}$$

Evaluate the square root: $= \frac{-7 \pm 23}{12}$

$$x = \frac{-7 - 23}{12} \quad \text{or} \quad x = \frac{-7 + 23}{12}$$

$$= -\frac{30}{12} \quad \text{or} \quad = \frac{16}{12}$$

$$= -\frac{5}{2} \quad \text{or} \quad = \frac{4}{3}$$

Work Smart
The quadratic equation solved in Example 2 could also be solved by factoring. Try it for yourself. Which method do you prefer? Why?

The solution set is $\left\{-\frac{5}{2}, \frac{4}{3}\right\}$. We leave it to you to verify the solutions.

QUICK ✓ *Solve each equation using the quadratic formula.*

3. $6x^2 + 6 = -13x$ **4.** $8x^2 + 6x = 5$

Not every quadratic equation has rational solutions.

Classroom Example
Solve: $3p^2 = 6p - 1$

Answer: $\left\{\frac{3 - \sqrt{6}}{3}, \frac{3 + \sqrt{6}}{3}\right\}$

EXAMPLE 3 Solving a Quadratic Equation Having Irrational Solutions

Solve: $4z^2 + 1 = 8z$

Solution

Once again, we first write the equation in standard form by setting the equation equal to 0.

$$4z^2 + 1 = 8z$$

Subtract $8z$ from both sides: $4z^2 - 8z + 1 = 8z - 8z$

$$4z^2 - 8z + 1 = 0$$

We see that $a = 4$, $b = -8$, and $c = 1$. Now we use the quadratic formula.

Work Smart
Notice we write the quadratic formula as $z =$ because the variable in the equation is z.

Write the quadratic formula: $z = \frac{-b \pm \sqrt{b^2 - 4ac}}{2a}$

Substitute $a = 4$, $b = -8$, $c = 1$: $z = \frac{-(-8) \pm \sqrt{(-8)^2 - 4(4)(1)}}{2(4)}$

$$= \frac{8 \pm \sqrt{64 - 16}}{8}$$

$$= \frac{8 \pm \sqrt{48}}{8}$$

Simplify the square root: $= \frac{8 \pm \sqrt{16 \cdot 3}}{8}$

$$= \frac{8 \pm 4\sqrt{3}}{8}$$

Factor the numerator: $= \frac{4(2 \pm \sqrt{3})}{8}$

Divide out common factors: $= \frac{2 \pm \sqrt{3}}{2}$

Work Smart
We could also simplify $z = \frac{8 \pm 4\sqrt{3}}{8}$ as follows:

$$z = \frac{8 \pm 4\sqrt{3}}{8}$$

$$= \frac{8}{8} \pm \frac{4\sqrt{3}}{8}$$

$$= 1 \pm \frac{\sqrt{3}}{2}$$

These answers are equivalent to the ones given in Example 3.

The solution set is $\left\{\frac{2-\sqrt{3}}{2}, \frac{2+\sqrt{3}}{2}\right\}$. We leave it to you to check the solutions.

QUICK ✓ *Solve each equation using the quadratic formula.*

5. $y^2 - 2 = -4y$

6. $16x^2 - 24x + 7 = 0$

When fractions appear as coefficients in a quadratic equation, we can multiply both sides of the equation by the least common denominator of the coefficients to clear the fractions. Then we use can use the quadratic formula without fractional values for a, b, and c.

Classroom Example
Solve:

$$\frac{2}{3}x^2 + \frac{1}{6}x - \frac{1}{2} = 0$$

Answer: $\left\{-1, \frac{3}{4}\right\}$

EXAMPLE 4 Solving a Rational Equation That Leads to a Quadratic Equation

Solve: $\frac{3}{4} + \frac{5}{8}x - \frac{1}{2}x^2 = 0$

Solution

Because the coefficients are fractions, we multiply both sides of the equation by 8, the least common denominator of the coefficients.

$$\frac{3}{4} + \frac{5}{8}x - \frac{1}{2}x^2 = 0$$

Multiply both sides by 8: $$8\left(\frac{3}{4} + \frac{5}{8}x - \frac{1}{2}x^2\right) = 8 \cdot 0$$

Apply the Distributive Property: $$8 \cdot \frac{3}{4} + 8 \cdot \frac{5}{8}x - 8 \cdot \frac{1}{2}x^2 = 8 \cdot 0$$

$$6 + 5x - 4x^2 = 0$$

Rearrange terms: $$-4x^2 + 5x + 6 = 0$$

Because the coefficient of the square term is negative, we multiply both sides by -1.

$$-1 \cdot (-4x^2 + 5x + 6) = -1 \cdot 0$$

Distribute: $$4x^2 - 5x - 6 = 0$$

The equation is now in standard form, with $a = 4$, $b = -5$, and $c = -6$.

Work Smart
Although not necessary, writing a quadratic equation in standard form with a leading coefficient that is positive is a good idea because it makes computation easier later on.

Write the quadratic formula: $$x = \frac{-b \pm \sqrt{b^2 - 4ac}}{2a}$$

Substitute $a = 4$, $b = -5$, $c = -6$: $$x = \frac{-(-5) \pm \sqrt{(-5)^2 - 4(4)(-6)}}{2(4)}$$

$$= \frac{5 \pm \sqrt{25 - (-96)}}{8}$$

$$= \frac{5 \pm \sqrt{121}}{8}$$

Simplify the radicand: $$= \frac{5 \pm 11}{8}$$

$$x = \frac{5 - 11}{8} \quad \text{or} \quad x = \frac{5 + 11}{8}$$

$$= \frac{-6}{8} \quad \text{or} \quad = \frac{16}{8}$$

Simplify: $$= -\frac{3}{4} \quad \text{or} \quad = 2$$

The solution set is $\left\{-\frac{3}{4}, 2\right\}$. We leave the check to you.

QUICK ✓ *Solve the equation using the quadratic formula.*

7. $x - \frac{3}{8} = \frac{1}{4}x^2$

Classroom Example ➤
Solve:

$3y(3y + 2) + 1 = 0$

Answer: $\left\{-\frac{1}{3}\right\}$

EXAMPLE 5 Solving a Quadratic Equation Having A Repeated Solution

Solve $4x(x - 3) + 9 = 0$ using the quadratic formula.

Solution

What's the first step? Write the equation in standard form!

$$4x(x - 3) + 9 = 0$$

Use the Distributive Property: $$4x^2 - 12x + 9 = 0$$

We see that $a = 4$, $b = -12$, and $c = 9$.

Teaching Tip
After you solve Example 5 using the quadratic formula, you may want to reinforce the idea of a double root by solving the equation by factoring.

Write the quadratic formula: $$x = \frac{-b \pm \sqrt{b^2 - 4ac}}{2a}$$

Substitute $a = 4$, $b = -12$, $c = 9$: $$x = \frac{-(-12) \pm \sqrt{(-12)^2 - 4(4)(9)}}{2(4)}$$

$$= \frac{12 \pm \sqrt{144 - 144}}{8}$$

$$= \frac{12 \pm \sqrt{0}}{8}$$

$$= \frac{12 \pm 0}{8}$$

The numerator is 12 ± 0, but $12 - 0 = 12$ and $12 + 0 = 12$. So the numerator is 12, and we have the solution: $x = \frac{12}{8} = \frac{3}{2}$. The solution $x = \frac{3}{2}$ is called a **repeated solution** or **double root** because it appears twice, once in the form $x = \frac{12 - 0}{8} = \frac{3}{2}$ and once in the form $x = \frac{12 + 0}{8} = \frac{3}{2}$. The solution set is $\left\{\frac{3}{2}\right\}$.

Work Smart
Did you notice that the quadratic equation solved in Example 5 can be solved by factoring? Try solving it this way. Which method do you prefer?

QUICK ✓ *Solve the equation using the quadratic formula.*

8. $16n^2 - 40n = -25$

When we solved linear equations, we found that some linear equations have no real solution. We also saw that some radical equations have no real solution. The same holds for quadratic equations.

Classroom Example Solve:

$$\frac{4}{3}x^2 - \frac{1}{6}x + 2 = 0$$

Answer: no real solution

EXAMPLE 6 Solving a Quadratic Equation Having No Real Solution

Solve: $\frac{1}{2}m^2 + \frac{1}{2}m + 4 = 0$

Solution

Is the equation is in standard form? Yes, but we have fractional coefficients. To clear the fractions, we multiply both sides of the equation by the LCD, 2, so working with the quadratic formula will be easier.

$$\frac{1}{2}m^2 + \frac{1}{2}m + 4 = 0$$

Clear fractions: $$2\left(\frac{1}{2}m^2 + \frac{1}{2}m + 4\right) = 2 \cdot 0$$

Distribute and simplify: $$m^2 + m + 8 = 0$$

The equation is now in standard form, with $a = 1$, $b = 1$, and $c = 8$.

Write the quadratic formula: $$m = \frac{-b \pm \sqrt{b^2 - 4ac}}{2a}$$

Substitute: $$m = \frac{-1 \pm \sqrt{1^2 - 4(1)(8)}}{2(1)}$$

$$= \frac{-1 \pm \sqrt{1 - 32}}{2}$$

$$= \frac{-1 \pm \sqrt{-31}}{2}$$

Because $\sqrt{-31}$ is not a real number, the equation has no real solution.

QUICK ✓ *Solve the equation using the quadratic formula.*

9. $4n^2 - 3n + 5 = 0$

Classroom Example Use Example 7 to find the number of pairs of sunglasses required to have a weekly revenue of \$6000 with $R = -0.1x^2 + 50x$.

Answer: 300 pairs of sunglasses

EXAMPLE 7 Revenue from Selling Sunglasses

A company uses the equation $R = -0.1x^2 + 70x$ to model the revenue received from selling x pairs of sunglasses per week. If the company always sells at least 250 pairs of sunglasses per week, find the number of pairs of sunglasses required to have weekly revenue of \$10,000.

Solution

Because we want to find the value of x for which revenue is \$10,000, we substitute $R = 10{,}000$ into the equation $R = -0.1x^2 + 70x$ and obtain $10{,}000 = -0.1x^2 + 70x$. We can use the quadratic formula to solve the equation for x by writing the equation in standard form, $ax^2 + bx + c = 0$.

Work Smart
Multiply both sides of the equation by -10 to clear decimals and make the coefficient of x^2 positive.

$$-0.1x^2 + 70x = 10{,}000$$

Multiply both sides by -10 to clear decimal: $$-10(-0.1x^2 + 70x) = -10 \cdot 10{,}000$$

Distribute: $$x^2 - 700x = -100{,}000$$

Add 100,000 to both sides: $x^2 - 700x + 100{,}000 = -100{,}000 + 100{,}000$

$$x^2 + 700x + 100{,}000 = 0$$

We see that $a = 1$, $b = 700$, and $c = 100{,}000$.

Write the quadratic formula: $x = \dfrac{-b \pm \sqrt{b^2 - 4ac}}{2a}$

Substitute: $x = \dfrac{-700 \pm \sqrt{700^2 - 4(1)(100{,}000)}}{2(1)}$

$$= \frac{-700 \pm \sqrt{490{,}000 - 400{,}000}}{2}$$

$$= \frac{700 \pm \sqrt{90{,}000}}{2}$$

Evaluate the square root: $= \dfrac{700 \pm 300}{2}$

$$x = \frac{700 - 300}{2} \quad \text{or} \quad x = \frac{700 + 300}{2}$$

$$= 200 \quad \text{or} \quad = 500$$

We are told that the company sells more than 250 pairs of sunglasses per week, so we conclude that 500 pairs of sunglasses must be sold to have revenue of $10,000. ■

QUICK ✓

10. The daily revenue received from a company selling x all-day passes to a theme park is given by the equation $R = -0.02x^2 + 24x$. Solve the equation $4000 = -0.02x^2 + 24x$ to determine the number of all-day passes that must be sold for the company to have daily revenue of $4000. Past records of the company show that the park sells more than 400 all-day passes daily.

2 Use the Discriminant to Determine Which Method to Use When Solving a Quadratic Equation

Teaching Tip
The material in Objective 2 is optional and can be skipped without loss of continuity.

We have now introduced four methods for solving quadratic equations:

1. Factoring
2. Square Root Property
3. Completing the square
4. The quadratic formula

Questions that you may be asking yourself are, "How do I know which method to use?" or "Is one method 'better' than another?" The answer is, "It depends." Answering the questions "Is the expression in the quadratic equation factorable? "or" Is the equation of the form $x^2 = p$?" will help you to decide the most efficient method of solving the equation. Another strategy to use when determining which method is most efficient is to find the value of the *discriminant.*

In Words
The discriminant is the radicand of the quadratic formula.

DEFINITION

For a quadratic equation $ax^2 + bx + c = 0$, the expression $\boldsymbol{b^2 - 4ac}$ is called the **discriminant.**

Why is the discriminant important? The following supplies the answer.

THE DISCRIMINANT AND SOLVING A QUADRATIC EQUATION

For a quadratic equation $ax^2 + bx + c = 0$, the discriminant is $b^2 - 4ac$.

1. If $b^2 - 4ac > 0$, and

- $b^2 - 4ac$ is a perfect square, the quadratic equation may be solved by factoring.
- $b^2 - 4ac$ is not a perfect square, the quadratic equation is not factorable over the integers. Completing the square or the quadratic formula should be used to solve the equation.

2. $b^2 - 4ac = 0$, the equation may be solved by factoring.

3. $b^2 - 4ac < 0$, the equation has no real solution.

Teaching Tip
Emphasize that the discriminant is the radicand of the quadratic formula—no square root!

Classroom Example
Use Example 8 with $x^2 - 5x + 2 = 0$.

Answer: $b^2 - 4ac = 17$; use the quadratic formula to obtain $\left\{\frac{5 - \sqrt{17}}{2}, \frac{5 + \sqrt{17}}{2}\right\}$.

EXAMPLE 8 Using the Discriminant to Determine a Method to Solve a Quadratic Equation

For the quadratic equation $x^2 - 5x + 3 = 0$, determine the discriminant. Then use the value of the discriminant to determine the most efficient method to solve the quadratic equation, and solve.

Solution

We compare $x^2 - 5x + 3 = 0$ to the standard form $ax^2 + bx + c = 0$.

$$\underset{a}{x^2} - \underset{b}{5}x + \underset{c}{3} = 0$$

We have $a = 1$, $b = -5$, and $c = 3$. Substituting these values into the formula for the discriminant, $b^2 - 4ac$, we obtain

$$\begin{aligned} b^2 - 4ac &= (-5)^2 - 4(1)(3) \\ &= 25 - 12 \\ &= 13 \end{aligned}$$

Because $b^2 - 4ac = 13$ is positive but not a perfect square, we use the quadratic formula to solve the equation.

$$x = \frac{-b \pm \sqrt{b^2 - 4ac}}{2a}$$

Substitute $a = 1$, $b = -5$, $c = 3$, and $b^2 - 4ac = 13$:

$$= \frac{-(-5) \pm \sqrt{13}}{2(1)}$$

$$= \frac{5 \pm \sqrt{13}}{2}$$

We leave the check to you. The solution set is $\left\{\frac{5 - \sqrt{13}}{2}, \frac{5 + \sqrt{13}}{2}\right\}$.

Classroom Example
Use Example 9 with
(a) $2x^2 + 7x - 15 = 0$
(b) $3x^2 + 2x - 4 = 0$

Answer:
(a) $b^2 - 4ac = 169$; use factoring to obtain $\left\{-5, \frac{3}{2}\right\}$.
(b) $b^2 - 4ac = 52$; use quadratic formula to obtain $\left\{\frac{-1 \pm \sqrt{13}}{3}\right\}$.

EXAMPLE 9 Using the Discriminant to Determine a Method to Solve a Quadratic Equation

Determine the discriminant of the quadratic equation $2x^2 - 5x - 12 = 0$. Then use the value of the discriminant to determine the most efficient method to solve the quadratic equation, and solve.

Solution

The equation is in standard form. The value of a is 2, b is -5, and c is -12. Now, we determine the value of the discriminant, $b^2 - 4ac$.

$$\begin{aligned} b^2 - 4ac &= (-5)^2 - 4(2)(-12) \\ &= 25 + 96 \\ &= 121 \end{aligned}$$

Since the discriminant is positive and a perfect square number, the quadratic equation $2x^2 - 5x - 12 = 0$ can be solved by factoring.

$$\begin{aligned} 2x^2 - 5x - 12 &= 0 \\ (2x + 3)(x - 4) &= 0 \\ 2x + 3 = 0 \quad &\text{or} \quad x - 4 = 0 \\ 2x = -3 \quad &\text{or} \quad x = 4 \\ x = -\frac{3}{2} \end{aligned}$$

We leave the check to you. The solution set is $\left\{-\frac{3}{2}, 4\right\}$.

QUICK ✓ *Determine the value of the discriminant. Use the discriminant to determine the most efficient method to solve the quadratic equation, and then solve.*

11. $9q^2 - 6q + 1 = 0$ **12.** $3w^2 = 5w - 2$ **13.** $4z^2 - 2z = 1$

Table 1 summarizes all four methods for solving a quadratic equation, guidance as to when to use each one, and an example of each.

Table 1

Method	When to Use	Example
Square Root Property	When the quadratic equation is written in the form $x^2 = p$ where p is any real number	$x^2 = 45$ Square Root Property: $x = \pm\sqrt{45}$ $= \pm 3\sqrt{5}$
Square Root Property	When the quadratic equation is written in the form $ax^2 - c = 0$	$3p^2 - 12 = 0$ Add 12 to both sides: $3p^2 = 12$ Divide both sides by 3: $p^2 = 4$ Square Root Property: $p = \pm\sqrt{4}$ $p = \pm 2$
Square Root Property	When the quadratic equation is written in the form $(ax + b)^2 = p$ where p is any real number	$(4n - 3)^2 = 12$ Square Root Property: $4n - 3 = \pm\sqrt{12}$ Simplify the radicand: $4n - 3 = \pm\sqrt{4 \cdot 3}$ $4n - 3 = \pm 2\sqrt{3}$ Add 3 to both sides: $4n = 3 \pm 2\sqrt{3}$ Divide both sides by 4: $n = \frac{3 \pm 2\sqrt{3}}{4}$

Method	When to Use	Example
Factoring or the Quadratic Formula	When the quadratic equation is written in the form $ax^2 + bx + c = 0$ where $b^2 - 4ac$ is a perfect square. Use factoring if the quadratic expression is easy to factor. Otherwise use the quadratic formula.	$2m^2 + m - 10 = 0$ $a = 2, b = 1, c = -10$ $b^2 - 4ac = 1^2 - 4(2)(-10) = 1 + 80 = 81$ 81 is a perfect square, so we can use factoring: $2m^2 + m - 10 = 0$ $(2m + 5)(m - 2) = 0$ $2m + 5 = 0$ or $m - 2 = 0$ $m = -\frac{5}{2}$ or $m = 2$
Quadratic Formula	When the quadratic equation is written in the form $ax^2 + bx + c = 0$ where $b^2 - 4ac$ is positive, but not a perfect square	$2x^2 + 4x - 1 = 0$ $a = 2, b = 4, c = -1$ $b^2 - 4ac = 4^2 - 4(2)(-1) = 16 + 8 = 24$ 24 is positive, but is not a perfect square, so we use the quadratic formula to solve: $x = \frac{-b \pm \sqrt{b^2 - 4ac}}{2a}$ $a = 2, b = 4, b^2 - 4ac = 24$: $= \frac{-4 \pm \sqrt{24}}{2(2)}$ $= \frac{-4 \pm 2\sqrt{6}}{4}$ $= \frac{-2 \pm \sqrt{6}}{2}$
Quadratic Formula	When the quadratic equation is written in the form $ax^2 + bx + c = 0$ where $b^2 - 4ac$ is negative. The equation has no real solution.	$2n^2 + 5n + 8 = 0$ $a = 2, b = 5, c = 8$ $b^2 - 4ac = 5^2 - 4(2)(8) = 25 - 64 = -39$ The value of the discriminant is a negative number, so the quadratic equation has no real solution. We discuss this type of equation in more detail in Section 7.5.

You may have noticed that we did not recommend completing the square as one of the methods to use in solving the quadratic equation. This is because the quadratic formula was developed by completing the square of $ax^2 + bx + c = 0$. Besides, completing the square is a cumbersome task, whereas the quadratic formula is fairly straightforward to use. We did not waste your time discussing completing the square, however, because it is needed to present a discussion of the quadratic formula. Additionally, completing the square is a skill that you will need in future math courses.

QUICK ✓ *Solve each quadratic equation using any method you wish.*

14. $4n^2 - 24 = 0$ **15.** $2y^2 = 3y + 35$ **16.** $1 = 3q^2 + 4q$

7.3 Exercises

1. standard
2. $b^2 - 4ac$; discriminant
3. (a) factoring
   (b) factoring
   (c) the quadratic formula
4. True
5. False
6. False
7. Answers may vary.
8. Answers may vary.
9. $-2x^2 + x + 4 = 0$; $a = -2, b = 1, c = 4$
10. $2x^2 + x - 1 = 0$; $a = 2, b = 1, c = -1$
11. $x^2 + 0x + 4 = 0$; $a = 1, b = 0, c = 4$
12. $3x^2 + 0x - 6 = 0$; $a = 3, b = 0, c = -6$
13. $\frac{3}{2}x^2 - \frac{6}{5}x - 2 = 0$; $a = \frac{3}{2}, b = -\frac{6}{5}, c = -2$ or $15x^2 - 12x - 20 = 0$; $a = 15, b = -12, c = -20$
14. $\frac{3}{7}x^2 - x - \frac{1}{2} = 0$; $a = \frac{3}{7}, b = -1, c = -\frac{1}{2}$ or $6x^2 - 14x - 7 = 0$; $a = 6, b = -14, c = -7$
15. $0.5x^2 - x + 3 = 0$; $a = 0.5, b = -1, c = 3$
16. $x^2 - 0.1x + 2 = 0$; $a = 1, b = -0.1, c = 2$
17. -1 or 3
18. $-\frac{5}{2}$ or 1
19. 2
20. -4
21. $-2 - \sqrt{6}$ or $-2 + \sqrt{6}$
22. $-4 - 3\sqrt{2}$ or $-4 + 3\sqrt{2}$
23. $\frac{1 \pm \sqrt{-15}}{4}$
24. $\frac{5 \pm \sqrt{-35}}{6}$

For Extra Help:

Student Solutions Manual

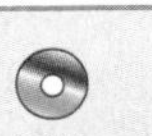
CD Video

PH Math/Tutor Center

MathXL Tutorials on CD

MathXL®

MyMathLab

Concepts and Vocabulary

In Problems 1–3, fill in the blanks.

1. Before using the quadratic formula, it is important that the quadratic equation be written in ________ form.

2. The radicand of the quadratic formula is the algebraic expression ________ and is called the ________

3. Fill in the blanks based on the value of the discriminant.

(a) If the discriminant of a quadratic equation has a value of zero, the most efficient way of solving the quadratic equation is ________.

(b) If the value of the discriminant of a quadratic equation is a perfect square, the most efficient way of solving the quadratic equation is ________.

(c) If the value of the discriminant of a quadratic equation is a positive number that is not a perfect square, the most efficient way of solving the quadratic equation is ________.

In Problems 4–6, answer True or False to each statement.

4. Using the quadratic formula to solve $\frac{1}{2}x^2 + \frac{2}{3}x + 1 = 0$, we can use either $a = \frac{1}{2}, b = \frac{2}{3}, c = 1$ or $a = 3, b = 4, c = 6$.

5. To solve $2x^2 + 5x + 3 = 0$ using the quadratic formula, we write

$$x = -5 \pm \frac{\sqrt{5^2 - 4(2)(3)}}{2(2)}.$$

6. The solution $x = \frac{3 \pm \sqrt{72}}{12}$ is in simplest form.

7. Explain how you can use the discriminant to tell if a quadratic expression is factorable.

8. Explain why completing the square as a method for solving quadratic equations did not appear in Table 1 of this section.

Building Skills

In Problems 9–16, write the quadratic equation in standard form and then identify the values assigned to a, b, and c. Do not solve the equation.

9. $x - 2x^2 = -4$ **10.** $2x^2 = 1 - x$ **11.** $x^2 + 4 = 0$ **12.** $3x^2 - 6 = 0$

13. $2 + \frac{6}{5}x = \frac{3}{2}x^2$ **14.** $\frac{3}{7}x^2 = x + \frac{1}{2}$ **15.** $0.5x^2 = x - 3$ **16.** $0.1x - x^2 = 2$

In Problems 17–24, simplify each expression, if possible.

17. $\frac{6 \pm \sqrt{(-6)^2 - 4(3)(-9)}}{2(3)}$

18. $\frac{-3 \pm \sqrt{3^2 - 4(2)(-5)}}{2(2)}$

19. $\frac{-12 \pm \sqrt{12^2 - 4(-3)(-12)}}{2(-3)}$

20. $\frac{8 \pm \sqrt{(-8)^2 - 4(-1)(-16)}}{2(-1)}$

21. $\frac{-4 \pm \sqrt{4^2 - 4(1)(-2)}}{2(1)}$

22. $\frac{-8 \pm \sqrt{8^2 - 4(1)(-2)}}{2(1)}$

23. $\frac{1 \pm \sqrt{(-1)^2 - 4(2)(2)}}{2(2)}$

24. $\frac{5 \pm \sqrt{(-5)^2 - 4(3)(5)}}{2(3)}$

25. $\{-8, 3\}$ 26. $\{2 - \sqrt{10}, 2 + \sqrt{10}\}$

27. $\left\{-4, -\frac{3}{2}\right\}$ 28. $\left\{-5, \frac{1}{2}\right\}$

29. $\left\{\frac{1 - \sqrt{21}}{2}, \frac{1 + \sqrt{21}}{2}\right\}$

30. $\left\{\frac{7 - \sqrt{61}}{2}, \frac{7 + \sqrt{61}}{2}\right\}$

31. $\left\{\frac{-2 - \sqrt{7}}{3}, \frac{-2 + \sqrt{7}}{3}\right\}$

32. $\left\{\frac{-1 - \sqrt{11}}{2}, \frac{-1 + \sqrt{11}}{2}\right\}$

33. $\left\{\frac{1}{2}, \frac{3}{2}\right\}$ 34. $\left\{-\frac{1}{2}, \frac{2}{5}\right\}$

35. $\left\{\frac{-7 - 3\sqrt{5}}{2}, \frac{-7 + 3\sqrt{5}}{2}\right\}$

36. $\left\{\frac{3 - 3\sqrt{5}}{2}, \frac{3 + 3\sqrt{5}}{2}\right\}$

37. $\left\{\frac{5 - \sqrt{17}}{4}, \frac{5 + \sqrt{17}}{4}\right\}$

38. $\left\{\frac{7 - \sqrt{13}}{6}, \frac{7 + \sqrt{13}}{6}\right\}$

39. $\{3, 4\}$ 40. $\{-7, -2\}$

41. $\left\{-\frac{2}{3}, 2\right\}$ 42. $\left\{\frac{5}{2}, 3\right\}$

43. $\{-3 - \sqrt{5}, -3 + \sqrt{5}\}$

44. $\{1 - \sqrt{3}, 1 + \sqrt{3}\}$

45. $\left\{1 - \frac{\sqrt{2}}{2}, 1 + \frac{\sqrt{2}}{2}\right\}$

46. $\left\{\frac{-2 - \sqrt{10}}{3}, \frac{-2 + \sqrt{10}}{3}\right\}$

47. $\left\{\frac{-5 - \sqrt{33}}{4}, \frac{-5 + \sqrt{33}}{4}\right\}$

48. $\left\{\frac{-3 - \sqrt{41}}{4}, \frac{-3 + \sqrt{41}}{4}\right\}$

49. $\left\{\frac{1}{5}, \frac{1}{3}\right\}$ 50. $\left\{-\frac{5}{2}, -\frac{1}{2}\right\}$

51. $\left\{-\frac{3}{2}\right\}$ 52. $\left\{\frac{2}{3}\right\}$

53. $\{-6, 3\}$ 54. $\{-4, 6\}$

55. $\left\{\frac{4 - \sqrt{10}}{3}, \frac{4 + \sqrt{10}}{3}\right\}$

56. $\left\{\frac{-1 - \sqrt{7}}{2}, \frac{-1 + \sqrt{7}}{2}\right\}$

57. $\{3\}$ 58. $\{4\}$

59. $\left\{\frac{-1 - \sqrt{61}}{2}, \frac{-1 + \sqrt{61}}{2}\right\}$

60. $\left\{\frac{-3 - \sqrt{41}}{2}, \frac{-3 + \sqrt{41}}{2}\right\}$

61. no real solution

62. no real solution

63. $\{-2\sqrt{3}, 2\sqrt{3}\}$

64. $\{-4\sqrt{2}, 4\sqrt{2}\}$

65. $\left\{-\frac{3}{2}, 6\right\}$ 66. $\left\{-3, -\frac{1}{3}\right\}$

67. $\{0, 5\}$ 68. $\{0, 3\}$

69. -39; no real solution

70. -47; no real solution

71. 25; factoring

72. 100; factoring

73. 0; factoring

74. 0; factoring

75. 40; quadratic formula

In Problems 25–68, solve each equation using the quadratic formula.

25. $x^2 + 5x - 24 = 0$ **26.** $x^2 - 4x - 6 = 0$ **27.** $2x^2 + 11x + 12 = 0$

28. $2x^2 + 9x - 5 = 0$ **29.** $x^2 - x - 5 = 0$ **30.** $q^2 - 7q - 3 = 0$

31. $3r^2 + 4r - 1 = 0$ **32.** $2s^2 + 2s - 5 = 0$ **33.** $2t^2 - 4t + \frac{3}{2} = 0$

34. $5a^2 + \frac{1}{2}a - 1 = 0$ **35.** $z^2 = -1 - 7z$ **36.** $x^2 = 3x + 9$

37. $2x^2 + 1 = 5x$ **38.** $3x^2 + 3 = 7x$ **39.** $x^2 - 7x = -12$

40. $x^2 + 9x = -14$ **41.** $3z^2 - 4z = 4$ **42.** $2y^2 + 15 = 11y$

43. $\frac{3}{2}n + 1 + \frac{1}{4}n^2 = 0$ **44.** $0 = \frac{1}{2}v + \frac{1}{2} - \frac{1}{4}v^2$ **45.** $2k^2 + 1 - 4k = 0$

46. $3x^2 - 2 + 4x = 0$ **47.** $x(2x + 5) - 1 = 0$ **48.** $x(2x + 3) - 4 = 0$

49. $15m^2 - 8m = -1$ **50.** $4n^2 + 12n = -5$ **51.** $4x^2 + 18x + 9 = 6x$

52. $9x^2 - 12x + 6 = 2$ **53.** $\frac{k^2}{3} - 6 = -k$ **54.** $\frac{1}{12}p^2 = \frac{1}{6}p + 2$

55. $4z = \frac{3}{2}z^2 + 1$ **56.** $n^2 + n = \frac{3}{2}$ **57.** $x = \frac{x^2}{6} + \frac{3}{2}$

58. $\frac{x^2}{6} - \frac{4}{3}x + \frac{8}{3} = 0$ **59.** $(x + 3)(x - 2) = 9$ **60.** $(x + 4)(x - 1) = 4$

61. $t(1 - 2t) = 1$ **62.** $2(t - t^2) = 3$ **63.** $3w^2 = 36$

64. $2q^2 = 64$ **65.** $\frac{2}{9}w^2 - w - 2 = 0$ **66.** $v^2 + \frac{10}{3}v + 1 = 0$

67. $8x = 2x^2 - 2x$ **68.** $5x = 3x^2 - 4x$

In Problems 69–80, determine the discriminant and then use this value to determine the most efficient way of solving the quadratic equation. DO NOT SOLVE THE EQUATION.

69. $2p^2 - p + 5 = 0$ **70.** $3s^2 - s + 4 = 0$ **71.** $0 = 2x^2 + 3x - 2$

72. $0 = 3x^2 - 8x - 3$ **73.** $4x(x + 3) = -9$ **74.** $2x = 1 + x^2$

75. $y = y^2 + 7y - 1$ **76.** $n(2 + n) = 7$ **77.** $0 = -20z - 4z^2 - 25$

78. $0 = 4z - 1 - 4z^2$ **79.** $\frac{x^2}{2} - \frac{x}{3} + \frac{5}{6} = 0$ **80.** $\frac{x^2}{2} - \frac{x}{2} + \frac{1}{5} = 0$

Mixed Practice

In Problems 81–98, solve each quadratic equation using any method you wish.

81. $x^2 - 2x + \frac{2}{3} = 0$ **82.** $x^2 - 2x - \frac{5}{2} = 0$ **83.** $(3p + 1)^2 = 16$

84. $(2m - 3)^2 = 25$ **85.** $(y + 1)(y + 5) + 9 = 0$

86. $(r - 3)(r - 2) + 4 = 0$ **87.** $x + 1 = \frac{-1}{4x}$ **88.** $x - 5 = \frac{-25}{4x}$

89. $2u(u + 5) = -10(u + 1)$ **90.** $6v(v - 5) = -3(4v + 1)$

91. $2n^2 = -12 + 6(n + 2)$ **92.** $3k^2 = -24 + 12(k + 2)$

93. $(2x + 7)^2 - 9 = 0$ **94.** $(3x - 2)^2 - 16 = 0$ **95.** $x + \frac{7}{2} - \frac{2}{x} = 0$

96. $\frac{x}{4} - 1 - \frac{3}{x} = 0$ **97.** $w^2 + 10 = 2$ **98.** $q^2 + 23 = 5$

Applying the Concepts

99. Fun with Numbers Find the number or numbers such that the square of three more than the number is 18.

76. 32; quadratic formula
77. 0; factoring 78. 0; factoring
79. −56; no real solution
80. −15; no real solution
81. $\left\{1 - \frac{\sqrt{3}}{3}, 1 + \frac{\sqrt{3}}{3}\right\}$
82. $\left\{1 - \frac{\sqrt{14}}{2}, 1 + \frac{\sqrt{14}}{2}\right\}$
83. $\left\{-\frac{5}{3}, 1\right\}$ 84. $\{-1, 4\}$
85. no real solution 86. no real solution
87. $\left\{-\frac{1}{2}\right\}$ 88. $\left\{\frac{5}{2}\right\}$
89. $\{-5 - 2\sqrt{5}, -5 + 2\sqrt{5}\}$
90. $\left\{\frac{3 - \sqrt{7}}{2}, \frac{3 + \sqrt{7}}{2}\right\}$
91. $\{0, 3\}$ 92. $\{0, 4\}$ 93. $\{-5, -2\}$
94. $\left\{-\frac{2}{3}, 2\right\}$ 95. $\left\{-4, \frac{1}{2}\right\}$ 96. $\{-2, 6\}$
97. no real solution 98. no real solution
99. $-3 - 3\sqrt{2}$ or $-3 + 3\sqrt{2}$
100. $5 - \sqrt{13}$ or $5 + \sqrt{13}$
101. −1 or 5
102. $-\frac{11}{4}$ or -2
103. (a) \$1515
(b) 996 or 2004
104. (a) \$1012.50
(b) 50
105. (a) 43
(b) 37
(c) 33
(d) 2029
106. (a) 50
(b) 43
(c) 46
(d) 2008
107. $\{-4\sqrt{2}, \sqrt{2}\}$
108. $\{-2\sqrt{3} - 3\sqrt{2}, -2\sqrt{3} + 3\sqrt{2}\}$
109. $\left\{\frac{2\sqrt{3}}{3}, \sqrt{3}\right\}$
110. $\left\{\frac{\sqrt{6} - 3\sqrt{2}}{2}, \frac{\sqrt{6} + 3\sqrt{2}}{2}\right\}$

100. Fun with Numbers Find the number or numbers such that the square of five less than the number is 13.

101. Fun with Numbers Four times the sum of a number and two is the same as three more than the square of the number. Find the number(s).

102. Fun with Numbers The square of five more than twice a number is the same as three more than the number. Find the number.

103. Self-Help Magazine The annual profit, P, from selling x subscriptions to *Enjoying Life While Studying Algebra* is given by the equation $P = 40 + 30x - 0.01x^2$.

(a) What is the annual profit from selling 50 subscriptions of this popular magazine?

(b) Approximately how many subscriptions would have to be sold if the publisher wants annual profit of \$20,000?

104. Heavenly Spa The monthly revenue from selling x passes to Heavenly Spa is given by the equation $R = 0.02x^2 + 40x$.

(a) What is the revenue during a month that the Spa sells 25 passes?

(b) How many passes does Heavenly Spa need to sell to have a monthly income of \$2050?

105. Engineering Majors The number of students graduating from a local high school and choosing to enter college as an engineering major is given by the equation $N = 0.05x^2 - 1.5x + 43$, where x is the number of years after 1990.

(a) How many students chose engineering as a major in 1990?

(b) Approximately how many students chose engineering as a major 5 years later?

(c) How many students chose engineering as a major 2000?

(d) In what year will 60 students choose engineering as a major?

106. Preschool Attendance The number of children attending Happy Days Preschool in a given year is given by the equation $N = 0.3x^2 - 3.2x + 50$, where x is the number of years after 1995.

(a) How many children attended this school in 1995?

(b) Approximately how many children attended Happy Days Preschool 3 years later?

(c) Approximately how many children attended this school in 2004?

(d) In what year will 60 students attend Happy Days Preschool?

Extending the Concepts

Up to this point, we have seen values of a, b, and c that were rational (that is, positive or negative whole numbers, fractions, or terminating or repeating decimals). It is possible to have quadratic equations that have irrational coefficients and constants. In Problems 107–110, use the quadratic formula to solve each equation. Find only the exact solutions expressed in simplest form.

107. $x^2 + 3\sqrt{2}x - 8 = 0$ **108.** $x^2 + 4\sqrt{3}x - 6 = 0$

109. $\sqrt{3}x^2 - 5x + \sqrt{12} = 0$ **110.** $\sqrt{2}x^2 - \sqrt{12}x - \sqrt{18} = 0$

PUTTING THE CONCEPTS TOGETHER (SECTIONS 7.1–7.3)

These problems cover important concepts from Sections 7.1 to 7.3. We designed these problems so that you can review the chapter so far and show your mastery of the concepts. Take time to work these problems before proceeding with the next section. The answers to these problems are located at the back of the text starting on page AN-15.

1. (a) $\left\{-\frac{\sqrt{5}}{2}, \frac{\sqrt{5}}{2}\right\}$
(b) $\{2 - 2\sqrt{3}, 2 + 2\sqrt{3}\}$
(c) $\left\{\frac{-3 - \sqrt{3}}{2}, \frac{-3 + \sqrt{3}}{2}\right\}$

1. *Solve using the Square Root Method:*

(a) $w^2 = \frac{5}{4}$ **(b)** $(y - 2)^2 = 12$ **(c)** $\left(y + \frac{3}{2}\right)^2 = \frac{3}{4}$

2. (a) add 81;
$z^2 - 18z + 81 = (z - 9)^2$
(b) add $\frac{81}{4}$;
$y^2 + 9y + \frac{81}{4} = \left(y + \frac{9}{2}\right)^2$
(c) add $\frac{4}{25}$;
$m^2 + \frac{4}{5}m + \frac{4}{25} = \left(m + \frac{2}{5}\right)^2$

3. (a) $\{-7, -5\}$
(b) no real solution
(c) $\{4 - 3\sqrt{2}, 4 + 3\sqrt{2}\}$

4. (a) $\{-4 - 2\sqrt{5}, -4 + 2\sqrt{5}\}$
(b) $\left\{-\frac{3}{2}, 4\right\}$
(c) $\left\{-\frac{9}{2}\right\}$

5. $\{-2, 3\}$

6. $\left\{\frac{-1 - \sqrt{11}}{2}, \frac{-1 + \sqrt{11}}{2}\right\}$

7. $\left\{-1 - \frac{\sqrt{3}}{2}, -1 + \frac{\sqrt{3}}{2}\right\}$ or $\left\{\frac{-2 - \sqrt{3}}{2}, \frac{-2 + \sqrt{3}}{2}\right\}$

8. (a) discriminant $= 64$
(b) factoring
(c) Answers may vary.

9. (a) $5\sqrt{15}$ feet
(b) 19.4 feet

10. (a) $12,000 revenue from selling 100 calculus texts
(b) 60 or 160 texts must be sold for the revenue to be $9600.

2. *Determine the number that must be added to the expression to make it a perfect square trinomial. Then factor the expression.*

(a) $z^2 - 18z$ **(b)** $y^2 + 9y$ **(c)** $m^2 + \frac{4}{5}m$

3. *Solve by completing the square:*

(a) $x^2 + 12x + 35 = 0$ **(b)** $z^2 - 4z + 9 = 0$ **(c)** $m^2 = 8m + 2$

4. Solve using the quadratic formula:

(a) $x^2 + 8x - 4 = 0$ **(b)** $2y^2 = 5y + 12$ **(c)** $4p^2 + 36p + 81 = 0$

In Problems 5–7, solve the quadratic equation using any appropriate method you wish.

5. $(x + 3)(x - 4) = -6$ **6.** $a + 1 = \frac{5}{2a}$ **7.** $\frac{1}{2}z^2 + \frac{1}{8} = -z$

8. Given the quadratic equation $2x = -5 + 3x^2$,

(a) determine the value of the discriminant.
(b) determine which method you would choose to solve the equation.
(c) explain why this method is appropriate for this equation.

9. Painter's Ladder A painter has a ladder that can be extended to 20 feet. The painter decides to place the bottom of the ladder 5 feet from the house to be painted.

(a) Exactly how far up the house will the ladder reach?
(b) Approximate this height to nearest tenth of a foot.

10. Bookstore Sales Rock Bottom Discount Bookstore determines that the revenue from selling x calculus texts can be given by the equation $R = -x^2 + 220x$.

(a) What is the revenue from selling 100 calculus texts?
(b) How many texts must the bookstore sell to have revenue of $9600?

7.4 Problem Solving Using Quadratic Equations

OBJECTIVES

1 Model and Solve Direct Translation Problems

2 Solve Problems Modeled by Quadratic Equations

Preparing for Problem Solving Using Quadratic Equations

Before getting started, take the following readiness quiz. If you get a problem wrong, go back to the section cited and review the material.

1. Use the Pythagorean Theorem to find a when $b = 7$ and $c = 25$. [Section 4.7, pp. 293–295]

Many applied problems require solving quadratic equations. For example, finding the length and width of a rectangle whose area is known requires solving a quadratic equation. As always, we will rely on the problem-solving strategy first presented in Section 2.5.

1 Model and Solve Direct Translation Problems

We begin by looking at a problem that involves geometry.

Classroom Example
Use Example 1 with area = 192 sq. inches and the length of the rectangle is 4 inches more than the width.

Answer: The dimensions are width = 12 inches and length = 16 inches.

EXAMPLE 1 Finding the Dimensions of a Rectangle from Its Area

The area of a rectangle is 112 square inches. The length of the rectangle is 6 inches more than the width. What are the dimensions of the rectangle?

Preparing for...Answer **1.** 24

Work Smart

Generally, we let the variable represent the quantity we know *least* about. In this problem, we know that the length is 6 inches more than the width, so we let w represent the width.

Figure 5

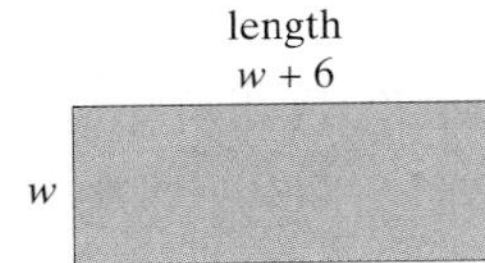

Solution

Step 1: Identify To find the dimensions of the rectangle, we need to know the length and width of the rectangle.

Step 2: Name The length of the rectangle is expressed in terms of the width, so we let w represent the width of the rectangle. The problem stated that the length is 6 inches more than the width so we let $w + 6$ represent the length.

Step 3: Translate Figure 5 illustrates the situation. We know that the area of a rectangle is found by multiplying the length by the width. That is, $A = l \cdot w$, so we develop the model by replacing $A = 112$ and $l = w + 6$.

$$A = l \cdot w$$
$$112 = (w + 6)w \quad \text{The Model}$$

Let's multiply out the right side.

$$112 = w^2 + 6w$$

We can now see that this is a quadratic equation, so we write the equation in standard form, $ax^2 + bx + c = 0, a \neq 0$.

$$112 = w^2 + 6w$$
$$0 = w^2 + 6w - 112$$

Step 4: Solve We can solve this equation by factoring.

$$w^2 + 6w - 112 = 0$$
$$(w + 14)(w - 8) = 0$$
$$w + 14 = 0 \quad \text{or} \quad w - 8 = 0$$
$$w = -14 \quad \text{or} \quad w = 8$$

Step 5: Check We disregard the solution $w = -14$ because w represents the width of the rectangle and cannot be negative. A rectangle whose dimensions are 8 inches by $8 + 6 = 14$ inches has an area that is 112 square inches.

$$(8)(14) \stackrel{?}{=} 112$$
$$112 = 112 \quad \text{True}$$

Step 6: Answer The dimensions of the rectangle are 8 inches by 14 inches.

QUICK ✓

1. The height of a triangle is two inches more than the base and the area of the triangle is 40 square inches. Find the measurements of the base and the height of the triangle.

Classroom Example

Use Example 2 with hypotenuse measuring 9 feet.

Answer: The lengths of the sides are approximately 5.3 feet and 7.3 feet.

EXAMPLE 2 Putting in a Flower Garden

Melissa plans to plant a flower garden at the corner of her lot. The garden is in the shape of a right triangle with the hypotenuse measuring 12 feet as shown in Figure 6. Because of the location of a dogwood tree, one leg of the triangle will be two feet longer than the other leg. Find the approximate lengths of the sides of the triangular flower bed, to the nearest tenth of a foot.

Figure 6

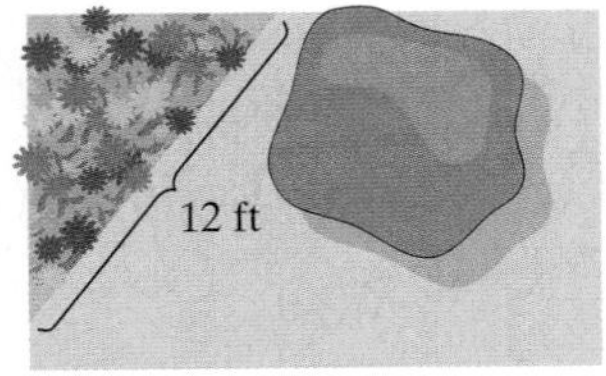

Solution

Step 1: Identify To find the lengths of the two sides of Melissa's garden, we must find the length of each leg of the right triangle that is formed.

Step 2: Name We let x represent the length of the shorter leg of the right triangle. Because the length of the longer leg is described in the problem as "one leg of the triangle will be two feet longer than the other leg," $x + 2$ represents the length of the longer leg.

Step 3: Translate We develop the model by stating the relationship among the three sides of the right triangle using the Pythagorean Theorem with $a = x$ and $b = x + 2$.

Teaching Tip
Remind students that the hypotenuse is the longest side of the triangle.

$$a^2 + b^2 = c^2$$
$$x^2 + (x + 2)^2 = 12^2$$
$(a + b)^2 = a^2 + 2ab + b^2$: $$x^2 + (x^2 + 4x + 4) = 144$$
Write in standard form: $$2x^2 + 4x - 140 = 0$$
Divide each side by 2: $$\frac{2x^2 + 4x - 140}{2} = \frac{0}{2}$$
$$x^2 + 2x - 70 = 0$$

Step 4: Solve We will solve the quadratic equation using the quadratic formula. Do you know why? We see that $a = 1, b = 2$, and $c = -70$.

$$x^2 + 2x - 70 = 0$$
$$x = \frac{-b \pm \sqrt{b^2 - 4ac}}{2a}$$
Use the quadratic formula: $$x = \frac{-2 \pm \sqrt{2^2 - 4(1)(-70)}}{2(1)}$$
$$= \frac{-2 \pm \sqrt{284}}{2}$$
$$= \frac{-2 \pm \sqrt{4 \cdot 71}}{2}$$
$$= \frac{-2 \pm 2\sqrt{71}}{2}$$
$$= -\frac{2}{2} \pm \frac{2\sqrt{71}}{2}$$
$$= -1 \pm \sqrt{71}$$

The two solutions are $x = -1 - \sqrt{71} \approx -9.4$ and $x = -1 + \sqrt{71} \approx 7.4$. We disregard the solution $x = -1 - \sqrt{71} \approx -9.4$ because x represents the length of a side of the garden. One leg is approximately 7.4 feet long. The other leg is $x + 2 = \left(-1 + \sqrt{71}\right) + 2 = 1 + \sqrt{71} \approx 9.4$ feet long.

Step 5: Check Let's verify our answers. Does $\left(-1 + \sqrt{71}\right)^2 + \left(1 + \sqrt{71}\right)^2 = 144$? Because $\left(1 - 2\sqrt{71} + 71\right) + \left(1 + 2\sqrt{71} + 71\right) = 144$, our answer checks.

Step 6: Answer The lengths of the sides of the triangular flowerbed are approximately 7.4 feet and 9.4 feet.

QUICK ✓

2. A rectangular plot of land is designed so that its length is 4 meters more than its width. The diagonal of the land is known to be 14 meters. To the nearest tenth of a meter, what are the dimensions of the land?

Classroom Example ➤
Use Example 3 with 10 cubic feet of potting soil.

Answer: The box should be approximately 4.3 feet long.

EXAMPLE 3 Building a Flower Box

Jonathan is building a 1-foot-tall flower box for his patio. In addition, the flower box is to be 2 feet longer than it is wide. Jonathan wants to fill the planter with 6 cubic feet of potting soil. How long, to the nearest tenth of a foot, should he make the flower box?

Solution

Step 1: Identify We wish to know the length of the flower box.

Figure 7

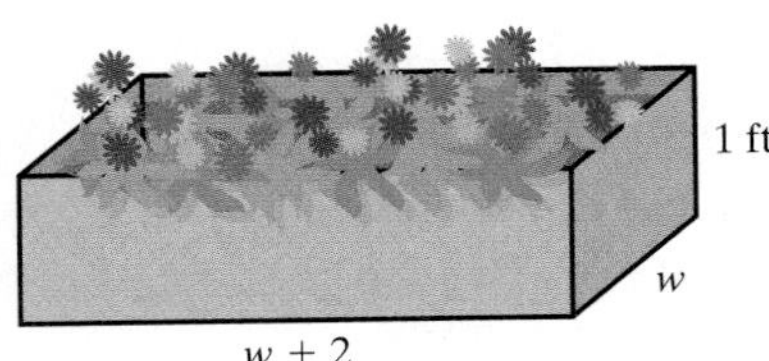

Step 2: Name We know that the box is to be 2 feet longer than it is wide, so we let w represent the width of the flower box. We'll let $w + 2$ represent the length of the box as shown in Figure 7.

Step 3: Translate The volume of the rectangular box is $V = l \cdot w \cdot h$. We know that the volume is 6 cubic feet, and the height is 1 foot, so we substitute the values $V = 6, l = w + 2$, and $h = 1$ into the equation to develop the model.

$$V = lwh$$
$$6 = (w + 2)(w)(1) \quad \text{The Model}$$

Simplify the right side of the equation.

Use the Distributive Property: $6 = w^2 + 2w$

Write in standard form: $0 = w^2 + 2w - 6$

Step 4: Solve We will solve the quadratic equation using the quadratic formula. In the model, $a = 1, b = 2, c = -6$.

$$w^2 + 2w - 6 = 0$$

$$w = \frac{-2 \pm \sqrt{2^2 - 4(1)(-6)}}{2(1)}$$

Use the quadratic formula: $= \frac{-2 \pm \sqrt{28}}{2}$

$$= \frac{-2 \pm \sqrt{4 \cdot 7}}{2}$$

Simplify the radical: $= \frac{-2 \pm 2\sqrt{7}}{2}$

$$= \frac{2\left(-1 \pm \sqrt{7}\right)}{2}$$

$$= \frac{\cancel{2}\left(-1 \pm \sqrt{7}\right)}{\cancel{2}}$$

$$= -1 \pm \sqrt{7}$$

The two exact solutions are $w = -1 - \sqrt{7}$ feet and $w = -1 + \sqrt{7}$ feet. The approximate values of the two solutions are $w = -1 - \sqrt{7} \approx -3.6$ and $w = -1 + \sqrt{7} \approx 1.6$. We disregard the solution $w = -1 - \sqrt{7} \approx -3.6$ because w represents the width of the flower box and cannot be negative. The flower box should be $\left(-1 + \sqrt{7}\right) + 2 = 1 + \sqrt{7} \approx 3.6$ feet long.

Step 5: Check Let's verify our answers. If we multiply length · width · height, we should obtain volume $= 6$ cubic feet. $\left(1 + \sqrt{7}\right)\left(-1 + \sqrt{7}\right)(1) = -1 + 7 = 6$, so our dimensions are correct.

Step 6: Answer Jonathan should construct the flower box to be approximately 3.6 feet long.

QUICK ✓

3. The compost bin that Alyssa is constructing must be 4 feet high and have a length that is 1 foot more than the width. The bin should hold 50 cubic feet of compost. To the nearest tenth of a foot, what should the length and width of the compost bin be?

2 Solve Problems Modeled by Quadratic Equations

Numerous problems from the physical, behavioral, and social sciences require solving a quadratic equation.

Classroom Example ➤
Use Example 4 to determine the year in which the number of fatalities was 750.

Answer: There were 750 fatalities in the year 2005.

EXAMPLE 4 Assessing Highway Worker Fatalities

The number of fatalities of highway workers peaked at 1314 in 1999 and has declined in recent years. The equation $N = -11x^2 + 75x + 1134$ models the number of fatalities of private industry highway workers per year, where x represents the number of years since 1995. Assuming that this trend is still true today, use the model to determine the year in which the number of fatalities was 1000.

Solution

We want to find the year in which the number of fatalities was 1000. We let $N = 1000$ in the equation $N = -11x^2 + 75x + 1134$.

$$1000 = -11x^2 + 75x + 1134$$

Subtract 1000 from each side: $$0 = -11x^2 + 75x + 134$$

We'll use the quadratic formula with $a = -11$, $b = 75$, and $c = 134$ to solve the equation.

Quadratic formula: $$x = \frac{-b \pm \sqrt{b^2 - 4ac}}{2a}$$

Substitute: $$x = \frac{-75 \pm \sqrt{75^2 - 4(-11)(134)}}{2(-11)}$$

$$= \frac{-75 \pm \sqrt{5625 + 5896}}{-22}$$

$$= \frac{-75 \pm \sqrt{11{,}521}}{-22}$$

$$= \frac{-75 - \sqrt{11{,}521}}{-22} \quad \text{or} \quad = \frac{-75 + \sqrt{11{,}521}}{-22}$$

$$\approx 8.3 \quad \text{or} \quad \approx -1.5$$

Since x represents the number of years since 1995, we disregard the solution $x \approx -1.5$. We round the solution $x \approx 8.3$ to the nearest whole number because it represents years, and we determine that in the year $1995 + 8 = 2003$, 1000 highway workers were killed in accidents.

QUICK ✓

4. A part of a theory from economics states that a worker's income depends on his or her age. The equation $I = -55a^2 + 5119a - 54{,}448$ represents the relationship between average annual income I and age a. For what age does average income I equal \$50,000? Round your answer to the nearest year.

Classroom Example
Use Example 5 and determine the time the rocket will be 60 feet above the ground.

Answer: t is approximately 0.88 second or 4.12 seconds

EXAMPLE 5 Launching a Toy Rocket

The height of a toy rocket after t seconds when fired straight up with an initial speed of 80 feet per second from an initial height of 2 feet can be modeled by the equation $h = -16t^2 + 80t + 2$, where t is time in seconds. After how many seconds will the rocket be 82 feet above the ground?

Solution

Substitute 82 for h in the equation $h = -16t^2 + 80t + 2$ and then solve for t.

$$-16t^2 + 80t + 2 = 82$$

Subtract 82 from each side: $-16t^2 + 80t - 80 = 0$

We can factor out -16 from the equation.

$$-16(t^2 - 5t + 5) = 0$$

Divide both sides by -16: $t^2 - 5t + 5 = 0$

We see that $a = 1, b = -5$, and $c = 5$. Because $b^2 - 4ac = (-5)^2 - 4(1)(5) = 25 - 20 = 5$, we will use the quadratic formula to solve the equation.

Quadratic formula: $t = \dfrac{-b \pm \sqrt{b^2 - 4ac}}{2a}$

Substitute: $t = \dfrac{-(-5) \pm \sqrt{5}}{2(1)}$

$$= \frac{5 \pm \sqrt{5}}{2}$$

$$t = \frac{5 - \sqrt{5}}{2} \quad \text{or} \quad t = \frac{5 + \sqrt{5}}{2}$$

$$\approx 1.38 \quad \text{or} \quad \approx 3.62$$

Teaching Tip
Pose this question: Why are there two possible answers that make sense for this problem?

So t is approximately 1.38 seconds or t is approximately 3.62 seconds. Both answers are positive, so both are solutions. Do you understand why both answers are possible?

QUICK ✓

5. The height of a cannon ball shot upward from a height of 3 feet above the ground with an initial velocity of 100 feet/second is given by the equation $h = -16t^2 + 100t + 3$. After how many seconds will the cannon ball be 50 feet above ground? Round your answer to the nearest hundredth of a second.

7.4 Exercises

For Extra Help:

Student Solutions Manual

CD Video

PH Math/Tutor Center

MathXL Tutorials on CD

MathXL®

MyMathLab

Concepts and Vocabulary

In Problems 1–3, fill in the blanks.

1. $a^2 + b^2 = c^2$
2. positive
3. quadratic formula

1. The Pythagorean Theorem tells the relationship between the lengths of the legs and the hypotenuse of a right triangle and is given by the formula ________.
2. When solving an applied problem in which the variable represents some measurable quantity such as a length, the result should always be ________.
3. A real-world situation may be modeled by a quadratic equation. This equation may contain large numbers or decimals, so when solving a real world problem that is modeled by a quadratic equation it is likely that the ________ ________ will be used to solve the equation.

4. False
5. True
6. False
7. reaches height twice, once on way up and once on way down
8. Answers may vary.
9. $-17, -16$ and $16, 17$
10. $-14, -13$ and $13, 14$
11. 14, 15
12. $-11, -10$ and $10, 11$
13. $-1 - 2\sqrt{6}$ or $-1 + 2\sqrt{6}$
14. $2 - 3\sqrt{2}$ or $2 + 3\sqrt{2}$
15. $5 - 2\sqrt{3}$ or $5 + 2\sqrt{3}$
16. $2 - \sqrt{3}$ or $2 + \sqrt{3}$
17. width = 6 m, length = 15 m
18. width = 10 ft; length = 15 ft
19. width = 3 in.; length = 12 in.
20. width = 12 cm; length = 10 cm
21. height = 5 cm; base = 8 cm
22. height = 15 in.; base = 8 in.
23. height = 11 ft; base = 12 ft
24. height = 18 m; base = 11 m

In Problems 4–6, answer True or False to each statement.

4. In a rectangle, the width is five feet less than the length. If l represents the length, then the width is $5 - l$.

5. In a rectangle, the width is two times the sum of 3 and the length. If l represents the length, then the width is $2(l + 3)$.

6. The length of a rectangular box could be either $2 + \sqrt{7}$ feet or $2 - \sqrt{7}$ feet long.

7. Example 5 asked you find the times a rocket will be 82 feet above the ground. There were two answers: 1.38 seconds and 3.62 seconds after the rocket was fired. Explain why there are two answers and why we are not discarding one from the solution set.

8. Explain why we see quadratic equations in problems about the area of a figure but we see linear equations in problems about perimeter. Make up an example where you need to find both the area and perimeter of a rectangle.

Building Skills

In Problems 9–16, write an equation to represent the unknown number and then solve.

9. The product of two consecutive integers is 272. Find two pairs of integers that have this product.

10. The product of two consecutive integers is 182. Find two pairs of integers that have this product.

11. Consider two consecutive integers. The product of twice the smaller integer and 10 less than the larger integer is 140. Find the positive integers that satisfy these conditions.

12. Consider two consecutive integers. The product of the larger and half of the smaller is 55. Find the positive integers that satisfy these conditions.

13. Consider any real number such that its square increased by twice the number is 23. Find the exact value of the number(s).

14. Consider any real number such that its square decreased by four times the number is 14. Find the exact value of the number(s).

15. Consider any real number such that its square is the same as thirteen less than ten times the number. Find the exact value of the number(s).

16. Consider any real number such that the square of the number decreased by four times the number is negative one. Find the exact value of the number(s).

In Problems 17–20, find the dimensions of each rectangle.

△ **17.** The length of a rectangle is three more than twice the width. If the area of the rectangle is 90 square meters, find the dimensions.

△ **18.** The length of a rectangle is five less than twice the width. If the area of the rectangle is 150 square feet, find the dimensions.

△ **19.** The width of a rectangle is three less than half of the length. If the area of the rectangle is 36 square inches, find the dimensions.

△ **20.** The width of a rectangle is seven more than half of the length. If the area of the rectangle is 120 square centimeters, find the dimensions.

In Problems 21–24, find the unknown values in each triangle.

△ **21.** In a triangle, the base is three centimeters longer than the height. If the area of the triangle is 20 square centimeters, find the base and height of the triangle.

△ **22.** In a triangle, the height is one inch less than twice the base. If the area of the triangle is 60 square inches, find the base and height of the triangle.

△ **23.** In a triangle, the height is five feet more than half of the base. If the area of the triangle is 66 square feet, find the base and height of the triangle.

△ **24.** In a triangle, the base is seven meters less than the height. If the area of the triangle is 99 square meters, find the base and height of the triangle.

In Problems 25–28, use the Pythagorean Theorem to find the unknown values.

△ **25.** In a right triangle, one leg is 3 inches shorter than the other. If the length of the hypotenuse is $3\sqrt{2}$ inches, find exact length of each leg. Then approximate these lengths to the nearest tenth of an inch.

△ **26.** In a right triangle, one leg is five less than the other. If the length of the hypotenuse is 10 kilometers, find exact length of each leg. Then approximate these lengths to the nearest tenth of a kilometer.

△ **27.** In a right triangle, the hypotenuse is twice as long as one leg. If the length of one leg is 9 meters, find the exact length of the remaining leg and the hypotenuse. Then approximate these lengths to the nearest tenth of a meter.

△ **28.** In a right triangle, the hypotenuse is one less than twice one of the legs. If the length of one of the legs is 3 yards, find exact length of the other leg and the hypotenuse. Then approximate these lengths to the nearest tenth of a yard.

Applying the Concepts

29. Model Rocket A model rocket is launched from the ground with an initial speed of 90 feet per second. The equation that models its height, h (in feet), t seconds after it was fired is $h = -16t^2 + 90t$.

(a) How high is the rocket 0.5 second after it was fired?
(b) How high is the rocket 2 seconds after it was fired?
(c) How long will it take the rocket to reach an altitude of 100 feet?
(d) How long will it take the rocket to return to Earth? (**Hint:** It is on the ground when $h = 0$.)

30. Projectile Motion Omar shot a projectile up vertically with a initial speed of 92 feet per second. If he was standing on a building 25 feet high, the equation that models the height of the projectile is $h = -16t^2 + 92t + 25$.

(a) How high was the projectile $\frac{3}{4}$ of a second after it was launched?
(b) How high is it after three seconds?
(c) How long does it take for the projectile to reach an altitude of 75 feet?
(d) How long will it take to hit the ground? (**Hint:** It is on the ground when $h = 0$.)

31. Fruit Fly Experiment Ruben is conducting an experiment with fruit flies in a Petri dish. Ruben discovers the population of fruit flies in the Petri dish can be modeled by the equation $P = \frac{4000t}{4t^2 + 90}$, where t represents the number of hours after starting the experiment.

(a) How many fruit flies are in the Petri dish after 1 hour?
(b) How many fruit flies are in the Petri dish after 2 hours?
(c) How long does it take the population to reach 80 flies?

32. Another Experiment Tuan is doing the same experiment as Ruben (Problem 31) but with a different variety of fruit flies. The number of fruit flies in his Petri dish is given by the equation $P = \frac{4000t}{4t^2 + 50}$.

(a) How many fruit flies are in the Petri dish after 1 hour?
(b) How many fruit flies are in the Petri dish after 2 hours?
(c) How long does it take the population to reach 100 flies?

△ **33. Diagonals of a Polygon** In geometry, a convex polygon is a many-sided closed figure, with no sides collapsing in towards the middle. The points where the sides intersect are called vertices and when you join any two vertices, this line segment is called a diagonal of the polygon. For example, a rectangle has two diagonals. The number of diagonals of other polygons is given by the formula $D = \frac{n(n-3)}{2}$, where n is the number of sides.

---

25. Longer leg is exactly $\frac{3 + 3\sqrt{3}}{2}$ inches long; shorter leg is exactly $\frac{-3 + 3\sqrt{3}}{2}$ inches. Approximate lengths are: longer leg 4.1 inches and shorter leg 1.1 inches.
26. $\frac{5 + 5\sqrt{7}}{2} \approx 9.1$ km; $\frac{-5 + 5\sqrt{7}}{2} \approx 4.1$ km
27. leg $= 3\sqrt{3} \approx 5.2$ m; hypotenuse $= 6\sqrt{3} \approx 10.4$ m
28. leg $\frac{2 + 2\sqrt{7}}{3} \approx 2.4$ yd; hypotenuse $= \frac{1 + 4\sqrt{7}}{3} \approx 3.9$ yd
29. (a) 41 ft (b) 116 ft (c) 1.52 s and 4.10 s (d) 5.625 s
30. (a) 85 ft (b) 157 ft (c) 0.61 s and 5.14 s (d) 6.01 s
31. (a) 43 fruit flies (b) 75 fruit flies (c) 2.18 hours
32. (a) 74 fruit flies (b) 121 fruit flies (c) 1.5 hours and 8.5 hours
33. (a) 2 (b) 20 (c) 15

34. (a) 14
(b) 6
35. (a) $(180 - x)°$
(b) 30° or 125°
36. (a) $(90 - y)°$
(b) 20° or 70°
37. (a) $146.40
(b) $992.40
(c) approximately 3 months
(d) approximately 13 months
38. (a) $333.33
(b) $1714
(c) about 9 years

(a) Use this formula to calculate the number of diagonals in a rectangle.
(b) How many diagonals does an octagon have? (**Hint:** An octagon has 8 sides.)
(c) If a polygon has 90 diagonals, how many sides does it have?

△ **34. Diagonals of a Polygon** Using the formula given in Problem 33, find the following.

(a) How many diagonals does have heptagon have? (**Hint:** A heptagon has 7 sides.)
(b) If a polygon has 9 diagonals, how many sides does it have?

△ **35. Supplementary Angles** Two angles are supplementary if the sum of their measures is 180°.

(a) If the measure of one angle is $x°$, what is the measure of its supplement?
(b) One-third of the product of the measure of an angle and 25° less than its supplement is 1250°. Find the measure of the angle.

△ **36. Complementary Angles** Two angles are complementary if the sum of their measures is 90°.

(a) If the measure of one angle is $y°$, what is the measure of its complement?
(b) Half of the product of the measure of an angle and its complement is 700°. Find the measure of the angle.

37. Profit from Jewelry The monthly profit, P, that Dorothy will earn from her jewelry business is given by the equation $P = -0.1x^2 + 50x - 150$, where x is the number of months that she has been in business.

(a) What will be her monthly profit after 6 months?
(b) What will be her monthly profit after 2 years?
(c) How long does Dorothy have to stay in business before she breaks even? The break-even point occurs when $P = 0$.
(d) How many months will it take Dorothy to get to the point where she is earning $500 per month?

38. Advertising Costs Joel has started a catering business for which he buys advertising. He predicts that as time goes by, this expense will decrease because he will have all the business he can handle. A model that predicts the monthly advertising expenditures N (in dollars) x years after he opens his business is given by $N = \dfrac{2000}{x^2 + 5}$.

(a) How much did Joel spend on advertising each month during the first year he was open?
(b) What was his annual expenditure for advertising during the third year of operation?
(c) After how many years will Joel be spending $25 per month on advertising?

Extending the Concepts

Problems 39–48 use the following discussion. Two right triangles have special properties worth noting. One of these is the isosceles right triangle, sometimes called the 45-45-90° right triangle, in which two angles have a measure of 45°. The other is the 30-60-90° triangle, in which one of the angles measures 30° and the other angle measures 60°. The relationship between the sides for each triangle is given.

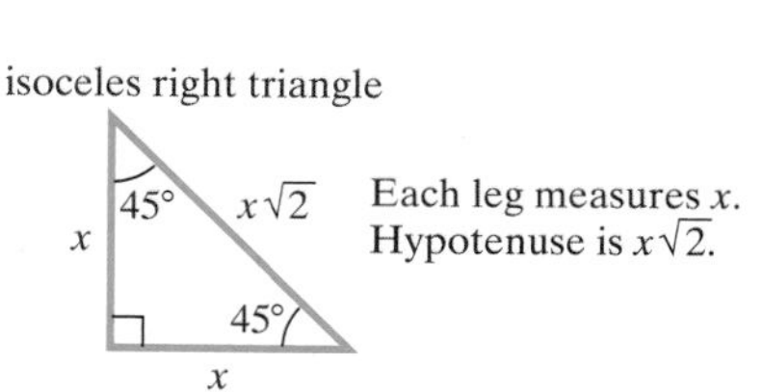

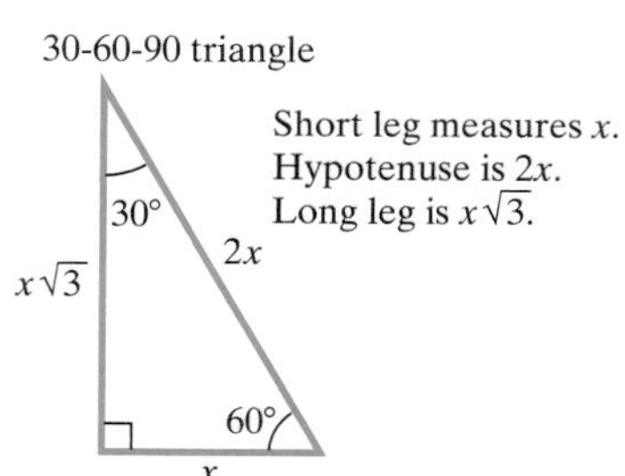

39. leg = 1, hypotenuse = $\sqrt{2}$
40. hypotenuse = 2, long leg = $\sqrt{3}$
41. leg = 3, hypotenuse = $3\sqrt{2}$
42. hypotenuse = 10, long leg = $5\sqrt{3}$
43. leg = $\sqrt{6}$, hypotenuse = $2\sqrt{3}$
44. short leg = 8, long leg = $8\sqrt{3}$
45. leg = $6\sqrt{2}$, leg = $6\sqrt{2}$
46. short leg = $3\sqrt{3}$, hypotenuse = $6\sqrt{3}$
47. leg = 2, leg = 2
48. short leg = $\sqrt{2}$, hypotenuse = $2\sqrt{2}$

△ *Use the relationships regarding the sides of isosceles and 30-60-90 triangles to find the missing lengths of the sides of each triangle.*

	Isosceles Right Triangle		
	Leg	**Leg**	**Hypotenuse**
	x	x	$x\sqrt{2}$
39.	1	?	?
41.	?	3	?
43.	$\sqrt{6}$	?	?
45.	?	?	12
47.	?	?	$\sqrt{8}$

	30-60-90° Right Triangle		
	Short Leg	**Long Leg**	**Hypotenuse**
	x	$x\sqrt{3}$	$2x$
40.	1	?	?
42.	5	?	?
44.	?	?	16
46.	?	9	?
48.	?	$\sqrt{6}$	?

7.5 The Complex Number System

OBJECTIVES

1. Evaluate the Square Root of Negative Real Numbers
2. Add or Subtract Complex Numbers
3. Multiply Complex Numbers
4. Divide Complex Numbers
5. Solve Quadratic Equations with Complex Solutions

Preparing for the Complex Number System

Before getting started, take the following readiness quiz. If you get a problem wrong, go back to the section cited and review the material.

1. List the numbers in the set $\left\{6, -\frac{5}{6}, -14, 0, \sqrt{3}, 1.\overline{65}, -\frac{18}{3}, \sqrt{-2}\right\}$ that are [Section 1.2, pp. 9–12]

(a) natural numbers **(b)** whole numbers
(c) integers **(d)** rational numbers
(e) irrational numbers **(f)** real numbers

2. Distribute: $2x(3x - 4)$ [Section 3.3, pp. 190–191]

3. Multiply: $(m + 2)(3m - 2)$ [Section 3.3, pp. 191–193]

4. Multiply: $(6y + 7)(6y - 7)$ [Section 3.3, p. 194]

Each time we encounter a situation where a number system can't handle a problem, we expand the number system. For example, if we considered only the whole numbers, we could not describe a negative balance in a checking account, so we introduce integers. But if the world could only be described by integers, then we could not talk about parts of a whole such as $\frac{1}{2}$ a pizza or $\frac{3}{4}$ of a dollar, so we introduce rational numbers. If we considered only rational numbers, then we wouldn't be able to find a number whose square is 2, so we introduce the irrational numbers to show that $(\sqrt{2})^2 = 2$. By combining the rational numbers with the irrational numbers, we created the real number system. The real number system is usually sufficient for solving most problems in mathematics, but not for all problems.

For example, suppose we wanted to determine a number whose square is -1. We know that when we square any real number, the result is never negative. We call this property of real numbers, the *Nonnegativity Property.*

NONNEGATIVITY PROPERTY OF REAL NUMBERS

For any real number a, $a^2 \geq 0$.

Because the square of any real number is never negative, there is no real number x for which

$$x^2 = -1$$

The above equation does not have a solution that is a real number. To remedy this situation, we introduce a new number.

Preparing for...Answers **1. (a)** 6 **(b)** 6, 0 **(c)** 6, -14, 0, $-\frac{18}{3}$ **(d)** 6, $-\frac{5}{6}$, -14, 0, $-\frac{18}{3}$, $1.\overline{65}$ **(e)** $\sqrt{3}$ **(f)** 6, $-\frac{5}{6}$, -14, 0, $\sqrt{3}$, $1.\overline{65}$, $-\frac{18}{3}$ **2.** $6x^2 - 8x$ **3.** $3m^2 + 4m - 4$ **4.** $36y^2 - 49$

DEFINITION: IMAGINARY UNIT

The **imaginary unit,** denoted by $\boldsymbol{i}$, is the number whose square is -1. That is,

$$i^2 = -1$$

If we take the square root of both sides of $i^2 = -1$, we find that

$$i = \sqrt{-1}$$

If we look back at the development of the real number system, we see that each time a new problem is encountered that cannot be solved using the existing number system, a new number system is developed. Further, each new number system contains the earlier number system as a subset. By introducing the number i, we now have a new number system called the **complex number system.**

In Words
The real number system is a subset of the complex number system. This means that all real numbers are a type of complex numbers.

DEFINITION

Complex numbers are numbers of the form $\boldsymbol{a + bi}$, where a and b are real numbers. The real number a is called the **real part** of the number $a + bi$; the real number b is called the **imaginary part** of $a + bi$.

For example, the complex number $3 + 4i$ has the real part 3 and the imaginary part 4. The complex number $6 - 2i = 6 + (-2)i$ has the real part 6 and the imaginary part -2.

When a complex number is written in the form $a + bi$, where a and b are real numbers, we say that it is in **standard form.** The complex number $a + 0i$ is typically written as a. This serves as a reminder that the real number system is a subset of the complex number system. The complex number $0 + bi$ is usually written as bi. Any number of the form bi is called a **pure imaginary number.** Figure 8 shows the relation between the number systems.

Figure 8
The complex number system.

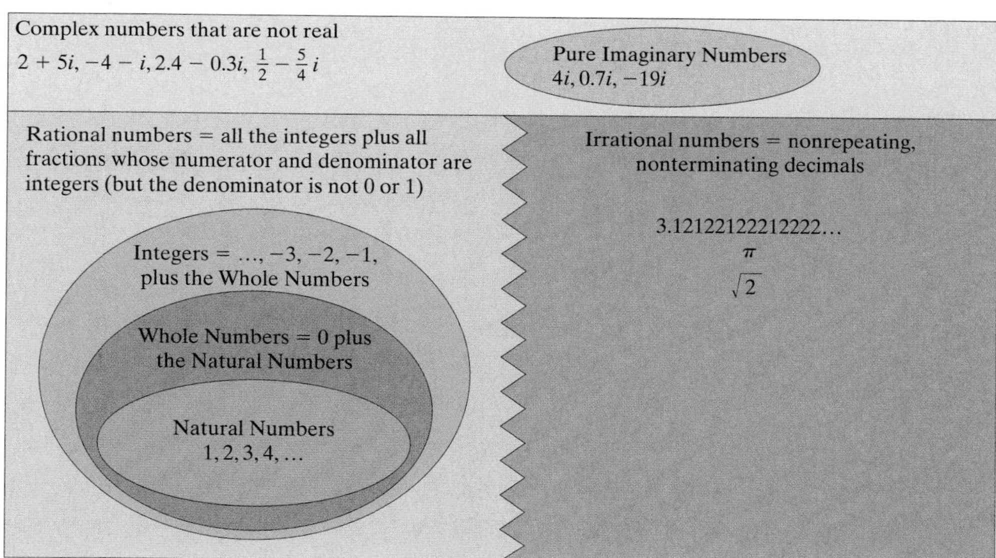

Complex numbers that are not real
$2 + 5i, -4 - i, 2.4 - 0.3i, \frac{1}{2} - \frac{5}{4}i$
Pure Imaginary Numbers
$4i, 0.7i, -19i$
Rational numbers = all the integers plus all fractions whose numerator and denominator are integers (but the denominator is not 0 or 1)
Irrational numbers = nonrepeating, nonterminating decimals
3.12122122212222...
π
$\sqrt{2}$
Integers = ..., −3, −2, −1, plus the Whole Numbers
Whole Numbers = 0 plus the Natural Numbers
Natural Numbers
1, 2, 3, 4, ...

1 Evaluate the Square Root of Negative Real Numbers

By using the definition of i along with the fact that $\sqrt{ab} = \sqrt{a} \cdot \sqrt{b}$, we now have the ability to simplify square roots of negative numbers.

EVALUATING SQUARE ROOTS OF NEGATIVE NUMBERS

If N is a positive real number, we define the **principal square root of $-N$,** denoted by $\sqrt{-N}$, as

$$\sqrt{-N} = \sqrt{N}i$$

where $i = \sqrt{-1}$.

Classroom Example Write each as a pure imaginary number.

(a) $\sqrt{-64}$ (b) $\sqrt{-20}$

Answer:

(a) $8i$ (b) $2\sqrt{5}i$

EXAMPLE 1 Evaluating the Square Root of a Negative Number

Write each of the following as a pure imaginary number.

(a) $\sqrt{-25}$ **(b)** $\sqrt{-12}$

Solution

(a)
$$\begin{aligned} \sqrt{-25} &= \sqrt{25\cdot(-1)} \\ \sqrt{ab} = \sqrt{a}\cdot\sqrt{b}: \quad &= \sqrt{25}\cdot\sqrt{-1} \\ &= 5i \end{aligned}$$

(b)
$$\begin{aligned} \sqrt{-12} &= \sqrt{12\cdot(-1)} \\ &= \sqrt{12}\cdot\sqrt{-1} \\ &= 2\sqrt{3}i \end{aligned}$$

QUICK ✓ *Write each radical as a pure imaginary number.*

1. $\sqrt{-16}$ **2.** $\sqrt{-2}$ **3.** $\sqrt{-24}$

Classroom Example Write each of the following in standard form:

(a) $2-\sqrt{-25}$ (b) $5+\sqrt{-12}$.

Answer:

(a) $2-5i$ (b) $5+2\sqrt{3}i$

EXAMPLE 2 Writing Complex Numbers in Standard Form

Write each of the following in standard form.

(a) $3-\sqrt{-100}$ **(b)** $5+\sqrt{-28}$

Solution

(a)
$$\begin{aligned} 3-\sqrt{-100} &= 3-\sqrt{100\cdot -1} \\ \sqrt{ab} = \sqrt{a}\cdot\sqrt{b}: \quad &= 3-\sqrt{100}\cdot\sqrt{-1} \\ &= 3-10i \end{aligned}$$

(b)
$$\begin{aligned} 5+\sqrt{-28} &= 5+\sqrt{28}\cdot\sqrt{-1} \\ \sqrt{ab} = \sqrt{a}\cdot\sqrt{b}: \quad &= 5+\sqrt{4}\cdot\sqrt{7}\cdot\sqrt{-1} \\ &= 4+2\sqrt{7}i \end{aligned}$$

Classroom Example Write in standard form: $\frac{4-\sqrt{-12}}{2}$

Answer: $2-\sqrt{3}i$

EXAMPLE 3 Writing a Complex Number in Standard Form

Write the following in standard form: $\frac{4-\sqrt{-72}}{2}$

Solution

$$\begin{aligned} \frac{4-\sqrt{-72}}{2} &= \frac{4-\sqrt{72\cdot(-1)}}{2} \\ &= \frac{4-\sqrt{72}\cdot\sqrt{-1}}{2} \\ \sqrt{72} = \sqrt{36}\cdot\sqrt{2} = 6\sqrt{2}: \quad &= \frac{4-6\sqrt{2}i}{2} \\ \text{Factor out 2:} \quad &= \frac{2(2-3\sqrt{2}i)}{2} \\ \text{Divide out the 2s:} \quad &= \frac{\cancel{2}(2-3\sqrt{2}i)}{\cancel{2}} \\ &= 2-3\sqrt{2}i \end{aligned}$$

QUICK ✓ *Write each expression in the standard form of a complex number, $a + bi$.*

4. $9+\sqrt{-144}$ **5.** $-7-\sqrt{-20}$ **6.** $\frac{8-\sqrt{-32}}{4}$

2 Add or Subtract Complex Numbers

Throughout your math career, whenever you were introduced to a new number system, the four binary operations of addition, subtraction, multiplication, and division using that number system were reviewed. In keeping with that practice, we are now going to examine how to add and subtract complex numbers. Then we will discuss multiplying and dividing complex numbers.

Two complex numbers are added by adding the real parts and then adding the imaginary parts.

In Words
To add two complex numbers, add the real parts, and then add the imaginary parts. To subtract two complex numbers, subtract the real parts, and then subtract the imaginary parts.

THE SUM OF COMPLEX NUMBERS

$$(a + bi) + (c + di) = (a + c) + (b + d)i$$

To subtract two complex numbers, we use this rule:

THE DIFFERENCE OF COMPLEX NUMBERS

$$(a + bi) - (c + di) = (a - c) + (b - d)i$$

Classroom Example
Add:
(a) $(6 - 2i) + (-4 + 7i)$
(b) $(4 + \sqrt{-25}) + (6 - \sqrt{-16})$
Answer:
(a) $2 + 5i$ (b) $10 + i$

EXAMPLE 4 Adding Complex Numbers

Add:

(a) $(4 + 6i) + (-3 + 5i)$ **(b)** $(8 + \sqrt{-9}) + (2 - \sqrt{-64})$

Solution

(a) $(4 + 6i) + (-3 + 5i) = [4 + (-3)] + (6 + 5)i$
$= 1 + 11i$

(b) $\sqrt{-9} = 3i$; $\sqrt{-64} = 8i$ $(8 + \sqrt{-9}) + (2 - \sqrt{-64}) = (8 + 3i) + (2 - 8i)$
$= (8 + 2) + (3 - 8)i$
$= 10 + (-5)i$
$= 10 - 5i$

Work Smart
Notice how adding complex numbers is like combining like terms. For example,

$(4 + 6x) + (-3 + 5x)$
$= [4 + (-3)] + (6 + 5)x$
$= 1 + 11x$

and

$(4 + 6i) + (-3 + 5i)$
$= [4 + (-3)] + (6 + 5)i$
$= 1 + 11i$

Classroom Example
Subtract:
(a) $(-11 + 3i) - (8 - 2i)$
(b) $(-1 + \sqrt{-48}) - (-3 - \sqrt{-12})$
Answer:
(a) $-19 + 5i$ (b) $2 + 6\sqrt{3}i$

EXAMPLE 5 Subtracting Complex Numbers

Subtract:

(a) $(4 - 2i) - (-3 + 7i)$ **(b)** $(4 - \sqrt{-4}) - (-7 + \sqrt{-9})$

Solution

(a) $(4 - 2i) - (-3 + 7i) = (4 - (-3)) + (-2 - 7)i$
$= 7 + (-9)i$
$= 7 - 9i$

(b) $\sqrt{-4} = 2i$; $\sqrt{-9} = 3i$: $(4 - \sqrt{-4}) - (-7 + \sqrt{-9}) = (4 - 2i) - (-7 + 3i)$
$= (4 - (-7)) + (-2 - 3)i$
$= (4 + 7) + (-2 - 3)i$
$= 11 + (-5i)$
$= 11 - 5i$

Work Smart
Always write a number in standard form before you add or subtract.

QUICK ✓ *Add or subtract as indicated.*

7. $(4 - 3i) + (-2 + 5i)$ **8.** $(-3 + 7i) - (5 - 4i)$
9. $(3 + \sqrt{-12}) - (-2 - \sqrt{-27})$

3 Multiply Complex Numbers

To multiply complex numbers we apply the same ideas that we used to multiply polynomials.

Classroom Example Multiply and write the answer in standard form: $2i(5 - 3i)$

Answer: $6 + 10i$

EXAMPLE 6 Using the Distributive Property to Multiply Complex Numbers

Multiply and write the answer in standard form: $3i(5 - 4i)$

Solution

We distribute the $3i$ into each term in the parentheses.

$$\begin{aligned} 3i(5 - 4i) &= 3i \cdot 5 - 3i \cdot 4i \\ &= 15i - 12i^2 \\ i^2 = -1: \quad &= 15i - 12(-1) \\ &= 12 + 15i \end{aligned}$$

Classroom Example Multiply and write the answer in standard form: $(-3 + 2i)(5 - i)$

Answer: $-13 + 13i$

EXAMPLE 7 Using the FOIL Method to Multiply Complex Numbers

Multiply and write the answer in standard form: $(-2 + 5i)(4 - 2i)$

Solution

$$\begin{aligned} (-2 + 5i)(4 - 2i) &= -2 \cdot 4 - 2 \cdot (-2i) + 5i \cdot 4 + 5i \cdot (-2i) \\ &= -8 + 4i + 20i - 10i^2 \\ \text{Combine like terms; } i^2 = -1: \quad &= -8 + 24i - 10(-1) \\ &= -8 + 24i + 10 \\ &= 2 + 24i \end{aligned}$$

QUICK ✓ *Multiply.*

10. $4i(3 - 6i)$

11. $(-2 + 4i)(3 - i)$

Work Smart

$\sqrt{a \cdot b} = \sqrt{a} \cdot \sqrt{b}$ only if a and b are nonnegative (zero or positive) real numbers.

Look back at the Product Property for Square Roots in Section 6.4. You should notice that the property only applies when $\sqrt{a}$ and $\sqrt{b}$ are real numbers. This means that

$$\sqrt{a} \cdot \sqrt{b} \neq \sqrt{ab} \quad \text{if} \quad a < 0 \text{ or } b < 0$$

So how do we perform this multiplication? We write the radical as a complex number, using the fact that $\sqrt{-N} = \sqrt{N}i$, and then perform the multiplication.

Classroom Example Multiply: $\sqrt{-49} \cdot \sqrt{-4}$

Answer: -14

EXAMPLE 8 Multiplying Square Roots of Negative Numbers

Multiply: $\sqrt{-9} \cdot \sqrt{-36}$

Solution

Work Smart

$\sqrt{-9} \cdot \sqrt{-36} \neq \sqrt{(-9)(-36)}$

because the Product Property of Square Roots only applies when the radical is a real number.

We cannot use the Product Property of Radicals to multiply these radicals because neither $\sqrt{-9}$ nor $\sqrt{-36}$ is a real number. Therefore, we express the radicals as pure imaginary numbers and then multiply.

$$\begin{aligned} \sqrt{-9} \cdot \sqrt{-36} &= 3i \cdot 6i \\ &= 18i^2 \\ i^2 = -1: \quad &= 18(-1) \\ &= -18 \end{aligned}$$

Therefore, $\sqrt{-9} \cdot \sqrt{-36} = -18$.

Classroom Example ➤
Multiply:
$\left(3 + \sqrt{-25}\right)\left(1 - \sqrt{-9}\right)$

Answer: $18 - 4i$

EXAMPLE 9 Multiplying Square Roots of Negative Numbers

Multiply: $\left(2 - \sqrt{-36}\right)\left(4 - \sqrt{-25}\right)$

Solution

First, we rewrite each expression as a complex number in standard form.

$$\left(2 - \sqrt{-36}\right)\left(4 - \sqrt{-25}\right) = (2 - 6i)(4 - 5i)$$

$$\begin{aligned} \text{FOIL:} \quad &= 2\cdot 4 + 2\cdot(-5i) - 6i\cdot 4 - 6i\cdot(-5i) \\ &= 8 - 10i - 24i + 30i^2 \\ \text{Combine like terms; } i^2 = -1\text{:} \quad &= 8 - 34i + 30(-1) \\ &= 8 - 34i - 30 \\ &= -22 - 34i \end{aligned}$$

So $\left(2 + \sqrt{-36}\right)\left(4 - \sqrt{-25}\right) = -22 - 34i.$

QUICK ✓ *Multiply.*

12. $\sqrt{-25}\cdot\sqrt{-4}$

13. $\left(2 - \sqrt{-16}\right)\left(1 - \sqrt{-4}\right)$

Conjugates

We now introduce a special product that involves the *conjugate* of a complex number.

In Words
To find the complex conjugate of $a + bi$, change the sign from "+" to "−" or "−" to "+" between the "a" and "b" in the complex number.

THE CONJUGATE OF A COMPLEX NUMBER

If $a + bi$ is a complex number, then its **conjugate** is defined as $a - bi$. The expression $a - bi$ is called the *complex conjugate* of $a + bi$.

For example,

Complex Number	Complex Conjugate
$2 + 3i$	$2 - 3i$
$-12 - 7i$	$-12 + 7i$
$5i$	$-5i$

The next example illustrates what happens when we multiply a complex number and its conjugate.

Classroom Example ➤
Find the product of $4 + 3i$ and its conjugate, $4 - 3i$.

Answer: 25

EXAMPLE 10 Multiplying a Complex Number by Its Conjugate

Find the product of $5 + 2i$ and its conjugate, $5 - 2i$.

Solution

$$\begin{aligned} (5 + 2i)(5 - 2i) &= 5\cdot 5 + 5\cdot(-2i) + 2i\cdot 5 + 2i\cdot(-2i) \\ &= 25 - 10i + 10i - 4i^2 \\ &= 25 - 4(-1) \\ &= 25 + 4 \\ &= 29 \end{aligned}$$

Teaching Tip
Have students compare the products

$$(3 + 2x)(3 - 2x) = 9 - 4x^2$$

and

$$(3 + 2i)(3 - 2i) = 9 - 4i^2 = 9 + 4 = 13$$

Work Smart

$(a + b)(a - b) = a^2 - b^2$

but

$(a + bi)(a - bi) = a^2 + b^2$

Wow! The product of the complex number $5 + 2i$ and its conjugate $5 - 2i$ is 29, a real number. In fact, the results of Example 10 are true in general.

THE PRODUCT OF A COMPLEX NUMBER AND ITS CONJUGATE

The product of a complex number and its conjugate is a nonnegative real number. That is,

$$(a + bi)(a - bi) = a^2 + b^2$$

Perhaps you noticed that multiplying a complex number and its conjugate is similar to multiplying the sum and difference of the same two terms. But be careful: $(a + b)(a - b) = a^2 - b^2$, whereas $(a + bi)(a - bi) = a^2 + b^2$.

QUICK ✓ *Multiply.*

14. $(2 - 7i)(2 + 7i)$ **15.** $(-4 + 9i)(-4 - 9i)$

4 Divide Complex Numbers

Now that we understand that the product of a complex number and its conjugate is a nonnegative real number, we can illustrate how to divide complex numbers.

EXAMPLE 11 How to Divide Complex Numbers

Divide: $\dfrac{15}{2 + i}$

Classroom Example
Divide: $\dfrac{58}{5 + 2i}$
Answer: $10 - 4i$

Step-by-Step Solution

Step 1: Write the numerator and denominator in standard form, $a + bi$.

The denominator is in standard form. $\dfrac{15}{2 + i}$

Step 2: Multiply the numerator and denominator by the complex conjugate of the denominator.

Multiply by $1 = \dfrac{2 - i}{2 - i}$: $\dfrac{15}{2 + i} \cdot \dfrac{2 - i}{2 - i} = \dfrac{15(2 - i)}{(2 + i)(2 - i)}$

$(a + bi)(a - bi) = a^2 + b^2$: $= \dfrac{15(2 - i)}{2^2 + 1^2}$

Divide out like factors: $= \dfrac{\overset{3}{\cancel{15}}(2 - i)}{\underset{1}{\cancel{5}}}$

$= 3(2 - i)$

Step 3: Simplify by writing the quotient in standard form, $a + bi$.

$= 6 - 3i$

So $\dfrac{15}{2 + i} = 6 - 3i$.

Work Smart
Dividing complex numbers is a lot like rationalizing a denominator.

$$\frac{15}{2 + \sqrt{3}} = \frac{15}{2 + \sqrt{3}} \cdot \frac{2 - \sqrt{3}}{2 - \sqrt{3}}$$
$$= \frac{15(2 - \sqrt{3})}{4 - 2\sqrt{3} + 2\sqrt{3} - \sqrt{9}}$$
$$= \frac{15(2 - \sqrt{3})}{4 - 3}$$
$$= 15(2 - \sqrt{3})$$
$$= 30 - 15\sqrt{3}$$

The following steps can be used to divide complex numbers.

Steps to Divide Complex Numbers

Step 1: Write the numerator and denominator in standard form, $a + bi$.

Step 2: Multiply the numerator and denominator by the complex conjugate of the denominator.

Step 3: Simplify by writing the quotient in standard form, $a + bi$.

QUICK ✓ *Divide.*

16. $\dfrac{26}{-2 + 3i}$

17. $\dfrac{-12}{4 - i}$

Classroom Example ➤

Divide: $\dfrac{3 + 4i}{\sqrt{-4}}$

Answer: $2 - \dfrac{3}{2}i$

Teaching Tip

Consider working the classroom example showing Method 1 and Method 2 side by side.

EXAMPLE 12 Dividing Complex Numbers

Divide: $\dfrac{8 + 5i}{\sqrt{-4}}$

Solution

There are two approaches that we can take to performing this division. We will show both and leave it to you to decide which method you like better. Both approaches require that we write the numerator and denominator of the expression in standard form. Because $\sqrt{-4} = 2i$, we have

$$\frac{8 + 5i}{\sqrt{-4}} = \frac{8 + 5i}{2i}$$

Now let's look at each approach.

METHOD 1:

Because $2i$ can be written as $0 + 2i$, we have that the complex conjugate of the denominator, $0 + 2i$, is $0 - 2i$. Because $0 - 2i = -2i$, we multiply the numerator and denominator by $-2i$.

Multiply by $\dfrac{-2i}{-2i}$: $\dfrac{8 + 5i}{2i} = \dfrac{8 + 5i}{2i} \cdot \dfrac{-2i}{-2i}$

$= \dfrac{(8 + 5i)(-2i)}{(2i)(-2i)}$

Distribute $-2i$: $= \dfrac{8(-2i) + 5i(-2i)}{-4i^2}$

$= \dfrac{-16i - 10i^2}{-4i^2}$

$i^2 = -1$: $= \dfrac{-16i - (10)(-1)}{-4(-1)}$

$= \dfrac{-16i + 10}{4}$

Divide each term in the numerator by 4: $= \dfrac{-16i}{4} + \dfrac{10}{4}$

$= -4i + \dfrac{5}{2}$

Write in standard form: $= \dfrac{5}{2} - 4i$

METHOD 2:

We see that the denominator has only an imaginary part. We use the fact that $\dfrac{A + B}{C} = \dfrac{A}{C} + \dfrac{B}{C}$ and divide $2i$ into real part and imaginary part of the numerator.

$\dfrac{8 + 5i}{2i} = \dfrac{8}{2i} + \dfrac{5i}{2i}$

Simplify: $= \dfrac{4}{i} + \dfrac{5}{2}$

Multiply $\dfrac{4}{i}$ by $1 = \dfrac{i}{i}$: $= \dfrac{4}{i} \cdot \dfrac{i}{i} + \dfrac{5}{2}$

$= \dfrac{4i}{i^2} + \dfrac{5}{2}$

$i^2 = -1$: $= \dfrac{4i}{-1} + \dfrac{5}{2}$

Write in standard form: $= -4i + \dfrac{5}{2}$

$= \dfrac{5}{2} - 4i$

Regardless of the method we choose, the quotient $\dfrac{8 + 5i}{\sqrt{-4}}$ is equal to $\dfrac{5}{2} - 4i$.

QUICK ✓ *Divide.*

18. $\dfrac{8 + 3i}{3i}$

19. $\dfrac{1 + 5i}{2i}$

5 Solve Quadratic Equations with Complex Solutions

Work Smart
Just as all integers are rational numbers, or all rational numbers are real numbers, all real numbers are complex numbers. For example -2 can be written $-2 + 0i$.

Equations such as $x^2 = -9$ and $(y + 2)^2 + 3 = 1$ have no real solution, but they do have solutions. The solutions of the equations $x^2 = -9$ and $(y + 2)^2 + 3 = 1$ are complex numbers that are not real. Before we solve a quadratic equation having solutions that are nonreal complex numbers, let's see how to determine whether or not a given complex number is a solution of a quadratic equation.

Classroom Example
Is $x = 2 + 2i$ a solution of the equation
$x^2 - 4x + 8 = 0$?
Answer: yes

EXAMPLE 13 Determining Whether or Not a Complex Number Is a Solution of a Quadratic Equation

Is $x = 2 + i$ a solution of the equation $x^2 - 4x + 5 = 0$?

Solution

A number is a solution of an equation if it satisfies the equation. Does $x = 2 + i$ satisfy $x^2 - 4x + 5 = 0$? Let's see.

$$x^2 - 4x + 5 = 0$$

Substitute $2 + i$ for x: $$(2 + i)^2 - 4(2 + i) + 5 \stackrel{?}{=} 0$$

$$2^2 + 2(2)(i) + (i)^2 - 4 \cdot 2 - 4 \cdot i + 5 \stackrel{?}{=} 0$$

$$4 + 4i + i^2 - 8 - 4i + 5 \stackrel{?}{=} 0$$

$$4 + 4i - 1 - 8 - 4i + 5 \stackrel{?}{=} 0$$

$$0 = 0 \quad \text{True}$$

We see that we have a true statement, so $x = 2 + i$ is a solution of $x^2 - 4x + 5 = 0$.

Now let's solve a quadratic equation that has complex solutions that are not real numbers.

Classroom Example
Solve the quadratic equation:
$(2x + 5)^2 + 7 = 6$
Answer: $\left\{-\frac{5}{2} - \frac{1}{2}i, -\frac{5}{2} + \frac{1}{2}i\right\}$

EXAMPLE 14 Solving a Quadratic Equation Having Complex Solutions

Solve the quadratic equation: $(x - 3)^2 + 9 = 5$

Solution

Let's examine the appearance of this equation. In just one step, we can write it in the form $(ax + b)^2 = p$ and then use the Square Root Property.

Work Smart
A quadratic equation may be solved by factoring, using the Square Root Property, or using the quadratic formula. Study the form of the quadratic equation to see which method is most efficient.

$$(x - 3)^2 + 9 = 5$$

Isolate the squared term: $$(x - 3)^2 + 9 - 9 = 5 - 9$$

$$(x - 3)^2 = -4$$

Use the Square Root Property: $$x - 3 = \pm\sqrt{-4}$$

$\sqrt{-4} = \sqrt{4 \cdot -1} = 2i$: $$x - 3 = \pm 2i$$

Add 3 to both sides: $$x - 3 + 3 = \pm 2i + 3$$

$$x = 3 \pm 2i$$

Let's check these solutions.

$x = 3 - 2i$:
$$(x - 3)^2 + 9 = 5$$
$$[(3 - 2i) - 3]^2 + 9 \stackrel{?}{=} 5$$
$$[3 - 2i - 3]^2 + 9 \stackrel{?}{=} 5$$
$$[-2i]^2 + 9 \stackrel{?}{=} 5$$
$$4i^2 + 9 \stackrel{?}{=} 5$$
$$-4 + 9 \stackrel{?}{=} 5$$
$$5 = 5 \quad \text{True}$$

$x = 3 + 2i$:
$$(x - 3)^2 + 9 = 5$$
$$[(3 + 2i) - 3]^2 + 9 \stackrel{?}{=} 5$$
$$[3 + 2i - 3]^2 + 9 \stackrel{?}{=} 5$$
$$[2i]^2 + 9 \stackrel{?}{=} 5$$
$$4i^2 + 9 \stackrel{?}{=} 5$$
$$-4 + 9 \stackrel{?}{=} 5$$
$$5 = 5 \quad \text{True}$$

Our answers both check, so the solution set is $\{3 - 2i, 3 + 2i\}$.

Classroom Example
Solve the quadratic equation:

$$x^2 - 4x + 13 = 0$$

Answer: $\{2 - 3i, 2 + 3i\}$

Teaching Tip
Remind students that they're putting together two concepts that they already know: the quadratic formula and complex numbers.

EXAMPLE 15 Solving a Quadratic Equation Having Complex Solutions

Solve the quadratic equation: $x^2 - 2x + 7 = 0$

Solution

We'll use the quadratic formula to solve $x^2 - 2x + 7 = 0$.

Quadratic formula: $x = \dfrac{-b \pm \sqrt{b^2 - 4ac}}{2a}$

Substitute $a = 1, b = -2, c = 7$: $x = \dfrac{-(-2) \pm \sqrt{(-2)^2 - 4(1)7}}{2(1)}$

$$= \frac{2 \pm \sqrt{4 - 28}}{2}$$

$$= \frac{2 \pm \sqrt{-24}}{2}$$

$$= \frac{2 \pm \sqrt{4 \cdot 6 \cdot (-1)}}{2}$$

Simplify the square root: $= \dfrac{2 \pm 2\sqrt{6}i}{2}$

Factor out 2: $= \dfrac{2\left(1 \pm \sqrt{6}i\right)}{2}$

Divide out the 2s: $= \dfrac{\not{2}\left(1 \pm \sqrt{6}i\right)}{\not{2}}$

$$= 1 \pm \sqrt{6}i$$

Teaching Tip
Remind students that the expression $\dfrac{2 \pm 2\sqrt{6}i}{2}$ can also be simplified using $\dfrac{A + B}{C} = \dfrac{A}{C} + \dfrac{B}{C}$.

We leave the check to you. The solution set is $\left\{1 - \sqrt{6}i, 1 + \sqrt{6}i\right\}$.

QUICK ✓ *Solve the quadratic equation.*

20. $2w^2 + 5 = 0$ **21.** $n^2 + 6n = -17$

7.5 Exercises

For Extra Help: Student Solutions Manual · CD Video · PH Math/Tutor Center · MathXL Tutorials on CD · MathXL® · MyMathLab

Concepts and Vocabulary

1. $\sqrt{-1}$
2. real, imaginary
3. complex conjugates, real
4. False
5. True
6. False

In Problems 1–3, fill in the blanks.

1. The imaginary unit i equals ________.

2. In the complex number $-3 + 4i$, -3 is called the ________ part and 4 is called the ________ part.

3. The numbers $(3 + 4i)$ and $(3 - 4i)$ are called ________ ________ and the product of two numbers of this type is always a ________ number.

In Problems 4–6, answer True or False to each statement.

4. The product of two non-real complex numbers is always a non-real complex number.

5. Every real number is also a complex number but not every complex number is a real number.

6. The following product is correct:

$$(3 - 2i)(3 + 2i) = 3^2 + (2i)^2 = 9 + 4(-1) = 5$$

7. complex solutions; quadratic formula
8. No
9. $25i$ 10. $13i$
11. $-2\sqrt{13}i$ 12. $-3\sqrt{6}i$
13. $5 + 2i$ 14. $7 + 6i$
15. $12 - 3\sqrt{3}i$ 16. $15 - 3\sqrt{5}i$
17. $-\frac{3}{2} + \frac{\sqrt{2}}{2}i$ 18. $-3 + \sqrt{3}i$
19. $-\frac{1}{2} + \sqrt{3}i$ 20. $-2 + \sqrt{2}i$
21. $1 + 3i$ 22. $-1 - 11i$
23. -10 24. -12
25. $-3 + 21i$ 26. $4 + i$
27. $13 - 10i$ 28. $5 + i$
29. $-8 - i$ 30. $5 - 18i$
31. $12 + \sqrt{2}i$
32. $-6 + \sqrt{6}i$
33. $-5\sqrt{10}i$
34. $-1 + 4\sqrt{2}i$
35. $15 + 12i$
36. $-35 - 10i$
37. $-10 - 10i$
38. $3 - 5i$
39. $-3 - 16i$
40. $-9 + 3i$
41. -256
42. $-10{,}000$
43. $24 - 42i$
44. $10 + 24i$
45. $7 - 24i$
46. $28 - 96i$
47. -20 48. -2
49. $-6\sqrt{2}$ 50. $-3\sqrt{5}$
51. $1 - 31i$ 52. $16 + 4i$
53. $-3 - 2i$; 13
54. $4 + 5i$; 41
55. $6 - 3\sqrt{2}i$; 54
56. $2 + 2\sqrt{3}i$; 16
57. $4 + 8i$ 58. $3 + 6i$
59. $3 + 2i$ 60. $-5 - 2i$
61. $\frac{3}{2} + \frac{3}{2}i$
62. $-3 - 3i$
63. $-1 + \frac{1}{2}i$
64. $-\frac{27}{20} - \frac{9}{20}i$
65. $-\frac{27}{4} + 2i$
66. $\frac{4}{5} - i$
67. $-\frac{3}{13} + \frac{2}{13}i$
68. $\frac{3}{2} + \frac{1}{2}i$
69. $-6 + \sqrt{3}i$
70. $2 + 9\sqrt{2}i$
71. $-5 + 12i$
72. $24 + 10i$
73. $23 - 10i$
74. $-2 - 16i$

7. Suppose you computed the discriminant of a quadratic equation and found it to be a negative number. What kind of solutions does the equation have? Which method would be most efficient to solve the equation?

8. Is it possible to have a solution set to a quadratic equation that contains one real number and one complex number that is not real? Explain your response.

Building Skills

In Problems 9–12, write each of the following as a pure imaginary number.

9. $\sqrt{-625}$ **10.** $\sqrt{-169}$ **11.** $-\sqrt{-52}$ **12.** $-\sqrt{-54}$

In Problems 13–20, write each of the following in standard form.

13. $5 + \sqrt{-4}$ **14.** $7 + \sqrt{-36}$ **15.** $12 - \sqrt{-27}$ **16.** $15 - \sqrt{-45}$

17. $\dfrac{-9 + \sqrt{-18}}{6}$ **18.** $\dfrac{-6 + \sqrt{-12}}{2}$ **19.** $\dfrac{2 - \sqrt{-48}}{-4}$ **20.** $\dfrac{10 - \sqrt{-50}}{-5}$

In Problems 21–34, add or subtract as indicated. Express your answer in standard form.

21. $(3 - 2i) + (-2 + 5i)$ **22.** $(-5 - 2i) + (4 - 9i)$

23. $(-4 + 6i) - (6 + 6i)$ **24.** $(-5 + 4i) - (7 + 4i)$

25. $9i - (3 - 12i)$ **26.** $(4 + 2i) - i$

27. $\left(10 - \sqrt{-4}\right) - \left(-3 + \sqrt{-64}\right)$ **28.** $\left(8 + \sqrt{-25}\right) + \left(-3 - \sqrt{-16}\right)$

29. $\left(-6 + \sqrt{-64}\right) + \left(-2 - \sqrt{-81}\right)$ **30.** $\left(13 - \sqrt{-121}\right) - \left(8 + \sqrt{-49}\right)$

31. $\left(11 + \sqrt{-18}\right) + \left(1 - \sqrt{-8}\right)$ **32.** $\left(-4 - \sqrt{-24}\right) + \left(-2 + \sqrt{-54}\right)$

33. $\left(-1 - \sqrt{-40}\right) - \left(-1 + \sqrt{-90}\right)$ **34.** $\left(1 + \sqrt{-18}\right) - \left(2 - \sqrt{-2}\right)$

In Problems 35–52, multiply. Express your answer in standard form.

35. $3i(4 - 5i)$ **36.** $5i(-2 + 7i)$ **37.** $(1 - i)(-10i)$

38. $(5 + 3i)(-i)$ **39.** $(-2 + 7i)(-2 + i)$ **40.** $(-3 - 3i)(1 - 2i)$

41. $(16i)^2$ **42.** $(100i)^2$ **43.** $(6 + 12i)(-2 - 3i)$

44. $(6 + 4i)(3 + 2i)$ **45.** $(4 - 3i)^2$ **46.** $(-8 + 6i)^2$

47. $\sqrt{-25} \cdot \sqrt{-16}$ **48.** $\sqrt{-4} \cdot \sqrt{-1}$

49. $\sqrt{-6} \cdot \sqrt{-12}$ **50.** $\sqrt{-3} \cdot \sqrt{-15}$

51. $\left(-5 - \sqrt{-1}\right)\left(1 + \sqrt{-36}\right)$ **52.** $\left(-2 + \sqrt{-4}\right)\left(-3 - \sqrt{-25}\right)$

In Problems 53–56, write the complex conjugate of the given complex number and then find the product of the complex number and its conjugate.

53. $-3 + 2i$ **54.** $4 - 5i$ **55.** $6 + \sqrt{-18}$ **56.** $2 - \sqrt{-12}$

In Problems 57–68, divide. Express your answer in standard form.

57. $\dfrac{20}{1 - 2i}$ **58.** $\dfrac{15}{1 - 2i}$ **59.** $\dfrac{-4 + 6i}{2i}$ **60.** $\dfrac{2 - 5i}{i}$

61. $\dfrac{3}{1 - i}$ **62.** $\dfrac{-6i}{1 + i}$ **63.** $\dfrac{-5i}{-2 + 4i}$ **64.** $\dfrac{9i}{-2 - 6i}$

65. $\dfrac{8 + 27i}{-4i}$ **66.** $\dfrac{-5 - 4i}{-5i}$ **67.** $\dfrac{-1}{3 + 2i}$ **68.** $\dfrac{-5}{-3 + i}$

Mixed Practice

In Problems 69–80, perform the indicated operation and express your answer in standard form.

69. $\left(-3 + \sqrt{-27}\right) + \left(-3 - \sqrt{-12}\right)$ **70.** $\left(-4 + \sqrt{-32}\right) + \left(6 + \sqrt{-50}\right)$

71. $(-2 - 3i)^2$ **72.** $(5 + i)^2$

73. $(20 - 3i) - (-3 + 7i)$ **74.** $(6 - 12i) - (8 + 4i)$

75. $-\frac{5}{2}+\frac{5}{2}i$
76. $\frac{8}{5}-\frac{4}{5}i$
77. $\frac{2}{3}+\frac{5}{2}i$ 78. $-2-\frac{4}{3}i$
79. $16\sqrt{3}-6+\left(-6\sqrt{3}-16\right)i$
80. $26+32i$
81. $\left\{1-\frac{\sqrt{6}}{2}i, 1+\frac{\sqrt{6}}{2}i\right\}$
82. $\left\{\frac{3}{4}-\frac{\sqrt{39}}{4}i, \frac{3}{4}+\frac{\sqrt{39}}{4}i\right\}$
83. $\{-1-2\sqrt{3}i, -1+2\sqrt{3}i\}$
84. $\{2-3\sqrt{2}i, 2+3\sqrt{2}i\}$
85. $\{5-2i, 5+2i\}$
86. $\{-6-3i, -6+3i\}$
87. $\{-2-\sqrt{5}i, -2+\sqrt{5}i\}$
88. $\{-2-\sqrt{2}i, -2+\sqrt{2}i\}$
89. $\left\{-\frac{3}{2}, 8\right\}$
90. $\left\{\frac{1}{3}, -5\right\}$
91. $\{-2\sqrt{2}, 2\sqrt{2}\}$
92. $\{-3\sqrt{5}, 3\sqrt{5}\}$
93. $\{1-\sqrt{2}i, 1+\sqrt{2}i\}$
94. $\left\{\frac{1}{4}-\frac{3\sqrt{3}}{4}i, \frac{1}{4}+\frac{3\sqrt{3}}{4}i\right\}$
95. $\left\{\frac{1-\sqrt{105}}{4}, \frac{1+\sqrt{105}}{4}\right\}$
96. $\{-3-3\sqrt{2}, -3+3\sqrt{2}\}$
97. $\frac{1}{3}-\frac{\sqrt{14}}{3}i$ or $\frac{1}{3}+\frac{\sqrt{14}}{3}i$
98. $1-\sqrt{2}i$ or $1+\sqrt{2}i$
99. $1-3i$ or $1+3i$
100. $1-\frac{\sqrt{2}}{2}i$ or $1+\frac{\sqrt{2}}{2}i$
101. $5-\sqrt{15}i$ and $5+\sqrt{15}i$

75. $\dfrac{-15i}{-3+3i}$ **76.** $\dfrac{-8i}{2-4i}$

77. $\dfrac{-15+4i}{6i}$ **78.** $\dfrac{16-24i}{12i}$

79. $\left(-3-\sqrt{-64}\right)\left(2+\sqrt{-12}\right)$ **80.** $\left(1-\sqrt{-16}\right)\left(-6+\sqrt{-64}\right)$

In Problems 81–96, solve the quadratic equation by any method.

81. $2x^2-4x+5=0$ **82.** $2x^2-3x+6=0$ **83.** $(n+1)^2+12=0$

84. $(n-2)^2+18=0$ **85.** $(w-5)^2+6=2$ **86.** $(a+6)^2+13=4$

87. $z^2+6z+9=2z$ **88.** $r^2+3r+6=-r$ **89.** $2x^2-13x=24$

90. $3x^2+14x=5$ **91.** $2x^2-16=0$ **92.** $2x^2-90=0$

93. $3=2t-t^2$ **94.** $2p=4p^2+7$

95. $13=k(2k-1)$ **96.** $x(x+6)=9$

Applying the Concepts

97. Fun with Numbers The square of one more than a number is the same as the square of twice the number increased by six. Find the number(s).

98. Fun with Numbers Three times one less than a number is the same as the sum of the number and its square. Find the number(s).

99. Fun with Numbers The square of a number is the same as ten less than twice the number. Find the number(s).

100. Fun with Numbers Three times the reciprocal of a number is the same as four decreased by twice the number. Find the number(s).

Extending the Concepts

Complex numbers have not always enjoyed the acceptance that they have today. In 1545, Girolamo Cardano published a mathematical treatise on solutions to higher-degree equations. In this work, he recognized the existence of negative roots but claimed them to be "fictitious." Complex numbers became more accepted as established mathematicians such as Descartes, Euler, and Gauss used them in their work. Descartes originated the terms "real" and "imaginary" parts (in 1637), Euler is responsible for the notation $i=\sqrt{-1}$ (in 1777), and Gauss coined the term "complex" (in 1831).

101. Part of Cardano's gradual acceptance of the notion of negative roots stemmed from his investigation into the following problem: "Divide 10 into two parts such that the product of one part times the remainder is 40." Here "remainder" is not a part of a division problem but rather another way of saying a number, x, and its remainder $(10-x)$ sum to 10. Can you help him find an answer to this puzzle?

CHAPTER 7 ACTIVITY: The Math Game

1. $\left\{-5, \frac{1}{2}\right\}$ 2. $\{-5, 5\}$
3. $\left\{\frac{-2-\sqrt{10}}{3}, \frac{-2+\sqrt{10}}{3}\right\}$
4. $\{-1-i, -1+i\}$

Focus: Solving quadratic equations

Time: 15 minutes

Group size: 2–4

- The instructor will announce when the groups may begin solving the problems below.
- When your group has completed all of the problems, ask the instructor to check the answers. The instructor will tell you how many answers are correct, but not which ones.
- The first group to complete all of the problems correctly will win a prize, as determined by the instructor.

5. $\left\{\frac{3}{2}-\frac{3\sqrt{3}}{2}i, \frac{3}{2}+\frac{3\sqrt{3}}{2}i\right\}$
6. $\left\{-\frac{1}{3}, 1\right\}$
7. $\left\{\frac{5-\sqrt{15}}{5}, \frac{5+\sqrt{15}}{5}\right\}$
8. $\left\{\frac{5-\sqrt{17}}{4}, \frac{5+\sqrt{17}}{4}\right\}$

Solve the following quadratic equations. Express radicals in simplest form, and express complex answers in standard form.

1. $2x^2 + 9x - 5 = 0$ **2.** $x^2 - 20 = 5$ **3.** $3x^2 - 2 = -4x$

4. $3x^2 + 6x + 6 = 0$ **5.** $x^2 - 3x + 9 = 0$ **6.** $3x^2 - 1 = 2x$

7. $5(x-1)^2 + 3 = 6$ **8.** $5x = 2x^2 + 1$

CHAPTER 7 REVIEW

Section 7.1 Solving Quadratic Equations Using the Square Root Property

KEY CONCEPT

- **The Square Root Method**
  If $x^2 = p$, where $p \geq 0$, then $x = \sqrt{p}$ or $x = -\sqrt{p}$.

YOU SHOULD BE ABLE TO . . .	EXAMPLE	REVIEW EXERCISES
(1) Solve quadratic equations using the Square Root Property (p. 458)	Examples 1 through 6	1–14
(2) Solve problems using the Pythagorean Theorem (p. 463)	Examples 7 and 8	15–20

1. $\{-5, 5\}$ 2. $\{-12, 12\}$
3. $\left\{-\frac{3}{2}, \frac{3}{2}\right\}$ 4. $\left\{-\frac{2}{5}, \frac{2}{5}\right\}$
5. $\{-2\sqrt{3}, 2\sqrt{3}\}$ 6. $\{-3\sqrt{2}, 3\sqrt{2}\}$
7. $\left\{-\frac{5}{3}, 1\right\}$ 8. $\{-1, 6\}$
9. $\{-5-\sqrt{2}, -5+\sqrt{2}\}$
10. $\{2-\sqrt{2}, 2+\sqrt{2}\}$
11. no real solution
12. no real solution
13. $\{2-2\sqrt{6}, 2+2\sqrt{6}\}$
14. $\{-4-2\sqrt{7}, -4+2\sqrt{7}\}$

In Problems 1–14, solve each quadratic equation using the Square Root Property. Express radicals in simplest form.

1. $x^2 - 25 = 0$ **2.** $0 = x^2 - 144$ **3.** $4p^2 = 9$

4. $25t^2 = 4$ **5.** $3x^2 - 10 = 26$ **6.** $2x^2 + 5 = 41$

7. $(3x+1)^2 = 16$ **8.** $(2x-5)^2 = 49$ **9.** $9 = (2n+10)^2 + 1$

10. $25 = (4d-8)^2 - 7$ **11.** $(2x+7)^2 + 6 = 2$ **12.** $(5x-1)^2 - 5 = -30$

13. $x^2 - 4x + 4 = 24$ **14.** $x^2 + 8x + 16 = 28$

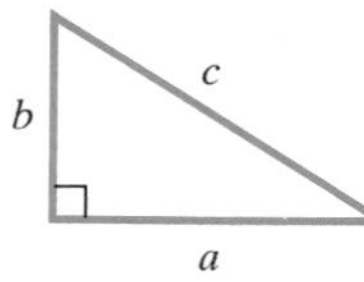

In Problems 15–18, use the right triangle shown in the margin and find the missing length. Give exact answers and decimal approximations correct to two decimal places.

15. $a = 12, b = 8$ **16.** $a = 6, b = 12$

17. $b = 8, c = 14$ **18.** $a = 4, c = 12$

15. $c = 4\sqrt{13} \approx 14.42$
16. $c = 6\sqrt{5} \approx 13.42$
17. $a = 2\sqrt{33} \approx 11.49$
18. $b = 8\sqrt{2} \approx 11.31$
19. (a) $3\sqrt{2}$ ft by $3\sqrt{2}$ ft (b) 4.24 ft by 4.24 ft
20. (a) $10\sqrt{2}$ ft (b) 14.1 ft

19. Picture Window A square window in the front of the Browns' house measures 6 feet on the diagonal.
(a) Exactly how wide and how tall is the window?
(b) Approximate the dimensions of the window rounded to 2 decimal places.

20. Painter's Ladder A painter has a ladder that can be extended to 15 feet. The painter decides to place the bottom of the ladder 5 feet from the house to be painted.
(a) Exactly how far up the house will the ladder reach?
(b) Approximate this height to nearest tenth of a foot.

Section 7.2 Solving Quadratic Equations by Completing the Square

KEY CONCEPT	KEY TERM
• Steps to Solve a Quadratic Equation by Completing the Square. See p. 471.	Complete the square

YOU SHOULD BE ABLE TO . . .	EXAMPLE	REVIEW EXERCISES
(1) Complete the square in one variable (p. 469)	Example 1	21–26
(2) Solve quadratic equations by completing the square (p. 470)	Examples 2 through 6	27–40

21. 36; $(x-6)^2$ 22. 25; $(x+5)^2$
23. $\frac{1}{9}$; $\left(a+\frac{1}{3}\right)^2$ 24. $\frac{4}{25}$; $\left(k-\frac{2}{5}\right)^2$
25. $\frac{1}{16}$; $\left(n-\frac{1}{4}\right)^2$ 26. $\frac{1}{36}$; $\left(t+\frac{1}{6}\right)^2$
27. $\{-1, 7\}$ 28. $\{1, 11\}$
29. $\{-4, 1\}$ 30. $\{-7, -2\}$
31. $\{2-2\sqrt{5}, 2+2\sqrt{5}\}$
32. $\{1-2\sqrt{2}, 1+2\sqrt{2}\}$
33. $\left\{-\frac{5}{2}, 2\right\}$ 34. $\left\{-3, -\frac{2}{3}\right\}$
35. $\left\{\frac{3-\sqrt{53}}{2}, \frac{3+\sqrt{53}}{2}\right\}$
36. $\left\{\frac{-1-\sqrt{33}}{2}, \frac{-1+\sqrt{33}}{2}\right\}$
37. $\{5-3\sqrt{5}, 5+3\sqrt{5}\}$
38. $\{4-2\sqrt{6}, 4+2\sqrt{6}\}$
39. $-\frac{5}{2}$ or $\frac{1}{2}$ 40. -8 or 2

In Problems 21–26, determine the number that must be added to the expression to make it a perfect square trinomial. Then factor the expression.

21. $x^2 - 12x$ **22.** $x^2 + 10x$ **23.** $a^2 + \frac{2}{3}a$

24. $k^2 - \frac{4}{5}k$ **25.** $n^2 - \frac{n}{2}$ **26.** $t^2 + \frac{t}{3}$

In Problems 27–38, solve the quadratic equation by completing the square.

27. $x^2 - 6x = 7$ **28.** $x^2 - 12x = -11$ **29.** $x^2 = 4 - 3x$

30. $x^2 = -14 - 9x$ **31.** $\frac{1}{4}x^2 - x - 4 = 0$ **32.** $\frac{1}{2}x^2 - x - \frac{7}{2} = 0$

33. $2a^2 + a - 10 = 0$ **34.** $3n^2 + 11n + 6 = 0$ **35.** $(x-5)(x+2) = 1$

36. $(x+3)(x-2) = 2$ **37.** $x(x-10) - 15 = 5$ **38.** $x(x-8) + 2 = 10$

39. Fun with Numbers The product of a number and 2 more than the number is $\frac{5}{4}$. Find the pair(s) of numbers that satisfy this condition.

40. Fun with Numbers The sum of the square of a number and 6 times this number is 16. Find the two numbers that satisfy this condition.

Section 7.3 Solving Quadratic Equations Using the Quadratic Formula

KEY CONCEPTS	KEY TERMS
• The solution(s) to the quadratic equation $ax^2 + bx + c = 0$, $a \neq 0$, are given by $x = \frac{-b \pm \sqrt{b^2 - 4ac}}{2a}$ • Steps to Solve a Quadratic Equation Using the Quadratic Formula. See p. 479. • The discriminant of a quadratic equation is the expression $b^2 - 4ac$.	Quadratic formula Repeated solution Discriminant

YOU SHOULD BE ABLE TO . . .	EXAMPLE	REVIEW EXERCISES
(1) Solve quadratic equations using the quadratic formula (p. 477)	Examples 1 through 7	41–56
(2) Use the discriminant to determine which method to use when solving a quadratic equation (p. 484)	Examples 8 and 9; Table 1	57–62

41. $2x^2 - x + 3 = 0$; $a = 2, b = -1, c = 3$
42. $\frac{1}{2}x^2 + x - \frac{3}{4}$; $a = \frac{1}{2}, b = 1, c = -\frac{3}{4}$ or $2x^2 + 4x - 3 = 0$; $a = 2, b = 4, c = -3$
43. -1 or $-\frac{1}{2}$ 44. $-\frac{2}{3}$ or 1

In Problems 41 and 42, write the quadratic equation in standard form and then identify the values assigned to a, b, and c. Do not solve the equation.

41. $3 = x - 2x^2$ **42.** $\frac{1}{2}x^2 = \frac{3}{4} - x$

In Problems 43 and 44, simplify each expression.

43. $\frac{-3 \pm \sqrt{3^2 - 4(2)(1)}}{2(2)}$ **44.** $\frac{1 \pm \sqrt{(-1)^2 - 4(3)(-2)}}{2(3)}$

In Problems 45–54, solve each equation using the quadratic formula.

45. $2x^2 - 13x + 15 = 0$ **46.** $4x^2 + 8x + 3 = 0$ **47.** $x^2 - 12x = -5$

48. $x^2 - 3x = 9$ **49.** $p(3p + 5) - 8 = 0$ **50.** $v(2v + 7) - 4 = 0$

51. $1 + \frac{1}{x} = \frac{3}{2x^2}$ **52.** $1 + \frac{4}{x} = \frac{2}{x^2}$

53. $\frac{b}{3} + \frac{1}{12b} = 1$ **54.** $\frac{3n}{2} + \frac{1}{n} = 4$

55. Bookstore Sales Dirt Cheap Discount Bookstore determines that the revenue R from selling x algebra texts is be given by the equation $R = -\frac{1}{2}x^2 + 130x$.
(a) What is the revenue from selling 150 algebra texts?
(b) How many texts must the bookstore sell to have revenue of $8000?

56. Fancy Pet Grooming Martin owns a dog grooming business and has determined that his profit P from grooming x dogs is given by the equation $P = x^2 + 30x - 60$.
(a) What is Martin's profit when he grooms 8 dogs?
(b) How many dogs must he groom to have a profit of $615?

In Problems 57–60, determine the discriminant and then use this value to determine the most efficient way of solving the quadratic equation. Lastly, solve the quadratic equation by any method you wish.

57. $5t^2 + 2 = 8t$ **58.** $x^2 - 5x = 6$ **59.** $3x^2 + 7x = 10$

60. $n^2 - 2 = -3n$ **61.** $2 + 3d^2 = d$ **62.** $3(p^2 + 1) + 5p = 0$

45. $\left\{\frac{3}{2}, 5\right\}$ 46. $\left\{-\frac{3}{2}, -\frac{1}{2}\right\}$
47. $\{6 - \sqrt{31}, 6 + \sqrt{31}\}$
48. $\left\{\frac{3 - 3\sqrt{5}}{2}, \frac{3 + 3\sqrt{5}}{2}\right\}$
49. $\left\{-\frac{8}{3}, 1\right\}$ 50. $\left\{-4, \frac{1}{2}\right\}$
51. $\left\{\frac{-1 - \sqrt{7}}{2}, \frac{-1 + \sqrt{7}}{2}\right\}$
52. $\{-2 - \sqrt{6}, -2 + \sqrt{6}\}$
53. $\left\{\frac{3 - 2\sqrt{2}}{2}, \frac{3 + 2\sqrt{2}}{2}\right\}$
54. $\left\{\frac{4 + \sqrt{10}}{3}, \frac{4 - \sqrt{10}}{3}\right\}$
55. (a) $8250 (b) 100 or 160 texts
56. (a) $244 (b) 15 dogs
57. $b^2 - 4ac = 24$; quadratic formula; $\left\{\frac{4 - \sqrt{6}}{5}, \frac{4 + \sqrt{6}}{5}\right\}$
58. $b^2 - 4ac = 49$; factoring; $\{-1, 6\}$
59. $b^2 - 4ac = 169$; factoring; $\left\{-\frac{10}{3}, 1\right\}$
60. $b^2 - 4ac = 17$; quadratic formula; $\left\{\frac{-3 - \sqrt{17}}{2}, \frac{-3 + \sqrt{17}}{2}\right\}$
61. $b^2 - 4ac = -23$; no real solution
62. $b^2 - 4ac = -11$; no real solution

Section 7.4 Problem Solving Using Quadratic Equations

YOU SHOULD BE ABLE TO ...	EXAMPLE	REVIEW EXERCISES
1 Model and solve direct translation problems (p. 491)	Examples 1 through 3	63–72
2 Solve problems modeled by quadratic equations (p. 495)	Examples 4 and 5	73–76

63. Consecutive Integers The product of two consecutive even integers is 288. Find both pairs of integers that have this product.

64. Consecutive Integers The sum of a number and its reciprocal is $\frac{13}{6}$. Find the number and the reciprocal.

△ **65. A Rectangle** The length of a rectangle is two meters less than twice the width. If the area of the rectangle is 112 square meters, find the dimensions of the rectangle.

△ **66. A Rectangle** The width of a rectangle is one yard more than one-third of the length. If the area of the rectangle is 60 square yards, find the dimensions of the rectangle.

△ **67. A Triangle** In a right triangle, one leg is double the length of the other. If the length of the hypotenuse is $4\sqrt{5}$ feet, find the length of each of the legs of the triangle.

△ **68. A Triangle** In a right triangle, the hypotenuse is 3 centimeters more than twice one leg. If the length of the other leg is 5 centimeters, find the exact lengths of each of the unknown sides of the triangle.

△ **69. A Triangle** In a triangle, the base is 5 meters more than the height. If the area of the triangle is 42 square meters, find the base and height of the triangle.

△ **70. A Triangle** In a triangle, the base is one and a half inches more than the height. If the area of the triangle is $\frac{9}{4}$ square inches, find the base and height of the triangle.

63. $-18, -16$ and $16, 18$
64. $\frac{3}{2}; \frac{2}{3}$
65. width = 8 m, length = 14 m
66. width = 5 yd, length = 12 yd
67. 4 ft, 8 ft
68. leg $= \frac{-6 + 2\sqrt{21}}{3}$ cm, hypotenuse $= \frac{-3 + 4\sqrt{21}}{3}$ cm
69. base = 12 m, height = 7 m
70. base = 3 in., height $= \frac{3}{2}$ in.

71. length = 10 in., width = 4 in.
72. length = 15 ft., width = 8 ft.
73. $\frac{5 + \sqrt{26}}{2} \approx 5.0$ s
74. $\frac{15 - \sqrt{145}}{4} \approx 0.7$ s and $\frac{15 + \sqrt{145}}{4} \approx 6.8$ s
75. 5 or 25 units
76. $-5(-5 + \sqrt{11}) \approx 8$ units or $5(5 + \sqrt{11}) \approx 42$ units

△ **71. Jewelry Box** Kaysie is taking a jewelry making class and is building a box to hold her treasures. The height of her box is to be 5 inches and the volume will be 200 cubic inches. The base of the box is a rectangle whose width is one inch less than half the length. Find the dimensions of the box.

△ **72. Harold's Driveway** Harold is ordering cement for his new driveway. He orders 60 cubic feet of cement that will fill his rectangular driveway 6 inches deep. If the length of the driveway is one foot less than twice the width, find the dimensions of his driveway.

73. Projectile Motion Ron and Ellen are shooting a toy rocket from a sling shot. The rocket leaves their hand 4 feet above the ground with an initial velocity of 80 feet per second. The equation that models the height of the rocket h feet above the ground t seconds after it was fired is $h = -16t^2 + 80t + 4$. How long does it take for the rocket to hit the ground? Find the exact answer and then approximate to the nearest tenth of a second.

74. Model Rocket A model rocket is launched from the ground with an initial speed of 120 feet per second. The equation that models its height, h (in feet), off the ground t seconds after it was fired is $h = -16t^2 + 120t$. How long will it take for the rocket to be 80 feet off the ground? Find the exact answer and then approximate to the nearest tenth of a second.

The point at which a company's profit equal zero is called the company's ***break-even point.*** *In Problems 75 and 76, let R represent a company's revenue, let C represent the company's costs, and let x represent the number of units produced and sold. Find the company's break-even point; that is, find x so that* $R - C = 0$. *Give an exact answer and, if necessary, approximate this value to the nearest whole number.*

75. $R = -\frac{x^2}{5} + 10x$
$C = 4x + 25$

76. $R = -\frac{x^2}{10} + 8x$
$C = 3x + 35$

Section 7.5 The Complex Number System

KEY CONCEPTS	KEY TERMS
• For any real number a, $a^2 \geq 0$ • $i^2 = -1; i = \sqrt{-1}$ • $\sqrt{-N} = \sqrt{N}i$ • $(a + bi) + (c + di) = (a + c) + (b + d)i$ • $(a + bi) - (c + di) = (a - c) + (b - d)i$ • $(a + bi)(a - bi) = a^2 + b^2$	Imaginary unit Complex number Real part Imaginary part Standard form Pure imaginary number Principal square root of $-N$ Complex conjugate

YOU SHOULD BE ABLE TO . . .	EXAMPLE	REVIEW EXERCISES
1 Evaluate the square root of negative real numbers (p. 501)	Examples 1 through 3	77–82
2 Add or subtract complex numbers (p. 503)	Examples 4 and 5	83–88
3 Multiply complex numbers (p. 504)	Examples 6 through 10	89–98
4 Divide complex numbers (p. 506)	Examples 11 and 12	99–104
5 Solve quadratic equations with complex solutions (p. 508)	Examples 13 through 15	105–108

77. $-10i$ 78. $-7i$
79. $5 + 2\sqrt{5}i$ 80. $4 - 2\sqrt{15}i$
81. $-1 - \sqrt{3}i$
82. $-\frac{1}{2} + \frac{\sqrt{3}}{2}i$

In Problems 77–82, write each complex number in standard form.

77. $-\sqrt{-100}$ **78.** $-\sqrt{-49}$ **79.** $5 + \sqrt{-20}$

80. $4 - \sqrt{-60}$ **81.** $\frac{-3 - \sqrt{-27}}{3}$ **82.** $\frac{-5 + \sqrt{-75}}{10}$

83. $1 + 8i$ 84. $11i$
85. $-16 + 10i$
86. $-26 + 13i$
87. $-10 + 7i$
88. $14 - 4i$
89. $-10 + 12i$
90. $-3 - 3i$
91. $2 - 23i$
92. $27 - 34i$
93. $-5 - 12i$
94. $-16 + 30i$
95. $-36\sqrt{2}$
96. $-48\sqrt{5}$
97. $5 - 2i$; 29
98. $3 + 3i$; 18
99. $-\frac{1}{3} - i$ 100. $2 + 3i$
101. $\frac{3}{5} + \frac{4}{5}i$ 102. $\frac{6}{5} - \frac{3}{5}i$
103. $-\frac{5}{2} + \frac{5}{2}i$ 104. $\frac{5}{17} + \frac{3}{17}i$
105. $\left\{-\frac{1}{2} - \frac{3}{2}i, -\frac{1}{2} + \frac{3}{2}i\right\}$
106. $\left\{\frac{3}{2} - \frac{3}{2}i, \frac{3}{2} + \frac{3}{2}i\right\}$
107. $\{2 - 2\sqrt{2}i, 2 + 2\sqrt{2}i\}$
108. $\{-3 - 4\sqrt{2}i, -3 + 4\sqrt{2}i\}$

In Problems 83–96, perform the indicated operation and express your answer in standard form.

83. $(4 + 10i) + (-3 - 2i)$
84. $(-3 + 5i) + (3 + 6i)$
85. $(-8 + 5i) - (8 - 5i)$
86. $(-13 + i) - (13 - 12i)$
87. $\left(2 + \sqrt{-9}\right) - \left(12 - \sqrt{-16}\right)$
88. $\left(12 - \sqrt{-25}\right) + \left(2 + \sqrt{-1}\right)$
89. $2i(6 + 5i)$
90. $-3i(1 - i)$
91. $(4 - 5i)(3 - 2i)$
92. $(12 + i)(2 - 3i)$
93. $(2 - 3i)^2$
94. $(3 + 5i)^2$
95. $3\sqrt{-12} \cdot 2\sqrt{-6}$
96. $4\sqrt{-10} \cdot 3\sqrt{-8}$

In Problems 97 and 98, write the conjugate of the given complex number and then find the product of the complex number and its conjugate.

97. $5 + \sqrt{-4}$
98. $3 - \sqrt{-9}$

In Problems 99–104, divide and express your answer in standard form.

99. $\dfrac{9 - 3i}{9i}$
100. $\dfrac{-6 + 4i}{2i}$
101. $\dfrac{5}{3 - \sqrt{-16}}$
102. $\dfrac{6}{4 + \sqrt{-4}}$
103. $\dfrac{5i}{1 - i}$
104. $\dfrac{2i}{3 + 5i}$

In Problems 105–108, solve each quadratic equation by any method. Express your answer in standard form.

105. $2x^2 + 2x + 5 = 0$
106. $2x^2 - 6x + 9 = 0$
107. $(x - 2)^2 + 12 = 4$
108. $(x + 3)^2 + 9 = -23$

CHAPTER 7 TEST

Remember to use your Chapter Test Prep Video CD to see fully worked-out solutions to any of these problems you would like to review.

Note to Instructor: A special file in TestGen provides algorithms specifically matched to the problems in this Chapter Test for easy-to-replicate practice or assessment purposes.

1. $\{-3, 3\}$
2. $\{-10, 4\}$
3. $\{-3 - 2\sqrt{2}, -3 + 2\sqrt{2}\}$
4. $\{-2, 3\}$
5. $\left\{\frac{2 - \sqrt{10}}{2}, \frac{2 + \sqrt{10}}{2}\right\}$
6. $\left\{\frac{3 - \sqrt{13}}{4}, \frac{3 + \sqrt{13}}{4}\right\}$
7. (a) 61
(b) quadratic formula
(c) Answers may vary.
8. $-\frac{1}{2} - \frac{3}{2}i$
9. $-14 - 4i$
10. $1 + 3i$
11. -36
12. $-5 - 14i$
13. $-20 - 48i$
14. $4 - \frac{4}{3}i$
15. $\frac{15}{17} - \frac{9}{17}i$

1. Solve using the Square Root Method: $4x^2 - 12 = 24$

2. Solve by completing the square: $x^2 + 6x - 40 = 0$

3. Solve using the quadratic formula: $x^2 + 6x + 1 = 0$

In Problems 4–6, solve the quadratic equation by any method.

4. $(x - 2)(x + 1) = 4$
5. $2n = 4 + \dfrac{3}{n}$
6. $\dfrac{1}{2}t^2 = \dfrac{1}{8} + \dfrac{3}{4}t$

7. Given the quadratic equation $x = 5 - 3x^2$,
(a) determine the value of the discriminant.
(b) determine which method you would choose to solve the equation.
(c) explain why the method you chose is appropriate for this equation.

8. Write the complex number in standard form: $\dfrac{-3 - \sqrt{-81}}{6}$

In Problems 9–15, perform the indicated operation and write your answer in standard form.

9. $(-12 + 4i) + (-2 - 8i)$
10. $\left(3 - \sqrt{-4}\right) - \left(2 - \sqrt{-25}\right)$
11. $2\sqrt{-3} \cdot 3\sqrt{-12}$
12. $(1 - 4i)(3 - 2i)$
13. $(4 - 6i)^2$
14. $\dfrac{4 + 12i}{3i}$
15. $\dfrac{6}{5 + 3i}$

16. $\{2 - i, 2 + i\}$

17. $\left\{-2 - \frac{\sqrt{14}}{2}i, -2 + \frac{\sqrt{14}}{2}i\right\}$

18. (a) width $= -1 + \sqrt{127}$ ft, length $= 1 + \sqrt{127}$ ft
(b) width ≈ 10.3 ft, length ≈ 12.3 ft

19. (a) $\frac{25 + \sqrt{785}}{8}$ s
(b) 6.6 s

20. (a) width $= 25 + 25\sqrt{161}$ yards, length $= -50 + 50\sqrt{161}$ yards
(b) width ≈ 342.21 yards, length ≈ 584.43 yards

In Problems 16 and 17, solve the quadratic equation by any method. Express complex answers in standard form.

16. $x^2 - 4x + 5 = 0$ **17.** $2p^2 + 8p + 15 = 0$

18. Sibling Rivalry Cybil and her sister, Kara, share their bedroom. During a bout of sibling rivalry, they decide they are going to split the bedroom into two parts by installing a room divider along the diagonal of the room, creating two triangular spaces. The divider is 16 feet long and the room is 2 feet longer than it is wide.

(a) Find the exact dimensions of Cybil and Kara's bedroom.
(b) Approximate these dimensions to the nearest tenth of a foot.

19. Pirate Battle Pirates protecting their harbor fire a cannonball towards an intruder's sailing ship. The cannon sits on the cliffs above the harbor and the pirates use the equation $h = -16t^2 + 100t + 40$ to calculate how long it will take for the cannonball to strike the ship, where h is the height of the cannonball above the water in feet, and t is the time, in seconds, the ball travels.

(a) If the cannonball strikes the ship at the water line (that is, $h = 0$), calculate exactly how long the cannonball was in the air.
(b) Approximate this time to the nearest tenth of a second.

20. Cornfield Farmer Green is planting this season's cornfield. The length of the rectangular field is 100 yards less than twice the width and it has area of 200,000 square yards.

(a) Find the exact dimensions of the field.
(b) Approximate these dimensions to the nearest hundredth of a yard.

CUMULATIVE REVIEW CHAPTERS 1–7

1. undefined
2. 0
3. $-\frac{1}{2}$
4. $\frac{27x^{12}}{y^2}$
5. $\frac{9}{x^2y^6}$
6. (a) -1
(b) -2
(c) 2
(d) trinomial
7. (a) $y = \frac{3}{2}x - 6$
(b) $\frac{3}{2}$
(c) -6

In Problems 1–3, perform the indicated operations.

1. $\dfrac{-8 + 4(-2)^2}{-1 - (-1)}$ **2.** $\dfrac{\frac{1}{2}\cdot\frac{8}{3} - \frac{3}{4}\cdot\frac{16}{9}}{\left(\frac{1}{2}\right)^2}$ **3.** $\dfrac{4 - 7}{-3 - (-9)}$

In Problems 4 and 5, simplify each expression. Write your answers with positive exponents only.

4. $(3x^2y^{-2})^3(-x^3y^2)^2$ **5.** $\left(\dfrac{2x^2y}{6xy^{-2}}\right)^{-2}$

6. Given the polynomial $-2x^2 - 3x + 1$,

(a) Evaluate the polynomial for $x = -2$.
(b) What is the leading coefficient?
(c) What is the degree of the polynomial?
(d) Is the polynomial a monomial, binomial, trinomial, or none of these?

7. Given the equation $3x - 2y = 12$,

(a) Solve for y.
(b) When the equation is solved for y, identify the coefficient of x.
(c) When the equation is solved for y, identify the constant.

In Problems 8–11, solve each equation.

8. $3 - 5(x + 2) = -3(x - 7)$

9. $\dfrac{x - 1}{4} = \dfrac{2x + 1}{5}$

10. $\dfrac{x^2}{3} + \dfrac{x}{6} = 1$

11. $x(x - 2) = 3$

In Problems 12–14, factor completely, if possible.

12. $6h^2 - 11h - 10$

13. $3x^3 + 24$

14. $xy + 2x - 3y - 6$

In Problems 15 and 16, perform the indicated operations. Write your answer in lowest terms, if possible.

15. $\dfrac{6x^3 + 5x^2 - x + 3}{2x + 1}$

16. $\dfrac{3x^2 + 6x - 4}{2x}$

In Problems 17 and 18, evaluate each expression.

17. $-8^{2/3}$

18. $9^{-\frac{3}{2}}$

19. Write in scientific notation: 34,000,000,000

20. Write in decimal notation: 3.04×10^{-4}

In Problems 21–23, perform the indicated operation and write your answer in standard form.

21. $(-3 + 2i)^2$

22. $2\sqrt{-8} \cdot 4\sqrt{-3}$

23. $\dfrac{4 + 3i}{6i}$

24. Old Globe Tickets All 580 seats to the opening night performance of *Twelfth Night* at the Old Globe Theatre were sold. Tickets were sold at two different rates, \$20 for general seating and \$50 for reserved seats. If the Old Globe Theatre took in \$17,300 for the night, how many of each type of ticket were sold?

25. Pole Vault Misha is practicing pole vault for the upcoming track meet. The height of his jump, h (in feet), can be approximated by the equation $h = -2t^2 + 9t + 6$, where t is the number of seconds after takeoff.

(a) How high is Misha 4 seconds after he plants his vaulting pole?
(b) Exactly how long does it take for him to land on the mat in the pit?
(c) Approximate the time found in part (b) to the nearest hundredth of a second.

8. $\{-14\}$
9. $\{-3\}$
10. $\left\{-2, \frac{3}{2}\right\}$
11. $\{-1, 3\}$
12. $(2h - 5)(3h + 2)$
13. $3(x + 2)(x^2 - 2x + 4)$
14. $(y + 2)(x - 3)$
15. $3x^2 + x - 1 + \dfrac{4}{2x + 1}$
16. $\dfrac{3}{2}x + 3 - \dfrac{2}{x}$
17. -4
18. $\dfrac{1}{27}$
19. 3.4×10^{10}
20. 0.000304
21. $5 - 12i$
22. $-16\sqrt{6}$
23. $\dfrac{1}{2} - \dfrac{2}{3}i$
24. There were 390 general seating tickets and 190 reserved tickets sold.
25. (a) 10 feet
(b) $\dfrac{9 + \sqrt{129}}{4}$ seconds
(c) 5.09 seconds

CHAPTER

8 Introduction to Graphing and Equations of Lines

Your electric bill arrives. How do electric companies compute the amount owed? They use linear equations! See Example 7 in Section 8.1.

OUTLINE

The Big Picture: Putting It Together

It is now time to switch gears. In Chapters 2–7, we simplified expressions and solved equations involving one unknown. We are now going to focus our attention on equations and inequalities involving two unknowns.

When we dealt with a single unknown, we could represent solutions to equations or inequalities graphically using a real number line. In that case, we only needed one dimension to represent the solution. When we have two unknowns, we need to work in two dimensions. This is accomplished through the *rectangular coordinate system*.

8.1 The Rectangular Coordinate System and Equations in Two Variables

OBJECTIVES

1. Plot Points in the Rectangular Coordinate System
2. Determine If an Ordered Pair Satisfies an Equation
3. Create a Table of Values That Satisfies an Equation

Preparing for the Rectangular Coordinate System and Equations in Two Variables

Before getting started, take the following readiness quiz. If you get a problem wrong, go back to the section cited and review the material.

1. Plot the following points on the real number line: [Section 1.2, pp. 12–13]

$$4, -3, \frac{1}{2}, 5.5$$

2. Evaluate $3x + 5$ for (a) $x = 4$ (b) $x = -1$ [Section 1.7, pp. 57–58]

3. Evaluate $2x - 5y$ for (a) $x = 3, y = 2$ (b) $x = 1, y = -4$ [Section 1.7, pp. 57–58]

4. Solve: $3x + 5 = 14$ [Section 2.2, pp. 84–86]

5. Solve: $5(x - 3) - 2x = 3x + 12$ [Section 2.2, pp. 86–89]

1 Plot Points in the Rectangular Coordinate System

Figure 1

We have all heard the saying, "A picture is worth a thousand words." Because pictures allow individuals to visualize ideas, they are typically more powerful than any other form of printed communication. Consider the picture shown in Figure 1, which shows the results of the Manhattan Project from the test conducted July 16, 1945. It illustrates the power of the atom in a way that words never could.

While the pictures that we use in mathematics might not deliver as powerful a message as the picture in Figure 1, they are powerful nonetheless. To draw pictures of mathematical relationships, we need a "canvas." The "canvas" that we use in this chapter is the *rectangular coordinate system*.

In Section 1.2, we learned how to plot points on the real number line. We locate a point on the real number line by assigning it a single real number, called the *coordinate of the point*. We can think of this as plotting in one dimension. In this chapter, we use the *rectangular coordinate system,* a system that allows us to plot points in two dimensions.

We begin by drawing two real number lines that intersect at right (90°) angles. One of the real number lines is drawn horizontally, while the other is drawn vertically. We call the horizontal real number line the ***x*-axis,** and the vertical real number line is called the ***y*-axis.** The point where the x-axis and y-axis intersect is called the **origin, *O*.** See Figure 2.

Figure 2
The rectangular coordinate system

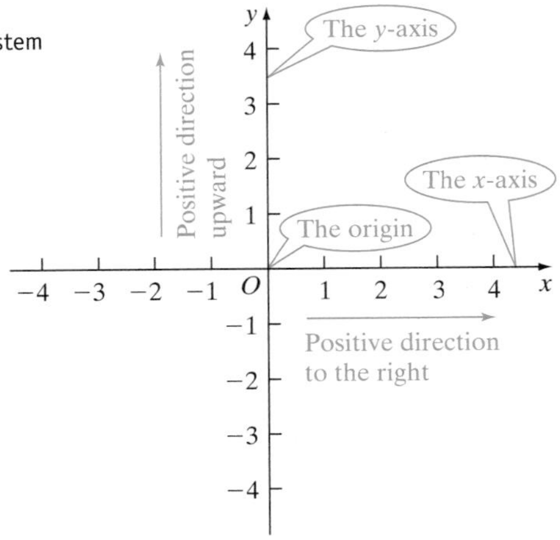

The origin O has a value of 0 on the x-axis and on the y-axis. Points on the x-axis to the right of O represent positive real numbers; points on the x-axis to the left of O represent negative real numbers. Points on the y-axis that are above O represent positive real numbers. Points on the y-axis that are below O represent negative real numbers.

Preparing for...Answers

1.

2. (a) 17 **(b)** 2 **3. (a)** −4 **(b)** 22 **4.** {3} **5.** ∅

Notice in Figure 2 that we label the x-axis "x" and the y-axis "y." An arrow is used at the end of each axis to denote the positive direction.

The coordinate system presented in Figure 2 is called a **rectangular** or **Cartesian coordinate system,** named after René Descartes (1596–1650), a French mathematician, philosopher, and theologian. The plane formed by the x-axis and y-axis is often referred to as the ***xy*-plane,** and the x-axis and y-axis are called the **coordinate axes.**

We can represent any point P in the rectangular coordinate system by using an **ordered pair (x, y)** of real numbers. We say that x represents the distance that P is from the y-axis. If $x > 0$ (that is, if x is positive), we travel x units to the right of the y-axis. If $x < 0$, we travel $|x|$ units to the left of the y-axis. We say that y represents the distance that P is from the x-axis. If $y > 0$, we travel y units above the x-axis. If $y < 0$, we travel $|y|$ units below the x-axis. The ordered pair (x, y) is also called the **coordinates** of P. For example, to plot the point (2, 5), from the origin we would travel 2 units to the right along the x-axis and then 5 units up as shown in Figure 3(a).

Figure 3

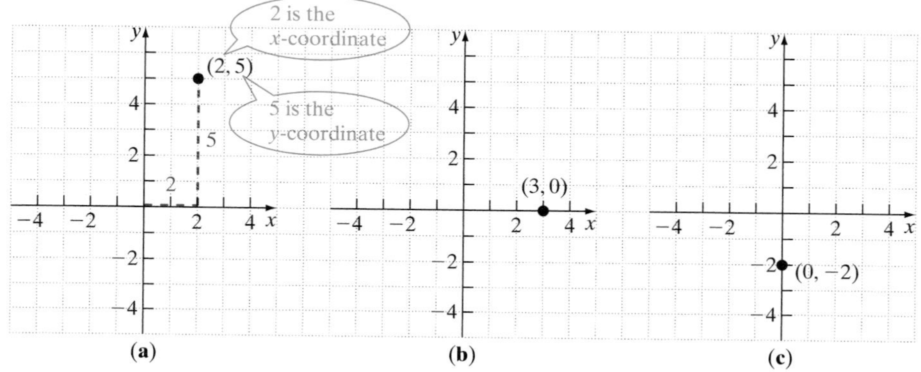

The origin O has coordinates (0, 0). Any point on the x-axis has coordinates of the form $(x, 0)$, and any point on the y-axis has coordinates of the form $(0, y)$. For example, the point whose coordinates are (3, 0) is 3 units to the right of the y-axis on the x-axis. See Figure 3(b). The point whose coordinates are $(0, -2)$ is 2 units below the x-axis on the y-axis. See Figure 3(c).

Work Smart

Be careful: The order in which numbers appear in the ordered pairs matters. For example, (3, 2) represents a different point from (2, 3).

If (x, y) are the coordinates of a point P, then x is called the ***x*-coordinate** of P and y is called the ***y*-coordinate** of P.

If you look back at Figure 2, you should notice that the x- and y-axes divide the plane into four separate regions or **quadrants.** In quadrant I, both the x-coordinate and y-coordinate are positive. In quadrant II, x is negative and y is positive. In quadrant III, both x and y are negative. In quadrant IV, x is positive and y is negative. Points on the coordinate axes do not belong to a quadrant. See Figure 4.

Figure 4

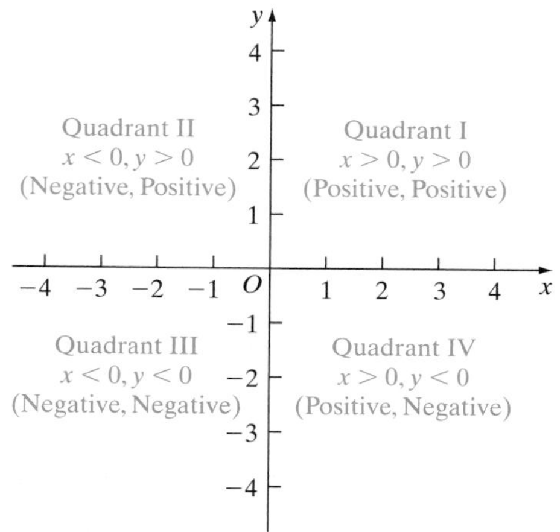

Let's summarize what we've learned so far.

SUMMARY: The Rectangular Coordinate System

- Composed of two real number lines—one horizontal (the x-axis) and one vertical (the y-axis). The x- and y-axes intersect at the origin.
- Also called the Cartesian coordinate system or xy-plane.
- Points in the rectangular coordinate system are denoted (x, y) and are called the coordinates of the point. We call x the x-coordinate and y the y-coordinate.
- If both x and y are positive, the point lies in quadrant I; if x is negative, but y is positive, the point lies in quadrant II; if x is negative and y is negative, the point lies in quadrant III; if x is positive and y is negative, the point lies in quadrant IV.
- Points on the x-axis have a y-coordinate of 0; points on the y-axis have an x-coordinate of 0.

Classroom Example Plot the following ordered pairs in the rectangular coordinate system. Tell which quadrant each point lies in.

(a) $A(4, 2)$ (b) $B(-2, 5)$
(c) $C(3, -6)$ (d) $D(2, 0)$
(e) $E(0, 7)$ (f) $F\left(-\frac{3}{2}, -\frac{7}{2}\right)$

Answer:

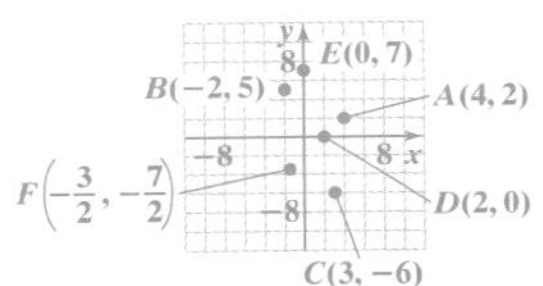

Point A is in quadrant I
Point B is in quadrant II
Point C is in quadrant IV
Point D is on the x-axis
Point E is on the y-axis
Point F is in quadrant III

EXAMPLE 1 Plotting Points in the Rectangular Coordinate System

Plot the following ordered pairs in the rectangular coordinate system. Tell which quadrant each point lies in or state that the point lies on the x- or y-axis.

(a) $A(3, 1)$ **(b)** $B(-4, 2)$ **(c)** $C(3, -5)$

(d) $D(4, 0)$ **(e)** $E(0, -3)$ **(f)** $F\left(\frac{5}{2}, -\frac{1}{2}\right)$

Solution

Before we can plot the points, we draw a rectangular or Cartesian coordinate system. See Figure 5(a). We now plot the points.

Figure 5

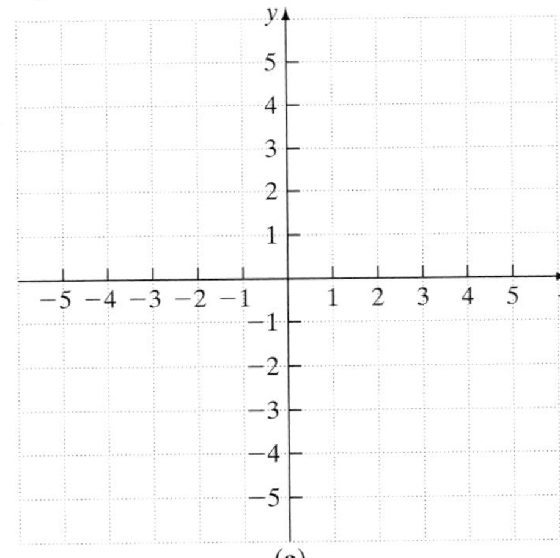
(a)

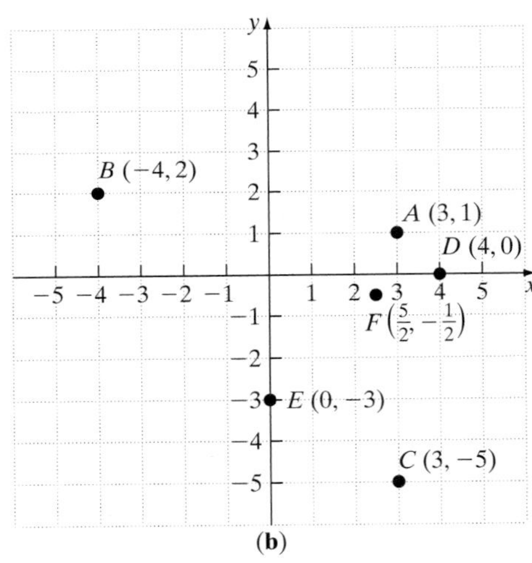

(b)

(a) To plot $A(3, 1)$, from the origin O, we travel 3 units to the right and then 1 unit up. Label the point A. Point A lies in quadrant I because both x and y are positive. See Figure 5(b).

(b) To plot $B(-4, 2)$, from the origin O, we travel 4 units to the left and then 2 units up. Label the point B. See Figure 5(b). Point B lies in quadrant II.

(c) See Figure 5(b). Point C lies in quadrant IV.

(d) See Figure 5(b). Point D does not lie in a quadrant because it lies on the x-axis.

(e) See Figure 5(b). Point E does not lie in a quadrant because it lies on the y-axis.

(f) It is helpful to convert the fractions to decimals, so $F\left(\frac{5}{2}, -\frac{1}{2}\right) = F(2.5, -0.5)$. We see that the x-coordinate of the point is halfway between 2 and 3; the y-coordinate of the point is halfway between -1 and 0. See Figure 5(b). Point F lies in quadrant IV.

QUICK ✓ *Plot the ordered pairs in the rectangular coordinate system. Tell which quadrant each point lies in or state that the point lies on the x-axis or y-axis.*

1. **(a)** $(5, 2)$ **(b)** $(-4, -3)$ **(c)** $(1, -3)$ **(d)** $(-2, 0)$ **(e)** $(0, 6)$ **(f)** $\left(-\frac{3}{2}, \frac{5}{2}\right)$

2. **(a)** $(-6, 2)$ **(b)** $(1, 7)$ **(c)** $(-3, -2)$ **(d)** $(4, 0)$ **(e)** $(0, -1)$ **(f)** $\left(\frac{3}{2}, -\frac{7}{2}\right)$

Classroom Example ➤
Use Example 2.

EXAMPLE 2 Identifying Points in the Rectangular Coordinate System

Identify the coordinates of each point labeled in Figure 6.

Figure 6

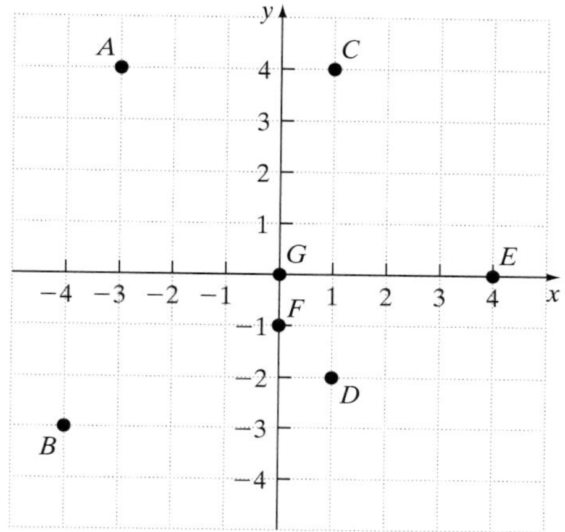

Solution

In an ordered pair (x, y), remember that x represents the position of the point left or right of the y-axis, while y represents the position of the point above or below the x-axis.

Since the point A is 3 units left of the y-axis, it has an x-coordinate of -3; since the point A is 4 units above the x-axis, it has a y-coordinate of 4. The ordered pair corresponding to the point A is $(-3, 4)$. We find the remaining coordinates in a similar fashion.

Point	Position	Ordered Pair
A	3 units left of the y-axis, 4 units above the x-axis	$(-3, 4)$
B	4 units left of the y-axis, 3 units below the x-axis	$(-4, -3)$
C	1 unit right of the y-axis, 4 units above the x-axis	$(1, 4)$
D	1 unit right of the y-axis, 2 units below the x-axis	$(1, -2)$
E	4 units right of the y-axis, on the x-axis	$(4, 0)$
F	On the y-axis, 1 unit below the x-axis	$(0, -1)$
G	On the x-axis; on the y-axis	$(0, 0)$

QUICK ✓ *Identify the coordinates of each point labeled in the figure below.*

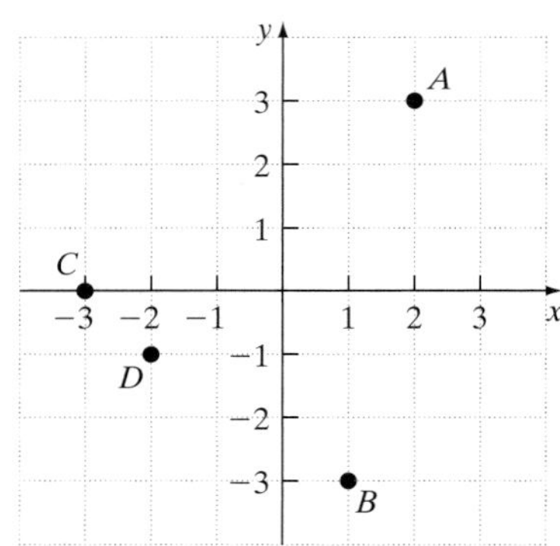

3. A **4.** B **5.** C **6.** D

2 Determine If an Ordered Pair Satisfies an Equation

In Sections 2.1–2.3, we solved linear equations in one variable. Recall that the solution of a linear equation is either a single value of the variable (conditional equation), the empty set (contradiction), or all real numbers (identity). Table 1 illustrates the three categories of linear equations in one variable.

Table 1

Conditional Equation	Contradiction	Identity
$3x + 2 = 11$	$4(x - 2) - x = 2(x + 1) + x$	$-2(x + 3) + 4x = 2(x - 3)$
$3x = 9$	$4x - 8 - x = 2x + 2 + x$	$-2x - 6 + 4x = 2x - 6$
$x = 3$	$3x - 8 = 3x + 2$	$2x - 6 = 2x - 6$
	$-8 = 2$	$0 = 0$
	no solution	solution is all real numbers

We will now look at equations in two variables.

DEFINITION

An **equation in two variables,** x and y, is a statement in which the algebraic expressions involving x and y are equal. The expressions are called **sides** of the equation.

Teaching Tip
Start with a linear equation such as $x + y = 6$ and have students come up with as many solutions as possible. Don't forget those such as $\left(\frac{1}{2}, \frac{11}{2}\right)$ and $(-1, 7)$, and so on. Emphasize that there are an infinite number of solutions of the linear equation.

For example, the following are all equations in two variables.

$$x + 2y = 6 \qquad y = -3x + 7 \qquad y = x^2 + 3$$

Since an equation is a statement, it may be true or false, depending upon the values of the variables. Any values of the variable that make the equation a true statement are said to **satisfy** the equation.

The first equation $x + 2y = 6$ is satisfied when $x = 4$ and $y = 1$. Why? If we substitute 4 for x and 1 for y into the equation $x + 2y = 6$, we obtain a true statement.

$$x + 2y = 6$$

$$x = 4, y = 1: \quad 4 + 2(1) \stackrel{?}{=} 6$$

$$6 = 6 \quad \text{True}$$

Rather than saying the equation is satisfied when $x = 4$ and $y = 1$, we can say that the ordered pair $(4, 1)$ satisfies the equation. Does $x = 8$ and $y = -1$ satisfy the equation $x + 2y = 6$?

$$x + 2y = 6$$
$$x = 8, y = -1: \quad 8 + 2(-1) \stackrel{?}{=} 6$$
$$6 = 6 \quad \text{True}$$

So the ordered pair $(8, -1)$ satisfies the equation as well. In fact, there are infinitely many choices of x and y that satisfy the equation $x + 2y = 6$. However, there are some choices of x and y that do not satisfy the equation $x + 2y = 6$. For example, $x = 3$ and $y = 4$ do not satisfy the equation $x + 2y = 6$.

$$x + 2y = 6$$
$$x = 3, y = 4: \quad 3 + 2(4) \stackrel{?}{=} 6$$
$$11 = 6 \quad \text{False}$$

Classroom Example Determine if the following ordered pairs satisfy the equation $2x + y = 5$.

(a) $(-2, 9)$

(b) $(4, -1)$

(c) $(6, -7)$

Answer:

(a) yes (b) no (c) yes

EXAMPLE 3 Determining Whether an Ordered Pair Satisfies an Equation

Determine if the following ordered pairs satisfy the equation $x + 2y = 8$.

(a) $(2, 3)$ **(b)** $(6, -1)$ **(c)** $(-2, 5)$

Solution

(a) For the ordered pair $(2, 3)$, we check to see if $x = 2$, $y = 3$ satisfies the equation $x + 2y = 8$.

$$x + 2y = 8$$
$$\text{Let } x = 2, y = 3: \quad 2 + 2(3) \stackrel{?}{=} 8$$
$$2 + 6 \stackrel{?}{=} 8$$
$$8 = 8 \quad \text{True}$$

The statement is true, so $(2, 3)$ satisfies the equation $x + 2y = 8$.

(b) For the ordered pair $(6, -1)$, we have

$$x + 2y = 8$$
$$\text{Let } x = 6, y = -1: \quad 6 + 2(-1) \stackrel{?}{=} 8$$
$$6 - 2 \stackrel{?}{=} 8$$
$$4 = 8 \quad \text{False}$$

The statement $4 = 8$ is false, so $(6, -1)$ does not satisfy the equation $x + 2y = 8$.

(c) For the ordered pair $(-2, 5)$, we have

$$x + 2y = 8$$
$$\text{Let } x = -2, y = 5: \quad -2 + 2(5) \stackrel{?}{=} 8$$
$$-2 + 10 \stackrel{?}{=} 8$$
$$8 = 8 \quad \text{True}$$

The statement is true, so $(-2, 5)$ satisfies the equation $x + 2y = 8$.

QUICK ✓

7. Determine if the following ordered pairs satisfy the equation $x + 4y = 12$.

(a) $(4, 2)$ **(b)** $(-2, 4)$ **(c)** $(1, 8)$

8. Determine if the following ordered pairs satisfy the equation $y = 4x + 3$.

(a) $(1, 3)$ **(b)** $(-2, -5)$ **(c)** $\left(-\frac{3}{2}, -3\right)$

3 Create a Table of Values That Satisfies an Equation

In Example 3 we learned how to determine whether a given ordered pair satisfies an equation. However, we do not yet know how to find an ordered pair that satisfies an equation.

EXAMPLE 4 How to Determine an Ordered Pair That Satisfies an Equation

Find an ordered pair that satisfies the equation $3x + y = 5$.

Classroom Example
Find an ordered pair that satisfies the equation $2x + y = 7$.

Answer: Many possible answers, such as $(-2, 11)$, $(-1, 9)$, $(0, 7)$, $(1, 5)$, and $(2, 3)$.

Step-by-Step Solution

Step 1: Choose any value for one of the variables in the equation.

You may choose any value of x or y that you wish. In this example, we will let $x = 2$.

Step 2: Substitute the value of the variable chosen in Step 1 into the equation and then use the techniques learned in Chapter 2 to solve for the remaining variable.

Teaching Tip
Remind students that the solutions of a linear equation in two variables are those **ordered pairs** that make the equation true. In an ordered pair, x comes first, followed by y.

Substitute 2 for x in the equation $3x + y = 5$ and then solve for y.

$$3x + y = 5$$

Let $x = 2$: $3(2) + y = 5$

Simplify: $6 + y = 5$

Subtract 6 from both sides: $6 - 6 + y = 5 - 6$

$$y = -1$$

The ordered pair that satisfies the equation is $(2, -1)$.

QUICK ✓ *Determine an ordered pair that satisfies the given equation by substituting the given value of the variable into the equation.*

9. $2x + y = 10$; $x = 3$

10. $-3x + 2y = 11$; $y = 1$

Look again at Example 3. Did you notice that two different ordered pairs satisfy the equation? In fact, there are an infinite number of ordered pairs that satisfy the equation $x + 2y = 8$ because for any real number y we can find a value of x that makes the equation a true statement. One approach to find some of the solutions of an equation in two variables is to create a table of values that satisfy the equation. The table is created by choosing values of x and using the equation to find the value of y (as we did in Example 4) or choosing a value of y and using the equation to find the value of x.

Classroom Example
Use the equation $y = -3x + 1$ to complete the ordered pairs. Use the table format, as in Example 5.

$(-2, \_\_\_)$, $(0, \_\_\_)$, $(1, \_\_\_)$

Answer:

(a) $y = 7$; $(-2, 7)$

(b) $y = 1$; $(0, 1)$

(c) $y = -2$; $(1, -2)$

EXAMPLE 5 Creating a Table of Values That Satisfies an Equation

Use the equation $y = -2x + 5$ to complete Table 2 and list the ordered pairs that satisfy the equation.

Table 2

x	y	(x, y)
−2		
0		
1		

Solution

The first entry in the table is $x = -2$. We substitute -2 for x and use the equation $y = -2x + 5$ to find y.

$$\begin{aligned} & y = -2x + 5 \\ x = -2: \quad & y = -2(-2) + 5 \\ & y = 4 + 5 \\ & y = 9 \end{aligned}$$

Now substitute 0 for x in the equation $y = -2x + 5$.

$$\begin{aligned} & y = -2x + 5 \\ x = 0: \quad & y = -2(0) + 5 \\ & y = 0 + 5 \\ & y = 5 \end{aligned}$$

Finally, substitute 1 for x in the equation $y = -2x + 5$.

$$\begin{aligned} & y = -2x + 5 \\ x = 1: \quad & y = -2(1) + 5 \\ & y = -2 + 5 \\ & y = 3 \end{aligned}$$

The completed table is shown as Table 3.

Table 3

x	y	(x, y)
−2	9	(−2, 9)
0	5	(0, 5)
1	3	(1, 3)

The ordered pairs that satisfy the equation are $(-2, 9)$, $(0, 5)$, and $(1, 3)$.

QUICK ✓ *Use the equation to complete the table. Use the table to list some of the ordered pairs that satisfy the equation.*

11. $y = 5x - 2$

x	y	(x, y)
−2		
0		
1		

12. $y = -3x + 4$

x	y	(x, y)
−1		
2		
5		

Classroom Example Use the equation $3x - 2y = -6$ to complete the ordered pairs:

(a) $x = -4$ (b) $y = 6$ (c) $x = -2$

Use the table format, as in Example 6.

Answer:

(a) $y = -3$; $(-4, -3)$

(b) $x = 2$; $(2, 6)$

(c) $y = 0$; $(-2, 0)$

EXAMPLE 6 Creating a Table of Values That Satisfies an Equation

Use the equation $2x - 3y = 12$ to complete Table 4 and list the ordered pairs that satisfy the equation.

Table 4

x	y	(x, y)
−3		
	−4	
6		

Solution

The first entry in the table is $x = -3$. We substitute -3 for x and use the equation $2x - 3y = 12$ to find y.

$$2x - 3y = 12$$

$x = -3$: $$2(-3) - 3y = 12$$
$$-6 - 3y = 12$$
Add 6 to both sides: $$-3y = 18$$
Divide both sides by -3: $$y = -6$$

Substitute -4 for y in the equation $2x - 3y = 12$.

$$2x - 3y = 12$$
$y = -4$: $$2x - 3(-4) = 12$$
$$2x + 12 = 12$$
Subtract 12 from both sides: $$2x = 0$$
Divide both sides by 2: $$x = 0$$

Substitute 6 for x in the equation $2x - 3y = 12$.

$$2x - 3y = 12$$
$x = 6$: $$2(6) - 3y = 12$$
$$12 - 3y = 12$$
Subtract 12 from both sides: $$-3y = 0$$
Divide both sides by -3: $$y = 0$$

The completed table is shown in Table 5.

Table 5

x	y	(x, y)
−3	−6	(−3, −6)
0	−4	(0, −4)
6	0	(6, 0)

The ordered pairs that satisfy the equation are $(-3, -6)$, $(0, -4)$, and $(6, 0)$.

QUICK ✓ *Use the equation to complete the table. Use the table to list some of the ordered pairs that satisfy the equation.*

13. $2x + y = -8$

x	y	(x, y)
−5		
	−4	
2		

14. $2x - 5y = 18$

x	y	(x, y)
−6		
	−4	
2		

When we are working with equations that model a situation, we often do not use the variables x and y. Instead, we use variables that remind us of what they represent. For example, we might use the equation $C = 1.20m + 3$ to represent the cost of taking a taxi. Here C represents the cost (in dollars) and m represents the number of miles driven.

Recall from Section 1.2 that the scale of a number line refers to the distance between tick marks on the number line. Up to this point we have been using a scale of 1 on both the x-axis and y-axis. In applications, a different scale is often used to accomodate the ordered pairs that need to be plotted.

Classroom Example Use Example 7 with

(a) $x = 25$ kWh

(b) $x = 175$ kWh

(c) $x = 200$ kWh

Plot the ordered pairs in the rectangular coordinate system.

Answer:

(a) \$9.53

(b) \$19.51

(c) \$21.17

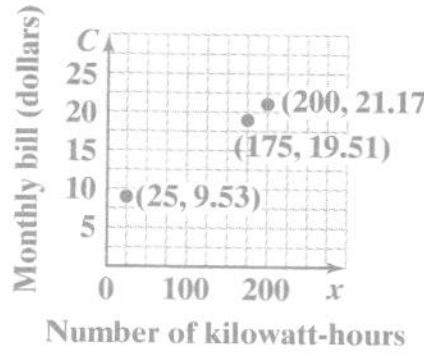

EXAMPLE 7 An Electric Bill

In North Carolina, Duke Power determines that the monthly electric bill for a household will be C dollars for using x kilowatt-hours (kWh) of electricity using the formula

$$C = 7.87 + 0.066508x$$

where $x \leq 300$ kWh. (*Source:* Duke Power)

(a) Complete Table 6 and use the table to list ordered pairs that satisfy the equation. Express answers rounded to the nearest penny.

Table 6

x (kWh)	50 kWh	100 kWh	250 kWh
C (\$)			

(b) Plot the ordered pairs (x, C) found in part (a) in a rectangular coordinate system.

Solution

(a) The first entry in the table is $x = 50$. We substitute 50 for x and use the equation $C = 7.87 + 0.066508x$ to find C. When we find C we will round our answer to two decimal places because C represents the cost, so our answer is correct to the nearest penny.

$$C = 7.87 + 0.066508x$$

$x = 50$: $\quad C = 7.87 + 0.066508(50)$

Use a calculator: $\quad C = 11.20$

Now substitute 100 for x and use the equation $C = 7.87 + 0.066508x$ to find C.

$$C = 7.87 + 0.066508x$$

$x = 100$: $\quad C = 7.87 + 0.066508(100)$

Use a calculator: $\quad C = 14.52$

Now substitute 250 for x and use the equation $C = 7.87 + 0.066508x$ to find C.

$$C = 7.87 + 0.066508x$$

$x = 250$: $\quad C = 7.87 + 0.066508(250)$

Use a calculator: $\quad C = 24.50$

Table 7 shows the completed table.

Table 7

x (kWh)	50 kWh	100 kWh	250 kWh
C (\$)	\$11.20	\$14.52	\$24.50

The ordered pairs that satisfy the equation are (50, 11.20), (100, 14.52), and (250, 24.50).

(b) Remember that x represents the number of kilowatt-hours used, so we label the horizontal axis x. Also, recall that C represents the bill, so we label the vertical axis C. Figure 7 shows a rectangular coordinate system with the ordered pairs found in part (a) plotted. Notice we use a different scale on the horizontal and vertical axis.

Figure 7

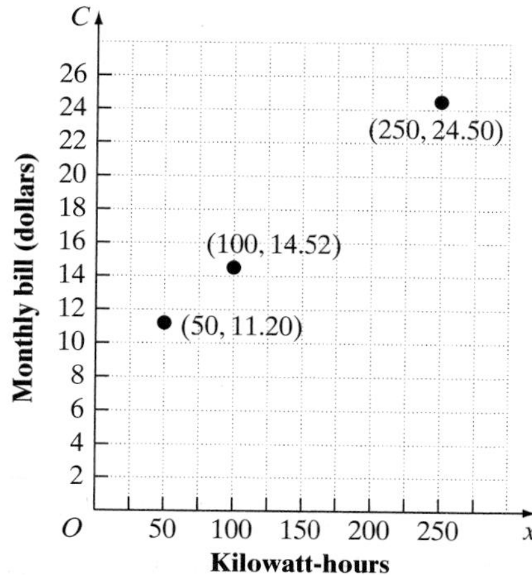

Notice in Figure 7 that we labeled the horizontal and vertical axis so that it is clear what they represent. Labeling the axes is a good practice to follow whenever you are drawing a graph.

Quick ✓

15. Cinergy Corporation charges its customers in Cincinnati, Ohio, a monthly fee C dollars for using x cubic feet of natural gas using the formula

$$C = 5.50 + 0.666652x$$

(*Source*: Cinergy Corp.)

(a) Complete the table and use the results to list ordered pairs (x, C) that satisfy the equation. Express answers rounded to the nearest penny.

x (ft³)	50 ft³	100 ft³	150 ft³
C ($)			

(b) Plot the ordered pairs found in part (a) in a rectangular coordinate system.

8.1 Exercises

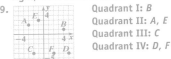

1. x-axis, y-axis, origin
2. x-coordinate, y-coordinate
3. solution
4. False
5. True
6. False
7. Answers may vary.
8. From the origin, count 3 units to the right, then down 5 units.
9. Quadrant I: B; Quadrant II: A, E; Quadrant III: C; Quadrant IV: D, F
10. Quadrant I: R; Quadrant II: S; Quadrant III: P, T; Quadrant IV: Q, U
11. Quadrant I: C, E; Quadrant III: F; Quadrant IV: B; x-axis: A, G; y-axis: D, G
12. Quadrant I: U; Quadrant III: T; Quadrant IV: P, V; x-axis: R, S; y-axis: Q, S
13. Positive x-axis: A; Negative x-axis: D; Positive y-axis: C; Negative y-axis: B
14. Positive x-axis: S; Negative x-axis: Q; Positive y-axis: R; Negative y-axis: P

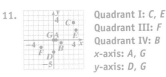

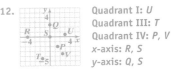

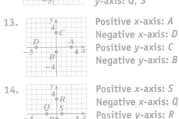

Concepts and Vocabulary

In Problems 1–3, fill in the blanks.

1. In the rectangular coordinate system, we call the horizontal real number line the __________ and we call the vertical real number line the __________. The point where these two axes intersect is called the __________.

2. If (x, y) are the coordinates of a point P, then x is called the __________ of P and y is called the __________ of P.

3. A(n) __________ to an equation in two variables x and y is an ordered pair, (x, y), that satisfies the equation.

In Problems 4–6, answer True or False to each statement.

4. The point whose ordered pair is $(-2, 4)$ is located in quadrant IV.

5. An equation in two variables can have more than one solution.

6. The ordered pairs $(3, 2)$ and $(2, 3)$ represent the same point in the Cartesian plane.

7. Describe how the quadrants in the rectangular coordinate system are labeled and how you can determine the quadrant in which a point lies. Describe the characteristics of a point that lies on either the x- or y-axis.

8. Describe how to plot the ordered pair $(3, -5)$.

Building Skills

In Problems 9–12, plot the following ordered pairs in the rectangular coordinate system. Tell which quadrant each point lies in or state that the point lies on the x-axis or y-axis.

9. $A(-3, 2)$; $B(4, 1)$; $C(-2, -4)$; $D(5, -4)$; $E(-1, 3)$; $F(2, -4)$

10. $P(-3, -2)$; $Q(2, -4)$; $R(4, 3)$; $S(-1, 4)$; $T(-2, -4)$; $U(3, -3)$

11. $A\left(\frac{1}{2}, 0\right)$; $B\left(\frac{3}{2}, -\frac{1}{2}\right)$; $C\left(4, \frac{7}{2}\right)$; $D\left(0, -\frac{5}{2}\right)$; $E\left(\frac{9}{2}, 2\right)$; $F\left(-\frac{5}{2}, -\frac{3}{2}\right)$; $G(0, 0)$

12. $P\left(\frac{3}{2}, -2\right)$; $Q\left(0, \frac{5}{2}\right)$; $R\left(-\frac{9}{2}, 0\right)$; $S(0, 0)$; $T\left(-\frac{3}{2}, -\frac{9}{2}\right)$; $U\left(3, \frac{1}{2}\right)$; $V\left(\frac{5}{2}, -\frac{7}{2}\right)$

In Problems 13 and 14, plot the following ordered pairs in the rectangular coordinate system. Tell the location of each point: positive x-axis, negative x-axis, positive y-axis, or negative y-axis.

13. $A(3, 0)$; $B(0, -1)$; $C(0, 3)$; $D(-4, 0)$

14. $P(0, -1)$; $Q(-2, 0)$; $R(0, 3)$; $S(1, 0)$

15. A(4,0); positive x-axis
B(−3, 2); quadrant II
C(1, −4); quadrant IV
D(−2, −4); quadrant III
E(3, 5); quadrant I
F(0, −3); negative y-axis
16. P(0, 2); positive y-axis
Q(−3, −1); quadrant III
R(−2, 4); quadrant II
S(2, 2); quadrant I
T(3, −2); quadrant IV
U(−5, 0); negative x-axis
17. A No B Yes C Yes
18. A Yes B No C Yes
19. A Yes B No C Yes
20. A No B No C Yes
21. A Yes B No C Yes
22. A Yes B Yes C No
23. (4, 1) 24. (2, 5) 25. (5, −1)
26. (−3, 7) 27. (−3, 3) 28. (4, 3)

29.

x	y	(x, y)
−3	3	(−3, 3)
0	0	(0, 0)
1	−1	(1, −1)

30.

x	y	(x, y)
−4	−4	(−4, −4)
0	0	(0, 0)
2	2	(2, 2)

31.

x	y	(x, y)
−2	7	(−2, 7)
−1	4	(−1, 4)
4	−11	(4, −11)

32.

x	y	(x, y)
−3	−17	(−3, −17)
1	−1	(1, −1)
2	3	(2, 3)

33.

x	y	(x, y)
−1	8	(−1, 8)
2	2	(2, 2)
3	0	(3, 0)

34.

x	y	(x, y)
−2	2	(−2, 2)
2	−1	(2, −1)
4	$-\frac{5}{2}$	$\left(4, -\frac{5}{2}\right)$

In Problems 15 and 16, identify the coordinates of each point labeled in the figure. Name the quadrant in which each point lies or state that the point lies on the x- or y-axis.

15.

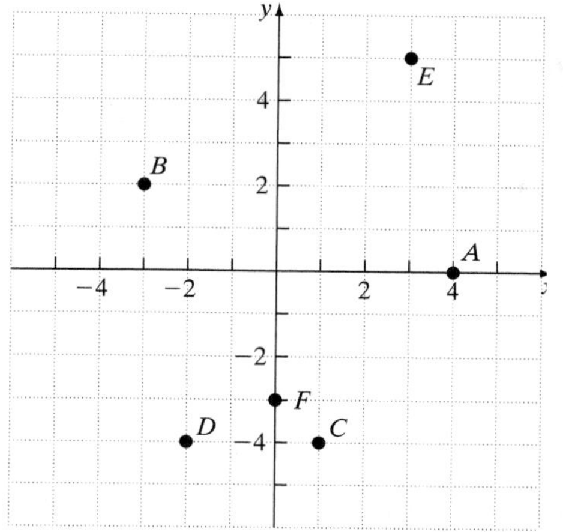

16.

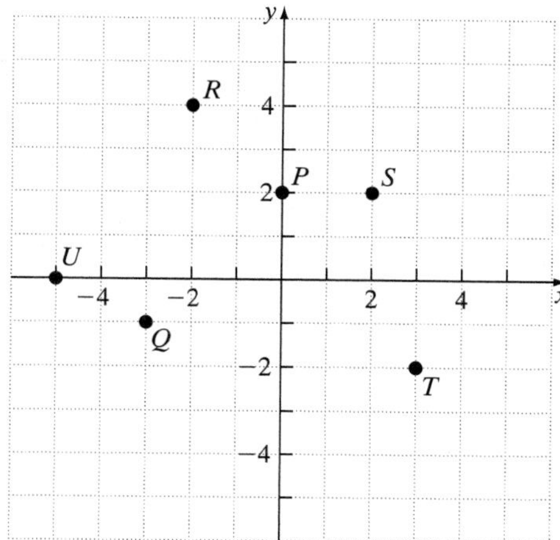

In Problems 17–22, determine whether or not the ordered pair is a solution to the equation.

17. $y = -3x + 5$
$A(-2, -1)$
$B(2, -1)$
$C\left(\frac{1}{3}, 4\right)$

18. $y = 2x - 3$
$A(-1, -5)$
$B(4, -5)$
$C(-2, -7)$

19. $3x + 2y = 4$
$A(0, 2)$
$B(1, 0)$
$C(4, -4)$

20. $5x - y = 12$
$A(2, 0)$
$B(0, 12)$
$C(-2, -22)$

21. $\frac{4}{3}x + y - 1 = 0$
$A(3, -3)$
$B(-6, -9)$
$C\left(\frac{3}{4}, 0\right)$

22. $\frac{3}{4}x + 2y = 0$
$A\left(-4, \frac{3}{2}\right)$
$B(0, 0)$
$C\left(1, -\frac{3}{2}\right)$

23. Find an ordered pair that satisfies the equation $x + y = 5$ by letting $x = 4$.

24. Find an ordered pair that satisfies the equation $x + y = 7$ by letting $x = 2$.

25. Find an ordered pair that satisfies the equation $2x + y = 9$ by letting $y = -1$.

26. Find an ordered pair that satisfies the equation $-4x - y = 5$ by letting $y = 7$.

27. Find an ordered pair that satisfies the equation $-3x + 2y = 15$ by letting $x = -3$.

28. Find an ordered pair that satisfies the equation $5x - 3y = 11$ by letting $y = 3$.

In Problems 29–42, use the equation to complete the table. Use the table to list some of the ordered pairs that satisfy the equation.

29. $y = -x$

x	y	(x, y)
−3		
0		
1		

30. $y = x$

x	y	(x, y)
−4		
0		
2		

31. $y = -3x + 1$

x	y	(x, y)
−2		
−1		
4		

32. $y = 4x - 5$

x	y	(x, y)
−3		
1		
2		

33. $2x + y = 6$

x	y	(x, y)
−1		
2		
3		

34. $3x + 4y = 2$

x	y	(x, y)
−2		
2		
4		

35. $y = 6$

x	y	(x, y)
−4		
1		
12		

36. $x = 2$

x	y	(x, y)
	−4	
	0	
	8	

37. $x - 2y + 6 = 0$

x	y	(x, y)
1		
	1	
−2		

38. $2x + y - 4 = 0$

x	y	(x, y)
−1		
	−4	
	2	

39. $y = 5 + \frac{1}{2}x$

x	y	(x, y)
	7	
−4		
	2	

40. $y = 8 - \frac{1}{3}x$

x	y	(x, y)
	10	
9		
	27	

41. $\frac{x}{2} + \frac{y}{3} = -1$

x	y	(x, y)
0		
	0	
	−6	

42. $\frac{x}{5} - \frac{y}{2} = 1$

x	y	(x, y)
	0	
0		
−5		

In Problems 43–54, for each equation find the missing value in the ordered pair.

43. $y = -3x - 10$
$A(\_\_\_, -16)$
$B(-3, \_\_\_)$
$C(\_\_\_, -9)$

44. $y = 5x - 4$
$A(-1, \_\_\_)$
$B(\_\_\_, 31)$
$C\left(-\frac{2}{5}, \_\_\_\right)$

45. $x = -\frac{1}{3}y$
$A(2, \_\_\_)$
$B(\_\_\_, 0)$
$C\left(\_\_\_, -\frac{1}{2}\right)$

46. $y = \frac{2}{3}x$
$A\left(\_\_\_, -\frac{8}{3}\right)$
$B(0, \_\_\_)$
$C\left(-\frac{5}{6}, \_\_\_\right)$

35.

x	y	(x, y)
−4	6	(−4, 6)
1	6	(1, 6)
12	6	(12, 6)

36.

x	y	(x, y)
2	−4	(2, −4)
2	0	(2, 0)
2	8	(2, 8)

37.

x	y	(x, y)
1	$\frac{7}{2}$	$\left(1, \frac{7}{2}\right)$
−4	1	(−4, 1)
−2	2	(−2, 2)

38.

x	y	(x, y)
−1	6	(−1, 6)
4	−4	(4, −4)
1	2	(1, 2)

39.

x	y	(x, y)
4	7	(4, 7)
−4	3	(−4, 3)
−6	2	(−6, 2)

40.

x	y	(x, y)
−6	10	(−6, 10)
9	5	(9, 5)
−57	27	(−57, 27)

41.

x	y	(x, y)
0	−3	(0, −3)
−2	0	(−2, 0)
2	−6	(2, −6)

42.

x	y	(x, y)
5	0	(5, 0)
0	−2	(0, −2)
−5	−4	(5, −4)

43. $A(2, -16)$ $B(-3, -1)$ $C\left(-\frac{1}{3}, -9\right)$

44. $A(-1, -9)$ $B(7, 31)$ $C\left(-\frac{2}{5}, -6\right)$

45. $A(2, -6)$ $B(0, 0)$ $C\left(\frac{1}{6}, -\frac{1}{2}\right)$

46. $A\left(-4, -\frac{8}{3}\right)$ $B(0, 0)$ $C\left(-\frac{5}{6}, -\frac{5}{9}\right)$

47. $A(4, -8)$ $B(4, -19)$ $C(4, 5)$
48. $A(6, -1)$ $B(-1, -1)$ $C(0, -1)$
49. $A(3, 4)$ $B(-6, -2)$ $C\left(\frac{1}{2}, \frac{7}{3}\right)$
50. $A(4, -6)$ $B(-8, 9)$ $C\left(\frac{2}{3}, -\frac{11}{6}\right)$
51. $A\left(-4, -\frac{4}{3}\right)$ $B(-2, -1)$ $C\left(-\frac{2}{3}, -\frac{7}{9}\right)$
52. $A\left(-4, \frac{1}{6}\right)$ $B\left(\frac{3}{2}, -\frac{3}{4}\right)$ $C\left(0, -\frac{1}{2}\right)$
53. $A(20, 23)$ $B(-4, -17)$ $C(2.6, -6)$
54. $A(-10, -110)$ $B(40, 315)$ $C(2.4, -4.6)$
55. (a) \$24.85 (b) \$54.70 (c) 6 CDs (d) It costs \$34.80 to order 3 CDs.
56. (a) \$11.70 (b) \$41.70 (c) 15.5 mi (d) It costs \$29.70 to take a taxi 14 miles.
57. (a) \$1.15 (b) \$1.23 (c) \$1.60 (d) 1996 (e) Answers may vary.
58. (a) 67.895 or 68 yr (b) 73.385 or 73 yr (c) 80.705 or 81 yr (d) 2000 (e) Answers may vary.
59. $k = 4$
60. $k = 13$
61. $k = 2$
62. $k = 3$
63. $k = \frac{1}{2}$
64. $k = \frac{1}{3}$

65.

a	b	(a, b)
2	−8	(2, −8)
0	−4	(0, −4)
−5	6	(−5, 6)

66.

r	s	(r, s)
6	3	(6, 3)
0	−1	(0, −1)
3	−3	(3, −3)

47. $x = 4$
$A(\_\_\_, -8)$
$B(\_\_\_, -19)$
$C(\_\_\_, 5)$

48. $y = -1$
$A(6, \_\_\_)$
$B(-1, \_\_\_)$
$C(0, \_\_\_)$

49. $y = \frac{2}{3}x + 2$
$A(\_\_\_, 4)$
$B(-6, \_\_\_)$
$C\left(\frac{1}{2}, \_\_\_\right)$

50. $y = -\frac{5}{4}x - 1$
$A(\_\_\_, -6)$
$B(-8, \_\_\_)$
$C\left(\_\_\_, -\frac{11}{6}\right)$

51. $\frac{1}{2}x - 3y = 2$
$A(-4, \_\_\_)$
$B(\_\_\_, -1)$
$C\left(-\frac{2}{3}, \_\_\_\right)$

52. $\frac{1}{3}x + 2y = -1$
$A(-4, \_\_\_)$
$B\left(\_\_\_, -\frac{3}{4}\right)$
$C(0, \_\_\_)$

53. $0.5x - 0.3y = 3.1$
$A(20, \_\_\_)$
$B(\_\_\_, -17)$
$C(2.6, \_\_\_)$

54. $-1.7x + 0.2y = -5$
$A(\_\_\_, -110)$
$B(40, \_\_\_)$
$C(2.4, \_\_\_)$

Applying the Concepts

55. Mail-Order CDs The cost of ordering CDs from a mail-order house is \$9.95 per CD plus \$4.95 for shipping and handling per order. The equation $C = 9.95n + 4.95$ represents the total cost, C, of ordering n CDs.

(a) How much will it cost to order 2 CDs?
(b) How much will it cost to order 5 CDs?
(c) If you have \$64.65 to spend, how many CDs can you order?
(d) If (n, C) represents any ordered pair that satisfies $C = 9.95n + 4.95$, interpret the meaning of $(3, 34.8)$ in the context of this problem.

56. Taxi Ride The cost to take a taxi is \$1.70 plus \$2.00 per mile for each mile driven. The total cost, C, is given by the equation $C = 1.7 + 2m$, where m represents the total miles driven.

(a) How much will it cost to take a taxi 5 miles?
(b) How much will it cost to take a taxi 20 miles?
(c) If you spent \$32.70 on cab fare, how far was your trip?
(d) If (m, C) represents any ordered pair that satisfies $C = 1.7 + 2m$, interpret the meaning of $(14, 29.7)$ in the context of this problem.

57. Cost of a Gallon of Gasoline An equation to approximate the future cost of a gallon of gasoline can be given by the model $C = \$0.0375n + \$1.15$, where n is the number of years after 1992.

(a) According to the model, what was the cost of a gallon of gas in 1992 ($n = 0$)?
(b) According to the model, what was the approximate cost of a gallon of gas in 1994 ($n = 2$)?
(c) According to the model, what was the approximate cost of a gallon of gas in 2004 ($n = 12$)?
(d) In what year was the cost of a gallon of gas \$1.30?
(e) The price will not increase at a constant rate forever. What factors do you think will affect the cost of gasoline in the future?

58. Life Expectancy The model $A = 0.183n + 67.895$ is used to estimate the life expectancy A of residents of the United States n years after 1950.

(a) According to the model, what is the life expectancy for a person born in 1950?
(b) According to the model, what is the life expectancy for a person born in 1980 ($n = 30$)?

67.

p	q	(p, q)
0	$\frac{10}{3}$	$\left(0, \frac{10}{3}\right)$
$\frac{5}{2}$	0	$\left(\frac{5}{2}, 0\right)$
-10	$\frac{50}{3}$	$\left(-10, \frac{50}{3}\right)$

68.

a	b	(a, b)
$-\frac{3}{4}$	0	$\left(-\frac{3}{4}, 0\right)$
0	$-\frac{5}{2}$	$\left(0, -\frac{5}{2}\right)$
$-\frac{15}{4}$	10	$\left(-\frac{15}{4}, 10\right)$

69. Points may vary; line

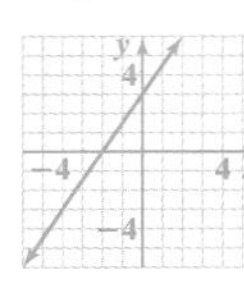

70. Points may vary; line

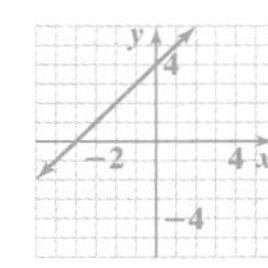

71.

x	y	(x, y)
-2	0	$(-2, 0)$
-1	-3	$(-1, -3)$
0	-4	$(0, -4)$
1	-3	$(1, -3)$
2	0	$(2, 0)$

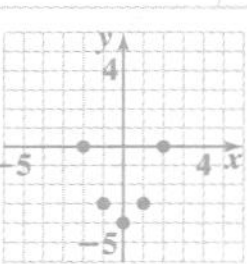

72.

x	y	(x, y)
-2	-1	$(-2, -1)$
-1	2	$(-1, 2)$
0	3	$(0, 3)$
1	2	$(1, 2)$
2	-1	$(2, -1)$

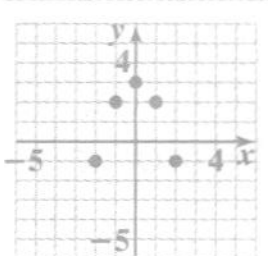

73.

x	y	(x, y)
-2	10	$(-2, 10)$
-1	3	$(-1, 3)$
0	2	$(0, 2)$
1	1	$(1, 1)$
2	-6	$(2, -6)$

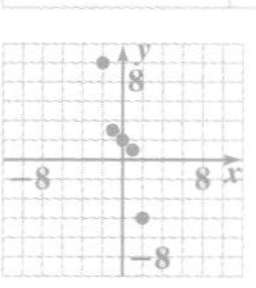

(c) If the model holds true for future generations, what is the life expectancy for a person born in 2020?

(d) If a person has a life expectancy of 77 years, to the nearest year, when was the person born?

(e) Do you think life expectancy will continue to increase in the future? What could happen that would change this model?

In Problems 59–64, determine the value of k so that the given ordered pair is a solution to the equation.

59. Find the value of k for which $(1, 2)$ is a solution of $y = -2x + k$.

60. Find the value of k for which $(-1, 10)$ is a solution of $y = 3x + k$.

61. Find the value of k for which $(2, 9)$ is a solution of $7x - ky = -4$.

62. Find the value of k for which $(3, -1)$ is a solution of $4x + ky = 9$.

63. Find the value of k for which $\left(-8, -\frac{5}{2}\right)$ is a solution of $kx - 4y = 6$.

64. Find the value of k for which $\left(-9, \frac{1}{2}\right)$ is a solution of $kx + 2y = -2$.

In Problems 65–68, use the equation to complete the table. Use the table to list some of the ordered pairs that satisfy the equation.

65. $4a + 2b = -8$

a	b	(a, b)
2		
	-4	
	6	

66. $2r - 3s = 3$

r	s	(r, s)
	3	
	-1	
-3		

67. $\dfrac{2p}{5} + \dfrac{3q}{10} = 1$

p	q	(p, q)
0		
	0	
-10		

68. $\dfrac{4a}{3} + \dfrac{2b}{5} = -1$

a	b	(a, b)
	0	
0		
	10	

Extending the Concepts

In Problems 69 and 70, use the equation to complete the table. Choose any value for x and then solve the resulting equation to find the corresponding value for y. Then plot these ordered pairs in a rectangular coordinate system. Connect the points and describe the figure.

69. $3x - 2y = -6$

x	y	(x, y)

70. $-x + y = 4$

x	y	(x, y)

74.

x	y	(x, y)
−2	−17	(−2, −17)
−1	−3	(−1, −3)
0	−1	(0, −1)
1	1	(1, 1)
2	15	(2, 15)

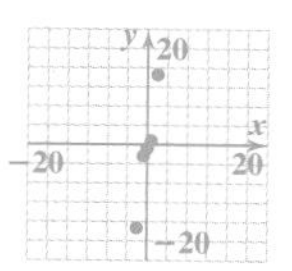

75.

X	Y1
-3	-15
-2	-13
-1	-11
0	-9
1	-7
2	-5
3	-3

Y1 = 2X−9

76.

X	Y1
-3	17
-2	14
-1	11
0	8
1	5
2	2
3	-1

Y1 = -3X+8

77.

X	Y1
-3	11
-2	10
-1	9
0	8
1	7
2	6
3	5

Y1 = -X+8

78.

X	Y1
-3	-10
-2	-8
-1	-6
0	-4
1	-2
2	0
3	2

Y1 = 2X−4

79.

X	Y1
-3	19
-2	17
-1	15
0	13
1	11
2	9
3	7

Y1 = 13−2X

80.

X	Y1
-3	-18
-2	-17
-1	-16
0	-15
1	-14
2	-13
3	-12

Y1 = -15+X

81.

X	Y1
-3	-53
-2	-23
-1	-5
0	1
1	-5
2	-23
3	-53

Y1 = -6X²+1

82.

X	Y1
-3	-18
-2	-10
-1	-4
0	0
1	2
2	2
3	0

Y1 = -X²+3X

In Problems 71–74, use the equation to complete the table. Then plot the points in a rectangular coordinate system.

71. $y = x^2 - 4$

x	y	(x, y)
−2		
−1		
0		
1		
2		

72. $y = -x^2 + 3$

x	y	(x, y)
−2		
−1		
0		
1		
2		

73. $y = -x^3 + 2$

x	y	(x, y)
−2		
−1		
0		
1		
2		

74. $y = 2x^3 - 1$

x	y	(x, y)
−2		
−1		
0		
1		
2		

The Graphing Calculator

Graphing calculators can also create tables of values that satisfy an equation. To do this, we first solve the equation for y. For example, to obtain a table of values that satisfy the equation $2x - 3y = 12$ (Example 6), we solve for y as follows:

$$2x - 3y = 12$$

Subtract $2x$ from both sides: $$-3y = -2x + 12$$

Divide both sides by −3: $$y = \frac{-2x + 12}{-3}$$

Divide −3 into each term in the numerator: $$y = \frac{-2x}{-3} + \frac{12}{-3}$$

Simplify: $$y = \frac{2}{3}x - 4$$

We now enter the equation $y = \frac{2}{3}x - 4$ into the calculator and create the table shown below.

X	Y1
-3	-6
-2	-5.333
-1	-4.667
0	-4
1	-3.333
2	-2.667
3	-2

Y1 = (2/3)X−4

In Problems 75–82, use a graphing calculator to create a table of values that satisfy each equation. Have the table begin at −3 and increase by 1.

75. $y = 2x - 9$ **76.** $y = -3x + 8$ **77.** $y = -x + 8$ **78.** $y = 2x - 4$

79. $y + 2x = 13$ **80.** $y - x = -15$ **81.** $y = -6x^2 + 1$ **82.** $y = -x^2 + 3x$

8.2 Graphing Equations in Two Variables

OBJECTIVES

1. Graph a Line by Plotting Points
2. Graph a Line Using Intercepts
3. Graph Vertical and Horizontal Lines

Preparing for Graphing Equations in Two Variables

Before getting started, take the following readiness quiz. If you get a problem wrong, go back to the section cited and review the material.

1. Solve: $4x = 24$ [Section 2.1, pp. 78–80]

2. Solve: $-3y = 18$ [Section 2.1, pp. 78–80]

3. Solve: $2x + 5 = 13$ [Section 2.2, pp. 84–86]

1 Graph a Line by Plotting Points

In Words
The graph of an equation is a geometric way of representing the set of all ordered pairs that make the equation a true statement. Think of the graph as a picture of the solution set.

In the previous section, we found values of x and y that satisfy an equation. What does this mean? Well, it means that the ordered pair (x, y) is a point on the graph of the equation.

DEFINITION
The **graph of an equation in two variables** x and y is the set of points (x, y) in the xy-plane whose coordinates (x, y) satisfy the equation.

But how do we obtain the graph of an equation? One method for graphing an equation is the **point-plotting method.**

EXAMPLE 1 How to Graph an Equation Using the Point-Plotting Method

Graph the equation $y = 2x - 3$ using the point-plotting method.

Step-by-Step Solution

Step 1: We find the ordered pairs that satisfy the equation by choosing some values of x and using the equation to find the corresponding values of y. See Table 8.

Classroom Example
Graph the equation $y = 3x + 2$ using the point-plotting method. Use the values $x = -2, -1, 0, 1, 2$ to develop a table, as in Example 1.

Answer: $(-2, -4)$, $(-1, -1)$, $(0, 2)$, $(1, 5)$, $(2, 8)$

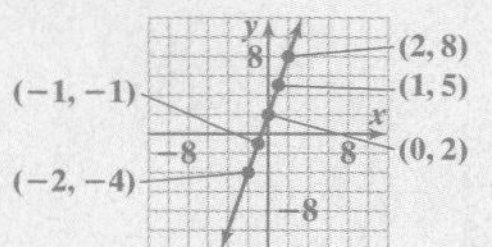

Table 8

x	y	(x, y)
-2	$y = 2(-2) - 3$ $= -4 - 3$ $= -7$	$(-2, -7)$
-1	$y = 2(-1) - 3$ $= -5$	$(-1, -5)$
0	$y = 2(0) - 3$ $= -3$	$(0, -3)$
1	$y = 2(1) - 3$ $= -1$	$(1, -1)$
2	$y = 2(2) - 3$ $= 1$	$(2, 1)$

Preparing for...Answers
1. $\{6\}$ **2.** $\{-6\}$ **3.** $\{4\}$

Step 2: Plot the points found in Step 1 in a rectangular coordinate system. See Figure 8.

Figure 8

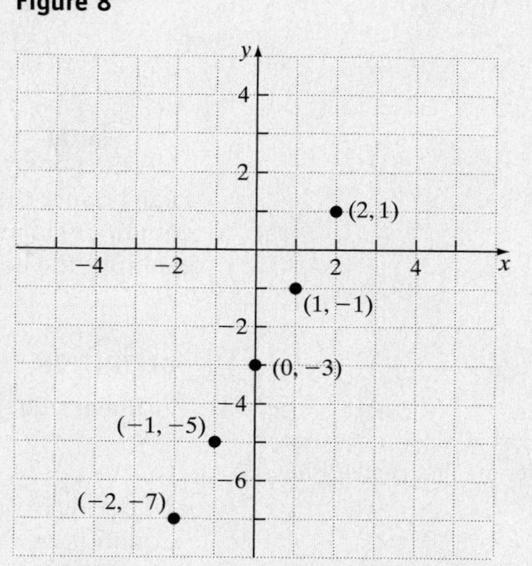

Step 3: Connect the points in a straight line. See Figure 9.

Figure 9

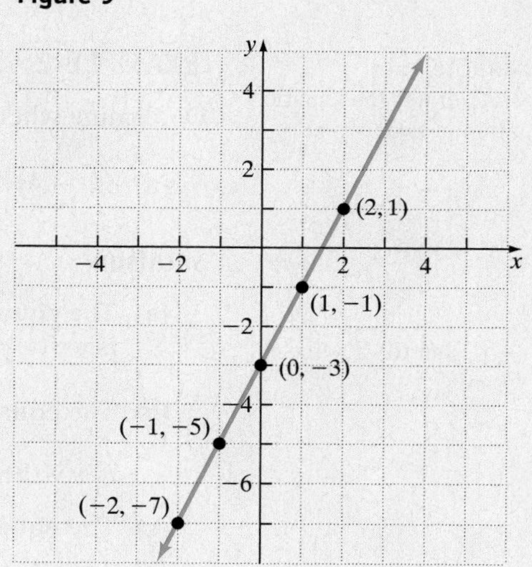

Work Smart

Remember, all coordinates of the points on the graph shown in Figure 9 satisfy the equation $y = 2x - 3$.

The graph of the equation shown in Figure 9 does not show all the points that satisfy the equation. For example, in Figure 9 the point (5, 7) is a part of the graph of $y = 2x - 3$, but it is not shown. Since the graph of $y = 2x - 3$ could be extended as far as we please, we use arrows to indicate that the pattern shown continues. It is important to show enough of the graph so that anyone who is looking at it will "see" the rest of it as an obvious continuation of what is there. This is called a **complete graph.**

We summarize the steps for graphing an equation using the point-plotting method below.

Teaching Tip

Emphasize that the graph of an equation does not show all the solutions of an equation in two variables. We use arrows to indicate that the pattern shown continues.

Graphing an Equation Using the Point-Plotting Method

Step 1: Find several ordered pairs that satisfy the equation.

Step 2: Plot the points found in Step 1 in a rectangular coordinate system.

Step 3: Connect the points in a smooth curve or line.

QUICK ✓ *Draw a complete graph of each equation using point plotting.*

1. $y = 3x - 2$ **2.** $y = -4x + 8$

A question you may be asking yourself is, "How many points do I need to find before I can be sure that I have a complete graph?" The answer is that it depends on the type of equation you are graphing. In mathematics, we classify equations as different types. For example, the equation that we graphed in Example 1 is called a *linear equation*.

Teaching Tip
Draw on students' knowledge of degree. Both variables in a linear equation in two variables are first degree.

DEFINITION

A **linear equation in two variables** is an equation of the form

$$Ax + By = C$$

where A, B, and C are real numbers. A and B cannot both be 0. When a linear equation is written in the form $Ax + By = C$, we say that the linear equation is in **standard form.**

Classroom Example ➤
Determine whether or not the equation is a linear equation in two variables.
(a) $2x + y = 7$
(b) $\frac{5}{3}x - \frac{2}{5}y = 1$
(c) $x + 2y^2 = 3$
(d) $-4x = 9$
Answer: (a), (b), and (d) are linear equations; (c) is not

EXAMPLE 2 Identifying Linear Equations in Two Variables

Determine whether or not the equation is a linear equation in two variables.

(a) $3x - 4y = 9$ **(b)** $\frac{1}{2}x + \frac{2}{3}y = 4$ **(c)** $x^2 + 5y = 10$ **(d)** $-2y = 5$

Solution

(a) The equation $3x - 4y = 9$ is a linear equation in two variables because it is written in the form $Ax + By = C$ with $A = 3$, $B = -4$, and $C = 9$.

(b) The equation $\frac{1}{2}x + \frac{2}{3}y = 4$ is a linear equation in two variables because it is written in the form $Ax + By = C$ with $A = \frac{1}{2}$, $B = \frac{2}{3}$, and $C = 4$.

(c) The equation $x^2 + 5y = 10$ is not a linear equation because x is squared.

(d) The equation $-2y = 5$ is a linear equation in two variables because it is written in the form $Ax + By = C$ with $A = 0$, $B = -2$, and $C = 5$.

QUICK ✓ *Determine whether or not the equation is a linear equation in two variables.*

3. $4x - y = 12$ **4.** $5x - y^2 = 10$ **5.** $\frac{1}{2}x + \frac{2}{3}y = 4$ **6.** $5x = 20$

Work Smart
When graphing a line, be sure to find three points—just to be safe!

For the remainder of the text, we will refer to linear equations in two variables as **linear equations.** The graph of a linear equation is a **line.** To graph a linear equation requires only two points; however, we recommend that you find a third point as a check.

Classroom Example ➤
Graph the linear equation $2x + y = 1$.
Answer:

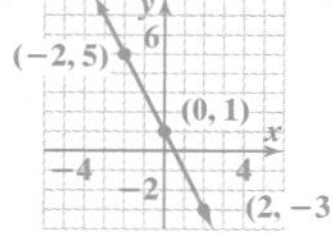

EXAMPLE 3 Graphing a Linear Equation Using the Point-Plotting Method

Graph the linear equation $2x + y = 4$.

Work Smart

Choose values of x (or y) that make the algebra easy.

Solution

We need to find ordered pairs that satisfy the equation. Because the coefficient of y is 1, it is easier to choose values of x and find the corresponding values of y. We will determine the value of y for $x = -2, 0,$ and 2. There is nothing magical about these choices. Any three different values of x will give us the results we want.

$x = -2$:

$$2x + y = 4$$

Let $x = -2$: $2(-2) + y = 4$

$$-4 + y = 4$$

Add 4 to both sides: $y = 8$

$x = 0$:

$$2x + y = 4$$

Let $x = 0$: $2(0) + y = 4$

$$y = 4$$

$x = 2$:

$$2x + y = 4$$

Let $x = 2$: $2(2) + y = 4$

$$4 + y = 4$$

Subtract 4 from both sides: $y = 0$

Table 9 summarizes the results. The ordered pairs $(-2, 8)$, $(0, 4)$, and $(2, 0)$ represent points that are on the graph of the equation. We plot these points in Figure 10(a). After connecting the points in a straight line we obtain the graph in Figure 10(b).

Table 9

x	y	(x, y)
−2	8	(−2, 8)
0	4	(0, 4)
2	0	(2, 0)

Figure 10

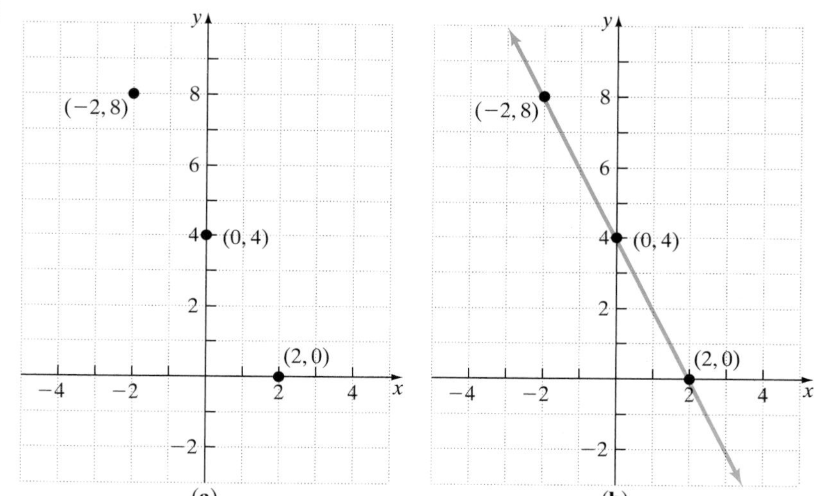

QUICK ✓ *Graph each linear equation using point plotting.*

7. $-3x + y = -6$

8. $2x + 3y = 12$

Classroom Example

Use Example 4 with $m = 0$, 75, and 150.

Answer: (0, 30), (75, 45), (150, 60)

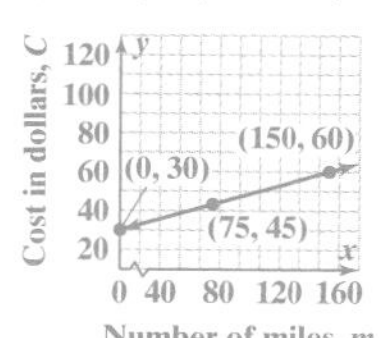

EXAMPLE 4 Cost of Renting a Car

Your favorite car-rental agency quotes you the cost of renting a car in Washington, D.C., as \$30 per day plus \$0.20 per mile. The linear equation $C = 0.20m + 30$ models the cost, where C represents total cost and m represents the number of miles that were traveled.

(a) Complete Table 10 and use the results to list ordered pairs that satisfy the equation. Express answers rounded to the nearest penny.

(b) Graph the linear equation $C = 0.20m + 30$ using the points obtained in part (a).

Table 10

m	C	(m, C)
0		
50		
100		

Teaching Tip
Tell students that the variable m "acts like" x and the variable C "acts like" y in setting up ordered pair solutions.

Solution

(a) The first entry in the table is $m = 0$. We substitute 0 for m and use the equation $C = 0.20m + 30$ to find C.

$$C = 0.20m + 30$$
$$m = 0: \quad C = 0.20(0) + 30$$
$$C = 30$$

Now substitute 50 for m and use the equation $C = 0.20m + 30$ to find C.

$$C = 0.20m + 30$$
$$m = 50: \quad C = 0.20(50) + 30$$
$$C = 40$$

Now substitute 100 for m and use the equation $C = 0.20m + 30$ to find C.

$$C = 0.20m + 30$$
$$m = 100: \quad C = 0.20(100) + 30$$
$$C = 50$$

Table 11

m	C	(m, C)
0	30	(0, 30)
50	40	(50, 40)
100	50	(100, 50)

Table 11 shows the completed table.

The ordered pairs that satisfy the equation are (0, 30), (50, 40), and (100, 50). The ordered pair (50, 40) means if a car is driven 50 miles, the cost of renting the car will be \$40.

(b) Remember that m represents the number miles driven, so we label the horizontal axis m. Also, recall that C represents the cost of renting the car, so we label the vertical axis C. When drawing the horizontal axis, we set the scale to 10, which means each tick mark represents 10 miles. We scale the vertical axis to 5, which means each tick mark represents 5 dollars. Scaling in this way makes it easier to plot the ordered pairs. We plot the points found in part (a) in the rectangular coordinate system and then draw the line. See Figure 11.

Teaching Tip
Ask students to explain why we do not graph in quadrants II or III for the graph shown in Figure 11.

Figure 11

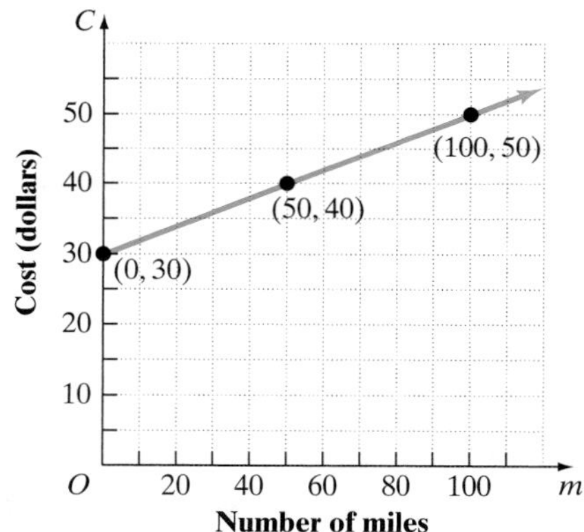

QUICK ✓

9. Michelle sells computers. Her monthly salary is \$3000 plus 8% of total sales. The linear equation $S = 0.08x + 3000$ models Michelle's monthly salary, S, where x represents her total sales in the month.

(a) Complete the table and use the results to list ordered pairs that satisfy the equation. Express answers rounded to the nearest penny.

x	S	(x, S)
0		
10,000		
25,000		

(b) Graph the linear equation $S = 0.08x + 3000$ using the points obtained in part (a).

2 Graph a Line Using Intercepts

Intercepts should always be displayed in a complete graph.

DEFINITIONS

The **intercepts** are the points, if any, where a graph crosses or touches the coordinate axes. The x-coordinate of a point at which the graph crosses or touches the x-axis is an ***x*-intercept,** and the y-coordinate of a point at which the graph crosses or touches the y-axis is a ***y*-intercept.**

Figure 12

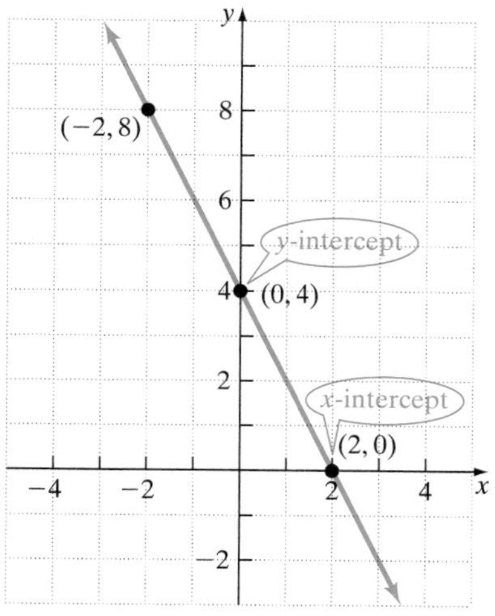

In Words

An x-intercept exists when $y = 0$.
A y-intercept exists when $x = 0$.

See Figure 12 for an illustration. The graph in Figure 12 is the graph obtained in Example 3.

EXAMPLE 5 Finding Intercepts from a Graph

Find the intercepts of the graphs shown in Figures 13(a) and 13(b). What are the x-intercepts? What are the y-intercepts?

Figure 13

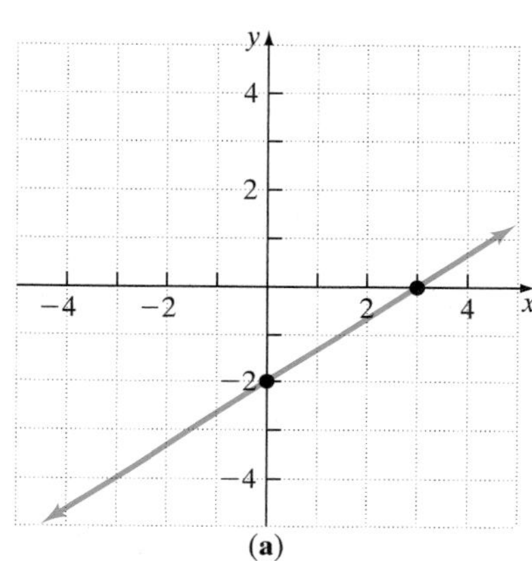
(a)

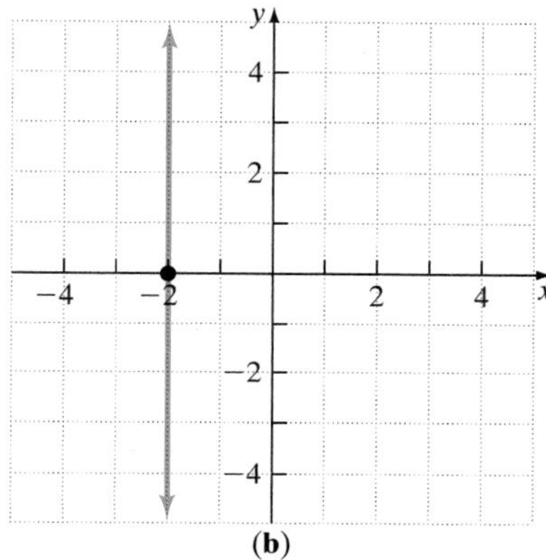
(b)

Classroom Example
Find the intercepts of the graphs shown. What are the x-intercepts? What are the y-intercepts?

Answer:

(a) $(0, -3)$, $(2, 0)$; x-intercept $= 2$, y-intercept $= -3$

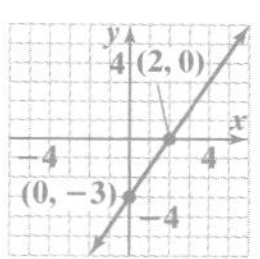

(b) $(0, 3)$; no x-intercept; y-intercept $= 3$

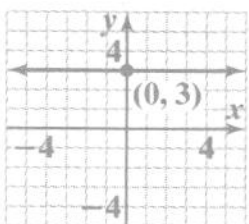

Solution

(a) The intercepts of the graph in Figure 13(a) are the points $(0, -2)$ and $(3, 0)$. The x-intercept is 3. The y-intercept is -2.

(b) The intercept of the graph in Figure 13(b) is the point $(-2, 0)$. The x-intercept is -2. There are no y-intercepts.

In Example 5, you should notice the following: If we do not specify the type of intercept (x-intercept versus y-intercept), we report the intercept as an ordered pair. However, if we specify the type of intercept, then we only need report the coordinate of the intercept. For example, we would say $(4, 0)$ is an intercept, while we would say 4 is an x-intercept.

QUICK ✓ *Find the intercepts of the graph shown in the figure. What are the x-intercepts? What are the y-intercepts?*

10.

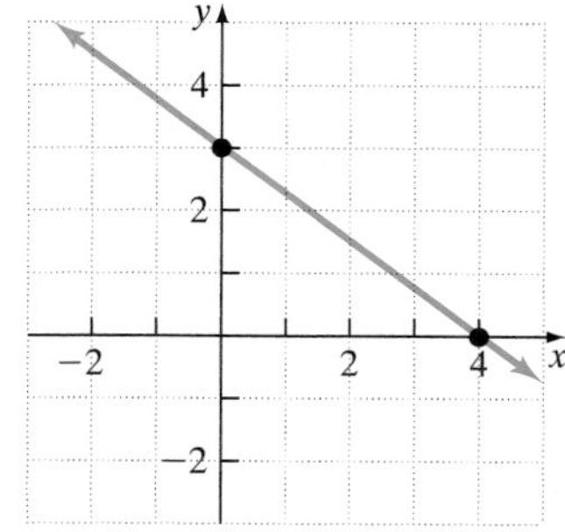

11.

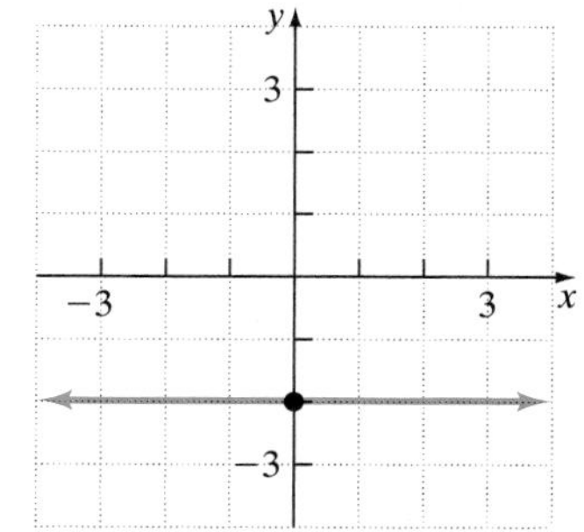

Teaching Tip
Emphasize the content of the following Work Smart. Students may want to just memorize the rule instead of understanding it.

Work Smart
Every point on the x-axis has a y-coordinate of 0. That's why we set $y = 0$ to find the x-intercept. Likewise, every point on the y-axis has an x-coordinate of 0. That's why we set $x = 0$ to find the y-intercept.

Now we will explain how to find the intercepts algebraically. From Figure 12 it should be apparent that an x-intercept exists when the value of y is 0 and that a y-intercept exists when the value of x is 0. This leads to the following procedure for finding intercepts.

Procedure for Finding Intercepts

1. To find the x-intercept(s), if any, of the graph of an equation, let $y = 0$ in the equation and solve for x.
2. To find the y-intercept(s), if any, of the graph of an equation, let $x = 0$ in the equation and solve for y.

Because this method for graphing a linear equation results in only two points, we find a third point so that we can check our work.

EXAMPLE 6 How to Graph a Linear Equation by Finding Its Intercepts

Graph the linear equation $4x - 3y = 24$ by finding its intercepts.

Classroom Example
Graph the linear equation $2x - 3y = 6$ by finding its intercepts.

Answer: x-intercept $= 3$; y-intercept $= -2$; additional point $(-3, -4)$

Step-by-Step Solution

Step 1: Find the y-intercept by letting $x = 0$ and solving the equation for y.

$$4x - 3y = 24$$
Let $x = 0$: $$4(0) - 3y = 24$$
$$0 - 3y = 24$$
$$-3y = 24$$
Divide both sides by -3: $$y = -8$$

The y-intercept is -8, so the point $(0, -8)$ is on the graph of the equation.

Step 2: Find the x-intercept by letting $y = 0$ and solving the equation for x.

$$4x - 3y = 24$$
Let $y = 0$: $$4x - 3(0) = 24$$
$$4x - 0 = 24$$
$$4x = 24$$
Divide both sides by 4: $$x = 6$$

The x-intercept is 6, so the point $(6, 0)$ is on the graph of the equation.

Step 3: Find one additional point on the graph by choosing any value of x that is convenient and solving the equation for y.

We will let $x = 3$ and solve the equation $4x - 3y = 24$ for y.

Let $x = 3$: $$4(3) - 3y = 24$$
$$12 - 3y = 24$$
Subtract 12 from both sides: $$-3y = 12$$
Divide both sides by -3: $$y = -4$$

The point $(3, -4)$ is on the graph of the equation.

Step 4: Plot the points found in Steps 1–3 and draw in the line.

We plot the points $(0, -8)$, $(6, 0)$, and $(3, -4)$. Connect the points in a straight line and obtain the graph in Figure 14.

Figure 14

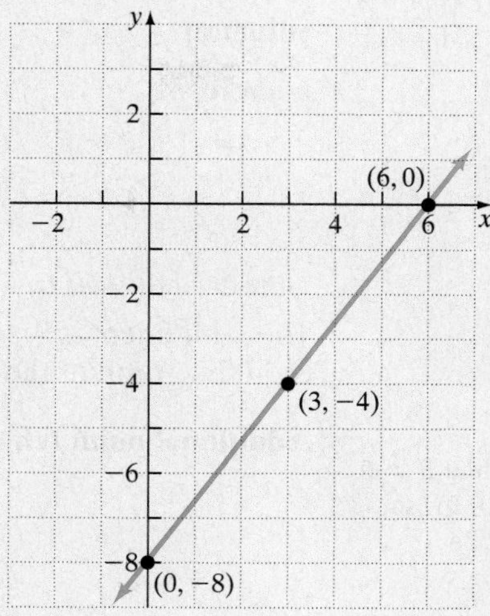

Classroom Example
Graph the linear equation $2x - \frac{1}{2}y = 3$ by finding its intercepts.

Answer: x-intercept $= \frac{3}{2}$; y-intercept $= -6$; additional point $(2, 2)$

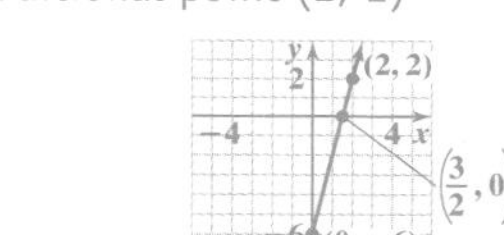

EXAMPLE 7 Graphing a Linear Equation by Finding Its Intercepts

Graph the linear equation $\frac{1}{2}x - 2y = 3$ by finding its intercepts.

Solution

***x*-intercept:**

$$\frac{1}{2}x - 2y = 3$$

Let $y = 0$: $\frac{1}{2}x - 2(0) = 3$

$$\frac{1}{2}x = 3$$

Multiply both sides by 2: $x = 6$

***y*-intercept:**

$$\frac{1}{2}x - 2y = 3$$

Let $x = 0$: $\frac{1}{2}(0) - 2y = 3$

$$-2y = 3$$

Divide both sides by -2: $y = -\frac{3}{2}$

Additional point (choose $x = 2$):

$$\frac{1}{2}x - 2y = 3$$

Let $x = 2$: $\frac{1}{2}(2) - 2y = 3$

$$1 - 2y = 3$$

Subtract 1 from both sides: $-2y = 2$

Divide both sides by -2: $y = -1$

Plot the points $(6, 0)$, $\left(0, -\frac{3}{2}\right)$, and $(2, -1)$. Connect the points in a straight line. See Figure 15.

Figure 15

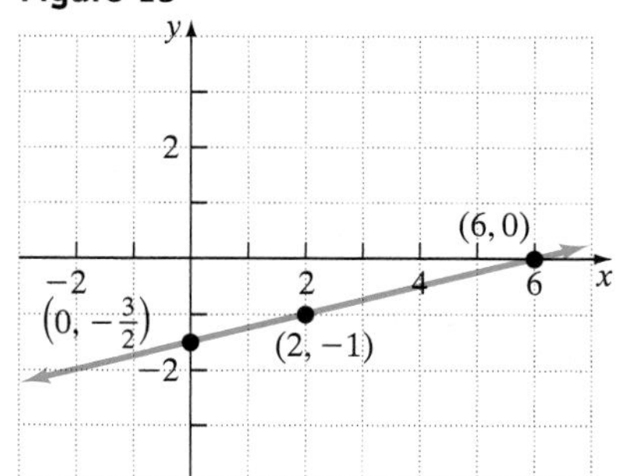

QUICK ✓ *Graph each linear equation by finding its intercepts.*

12. $x + y = 3$ **13.** $2x - 5y = 20$ **14.** $-3x + 4y = 9$ **15.** $\frac{3}{2}x - 2y = 9$

Classroom Example Graph the linear equation $3x + 4y = 0$ by finding its intercepts.

Answer: x-intercept = y-intercept = 0; use $(-4, 3)$ and $(4, -3)$ as additional points.

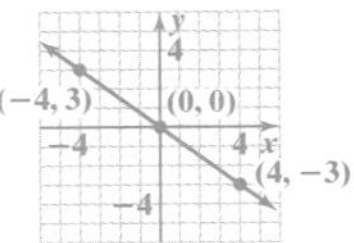

EXAMPLE 8 Graphing a Linear Equation of the Form $Ax + By = 0$

Graph the linear equation $2x + 3y = 0$ by finding its intercepts.

Solution

x-intercept:

$$2x + 3y = 0$$

Let $y = 0$: $2x + 3(0) = 0$

$$2x + 0 = 0$$

Divide both sides by 2: $x = 0$

The x-intercept is 0, so the point $(0, 0)$ is on the graph of the equation.

y-intercept:

$$2x + 3y = 0$$

Let $x = 0$: $2(0) + 3y = 0$

$$3y = 0$$

Divide both sides by 3: $y = 0$

The y-intercept is 0, so the point $(0, 0)$ is on the graph of the equation.

Work Smart

Linear equations of the form $Ax + By = 0$, where $A \neq 0$ and $B \neq 0$ have only one intercept at $(0, 0)$, so two additional points should be plotted to obtain the graph.

Additional point (choose $x = 3$):

$$2x + 3y = 0$$

Let $x = 3$: $2(3) + 3y = 0$

$$6 + 3y = 0$$

Subtract 6 from both sides: $3y = -6$

Divide both sides by 3: $y = -2$

The point $(3, -2)$ is on the graph of the equation. Because both the x- and y-intercepts are 0, we find *two* additional points on the graph of the equation. We already have one additional point, $(3, -2)$, so we need one more. By letting $x = -3$, we find that $y = 2$. We plot the points $(0, 0)$, $(-3, 2)$, and $(3, -2)$. Connect the points in a straight line and obtain the graph in Figure 16.

Figure 16

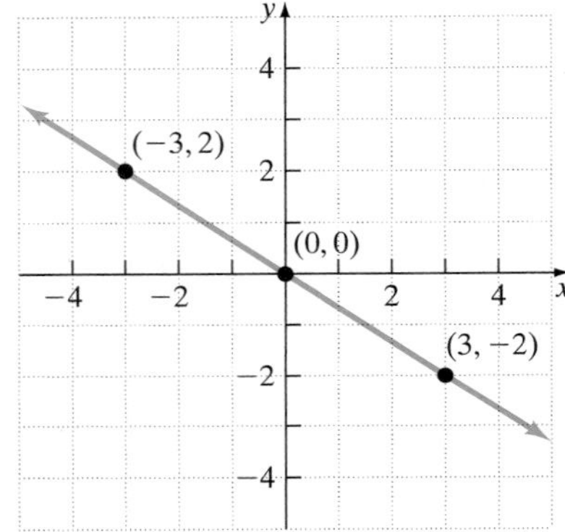

QUICK ✓ *Graph the equation by finding its intercepts.*

16. $y = \frac{1}{2}x$

17. $4x + y = 0$

3 Graph Vertical and Horizontal Lines

In the equation of a line, $Ax + By = C$, we said that A and B cannot both be zero. But what if $A = 0$ or $B = 0$? We find that this leads to special types of lines called *vertical lines* (when $B = 0$) and *horizontal lines* (when $A = 0$).

Classroom Example Graph the equation $x = -4$ using the point-plotting method.

Answer:

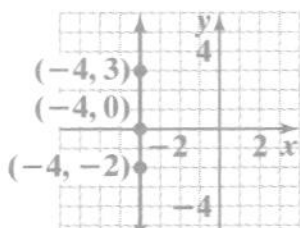

EXAMPLE 9 Graphing a Vertical Line

Graph the equation $x = 3$ using the point-plotting method.

Solution

Because the equation $x = 3$ can be written as $1x + 0y = 3$, we know that the graph is a line. When you look at the equation $x = 3$, notice that no matter what value of y we choose, the corresponding value of x is going to be 3. For example, if $y = -1$, then

$$1x + 0(-1) = 3$$
$$x = 3$$

See Table 12 for other choices for y. We see that the points $(3, -2)$, $(3, -1)$, $(3, 0)$, $(3, 1)$, and $(3, 2)$ are all points on the line. See Figure 17.

Teaching Tip
Tell students that an equation with only the variable x will intersect the x-axis: a vertical line. An equation with only the variable y will intersect the y-axis: a horizontal line.

Table 12

x	y	(x, y)
3	−2	(3, −2)
3	−1	(3, −1)
3	0	(3, 0)
3	1	(3, 1)
3	2	(3, 2)

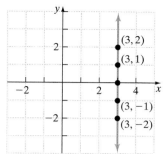

Figure 17

Based on the results of Example 9, we can write a definition of a vertical line:

DEFINITION: EQUATION OF A VERTICAL LINE
A vertical line is given by an equation of the form

$$x = a$$

where a is the x-intercept.

Now let's look at equations that lead to graphs that are horizontal lines.

Classroom Example
Graph the linear equation $y = 2$ using the point-plotting method.

Answer:

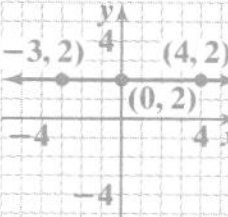

EXAMPLE 10 Graphing a Horizontal Line

Graph the equation $y = -2$ using the point-plotting method.

Solution

Because the equation $y = -2$ can be written as $0x + 1y = -2$, we know that the graph is a line. In looking at the equation $y = -2$, notice that no matter what value of x we choose, the corresponding value of y is going to be -2. For example, if $x = -2$, then

$$0(-2) + 1y = -2$$
$$y = -2$$

See Table 13 for other choices of x. Therefore, the points $(-2, -2)$, $(-1, -2)$, $(0, -2)$, $(1, -2)$, and $(2, -2)$ are all points on the line. See Figure 18.

Table 13

x	y	(x, y)
−2	−2	(−2, −2)
−1	−2	(−1, −2)
0	−2	(0, −2)
1	−2	(1, −2)
2	−2	(2, −2)

Figure 18

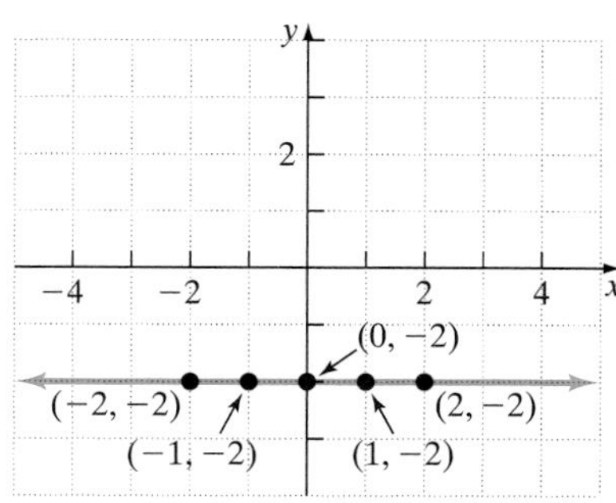

Based on the results of Example 10, we can generalize:

DEFINITION: EQUATION OF A HORIZONTAL LINE
A horizontal line is given by an equation of the form

$$y = b$$

where b is the y-intercept.

QUICK ✓ *Graph each equation.*

18. $x = -5$ **19.** $y = -4$ **20.** $x - 4 = 0$

We covered a lot of material in this section. We present a summary below to help you organize the information presented.

SUMMARY: Intercepts and Equations of Lines

Topic	Comments
Intercepts: Points where the graph crosses or touches a coordinate axis. **x-intercept:** Point where the graph crosses or touches the x-axis. Found by letting $y = 0$ in the equation. **y-intercept:** Point where the graph crosses or touches the y-axis. Found by letting $x = 0$ in the equation.	Intercepts need to be shown for a graph to be complete.
Standard Form of an Equation of a Line: $Ax + By = C$, where A and B are not both zero	Can be graphed using point-plotting or intercepts.
Equation of a Vertical Line: $x = a$	Graph is a vertical line whose x-intercept is a.
Equation of a Horizontal Line: $y = b$	Graph is a horizontal line whose y-intercept is b.

8.2 Exercises

1. linear, standard form 2. line
3. intercepts 4. False 5. False 6. True
7. Two; three is recommended for a check.
8. To graph $Ax + By = C$, where A, B, and C are not equal to zero, to find the x-intercept let $y = 0$, and to find the y-intercept, let $x = 0$. When $C = 0$, both the x- and the y-intercepts are 0, so find two additional points; answers may vary.
9. linear 10. nonlinear 11. nonlinear
12. linear 13. nonlinear 14. linear
15. linear 16. nonlinear

17. $y = 2x$

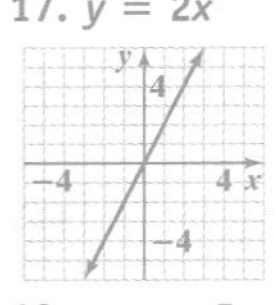

18. $y = 3x$

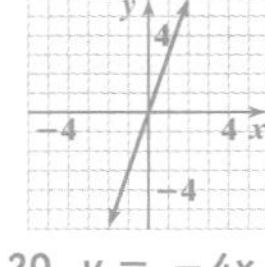

19. $y = -5x$ 20. $y = -4x$

21. $y = 4x - 2$ 22. $y = -3x - 1$

For Extra Help:

Student Solutions Manual

CD Video

PH Math/Tutor Center

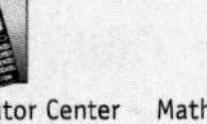
MathXL Tutorials on CD

MathXL®

MyMathLab

Concepts and Vocabulary

In Problems 1–3, fill in the blanks.

1. A(n) _________ equation is an equation of the form $Ax + By = C$, where A, B, and C are real numbers, and A and B are not both zero. Equations written in this form are said to be in _________ _________.

2. The graph of a linear equation is a(n) _________.

3. The _________ are the points, if any, where a graph crosses or touches the coordinate axes.

In Problems 4–6, answer True or False to each statement.

4. To find the y-intercept(s), if any, of the graph of an equation, let $y = 0$ in the equation and solve for x.

5. All linear equations have exactly one x-intercept and one y-intercept.

6. A horizontal line can be represented by the equation $y = b$, where b is the y-intercept of the graph of the equation.

7. How many points are required to graph a line? Explain your reasoning and why you might include additional point(s) when graphing a line.

8. Explain how to use the intercepts to graph the equation $Ax + By = C$, where A, B, and C are not equal to zero. Explain how to graph the same equation when C is equal to zero. Can you use the same techniques for both equations? Why or why not?

Building Skills

In Problems 9–16, determine whether or not the equation is a linear equation in two variables.

9. $2x - 5y = 10$ **10.** $y^2 = 2x + 3$ **11.** $\sqrt{x} + y = 1$ **12.** $y - 2x = 9$

13. $y = \dfrac{4}{x}$ **14.** $x - 8 = 0$ **15.** $y - 1 = 0$ **16.** $y = \dfrac{-2}{x}$

In Problems 17–36, graph each linear equation using the point-plotting method.

17. $y = 2x$ **18.** $y = 3x$ **19.** $y = -5x$ **20.** $y = -4x$

21. $y = 4x - 2$ **22.** $y = -3x - 1$ **23.** $y = -2x + 5$ **24.** $y = x - 6$

25. $x + y = 5$ **26.** $x - y = 6$ **27.** $-2x + y = 6$ **28.** $5x - 2y = -10$

29. $4x - 2y = -8$ **30.** $x + 3y = 6$ **31.** $x = -4y$ **32.** $x = \dfrac{1}{2}y$

33. $y + 7 = 0$ **34.** $x - 6 = 0$

35. $y - 2 = 3(x + 1)$ **36.** $y + 3 = -2(x - 2)$

In Problems 37–44, find the intercepts of each graph.

37.

38.

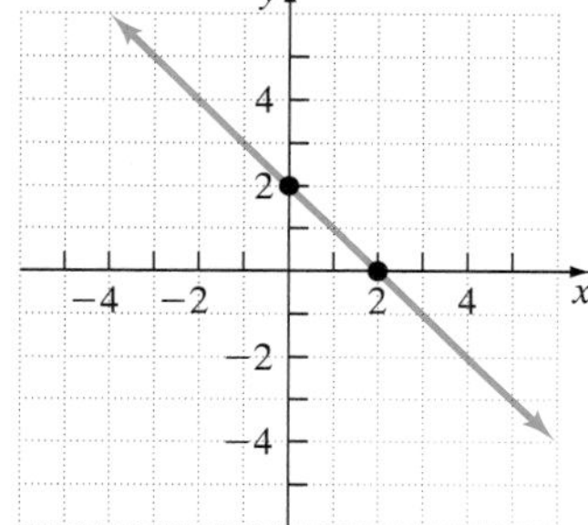

39.

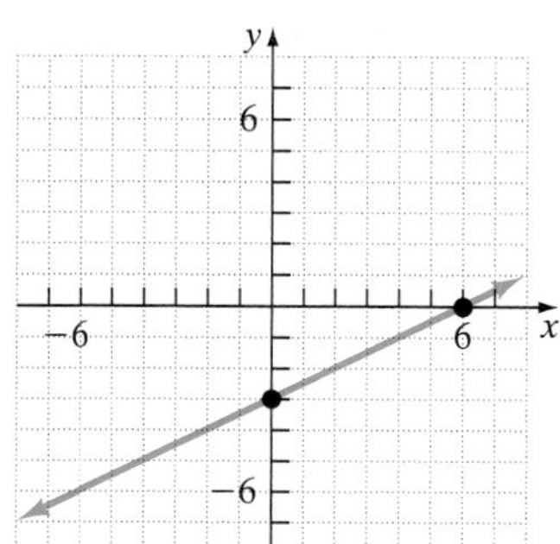

40.

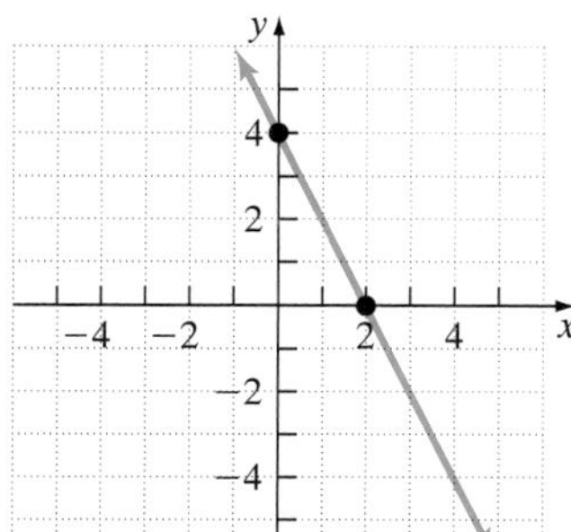

41.

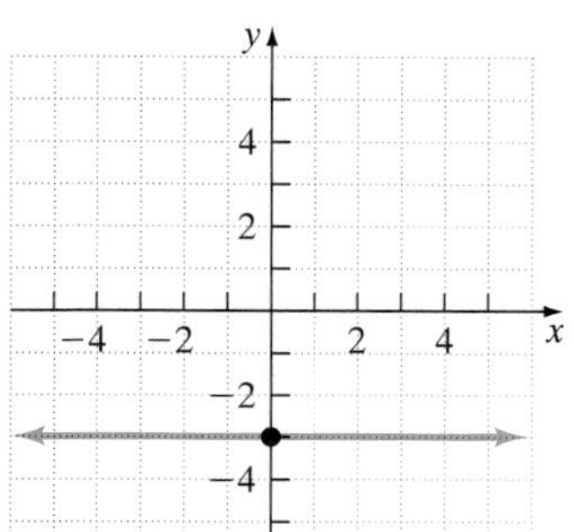

42.

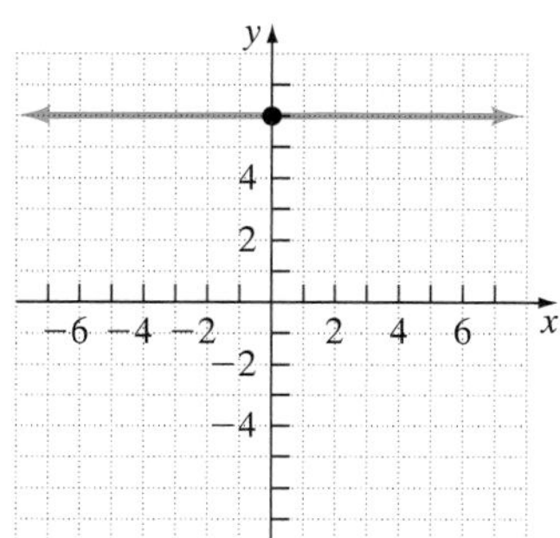

23. $y = -2x + 5$ 24. $y = x - 6$

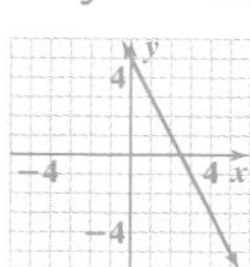

25. $x + y = 5$ 26. $x - y = 6$

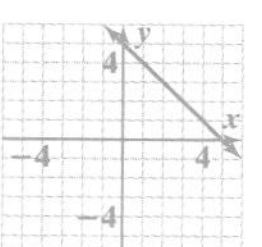

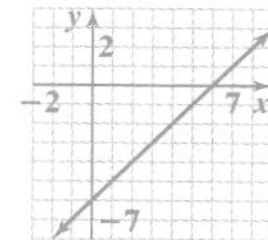

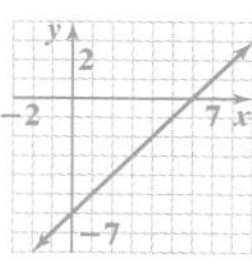

27. $-2x + y = 6$ 28. $5x - 2y = -10$

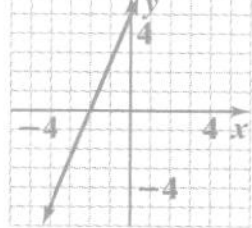

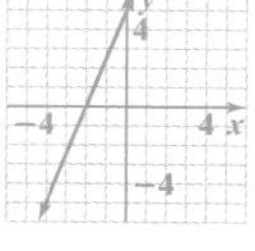

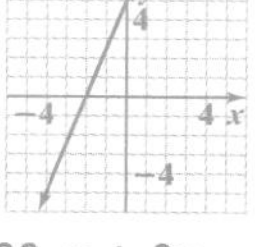

29. $4x - 2y = -8$ 30. $x + 3y = 6$

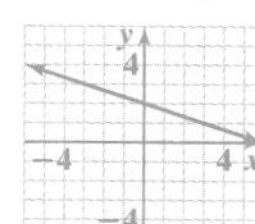

31. $x = -4y$ 32. $x = \frac{1}{2}y$

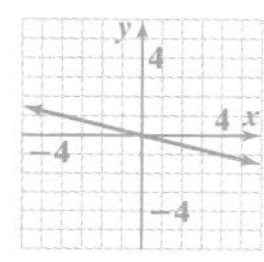

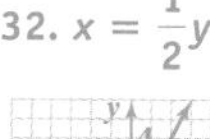

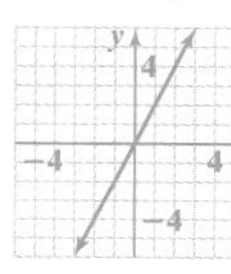

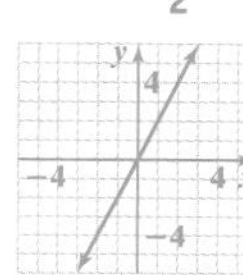

33. $y + 7 = 0$ 34. $x - 6 = 0$

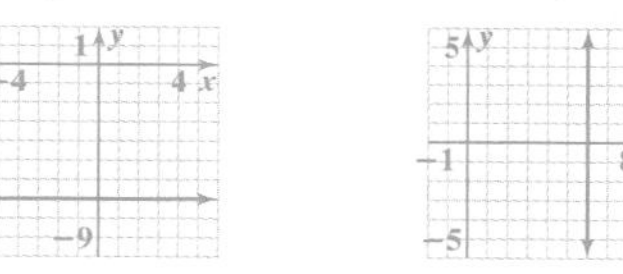

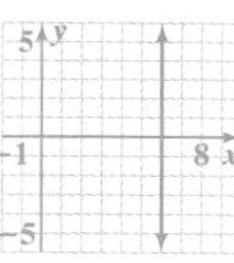

35. $y - 2 = 3(x + 1)$ 36. $y + 3 = -2(x - 2)$

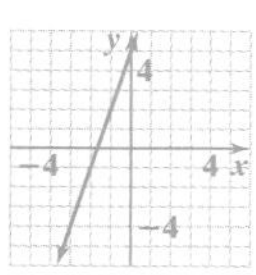

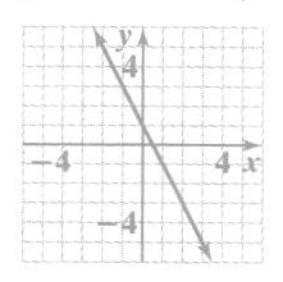

37. $(0, -5)$, $(5, 0)$ 38. $(0, 2)$, $(2, 0)$
39. $(0, -3)$, $(6, 0)$ 40. $(0, 4)$, $(2, 0)$
41. $(0, -3)$ 42. $(0, 6)$ 43. $(-5, 0)$
44. $(7, 0)$ 45. $(0, -4)$, $(-6, 0)$
46. $(0, -6)$, $(10, 0)$ 47. $(0, 0)$ 48. $(0, 0)$
49. $(0, -5)$, $(5, 0)$ 50. $(0, 7)$, $(7, 0)$
51. $(0, 8)$, $(6, 0)$ 52. $(0, -8)$, $(2, 0)$
53. $(4, 0)$ 54. $(0, 6)$ 55. $(0, -2)$
56. $(-8, 0)$
57. $3x + 6y = 18$ 58. $4x - 2y = -8$

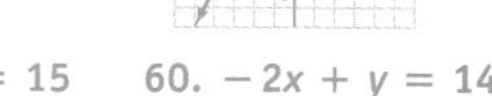

59. $-x + 5y = 15$ 60. $-2x + y = 14$

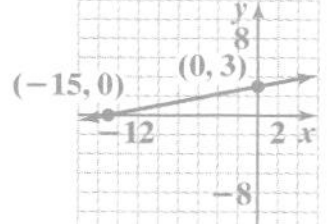

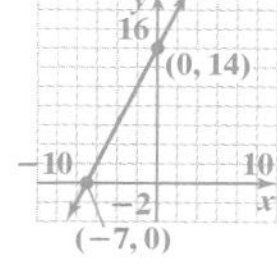

61. $\frac{1}{2}x = y + 3$ 62. $\frac{4}{3}x = -y + 1$ 63. $9x - 2y = 0$ 64. $\frac{1}{3}x - y = 0$

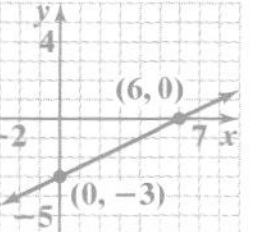

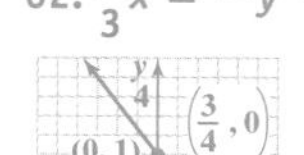

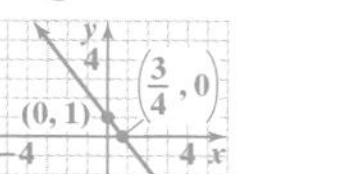

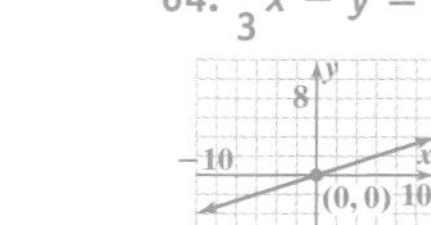

43.

44.

In Problems 45–56, find the intercepts of each equation.

45. $2x + 3y = -12$ **46.** $3x - 5y = 30$ **47.** $x = -6y$ **48.** $y = 10x$

49. $y = x - 5$ **50.** $y = -x + 7$ **51.** $\frac{x}{6} + \frac{y}{8} = 1$ **52.** $\frac{x}{2} - \frac{y}{8} = 1$

53. $x = 4$ **54.** $y = 6$ **55.** $y = -2$ **56.** $x = -8$

In Problems 57–72, graph each linear equation by finding its intercepts.

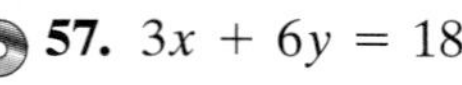

57. $3x + 6y = 18$ **58.** $4x - 2y = -8$ **59.** $-x + 5y = 15$ **60.** $-2x + y = 14$

61. $\frac{1}{2}x = y + 3$ **62.** $\frac{4}{3}x = -y + 1$ **63.** $9x - 2y = 0$ **64.** $\frac{1}{3}x - y = 0$

65. $y = -\frac{1}{2}x + 3$ **66.** $y = \frac{2}{3}x - 3$ **67.** $\frac{1}{3}y + 2 = 2x$ **68.** $\frac{1}{2}x - 3 = 3y$

69. $\frac{x}{2} + \frac{y}{3} = 1$ **70.** $\frac{y}{4} - \frac{x}{3} = 1$

71. $4y - 2x + 1 = 0$ **72.** $2y - 3x + 2 = 0$

In Problems 73–80, graph each horizontal or vertical line.

73. $x = 5$ **74.** $x = -7$ **75.** $y = -6$ **76.** $y = 2$

77. $y - 12 = 0$ **78.** $y + 3 = 0$ **79.** $3x - 5 = 0$ **80.** $2x - 7 = 0$

Mixed Practice

In Problems 81–92, graph each linear equation by any method.

81. $y = 2x - 5$ **82.** $y = -3x + 2$ **83.** $y = -5$ **84.** $x = 2$

85. $2x + 5y = -20$ **86.** $3x - 4y = 12$ **87.** $2x = -6y + 4$ **88.** $5x = 3y - 10$

89. $x - 3 = 0$ **90.** $y + 4 = 0$ **91.** $3y - 12 = 0$ **92.** $-4x + 8 = 0$

Applying the Concepts

93. Plot the points $(3, 5)$ and $(-2, 5)$ and draw a line through the points. What is the equation of this line?

94. Plot the points $(-1, 2)$ and $(5, 2)$ and draw a line through the points. What is the equation of this line?

95. Plot the points $(-2, -4)$ and $(-2, 1)$ and draw a line through the points. What is the equation of this line?

96. Plot the points $(3, -1)$ and $(3, 2)$ and draw a line through the points. What is the equation of this line?

65. $y = -\frac{1}{2}x + 3$

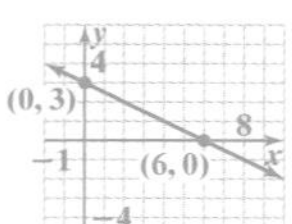

66. $y = \frac{2}{3}x - 3$

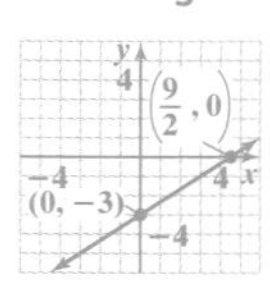

67. $\frac{1}{3}y + 2 = 2x$

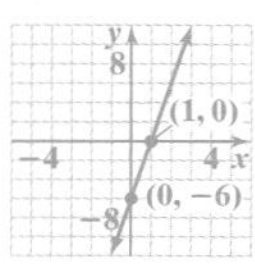

68. $\frac{1}{2}x - 3 = 3y$

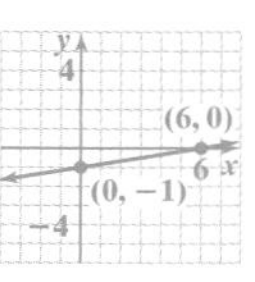

69. $\frac{x}{2} + \frac{y}{3} = 1$

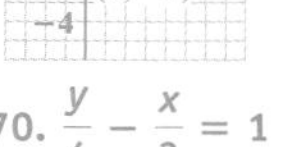

70. $\frac{y}{4} - \frac{x}{3} = 1$

71. $4y - 2x + 1 = 0$

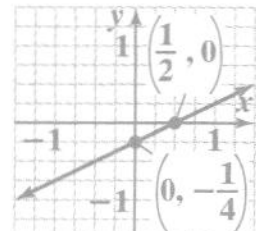

72. $2y - 3x + 2 = 0$

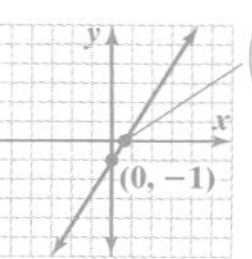

73. $x = 5$

74. $x = -7$

75. $y = -6$

76. $y = 2$

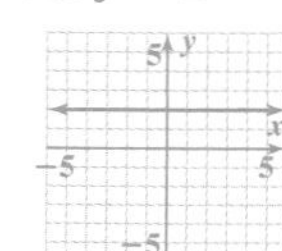

77. $y - 12 = 0$

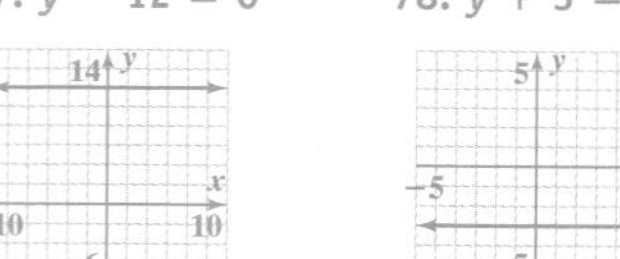

78. $y + 3 = 0$

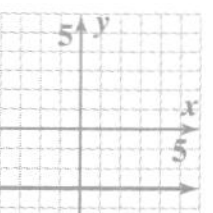

79. $3x - 5 = 0$

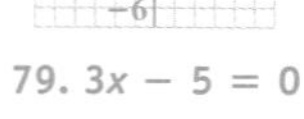

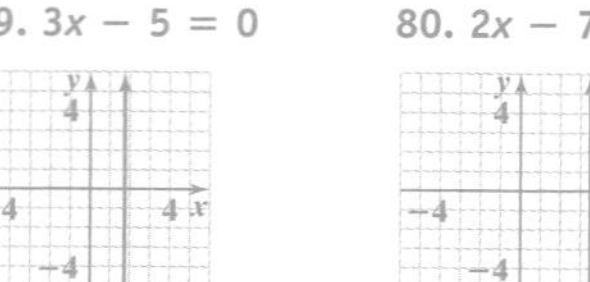

80. $2x - 7 = 0$

81. $y = 2x - 5$

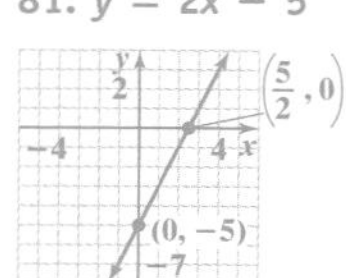

82. $y = -3x + 2$

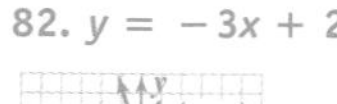

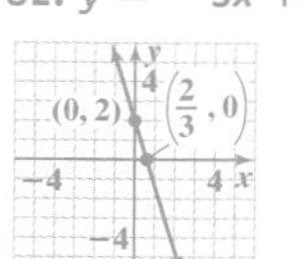

83. $y = -5$

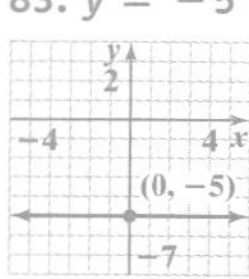

84. $x = 2$

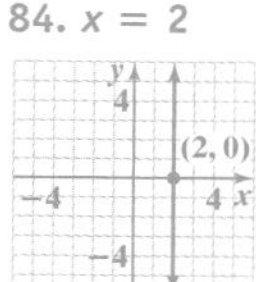

85. $2x + 5y = -20$

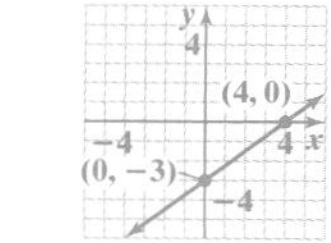

86. $3x - 4y = 12$

87. $2x = -6y + 4$

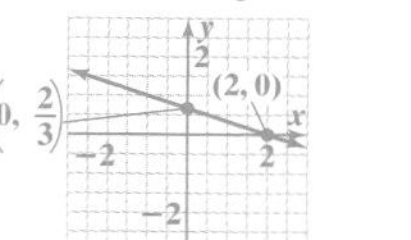

88. $5x = 3y - 10$

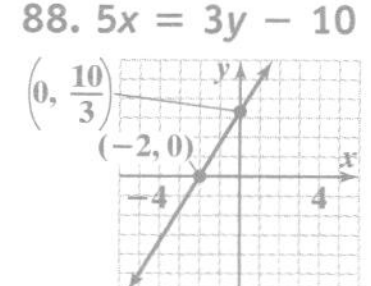

89. $x - 3 = 0$

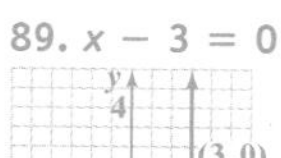
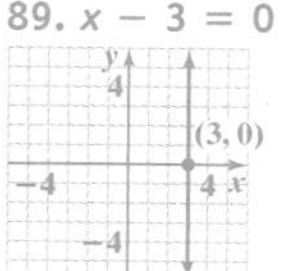

90. $y + 4 = 0$

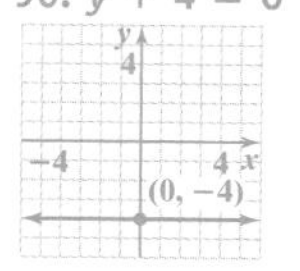

91. $3y - 12 = 0$

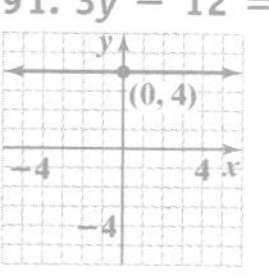

92. $-4x + 8 = 0$

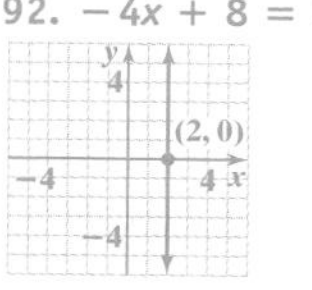

93. $y = 5$

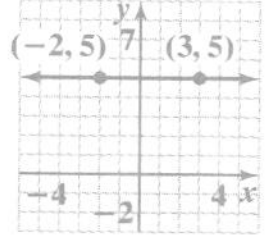

94. $y = 2$

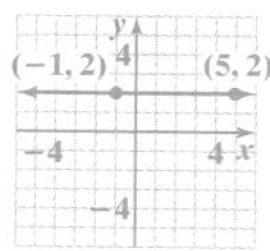

95. $x = -2$

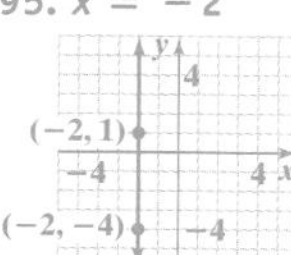

96. $x = 3$

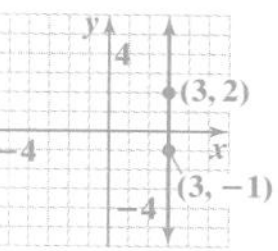

97. $y = 4$ 98. $y = 8$ 99. $x = -9$
100. $x = 2$ 101. $x = 2y$ 102. $y = 2x$
103. $y = x + 2$ 104. $x = y - 3$
105. $y = 2$ 106. $y = -11$
107. $x = 7$ 108. $x = -\frac{1}{2}$

109. (a) (0, 500), (4, 900), (10, 1500)
(b)

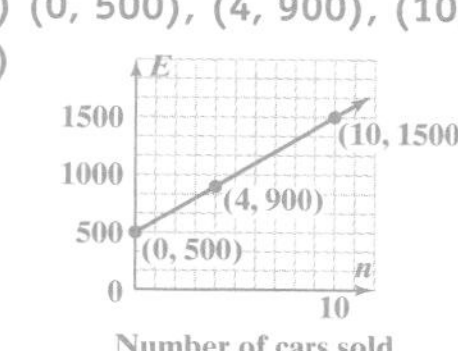

(c) If she sells 0 cars, her earnings are $500.

110. (a) (1000, 150), (2000, 250), (2500, 300)
(b)

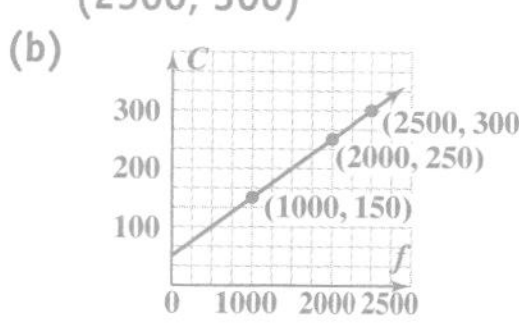

(c) If there is 0 square feet of carpet, the charge is $50.

111. The "steepness" of the lines is the same.

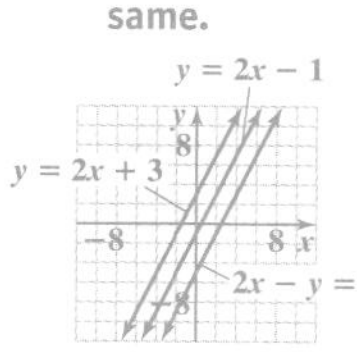

112. The lines are all the same.

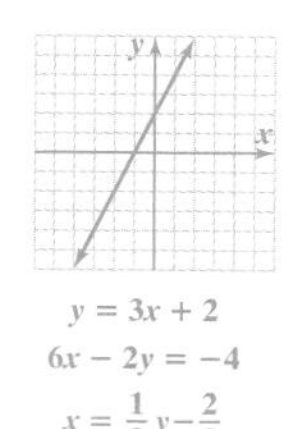

113. The lines get more steep as the coefficient of x gets larger.

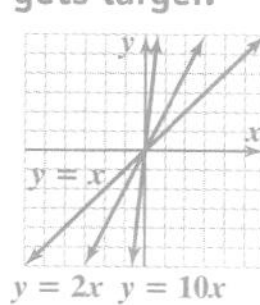

114. The lines get less steep as the coefficient of x gets smaller.

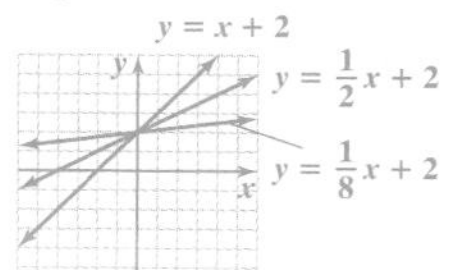

In Problems 97–100, find the equation of each line.

97.

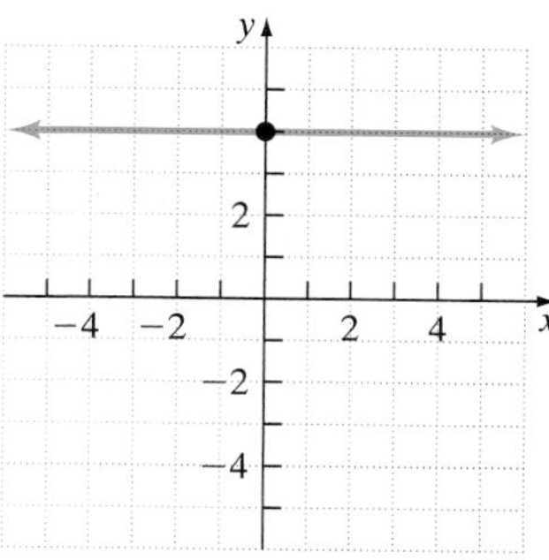

98.

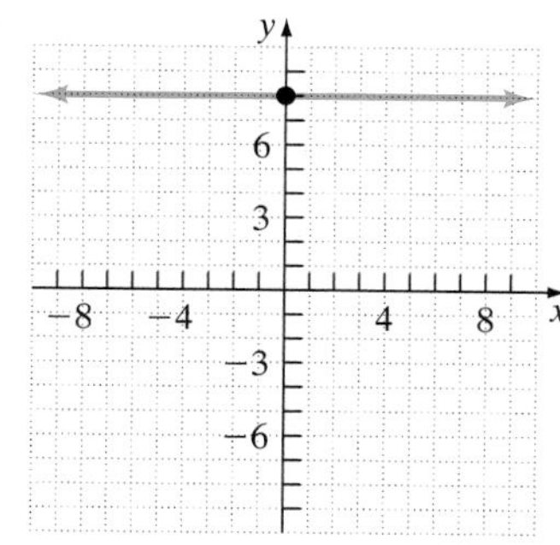

99.

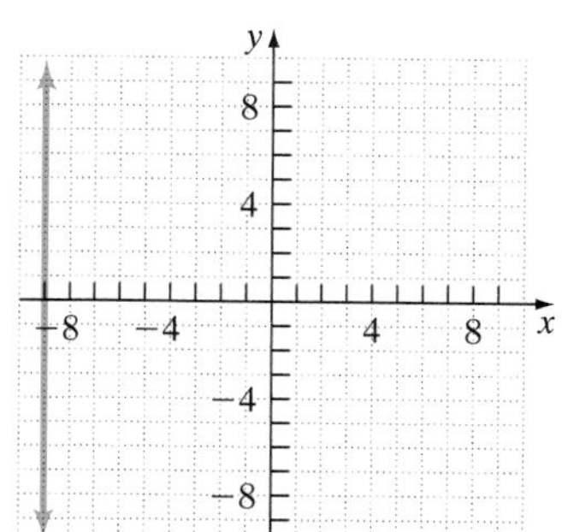

100.

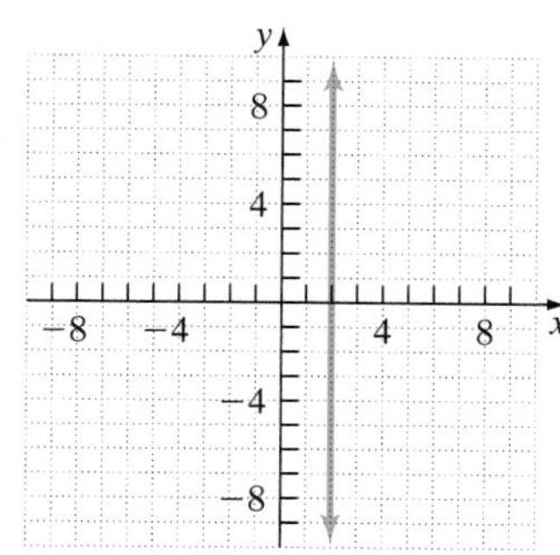

101. Create a set of ordered pairs in which the x-coordinate is twice the y-coordinate. What is the equation of this line?

102. Create a set of ordered pairs in which the y-coordinate is twice the x-coordinate. What is the equation of this line?

103. Create a set of ordered pairs in which the y-coordinate is 2 more than the x-coordinate. What is the equation of this line?

104. Create a set of ordered pairs in which the x-coordinate is 3 less than the y-coordinate. What is the equation of this line?

105. If $(3, y)$ is a point on the graph of $4x + 3y = 18$, find y.

106. If $(-4, y)$ is a point on the graph of $3x - 2y = 10$, find y.

107. If $(x, -2)$ is a point on the graph of $3x + 5y = 11$, find x.

108. If $(x, -3)$ is a point on the graph of $4x - 7y = 19$, find x.

109. Calculating Wages Marta earns $500 per week plus $100 in commission for every car she sells. The linear equation that calculates her weekly earnings is $E = 100n + 500$, where E represents her weekly earnings in dollars and n represents the number of cars she sold during the week.

(a) Create a set of ordered pairs (n, E) if, in three consecutive weeks, she sold 0 cars, 4 cars, and 10 cars.

(b) Graph the linear equation $E = 100n + 500$ using the ordered pairs obtained in part (a). Be sure to label the axes appropriately.

(c) Explain the meaning of the E-intercept.

110. Carpet Cleaning Harry's Carpet Cleaning charges a $50 service charge plus $0.10 for each square foot of carpeting to be cleaned. The linear equation that calculates the total cost to clean a carpet is $C = 0.1f + 50$, where C is the total cost in dollars and f is the number of square feet of carpet.

(a) Create a set of ordered pairs (f, C) for the following number of square feet to be cleaned: 1000 sq ft, 2000 sq ft, 2500 sq ft.

(b) Graph the linear equation $C = 0.1f + 50$ using the ordered pairs obtained in part (a). Be sure to label the axes appropriately.

(c) Explain the meaning of the C-intercept.

115. (0, −6), (−2, 0), (3, 0)
116. (0, 12), (−2, 0), (2, 0)
117. (0, 14), (−3, 0), (2, 0), (5, 0)
118. (0, 12), (−5, 0), (−2, 0), (2, 0)
119. $y = 2x - 9$

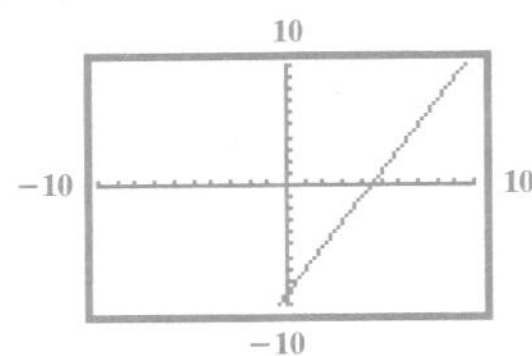

120. $y = -3x + 8$

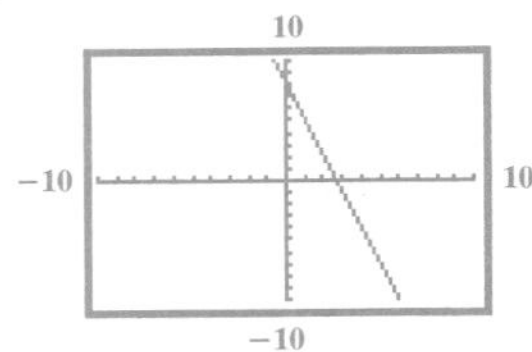

121. $y = -x + 8$

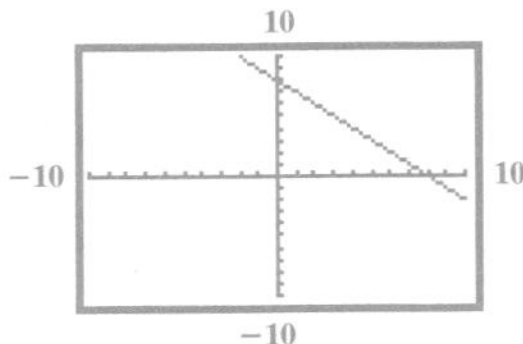

122. $y = 2x - 4$

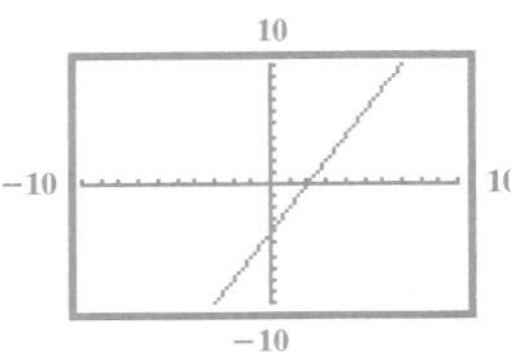

123. $y + 2x = 13$ or $y = -2x + 13$

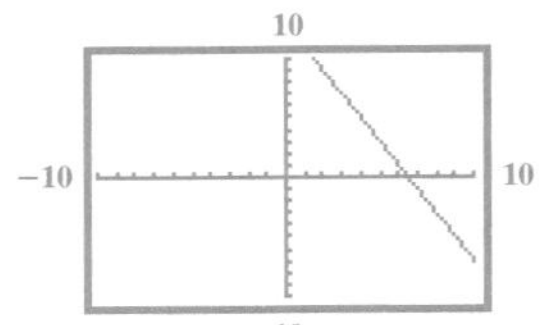

124. $y - x = -15$ or $y = x - 15$

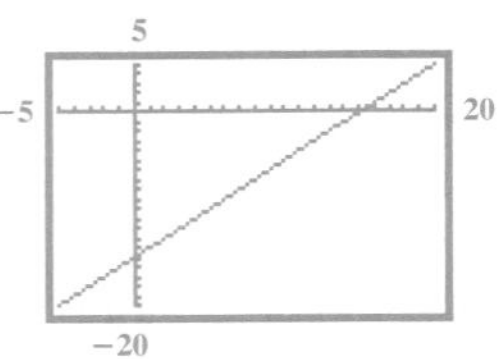

125. $y = -6x^2 + 1$

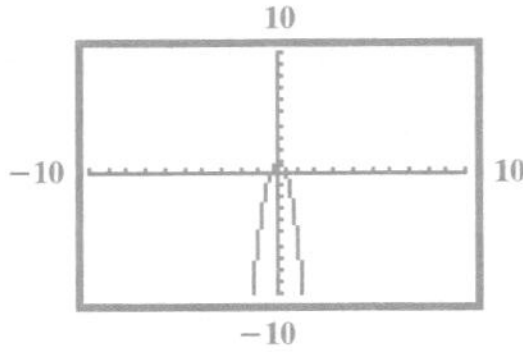

126. $y = -x^2 + 3x$

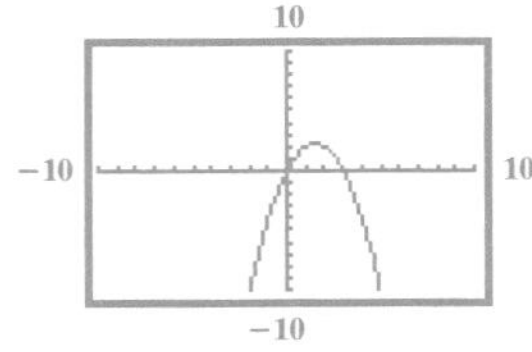

Extending the Concepts

111. Graph each of the following linear equations in the same xy-plane. What can you infer from these graphs?

$$y = 2x - 1 \qquad y = 2x + 3 \qquad 2x - y = 5$$

112. Graph each of the following linear equations in the same xy-plane. What can you infer from these graphs?

$$y = 3x + 2 \qquad 6x - 2y = -4 \qquad x = \frac{1}{3}y - \frac{2}{3}$$

113. Graph each of the following linear equations in the same xy-plane. What statement can you make about the steepness of the line as the coefficient of x gets larger?

$$y = x \qquad y = 2x \qquad y = 10x$$

114. Graph each of the following linear equations in the same xy-plane. What statement can you make about the steepness of the line as the coefficient of x gets smaller?

$$y = x + 2 \qquad y = \frac{1}{2}x + 2 \qquad y = \frac{1}{8}x + 2$$

In Problems 115–118, find the intercepts of each graph.

115.

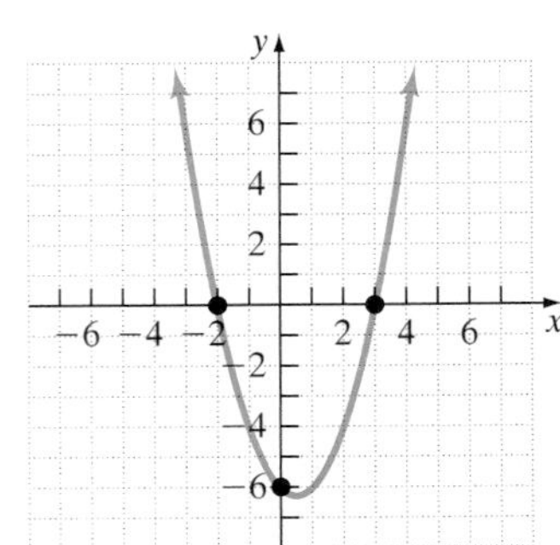

116.

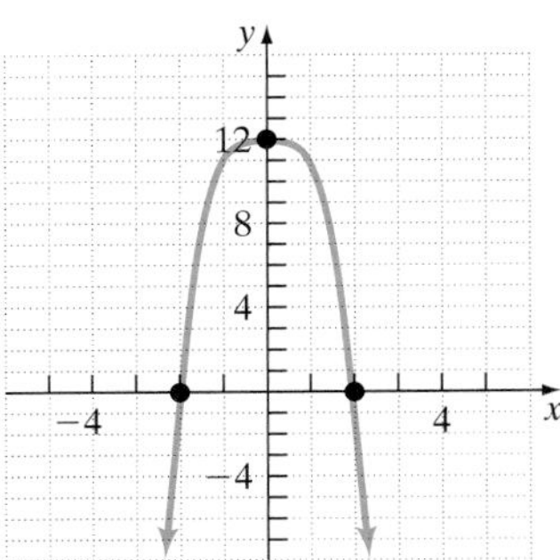

117.

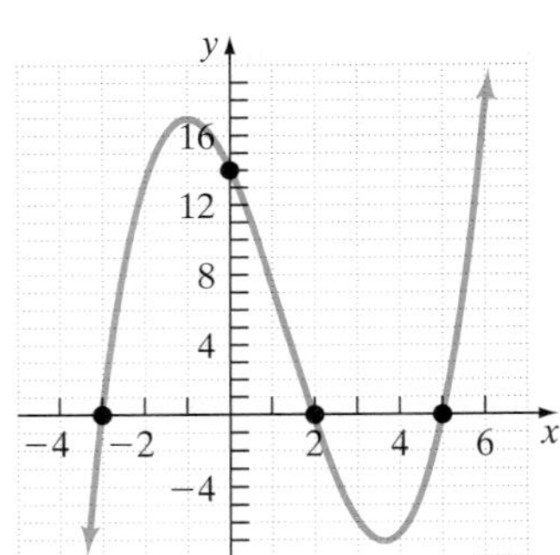

118.

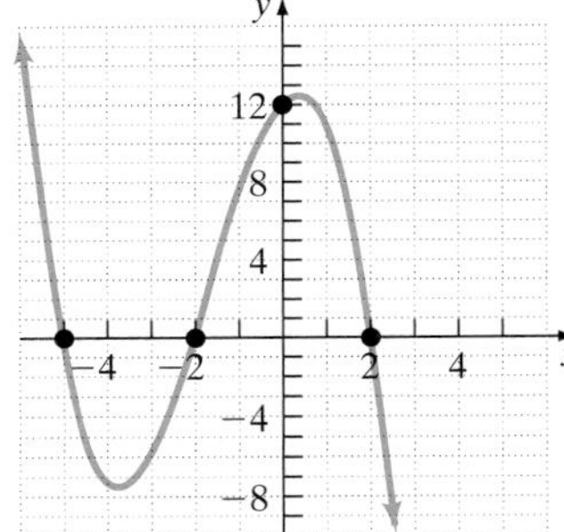

The Graphing Calculator

Graphing calculators can graph equations. In fact, graphing calculators also use the point-plotting method to obtain the graph by choosing 95 values of x and using the equation to find the corresponding value of y. As with creating tables, we first solve the equation for y. For example, to obtain a table of values that satisfy the equation $2x - 3y = 12$ (Example 6 from Section 8.1), we must solve for y and we obtain $y = \frac{2}{3}x - 4$. We enter the equation $y = \frac{2}{3}x - 4$ into the calculator and create the graph shown in the right.

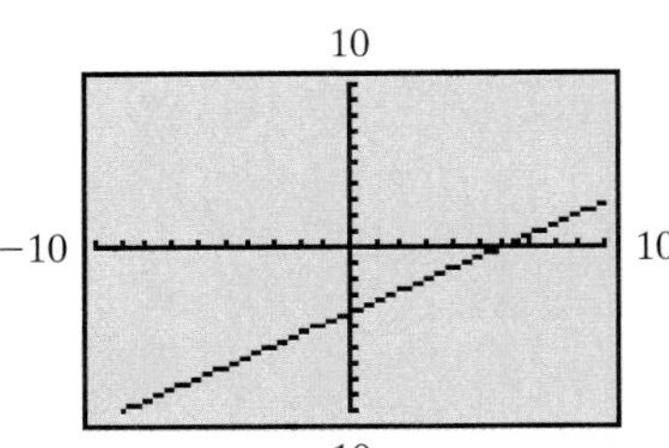

In Problems 119–126, use a graphing calculator to graph each equation.

119. $y = 2x - 9$ **120.** $y = -3x + 8$ **121.** $y = -x + 8$ **122.** $y = 2x - 4$

123. $y + 2x = 13$ **124.** $y - x = -15$ **125.** $y = -6x^2 + 1$ **126.** $y = -x^2 + 3x$

8.3 Slope

OBJECTIVES

1. Find the Slope of a Line Given Two Points
2. Find the Slope of Vertical and Horizontal Lines
3. Graph a Line Using Its Slope and a Point on the Line
4. Work with Applications of Slope

Preparing for Slope

Before getting started, take the following readiness quiz. If you get a problem wrong, go back to the section cited and review the material.

1. Evaluate: $\dfrac{5-2}{8-7}$ [Section 1.6, p. 52]

2. Evaluate: $\dfrac{3-7}{9-3}$ [Section 1.6, p. 52]

3. Evaluate: $\dfrac{-3-4}{6-(-1)}$ [Section 1.6, p. 52]

Figure 19

Pretend you are on snow skis for the first time in your life. The ski resort that you are visiting has two hills available to beginning skiers. The profile of each hill is shown in Figure 19. Which hill would you prefer to go down? Why?

It is clear from the figure that the hill in Figure 19(a) is not as steep as the hill in Figure 19(b). One of the things that mathematicians like to do is give numerical descriptions to situations such as the steepness of a hill. Measuring the steepness of each hill allows for them to be compared more easily. The numerical measure that we could use to describe the steepness of a hill is its *slope*.

1 Find the Slope of a Line Given Two Points

Consider the staircase drawn in Figure 20(a). If we draw a line through the top of each riser on the staircase (in blue), we can see that each step contains exactly the same horizontal change (or **run**) and the same vertical change (or **rise**).

Figure 20

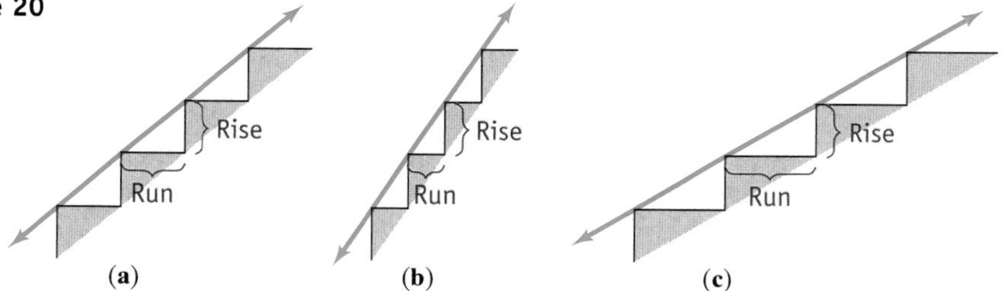

DEFINITION

The **slope** of a line, denoted by the letter m, is the ratio of the rise to the run. That is,

$$\text{Slope} = m = \frac{\text{Rise}}{\text{Run}}$$

Slope is a numerical measure of the steepness of the line. For example, if the run is decreased and the rise remains the same, then the staircase becomes steeper. See Figure 20(b). If the run is increased and the rise remains the same, then the staircase becomes less steep. See Figure 20(c).

Suppose that the staircase in Figure 20(a) has a rise of 6 inches and a run of 7 inches. Then the slope of the line is

$$m = \frac{\text{rise}}{\text{run}}$$
$$= \frac{6 \text{ inches}}{7 \text{ inches}}$$

Preparing for...Answers **1.** 3 **2.** $-\frac{2}{3}$ **3.** -1

If the rise of the stair is increased to 9 inches, then the slope of the line is

$$m = \frac{\text{rise}}{\text{run}}$$
$$= \frac{9 \text{ inches}}{7 \text{ inches}}$$

The main idea is the steeper the line, the larger the slope. We can define the slope of a line using rectangular coordinates.

Work Smart
The subscripts 1 and 2, on x_1, x_2, y_1, and y_2 do not represent a computation (as superscripts do in x^2). Instead, they are used to indicate that the values of the variable x_1 may be different from x_2, and y_1 may be different from y_2.

DEFINITION
If $x_1 \neq x_2$, the **slope *m*** of the line containing the points (x_1, y_1) and (x_2, y_2) is defined by the formula

$$m = \frac{y_2 - y_1}{x_2 - x_1}, \quad x_1 \neq x_2$$

Figure 21 provides an illustration of the slope of a line.

In Words
The accepted symbol for the slope of a line is m. It comes from the French word *monter*, which means "to go up, ascend, or climb."

Figure 21

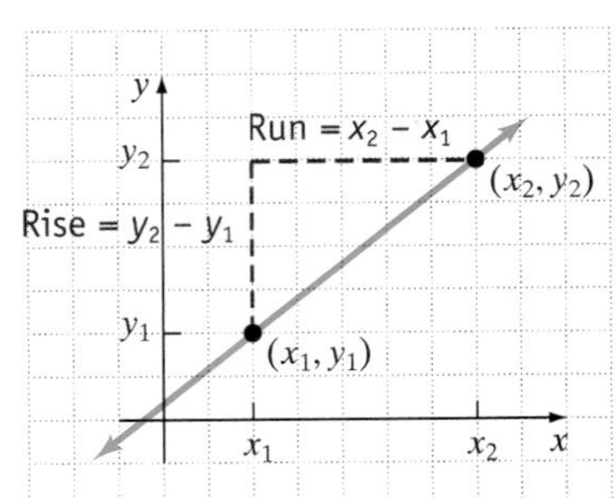

From Figure 21 we can see that the slope m of a line may be viewed as

$$m = \frac{\text{rise}}{\text{run}} = \frac{y_2 - y_1}{x_2 - x_1}$$

In Words
Slope is rise over run, or the change in y divided by the change in x.

We can also write the slope m of a line as

$$m = \frac{y_2 - y_1}{x_2 - x_1} = \frac{\text{change in } y}{\text{change in } x} = \frac{\Delta y}{\Delta x}$$

Work Smart
The symbol Δ comes from the first letter of the Greek word *dunamis*, which means "change."

The symbol Δ is the Greek letter "delta." In mathematics, we read the symbol Δ as "change in." So the notation $\frac{\Delta y}{\Delta x}$ is read "change in y divided by change in x."

EXAMPLE 1 How to Find the Slope of a Line

Find the slope of the line drawn in Figure 22.

Classroom Example
Find the slope of the given line.

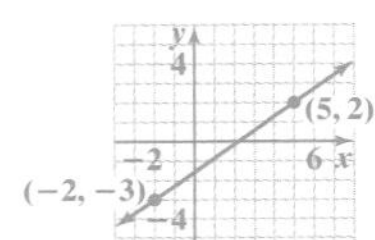

Answer: $m = \frac{5}{7}$

Figure 22

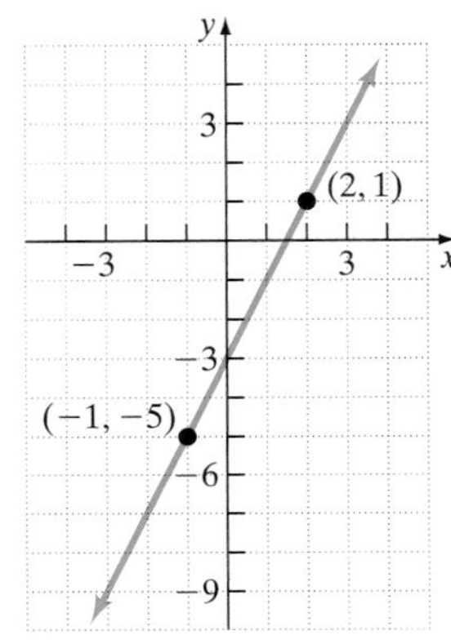

Step-by-Step Solution

Step 1: Let one of the points be (x_1, y_1) and the other point be (x_2, y_2).

Let's say that $(x_1, y_1) = (-1, -5)$ and $(x_2, y_2) = (2, 1)$.

Step 2: Find the slope by evaluating

$$m = \frac{y_2 - y_1}{x_2 - x_1} = \frac{\text{Change in } y}{\text{Change in } x} = \frac{\Delta y}{\Delta x}$$

$$m = \frac{y_2 - y_1}{x_2 - x_1}$$

$x_1 = -1, y_1 = -5; x_2 = 2, y_2 = 1$:

$$= \frac{1 - (-5)}{2 - (-1)} = \frac{6}{3} = 2$$

Work Smart

It doesn't matter which point is called (x_1, y_1) and which is called (x_2, y_2). The answer will be the same. In Example 1, if we let $(x_1, y_1) = (2, 1)$ and $(x_2, y_2) = (-1, -5)$, we obtain

$$m = \frac{y_2 - y_1}{x_2 - x_1} = \frac{-5 - 1}{-1 - 2} = \frac{-6}{-3} = 2$$

Figure 23

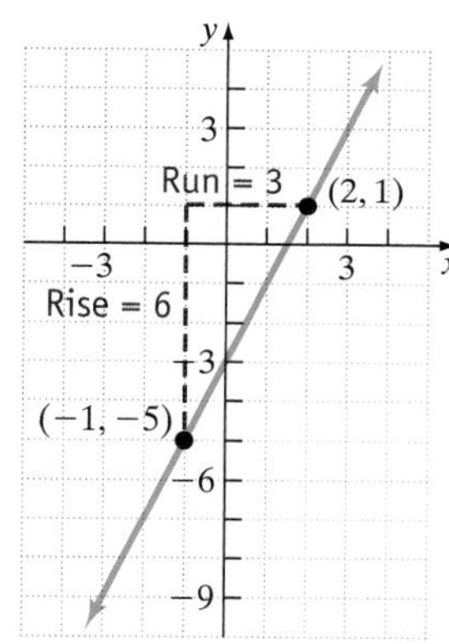

Remember—we said that the slope of the line can be thought of as "rise divided by run." This description of the slope of a line is illustrated in Figure 23.

We interpret the slope of the line drawn in Figure 23 as follows: "For every 6-unit increase in y, the values of x increase by 3 units." Or because $\frac{6}{3} = 2 = \frac{2}{1}$ "for every 2-unit increase in y, the value of x will increase by 1." Both interpretations are acceptable.

Classroom Example ➤
Plot the points (3, 2) and (−1, 10) and draw a line through the points. Find and interpret the slope of the line.

Answer: The slope = −2. The value of y decreases by 2 units when x increases by 1 unit

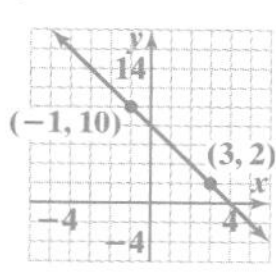

EXAMPLE 2 Finding and Interpreting the Slope of a Line

Plot the points (−1, 3) and (2, −2) in a rectangular coordinate system. Then draw a line through the two points. Find and interpret the slope of the line.

Solution

We plot the points $(x_1, y_1) = (-1, 3)$ and $(x_2, y_2) = (2, -2)$ in the rectangular coordinate system and draw a line through the two points. See Figure 24. The slope of the line drawn in Figure 24 is

$$m = \frac{y_2 - y_1}{x_2 - x_1} = \frac{-2 - 3}{2 - (-1)} = \frac{-5}{3} = -\frac{5}{3}$$

Figure 24

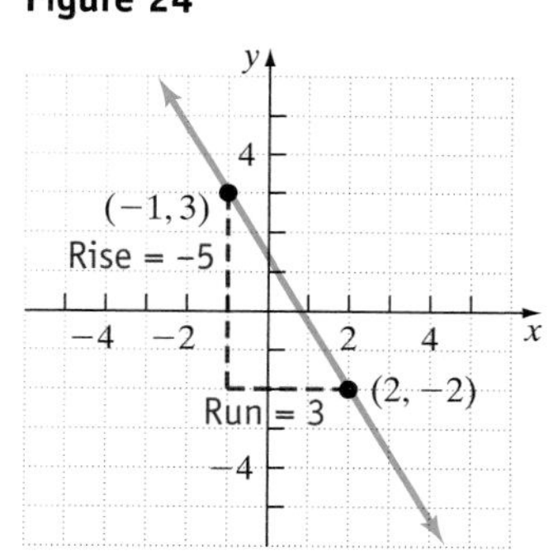

You can interpret a slope of $-\frac{5}{3} = \frac{-5}{3}$ this way: The value of y will go down 5 units whenever x increases by 3 units. Because $-\frac{5}{3} = \frac{5}{-3} = \frac{\text{rise}}{\text{run}}$, a second interpretation is as follows: The value of y will increase by 5 units whenever x decreases by 3 units.

Notice that the line drawn in Figure 23 goes up and to the right and the slope is positive, while the line drawn in Figure 24 goes down and to the right and slope is negative. In general, a line that goes down and to the right will have negative slope and a line that goes up and to the right will have positive slope. We illustrate this idea in Figure 25.

Work Smart
We read a graph from left to right, just like we read a book.

Figure 25

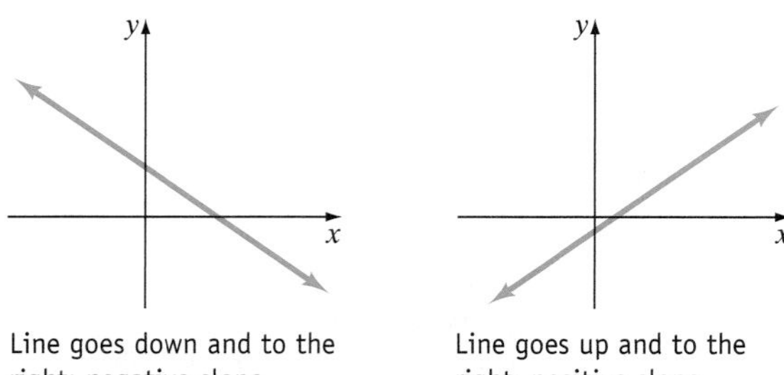

Line goes down and to the right: negative slope

Line goes up and to the right: positive slope

Quick ✓ *Find and interpret the slope of the line containing the points.*

1. (0, 4); (2, 10) **2.** (−2, 3); (2, −7)

2 Find the Slope of Vertical and Horizontal Lines

Did you notice in the definition of slope, $m = \frac{y_2 - y_1}{x_2 - x_1}$, we have the restriction that $x_1 \neq x_2$? This means that the formula does not apply if the x-coordinates of the two points are the same. Why? Let's look at the following example.

Classroom Example Plot the points (4, 3) and (4, −4); draw a line through the points. Find and interpret the slope of the line.

Answer: Undefined slope; when y increases by 1, there is no change in x.

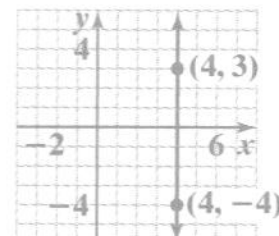

EXAMPLE 3 The Slope of a Vertical Line

Plot the points (2, −1) and (2, 3) in a rectangular coordinate system. Then draw a line through the two points. Find and interpret the slope of the line.

Solution

We plot the points $(x_1, y_1) = (2, -1)$ and $(x_2, y_2) = (2, 3)$ in the rectangular coordinate system and draw a line through the two points. See Figure 26. The slope of the line drawn in Figure 26 is

$$m = \frac{y_2 - y_1}{x_2 - x_1} = \frac{3 - (-1)}{2 - 2}$$
$$= \frac{4}{0}$$

Because division by 0 is undefined, we say that the slope of the line is undefined. When y increases by 1, there is no change in x.

Figure 26

(2, 3)
(2, −1)

Let's generalize the results of Example 3. Let (x_1, y_1) and (x_2, y_2) be two distinct points. If $x_1 = x_2$, then we have a **vertical line** whose slope m is **undefined** (since this results in division by 0). Figure 27 illustrates a vertical line.

Figure 27
Slope of a vertical line.

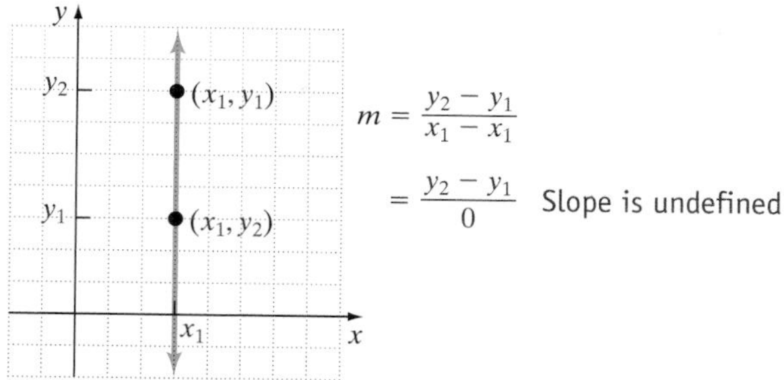

Okay, but what if $y_1 = y_2$?

Classroom Example ➤
Plot the points $(-5, -2)$ and $(3, -2)$; draw a line through the points. Find and interpret the slope of the line.

Answer: There is no change in y when x increases by 1 unit.

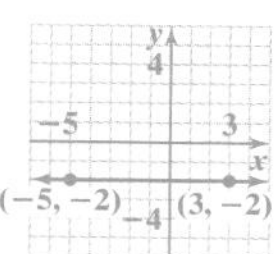

EXAMPLE 4 The Slope of a Horizontal Line

Plot the points $(-2, 4)$ and $(3, 4)$ in a rectangular coordinate system. Then draw a line through the two points. Finally, find and interpret the slope of the line.

Solution

We plot the points $(x_1, y_1) = (-2, 4)$ and $(x_2, y_2) = (3, 4)$ in the rectangular coordinate system and draw a line through the two points. See Figure 28. The slope of the line drawn in Figure 28 is

$$m = \frac{y_2 - y_1}{x_2 - x_1} = \frac{4 - 4}{3 - (-2)}$$
$$= \frac{0}{5}$$
$$= 0$$

Figure 28

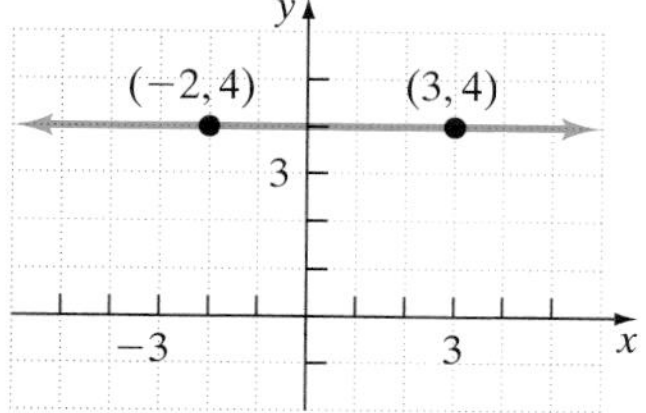

The slope of the line is 0. A slope of 0 can be interpreted as: There is no change in y when x increases by 1 unit.

Let's generalize the results of Example 4. Let $P = (x_1, y_1)$ and $Q = (x_2, y_2)$ be two distinct points. If $y_1 = y_2$, then we have a **horizontal line** whose slope m is 0. Figure 29 illustrates a horizontal line.

Figure 29

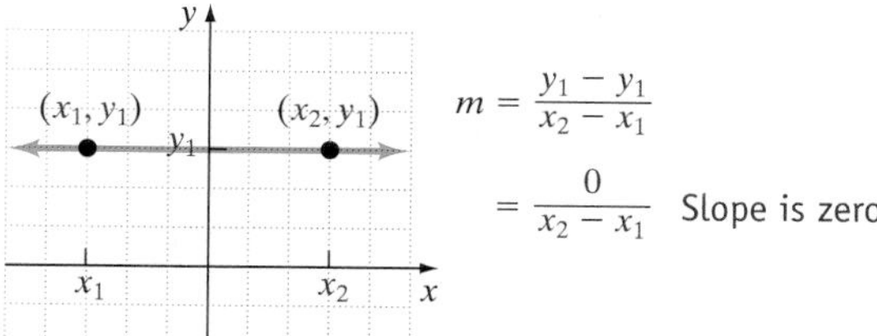

QUICK ✓ *Plot the given points in a rectangular coordinate system. Then draw a line through the two points. Finally, find and interpret the slope of the line.*

3. $(2, 5), (2, -1)$ **4.** $(2, 5), (6, 5)$ **5.** $(-3, -2), (1, -2)$ **6.** $\left(\frac{3}{2}, \frac{4}{3}\right), \left(\frac{3}{2}, \frac{8}{3}\right)$

SUMMARY: The Slope of a Line

Figure 30 illustrates the four possibilities for the slope of a line. Remember, just as we read a text from left to right, we also read graphs from left to right.

Figure 30

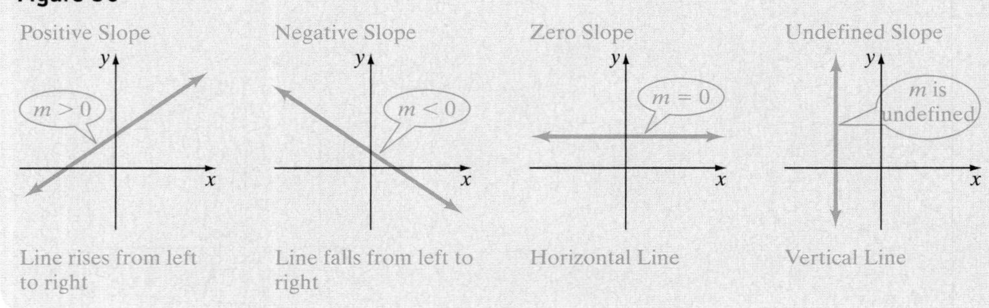

3 Graph a Line Using Its Slope and a Point on the Line

We now illustrate how to use the slope of a line to graph lines.

Classroom Example Draw a graph of the line that contains the point $(-1, 3)$ and has slope of 2.

Answer:

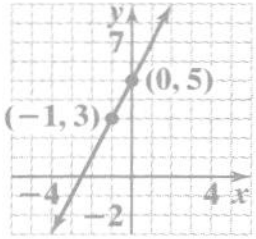

In Words
You can graph a line if you have a single point on the line and the slope of the line.

Teaching Tip
Point out that it takes two points to draw a line. One point is given and the slope acts like "driving directions" to arrive at the second point on the line.

EXAMPLE 5 Graphing a Line Given a Point and Its Slope

Draw a graph of the line that contains the point $(1, 3)$ and has a slope of 2.

Solution

Because slope $= \dfrac{\text{rise}}{\text{run}}$, we have that $2 = \dfrac{2}{1} = \dfrac{\text{rise}}{\text{run}}$. This means that y will increase by 2 units (the rise), when x increases by 1 unit (the run). So if we start at $(1, 3)$ and move 2 units up and then 1 unit to the right, we end up at the point $(2, 5)$. We then draw a line through the points $(1, 3)$ and $(2, 5)$ to obtain the graph of the line. See Figure 31.

Figure 31

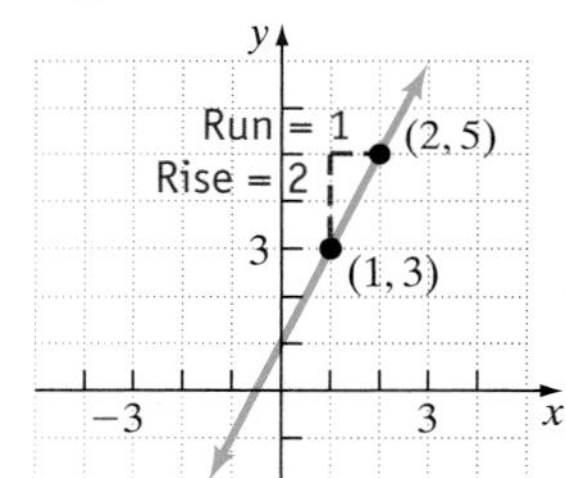

Classroom Example Draw a graph of the line that contains the point $(-1, 3)$ and has slope of $-\dfrac{2}{3}$.

Answer:

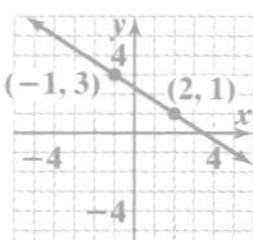

Work Smart
If the "rise" is positive, we go up. If the "rise" is negative, we go down. Similarly, if the "run" is positive, then we go to the right. If the "run" is negative, then we go to the left.

EXAMPLE 6 Graphing a Line Given a Point and Its Slope

Draw a graph of the line that contains the point $(1, 3)$ and has a slope of $-\dfrac{2}{3}$.

Solution

Because slope $= \dfrac{\text{rise}}{\text{run}}$, we have $-\dfrac{2}{3} = \dfrac{-2}{3} = \dfrac{\text{rise}}{\text{run}}$. This means that y will decrease by 2 units when x increases by 3 units. If we start at $(1, 3)$ and move 2 units down and then 3 units to the right, we end up at the point $(4, 1)$. We then draw a line through the points $(1, 3)$ and $(4, 1)$ to obtain the graph of the line. See Figure 32.

Figure 32

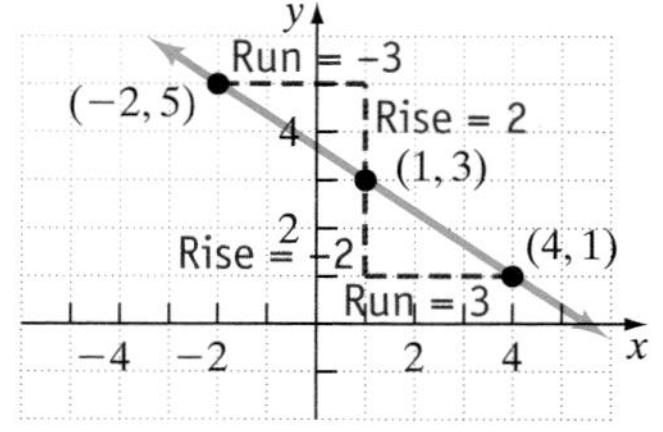

It is perfectly acceptable to set $\dfrac{\text{rise}}{\text{run}} = -\dfrac{2}{3} = \dfrac{2}{-3}$ so that we move 2 units up from $(1, 3)$ and then 3 units to the left. We would then end up at $(-2, 5)$, which is also on the graph of the line as indicated in Figure 32.

QUICK ✓

7. Draw a graph of the line that contains the point $(1, 2)$ and has a slope of

(a) $\frac{1}{2}$ **(b)** -3 **(c)** 0

4 Work with Applications of Slope

In its simplest form, slope is a ratio of rise over run. For example, if we are climbing a hill whose grade is 5% $\left(= 0.05 = \frac{5}{100}\right)$, then we go up 5 feet (the rise) for every 100 feet we travel horizontally (the run). See Figure 33.

Figure 33

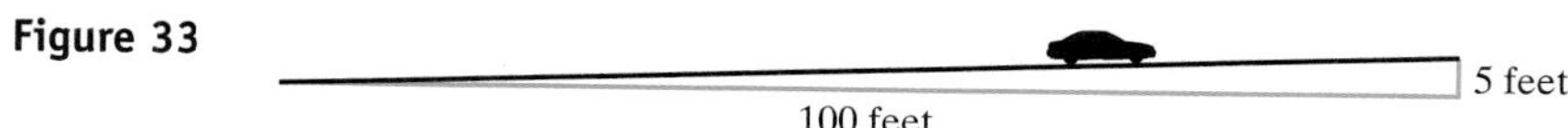

Or consider the pitch of a roof. If a roof's pitch is $\frac{5}{12}$, then every 5-foot measurement downward will result in a horizontal measurement of 12 feet. See Figure 34.

Work Smart
The pitch of a roof or grade of a road is always represented as a positive number.

Figure 34

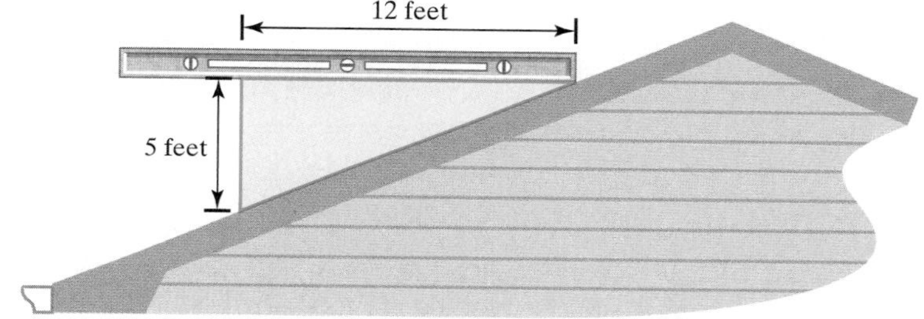

Classroom Example
A wheelchair ramp has a rise of 1 foot for every 12 feet of length. Tell the slope of the wheelchair ramp.

Answer: $\frac{1}{12}$

EXAMPLE 7 Finding the Grade of a Road

In Heckman Pass, British Columbia, there is a road that rises 9 feet for every 50 feet of horizontal distance covered. What is the grade of the road?

Solution

The grade of the road is given by $\frac{\text{rise}}{\text{run}}$. Since a rise of 9 feet is accompanied by a run of 50 feet, the grade of the road is $\frac{9 \text{ feet}}{50 \text{ feet}} = 0.18 = 18\%$.

The slope m of a line measures the amount that y changes as x changes from x_1 to x_2. The slope of a line is also called the **average rate of change** of y with respect to x.

In applications, we are often interested in knowing how the change in one variable might impact some other variable. For example, if your income increases by $1,000, how much will your spending (on average) change? Or, if the speed of your car increases by 10 miles per hour, how much (on average) will your car's gas mileage change?

Classroom Example
Use Example 8 with a 1750-square-foot house that costs $275,000, and a 1875-square-foot house that costs $284,000.

Answer: $m = 72$. Between 1750 and 1875 square feet, the price increases by $72 per square foot.

EXAMPLE 8 Slope as an Average Rate of Change

In Naples, Florida, the price of a new three-bedroom house that is 1828 square feet costs $280,000. The price of a new three-bedroom home that is 1987 square feet costs $296,000. Find and interpret the slope of the line joining the points $(1828, 280000)$ and $(1987, 296000)$.

Solution

Let x represent the square footage of the house and y represent the price. Let $(x_1, y_1) = (1828, 280000)$ and $(x_2, y_2) = (1987, 296000)$ and compute the slope as

$$m = \frac{y_2 - y_1}{x_2 - x_1} = \frac{296{,}000 - 280{,}000}{1987 - 1828} = \frac{16{,}000}{159} = 100.63$$

The unit of measure of y is dollars while the unit of measure for x is square feet. So, the slope can be interpreted as follows: Between 1828 and 1987 square feet, the price increases by \$100.63 per square foot, on average.

QUICK ✓

8. A road rises 4 feet for every 50 feet of horizontal distance covered. What is the grade of the road?
9. A roof is pitched so that the vertical drop is 2 inches for every horizontal measurement of 12 inches. What is the pitch of the roof?
10. The annual cost of gasoline and maintenance on a Chevy Cobalt is \$1370 when it is driven 10,000 miles. The annual cost of gasoline and maintenance on a Chevy Cobalt is \$1850 when it is driven 14,000 miles. Find and interpret the slope of the line joining (10000, 1370) and (14000, 1850).

8.3 Exercises

For Extra Help: Student Solutions Manual | CD Video | PH Math/Tutor Center | MathXL Tutorials on CD | MathXL® | MyMathLab

1. $\frac{6}{10}$ or $\frac{3}{5}$
2. 0; undefined
3. positive
4. False
5. True
6. True
7. vertical line; answers may vary
8. horizontal line; answers may vary

Concepts and Vocabulary

In Problems 1–3, fill in the blanks.

1. If the run of a line is 10 and its rise is 6, then its slope is __________.
2. The slope of a horizontal line is __________, while the slope of a vertical line is __________.
3. If the graph of a line goes up as you move to the right, then the slope of this line must be __________.

In Problems 4–6, answer True or False to each statement.

4. If $P = (x_1, y_1)$ and $Q = (x_2, y_2)$ are two distinct points with $y_1 \neq y_2$, the slope m of the line that contains points P and Q is defined by the formula $m = \frac{x_2 - x_1}{y_2 - y_1}$.
5. The slope of a line is also called the average rate of change of y with respect to x.
6. If the slope of a line is $\frac{3}{2}$, then y will increase by 3 units when x increases by 2 units.
7. Describe a line that has one x-intercept but no y-intercept. Give two ordered pairs that could lie on this line and then describe how to find its slope.
8. Describe a line that has one y-intercept but no x-intercept. Give two ordered pairs that could lie on this line and then describe how to find its slope.

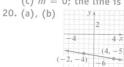

Building Skills

In Problems 9–16, find the slope of the line whose graph is given.

9.

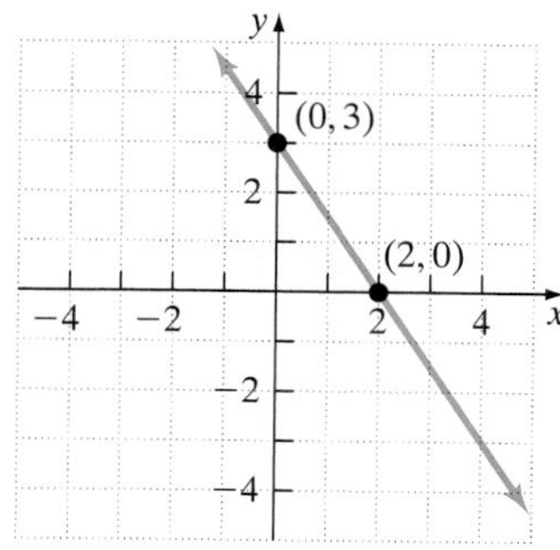

10.

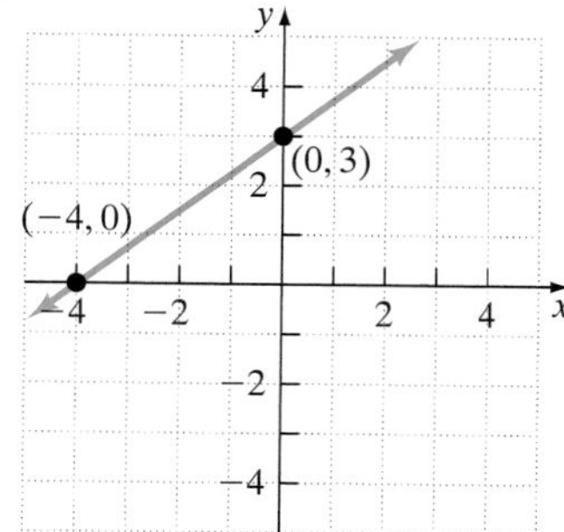

11.

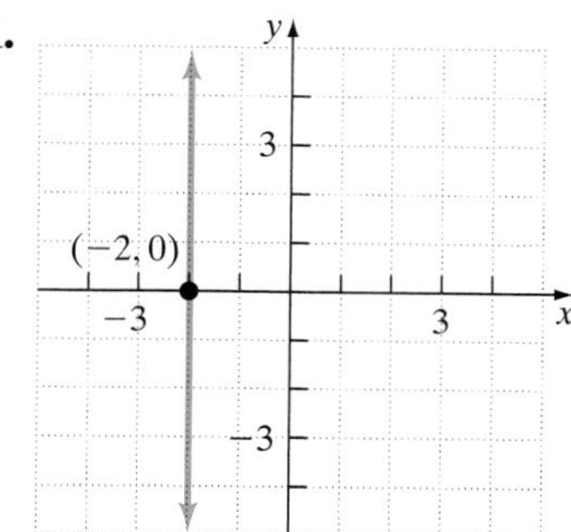

12.

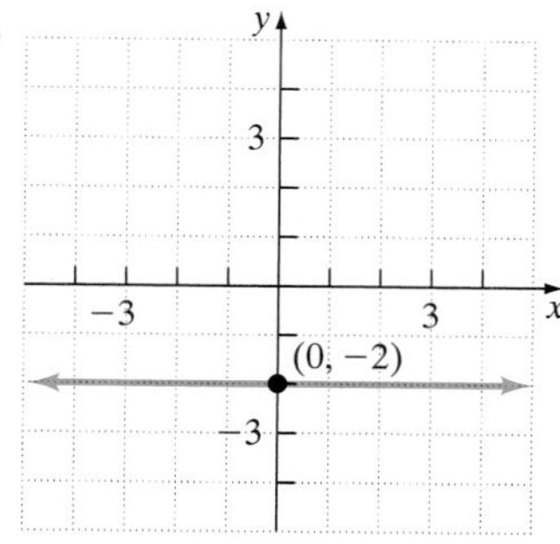

13.

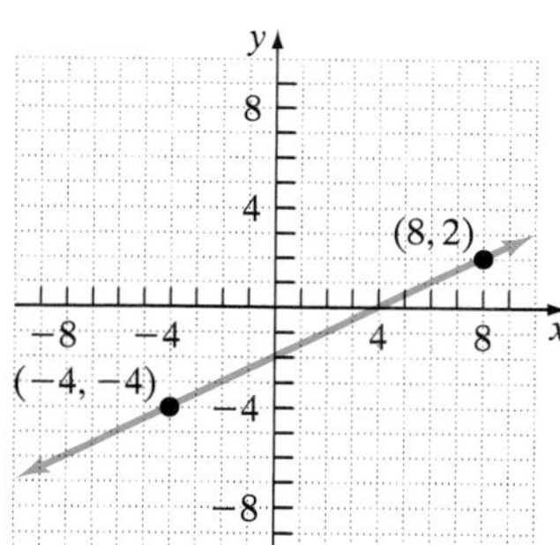

14.

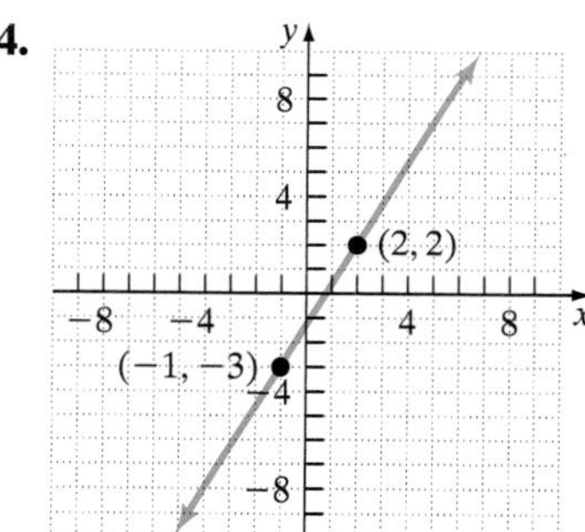

15.

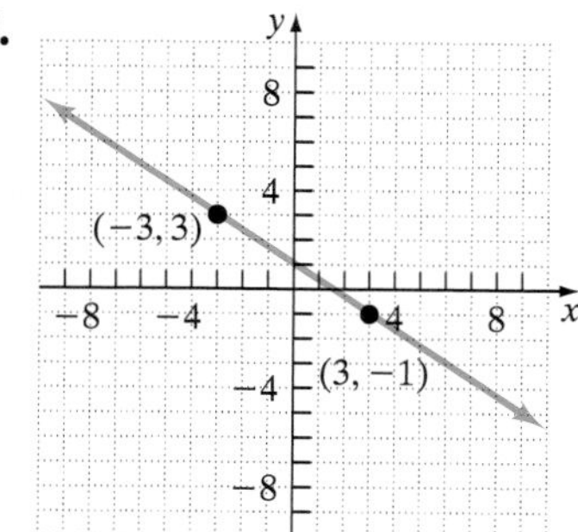

16.

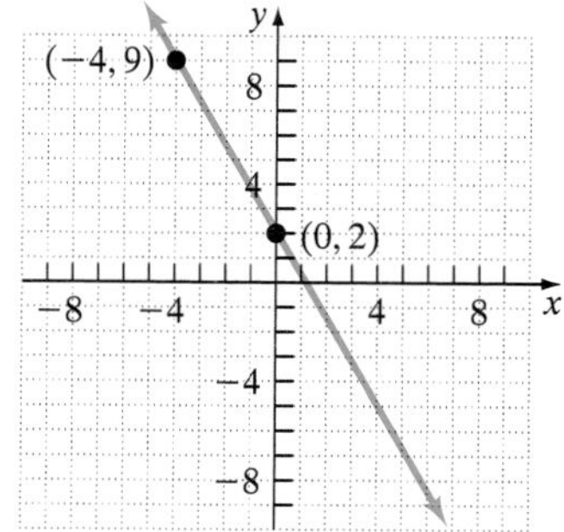

In Problems 17–20, (a) plot the points in a rectangular coordinate system, (b) draw a line through the points, (c) find and interpret the slope of the line.

17. $(-3, 2)$ and $(3, 5)$ **18.** $(-2, 6)$ and $(-2, -4)$

19. $(3, -1)$ and $(-2, -1)$ **20.** $(4, -5)$ and $(-2, -4)$

In Problems 21–40, find and interpret the slope of the line containing the given points.

21. $(10, 4)$ and $(6, 12)$ **22.** $(7, 3)$ and $(0, -11)$ **23.** $(4, -4)$ and $(12, -12)$

24. $(-3, 2)$ and $(2, -3)$ **25.** $(7, -2)$ and $(4, 3)$ **26.** $(-8, -1)$ and $(2, 3)$

9. $-\frac{3}{2}$
10. $\frac{3}{4}$
11. undefined
12. 0
13. $\frac{1}{2}$
14. $\frac{5}{3}$
15. $-\frac{2}{3}$
16. $-\frac{7}{4}$
17. (a), (b) (c) $m = \frac{1}{2}$; the value of y increases by 1 unit when x increases by 2 units.
18. (a), (b) (c) m is undefined; the line is vertical.
19. (a), (b) (c) $m = 0$; the line is horizontal.
20. (a), (b) (c) $m = -\frac{1}{6}$; the value of y decreases by 1 unit when x increases by 6 units.
21. $m = -2$; the value of y decreases by 2 units when x increases by 1 unit.
22. $m = 2$; the value of y increases by 2 units when x increases by 1 unit.
23. $m = -1$; the value of y decreases by 1 unit when x increases by 1 unit.
24. $m = -1$; the value of y decreases by 1 unit when x increases by 1 unit.
25. $m = -\frac{5}{3}$; the value of y decreases by 5 units when x increases by 3 units.
26. $m = \frac{2}{5}$; the value of y increases by 2 units when x increases by 5 units.
27. $m = \frac{3}{2}$; the value of y increases by 3 units when x increases by 2 units.
28. $m = \frac{3}{5}$; the value of y increases by 3 units when x increases by 5 units.
29. $m = 0$; there is no change in the y-values. The line is horizontal.
30. m is undefined; the line is vertical.
31. $m = \frac{2}{3}$; the value of y increases by 2 units when x increases by 3 units.
32. $m = \frac{1}{3}$; the value of y increases by 1 unit when x increases by 3 units.

27. $(0, 6)$ and $(-4, 0)$ **28.** $(-5, 0)$ and $(0, 3)$ **29.** $(4, -6)$ and $(-1, -6)$

30. $(-1, -3)$ and $(-1, 2)$ **31.** $(-4, -1)$ and $(2, 3)$ **32.** $(5, 1)$ and $(-1, -1)$

33. $(3, 9)$ and $(3, -2)$ **34.** $(5, 1)$ and $(-2, 1)$ **35.** $\left(\frac{1}{2}, \frac{3}{4}\right)$ and $\left(-\frac{5}{2}, -\frac{1}{4}\right)$

36. $\left(-\frac{1}{3}, \frac{2}{5}\right)$ and $\left(\frac{2}{3}, -\frac{3}{5}\right)$ **37.** $\left(\frac{1}{3}, \frac{4}{9}\right)$ and $\left(-\frac{1}{3}, \frac{2}{9}\right)$ **38.** $\left(\frac{1}{4}, -\frac{4}{3}\right)$ and $\left(-\frac{5}{4}, \frac{1}{3}\right)$

39. $\left(\frac{1}{2}, \frac{1}{3}\right)$ and $\left(\frac{3}{4}, \frac{5}{6}\right)$ **40.** $\left(\frac{1}{3}, \frac{3}{4}\right)$ and $\left(-\frac{2}{5}, \frac{1}{2}\right)$

In Problems 41–58, draw a graph of the line that contains the given point and has the given slope.

41. $(4, 2)$; $m = 1$ **42.** $(3, -1)$; $m = -1$ **43.** $(0, 6)$; $m = -2$

44. $(-1, 3)$; $m = 3$ **45.** $(-1, 0)$; $m = \frac{1}{4}$ **46.** $(5, 2)$; $m = -\frac{1}{2}$

47. $(2, -3)$; $m = 0$ **48.** $(1, 4)$; $m =$ undefined **49.** $(2, 1)$; $m = \frac{2}{3}$

50. $(-2, -3)$; $m = \frac{5}{2}$ **51.** $(-1, -4)$; $m = \frac{5}{3}$ **52.** $(0, -2)$; $m = -\frac{3}{2}$

53. $(0, 0)$; $m =$ undefined **54.** $(3, -1)$; $m = 0$ **55.** $(0, 2)$; $m = -4$

56. $(0, 0)$; $m = \frac{1}{5}$ **57.** $(2, -3)$; $m = \frac{3}{4}$ **58.** $(-3, 0)$; $m = -3$

Applying the Concepts

In Problems 59–70, determine the missing value so that the line containing the two points will have the required slope.

59. $(3, 7)$ and $(x, 2)$; $m = 5$ **60.** $(-2, 6)$ and $(x, 4)$; $m = -2$

61. $(-1, 3)$ and $(9, y)$; $m = -\frac{1}{2}$ **62.** $(-1, -6)$ and $(4, y)$; $m = 2$

63. $(-2, 4)$ and $(x, 5)$; $m =$ undefined **64.** $(0, 4)$ and $(5, y)$; $m = 0$

65. $(x, -8)$ and $(-2, -3)$; $m = -1$ **66.** $(-12, y)$ and $(3, -1)$; $m = -\frac{1}{3}$

67. $(x, -3)$ and $(-6, 0)$; $m = -\frac{1}{2}$ **68.** $(4, y)$ and $(-3, -1)$; $m = \frac{1}{7}$

69. $\left(\frac{1}{4}, \frac{1}{3}\right)$ and $\left(\frac{3}{4}, y\right)$; $m = 2$ **70.** $\left(\frac{1}{2}, \frac{3}{5}\right)$ and $\left(x, \frac{1}{5}\right)$; $m = \frac{2}{15}$

In Problems 71–74, draw the graph of the two lines with the given properties on the same rectangular coordinate system.

71. Both lines pass through the point $(2, -1)$. One has slope of 2 and the other has slope of $-\frac{1}{2}$.

72. Both lines pass through the point $(3, 0)$. One has slope of $\frac{2}{3}$ and the other has slope of $-\frac{3}{2}$.

73. Both lines have a slope of $\frac{3}{4}$. One passes through the point $(-1, -2)$, and the other passes through the point $(2, 1)$.

33. m is undefined; the line is vertical.
34. $m = 0$; there is no change in the y-values. The line is horizontal.
35. $m = \frac{1}{3}$; the value of y increases by 1 unit when x increases by 3 units.
36. $m = -1$; the value of y decreases by 1 unit when x increases by 1 unit.
37. $m = \frac{1}{3}$; the value of y increases by 1 unit when x increases by 3 units.
38. $m = -\frac{10}{9}$; the value of y decreases by 10 units when x increases by 9 units.
39. $m = 2$; the value of y increases by 2 units when x increases by 1 unit.
40. $m = \frac{15}{44}$; the value of y increases by 15 units when x increases by 44 units.

41. 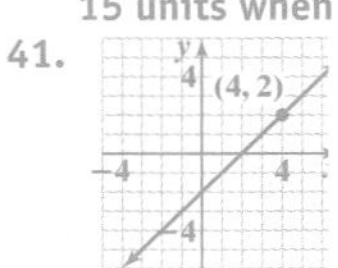

42. 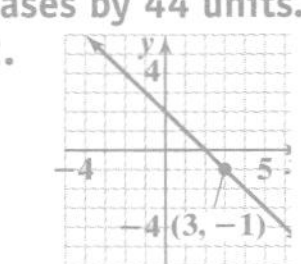

43. 45. 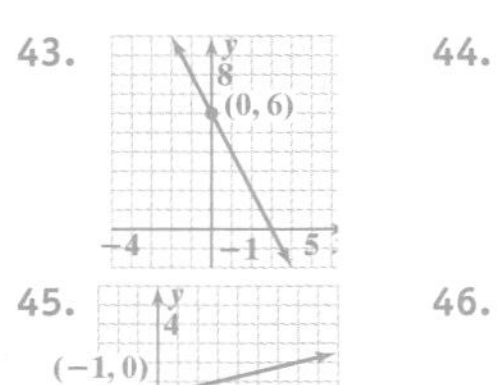

44. 46. 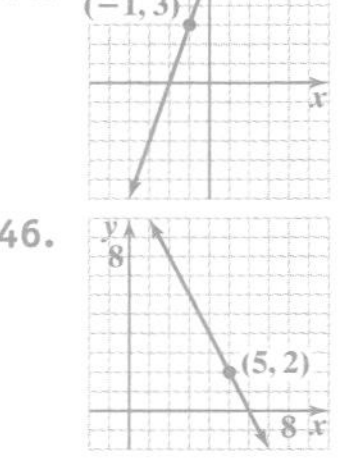

47. 49. 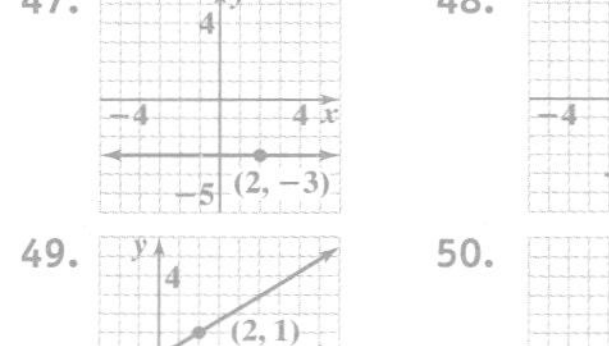

48. 50. 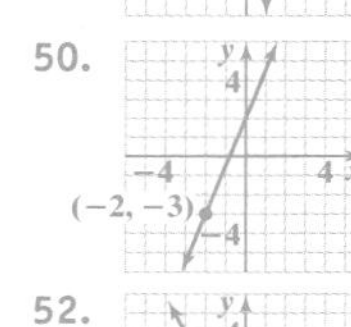

51. 53. 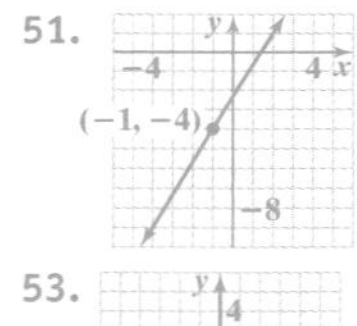

52. 54. 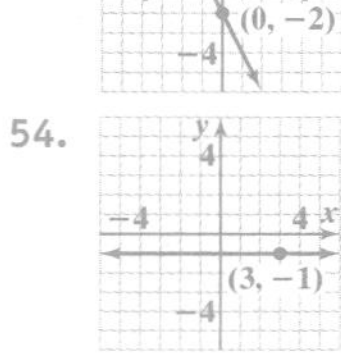

55. 56. 57. 58. 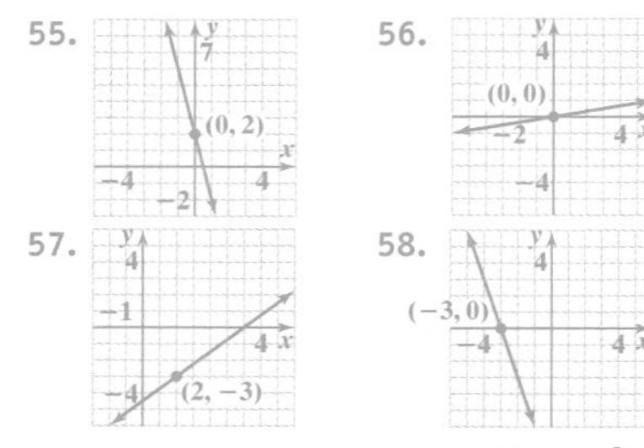

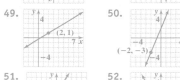

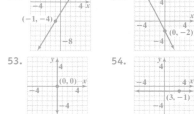

59. $x = 2$ 60. $x = -1$ 61. $y = -2$
62. $y = 4$ 63. $x = -2$ 64. $y = 4$
65. $x = 3$ 66. $y = 4$ 67. $x = 0$
68. $y = 0$ 69. $y = \frac{4}{3}$ 70. $x = -\frac{5}{2}$

71.

72.
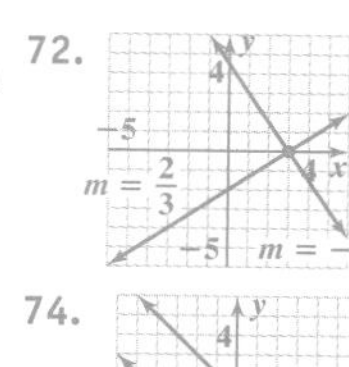

73.
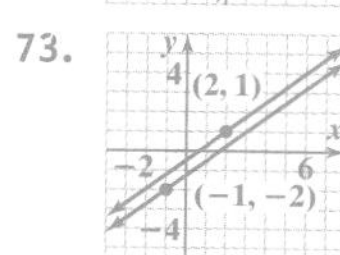

74.

75. $\frac{1}{3}$

76. $\frac{4}{45}$

77. 12 in. or 1 ft

78. $1\frac{3}{4}$ ft

79. 16% 80. 15%

81. $m = 2.26$ million; the population is increasing at an average rate of about 2.26 million people per year.

82. $m = 0.225$ million; the income increases by 0.225 million dollars or $225,000 for each year of college.

83. Points may vary. $(-2, 1)$, $(0, -5)$ $m = -3$

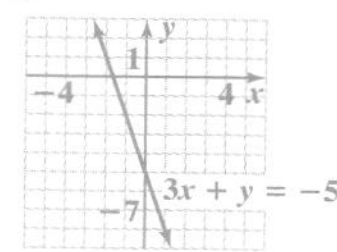

84. Points may vary. $(1, 2)$, $(-4, 4)$ $m = \frac{-2}{5}$

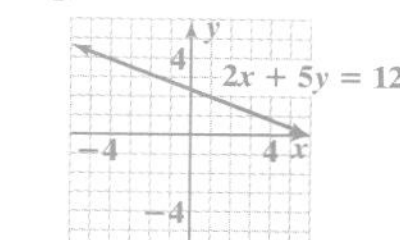

85. Points may vary. $(4, -10)$, $(-2, -10)$ $m = 0$

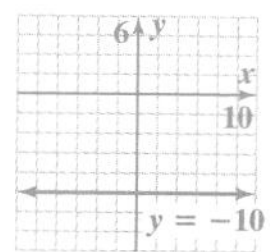

86. Points may vary. $(7, 0)$, $(7, 1)$ m is undefined.

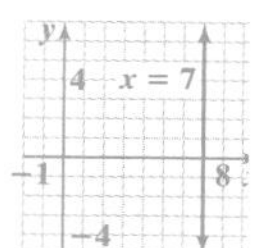

87. Points may vary. $(-2, -2)$, $(0, 4)$ $m = 3$

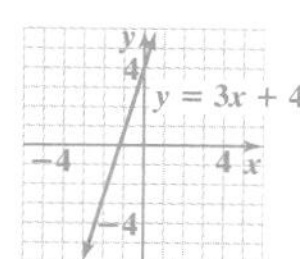

88. Points may vary. $(-2, -4)$, $(-4, -2)$ $m = -1$

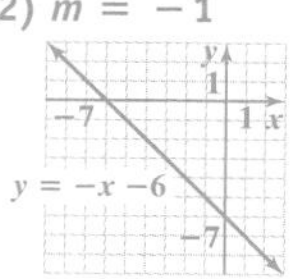

74. Both lines have a slope of -1. One passes through the point $(0, -3)$, and the other passes through the point $(2, -1)$.

75. Roof Pitch A carpenter who was installing a new roof on a garage noticed that for every one-foot horizontal run, the roof was elevated by 4 inches. What is the pitch of this roof?

76. Roof Pitch A canopy is set up on the football field. On the 45-yard line, the height of the canopy is 68 inches. The peak of the canopy is at the 50-yard marker where the height is 84 inches. What is the pitch of the roof of the canopy?

77. Building a Roof To build a shed in his back yard, Moises has decided to use a pitch of $\frac{2}{5}$ for his roof. The shed measures 30" from the side to the center. How much height should he add to the roof to get the desired pitch?

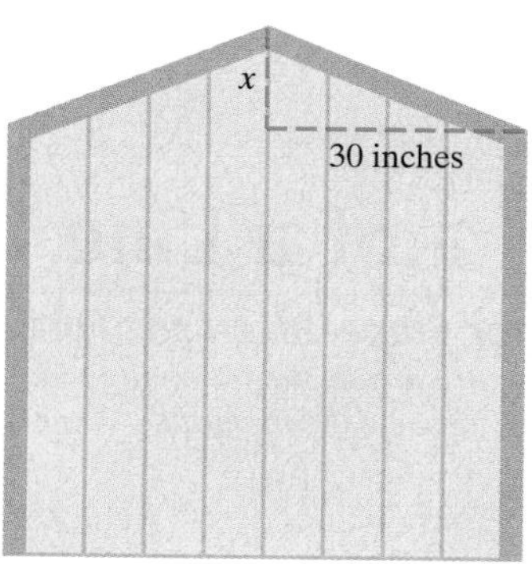

78. Building a Roof The design for the bedroom of a house requires a roof pitch of $\frac{7}{20}$. If the room measures 5 feet from the wall to the center, how high above the ceiling is the peak of the roof?

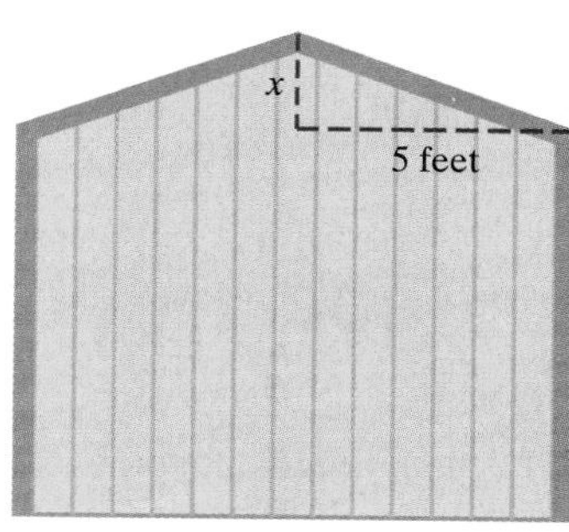

79. Road Grade Fall River Road was completed in 1920 and was the first road built through the Rocky Mountains in Colorado. It was so steep that sometimes the early model cars had to drive up the hill in reverse to maximize their weak engines and fuel systems. If the road rises 200 feet for every 1250 feet of horizontal change, in percent, what is the grade of this road?

80. Road Grade Barbara decided to take a bicycle trip up to the observatory on Mauna Kea on the island of Hawaii. The road has a vertical rise of 120 feet for every 800 feet of horizontal change. In percent, what is the grade of this road?

81. Population Growth The population of the United States was 123,202,624 in 1930 and 281,421,906 in 2000. Use the ordered pairs (0, 123 million) and (70, 281 million) to find and interpret the slope of line representing the average rate of change in the population of the United States.

82. Earning Potential On average, a person who graduates from high school can expect to have lifetime earnings of 1.2 million dollars. It takes four years to earn a bachelor's degree, but the lifetime earnings will increase to 2.1 million dollars. Use the ordered pairs (0, 1.2 million) and (4, 2.1 million) to find and interpret the slope of the line representing the increase in earnings due to finishing college.

Extending the Concepts

In Problems 83–88, find any two ordered pairs that lie on the given line. Graph the line and then determine the slope of the line.

83. $3x + y = -5$ **84.** $2x + 5y = 12$ **85.** $y = -10$

86. $x = 7$ **87.** $y = 3x + 4$ **88.** $y = -x - 6$

89. $m = -2$
90. $m = -\frac{1}{2}$
91. $m = \frac{q}{p}$
92. $m = -\frac{q}{p}$
93. $m = \frac{6}{a-6}$
94. $m = \frac{4b-5}{2a+10}$
95. $MR = 2$; for every hot dog sold, revenue increases by $2.
96. $MR = 5$; for every CD sold, revenue increases by $5.

In Problems 89–94, find the slope of the line containing the given points.

89. $(2a, a)$ and $(3a, -a)$

90. $(4p, 2p)$ and $(-2p, 5p)$

91. $(2p + 1, q - 4)$ and $(3p + 1, 2q - 4)$

92. $(3p + 1, 4q - 7)$ and $(5p + 1, 2q - 7)$

93. $(a + 1, b - 1)$ and $(2a - 5, b + 5)$

94. $(2a - 3, b + 4)$ and $(4a + 7, 5b - 1)$

*In economics, **marginal revenue** is a rate of change defined as the change in total revenue divided by the change in output. If Q_1 represents the number of units sold, then the total revenue from selling these goods is represented by R_1. If Q_2 represents a different number of units sold, then the total revenue from this sale is represented by R_2. We compute marginal revenue as $MR = \frac{R_2 - R_1}{Q_2 - Q_1}$.*

So marginal revenue is a rate of change or slope. Marginal revenue is important in economics because it is used to determine the level of output that maximizes profits for a company. Use the marginal revenue formula to solve Problems 95 and 96.

95. Determine and interpret marginal revenue if total revenue is $1000 when 400 hot dogs are sold at a baseball game and total revenue is $1200 when 500 hot dogs are sold.

96. Determine and interpret marginal revenue if total revenue is $300 when 30 compact disks are sold and total revenue is $400 when 50 compact disks are sold.

8.4 Slope-Intercept Form of a Line

OBJECTIVES

1. Use the Slope-Intercept Form to Identify the Slope and y-Intercept of a Line
2. Graph a Line Whose Equation Is in Slope-Intercept Form
3. Graph a Line Whose Equation Is in the Form $Ax + By = C$
4. Find the Equation of a Line Given Its Slope and y-Intercept
5. Work with Linear Models in Slope-Intercept Form

Preparing for Slope-Intercept Form of a Line

Before getting started, take the following readiness quiz. If you get a problem wrong, go back to the section cited and review the material.

1. Solve $4x + 2y = 10$ for y. [Section 2.4, pp. 108–111]

2. Solve: $10 = 2x - 8$ [Section 2.2, pp. 84–86]

1 Use the Slope-Intercept Form to Identify the Slope and y-Intercept of a Line

We have defined a linear equation as an equation of the form $Ax + By = C$, where A and B are not both zero. So far, we have graphed equations of this form using point plotting or by finding its intercepts. From the previous section, we know the slope can be used to help us graph a line.

In this section, we use the slope and y-intercept to graph a line. This method for graphing will be more efficient than plotting points. Why? Well, suppose we wish to graph the equation $-2x + y = 5$ by plotting points. To do this, we will first solve the equation for y (get y by itself), then choose values of x and use the equation to find the corresponding value of y. The reason for doing this will become clear soon. We solve the equation $-2x + y = 5$ for y by adding $2x$ to both sides of the equation.

$$-2x + y = 5$$

Add $2x$ to both sides: $y = 2x + 5$

Table 14

x	$y = 2x + 5$	(x, y)
-2	$2(-2) + 5 = 1$	$(-2, 1)$
-1	$2(-1) + 5 = 3$	$(-1, 3)$
0	$2(0) + 5 = 5$	$(0, 5)$

We now create Table 14, which gives us points on the graph of the equation. Figure 35 shows the graph of the line.

Notice two things about the line in Figure 35. First, the slope is $m = 2$. Second, the y-intercept is 5. If you look back at the form of the equation $-2x + y = 5$ after we solved for y, namely, $y = 2x + 5$, you should notice that the coefficient of the variable x is 2 and the constant is 5. This is no coincidence!

Figure 35

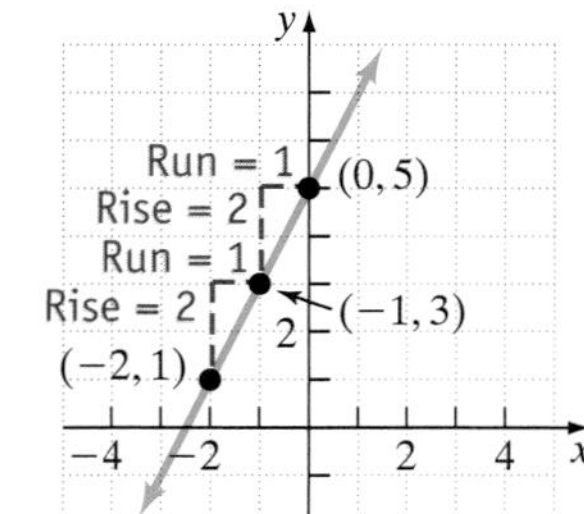

Preparing for...Answers
1. $y = -2x + 5$ **2.** $\{9\}$

SLOPE-INTERCEPT FORM OF AN EQUATION OF A LINE

An equation of a line with slope m and y-intercept b is

$$y = mx + b$$

Classroom Example ➤
Find the slope and y-intercept of the line whose equation is $y = 4x + 5$.

Answer: $m = 4$, y-intercept is 5

EXAMPLE 1 Finding the Slope and y-Intercept of a Line

Find the slope and y-intercept of the line whose equation is $y = -3x + 1$.

Solution

We compare the equation $y = -3x + 1$ to the slope-intercept form of a line $y = mx + b$ and find that the coefficient of x, -3, is the slope and the constant, 1, is the y-intercept.

Classroom Example ➤
Find the slope and y-intercept of the line whose equation is $4x - 3y = 12$.

Answer: $m = \frac{4}{3}$, y-intercept is $b = -4$

EXAMPLE 2 Finding the Slope and y-Intercept of a Line Whose Equation Is in Standard Form

Find the slope and y-intercept of the line whose equation is $3x + 2y = 6$.

Solution

First, we must rewrite the equation $3x + 2y = 6$ so that it is of the form $y = mx + b$. That is, we want to solve the equation for y.

$$3x + 2y = 6$$

Subtract $3x$ from both sides: $2y = -3x + 6$

Divide both sides by 2: $y = \dfrac{-3x + 6}{2}$

$\dfrac{a+b}{c} = \dfrac{a}{c} + \dfrac{b}{c}$: $y = \dfrac{-3}{2}x + \dfrac{6}{2}$

Simplify: $y = -\dfrac{3}{2}x + 3$

Work Smart

Notice that after we subtracted $3x$ from both sides, we wrote the equation as $2y = -3x + 6$ rather than $2y = 6 - 3x$. This is because we want to get the equation in the form $y = mx + b$, so the term involving x should be first.

Now we compare the equation $y = -\frac{3}{2}x + 3$ to the slope-intercept form of a line $y = mx + b$ and find that the coefficient of x, $-\frac{3}{2}$, is the slope and the constant, 3, is the y-intercept. So for the line $3x + 2y = 6$, $m = -\frac{3}{2}$, and the y-intercept is $b = 3$.

Teaching Tip

Emphasize to students that the slope of the line is the **numerical** coefficient of x. It's incorrect to say the slope of a line is $\frac{3}{2}x$.

QUICK ✓ *Find the slope and y-intercept of the line whose equation is given.*

1. $y = 4x - 3$ **2.** $3x + y = 7$ **3.** $2x + 5y = 15$

(2) Graph a Line Whose Equation Is in Slope-Intercept Form

In the previous section, we graphed an equation of a line using a point on the line and its slope. If an equation is in slope-intercept form, we can graph the line by plotting the y-intercept and using the slope to find another point on the line.

EXAMPLE 3 How to Graph a Line Whose Equation Is in Slope-Intercept Form

Graph the line $y = 3x - 1$ using the slope and y-intercept.

Step-by-Step Solution

Step 1: Identify the slope and y-intercept of the line.

$$y = 3x - 1$$
$$y = 3x + (-1)$$

$m = 3$ $b = -1$

The slope is $m = 3$ and the y-intercept is $b = -1$.

Step 2: Plot the y-intercept and then use the slope to find a second point on the graph. Draw a straight line through the points.

Plot the y-intercept at $(0, -1)$. Use the slope $m = \frac{3}{1} = \frac{\text{rise}}{\text{run}}$ to find a second point on the graph. See Figure 36.

Figure 36

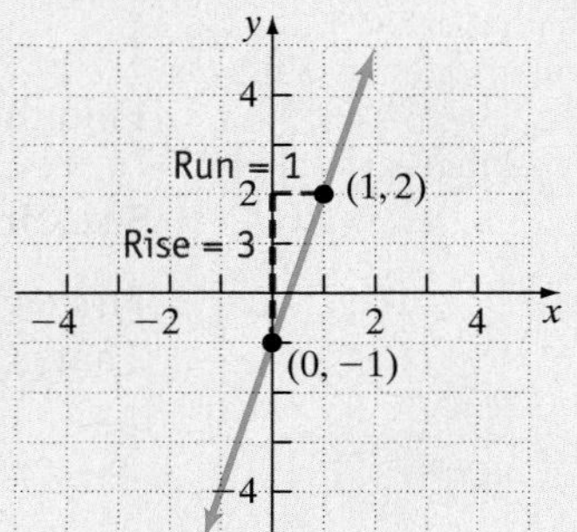

Classroom Example ▲
Graph the line $y = 2x + 2$ using the slope and y-intercept.

Answer: $m = 2$, y-intercept is 2

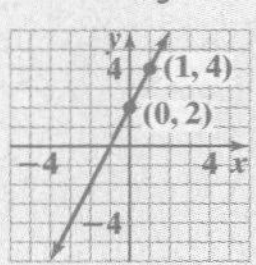

QUICK ✓ *Graph the line using the slope and y-intercept.*

4. $y = 2x - 5$

5. $y = \frac{1}{2}x - 5$

Classroom Example ▼
Graph the line $y = -\frac{2}{3}x + 1$ using the slope and y-intercept.

Answer: $m = -\frac{2}{3}$, $b = 1$

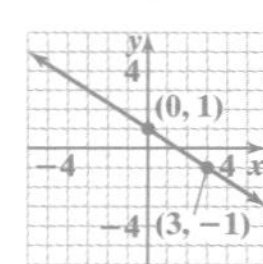

EXAMPLE 4 Graphing a Linear Equation Whose Equation Is in Slope-Intercept Form

Graph the line $y = -\frac{4}{3}x + 2$ using the slope and y-intercept.

Solution

First, we determine the slope and y-intercept.

$$y = -\frac{4}{3}x + 2$$

$m = -\frac{4}{3}$ $b = 2$

The slope is $m = -\frac{4}{3}$ and the y-intercept is $b = 2$. We plot the point $(0, 2)$. Now use the slope $m = -\frac{4}{3} = \frac{-4}{3} = \frac{\text{rise}}{\text{run}}$ to find a second point on the graph. We then draw a line through these two points. See Figure 37.

Figure 37

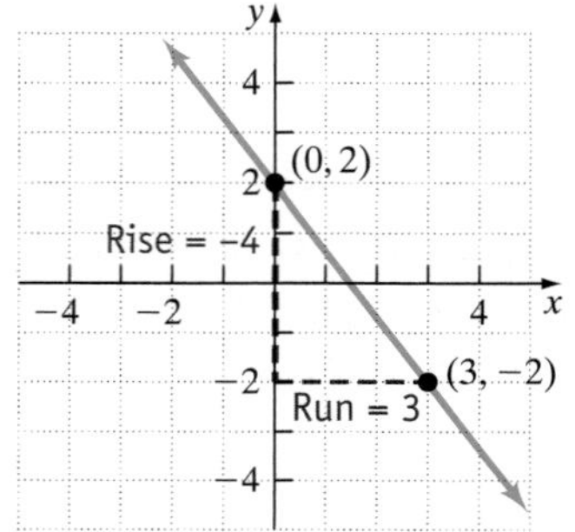

Classroom Example Graph the line $2x - 3y = -6$ using the slope and y-intercept.

Answer: $m = \frac{2}{3}, b = 2$

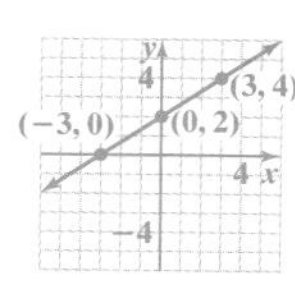

QUICK ✓ *Graph the line using the slope and y-intercept.*

6. $y = -3x + 1$ **7.** $y = -\frac{3}{2}x + 4$

3 Graph a Line Whose Equation Is in the Form $Ax + By = C$

If a linear equation is written in standard form $Ax + By = C$, we can still use the slope and y-intercept to obtain the graph of the equation. Let's see how.

EXAMPLE 5 How to Graph a Line Whose Equation Is in the Form $Ax + By = C$

Graph the line $8x + 2y = 10$ using the slope and y-intercept.

Step-by-Step Solution

Step 1: Solve the equation for y to put it in the form $y = mx + b$.

Teaching Tip Remind students that the slope of a line is the coefficient of x only when the equation is in slope-intercept form: $y = mx + b$.

$$8x + 2y = 10$$

Subtract $8x$ from both sides: $2y = -8x + 10$

Divide both sides by 2: $y = \frac{-8x + 10}{2}$

Simplify: $y = -4x + 5$

Step 2: Identify the slope and y-intercept of the line.

The slope is -4 and the y-intercept is 5.

Step 3: Plot the y-intercept and then use the slope to find a second point on the graph. Draw a straight line through the points.

Plot the point $(0, 5)$ and use the slope $m = -4 = \frac{-4}{1} = \frac{\text{rise}}{\text{run}}$ to find a second point on the graph. See Figure 38.

Figure 38

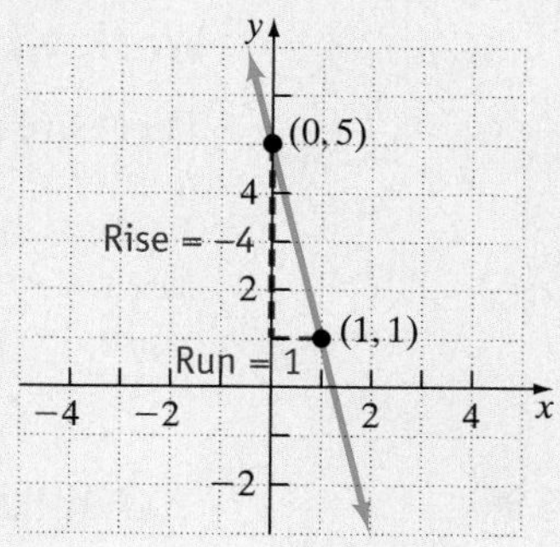

Work Smart
An alternative to graphing the equation in Example 5 using the slope and y-intercept would be to graph the line using intercepts.

QUICK ✓ *Graph each line using the slope and y-intercept.*

8. $-2x + y = -3$ **9.** $6x - 2y = 2$ **10.** $3x + 5y = 0$

4 Find the Equation of a Line Given Its Slope and y-Intercept

Teaching Tip This concept is just reversing the process of finding the slope and y-intercept from the equation of a line.

Up to now, we have identified the slope and y-intercept from its equation. We will now reverse the process and find the equation of a line whose slope and y-intercept are given. This is a fairly straightforward process—replace m with the given slope and b with the y-intercept.

Classroom Example ➤
Find the equation of a line whose slope is $\frac{2}{5}$ and whose y-intercept is -2.

Answer: $y = \frac{2}{5}x - 2$

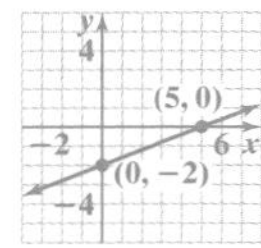

EXAMPLE 6 Finding the Equation of a Line Given Its Slope and y-Intercept

Find the equation of a line whose slope is $\frac{3}{8}$ and whose y-intercept is -4. Graph the line.

Solution

The slope is $m = \frac{3}{8}$ and the y-intercept is $b = -4$. Substitute $\frac{3}{8}$ for m and -4 for b in the slope-intercept form of a line $y = mx + b$ to obtain

$$y = \frac{3}{8}x - 4$$

Figure 39 shows the graph of the equation.

Figure 39

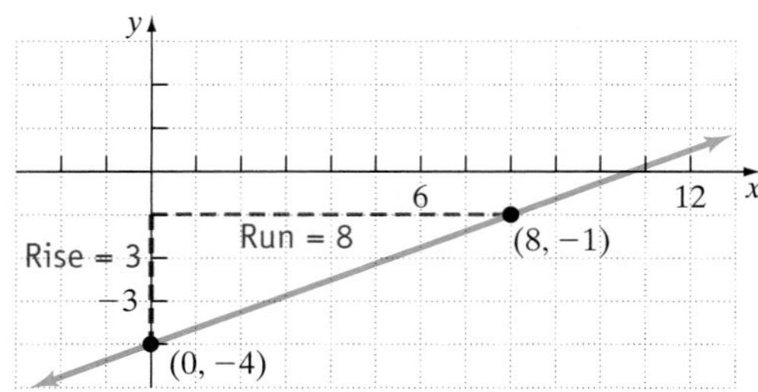

QUICK ✓ *Find the equation of the line whose slope and y-intercept are given. Graph the line.*

11. $m = 3, b = -2$ **12.** $m = -\frac{1}{4}, b = 3$ **13.** $m = 0, b = -1$

5 Work with Linear Models in Slope-Intercept Form

There are many situations where we can use a linear equation to describe the relationship that exists between two variables. For example, your long-distance phone bill depends linearly upon the number of minutes used or the cost of renting a moving truck depends linearly on the number of miles driven. Let's look at a couple of examples.

Classroom Example ➤
Use Example 7 except (a) predict the total cholesterol of a woman who is 50 years old. Use (b), (c), and (d) as given.

Answer: (a) Total cholesterol for a 50-year-old woman is predicted to be 212 mg/dL.

EXAMPLE 7 A Model for Total Cholesterol

When you have a physical exam, your doctor draws blood for your "cholesterol test." Your total cholesterol count is measured in milligrams per deciliter (mg/dL). It is the sum of low-density lipoprotein cholesterol (LDL)—sometimes called "bad cholesterol"—and high-density lipoprotein cholesterol (HDL)—sometimes called your "good cholesterol." Based on data from the National Center for Health Statistics, a woman's total cholesterol y is related to her age x by the following linear equation:

$$y = 1.1x + 157$$

(a) Use the equation to predict the total cholesterol of a woman who is 40 years old.

(b) Determine and interpret the slope of the equation.

(c) Determine and interpret the y-intercept of the equation.

(d) Graph the equation in a rectangular coordinate system.

Solution

(a) Because x represents the woman's age, we substitute 40 for x in the equation $y = 1.1x + 157$ to find the total cholesterol y.

$$\begin{aligned} & y = 1.1x + 157 \\ x = 40: \quad & y = 1.1(40) + 157 \\ & = 201 \end{aligned}$$

We predict that a 40-year-old woman will have a total cholesterol of 201 mg/dL.

(b) The slope of the equation $y = 1.1x + 157$ is 1.1. Because slope equals $\frac{\text{rise}}{\text{run}} = \frac{1.1 \text{ mg/dL}}{1 \text{ year}}$, we interpret the slope as follows: The total cholesterol of a female increases by 1.1 mg/dL as age increases by 1 year."

(c) The y-intercept of the equation $y = 1.1x + 157$ is 157. The y-intercept is the value of total cholesterol, y, when $x = 0$. Since x represents age, we interpret the y-intercept as follows: "The total cholesterol of a newborn girl is 157 mg/dL."

(d) Figure 40 shows the graph of the equation. Because it does not make sense for x to be less than 0, we only graph the equation in quadrant I.

Work Smart

Notice that we did not use the slope to obtain an additional point on the graph of the equation. It would be difficult to find an additional point with a slope of 1.1. For example, from the y-intercept, we would go up 1.1 mg/dL and right 1 year and end up at (1, 158.1). It would be hard to draw the line through these two points!

Figure 40

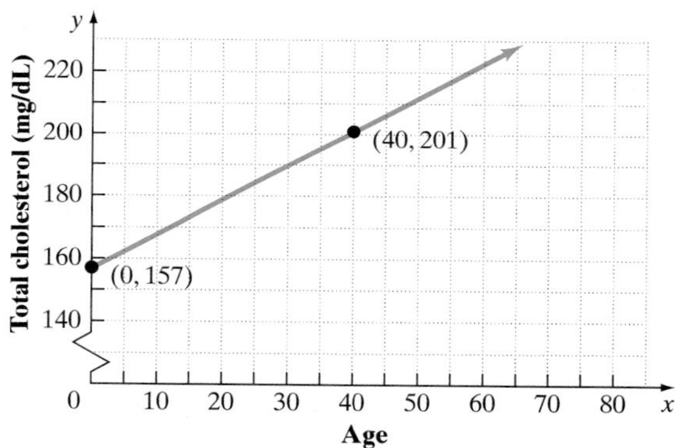

In Figure 40, notice the "broken line" (⌇) on the y-axis near the origin. We include this in the graph to indicate that a portion of the graph has been removed. This is done so that we do not have to start the y-axis at 0 and work our way higher. This avoids a lot of white space in the graph. Whenever you are reading a graph, always look carefully at how the axes are labeled and the units on each axis.

QUICK ✓

14. Based on data obtained from the National Center for Health Statistics, the birth weight y of a baby, measured in grams, is linearly related to gestation period x (in weeks) according to the equation

$$y = 143x - 2215$$

(a) Use the equation to predict the birth weight of a baby if the gestation period is 30 weeks.

(b) Use the equation to predict the birth weight of a baby if the gestation period is 36 weeks.

(c) Determine and interpret the slope of the equation.

(d) Explain why it does not make sense to interpret the y-intercept of the equation.

(e) Graph the equation in a rectangular coordinate system for $28 \leq x \leq 43$.

We know that the slope can be interpreted as a rate of change. For this reason, when information in a problem is given as a rate of change as in miles per gallon or dollars per pound, the rate of change will represent the slope in a linear model.

Classroom Example Use Example 8 with average cost = \$0.38 per mile.

Answer:

(a) $y = 0.38x + 3000$

(b) \$7180

(c)

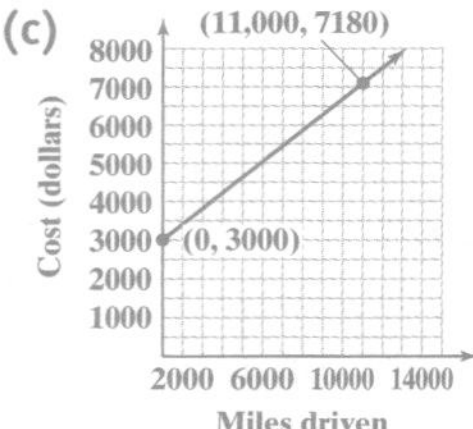

EXAMPLE 8 Cost of Owning and Operating a Car

There are many costs that factor into owning a car including gas, maintenance, and insurance. Some of these costs are affected by the number of miles that are driven (gas and maintenance), while others are not (comprehensive insurance, license plates, depreciation). Suppose the annual cost of operating a Chevy Cobalt is \$0.25 per mile plus \$3000.

(a) Write a linear equation that relates the annual cost of operating the car y to the number of miles driven in a year x.

(b) What is the annual cost of driving 11,000 miles?

(c) Graph the equation in a rectangular coordinate system.

Solution

(a) The rate of change in the problem is \$0.25 per mile. We can express this as $\frac{\$0.25}{1 \text{ mile}}$, which is the slope m of the linear equation. The cost of \$3000 is a cost that does not change with the number of miles driven. Put another way, if we drive 0 miles, the cost will be \$3000, so this value represents the y-intercept, b. The linear equation that relates cost y to the number of miles driven x is

$$y = 0.25x + 3000$$

(b) Let $x = 11{,}000$ in the equation $y = 0.25x + 3000$.

$$\begin{aligned} y &= 0.25(11{,}000) + 3000 \\ &= 2750 + 3000 \\ &= \$5750 \end{aligned}$$

The cost of driving 11,000 miles in a year is \$5750. Remember, this cost includes gas, insurance, maintenance, and depreciation in the value of the vehicle!

(c) See Figure 41.

Figure 41

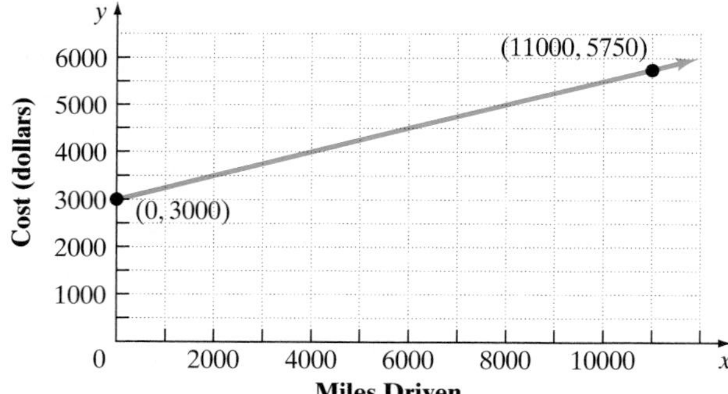

QUICK ✓

15. The daily cost, y, of renting a 16-foot moving truck for a day is \$50 plus \$0.38 per mile driven, x.

(a) Write a linear equation relating the daily cost y to the number of miles driven, x.

(b) Determine the cost of renting the truck if the truck is driven 75 miles.

(c) If the cost of renting the truck is \$84.20, how many miles were driven?

(d) Graph the linear equation.

8.4 Exercises

1. 3; 7 2. y
3. plotting points, using intercepts, using the slope and a point.
4. False 5. False 6. False
7. (a), (e)
8. (a) rises to the right
(b) falls to the right
(c) rises to the right
(d) horizontal line
9. $m = 5; b = 2$ 10. $m = 7; b = 1$
11. $m = 2; b = -9$ 12. $m = 3; b = -7$
13. $m = -10; b = 7$ 14. $m = -6; b = 2$
15. $m = -1; b = -9$
16. $m = -1; b = -12$
17. $m = 0; b = -5$
18. $m = \frac{5}{4}; b = 2$ 19. $m = \frac{2}{3}; b = -4$
20. $m = 0; b = 3$ 21. $m = 2; b = -4$
22. $m = 3; b = -9$
23. $m = -\frac{2}{3}; b = 8$ 24. $m = \frac{3}{4}; b = 3$
25. $m = \frac{5}{3}; b = -\frac{1}{3}$ 26. $m = -\frac{1}{3}; b = \frac{4}{3}$
27. $m = \frac{1}{2}; b = -\frac{5}{2}$
28. $m = -\frac{1}{5}; b = -\frac{3}{5}$
29. m is undefined; no y-intercept
30. m is undefined; no y-intercept

31. 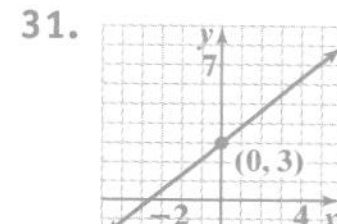
32. 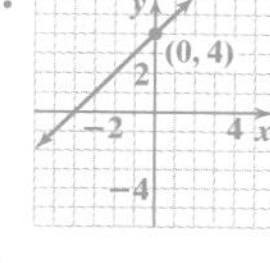
33.
34. 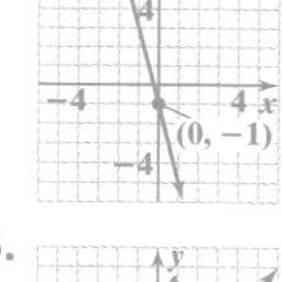
35. 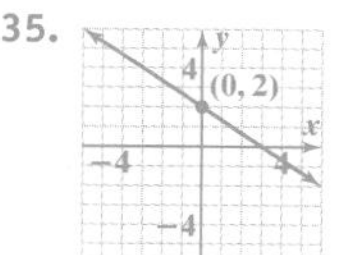
36.
37.
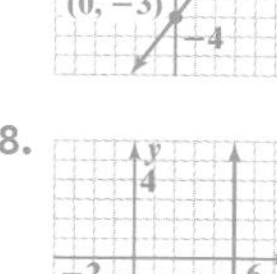
38.
39. 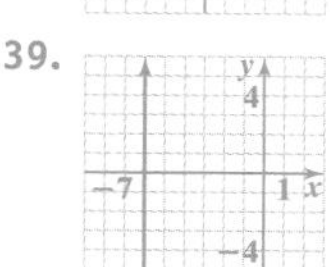
40.
41. 42. 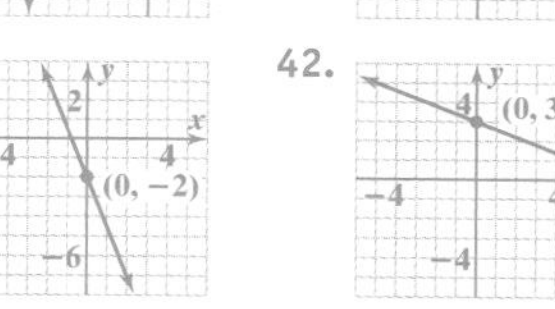
43. 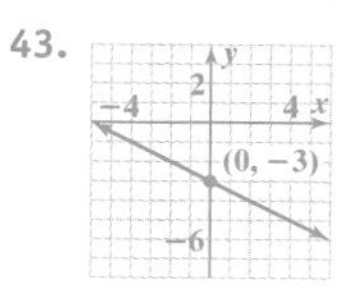
44. 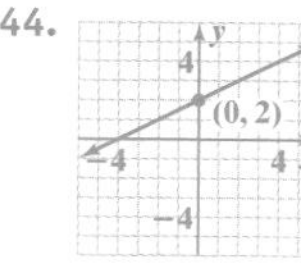

For Extra Help:

Student Solutions Manual  CD Video  PH Math/Tutor Center  MathXL Tutorials on CD  MathXL® MyMathLab

Concepts and Vocabulary

In Problems 1–3, fill in the blanks.

1. The graph of the line whose equation is $y = 3x + 7$ has slope of __________ and y-intercept at __________.
2. To identify the slope and y-intercept in the equation $4x - 2y = 3$, you must first solve the equation for __________.
3. List three techniques that can be used to graph a line: __________, __________, __________.

In Problems 4–6, answer True or False to each statement.

4. The slope of the line $2y = 3x - 4$ is 3.
5. Every linear equation can be written in both standard form and slope-intercept form.
6. The graph below shows the graph of the linear equation $y = x + 2$.

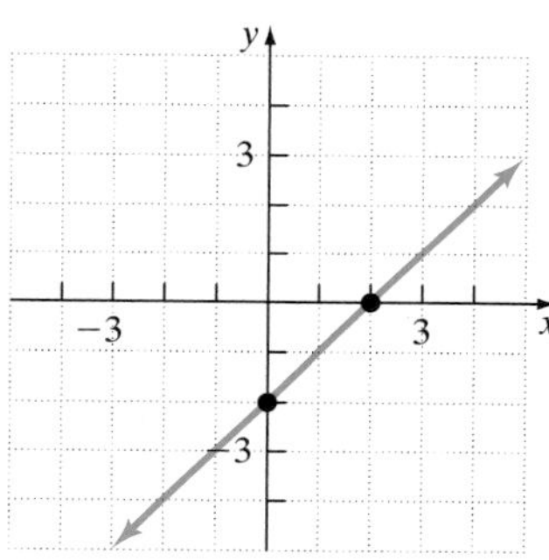

7. Which of the following equations could have the graph that is shown?

(a) $y = 3x - 2$ (b) $y = -2x + 5$
(c) $y = 3$ (d) $2x + 3y = 6$
(e) $3x - 2y = 8$ (f) $4x - y = -4$
(g) $-5x + 2y = 12$ (h) $x - y = -3$

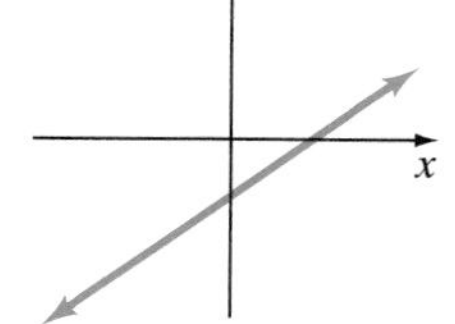

8. Without graphing, describe the orientation of each line (rises to the right, and so on). Explain how you came to this conclusion.

(a) $y = 4x - 3$ (b) $y = -2x + 5$ (c) $y = x$ (d) $y = 4$

Building Skills

In Problems 9–30, find the slope and y-intercept of the line whose equation is given.

9. $y = 5x + 2$
10. $y = 7x + 1$
11. $y = 2x - 9$
12. $y = 3x - 7$
13. $y = -10x + 7$
14. $y = -6x + 2$
15. $y = -x - 9$
16. $y = -x - 12$
17. $y = -5$
18. $y = \frac{5}{4}x + 2$
19. $y = \frac{2}{3}x - 4$
20. $y = 3$
21. $2x - y = 4$
22. $3x - y = 9$
23. $2x + 3y = 24$
24. $6x - 8y = -24$
25. $5x - 3y = 1$
26. $2x + 6y = 8$
27. $x - 2y = 5$
28. $-x - 5y = 3$
29. $x = 6$
30. $x = -2$

In Problems 31–60, use the slope and y-intercept, if possible, to graph each line whose equation is given.

31. $y = x + 3$ **32.** $y = x + 4$ **33.** $y = -2x - 3$ **34.** $y = -4x - 1$

35. $y = -\frac{2}{3}x + 2$ **36.** $y = \frac{4}{3}x - 3$ **37.** $y = 0$ **38.** $x = 5$

39. $x = -6$ **40.** $y = 4$ **41.** $y = -\frac{5}{2}x - 2$ **42.** $y = -\frac{2}{5}x + 3$

43. $x + 2y = -6$ **44.** $x - 2y = -4$ **45.** $3x - 2y = 10$ **46.** $4x + 3y = -6$

47. $6x + 3y = -15$ **48.** $5x - 2y = 6$ **49.** $y = \frac{x}{3}$ **50.** $x = \frac{5}{2}$

51. $2x = -8y$ **52.** $y = -\frac{x}{4}$ **53.** $y = -\frac{4}{3}$ **54.** $-3x = 5y$

55. $y = -\frac{2x}{3} + 1$ **56.** $y = \frac{3x}{2} - 4$ **57.** $x + 2 = -7$ **58.** $y - 4 = -1$

59. $5x + y + 1 = 0$ **60.** $2x - y + 4 = 0$

In Problems 61–72, find the equation of the line with the given slope and intercept.

61. slope is -1; y-intercept is 8

62. slope is 1; y-intercept is 10

63. slope is $\frac{6}{7}$; y-intercept is -6

64. slope is $\frac{4}{7}$; y-intercept is -9

65. slope is $-\frac{1}{3}$; y-intercept is $\frac{2}{3}$

66. slope is $\frac{1}{4}$; y-intercept is $\frac{3}{8}$

67. slope is undefined; x-intercept is -5

68. slope is 0; y-intercept is -2

69. slope is 0; y-intercept is 3

70. slope is undefined; x-intercept is 4

71. slope is 5; y-intercept is 0

72. slope is -3; y-intercept is 0

Mixed Practice

In Problems 73–84, graph each equation using any method you wish.

73. $y = 2x - 7$ **74.** $y = -4x + 1$ **75.** $3x - 2y = 24$ **76.** $2x + 5y = 30$

77. $y = -5$ **78.** $x = -3$ **79.** $6x - 4y = 0$ **80.** $3x + 8y = 0$

81. $y = -\frac{5}{3}x + 6$ **82.** $y = -\frac{3}{5}x + 4$ **83.** $2y = x + 4$ **84.** $3y = x - 9$

Applying the Concepts

85. Weekly Salary Dien is paid a salary of \$400 per week plus an 8% commission on all sales he makes during the week.

(a) Write a linear equation that calculates his weekly income, where y represents his income and x represents the amount of sales.

(b) What is Dien's weekly income if he sold \$1200 worth of merchandise?

(c) Graph the equation in a rectangular coordinate system. Label the axes appropriately.

86. Car Rental To rent a car for a day, Gloria pays \$75 plus \$0.10 per mile.

(a) Write a linear equation that calculates the daily cost, y, to rent a car which will be driven x miles.

(b) What is the cost to drive this car for 200 miles?

(c) If Gloria paid \$87.50, how many miles did she drive?

(d) Graph the equation in a rectangular coordinate system. Label the axes appropriately.

45.
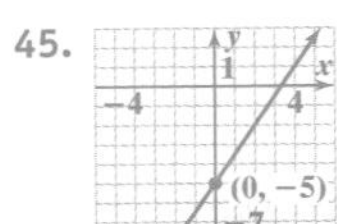

46.
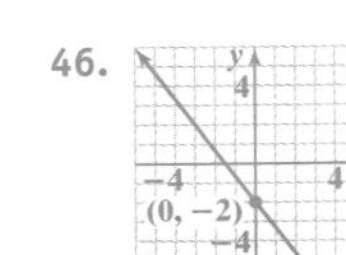

47.
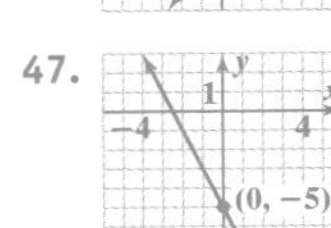

48.
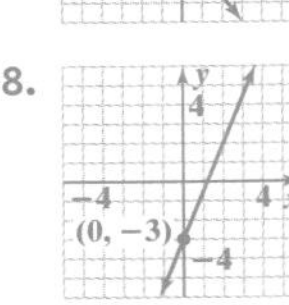

49.
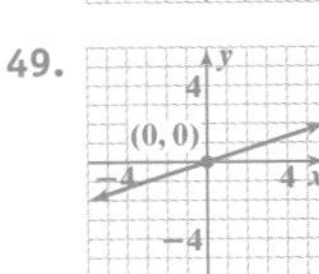

50.
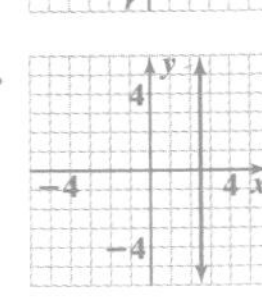

51.
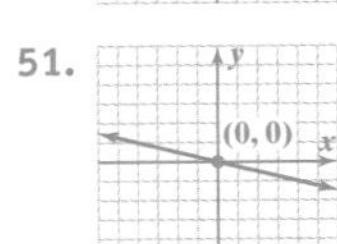

52.
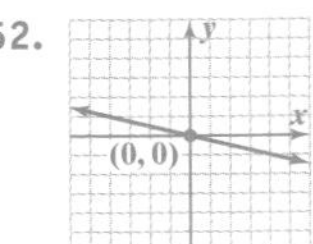

53.
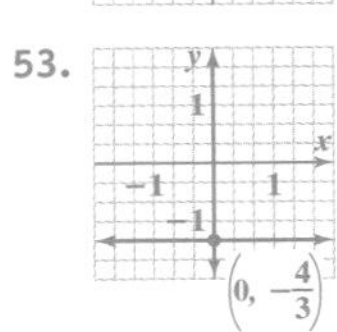

54.
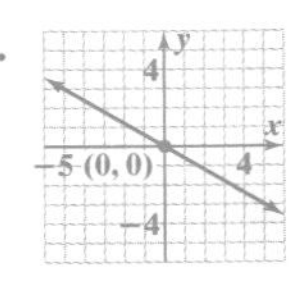

55.
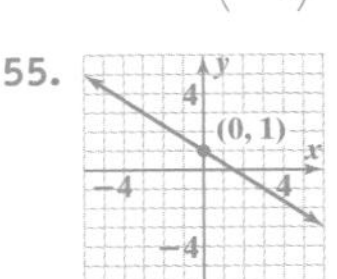

56.
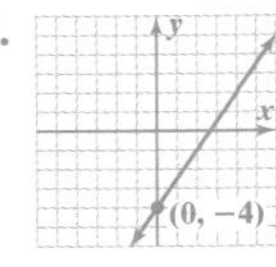

57.
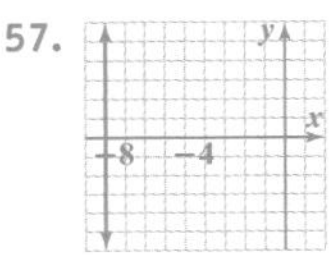

58.

59.

60.
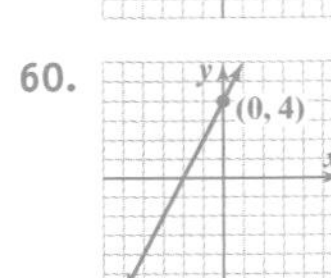

61. $y = -x + 8$ 62. $y = x + 10$

63. $y = \frac{6}{7}x - 6$ 64. $y = \frac{4}{7}x - 9$

65. $y = -\frac{1}{3}x + \frac{2}{3}$ 66. $y = \frac{1}{4}x + \frac{3}{8}$

67. $x = -5$ 68. $y = -2$

69. $y = 3$ 70. $x = 4$

71. $y = 5x$ 72. $y = -3x$

73. 74.
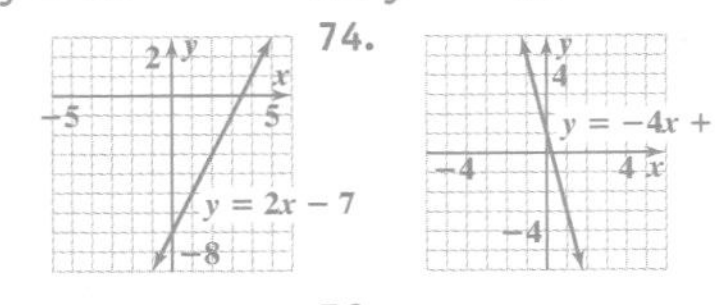

75. 76.
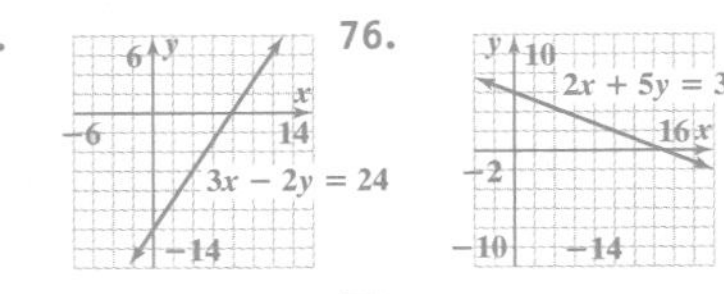

77. 78.
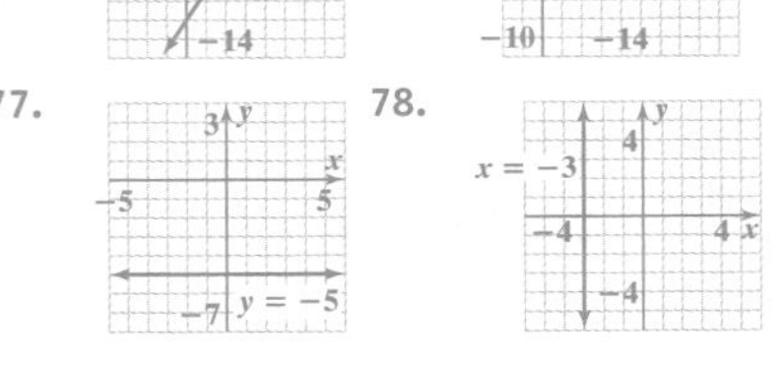

79.
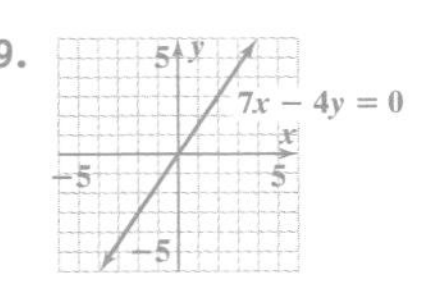

80.
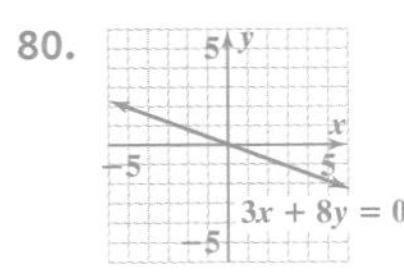

81.
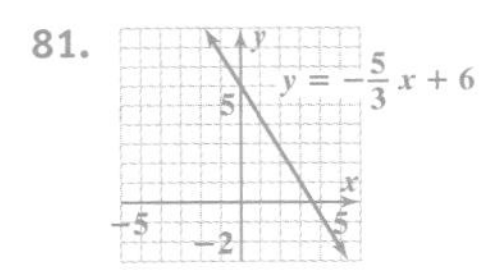

82.
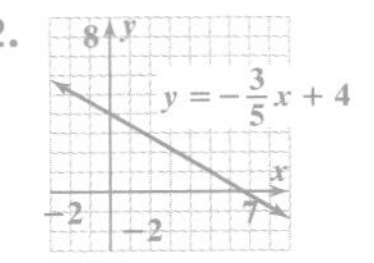

83. 84.

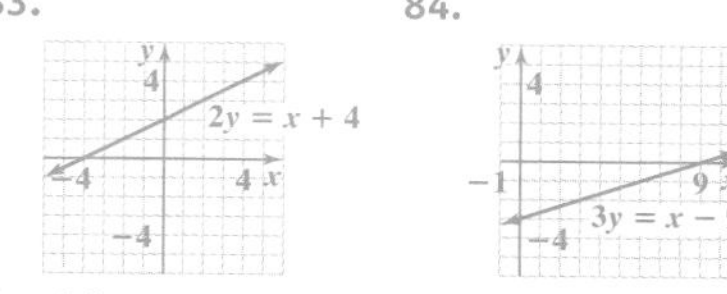

85. (a) $y = 0.08x + 400$
(b) $496 (c)

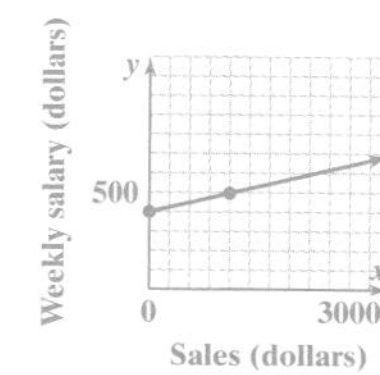

86. (a) $y = 0.1x + 75$ (b) $95
(c) 125 mi (d)

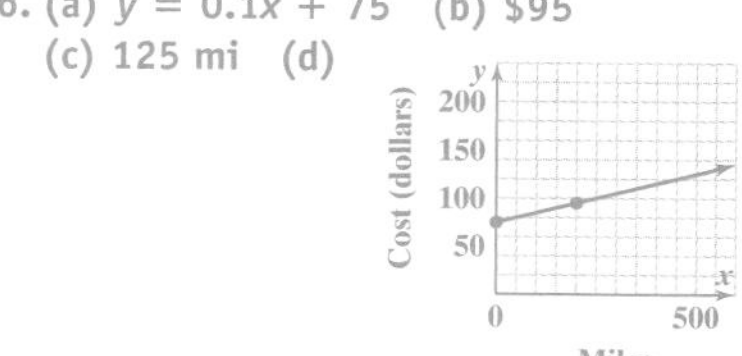

87. (a) 34.5¢ (b) 2001
(c) The cost per minute is decreasing by 5.375¢ per year.
(d) No, the cost will never be 0 or negative.
(e)

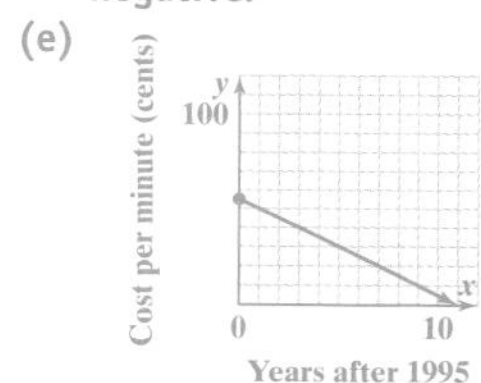

88. (a) 2625 calories (b) 9 years old
(c) The recommended caloric intake increased by 125 calories per year.
(d) Age 3 is outside the given age range for the equation.
(e)

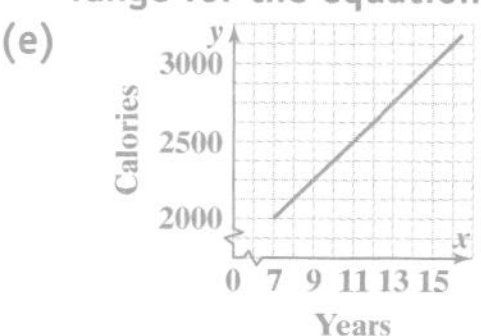

89. $B = -4$ 90. $A = -3$ 91. $A = -4$
92. $B = 3$ 93. $B = -3$ 94. $B = 2$
95. (a) $y = 40x + 4000$
(b) $24,000 (c) 375 calculators
(d)

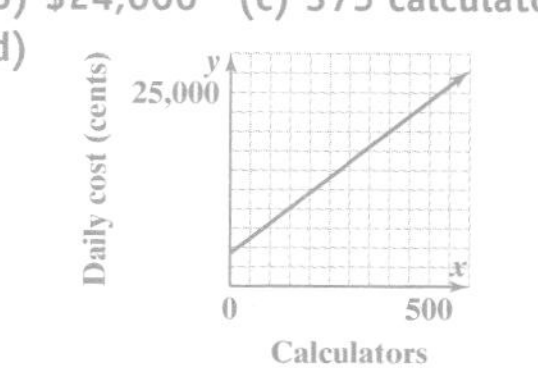

96. (a) $y = 35x + 3600$
(b) $17,600
(c) 275 cellular telephones
(d)

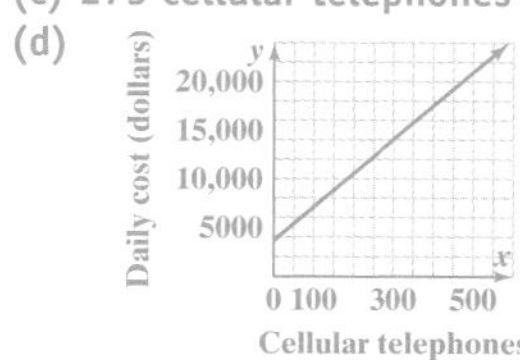

87. Cell Phone Costs The cost per minute for cell phone users has gone down over the years. In 1995, cell phone users paid, on the average, 56¢ per minute. In 2003, they paid 13¢ per minute. Assuming that the rate of decline of the cost per minute was constant, the cost per minute can be calculated by the equation $y = -5.375x + 56$, where x represents the number of years after 1995 and y represents the cost per minute of cell phone usage in cents.

(a) What was the cost per minute for a cell phone user in 1999?
(b) In which year did a cell phone user pay 23.75¢ per minute?
(c) Interpret the slope of $y = -5.375x + 56$.
(d) Can this trend continue indefinitely?
(e) Graph the equation in a rectangular coordinate system. Label the axes appropriately.

88. Counting Calories According to a 1989 National Academy of Sciences Report, the recommended daily intake of calories for males between the ages of 7 and 15 can be calculated by the equation $y = 125x + 1125$, where x represents the boy's age and y represents the recommended calorie intake.

(a) What is the recommended caloric intake for a 12-year-old boy?
(b) What is the age of a boy whose recommended caloric intake is 2250 calories?
(c) Interpret the slope of $y = 125x + 1125$.
(d) Why would this equation not be accurate for a 3-year-old male?
(e) Graph the equation in a rectangular coordinate system. Label the axes appropriately.

Extending the Concepts

In Problems 89–94, find the value of the missing coefficient so that the line will have the given property.

89. $2x + By = 12$; slope is $\frac{1}{2}$

90. $Ax + 2y = 5$; slope is $\frac{3}{2}$

91. $Ax - 2y = 10$; slope is -2

92. $12x + By = -1$; slope is -4

93. $x + By = \frac{1}{2}$; y-intercept is $-\frac{1}{6}$

94. $4x + By = \frac{4}{3}$; y-intercept is $\frac{2}{3}$

In Problems 95 and 96, use the following information. In business, a cost equation relates the total cost of producing a product or good such as a refrigerator, rug, or blender to the number of goods produced. The simplest cost model is the linear cost model. In the linear cost model, the slope of the linear equation represents the variable cost of producing a good—variable costs are costs that change with the level of output. Variable cost is reported as a rate of change, such as $40 per calculator. Examples of variable costs would be labor costs and materials. The y-intercept of the linear equation represents the fixed cost of production—these are costs that exist regardless of the level of production. Fixed costs would be cost of the manufacturing facility and insurance.

95. Cost Equations Suppose the variable cost of manufacturing a graphing calculator is $40 per calculator while the daily fixed cost is $4000.

(a) Write a linear equation that relates cost y to the number of calculators manufactured x.
(b) What is the daily cost of manufacturing 500 calculators?
(c) One day, the total cost was $19,000. How many calculators were manufactured?
(d) Graph the equation relating cost and number of calculators manufactured.

96. Cost Equations Suppose the variable cost of manufacturing a cellular telephone is $35 per phone, and the daily fixed cost is $3600.

(a) Write a linear equation that relates the daily cost y to the number of cellular telephones manufactured x.
(b) What is the daily cost of manufacturing 400 cellular phones?
(c) One day, the total cost was $13,225. How many cellular phones were manufactured?
(d) Graph the equation relating cost and number of cellular telephones manufactured.

8.5 Point-Slope Form of a Line

OBJECTIVES

1. Find the Equation of a Line Given a Point and a Slope
2. Find the Equation of a Line Given Two Points
3. Build Linear Models Using the Point-Slope Form of a Line

Preparing for Point-Slope Form of a Line

Before getting started, take the following readiness quiz. If you get a problem wrong, go back to the section cited and review the material.

1. Solve $y - 3 = 2(x + 1)$ for y. [Section 2.4, pp. 109–110]

2. Evaluate $\dfrac{7 - 3}{4 - 2}$. [Section 1.6, p. 52]

1 Find the Equation of a Line Given a Point and a Slope

We now have two forms for the equation of a line. We have the standard equation of a line $Ax + By = C$, where A and B are not both zero, and the slope-intercept form of a line $y = mx + b$, where m is the slope and b is the y-intercept. We now introduce another form for the equation of a line.

Suppose that we have a nonvertical line with slope m containing the point (x_1, y_1). For any other point (x, y) on the line, we know from the formula for the slope of a line that

$$m = \frac{y - y_1}{x - x_1}$$

See Figure 42. Multiplying both sides by $x - x_1$, we can rewrite this expression as

$$m(x - x_1) = y - y_1 \quad \text{or} \quad y - y_1 = m(x - x_1)$$

Figure 42

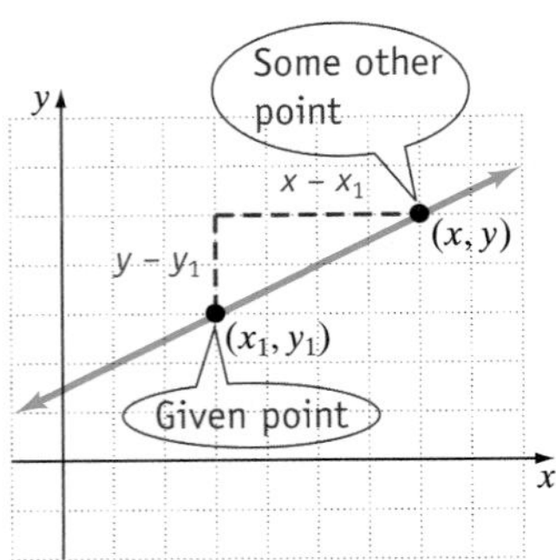

POINT-SLOPE FORM OF AN EQUATION OF A LINE

An equation of a nonvertical line of slope m that contains the point (x_1, y_1) is

Slope ↓

$$y - y_1 = m(x - x_1)$$

↑ Given point ↑

The point-slope form of a line can be used to write an equation in either slope-intercept form $(y = mx + b)$ or standard form $(Ax + By = C)$.

Classroom Example Use Example 1 with a line whose slope is 2 and contains the point $(-2, 1)$.

Answer: $y = 2x + 5$

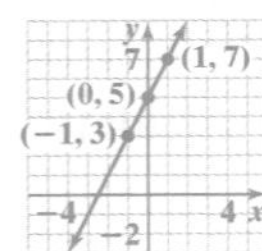

EXAMPLE 1 Using the Point-Slope Form of an Equation of a Line—Positive Slope

Find the equation of a line whose slope is 3 and contains the point $(-1, 3)$. Write the equation in slope-intercept form. Graph the line.

Solution

Because we are given the slope and a point on the line, we use the point-slope form of a line with $m = 3$ and $(x_1, y_1) = (-1, 3)$.

$$y - y_1 = m(x - x_1)$$

$m = 3, x_1 = -1, y_1 = 3$: $\quad y - 3 = 3(x - (-1))$

$$y - 3 = 3(x + 1)$$

To put the equation in slope-intercept form, $y = mx + b$, we solve the equation for y.

Distribute: $\quad y - 3 = 3x + 3$

Add 3 to both sides: $\quad y = 3x + 6$

See Figure 43 for a graph of the line.

Figure 43

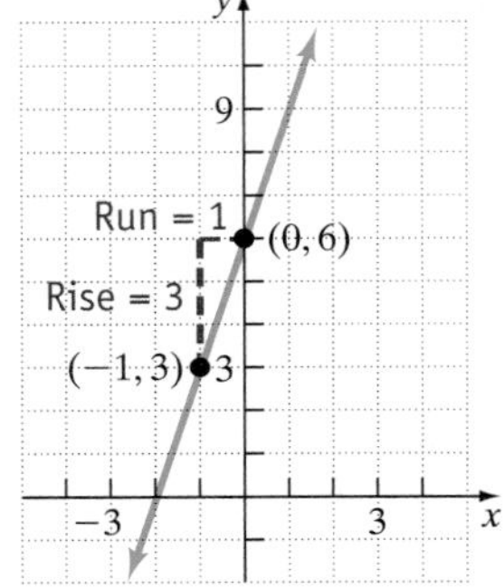

Preparing for...Answers **1.** $y = 2x + 5$ **2.** 2

QUICK ✓ *Find an equation of the line with the given properties. Write the equation in slope-intercept form. Graph the line.*

1. $m = 3$ containing $(x_1, y_1) = (2, 1)$ **2.** $m = \frac{1}{3}$ containing $(x_1, y_1) = (3, -4)$

Classroom Example ➤
Use Example 2 with a line whose slope is $-\frac{3}{2}$ and contains the point $(-2, 4)$.

Answer: $y = -\frac{3}{2}x + 1$

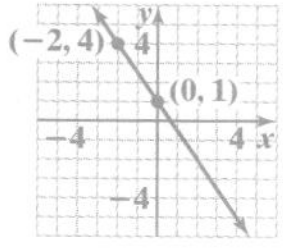

EXAMPLE 2 Using the Point-Slope Form of an Equation of a Line—Negative Slope

Find the equation of a line whose slope is $-\frac{3}{4}$ and contains the point $(-4, 3)$. Write the equation in slope-intercept form. Graph the line.

Solution

Because we are given the slope and a point on the line, we use the point-slope form of a line with $m = -\frac{3}{4}$ and $(x_1, y_1) = (-4, 3)$.

$$y - y_1 = m(x - x_1)$$

$m = -\frac{3}{4}, x_1 = -4, y_1 = 3$: $\quad y - 3 = -\frac{3}{4}(x - (-4))$

Simplify: $\quad y - 3 = -\frac{3}{4}(x + 4)$

Distribute: $\quad y - 3 = -\frac{3}{4}x - 3$

Add 3 to both sides: $\quad y = -\frac{3}{4}x$

See Figure 44 for a graph of the line.

Figure 44

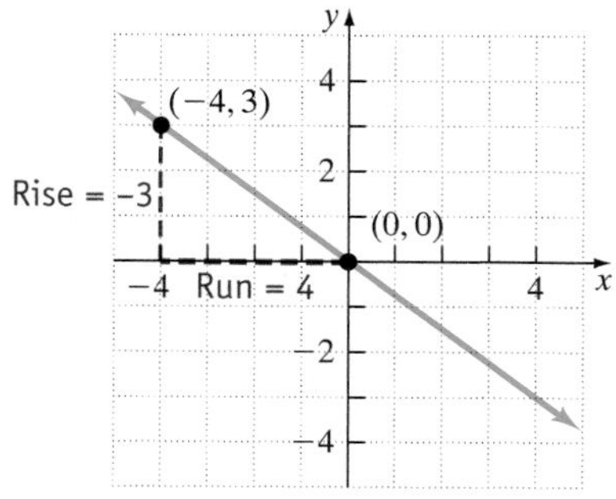

QUICK ✓ *Find an equation of the line with the given properties. Write the equation in slope-intercept form. Graph the line.*

3. $m = -4$ containing $(x_1, y_1) = (-2, 5)$ **4.** $m = -\frac{5}{2}$ and $(x_1, y_1) = (-4, 5)$

Classroom Example ➤
Use Example 3 with the point $(5, -4)$.

Answer: $y = -4$

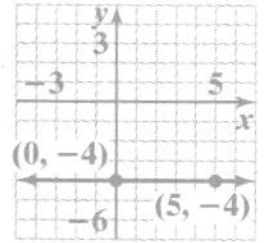

Work Smart
When the slope of a line is 0, the equation of the line will always be in the form "y = some number."

EXAMPLE 3 Finding the Equation of a Horizontal Line

Find the equation of a horizontal line that contains the point $(-4, 2)$. Write the equation of the line in slope-intercept form. Graph the line.

Solution

The line is a horizontal line, so the slope of the line is 0. Because we know the slope and a point on the line, we use the point-slope form of a line with $m = 0$, $x_1 = -4$, and $y_1 = 2$.

$$y - y_1 = m(x - x_1)$$

$m = 0, x_1 = -4, y_1 = 2$: $\quad y - 2 = 0(x - (-4))$

$$y - 2 = 0$$

To put the equation of the line in slope-intercept form, we add 2 to both sides.

Add 2 to both sides: $\quad y = 2 \quad$ Slope-intercept form

See Figure 45 for a graph of the line.

Figure 45

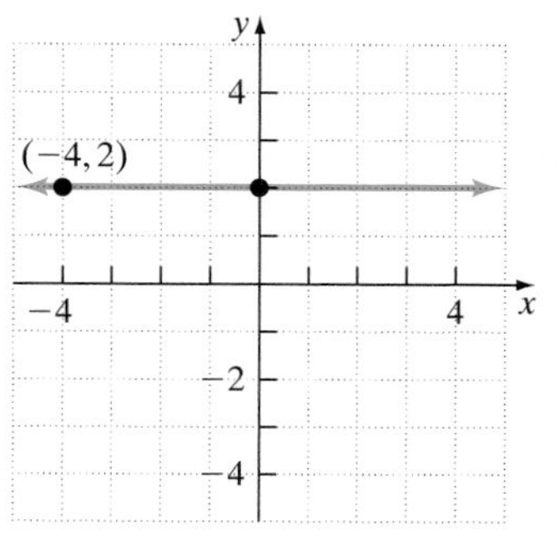

QUICK ✓

5. Find the equation of a horizontal line that contains the point $(-2, 3)$. Write the equation of the line in slope-intercept form. Graph the line.

2 Find the Equation of a Line Given Two Points

From Section 8.2, we know that two points are all that is needed to graph a line. If we are given two points, we can find an equation of the line through the points by first finding the slope of the line and then using the point-slope form of a line.

EXAMPLE 4 How to Find the Equation of a Line from Two Points

Find the equation of a line through the points $(1, 3)$ and $(4, 9)$. Write the equation in slope-intercept form. Graph the line.

Step-by-Step Solution

Step 1: Find the slope of the line containing the points.

Let $(x_1, y_1) = (1, 3)$ and $(x_2, y_2) = (4, 9)$. Substitute these values into the formula for the slope of a line.

$$m = \frac{y_2 - y_1}{x_2 - x_1} = \frac{9 - 3}{4 - 1} = \frac{6}{3} = 2$$

Step 2: Substitute the slope found in Step 1 and either point into the point-slope form of a line to find the equation.

With $m = 2$, $x_1 = 1$, and $y_1 = 3$, we have

$$y - y_1 = m(x - x_1)$$

$m = 2, x_1 = 1, y_1 = 3$: $\quad y - 3 = 2(x - 1)$

Step 3: Solve the equation for y.

Distribute the 2: $\quad y - 3 = 2x - 2$

Add 3 to both sides: $\quad y = 2x + 1$

The slope-intercept form of the equation is $y = 2x + 1$. The slope of the line is 2 and the y-intercept is 1. See Figure 46 for the graph.

Classroom Example Use Example 4 with the points (1, 2) and (3, 8).

Answer: $y = 3x - 1$

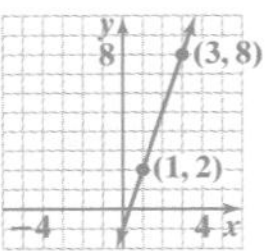

Figure 46

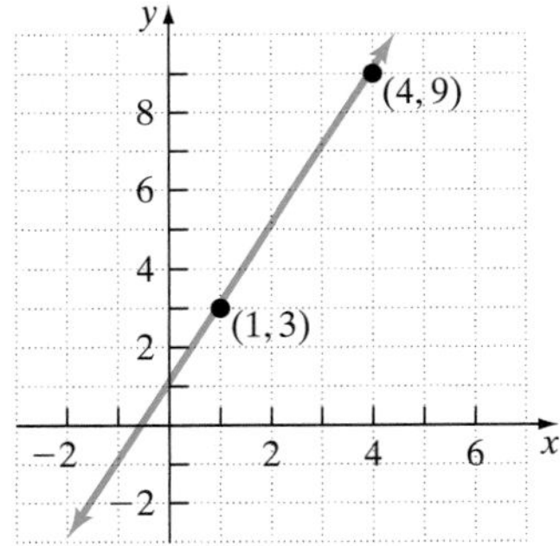

QUICK ✓ *Find the equation of the line containing the given points. Write the equation in slope-intercept form. Graph the line.*

6. $(0, 2); (3, 5)$

7. $(-1, 4); (1, -2)$

Work Smart: Study Skills

To write the equation of a nonvertical line, we must know either the slope of the line along with a point on the line or two points on the line.

- If the **slope** and the **y-intercept** are known, use the slope-intercept form, $y = mx + b$.
- If the **slope** and a **point** which is not the y-intercept are known, use the **point-slope** form, $y - y_1 = m(x - x_1)$.
- If **two points** are known, first find the **slope,** then use that slope, along with one of the **points** in the **point-slope** formula, $y - y_1 = m(x - x_1)$.

Classroom Example
Use Example 5 with the points $(4, -2)$ and $(4, 5)$.

Answer: $x = 4$

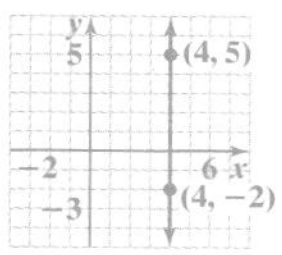

Work Smart

The equation of a vertical line cannot be written in slope-intercept form.

EXAMPLE 5 Finding an Equation of a Vertical Line from Two Points

Find the equation of a line through the points $(-3, 2)$ and $(-3, -4)$. Write the equation in slope-intercept form. Graph the line.

Solution

Let $(x_1, y_1) = (-3, 2)$ and $(x_2, y_2) = (-3, -4)$. Substitute these values into the formula for the slope of a line.

$$m = \frac{y_2 - y_1}{x_2 - x_1} = \frac{-4 - 2}{-3 - (-3)} = \frac{-6}{0}$$

The slope is undefined, so the line is vertical. No matter what value of y we choose, the x-coordinate of the point on the line will be -3. For this reason, the equation of the line is $x = -3$. See Figure 47 for the graph.

Figure 47

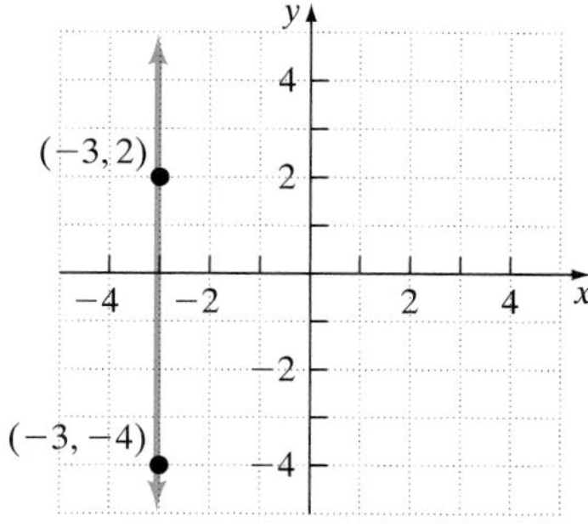

QUICK ✓

8. Find an equation of the line containing the points $(3, 2)$ and $(3, -4)$. If possible, write the answer in slope-intercept form. Graph the line.

SUMMARY: Equations of Lines

Form of Line	Formula	Comments
Horizontal line	$y = b$	Graph is a horizontal line (slope is 0) with y-intercept b.
Vertical line	$x = a$	Graph is a vertical line (undefined slope) with x-intercept a.
Point-slope	$y - y_1 = m(x - x_1)$	Useful for finding the equation of a line, given a point and a slope, or two points.
Slope-intercept	$y = mx + b$	Useful for finding the equation of a line, given the slope and y-intercept, or for quickly determining the slope and y-intercept of the line, given the equation of the line.
Standard form	$Ax + By = C$	Straightforward to find the x- and y-intercepts.

3 Build Linear Models Using the Point-slope Form of a Line

We can use the point-slope form of a line to build linear models from data.

Classroom Example ➤
Use Example 6.

EXAMPLE 6 Building a Linear Model from Data

Healthcare costs are skyrocketing. For individuals 20 years of age or older, the percentage of total income y that an individual spends on healthcare increases linearly with age x. According to data obtained from the Bureau of Labor Statistics, a 35-year-old spends about 4.0% of income on healthcare, while a 65-year-old spends about 11.2% of income on healthcare.

(a) Plot the points (35, 4.0) and (65, 11.2) in a rectangular coordinate system and graph the line. Find the linear equation in slope-intercept form that relates the percent of income spent on healthcare y to the age x.

(b) Use the equation found in part (a) to predict the percentage of income that a 50-year-old spends on healthcare.

(c) Interpret the slope.

Solution

(a) We plot the ordered pairs (35, 4.0) and (65, 11.2) and draw a line through the points. See Figure 48.

Figure 48

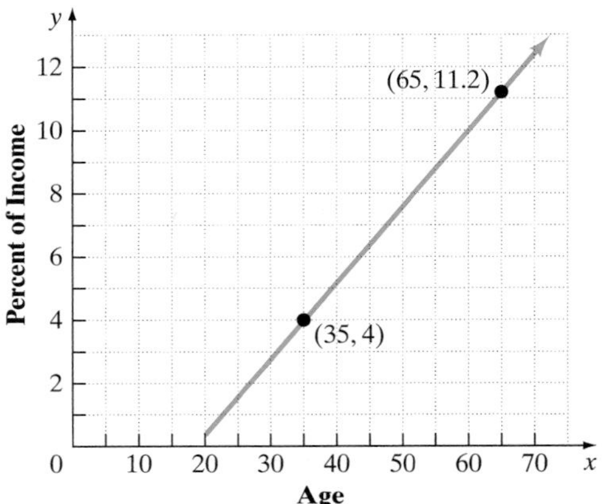

Because we know two points on the line, we will use the point-slope form of a line to find the equation of the line.

First, we must find the slope of the line:

$$m = \frac{y_2 - y_1}{x_2 - x_1} = \frac{11.2 - 4.0}{65 - 35} = \frac{7.2}{30} = 0.24$$

We use the point-slope form of a line with $m = 0.24$, $x_1 = 35$, and $y_1 = 4.0$:

$$y - y_1 = m(x - x_1)$$

$m = 0.24$, $x_1 = 35$, and $y_1 = 4.0$: $$y - 4.0 = 0.24(x - 35)$$

$$y - 4.0 = 0.24x - 8.4$$

Add 4.0 to both sides of the equation: $$y = 0.24x - 4.4$$

The equation $y = 0.24x - 4.4$ relates the percent of income spent on healthcare y to the age x.

(b) We substitute 50 for x in the equation found in part (a), so the predicted percentage of income spent on health care for a 50-year old is

$$y = 0.24x - 4.4$$
$$\text{Let } x = 50: \quad = 0.24(50) - 4.4$$
$$= 7.6$$

We predict that 7.6% of a 50-year-old's income is spent on healthcare.

(c) The slope is 0.24. The percentage of income spent on healthcare for the individual increases by 0.24% as an individual ages by one year.

QUICK ✓

9. Armando owns a gas station. He has found that when the price of regular unleaded gasoline is \$2.20, he sells 400 gallons of gasoline between the hours of 7:00 A.M. and 8:00 A.M. When the price of regular unleaded gasoline is \$2.40, he sells 380 gallons of gasoline between the hours of 7:00 A.M. and 8:00 A.M. Suppose that the relation between the quantity of gasoline sold and price is linear.

(a) Plot the points in a rectangular coordinate system and graph the line. Find the linear equation in slope-intercept form that relates quantity of gasoline sold y to the price x.

(b) Use the equation found in part (a) to predict the number of gallons of gasoline sold if the price is \$2.30.

(c) Interpret the slope.

8.5 Exercises

1. $y - y_1 = m(x - x_1)$ 2. $y = mx + b$
3. standard form, slope-intercept form, point-slope form; $x = a$; $y = b$
4. False 5. True 6. False
7. Yes; Answers may vary.
8. Answers may vary.
9. $y = 3x - 1$ 10. $y = 6x - 23$

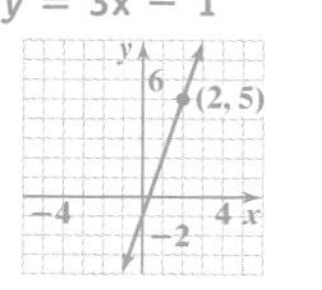

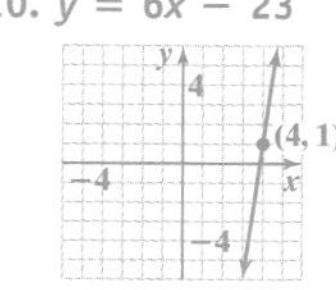

11. $y = -2x$ 12. $y = -5x + 27$

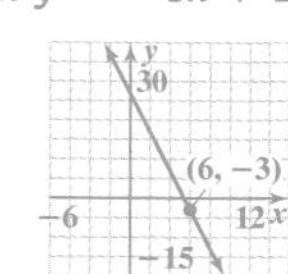

13. $y = \frac{1}{4}x - 3$ 14. $y = -\frac{1}{2}x - 2$

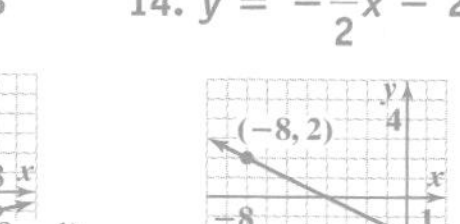

15. $y = -6x + 13$ 16. $y = 9x - 4$

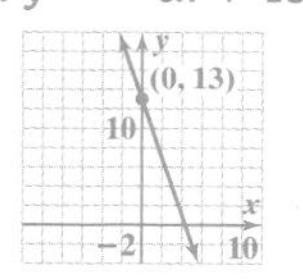

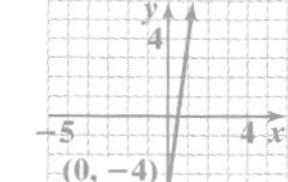

For Extra Help:

Student Solutions Manual | CD Video | PH Math/Tutor Center | MathXL Tutorials on CD | MathXL® | MyMathLab

Concepts and Vocabulary

In Problems 1–3, fill in the blanks.

1. The point-slope form of a non-vertical line whose slope is m that contains the point (x_1, y_1) is ________.

2. The slope-intercept form of a non-vertical line whose slope is m and y-intercept is b is ________.

3. List the five forms that are used when writing the equation of a line: ________, ________, ________, ________, ________.

In Problems 4–6, answer True or False to each statement.

4. The line $y = 3$ is a vertical line.

5. The slope of the line $y - 3 = 4(x - 1)$ is 4.

6. The y-intercept of the line $y - 7 = 4(x - 1)$ is -1.

7. You are asked to write the equation of the line through the points (3, 1) and (4, 7) in slope-intercept form. After calculating the slope of the line, you choose the point-slope form and assign $x_1 = 3$ and $y_1 = 1$. Your friend lets $x_1 = 4$ and $y_1 = 7$. Will you and your friend obtain the same answer? Explain why or why not.

8. You are asked to write the equation of the line through $(-1, 3)$ and $(0, 4)$. Which form of a line would you choose to find the equation? Explain why you chose this form. Could you also use one of the other forms?

17. $y = -7$

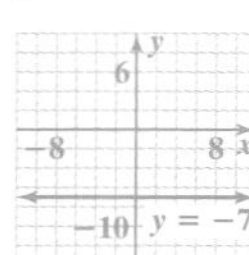

18. $x = 3$

19. $x = -4$

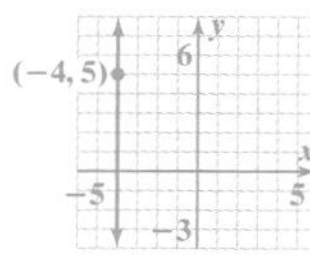

20. $y = -1$

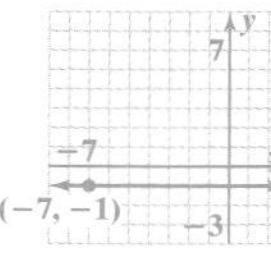

21. $y = \frac{2}{3}x + 2$

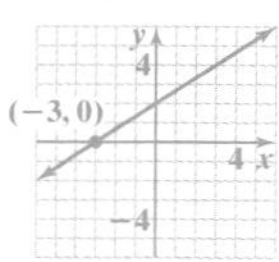

22. $y = -\frac{4}{5}x - 8$

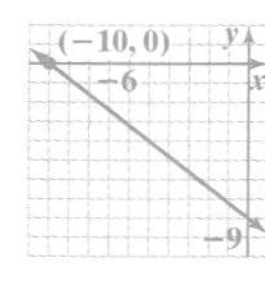

23. $y = -\frac{3}{4}x - \frac{39}{4}$

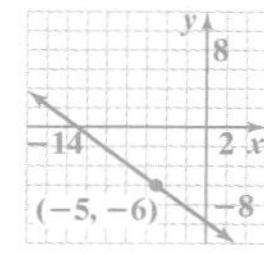

24. $y = \frac{3}{2}x + \frac{5}{2}$

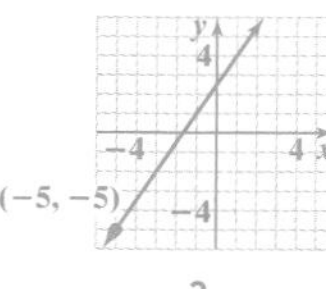

25. $y = \frac{1}{2}x - 5$

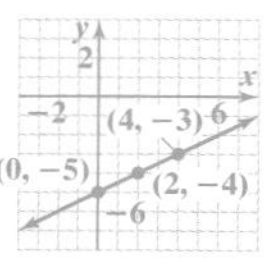

26. $y = \frac{3}{2}x - 8$

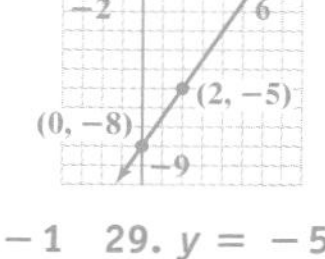

27. $x = -3$ 28. $y = -1$ 29. $y = -5$
30. $x = 4$ 31. $y = -4.3$ 32. $x = 3.5$
33. $x = \frac{1}{2}$ 34. $y = \frac{9}{4}$ 35. $y = 2x + 4$
36. $y = -\frac{1}{2}x + 3$ 37. $y = -4x + 6$
38. $y = -2x + 8$ 39. $y = -\frac{3}{2}x - \frac{5}{2}$
40. $y = -\frac{3}{2}x + 1$ 41. $y = 2x - 5$
42. $y = 3x + 6$ 43. $y = -3$
44. $y = 5$ 45. $x = 2$ 46. $x = -3$
47. $y = 0.25x + 0.575$
48. $y = 0.8x + 0.24$
49. $y = x - \frac{11}{4}$ 50. $y = -2x + \frac{46}{15}$

51. $y = 5x - 22$ 52. $y = 4x - 10$

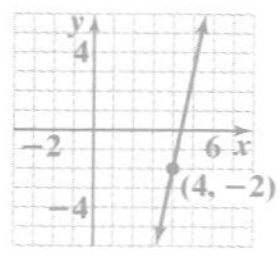

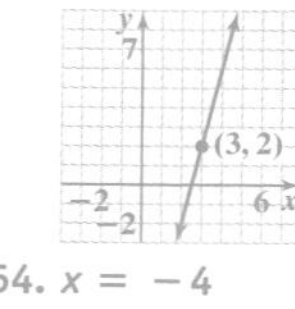

53. $y = 5$ 54. $x = -4$

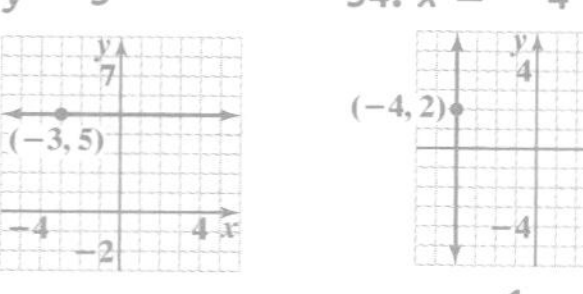

55. $y = x + 2$ 56. $y = \frac{1}{2}x - 7$

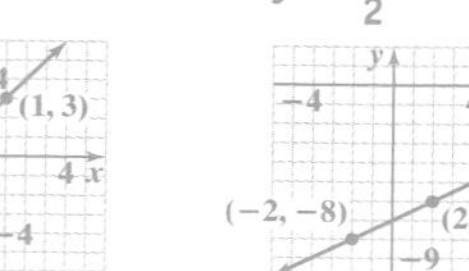

Building Skills

In Problems 9–26, find the equation of the line that contains the given point and slope. Write the equation in slope-intercept form and graph the line.

9. $(2, 5)$; slope $= 3$
10. $(4, 1)$; slope $= 6$
11. $(-1, 2)$; slope $= -2$
12. $(6, -3)$; slope $= -5$
13. $(8, -1)$; slope $= \frac{1}{4}$
14. $(-8, 2)$; slope $= -\frac{1}{2}$
15. $(0, 13)$; slope $= -6$
16. $(0, -4)$; slope $= 9$
17. $(5, -7)$; slope $= 0$
18. $(3, 12)$; undefined slope
19. $(-4, 5)$; undefined slope
20. $(-7, -1)$; slope $= 0$
21. $(-3, 0)$; slope $= \frac{2}{3}$
22. $(-10, 0)$; slope $= -\frac{4}{5}$
23. $(-5, -6)$; slope $= -\frac{3}{4}$
24. $(-5, -5)$; slope $= \frac{3}{2}$
25. $(4, -3)$; slope $= \frac{1}{2}$
26. $(2, -5)$; slope $= \frac{3}{2}$

In Problems 27–34, find the equation of the line that contains the given point and satisfies the given information. Write the equation in slope-intercept form, if possible.

27. Vertical line that contains $(-3, 10)$
28. Horizontal line that contains $(-6, -1)$
29. Horizontal line that contains $(-1, -5)$
30. Vertical line that contains $(4, -3)$
31. Horizontal line that contains $(0.2, -4.3)$
32. Vertical line that contains $(3.5, 2.4)$
33. Vertical line that contains $\left(\frac{1}{2}, \frac{7}{4}\right)$
34. Horizontal line that contains $\left(\frac{3}{2}, \frac{9}{4}\right)$

In Problems 35–50, find the equation of the line that contains the given points. Write the equation in slope-intercept form, if possible.

35. $(0, 4)$ and $(-2, 0)$
36. $(0, 3)$ and $(6, 0)$
37. $(1, 2)$ and $(0, 6)$
38. $(2, 4)$ and $(0, 8)$
39. $(-3, 2)$ and $(1, -4)$
40. $(-2, 4)$ and $(2, -2)$
41. $(-3, -11)$ and $(2, -1)$
42. $(4, 18)$ and $(-1, 3)$
43. $(4, -3)$ and $(-3, -3)$
44. $(-6, 5)$ and $(7, 5)$
45. $(2, -1)$ and $(2, -9)$
46. $(-3, 8)$ and $(-3, 1)$
47. $(0.1, 0.6)$ and $(0.5, 0.7)$
48. $(0.7, 0.8)$ and $(0.2, 0.4)$
49. $\left(\frac{1}{2}, -\frac{9}{4}\right)$ and $\left(\frac{5}{2}, -\frac{1}{4}\right)$
50. $\left(\frac{1}{3}, \frac{12}{5}\right)$ and $\left(\frac{4}{3}, \frac{2}{5}\right)$

Mixed Practice

In Problems 51–70, find the equation of the line described. Write the equation in slope-intercept form. Graph the line.

51. Contains $(4, -2)$ with slope $= 5$
52. Contains $(3, 2)$ with slope $= 4$
53. Horizontal line that contains $(-3, 5)$
54. Vertical line that contains $(-4, 2)$
55. Contains $(1, 3)$ and $(-4, -2)$
56. Contains $(-2, -8)$ and $(2, -6)$
57. Contains $(-2, 3)$ with slope $= \frac{1}{2}$
58. Contains $(-8, 3)$ with slope $= \frac{1}{4}$
59. Vertical line that contains $(5, 2)$
60. Horizontal line that contains $(-2, -6)$
61. Contains $(3, -19)$ and $(-1, 9)$
62. Contains $(-3, 13)$ and $(4, -22)$
63. Contains $(6, 3)$ with slope $= -\frac{2}{3}$
64. Contains $(-6, 3)$ with slope $= -\frac{2}{3}$

65. Contains $(-2, 3)$ and $(4, -6)$

66. Contains $(5, -3)$ and $(-3, 3)$

67. Contains $(-3, 0)$ with slope $= \frac{4}{5}$

68. Contains $(-10, 0)$ with slope $= -\frac{5}{4}$

69. Contains $(-3, 4)$ and $(6, 0)$

70. Contains $(8, -11)$ and $(-4, 4)$

Applying the Concepts

71. Shipping Packages The shipping department for a warehouse has noted that if 60 packages are shipped during a month, the total expenses for the department are \$1635. If 120 packages are shipped during a month, the total expenses for the shipping department are \$1770. Let x represent the number of packages and y represent the total expenses for the shipping department.

(a) Interpret the meaning of the point $(60, 1635)$ in the context of this problem.
(b) Plot the ordered pairs $(60, 1635)$ and $(120, 1770)$ in a rectangular coordinate system and graph the line through the points.
(c) Find the linear equation in slope-intercept form that relates the total expenses for the shipping department, y, to the number of packages sent, x.
(d) Use the equation found in part (c) to find the total expenses during a month when 200 packages were sent.
(e) Interpret the slope.

72. Retirement Plans Based on the retirement plan available by his employer, Kei knows that if he retires after 20 years, his monthly retirement income will be \$3150. If he retires after 30 years, his monthly income increases to \$3600. Let x represent the number of years of service and y represent the monthly retirement income.

(a) Interpret the meaning of the point $(30, 3600)$ in the context of this problem.
(b) Plot the ordered pairs $(20, 3150)$ and $(30, 3600)$ in a rectangular coordinate system and graph the line through the points.
(c) Find the linear equation in slope-intercept form that relates the monthly retirement income, y, to the number of years of service, x.
(d) Use the equation found in part (c) to find the monthly income for 15 years of service.
(e) Interpret the slope.

73. Traffic Fatalities In 1975, one region of the country had 181 traffic fatalities. In 2002, the same region had 276 fatalities. Let x represent the number of years after 1975 and let y represent the number of traffic fatalities.

(a) Fill in the ordered pairs: (_______, 181); (_______, 276).
(b) Plot the ordered pairs from part (a) in a rectangular coordinate system and graph the line through the points.
(c) Find the linear equation in slope-intercept form that relates the number of traffic fatalities, y, to the number of years after 1975, x.
(d) Use the equation found in part (c) to predict the number of traffic fatalities in 1990.
(e) Interpret the slope.

74. U.S. Traffic Fatalities Nationwide, the statistics for traffic fatalities show a decline. In 1975, the United States had 39,161 fatal crashes while in 2002 the number dropped to 38,309. Let y represent the number of traffic fatalities, and x represent the number of years after 1975.

(a) Fill in the ordered pairs: (_______, 39,161); (_______, 38,309).
(b) Plot the ordered pairs from part (a) in a rectangular coordinate system and graph the line through the points.

57. $y = \frac{1}{2}x + 4$

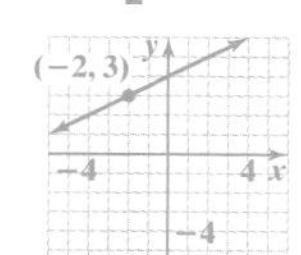

58. $y = \frac{1}{4}x + 5$

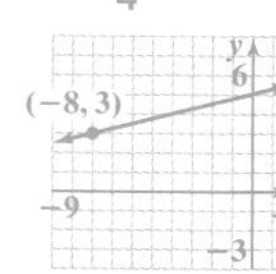

59. $x = 5$

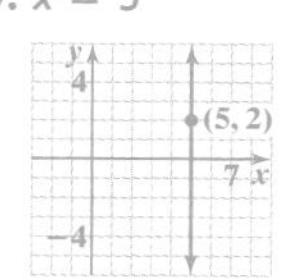

60. $y = -6$

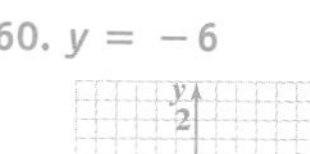

61. $y = -7x + 2$ 62. $y = -5x - 2$

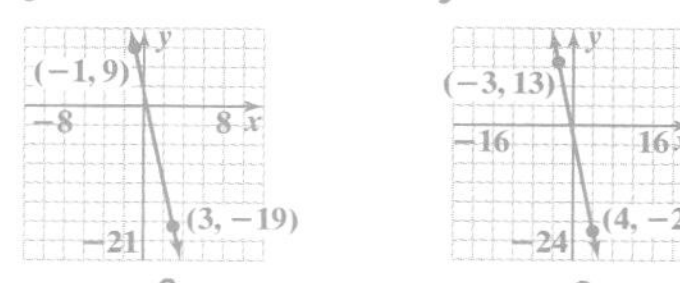

63. $y = -\frac{2}{3}x + 7$

64. $y = -\frac{2}{3}x - 1$

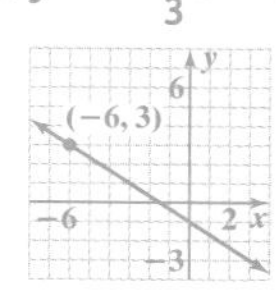

65. $y = -\frac{3}{2}x$

66. $y = -\frac{3}{4}x + \frac{3}{4}$

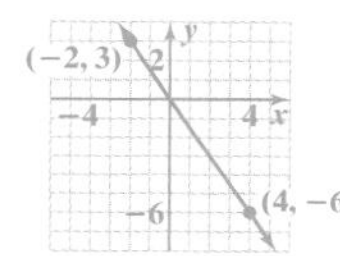

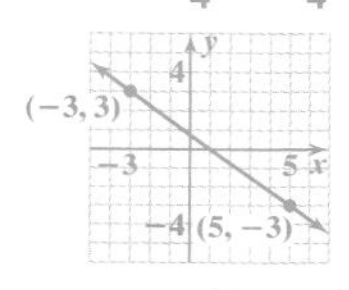

67. $y = \frac{4}{5}x + \frac{12}{5}$

68. $y = -\frac{5}{4}x - \frac{25}{2}$

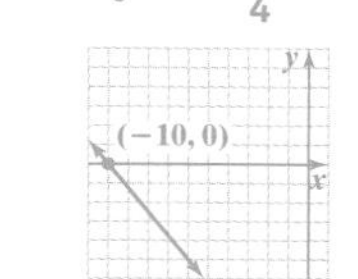

69. $y = -\frac{4}{9}x + \frac{8}{3}$

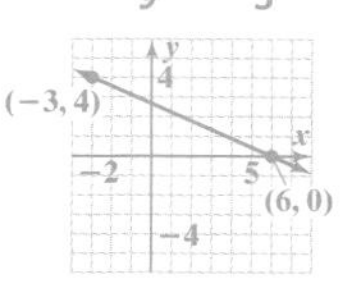

70. $y = -\frac{5}{4}x - 1$

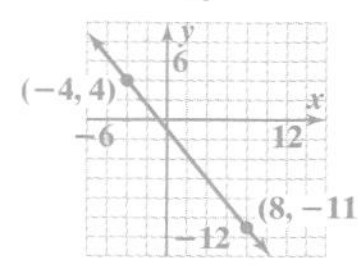

71. (a) When 60 packages are shipped, the expenses are \$1635.
(b)

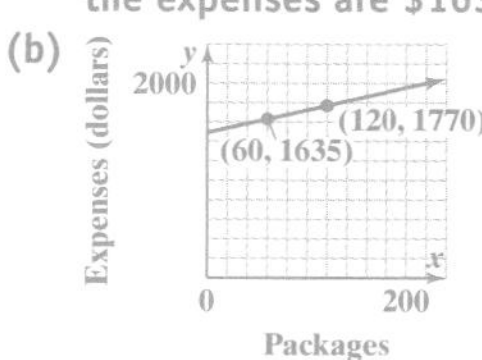

(c) $y = \frac{9}{4}x + 1500$
(d) \$1950
(e) Expenses increase by \$2.25 for each additional package.

72. (a) After 30 years, monthly income is \$3600.
(b)

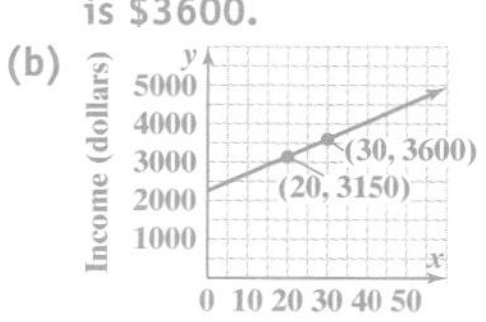

(c) $y = 45x + 2250$ (d) \$2925 (e) Monthly income increases by \$45 for each additional year of service.

73. (a) (0, 181), (27, 276)
(b)
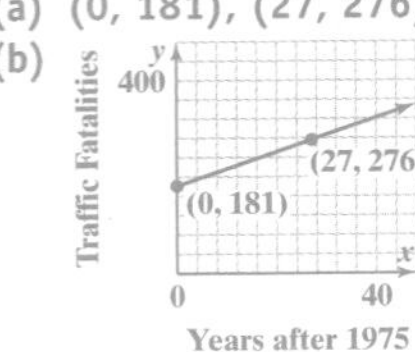

(c) $y = \frac{95}{27}x + 181$
(d) about 234
(e) The number of traffic fatalities in this region increases by about 3.5 each year.

74. (a) (0, 39161), (27, 38309)
(b)
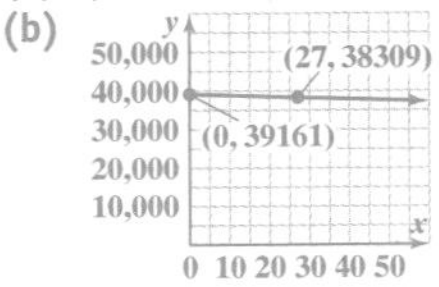

(c) $y = -\frac{284}{9}x + 39{,}161$
(d) about 38,782
(e) The number of traffic fatalities in this region decreases by about 31.6 each year.
(f) Answers may vary.

75. $y = 3x + 14$ 76. $y = 4x - 22$

77. $y = -2x - 2$ 78. $y = -5x + 2$

79. $y = 6x - \frac{7}{2}$ 80. $y = -9x + \frac{21}{2}$

81. $y = -x - 7$ 82. $y = x + 5$

83. $y = \frac{1}{5}x - 2$ 84. $y = \frac{1}{3}x - 1$

85. $y = -\frac{3}{2}x + 2$ 86. $y = -\frac{5}{2}x + 1$

(c) Find the linear equation in slope-intercept form that relates the number of traffic fatalities, y, to the number of years after 1975, x.

(d) Use the equation found in part (c) to find the number of traffic fatalities in 1987.

(e) Interpret the slope.

(f) Do you think that the slope is misleading in terms of describing the decline in traffic fatalities? Why?

Extending the Concepts

Up to this point, when we knew the slope of a line and a point on the line we found the equation of the line using point-slope form. We could also use the slope-intercept form to find this equation.

For example, suppose that we were asked to find the equation of the line whose slope is 6 and that passes through the point $(2, -5)$. *Let's use the slope-intercept form,* $y = mx + b$, *to write the equation of this line. We know* $m = 6$, *so we have* $y = 6x + b$. *We also know that* $y = -5$ *when* $x = 2$. *So we substitute 2 for x and −5 for y into the equation and solve for b.*

$$y = 6x + b$$

Let $x = 2$ and $y = -5$: $\quad -5 = 6(2) + b$

Multiply: $\quad -5 = 12 + b$

Subtract 12 from each side: $\quad -5 - 12 = 12 - 12 + b$

$$-17 = b$$

We now know that $b = -17$ *and because we also know that* $m = 6$, *the equation of the line is* $y = 6x - 17$. *Use this technique to find the slope-intercept form of the line for Problems 75–86.*

75. $(-4, 2)$; slope $= 3$ **76.** $(5, -2)$; slope $= 4$ **77.** $(3, -8)$; slope $= -2$

78. $(-1, 7)$; slope $= -5$ **79.** $\left(\frac{2}{3}, \frac{1}{2}\right)$; slope $= 6$ **80.** $\left(\frac{4}{3}, -\frac{3}{2}\right)$; slope $= -9$

81. $(6, -13)$ and $(-2, -5)$ **82.** $(-10, -5)$ and $(2, 7)$ **83.** $(5, -1)$ and $(-10, -4)$

84. $(-6, -3)$ and $(9, 2)$ **85.** $(-4, 8)$ and $(2, -1)$ **86.** $(4, -9)$ and $(8, -19)$

8.6 Parallel and Perpendicular Lines

OBJECTIVES

1. Determine Whether Two Lines Are Parallel
2. Find the Equation of a Line Parallel to a Given Line
3. Determine Whether Two Lines Are Perpendicular
4. Find the Equation of a Line Perpendicular to a Given Line

Preparing for Parallel and Perpendicular Lines

Before getting started, take the following readiness quiz. If you get a problem wrong, go back to the section cited and review the material.

1. Determine the reciprocal of 3. [Section 1.4, pp. 24–25]

2. Determine the reciprocal of $-\frac{3}{5}$. [Section 1.4, pp. 24–25]

1 Determine Whether Two Lines Are Parallel

When two lines in the rectangular coordinate system do not intersect (that is, they have no points in common), they are said to be *parallel.* However, rather than looking at graphs of linear equations to determine whether they are parallel or not, we can look at the equations themselves to determine whether two lines might be parallel.

DEFINITION

Two nonvertical lines are **parallel** if and only if their slopes are equal and they have different y-intercepts. Vertical lines are parallel if they have different x-intercepts.

Preparing for...Answers **1.** $\frac{1}{3}$ **2.** $-\frac{5}{3}$

Work Smart

The use of the words "if and only if" given in the definition of parallel lines means that there are two statements being made:

If two nonvertical lines are parallel, then their slopes are equal and they have different y-intercepts.

If two nonvertical lines have equal slopes and different y-intercepts, then they are parallel.

Figure 49(a) shows nonvertical parallel lines. Figure 49(b) shows vertical parallel lines.

Figure 49
Parallel lines.

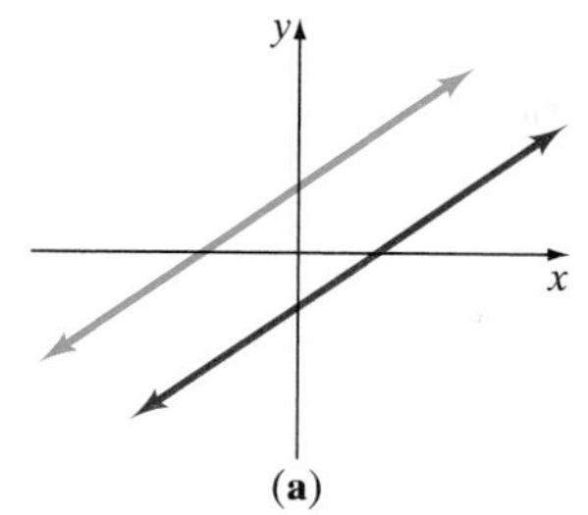

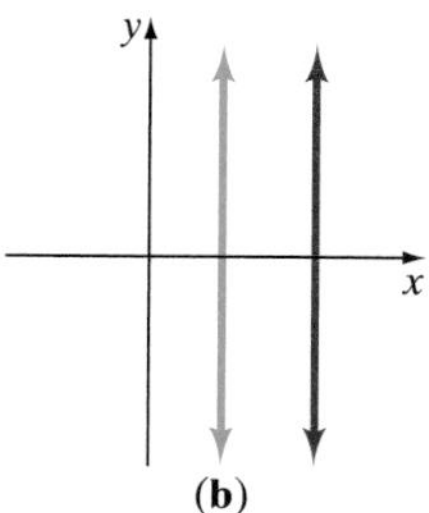

To determine whether two lines are parallel, we find the slope and y-intercept of each line by putting the equation of the line in slope-intercept form. If the slopes are the same, but the y-intercepts are different, then the lines are parallel.

Classroom Example Determine whether the line $y = 2x - 4$ is parallel to $y = 3x + 1$.

Answer: The lines are not parallel.

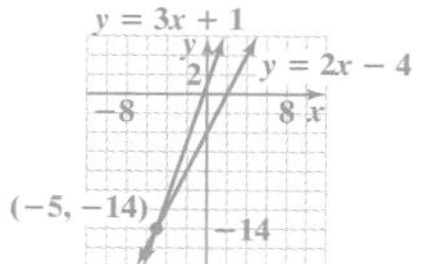

EXAMPLE 1 Determining Whether Two Lines Are Parallel

Determine whether the line $y = 4x - 5$ is parallel to $y = 3x - 2$. Graph the lines to confirm your results.

Figure 50

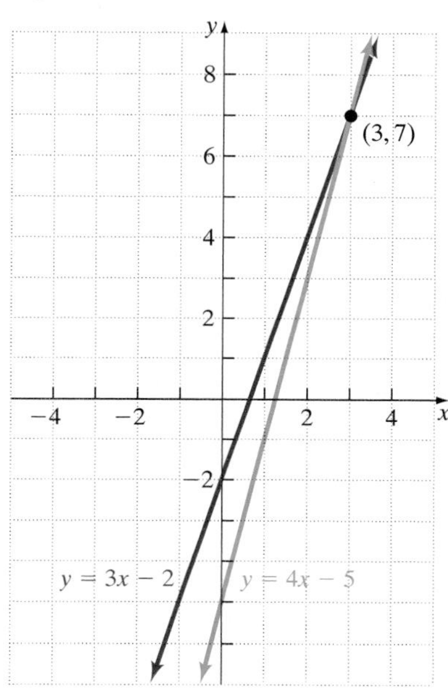

Solution

For the line $y = 4x - 5$, the slope is 4 and the y-intercept is -5. For the line $y = 3x - 2$, the slope is 3 and the y-intercept is -2. Because the lines have different slopes, they are not parallel. Figure 50 shows the graph of the two lines.

We can see from the graph in Figure 50 that the lines intersect at $(3, 7)$. Therefore, the lines are not parallel.

Classroom Example Determine whether the line $2x - y = 5$ is parallel to $-4x + 2y = 7$.

Answer: The lines are parallel.

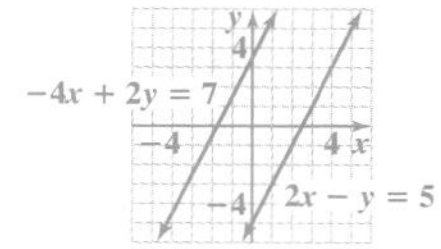

EXAMPLE 2 Determining Whether Two Lines Are Parallel

Determine whether the line $-3x + y = 5$ is parallel to $6x - 2y = -2$. Graph the lines to confirm your results.

Solution

To find the slope and y-intercept, we solve each equation for y so that each is in slope-intercept form.

$$-3x + y = 5$$

Add $3x$ to both sides: $$y = 3x + 5$$

The slope of the line $-3x + y = 5$ is 3 and the y-intercept is 5.

$$6x - 2y = -2$$

Subtract $6x$ from both sides: $$-2y = -6x - 2$$

Divide both sides by -2: $$y = \frac{-6x - 2}{-2}$$

Divide each term in the numerator by -2: $$y = 3x + 1$$

The slope of $6x - 2y = -2$ is 3 and the y-intercept is 1.

Because the lines have the same slope, 3, but different y-intercepts, the lines are parallel. Figure 51 shows a graph of the two lines.

Work Smart

Make sure both criteria for parallel lines are satisfied.

1. Same slope
2. Different y-intercepts

Lines with the same slope and same y-intercept are called *coincident lines.*

Figure 51

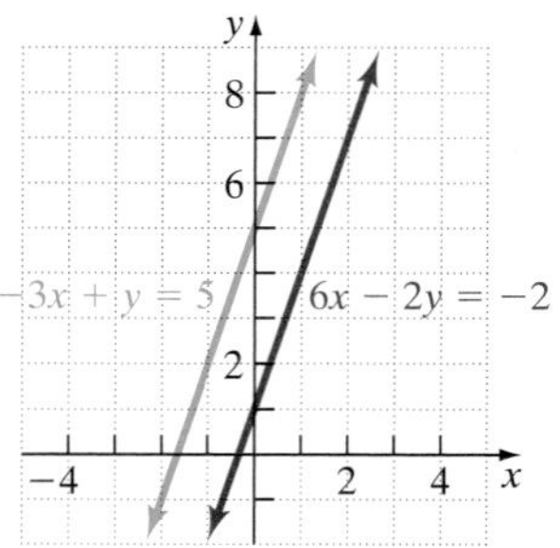

QUICK ✓ *Determine whether the two lines are parallel. Graph the lines to confirm your results.*

1. $y = 2x + 1$
   $y = -2x - 3$
2. $6x + 3y = 3$
   $10x + 5y = 10$
3. $4x + 5y = 10$
   $8x + 10y = 20$

Classroom Example

Find an equation for the line that is parallel to $2x + y = 5$ that contains the point $(2, -2)$.

Answer: $y = -2x + 2$

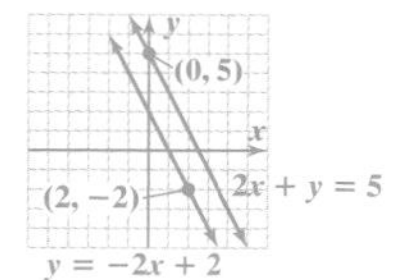

2 Find the Equation of a Line Parallel to a Given Line

Now that we know how to identify parallel lines, we can find the equation of a line that is parallel to a given line.

EXAMPLE 3 How to Find an Equation of a Line That Is Parallel to a Given Line

Find an equation for the line that is parallel to $2x + y = 5$ and contains the point $(-1, 3)$. Write the equation of the line in slope-intercept form. Graph the lines.

Step-by-Step Solution

Step 1: Find the slope of the given line.

$$2x + y = 5$$

Subtract $2x$ from both sides: $y = -2x + 5$

The slope of the line is -2, so the slope of the parallel line is also -2.

Step 2: Use the point-slope form of a line with the given point and the slope found in Step 1 to find the equation of the parallel line.

$$y - y_1 = m(x - x_1)$$

$m = -2, x_1 = -1, y_1 = 3$: $y - 3 = -2(x - (-1))$

Step 3: Put the equation in slope-intercept form by solving for y.

$$y - 3 = -2(x + 1)$$

Distribute the -2: $y - 3 = -2x - 2$

Add 3 to both sides: $y = -2x + 1$

Teaching Tip

Remind students that they are reducing this problem to one they already know how to do. Finding the slope of the line parallel to the given line is the only "layer" that is added to previously learned concepts.

The equation of the line parallel to $2x + y = 5$ is $y = -2x + 1$. Figure 52 shows the graph of the parallel lines.

Figure 52

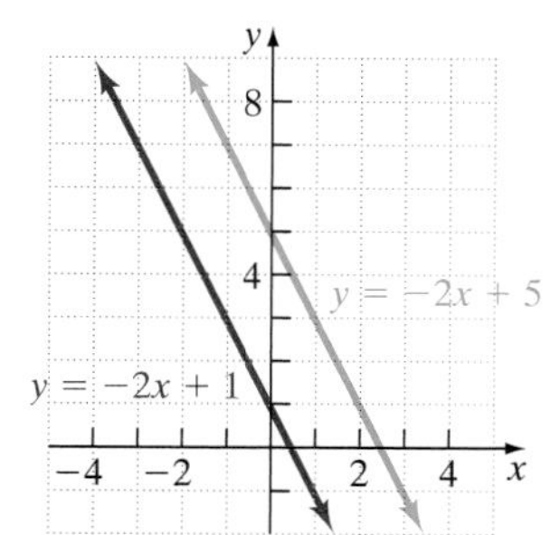

QUICK ✓ *Find the equation of the line that contains the given point and is parallel to the given line. Write the line in slope-intercept form. Graph the lines.*

4. $y = 2x + 1$ containing $(2, 3)$ **5.** $3x + 2y = 4$ containing $(-2, 3)$

Classroom Example
Find an equation for the line that is parallel to $x = -1$ and contains the point $(5, 2)$.

Answer: $x = 5$

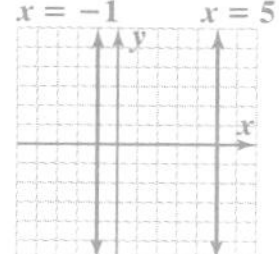

EXAMPLE 4 Finding an Equation of a Line That Is Parallel to a Given Line

Find an equation for the line that is parallel to $x = 3$ and contains the point $(1, 5)$. Graph the lines.

Solution

The equation of the given line is $x = 3$. Because this is the equation of a vertical line, the line parallel to it will also be vertical. Vertical lines have equations of the form $x = a$. Since the line parallel to $x = 3$ contains the point $(1, 5)$, the equation of the parallel line is $x = 1$. Figure 53 shows the graphs of the lines $x = 3$ and $x = 1$.

Figure 53

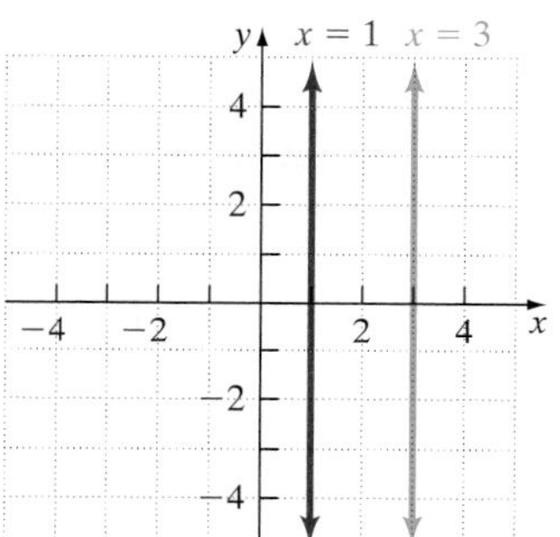

QUICK ✓ *Find the equation of the line that contains the given point and is parallel to the given line. Write the line in slope-intercept form, if possible. Graph the lines.*

6. $x = -2$ containing $(3, 1)$ **7.** $y + 3 = 0$ containing $(-2, 5)$

3 Determine Whether Two Lines Are Perpendicular

When two lines intersect at a right (90°) angle, they are said to be **perpendicular.** See Figure 54.

Just as we use the slopes of lines to determine whether two lines are parallel, we also use slopes of lines to determine whether two lines are perpendicular.

Figure 54
Perpendicular lines.

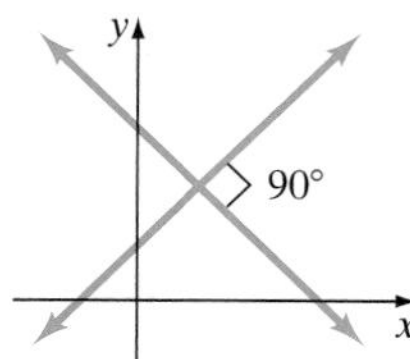

Work Smart
If m_1 and m_2 are negative reciprocals of each other, then $m_1 = \frac{-1}{m_2}$. For example, the numbers 4 and $-\frac{1}{4}$ are negative reciprocals. Watch out though, because the use of the word "*negative*" does not mean that the slope of the perpendicular line must be negative, it means that the nonvertical lines have slopes that are opposite in sign. One is positive and one is negative.

DEFINITION

Two nonvertical lines are **perpendicular** if and only if the product of their slopes is -1. Put another way, two nonvertical lines are perpendicular if their slopes are negative reciprocals of each other. Any vertical line is perpendicular to any horizontal line.

Classroom Example Find the slope of a line perpendicular to a line whose slope is

(a) 3 (b) $-\frac{2}{7}$

Answer:

(a) $-\frac{1}{3}$ (b) $\frac{7}{2}$

EXAMPLE 5 Finding the Slope of a Line Perpendicular to a Given Line

Find the slope of a line perpendicular to a line whose slope is (a) 5 (b) $-\frac{2}{3}$.

Solution

(a) To find the slope of a line perpendicular to a given line, we determine the negative reciprocal of the slope of the given line. The negative reciprocal of 5 is $\frac{-1}{5} = -\frac{1}{5}$. Any line whose slope is $-\frac{1}{5}$ will be perpendicular to the line whose slope is 5.

(b) The negative reciprocal of $-\frac{2}{3}$ is $\frac{-1}{-\frac{2}{3}} = \frac{3}{2}$. Any line whose slope is $\frac{3}{2}$ will be perpendicular to the line whose slope is $-\frac{2}{3}$.

QUICK ✓ *Find the slope of a line perpendicular to the line whose slope is given.*

8. -4 **9.** $\frac{5}{4}$ **10.** $-\frac{1}{5}$

Classroom Example Determine whether the line $y = 2x - 5$ is perpendicular to $y = \frac{1}{2}x + 3$.

Answer: The lines are not perpendicular.

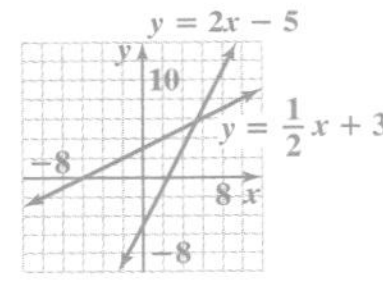

EXAMPLE 6 Determining Whether Two Lines Are Perpendicular

Determine whether the line $y = 3x - 2$ is perpendicular to $y = \frac{1}{3}x + 1$. Graph the lines to confirm your results.

Solution

To determine whether the lines are perpendicular, we first need to find the slope of each line. If the product of the slopes is -1 (the slopes are negative reciprocals of each other), then the lines are perpendicular. The slope of $y = 3x - 2$ is $m_1 = 3$. The slope of $y = \frac{1}{3}x + 1$ is $m_2 = \frac{1}{3}$. Because the product of the slopes, $m_1 \cdot m_2 = 3 \cdot \left(\frac{1}{3}\right) = 1 \neq -1$, the lines are not perpendicular. Notice that the slopes are reciprocals of each other, but are not *negative* reciprocals of each other. Figure 55 shows the graph of the two lines.

Figure 55

Classroom Example Determine whether the line $4x - 3y = 7$ is perpendicular to $6x + 8y = -5$.

Answer: The lines are perpendicular.

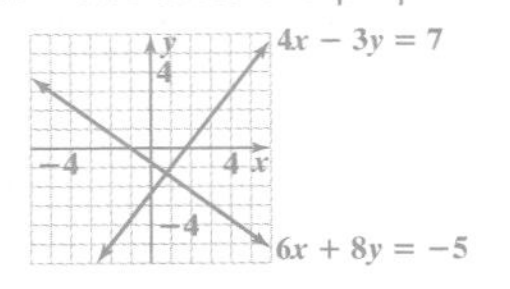

EXAMPLE 7 Determining Whether Two Lines Are Perpendicular

Determine whether the line $2x + 3y = -6$ is perpendicular to $3x - 2y = 2$. Graph the lines to confirm your results.

Solution

To find the slopes of the two lines, we write the equations of the lines in slope-intercept form.

$$2x + 3y = -6$$

Subtract $2x$ from both sides: $3y = -2x - 6$

Divide both sides by 3: $y = \frac{-2x - 6}{3}$

Divide 3 into each term in the numerator: $y = -\frac{2}{3}x - 2$

The slope of $2x + 3y = -6$ is $m_1 = -\frac{2}{3}$.

$$3x - 2y = 2$$

Subtract $3x$ from both sides: $-2y = -3x + 2$

Divide both sides by -2: $y = \frac{-3x + 2}{-2}$

Divide -2 into each term in the numerator: $y = \frac{3}{2}x - 1$

The slope of $3x - 2y = 2$ is $m_2 = \frac{3}{2}$.

The product of the slopes is $m_1 \cdot m_2 = -\frac{2}{3} \cdot \frac{3}{2} = -1$, so the lines are perpendicular. Put another way, because the slopes are negative reciprocals of each other, the lines are perpendicular. See Figure 56 for the graph of the two lines.

Figure 56

$3x - 2y = 2$

$2x + 3y = -6$

QUICK ✓ *Determine whether the given lines are perpendicular. Graph the lines.*

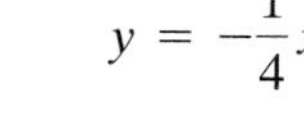

11. $y = 4x - 3$
$y = -\frac{1}{4}x - 4$

12. $2x - y = 3$
$x - 2y = 2$

13. $5x + 2y = 8$
$2x - 5y = 10$

Classroom Example
Use Example 8 with the point (6, 2) and the line $y = 3x - 2$.

Answer: $y = -\frac{1}{3}x + 4$

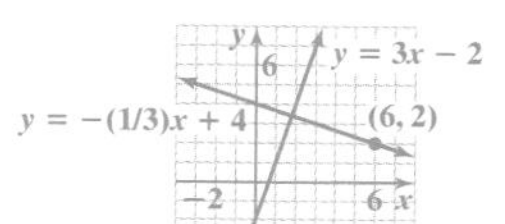

4 Find the Equation of a Line Perpendicular to a Given Line

Now that we know how to find the slope of a line perpendicular to a second line, we can find the equation of a line that is perpendicular to a given line.

EXAMPLE 8 How to Find the Equation of a Line Perpendicular to a Given Line

Find an equation of the line that is perpendicular to the line $y = 3x - 2$ and contains the point $(3, -1)$. Write the equation of the line in slope-intercept form. Graph the two lines.

Step-by-Step Solution

Step 1: Find the slope of the given line.

$$y = 3x - 2$$

The slope of the line is 3.

Step 2: Find the slope of the perpendicular line.

The slope of the perpendicular line is the negative reciprocal of 3, which is $-\frac{1}{3}$.

Step 3: Use the point-slope form of a line with the given point and the slope found in Step 2 to find the equation of the perpendicular line.

$$y - y_1 = m(x - x_1)$$

$m = -\frac{1}{3}, x_1 = 3, y_1 = -1$: $y - (-1) = -\frac{1}{3}(x - 3)$

Step 4: Put the equation in slope-intercept form by solving for y.

$$y + 1 = -\frac{1}{3}(x - 3)$$

Distribute the $-\frac{1}{3}$: $y + 1 = -\frac{1}{3}x + 1$

Subtract 1 from both sides: $y = -\frac{1}{3}x$

The equation of the line perpendicular to $y = 3x - 2$ is $y = -\frac{1}{3}x$.

Teaching Tip
Remind students of the appearance of perpendicular lines. A line with positive slope rises from left to right, so a line that is perpendicular to that line must have a negative slope: It will fall from left to right, as in Figure 57.

(continued)

Figure 57

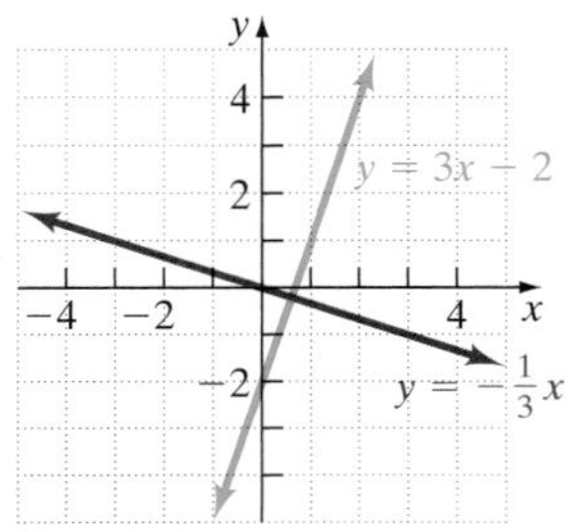

Figure 57 shows the graphs of the two lines.

QUICK ✓ *Find the equation of the line that contains the given point and is perpendicular to the given line. Write the line in slope-intercept form. Graph the lines.*

14. $(-4, 2); y = 2x + 1$ **15.** $(-2, -1); y = -\frac{2}{3}x + 1$

Classroom Example ➔
Use Example 9 with the point $(-3, 5)$.
Answer: $y = 5$

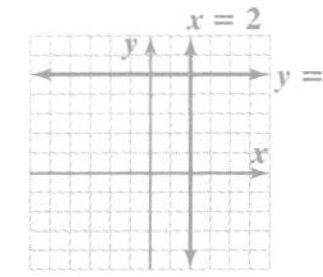

EXAMPLE 9 Finding the Equation of a Line Perpendicular to a Given Line

Find an equation of the line that is perpendicular to the line $x = 2$ and contains the point $(-4, 3)$. Write the equation of the line in slope-intercept form, if possible. Graph the two lines.

Solution

The line $x = 2$ is the equation of a vertical line. Therefore, the line perpendicular will be horizontal. Horizontal lines have a slope equal to 0. To find the equation of the perpendicular line we use the point-slope formula with $m = 0$, $x_1 = -4$, and $y_1 = 3$.

$$y - y_1 = m(x - x_1)$$

$m = 0, x_1 = -4, y_1 = 3$: $$y - 3 = 0(x - (-4))$$

$$y - 3 = 0$$

Add 3 to both sides: $$y = 3$$

The equation of the line perpendicular to $x = 2$ through the point $(-4, 3)$ is $y = 3$. See Figure 58 for the graph of the two lines.

Work Smart
The line perpendicular to a vertical line is horizontal and vice versa.

Figure 58

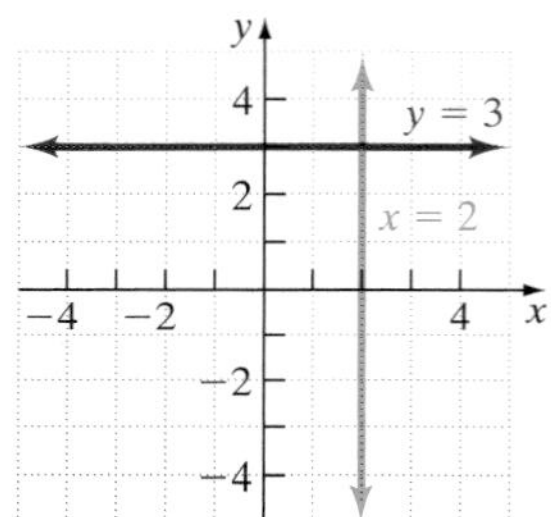

QUICK ✓ *Find the equation of the line that contains the given point and is perpendicular to the given line. Write the line in slope-intercept form, if possible. Graph the lines.*

16. $x = -4$ containing $(-1, -5)$ **17.** $y + 2 = 0$ containing $(3, -2)$

8.6 Exercises

For Extra Help:

Student Solutions Manual

CD Video

PH Math/Tutor Center

MathXL Tutorials on CD

MathXL® MyMathLab

Concepts and Vocabulary

In Problems 1–3, fill in the blanks.

1. Given any two nonvertical lines, if the product of their slopes is -1, then the lines are ________.

2. A line has a slope of 4. The slope of the line parallel to this line is ________, while the slope of the line perpendicular to this line is ________.

3. If m_1 and m_2 are the slopes of two non-vertical lines with $m_1 = \frac{-1}{m_2}$, then the lines are ________.

In Problems 4–6, answer True or False to each statement.

4. Two different lines L_1 and L_2 have slopes $m_1 = 4$ and $m_2 = -4$, so L_1 is perpendicular to L_2.

5. Given any two lines, L_1 and L_2, if the slopes are equal (that is $m_1 = m_2$), then the lines must be parallel.

6. Lines in the same plane that never intersect are called parallel.

1. perpendicular
2. 4, $-\frac{1}{4}$
3. perpendicular
4. False 5. False 6. True

7. L_1 could be parallel to L_2. L_1 could be perpendicular to L_2. L_1 and L_2 could intersect, but not at right angles. L_1 could be coincident with L_2.
8. Answers may vary.
9.–16. See table.
17. perpendicular
18. neither 19. parallel 20. parallel
21. perpendicular 22. perpendicular
23. perpendicular 24. parallel
25. perpendicular 26. neither
27. parallel 28. perpendicular
29. parallel 30. perpendicular
31. (a) $m_1 = 3$, $m_2 = -3$ (b) neither
32. (a) $m_1 = 4$, $m_2 = 4$ (b) parallel
33. (a) $m_1 = 2$, $m_2 = 2$ (b) parallel
34. (a) $m_1 = -1$, $m_2 = 1$ (b) perpendicular
35. (a) $m_1 = \frac{1}{2}$, $m_2 = \frac{1}{2}$ (b) parallel
36. (a) $m_1 = 8$, $m_2 = -8$ (b) neither
37. (a) $m_1 = \frac{5}{3}$, $m_2 = -\frac{3}{5}$ (b) perpendicular
38. (a) $m_1 = \frac{3}{4}$, $m_2 = \frac{3}{4}$ (b) parallel
39. $y = 3x - 14$
40. $y = 2x - 19$
41. $y = -4x - 4$
42. $y = -5x - 4$
43. $y = -7$
44. $x = -4$
45. $x = -1$
46. $y = -8$
47. $y = \frac{3}{2}x - 13$
48. $y = -\frac{2}{3}x + 11$
49. $y = -\frac{1}{2}x - \frac{21}{2}$
50. $y = \frac{2}{5}x - \frac{19}{5}$
51. $y = -2x + 11$
52. $y = -3x + 19$
53. $y = \frac{1}{4}x$
54. $y = \frac{1}{2}x - 4$
55. $x = -2$
56. $y = -6$
57. $y = 5$
58. $x = 11$
59. $y = \frac{5}{2}x$
60. $y = \frac{2}{3}x$
61. $y = -\frac{3}{5}x - 9$
62. $y = -\frac{5}{3}x$
63. $y = 7x - 26$ 64. $y = 3x + 16$

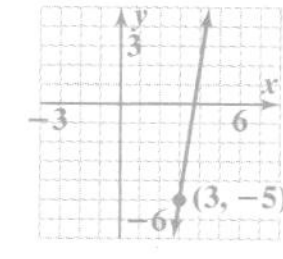

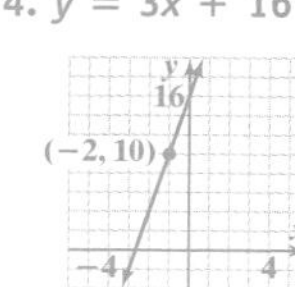

7. Describe the four possible relationships of the graphs of two lines in a rectangular coordinate system.

8. You are asked to determine if the following lines are parallel, perpendicular or neither: L_1 contains the points $(1, 1)$ and $(3, 5)$ and L_2 contains the points $(-1, -1)$ and $(4, 9)$. List the steps you would follow to make this determination, and describe the criteria you would use to answer the question.

Building Your Skills

In Problems 9–16, fill in the chart with the missing slopes.

	Slope of the Given Line	Slope of a Line Parallel to the Given Line	Slope of a Line Perpendicular to the Given Line
9.	$m = -3$	-3	$\frac{1}{3}$
10.	$m = 4$	4	$-\frac{1}{4}$
11.	$m = \frac{1}{2}$	$\frac{1}{2}$	-2
12.	$m = -\frac{1}{8}$	$-\frac{1}{8}$	8
13.	$m = -\frac{4}{9}$	$-\frac{4}{9}$	$\frac{9}{4}$
14.	$m = \frac{5}{2}$	$\frac{5}{2}$	$-\frac{2}{5}$
15.	$m = 0$	0	undefined
16.	$m =$ undefined	undefined	0

In Problems 17–30, determine if the lines are parallel, perpendicular, or neither.

17. $L_1: y = x - 3$
$L_2: y = 1 - x$

18. $L_1: y = -4x + 3$
$L_2: y = 4x - 1$

19. $L_1: y = \frac{3}{4}x + 2$
$L_2: y = 0.75x - 1$

20. $L_1: y = 0.8x + 6$
$L_2: y = \frac{4}{5}x + \frac{19}{3}$

21. $L_1: y = -\frac{5}{3}x - 6$
$L_2: y = \frac{3}{5}x - 1$

22. $L_1: y = 3x - 1$
$L_2: y = 6 - \frac{x}{3}$

23. $L_1: x + y = -3$
$L_2: y - x = 1$

24. $L_1: x - 4y = 24$
$L_2: 2x - 8y = -8$

25. $L_1: 2x - 5y = 5$
$L_2: 5x + 2y = 4$

26. $L_1: x - 2y = -8$
$L_2: x + 2y = 2$

27. $L_1: 4x - 5y - 15 = 0$
$L_2: 8x - 10y + 5 = 0$

28. $L_1: x + y = 6$
$L_2: x - y = -2$

29. $L_1: 4x = 3y + 3$
$L_2: 6y = 8x + 36$

30. $L_1: 2x - 5y - 45 = 0$
$L_2: 5x + 2y - 8 = 0$

In Problems 31–38, each line contains the given points. (a) Find the slope of each line. (b) Determine if the lines are parallel, perpendicular, or neither.

31. L_1: $(0, -1)$ and $(-2, -7)$
L_2: $(-1, 5)$ and $(2, -4)$

32. L_1: $(-3, -14)$ and $(1, 2)$
L_2: $(0, 2)$ and $(-3, -10)$

33. L_1: $(2, 8)$ and $(7, 18)$
L_2: $(-2, -3)$ and $(6, 13)$

34. L_1: $(6, 0)$ and $(-2, 8)$
L_2: $(4, 1)$ and $(-6, -9)$

65. $y = \frac{1}{5}x + \frac{43}{5}$

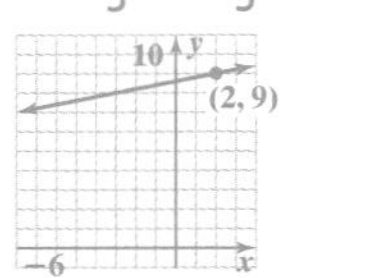

66. $y = -4x + 35$

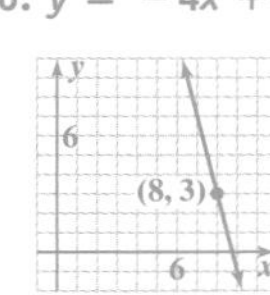

67. $y = -7x + 41$

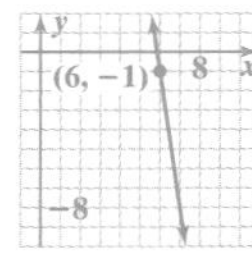

68. $y = -\frac{1}{4}x - \frac{3}{4}$

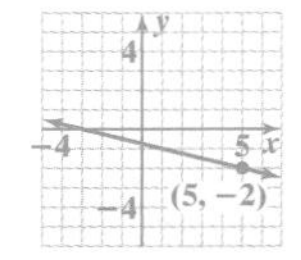

69. $y = -2x - 10$

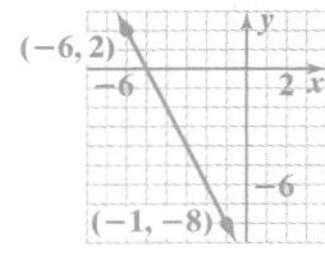

70. $y = -4x - 17$

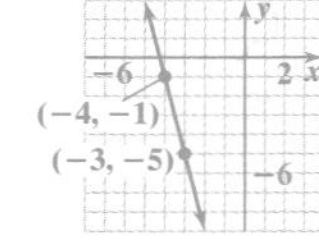

71. $y = 3x - 2$

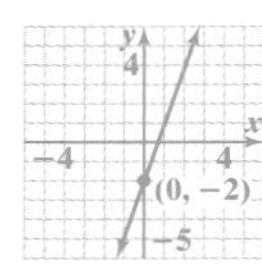

72. $y = -2x + 7$

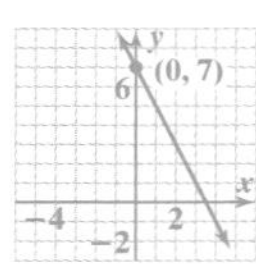

73. $x = 5$

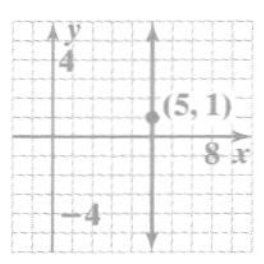

74. $y = 7$

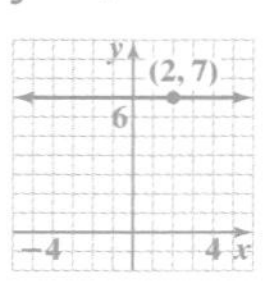

75. $y = -\frac{4}{3}x + 2$

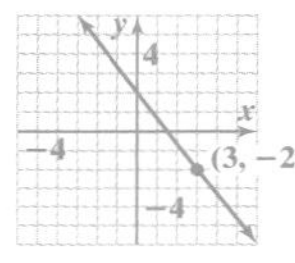

76. $y = -\frac{1}{3}x - 3$

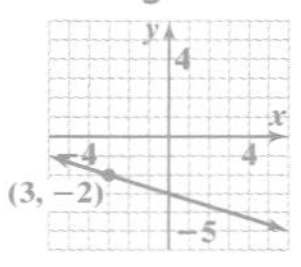

77. $y = -2x - 5$

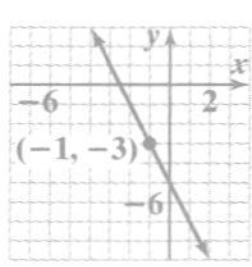

78. $y = -\frac{1}{4}x + \frac{1}{4}$

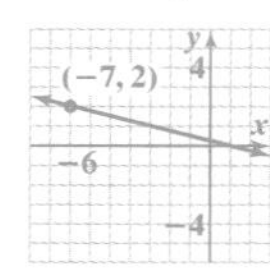

79.

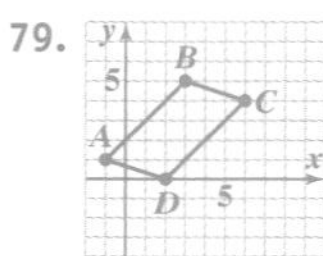

parallelogram

80.

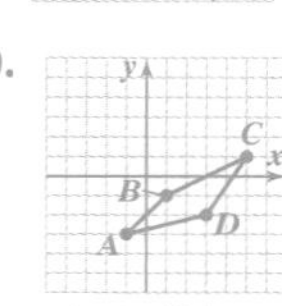

not a parallelogram

81.

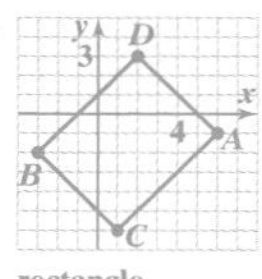

rectangle

82.

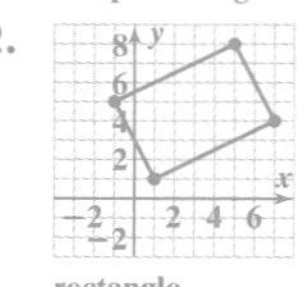

rectangle

83.

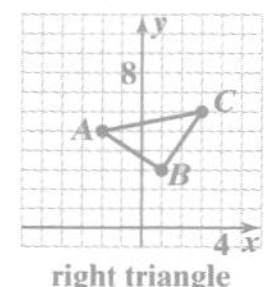

right triangle

84.

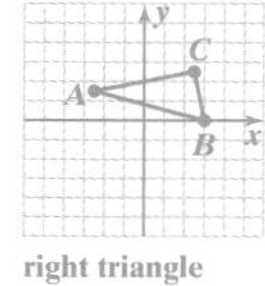

right triangle

35. L_1: $(-2, -5)$ and $(4, -2)$
L_2: $(-8, -5)$ and $(0, -1)$

36. L_1: $(1, 6)$ and $(-1, -10)$
L_2: $(0, 1)$ and $(-2, 17)$

37. L_1: $(-6, -9)$ and $(3, 6)$
L_2: $(10, -8)$ and $(-5, 1)$

38. L_1: $(-8, -8)$ and $(4, 1)$
L_2: $(12, 8)$ and $(-4, -4)$

In Problems 39–50, find the equation of the line that contains the given point and is parallel to the given line. Write the equation in slope-intercept form, if possible.

39. $(4, -2)$; $y = 3x - 1$
40. $(7, -5)$; $y = 2x + 6$
41. $(-3, 8)$; $y = -4x + 5$
42. $(-2, 6)$; $y = -5x - 2$
43. $(3, -7)$; $y = 4$
44. $(-4, 5)$; $x = -3$
45. $(-1, 10)$; $x = 10$
46. $(4, -8)$; $y = -1$
47. $(10, 2)$; $3x - 2y = 5$
48. $(6, 7)$; $2x + 3y = 9$
49. $(-1, -10)$; $x + 2y = 4$
50. $(-3, -5)$; $2x - 5y = 6$

In Problems 51–62, find the equation of the line that contains the given point and is perpendicular to the given line. Write the equation in slope-intercept form, if possible.

51. $(3, 5)$; $y = \frac{1}{2}x - 2$
52. $(4, 7)$; $y = \frac{1}{3}x - 3$
53. $(-4, -1)$; $y = -4x + 1$
54. $(-2, -5)$; $y = -2x + 5$
55. $(-2, 1)$; x-axis
56. $(3, -6)$; y-axis
57. $(7, 5)$; y-axis
58. $(11, -6)$; x-axis
59. $(0, 0)$; $2x + 5y = 7$
60. $(0, 0)$; $6x + 4y = 3$
61. $(-10, -3)$; $5x - 3y = 4$
62. $(-6, 10)$; $3x - 5y = 2$

Mixed Practice

In Problems 63–78, find the equation of the line that has the given properties. Write the equation in slope-intercept form, if possible. Graph each line.

63. Contains $(3, -5)$; slope $= 7$
64. Contains $(-2, 10)$; slope $= 3$
65. Contains $(2, 9)$; perpendicular to the line $y = -5x + 3$
66. Contains $(8, 3)$; parallel to the line $y = -4x + 2$
67. Contains $(6, -1)$; parallel to the line $y = -7x + 2$
68. Contains $(5, -2)$; perpendicular to the line $y = 4x + 3$
69. Contains $(-6, 2)$ and $(-1, -8)$
70. Contains $(-4, -1)$ and $(-3, -5)$
71. Slope $= 3$, y-intercept $= -2$
72. Slope $= -2$, y-intercept $= 7$
73. Contains $(5, 1)$; parallel to the line $x = -6$
74. Contains $(2, 7)$; perpendicular to the line $x = -8$
75. Contains $(3, -2)$; parallel to the line $4x + 3y = 9$
76. Contains $(-3, -2)$; perpendicular to the line $6x - 2y = 1$
77. Contains $(-1, -3)$; perpendicular to the line $x - 2y = -10$
78. Contains $(-7, 2)$; parallel to the line $x + 4y = 2$

Applying the Concepts

A parallelogram is a quadrilateral in which both pairs of opposite sides are parallel. In Problems 79 and 80, plot the following points, draw the figure, and then use slope to determine if the figure is a parallelogram.

△ **79.** $A(-1, 1)$; $B(3, 5)$; $C(6, 4)$; $D(2, 0)$
△ **80.** $A(-1, -3)$; $B(1, -1)$; $C(5, 1)$; $D(3, -2)$

A rectangle is a parallelogram that contains one right angle. That is, one pair of sides is perpendicular. In Problems 81 and 82, plot the following points, draw the figure, and then use slope to determine if the figure is a rectangle.

△ **81.** $A(6, -1)$; $B(-3, -2)$; $C(1, -6)$; $D(2, 3)$
△ **82.** $A(1, 1)$; $B(-1, 5)$; $C(5, 8)$; $D(7, 4)$

85. 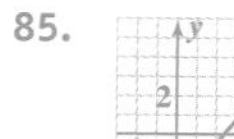 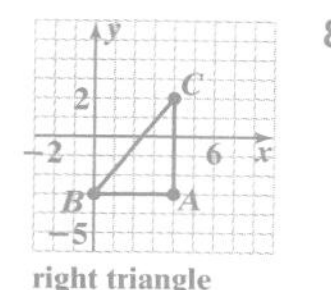
right triangle

86. 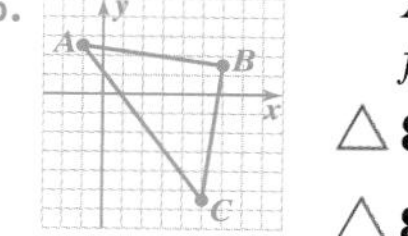
right triangle

87. $B = 2$ 88. $A = -6$
89. $A = 4$ 90. $B = 2$

91. 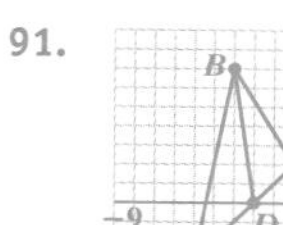
not an altitude

92. slope $\overline{AC} = \frac{5}{4}$

slope $\overline{BD} = -\frac{4}{5}$

$\frac{5}{4}\left(-\frac{4}{5}\right) = -1$

Diagonals are perpendicular.

A right triangle is a triangle that contains one right angle. In Problems 83–86, plot each point and form the triangle ABC. Verify using slope that the triangle is a right triangle.

△ **83.** $A(-2, 5)$; $B(1, 3)$; $C(3, 6)$ △ **84.** $A(-5, 3)$; $B(6, 0)$; $C(5, 5)$

△ **85.** $A(4, -3)$; $B(0, -3)$; $C(4, 2)$ △ **86.** $A(-2, 5)$; $B(12, 3)$; $C(10, -11)$

Extending the Concepts

In Problems 87 and 88, find the missing coefficient so that the lines are parallel.

87. $-3y = 6x - 12$ and $4x + By = -2$

88. $Ax + 2y = 4$ and $15x = 5y + 20$

In Problems 89 and 90, find the missing coefficient so that the lines are perpendicular.

89. $Ax + 6y = -6$ and $12 - 6y = -9x$

90. $x - By = 10$ and $3y = -6x + 9$

△ **91.** The altitude of a triangle is a line segment drawn from a vertex of the triangle perpendicular to the opposite side. Plot the following points, draw triangle ABC and the segment joining points B and D, and then determine if $\overline{BD}$ is an altitude of the triangle. $A(-6, -3)$; $B(-4, 7)$; $C(-1, 2)$; $D(-3, 0)$

△ **92.** The coordinates of the vertices of a quadrilateral are $A(2, 1)$, $B(4, 6)$, $C(6, 6)$, and $D(9, 2)$. Use slopes to show that the diagonals of the quadrilateral, $\overline{AC}$ and $\overline{BD}$, are perpendicular to each other.

PUTTING THE CONCEPTS TOGETHER (SECTIONS 8.1–8.6)

1. yes, $(1, -2)$ is a solution

2. 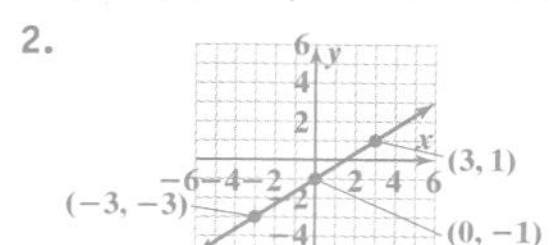

3. 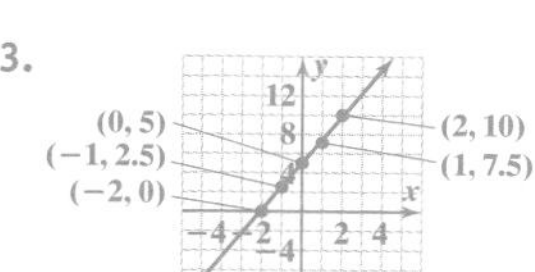

4. (a) x-intercept is $-\frac{3}{4}$
(b) y-intercept is 3

5. (a) x-intercept is $\frac{3}{2}$
(b) y-intercept is 2
(c) 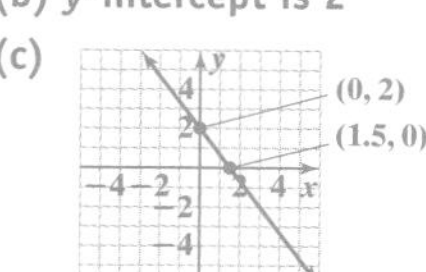

6. (a) slope $= -\frac{2}{3}$
(b) y-intercept $= -\frac{4}{3}$

7. slope $= -\frac{1}{3}$

These problems cover important concepts from Sections 8.1 to 8.6. We designed these problems so that you can review the chapter so far and show your mastery of the concepts. Take time to work these problems before proceeding with the next section. The answers to these problems are located at the back of the text starting on page AN-21.

1. Given the linear equation $4x - 3y = 10$, determine whether the ordered pair $(1, -2)$ is a solution to the equation.

In Problems 2 and 3, graph each equation using the point-plotting method.

2. $y = \frac{2}{3}x - 1$ **3.** $-5x + 2y = 10$

4. Given the equation $-8x + 2y = 6$, determine
(a) the x-intercept **(b)** the y-intercept

5. Graph the equation $4x + 3y = 6$ by finding the x-intercept and the y-intercept.

6. Given the equation $6x + 9y = -12$, determine
(a) the slope **(b)** the y-intercept

7. Find the slope of the line that contains the points $(3, -5)$ and $(-6, -2)$.

8. Given the linear equation $2y = -5x - 4$, find
(a) the slope of the line perpendicular to the given line
(b) the slope of the line parallel to the given line

9. Determine whether the lines are parallel, perpendicular, or neither. Explain how you came to your conclusion.

L_1: $10x + 5y = 2$ L_2: $y = -2x + 3$

8. (a) $m = \frac{2}{5}$ (b) $m = -\frac{5}{2}$
9. The lines are parallel. Answers may vary.
10. $y = 3x + 1$
11. $y = -6x - 2$
12. $y = -2x + 7$
13. $y = -\frac{5}{2}x - 20$
14. $y = \frac{1}{4}x + 5$
15. $y = -8$
16. $x = 2$
17. $m = 11$. Every package increases expenses by \$11.
18. (a) $y = 8350x - 2302$, where x is the weight, in carats, and y is the price.
(b) For every 1-carat increase in weight, the cost increases by \$8350.
(c) \$4044

In Problems 10–16, write the equation of the line that satisfies the given conditions. Write the equation in slope-intercept form, if possible.

10. slope $= 3$ and y-intercept is 1

11. slope $= -6$ and passes through $(-1, 4)$

12. through $(4, -1)$ and $(-2, 11)$

13. through $(-8, 0)$ and perpendicular to $y = \frac{2}{5}x - 5$

14. through $(-8, 3)$ and parallel to $-8y + 2x = -1$

15. horizontal line through $(-6, -8)$

16. through $(2, 6)$ with undefined slope

17. Shipping Expenses The shipping department records indicate that during a week when 80 packages were shipped, the total expenses recorded for the shipping department were \$1180. During a different week 50 packages were shipped and the expenses recorded were \$850. Use the ordered pairs (80, 1180) and (50, 850) to determine the average rate to ship an additional package.

18. Diamonds The relation between the cost of a diamond and its weight is linear. In looking at two diamonds, we find that one of the diamonds weighs 0.7 carats and costs \$3543, while the other diamond weighs 0.8 carats and costs \$4378. (*Source:* diamonds.com)

(a) Use the ordered pairs (0.7, 3543) and (0.8, 4378) to find a linear equation that relates the price of a diamond to its weight.
(b) Interpret the slope.
(c) Predict the price of a diamond that weighs 0.76 carats.

8.7 Variation

OBJECTIVES

1. Model and Solve Direct Variation Problems
2. Model and Solve Inverse Variation Problems

Preparing for Variation

Before getting started, take the following readiness quiz. If you get a problem wrong, go back to the section cited and review the material.

1. Solve $10 = 2k$ for k. [Section 2.1, pp. 78–81]

2. Solve $3 = \frac{k}{5}$ for k. [Section 2.1, pp. 78–81]

1 Model and Solve Direct Variation Problems

Teaching Tip
Have students develop some examples of direct variation. For example, the more hours a student studies, the higher the test score.

Often two variables are related as proportions. For example, we say, "Revenue is proportional to sales" or "Force is proportional to acceleration." When we say that one variable is proportional to another variable, we are talking about *variation*. **Variation** describes how one quantity changes in relation to another quantity. In this text, we will discuss two types of variation: *direct* and *inverse*. We will discuss direct variation first.

In Words
When y is directly proportional to x, then y and x are related through a linear equation whose y-intercept is 0 and slope is k, the constant of proportionality.

DEFINITION

If x and y represent two quantities, we say that y **varies directly** with x, or y is **directly proportional to** x, if there is a nonzero number k such that

$$y = kx$$

The number k is called the **constant of proportionality,** or the **constant of variation.**

Preparing for...Answers **1.** $\{5\}$ **2.** $\{15\}$

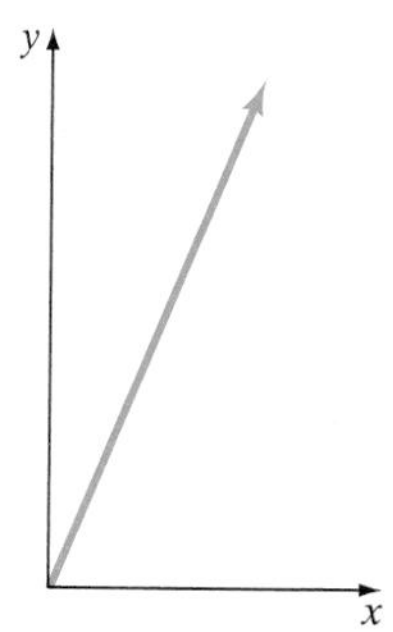

Figure 59
$y = kx,\ k > 0,\ x \geq 0$

The graph in Figure 59 illustrates the relationship between y and x if y varies directly with x and $k > 0$, $x \geq 0$. Notice that the constant of proportionality k is the slope of the line.

If we know that two quantities vary directly, then knowing the value of each quantity in one instance allows us to write a formula that is true in all cases. Although the definition for direct variation uses y and x as the variables, any variables can be used.

EXAMPLE 1 Direct Variation

Suppose that y varies directly with x for $x \geq 0$. Find an equation that relates y and x if it is known that $y = 20$ when $x = 4$. Graph the equation.

Classroom Example
Use Example 1 with $y = 12$ when $x = 3$.

Answer: $y = 4x$

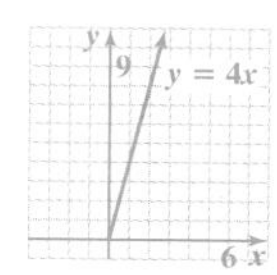

Solution

Because y varies directly with x, we know that $y = kx$. Now we can use the given information to find k, the constant of proportionality.

$$y = kx$$

$y = 20,\ x = 4$: $\quad 20 = k(4)$

Divide both sides by 4: $\quad 5 = k$

Symmetric Property: $\quad k = 5$

Work Smart
Recall that the Symmetric Property states that if $a = b$, then $b = a$.

With $k = 5$, we have that $y = 5x$. Because $x \geq 0$, we only need to graph the equation in quadrant I. See Figure 60.

Figure 60

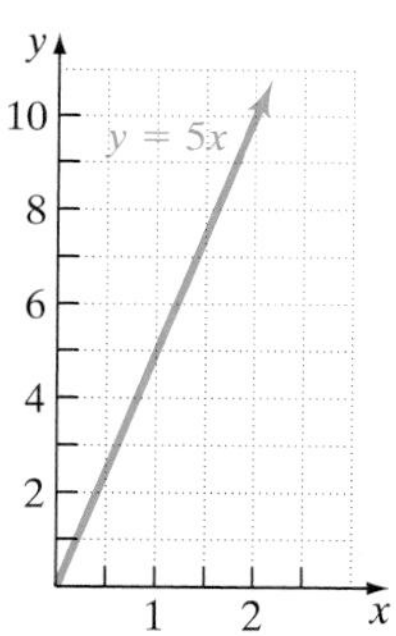

Work Smart
The constant of variation, k, is the slope of the line.

QUICK ✓

1. Suppose that y varies directly with x for $x \geq 0$. Find an equation that relates y and x if it is known that $y = 15$ when $x = 5$. Graph the equation.
2. Suppose that y varies directly with x for $x \geq 0$. Find an equation that relates y and x if it is known that $y = 6$ when $x = 18$. Graph the equation.
3. Suppose that q varies directly with w for $w \geq 0$.
   (a) Find an equation that relates q and w if it is known that $q = 10$ when $w = 40$.
   (b) Use the equation found in part (a) to determine q when $w = 60$.
   (c) Graph the equation found in part (a).

EXAMPLE 2 Car Payments

Classroom Example
Use Example 2.

Brandon just bought a used car for \$12,000. He decides to put \$2000 down on the car and borrow the remaining \$10,000. The bank lends Brandon \$10,000 at 5.9% interest for 48 months. His payments are \$234.39. The monthly payment p on a car varies directly with the amount borrowed b. If Brandon puts \$3000 down on the car instead, what would his monthly payment be?

Teaching Tip
Point out that solving direct variation problems require two steps: (1) finding the constant of variation and (2) solving the problem for the unknown quantity.

Solution

Step 1: Identify We want to know the monthly payment if Brandon puts \$3000 down. The model will involve variables that are directly related.

Step 2: Name The variables have been named already: p is the monthly payment and b is the amount borrowed.

Step 3: Translate The monthly payment p varies directly with the amount borrowed b. So

$$p = kb$$

Since $p = 234.39$ when $b = 10{,}000$, we have that

$$234.39 = k(10{,}000)$$

Work Smart
If possible, find the exact value of k. If you cannot avoid approximate values, do not round the value of k to fewer than 5 decimal places.

Divide both sides by 10,000 to obtain $k = 0.023439$ and we have our model:

$$p = 0.023439b$$

Step 4: Solve If Brandon puts \$3000 down, he will need to borrow \$9000. We let $b = 9000$ in the model to determine the monthly payment.

$$p = 0.023439b$$

Let $b = 9000$: $\quad p = 0.023439(9000)$

Evaluate: $\quad p \approx 210.95$

Step 5: Check We can check the reasonableness of our answer. Brandon's payments are \$234.39 when he borrows \$10,000. It seems reasonable that his payments will decrease a little when he puts more money down. With \$3000 down, we determine his payment will be \$210.95. This seems reasonable.

Step 6: Answer When \$3000 is put down, Brandon's monthly payment will be \$210.95.

QUICK ✓

4. The cost of gas C varies directly with the number of gallons pumped, g. Suppose that the cost of pumping 8 gallons of gas is \$20.80. If 6.8 gallons are pumped into your car, what would the cost be?

Teaching Tip
Have students develop some examples of inverse variation. For example, if you increase your rate of speed while driving, the time it takes to reach your destination decreases.

2 Model and Solve Inverse Variation Problems

We now turn our attention to another kind of variation. This second type of variation is used to describe relations between two variables when an increase in one variable results in a decrease of a second variable. For example, as the price for a product increases, the quantity demanded of the product decreases.

Figure 61
$y = \frac{k}{x}, k > 0, x > 0$

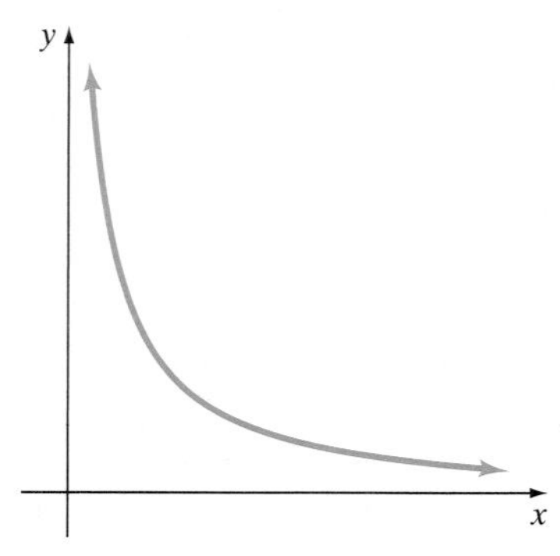

DEFINITION

If x and y represent two quantities, we say that y **varies inversely** with x, or y is **inversely proportional to** x, if there is a nonzero number k such that

$$y = \frac{k}{x}$$

where k is the **constant of variation.**

The graph in Figure 61 illustrates the relationship between y and x if y varies inversely with x with $k > 0$ and $x > 0$. Notice from the graph that as x increases the values of y decrease.

Classroom Example ➤
Use Example 3 with $y = 24$ when $x = 6$. Then find y when $x = 9$.

Answer:
(a) $y = \frac{144}{x}$
(b) $y = 16$

EXAMPLE 3 **Inverse Variation**

Suppose that y varies inversely with x for $x > 0$.

(a) Find an equation that relates y and x if it is known that $y = 20$ when $x = 3$.
(b) Use the equation found in part (a) to determine y when $x = 5$.

Solution

(a) Because y varies inversely with x, we know that $y = \frac{k}{x}$. Now we can use the given information to find k.

$$y = \frac{k}{x}$$

$y = 20, x = 3$: $\quad 20 = \frac{k}{3}$

Multiply both sides by 3: $\quad 60 = k$

Symmetric Property: $\quad k = 60$

With $k = 60$, we have that $y = \frac{60}{x}$.

(b) We can let $x = 5$ in the equation $y = \frac{60}{x}$ to find y. We obtain

$$y = \frac{60}{5} = 12$$

When $x = 5$, $y = 12$.

QUICK ✓

5. Suppose that y varies inversely with x for $x \geq 0$.
   (a) Find an equation that relates y and x if it is known that $y = 6$ when $x = 2$.
   (b) Use the equation found in part (a) to determine y when $x = 4$.
6. Suppose that p varies inversely with q for $q \geq 0$.
   (a) Find an equation that relates p and q if it is known that $p = 10$ when $q = 4$.
   (b) Use the equation found in part (a) to determine p when $q = 8$.

Classroom Example ➤
Use Example 4 with these changes: It takes 8 hours using 12 employees. How long will it take to clean the stadium with 20 people?

Answer: $T = \frac{96}{n}$; $T = 4.8$ hours

EXAMPLE 4 **Cleaning a Stadium**

The amount of time T that it takes to clean a stadium after a game varies inversely with the number of employees cleaning, n. Suppose that it takes 5 hours to clean a stadium using 20 employees. How long will it take to clean the stadium using 16 employees?

Solution

Step 1: Identify We want to know the amount of time it will take to clean the stadium using 16 employees. In addition, we know that the model will involve variables that are inversely related.

Step 2: Name The variables have been named already: T is the time it takes to clean the stadium and n is the number of employees.

Step 3: Translate The time to clean T varies inversely with the number of employees n. So

$$T = \frac{k}{n}$$

Since $T = 5$ when $n = 20$, we have that

$$5 = \frac{k}{20}$$

Multiply both sides by 20 to obtain $k = 100$.

A model for the time T to clean the stadium with n employees is given by

$$T = \frac{100}{n}$$

Step 4: Solve We let $n = 16$ in the model to predict the time to clean the stadium.

$$T = \frac{100}{n}$$

Let $n = 16$: $T = \frac{100}{16}$

Evaluate: $T = 6.25$

Step 5: Check We can check the reasonableness of our answer. Because it takes 5 hours to clean the stadium when 20 employees are used, it seems reasonable that it will take more time with 16 employees (but not that much more). With 16 employees, we are predicting it will take 6.25 hours. This seems reasonable.

Step 6: Answer When 16 employees are used, we predict it will take 6.25 hours to clean the stadium.

QUICK ✓

7. The rate of vibration (in oscillations per second) V of a string under constant tension varies inversely with the length l. If a string is 30 inches long and vibrates 20 times per second, determine the rate of vibration of a string that is 24 inches long.

8.7 Exercises

Concepts and Vocabulary

In Problems 1–3, fill in the blanks.

1. The statement t varies directly as s, is given by the equation ________.
2. The statement f varies inversely as d, is given by the equation ________.
3. In the equation $y = kx$, k is called the ________ ________ ________ and represents the rate of change between the variables x and y.

In Problems 4–6, answer True or False to each statement.

4. It is possible to solve a variation problem without knowing the constant of proportionality.
5. The constant of proportionality k in the direct variation model $y = kx$, is the slope.
6. If your employer reimburses you 36¢ per mile for driving your car on business trips, the amount you are reimbursed is inversely proportional to the miles you drive on business.
7. List 3 everyday occurrences of direct variation. Make up an example for each, and explain how to use this equation to acquire information about the situation.
8. Describe the conditions that are necessary for two variables to be directly proportional. Also describe the conditions necessary for two variables to be inversely proportional.

Building Skills

In Problems 9–14, suppose y varies directly with x. Find an equation that relates x and y given the following.

9. $y = 3$ when $x = 6$
10. $y = 15$ when $x = 45$
11. $x = 12$ when $y = 24$
12. $x = 2$ when $y = 8$
13. $y = -6$ when $x = 8$
14. $y = -10$ when $x = 24$

1. $t = ks$
2. $f = \frac{k}{d}$
3. constant of proportionality or constant of variation
4. False
5. True
6. False
7. Answers may vary.
8. Answers may vary.
9. $y = \frac{1}{2}x$
10. $y = \frac{1}{3}x$
11. $y = 2x$
12. $y = 4x$
13. $y = -\frac{3}{4}x$
14. $y = -\frac{5}{12}x$

15. $y = \frac{6}{x}$
16. $y = \frac{70}{x}$
17. $y = \frac{12}{x}$
18. $y = \frac{16}{x}$
19. $y = \frac{2}{7x}$
20. $y = -\frac{1}{6x}$
21. direct variation; $\frac{2}{3}$
22. neither
23. neither
24. direct variation; 4
25. inverse variation; 9
26. direct variation; $\frac{2}{5}$
27. direct variation; 2
28. inverse variation; 4
29. (a) $p = \frac{1}{3}g$ (b) $p = 3$
30. (a) $d = 40t$ (b) $d = 200$
31. (a) $y = -\frac{3}{4}x$ (b) $x = -\frac{5}{3}$
32. (a) $n = \frac{3}{4}m$ (b) $m = -\frac{32}{15}$
33. (a) $e = \frac{8}{n}$ (b) $e = \frac{1}{2}$
34. (a) $y = \frac{12}{x}$ (b) $y = \frac{2}{3}$
35. (a) $b = \frac{3}{4a}$ (b) $a = \frac{5}{6}$
36. (a) $f = \frac{5}{3d}$ (b) $d = \frac{4}{9}$
37. $\frac{9}{2}$
38. $p = 16$
39. $P = 75$
40. $B = 80$
41. $t = 42$
42. $r = 64$
43. $b = 18$
44. $x = 84$

In Problems 15–20, suppose y varies inversely with x. Find an equation that relates x and y given the following.

15. $x = 2$ when $y = 3$ **16.** $x = 7$ when $y = 10$ **17.** $y = -3$ when $x = -4$

18. $y = -2$ when $x = -8$ **19.** $x = -\frac{1}{2}$ when $y = -\frac{4}{7}$ **20.** $x = -\frac{3}{4}$ when $y = \frac{2}{9}$

In Problems 21–28, indicate whether the equation represents direct variation, inverse variation, or neither. If it is a variation equation, identify the constant of proportionality.

21. $y = \frac{2x}{3}$ **22.** $x = 4$ **23.** $y = \frac{1}{2}$

24. $4x = y$ **25.** $xy = 9$ **26.** $5y = 2x$

27. $6x = 3y$ **28.** $xy = 4$

In Problems 29–32, use the direct variation model to find each of the following.

29. Suppose p varies directly with g.
(a) Find an equation that relates p and g if it is known that $p = 12$ when $g = 36$.
(b) Use this equation to determine p if g is 9.

30. Suppose d varies directly with t.
(a) Find an equation that relates d and t if it is known that $d = 320$ when $t = 8$.
(b) Use this equation to determine d if t is 5.

31. Suppose y varies directly with x.
(a) Find an equation that relates y and x if it is known that $x = 12$ when $y = -9$.
(b) Use this equation to determine x if y is $\frac{5}{4}$.

32. Suppose n varies directly with m.
(a) Find an equation that relates n and m if it is known that $m = 12$ when $n = 9$.
(b) Use this equation to determine m if n is $-\frac{8}{5}$.

In Problems 33–36, use the inverse variation model to find each of the following.

33. Suppose e varies inversely with n.
(a) Find an equation that relates e and n if it is known that $e = 4$ when $n = 2$.
(b) Use this equation to determine e if n is 16.

34. Suppose y varies inversely with x.
(a) Find an equation that relates y and x if it is known that $y = 4$ when $x = 3$.
(b) Use this equation to determine y if x is 18.

35. Suppose b varies inversely with a.
(a) Find an equation that relates b and a if it is known that $b = \frac{3}{2}$ when $a = \frac{1}{2}$.
(b) Use this equation to determine a if b is $\frac{9}{10}$.

36. Suppose f varies inversely with d.
(a) Find an equation that relates f and d if it is known that $f = \frac{2}{9}$ when $d = \frac{15}{2}$.
(b) Use this equation to determine d if f is $\frac{15}{4}$.

Mixed Practice

In Problems 37–44, find the quantity indicated.

37. x varies inversely with y. If $x = 6$ when $y = 3$, find x when $y = 4$.

38. p varies inversely with v. If $p = 50$ when $v = 24$, find p when $v = 75$.

39. J is directly proportional to P. If $J = 80$ when $P = 120$, find P when $J = 50$.

40. A is directly proportional to B. If $A = 360$ when $B = 72$, find B when $A = 400$.

41. s varies directly with t. If $t = 18$ when $s = 21$, find t when $s = 49$.

42. m varies directly with r. If $r = 24$ when $m = 9$, find r when $m = 24$.

43. b is inversely proportional to a. If $a = 4$ when $b = 6$, find b when $a = \frac{4}{3}$.

44. x is inversely proportional to y. If $y = 14$ when $x = 4$, find x when $y = \frac{2}{3}$.

45. 10 representatives
46. 25 representatives
47. 120 board feet
48. \$80.50
49. 6.25 atmospheres
50. 13.5 meters
51. 150 bags
52. 28 min
53. $m = 90$
54. $m = 36$
55. $r = 5$
56. $r = \frac{2}{3}$
57. (a) $e = \frac{1.989 \times 10^{-25}}{\lambda}$
(b) 3.978×10^{-19} joules
58. 3.315×10^{-19} joules

Applying the Concepts

45. House of Representatives Apportionment The number of representatives that each state has in the U.S. House of Representatives is directly proportional to the state's population. According to the 2000 Census, Ohio's 18 delegates represent a state whose population is 11.375 million. Find the number of representatives from Massachusetts if its population is 6.356 million.

46. House of Representatives Apportionment The number of representatives that each state has in the U.S. House of Representatives is directly proportional to the state's population. According to the 2000 Census, California's 53 delegates represent a state whose population is 33.931 million. Find the number of representatives from Florida if its population is 16.029 million.

47. Buying Lumber The cost of a certain type of lumber at Beechwold Lumber varies directly with the number of board feet purchased. If 70 board feet of this type of lumber cost Christine \$717.50, find the number of board feet in Elizabeth's order of \$1230.

48. Buying Gasoline The cost to purchase a tank of gasoline varies directly with the number of gallons purchased. You notice that the person in front of you spent \$34.50 on 15 gallons of gas. If your SUV needs 35 gallons of gas, how much will you spend?

49. Measuring Pressure A fixed amount of gas is placed in a collapsible cylinder. The pressure in the cylinder, P, varies inversely with the volume V. If the pressure is measured at 5 atmospheres when the volume is 2.5 liters, what will the pressure be when the cylinder is collapsed to 2 liters?

50. Measuring Sound The frequency of sound varies inversely with the wavelength. If a radio station broadcasts at a frequency of 90 megahertz, the wave length of the sound is approximately 3 meters. Find the wave length of a radio broadcast at 20 megahertz.

51. Demand Suppose that the demand D for candy at the movie theater is inversely related to the price p. When the price of candy is \$2.50 per bag, the theater sells 180 bags of candy. Determine the number of bags of candy that will be sold if the price is raised to \$3 a bag.

52. Driving to School The time t that it takes to drive to school varies inversely with your average speed s. It takes you 24 minutes to drive to school when your average speed is 35 miles per hour. Suppose that your average speed to school yesterday was 30 miles per hour. How long did it take you to get to school?

Extending the Concepts

Joint variation *is a model in which a variable is proportional to the product of two or more variables. For example, we may say that y varies jointly as x and z, or* $y = kxz$. *Use joint variation to solve Problems 53–56.*

53. m varies jointly as r and s. If $m = 12$ when $r = 0.5$ and $s = 4$, find m when $s = 9$ and $r = \frac{5}{3}$.

54. m varies jointly as r and s. If $m = 9$ when $r = 6$ and $s = 2$, find m when $s = 3$ and $r = 16$.

55. p varies jointly as q and the square of r. If $p = 162$ when $q = 9$ and $r = 6$, find r when $p = 300$ and $q = 24$.

56. p varies jointly as q and the square of r. If $p = 6$ when $q = \frac{1}{8}$ and $r = 4$, find r when $p = 40$ and $q = 30$.

The energy produced by a photon of light, e, (in joules) is inversely proportional to the wavelength of the light, λ (in meters). The constant of variation is the product of Planck's Constant (6.63×10^{-34}) *and the speed of light* (3×10^8).

57. (a) Write an equation to calculate the energy produced by a photon with a wavelength, λ.
(b) Calculate the energy produced when the wavelength is 500×10^{-9} meters.

58. Use the equation from Problem 57(a) to calculate the energy produced when the wavelength is 600 nanometers. (1 nanometer $= 10^{-9}$ meters)

8.8 Linear Inequalities in Two Variables

OBJECTIVES

1. Determine Whether an Ordered Pair Is a Solution to a Linear Inequality
2. Graph Linear Inequalities
3. Solve Problems Involving Linear Inequalities

Preparing for Linear Inequalities in Two Variables

Before getting started, take the following readiness quiz. If you get a problem wrong, go back to the section cited and review the material.

1. Solve: $x - 4 > 5$ [Section 2.8, p. 153]

2. Solve: $3x + 1 \le 10$ [Section 2.8, pp. 155–156]

3. Solve: $2(x + 1) - 6x > 18$ [Section 2.8, pp. 155–157]

1 Determine Whether an Ordered Pair Is a Solution to a Linear Inequality

In Chapter 2, we solved inequalities in one variable. In this section, we discuss linear inequalities in two variables.

> **DEFINITION**
>
> **Linear inequalities in two variables** are inequalities in one of the forms
>
> $$Ax + By < C \qquad Ax + By > C \qquad Ax + By \le C \qquad Ax + By \ge C$$
>
> where A and B are not both zero. A linear inequality in two variables x and y is **satisfied** by an ordered pair (a, b) if a true statement results when x is replaced by a and y is replaced by b.

Classroom Example Determine which of the following points are solutions to the linear inequality $x + 3y \le 6$.

(a) $(1, 2)$
(b) $(-5, 3)$
(c) $(0, 2)$

Answer:

(a) not a solution
(b) a solution
(c) a solution

EXAMPLE 1 Determining Whether an Ordered Pair Is a Solution to a Linear Inequality in Two Variables

Determine which of the following points are solutions to the linear inequality $2x + y \le 9$.

(a) $(3, 5)$ **(b)** $(1, 3)$ **(c)** $(3, -2)$

Solution

(a) Let $x = 3$ and $y = 5$ in the inequality. If a true statement results, then $(3, 5)$ is a solution to the inequality.

$$2x + y \le 9$$

$$x = 3, y = 5: \quad 2(3) + 5 \overset{?}{\le} 9$$

$$6 + 5 \overset{?}{\le} 9$$

$$11 \le 9 \quad \text{False}$$

The statement $11 \le 9$ is false, so $(3, 5)$ is not a solution to the inequality.

(b) Let $x = 1$ and $y = 3$ in the inequality. If a true statement results, then $(1, 3)$ is a solution to the inequality.

$$2x + y \le 9$$

$$x = 1, y = 3: \quad 2(1) + 3 \overset{?}{\le} 9$$

$$2 + 3 \overset{?}{\le} 9$$

$$5 \le 9 \quad \text{True}$$

The statement $5 \le 9$ is true, so $(1, 3)$ is a solution to the inequality.

Preparing for...Answers **1.** $\{x|x > 9\}$ **2.** $\{x|x \le 3\}$ **3.** $\{x|x < -4\}$

(c) Let $x = 3$ and $y = -2$ in the inequality. If a true statement results, then $(3, -2)$ is a solution to the inequality.

$$2x + y \leq 9$$

$$x = 3, y = -2: \quad 2(3) + (-2) \stackrel{?}{\leq} 9$$

$$6 - 2 \stackrel{?}{\leq} 9$$

$$4 \leq 9 \quad \text{True}$$

The statement $4 \leq 9$ is true, so $(3, -2)$ is a solution to the inequality.

QUICK ✓ *Determine which of the following points are solutions to the given linear inequality.*

(a) $(2, 1)$ **(b)** $(3, 4)$ **(c)** $(-1, 10)$

1. $2x + y > 7$ **2.** $-3x + 2y \leq 8$

2 Graph Linear Inequalities

Teaching Tip
Point out that our definition of a solution continues to be true here: the solutions of a linear inequality consist of the ordered pairs that make the inequality true.

Now that we know how to determine whether a point is a solution to a linear inequality in two variables, we are prepared to graph linear inequalities in two variables. A **graph of a linear inequality in two variables** x and y consists of all points (x, y) whose coordinates satisfy the inequality.

If we replace the inequality symbol with an equal sign in a linear inequality of the form

$$Ax + By < C \qquad Ax + By > C \qquad Ax + By \leq C \qquad Ax + By \geq C$$

we obtain the equation of a line, $Ax + By = C$. This **boundary line** separates the xy-plane into two regions, called **half-planes.** See Figure 62.

Work Smart
We graph the solutions of a linear inequality in one variable on the real number line. We graph the solutions of a linear inequality in two variables in the xy-plane.

Let's consider the linear inequality $2x + y \leq 9$ in Example 1. Figure 63 shows the graph of the equation $2x + y = 9$ and the three points $(3, 5)$, $(1, 3)$, and $(3, -2)$ examined in Example 1.

Figure 62

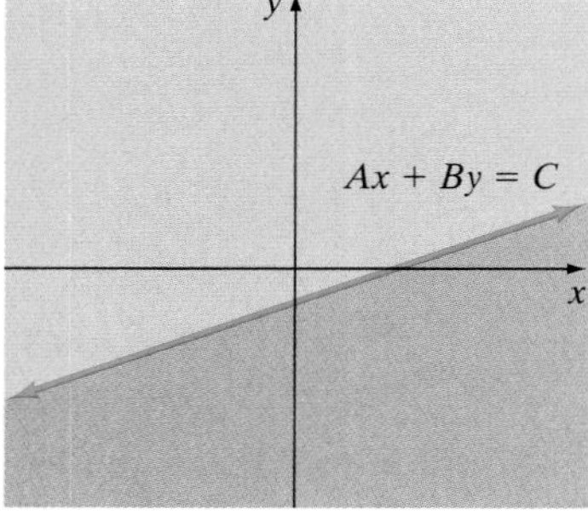

Figure 63

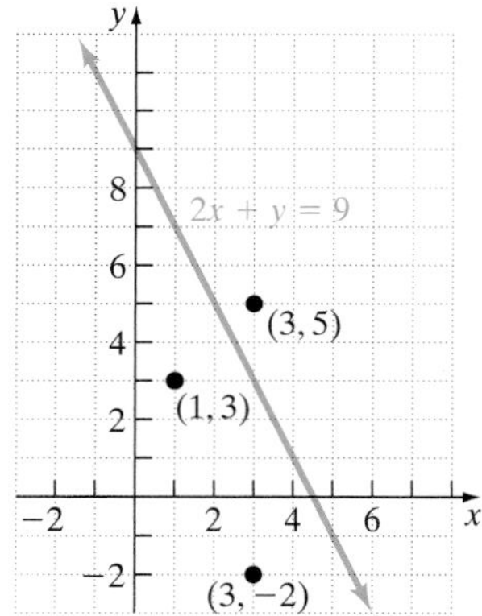

From the results of Example 1, we know that the two points below the line, $(1, 3)$ and $(3, -2)$, satisfy the inequality $2x + y \leq 9$ while $(3, 5)$ does not. In fact, all points below the line will satisfy the inequality and all points above the inequality will not satisfy the inequality. What can we learn from this? If you find a point that satisfies a linear

Classroom Example

Graph the linear inequality $y < \frac{1}{2}x + 3$.

Answer:

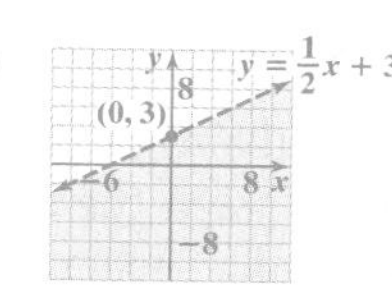

inequality in two variables, then all points in the half-plane containing the point will satisfy the inequality. If you find a point that does not satisfy a linear inequality in two variables, then all points in the opposite half-plane will satisfy the inequality. For this reason, the use of a single **test point** is all that is required to obtain the graph of a linear inequality in two variables.

EXAMPLE 2 How to Graph a Linear Inequality in Two Variables

Graph the linear inequality $y < 2x - 5$.

Step-by-Step Solution

Step 1: Replace the inequality symbol with an equal sign and graph the resulting equation. If the inequality is strict ($<$ or $>$), use dashes to graph the line; if the inequality is non-strict ($\leq$ or $\geq$) use a solid line.

$$y < 2x - 5$$

We graph the line $y = 2x - 5$ whose slope is 2 and y-intercept is -5 using a dashed line because the inequality is strict ($<$). See Figure 64(a).

Work Smart

A strict inequality uses the symbol $<$ or $>$; a nonstrict inequality uses $\leq$ or $\geq$.

Work Smart

When the line does not contain the origin, it is usually easiest to choose the origin, (0, 0), as the test point.

Figure 64 (a)

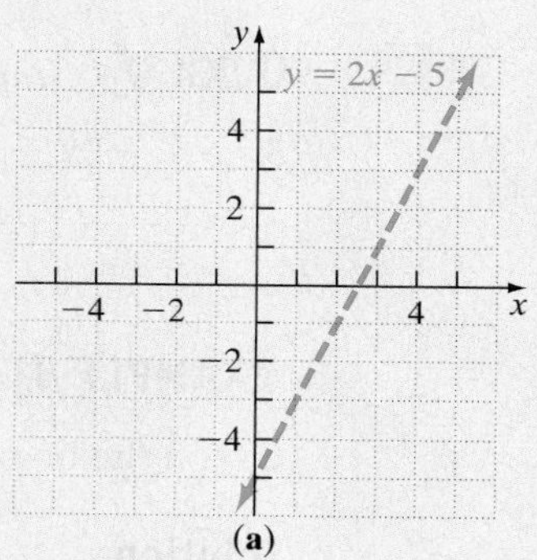

(a)

Step 2: We select any test point that is not on the line and determine whether the test point satisfies the inequality. We will use (0, 0) as the test point. If (0, 0) satisfies the inequality, we shade the half-plane containing (0, 0); otherwise shade the half-plane opposite (0, 0).

$$y < 2x - 5$$

$$x = 0, y = 0: \quad 0 \stackrel{?}{<} 2(0) - 5$$

$$0 \stackrel{?}{<} 0 - 5$$

$$0 < -5 \quad \text{False}$$

Because (0, 0) does not satisfy the inequality $y < 2x - 5$, we shade the half-plane opposite the half-plane containing (0, 0). See Figure 64(b).

Work Smart

To double check your results, choose a point, such as (4, 0), in the shaded region. Does (4, 0) satisfy $y < 2x - 5$?

Figure 64 (b)

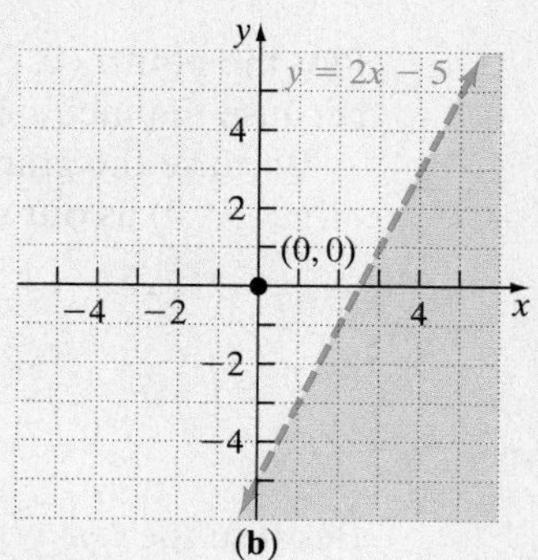

(b)

The shaded region represents the solution to the linear inequality. Because the inequality is strict, points on the line $y = 2x - 5$ do not satisfy the inequality.

We summarize the steps for graphing a linear inequality in two variables.

Steps for Graphing a Linear Inequality in Two Variables

Step 1: Replace the inequality symbol with an equal sign and graph the resulting equation. If the inequality is strict ($<$ or $>$), use dashes to graph the line; if the inequality is nonstrict ($\leq$ or $\geq$), use a solid line. The graph separates the *xy*-plane into two half-planes.

Step 2: Select a test point *P* that is not on the line (that is, select a test point in one of the half-planes).

(a) If the coordinates of *P* satisfy the inequality, then shade the half-plane containing *P*.

(b) If the coordinates of *P* do not satisfy the inequality, then shade the half-plane that does not contain *P*.

Note: If the inequality is nonstrict, then all points on the line also satisfy the inequality. If the inequality is strict, then all points on the line *do not* satisfy the inequality.

QUICK ✓ *Graph each linear inequality.*

3. $y < -2x + 1$ **4.** $y \geq 3x + 2$

Classroom Example ➤
Graph the linear inequality $2x - 5y \leq 10$.

Answer:

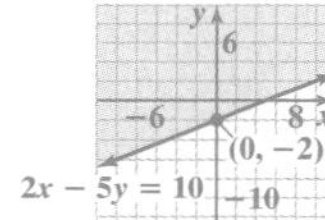

EXAMPLE 3 Graphing a Linear Inequality in Two Variables

Graph the linear inequality $5x + 2y \leq 10$.

Solution

First, we need to graph the equation $5x + 2y = 10$. We will do this by finding the intercepts of the equation. To find the x-intercept, let $y = 0$; to find the y-intercept, let $x = 0$.

$$5x + 2y = 10$$
Let $x = 0$: $$5(0) + 2y = 10$$
$$2y = 10$$
Divide both sides by 2: $$y = 5$$

$$5x + 2y = 10$$
Let $y = 0$: $$5x + 2(0) = 10$$
$$5x = 10$$
Divide both sides by 5: $$x = 2$$

Plot the points $(0, 5)$ and $(2, 0)$ and draw the equation of the line as a solid line because the inequality is nonstrict ($\leq$). See Figure 65(a).

Because the graph of the equation does not contain the origin, we will select the origin $(0, 0)$ as our test point.

$$5x + 2y \leq 10$$
Let $x = 0$, let $y = 0$: $$5(0) + 2(0) \overset{?}{\leq} 10$$
$$0 \leq 10 \quad \text{True}$$

Because the test point $(0, 0)$ results in a true statement, we shade the half-plane that contains $(0, 0)$. See Figure 65(b). The shaded region and all points on the line $5x + 2y = 10$ represents the solution set.

Work Smart
The test point (0, 0) is a solution of the linear inequality, so all other points in that half-plane are also solutions. That's why we shade the half-plane that contains the origin.

Figure 65

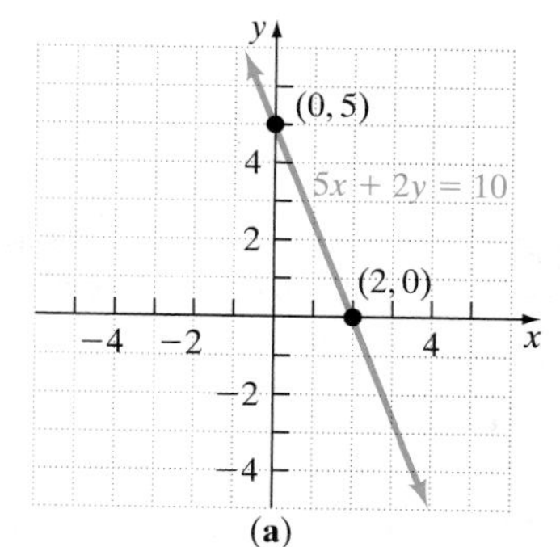

(a)

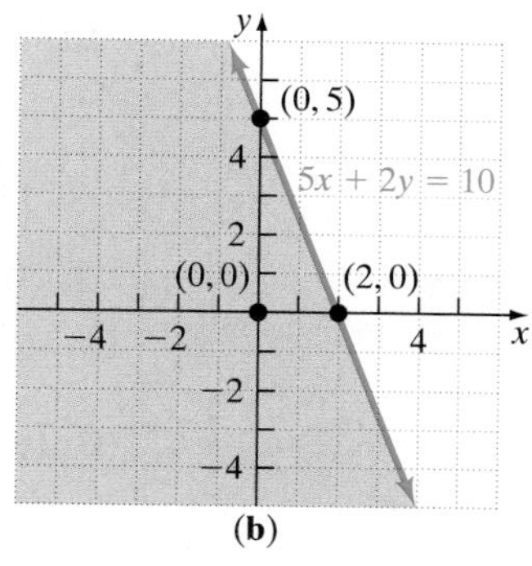

(b)

QUICK ✓ *Graph each linear inequality.*

5. $2x + 3y \leq 6$

6. $4x - 6y > 12$

Classroom Example
Graph the linear inequality $3x - 4y \geq 0$.

Answer:

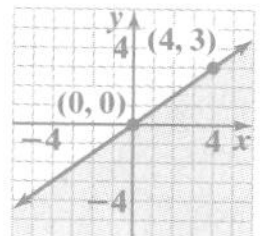

EXAMPLE 4 Graphing a Linear Inequality where the Line Goes through the Origin

Graph the linear inequality $-4x + 3y > 0$.

Solution

We graph the equation $-4x + 3y = 0$ using a dashed line. Because this is an equation of the form $Ax + By = 0$, we know that the graph will pass through the origin. For this reason, we write the equation in slope-intercept form.

$$-4x + 3y = 0$$

Add $4x$ to both sides: $$3y = 4x$$

Divide both sides by 3: $$y = \frac{4}{3}x$$

The line has slope $\frac{4}{3}$ and y-intercept 0. See Figure 66(a).

Work Smart
Because the boundary line contains the origin, the test point cannot be the origin.

The graph of the equation contains the origin, so we will use (1, 3) as our test point.

$$-4x + 3y > 0$$

Let $x = 1$, let $y = 3$: $$-4(1) + 3(3) \stackrel{?}{>} 0$$

$$-4 + 9 \stackrel{?}{>} 0$$

$$5 > 0 \quad \text{True}$$

Because the test point (1, 3) results in a true statement, we shade the half-plane that contains (1, 3). See Figure 66(b). The shaded region represents the solution set.

Figure 66

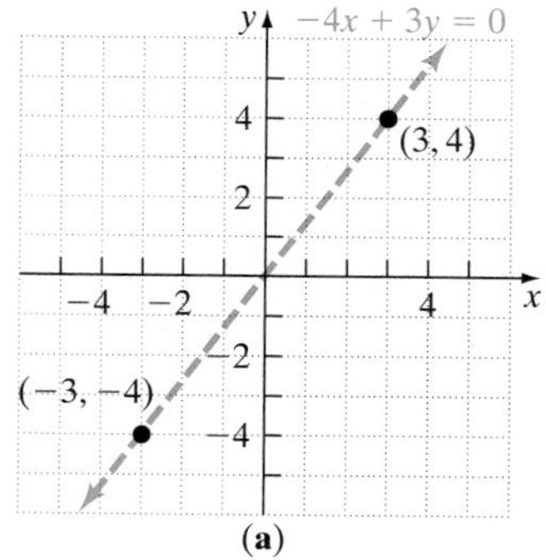

(a)

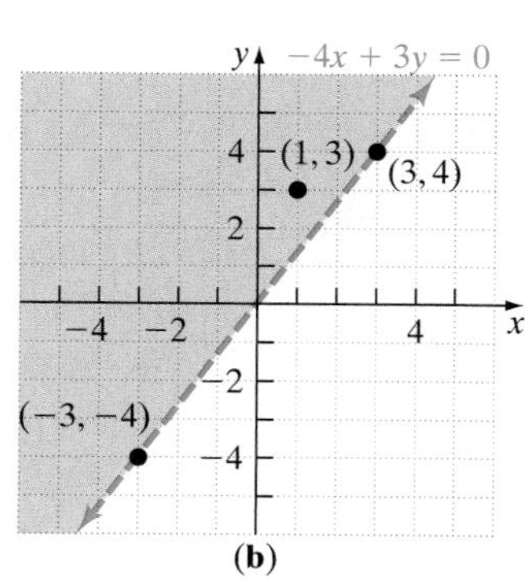

(b)

QUICK ✓ *Graph each linear inequality.*

7. $3x + y < 0$ **8.** $2x - 5y \leq 0$

Classroom Example ➤
Graph each linear inequality:
(a) $y > 4$ (b) $x < -1$.

Answer:

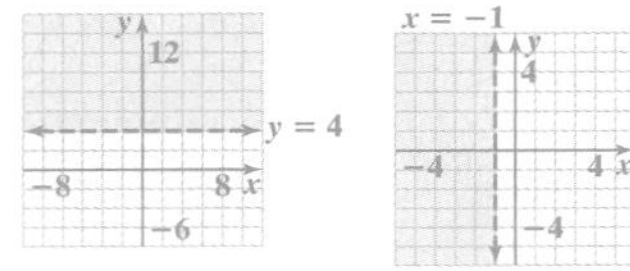

EXAMPLE 5 Graphing a Linear Inequality Involving a Horizontal Line

Graph the linear inequality $y < 2$.

Solution

We begin by graphing the line horizontal line $y = 2$ using a dashed line (because the inequality is strict). See Figure 67(a). Next we select a test point. We can select the origin (0, 0) as our test point because the line does not contain (0, 0).

$$y < 2$$

Test point (0, 0): $0 \stackrel{?}{<} 2$ True

Because the test point satisfies the inequality, we shade the half-plane that contains (0, 0) as shown in Figure 67(b). The shaded region represents the solution to the linear inequality. Doesn't it seem intuitive that we shade below the line since these are the y-values that are less than 2?

Work Smart: Study Skills
Have you noticed the patterns?
If $y > mx + b$, shade above.
If $y < mx + b$, shade below.
If $y > b$, shade above.
If $y < b$, shade below.
If $x > a$, shade to the right.
If $x < a$, shade to the left.

Figure 67

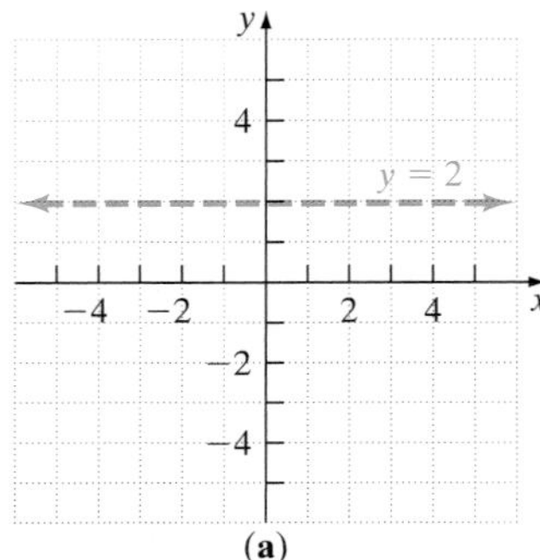

(a)

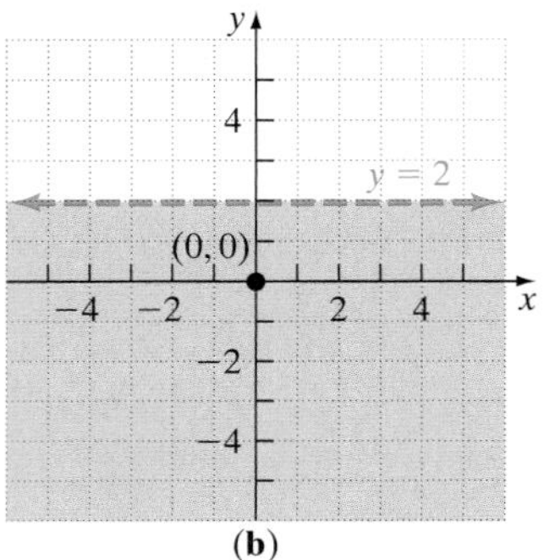

(b)

The process for graphing a linear inequality involving a vertical line parallels the process for graphing a linear inequality involving a horizontal line. That is, graph the vertical boundary line, choose a test point, and shade the half-plane containing the points that satisfy the inequality.

QUICK ✓ *Graph each linear inequality.*

9. $y < -1$ **10.** $3y - 9 \geq 0$ **11.** $x > 6$

3 Solve Problems Involving Linear Inequalities

Many applications of linear inequalities that involve two variables can be used to solve problems in areas such as nutrition, manufacturing, or sales. Linear inequalities can also be used to describe weight loads in elevators! Let's see how.

Classroom Example ➤
Use Example 6. (a) same (b) Can 8 men and 10 women ride safely? (c) Can 9 men and 10 women ride safely?

Answer:
(a) $180m + 150f \leq 3000$
(b) yes
(c) no

EXAMPLE 6 Elevator Capacity

An elevator designed and manufactured by Otis Elevator Company has a capacity of 3000 pounds (*Source:* otis.com). According to the National Center for Health Statistics, the average man 20 years or older weighs 180 pounds and the average woman 20 years or older weighs 150 pounds.

(a) Write a linear inequality that describes the various combinations of men and women who can ride this elevator.

(b) Can 10 men and 9 women ride in this elevator safely?

(c) Can 7 men and 11 women ride in this elevator safely?

Solution

(a) We are going to use the first three steps in problem solving strategy given in Chapter 2 on page 122 to help us develop the linear inequality.

Step 1: Identify We want to determine the number of men and women who can ride in the elevator without going over 3000 pounds.

Step 2: Name Unknowns Let m represent the number of males and f represent the number of females on the elevator.

Step 3: Translate For the sake of simplicity, assume that all the people who get on the elevator are "average." If there is one male on the elevator, the weight on the elevator will be 180 pounds. If there are two males on the elevator, the weight will be $2(180) = 360$ pounds. In general, if there are m males on the elevator, the weight will be $180m$. Similar logic for the females tells us that if there are f females on the elevator, the weight will be $150f$. Because the capacity of the elevator is 3000 pounds, we use a less than or equal ($\leq$) inequality. A linear inequality that describes the weight limitations on the elevator is

$$180m + 150f \leq 3000 \quad \text{The Model}$$

(b) Letting $m = 10$ and $f = 9$, we obtain

$$180(10) + 150(9) \stackrel{?}{\leq} 3000$$
$$3150 \leq 3000 \quad \text{False}$$

Because the inequality is false, 10 men and 9 women cannot ride the elevator safely.

(c) Letting $m = 7$ and $f = 11$, we obtain

$$180(7) + 150(11) \stackrel{?}{\leq} 3000$$
$$2910 \leq 3000 \quad \text{True}$$

Because the inequality is true, 7 men and 11 women can ride the elevator safely.

QUICK ✓

12. Kevin just received $2 from his grandma. He goes to a candy store where each sucker sells for $0.20 and each taffy stick sells for $0.25.

(a) Write a linear inequality that describes the various combinations of suckers, s, and taffy sticks, t, Kevin can buy.

(b) Can Kevin buy 6 suckers and 3 taffy sticks?

(c) Can Kevin buy 5 suckers and 5 taffy sticks?

8.8 Exercises

1. solid 2. dashed 3. half-planes 4. False 5. True 6. False 7. Answers may vary. 8. Answers may vary. 9. A is a solution. 10. None 11. C is a solution. 12. C is a solution. 13. A and C are solutions. 14. A and B are solutions. 15. B is a solution. 16. A, B, C are solutions. 17. B is a solution. 18. A and B are solutions. 19. A is a solution. 20. B is a solution. 21. $y > 3x - 2$ 22. $y \leq 4x + 2$

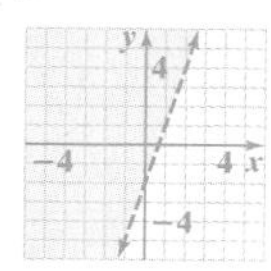

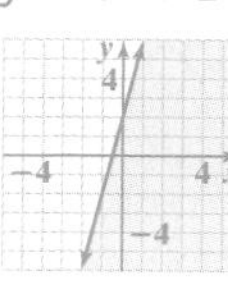

For Extra Help: Student Solutions Manual | CD Video | PH Math/Tutor Center | MathXL Tutorials on CD | MathXL® | MyMathLab

Concepts and Vocabulary

In Problems 1–3, fill in the blanks.

1. When drawing the boundary line for the graph of $Ax + By \geq C$, we use a __________ line.

2. When drawing the boundary line for the graph of $Ax + By < C$, we use a __________ line.

3. The boundary line separates the xy-plane into two regions, called __________.

23. $y \le -x + 1$ 24. $y > x - 3$

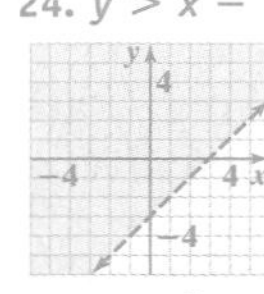

25. $y < \frac{x}{2}$

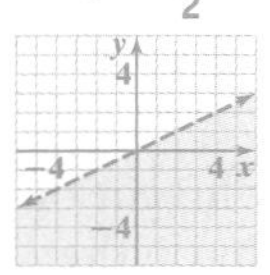

26. $y > \frac{x}{3}$

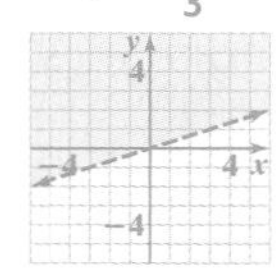

27. $y > 5$

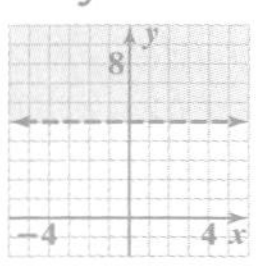

28. $y < 4$

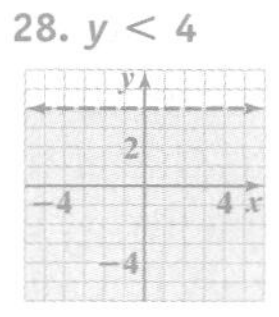

29. $y \le \frac{2}{5}x + 3$

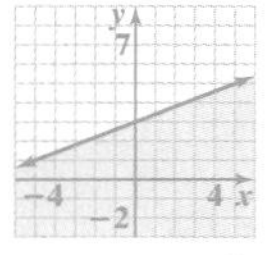

30. $y \ge \frac{3}{2}x - 2$

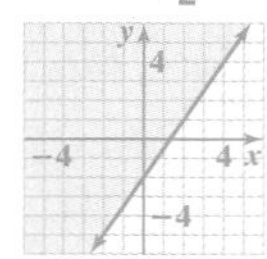

31. $y \ge -\frac{4}{3}x + 2$

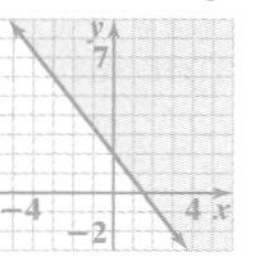

32. $y \le -\frac{3}{4}x + 1$

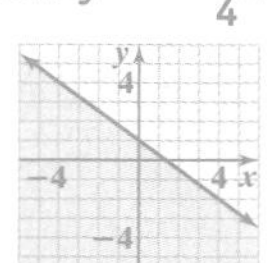

33. $x < 2$ 34. $x > -1$

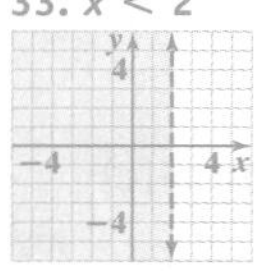

35. $3x - 4y < 12$ 36. $2x + 6y < -6$

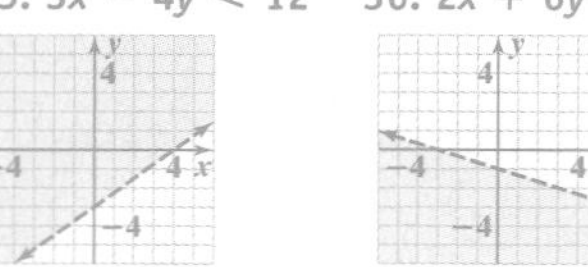

37. $2x + y \ge -4$ 38. $6x - 8y \ge 24$

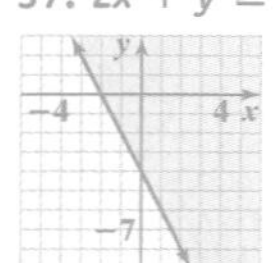

39. $x + y > 0$ 40. $x - y > 0$

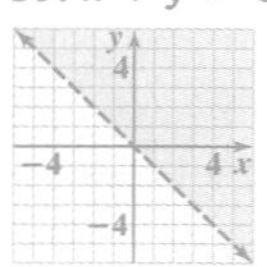

41. $5x - 2y < -8$ 42. $3x + 2y \le -9$

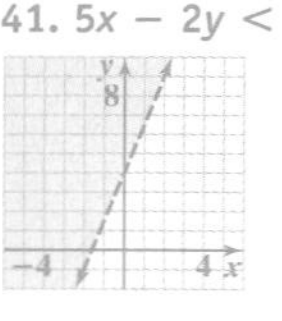

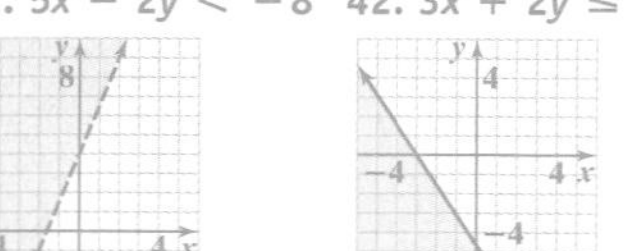

In Problems 4–6, answer True or False to each statement.

4. After plotting the boundary line for the graph of $3x \le 2y$, we can use $(0, 0)$ as a test point to decide which half-plane contains the solutions.

5. After drawing the graph of a linear inequality in two variables, it is a good idea to pick a test point in the shaded region to verify that the correct half-plane has been shaded.

6. The boundary line for the solution of the inequality $3x - 2y > -2$ is solid.

7. Describe how you can tell if a point (a, b) is a solution to a linear inequality in two variables. How can you decide if (a, b) lies on the boundary line?

8. Explain why $(0, 0)$ is generally a good test point. Describe how you can decide when this is a good test point to use and when you should choose a different test point.

Building Skills

In Problems 9–20, determine which of the following points, if any, are solutions to the given linear inequality in two variables.

9. $y > -x + 2$
$A(2, 4)$ $B(3, -6)$ $C(0, 0)$

10. $y \le x - 5$
$A(-4, -6)$ $B(0, 0)$ $C(3, 10)$

11. $y \le 3x - 1$
$A(-6, -15)$ $B(0, 0)$ $C(-1, -6)$

12. $y > -2x + 1$
$A(0, 0)$ $B(-2, 3)$ $C(2, -1)$

13. $3x \ge 2y$
$A(-8, -12)$ $B(3, 5)$ $C(-5, -8)$

14. $4y < 5x$
$A(-4, -6)$ $B(5, 6)$ $C(-12, -15)$

15. $2x - 3y < -6$
$A(2, -1)$ $B(4, 8)$ $C(-3, 0)$

16. $3x + 5y \ge 4$
$A(2, 0)$ $B(-3, 4)$ $C(6, -2)$

17. $x \le 2$
$A(7, 2)$ $B(2, 5)$ $C(4, 2)$

18. $y \le -3$
$A(-3, -3)$ $B(-1, -4)$ $C(-6, -1)$

19. $y > -1$
$A(-1, 1)$ $B(3, -1)$ $C(4, -2)$

20. $x > 10$
$A(3, 12)$ $B(11, -6)$ $C(10, 16)$

In Problems 21–50, graph each linear inequality.

21. $y > 3x - 2$ **22.** $y \le 4x + 2$ **23.** $y \le -x + 1$

24. $y > x - 3$ **25.** $y < \frac{x}{2}$ **26.** $y > \frac{x}{3}$

27. $y > 5$ **28.** $y < 4$ **29.** $y \le \frac{2}{5}x + 3$

30. $y \ge \frac{3}{2}x - 2$ **31.** $y \ge -\frac{4}{3}x + 2$ **32.** $y \le -\frac{3}{4}x + 1$

33. $x < 2$ **34.** $x > -1$ **35.** $3x - 4y < 12$

36. $2x + 6y < -6$ **37.** $2x + y \ge -4$ **38.** $6x - 8y \ge 24$

39. $x + y > 0$ **40.** $x - y > 0$ **41.** $5x - 2y < -8$

42. $3x + 2y \le -9$ **43.** $x > -1$ **44.** $y < -3$

45. $y \le 4$ **46.** $x \ge 2$ **47.** $\frac{x}{3} - \frac{y}{5} \ge 1$

48. $\frac{x}{4} + \frac{y}{2} < 1$ **49.** $-3 \ge x - y$ **50.** $-2 > x + y$

43. $x > -1$

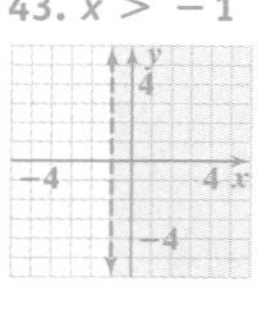

44. $y < -3$

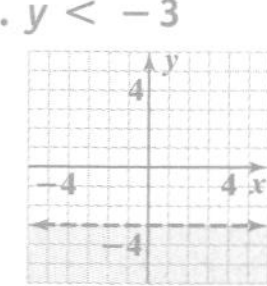

45. $y \le 4$ 46. $x \ge 2$

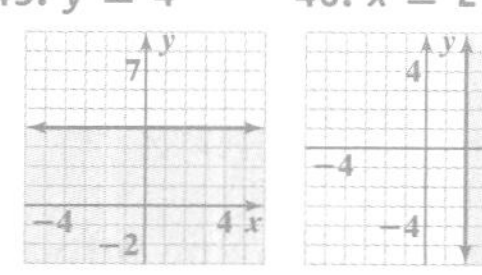

47. $\frac{x}{3} - \frac{y}{5} \ge 1$

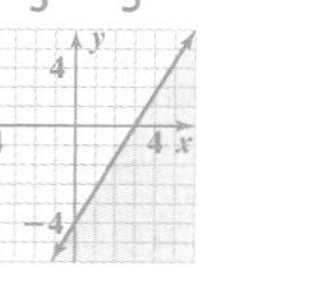

48. $\frac{x}{4} + \frac{y}{2} < 1$ 49. $-3 \ge x - y$

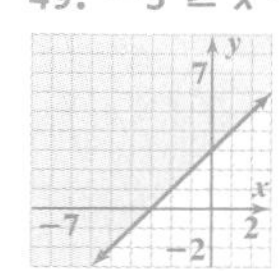

50. $-2 > x + y$

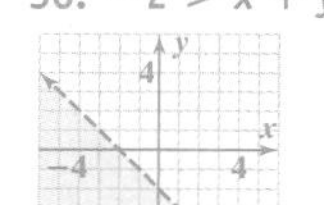

51. $x + y \geq 26$

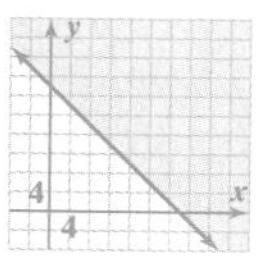

52. $\frac{x}{3} < -2$

53. $\frac{y}{-2} \leq 4$

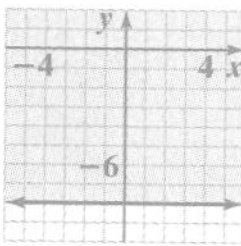

54. $x \geq y + 12$

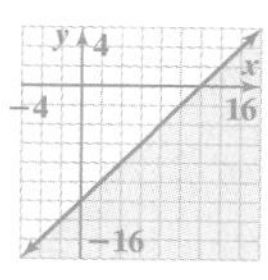

55. $x \leq y - 3$

56. $x - \frac{1}{2}y > 0$

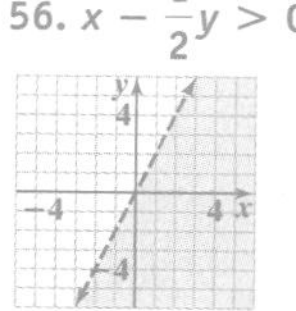

57. $x + 3y < 0$

58. $x + y \geq -3$

59. $2x - \frac{1}{2}y \geq 5$

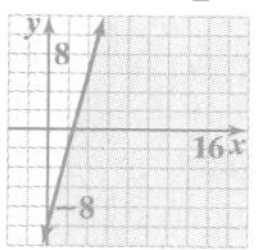

60. $2y \leq -10$

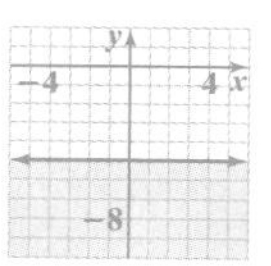

61. $-2x > -1$

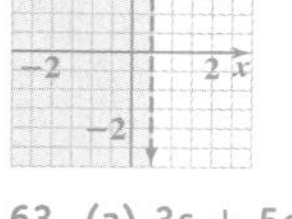

62. $\frac{1}{2}x + 2y \geq 12$

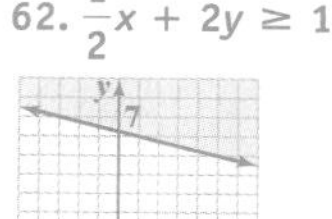

63. (a) $3s + 5a \leq 120$ (b) No (c) Yes
64. (a) $160a + 75c \leq 500$ (b) Yes (c) Yes
65. (a) $3.1s + 5.7b \leq 22$ (b) No (c) Yes
66. (a) $5m + 2n \geq 40$ (b) Yes (c) Yes

67. $3x - 2y > 6$ and $x + y < 2$

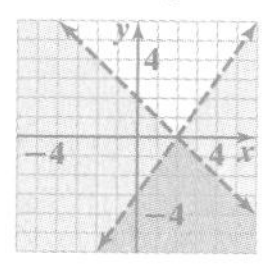

68. $2x - 4y \leq 4$ and $3x + 2y \geq 6$

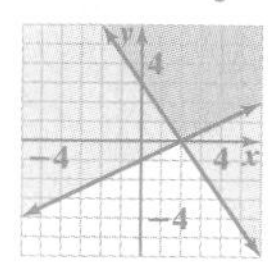

69. $y > \frac{3}{4}x - 1$ and $x \geq 0$

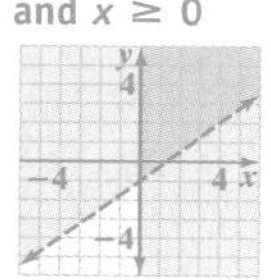

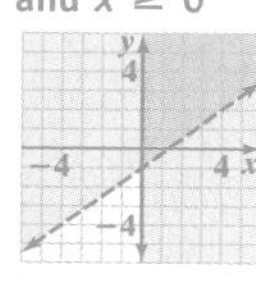

70. $y > \frac{2}{3}x + 3$ and $y \geq 0$

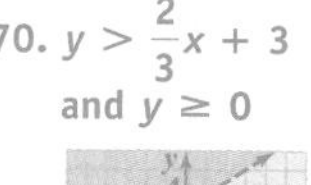

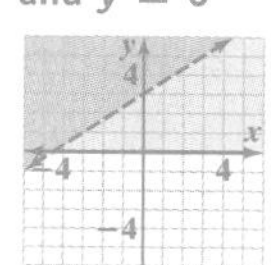

71. $x < -3$ and $y \leq 4$

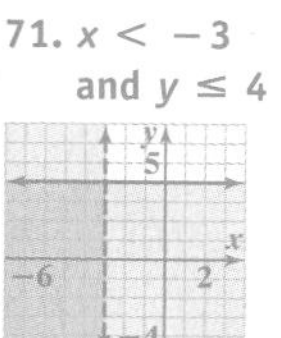

72. $y < -1$ and $x \geq -2$

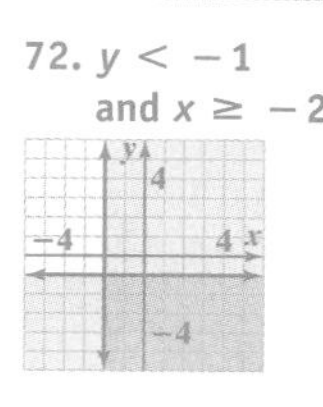

Applying the Concepts

In Problems 51–62, translate each statement into a linear inequality. Then graph the inequality.

51. The sum of two numbers, x and y, is at least 26.

52. The ratio of a number, x, and 3 is less than -2.

53. The ratio of a number, y, and -2 is at most 4.

54. One number, x, is at least 12 more than a second number, y.

55. One number, x, is no more than 3 less than a second number, y.

56. The difference of a number, x, and half a second number, y, is positive.

57. The sum of a number, x, and 3 times a second number, y, is negative.

58. The sum of two numbers, x and y, is at least -3.

59. The difference of twice a number, x, and half a second number, y, is at least 5.

60. The product of a number, y, and 2 is at most -10.

61. The product of a number, x, and -2 is more than -1.

62. The sum of half a number, x, and twice a second number, y, is at least 12.

63. Trip to the Aquarium A kindergarten class has a maximum of \$120 to spend on a trip to the aquarium. The cost of admission for students is \$3, while adults must pay \$5.

(a) Write a linear inequality that describes the various combinations of the number of students s and adults a that can go on the field trip.
(b) Is there enough money to pay for 32 students and 6 adults?
(c) Is there enough money to pay for 29 students and 4 adults?

64. Fishing Trip Patrick's row boat can hold a maximum of 500 pounds. In Patrick's circle of friends, the average adult weighs 160 pounds and the average child weighs 75 pounds.

(a) Write a linear inequality that describes the various combinations of the number of adults a and children c that can go on a fishing trip in the row boat.
(b) Will the boat sink with 2 adults and 4 children?
(c) Will the boat sink with 1 adult and 5 children?

65. Backpacking Trip Scott's dad decided he can carry no more than 22 pounds in his backpack on the next group trip. He carries only camping stoves, which weigh 3.1 pounds, and sleeping bags, which weigh 5.7 pounds.

(a) Write a linear inequality that describes the various combinations of the number of stoves s and sleeping bags b that he will carry.
(b) Will his pack be too heavy if he carries 3 stoves and 2 sleeping bags?
(c) Will his pack be too heavy if he carries 2 stoves and 3 sleeping bags?

66. School Fundraiser For the Clinton Elementary School fundraiser, Guillermo earns 5 points for each magazine subscription he sells and 2 points for each novelty. It takes at least 40 points to earn the portable CD player that Guillermo wants.

(a) Write a linear inequality that describes the various combinations of subscriptions m and novelties n that Guillermo can sell to earn enough points for the portable CD player.
(b) Will he earn the CD player if he sells 8 novelties and 5 subscriptions?
(c) Will he earn the CD player if he sells 20 novelties and 2 subscriptions?

73. $y > 3$

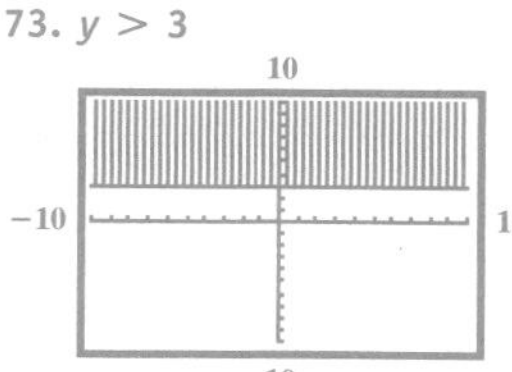

74. $y < -2$

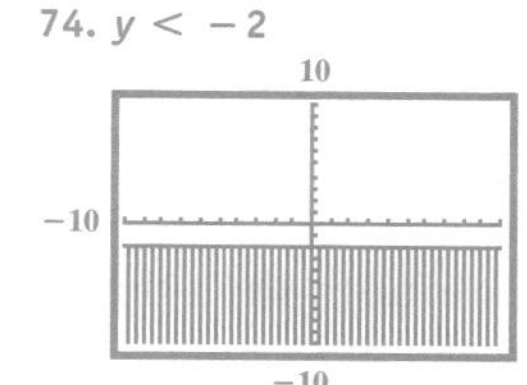

75. $y < 5x$

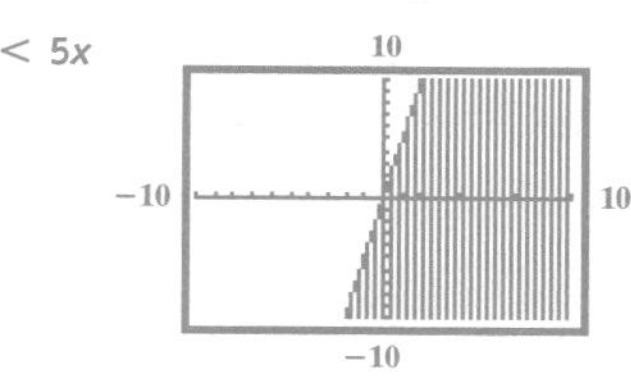

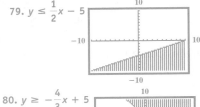

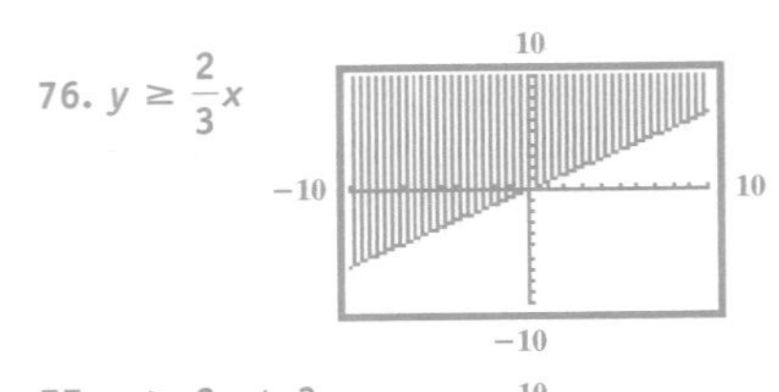

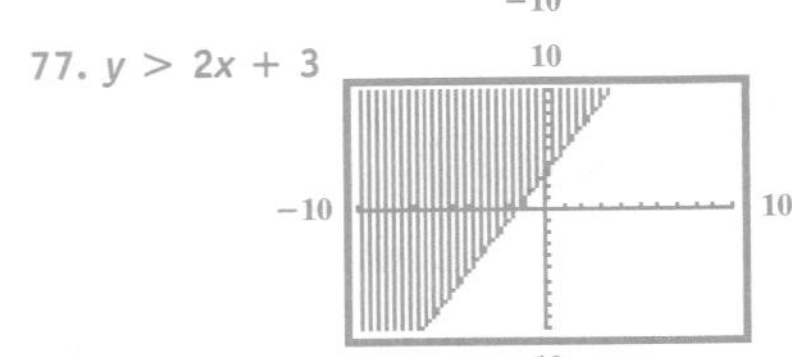

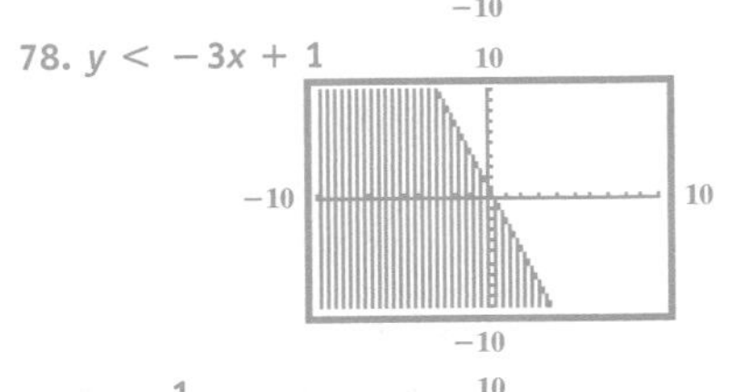

79. $y \leq \frac{1}{2}x - 5$

10 -10 10 -10

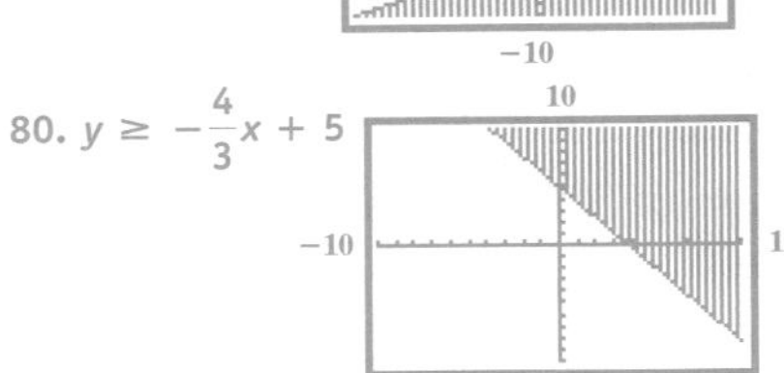

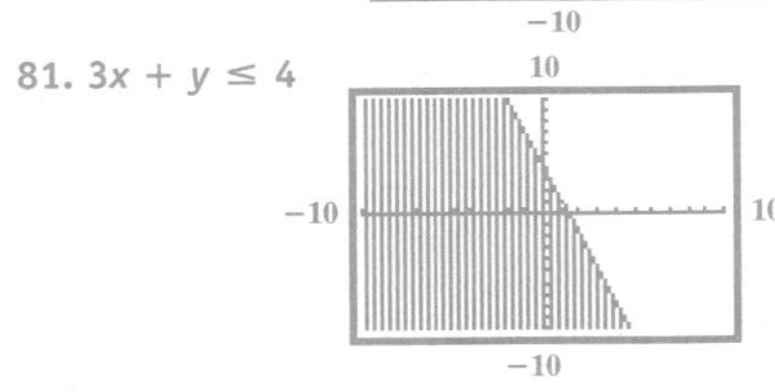

Extending the Concepts

Two inequalities joined by the word "and" or "or" form a ***compound inequality.*** *The directions "Solve $3x - 2y > 6$ and $x + y < 2$" mean to find the ordered pairs that satisfy both inequalities at the same time. The solution of a compound inequality of this type is found by graphing the two linear inequalities in the same Cartesian coordinate plane and then shading the area where the graphs overlap. Solve the compound inequalities in Problems 67–72.*

67. $3x - 2y > 6$ and $x + y < 2$

68. $2x - 4y \leq 4$ and $3x + 2y \geq 6$

69. $y > \frac{3}{4}x - 1$ and $x \geq 0$

70. $y > \frac{2}{3}x + 3$ and $y \geq 0$

71. $x < -3$ and $y \leq 4$

72. $y < -1$ and $x \geq -2$

The Graphing Calculator

Graphing calculators can be used to graph linear inequalities in two variables. The figure to the right shows the graph of $3x + y < 7$ using a TI-84 Plus graphing calculator. To obtain the graph, we solve the inequality for y and obtain $y < -3x + 7$. Graph the line $y = -3x + 7$ and shade below. Consult your owner's manual for specific keystrokes.

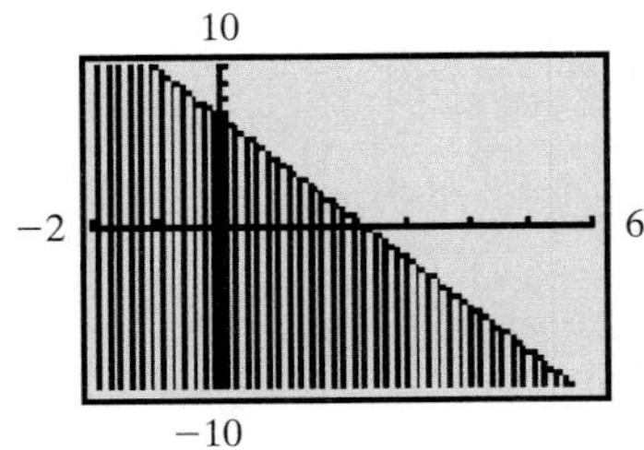

In Problems 73–84, graph the inequalities using a graphing calculator.

73. $y > 3$

74. $y < -2$

75. $y < 5x$

76. $y \geq \frac{2}{3}x$

77. $y > 2x + 3$

78. $y < -3x + 1$

79. $y \leq \frac{1}{2}x - 5$

80. $y \geq -\frac{4}{3}x + 5$

81. $3x + y \leq 4$

82. $-4x + y \geq -5$

83. $2x + 5y \leq -10$

84. $3x + 4y \geq 12$

82. $-4x + y \geq -5$

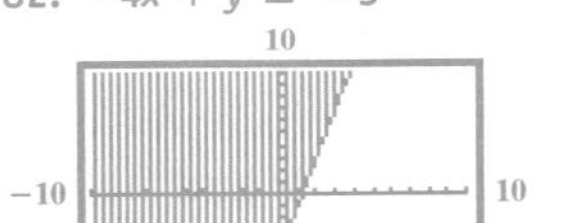

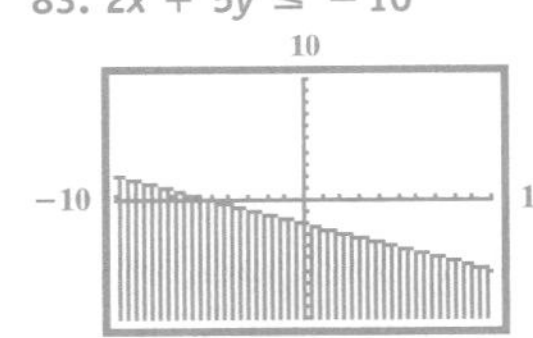

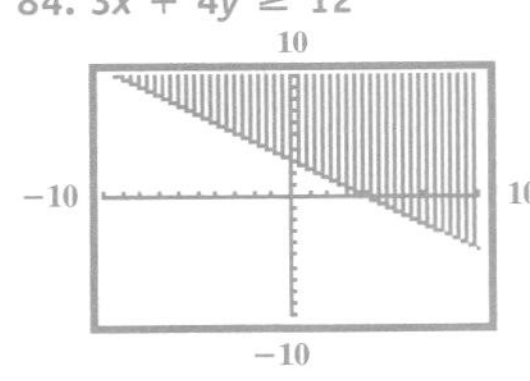

CHAPTER 8 ACTIVITY: GRAPHING PRACTICE

1. (a)

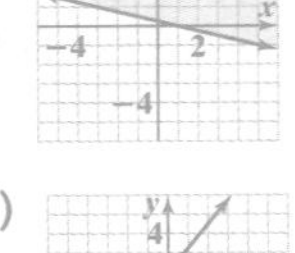

(b)

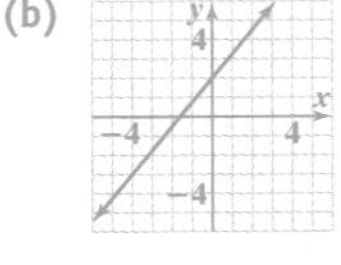

(c)

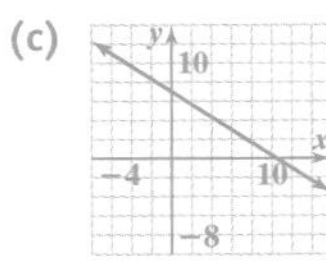

(d)

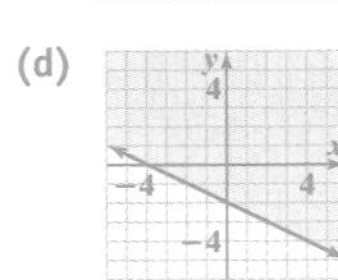

Focus: Graphing and identifying the graphs of linear equations and inequalities.

Time: 15–20 minutes

Group size: 3–4

Materials needed: Blank piece of paper and graph paper for each group member.

1. On each of four pieces of paper, write one of the following equations or inequalities.

   (a) $x + 3 \geq -4(y - 1)$

   (b) $4(y - 2) + 5 = 5(x + 1)$

   (c) $3(y - 7) + 4 = -2(x - 3) - 2$

   (d) $\frac{2}{3}(x + 2y) + \frac{8}{3} \geq 0$

2. Place the four papers face down on the desk and mix them up. One by one, each group member should choose a piece of paper. Do not show your choice to the other members of the group.

3. Carefully graph that equation/inequality on your graph paper. Do not label your graphing.

4. When finished, place the unlabeled graphs in a pile.
5. As a group, with no help from the member who drew the graph, write the equation or inequality that is drawn.
6. As a group, discuss the outcome.

Chapter 8 Review

Section 8.1	The Rectangular Coordinate System and Equations in Two Variables	
KEY TERMS		
x-axis y-axis Origin Rectangular or Cartesian coordinate system xy-plane Coordinate axes	Ordered pair (x, y) Coordinates x-coordinate y-coordinate	Quadrants Equation in two variables Sides Satisfy

YOU SHOULD BE ABLE TO . . .	EXAMPLE	REVIEW EXERCISES
(1) Plot points in the rectangular coordinate system (p. 520)	Examples 1 and 2	1–6
(2) Determine if an ordered pair satisfies an equation (p. 524)	Example 3	7, 8
(3) Create a table of values that satisfies an equation (p. 526)	Examples 4 through 7	9–16

1–4. $A(3, -2)$ quadrant IV
$B(-1, -3)$ quadrant III
$C(-4, 0)$ negative x-axis
$D(0, 2)$ positive y-axis

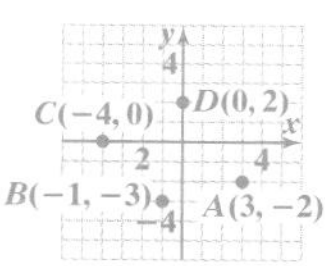

5. $A(1, 4)$; quadrant I
$B(-3, 0)$; x-axis
6. $A(0, 1)$; y-axis
$B(-2, 2)$; quadrant II
7. A. Yes B. No
8. A. Yes B. No
9. (a) $\left(4, -\frac{4}{3}\right)$ (b) $(-6, 2)$
10. (a) $(4, 4)$ (b) $(-2, -2)$

In Problems 1–4, plot the following points in the rectangular coordinate system. Tell which quadrant each point belongs to or on which axis the point lies.

1. $A(3, -2)$ **2.** $B(-1, -3)$ **3.** $C(-4, 0)$ **4.** $D(0, 2)$

In Problems 5 and 6, identify the coordinates of each point labeled in the figure. Tell which quadrant each point belongs to (or on which axis the point lies).

5.

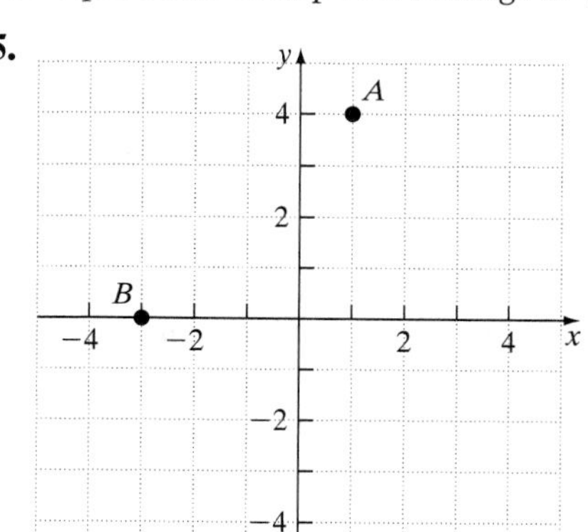

6.

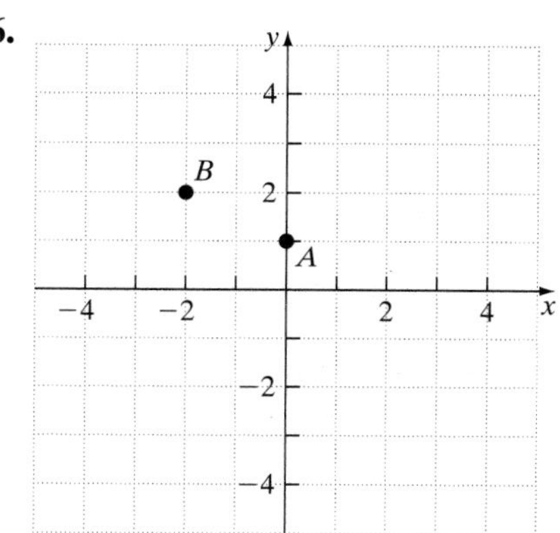

In Problems 7 and 8, determine whether the ordered pairs satisfy the given equation.

7. $y = 3x - 7$
$A(-1, -10)$
$B(-7, 0)$

8. $4x - 3y = 2$
$A(2, 2)$
$B(6, 5)$

9. Find an ordered pair that satisfies the equation $-x = 3y$ when
(a) $x = 4$
(b) $y = 2$

10. Find an ordered pair that satisfies the equation $x - y = 0$ when
(a) $x = 4$
(b) $y = -2$

In Problems 11–14, use the equation to complete the table. Use the table to list some of the ordered pairs that satisfy the equation.

11. $3x - 2y = 10$

x	y	(x, y)
−2		
0		
4		

12. $y = -x + 2$

x	y	(x, y)
−3		
2		
4		

13. $y = -\frac{1}{3}x - 4$

x	y	(x, y)
	2	
−6		
	−12	

14. $3x - 2y = 7$

x	y	(x, y)
	−8	
3		
	4	

15. Mail Order Shipping Martha purchases some clothes from a mail order catalogue. The shipping costs added to her order will be \$5.00 plus \$2.00 per item. The equation that calculates her total shipping cost, C, is given by the equation $C = 5 + 2x$, where x is the number of items purchased. Complete the table below and then graph the ordered pairs in a rectangular coordinate system.

x	C	(x, C)
1		
2		
3		

16. Department Store Wages Isabel works in a department store where her monthly earnings are \$1000 plus 10% commission on her sales. The equation that calculates her earnings, E, is given by the equation: $E = 1000 + 0.10x$, where x is Isabel's sales for the month. Complete the table below and then graph the ordered pairs in a rectangular coordinate system.

x	E	(x, E)
500		
1000		
2000		

11.

x	y	(x, y)
−2	−8	(−2, −8)
0	−5	(0, −5)
4	1	(4, 1)

12.

x	y	(x, y)
−3	5	(−3, 5)
2	0	(2, 0)
4	−2	(4, −2)

13.

x	y	(x, y)
−18	2	(−18, 2)
−6	−2	(−6, −2)
24	−12	(24, −12)

14.

x	y	(x, y)
−3	−8	(−3, −8)
3	1	(3, 1)
5	4	(5, 4)

15.

x	y	(x, y)
1	7	(1, 7)
2	9	(2, 9)
3	11	(3, 11)

16.

x	E	(x, E)
500	1050	(500, 1050)
1000	1100	(1000, 1100)
2000	1200	(2000, 1200)

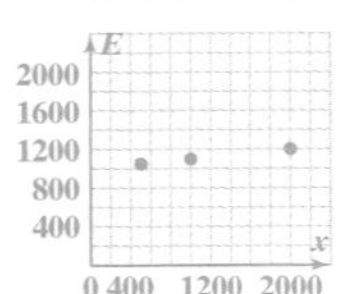

Section 8.2 Graphing Equations in Two Variables

KEY CONCEPTS	KEY TERMS
• **Linear Equation** • A linear equation in two variables is an equation of the form $Ax + By = C$, where A, B, and C are real numbers. A and B cannot both be zero. • **Procedure for Finding Intercepts** **1.** To find the x-intercept(s), if any, of the graph of an equation, let $y = 0$ in the equation and solve for x. **2.** To find the y-intercept(s), if any, of the graph of an equation, let $x = 0$ in the equation and solve for y. • **Vertical Line** • A vertical line is given by the equation $x = a$, where a is the x-intercept. • **Horizontal Line** • A horizontal line is given by the equation $y = b$ where b is the y-intercept.	Graph of an equation in two variables Point-plotting method Complete graph Linear equation in two variables Standard form of an equation of a line Line Intercepts x-intercept y-intercept

YOU SHOULD BE ABLE TO . . .	EXAMPLE	REVIEW EXERCISES
1 Graph a line by plotting points (p. 536)	Examples 1, 3, and 4	17–22
2 Graph a line using intercepts (p. 541)	Examples 5 through 8	23–32
3 Graph vertical and horizontal lines (p. 544)	Examples 9 and 10	33–36

17. $y = -2x$

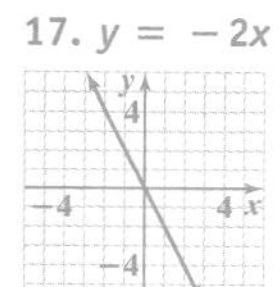

18. $y = x$

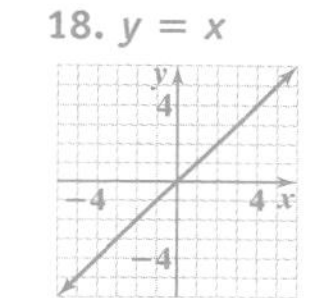

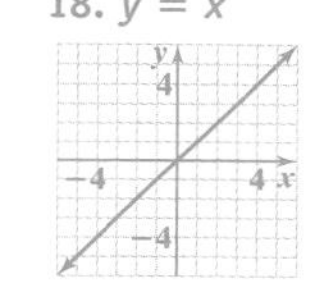

19. $4x + y = -2$

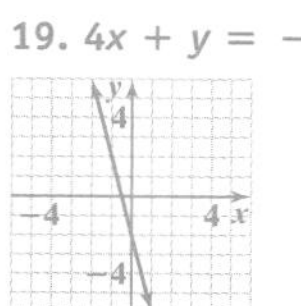

20. $3x - y = -1$

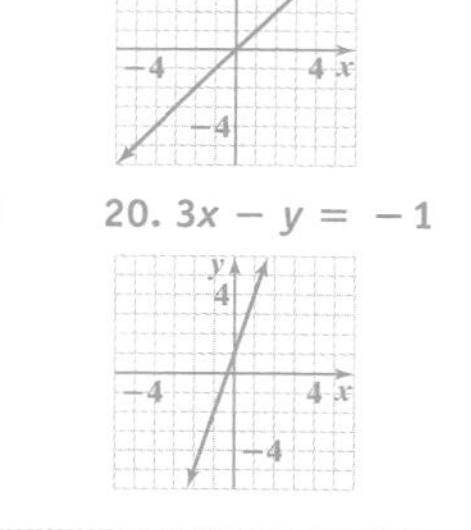

21.

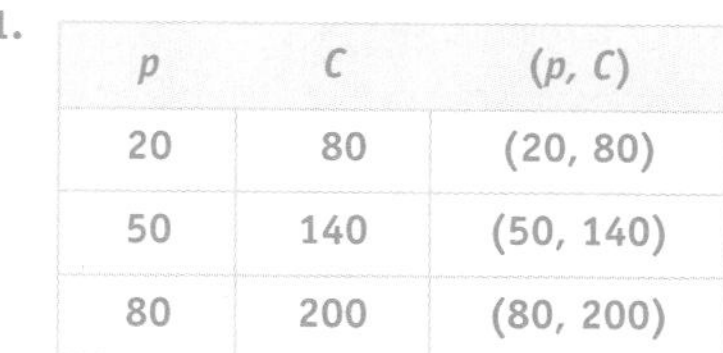

p	C	(p, C)
20	80	(20, 80)
50	140	(50, 140)
80	200	(80, 200)

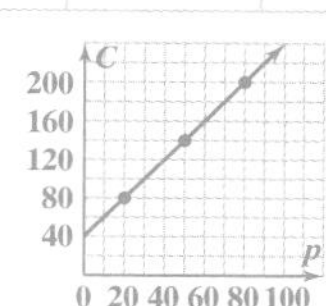

22.

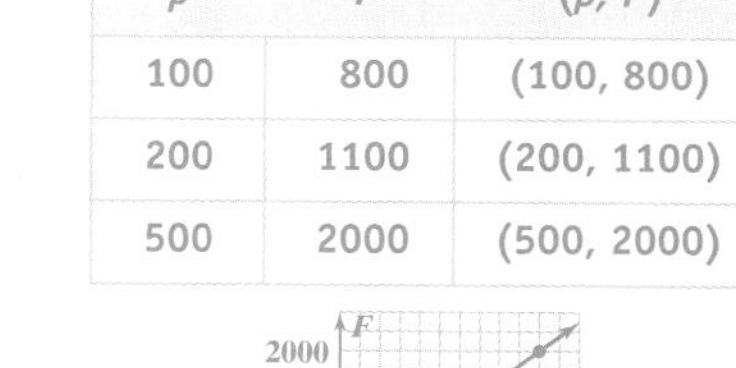

p	F	(p, F)
100	800	(100, 800)
200	1100	(200, 1100)
500	2000	(500, 2000)

In Problems 17–20, graph each linear equation using the point-plotting method.

17. $y = -2x$ **18.** $y = x$ **19.** $4x + y = -2$ **20.** $3x - y = -1$

21. Printing Cost The cost to print pamphlets for a new healthcare clinic is a \$40 set-up fee plus \$2.00 per pamphlet that will be printed. The equation that calculates the total cost for printing, C, is $C = 40 + 2p$, where p is the number of pamphlets to be printed. Complete the table below and then graph the equation.

p	C	(p, C)
20		
50		
80		

22. Performance Fees The Crickets are the newest band in town and have a gig performing at a local concert. They agree that their total fee will be \$500 plus an additional \$3.00 per person who attends the concert. The equation that calculates the total fee for performing, F, is $F = 500 + 3p$, where p is the number of people in attendance. Complete the table below and then graph the equation.

p	F	(p, F)
100		
200		
500		

23. $(-2, 0), (0, -4)$ 24. $(0, 1)$
25. $(0, 9), (-3, 0)$ 26. $(0, -6), (3, 0)$
27. $(3, 0)$ 28. $\left(0, -\frac{2}{5}\right), (1, 0)$

29. $y - 3x = 3$ 30. $2x + 5y = 0$

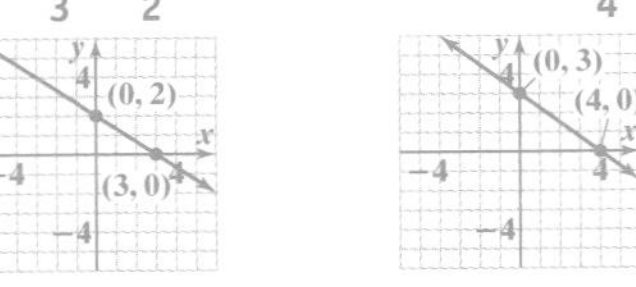

31. $\frac{x}{3} + \frac{y}{2} = 1$ 32. $y = -\frac{3}{4}x + 3$

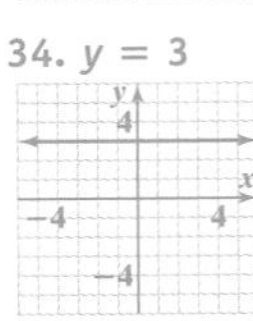

33. $x = -2$ 34. $y = 3$

In Problems 23 and 24, find the intercepts of each graph.

23.

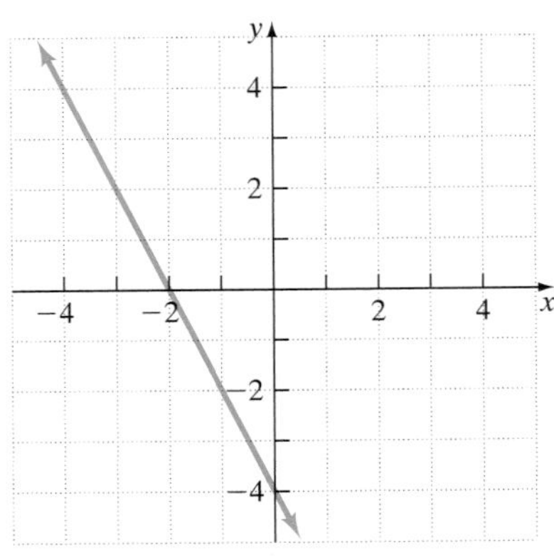

24.

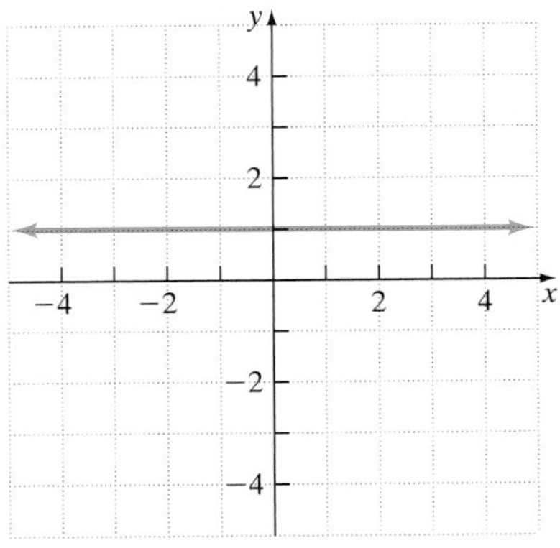

In Problems 25–28, find the x-intercept and y-intercept of each equation.

25. $-3x + y = 9$ **26.** $y = 2x - 6$ **27.** $x = 3$ **28.** $2x - 5y = 2$

In Problems 29–32, graph each linear equation by finding its intercepts.

29. $y - 3x = 3$ **30.** $2x + 5y = 0$ **31.** $\frac{x}{3} + \frac{y}{2} = 1$ **32.** $y = -\frac{3}{4}x + 3$

In Problems 33–36, graph each vertical or horizontal line.

33. $x = -2$ **34.** $y = 3$ **35.** $y = -4$ **36.** $x = 1$

Section 8.3 Slope

KEY CONCEPT	KEY TERMS
• **Slope** If $x_1 \neq x_2$, the **slope** m of the line containing (x_1, y_1) and (x_2, y_2) is defined by the formula $m = \frac{y_2 - y_1}{x_2 - x_1}$. The slope of vertical line is undefined. The slope of a horizontal line is 0.	Run Rise Slope Average rate of change

YOU SHOULD BE ABLE TO . . .	EXAMPLE	REVIEW EXERCISES
(1) Find the slope of a line given two points (p. 551)	Examples 1 and 2	37–42
(2) Find the slope of vertical and horizontal lines (p. 554)	Examples 3 and 4	43–46
(3) Graph a line using its slope and a point on the line (p. 556)	Examples 5 and 6	47–50
(4) Work with applications of slope (p. 557)	Examples 7 and 8	51, 52

35. $y = -4$ 36. $x = 1$

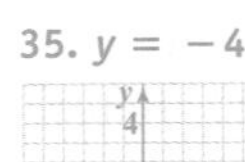

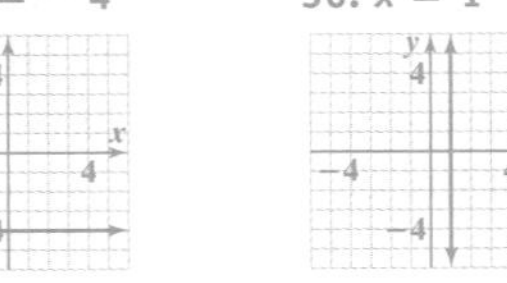

37. $\frac{4}{3}$ 38. -2

In Problems 37 and 38, find the slope of the line whose graph is shown.

37.

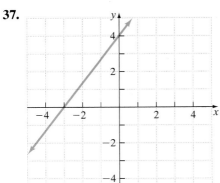

38.

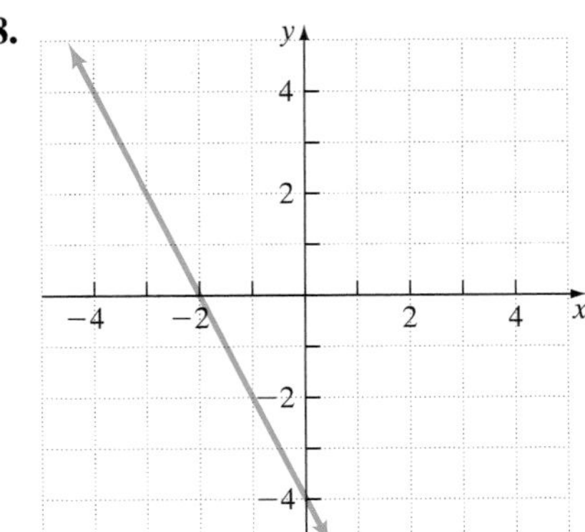

39. −8 40. 2

41. $\frac{1}{4}$ 42. $-\frac{1}{6}$ 43. undefined

44. 0 45. 0 46. undefined

47.

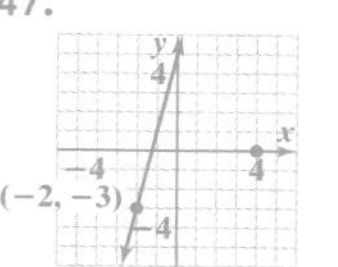

48.

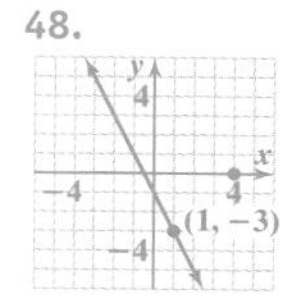

49.

50.

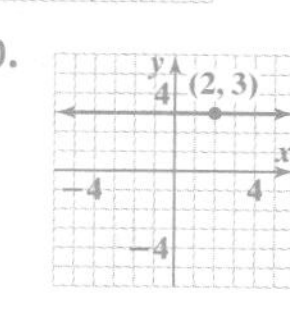

51. $m = 45$; the cost to produce 1 additional bicycle is \$45.

52. 5%

53. $m = -1$; $b = \frac{1}{2}$ 54. $m = 1$; $b = -\frac{3}{2}$

55. $m = \frac{3}{4}$; $b = 1$ 56. $m = -\frac{2}{5}$; $b = \frac{8}{5}$

In Problems 39–46, find the slope of the line containing the given points.

39. $(-4, 6)$ and $(-3, -2)$ **40.** $(4, 1)$ and $(0, -7)$

41. $\left(\frac{1}{2}, -\frac{3}{4}\right)$ and $\left(\frac{5}{2}, -\frac{1}{4}\right)$ **42.** $\left(-\frac{1}{2}, \frac{2}{3}\right)$ and $\left(\frac{3}{2}, \frac{1}{3}\right)$

43. $(-3, -6)$ and $(-3, -10)$ **44.** $(-5, -1)$ and $(-1, -1)$

45. $\left(\frac{3}{4}, \frac{1}{2}\right)$ and $\left(-\frac{1}{4}, \frac{1}{2}\right)$ **46.** $\left(\frac{1}{3}, -\frac{3}{5}\right)$ and $\left(\frac{3}{9}, -\frac{1}{5}\right)$

In Problems 47–50, graph the line that contains the given point and has the given slope.

47. $(-2, -3)$; $m = 4$ **48.** $(1, -3)$; $m = -2$

49. $(0, 1)$; $m = -\frac{2}{3}$ **50.** $(2, 3)$; $m = 0$

51. Production Cost The total cost to produce 20 bicycles is \$1400 and the total cost to produce 50 bicycles is \$2750. Find and interpret the slope of the line containing the points (20, 1400) and (50, 2750).

52. Road Grade Scott is driving to the river on a road which falls 5 feet for every 100 feet of horizontal distance. What is the grade of this road? Express your answer as a percent.

Section 8.4 Slope-Intercept Form of a Line

KEY CONCEPT

- **Slope-intercept form of an equation of a line**
  - An equation of a line with slope m and y-intercept b is $y = mx + b$.

YOU SHOULD BE ABLE TO . . .	EXAMPLE	REVIEW EXERCISES
① Use the slope-intercept form to identify the slope and y-intercept of a line (p. 562)	Examples 1 and 2	53–56
② Graph a line whose equation is in slope-intercept form (p. 563)	Examples 3 and 4	57–62
③ Graph a line whose equation is in the form $Ax + By = C$ (p. 565)	Example 5	63, 64
④ Find the equation of a line given its slope and y-intercept (p. 565)	Example 6	65–70
⑤ Work with linear models in slope-intercept form. (p. 566)	Examples 7 and 8	71, 72

57. $y = \frac{1}{3}x + 1$

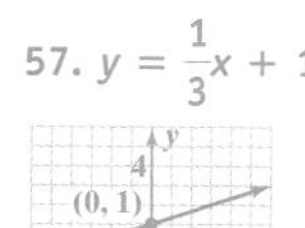

58. $y = -\frac{x}{2} - 1$

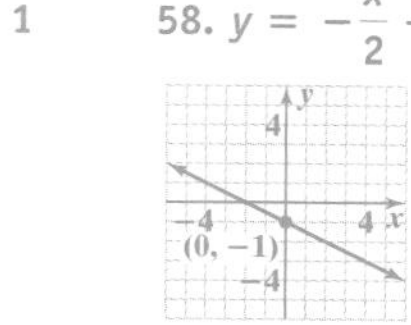

59. $y = -\frac{2x}{3} - 2$

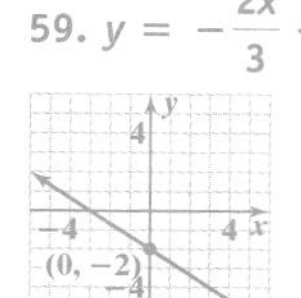

60. $y = \frac{3x}{4} + 3$

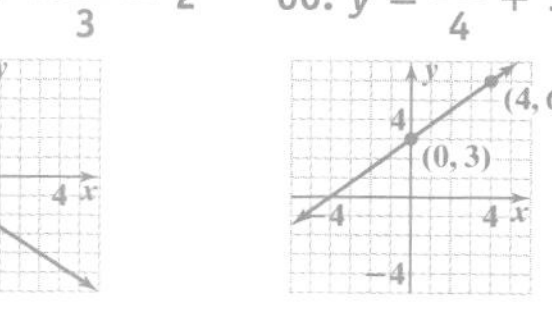

61. $y = x$

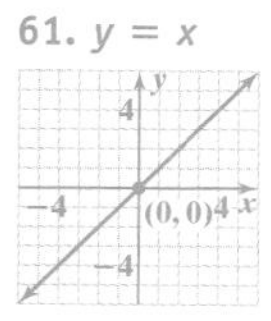

62. $y = -2x$

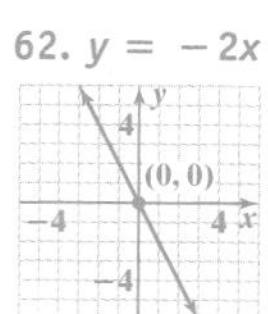

In Problems 53–56, find the slope and the y-intercept of the line whose equation is given.

53. $y = -x + \frac{1}{2}$ **54.** $y = x - \frac{3}{2}$ **55.** $3x - 4y = -4$ **56.** $2x + 5y = 8$

In Problems 57–64, use the slope and y-intercept to graph each line whose equation is given.

57. $y = \frac{1}{3}x + 1$ **58.** $y = -\frac{x}{2} - 1$ **59.** $y = -\frac{2}{3}x - 2$ **60.** $y = \frac{3x}{4} + 3$

61. $y = x$ **62.** $y = -2x$ **63.** $2x - y = -4$ **64.** $-4x + 2y = 2$

In Problems 65–70, find the equation of the line with the given slope and intercept.

65. slope is $-\frac{3}{4}$; y-intercept is $\frac{2}{3}$ **66.** slope is $\frac{1}{5}$; y-intercept is 10

67. slope is undefined; x-intercept is -12 **68.** slope is 0; y-intercept is -4

69. slope is 1; y-intercept is -20 **70.** slope is -1; y-intercept is -8

63. $2x - y = -4$

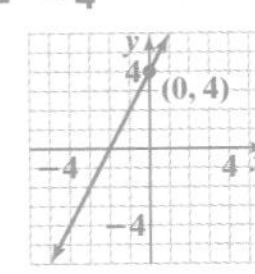

64. $-4x + 2y = 2$

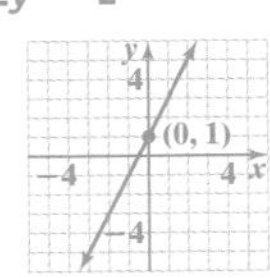

65. $y = -\frac{3}{4}x + \frac{2}{3}$ 66. $y = \frac{1}{5}x + 10$

67. $x = -12$ 68. $y = -4$

69. $y = x - 20$ 70. $y = -x - 8$

71. Car Rentals Hot Rod Car Rentals rents sports cars for \$120 plus \$80 per day. The equation that calculates the total cost C to rent a sports car is $C = 120 + 80d$, where d is the number of days that the car is used.

(a) How much will is cost to rent the car for 3 days?

(b) If the bill came to \$680, how many days was the sports car rented out?

(c) Graph the equation in a rectangular coordinate system.

72. Computer Rentals Anthony plans to rent a computer to finish his Master's thesis. There is no fixed fee, only a charge for each day the computer is checked out. If he keeps the computer for 22 days, he will pay \$418. If his thesis advisor asks him to redo a section and he finds that he must keep the computer for 35 days, it will cost him \$665.

(a) Write two ordered pairs (d, C), where d represents the days the computer is rented and C represents the cost. Calculate the slope of the line that contains these two points.

(b) Interpret the meaning of the slope of this line.

(c) Write an equation that represents the total cost for Anthony to rent a computer.

(d) Calculate the cost to rent a computer for 8 days.

Section 8.5 Point-Slope Form of a Line

KEY CONCEPT

- **Point-Slope Form of an Equation of a Line**
  An equation of a nonvertical line of slope m that contains the point (x_1, y_1) is $y - y_1 = m(x - x_1)$.

YOU SHOULD BE ABLE TO . . .	EXAMPLE	REVIEW EXERCISES
1 Find the equation of a line given a point and a slope (p. 572)	Examples 1 through 3	73–80
2 Find the equation of a line given two points (p. 574)	Examples 4 and 5	81–84
3 Build linear models using the point-slope form of a line (p. 576)	Example 6	85, 86

In Problems 73–80, find the equation of the line that contains the given point and the given slope. Write the equation in slope-intercept form, if possible.

73. $(0, -3)$; slope $= 6$

74. $(4, 0)$; slope $= -2$

75. $(3, -1)$; slope $= -\frac{1}{2}$

76. $(-1, -3)$; slope $= \frac{2}{3}$

77. $\left(-\frac{4}{3}, -\frac{1}{2}\right)$; horizontal

78. $\left(-\frac{4}{7}, \frac{8}{5}\right)$; vertical

79. $(-5, 2)$; slope is undefined

80. $(6, 0)$; slope $= 0$

In Problems 81–84, find the equation of the line that contains the given points. Write the equation of the line in slope-intercept form, if possible.

81. $(-7, 0)$ and $(0, 8)$

82. $(0, -6)$ and $(4, 0)$

83. $(3, 5)$ and $(-2, -10)$

84. $(-15, 1)$ and $(-5, -3)$

85. Harvesting Hay Farmer Myers noted that at the end of 3 days of harvesting 12 acres of hay remained in his field and at the end of 5 days, 4 acres remained. Use the ordered pairs $(3, 12)$ and $(5, 4)$ to write an equation that calculates how many acres of hay, A, are left to be harvested after d days.

86. Fish Population Pat works at Scripps Institute of Oceanography and is responsible for monitoring the fish population in various waters around the world. In one bay, the sunfish population was declining at a constant rate. His initial sample netted 15 sunfish and six months later, the same location netted 13 sunfish. Use the ordered pairs $(0, 15)$ and $(6, 13)$ to write an equation that will predict how many sunfish, F, the sample will yield after m months.

71. (a) \$360 (b) 7 days
(c)

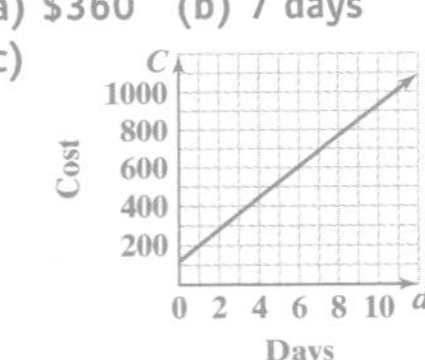

72. (a) $(22, 418)$, $(35, 665)$, $m = 19$
(b) It costs \$19 more for each additional day.
(c) $C = 19d$ (d) \$152

73. $y = 6x - 3$ 74. $y = -2x + 8$

75. $y = -\frac{1}{2}x + \frac{1}{2}$ 76. $y = \frac{2}{3}x - \frac{7}{3}$

77. $y = -\frac{1}{2}$ 78. $x = -\frac{4}{7}$

79. $x = -5$ 80. $y = 0$

81. $y = \frac{8}{7}x + 8$ 82. $y = \frac{3}{2}x - 6$

83. $y = 3x - 4$ 84. $y = -\frac{2}{5}x - 5$

85. $A = -4d + 24$ 86. $F = -\frac{1}{3}m + 15$

Section 8.6 Parallel and Perpendicular Lines

KEY TERMS

Parallel
Perpendicular

YOU SHOULD BE ABLE TO . . .	EXAMPLE	REVIEW EXERCISES
1 Determine whether two lines are parallel (p. 580)	Examples 1 and 2	87, 88
2 Find the equation of a line parallel to a given line (p. 582)	Examples 3 and 4	89–94
3 Determine whether two lines are perpendicular (p. 583)	Examples 5 through 7	95–98
4 Find the equation of a line perpendicular to a given line (p. 585)	Examples 8 and 9	99–102

In Problems 87 and 88, determine if the two lines are parallel.

87. $y = -\frac{1}{3}x + 2$
$x - 3y = 3$

88. $y = \frac{1}{2}x - 4$
$x - 2y = 6$

In Problems 89–94, find the equation of the line that contains the given point and is parallel to the given line. Write the equation in slope-intercept form, if possible.

89. $(3, -1)$; $y = -x + 5$

90. $(-2, 4)$; $y = 2x - 1$

91. $(-1, 10)$; $3x + y = -7$

92. $(4, -5)$; $6x + 2y = 5$

93. $(5, 19)$; y-axis

94. $(-1, -12)$; x-axis

In Problems 95 and 96, determine the slope of the line perpendicular to the given line.

95. $3x - 2y = 5$

96. $4x - 9y = 1$

In Problems 97 and 98, determine if the two lines are perpendicular.

97. $x + 3y = 3$
$y = 3x + 1$

98. $5x - 2y = 2$
$y = \frac{2}{5}x + 12$

In Problems 99–102, write the equation of the line that contains the given point and is perpendicular to the given line. Write the equation in slope-intercept form, if possible.

99. $(-3, 4)$; $y = -3x + 1$

100. $(4, -1)$; $y = 2x - 1$

101. $(1, -3)$; $2x - 3y = 6$

102. $\left(-\frac{3}{5}, \frac{2}{5}\right)$; $x + y = -7$

87. not parallel 88. parallel
89. $y = -x + 2$ 90. $y = 2x + 8$
91. $y = -3x + 7$ 92. $y = -3x + 7$
93. $x = 5$ 94. $y = -12$
95. $-\frac{2}{3}$ 96. $-\frac{9}{4}$
97. perpendicular 98. not perpendicular
99. $y = \frac{1}{3}x + 5$
100. $y = -\frac{1}{2}x + 1$
101. $y = -\frac{3}{2}x - \frac{3}{2}$
102. $y = x + 1$

Section 8.7 Variation

KEY TERMS

Variation
Varies directly
Directly proportional to
Constant of proportionality
Constant of variation
Varies inversely
Inversely proportional

YOU SHOULD BE ABLE TO . . .	EXAMPLE	REVIEW EXERCISES
1 Model and solve direct variation problems (p. 590)	Examples 1 and 2	103–108
2 Model and solve inverse variation problems (p. 592)	Examples 3 and 4	109–114

103. $y = \frac{1}{4}x$ 104. $y = 6x$
105. $f = 16$ 106. $p = 110$
107. $507.50 108. 160 mi
109. $y = \frac{48}{x}$ 110. $y = \frac{54}{x}$
111. $r = 9$ 112. $s = \frac{1}{4}$
113. 78 mph 114. 25 liters
115. *A*, *B* are solutions.
116. *C* is a solution.
117. $y < -\frac{1}{4}x + 2$ 118. $y > 2x - 1$

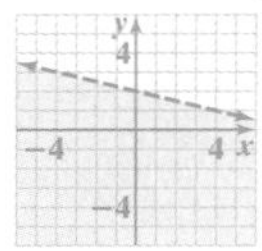
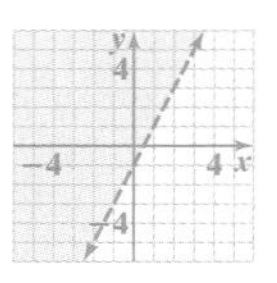

119. $3x + 2y \geq -6$ 120. $-2x + y \geq 4$

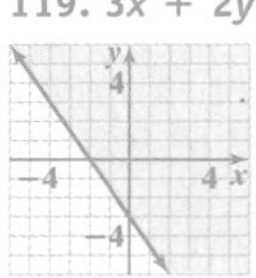
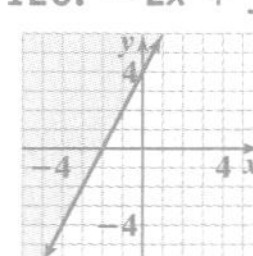

In Problems 103 and 104, assume that y varies directly with x. Find an equation that relates x and y given the following:

103. $x = 12$ when $y = 3$ **104.** $x = 3$ when $y = 18$

105. Suppose that f varies directly with g. If $f = 12$ when $g = 18$, find f when $g = 24$.

106. Suppose that p varies directly with q. If $p = 25$ when $q = 20$, find p when $q = 88$.

107. CD Sales The income from CD sales varies directly with the number of customers that enter the store. If 20 customers produce CD sales of $290, how much in sales will 35 customers produce?

108. Map Distances The distance between two points on a map varies directly with the actual distance between the cities. If two cities are 3 cm apart on a map and their actual distance is 60 miles, find the actual distance for two cities which measure 8 cm apart on the same map.

In Problems 109 and 110, suppose that y varies inversely with x. Find an equation that relates x and y given the following:

109. $x = 8$ when $y = 6$ **110.** $x = 6$ when $y = 9$

111. Suppose that r varies inversely with s. If $r = 18$ when $s = 2$, find r when $s = 4$.

112. Suppose s varies inversely with u. If $s = 3$ when $u = 2$, find s when $u = 24$.

113. Road Trip The time that it takes to drive from Los Angeles to San Francisco is inversely proportional to the speed the car is driven. If the trip takes 6.5 hours at 60 mph, what speed is necessary to make the trip in 5 hours?

114. Measuring Pressure The volume of a gas varies inversely with the pressure. If the volume is 8 liters when the pressure is 100 grams per liter, what is the volume of this gas if the pressure is 32 grams per liter?

Section 8.8 Linear Inequalities in Two Variables

KEY TERMS

Satisfied	Half-planes
Graph of a linear inequality in two variables	Test point

YOU SHOULD BE ABLE TO . . .	EXAMPLE	REVIEW EXERCISES
1 Determine whether an ordered pair is a solution of a linear inequality (p. 597)	Example 1	115, 116
2 Graph linear inequalities (p. 598)	Examples 2 through 5	117–124
3 Solve problems involving linear inequalities (p. 602)	Example 6	125, 126

121. $x - 3y \leq 0$ 122. $x - 4y \geq 4$

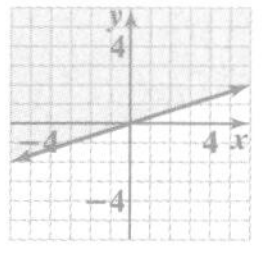
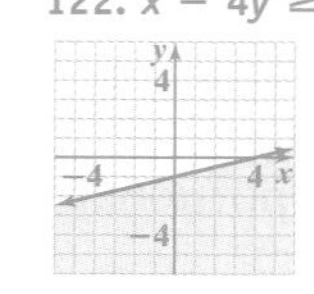

123. $x < -3$ 124. $y > 2$

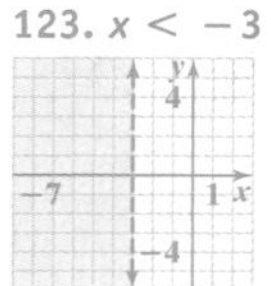
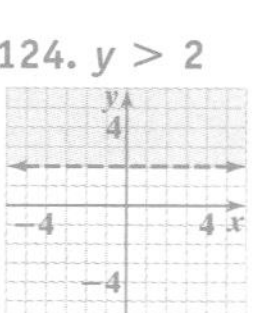

In Problems 115 and 116, determine which of the following points, if any, are solutions to the given linear inequality in two variables.

115. $y \leq 3x + 4$

$A(2, 0)$ $B(-4, -8)$ $C(7, 26)$

116. $y > \frac{1}{3}x + 4$

$A(6, -2)$ $B(0, 4)$ $C(-18, -1)$

In Problems 117–124, graph each linear inequality.

117. $y < -\frac{1}{4}x + 2$ **118.** $y > 2x - 1$ **119.** $3x + 2y \geq -6$

120. $-2x + y \geq 4$ **121.** $x - 3y \leq 0$ **122.** $x - 4y \geq 4$

123. $x < -3$ **124.** $y > 2$

125. $0.25x + 0.1y \geq 12$

126. $2x - \frac{1}{2}y \leq 10$

125. Coin Jar A jar holding only quarters and dimes contains at least \$12. If there are x quarters and y dimes in the jar, write a linear inequality in two variables that describes how many of each type of coin are in the jar.

126. Number The difference between twice a number, x, and half of a second number, y, is at most 10. Write a linear inequality in two variables that describes these two numbers.

CHAPTER 8 TEST

Remember to use your Chapter Test Prep Video CD to see fully worked-out solutions to any of these problems you would like to review.

Note to Instructor: A special file in TestGen provides algorithms specifically matched to the problems in this Chapter Test for easy-to-replicate practice or assessment purposes.

1. No
2. (a) $(4, 0)$
   (b) $\left(0, -\frac{4}{3}\right)$
3. (a) $m = \frac{4}{3}$
   (b) $b = 8$
4. $y = -\frac{3}{4}x + 2$
5. $3x - 6y = -12$

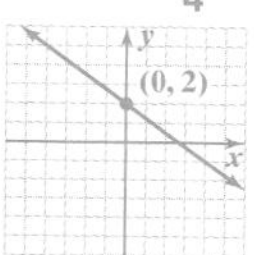

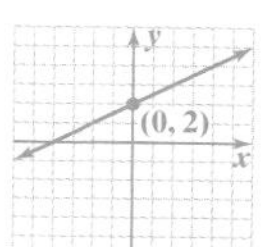

6. $-\frac{1}{6}$
7. (a) $-\frac{3}{2}$
   (b) $\frac{2}{3}$
8. neither; Answers may vary.
9. $y = -4x - 15$
10. $y = 2x + 14$
11. $y = -3x - 11$
12. $y = \frac{1}{2}x - 2$
13. $y = -\frac{3}{2}x + 8$
14. $y = 5$
15. $x = -2$
16. $m = 30$
17. \$8 per package
18. $y \geq x - 3$
19. $-2x - 4y < 8$

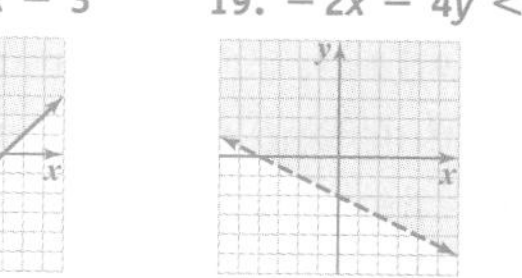

20. $x \leq -4$

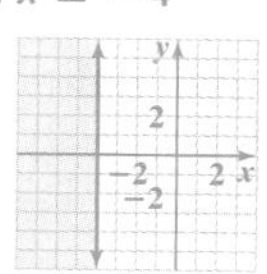

1. Determine whether the ordered pair $(-3, -2)$ is a solution to the equation $3x - 4y = -17$.

2. Given the equation $3x - 9y = 12$, determine
(a) the x-intercept **(b)** the y-intercept

3. Given the equation $4x - 3y = -24$, determine
(a) the slope **(b)** the y-intercept

In Problems 4 and 5, graph each linear equation.

4. $y = -\frac{3}{4}x + 2$ **5.** $3x - 6y = -12$

6. Find the slope of the line that contains the points $(2, -2)$ and $(-4, -1)$.

7. Given the linear equation $3y = 2x - 1$, find
(a) the slope of a line perpendicular to the given line.
(b) the slope of a line parallel to the given line.

8. Determine whether the lines are parallel, perpendicular, or neither. Explain how you came to your conclusion.

$$L_1\colon 3x - 7y = 2$$
$$L_2\colon y = \frac{7}{3}x + 4$$

In Problems 9–15, write the equation of the line that satisfies the given conditions. Write the equation in slope-intercept form, if possible.

9. slope $= -4$ and y-intercept is -15

10. slope $= 2$ and contains $(-3, 8)$

11. contains $(-3, -2)$ and $(-4, 1)$

12. contains $(4, 0)$ and parallel to $y = \frac{1}{2}x + 2$

13. contains $(4, 2)$ and perpendicular to $4x - 6y = 5$

14. horizontal line through $(3, 5)$

15. contains $(-2, -1)$ with undefined slope

16. m varies directly with n. If $m = 12$ when $n = 8$, find m when $n = 20$.

17. Shipping Packages The shipping department records indicate that during a week when 20 packages were shipped, the total expenses recorded for the shipping department were \$560. During a different week when 30 packages were shipped, the expenses recorded were \$640. Use the ordered pairs $(20, 560)$ and $(30, 640)$ to determine the average rate to ship a package.

In Problems 18–20, graph each inequality.

18. $y \geq x - 3$ **19.** $-2x - 4y < 8$ **20.** $x \leq -4$

CHAPTER

9 Systems of Linear Equations and Inequalities

It is well documented that American women can expect to live longer than American men. But is there ever going to be a time when men can expect to live as long as women? See Example 6 in Section 9.2.

OUTLINE

The Big Picture: Putting It Together

In Chapter 2, we solved linear equations and inequalities in one variable. Recall that linear equations in one variable can have no solution (a contradiction), one solution, or infinitely many solutions (an identity). In Chapter 8, we learned how to graph both linear equations and linear inequalities in two variables. Remember, the graph of a linear equation in two variables represents the set of all points whose ordered pairs satisfy the equation. The graph of a linear equation will be used in this chapter to help us visualize results.

In this chapter, we will discuss methods for finding solutions that simultaneously satisfy two linear equations involving two variables. We are going to learn a graphical method for finding the solution and two algebraic methods. These systems can have no solution, one solution, or infinitely many solutions, just like linear equations in one variable. We conclude the chapter by looking at systems of linear inequalities. These systems require us to determine the region that satisfies two or more linear inequalities simultaneously.

9.1 Solving Systems of Linear Equations by Graphing

OBJECTIVES

1. Determine If an Ordered Pair Is a Solution of a System of Linear Equations
2. Solve a System of Linear Equations by Graphing
3. Classify Systems of Linear Equations as Consistent or Inconsistent
4. Solve Applied Problems Involving Systems of Linear Equations

Preparing for Solving Systems of Linear Equations by Graphing

Before getting started, take the following readiness quiz. If you get a problem wrong, go back to the section cited and review the material.

1. Graph: $y = 2x - 3$ [Section 8.4, pp. 563–565]

2. Graph: $3x + 4y = 12$ [Section 8.2, pp. 541–544]

3. Determine whether $2x + 6y = 12$ is parallel to $-3x - 9y = 18$. [Section 8.6, pp. 580–582]

In Section 8.2, we learned that an equation in two variables is linear provided that it can be written in the form $Ax + By = C$, where A and B cannot both be zero. We now learn methods for finding ordered pairs that satisfy two linear equations at the same time.

DEFINITION

A **system of linear equations** is a grouping of two or more linear equations where each equation contains one or more variables.

In this book, we will discuss only systems of linear equations containing two variables and two equations. Example 1 gives some examples of systems of linear equations.

Classroom Example
Examples of systems of linear equations:

(a) $\begin{cases} x + 3y = 9 \\ 2x - 4y = -1 \end{cases}$

(b) $\begin{cases} -5a + b = 8 \\ a - 6b = 0 \end{cases}$

EXAMPLE 1 Examples of Systems of Linear Equations

(a) $\begin{cases} 2x + y = 5 \\ x - 5y = -10 \end{cases}$ Two equations containing two variables, x and y

(b) $\begin{cases} a + b = 5 \\ 2a - 3b = -3 \end{cases}$ Two equations containing two variables, a and b

Teaching Tip
Remind students that a solution is the value of the variable(s) that make a true statement—in this case, both statements must be satisfied.

We use a brace, as shown in the systems in Example 1, to remind us that we are dealing with a system of equations.

Preparing for...Answers

1.

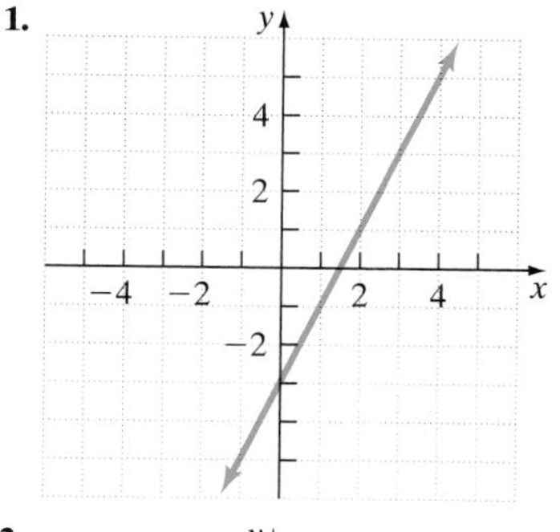

2.

3. Parallel

1 Determine If an Ordered Pair Is a Solution of a System of Linear Equations

DEFINITION

A **solution** of a system of equations consists of values for the variables that satisfy each equation of the system. When we are solving systems of two linear equations containing two unknowns, we represent the solution as an ordered pair, (x, y).

So to determine whether an ordered pair is a solution of a system of equations, we replace each variable with its value in each equation. If *both* the equations in the system are satisfied, then the ordered pair is a solution.

EXAMPLE 2 Determining Whether an Ordered Pair Is a Solution of a System of Linear Equations

Determine whether the given ordered pairs are solutions of the system of equations.

$$\begin{cases} x + 3y = 9 \\ 4x - 2y = 8 \end{cases}$$

(a) $(-3, 4)$ **(b)** $(3, 2)$

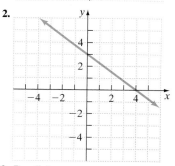

Classroom Example
Determine whether the given ordered pairs are solutions to the system of equations.

$$\begin{cases} 2x + 4y = -2 \\ x - 2y = -7 \end{cases}$$

(a) $(3, -2)$ (b) $\left(-4, \frac{3}{2}\right)$

Answer:

(a) no (b) yes

Solution

(a) Let $x = -3$ and $y = 4$ in both equations. If both equations are true, then $(-3, 4)$ is a solution.

First equation: $x + 3y = 9$

$x = -3, y = 4$: $-3 + 3(4) \stackrel{?}{=} 9$

$-3 + 12 \stackrel{?}{=} 9$

$9 = 9$ True

Second equation: $4x - 2y = 8$

$4(-3) - 2(4) \stackrel{?}{=} 8$

$-12 - 8 \stackrel{?}{=} 8$

$-20 = 8$ False

Although $(-3, 4)$ satisfies the first equation, it does not satisfy the second equation. Because the values $x = -3$ and $y = 4$ do not satisfy both equations, $(-3, 4)$ is not a solution of the system.

(b) Let $x = 3$ and $y = 2$ in both equations. If both equations are true, then $(3, 2)$ is a solution.

First equation: $x + 3y = 9$

$x = 3, y = 2$: $3 + 3(2) \stackrel{?}{=} 9$

$3 + 6 \stackrel{?}{=} 9$

$9 = 9$ True

Second equation: $4x - 2y = 8$

$4(3) - 2(2) \stackrel{?}{=} 8$

$12 - 4 \stackrel{?}{=} 8$

$8 = 8$ True

Because the values $x = 3$ and $y = 2$ satisfy both equations, the ordered pair $(3, 2)$ is a solution of the system.

QUICK ✓ *Determine whether the given set of ordered pairs is a solution of the system of equations.*

1. $\begin{cases} 2x + 3y = 7 \\ 3x + y = -7 \end{cases}$

(a) $(3, 1)$ **(b)** $(-4, 5)$ **(c)** $(-2, -1)$

2. $\begin{cases} 3x - 6y = 6 \\ -2x + 4y = -4 \end{cases}$

(a) $(2, 0)$ **(b)** $(0, -1)$ **(c)** $(4, 1)$

2 Solve a System of Linear Equations by Graphing

Teaching Tip
Review how to write the equation in slope-intercept form, $y = mx + b$: Solve for y and then use the slope and the y-intercept to graph the line. Also review finding intercepts: To find the x-intercept, let $y = 0$, and to find the y-intercept, let $x = 0$.

It is useful to visualize the problem of solving a system of two linear equations containing two variables as a geometry problem. The graph of each equation in the system is a line. So a system of two equations containing two variables represents a pair of lines. We know from Chapter 8 that the graph of an equation represents the set of all ordered pairs that make the equation a true statement. Therefore, **the point, or points, at which two lines intersect, if any, represents the solution of the system of equations.**

Let's look at an example to learn how to solve a system of linear equations by graphing.

EXAMPLE 3 How to Solve a System of Linear Equations by Graphing

Solve the following system by graphing: $\begin{cases} 3x + y = 7 \\ -2x + 3y = -12 \end{cases}$

Classroom Example
Solve the system by graphing:

$$\begin{cases} 2x + y = 1 \\ -x + 2y = 12 \end{cases}$$

Answer: $(-2, 5)$

Step-by-Step Solution

Step 1: *Graph the first equation in the system.*

We graph the equation $3x + y = 7$ by putting the equation in slope-intercept form, $y = mx + b$.

$$3x + y = 7$$
$$y = -3x + 7$$

We graph the line whose slope is -3 and y-intercept is 7. See Figure 1(a).

Figure 1

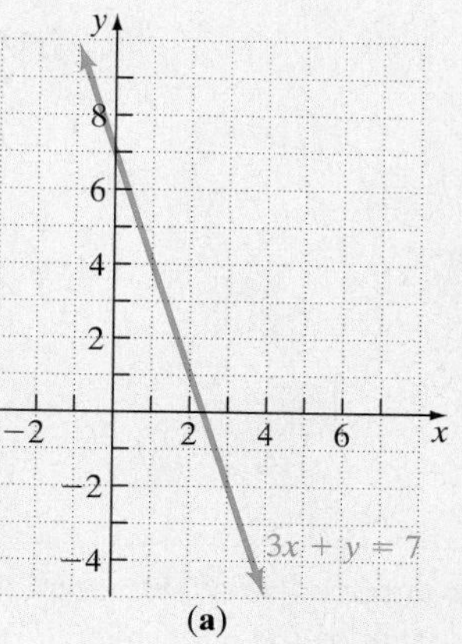

(a)

Step 2: Graph the second equation in the system.

Work Smart
We graph the line $-2x + 3y = -12$ using intercepts because the equation is in standard form and we can find the intercepts easily.

Teaching Tip
In Example 3 we graphed the equation $3x + y = 7$ using the slope-intercept form and we graphed the equation $-2x + 3y = -12$ using intercepts. Point out to students that the intercept method works well if both intercepts are integers; otherwise the slope-intercept method is likely to be better.

Graph the equation $-2x + 3y = -12$ using intercepts.

x-intercept: Let $y = 0$ and solve for x.

$$\begin{aligned} -2x + 3y &= -12 \\ -2x + 3(0) &= -12 \\ -2x &= -12 \\ x &= 6 \end{aligned}$$

y-intercept: Let $x = 0$ and solve for y.

$$\begin{aligned} -2x + 3y &= -12 \\ -2(0) + 3y &= -12 \\ 3y &= -12 \\ y &= -4 \end{aligned}$$

Plot the points $(6, 0)$ and $(0, -4)$ and draw a line through these points on the same rectangular coordinate system that we graphed the line $3x + y = 7$. See Figure 1(b).

Figure 1

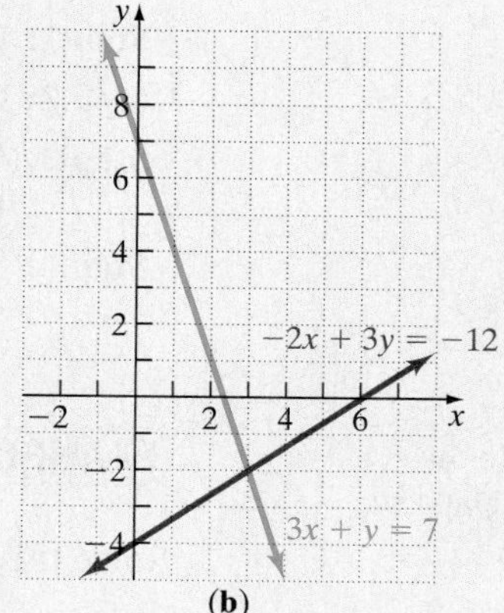

(b)

Step 3: Determine the point of intersection of the two lines, if any.

From the graphs in Figure 1(b), we can see that the lines appear to intersect at $(3, -2)$. We label this point in Figure 1(c).

(continued)

Figure 1 (*Continued*)

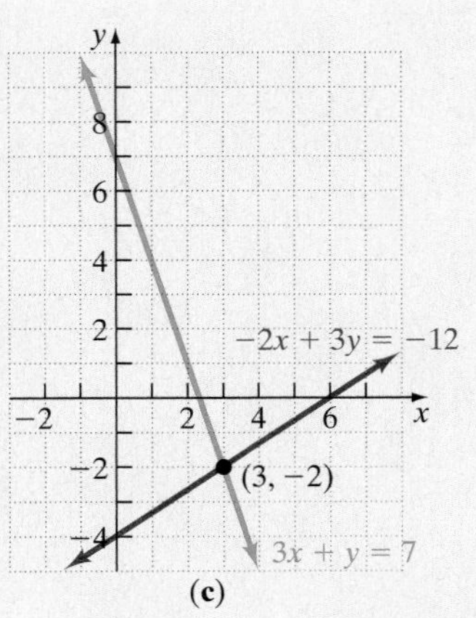

(c)

Step 4: Verify that the point of intersection determined in Step 3 is a solution of the system.

Let $x = 3$ and $y = -2$ in both equations.

First equation:
$$
\begin{aligned}
3x + y &= 7 \\
3(3) + (-2) &\stackrel{?}{=} 7 \\
9 - 2 &\stackrel{?}{=} 7 \\
7 &= 7 \quad \text{True}
\end{aligned}
$$

Second equation:
$$
\begin{aligned}
-2x + 3y &= -12 \\
-2(3) + 3(-2) &\stackrel{?}{=} -12 \\
-6 - 6 &= -12 \\
-12 &= -12 \quad \text{True}
\end{aligned}
$$

The solution of the system $\begin{cases} 3x + y = 7 \\ -2x + 3y = -12 \end{cases}$ is $(3, -2)$.

Below, we summarize the steps for obtaining the solution of a system of linear equations by graphing.

Steps for obtaining the solution of a system of linear equations by graphing

Step 1: Graph the first equation in the system.

Step 2: Graph the second equation in the system.

Step 3: Determine the point of intersection, if any. The point of intersection of the lines is the solution of the system.

Step 4: Verify that the point of intersection determined in Step 3 is a solution of the system.

Classroom Example Solve the system by graphing:

$$\begin{cases} 2x - 3y = -12 \\ 2x + 3y = 0 \end{cases}$$

Answer: $(-3, 2)$

EXAMPLE 4 Solving a System of Linear Equations Graphically

Solve the following system by graphing: $\begin{cases} 3x - 2y = -16 \\ 5x + 2y = 0 \end{cases}$

Solution

We graph both equations in the system by putting each equation in slope-intercept form.

$$
\begin{aligned}
\text{First equation:} \quad 3x - 2y &= -16 \\
\text{Subtract } 3x \text{ from both sides:} \quad -2y &= -3x - 16 \\
\text{Divide both sides by } -2\text{:} \quad y &= \frac{3}{2}x + 8
\end{aligned}
$$

Figure 2

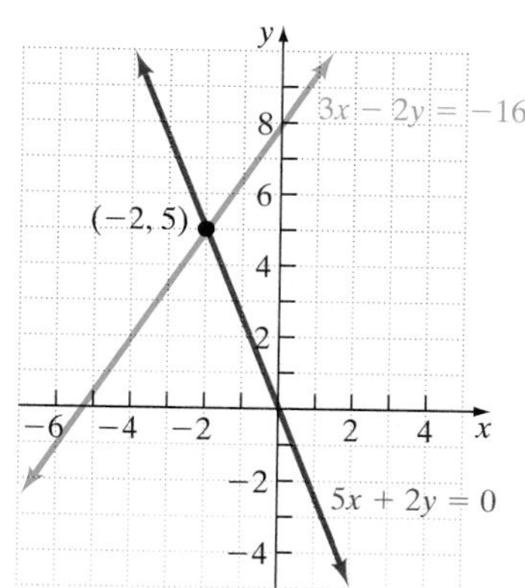

The slope of the first equation is $\frac{3}{2}$ and the y-intercept is 8.

$$\begin{aligned} \text{Second equation:} \quad 5x + 2y &= 0 \\ \text{Subtract } 5x \text{ from both sides:} \quad 2y &= -5x \\ \text{Divide both sides by 2:} \quad y &= -\frac{5}{2}x \end{aligned}$$

The slope of the second equation is $-\frac{5}{2}$ and the y-intercept is 0. Figure 2 shows the graph of each equation. The point of intersection is $(-2, 5)$. Now, we check to see if the solution is $x = -2$, $y = 5$.

Check

$$\begin{aligned} \text{First equation:} \quad 3x - 2y &= -16 \\ x = -2, y = 5: \quad 3(-2) - 2(5) &\stackrel{?}{=} -16 \\ -6 - 10 &\stackrel{?}{=} -16 \\ -16 &= -16 \quad \text{True} \end{aligned}$$

$$\begin{aligned} \text{Second equation:} \quad 5x + 2y &= 0 \\ x = -2, y = 5: \quad 5(-2) + 2(5) &\stackrel{?}{=} 0 \\ -10 + 10 &\stackrel{?}{=} 0 \\ 0 &= 0 \quad \text{True} \end{aligned}$$

Both equations check, so the solution is $(-2, 5)$.

QUICK ✓ *Solve each system by graphing.*

3. $\begin{cases} y = -2x + 9 \\ y = 3x - 11 \end{cases}$ 4. $\begin{cases} 4x + y = -3 \\ 3x - y = -11 \end{cases}$ 5. $\begin{cases} 2x + 3y = 1 \\ -4x + 2y = -26 \end{cases}$

Sometimes the lines that make up a system of linear equations in two variables do not intersect at all, as shown in Example 5.

Classroom Example
Solve the system by graphing:

$\begin{cases} 3x - y = 5 \\ -6x + 2y = 4 \end{cases}$

Answer: No solution or $\varnothing$

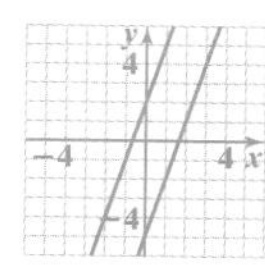

EXAMPLE 5 Solving a System of Linear Equations That Have No Point of Intersection

Solve the following system by graphing: $\begin{cases} 3x - y = -4 \\ -6x + 2y = 2 \end{cases}$

Solution

We graph both equations in the system by putting each equation in slope-intercept form.

$$\begin{aligned} \text{First equation:} \quad 3x - y &= -4 \\ \text{Subtract } 3x \text{ from both sides:} \quad -y &= -3x - 4 \\ \text{Divide both sides by } -1: \quad y &= 3x + 4 \end{aligned}$$

The slope of the first equation is 3 and the y-intercept is 4.

$$\begin{aligned} \text{Second equation:} \quad -6x + 2y &= 2 \\ \text{Add } 6x \text{ to both sides:} \quad 2y &= 6x + 2 \\ \text{Divide both sides by 2:} \quad y &= 3x + 1 \end{aligned}$$

The slope of the second equation is 3 and the y-intercept is 1. Figure 3 shows the graph of each equation. The lines are parallel and therefore do not intersect anywhere. The system of equations has no solution. We could also say the solution set is the empty set, $\varnothing$.

Work Smart
Notice that the slopes of the two lines in Example 5 are the same, and the y-intercepts are different.

Figure 3

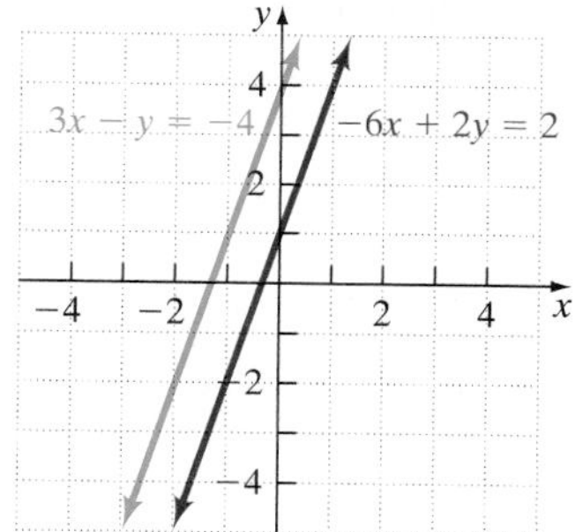

QUICK ✓ *Solve each system by graphing.*

6. $\begin{cases} y = 2x + 4 \\ y = 2x - 1 \end{cases}$

7. $\begin{cases} 2x - 3y = -3 \\ -4x + 6y = -12 \end{cases}$

Another special case with systems of linear equations in two variables occurs when the graph of each of the equations in the system results in the same line.

Classroom Example ➤
Solve the system by graphing:

$$\begin{cases} 3x + 2y = 2 \\ -6x - 4y = -4 \end{cases}$$

Answer: Infinitely many solutions

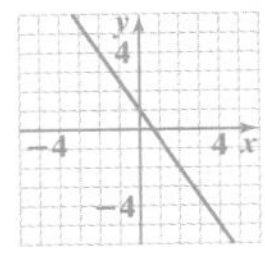

EXAMPLE 6 Solving a System of Linear Equations where the Graph of Each Equation Is the Same Line

Solve the following system by graphing: $\begin{cases} 5x + 2y = -4 \\ 10x + 4y = -8 \end{cases}$

Solution

We graph both equations in the system by putting each equation in slope-intercept form.

First equation: $5x + 2y = -4$

Subtract $5x$ from both sides: $2y = -5x - 4$

Divide both sides by 2: $y = -\frac{5}{2}x - 2$

The slope of the first equation is $-\frac{5}{2}$ and the y-intercept is -2.

Second equation: $10x + 4y = -8$

Subtract $10x$ from both sides: $4y = -10x - 8$

Divide both sides by 4: $y = -\frac{5}{2}x - 2$

The slope of the second equation is $-\frac{5}{2}$ and the y-intercept is -2. Figure 4 shows the graph of each equation. The equations are the same so the lines coincide. The system of equations has infinitely many solutions. ■

Work Smart

Notice that the slopes and y-intercepts of the two lines in Example 6 are the same.

Figure 4

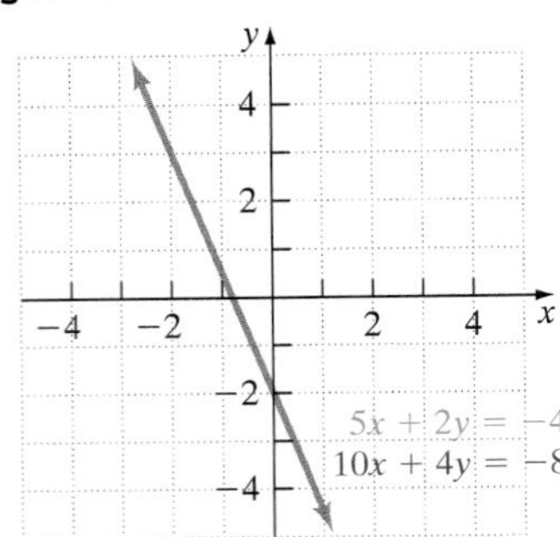

QUICK ✓ *Solve each system by graphing.*

8. $\begin{cases} y = 3x + 2 \\ -6x + 2y = 4 \end{cases}$

9. $\begin{cases} 7x + 3y = -6 \\ -14x - 6y = 12 \end{cases}$

3 Classify Systems of Linear Equations as Consistent or Inconsistent

The results of Examples 3–6 tell us that there are three possibilities for the solution of a system of linear equations (one solution, no solution, or infinitely many solutions). In Examples 3 and 4, the system of linear equations had exactly one solution. In Example 5, the system of linear equations had no solution. In Example 6, the system of equations had infinitely many solutions. We classify systems of equations based on the number of solutions the system has.

DEFINITION

A **consistent system** is a system of equations that has at least one solution.

An **inconsistent system** is a system of equations that has no solution.

We summarize these possibilities next.

CLASSIFYING A SYSTEM OF EQUATIONS GRAPHICALLY

Intersect: If the lines intersect, then the system of equations has one solution given by the point of intersection. We say that the system is **consistent** and the equations are **independent.** See Figure 5(a).

Parallel: If the lines are parallel, then the system of equations has no solution because the lines never intersect. We say that the system is **inconsistent.** See Figure 5(b).

Coincident: If the lines lie on top of each other (are coincident), then the system of equations has infinitely many solutions. The solution set is the set of all points on the line. The system is **consistent** and the equations are **dependent.** See Figure 5(c).

Figure 5

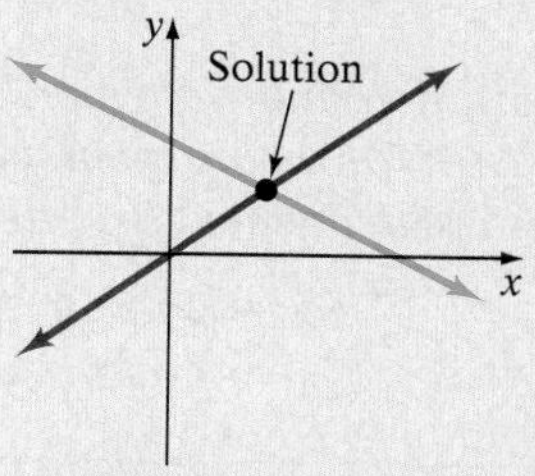

(a) Intersecting lines; system has exactly one solution

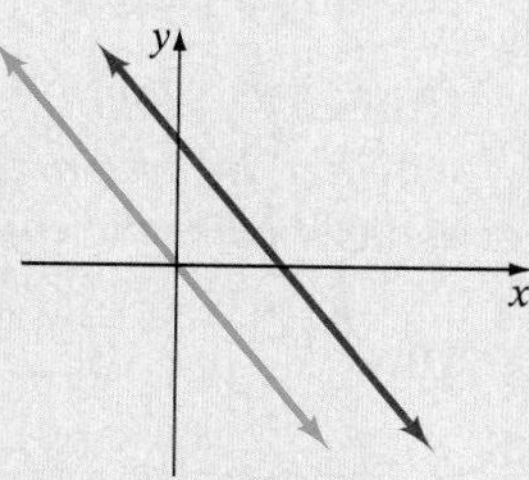

(b) Parallel lines; system has no solution

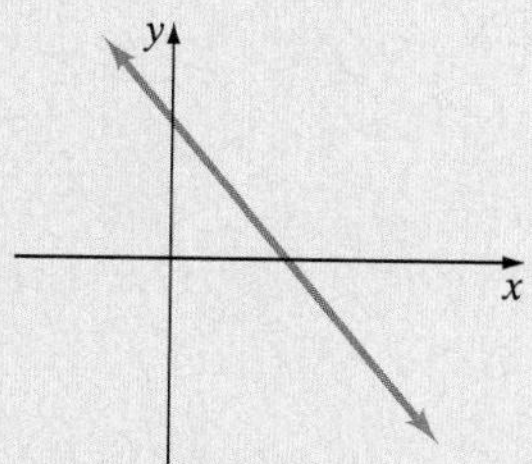

(c) Coincident lines; system has infinitely many solutions

Rather than graphing the lines in a system to classify the system, we can also classify a system algebraically.

Classifying a System of Equations Algebraically

Step 1: Write each equation in the system in slope-intercept form.

Step 2: **(a)** If the equations of the lines in the system have different slopes, then the lines will intersect. The point of intersection represents the solution. We say that the system is consistent and the equations are independent. See Examples 3 and 4.

(b) If the equations of the lines have the same slope, but different y-intercepts, then the lines are parallel. We say that the system is inconsistent. See Example 5.

(c) If the equations of the lines have the same slope and the same y-intercept, then the lines are coincident. We say that the system is consistent and the equations are dependent. See Example 6.

Classroom Example
Without graphing, determine the number of solutions of each system. State whether the system is consistent or inconsistent. If the system is consistent, state whether the equations are dependent or independent.

(a) $\begin{cases} -x + 2y = 4 \\ 2x - 4y = -8 \end{cases}$

(b) $\begin{cases} 3x + 4y = 4 \\ 2x - y = 2 \end{cases}$

Answer:

(a) Infinitely many solutions; consistent, dependent

(b) One solution; consistent; independent

EXAMPLE 7 Determining the Number of Solutions a System Has

Without graphing, determine the number of solutions of the system:

$$\begin{cases} 12x + 4y = -16 \\ -9x - 3y = 3 \end{cases}$$

State whether the system is consistent or inconsistent. If the system is consistent, state whether the equations are dependent or independent.

Solution

We write the equations in slope-intercept form. We then determine the slope and y-intercept to determine the number of solutions.

$$\begin{aligned} \text{First equation:} &\quad 12x + 4y = -16 \\ \text{Subtract } 12x \text{ from both sides:} &\quad 4y = -12x - 16 \\ \text{Divide both sides by 4:} &\quad y = -3x - 4 \end{aligned}$$

$$\begin{aligned} \text{Second equation:} &\quad -9x - 3y = 3 \\ \text{Add } 9x \text{ to both sides:} &\quad -3y = 9x + 3 \\ \text{Divide both sides by } -3\text{:} &\quad y = -3x - 1 \end{aligned}$$

The slope of both equations is -3. The y-intercept of the first equation is -4, while the y-intercept of the second equation is -1. The equations have the same slope, but different y-intercepts, so the lines are parallel. Therefore, the system has no solution and is inconsistent.

QUICK ✓ *Without graphing, determine the number of solutions of each system. State whether the system is consistent or inconsistent. For those systems that are consistent, state whether the equations are dependent or independent.*

10. $\begin{cases} 7x - 2y = 4 \\ 2x + 7y = 7 \end{cases}$ **11.** $\begin{cases} 6x + 4y = 4 \\ -12x - 8y = -8 \end{cases}$ **12.** $\begin{cases} 3x - 4y = 8 \\ -6x + 8y = 8 \end{cases}$

4 Solve Applied Problems Involving Systems of Linear Equations

We have seen in Chapter 8 that numerous situations, such as the cost of owning a car, can be modeled by linear equations. Many times we can use a system of two linear equations with two variables to answer interesting questions such as, "Which long-distance telephone plan should I choose?"

Classroom Example ➔
Bob's Truck Rental has two rental programs for its 20-foot truck. Program A charges a daily fee of \$10 plus \$0.25 per mile driven. Program B charges a daily fee of \$20 plus \$0.15 per mile driven. For how many miles is the cost the same for the two programs?

Answer: 100 miles

EXAMPLE 8 SBC Long Distance

SBC, a company that offers phone service in many states, has two domestic long-distance phone plans of interest. The "Just Call" plan charges \$3.00 per month plus \$0.03 per minute. The "Value Plus Flat Rate" plan has no monthly service fee and charges \$0.05 per minute. Determine the number of minutes of long-distance phone calls that can be used each month in order for the two plans to charge the same fee.

Solution

Step 1: Identify We want to know the number of minutes of long-distance phone calls that can be used such that the two plans have the same fee.

Step 2: Name Let m represent the number of minutes used and let C represent the monthly cost.

Step 3: Translate For the "Just Call" plan, talking for 0 minutes results in a cost of \$3.00; talking for 1 minute results in a cost of \$3.00 plus \$0.03; talking for two minutes results in a cost of \$3.00 plus 2(\$0.03), or \$3.06. To generalize, the monthly cost for the "Just Call" plan is \$3.00 plus \$0.03 times the number of minutes used. This leads to the following equation:

$$C = 0.03m + 3.00 \quad \text{Equation (1)}$$

For the "Value Plus Flat Rate" plan, talking for 0 minutes results in a cost of \$0; talking for 1 minute results in a cost of \$0.05; talking for two minutes results in a cost of \$0.10. To generalize, the monthly cost for the "Value Plus Flat Rate" plan is \$0.05 times the number of minutes used. This leads to the following equation:

$$C = 0.05m \quad \text{Equation (2)}$$

We use equations (1) and (2) to form the following system:

$$\begin{cases} C = 0.03m + 3.00 \\ C = 0.05m \end{cases} \quad \text{The Model}$$

Step 4: Solve To obtain the graph of each equation in the system, we create Table 1, which shows the monthly fee for various minutes used for each plan. Figure 6 shows the graph of the two equations in the system with the horizontal axis labeled m and the vertical axis labeled C.

Table 1

m	Cost for "Just Call" Plan	Cost for "Value Plus Flat Rate" Plan
0	$3.00	$ 0
100	$6.00	$ 5.00
200	$9.00	$10.00

Figure 6

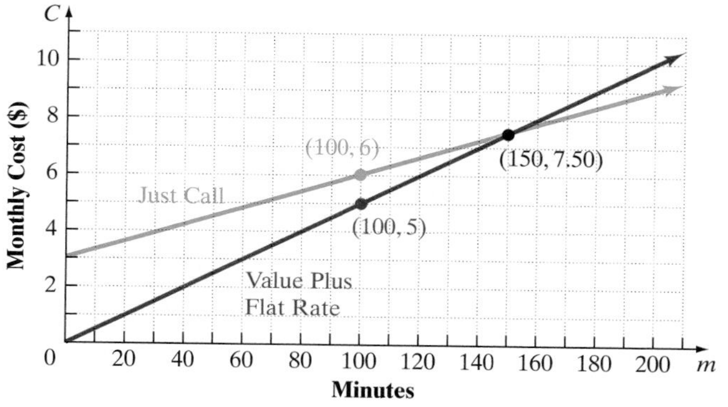

Work Smart

Use the graphical method to solve a system of equations when it is helpful to visualize the solutions.

From the graph, we see that the monthly charge will be the same if 150 minutes are used in a month.

Step 5: Check For the "Just Call" plan, the bill is $3.00 plus $0.03 times the number of minutes used. If 150 minutes are used, the bill is $3.00 plus $0.03 times 150 or $3.00 plus $4.50 for a total bill of $7.50. For the "Value Plus Flat Rate Plan," the bill is $0.05 times the number of minutes used. If 150 minutes are used, the bill is $0.05 times 150 or $7.50. Both bills are the same. Our answer checks.

Step 6: Answer The two plans result in the same monthly bill when 150 minutes are used. If you use the phone for fewer than 150 minutes, we can see that the "Value Plus Flat Rate" plan is the better deal; otherwise the "Just Call" plan is the better deal. ■

QUICK ✓

13. John needs to rent a 12-foot moving truck for a day. He calls two rental companies to determine their fee structure. EZ-Rental charges $20 plus $0.40 per mile, while U-Move-It charges $35 plus $0.25 per mile. After how many miles will the cost of renting the truck be the same? What is the cost at this mileage?

9.1 Exercises

For Extra Help: Student Solutions Manual · CD Video · PH Math/Tutor Center · MathXL Tutorials on CD · MathXL® · MyMathLab

Concepts and Vocabulary

In Problems 1–3, fill in the blanks.

1. intersect in one point
2. consistent, independent, dependent
3. inconsistent

1. In order for a system of linear equations to have exactly one solution, the graphs of the equations must __________.

2. If a system of linear equations has at least one solution, the system is called __________. In addition, we call this system __________ if it has exactly one solution and __________ if it has more than one solution.

3. A system with no solutions is called __________.

4. True
5. False
6. False
7. Answers may vary.
8. Answers may vary.
9. (a) No (b) Yes (c) No
10. (a) Yes (b) No (c) No
11. (a) No (b) Yes (c) Yes
12. (a) No (b) No (c) No
13. (a) No (b) No (c) No
14. (a) No (b) No (c) No

15. 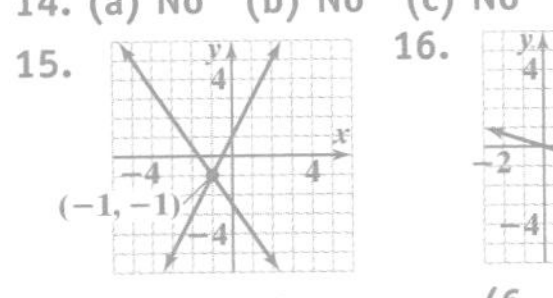

(−1, −1)

16. 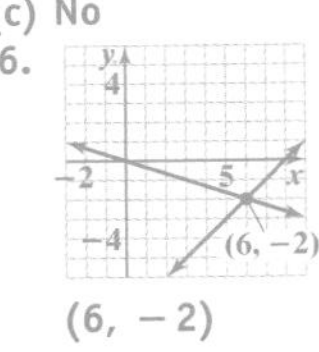

(6, −2)

17. 18. 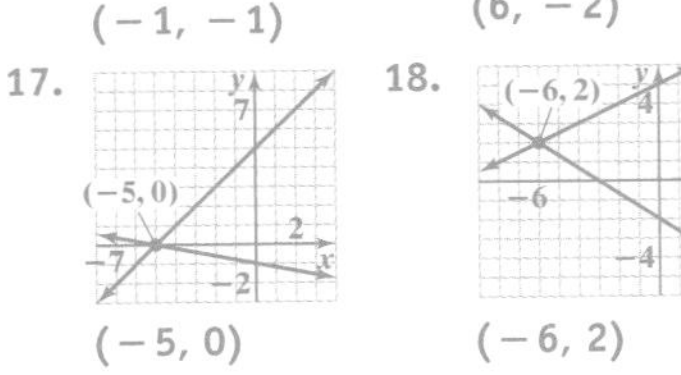

(−5, 0) (−6, 2)

19. 20. 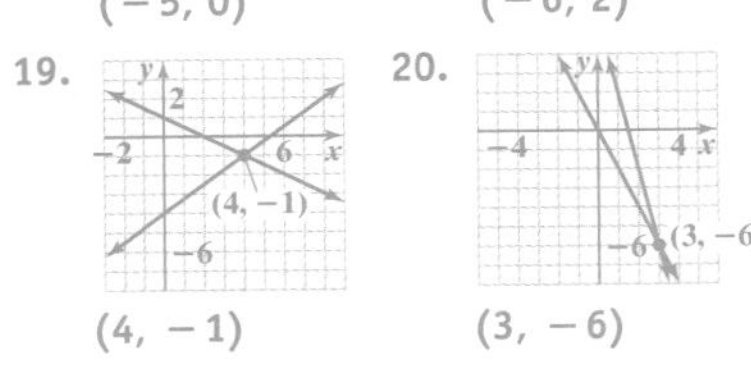

(4, −1) (3, −6)

21. 22. 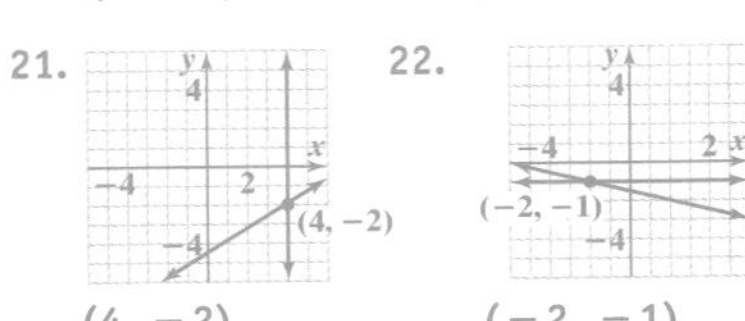

(4, −2) (−2, −1)

23. 24. 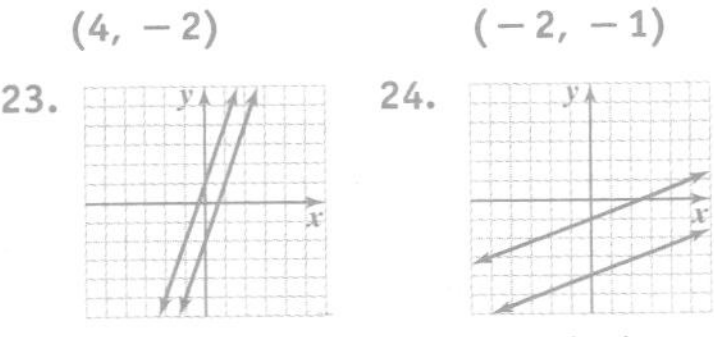
no solution no solution

25. 26. 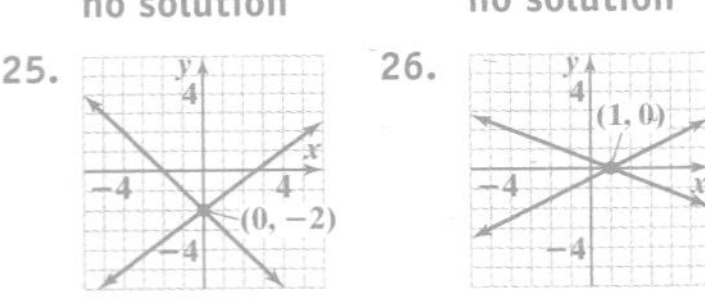

(0, −2) (1, 0)

27. 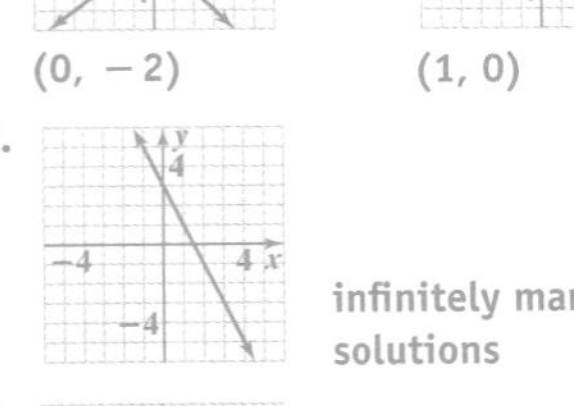
infinitely many solutions

28. 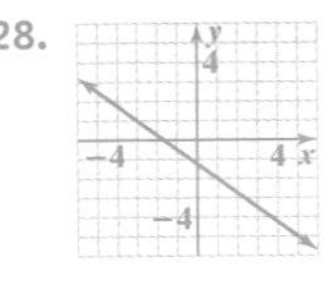
infinitely many solutions

29. 30. 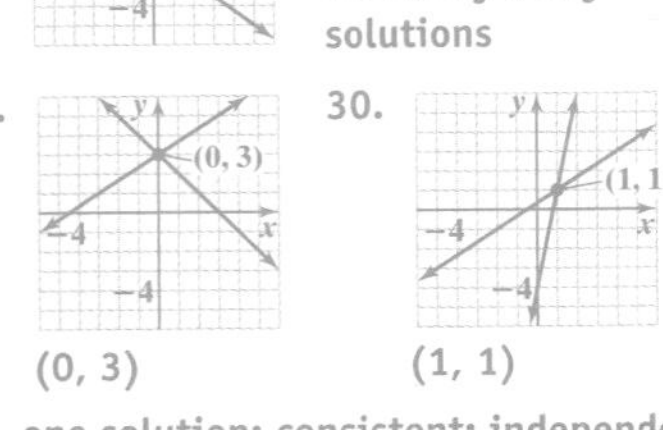

(0, 3) (1, 1)

In Problems 4–6, answer True or False to each statement.

4. If a solution of a system of linear equations has more than one solution, it must have infinitely many solutions.

5. A solution of a system of two linear equations is an ordered pair that satisfies one of the equations in the system.

6. It is always possible to tell the exact solution of a system of linear equations by the graphing method.

7. The graphing method for solving a system of equations depends on your ability to graph a line. List all of the methods to graph a line discussed in Chapter 8 and then briefly describe the benefits and drawbacks to each method in the context of solving a system of linear equations. Can you think of any drawbacks to solving a system using the graphing method?

8. When solving a system of linear equations using the graphing method, you find that the solution appears to be (3, 4). How do you verify that (3, 4) is the solution? Is it possible that there are exactly two points that satisfy the given system? Why or why not? Make up a system of equations for which (3, 4) is a solution.

Building Skills

In Problems 9–14, determine whether the ordered pair is a solution of the system of equations.

9. $\begin{cases} x - y = -4 \\ 3x + y = -4 \end{cases}$
(a) $(2, 6)$ **(b)** $(-2, 2)$ **(c)** $(2, -2)$

10. $\begin{cases} 2x + 5y = 0 \\ x - 3y = 11 \end{cases}$
(a) $(5, -2)$ **(b)** $(-5, 2)$ **(c)** $(2, -3)$

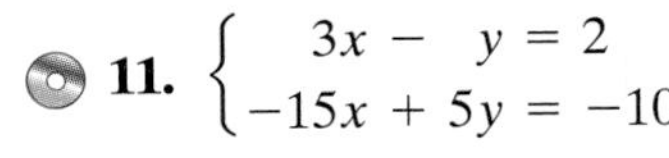
11. $\begin{cases} 3x - y = 2 \\ -15x + 5y = -10 \end{cases}$
(a) $(1, -1)$ **(b)** $(-2, -8)$ **(c)** $(0, -2)$

12. $\begin{cases} x - 4y = 2 \\ 5x - 8y = -6 \end{cases}$
(a) $(-8, -2)$ **(b)** $(2, 2)$ **(c)** $(-6, -3)$

13. $\begin{cases} 6x - 2y = 1 \\ y = 3x + 2 \end{cases}$
(a) $\left(0, -\dfrac{1}{2}\right)$ **(b)** $(-2, -4)$ **(c)** $(0, 2)$

14. $\begin{cases} y = -4x - 1 \\ 2y + 8x = 2 \end{cases}$
(a) $\left(-\dfrac{1}{2}, 0\right)$ **(b)** $(0, 1)$ **(c)** $(-1, 0)$

In Problems 15–30, solve each system of equations by graphing.

15. $\begin{cases} 2x - y = -1 \\ 3x + 2y = -5 \end{cases}$

16. $\begin{cases} x + 3y = 0 \\ x - y = 8 \end{cases}$

17. $\begin{cases} y = x + 5 \\ y = -\dfrac{1}{5}x - 1 \end{cases}$

18. $\begin{cases} y = -\dfrac{2}{3}x - 2 \\ y = \dfrac{1}{2}x + 5 \end{cases}$

19. $\begin{cases} y = \dfrac{3}{4}x - 4 \\ y = -\dfrac{1}{2}x + 1 \end{cases}$

20. $\begin{cases} y = -4x + 6 \\ y = -2x \end{cases}$

21. $\begin{cases} x - 4 = 0 \\ 3x - 5y = 22 \end{cases}$

22. $\begin{cases} y + 1 = 0 \\ x + 4y = -6 \end{cases}$

23. $\begin{cases} 3x - y = -1 \\ -6x + 2y = -4 \end{cases}$

24. $\begin{cases} -2x + 5y = -20 \\ 4x - 10y = 10 \end{cases}$

25. $\begin{cases} x + y = -2 \\ 3x - 4y = 8 \end{cases}$

26. $\begin{cases} 2x + 5y = 2 \\ -x + 2y = -1 \end{cases}$

27. $\begin{cases} 2y = 6 - 4x \\ 6x = 9 - 3y \end{cases}$

28. $\begin{cases} 3y = -4x - 4 \\ 8x + 2 = -6y - 6 \end{cases}$

29. $\begin{cases} y = -x + 3 \\ 3y = 2x + 9 \end{cases}$

30. $\begin{cases} 3y = 2x + 1 \\ y = 5x - 4 \end{cases}$

31. one solution; consistent; independent; (3, 2) 32. no solution; inconsistent 33. no solution; inconsistent 34. one solution; consistent; independent; (−2, 1) 35. infinitely many solutions; consistent; dependent 36. one solution; consistent; independent; (1, −3) 37. one solution; consistent; independent; (−1, −2) 38. infinitely many solutions; consistent; dependent 39. one solution; consistent; independent

40. one solution; consistent; independent
41. no solution; inconsistent
42. no solution; inconsistent
43. infinitely many solutions; consistent; dependent
44. infinitely many solutions; consistent; dependent
45. one solution; consistent; independent
46. one solution; consistent; independent
47. no solution; inconsistent
48. infinitely many solutions; consistent; dependent
49. infinitely many solutions; consistent; dependent
50. no solution; inconsistent
51. consistent; independent; $(-3, 3)$
52. consistent; independent; $(4, 7)$
53. inconsistent; no solution
54. inconsistent; no solution
55. consistent; independent; $(2, 4)$
56. consistent; independent; $(-2, 5)$
57.

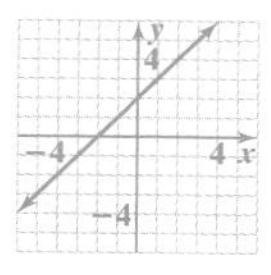

consistent; dependent; infinitely many solutions

58.

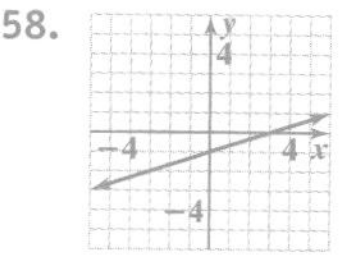

consistent; dependent; infinitely many solutions

59. consistent; dependent; infinitely many solutions
60. consistent; dependent; infinitely many solutions

In Problems 31–38, determine the number of solutions of each system. State whether the system is consistent or inconsistent. For those systems that are consistent, state whether the equations are dependent or independent. State the solution of each system.

31.

32.

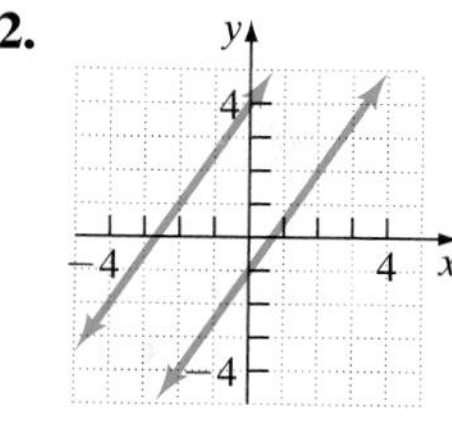

33.

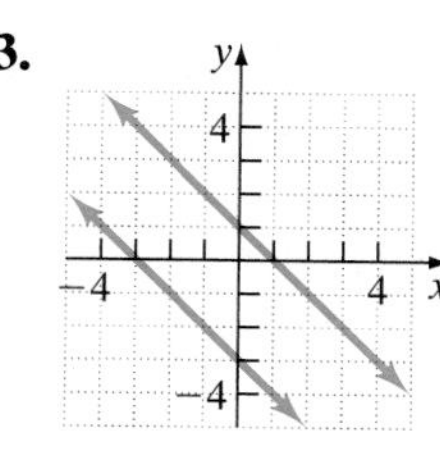

34.

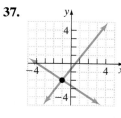

35.

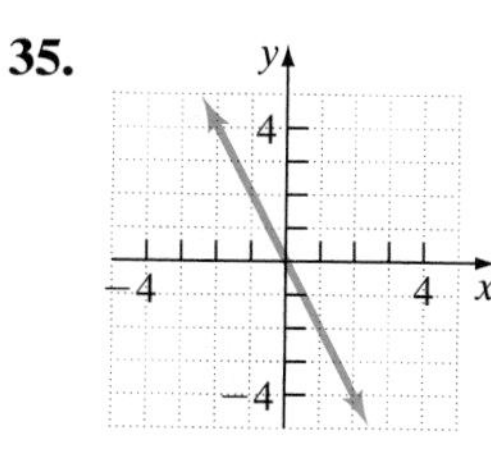

36.

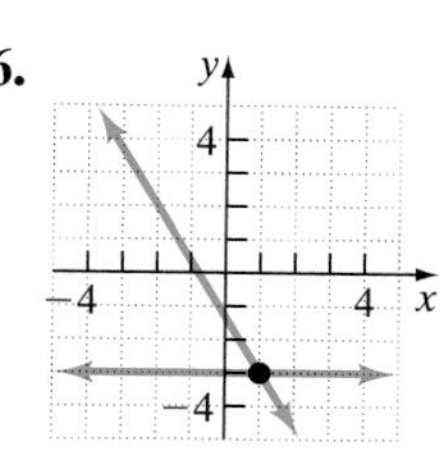

37.

38.

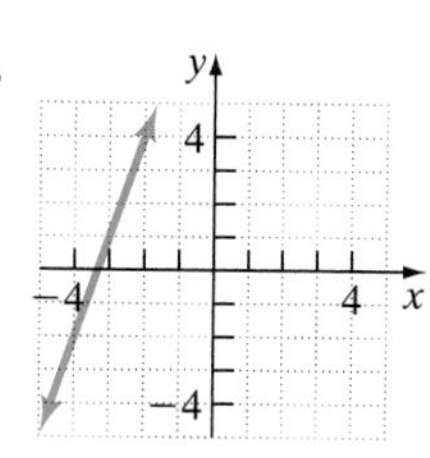

In Problems 39–50, without graphing, determine the number of solutions of each system of equations. State whether the system is consistent or inconsistent. For those systems that are consistent, state whether the equations are dependent or independent.

39. $\begin{cases} y = 2x + 3 \\ y = -2x + 3 \end{cases}$

40. $\begin{cases} y = -x + 5 \\ y = x - 5 \end{cases}$

41. $\begin{cases} 6x + 2y = 12 \\ 3x + y = 12 \end{cases}$

42. $\begin{cases} 2x + y = -3 \\ -2x - y = 6 \end{cases}$

43. $\begin{cases} x + 2y = 2 \\ 2x + 4y = 4 \end{cases}$

44. $\begin{cases} -x + y = 5 \\ x - y = -5 \end{cases}$

45. $\begin{cases} y = 4 \\ x = 4 \end{cases}$

46. $\begin{cases} y - 3 = 0 \\ x - 3 = 0 \end{cases}$

47. $\begin{cases} x - 2y = -4 \\ -x + 2y = -4 \end{cases}$

48. $\begin{cases} 5x + y = -1 \\ 10x + 2y = -2 \end{cases}$

49. $\begin{cases} x + 2y = 4 \\ x + 1 = 5 - 2y \end{cases}$

50. $\begin{cases} 2y + 6 = 2x \\ x + 4 = y - 7 \end{cases}$

Mixed Practice

In Problems 51–62, solve each system of equations by graphing. Based on the graph, state whether the system is consistent or inconsistent. For those systems that are consistent, state whether the equations are dependent or independent.

51. $\begin{cases} y = 2x + 9 \\ y = -3x - 6 \end{cases}$

52. $\begin{cases} y = 3x - 5 \\ y = -x + 11 \end{cases}$

53. $\begin{cases} x - 2y = 6 \\ 2x - 4y = 0 \end{cases}$

54. $\begin{cases} 2x + 3y = -3 \\ 4x + 6y = 6 \end{cases}$

55. $\begin{cases} 3x - 2y = -2 \\ 2x + y = 8 \end{cases}$

56. $\begin{cases} 3x - y = -11 \\ -2x + y = 9 \end{cases}$

57. $\begin{cases} y = x + 2 \\ 3x - 3y = -6 \end{cases}$

58. $\begin{cases} -2x + 6y = -6 \\ 3x - 9y = 9 \end{cases}$

59. $\begin{cases} -5x - 2y = -2 \\ 10x + 4y = 4 \end{cases}$

60. $\begin{cases} 3x + y = -2 \\ 6x + 2y = -4 \end{cases}$

61. $\begin{cases} 3x = 4y - 12 \\ 2x = -4y - 8 \end{cases}$

62. $\begin{cases} 2x = -5y - 25 \\ -3x = 10y + 50 \end{cases}$

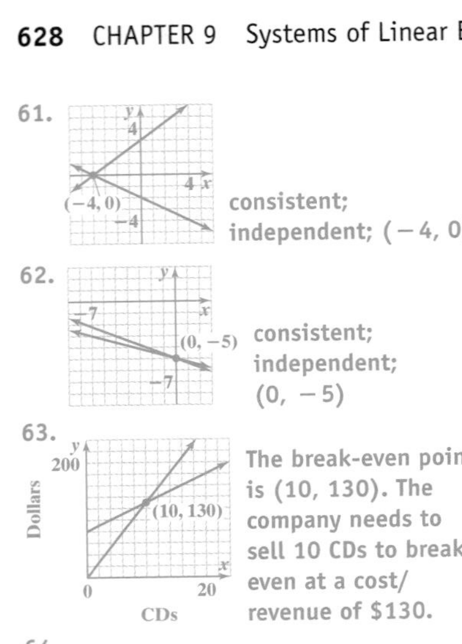

61. consistent; independent; $(-4, 0)$

62. consistent; independent; $(0, -5)$

63. The break-even point is (10, 130). The company needs to sell 10 CDs to break even at a cost/revenue of \$130.

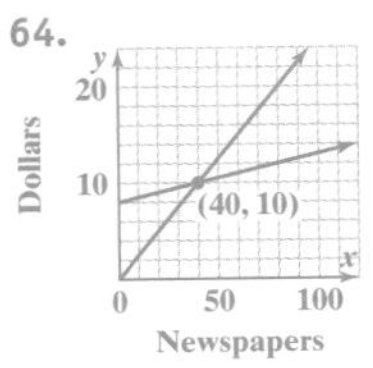

64. The break-even point is (40, 10). The company needs to sell 40 newspapers to break even at a cost/revenue of \$10.

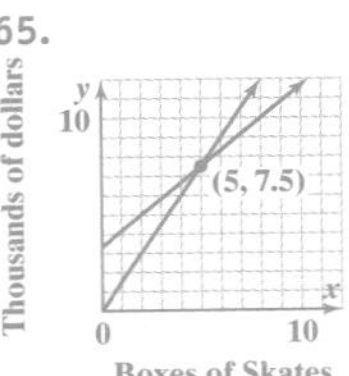

65. The break-even point is (5, 7.5). The company needs to sell 5 boxes of skates to break even at a cost/revenue of \$7500.

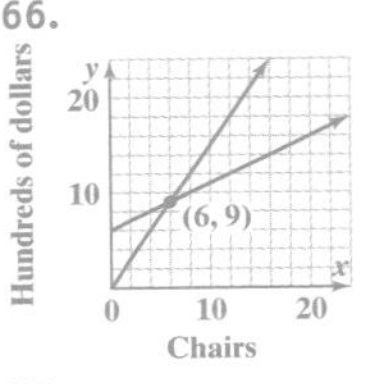

66. The break-even point is (6, 9). The company needs to sell 6 chairs in order to break even at a cost/revenue of \$900.

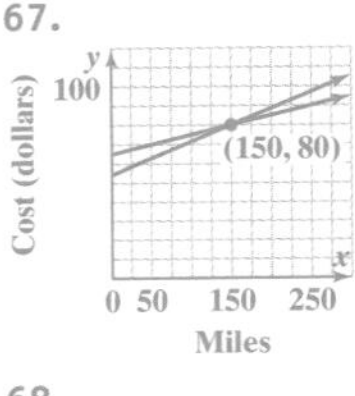

67. The cost for driving 150 miles is the same for both companies (\$80). Choose Acme to drive 200 miles.

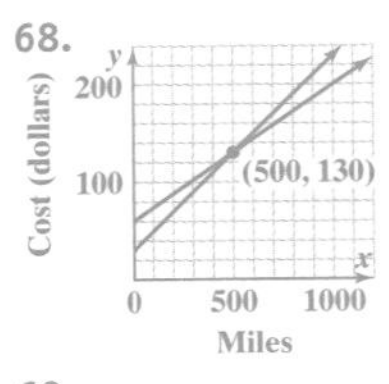

68. The cost for driving 500 miles is the same for both companies (\$130). Choose Slow-but-Cheap to drive 400 miles.

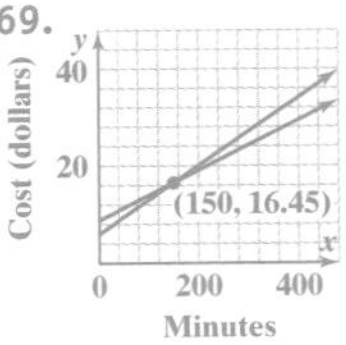

69. The cost is the same (\$16.45) when 150 minutes are used. Choose Plan B for 100 long-distance minutes.

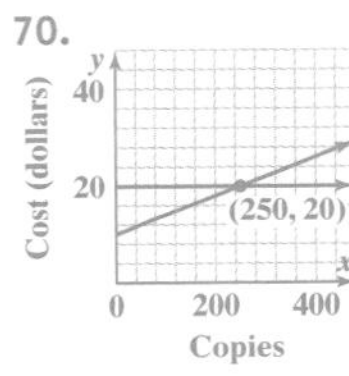

70. The cost is the same (\$20) when 250 copies are made.

Applying the Concepts

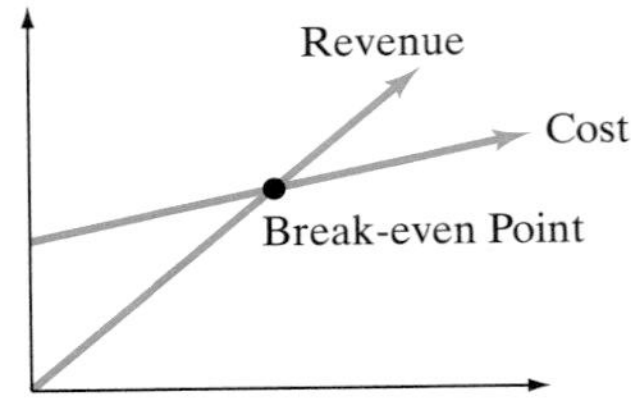

*In business, a company's **break-even point** is the number of units of a product that is produced so that the costs of production will be equal to the amount of revenue from the sale of the product. In Problems 63–66, solve the system of equations by graphing to find the number of units that need to be produced for the company to break even.*

63. Producing CDs California Dreamin' produces CDs of local talent and sells the CDs to neighborhood music stores. The cost y to produce x CDs can be determined by the equation $y = 5x + 80$ and the revenue y from the sale of these CDs is given by the equation $y = 13x$. Find the break-even point: the point where the line that represents the cost, $y = 5x + 80$, intersects the revenue line, $y = 13x$. Interpret the x-coordinate and the y-coordinate of the break-even point.

64. School Newspaper School Printers, Inc. prints and sells the school newspaper. The cost y to produce x newspapers is determined by the equation $y = 0.05x + 8$ and the revenue y from the sale of these newspapers is given by the equation $y = 0.25x$. Find the break-even point: the point where the line that represents the cost, $y = 0.05x + 8$, intersects the revenue line, $y = 0.25x$. Interpret the x-coordinate and the y-coordinate of the break-even point.

65. Hockey Skates The Ice Blades Manufacturing Company produces and sells hockey skates to discount stores. The skates come in boxes of 10 pairs of skates. The cost y (in thousands of dollars) to produce x boxes of skates can be determined by the equation $y = 0.8x + 3.5$ and the revenue y (in thousands of dollars) from the sale of these skates is given by the equation $y = 1.5x$. Find the break-even point.

66. Patio Chairs Maumee Lumber Company produces wood patio chairs that they sell to local hardware and home improvement stores. The cost y (in hundreds of dollars) to produce x patio chairs can be determined by the equation $y = 0.5x + 6$. The revenue y (in hundreds of dollars) from the sale of x patio chairs is given by the equation $y = 1.5x$. Find the break-even point.

In Problems 67–70, set up a system of linear equations in two variables that models the problem. Then solve the system of linear equations.

67. Car Rental The Wheels-to-Go car rental agency rents cars for \$50 daily plus \$0.20 per mile. Acme Rental agency will rent the same car for \$62 daily plus \$0.12 per mile. On Liza's trip to Houston, she decides to rent a car but is not sure how far she will need to drive. Determine the number of miles for which the cost of the car rental will the same for both companies. If Liza plans to drive the car for 200 miles, which company should she use?

68. Car Rental The Speedy Car Rental agency charges \$60 per day plus \$0.14 per mile while the Slow-but-Cheap Car Rental agency charges \$30 per day plus \$0.20 per mile. Determine the number of miles for which the cost of the car rental will the same for both companies. If you plan to drive the car for 400 miles, which company should you use?

69. Phone Charges Sprint has two different long-distance phone plans. Plan A charges a monthly fee of \$8.95 plus \$0.05 per minute. Plan B charges a monthly fee of \$5.95 plus \$0.07 per minute. Determine the number of minutes for which the cost of each plan will be the same. If you typically use 100 long-distance minutes each month, which plan should you choose? (*Source:* Sprint.com)

70. Political Flyers A politician running for city council has found that PrintQuick can print pamphlets for her campaign for \$0.04 per copy plus a one-time set-up fee of \$10. She has also learned that Print-A-Lot will print the same pamphlet for a flat fee of \$20. Determine the number of political pamphlets that can be printed for the cost at PrintQuick to be the same as at Print-A-Lot.

71. $c = 3$ 72. $c = \frac{1}{2}$

73. $c = 1$ 74. $c = 6$

75.

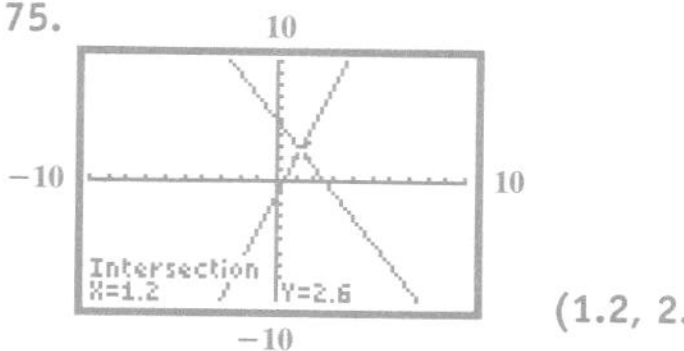

(1.2, 2.6)

76.

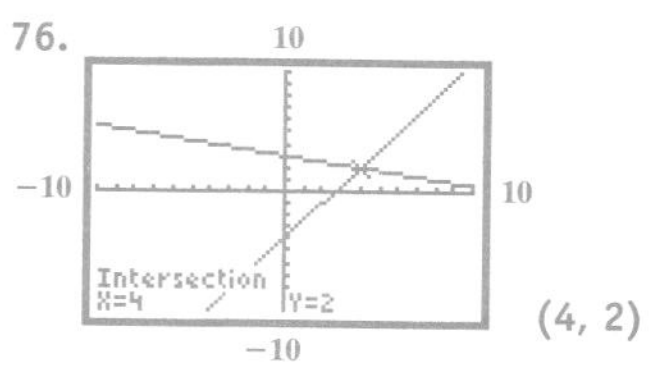

(4, 2)

77.

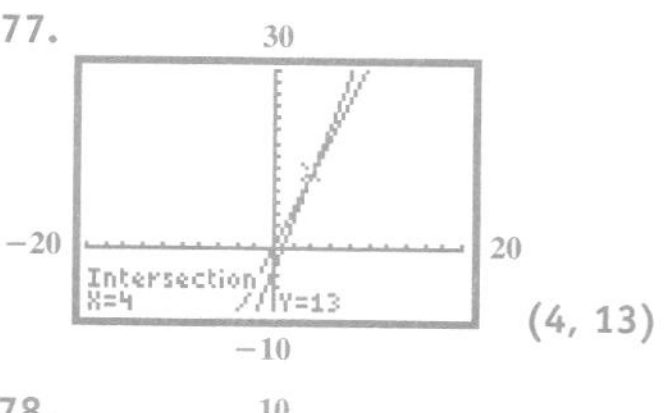

(4, 13)

78.

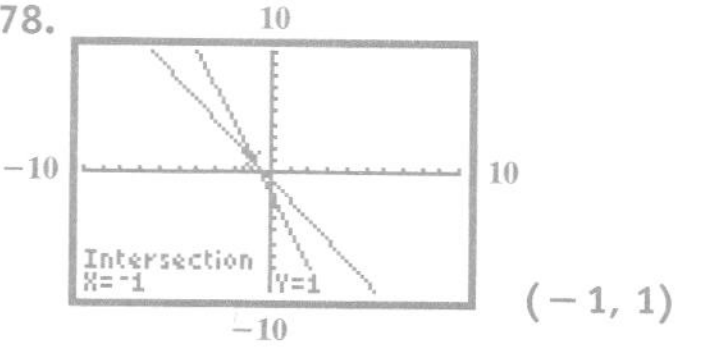

(−1, 1)

79.

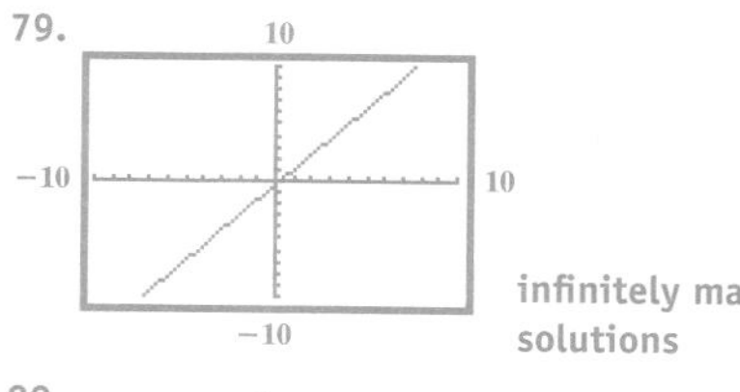

infinitely many solutions

80.

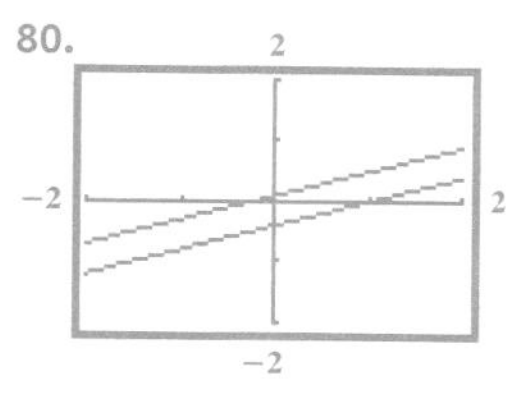

no solution

Extending the Concepts

In Problems 71 and 72, determine the value of the coefficient, c, so that the given system of equations is dependent.

71. $\begin{cases} 3x - y = -4 \\ y = cx + 4 \end{cases}$

72. $\begin{cases} 2(x + 4) = 4y + 4 \\ y = cx + 1 \end{cases}$

In Problems 73 and 74, determine the value of c so that the given system of equations is dependent.

73. $\begin{cases} x + 3 = 3(x - y) \\ 2y + 3 = 2cx - y \end{cases}$

74. $\begin{cases} 5x + 2y = 3 \\ 4y = -10x + c \end{cases}$

The Graphing Calculator

A graphing calculator can be used to approximate the point of intersection between two equations using its INTERSECT command. We illustrate this feature of the graphing calculator by solving the system $\begin{cases} x + y = -1 \\ -2x + y = -7 \end{cases}$. Start by graphing each equation in the system as shown in Figure 7(a). Then use the INTERSECT command and find that the lines intersect at $x = 2$, $y = -3$. See Figure 7(b) The solution is the ordered pair $(2, -3)$.

Figure 7

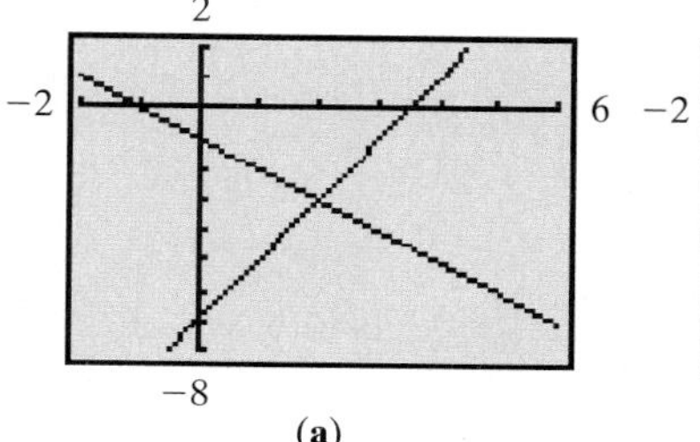

(a)

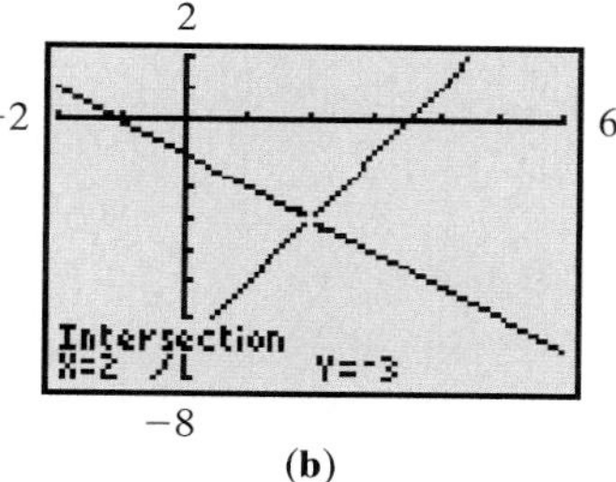

(b)

In Problems 75–80, use a graphing calculator to solve each system of equations. If necessary, express your solution rounded to two decimal places.

75. $\begin{cases} y = 3x - 1 \\ y = -2x + 5 \end{cases}$

76. $\begin{cases} y = \frac{3}{2}x - 4 \\ y = -\frac{1}{4}x + 3 \end{cases}$

77. $\begin{cases} 3x - y = -1 \\ -4x + y = -3 \end{cases}$

78. $\begin{cases} -6x - 2y = 4 \\ 5x + 3y = -2 \end{cases}$

79. $\begin{cases} 4x - 3y = 1 \\ -8x + 6y = -2 \end{cases}$

80. $\begin{cases} -2x + 5y = -2 \\ 4x - 10y = 1 \end{cases}$

9.2 Solving Systems of Linear Equations Using Substitution

OBJECTIVES

1. Solve a System of Linear Equations Using the Substitution Method
2. Solve Applied Problems Involving Systems of Linear Equations

Preparing for Solving Systems of Linear Equations Using Substitution

Before getting started, take the following readiness quiz. If you get a problem wrong, go back to the section cited and review the material.

1. Solve $3x - y = 2$ for y. [Section 2.4, pp. 108–111]

2. Solve $2x + 5y = 8$ for x. [Section 2.4, pp. 108–111]

3. Solve: $3x - 2(5x + 1) = 12$ [Section 2.2, pp. 88–89]

Preparing for...Answers

1. $y = 3x - 2$ **2.** $x = -\frac{5}{2}y + 4$ **3.** $\{-2\}$

1 Solve a System of Linear Equations Using the Substitution Method

If the x- and y-coordinates of the point of intersection of two lines are not integers, then obtaining an exact result using the graphing method can be difficult. For example,

Figure 8

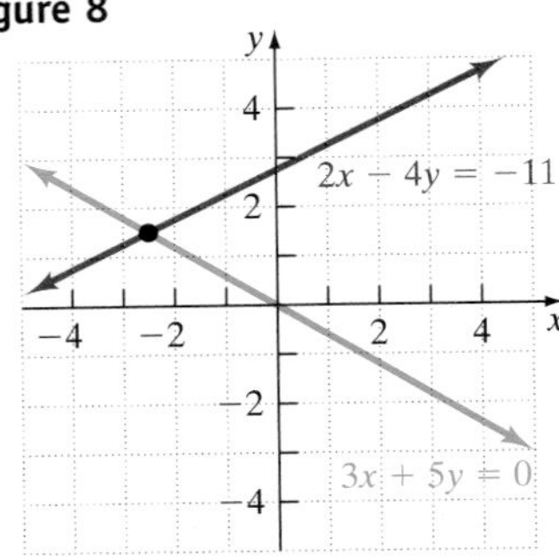

consider the following system of equations:

$$\begin{cases} 3x + 5y = 0 \\ 2x - 4y = -11 \end{cases}$$

Figure 8 shows the graph of the two equations in the system. The lines intersect at $\left(-\frac{5}{2}, \frac{3}{2}\right)$. Because the lines do not intersect at integer values, it is difficult to determine the solution of the system graphically. Therefore, rather than using graphical methods to obtain solutions of some systems of two linear equations, we prefer to use algebraic methods. The first algebraic method that we present is the *method of substitution*. Let's see how to solve a system of equations using this technique.

EXAMPLE 1 How to Solve a System of Two Equations in Two Variables by Substitution

Solve the following system by substitution: $\begin{cases} y = -2x + 10 \\ 3x + 5y = 8 \end{cases}$

Step-by-Step Solution

First, we will call the equation $y = -2x + 10$ equation (1) and $3x + 5y = 8$ equation (2), so that the system is displayed as follows:

$$\begin{cases} y = -2x + 10 & (1) \\ 3x + 5y = 8 & (2) \end{cases}$$

Step 1: Solve one of the equations for one of the unknowns.

Equation (1) is already solved for y: $y = -2x + 10$

Step 2: Substitute $y = -2x + 10$ into equation (2).

$3x + 5y = 8$

Equation (2): $3x + 5(-2x + 10) = 8$

Step 3: Solve the equation for x.

Distribute the 5: $3x - 10x + 50 = 8$

Combine like terms: $-7x + 50 = 8$

Subtract 50 from both sides: $-7x = -42$

Divide both sides by -7: $x = 6$

Step 4: Let $x = 6$ in equation (1) to find the value of y.

Equation (1): $y = -2x + 10$

$x = 6$: $y = -2(6) + 10$

$y = -12 + 10$

$y = -2$

Step 5: Verify that $x = 6$ and $y = -2$ is the solution.

Equation (1):
$$\begin{aligned} y &= -2x + 10 \\ -2 &\stackrel{?}{=} -2(6) + 10 \\ -2 &\stackrel{?}{=} -12 + 10 \\ -2 &= -2 \quad \text{True} \end{aligned}$$

Equation (2):
$$\begin{aligned} 3x + 5y &= 8 \\ 3(6) + 5(-2) &\stackrel{?}{=} 8 \\ 18 - 10 &\stackrel{?}{=} 8 \\ 8 &= 8 \quad \text{True} \end{aligned}$$

Classroom Example
Solve:

$$\begin{cases} y = 2x + 7 \\ x - 3y = 4 \end{cases}$$

Answer: $(-5, -3)$

Both equations are satisfied so the solution of the system $\begin{cases} y = -2x + 10 \\ 3x + 5y = 8 \end{cases}$ is the ordered pair $(6, -2)$.

Figure 9 shows the graph of each equation in the system from Example 1. The point of intersection is $(6, -2)$, as we would expect.

Notice that we named each equation in the system in Example 1 before going through the steps for solving the system. For convenience, we shall name the equations in the problem for the remainder of the text.

Figure 9

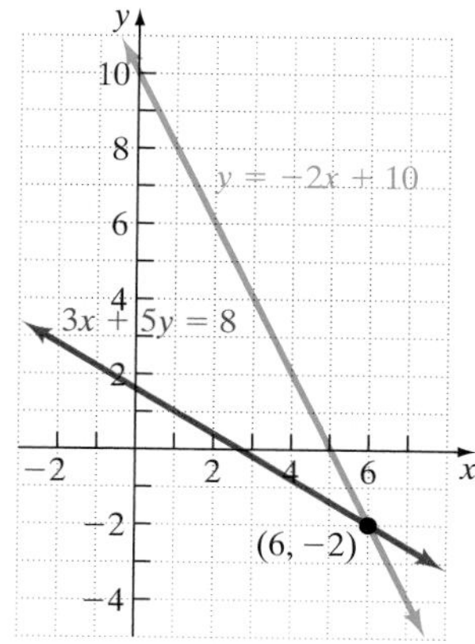

Below we summarize the steps for solving a system using substitution.

> **Steps for Solving a System of Two Linear Equations Containing Two Unknowns by Substitution**
>
> **Step 1:** Solve one of the equations for one of the unknowns.
>
> **Step 2:** Substitute the expression solved for in Step 1 into the *other* equation. The result will be a single linear equation in one unknown.
>
> **Step 3:** Solve the linear equation in one unknown found in Step 2.
>
> **Step 4:** Substitute the value of the variable found in Step 3 into one of the *original* equations to find the value of the other variable.
>
> **Step 5:** Check your answer.

Classroom Example Solve:

$$\begin{cases} x + 5y = -2 \\ 3x + 8y = 8 \end{cases}$$

Answer: $(8, -2)$

EXAMPLE 2 Solving a System of Equations by Substitution—First Solving for One Variable

Solve the following system by substitution: $\begin{cases} 2x - y = -15 & (1) \\ 4x + 3y = 5 & (2) \end{cases}$

Solution

When using substitution, solve for the variable whose coefficient is 1 or -1, if possible, to simplify the algebra. We will solve equation (1) for y.

Equation (1):	$2x - y = -15$
Subtract $2x$ from both sides:	$-y = -2x - 15$
Multiply both sides by -1:	$y = 2x + 15$

Work Smart
If you solve for a variable in equation (1), be sure to substitute into equation (2). If you solve for a variable in equation (2), be sure to substitute into equation (1).

We substitute $2x + 15$ for y in equation (2).

Equation (2):	$4x + 3y = 5$
Let $y = 2x + 15$:	$4x + 3(2x + 15) = 5$

Now we solve for x.

Distribute to remove parentheses:	$4x + 6x + 45 = 5$
Combine like terms:	$10x + 45 = 5$
Subtract 45 from both sides:	$10x = -40$
Divide both sides by 10:	$x = -4$

Let $x = -4$ in equation (1) and solve for y.

Equation (1):	$2x - y = -15$
Let $x = -4$:	$2(-4) - y = -15$
	$-8 - y = -15$
Add 8 to both sides:	$-y = -7$
Multiply both sides by -1:	$y = 7$

Check

Equation (1):	$2x - y = -15$	Equation (2):	$4x + 3y = 5$
Let $x = -4, y = 7$:	$2(-4) - 7 \stackrel{?}{=} -15$	Let $x = -4, y = 7$:	$4(-4) + 3(7) \stackrel{?}{=} 5$
	$-8 - 7 \stackrel{?}{=} -15$		$-16 + 21 \stackrel{?}{=} 5$
	$-15 = -15$ True		$5 = 5$ True

The solution of the system $\begin{cases} 2x - y = -15 \\ 4x + 3y = 5 \end{cases}$ is $(-4, 7)$.

QUICK ✓ *Solve the system using substitution.*

1. $\begin{cases} y = 3x - 2 \\ 2x - 3y = -8 \end{cases}$

2. $\begin{cases} 2x + y = -1 \\ 4x + 3y = 3 \end{cases}$

Classroom Example ➤
Solve:

$\begin{cases} 2x + 4y = 11 \\ 3x + y = -6 \end{cases}$

Answer: $\left(-\frac{7}{2}, \frac{9}{2}\right)$

EXAMPLE 3 Solving a System of Equations by Substitution—First Solving for One Variable

Solve the following system by substitution: $\begin{cases} 2x + 3y = 9 & (1) \\ x + 2y = \dfrac{13}{2} & (2) \end{cases}$

Solution

In this problem, we will solve equation (2) for x. Why did we choose this approach? Because the coefficient of x in equation (2) is one. That will make solving for x easier. If we try to solve for x or y in equation (1), or solve for y in equation (1), we will end up with an equation that contains a variable with a fractional coefficient. We'd like to avoid that situation, if possible.

Equation (2): $x + 2y = \dfrac{13}{2}$

Subtract $2y$ from both sides: $x = -2y + \dfrac{13}{2}$

We substitute $-2y + \dfrac{13}{2}$ for x in equation (1).

Equation (1): $2x + 3y = 9$

Let $x = -2y + \dfrac{13}{2}$: $2\left(-2y + \dfrac{13}{2}\right) + 3y = 9$

Now we solve this equation for y.

Distribute the 2: $2(-2y) + 2\left(\dfrac{13}{2}\right) + 3y = 9$

$-4y + 13 + 3y = 9$

Combine like terms: $-y + 13 = 9$

Subtract 13 from each side: $-y = -4$

Multiply both sides by -1: $y = 4$

Let $y = 4$ in equation (1) and solve for x.

Equation (1): $2x + 3y = 9$

Let $y = 4$: $2x + 3(4) = 9$

$2x + 12 = 9$

Subtract 12 from both sides: $2x = -3$

Divide both sides by 2: $\dfrac{2x}{2} = -\dfrac{3}{2}$

$x = -\dfrac{3}{2}$

Check

Equation (1): $2x + 3y = 9$

Let $x = -\frac{3}{2}, y = 4$: $2\left(-\frac{3}{2}\right) + 3(4) \stackrel{?}{=} 9$

$-3 + 12 \stackrel{?}{=} 9$

$9 = 9$ True

Equation (2): $x + 2y = \frac{13}{2}$

Let $x = -\frac{3}{2}, y = 4$: $-\frac{3}{2} + 2(4) \stackrel{?}{=} \frac{13}{2}$

$-\frac{3}{2} + 8 \stackrel{?}{=} \frac{13}{2}$

$\frac{13}{2} = \frac{13}{2}$ True

The solution of the system $\begin{cases} 2x + 3y = 9 \\ x + 2y = \frac{13}{2} \end{cases}$ is $\left(-\frac{3}{2}, 4\right)$.

Work Smart

Solving for x or for y in the first equation in Example 3 is possible, but will lead to fractions, which will make substitution more difficult. For example, if we solve for y in the first equation we obtain

$$2x + 3y = 9$$
$$3y = -2x + 9$$
$$y = -\frac{2}{3}x + 3$$

Then

$$x + 2\left(-\frac{2}{3}x + 3\right) = \frac{13}{2}$$
$$x - \frac{4}{3}x + 6 = \frac{13}{2}$$
$$-\frac{1}{3}x + 6 = \frac{13}{2}$$
$$-\frac{1}{3}x = \frac{1}{2}$$
$$x = -\frac{3}{2}$$

Whew!!!

QUICK ✓ *Solve the system using substitution.*

3. $\begin{cases} y = 4x + 1 \\ 8x - y = 5 \end{cases}$

4. $\begin{cases} 3x + 2y = -3 \\ -x + y = \frac{11}{6} \end{cases}$

Teaching Tip
Ask students to solve $3x - 2 = 3x + 1$. What happens? Ask students to graph $y = 3x - 2$ and $y = 3x + 1$. What happens?

Recall that a system of two linear equations containing two unknowns can be inconsistent. This means that the two lines in the system are parallel. What happens when we use the method of substitution on a system that is inconsistent? Let's find out!

Classroom Example
Solve:

$\begin{cases} x + 5y = 10 \\ -3x - 15y = 5 \end{cases}$

Answer: Inconsistent; no solution

EXAMPLE 4 Solving an Inconsistent System Using Substitution

Solve the following system by substitution: $\begin{cases} x - 3y = 5 & (1) \\ -2x + 6y = 3 & (2) \end{cases}$

Work Smart
Solve for the variable whose coefficient is 1 or -1 to simplify the algebra.

Solution

Is the coefficient of one of the variables 1 or -1? Yes, the coefficient of x in equation (1) is 1, so we solve for x in that equation.

Equation (1):	$x - 3y = 5$
Add $3y$ to both sides:	$x = 3y + 5$
In equation (2), replace x by $3y + 5$:	$-2(3y + 5) + 6y = 3$
Distribute to remove parentheses:	$-6y - 10 + 6y = 3$
Combine like terms:	$-10 = 3$

Notice that the variable, y, was eliminated from the equation and that the last equation, $-10 = 3$, is a false statement. We conclude that the system $\begin{cases} x - 3y = 5 \\ -2x + 6y = 3 \end{cases}$ is inconsistent and has no solution. ■

So how do we determine if a system is inconsistent algebraically? If we end up with a false statement such as $-10 = 3$ or $-13 = 0$, we conclude the system is inconsistent.

How do we determine if a system is consistent, but dependent? That is, how do we determine if a system has infinitely many solutions? Example 5 supplies the answer.

Classroom Example ➤
Solve:

$$\begin{cases} 8x - 4y = -2 \\ -2x + y = \frac{1}{2} \end{cases}$$

Answer: Consistent and dependent; infinitely many solutions.

EXAMPLE 5 Solving a System with Infinitely Many Solutions Using Substitution

Solve the following system by substitution: $\begin{cases} 6x - 2y = -4 & (1) \\ -3x + y = 2 & (2) \end{cases}$

Solution

It is easiest to solve equation (2) for y.

Equation (2):	$-3x + y = 2$
Add $3x$ to both sides:	$y = 3x + 2$
Let $y = 3x + 2$ in equation (1):	$6x - 2(3x + 2) = -4$
Distribute to remove parentheses:	$6x - 6x - 4 = -4$
Combine like terms:	$-4 = -4$

Notice that the variable x has been eliminated from the equation and the last equation, $-4 = -4$, is a true statement. We conclude that the system $\begin{cases} 6x - 2y = -4 \\ -3x + y = 2 \end{cases}$ is consistent, but dependent. The system has infinitely many solutions. ■

Teaching Tip
Remind students what a linear equation in one variable with infinitely many solutions is. Ask students to find an equation that is equivalent to $y = 2x - 3$ by multiplying both sides by some nonzero number k. Now ask students to solve the equation formed by the system.

Work Smart
If a true statement with no variables occurs after the substitution (such as $-4 = -4$), then the system has infinitely many solutions. If a false statement with no variables occurs after the substitution (such as $-10 = 3$), the system has no solution.

Notice that for both the system that has no solution and for the system that has infinitely many solutions, the variables dropped out. The difference between the systems is that the system with infinitely many solutions ends up with a statement that is true (such as $6 = 6$), while the system with no solution ends up with a statement that is false (such as $0 = 5$).

QUICK ✓ *Use the method of substitution to determine whether the system has infinitely many solutions or no solution.*

5. $\begin{cases} y = 5x + 2 \\ -10x + 2y = 4 \end{cases}$ **6.** $\begin{cases} 3x - 2y = 0 \\ -9x + 6y = 5 \end{cases}$ **7.** $\begin{cases} 2x - 6y = 2 \\ -3x + 9y = 4 \end{cases}$

2 Solve Applied Problems Involving Systems of Linear Equations

Now let's look at a problem that can be solved using the method of substitution.

Classroom Example ➤
The number of food processors supplied S is given by the equation $S = 60p - 900$, while the quantity of food processors demanded D is $D = -15p + 2850$, where p is the price of the food processor, in dollars. Find the equilibrium price, which occurs when $S = D$.

Answer: $50

EXAMPLE 6 Life Expectancy of Men versus Women

It is well documented that American women can expect to live longer than American men. But is there ever going to be a time when men can expect to live as long as women? The data given in Table 2 show the life expectancy at birth for both men and women from 1990 to 2001. The data are from the Centers for Disease Control and Prevention. Using techniques from statistics, we can determine an equation that models the life expectancy of Americans of each gender. These techniques indicate that the life expectancy E of women is given by $E = 0.09t - 96.63$, where t is the year. The life expectancy E of men is given by $E = 0.25t - 434.22$. The graphs in Figure 10 plot the data given in Table 2.

Table 2

Year	Life Expectancy—Women	Life Expectancy—Men
1990	78.8	71.8
1991	78.9	72.0
1992	79.1	72.3
1993	78.8	72.2
1994	79.1	72.4
1995	78.8	72.5
1996	79.0	73.1
1997	78.9	73.6
1998	79.1	73.8
1999	79.4	73.9
2000	79.7	74.3
2001	79.8	74.4

Figure 10

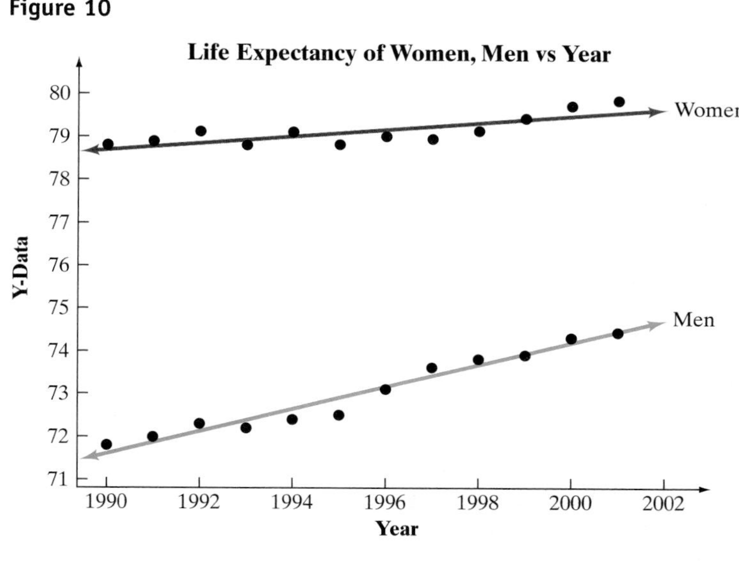

Projecting forward from these models, in what year, if ever, can men expect to live as long as women?

Solution

Step 1: Identify We want to know the year in which American men can expect to live as long as American women.

Step 2: Name The variables are E, which represents the life expectancy, and t, which represents the year.

Step 3: Translate The life expectancy for American women is given by the model $E = 0.09t - 96.63$. We call this equation (1). The life expectancy for American men is given by the model $E = 0.25t - 434.22$. We call this equation (2). We combine equations (1) and (2) to form the following system:

$$\begin{cases} E = 0.09t - 96.63 & (1) \\ E = 0.25t - 434.22 & (2) \end{cases} \quad \text{The Model}$$

Step 4: Solve We will solve this system of equations by substituting $0.09t - 96.63$ for E into equation (2). This leads to the following equation.

	$0.09t - 96.63 = 0.25t - 434.22$
Add 96.63 to both sides:	$0.09t = 0.25t - 337.59$
Subtract $0.25t$ from both sides:	$-0.16t = -337.59$
Divide both sides by -0.16:	$t \approx 2109.9$

We round this result to the nearest year, 2110.

Step 5: Check For American women, the life expectancy in 2110 is predicted to be $E = 0.09(2110) - 96.63 = 93.27$ years. For American men, the life expectancy in 2110 is predicted to be $E = 0.25(2110) - 434.22 = 93.28$ years. The answers differ slightly because of rounding, but our answer checks.

In Words
Caveat (pronounced kä-vē-at) is Latin for "warning" or "beware!"

Step 6: Answer According to these models, the life expectancy of American men will be the same as American women in 2110. One caveat to this prediction is that we cannot be sure that the trend that we see for the years 1990–2001 will continue for the next century.

QUICK 

8. The participation rate in the workforce is defined by economists to be the number of individuals employed divided by the number of individuals 16 years of age or older. The participation rate P for men is given by $P = -0.08t + 243.96$, where t is the year. The participation rate P for women is given by $P = 0.24t - 415.10$, where t is the year. Based on these models, when, if ever, can we expect the participation rate of women in the workforce to equal the participation rate of men?

9.2 Exercises

For Extra Help: Student Solutions Manual | CD Video | PH Math/Tutor Center | MathXL Tutorials on CD | MathXL® | MyMathLab

Concepts and Vocabulary

In Problems 1–3, fill in the blanks.

1. When solving a system of equations by substitution, whenever possible solve one of the equations for the variable that has a coefficient of ________ or ________.

2. While solving a system of equations by substitution, the variable expression has been eliminated and a *false* statement such as $-3 = 0$ results. This means that the solution of the system is ________.

3. While solving a system of equations by substitution, the variable expression has been eliminated and a *true* statement such as $0 = 0$ results. This means that the system has ________ ________ ________.

1. 1, -1
2. no solution or ∅
3. infinitely many solutions
4. False

In Problems 4–6, answer True or False to each statement.

4. When solving a system of equations by substitution, you solve equation (1) for x and then you substitute this expression back into equation (1), resulting in the statement $5 = 5$. This means that there are infinitely many solutions of the system.

5. True
6. False
7. Answers may vary.
8. Answers may vary.
9. $(4, -1)$
10. $(3, -2)$
11. $(-1, 1)$
12. $(2, -5)$
13. $(-3, -4)$
14. $(-3, 5)$
15. $(-4, -7)$
16. $(18, 13)$
17. $\left(-\frac{2}{3}, 2\right)$
18. $\left(-\frac{2}{5}, -5\right)$
19. $\left(2, -\frac{1}{3}\right)$
20. $\left(-\frac{1}{2}, -2\right)$
21. $\left(\frac{1}{4}, \frac{1}{6}\right)$
22. $\left(-\frac{3}{2}, \frac{2}{3}\right)$
23. inconsistent; no solution
24. inconsistent; no solution
25. dependent; infinitely many solutions
26. dependent; infinitely many solutions
27. inconsistent; no solution
28. inconsistent; no solution
29. dependent; infinitely many solutions
30. inconsistent; no solution
31. $(-2, 0)$
32. $(-6, -1)$
33. $\left(2, -\frac{1}{3}\right)$

5. It is possible to solve a system of equations such as $\begin{cases} x = 10 \\ x - y = -2 \end{cases}$ by substitution.

6. A system of equations can have one solution or no solution, but not more than one solution.

7. A test question asks you to solve the following system of equations by substitution. What would be a reasonable first step? Explain which variable you would solve for and why.

$$\begin{cases} \frac{1}{3}x - \frac{1}{6}y = \frac{2}{3} \\ \frac{3}{2}x + \frac{1}{2}y = -\frac{5}{4} \end{cases}$$

8. List two advantages of solving a system of linear equations by substitution rather than by the graphing method.

Building Skills

In Problems 9–22, solve each system of equations using substitution.

9. $\begin{cases} x + 2y = 2 \\ y = 2x - 9 \end{cases}$ **10.** $\begin{cases} y = 3x - 11 \\ -x + 4y = -11 \end{cases}$ **11.** $\begin{cases} -2x + 5y = 7 \\ x = 3y - 4 \end{cases}$

12. $\begin{cases} -4x - y = -3 \\ x = y + 7 \end{cases}$ **13.** $\begin{cases} x + y = -7 \\ 2x - y = -2 \end{cases}$ **14.** $\begin{cases} 5x + 2y = -5 \\ 3x - y = -14 \end{cases}$

15. $\begin{cases} y = \frac{1}{2}x - 5 \\ y = -\frac{3}{4}x - 10 \end{cases}$ **16.** $\begin{cases} y = \frac{2}{3}x + 1 \\ y = -\frac{3}{2}x + 40 \end{cases}$ **17.** $\begin{cases} y = 3x + 4 \\ y = -\frac{1}{2}x + \frac{5}{3} \end{cases}$

18. $\begin{cases} y = 5x - 3 \\ y = 2x - \frac{21}{5} \end{cases}$ **19.** $\begin{cases} x = -6y \\ x - 3y = 3 \end{cases}$ **20.** $\begin{cases} y = 4x \\ 2x - 3y = 5 \end{cases}$

21. $\begin{cases} 2x - 3y = 0 \\ 8x + 6y = 3 \end{cases}$ **22.** $\begin{cases} 2x + 3y = -1 \\ 2x - 9y = -9 \end{cases}$

In Problems 23–30, determine if the given system is a dependent or inconsistent system. State the solution to each system.

23. $\begin{cases} x + 3y = -12 \\ x + 3y = 6 \end{cases}$ **24.** $\begin{cases} 2x - y = 3 \\ y - 2x = 3 \end{cases}$ **25.** $\begin{cases} y = 4x - 1 \\ 8x - 2y = 2 \end{cases}$

26. $\begin{cases} y = -x + 1 \\ x + y = 1 \end{cases}$ **27.** $\begin{cases} x + 2y = 6 \\ x = 3 - 2y \end{cases}$ **28.** $\begin{cases} 4 + y = 3x \\ 6x - 2y = -2 \end{cases}$

29. $\begin{cases} 5x + 6 = 2 - y \\ y = -5x - 4 \end{cases}$ **30.** $\begin{cases} x + y = 2 \\ 2x + 2y = 2 \end{cases}$

Mixed Practice

In Problems 31–50, solve each system of equations using substitution.

31. $\begin{cases} x - 3y = -2 \\ x = -2 \end{cases}$ **32.** $\begin{cases} y = -1 \\ -2x + 9y = 3 \end{cases}$ **33.** $\begin{cases} x = 2 \\ x - 6y = 4 \end{cases}$

34. $\left(\frac{5}{2}, 3\right)$
35. no solution
36. $\left(2, \frac{3}{2}\right)$
37. infinitely many solutions
38. infinitely many solutions
39. $\left(\frac{5}{3}, -\frac{1}{2}\right)$
40. $\left(\frac{19}{21}, -\frac{6}{7}\right)$
41. no solution
42. no solution
43. infinitely many solutions
44. infinitely many solutions
45. $\left(\frac{5}{4}, -\frac{7}{4}\right)$
46. $\left(-\frac{1}{3}, \frac{1}{6}\right)$
47. infinitely many solutions
48. infinitely many solutions
49. $\left(-\frac{5}{2}, 4\right)$
50. $\left(\frac{1}{2}, \frac{1}{2}\right)$
51. 5 and 12
52. 11 and 14
53. length 10 ft; width 7 ft

34. $\begin{cases} 2x + y = 8 \\ y = 3 \end{cases}$

35. $\begin{cases} y = \frac{1}{2}x \\ x = 2(y + 1) \end{cases}$

36. $\begin{cases} y = \frac{3}{4}x \\ x = 4(y - 1) \end{cases}$

37. $\begin{cases} y = 2x - 8 \\ x - \frac{1}{2}y = 4 \end{cases}$

38. $\begin{cases} x = 3y + 6 \\ y - \frac{1}{3}x = -2 \end{cases}$

39. $\begin{cases} 3x + 2y = 4 \\ 3x + y = \frac{9}{2} \end{cases}$

40. $\begin{cases} 3x + 2y = 1 \\ x + 3y = -\frac{5}{3} \end{cases}$

41. $\begin{cases} x - 5y = 3 \\ -2x + 10y = 8 \end{cases}$

42. $\begin{cases} -4x + y = 3 \\ 8x - 2y = 1 \end{cases}$

43. $\begin{cases} 3x - 2y = 1 \\ -6x + 4y = -2 \end{cases}$

44. $\begin{cases} -x + 3y = 4 \\ 2x - 6y = -8 \end{cases}$

45. $\begin{cases} 4x + 8y = -9 \\ 8x + 4y = 3 \end{cases}$

46. $\begin{cases} 18x - 6y = -7 \\ x + 2y = 0 \end{cases}$

47. $\begin{cases} \frac{3}{2}x - y = 1 \\ 3x - 2y = 2 \end{cases}$

48. $\begin{cases} \frac{2}{3}x + \frac{1}{3}y = 1 \\ -4x - 2y = -6 \end{cases}$

49. $\begin{cases} \frac{x}{2} + \frac{y}{3} = \frac{1}{12} \\ \frac{5x}{4} + \frac{2y}{3} = -\frac{11}{24} \end{cases}$

50. $\begin{cases} \frac{x}{4} + \frac{y}{2} = \frac{3}{8} \\ \frac{x}{5} - \frac{3y}{4} = -\frac{11}{40} \end{cases}$

Applying the Concepts

Use substitution to solve each system of linear equations in two variables.

51. Fun with Numbers The sum of two numbers is 17, and their difference is 7. Determine the numbers by solving the following system of equations, using x and y to represent the unknown numbers.

$$\begin{cases} x + y = 17 & (1) \\ x - y = 7 & (2) \end{cases}$$

52. Fun with Numbers The sum of two numbers is 25, and their difference is 3. Determine the numbers by solving the following system of equations, using x and y to represent the unknown numbers.

$$\begin{cases} x + y = 25 & (1) \\ x - y = 3 & (2) \end{cases}$$

△ **53. Dimensions of a Garden** The perimeter of a rectangular garden is 34 feet. The length of the garden is 3 feet more than the width. Determine the dimensions of the garden by solving the following system of equations, where l and w represent the length and width of the garden.

$$\begin{cases} 2l + 2w = 34 & (1) \\ l = w + 3 & (2) \end{cases}$$

54. length 15 in.; width 11 in.
55. $8000 in money market; $4000 in international fund
56. $8000 in international stock fund; $16,000 in domestic growth fund
57. 2003
58. 2002
59. $A = \frac{7}{6}$; $B = -\frac{1}{2}$
60. Answers may vary.
61. Answers may vary.
62. Answers may vary.
63. Answers may vary.

△ **54. Dimensions of a Rectangle** The perimeter of a rectangle is 52 inches. The length of the rectangle is 4 inches more than the width. Determine the dimensions of the rectangle by solving the following system of equations, where l and w represent the length and width of the rectangle.

$$\begin{cases} 2l + 2w = 52 & (1) \\ l = w + 4 & (2) \end{cases}$$

55. Investment Paul wants to invest part of his Ohio Lottery winnings in a safe money-market fund that earns 2.5% annual interest and the rest in a risky international fund that expects to yield of 9% annual interest. The amount of money invested in the money-market fund, x, is to be exactly twice the amount, y, invested in the international fund. Use the system of equations to determine the amount to be invested in each fund if a total of $560 is to be earned at the end of one year.

$$\begin{cases} x = 2y & (1) \\ 0.025x + 0.09y = 560 & (2) \end{cases}$$

56. Investment Elaine wants to invest part of her $24,000 Virginia Lottery winnings in an international stock fund that yields 4% annual interest and the remainder in a domestic growth fund that yields 6.5% annually. Use the system of equations to determine the amount Elaine should invest in each account to earn $1360 interest at the end of one year, where x represents the amount invested in the international stock fund and y represents the amount invested in the domestic growth fund.

$$\begin{cases} x + y = 24{,}000 & (1) \\ 0.04x + 0.065y = 1360 & (2) \end{cases}$$

57. Sales of Music Videos According to the Recording Industry of America, the number of music videos sold in the U.S. annually since 1998 has been decreasing according to the equation $y = -2.17x + 24.94$, where y is the number of music videos sold (in millions) and x is the number of years since 1998. The sales of DVD videos have been increasing over this same period, according to the equation $y = 2.58x - 0.18$, where y is the number of DVD videos, in millions, and x is the number of years since 1998. Find, to the nearest year, when the sales of DVD videos will be equal to the sales of music videos. (*Source:* Recording Industry of America)

58. Sales of Music Videos According to the Recording Industry of America, the total value of DVD videos has increased since 1998, according to the equation $y = 57.2x + 2.7$, where y is the dollar value of U.S. sales and x is the number of years since 1998. During this same period, the dollar value of music videos has declined according to the equation $y = -48.8x + 454.4$, where y is the dollar value of U.S. music video sales and x is the number of years since 1998. Find, to the nearest year, when the dollar value of sales of music videos and DVD videos was the same. (*Source:* Recording Industry of America)

Extending the Concepts

59. For the system $\begin{cases} Ax + 3By = 2 \\ -3Ax + By = -11 \end{cases}$, find A and B such that $x = 3$, $y = 1$ is a solution.

60. Write a system of equations that has $(3, 5)$ as a solution.

61. Write a system of equations that has $(-1, 4)$ as a solution.

62. Write a system of equations that has infinitely many solutions.

63. Write a system of equations that has no solution.

9.3 Solving Systems of Linear Equations Using Elimination

OBJECTIVES

1. Solve a System of Linear Equations Using the Elimination Method
2. Solve Applied Problems Involving Systems of Linear Equations

Preparing for Solving Systems of Linear Equations Using Elimination

Before getting started, take the following readiness quiz. If you get a problem wrong, go back to the section cited and review the material.

1. What is the additive inverse of 5? [Section 1.3, p. 21]

2. What is the additive inverse of -8? [Section 1.3, p. 21]

3. Distribute: $\frac{2}{3}(3x - 9y)$ [Section 1.7, p. 61]

4. Solve: $2y - 5y = 12$ [Section 2.2, p. 86]

1 Solve a System of Linear Equations Using the Elimination Method

We have seen how to use the method of graphing and the method of substitution to solve a system of linear equations containing two unknowns. The method of graphing is nice because we can visualize the solutions, but it is not always easy to find the exact coordinates of the point of intersection. The method of substitution has the advantage of finding an exact solution because it is an algebraic method, but it may require that we introduce fractions to the solution process. We would rather avoid introducing fractions, if possible. For this reason, we introduce a second algebraic method that can be used to solve a system of equations called the *method of elimination*. This method is usually preferred over the method of substitution when substitution leads to fractions.

Work Smart

Use the method of substitution if it is easy to solve for one of the variables in the system. Use elimination if substitution will lead to fractions or solving for a variable is not easy.

Remember that the Addition Property of Equality states that if we add the same quantity to both sides of an equation, we obtain an equivalent equation. Solving a system of linear equations using the elimination method takes this principle a step further. Adding equal quantities to both sides of an equation also results in a true statement.

Classroom Example Solve the system

$$\begin{cases} x + 2y = 11 \\ -x + 5y = 10 \end{cases}$$

Answer: $(5, 3)$

EXAMPLE 1 Solving a System of Linear Equations Using Elimination

Solve the system $\begin{cases} x + y = 13 & (1) \\ 2x - y = -7 & (2) \end{cases}$

Solution

Equation (2) says that $2x - y$ equals -7. So, if we add $2x - y$ to the left side of equation (1) and -7 to the right side of equation (1), we are adding to the same value to each side of equation (1). We will perform this addition vertically.

$$\begin{array}{rl} x + y = 13 & (1) \\ \underline{2x - y = -7} & (2) \\ 3x + 0 = 6 & \end{array}$$

Work Smart

This method is called "elimination" because one of the variables is eliminated through the process of addition. The elimination method is also sometimes referred to as the **addition method.**

Notice that the variable y has been eliminated, and we can solve for x.

$$3x + 0 = 6$$

Divide each side by 3: $3x = 6$

$$x = 2$$

Now that we have the value of x, we can substitute $x = 2$ into either equation (1) or equation (2) to find the value of y. Let's substitute $x = 2$ into equation (1).

$$x + y = 13$$

Substitute 2 for x: $2 + y = 13$

Subtract 2 from each side: $y = 11$

Preparing for...Answers **1.** -5 **2.** 8 **3.** $2x - 6y$ **4.** $\{-4\}$

We have $x = 2$ and $y = 11$. Let's check this solution in both equations.

Check

Equation (1): $x + y = 13$

$x = 2, y = 11$: $2 + 11 \stackrel{?}{=} 13$

$13 = 13$ True

Equation (2): $2x - y = -7$

$x = 2, y = 11$: $2(2) - 11 \stackrel{?}{=} -7$

$4 - 11 \stackrel{?}{=} -7$

$-7 = -7$ True

The solution of the system $\begin{cases} x + y = 13 \\ 2x - y = -7 \end{cases}$ is $(2, 11)$. ■

The idea in using the elimination method is to get the coefficients of one of the variables to be additive inverses. In Example 1, the coefficients of the variable y were opposites: 1 and -1. Unfortunately, having linear equations in which one of the variables has opposite coefficients is uncommon, so we need to develop a strategy to solve systems in which the coefficients of one of the variables are not opposites.

EXAMPLE 2 How to Solve a System of Linear Equations by Elimination—Multiply One Equation by a Number to Create Additive Inverses

Solve the following system by elimination: $\begin{cases} 2x + 3y = -6 & (1) \\ -4x + 5y = -21 & (2) \end{cases}$

Classroom Example Solve the system $\begin{cases} 2x + 6y = 5 \\ -4x + 3y = -15 \end{cases}$

Answer: $\left(\dfrac{7}{2}, -\dfrac{1}{3}\right)$

Step-by-Step Solution

Step 1: We need to get the coefficients of one of the variables to be additive inverses. In looking at the system, this can be accomplished by multiplying both sides of equation (1) by 2. Do you see why?

$$\begin{cases} 2x + 3y = -6 & (1) \\ -4x + 5y = -21 & (2) \end{cases}$$

Multiply (1) by 2: $\begin{cases} 2(2x + 3y) = 2(-6) & (1) \\ -4x + 5y = -21 & (2) \end{cases}$

Distribute the 2 in (1): $\begin{cases} 4x + 6y = -12 & (1) \\ -4x + 5y = -21 & (2) \end{cases}$

Step 2: Notice that the coefficients of the variable x are additive inverses. We now add equations (1) and (2) to eliminate the variable x and then solve for y.

$$\begin{cases} 4x + 6y = -12 & (1) \\ -4x + 5y = -21 & (2) \end{cases}$$

Add (1) and (2): $11y = -33$

Divide both sides by 11: $y = -3$

Step 3: We let $y = -3$ in either equation (1) or (2). Since equation (1) looks a little easier to work with, we will substitute $y = -3$ into equation (1) and solve for x.

Equation (1): $2x + 3y = -6$

Let $y = -3$: $2x + 3(-3) = -6$

$2x - 9 = -6$

Add 9 to both sides: $2x = 3$

Divide both sides by 2: $x = \dfrac{3}{2}$

We have that $x = \dfrac{3}{2}$, $y = -3$.

(continued)

Step 4: Check

Equation (1): $2x + 3y = -6$

$x = \frac{3}{2}, y = -3$: $2\left(\frac{3}{2}\right) + 3(-3) \stackrel{?}{=} -6$

$3 + (-9) \stackrel{?}{=} -6$

$-6 = -6$ True

Equation (2): $-4x + 5y = -21$

$x = \frac{3}{2}, y = -3$: $-4\left(\frac{3}{2}\right) + 5(-3) \stackrel{?}{=} -21$

$-6 + (-15) \stackrel{?}{=} -21$

$-21 = -21$ True

Both equations check, so the solution of the system $\begin{cases} 2x + 3y = -6 \\ -4x + 5y = -21 \end{cases}$ is $\left(\frac{3}{2}, -3\right)$.

The following summarizes the steps just used in Example 2.

Work Smart
The idea behind the elimination method is to get the coefficients of one of the variables to be additive inverses (opposites), so that one of the variables will be eliminated.

Steps for Solving a System of Linear Equations by Elimination

Step 1: Make sure that the coefficients on one of the variables are additive inverses by multiplying (or dividing) both sides of one (or both) equation(s) by a nonzero constant.

Step 2: Add the equations to eliminate the variable whose coefficients are now additive inverses. Solve the resulting equation for the remaining unknown.

Step 3: Substitute the value of the variable found in Step 2 into one of the *original* equations to find the value of the remaining variable.

Step 4: Check your answer.

QUICK ✓ *Solve the system using elimination.*

1. $\begin{cases} x - 3y = 2 \\ 2x + 3y = -14 \end{cases}$

2. $\begin{cases} x - 2y = 2 \\ -2x + 5y = -1 \end{cases}$

Classroom Example ➤
Solve the system
$\begin{cases} 2x + 5y = -11 \\ 3x + 2y = 11 \end{cases}$
Answer: $(7, -5)$

EXAMPLE 3 Solve a System of Equations by Elimination—Multiply Both Equations by a Number to Create Additive Inverses

Solve the following system by elimination: $\begin{cases} 2x + 3y = 3 & (1) \\ 3x + 5y = 7 & (2) \end{cases}$

Solution

We want the coefficients of one of the variables to be additive inverses. We cannot do this by multiplying a single equation by some nonzero constant, so we will need to multiply both equations by some nonzero constant. For example, we can multiply equation (1) by 3 and equation (2) by -2 so that the coefficients on x are additive inverses.

Work Smart
There is no single right way to multiply the equations in a system by a nonzero constant to get coefficients to be additive inverses. For example, we could also multiply equation (1) by 5 and equation (2) by -3. Do you see why this also works?

$$\begin{cases} 2x + 3y = 3 & (1) \\ 3x + 5y = 7 & (2) \end{cases}$$

Multiply equation (1) by 3:
Multiply equation (2) by -2:
$$\begin{cases} 3(2x + 3y) = 3 \cdot 3 & (1) \\ -2(3x + 5y) = -2 \cdot 7 & (2) \end{cases}$$

Distribute: $\begin{cases} 6x + 9y = 9 & (1) \\ -6x - 10y = -14 & (2) \end{cases}$

Add (1) and (2): $-y = -5$

Multiply both sides by -1: $y = 5$

Now let $y = 5$ in equation (1) to determine the value of x.

Equation (1): $2x + 3y = 3$

Let $y = 5$ in (1) and find x: $2x + 3(5) = 3$

Simplify: $2x + 15 = 3$

Subtract 15 from both sides: $2x = -12$

Divide both sides by 2: $x = -6$

We have that $x = -6$ and $y = 5$. Now we check our answer.

Work Smart: Study Skills
At this point you should be getting a feel as to when each method for solving a system of equations can be used. If not, start creating a list that describes each method, its benefits, and when it should be used.

Check

Equation (1): $2x + 3y = 3$

$x = -6, y = 5$: $2(-6) + 3(5) \stackrel{?}{=} 3$

$-12 + 15 \stackrel{?}{=} 3$

$3 = 3$ True

Equation (2): $3x + 5y = 7$

$x = -6, y = 5$: $3(-6) + 5(5) \stackrel{?}{=} 7$

$-18 + 25 \stackrel{?}{=} 7$

$7 = 7$ True

The solution to the system $\begin{cases} 2x + 3y = 3 \\ 3x + 5y = 7 \end{cases}$ is $(-6, 5)$.

QUICK ✓ *Solve the system using elimination.*

3. $\begin{cases} 5x + 4y = 10 \\ -2x + 3y = -27 \end{cases}$

4. $\begin{cases} 6x + 4y = 30 \\ 7x + 10y = 35 \end{cases}$

Now let's look at the results we obtain when we have systems that have no solution or infinitely many solutions.

Classroom Example
Solve the following system by elimination:

$\begin{cases} 2x + 6y = 3 \\ 3x + 9y = 2 \end{cases}$

Answer: Inconsistent; no solution

EXAMPLE 4 Solving a System of Equations with No Solution Using Elimination

Solve the following system by elimination: $\begin{cases} 4x - 6y = -5 & (1) \\ 6x - 9y = 2 & (2) \end{cases}$

Solution

$\begin{cases} 4x - 6y = -5 & (1) \\ 6x - 9y = 2 & (2) \end{cases}$

To eliminate x,
multiply equation (1) by -3:
multiply equation (2) by 2:
$\begin{cases} -3(4x - 6y) = -3 \cdot (-5) & (1) \\ 2(6x - 9y) = 2 \cdot 2 & (2) \end{cases}$

Distribute: $\begin{cases} -12x + 18y = 15 & (1) \\ 12x - 18y = 4 & (2) \end{cases}$

Add (1) and (2): $0 = 19$

The statement $0 = 19$ is false. Therefore, the system $\begin{cases} 4x - 6y = -5 \\ 6x - 9y = 2 \end{cases}$ is inconsistent and has no solution.

Classroom Example ➤
Solve the following system by elimination:

$$\begin{cases} 3x - 9y = 3 \\ 2x - 6y = 2 \end{cases}$$

Answer: Dependent; Infinitely many solutions

EXAMPLE 5 Solving a System of Equations with Infinitely Many Solutions Using Elimination

Solve the following system by elimination: $\begin{cases} 3x + 4y = -6 & (1) \\ \frac{3}{2}x + 2y = -3 & (2) \end{cases}$

Solution

$$\begin{cases} 3x + 4y = -6 & (1) \\ \frac{3}{2}x + 2y = -3 & (2) \end{cases}$$

To eliminate y, multiply equation (2) by -2:

$$\begin{cases} 3x + 4y = -6 & (1) \\ -2\left(\frac{3}{2}x + 2y\right) = -2(-3) & (2) \end{cases}$$

Distribute:

$$\begin{cases} 3x + 4y = -6 & (1) \\ -3x - 4y = 6 & (2) \end{cases}$$

Add (1) and (2):

$$0 = 0$$

Notice that both the variables were eliminated. However, the statement $0 = 0$ is true. Therefore, the system $\begin{cases} 3x + 4y = -6 \\ \frac{3}{2}x + 2y = -3 \end{cases}$ is consistent but dependent. The system has infinitely many solutions.

QUICK ✓ *Use elimination to determine whether the system has no solution or infinitely many solutions.*

5. $\begin{cases} 2x - 6y = 10 \\ 5x - 15y = 4 \end{cases}$

6. $\begin{cases} -x + 3y = 2 \\ 3x - 9y = -6 \end{cases}$

7. $\begin{cases} -4x + 8y = 4 \\ 3x - 6y = -3 \end{cases}$

We summarize the three methods that we have discussed for solving a system of two linear equations containing two unknowns.

SUMMARY: Which Method Should I Use?

We have presented three methods for solving systems of linear equations. Below, we present a summary of the questions you should ask before solving a system of linear equations. The answers to these questions tell us the most appropriate method. We also present an example of the situation and the advantages and disadvantages of each method.

Question	Method to Use	Example	Advantages/Disadvantages
Would it be beneficial to see the solutions visually?	Graphical	Find the break-even point of the cost equation $y = 20 + 4x$ and the revenue equation $y = 9x$, where x represents the number of calculators produced and sold.	Allows us to "see" the answer, but if the coordinates of the intersection point are not integers, it can be difficult to determine the answer.
Is one of the coefficients of the variables 1 or -1?	Substitution	$\begin{cases} -x + 2y = 7 \\ 2x - 3y = -15 \end{cases}$	Gives exact solutions. The algebra can be easy provided one of the variables has a coefficient of 1 or -1. If none of the coefficients are 1 or -1, the algebra can get messy.

Question	Method to Use	Example	Advantages/Disadvantages
Are the coefficients of one of the variables additive inverses? If the coefficients are not additive inverses, is it easy to get the coefficients to be additive inverses?	Elimination	$\begin{cases} 2x + 3y = -5 \\ -2x + y = -3 \end{cases}$	Gives exact solutions. It is easy to use when neither of the variables has a coefficient of 1 or -1.

2 Solve Applied Problems Involving Systems of Linear Equations

We saw in the last section that applied problems are easily solved using substitution when at least one of the variables is isolated (that is, by itself). But many mathematical models that involve two equations containing two unknowns do not have one of the equations solved for one of the unknowns. If this occurs, it is better to use elimination to solve the problem.

Classroom Example Mark and John attend a baseball game with their kids. Mark buys dinner. He gets 9 hot dogs and 5 drinks for \$55.50. After the fifth inning everyone is hungry again, so John buys 5 hot dogs and 3 drinks for \$31.50. We can determine the price of a hot dog and the price of a Pepsi by solving the system of equations

$$\begin{cases} 9h + 5d = 55.50 \\ 5h + 3d = 31.50 \end{cases}$$

where h represents the price of a hot dog and d represents the price of a drink. How much does a hot dog cost at the ball park? How much does a drink cost?

Answer: A hot dog costs \$4.50 and a drink costs \$3.00.

EXAMPLE 6 Hot Dogs and Soda at the Game

Bill and Roger attend a baseball game with their kids. They get to the game early to watch batting practice and have dinner. Roger says that he will buy dinner. He gets 7 hot dogs and 5 Pepsis for \$38.25. After the sixth inning everyone is hungry again, so Bill buys 5 hot dogs and 4 Pepsis for \$28.50. We can determine the price of a hot dog and the price of a Pepsi by solving the system of equations

$$\begin{cases} 7h + 5p = 38.25 \\ 5h + 4p = 28.50 \end{cases}$$

where h represents the price of a hot dog and p represents the price of a Pepsi. How much does a hot dog cost at the ballpark? How much does a Pepsi cost?

Solution

We name the equation $7h + 5p = 38.25$ equation (1) and name $5h + 4p = 28.50$ equation (2). Now let's solve the system.

$$\begin{cases} 7h + 5p = 38.25 & (1) \\ 5h + 4p = 28.50 & (2) \end{cases}$$

Let's eliminate p. Multiply (1) by -4: Multiply (2) by 5:
$$\begin{cases} -4(7h + 5p) = -4 \cdot 38.25 & (1) \\ 5(5h + 4p) = 5 \cdot 28.50 & (2) \end{cases}$$

Distribute:
$$\begin{cases} -28h - 20p = -153 & (1) \\ 25h + 20p = 142.50 & (2) \end{cases}$$

Add (1) and (2): $-3h = -10.50$

Divide both sides by -3: $h = 3.50$

Let $h = 3.50$ in (2): $5(3.50) + 4p = 28.50$

$17.50 + 4p = 28.50$

Subtract 17.50 from both sides: $4p = 11$

Divide both sides by 4: $p = 2.75$

We leave it to you to verify the answer. A hot dog costs \$3.50 and a Pepsi costs \$2.75 at the ballpark.

QUICK ✓

8. At a fast food joint, 5 cheeseburgers and 3 medium shakes cost \$15.50. At the same fast food joint, 3 cheeseburgers and 2 medium shakes cost \$9.75. We can determine the price of a cheeseburger and the price of a shake by solving the system of equations

$$\begin{cases} 5c + 3s = 15.50 \\ 3c + 2s = 9.75 \end{cases}$$

where c represents the price of a cheeseburger and s represents the price of a shake. How much does a cheeseburger cost at the fast food joint? How much does a shake cost?

9.3 Exercises

For Extra Help: Student Solutions Manual | CD Video | PH Math/Tutor Center | MathXL Tutorials on CD | MathXL® | MyMathLab

1. graphing, substitution, elimination
2. dependent, infinitely
3. additive inverses
4. True
5. True
6. True
7. Answers may vary.
8. Answers may vary.

Concepts and Vocabulary

In Problems 1–3, fill in the blanks.

1. List the three methods presented in this chapter for solving systems of equations: __________, __________, __________.

2. When using the elimination method to solve a system of equations, you add equation (1) and equation (2), resulting in the statement $-50 = -50$. This means that the system is __________ and has __________ many solutions.

3. The basic idea in using the elimination method is to get the coefficients of one of the variables to be __________ __________, such as 3 and -3.

In Problems 4–6, answer True or False to each statement.

4. It is possible to solve a system of equations using the elimination method without multiplying either equation by some nonzero constant.

5. Consider the following system of equations. This system could be solved by all of the following:

$$\begin{cases} 6x + 3y = 3 & (1) \\ -4x - y = 1 & (2) \end{cases}$$

(a) Multiply equation (1) by 4 and equation (2) by 6, then add.
(b) Multiply equation (1) by 2 and equation (2) by 3, then add.
(c) Multiply equation (2) by 3, then add.
(d) Multiply equation (1) one-third, then add.
(e) Solve equation (2) for y and substitute this expression into equation (1).

6. If an equation contains fractions, it is a good idea to multiply both sides of the equation by the LCD to clear the denominators before deciding which variable to eliminate.

✎ **7.** Suppose you are given the system of equations $\begin{cases} x - 3y = 6 \\ 2x + y = 5 \end{cases}$. Which variable is easier to eliminate? Why? List the steps you would use to solve this system.

✎ **8.** One of the methods for solving a system of equations that you have studied in this section has two commonly used titles: elimination method or addition method. Explain why either title would be appropriate for this process.

9. $(2, -1)$
10. $(-5, -15)$
11. $(6, 4)$
12. $(-3, 2)$
13. $\left(-\frac{11}{4}, \frac{1}{2}\right)$
14. $\left(\frac{4}{3}, \frac{3}{2}\right)$
15. $(-1, -3)$
16. $(4, -2)$
17. inconsistent; 0
18. inconsistent; 0
19. dependent; infinitely many
20. inconsistent; 0
21. dependent; infinitely many
22. dependent; infinitely many
23. inconsistent; 0
24. inconsistent; 0
25. $(-5, 8)$
26. $(-7, 10)$
27. $\left(\frac{3}{2}, -\frac{3}{4}\right)$
28. $\left(-\frac{2}{3}, -\frac{5}{6}\right)$
29. no solution
30. no solution
31. $(12, -9)$
32. $(-14, 10)$
33. no solution
34. no solution
35. $\left(\frac{6}{29}, -\frac{8}{29}\right)$
36. $\left(\frac{95}{29}, -\frac{43}{29}\right)$
37. $\left(\frac{1}{2}, 4\right)$
38. $\left(-\frac{20}{39}, -\frac{51}{13}\right)$
39. $(-1, 2.1)$
40. $(-0.1, -0.2)$
41. $(-2, -6)$
42. $\left(\frac{8}{7}, -\frac{9}{7}\right)$
43. $(50, 30)$
44. $(-15, 25)$
45. $(1, 5)$
46. $(-2, -1)$
47. $(5, 2)$
48. $\left(\frac{1}{2}, -\frac{5}{2}\right)$
49. $\left(-\frac{9}{17}, \frac{31}{17}\right)$
50. $(1, 4)$
51. $\left(\frac{8}{3}, -\frac{4}{7}\right)$
52. $(8, 17)$
53. infinitely many solutions

Building Skills

In Problems 9–16, solve each system of equations using elimination.

9. $\begin{cases} 2x + y = 3 \\ 5x - y = 11 \end{cases}$ **10.** $\begin{cases} x + y = -20 \\ x - y = 10 \end{cases}$ **11.** $\begin{cases} 3x - 2y = 10 \\ -3x + 12y = 30 \end{cases}$

12. $\begin{cases} 2x + 6y = 6 \\ -2x + y = 8 \end{cases}$ **13.** $\begin{cases} 2x + 3y = -4 \\ -2x + y = 6 \end{cases}$ **14.** $\begin{cases} 3x + 2y = 7 \\ -3x + 4y = 2 \end{cases}$

15. $\begin{cases} 6x - 2y = 0 \\ -9x - 4y = 21 \end{cases}$ **16.** $\begin{cases} 2x + 5y = -2 \\ -3x + 10y = -32 \end{cases}$

In Problems 17–24, determine if the system is dependent or inconsistent. State the number of solutions the system has.

17. $\begin{cases} 2x + 2y = 1 \\ -2x - 2y = 1 \end{cases}$ **18.** $\begin{cases} 3x - 2y = 4 \\ -3x + 2y = 4 \end{cases}$ **19.** $\begin{cases} 3x + y = -1 \\ 6x + 2y = -2 \end{cases}$

20. $\begin{cases} x - y = -4 \\ -2x + 2y = -8 \end{cases}$ **21.** $\begin{cases} 2x - 3y = 10 \\ -4x + 6y = -20 \end{cases}$ **22.** $\begin{cases} 2x + 5y = 15 \\ -6x - 15y = -45 \end{cases}$

23. $\begin{cases} -4x + 8y = 1 \\ 3x - 6y = 1 \end{cases}$ **24.** $\begin{cases} 4x + 6y = -10 \\ 9x + \frac{27}{2}y = 3 \end{cases}$

In Problems 25–44, solve each system of equations using elimination.

25. $\begin{cases} 2x + 3y = 14 \\ -3x + y = 23 \end{cases}$ **26.** $\begin{cases} 2x + y = -4 \\ 3x + 5y = 29 \end{cases}$ **27.** $\begin{cases} 2x + 4y = 0 \\ 5x + 2y = 6 \end{cases}$

28. $\begin{cases} -2x - 2y = 3 \\ x - 2y = 1 \end{cases}$ **29.** $\begin{cases} x - 3y = 4 \\ -2x + 6y = 3 \end{cases}$ **30.** $\begin{cases} 5x - y = 3 \\ -10x + 2y = 2 \end{cases}$

31. $\begin{cases} 2x + 3y = -3 \\ 3x + 5y = -9 \end{cases}$ **32.** $\begin{cases} 2x + 3y = 2 \\ 5x + 7y = 0 \end{cases}$ **33.** $\begin{cases} 5y = 2x - 5 \\ 2x - 5y = 1 \end{cases}$

34. $\begin{cases} x + 3y = 6 \\ 3y = -x + 12 \end{cases}$ **35.** $\begin{cases} 4x + 3y = 0 \\ 3x - 5y = 2 \end{cases}$ **36.** $\begin{cases} 5x + 7y = 6 \\ 2x - 3y = 11 \end{cases}$

37. $\begin{cases} 4x - 3y = -10 \\ -\frac{2}{3}x + y = \frac{11}{3} \end{cases}$ **38.** $\begin{cases} 12x + 15y = -65 \\ \frac{1}{2}y + 3x = -\frac{7}{2} \end{cases}$

39. $\begin{cases} 1.5x + 0.5y = -0.45 \\ -0.3x - 0.4y = -0.54 \end{cases}$ **40.** $\begin{cases} -2.4x - 0.4y = 0.32 \\ 4.2x + 0.6y = -0.54 \end{cases}$

41. $\begin{cases} \frac{1}{2}x + \frac{2}{3}y = -5 \\ \frac{5}{2}x + \frac{5}{6}y = -10 \end{cases}$ **42.** $\begin{cases} x + \frac{5}{3}y = -1 \\ \frac{1}{2}x + \frac{1}{4}y = \frac{1}{4} \end{cases}$

43. $\begin{cases} 0.05x + 0.10y = 5.50 \\ x + y = 80 \end{cases}$ **44.** $\begin{cases} -1.8x + 1.5y = 64.50 \\ 0.8x - 0.6y = -27 \end{cases}$

Mixed Practice

In Problems 45–68, solve by any method: graphing, substitution, or elimination.

45. $\begin{cases} x - y = -4 \\ 3x + y = 8 \end{cases}$ **46.** $\begin{cases} x - 2y = 0 \\ 3x + 5y = -11 \end{cases}$ **47.** $\begin{cases} 3x - 10y = -5 \\ 6x - 8y = 14 \end{cases}$

48. $\begin{cases} 3x - 5y = 14 \\ -2x + 6y = -16 \end{cases}$ **49.** $\begin{cases} x - 3y = -6 \\ 4x + 5y = 7 \end{cases}$ **50.** $\begin{cases} 2x - 3y = -10 \\ y - 3x = 1 \end{cases}$

51. $\begin{cases} 0.3x - 0.7y = 1.2 \\ 1.2x + 2.1y = 2 \end{cases}$ **52.** $\begin{cases} 0.25x + 0.10y = 3.70 \\ x + y = 25 \end{cases}$ **53.** $\begin{cases} x = 3y - 1 \\ 2x - 6y = -2 \end{cases}$

54. infinitely many solutions
55. infinitely many solutions
56. infinitely many solutions
57. $(1, -3)$
58. $(14, -2)$
59. $(-4, 0)$
60. $(3, 5)$
61. $\left(-\frac{8}{5}, \frac{27}{10}\right)$
62. $\left(\frac{1}{5}, 4\right)$
63. $(-4, -5)$
64. $(5, 2)$
65. no solution
66. no solution
67. $\left(\frac{106}{37}, \frac{16}{37}\right)$
68. $\left(\frac{1}{2}, \frac{1}{3}\right)$
69. hamburger 280 cal; Coke 210 cal
70. biscuit 28 g; orange juice 42 g
71. tomatoes 10 hr; zucchini 15 hr; no

54. $\begin{cases} y = -4x + 3 \\ 8x + 2y = 6 \end{cases}$

55. $\begin{cases} 3x - 2y = 6 \\ \frac{3}{2}x - y = 3 \end{cases}$

56. $\begin{cases} 4x + 3y = 9 \\ \frac{4}{3}x + y = 3 \end{cases}$

57. $\begin{cases} y = -\frac{2}{3}x - \frac{7}{3} \\ y = \frac{3}{4}x - \frac{15}{4} \end{cases}$

58. $\begin{cases} y = \frac{1}{7}x - 4 \\ y = -\frac{3}{2}x + 19 \end{cases}$

59. $\begin{cases} \frac{x}{2} + \frac{y}{4} = -2 \\ \frac{3x}{2} + \frac{y}{5} = -6 \end{cases}$

60. $\begin{cases} \frac{x}{3} + \frac{y}{5} = 2 \\ \frac{x}{3} - \frac{2y}{5} = -1 \end{cases}$

61. $\begin{cases} x - 2y = -7 \\ 3x + 4y = 6 \end{cases}$

62. $\begin{cases} -5x + y = 3 \\ 10x + 3y = 14 \end{cases}$

63. $\begin{cases} 6x - 5y = 1 \\ 8x - 2y = -22 \end{cases}$

64. $\begin{cases} -3x - 2y = -19 \\ 4x + 5y = 30 \end{cases}$

65. $\begin{cases} y = 2x - 4y \\ 4x + 1 = 10y + 3 \end{cases}$

66. $\begin{cases} 12y = 8x + 3 \\ -2 + 4(3y - 2x) = 5 \end{cases}$

67. $\begin{cases} \frac{x}{2} - y = 1 \\ \frac{x}{5} + \frac{5y}{6} = \frac{14}{15} \end{cases}$

68. $\begin{cases} \frac{x}{2} + \frac{3y}{4} = \frac{1}{2} \\ -\frac{3x}{5} + \frac{3y}{4} = -\frac{1}{20} \end{cases}$

Applying the Concepts

69. Counting Calories Suppose that Kristin ate two McDonald's hamburgers and drank one medium Coke, for a total of 770 calories. Kristin's friend, Jack, ate three hamburgers and drank two medium Cokes (Jack takes advantage of free refills) for a total of 1260 calories. How many calories are in a McDonald's hamburger? How many calories are in a medium Coke? Solve the system

$$\begin{cases} 2h + c = 770 & (1) \\ 3h + 2c = 1260 & (2) \end{cases}$$

where h represents the number of calories in a hamburger and c represents the number of calories in a medium Coke to find the answers.

70. Carbs Yvette and José go to McDonald's for breakfast. Yvette orders two sausage biscuits and one 16-ounce orange juice. The entire meal had 98 grams of carbohydrates. José orders three sausage biscuits and two 16-ounce orange juices and his meal had 168 grams of carbohydrates. How many grams of carbohydrates are in a sausage biscuit? How many grams of carbohydrates are in a 16-ounce orange juice? Solve the system

$$\begin{cases} 2b + u = 98 & (1) \\ 3b + 2u = 168 & (2) \end{cases}$$

where b represents the number of grams of carbohydrates in a sausage biscuit and u represents the number of grams of carbohydrates in an orange juice to find the answers.

71. Planting Crops Farmer Green runs an organic farm and is planting his fields. He remembers from previous years that he planted 2 acres of tomatoes and 3 acres of zucchini in 65 hours. A different year he planted 3 acres of tomatoes and 4 acres of zucchini in 90 hours. If t represents the number of hours it takes to plant an acre of tomatoes, and z represents the number of hours it takes to plant one acre of zucchini, determine how long it takes Farmer Green to plant an acre of each crop by solving the following system of equations.

$$\begin{cases} 2t + 3z = 65 & (1) \\ 3t + 4z = 90 & (2) \end{cases}$$

If he wants to plant 5 acres of tomatoes and 2 acres of zucchini, will he get the crops in if the television meteorologist predicts rain in 72 hours?

72. corn 8 hr; lettuce 12 hr; $3\frac{1}{3}$ acres
73. 60 lb Arabica; 40 lb Robusta
74. 40 lb almonds; 10 lb peanuts
75. 70° and 20°
76. 30° and 150°
77. $\left(-\frac{3}{a}, 1\right)$
78. $\left(-1, \frac{4}{b}\right)$

72. Farmer's Daughter Farmer Green has talked his daughter, Haydee, into helping with some of the planting in the scenario in Problem 71. Haydee's specialty is planting corn and lettuce, which she has done for the past several years. One year it took her 48 hours to plant 3 acres of corn and 2 acres of lettuce. Another year it took her 68 hours to plant 4 acres of corn and 3 acres of lettuce. If c represents the number of hours to plant one acre of corn and l represents the number of hours required to plant one acre of lettuce, determine how long it takes Haydee to plant an acre of each crop by solving the following system of equations.

$$\begin{cases} 3c + 2l = 48 & (1) \\ 4c + 3l = 68 & (2) \end{cases}$$

If Farmer Green wants to have 4 acres of corn planted before the rains begin in 72 hours, how many acres of lettuce can Haydee expect to plant?

73. Making Coffee Suppose that you want to blend two coffees in order to obtain a new blend. The blend will be made with the best Arabica beans that sell for \$9.00 per pound and select African Robustas that sell for \$11.50 per pound to obtain 100 pounds of the new blend that will sell for \$10.00 per pound. How many pounds of the Arabica and Robusta beans are required? Solve the system

$$\begin{cases} a + r = 100 & (1) \\ 9a + 11.50r = 1000 & (2) \end{cases}$$

where a represents the number of pounds of Arabica beans and r represents the number of pounds of African Robustas beans to determine the answer.

74. Candy A candy store sells chocolate-covered almonds for \$6.50 per pound and chocolate-covered peanuts for \$4.00 per pound. The manager decides to make a bridge mix that combines the almonds with the peanuts. She wants the bridge mix to sell for \$6.00 per pound. How many pounds of chocolate-covered almonds and chocolate-covered peanuts are required to create 50 pounds of bridge mix? Solve the system

$$\begin{cases} a + p = 50 & (1) \\ 6.50a + 4.00p = 300 & (2) \end{cases}$$

where a represents the number of pounds of chocolate-covered almonds and p represents the number of pounds of chocolate-covered peanuts to determine the answer.

△ **75. Complementary Angles** Two angles are complementary if the sum of their measures is 90°. The measure of one angle is 10° more than three times the measure of its complement. If A and B represent the measures of the two complementary angles, determine the measure of each of the angles by solving the following system of equations.

$$\begin{cases} A + B = 90 & (1) \\ A = 10 + 3B & (2) \end{cases}$$

△ **76. Supplementary Angles** Two angles are supplementary if the sum of their measures is 180°. One-half the measure of one angle is 45° more than the measure of its supplement. If A and B represent the measures of the two supplementary angles, determine the measure of each of the angles by solving the following system of equations.

$$\begin{cases} A + B = 180 & (1) \\ \dfrac{A}{2} = 45 + B & (2) \end{cases}$$

Extending the Concepts

In Problems 77 and 78, solve the system of equations for x and y using the elimination method.

77. $\begin{cases} ax + 4y = 1 \\ -2ax - 3y = 3 \end{cases}$

78. $\begin{cases} 4x + 2by = 4 \\ 5x + 3by = 7 \end{cases}$

79. $\left(-3a - b, -\frac{3}{2}a - \frac{3}{2}b\right)$
80. $(-3a + 2b, 7a - 6b)$

In Problems 79 and 80, solve the system of equations for x and y using the elimination method.

79. $\begin{cases} -3x + 2y = 6a \\ x - 2y = 2b \end{cases}$

80. $\begin{cases} -7x - 3y = 4b \\ 3x + y = -2a \end{cases}$

PUTTING THE CONCEPTS TOGETHER (SECTIONS 9.1–9.3)

These problems cover important concepts from Sections 9.1 to 9.3. We designed these problems so that you can review the chapter so far and show your mastery of the concepts. Take time to work these problems before proceeding with the next section. The answers to these problems are located at the back of the text starting on page AN-25.

1. (a) No
   (b) Yes
   (c) No
2. (a) infinitely many
   (b) consistent
   (c) dependent
3. (a) one
   (b) consistent
   (c) independent
4. (1, 1)
5. (5, −1)
6. (1, −5)
7. (−1, 0)
8. (−4, 5)
9. (−3, −1)
10. $\left(-\frac{1}{3}, 3\right)$
11. infinitely many solutions
12. (2.5, 3)
13. no solution

1. Determine whether the given ordered pairs are solutions of the given system of equations.

$$\begin{cases} 4x + y = -20 \\ y = -\frac{1}{6}x + 3 \end{cases}$$

(a) $\left(3, \frac{5}{2}\right)$ **(b)** $(-6, 4)$ **(c)** $(-4, -4)$

2. Suppose that you begin to solve a system of linear equations where each equation is written in slope-intercept form. You notice that the slopes and *y*-intercepts of the two lines are equal.
(a) How many solutions exist?
(b) State whether the system is consistent or inconsistent.
(c) If the system is consistent, state whether the system in dependent or independent.

3. Suppose that you begin to solve a system of linear equations where each equation is written in slope-intercept form. The slopes of the two lines are not equal, but the *y*-intercepts are the same.
(a) How many solutions exist?
(b) State whether the system is consistent or inconsistent.
(c) If the system is consistent, state whether the system in dependent or independent.

In Problems 4–13, solve each system of equations using any method that you wish.

4. $\begin{cases} 4x + y = 5 \\ -x + y = 0 \end{cases}$

5. $\begin{cases} y = -\frac{2}{5}x + 1 \\ y = -x + 4 \end{cases}$

6. $\begin{cases} x = 2y + 11 \\ 3x - y = 8 \end{cases}$

7. $\begin{cases} 4x + 3y = -4 \\ x + 5y = -1 \end{cases}$

8. $\begin{cases} y = -2x - 3 \\ y = \frac{1}{2}x + 7 \end{cases}$

9. $\begin{cases} -2x + 4y = 2 \\ 3x + 5y = -14 \end{cases}$

10. $\begin{cases} -\frac{3}{4}x + \frac{2}{3}y = \frac{9}{4} \\ 3x - \frac{1}{2}y = -\frac{5}{2} \end{cases}$

11. $\begin{cases} 3(y + 3) = 1 + 4(3x - 1) \\ x = \frac{y}{4} + 1 \end{cases}$

12. $\begin{cases} 0.4x - 2.5y = -6.5 \\ x + y = 5.5 \end{cases}$

13. $\begin{cases} 2(y + 1) = 3x + 4 \\ x = \frac{2}{3}y - 3 \end{cases}$

9.4 Solving Direct Translation, Geometry, and Uniform Motion Problems Using Systems of Linear Equations

OBJECTIVES

1. Model and Solve Direct Translation Problems
2. Model and Solve Geometry Problems
3. Model and Solve Uniform Motion Problems

Preparing for Solving Direct Translation, Geometry, and Uniform Motion Problems Using Systems of Linear Equations

Before getting started, take the following readiness quiz. If you get a problem wrong, go back to the section cited and review the material.

1. If you travel at an average speed of 45 miles per hour for 3 hours, how far will you travel? [Section 2.7, pp. 143–145]

2. Suppose that you have \$3600 in a savings account. The bank pays 1.5% annual simple interest. What is the interest paid after 1 month? [Section 2.4, pp. 104–105]

Teaching Tip
Remind students of key phrases that directly translate to math. For example, "more than" means "addition." While going over this concept, ask students what it would mean if Bob has \$5 more than Amy. Bob's amount = Amy's amount + \$5.

Recall, in Sections 2.5–2.7 we modeled and solved problems using a linear equation with a single variable. In this section, we learn how to model problems using a system of equations. The problems that we present in this section will have two unknowns, and the unknowns will be related through a system of equations. This is different from the problems in Sections 2.5–2.7 because the problems in Sections 2.5–2.7 had a single unknown whose value was determined from a single equation. If you haven't done so already, go back to Section 2.5 and review the problem-solving strategy given on page 122.

1 Model and Solve Direct Translation Problems

One class of problems that we discussed in Chapter 2 was direct translation problems. Let's look at a couple of examples where direct translation results in a system of two linear equations containing two unknowns.

Classroom Example ➤
The sum of two numbers is 62. The second number is 50 less than 3 times the first number. Find the numbers.
Answer: 28 and 34

EXAMPLE 1 Fun with Numbers

The sum of two numbers is 45. Twice the first number minus the second number is 27. Find the numbers.

Solution

Step 1: Identify We are looking for two unknown numbers.

Step 2: Name Let x represent the first number and y represent the second number.

Step 3: Translate The sum of the two numbers is 45, so we know that

$$x + y = 45 \quad \text{Equation (1)}$$

Twice the first number minus the second number is 27. So

$$2x - y = 27 \quad \text{Equation (2)}$$

Equations (1) and (2) are combined to form the following system:

$$\begin{cases} x + y = 45 & (1) \\ 2x - y = 27 & (2) \end{cases} \quad \text{The Model}$$

Step 4: Solve We will use the method of elimination since the coefficients of y are opposites.

$$\begin{cases} x + y = 45 & (1) \\ 2x - y = 27 & (2) \end{cases}$$

$$\text{Add equations (1) and (2):} \quad 3x = 72$$

$$\text{Divide both sides by 3:} \quad x = 24$$

Preparing for...Answers **1.** 135 miles **2.** \$4.50

Let $x = 24$ in equation (1) and solve for y.

Equation (1):	$x + y = 45$
Let $x = 24$:	$24 + y = 45$
Subtract 24 from both sides:	$y = 21$

Step 5: Check The sum of 24 and 21 is 45. Twice 24 minus 21 is 48 minus 21, which equals 27.

Step 6: Answer The two numbers are 24 and 21.

QUICK ✓

1. The sum of two numbers is 104. The second number is 25 less than twice the first number. Find the numbers.

2 Model and Solve Geometry Problems

Teaching Tip
Ask students to review the geometry formulas on pages 105–106.

Formulas from geometry are often needed to solve certain types of problems. The formula that you will need is determined by the problem. Remember that a formula is a model—and in geometry, formulas describe relationships and shapes.

Classroom Example Pedro needs 320 feet of fence to enclose a rectangular lot. The length of the lot is 40 feet more than the width. What are the dimensions of the lot?

Answer: 60 feet by 100 feet

EXAMPLE 2 Enclosing a Yard with a Fence

The Freese Family just bought a wooded lot that is on a lake. They want to enclose the lot with a fence but do not want to put up a fence along the lake. Dave Freese determined that he will need 240 feet of fence. He also knows that the width of the lot is 30 feet more than the length. See Figure 11. What are the dimensions of the lot?

Figure 11

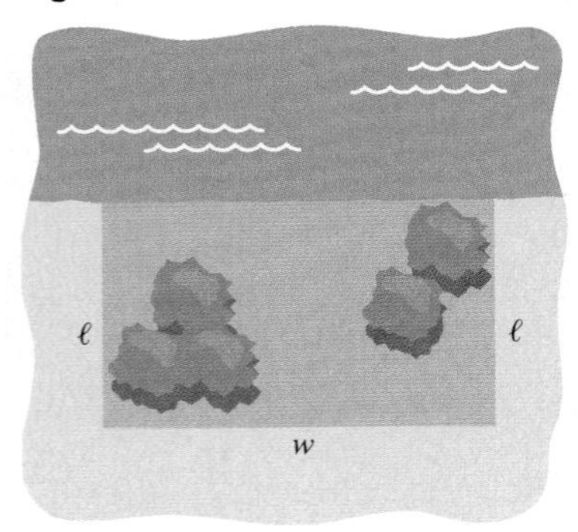

Solution

Step 1: Identify We are looking for the length and the width of the lot.

Step 2: Name Let l represent the length and w represent the width of the lot.

Step 3: Translate The perimeter P of a rectangle excluding the waterfront is $P = 2l + w$, where l is the length and w is the width. So we know that

$$2l + w = 240 \quad \text{Equation (1)}$$

In addition, the width is 30 feet more than the length.

$$w = l + 30 \quad \text{Equation (2)}$$

Equations (1) and (2) are combined to form the following system:

$$\begin{cases} 2l + w = 240 & (1) \\ w = l + 30 & (2) \end{cases} \quad \text{The Model}$$

Step 4: Solve We will use the method of substitution since equation (2) is already solved for w.

Equation (2):	$w = l + 30$
Let $w = l + 30$ in equation (1):	$2l + (l + 30) = 240$
Combine like terms:	$3l + 30 = 240$
Subtract 30 from both sides:	$3l = 210$
Divide both sides by 3:	$l = 70$

Let $l = 70$ in equation (2) and solve for w.

Equation (2): $w = l + 30$

Let $l = 70$: $w = 70 + 30$

Simplify: $w = 100$

Step 5: Check With $l = 70$ and $w = 100$, the perimeter (excluding the waterfront) would be $2(70) + 100 = 240$ feet. The width (100 feet) is 30 feet more than the length (70 feet).

Step 6: Answer The length of the backyard is 70 feet and the width is 100 feet.

QUICK ✓

2. A rectangular field has a perimeter of 400 yards. The length of the yard is three times the width. What is the length of the yard? What is the width of the yard?

Recall from Section 2.7 that complementary angles are two angles whose measures sum to 90°. Each angle is called the *complement* of the other. For example, the angles shown in Figure 12(a) below are complements because their measures sum to 90°. Supplementary angles are two angles whose measures sum to 180°. Each angle is called the *supplement* of the other. The angles shown in Figure 12(b) are supplementary.

Figure 12

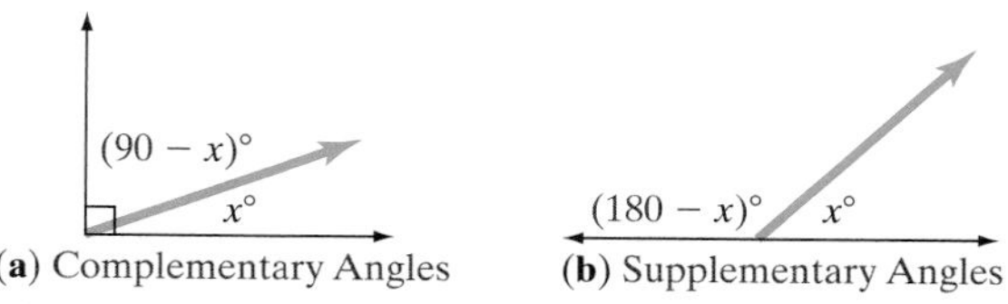

(a) Complementary Angles **(b)** Supplementary Angles

The next example was first presented in Section 2.7. In Section 2.7 we solved the problem by developing a model that involved only one unknown. We now show how the same problem can be solved by developing a model of two equations containing two unknowns.

Classroom Example Find the measure of two supplementary angles such that the measure of the larger angle is 32° greater than the measure of the smaller angle.

Answer: 74° and 106°

EXAMPLE 3 Solve a Complementary Angle Problem

Find the measure of two complementary angles such that the measure of the larger angle is 6° greater than twice the measure of the smaller angle.

Solution

Step 1: Identify This is a complementary angle problem. We are looking for the measure of the two angles whose sum is 90°.

Step 2: Name Let x represent the measure of the smaller angle and y represent the measure of the other angle.

Step 3: Translate Because these are complementary angles, we know that the sum of the measures of the angles must be 90°. So we have

$$x + y = 90 \quad \text{Equation (1)}$$

The measure of the larger angle y is 6° more than twice the measure of the smaller angle, x. This leads to the following equation.

$$y = 2x + 6 \quad \text{Equation (2)}$$

Using equations (1) and (2), we obtain the following system.

$$\begin{cases} x + y = 90 & (1) \\ y = 2x + 6 & (2) \end{cases} \quad \text{The Model}$$

Step 4: Solve Since equation (2) is already solved for y, we let $y = 2x + 6$ in equation (1).

Equation (1):	$x + y = 90$
Let $y = 2x + 6$:	$x + (2x + 6) = 90$
Combine like terms:	$3x + 6 = 90$
Subtract 6 from each side of the equation:	$3x = 84$
Divide each side by 3:	$x = 28$

Now we let $x = 28$ in equation (2), $y = 2x + 6$, to find the measure of the larger angle, y.

$$y = 2x + 6$$

$$x = 28: \quad y = 2(28) + 6$$

$$y = 62$$

Step 5: Check The measure of the smaller angle, x, is 28°. The measure of the larger angle, y, is 62°. Is the sum of the measures of these angles 90°? Yes, since $28° + 62° = 90°$. Is the measure of the larger angle 6° more than twice the measure of the smaller angle? Yes, since 2 times 28° is 56°. Six more than this is $56° + 6° = 62°$. The answers check!

Step 6: Answer The two complementary angles measure 28° and 62°.

QUICK ✓ *Solve each problem for the unknown angle measures.*

3. Find two complementary angles such that the measure of the larger angle is 18° more than the measure of the smaller angle.
4. Find two supplementary angles such that the measure of the larger angle is 16° less than three times the measure of the smaller angle.

Teaching Tip
To review $d = rt$, ask students the following: If Kathy rides her bicycle at an average speed of 18 miles per hour for 2 hours, how far will she travel?

3 Model and Solve Uniform Motion Problems

Let's now look at a problem involving uniform motion. Remember, these problems use the fact that distance equals rate times time ($d = rt$).

Classroom Example
Suppose a plane flying west a distance of 600 miles takes 6 hours. The return trip takes 5 hours. Find the airspeed of the plane and the effect wind resistance has on the plane.

Answer: Airspeed: 110 miles per hour; effect of wind: 10 miles per hour

EXAMPLE 4 Uniform Motion—Flying a Piper Aircraft

The airspeed of a plane is its speed through the air. This speed is different from the plane's groundspeed—its speed relative to the ground. The groundspeed of an airplane is affected by the speed of the wind. Suppose that a Piper aircraft flying west a distance of 500 miles takes 5 hours. The return trip takes 4 hours. Find the airspeed of the plane and the effect wind resistance has on the plane.

Solution

Step 1: Identify This is a uniform motion problem. We want to determine the airspeed of the plane and the effect of wind resistance.

Step 2: Name There are two unknowns in the problem—the airspeed of the plane and the impact of wind resistance on the plane. We will let a represent the airspeed of the plane and w represent the impact of wind resistance.

Step 3: Translate Going west, the plane is flying into the jet stream, so that the plane is slowed down by the wind. Therefore, the groundspeed of the plane will be $a - w$.

Going east, the wind is helping the plane. Therefore, the groundspeed of the plane will be $a + w$. We set up Table 3.

Table 3

	Distance	Rate	Time
With Wind (East)	500	$a + w$	4
Against Wind (West)	500	$a - w$	5

Work Smart

Instead of distributing 4 into the equation $4(a + w) = 500$, we could divide both sides by 4. The same approach can be taken with the equation $5(a - w) = 500$—divide both sides of the equation by 5. Try solving the system this way. Is it easier?

Going with the wind, we have the equation

$$4(a + w) = 500 \quad \text{or} \quad 4a + 4w = 500 \quad (1)$$

Going against the wind, we have the equation

$$5(a - w) = 500 \quad \text{or} \quad 5a - 5w = 500 \quad (2)$$

Combining equations (1) and (2), we form a system of two linear equations containing two unknowns.

$$\begin{cases} 4a + 4w = 500 & (1) \\ 5a - 5w = 500 & (2) \end{cases} \quad \text{The Model}$$

Step 4: Solve We will use the elimination method by multiplying equation (1) by 5 and equation (2) by 4 and then adding equations (1) and (2).

$$\begin{cases} 5(4a + 4w) = 5(500) & (1) \\ 4(5a - 5w) = 4(500) & (2) \end{cases}$$

$$\text{Distribute:} \quad \begin{cases} 20a + 20w = 2500 & (1) \\ 20a - 20w = 2000 & (2) \end{cases}$$

$$\text{Add:} \quad 40a = 4500$$

$$\text{Divide both sides by 40:} \quad a = 112.5$$

We use equation (1) with $a = 112.5$ to find the effect of wind resistance.

$$\begin{aligned} \text{Equation (1):} \quad 4a + 4w &= 500 \\ a = 112.5: \quad 4(112.5) + 4w &= 500 \\ 450 + 4w &= 500 \\ \text{Subtract 450 from both sides:} \quad 4w &= 50 \\ \text{Divide both sides by 4:} \quad w &= 12.5 \end{aligned}$$

Step 5: Check Flying west, the groundspeed of the plane is $112.5 - 12.5 = 100$ miles per hour. Flying west, the plane flies 500 miles in 5 hours for an average speed of 100 miles per hour. Flying east, the groundspeed of the plane is $112.5 + 12.5 = 125$ miles per hour. Flying east, the plane flies 500 miles in 4 hours for an average speed of 125 miles per hour. Everything checks!

Step 6: Answer The airspeed of the plane is 100 miles per hour. The impact of wind resistance on the plane is 12.5 miles per hour.

QUICK ✓

5. Suppose that a plane flying 1200 miles west requires 4 hours and flying 1200 miles east requires 3 hours. Find the airspeed of the plane and the effect wind resistance has on the plane.

9.4 Exercises

For Extra Help: Student Solutions Manual  CD Video  PH Math/Tutor Center  MathXL Tutorials on CD MathXL® MyMathLab

1. perimeter; $P = 2l + 2w$
2. $2y + 3$
3. 180°; 90°
4. True
5. True
6. True
7. No; Answers may vary.
8. No; Answers may vary.
9. $2l + 2w$
10. $l = 3w - 8$
11. 33, 49
12. 19, 36
13. 14, 37
14. 14, 18
15. Thursday 35,000; Friday 42,000
16. 60 win; 102 lose

Concepts and Vocabulary

In Problems 1–3, fill in the blanks.

1. The distance around the outside of a rectangle is called the ________ and uses the formula ________.

2. If a number x is three more than twice a number y, then $x =$ ________.

3. Supplementary angles are two angles whose measures sum to ________, while the measures of two angles that are complementary sum to ________.

In Problems 4–6, answer True or False to each statement.

4. If the sum of a number x and three times a second number y is 45, then $x + 3y = 45$.

5. Complementary angles are angles whose measures sum to 90°.

6. In uniform motion problems we use the equation $d = rt$, where d is distance, r is rate, and t is time.

7. While solving a supplementary angle problem, you find that one of the angles has a measure of 190°. Is this possible? Why or why not?

8. While solving a uniform motion problem, you find that the average speed of a boat in still water is −12 miles per hour. Is this possible? Why or why not?

Building Skills

△ **9.** The perimeter of a rectangle is 59 inches. The length is 5 inches less than twice the width. Let l represent the length of the rectangle and let w represent the width of the rectangle. Complete the system of equations.

$$\begin{cases} l = 2w - 5 \\ \_\_\_\_\_\_\_\_ = 59 \end{cases}$$

△ **10.** The perimeter of a rectangle is 212 centimeters. The length is 8 centimeters less than three times the width. Let l represent the length of the rectangle and let w represent the width of the rectangle. Complete the system of equations.

$$\begin{cases} 2w + 2l = 212 \\ \_\_\_\_\_\_\_\_ \end{cases}$$

Applying the Concepts

11. Fun with Numbers Find two numbers whose sum is 82 and whose difference is 16.

12. Fun with Numbers Find two numbers whose sum is 55 and whose difference is 17.

13. Fun with Numbers Two numbers sum to 51. Twice the first subtracted from the second is 9. Find the numbers.

14. Fun with Numbers Two numbers sum to 32. Twice the first subtracted from the second is −22. Find the numbers.

15. Oakland Baseball The attendance at the games on two successive nights of Oakland A's baseball was 77,000. The attendance on Thursday's game was 7000 more than two-thirds of the attendance at Friday night's game. How many people attended the baseball game each night?

16. Winning Baseball The number of games the Oakland A's are expected to win this year is 8 less than two-thirds of the number that they are expected to lose. If there are 162 games in a season, how many games are the A's expected to win?

17. $12,000 in stocks; $9000 in bonds
18. $16,000 in stocks; $24,000 in bonds
19. length 12 ft; width 9 ft
20. length 36.25 ft; width 26.25 ft
21. length 25 m; width 10 m
22. height 36 in.; width 45 in.
23. 40°, 50°
24. 25°, 65°
25. 22.5°, 157.5°
26. 96°, 84°
27. current 0.4 mph; still-water 3.9 mph
28. airplane 485 mph; wind 30 mph
29. bike 11 mph; wind 1 mph
30. still-water rate 5 mph; current speed 1 mph
31. northbound 60 mph; southbound 72 mph
32. Gabriella's horse 2 mph; Monica's horse 6 mph

17. Investments Suppose that you received an unexpected inheritance of $21,000. You have decided to invest the money by placing some of the money in stocks and the remainder in bonds. To diversify, you decide that four times the amount invested in bonds should equal three times the amount invested in stocks. How much should be invested in stocks? How much should be invested in bonds?

18. Investments Marge and Homer have $40,000 to invest. Their financial advisor has recommended that they diversify by placing some of the money in stocks and the remainder in bonds. Based upon current market conditions, he has recommended that two times the amount in bonds should equal three times the amount invested in stocks. How much should be invested in stocks? How much should be invested in bonds?

△ **19. Fencing a Garden** Melody Jackson wishes to enclose a rectangular garden with fencing, using the side of her garage as one side of the rectangle. A neighbor gave her 30 feet of fencing, and Melody wants the length of the garden along the garage to be 3 feet more than the width. What are the dimensions of the garden?

△ **20. Perimeter of a Parking Lot** A rectangular parking lot has a perimeter of 125 feet. The length of the parking lot is 10 feet more than the width. What is the length of the parking lot? What is the width?

△ **21. Perimeter** The perimeter of a rectangle is 70 meters. If the width is 40% of the length, find the dimensions of the rectangle.

△ **22. Window Dimensions** The perimeter of a rectangular window is 162 inches. If the height of the window is 80% of the width, find the dimensions of the window.

△ **23. Working with Complements** The measure of one angle is 15° more than half the measure of its complement. Find the measures of the two angles.

△ **24. Working with Complements** The measure of one angle is 10° less than the measure of three times its complement. Find the measures of the two angles.

△ **25. Finding Supplements** The measure of one angle is 30° less than one-third the measure of its supplement. Find the measures of the two angles.

△ **26. Finding Supplements** The measure of one angle is 20° more than two-thirds the measure of its supplement. Find the measures of the two angles.

27. Kayaking Michael is kayaking in the Chesapeake Bay. He can paddle 3.5 mph against the current and 4.3 mph when he rows with the current. Find the speed of the current and the speed Michael can paddle in still water.

28. Southwest Airlines Plane A Southwest Airlines plane can fly 455 mph against the wind and 515 mph when it flies with the wind. Find the speed of the wind and the groundspeed of the airplane.

29. Biking Suppose that Jose bikes into the wind for 60 miles for 6 hours. After a long rest, he returns (with the wind at his back) in 5 hours. Determine the speed at which Jose can ride his bike in still air and determine the impact that the wind had on his speed.

30. Rowing On Monday afternoon, you rowed your boat with the current for 4.5 hours and covered 27 miles, stopping in the evening at a campground. On Tuesday morning you returned to your starting point against the current in 6.75 hours. Find the speed of the current and the rate at which you row in still water.

31. Outbound from Chicago Two trains leave Chicago going opposite directions, one going north and the other going south. The northbound train is traveling 12 mph slower than the southbound train. After 4 hours the trains are 528 miles apart. Find the speed of each train.

32. Horseback Riding Monica and Gabriella enjoy riding horses at a dude ranch in Colorado. They decide to go down different trails, agreeing to meet back at the ranch later in the day. Monica's horse is going 4 mph faster than the Gabriella's, and after 2.5 hours, they are 20 miles apart. Find the speed of each horse.

33. 4 hr
34. 1 hr, 15 min
35. wind 25 mph; Piper 175 mph
36. 30 mph
37. Sam 24 hr; Diane 8 hr

33. Riding Bikes Vanessa and Richie are riding their bikes down a trail to the next campground. Vanessa rides at 10 mph while Richie rides at 7 mph. Since Vanessa is a little speedier, she stays behind and cleans up camp for 30 minutes before leaving. How long has Richie been riding when Vanessa is 7 miles ahead of Richie?

34. Running a Marathon Rafael and Edith are running in a marathon to raise money for breast cancer. Rafael can run at 12 mph and Edith runs at 10 mph. Unfortunately, Rafael lost his car keys and went back to look for them while Edith started down the course. If Rafael was back at the starting line 15 minutes after Edith left, how long will it take him to catch up to her?

35. Computing Wind Speed With a tail wind, a small Piper aircraft can fly 600 miles in 3 hours. Against this same wind, the Piper can fly the same distance in 4 hours. Find the average wind speed and the average airspeed of the Piper.

36. Computing Wind Speed The average airspeed of a single engine aircraft is 150 miles per hour. If the aircraft flew the same distance in 2 hours with the wind as it flew in 3 hours against the wind, what was the wind speed?

Extending the Concepts

37. Painting the House Sam and Diane, working together, can paint a garage in 6 hours. When working alone, Diane paints a similar garage three times faster than her apprentice, Sam. Assuming no gain or loss of efficiency, how long should it take each person to complete such a job working alone?

9.5 Solving Mixture Problems Using Systems of Linear Equations

OBJECTIVES

1. Draw Up a Plan for Modeling Mixture Problems
2. Set Up and Solve Money Problems Using the Mixture Model
3. Set Up and Solve Dry Mixture and Percent Mixture Problems

Preparing for Solving Mixture Problems Using Systems of Linear Equations

Before getting started, take the following readiness quiz. If you get a problem wrong, go back to the section cited and review the material.

1. Suppose that Roberta has a credit card balance of \$1200. Each month, the credit card charges 14% annual simple interest on any outstanding balances. What is the interest that Roberta will be charged on this loan after one month? What is Roberta's credit card balance after one month? [Section 2.4, pp. 104–105]

1 Draw Up a Plan for Modeling Mixture Problems

Sometimes ingredients must be combined to form a new mixture. Problems that involve mixing two or more substances are called **mixture problems.** They can be set up using the model *number of units of the same kind* · *rate* = *amount*. The *rates* that we use in mixture problems are items such as cost per person, interest rates, or cost per pound. Using a chart such as the one below helps to organize the information given in the problem.

Work Smart
Use a chart to organize your thoughts and keep track of information.

	Number of Units	·	Rate	=	Amount
Item 1					
Item 2					
Total					

Because each problem will be slightly different, we will adjust the titles of the categories, but the general equation *number of units of the same kind* · *rate* = *amount* will remain constant. Let's begin by practicing filling in the chart.

Preparing for...Answer **1.** \$14; \$1214

Classroom Example ➤
A group of neighbors rented a bus and traveled to a play. There were 10 more adults than children and total admission was $620. If the adults paid $12 each and the children paid $8 each, how many adults and how many children went to the play? Fill in a chart that summarizes the information in the problem. Do not solve the problem.

Answer:

	Number	Cost	Amount
Adults	a	12	$12a$
Children	c	8	$8c$
Total			620

Work Smart
Not every entry in the chart must be filled.

EXAMPLE 1 Set Up a Chart for Ticket Sales Problem Using the Mixture Model

A class of school children and their adult chaperones took a field trip to the San Diego Zoo. There were 20 more children than adults and total admission to the zoo was $230. If the children paid $5 each and the adults paid $8 each, how many adults and how many children went on the trip? Fill in a chart that summarizes the information in the problem. Do not solve the problem.

Solution

Both the number of children and the number of adults are unknown. We let a represent the number of adults on the field trip and c represent the number of children. We can now fill in Table 4 with the quantities that we know.

Table 4

	Number	•	Cost per Person	=	Amount
Adults	a		8		$8a$
Children	c		5		$5c$
Total					230

QUICK ✓

1. A one-day admission ticket to Cedar Point Amusement Park costs $36.95 for adults and $21.95 for children. Two families purchased five more adult tickets than children's tickets and spent $302.55 for the tickets. How many adult tickets did the families purchase? Fill in a chart that summarizes the information in the problem. Do not solve the problem.

2 Set Up and Solve Money Problems Using the Mixture Model

We are now ready to solve problems using the mixture model. We will set up a table to organize the given information, and then use the table to develop a model (equation). We continue to use the six-step procedure that was introduced in Section 2.5.

Classroom Example ➤
A fourth-grade class contributed $17.70 in dimes and quarters to the Red Cross. In all there were 96 coins. Find the number of dimes and quarters that the children contributed to the Red Cross.

Answer: 42 dimes and 54 quarters

Work Smart
It's always a good idea to name your variable so that it reminds you of what it represents, as in q for the number of quarters.

EXAMPLE 2 Solve a Coin Problem Using the Mixture Model

A third-grade class contributed $9.20 in dimes and quarters to the Red Cross. In all there were 56 coins. Find the number of dimes and the number of quarters that the children contributed to the Red Cross.

Solution

Step 1: Identify This is a money problem and we can use the mixture model to solve it. We want to know the number of dimes and the number of quarters that the children contributed, and we know that $9.20 was given.

Step 2: Name Let q represent the number of quarters and let d represent the number of dimes.

Step 3: Translate We fill in Table 5 with the information that we know.

Table 5

	Number of Coins	·	Value per Coin in Dollars	=	Total Value
Quarters	q		0.25		$0.25q$
Dimes	d		0.10		$0.10d$
Total	56				9.20

The total value of the quarters plus the total value of the dimes equals \$9.20. Based on the information in Table 5, we have that

$$\overbrace{0.25q}^{\text{value of quarters}} + \overbrace{0.10d}^{\text{value of dimes}} = \overbrace{9.20}^{\text{total value of coins}} \qquad \text{Equation (1)}$$

There are a total of 56 coins, so

$$q + d = 56 \qquad \text{Equation (2)}$$

We use equations (1) and (2) to form a system of equations.

$$\begin{cases} 0.25q + 0.1d = 9.20 & (1) \\ q + d = 56 & (2) \end{cases} \quad \text{The Model}$$

Step 4: Solve We will use the method of elimination.

Multiply equation (2) by -0.1:
$$\begin{cases} 0.25q + 0.1d = 9.20 & (1) \\ -0.1(q + d) = -0.1(56) & (2) \end{cases}$$

Distribute -0.1 in equation (2):
$$\begin{cases} 0.25q + 0.1d = 9.20 & (1) \\ -0.1q - 0.1d = -5.6 & (2) \end{cases}$$

Add equations (1) and (2): $0.15q = 3.6$

Divide both sides by 0.15: $q = 24$

Now we need to determine the value of d. We let $q = 24$ in equation (2) and solve for d.

Equation (2): $q + d = 56$

$q = 24$: $24 + d = 56$

Subtract 24 from both sides: $d = 32$

Step 5: Check Because q represents the number of quarters, we know that the children collected 24 quarters. They collected 32 dimes. Is the total value of the dimes and quarters collected equal to \$9.20? Because $24(\$0.25) + 32(\$0.10) = \$6.00 + \$3.20 = \$9.20$, our answer is correct.

Step 6: Answer The children collected 24 quarters and 32 dimes for the Red Cross.

QUICK ✓

2. You have a piggy bank containing a total of 85 coins in dimes and quarters. If the piggy bank contains \$14.50, how many dimes are there in the piggy bank?

Teaching Tip
Review the simple interest formula $I = Prt$. Do an example with $P = \$1000$, $r = 0.08$, and $t = \frac{1}{12}$. This might be the monthly interest on a credit card.

Recall, in Section 2.4 we introduced the simple interest formula $I = Prt$, where I is interest, P is principal (either an amount borrowed or deposited), r is the annual interest rate (expressed as a decimal), and t is time (measured in years). When solving interest problems using the mixture model in this section we will consider only the situation in which time is 1 year, so the simple interest formula reduces to *interest* = *principal* · *rate* · 1, or *interest* = *principal* · *rate*.

Classroom Example ➔
You invested \$8000 in a stock fund that payed 10.5% annual interest and a bond fund that payed 7% annual interest. You earned \$735 in interest at the end of one year. How much was invested in each account?

Answer: \$5000 in stocks, \$3000 in bonds

EXAMPLE 3 Solve an Interest Problem Using the Mixture Model

You have \$12,000 invested in a money market account that payed 6% annual interest and a stock fund that payed 8.5% annual interest. Suppose that you earned \$807.50 in interest at the end of one year. How much was invested in each account?

Solution

Step 1: Identify This is a mixture problem involving simple interest. We need to know how much was invested in the money market account and how much was invested in the stock fund to earn \$807.50 in interest.

Step 2: Name Let m represent the amount invested in the money market account and let s represent the amount invested in stocks.

Step 3: Translate We organize the information given in Table 6.

Table 6

	Principal \$	·	Rate %	=	Interest \$
Money Market	m		0.06		$0.06m$
Stock Fund	s		0.085		$0.085s$
Total	12,000				807.50

You earned interest of \$807.50 on your principal of \$12,000. The total interest is the sum of the interest from the money market account and the stock fund.

Interest from money market account + Interest from stock fund = \$807.50

Work Smart
We use the simple interest formula $I = Prt$, where P is the principal, r is the interest rate (as a decimal), and t is time (in years).

Since $t = 1$ year, we have

$$\underbrace{0.06m}_{\text{Interest earned from money market}} + \underbrace{0.085s}_{\text{Interest earned from stock fund}} = \underbrace{807.50}_{\text{Total interest earned}} \quad \text{Equation (1)}$$

In addition, the total investment is to be \$12,000 so that the amount invested in the money market account plus the amount invested in the stock fund equals \$12,000:

$$m + s = 12{,}000 \quad \text{Equation (2)}$$

We use equations (1) and (2) to form a system of equations.

$$\begin{cases} 0.06m + 0.085s = 807.50 & (1) \\ m + s = 12{,}000 & (2) \end{cases} \quad \text{The Model}$$

Work Smart
In Step 4, we could have multiplied both sides of equation (1) by 1000 to eliminate decimals and then solved the system

$$\begin{cases} 60m + 85s = 807{,}500 & (1) \\ m + s = 12{,}000 & (2) \end{cases}$$

Which method do you prefer?

Step 4: Solve We will use the method of elimination.

Multiply equation (2) by -0.06:
$$\begin{cases} 0.06m + 0.085s = 807.50 & (1) \\ -0.06(m + s) = -0.06(12{,}000) & (2) \end{cases}$$

Distribute -0.06 in equation (2):
$$\begin{cases} 0.06m + 0.085s = 807.50 & (1) \\ -0.06m - 0.06s = -720 & (2) \end{cases}$$

Add equations (1) and (2): $0.025s = 87.50$

Divide both sides by 0.025: $s = 3500$

Now we need to determine the value of m. We let $s = 3500$ in equation (2) and solve for m.

Equation (2): $m + s = 12{,}000$

$s = 3500$: $m + 3500 = 12{,}000$

Subtract 3500 from both sides: $m = 8500$

Step 5: Check The simple interest earned each year on the money market account is $(\$8500)(0.06)(1) = \$510$. The simple interest earned each year on the stock fund is $(\$3500)(0.085)(1) = \$297.50$. The total interest earned is $\$510 + \$297.50 = \$807.50$. You earned \$807.50 per year in interest, so this agrees with the information presented in the problem. In addition, the total amount invested is $\$8500 + \$3500 = \$12{,}000$.

Step 6: Answer You invested \$8500 in the money market account and \$3500 in the stock fund.

QUICK ✓

3. Faye has recently retired and requires an extra \$5500 per year in income. She has \$90,000 to invest and can invest in either an Aa-bond that pays 5% per annum or a B-rated bond paying 7% annually. How much should be placed in each investment for Faye to achieve her goal?

3 Set Up and Solve Dry Mixture and Percent Mixture Problems

Mixture of Two Substances—Dry Mixture Problems

Often, new blends are created by mixing two quantities. For example, a chef might mix buckwheat flour with wheat flour to make buckwheat pancakes. Or a coffee shop might mix two different types of coffee to create a new coffee blend.

Classroom Example ➤
A candy store sells chocolate-covered almonds for \$7 per pound and chocolate-covered peanuts for \$3 per pound. The manager decides to make a bridge mix that combines the almonds with the peanuts. She wants the bridge mix to sell for \$4 per pound. How many pounds of chocolate-covered almonds and chocolate-covered peanuts are required to create 20 pounds of bridge mix so that the value of the mix is the same as the value of the individual items?

Answer: 5 pounds of almonds and 15 pounds of peanuts.

EXAMPLE 4 Solve a Dry Mixture Problem Using the Mixture Model

A store manager mixed together nuts and M&M's to make a trail mix. The manager wants to create 10 pounds of the mix and sell it for \$5.10 per pound. If the price of the nuts was \$3 per pound and the price of the M&M's was \$6 per pound, how many pounds of each type did he use so that the value of the mix is the same as the value of the individual items?

Solution

Step 1: Identify This is a mixture problem. We want to know the number of pounds of nuts and the number of pounds of M&M's that are required in the trail mix.

Step 2: Name Let n represent the number of pounds of nuts and m represent the number of pounds of M&M's that are required.

Step 3: Translate We are told that there is to be no difference in revenue between selling the nuts and M&M's separately versus the blend. This means that if the blend contains one pound of nuts and one pound of M&M's, we should collect $\$3(1) + \$6(1) = \$9$ because that is how much we would collect if we sold the nuts and M&M's separately.

We set up Table 7.

Table 7

	Price \$/Pound	·	Number of Pounds	=	Revenue
Nuts	3		n		$3n$
M&M's	6		m		$6m$
Blend	5.10		10		$5.10(10) = 51$

In general, if the mixture contains n pounds of nuts, we should collect $\$3n$. If the mixture contains m pounds of M&M's, we should collect $\$6m$. If the mixture sells for \$5.10 per pound and we make 10 pounds of the blend, then we should collect $\$5.10(10) = \$51$ for the blend.

$$\binom{\text{Price per pound}}{\text{of nuts}}\binom{\text{Pounds of}}{\text{nuts}} + \binom{\text{Price per pound}}{\text{of M\&M's}}\binom{\text{Pounds of}}{\text{M\&M's}} = \binom{\text{Price per pound}}{\text{of blend}}\binom{\text{Pounds of}}{\text{blend}}$$

$$\$3 \cdot n + \$6 \cdot m = \$5.10 \cdot (10)$$

We have the equation

$$3n + 6m = 51 \quad \text{Equation (1)}$$

Work Smart
Remember that mixtures can include interest (money), solids (nuts), liquids (chocolate milk), and even gases (Earth's atmosphere).

The number of pounds of nuts plus the number of M&M's should equal 10 pounds. We have the equation

$$n + m = 10 \quad \text{Equation (2)}$$

We use equations (1) and (2) to form a system of equations.

$$\begin{cases} 3n + 6m = 51 & (1) \\ n + m = 10 & (2) \end{cases} \quad \text{The Model}$$

Work Smart
The system $\begin{cases} 3n + 6m = 51 \\ n + m = 10 \end{cases}$ could also have been solved by elimination. Which method do you prefer?

Step 4: Solve We solve the system by substitution by solving equation (2) for n.

Equation (2):	$n + m = 10$
Subtract m from both sides:	$n = 10 - m$
Let $n = 10 - m$ in equation (1):	$3(10 - m) + 6m = 51$
Use the Distributive Property to remove the parentheses:	$30 - 3m + 6m = 51$
Combine like terms:	$30 + 3m = 51$
Subtract 30 from both sides:	$3m = 21$
Divide both sides by 3:	$m = 7$

With $m = 7$ pounds, we have that $n = 10 - m = 10 - 7 = 3$ pounds.

Step 5: Check It appears that we should mix 3 pounds of nuts with 7 pounds of M&M's. The 7 pounds of M&M's would sell for $\$6(7) = \$42$ and the 3 pounds of nuts would sell for $\$3(3) = \$9$; the total revenue would be $\$42 + \$9 = \$51$, which equals the revenue obtained from selling the blend. This checks with the information presented in the problem. In addition, the blend weighs 7 pounds + 3 pounds = 10 pounds, as required.

Step 6: Answer The manager should mix 3 pounds of nuts with 7 pounds of M&M's to make the blend.

QUICK ✓

4. A coffee house has Brazilian coffee that sells for $6 per pound and Colombian coffee that sells for $10 per pound. How many pounds of each coffee should be mixed to obtain 20 pounds of a blend that costs $9 per pound?

Classroom Example
You work in the chemistry stock room and your instructor has asked you to prepare 6 liters of 10% hydrochloric acid (HCl). Looking through the supply room, you see that there is a bottle of 8% HCl and another of 14% HCl. How much of each should you mix so that your instructor has the required solution?

Answer: 4 liters of 8% HCl and 2 liters of 14% HCl.

Mixture of Two Substances—Percent Mixture Problems

The last example dealt with mixing two dry substances. We now turn our attention to mixing two liquids. These problems require using percents.

EXAMPLE 5 Solve a Percent Mixture Problem Using the Mixture Model

You work in the chemistry stock room and your instructor has asked you to prepare 4 liters of 15% hydrochloric acid (HCl). Looking through the supply room you see that there is a bottle of 12% HCl and another of 20% HCl. How much of each should you mix so that your instructor has the required solution?

Solution

Step 1: Identify This is a percent mixture problem so we use the mixture model. We want to know the number of liters of 12% HCl that must be mixed with a 20% HCl solution to prepare 4 liters of 15% HCl. We also know that we need 4 liters of the solution.

Step 2: Name We let x represent the number of liters of the 12% HCl solution and let y represent the number of liters of 20% HCl.

Step 3: Translate We fill in Table 8 with the information that we know.

Table 8

	Number of Liters	•	Concentration (part of solution that is pure HCl per liter)	=	Amount of Pure HCl
12% HCl Solution	x		0.12		$0.12x$
20% HCl Solution	y		0.20		$0.20y$
Total	4		0.15		(0.15)(4)

The amount of 12% HCl solution plus the amount of 20% HCl solution should yield 4 liters of 15% HCl solution. This leads to the following equation:

$$\overbrace{0.12x}^{\text{part that is 12\% HCl}} + \overbrace{0.20y}^{\text{part that is 20\% HCl}} = \overbrace{(0.15)(4)}^{\text{part of the total that is HCl}} \quad \text{Equation (1)}$$

The number of liters of HCl solution should equal 4 liters. We have the equation

$$x + y = 4 \quad \text{Equation (2)}$$

We use equations (1) and (2) to form a system of equations.

$$\begin{cases} 0.12x + 0.20y = 0.6 & (1) \\ x + y = 4 & (2) \end{cases} \quad \text{The Model}$$

Step 4: Solve We solve the system by substitution by solving equation (2) for y.

Equation (2):	$x + y = 4$
Subtract x from both sides:	$y = 4 - x$
Substitute $y = 4 - x$ in equation (1):	$0.12x + 0.20(4 - x) = 0.6$
Use the Distributive Property:	$0.12x + 0.8 - 0.20x = 0.6$
Combine like terms:	$-0.08x + 0.8 = 0.6$
Subtract 0.8 from each side of the equation:	$-0.08x = -0.2$
Divide each side by -0.08:	$x = 2.5$

Because x represents the number of liters of 12% HCl solution, $x = 2.5$ liters of 12% HCl solution is required. We need $4 - x = 4 - 2.5 = 1.5$ liters of 20% HCl.

Step 5: Check The quantity of HCl in the 2.5 liters of 12% HCl is $0.12(2.5) = 0.3$ liters. The quantity of HCl in 1.5 liters of 20% HCl is $0.20(1.5) = 0.3$ liters. Combined, these solutions give us 0.6 liters of pure HCl. The total amount of HCl in 4 liters of 15% HCl is $0.15(4) = 0.6$ liters. Our answer is correct.

Step 6: Answer 2.5 liters of 12% HCl must be mixed with 1.5 liters of 20% HCl to form the required 4 liters of 15% HCl. ■

QUICK ✓

5. You are a wine maker and decide that wine with 9% alcohol is the best to sell. How many gallons of wine with 5% alcohol should be mixed with wine that is 15% alcohol to make 200 gallons of the desired blend?

9.5 Exercises

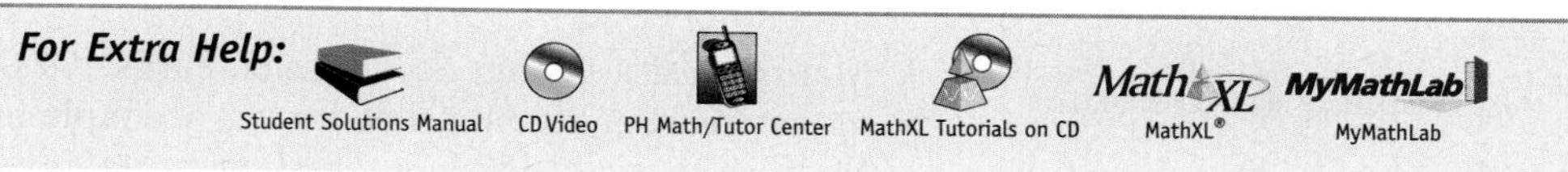

1. mixture problems
2. number of units of the same kind, rate, amount
3. Principal, rate, time
4. True
5. False
6. Answers may vary.
7.

	Number	Cost	Total
Adult	a	4	$4a$
Student	s	1.5	$1.5s$
Total	215		580

$$\begin{cases} a + s = 215 \\ 4a + 1.5s = 580 \end{cases}$$

Concepts and Vocabulary

In Problems 1–3, fill in the blanks.

1. Problems that involve mixing two or more substances are called __________ __________.

2. The general equation we use to solve mixture problems is __________ · __________ = __________.

3. The simple interest formula states that interest = __________ · __________ · __________.

In Problems 4 and 5, answer True or False to each statement.

4. When we solve a simple interest problem using the mixture model, we assume that $t = 1$.

5. The system $\begin{cases} a + b = 700 \\ 0.05a + 0.1b = 10{,}000 \end{cases}$ correctly models the following problem.

   Molly invests $10,000 in two accounts, one paying 5% annual interest and the other paying 10% annual interest, and she earns $700 in interest at the end of the year on the two accounts. Let a represent the amount of money she invests in the account that pays 5% annual interest and b represent the amount of money she invests in the account that pays 10%. How much is invested in each account?

6. Without solving, explain what is wrong with the following mixture problem: How many liters of 25% ethanol should be added to 20 liters of 48% ethanol to obtain a solution of 58% ethanol?

Building Skills

In Problems 7–12, fill in the table from the information given. Then write the system that models the problem. DO NOT SOLVE.

7. The PTA had an ice cream social and sold adult tickets for $4 and student tickets for $1.50. The PTA treasurer found that 215 tickets had been sold and the receipts were $580.

	Number	·	Cost per Person	=	Total Value
Adults' Tickets	?		?		?
Students' Tickets	?		?		?
Total	?				?

8.

	Number ·	Cost =	Total
Bracelets	b	10	$10b$
Necklaces	n	15	$15n$
Total	69		895

$$\begin{cases} b + n = 69 \\ 10b + 15n = 895 \end{cases}$$

9.

	P ·	r =	I
Savings	s	0.05	$0.05s$
Money Market	m	0.03	$0.03m$
Total	1600		50

$$\begin{cases} s + m = 1600 \\ 0.05s + 0.03m = 50 \end{cases}$$

10.

	P ·	r =	I
Savings	s	0.0275	$0.0275s$
CD	c	0.02	$0.02c$
Total	1700		37.75

$$\begin{cases} s + c = 1700 \\ 0.0275s + 0.02c = 37.75 \end{cases}$$

11.

	lb ·	Price =	Total
Mild	m	7.5	$7.5m$
Robust	r	10	$10r$
Total	12	8.75	8.75(12)

$$\begin{cases} m + r = 12 \\ 7.5m + 10r = 8.75(12) \end{cases}$$

12.

	lb ·	Price =	Total
Peanuts	p	5	$5p$
Trail Mix	t	2	$2t$
Total	40	3	3(40)

$$\begin{cases} p + t = 40 \\ 5p + 2t = 3(40) \end{cases}$$

8. John Murphy sells jewelry at art shows. He sells bracelets for \$10 and necklaces for \$15. At the end of one day John found that he had sold 69 pieces of jewelry and had receipts of \$895.

	Number	•	Cost per Item	=	Total Value
Bracelets	?		?		?
Necklaces	?		?		?
Total	?				?

9. Maurice has a savings account that earns 5% simple interest per year and a money market account that earns 3% simple interest. At the end of one year, Maurice received \$50 in interest on a total investment of \$1600 in the two accounts.

	Principal	•	Rate	=	Interest
Savings Account	?		?		?
Money Market	?		?		?
Total	?				?

10. Sherry has a savings account that earns 2.75% simple interest per year and a certificate of deposit (CD) that earns 2% simple interest annually. At the end of one year, Sherry received \$37.75 in interest on a total investment of \$1700 in the two accounts.

	Principal	•	Rate	=	Interest
Savings Account	?		?		?
Certificate of Deposit	?		?		?
Total	?				?

11. A coffee shop wishes to blend two types of coffee to create a breakfast blend. They mix a mild coffee that sells for \$7.50 per pound with a robust coffee that sells for \$10.00 per pound. The owner wants to make 12 pounds of breakfast blend that will sell for \$8.75 per pound.

	Number of Pounds	•	Price per Pound	=	Total Value
Mild Coffee	?		?		?
Robust Coffee	?		?		?
Total	?		?		?

12. A merchant wishes to mix peanuts worth \$5 per pound and trail mix worth \$2 per pound to yield 40 pounds of a nutty mixture that will sell for \$3 per pound.

	Number of Pounds	•	Price per Pound	=	Total Value
Peanuts	?		?		?
Trail Mix	?		?		?
Total	?		?		?

13. $32a + 24c$
14. $15a + 7c$
15. $0.1A + 0.07B$
16. 2560; $0.05A + 0.065B$
17. $5.85r + 4.20y = 128.85$
18. 143.45; $2W - 2$

In Problems 13–18, complete the system of linear equations to solve the problem. Do not solve.

13. A family of 11 decides to take a trip to Water World Water Park. The total cost of admission for the family is \$296. If adult tickets cost \$32 and children's tickets cost \$24, how many adults and how many children went to Water World? Let a represent the number of adult tickets purchased and let c represent the number of children's tickets purchased.

$$\begin{cases} a + c = 11 \\ \_ + \_ = 296 \end{cases}$$

14. On a school field trip, 22 people attended a dress rehearsal of the Broadway play, "Avenue Q." They paid \$274 for the tickets which cost \$15 for each adult and \$7 for each child. How many adults and how many children attended "Avenue Q"? Let a represent the number of adult tickets purchased and let c represent the number of children's tickets purchased.

$$\begin{cases} a + c = 22 \\ \_ + \_ = 274 \end{cases}$$

15. You have a total of \$2250 to invest. Account A pays 10% annual interest and account B pays 7% annual interest. How much should you invest in each account if you would like the investment to earn \$195 at the end of one year? Let A represent the amount of money invested in the account that earns 10% annual interest and let B represent the amount of money invested in the account that earns 7% annual interest.

$$\begin{cases} A + B = 2250 \\ \_ + \_ = 195 \end{cases}$$

16. You have a total of \$2650 to invest. Account A pays 5% annual interest and account B pays 6.5% annual interest. How much should you invest in each account if you would like the investment to earn \$155 at the end of one year? Let A represent the amount of money invested in the account that earns 5% annual interest and let B represent the amount of money invested in the account that earns 6.5% annual interest.

$$\begin{cases} A + B = \_\_\_ \\ \_ + \_ = 155 \end{cases}$$

17. Flowers on High sells flower bouquets at different prices depending on color: red for \$5.85 per bouquet or yellow for \$4.20 per bouquet. One day the number of red bouquets sold was 3 more than twice the number of yellow bouquets sold, and revenue from selling the bouquets was \$128.85. How many of each color were sold? Let r represent the number of red bouquets sold and let y represent the number of yellow bouquets sold.

$$\begin{cases} r = 3 + 2y \\ \_ + \_ = \_\_\_ \end{cases}$$

18. The Latte Shoppe sells Bold Breakfast coffee for \$8.60 per pound and Wake-Up coffee for \$5.75 per pound. One day, the amount of Bold Breakfast coffee sold was 2 pounds less than twice the amount of Wake-Up coffee and the revenue received from selling both types of coffee was \$143.45. How many pounds of each type of coffee were sold that day? Let B represent the number of pounds of Bold Breakfast coffee sold and let W represent the number of pounds of Wake-Up sold. Complete the system of equations:

$$\begin{cases} 8.60B + 5.75W = \_\_\_ \\ B = \_\_\_ - \_\_\_ \end{cases}$$

19. 315
20. adult $4.50; child $3.25
21. 400
22. 5
23. 60 nickels, 90 dimes
24. 42 ones, 26 fives
25. first-class $0.37; postcard $0.19
26. 8 30-second spots; 5 one-minute spots
27. $7500 in 5% account; $2500 in 8% account
28. $1500 in 4.5% account; $3500 in 9% account
29. $3200 in risky plan; $1800 in safer plan
30. $3500 mall; $8500 suburban office
31. 2 lb arbequina; 3 lb green

Applying the Concepts

19. **Theater** Tickets to a student theater production cost $8 for students and $10 for nonstudents. The receipts for opening night came to $3270 from selling 390 tickets. How many student tickets were sold?

20. **Ticket Pricing** A ticket on the roller coaster is priced differently for adults and children. One day there were 5 adults and 8 children in a group and the cost of their tickets was $48.50. Another group of 4 adults and 12 children paid $57. What is the price of each type of ticket?

21. **Bedding Plants** A girls' softball team is raising money for uniforms and travel expenses. They are selling flats of bedding plants for $13.00 and hanging baskets for $18.00. They hope to sell twice as many flats of bedding plants than hanging baskets. If they do, their total revenue will be $8800. How many flats of bedding plants do they hope to sell?

22. **Amusement Park** An adult discount ticket to King's Island Amusement Park costs $26, and a child's discount ticket to King's Island Amusement Park costs $24.50. A group of 13 friends purchased adult and children's tickets and paid $330.50. How many discount tickets for children were purchased?

23. **Tip Jar** The tip jar next to the cash register has 150 coins, all in nickels and dimes. If the total value of the tips is $12, how many of each coin are in the jar?

24. **Bigger Tips** Christa is a waitress and collects her tips at the table. At the end of the shift she has 68 bills in her tip wallet, all ones and fives. If the total value of her tips is $172, how many of each bill does she have?

25. **Buying Stamps** Rosemarie can buy stamps at two different prices. One day she bought 20 first-class stamps and 10 postcard stamps for $9.30. Another day she spent $39.10 on 80 first-class stamps and 50 postcard stamps. Find the cost of each type of stamp.

26. **TV Commercials** Advertising on television can be expensive depending on the expected number of viewers. The Fox network sells 30-second spots for $175,000 and one-minute spots for $250,000 during the basketball playoffs. If they sold 13 commercial spots and earned $2,650,000, how many of each type did they sell?

27. **Investments** Harry has $10,000 to invest. He invests in two different accounts, one expected to return 5% and the other expected to return 8%. If he wants to earn $575 for the year, how much should he invest at each rate?

28. **Investments** Ann has $5000 to invest. She invests in two different accounts, one expected to return 4.5% and the other expected to return 9%. In order to earn $382.50 for the year, how much should she invest at each rate?

29. **Stock Return** Esmeralda is investing $5000 in two stock plans. On the risky plan she hopes to earn 12% annual interest and on a safer plan she expects to earn 8% annual interest. If Esmeralda saw a return on her investment of $528 last year, how much did she invest in each of the stock plans?

30. **Real Estate Return** Victor, Esmeralda's wealthy brother, is also investing his money. He thinks he will do better if he invests his money in real estate partnerships. He has $12,000 to invest and has two groups in mind. One of the groups is promoting a 13% annual return on a downtown mall the other 10.5% annual return on a suburban office center. If Victor saw a profit of $1347.50 on his investment last year, how much did he invest with each group?

31. **Olive Blend** Juana works at a delicatessen, and it is her turn to make the zesty olive blend. If arbequina olives are $9 per pound and green olives are $4 per pound, how much of each kind of olive should she mix to get five pounds of zesty olive blend that will sell for $6 a pound?

32. **Churrascaria Platter** Churrascaria is a name that comes from the Brazilian gauchos of the 1800s. To celebrate, the cowboys would take a large variety of different meats and barbeque them. Today, some specialty restaurants still serve Brazilian barbeque in the Churrascaria style. Elis da Silva's Restaurant wants to offer a one-pound Churrascaria platter for \$12. If it costs the restaurant \$8 per pound to prepare grilled pork and \$14 per pound to make barbecued flank steak, how many pounds of each meat can you serve on the platter so that the price is still \$12?

33. **Blending Coffee** A coffee manufacturer wants to market a new blend of coffee that will cost \$3.90 per pound by mixing two coffees that sell for \$2.75 per pound and \$5 per pound, respectively. What amounts of each coffee should be blended to obtain 100 pounds of the desired mixture?

34. **Blending Nuts** The Nutty Professor store wishes to mix peanuts that sell for \$2 per pound with cashews that sell for \$6 per pound. The store plans to use five pounds more of peanuts than cashews and sell the mixture for \$26. How many pounds of each nut should be used in the mixture?

35. **Grass Seed** A nursery decides to make a grass-seed blend by mixing rye seed that sells for \$4.20 per pound with blue grass seed that sells for \$3.75 per pound. The final mix is 180 pounds and sells at \$3.95. How much of each type was used?

36. **Stock Account** A stock portfolio is currently valued at \$5872.44. IBM is currently selling at \$82.01 per share and Microsoft is currently selling at \$26.52 per share. If the number of IBM shares is 25 less than the number of Microsoft shares, find the number of shares of each.

37. **Saline Solutions** A lab technician needs 60 ml of a 50% saline solution. How many ml of 30% saline solution should she add to a 60% saline solution to obtain the required mixture?

38. **Alcohol** A laboratory assistant is asked to mix a 30% alcohol solution with 21 liters of an 80% alcohol solution to make a 60% alcohol solution. How many liters of the 30% alcohol solution should be used?

39. **Silver Alloy** How many liters of 10% silver must be added to 70 liters of 50% silver to make an alloy that is 30% silver?

40. **Paint** Four gallons of paint contain 2.5% pigment. How many gallons of paint that is 6% pigment must be added to make a paint that is 4% pigment?

Extending the Concepts

41. **Antifreeze** A radiator holds 3 gallons. How much of the 25% antifreeze solution should be drained and replaced with pure water to reduce the solution to 15% antifreeze?

42. **Antifreeze** A radiator holds 6 liters. How much of the 20% antifreeze solution should be drained and replaced with pure antifreeze to bring the solution up to 50% antifreeze?

43. **Finance** Jim Davidson invested \$10,000 in two businesses. One business earned a profit of 5% for the year, while the other lost 7.5%. Find the amount he invested in each business if he earned \$25 for the year.

44. **Finance** Marco invested money in two stocks. The stock that made a profit of 12% had \$5000 less than the stock that lost 5.5%. Find the amount he invested in each if there was a net loss of \$80 on his portfolio.

---

32. $\frac{1}{3}$ lb pork; $\frac{2}{3}$ lb steak
33. 48.9 lb of the \$2.75 per pound coffee; 51.1 lb of the \$5 per pound coffee
34. 7 lb peanuts; 2 lb cashews
35. 80 lb rye; 100 lb blue grass
36. 48 shares IBM, 73 shares Microsoft
37. 20 ml 30% saline solution; 40 ml 60% saline solution
38. 14 l
39. 70 l
40. 3 gal
41. 1.2 gal
42. 2.25 l
43. \$6200 at 5%; \$3800 at $7\frac{1}{2}$% loss
44. \$3000 at 12%; \$8000 at 5.5% loss

9.6 Systems of Linear Inequalities

OBJECTIVES

1. Determine Whether an Ordered Pair Is a Solution of a System of Linear Inequalities
2. Graph a System of Linear Inequalities
3. Solve Applied Problems Involving Systems of Linear Inequalities

Preparing for Systems of Linear Inequalities

Before getting started, take the following readiness quiz. If you get a problem wrong, go back to the section cited and review the material.

1. Solve: $3x - 2 \geq 7$ [Section 2.8, pp. 155–157]
2. Solve: $4(x - 1) < 6x + 4$ [Section 2.8, pp. 155–157]
3. Graph: $y > 2x - 5$ [Section 8.8, pp. 598–602]
4. Graph: $2x + 3y \leq 9$ [Section 8.8, pp. 598–602]

In Section 8.8, we graphed a single linear inequality in two variables. In this section, we discuss how to graph a system of linear inequalities in two variables.

1 Determine Whether an Ordered Pair Is a Solution of a System of Linear Inequalities

An ordered pair **satisfies** a system of linear inequalities if it makes each inequality in the system a true statement.

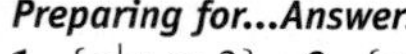
Classroom Example
Which of the following points, if any, satisfies the system of linear inequalities?

$$\begin{cases} 3x + y \leq 9 \\ 5x - 3y \geq 2 \end{cases}$$

(a) (3, 4)
(b) (2, 1)
Answer: (a) No (b) Yes

EXAMPLE 1 Determining Whether an Ordered Pair Is a Solution of a System of Linear Inequalities

Which of the following points, if any, satisfies the system of linear inequalities?

$$\begin{cases} 2x + y \leq 7 \\ 7x - 2y \geq 4 \end{cases}$$

(a) (2, 1) **(b)** (−2, 5)

Solution

(a) Let $x = 2$ and $y = 1$ in each inequality in the system. If each statement is true, then (2, 1) is a solution of the inequality.

$$2x + y \leq 7 \qquad\qquad 7x - 2y \geq 4$$

$$x = 2, y = 1: \quad 2(2) + 1 \overset{?}{\leq} 7 \qquad 7(2) - 2(1) \overset{?}{\geq} 4$$

$$4 + 1 \overset{?}{\leq} 7 \qquad 14 - 2 \overset{?}{\geq} 4$$

$$5 \leq 7 \text{ True} \qquad 12 \geq 4 \text{ True}$$

Both inequalities are true when $x = 2$ and $y = 1$, so (2, 1) is a solution of the system of inequalities.

(b) Let $x = -2$ and $y = 5$ in each inequality in the system. If each statement is true, then (−2, 5) is a solution of the inequality.

$$2x + y \leq 7 \qquad\qquad 7x - 2y \geq 4$$

$$x = -2, y = 5: \quad 2(-2) + 5 \overset{?}{\leq} 7 \qquad 7(-2) - 2(5) \overset{?}{\geq} 4$$

$$-4 + 5 \overset{?}{\leq} 7 \qquad -14 - 10 \overset{?}{\geq} 4$$

$$1 \leq 7 \text{ True} \qquad -24 \geq 4 \text{ False}$$

The inequality $7x - 2y \geq 4$ is not true when $x = -2$ and $y = 5$, so (−2, 5) is not a solution of the system of inequalities.

Preparing for...Answers
1. $\{x \mid x \geq 3\}$ **2.** $\{x \mid x > -4\}$
3.

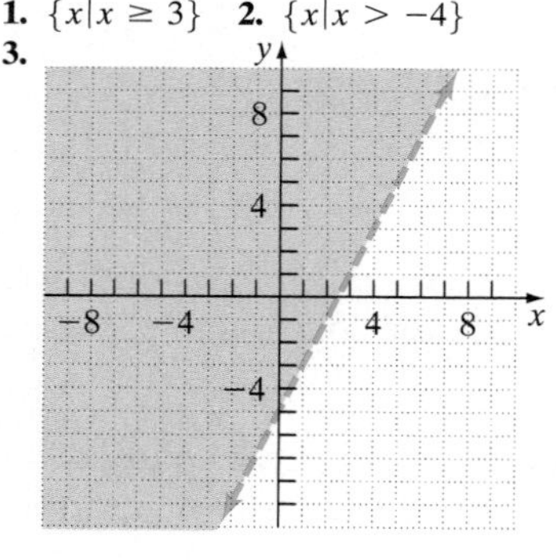

4.

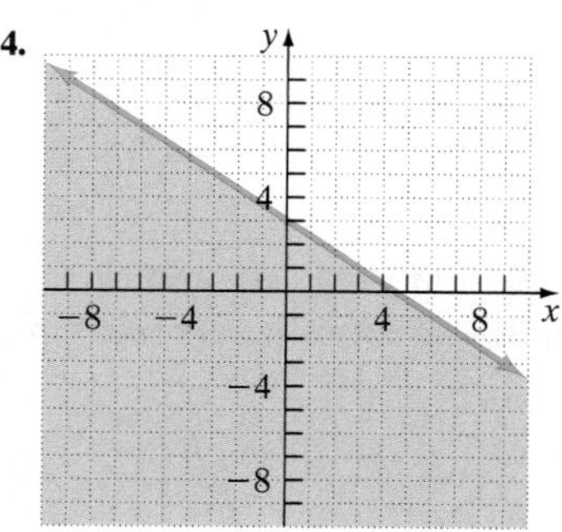

QUICK ✓

1. Determine which of the following points is a solution of the system of linear inequalities.

$$\begin{cases} 4x + y \leq 6 \\ 2x - 5y < 10 \end{cases}$$

(a) $(1, 2)$ **(b)** $(-1, -3)$

2 Graph a System of Linear Inequalities

Work Smart

Don't forget that we use a solid line when the inequality is nonstrict ($\leq$ or $\geq$) and a dashed line when the inequality is strict ($<$ or $>$).

The graph of a system of inequalities in two variables x and y is the set of all points (x, y) that simultaneously satisfy *each* of the inequalities in the system. The graph of a system of linear inequalities can be obtained by graphing each linear inequality individually and then determining where, if at all, they intersect. The ONLY way we show the solution of a system of linear inequalities is by its graphical representation.

EXAMPLE 2 How to Graph a System of Linear Inequalities

Graph the system: $\begin{cases} y \geq 3x - 5 \\ y \leq -2x + 5 \end{cases}$

Step-by-Step Solution

Step 1: We graph the first inequality in the system $y \geq 3x - 5$.

Classroom Example

Graph the system: $\begin{cases} y \geq 2x - 3 \\ y \leq -3x + 1 \end{cases}$

Answer:

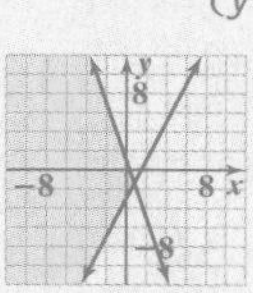

Graph $y \geq 3x - 5$ by graphing the line $y = 3x - 5$ (slope $= 3$, y-intercept $= -5$) with a solid line because the inequality is nonstrict ($\geq$). We choose to use the test point $(0, 0)$. Since $(0, 0)$ makes the inequality true $(0 \geq 3(0) - 5)$, we shade the half-plane containing $(0, 0)$. See Figure 13.

Figure 13

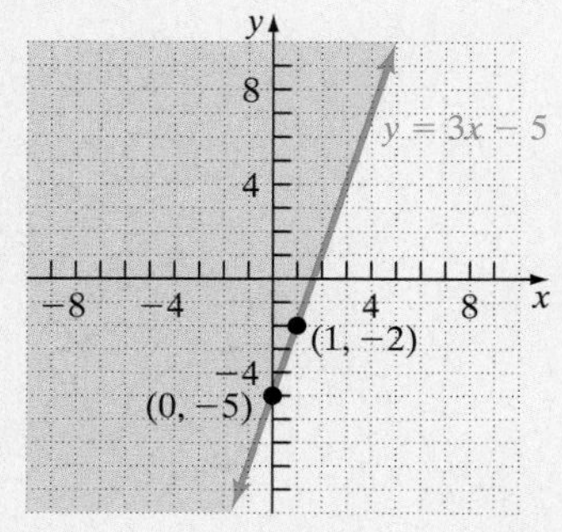

Step 2: Graph the second inequality in the system, $y \leq -2x + 5$.

Work Smart

We can also graph inequalities by solving the inequality for y. Then, if the inequality is of the form $y >$ or $y \geq$, we shade above the line. If the inequality is of the form $y <$ or $y \leq$, shade below the line.

Graph $y \leq -2x + 5$ by graphing the line $y = -2x + 5$ (slope $= -2$, y-intercept $= 5$) with a solid line because the inequality is nonstrict ($\leq$). Again, we choose to use the test point $(0, 0)$. Since $(0, 0)$ makes the inequality true $(0 \leq -2(0) + 5)$, we shade the half-plane containing $(0, 0)$. See Figure 14.

Figure 14

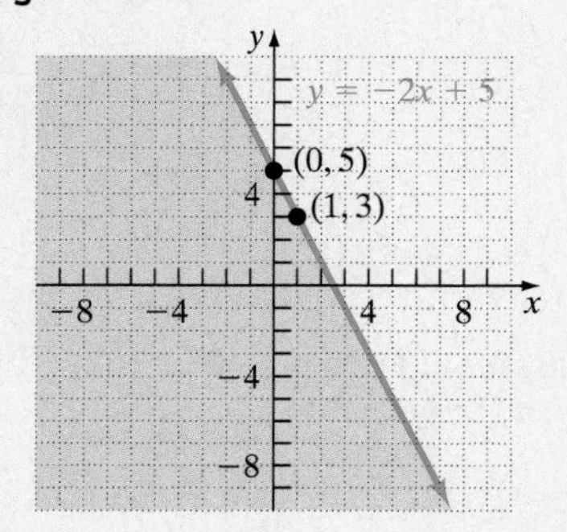

(continued)

Step 3: Combine the graphs in Steps 1 and 2
The overlapping shaded region is the solution of the system of linear inequalities.

See Figure 15. Note: The ordered pair (2, 1) represents the solution to the system $\begin{cases} y = 3x - 5 \\ y = -2x + 5 \end{cases}$.

Figure 15

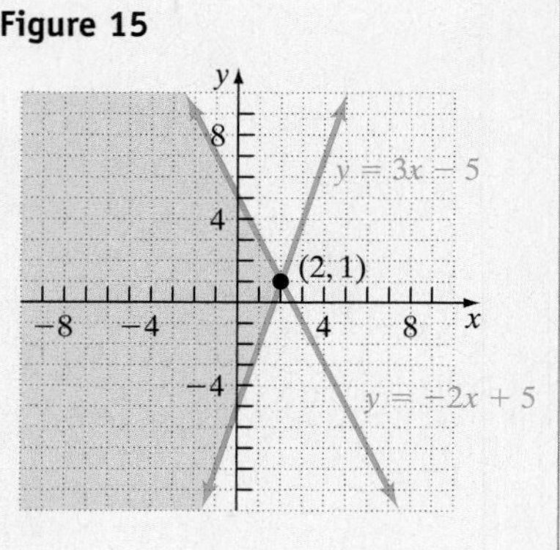

Classroom Example
Graph the system:

$$\begin{cases} 2x + y > 7 \\ -3x + 2y > 0 \end{cases}$$

Answer:

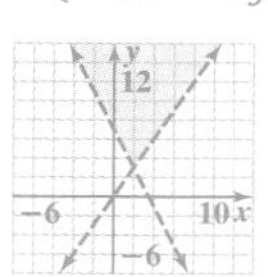

EXAMPLE 3 Graphing a System of Linear Inequalities

Graph the system: $\begin{cases} 2x + y < 7 \\ 7x - 2y > 4 \end{cases}$

Solution

We graph the inequality $2x + y < 7$ $(y < -2x + 7)$. See Figure 16(a). We then graph the inequality $7x - 2y > 4\left(y < \frac{7}{2}x - 2\right)$. See Figure 16(b). Don't forget to use dashed lines!

Now we combine the graphs in Figures 16(a) and (b). The overlapping shaded region is the solution to the system of linear inequalities. See Figure 16(c). The points on the two boundary lines are not solutions of the system. Do you know why?

Figure 16

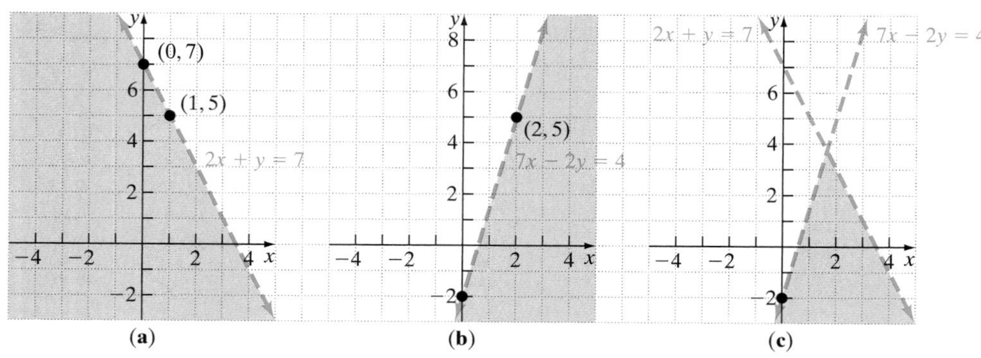

QUICK ✓ *Graph each system of linear inequalities.*

2. $\begin{cases} y \geq -3x + 8 \\ y \geq 2x - 7 \end{cases}$

3. $\begin{cases} 4x + 2y < -9 \\ x + 3y < -1 \end{cases}$

Rather than use multiple graphs to obtain the solution set to the system of linear inequalities, we can use a single graph to determine the overlapping region.

Classroom Example
Graph the system:

$$\begin{cases} 2x + y \geq -2 \\ -5x + 3y \leq 27 \end{cases}$$

Answer:

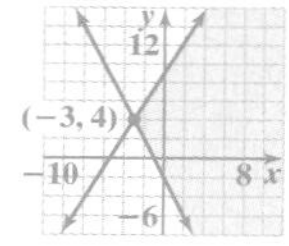

EXAMPLE 4 Graphing a System of Linear Inequalities

Graph the system: $\begin{cases} 2x + y \geq 3 \\ 3x - 2y < 8 \end{cases}$

Work Smart

$$3x - 2y < 8$$
$$-2y < -3x + 8$$
$$y > \frac{3}{2}x - 4$$

Remember to reverse the inequality symbol! Now graph $y = \frac{3}{2}x - 4$ using a dashed line and shade above.

Work Smart

The points on the solid boundary lines are also solutions of the system of linear inequalities.

Solution

We graph the inequality $2x + y \geq 3$ $(y \geq -2x + 3)$. On the same rectangular coordinate system, we graph the inequality $3x - 2y < 8$ $\left(y > \frac{3}{2}x - 4\right)$ using a different shading pattern.

The overlapping shaded region represents the solution. Figure 17 shows the solution. The points on the solid boundary line are solutions of the system, but the points on the dotted boundary line are not solutions.

Figure 17

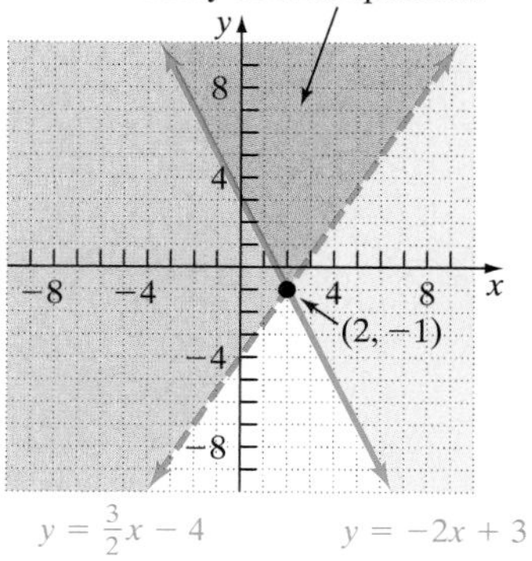

QUICK ✓ *Graph the system of linear inequalities.*

4. $\begin{cases} x + y \leq 4 \\ -x + y \geq -4 \end{cases}$

5. $\begin{cases} 3x + y > -5 \\ x + 2y < 0 \end{cases}$

3 Solve Applied Problems Involving Systems of Linear Inequalities

Now let's look at some problems that lead to systems of linear inequalities. As always, we use the problem-solving strategy presented in Section 2.5.

Classroom Example Jacob recently retired and has up to \$70,000 to invest. His financial advisor has recommended that he place no more than \$45,000 in corporate bonds and at least \$20,000 in Treasury notes. A system of linear inequalities that models this situation is given by

$$\begin{cases} c + t \leq 70{,}000 \\ c \leq 45{,}000 \\ t \geq 20{,}000 \end{cases}$$

where c represents the amount invested in corporate bonds and t represents the amount invested in Treasury notes and.

(a) Graph the system.

(b) Can Jacob put \$25,000 in corporate bonds and \$35,000 in Treasury notes?

(c) Can Jacob put \$40,000 in corporate bonds and \$15,000 in Treasury notes?

Answer:

(a)

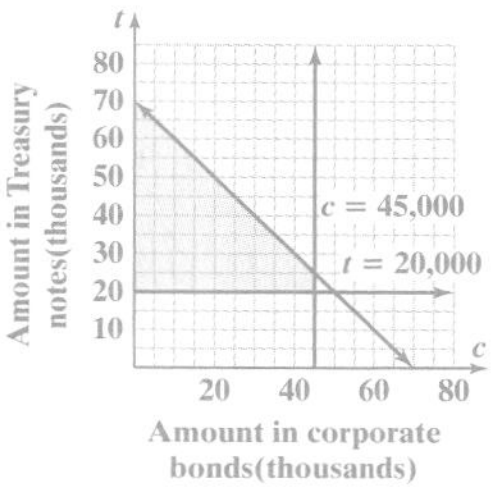

(b) Yes (c) No

EXAMPLE 5 Financial Planning

Maurice recently retired and has up to \$50,000 to invest. His financial advisor has recommended that he place at least \$10,000 in Treasury notes and no more than \$35,000 in corporate bonds. A system of linear inequalities that models this situation is given by

$$\begin{cases} c + t \leq 50{,}000 \\ c \leq 35{,}000 \\ t \geq 10{,}000 \end{cases}$$

where c represents the amount invested in corporate bonds and t represents the amount invested in Treasury notes.

(a) Graph the system.

(b) Can Maurice put \$20,000 in corporate bonds and \$30,000 in Treasury notes?

(c) Can Maurice put \$40,000 in corporate bonds and \$10,000 in Treasury notes?

Solution

(a) Draw a rectangular coordinate system with the horizontal axis labeled c (for corporate bonds) and the vertical axis labeled t (for Treasury notes). Each axis will be in thousands, so we draw the line $t + c = 50$ and shade below. We draw the line $t = 10$ and shade above. We draw the line $c = 35$ and shade to the left. Figure 18 shows the graph of the system of linear inequalities.

Figure 18

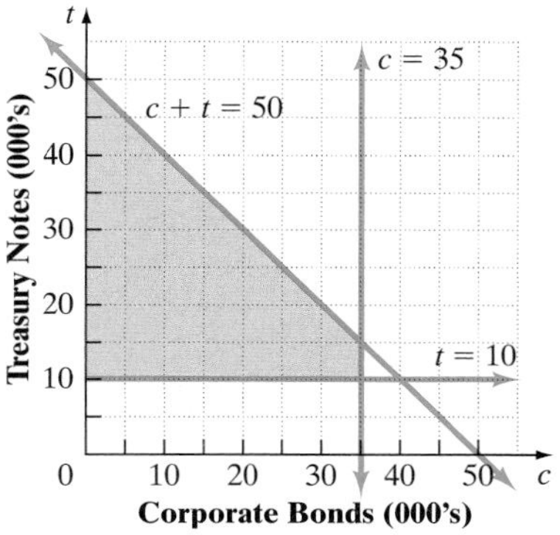

(b) Yes, Maurice can put \$20,000 in corporate bonds and \$30,000 in Treasury notes because these values lie within the shaded region. Put another, way $c = 20$ and $t = 30$ satisfies all three inequalities.

(c) No, Maurice cannot put \$40,000 in corporate bonds and \$10,000 in Treasury notes because these values do not lie within the shaded region. Put another way, $c = 40$ and $t = 10$ does not satisfy the inequality $c \leq 35{,}000$.

QUICK ✓

6. Jack and Mary recently retired and have up to \$75,000 to invest. Their financial advisor has recommended that they place no more than \$50,000 in corporate bonds and at least \$25,000 in Treasury notes. A system of linear inequalities that models this situation is given by

$$\begin{cases} c + t \leq 75{,}000 \\ c \qquad \leq 50{,}000 \\ \qquad t \geq 25{,}000 \end{cases}$$

where c represents the amount invested in corporate bonds and t represents the amount invested in Treasury notes.

(a) Graph the system.
(b) Can Jack and Mary invest \$30,000 in corporate bonds and \$35,000 in Treasury notes?
(c) Can Jack and Mary invest \$60,000 in corporate bonds and \$15,000 in Treasury notes?

9.6 Exercises

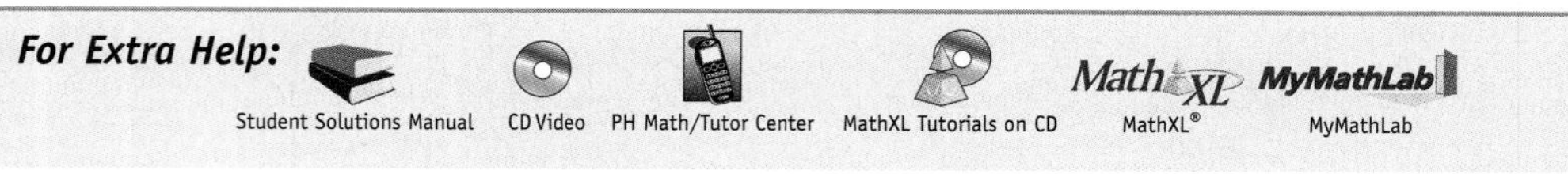

1. solution
2. satisfies
3. dashed; solid
4. False
5. True
6. True
7. (a) $\begin{cases} y \geq x \\ y \geq -2x \end{cases}$
(b) $\begin{cases} y \leq x \\ y \geq -2x \end{cases}$
(c) $\begin{cases} y \geq x \\ y \leq -2x \end{cases}$
(d) $\begin{cases} y \leq x \\ y \leq -2x \end{cases}$
8. Answers may vary.
9. (a) Yes (b) Yes (c) Yes
10. (a) Yes (b) No (c) No
11. (a) Yes (b) No (c) No
12. (a) No (b) No (c) Yes
13. $\begin{cases} x > 2 \\ y \leq -1 \end{cases}$

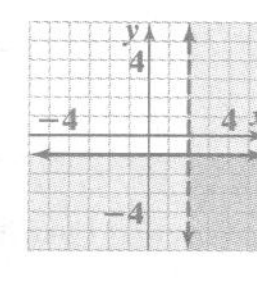

Concepts and Vocabulary

In Problems 1–3, fill in the blanks.

1. An ordered pair is a ________ of a system of linear inequalities if it makes each inequality in the system a true statement.

2. An ordered pair ________ a system of linear inequalities if it makes each inequality in the system a true statement.

3. When graphing linear inequalities, we use a ________ line when graphing strict inequalities ($>$ or $<$) and a ________ line when graphing nonstrict inequalities ($\geq$ or $\leq$).

In Problems 4–6, answer True or False to each statement.

4. We can list every ordered pair that satisfies a system of linear inequalities.

5. It is possible to have the system of linear inequalities graphed correctly and *not* have an overlapping shaded region to represent the solution.

6. Every point that is included in an overlapping shaded region on the graph of a system of linear inequalities satisfies all of the inequalities and therefore is a solution to the system.

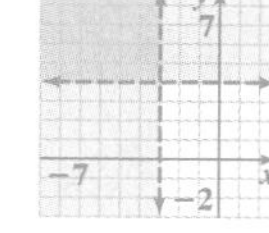

7. Each of the graphs below uses two of the following lines:

$$l_1: y \geq -2x \qquad l_2: y \leq -2x \qquad l_3: y \geq x \qquad l_4: y \leq x$$

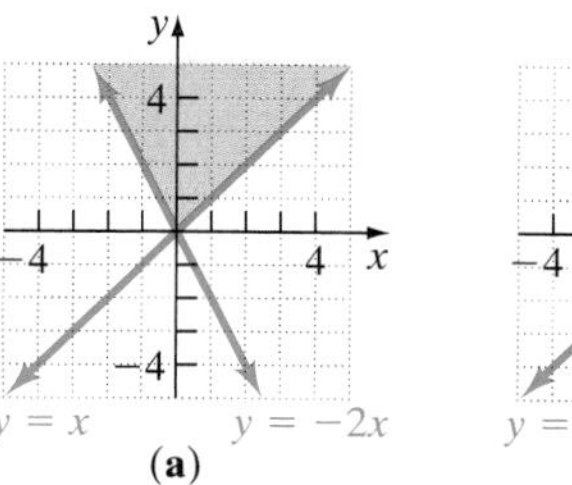

(a)

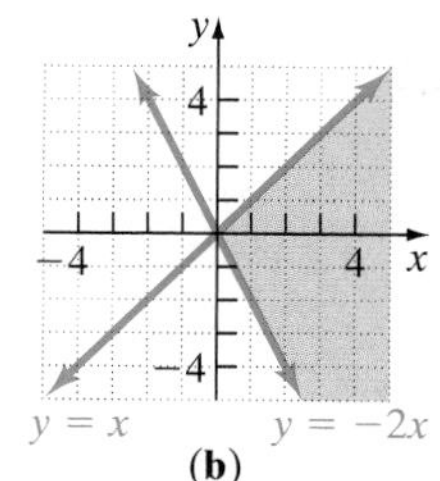

(b)

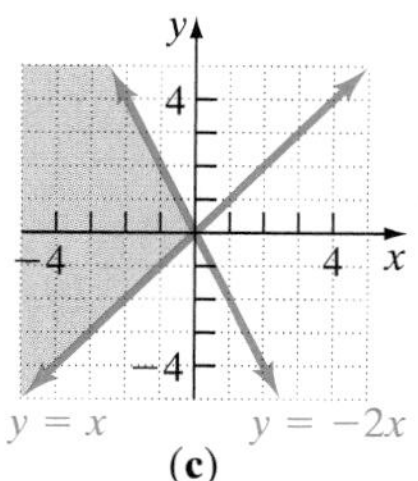

(c)

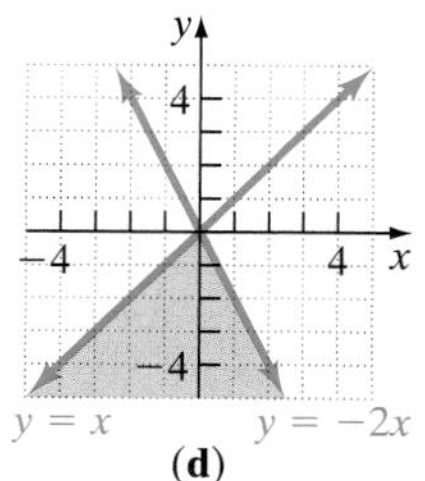

(d)

Write the system of linear inequalities for each graph and explain how you came to your conclusions.

8. When solving application problems which require solving a system of linear inequalities, typically the graph appears only in the first quadrant. Explain why this is so. Can you think of a situation when it would not be the case?

Building Skills

In Problems 9–12, determine which point(s), if any, is a solution of the system of linear inequalities.

9. $\begin{cases} x \geq 5 \\ y < -\frac{1}{2}x + 3 \end{cases}$

(a) $(5, -2)$
(b) $(10, -4)$
(c) $(8, -3)$

10. $\begin{cases} 2x - 3y < 3 \\ 2x + y < -5 \end{cases}$

(a) $(-4, 1)$
(b) $\left(-\frac{3}{2}, -2\right)$
(c) $(-1, -2)$

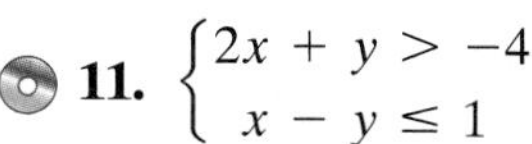

11. $\begin{cases} 2x + y > -4 \\ x - y \leq 1 \end{cases}$

(a) $(-2, 1)$
(b) $(-1, -2)$
(c) $(2, -3)$

12. $\begin{cases} x - 2y > 2 \\ -3x - 2y \leq 6 \end{cases}$

(a) $\left(-1, -\frac{3}{2}\right)$
(b) $(0, -4)$
(c) $(4, -1)$

In Problems 13–38, graph each system of linear inequalities.

13. $\begin{cases} x > 2 \\ y \leq -1 \end{cases}$

14. $\begin{cases} y > 4 \\ x < -3 \end{cases}$

15. $\begin{cases} y > -2 \\ x > -3 \end{cases}$

16. $\begin{cases} x \leq 3 \\ y < 1 \end{cases}$

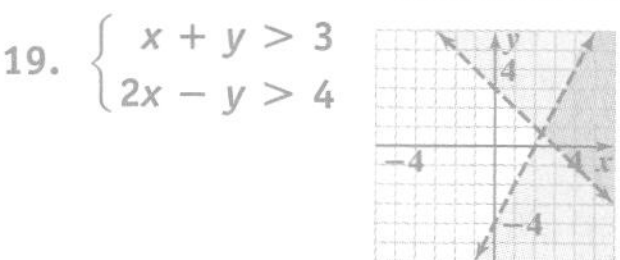

17. $\begin{cases} x + y < 3 \\ x - y > 5 \end{cases}$

18. $\begin{cases} x - y > 2 \\ x + y \geq -2 \end{cases}$

19. $\begin{cases} x + y > 3 \\ 2x - y > 4 \end{cases}$

20. $\begin{cases} x - y \geq -1 \\ x + 2y > 4 \end{cases}$

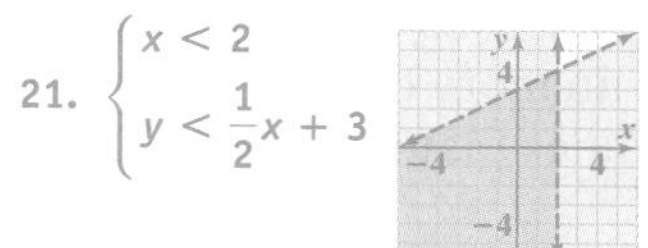

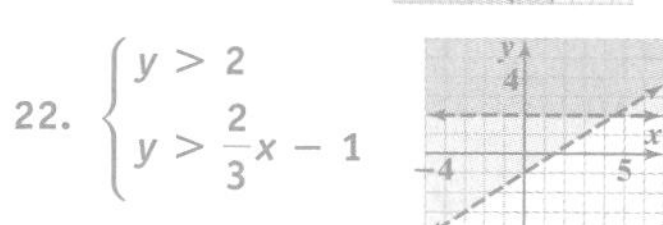

21. $\begin{cases} x < 2 \\ y < \frac{1}{2}x + 3 \end{cases}$

22. $\begin{cases} y > 2 \\ y > \frac{2}{3}x - 1 \end{cases}$

23. $\begin{cases} x \geq -2 \\ y < 2x + 3 \end{cases}$

24. $\begin{cases} y > 2 \\ 2x - y \leq 1 \end{cases}$

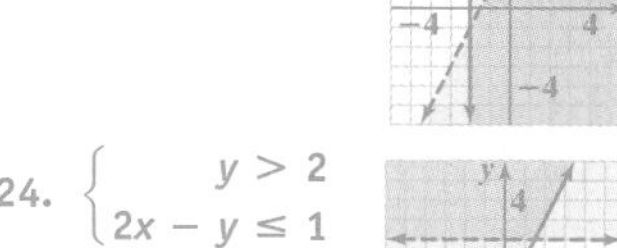

25. $\begin{cases} x > 0 \\ y \leq \frac{2}{5}x - 1 \end{cases}$

26. $\begin{cases} y < -2 \\ y > 2x - 1 \end{cases}$

27. $\begin{cases} -y \leq x \\ 3x - y \geq -5 \end{cases}$

28. $\begin{cases} x + y < 2 \\ 3x - 5y \geq 0 \end{cases}$

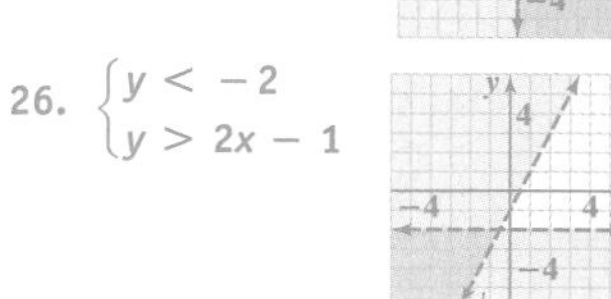

29. $\begin{cases} x + y \leq -2 \\ y \geq x + 3 \end{cases}$

30. $\begin{cases} 3x + y > 3 \\ y > 4x - 2 \end{cases}$

31. $\begin{cases} x + 3y \geq 0 \\ 2y < x + 1 \end{cases}$

32. $\begin{cases} -y < \dfrac{2}{3}x + 1 \\ -3x + y \le 2 \end{cases}$

33. $\begin{cases} x + y \ge 0 \\ x < 2y + 4 \end{cases}$

34. $\begin{cases} 2x - 3y \le 9 \\ y < -2x + 3 \end{cases}$

35. $\begin{cases} x + 3y > 6 \\ 2x - y \le 4 \end{cases}$

36. $\begin{cases} x - y \le 3 \\ 2x + 3y \le -9 \end{cases}$

37. $\begin{cases} -y \le 3x - 4 \\ 2x + 3y \ge -3 \end{cases}$

38. $\begin{cases} x + 4y \le 4 \\ 2x + 3y \ge 6 \end{cases}$

Applying the Concepts

39. **House Blend** The Coffee Cup coffee shop is experimenting with blending the "house" coffee. It will be made up of two varieties of coffee, French Roast and Hazelnut. The management has decided that they will make at most 30 pounds of "house" coffee in a day. The tasters have determined that the blend should be mixed so that the amount of French Roast coffee is at least twice the amount of Hazelnut coffee. A system of linear inequalities that models this situation is given by

$$\begin{cases} f + h \le 30 \\ f \ge 2h \end{cases}$$

where f represents the number of pounds of French Roast coffee and h represents the number of pounds of Hazelnut coffee.

(a) Graph the system of linear inequalities.
(b) Is it possible to use 18 pounds of French Roast coffee and 11 pounds of Hazelnut coffee in the house blend?
(c) Is it possible to use 8 pounds of French Roast coffee and 4 pounds of Hazelnut coffee in the house blend?

40. **Party Food** Steven and Christopher are planning a party. They plan to buy bratwurst for \$4.00 per pound and hamburger patties that cost \$3.00 per pound. They can spend at most \$70 and think they should have no more than 20 pounds of bratwurst and hamburger patties. A system of inequalities that models the situation is given by

$$\begin{cases} 4b + 3h \le 70 \\ b + h \le 20 \end{cases}$$

where b represents the number of pounds of bratwurst and h represents the number of pounds of hamburger patties.

(a) Graph the system of linear inequalities.
(b) Is it possible to purchase 10 pounds of bratwurst and 10 pounds of hamburger patties?
(c) Is it possible to purchase 15 pounds of bratwurst and 5 pounds of hamburger patties?

41. **Auto Manufacturing** A manufacturing plant has 360 hours of machine time budgeted for the production of cars and trucks on a given day. It takes 12 hours to build a car and 18 hours to build a truck. Based on experience, the plant manager knows that the number of trucks is fewer than 20 less than twice the number of cars produced. Due to limitations in the machinery, the maximum number of cars that can be produced in a day is 24. A system of linear inequalities that models this situation is given by

$$\begin{cases} 12c + 18t \le 360 \\ t < 2c - 20 \\ c \le 24 \end{cases}$$

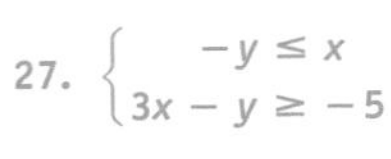
27. $\begin{cases} -y \le x \\ 3x - y \ge -5 \end{cases}$

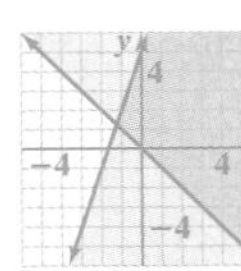

28. $\begin{cases} x + y < 2 \\ 3x - 5y \ge 0 \end{cases}$

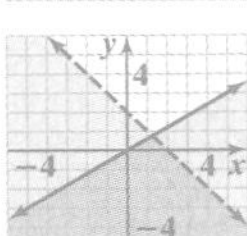

29. $\begin{cases} x + y \le -2 \\ y \ge x + 3 \end{cases}$

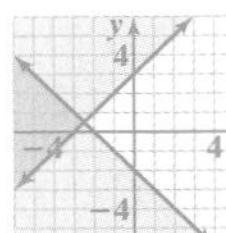

30. $\begin{cases} 3x + y > 3 \\ y > 4x - 2 \end{cases}$

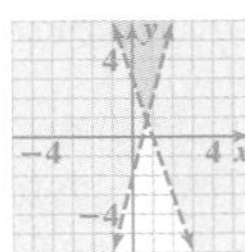

31. $\begin{cases} x + 3y \ge 0 \\ 2y < x + 1 \end{cases}$

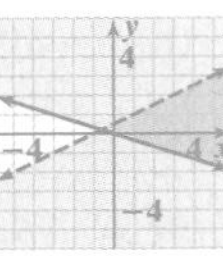

32. $\begin{cases} -y < \dfrac{2}{3}x + 1 \\ -3x + y \le 2 \end{cases}$

33. $\begin{cases} x + y \ge 0 \\ x < 2y + 4 \end{cases}$

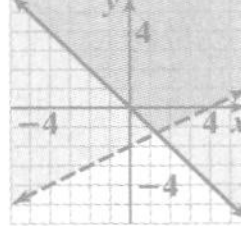

34. $\begin{cases} 2x - 3y \le 9 \\ y < -2x + 3 \end{cases}$

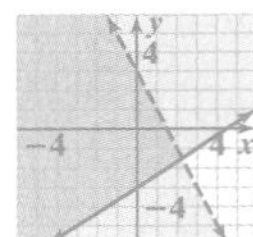

35. $\begin{cases} x + 3y > 6 \\ 2x - y \le 4 \end{cases}$

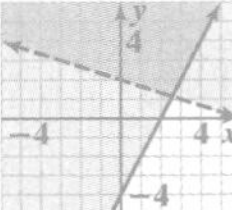

36. $\begin{cases} x - y \le 3 \\ 2x + 3y \le -9 \end{cases}$

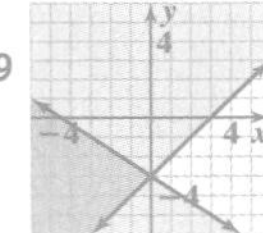

37. $\begin{cases} -y \le 3x - 4 \\ 2x + 3y \ge -3 \end{cases}$

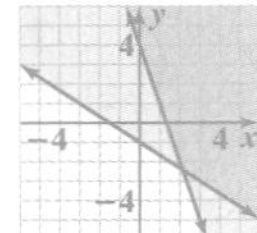

38. $\begin{cases} x + 4y \le 4 \\ 2x + 3y \ge 6 \end{cases}$

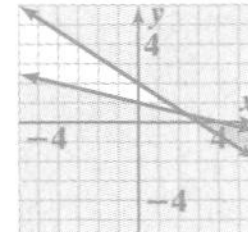

39. (a)

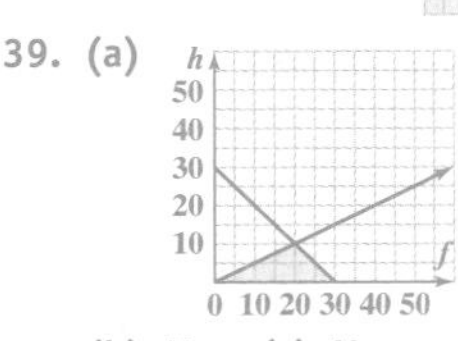

(b) No (c) Yes

40. (a)
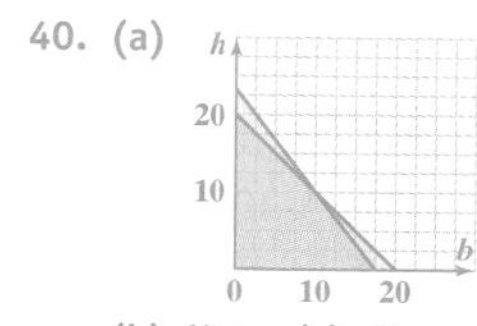
(b) Yes (c) No

41. (a) (b) No (c) No

42. (a) (b) Yes (c) Yes

43. $\begin{cases} \frac{y}{2} - \frac{x}{6} \geq 1 \\ \frac{x}{3} - \frac{y}{1} \geq 1 \end{cases}$
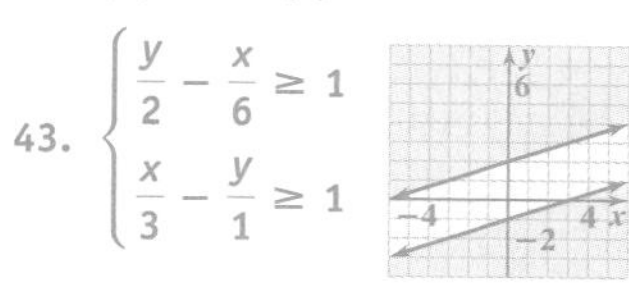

44. $\begin{cases} \frac{x}{2} + \frac{y}{4} > 1 \\ -\frac{x}{4} - \frac{y}{8} > 1 \end{cases}$
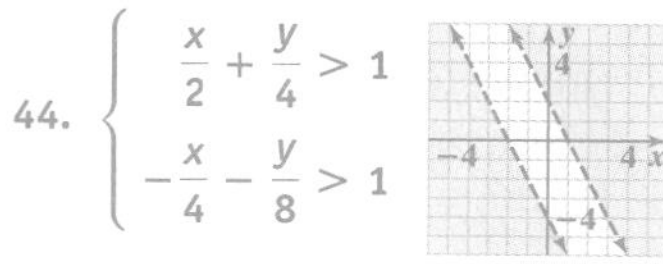

45. $\begin{cases} x < \frac{3}{2}y + \frac{9}{2} \\ -2x < 3(y + 2) \end{cases}$
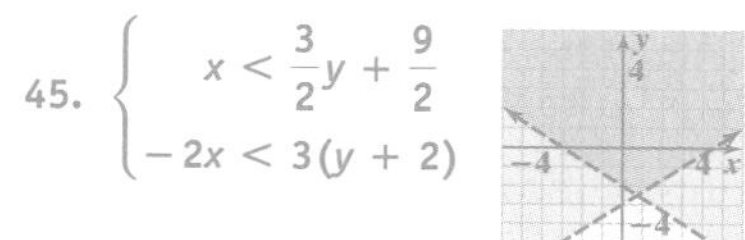

46. $\begin{cases} x > \frac{5}{4}y - \frac{1}{2} \\ 4x < 5(y + 3) \end{cases}$
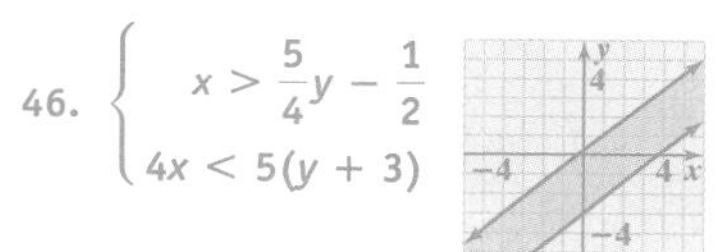

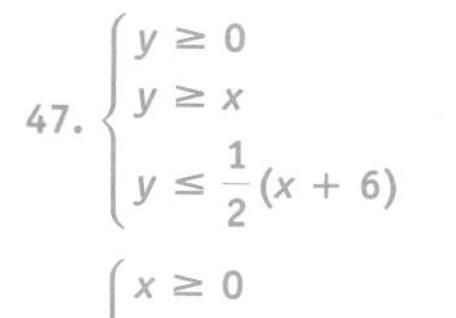

47. $\begin{cases} y \geq 0 \\ y \geq x \\ y \leq \frac{1}{2}(x + 6) \end{cases}$

48. $\begin{cases} x \geq 0 \\ y \leq -x + 4 \\ y \geq 2x - 2 \end{cases}$

where c represents the number of cars produced and t represents the number of trucks built.

(a) Graph the system of linear inequalities.
(b) Is it possible to build 17 cars and 9 trucks at this plant?
(c) Is it possible to build 12 cars and 11 trucks at this plant?

42. Breakfast at Burger King Aman decides to eat breakfast at Burger King. He is a big fan of their French toast sticks and he likes orange juice. He wants to eat no more than 500 calories and consume no more than 425 mg of sodium. Each French toast stick (with syrup) has 90 calories and 100 mg of sodium. Each small orange juice has 140 calories and 25 mg of sodium. The system of linear inequalities that represents the possible combination of French toast sticks and orange juice that Aman can consume is

$$\begin{cases} 90x + 140y \leq 500 \\ 100x + 25y \leq 425 \\ x \geq 0 \\ y \geq 0 \end{cases}$$

where x represents the number of Burger King French toast sticks and y represents the number of containers of orange juice.

(a) Graph the system of linear inequalities.
(b) Is it possible for Aman to consume 3 French toast sticks and 1 container of orange juice and stay within the allowance for calories and sodium?
(c) Is it possible for Aman to consume 2 French toast sticks and 2 containers of orange juice and stay within the allowance for calories and sodium?

Extending the Concepts

In Problems 43–46, the shaded region for the solution to each system of linear inequalities below differs from what we have seen previously. Graph and then state the solution of each system.

43. $\begin{cases} \frac{y}{2} - \frac{x}{6} \geq 1 \\ \frac{x}{3} - \frac{y}{1} \geq 1 \end{cases}$

44. $\begin{cases} \frac{x}{2} + \frac{y}{4} > 1 \\ -\frac{x}{4} - \frac{y}{8} > 1 \end{cases}$

45. $\begin{cases} x < \frac{3}{2}y + \frac{9}{2} \\ -2x < 3(y + 2) \end{cases}$

46. $\begin{cases} x > \frac{5}{4}y - \frac{1}{2} \\ 4x < 5(y + 3) \end{cases}$

In Problems 47 and 48, write a system of linear inequalities that will produce each shaded region as its solution.

47.
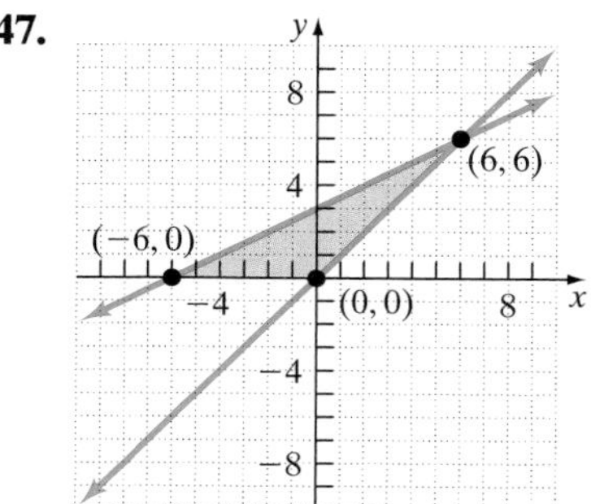

48.
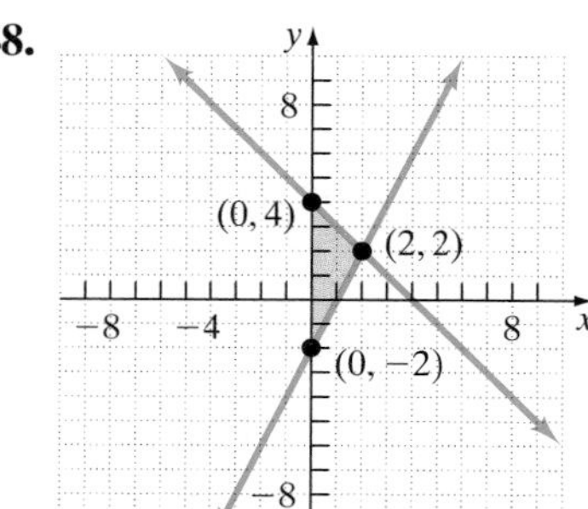

CHAPTER 9 ACTIVITY: FIND THE NUMBERS

Focus: Solving systems of equations
Time: 15 minutes
Group size: 2

Consider the following dialogue between two students:

Ryan: Think of two numbers between 1 and 10, and don't tell me what they are.

Melissa: Ok, I've thought of two numbers.

Ryan: Now tell me the sum of the two numbers and the difference of the two numbers, and I'll tell you what your two numbers are.

Melissa: Their sum is 14 and their difference is 6.

Ryan: Your numbers are 10 and 4.

Melissa: That's right! How did you do that?

Ryan: I set up a system of equations using the sum and difference that you gave me, along with the variables x and y.

1. Answers will vary.
2. Answers will vary.
3. yes

1. Each group member should set up and solve the system of equations described by Ryan. Discuss your results and be sure that you both arrive at the solutions 10 and 4.
2. Now each of you will think of two new numbers and the other will try to find the numbers by solving a system of equations. But this time the system will be a bit trickier! Give each other the following information about your two numbers, and then figure out each other's numbers:

   six more than three times the sum of the numbers

   two less than four times the difference of the numbers
3. Would the systems in this activity work if negative numbers were used? Try it and see!

CHAPTER 9 REVIEW

Section 9.1 Solve Systems of Linear Equations by Graphing

KEY CONCEPT	KEY TERMS
• **Recognizing Solutions of Systems of Two Linear Equations with Two Unknowns** • If the lines in a system of two linear equations containing two unknowns intersect, then the point of intersection is the solution and the system is consistent and independent. • If the lines in a system of two linear equations containing two unknowns are parallel, then the system has no solution and the system is inconsistent. • If the lines in a system of two linear equations containing two unknowns lie on top of each other, then the system has infinitely many solutions. The solution set is the set of all points on the line and the system is consistent, but dependent.	System of linear equations Solution Consistent and independent Inconsistent Consistent and dependent

YOU SHOULD BE ABLE TO . . .	EXAMPLE	REVIEW EXERCISES
(1) Determine if an ordered pair is a solution of a system of linear equations (p. 617)	Example 2	1–4
(2) Solve a system of linear equations by graphing (p. 618)	Examples 3 through 6	5–12
(3) Classify systems of linear equations as consistent or inconsistent (p. 622)	Example 7	13–18
(4) Solve applied problems involving systems of linear equations (p. 624)	Example 8	19, 20

1. (a) No (b) Yes (c) No
2. (a) No (b) No (c) Yes
3. (a) Yes (b) Yes (c) Yes
4. (a) No (b) No (c) No
5. 6.

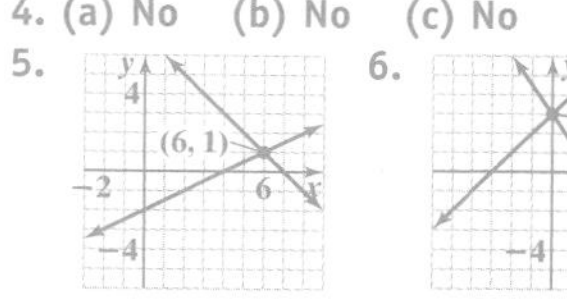

(6, 1) (0, 3)

7. 8.

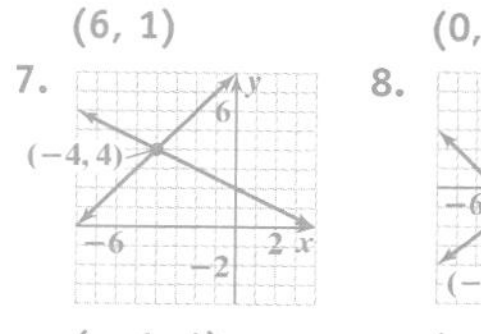

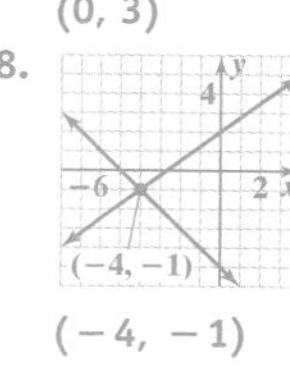

(−4, 4) (−4, −1)

9. 10.

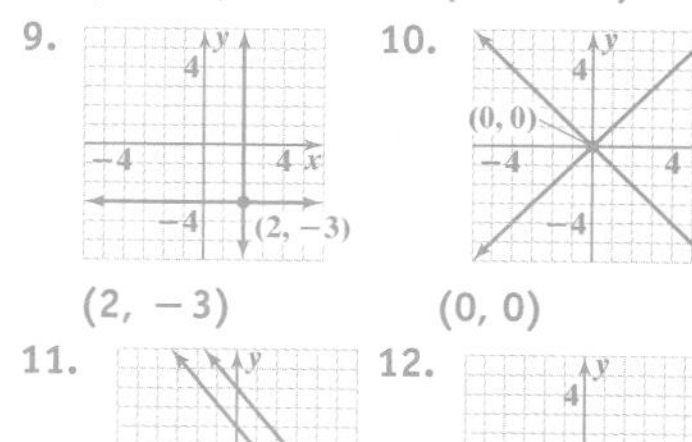

(2, −3) (0, 0)

11. 12.

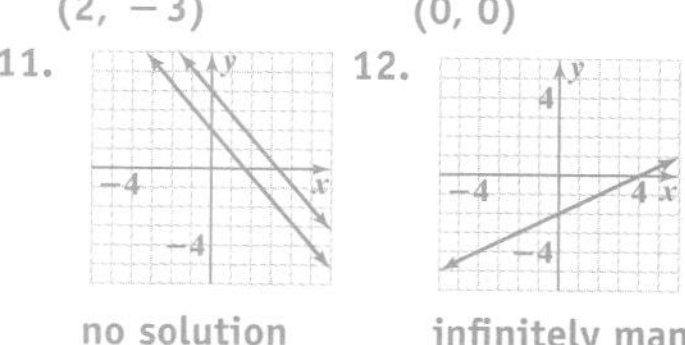

no solution infinitely many solutions

13. none; inconsistent
14. none; inconsistent
15. infinitely many; consistent; dependent
16. infinitely many; consistent; dependent
17. one; consistent; independent
18. one; consistent; independent
19. (a) $\begin{cases} y = 70 + 0.1x \\ y = 100 + 0.04x \end{cases}$
(b) 500 fliers

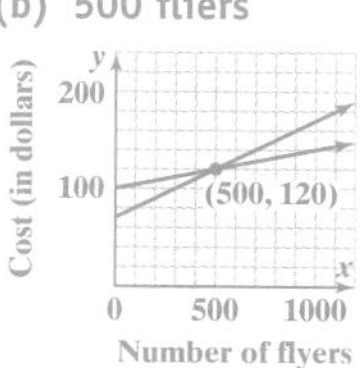

(c) Printer A

20. (a) $\begin{cases} y = 50 + 40x \\ y = 350 + 10x \end{cases}$
(b) 10 sq yd

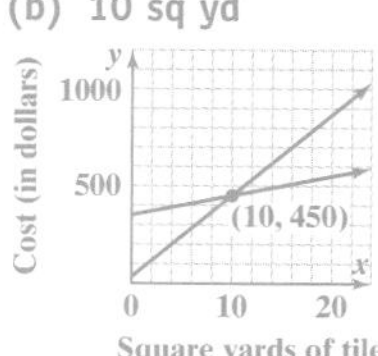

(c) tile

In Problems 1–4, determine whether the ordered pair is a solution to the system of equations.

1. $\begin{cases} x + 2y = 6 \\ 3x - y = -10 \end{cases}$

(a) $(3, -1)$ **(b)** $(-2, 4)$ **(c)** $(4, 1)$

2. $\begin{cases} y = 3x - 5 \\ 3y = 6x - 5 \end{cases}$

(a) $(2, 1)$ **(b)** $\left(0, -\frac{5}{3}\right)$ **(c)** $\left(\frac{10}{3}, 5\right)$

3. $\begin{cases} 3x - 4y = 2 \\ 20y = 15x - 10 \end{cases}$

(a) $\left(\frac{1}{2}, -\frac{1}{8}\right)$ **(b)** $(6, 4)$
(c) $(0.4, -0.2)$

4. $\begin{cases} x = -4y + 2 \\ 2x + 8y = 12 \end{cases}$

(a) $(10, -1)$ **(b)** $\left(-\frac{1}{2}, \frac{1}{2}\right)$
(c) $(2, 1)$

In Problems 5–12, solve each system of equations by graphing.

5. $\begin{cases} 2x - 4y = 8 \\ x + y = 7 \end{cases}$

6. $\begin{cases} x - y = -3 \\ 3x + 2y = 6 \end{cases}$

7. $\begin{cases} y = -\dfrac{x}{2} + 2 \\ y = x + 8 \end{cases}$

8. $\begin{cases} y = -x - 5 \\ y = \dfrac{3x}{4} + 2 \end{cases}$

9. $\begin{cases} 4x - 8 = 0 \\ 3y + 9 = 0 \end{cases}$

10. $\begin{cases} x = y \\ x + y = 0 \end{cases}$

11. $\begin{cases} 0.6x + 0.5y = 2 \\ 10y = -12x + 20 \end{cases}$

12. $\begin{cases} \dfrac{1}{4}x - \dfrac{1}{2}y = 1 \\ 3x - 6y = 12 \end{cases}$

In Problems 13–18, without graphing, determine the number of solutions to each system of equations. State whether the system is consistent or inconsistent. For those systems that are consistent, state whether the equations are dependent or independent.

13. $\begin{cases} 3x = y + 4 \\ 3x - y = -4 \end{cases}$

14. $\begin{cases} 4y = 2x - 8 \\ x - 2y = -4 \end{cases}$

15. $\begin{cases} -3x + 3y = -3 \\ \dfrac{1}{2}x - \dfrac{1}{2}y = 0.5 \end{cases}$

16. $\begin{cases} x - 2 = -\dfrac{2}{3}y - \dfrac{2}{3} \\ 3x = 4 - 2y \end{cases}$

17. $\begin{cases} 3 - 2x = y \\ \dfrac{y}{2} = x + 1.5 \end{cases}$

18. $\begin{cases} \dfrac{y}{2} = \dfrac{x}{4} + 2 \\ \dfrac{x}{8} + \dfrac{y}{4} = -1 \end{cases}$

19. Printing Costs Monique is creating a flyer to give to local businesses to advertise her new vintage clothing store. She is trying to decide between two quotes for the printing. Printer A has given her a quote of \$70 setup fee plus \$0.10 per flyer printed. Printer B has given her a quote of \$100 plus \$0.04 per flyer.

(a) Write a system of linear equations that models the problem.
(b) Graph the system of equations to determine how many fliers she needs to have printed in order for the cost to be the same at each printer.
(c) If 400 fliers are printed, which printer should she choose to have the lower cost?

20. Flooring Installation To install carpeting in Juan's bedroom, it costs \$50 for installation plus \$40 per square yard of carpet. In the same room, Juan can install ceramic tile for a cost of \$350 for installation plus \$10 per square yard of tile.

(a) Write a system of linear equations that models the problem.
(b) Graph the system of equations to determine how many square yards of flooring he needs to have installed for the cost to be the same.
(c) If the area of Juan's bedroom is 15 square yards, which material would be cheaper?

Section 9.2 Solving Systems of Linear Equations in Two Variables Using Substitution

KEY CONCEPT

- **Solving Systems of Linear Equations Using Substitution**

  Step 1: Solve one of the equations for one of the unknowns. For example, we might solve equation (1) for y in terms of x.

  Step 2: Substitute the expression solved for in Step 1 into the other equation. The result will be a single linear equation in one unknown. For example, if we solved equation (1) for y in terms of x in Step 1, then we would replace y in equation (2) with the expression in x.

  Step 3: Solve the linear equation in one unknown found in Step 2.

  Step 4: Substitute the value of the variable found in Step 3 into one of the original equations to find the value of the other variable.

  Step 5: Check your answer.

YOU SHOULD BE ABLE TO . . .	EXAMPLE	REVIEW EXERCISES
1 Solve a system of linear equations using the substitution method (p. 629)	Examples 1 through 5	21–34
2 Solve applied problems involving systems of linear equations (p. 635)	Example 6	35, 36

21. (2, 1)
22. (1, −1)
23. (7, −2)
24. (4, −2)
25. (18, 11)
26. (−6, 8)
27. (2, −2)
28. (3, −1)
29. infinitely many solutions
30. infinitely many solutions
31. no solution
32. infinitely many solutions
33. $\left(\frac{3}{2}, 1\right)$
34. $\left(\frac{1}{3}, -1\right)$
35. width 125 m; length 200 m
36. −3

In Problems 21–34, solve each system of equations using substitution.

21. $\begin{cases} x + 4y = 6 \\ y = 2x - 3 \end{cases}$ **22.** $\begin{cases} 7x - 3y = 10 \\ y = 3x - 4 \end{cases}$ **23.** $\begin{cases} 2x + 5y = 4 \\ x = 3 - 2y \end{cases}$ **24.** $\begin{cases} 3x + y = 10 \\ x = 8 + 2y \end{cases}$

25. $\begin{cases} y = \frac{2}{3}x - 1 \\ y = \frac{1}{2}x + 2 \end{cases}$ **26.** $\begin{cases} y = -\frac{5}{6}x + 3 \\ y = -\frac{4}{3}x \end{cases}$ **27.** $\begin{cases} 2x - y = 6 \\ 4x + 3y = 2 \end{cases}$ **28.** $\begin{cases} 5x + 2y = 13 \\ x + 4y = -1 \end{cases}$

29. $\begin{cases} 6x + 3y = 12 \\ y = -2x + 4 \end{cases}$ **30.** $\begin{cases} x = 4y - 2 \\ 8y - 2x = 4 \end{cases}$ **31.** $\begin{cases} -6 - 2(3x - 6y) = 0 \\ 6 - 12(x - 2y) = 0 \end{cases}$

32. $\begin{cases} 6 - 2(3y + 4x) = 0 \\ 9(y - 1) + 12x = 0 \end{cases}$ **33.** $\begin{cases} \frac{1}{2}x - \frac{1}{4}y = \frac{1}{2} \\ \frac{1}{3}x - \frac{3}{4}y = -\frac{1}{4} \end{cases}$ **34.** $\begin{cases} -\frac{5x}{4} + \frac{y}{6} = -\frac{7}{12} \\ \frac{3x}{2} - \frac{y}{10} = \frac{3}{5} \end{cases}$

△ **35. Walk around the Park** A rectangular park is surrounded by a walking path. The park is 75 meters longer than it is wide. If a person walks completely around the park and covers a distance of 650 meters, find the length and width of the park by solving the following system of equations where l is the length and w is the width.

$$\begin{cases} 2l + 2w = 650 & (1) \\ l = w + 75 & (2) \end{cases}$$

36. Fun with Numbers The sum of two numbers is 12. If twice the smaller number is subtracted from the larger number, the difference is 21. Determine the smaller of the two numbers by solving the following system of equations where x and y represent the unknown numbers.

$$\begin{cases} x + y = 12 & (1) \\ x - 2y = 21 & (2) \end{cases}$$

Section 9.3 Solving Systems of Linear Equations Using Elimination

KEY CONCEPT

- **Solving Systems of Linear Equations Using Elimination**

  Step 1: Make sure that the coefficients on one of the variables are additive inverses by multiplying (or dividing) both sides of one (or both) equation(s) by a nonzero constant.

  Step 2: Add the equations to eliminate the variable whose coefficients are now additive inverses. Solve the resulting equation for the remaining unknown.

Step 3: Substitute the value of the variable found in Step 2 into one of the original equations to find the value of the remaining variable.

Step 4: Check your answer.

YOU SHOULD BE ABLE TO . . .	EXAMPLE	REVIEW EXERCISES
1 Solve a system of linear equations using the elimination method (p. 640)	Examples 1 through 5	37–52
2 Solve applied problems involving systems of linear equations (p. 645)	Example 6	53, 54

37. $(0, -12)$
38. $(-3, 7)$
39. $(-3, 4)$
40. $(-3, 0)$
41. $(-1, -1)$
42. $(5, 2)$
43. $(-2, 2)$
44. $\left(\frac{202}{55}, -\frac{59}{11}\right)$
45. $\left(\frac{1}{2}, -2\right)$
46. $\left(-\frac{3}{2}, 1\right)$
47. infinitely many solutions
48. $\left(\frac{6}{7}, \frac{5}{2}\right)$
49. $(-20, -13)$
50. $\left(\frac{1}{3}, \frac{2}{3}\right)$
51. no solution
52. $\left(-\frac{1}{7}, 0\right)$
53. 2.4 lb cookies; 1.6 lb chocolates
54. 400 individual tickets; 325 block tickets

In Problems 37–44, solve each system of equations using elimination.

37. $\begin{cases} 4x - y = 12 \\ 2x + y = -12 \end{cases}$ **38.** $\begin{cases} -2x + 3y = 27 \\ 2x - 5y = -41 \end{cases}$ **39.** $\begin{cases} -3x + 4y = 25 \\ x - 5y = -23 \end{cases}$

40. $\begin{cases} 5x + 8y = -15 \\ -2x + y = 6 \end{cases}$ **41.** $\begin{cases} 4x - 3y = -1 \\ 2x - 5y = 3 \end{cases}$ **42.** $\begin{cases} -2x + 5y = 0 \\ -3x - 2y = -19 \end{cases}$

43. $\begin{cases} 1.3x - 0.2y = -3 \\ -0.1x + 0.5y = 1.2 \end{cases}$ **44.** $\begin{cases} 2.5x + 0.5y = 6.5 \\ -0.5x - 1.2y = 4.6 \end{cases}$

In Problems 45–52, solve each system of equations by any method.

45. $\begin{cases} 2x + y = -1 \\ -6x - 8y = 13 \end{cases}$ **46.** $\begin{cases} \dfrac{x+9}{6} = \dfrac{9-4y}{4} \\ \dfrac{2y-8}{3} = 4x + 4 \end{cases}$ **47.** $\begin{cases} y + 5 = \dfrac{2}{3}x + 3 \\ \dfrac{2x+3}{6} = \dfrac{y+3}{2} \end{cases}$

48. $\begin{cases} \dfrac{1}{14} - \dfrac{x}{2} = -\dfrac{y}{7} \\ y = \dfrac{1}{2} + \dfrac{7x}{3} \end{cases}$ **49.** $\begin{cases} -x + y = 7 \\ -3x + 4y = 8 \end{cases}$ **50.** $\begin{cases} 4x + y = 2 \\ 9y - 3x = 5 \end{cases}$

51. $\begin{cases} 4x + 6 = 3y + 5 \\ 4(-2x - 4) = 6(-y - 3) \end{cases}$ **52.** $\begin{cases} 3y + 2x = 16x + 2 \\ x = -\dfrac{3}{14}y - \dfrac{1}{7} \end{cases}$

53. Valentine Gift The specialty bakery around the corner is putting together a Valentine's Day gift box to sell. The gift box is going to contain two items, heart-shaped cookies and chocolate candies. If the cookies sell for \$1.50 per pound and the chocolates sell for \$7.75 per pound, how many pounds of each type should be included in a four-pound box that sells for \$4.00 per pound? Solve the system

$$\begin{cases} h + c = 4 & (1) \\ 1.50h + 7.75c = 16 & (2) \end{cases}$$

where h represents the number of pounds of heart-shaped cookies and c represents the number of pounds of candies to find the answers.

54. Raffle Tickets The parent booster club is selling 50–50 raffle tickets to raise money for the upcoming hockey banquet. Raffle tickets to win half the money collected are sold in two ways: Individual chances are sold for \$1.00 each or a ticket with a block of 6 chances can be purchased for \$5.00. On one particular night the booster club brought in \$2025 from the sale of 725 raffle tickets. How many of each type of raffle ticket was sold? Solve the system

$$\begin{cases} i + b = 725 & (1) \\ 1i + 5b = 2025 & (2) \end{cases}$$

where i is the number of individual tickets and b is the number of block tickets to find the answers.

Section 9.4 Solving Direct Translation, Geometry, and Uniform Motion Problems Using Systems of Linear Equations		
YOU SHOULD BE ABLE TO . . .	**EXAMPLE**	**REVIEW EXERCISES**
1 Model and solve direct translation problems (p. 651)	Example 1	55–58
2 Model and solve geometry problems (p. 652)	Examples 2 and 3	59–62
3 Model and solve uniform motion problems (p. 654)	Example 4	63–66

55. $\frac{3}{8}, \frac{1}{3}$
56. 2.6, 3.2
57. $32,000 in stocks, $18,000 in bonds
58. 14 notebooks, 10 calculators
59. 77.5°, 102.5°
60. 40°, 50°
61. 15 m by 22 m
62. 110°, 40°
63. plane 450 mph, wind 50 mph
64. current 2 mph, paddling 6 mph
65. cyclist 10 mph, wind 2 mph
66. faster 8 mph; slower 6 mph

55. Fun with Numbers Find two numbers whose difference is $\frac{1}{24}$ and whose sum is $\frac{17}{24}$.

56. Fun with Numbers The sum of two numbers is 5.8. If twice the smaller number is subtracted from the larger number, the difference is -2. Find the smaller number.

57. Investments Aaron has $50,000 to invest. His financial advisor recommends he invest in stocks and bonds with the amount invested in stocks equaling $4000 less than twice the amount invested in bonds. How much should be invested in stocks? How much should be invested in bonds?

58. Bookstore Sales A register at the bookstore sold only notebooks and scientific calculators. The cost of a notebook is $\frac{1}{3}$ of the cost of the calculator. If the calculator sells for $7.50 and the register tape shows that 24 items resulted in revenue of $110, how many of each was sold?

△ **59. Geometry** The measure of one angle is 25° less than the measure of its supplement. Find the measures of the two angles.

△ **60. Geometry** The measure of one angle is 15° more than half the measure of its complement. Find the measures of the two angles.

△ **61. Horse Corral** The O'Connells are planning to build a rectangular corral for their horses using the river on their property as one of the sides. On the other three sides they are installing 52 meters of fencing. The longest side of the corral is opposite the water and is 8 meters less than twice the short side. Find the dimensions of the corral.

△ **62. Geometry** In a triangle, the measure of the first angle is 10° less than the three times the measure of the second. If the measure of the third angle is 30°, find the measure of the two unknown angles of the triangle.

63. Wind Resistance A plane can fly 2000 miles with a tailwind in 4 hours. It can make the return trip in 5 hours. Find the speed of the plane in still air and the speed of the wind.

64. Fishing Spot Dayle is paddling his canoe upstream, against the current, to a fishing spot 10 miles away. If he paddles upstream for 2.5 hours and his return trip takes 1.25 hours, find the speed of the current and his paddling speed in still water.

65. Cycling in the Wind A cyclist can go 36 miles with the wind blowing at her back in 3 hours. On the return trip, after 4 hours, the cyclist still has 4 miles remaining to return to the starting point. Find the speed of the cyclist and the speed of the wind.

66. Out for a Run One speedy jogger can run 2 mph faster than his running buddy. In the last race they entered, the faster runner covered 12 miles in the same time that it took for the slower jogger to cover 9 miles. Find the speed of each of the runners.

Section 9.5 Solving Mixture Problems Using Systems of Linear Equations in Two Variables		
YOU SHOULD BE ABLE TO . . .	**EXAMPLE**	**REVIEW EXERCISES**
(1) Draw up a plan for modeling mixture problems (p. 658)	Example 1	67, 68
(2) Set up and solve money problems using the mixture model (p. 659)	Examples 2 and 3	69–72
(3) Set up and solve dry mixture and percent mixture problems (p. 662)	Examples 4 and 5	73–76

In Problems 67 and 68, fill in the table from the information given. DO NOT SOLVE.

67. Brendan has twice as many nickels as dimes. He has a total of $2.25 in change.

	Number of Coins	· Value of Each Coin	= Total Value
Dimes	?	?	?
Nickels	?	?	?
Total			?

68. Belinda invested part of her $15,000 inheritance in a savings account with an annual rate of 6.5% simple interest and the rest of her inheritance in a mutual fund with a rate of 8% simple interest. Her total yearly interest income from both accounts was $1050.

	Principal	Rate	Interest
Savings Account	?	?	?
Mutual Fund	?	?	?
Total	?		?

69. Movie Tickets A movie theater sells three types of tickets: children $4.00 each, adults $6.25 each, and seniors $5.00 each. At one particular show, the same number of children and adult tickets were sold. If a total of 300 tickets brought in $1531.25 in sales, how many of each type were sold?

70. Coins Sharona has $1.70 in change consisting of three more dimes than quarters. Find the number of quarters she has.

71. Investment Carlos Singer has some money invested at 5%, and $5000 more than that invested at 9%. His total annual interest income is $1430. Find the amount Carlos has invested at each rate.

72. Investment Hilda invested part of her $25,000 advance in savings bonds at 7% annual simple interest and the rest in a stock portfolio yielding 8% annual simple interest. If her total yearly interest income is $1900, find the amount Hilda has invested at each rate.

73. Sugar A baker wants to mix a 60% sugar solution with a 30% sugar solution to obtain 10 quarts of a 51% sugar solution. How much of the 30% sugar solution will the baker use?

74. Peroxide How many pints of a 25% peroxide solution should be added to 10 pints of a 60% peroxide solution to obtain a 30% peroxide solution?

75. Almond-Peanut Butter Joni works at a health food store and is planning to grind a special blend of peanut butter by mixing peanuts, which sell for $4.00 per pound, and almonds, which sell for $6.50 per pound. How many pounds of each type should she use if she needs 5 pounds of almond-peanut butter, which sells for $5.00 per pound?

76. Mixing Chemicals Harold needs 20 liters of 55% acid. If he has 35% acid and 60% acid that he can mix together, how many liters of each should he use to obtain the desired mixture?

67.

	Number	· Value	= Total Value
Dimes	d	0.10	$0.10d$
Nickels	n	0.05	$0.05n$
Total			2.25

68.

	P	· r	= I
Savings	s	0.065	$0.065s$
Mutual Fund	m	0.08	$0.08m$
Total	15,000		1050

69. 125 children; 125 adult; 50 senior
70. 7 dimes; 4 quarters
71. $7000 at 5%, $12,000 at 9%
72. $10,000 bonds, $15,000 stocks
73. 7 quarts 60% sugar solution; 3 quarts 30% sugar solution
74. 60 pints
75. 3 lb peanuts; 2 lb almonds
76. 4 l 35% acid; 16 l 60% acid
77. (a) Yes (b) Yes (c) No
78. (a) No (b) Yes (c) Yes
79. (a) Yes (b) No (c) No
80. (a) No (b) Yes (c) Yes
81. (a) No (b) No (c) No
82. (a) No (b) Yes (c) No
83. $\begin{cases} x > -2 \\ y > 1 \end{cases}$

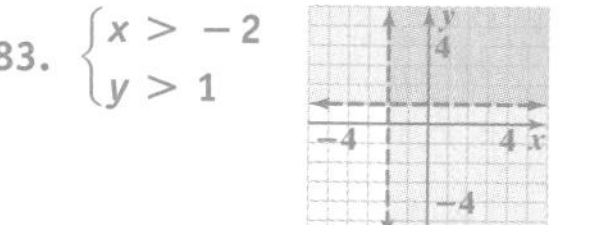

84. $\begin{cases} x \le 3 \\ y > -1 \end{cases}$

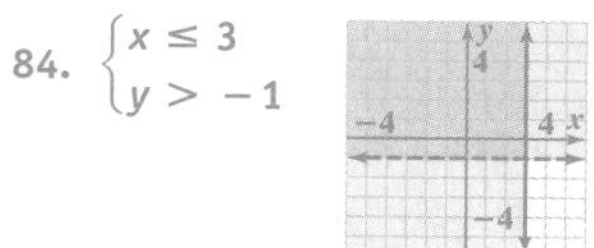

85. $\begin{cases} x + y \ge -2 \\ 2x - y \le -4 \end{cases}$

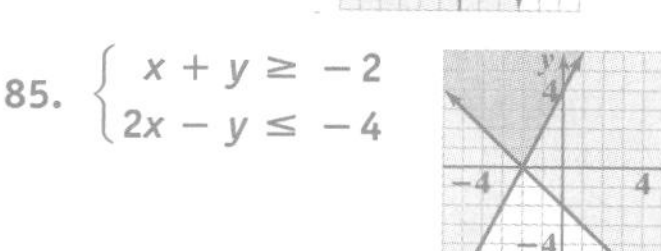

Section 9.6 Solving Systems of Linear Inequalities

KEY TERM

Satisfies

YOU SHOULD BE ABLE TO . . .	EXAMPLE	REVIEW EXERCISES
(1) Determine whether an ordered pair is a solution of a system of linear inequalities (p. 670)	Example 1	77–82
(2) Graph a system of linear inequalities (p. 671)	Examples 2 through 4	83–94
(3) Solve applied problems involving systems of linear inequalities (p. 673)	Example 5	95

In Problems 77–82, determine which point(s), if any, is a solution to the system of linear inequalities.

77. $\begin{cases} x + y \le 2 \\ 3x - 2y > 6 \end{cases}$
(a) $(-1, -5)$ **(b)** $(3, -1)$ **(c)** $(4, 1)$

78. $\begin{cases} y \ge 3x + 5 \\ y \ge -2x \end{cases}$
(a) $(-1, 0)$ **(b)** $(-2, 4)$ **(c)** $(1, 8)$

79. $\begin{cases} x > 5 \\ y < -2 \end{cases}$
(a) $(10, -10)$ **(b)** $(5, -3)$ **(c)** $(7, -2)$

80. $\begin{cases} y > x \\ 2x - y \le 3 \end{cases}$
(a) $(4, 3)$ **(b)** $(-3, -2)$ **(c)** $(-1, 4)$

81. $\begin{cases} x + 2y < 6 \\ 4y - 2x > 16 \end{cases}$
(a) $(-4, 2)$ **(b)** $(4, 5)$ **(c)** $(-2, -3)$

82. $\begin{cases} 2x - y \ge 3 \\ y \le 2x + 1 \end{cases}$
(a) $(1, 2)$ **(b)** $(3, -2)$ **(c)** $(-1, 4)$

In Problems 83–94, graph the solution set of each system of linear inequalities.

83. $\begin{cases} x > -2 \\ y > 1 \end{cases}$

84. $\begin{cases} x \le 3 \\ y > -1 \end{cases}$

85. $\begin{cases} x + y \ge -2 \\ 2x - y \le -4 \end{cases}$

86. $\begin{cases} 3x + 2y < -6 \\ x - y < 2 \end{cases}$

87. $\begin{cases} x > 0 \\ y \le \dfrac{3}{4}x + 1 \end{cases}$

88. $\begin{cases} y \le 0 \\ y \le -\dfrac{1}{2}x - 3 \end{cases}$

89. $\begin{cases} -y \ge x \\ 4x - 3y \ge -12 \end{cases}$

90. $\begin{cases} -y \le x + 2 \\ 2x + 2y \ge -9 \end{cases}$

91. $\begin{cases} 2x + 3y > -3 \\ y > -\dfrac{2}{3}x + 2 \end{cases}$

92. $\begin{cases} x + 4y \le -4 \\ y \ge \dfrac{1}{4}x + 3 \end{cases}$

93. $\begin{cases} y > 2x - 5 \\ y - 2x \le 0 \end{cases}$

94. $\begin{cases} y > x + 2 \\ y < x - 4 \end{cases}$

95. Party Planning Alexis and Sarah are planning a barbeque for their friends. They plan to serve grilled fish and carne asada and want to spend at most \$40 on the meat. The fish sells for \$8 per pound and the carne asada is \$5 per pound. Since most of their friends do not eat red meat, they plan to buy at least twice as much fish as carne asada. A system of linear inequalities that models this situation is given by

$$\begin{cases} 8x + 5y \le 40 \\ x \ge 2y \\ x \ge 0 \\ y \ge 0 \end{cases}$$

where x represents the number of pounds of fish and y represents the number of pounds of carne asada. Graph the system of linear inequalities that shows how many pounds of each they can buy and stay within their budget.

86. $\begin{cases} 3x + 2y < -6 \\ x - y < 2 \end{cases}$

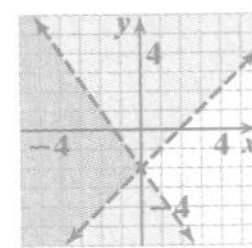

87. $\begin{cases} x > 0 \\ y \le \frac{3}{4}x + 1 \end{cases}$

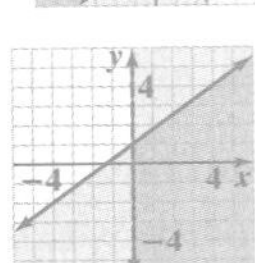

88. $\begin{cases} y \le 0 \\ y \le -\frac{1}{2}x - 3 \end{cases}$

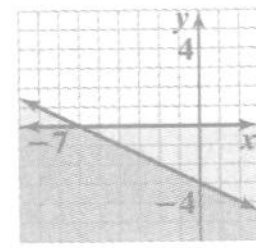

89. $\begin{cases} -y \ge x \\ 4x - 3y > -12 \end{cases}$

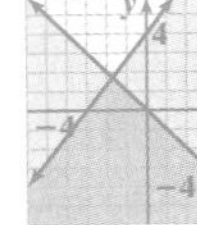

90. $\begin{cases} -y < x + 2 \\ 2x + 2y \ge -9 \end{cases}$

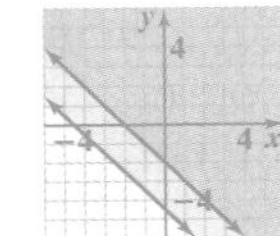

91. $\begin{cases} 2x + 3y > -3 \\ y > -\frac{2}{3}x + 2 \end{cases}$

92. $\begin{cases} x + 4y \le -4 \\ y \ge \frac{1}{4}x + 3 \end{cases}$

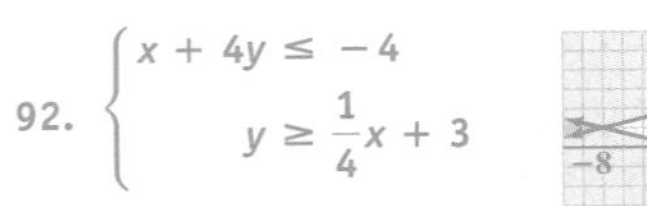

93. $\begin{cases} y > 2x - 5 \\ y - 2x \le 0 \end{cases}$

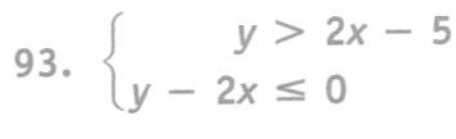

94. $\begin{cases} y > x + 2 \\ y < x - 4 \end{cases}$

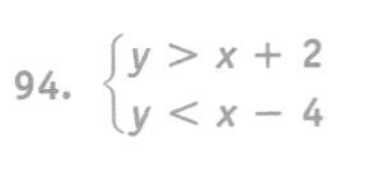

95. x = fish, y = carne asada

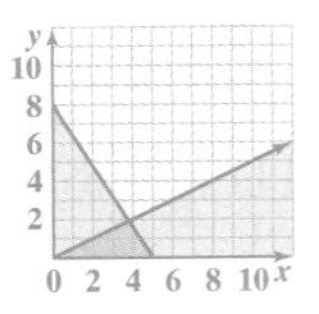

CHAPTER 9 TEST

Remember to use your Chapter Test Prep Video CD to see fully worked-out solutions to any of these problems you would like to review.

Note to Instructor: A special file in TestGen provides algorithms specifically matched to the problems in this Chapter Test for easy-to-replicate practice or assessment purposes.

1. (a) No
   (b) No
   (c) Yes
2. (a) Yes
   (b) No
   (c) No
3. (a) one
   (b) consistent
   (c) independent
4. (a) none
   (b) inconsistent
   (c) not applicable
5. $\begin{cases} 2x + 3y = 0 \\ x + 4y = 5 \end{cases}$

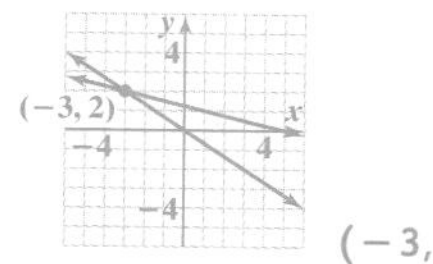

$(-3, 2)$

6. $\begin{cases} y = 2x - 6 \\ y = -\dfrac{1}{4}x + 3 \end{cases}$

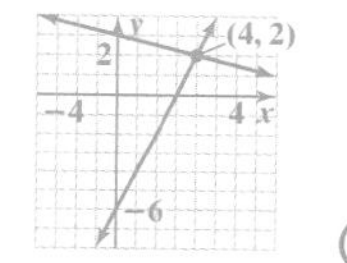

$(4, 2)$

7. $(0, -3)$
8. $(-12, -1)$
9. $(-3, 3)$
10. $(-2, -4)$
11. $\left(3, -\dfrac{1}{2}\right)$
12. $(2.5, 3)$
13. infinitely many solutions
14. airplane 200 mph; wind 25 mph

In Problems 1 and 2, determine which of the following is a solution to the system of linear equations or inequalities.

1. $\begin{cases} 3x - y = -5 \\ y = \dfrac{2}{3}x - 2 \end{cases}$

(a) $(-1, 2)$
(b) $(-9, -8)$
(c) $(-3, -4)$

2. $\begin{cases} 2x + y \geq 10 \\ y < \dfrac{x}{2} + 1 \end{cases}$

(a) $(5, 3)$
(b) $(-2, 14)$
(c) $(-3, -1)$

In Problems 3 and 4, you begin to solve a system of linear equations in two variables. Before you draw the graphs you note the given conditions.

3. The slopes of the two lines are negative reciprocals and the y-intercepts are the same.

(a) Tell how many solutions exist.
(b) State whether the system is consistent or inconsistent.
(c) If the system is consistent, state whether the system is dependent or independent.

4. The slopes of the two lines are the same and the y-intercepts are different.

(a) Tell how many solutions exist.
(b) State whether the system is consistent or inconsistent.
(c) If the system is consistent, state whether the system is dependent or independent.

In Problems 5 and 6, solve each system of equations by graphing.

5. $\begin{cases} 2x + 3y = 0 \\ x + 4y = 5 \end{cases}$

6. $\begin{cases} y = 2x - 6 \\ y = -\dfrac{1}{4}x + 3 \end{cases}$

In Problems 7 and 8, solve each system of equations using substitution.

7. $\begin{cases} 3x - y = 3 \\ 4x + 5y = -15 \end{cases}$

8. $\begin{cases} y = \dfrac{2}{3}x + 7 \\ y = \dfrac{1}{4}x + 2 \end{cases}$

In Problems 9 and 10, solve each system of equations using elimination.

9. $\begin{cases} 3x + 2y = -3 \\ 5x - y = -18 \end{cases}$

10. $\begin{cases} 4x - 5y = 12 \\ 3x + 4y = -22 \end{cases}$

In Problems 11–13, solve by any method.

11. $\begin{cases} \dfrac{2}{3}x - \dfrac{1}{6}y = \dfrac{25}{12} \\ -\dfrac{1}{4}x = \dfrac{3}{2}y \end{cases}$

12. $\begin{cases} 0.4x - 2.5y = -6.5 \\ x + y = 5.5 \end{cases}$

13. $\begin{cases} 4 + 3(y - 3) = -2(x + 1) \\ x + \dfrac{3y}{2} = \dfrac{3}{2} \end{cases}$

14. Plane Trip A small plane can fly 1575 miles in 7 hours with a tailwind. On the return trip, with a headwind, the same trip takes 9 hours. Find the speed of the plane in still air and the speed of the wind.

15. 7 containers vanilla; 4 containers peach
16. 52.5°, 127.5°
17. 15 basketballs, 25 volleyballs
18. $\begin{cases} 4x - 2y < 8 \\ x + 3y < 6 \end{cases}$

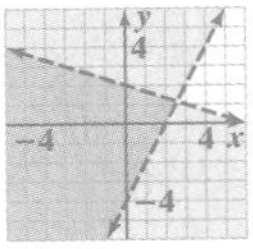

19. $\begin{cases} x \le -2 \\ -2x - 4y \le 8 \end{cases}$

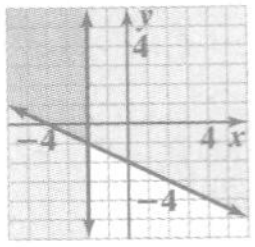

20. $\begin{cases} y \le \frac{2}{3}x + 4 \\ -2x + 3y > -3 \end{cases}$

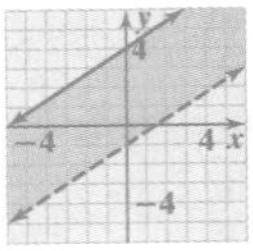

15. Ice Cream Manny is blending a peach swirl ice cream. He mixes together vanilla ice cream, which sells for \$6 per container, and peach ice cream, which sells for \$11.50 per container. How many containers of each should he use if he wants to have 11 containers of swirl ice cream, which he can sell for \$8 each?

△ **16. Supplementary Angles** The measure of one angle is 30° less than three times the measure of its supplement. Find the measures of the two angles.

17. Playground Equipment Moises is distributing new playground equipment to several elementary schools. Each school will receive some basketballs and some volleyballs. The cost of a basketball is \$25 and the cost of a volleyball is \$15. If each school needs 40 balls and he can spend \$750 per school, how many of each type will a school receive?

In Problems 18–20, graph the solution set of each system of inequalities.

18. $\begin{cases} 4x - 2y < 8 \\ x + 3y < 6 \end{cases}$ **19.** $\begin{cases} x \le -2 \\ -2x - 4y \le 8 \end{cases}$ **20.** $\begin{cases} y \le \frac{2}{3}x + 4 \\ -2x + 3y > -3 \end{cases}$

CUMULATIVE REVIEW CHAPTERS 1–9

1. $8x + 23$
2. $\dfrac{y^4}{81x^2}$
3. $\dfrac{x^2 + 2x + 4}{x + 2}$
4. $34 - 24\sqrt{2}$
5. $\dfrac{x - 6}{5 - 4x}$
6. 9
7. $\dfrac{1}{\sqrt{-729}}$ is not a real number.
8. $-\dfrac{1}{27}$
9. $\left\{\dfrac{5}{2}, -5\right\}$
10. $\{-2, 0, 2\}$
11. no solution
12. $\{6\}$
13. $\{x \mid x \le 5\}$; $(-\infty, 5]$

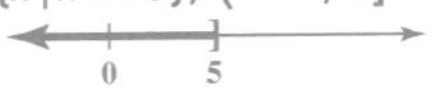

14. $q = \dfrac{3pr}{2r - 6p}$
15. 80%
16. $2\sqrt{6}$

In Problems 1–5, simplify each expression.

1. $5[3 - 2(-2x + 1)] - 6(2x - 3)$ **2.** $\dfrac{(3x^2y^{-2})^{-3}}{3x^{-4}y^2}$ **3.** $\dfrac{x^3 - 8}{x^2 - 4}$

4. $(3\sqrt{2} - 4)^2$ **5.** $\dfrac{\frac{1}{2x} - \frac{3}{x^2}}{\frac{10}{4x^2} - \frac{2}{x}}$

In Problems 6–8, evaluate each expression. If the expression does not represent a real number, so state.

6. $\dfrac{-3x^2 - y^2}{z}$ when $x = -3$; $y = -3$; $z = -4$ **7.** $(-9)^{-\frac{3}{2}}$ **8.** $-9^{-\frac{3}{2}}$

In Problems 9–12, solve each equation, if possible.

9. $2x^2 + 5x - 25 = 0$ **10.** $x^3 - 4x = 0$

11. $\sqrt{3x + 4} = \sqrt{x} - 2$ **12.** $\dfrac{3x - 1}{2} - 4 = \dfrac{2x}{3} + \dfrac{3}{6}$

13. Solve and graph the following inequality: $3x + 4 \ge 5x - 6$

14. Solve the equation $\dfrac{2}{p} - \dfrac{3}{q} = \dfrac{6}{r}$ for q.

15. 36 is what percent of 45?

16. Use the Pythagorean Theorem to find the missing side of the triangle:

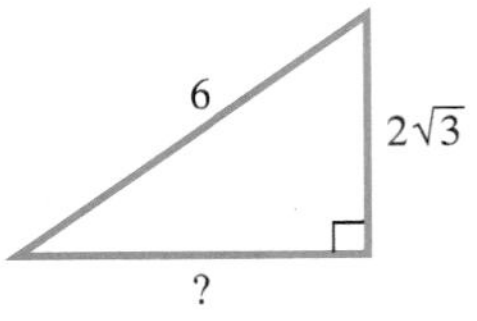

17. $-\frac{7}{6}$
18. (a) $\frac{4}{3}$
(b) (0, 2)
(c) $\left(-\frac{3}{2}, 0\right)$
19. $y = -\frac{3}{2}x - 8$
20. $(a^8 + 4b^2)(a^4 + 2b)(a^4 - 2b)$
21. $2a(2a + 5)(a - 3)$
22. 4.32×10^2
23. $\{2 - \sqrt{5}i, 2 + \sqrt{5}i\}$
24. −7, −5, −3
25. 5 and 3

17. Find the slope of line shown:

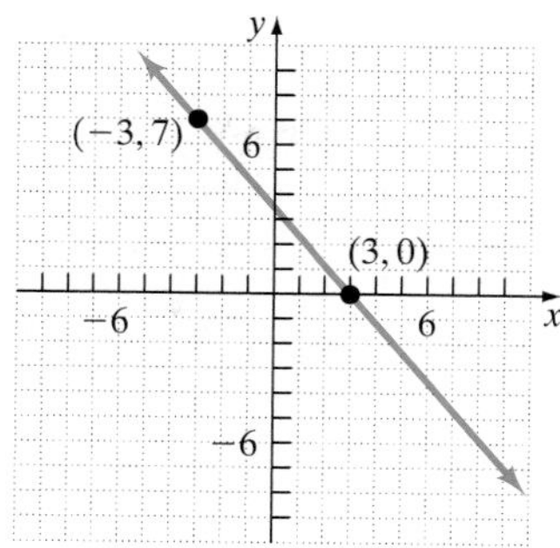

18. Consider the graph of $-4x + 3y = 6$.

(a) Find the slope of the line.
(b) Find the y-intercept.
(c) Find the x-intercept.

19. Write the equation of the line whose slope is $-\frac{3}{2}$ and passes through the point $(-4, -2)$.

In Problems 20 and 21, factor completely, if possible.

20. $a^{16} - 16b^4$

21. $4a^3 - 2a^2 - 30a$

22. Perform the indicated operation: $(1.2 \times 10^6) \cdot (3.6 \times 10^{-4})$

23. Solve: $x^2 - 4x + 9 = 0$

24. Consecutive Integers Find three consecutive odd integers such that three times the middle number, increased by 5, is the same as the sum of the first and last integer.

25. Missing Numbers Find two numbers such that twice the first decreased by the second is 7 and the difference between the first and 3 times the second is −4.

CHAPTER

10 Graphs of Quadratic Equations in Two Variables and an Introduction to Functions

One of the oldest competitions in the Summer Olympics is the shot-put competition. Did you know that the distance a shot-putter can throw the 7.26-kilogram metal ball can be described mathematically? See Problem 63 in Section 10.1.

OUTLINE

The Big Picture: Putting It All Together

We have reached the last chapter of the text! Congratulations! This chapter begins with an introduction to graphing quadratic equations in two variables. In Chapter 8, we learned that we can graph linear equations in two variables by plotting points. However, it is usually more efficient to use the properties of linear equations (such as slope and intercepts) to obtain the graph. The same is true for finding the graphs of quadratic equations in two variables. We will learn how to use the properties of quadratic equations in two variables to obtain the graph of the quadratic equation. Just as linear equations describe everyday phenomena, quadratic equations and their graphs are all around us. You can "see" quadratic equations in the path followed by the shot hurled by a shot-putter, in car headlights, in satellite dishes, and even in the wok in your kitchen.

The chapter ends with two sections that present an introduction to functions. For those of you continuing your study of algebra, you will see functions often. So it is worthwhile to start learning about functions now so that concept will not be entirely new in your next class.

10.1 Quadratic Equations in Two Variables

OBJECTIVES

1. Graph a Quadratic Equation by Plotting Points
2. Find the Intercepts of the Graph of a Quadratic Equation
3. Find the Vertex and Axis of Symmetry of the Graph of a Quadratic Equation
4. Graph a Quadratic Equation Using Its Vertex, Intercepts, and Axis of Symmetry
5. Determine the Maximum or Minimum Value of a Quadratic Equation

Preparing for Quadratic Equations in Two Variables

Before getting started, take the following readiness quiz. If you get a problem wrong, go back to the section cited and review the material.

1. Graph: $y = 3x - 8$ [Section 8.4, pp. 561–563]
2. Graph: $x = -1$ [Section 8.2, pp. 542–543]
3. Solve: $x^2 - 6x + 8 = 0$ [Section 4.6, pp. 281–287]
4. Solve: $2x^2 + x - 5 = 0$ [Section 7.3, pp. 477–487]
5. Find the intercepts of $3x - 5y = 30$. [Section 8.2, pp. 539–542]

In Section 8.1, we first graphed linear equations in two variables by plotting points. We then learned that the graph of a linear equation can also be obtained by using its properties such as slope and y-intercept.

We now extend this idea to graphing a second type of equation, called a *quadratic equation.*

DEFINITION

A **quadratic equation** in two variables is an equation of the form

$$y = ax^2 + bx + c$$

where a, b, and c are real numbers and $a \neq 0$.

Classroom Example
Graph $y = x^2 - 4x - 5$ by plotting points.

Answer:

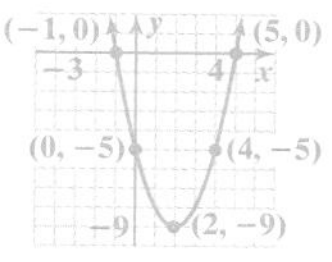

Just as we did for linear equations, we begin by graphing quadratic equations by plotting points.

1 Graph a Quadratic Equation by Plotting Points

Recall that the point-plotting method for obtaining the graph of an equation requires that we choose certain x-values and then use the equation to find the corresponding y-values. We plot the ordered pairs and connect the points in a smooth curve.

Preparing for...Answers

1.

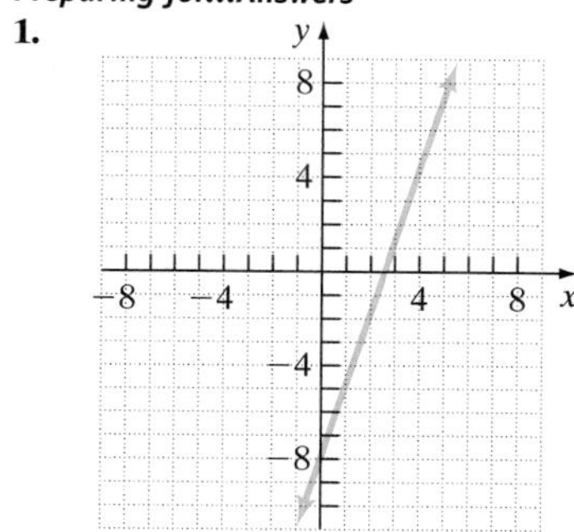

2.

3. $\{2, 4\}$

4. $\left\{\dfrac{-1 - \sqrt{41}}{4}, \dfrac{-1 + \sqrt{41}}{4}\right\}$

5. $(10, 0), (0, -6)$

EXAMPLE 1 Graphing a Quadratic Equation in Two Variables by Plotting Points

Graph $y = x^2 - 2x - 3$ by plotting points.

Solution

We begin by creating a table of values. Without knowing much about the graph of the equation, it is hard to tell which x-values to choose. We will start with the values shown in Table 1. We then use the equation to determine the corresponding values of y shown in the second column of Table 1.

Table 1

x	y	(x, y)
−2	$y = (-2)^2 - 2(-2) - 3 = 5$	$(-2, 5)$
−1	$y = (-1)^2 - 2(-1) - 3 = 0$	$(-1, 0)$
0	$y = (0)^2 - 2(0) - 3 = -3$	$(0, -3)$
1	$y = (1)^2 - 2(1) - 3 = -4$	$(1, -4)$
2	$y = (2)^2 - 2(2) - 3 = -3$	$(2, -3)$
3	$y = (3)^2 - 2(3) - 3 = 0$	$(3, 0)$
4	$y = (4)^2 - 2(4) - 3 = 5$	$(4, 5)$

Figure 1

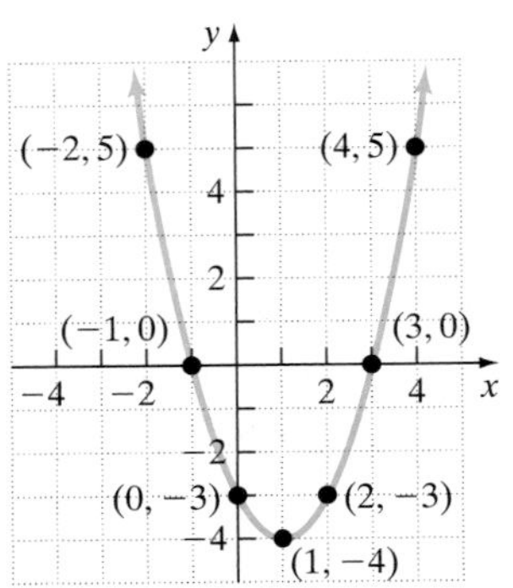

We plot the ordered pairs in the third column of Table 1 and connect the points in a smooth curve. See Figure 1.

Now let's look at another example.

EXAMPLE 2 Graphing a Quadratic Equation by Plotting Points

Graph $y = -x^2 + 6x - 8$ by plotting points.

Solution

We begin by creating a table of values. As in Example 1, without knowing much about the graph of the equation, it is difficult to determine which x-values to choose. We will start with the values shown in Table 2. We then use the equation to determine the corresponding values of y shown in the second column of Table 2.

We plot the ordered pairs in the third column of Table 2 and connect the points in a smooth curve. See Figure 2.

Classroom Example
Graph $y = -x^2 + 3x + 4$ by plotting points.

Answer:

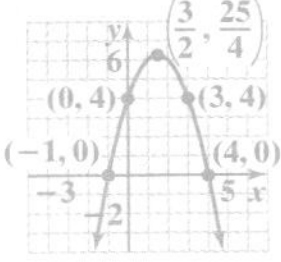

Table 2

x	y	(x, y)
0	$y = -(0)^2 + 6(0) - 8 = -8$	$(0, -8)$
1	$y = -(1)^2 + 6(1) - 8 = -3$	$(1, -3)$
2	$y = -(2)^2 + 6(2) - 8 = 0$	$(2, 0)$
3	$y = -(3)^2 + 6(3) - 8 = 1$	$(3, 1)$
4	$y = -(4)^2 + 6(4) - 8 = 0$	$(4, 0)$
5	$y = -(5)^2 + 6(5) - 8 = -3$	$(5, -3)$
6	$y = -(6)^2 + 6(6) - 8 = -8$	$(6, -8)$

Figure 2

We stated in Examples 1 and 2 that choosing the x-values to use to obtain the graph of the quadratic equation is difficult. For this reason, we will supply the x-values to use—your job will be to find the corresponding y-values, plot the points, and connect the points in a smooth curve.

QUICK ✓ *Graph each quadratic equation by filling in the table and plotting points.*

1. $y = x^2 - 3$

x	y
−3	
−2	
−1	
0	
1	
2	
3	

2. $y = x^2 + 4x - 5$

x	y
−5	
−4	
−3	
−2	
−1	
0	
1	

3. $y = -x^2 + 8x - 15$

x	y
1	
2	
3	
4	
5	
6	
7	

2 Find the Intercepts of the Graph of a Quadratic Equation

Figure 3

Figure 3 shows the graph of the quadratic equation $y = x^2 - 2x - 3$, found in Example 1. We have labeled the x- and y-intercepts.

The y-intercept is the value of the quadratic equation $y = ax^2 + bx + c$ at $x = 0$.

The x-intercepts, if there are any, are found by letting $y = 0$ in the quadratic equation $y = ax^2 + bx + c$ and solving the equation

$$ax^2 + bx + c = 0$$

In Section 7.3, we learned how to use the discriminant to determine whether a quadratic equation was factorable. We also can use the discriminant to determine the number of solutions a quadratic equation has. Recall the solutions to the equation

$$ax^2 + bx + c = 0 \qquad \text{are given by} \qquad x = \frac{-b \pm \sqrt{b^2 - 4ac}}{2a}$$

The discriminant, $b^2 - 4ac$, is the radicand of the square root. Recall from our study of square roots that

- The square root of a positive number (a number greater than 0) is a real number.
- The square root of 0 is 0.
- The square root of a negative number is not a real number.

This leads to the following result.

> **Finding the x-Intercepts of the Graph of a Quadratic Equation**
>
> **1.** If the discriminant $b^2 - 4ac > 0$, the graph of $y = ax^2 + bx + c$ has two different x-intercepts. The graph will cross the x-axis at the solutions to the equation $ax^2 + bx + c = 0$.
>
> **2.** If the discriminant $b^2 - 4ac = 0$, the graph of $y = ax^2 + bx + c$ has one x-intercept. The graph will touch the x-axis at the solution to the equation $ax^2 + bx + c = 0$.
>
> **3.** If the discriminant $b^2 - 4ac < 0$, the graph of $y = ax^2 + bx + c$ has no x-intercepts. This means that the graph will not cross or touch the x-axis.

As we can see from Examples 1 and 2, the graph of a quadratic equation can either *open up* or *down*. Figure 4 illustrates the possibilities for x-intercepts for graphs that open up.

Figure 4

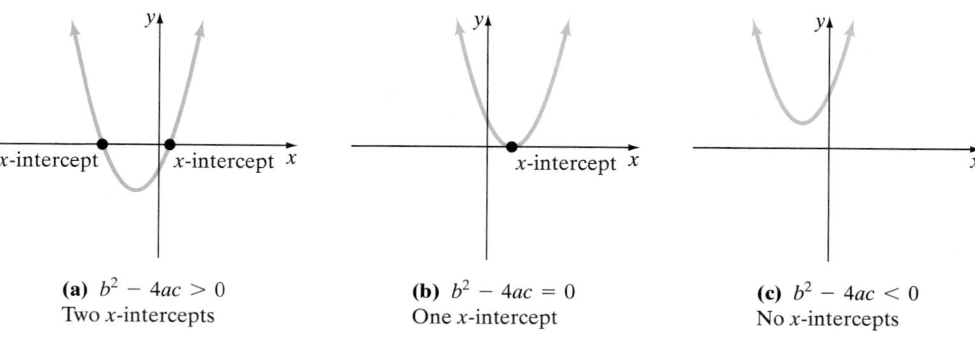

(a) $b^2 - 4ac > 0$
Two x-intercepts

(b) $b^2 - 4ac = 0$
One x-intercept

(c) $b^2 - 4ac < 0$
No x-intercepts

The intercepts of the graph of an equation represent one of the "interesting" features of the graph. For this reason, all graphs should have the intercepts labeled.

Classroom Example Find the intercepts of the graph of $y = x^2 + x - 20$.

Answer: The x-intercepts are -5 and 4. The y-intercept is -20. The graph passes through the points $(0, -20)$, $(-5, 0)$, and $(4, 0)$.

EXAMPLE 3 Finding the Intercepts of the Graph of a Quadratic Equation

Find the intercepts of the graph of $y = x^2 - 2x - 15$.

Solution

To find the y-intercept, we let $x = 0$ in the equation $y = x^2 - 2x - 15$.

$$\begin{aligned} y &= 0^2 - 2(0) - 15 \\ &= -15 \end{aligned}$$

The y-intercept is -15. Before we find the x-intercepts, we first determine the value of the discriminant, $b^2 - 4ac$. We compare $y = x^2 - 2x - 15$ with $y = ax^2 + bx + c$ to identify the values of a, b, and c. We find that $a = 1$, $b = -2$, and $c = -15$. So

$$\begin{aligned} b^2 - 4ac &= (-2)^2 - 4(1)(-15) \\ &= 4 + 60 \\ &= 64 \end{aligned}$$

Because the discriminant is positive $(64 > 0)$, we know that the graph of the equation will have two x-intercepts. We find the x-intercepts by letting $y = 0$ in the equation $y = x^2 - 2x - 15$.

$$y = x^2 - 2x - 15$$

Let $y = 0$: $\quad 0 = x^2 - 2x - 15$

Factor: $\quad 0 = (x - 5)(x + 3)$

Zero-Product Property: $\quad x - 5 = 0$ or $x + 3 = 0$

$$x = 5 \quad \text{or} \quad x = -3$$

The x-intercepts are -3 and 5. The graph passes through the points $(0, -15)$, $(-3, 0)$, and $(5, 0)$.

Classroom Example Find the intercepts of the graph of $y = 3x^2 + 4x + 4$.

Answer: The y-intercept is 4; there are no x-intercepts.

EXAMPLE 4 Finding the Intercepts of the Graph of a Quadratic Equation

Find the intercepts of the graph of $y = -2x^2 + x - 3$.

Solution

To find the y-intercept, we let $x = 0$ in the equation $y = -2x^2 + x - 3$.

$$\begin{aligned} y &= -2(0)^2 + 0 - 3 \\ &= -3 \end{aligned}$$

The y-intercept is -3. Before we find the x-intercepts, we first determine the value of the discriminant, $b^2 - 4ac$. We compare $y = -2x^2 + x - 3$ with $y = ax^2 + bx + c$ to identify the values of a, b, and c. We find that $a = -2$, $b = 1$, and $c = -3$. So

$$\begin{aligned} b^2 - 4ac &= (1)^2 - 4(-2)(-3) \\ &= 1 - 24 \\ &= -23 \end{aligned}$$

Because the discriminant is negative $(-23 < 0)$, we know that the graph of the equation will have no x-intercepts. The graph passes through the point $(0, -3)$.

QUICK ✓ *Determine the intercepts of the graph of each quadratic equation.*

4. $y = x^2 + 5x + 6$ **5.** $y = -3x^2 + 4x - 5$ **6.** $y = 4x^2 + 4x + 1$

3 Find the Vertex and Axis of Symmetry of the Graph of a Quadratic Equation

The graphs obtained in Examples 1 and 2 are typical of graphs of all quadratic equations. We call the graph of a quadratic equation a **parabola** (pronounced puh-rab-ō-luh). Refer to Figure 5, where two parabolas are shown. The shape of parabolas should be familiar to you—you see them in satellite dishes, car headlights, and some bridges.

Figure 5

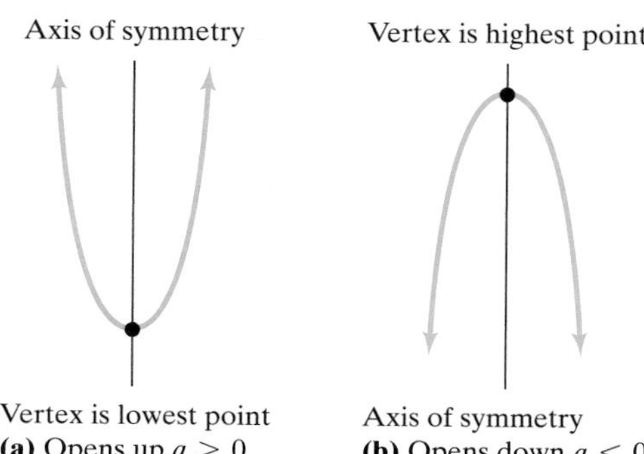

In Words
The original meaning of "vertex" is "a turning point."

Symmetry is all around us—do you see the symmetry in the butterfly?

The parabola in Figure 5(a) **opens up** and has a lowest point. The parabola in Figure 5(b) **opens down** and has a highest point. The lowest or highest point of a parabola is called the **vertex.** The vertical line passing through the vertex in each parabola in Figure 5 is called the **axis of symmetry** of the parabola. If we were to take the portion of the parabola to the right of the vertex and fold it over the axis of symmetry, it would lie directly on top of the portion of the parabola to the left of the vertex. Therefore, we say that the parabola is symmetric about its axis of symmetry. It is important to note that the axis of symmetry is not part of the graph of the quadratic equation.

How do we know whether a parabola opens up or down? The results of Examples 1 and 2 provide the answer. In Example 1, the coefficient on the squared term, a, is 1. The parabola in Example 1 opens up. In Example 2, $y = -x^2 + 6x - 8$, so $a = -1$. The parabola in Example 2 opens down. This leads us to the following conclusion.

Work Smart
The axis of symmetry is *not* part of the graph of the parabola. However, it serves as a guide in graphing the parabola.

> The graph of the quadratic equation $y = ax^2 + bx + c$ will open up if $a > 0$ and open down if $a < 0$.

In Words
If the leading coefficient is positive, the graph opens up, and if the leading coefficient is negative, it opens down.

Now let's learn how to find the vertex and axis of symmetry of a parabola. The x-intercepts of the graph of $y = ax^2 + bx + c$ are found by solving the equation $ax^2 + bx + c = 0$. The solution(s) to this equation using the quadratic formula are given by

$$x = \frac{-b - \sqrt{b^2 - 4ac}}{2a} \quad \text{and} \quad x = \frac{-b + \sqrt{b^2 - 4ac}}{2a}$$

The x-coordinate of the vertex of a parabola is halfway between its x-intercepts. Put another way, if we find the average of the x-intercepts, we obtain the x-coordinate of the vertex.

$$x = \frac{\frac{-b - \sqrt{b^2 - 4ac}}{2a} + \frac{-b + \sqrt{b^2 - 4ac}}{2a}}{2}$$

Write numerator over common denominator: $$= \frac{\frac{-b - \sqrt{b^2 - 4ac} - b + \sqrt{b^2 - 4ac}}{2a}}{2}$$

Combine like terms: $$= \frac{\frac{-2b}{2a}}{2}$$

$\frac{\frac{-2b}{2a}}{2} = \frac{\frac{-2b}{2a}}{\frac{2}{1}} = \frac{-2b}{2a} \cdot \frac{1}{2} = \frac{-b}{2a}$: $$= \frac{-b}{2a}$$

THE VERTEX AND AXIS OF SYMMETRY OF A PARABOLA

Any quadratic equation $y = ax^2 + bx + c$, $a \neq 0$, has a vertex whose x-coordinate is

$$-\frac{b}{2a}$$

The y-coordinate of the vertex is found by evaluating the equation at the x-coordinate of the vertex.

The axis of symmetry of a parabola is given by the equation

$$x = -\frac{b}{2a}$$

Classroom Example
For the quadratic equation $y = 2x^2 + 8x - 1$,
(a) Determine whether the parabola opens up or down.
(b) Determine the vertex of the parabola.
(c) Determine the axis of symmetry of the parabola.
Answer:
(a) Opens up
(b) $(-2, -9)$
(c) $x = -2$

EXAMPLE 5 Determining If a Parabola Opens Up or Down and Finding Its Vertex and Axis of Symmetry

For the quadratic equation $y = -2x^2 + 8x - 1$,

(a) Determine whether the parabola opens up or down.
(b) Determine the vertex of the parabola.
(c) Determine the axis of symmetry of the parabola.

Solution

We compare $y = -2x^2 + 8x - 1$ to $y = ax^2 + bx + c$ and find that $a = -2$, $b = 8$, and $c = -1$.

(a) Because a is negative, the parabola will open down.

(b) To find the x-coordinate of the vertex, we substitute $a = -2$ and $b = 8$ into the formula $x = -\frac{b}{2a}$.

$$x = -\frac{b}{2a}$$

$a = -2$, $b = 8$: $$= -\frac{8}{2(-2)} = -\frac{8}{-4}$$

$$x = 2$$

We now know that the x-coordinate of the vertex is 2. To find the y-coordinate of the vertex, let $x = 2$ in the equation $y = -2x^2 + 8x - 1$.

$$
\begin{aligned}
& y = -2x^2 + 8x - 1 \\
x = 2: \quad & = -2(2)^2 + 8(2) - 1 \\
& = -2(4) + 16 - 1 \\
& = 7
\end{aligned}
$$

Work Smart

Do not write that the axis of symmetry is 2. The axis of symmetry is a vertical line and must be written in the form $x = -\frac{b}{2a}$. Don't forget the "$x =$" part!

The vertex is $(2, 7)$.

(c) The axis of symmetry is given by $x = -\frac{b}{2a}$. Since $-\frac{b}{2a} = 2$, we have the axis of symmetry is $x = 2$.

QUICK ✓ *For each quadratic equation, (a) determine whether the parabola opens up or down, (b) find the vertex of the parabola, (c) find the axis of symmetry.*

7. $y = 3x^2 - 6x + 2$ **8.** $y = -2x^2 - 12x - 7$

Classroom Example

Graph $y = x^2 + 4x - 12$ using its properties.

Answer:

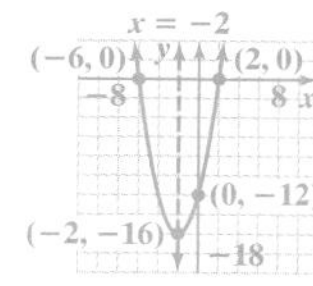

4 Graph a Quadratic Equation Using Its Vertex, Intercepts, and Axis of Symmetry

We now know how to determine whether the graph of a quadratic equation opens up or down, how to find its intercepts, how to find its vertex, and how to find its axis of symmetry. We can use this information to draw the graph of the parabola. Example 6 illustrates how to graph a quadratic equation using its properties.

EXAMPLE 6 How to Graph a Quadratic Equation Using Its Properties

Graph $y = x^2 + 2x - 8$ using its properties.

Step-by-Step Solution

Step 1: Determine whether the parabola opens up or down.

We compare $y = x^2 + 2x - 8$ to $y = ax^2 + bx + c$ and see that $a = 1$, $b = 2$, and $c = -8$. The parabola opens up because a is positive.

Step 2: Determine the vertex and axis of symmetry.

The x-coordinate of the vertex is

$$x = -\frac{b}{2a} = -\frac{2}{2(1)} = -1$$

The y-coordinate of the vertex is found by letting $x = -1$ in the equation $y = x^2 + 2x - 8$.

$$
\begin{aligned}
y &= (-1)^2 + 2(-1) - 8 \\
&= 1 - 2 - 8 \\
&= -9
\end{aligned}
$$

The vertex is $(-1, -9)$.
The axis of symmetry is the line

$$x = -\frac{b}{2a} = -1$$

Step 3: Determine the y-intercept by letting $x = 0$ in the equation.

$$
\begin{aligned}
y &= 0^2 + 2(0) - 8 \\
&= -8
\end{aligned}
$$

(continued)

Step 4: Determine the discriminant, $b^2 - 4ac$, to determine the number of x-intercepts. Then determine the x-intercepts, if any.

We have that $a = 1$, $b = 2$, and $c = -8$, so $b^2 - 4ac = (2)^2 - 4(1)(-8) = 36$. Because the discriminant is positive, the parabola will have two different x-intercepts. We find the x-intercepts by letting $y = 0$ and solving the resulting equation

$$x^2 + 2x - 8 = 0$$
$$(x + 4)(x - 2) = 0$$
$$x + 4 = 0 \quad \text{or} \quad x - 2 = 0$$
$$x = -4 \quad \text{or} \quad x = 2$$

Step 5: Plot the points and draw the graph of the quadratic equation. Use the axis of symmetry to find an additional point.

See Figure 6. Notice how we use the axis of symmetry to find the additional point $(-2, -8)$. The y-intercept, $(0, -8)$, is 1 unit to the right of the axis of symmetry, so there must be a point 1 unit to the left of the axis of symmetry, $(-2, -8)$.

Figure 6

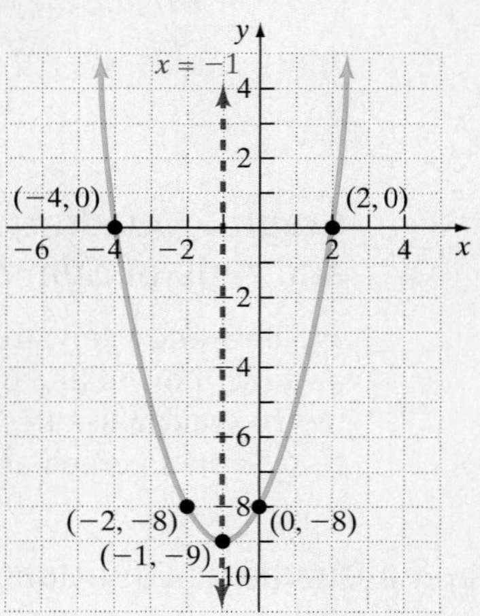

We summarize the steps used in Example 6.

Graphing a Quadratic Equation Using Its Properties

To graph any quadratic equation of the form $y = ax^2 + bx + c$, $a \neq 0$, we use the following steps:

Step 1: Determine whether the parabola opens up or down.

Step 2: Determine the vertex and axis of symmetry.

Step 3: Determine the y-intercept by evaluating the equation at $x = 0$.

Step 4: Determine the x-intercept(s), if any, by first determining the discriminant, $b^2 - 4ac$.

- **(a)** If $b^2 - 4ac > 0$, then the parabola has two x-intercepts, which are found by letting $y = 0$ in the equation and solving the resulting equation $(ax^2 + bx + c = 0)$.
- **(b)** If $b^2 - 4ac = 0$, the vertex is the x-intercept.
- **(c)** If $b^2 - 4ac < 0$, there are no x-intercepts.

Step 5: Plot the points and draw the graph of the quadratic equation. Use the axis of symmetry to find an additional point.

QUICK ✓ *Graph each quadratic equation using its properties.*

9. $y = x^2 - 2x - 15$ **10.** $y = -x^2 - 2x + 3$

In Example 6, we were able to solve the quadratic equation by factoring, so the x-intercepts were rational numbers. When the quadratic equation cannot be solved by factoring, we can use the quadratic formula to find the x-intercepts. For the purpose of graphing the quadratic equation, we will approximate the x-intercepts rounded to two decimal places.

Classroom Example ➤ Graph $y = -2x^2 - 4x + 3$ using its properties.

Answer:

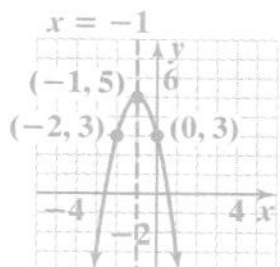

EXAMPLE 7 Graphing a Quadratic Equation Using Its Properties

Graph $y = -2x^2 + 4x + 3$ using its properties.

Solution

We compare $y = -2x^2 + 4x + 3$ to $y = ax^2 + bx + c$ and see that $a = -2, b = 4$, and $c = 3$. Because a is negative ($a = -2 < 0$), the parabola opens down. The x-coordinate of the vertex is

$$x = -\frac{b}{2a}$$

$$a = -2,\ b = 4: \quad = -\frac{4}{2(-2)}$$

$$= 1$$

We find the y-coordinate of the vertex by evaluating $y = -2x^2 + 4x + 3$ at $x = 1$.

$$y = -2(1)^2 + 4(1) + 3$$
$$= -2 + 4 + 3$$
$$= 5$$

The vertex is at $(1, 5)$. The axis of symmetry is the line $x = 1$. The y-intercept is found by evaluating $y = -2x^2 + 4x + 3$ at $x = 0$.

$$y = -2(0)^2 + 4(0) + 3$$
$$= 3$$

The y-intercept is 3. The x-intercept is found by letting $y = 0$ in the equation $y = -2x^2 + 4x + 3$ and solving for x.

$$0 = -2x^2 + 4x + 3$$

The discriminant is $b^2 - 4ac = 4^2 - 4(-2)(3) = 40$. Because the discriminant is positive, we know the parabola will have two x-intercepts. The equation cannot be solved by factoring because the discriminant is not a perfect square. We will find the x-intercepts using the quadratic formula.

$$x = \frac{-b \pm \sqrt{b^2 - 4ac}}{2a}$$

$$a = -2,\ b = 4,\ b^2 - 4ac = 40: \quad = \frac{-4 \pm \sqrt{40}}{2(-2)}$$

$$\sqrt{40} = \sqrt{4 \cdot 10} = 2\sqrt{10}: \quad = \frac{-4 \pm 2\sqrt{10}}{-4}$$

$$\frac{a+b}{c} = \frac{a}{c} + \frac{b}{c}: \quad = \frac{-4}{-4} \pm \frac{2\sqrt{10}}{-4}$$

$$\text{Simplify:} \quad = 1 \pm \frac{\sqrt{10}}{2}$$

Figure 7

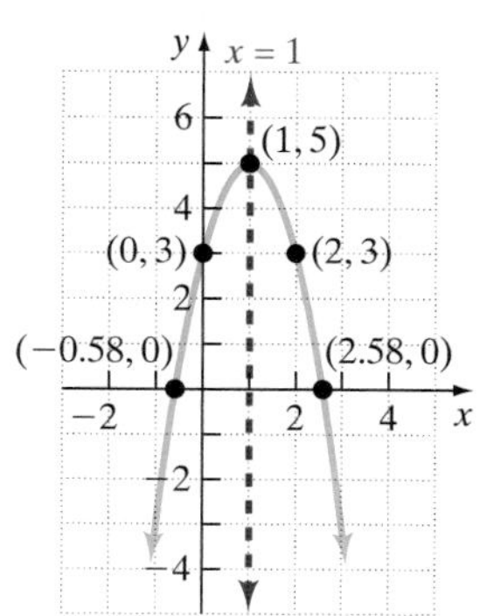

Using a calculator, we find that the x-intercepts are $1 - \frac{\sqrt{10}}{2} \approx -0.58$ and $1 + \frac{\sqrt{10}}{2} \approx 2.58$. We plot the vertex, intercepts and axis of symmetry and trace in a smooth curve as shown in Figure 7. Notice how the axis of symmetry was used to find the additional point $(2, 3)$.

QUICK ✓ *Graph each quadratic equation using its properties.*

11. $y = 2x^2 + 8x - 1$ **12.** $y = -3x^2 - 12x + 5$

5 Determine the Maximum or Minimum Value of a Quadratic Equation

Recall that the graph of a quadratic equation $y = ax^2 + bx + c$ is a parabola that opens up (if $a > 0$) or down (if $a < 0$). The vertex of the parabola will be the highest point on the graph if $a < 0$ and the lowest point on the graph if $a > 0$. The terms *maximum* and *minimum* apply to the highest point and the lowest point on a parabola.

DEFINITION

If the vertex is the highest point on the graph of a parabola ($a < 0$), then the y-coordinate of the vertex represents the **maximum value** of the equation.

If the vertex is the lowest point on the graph of a parabola ($a > 0$), then the y-coordinate of the vertex represents the **minimum value** of the equation.

Classroom Example Determine whether the quadratic equation $y = -4x^2 + 16x - 5$ has a maximum or minimum value. Then find the maximum or minimum value of the equation.

Answer: Maximum; 11

EXAMPLE 8 Finding the Maximum or Minimum Value of a Quadratic Equation

Determine whether the quadratic equation

$$y = 3x^2 + 12x - 1$$

has a maximum or minimum value. Then find the maximum or minimum value of the equation.

Solution

Work Smart
The maximum or the minimum value of the quadratic equation is the y-coordinate of the vertex.

If we compare $y = 3x^2 + 12x - 1$ to $y = ax^2 + bx + c$, we find that $a = 3, b = 12$, and $c = -1$. Because a is positive, we know that the graph of the quadratic equation will open up, so the equation has a minimum value. The minimum value of the equation occurs at

$$x = -\frac{b}{2a} = -\frac{12}{2(3)} = -2$$

The minimum value of the equation is found by evaluating the equation at $x = -2$:

$$\begin{aligned} y &= 3(-2)^2 + 12(-2) - 1 \\ &= 3(4) - 24 - 1 \\ &= -13 \end{aligned}$$

So the minimum value of the equation is -13 and occurs at $x = -2$.

QUICK ✓ *Determine whether the quadratic equation has a maximum or minimum value, then find the maximum or minimum value of the equation.*

13. $y = 2x^2 - 8x + 1$ **14.** $y = -x^2 + 10x + 8$

Classroom Example ➤
Adam fires a rocket straight up with an initial speed of 128 feet per second from an initial height of 2 feet. The height h of the rocket after t seconds can be modeled by the equation

$$h = -16t^2 + 128t + 2$$

Use parts (a) and (b) of Example 9.

Answer:
(a) 4 seconds
(b) 258 feet

EXAMPLE 9 Projectile Motion: Firing a Rocket

Kevin just received a "sling-shot rocket" for his birthday. If Kevin fires the rocket straight up with an initial speed of 64 feet per second from an initial height of 4 feet, then the height h of the rocket after t seconds can be modeled by the equation

$$h = -16t^2 + 64t + 4$$

(a) Determine the time at which the rocket is at a maximum height by finding $t = -\frac{b}{2a}$.

(b) Determine the maximum height of the rocket by evaluating the equation at the value of t found in part (a).

Solution

Notice that the quadratic equation has a graph that opens down, because a is negative. Therefore, the vertex will be the high point of the graph.

(a) Let $a = -16$ and $b = 64$ in the expression $t = -\frac{b}{2a}$.

$$t = -\frac{64}{2(-16)} = 2$$

The rocket will be at a maximum height after 2 seconds.

(b) We evaluate the equation at $t = 2$.

$$\begin{aligned} h &= -16t^2 + 64t + 4 \\ \text{Let } t = 2: \quad &= -16(2)^2 + 64(2) + 4 \\ &= -16(4) + 128 + 4 \\ &= -64 + 128 + 4 \\ &= 68 \end{aligned}$$

The rocket will have a maximum height of 68 feet.

QUICK ✓

15. The height h of a ball fired straight up with an initial speed of 128 feet per second from an initial height of 15 feet after t seconds is given by the equation

$$h = -16t^2 + 128t + 15$$

(a) Determine the time at which the ball is highest by finding $t = -\frac{b}{2a}$.

(b) Determine the maximum height of the ball by evaluating the equation at the time found in part (a).

10.1 Exercises

For Extra Help: Student Solutions Manual  CD Video  PH Math/Tutor Center MathXL Tutorials on CD MathXL® MyMathLab

1. parabola 2. up, down
3. vertex; $x = -\frac{b}{2a}$
4. True 5. False 6. True
7. the discriminant; 1
8. Answers may vary.
9. $y = x^2$ 10. $y = -x^2$

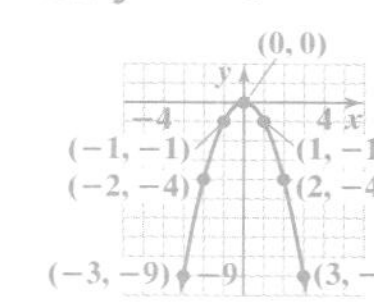

Concepts and Vocabulary

In Problems 1–3, fill in the blanks.

1. The graph of a quadratic equation in the form $y = ax^2 + bx + c$ is a ________.

2. When graphing a quadratic equation, if the leading coefficient is positive, the graph will open ________ and if the leading coefficient is negative, the graph will open ________.

3. The lowest or highest point of the graph of $y = ax^2 + bx + c$ is called the ________. Its x-coordinate can be determined using the formula ________.

11. $y = -x^2 + 4$ 12. $y = x^2 - 2$

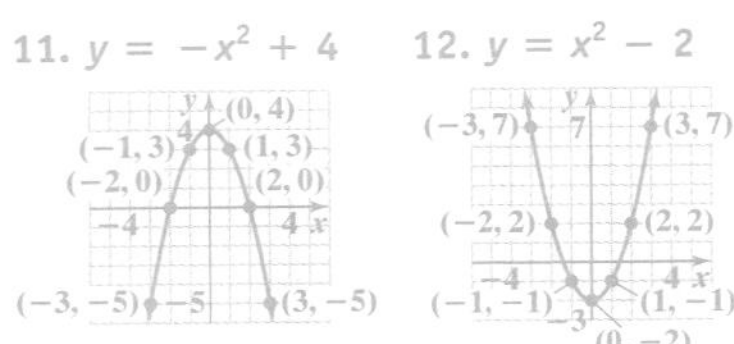

13. $y = -x^2 + 3x + 4$

14. $y = x^2 - 5x + 6$

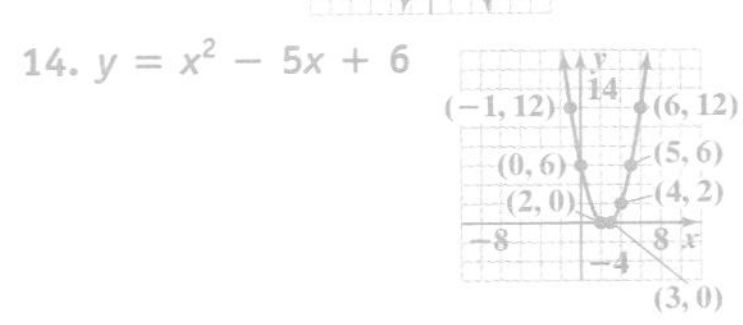

15. $y = 2x^2 + 5x - 3$

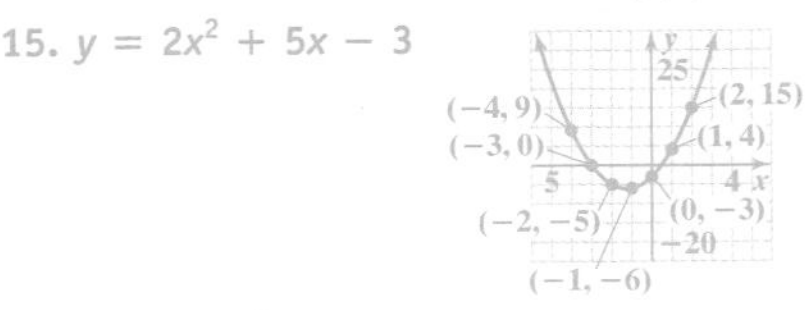

16. $y = -2x^2 - x + 3$

17. x-intercepts: 2, 4; y-intercept: -8
18. x-intercepts: -1, 3; y-intercept: -3
19. x-intercepts: $-\frac{1}{3}, \frac{1}{2}$; y-intercept: -1
20. x-intercepts: $-\frac{2}{3}$, -1; y-intercept: -2
21. x-intercepts: -3, 0; y-intercept: 0
22. x-intercepts: -6, 0; y-intercept: 0
23. x-intercepts: none; y-intercept: -5
24. x-intercepts: $-\sqrt{2} \approx -1.41$, $\sqrt{2} \approx 1.41$; y-intercept: -2
25. x-intercepts: $\frac{1-\sqrt{17}}{4} \approx -0.78$, $\frac{1+\sqrt{17}}{4} \approx 1.28$; y-intercept: 2
26. x-intercepts: $\frac{-5-\sqrt{13}}{6} \approx -1.43$, $\frac{-5+\sqrt{13}}{6} \approx -0.23$; y-intercept: 1
27. x-intercepts: $-2-\sqrt{2} \approx -3.41$, $-2+\sqrt{2} \approx -0.59$; y-intercept: 2
28. x-intercepts: $\frac{-1-\sqrt{7}}{2} \approx -1.82$, $\frac{-1+\sqrt{7}}{2} \approx 0.82$; y-intercept: 3
29. (a) up (b) $(0, 6)$ (c) $x = 0$
30. (a) down (b) $(0, 3)$ (c) $x = 0$
31. (a) down (b) $(4, 1)$ (c) $x = 4$
32. (a) up (b) $(1, -9)$ (c) $x = 1$
33. (a) down (b) $\left(\frac{3}{4}, \frac{49}{8}\right)$ (c) $x = \frac{3}{4}$
34. (a) up (b) $\left(\frac{9}{4}, -\frac{49}{8}\right)$ (c) $x = \frac{9}{4}$
35. (a) up (b) $\left(\frac{4}{3}, -\frac{4}{3}\right)$ (c) $x = \frac{4}{3}$
36. (a) down (b) $\left(\frac{1}{6}, \frac{7}{6}\right)$ (c) $x = \frac{1}{6}$

In Problems 4–6, answer True or False to each statement.

4. If a parabola has a vertex at $(-3, 4)$ and opens down, it must have two x-intercepts.

5. The parabola determined by the equation $y = 3x^2 + 2x - 1$ has a maximum value.

6. Every parabola has an axis of symmetry that contains either the maximum or minimum value.

7. Given a quadratic equation, without graphing, explain how you can tell how many x-intercepts the graph has and how you find them. How many y-intercepts can the graph of a quadratic equation have and how can you find them?

8. Describe what the axis of symmetry tells you about the graph of a quadratic equation. How do you find the axis of symmetry? How many different parabolas have a given vertex and axis of symmetry?

Building Skills

In Problems 9–16, graph each quadratic equation by filling in the table and plotting the points.

9. $y = x^2$

x	y
−3	
−2	
−1	
0	
1	
2	
3	

10. $y = -x^2$

x	y
−3	
−2	
−1	
0	
1	
2	
3	

11. $y = -x^2 + 4$

x	y
−3	
−2	
−1	
0	
1	
2	
3	

12. $y = x^2 - 2$

x	y
−3	
−2	
−1	
0	
1	
2	
3	

13. $y = -x^2 + 3x + 4$

x	y
−2	
−1	
0	
1	
2	
3	
4	

14. $y = x^2 - 5x + 6$

x	y
−1	
0	
1	
2	
3	
4	
5	

15. $y = 2x^2 + 5x - 3$

x	y
−4	
−3	
−2	
−1	
0	
1	
2	

16. $y = -2x^2 - x + 3$

x	y
−3	
−2	
−1	
0	
1	
2	
3	

37. $y = x^2 + 2x - 3$
opens up;
x-intercepts: $-3, 1$
y-intercept: -3
axis of symmetry: $x = -1$

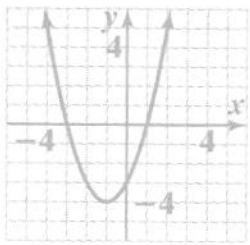

38. $y = x^2 - 4x - 5$
opens up;
x-intercepts: $-1, 5$
y-intercept: -5
axis of symmetry: $x = 2$

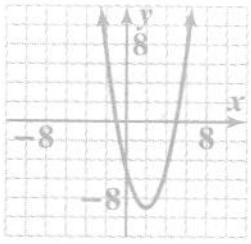

39. $y = -x^2 - 2x$
opens down;
x-intercepts: $-2, 0$
y-intercept: 0
axis of symmetry: $x = -1$

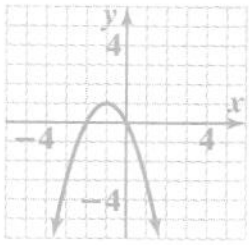

40. $y = -x^2 + 4x$
opens down;
x-intercepts: 0, 4
y-intercept: 0
axis of symmetry: $x = 2$

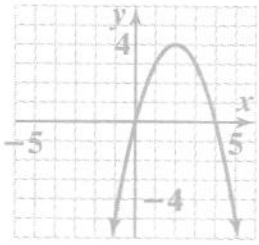

41. $y = x^2 - 8x + 12$
opens up;
x-intercepts: 2, 6
y-intercept: 12
axis of symmetry: $x = 4$

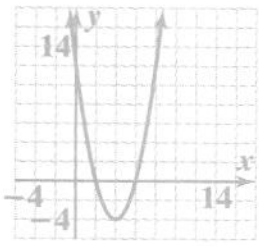

42. $y = x^2 + 8x + 15$
opens up;
x-intercepts: $-5, -3$
y-intercept: 15
axis of symmetry: $x = -4$

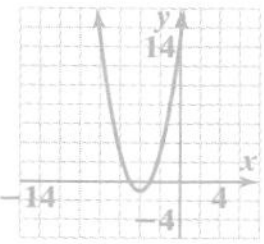

43. $y = -x^2 - 8x - 9$
opens down;
x-intercepts: $-4 - \sqrt{7}, -4 + \sqrt{7}$
y-intercept: -9
axis of symmetry: $x = -4$

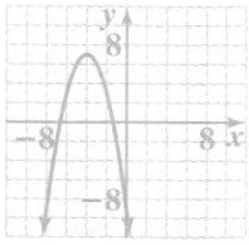

44. $y = -x^2 - 6x + 16$
opens down;
x-intercepts: $-8, 2$
y-intercept: 16
axis of symmetry: $x = -3$

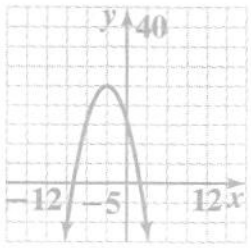

45. $y = x^2 - 9$
opens up;
x-intercepts: $-3, 3$
y-intercept: -9
axis of symmetry: $x = 0$

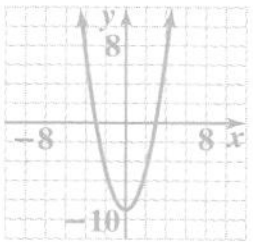

46. $y = -x^2 + 25$
opens down;
x-intercepts: $-5, 5$
y-intercept: 25
axis of symmetry: $x = 0$

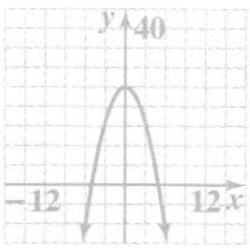

47. $y = x^2 - 7x + 10$
opens up;
x-intercepts: 2, 5
y-intercept: 10
axis of symmetry: $x = \frac{7}{2}$

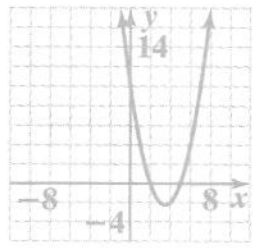

48. $y = -x^2 + 5x - 4$
opens down;
x-intercepts: 1, 4
y-intercept: -4
axis of symmetry: $x = \frac{5}{2}$

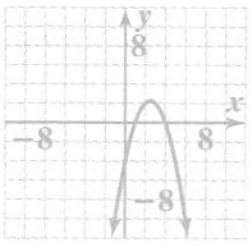

In Problems 17–28, determine the x- and y-intercepts of the graph of each quadratic equation. If an intercept is an irrational number, give both the exact and approximate value.

17. $y = -x^2 + 6x - 8$
18. $y = x^2 - 2x - 3$
19. $y = 6x^2 - x - 1$
20. $y = -3x^2 - 5x - 2$
21. $y = x^2 + 3x$
22. $y = x^2 + 6x$
23. $y = -x^2 - 5$
24. $y = x^2 - 2$
25. $y = -2x^2 + x + 2$
26. $y = 3x^2 + 5x + 1$
27. $y = x^2 + 4x + 2$
28. $y = -2x^2 - 2x + 3$

In Problems 29–36, for each quadratic equation, (a) determine whether the parabola opens up or down, (b) find the vertex of the parabola, and (c) find the axis of symmetry.

29. $y = x^2 + 6$
30. $y = -x^2 + 3$
31. $y = -x^2 + 8x - 15$
32. $y = x^2 - 2x - 8$
33. $y = -2x^2 + 3x + 5$
34. $y = 2x^2 - 9x + 4$
35. $y = 3x^2 - 8x + 4$
36. $y = -6x^2 + 2x + 1$

In Problems 37–54, graph each quadratic equation by determining whether the parabola opens up or down, finding the intercepts, and the axis of symmetry.

37. $y = x^2 + 2x - 3$
38. $y = x^2 - 4x - 5$
39. $y = -x^2 - 2x$
40. $y = -x^2 + 4x$
41. $y = x^2 - 8x + 12$
42. $y = x^2 + 8x + 15$
43. $y = -x^2 - 8x - 9$
44. $y = -x^2 - 6x + 16$
45. $y = x^2 - 9$
46. $y = -x^2 + 25$
47. $y = x^2 - 7x + 10$
48. $y = -x^2 + 5x - 4$
49. $y = -3x^2 + 4x + 4$
50. $y = 2x^2 - x - 3$
51. $y = -2x^2 - 4x + 5$
52. $y = \frac{1}{2}x^2 - 4x + 6$
53. $y = x^2 - 2x - 5$
54. $y = x^2 - 4x + 1$

In Problems 55–62, determine whether the quadratic equation has a maximum or minimum value, then find the maximum or minimum value of the equation.

55. $y = x^2 - 6x + 5$
56. $y = x^2 + 8x - 2$
57. $y = 2x^2 + 8x + 15$
58. $y = 4x^2 - 16x + 19$
59. $y = -2x^2 + 6x + 3$
60. $y = -2x^2 - 4x + 5$
61. $y = -x^2 - 3x - 7$
62. $y = -x^2 + 2x - 6$

Applying the Concepts

63. Height of a Shot-Put A track and field athlete throws a shot-put, and during one competition, the path of the shot-put could be approximated by the quadratic equation

$$h = -0.004x^2 + x + 6$$

where h is the height (in feet) of the shot-put at the instant it has traveled x feet horizontally.

(a) Determine the distance x at which the height of the shot-put is a maximum by finding $x = -\frac{b}{2a}$.

(b) Determine the maximum height of the shot-put by evaluating the equation at the value of x found in part (a).

64. Maximizing Miles per Gallon An engineer has determined that the miles per gallon a Ford Taurus will get can be calculated at various driving speeds by the quadratic equation

$$m = -0.02s^2 + 2s - 25$$

where m represents the miles per gallon and s represents the speed in miles per hour.

49. $y = -3x^2 + 4x + 4$
opens down;
x-intercepts: $-\frac{2}{3}, 2$
y-intercept: 4
axis of symmetry: $x = \frac{2}{3}$

50. $y = 2x^2 - x - 3$
opens up;
x-intercepts: $-1, \frac{3}{2}$
y-intercept: -3
axis of symmetry: $x = \frac{1}{4}$

51. $y = -2x^2 - 4x + 5$
opens down;
x-intercepts: $\frac{-2 - \sqrt{14}}{2}, \frac{-2 + \sqrt{14}}{2}$
y-intercept: 5
axis of symmetry: $x = -1$

52. $y = \frac{1}{2}x^2 - 4x + 6$
opens up;
x-intercepts: 2, 6
y-intercept: 6
axis of symmetry: $x = 4$

53. $y = x^2 - 2x - 5$
opens up;
x-intercepts: $1 - \sqrt{6}, 1 + \sqrt{6}$
y-intercept: -5
axis of symmetry: $x = 1$

54. $y = x^2 - 4x + 1$
opens up;
x-intercepts: $2 - \sqrt{3}, 2 + \sqrt{3}$
y-intercepts: 1
axis of symmetry: $x = 2$

55. minimum; -4 at $x = 3$
56. minimum; -18 at $x = -4$
57. minimum; 7 at $x = -2$
58. minimum; 3 at $x = 2$
59. maximum; $\frac{15}{2}$ at $x = \frac{3}{2}$
60. maximum; 7 at $x = -1$
61. maximum; $-\frac{19}{4}$ at $x = -\frac{3}{2}$
62. maximum; -5 at $x = 1$
63. (a) 125 ft (b) 68.5 ft
64. (a) 50 mph (b) 25 miles per gallon
65. (a) 25 yd (b) 625 yd^2
66. (a) 15 ft (b) 225 ft^2
67. (a) 50 ft (b) 5000 ft^2
68. (a) 12.5 ft (b) 312.5 ft^2

69. $y = x^2 + 6x + 10$
opens up;
vertex $(-3, 1)$
axis of symmetry $x = -3$
y-intercept: 10
x-intercepts: none

70. $y = x^2 + 4x + 5$
opens up;
vertex $(-2, 1)$
axis of symmetry $x = -2$
y-intercept: 5
x-intercepts: none

(a) Determine the speed s that maximizes the miles per gallon by finding $s = -\frac{b}{2a}$.

(b) Determine the miles per hour that a Ford Taurus should be driven to maximize miles per gallon by evaluating the equation at the value of s found in part (a).

△ **65. Maximizing an Enclosed Area** A farmer has 100 yards of fence and wishes to enclose a rectangular field for grazing cattle. The area A of the enclosed field is given by

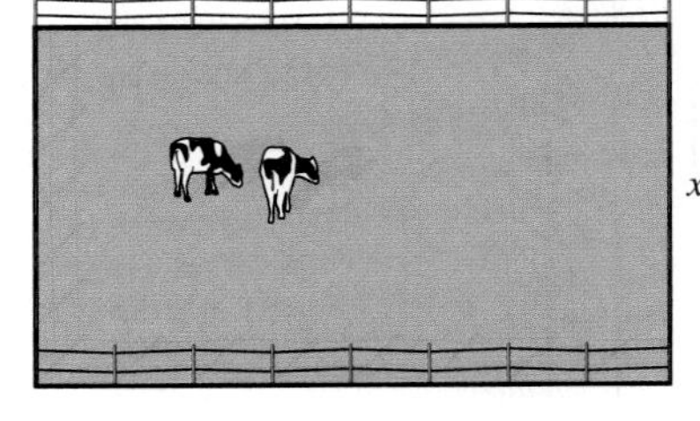

$$A = -x^2 + 50x$$

where x is the length of one of the sides of the enclosed field. See the figure.

(a) Determine the length of the side x that maximizes the enclosed area by finding $x = -\frac{b}{2a}$.

(b) Determine the maximum enclosed area by evaluating the equation at the value of x found in part (a).

△ **66. Maximizing an Enclosed Area** Mayra is planning to provide child care in her home. She needs to enclose an area in her backyard so that the children will be safe when they play. She has 60 feet of chain link fence that she can use to make the play area. The area A of the enclosed region is given by

$$A = -x^2 + 30x$$

where x is the length of one of the sides of the enclosed region.

(a) Determine the length of the side x that maximizes the enclosed area by finding $x = -\frac{b}{2a}$.

(b) Determine the maximum enclosed area by evaluating the equation at the value of x found in part (a).

△ **67. Maximizing Corral Area** Kevin is adding a corral for his horses on his property. There is a river on one side, so he only has to put the fencing on three sides, enclosing the corral. He has 200 feet of fencing already on his property that he can use. The maximum area he can enclose can be determined from the equation

$$A = -2x^2 + 200x$$

where x is the length of one of the sides of the enclosed region. See the figure.

(a) Determine the length of the side x that maximizes the enclosed area by finding $x = -\frac{b}{2a}$.

(b) Determine the maximum enclosed area by evaluating the equation at the value of x found in part (a).

△ **68. Maximizing a Dog Pen** Martin is making a dog run in which he will keep his show dogs and decides to enclose an area of his backyard. He uses one side of his house for the pen and encloses the other three sides with fencing. If he has 50 feet of fencing he plans to use, the maximum area he can enclose can be determined from the equation

$$A = -2x^2 + 50x$$

where x is the length of one of the sides of the enclosed region.

71. $y = -2x^2 + 6x - 7$
opens down; vertex $\left(\frac{3}{2}, -\frac{5}{2}\right)$
axis of symmetry $x = \frac{3}{2}$
y-intercept: -7; x-intercepts: none

72. $y = -3x^2 + 4x - 2$
opens down; vertex $\left(\frac{2}{3}, -\frac{2}{3}\right)$
axis of symmetry $x = \frac{2}{3}$
y-intercept: -2; x-intercepts: none

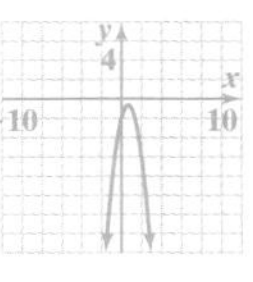

73. $y = 4x^2 - 4x + 1$
opens up;
vertex $\left(\frac{1}{2}, 0\right)$
axis of symmetry $x = \frac{1}{2}$
y-intercept: 1; x-intercept: $\frac{1}{2}$

74. $y = 9x^2 + 12x + 4$
opens up
vertex $\left(-\frac{2}{3}, 0\right)$
axis of symmetry $x = -\frac{2}{3}$
y-intercept: 4
x-intercept: $\frac{-2}{3}$

(a) Determine the length of the side x that maximizes the enclosed area by finding $x = -\frac{b}{2a}$.

(b) Determine the maximum enclosed area by evaluating the equation at the value of x found in part (a).

Extending the Concepts

In Problems 69–76, graph each quadratic equation using its properties.

69. $y = x^2 + 6x + 10$ **70.** $y = x^2 + 4x + 5$ **71.** $y = -2x^2 + 6x - 7$

72. $y = -3x^2 + 4x - 2$ **73.** $y = 4x^2 - 4x + 1$ **74.** $y = 9x^2 + 12x + 4$

10.2 Relations

OBJECTIVES

1. Define Relations
2. Find the Domain and the Range of a Relation
3. Graph a Relation Defined by an Equation

Preparing for Relations

Before getting started, take this readiness quiz. If you get a problem wrong, go back to the section cited and review the material.

1. Plot the points $(1, 5)$, $(-4, 0)$, and $(3, -1)$ in the rectangular coordinate system. [Section 8.1, pp. 518–521]

2. Graph: $y = -2x + 6$ [Section 8.4, pp. 561–563]

1 Define Relations

Frequently the value of one variable is somehow linked to the value of some other variable. For example, the total revenue of Brownie Troop 45 depends on the number of boxes of Girl Scout cookies they sell. Or the total cholesterol of an individual depends on the amount of saturated fat consumed.

Preparing for...Answers

1.

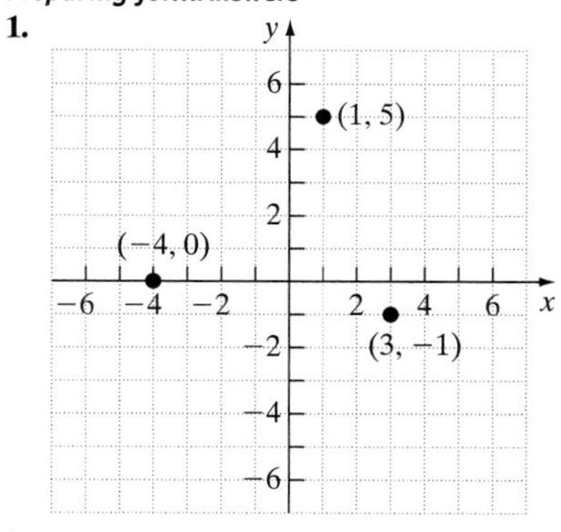

2.

DEFINITION

When the elements in one set are linked to elements in a second set, we have a **relation.** If x is an element in the first set and y is an element in the second set and if a relation exists between x and y, then we say that x **corresponds to** y or that y **depends on** x, and write $x \rightarrow y$. We may also write y depends on x as an ordered pair (x, y).

EXAMPLE 1 Illustrating a Relation

Consider the information presented in Figure 8, where we have a correspondence between states and number of representatives in the U.S. House of Representatives. By representing the relation as we do in Figure 8, we are using a **map,** in which we draw an arrow from an element from the first set (State) to an element in the second set (Number of Representatives). We could also represent the relation in Figure 8 using ordered pairs. The first value in the ordered pair represents the state and the second value in the ordered pair represents the number of representatives.

Figure 8

State	Number of Representatives
Alabama	7
Colorado	7
Georgia	13
Massachusetts	10
New York	29

$$\{(\text{Alabama}, 7), (\text{Colorado}, 7), (\text{Georgia}, 13), (\text{Massachusetts}, 10), (\text{New York}, 29)\}.$$

QUICK ✓

1. Use the map to represent the relation as a set of ordered pairs.

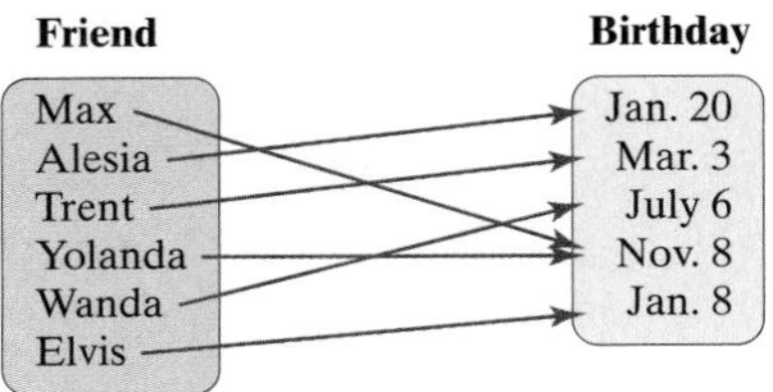

2. Use the set of ordered pairs to represent the relation as a map.

$$\{(1, 5), (2, 4), (3, 7), (8, 12)\}$$

Classroom Example Consider the relation presented where there is a correspondence between state and number of representatives. Write the relation as a set of ordered pairs.

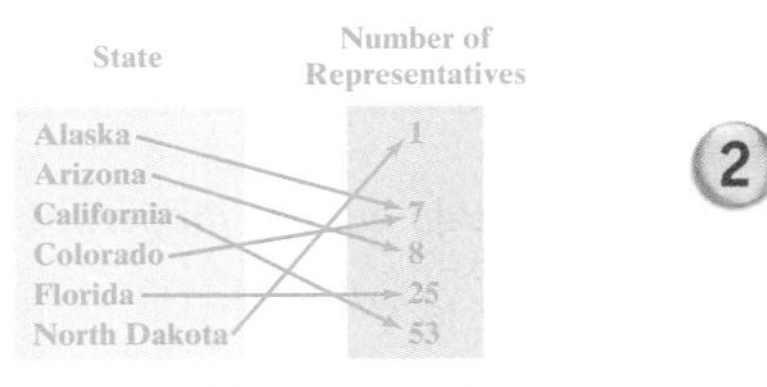

Answer: {(Alaska, 7), (Arizona, 8), (California, 53), (Colorado, 7), (Florida, 25), (North Dakota, 1)}

2 Find the Domain and the Range of a Relation

In a relation we say that y depends on x, and we can write the relation as a set of ordered pairs (x, y). We can think of the set of all x as the **inputs** of the relation. The set of all y can be thought of as the **outputs** of the relation. We use this interpretation of a relation to define *domain* and *range*.

DEFINITION

The **domain** of a relation is the set of all inputs of the relation. The **range** is the set of all outputs of the relation.

Classroom Example Find the domain and the range of the relation presented for Classroom Example 1.

Answer: Domain: {Alaska, Arizona, California, Colorado, Florida, North Dakota}; Range: {1, 7, 8, 25, 53}

EXAMPLE 2 Finding the Domain and the Range of a Relation

Find the domain and the range of the relation presented in Figure 8.

Solution

The domain is the set of all inputs and the range is the set of all outputs. The inputs and, therefore, the domain of the relation is

$$\{\text{Alabama, Colorado, Georgia, Massachusetts, New York}\}$$

The outputs and, therefore, the range of the relation is

$$\{7, 13, 10, 29\}$$

Did you notice that we did not list 7 twice in the range? The domain and the range are sets, and we never list elements in a set more than once.

Work Smart

Never list elements in the domain or range more than once.

QUICK ✓

3. State the domain and the range of the relation.

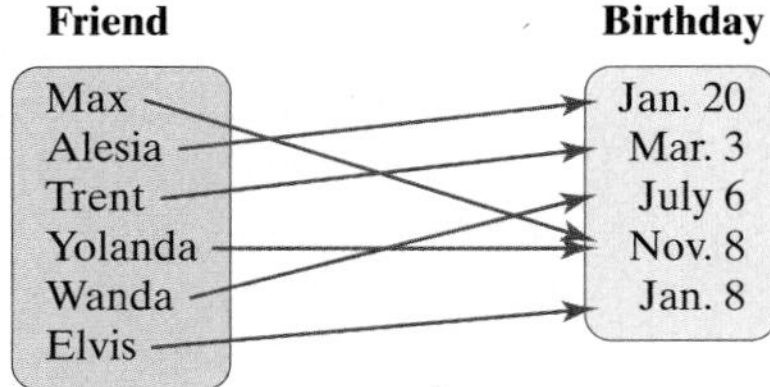

4. State the domain and the range of the relation.

$$\{(1, 5), (2, 4), (3, 7), (8, 7)\}$$

Classroom Example
Identify the domain and the range of the relation.

Answer: Domain: $\{-2, 0, 1, 2\}$; Range: $\{-2, -1, 1, 3\}$

Relations can also be represented by plotting a set of ordered pairs. The set of all x-coordinates represents the domain of the relation, and the set of all y-coordinates represents the range of the relation.

EXAMPLE 3 Finding the Domain and Range of a Relation

Figure 9 shows the graph of a relation. Identify the domain and the range of the relation.

Figure 9

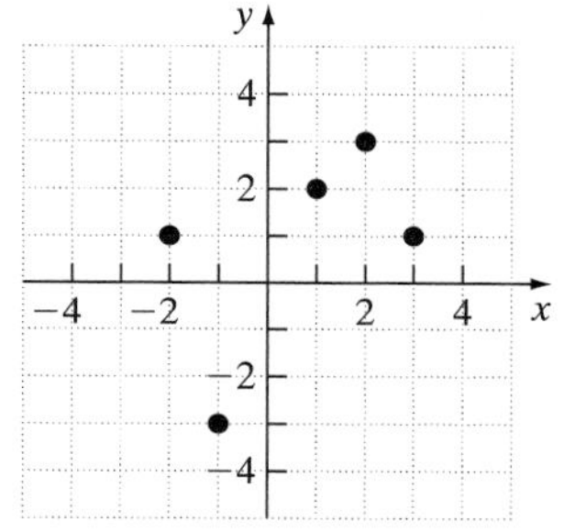

Solution

First, we notice that the ordered pairs of the points in the graph are $(-2, 1)$, $(-1, -3)$, $(1, 2)$, $(2, 3)$, and $(3, 1)$. The domain is the set of all x-coordinates: $\{-2, -1, 1, 2, 3\}$. The range is the set of all y-coordinates: $\{-3, 1, 2, 3\}$.

Work Smart
First write the points as ordered pairs to assist in finding the domain and range.

Quick ✓

5. Identify the domain and the range of the relation shown in the figure.

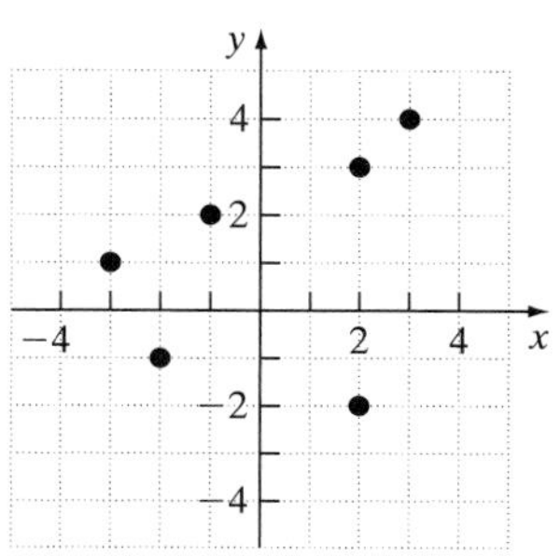

Work Smart
The order in which we list elements in the domain or the range does not matter.

To identify the domain and the range from a graph, we need to remember that a graph is the set of all points whose ordered pairs (x, y) that make the equation a true statement. If the graph exists for some ordered pair (x, y), then the x-coordinate is in the domain and the y-coordinate is in the range.

Classroom Example
Determine the domain and the range of the relation.

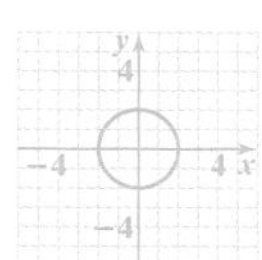

Answer: Domain: $\{x|-2 \le x \le 2\}$ Range: $\{y|-2 \le y \le 2\}$

EXAMPLE 4 Identifying the Domain and Range of a Relation from Its Graph

Figure 10 shows the graph of a relation. Determine the domain and range of the relation.

Figure 10

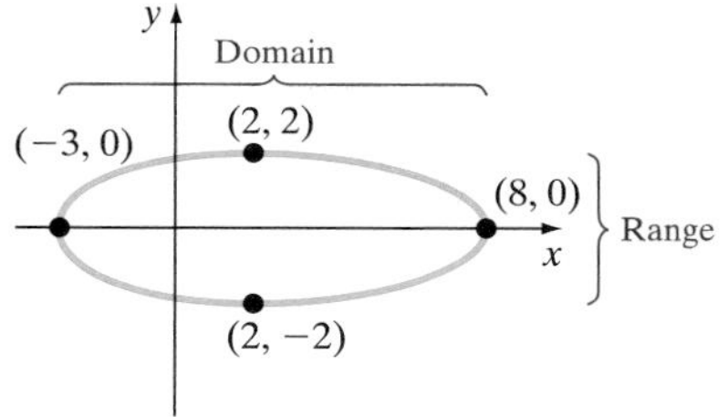

Solution

To find the domain of the relation, we determine the x-coordinates for which the graph exists. The graph exists for $-3 \le x \le 8$. Therefore, the domain is $\{x|-3 \le x \le 8\}$.

To find the range of the relation, we determine the y-coordinates for which the graph exists. The graph exists for $-2 \le y \le 2$. Therefore, the range is $\{y|-2 \le y \le 2\}$.

Quick ✓ *Identify the domain and the range of the relation from its graph.*

6.

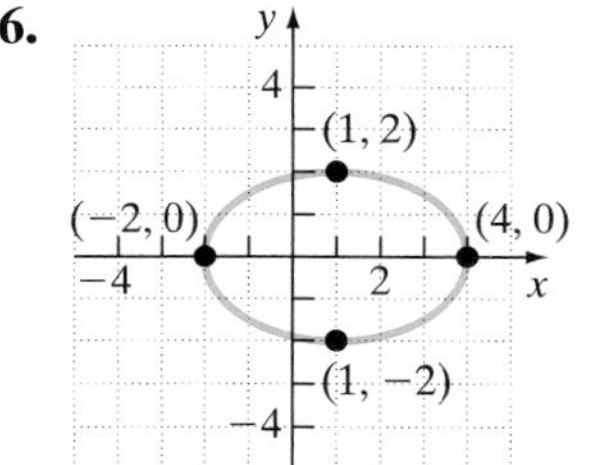

7.

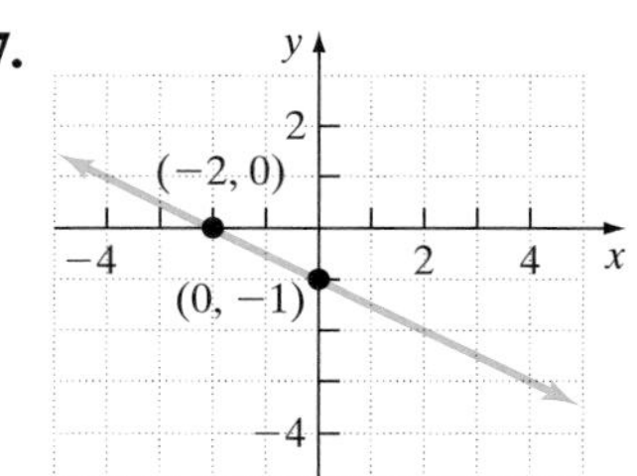

3 Graph a Relation Defined by an Equation

Work Smart

We can define relations four ways:

1. Maps
2. Sets of ordered pairs
3. Graphs
4. Equations

Another approach to defining a relation (instead of a map, a set of ordered pairs, or a graph) is to define relations through equations such as $x + y = 4$ or $x = y^2$. When relations are defined by equations, we typically graph the relation so that we can see how y depends upon x. The graph of the relation is also useful for helping us to identify the domain and the range of the relation.

In Chapter 8, we learned how to graph equations using the point-plotting method. We will use the same technique here.

Classroom Example Graph the relation $y = -x^2 + 1$. Use the graph to determine the domain and the range of the relation.

Answer:

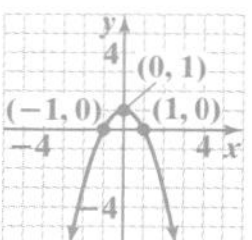

Domain: $\{x|x \text{ is a real number}\}$
Range: $\{y|y \le 1\}$

EXAMPLE 5 Relations Defined by Equations

Graph the relation $y = x^2 + 1$. Use the graph to determine the domain and the range of the relation.

Solution

We use the point-plotting method to graph the relation. Table 3 shows some points on the graph. Figure 11 shows a graph of the relation.

Table 3

x	$y = x^2 + 1$	(x, y)
−3	$(-3)^2 + 1 = 10$	(−3, 10)
−2	$(-2)^2 + 1 = 5$	(−2, 5)
−1	2	(−1, 2)
0	1	(0, 1)
1	2	(1, 2)
2	5	(2, 5)
3	10	(3, 10)

Figure 11

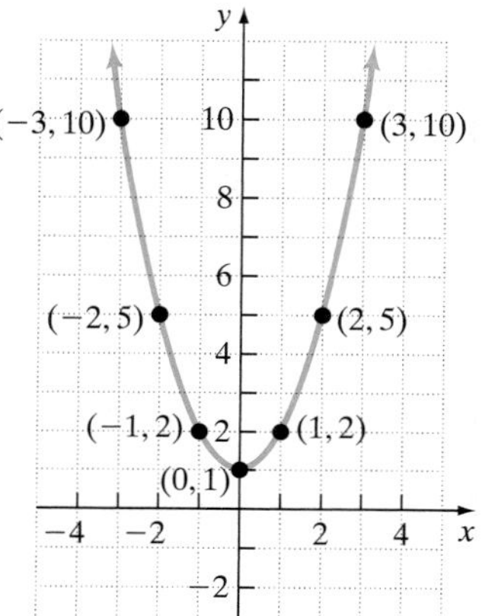

We can see that the graph extends indefinitely to the left and to the right (that is, the graph exists for all x-values), so the domain of the relation is the set of all real numbers. That is, the domain is $\{x|x \text{ is a real number}\}$. We also notice from the graph that there are no y-values less than 1, but the graph exists everywhere for y-values greater than or equal to 1. Therefore, the range of the relation is $\{y|y \ge 1\}$. ■

QUICK ✓ *Graph each relation. Use the graph to identify the domain and the range of the relation.*

8. $y = 2x - 5$ **9.** $y = x^2 - 3$

10.2 Exercises

For Extra Help: Student Solutions Manual | CD Video | PH Math/Tutor Center | MathXL Tutorials on CD | MathXL® | MyMathLab

Concepts and Vocabulary

In Problems 1–3, fill in the blanks.

1. depends on

1. If a relation exists between x and y, then we say that y __________ on x. The notation we use to represent this situation is $x \rightarrow y$ or (x, y).

2. domain; range
3. relation
4. True
5. False
6. True
7. Answers may vary.
8. Answers may vary.
9. {(Physician, 120), (Teacher, 37), (Mathematician, 64), (Computer Engineer, 64), (Lawyer, 82)}
Domain: {(Physician, Teacher, Mathematician, Computer Engineer, Lawyer}
Range: {120, 37, 64, 82}
10. {(Sandy, 426–555), (Kathleen, 278–896), (Shawn, 377–204), (Roberto, 368–110)}
Domain: {Sandy, Kathleen, Shawn, Roberto}
Range: {426–555, 278–896, 377–204, 368–110}
11. {(U.S., 97), (Great Britain, 28), (Korea, 28), (Poland, 14), (Canada, 14)}
Domain: {U.S., Great Britain, Korea, Poland, Canada}
Range: {97, 28, 14}
12. {(Christian, 8), (Maria, 11), (Irma, 5), (Dwayne, 11), (Eleazar, 20)}
Domain: {Christian, Maria, Irma, Dwayne, Eleazar}
Range: {5, 8, 11, 20}

13.
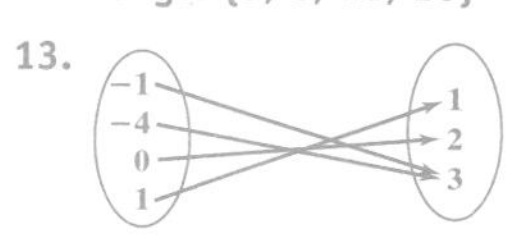

Domain: {−1, −4, 0, 1}
Range: {3, 2, 1}

14.
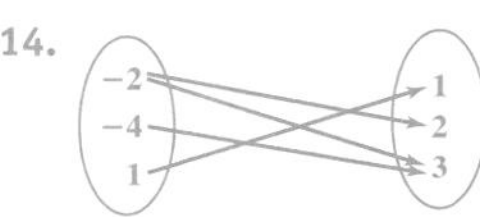

Domain: {−2, −4, 1}
Range: {3, 2, 1}

15.
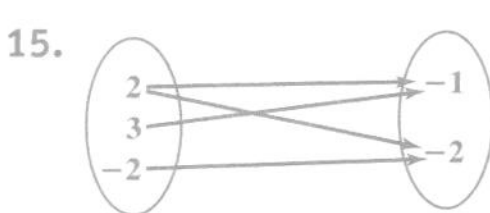

Domain: {2, 3, −2}
Range: {−1, −2}

16.
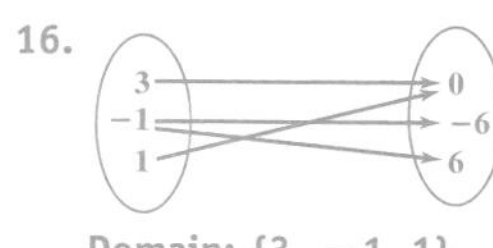

Domain: {3, −1, 1}
Range: {0, −6, 6}

17.
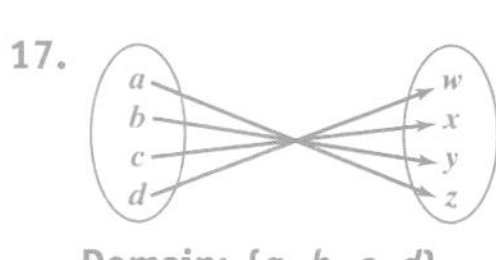

Domain: {a, b, c, d}
Range: {w, x, y, z}

2. The __________ of a relation is the set of all inputs to the relation. The __________ is the set of all outputs of the relation.

3. When the elements in one set of data are linked to elements in a second set of data, we have a __________.

In Problems 4–6, answer True or False to each statement.

4. If the graph of a relation does not exist at $x = 3$, then 3 is not an element in the domain of the relation.

5. Relations can only exist between real numbers.

6. For the relation $(1, 3)$, the input is 1 and the output is 3.

7. In your own words, explain what a relation is. Be sure to include an explanation of *domain* and *range*.

8. State the four methods presented in this section to describe a relation. When is using ordered pairs most appropriate? When is using a graph most appropriate? Support your opinion with an example.

Building Skills

In Problems 9–12, represent each relation as a set of ordered pairs. Then state the domain and the range of each relation.

9.
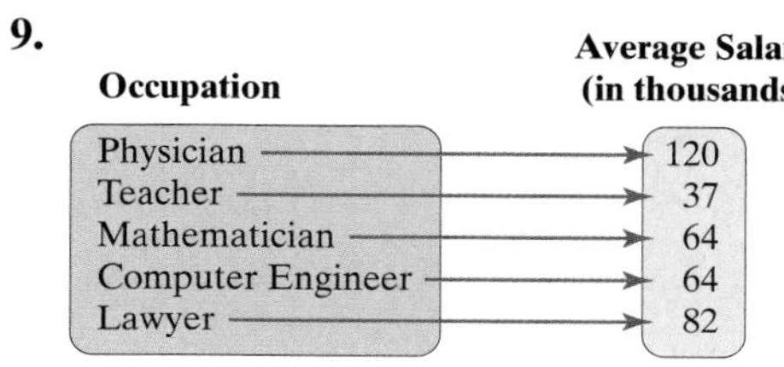

10.
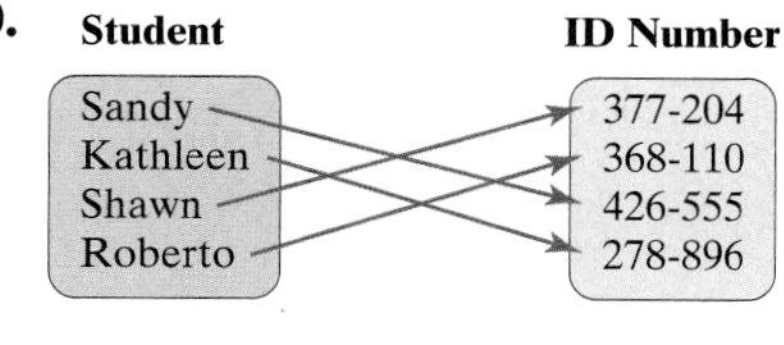

11.
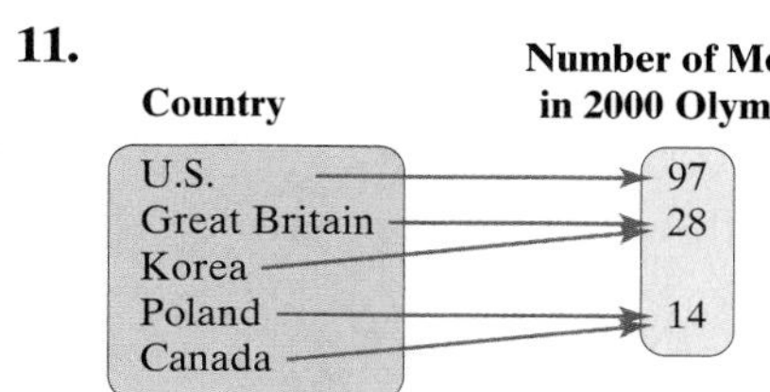

12.
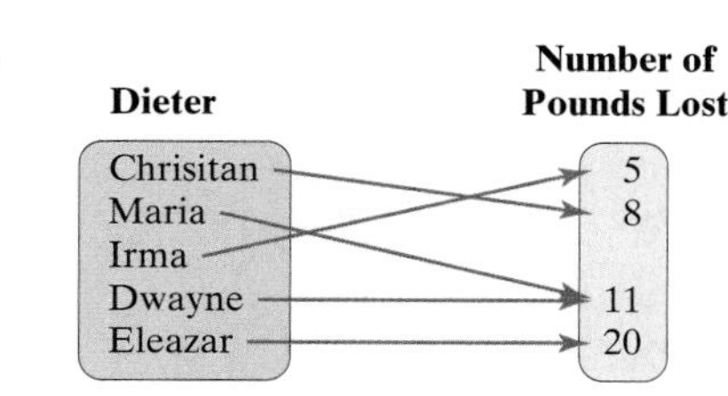

In Problems 13–18, use the set of ordered pairs to represent the relation as a map. Then state the domain and the range of each relation.

13. $\{(-1, 3); (-4, 3); (0, 2); (1, 1)\}$

14. $\{(-2, 3); (-4, 3); (-2, 2); (1, 1)\}$

15. $\{(2, -1); (2, -2); (3, -1); (-2, -2)\}$

16. $\{(3, 0); (-1, -6); (-1, 6); (1, 0)\}$

17. $\{(a, z); (b, y); (c, x); (d, w)\}$

18. $\{(b, o); (o, k); (b, a); (g, s)\}$

In Problems 19–34, identify the domain and the range of the relation shown in the figure.

19.
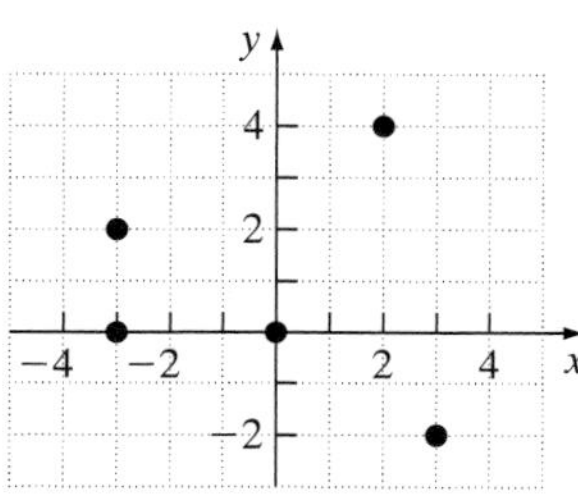

20.
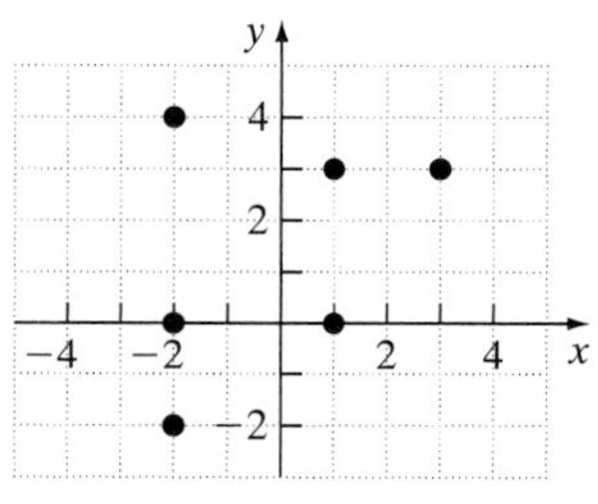

21.
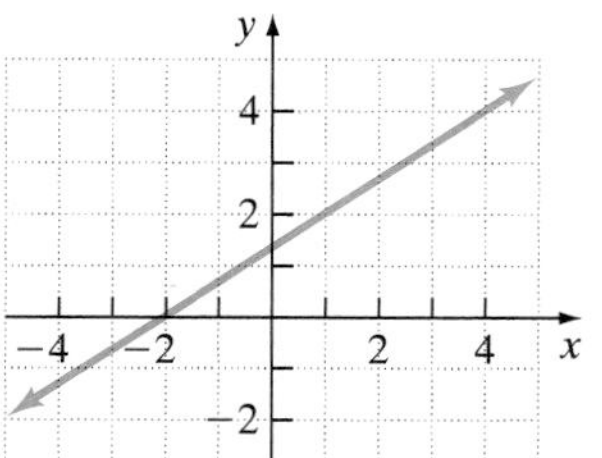

18. b, o, g → o, k, a, s
Domain: {b, o, g}
Range: {o, k, a, s}

19. Domain: {−3, 0, 2, 3}
Range: {−2, 0, 2, 4}
20. Domain: {−2, 1, 3}
Range: {−2, 0, 3, 4}

21. Domain: $\{x \mid x \text{ is a real number}\}$
Range: $\{y \mid y \text{ is a real number}\}$

22. Domain: $\{x \mid x \text{ is a real number}\}$
Range: $\{y \mid y \text{ is a real number}\}$
23. Domain: $\{3\}$
Range: $\{y \mid y \text{ is a real number}\}$
24. Domain: $\{x \mid x \text{ is a real number}\}$
Range: $\{-2\}$
25. Domain: $\{x \mid x \text{ is a real number}\}$
Range: $\{y \mid y \leq 0\}$
26. Domain: $\{x \mid x \text{ is a real number}\}$
Range: $\{y \mid y \leq 3\}$
27. Domain: $\{x \mid x \text{ is a real number}\}$
Range: $\{y \mid y \geq -2\}$
28. Domain: $\{x \mid x \text{ is a real number}\}$
Range: $\{y \mid y \geq 0\}$
29. Domain: $\{x \mid x \geq -1\}$
Range: $\{y \mid y \text{ is a real number}\}$
30. Domain: $\{x \mid x \leq 3\}$
Range: $\{y \mid y \text{ is a real number}\}$
31. Domain: $\{x \mid -1 \leq x \leq 5\}$
Range: $\{y \mid -1 \leq y \leq 5\}$
32. Domain: $\{x \mid -3 \leq x \leq 1\}$
Range: $\{y \mid -1 \leq y \leq 3\}$
33. Domain: $\{x \mid -4 \leq x \leq 4\}$
Range: $\{y \mid -3 \leq y \leq 3\}$
34. Domain: $\{x \mid -2 \leq x \leq 2\}$
Range: $\{y \mid -4 \leq y \leq 4\}$

35. $y = -2x + 1$

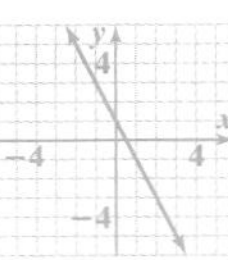

Domain: $\{x \mid x \text{ is a real number}\}$
Range: $\{y \mid y \text{ is a real number}\}$

36. $y = x + 4$

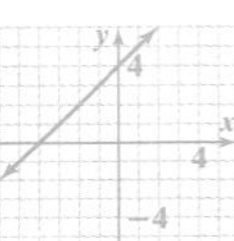

Domain: $\{x \mid x \text{ is a real number}\}$
Range: $\{y \mid y \text{ is a real number}\}$

37. $y = -\frac{2}{3}x - 1$

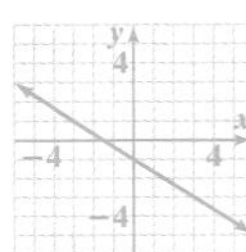

Domain: $\{x \mid x \text{ is a real number}\}$
Range: $\{y \mid y \text{ is a real number}\}$

38. $y = 3x - 2$

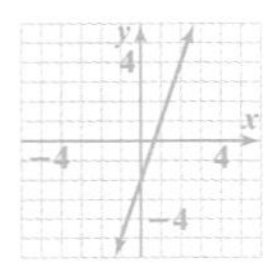

Domain: $\{x \mid x \text{ is a real number}\}$
Range: $\{y \mid y \text{ is a real number}\}$

39. $y = 4x - 2x^2$

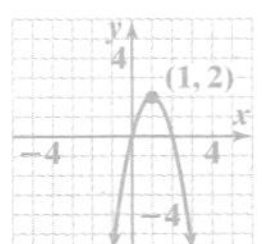

Domain: $\{x \mid x \text{ is a real number}\}$
Range: $\{y \mid y \leq 2\}$

40. $y = -x^2 + 4x - 4$

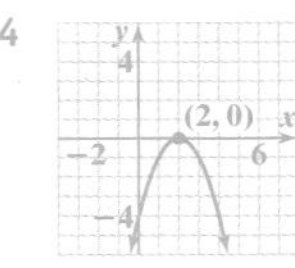

Domain: $\{x \mid x \text{ is a real number}\}$
Range: $\{y \mid y \leq 0\}$

22.

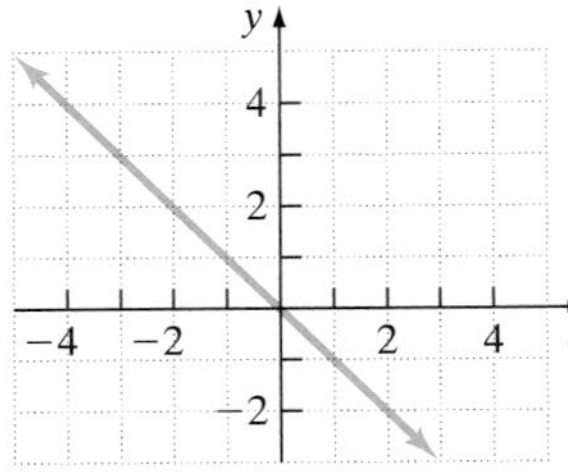

23.

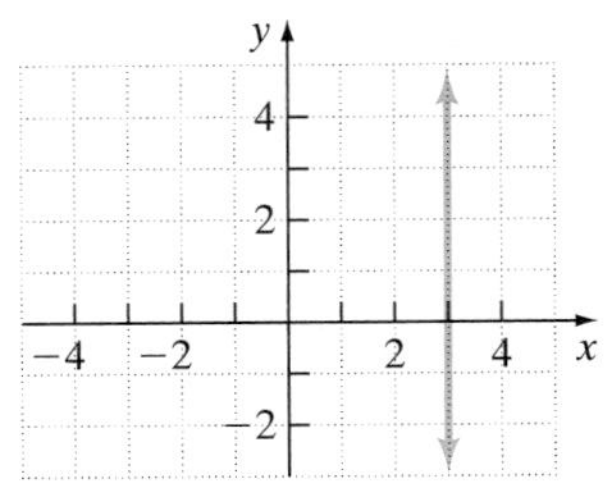

24.

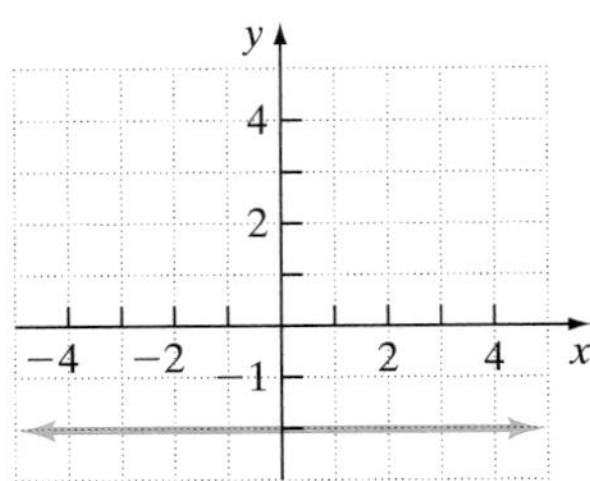

25.

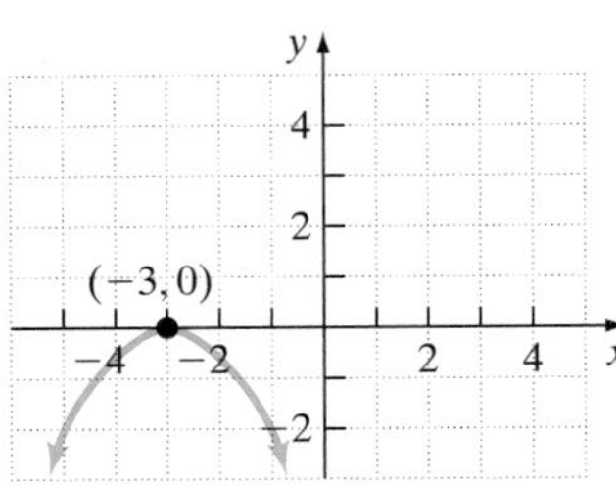

26.

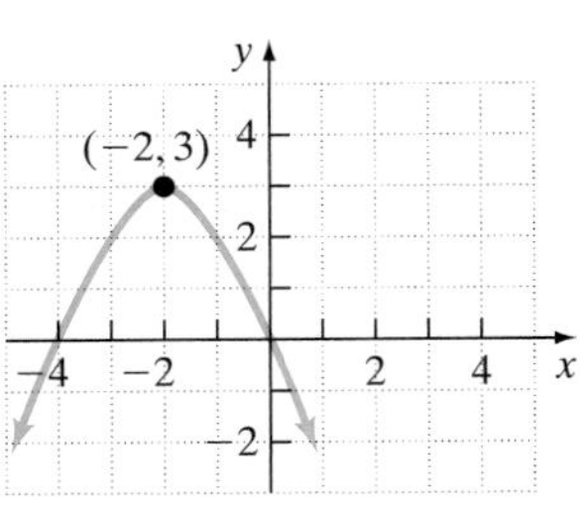

27.

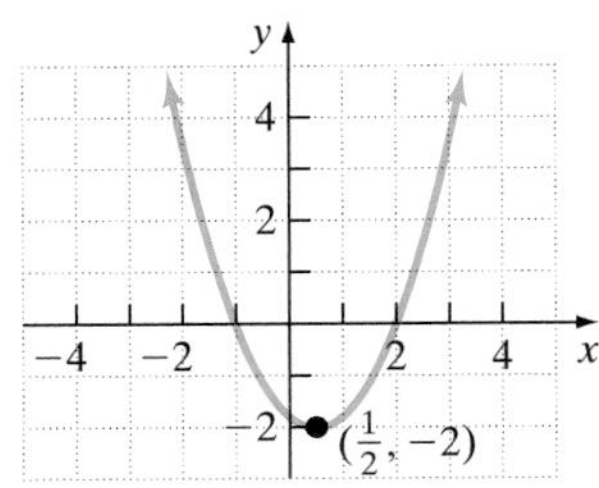

28.

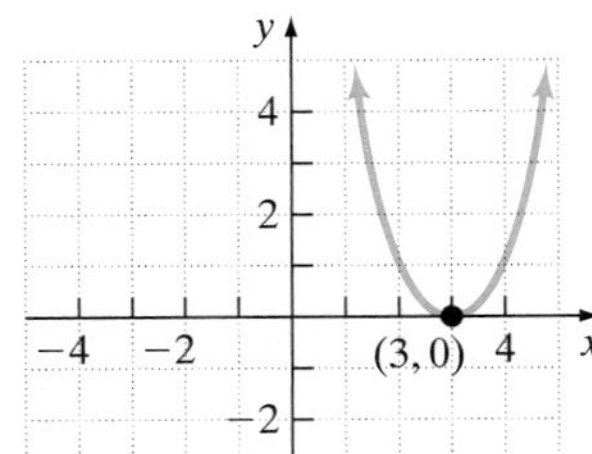

29.

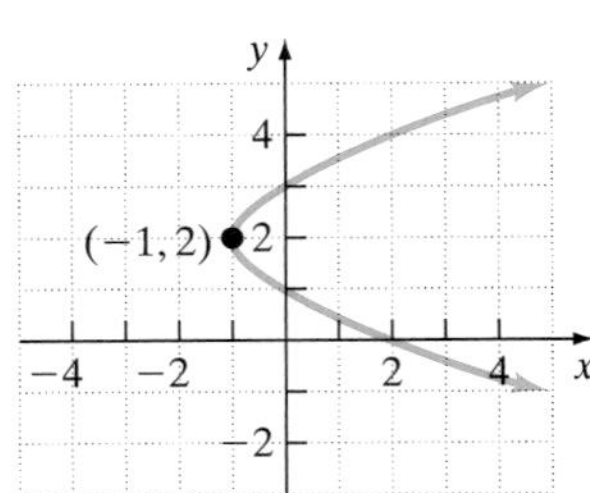

30.

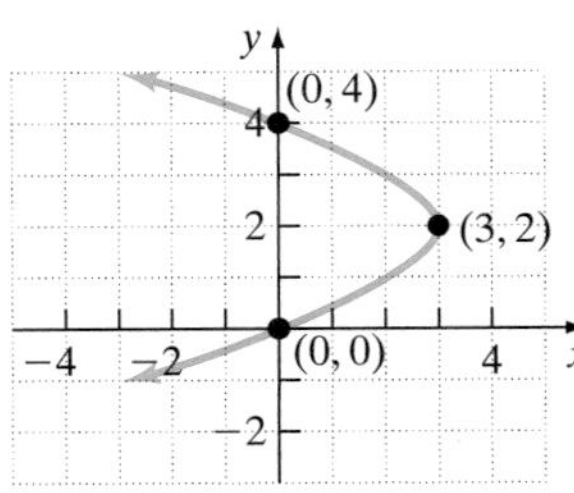

31.

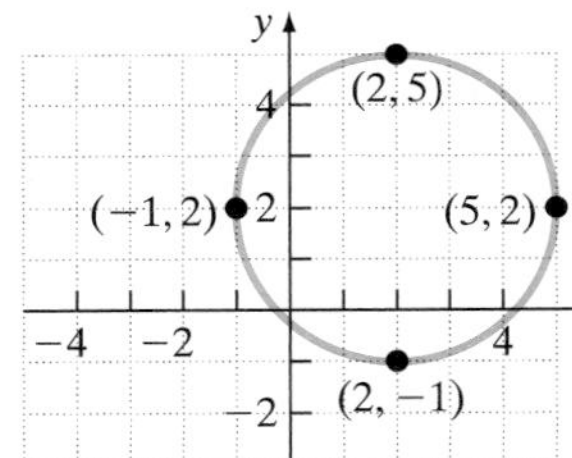

32.

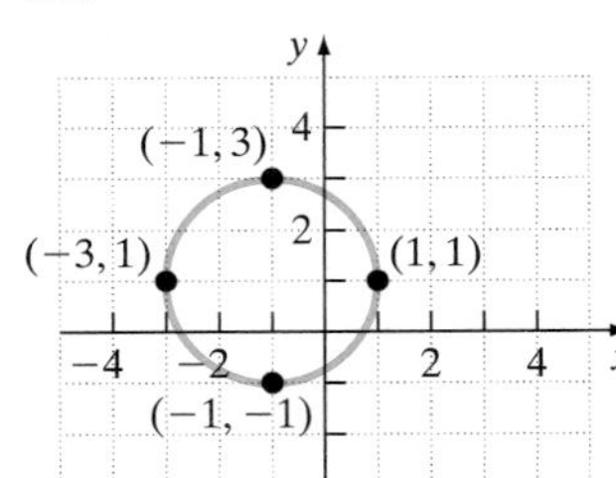

33.

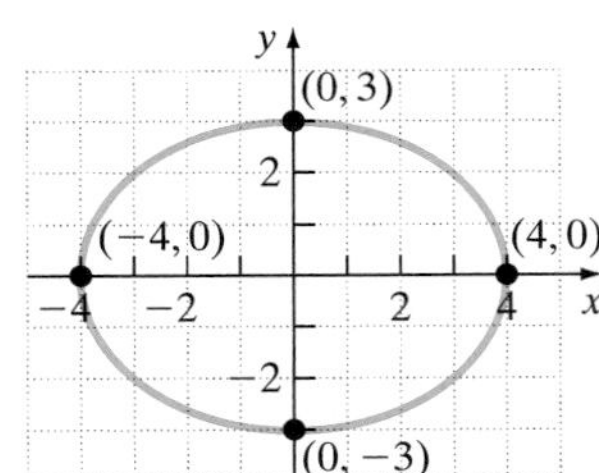

34.

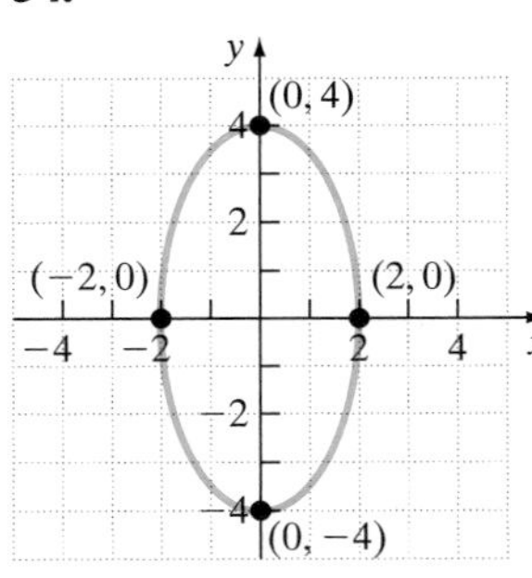

In Problems 35–50, graph each relation. Use the graph to identify the domain and the range of the relation.

35. $y = -2x + 1$ **36.** $y = x + 4$ **37.** $y = -\frac{2}{3}x - 1$ **38.** $y = 3x - 2$

39. $y = 4x - 2x^2$ **40.** $y = -x^2 + 4x - 4$ **41.** $y = x^2 - 7x + 10$

42. $y = 3x^2 + 6x$ **43.** $y = 4x$ **44.** $3x + 3y = 2$

45. $y = 2x^2 - 5$ **46.** $y = -x^2 + 5x + 2$ **47.** $y = -x^2 + 4x + 2$

48. $y = x^2 - 5x + 1$ **49.** $3x + 4y = 8$ **50.** $2 = x - 2y$

41. $y = x^2 - 7x + 10$

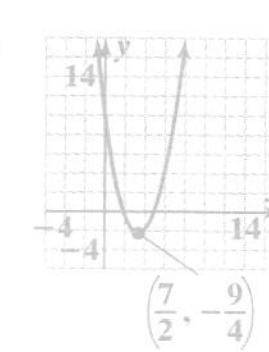

Domain: $\{x \mid x \text{ is a real number}\}$
Range: $\left\{y \mid y \geq -\frac{9}{4}\right\}$

42. $y = 3x^2 + 6x$

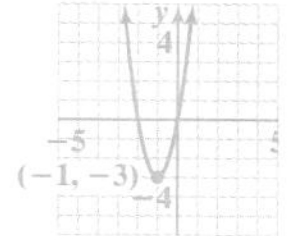

Domain: $\{x \mid x \text{ is a real number}\}$
Range: $\{y \mid y \geq -3\}$

43. $y = 4x$

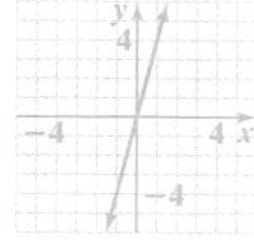

Domain: $\{x \mid x \text{ is a real number}\}$
Range: $\{y \mid y \text{ is a real number}\}$

44. $3x + 3y = 2$

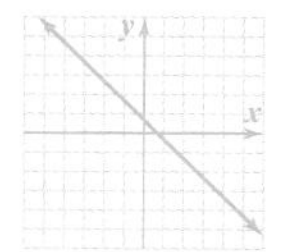

Domain: $\{x \mid x \text{ is a real number}\}$
Range: $\{y \mid y \text{ is a real number}\}$

45. $y = 2x^2 - 5$

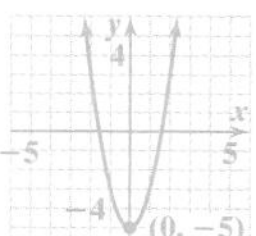

Domain: $\{x \mid x \text{ is a real number}\}$
Range: $\{y \mid y \geq -5\}$

46. $y = -x^2 + 5x + 2$

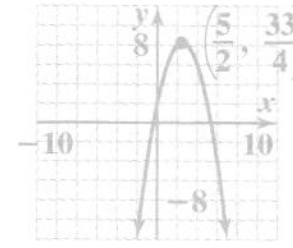

Domain: $\{x \mid x \text{ is a real number}\}$
Range: $\left\{y \mid y \leq \frac{33}{4}\right\}$

47. $y = -x^2 + 4x + 2$

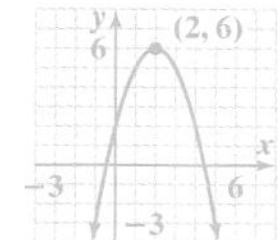

Domain: $\{x \mid x \text{ is a real number}\}$
Range: $\{y \mid y \leq 6\}$

48. $y = x^2 - 5x + 1$

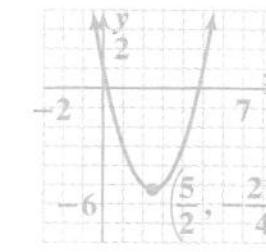

Domain: $\{x \mid x \text{ is a real number}\}$
Range: $\left\{y \mid y \geq -\frac{21}{4}\right\}$

Applying the Concepts

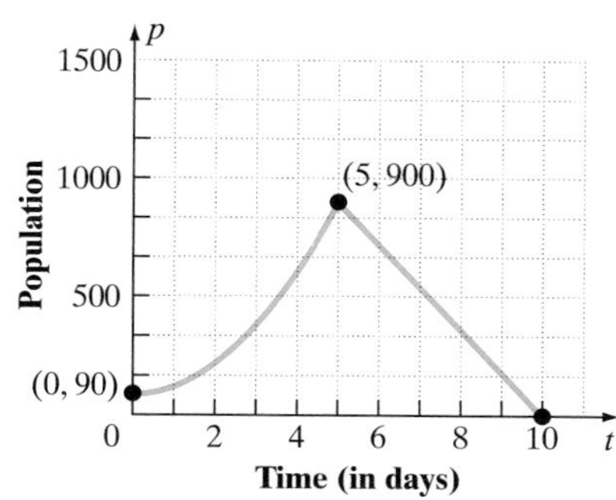

51. **Bacteria Population** When you are ill, there may be bacteria growing in your body. Once you take an antibiotic, the population of bacteria will decline (hopefully, quickly). The graph shows the population of bacteria after t days.
    (a) Find the domain and the range of this function.
    (b) On what day did the patient take the antibiotic?
    (c) How many days did it take to be rid of this colony of bacteria?

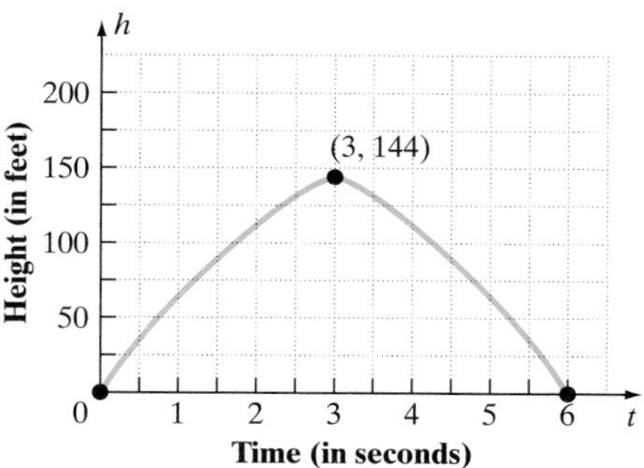

52. **Ball Launch** A ball is launched into the air from the ground with an initial velocity of 96 feet per second. The graph shows the height, h, of the ball after t seconds.
    (a) Find the domain and the range of this function.
    (b) How many seconds does it take for the ball to get to its highest point?
    (c) What is the maximum height of the ball?

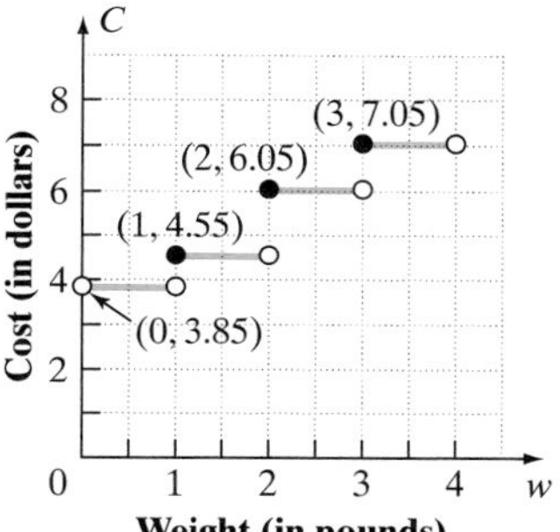

53. **Mailing a Package** When you send a package in the mail, you pay according the weight of the package. All packages weighing within a certain number of ounces pay one rate. Once you go over that weight, the price jumps to the next higher rate (the open circle is the graph is the point where the jump occurs). The graph shows the cost to send a package by priority mail, where C represents the cost and w represents the weight in pounds.
    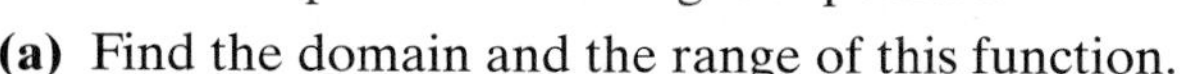
    (a) Find the domain and the range of this function.
    (b) How much does it cost to send a 10-ounce package?
    (c) How much does it cost to send a 20-ounce package?

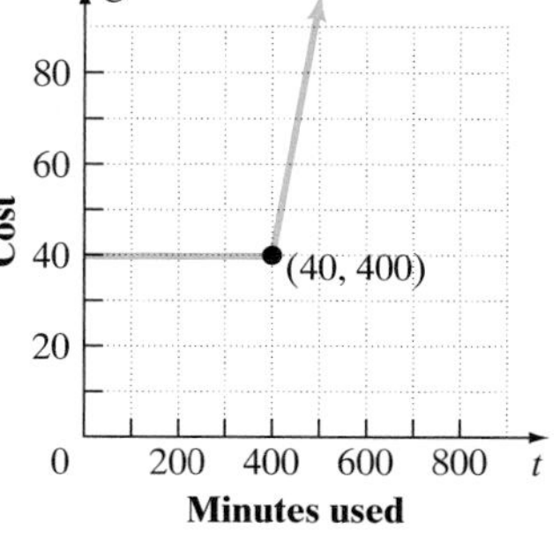

54. **Cell Phone Expenses** Cell phone users pay according to the number of minutes of air time they use. There is a fixed rate for a specified number of minutes, and after that you are charged for each additional minute. The graph shows the rates for one of Verizon's phone plans, where t is the time in minutes and C is the cost in dollars.
    (a) Find the domain and the range of this function.
    (b) How much air time does the fixed plan allot?

Extending the Concepts

55. Make up an example of a relation in which each element of the domain corresponds to a single element in the range.

56. Make up an example of a relation in which each element in the range corresponds to a single element in the domain.

49. $3x + 4y = 8$

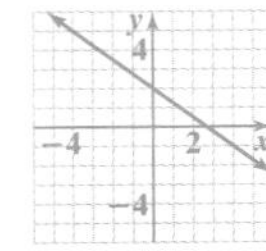

Domain: $\{x \mid x \text{ is a real number}\}$
Range: $\{y \mid y \text{ is a real number}\}$

50. $2 = x - 2y$

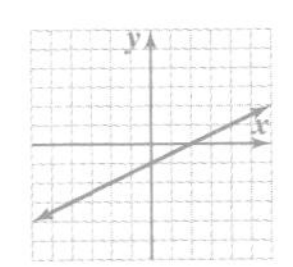

Domain: $\{x \mid x \text{ is a real number}\}$
Range: $\{y \mid y \text{ is a real number}\}$

51. (a) Domain: $\{t \mid 0 \le t \le 10\}$
Range: $\{p \mid 0 \le p \le 900\}$
(b) day 5 (c) 5 days

52. (a) Domain: $\{t \mid 0 \le t \le 6\}$
Range: $\{h \mid 0 \le h \le 144\}$
(b) 3 sec (c) 144 ft

53. (a) Domain: $\{w \mid 0 < w < 4\}$
Range: $\{3.85, 4.55, 6.05, 7.05\}$
(b) \$3.85 (c) \$4.55

54. (a) Domain: $\{m \mid m \ge 0\}$
Range: $\{c \mid c \ge 40\}$
(b) 400 minutes

55. Answers may vary.

56. Answers may vary.

57. Johann Dirichlet, born in the French Empire in 1805 (in a region now in Germany), developed a passion for mathematics at an early age and spent his pocket money buying math books. One of the things he is famous for is the following mapping:

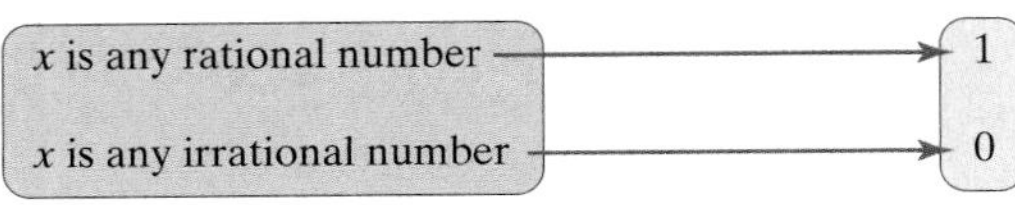

(a) Is it possible to list every ordered pair represented in Dirichlet's mapping?
(b) Show six ordered pairs that demonstrate how this relation maps all real numbers.
(c) Could you make a graph of this mapping similar to those in Problems 51–54? Why or why not?

58. Draw the graph of a relation whose domain is all real numbers but whose range is a single real number. Compare your graph with those of your classmates. How are they similar?

59. Draw the graph of a relation whose domain is a single real number but whose range is all real numbers. Compare your graph with those of your classmates. How are they similar?

60. Draw the graph of a relation that contains all ordered pairs (x, y) such that $y = x$ and x is any real number.

57. (a) No (b) Answers may vary. (c) No; answers may vary.

58. Answers may vary.

59. Answers may vary.

60. $y = x$

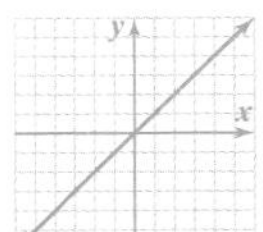

PUTTING THE CONCEPTS TOGETHER (SECTIONS 10.1 AND 10.2)

These problems cover important concepts from Sections 10.1 and 10.2. We designed these problems so that you can review the chapter so far and show your mastery of the concepts. Take time to work these problems to make sure you understand these concepts before proceeding with the next section. The answers to these problems are located at the back of the text starting on page AN-29.

1. For the quadratic equation $y = x^2 - 2x - 3$.

(a) Determine whether the parabola opens up or down.
(b) Find the vertex of the parabola.
(c) Find the equation of the axis of symmetry.
(d) Find all intercepts.

In Problems 2–4, graph each quadratic equation using its properties. Label the vertex, intercepts, and axis of symmetry. Based on the graph determine the domain and the range of the relation.

2. $y = x^2 - 6x$ **3.** $y = x^2 + 4x - 1$ **4.** $y = -x^2 + 9$

5. For the quadratic equation $y = 2x^2 + 12x - 3$,

(a) Determine whether the equation has a maximum or minimum value.
(b) Find the maximum or minimum value of the equation.

In Problems 6–9, determine (a) the domain and (b) the range of the given relation.

6. $\{(a, 1), (b, 2), (c, 1)\}$ **7.** $\{(1, -1), (3, -3), (3, -1)\}$

1. (a) Opens up (b) $(1, -4)$
(c) $x = 1$
(d) $(0, -3)$, $(-1, 0)$, $(3, 0)$

2.

$x = 3$
$(6, 0)$
-10 $(0, 0)$ 10
-8 $(3, -9)$
8

Domain: $\{x \mid x \text{ is a real number}\}$
Range: $\{y \mid y \ge -9\}$

3.

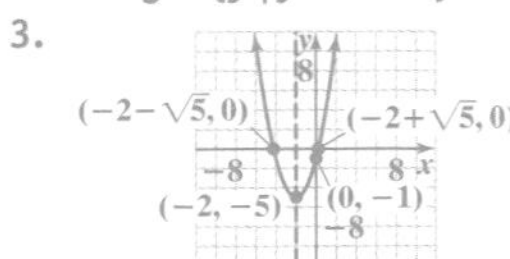

Domain: $\{x \mid x \text{ is a real number}\}$
Range: $\{y \mid y \ge -5\}$

4.

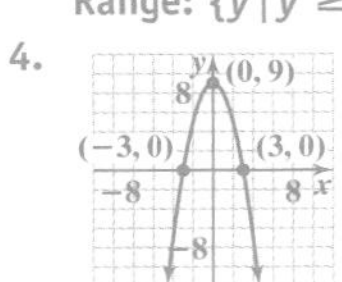

Domain: $\{x \mid x \text{ is a real number}\}$
Range: $\{y \mid y \le 9\}$

5. (a) Minimum (b) -21

6. Domain: $\{a, b, c\}$
Range: $\{1, 2\}$

7. Domain: $\{1, 3\}$
Range: $(-1, -3\}$

8. Domain: $\{x \mid x \geq -4\}$
Range: $\{y \mid y \text{ is a real number}\}$
9. Domain: $\{x \mid 1 \leq x \leq 7\}$
Range: $\{y \mid -1 \leq y \leq 5\}$

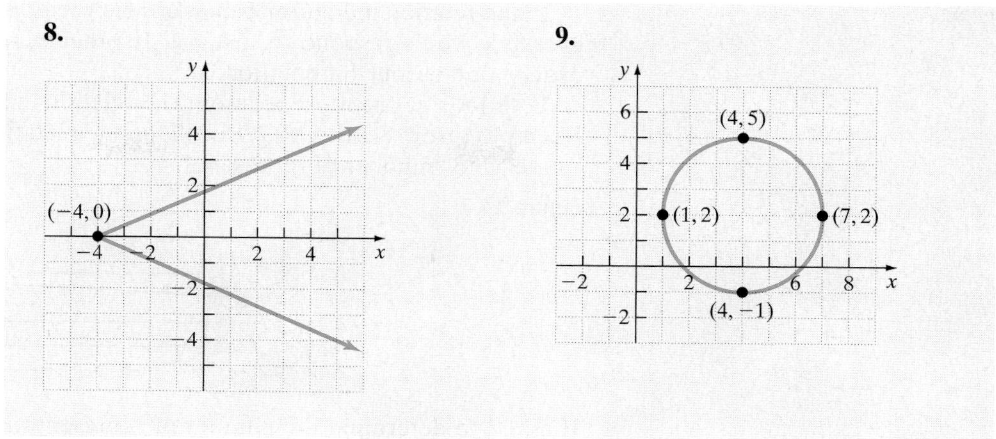

10.3 An Introduction to Functions

OBJECTIVES

1. Determine Whether or Not a Relation Expressed as a Map or Ordered Pairs Represents a Function
2. Determine Whether or Not a Relation Expressed as an Equation Represents a Function
3. Identify the Graph of a Function Using the Vertical Line Test
4. Find the Value of a Function

Preparing for An Introduction to Functions

Before getting started, take this readiness quiz. If you get a problem wrong, go back to the section cited and review the material.

1. Evaluate the expression $2x^2 - 5x$ for (a) $x = 1$, (b) $x = 4$, and (c) $x = -3$. [Section 1.7, pp. 57–58]

1 Determine Whether or Not a Relation Expressed as a Map or Ordered Pairs Represents a Function

We now present one of the most important concepts in algebra—the *function*. A function is a special type of relation. To understand the idea behind a function, consider the relation presented in Figure 12. This is a correspondence between people and their telephone numbers.

Work Smart
Not all relations are functions!

Figure 12

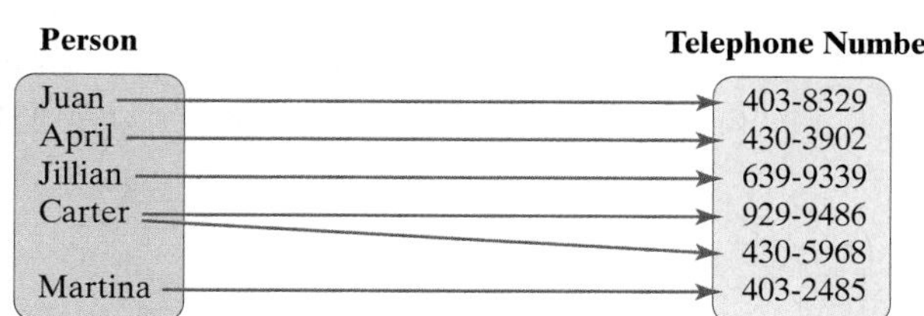

For the relation in Figure 12, if someone were asked to provide the phone number of Carter, some would respond "929-9486," while others would respond "430-5968." In other words, the input "person" does not correspond to a single output "telephone number."

Let's consider a second relation where we have a correspondence between states and their populations presented in Figure 13.

Figure 13

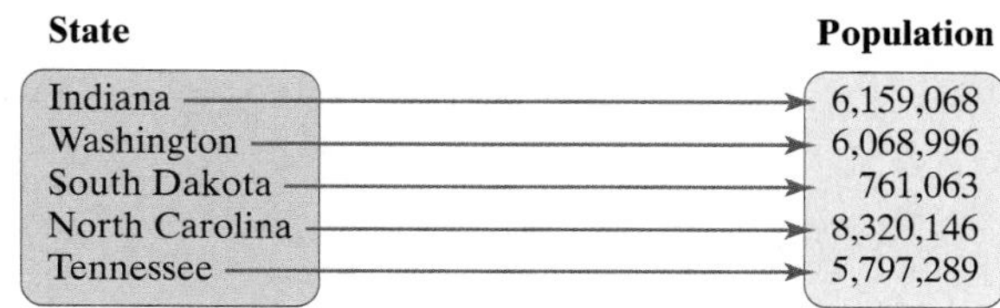

Preparing for...Answer
1. (a) −3 **(b)** 12 **(c)** 33

In the relation in Figure 13, if asked for the population that corresponds to Indiana, everyone would respond "6,159,068." In other words, each input "state" corresponds to exactly one output "population."

Let's look at one more relation. The relation in Figure 14 is the relation presented in Example 1 from Section 10.2. Recall that this relation shows a correspondence between "state" and "number of representatives."

Figure 14

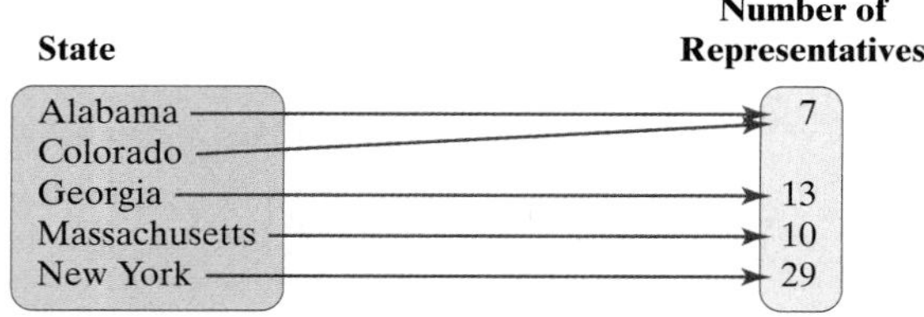

If asked to determine the number of representatives from Alabama, we would all respond "7." If asked to determine the number of representatives from Colorado, we would all respond "7."

Looking carefully at the three relations, we should notice that the relations presented in Figures 13 and 14 have something in common. Can you tell what it is? The common link between these two relations is that each input corresponds to only one output. Remember the set of all inputs is called the domain of the relation and the set of all outputs is called the range of the relation. This leads to the definition of a *function*.

In Words
For a relation to be classified as a function, each input may have only one output.

DEFINITION

A **function** is a relation in which each element in the domain of the relation corresponds to exactly one element in the range of the relation.

In Figure 12, Carter has two telephone numbers (two outputs). According to the definition, the relation in Figure 12 is not a function.

Classroom Example
Determine whether the following relations represent functions or not. If the relation is a function, then state its domain and range.

(a)

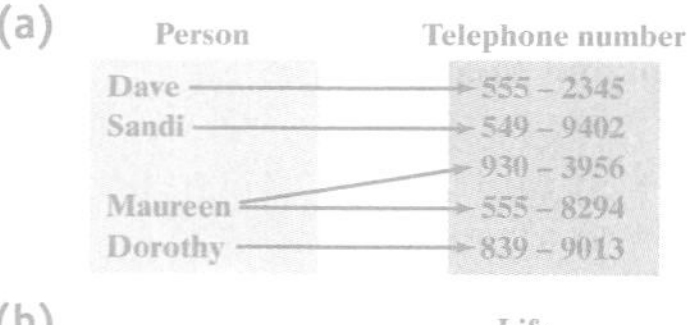

(b)

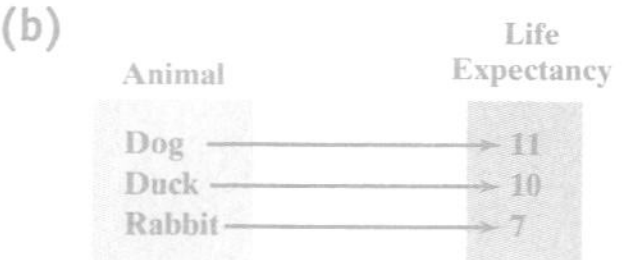

Answer:

(a) Not a function

(b) Function; Domain: {Dog, Duck, Rabbit}; Range: {11, 10, 7}

EXAMPLE 1 Determining If a Relation Defined by a Map Represents a Function

Determine whether the following relations represent functions. If the relation is a function, then state its domain and range.

(a) For this relation, the domain represents the length (mm) of the right humerus and the range represents the length (mm) of the right tibia for each of five rats sent into space. The lengths were measured once the rats returned from their trip.

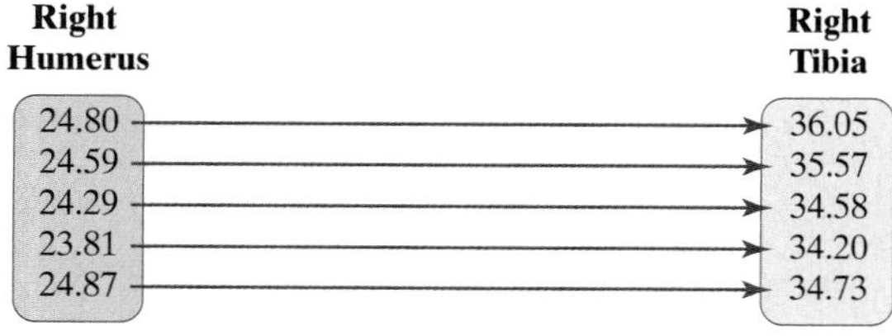

Source: Nasa Life Sciences Data Archive

(b) For this relation, the domain represents the state and the range represents the percent of the population living in poverty in 2002.

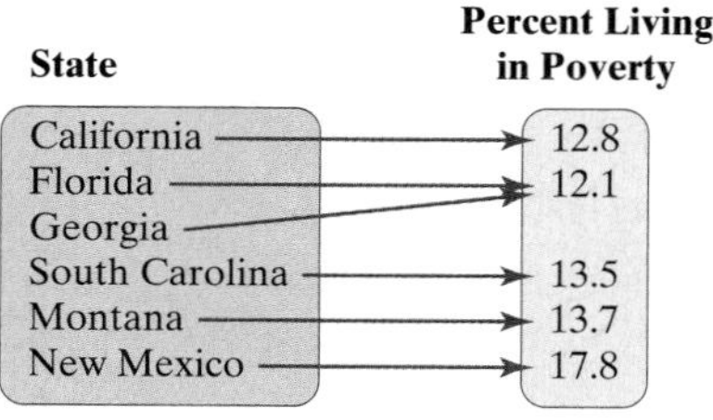

Source: United States Census Bureau

(c) For the following relation, the domain represents the average total cholesterol (mg/dL) and the range represents the age category.

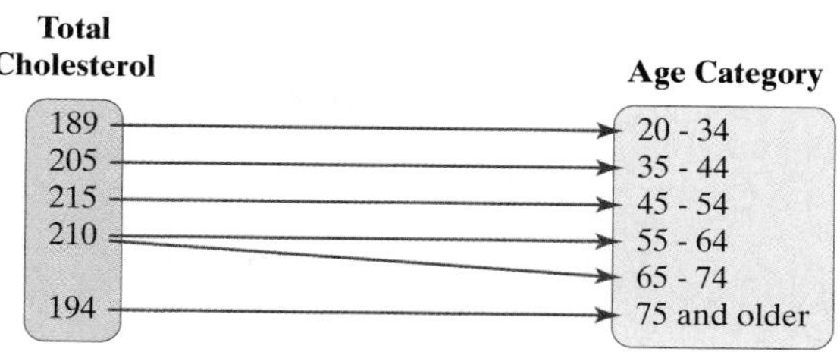

Source: Centers for Disease Control and Prevention

Solution

(a) The relation is a function because each element in the domain corresponds to exactly one element in the range. The domain of the function is {24.80, 24.59, 24.29, 23.81, 24.87}. The range of the function is {36.05, 35.57, 34.58, 34.20, 34.73}.

(b) The relation is a function because each element in the domain corresponds to exactly one element in the range. The domain of the function is {California, Florida, Georgia, South Carolina, Montana, New Mexico}. The range of the function is {12.8, 12.1, 13.5, 13.7, 17.8}.

(c) The relation is not a function because there is an element in the domain, 210, that corresponds to more than one element the range. If asked which age group has an average total cholesterol of 210, some would respond 55–64 and others would respond 65–74.

QUICK ✓ *Determine whether or not the relation represents a function. If the relation is a function, then state its domain and range.*

1.

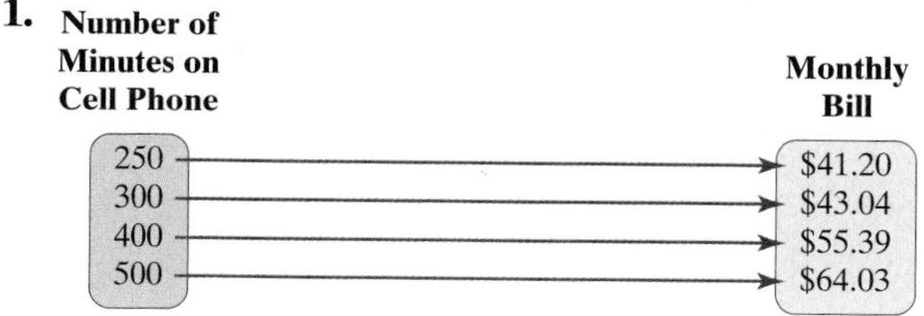

2.

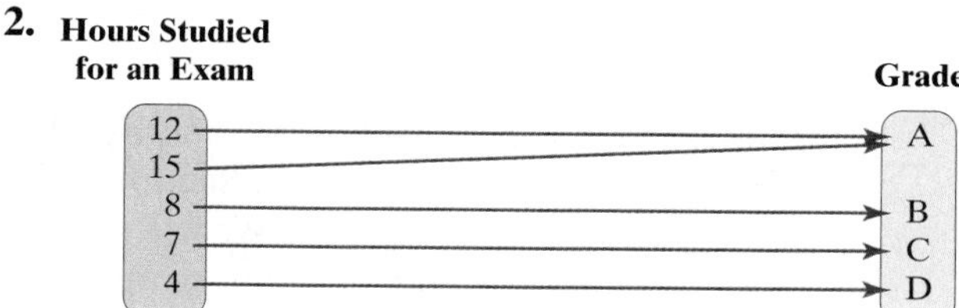

3.

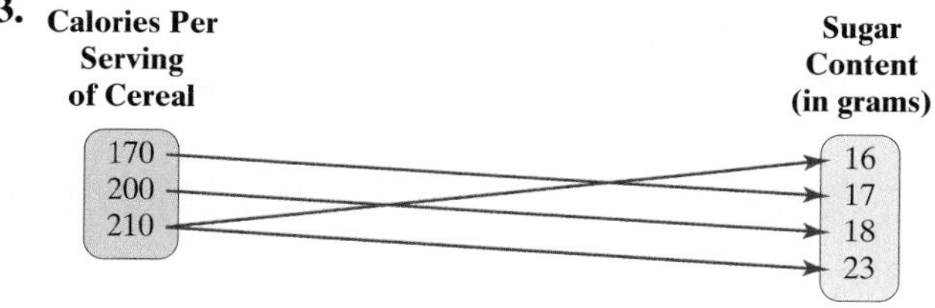

We may also think of a function as a set of ordered pairs (x, y) in which no two ordered pairs have the same first coordinate but different second coordinates.

Classroom Example Determine whether or not each relation represents a function. If the relation is a function, then state its domain and range.

(a) $\{(2, 3), (3, 3), (5, 7)\}$

(b) $(-1, 4), (0, 8), (-1, 9)\}$

Answer:

(a) Function; Domain: $\{2, 3, 5\}$ Range: $\{3, 7\}$

(b) Not a function

EXAMPLE 2 Determining If a Relation Written as Ordered Pairs Represents a Function

Determine whether or not each relation represents a function. If the relation is a function, then state its domain and range.

(a) $\{(1, 3), (-1, 4), (0, 5), (2, 7)\}$ **(b)** $\{(-2, 8), (-1, 4), (0, 1), (1, 6), (2, 8)\}$

(c) $\{(0, 3), (1, 4), (4, 5), (9, 5), (4, 1)\}$

Solution

(a) This relation is a function because there are no ordered pairs with the same first coordinate but different second coordinates. The domain of the function is the set of all first coordinates, $\{-1, 0, 1, 2\}$; the range of the function is the set of all second coordinates, $\{3, 4, 5, 7\}$.

(b) This relation is a function because there are no ordered pairs with the same first coordinate but different second coordinates. The domain of the function is the set of all first coordinates, $\{-2, -1, 0, 1, 2\}$; the range of the function is the set of all second coordinates, $\{1, 4, 6, 8\}$. Remember that it is okay for two different inputs to correspond to the same output.

(c) This relation is not a function because there are two ordered pairs, $(4, 5)$ and $(4, 1)$, with the same first coordinate but different second coordinates. In other words, this is not a function because an element in the domain, 4, corresponds to two different elements in the range, 1 and 5. ■

Teaching Tip At this point, make sure that students understand that all functions are relations, but all relations are not functions.

In Example 2(b), notice that -2 and 2 in the domain each correspond to 8 in the range. This does not violate the definition of a function—two different first coordinates can have the same second coordinate. A violation of the definition occurs when two ordered pairs have the same first coordinate and different second coordinates as in Example 2(c).

QUICK ✓ *Determine whether each relation represents a function. If the relation is a function, then state its domain and range.*

4. $\{(-3, 3), (-2, 2), (-1, 1), (0, 0), (1, 1)\}$

5. $\{(-3, 2), (-2, 5), (-1, 8), (-3, 6)\}$

6. $\{(3, 8), (4, 10), (5, 11), (6, 8)\}$

2 Determine Whether or Not a Relation Expressed as an Equation Represents a Function

At this point, we have shown how to identify when a relation is a function for relations defined by maps or ordered pairs. In Section 10.2, we learned that relations can also be expressed as equations and graphs. We will now address the circumstances under which equations are functions and, in the next objective, look at instances for which graphs are functions.

To determine whether an equation, where y depends upon x, is a function, it is often easiest to solve the equation for y. If a value of x corresponds to exactly one y, the equation defines a function. Otherwise, it does not define a function.

Classroom Example Determine whether the equation $y = -3x + 5$ is a function.

Answer: A function

EXAMPLE 3 Determining If a Linear Equation Represents a Function

Determine whether or not the equation $y = 2x + 7$ is a function.

Solution

The rule for getting from x to y is to multiply x by 2 and then add 7. Since there is only one output y that can result by performing these operations on any given input x, the equation is a function. ■

Classroom Example Determine whether the equation $y = \pm 5x$ is a function.

Answer: Not a function

EXAMPLE 4 Determining If an Equation Represents a Function

Determine whether or not the equation $y = \pm\sqrt{x}$ is a function.

Solution

Notice that for any single value of x greater than 0, two values of y result. For example, if $x = 4$, then $y = \pm 2$. Because a single x corresponds to more than one y, the equation is not that of a function.

QUICK ✓ *Determine whether each equation is that of a function.*

7. $y = -5x + 1$ **8.** $y = \pm 6x$ **9.** $y = x^2 + 3x$ **10.** $x^2 + y^2 = 4$

> **Work Smart: Study Skills**
> Here are two student definitions of a function:
>
> 1. A **function** is a relation in which each input matches up with only one output.
> 2. A **function** is a relation in which every input appears only once in the correspondence.
>
> Each of these definitions is correct. How do these definitions compare with the one presented in this section? How will they help you remember the definition of a function?

3 Identify the Graph of a Function Using the Vertical Line Test

The graph of an equation is the set of all ordered pairs (x, y) that satisfy the equation. Not all graphs represent functions. In a function, each number x in the domain corresponds to exactly one number y in the range. This means that the graph of an equation will *not* represent a function if two points with the same x-coordinate have different y-coordinates.

In Words
If any vertical line intersects a graph at more than one point, the graph does not describe a function.

> **VERTICAL LINE TEST**
> A set of points in the coordinate plane is the graph of a function if and only if every vertical line intersects the graph in at most one point.

Classroom Example Which of the graphs below are functions?

(a)

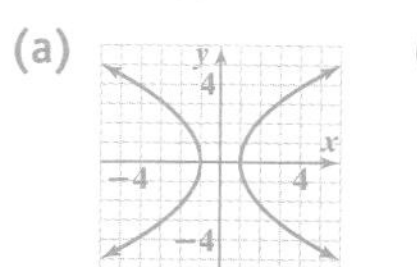

(b)

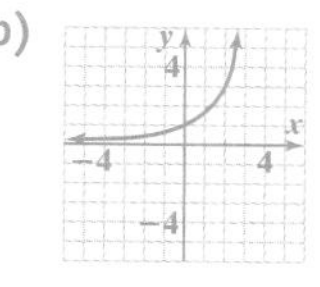

Answer:
(a) Not a function (b) Function

EXAMPLE 5 Using the Vertical Line Test to Identify Graphs of Functions

Which of the graphs in Figure 15 are graphs of functions?

Figure 15

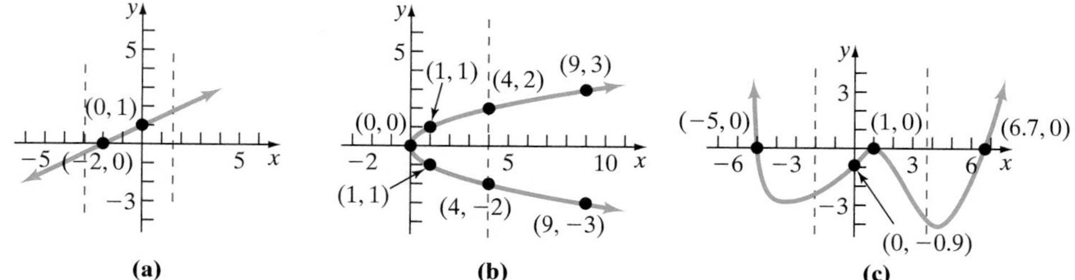

Solution

The graph in Figure 15(a) is a function, because every vertical line intersects the graph in at most one point. The graph in Figure 15(b) is not a function, because a vertical line intersects the graph in more than one point. The graph in Figure 15(c) is a function, because a vertical line intersects the graph in at most one point.

QUICK ✓ *Use the vertical line test to determine whether the graph is that of a function.*

11.

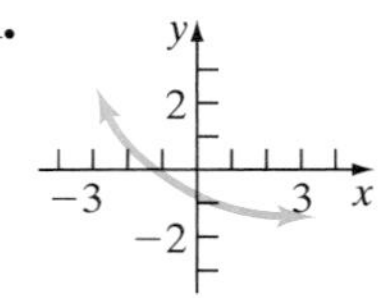

12.

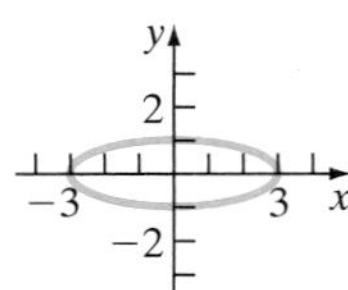

4 Find the Value of a Function

A relation is a function if each input corresponds to exactly one output. A common way of expressing functions is by an equation that describes the relationship between two variables. When a function is written as an equation, we use a special notation, called *function notation,* to represent the function. The following table shows the similarity between an equation in two variables that is solved for y and function notation.

Equation in Two Variables	Function Notation
$y = \frac{1}{2}x + 5$	$f(x) = \frac{1}{2}x + 5$
$y = -3x + 1$	$f(x) = -3x + 1$
$y = 2x^2 - 3x$	$f(x) = 2x^2 - 3x$

Teaching Tip
Be sure to emphasize that $f(x)$ does not mean "f times x."

Observe that the expression $f(x)$ replaces the variable y. Functions are often denoted by letters such as f, F, g, G and so on. If f is a function, then for each number x in its domain, the corresponding value in the range is denoted $f(x)$, read as "f of x" or as "f at x." Note that $f(x)$ does not mean "f times x."

We call $f(x)$ the *value of f at the number x*—$f(x)$ is the number that results when the function is applied to x. Finding the value of a function parallels finding the value of y in an equation in two variables.

Equation in Two Variables	Function Notation
Given the equation $y = \frac{1}{2}x + 5$, find the value of y when $x = -2$ and $x = 4$.	Given the function $f(x) = \frac{1}{2}x + 5$, evaluate $f(-2)$ and $f(4)$.
$y = \frac{1}{2}(-2) + 5 = -1 + 5 = 4$ $y = \frac{1}{2}(4) + 5 = 2 + 5 = 7$	$f(-2) = \frac{1}{2}(-2) + 5 = -1 + 5 = 4$ $f(4) = \frac{1}{2}(4) + 5 = 2 + 5 = 7$

In the first column, we say "the value of y when $x = -2$ is 4." In the second column, we say "f of -2 equals 4." In both cases, we replace the value of x with the indicated number, and evaluate.

Classroom Example
For the function defined by $f(x) = 2x - 5$, evaluate:
(a) $f(3)$ (b) $f(-2)$
Answer:
(a) 1 (b) -9

EXAMPLE 6 Finding Values of a Function Defined by a Linear Equation

For the function defined by $G(x) = 3x - 8$, evaluate:

(a) $G(4)$ **(b)** $G(-3)$

Solution

(a) Wherever we see an x in the equation defining the function G, we substitute 4 to get

$$\begin{aligned} G(4) &= 3(4) - 8 \\ &= 12 - 8 \\ &= 4 \end{aligned}$$

(b) We substitute -3 for x in the expression $3x - 8$ to get

$$\begin{aligned} G(-3) &= 3(-3) - 8 \\ &= -9 - 8 \\ &= -17 \end{aligned}$$

Classroom Example ➔
For the function defined by $f(x) = x^2 - 5x$, evaluate:
(a) $f(4)$ (b) $f(-3)$
Answer:
(a) -4 (b) 24

EXAMPLE 7 Finding Values of a Function

For the function defined by $f(x) = x^2 + 2x$, evaluate:

(a) $f(3)$ **(b)** $f(-2)$

Solution

(a) Wherever we see an x in the equation defining the function f, we substitute 3 to get

$$\begin{aligned} f(3) &= 3^2 + 2(3) \\ &= 9 + 6 \\ &= 15 \end{aligned}$$

(b) We substitute -2 for x in the expression $x^2 + 2x$ to get

$$\begin{aligned} f(-2) &= (-2)^2 + 2(-2) \\ &= 4 + (-4) \\ &= 0 \end{aligned}$$

QUICK ✓ *Let $f(x) = 4x + 1$ and $g(x) = 2x^2 + 5x - 3$ to evaluate each function.*

13. $f(4)$ **14.** $f(-2)$ **15.** $g(2)$ **16.** $g(-5)$

SUMMARY: Important Facts about Functions

1. For each x in the domain, there corresponds exactly one y in the range.
2. We use $f(x)$ to denote the function. It represents the value of the function at x.

10.3 Exercises

For Extra Help: Student Solutions Manual | CD Video | PH Math/Tutor Center | MathXL Tutorials on CD | MathXL® | MyMathLab

Concepts and Vocabulary

In Problems 1–3, fill in the blanks.

1. function
2. vertical line test
3. f of x
4. False
5. True
6. True

1. A ________ is a relation in which each element in the domain of the relation corresponds to exactly one element in the range of the relation.

2. The ________ ________ ________ states that a set of points in the coordinate plane is the graph of a function if and only if every vertical line intersects the graph in at most one point.

3. The notation $f(x)$ is read ________ ________ ________.

In Problems 4–6, answer True or False to each statement.

4. Every relation is a function.

5. Graphs that represent functions must have an imaginary vertical line that intersects the graph in at most one point.

6. If all of the elements of the domain of a relation correspond to the same element in the range of the relation, then that relation is a function.

7. Maps, ordered pairs, equations, graphs; answers may vary.
8. Answers may vary.
9. not a function
10. function
Domain: {English, Math, Psychology, History}
Range: {125, 85, 120}
11. function
Domain: {Apples, Pears, Oranges, Grapes}
Range: {$0.59, $0.39, $1.69}
12. not a function
13. function
Domain: $\{-1, 0, 1, 2\}$
Range: $\{-1, 0, 1, 2\}$
14. function
Domain: $\{3, 2, 0, -1\}$
Range: $\{1\}$
15. not a function
16. not a function
17. function
Domain: $\{a, b, c, d\}$
Range: $\{a, b\}$
18. function
Domain: $\{a, b, c, d\}$
Range: $\{z\}$
19. function
20. function
21. not a function
22. not a function
23. function
24. function
25. function
26. function
27. not a function
28. function
29. not a function
30. not a function
31. not a function
32. not a function
33. not a function
34. not a function
35. function

7. We described four ways to represent a function. What are they? Give an example of each.

8. In your own words, explain why the vertical line test can be used to identify the graph of a function. Show two cases of graphs that are functions and two cases of graphs that are not functions.

Building Skills

In Problems 9–12, determine whether each relation represents a function. If the relation is a function, state its domain and range.

9.

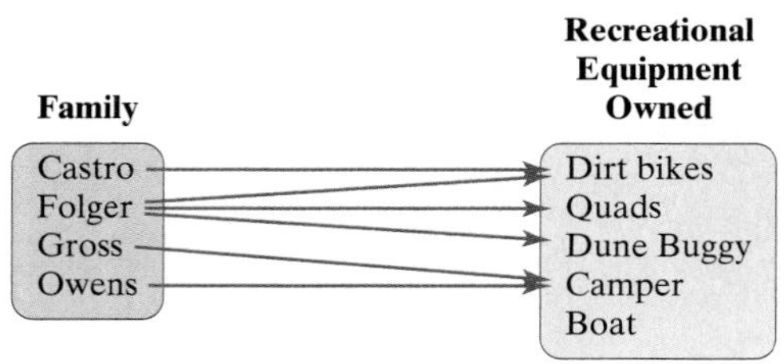

10.

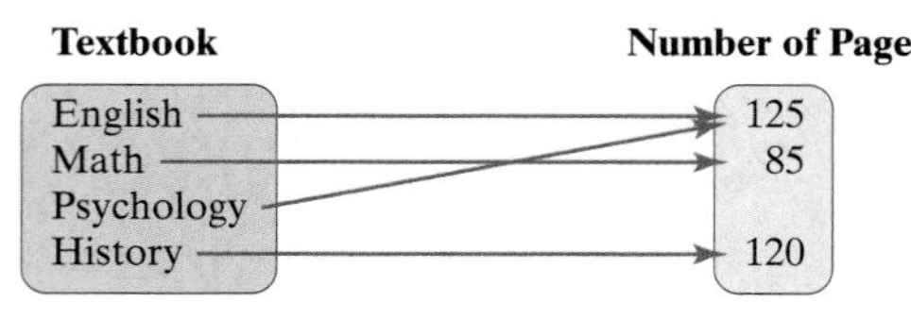

11.

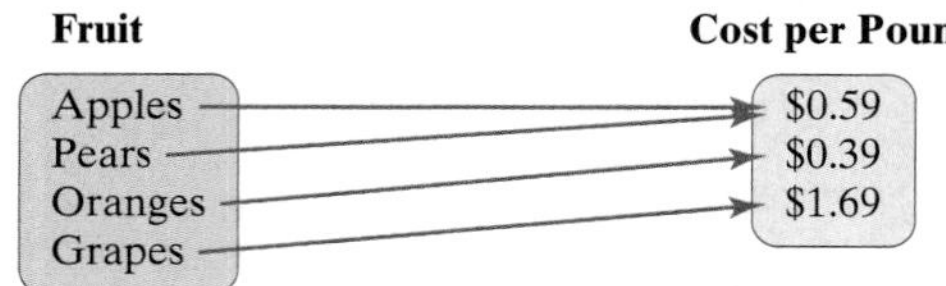

12.

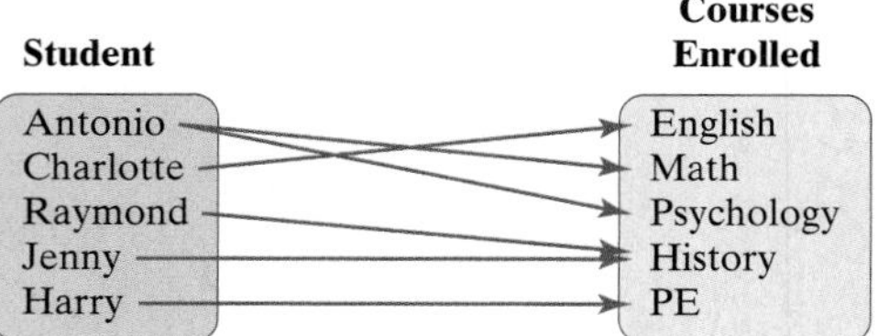

In Problems 13–18, determine whether each relation represents a function. If the relation is a function, state its domain and range.

13. $\{(-1, -1); (0, 0); (1, 1); (2, 2)\}$

14. $\{(3, 1); (2, 1); (0, 1); (-1, 1)\}$

15. $\{(-1, 1); (0, 0); (1, -1); (-1, -2)\}$

16. $\{(5, 0); (3, 1); (-1, 2); (3, 2)\}$

17. $\{(a, a); (b, a); (c, b); (d, b)\}$

18. $\{(a, z); (b, z); (c, z); (d, z)\}$

In Problems 19–32, determine whether each equation represents a function.

19. $3x - y = 2$

20. $2x + 3y = 1$

21. $x = 3$

22. $x = -4$

23. $y = -1$

24. $y = 0$

25. $y = -x^2 + 2x - 3$

26. $y = x^2 - 3x + 2$

27. $x = y^2$

28. $y = \sqrt{x}$

29. $y = \pm x$

30. $y = \pm\sqrt{x}$

31. $x^2 + y^2 = 9$

32. $x^2 - y^2 = 1$

In Problems 33–46, determine whether the graph is that of a function.

33.

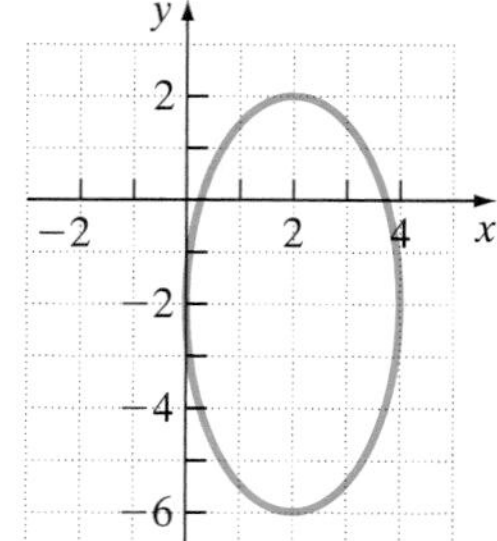

34.

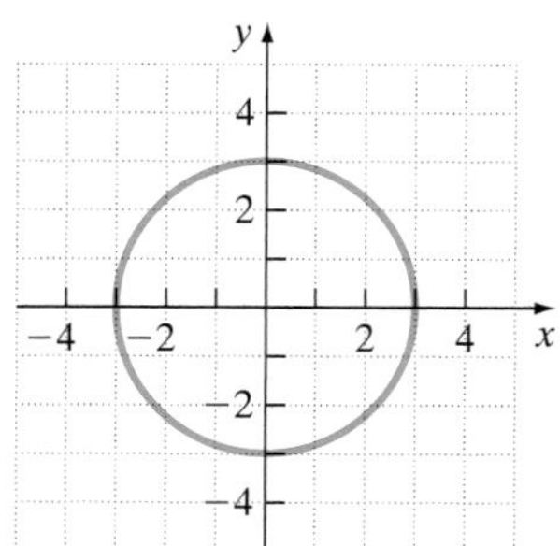

35.

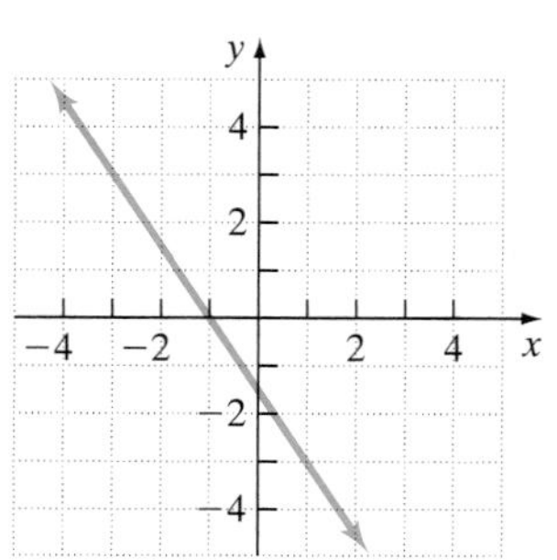

36. function
37. function
38. function
39. not a function
40. function
41. function
42. not a function
43. not a function
44. function
45. function
46. function
47. (a) -3 (b) -6 (c) -1
48. (a) 4 (b) 10 (c) 0
49. (a) 4 (b) -2 (c) 8
50. (a) 1 (b) -2 (c) 3
51. (a) 3 (b) 3 (c) 3
52. (a) -1 (b) -1 (c) -1
53. (a) -3 (b) $-\frac{3}{2}$ (c) -4
54. (a) 4 (b) 2 (c) $\frac{16}{3}$
55. (a) 3 (b) -12 (c) 13
56. (a) -3 (b) 9 (c) -11
57. (a) 0 (b) 0 (c) 10
58. (a) 4 (b) 13 (c) 8
59. (a) 1 (b) -32 (c) 3
60. (a) -3 (b) -18 (c) -13
61. 18
62. 0
63. -8
64. -2
65. 0
66. 15
67. 2
68. 3
69. $-\frac{1}{2}$
70. $\frac{2}{3}$

36.

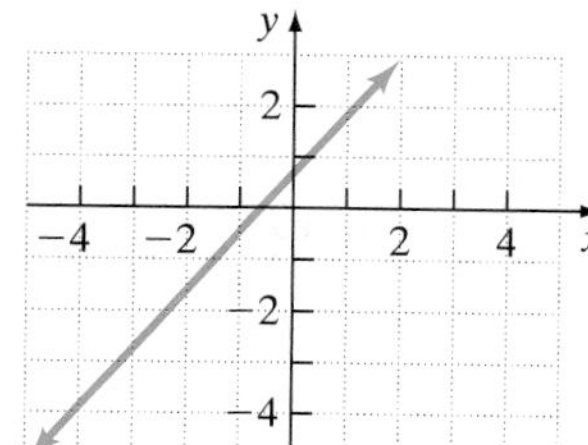

37.

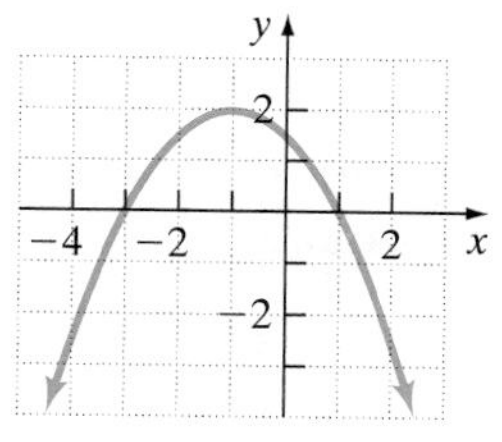

38.

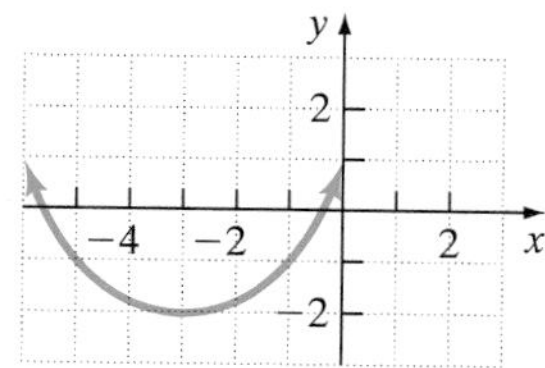

39.

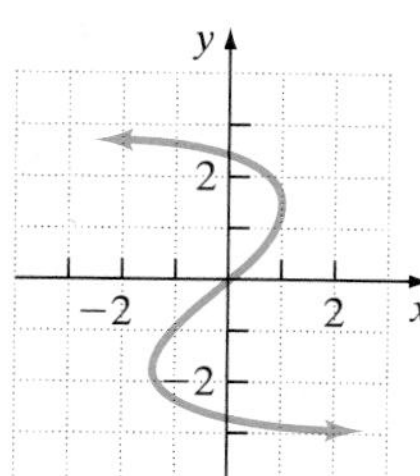

40.

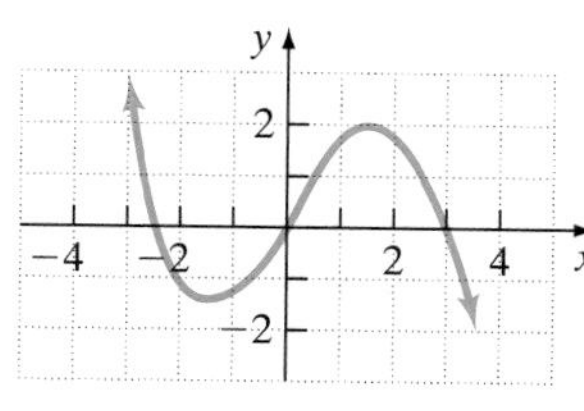

41.

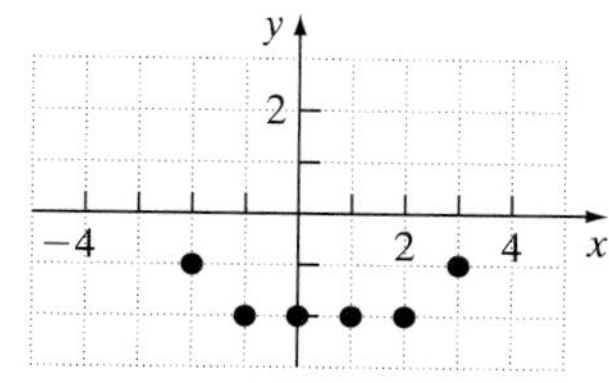

42.

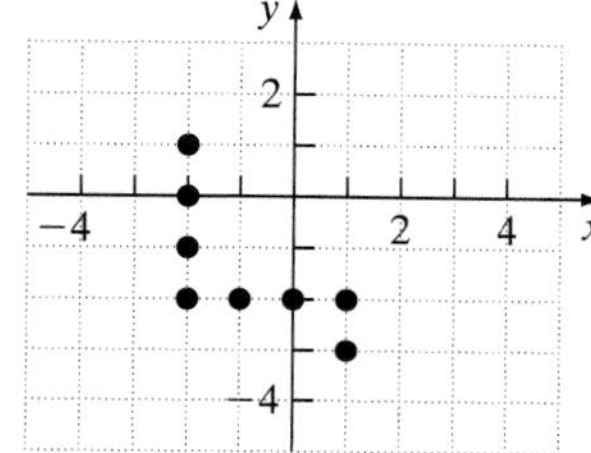

43.

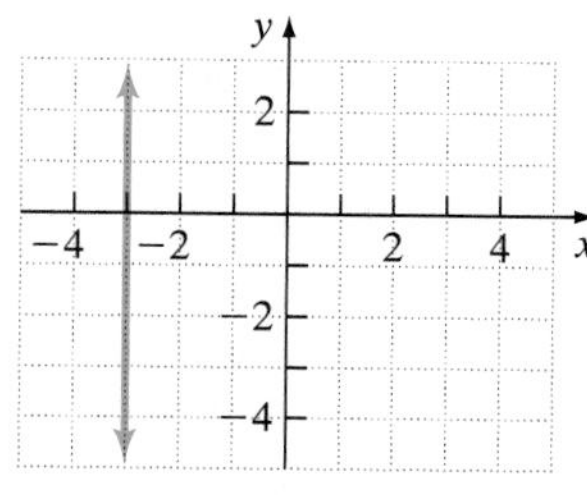

44.

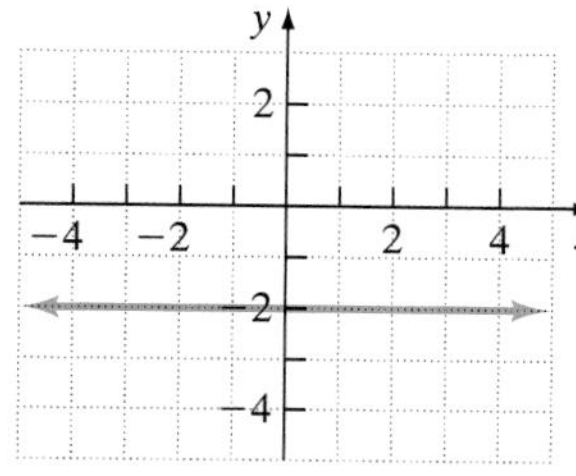

45.

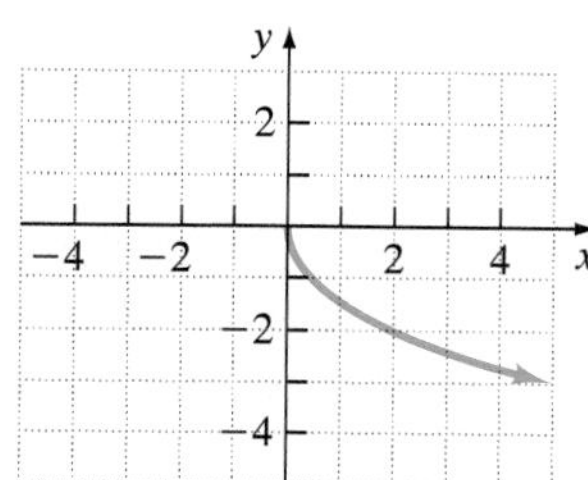

46.

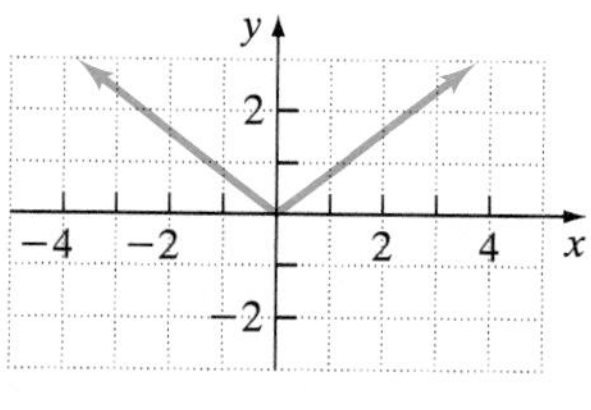

In Problems 47–60, find the following values for each function:

(a) $f(0)$ **(b)** $f(3)$ **(c)** $f(-2)$

47. $f(x) = -x - 3$
48. $f(x) = 2x + 4$
49. $f(x) = 4 - 2x$
50. $f(x) = 1 - x$
51. $f(x) = 3$
52. $f(x) = -1$
53. $f(x) = \frac{1}{2}x - 3$
54. $f(x) = -\frac{2}{3}x + 4$
55. $f(x) = 3 - 5x$
56. $f(x) = -3 + 4x$
57. $f(x) = x^2 - 3x$
58. $f(x) = x^2 + 4$
59. $f(x) = -2x^2 - 5x + 1$
60. $f(x) = -2x^2 + x - 3$

In Problems 61–74, find the value of each function.

61. $f(x) = x^2 - 3x; f(-3)$
62. $f(x) = -x^2 + 4x; f(4)$
63. $g(x) = -2x^2 + x - 5; g(-1)$
64. $g(x) = 3x^2 - 2; g(0)$
65. $a(x) = 2 + x - x^2; a(-1)$
66. $a(x) = 3 - x + x^2; a(-3)$
67. $h(x) = \sqrt{x + 2}; h(2)$
68. $h(x) = \sqrt{2x - 1}; h(5)$
69. $h(x) = \dfrac{-2}{2x + 3}; h\left(\dfrac{1}{2}\right)$
70. $h(x) = \dfrac{-4}{5x - 1}; h(-1)$

71. 7
72. 0
73. $2\sqrt{3}$
74. 4
75. $C = 5$
76. $C = 4$
77. $C = 3$
78. $C = -6$
79. $A = 5$
80. $B = 5$
81. $G(h) = 12h$; \$240
82. $G(p) = 0.10p + 350$; \$500
83. (a) $h(30) = 8.132$; in 1930, there were 8.132 homicides per 100,000 people.
(b) $h(50) = 6.632$; in 1950, there were 6.632 homicides per 100,000 people.
84. (a) $p(30) = 12.408$; 12.4% of the workers earn \$30,000.
(b) $p(60) = 3.795$; 3.8% of the workers earn \$60,000.
(c) $p(90) = 0.78$; 0.78% of the workers earn \$90,000.
85. $A(r) = \pi r^2$; $A(3) = 9\pi$ in.^{2}
86. $A(h) = \frac{5}{2}h$; $A(8) = 20$ cm^2
87. $x = 2$
88. $x = -\frac{2}{3}$
89. $x = 2$
90. $x = -18$
91. $x = -4$
92. $x = -8$
93. (a) -5 (b) $1 - 2h$ (c) $-5 - 2h$ (d) $-4 - 2h$ (e) $-2h$
94. (a) -1 (b) $5 - 3h$ (c) $-1 - 3h$ (d) $4 - 3h$ (e) $-3h$

71. $f(x) = |3x^2 - 10|; f(-1)$

72. $f(x) = |2x^2 - 8|; f(-2)$

73. $g(x) = \sqrt{16 - x^2}; g(-2)$

74. $g(x) = \sqrt{25 - x^2}; g(3)$

Applying the Concepts

75. If $f(x) = 2x + C$ and $f(3) = 11$, what is the value of C?

76. If $f(x) = 3x - C$ and $f(-2) = -10$, what is the value of C?

77. If $f(x) = -2x^2 + 5x + C$ and $f(-2) = -15$, what is the value of C?

78. If $f(x) = 3x^2 - x + C$ and $f(3) = 18$, what is the value of C?

79. If $f(x) = \dfrac{2x + 5}{x - A}$ and $f(0) = -1$, what is the value of A?

80. If $f(x) = \dfrac{-x + B}{x - 5}$ and $f(3) = -1$, what is the value of B?

81. Salary Express the gross salary, G, of Jackie, who earns \$12 per hour as a function of the number of hours worked, h. Determine the gross salary of Jackie if she works 20 hours. That is, find $G(20)$.

82. Commission Roberta is a commissioned salesperson. She earns a base weekly salary of \$350 per week plus a 10% commission on her sales. Express her gross salary, G, as a function of the price, p, of the items sold. Determine the gross salary of Roberta if the value of the items sold is \$1500. That is, find $G(1500)$.

83. Homicide Rate The number of homicides in the United States can be modeled by the function $h(x) = -0.00535x^2 + 0.353x + 2.357$, where x is the number of years after 1900 and h is the number of homicides per 100,000 people.

(a) Find $h(30)$ and interpret your results.
(b) Find $h(50)$ and interpret your results.

84. Salary The percent of workers, p, who earn a given salary, x (in thousands of dollars), can be modeled by the function $p(x) = 0.00311x^2 - 0.567x + 26.619$.

(a) Find $p(30)$ and interpret your results.
(b) Find $p(60)$ and interpret your results.
(c) Find $p(90)$ and interpret your results.

△ **85. Area of a Circle** Express the area, A, of a circle as a function of its radius, r. Determine the area of a circle whose radius is 3 inches. That is, find $A(3)$.

△ **86. Area of a Triangle** Express the area, A, of a triangle as a function of its height, h, assuming that the length of the base of the triangle is 5 centimeters. Determine the area of this triangle if its height is 8 centimeters. That is, find $A(8)$.

Extending the Concepts

In Problems 87–92, the value of a function is given. Find the value of x that corresponds to the given function value.

87. $f(x) = 2x + 5$; find x if $f(x) = 9$.

88. $f(x) = 3x - 8$; find x if $f(x) = -10$.

89. $f(x) = 2 - \frac{1}{2}x$; find x if $f(x) = 1$.

90. $f(x) = -4 - \frac{2}{3}x$; find x if $f(x) = 8$.

91. $f(x) = \dfrac{12}{x + 2}$; find x if $f(x) = -6$.

92. $f(x) = \dfrac{x}{x + 4}$; find x if $f(x) = 2$.

In Problems 93 and 94, find the values for the given function.

93. $f(x) = 1 - 2x$

(a) $f(3)$
(b) $f(h)$
(c) $f(h + 3)$
(d) $f(h) + f(3)$
(e) $f(h + 3) - f(3)$

94. $f(x) = 5 - 3x$

(a) $f(2)$
(b) $f(h)$
(c) $f(h + 2)$
(d) $f(h) + f(2)$
(e) $f(h + 2) - f(2)$

CHAPTER 10 ACTIVITY: DISCOVERING SHIFTING

2.

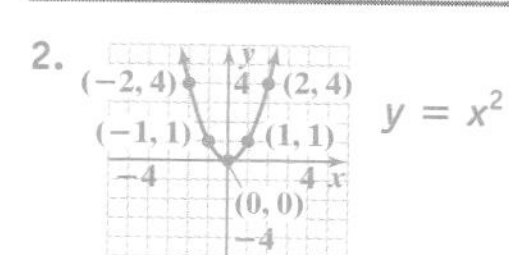

$y = x^2$

3. (a) $y = x^2 + 4$ (b) $y = x^2 - 4$

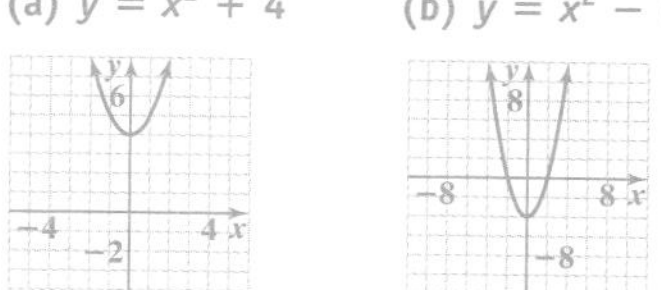

(c) $y = (x - 4)^2$ (d) $y = (x + 4)^2$

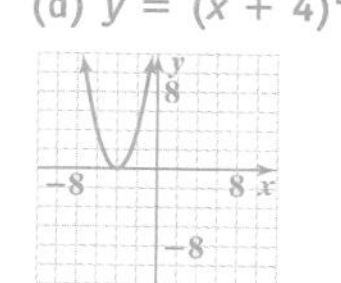

5. (a) $y = (x - 2)^2$ (b) $y = x^2 - 2$

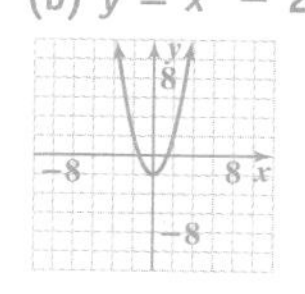

(c) $y = (x + 2)^2$ (d) $y = x^2 + 2$

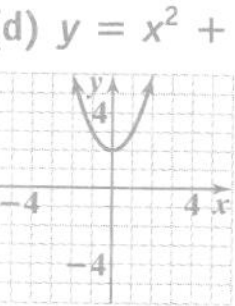

Focus: Discovering the rules for graphing quadratic equations by shifting

Time: 20–30 minutes

Group size: 3–4

Materials needed: Graph paper

1. Each member of the group needs to draw a coordinate plane and label the axes.
2. Each member of the group needs to graph the primary equation $y = x^2$.
3. On the same coordinate plane, each group member should choose one of the following quadratic equations and graph by plotting points. Be sure your graph of the primary equation and one of the equations (a)–(d) are on the same coordinate plane.

   (a) $y = x^2 + 4$ **(b)** $y = x^2 - 4$ **(c)** $y = (x - 4)^2$ **(d)** $y = (x + 4)^2$
4. As a group, discuss the following:

   (a) What shape are the graphs?

   (b) Each member of the group, share the difference between your graph and the primary graph.

   (c) As a group, can you develop possible rules for these differences?
5. With these rules in mind, repeat the above procedure with the primary equation $y = x^2$ and one of the following equations. As a group, decide whether your rules held.

   (a) $y = (x - 2)^2$ **(b)** $y = x^2 - 2$ **(c)** $y = (x + 2)^2$ **(d)** $y = x^2 + 2$

CHAPTER 10 REVIEW

Section 10.1 Quadratic Equations in Two Variables

KEY CONCEPTS	KEY TERMS
• Any quadratic equation $y = ax^2 + bx + c, a \neq 0$, has a vertex whose x-coordinate is $-\frac{b}{2a}$ • The y-coordinate of the vertex is found by evaluating the equation at the x-coordinate of the vertex. • The axis of symmetry of a parabola is given by the equation $x = -\frac{b}{2a}$ **The x-intercepts of the Graph of a Quadratic Equation** **1.** If the discriminant $b^2 - 4ac > 0$, the graph of $y = ax^2 + bx + c$ has two different x-intercepts. The graph will cross the x-axis at the solutions to the equation $ax^2 + bx + c = 0$. **2.** If the discriminant $b^2 - 4ac = 0$, the graph of $y = ax^2 + bx + c$ has one x-intercept. The graph will touch the x-axis at the solution to the equation $ax^2 + bx + c = 0$. **3.** If the discriminant $b^2 - 4ac < 0$, the graph of $y = ax^2 + bx + c$ has no x-intercepts. The graph will not cross or touch the x-axis.	Quadratic equation in two variables Parabola Opens up Opens down Vertex Axis of symmetry Maximum value Minimum value

(continued)

YOU SHOULD BE ABLE TO . . .	EXAMPLE	REVIEW EXERCISES
(1) Graph a quadratic equation by plotting points (p. 687)	Examples 1 and 2	1–6
(2) Find the intercepts of the graph of a quadratic equation (p. 689)	Examples 3 and 4	7–14
(3) Find the vertex and axis of symmetry of the graph of a quadratic equation (p. 691)	Example 5	15–22
(4) Graph a quadratic equation using its vertex, intercepts, and axis of symmetry (p. 693)	Examples 6 and 7	23–28
(5) Determine the maximum or minimum value of a quadratic equation (p. 696)	Examples 8 and 9	29–34

In Problems 1–6, graph each quadratic equation by making a table of values and plotting the points.

1. $y = -x^2$ **2.** $y = x^2$ **3.** $y = 2x^2 - 3$

4. $y = -\frac{1}{2}x^2 + 4$ **5.** $y = x^2 - 4x - 3$ **6.** $y = -2x^2 + 4x + 1$

In Problems 7–14, find the intercepts of the graph.

7. $y = x^2 + x - 6$ **8.** $y = x^2 + 6x + 8$ **9.** $y = -4x^2 + 12x - 9$

10. $y = -9x^2 - 6x - 1$ **11.** $y = -4x^2 + 8x - 5$ **12.** $y = -3x^2 + 2x - 1$

13. $y = x^2 - 4x + 2$ **14.** $y = -x^2 - 4x + 14$

In Problems 15–22, for each quadratic equation, (a) determine whether the parabola opens up or down, (b) find the vertex of the parabola, (c) find the axis of symmetry.

15. $y = x^2 + 5$ **16.** $y = -x^2 - 4$ **17.** $y = -2x^2 + 4x - 1$

18. $y = 3x^2 - 12x + 4$ **19.** $y = \frac{3}{2}x^2 + 9x$ **20.** $y = -\frac{3}{2}x^2 - 6x$

21. $y = -x^2 + 3x - 4$ **22.** $y = x^2 + 5x + 5$

In Problems 23–28, graph each quadratic equation by using its properties.

23. $y = x^2 + 2x - 3$ **24.** $y = -x^2 + 6x - 8$ **25.** $y = -x^2 + 4$

26. $y = x^2 - 9$ **27.** $y = -x^2 + 3x + 4$ **28.** $y = x^2 - 5x + 4$

In Problems 29–32, determine whether the quadratic equation has a maximum or minimum value, then find the maximum or minimum value of the equation.

29. $y = -2x^2 - 8x + 7$ **30.** $y = 3x^2 + 12x - 4$

31. $y = x^2 - 4x - 6$ **32.** $y = -x^2 - 10x - 15$

33. Throwing a Baseball Vladimir Guerrero throws a baseball from right field to home plate. The height of the path of the baseball can be approximated by the following quadratic equation:

$$h = -0.00175x^2 + 0.325x + 5$$

where h is the height, in feet, of the baseball at any time during the throw and x is the distance, in feet, the ball has traveled.

(a) Determine the distance x (to the nearest tenth of a foot) that maximizes the height of the throw.
(b) Determine the maximum height (to the nearest tenth of a foot) of the path of the baseball.
(c) How far (to the nearest tenth of a foot) from Vladimir does the baseball land?

1. $y = -x^2$

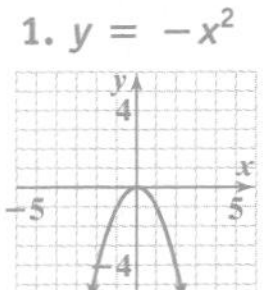

2. $y = x^2$

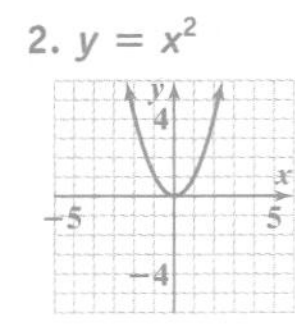

3. $y = 2x^2 - 3$

4. $y = -\frac{1}{2}x^2 + 4$

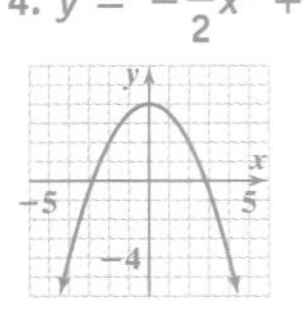

5. $y = x^2 - 4x - 3$

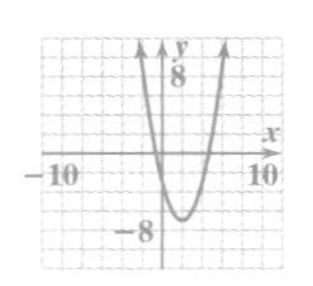

6. $y = -2x^2 + 4x + 1$

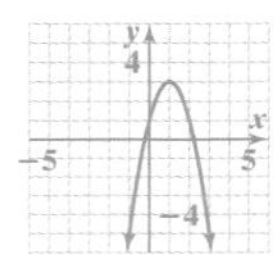

7. x-intercepts: $-3, 2$; y-intercept: -6
8. x-intercepts: $-2, -4$; y-intercept: 8
9. x-intercept: $\frac{3}{2}$; y-intercept: -9
10. x-intercept: $-\frac{1}{3}$; y-intercept: -1
11. x-intercepts: none; y-intercept: -5
12. x-intercepts: none; y-intercept: -1
13. x-intercepts: $2 - \sqrt{2}, 2 + \sqrt{2}$; y-intercept: 2
14. x-intercepts: $-2 - 3\sqrt{2}, -2 + 3\sqrt{2}$; y-intercept: 14
15. (a) up (b) $(0, 5)$ (c) $x = 0$
16. (a) down (b) $(0, -4)$ (c) $x = 0$
17. (a) down (b) $(1, 1)$ (c) $x = 1$
18. (a) up (b) $(2, -8)$ (c) $x = 2$
19. (a) up (b) $\left(-3, -\frac{27}{2}\right)$ (c) $x = -3$
20. (a) down (b) $(-2, 6)$ (c) $x = -2$
21. (a) down (b) $\left(\frac{3}{2}, -\frac{7}{4}\right)$ (c) $x = \frac{3}{2}$
22. (a) up (b) $\left(-\frac{5}{2}, -\frac{5}{4}\right)$ (c) $x = -\frac{5}{2}$
23. $y = x^2 + 2x - 3$ opens up; vertex $(-1, -4)$ axis of symmetry $x = -1$ x-intercepts: $-3, 1$ y-intercept: -3

24. $y = -x^2 + 6x - 8$
opens down; vertex (3, 1)
axis of symmetry $x = 3$
x-intercepts: 2, 4
y-intercept: -8

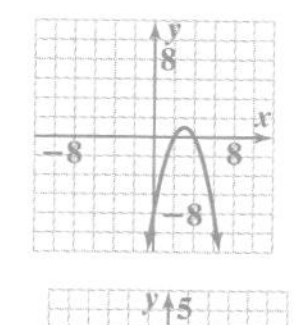

25. $y = -x^2 + 4$
opens down; vertex (0, 4)
axis of symmetry $x = 0$
x-intercepts: -2, 2
y-intercept: 4

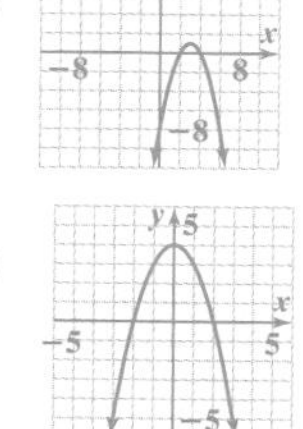

34. Enclosing a Pasture A farmer has 400 meters of fencing to enclose a rectangular pasture. The area, A, of the pasture is given by the quadratic equation

$$A = -x^2 + 200x$$

where x is the length of the pasture (in meters).

(a) Determine the length of the side x that maximizes the enclosed area.
(b) Determine the maximum area of the pasture that can be enclosed with 400 meters of fence.

Section 10.2 Relations

KEY CONCEPT

- **Relation**
A correspondence between two variables x and y where y depends on x. Relations can be represented through maps, sets of ordered pairs, equations, or graphs.

KEY TERMS

Relation
Corresponds to
Depends on
Map
Input
Output
Domain
Range

YOU SHOULD BE ABLE TO . . .	EXAMPLE	REVIEW EXERCISES
1 Define relations (p. 701)	Example 1	35, 36
2 Find the domain and the range of a relation (p. 702)	Examples 2 through 4	37–50
3 Graph a relation defined by an equation (p. 704)	Example 5	51–54

26. $y = x^2 - 9$
opens up;
vertex (0, -9)
axis of symmetry $x = 0$
x-intercepts: -3, 3
y-intercept: -9

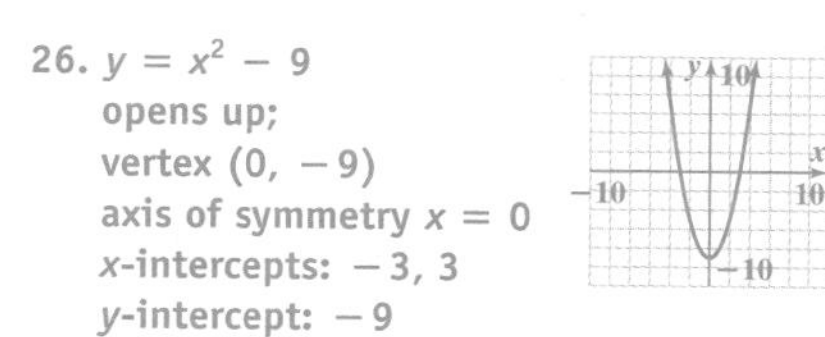

27. $y = -x^2 + 3x + 4$
opens down;
vertex $\left(\frac{3}{2}, \frac{25}{4}\right)$
axis of symmetry $x = \frac{3}{2}$
x-intercepts: -1, 4; y-intercept: 4

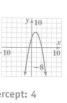

28. $y = x^2 - 5x + 4$
opens up;
vertex $\left(\frac{5}{2}, -\frac{9}{4}\right)$
axis of symmetry $x = \frac{5}{2}$
x-intercepts: 1, 4; y-intercept: 4

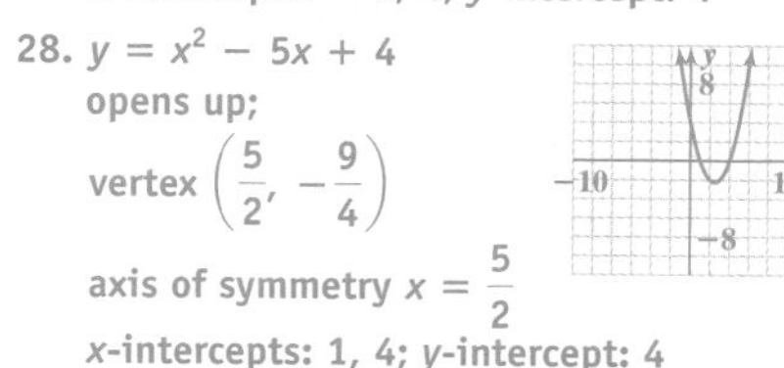

29. maximum; 15 at $x = -2$
30. minimum; -16 at $x = -2$
31. minimum; -10 at $x = 2$
32. maximum; 10 at $x = -5$
33. (a) 92.9 ft (b) 20.0 ft (c) 200 ft
34. (a) 100 m (b) 10,000 m^2
35. {(Alabama, 38), (California, 54), (Florida, 45), (Nebraska, 41), (New York, 55), (Ohio, 41), (South Dakota, 37)}
36. {(History, 400), (Math, 400), (Music, 200), (Psychology, 225)}
37. Domain: {Alabama, California, Florida, Nebraska, New York, Ohio, South Dakota}
Range: {37, 38, 41, 45, 54, 55}

In Problems 35 and 36, represent each relation as a set of ordered pairs.

35.

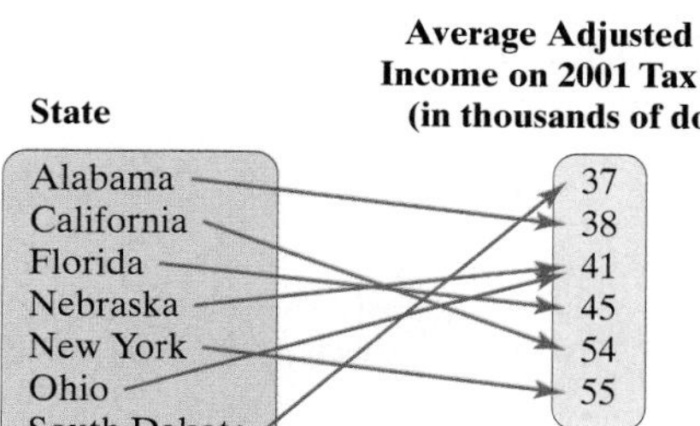

36.

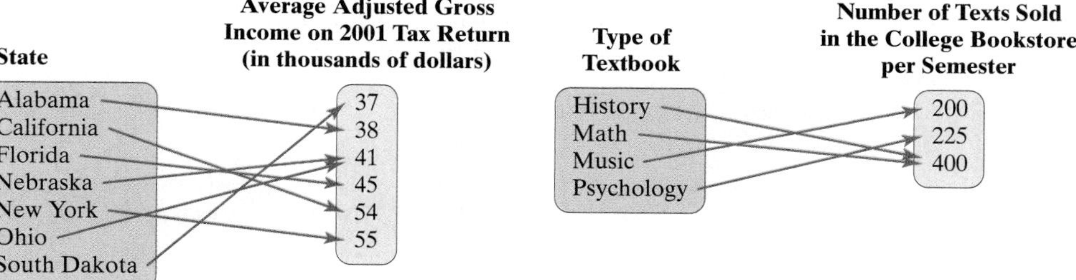

37. State the domain and range of the relation shown in Problem 35.

38. State the domain and range of the relation shown in Problem 36.

In Problems 39–50, identify the domain and range of each relation.

39. $\{(-1, 3); (-1, 5); (0, -2); (0, -4)\}$

40. $\{(a, x^2); (b, x^3); (c, x^4); (d, x^5)\}$

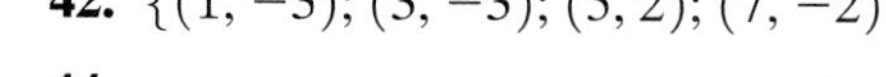

41. $\{(a_1, 1); (a_2, 4); (a_3, 9); (a_4, 16)\}$

42. $\{(1, -3); (3, -3); (5, 2); (7, -2)\}$

43.

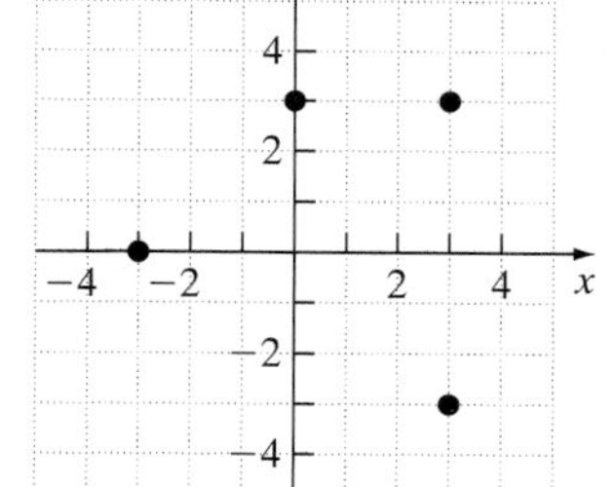

44.

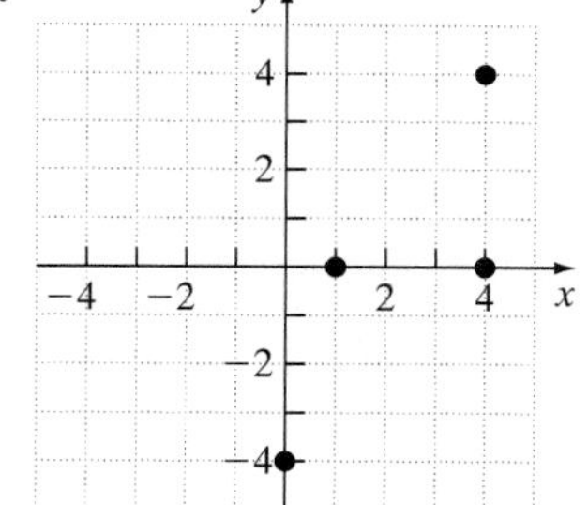

38. Domain: {History, Math, Music, Psychology}
Range: {200, 225, 400}
39. Domain: $\{-1, 0\}$
Range: $\{3, 5, -2, -4\}$
40. Domain: $\{a, b, c, d\}$
Range: $\{x^2, x^3, x^4, x^5\}$
41. Domain: $\{a_1, a_2, a_3, a_4\}$
Range: {1, 4, 9, 16}
42. Domain: {1, 3, 5, 7}
Range: $\{-3, 2, -2\}$
43. Domain: $\{-3, 0, 3\}$
Range: $\{-3, 0, 3\}$
44. Domain: {0, 1, 4}
Range: $\{-4, 0, 4\}$
45. Domain: $\{x \mid x \text{ is a real number}\}$
Range: $\{y \mid y \text{ is a real number}\}$
46. Domain: $\{x \mid x \text{ is a real number}\}$
Range: {4}
47. Domain: {5}
Range: $\{y \mid y \text{ is a real number}\}$
48. Domain: $\{x \mid x \text{ is a real number}\}$
Range: $\{y \mid y \text{ is a real number}\}$
49. Domain: $\{x \mid 0 \leq x \leq 6\}$
Range: $\{y \mid 1 \leq y \leq 5\}$
50. Domain: $\{x \mid -4 \leq x \leq 0\}$
Range: $\{y \mid -4 \leq x \leq 3\}$
51. $y = \dfrac{-3x}{2} + 4$

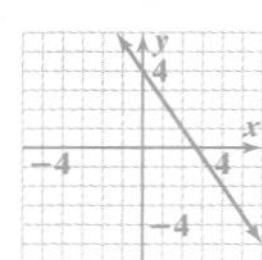

Domain: $\{x \mid x \text{ is a real number}\}$
Range: $\{y \mid y \text{ is a real number}\}$

52. $y = \dfrac{4x}{3} - 5$

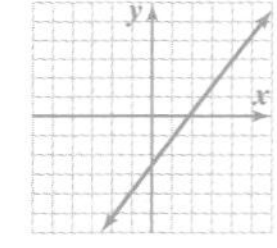

Domain: $\{x \mid x \text{ is a real number}\}$
Range: $\{y \mid y \text{ is a real number}\}$

45.

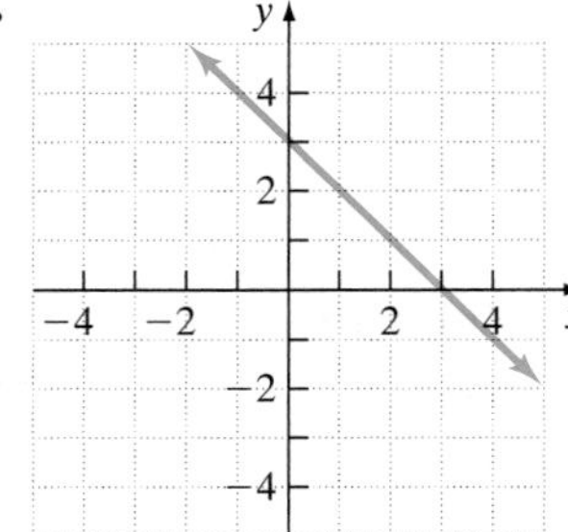

46.

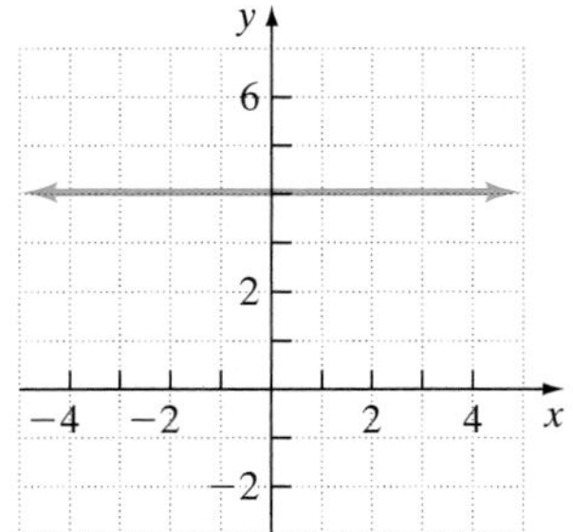

47.

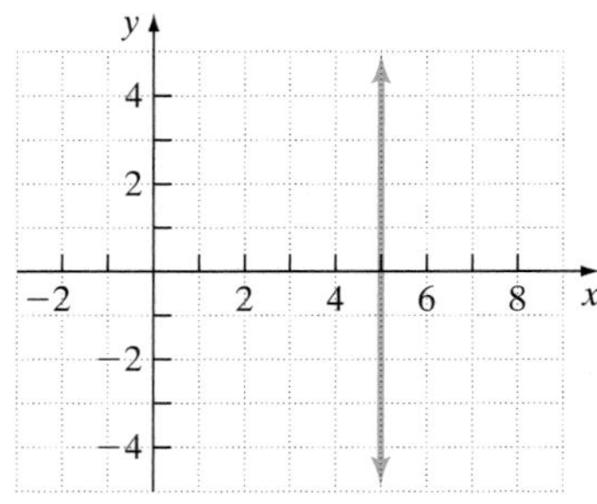

48.

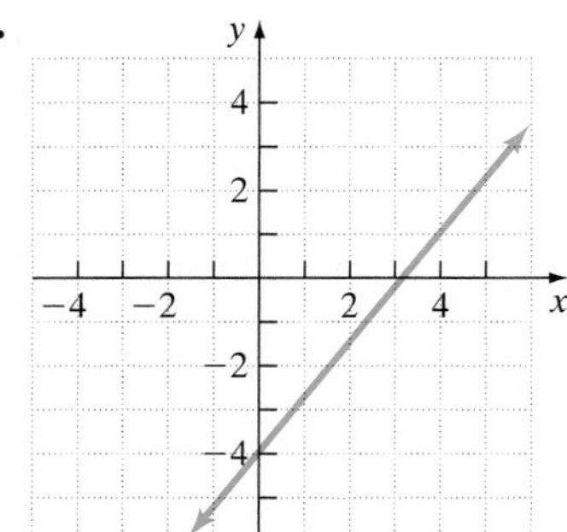

49.

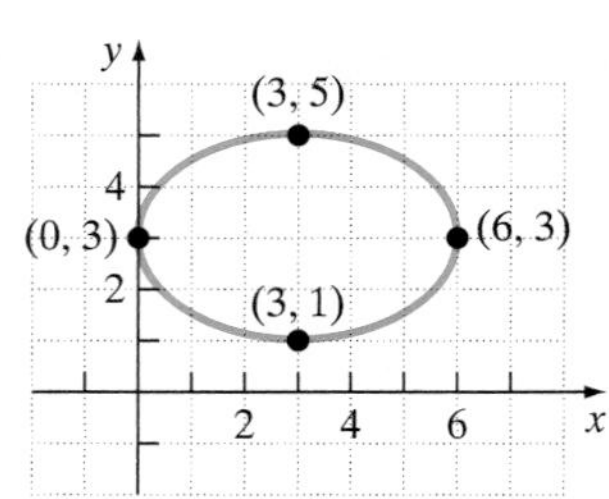

50.

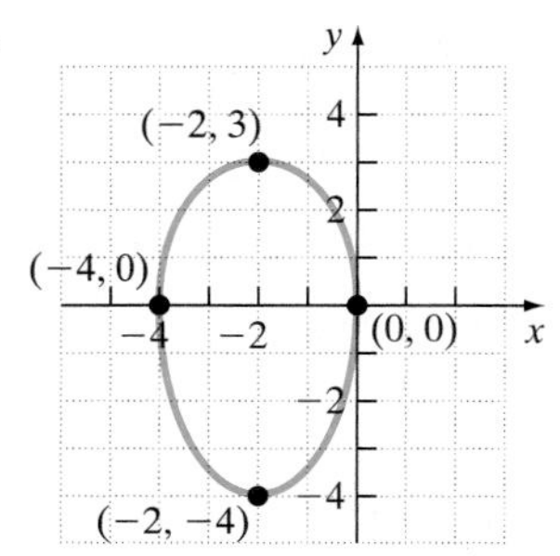

In Problems 51–54, graph each relation. Use the graph to identify the domain and the range of the relation.

51. $y = \dfrac{-3x}{2} + 4$ **52.** $y = \dfrac{4x}{3} - 5$ **53.** $y = 8x - 4x^2$ **54.** $y = x^2 - 4x + 3$

Section 10.3 An Introduction to Functions

KEY CONCEPTS	KEY TERMS
• **Function** A relation in which each element in the domain of the relation corresponds to exactly one element in the range. Functions can be represented through maps, sets of ordered pairs, equations, or graphs. • **Vertical Line Test** A set of points in the xy-plane is the graph of a function if and only if every vertical line intersects the graph in at most one point.	Function Vertical line test Value of f at the number x

YOU SHOULD BE ABLE TO . . .	EXAMPLE	REVIEW EXERCISES
(1) Determine whether or not a relation expressed as a map or ordered pairs represents a function (p. 709)	Examples 1 and 2	55–60
(2) Determine whether or not a relation expressed as an equation represents a function (p. 712)	Examples 3 and 4	61–70
(3) Identify the graph of a function using the vertical line test (p. 713)	Example 5	71–78
(4) Find the value of a function (p. 714)	Examples 6 and 7	79–87

53. $y = 8x - 4x^2$

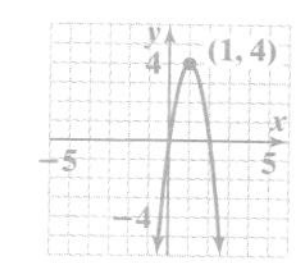

Domain: $\{x \mid x \text{ is a real number}\}$
Range: $\{y \mid y \leq 4\}$

54. $y = x^2 - 4x + 3$

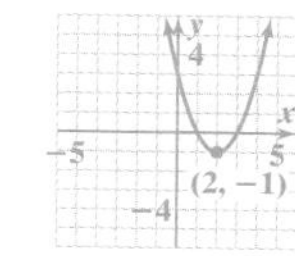

Domain: $\{x \mid x \text{ is a real number}\}$
Range: $\{y \mid y \geq -1\}$

55. not a function

56. function; Domain: {Less than High School Degree, High School Graduate, Some College/AA Degree, Bachelor's Degree and more}
Range: {3.77, 3.88, 4.22}

57. function; Domain: $\{-4, -3, -2, -1\}$
Range: {0, 1}

58. not a function 59. not a function

60. function; Domain: $\{5, 3, 1, -1\}$;
Range: {x, z}

61. function 62. not a function
63. not a function 64. function
65. not a function 66. function
67. function 68. not a function
69. not a function 70. function
71. not a function 72. function
73. function 74. not a function
75. function 76. function

In Problems 55–60, determine whether each relation represents a function. If the relation is a function, state its domain and range.

55.

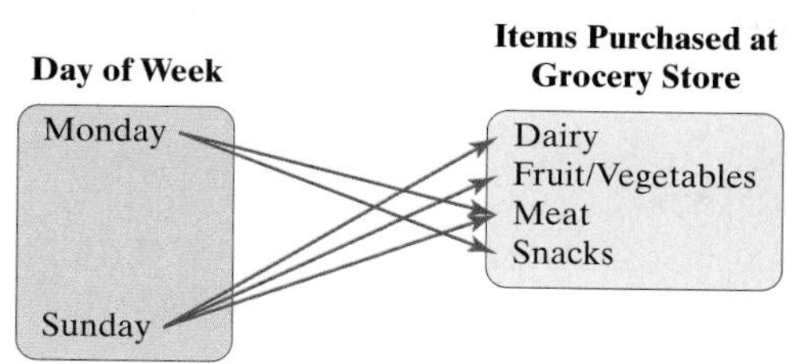

56.

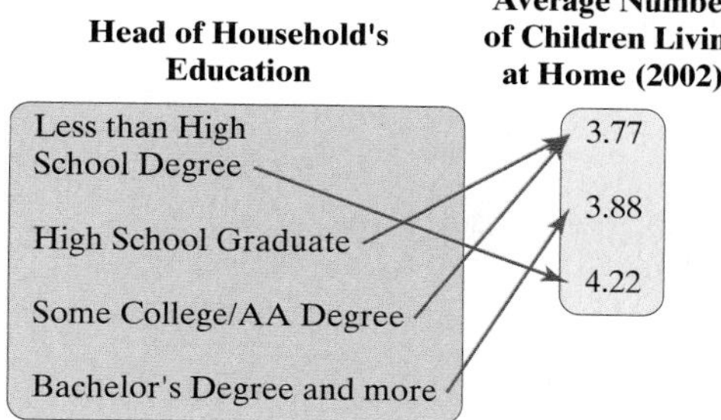

57. $\{(-4, 0); (-3, 0); (-2, 1); (-1, 1)\}$

58. $\{(1, 4); (3, 3); (1, 2); (-1, 1)\}$

59. $\{(a, 9); (b, 8); (b, 7); (a, 6)\}$

60. $\{(5, z); (3, x); (1, x); (-1, z)\}$

In Problems 61–70, determine whether each equation represents a function.

61. $x = y + 2$ **62.** $x = y^2$ **63.** $x = -3$ **64.** $y = 2x$

65. $y = \pm\sqrt{x}$ **66.** $y = 7$ **67.** $y = x^2 + 4x$ **68.** $x^2 + y^2 = 25$

69. $x^2 - y^2 = 16$ **70.** $x^2 = y - 1$

In Problems 71–78, use the vertical line test to determine whether the graph is that of a function.

71.

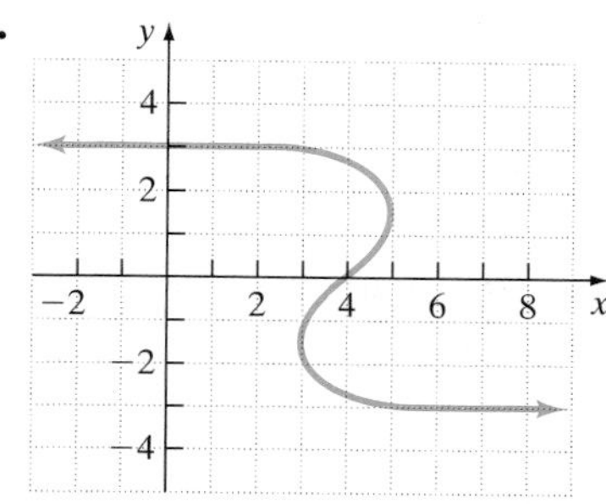

72.

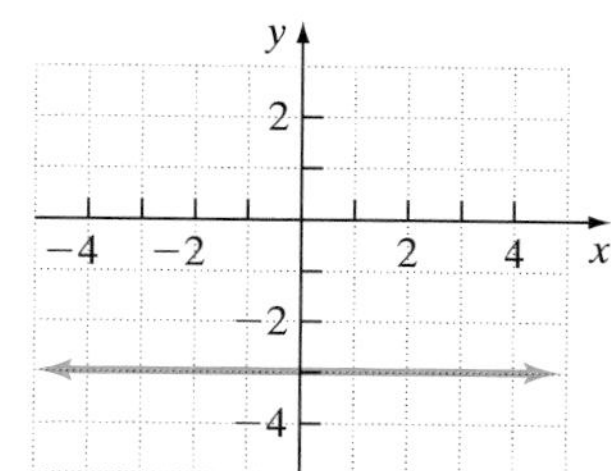

73.

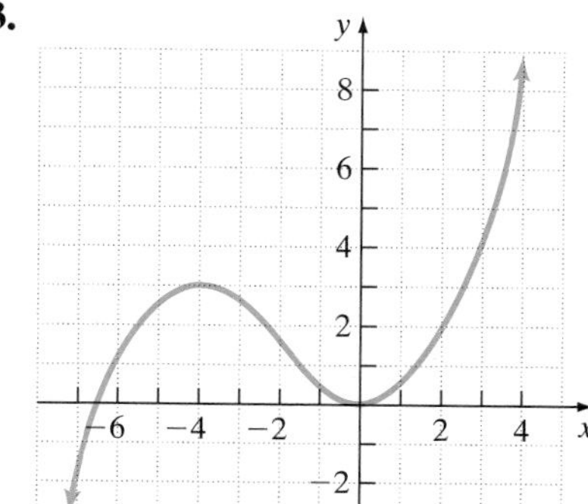

74.

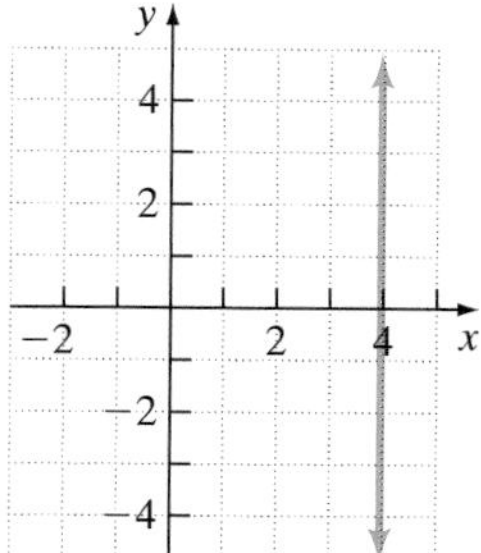

75.

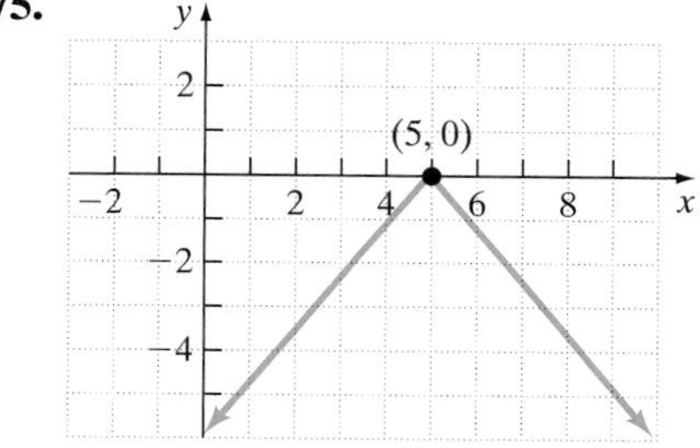

76.

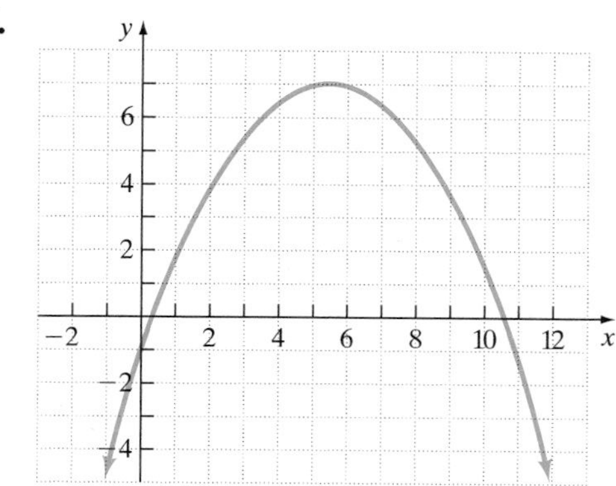

77. not a function
78. not a function
79. 1
80. 18
81. -3
82. -10
83. 14
84. 7
85. $2\sqrt{6}$
86. 5
87. (a) $C(m) = 0.10m + 75$
(b) \$88

77.

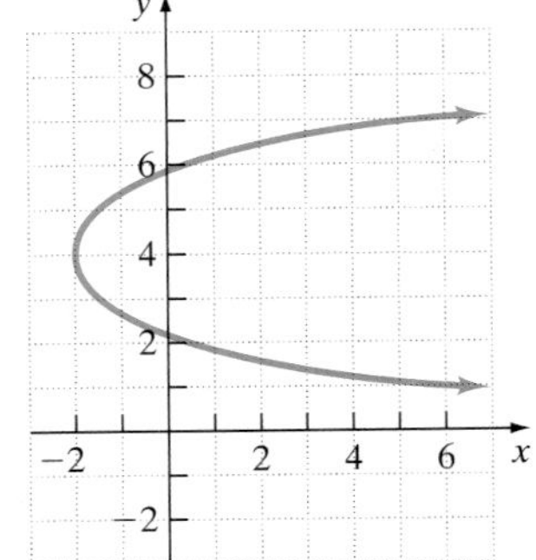

78.

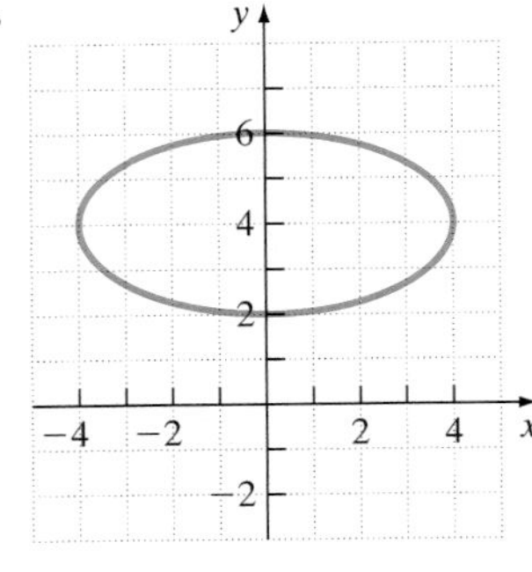

In Problems 79–86, find the following values for each function.

79. $f(x) = -3x - 2; f(-1)$

80. $g(x) = 6 - 4x; g(-3)$

81. $h(x) = -3; h(2)$

82. $F(x) = -10; F(-2)$

83. $G(x) = x^2 - 3x + 4; G(-2)$

84. $H(x) = -x^2 + 6x - 1; H(4)$

85. $A(x) = \sqrt{2x + x^2}; A(4)$

86. $B(x) = \left|\frac{3x + 2}{2}\right|; B(-4)$

87. Car Rental The cost to rent a car for a week is \$75 plus \$0.10 per mile driven.
(a) Express the cost, C, to rent a car as a function of the number of miles, m, driven.
(b) Determine the total cost to rent a car and drive 130 miles in a week.

CHAPTER 10 TEST

Remember to use your Chapter Test Prep Video CD to see fully worked-out solutions to any of these problems you would like to review.

Note to Instructor: A special file in TestGen provides algorithms specifically matched to the problems in this Chapter Test for easy-to-replicate practice or assessment purposes.

1. (a) up (b) $(-2, -9)$
(c) $x = -2$
(d) x-intercepts: $-5, 1$
y-intercept: -5
2. $y = -x^2 + 6x - 9$
3. $y = x^2 - 4$
4. $y = x^2 - 4x - 3$
5. (a) maximum
(b) 7 at $x = -1$
6. Domain: $\{1\}$
Range: $\{-1, -2, -3\}$
7. Domain: $\{-2, -3\}$
Range: $\{a, b, c\}$
8. Domain: $\{x \mid x \text{ is a real number}\}$
Range: $\{y \mid y \geq -2\}$

1. For the quadratic equation $y = x^2 + 4x - 5$,
(a) Determine whether the parabola opens up or down.
(b) Find the vertex of the parabola.
(c) Find the equation of the axis of symmetry.
(d) Find all intercepts.

In Problems 2–4, graph each quadratic equation using its properties. Label the vertex, intercepts, and axis of symmetry.

2. $y = -x^2 + 6x - 9$ **3.** $y = x^2 - 4$ **4.** $y = x^2 - 4x - 3$

5. For the quadratic equation $y = -2x^2 - 4x + 5$,
(a) Determine whether the equation has a maximum or minimum value.
(b) Find the maximum or minimum value of the equation.

In Problems 6–9, determine (a) the domain and (b) the range of the given relation.

6. $\{(1, -1), (1, -2), (1, -3)\}$

7. $\{(-2, a), (-3, b), (-3, c)\}$

8.

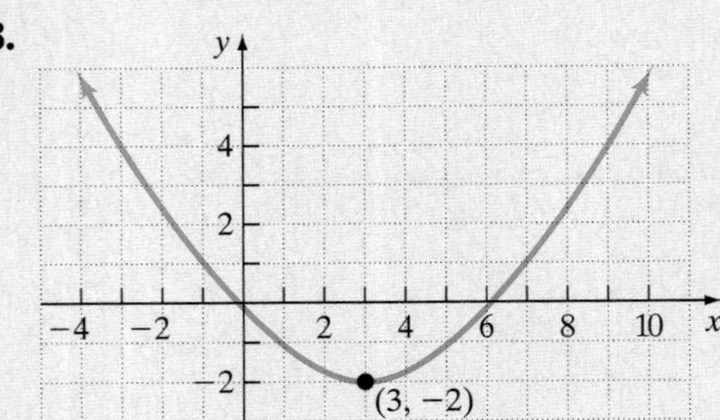

9.

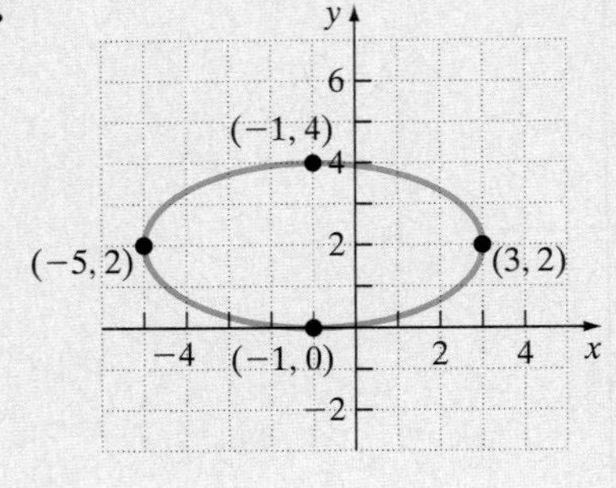

9. Domain: $\{x \mid -5 \le x \le 3\}$
Range: $\{y \mid 0 \le y \le 4\}$
10. not a function
11. function
12. function
13. function
14. not a function
15. function
16. -5
17. $\frac{5}{16}$
18. 8
19. (a) $\frac{3}{2}$ sec
(b) $\frac{27}{2}$ ft
20. (a) \$3
(b) 9.8 years
(c) \$35
(d) \$27

In Problems 10–15, determine whether each of the following represents a function.

10. $\{(-5, 10), (-5, 9), (-5, 8)\}$

11. $\{(4, -2), (3, -2), (2, -2)\}$

12. $x = \sqrt{y + 2}$

13. $x = 4 + y$

14.

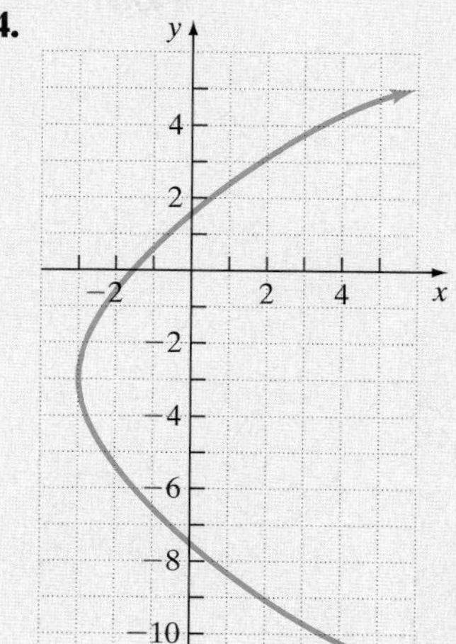

15.

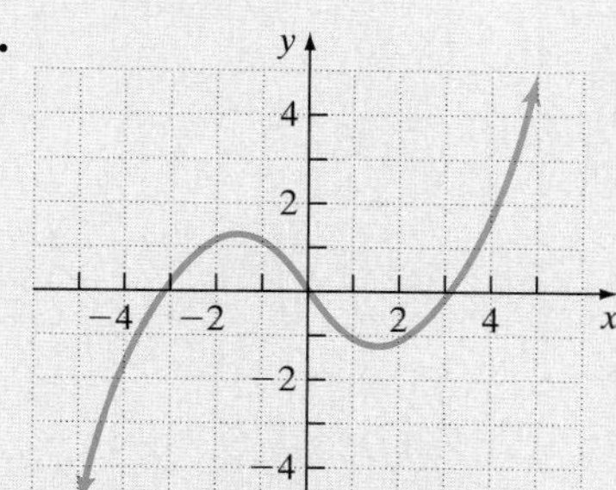

In Problems 16–18, find the value for each function.

16. $f(x) = -x^2 + 3x - 1; f(-1)$

17. $g(x) = \dfrac{2x + 4x^2}{8x^3}; g(2)$

18. $h(x) = 2\sqrt{x + 4}; h(12)$

19. Height of a Ball The height, h, in feet, of the path of a ball can be approximated by the quadratic equation $h(t) = -6t^2 + 18t$, where t is the number of seconds after the ball was thrown.

(a) How long does it take for the ball to reach its maximum height?
(b) What is the maximum height of the ball?

20. Oil Price Assume the price per barrel of crude oil rose sharply and then came back down between 1969 and 1989. The equation that can be used to approximate the cost of a barrel of crude oil can be given by the function $P(t) = -0.33t^2 + 6.5t + 3$, where P is the price of a barrel of crude oil (not adjusted for inflation) and t is the number of years after 1969.

(a) What was the price per barrel of crude oil in 1969?
(b) How many years, to the nearest tenth, after 1969 did the price reach its maximum?
(c) What was the highest price per barrel, to the nearest dollar, during this time?
(d) Find the price per barrel of crude oil, to the nearest dollar, in 1974.

Fractions, Decimals, and Percents

A.1 Fractions

OBJECTIVES

1. Factor a Number as a Product of Prime Factors
2. Find the Least Common Multiple of Two or More Numbers
3. Write Equivalent Fractions
4. Write a Fraction in Lowest Terms

We begin this section by reviewing two concepts from arithmetic: factoring a number as a product of primes and finding the least common multiple of a list of numbers. We base our discussion in this section on natural numbers.

1 Factor a Number as a Product of Prime Factors

When we multiply, the numbers that are multiplied together are called **factors** and the answer is called the **product.**

$$\underbrace{7}_{\text{factor}} \cdot \underbrace{5}_{\text{factor}} = \underbrace{35}_{\text{product}}$$

Teaching Tip
Emphasize vocabulary: factor, product. These terms are used throughout the text.

When we write a number as a product, we say that we **factor** the number. For example, when we write 20 as the product $10 \cdot 2$, we say that we have factored 20.

Some natural numbers are prime numbers and others are composite.

Work Smart
Remember, the natural numbers are 1, 2, 3, ...

DEFINITION

A natural number is **prime** if its factors are only one and itself. Natural numbers that are not prime are called **composite.** The number 1 is neither prime nor composite.

Work Smart
The first six primes are 2, 3, 5, 7, 11, and 13.

Examples of prime numbers are 2, 3, 5, 7, 11, and 13. We use a *factor tree* to find the prime factorization of a number. The process begins with finding two factors of the given number. Continue to factor until all factors are primes.

When a composite number is written as the product of prime numbers, we say that we are writing the **prime factorization** of the number. One technique that may be used to write the prime factorization of a number is shown below.

Classroom Example ➤
Write the prime factorization of 80 using the method in Example 1.
Answer: $2^4 \cdot 5$

EXAMPLE 1 Finding the Prime Factorization

Write the prime factorization of 24.

Solution

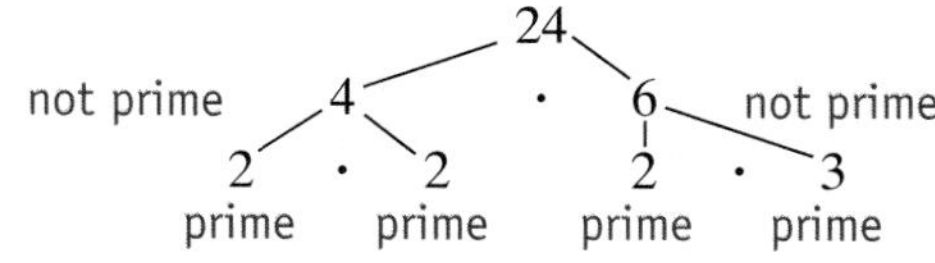

Work Smart
We could have begun the factorization of 24 with the factors 8 and 3 instead of 4 and 6. Try this for yourself.

All the numbers are prime, so we are done. The prime factorization of 24 is $2 \cdot 2 \cdot 2 \cdot 3$. Order is not important in multiplying factors. The product could also be written as $3 \cdot 2 \cdot 2 \cdot 2$ or $2 \cdot 3 \cdot 2 \cdot 2$.

QUICK ✓ *Find the prime factorization of each of the following:*

1. 12 **2.** 18 **3.** 75 **4.** 120 **5.** 131 **6.** 459

2 Find the Least Common Multiple of Two or More Numbers

A **multiple** of a number is the product of that number and any natural number. For example, the multiples of 2 are

$$2 \cdot 1 = 2, 2 \cdot 2 = 4, 2 \cdot 3 = 6, 2 \cdot 4 = 8, 2 \cdot 5 = 10, 2 \cdot 6 = 12, \text{ and so on.}$$

Multiples of 3 are

$$3 \cdot 1 = 3, 3 \cdot 2 = 6, 3 \cdot 3 = 9, 3 \cdot 4 = 12, 3 \cdot 5 = 15, 3 \cdot 6 = 18, \text{ and so on.}$$

Notice that the numbers 2 and 3 share two common multiples in this list: 6 and 12. The smallest common multiple, called the least common multiple, of 2 and 3 is 6.

DEFINITION

The **least common multiple (LCM)** of two or more natural numbers is the smallest number that is a multiple of each of the numbers.

Classroom Example
Find the least common multiple of the numbers 18 and 24.

Answer: 72

EXAMPLE 2 How to Find the Least Common Multiple

Find the least common multiple of the numbers 6 and 15.

Step-by-Step Solution

Step 1: Write each number as the product of prime factors, aligning common factors vertically.

Arrange the common factor of 3 in its own column

$$6 = 2 \cdot 3$$
$$15 = \quad 3 \cdot 5$$

Step 2: Write down the factor(s) that the numbers share, if any. Then write down the remaining factors the greatest number of times that the factors appear in any number.

The common factor is 3.
The remaining factors are 2 and 5.

Step 3: Multiply the factors listed in Step 2. The product is the least common multiple (LCM).

The LCM is $2 \cdot 3 \cdot 5 = 30$

The least common multiple of 6 and 15 is 30. We see that 30 is a multiple of 6 because $6 \cdot 5 = 30$, and 30 is a multiple of 15 because $15 \cdot 2 = 30$.

Work Smart
To see the individual factors more clearly, arrange the factors vertically.

We could also find the least common multiple by listing the factors of each number until we find the smallest common multiple as follows:

Multiples of 6:	6, 12, 18, 24, 30, 36, 42, ...
Multiples of 15:	15, 30, 45, 60, ...

Teaching Tip
Suggest that students use the technique used in Example 2: Arrange the factors vertically.

The least common multiple is 30 (indicated in red). This approach works just fine for numbers, but does not work for algebra. For this reason, it is recommended that you follow the steps used in Example 2 to find the least common multiple of two or more numbers so that you are better prepared when we discuss the least common multiple again later in the text.

Classroom Example Write the LCM of the numbers 24 and 30.

Answer: 120

EXAMPLE 3 Finding the Least Common Multiple

Find the LCM of the numbers 18 and 15.

Solution

We first write each number as the product of prime factors.

$$18 = 2 \cdot 3 \cdot 3$$
$$15 = \quad 3 \cdot \quad 5$$
$$\downarrow \; \downarrow \; \downarrow \; \downarrow$$
$$2 \cdot 3 \cdot 3 \cdot 5$$

Write the factors in each column that the numbers share, if any. Then write down the remaining factors the greatest number of times that the factors appear in any number.

Find the product of the factors. The LCM is $2 \cdot 3 \cdot 3 \cdot 5 = 90$.

QUICK ✓ *Find the LCM of the numbers.*

7. 6 and 8 **8.** 5 and 10 **9.** 45 and 72 **10.** 7 and 3 **11.** 12, 18, and 30

3 Write Equivalent Fractions

A fraction represents a part of a whole. For example, the fraction $\frac{5}{8}$ means "5 parts out of 8 parts." A fraction also indicates division: $\frac{5}{8}$ means "five divided by 8." Since $\frac{5}{8}$ indicates division, $\frac{5}{8}$ may be written as $8\overline{)5}$. Figure 1 shows the fraction $\frac{5}{8}$ visually.

Figure 1

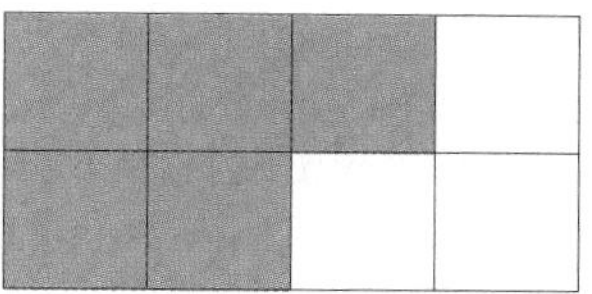

In the fraction $\frac{5}{8}$, the number 5 is called the **numerator** and the number 8 is called the **denominator.** The denominator tells the number of equal parts that the whole is divided into, and the numerator tells the number of equal parts that are shaded. For example, in Figure 1 the box is divided into 8 equal parts, 5 of which are shaded.

We use whole numbers (0, 1, 2, 3, . . .) for the numerators of fractions and natural numbers (1, 2, 3, 4, . . .) for denominators.

Work Smart
The denominator of a number such as 7 is 1 because $7 = \frac{7}{1}$.

Fractions without common denominators can be rewritten in an equivalent form so they have the same denominator.

DEFINITION
Equivalent fractions are fractions that represent the same part of a whole.

Figure 2

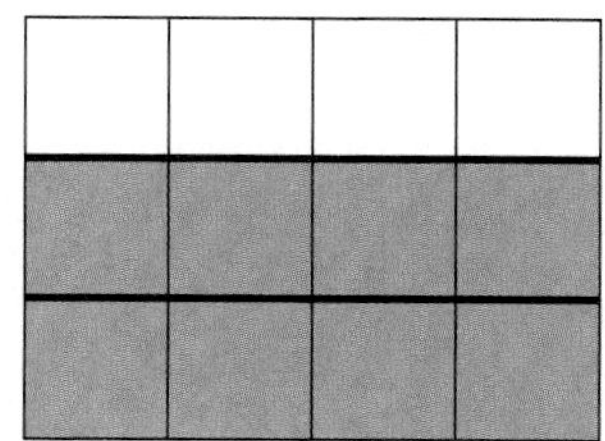

For example, $\frac{2}{3}$ and $\frac{8}{12}$ are equivalent fractions. To understand why, consider Figure 2. If we break the whole into 12 parts and shade 8 of these parts, the shaded region represents $\frac{8}{12}$ of the rectangle. If we only consider the 3 parts separated by the thick black lines, we can see that 2 parts are shaded for a fraction of $\frac{2}{3}$. In each case, the same portion of the rectangle is shaded, so $\frac{2}{3}$ and $\frac{8}{12}$ are equivalent fractions.

Now the question becomes, how do we obtain equivalent fractions? The answer lies in the following property.

In Words
We can obtain an equivalent fraction by multiplying the numerator and denominator of the fraction by the same nonzero number.

If a, b, and c are whole numbers, then

$$\frac{a}{b} = \frac{a \cdot c}{b \cdot c} \quad \text{if } b \neq 0, c \neq 0$$

Let's see how this property works.

Classroom Example

Write the fraction $\frac{2}{5}$ as an equivalent fraction with a denominator of 15.

Answer: $\frac{6}{15}$

EXAMPLE 4 Writing an Equivalent Fraction

Write the fraction $\frac{3}{4}$ as an equivalent fraction with a denominator of 20.

Solution

We want to know $\frac{3}{4}$ equals "what" over 20, or $\frac{3}{4} = \frac{?}{20}$. To write the fraction $\frac{3}{4}$ with a denominator of 20, we multiply the numerator and denominator of $\frac{3}{4}$ by 5. Do you see why?

$$\frac{3}{4} = \frac{3 \cdot 5}{4 \cdot 5} = \frac{15}{20}$$

QUICK ✓ *Rewrite each fraction with the denominator indicated.*

12. $\frac{1}{2}$; 10

13. $\frac{5}{8}$; 48

It is sometimes necessary to rewrite two or more fractions so that each has the same denominator. For example, we could rewrite the fractions $\frac{5}{6}$ and $\frac{3}{8}$ so that they have a common denominator of 24, 48, 96 and so on because these are common multiples of the denominators 6 and 8. Notice that 24 is the least common multiple of 6 and 8. When talking about the least common multiple as it applies to denominators of fractions, we use the phrase least common denominator.

Classroom Example

Write $\frac{3}{14}$ and $\frac{2}{21}$ as equivalent fractions with the least common denominator.

Answer: $\frac{3}{14} = \frac{9}{42}$ and $\frac{2}{21} = \frac{4}{42}$

DEFINITION

The **least common denominator (LCD)** is the least common multiple of the denominators of a group of fractions.

EXAMPLE 5 How to Write Two Fractions as Equivalent Fractions with the LCD

Write $\frac{5}{8}$ and $\frac{9}{20}$ as equivalent fractions with the least common denominator.

Step-by-Step Solution

Step 1: Find the least common denominator of the fractions.

The denominators of $\frac{5}{8}$ and $\frac{9}{20}$ are 8 and 20.

Write each denominator as the product of prime factors:

$$8 = 2 \cdot 2 \cdot 2$$
$$20 = 2 \cdot 2 \cdot \quad 5$$
$$\downarrow \; \downarrow \; \downarrow \; \downarrow$$
$$\text{LCD} = 2 \cdot 2 \cdot 2 \cdot 5 = 40$$

(continued)

Step 2: Rewrite each fraction with the least common denominator.

Multiply the numerator and denominator of $\frac{5}{8}$ by 5: $\frac{5}{8} = \frac{5 \cdot 5}{8 \cdot 5} = \frac{25}{40}$

Multiply the numerator and denominator of $\frac{9}{20}$ by 2: $\frac{9}{20} = \frac{9 \cdot 2}{20 \cdot 2} = \frac{18}{40}$

So $\frac{5}{8} = \frac{25}{40}$ and $\frac{9}{20} = \frac{18}{40}$.

QUICK  *Write the equivalent fractions with the least common denominator.*

14. $\frac{1}{4}$ and $\frac{5}{6}$ **15.** $\frac{5}{12}$ and $\frac{4}{15}$ **16.** $\frac{9}{20}$ and $\frac{11}{16}$

4 Write a Fraction in Lowest Terms

In Words
To write a fraction in lowest terms, find a common factor between the numerator and denominator and divide out the common factors.

DEFINITION
A fraction is written in lowest terms if the numerator and the denominator share no common factor other than 1.

We can write fractions in lowest terms using the fact that

$$\frac{a \cdot c}{b \cdot c} = \frac{a}{b}$$

So, to write a fraction in lowest terms, we write the numerator and the denominator as a product of primes, and then divide out common factors.

Classroom Example
Write $\frac{15}{55}$ in lowest terms.

Answer: $\frac{3}{11}$

Teaching Tip
In Example 6, show the use of different slash marks for each factor, so students can keep track of factors that have divided out. Remind students that each quotient equals 1. You may even want to show the "1" in each quotient.

EXAMPLE 6 Writing a Fraction in Lowest Terms

Write $\frac{24}{40}$ in lowest terms.

Solution

Write the numerator and the denominator as the product of primes and divide out common factors.

$$\frac{24}{40} = \frac{2 \cdot 2 \cdot 2 \cdot 3}{2 \cdot 2 \cdot 2 \cdot 5}$$

Divide out common factors: $= \frac{\cancel{2} \cdot \cancel{2} \cdot \cancel{2} \cdot 3}{\cancel{2} \cdot \cancel{2} \cdot \cancel{2} \cdot 5}$

$$= \frac{3}{5}$$

So $\frac{24}{40} = \frac{3}{5}$.

Work Smart
Use different slash marks to keep track of factors that have divided out. Also, you may wish to use nonprime factors when writing a fraction in lowest terms. In Example 6 we could have written $\frac{24}{40}$ in lowest terms as follows:

$$\frac{24}{40} = \frac{8 \cdot 3}{8 \cdot 5} = \frac{3}{5}$$

QUICK ✓ *Write each fraction in lowest terms if possible.*

17. $\frac{3}{6}$ **18.** $\frac{4}{9}$ **19.** $\frac{45}{80}$ **20.** $\frac{20}{50}$ **21.** $\frac{30}{105}$

A.1 Exercises

For Extra Help: Student Solutions Manual | CD Video | PH Math/Tutor Center | MathXL Tutorials on CD | MathXL® | MyMathLab

Concepts and Vocabulary

In Problems 1–3, fill in the blanks.

1. In the statement $6 \cdot 8 = 48$, 6 and 8 are called ________, and 48 is called the ________.

2. The ________ ________ ________ of two natural numbers is the smallest number that is a multiple of the numbers.

3. Fractions which represent the same portion of a whole are called ________ ________.

In Problems 4–6, answer True or False to each statement.

4. The number 42 is a natural number.

5. The LCM of 5 and 15 is 75.

6. When finding the least common multiple of two numbers, list only the common or shared factors of the numbers.

7. You are explaining how to find the LCM to the members of your study group. In your own words, explain how you would find the LCM of 12, 18, and 45.

8. How can you tell if the number 91 is prime? In your own words, write a definition for *prime number*.

Building Skills

In Problems 9–32, find the prime factorization of each number.

9. 25	**10.** 9	**11.** 28	**12.** 100	**13.** 21
14. 35	**15.** 36	**16.** 54	**17.** 20	**18.** 63
19. 30	**20.** 45	**21.** 50	**22.** 70	**23.** 53
24. 79	**25.** 252	**26.** 315	**27.** 256	**28.** 243
29. 1300	**30.** 693	**31.** 2275	**32.** 6699	

In Problems 33–48, find the LCM of each set of numbers.

33. 6 and 15	**34.** 10 and 14	**35.** 12 and 10	**36.** 21 and 18
37. 15 and 14	**38.** 55 and 6	**39.** 30 and 45	**40.** 8 and 70
41. 42 and 14	**42.** 112 and 28	**43.** 5, 6 and 12	**44.** 9, 15 and 20
45. 3, 8 and 9	**46.** 4, 18 and 20	**47.** 4, 8, 12 and 18	**48.** 3, 6, 15 and 21

In Problems 49–52, write the fraction that is indicated by the shaded region.

49.

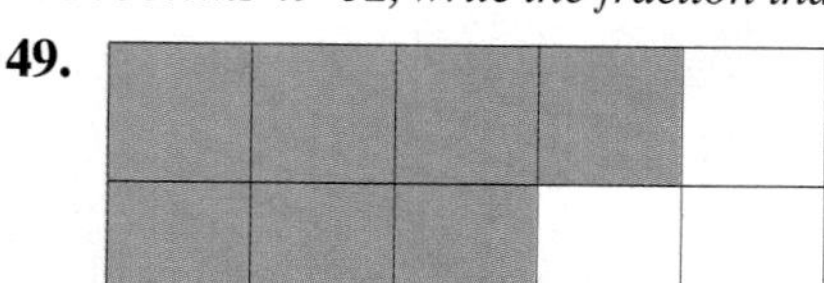

50.

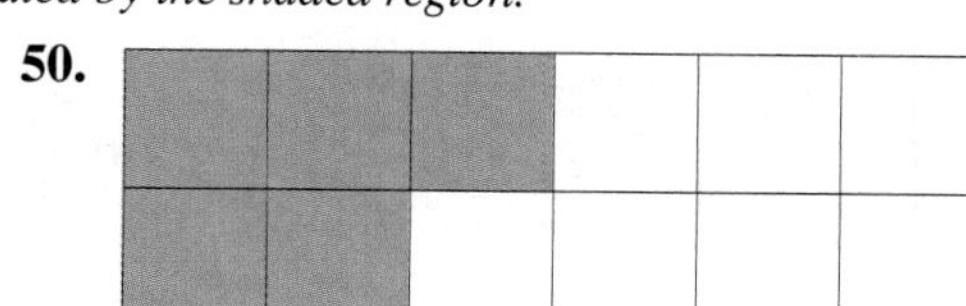

1. factors, product
2. least common multiple
3. equivalent fractions
4. True
5. False
6. False
7. Answers will vary.
8. Answers will vary.
9. $5 \cdot 5$
10. $3 \cdot 3$
11. $2 \cdot 2 \cdot 7$
12. $2 \cdot 2 \cdot 5 \cdot 5$
13. $3 \cdot 7$
14. $5 \cdot 7$
15. $2 \cdot 2 \cdot 3 \cdot 3$
16. $2 \cdot 3 \cdot 3 \cdot 3$
17. $2 \cdot 2 \cdot 5$
18. $3 \cdot 3 \cdot 7$
19. $2 \cdot 3 \cdot 5$
20. $3 \cdot 3 \cdot 5$
21. $2 \cdot 5 \cdot 5$
22. $2 \cdot 5 \cdot 7$
23. 53 is prime.
24. 79 is prime.
25. $2 \cdot 2 \cdot 3 \cdot 3 \cdot 7$
26. $3 \cdot 3 \cdot 5 \cdot 7$
27. $2 \cdot 2 \cdot 2 \cdot 2 \cdot 2 \cdot 2 \cdot 2 \cdot 2$
28. $3 \cdot 3 \cdot 3 \cdot 3 \cdot 3$
29. $2 \cdot 2 \cdot 5 \cdot 5 \cdot 13$
30. $3 \cdot 3 \cdot 7 \cdot 11$
31. $5 \cdot 5 \cdot 7 \cdot 13$
32. $3 \cdot 7 \cdot 11 \cdot 29$
33. 30
34. 70
35. 60
36. 126
37. 210
38. 330
39. 90
40. 280
41. 42
42. 112
43. 60
44. 180
45. 72
46. 180
47. 72
48. 210
49. $\frac{7}{10}$
50. $\frac{5}{12}$

51. $\frac{5}{6}$
52. $\frac{1}{6}$
53. $\frac{8}{12}$
54. $\frac{12}{15}$
55. $\frac{18}{24}$
56. $\frac{10}{28}$
57. $\frac{21}{3}$
58. $\frac{40}{10}$
59. $\frac{4}{8}$ and $\frac{3}{8}$
60. $\frac{9}{12}$ and $\frac{5}{12}$
61. $\frac{9}{15}$ and $\frac{10}{15}$
62. $\frac{9}{36}$ and $\frac{8}{36}$
63. $\frac{20}{24}$ and $\frac{15}{24}$
64. $\frac{3}{36}$ and $\frac{10}{36}$
65. $\frac{44}{60}$ and $\frac{21}{60}$
66. $\frac{25}{60}$ and $\frac{28}{60}$
67. $\frac{27}{48}$ and $\frac{14}{48}$
68. $\frac{138}{168}$ and $\frac{147}{168}$
69. $\frac{12}{96}$ and $\frac{21}{96}$
70. $\frac{51}{90}$ and $\frac{38}{90}$
71. $\frac{20}{90}$ and $\frac{35}{90}$ and $\frac{21}{90}$
72. $\frac{42}{60}$ and $\frac{15}{60}$ and $\frac{50}{60}$
73. $\frac{2}{3}$ 74. $\frac{3}{5}$ 75. $\frac{19}{9}$
76. $\frac{9}{4}$ 77. $\frac{1}{2}$ 78. $\frac{8}{9}$
79. $\frac{4}{5}$ 80. $\frac{7}{9}$ 81. $\frac{3}{5}$
82. $\frac{2}{3}$ 83. $\frac{2}{3}$ 84. $\frac{12}{13}$
85. 84 months
86. 630 laps
87. 20 days

51.

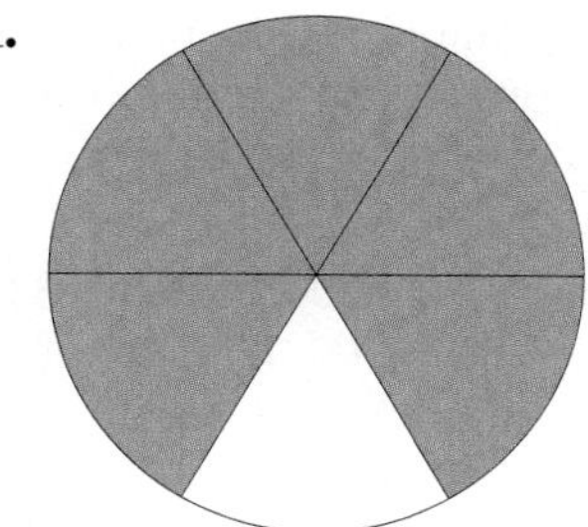

52.

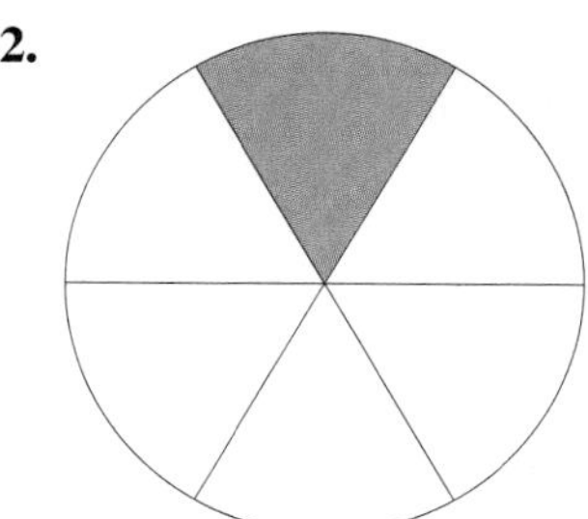

In Problems 53–58, write each fraction with the given denominator.

53. Write $\frac{2}{3}$ with denominator 12.

54. Write $\frac{4}{5}$ with denominator 15.

55. Write $\frac{3}{4}$ with denominator 24.

56. Write $\frac{5}{14}$ with denominator 28.

57. Write 7 with denominator 3.

58. Write 4 with denominator 10.

In Problems 59–72, write the equivalent fractions with the least common denominator.

59. $\frac{1}{2}$ and $\frac{3}{8}$ **60.** $\frac{3}{4}$ and $\frac{5}{12}$ **61.** $\frac{3}{5}$ and $\frac{2}{3}$ **62.** $\frac{1}{4}$ and $\frac{2}{9}$

63. $\frac{5}{6}$ and $\frac{5}{8}$ **64.** $\frac{1}{12}$ and $\frac{5}{18}$ **65.** $\frac{11}{15}$ and $\frac{7}{20}$ **66.** $\frac{5}{12}$ and $\frac{7}{15}$

67. $\frac{9}{16}$ and $\frac{7}{24}$ **68.** $\frac{23}{28}$ and $\frac{21}{24}$ **69.** $\frac{3}{24}$ and $\frac{7}{32}$ **70.** $\frac{17}{30}$ and $\frac{19}{45}$

71. $\frac{2}{9}$ and $\frac{7}{18}$ and $\frac{7}{30}$ **72.** $\frac{7}{10}$ and $\frac{1}{4}$ and $\frac{5}{6}$

In Problems 73–84, write each fraction in lowest terms if possible.

73. $\frac{14}{21}$ **74.** $\frac{9}{15}$ **75.** $\frac{38}{18}$ **76.** $\frac{81}{36}$

77. $\frac{22}{44}$ **78.** $\frac{24}{27}$ **79.** $\frac{32}{40}$ **80.** $\frac{49}{63}$

81. $\frac{15}{25}$ **82.** $\frac{34}{51}$ **83.** $\frac{150}{225}$ **84.** $\frac{144}{156}$

Applying the Concepts

85. Planets in Our Solar System At a certain point, Mercury, Venus, and Earth lie on a straight line. If it takes each planet 3, 7, and 12 months, respectively, to revolve around the sun, what is the fewest number of months until they align this way again?

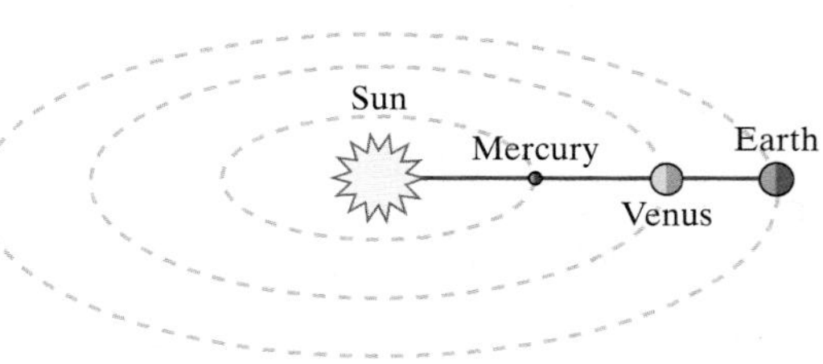

86. Talladega Raceway At Talladega one of the crew chiefs discovered that in a given time interval, Jeff Gordon completed 21 laps, Dale Earnhardt Jr. completed 18 laps, and Robby Gordon completed 15 laps. How many laps would a race have to have so that all three drivers were at the finish line at exactly the same time?

87. Sam's Medication Bob gives his dog Sam one type of medication every 4 days, and a second type of medication every 10 days. How often does Bob give Sam both types of medication on the same day?

88. 140 days
89. $\frac{13}{20}$
90. $\frac{3}{5}$
91. $\frac{3}{5}$
92. $\frac{3}{5}$
93. $\frac{2}{3}$
94. $\frac{5}{6}$
95. 2, 3, 5, 7, 11, 13, 17, 19, 23, 29, 31, 37, 41, 43, 47, 53, 59, 61, 67, 71, 73, 79, 83, 89, 97

88. Visiting Columbus Pamela and Geoff both visit Columbus on business. Pamela flies to Columbus from Atlanta about every 14 days, and Geoff takes the train to Columbus from Cincinnati every 20 days. How often are both Pamela and Geoff in Columbus on business?

89. Survey Data In a survey of 500 students, 325 stated that they work at least 25 hours per week. Express the fraction of students that work at least 25 hours per week as a fraction in lowest terms.

90. Survey Data In a survey of 750 students, 450 stated that they are enrolled in 15 or more semester hours. Express the fraction of students that are enrolled in 15 or more semester hours as a fraction in lowest terms.

91. Quiz Score A student earns 12 points out of a total of 20 points on a quiz. Express this score as a fraction in lowest terms.

92. Quiz Score A student earns 15 points out of a total of 25 points on a quiz. Express this score as a fraction in lowest terms.

93. Pizza Marko orders a pizza that is cut into 12 equal pieces. He eats 8 of the pieces. Write the amount of the pizza that Marko ate as a fraction in lowest terms.

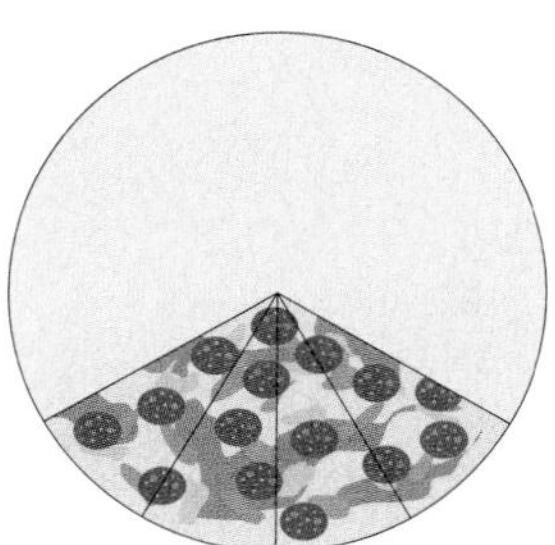

94. Sub Sandwich Aisha orders a 12-foot sub sandwich that is cut into 24 equal pieces for a party. After the party, she notices that 4 pieces remain. Write the amount of the sub sandwich that was eaten as a fraction in lowest terms.

95. The Sieve of Eratosthenes Eratosthenes (276 B.C.–194 B.C.) was born in Cyrene, which is now in Libya in North Africa. He devised an algorithm (a series of steps that are followed to solve a problem) for identifying prime numbers. The algorithm works as follows:

Step 1: List all the natural numbers that are greater than or equal to 2.

Step 2: The first number in the list, 2, is prime. Cross out all multiples of 2. For example, cross out 2, 4, 6, ...

Step 3: Identify the next number in the list after the most recently identified prime number. For example, we already know 2 is a prime number, so the next number in the list, 3, is also prime. Cross out all multiples of this number.

Step 4: Repeat Step 3.

Use the algorithm to find all the prime numbers less than 100.

A.2 Decimals and Percents

OBJECTIVES

1. Use Place Value
2. Round Decimals
3. Convert a Fraction to a Decimal and a Decimal to a Fraction
4. Convert a Percent to a Decimal and a Decimal to a Percent

Decimals and percentages commonly occur in everyday life. You received a 92% on your test, there is a 10% discount on jeans, we pay 7.75% in sales tax, 45% of the people polled support a proposition. Before we discuss decimals and percents, let us consider place value.

1 Use Place Value

Figure 3 shows how we interpret the place value of each digit in the number 9186.347. For example, the 7 is in the "thousandths" position, 3 is in the "tenths" position, and the 8 is in the "tens" position.

Figure 3

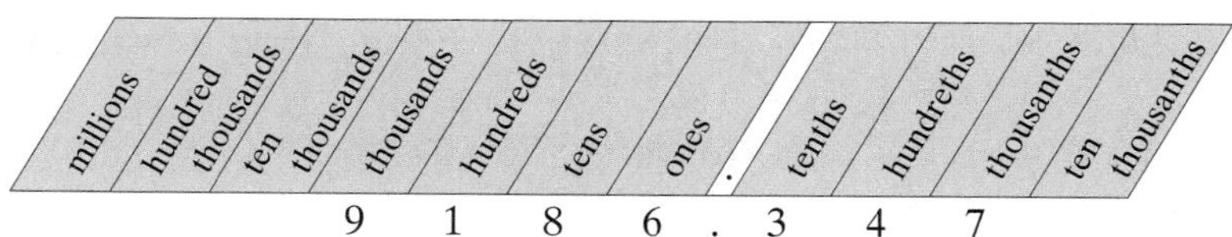

The number 9186.347 is read "nine thousand, one hundred eighty-six and three hundred forty-seven thousandths."

QUICK ✓ *Tell the place value of the digit in the given number.*

1. 235.71; the 1
2. 56,701.28; the 2
3. 278,403.95; the 8
4. 0.189; the 9
5. 3.590; the 3
6. 9,021,458.5; the 2

2 Round Decimals

We round decimals in the same way we round whole numbers. First, identify the specified place value in the decimal. If the digit to the right is 5 or more, add 1 to the digit; if the next digit is 4 or less, leave the digit as it is. Then drop the digits to the right of the rounding number.

Classroom Example ➤
Round 4.359 to the nearest hundredth.
Answer: 4.36

EXAMPLE 1 Rounding a Decimal Number

Round 8.726 to the nearest hundredth.

Solution

To round to the nearest hundredth, we determine that the number 2 is in the hundredths place: 8.7<u>2</u>6. The number to the right of 2 is 6. Since 6 is greater than 5, we round 8.726 to 8.73. ■

Classroom Example ➤
Round 0.3692 to the nearest thousandth.
Answer: 0.369

EXAMPLE 2 Rounding a Decimal Number

Round 0.9451 to the nearest thousandth.

Solution

To round 0.9451 to the nearest thousandth, we see that the number 5 is in the thousandths place: 0.94<u>5</u>1. The number to the right of 5 is 1. Since 1 is less than 5, we round 0.9451 to 0.945. ■

QUICK ✓ *Round each number to the given decimal place.*

7. 0.17 to the nearest tenth
8. 0.932 to the nearest hundredth
9. 1.396 to the nearest hundredth
10. 14.3983 to the nearest thousandth
11. 690.004 to the nearest hundredth
12. 59.98 to the nearest tenth

3 Convert a Fraction to a Decimal and a Decimal to a Fraction

Convert a Fraction to a Decimal

To convert a fraction to a decimal, divide the numerator of the fraction by the denominator of the fraction until the remainder is 0 or the remainder repeats.

Classroom Example

Convert $\frac{7}{25}$ to a decimal.

Answer: 0.28

EXAMPLE 3 Converting a Fraction to a Decimal

Convert $\frac{9}{20}$ to a decimal.

Solution

$$\frac{9}{20} = 20\overline{)9.00} = 0.45$$

$$\begin{array}{r} 0.45 \\ 20\overline{)9.00} \\ \underline{8\,0} \\ 100 \\ \underline{100} \\ 0 \end{array}$$

Therefore, $\frac{9}{20} = 0.45$.

Classroom Example

Convert $\frac{7}{9}$ to a decimal.

Answer: $0.\overline{7}$ or 0.777 . . .

EXAMPLE 4 Converting a Fraction to a Decimal

Convert $\frac{2}{3}$ to a decimal.

Solution

$$\frac{2}{3} = \begin{array}{r} 0.666 \\ 3\overline{)2.000} \\ \underline{1\,8} \\ 20 \\ \underline{18} \\ 20 \\ \underline{18} \\ 2 \end{array}$$

Notice that the remainder, 2, repeats. So $\frac{2}{3} = 0.666 \ldots.$

In Example 3, the decimal 0.45 is called a **terminating decimal** because the decimal stops after the 5. In Example 4, the number 0.666 . . . is called a **repeating decimal** because the 6 continues repeating indefinitely. The decimal 0.666 . . . can also be written as $0.\overline{6}$. The bar over the 6 means the 6 repeats.

QUICK ✓ *Write the fraction as a decimal.*

13. $\frac{2}{5}$ **14.** $\frac{3}{7}$ **15.** $\frac{11}{8}$ **16.** $\frac{5}{6}$ **17.** $\frac{5}{9}$

Based on Examples 3 and 4 and Quick Check Problems 13–17, you should notice that **every fraction has a decimal representation that either terminates or repeats.**

Convert a Decimal to a Fraction

To convert a decimal to a fraction, identify the place value of the denominator, write the decimal as a fraction using the given denominator, and reduce.

Classroom Example

Convert each decimal to a fraction, and write in lowest terms, if possible.

(a) 0.4 **(b)** 0.53

Answer:

(a) $\frac{2}{5}$ **(b)** $\frac{53}{100}$

EXAMPLE 5 Writing a Decimal as a Fraction

Convert each decimal to a fraction and write in lowest terms, if possible.

(a) 0.8 **(b)** 0.77

Solution

(a) 0.8 is equivalent to 8 tenths, or $\frac{8}{10}$. Because, $\frac{8}{10} = \frac{4 \cdot \cancel{2}}{5 \cdot \cancel{2}} = \frac{4}{5}$, we write $0.8 = \frac{4}{5}$.

(b) 0.77 is equivalent to 77 hundredths, or $\frac{77}{100}$.

QUICK ✓ *Write the decimal as a fraction and write in lowest terms, if possible.*

18. 0.65 **19.** 0.2 **20.** 0.625

4 Convert a Percent to a Decimal and a Decimal to a Percent

Teaching Tip
Students may not be aware that the word "per" means "divided by." "Miles/hour" means "miles divided by hours." "Miles/gallon" means "miles divided by gallons," and so on.

When computing with percents, it is convenient to write percents as decimals. How is a percent converted to a decimal? Let's see.

Convert a Percent to a Decimal

DEFINITION
The word **percent** means **parts per hundred** or **parts out of one hundred.**

So 25% means 25 parts out of 100 parts. Therefore, $25\% = \frac{25}{100} = \frac{1 \cdot \cancel{25}}{4 \cdot \cancel{25}} = \frac{1}{4}$.

Since the word percent means "parts per hundred," we have that 100% means 100 "parts per 100," so 100% = 1. Therefore, to convert from a percent to a decimal, multiply the percent by $\frac{1}{100\%}$.

Classroom Example ➤
Write the following percents as decimals.
(a) 3% (b) 212% (c) 18.5%
Answer:
(a) 0.03 (b) 2.12 (c) 0.185

Work Smart
To convert from a percent to a decimal, move the decimal point two places to the left and drop the % symbol.

EXAMPLE 6 Writing a Percent as a Decimal

Write the following percents as decimals:

(a) 17% **(b)** 150%

Solution

(a)
$$17\% = 17\% \cdot \frac{1}{100\%} = \frac{17}{100} = 0.17$$

(b)
$$150\% = 150\% \cdot \frac{1}{100\%} = \frac{150}{100} = 1.5$$

QUICK ✓ *Write the percent as a decimal.*

21. 23% **22.** 1% **23.** 72.4% **24.** 127% **25.** 89.26%

Convert a Decimal to a Percent

Because 100% = 1, to convert a decimal to a percent, multiply the decimal by $\frac{100\%}{1}$.

Classroom Example ➢
Write the following decimals in percent form:
(a) 0.326 (b) 0.07 (c) 1.06
Answer:
(a) 32.6% (b) 7% (c) 106%

EXAMPLE 7 Writing a Decimal as a Percent

Write the following decimals in percent form:

(a) 0.445 **(b)** 1.42

Solution

(a) $0.445 = 0.445 \cdot \frac{100\%}{1} = 44.5\%$

(b) $1.42 = 1.42 \cdot \frac{100\%}{1} = 142\%$

Work Smart
To convert from a decimal to a percent, move the decimal point two places to the right and add the % symbol.

QUICK ✓ *Write the decimal as a percent.*

26. 0.15 **27.** 0.8 **28.** 1.3 **29.** 0.398 **30.** 0.004

A.2 Exercises

For Extra Help: Student Solutions Manual, CD Video, PH Math/Tutor Center, MathXL Tutorials on CD, MathXL®, MyMathLab

1. terminating
2. repeating
3. ones
4. True
5. True
6. True
7. Answers will vary.
8. Answers will vary.
9. hundredths place
10. hundreds place
11. tens place
12. tenths place
13. thousandths place
14. ones place
15. 578.2
16. 7300
17. 350
18. 543
19. 3682.010
20. 683.10
21. 30
22. 37.4

Concepts and Vocabulary

In Problems 1–3, fill in the blanks.

1. The decimal 0.492 is called a _________ decimal.

2. The decimal $8.\overline{34}$ is called a _________ decimal.

3. In the number 465.39, the digit 5 is in the _________ place.

In Problems 4–6, answer True or False to each statement.

4. The word percent means parts per hundred.

5. When converting 9.2% to a decimal, the decimal point will shift 2 places to the left.

6. Every fraction is equivalent to either a terminating or a repeating decimal.

7. Your teacher called you to the board to solve the following problem: convert $\frac{7}{8}$ to a percent. As you walk by, your friend whispers, "make $\frac{7}{8}$ a decimal first." With a sigh of relief, you begin. Write out your explanation.

8. Explain why multiplying by the fraction $\frac{1}{100\%}$ correctly converts a percent to a decimal.

Building Skills

In Problems 9–14, tell the place value of the digit in the given number.

9. 3465.902; the 0 **10.** 549,813.0267; the 8 **11.** 357.469; the 5

12. 9124.786; the 7 **13.** 2018.3764; the 6 **14.** 539.016; the 9

In Problems 15–22, round each number to the given place.

15. 578.206 to the nearest tenth

16. 7298.0845 to the nearest hundred

17. 354.678 to the nearest ten

18. 543.06 to the nearest unit

19. 3682.0098 to the nearest thousandth

20. 683.098 to the nearest hundredth

21. 29.96 to the nearest unit

22. 37.439 to the nearest tenth

In Problems 23–32, write each fraction as a terminating or repeating decimal.

23. $\frac{5}{8}$ **24.** $\frac{3}{4}$ **25.** $\frac{2}{7}$ **26.** $\frac{5}{6}$ **27.** $\frac{5}{16}$

28. $\frac{11}{32}$ **29.** $\frac{3}{13}$ **30.** $\frac{6}{13}$ **31.** $\frac{29}{25}$ **32.** $\frac{57}{50}$

In Problems 33–40, write each fraction as a decimal, rounded to the indicated place.

33. $\frac{13}{6}$ to the nearest tenth

34. $\frac{15}{8}$ to the nearest tenth

35. $\frac{6}{21}$ to the nearest hundredth

36. $\frac{24}{28}$ to the nearest hundredth

37. $\frac{8}{3}$ to the nearest hundredth

38. $\frac{9}{7}$ to the nearest hundredth

39. $\frac{14}{27}$ to the nearest thousandth

40. $\frac{18}{31}$ to the nearest thousandth

In Problems 41–48, write each decimal as a fraction, written in lowest terms.

41. 0.75 **42.** 0.25 **43.** 0.9 **44.** 0.4

45. 0.982 **46.** 0.358 **47.** 0.2525 **48.** 0.3334

In Problems 49–54, write each percent as a decimal.

49. 37% **50.** 59% **51.** 6.02% **52.** 8.25% **53.** 0.1% **54.** 0.5%

In Problems 55–60, write each decimal as a percent.

55. 0.2 **56.** 0.5 **57.** 0.275 **58.** 0.349 **59.** 2 **60.** 1

Applying the Concepts

61. Eating Healthy? In a poll conducted by Zogby International of 1200 adult Americans, 840 stated that they believe that they eat healthy. What percentage of adult Americans believe that they eat healthy?

62. Ghosts In a survey of 1100 adult women conducted by Harris Interactive, it was determined that 640 believe in ghosts. What percentage of adult women believe in ghosts?

63. Test Score A student earns 85 points out of a total of 110 points on an exam. Express this score as a percent.

64. Test Score A student earns 80 points out of a total of 115 points on an exam. Express this score as a percent.

65. Time Utilization In a 24-hour day, Jackson sleeps for 8 hours, works for 4 hours, and goes to school and studies for 6 hours.

(a) What percent of the time does Jackson sleep?
(b) What percent of the time does Jackson work?
(c) What percent of the time does Jackson go to school and study?

66. United States Senate In 2004, the United States Senate was comprised of 48 Democrats, 51 Republicans, and 1 Independent.

(a) What percent of the U.S. Senate was Democrat in 2004?
(b) What percent of the U.S. Senate was Republican in 2004?
(c) What percent of the U.S. Senate was Independent in 2004?

67. Cashews A single serving of cashews contains 14 grams of fat. Of this, 3 grams is saturated fat. What percentage of fat grams is saturated fat in a single serving of cashews?

68. Cheese Pizza A single serving of cheese pizza contains 11 grams of fat. Of this, 5 grams are saturated fat. What percentage of fat grams is saturated fat in a single serving of cheese pizza?

23. 0.625
24. 0.75
25. $0.\overline{285714}$
26. $0.8\overline{3}$
27. 0.3125
28. 0.34375
29. $0.\overline{230769}$
30. $0.\overline{461538}$
31. 1.16
32. 1.14
33. 2.2
34. 1.9
35. 0.29
36. 0.86
37. 2.67
38. 1.29
39. 0.519
40. 0.581
41. $\frac{3}{4}$
42. $\frac{1}{4}$
43. $\frac{9}{10}$
44. $\frac{2}{5}$
45. $\frac{491}{500}$
46. $\frac{179}{500}$
47. $\frac{101}{400}$
48. $\frac{1667}{5000}$
49. 0.37
50. 0.59
51. 0.0602
52. 0.0825
53. 0.001
54. 0.005
55. 20%
56. 50%
57. 27.5%
58. 34.9%
59. 200%
60. 100%
61. 70%
62. 58.18%
63. 77.27%
64. 69.57%
65. (a) 33.33% (b) 16.67% (c) 25%
66. (a) 48% (b) 51% (c) 1%
67. 21.43%
68. 45.45%

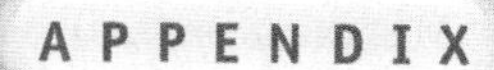

B Table of Square Roots

n	$\sqrt{n}$	n	$\sqrt{n}$	n	$\sqrt{n}$	n	$\sqrt{n}$
1	1	26	5.09902	51	7.14143	76	8.71780
2	1.41421	27	5.19615	52	7.21110	77	8.77496
3	1.73205	28	5.29150	53	7.28011	78	8.83176
4	2	29	5.38516	54	7.34847	79	8.88819
5	2.23607	30	5.47723	55	7.41620	80	8.94427
6	2.44949	31	5.56776	56	7.48331	81	9
7	2.64575	32	5.65685	57	7.54983	82	9.05539
8	2.82843	33	5.74456	58	7.61577	83	9.11043
9	3	34	5.83095	59	7.68115	84	9.16515
10	3.16228	35	5.91608	60	7.74597	85	9.21954
11	3.31662	36	6	61	7.81025	86	9.27362
12	3.46410	37	6.08276	62	7.87401	87	9.32738
13	3.60555	38	6.16441	63	7.93725	88	9.38083
14	3.74166	39	6.24500	64	8	89	9.43398
15	3.87298	40	6.32456	65	8.06226	90	9.48683
16	4	41	6.40312	66	8.12404	91	9.53939
17	4.12311	42	6.48074	67	8.18535	92	9.59166
18	4.24264	43	6.55744	68	8.24621	93	9.64365
19	4.35890	44	6.63325	69	8.30662	94	9.69536
20	4.47214	45	6.70820	70	8.36660	95	9.74679
21	4.58258	46	6.78233	71	8.42615	96	9.79796
22	4.69042	47	6.85565	72	8.48528	97	9.84886
23	4.79583	48	6.92820	73	8.54400	98	9.89949
24	4.89898	49	7	74	8.60233	99	9.94987
25	5	50	7.07107	75	8.66025	100	10

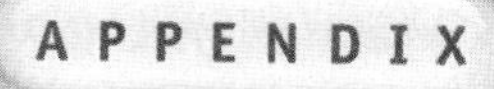

Geometry Review

C.1 Lines and Angles

OBJECTIVES

1. Understand the Terms Point, Line, and Plane
2. Work with Angles
3. Find the Measures of Angles Formed by Parallel Lines

The word geometry comes from the Greek words "geo," meaning "earth" and "metra," meaning "measure." The Greek scholar Euclid collected and organized the geometry known in his day into a logical system more than two thousand years ago. Euclid's system forms the basis of the geometry we still study today.

1 Understand the Terms Point, Line, and Plane

A **point** has no size, only position, and is usually designated by a capital letter as shown below.

P •

A **line** is a set of points extending infinitely far in opposite directions. A line has no width or height, just length and is uniquely determined by two points. For example, the line in Figure 1 is passing through the points A and B. The notation for the line shown is $\overleftrightarrow{AB}$.

Figure 1

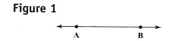

A **ray** is a half-line with one **endpoint,** which extends infinitely far in one direction. See Figure 2. The notation for the ray shown is $\overrightarrow{AB}$.

Figure 2

A B

A **line segment** is a portion of a line that has a beginning and an end. See Figure 3. If two line segments have the same length, they are said to be **congruent.** Notation for the congruent line segments shown are $\overline{AB}$ and $\overline{CD}$.

Figure 3

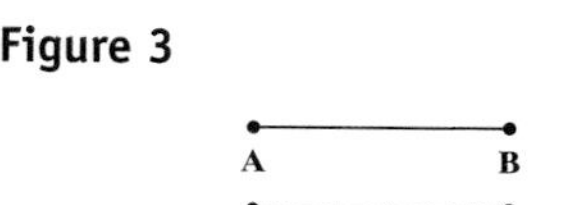

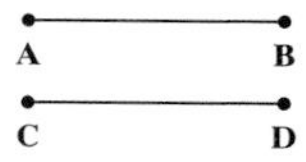

A **plane** is the set of points that forms a flat surface that extends indefinitely. A plane has no thickness. See Figure 4. The arrows indicate that the plane extends indefinitely in each direction.

Figure 4

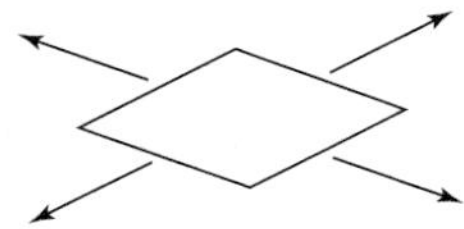

2 Work with Angles

Figure 5

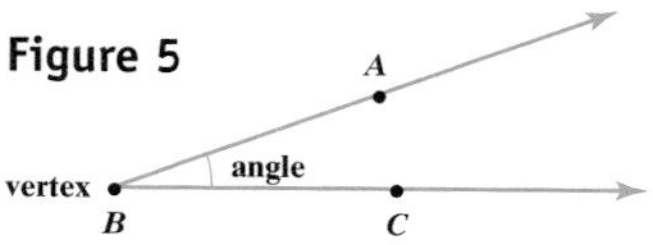

vertex B, angle, A, C

Suppose we draw two rays with a common endpoint as shown in Figure 5. The amount of rotation from one ray to the second ray is called the **angle** between the rays. The common endpoint is called the **vertex.** In Figure 5, the name of the angle is $\angle ABC$, $\angle CBA$ or $\angle B$. Angles are measured in **degrees,** which is symbolized °. One full rotation represents 360°. The notation $m\angle A = 60°$ means "the measure of angle A is 60 degrees." Because 60° is $\frac{1}{6}$ of 360°, an angle whose measure is 60° is $\frac{1}{6}$ of a full rotation. If two angles have the same measure, they are called **congruent.**

Work Smart

One full rotation is represented below.

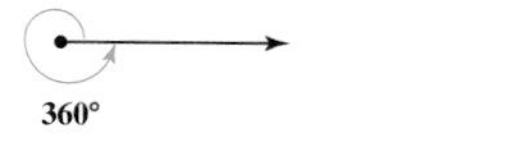

DEFINITION

An angle that measures 90° is called a **right angle.** The symbol ⏋ is used to denote a right angle. A right angle has $\frac{1}{4}$ of a full rotation.

An angle whose measure is between 0° and 90° is called an **acute angle.**

An angle whose measure is between 90° and 180° is called an **obtuse angle.**

An angle whose measure is 180° is called a **straight angle.** A straight angle has $\frac{1}{2}$ of a full rotation.

Figure 6

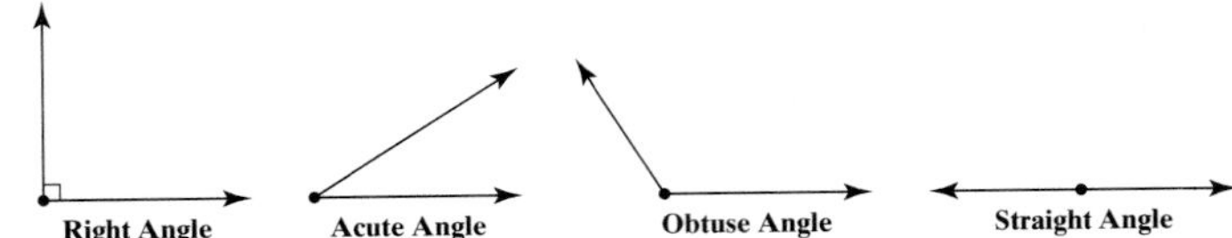

QUICK ✓ *Classify each angle as right, acute, obtuse, or straight.*

1. **2.** 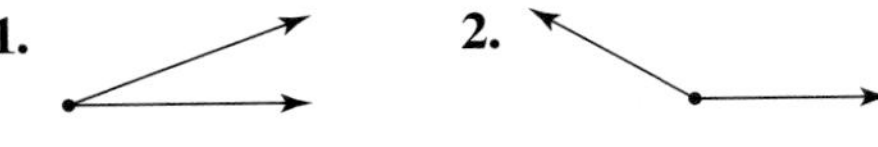 **3.**  **4.** 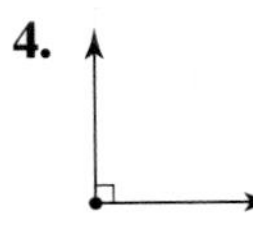

DEFINITION

Two angles whose measures sum to 90° are called **complementary** angles. Each angle is the **complement** of the other. Two angles whose measures sum to 180° are called supplementary angles. Each angle is the **supplement** of the other. See Figure 7.

Figure 7

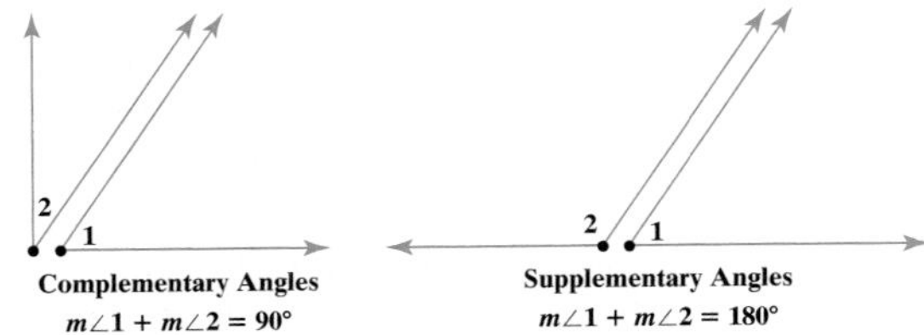

Classroom Example ➔ Find the complement of an angle whose measure is 47°.

Answer: 43°

EXAMPLE 1 Finding the Complement of an Angle

Find the complement of an angle whose measure is 18°.

Solution

Two angles are complementary if their sum is 90°. The measure of an angle that is complementary to an angle whose measure is 18° is $90° - 18° = 72°$. ■

Classroom Example ➔ Find the supplement of an angle whose measure is 62°.

Answer: 118°

EXAMPLE 2 Finding the Supplement of an Angle

Find the supplement of an angle whose measure is 97°.

Solution

Two angles are supplementary if their sum is 180°. The measure of an angle that is supplementary to an angle whose measure is 97° is $180° - 97° = 83°$. ■

QUICK ✓ *Find the complement and the supplement of each angle.*

5. 15° **6.** 60°

3 Find the Measures of Angles Formed by Parallel Lines

Lines that lie in the same plane are called **coplanar.**

DEFINITION

Parallel lines are lines in the same plane that never meet. **Intersecting lines** meet or cross in one point. Two lines that intersect to form right (90°) angles are called **perpendicular lines.**

Figure 8

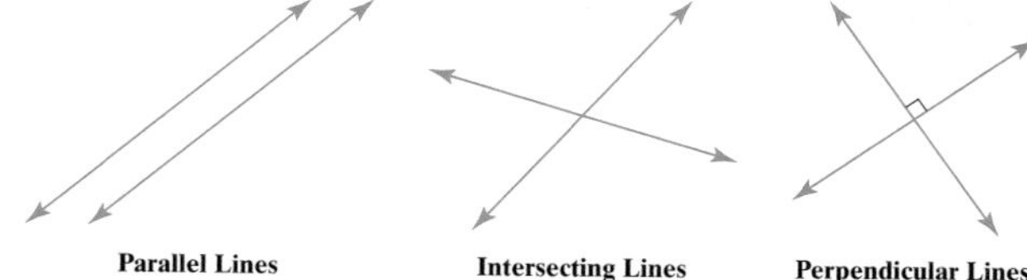

Figure 9
$m\angle 1 = m\angle 3$
$m\angle 2 = m\angle 4$

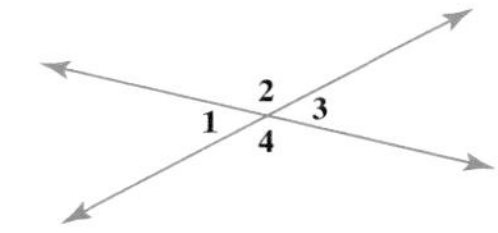

Figure 10

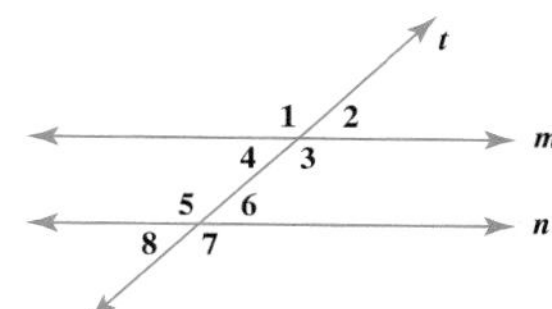

When two lines intersect they form four angles. Two angles that are opposite each other are called **vertical angles.** Vertical angles have equal measures. **Adjacent angles** have the same vertex and share a side. In Figure 9, angles 1 and 3 are vertical angles and angles 1 and 2 are adjacent angles. Angles 1 and 2 are also supplementary angles. Other pairs of adjacent angles are angles 2 and 3, angles 3 and 4, and angles 1 and 4. A line that cuts two parallel lines is called a **transversal.** In Figure 10, lines m and n are parallel and the transversal is labeled t. In this figure, there are certain angles with special names.

- There are 4 pairs of **corresponding angles:**

$$\angle 1 \text{ and } \angle 5, \quad \angle 2 \text{ and } \angle 6, \quad \angle 3 \text{ and } \angle 7, \quad \angle 4 \text{ and } \angle 8$$

- There are 2 pairs of **alternate interior angles:**

$$\angle 3 \text{ and } \angle 5, \quad \angle 4 \text{ and } \angle 6$$

Parallel lines and these angles are related in the following way.

PARALLEL LINES CUT BY A TRANSVERSAL

If two parallel lines are cut by a transversal, then

- Corresponding angles are equal in measure.
- Alternate interior angles are equal in measure.

Classroom Example ➤
Work Example 3, substituting 55° for angle 1.

Answer: $m\angle 2 = 180° - 55° = 125°$
$m\angle 3 = 55°$
$m\angle 4 = 180° - 55° = 125°$
$m\angle 5 = 55°$
$m\angle 6 = 55°$

EXAMPLE 3 Finding the Measure of Corresponding and Alternate Interior Angles

Given that lines m and n are parallel, t is a transversal, and the measure of angle 1 is 85°, find the measure of angles 2, 3, 4, 5 and 6.

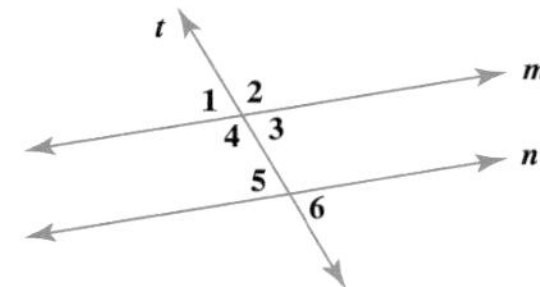

Solution

$m\angle 2 = 180° - 85° = 95°$ because $\angle 1$ and $\angle 2$ are supplementary angles.
$m\angle 3 = 85°$ because $\angle 1$ and $\angle 3$ are vertical angles.
$m\angle 4 = 180° - 85° = 95°$ because $\angle 3$ and $\angle 4$ are supplementary angles.
$m\angle 5 = 85°$ because $\angle 1$ and $\angle 5$ are corresponding angles.
$m\angle 6 = 85°$ because $\angle 1$ and $\angle 6$ are alternate interior angles.

QUICK ✓

7. Find the measure of angles 1–7 given that lines m and n are parallel and t is a transversal.

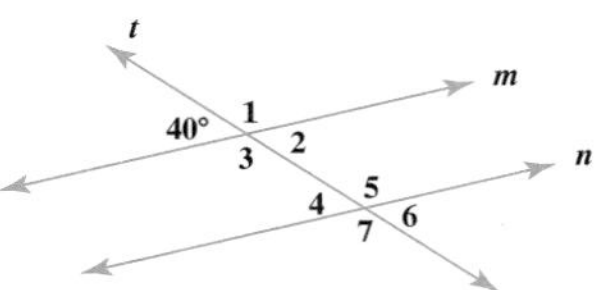

C.1 Exercises

For Extra Help: Student Solutions Manual | CD Video | PH Math/Tutor Center | MathXL Tutorials on CD | MathXL® | MyMathLab

1. congruent
2. angle
3. right
4. acute
5. supplementary
6. True
7. False
8. False
9. acute
10. right
11. right
12. obtuse
13. straight
14. acute
15. obtuse
16. straight
17. 58°
18. 71°
19. 17°
20. 39°
21. 113°
22. 35°
23. 172°
24. 74°
25. $m\angle 1 = 130°$; $m\angle 2 = 50°$; $m\angle 3 = 130°$; $m\angle 4 = 50°$; $m\angle 5 = 130°$; $m\angle 6 = 50°$; $m\angle 7 = 130°$
26. $m\angle 1 = 50°$; $m\angle 2 = 130°$; $m\angle 3 = 50°$; $m\angle 4 = 130°$; $m\angle 5 = 130°$; $m\angle 6 = 50°$; $m\angle 7 = 130°$

Concepts and Vocabulary

In Problems 1–5, fill in the blanks.

1. If two line segments have the same length, they are __________.
2. Two rays that have a common endpoint form a(n) __________.
3. An angle that measures 90° is called a __________ angle.
4. An angle whose measure is between 0° and 90° is called a(an) __________ angle.
5. Two angles whose measures sum to 180° are called __________ angles.

In Problems 6–8, answer True or False to each statement.

6. An angle whose measure is between 90° and 180° is called an obtuse angle.
7. Two intersecting lines form four angles. The two angles that are opposite each other are supplementary.
8. If two parallel lines are cut by a transversal, the corresponding angles are supplementary.

Building Skills

In Problems 9–16, classify each angle as right, acute, obtuse, or straight.

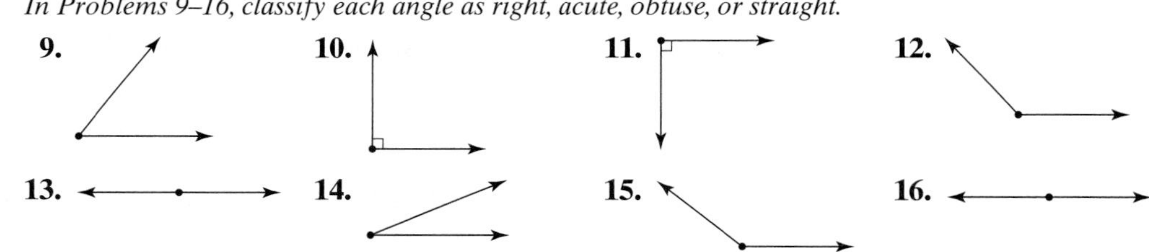

In Problems 17–20, find the complement of each angle.

17. 32° **18.** 19° **19.** 73° **20.** 51°

In Problems 21–24, find the supplement of each angle.

21. 67° **22.** 145° **23.** 8° **24.** 106°

25. Find the measure of angles 1–7 given that lines m and n are parallel and t is a transversal.

26. Find the measure of angles 1–7 given that lines m and n are parallel and t is a transversal.

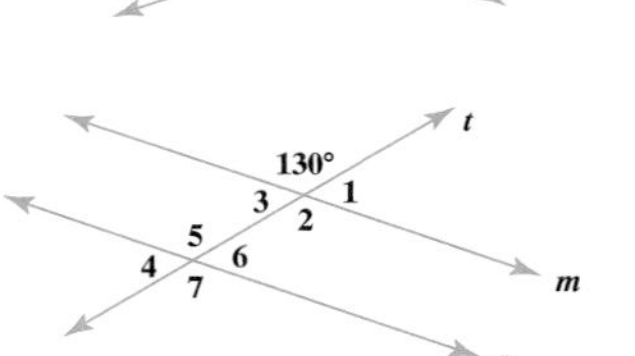

C.2 Polygons

OBJECTIVES

1. Define Polygon
2. Work with Triangles
3. Identify Quadrilaterals
4. Work with Circles

1 Define Polygon

A **polygon** is a closed figure in a plane consisting of line segments that meet at the **vertices.** A **regular polygon** is a polygon in which the sides are congruent and the angles are congruent. Figure 11 shows four regular polygons.

In Words
Vertices is the plural form of the word "vertex." The word polygon comes from the Greek words "poly," which means "many," and "gon," which means "angle."

Figure 11
Regular polygons: All the sides are the same length; all the angles have the same measure.

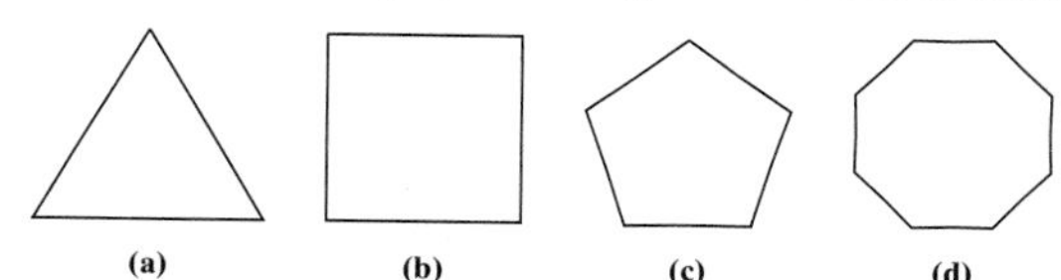

A polygon is named according to the number of sides. Table 1 summarizes the names of the polygons with 3 to 10 sides. A **triangle** is a polygon with three sides. Figure 11(a) is a regular triangle. A **quadrilateral** is a polygon with four sides. Figure 11(b) is a regular quadrilateral (also known as a *square*). A **pentagon** is a polygon with five sides. Figure 5(c) is a regular pentagon. An **octagon** is a polygon with eight sides. Figure 5(d) shows a regular octagon.

Table 1
Polygons

Number of Sides	Name of Polygon
3	Triangle
4	Quadrilateral
5	Pentagon
6	Hexagon
7	Heptagon
8	Octagon
9	Nonagon
10	Decagon

2 Work with Triangles

Figure 12 shows triangle ABC. In triangle ABC angles A, B and C are called **interior angles.** The sum of the measures of the interior angles of a triangle is 180°. If x, y, and z represent the measures of angles A, B, and C, respectively, then

$$x + y + z = 180°$$

Figure 12

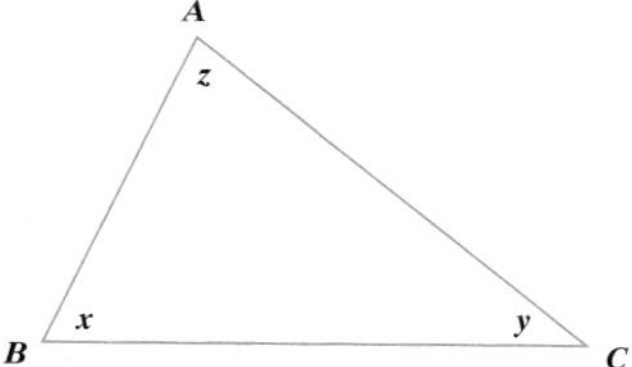

Classroom Example ➤
Find the measure of angle B in the triangle

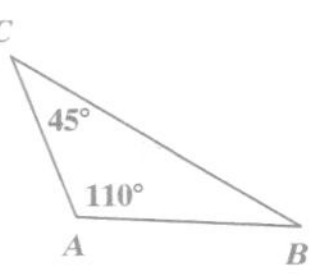

Answer: $m\angle B = 25°$

EXAMPLE 1 Finding the Measure of an Interior Angle of a Triangle

Find the measure of angle C in the triangle.

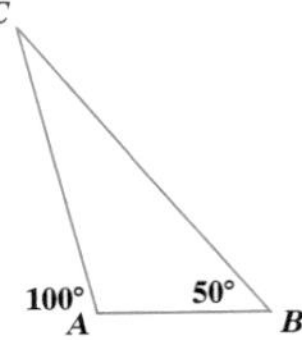

Solution

The measure of angle A is 100° and the measure of angle B is 50°. We compute the measure of angle C as

$$m\angle C = 180° - 50° - 100° = 30°$$

We classify triangles by the lengths of their sides. A triangle in which all three sides are congruent is called an **equilateral** triangle. A triangle in which two sides are congruent is called an **isosceles** triangle, and a triangle in which none of the sides are congruent is called a **scalene** triangle. See Figure 13.

Figure 13

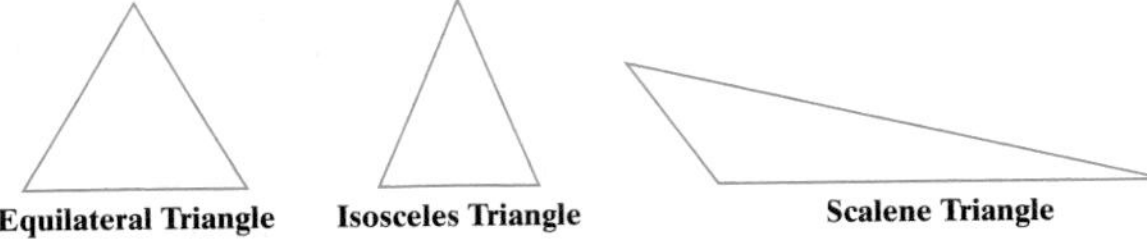

A **right triangle** is a triangle that contains a right (90°) angle. In a right triangle, the longest side is called the hypotenuse, and the remaining two sides are called **legs.** See Figure 14.

Figure 14

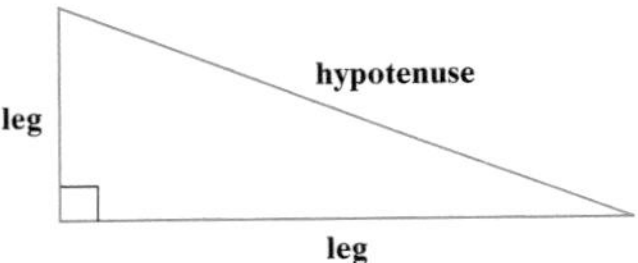

Classroom Example ➤
Find the measure of angle C in the triangle

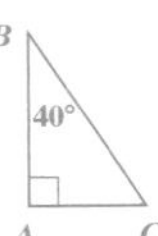

Answer: $m\angle C = 50°$

EXAMPLE 2 Finding the Measure of an Angle of a Right Triangle

Find the measure of angle B in the right triangle.

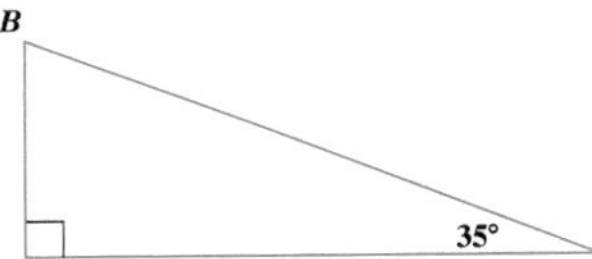

Solution

There are 180° degrees in a triangle and the right angle measures 90° so

$$m\angle B = 180° - 90° - 35° = 55°$$

QUICK ✓ *Find the measure of each angle B in the triangle.*

1.
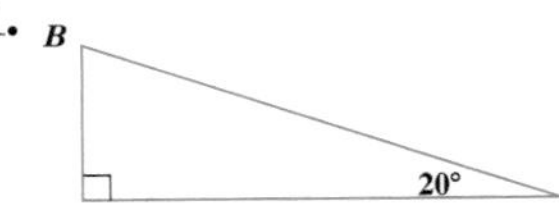

2.
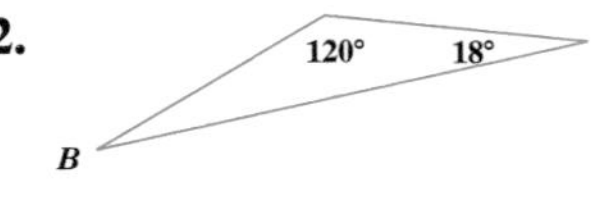

Congruent Triangles

Two triangles are congruent if the corresponding angles have the same measure and the corresponding sides have the same length. See Figures 15(a) and (b). We see that the corresponding angles in triangle (a) and (b) are equal. Also, the lengths of the corresponding sides are equal: $a = d, b = e$, and $c = f$.

Figure 15

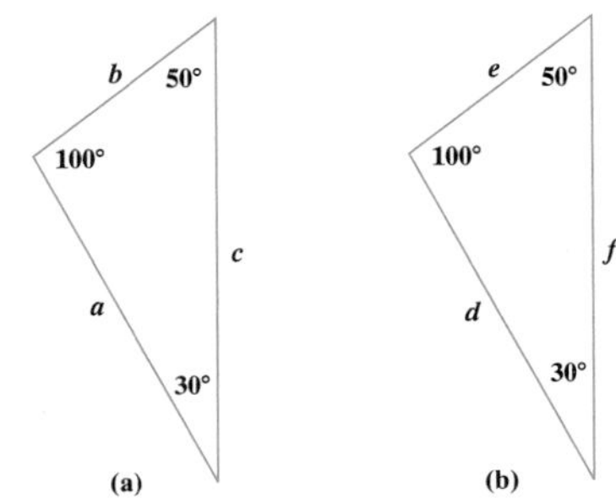

It is not necessary to verify that all three angles and all three sides are the same measure to determine whether two triangles are congruent.

Determining Congruent Triangles

1. Two triangles are congruent if two of the angles are equal and the lengths of the corresponding sides between the two angles are equal. See Figure 16(a).
2. Two triangles are congruent if the lengths of the corresponding sides of the triangle are equal. See Figure 16(b).
3. Two triangles are congruent if the lengths of two corresponding sides are equal and the measures of the angles between the two sides are equal. See Figure 16(c).

Figure 16

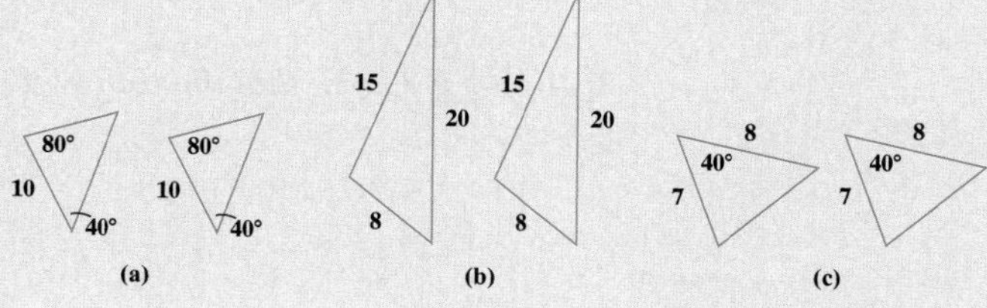

Similar Triangles

Figure 17

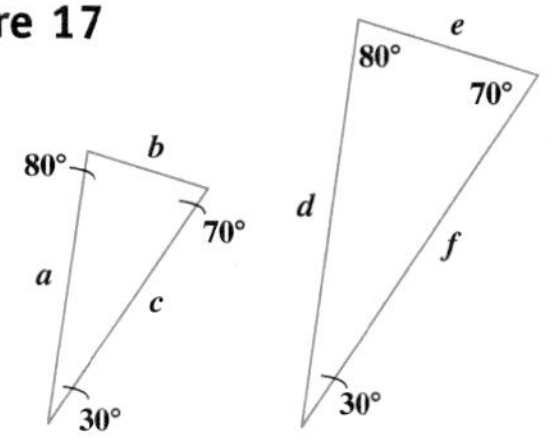

Two triangles are **similar** if they have the same shape. That is, the triangles are similar if the corresponding angles are equal and the lengths of the corresponding sides are proportional. In Figure 17 the triangles are similar because the corresponding angles are equal and the corresponding sides are proportional: $\frac{d}{a} = \frac{e}{b} = \frac{f}{c}$. It is not necessary to verify that all three angles are congruent and all three sides are proportional to determine whether two triangles are similar.

Determining Similar Triangles

1. Two triangles are similar if two of the corresponding angles are equal. See Figure 18(a).
2. Two triangles are similar if the lengths of the all three sides of each triangle are proportional. See Figure 18(b).
3. Two triangles are similar if two corresponding sides are proportional and the angles between the two sides are congruent. See Figure 18(c).

Figure 18

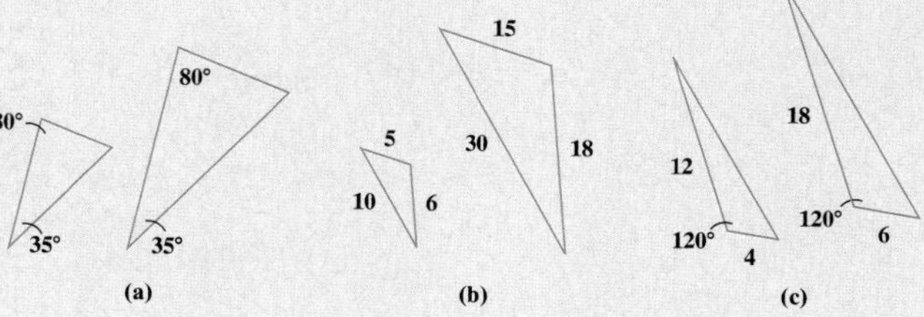

Classroom Example ➔
Find the missing length in these two similar triangles.

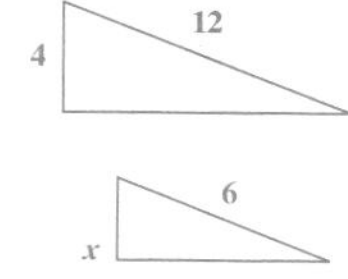

Answer: $x = 2$ units

EXAMPLE 3

Given that the triangles in Figure 19 are similar, find the missing length.

Figure 19

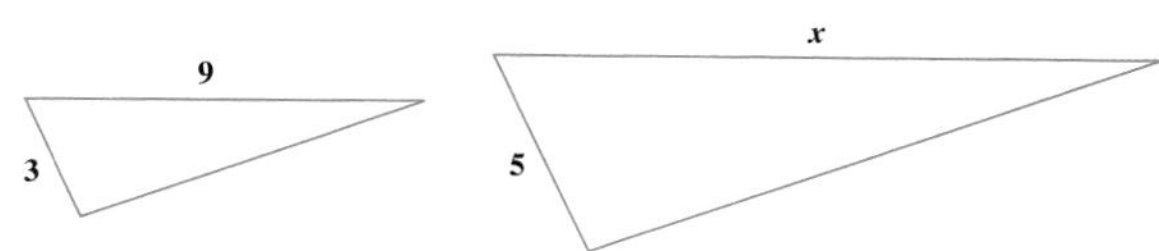

Solution

Because the triangles are similar, the corresponding sides are proportional. That is, $\frac{3}{5} = \frac{9}{x}$. We solve this equation for x.

$$\frac{3}{5} = \frac{9}{x}$$

Multiply both sides by the LCD, $5x$: $$5x \cdot \left(\frac{3}{5}\right) = 5x \cdot \left(\frac{9}{x}\right)$$

Simplify: $$3x = 45$$

Divide both sides by 3: $$x = 15$$

The missing length is 15 units.

QUICK ✓ *Given that the following triangles are similar, find the missing length.*

3.

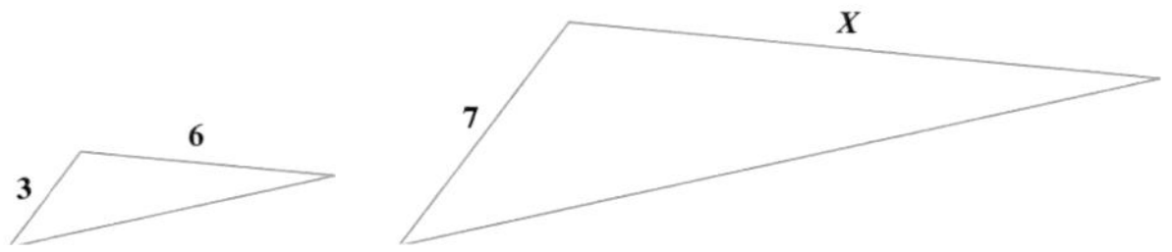

3 Identify Quadrilaterals

A **quadrilateral** is a polygon with four sides. A **parallelogram** is a quadrilateral in which both pairs of opposite sides are parallel. A **rectangle** is a parallelogram that contains a right angle. A **square** is a rectangle with all sides of equal length. A **rhombus** is a parallelogram that has all sides equal in length. A **trapezoid** is a quadrilateral with exactly one pair of opposite sides that are parallel.

Teaching Tip
Students should be able to identify the difference between a square and a rhombus. A square must have right angles (because it's a rectangle), but a rhombus doesn't.

Figure	Sketch
Parallelogram	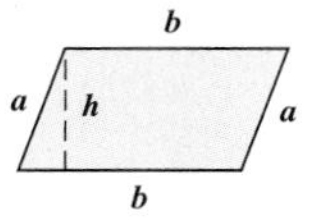
Rectangle	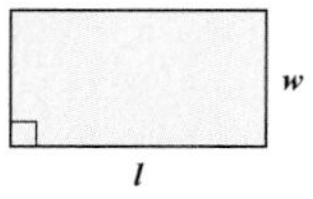
Square	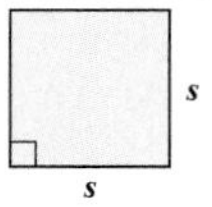
Rhombus	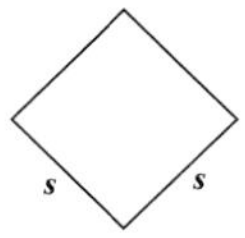
Trapezoid	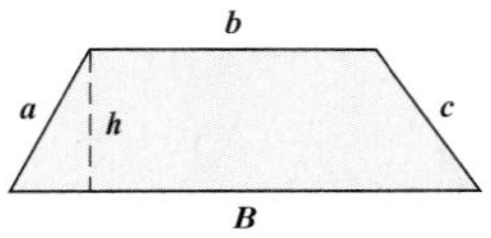

4 Work with Circles

Figure 20

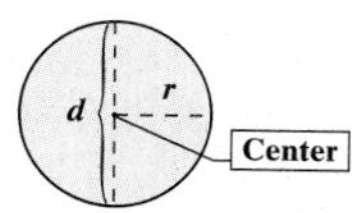

A *circle* is also a plane figure. A **circle** is a figure made up of all points in the plane that are a fixed distance from a point called the **center.** The **radius** of the circle is the line segment drawn from the center of the circle to any point on the circle. The **diameter** of the circle is the line segment which has endpoints on the circle. The diameter passes through the center of the circle. See Figure 20. Notice that the diameter d of a circle is twice the radius r. That is, $d = 2r$.

Classroom Example ➢
Find the length of the diameter of the circle with radius 12 inches.
Answer: 24 inches

EXAMPLE 4 Finding the Length of the Diameter of a Circle

Find the length of the diameter of a circle with radius 4 cm.

Solution

The length of the diameter is twice the length of the radius.

$$d = 2 \cdot r$$
$$d = 2 \cdot 4 \text{ cm}$$
$$d = 8 \text{ cm}$$

The diameter of the circle is 8 cm.

Classroom Example ➢
Find the length of the radius of the circle with diameter 6 feet.
Answer: 3 feet

EXAMPLE 5 Finding the Length of the Radius of a Circle

Find the length of the radius of the circle with diameter 18 yards.

Solution

The length of the radius is one-half the length of the diameter.

$$r = \frac{1}{2} \cdot d$$
$$r = \frac{1}{2} \cdot 18 \text{ yards}$$
$$r = 9 \text{ yards}$$

The radius of the circle is 9 yards.

QUICK ✓ *Find the length of the radius or diameter of each circle.*

4. $d = 15$ inches, find r.
5. $d = 24$ feet, find r.
6. $r = 3.6$ yards, find d.
7. $r = 9$ cm, find d.

C.2 Exercises

Concepts and Vocabulary

In Problems 1–4, fill in the blanks.

1. regular
2. 180
3. isosceles
4. similar

1. A _______ polygon is one in which the sides are congruent and the angles are congruent.

2. The sum of the measures of the interior angles of a triangle is _______ degrees.

3. A(n) _______ triangle is a triangle that has two congruent sides.

4. Two triangles are _______ if the measures of the angles of the triangle are equal and the sides are proportional.

5. True
6. False
7. True
8. True
9. 55°
10. 83°
11. 48°
12. 46°
13. 4 units
14. 8 units
15. 67.5 units
16. 40 units
17. 10 inches
18. 32 feet
19. 5 cm
20. 11.8 in.
21. 7 cm
22. 29 inches
23. $\frac{11}{2}$ yards or 5.5 yards
24. $\frac{27}{2}$ feet or 13.5 feet

In Problems 5–8, answer True or False to each statement.

5. A rhombus is a parallelogram that has all sides equal in length.

6. A heptagon is a polygon with six sides.

7. The diameter of a circle is exactly twice the length of the radius.

8. In triangle ABC, if $m\angle A = 83°$ and $m\angle B = 47°$, then $m\angle C = 50°$.

Building Skills

In Problems 9–12, find the measure of the missing angle of the triangle.

9.

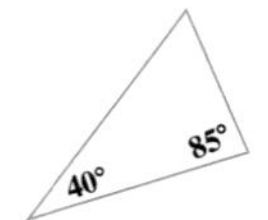

10.

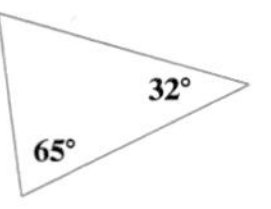

11.

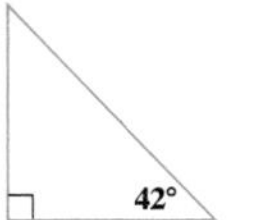

12.

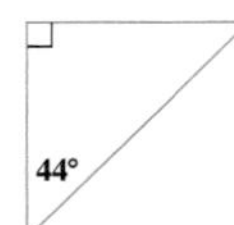

In Problems 13–16, determine the length of the missing side of the triangle. (These are similar triangles.)

13.

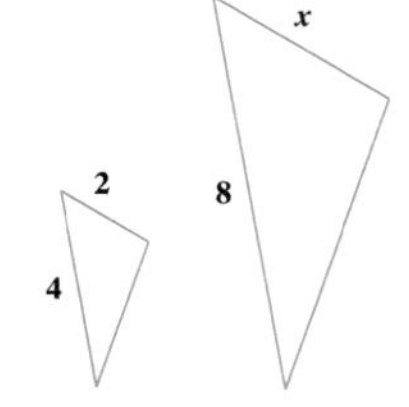

14.

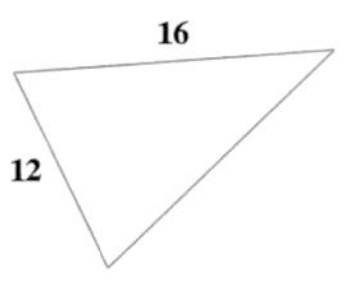

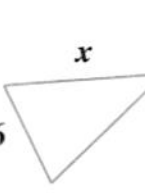

15.

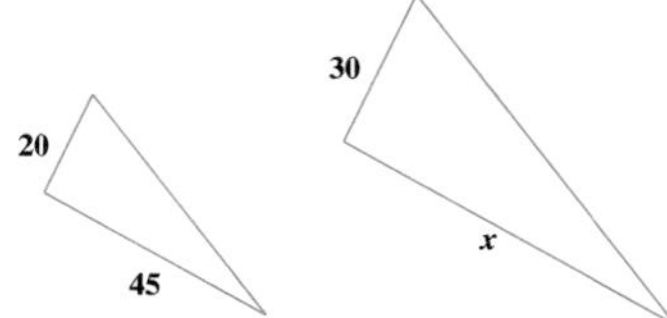

16.

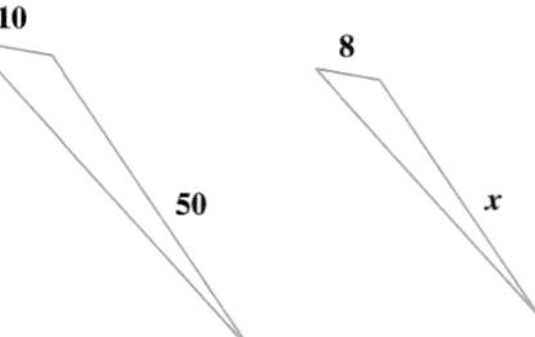

In Problems 17–20, find the length of the diameter of the circle.

17. $r = 5$ in **18.** $r = 16$ feet **19.** $r = 2.5$ cm **20.** $r = 5.9$ in

In Problems 21–24, find the length of the radius of the circle.

21. $d = 14$ cm **22.** $d = 58$ inches **23.** $d = 11$ yards **24.** $d = 27$ feet

C.3 Perimeter and Area of Polygons and Circles

OBJECTIVES

1. Find the Perimeter and Area of a Rectangle and a Square
2. Find the Perimeter and Area of a Parallelogram and a Trapezoid
3. Find the Perimeter and Area of a Triangle
4. Find the Circumference and Area of a Circle

The first concept we introduce in this section is the concept of *perimeter*. The **perimeter** of a polygon is the distance around the polygon. Put another way, the perimeter of a polygon is the sum of the lengths of the sides.

1 Find the Perimeter and Area of a Rectangle and a Square

A rectangle is a polygon, so the perimeter of a rectangle is the sum of the lengths of the sides.

Classroom Example ➤
Find the perimeter of the rectangle.

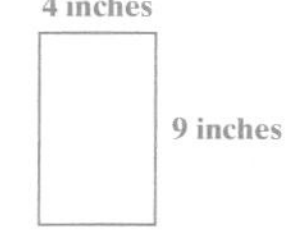

Answer: 26 inches

EXAMPLE 1 **Finding the Perimeter of a Rectangle**

Find the perimeter of the rectangle in Figure 21.

Figure 21

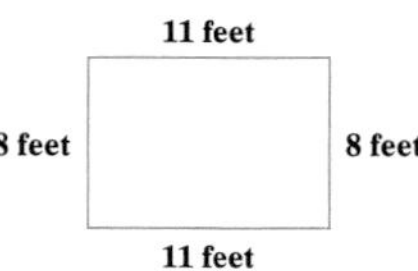

Solution

The perimeter of the rectangle is the sum of the lengths of the sides, so

$$\begin{aligned}\text{perimeter} &= 11\text{ feet} + 8\text{ feet} + 11\text{ feet} + 8\text{ feet}\\ &= 38\text{ feet}\end{aligned}$$

Did you notice that the perimeter from Example 1 can also be written as

$$\text{perimeter} = 2 \cdot 11\text{ feet} + 2 \cdot 8\text{ feet?}$$

In general, the perimeter of a rectangle is written $P = 2l + 2w$.

A different measure of a polygon is its *area*. The **area** of a polygon is the amount of surface the polygon covers. Consider the rectangle shown in Figure 22. If we count the number of 1-unit by 1-unit squares within the rectangle, we see that the area of the rectangle is 6 square units. The area can also be found by multiplying the number of units of length by the number of units of width.

Figure 22

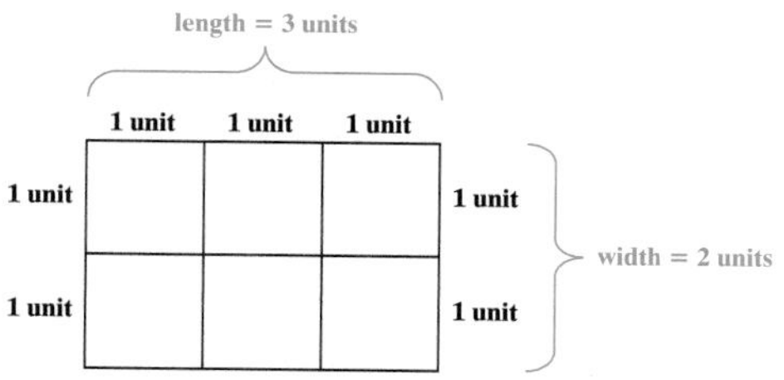

We can generalize this result to say that the area of a rectangle is the product of its length and width.

Classroom Example Find the area of the rectangle in Classroom Example 1.

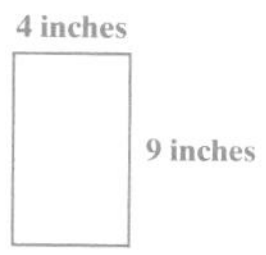

Answer: 36 square inches

EXAMPLE 2 Finding the Area of a Rectangle

Find the area of the rectangle in Figure 23.

Figure 23

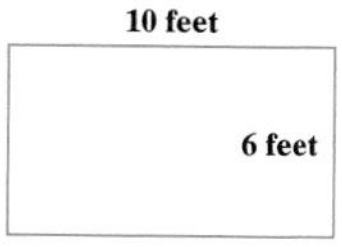

Solution

The area of the rectangle is the product of the length and the width, so

$$\begin{aligned}\text{area} &= 6\text{ feet} \cdot 10\text{ feet}\\ &= 60\text{ square feet}\end{aligned}$$

We give a summary of how to find the perimeter and area of a rectangle.

Figure	Sketch	Perimeter	Area
Rectangle	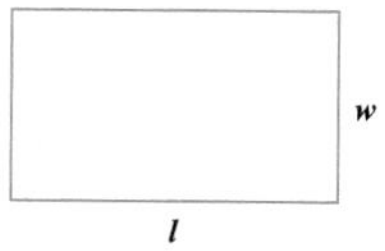	$P = 2l + 2w$	$A = lw$

Classroom Example Find the (a) perimeter and (b) area of the rectangle.

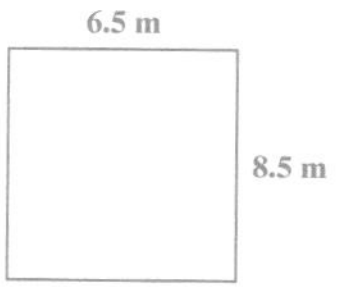

Answer:

(a) 30 m **(b)** 55.25 square m

EXAMPLE 3 Finding the Perimeter and Area of a Rectangle

Find (a) the perimeter and (b) the area of the rectangle shown in Figure 24.

Figure 24

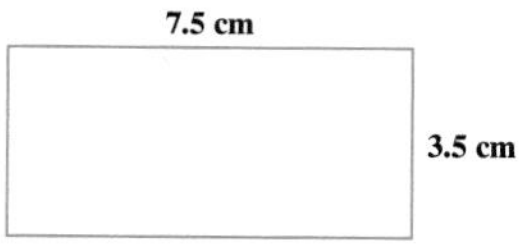

Solution

(a) The perimeter of the rectangle is

$$\begin{aligned} P &= 2l + 2w \\ P &= 2 \cdot 7.5 \text{ cm} + 2 \cdot 3.5 \text{ cm} \\ &= 15 \text{ cm} + 7 \text{ cm} \\ &= 22 \text{ cm} \end{aligned}$$

The perimeter of the rectangle is 22 cm.

(b) The area of the rectangle is

$$\begin{aligned} A &= lw \\ &= 7.5 \text{ cm} \cdot 3.5 \text{ cm} \\ &= 26.25 \text{ square cm} \end{aligned}$$

The area of the given rectangle is 26.25 square cm. ■

QUICK ✓ *Find the perimeter and area of each rectangle.*

1. **2.**

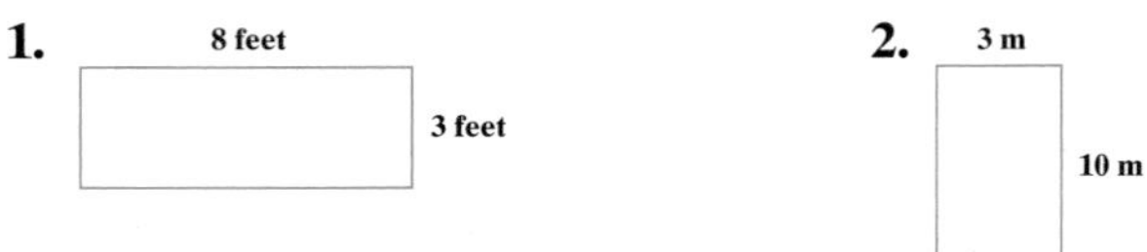

We can find the perimeter and area of a square using the same methods we use to find the perimeter and area of a rectangle. A square is a rectangle that has four congruent sides, so to find the perimeter of a square we use

$$\text{perimeter} = \text{side} + \text{side} + \text{side} + \text{side} = 4 \cdot \text{side} = 4s$$

where s is the length of a side.

Classroom Example ➤
Find the perimeter of square with sides that measure 5 m.
Answer: 20 m

EXAMPLE 4 Finding the Perimeter of a Square

Find the perimeter of the square in Figure 25.

Figure 25

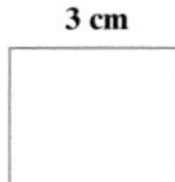

Solution

The perimeter of the square is

$$\begin{aligned} \text{perimeter} &= 4 \cdot s \\ &= 4 \cdot 3 \text{ cm} \\ &= 12 \text{ cm} \end{aligned}$$ ■

We know that the area of a rectangle is found by finding the product of the length and the width. In a square, the sides are congruent, so

$$\text{area} = \text{side} \cdot \text{side} = \text{side}^2$$

Classroom Example ➤
Find the area of the square with sides that measure 9 feet.
Answer: 81 square feet

EXAMPLE 5 Finding the Area of a Square

Find the area of the square in Figure 26.

Figure 26

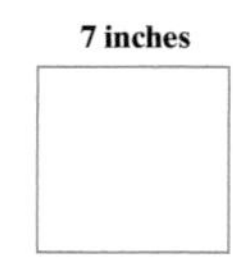

Solution

The area of the square is

$$\begin{aligned} \text{area} &= \text{side}^2 \\ &= (7 \text{ inches})^2 \\ &= 49 \text{ square inches} \end{aligned}$$ ■

We summarize the formulas for the perimeter and area of a square in the following table.

Figure	Sketch	Perimeter	Area
Square	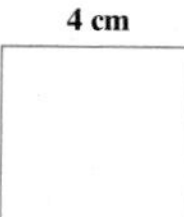	$P = 4s$	$A = s^2$

QUICK ✓ *Find the perimeter and area of each square.*

3. 4 cm

4. 1.5 yards

Classroom Example ➤
Find the (a) perimeter and (b) area of the figure:

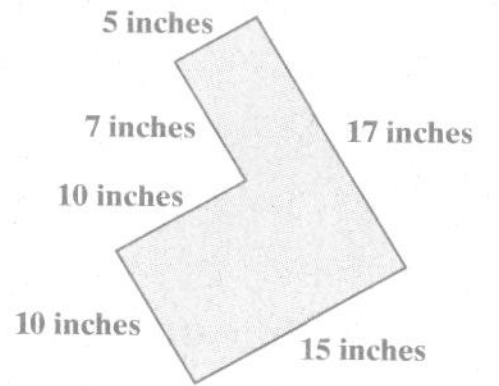

Answer:
(a) 61 inches **(b)** 185 inches^2

EXAMPLE 6 Finding the Perimeter and Area of a Geometric Figure

Find (a) the perimeter and (b) the area of the region shown in Figure 27.

Figure 27

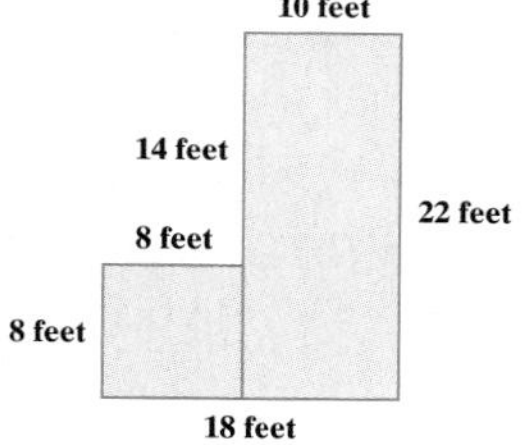

Solution

(a) The perimeter is the distance around the polygon. So,

$$\begin{aligned}\text{Perimeter} &= 8 \text{ feet} + 8 \text{ feet} + 14 \text{ feet} + 10 \text{ feet} + 22 \text{ feet} + 18 \text{ feet}\\ &= 80 \text{ feet}\end{aligned}$$

(b) Notice the region can be divided into an 8-foot by 8-foot square plus a 10-foot by 22-foot rectangle. To find the area, we find the area of the square and add it to the area of the rectangle.

$$\begin{aligned}\text{Area} &= \text{Area of square} + \text{Area of rectangle}\\ &= (8 \text{ feet})^2 + (10 \text{ feet})(22 \text{ feet})\\ &= 64 \text{ feet}^2 + 220 \text{ feet}^2\\ &= 284 \text{ feet}^2\end{aligned}$$

QUICK ✓ *Find the perimeter and area of the figure.*

5.

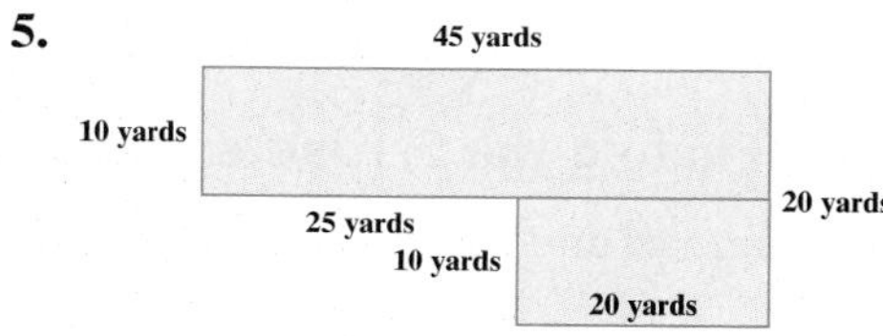

Classroom Example ➤
Use the same scenario as in Example 7. (a) Find the area of a rectangular room that is 11 feet long by 13 feet wide with 9 foot high ceilings. (b) How many gallons of paint are needed?

Answer:
(a) 432 square feet (b) 2 gallons

EXAMPLE 7 Painting a Room

You've decided to paint your rectangular bedroom. Two walls are 14 feet long and 7 feet high, and the other two walls are 10 feet long and 7 feet high.

(a) Ignoring the window and door openings in the bedroom, what is the area of the walls in your bedroom?

(b) You know that one gallon of paint will cover 300 square feet. A one-gallon can of paint costs \$21.99. How many one-gallon cans of paint must you purchase to paint your bedroom?

Solution

(a) The bedroom consists of two walls that are 10 feet long and 7 feet high. The area of these two walls is

$$\begin{aligned} \text{area} &= l \cdot w \cdot 2 \\ \text{area} &= 14 \text{ feet} \cdot 7 \text{ feet} \cdot 2 \\ &= 196 \text{ square feet} \end{aligned}$$

The area of the other two walls is

$$\begin{aligned} \text{area} &= l \cdot w \cdot 2 \\ \text{area} &= 10 \text{ feet} \cdot 7 \text{ feet} \cdot 2 \\ &= 140 \text{ square feet} \end{aligned}$$

The total area to be painted is

$$\begin{aligned} \text{area} &= 196 \text{ square feet} + 140 \text{ square feet} \\ &= 336 \text{ square feet} \end{aligned}$$

(b) A one-gallon can of paint covers 300 square feet. You have 336 square feet, so you will need to purchase 2 gallons of paint.

2 Find the Perimeter and Area of a Parallelogram and a Trapezoid

Recall that a parallelogram is a quadrilateral in which opposite sides are parallel. A trapezoid is a quadrilateral with exactly one pair of opposite sides that are parallel. The following table gives the formulas for the perimeter and area of parallelograms and trapezoids.

Figure	Sketch	Perimeter	Area
Parallelogram		$P = 2a + 2b$	$A = b \cdot h$
Trapezoid		$P = a + b + c + B$	$A = \frac{1}{2}h(b + B)$

Classroom Example ➤
Find (a) the perimeter and (b) the area of parallelogram.

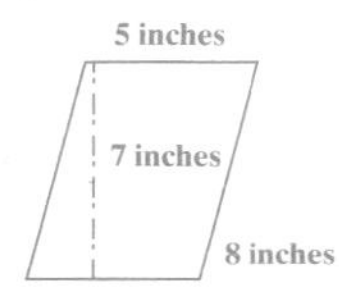

Answer:
(a) 26 inches
(b) 35 square inches

EXAMPLE 8 Finding the Perimeter and Area of a Parallelogram

Find (a) the perimeter and (b) the area of parallelogram shown in Figure 28.

Figure 28

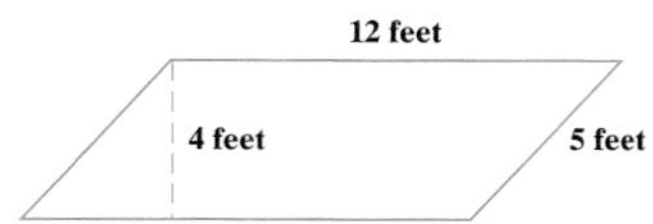

Solution

(a) The perimeter of the parallelogram is

$$\begin{aligned} P &= 2l + 2w \\ &= 2 \cdot 12 \text{ feet} + 2 \cdot 5 \text{ feet} \\ &= 24 \text{ feet} + 10 \text{ feet} \\ &= 34 \text{ feet} \end{aligned}$$

The perimeter of the parallelogram is 34 feet.

(b) The area of the parallelogram is

$$\begin{aligned} A &= b \cdot h \\ &= 12 \text{ feet} \cdot 4 \text{ feet} \\ &= 48 \text{ square feet} \end{aligned}$$

The area of the parallelogram is 48 square feet.

Classroom Example Find (a) the perimeter and (b) the area of trapezoid.

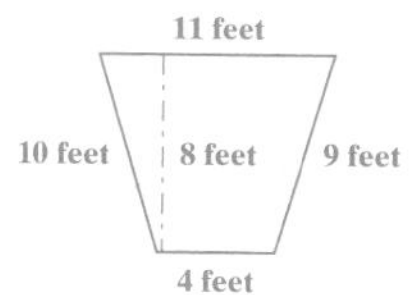

Answer:
(a) 34 feet (b) 60 square feet

EXAMPLE 9 Finding the Perimeter and Area of a Trapezoid

Find (a) the perimeter and (b) the area of trapezoid shown in Figure 29.

Figure 29

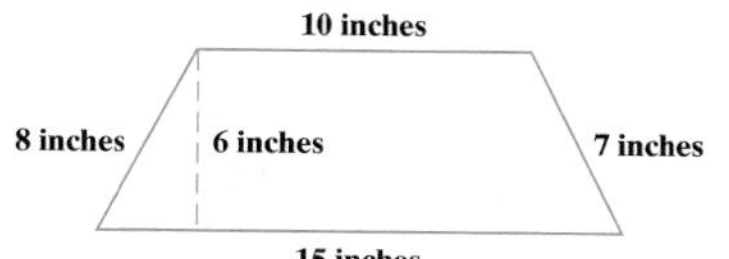

Solution

(a) The perimeter of the trapezoid is

$$\begin{aligned} \text{Perimeter} &= 7 \text{ inches} + 8 \text{ inches} + 10 \text{ inches} + 15 \text{ inches} \\ &= 40 \text{ inches} \end{aligned}$$

The perimeter of the trapezoid is 40 inches.

(b) The area of the trapezoid is

$$\begin{aligned} A &= \frac{1}{2}h(b + B) \\ &= \frac{1}{2} \cdot 6 \text{ inches} \cdot (7 \text{ inches} + 10 \text{ inches}) \\ &= \frac{1}{2} \cdot 6 \text{ inches} \cdot 17 \text{ inches} \\ &= 51 \text{ square inches} \end{aligned}$$

The area of the trapezoid is 51 square inches.

QUICK ✓ *Find the perimeter and area of each figure.*

6.

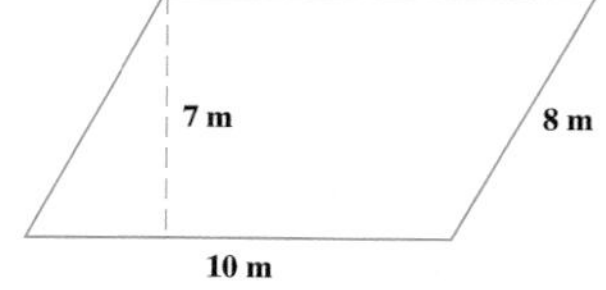

7.

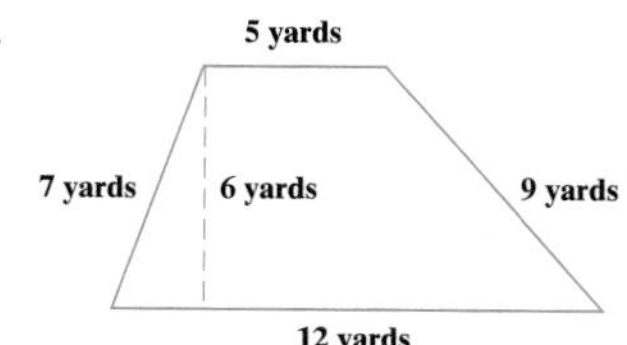

3 Find the Perimeter and Area of a Triangle

Recall that a triangle is a polygon with three sides. The following table gives the formulas for the perimeter and area of a triangle.

Figure	Sketch	Perimeter	Area
Triangle	a, h, c, b	$P = a + b + c$	$A = \frac{1}{2}bh$

Classroom Example ➤
Find (a) the perimeter and (b) the area of trapezoid.

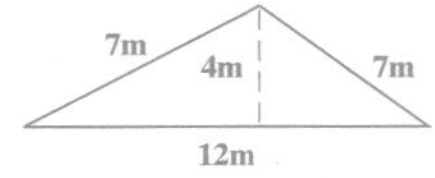

Answer:
(a) 26 m (b) 24 m^2

EXAMPLE 10 Finding the Perimeter and Area of a Triangle

Find (a) the perimeter and (b) the area of the triangle shown in Figure 30.

Figure 30

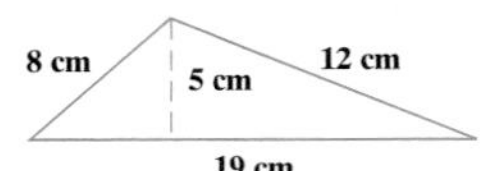

Solution

(a) To find the perimeter of the triangle, add the lengths of the three sides of the triangle.

$$\begin{aligned}\text{perimeter} &= a + b + c \\ &= 8\text{ cm} + 12\text{ cm} + 19\text{ cm} \\ &= 39\text{ cm}\end{aligned}$$

The perimeter of the triangle shown in Figure 30 is 39 cm.

(b) We use $A = \frac{1}{2}bh$ with base $= b = 19$ cm and height $= h = 5$ cm.

$$A = \frac{1}{2}bh = \frac{1}{2} \cdot 19\text{ cm} \cdot 5\text{ cm} = 47.5\text{ square cm}$$

The area of the triangle in Figure 30 is 47.5 square cm. ■

QUICK ✓ *Find the perimeter and area of each triangle.*

8.

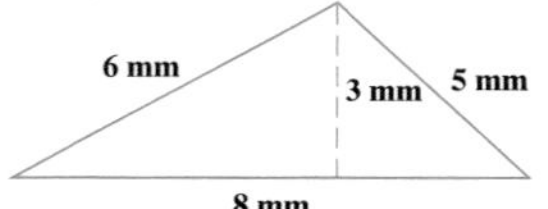

9.

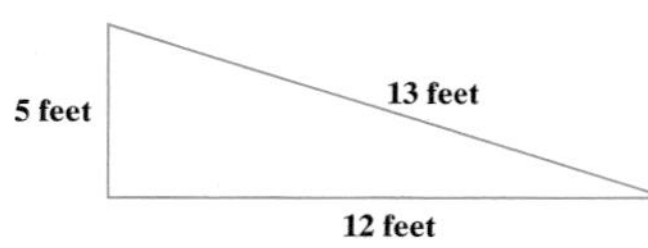

4 Find the Circumference and Area of a Circle

The **circumference** of a circle is the distance around a circle. We use the diameter or the radius of the circle to find the circumference of a circle according to the formulas given below. We also give the formula for the area of a circle.

Figure	Sketch	Perimeter	Area
Circle	d, r, Center	$C = \pi d$ where d is the diameter $C = 2\pi r$ where r is the radius	$A = \frac{1}{2}\pi r^2$

Classroom Example

Find the circumference of the circles with (a) radius of 5 inches and (b) diameter of 15 cm.

Answer:

(a) 10π inches; 31.42 inches

(b) 15π cm; 47.12 cm

Figure 31

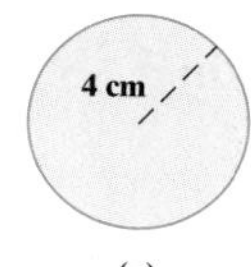

(a)

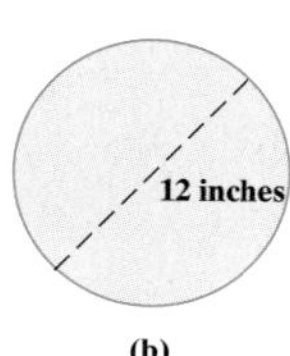

(b)

EXAMPLE 11 Finding the Circumference of a Circle

Find the circumference of the circles in Figure 31.

Solution

(a) We know the length of the radius is 4 cm so we use the formula $C = 2\pi r$.

$$C = 2\pi r$$

$r = 4$ cm: $= 2 \cdot \pi \cdot 4$ cm

$= 8 \cdot \pi$ cm

Use a calculator: ≈ 25.13 cm

The exact circumference of the circle is 8π cm, and 25.13 cm is the approximate circumference, to the nearest hundredth of a centimeter.

(b) The length of the diameter is given to be 12 inches, so we use $C = \pi d$.

$$C = \pi d$$

$d = 12$ inches: $= \pi \cdot 12$ inches

$= 12\ \pi$ inches

Use a calculator: ≈ 37.70 inches

The exact circumference of the circle is 12π in., and 37.70 in. is the approximate circumference, to the nearest hundredth of an inch.

Classroom Example

Find the area of a circle with radius of 3 inches.

Answer: 9π square inches; 28.27 square inches

Figure 32

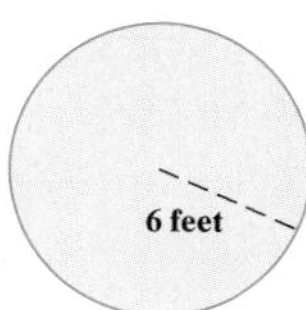

EXAMPLE 12 Finding the Area of a Circle

Find the area of the circle in Figure 32.

Solution

The circle in Figure 32 has radius 6 feet, so we substitute $r = 6$ in the equation $A = \frac{1}{2}\pi r^2$.

$$A = \frac{1}{2}\pi r^2$$

$$= \frac{1}{2} \cdot \pi \cdot (6 \text{ feet})^2$$

$$= \frac{1}{2} \cdot \pi(36) \text{ square feet}$$

$$= 18\pi \text{ square feet}$$

Use a calculator: ≈ 56.55 square feet

The area of the circle is exactly 18π square feet or approximately 56.55 square feet.

QUICK ✓ *Find the circumference and area of each circle.*

10.

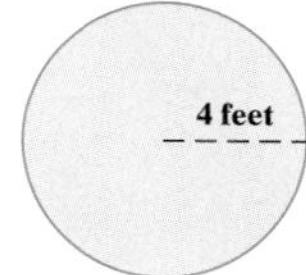

11.

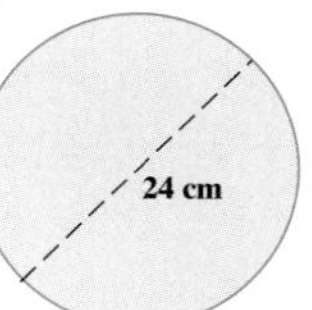

C.3 Exercises

For Extra Help:  Student Solutions Manual  CD Video  PH Math/Tutor Center  MathXL Tutorials on CD | MathXL® | MyMathLab

1. perimeter
2. circumference
3. $\frac{1}{2}h(b + B)$; h; b; B
4. False
5. False
6. False
7. Perimeter: 28 feet; Area: 40 square feet
8. Perimeter: 40 miles; Area: 96 square miles
9. Perimeter: 40 m; Area: 75 m^2
10. Perimeter: 44 cm; Area: 40 cm^2
11. Perimeter: 24 km; Area: 36 km^2
12. Perimeter: 52 yards; Area: 169 square yards
13. Perimeter: 72 feet; Area: 218 square feet
14. Perimeter: 26 m; Area: 30 m^2
15. Perimeter: 54 m; Area: 62 m^2
16. Perimeter: 120 yards; Area: 425 square yards

Concepts and Vocabulary

In Problems 1–3, fill in the blanks.

1. The __________ of a polygon is the distance around the polygon.

2. The __________ of a circle is the distance around the circle.

3. To find the area of a trapezoid, we use the formula A = __________, where __________ is the height of the trapezoid and the bases have lengths __________ and __________.

In Problems 4–6, answer True or False to each statement.

4. The area of a circle is $A = \frac{1}{2}\pi d^2$.

5. The length of the radius of a circle is twice the length of the diameter.

6. The area of a square is the sum of the lengths of the sides.

Building Skills

In Problems 7–10, find the perimeter and area of each rectangle.

7.

10 feet

4 feet

8.

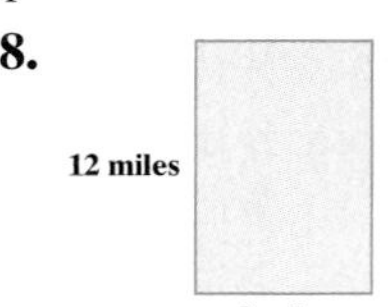

9.

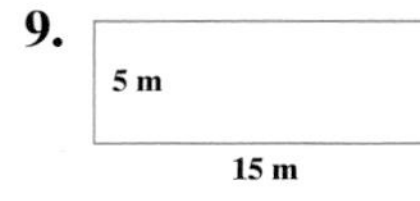

10.

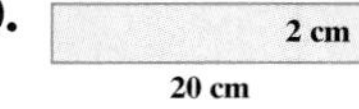

In Problems 11 and 12, find the perimeter and area of each square.

11.

6 km

6 km

12.

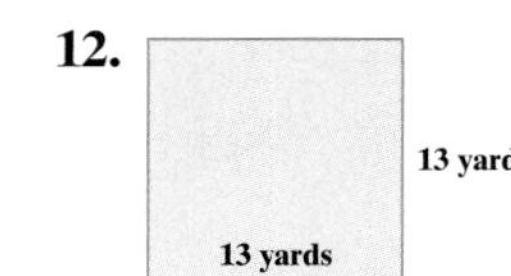

In Problems 13–16, find the perimeter and area of each figure.

13.

7 feet

6 feet

15 feet

14 feet

8 feet

22 feet

14.

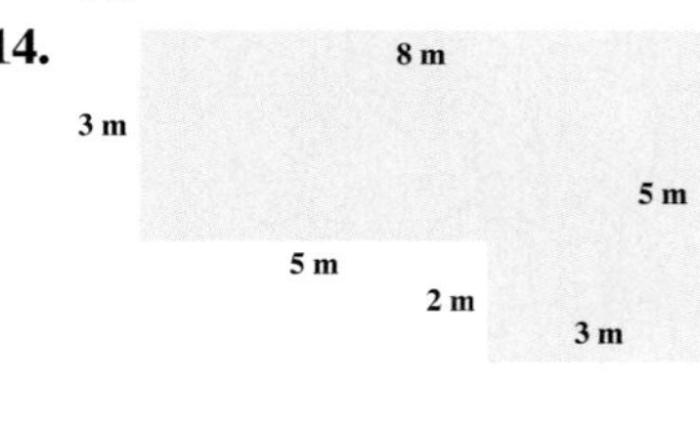

15.

13 m

2 m

8 m

2 m

8 m

2 m

16.

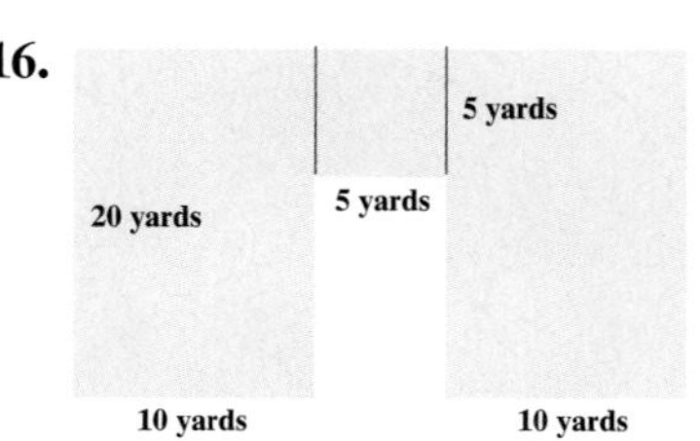

17. Perimeter: 30 feet; Area: 45 square feet
18. Perimeter: 36 cm; Area: 28 cm^2
19. Perimeter: 28 mm; Area: 36 mm^2
20. Perimeter: 56 in.; Area: 152 $in.^2$
21. Perimeter: 40 in.; Area: 84 $in.^2$
22. Perimeter: 36 m; Area: 60 m^2
23. Perimeter: 45 cm; Area: 94.5 cm^2
24. Perimeter: 45 feet; 77.5 square feet
25. Perimeter: 32 m; Area: 42 m^2
26. Perimeter: 28 in.; Area: 36 $in.^2$
27. Perimeter: 32 ft; Area: 24 ft^2
28. Perimeter: 46 km; Area: 48 km^2
29. Circumference: 32π in. ≈ 100.53 in.; Area: 256π $in.^2 \approx 804.25$ $in.^2$
30. Circumference: 18π mm ≈ 56.55 mm; Area: 81π $mm^2 \approx 254.47$ mm^2
31. Circumference: 20π cm ≈ 62.83 cm; Area: 100π $cm^2 \approx 314.16$ cm^2
32. Circumference: 3π miles ≈ 9.42 miles; Area: 2.25π square miles ≈ 7.07 square miles

In Problems 17–24, find the perimeter and area of each quadrilateral.

17. 6 feet, 5 feet, 9 feet

18.

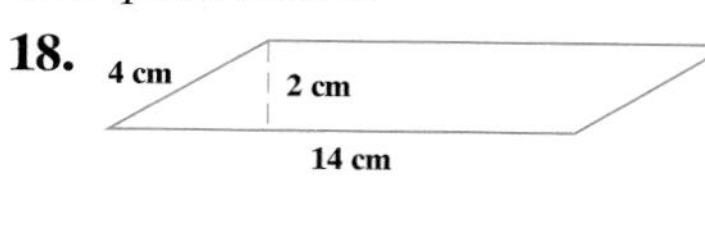

19. 9 mm, 10 mm, 4 mm

20.

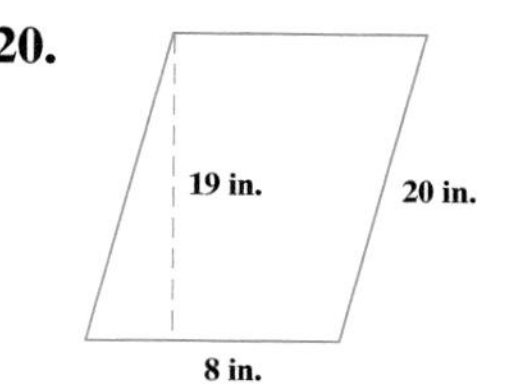

21. 8 in., 8 in., 7 in., 8 in., 16 in.

22.

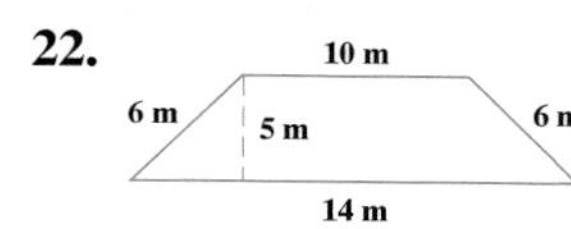

23. 8 cm, 8 cm, 7 cm, 10 cm, 19 cm

24.

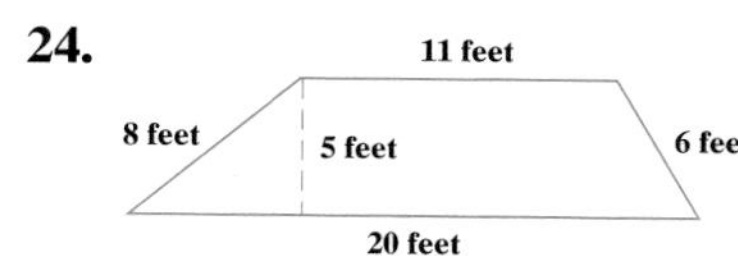

In Problems 25–28, find the perimeter and area of each triangle.

25.

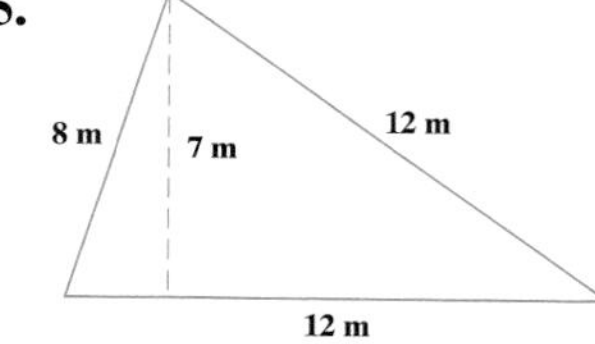

26.

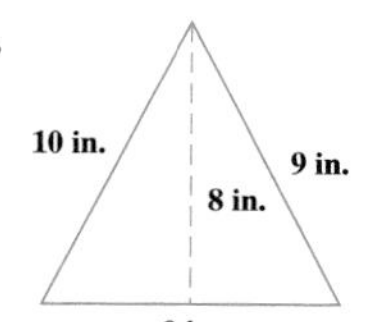

27.

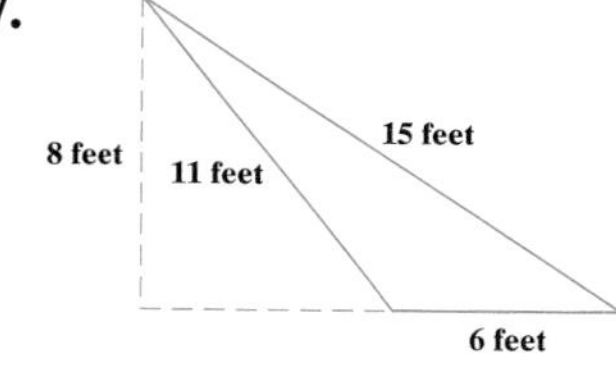

28.

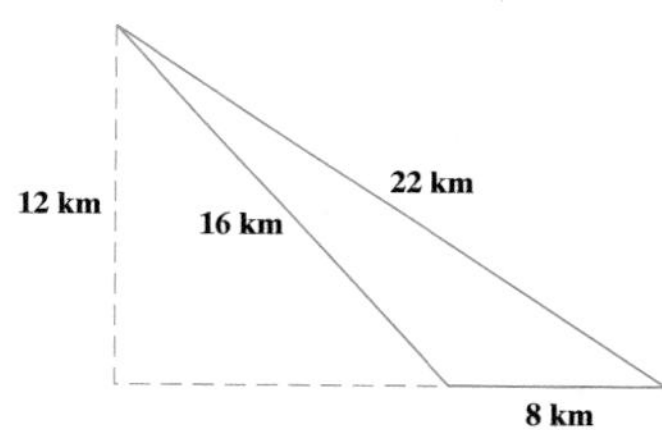

In Problems 29–32, find (a) the circumference and (b) the area of each circle. For both the circumference and area, provide exact answers and approximate answers rounded to the nearest hundredth.

29.

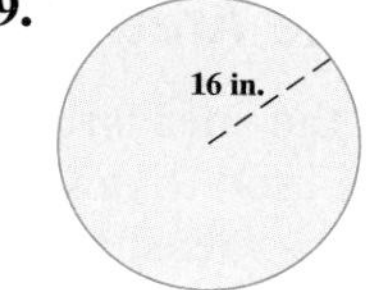

30.

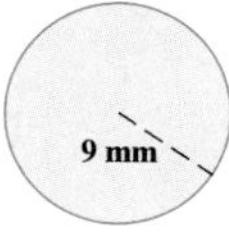

31.

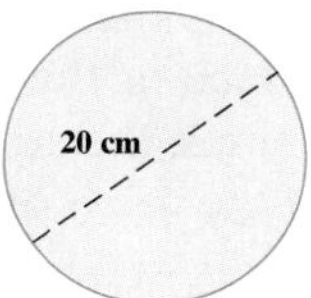

32.

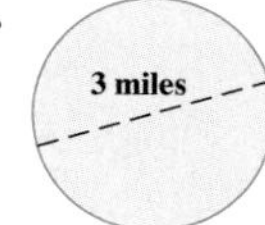

33. π square units
34. $4 - \pi$ square units
35. about 26.18 feet
36. about 14.14 feet

Applying the Concepts

In Problems 33–34, find the area of the shaded region.

33.

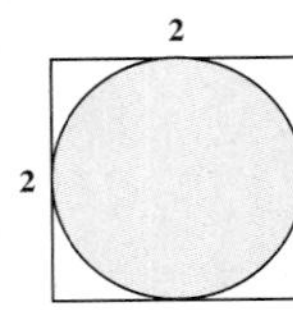

34.

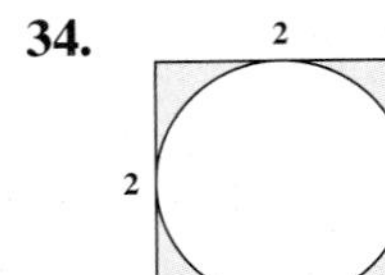

35. How many feet will a wheel with a diameter of 20 inches travel after 5 revolutions?

36. How many feet will a wheel with a diameter of 18 inches travel after 3 revolutions?

C.4 Volume and Surface Area

OBJECTIVES

1. Identify Solid Figures
2. Find the Volume and Surface Area of Solid Figures

In Words
Polyhedra is the plural form of the Greek word polyhedron.

1 Identify Solid Figures

A **geometric solid** is a three-dimensional region of space enclosed by planes and curved surfaces. Examples of geometric solids are cubes, pyramids, spheres, cylinders, and cones.

There are some geometric solids that you see and use every day. When you grab a box of cereal, you are holding a rectangular solid. When you reach for a can of soup, you are holding a circular cylinder.

We first discuss *polyhedra*. A **polyhedron** is a three-dimensional solid formed by connecting polygons. Figure 33 shows an example of a polyhedron called a hexagonal dipyramid. Notice that the top and bottom are formed by six connected triangles.

Figure 33
Hexagonal Dipyramid

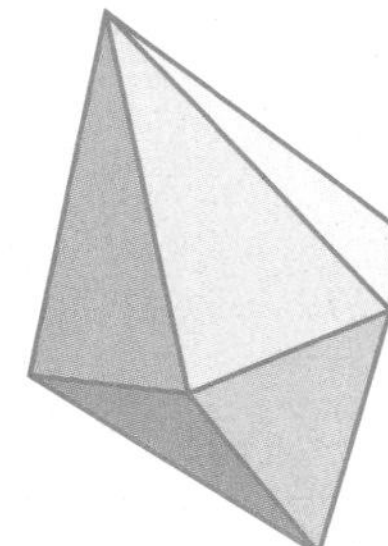

Each of the planes of the polyhedron is called a **face.** The line segment which is the intersection of any two faces of a polyhedron is called an **edge.** The point of intersection of three or more edges is called a **vertex.**

2 Find the Volume and Surface Area of Solid Figures

Figure 34

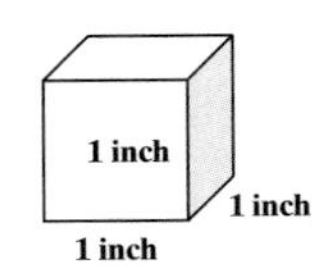

Work Smart
Volume is measured in cubic units. Surface area is measured in square units.

The **volume** of a polyhedron is the measure of the number of units of space contained in the solid. Volume can be used to describe the amount of soda in a can, or the amount of cereal in a box. Volume is measured in cubic units. For example, the cube in Figure 34 represents 1 cubic inch.

The **surface area** of a polyhedron is the sum of the areas of the faces of the polyhedron. For example, because each face in the cube shown in Figure 34 has an area of 1 square inch, and a cube has six sides, the surface area of the cube is 6 square inches. Because surface area is the sum of the areas of each polygon in the polyhedron, surface area is measured in square units.

Table 2 shows some common geometric solids, and formulas to find their volume and surface area.

Table 2

Solids	Formulas
Cube	**Volume:** $V = s^3$ **Surface Area:** $S = 6s^2$
Rectangular Solid	**Volume:** $V = lwh$ **Surface Area:** $S = 2lw + 2lh + 2wh$
Sphere	**Volume:** $V = \frac{4}{3}\pi r^3$ **Surface Area:** $S = 4\pi r^2$
Right Circular Cylinder	**Volume:** $V = \pi r^2 h$ **Surface Area:** $S = 2\pi r^2 + 2\pi rh$
Cone	**Volume:** $V = \frac{1}{3}\pi r^2 h$
Square Pyramid	**Volume:** $V = \frac{1}{3}b^2 h$ **Surface Area:** $S = b^2 + 2bs$

Classroom Example ➤
Find (a) the volume and (b) the surface area of the rectangular solid.

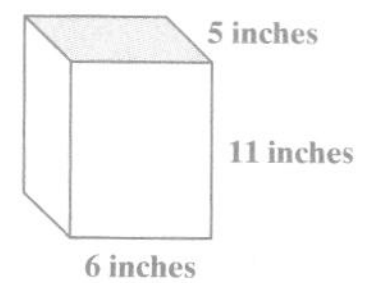

Answer:
(a) 330 cubic inches
(b) 302 square inches

EXAMPLE 1 Finding the Volume and Surface Area of a Rectangular Solid

Find the volume and surface area of the rectangular solid shown in Figure 35.

Figure 35

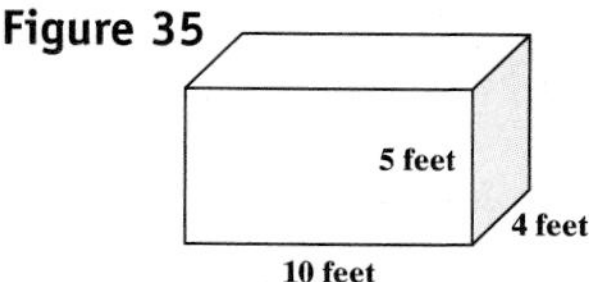

Solution

We have that $l = 10$ feet, $h = 5$ feet, and $w = 4$ feet. The volume of the rectangular solid is

$$\begin{aligned} V &= lwh \\ &= (10 \text{ feet})(4 \text{ feet})(5 \text{ feet}) \\ &= 200 \text{ cubic feet} \end{aligned}$$

The volume of the rectangular solid is 200 cubic feet.

The surface area of the rectangular solid is

$$\begin{aligned} S &= 2lw + 2lh + 2wh \\ &= 2(10 \text{ feet})(4 \text{ feet}) + 2(10 \text{ feet})(5 \text{ feet}) + 2(4 \text{ feet})(5 \text{ feet}) \\ &= 220 \text{ square feet} \end{aligned}$$

The surface area of the rectangular solid is 220 square feet. ■

Classroom Example ➤
Find (a) the volume and (b) the surface area of the right circular cylinder.

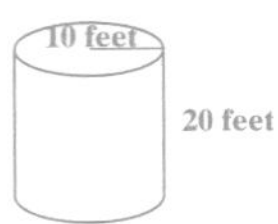

Answer:

(a) 2000π cubic feet; approximately 6283.19 cubic feet

(b) 600π square feet; 1884.96 square feet

EXAMPLE 2 Finding the Volume of a Right Circular Cylinder

Find the volume and surface area of the right circular cylinder shown in Figure 36.

Solution

We have that $h = 10$ inches and $r = 3$ inches. The volume of the right circular cylinder is

$$\begin{aligned} V &= \pi r^2 h \\ &= \pi(3 \text{ in.})^2(10 \text{ in.}) \\ &= 90\pi \text{ in.}^3 \\ &\approx 282.74 \text{ in.}^3 \end{aligned}$$

Figure 36

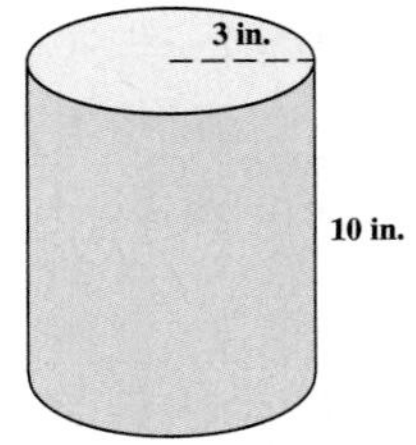

The volume of the right circular cylinder is 90π cubic inches exactly; and 282.74 cubic inches approximately.

The surface area of the right circular cylinder is

$$\begin{aligned} V &= 2\pi r^2 + 2\pi rh \\ &= 2\pi(3 \text{ in.})^2 + 2\pi(3 \text{ in.})(10 \text{ in.}) \\ &= 18\pi \text{ in.}^2 + 60\pi \text{ in.}^2 \\ &= 78\pi \text{ in.}^2 \\ &\approx 245.04 \text{ in.}^2 \end{aligned}$$

The surface area of the right circular cylinder is 78π square inches exactly; and 245.04 square inches approximately. ■

QUICK ✓ *Find the volume and surface area of the following polyhedra.*

1.

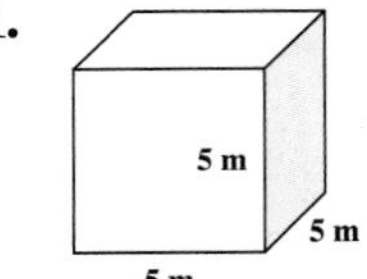

2.

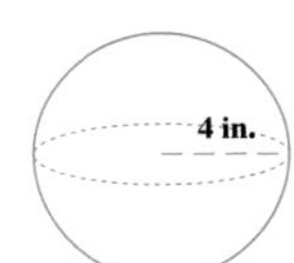

C.4 Exercises

For Extra Help: Student Solutions Manual | CD Video | PH Math/Tutor Center | MathXL Tutorials on CD | MathXL® | MyMathLab

Concepts and Vocabulary

In Problems 1–3, fill in the blanks.

1. A __________ is a three-dimensional solid formed by connecting polygons.

2. The __________ of a polyhedron is the measure of the number of units of space contained in the solid.

3. The __________ __________ of a polyhedron is the sum of the areas of the faces of the polyhedron.

In Problems 4–6, answer True or False to each statement.

4. Each of the planes of the polyhedron is called a face.

5. The volume of a polyhedron is measured in square units.

6. The volume of a rectangular solid is the product of the length, width, and height.

1. polyhedron
2. volume
3. surface area
4. True
5. False
6. True

7. Volume: 600 cubic feet; Surface Area: 460 square feet
8. Volume: 108 cubic meters; Surface Area: 168 square meters
9. Volume: 288π cubic centimeters ≈ 904.78 cubic centimeters; Surface Area: 144π square centimeters ≈ 452.39 square centimeters
10. Volume: $\frac{4000}{3}\pi$ cubic inches ≈ 4188.79 cubic inches; Surface Area: 400π square inches ≈ 1256.64 square inches
11. Volume: 32π cubic inches ≈ 100.53 cubic inches; Surface Area: 40π square inches ≈ 125.66 square inches
12. Volume: 54π cubic inches ≈ 169.65 cubic inches; Surface Area: 54π square inches ≈ 169.65 square inches
13. Volume: $\frac{800}{3}\pi$ cubic millimeters ≈ 837.76 cubic millimeters
14. Volume: 4000π cubic feet $\approx 12{,}566.37$ cubic feet
15. Volume: $\frac{640}{3}$ cubic feet; Surface Area: 256 square feet
16. Volume: $\frac{500}{3}$ cubic meters; Surface Area: 260 square meters
17. 1728 cubic inches
18. 60 cubic feet
19. 75.40 cubic inches; 251.86 square inches
20. 502.65 cubic inches; 100.53 square inches
21. 268.08 cubic centimeters
22. 52.36 cubic inches

Building Skills

7. Find the volume V and surface area S of a rectangular box with length 10 feet, width 5 feet, and height 12 feet.

8. Find the volume V and surface area S of a rectangular box with length 2 meters, width 6 meters, and height 9 meters.

9. Find the volume V and surface area S of a sphere of radius 6 centimeters.

10. Find the volume V and surface area S of a sphere of radius 10 inches.

11. Find the volume V and surface area S of a right circular cylinder of radius 2 inches and height 8 inches.

12. Find the volume V and surface area S of a right circular cylinder of radius 3 inches and height 6 inches.

13. Find the volume V of a cone of radius 10 mm and height 8 mm.

14. Find the volume V of a cone of radius 20 feet and height 30 feet.

15. Find the volume V and surface area S of a square pyramid of height 10 feet, slant 12 feet, and base 8 feet.

16. Find the volume V and surface area S of a square pyramid of height 5 m, slant 8 m, and base 10 m.

Applying the Concepts

17. Rain Gutter A rain gutter is in the shape of a rectangular solid. How much water can the gutter hold if it is 4 inches in height, 3 inches wide and 12 feet long?

18. Water for the Horses A trough for horses is 10 feet long, 2 feet wide, and 3 feet deep. How much water can the trough hold?

19. A Can of Peaches A can of peaches is in the shape of a right circular cylinder. The can has a 4 inch diameter and is 6 inches tall. What is the volume of the can? What is the surface area of the can? Express your answer as a decimal rounded to the nearest hundredth.

20. Coffee Can A coffee can is in the shape of a right circular cylinder. The can has an 8-inch diameter and is 10 inches tall. What is the volume of the can? What is the surface area of the can? Express your answer as a decimal rounded to the nearest hundredth.

21. Ice Cream Cone A waffle cone for ice cream has a diameter of 8 cm and a height of 16 cm. How much ice cream can the cone hold if the ice cream is flush with the top of the cone? Express your answer as a decimal rounded to the nearest hundredth.

22. Water Cooler The cups at the water cooler are in the shape of a cone. How much water can a cup hold if it has a 5 inch diameter and is 8 inches in height? Express your answer as a decimal rounded to the nearest hundredth.

Answers to Quick ✔ Exercises

Chapter 1

Section 1.2

1. $\{1, 3, 5, 9\}$ **2.** {Alabama, Alaska, Arkansas, Arizona} **3.** $\emptyset$ or $\{\ \}$ **4.** 12 **5.** 12, 0 **6.** $-5, 12, 0$ **7.** $\frac{11}{5}, -5, 12, 2.\overline{76}, 0, \frac{18}{4}$

8. π **9.** All **10.** 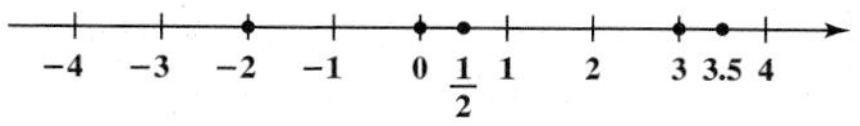 **11.** $<$ **12.** $<$ **13.** $>$ **14.** $>$ **15.** $=$ **16.** 15 **17.** $\frac{3}{4}$

Section 1.3

1. 27 **2.** -8 **3.** 14 **4.** -7 **5.** -17 **6.** 3 **7.** -1 **8.** 2 **9.** -4 **10.** -20 **11.** -6 **12.** 10 **13.** -4 **14.** -3 **15.** -24 **16.** -56 **17.** -7 **18.** $-\frac{3}{7}$ **19.** 21 **20.** $\frac{8}{5}$ **21.** 5.75 **22.** 80 **23.** -178 **24.** -18 **25.** -2572 **26.** -21 **27.** -58 **28.** 12 **29.** 550 **30.** -21 **31.** -52 **32.** 80 **33.** 108 **34.** 325 **35.** 108 **36.** 360 **37.** $\frac{1}{6}$ **38.** $-\frac{1}{2}$ **39.** -5 **40.** -7 **41.** 9 **42.** $-\frac{27}{2}$

Section 1.4

1. $-\frac{2}{7}$ **2.** $-\frac{3}{5}$ **3.** -6 **4.** $\frac{27}{32}$ **5.** $-\frac{8}{3}$ **6.** $-\frac{6}{25}$ **7.** $\frac{3}{4}$ **8.** $-\frac{3}{22}$ **9.** $\frac{1}{12}$ **10.** $\frac{5}{7}$ **11.** -4 **12.** $-\frac{20}{31}$ **13.** $\frac{50}{49}$ **14.** $-\frac{3}{8}$ **15.** $-\frac{16}{7}$ **16.** $\frac{5}{27}$ **17.** $-\frac{6}{5}$ **18.** $\frac{10}{11}$ **19.** $-\frac{3}{7}$ **20.** $\frac{1}{7}$ **21.** $\frac{5}{36}$ **22.** $\frac{29}{42}$ **23.** $-\frac{13}{4}$ **24.** $\frac{13}{55}$ **25.** $-\frac{25}{16}$ **26.** $\frac{15}{4}$ **27.** 21.014 **28.** 64.57 **29.** 22.368 **30.** 337.5788 **31.** -71.412 **32.** -89.112 **33.** 4.78 **34.** 65.884 **35.** -0.1035 **36.** -899.5 **37.** 0.0135 **38.** 0.0198 **39.** 0.25 **40.** 8.36 **41.** -0.094 **42.** -0.03

Section 1.5

1. 8 feet **2.** 8 hours, 20 minutes **3.** 5 pounds, 8 ounces **4.** 22 **5.** $\frac{3}{20}$ **6.** 11.98 **7.** -78 **8.** $-\frac{36}{331}$ **9.** 349 **10.** 14 **11.** 14 **12.** -24.2 **13.** $\frac{50}{13}$ **14.** 0 **15.** undefined **16.** 0 **17.** undefined

Section 1.6

1. 11^5 **2.** 7^8 **3.** $(-2)^3$ **4.** 16 **5.** 64 **6.** $-\frac{1}{216}$ **7.** 0.81 **8.** -16 **9.** 16 **10.** 15 **11.** -31 **12.** 12 **13.** 31 **14.** -3 **15.** 19 **16.** -63 **17.** 4 **18.** $-\frac{8}{7}$ **19.** $\frac{4}{7}$ **20.** $\frac{5}{3}$ **21.** 20 **22.** 40 **23.** -9 **24.** 48 **25.** $-\frac{5}{11}$ **26.** 41 **27.** 12 **28.** -108 **29.** 10

Section 1.7

1. -7 **2.** 9 **3.** 2 **4.** \$220 **5.** $5x^2$; $3xy$ **6.** $9ab$; $-3bc, 5ac$; $-ac^2$ **7.** $\frac{2mn}{5}$; $\frac{-3n}{7}$ **8.** $\frac{m^2}{3}$; -8 **9.** 2 **10.** 1 **11.** -1 **12.** 5 **13.** $-\frac{2}{3}$ **14.** $\frac{1}{6}$ **15.** like **16.** like **17.** unlike **18.** unlike **19.** $6x + 12$ **20.** $-5x - 10$ **21.** $-2k + 14$ **22.** $6x + 9$ **23.** $-5x$ **24.** $-4x^2$ **25.** $-8x + 3$ **26.** $-8x + 14$ **27.** $-2a + 9b - 4$ **28.** $12ac - 5a + b$ **29.** $8ab^2 - a^2b$ **30.** $2rs - \frac{3}{2}r^2 - 5$ **31.** $-2x - 1$ **32.** $-2m - n - 7$ **33.** $a - 11b$ **34.** $6x - 2$

Chapter 2

Section 2.1

1. yes **2.** no **3.** yes **4.** no **5.** $\{32\}$ **6.** $\{14\}$ **7.** $\{12\}$ **8.** $\{-15\}$ **9.** $\left\{\frac{7}{3}\right\}$ **10.** $\{-1\}$ **11.** $\left\{\frac{5}{8}\right\}$ **12.** $\left\{-\frac{13}{12}\right\}$ **13.** $p = \$12{,}455$ **14.** $\{2\}$ **15.** $\{-2\}$ **16.** $\left\{\frac{5}{2}\right\}$ **17.** $\left\{-\frac{7}{3}\right\}$ **18.** $\{9\}$ **19.** $\{-9\}$ **20.** $\left\{-\frac{10}{3}\right\}$ **21.** $\{6\}$ **22.** $\left\{\frac{8}{3}\right\}$ **23.** $\left\{-\frac{5}{14}\right\}$

Section 2.2

1. $\{3\}$ **2.** $\{2\}$ **3.** $\{18\}$ **4.** $\left\{\frac{3}{2}\right\}$ **5.** $\{2\}$ **6.** $\{-18\}$ **7.** $\left\{\frac{5}{2}\right\}$ **8.** $\{-8\}$ **9.** $\{2\}$ **10.** $\left\{\frac{4}{3}\right\}$ **11.** $\{13\}$ **12.** $\{6\}$ **13.** $\{6\}$ **14.** $\left\{-\frac{7}{2}\right\}$ **15.** $\{1\}$ **16.** $\left\{-\frac{5}{3}\right\}$ **17.** 52 hours

Section 2.3

1. $\{10\}$ **2.** $\left\{-\frac{5}{18}\right\}$ **3.** $\{-14\}$ **4.** $\left\{\frac{35}{3}\right\}$ **5.** $\{100\}$ **6.** $\{50\}$ **7.** $\{50\}$ **8.** $\{160\}$ **9.** $\{4\}$ **10.** $\{3000\}$ **11.** $\emptyset$ **12.** all real numbers **13.** all real numbers **14.** $\emptyset$ **15.** identity; all real numbers **16.** conditional; $\{4\}$ **17.** contradiction; $\emptyset$ **18.** contradiction; $\emptyset$ **19.** \$500

Section 2.4

1. 59° **2.** size 40 **3.** \$312.50 **4.** 5936 persons **5.** \$50 interest; \$2550 total **6.** 36 square inches **7. (a)** 187 square feet **(b)** \$46.75 **8.** 28.27 sq feet **9.** The extra large pizza is the better buy. **10.** $C = \frac{5}{9}(F - 32)$ **11.** $h = \frac{A - 2\pi r^2}{2\pi r}$ **12.** $y = \frac{7 - x}{2}$ **13.** $y = \frac{15 - 5x}{-3}$ **14.** $b = \frac{28 - 3a}{8}$ **15.** $t = 24 - 6rs$ **16.** $\frac{d}{r} = t; t = 9\frac{1}{6}$ hours, or 9 hours 10 min. **17.** $\frac{I}{Pr} = t; t = 0.5$ year

Section 2.5

1. $5 + 17$ **2.** $-2 \cdot 6$ **3.** $\frac{25}{3}$ **4.** $7 - 4$ **5.** $2a - 2$ **6.** $3 + \frac{z}{4}$ **7.** $z + 50$ **8.** $x - 15$ **9.** $75 - d$ **10.** $3l - 2$ **11.** $2q + 3$ **12.** $3b - 5$ **13.** $3y = 21$ **14.** $3 + x = 5x$ **15.** $x - 10 = \frac{x}{2}$ **16.** $y - 3 = 5y$ **17.** $s + \frac{2}{3}s = 15$; Sean pays \$9 and Connor pays \$6 **18.** $n + (n + 2) + (n + 4) = 270$; 88, 90, 92 **19.** $76 = x + (x + 24) + \frac{1}{2}(x + 24)$, where x = length of smallest piece of ribbon; 16 inches, 40 inches, 20 inches **20.** $x + 2x = 18{,}000$, where x = amount invested in stocks; \$6,000 in stocks; \$12,000 in bonds **21.** 150 miles

Section 2.6

1. 801 **2.** 2.52 **3.** 36 **4.** 3.5 **5.** 40% **6.** 37.5% **7.** 20.5% **8.** 110% **9.** 50 **10.** 150 **11.** 80 **12.** 75 **13.** 31,450,000 residents **14.** \$39,975 **15.** \$7300 **16.** \$1.25 per gallon. **17.** \$659

Section 2.7

1. 39° and 51° **2.** 70° and 110° **3.** 35°, 40°, and 105° **4.** width is $1\frac{1}{2}$ feet; length is 3 feet **5.** 5 feet **6.** José's speed is 13 mph and Luis' speed is 8 mph. **7.** It takes $\frac{1}{2}$ hour to catch up to Tanya. Each of you has traveled 20 miles.

Section 2.8

1. $n \geq 8$

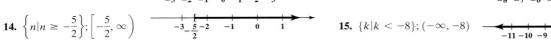

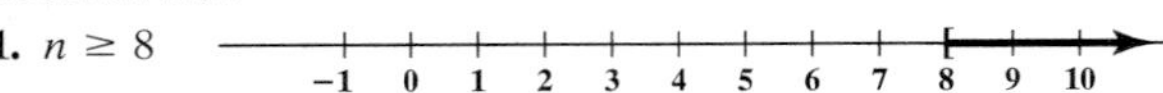

2. $a < -6$

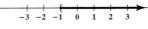

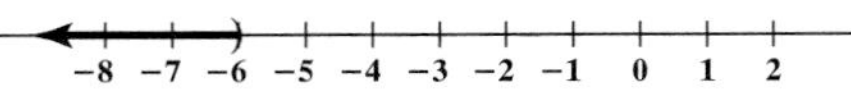

3. $x > -1$ **4.** $p \leq 0$ **5.** $[-3, \infty)$ **6.** $(-\infty, 12)$ **7.** $(-\infty, 2.5]$ **8.** $(125, \infty)$ **9.** $\{n|n > 3\}; (3, \infty)$ **10.** $\{x|x < 4\}; (-\infty, 4)$

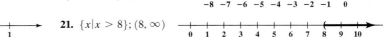

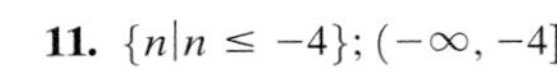

11. $\{n|n \leq -4\}; (-\infty, -4]$ **12.** $\{x|x > -1\}; (-1, \infty)$ **13.** $\{k|k < -6\}; (-\infty, -6)$

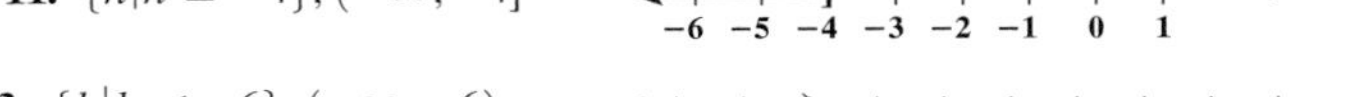

14. $\left\{n|n \geq -\frac{5}{2}\right\}; \left[-\frac{5}{2}, \infty\right)$ **15.** $\{k|k < -8\}; (-\infty, -8)$ **16.** $\left\{p|p \geq \frac{3}{5}\right\}; \left[\frac{3}{5}, \infty\right)$ **17.** $\{x|x > 7\}; (7, \infty)$

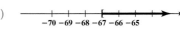

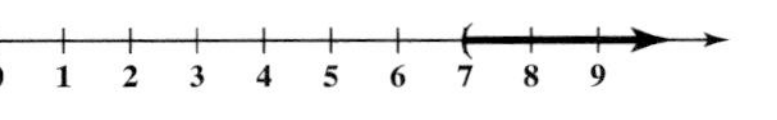

18. $\{n|n > -3\}; (-3, \infty)$ **19.** $\{x|x > -4\}; (-4, \infty)$ **20.** $\{x|x \leq -6\}; (-\infty, -6]$ **21.** $\{x|x > 8\}; (8, \infty)$ **22.** $\left\{x|x \leq \frac{19}{8}\right\}; \left(-\infty, \frac{19}{8}\right]$ **23.** $\{x|x \geq -67\}; [-67, \infty)$ **24.** Ø or { } **25.** $\{x|x > 4\}; (4, \infty)$

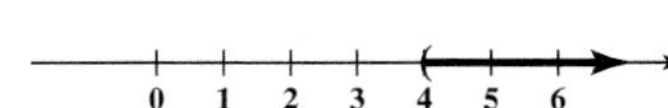

26. $\{x|x$ is any real number$\}; (-\infty, \infty)$ **27.** at most 20 boxes

Chapter 3

Section 3.1

1. Monomial, Coefficient = 12, Degree = 6 **2.** Not a monomial **3.** Monomial, Coefficient = 10, Degree = 0 **4.** Not a monomial **5.** Coefficient: 3, Degree: 7 **6.** Coefficient: −2, Degree: 4 **7.** Not a monomial **8.** Coefficient: −1, Degree: 2 **9.** Polynomial; degree: 3 **10.** Not a polynomial **11.** Not a polynomial **12.** Polynomial; degree: 5 **13.** Polynomial; degree: 4 **14.** $12x^2 + 3x + 3$ **15.** $2z^4 + 4z^3 - z^2 + z - 1$ **16.** $5x^2y + 6x^2y^2 - xy^2$ **17.** $5x^3 - 15x^2 + 3x + 7$ **18.** $8y^3 - 3y^2 - 3y - 6$ **19.** $11x^2y + 2x^2y^2 - 9xy^2$ **20.** $7a^2 - 4ab - 12b^2$ **21. (a)** 1 **(b)** −214 **(c)** 101 **22. (a)** 40 **(b)** −20 **23.** \$1875

Section 3.2

1. 27 **2.** -3125 **3.** c^8 **4.** y^9 **5.** a^9b^6 **6.** 2^8 **7.** $(-3)^6$ **8.** b^{10} **9.** z^{20} **10.** $8n^3$ **11.** $36y^2$ **12.** $-125x^{12}$ **13.** $49a^6b^2$ **14.** $8a^{11}$ **15.** $-12p^7$ **16.** $-18m^3n^9$ **17.** $\frac{2}{3}x^3y^3$

Section 3.3

1. $3x^3 - 6x^2 + 12x$ **2.** $-2a^4b^3 + 4a^3b^3 - 3a^3b^2$ **3.** $3n^5 - \frac{9}{8}n^4 - \frac{6}{7}n^3$ **4.** $x^2 + 9x + 14$ **5.** $2n^2 - 5n - 3$ **6.** $6p^2 - 7p + 2$ **7.** $x^2 + 7x + 12$ **8.** $y^2 + 2y - 15$ **9.** $a^2 - 6a + 5$ **10.** $2x^2 + 11x + 12$ **11.** $6y^2 + y - 15$ **12.** $6a^2 - 5a - 4$ **13.** $15x^2 - 14xy - 8y^2$ **14.** $a^2 - 16$ **15.** $9w^2 - 49$ **16.** $36b^2 - 4c^2$ **17.** $x^2 - 4y^6$ **18.** $z^2 - 18z + 81$ **19.** $p^2 + 2p + 1$ **20.** $16 - 8a + a^2$ **21.** $w^2 + 2wy + y^2$ **22.** $9z^2 - 24z + 16$ **23.** $25p^2 + 10p + 1$ **24.** $16a^2 - 40a + 25$ **25.** $4w^2 + 28wy + 49y^2$ **26.** $x^3 - 8$ **27.** $3y^3 + 4y^2 + 8y - 8$ **28.** $x^3 - 8$ **29.** $3y^3 + 4y^2 + 8y - 8$ **30.** $-24a^3 - 34a^2 + 10a$ **31.** $15x^2 + 54x - 24$

Section 3.4

1. 9 **2.** y^6 **3.** $\frac{7}{5}c$ **4.** $-\frac{3}{2}wz^7$ **5.** $\frac{p^4}{16}$ **6.** $\frac{b^3}{27}$ **7.** $\frac{a^6}{b^8}$ **8.** $-\frac{8a^6}{b^{12}}$ **9.** 1 **10.** 4 **11.** -1 **12.** $\frac{1}{z^5}$ **13.** $\frac{1}{p^4}$ **14.** $-\frac{1}{x^3}$ **15.** $\frac{1}{16}$ **16.** $-\frac{1}{49}$ **17.** $\frac{1}{49}$ **18.** $\frac{1}{8}$ **19.** 9 **20.** v **21.** -100 **22.** $\frac{5}{2}z^2$ **23.** $\frac{8}{7}$ **24.** -64 **25.** $\frac{25}{9a^2}$ **26.** $-\frac{27}{8}n^{12}$ **27.** $\frac{-20}{a^2}$ **28.** $\frac{2}{mn}$ **29.** $\frac{8}{m^5}$ **30.** $-\frac{4}{3}a^3b^3$ **31.** $\frac{9x^2}{7y^3}$ **32.** $\frac{z^6}{9y^4}$ **33.** $\frac{4w^4}{49z^6}$ **34.** $-\frac{2}{p^{11}}$ **35.** $\frac{-4}{k}$

Section 3.5

1. $2n^2 - 4n + 1$ **2.** $6k^2 - 9 + \frac{5}{2k^2}$ **3.** $\frac{xy^3}{4} + \frac{2y}{x} - \frac{1}{x^2}$ **4.** $x - 8$ **5.** $x - 4$ **6.** $4x + 5 + \frac{6}{x + 3}$ **7.** $4x^2 - 11x + 23 - \frac{45}{x + 2}$ **8.** $x^2 + 4x + 10 + \frac{60}{2x - 5}$ **9.** $4x - 3 + \frac{-7x + 7}{x^2 + 2}$

Section 3.6

1. 4.32×10^2 **2.** 1.0302×10^4 **3.** 5.432×10^6 **4.** 9.3×10^{-2} **5.** 4.59×10^{-5} **6.** 8×10^{-8} **7.** 310 **8.** 0.901 **9.** 170,000 **10.** 7 **11.** 0.00089 **12.** 6×10^7 **13.** 8×10^{-3} **14.** 1.5×10^4 **15.** 2.8×10^{-5} **16.** 4×10^5 **17.** 2×10^{-4} **18.** 5×10^3 **19.** 6.25×10^{-5} **20.** $2.604 \times 10^8 = 260{,}400{,}000$ barrels

Chapter 4

Section 4.1

1. 8 **2.** 5 **3.** 3 **4.** 7 **5.** $7y^2$ **6.** $2z$ **7.** $4xy^3$ **8.** $(2x + 3)$ **9.** $3(k + 8)$ **10.** $5z(z - 6)$ **11.** $6p^2(2 + 5p^2)$ **12.** $4y(4y^2 - 3y + 1)$ **13.** $2m^2n^2(3m^2 + 9mn^2 - 11n^3)$ **14.** $-4y(y - 2)$ **15.** $-3a(2a^2 - 4a + 1)$ **16.** $(a - 5)(2a + 3)$ **17.** $(z + 5)(7z - 4)$ **18.** $(x + y)(4 + b)$ **19.** $(2a - 3b)(3z - 1)$ **20.** $(2b + 1)(4 - 5a)$ **21.** $3(z + 4)(z^2 + 2)$ **22.** $2n(n + 1)(n^2 - 2)$

Section 4.2

1. $(y + 5)(y + 4)$ **2.** $(w + 6)(w + 4)$ **3.** $(q - 3)(q - 2)$ **4.** $(z - 7)(z - 2)$ **5.** $(y - 5)(y + 3)$ **6.** $(w - 3)(w + 4)$ **7.** $(q - 6)(q + 2)$ **8.** $(n + 6)(n - 1)$ **9.** Prime **10.** $(q + 9)(q - 5)$ **11.** $(x + 4y)(x + 5y)$ **12.** $(m + 7n)(m - 6n)$ **13.** $(n + 8)(n - 7)$ **14.** $(y - 7)(y - 5)$ **15.** $4(m + 3)(m - 7)$ **16.** $3z(z - 1)(z + 5)$ **17.** $-1(w + 5)(w - 2)$ **18.** $-2(a + 6)(a - 2)$

Section 4.3

1. $(3x + 2)(x + 1)$ **2.** $(7y + 1)(y + 3)$ **3.** $(3x - 4)(x - 3)$ **4.** $(5p - 1)(p - 4)$ **5.** $(2n + 1)(n - 9)$ **6.** $(4b + 3)(b - 2)$ **7.** $(3x + 2)(4x + 3)$ **8.** $(6y - 5)(2y + 7)$ **9.** $3(6x - y)(5x + 2y)$ **10.** $-1(6y + 1)(y - 4)$ **11.** $-1(3x + 2y)(3x + 5y)$ **12.** $-2(2x - 1)(3x + 4)$ **13.** $-3(2x - 5)(x + 3)$ **14.** $(3x + 4)(x - 2)$ **15.** $(5z + 3)(2z + 3)$ **16.** $3(4x + 3)(2x - 1)$ **17.** $-(5n - 1)(2n - 3)$

Section 4.4

1. $(x - 6)^2$ **2.** $(4x + 5)^2$ **3.** $(3a - 10b)^2$ **4.** $(z - 4)^2$ **5.** $(2n + 3)^2$ **6.** prime **7.** $4(z + 3)^2$ **8.** $2a(5a + 4)^2$ **9.** $2(5a^2 + 2b)^2$ **10.** $(z - 5)(z + 5)$ **11.** $(9m - 4n)(9m + 4n)$ **12.** $(4a - \frac{2}{3}b)(4a + \frac{2}{3}b)$ **13.** $(6b^2 - c)(6b^2 + c)$ **14.** $(10k^2 - 9w)(10k^2 + 9w)$ **15.** $(x^2 + 4)(x - 2)(x + 2)$ **16.** $3(7x - 4)(7x + 4)$ **17.** $-3ab(3a - 5b)(3a + 5b)$ **18.** $(z + 5)(z^2 - 5z + 25)$ **19.** $(2p - 3q^2)(4p^2 + 6pq^2 + 9q^4)$ **20.** $2a(3 - 2a)(9 + 6a + 4a^2)$ **21.** $-3(5b - 1)(25b^2 + 5b + 1)$

Section 4.5

1. $2(p - 5)(p + 9)$ **2.** $-3(3x + y)(5x - 2y)$ **3.** $(10x - 9y)(10x + 9y)$ **4.** $2a(b - 11)(b + 11)$ **5.** $(p - 6q)^2$ **6.** $3(5x + 3)^2$ **7.** $(5y - 4)(25y^2 + 20y + 16)$ **8.** $-3(2a - b)(4a^2 + 2ab + b^2)$ **9.** $(x^2 + 1)(x - 1)(x + 1)$ **10.** $-4y(3x - 2)(3x + 2)$ **11.** $(x^2 + 2)(2x + 3)$ **12.** $3(2x + 3)(x - 1)(x + 1)$ **13.** $-3(z^2 - 3z + 7)$ **14.** $3x(2y^2 + 5x^2)$

Section 4.6

1. $\{-3, 0\}$ **2.** $\left\{-\frac{5}{4}, 2\right\}$ **3.** $\{2, 4\}$ **4.** $\left\{-\frac{1}{2}, 3\right\}$ **5.** $\left\{-\frac{5}{2}, 1\right\}$ **6.** $\{-5, -4\}$ **7.** $\left\{-1, \frac{5}{2}\right\}$ **8.** $\left\{-4, \frac{1}{3}\right\}$ **9.** $\{-6, 4\}$ **10.** $\left\{-\frac{3}{5}, 1\right\}$ **11.** $\left\{\frac{4}{3}\right\}$ **12.** $\{-6, 3\}$ **13.** $\{-2, 3\}$ **14.** After 1 second or 4 seconds **15.** $\left\{-3, 3, \frac{5}{4}\right\}$ **16.** $\{-2, -1, 0\}$

Section 4.7

1. (a) After 4 seconds and after 6 seconds (b) After 10 seconds 2. 8 km by 13 km 3. base = 12 yards, height = 8 yards
4. $x = 12, x - 3 = 9$ 5. 6 inches by 8 inches

Chapter 5

Section 5.1

1. (a) $-\frac{1}{8}$ (b) $\frac{3}{16}$ 2. (a) -1 (b) 9 3. $\frac{11}{2}$ 4. undefined 5. $x = -7$ 6. $n = -\frac{5}{3}$ 7. $x = -3$ or $x = 1$ 8. $k = -3$ or $k = 3$ 9. $\frac{1}{2}$
10. $\frac{5}{2p - 1}$ 11. $-\frac{z + 1}{2}$ 12. $\frac{a + 7}{2a + 7}$ 13. $\frac{x - y}{z - 2}$ 14. $\frac{2k + 1}{2k - 1}$ 15. -7 16. $\frac{-4}{4x - 1}$ 17. $\frac{-(5z + 1)}{3}$

Section 5.2

1. $\frac{5p(p + 3)}{2}$ 2. $\frac{7}{3(x + 5)}$ 3. $\frac{2(x + 3)}{x + 5}$ 4. $\frac{-3a}{7}$ 5. $\frac{1}{3}$ 6. $\frac{6(x + 1)}{x(2x - 1)}$ 7. $\frac{(x - 3)^2}{(x - 4)^2}$ 8. $\frac{q + 1}{(q - 5)(q + 5)}$ 9. $\frac{7x}{4(x - 2)}$
10. $\frac{6m}{m + 2n}$

Section 5.3

1. $\frac{4}{x - 2}$ 2. $x + 1$ 3. $\frac{11x - 3}{6x - 5}$ 4. $\frac{1}{x - 5}$ 5. $\frac{2(4y - 3)}{2y - 5}$ 6. $\frac{z + 1}{2z}$ 7. $\frac{x + 8}{3}$ 8. $\frac{x + 2}{x - 7}$ 9. $\frac{3x - 1}{x - 5}$ 10. 2 11. $\frac{2}{n - 3}$
12. $\frac{2k + 3}{2(k - 1)}$

Section 5.4

1. $24x^2y^3$ 2. $42a^3b^3$ 3. $15z(z + 1)$ 4. $(x - 1)(x + 5)^2$ 5. $-3(x + 7)(x - 7)$ 6. $\frac{12p^2}{16p^3(p - 2)}$
7. $\frac{2x - 8}{(x + 2)(x - 4)(x + 3)}$ 8. LCD $= (x - 5)(x - 2)(x + 1)$; $\frac{5}{x^2 - 4x - 5} = \frac{5x - 10}{(x - 5)(x - 2)(x + 1)}$; $\frac{-3}{x^2 - 7x + 10} = \frac{-3x - 3}{(x - 5)(x - 2)(x + 1)}$

Section 5.5

1. $\frac{25}{36}$ 2. $\frac{7}{2}$ 3. $\frac{11}{4}$ 4. $\frac{7}{30}$ 5. $\frac{3b + 10a^2}{24a^3b^2}$ 6. $\frac{6x^2 + 35y}{90x^3y^2}$ 7. $\frac{2(4x - 1)}{(x - 4)(x + 2)}$ 8. $\frac{3n - 13}{(n - 3)(n + 1)}$ 9. $\frac{2}{x - 2}$ 10. $\frac{1}{z + 2}$
11. $\frac{2}{(x + 5)(x - 5)}$ 12. $\frac{-16ab - 15}{20a^2b^3}$ 13. $\frac{2(x - 10)}{x(x - 4)}$ 14. $\frac{x - 11}{(x - 5)(x + 4)(x - 1)}$ 15. $\frac{(x + 8)(x - 1)}{4x(x - 3)}$ 16. $\frac{14 - 5p}{6p(p - 1)}$
17. $\frac{z - 2}{z - 5}$ 18. $\frac{2x - 5}{x - 1}$ 19. $\frac{4x}{(x + 1)(x - 1)}$ 20. $\frac{2}{x - 2}$

Section 5.6

1. $\frac{k + 1}{2}$ 2. $\frac{1}{4n}$ 3. $\frac{2}{3x}$ 4. $\frac{1}{3 + y}$ 5. $\frac{3}{x + 5}$ 6. $\frac{-2}{y + 4}$ 7. $\frac{y + 2x}{2y - x}$

Section 5.7

1. $\left\{\frac{2}{3}\right\}$ 2. $\{-18\}$ 3. $\{-4\}$ 4. $\{-6\}$ 5. $\{4\}$ 6. $\{-5, 6\}$ 7. $\{5\}$ 8. $\{\ \}$ or $\varnothing$ 9. After $2\frac{1}{2}$ hours and 10 hours
10. $x = \frac{4g}{R}$ 11. $r = 1 - \frac{a}{S}$ or $r = \frac{S - a}{S}$ 12. $p = \frac{fq}{q - f}$

Section 5.8

1. $\{-12\}$ 2. $\left\{\frac{13}{3}\right\}$ 3. $\left\{-\frac{2}{3}\right\}$ 4. $\{0, 3\}$ 5. about 111 minutes 6. \$233.38 7. 210 miles 8. $XY = 8$ 9. 6 feet
10. 2 hours 11. $3\frac{1}{3}$ hours 12. 20 mph 13. 6 mph

Chapter 6

Section 6.1

1. -8 and 8 2. $-\frac{5}{7}$ and $\frac{5}{7}$ 3. -0.6 and 0.6 4. 10 5. -3 6. $\frac{5}{7}$ 7. 0.6 8. 12 9. 13 10. 17 11. 7 12. 5.92
13. -2.45 14. rational; 7 15. irrational; ≈ 8.43 16. not a real number 17. rational; -4 18. $|b|$ 19. $4|p|$ 20. $|w + 3|$
21. $5b$ 22. $z - 6$ 23. $2h - 5$

Section 6.2

1. $2\sqrt{5}$ **2.** $5\sqrt{2}$ **3.** $2\sqrt{6}$ **4.** $6\sqrt{2}$ **5.** $2+\sqrt{5}$ **6.** $\frac{-1+2\sqrt{2}}{2}$ or $-\frac{1}{2}+\sqrt{2}$ **7.** $6y^4$ **8.** $2n^3\sqrt{3}$ **9.** $2ab^5\sqrt{7}$ **10.** $3x^6\sqrt{5}$ **11.** $5y^5\sqrt{y}$ **12.** $11a^2\sqrt{a}$ **13.** $4b^4\sqrt{2ab}$ **14.** $2x^3y^6\sqrt{3xy}$ **15.** $\frac{8}{5}$ **16.** $\frac{7a\sqrt{a}}{10}$ **17.** $\frac{\sqrt{11}}{4m}$ **18.** 5 **19.** $3b^2\sqrt{b}$

Section 6.3

1. $14\sqrt{3}$ **2.** $-\sqrt{11}+2\sqrt{10}$ **3.** $-8\sqrt{13}+\sqrt{6}$ **4.** $12\sqrt{5x}$ **5.** $9\sqrt{2}$ **6.** $-2\sqrt{3}$ **7.** $5\sqrt{5}-3\sqrt{2}$ **8.** $8\sqrt{3z}$ **9.** $-8a\sqrt{3a}$

Section 6.4

1. $\sqrt{15}$ **2.** $\sqrt{66}$ **3.** $\sqrt{30c}$ **4.** $\sqrt{14z}$ **5.** $5b\sqrt{3}$ **6.** $6k^2\sqrt{2}$ **7.** $2z^3\sqrt{5z}$ **8.** 11 **9.** 31 **10.** 21 **11.** −24 **12.** $-30\sqrt{2}$ **13.** 44 **14.** $-15+3\sqrt{7}$ **15.** $3\sqrt{5}-5$ **16.** $-4\sqrt{2}+2\sqrt{5}$ **17.** $-4+4\sqrt{3}$ **18.** $19+7\sqrt{7}$ **19.** $42+\sqrt{5n}-5n$ **20.** $-2+4\sqrt{3}$ **21.** $-18+14\sqrt{3}$ **22.** $-32-11\sqrt{7}$ **23.** $67+16\sqrt{3}$ **24.** $4-4\sqrt{7z}+7z$ **25.** $49+12\sqrt{5}$ **26.** $51+10\sqrt{2}$ **27.** 33 **28.** −47 **29.** $1-80n$

Section 6.5

1. 3 **2.** $\sqrt{7}$ **3.** $\frac{4}{5}$ **4.** $\sqrt{6}-\sqrt{15}$ **5.** $\frac{\sqrt{2}}{2}$ **6.** $-\frac{\sqrt{6}}{15}$ **7.** $\frac{\sqrt{5}}{2}$ **8.** $4-2\sqrt{3}$ **9.** $9+3\sqrt{5}$ **10.** $\frac{4\sqrt{3}-\sqrt{21}}{9}$ **11.** $\frac{3\sqrt{3}+\sqrt{3x}}{9-x}$ **12.** $-\sqrt{6}+2$

Section 6.6

1. yes **2.** no **3.** yes **4.** {14} **5.** {13} **6.** {36} **7.** {12} **8.** {3} **9.** {4} **10.** {6} **11.** {4} **12.** ∅ **13.** {6} **14.** {−1, 10} **15.** {4} **16.** {16} **17.** 49

Section 6.7

1. 2 **2.** −3 **3.** not a real number **4.** 2 **5.** 7 **6.** w **7.** $12p$ **8.** $-2\sqrt[3]{4}$ **9.** $5n\sqrt[3]{2}$ **10.** $3b^2$ **11.** 3 **12.** $4m^3\sqrt[3]{2}$ **13.** 7 **14.** −4 **15.** −10 **16.** not a real number **17.** $n^{1/3}$ **18.** $d^{3/4}$ **19.** $y^{8/5}$ **20.** $\sqrt[5]{z^4}=(\sqrt[5]{z})^4$ **21.** $2\sqrt[3]{n^2}=2(\sqrt[3]{n})^2$ **22.** $\sqrt[3]{(2n)^2}=(\sqrt[3]{2n})^2$ **23.** 125 **24.** 16 **25.** −8 **26.** 16 **27.** not a real number **28.** $\frac{1}{6}$ **29.** $\frac{1}{16}$ **30.** $-\frac{1}{125}$ **31.** $7^{1/6}$ **32.** $11^{3/5}$ **33.** 4 **34.** $a^{11/2}$

Chapter 7

Section 7.1

1. $\{-6, 6\}$ **2.** $\{-2\sqrt{5}, 2\sqrt{5}\}$ **3.** $\{-5, 5\}$ **4.** $\{-3\sqrt{2}, 3\sqrt{2}\}$ **5.** $\left\{-\frac{\sqrt{2}}{2}, \frac{\sqrt{2}}{2}\right\}$ **6.** $\left\{-\frac{4\sqrt{5}}{5}, \frac{4\sqrt{5}}{5}\right\}$ **7.** $\{-8, 14\}$ **8.** $\left\{\frac{-1-2\sqrt{7}}{5}, \frac{-1+2\sqrt{7}}{5}\right\}$ **9.** $\{4, -1\}$ **10.** $\{-4-2\sqrt{6}, -4+2\sqrt{6}\}$ **11.** no real solution **12.** no real solution **13.** $5\sqrt{10}$, 15.8 **14.** $6\sqrt{2}$, 8.5 **15.** Yes. The ladder reaches 28.9 feet up on the wall.

Section 7.2

1. 9; $(p+3)^2$ **2.** 49; $(z-7)^2$ **3.** 81; $(b-9)^2$ **4.** $\frac{9}{4}$; $\left(w+\frac{3}{2}\right)^2$ **5.** $\{-7, -1\}$ **6.** $\{-8, 4\}$ **7.** $\{-4-\sqrt{7}, -4+\sqrt{7}\}$ **8.** $\{5-\sqrt{6}, 5+\sqrt{6}\}$ **9.** $\{-9, 1\}$ **10.** $\left\{\frac{-5-\sqrt{65}}{2}, \frac{-5+\sqrt{65}}{2}\right\}$ **11.** $\left\{-2, \frac{5}{2}\right\}$ **12.** $\left\{\frac{2-\sqrt{10}}{3}, \frac{2+\sqrt{10}}{3}\right\}$ **13.** no real solution **14.** $\{-7, 6\}$

Section 7.3

1. $\left\{-\frac{4}{3}, -1\right\}$ **2.** $\left\{-6, -\frac{1}{2}\right\}$ **3.** $\left\{-\frac{2}{3}, -\frac{3}{2}\right\}$ **4.** $\left\{-\frac{5}{4}, \frac{1}{2}\right\}$ **5.** $\{-2-\sqrt{6}, -2+\sqrt{6}\}$ **6.** $\left\{\frac{3-\sqrt{2}}{4}, \frac{3+\sqrt{2}}{4}\right\}$ **7.** $\left\{\frac{4-\sqrt{10}}{2}, \frac{4+\sqrt{10}}{2}\right\}$ **8.** $\left\{\frac{5}{4}\right\}$ **9.** no real solution **10.** 1000 **11.** 0; Factoring; $\left\{\frac{1}{3}\right\}$ **12.** 1; Factoring; $\left\{\frac{2}{3}, 1\right\}$ **13.** 20; quadratic formula; $\left\{\frac{1 \pm \sqrt{5}}{4}\right\}$ **14.** $\{-\sqrt{6}, \sqrt{6}\}$ **15.** $\left\{-\frac{7}{2}, 5\right\}$ **16.** $\left\{\frac{-2-\sqrt{7}}{3}, \frac{-2+\sqrt{7}}{3}\right\}$

Section 7.4

1. base = 8 inches; height = 10 inches **2.** 7.7 m × 11.7 m **3.** length = 4.1 feet; width = 3.1 feet **4.** ages 30 or 63 years **5.** The cannon ball will be 50 feet above the ground after 0.51 s and after 5.74 s.

Section 7.5

1. $4i$ **2.** $\sqrt{2}i$ **3.** $2\sqrt{6}i$ **4.** $9+12i$ **5.** $-7-2\sqrt{5}i$ **6.** $2-\sqrt{2}i$ **7.** $2+2i$ **8.** $-8+11i$ **9.** $5+5\sqrt{3}i$ **10.** $24+12i$ **11.** $-2+14i$ **12.** −10 **13.** $-6-8i$ **14.** 53 **15.** 97 **16.** $-4-6i$ **17.** $-\frac{48}{17}-\frac{12}{17}i$ **18.** $1-\frac{8}{3}i$ **19.** $\frac{5}{2}-\frac{1}{2}i$ **20.** $\left\{\frac{-\sqrt{10}i}{2}, \frac{\sqrt{10}i}{2}\right\}$ **21.** $\{-3-2\sqrt{2}i, -3+2\sqrt{2}i\}$

Chapter 8

Section 8.1

1. 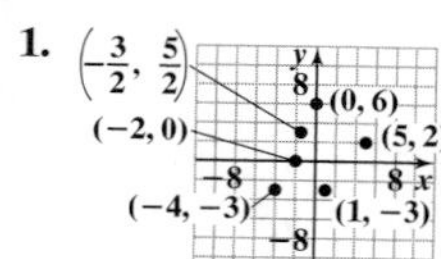

(a) I, **(b)** III, **(c)** IV, **(d)** x-axis, **(e)** y-axis, **(f)** II

2. (1, 7), (0, −1), (4, 0), (−6, 2), (−3, −2), $\left(\frac{3}{2}, -\frac{7}{2}\right)$

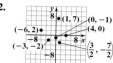

(a) II, **(b)** I, **(c)** III, **(d)** x-axis, **(e)** y-axis, **(f)** IV

3. (2, 3) **4.** (1, −3) **5.** (−3, 0) **6.** (−2, −1) **7. (a)** Yes **(b)** No **(c)** No
8. (a) No **(b)** Yes **(c)** Yes **9.** (3, 4) **10.** (−3, 1)

11.

x	y	(x, y)
−2	−12	(−2, −12)
0	−2	(0, −2)
1	3	(1, 3)

12.

x	y	(x, y)
−1	7	(−1, 7)
2	−2	(2, −2)
5	−11	(5, −11)

13.

x	y	(x, y)
−5	2	(−5, 2)
−2	−4	(−2, −4)
2	−12	(2, −12)

14.

x	y	(x, y)
−6	−6	(−6, −6)
−1	−4	(−1, −4)
2	$-\frac{14}{5}$	$\left(2, -\frac{14}{5}\right)$

15. (a)

x (ft³)	50 ft³	100 ft³	150 ft³
C ($)	$38.83	$72.17	$105.50

(50, 38.83), (100, 72.17), (150, 105.50)

(b) 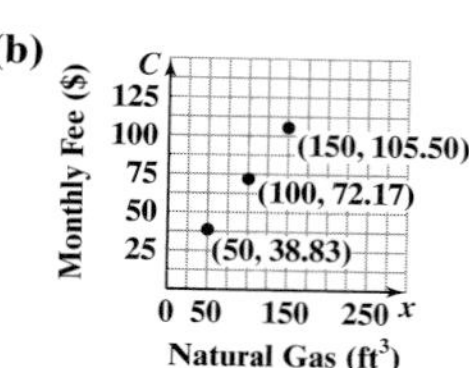

Section 8.2

1. 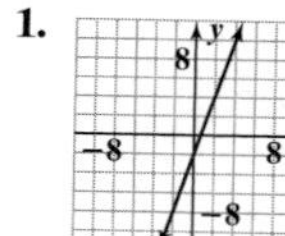

2. 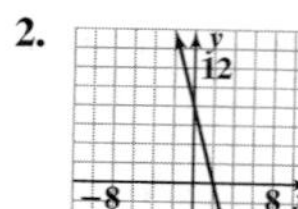

3. Linear **4.** Not linear **5.** Linear **6.** Linear

7. 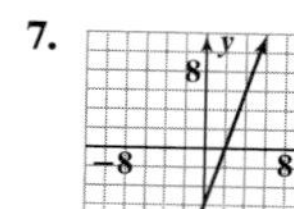

8. 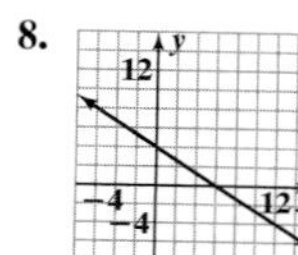

9. (a) (0, 3000), (10000, 3800), (25000, 5000) **(b)** 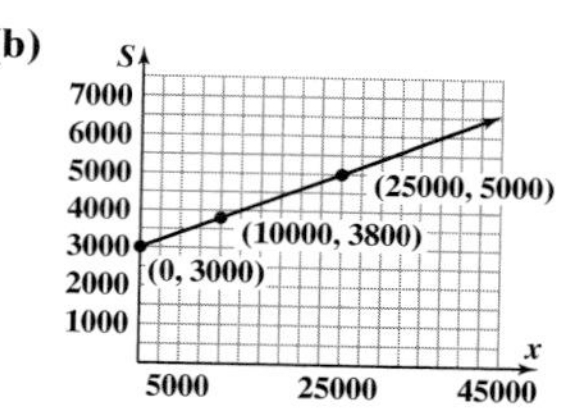

10. Intercepts: (0, 3), (4, 0); x-intercept: 4; y-intercept: 3
11. Intercept: (0, −2); y-intercept: −2

12. 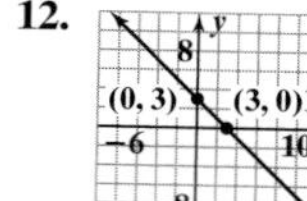
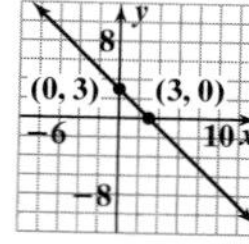

13. 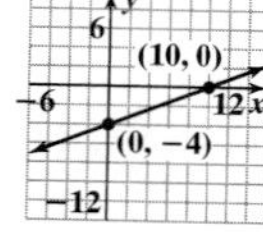

14. 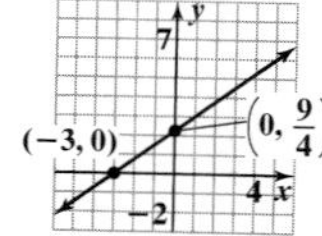

15. **16.** 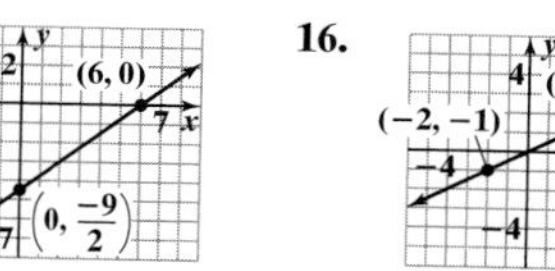

17. 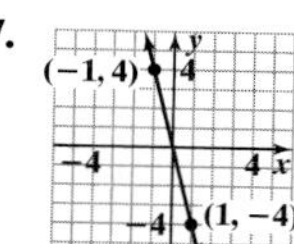

18. 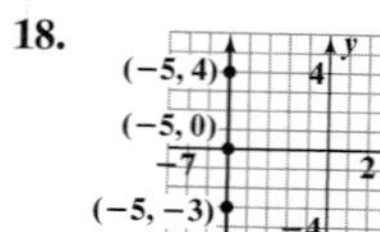

19. 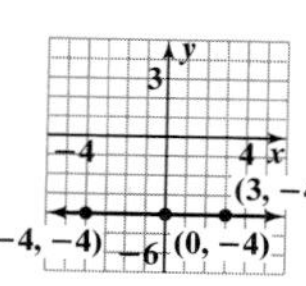

20. 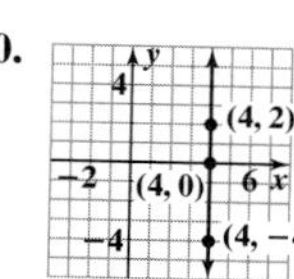

Section 8.3

1. 3; The value of y increases by 3 when x increases by 1. **2.** $-\frac{5}{2}$; The value of y decreases by 5 when x increases by 2.

3. Slope undefined; when y increases by 1, there is no change in x.

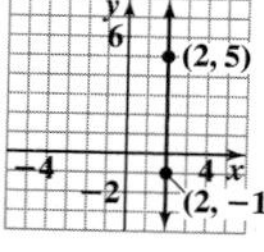

4. $m = 0$; there is no change in y when x increases by 1 unit.

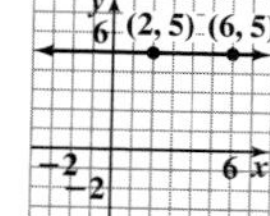

5. $m = 0$; there is no change in y when x increases by 1 unit.

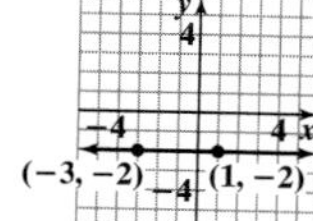

6. Slope undefined; when y increases by 1, there is no change in x.

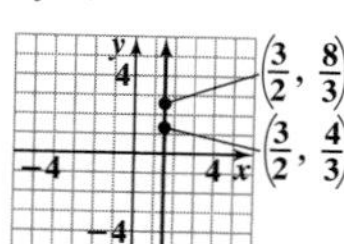

7. (a) 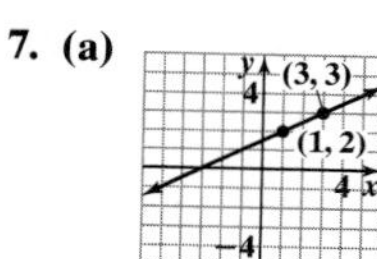
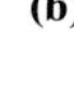

(b) 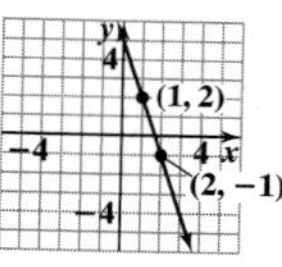
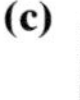

(c) 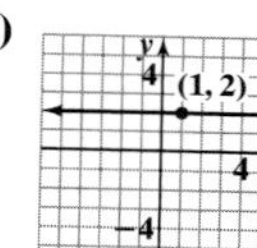

8. 8% **9.** $\frac{1}{6}$ **10.** 0.12; between 10,000 and 14,000 miles driven, the annual cost of gasoline and maintenance on a Chevy Cavalier is $0.12 per mile.

Section 8.4

1. slope: 4, y-intercept: -3 **2.** slope: -3, y-intercept: 7 **3.** slope: $-\frac{2}{5}$, y-intercept: 3

4. 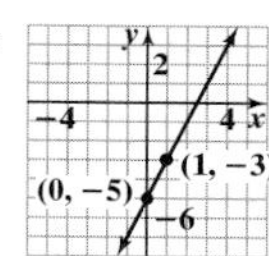
5. 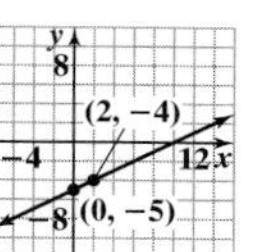
6. 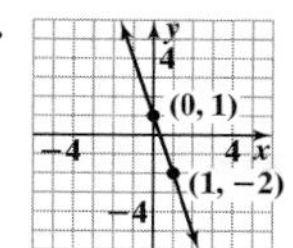
7. 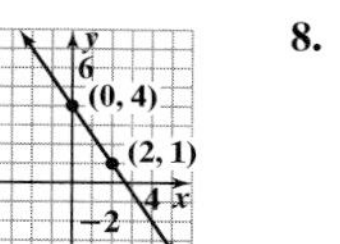
8. 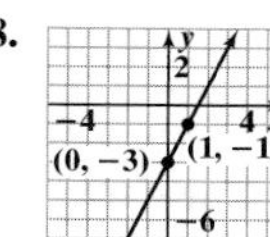
9. **10.** 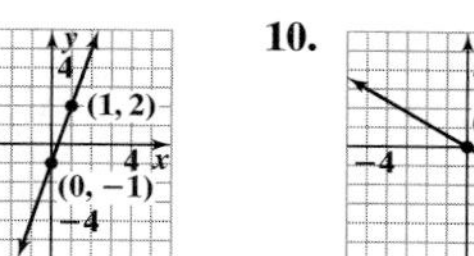

11. $y = 3x - 2$ 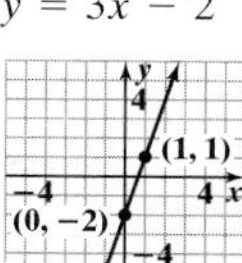

12. $y = -\frac{1}{4}x + 3$

13. $y = -1$ 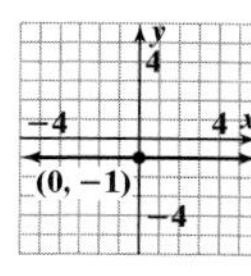

14. **(a)** 2075 grams **(b)** 2933 grams **(c)** slope = 143. Birth weight increases by 143 grams for each additional week of pregnancy.
(d) A gestation period of 0 weeks does not make sense.
(e) 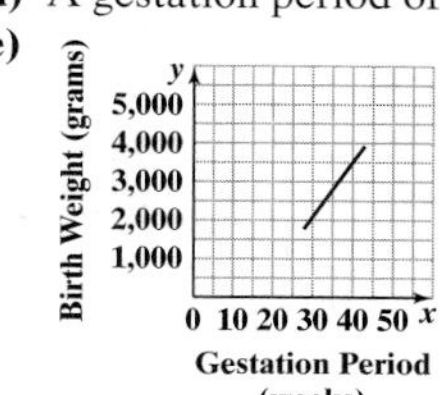

15. **(a)** $y = 0.38x + 50$ **(b)** \$78.50 **(c)** 90 miles
(d) 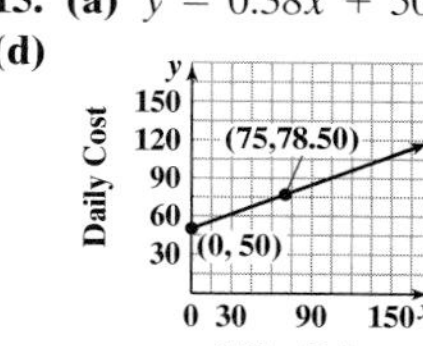
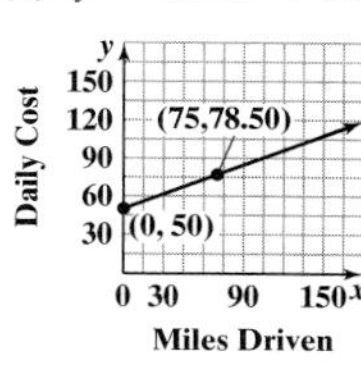

Section 8.5

1. $y = 3x - 5$ 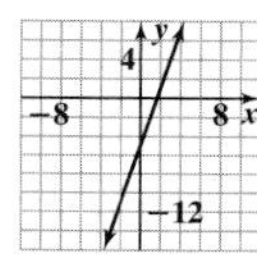

2. $y = \frac{1}{3}x - 5$ 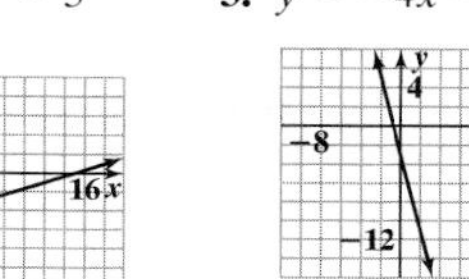

3. $y = -4x - 3$ 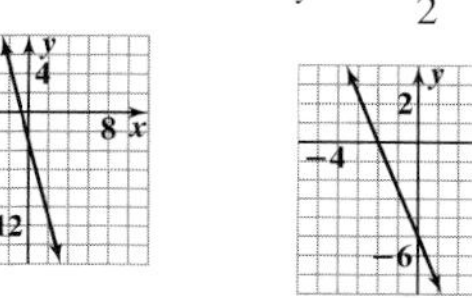

4. $y = -\frac{5}{2}x - 5$

5. $y = 3$ 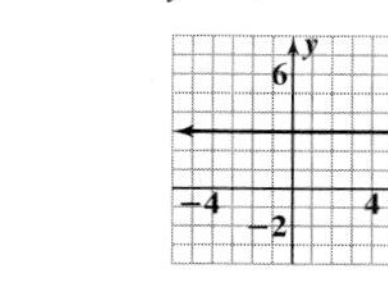

6. $y = x + 2$

7. $y = -3x + 1$ 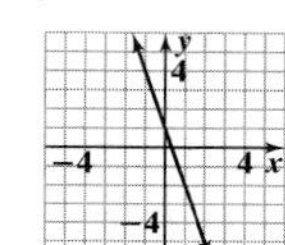

8. $x = 3$ 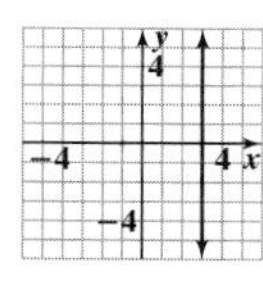

9. **(a)** $y = -100x + 620$ 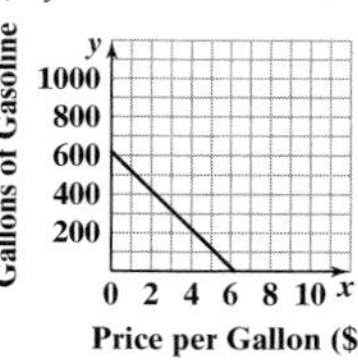
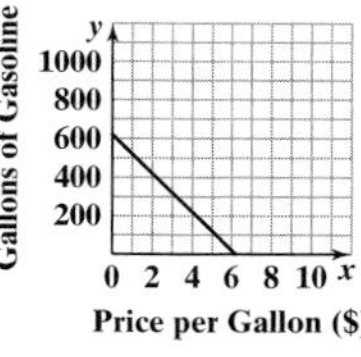

(b) 390 gallons **(c)** The number of gallons sold will decrease by 100 if the price per gallon of gasoline increases by \$1.

Section 8.6

1. Not parallel

2. Parallel 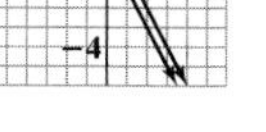

3. Not parallel 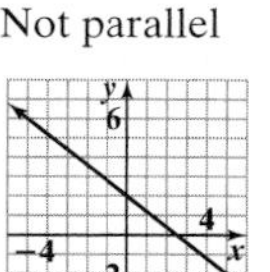
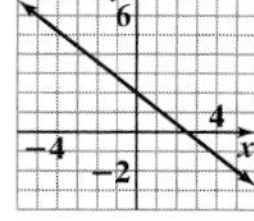

4. $y = 2x - 1$ 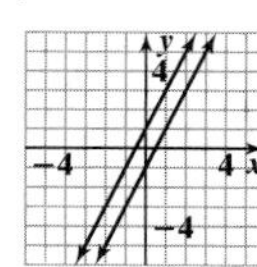
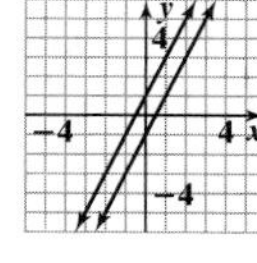

5. $y = -\frac{3}{2}x$

6. $x = 3$ 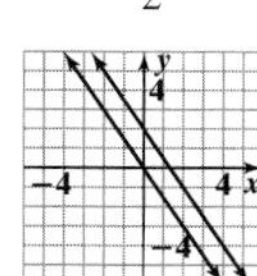

7. $y = 5$ 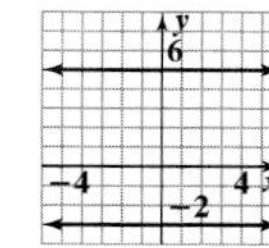

8. $\frac{1}{4}$ 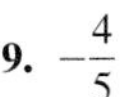  **9.** $-\frac{4}{5}$ **10.** 5

11. Perpendicular 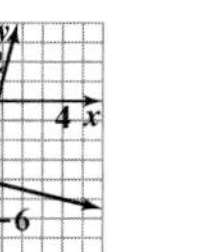

12. Not perpendicular 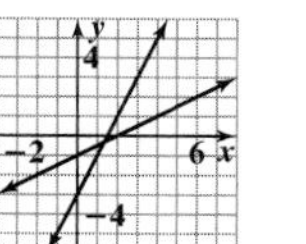

13. Perpendicular 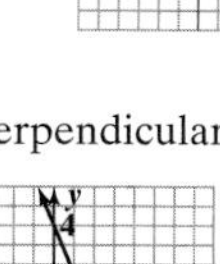
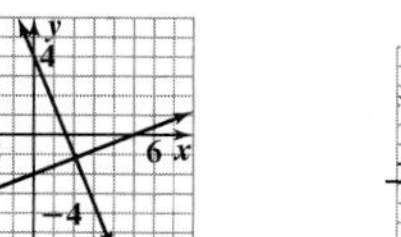

14. $y = -\frac{1}{2}x$ 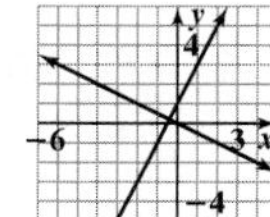

15. $y = \frac{3}{2}x + 2$ 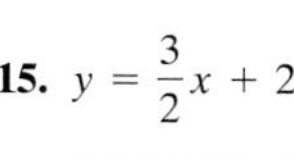
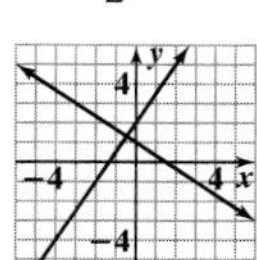

16. $y = -5$ 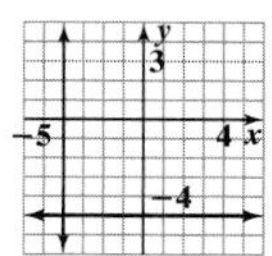

17. $x = 3$ 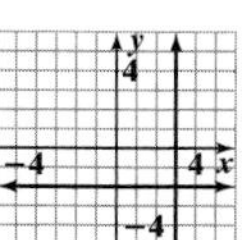

Section 8.7

1. $y = 3x$ **2.** $y = \frac{1}{3}x$ **3. (a)** $q = \frac{1}{4}w$ **(b)** 15 **(c)**

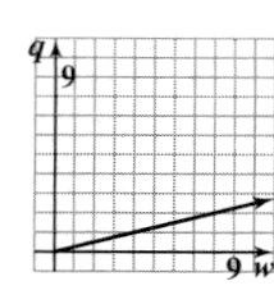

4. \$17.68 **5. (a)** $y = \frac{12}{x}$ **(b)** 3 **6. (a)** $p = \frac{40}{q}$ **(b)** 5 **7.** 25 times per second

Section 8.8

1. (a) No **(b)** Yes **(c)** Yes **2. (a)** Yes **(b)** Yes **(c)** No

3. **4.** **5.** **6.** **7.**

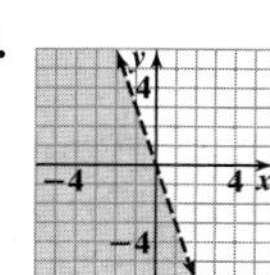

8.

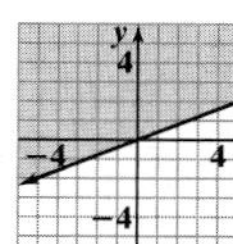

9.

10.

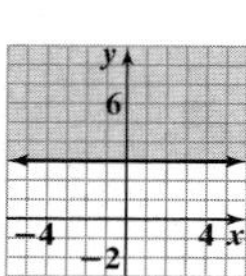

11.

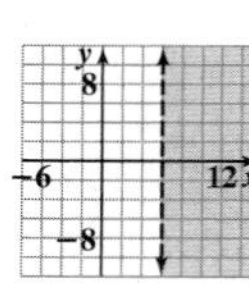

12. (a) $0.2s + 0.25t \le 2$ **(b)** Yes **(c)** No

Chapter 9

Section 9.1

1. (a) No **(b)** Yes **(c)** No **2. (a)** Yes **(b)** Yes **(c)** Yes **3.** $(4, 1)$ **4.** $(-2, 5)$ **5.** $(5, -3)$ **6.** No solution; { } or ∅ **7.** No solution; { } or ∅ **8.** Infinitely many solutions **9.** Infinitely many solutions **10.** One solution; consistent; independent **11.** Infinitely many solutions; consistent; dependent **12.** No solution; inconsistent **13.** After 100 miles the cost is \$60.

Section 9.2

1. $(2, 4)$ **2.** $(-3, 5)$ **3.** $\left(\frac{3}{2}, 7\right)$ **4.** $\left(-\frac{4}{3}, \frac{1}{2}\right)$ **5.** Infinitely many solutions **6.** No solution **7.** No solution **8.** During the year 2060

Section 9.3

1. $(-4, -2)$ **2.** $(8, 3)$ **3.** $(6, -5)$ **4.** $(5, 0)$ **5.** No solution **6.** Infinitely many solutions **7.** Infinitely many solutions **8.** Cheeseburger: \$1.75; shake: \$2.25

Section 9.4

1. 43 and 61 **2.** Width: 50 yards; length: 150 yards **3.** 36° and 54° **4.** 49° and 131° **5.** Airspeed: 350 mph; wind resistance: 50 mph

Section 9.5

1.

	Number	·	Cost per Person	=	Amount
Adults	a		36.95		$36.95a$
Children	c		21.95		$21.95c$
Total					302.55

2. 45 dimes **3.** \$40,000 in Aa-rated bonds: \$50,000 in B-rated bond **4.** 5 pounds of Brazilian coffee and 15 pounds of Colombian coffee **5.** Mix 120 gallons of the wine with 5% alcohol and 80 gallons of the wine with 15% alcohol.

Section 9.6

1. (a) solution **(b)** not a solution

2.

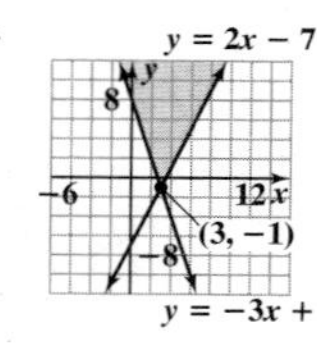

3.

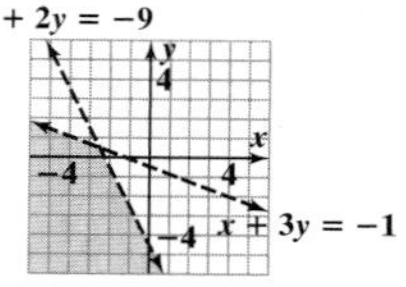

4.

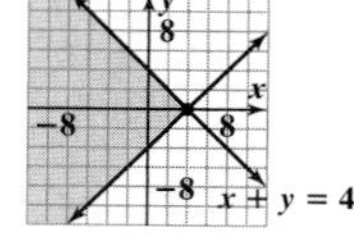

5.

6. (a)

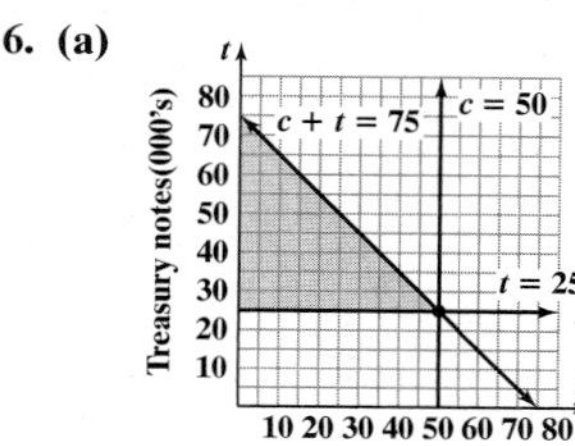

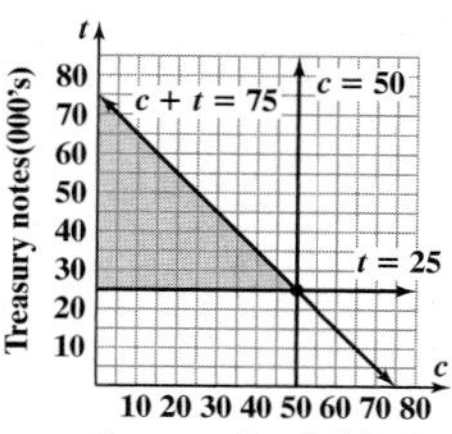

(b) Yes
(c) No

Chapter 10

Section 10.1

1. 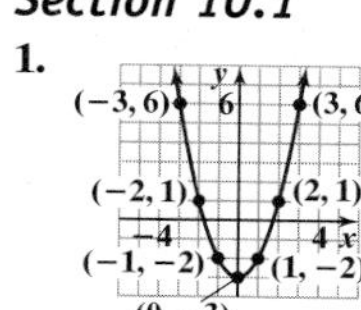

2. 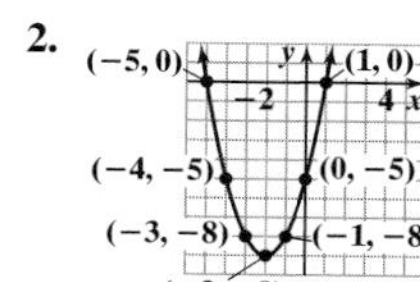

3. 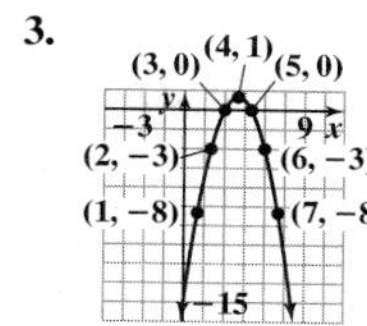

4. y-intercept: 6, x-intercepts: $(-3, 0)$, $(-2, 0)$
5. y-intercept: -5, no x-intercepts
6. y-intercept: 1; x-intercept: $\left(-\frac{1}{2}, 0\right)$
7. **(a)** Opens up **(b)** $(1, -1)$ **(c)** $x = 1$
8. **(a)** Opens down **(b)** $(-3, 11)$ **(c)** $x = -3$

9. 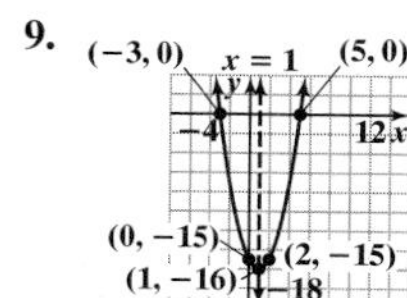

10. 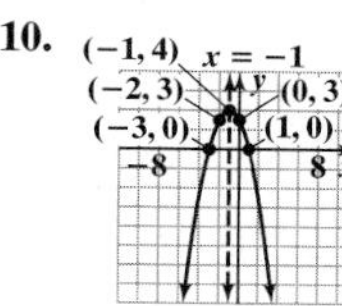

11. 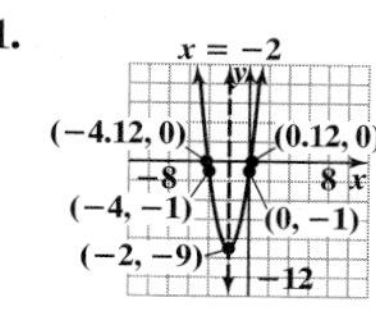

12. 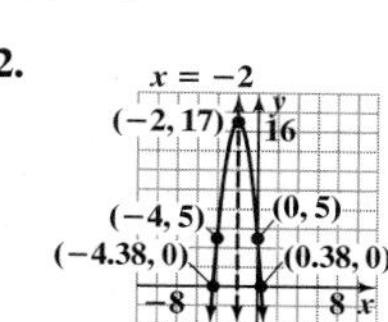

13. Minimum; -7 **14.** Maximum; 33
15. **(a)** After 4 seconds **(b)** 271 feet

Section 10.2

1. {(Max, November 8), (Alesia, January 20), (Trent, March 3), (Yolanda, November 8), (Wanda, July 6), (Elvis, January 8)}

2. 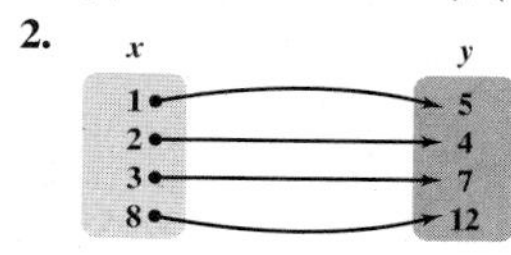

3. Domain: {Max, Alesia, Trent, Yolanda, Wanda, Elvis}; Range: {January 20, March 3, July 6, November 8, January 8}
4. Domain: {1, 2, 3, 8}; Range: {5, 4, 7} **5.** Domain: $\{-3, -2, -1, 2, 3\}$; Range: $\{-2, -1, 1, 2, 3, 4\}$
6. Domain: $\{x \mid -2 \le x \le 4\}$; Range: $\{y \mid -2 \le y \le 2\}$
7. Domain: $\{x \mid x$ is a real number$\}$; Range: $\{y \mid y$ is a real number$\}$

8. 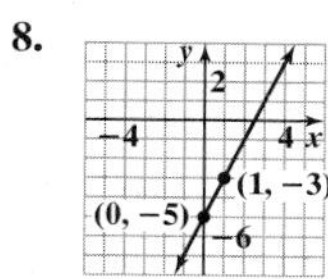

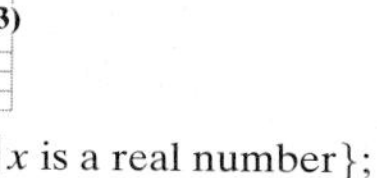

Domain: $\{x \mid x$ is a real number$\}$;
Range: $\{y \mid y$ is a real number$\}$

9. 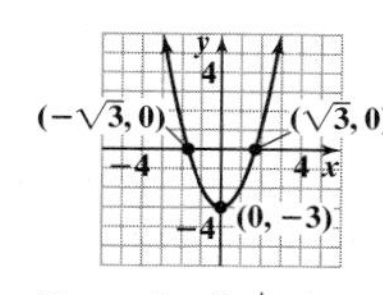

Domain: $\{x \mid x$ is a real number$\}$;
Range: $\{y \mid y \ge -3\}$

Section 10.3

1. Function; Domain: {250, 300, 400, 500}; Range: {\$41.20, \$43.04, \$55.39, \$64.03}
2. Function; Domain: {12, 15, 8, 7, 4}; Range: {A, B, C, D} **3.** not a function **4.** Function; Domain: $\{-3, -2, -1, 0, 1\}$; Range: {3, 2, 1, 0}
5. not a function **6.** Function; Domain: {3, 4, 5, 6}; Range: {8, 10, 11} **7.** function **8.** not a function **9.** function **10.** not a function
11. function **12.** not a function **13.** 17 **14.** -7 **15.** 15 **16.** 22

Appendix A

Section A.1

1. $2 \cdot 2 \cdot 3$ **2.** $2 \cdot 3 \cdot 3$ **3.** $3 \cdot 5 \cdot 5$ **4.** $2 \cdot 2 \cdot 2 \cdot 3 \cdot 5$ **5.** prime **6.** $3 \cdot 3 \cdot 3 \cdot 17$ **7.** 24 **8.** 10 **9.** 360 **10.** 21 **11.** 180
12. $\frac{5}{10}$ **13.** $\frac{30}{48}$ **14.** $\frac{1}{4} = \frac{3}{12}; \frac{5}{6} = \frac{10}{12}$ **15.** $\frac{5}{12} = \frac{25}{60}; \frac{4}{15} = \frac{16}{60}$ **16.** $\frac{9}{20} = \frac{36}{80}; \frac{11}{16} = \frac{55}{80}$ **17.** $\frac{1}{2}$ **18.** $\frac{4}{9}$ **19.** $\frac{9}{16}$ **20.** $\frac{2}{5}$ **21.** $\frac{2}{7}$

Section A.2

1. hundredths **2.** tenths **3.** thousands **4.** thousandths **5.** ones **6.** ten-thousands **7.** 0.2 **8.** 0.93 **9.** 1.40 **10.** 14.398
11. 690.00 **12.** 60.0 **13.** 0.4 **14.** $0.\overline{428571}$ **15.** 1.375 **16.** 0.833... **17.** $0.\overline{5}$ **18.** $\frac{13}{20}$ **19.** $\frac{1}{5}$ **20.** $\frac{5}{8}$ **21.** 0.23 **22.** 0.01
23. 0.724 **24.** 1.27 **25.** 0.8926 **26.** 15% **27.** 80% **28.** 130% **29.** 39.8% **30.** 0.4%

Appendix C

Section C.1

1. Acute **2.** Obtuse **3.** Straight **4.** Right **5.** Complement: 75°; Supplement: 165° **6.** Complement: 30°; Supplement: 120°
7. $m\angle 1 = 140°$; $m\angle 2 = 40°$; $m\angle 3 = 140°$; $m\angle 4 = 40°$; $m\angle 5 = 140°$; $m\angle 6 = 40°$; $m\angle 7 = 140°$

Section C.2

1. 70° **2.** 42° **3.** 14 units **4.** $\frac{15}{2}$ inches or 7.5 inches **5.** 12 feet **6.** 7.2 yards **7.** 18 cm

Section C.3

1. Perimeter: 22 feet; Area: 24 square feet **2.** Perimeter: 26 m; Area: 30 m^2 **3.** Perimeter: 16 cm; Area 16 cm^2 **4.** Perimeter: 6 yards; 2.25 square yards **5.** Perimeter: 130 yards; 650 square yards **6.** Perimeter: 36 m; Area: 70 m^2 **7.** Perimeter: 33 yards; Area: 51 square yards
8. Perimeter: 19 mm; Area: 12 mm^2 **9.** Perimeter: 30 feet; Area: 30 square feet **10.** Circumference: 8π feet $\approx$ 25.13 feet; Area: 16π square feet $\approx$ 50.27 square feet **11.** Circumference: 24π cm $\approx$ 75.40 cm; Area: 144π $cm^2 \approx 452.39$ cm^2

Section C.4

1. Volume: 125 m^3; Surface area: 150 m^2 **2.** Volume: $\frac{256}{3}\pi$ in.$^3 \approx 268.08$ in.^{3}; Surface area: 64π in.$^2 \approx 201.06$ in.^{2}

Answers to Selected Exercises

Chapter 1

Section 1.1
Answers will vary.

Section 1.2
1. rational **3.** absolute value **5.** True **7.** Answers may vary. **9.** $A = \{0, 1, 2, 3, 4\}$ **11.** $D = \{1, 2, 3, 4\}$ **13.** $E = \{6, 8, 10, 12, 14\}$
15. 3 **17.** $-4, 3, 0$ **19.** 2.303003000... **21.** All numbers listed **23.** π **25.** $\frac{5}{5} = 1$ **27.**

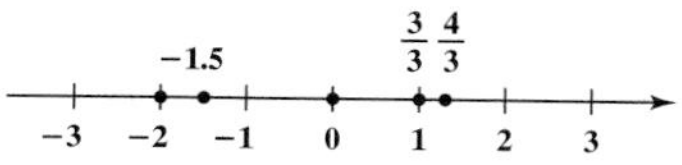

29. 12 **31.** 4 **33.** $\frac{3}{8}$ **35.** 2.1 **37.** True **39.** True **41.** True **43.** False **45.** $<$ **47.** $>$ **49.** $>$ **51.** $=$
53. **(a)**

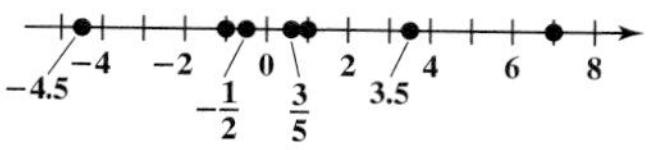

(b) $-4.5, -1, -\frac{1}{2}, \frac{3}{5}, 1, 3.5, |-7| = 7$ **(c)** **(i)** $-1, 1, |-7| = 7$ **(ii)** All numbers listed
55. -100, Integers, Rational, Real **57.** -10.5 Rational, Real **59.** $\frac{75}{25}$, Natural, Whole, Integers, Rational, Real **61.** 7.56556555..., Irrational, Real **63.** True **65.** False **67.** True **69.** True **71.** True **73.** Irrational numbers **75.** Real numbers **77.** 0 **79.** True **81.** True **83.** $\{7, 8, 9, 10, 11, 12, 13, 14, 15\}$ **85.** $\{10, 11, 12, 13, 14, 15\}$ **87.** $\{11, 12\}$ **89.** $\{2, 4, 6, 8, 10\}$ **91.** **(a)** $\{1\}, \{2\}, \{3\}, \{4\}, \{1, 2\}, \{1, 3\}, \{1, 4\}, \{2, 3\}, \{2, 4\}, \{3, 4\}, \{1, 2, 3\}, \{1, 2, 4\}, \{1, 3, 4\}, \{2, 3, 4\}, \{1, 2, 3, 4\}, \emptyset$ **(b)** 16

Section 1.3
1. sum **3.** product **5.** False **7.** Answers may vary. **9.** 15 **11.** 4 **13.** 4 **15.** -19 **17.** 21 **19.** -328 **21.** -11 **23.** 43 **25.** 325 **27.** -125 **29.** 11 **31.** -8 **33.** -28 **35.** 54 **37.** 0 **39.** -41 **41.** -31 **43.** 172 **45.** 40 **47.** -56 **49.** 0 **51.** 144 **53.** -126 **55.** -90 **57.** 210 **59.** 120 **61.** $\frac{1}{8}$ **63.** $-\frac{1}{4}$ **65.** 1 **67.** 5 **69.** 7 **71.** -15 **73.** $\frac{7}{2}$ **75.** $-\frac{10}{7}$ **77.** $\frac{35}{4}$ **79.** -72 **81.** 60 **83.** 171 **85.** -15 **87.** -42 **89.** $\frac{15}{4}$ **91.** 40 **93.** -238 **95.** $28 + (-21) = 7$ **97.** $-21 - 47 = -68$ **99.** $-12 \cdot 18 = -216$ **101.** $-36 \div (-108)$ or $\frac{-36}{-108} = \frac{1}{3}$ **103.** -3.25 points **105.** -6 yards **107.** $-\$48$ **109.** 14 miles **111.** \$655 **113.** No; -125 cases **115.** 25,725 feet **117.** $-3, -5$ **119.** $-12, 2$ **121.** **(a)** $1, 2, 1.5, 1.\overline{6}, 1.6, 1.625, 1.615, 1.619, 1.618, \ldots$ **(b)** 1.618 **(c)** Answers may vary.

Section 1.4
1. reciprocals, multiplicative inverses **3.** $\frac{a - b}{c}$ **5.** False **7.** Answers may vary. **9.** $\frac{2}{3}$ **11.** $-\frac{19}{9}$ **13.** $-\frac{1}{2}$ **15.** $\frac{4}{5}$ **17.** $-\frac{3}{5}$ **19** $\frac{2}{3}$ **21.** $\frac{12}{25}$ **23.** -25 **25.** $-\frac{2}{3}$ **27.** 8 **29.** $\frac{31}{3}$ **31.** $\frac{6}{11}$ **33.** $\frac{5}{3}$ **35.** $-\frac{1}{5}$ **37.** $\frac{5}{6}$ **39.** $-\frac{1}{9}$ **41.** $-\frac{2}{3}$ **43.** $\frac{9}{11}$ **45.** -12 **47.** 32 **49.** $\frac{3}{2}$ **51.** $\frac{1}{2}$ **53.** 2 **55.** $\frac{1}{3}$ **57.** $\frac{5}{2}$ **59.** $-\frac{13}{12}$ **61.** $\frac{1}{4}$ **63.** $\frac{9}{5}$ **65.** $-\frac{1}{6}$ **67.** $\frac{33}{40}$ **69.** $\frac{139}{36}$ **71.** $-\frac{7}{18}$ **73.** -6.5 **75.** 8.4 **77.** 55.92 **79.** 1.49 **81.** 42.55 **83.** 24.94 **85.** 9 **87.** 24.3 **89.** 490 **91.** 0.95 **93.** $-\frac{11}{30}$ **95.** $-\frac{4}{3}$ **97.** $-\frac{2}{3}$ **99.** $-\frac{1}{4}$ **101.** $-\frac{129}{35}$ **103.** 1.6 **105.** $-\frac{2}{21}$ **107.** -50.526 **109.** -16 **111.** -15 **113.** -6.58 **115.** -7.9 **117.** $-\frac{25}{14}$ **119.** 58.39 **121.** 30.96 **123.** $\frac{1}{8}$ **125.** 21 hours **127.** 18 students **129.** $-\$68.79$ **131.** \$2.40 **133.** 13.2 **135.** $\frac{86}{15}$

Putting the Concepts Together
1. **(a)** $-12, -\frac{14}{7} = -2, 0, 3$ **(b)** $-12, -\frac{14}{7} = -2, -1.25, 0, 3, 11.2$ **(c)** $\sqrt{2}$ **(d)** All the numbers in the set are real numbers. **2.** $<$ **3.** -11 **4.** -65 **5.** -27 **6.** 27 **7.** -5.5 **8.** -10 **9.** -100 **10.** 72 **11.** -5 **12.** 16 **13.** -9 **14.** -3 **15.** $\frac{31}{5}$ **16.** $\frac{31}{36}$ **17.** $-\frac{17}{36}$ **18.** $\frac{9}{5}$ **19.** $-\frac{1}{28}$ **20.** 0 **21.** 10.76 **22.** 7.646 **23.** 1.46 **24.** 22.232

Section 1.5
1. Identity Property; Addition **3.** undefined **5.** True **7.** Answers may vary. **9.** Additive Inverse Property **11.** Multiplicative Identity Property **13.** Multiplicative Inverse Property **15.** Additive Inverse Property **17.** Commutative Property of Multiplication **19.** Commutative Property of Addition **21.** 156 inches **23.** 45 meters **25.** $10\frac{1}{2}$ gallons **27.** $11\frac{1}{4}$ pounds **29.** $4\frac{1}{2}$ hours = 4 hours, 30 minutes **31.** 29 **33.** 18 **35.** -65 **37.** 347 **39.** -90 **41.** undefined **43.** -34 **45.** 0 **47.** 0 **49.** $-\frac{20}{3}$ **51.** \$203.16 **53.** Answers may vary. **55.** Answers may vary. **57.** $-6 - (4 + 10)$ **59.** $25 - (6 - 10) - 1$ **61.** $58\frac{2}{3}$ feet per second

Section 1.6

1. base **3.** grouping symbols **5.** False **7.** Answers may vary. **9.** 5^2 **11.** $\left(\frac{3}{5}\right)^3$ **13.** 64 **15.** 64 **17.** 1000 **19.** $\frac{27}{64}$ **21.** 2.25 **23.** -9 **25.** -1 **27.** 0 **29.** $\frac{1}{64}$ **31.** $-\frac{1}{27}$ **33.** 14 **35.** -3 **37.** 2500 **39.** 160 **41.** 20 **43.** 4 **45.** $\frac{3}{5}$ **47.** -1 **49.** 42 **51.** 5 **53.** -4 **55.** 115 **57.** -5 **59.** $-\frac{65}{4}$ **61.** -24 **63.** $-\frac{13}{12}$ **65.** 0.5 **67.** 12 **69.** $-\frac{1}{2}$ **71.** $\frac{3}{4}$ **73.** 1 **75.** 24 **77.** $\frac{133}{8}$ **79.** $\frac{2}{5}$ **81.** 3 **83.** $\frac{3}{2}$ **85.** $\frac{64}{27}$ **87.** $-\frac{1}{6}$ **89.** $2^3 \cdot 3^2$ **91.** $2^4 \cdot 3$ **93.** $(4 \cdot 3 + 6) \cdot 2$ **95.** $(4 + 3) \cdot (4 + 2)$ **97.** $(6 - 4) + (3 - 1)$ **99.** \$514.93 **101.** 603.19 in.^{2} **103.** \$1060.90 in.^{2} **105.** 115.75°

Section 1.7

1. Distributive **3.** -1 **5.** False **7.** Answers may vary. **9.** $2x^3, \frac{x^2}{4}, -x, 6; 2, \frac{1}{4}, -1, 6$ **11.** $z^2, \frac{2y}{3}; 1, \frac{2}{3}$ **13.** 13 **15.** 17 **17.** -21 **19.** 104 **21.** $\frac{17}{5}$ **23.** 225 **25.** 153 **27.** unlike **29.** like **31.** like **33.** unlike **35.** $3m + 6n$ **37.** $18n^2 + 12n - 6$ **39.** $-x + y$ **41.** $-4x + 3y$ **43.** $3x$ **45.** $6z$ **47.** $10m + 10n$ **49.** $2.2x^7$ **51.** $10y^6$ **53.** $-6w - 12y + 13z$ **55.** $-3k + 15$ **57.** $4n - 8$ **59.** $-4n + 20$ **61.** $\frac{5}{6}x$ **63.** $-\frac{11}{2}$ **65.** $-3.5x - 6$ **67.** $6x - 0.06$ **69.** 32 **71.** 27 **73.** 0 **75.** -13 **77.** -7 **79.** -12 **81.** 44 **83.** $-\frac{3}{2}$ **85.** 36 **87.** \$78.70 **89.** \$4819 **91. (a)** $8w - 8$ **(b)** 32 yards **93.** \$228.88 **95.** Answers may vary; $-3x^2 + 7x - 3$

Chapter 1 Review

1. $A = \{0, 1, 2, 3, 4, 5, 6\}$ **2.** $B = \{1, 2, 3\}$ **3.** $C = \{-2, -1, 0, 1, 2, 3, 4, 5\}$ **4.** $D = \{-2, -1, 0, 1, 2, 3\}$ **5.** $\frac{9}{3} = 3, 11$ **6.** $0, \frac{9}{3} = 3, 11$ **7.** $-6, 0, \frac{9}{3} = 3, 11$ **8.** $-6, -3.25, 0, \frac{9}{3} = 3, 11, \frac{5}{7}$ **9.** 5.030030003... **10.** All numbers listed

11. 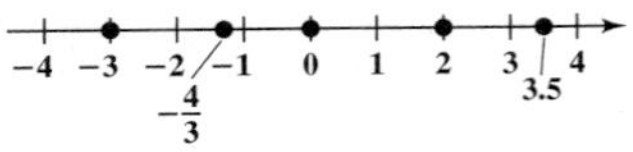

12. 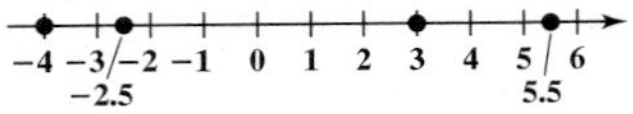

13. False **14.** True **15.** True **16.** True **17.** $-\frac{1}{2}$ **18.** 7 **19.** -6 **20.** -8.2 **21.** = **22.** < **23.** > **24.** > **25.** > **26.** < **27.** Answers may vary. **28.** natural numbers **29.** 7 **30.** -4 **31.** -34 **32.** -95 **33.** -4 **34.** 47 **35.** -53 **36.** -115 **37.** -22 **38.** -7 **39.** 21 **40.** 67 **41.** 26 **42.** -36 **43.** 12 **44.** -40 **45.** -1118 **46.** -8037 **47.** -715 **48.** $-11{,}130$ **49.** 5 **50.** -12 **51.** 5 **52.** -25 **53.** -8 **54.** $-\frac{16}{5}$ **55.** $-\frac{10}{3}$ **56.** $-\frac{30}{7}$ **57.** -13 **58.** 45 **59.** $-43 + 101 = 58$ **60.** $45 + (-28) = 17$ **61.** $-10 - (-116) = 106$ **62.** $74 - 56 = 18$ **63.** $13 + (-8) = 5$ **64.** $-60 - (-10) = -50$ **65.** $-21 \cdot (-3) = 63$ **66.** $54 \cdot (-18) = -972$ **67.** $-34 \div (-2)$ or $\frac{-34}{-2} = 17$ **68.** $-49 \div 14$ or $\frac{-49}{14} = -\frac{7}{2}$ **69.** 26 yards **70.** -3°F **71.** 24°F **72.** 87 points **73.** $\frac{1}{2}$ **74.** $-\frac{1}{3}$ **75.** $-\frac{2}{3}$ **76.** $-\frac{7}{5}$ **77.** $\frac{5}{4}$ **78.** $-\frac{5}{28}$ **79.** $-\frac{1}{20}$ **80.** $-\frac{3}{2}$ **81.** $\frac{4}{17}$ **82.** $-\frac{2}{3}$ **83.** $-\frac{3}{10}$ **84.** -32 **85.** $\frac{1}{3}$ **86.** $-\frac{2}{5}$ **87.** $\frac{3}{7}$ **88.** 3 **89.** $\frac{7}{20}$ **90.** $\frac{31}{36}$ **91.** $-\frac{59}{245}$ **92.** $\frac{13}{12}$ **93.** $-\frac{19}{12}$ **94.** $-\frac{11}{4}$ **95.** 0 **96.** $-\frac{23}{24}$ **97.** 48.5 **98.** -24.66 **99.** 82.98 **100.** -53.74 **101.** 0.0804 **102.** -260.154 **103.** 18.4 **104.** -25.79 **105.** 2.3 **106.** -22.9 **107.** -5.418 **108.** -0.732 **109.** $-\$98.93$ **110.** 24 friends **111.** $\frac{23}{2}$ or $11\frac{1}{2}$ inches **112.** \$186.81 **113.** Associative Property of Multiplication **114.** Multiplicative Inverse Property **115.** Multiplicative Inverse Property **116.** Commutative Property of Multiplication **117.** Commutative Property of Multiplication **118.** Additive Inverse Property **119.** Identity Property of Addition **120.** Identity Property of Addition **121.** Commutative Property of Addition **122.** Multiplicative Identity Property **123.** Multiplication Property of Zero **124.** Associative Property of Addition **125.** 29 **126.** 99 **127.** 18 **128.** 121 **129.** 3.4 **130.** 5.3 **131.** -33 **132.** 6 **133.** undefined **134.** 0 **135.** -334 **136.** 2 **137.** 0 **138.** 1 **139.** 0 **140.** 130 **141.** $-\frac{5}{3}$ **142.** $-\frac{150}{13}$ **143.** 3^4 **144.** $\left(\frac{2}{3}\right)^3$ **145.** $(-4)^2$ **146.** $(-3)^3$ **147.** 125 **148.** 32 **149.** 81 **150.** -64 **151.** -81 **152.** $\frac{1}{64}$ **153.** -4 **154.** -76 **155.** 206 **156.** 32 **157.** 2 **158.** $\frac{1}{2}$ **159.** $\frac{6}{5}$ **160.** $\frac{7}{5}$ **161.** 21 **162.** -18 **163.** -729 **164.** -3 **165.** $3x^2, -x, 6; 3, -1, 6$ **166.** $2x^2y^3, -\frac{y}{5}; 2, -\frac{1}{5}$ **167.** like **168.** unlike **169.** unlike **170.** like **171.** $-3x$ **172.** $-4x - 15$ **173.** $-4.1x^4 + 0.3x^3$ **174.** $-3x^4 + 6x^2 + 12$ **175.** $18 - x$ **176.** $4x - 18$ **177.** $9x - 4$ **178.** -1 **179.** \$98.70

Chapter 1 Test

1. $\frac{1}{3}$ **2.** $\frac{9}{4}$ **3.** $-\frac{320}{3}$ **4.** -102 **5.** -12.16 **6.** -20 **7.** undefined **8.** -14 **9.** 55 **10. (a)** 6 **(b)** 0, 6 **(c)** $-2, 0, 6$ **(d)** $-2, -\frac{1}{2}, 0, 2.5, 6$ **(e)** none **(f)** All those listed. **11.** < **12.** = **13.** -7 **14.** 15 **15.** -102 **16.** -343 **17.** $-16x - 28$ **18.** $-4x^2 + 5x + 4$ **19.** \$531.85 **20.** $4x + 10$

Chapter 2

Section 2.1

1. Addition Property of Equality **3.** subtract 9 **5.** False **7.** Answers may vary. **9.** Answers may vary. **11.** Yes **13.** No **15.** Yes **17.** Yes **19.** $\{20\}$ **21.** $\{-12\}$ **23.** $\{19\}$ **25.** $\{-13\}$ **27.** $\{2\}$ **29.** $\left\{\frac{1}{4}\right\}$ **31.** $\left\{\frac{19}{24}\right\}$ **33.** $\{-6.1\}$ **35.** $\{5\}$ **37.** $\{-4\}$ **39.** $\left\{\frac{7}{2}\right\}$ **41.** $\left\{-\frac{5}{2}\right\}$ **43.** $\{21\}$ **45.** $\{121\}$ **47.** $\left\{\frac{3}{5}\right\}$ **49.** $\left\{-\frac{1}{3}\right\}$ **51.** $\{9\}$ **53.** $\left\{-\frac{4}{9}\right\}$ **55.** $\left\{-\frac{5}{3}\right\}$ **57.** $\{2\}$ **59.** $\{-3\}$ **61.** $\left\{\frac{2}{3}\right\}$ **63.** $\{-6\}$ **65.** $\{19\}$ **67.** $\{283\}$ **69.** $\{-50\}$ **71.** $\{41.1\}$ **73.** $\left\{\frac{20}{3}\right\}$ **75.** $\{-4\}$ **77.** $\{-12\}$ **79.** $\left\{\frac{1}{2}\right\}$ **81.** $\left\{\frac{3}{16}\right\}$ **83.** $\left\{-\frac{5}{4}\right\}$ **85.** $18,499.80 **87.** $68 **89.** 12 **91.** 0.18 or 18% **93.** $x = 48 - \lambda$ **95.** $x = \frac{14}{\theta}$ **97.** $\lambda = \frac{50}{9}$ **99.** $\theta = -\frac{6}{7}$

Section 2.2

1. add 8 **3.** add x **5.** True **7.** Answers may vary. **9.** Answers may vary. **11.** $\{1\}$ **13.** $\{-2\}$ **15.** $\{-3\}$ **17.** $\left\{\frac{3}{2}\right\}$ **19.** $\{-3\}$ **21.** $\{12\}$ **23.** $\{4\}$ **25.** $\{-2\}$ **27.** $\{-5\}$ **29.** $\{-8\}$ **31.** $\{-5\}$ **33.** $\{-8\}$ **35.** $\{3\}$ **37.** $\left\{-\frac{7}{2}\right\}$ **39.** $\{7\}$ **41.** $\{-18\}$ **43.** $\left\{\frac{7}{2}\right\}$ **45.** $\left\{-\frac{27}{16}\right\}$ **47.** $\{2\}$ **49.** $\left\{-\frac{3}{4}\right\}$ **51.** $\left\{\frac{1}{3}\right\}$ **53.** $\left\{\frac{1}{16}\right\}$ **55.** $\left\{\frac{15}{2}\right\}$ **57.** $\left\{\frac{1}{2}\right\}$ **59.** $\{-7\}$ **61.** $\left\{\frac{51}{7}\right\}$ **63.** McDonald's: 23 g; Burger King: 27 g **65.** width: $\frac{13}{3}$ or $4\frac{1}{3}$ feet; length: $\frac{32}{3}$ or $10\frac{2}{3}$ feet **67.** $8 **69.** Yes **71.** $\{2.47\}$ **73.** $\{2.6\}$ **75.** $\frac{20}{3}$ **77.** $-\frac{5}{4}$

Section 2.3

1. conditional equation **3.** 100 **5.** False **7.** Answers may vary. **9.** $\left\{\frac{9}{2}\right\}$ **11.** $\{-2\}$ **13.** $\{-6\}$ **15.** $\left\{\frac{2}{3}\right\}$ **17.** $\{30\}$ **19.** $\{-4\}$ **21.** $\{50\}$ **23.** $\{4.8\}$ **25.** $\{150\}$ **27.** $\{6\}$ **29.** $\left\{\frac{25}{8}\right\}$ **31.** $\{-20\}$ **33.** $\{1\}$ **35.** $\{-12\}$ **37.** $\{60\}$ **39.** $\{5\}$ **41.** $\{2\}$ **43.** $\{75\}$ **45.** contradiction; $\emptyset$ or $\{\ \}$ **47.** identity; all real numbers **49.** conditional equation; $\left\{-\frac{1}{2}\right\}$ **51.** contradiction; $\emptyset$ or $\{\ \}$ **53.** contradiction; $\emptyset$ or $\{\ \}$ **55.** identity; all real numbers **57.** $\left\{-\frac{3}{4}\right\}$ **59.** identity; all real numbers **61.** $\left\{-\frac{1}{3}\right\}$ **63.** $\{-20\}$ **65.** $\left\{\frac{7}{2}\right\}$ **67.** contradiction; $\emptyset$ or $\{\ \}$ **69.** $\{-2\}$ **71.** $\{3\}$ **73.** contradiction; $\emptyset$ or $\{\ \}$ **75.** $\{-4\}$ **77.** $\{39\}$ **79.** $\{0\}$ **81.** $\left\{\frac{2}{3}\right\}$ **83.** -1.70 **85.** -13.2 **87.** $50 **89.** $18,000 **91.** $8.50 **93.** $95 **95.** 15 quarters **97.** 6 units **99.** $12,000 **101.** Answers may vary.

Section 2.4

1. formula **3.** subtract **5.** False **7.** Answers may vary. **9.** Rogan 1099.14 feet; Nurek 984.3 feet **11.** $19.20 **13.** $650 **15.** 20°C **17.** $3 **19. (a)** 50 units **(b)** 144 square units **21. (a)** 36.2 meters **(b)** 70 square meters **23. (a)** 36 units **(b)** 81 square units **25. (a)** 31.4 cm **(b)** 78.5 cm^2 **27.** 68.44 square inches **29.** $t = \frac{d}{r}$ **31.** $d = \frac{C}{\pi}$ **33.** $r = \frac{I}{Pt}$ **35.** $h = \frac{2A}{b}$ **37.** $a = P - b - c$ **39.** $t = \frac{A - P}{Pr}$ **41.** $b = \frac{2A}{h} - B$ **43.** $y = -3x + 12$ **45.** $y = 2x - 5$ **47.** $y = \frac{-4x + 13}{3}$ **49.** $y = 3x - 12$ **51. (a)** $C = R - P$ **(b)** $450 **53. (a)** $t = \frac{I}{Pr}$ **(b)** 2 years **55. (a)** $m = \frac{2K}{v^2}$ **(b)** 16 **57. (a)** $C = \frac{5}{9}(F - 32)$ **(b)** 15°C **59. (a)** $r = \frac{A - P}{Pt}$ **(b)** 0.04 or 4% **61. (a)** $h = \frac{V}{\pi r^2}$ **(b)** 5 mm **63. (a)** $b = \frac{2A}{h}$ **(b)** 18 ft **65.** 1834.05 **67. (a)** $h = \frac{S - 2\pi r^2}{2\pi r}$ **(b)** 1.25 inches **69.** medium 12" **71. (a)** 12 hours **(b)** $336 **73.** 50.5 square inches **75.** $96\pi\ \text{cm}^3 \approx 301.59\ \text{cm}^3$ **77. (a)** $I = \frac{D - P}{0.03} + 137{,}300$ **(b)** $158,000 **79. (a)** 62 **(b)** $372 **(c)** Yes **81. (a)** 4948 ft^2 **(b)** $1237 **83. (a)** 1080 in.^2 **(b)** 7.5 ft^2 **85.** Multiply by $\frac{1\ \text{ft}^2}{144\ \text{in}^2}$.

Putting the Concepts Together

1. (a) Yes **(b)** No **2. (a)** No **(b)** Yes **3.** $\left\{-\frac{2}{3}\right\}$ **4.** $\{-40\}$ **5.** $\{-6\}$ **6.** $\{3\}$ **7.** $\{-6\}$ **8.** $\left\{-\frac{7}{3}\right\}$ **9.** $\left\{-\frac{57}{2}\right\}$ **10.** $\{25\}$ **11.** $\{6\}$ **12.** $\left\{\frac{74}{5}\right\}$ or $\{14.8\}$ **13.** contradiction; $\emptyset$ or $\{\ \}$ **14.** identity; all real numbers **15.** $5000 **16. (a)** $b = \frac{2A}{h} - B$ **(b)** 6 in. **17. (a)** $h = \frac{V}{\pi r^2}$ **(b)** 13 in. **18.** $y = -\frac{3}{2}x + 7$

Section 2.5

1. mathematical modeling **3.** False **5.** Answers may vary. **7.** Answers may vary. **9.** $-5 + x$ **11.** $x\left(\frac{2}{3}\right)$ or $\frac{2}{3}x$ **13.** $\frac{1}{2}x$ **15.** $x - (-25)$ **17.** $\frac{x}{3}$ **19.** $x + \frac{1}{2}$ **21.** $6x + 9$ **23.** $2(13.7 + x)$ **25.** $2x + 31$ **27.** $x + 15 = -34$ **29.** $35 = 3x - 7$ **31.** $\frac{x}{-4} + 5 = 36$ **33.** $2[x + 6] = x + 3$ **35.** Braves: r; Clippers: $r + 5$ **37.** Bill's amount: b; Jan's amount: $b + 0.55$ **39.** Janet's share: j; Kathy's share: $200 - j$ **41.** number adults: a; number children: $1433 - a$ **43.** 83 **45.** −14 **47.** 54, 55, 56 **49.** Verrazano-Narrows Bridge: 4260 ft; Golden Gate Bridge: 4200 ft **51.** $11,215 **53.** CD: $8500; Bonds: $11,500 **55.** Stocks: $20,000; Bonds: $12,000 **57.** Smart Start: 2 g; Go Lean: 8 g **59.** $29,140 **61.** 150 miles **63.** 2000 pages **65.** Jensen: $36,221; Maureen: $35,972 **67.** 5, 6, 7, and 8 **69.** Answers may vary. **71.** Answers may vary. **73.** 20°, 40°, 120°

Section 2.6

1. 0.032 **3.** True **5.** Answers may vary. **7.** 80 **9.** 14 **11.** 4.8 **13.** 210 **15.** 72 **17.** $8\frac{1}{3}$ **19.** 40% **21.** 7.5% **23.** 200% **25.** $54 **27.** $102,000 **29.** $25,000 **31.** $68 **33.** $570 **35.** $400 **37.** winner: 530; loser: 318 **39.** $285,700 **41.** 597 **43.** 32.4 million **45.** 70.2% **47.** 24.6% **49.** $24 **51.** 28.6% **53. (a)** $31,314.38 **(b)** 43.7% **55.** 21.2%

Section 2.7

1. 90 **3.** True **5.** False **7.** 20°, 70° **9.** 50°, 130° **11.** 60°, 75°, 45° **13.** 42°, 44°, 94° **15.** 47.5°, 132.5° **17.** 52.5°, 37.5° **19.** 44°, 46° **21.** 18°, 72°, 90° **23.** $l = 32$ feet; $w = 12$ feet **25.** 42 ft by 84 ft **27.** 26.5 in. **29.** 49°, 49°, 82° **31. (a)** $62t$ **(b)** $68t$ **(c)** $62t + 68t$ **(d)** $62t + 68t = 585$ **33.** $528(t + 10) = 880t$ **35.** length = 14 in.; width = 4 in. **37.** 40 ft **39.** 40 ft; 32 ft **41. (a)** length = 11 ft; width = 19 ft **(b)** 209 ft^2 **43.** 13 hours **45.** fast car: 60 mph; slow car: 48 mph **47.** freeway: 4.5 hours; 2-lane: 1.5 hours **49.** 6 mph **51.** $x = 3$ **53.** $x = 32.5$ **55.** $x = 18$

Section 2.8

1. solve **3.** $\{x|x$ is a real number$\}$ **5.** True **7.** A left parenthesis is used when the endpoint is not included. A left bracket is used when the endpoint is included. **9.** $x \geq 16{,}000$ **11.** $x \leq 20{,}000$ **13.** $x > 12{,}000$ **15.** $x > 0$ **17.** $x \leq 0$ **19.** $(2, \infty)$

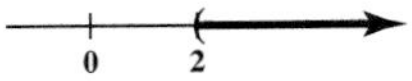

21. $(-\infty, -1]$

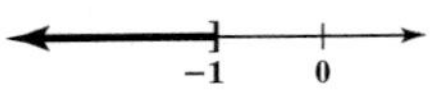

23. $[-3, \infty)$

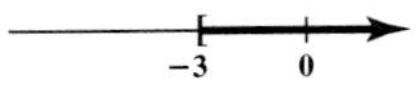

25. $(-\infty, 4)$

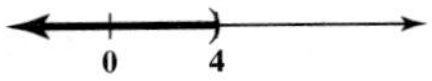

27. $(-\infty, 2)$

29. $\emptyset$ or { } **31.** $(-\infty, \infty)$ **33.** $\{x|x < 4\}; (-\infty, 4)$

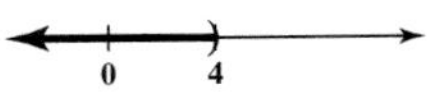

35. $\{x|x \geq 2\}; [2, \infty)$

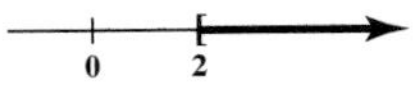

37. $\{x|x \leq 5\}; (-\infty, 5]$

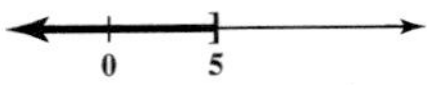

39. $\{x|x > -7\}; (-7, \infty)$

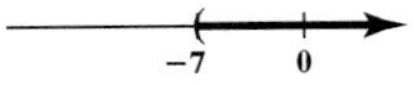

41. $\{x|x > 3\}; (3, \infty)$

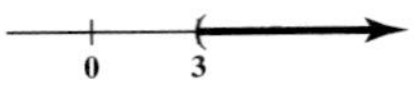

43. $\{x|x \geq 2\}; [2, \infty)$

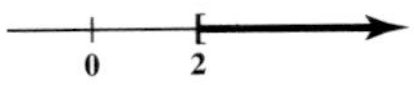

45. $\{x|x \geq -1\}; [-1, \infty)$

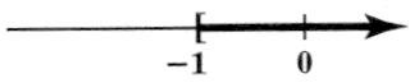

47. $\{x|x > -7\}; (-7, \infty)$

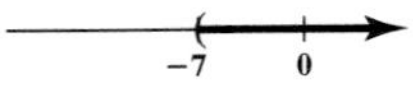

49. $\left\{x|x \leq \frac{2}{3}\right\}; \left(-\infty, \frac{2}{3}\right]$

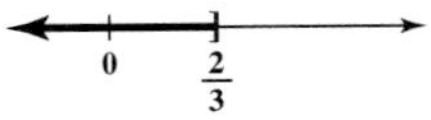

51. $\{x|x < -20\}; (-\infty, -20)$

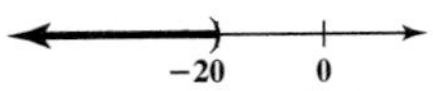

53. $\emptyset$ or { }

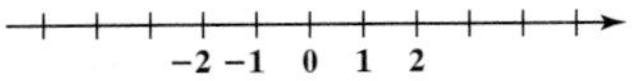

55. $\{n|n$ is any real number$\}; (-\infty, \infty)$

57. $\{n|n > 5\}; (5, \infty)$

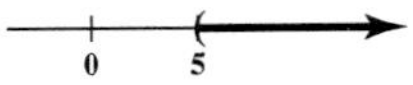

59. $\{w|w$ is any real number$\}; (-\infty, \infty)$

61. $\left\{y|y < -\frac{3}{2}\right\}; \left(-\infty, -\frac{3}{2}\right)$

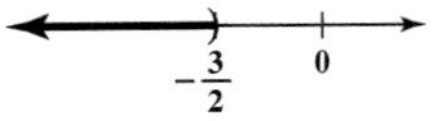

63. $\{x|x > 4\}; (4, \infty)$

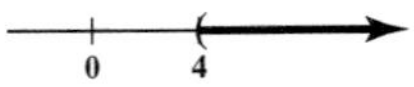

65. $\left\{x|x < \frac{3}{4}\right\}; \left(-\infty, \frac{3}{4}\right)$

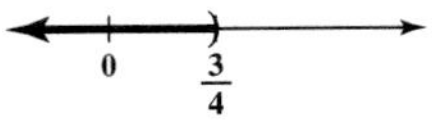

67. $\{x|x$ is any real number$\}; (-\infty, \infty)$

69. $\{a|a < -1\}; (-\infty, -1)$

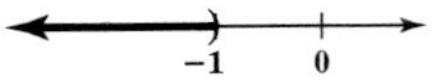

71. $\{n|n$ is any real number$\}; (-\infty, \infty)$

73. $\left\{x|x \geq \frac{4}{3}\right\}; \left[\frac{4}{3}, \infty\right)$

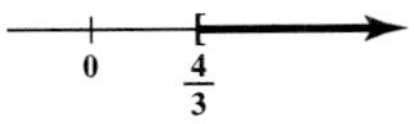

75. $\{x|x < 25\}; (-\infty, 25)$

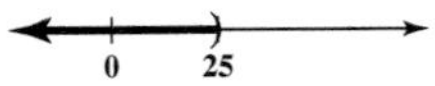

77. $\emptyset$ or { }

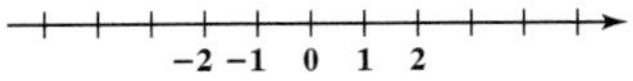

79. $\{x|x > 5.9375\}; (5.9375, \infty)$

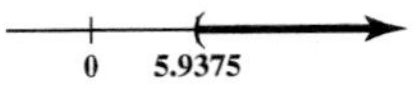

81. at most 1250 miles **83.** 32 **85.** more than 400 minutes **87.** greater than $50,361.11 **89.** at least 74 **91.** $\{x|-33 < x < -14\}$ **93.** $\{x|-4 \leq x \leq 6\}$ **95.** $\{x|-2 \leq x \leq 9\}$ **97.** $\{x|-8 \leq x < 4\}$

Chapter Review

1. No **2.** No **3.** No **4.** Yes **5.** $\{16\}$ **6.** $\{20\}$ **7.** $\{-16\}$ **8.** $\{-7\}$ **9.** $\{-95\}$ **10.** $\{50\}$ **11.** $\{24\}$ **12.** $\{80\}$ **13.** $\{-6\}$ **14.** $\{5\}$ **15.** $\left\{-\frac{1}{3}\right\}$ **16.** $\left\{\frac{3}{8}\right\}$ **17.** $\{4\}$ **18.** $\{5\}$ **19.** \$20,100 **20.** \$2.55 **21.** $\{-4\}$ **22.** $\{4\}$ **23.** $\{9\}$ **24.** $\{-21\}$ **25.** $\{-4\}$ **26.** $\{-6\}$ **27.** $\{4\}$ **28.** $\{-5\}$ **29.** $\{6\}$ **30.** $\{-6\}$ **31.** $\left\{\frac{4}{3}\right\}$ **32.** $\{5\}$ **33.** $\{2\}$ **34.** $\{7\}$ **35.** 14 years **36.** width = 19 yards; length = 29 yards **37.** $\left\{-\frac{35}{12}\right\}$ **38.** $\left\{-\frac{62}{3}\right\}$ **39.** $\{-2\}$ **40.** $\left\{\frac{6}{5}\right\}$ **41.** $\left\{\frac{3}{2}\right\}$ **42.** $\{3\}$ **43.** $\{4\}$ **44.** $\{-1\}$ **45.** $\left\{-\frac{7}{2}\right\}$ **46.** $\{-3\}$ **47.** $\{58\}$ **48.** $\{-10.5\}$ **49.** contradiction; $\varnothing$ or $\{\ \}$ **50.** contradiction; $\varnothing$ or $\{\ \}$ **51.** conditional equation; $\{0\}$ **52.** conditional equation; $\{0\}$ **53.** identity; all real numbers **54.** identity; all real numbers **55.** \$15.75 **56.** 3 dimes **57.** 48 in.^{2} **58.** 64 cm **59.** $\frac{3}{2}$ yards **60.** 15 mm **61.** $H = \frac{V}{LW}$ **62.** $P = \frac{I}{rt}$ **63.** $W = \frac{S - 2LH}{2L + 2H}$ **64.** $M = \frac{\rho - mv}{V}$ **65.** $y = \frac{-2x + 10}{3}$ **66.** $x = \frac{14 + 7y}{6}$ **67. (a)** $P = \frac{A}{(1 + r)^t}$ **(b)** \$2238.65 **68. (a)** $h = \frac{A - 2\pi r^2}{2\pi r}$ **(b)** 5 cm **69.** \$11.25 **70.** $\frac{9}{4}\pi$ ft$^2 \approx 7.1$ ft^2 **71.** $x - 6$ **72.** $x - 8$ **73.** $-8x$ **74.** $\frac{x}{10}$ **75.** $2(6 + x)$ **76.** $4(5 - x)$ **77.** $6 + x = 2x + 5$ **78.** $6x - 10 = 2x + 1$ **79.** $x - 8 = \frac{1}{2}x$ **80.** $\frac{6}{x} = 10 + x$ **81.** $4(2x + 8) = 16$ **82.** $5(2x - 8) = -24$ **83.** Sarah's age: s; Jacob's age: $s + 7$ **84.** Consuelo's speed: c; Jose's speed: $2c$ **85.** Max's amount: m; Irene's amount: $m - 6$ **86.** Victor's amount: v; Larry's amount: $350 - v$ **87.** 153 pounds **88.** 12, 13, 14 **89.** Juan: \$11,000; Roberto: \$9000 **90.** 100 miles **91.** 5.2 **92.** 60 **93.** 13% **94.** 50 **95.** \$18.50 **96.** \$30 **97.** \$40 **98.** \$200 **99.** \$125,000 **100.** winner: 500 votes; loser: 400 votes **101.** 10°, 80° **102.** 80°, 100° **103.** 30°, 60°, 90° **104.** 50°, 45°, 85° **105.** length = 31 in.; width = 8 in. **106.** length = 28 cm; width = 7 cm **107. (a)** length = 20 ft; width = 40 ft **(b)** 800 ft^2 **108.** 50 ft; 40 ft **109.** 5 hours **110.** 65 mph **111.** 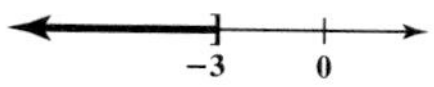 **112.** 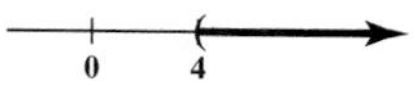 **113.** 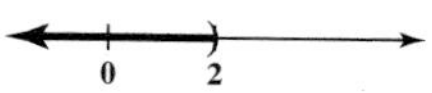 **114.** 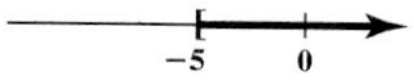 **115.** 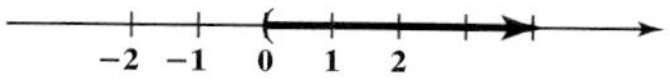 **116.** 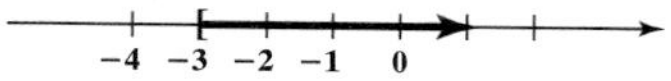 **117.** $(-\infty, -4)$ **118.** $[7, \infty)$ **119.** $[2, \infty)$ **120.** $(-\infty, 3)$ **121.** $\left\{x \mid x < -\frac{13}{2}\right\}$; $\left(-\infty, -\frac{13}{2}\right)$ **122.** $\left\{x \mid x \ge -\frac{7}{3}\right\}$; $\left[-\frac{7}{3}, \infty\right)$ **123.** $\left\{x \mid x \ge -\frac{4}{5}\right\}$; $\left[-\frac{4}{5}, \infty\right)$ **124.** $\{x \mid x > -12\}$; $(-12, \infty)$ **125.** $\varnothing$ or $\{\ \}$ **126.** $\{x \mid x \text{ is any real number}\}$, $(-\infty, \infty)$ **127.** $\{x \mid x > 3\}$; $(3, \infty)$ **128.** $\left\{x \mid x < -\frac{38}{5}\right\}$; $\left(-\infty, -\frac{38}{5}\right)$ **129.** at most 65 miles **130.** more than 150

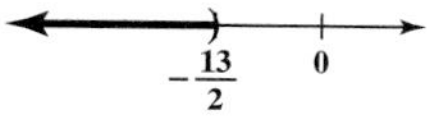

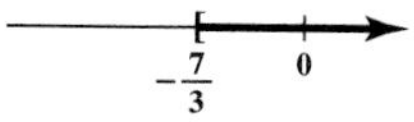

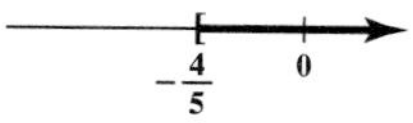

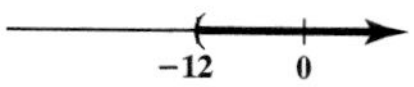

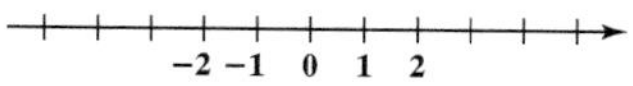

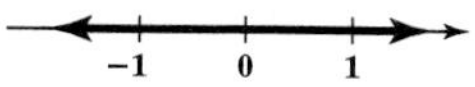

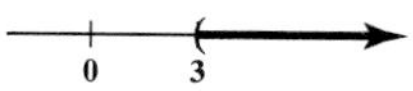

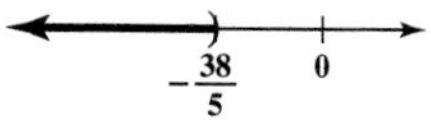

Chapter 2 Test

1. $\{-17\}$ **2.** $\left\{-\frac{4}{9}\right\}$ **3.** $\{4\}$ **4.** $\left\{\frac{10}{13}\right\}$ **5.** $\left\{\frac{5}{8}\right\}$ **6.** $\{5\}$ **7.** $\varnothing$ or $\{\ \}$ **8.** all real numbers **9. (a)** $l = \frac{V}{wh}$ **(b)** 9 in. **10. (a)** $y = -\frac{2}{3}x + 4$ **(b)** $y = -\frac{4}{3}$ **11.** $6(x - 8) = 2x - 5$ **12.** 60 **13.** 15, 16, 17 **14.** 10 in., 24 in., 26 in. **15.** 3.5 hours **16.** shorter piece is 5 feet; longer is 16 feet **17.** \$36 **18.** $\{x \mid x \le 6\}$; $(-\infty, 6]$ 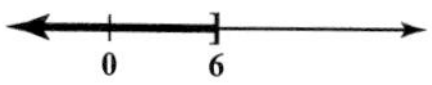 **19.** $\left\{x \mid x > \frac{5}{4}\right\}$; $\left(\frac{5}{4}, \infty\right)$ 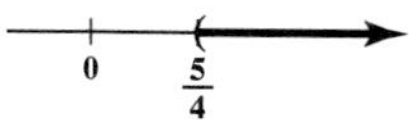 **20.** at most 200 minutes

Chapter 3

Section 3.1

1. monomial **3.** 5 **5.** True **7.** Answers may vary. **9.** Yes; coefficient $\frac{1}{2}$; degree 3 **11.** Yes; coefficient $\frac{1}{7}$; degree 2 **13.** No **15.** Yes; coefficient 12; degree 5 **17.** No **19.** Yes; coefficient 4; degree 0 **21.** $x^3 + 4x - 5$ **23.** $-y^4 + y^2 - 3y + 2$ **25.** Yes; $6x^2 - 10$; degree 2; binomial **27.** No **29.** No **31.** Yes; $\frac{1}{8}$; degree 0; monomial **33.** Yes; $-\frac{1}{2}t^4 + 3t^2 + 6t$; degree 4; trinomial **35.** No **37.** Yes; $5z^3 - 10z^2 + z + 12$; degree 3; polynomial **39.** Yes; $2xy^4 + 3x^2y^2 + 4$; degree 5; trinomial **41.** Yes; $-3y^5 + 4x^3$; degree 5; binomial **43.** No **45.** $7x - 10$ **47.** $-2m^2 + 5$ **49.** $-2p^2 + p + 8$ **51.** $-3y^2 - 2y - 4$ **53.** $\frac{5}{4}p^2 + \frac{1}{6}p - 3$

55. $6m^2 - 4mn - n^2$ **57.** $4n^2 - 8n + 5$ **59.** $-x - 16$ **61.** $14x^2 - 3x - 5$ **63.** $4y^3 - 3y - 4$ **65.** $2y^3 - y^2 - 4y$ **67.** $\frac{14}{9}q^2 - \frac{23}{8}q + 2$ **69.** $-8m^2n^2 - 4mn - 7$ **71.** $-4x - 5$ **73. (a)** 3 **(b)** 48 **(c)** 13 **75. (a)** -2 **(b)** $\frac{3}{4}$ **(c)** 4.75 **77.** 45 **79.** -328 **81.** $3t - 3$ **83.** $3x^2 - 3x - 3$ **85.** $9xy^2 + 1$ **87.** $-y^2 + 15y + 1$ **89.** $\frac{7}{3}q^2 + \frac{5}{3}$ **91.** $13d^2 + d + 10$ **93.** $6a^2 + 5a - 3$ **95.** $-2x^2 - 2xy + y^2$ **97.** $-5x + 12$ **99.** $3x^2 + 4x - 3$ **101.** $-14x^2 + 4x - 13$ **103.** $-3x + 5$ **105.** 26 ft **107. (a)** \$22,579.65 **(b)** \$43,793.85 **109. (a)** $-2.125x^2 + 105x$ **(b)** \$1250 **111. (a)** $10x$ **(b)** \$500 **(c)** \$26,000 **113.** $16x - 12$ **115.** $7x + 18$ **117.** $2x - 5$ **119.** $-x^2 - x + 15$ **121.** $2p^2 + 10p + 16$ **123.** $y^3 - 2y^2 + 11y - 14$

Section 3.2

1. a^{mn} **3.** expanded form **5.** False **7.** Answers may vary. **9.** 1024 **11.** -128 **13.** m^9 **15.** b^{20} **17.** x^8 **19.** p^9 **21.** $(-n)^7$ **23.** a^{17} **25.** 64 **27.** z^{10} **29.** $(-m)^{18}$ **31.** m^{14} **33.** $(-b)^{20}$ **35.** 225 **37.** $27x^6$ **39.** $81p^{28}q^8$ **41.** $-8m^6n^3$ **43.** $25x^2y^4z^6$ **45.** $12x^5$ **47.** $-40a^{10}$ **49.** $-7m^4$ **51.** $6x^7$ **53.** $6x^6y^4$ **55.** $-5m^2n^4$ **57.** $4x^3$ **59.** b^6 **61.** $81b^4$ **63.** $9p^5$ **65.** $25x^8$ **67.** $-2x^7y^6$ **69.** $72x^{14}$ **71.** $4q^6$ **73.** x^6 **75.** $48x^2$ **77.** $16x^{14}$ **79.** $-\frac{24}{5}m^{17}n^9$

Section 3.3

1. $a^2 - b^2$ **3.** perfect square trinomial **5.** False **7.** Yes; $x = 0, y = 0$ **9.** $6x^2 - 10x$ **11.** $-7b^2 + 35b$ **13.** $2n^2 - 3n$ **15.** $12n^4 + 6n^3 - 15n^2$ **17.** $12m^3n - 4m^2n^2 + 2mn^3$ **19.** $4x^4y^2 - 3x^3y^3$ **21.** $x^2 + 5x + 6$ **23.** $q^2 - 13q + 42$ **25.** $y^2 - y - 20$ **27.** $x^4 + 4x^2 + 3$ **29.** $42 - 13x + x^2$ **31.** $-n^2 + 4n - 3$ **33.** $3a^2 + 5a + 2$ **35.** $10u^2 + 17uv + 6v^2$ **37.** $4n^2 - 9p^2$ **39.** $x^3 + x^2 - 5x - 2$ **41.** $2y^3 - 12y^2 + 19y - 3$ **43.** $2x^3 - 7x^2 + 4x + 3$ **45.** $6p^3 - 13p^2 + 4p + 3$ **47.** $2b^3 + 2b^2 - 24b$ **49.** $-x^3 + 9x$ **51.** $2x^4 - 5x^3 + 10x^2 - 11x - 4$ **53.** $10y^5 + 3y^4 + 4y^3 + 3y^2 + 2y + 2$ **55.** $x^2 - 4x - 21$ **57.** $z^2 - 7z - 30$ **59.** $6x^2 + 7x - 3$ **61.** $7k^2 + 39k - 18$ **63.** $14x^2 + 41xy + 15y^2$ **65.** $10a^2 - ab - 2b^2$ **67.** $x^2 - 9$ **69.** $4z^2 - 25$ **71.** $16x^4 - 1$ **73.** $4x^2 - 9y^2$ **75.** $x^2 - \frac{1}{9}$ **77.** $x^6 - y^4$ **79.** $x^2 - 4x + 4$ **81.** $x^2 + 4xy + 4y^2$ **83.** $4a^2 - 12ab + 9b^2$ **85.** $16x^2 + 24xy + 9y^2$ **87.** $25a^2 - 20ab^2 + 4b^4$ **89.** $x^2 + x + \frac{1}{4}$ **91.** $3x^3 - 6x^2 + 1$ **93.** $-5x^7 + 4x^6 - 6x^5$ **95.** $5x^3 + 21x^2 + 2x$ **97.** $-3w^3 + 3w^2 + 36w$ **99.** $3a^3 + 24a^2 + 48a$ **101.** $2n^2 + 6n$ **103.** $24ab$ **105.** $x^2 - 0.6x + 0.09$ **107.** $\frac{4}{9}b^2 - \frac{25}{81}$ **109.** $-x^2 + 3x + 2$ **111.** $4x^2 + 4x + 1$ **113.** $x^3 - 3x^2 + 3x - 1$ **115.** $2x^2 - 9$ **117.** $2x^2 + 7x - 15$ **119.** $x^2 + 10x + 25$ **121.** $9x^2 - 25$ **123.** $2x^2 + x - 1$ **125.** $x^2 + 12x + 20$ **127.** $x^2 + 3x + 2$ **129.** $40.8 + 0.4x$ **131.** $\pi x^2 + 4\pi x + 4\pi$ feet **133. (a)** a^2, ab, ab, b^2 **(b)** $a^2 + 2ab + b^2$ **(c)** length: $a + b$, width: $a + b$, area: $(a + b)^2$ **135.** $x^3 + 2x^2 - 4x - 8$ **137.** $x^2 + 14x + 49$ **139.** $x^2 - 2xy + y^2 - 9$ **141.** $x^3 + 6x^2 + 12x + 8$ **143.** $z^4 + 12z^3 + 54z^2 + 108z + 81$

Section 3.4

1. a^{m-n} **3.** $\frac{1}{a^n}$ **5.** False **7.** Answers may vary. **9.** 27 **11.** 16 **13.** x^9 **15.** a^{12} **17.** $4y^3$ **19.** $-\frac{2}{3}m^7$ **21.** $2m^8n^2$ **23.** $\frac{11}{8}b^2$ **25.** $\frac{27}{8}$ **27.** $\frac{x^{15}}{27}$ **29.** $\frac{x^{20}}{y^{28}}$ **31.** $\frac{49a^4b^2}{c^6}$ **33.** 1 **35.** -1 **37.** 18 **39.** $\frac{1}{3}$ **41.** $\frac{1}{1000}$ **43.** $\frac{1}{m^2}$ **45.** $\frac{4}{a^2}$ **47.** $-\frac{4}{y^3}$ **49.** $\frac{11}{18}$ **51.** 2 **53.** $\frac{25}{4}$ **55.** $\frac{z^2}{3}$ **57.** $-\frac{m^6}{8n^3}$ **59.** 16 **61.** $6x^4$ **63.** 4 **65.** $\frac{1}{8}$ **67.** 81 **69.** $\frac{1}{x^9}$ **71.** x^{13} **73.** $\frac{4}{x^3}$ **75.** $-\frac{3z^3}{2x^3}$ **77.** $\frac{4a}{5b}$ **79.** $\frac{1}{4x^3y^9}$ **81.** $-\frac{1}{a^{12}}$ **83.** $\frac{27}{m^6}$ **85.** $\frac{x^4y^6}{9}$ **87.** $\frac{2}{p^7}$ **89.** 12 **91.** $\frac{2x^8}{3}$ **93.** $\frac{16y}{27x^{18}}$ **95.** $\frac{9y}{2z^3}$ **97.** $216x^3y^6$ **99.** $\frac{5ab^7}{3}$ **101.** $\frac{2x^6}{5}$ **103.** $x^3 + 6x^2 + 12x + 8$ cubic feet **105.** $3x^3$ cubic meters **107.** $V = \frac{\pi d^2h}{4}$ **109.** $x^5 + 3x^4 - 3x^3 - 9x^2$ cubic cm **111.** $9\pi x^3$ square units **113.** $\frac{1}{x^n}$ **115.** $x^{6n-2}y^2$ **117.** $\frac{x^{6a^2}y^{3ab}}{z^{3ac}}$ **119.** $\frac{9}{8a^{10n}}$

Chapter 3 Putting the Concepts Together

1. Yes; degree 6; trinomial **2. (a)** 0 **(b)** -4 **(c)** 2 **3.** $8x^4 - 4x^2 + 14$ **4.** $5x^2y + 3xy + 2y^2$ **5.** $-6m^3n^2 + 2m^2n^4$ **6.** $5x^2 - 17x - 12$ **7.** $8x^2 - 26xy + 21y^2$ **8.** $25x^2 - 64$ **9.** $4x^2 + 12xy + 9y^2$ **10.** $6a^3 + 22a^2 - 40a$ **11.** $16m^4 + 4m^3 + 10m^2 - 20m - 24$ **12.** $-8x^8z^9$ **13.** $-\frac{15}{mn}$ **14.** $-3a^3b^2$ **15.** $\frac{1}{2x^2y^2}$ **16.** $\frac{y^4}{16z^6}$ **17.** $\frac{8}{27r^6}$ **18.** $\frac{q^{56}}{r^{40}t^{64}}$ **19.** $\frac{1}{128x^6y^2}$ **20.** $-\frac{48y^6}{x^{12}}$

Section 3.5

1. standard **3.** quotient; remainder; dividend **5.** False **7.** Answers may vary. **9.** $2x - 1$ **11.** $3a + 9 - \frac{1}{a^2}$ **13.** $\frac{n^4}{5} - \frac{2n^2}{5} - 1$ **15.** $\frac{5r^2}{3} - 3$ **17.** $-x^2 + 3x - \frac{9}{x^2}$ **19.** $\frac{3}{8} + \frac{z^2}{2} - \frac{z}{4}$ **21.** $x + \frac{3x^2}{2} + \frac{2}{3x}$ **23.** $-7x + 5y$ **25.** $-\frac{6y}{x} + 15$ **27.** $-\frac{5}{a} - \frac{2c^2}{a^2b}$ **29.** $x - 7$ **31.** $x - 5$ **33.** $x^2 + 6x - 3$ **35.** $x^3 - 3x^2 + 6x - 2$ **37.** $x^2 - 2x + 3 + \frac{5}{x + 1}$ **39.** $2x + 3$ **41.** $x^2 + 6x + 7 + \frac{16}{x - 2}$ **43.** $2x^3 + 5x^2 + 9x - 4 + \frac{-17}{x - 4}$ **45.** $2x - 6$ **47.** $x^2 + 4x - 3 + \frac{2}{2x - 1}$ **49.** $x - 4 + \frac{-4}{5 + x}$ **51.** $2x - 1 + \frac{6}{1 + 2x}$ **53.** $x + 2 + \frac{2x - 4}{x^2 - 2}$ **55.** $x^2 + 3x - 6$ **57.** $a^2 + a - 30$ **59.** $-x^2 - x - 6$ **61.** $3x + 1$ **63.** $-2ab + 2b^2$ **65.** $x - 2 + \frac{3}{x}$

67. $\frac{2}{3}x^7y^8z^{12}$ **69.** $n^2 - 6n + 9$ **71.** $-10x^5 - 10x^3 + 40x^6$ **73.** $6pq - 2q^2 - p^2$ **75.** $4n^2 + 10nm - 6m^2$ **77.** $x - 3 + \frac{-10}{x - 5}$ **79.** $x^3 + 6x^2 + 4x - 5$ **81.** $15rs^2 - 4r^2s$ **83.** $-3n - \frac{8}{3}m + 4$ **85.** $2x^6 - 4x^5 + 6x^4$ **87.** $-x + \frac{2}{x^2} - \frac{4}{x}$ **89.** $-\frac{1}{2x} - \frac{2}{x^2} + \frac{1}{x^3}$ **91.** x **93.** $z + 3$ **95.** $4x + 4$ **97.** \$182.11 per computer **99.** 6

Section 3.6

1. scientific notation **3.** positive **5.** False **7.** Answers may vary. **9.** 3×10^5 **11.** 6.4×10^7 **13.** 5.1×10^{-4} **15.** 1×10^{-9} **17.** 8.007×10^9 **19.** 3.09×10^{-5} **21.** 6.2×10^2 **23.** 4×10^0 **25.** 420,000 **27.** 100,000,000 **29.** 0.0039 **31.** 0.4 **33.** 3760 **35.** 0.0082 **37.** 0.00006 **39.** 7,050,000 **41.** 5×10^{-4} **43.** 142,000 **45.** 0.008 **47.** 6.9×10^4 **49.** 6.415×10^9 **51.** 1.46×10^9 **53.** 3×10^{-5} **55.** 2,158,000 **57.** 0.000000000000001 **59.** 0.00225 **61.** 3×10^9 **63.** 8.4×10^{-3} **65.** 1.05×10^{-3} **67.** 3×10^8 **69.** 2.5×10^1 **71.** 1.8×10^8 **73.** 8×10^8 **75.** 9×10^{15} **77.** 3×10^{-14} **79.** 2.11×10^5 mi **81.** 4.87×10^8 km **83.** 4.56×10^8 **85.** 2.5×10^{-8} M **87.** $\approx 1.65 \times 10^3$ lb/person **89.** 1.116×10^7 miles **91.** 2.5×10^{-4} m **93.** 8×10^{-10} m **95.** 7.15×10^{-8} m **97.** $1.2348\pi \times 10^{-14}$ m^3 **99.** $2.88\pi \times 10^{-25}$ m^3

Chapter Review

1. Yes; degree 3; coefficient 4 **2.** No **3.** No **4.** Yes; degree 3; coefficient 1 **5.** No **6.** No **7.** Yes; degree 0; monomial **8.** Yes; degree 5; binomial **9.** Yes; degree 6; trinomial **10.** Yes; 10; trinomial **11.** $9x^2 + 8x - 2$ **12.** $m^3 + mn - 5m$ **13.** $-x^2y + 12x$ **14.** $-7y^2 - 6yz + 9z^2$ **15.** $-2x^2 - 2$ **16.** $-20y^2 - 8y + 5$ **17. (a)** 0 **(b)** 8 **(c)** 2 **18. (a)** 3 **(b)** 2 **(c)** $\frac{11}{4}$ **19.** 0 **20.** 29 **21.** 279,936 **22.** $-\frac{1}{243}$ **23.** 16,777,216 **24.** 1 **25.** x^{13} **26.** m^6 **27.** r^{12} **28.** m^{24} **29.** $1024x^5$ **30.** $64n^6$ **31.** $81x^8y^4$ **32.** $4x^6y^8$ **33.** $15x^6$ **34.** $-36a^4$ **35.** $16y^5$ **36.** $-12p^6$ **37.** $12w^4$ **38.** $-\frac{3}{4}z^3$ **39.** $108x^8$ **40.** $80a^6$ **41.** $-8x^5 + 6x^4 - 2x^3$ **42.** $2x^7 + 4x^6 - x^4$ **43.** $6x^2 - 7x - 5$ **44.** $4x^2 - 5x - 6$ **45.** $x^2 - 3x - 40$ **46.** $w^2 + 9w - 10$ **47.** $24m^3 - 22m^2 + 7m - 3$ **48.** $8y^5 + 12y^4 + 4y^3 + 6y^2 - 6y - 9$ **49.** $x^2 + 8x + 15$ **50.** $2x^2 - 17x + 8$ **51.** $6m^2 + 17m - 14$ **52.** $48m^2 - 26m - 4$ **53.** $21x^2 + 5xy - 6y^2$ **54.** $20x^2 + 7xy - 3y^2$ **55.** $x^2 - 16$ **56.** $4x^2 - 25$ **57.** $4x^2 + 12x + 9$ **58.** $49x^2 - 28x + 4$ **59.** $9x^2 - 16y^2$ **60.** $64m^2 - 36n^2$ **61.** $25x^2 - 20xy + 4y^2$ **62.** $4a^2 + 12ab + 9b^2$ **63.** $x^2 - 0.25$ **64.** $r^2 - 0.0625$ **65.** $y^2 + \frac{4}{3}y + \frac{4}{9}$ **66.** $y^2 - y + \frac{1}{4}$ **67.** 36 **68.** $\frac{1}{343}$ **69.** x^4 **70.** $\frac{1}{x^8}$ **71.** 1 **72.** -1 **73.** 1 **74.** -1 **75.** $\frac{5x^2}{2y^3}$ **76.** $\frac{x^2}{3y^8}$ **77.** $\frac{x^{15}}{y^{10}}$ **78.** $\frac{343}{x^6}$ **79.** $\frac{8m^6n^3}{p^{12}}$ **80.** $\frac{81m^4n^8}{p^{20}}$ **81.** $-\frac{1}{25}$ **82.** 64 **83.** $\frac{81}{16}$ **84.** 27 **85.** $\frac{7}{12}$ **86.** $\frac{1}{8}$ **87.** $\frac{2x^3y^5}{3}$ **88.** $\frac{3}{7xy^{10}}$ **89.** $\frac{9m^4}{16}$ **90.** $\frac{64}{9m^6n^6}$ **91.** $\frac{8r^4s^6}{9}$ **92.** $\frac{32}{3r^{10}s^{15}}$ **93.** $6x^5 - 4x^4 + 5$ **94.** $3x^4 + 5x^2 - 6x$ **95.** $4n^3 + 1 - \frac{5}{2n^4}$ **96.** $6n - 4 - \frac{16}{5n^2}$ **97.** $-\frac{p^3}{8} - \frac{1}{4} + \frac{1}{2p^2}$ **98.** $-\frac{p^2}{2} + 1 - \frac{3}{2p^2}$ **99.** $4x - 7$ **100.** $x + 6$ **101.** $x^2 + 7x + 5 + \frac{6}{x - 1}$ **102.** $2x^2 - 3x - 12 + \frac{-16}{x - 2}$ **103.** $x^2 - 2x + 4$ **104.** $3x^2 + 15x + 77 + \frac{378}{x - 5}$ **105.** 2.7×10^7 **106.** 1.23×10^9 **107.** 6×10^{-5} **108.** 3.05×10^{-6} **109.** 3×10^0 **110.** 8×10^0 **111.** 0.0006 **112.** 0.00125 **113.** 613,000 **114.** 80,000 **115.** 0.37 **116.** 54,000,000 **117.** 6×10^3 **118.** 4.2×10^{-8} **119.** 2×10^2 **120.** 2×10^8 **121.** 4×10^{15} **122.** 2×10^2

Chapter 3 Test

1. Yes; degree 5; binomial **2. (a)** 5 **(b)** 21 **(c)** 26 **3.** $7x^2y^2 - 6x - 3y$ **4.** $10m^3 + m^2 - 6$ **5.** $-6x^5 + 18x^4 - 15x^3$ **6.** $2x^2 - 3x - 35$ **7.** $4x^2 - 28x + 49$ **8.** $16x^2 - 9y^2$ **9.** $6x^3 + x^2 - 25x + 8$ **10.** $2x - \frac{8}{3} + \frac{3}{x^3}$ **11.** $3x^2 - 11x + 33 + \frac{-94}{x + 3}$ **12.** $-12x^4y^6$ **13.** $\frac{2m^3}{3n^5}$ **14.** $\frac{n^{24}}{m^{30}}$ **15.** $\frac{x^{22}}{y^{14}}$ **16.** $\frac{8m^7}{n^7}$ **17.** 1.2×10^{-5} **18.** 210,100 **19.** 3.57×10^4 **20.** 2×10^{-7}

Chapters 1–3 Cumulative Review

1. (a) $\{\sqrt{25}\}$ **(b)** $\{0, \sqrt{25}\}$ **(c)** $\left\{-6, -\frac{4}{2}, 0, \sqrt{25}\right\}$ **(d)** $\left\{-6, -\frac{4}{2}, 0, 1.4, \sqrt{25}\right\}$ **(e)** $\{\sqrt{7}\}$ **(f)** $\left\{-6, -\frac{4}{2}, 0, 1.4, \sqrt{7}, \sqrt{25}\right\}$ **2.** $-\frac{4}{9}$ **3.** -19 **4.** $6x^3 + 5x^2 - 3x$ **5.** $-18x + 8$ **6.** $\{1\}$ **7.** $4(x - 5) = 2x + 10$ **8.** \$640 **9.** 45 mi/h **10.** $\{x | x < -3\}$ or $(-\infty, -3)$;

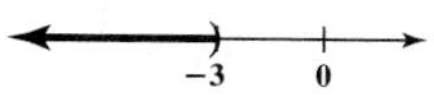

11. $-\frac{1}{2}$ **12.** $6x^3 - 3x^2 + 7x$ **13.** $28m^2 - 13m - 6$ **14.** $9m^2 - 4n^2$ **15.** $49x^2 + 14xy + y^2$ **16.** $4m^3 - 19m + 15$ **17.** $\frac{2}{x} + \frac{1}{y}$ **18.** $x^2 + 3x + 9 + \frac{54}{x - 3}$ **19.** $-24n^4$ **20.** $-\frac{5n^8}{2m^2}$ **21.** $\frac{1}{64x^6y^{24}z^{12}}$ **22.** $\frac{1}{2x^{12}y^3}$ **23.** 6.05×10^{-5} **24.** 2,175,000 **25.** 7.14×10^5

Chapter 4

Section 4.1

1. greatest common factor **3.** Distributive **5.** true **7.** false **9.** Answers may vary **11.** 2 **13.** 1 **15.** 5 **17.** 26 **19.** 4 **21.** 18 **23.** x^2 **25.** $7x$ **27.** m^2n^2 **29.** $15ab^2$ **31.** $2abc^2$ **33.** 3 **35.** $2(x - 4)^2$ **37.** $6(x - 3)^2$ **39.** $4x^3$ **41.** $-3s$ **43.** mn^5 **45.** $6(2x - 3)$ **47.** $5x^2y(1 - 3xy)$ **49.** $3x(x^2 + 2x - 1)$ **51.** $3(3m^5 - 6m^3 - 4m^2 + 27)$ **53.** $15ab^2(ab^2 - 4b + 3a^2)$ **55.** $(x - 3)(x - 5)$ **57.** $(x - 1)(x^2 + y^2)$ **59.** $(4x + 1)(x^2 + 2x + 5)$ **61.** $-x(3x - 4)$ **63.** $-5x(x^2 - 2x + 3)$

65. $-4z(3z^2 - 4z + 2)$ **67.** $-5(3b^3 + b - 2)$ **69.** $(x + 3)(y + 4)$ **71.** $(y + 1)(z - 1)$ **73.** $(x - 1)(x^2 + 2)$ **75.** $(2t - 1)(t^2 - 1)$ **77.** $(2t - 1)(t^3 - 1)$ **79.** $4(y - 5)$ **81.** $7m(4m^2 + m + 9)$ **83.** $6m^2n(2mnp - 3)$ **85.** $3(p + 3)(3p + 1)$ **87.** $3(2x - y)(3a - 2b)$ **89.** $(x - 2)(x - 2)$ or $(x - 2)^2$ **91.** $3x(5x - 2)(x^2 + 2)$ **93.** $-3x(x^2 - 2x + 3)$ **95.** $-4b(3 - 4b)$ **97.** $(4y + 3)(3x - 2)$ **99.** $\frac{1}{3}x^2\left(x - \frac{2}{3}\right)$ **101.** Answers may vary. **103.** Answers may vary. **105.** $-2t(8t - 75)$ **107.** $150(140 - p)$ **109.** $4x^3(2x^2 - 7)$ **111.** $2\pi r(r + 4)$ **113.** $x - 2$ **115.** $4x^{2n} + 5$ **117.** $3x^3 - 4x^2 + 2$ **119.** $x^4 - 3x + 2$ **121.** $\frac{3}{5}x^3 - \frac{1}{2}x + 1$

Section 4.2

1. quadratic trinomial **3.** opposite **5.** false **7.** Answers may vary. **9.** $(x + 2)(x + 3)$ **11.** $(m + 3)(m + 6)$ **13.** $(x - 3)(x - 12)$ **15.** $(a - 2)(a - 6)$ **17.** $(x - 4)(x + 3)$ **19.** $(x - 4)(x + 5)$ **21.** $(z - 3)(z + 15)$ **23.** prime **25.** $(x - 2y)(x - 3y)$ **27.** $(r - 2s)(r + 3s)$ **29.** $(x - 4y)(x + y)$ **31.** prime **33.** $3(x + 2)(x - 1)$ **35.** $3a(a - 3)(a - 5)$ **37.** $5z(x - 8)(x + 4)$ **39.** $-2(y - 2)^2$ **41.** $(x - 6)(x + 3)$ **43.** $(x + 8)(x - 4)$ **45.** $x(x + 5)(x - 3)$ **47.** $(a + 4b)(a - 7b)$ **49.** prime **51.** $(k - 5)(k + 4)$ **53.** $(x - 5y)(x + 6y)$ **55.** $(st - 3)(st - 5)$ **57.** $-3c(c - 2)(c + 1)$ **59.** $-(x + 3)(x - 2)$ **61.** prime **63.** $n^2(n - 6)(n + 5)$ **65.** $(m + 5n)(m - 3n)$ **67.** $(a + 2)(b - 4)$ **69.** $(s + 5)(s + 7)$ **71.** $3xy(x^2 - 2x - 2)$ **73.** prime **75.** $(ab - 6)(ab - 2)$ **77.** $-x(x - 14)(x + 2)$ **79.** $(y - 6)^2$ **81.** $(2r - 3)(3s - 2)$ **83.** $-2x(x - 3)(x - 6)$ **85.** $-7xy(3x^2 + 2y)$ **87.** $(x - a)(x + b)$ **89.** $-16(t + 1)(t - 5)$ **91.** $(x + 6)$ and $(x + 3)$ **93.** $(x + 5)$ and $(x - 3)$ **95.** $-7, -5, 5, 7$ **97.** $-20, -4, 4, 20$ **99.** 1 **101.** 6, 10, 12

Section 4.3

1. common factor **3.** close **5.** false **7.** Answers may vary. **9.** $(2n - 3)(3n - 1)$ **11.** $(2w - 5)(2w + 1)$ **13.** prime **15.** $(2x - 1)(2x + 3)$ **17.** $(9z - 2)(3z + 1)$ **19.** $(3x - 1)^2$ **21.** $(2x + 3)(x + 1)$ **23.** $(6n - 5)(n - 1)$ **25.** $2(2x - y)(3x + 2y)$ **27.** prime **29.** $(3m - 4)(2m + 1)$ **31.** $-3(3n - 2)(2n - 3)$ **33.** $(4a + 5)(3a + 1)$ **35.** $(2x - 3)(x - 1)$ **37.** $(3x - 4)(5x - 1)$ **39.** $-(a + 4)(4a - 3)$ **41.** $2(x - 2y)(5x + 6y)$ **43.** $(3x - 4)(3x - 2)$ **45.** $-3(4x + 3)(2x - 1)$ **47.** $4x^2y^2(x - 2y - 1)$ **49.** $(x - 6y)(x - 7y)$ **51.** $(m + 3n)(4m + n)$ **53.** $-(x - 12)(x + 2)$ **55.** $(n - 1)(n^2 + 1)$ **57.** prime **59.** $6(2x + 5y)(2x - y)$ **61.** $4(a + 2)(a - 2)$ **63.** $3(3a + 8b)(2a - b)$ **65.** $(2ab^2 - c)(a + b)$ **67.** $-2x^2(2x + 5)(x - 1)$ **69.** $(2x + 7)$ m and $(3x - 4)$ m **71.** $2x - 5$ **73.** $-2x(4x + 3)(3x - 5)$ **75.** $(9z^2 + 8)(3z^2 + 2)$ **77.** $(3x^n + 1)(x^n + 6)$ **79.** $(x^2 + 1)(2x - 7)(3x - 2)$ **81.** ± 2 or ± 14

Section 4.4

1. sum, two cubes **3.** difference, two squares **5.** false **7.** Answers may vary. **9.** $(x - 3)^2$ **11.** $(x + 5)^2$ **13.** $(2p - 1)^2$ **15.** $(4x + 3)^2$ **17.** $(x - 2y)^2$ **19.** $(2z - 3)^2$ **21.** $(x - 11)(x + 11)$ **23.** $(4x - 7y)(4x + 7y)$ **25.** $(2x - 5)(2x + 5)$ **27.** $(10n^4 - 9p^2)(10n^4 + 9p^2)$ **29.** $(k - 2)(k + 2)(k^2 + 4)(k^4 + 16)$ **31.** $(5p^2 - 7q)(5p^2 + 7q)$ **33.** $(1 + x)(1 - x + x^2)$ **35.** $(2x - 3y)(4x^2 + 6xy + 9y^2)$ **37.** $(x^2 - 2y)(x^4 + 2x^2y + 4y^2)$ **39.** $(3c + 4d^3)(9c^2 - 12cd^3 + 16d^6)$ **41.** prime **43.** $\left(x - \frac{1}{3}\right)^2$ **45.** $(4m + 5n)^2$ **47.** $2(x - 3)^2$ **49.** $3a(4a + 3)^2$ **51.** $\left(9p - \frac{1}{2}\right)\left(9p + \frac{1}{2}\right)$ **53.** $x^2y^2(x - y)(x + y)$ **55.** $2t(t - 3)(t^2 + 3t + 9)$ **57.** $3s(s^2 + 2)(s^4 - 2s^2 + 4)$ **59.** $4x(1 - x)(1 + x)$ **61.** $xy(x - y)(x + y)$ **63.** $(xy + 1)(x^2y^2 - xy + 1)$ **65.** $(x^4 - 5y^5)(x^4 + 5y^5)$ **67.** $\left(x - \frac{y^2}{4}\right)\left(x^2 + \frac{xy^2}{4} + \frac{y^4}{16}\right)$ **69.** $2(x - 2)^2$ **71.** $-2n(2n - 1)^2(2n + 1)^2$ **73.** $-2x(1 + 2x)^2$ **75.** $(x^2 - y^3)(x^4 + x^2y^3 + y^6)$ **77.** prime **79.** $(x - 2)(x + 2)(2x + 5)$ **81.** $(2y + 5)(y - 4)(y + 4)$ **83.** $(x - 2)(x + 2)(x^2 - 2x + 4)$ **85.** $2x + 5$ **87.** $(x - 3)(x + 1)$ **89.** $-3(2x - 1)$ **91.** $x(x^2 - 3xy + 3y^2)$ **93.** $(x - 2)(x + 4)$ **95.** $(x + 1)(2a - 5)(a - 6)$ **97.** $x(5x + 13)$

Section 4.5

1. perfect square trinomial, perfect squares **3.** greatest common factor **5.** false **7.** Answers may vary. **9.** $(x - 10)(x + 10)$ **11.** $(t + 3)(t - 2)$ **13.** $(x + y)(1 + 2a)$ **15.** $(a - 2)(a^2 + 2a + 4)$ **17.** $(a - 3b)(a + 2b)$ **19.** $(2x - 7)(x + 1)$ **21.** $2(x - 5y)(x + 2y)$ **23.** $(3 - a)(3 + a)$ **25.** $(u - 11)(u - 3)$ **27.** $(x - a)(y - b)$ **29.** $(w + 2)(w + 4)$ **31.** $(6a - 7b^2)(6a + 7b^2)$ **33.** $(x + 4m)(x - 2m)$ **35.** $(3xy - 2)(2xy - 3)$ **37.** $(x + 1)(x^2 + 1)$ **39.** $6(2z^2 + 2z + 3)$ **41.** $(7c - 1)(2c + 3)$ **43.** $(3m + 4n^2)(9m^2 - 12mn^2 + 16n^4)$ **45.** $2j^2(j - 1)(j + 1)(j^2 + 1)$ **47.** $(4a - b)(2a + 5b)$ **49.** $2a(a^2 + 3)$ **51.** $3(2z - 1)(2z + 1)$ **53.** prime **55.** $2a(2a + b)(4a^2 - 2ab + b^2)$ **57.** $(pq + 7)(pq - 1)$ **59.** $(s + 2)^2(s - 2)$ **61.** $-2x(3x + 1)(2x - 1)$ **63.** $(5v + 2)(2v - 1)$ **65.** $-n^2(n - 4)(n + 1)$ **67.** $-2ab(2a^2 - a + 1)$ **69.** $-p(p - 4)(p + 3)$ **71.** $-8x(2x - 3y)(2x + 3y)$ **73.** $2(n - 5)(n^2 - 3)$ **75.** $4(4x - 3)(x + 1)$ **77.** $(3x^2 + 2)(x^2 + 4)$ **79.** $-xy(2x - 3y)(x + y)$ **81.** $x(x - 5)(x - 3)$ **83.** $2x, 2x + 1, x - 3$ **85.** $(2m - n + p)(2m - n - p)$ **87.** $(x + y - z)(x - y + z)$ **89.** $(x - y - 4)(x - y - 2)$ **91.** $(x + y - a + b)(x + y + a - b)$

Putting the Concepts Together (Sections 4.1–4.5)

1. $5x^2y$ **2.** $(x - 4)(x + 1)$ **3.** $(x^2 - 3)(x^4 + 3x^2 + 9)$ **4.** $(2x + 1)(6x + 5z)$ **5.** $(x + 6y)(x - y)$ **6.** $(x + 4)(x^2 - 4x + 16)$ **7.** prime **8.** $3(x + 6y)(x - 2y)$ **9.** $4z^2(3z^3 - 11z - 6)$ **10.** prime **11.** $m^2(m + 2)(4m - 3)$ **12.** $(5p - 2)(p - 3)$ **13.** $(2m + 5)(5m - 3)$ **14.** $6(3m - 1)(2m + 1)$ **15.** $(2m - 5)^2$ **16.** $(5x + 4y)(x - y)$ **17.** $S = 2\pi r(h + r)$ **18.** $h = 16t(3 - t)$

Section 4.6

1. quadratic **3.** second **5.** false **7.** Answers may vary. **9.** linear **11.** quadratic **13.** $\{-4, 0\}$ **15.** $\{-3, 9\}$ **17.** $\left\{-\frac{1}{3}, 5\right\}$ **19.** $\left\{-\frac{3}{5}, \frac{2}{7}\right\}$ **21.** $\{-1, 4\}$ **23.** $\{-7, -2\}$ **25.** $\left\{-\frac{1}{2}, 0\right\}$ **27.** $\left\{-\frac{1}{2}, 2\right\}$ **29.** $\{3\}$ **31.** $\{0, 6\}$ **33.** $\{-2, 3\}$ **35.** $\{2, 9\}$

37. $\left\{\frac{1}{4}, 1\right\}$ **39.** $\{-4, 6\}$ **41.** $\{-5, 10\}$ **43.** $\{-5, 1\}$ **45.** $\{-3, 0, 2\}$ **47.** $\{-3, -2, 2\}$ **49.** $\left\{-\frac{3}{2}, 2, -2\right\}$ **51.** $\left\{-\frac{3}{5}, 4\right\}$ **53.** $\{-4, 5\}$ **55.** $\{-5\}$ **57.** $\left\{-\frac{3}{4}, 7\right\}$ **59.** $\{-4, 2\}$ **61.** $\left\{-4, -\frac{1}{2}, 4\right\}$ **63.** $\{20\}$ **65.** $\{-10, 10\}$ **67.** $\left\{-3, -\frac{2}{3}, 5\right\}$ **69.** $\left\{-\frac{1}{2}, \frac{3}{2}, 5\right\}$ **71.** $\{0, 8\}$ **73.** $\left\{-4, \frac{3}{2}\right\}$ **75.** $\left\{-6, \frac{1}{2}\right\}$ **77.** 1 sec, 3 sec **79.** 1 sec **81.** -4 and -3 or 3 and 4 **83.** -12, -10, and -8, or 8, 10, and 12 **85.** 15 by 17 **87.** 8 teams **89.** $x(x - 3)(x + 5) = 0$ **91.** $(z - 6)(z - 6) = 0$ or $(z - 6)^2 = 0$ **93.** $(a - 4)(2a + 1)(3a - 2) = 0$ **95.** Student divided by a variable expression; $\left\{0, \frac{1}{3}\right\}$ **97.** $\{a, -b\}$ **99.** $\{0, 2a\}$

Section 4.7

1. hypotenuse **3.** $a^2 + b^2 = c^2$ **5.** false **7.** Answers may vary. **9.** 4, 14 **11.** 2, 9 **13.** base = 26; height = 8 **15.** base = 18; height = 16 **17.** base = 13; height = 11 **19.** $B = 53$; $b = 43$ **21.** 9, 12 **23.** 12, 5 **25.** 256, 252, 240, 220, 192, 156, 112, 60, 0 feet **27.** 6 sec **29.** width = 4 m; length = 12 m **31.** base = 6 ft; height = 18 ft **33.** 30 inches **35.** width = 12 mm; length = 25 mm **37.** width = 4 cm; length = 25 cm **39. (a)** width = 25 in.; length = 42 in. **(b)** 28 in. by 45 in. **41.** 6 ft **43.** width = 7 ft; length = 14 feet

Chapter 4 Review

1. 12 **2.** 27 **3.** 10 **4.** 4 **5.** x^2 **6.** m **7.** $15ab^2$ **8.** $6x^3y^2z$ **9.** $2(2a + 1)^2$ **10.** $9(x - y)$ **11.** $-6a^2(3a + 4)$ **12.** $-3x(3x - 4)$ **13.** $5y^2z(3 + y^5 + 4y)$ **14.** $7xy(x^2 - 3xy + 2y^2)$ **15.** $(5 - y)(x + 2)$ **16.** $(a + b)(z + y)$ **17.** $(5m + 2n)(m + 3n)$ **18.** $(2x + y)(y + x)$ **19.** $(x + 2)(8 - y)$ **20.** $(y^2 + 1)(x - 3)$ **21.** $(x + 3)(x + 2)$ **22.** $(x + 2)(x + 4)$ **23.** $(x - 7)(x + 3)$ **24.** $(x + 5)(x - 2)$ **25.** prime **26.** prime **27.** $(x - 3y)(x - 5y)$ **28.** $(m + 5n)(m - n)$ **29.** $-(p + 6)(p + 5)$ **30.** $-(y - 5)(y + 3)$ **31.** $3x(x^2 + 11x + 12)$ **32.** $4(x + 1)(x + 8)$ **33.** $2(x - 7y)(x + 6y)$ **34.** $4y(y + 5)(y - 2)$ **35.** $(5y - 6)(y + 4)$ **36.** $(y - 7)(6y + 1)$ **37.** $(2x - 3)(x - 1)$ **38.** $(2x + 7)(3x + 1)$ **39.** prime **40.** $(4m + 3)(2m + 3)$ **41.** $3m(m + n)(3m + 7n)$ **42.** $2(7m + n)(m + n)$ **43.** $x(5x + 2)(3x - 1)$ **44.** $p^2(3p - 1)(2p + 1)$ **45.** $(2x - 3)^2$ **46.** $(x - 5)^2$ **47.** $(x + 3y)^2$ **48.** prime **49.** $2(2m + 1)^2$ **50.** $2(m - 6)^2$ **51.** $(2x - 5y)(2x + 5y)$ **52.** $(7x - 6y)(7x + 6y)$ **53.** prime **54.** prime **55.** $(x - 3)(x + 3)(x^2 + 9)$ **56.** $(x - 5)(x + 5)(x^2 + 25)$ **57.** $(m + 3)(m^2 - 3m + 9)$ **58.** $(m + 5)(m^2 - 5m + 25)$ **59.** $(3p - 2)(9p^2 + 6p + 4)$ **60.** $(4p - 1)(16p^2 + 4p + 1)$ **61.** $(y^3 + 4z^2)(y^6 - 4y^3z^2 + 16z^4)$ **62.** $(2y + 3z^2)(4y^2 - 6yz^2 + 9z^4)$ **63.** $(5a - 2b)(3a^2 - 5b^2)$ **64.** $(4a - 3b)(3a + b)$ **65.** prime **66.** $(x - 12y)(x + 2y)$ **67.** $x(x - 7)(x + 6)$ **68.** $3x^4(x - 7)(x - 3)$ **69.** $(3x + 1)(2x + 3)$ **70.** $(2z + 3)(5z - 3)$ **71.** $(3x + 2)(9x^2 - 6x + 4)$ **72.** $(2z - 1)(4z^2 + 2z + 1)$ **73.** $2(2y - 1)(y + 5)$ **74.** $xy^2(5x - 3)(x - 1)$ **75.** $(5k - 9m)(5k + 9m)$ **76.** $(x^2 - 3)(x^2 + 3)$ **77.** prime **78.** prime **79.** $\left\{\frac{3}{2}, 4\right\}$ **80.** $\left\{-7, -\frac{1}{2}\right\}$ **81.** $\{-3, 15\}$ **82.** $\{2, 5\}$ **83.** $\{0, -2\}$ **84.** $\left\{-\frac{9}{2}, 0\right\}$ **85.** $\{-1, 3\}$ **86.** $\{-3, -1\}$ **87.** $\{-3, -1\}$ **88.** $\left\{-\frac{1}{2}, \frac{3}{2}\right\}$ **89.** $\{-14, 0, 3\}$ **90.** $\{-3, 0, 6\}$ **91.** $\left\{-3, \frac{4}{3}, 3\right\}$ **92.** $\left\{-\frac{5}{2}, -2\right\}$ **93.** 5 sec **94.** 2 sec or 3 sec **95.** 9 ft, 6 ft **96.** 3 yd, 5 yd **97.** 8 ft, 6 ft **98.** 24 ft, 10 ft

Chapter 4 Test

1. $4x^4y^2$ **2.** $(x + 3)(x - 3)(x^2 + 9)$ **3.** $9x(2x^2 - 3)(x^2 + 1)$ **4.** $(x - 7)(y - 4)$ **5.** $(3x + 5)(9x^2 - 15x + 25)$ **6.** $(y - 12)(y + 4)$ **7.** $(6m + 5)(m - 1)$ **8.** prime **9.** $(x - 5)(4 + y)$ **10.** $3y(x - 7)(x + 2)$ **11.** prime **12.** $3(x - 1)(x + 1)(x^2 + x + 1)(x^2 - x + 1)$ **13.** $3x(3x + 1)(x + 4)$ **14.** $(3m + 2)(2m + 1)$ **15.** $2(2m^2 - 3mn + 2)$ **16.** $(5x + 7y)^2$ **17.** $\left\{-\frac{1}{5}, -3\right\}$ **18.** $\{0, 2\}$ **19.** length = 7 in.; width = 5 in. **20.** legs: 12 in., 5 in., hypotenuse: 13 in.

Chapter 5

Section 5.1

1. rational expression **3.** simplify **5.** True **7.** Answers may vary. **9. (a)** 2 **(b)** $\frac{1}{2}$ **(c)** undefined **11. (a)** undefined **(b)** 3 **(c)** $\frac{5}{3}$ **13. (a)** 3 **(b)** undefined **(c)** 0 **15. (a)** -3 **(b)** 0 **(c)** $-\frac{15}{7}$ **17. (a)** 0 **(b)** undefined **(c)** undefined **19.** 0 **21.** 5 **23.** $\frac{3}{2}$ **25.** $-6, 6$ **27.** 2, 5 **29.** $-2, 0, 3$ **31.** $\frac{3}{x - 2}$ **33.** $\frac{(1 + z)(1 - z)}{3}$ **35.** $\frac{1}{p + 2}$ **37.** -1 **39.** $-2k$ **41.** $\frac{x - 1}{x + 4}$ **43.** $\frac{5}{9}$ **45.** $\frac{10}{11}$ **47.** $\frac{1}{2x^3}$ **49.** $\frac{12}{a^2}$ **51.** -3 **53.** $\frac{b - 5}{4}$ **55.** $\frac{x + 3}{x - 3}$ **57.** $\frac{x - 3}{x - 5}$ **59.** $-(x + y)$ **61.** $\frac{4}{a + b}$ **63.** $\frac{x}{x - 4}$ **65.** $\frac{-(4 + c)}{c - 4}$ **67.** $\frac{2(x - 2)}{3(x - 5)}$ **69.** $\frac{-(x - 3)}{x - 2}$ **71.** $\frac{2}{t^2 + 9}$ **73.** $\frac{-3}{w - 1}$ **75. (a)** 5 mg/mL **(b)** 4 mg/mL **77.** Yes; BMI = 27.5 **79.** \$15,600 **81.** $\frac{(c^2 - c + 1)(c^2 + c + 1)}{c^2 + 1}$ **83.** $\frac{(x^2 + 4)(x^2 - 3)}{(x + 2)(x^3 - 9)}$ **85.** $\frac{(t^2 + 4)(t + 2)}{t^2 - 2t + 4}$

Section 5.2

1. quotient, product **3.** factor, common factors **5.** False **7.** Answers may vary. **9.** $\frac{16}{63x^6}$ **11.** $-\frac{5a^4}{8}$ **13.** $\frac{x}{(x + 1)^2}$ **15.** $p - 1$ **17.** $\frac{18}{25z}$ **19.** $-\frac{3m^5}{10}$ **21.** x **23.** 1 **25.** $\frac{10}{3}$ **27.** 1 **29.** $-\frac{4}{11}$ **31.** $\frac{4}{3y(y - 3)}$ **33.** $\frac{a}{3}$ **35.** $\frac{3x}{x - 4}$ **37.** $\frac{4 - w}{(w + 2)(w + 4)}$

39. $-\frac{(x-y)^2}{x}$ **41.** $-\frac{1}{2}$ **43.** 1 **45.** $\frac{3(2n+3)(n-3)}{2(n+4)}$ **47.** $\frac{x}{4y^2(x-2y)}$ **49.** $\frac{2a-b}{2a(a+b)}$ **51.** $\frac{(t+3)(t-2)}{3}$ **53.** -1
55. $-\frac{x(x+3)}{(x+1)^2}$ **57.** $\frac{a+b}{2}$ **59.** $\frac{1}{3}$ **61.** $\frac{1}{3(x-y)}$ **63.** $\frac{9}{8}$ **65.** $\frac{x-y}{(x+y)^3}$ **67.** $\frac{4x}{(x-2)(x-3)}$ **69.** $\frac{1}{x}$ square feet
71. $\frac{2(x+2)}{x-3}$ square inches **73.** $\frac{(x-a)^2}{2(x+a)}$ **75.** $\frac{1-x}{4}$ **77.** $\frac{2a^2}{3}$ **79.** $\frac{(x+y)^2}{(x^2-xy+y^2)(x+2)}$ **81.** $3p^5q^2(p^2+3pq+9q^2)(p-q)$
83. $6x+12$

Section 5.3

1. numerators **3.** $a+b$; c **5.** True **7.** Incorrect; Answers may vary. **9.** $\frac{7p}{4}$ **11.** $2n$ **13.** $\frac{2a-1}{a}$ **15.** $\frac{2(5c-2)}{c-1}$ **17.** 2
19. $\frac{7x(1+x)}{x+2}$ **21.** 2 **23.** $2x$ **25.** $x-3$ **27.** $-\frac{x}{2}$ **29.** $-\frac{c+3}{c}$ **31.** 3 **33.** -1 **35.** $x-1$ **37.** $-\frac{1}{x}$ **39.** -1 **41.** 0
43. $n-1$ **45.** $\frac{3}{v+3}$ **47.** $\frac{x+1}{x-1}$ **49.** $\frac{x^2+3}{x(x-3)}$ **51.** $\frac{1}{a-5}$ **53.** 1 **55.** $\frac{2(2p+3q)}{p-q}$ **57.** $\frac{2p^2+2p+1}{p-1}$ **59.** $\frac{4b}{a-b}$ **61.** -1
63. $\frac{2}{x-y}$ **65.** $2(x-1)$ **67.** $\frac{3x+3y-2}{x-y}$ **69.** $\frac{p+2q}{p^2-6q^2}$ **71.** $\frac{2g-7}{g-3}$ **73.** $12a$ **75.** $\frac{n-3}{n+1}$ **77.** $\frac{4}{n}$ **79.** $\frac{x}{x+3}$
81. $\frac{4x-2}{x+2}$ **83.** $\frac{2(6x-1)}{x}$ cm **85.** $\frac{-x(x-1)}{(3x-1)(x+2)}$ **87.** $\frac{-5x+12}{x-2}$ **89.** $-4n-6$

Section 5.4

1. least common denominator **3.** 1 **5.** False **7.** Answers may vary. **9.** 36 **11.** 60 **13.** $5x^2$ **15.** $60x^3y^2$ **17.** $x(x+1)$
19. $2x+1$ **21.** $4(b-3)$ **23.** $p(p+1)(p-2)$ **25.** $(r+2)^2(r+1)(r-2)$ **27.** $-(x-4)$ **29.** $-2(x+3)(x-3)$
31. $-(x-1)(x+2)$ **33.** $p^2(p-1)(p-3)$ **35.** $\frac{12x^2}{3x^3}$ **37.** $\frac{3c+c^2}{a^2b^2c^2}$ **39.** $\frac{(x-4)^2}{x^2-16}$ **41.** $\frac{9n^2-9n}{6n^2-6}$ **43.** $\frac{4t^2-4t}{t-1}$ **45.** $\frac{21}{36}; \frac{-32}{36}$
47. $\frac{7}{15}; \frac{30}{15}$ **49.** $\frac{6xy}{9y^2}; \frac{4}{9y^2}$ **51.** $\frac{6a+2}{4a^3}; \frac{4a^3-a^2}{4a^3}$ **53.** $\frac{2m+2}{m(m+1)}; \frac{3m}{m(m+1)}$ **55.** $\frac{2y^2-5y+2}{4y(2y-1)}; \frac{y^2}{4y(2y-1)}$
57. $\frac{-1}{-(x-1)}; \frac{2x}{-(x-1)}$ **59.** $\frac{-3x}{-(x-7)}; \frac{-5}{-(x-7)}$ **61.** $\frac{4x}{(x+2)(x-2)}; \frac{2x-4}{(x+2)(x-2)}$
63. $\frac{x^2+2x+1}{(x+3)(x-3)(x+1)}; \frac{x^2+5x+6}{(x+3)(x-3)(x+1)}$ **65.** $\frac{6x-3}{(x+4)(2x-1)}; \frac{2x-1}{(x+4)(2x-1)}$ **67.** $3x; \frac{3}{3x}; \frac{1}{3x}$
69. $(r-4)(r+4); \frac{12r+48}{(r-4)(r+4)}; \frac{12r-48}{(r-4)(r+4)}$ **71.** $(x+2)(x-2)(x^2-2x+4)(x^2+2x+4)$
73. $4a^2b^2(a+b)(a-b)(a^2+ab+b^2)$

Section 5.5

1. least common denominator **3.** denominator **5.** True **7.** Answers may vary. **9.** $-\frac{5}{6}$ **11.** $\frac{5}{3x}$ **13.** $\frac{2a^2+7a-3}{(2a-1)(2a+1)}$
15. $\frac{-2}{x-4}$ **17.** $\frac{2a^2-3a-12}{4a(a-5)}$ **19.** $\frac{5}{x}$ **21.** $\frac{8}{75}$ **23.** $\frac{(m-4)(m+4)}{m}$ **25.** $\frac{2(4x-3)}{(x-3)(x+3)}$ **27.** $\frac{6(x+1)}{(x+3)^2}$ **29.** $\frac{-4(a-1)}{a+3}$
31. $\frac{x}{x-2}$ **33.** $\frac{5}{4}$ **35.** $\frac{20-9y}{12y^2}$ **37.** $\frac{6}{5x}$ **39.** $\frac{7y+4x}{2x^2y^2}$ **41.** $\frac{-3}{2x+3}$ **43.** $\frac{2(n^2-2)}{n(n-2)}$ **45.** $\frac{2x^2-5x+15}{x(x-3)}$ **47.** $\frac{-1}{a(a-1)}$
49. $\frac{2n+5}{2n+1}$ **51.** $\frac{4x-5}{x-2}$ **53.** $\frac{3x-28}{x-9}$ **55.** $\frac{4n^2-7n+9}{n^2(n-3)(n-1)}$ **57.** $\frac{-13}{2a(a-2)}$ **59.** $\frac{n^2-n-3}{-(n+2)(n-2)(n+3)}$ **61.** $\frac{2x^2+x+1}{(x+1)(x-1)^2}$
63. $\frac{-3n+5}{(n+3)(n-2)}$ **65.** $\frac{x}{x-2}$ **67.** $\frac{x-4}{x}$ **69.** $\frac{6}{m(m-2)}$ **71.** $\frac{a+2}{a(a-1)^2}$ **73.** $\frac{x^2+4}{(x+2)(x+3)}$ **75.** $\frac{x+3}{(x+1)(x-1)}$
77. $\frac{2x^2-x+10}{(x+3)(x-2)(x+2)}$ **79.** $\frac{2x^2-5x+3}{(x+6)(x-7)}$ **81.** $\frac{x-4}{8}$ square units **83.** $\frac{2x^2-9x-18}{6}$ square units **85.** $\frac{x^3-2x^2+4x+3}{x^2(x+1)(x-1)}$
87. $\frac{1}{a+b}$ **89.** $\frac{x^2+2x+5}{(x-4)(x+1)}$ **91.** $\frac{-x^3+6x^2-11x+10}{x-2}$

Section 5.6

1. complex rational expression **3.** reciprocal, multiply **5.** False **7.** Answers may vary. **9.** -5 **11.** $\frac{2}{17}$ **13.** -24 **15.** $\frac{3(x-2)}{2}$
17. $\frac{9}{x-3}$ **19.** $\frac{n(m+2n)}{2m}$ **21.** $\frac{y-3x}{4y+x}$ **23.** $\frac{1}{2}$ **25.** $\frac{b-7}{b}$ **27.** $\frac{x+4}{x}$ **29.** $\frac{-(x+y)}{x^2y^2}$ **31.** $\frac{x^2+2xy-y^2}{xy}$ **33.** $\frac{-1}{2x+7}$
35. $-b$ **37.** $\frac{11(2n+3)}{72}$ **39.** $\frac{-x^3}{x+3}$ **41. (a)** $R=\frac{R_1R_2}{R_2+R_1}$ **(b)** $\frac{15}{4}$ ohms **43.** $\frac{2x+1}{x+1}$ **45.** $\frac{5-2x}{2-x}$

Putting the Concepts Together

1. -35 **2. (a)** The expression is undefined for $a = 6$. **(b)** The expression is undefined for $y = 0$ or $y = -4$. **3. (a)** $\frac{a-4b}{2}$ **(b)** $\frac{-(x+2)}{x+1}$ **4.** The LCD is $8(a+2b)(a-4b)$. **5.** $\frac{7}{3x^2-x} = \frac{35x}{5x^2(3x-1)}$ **6.** $\frac{-2y^2}{y+1}$ **7.** $\frac{2(m-1)^2}{m^2}$ **8.** $\frac{2(y-1)}{5(y-3)}$ **9.** $4x+1$ **10.** $\frac{1}{x+1}$ **11.** $\frac{1}{2x-1}$ **12.** $\frac{15m+17}{(m+3)(3m+2)}$ **13.** $\frac{m}{(m+5)(m+4)}$ **14.** $-\frac{1}{m-1}$ **15.** $\frac{a(24+a)}{3(6+a^2)}$

Section 5.7

1. least common denominator **3.** undefined **5.** True **7.** Answers may vary. **9.** $\{18\}$ **11.** $\{-3\}$ **13.** $\{2\}$ **15.** $\{6\}$ **17.** $\left\{\frac{5}{2}\right\}$ **19.** $\left\{\frac{31}{2}\right\}$ **21.** $\{-7\}$ **23.** $\left\{\frac{1}{2}\right\}$ **25.** $\{7\}$ **27.** $\{\ \}$ or $\emptyset$ **29.** $\{-3\}$ **31.** $\left\{\frac{3}{4}\right\}$ **33.** $\left\{-\frac{5}{6}\right\}$ **35.** $\{-4\}$ **37.** $\{-3, -2\}$ **39.** $\{-1\}$ **41.** $\{-9, 4\}$ **43.** $\left\{-\frac{2}{3}, \frac{1}{2}\right\}$ **45.** $\{\ \}$ or $\emptyset$ **47.** $\{-5, 5\}$ **49.** $\{-1, 5\}$ **51.** $\{-6\}$ **53.** $\{-5\}$ **55.** $\{\ \}$ or $\emptyset$ **57.** $y = \frac{2}{x}$ **59.** $R = \frac{E}{I}$ **61.** $b = \frac{2A}{h} - B$ **63.** $y = \frac{x}{z} - 3$ **65.** $S = \frac{RT}{T-R}$ **67.** $y = \frac{an-p}{am}$ **69.** $x = -\frac{Ay}{y-A}$ **71.** $y = \frac{xz}{2z-6x}$ **73.** $x = \frac{cy}{y-1}$ **75.** $\frac{4x+5}{x(x+5)}$ **77.** $\{-2, 3\}$ **79.** $\frac{x+7}{2}$ **81.** $\left\{-\frac{1}{6}, 1\right\}$ **83.** $\frac{x^3-x^2+4}{x(x+2)(x-2)}$ **85.** 1 hour and 9 hours **87.** 40 or 125 bicycles **89.** $k = 33$

Section 5.8

1. extremes, means **3.** $d = rt$ **5.** True **7.** Answers may vary. **9.** $\{12\}$ **11.** $\left\{\frac{18}{7}\right\}$ **13.** $\{16\}$ **15.** $\left\{\frac{24}{5}\right\}$ **17.** $\left\{-\frac{4}{5}\right\}$ **19.** $\{-2\}$ **21.** $\left\{\frac{5}{7}\right\}$ **23.** $\{-3, 2\}$ **25.** $\{2, 5\}$ **27.** $\{-2, 9\}$ **29.** $XZ = 12$ **31.** $n = 3$; $ZY = 5$ **33.** $x = 30$; $AB = 28$ **35.** 2x **37.** $t - 3$ **39.** $r - 2$ **41.** $\frac{1}{5} + \frac{1}{3} = \frac{1}{t}$ **43.** $\frac{1}{b+4} + \frac{1}{b} = \frac{1}{7}$ **45.** $\frac{4}{14-c} = \frac{7}{14+c}$ **47.** 145 miles **49.** $8\frac{1}{3}$ lb **51.** 37,245 rubles **53.** 24 ft **55.** $1\frac{7}{8}$ hr **57.** $3\frac{3}{7}$ minutes **59.** $5\frac{1}{7}$ hr **61.** 15 hr **63.** 24 hr **65.** $\frac{14}{3}$ km per hour **67.** 1 hour, 40 minutes **69.** 40 mph **71.** 35 mph **73.** $x = 6$ **75.** $x = \frac{98}{5}$ **77.** $x = \frac{72}{5}$

Chapter 5 Review

1. (a) $\frac{1}{2}$ **(b)** undefined **(c)** 1 **2. (a)** 0 **(b)** -1 **(c)** undefined **3. (a)** undefined **(b)** 9 **(c)** 49 **4. (a)** -1 **(b)** undefined **(c)** 42 **5.** $\frac{7}{3}$ **6.** $\frac{1}{2}$ **7.** none **8.** none **9.** -10 or -2 **10.** -1 or 4 **11.** $\frac{4x^4}{5y^4}$ **12.** $\frac{3}{4x^2y^4}$ **13.** $\frac{3}{k+2}$ **14.** $-\frac{2}{x-3}$ **15.** $\frac{x+5}{2x-1}$ **16.** $\frac{3x-1}{2(x^2-2x+4)}$ **17.** $\frac{2m^2}{n^2}$ **18.** $\frac{y^3}{6x}$ **19.** $\frac{14}{9n^2}$ **20.** $\frac{1}{ab^7}$ **21.** $-\frac{5(x-2)}{x+3}$ **22.** $-\frac{4(x-3)(x+3)}{(x-9)(x-9)}$ **23.** $\frac{x-2}{4(x-3)}$ **24.** $15x^3$ **25.** $\frac{x+5}{x+6}$ **26.** $\frac{y+1}{y-3}$ **27.** $\frac{x^2}{(x+1)(x+9)}$ **28.** $\frac{y-2}{3y+1}$ **29.** $\frac{9}{x-3}$ **30.** $\frac{10}{x+4}$ **31.** m **32.** $\frac{1}{m}$ **33.** $\frac{1}{m-2}$ **34.** $2m+3$ **35.** $\frac{1}{3m}$ **36.** $\frac{2}{b}$ **37.** $y+7$ **38.** $\frac{1}{m-6}$ **39.** $-\frac{7}{x-y}$ **40.** $\frac{4}{a-b}$ **41.** $\frac{3x-5}{x^2-25}$ **42.** $\frac{x+4}{x-3}$ **43.** $12x^4y^7$ **44.** $60a^3b^4c^7$ **45.** $8a(a+2)$ **46.** $5a(a+6)$ **47.** $4(x-3)(x+1)$ **48.** $x(x-7)(x+7)$ **49.** $\frac{6xy^6}{x^4y^7}$ **50.** $\frac{11a^4b^3}{a^7b^5}$ **51.** $\frac{(x-1)(x+2)}{(x-2)(x+2)}$ **52.** $\frac{(m+2)(m-2)}{(m+7)(m-2)}$ **53.** $\frac{20x^2y}{24x^5}; \frac{21}{24x^5}$ **54.** $\frac{12}{10a^3}; \frac{11a^2b}{10a^3}$ **55.** $\frac{-4x}{-(x-2)}; \frac{6}{-(x-2)}$ **56.** $\frac{-3}{-(m-5)}; \frac{-2m}{-(m-5)}$ **57.** $\frac{2m+4}{(m+7)(m-2)(m+2)}; \frac{m^2-m-2}{(m+7)(m-2)(m+2)}$ **58.** $\frac{n^2+3n-10}{n(n-5)(n+5)}; \frac{n^2}{n(n-5)(n+5)}$ **59.** $\frac{4xz+8y^4}{x^2y^3z}$ **60.** $\frac{5xy^2+x^2y}{10x^3y^3}$ **61.** $\frac{x^2-5x+14}{(x+7)(x-7)}$ **62.** $\frac{2x^2+13x-9}{(2x+3)(2x-3)}$ **63.** $\frac{-4x-10}{x(x-2)}$ **64.** $\frac{2x^2-6x-5}{x(x+5)}$ **65.** $\frac{3x-1}{4x+1}$ **66.** $\frac{3x-4}{(x-1)(x-2)}$ **67.** $\frac{-(4m-9)(m-4)}{(m+3)(m-3)}$ **68.** $\frac{2m^2+5m-2}{(m-5)(m+2)}$ **69.** $\frac{4}{m-2}$ **70.** $\frac{m+n}{m-n}$ **71.** $\frac{5x+12}{x+3}$ **72.** $\frac{7x+13}{x+2}$ **73.** $-\frac{3}{23}$ **74.** $\frac{18}{11}$ **75.** $\frac{2m^2-10m}{m^2+10}$ **76.** $\frac{6+4m^2}{6m-5m^2}$ **77.** $\frac{1}{6}$ **78.** $\frac{56}{9}$ **79.** $\frac{y+8}{-y+2}$ **80.** $\frac{5y+50}{-5y-22}$ **81.** $x = -5$ or $x = 0$ **82.** $x = -6$ or $x = 6$ **83.** $\{-4\}$ **84.** $\left\{\frac{2}{3}\right\}$ **85.** $\{5\}$ **86.** $\left\{\frac{14}{3}\right\}$ **87.** $\{38\}$ **88.** $\left\{\frac{9}{7}\right\}$ **89.** $\{\ \}$ or $\emptyset$ **90.** $\{\ \}$ or $\emptyset$ **91.** $\left\{-\frac{5}{2}\right\}$ **92.** $\{24\}$ **93.** $k = \frac{4}{y}$ **94.** $y = \frac{x}{6}$ **95.** $y = \frac{xz}{x-z}$ **96.** $z = \frac{xy}{x+y}$ **97.** $\{4\}$ **98.** $\{5\}$ **99.** $\{1\}$ **100.** $\{3\}$ **101.** $x = 15$ **102.** $x = 3.5$ **103.** 8 tanks **104.** \$26.25 **105.** $1\frac{1}{5}$ hours **106.** $1\frac{7}{8}$ hours **107.** 15 min **108.** $9\frac{1}{3}$ hours **109.** 9 mph **110.** 55 mph **111.** 70 mph **112.** 45 mph

Chapter 5 Test

1. 2 **2.** $x = 5$ or $x = -2$ **3.** $-\frac{x+3}{2}$ **4.** $\frac{14}{5x^2}$ **5.** $\frac{25}{9}$ **6.** $\frac{x-y}{2x-3y}$ **7.** y **8.** $\frac{x-5}{x+3}$ **9.** $\frac{-1}{y-z}$ or $\frac{1}{z-y}$
10. $\frac{2(x+3)(x-1)}{(x-2)(2x+1)}$ **11.** $\frac{x^2-5x-2}{(x+2)(x+3)(x-1)}$ **12.** $\frac{y-3}{3y}$ **13.** $\{-20, 5\}$ **14.** $\{2\}$ **15.** $x = \frac{yz}{y-z}$ **16.** $\{5\}$ **17.** $x = \frac{24}{5}$
18. \$33 **19.** 36 min **20.** 4 mph

Chapters 1–5 Cumulative Review

1. -144 **2.** $9x - 11$ **3.** $\left\{\frac{19}{6}\right\}$ **4.** all real numbers **5.** $\{4\}$ **6.** 7 feet **7.** 44, 46, 48
8. $x \geq 1$ (number line: 0, 1) **9.** $9x^2 - 12xy + 4y^2$ **10.** $\frac{16}{a^{16}b^{12}}$ **11.** $\frac{27z^9}{y^{12}}$ **12.** $2b(4a-7)(a+6)$ **13.** $\left\{0, \frac{1}{6}, 5\right\}$
14. 10 cm **15.** 0 **16.** $-\frac{3(x+1)}{5x(2x-3)}$ **17.** $\frac{(x-2)(x+1)}{4(x+2)}$ **18.** $\frac{1}{x-6}$ **19.** $\frac{17x+30}{2(x-2)(x+2)}$ **20.** $\frac{4(10-3x)}{85x}$
21. $\{\ \}$ or $\varnothing$ **22.** $\{-16\}$ **23.** 10 inches **24.** 9 hr **25.** 7 mph

Chapter 6

Section 6.1

1. square roots **3.** radical, principal (or nonnegative) **5.** True **7.** Answers may vary. **9.** 16 **11.** $-2, 2$ **13.** $-\frac{1}{3}, \frac{1}{3}$ **15.** $\frac{1}{81}$
17. 12 **19.** -3 **21.** 6 **23.** 15 **25.** $\frac{1}{11}$ **27.** 0.2 **29.** -18 **31.** $\frac{6}{9}$ **33.** 0.04 **35.** 35 **37.** 2.828 **39.** 5.48 **41.** 7.5
43. not a real number **45.** rational; 20 **47.** rational; $\frac{1}{2}$ **49.** irrational; 7.35 **51.** irrational; 7.07 **53.** not a real number **55.** d
57. $|x-9|$ **59.** $2m - n$ **61.** 1.73 **63.** 0.63 **65.** not a real number **67.** 10 **69.** 14 **71.** 6 **73.** -8 **75.** not a real number
77. 8 **79.** 6.63 **81.** 5 **83.** 6.71 **85.** 25 ft **87.** 16 km **89.** 7 m **91.** 14 in. **93.** 398 m **95.** 2 **97.** 5

Section 6.2

1. $\sqrt{a} \cdot \sqrt{b}$ **3.** $\frac{\sqrt{a}}{\sqrt{b}}$ **5.** True **7.** No, not simplified. Answers may vary. **9.** $2\sqrt{2}$ **11.** $2\sqrt{10}$ **13.** $3\sqrt{2}$ **15.** $\sqrt{33}$ **17.** $3\sqrt{5}$
19. $5\sqrt{5}$ **21.** $\sqrt{42}$ **23.** $7\sqrt{2}$ **25.** $3\sqrt{30}$ **27.** $2\sqrt{33}$ **29.** x^5 **31.** n^{72} **33.** $3y^2$ **35.** $2z^7\sqrt{6}$ **37.** $7x^3\sqrt{x}$
39. $5c^2d\sqrt{3}$ **41.** $\frac{5}{6}$ **43.** $\frac{1}{14}$ **45.** $\frac{\sqrt{11}}{3}$ **47.** $\frac{y^8}{11}$ **49.** $\frac{\sqrt{3}}{a^2}$ **51.** $\frac{x^2\sqrt{x}}{y^4}$ **53.** $\frac{2xy^2\sqrt{5y}}{z^2}$ **55.** $\frac{\sqrt{15}}{x^2}$ **57.** $\frac{5a}{2}$ **59.** $\frac{8n^2\sqrt{n}}{3}$
61. $3\sqrt{7}$ **63.** $\frac{2\sqrt{3}}{5}$ **65.** $12a^3$ **67.** $\frac{x^2\sqrt{x}}{3}$ **69.** $2a^2bc^3\sqrt{3}$ **71.** $q^2r^2\sqrt{15pq}$ **73.** $\frac{2x^3\sqrt{x}}{y^6}$ **75.** $3a^{12}\sqrt{3a}$ **77.** 5 **79.** $3 + \sqrt{2}$
81. $\frac{-4-9\sqrt{2}}{6}$ **83.** $\frac{1-\sqrt{2}}{2}$ **85.** $x = 90\sqrt{2}$ **87.** $x = 7\sqrt{5}$ **89.** $x = 2\sqrt{34}$ **91.** $2\sqrt{73}$ **93.** **(a)** $\frac{\sqrt{6}}{\pi}$ **(b)** 0.78
95. x^{24} **97.** $16s^{18n}$

Section 6.3

1. radicands **3.** 1 **5.** False **7.** Answers may vary. **9.** $2\sqrt{3}$ **11.** $2a\sqrt{2a}$ **13.** $\sqrt{15} + \sqrt{7}$ **15.** 0 **17.** $-\sqrt{2}$
19. $-3\sqrt{6} - 6\sqrt{11}$ **21.** $2\sqrt{22}$ **23.** $7\sqrt{14}$ **25.** 0 **27.** $30\sqrt{7}$ **29.** 0 **31.** $-5\sqrt{y}$ **33.** $\sqrt{6} + 9\sqrt{5}$ **35.** $-13\sqrt{qr}$
37. $9\sqrt{2x} - 5\sqrt{2}$ **39.** $-2n\sqrt{5n}$ **41.** $-\sqrt{3}$ **43.** $21\sqrt{15}$ **45.** $3\sqrt{5}$ **47.** $18\sqrt{3}$ **49.** $11\sqrt{5} - 4\sqrt{6}$ **51.** $18\sqrt{2} - 36\sqrt{3}$
53. $-\sqrt{6x}$ **55.** 0 **57.** $5x\sqrt{5y} - 5x\sqrt{5}$ **59.** $a^4\sqrt{10a} + 6a\sqrt{3a}$ **61.** $7 + 13\sqrt{2}$ **63.** $-1 + 5\sqrt{3}$ **65.** $\frac{7}{6}\sqrt{3}$ **67.** $-\frac{16}{5}\sqrt{5}$
69. $\frac{9}{20}\sqrt{3}$ **71.** $\frac{5}{6}\sqrt{15}$ **73.** $\frac{11}{3}\sqrt{6}$ **75.** $24\sqrt{3}$ cm **77.** $(18\sqrt{2} + 32\sqrt{3})$ inches **79.** 23.3 ft **81.** $13 + 8\sqrt{3}$ **83.** Answers may vary.

Section 6.4

1. FOIL **3.** Distributive **5.** False **7.** It is true that $(\sqrt{-4})^2 = -4$, but not because $\sqrt{a^2} = a$. The property $\sqrt{a^2} = a$ is only true if $a \geq 0$. The expression $(\sqrt{-4})^2$ cannot be simplified using $(\sqrt{a})^2 = \sqrt{a} \cdot \sqrt{a}$ since this requires $a \geq 0$. **9.** 3 **11.** $7x$ **13.** $\sqrt{39}$ **15.** $\sqrt{14a}$
17. $-2\sqrt{15}$ **19.** $24\sqrt{3}$ **21.** $-9x^2$ **23.** $72\sqrt{6}$ **25.** $270\sqrt{6}$ **27.** $-45\sqrt{6}$ **29.** $32x^2\sqrt{3x}$ **31.** $42y^7\sqrt{3}$ **33.** 37 **35.** $7x$
37. $24 + 8\sqrt{5}$ **39.** $12 - 3\sqrt{6}$ **41.** $21\sqrt{5} + 4\sqrt{3}$ **43.** $a\sqrt{2} + 2\sqrt{a}$ **45.** $33 + 11\sqrt{3}$ **47.** $3 - 14\sqrt{x} + 8x$
49. $20 - 15\sqrt{2} + 8\sqrt{3} - 6\sqrt{6}$ **51.** $8x\sqrt{3} - 7x$ **53.** $19 + 8\sqrt{3}$ **55.** $9x - 12\sqrt{x} + 4$ **57.** 1 **59.** $x - 9y$ **61.** $3m^3\sqrt{5}$
63. $-5\sqrt{2}$ **65.** $12\sqrt{2n}$ **67.** 32 **69.** $a + 2\sqrt{ab} + b$ **71.** $27 + 2\sqrt{30}$ **73.** $m\sqrt{2} - m$ **75.** $1 + 2\sqrt{6}$ **77.** $-\frac{\sqrt{3}}{12}$
79. $9s - 12\sqrt{s} + 4$ **81.** -120 **83.** $6a^2\sqrt{a}$ **85.** $-\frac{9}{2}\sqrt{3}$ **87.** 4 **89.** 9 **91.** $126\sqrt{3}$ square units **93.** $112\sqrt{7}$ square units
95. $(12 + 18\sqrt{3})$ square units **97.** 147π square meters **99.** 126π square inches **101.** $12\sqrt{30}$ square units **103.** $50\sqrt{14}$ square units

Section 6.5

1. rationalize the denominator **3.** $\frac{\sqrt{2}}{\sqrt{2}}; \frac{1+\sqrt{2}}{1+\sqrt{2}}$ **5.** True **7.** Neither is correct. The correct approach is $\frac{3}{\sqrt{2}+\sqrt{3}} \cdot \frac{\sqrt{2}-\sqrt{3}}{\sqrt{2}-\sqrt{3}}$. **9.** 4 **11.** $-\frac{1}{5}$ **13.** $3\sqrt{2}$ **15.** $2\sqrt{7}$ **17.** $-5\sqrt{y}$ **19.** 8 **21.** $\frac{3\sqrt{5}}{5}$ **23.** $\frac{10\sqrt{6}}{3}$ **25.** $-\frac{8\sqrt{2}}{3}$ **27.** $-\frac{2\sqrt{x}}{x}$ **29.** $\frac{9\sqrt{2y}}{2}$ **31.** $\frac{\sqrt{3}}{2}$ **33.** $\frac{3\sqrt{2n}}{2n}$ **35.** $\frac{4\sqrt{3}}{9}$ **37.** $\frac{\sqrt{6ab}}{21b}$ **39.** $-\frac{7\sqrt{10}}{12}$ **41.** $\frac{\sqrt{2}}{2}$ **43.** $-\frac{\sqrt{6}}{3}$ **45.** $\frac{\sqrt{30}}{10}$ **47.** $\frac{\sqrt{7xy}}{7x^2y}$ **49.** $2-\sqrt{3}; 1$ **51.** $\sqrt{6}+1; 5$ **53.** $-1+3\sqrt{14}; -125$ **55.** $\sqrt{p}-3\sqrt{q}; p-9q$ **57.** $\frac{3-\sqrt{5}}{4}$ **59.** $\frac{27+3\sqrt{3}}{26}$ **61.** $\frac{24-4\sqrt{7}}{29}$ **63.** $\frac{3\sqrt{y}+18}{y-36}$ **65.** $32\sqrt{2}+48$ **67.** $\frac{6\sqrt{2}+\sqrt{42}}{15}$ **69.** $\frac{n\sqrt{n}-n}{n-1}$ **71.** $\frac{2\sqrt{x}-3x}{4-9x}$ **73.** $\frac{\sqrt{55}+\sqrt{30}}{5}$ **75.** $\frac{\sqrt{3}+1}{2}$ **77.** $\sqrt{3}-\sqrt{6}$ **79.** $\frac{3\sqrt{5}}{10}-\frac{\sqrt{2}}{2}$ **81.** $7\sqrt{2}$ **83.** $\frac{5\sqrt{7}}{98}$ **85.** $2\sqrt{2x}$ **87.** $\frac{2\sqrt{3}}{3}$ **89.** $-8\sqrt{2}+6\sqrt{5}$ **91.** $4-2\sqrt{3}$ **93.** $\frac{\sqrt{2}}{10}$ **95.** $11-6\sqrt{2}$ **97. (a)** 0.2679 **(b)** $2-\sqrt{3}$ **(c)** 0.2679 **(d)** They are the same. **99. (a)** 2.0291 **(b)** $\frac{10\sqrt{3}+5}{11}$ **(c)** 2.0291 **(d)** They are the same. **101. (a)** -1.7836 **(b)** $-\frac{4+6\sqrt{2}}{7}$ **(c)** -1.7836 **(d)** They are the same. **103. (a)** 0.2101 **(b)** $\frac{5\sqrt{2}-\sqrt{6}}{22}$ **(c)** 0.2101 **(d)** They are the same. **105.** $-2-\sqrt{3}$ **107.** $-2\sqrt{3}-3\sqrt{2}+\sqrt{6}+3$ **109.** $\frac{2x-7\sqrt{xy}-4y}{x-16y}$

Putting the Concepts Together

1. (a) irrational; $\sqrt{98} \approx 9.90$ **(b)** rational; $\sqrt{361} = 19$ **2.** $-6\sqrt{2}$ **3.** $4xy^3\sqrt{2x}$ **4.** $\frac{5}{2x}$ **5.** 7 **6.** $3m^3$ **7.** $3\sqrt{3}$ **8.** $3\sqrt{2}$ **9.** $6\sqrt{6}$ **10.** -42 **11.** $60\sqrt{2}+18\sqrt{5}$ **12.** $24-26\sqrt{3}$ **13.** -11 **14.** $11-4\sqrt{7}$ **15.** $-2+\sqrt{7}$ **16.** $\frac{3\sqrt{2}}{x^2}$ **17.** $\frac{2\sqrt{6}}{3}$ **18.** $-2\sqrt{3}+6$

Section 6.6

1. radical equation **3.** isolate **5.** False **7.** Incorrect; answers may vary. **9.** Yes **11.** No **13.** Yes **15.** No **17.** $\{16\}$ **19.** $\{-4\}$ **21.** $\{7\}$ **23.** $\{4\}$ **25.** $\{9\}$ **27.** $\emptyset$ **29.** $\{25\}$ **31.** $\{3, 6\}$ **33.** $\{1\}$ **35.** $\emptyset$ **37.** $\{9\}$ **39.** $\{6, 9\}$ **41.** $\{9\}$ **43.** $\{6\}$ **45.** $\{1\}$ **47.** $\{0\}$ **49.** $\{-5\}$ **51.** $\{5\}$ **53.** $\emptyset$ **55.** $\{6\}$ **57.** $\left\{\frac{13}{4}\right\}$ **59.** $\{23\}$ **61.** $\{20\}$ **63.** $\{-1\}$ **65.** $\{16\}$ **67.** $\{2\}$ **69.** $\left\{\frac{1}{4}\right\}$ **71.** $\{-4, 7\}$ **73.** $\{4\}$ **75.** $\emptyset$ **77.** $\emptyset$ **79.** 11 **81.** 3 **83. (a)** 8 seconds **(b)** $\frac{5\sqrt{3}}{2}$ seconds **85.** about 28 ft **87. (a)** about 42 ft **(b)** Yes. **89.** $A = \pi r^2$ **91.** $h = \frac{3V}{S^2}$ **93.** $a = \frac{b^2-b}{2}$ **95.** $n = \frac{1}{2}m$

Section 6.7

1. radicand, index **3.** even, negative **5.** True **7.** First answer incorrect; second answer correct; answers may vary **9.** 8^3 or 512 **11.** $\sqrt[3]{-1}$ or -1 **13.** $\sqrt[4]{16}$ or 2 **15.** $(-1)^4$ or 1 **17.** 3 **19.** not a real number **21.** -5 **23.** 5 **25.** 0 **27.** 3 **29.** x **31.** $2\sqrt[3]{2}$ **33.** not a real number **35.** $-2\sqrt[3]{5}$ **37.** 3 **39.** $-2n^2\sqrt[4]{2}$ **41.** $10\sqrt[4]{2}$ **43.** $4\sqrt[3]{2}$ **45.** $3x\sqrt[3]{3}$ **47.** $2b$ **49.** $c^{1/2}$ **51.** $x^{2/3}$ **53.** $p^{5/4}$ **55.** $\sqrt[3]{u^2} = (\sqrt[3]{u})^2$ **57.** $\sqrt{2a}$ **59.** $4\sqrt[3]{x^2} = 4(\sqrt[3]{x})^2$ **61.** -2 **63.** -2 **65.** 512 **67.** $\frac{1}{9}$ **69.** -6 **71.** 16 **73.** not a real number **75.** $-\frac{1}{27}$ **77.** $\frac{1}{343}$ **79.** not a real number **81.** 16 **83.** $\frac{1}{36}$ **85.** x^6 **87.** $4\sqrt[3]{n}$ **89.** $x^{1/4}$ or $\sqrt[4]{x}$ **91.** $\frac{1}{2x}$ **93.** -125 **95.** $\frac{1}{6}$ **97.** $-\frac{1}{25}$ **99.** 127 **101.** 5 **103.** $\frac{125}{8}$ **105.** $\frac{1}{y^2}$ **107.** $\frac{1}{4}$ **109.** 6 **111.** $\frac{1}{\sqrt[4]{x}}$ **113.** $-2\sqrt{a}$ **115.** $81x\sqrt[6]{x^5}$

Chapter 6 Review

1. $-2, 2$ **2.** $-9, 9$ **3.** -1 **4.** -5 **5.** 0.4 **6.** 0.2 **7.** $\frac{5}{4}$ **8.** 6 **9.** 4 **10.** 12 **11.** 21 **12.** 11 **13.** rational; -3 **14.** irrational; -3.46 **15.** irrational; 3.74 **16.** not a real number **17.** $|4x-9|$ **18.** $|16m-25|$ **19.** $2\sqrt{7}$ **20.** $3\sqrt{5}$ **21.** $10\sqrt{2}$ **22.** $5\sqrt{6}$ **23.** $-1+\sqrt{2}$ **24.** $\frac{-1+\sqrt{3}}{2}$ **25.** a^{18} **26.** x^8 **27.** $4x^5$ **28.** $7a^6$ **29.** $3n^4\sqrt{2n}$ **30.** $2y^{12}\sqrt{2y}$ **31.** $\frac{3}{x^4}$ **32.** $\frac{2}{x^2}$ **33.** $\frac{9y^2}{5}$ **34.** $\frac{6}{11n^4}$ **35.** $-2\sqrt{7}$ **36.** $3\sqrt{11}$ **37.** $\sqrt{x}$ **38.** $-\sqrt{n}$ **39.** $25a\sqrt{ab}$ **40.** $3x\sqrt{3xy}$ **41.** $5\sqrt{3}$ **42.** $8\sqrt{2}$ **43.** $4+\sqrt{5}$ **44.** $6-2\sqrt{3}$ **45.** $19n^2\sqrt{2n}$ **46.** $9a^2\sqrt{14a}$ **47.** $2\sqrt{2}+2\sqrt{3}$ **48.** $\frac{9}{2}\sqrt{6}+\frac{2}{3}\sqrt{3}$ **49.** $\frac{11}{20}\sqrt{3}$ **50.** $-\frac{8}{5}\sqrt{2}$ **51.** $4\sqrt{6}$ **52.** $4\sqrt{15}$ **53.** $7x^2\sqrt{6x}$ **54.** $6y^2\sqrt{5y}$ **55.** $-128\sqrt{10}$ **56.** -200 **57.** $8a$ **58.** $45n$ **59.** -24 **60.** -154 **61.** $3\sqrt{a}+3\sqrt{5a}$ **62.** $2x+2x\sqrt{2}$ **63.** $6+12\sqrt{3}+4\sqrt{6}+24\sqrt{2}$ **64.** $15+10\sqrt{6}-9\sqrt{3}-18\sqrt{2}$ **65.** $4\sqrt{y}+12-2y\sqrt{2}-6\sqrt{2y}$ **66.** $\sqrt{x}+9+3x\sqrt{3}+27\sqrt{3x}$ **67.** $11-6\sqrt{2}$ **68.** $21-8\sqrt{5}$ **69.** $11x+6x\sqrt{2}$

70. $13n + 4n\sqrt{3}$ **71.** $9 - 25x$ **72.** $4x - 16$ **73.** $3\sqrt{6}$ **74.** $3\sqrt{5}$ **75.** $\frac{\sqrt{6}}{2x}$ **76.** $\frac{x\sqrt{15}}{6}$ **77.** $\sqrt{5} + \sqrt{6}$ **78.** $\sqrt{6} - \sqrt{3}$ **79.** $2\sqrt{7}$ **80.** $5\sqrt{5}$ **81.** $\frac{\sqrt{6}}{4}$ **82.** $\frac{\sqrt{10}}{25}$ **83.** $-2 - 3\sqrt{5}; -41$ **84.** $-3 + 2\sqrt{7}; -19$ **85.** $4 + 2\sqrt{2}$ **86.** $\frac{6 - \sqrt{3}}{11}$ **87.** $2\sqrt{2} - 2$ **88.** $\sqrt{3} + 1$ **89.** Yes **90.** Yes **91.** $\{-13\}$ **92.** $\{13\}$ **93.** $\{2\}$ **94.** $\{9\}$ **95.** $\{3, 5\}$ **96.** $\{6\}$ **97.** $\left\{\frac{1}{4}\right\}$ **98.** $\{-3\}$ **99.** $\{3\}$ **100.** $\{2\}$ **101.** $\varnothing$ **102.** $\varnothing$ **103.** 18π square cm **104.** 6 in. **105.** -2 **106.** not a real number **107.** $2x\sqrt[4]{3}$ **108.** $5x\sqrt[3]{5}$ **109.** $2n\sqrt[3]{3n}$ **110.** $3z\sqrt[4]{2z^2}$ **111.** -2 **112.** not a real number **113.** $x^{3/4}$ **114.** $z^{1/3}$ **115.** $2\sqrt{x^3} = 2x\sqrt{x}$ **116.** $\sqrt[3]{3v}$ **117.** 9 **118.** 32 **119.** $-\frac{1}{64}$ **120.** $\frac{1}{8}$ **121.** 4 **122.** 36 **123.** $\frac{1}{x^{3/4}}$ **124.** $x^{1/5}$

Chapter 6 Test

1. $-6\sqrt{3}$ **2.** $5xy\sqrt{2y}$ **3.** $\frac{4x}{3}$ **4.** 7 **5.** $\sqrt{2}$ **6.** $17\sqrt{3}$ **7.** $15\sqrt{3}$ **8.** $12\sqrt{3} - 4\sqrt{2}$ **9.** $19 - 8\sqrt{3}$ **10.** $2x\sqrt{3x}$ **11.** $2 - 10\sqrt{2}$ **12.** $\frac{-1 + 3\sqrt{2}}{2}$ **13.** $\frac{6\sqrt{5}}{5}$ **14.** $-3\sqrt{2} + 6$ **15.** $\{5\}$ **16.** $\{5\}$ **17.** $\{7, -3\}$ **18.** $3x\sqrt[3]{4x^2}$ **19.** $\frac{1}{9}$ **20.** $3\sqrt{5} \approx 6.7$ feet

Chapter 7

Section 7.1

1. $\sqrt{p}; -\sqrt{p}$ **3.** two **5.** False **7.** Answers may vary. **9.** $\{-12, 12\}$ **11.** $\{0\}$ **13.** $\{-4\sqrt{2}, 4\sqrt{2}\}$ **15.** $\{-2\sqrt{3}, 2\sqrt{3}\}$ **17.** $\left\{-\frac{2}{3}, \frac{2}{3}\right\}$ **19.** $\left\{-\frac{2\sqrt{3}}{3}, \frac{2\sqrt{3}}{3}\right\}$ **21.** $\{-4\sqrt{2}, 4\sqrt{2}\}$ **23.** $\{-13, 13\}$ **25.** $\{-5\sqrt{2}, 5\sqrt{2}\}$ **27.** no real solution **29.** $\left\{-\frac{1}{3}, \frac{1}{3}\right\}$ **31.** $\{-8, 8\}$ **33.** $\{-6, 6\}$ **35.** $\left\{-\frac{5\sqrt{3}}{3}, \frac{5\sqrt{3}}{3}\right\}$ **37.** no real solution **39.** $\{-4\sqrt{2}, 4\sqrt{2}\}$ **41.** $\{-6, -2\}$ **43.** $\left\{\frac{2}{3}, \frac{4}{3}\right\}$ **45.** $\{-5 - \sqrt{6}, -5 + \sqrt{6}\}$ **47.** $\{6 - 2\sqrt{2}, 6 + 2\sqrt{2}\}$ **49.** $\left\{-\frac{5}{3}, 1\right\}$ **51.** $\left\{\frac{5 - \sqrt{15}}{2}, \frac{5 + \sqrt{15}}{2}\right\}$ **53.** $\{-3, 5\}$ **55.** $\left\{-\frac{1}{2}, -\frac{1}{4}\right\}$ **57.** $\left\{\frac{-2 - 3\sqrt{6}}{9}, \frac{-2 + 3\sqrt{6}}{9}\right\}$ **59.** no real solution **61.** $\{-9, 1\}$ **63.** $\{-3, 4\}$ **65.** $c = 10$ **67.** $c = 3\sqrt{2} \approx 4.24$ **69.** $a = 8\sqrt{2} \approx 11.31$ **71.** $\{4, 9\}$ **73.** $\left\{-1, \frac{1}{2}\right\}$ **75.** $\{-2\sqrt{2}, 2\sqrt{2}\}$ **77.** $\{-\sqrt{15}, \sqrt{15}\}$ **79.** $\left\{-\frac{\sqrt{15}}{5}, \frac{\sqrt{15}}{5}\right\}$ **81.** no real solution **83.** $\{-4, 3\}$ **85.** $\{-3\sqrt{3}, 3\sqrt{3}\}$ **87.** $\{5\}$ **89.** $\left\{\frac{-4 - \sqrt{10}}{2}, \frac{-4 + \sqrt{10}}{2}\right\}$ **91.** $\left\{-\frac{\sqrt{6}}{3}, \frac{\sqrt{6}}{3}\right\}$ **93.** $\left\{-\frac{3}{2}, 4\right\}$ **95.** $\{-2\sqrt{10}, 2\sqrt{10}\}$ **97.** no real solution **99.** $2\sqrt{10}$ **101.** **(a)** $2\sqrt{30}$ ft **(b)** 11.0 ft **103.** $20\sqrt{26}$ yd **105.** **(a)** 13.8 ft **(b)** 13 ft 9.6 in. **107.** **(a)** -2.08 or 0.08 **(b)** 8% **109.** $x^2 - 16 = 0$ **111.** $x^2 - 6 = 0$ **113.** $x^2 - 45 = 0$

Section 7.2

1. 25 **3.** $x + \frac{b}{2}$ **5.** False **7.** $\frac{b^2}{4a^2}$ **9.** $64; (x + 8)^2$ **11.** $36; (z - 6)^2$ **13.** $\frac{4}{9}; \left(k + \frac{2}{3}\right)^2$ **15.** $\frac{81}{4}; \left(y - \frac{9}{2}\right)^2$ **17.** $\frac{1}{4}; \left(t + \frac{1}{2}\right)^2$ **19.** $\frac{1}{16}; \left(s - \frac{1}{4}\right)^2$ **21.** $\{-3, 7\}$ **23.** $\{-2 - \sqrt{11}, -2 + \sqrt{11}\}$ **25.** $\left\{\frac{5 - \sqrt{65}}{2}, \frac{5 + \sqrt{65}}{2}\right\}$ **27.** $\{-3, 6\}$ **29.** $\{3, 5\}$ **31.** no real solution **33.** $\{-1, 4\}$ **35.** $\{4 - 2\sqrt{7}, 4 + 2\sqrt{7}\}$ **37.** $\left\{\frac{3 - 3\sqrt{13}}{2}, \frac{3 + 3\sqrt{13}}{2}\right\}$ **39.** $\{-5, -2\}$ **41.** $\left\{\frac{-7 - \sqrt{249}}{2}, \frac{-7 + \sqrt{249}}{2}\right\}$ **43.** no real solution **45.** $\left\{-1, \frac{1}{3}\right\}$ **47.** $\left\{-\frac{1}{2}, \frac{3}{2}\right\}$ **49.** $\left\{-3, \frac{1}{2}\right\}$ **51.** $\left\{\frac{-2 - \sqrt{10}}{3}, \frac{-2 + \sqrt{10}}{3}\right\}$ **53.** $\left\{\frac{-1 - \sqrt{6}}{2}, \frac{-1 + \sqrt{6}}{2}\right\}$ **55.** no real solution **57.** $\left\{\frac{3}{2}, \frac{5}{2}\right\}$ **59.** $\left\{\frac{3 - \sqrt{2}}{2}, \frac{3 + \sqrt{2}}{2}\right\}$ **61.** $\left\{\frac{6 - \sqrt{37}}{2}, \frac{6 + \sqrt{37}}{2}\right\}$ **63.** $\left\{\frac{9 - \sqrt{91}}{3}, \frac{9 + \sqrt{91}}{3}\right\}$ **65.** $\left\{\frac{3 - \sqrt{41}}{4}, \frac{3 + \sqrt{41}}{4}\right\}$ **67.** $\{-4, 6\}$ **69.** $\left\{\frac{1}{2}\right\}$ **71.** $\{-3 - 2\sqrt{3}, -3 + 2\sqrt{3}\}$ **73.** $\{-2\sqrt{5}, 2\sqrt{5}\}$ **75.** $\left\{\frac{-3 - \sqrt{15}}{3}, \frac{-3 + \sqrt{15}}{3}\right\}$ **77.** $\left\{\frac{-5 - \sqrt{97}}{2}, \frac{-5 + \sqrt{97}}{2}\right\}$ **79.** $\{-3, 3\}$ **81.** $\{-6, 2\}$ **83.** $\{-12, 2\}$ **85.** $\{-3\sqrt{2}, 3\sqrt{2}\}$ **87.** $\{-2\sqrt{3}, 2\sqrt{3}\}$ **89.** $\left\{\frac{-6 - 3\sqrt{10}}{2}, \frac{-6 + 3\sqrt{10}}{2}\right\}$ **91.** $\left\{\frac{2}{3}, 3\right\}$ **93.** $\{-1 - \sqrt{34}, -1 + \sqrt{34}\}$ **95.** -4 or 6 **97.** $6 - 3\sqrt{3}$ or $6 + 3\sqrt{3}$ **99.** $15\sqrt{2}$ **101.** 3 and 6 **103.** **(a)** 9 **(b)** $\{3, 6\}$ **(c)** $3 + 6 = 9$ **105.** **(a)** 2 **(b)** $\left\{1 - \frac{\sqrt{6}}{2}, 1 + \frac{\sqrt{6}}{2}\right\}$ **(c)** $1 - \frac{\sqrt{6}}{2} + 1 + \frac{\sqrt{6}}{2} = 2$ **107.** product $= -\frac{3}{4}; \left\{-\frac{3}{2}, \frac{1}{2}\right\}; \left(-\frac{3}{2}\right)\left(\frac{1}{2}\right) = -\frac{3}{4}$ **109.** product $= -7; \{1 - 2\sqrt{2}, 1 + 2\sqrt{2}\}; (1 - 2\sqrt{2})(1 + 2\sqrt{2}) = -7$ **111.** center $= (5, -3)$, radius $= 2$ **113.** center $= (10, 8)$, radius $= \sqrt{142}$

Section 7.3

1. standard **3. (a)** factoring **(b)** factoring **(c)** the quadratic formula **5.** False **7.** Answers may vary. **9.** $-2x^2 + x + 4 = 0; a = -2, b = 1, c = 4$ **11.** $x^2 + 0x + 4 = 0; a = 1, b = 0, c = 4$ **13.** $\frac{3}{2}x^2 - \frac{6}{5}x - 2 = 0; a = \frac{3}{2}, b = -\frac{6}{5}, c = -2$ or $15x^2 - 12x - 20 = 0; a = 15, b = -12, c = -20$ **15.** $0.5x^2 - x + 3 = 0; a = 0.5, b = -1, c = 3$ **17.** -1 or 3 **19.** 2 **21.** $-2 - \sqrt{6}$ or $-2 + \sqrt{6}$ **23.** $\frac{1 \pm \sqrt{-15}}{4}$ **25.** $\{-8, 3\}$ **27.** $\left\{-4, -\frac{3}{2}\right\}$ **29.** $\left\{\frac{1 - \sqrt{21}}{2}, \frac{1 + \sqrt{21}}{2}\right\}$ **31.** $\left\{\frac{-2 - \sqrt{7}}{3}, \frac{-2 + \sqrt{7}}{3}\right\}$ **33.** $\left\{\frac{1}{2}, \frac{3}{2}\right\}$ **35.** $\left\{\frac{-7 - 3\sqrt{5}}{2}, \frac{-7 + 3\sqrt{5}}{2}\right\}$ **37.** $\left\{\frac{5 - \sqrt{17}}{4}, \frac{5 + \sqrt{17}}{4}\right\}$ **39.** $\{3, 4\}$ **41.** $\left\{-\frac{2}{3}, 2\right\}$ **43.** $\{-3 - \sqrt{5}, -3 + \sqrt{5}\}$ **45.** $\left\{1 - \frac{\sqrt{2}}{2}, 1 + \frac{\sqrt{2}}{2}\right\}$ **47.** $\left\{\frac{-5 - \sqrt{33}}{4}, \frac{-5 + \sqrt{33}}{4}\right\}$ **49.** $\left\{\frac{1}{5}, \frac{1}{3}\right\}$ **51.** $\left\{-\frac{3}{2}\right\}$ **53.** $\{-6, 3\}$ **55.** $\left\{\frac{4 - \sqrt{10}}{3}, \frac{4 + \sqrt{10}}{3}\right\}$ **57.** $\{3\}$ **59.** $\left\{\frac{-1 - \sqrt{61}}{2}, \frac{-1 + \sqrt{61}}{2}\right\}$ **61.** no real solution **63.** $\{-2\sqrt{3}, 2\sqrt{3}\}$ **65.** $\left\{-\frac{3}{2}, 6\right\}$ **67.** $\{0, 5\}$ **69.** -39; no real solution **71.** 25; factoring **73.** 0; factoring **75.** 40; quadratic formula **77.** 0; factoring **79.** -56; no real solution **81.** $\left\{1 - \frac{\sqrt{3}}{3}, 1 + \frac{\sqrt{3}}{3}\right\}$ **83.** $\left\{-\frac{5}{3}, 1\right\}$ **85.** no real solution **87.** $\left\{-\frac{1}{2}\right\}$ **89.** $\{-5 - 2\sqrt{5}, -5 + 2\sqrt{5}\}$ **91.** $\{0, 3\}$ **93.** $\{-5, -2\}$ **95.** $\left\{-4, \frac{1}{2}\right\}$ **97.** no real solution **99.** $-3 - 3\sqrt{2}$ or $-3 + 3\sqrt{2}$ **101.** -1 or 5 **103. (a)** \$1515 **(b)** 996 or 2004 **105. (a)** 43 **(b)** 37 **(c)** 33 **(d)** 2029 **107.** $\{-4\sqrt{2}, \sqrt{2}\}$ **109.** $\left\{\frac{2\sqrt{3}}{3}, \sqrt{3}\right\}$

Putting the Concepts Together

1. (a) $\left\{\pm\frac{\sqrt{5}}{2}\right\}$ **(b)** $\{2 \pm 2\sqrt{3}\}$ **(c)** $\left\{\frac{-3 \pm \sqrt{3}}{2}\right\}$ **2. (a)** add 81; $z^2 - 18z + 81 = (z - 9)^2$ **(b)** add $\frac{81}{4}$; $y^2 + 9y + \frac{81}{4} = \left(y + \frac{9}{2}\right)^2$ **(c)** add $\frac{4}{25}$; $m^2 + \frac{4}{5}m + \frac{4}{25} = \left(m + \frac{2}{5}\right)^2$ **3. (a)** $\{-7, -5\}$ **(b)** no real solution **(c)** $\{4 \pm 3\sqrt{2}\}$ **4. (a)** $\{-4 \pm 2\sqrt{5}\}$ **(b)** $\left\{-\frac{3}{2}, 4\right\}$ **(c)** $\left\{-\frac{9}{2}\right\}$ **5.** $\{-2, 3\}$ **6.** $\left\{\frac{-1 \pm \sqrt{11}}{2}\right\}$ **7.** $\left\{-1 \pm \frac{1}{2}\sqrt{3}\right\}$ or $\left\{\frac{-2 \pm \sqrt{3}}{2}\right\}$ **8. (a)** discriminant $= 64$ **(b)** factoring **(c)** Answers may vary. **9. (a)** $5\sqrt{15}$ feet **(b)** 19.4 feet **10. (a)** \$12,000 revenue from selling 100 calculus texts **(b)** 60 or 160 texts must be sold for the revenue to be \$9600.

Section 7.4

1. $a^2 + b^2 = c^2$ **3.** quadratic formula **5.** True **7.** reaches height twice, once on way up and once on way down **9.** $-17, -16$ and $16, 17$ **11.** 14, 15 **13.** $-1 - 2\sqrt{6}$ or $-1 + 2\sqrt{6}$ **15.** $5 - 2\sqrt{3}$ or $5 + 2\sqrt{3}$ **17.** width $= 6$ m, length $= 15$ m **19.** width $= 3$ in.; length $= 12$ in. **21.** height $= 5$ cm; base $= 8$ cm **23.** height $= 11$ ft; base $= 12$ ft **25.** Longer leg is exactly $\frac{3 + 3\sqrt{3}}{2}$ inches; shorter leg is exactly $\frac{-3 + 3\sqrt{3}}{2}$ inches. Approximate lengths are longer leg 4.1 inches and shorter leg 1.1 inches. **27.** leg $= 3\sqrt{3} \approx 5.2$ m; hypotenuse $= 6\sqrt{3} \approx 10.4$ m **29. (a)** 41 ft **(b)** 116 ft **(c)** 1.52 s and 4.10 s **(d)** 5.625 s **31. (a)** 43 fruit flies **(b)** 75 fruit flies **(c)** 2.18 hours **33. (a)** 2 **(b)** 20 **(c)** 15 **35. (a)** $(180 - x)°$ **(b)** 30° or 125° **37. (a)** \$146.40 **(b)** \$992.40 **(c)** approximately 3 months **(d)** approximately 13 months **39.** leg $= 1$, hypotenuse $= \sqrt{2}$ **41.** leg $= 3$, hypotenuse $= 3\sqrt{2}$ **43.** leg $= \sqrt{6}$, hypotenuse $= 2\sqrt{3}$ **45.** leg $= 6\sqrt{2}$, leg $= 6\sqrt{2}$ **47.** leg $= 2$, leg $= 2$

Section 7.5

1. $\sqrt{-1}$ **3.** complex conjugates, real **5.** True **7.** complex solutions; quadratic formula **9.** $25i$ **11.** $-2\sqrt{13}i$ **13.** $5 + 2i$ **15.** $12 - 3\sqrt{3}i$ **17.** $-\frac{3}{2} + \frac{\sqrt{2}}{2}i$ **19.** $-\frac{1}{2} + \sqrt{3}i$ **21.** $1 + 3i$ **23.** -10 **25.** $-3 + 21i$ **27.** $13 - 10i$ **29.** $-8 - i$ **31.** $12 + \sqrt{2}i$ **33.** $-5\sqrt{10}i$ **35.** $15 + 12i$ **37.** $-10 - 10i$ **39.** $-3 - 16i$ **41.** -256 **43.** $24 - 42i$ **45.** $7 - 24i$ **47.** -20 **49.** $-6\sqrt{2}$ **51.** $1 - 31i$ **53.** $-3 - 2i$; 13 **55.** $6 - 3\sqrt{2}i$; 54 **57.** $4 + 8i$ **59.** $3 + 2i$ **61.** $\frac{3}{2} + \frac{3}{2}i$ **63.** $-1 + \frac{1}{2}i$ **65.** $-\frac{27}{4} + 2i$ **67.** $-\frac{3}{13} + \frac{2}{13}i$ **69.** $-6 + \sqrt{3}i$ **71.** $-5 + 12i$ **73.** $23 - 10i$ **75.** $-\frac{5}{2} + \frac{5}{2}i$ **77.** $\frac{2}{3} + \frac{5}{2}i$ **79.** $16\sqrt{3} - 6 + (-6\sqrt{3} - 16)i$ **81.** $\left\{1 - \frac{\sqrt{6}}{2}i, 1 + \frac{\sqrt{6}}{2}i\right\}$ **83.** $\{-1 - 2\sqrt{3}i, -1 + 2\sqrt{3}i\}$ **85.** $\{5 - 2i, 5 + 2i\}$ **87.** $\{-2 - \sqrt{5}i, -2 + \sqrt{5}i\}$ **89.** $\left\{-\frac{3}{2}, 8\right\}$ **91.** $\{-2\sqrt{2}, 2\sqrt{2}\}$ **93.** $\{1 - \sqrt{2}i, 1 + \sqrt{2}i\}$ **95.** $\left\{\frac{1 - \sqrt{105}}{4}, \frac{1 + \sqrt{105}}{4}\right\}$ **97.** $\frac{1}{3} - \frac{\sqrt{14}}{3}i$ or $\frac{1}{3} + \frac{\sqrt{14}}{3}i$ **99.** $1 - 3i$ or $1 + 3i$ **101.** $5 - \sqrt{15}i$ and $5 + \sqrt{15}i$

Chapter 7 Review

1. $\{-5, 5\}$ **2.** $\{-12, 12\}$ **3.** $\left\{-\frac{3}{2}, \frac{3}{2}\right\}$ **4.** $\left\{-\frac{2}{5}, \frac{2}{5}\right\}$ **5.** $\{-2\sqrt{3}, 2\sqrt{3}\}$ **6.** $\{-3\sqrt{2}, 3\sqrt{2}\}$ **7.** $\left\{-\frac{5}{3}, 1\right\}$ **8.** $\{-1, 6\}$ **9.** $\{-5 - \sqrt{2}, -5 + \sqrt{2}\}$ **10.** $\{2 - \sqrt{2}, 2 + \sqrt{2}\}$ **11.** no real solution **12.** no real solution **13.** $\{2 - 2\sqrt{6}, 2 + 2\sqrt{6}\}$ **14.** $\{-4 - 2\sqrt{7}, -4 + 2\sqrt{7}\}$ **15.** $c = 4\sqrt{13} \approx 14.42$ **16.** $c = 6\sqrt{5} \approx 13.42$ **17.** $a = 2\sqrt{33} \approx 11.49$ **18.** $b = 8\sqrt{2} \approx 11.31$ **19. (a)** $3\sqrt{2}$ ft by $3\sqrt{2}$ ft **(b)** 4.24 ft by 4.24 ft **20. (a)** $10\sqrt{2}$ ft **(b)** 14.1 ft **21.** 36; $(x - 6)^2$ **22.** 25; $(x + 5)^2$ **23.** $\frac{1}{9}; \left(a + \frac{1}{3}\right)^2$ **24.** $\frac{4}{25}; \left(k - \frac{2}{5}\right)^2$ **25.** $\frac{1}{16}; \left(n - \frac{1}{4}\right)^2$ **26.** $\frac{1}{36}; \left(t + \frac{1}{6}\right)^2$ **27.** $\{-1, 7\}$ **28.** $\{1, 11\}$ **29.** $\{-4, 1\}$ **30.** $\{-7, -2\}$ **31.** $\{2 - 2\sqrt{5}, 2 + 2\sqrt{5}\}$ **32.** $\{1 - 2\sqrt{2}, 1 + 2\sqrt{2}\}$ **33.** $\left\{-\frac{5}{2}, 2\right\}$ **34.** $\left\{-3, -\frac{2}{3}\right\}$ **35.** $\left\{\frac{3 - \sqrt{53}}{2}, \frac{3 + \sqrt{53}}{2}\right\}$ **36.** $\left\{\frac{-1 - \sqrt{33}}{2}, \frac{-1 + \sqrt{33}}{2}\right\}$ **37.** $\{5 - 3\sqrt{5}, 5 + 3\sqrt{5}\}$ **38.** $\{4 - 2\sqrt{6}, 4 + 2\sqrt{6}\}$ **39.** $-\frac{5}{2}$ or $\frac{1}{2}$ **40.** -8 or 2 **41.** $2x^2 - x + 3 = 0; a = 2, b = -1, c = 3$ **42.** $\frac{1}{2}x^2 + x - \frac{3}{4}; a = \frac{1}{2}, b = 1, c = -\frac{3}{4}$ or $2x^2 + 4x - 3 = 0; a = 2, b = 4, c = -3$ **43.** -1 or $-\frac{1}{2}$ **44.** $-\frac{2}{3}$ or 1 **45.** $\left\{\frac{3}{2}, 5\right\}$ **46.** $\left\{-\frac{3}{2}, -\frac{1}{2}\right\}$ **47.** $\{6 - \sqrt{31}, 6 + \sqrt{31}\}$ **48.** $\left\{\frac{3 - 3\sqrt{5}}{2}, \frac{3 + 3\sqrt{5}}{2}\right\}$ **49.** $\left\{-\frac{8}{3}, 1\right\}$ **50.** $\left\{-4, \frac{1}{2}\right\}$ **51.** $\left\{\frac{-1 - \sqrt{7}}{2}, \frac{-1 + \sqrt{7}}{2}\right\}$ **52.** $\{-2 - \sqrt{6}, -2 + \sqrt{6}\}$ **53.** $\left\{\frac{3 - 2\sqrt{2}}{2}, \frac{3 + 2\sqrt{2}}{2}\right\}$ **54.** $\left\{\frac{4 + \sqrt{10}}{3}, \frac{4 - \sqrt{10}}{3}\right\}$ **55. (a)** \$8250 **(b)** 100 or 160 texts **56. (a)** \$244 **(b)** 15 dogs **57.** $b^2 - 4ac = 24$; quadratic formula; $\left\{\frac{4 - \sqrt{6}}{5}, \frac{4 + \sqrt{6}}{5}\right\}$ **58.** $b^2 - 4ac = 49$; factoring; $\{-1, 6\}$ **59.** $b^2 - 4ac = 169$; factoring; $\left\{-\frac{10}{3}, 1\right\}$ **60.** $b^2 - 4ac = 17$; quadratic formula; $\left\{\frac{-3 - \sqrt{17}}{2}, \frac{-3 + \sqrt{17}}{2}\right\}$ **61.** $b^2 - 4ac = -23$; no real solution **62.** $b^2 - 4ac = -11$; no real solution **63.** $-18, -16$ and $16, 18$ **64.** $\frac{3}{2}; \frac{2}{3}$ **65.** width = 8 m, length = 14 m **66.** width = 5 yd, length = 12 yd **67.** 4 ft, 8 ft **68.** leg $= \frac{-6 + 2\sqrt{21}}{3}$ cm, hypotenuse $= \frac{-3 + 4\sqrt{21}}{3}$ cm **69.** base = 12 m, height = 7 m **70.** base = 3 in., height = $\frac{3}{2}$ in. **71.** length = 10 in., width = 4 in. **72.** length = 15 ft, width = 8 ft **73.** $\frac{5 + \sqrt{26}}{2} \approx 5.0$ s **74.** $\frac{15 - \sqrt{145}}{4} \approx 0.7$ s and $\frac{15 + \sqrt{145}}{4} \approx 6.8$ s **75.** 5 or 25 units **76.** $-5(-5 + \sqrt{11}) \approx 8$ units or $5(5 + \sqrt{11}) \approx 42$ units **77.** $-10i$ **78.** $-7i$ **79.** $5 + 2\sqrt{5}i$ **80.** $4 - 2\sqrt{15}i$ **81.** $-1 - \sqrt{3}i$ **82.** $-\frac{1}{2} + \frac{\sqrt{3}}{2}i$ **83.** $1 + 8i$ **84.** $11i$ **85.** $-16 + 10i$ **86.** $-26 + 13i$ **87.** $-10 + 7i$ **88.** $14 - 4i$ **89.** $-10 + 12i$ **90.** $-3 - 3i$ **91.** $2 - 23i$ **92.** $27 - 34i$ **93.** $-5 - 12i$ **94.** $-16 + 30i$ **95.** $-36\sqrt{2}$ **96.** $-48\sqrt{5}$ **97.** $5 - 2i$; 29 **98.** $3 + 3i$; 18 **99.** $-\frac{1}{3} - i$ **100.** $2 + 3i$ **101.** $\frac{3}{5} + \frac{4}{5}i$ **102.** $\frac{6}{5} - \frac{3}{5}i$ **103.** $-\frac{5}{2} + \frac{5}{2}i$ **104.** $\frac{5}{17} + \frac{3}{17}i$ **105.** $\left\{-\frac{1}{2} - \frac{3}{2}i, -\frac{1}{2} + \frac{3}{2}i\right\}$ **106.** $\left\{\frac{3}{2} - \frac{3}{2}i, \frac{3}{2} + \frac{3}{2}i\right\}$ **107.** $\{2 - 2\sqrt{2}i, 2 + 2\sqrt{2}i\}$ **108.** $\{-3 - 4\sqrt{2}i, -3 + 4\sqrt{2}i\}$

Chapter 7 Test

1. $\{-3, 3\}$ **2.** $\{-10, 4\}$ **3.** $\{-3 - 2\sqrt{2}, -3 + 2\sqrt{2}\}$ **4.** $\{-2, 3\}$ **5.** $\left\{\frac{2 - \sqrt{10}}{2}, \frac{2 + \sqrt{10}}{2}\right\}$ **6.** $\left\{\frac{3 - \sqrt{13}}{4}, \frac{3 + \sqrt{13}}{4}\right\}$ **7. (a)** 61 **(b)** quadratic formula **(c)** Answers may vary. **8.** $-\frac{1}{2} - \frac{3}{2}i$ **9.** $-14 - 4i$ **10.** $1 + 3i$ **11.** -36 **12.** $-5 - 14i$ **13.** $-20 - 48i$ **14.** $4 - \frac{4}{3}i$ **15.** $\frac{15}{17} - \frac{9}{17}i$ **16.** $\{2 - i, 2 + i\}$ **17.** $\left\{-2 - \frac{\sqrt{14}}{2}i, -2 + \frac{\sqrt{14}}{2}i\right\}$ **18. (a)** width $= -1 + \sqrt{127}$ ft, length $= 1 + \sqrt{127}$ ft **(b)** width ≈ 10.3 ft, length ≈ 12.3 ft **19. (a)** $\frac{25 + \sqrt{785}}{8}$ s **(b)** 6.6 s **20. (a)** width $= 25 + 25\sqrt{161}$ yards, length $= -50 + 50\sqrt{161}$ yards **(b)** width ≈ 342.21 yards, length ≈ 584.43 yards

Cumulative Review

1. undefined **2.** 0 **3.** $-\frac{1}{2}$ **4.** $\frac{27x^{12}}{y^2}$ **5.** $\frac{9}{x^2y^6}$ **6. (a)** -1 **(b)** -2 **(c)** 2 **(d)** trinomial **7. (a)** $y = \frac{3}{2}x - 6$ **(b)** $\frac{3}{2}$ **(c)** -6 **8.** $\{-14\}$ **9.** $\{-3\}$ **10.** $\left\{-2, \frac{3}{2}\right\}$ **11.** $\{-1, 3\}$ **12.** $(2h - 5)(3h + 2)$ **13.** $3(x + 2)(x^2 - 2x + 4)$ **14.** $(y + 2)(x - 3)$ **15.** $3x^2 + x - 1 + \frac{4}{2x + 1}$ **16.** $\frac{3}{2}x + 3 - \frac{2}{x}$ **17.** -4 **18.** $\frac{1}{27}$ **19.** 3.4×10^{10} **20.** 0.000304 **21.** $5 - 12i$ **22.** $-16\sqrt{6}$ **23.** $\frac{1}{2} - \frac{2}{3}i$ **24.** There were 390 general seating tickets and 190 reserved tickets sold. **25. (a)** 10 feet **(b)** $\frac{9 + \sqrt{129}}{4}$ s **(c)** 5.09 s

Chapter 8

Section 8.1

1. x-axis, y-axis, origin **3.** solution **5.** True **7.** Answers may vary.

9.

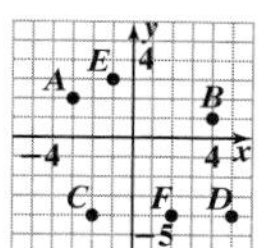

Quadrant I: B
Quadrant II: A, E
Quadrant III: C
Quadrant IV: D, F

11.

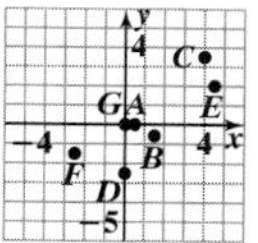

Quadrant I: C, E
Quadrant III: F
Quadrant IV: B
x-axis: A, G
y-axis: D, G

13.

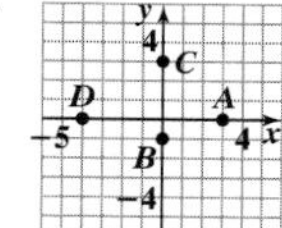

Positive x-axis: A
Negative x-axis: D
Positive y-axis: C
Negative y-axis: B

15. $A(4, 0)$; positive x-axis $B(-3, 2)$; quadrant II $C(1, -4)$; quadrant IV $D(-2, -4)$; quadrant III $E(3, 5)$; quadrant I $F(0, -3)$; negative y-axis

17. A No, B Yes, C Yes **19.** A Yes, B No, C Yes **21.** A Yes, B No, C Yes

23. $(4, 1)$ **25.** $(5, -1)$ **27.** $(-3, 3)$

29.

x	y	(x, y)
−3	3	$(-3, 3)$
0	0	$(0, 0)$
1	−1	$(1, -1)$

31.

x	y	(x, y)
−2	7	$(-2, 7)$
−1	4	$(-1, 4)$
4	−11	$(4, -11)$

33.

x	y	(x, y)
−1	8	$(-1, 8)$
2	2	$(2, 2)$
3	0	$(3, 0)$

35.

x	y	(x, y)
−4	6	$(-4, 6)$
1	6	$(1, 6)$
12	6	$(12, 6)$

37.

x	y	(x, y)
1	$\frac{7}{2}$	$\left(1, \frac{7}{2}\right)$
−4	1	$(-4, 1)$
−2	2	$(-2, 2)$

39.

x	y	(x, y)
4	7	$(4, 7)$
−4	3	$(-4, 3)$
−6	2	$(-6, 2)$

41.

x	y	(x, y)
0	−3	$(0, -3)$
−2	0	$(-2, 0)$
2	−6	$(2, -6)$

43. $A(2, -16)$ $B(-3, -1)$ $C\left(-\frac{1}{3}, -9\right)$ **45.** $A(2, -6)$ $B(0, 0)$ $C\left(\frac{1}{6}, -\frac{1}{2}\right)$

47. $A(4, -8)$ $B(4, -19)$ $C(4, 5)$ **49.** $A(3, 4)$ $B(-6, -2)$ $C\left(\frac{1}{2}, \frac{7}{3}\right)$ **51.** $A\left(-4, -\frac{4}{3}\right)$ $B(-2, -1)$ $C\left(-\frac{2}{3}, -\frac{7}{9}\right)$ **53.** $A(20, 23)$ $B(-4, -17)$ $C(2.6, -6)$

55. (a) \$24.85 **(b)** \$54.70 **(c)** 6 CDs **(d)** It costs \$34.80 to order 3 CDs. **57. (a)** \$1.15 **(b)** \$1.23 **(c)** \$1.60 **(d)** 1996 **(e)** Answers may vary. **59.** $k = 4$ **61.** $k = 2$ **63.** $k = \frac{1}{2}$

65.

a	b	(a, b)
2	−8	$(2, -8)$
0	−4	$(0, -4)$
−5	6	$(-5, 6)$

67.

p	q	(p, q)
0	$\frac{10}{3}$	$\left(0, \frac{10}{3}\right)$
$\frac{5}{2}$	0	$\left(\frac{5}{2}, 0\right)$
−10	$\frac{50}{3}$	$\left(-10, \frac{50}{3}\right)$

69. Points may vary; line

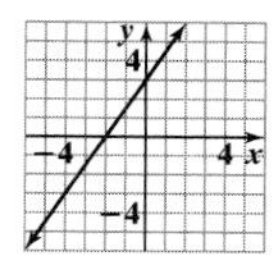

71.

x	y	(x, y)
−2	0	$(-2, 0)$
−1	−3	$(-1, -3)$
0	−4	$(0, -4)$
1	−3	$(1, -3)$
2	0	$(2, 0)$

73.

x	y	(x, y)
−2	10	$(-2, 10)$
−1	3	$(-1, 3)$
0	2	$(0, 2)$
1	1	$(1, 1)$
2	−6	$(2, -6)$

75.

X	Y1
-3	-15
-2	-13
-1	-11
0	-9
1	-7
2	-5
3	-3

Y1=2X−9

77.

X	Y1
-3	11
-2	10
-1	9
0	8
1	7
2	6
3	5

Y1=−X+8

79.

X	Y1
-3	19
-2	17
-1	15
0	13
1	11
2	9
3	7

Y1=13−2X

81.

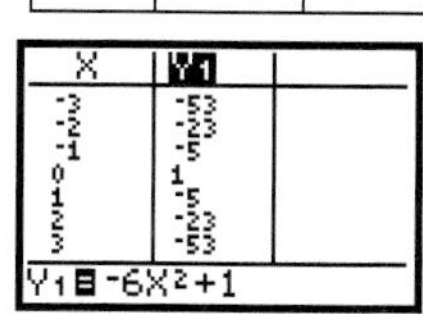

X	Y1
-3	-53
-2	-23
-1	-5
0	1
1	-5
2	-23
3	-53

Y1=−6X²+1

Section 8.2

1. linear, standard form **3.** intercepts **5.** False **7.** Two; three is recommended for a check. **9.** linear **11.** nonlinear **13.** nonlinear **15.** linear **17.** $y = 2x$

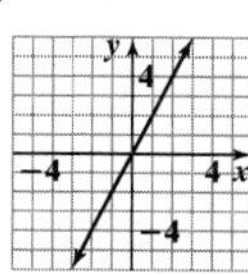

19. $y = -5x$

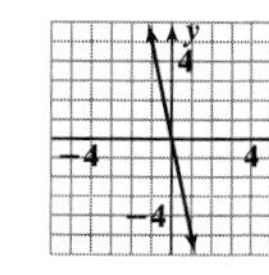

21. $y = 4x - 2$ **23.** $y = -2x + 5$

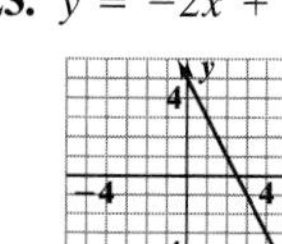

25. $x + y = 5$ **27.** $-2x + y = 6$

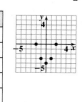

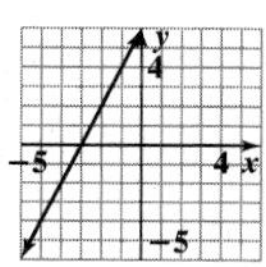

29. $4x - 2y = -8$

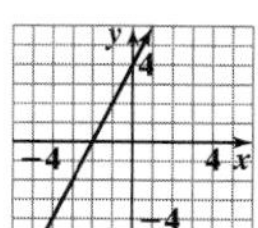

31. $x = -4y$ **33.** $y + 7 = 0$

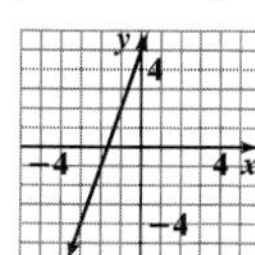

35. $y - 2 = 3(x + 1)$

37. $(0, -5), (5, 0)$ **39.** $(0, -3), (6, 0)$ **41.** $(0, -3)$ **43.** $(-5, 0)$ **45.** $(0, -4), (-6, 0)$ **47.** $(0, 0)$ **49.** $(0, -5), (5, 0)$ **51.** $(0, 8), (6, 0)$ **53.** $(4, 0)$ **55.** $(0, -2)$

57. $3x + 6y = 18$

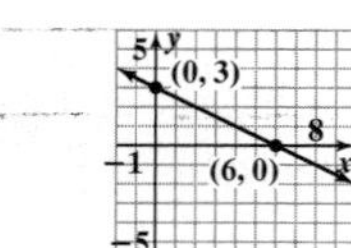

59. $-x + 5y = 15$

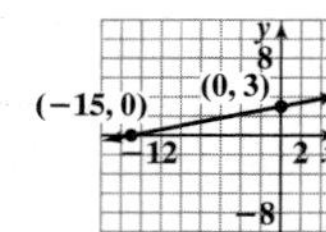

61. $\frac{1}{2}x = y + 3$

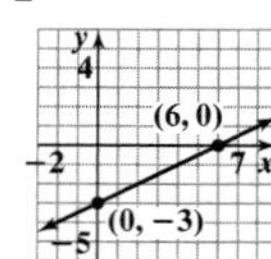

63. $9x - 2y = 0$

65. $y = -\frac{1}{2}x + 3$

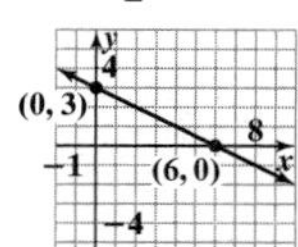

67. $\frac{1}{3}y + 2 = 2x$

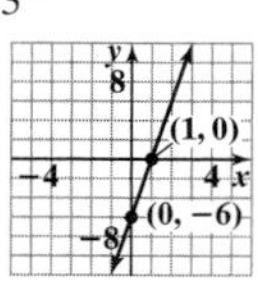

69. $\frac{x}{2} + \frac{y}{3} = 1$

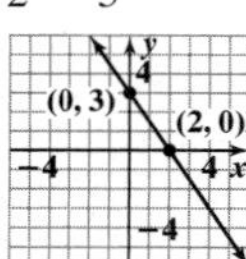

71. $4y - 2x + 1 = 0$

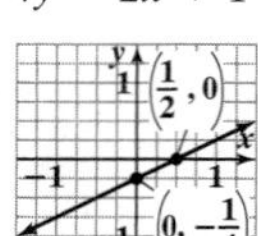

73. $x = 5$

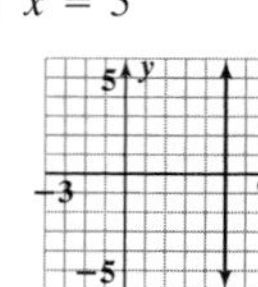

75. $y = -6$

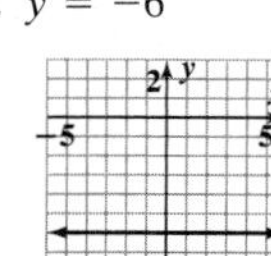

77. $y - 12 = 0$

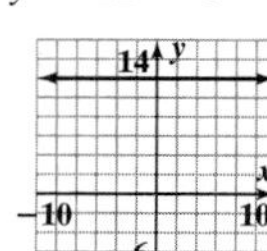

79. $3x - 5 = 0$

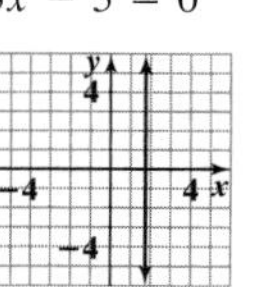

81. $y = 2x - 5$

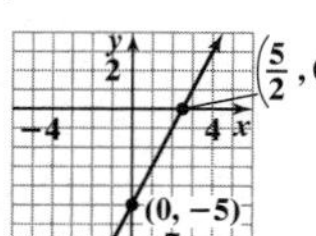

83. $y = -5$

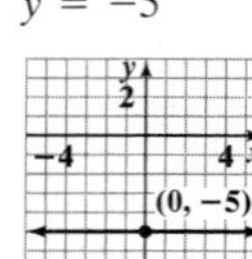

85. $2x + 5y = -20$

87. $2x = -6y + 4$

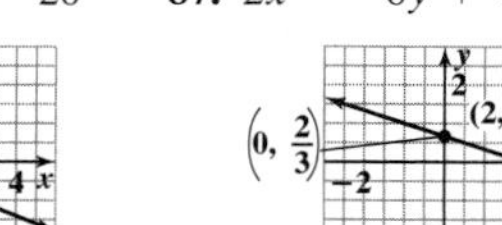

89. $x - 3 = 0$

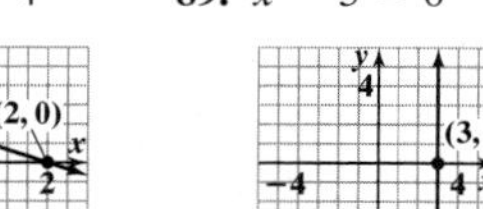

91. $3y - 12 = 0$

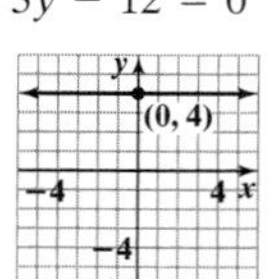

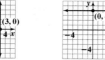

93. $y = 5$

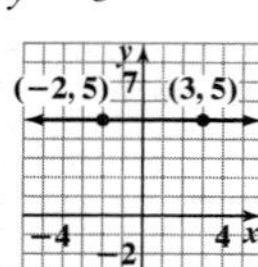

95. $x = -2$

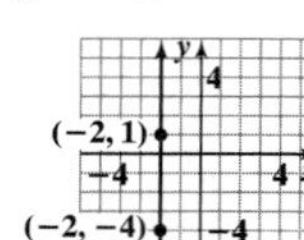

97. $y = 4$
99. $x = -9$
101. $x = 2y$
103. $y = x + 2$
105. $y = 2$
107. $x = 7$

109. **(a)** (0, 500), (4, 900), (10, 1500)
(b)

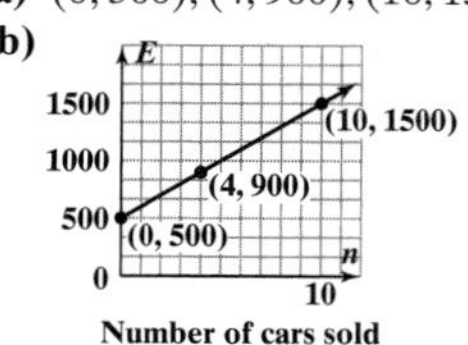

(c) If she sells 0 cars, her earnings are \$500.

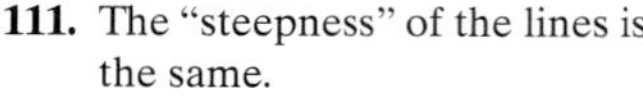

111. The "steepness" of the lines is the same.

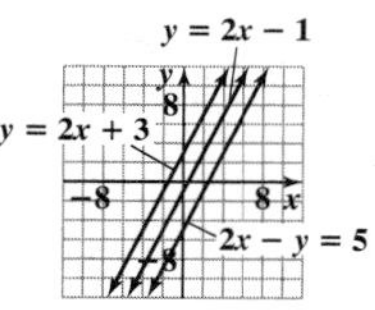

113. The lines get more steep as the coefficient of x gets larger.

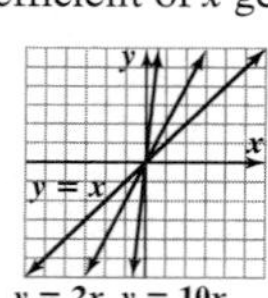

115. (0, −6), (−2, 0), (3, 0)
117. (0, 14), (−3, 0), (2, 0), (5, 0)

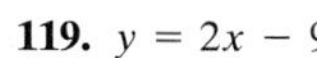

119. $y = 2x - 9$

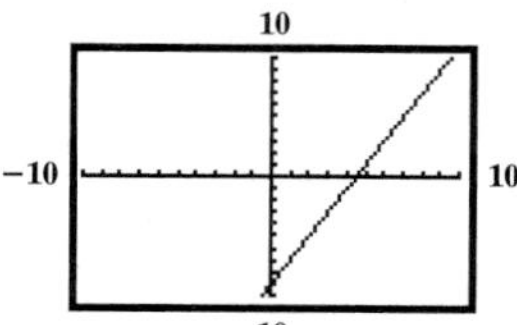

121. $y = -x + 8$

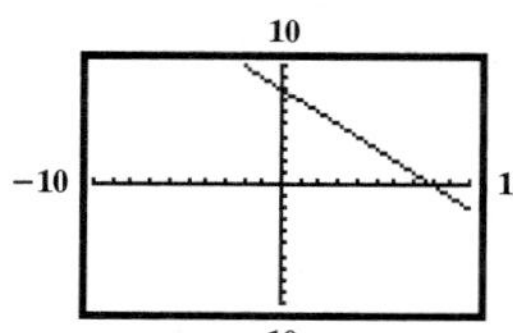

123. $y = 2x + 13$ or $y = -2x + 13$

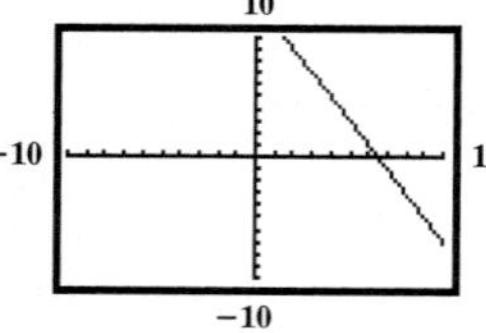

125. $y = -6x^2 + 1$

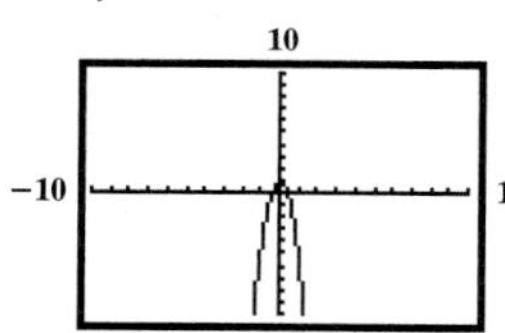

Section 8.3

1. $\frac{6}{10}$ or $\frac{3}{5}$ **3.** positive **5.** True **7.** vertical line; Answers may vary **9.** $-\frac{3}{2}$ **11.** undefined **13.** $\frac{1}{2}$ **15.** $-\frac{2}{3}$

17. **(a), (b)**

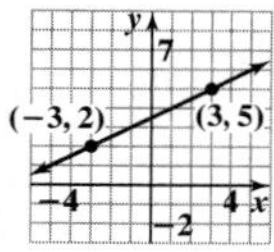

(c) $m = \frac{1}{2}$; for every 2-unit increase in x, there is a 1-unit increase in y.

19. **(a), (b)**

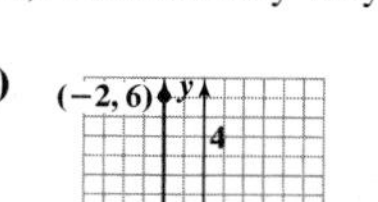

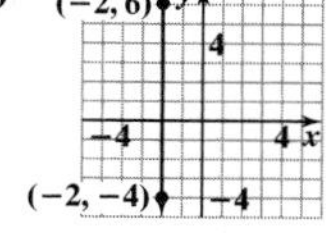

(c) $m = 0$; the line is horizontal

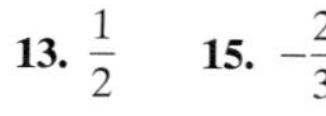

21. $m = -2$; for every 1-unit increase in x, there is a 2-unit decrease in y.

23. $m = -1$; for every 1-unit increase in x, there is a 1-unit decrease in y.

25. $m = -\frac{5}{3}$; for every 3-unit increase in x, there is a 5-unit decrease in y.

27. $m = \frac{3}{2}$; for every 2-unit increase in x, there is a 3-unit increase in y.

29. $m = 0$; there is no change in the y-values. The line is horizontal. **31.** $m = \frac{2}{3}$; for every 3-unit increase in x, there is a 2-unit increase in y.

33. m is undefined; the line is vertical. **35.** $m = \frac{1}{3}$; for every 3-unit increase in x, there is a 1-unit increase in y.

37. $m = \frac{1}{3}$; for every 3-unit increase in x, there is a 1-unit increase in y. **39.** $m = 2$; for every 1-unit increase in x, there is a 2-unit increase in y.

41. 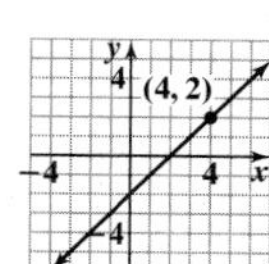

43. 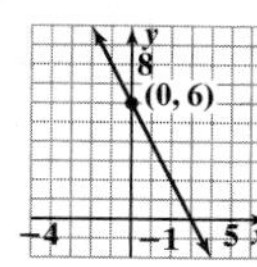

45. 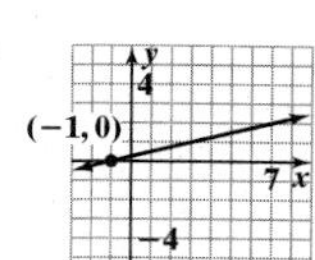

47. 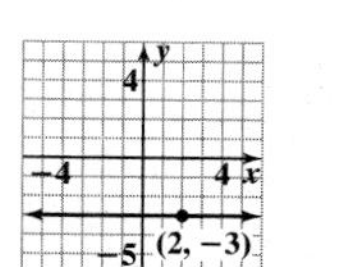

49. 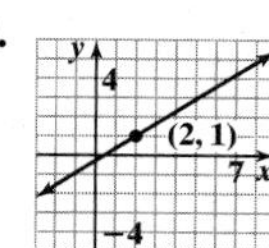

51. 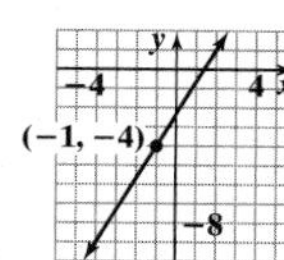

53. 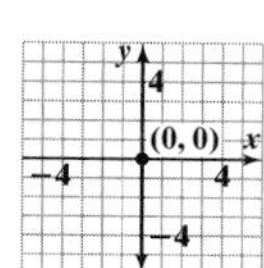

55. 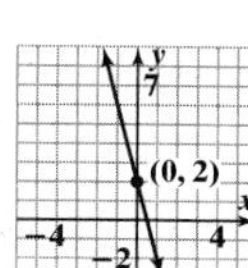

57. 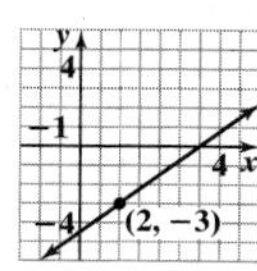

59. $x = 2$ **61.** $y = -2$ **63.** $x = -2$ **65.** $x = 3$ **67.** $x = 0$ **69.** $y = \frac{4}{3}$
71.
73. 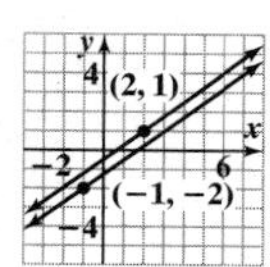

75. $\frac{1}{3}$ **77.** 12 in. or 1 ft **79.** 16% **81.** $m = 2.26$ million; the population is increasing at an average rate of about 2.26 million people per year.

83. Points may vary. $(-2, 1)$, $(0, -5)$ $m = -3$

85. Points may vary. $(4, -10)$, $(-2, -10)$ $m = 0$

87. Points may vary. $(-2, -2)$, $(0, 4)$ $m = 3$
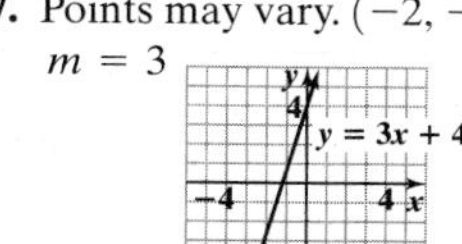

89. $m = -2$ **91.** $m = \frac{q}{p}$ **93.** $m = \frac{6}{a - 6}$ **95.** $MR = 2$; For every hot dog sold, revenue increases by \$2, on average.

Section 8.4

1. 3; 7 **3.** plotting points, using intercepts, using the slope and a point **5.** False **7. (a)** (a), (e)

9. $m = 5; b = 2$ **11.** $m = 2; b = -9$ **13.** $m = -10; b = 7$ **15.** $m = -1; b = -9$ **17.** $m = 0, b = -5$ **19.** $m = \frac{2}{3}; b = -4$

21. $m = 2; b = -4$ **23.** $m = -\frac{2}{3}; b = 8$ **25.** $m = \frac{5}{3}; b = -\frac{1}{3}$ **27.** $m = \frac{1}{2}; b = -\frac{5}{2}$ **29.** m is undefined; no y-intercept

31. 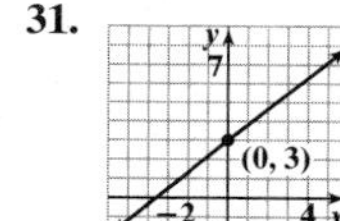

33. 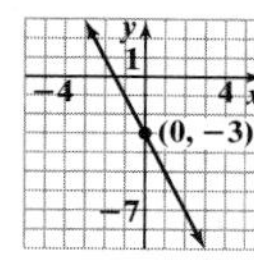

35. 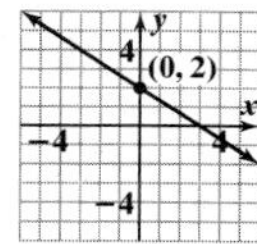

37. 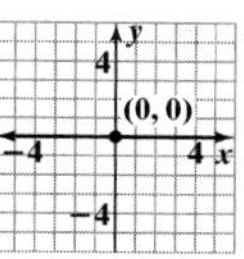

39. 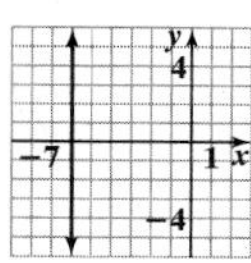
41. 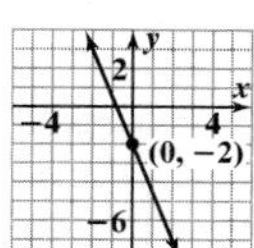

43. 
45. 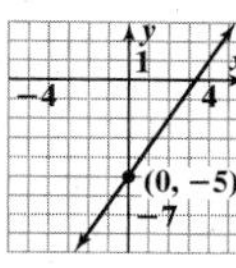

47. 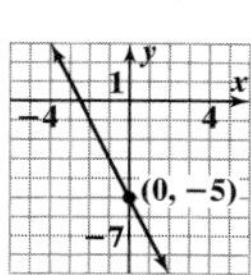

49. 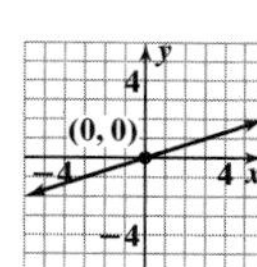

51. 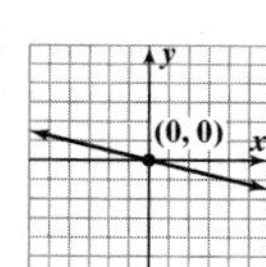

53. 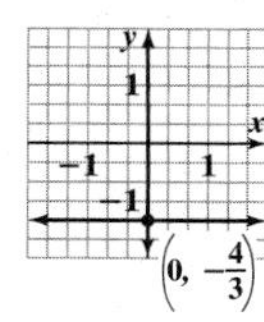

55.
57. 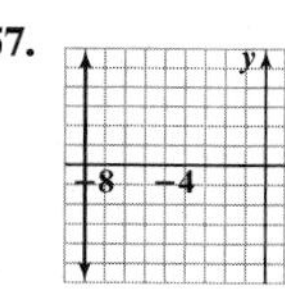
59. 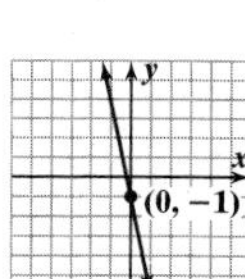

61. $y = -x + 8$ **63.** $y = \frac{6}{7}x - 6$ **65.** $y = -\frac{1}{3}x + \frac{2}{3}$ **67.** $x = -5$ **69.** $y = 3$ **71.** $y = 5x$

73. 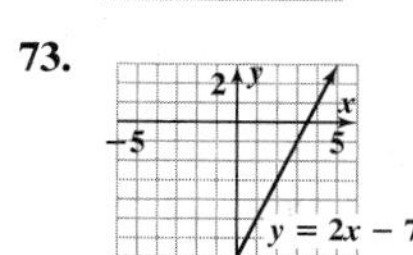

75. 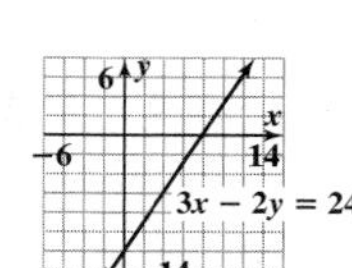

77. 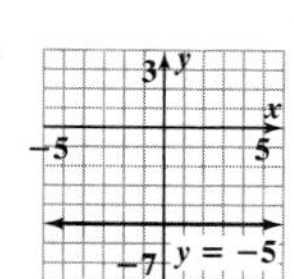

79. 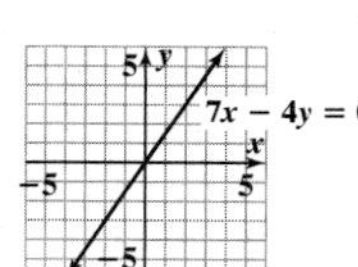

81. 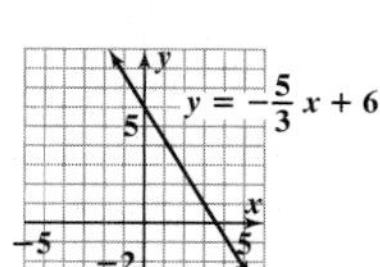

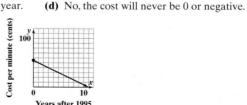

83.
85. (a) $y = 0.08x + 400$ **(b)** \$496 **(c)**
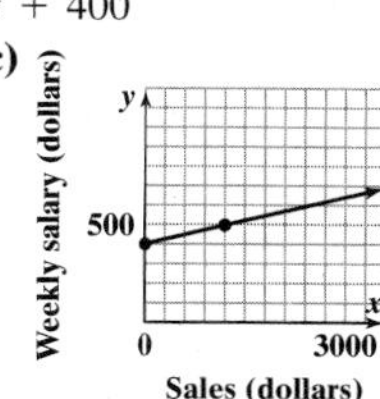

87. (a) 34.5¢ **(b)** 2001 **(c)** The cost per minute is decreasing by 5.375¢ per year. **(d)** No, the cost will never be 0 or negative. **(e)**
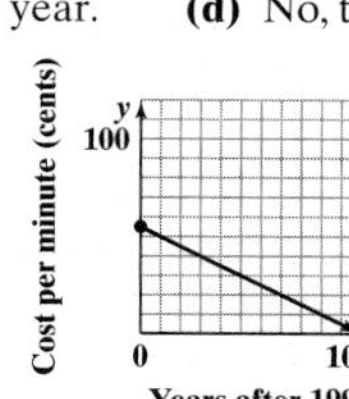

89. $B = -4$ **91.** $A = -4$ **93.** $B = -3$ **95. (a)** $y = 40x + 4000$ **(b)** \$24,000 **(c)** 375 calculators **(d)**
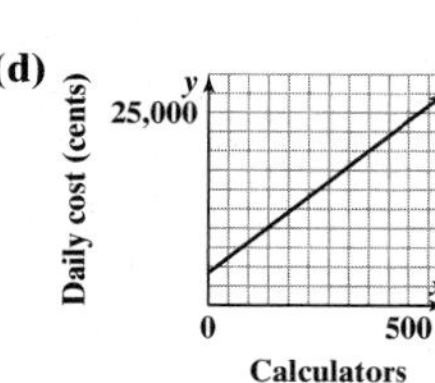

Section 8.5

1. $y - y_1 = m(x - x_1)$ **3.** standard form, slope-intercept form, point-slope form; $x = a$, $y = b$ **5.** True **7.** Yes; Answers may vary.

9. $y = 3x - 1$

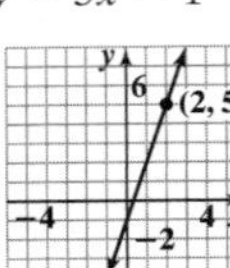

11. $y = -2x$

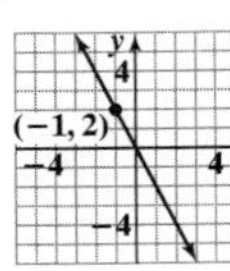

13. $y = \frac{1}{4}x - 3$

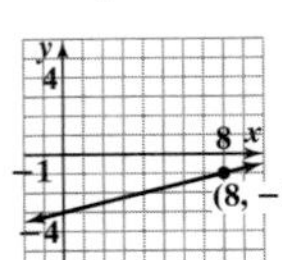

15. $y = -6x + 13$

17. $y = -7$

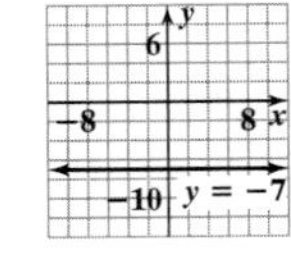

19. $x = -4$

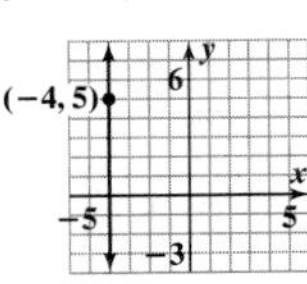

21. $y = \frac{2}{3}x + 2$ **23.** $y = -\frac{3}{4}x - \frac{39}{4}$

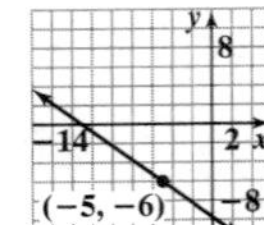

25. $y = \frac{1}{2}x - 5$ **27.** $x = -3$ **29.** $y = -5$ **31.** $y = -4.3$ **33.** $x = \frac{1}{2}$

35. $y = 2x + 4$ **37.** $y = -4x + 6$ **39.** $y = -\frac{3}{2}x - \frac{5}{2}$

41. $y = 2x - 5$ **43.** $y = -3$ **45.** $x = 2$

47. $y = 0.25x + 0.575$ **49.** $y = x - \frac{11}{4}$

51. $y = 5x - 22$

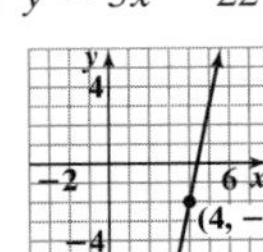

53. $y = 5$

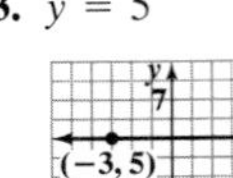

55. $y = x + 2$

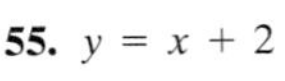

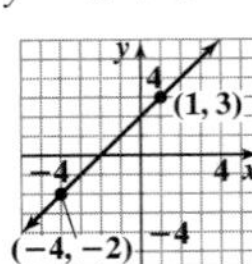

57. $y = \frac{1}{2}x + 4$

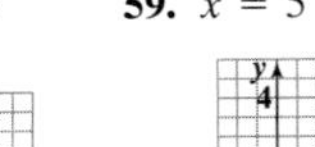

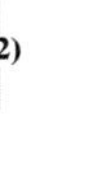

59. $x = 5$

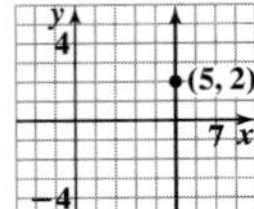

61. $y = -7x + 2$

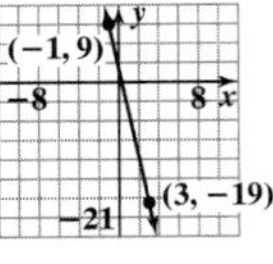

63. $y = -\frac{2}{3}x + 7$ **65.** $y = -\frac{3}{2}x$

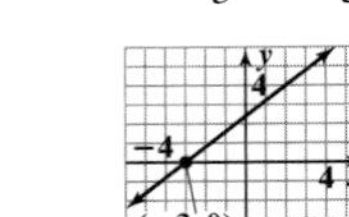

67. $y = \frac{4}{5}x + \frac{12}{5}$ **69.** $y = -\frac{4}{9}x + \frac{8}{3}$

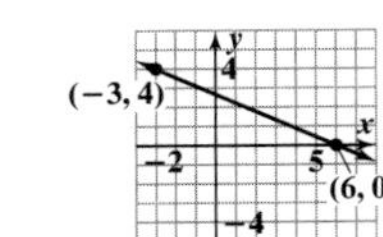

71. (a) When 60 packages are shipped, the expenses are \$1635.

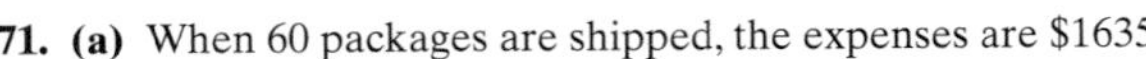

(b)

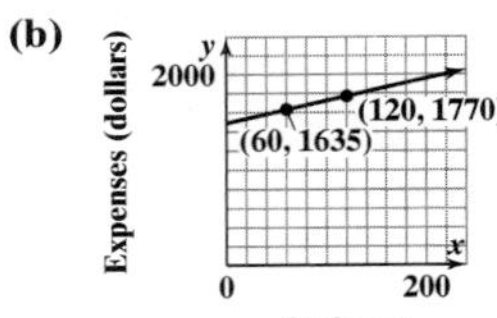

(c) $y = \frac{9}{4}x + 1500$ **(d)** \$1950

(e) Expenses increase by \$2.25 for each additional package.

73. (a) (0, 181), (27, 276) **(b)**

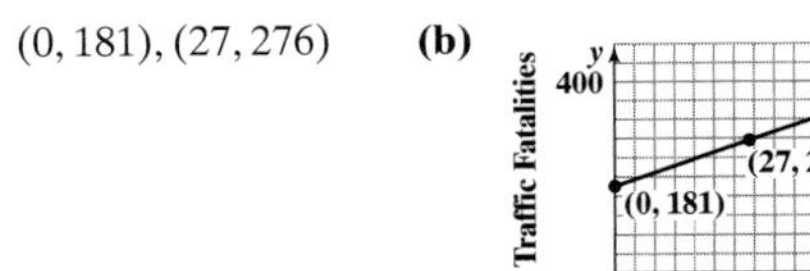

(c) $y = \frac{95}{27}x + 181$ **(d)** about 234

(e) The number of traffic fatalities in this region increases by about 3.5 each year.

75. $y = 3x + 14$ **77.** $y = -2x - 2$ **79.** $y = 6x - \frac{7}{2}$ **81.** $y = -x - 7$ **83.** $y = \frac{1}{5}x - 2$ **85.** $y = -\frac{3}{2}x + 2$

Section 8.6

1. perpendicular **3.** perpendicular **5.** False **7.** L_1 could be parallel to L_2. L_1 could be perpendicular to L_2. L_1 and L_2 could intersect, but not at right angles, L_1 could be coincident with L_2.

	Slope of the Given Line	Slope of a Line Parallel to the Given Line	Slope of a Line Perpendicular to the Given Line
9.	$m = -3$	$m_1 = -3$	$m_2 = \frac{1}{3}$
11.	$m = \frac{1}{2}$	$m_1 = \frac{1}{2}$	$m_2 = -2$
13.	$m = -\frac{4}{9}$	$m_1 = -\frac{4}{9}$	$m_2 = \frac{9}{4}$
15.	$m = 0$	$m_1 = 0$	$m_2 =$ undefined

17. perpendicular **19.** parallel **21.** perpendicular **23.** perpendicular **25.** perpendicular **27.** parallel **29.** parallel

31. (a) $m_1 = 3, m_2 = -3$ **(b)** neither **33. (a)** $m_1 = 2, m_2 = 2$ **(b)** parallel **35. (a)** $m_1 = \frac{1}{2}, m_2 = \frac{1}{2}$ **(b)** parallel

37. (a) $m_1 = \frac{5}{3}, m_2 = -\frac{3}{5}$ **(b)** perpendicular **39.** $y = 3x - 14$ **41.** $y = -4x - 4$ **43.** $y = -7$ **45.** $x = -1$ **47.** $y = \frac{3}{2}x - 13$

49. $y = -\frac{1}{2}x - \frac{21}{2}$ **51.** $y = -2x + 11$ **53.** $y = \frac{1}{4}x$ **55.** $x = -2$ **57.** $y = 5$ **59.** $y = \frac{5}{2}x$ **61.** $y = -\frac{3}{5}x - 9$

63. $y = 7x - 26$ **65.** $y = \frac{1}{5}x + \frac{43}{5}$ **67.** $y = -7x + 41$ **69.** $y = -2x - 10$ **71.** $y = 3x - 2$ **73.** $x = 5$

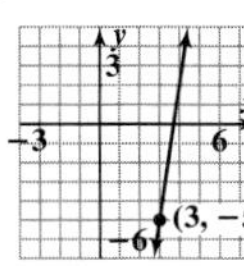
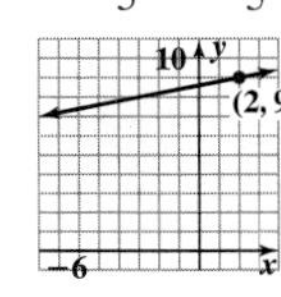
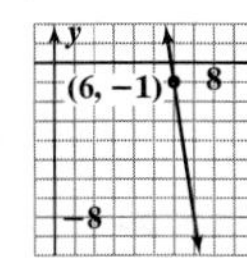
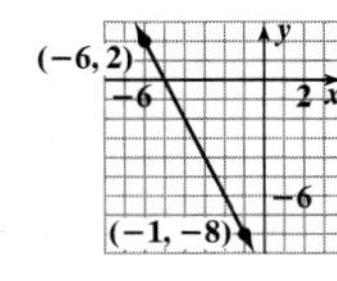
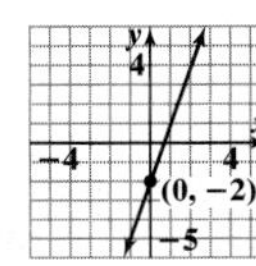
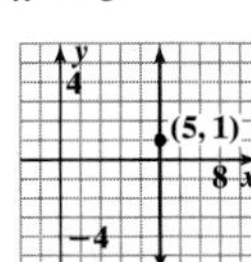

75. $y = -\frac{4}{3}x + 2$ **77.** $y = -2x - 5$ **79.** **81.** **83.** **85.**

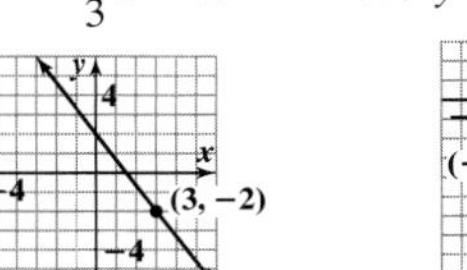
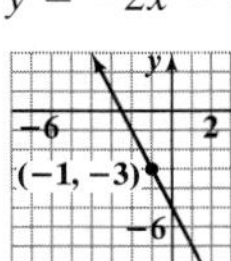
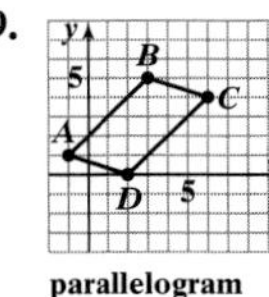
parallelogram
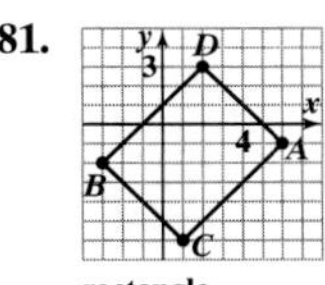
rectangle
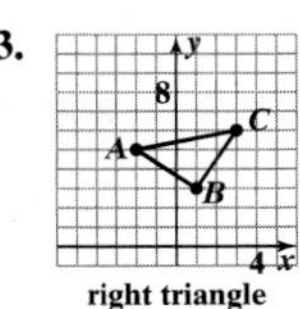
right triangle
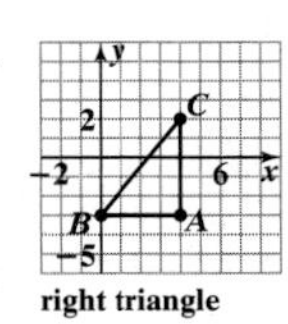
right triangle

87. $B = 2$ **89.** $A = 4$ **91.**

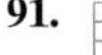
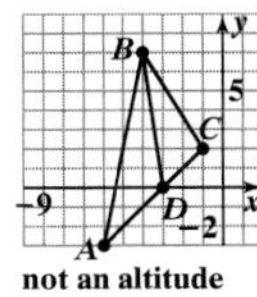
not an altitude

Putting the Concepts Together

1. yes, $(1, -2)$ is a solution **2.**

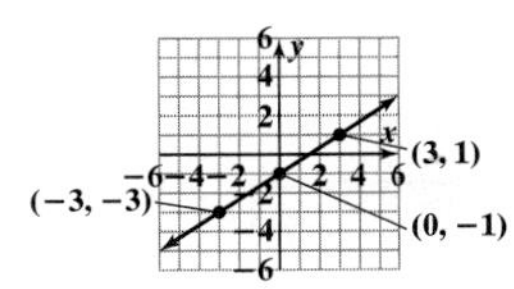

3.

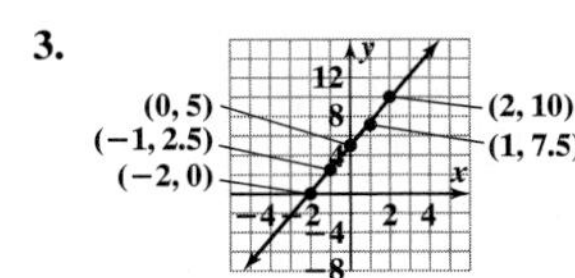

4. (a) x-intercept is $-\frac{3}{4}$ **(b)** y-intercept is 3

5. (a) x-intercept is $\frac{3}{2}$ **(b)** y-intercept is 2 **(c)**

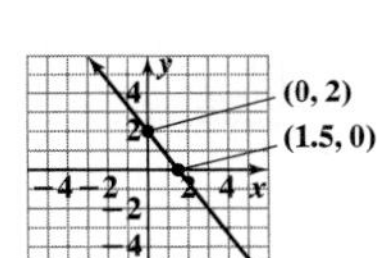

6. (a) slope $= -\frac{2}{3}$ **(b)** y-intercept $= -\frac{4}{3}$ **7.** slope $= -\frac{1}{3}$

8. (a) $m = \frac{2}{5}$ **(b)** $m = -\frac{5}{2}$ **9.** The lines are parallel. Answers may vary. **10.** $y = 3x + 1$ **11.** $y = -6x - 2$ **12.** $y = -2x + 7$

13. $y = -\frac{5}{2}x - 20$ **14.** $y = \frac{1}{4}x + 5$ **15.** $y = -8$ **16.** $x = 2$ **17.** $m = 11$. Every package increases expenses by $11.

18. (a) $y = 8350x - 2302$, where x is the weight, in carats, and y is the price. **(b)** For every 1-carat increase in weight, the cost increases by $8350. **(c)** $4044

Section 8.7

1. $t = ks$ **3.** constant of proportionality or constant of variation **5.** True **7.** Answers may vary. **9.** $y = \frac{1}{2}x$ **11.** $y = 2x$

13. $y = -\frac{3}{4}x$ **15.** $y = \frac{6}{x}$ **17.** $y = \frac{12}{x}$ **19.** $y = \frac{2}{7x}$ **21.** direct variation; $\frac{2}{3}$ **23.** neither **25.** inverse variation; 9

27. direct variation; 2 **29. (a)** $p = \frac{1}{3}g$ **(b)** $p = 3$ **31. (a)** $y = -\frac{3}{4}x$ **(b)** $x = -\frac{5}{3}$ **33. (a)** $e = \frac{8}{n}$ **(b)** $e = \frac{1}{2}$ **35. (a)** $b = \frac{3}{4a}$

(b) $a = \frac{5}{6}$ **37.** $\frac{9}{2}$ **39.** $P = 75$ **41.** $t = 42$ **43.** $b = 18$ **45.** 10 representatives **47.** 120 board feet **49.** 6.25 atmospheres

51. 150 bags **53.** $m = 90$ **55.** $r = 5$ **57. (a)** $e = \frac{1.989 \times 10^{-25}}{\lambda}$ **(b)** 3.978×10^{-19} joules

Section 8.8

1. solid **3.** half-planes **5.** True **7.** Answers may vary. **9.** A is a solution. **11.** C is a solution. **13.** A and C are solutions.
15. B is a solution. **17.** B is a solution. **19.** A is a solution.

21. $y > 3x - 2$ **23.** $y \le -x + 1$ **25.** $y < \frac{x}{2}$ **27.** $y > 5$ **29.** $y \le \frac{2}{5}x + 3$ **31.** $y \ge -\frac{4}{3}x + 2$ **33.** $x < 2$

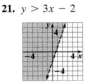

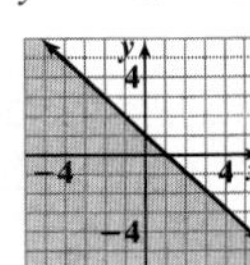
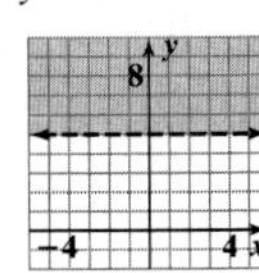
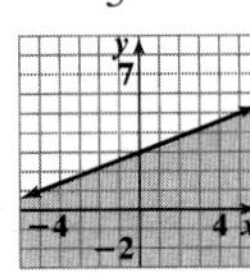
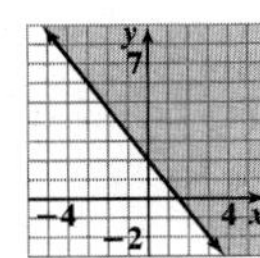
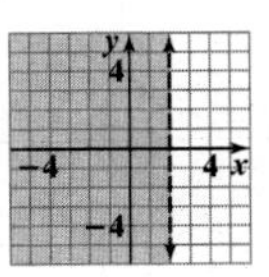

35. $3x - 4y < 12$

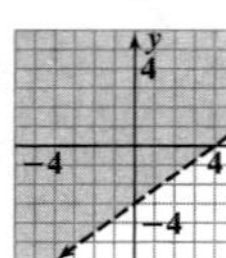

37. $2x + y \geq -4$

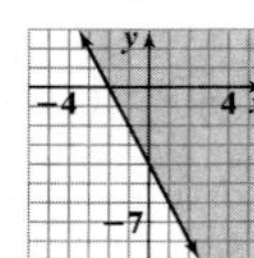

39. $x + y > 0$

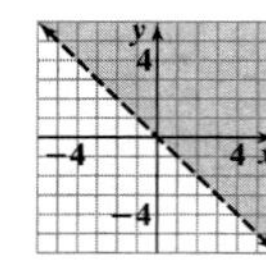

41. $5x - 2y < -8$

43. $x > -1$

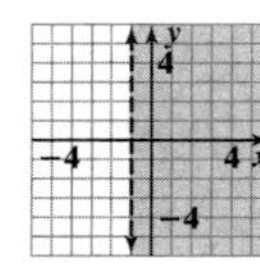

45. $y \leq 4$

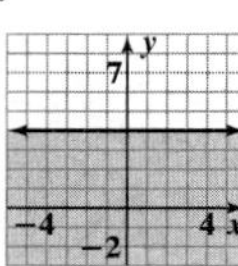

47. $\frac{x}{3} - \frac{y}{5} \geq 1$

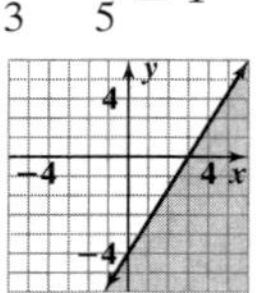

49. $-3 \geq x - y$

51. $x + y \geq 26$

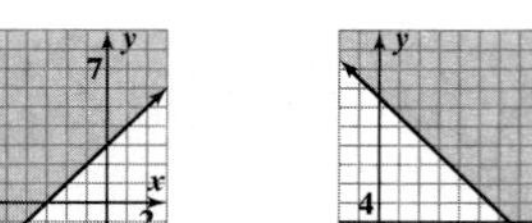

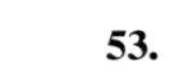

53. $\frac{y}{-2} \leq 4$

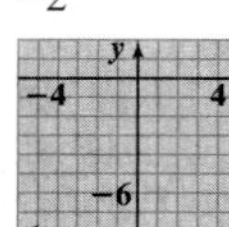

55. $x \leq y - 3$

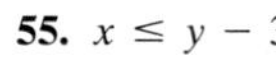

57. $x + 3y < 0$

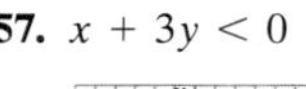

59. $2x - \frac{1}{2}y \geq 5$

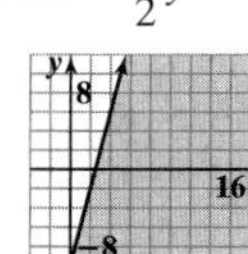

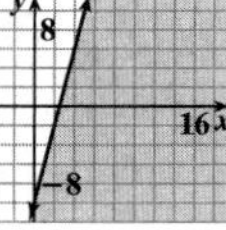

61. $-2x > -1$

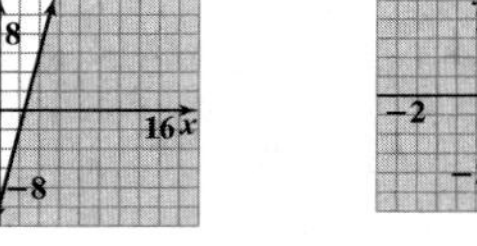

63. **(a)** $3s + 5a \leq 120$ **(b)** No **(c)** Yes

65. **(a)** $3.1s + 5.7b \leq 22$ **(b)** No **(c)** Yes

67. $3x - 2y > 6$ and $x + y < 2$

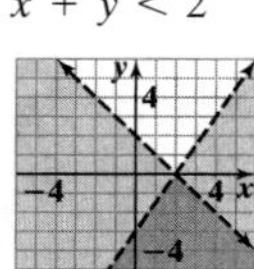

69. $y > \frac{3}{4}x - 1$ and $x \geq 0$

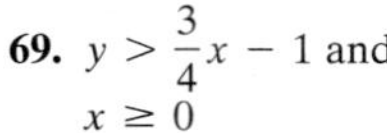

71. $x < -3$ and $y \leq 4$

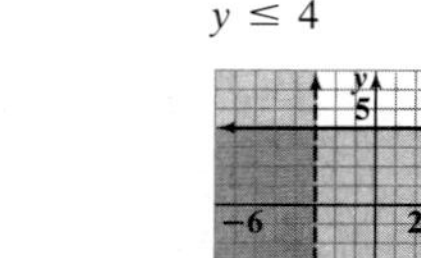

73. $y > 3$

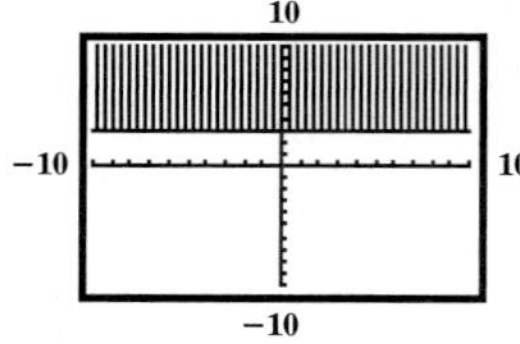

75. $y < 5x$

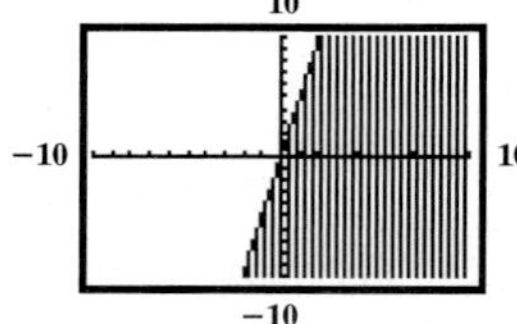

77. $y > 2x + 3$

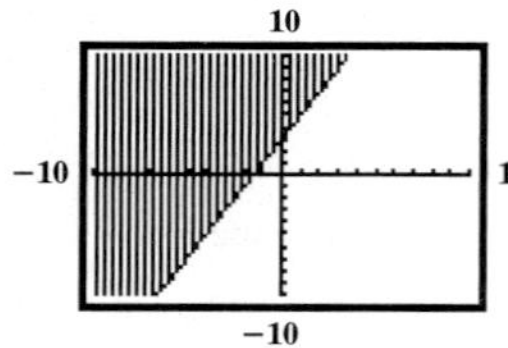

79. $y \leq \frac{1}{2}x - 5$

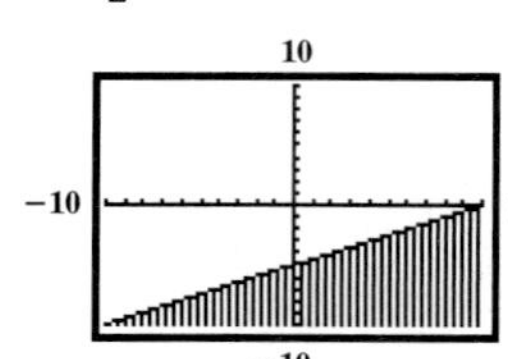

81. $3x + y \leq 4$

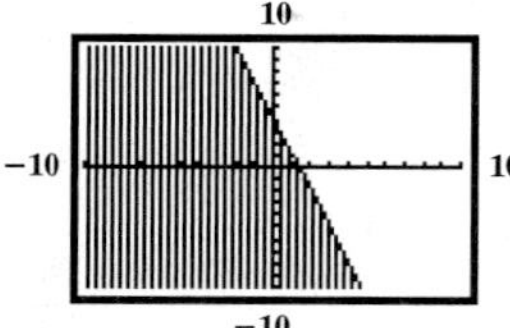

83. $2x + 5y \leq -10$

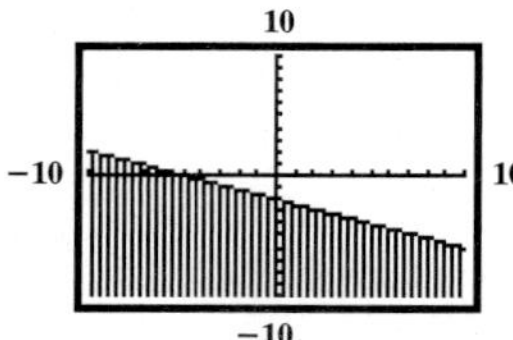

Chapter 8 Review

1–4.

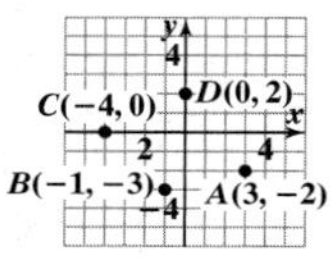

$A(3, -2)$ quadrant IV; $B(-1, -3)$ quadrant III; $C(-4, 0)$ negative x-axis; $D(0, 2)$ positive y-axis

5. $A(1, 4)$; quadrant I
$B(-3, 0)$; x-axis

6. $A(0, 1)$; y-axis
$B(-2, 2)$; quadrant II

7. A. Yes
B. No

8. A. Yes
B. No

9. **(a)** $\left(4, -\frac{4}{3}\right)$ **(b)** $(-6, 2)$

10. **(a)** $(4, 4)$ **(b)** $(-2, -2)$

11.

x	y	(x, y)
−2	−8	(−2, −8)
0	−5	(0, −5)
4	1	(4, 1)

12.

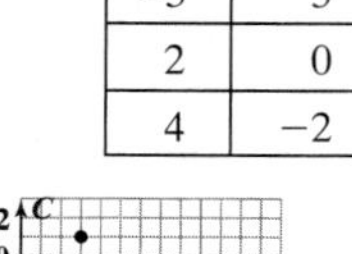

x	y	(x, y)
−3	5	(−3, 5)
2	0	(2, 0)
4	−2	(4, −2)

13.

x	y	(x, y)
−18	2	(−18, 2)
−6	−2	(−6, −2)
24	−12	(24, −12)

14.

x	y	(x, y)
−3	−8	(−3, −8)
3	1	(3, 1)
5	4	(5, 4)

15.

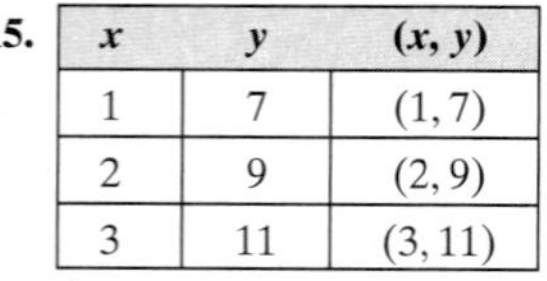

x	y	(x, y)
1	7	(1, 7)
2	9	(2, 9)
3	11	(3, 11)

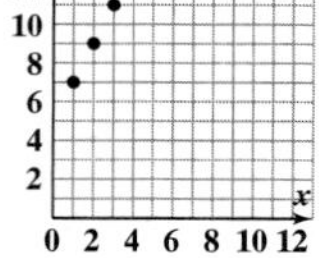

16.

x	E	(x, E)
500	1050	(500, 1050)
1000	1100	(1000, 1100)
2000	1200	(2000, 1200)

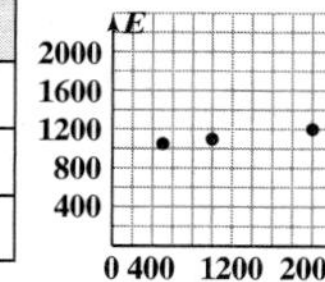

17. $y = -2x$

18. $y = x$

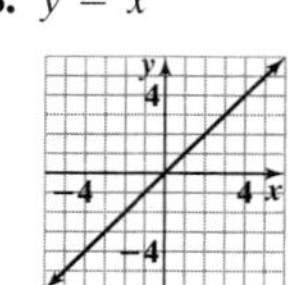

19. $4x + y = -2$

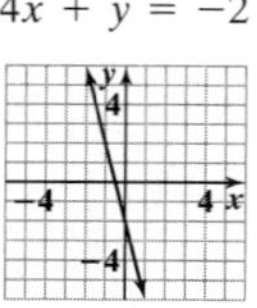

20. $3x - y = -1$

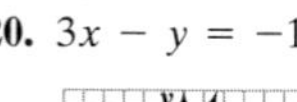

21.

p	C	(p, C)
20	80	(20, 80)
50	140	(50, 140)
80	200	(80, 200)

22.

p	F	(p, F)
100	800	(100, 800)
200	1100	(200, 1100)
500	2000	(500, 2000)

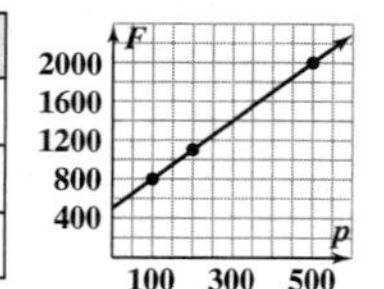

23. $(-2, 0), (0, -4)$ **24.** $(0, 1)$ **25.** $(0, 9), (-3, 0)$ **26.** $(0, -6), (3, 0)$ **27.** $(3, 0)$ **28.** $\left(0, -\frac{2}{5}\right), (1, 0)$

29. $y - 3x = 3$ **30.** $2x + 5y = 0$ **31.** $\frac{x}{3} + \frac{y}{2} = 1$ **32.** $y = -\frac{3}{4}x + 3$ **33.** $x = -2$ **34.** $y = 3$ **35.** $y = -4$

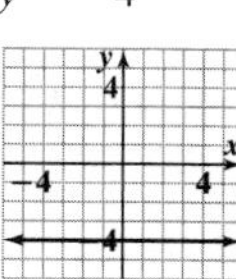

36. $x = 1$ **37.** $\frac{4}{3}$ **38.** -2 **39.** -8 **40.** 2 **41.** $\frac{1}{4}$ **42.** $-\frac{1}{6}$ **43.** undefined **44.** 0 **45.** 0 **46.** undefined

47. **48.** **49.** **50.** **51.** $m = 45$; the cost to produce 1 additional bicycle is \$45.

52. 5% **53.** $m = -1; b = \frac{1}{2}$ **54.** $m = 1; b = -\frac{3}{2}$ **55.** $m = \frac{3}{4}; b = 1$ **56.** $m = -\frac{2}{5}; b = \frac{8}{5}$

57. $y = \frac{1}{3}x + 1$ **58.** $y = -\frac{x}{2} - 1$ **59.** $y = -\frac{2}{3}x - 2$ **60.** $y = \frac{3x}{4} + 3$ **61.** $y = x$ **62.** $y = -2x$ **63.** $2x - y = -4$

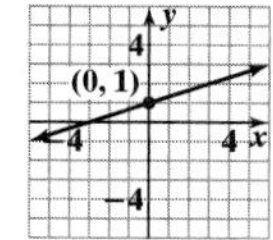
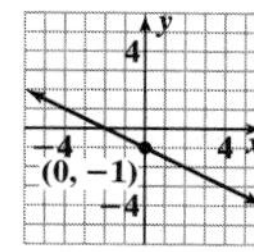
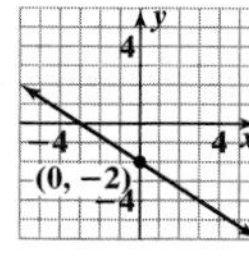
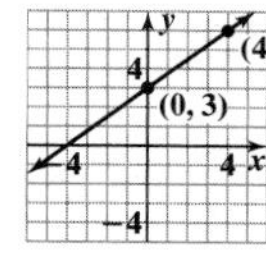
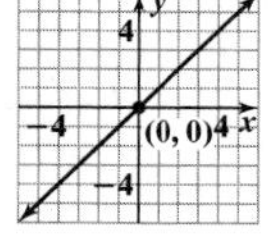
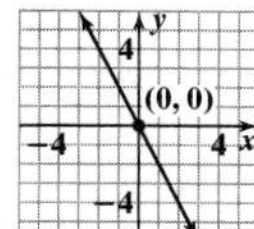
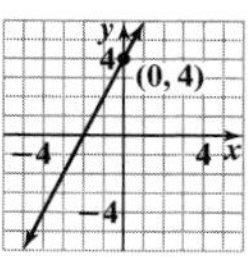

64. $-4x + 2y = 2$ **65.** $y = -\frac{3}{4}x + \frac{2}{3}$ **66.** $y = \frac{1}{5}x + 10$ **67.** $x = -12$ **68.** $y = -4$ **69.** $y = x - 20$ **70.** $y = -x - 8$

71. (a) \$360 **(b)** 7 days **(c)**

72. (a) $(22, 418), (35, 665), m = 19$
(b) It costs \$19 more for each additional day.
(c) $C = 19d$
(d) \$152

73. $y = 6x - 3$ **74.** $y = -2x + 8$ **75.** $y = -\frac{1}{2}x + \frac{1}{2}$ **76.** $y = \frac{2}{3}x - \frac{7}{3}$ **77.** $y = -\frac{1}{2}$ **78.** $x = -\frac{4}{7}$ **79.** $x = -5$ **80.** $y = 0$

81. $y = \frac{8}{7}x + 8$ **82.** $y = \frac{3}{2}x - 6$ **83.** $y = 3x - 4$ **84.** $y = -\frac{2}{5}x - 5$ **85.** $A = -4d + 24$ **86.** $F = -\frac{1}{3}m + 15$ **87.** not parallel

88. parallel **89.** $y = -x + 2$ **90.** $y = 2x + 8$ **91.** $y = -3x + 7$ **92.** $y = -3x + 7$ **93.** $x = 5$ **94.** $y = -12$ **95.** $-\frac{2}{3}$

96. $-\frac{9}{4}$ **97.** perpendicular **98.** not perpendicular **99.** $y = \frac{1}{3}x + 5$ **100.** $y = -\frac{1}{2}x + 1$ **101.** $y = -\frac{3}{2}x - \frac{3}{2}$ **102.** $y = x + 1$

103. $y = \frac{1}{4}x$ **104.** $y = 6x$ **105.** $f = 16$ **106.** $p = 110$ **107.** \$507.50 **108.** 160 mi **109.** $y = \frac{48}{x}$ **110.** $y = \frac{54}{x}$ **111.** $r = 9$

112. $s = \frac{1}{4}$ **113.** 78 mph **114.** 25 liters **115.** A, B are solutions. **116.** C is a solution.

117. $y < -\frac{1}{4}x + 2$ **118.** $y > 2x - 1$ **119.** $3x + 2y \geq -6$ **120.** $-2x + y \geq 4$ **121.** $x - 3y \leq 0$ **122.** $x - 4y \geq 4$

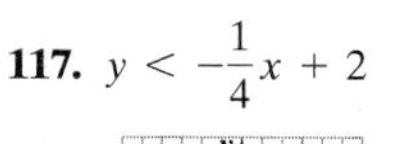

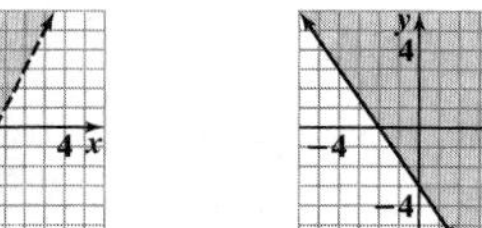
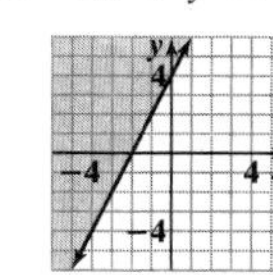
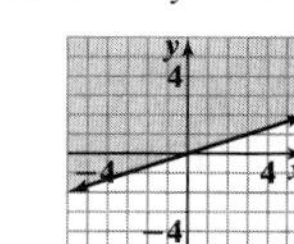
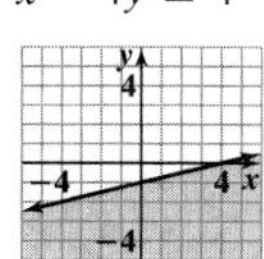

123. $x < -3$ **124.** $y > 2$ **125.** $0.25x + 0.1y \geq 12$ **126.** $2x - \frac{1}{2}y \leq 10$

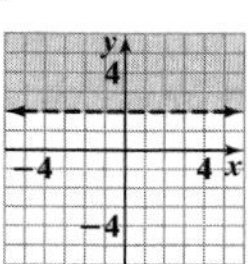

Chapter 8 Test

1. No **2. (a)** $(4, 0)$ **(b)** $\left(0, -\frac{4}{3}\right)$ **3. (a)** $m = \frac{4}{3}$ **(b)** $b = 8$

4. $y = -\frac{3}{4}x + 2$ **5.** $3x - 6y = -12$

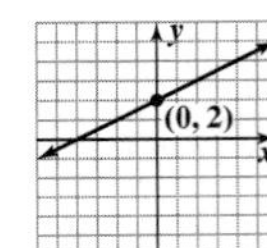

6. $-\frac{1}{6}$ **7. (a)** $-\frac{3}{2}$ **(b)** $\frac{2}{3}$ **8.** neither; Answers may vary. **9.** $y = -4x - 15$ **10.** $y = 2x + 14$

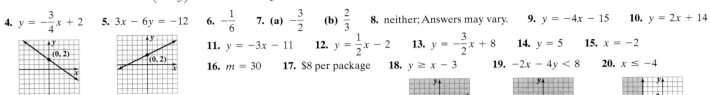

11. $y = -3x - 11$ **12.** $y = \frac{1}{2}x - 2$ **13.** $y = -\frac{3}{2}x + 8$ **14.** $y = 5$ **15.** $x = -2$

16. $m = 30$ **17.** $8 per package **18.** $y \geq x - 3$ **19.** $-2x - 4y < 8$ **20.** $x \leq -4$

Chapter 9

Section 9.1

1. intersect in one point **3.** inconsistent **5.** False **7.** Answers may vary. **9. (a)** No **(b)** Yes **(c)** Not

11. (a) No **(b)** Yes **(c)** Yes **13. (a)** No **(b)** No **(c)** No

15.

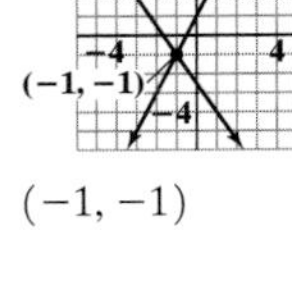

$(-1, -1)$

17.

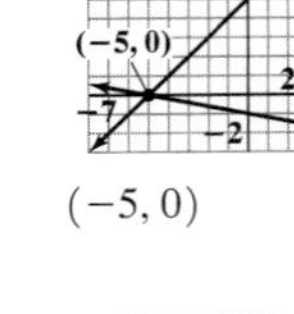

$(-5, 0)$

19.

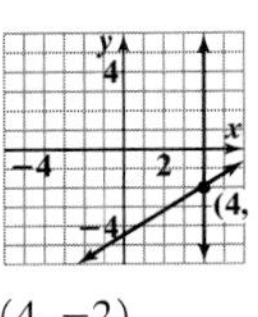

$(4, -1)$

21.

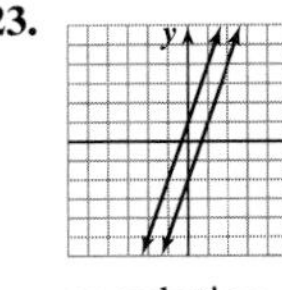

$(4, -2)$

23.

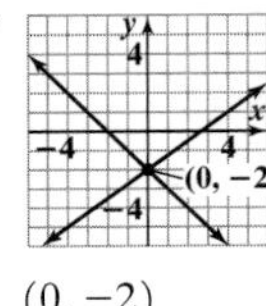

no solution

25.

$(0, -2)$

27.

infinitely many solutions

29.

$(0, 3)$

31. one solution; consistent; independent; $(3, 2)$ **33.** no solution; inconsistent **35.** infinitely many solutions; consistent; dependent **37.** one solution; consistent; independent; $(-1, -2)$ **39.** one solution; consistent; independent **41.** no solution; inconsistent **43.** infinitely many solutions; consistent; dependent **45.** one solution; consistent; independent **47.** no solution; inconsistent **49.** infinitely many solutions; consistent; dependent

51.

consistent; independent; $(-3, 3)$

53.

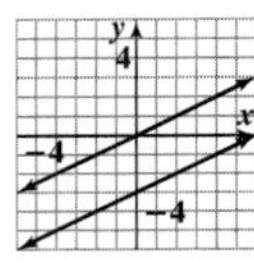

inconsistent; no solution

55.

consistent; independent; $(2, 4)$

57.

consistent; dependent; infinitely many solutions

59.

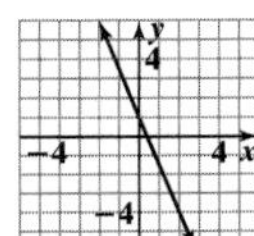

consistent; dependent; infinitely many solutions

61.

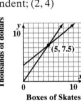

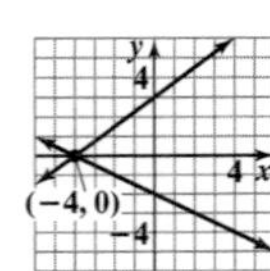

consistent; independent; $(-4, 0)$

63.

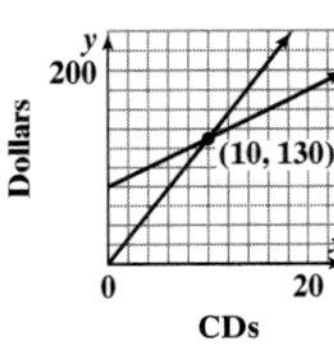

The break-even point is $(10, 130)$. The company needs to sell 10 CDs to break even at a cost/revenue of $130.

65.

The break-even point is $(5, 7.5)$. The company needs to sell 5 boxes of skates to break even at a cost/revenue of $7500.

67.

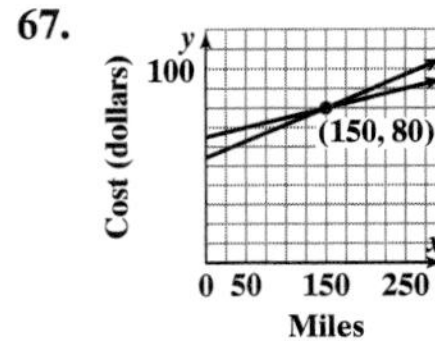

The cost for driving 150 miles is the same for both companies ($80). Choose Acme to drive 200 miles.

69.

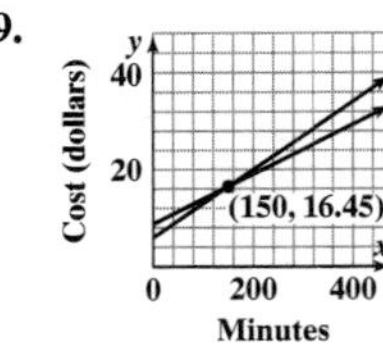

The cost is the same ($16.45) when 150 minutes are used. Choose Plan B for 100 long-distance minutes.

71. $c = 3$ **73.** $c = 1$ **75.**

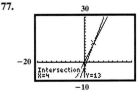

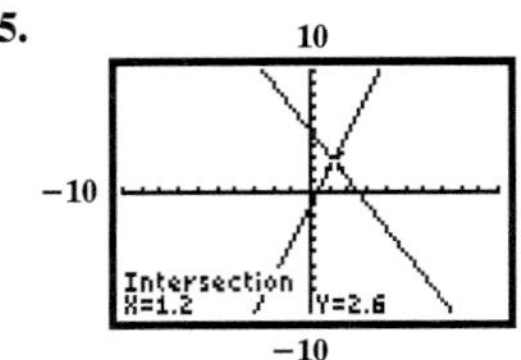

$(1.2, 2.6)$

77.

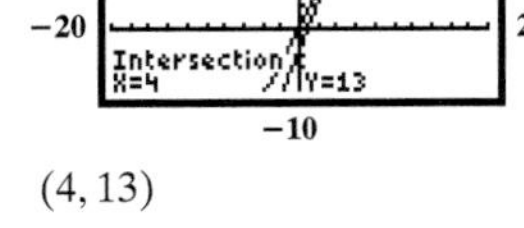

$(4, 13)$

79.

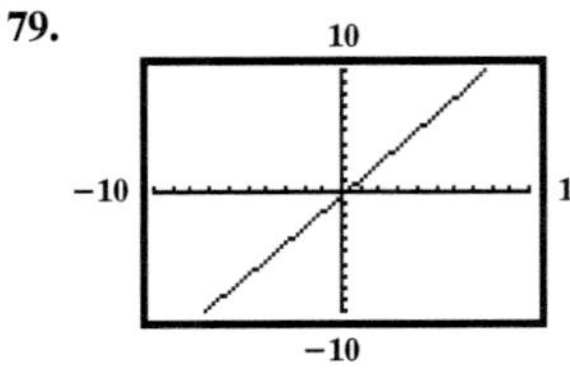

infinitely many solutions

Section 9.2

1. 1, −1 **3.** infinitely many solutions **5.** True **7.** Answers may vary. **9.** $(4, -1)$ **11.** $(-1, 1)$ **13.** $(-3, -4)$ **15.** $(-4, -7)$ **17.** $\left(-\frac{2}{3}, 2\right)$ **19.** $\left(2, -\frac{1}{3}\right)$ **21.** $\left(\frac{1}{4}, \frac{1}{6}\right)$ **23.** inconsistent; no solution **25.** dependent; infinitely many solutions **27.** inconsistent; no solution **29.** dependent; infinitely many solutions **31.** $(-2, 0)$ **33.** $\left(2, -\frac{1}{3}\right)$ **35.** no solution **37.** infinitely many solutions **39.** $\left(\frac{5}{3}, -\frac{1}{2}\right)$ **41.** no solution **43.** infinitely many solutions **45.** $\left(\frac{5}{4}, -\frac{7}{4}\right)$ **47.** infinitely many solutions **49.** $\left(-\frac{5}{2}, 4\right)$ **51.** 5 and 12 **53.** length 10 ft; width 7 ft **55.** \$8000 in money market; \$4000 in international fund **57.** 2003 **59.** $A = \frac{7}{6}; B = -\frac{1}{2}$ **61.** Answers may vary. **63.** Answers may vary.

Section 9.3

1. graphing, substitution, elimination **3.** additive inverses **5.** True **7.** Answers may vary. **9.** $(2, -1)$ **11.** $(6, 4)$ **13.** $\left(-\frac{11}{4}, \frac{1}{2}\right)$ **15.** $(-1, -3)$ **17.** inconsistent; 0 **19.** dependent; infinitely many **21.** dependent; infinitely many **23.** inconsistent; 0 **25.** $(-5, 8)$ **27.** $\left(\frac{3}{2}, -\frac{3}{4}\right)$ **29.** no solution **31.** $(12, -9)$ **33.** no solution **35.** $\left(\frac{6}{29}, -\frac{8}{29}\right)$ **37.** $\left(\frac{1}{2}, 4\right)$ **39.** $(-1, 2.1)$ **41.** $(-2, -6)$ **43.** $(50, 30)$ **45.** $(1, 5)$ **47.** $(5, 2)$ **49.** $\left(-\frac{9}{17}, \frac{31}{17}\right)$ **51.** $\left(\frac{8}{3}, -\frac{4}{7}\right)$ **53.** infinitely many solutions **55.** infinitely many solutions **57.** $(1, -3)$ **59.** $(-4, 0)$ **61.** $\left(-\frac{8}{5}, \frac{27}{10}\right)$ **63.** $(-4, -5)$ **65.** no solution **67.** $\left(\frac{106}{37}, \frac{16}{37}\right)$ **69.** hamburger 280 cal; Coke 210 cal **71.** tomatoes 10 hr; zucchini 15 hr; no **73.** 60 lb Arabica; 40 lb Robusta **75.** 70° and 20° **77.** $\left(-\frac{3}{a}, 1\right)$ **79.** $\left(-3a - b, -\frac{3}{2}a - \frac{3}{2}b\right)$

Putting the Concepts Together

1. (a) No **(b)** Yes **(c)** No **2. (a)** infinitely many **(b)** consistent **(c)** dependent **3. (a)** one **(b)** consistent **(c)** independent **4.** $(1, 1)$ **5.** $(5, -1)$ **6.** $(1, -5)$ **7.** $(-1, 0)$ **8.** $(-4, 5)$ **9.** $(-3, -1)$ **10.** $\left(-\frac{1}{3}, 3\right)$ **11.** infinitely many solutions **12.** $(2.5, 3)$ **13.** no solution

Section 9.4

1. perimeter; $P = 2l + 2w$ **3.** 180°; 90° **5.** True **7.** No; answers may vary. **9.** $2l + 2w$ **11.** 33, 49 **13.** 14, 37 **15.** Thursday 35,000; Friday 42,000 **17.** \$12,000 in stocks; \$9000 in bonds **19.** length 12 ft; width 9 ft **21.** length 25 m; width 10 m **23.** 40°, 50° **25.** 22.5°, 157.5° **27.** current 0.4 mph; still-water 3.9 mph **29.** bike 11 mph; wind 1 mph **31.** northbound 60 mph; southbound 72 mph **33.** 4 hr **35.** wind 25 mph; Piper 175 mph **37.** Sam 24 hr; Diane 8 hr

Section 9.5

1. mixture problems **3.** P, r, t **5.** False

7.

	Number	· Cost	= Total
Adult	a	4	$4a$
Student	s	1.5	$1.5s$
Total	215		580

$$\begin{cases} a + s = 215 \\ 4a + 1.5s = 580 \end{cases}$$

9.

	P	· r	= I
Savings	s	0.05	$0.05s$
Money Market	m	0.03	$0.03m$
Total	1600		50

$$\begin{cases} s + m = 1600 \\ 0.05s + 0.03m = 50 \end{cases}$$

11.

	lb	· Price	= Total
Mild	m	7.5	$7.5m$
Robust	r	10	$10r$
Total	12	8.75	8.75(12)

$$\begin{cases} m + r = 12 \\ 7.5m + 10r = 8.75(12) \end{cases}$$

13. $32a + 24c$ **15.** $0.1A + 0.07B$ **17.** $5.85r + 4.20y = 128.85$ **19.** 315 **21.** 400 **23.** 60 nickels, 90 dimes **25.** first-class \$0.37; postcard \$0.19 **27.** \$7500 in 5% account; \$2500 in 8% account **29.** \$3200 in risky plan; \$1800 in safer plan **31.** 2 lb arbequina; 3 lb green **33.** 48.9 lb of the \$2.75 per pound coffee; 51.1 lb of the \$5 per pound coffee **35.** 80 lb rye; 100 lb blue-grass **37.** 20 ml 30% saline solution; 40 ml 60% saline solution **39.** 70 l **41.** 1.2 gal **43.** \$6200 at 5%; \$3800 at $7\frac{1}{2}$% loss

Section 9.6

1. solution **3.** dashed; solid **5.** True **7. (a)** $\begin{cases} y \geq x \\ y \geq -2x \end{cases}$ **(b)** $\begin{cases} y \leq x \\ y \geq -2x \end{cases}$ **(c)** $\begin{cases} y \geq x \\ y \leq -2x \end{cases}$ **(d)** $\begin{cases} y \leq x \\ y \leq -2x \end{cases}$ **9. (a)** Yes **(b)** Yes **(c)** Yes **11. (a)** Yes **(b)** No **(c)** No

13. $\begin{cases} x > 2 \\ y \le -1 \end{cases}$ 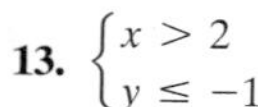

15. $\begin{cases} y > -2 \\ x > -3 \end{cases}$

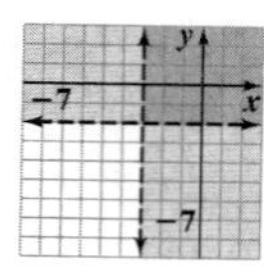

17. $\begin{cases} x + y < 3 \\ x - y > 5 \end{cases}$

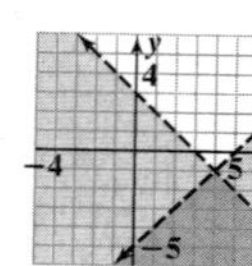

19. $\begin{cases} x + y > 3 \\ 2x - y > 4 \end{cases}$

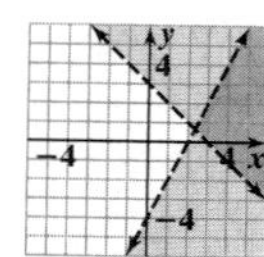

21. $\begin{cases} x < 2 \\ y < \frac{1}{2}x + 3 \end{cases}$

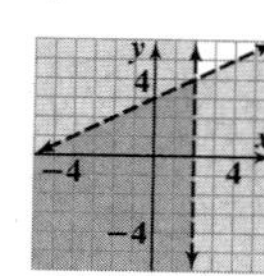

23. $\begin{cases} x \ge -2 \\ y < 2x + 3 \end{cases}$

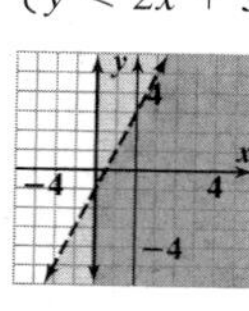

25. $\begin{cases} x > 0 \\ y \le \frac{2}{5}x - 1 \end{cases}$

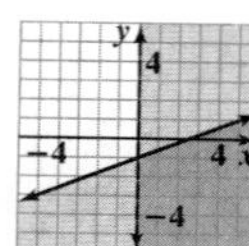

27. $\begin{cases} -y \le x \\ 3x - y \ge -5 \end{cases}$

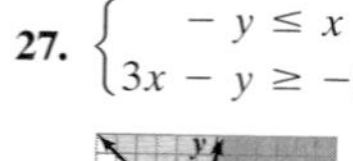

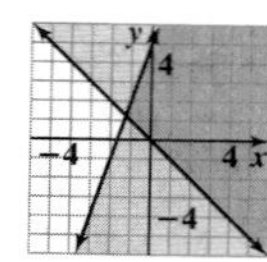

29. $\begin{cases} x + y \le -2 \\ y \ge x + 3 \end{cases}$

31. $\begin{cases} x + 3y \ge 0 \\ 2y < x + 1 \end{cases}$

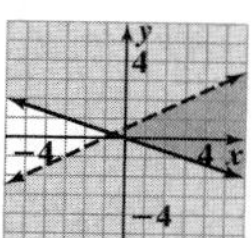

33. $\begin{cases} x + y \ge 0 \\ x < 2y + 4 \end{cases}$

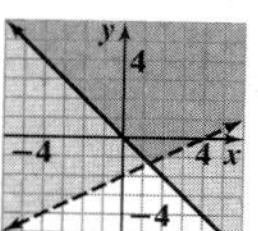

35. $\begin{cases} x + 3y > 6 \\ 2x - y \le 4 \end{cases}$

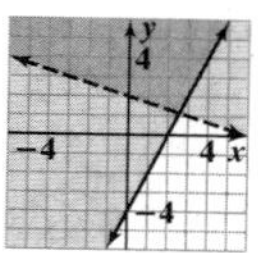

37. $\begin{cases} -y \le 3x - 4 \\ 2x + 3y \ge -3 \end{cases}$

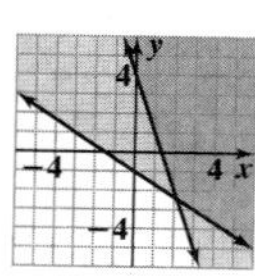

39. (a)

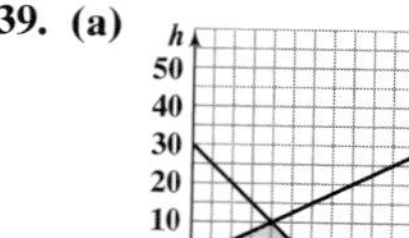

(b) No **(c)** Yes

41. (a)

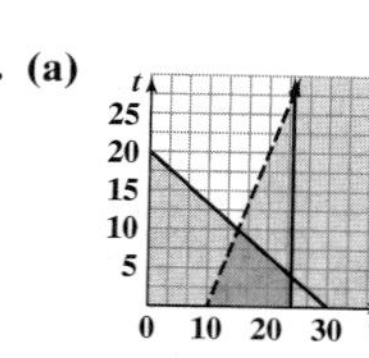

(b) No **(c)** No

43. $\begin{cases} \frac{y}{2} - \frac{x}{6} \ge 1 \\ \frac{x}{3} - \frac{y}{1} \ge 1 \end{cases}$

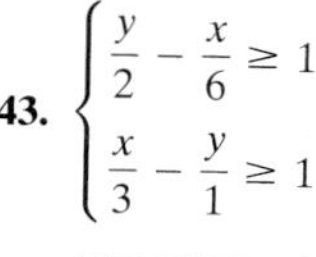

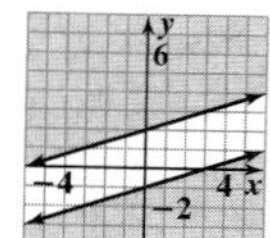

45. $\begin{cases} x < \frac{3}{2}y + \frac{9}{2} \\ -2x < 3(y + 2) \end{cases}$

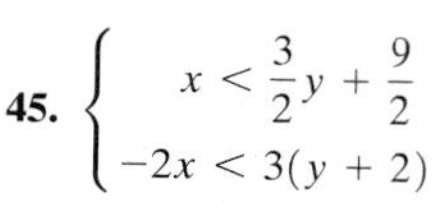

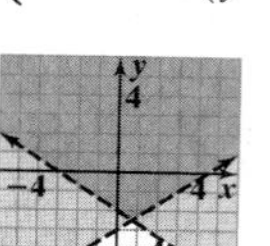

47. $\begin{cases} y \ge 0 \\ y \ge x \\ y \le \frac{1}{2}(x + 6) \end{cases}$

Chapter 9 Review

1. (a) No **(b)** Yes **(c)** No **2. (a)** No **(b)** No **(c)** Yes **3. (a)** Yes **(b)** Yes **(c)** Yes **4. (a)** No **(b)** No **(c)** No

5.

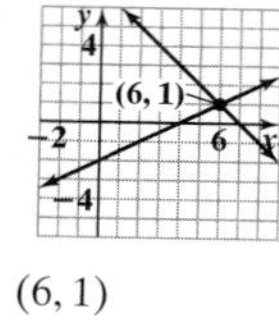

(6, 1)

6.

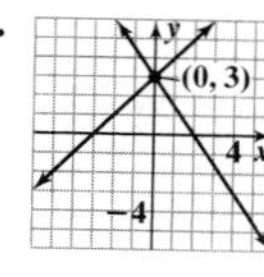

(0, 3)

7.

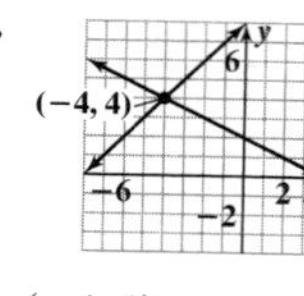

(−4, 4)

8.

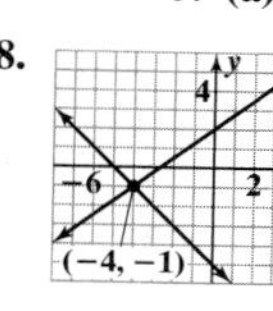

(−4, −1)

9.

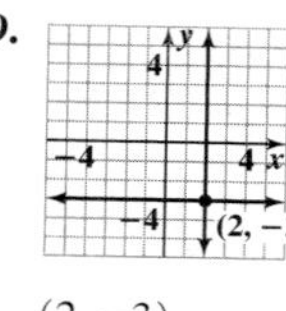

(2, −3)

10.

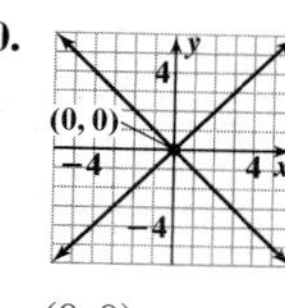

(0, 0)

11.

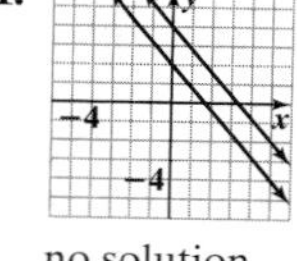

no solution

12. infinitely many solutions

13. none; inconsistent **14.** none; inconsistent **15.** infinitely many; consistent; dependent **16.** infinitely many; consistent; dependent **17.** one; consistent; independent **18.** one; consistent; independent

19. (a) $\begin{cases} y = 70 + 0.1x \\ y = 100 + 0.04x \end{cases}$ **(b)** 500 fliers **(c)** Printer A

20. (a) $\begin{cases} y = 50 + 40x \\ y = 350 + 10x \end{cases}$ **(b)** 10 sq yd

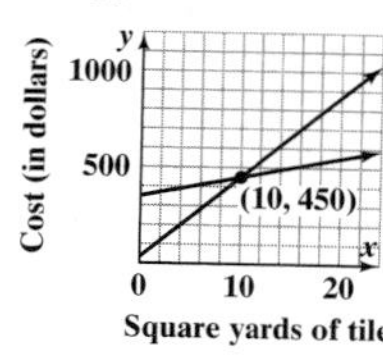

(c) tile

21. (2, 1) **22.** (1, −1) **23.** (7, −2) **24.** (4, −2) **25.** (18, 11) **26.** (−6, 8) **27.** (2, −2) **28.** (3, −1) **29.** infinitely many solutions **30.** infinitely many solutions **31.** no solution **32.** infinitely many solutions **33.** $\left(\frac{3}{2}, 1\right)$ **34.** $\left(\frac{1}{3}, -1\right)$ **35.** width 125 m; length 200 m **36.** −3

37. (0, −12) **38.** (−3, 7) **39.** (−3, 4) **40.** (−3, 0) **41.** (−1, −1) **42.** (5, 2) **43.** (−2, 2) **44.** $\left(\frac{202}{55}, -\frac{59}{11}\right)$ **45.** $\left(\frac{1}{2}, -2\right)$ **46.** $\left(-\frac{3}{2}, 1\right)$ **47.** infinitely many solutions **48.** $\left(\frac{6}{7}, \frac{5}{2}\right)$ **49.** (−20, −13) **50.** $\left(\frac{1}{3}, \frac{2}{3}\right)$ **51.** no solution **52.** $\left(-\frac{1}{7}, 0\right)$

53. 2.4 lb cookies; 1.6 lb chocolates **54. (a)** 400 individual tickets; 325 block tickets **55.** $\frac{3}{8}, \frac{1}{3}$ **56.** 2.6, 3.2 **57.** \$32,000 in stocks, \$18,000 in bonds **58.** 14 notebooks, 10 calculators **59.** 77.5°, 102.5° **60.** 40°, 50° **61.** 15 m by 22 m **62.** 110°, 40° **63.** plane 450 mph, wind 50 mph **64.** current 2 mph, paddling 6 mph **65.** cyclist 10 mph, wind 2 mph **66.** faster 8 mph; slower 6 mph

67.

	Number	· Value	= Total Value
Dimes	d	0.10	$0.10d$
Nickels	n	0.05	$0.05n$
Total			2.25

68.

	P	· r	= I
Savings	s	0.065	$0.065s$
Mutual Fund	m	0.08	$0.08m$
Total	15,000		1050

69. 125 children; 125 adult; 50 senior **70.** 7 dimes; 4 quarters **71.** \$7000 at 5%, \$12,000 at 9% **72.** \$10,000 bonds, \$15,000 stocks **73.** 7 quarts 60% sugar solution; 3 quarts 30% sugar solution

74. 60 pints **75.** 3 lb peanuts; 2 lb almonds **76.** 4 l 35% acid; 16 l 60% acid **77.** **(a)** Yes **(b)** Yes **(c)** No
78. **(a)** No **(b)** Yes **(c)** Yes **79.** **(a)** Yes **(b)** No **(c)** No **80.** **(a)** No **(b)** Yes **(c)** Yes
81. **(a)** No **(b)** No **(c)** No **82.** **(a)** No **(b)** Yes **(c)** No

83. $\begin{cases} x > -2 \\ y > 1 \end{cases}$ **84.** $\begin{cases} x \le 3 \\ y > -1 \end{cases}$ 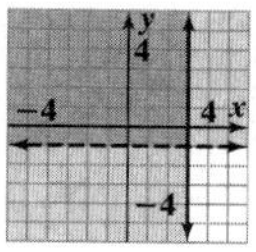 **85.** $\begin{cases} x + y \ge -2 \\ 2x - y \le -4 \end{cases}$ **86.** $\begin{cases} 3x + 2y < -6 \\ x - y < 2 \end{cases}$ **87.** $\begin{cases} x > 0 \\ y \le \frac{3}{4}x + 1 \end{cases}$ **88.** $\begin{cases} y \le 0 \\ y \le -\frac{1}{2}x - 3 \end{cases}$

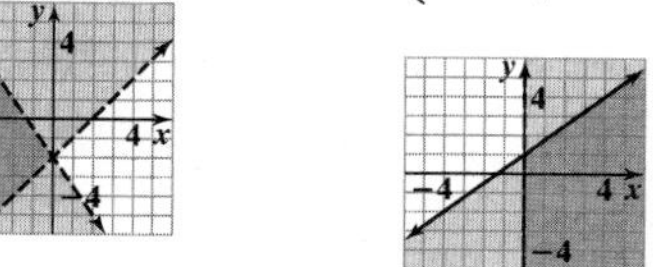
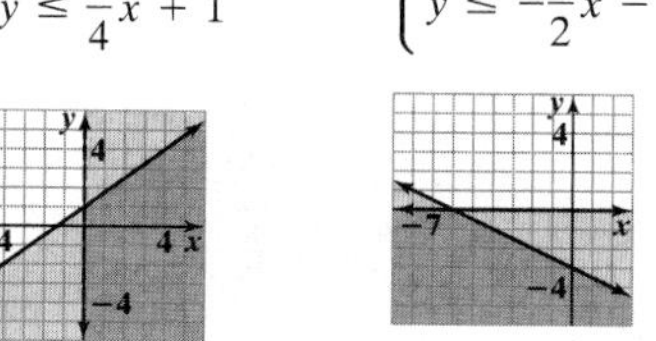

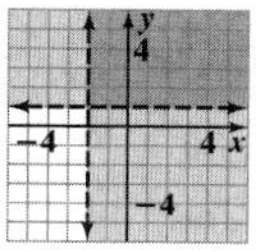

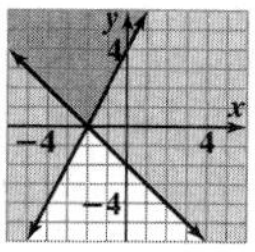

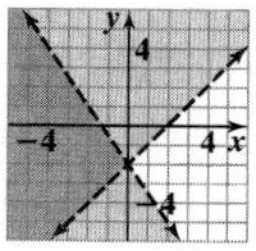

89. $\begin{cases} -y \ge x \\ 4x - 3y > -12 \end{cases}$ **90.** $\begin{cases} -y < x + 2 \\ 2x + 2y \ge -9 \end{cases}$ **91.** $\begin{cases} 2x + 3y > -3 \\ y > -\frac{2}{3}x + 2 \end{cases}$ **92.** $\begin{cases} x + 4y \le -4 \\ y \ge \frac{1}{4}x + 3 \end{cases}$ **93.** $\begin{cases} y > 2x - 5 \\ y - 2x \le 0 \end{cases}$

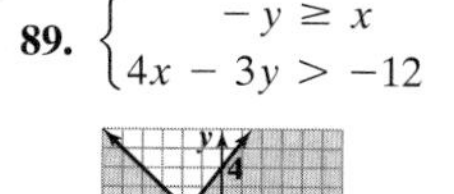
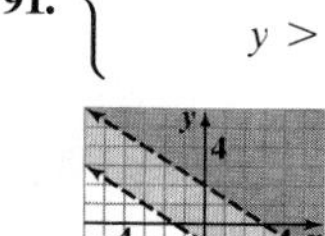
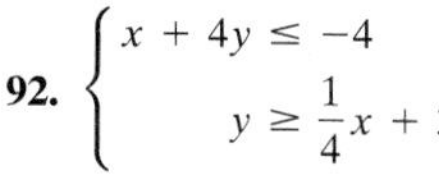
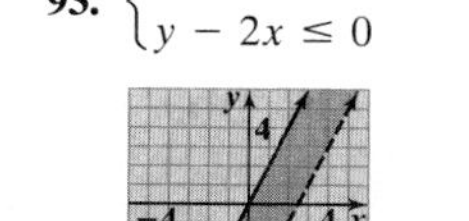
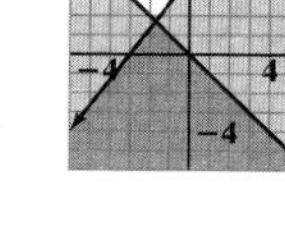
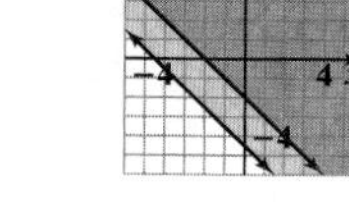
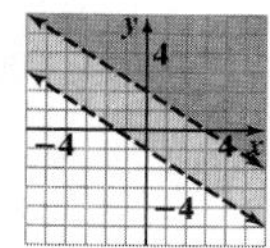
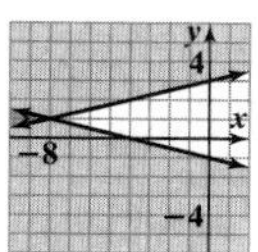

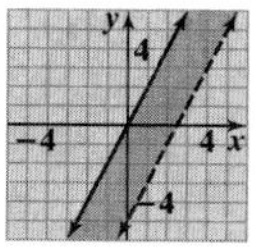

94. $\begin{cases} y > x + 2 \\ y < x - 4 \end{cases}$ **95.** x = fish, y = carne asada

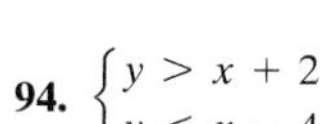
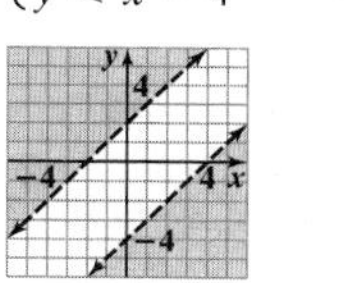
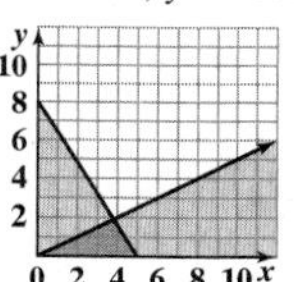

Chapter 9 Test

1. **(a)** No **(b)** No **(c)** Yes **2.** **(a)** Yes **(b)** No **(c)** No **3.** **(a)** one **(b)** consistent **(c)** independent
4. **(a)** none **(b)** inconsistent **(c)** not applicable

5. $\begin{cases} 2x + 3y = 0 \\ x + 4y = 5 \end{cases}$ $(-3, 2)$ **6.** $\begin{cases} y = 2x - 6 \\ y = -\frac{1}{4}x + 3 \end{cases}$ $(4, 2)$

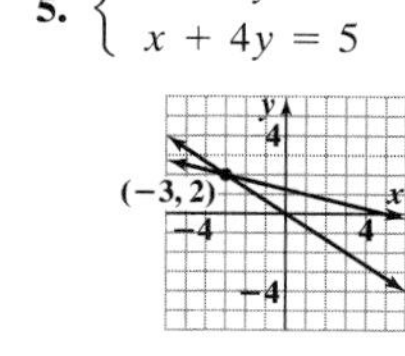
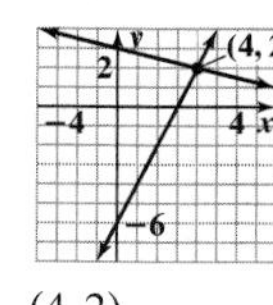

7. $(0, -3)$ **8.** $(-12, -1)$ **9.** $(-3, 3)$ **10.** $(-2, -4)$ **11.** $\left(3, -\frac{1}{2}\right)$ **12.** $(2.5, 3)$
13. infinitely many solutions **14.** airplane 200 mph; wind 25 mph
15. 7 containers vanilla; 4 containers peach **16.** 52.5°, 127.5° **17.** 15 basketballs, 25 volleyballs

18. $\begin{cases} 4x - 2y < 8 \\ x + 3y < 6 \end{cases}$ **19.** $\begin{cases} x \le -2 \\ -2x - 4y \le 8 \end{cases}$ **20.** $\begin{cases} y \le \frac{2}{3}x + 4 \\ -2x + 3y > -3 \end{cases}$

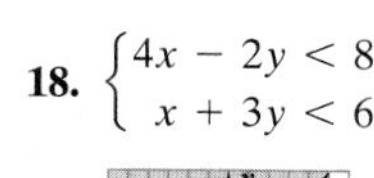
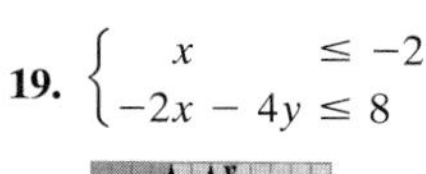
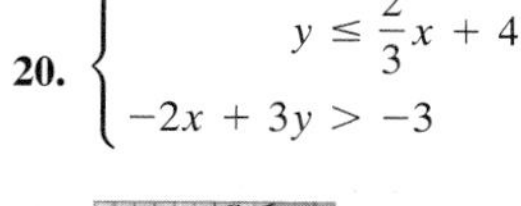
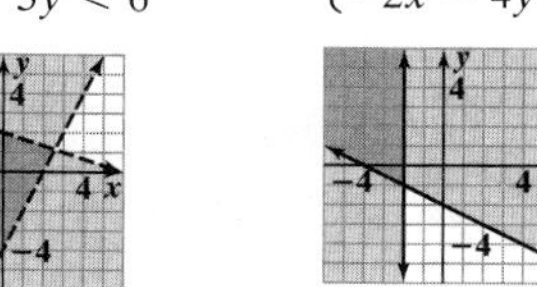
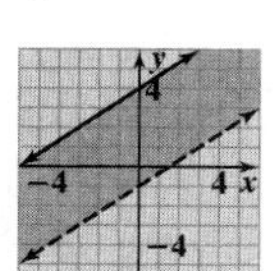

Chapters 1–9 Cumulative Review

1. $8x + 23$ **2.** $\frac{y^4}{81x^2}$ **3.** $\frac{x^2 + 2x + 4}{x + 2}$ **4.** $34 - 24\sqrt{2}$ **5.** $\frac{x - 6}{5 - 4x}$ **6.** 9 **7.** $\frac{1}{\sqrt{-729}}$ is not a real number. **8.** $-\frac{1}{27}$ **9.** $\left\{\frac{5}{2}, -5\right\}$

10. $\{-2, 0, 2\}$ **11.** no solution **12.** $\{6\}$ **13.** $\{x \mid x \le 5\}$; $(-\infty, 5]$ (number line: 0, 5) **14.** $q = \frac{3pr}{2r - 6p}$ **15.** 80% **16.** $2\sqrt{6}$

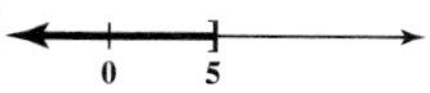

17. $-\frac{7}{6}$ **18.** **(a)** $\frac{4}{3}$ **(b)** $(0, 2)$ **(c)** $\left(-\frac{3}{2}, 0\right)$ **19.** $y = -\frac{3}{2}x - 8$ **20.** $(a^8 + 4b^2)(a^4 + 2b)(a^4 - 2b)$ **21.** $2a(2a + 5)(a - 3)$

22. 4.32×10^2 **23.** $\{2 - \sqrt{5}i, 2 + \sqrt{5}i\}$ **24.** $-7, -5, -3$ **25.** 5 and 3

Chapter 10

Section 10.1

1. parabola **3.** vertex; $x = -\dfrac{b}{2a}$ **5.** False **7.** the discriminant; 1

9. $y = x^2$

11. $y = -x^2 + 4$

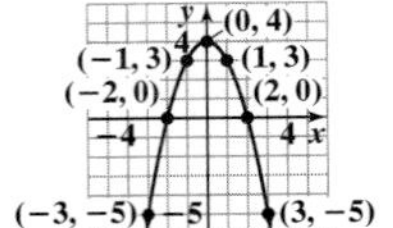

13. $y = -x^2 + 3x + 4$

15. $y = 2x^2 + 5x - 3$

17. x-intercepts: 2, 4; y-intercept: -8

19. x-intercepts: $-\dfrac{1}{3}, \dfrac{1}{2}$; y-intercept: -1

21. x-intercepts: $-3, 0$; y-intercept: 0

23. x-intercepts: none; y-intercept: -5

25. x-intercepts: $\dfrac{1 - \sqrt{17}}{4} \approx -0.78, \dfrac{1 + \sqrt{17}}{4} \approx 1.28$; y-intercepts: 2

27. x-intercepts: $-2 - \sqrt{2} \approx -3.41$, $-2 + \sqrt{2} \approx -0.59$; y-intercepts: 2

29. (a) up **(b)** $(0, 6)$ **(c)** $x = 0$

31. (a) down **(b)** $(4, 1)$ **(c)** $x = 4$

33. (a) down **(b)** $\left(\dfrac{3}{4}, \dfrac{49}{8}\right)$ **(c)** $x = \dfrac{3}{4}$

35. (a) up **(b)** $\left(\dfrac{4}{3}, -\dfrac{4}{3}\right)$ **(c)** $x = \dfrac{4}{3}$

37. $y = x^2 + 2x - 3$
opens up;
x-intercepts: $-3, 1$
y-intercept: -3
axis of symmetry: $x = -1$

39. $y = -x^2 - 2x$
opens down;
x-intercepts: $-2, 0$
y-intercept: 0
axis of symmetry: $x = -1$

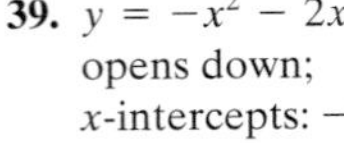

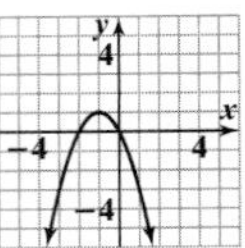

41. $y = x^2 - 8x + 12$
opens up;
x-intercepts: 2, 6
y-intercept: 12
axis of symmetry: $x = 4$

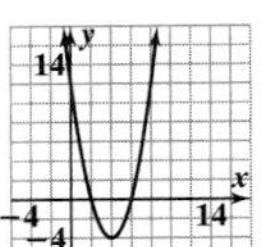

43. $y = -x^2 - 8x - 9$
opens down;
x-intercepts: $-4 - \sqrt{7}, -4 + \sqrt{7}$
y-intercept: -9
axis of symmetry: $x = -4$

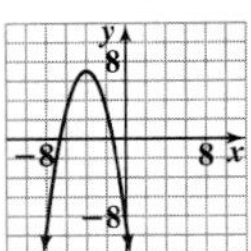

45. $y = x^2 - 9$
opens up;
x-intercepts: $-3, 3$
y-intercept: -9
axis of symmetry: $x = 0$

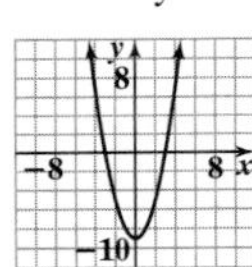

47. $y = x^2 - 7x + 10$
opens up;
x-intercepts: 2, 5
y-intercept: 10
axis of symmetry: $x = \dfrac{7}{2}$

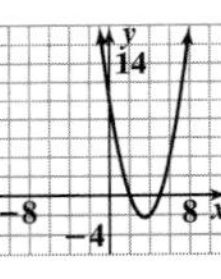

49. $y = -3x^2 + 4x + 4$
opens down;
x-intercepts: $-\dfrac{2}{3}, 2$
y-intercept: 4
axis of symmetry: $x = \dfrac{2}{3}$

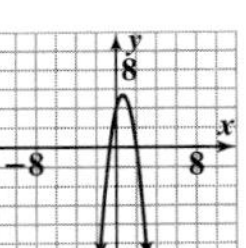

51. $y = -2x^2 - 4x + 5$
opens down;
x-intercepts: $\dfrac{-2 - \sqrt{14}}{2}, \dfrac{-2 + \sqrt{14}}{2}$
y-intercept: 5
axis of symmetry: $x = -1$

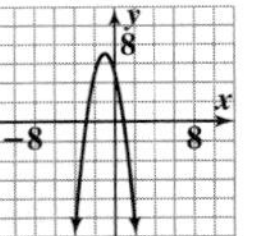

53. $y = x^2 - 2x - 5$
opens up;
x-intercepts: $1 - \sqrt{6}, 1 + \sqrt{6}$
y-intercept: -5
axis of symmetry: $x = 1$

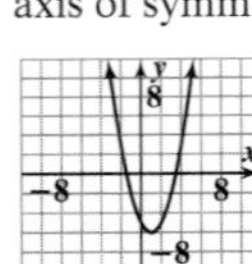

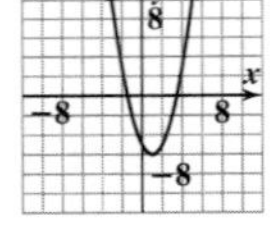

55. minimum; -4 at $x = 3$ **57.** minimum; 7 at $x = -2$ **59.** maximum; $\dfrac{15}{2}$ at $x = \dfrac{3}{2}$

61. maximum; $-\dfrac{19}{4}$ at $x = -\dfrac{3}{2}$

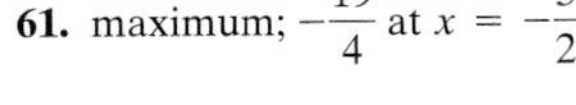

63. (a) 125 ft **(b)** 68.5 ft **65. (a)** 25 yd **(b)** 625 yd^2 **67. (a)** 50 ft **(b)** 5000 ft^2

69. $y = x^2 + 6x + 10$
opens up;
vertex $(-3, 1)$
axis of symmetry $x = -3$
y-intercept: 10
x-intercepts: none

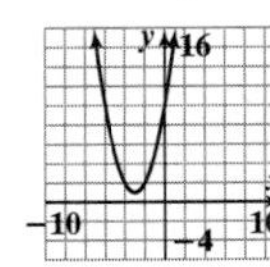

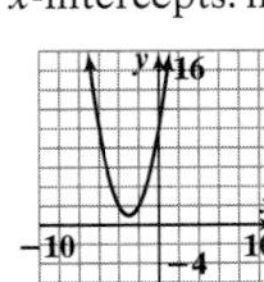

71. $y = -2x^2 + 6x - 7$
opens down;
vertex $\left(\dfrac{3}{2}, -\dfrac{5}{2}\right)$
axis of symmetry $x = \dfrac{3}{2}$
y-intercept: -7
x-intercepts: none

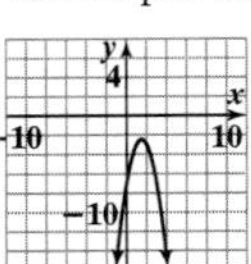

73. $y = 4x^2 - 4x + 1$
opens up;
vertex $\left(\dfrac{1}{2}, 0\right)$
axis of symmetry $x = \dfrac{1}{2}$
y-intercept: 1
x-intercept: $\dfrac{1}{2}$

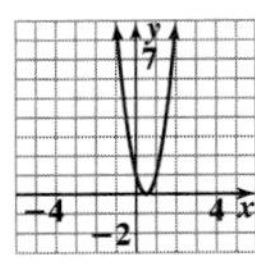

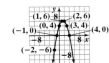

Section 10.2

1. depends on **3.** relation **5.** False **7.** Answers may vary. **9.** {(Physician, 120), (Teacher, 37), (Mathematician, 64), (Computer Engineer, 64), (Lawyer, 82)}; Domain: {(Physician, Teacher, Mathematician, Computer Engineer, Lawyer}; Range: {120, 37, 64, 82}

11. {(U.S., 97), (Great Britain, 28), (Korea, 28), (Poland, 14), (Canada, 14)}
Domain: {U.S., Great Britain, Korea, Poland, Canada}
Range: {97, 28, 14}

13.
Domain: $\{-1, -4, 0, 1\}$
Range: $\{3, 2, 1\}$

15.

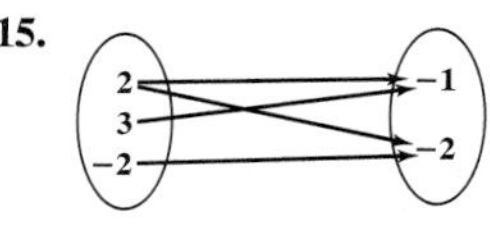

Domain: $\{2, 3, -2\}$
Range: $\{-1, -2\}$

17.

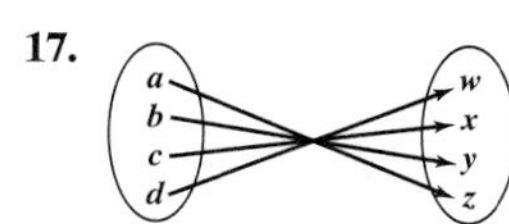

Domain: $\{a, b, c, d\}$
Range: $\{w, x, y, z\}$

19. Domain: $\{-3, 0, 2, 3\}$
Range: $\{-2, 0, 2, 4\}$

21. Domain: $\{x|x$ is a real number$\}$
Range: $\{y|y$ is a real number$\}$

23. Domain: $\{3\}$
Range: $\{y|y$ is a real number$\}$

25. Domain: $\{x|x$ is a real number$\}$
Range: $\{y|y \leq 0\}$

27. Domain: $\{x|x$ is a real number$\}$
Range: $\{y|y \geq -2\}$

29. Domain: $\{x|x \geq -1\}$
Range: $\{y|y$ is a real number$\}$

31. Domain: $\{x|-1 \leq x \leq 5\}$
Range: $\{y|-1 \leq y \leq 5\}$

33. Domain: $\{x|-4 \leq x \leq 4\}$

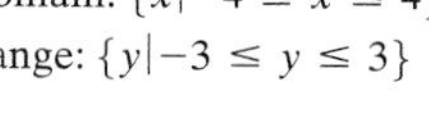

Range: $\{y|-3 \leq y \leq 3\}$

35. $y = -2x + 1$

Domain: $\{x|x$ is a real number$\}$
Range: $\{y|y$ is a real number$\}$

37. $y = -\frac{2}{3}x - 1$

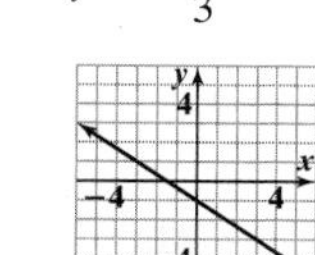

Domain: $\{x|x$ is a real number$\}$
Range: $\{y|y$ is a real number$\}$

39. $y = 4x - 2x^2$

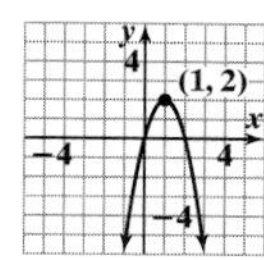

Domain: $\{x|x$ is a real number$\}$
Range: $\{y|y \leq 2\}$

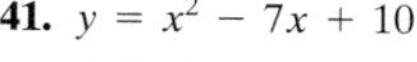

41. $y = x^2 - 7x + 10$

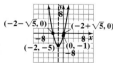

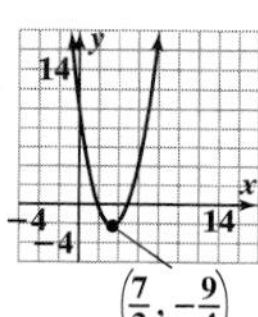

Domain: $\{x|x$ is a real number$\}$
Range: $\left\{y \mid y \geq -\frac{9}{4}\right\}$

43. $y = 4x$

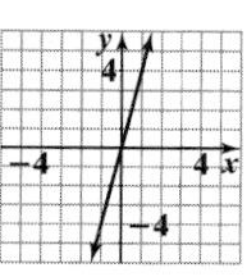

Domain: $\{x|x$ is a real number$\}$
Range: $\{y|y$ is a real number$\}$

45. $y = 2x^2 - 5$

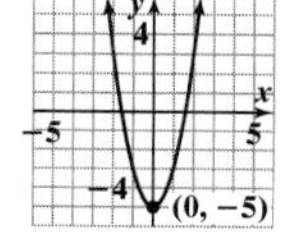

Domain: $\{x|x$ is a real number$\}$
Range: $\{y|y \geq -5\}$

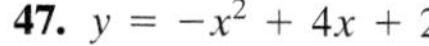

47. $y = -x^2 + 4x + 2$

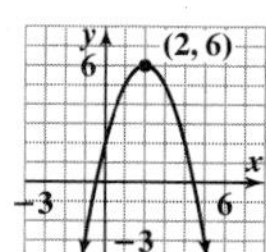

Domain: $\{x|x$ is a real number$\}$
Range: $\{y|y \leq 6\}$

49. $3x + 4y = 8$

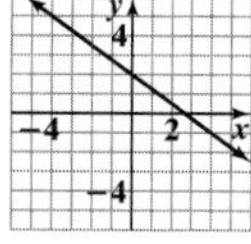

Domain: $\{x|x$ is a real number$\}$
Range: $\{y|y$ is a real number$\}$

51. (a) Domain: $\{t|0 \leq t \leq 10\}$
Range: $\{p|0 \leq p \leq 900\}$
(b) day 5 **(c)** 5 days

53. (a) Domain: $\{w|0 < w < 4\}$
Range: $\{3.85, 4.55, 6.05, 7.05\}$
(b) \$3.85 **(c)** \$4.55

55. Answers may vary. **57. (a)** No **(b)** Answers may vary. **(c)** No; answers may vary. **59.** Answers may vary.

Putting the Concepts Together

1. (a) Opens up **(b)** $(1, -4)$ **(c)** $x = 1$ **(d)** $(0, -3), (-1, 0), (3, 0)$

2.

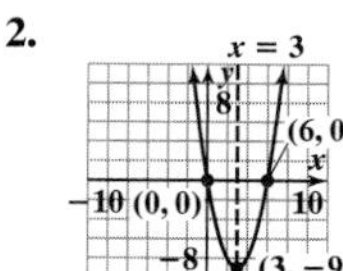

Domain: $\{x|x$ is a real number$\}$
Range: $\{y|y \geq -9\}$

3.
Domain: $\{x|x$ is a real number$\}$
Range: $\{y|y \geq -5\}$

4.

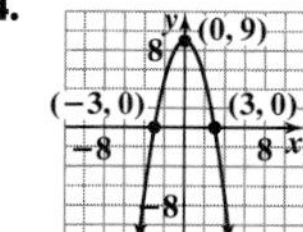

Domain: $\{x|x$ is a real number$\}$
Range: $\{y|y \leq 9\}$

5. (a) Minimum **(b)** -21
6. Domain: $\{a, b, c\}$
Range: $\{1, 2\}$
7. Domain: $\{1, 3\}$
Range: $\{-1, -3\}$
8. Domain: $\{x|x \geq -4\}$
Range: $\{y|y$ is a real number$\}$
9. Domain: $\{x|1 \leq x \leq 7\}$
Range: $\{y|-1 \leq y \leq 5\}$

Section 10.3

1. function **3.** *f* of *x* **5.** True **7.** Maps, ordered pairs, equations, graphs; answers may vary. **9.** not a function
11. function; Domain: {Apples, Pears, Oranges, Grapes} Range: {\$0.59, \$0.39, \$1.69} **13.** function; Domain: $\{-1, 0, 1, 2\}$ Range: $\{-1, 0, 1, 2\}$ **15.** not a function **17.** function; Domain: $\{a, b, c, d\}$ Range: $\{a, b\}$
19. function **21.** not a function **23.** function **25.** function **27.** not a function **29.** not a function **31.** not a function **33.** not a function **35.** function **37.** function **39.** not a function **41.** function **43.** not a function **45.** function **47.** **(a)** -3 **(b)** -6 **(c)** -1 **49.** **(a)** 4 **(b)** -2 **(c)** 8 **51.** **(a)** 3 **(b)** 3 **(c)** 3 **53.** **(a)** -3 **(b)** $-\frac{3}{2}$ **(c)** -4
55. **(a)** 3 **(b)** -12 **(c)** 13 **57.** **(a)** 0 **(b)** 0 **(c)** 10 **59.** **(a)** 1 **(b)** -32 **(c)** 3 **61.** 18 **63.** -8 **65.** 0 **67.** 2
69. $-\frac{1}{2}$ **71.** 7 **73.** $2\sqrt{3}$ **75.** $C = 5$ **77.** $C = 3$ **79.** $A = 5$ **81.** $G(h) = 12h$; \$240 **83.** **(a)** $h(30) = 8.132$; in 1930, there were 8.132 homicides per 100,000 people. **(b)** $h(50) = 6.632$; in 1950, there were 6.632 homicides per 100,000 people. **85.** $A(r) = \pi r^2$; $A(3) = 9\pi$ in.^{2}
87. $x = 2$ **89.** $x = 2$ **91.** $x = -4$ **93.** **(a)** -5 **(b)** $1 - 2h$ **(c)** $-5 - 2h$ **(d)** $-4 - 2h$ **(e)** $-2h$

Chapter 10 Review

1. $y = -x^2$ **2.** $y = x^2$ **3.** $y = 2x^2 - 3$ **4.** $y = -\frac{1}{2}x^2 + 4$ **5.** $y = x^2 - 4x - 3$ **6.** $y = -2x^2 + 4x + 1$

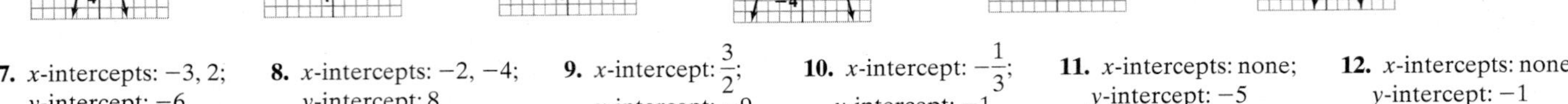

7. x-intercepts: $-3, 2$; y-intercept: -6 **8.** x-intercepts: $-2, -4$; y-intercept: 8 **9.** x-intercept: $\frac{3}{2}$; y-intercept: -9 **10.** x-intercept: $-\frac{1}{3}$; y-intercept: -1 **11.** x-intercepts: none; y-intercept: -5 **12.** x-intercepts: none; y-intercept: -1
13. x-intercepts: $2 - \sqrt{2}, 2 + \sqrt{2}$; y-intercept: 2 **14.** x-intercepts: $-2 - 3\sqrt{2}, -2 + 3\sqrt{2}$; y-intercept: 14 **15.** **(a)** up **(b)** $(0, 5)$ **(c)** $x = 0$ **16.** **(a)** down **(b)** $(0, -4)$ **(c)** $x = 0$
17. **(a)** down **(b)** $(1, 1)$ **(c)** $x = 1$ **18.** **(a)** up **(b)** $(2, -8)$ **(c)** $x = 2$ **19.** **(a)** up **(b)** $\left(-3, -\frac{27}{2}\right)$ **(c)** $x = -3$
20. **(a)** down **(b)** $(-2, 6)$ **(c)** $x = -2$ **21.** **(a)** down **(b)** $\left(\frac{3}{2}, -\frac{7}{4}\right)$ **(c)** $x = \frac{3}{2}$ **22.** **(a)** up **(b)** $\left(-\frac{5}{2}, -\frac{5}{4}\right)$ **(c)** $x = -\frac{5}{2}$

23. $y = x^2 + 2x - 3$ opens up; vertex $(-1, -4)$ axis of symmetry $x = -1$ x-intercepts: $-3, 1$ y-intercept: -3
24. $y = -x^2 + 6x - 8$ opens down; vertex $(3, 1)$ axis of symmetry $x = 3$ x-intercepts: 2, 4 y-intercept: -8
25. $y = -x^2 + 4$ opens down; vertex $(0, 4)$ axis of symmetry $x = 0$ x-intercepts: $-2, 2$ y-intercept: 4
26. $y = x^2 - 9$ opens up; vertex $(0, -9)$ axis of symmetry $x = 0$ x-intercepts: $-3, 3$ y-intercept: -9

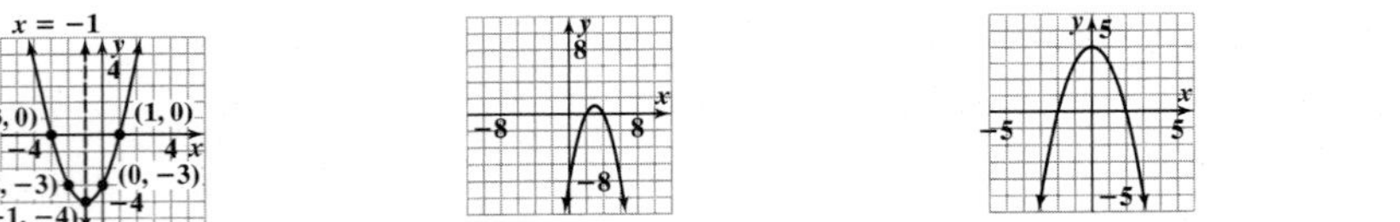

27. $y = -x^2 + 3x + 4$ opens down; vertex $\left(\frac{3}{2}, \frac{25}{4}\right)$ axis of symmetry $x = \frac{3}{2}$ x-intercepts: $-1, 4$ y-intercept: 4
28. $y = x^2 - 5x + 4$ opens up; vertex $\left(\frac{5}{2}, -\frac{9}{4}\right)$ axis of symmetry $x = \frac{5}{2}$ x-intercepts: 1, 4 y-intercept: 4
29. maximum; 15 at $x = -2$ **30.** minimum; -16 at $x = -2$ **31.** minimum; -10 at $x = 2$ **32.** maximum; 10 at $x = -5$ **33.** **(a)** 92.9 ft **(b)** 20.0 ft **(c)** 200 ft **34.** **(a)** 100 m **(b)** 10,000 m^2
35. {(Alabama, 38), (California, 54), (Florida, 45), (Nebraska, 41), (New York, 55), (Ohio, 41), (South Dakota, 37)} **36.** {(History, 400), (Math, 400), (Music, 200), (Psychology, 225)}
37. Domain: {Alabama, California, Florida, Nebraska, New York, Ohio, South Dakota} Range: {37, 38, 41, 45, 54, 55}
38. Domain: {History, Math, Music, Psychology} Range: {200, 225, 400}
39. Domain: $\{-1, 0\}$ Range: $\{3, 5, -2, -4\}$
40. Domain: $\{a, b, c, d\}$ Range: $\{x^2, x^3, x^4, x^5\}$ **41.** Domain: $\{a_1, a_2, a_3, a_4\}$ Range: {1, 4, 9, 16} **42.** Domain: {1, 3, 5, 7} Range: $\{-3, 2, -2\}$ **43.** Domain: $\{-3, 0, 3\}$ Range: $\{-3, 0, 3\}$ **44.** Domain: {0, 1, 4} Range: $\{-4, 0, 4\}$
45. Domain: $\{x \mid x$ is a real number$\}$ Range: $\{y \mid y$ is a real number$\}$ **46.** Domain: $\{x \mid x$ is a real number$\}$ Range: {4} **47.** Domain: {5} Range: $\{y \mid y$ is a real number$\}$

48. Domain: $\{x|x \text{ is a real number}\}$ Range: $\{y|y \text{ is a real number}\}$ **49.** Domain: $\{x|0 \le x \le 6\}$ Range: $\{y|1 \le y \le 5\}$ **50.** Domain: $\{x|-4 \le x \le 0\}$ Range: $\{y|-4 \le x \le 3\}$

51. $y = \frac{-3x}{2} + 4$

Domain: $\{x|x \text{ is a real number}\}$ Range: $\{y|y \text{ is a real number}\}$

52. $y = \frac{4x}{3} - 5$

Domain: $\{x|x \text{ is a real number}\}$ Range: $\{y|y \text{ is a real number}\}$

53. $y = 8x - 4x^2$

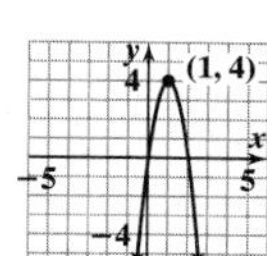

Domain: $\{x|x \text{ is a real number}\}$ Range: $\{y|y \le 4\}$

54. $y = x^2 - 4x + 3$

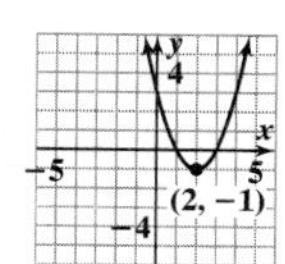

Domain: $\{x|x \text{ is a real number}\}$ Range: $\{y|y \ge -1\}$

55. not a function **56.** function; Domain: {Less than High School Degree, High School Graduate, Some College/AA Degree, Bachelor's Degree and more}; Range: $\{3.77, 3.88, 4.22\}$

57. function; Domain: $\{-4, -3, -2, -1\}$ Range: $\{0, 1\}$ **58.** not a function **59.** not a function **60.** function; Domain: $\{5, 3, 1, -1\}$ Range: $\{x, z\}$ **61.** function **62.** not a function **63.** not a function **64.** function

65. not a function **66.** function **67.** function **68.** not a function **69.** not a function **70.** function **71.** not a function **72.** function **73.** function **74.** not a function **75.** function **76.** function **77.** not a function **78.** not a function **79.** 1 **80.** 18 **81.** -3 **82.** -10 **83.** 14 **84.** 7 **85.** $2\sqrt{6}$ **86.** 5 **87. (a)** $C(m) = 0.10m + 75$ **(b)** \$88

Chapter 10 Test

1. (a) up **(b)** $(-2, -9)$ **(c)** $x = -2$ **(d)** x-intercepts: $-5, 1$ y-intercept: -5 **2.** $y = -x^2 + 6x - 9$ **3.** $y = x^2 - 4$ **4.** $y = x^2 - 4x - 3$

5. (a) maximum **(b)** 7 at $x = -1$ **6.** Domain: $\{1\}$ Range: $\{-1, -2, -3\}$ **7.** Domain: $\{-2, -3\}$ Range: $\{a, b, c\}$

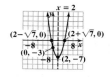

8. Domain: $\{x|x \text{ is a real number}\}$ Range: $\{y|y \ge -2\}$ **9.** Domain: $\{x|-5 \le x \le 3\}$ Range: $\{y|0 \le y \le 4\}$ **10.** not a function **11.** function **12.** function **13.** function **14.** not a function **15.** function **16.** -5 **17.** $\frac{5}{16}$ **18.** 8

19. (a) $\frac{3}{2}$ sec **(b)** $\frac{27}{2}$ ft **20. (a)** \$3 **(b)** 9.8 years **(c)** \$35 **(d)** \$27

Appendix A

Section A.1

1. factors, product **3.** equivalent fractions **5.** False **7.** Answers will vary. **9.** $5 \cdot 5$ **11.** $2 \cdot 2 \cdot 7$ **13.** $3 \cdot 7$ **15.** $2 \cdot 2 \cdot 3 \cdot 3$ **17.** $2 \cdot 2 \cdot 5$ **19.** $2 \cdot 3 \cdot 5$ **21.** $2 \cdot 5 \cdot 5$ **23.** 53 is prime. **25.** $2 \cdot 2 \cdot 3 \cdot 3 \cdot 7$ **27.** $2 \cdot 2 \cdot 2 \cdot 2 \cdot 2 \cdot 2 \cdot 2 \cdot 2$ **29.** $2 \cdot 2 \cdot 5 \cdot 5 \cdot 13$ **31.** $5 \cdot 5 \cdot 7 \cdot 13$ **33.** 30 **35.** 60 **37.** 210 **39.** 90 **41.** 42 **43.** 60 **45.** 72 **47.** 72 **49.** $\frac{7}{10}$ **51.** $\frac{5}{6}$ **53.** $\frac{8}{12}$ **55.** $\frac{18}{24}$ **57.** $\frac{21}{3}$ **59.** $\frac{4}{8}$ and $\frac{3}{8}$ **61.** $\frac{9}{15}$ and $\frac{10}{15}$ **63.** $\frac{20}{24}$ and $\frac{15}{24}$ **65.** $\frac{44}{60}$ and $\frac{21}{60}$ **67.** $\frac{27}{48}$ and $\frac{14}{48}$ **69.** $\frac{12}{96}$ and $\frac{21}{96}$ **71.** $\frac{20}{90}$ and $\frac{35}{90}$ and $\frac{21}{90}$ **73.** $\frac{2}{3}$ **75.** $\frac{19}{9}$ **77.** $\frac{1}{2}$ **79.** $\frac{4}{5}$ **81.** $\frac{3}{5}$ **83.** $\frac{2}{3}$ **85.** 84 months **87.** 20 days **89.** $\frac{13}{20}$ **91.** $\frac{3}{5}$ **93.** $\frac{2}{3}$ **95.** 2, 3, 5, 7, 11, 13, 17, 19, 23, 29, 31, 37, 41, 43, 47, 53, 59, 61, 67, 71, 73, 79, 83, 89, 97

Section A.2

1. terminating **3.** ones **5.** True **7.** Answers will vary. **9.** hundredths place **11.** tens place **13.** thousandths place **15.** 578.2 **17.** 350 **19.** 3682.010 **21.** 30 **23.** 0.625 **25.** $0.\overline{285714}$ **27.** 0.3125 **29.** $0.\overline{230769}$ **31.** 1.16 **33.** 2.2 **35.** 0.29 **37.** 2.67 **39.** 0.519 **41.** $\frac{3}{4}$ **43.** $\frac{9}{10}$ **45.** $\frac{491}{500}$ **47.** $\frac{101}{400}$ **49.** 0.37 **51.** 0.0602 **53.** 0.001 **55.** 20% **57.** 27.5% **59.** 200% **61.** 70% **63.** 77.27% **65. (a)** 33.33% **(b)** 16.67% **(c)** 25% **67.** 21.43%

Appendix C

Section C.1

1. congruent **3.** right **5.** supplementary **7.** False **9.** acute **11.** right **13.** straight **15.** obtuse **17.** 58° **19.** 17° **21.** 113° **23.** 172° **25.** $m\angle 1 = 130°$; $m\angle 2 = 50°$; $m\angle 3 = 130°$; $m\angle 4 = 50°$; $m\angle 5 = 130°$; $m\angle 6 = 50°$; $m\angle 7 = 130°$

Section C.2

1. regular **3.** isosceles **5.** True **7.** True **9.** 55° **11.** 48° **13.** 4 units **15.** 67.5 units **17.** 10 inches **19.** 5 cm **21.** 7 cm **23.** $\frac{11}{2}$ yards or 5.5 yards

Section C.3

1. perimeter **3.** $\frac{1}{2}h(b + B); h; b; B$ **5.** False **7.** Perimeter: 28 feet; Area: 40 square feet **9.** Perimeter: 40 m; Area: 75 m^2 **11.** Perimeter: 24 km; Area: 36 km^2 **13.** Perimeter: 72 feet; Area: 218 square feet **15.** Perimeter: 54 m; Area: 62 m^2 **17.** Perimeter: 30 feet; Area: 45 square feet **19.** Perimeter: 28 mm; Area: 36 mm^2 **21.** Perimeter: 40 in; Area: 84 $in.^2$ **23.** Perimeter: 45 cm; Area: 94.5 cm^2 **25.** Perimeter: 32 m; Area: 42 m^2 **27.** Perimeter: 32 ft; Area: 24 ft^2 **29.** Circumference: 32π in. $\approx$ 100.53 in; Area: 256π $in.^2 \approx 804.25$ $in.^2$ **31.** Circumference: 20π cm $\approx$ 62.83 cm; Area: 100π $cm^2 \approx 314.16$ cm^2 **33.** π square units **35.** about 26.18 feet

Section C.4

1. polyhedron **3.** surface area **5.** False **7.** Volume: 600 cubic feet; Surface Area: 460 square feet **9.** Volume: 288π cubic centimeters $\approx$ 904.78 cubic centimeters; Surface Area: 144π square centimeters $\approx$ 452.39 square centimeters **11.** Volume: 32π cubic inches $\approx$ 100.53 cubic inches; Surface Area: 40π square inches $\approx$ 125.66 square inches **13.** Volume: $\frac{800}{3}\pi$ cubic millimeters $\approx$ 837.76 cubic millimeters **15.** Volume: $\frac{640}{3}$ cubic feet; Surface Area: 256 square feet **17.** 1728 cubic inches **19.** 75.40 cubic inches; 100.53 square inches **21.** approximately 268.08 cubic centimeters

Applications Index

Subject Index

Photo Credits

Chapter 1	Pages 1, 28, Cosmo Condina/The Stock Connection
Chapter 2	Pages 74, 116, Randy M. Ury/CORBIS-NY
Chapter 3	Pages 173, 183, Getty Images-Digital Vision; Page 227, Karl Ronstrom/CORBIS/Reuters America LLC
Chapter 4	Page 237, Tim Boyle/Getty Images
Chapter 5	Pages 304, 364, Carlo Orlandi/AP Wide World Photos
Chapter 6	Pages 392, 439, Freudental Verhagen/Getty Images Inc.-Stone Allstock; Page 399, Norbert Schiller/The Image Works
Chapter 7	Pages 457, 495, Patrick Bennett/CORBIS-NY
Chapter 8	Page 519, Getty Images; Page 520, U.S. Air Force
Chapter 9	Page 616, Patrik Giardino/CORBIS/Bettmann
Chapter 10	Page 688, Michael Steele/Getty Images; Page 693, Larry L. Miller/Photo Researchers, Inc.

CHAPTER TEST PREP VIDEO CD
ELEMENTARY ALGEBRA

Michael Sullivan, III & Katherine R. Struve

ISBN 0-13-188784-X

CD License Agreement

YOU SHOULD CAREFULLY READ THE TERMS AND CONDITIONS BEFORE USING THE CD-ROM PACKAGE. USING THIS CD-ROM PACKAGE INDICATES YOUR ACCEPTANCE OF THESE TERMS AND CONDITIONS.

Pearson Education, Inc. provides this program and licenses its use. You assume responsibility for the selection of the program to achieve your intended results, and for the installation, use, and results obtained from the program. This license extends only to use of the program in the United States or countries in which the program is marketed by authorized distributors.

LICENSE GRANT
You hereby accept a nonexclusive, nontransferable, permanent license to install and use the program ON A SINGLE COMPUTER at any given time. You may copy the program solely for backup or archival purposes in support of your use of the program on the single computer. You may not modify, translate, disassemble, decompile, or reverse engineer the program, in whole or in part.

TERM
The License is effective until terminated. Pearson Education, Inc. reserves the right to terminate this License automatically if any provision of the License is violated. You may terminate the License at any time. To terminate this License, you must return the program, including documentation, along with a written warranty stating that all copies in your possession have been returned or destroyed.

LIMITED WARRANTY
THE PROGRAM IS PROVIDED "AS IS" WITHOUT WARRANTY OF ANY KIND, EITHER EXPRESSED OR IMPLIED, INCLUDING, BUT NOT LIMITED TO, THE IMPLIED WARRANTIES OF MERCHANTABILITY AND FITNESS FOR A PARTICULAR PURPOSE. THE ENTIRE RISK AS TO THE QUALITY AND PERFORMANCE OF THE PROGRAM IS WITH YOU. SHOULD THE PROGRAM PROVE DEFECTIVE, YOU (AND NOT PEARSON EDUCATION, INC. OR ANY AUTHORIZED DEALER) ASSUME THE ENTIRE COST OF ALL NECESSARY SERVICING, REPAIR, OR CORRECTION. NO ORAL OR WRITTEN INFORMATION OR ADVICE GIVEN BY PEARSON EDUCATION, INC., ITS DEALERS, DISTRIBUTORS, OR AGENTS SHALL CREATE A WARRANTY OR INCREASE THE SCOPE OF THIS WARRANTY.

SOME STATES DO NOT ALLOW THE EXCLUSION OF IMPLIED WARRANTIES, SO THE ABOVE EXCLUSION MAY NOT APPLY TO YOU. THIS WARRANTY GIVES YOU SPECIFIC LEGAL RIGHTS AND YOU MAY ALSO HAVE OTHER LEGAL RIGHTS THAT VARY FROM STATE TO STATE.

Pearson Education, Inc. does not warrant that the functions contained in the program will meet your requirements or that the operation of the program will be uninterrupted or error-free. However, Pearson Education, Inc. warrants the CD-ROM(s) on which the program is furnished to be free from defects in material and workmanship under normal use for a period of ninety (90) days from the date of delivery to you as evidenced by a copy of your receipt. The program should not be relied on as the sole basis to solve a problem whose incorrect solution could result in injury to person or property. If the program is employed in such a manner, it is at the user's own risk and Pearson Education, Inc. explicitly disclaims all liability for such misuse.

LIMITATION OF REMEDIES
Pearson Education, Inc.'s entire liability and your exclusive remedy shall be:
1. the replacement of any CD-ROM not meeting Pearson Education, Inc.'s "LIMITED WARRANTY" and that is returned to Pearson Education, or
2. if Pearson Education is unable to deliver a replacement CD-ROM that is free of defects in materials or workmanship, you may terminate this agreement by returning the program.

IN NO EVENT WILL PEARSON EDUCATION, INC. BE LIABLE TO YOU FOR ANY DAMAGES, INCLUDING ANY LOST PROFITS, LOST SAVINGS, OR OTHER INCIDENTAL OR CONSEQUENTIAL DAMAGES ARISING OUT OF THE USE OR INABILITY TO USE SUCH PROGRAM EVEN IF PEARSON EDUCATION, INC. OR AN AUTHORIZED DISTRIBUTOR HAS BEEN ADVISED OF THE POSSIBILITY OF SUCH DAMAGES, OR FOR ANY CLAIM BY ANY OTHER PARTY.

SOME STATES DO NOT ALLOW FOR THE LIMITATION OR EXCLUSION OF LIABILITY FOR INCIDENTAL OR CONSEQUENTIAL DAMAGES, SO THE ABOVE LIMITATION OR EXCLUSION MAY NOT APPLY TO YOU.

GENERAL
You may not sublicense, assign, or transfer the license of the program. Any attempt to sublicense, assign or transfer any of the rights, duties, or obligations hereunder is void.

This Agreement will be governed by the laws of the State of New York.

Should you have any questions concerning this Agreement, you may contact Pearson Education, Inc. by writing to:
ESM Media Development
Higher Education Division
Pearson Education, Inc.
1 Lake Street
Upper Saddle River, NJ 07458

Should you have any questions concerning technical support, you may write to:
New Media Production
Higher Education Division
Pearson Education, Inc.
1 Lake Street
Upper Saddle River, NJ 07458
YOU ACKNOWLEDGE THAT YOU HAVE READ THIS AGREEMENT, UNDERSTAND IT, AND AGREE TO BE BOUND BY ITS TERMS AND CONDITIONS. YOU FURTHER AGREE THAT IT IS THE COMPLETE AND EXCLUSIVE STATEMENT OF THE AGREEMENT BETWEEN US THAT SUPERSEDES ANY PROPOSAL OR PRIOR AGREEMENT, ORAL OR WRITTEN, AND ANY OTHER COMMUNICATIONS BETWEEN US RELATING TO THE SUBJECT MATTER OF THIS AGREEMENT.

System Requirements
-Windows
Pentium II 300-MHz processor-based computer or better
Windows 98, NT, 2000, ME, or XP
In addition to the minimum RAM required by the operating system, this CD-ROM requires 64 MB RAM (128 MB RAM required for Windows XP)
4.3 available hard drive space (optional—for minimum QuickTime installation)
800 × 600 resolution
8x or faster CD drive (required)
QuickTime 6.x (required). See Technical Support section below for information on QuickTime.
Sound Card
Internet Explorer 5.5, 6.0 or Netscape 6.23, 7.2.

-Macintosh
Power PC G3 233 MHz or better
Mac OS 9.x or 10.x
In addition to the minimum RAM required by the operating system, this CD-ROM requires 64 MB RAM
10 MB available hard drive space for Mac OS 9, 19 MB on OS X (optional—if QuickTime installation is needed)
800 × 600 resolution monitor
8x or faster CD drive (required)
QuickTime 6 is required. See Support section below for information on downloading QuickTime.
Internet browser - Mac 9.2.2: Internet Explorer 5.1 or Netscape 6.23
OSX 10.3.7: Internet Explorer 5.2, Safari 1.2, or Netscape 7.2

Support Information
If you are having problems with this software, please call our Media Support Line (800) 677-6337 Monday through Friday 8:00 a.m. to 8:00 p.m. and Sunday 5:00 p.m. to 12:00 a.m. EST.

You can also get support by filling out the web form located at:
http://247.prenhall.com/mediaform

Our technical staff will need to know certain things about your system in order to help us solve your problems more quickly and efficiently. If possible, please be at your computer when you call for support. You should have the following information ready:
- Textbook ISBN
- CD-ROM ISBN
- corresponding product and title
- computer make and model
- Operating System (Windows or Macintosh) and Version
- RAM available
- hard disk space available
- Sound card? Yes or No
- printer make and model
- network connection
- detailed description of the problem, including the exact wording of any error messages.

NOTE: Pearson does not support and/or assist with the following:
- third-party software (i.e. Microsoft including Microsoft Office Suite, Apple, Borland, etc.)
- homework assistance
- Textbooks and CD-ROM's purchased used are not supported and are non-replaceable. To purchase a new CD-ROM contact Pearson Individual Order Copies at 1-800-282-0693.

Note:
For updates, support, and installation instructions for QuickTime, please go to
http://www.info.apple.com/usen/quicktime/

QuickTime and the QuickTime logo are trademarks used under license.
The QuickTime logo is registered in the U.S. and other countries.

Macromedia Projector and the Macromedia logo are trademarks used under license.
The Macromedia logo is registered in the U.S. and other countries.